GEOCHIMICA ET COSMOCHIMICA ACTA

SUPPLEMENT 8

PROCEEDINGS
OF THE
EIGHTH LUNAR SCIENCE CONFERENCE

Houston, Texas, March 14–18, 1977

COVER ILLUSTRATIONS

OUR COMMON DESIRE TO SEE "what's on the other side of the hill" has been gratified in recent years by the planetary explorations of the American and Soviet space programs. This expansion of our physical horizons has been accompanied by an explosive increase in technical knowledge of our solar system, as evidenced by the wealth of new scientific data reported in these three volumes and in their twenty-one immediate predecessors. It seems appropriate to illustrate the covers of these three volumes with photographs which symbolize the accomplishments reported within them; therefore we have chosen six photographs which portray highlights of our exploration of the terrestrial planets.

Volume One, Front Cover. The surface of Mars in the early morning, taken by Viking Lander I. Accumulations of fine-grained sediment resemble terrestrial sand dunes, but appear partly to be drifts stabilized by the presence of large rocks. A portion of NASA photograph no. S76-28052.

Volume One, Back Cover. The surface of Mars, looking northeast, taken by Viking Lander II. A portion of NASA photograph no. S76-28780.

Volume Two, Front Cover. Earthrise over the farside of the moon, taken by the crew of Apollo 8. Man's first direct view of the Lunar farside and of Earthrise occurred on this trip. A portion of NASA photograph no. AS8-14-2383.

Volume Two, Back Cover. Close-up view of the surface of Earth's moon, showing the Apollo 14 Lunar Module and tracks left by the two-wheeled Modularized Equipment Transporter. A portion of NASA photograph no. AS14-67-9367.

Volume Three, Front Cover. Surface of Mars' moon, Phobos, northern hemisphere, taken from a distance of 120 km by Viking Orbiter I. The striations appear to be grooves, ranging between 100 and 200 m in width. NASA photograph no. S77-22262.

Volume Three, Back Cover. Surface of Venus, taken by Venera 9, October 22, 1975. Photograph courtesy of the U.S.S.R. Academy of Sciences.

GEOCHIMICA ET COSMOCHIMICA ACTA
Journal of The Geochemical Society and The Meteoritical Society
SUPPLEMENT 8

PROCEEDINGS OF THE EIGHTH LUNAR SCIENCE CONFERENCE

Houston, Texas, March 14–18, 1977

Compiled by
The Lunar Science Institute
Houston, Texas

VOLUME 2

PETROGENETIC STUDIES OF MARE AND HIGHLAND ROCKS

PERGAMON PRESS
NEW YORK · OXFORD · TORONTO · SYDNEY · FRANKFURT

Pergamon Press, Inc., Maxwell House, Fairview Park
Elmsford, N.Y. 10523 USA Teletype 137328

Pergamon Press, Ltd. Headington Hill Hall
Oxford OX3 0BW England

Pergamon of Canada Ltd., 75 The East Mall
Toronto M8Z 2Lq Canada

Pergamon GMBH, Postfach 6242 Kronberg/Ts, Germany

Pergamon House, 19a Boundary Street, Rushcutters Bay NSW2011, Australia

Type set by The European Printing Corporation, Ltd.
Dublin, Ireland

Printed by Publishers Production International
and bound by Arnold's Bindery
in the United States of America

Library of Congress Cataloging in Publication Data

Lunar Science Conference, 8th, Houston, Tex., 1977.
Proceedings of the Eighth Lunar Science Conference, Houston, Texas, March 14-18, 1977.

(Geochimica et cosmochimica acta : Supplement ; 8)
CONTENTS: v. 1. The moon and the inner solar system.--v. 2. Petrogenetic studies of mare and highland rocks.--v. 3. Planetary and lunar surfaces.
1. Moon--Congresses. 2. Solar system--Congresses. 3. Lunar petrology--Congresses. 4. Moon--Surface--Congresses. 5. Planets--Surfaces--Congresses.
I. Lunar Science Institute. II. Series.
QB581.L755 1977 559.9'1 77-17891
ISBN 0-08-022052-5

Supplement 8
GEOCHIMICA ET COSMOCHIMICA ACTA
Journal of The Geochemical Society and The Meteoritical Society

T. Maxwell, National Air and Space Museum, Smithsonian Institution, Washington, D.C. 20560.

H. O. A. Meyer, Department of Geosciences, Purdue University, West Lafayette, Indiana 47907.

Y. Nakamura, Geophysics Laboratory, University of Texas Marine Science Institute, Galveston, Texas 77550.

R. J. Phillips, Jet Propulsion Laboratory, California Institute of Technology, Pasadena, California 91103.

F. A. Podosek, McDonnell Center for the Space Sciences, Washington University, St. Louis, Missouri 63130.

G. Ryder, Center for Astrophysics, Harvard College Observatory and Smithsonian Astrophysical Observatory, Cambridge, Massachusetts 02138.

R. S. Saunders, Jet Propulsion Laboratory, California Institute of Technology, Pasadena, California 91103.

C. H. Simonds, The Lunar Science Institute, Houston, Texas 77058.

L. J. Srnka, The Lunar Science Institute, Houston, Texas 77058.

G. J. Taylor, Institute of Meteoritics, Department of Geology, University of New Mexico, Albuquerque, New Mexico 87131.

T. M. Usselman, Department of Geological Sciences, State University of New York at Buffalo, Amherst, New York 14226.

R. J. Williams, Geochemistry Branch, NASA Johnson Space Center, Houston, Texas 77058.

TECHNICAL EDITOR

EDITORIAL STAFF

Contents

(Volume 2)

Mare Basalts

Studies of the Apollo 12 and 17 Mare Basalts

Kinetics of Basalt Crystallization

Topical Studies of Mare Basalts

HIGHLAND ROCKS

Petrology of Highland Rocks

Proc. Lunar Sci. Conf. 8th (1977), p. 1305–1338.
Printed in the United States of America

Chemistry of Apollo 12 mare basalts: Magma types and fractionation processes

J. M. RHODES[1], D. P. BLANCHARD[2], M. A. DUNGAN[3], J. C. BRANNON[1], and K. V. RODGERS[1]

[1]Lockheed Electronics Co., Inc., Houston, Texas 77058
[2]NASA-Johnson Space Center, Houston, Texas 77058
[3]NRC Research Associate, Johnson Space Center, Houston, Texas 77058

Abstract—This paper presents major and trace element data for a large suite of petrographically diverse Apollo 12 mare basalts, and attempts to determine the number of magma types sampled at the Apollo 12 site, their compositional characteristics, their fractionation histories, and their spatial and temporal relationships. The new data confirm earlier classifications of these basalts into four compositional groups: the olivine, pigeonite, ilmenite, and feldspathic basalts respectively. Two of these groups, the olivine and pigeonite basalts, are shown to be co-magmatic, and related by olivine fractionation. The other groups differ in trace element and isotopic characteristics, and are derived from different sources within the lunar interior.

An important consequence of this study is the recognition that the ilmenite basalts are a major basalt type, indigenous to the Apollo 12 site, and with a compositional range equivalent to that of the olivine-pigeonite basalts. The spatial relationships between these main basalt types are discussed in terms of local cratering events. It is suggested that the flows are thicker than the majority of lunar flows, and that the younger ilmenite basalts overlie the olivine-pigeonite basalts. The chemical variation within these basalt groups is attributed to extensive, olivine-dominated, near-surface crystal fractionation (about 30%), from parental magmas with compositions equivalent to the vitrophyres in each group, and to the formation of complementary olivine cumulates. There is a systematic relationship between inferred cooling rate and the position of a sample in the fractionation sequence. The more slowly cooled samples depart farthest from the parental magma composition, through both differentiation and cumulate processes.

INTRODUCTION

THE APOLLO 12 MISSION provided the first lunar samples that were readily understandable in terms of common terrestrial magmatic processes. Initial studies (e.g., LSPET 1970; Kushiro and Haramura, 1971) immediately recognized the importance of olivine fractionation in this texturally and chemically diverse suite of basalts. However, in detail the relationships proved to be somewhat less straightforward. Three major basaltic types were identified, largely on the basis of petrography and major element chemistry, and the internal variations within these groups were ascribed to near-surface fractionation (James and Wright, 1972). However, the trace element and isotopic variability was greater than could be expected from the fractionation of these three major basalt groups (Compston *et al.*, 1971; Papanastassiou and Wasserburg, 1971), suggesting that multiple magma types, with similar major element chemistry but different trace element signatures, had been sampled.

Table 1. Major and trace element chemistry of mare basalts.

	Olivine basalts				
	12006,16	12012,15	12014,17	12076,18	12015,9-11
Major elements (wt.%)					
SiO_2	44.23	44.17	45.34	44.87	44.98
TiO_2	2.59	2.64	2.68	2.76	2.86
Al_2O_3	7.67	7.71	8.00	8.10	8.57
FeO	20.94	20.69	20.33	20.66	20.18
MnO	9.29	0.30	0.26	0.30	0.29
MgO	14.67	14.37	13.85	12.26	11.88
CaO	8.13	8.47	8.63	9.03	9.21
Na_2O	0.20	0.21	0.21	0.21	0.23
K_2O	0.05	0.06	0.06	0.06	0.06
P_2O_5	0.05	0.09	0.05	0.03	0.06
S	0.06	0.07	0.07	—	0.07
Cr_2O_3	0.91	0.69	0.64	0.68	0.68
Total	99.59	99.47	100.12	98.94	99.07
Mg′-value	55.5	55.3	54.8	51.4	51.2
Normative olivine or quartz	O 20.5	O 20.2	O 14.9	O 12.0	O 10.1
Trace elements (ppm)					
Sr	89	89	93	94	94
Ba	56	56	53	59	61
Ce	15.7	13.8	14.7		16.3
Sm	3.77	4.02	4.14	4.03	4.31
Eu	0.72	0.76	0.75		0.81
Tb	1.02	1.17	1.01		1.05
Yb	3.3	3.4	3.3	3.4	3.7
Lu	0.47	0.47	0.53	0.51	0.53
Y	31	33	32		35
Zr	97	99	101		110
Nb	6.4	6.6	6.8		6.6
Hf	3.0	3.4	3.3		3.5
Sc	40.1	41.9	42.8	46.4	46.1
Cr	6250	4780	4450	4640	4600
Co	60	56	54	54	51
Ni	110	60	40		50

Subsequent experience gained on mare basalts from other landing sites has shown that multiple magma types, erupted within narrow spatial and temporal limits, are a ubiquitous characteristic of mare volcanism (Chappell and Green, 1973; Rhodes and Hubbard, 1973; Helmke *et al.*, 1973; Rhodes *et al.*, 1976). Attempts to reconcile inferences based on major and trace element chemistry have met with varying success, but in general the trace element data imply greater complexity than is evident from petrography or fractionation trends derived from major element data.

Table 1. (*Continued*).

	Pigeonite basalts					
	12011,15	12043,2	12017,36	12055,25	12007,4	12039,26
Major elements (wt.%)						
SiO_2	46.63	46.77	47.27	47.00	46.42	46.09
TiO_2	3.29	3.38	3.37	3.52	3.90	4.46
Al_2O_3	9.77	10.09	10.02	10.15	11.28	10.52
FeO	19.53	19.50	19.72	19.54	19.05	20.32
MnO	9.29	0.29	0.29	0.29	0.28	0.29
MgO	8.26	7.68	7.63	7.46	5.86	5.75
CaO	10.63	10.96	10.97	11.10	11.52	11.67
Na_2O	0.25	0.27	0.27	0.27	0.32	0.29
K_2O	0.06	0.06	0.09	0.07	0.08	0.10
P_2O_5	0.07	0.06	0.09	0.07	0.10	0.09
S	0.06	0.07	0.03	0.07	0.10	0.11
Cr_2O_3	0.59	0.50	0.52	0.47	0.28	0.38
Total	99.42	99.63	100.47	100.01	99.19	100.07
Mg′-value	43.0	41.3	40.8	40.5	35.4	33.5
Normative olivine or quartz	Q 2.0	Q 2.4	Q 2.6	Q 2.8	Q 3.9	Q 3.5
Trace elements (ppm)						
Sr	113	117	118	121	142	122
Ba	71	73	75	69	91	88
Ce	19.9	17.7		18.2	23.6	25.7
Sm	5.0	5.25	5.1	5.25	6.4	6.55
Eu	0.95	1.00		0.95	1.20	1.18
Tb	1.06	1.25		1.02	1.48	1.66
Yb	4.2	4.4	4.4	4.4	5.3	5.5
Lu	0.62	0.63	0.66	0.67	0.77	0.81
Y	39	40		43	51	52
Zr	128	123		131	156	156
Nb	7.4	7.5		8.5	10.0	10.7
Hf	3.7	4.0		5.2	6.4	4.7
Sc	52.2	52.4	52.8	54.0	52.3	56.0
Cr	4050	3300	3550	3200	1980	2500
Co	39	37	32	38	26	28
Ni	n.d.	n.d.		n.d.	n.d.	n.d.

In this paper we present major and trace element chemistry for a large, texturally diverse suite of Apollo 12 mare basalts. We have attempted to integrate this information with petrographic inferences and with relevant experimental studies in an attempt to answer questions concerning the number of magma types at the Apollo 12 site, their spatial and temporal relationships, and their fractionation histories. In this attempt, we have benefited from a companion paper by Dungan and Brown (1977), which details the petrology and mineral chemistry of some of these samples. We have not concerned ourselves with the nature of the source of these basalts, or with its evolution. This is the

Table 1. (*Continued*).

	Feldspathic basalt	Ilmenite basalts			
	12031,23	12005,7	12036,24	12008,48	12016,9
Major elements (wt.%)					
SiO_2	46.97	41.56	43.11	42.75	42.78
TiO_2	2.88	2.76	3.20	4.45	4.02
Al_2O_3	12.63	5.30	6.16	7.98	7.23
FeO	16.78	22.27	21.82	21.94	22.64
MnO	0.26	0.30	0.30	0.30	0.30
MgO	7.13	19.97	16.71	12.33	12.65
CaO	12.25	6.31	7.46	8.97	8.42
Na_2O	0.33	0.16	0.18	0.25	0.22
K_2O	0.05	0.04	0.06	0.05	0.06
P_2O_5	0.05	0.04	0.02	0.07	0.08
S	0.05	0.04	0.07	0.08	0.08
Cr_2O_3	0.35	0.75	0.72	0.61	0.57
Total	99.73	99.50	99.81	99.78	99.05
Mg′-value	43.1	61.5	57.7	50.0	49.9
Normative olivine or quartz	Q 2.1	O 42.0	O 28.3	O 17.5	O 18.2
Trace elements (ppm)					
Sr	136	83	91	130	126
Ba	60	35	56	51	59
Ce	15.6	10.2	14.0	16.9	16.2
Sm	4.23	2.99	4.03	5.35	5.5
Eu	1.00	0.62	0.75	1.06	1.06
Tb	1.19	0.77	0.95	1.39	1.42
Yb	3.7	2.7	3.5	4.9	5.0
Lu	0.55	0.41	0.51	0.71	0.67
Y	35	28	36	45	45
Zr	100	66	97	117	117
Nb	7.0	4.3	6.6	5.9	6.1
Hf	3.3	2.4	4.7	3.8	6.3
Sc	48.9	37.1	42.6	52.4	49.4
Cr	2460	5200	4880	4200	3950
Co	26	71	63	51	54
Ni	n.d.	90	60	n.d.	25

subject of a companion paper by Nyquist *et al.* (1977), which relates Rb–Sr geochronology for the same suite of samples to the evolution of the lunar interior and the source of mare basalts.

Major and trace element data for 20 Apollo 12 mare basalt samples are given in Table 1, and their petrographic characteristics are briefly outlined in Table 2. In both tables, the data are arranged according to chemically defined magma types (as discussed below), and in decreasing atomic Mg/(Mg + Fe) values. For

Table 1. (*Continued*)

	Ilmenite basalts			
	12045,12	12056,10	12054,62	12047,3
Major elements (wt.%)				
SiO_2	42.30	43.44	45.86	45.13
TiO_2	4.78	5.07	4.63	5.20
Al_2O_3	8.06	8.82	10.47	10.10
FeO	22.09	21.60	19.51	20.50
MnO	0.29	0.29	0.29	0.29
MgO	11.63	9.30	6.76	6.59
CaO	9.09	10.21	11.93	11.32
Na_2O	0.26	0.29	0.31	0.31
K_2O	0.07	0.07	0.07	0.08
P_2O_5	0.09	0.07	0.06	0.08
S	0.09	0.10	0.09	0.12
Cr_2O_3	0.59	0.48	0.33	0.31
Total	99.34	99.74	100.31	100.02
Mg′-value	48.4	43.4	38.2	36.4
Normative olivine or quartz	O 16.3	O 6.9	Q 2.2	Q 2.2
Trace elements (ppm)				
Sr	136	159	162	171
Ba	52	62	64	69
Ce	17.4	20.2	18.8	20.1
Sm	5.6	6.4	6.0	6.5
Eu	1.19	1.31	1.27	1.36
Tb	1.51	1.73	1.85	1.80
Yb	5.1	6.0	5.8	5.9
Lu	0.73	0.82	0.78	0.89
Y	50	55	51	57
Zr	112	135	128	141
Nb	5.3	6.1	6.3	7.0
Hf	4.5	4.8	4.8	5.1
Sc	54.0	60.0	64.0	61.0
Cr	4060	3310	2300	2190
Co	52	42	31	32
Ni	40	n.d.	n.d.	n.d.

convenience, this ratio will be designated the Mg′-value throughout the paper. The major element data were obtained by X-ray fluorescence analysis following our customary procedures (Norrish and Hutton, 1969). Na, Hf, Sc, Cr, Co, Ni, and REE element abundances were determined by instrumental neutron activation analysis (Jacobs *et al.*, 1976), Sr, Y, Zr, and Nb by X-ray fluorescence analysis (Norrish and Chappel, 1967), and Ba by stable isotope dilution mass spectrometry (Gast *et al.*, 1970).

Table 2. Brief petrographic descriptions of Apollo 12 mare basalts analyzed in this study.

Olivine basalts

12006—Ophitic texture lacking olivine. Abundant silica. Petrography does not agree with chemical composition, or Apollo 12 catalogue description.

12012—Porphyritic-subophitic. Partly resorbed olivine phenocrysts in a subophitic groundmass.

12014—Similar to 12012.

12076—Porphyritic. Olivine and pigeonite phenocrysts in a variolitic groundmass.

12015—Olivine vitrophyre. Olivine phenocrysts and skeletal microphenocrysts in a rapidly quenched groundmass. Very similar to 12009.

Pigeonite basalts

12011—Porphyritic. Pyroxene and olivine phenocrysts in a variolitic groundmass.

12043—Porphyritic. Large pyroxene phenocrysts and minor olivine in a groundmass that varies in texture from subophitic to variolitic.

12017—Porphyritic. Pyroxene phenocrysts and very minor olivine in a subophitic groundmass.

12055—Porphyritic. Large pyroxene phenocrysts and minor olivine in a groundmass that varies from variolitic to subophitic.

12007—Microgabbro. Ophitic to graphic intergrowth of plagioclase and pyroxene plus acicular ilmenite and silica.

12039—Microgabbro. Similar to 12007.

Ilmenite basalts

12005—Coarse-grained heterogeneous cumulus texture includes large oikocrysts of pyroxene and an interstitial assemblace of olivine, plagioclase, and ilmenite.

12036—Coarse-grained cumulate dominated by equant pyroxene grains. Interstitial plagioclase, olivine, and ilmenite.

12008—Olivine vitrophyre. Olivine phenocrysts and skeletal microphenocrysts in a rapidly quenched groundmass. Similar to 12009.

12016—Medium-grained equigranular basalt. Abundant pyroxene plus partly resorbed olivine and interstitial plagioclase.

12045—Porphyritic. Olivine and augite phenocrysts in a variolitic groundmass.

12056—Medium-grained subophitic. Minor resorbed olivine.

12054—Subophitic. Lacks olivine. Heavily shocked.

12047—Equigranular ophitic. Lacks olivine.

Feldspathic basalt

12031—Coarse-grained graphic intergrowth of plagioclase and pyroxene plus needles of cristobalite and ilmenite.

THE APOLLO 12 BASALTS

James and Wright (1972) identified three major basalt groups at the Apollo 12 site: olivine-pigeonite basalts, ilmenite-bearing basalts, and feldspathic basalts. This scheme has been adopted in the more recent attempts at mare basalt classification, with the exception that the pigeonite and olivine basalts are considered separately in recognition of the distinct compositional hiatus between them (Rhodes *et al.*, 1975; Papike *et al.*, 1976).

This study has doubled the number of basalts from the Apollo 12 site for which comprehensive chemical information is available, and has essentially confirmed the four basalt groups recognized in the previous studies. In contrast with earlier work, we have used trace as well as major element data to discriminate between the various basalt types, using a combination of realistic crystal fractionation models and trace element signatures. As in our Apollo 17 mare basalt study (Rhodes *et al.*, 1976), we have emphasized fine-grained, rapidly chilled samples to establish discriminatory criteria, on the assumption that these samples most closely approach magma compositions and are less likely to be subject to sampling problems. Coarser-grained samples are subsequently assigned to the appropriate basalt type according to the criteria established for finer-grained samples. Similarly, we rely on magmaphile element ratios in preference to their absolute abundances as discriminants among basalt types. Since these elements have low crystal/liquid distribution coefficients, their ratios are unlikely to be grossly perturbed by fractionation processes or by unrepresentative sampling of coarse-grained basalts. The most useful of these ratios are Ba/Sm and Ce/Sm, both of which reflect the slope of the light rare-earth depletion, and also various combinations of such incompatible elements as Ba, Zr, Y, and Nb. Sc and Sr are useful discriminants, but their ratios to magmaphile elements tend to decrease in more evolved samples as a consequence of the onset of pyroxene and plagioclase fractionation. Table 3 summarizes some of the more useful discriminatory ratios for the four basalt groups, listing the "best estimate" for each group based on the most rapidly

Table 3. Trace element ratios in Apollo 12 mare basalts.

		Olivine basalts	Pigeonite basalts	Ilmenite basalt	12031 feldspathic basalt
Ba/Sm	(1)	14.1	14.2	9.5	14.2
	(2)	12.8–14.9	13.1–14.7	8.7–10.7	—
Ce/Sm	(1)	3.8	4.0	3.2	3.7
	(2)	3.4–4.2	3.4–4.0	2.9–3.4	—
Zr/Y	(1)	3.1	3.3	2.6	2.9
	(2)	3.0–3.2	3.0–3.3	2.4–2.6	—
Zr/Nb	(1)	17	17	20	14
	(2)	15–17	15–17	15–22	—
Zr/Sr	(1)	1.2	1.1	0.9	0.7
	(2)	1.1–1.2	1.1–1.3	0.8–0.9	—
Ba/Sr	(1)	0.65	0.63	0.39	0.44
	(2)	0.57–0.65	0.57–0.72	0.38–0.47	—
Ba/Sc	(1)	1.32	1.36	0.97	1.23
	(2)	1.24–1.40	1.28–1.74	0.94–1.19	—
Sc/Sm	(1)	10.7	10.4	9.8	11.6
	(2)	10.4–11.5	8.2–10.4	9.0–12.4	—

(1) "Best estimate" for group based on most rapidly cooled sample.
(2) Range.

chilled sample in the group, together with the range in the ratio found during this study. In order to minimize the effects of inter-laboratory bias we have included only data from this study in this compilation. Figure 1, a plot of the Ba/Sm ratio versus Mg′-value, serves to illustrate the relative constancy of such ratios within a basalt group over varying degrees of fractionation. An obvious feature of both this diagram and the ratios listed in Table 3 is the close correspondence in magmaphile element ratios of the olivine and pigeonite basalts: strong evidence for either a co-magmatic origin or derivation by different amounts of partial melting from a common source. The ilmenite basalts are distinctly different from both the olivine and pigeonite groups, and must be derived from a separate parental magma and from a different source in the lunar interior. The necessity

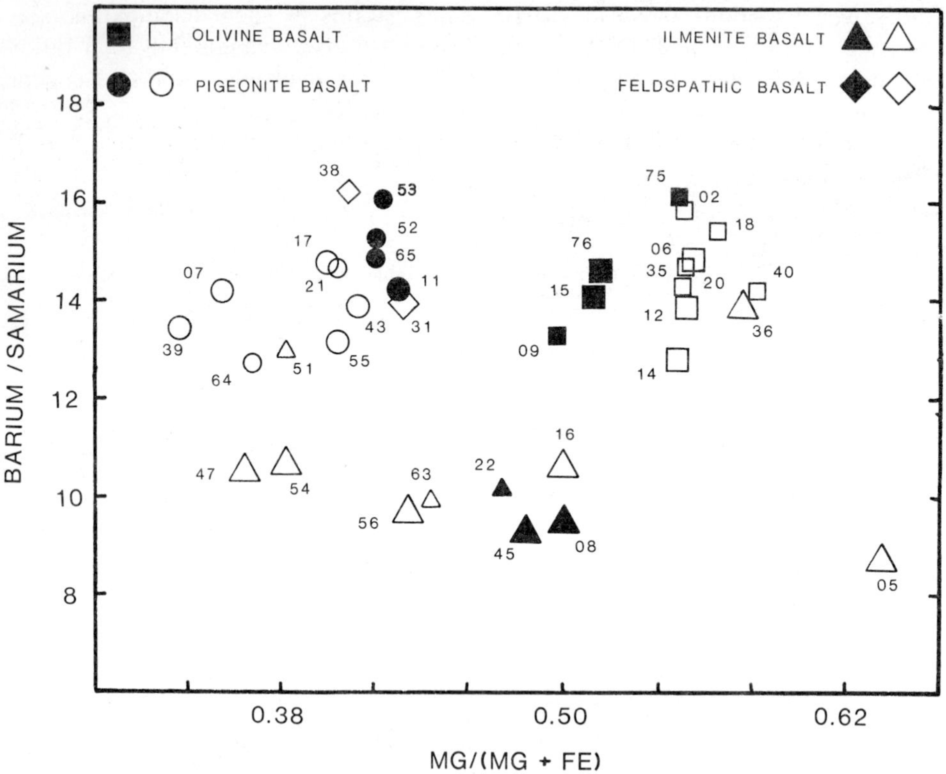

Fig. 1. Ba/Sm ratios versus atomic Mg/(Mg + Fe) values for Apollo 12 mare basalts. Rapidly cooled vitrophyres and basalts with a variolitic groundmass are represented by filled symbols, more slowly cooled samples with ophitic and gabbroic textures are represented by open symbols. Data from Table 1 are indicated by large symbols, that from other studies by small symbols (Compston *et al.*, 1971; Cuttita *et al.*, 1971; Engel *et al.*, 1971; Haskin *et al.*, 1971; Hubbard and Gast, 1971; Kushiro and Haramura, 1971; Maxwell and Wiik, 1971; Schnetzler and Philpotts, 1971; Scoon, 1971; Wänke *et al.*, 1971; Willis *et al.*, 1971).

for a separate source is further emphasized by higher initial Sr isotope ratios in the olivine and pigeonite basalts than those of the ilmenite basalts (Fig. 2). The isotopic data in this figure are taken from the companion paper of Nyquist *et al.* (1977) and from Papanastassiou and Wasserburg (1971), and are plotted against TiO_2 content, a measure of crystal fractionation in both basalt groups.

An important consequence of this study is the recognition that the ilmenite basalts are an abundant and significant rock type at the Apollo 12 site. Previously, only four chemically analyzed samples could be assigned to this group (Papike *et al.*, 1976), and one of these, 12064, should in fact be included with the pigeonite basalts on the basis of the criteria established in this paper. We now recognize ten samples of ilmenite basalts, spanning a wide compositional range from olivine-rich partial cumulates to highly evolved samples. In view of their

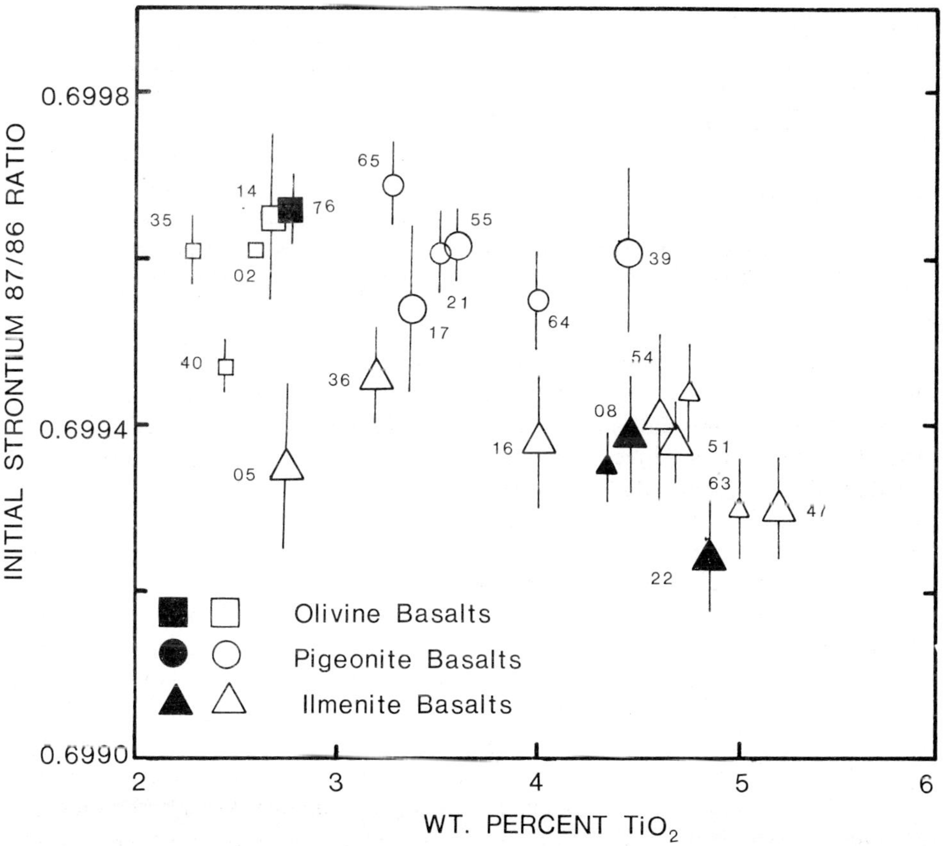

Fig. 2. Initial $^{87}Sr/^{86}Sr(I)$ versus TiO_2 for Apollo 12 mare basalts. Symbols are the same as in Fig. 1. Data from Nyquist *et al.* (1977) and Papanastassiou and Wasserburg (1971). The Cal-Tech data have been adjusted for inter-laboratory bias by the addition of 0.00012 in $^{87}Sr/^{86}Sr$.

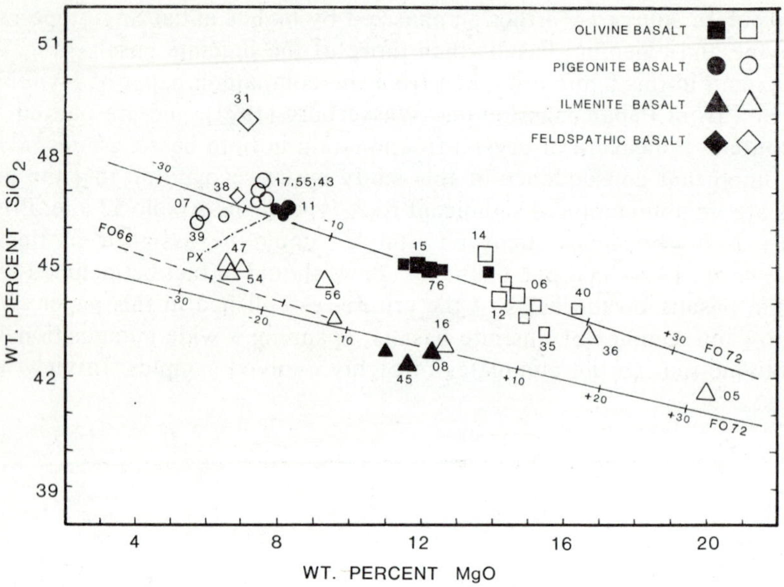

Fig. 3. MgO–SiO_2 relationships in Apollo 12 mare basalts. Symbols are the same as in Fig. 1. Olivine control lines (Fo_{72}) indicating the effects of olivine addition and subtraction are shown in solid lines; olivine (Fo_{66}) control in a broken line; and pigeonite control by a dot-dash line. These control lines are for reference purposes, they are not intended as lines of "best fit". The data sources are the same as in Fig. 1.

abundance and broad compositional range, it seems more probable that they are indigenous to the site, rather than exotic fragments thrown in from a nearby "unsampled" basaltic unit as has been suggested by Pieters and McCord (1976). The ilmenite basalts range from olivine-normative partial cumulates such as 12005, with the highest Mg′-value (0.61) of any mare basalt described to date, to highly evolved quartz-normative samples (12047, 12054) with Mg′-values as low as 0.36 (Table 1). In terms of major element chemistry they are readily distinguished from the olivine and pigeonite basalts by lower SiO_2 and higher FeO and TiO_2 abundances for a given MgO value (Figs. 3–5). Similarly, the ilmenite basalts are enriched relative to the olivine and pigeonite basalts in most magmaphile elements, with the exception of Ba, for a particular value of MgO (Figs. 6–8). In addition to a higher magmaphile element content, the ilmenite basalts display a pronounced depletion in the light rare earths, which is reflected in lower Ba/Sm (Fig. 1) and Ce/Sm ratios than in the olivine and pigeonite basalts (Table 3). Thus ilmenite basalts appear to possess chemical characteristics that are intermediate between those of typical low TiO_2 basalts and high TiO_2 basalts from the Apollo 17 and 11 sites. There are exceptions to this general rule; for example, the Zr/Nb ratio (about 20) is too high to fall between typical low TiO_2 basalts (about 17) and high TiO_2 basalts (10–15) (Duncan *et al.*, 1976). Sample 12036, although tentatively included with the ilmenite basalts, presents a special

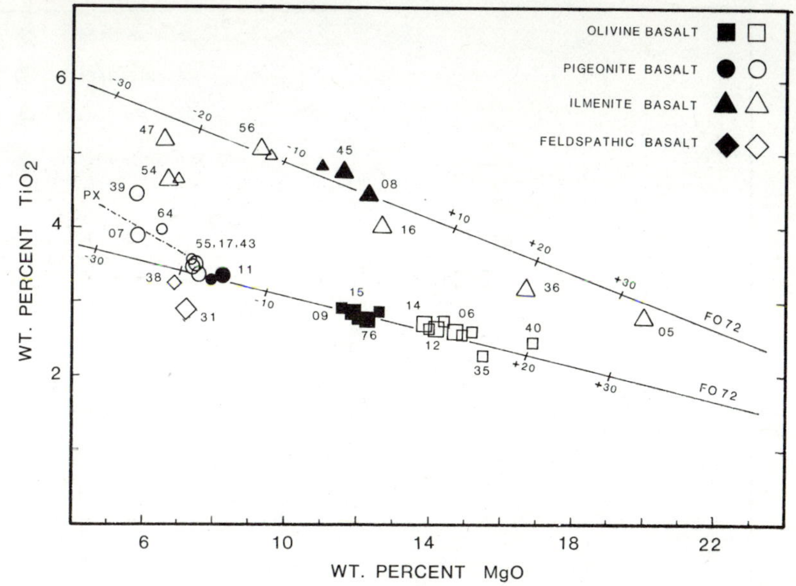

Fig. 4. MgO–TiO_2 relationships in Apollo 12 mare basalts. Symbols and data sources are the same as in Figs. 1 and 3.

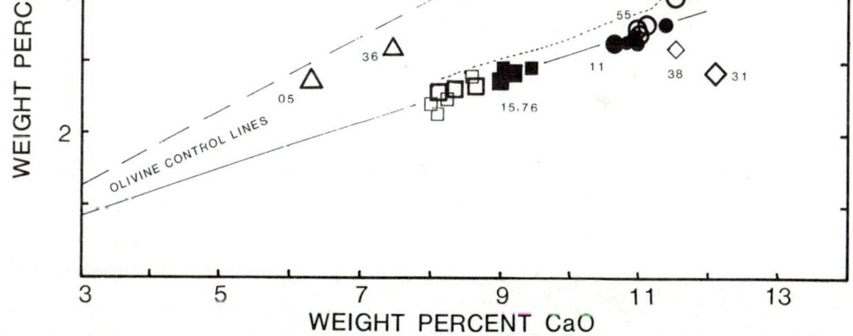

Fig. 5. TiO_2–CaO relationships in Apollo 12 mare basalts. Symbols and data sources are the same as in Figs. 1 and 3. Also shown in a dotted line is the experimentally determined liquid line of descent for 12002 (Walker *et al.*, 1976b).

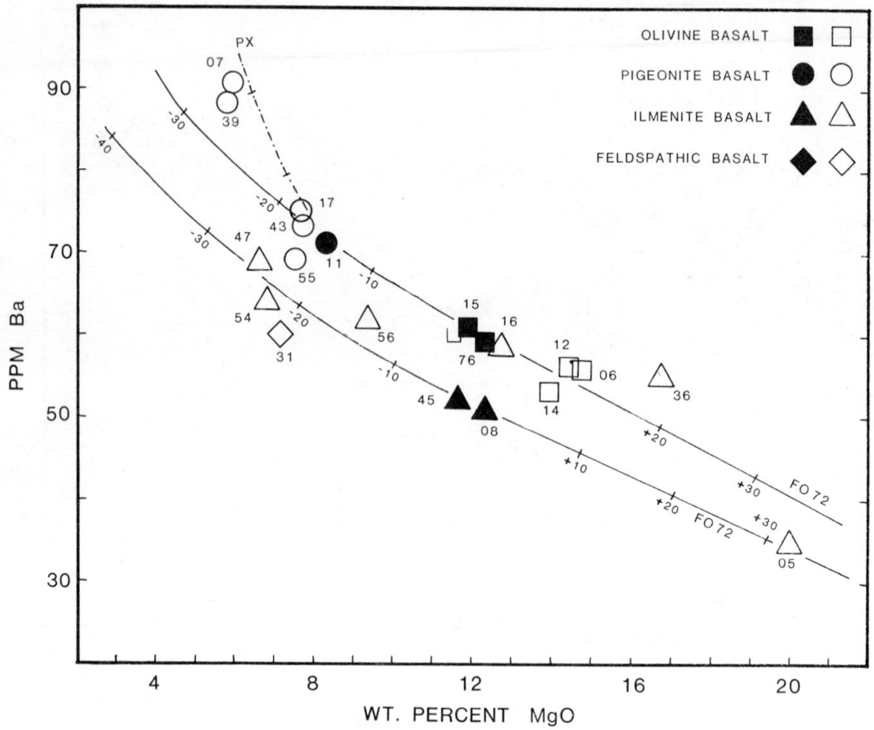

Fig. 6. MgO–Ba relationships in Apollo 12 mare basalts. Symbols are the same as in Figs. 1 and 3. Only data from this paper are shown.

classificatory problem. It is too high in TiO_2, Sc, and many magmaphile element abundances (Figs. 4–8) to be included with the olivine basalts, yet many of the element ratios (e.g., Ba/Sm + others) imply an affiliation with this group, whereas others (e.g., Ce/Sm) have values intermediate between the groups. Sr isotopic data (Nyquist *et al.*, 1977) do not resolve the problem, since these authors report an initial $Sr_{87/86}$ ratio of 0.69946, which is in between the majority of samples in the ilmenite and olivine-pigeonite groups (Fig. 2).

Our data for the olivine basalts fall within the range previously established for this group (James and Wright, 1972; Rhodes *et al.*, 1975; Papike *et al.*, 1976) and include an olivine vitrophyre (12015) which is both chemically and texturally similar to 12009, as well as several coarser-grained more olivine-rich samples. Sample 12006, which was classed as a feldspathic basalt by James and Wright (1972) solely on petrographic evidence, clearly belongs with these olivine-rich samples, on the basis of its composition, as its description in the Apollo 12 Lunar Sample Information Catalog (1970) would suggest. There is evidently a discrepancy for this sample between the thin section collection and the material allocated for our study.

All the olivine basalts contain both modal and normative olivine, and have

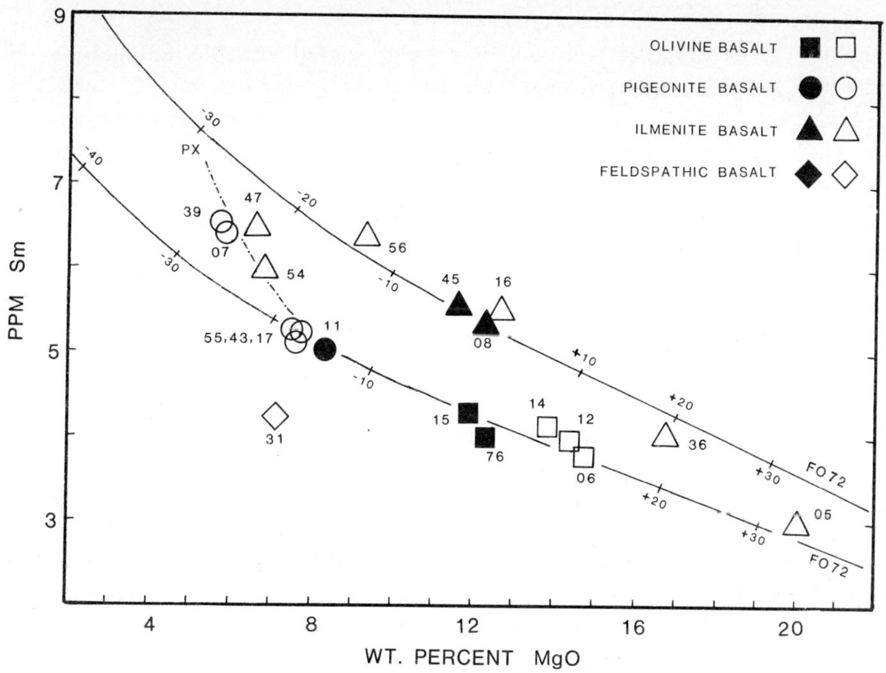

Fig. 7. MgO–Sm relationships for Apollo 12 mare basalts. Symbols are the same as in Figs. 1 and 3. Only data from this paper are shown.

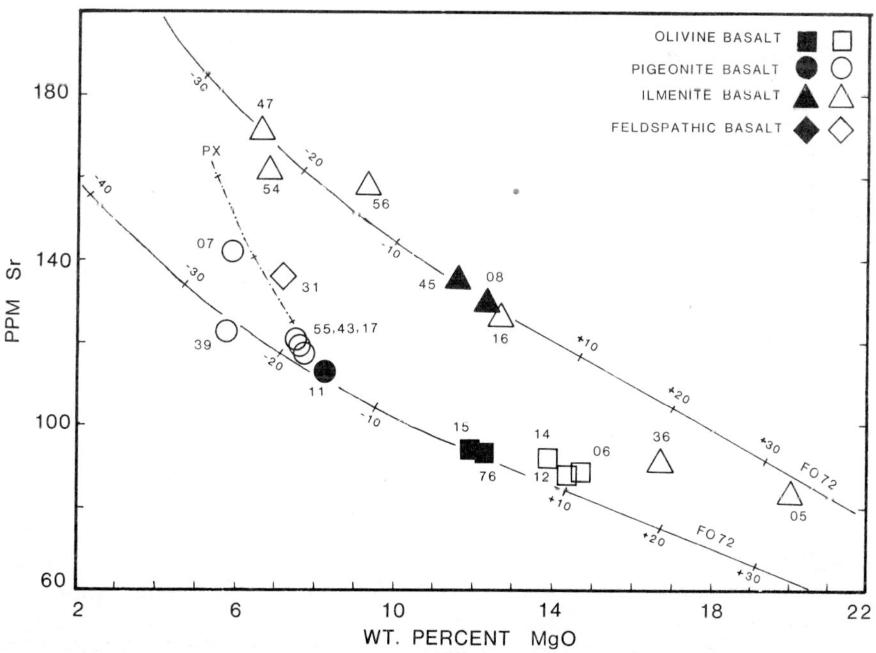

Fig. 8. MgO–Sr relationships for Apollo 12 mare basalts. Symbols are the same as in Figs. 1 and 3. Only data from this paper are shown.

high Mg′-values ranging from 0.49 in rapidly cooled vitrophyres such as 12009 and 12015 to 0.58 in more slowly cooled microgabbros such as 12035 and 12040. As noted by Walker *et al.* (1976a), there is a strong correlation between grain size and the normative and modal olivine content. The olivine basalts are separated from the pigeonite basalts by a distinct compositional hiatus (Figs. 3–5) with the olivine basalts containing lower SiO_2, TiO_2, Al_2O_3, CaO, and magmaphile element abundances and correspondingly higher amounts of MgO, Cr_2O_3, Co, and Ni. The magmaphile element ratios in the two groups are essentially the same (Table 3), and they have identical relative rare-earth patterns that are less depleted in light rare earths than the ilmenite basalt patterns.

Prior to this investigation, only four chemically analyzed pigeonite basalts were known (Papike *et al.*, 1976) and the compositional range was small. Our new data (Table 1) extend both the number of samples in the group and the chemical variation, including samples (e.g., 12007, 12039) that are substantially evolved from the main cluster of compositions (Figs. 3–5). All the pigeonite basalts are quartz normative, with Mg′-values ranging from 0.43 in 12011, a rapidly cooled sample, to 0.33 in 12039, a slowly cooled and evolved member of the group. The evolved pigeonite basalts have high TiO_2 contents, and in fact their major element compositions converge on those of low Mg′-value samples in the ilmenite group (Figs. 3–5) with which they are almost isochemical. They can, however, be readily resolved from the ilmenite basalts by several critical trace element ratios (Table 3) (e.g., Ba/Sm, Sc/Sm, Sr/Ba, Zr/Nb) and by initial Sr isotope ratios (Nyquist *et al.*, 1977). Sample 12064 has been assigned to the ilmenite basalts in previous studies on account of its high TiO_2 content of 3.99%. However, on the basis of high initial $Sr_{87/86}$ and Ba/Sm ratios it should be assigned to the pigeonite basalt group (Figs. 1–2). Similarly sample 12039, classed by James and Wright (1972) on petrographic evidence as an ilmenite basalt, should also be included with the pigeonite basalts.

Previously, only one sample of feldspathic basalt (12038) had been analyzed and its trace element characteristics (Compston *et al.*, 1971; Schnetzler and Philpotts, 1971) are sufficiently different from the other basalts to exclude it from the three major basalt types. An additional sample 12031, originally classed as an ilmenite basalt by James and Wright (1972), is very similar to 12038 in major element chemistry, having high amounts of Al_2O_3, CaO, and SiO_2 and correspondingly lower values of FeO and MgO. Both samples contain about 35–36% normative plagioclase, in contrast with the 27–29% common in the pigeonite basalts. Despite these major element similarities, 12031 differs from 12038 in its trace element characteristics and is similar to the olivine-pigeonite basalts (Table 3). It may be a plagioclase-enriched variant of the pigeonite basalts, either by plagioclase accumulation or by inadequate sampling of material that is heterogeneous at the hand-specimen level. However, the trace element ratios depart sufficiently from the "best estimates" for the pigeonite basalts to prevent unambiguous classification of this sample.

Table 4 assigns all currently analyzed Apollo 12 basalts to the appropriate

Table 4. Classification of chemically analyzed Apollo 12 mare basalts.

Olivine basalts	12009, 12015, 12075, 12076, 12004, 12012, 12006, 12002, 12020, 12014, 12018, 12040, 12035.
Pigeonite basalts	12011, 12053, 12052, 12065, 12043, 12055, 12017, 12021, 12007, 12039, 12064.
Ilmenite basalts	12008, 12045, 12022, 12051, 12047, 12063, 12056, 12054, 12016, 12036?, 12005.
Feldspathic basalts	12031?, 12038.

basalt group, ranking them in order of inferred cooling rate. The cooling rate estimates are based on a variety of criteria including overall grain size, plagioclase size and morphology, the nature of the groundmass, and comparisons with experimentally determined cooling-rate studies on rocks of similar composition (e.g., Lofgren *et al.*, 1974; Donaldson *et al.*, 1975).

Spatial Relationships

Since almost all "hand-specimen" sized basalts collected at the Apollo 12 site have now been analyzed in detail and most can be assigned with confidence to the major basalt groups, it is appropriate to consider the spatial aspects of these basalt types and to attempt to infer their stratigraphic and temporal relationships. Figure 9 is a sketch map of the Apollo 12 site, taken from Sutton and Schaber (1971), showing locations, where known, of all analyzed samples. It is readily apparent that the majority of the ilmenite basalts are located in the vicinity of Surveyor Crater in the SE part of the landing site, whereas the olivine basalts are largely confined to the NW section close to Middle Crescent Crater. The locations of several olivine basalts (12002, 12006, 12009, 12012, 12018, 12020) are not known with certainty, but they were all collected on EVA-1 somewhere between the LM and Middle Crescent Crater, four of them coming from the rim of the crater (Sutton and Schaber, 1971).

There is no apparent preferred distribution for the pigeonite basalts; they are found throughout the landing site. Sutton and Schaber (1971) show that continuous ejecta from the 400 m Middle Crescent Crater should cover the Apollo 12 landing site, and that this ejecta will have been reworked by the impacts creating Surveyor, Head, and Bench craters. Thus, the spatial distribution of samples may be depth related, controlled mainly by ejecta from Middle Crescent Crater, as originally suggested by Warner (1971). On this theory, the olivine basalts were excavated from the greatest depth and should be the oldest. The ilmenite basalts should be younger, overlying the olivine basalts, and the pigeonite basalts, if they are co-magmatic with the olivine basalts as the chemical data suggests, are probably sandwiched between the olivine and ilmenite basalts.

If one applies the crater ejecta model of McGetchin *et al.* (1973) to Middle Crescent Crater it is clear that its ejecta blanket will only be a few meters thick, ranging from 8–10 m at the rim to only 1–2 m in the vicinity of Surveyor Crater.

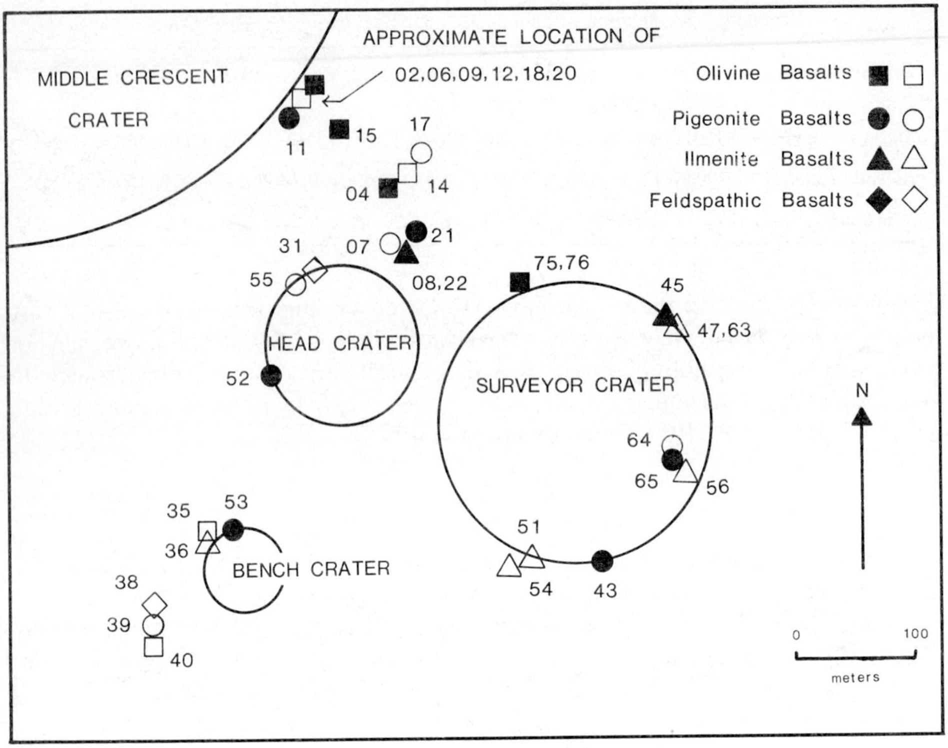

Fig. 9. Sample locations at the Apollo 12 landing site. Symbols are the same as in Fig. 1.

In view of the fact that Surveyor and Head craters probably excavated material from about 40 to 20 m, respectively, it seems unlikely that ejecta from Middle Crescent Crater will dominate the sample distribution throughout the site. The areal distribution can, however, be readily explained by assuming substantial thicknesses for the major basaltic types and that ilmenite basalts overlie older olivine basalts. Middle Crescent Crater, which probably excavated to a depth of about 80 m, sampled both olivine and ilmenite basalt. Olivine basalt, which is ubiquitous at the rim crest, will have been excavated from the greatest depth. On the other hand, Head and Surveyor craters, which penetrated only to depths of about 20 and 40 m, respectively, sampled both ilmenite and pigeonite basalt, but failed to penetrate the deeper olivine basalt. If this model is correct, it implies that the combined thickness of the ilmenite and pigeonite basalts must be substantial, possibly over 40 m thick. A similar thickness for the underlying, older olivine basalt is also required, probably between 30 and 50 m. Although lava flows up to 30 m thick have been described, apparently they are not common and have been ascribed to localized high rates of magma effusion (Schaber *et al.*, 1976). Flow thicknesses of about 10 m or less appear to be more common in the mare basins (Howard *et al.*, 1972; Schaber *et al.*, 1976).

On the basis of plagioclase grain size and thermal modeling, Walker *et al.* (1976a) arrive at cooling unit thicknesses for the olivine basalts similar to those suggested here. They suggest that some of the coarse-grained olivine basalts crystallized at a distance of 11 m from the flow base, and that a flow thickness of at least 30 m would be required to maintain the appropriate cooling history. Since the coarsest of the ilmenite basalts compare closely in texture with the olivine basalt samples (Dungan and Brown, 1977), a similar cooling history may be implied and a comparable thickness for the ilmenite basalt cooling unit is not unlikely.

Crystal Fractionation

It has been widely recognized that near-surface crystal fractionation has played a major role in the evolution of the Apollo 12 mare basalts (e.g., LSPET, 1970; Kushiro and Haramura, 1971; Compston *et al.*, 1971; Green *et al.*, 1971a; James and Wright, 1972; Walker *et al.*, 1976a). The olivine and pigeonite basalts were regarded as co-magmatic, the compositional variation being controlled primarily by olivine fractionation; but the relationships for the ilmenite basalts were obscure. The new data presented in this paper allow us to re-examine these relations in greater detail. Specifically, we are concerned with the nature and number of parental magmas involved; with the magmatic history of the ilmenite basalts; and with the fractionation history of the olivine basalts, and their relationship to the pigeonite basalts.

In a detailed chemical study, Compston *et al.* (1971) argued that the Apollo 12 basalts were derivatives from at least six parental magmas, each associated with a separate episode of partial melting in the lunar interior. Primary evidence for this was the wide variation in Rb/Sr ratios and also the variation in magmaphile element abundance that could not be quantitatively accounted for by crystal fractionation models consistent with the major element data. The wide spread in initial Sr isotope ratios also implied that multiple magma types had been sampled (Papanastassiou and Wasserburg, 1971). This conflict between large variances in the trace element data and the relatively simple crystal fractionation models implied by the covariance of the major element data, is a common problem in both lunar and terrestrial basalt petrogenesis. It has been widely interpreted as implying several parental basaltic magmas with distinctive trace element signatures, that follow essentially identical liquid lines of descent at the Apollo 15 (Helmke *et al.*, 1973) and at the Apollo 17 (Shih *et al.*, 1975) site, as well as at the Apollo 12 site. Several factors impede the resolution of this problem:

(1) Inter-laboratory bias in trace element determinations. This is probably the least troublesome.
(2) Most of the magmaphile elements used to distinguish between parental magmas and their associated derivatives are concentrated in the mesostasis. Consequently, their determination can be subject to severe samp-

ling problems. This is particularly true for coarse-grained rocks. For example, our measurements of rare-earth elements in 12039 differ from those reported by Nyquist *et al.* (1977), on a different subsample, by a factor of 1.2. This is not an inter-laboratory bias since Ba measurements on the two subsamples in one laboratory show the same discrepancy.

Clearly, it is dangerous to infer the existence of distinct magma types solely from differences in magmaphile element abundances. A safer approach is the use of element ratios, as discussed earlier, on the assumption that they are less likely to be disturbed by inadequate sampling than are the absolute abundances. The drawback to using element ratios is that although it is possible to distinguish magmas that are derived from different sources, it is difficult to resolve those magmas that are produced by varying amounts of melting from the same source, since the ratios will tend to remain the same. The only satisfactory solution to this dilemma is to rely heavily on fine-grained, rapidly chilled samples for evidence, or on suites of coarser-grained samples that exhibit systematic covariance with the major element data. Single coarse-grained samples that depart from such covariant trends are not sufficient evidence for the deduction of different magma types.

(3) By definition, magmaphile elements are excluded from the major rock-forming minerals and are concentrated in the residual liquids. It appears to us intrinsically probable that during the slow cooling of sizeable lava flows, this residual liquid will be mobilized, resulting in zones that are enriched or depleted in magmaphile elements, independent of the major element chemistry. Such a case has been documented in detail by Hart *et al.* (1971) for a single 10 m Icelandic lava flow, and Rhodes *et al.* (1976) explained the systematic magmaphile element depletion of several coarse-grained Apollo 17 basalts relative to fine-grained basalts having similar magmaphile element ratios, by this process. Again, the departure of absolute magmaphile element abundances from those predicted by major element fractionation trends may lead to erroneous conclusions concerning the number of magma types necessary to account for trace element variance, and the use of element ratios or data from rapidly cooled samples is to be preferred.

From the preceding discussion, it is clear that much of the diversity in trace element abundances in Apollo 12 basalts, and their departure from well-established major element fractionation trends, may result from a variety of causes. It is perhaps premature to advocate the existence of multiple magma types until all other possibilities can be excluded.

When we examine the Apollo 12 data in the light of the preceding discussion, emphasizing our own data to alleviate inter-laboratory bias, we arrive at the following conclusions:

(1) The major element data establish two well-defined fractionation trends for the olivine-pigeonite and ilmenite basalt groups. The feldspathic

basalts 12038 and 12031 may be related to these basalts by plagioclase accumulation processes, but the evidence is ambiguous.

(2) There are several rapidly chilled vitrophyres possessing almost identical Mg′-values and MgO, Cr_2O_3, and Co abundances (e.g., 12008, 12009, 12015, 12045). These samples are distinctly bi-modal in their abundances and ratios of many magamaphile elements, corresponding to the olivine (12009, 12015) and ilmenite (12008, 12045) basalt groups (Table 3, Fig. 1).

(3) The coarser-grained samples reflect the same bi-modal distributions in magmaphile element ratios (Table 3, Fig. 1), and although there is appreciable scatter, the magmaphile element abundances are broadly covariant with the major element trends (Figs. 6–8).

(4) The initial $Sr_{87/86}$ data are also bi-modal (Nyquist *et al.*, 1977) with the olivine-pigeonite basalts possessing higher ratios than the ilmenite basalts. There are some samples with intermediate values, and suggestions of subgroups within the data. However, the uncertainties are such that they cannot be resolved from the main groups.

Thus, with the exception of sample 12038, there is no compelling evidence for more than two major magma types at the Apollo 12 site: the olivine-pigeonite and the ilmenite basalts. With this established, we can discuss the nature of the parental magmas and their fractionation histories in more detail.

The olivine-pigeonite basalts

Olivine-controlled crystal fractionation has been extensively documented for the olivine and pigeonite basalts (e.g., Kushiro and Haramura, 1971; Compston *et al.*, 1971; James and Wright, 1972). The data in Table 1 confirm these observations, following well-defined trends of increasing SiO_2, TiO_2, Al_2O_3, and CaO with decreasing MgO and Cr_2O_3 (Figs. 3–5). Table 1 lists the data in order of decreasing Mg′-value, and it is readily seen that there are similar progressive and systematic changes in major and trace element abundances, and in normative olivine or quartz content. Such variations are consistent with crystal fractionation dominated by ferromagnesian minerals. Figure 10 attempts to summarize the relative positions of the samples in the fractionation sequence, and to integrate this with inferences concerning their cooling history. This diagram will be used as a framework for further discussion of the fractionation of these rocks. The Mg′-value is used as a fractionation index, and, within a particular interval, essentially isochemical samples are ranked according to their relative cooling rate. Samples that differ in composition but are inferred to have had similar cooling histories are placed at the same level. Cooling-rate estimates are based on examination of the curatorial thin section collection, and are influenced by a variety of criteria, including overall grain size, plagioclase size (Walker *et al.*, 1976a), and morphology (Donaldson *et al.*, 1975; Lofgren *et al.*, 1974).

Although necessarily subjective, our ranking for the olivine-normative basalts corresponds closely with similar cooling-rate evaluations of Donaldson *et al.*

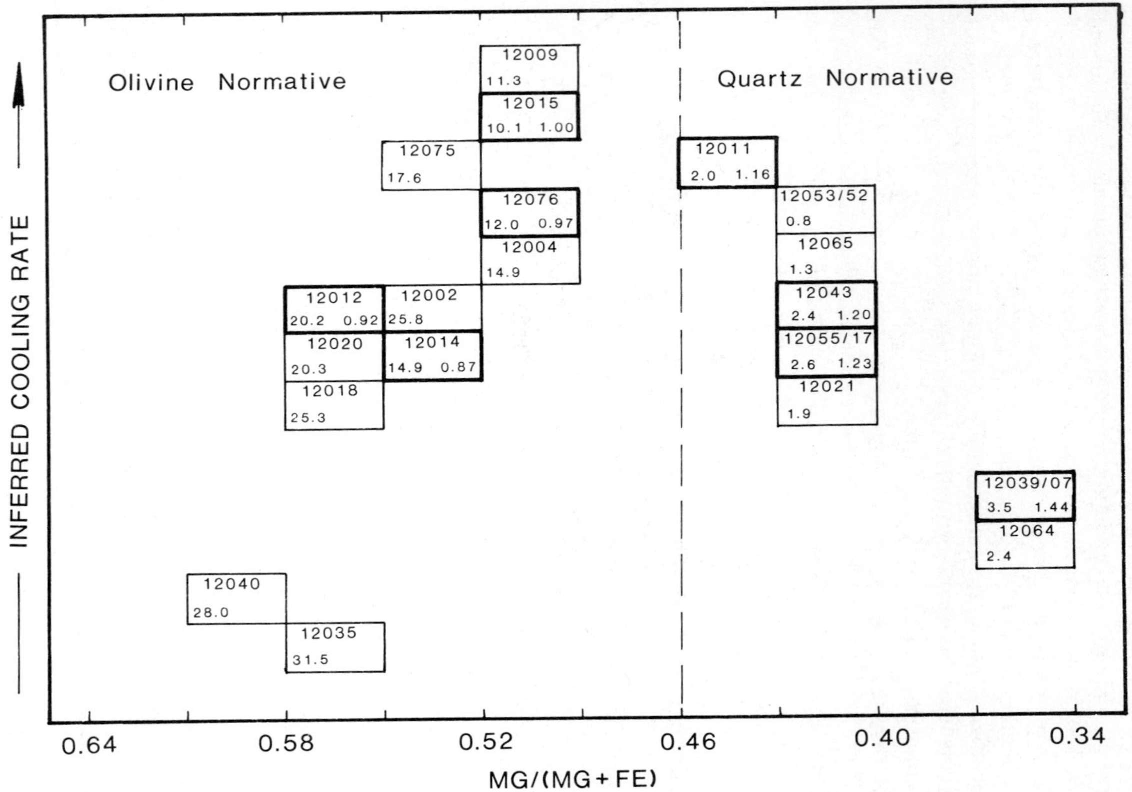

Fig. 10. Cooling rate–differentiation relationships for Apollo 12 olivine-pigeonite basalts. The Mg/(Mg + Fe) value provides a measure of the position of samples in the fractionation sequence relative to the inferred parental magma (12015). Also shown (lower left in each box) is the normative olivine or quartz content, and (lower right) the ratio of Ba_{sample}/Ba_{12015}. Both provide additional measures of the differentiation process. Samples studied in this paper are enclosed by heavier outlines. The source of data for Mg/(Mg + Fe) values is the same as in Fig. 1; only data from Table 1 are used for the Ba_{sample}/Ba_{12015} ratio.

(1975) and Walker *et al.* (1976a). In Fig. 10 samples discussed in this paper are included in a heavy outlined box. Included with the sample number is the normative olivine or quartz content and the ratio of Ba in the sample to its concentration in the inferred parental magma. Both factors provide independent estimates of the degree of fractionation, and for most samples there is good agreement between these estimates and the relative position in the sequence based on the Mg′-value.

It is evident from Fig. 10 that there are two contrasting cooling-rate trends within the olivine-pigeonite basalt series. The olivine basalts display an inverse relationship between cooling rate and Mg′-value: that is, the more slowly cooled samples are richer in normative olivine. This relationship has been discussed in detail by Walker *et al.* (1976a), and attributed to olivine accumulation at the base of a single flow. Conversely, there is a positive correlation between cooling rate and Mg′-value in the pigeonite basalts, fractionation having proceeded further in the more slowly cooled samples. Within the olivine basalts there are two subgroups, separated by a hiatus in the Mg′-value between 0.52 and 0.54 (Fig. 1). Samples with Mg′-values less than 0.52 (12009, 12015, 12004, 12076) have all cooled quickly and are closely comparable in composition. Two of these (12009, 12015) are vitrophyres, and must have been erupted onto the lunar surface as liquids, or liquids bearing a small percentage of olivine phenocrysts. Green (1971b) has argued that the composition of 12009 (Mg′-value 0.49) is close to that of the immediate parental magma for this suite of rocks. Several lines of evidence indicate that this is a realistic assumption, perhaps erring toward low Mg′-values:

(1) The most Mg-rich olivine composition reported for Apollo 12 mare basalts is Fo_{77} (Butler, 1972), which according to the Roeder and Emslie (1970) Fe/Mg distribution between olivine and melt, would indicate crystallization from a magma with an Mg′-value of 0.52.
(2) Donaldson *et al.* (1975) suggest, on the basis of experimentally determined textural evidence, that some of 12009 olivine crystallized slowly prior to eruption, and that 12009 was extruded onto the lunar surface bearing a small percentage of olivine phenocrysts.
(3) The hiatus in Mg′-values between 0.52 and 0.54 has been interpreted by Papike *et al.* (1976) as being indicative of the value for a magma parental to the Apollo 12 basalts.
(4) There is strong evidence, both textural and experimental (e.g., Green *et al.*, 1971a; Walker *et al.*, 1976a,b) that the samples with Mg′-values greater than 0.54 are olivine-enriched partial cumulates, thereby limiting the parental magma composition to a Mg′-value of less than 0.54.

In view of the close limits that can be placed on the composition of the inferred parental magma for the olivine basalts, we shall adopt the vitrophyre 12015 as the parental composition in the following discussion.

The suggestion that samples with Mg′-values higher than those of the vitrophyres must be olivine-enriched partial cumulates receives strong support from three areas: Firstly, it has been shown that the major element compositions of these samples can be modeled simply by adding olivine and minor Cr-spinel to the inferred magma composition (e.g., Kushiro and Haramura, 1971; Compston *et al.*, 1971); secondly, Green *et al.* (1971a) and Walker *et al.* (1976b) argue that the fact that the experimentally determined liquidus olivines in these samples are more forsteritic than the most magnesian cores in the basalts indicates olivine accumulation; thirdly, the inverse relationship between cooling rate and normative olivine content, or Mg′-value, seen in Fig. 10 has been shown by Walker *et al.* (1976a) to be consistent with olivine settling in the basal part of a 30 m cooling unit, where the amount of cumulus olivine increases upward with decreasing cooling rate and increasing grain size.

Starting with a parental magma of 12015 composition, basalts with Mg′-values greater than 0.54 can be derived by varying amounts of olivine (Fo_{72}) (±Cr-spinel) accumulation into the parental magma. Calculations indicate that sample 12014 requires about 6% olivine accumulation, whereas sample 12006 requires about 12%.

Inspection of Figs. 3–8 shows that there is reasonable consistency between major and trace element estimates in the amount of olivine accumulation required for these rocks. The "best-fit" olivine (Fo_{72}) required to account for these compositional variations is less forsteritic than experimentally determined liquidus olivines (Fo_{75-78}) in melts compositionally close to the inferred parental magma (Green 1971b; Walker *et al.*, 1976b) or than the most forsteritic cores (Fo_{77}) observed in rapidly chilled rocks. The majority of phenocryst cores are less forsteritic than either of these, about Fo_{69-71} (Butler, 1972). This is closer to the composition required to account for the fractionation trends, and presumably results from re-equilibration with the melt on cooling over a substantial temperature interval below the liquidus (Green, 1971b; Walker *et al.*, 1976b).

Major element data from the literature (Compston *et al.*, 1971; Kushiro and Haramura, 1971; Willis *et al.*, 1971) indicate that the majority of the other high Mg′-value samples have accumulated similar amounts of olivine, about 8–15%, together with minor spinel. Two samples, 12035 and 12040, are more extreme examples, requiring about 20% olivine accumulation (Figs. 3, 4). The trace element data for these coarse-grained olivine-enriched samples are highly variable, departing significantly from the calculated olivine accumulation trend. Nevertheless, because their trace element ratios and major element compositions are consistent with derivation from a magma of 12015 composition, we prefer to regard them as partial cumulates from this magma rather than derivatives from some unspecified and unsampled magma. These samples are all coarse grained, have cooled slowly (Fig. 10), and have had abundant opportunity for re-equilibration and migration of intergranular, trace element-enriched, residual fluids. Sample 12036 may be one extreme example of this type of process. It is clearly olivine enriched and could have been derived by about 20% olivine accumulation from a magma with an Mg′-value of 0.50. However, we are not clear from which, if any, of the two parental magma types it is derived. If it is derived from the 12015 composition, as some of its elemental ratios and Sr isotope ratios seem to imply (Figs. 1, 2), it has been enriched substantially in TiO_2 and other magmaphile elements.

If the cooling model proposed by Walker *et al.* (1976a) for the olivine basalts is essentiall correct and they have accumulated within a sizeable flow, complementary differentiates must exist within that cooling unit. The quartz-normative pigeonite basalts, with Mg′-values extending from 0.43 to 0.33, are obvious contenders for this role. We have seen previously that rapidly cooled members of the olivine and pigeonite basalt groups have almost identical magmaphile element ratios (Table 3), indicating either that they are co-magmatic or that they were derived from a common source. Earlier studies (e.g., Compston *et al.*, 1971; Kushiro and Haramura, 1971; James and Wright, 1972) have shown that the compositions of the few pigeonite basalts then analyzed fall on olivine control lines from the olivine basalts, indicating that both basalt types could have been derived from a common magma. Later reviews (Rhodes *et al.*, 1975; Papike *et al.*, 1976), impressed by the compositional hiatus between the two basalt types and the wide variations in trace element chemistry, have treated them separately although recognizing their potential co-magmatic nature.

Sample 12011 is the least evolved ($Mg' = 0.43$) of the pigeonite basalts. It is also the most rapidly cooled, with a fine-grained, variolitic, groundmass containing phenocrysts of pigeonite and skeletal microphenocrysts of olivine. It is an ideal sample to test for relationships between the olivine and pigeonite basalts, since sampling problems should be minimal, and there should have been no opportunity for migration of residual liquid. Least-squares calculations indicate that the 12011 major element composition can be very closely matched by the removal of 15% olivine (Fo_{72}) and 0.4% Cr-spinel from a 12015 magma composition. These are, of course, the petrographically observed and experimentally determined liquidus phases over a wide temperature interval in the olivine vitrophyres (Green, 1971b), and are the phases necessary to produce the olivine-enriched samples. Since the olivine/liquid distribution coefficients for the magmaphile trace elements are low (<0.02), the ratio of the concentration in the parent magma to its concentration in the derivative magma provides an estimate of the amount of liquid remaining (Anderson and Greenland, 1969). Using this relationship, we find that for most of these elements (e.g., Ba, Sr, Zr, Sm, Lu) the extent of fractionation varies between 14 and 16%, in excellent agreement with the independent estimate based on major element data. Similarly, by using $D^{ol/lq} = 3$ and an initial Co concentration of 51 ppm in a parental magma of 12015 composition, we can show that 15% olivine fractionation should decrease the Co concentration to 37 ppm, again, in excellent agreement with the 39 ppm found in 12011. Thus, there can be little doubt that the pigeonite basalts are co-magmatic, evolved derivatives of the olivine basalt.

Other, more evolved, members of the pigeonite basalt group (12043, 12017, 12055) continue this olivine fraction trend, requiring, respectively, the removal of 17.5, 17.7, and 18.4% olivine (Fo_{72}) and 0.5% Cr-spinel from a magma with the composition of 12015. The trace element data (Figs. 6–8) indicate similar amounts of olivine and spinel fractionation, up to 20%, but the internal consistency in these coarser-grained samples is not as satisfactory as that observed for the 12015–12011 relationship. After about 20% olivine + spinel fractionation, when the magma has reached an Mg′-value of about 0.40 olivine is no longer observed and pigeonite becomes the dominant fractionating phase. This results in marked departure from the olivine fractionation trends with an increase in TiO_2 and magmaphile elements and decrease in SiO_2, with relatively little change in the MgO content (Figs. 3–8). Similar trends are observed in the experimentally determined liquid line of descent for 12002 (Walker *et al.*, 1976b), at the onset of pigeonite crystallization. The three most evolved of the pigeonite basalts, 12064, 12007, and 12039, with Mg′-values of 0.37, 0.35, and 0.33, respectively, show evidence of substantial pigeonite fractionation. Attempts to model this stage in the fractionation process are not as satisfactory quantitatively as were the results for the earlier olivine fractionation, presumably because these are slowly cooled samples (Fig. 10), and will have had abundant opportunity for re-equilibration and migration of residual liquid, thus complicating the fractionation process. Furthermore, as we discussed earlier, sampling of coarse-grained samples, such as 12039, is a problem as evidenced by the 20% difference in rare-earth elements found in different subsamples of this rock. Nevertheless, there is sufficient consistency between the major and trace element data to indicate that these samples result from about 30 to 35% fractionation from a magma of 12015 composition, and that the later stages of this process are controlled by pigeonite. For example, major element calculations indicate that 12007 can be derived from a melt corresponding to the pigeonite basalt 12011 by removal of about 10% pigeonite and 3% olivine (Fo_{66}). Similarly, 12039 requires the removal of 17% pigeonite and only 1.5% olivine, but also 4% plagioclase (An_{95}). The magmaphile element abundances are generally supportive of these calculations, since, in Figs. 6–8, these samples plot closer to the pigeonite control line than to the olivine one. Furthermore, the amount of fractionation from the parental magma composition of 12015 indicated by the magmaphile elements is between 29 and 33% for 12007 and 29 and 35% for 12039, in comparison with 28 and 38% calculated from major element data.

In the experimental equilibrium crystallization studies on 12002 of Walker *et al.* (1976b), pigeonite is followed within 20°C by plagioclase in the crystallization sequence. Thus, we might expect to see the effects of plagioclase fractionation in samples that reflect more than 30% fractionation. Sample 12039 is depleted in Sr relative to 12007 and the pigeonite fractionation trend, which is qualitative support for the major element calculations that require some plagioclase removal to generate this composition from a magma comparable to 12011. Sample 12031 is remarkably similar to 12038 in major element chemistry, and both are enriched in plagioclase relative to other Apollo 12 basalts. However, 12038 is distinctly different from other Apollo 12 basalts in its trace element abundances and their ratios, whereas 12031 is not. Many of the magmaphile element ratios, with the exception of those that would be modified by plagioclase fractionation, are similar to those for the olivine and pigeonite basalts (Table 3). However, the SiO_2 content is far too high for this sample to be a plagioclase-enriched derivative of the pigeonite basalts. Its relationships are an enigma; perhaps it is another example of inadequate sampling?

In the preceding discussion, we have argued that the olivine and pigeonite basalts are co-magmatic; that fractional crystallization of a parental magma compositionally equivalent to the olivine vitrophyres gave rise to olivine-enriched samples, resulting from 10–20% olivine accumulation; that the same parental magma produced the evolved pigeonite basalts from 15% to more than 30% crystal fractionation, initially dominated by olivine, but subsequently by pigeonite and possibly plagioclase.

Although the genetic relationships can be firmly established, the spatial relationships present certain problems. In the case of a single differentiated cooling unit, the pigeonite basalts would be the complementary differentiates to the olivine-enriched basalts that are inferred to have formed at the base of a thick cooling unit (Walker *et al.*, 1976a). The marked relationship between cooling rate and degree of fractionation, evident in Fig. 10, for the pigeonite basalts, would suggest that these samples were from the upper half of the cooling unit, where such a relationship would be expected to hold. Although this model is appealing in its simplicity, two additional lines of evidence suggest that the relationships would be more complicated. First, the rapidly chilled pigeonite basalt 12011 must have crystallized very close to the surface of a cooling unit. This is unlikely in the case of a single differentiated flow, since olivine vitrophyres would crystallize at the upper surface. Secondly, the compositional hiatus between the olivine and pigeonite basalts should not exist in a single differentiated flow. It is in fact suggestive of a two-stage history involving subsurface crystallization of olivine basalt, followed by the subsequent eruption of a separate pigeonite basalt flow. In this case, the correlation between decreasing cooling rate and progressive fractionation in the pigeonite basalts would reflect sampling of the upper part of this flow. A possible way out of the dilemma, since there must be complementary differentiates to the olivine-enriched basalts, is to suggest that a single thick basalt flow, initially of olivine vitrophyre composition, fractured during differentiation and was breached by the evolved pigeonite basalt magma.

The ilmenite basalts

Prior to this study, only four ilmenite basalts had been chemically analyzed (Rhodes *et al.*, 1975; Papike *et al.*, 1976), and one of these (12064) is in fact an evolved pigeonite basalt. The seven additional samples of this basalt type, whose analyses are reported in Table 1, reveal a wide compositional range comparable to that of the olivine-pigeonite basalts, from an olivine-rich basalt with an Mg′-value of 0.61 to evolved quartz-normative basalts with Mg′-values as low as 0.36. As was the case with the olivine-pigeonite basalts, the ilmenite basalts follow well-defined trends of increasing SiO_2, Al_2O_3, CaO, and magmaphile element abundances with decreasing MgO and Cr_2O_3 content (Figs. 3–5). Table 1 lists the samples in order of decreasing Mg′-value, which generally corresponds with systematic changes in other element abundances, and in normative olivine and quartz content. These trends are indicative of substantial amounts of crystal fractionation involving ferromagnesian minerals. Figure 11 relates the inferred cooling history of the samples to their relative positions in the fractionation sequence. As was noted for the olivine-pigeonite basalts, there are two contrasting trends: the majority of samples reflect a correlation between cooling rate and the amount of differentiation, but the samples with high Mg′-values show an inverse relationship between cooling rate and position in the differentiation sequence. As is elaborated by Dungan and Brown (1977), these two contrasting trends can be, conceptually at least, related to complementary accumulation and differentiation processes occurring in a single cooling unit.

Sample 12008, a vitrophyre, is the most rapidly cooled of the ilmenite basalts. Texturally it resembles 12009, containing both equant and skeletal olivine phenocrysts. It can be presumed to have undergone a similar cooling history, and must have been erupted onto the lunar surface either as a liquid or as a liquid bearing a small percentage of pre-eruptive olivine phenocrysts (Dungan and Brown, 1977). Since it is also the least evolved of the ilmenite basalts (Mg′-value = 0.51), other than those of accumulative origin (e.g., 12005), it provides the best available estimate of the parental magma composition for the ilmenite basalts. This suggestion gains further support from two other areas: first, the most magnesian olivine recorded in the ilmenite basalts is Fo_{76} in 12022 (Weill *et al.*, 1971), which according to Fe/Mg partitioning between olivine and liquid (Roeder and Emslie, 1970) would be the liquidus olivine for a magma with an Mg′-value of 0.51, such as 12008; secondly, the Mg′-value (0.51), Cr_2O_3 (0.61%), and Co (51 ppm) content of 12008 are all indicative of the "primitive" nature of the sample and are remarkably similar to the values for these variables in 12015, the inferred parental magma composition for the olivine-pigeonite basalts. In view of this evidence, we have adopted the vitrophyre 12008 as the composition of the parental magma for the ilmenite basalts.

Inspection of Figs. 3–5 shows that, to a first approximation, much of the compositional variation can be accounted for by olivine (Fo_{72}) fractionation of a magma compositionally equivalent to 12008. Green *et al.* (1971a) show that for 12022, a rock whose composition is close to that of 12008, olivine is the liquidus phase over a wide temperature range, joined at 125°C below the liquidus by spinel, pigeonite, and plagioclase in that order. The majority of the olivine phenocrysts in 12008 fall within the narrow limits of Fo_{69-72} (Dungan and Brown, 1977), which is in the range Fo_{77-65} reported by Green *et al.* (1971a) for experimental studies on 12022 prior to the onset of pyroxene crystallization. Consequently, the fractionation trend, dominated by olivine (Fo_{72}) is consistent with what one would expect from experimental and petrographic observations.

Although there is a general correspondence between sample compositions and the predicted olivine fractionation trends, several of the more slowly cooled samples show significant correlated departures from these trends for TiO_2 and the magmaphile elements. We attribute this to a greater potential for re-equilibration and migration of trace element-enriched residual liquids in slowly

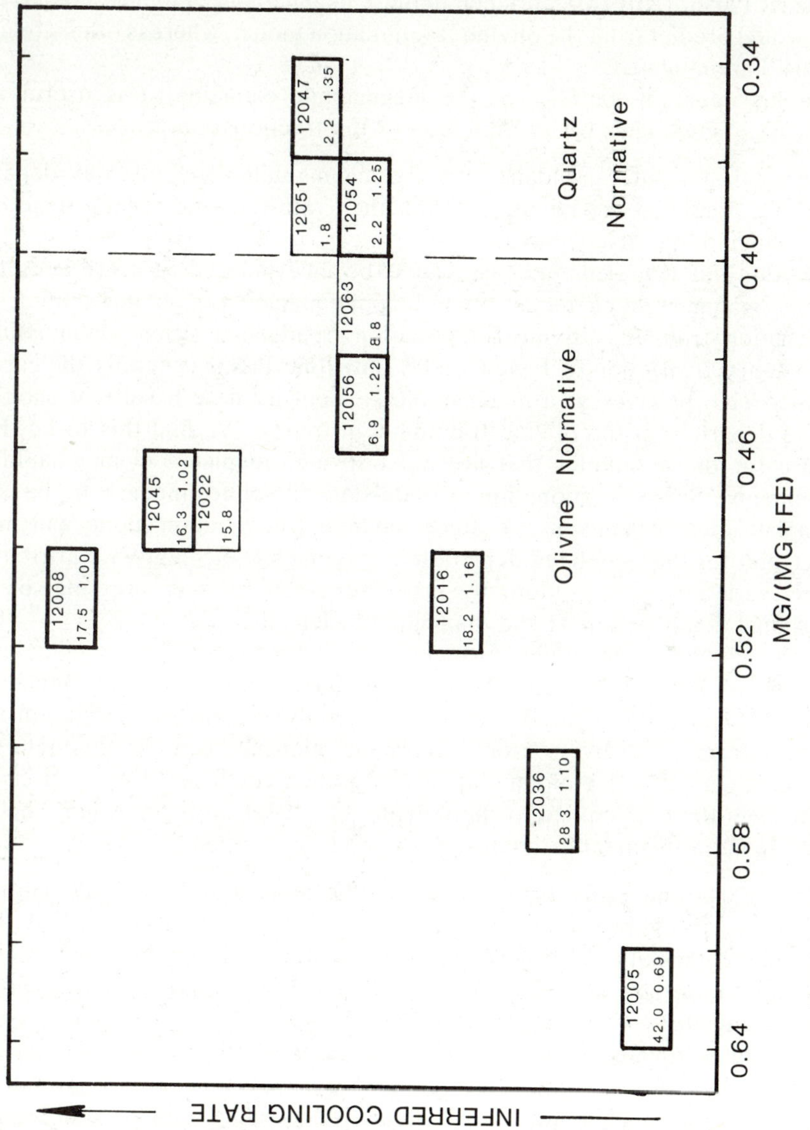

Fig. 11. Cooling rate–differentiation relationships for Apollo 12 ilmenite basalts. The conventions and sources of data are the same as in Fig. 10, except that the inferred parental magma is 12008.

cooled samples. A consequence of this process is that although magmaphile element ratios should remain essentially constant, and therefore characteristic of the ilmenite basalt group, their variable abundances will lead to inconsistencies in the amount of fractionation estimated from major and trace element data. Some samples (e.g., 12056, 12016) are enriched in these magmaphile elements relative to what one would predict from the olivine fractionation model, whereas others (e.g., 12054, 12047) are depleted.

Given this apparent mobility in the magmaphile elements, it is useful to establish certain guidelines for examination of the fractionation process:

(1) Although absolute abundances of the magmaphile elements may depart from a predicted fractionation trend, their ratios should reflect those of the ilmenite basalts (Table 3).
(2) Sr, Eu, and Ba, elements that would be incorporated in preference to other magmaphile elements in late-forming plagioclase, should show less deviation from the olivine fractionation trends than other magmaphile elements. Inspection of Figs. 6 and 8 show that this is generally the case.
(3) Since TiO_2 behaves as a magmaphile element in these basalts, it should also depart from the olivine fractionation trends. We find this to be the case for those samples that are most strongly depleted in magmaphile elements. Since fractionation of late-stage ilmenite appears to be an implausible mechanism, and since ilmenite fractionation alone will not account for the correlated depletion in magmaphile elements, we attribute this variation to migration, by processes such as filter pressing, of a residual liquid rich in Ti and magmaphile elements.
(4) Although the magmaphile elements cannot be used as an independent check of the major element modeling, ferromagnesian trace elements such as Co or Ni can. If we take the firmly established derivation of 12011 from 12015 by 15% olivine fractionation, we can use this data to calculate an effective olivine/liquid distribution coefficient for Co of 2.65. This can then be applied to the ilmenite basalts as an independent check of the major element estimates of olivine fractionation.

Samples 12045 and 12022 have cooled a little more slowly than 12008 and have slightly more evolved compositions (Fig. 11). Calculations show that the major element compositions of these two samples can be closely matched by the removal from a magma of 12008 composition of 2.6 and 7.0% olivine (Fo_{72}) respectively and 0.3% Cr-spinel (Figs. 3–5). The trace element data are consistent with this estimate, requiring between 2 and 4% fractionation, as can be seen in Figs. 6–8.

Sample 12056 has cooled more slowly than the preceding samples, and has as a consequence evolved to a greater extent. It can be derived from a magma of 12008 composition by removal of 12.5% olivine (Fo_{73}) and 0.3% Cr-spinel (Figs. 3–5). This sample is slightly enriched in all magmaphile elements relative to the calculated olivine fractionation trends (Figs. 6–8), with the result that predictions based on these elements would require greater degrees of olivine fractionation than is indicated by the major element data. Co provides an independent check on the major element estimate since by using $D^{ol/lq} = 2.65$, the calculated Co content of a magma derived from a magma of 12008 composition by 12.5% olivine fractionation would be 41 ppm. This is close to the observed abundance of 42 ppm in 12056.

The inferred migration of magmaphile elements is more pronounced in 12047 and 12054, the most evolved of the ilmenite basalts (Fig. 11). These two samples are depleted in Ti and most magmaphile elements, with the exception of Sr and Ba, relative to the theoretical fractionation trends (Figs. 4–8). The major element data can be modeled in a two-step process, by olivine (Fo_{72}) and spinel fractionation to the 12056 composition, and then by the removal of Fe-rich olivine (Fo_{60-66}), minor Cr-spinel and ilmenite. The necessity for fractionation involving more Fe-rich olivine at these evolved compositions is consistent with one atmosphere melting experiments on 12022 (Green *et al.*, 1971a) where the olivine at the olivine-pyroxene-plagioclase cotectic is Fo_{65}. The necessity for ilmenite fractionation is a consequence of the decrease in TiO_2 in these samples (Figs. 4, 5), which we attribute to movement of a Ti-rich residual liquid rather than the actual physical separation of ilmenite. If one accepts this interpretation, the major element data would then indicate that 12047 and 12054 resulted from 25 and 27% olivine and minor Cr-spinel fractionation from a magma of 12008 composition, and that the olivine removed became progressively less forsteritic (Fo_{72-60}) as fractionation progressed. Co abundances provide an independent check of this model, the calculated amounts remaining in the magma after 25 and 27% olivine fractionation being 31.6 and 30.1 ppm, respectively, compared to measured values of 32 and 31 ppm.

In summary, then, we propose that starting with a parental magma compositionally equivalent to 12008, varying amounts of olivine (Fo_{72-60}) and spinel fractionation have taken place, ranging from as little as 2% in rapidly chilled rocks (12045), up to 27% in more slowly cooled samples. Dungan and Brown (1977) argue that it is only in slowly cooled magmas that such extensive olivine fractionation can occur, and suggest that the unfractionated samples could have been derived from the upper surface of a single flow, with the more fractionated and slowly cooled samples located stratigraphically beneath those that are rapidly chilled. It follows, then, that at deeper levels in the flow olivine accumulation must occur. Two samples from the ilmenite basalt group, 12005 and possibly 12036 meet this requirement. Sample 12016, another slowly cooled sample, does not differ greatly in major element composition from 12008, but its trace element abundances are rather erratic. Dungan and Brown (1977) indicate that this sample has undergone substantial re-equilibration, and suggest that it has had a complex crystallization history, such as loss of olivine coupled with a compensatory gain of more Fe-rich olivine from higher levels in the cooling unit. The relationships for 12036 are even more nebulous. We have noted earlier that many of its magmaphile element ratios, suggest a genetic relationship to the olivine basalts (Fig. 1). On the other hand, it is too high in magmaphile elements to be related to these basalts without abundant enrichment in late-stage residual fluids. Furthermore, the major element composition can be derived successfully from 12008 by addition of olivine (14% Fo_{72}), pigeonite (8.7%), augite (3.5%), and minor Cr-spinel (0.3%). Similar calculations using 12015 as a parental liquid are not as successful. The necessity to include pyroxene as well as olivine in the calculations does not imply that both phases were co-precipitating, merely that substantial amounts of post-crystallization re-equilibration has taken place, involving cumulus olivine and the residual liquid.

Sample 12005 is the only clear example of olivine enrichment in the ilmenite basalts. It has an Mg′-value of 0.61, and is the most mafic of all lunar mare basalts. It has cooled very slowly (Fig. 11), commensurate with its high normative olivine content (42%) which is indicative of extensive olivine accumulation. By analogy with the models of Walker *et al.* (1976a), we attribute this to olivine settling near the base of the ilmenite basalt cooling unit. Calculations show that 12005 can be derived from a parental magma of 12008 composition by accumulation of 30% olivine (Fo_{72}), 3% pigeonite, 4% augite, and 0.5% Cr-spinel (Figs. 3–5). As discussed above for 12036, we attribute the necessity for pyroxene to reaction relationship between the cumulus olivine and the residual liquid.

CONCLUSIONS

(1) This study has doubled the amount of compositional data available for the Apollo 12 basalts. It confirms the four basalt types identified in previous studies: the olivine, pigeonite, ilmenite, and feldspathic basalts. The pigeonite basalts are shown to be co-magmatic with the olivine basalts, derived from a common parental magma by substantial amounts (>15%) of olivine fractionation. The ilmenite basalts are compositionally distinct from the olivine-pigeonite basalts, and derive from a separate parental magma. We infer, on the basis of initial Sr isotope ratios and significant differences in many magmaphile element ratios, that the parental magmas for these major basalt types are derived by partial melting of distinctly different source materials in the lunar interior. The relationship of the feldspathic basalts to the olivine-pigeonite and ilmenite basalts is enigmatic, and their paucity in the sample collection may imply that they are exotic to the area.

(2) The new data indicate that the ilmenite basalts are an abundant and important magma type at the Apollo 12 site. They are texturally and chemically varied, and, like the olivine-pigeonite basalts, reflect a wide range in cooling history and magmatic evolution. In both the olivine-pigeonite and ilmenite basalts this range includes rapidly cooled vitrophyres that we suggest are compositionally close to the parental magmas, and more slowly cooled evolved derivatives and complementary olivine-enriched accumulative basalts. There is a systematic relationship in both basalt groups between inferred cooling rate and position of a sample

within the fractionation sequence. Departure from the initial parental magma composition either through differentiation or cumulate processes is most pronounced in those samples that have cooled slowly. We deduce from this that the extensive fractionation displayed by both of these Apollo 12 basalt types is a consequence of crystallization in thick cooling units. The spatial relationships of the samples relative to local cratering events provide corroborative evidence for this assumption, and can be interpreted as indicating that a thick flow (40 m) of ilmenite basalt overlays an older flow of equally thick olivine-pigeonite basalt.

(3) The compositional variation within both the olivine-pigeonite basalts and the ilmenite basalts is extensive, and for each it can be modeled largely in terms of olivine fractionation within a single cooling unit. Olivine (Fo_{72}) plus minor Cr-spinel fractionation in the olivine basalt parental magma (equivalent in composition to 12015) has led to the formation of olivine-rich basalts, containing between 10 and 20% cumulus olivine, and complementary, evolved pigeonite basalt magmas. With further fractionation, pigeonite replaces olivine as the dominant fractionating phase, producing evolved magma compositions that result from removal of about 30% of the combined phenocryst phases. A similar pattern of fractionation is followed by the ilmenite basalts, producing olivine-rich variants, with up to 30% cumulus olivine, and evolved differentiates that result from over 30% olivine (Fo_{60-72}) plus minor spinel fractionation.

(4) In some of the more slowly cooled rocks magmaphile element abundances depart significantly from crystal/liquid fractionation trends based largely on major element data and involving petrographically observed and experimentally determined phenocryst phases. On the other hand, in rapidly chilled vitrophyres or variolitic samples, the major and trace element data provide self-consistent models of crystal/liquid fractionation that are realistic in terms of petrology and experimental studies. Sampling problems are a possible cause of the inconsistency in the coarse-grained samples. However, this inconsistency could also be produced by late-stage migration of magmaphile element-enriched residual liquids within slowly cooled magma bodies, resulting from mechanisms such as filter pressing. We believe that the mobilization of residual fluids is an intrinsically probable process that has been neglected in geochemical studies.

Acknowledgments—We wish to thank H. Wiesmann for providing isotope dilution barium determinations. Interchange of data and ideas with L. Nyquist proved to be most fruitful. Careful reviews by J. Green and G. Lofgren provided helpful advice and comments. Special thanks are due to L. Romero and C. Rhodes for assistance in preparation of the manuscript. This work was done in part under the auspices of the National Aeronautics and Space Administration, contract no. NAS 9-15200 to Lockheed Electronics Company, Inc. One of the authors (MAD) was a National Research Council post-Doctoral Fellow at the Johnson Space Center and is currently at the Lunar Science Institute (post-Doctoral Fellow, Basaltic Volcanism Project—contract no. NSR-09-051-001). This constitutes BV

Contribution No. 5 of the Basaltic Volcanism Study Project. The Project is organized and administered by the Lunar Science Institute/Universities Space Research Association under contract no. NSR-09-051-001 with the National Aeronautics and Space Administration.

REFERENCES

Anderson A. T. and Greenland L. P. (1969) Phosphorus fractionation diagram as a quantitative indicator of crystallization differentiation of basaltic liquids. *Geochim. Cosmochim. Acta* **33**, 493–505.

Apollo 12 Lunar Sample Information (1970) NASA publication TR R-353. 317 pp.

Apollo 12 Preliminary Science Report (1970) NASA publication SP-235. 227 pp.

Butler P., Jr. (1972) Compositional characteristics of olivines from Apollo 12 samples. *Geochim. Cosmochim. Acta* **36**, 773–785.

Chappell B. W. and Green D. H. (1973) Chemical composition and petrogenetic relationships in Apollo 15 mare basalts. *Earth Planet. Sci. Lett.* **18**, 237–246.

Compston W., Berry H., Vernon M. G., Chappell B. W., and Kaye M. (1971) Rubidium-strontium chronology and chemistry of lunar material from the Ocean of Storms. *Proc. Lunar Sci. Conf. 2nd*, p. 1471–1485.

Cuttitta F., Rose H. J., Annell C. S., Carron M. K., Christian R. P., Dwornik E. G., Greenland L. P., Helz A. W., and Ligon D. T. (1971) Elemental composition of some Apollo 12 lunar rocks and soil. *Proc. Lunar Sci. Conf. 2nd*, p. 1217–1229.

Donaldson C. H., Usselman T. M., Williams R. J., and Lofgren C. E. (1975) Experimental modeling of the cooling history of Apollo 12 olivine basalts. *Proc. Lunar Sci. Conf. 6th*, p. 843–869.

Duncan A. R., Erlank A. J., Sher M. K., Abraham Y. C., Willis J. P., and Ahrens L. H. (1976) Some trace element constraints on lunar basalt genesis. *Proc. Lunar Sci. Conf. 7th*, p. 1659–1671.

Dungan M. A. and Brown R. W. (1977) The petrology of the Apollo 12 ilmenite basalt suite. *Proc. Lunar Sci. Conf. 8th.* This volume.

Engel A. E. J., Engel C. G., Sutton A. L., and Myers A. T. (1971) Composition of five Apollo 11 and Apollo 12 rocks and one Apollo 11 soil and some petrogenic considerations. *Proc. Lunar Sci. Conf. 2nd*, p. 439–448.

Gast P. W., Hubbard N. J., and Wiesmann H. (1970) Chemical composition and petrogenesis of basalts from Tranquility base. *Proc. Apollo 11 Lunar Sci. Conf.*, p. 1143–1163.

Green D. H., Ringwood A. E., Ware N. G., Hibberson W. O., Major A., and Kiss E. (1971a) Experimental petrology and petrogenesis of Apollo 12 basalts. *Proc. Lunar Sci. Conf. 2nd*, p. 601–615.

Green D. H., Ware N. G., Hibberson W. O., and Major A. (1971b) Experimental petrology of Apollo 12 basalts, 1, Sample 12009. *Earth Planet. Sci. Lett.* **13**, 85–96.

Hart S. R., Gunn B. M., and Watkins N. D. (1971) Intralava variation of alkali elements in Icelandic Basalt. *Amer. J. Sci.* **270**, 315–318.

Haskin L. A., Helmke P. A., Allen R. O., Anderson M. R., Korotev R. L., and Zweifel K. A. (1971) Rare-earth elements in Apollo 12 lunar materials. *Proc. Lunar Sci. Conf. 2nd*, p. 1307–1317.

Helmke P. A., Blanchard D., Haskin L., Telander K., Weiss C., and Jacobs J. (1973) Major and trace elements in rocks from Apollo 15. *The Moon* **8**, 129–148.

Howard K. A., Head J. W., and Swann G. A. (1972) Geology of Hadley Rille. *Proc. Lunar Sci. Conf. 3rd*, p. 1–14.

Hubbard N. J. and Gast P. W. (1971) Chemical composition and origin of non-mare lunar basalts. *Proc. Lunar Sci. Conf. 2nd*, p. 999–1020.

Jacobs J. W., Korotev R. L., Blanchard D. P., and Haskin L. A. (1976) A well tested procedure for instrumental neutron activation analysis of silicate rocks and minerals. *J. Radioanal. Chem.* In press.

James O. B. and Wright T. L. (1972) Apollo 11 and 12 mare basalts and gabbros: Classification, compositional variations, and possible petrogenetic relations. *Bull. Geol. Soc. Am.* **83**, 2357–2382.

Kushiro I. and Haramura H. (1971) Major element variation and possible source materials of Apollo 12 crystalline rocks. *Science* **171**, 1235–1237.

Lofgren G., Donaldson C. H., Williams R. J., Mullins O., Jr., and Usselman T. M. (1974) Experimentally reproduced textures and mineral chemistry of Apollo 15 quartz normative basalts. *Proc. Lunar Sci. Conf. 5th*, p. 447–469.

LSPET (1970) Preliminary examination of lunar samples from Apollo 12. *Science* **167**, 1325–1339.

Maxwell J. A. and Wiik H. B. (1971) Chemical composition of Apollo 12 lunar samples 12004, 12033, 12051, 12052, 12065. *Earth Planet. Sci. Lett.* **10**, 285.

McGetchin T. R., Settle M., and Head J. W. (1973) Radial thickness variation in impact crater ejecta: Implications for lunar basin deposits. *Earth Planet. Sci. Lett.* **20**, 226–236.

Norrish K. and Chappell B. W. (1967) X-ray fluorescence spectrography. In *Physical Methods in Determinative Mineralogy* (J. Zussman, ed.), p. 161–214. Academic Press, New York.

Norrish K. and Hutton J. T. (1969) An accurate X-ray spectrographic method for the analysis of a wide range of geological samples. *Geochim. Cosmochim. Acta* **33**, 431–453.

Nyquist L. E., Bansal B. M., Wooden J. L., and Wiesmann H. (1977) Sr-isotopic constraints on the petrogenesis of Apollo 12 mare basalts. *Proc. Lunar Sci. Conf. 8th.* This volume.

Papanastassiou D. A. and Wasserburg G. J. (1971) Lunar chronology and evolution from Rb–Sr studies of Apollo 11 and 12 samples. *Earth Planet. Sci. Lett.* **11**, 37–62.

Papike J. G., Hodges F. N., Bence A. E., Cameron M., and Rhodes J. M. (1976) Mare basalts: Crystal chemistry, mineralogy, and petrology, *Rev. Geophys. Space Phys.* **14**, 475–540.

Pieters C. and McCord T. B. (1976) Characterization of lunar mare basalt types: I. A remote sensing study using reflection spectroscopy of surface soils. *Proc. Lunar Sci. Conf. 7th*, p. 2677–2690.

Rhodes J. M. and Hubbard N. J. (1973) Chemistry, classification and petrogenesis of Apollo 15 mare basalts. *Proc. Lunar Sci. Conf. 4th*, p. 1127–1148.

Rhodes J. M., Hodges F. N., and Papike J. J. (1975) Mare basalts: Major element composition and classification. In *Papers Presented to the Conference on Origins of Mare Basalts and Their Implications for Lunar Evolution*, p. 135–139. The Lunar Science Institute, Houston.

Rhodes J. M., Hubbard N. J., Wiesmann H., Rodgers K. V., Brannon J. C., and Bansal B. M. (1976) Chemistry, classification and petrogenesis of Apollo 17 mare basalts. *Proc. Lunar Sci. Conf. 7th*, p. 1467–1489.

Roeder P. L. and Emslie P. F. (1970) Olivine-liquid equilibrium. *Contrib. Mineral. Petrol.* **29**, 275–289.

Schaber G. G., Boyce J. M., and Moore H. J. (1976) The scarcity of mappable flow lobes on the lunar maria: Unique morphology of the Imbrium flows. *Proc. Lunar Sci. Conf. 7th*, p. 2783–2800.

Schnetzler C. C. and Philpotts J. A. (1971) Alkali, alkaline earth, and rare-earth element concentrations in some Apollo 12 soils, rocks, and separated phases. *Proc. Lunar Sci. Conf. 2nd*, p. 1101–1122.

Scoon G. H. (1971) Chemical analyses of lunar samples 12040 and 12064. *Proc. Lunar Sci. Conf. 2nd*, p. 1259–1260.

Shih C.-Y., Haskin L. A., Wiesmann H., Bansal B. M., and Brannon J. C. (1975) On the origin of high-Ti mare basalts. *Proc. Lunar Sci. Conf. 6th*, p. 1255–1285.

Sutton R. L. and Schaber G. G. (1971) Lunar locations and orientations of rock samples from Apollo missions 11 and 12. *Proc. Lunar Sci. Conf. 2nd*, p. 17–26.

Wanke H., Wlotzka F., Baddenhausen H., Balacescu A., Spettel B., Teschke F., Jagoutz E., Kruse H., Quijano-Rico M., and Rieder R. (1971) Apollo 12 samples: Chemical composition and its relation to sample locations and exposure ages, the two component origin of the various soil samples and studies on lunar metallic particles. *Proc. Lunar Sci. Conf. 2nd*, p. 1187–1208.

Walker D., Longhi J., Kirkpatrick R. J., and Hays J. F. (1976a) Differentiation of an Apollo 12 picrite magma. *Proc. Lunar Sci. Conf. 7th*, p. 1365–1389.

Walker D. R., Kirkpatrick R. J., Longhi J., and Hays J. F (1976b) Crystallization history of lunar picritic basalt sample 12002: Phase-equilibria and cooling-rate studies. *Bull. Geol. Soc. Am.* **87**, 646–656.

Warner J. L. (1971) Lunar crystalline rocks: Petrology and geology. *Proc. Lunar Sci. Conf. 2nd*, p. 469–480.

Weill D. F., Grieve R. A., McCallum I. S., Bottinga Y. (1971) Mineralogy-petrology of lunar samples. Microprobe studies of samples 12021 and 12022; viscosity of melts of selected lunar compositions. *Proc. Lunar Sci. Conf. 2nd*, p. 413–430.

Willis J. P., Ahrens L. H., Danchin R. V., Erlank A. J., Gurney J. J., Hofmeyr P. K., McCarthy T. S., and Orren M. J. (1971) Some interelement relationships between lunar rocks and fines, and stony meteorites. *Proc. Lunar Sci. Conf. 2nd*, p. 1123–1138.

Proc. Lunar Sci. Conf. 8th (1977), p. 1339–1381.
Printed in the United States of America

The petrology of the Apollo 12 ilmenite basalt suite

MICHAEL A. DUNGAN*

National Research Council-Resident Research Associate, NASA Johnson Space Center, Houston, Texas 77058

ROY W. BROWN

Lockheed Electronics Company, Inc., Houston, Texas 77058

Abstract—The Apollo 12 ilmenite basalt suite encompasses the range of variation in cooling history and chemical differentiation observed in low-Ti mare basalts. Numerous consistencies have emerged in the integration of combined mineral chemistry data and petrographic interpretations with fractionation modeling determined independently on the basis of bulk chemistry. The major conclusion we have derived from this petrologic synthesis is that the bulk chemical variations in the ilmenite suite can be reconciled with differentiation within a single cooling unit. The critical observation is that included within the suite there are (1) olivine porphyritic basalts, (2) medium-grained basalts with evolved compositions, and (3) coarse-grained olivine cumulates. Most of the chemical variation within the suite can be accounted for by fractionation of olivine equivalent to the composition of phenocrysts in vitrophyre 12008 which serves well as a parental magma for the other textural-chemical variants.

We have determined the relative cooling rate sequence within all three textural-chemical subgroups and found that mineral chemistry variations dependent on cooling rate (e.g., pyroxene compositions, spinel parageneses, and mineral reequilibration) are entirely consistent with this order. Furthermore, there are systematic correlations between the degree of olivine addition or subtraction and the relative cooling rate. The descending cooling rate sequence among the rapidly chilled porphyritic basalts corresponds to slightly increasing olivine removal. They are, therefore, believed to be representative of the cap of the flow. The medium-grained basalts include members with compositions indicative of variable degrees of olivine fractionation, with progressive removal correlating positively with increasing cooling rate. The cumulates represent the complementary derivatives of the evolved basalts and must underlie them in the stratigraphic sequence.

I. INTRODUCTION

DURING THE INITIAL STUDY of the Apollo 12 basalts, four magmatic groups were recognized on the basis of modal mineralogy and bulk chemistry. Several samples characterized by relatively high TiO_2, FeO, and low SiO_2 plus correspondingly abundant ilmenite were classified as the "ilmenite suite" basalts (ISB). Although the samples that are presently recognized as members of this group differ in part from earlier classifications (e.g., Warner, 1971; James and Wright, 1972) the general scheme remains valid and the terminology has been retained.

The Apollo 12 ilmenite basalts were the least well-characterized cogenetic group of mare basalts in the lunar sample collection prior to the current

*Present address: The Lunar Science Institute, Houston, Texas 77058.

re-examination of the Apollo 12 basalts by an informal JSC consortium (Rhodes *et al.*, 1977; Nyquist *et al.*, 1977; Gibson *et al.*, 1977). Some of the major results of these collaborative studies are that (1) the ilmenite suite basalts have been shown to comprise a significant proportion of the samples collected at the site and are, therefore, probably indigenous to it, (2) the compositional range of the ISBs extends from very magnesian cumulates to evolved quartz-normative basalts, and (3) they have been shown to be distinct from the other two major Apollo 12 basalt types, the olivine and pigeonite groups, on the basis of both major and trace elements (Rhodes *et al.*, 1977) and $^{87}Sr/^{86}Sr$ initial ratio (Nyquist *et al.*, 1977).

We have studied five key samples from the Apollo 12 ilmenite basalt suite (12008,20; 12045,6; 12056,7; 12016,14 and ,18; 12005,13) of varying petrography and bulk chemistry, reviewed the literature for available information on other well-studied samples and examined thin sections of all the samples in this suite. The purpose of the study has been to relate variations in texture, modal mineralogy and mineral chemistry among these samples to variations in post-eruption crystallization history.

II. ANALYTICAL TECHNIQUES

Microprobe analyses of spinels in 12005 and all reported analyses for plagioclase, olivine, and metal were performed on an ARL, EMX-SM electron microprobe using mineral and pure element standards at 15 kV acceleration potential and 20 nA sample current on clinopyroxene. Raw data were processed on a PDP 1120 computer for dead time, instrumental drift, background, and matrix corrections (Bence-Albee method for silicates and spinels). Matrix corrections for metal data were obtained using a version of the Mason and Frost program (MK5).

Zoned pyroxene, spinel, and ilmenite in samples 12045, 12008, 12016, 12005, and 12056 were analyzed on an automated (stage and spectrometers) MAC5 electron microprobe. Operating conditions were 15 kV acceleration potential and 40 nA sample current on benitoite. Mineral standards were used throughout the analysis. Raw data were corrected for dead time, background, and matrix corrected using a Bence–Albee routine. Structural formulae and relevant cation and oxide ratios for each sample point are printed out on line.

Modal analyses were accomplished on the automated MAC5 electron microprobe using a count rate-dependent sieving technique keyed to certain elements. Data are taken on an orthogonal grid with pre-determined point spacings using a sharply focused beam. A minimum of 2000 data points were classified during modal analysis. Initially, spectrometers are set for a "silicate loop" (Fe, Ca, Si). If the count rate permits classification of the data point as either olivine, plagioclase, low-Ca pyroxene, high-Ca pyroxene, or phosphate, the stage then moves to the next data point. If none of these phases can be identified, then depending on the count rate for Si a "K-loop" or "oxide-loop" or others #1 is chosen. K-loop-Si and K are set #2 and #3 spectrometers, respectively, classification is attempted as Si-rich mesostasis, silica phase or K-feldspar. In the oxide loop spectrometers #1 and #2 are set on Fe and Ti, respectively. Classification of the point as metal, troilite, spinel, ulvöspinel, ilmenite, armalcolite, or rutile is attempted. Points which cannot be classified as such are classified as others and are not used in the mode computation.

III. PETROGRAPHY

12008. This sample is the most fine grained and rapidly quenched of the porphyritic ISBs and is very similar in textures and mineralogy to the Apollo 12 olivine suite vitrophyres 12009 and 12015 (Fig. 1a,b). The liquidus phase, olivine,

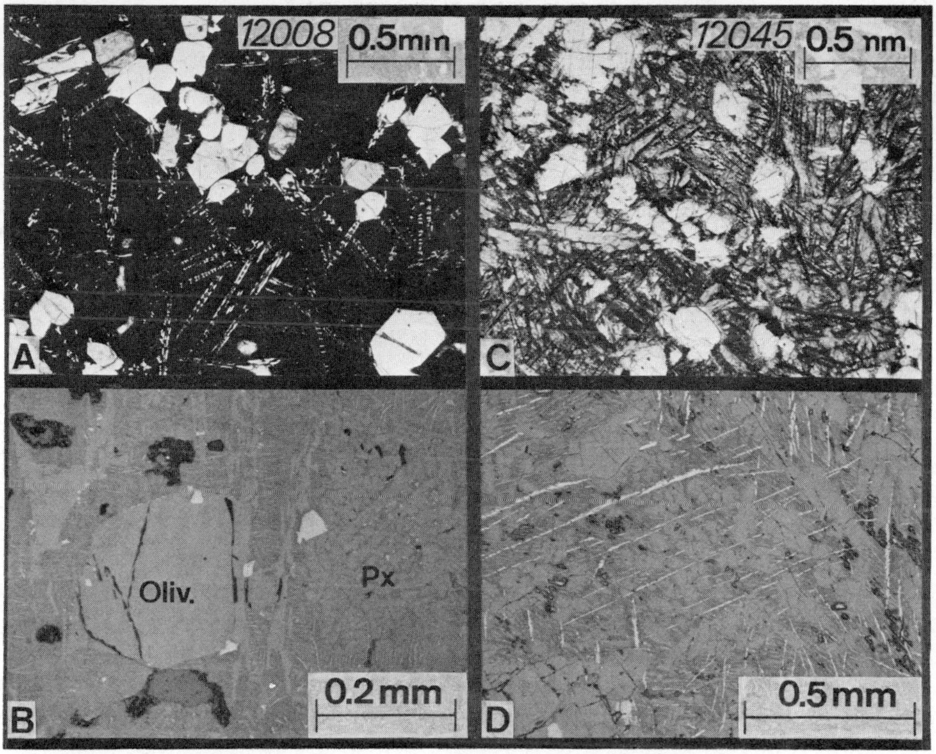

Fig. 1. Photomicrographs of porphritic basalts. (a) Vitrophyre 12008: Glomerophyric aggregates of olivine phenocrysts and skeletal chain-like olivine microphenocrysts set in a fine grained groundmass. (b) Reflected light detail of phenocryst and groundmass relations in 12008. *Left*—olivine phenocryst with small spinel octahedra included in rim. *Right*—feathery augite dendrite (px). Note very fine-grained filamental ilmenite in matrix. (c) 12045: Olivine phenocrysts (white) are rimmed by aluminous titanaugite (gray). Cigar-shaped phenocrysts of compositionally equivalent augite are set in a variolitic groundmass of pyroxene, plagioclase, a silica phase, and ilmenite. (d) Reflected light detail of augite phenocrysts and variolitic groundmass in 12045. Note parallelism of skeletal ilmenite plates.

is represented by ≈15–20% equant subhedral to euhedral phenocrysts (0.2–0.5 mm) heterogeneously distributed in the groundmass as glomerophyric aggregates. The most magnesian phenocryst core composition measured is Fo_{72} but most are Fo_{71-68} with narrow rims of Fo_{67-56}. Titanian chromian spinel is the second phase to crystallize. It occurs as small idiomorphic octahedra (0.01–0.05 mm) and as glomerophyric aggregates in the groundmass or as inclusions in the rims of olivine phenocrysts. Spherical metal droplets are typically associated with the spinel as inclusions or attached to grain margins. More rarely they are found as inclusions in olivine. Spherical melt inclusions are abundant in the cores of olivine phenocrysts in 12008 and in all the other samples studied.

A second generation of olivine is present as skeletal microphenocrysts in the groundmass and as ornamental overgrowths on the early equant grains. The dominant morphology of the second generation is chain-like (Donaldson, 1976).

The composition of these grains is the same as the phenocryst rims. Individual "chains" are up to 0.5 mm long and are typically 0.02–0.03 mm in width. The skeletal olvine precedes pyroxene, plagioclase, and ilmenite in the crystallization sequence. Olivine crystallization terminated at the appearance of these groundmass phases. Groundmass metal and troilite are also absent.

The first groundmass phase is aluminous titanaugite in the form of feathery dendritic crystals up to 0.5 mm across. Only a few of the dendrites in the section examined are coarse enough for microprobe analysis. Most merge with the extremely fine-grained to cryptocrystalline matrix consisting of ilmenite (less than 1 μm) and a silicate phase of lower reflectivity than pyroxene (plagioclase?). Ilmenite occurs as groups of sub-parallel wispy filaments and as tiny, scattered, equant cruciform grains.

12045. As in 12008, the early phenocryst phases are glomerophyric aggregates of equant olivine phenocrysts, spinel grains, and rare metal droplets. The chemistry of these phases is essentially identical to those in 12008 with the exception of the spinel which is zoned to ulvöspinel rims in 12045. This zoning only occurs on spinel grains not enclosed in olivine phenocrysts. Ulvöspinel microphenocrysts (some skeletal) are also present in the groundmass. The second generation of olivine in 12045 is less skeletal than in 12008. It generally forms ornamental overgrowths on phenocrysts in the form of acicular appendages.

Unlike 12008 the early pyroxene occurs as elongate non-skeletal phenocrysts and rims on olivine phenocrysts. They are zoned, brown titanaugite similar in composition to those in 12008. The phenocrysts are cigar-shaped grains which typically cluster in subparallel aggregates. The augite, spinel, and olivine phenocrysts are set in a variolitic groundmass of acicular pyroxene, plagioclase (An_{93-84}), a silica phase, ilmenite, troilite, and metal. The ilmenite is highly skeletal and occurs as parallel sets of thin platelets (0.2–0.7 mm $\times$ 10 μm) spaced at roughly 50 μm intervals. The parallelism of the individual platelets apparently reflects the presence of large framework grains that nucleated epitaxially on early pyroxene. Ilmenite appears to be the first groundmass phase to crystallize (Fig. 1c,d).

12022. Porphyritic basalts 12045 and 12022 are fundamentally similar in textures, mineralogy, mineral chemistry, and bulk composition (Weill *et al.*, 1971; see their Fig. 2). A less rapid cooling rate for 12022 is indicated by a coarsening of all the groundmass phases. Pyroxene phenocrysts in particular are larger and less elongate. Their compositions also vary in a way consistent with growth at lesser undercooling than in 12045 (see below).

12056. Sample 12056 is a medium-grained microgabbro with a seriate subophitic texture (Fig. 2d). The crystallization sequence is olivine, metal, and chromian spinel, followed by pyroxene in a peritectic reaction-relationship with olivine, ulvöspinel, plagioclase, ilmenite, and mesostasis consisting of a silica phase, Si + K-rich glass, fayalite, phosphate, metal, and troilite.

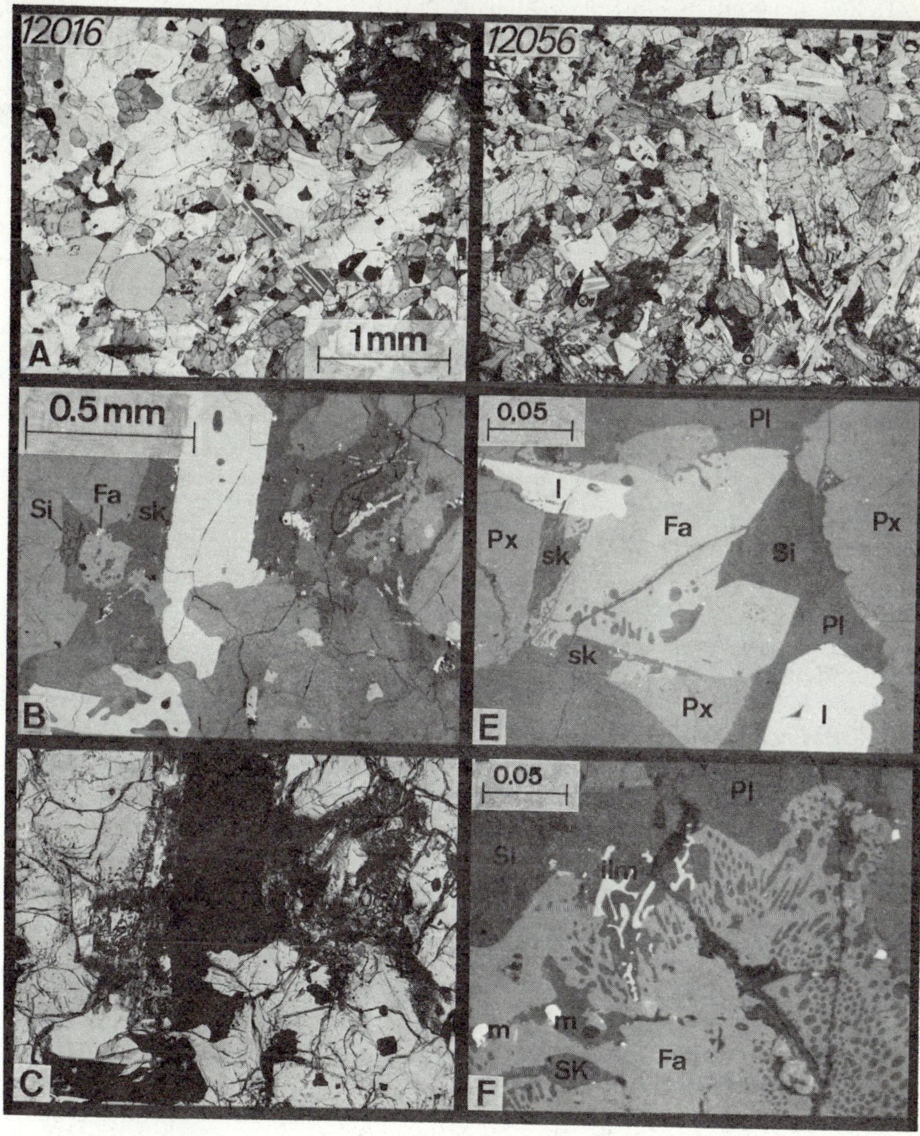

Fig. 2. Photomicrographs of medium-grained basalts. (a) 12016 (b,c) 12016—Reflected and transmitted light details of an anomalously large mesostasis segregation in upper right of (a). The large, light gray grains are ilmenite. Other well crystallized phases are irregular grains of fayalite (Fa) and silica (Si). Matrix consists of devitrified(?) Si + K-rich glass, phosphate, and feldspar. (d) 12056—Same scale as (a). (e,f) Reflected light details of mesostasis segregations in 12056.

Olivine is less abundant (Table 2) than in the porphyritic basalts (Dence *et al.*, 1971) and generally occurs as rounded, partly resorbed inclusions in larger pyroxenes. Small olivine grains included in plagioclase are subhedral rather than rounded and embayed. The rock is dominated modally by pyroxene. The cores of large grains are predominantly augite with minor intergrown low-Ca pyroxene. These larger grains tend to be elongate (1.0–1.5 × 0.3–0.6 mm). Textural evidence coupled with microprobe data indicate that the rims of these large grains and the finer-grained pyroxene in 12056 co-crystallized with plagioclase, ilmenite, and ulvöspinel. It is this phase of crystallization that gives rise to the subophitic texture and graphic intergrowths of plagioclase and pyroxene. Plagioclase morphology varies from equant to highly elongate. Early, equant plagioclase (0.3–0.5 mm) contains inclusions of pyroxene indicative of formation as cored grains, suggesting nucleation at substantial undercooling. The larger plagioclase laths, which are also in part cored, are up to 0.7 mm in length. Most of the lath-shaped grains range in length from 0.4 to 0.2 mm and are 0.05–0.2 mm wide. Ilmenite morphology is generally elongate. Ilmenite also tends to be skeletal and contain abundant silicate inclusions. Subsolidus reduction assemblages replacing ilmenite and ulvöspinel are rare.

12056 contains several percent of segregations of late-stage mesostasis which consist primarily of crystalline phases intimately intergrown in a symplectoid texture with minor Si + K-rich glass. The dominant phase is fayalitic olivine in which spheroidal blebs of the alkali-rich glass are included. The textural relationships between fayalite and the Si + K-rich glass are interpreted as evidence of immiscibility in the residual liquid. The phosphate is generally dispersed throughout as small elongate to acicular grains ($\approx 10\ \mu m$) whereas the silica phase occurs as a few anhedral interstitial grains intergrown with fayalite and the surrounding plagioclase and ferroaugite (Fig. 2e,f). Ilmenite, metal, and troilite are also invariably present in the mesostasis. Crystallization of the mesostasis into discrete phases with only the Si + K glass remaining is characteristic of most of the other evolved ISBs (e.g., 12051—Keil *et al.*, 1971; 12063—El Goresy *et al.*, 1971).

12051, 12054, 12047, and 12063. These four basalts are similar in petrography and modal mineralogy to each other and to 12056. They differ from 12056 in that they contain substantially less modal olivine. Olivine is absent in 12051 (Keil *et al.*, 1971; Busche *et al.*, 1972), 12054 (Schaal and Hörz, 1977) and 12047 with the exception of fayalite in the mesostasis. These three samples are chemically more evolved than 12056 and 12063 and contain substantial normative and modal silica (Rhodes *et al.*, 1977). The textures are medium-grained, equigranular-ophitic: lath-shaped plagioclase and equant to slightly elongate pyroxenes form an intergrowth suggestive of cotectic crystallization. The morphology of plagioclase tends to be highly elongate laths and differs significantly from 12056 and 12016 which contain more equant plagioclase. Elongate ilmenite crystals are a significant phase (up to 9%). Small segregations of symplectoid mesostasis consisting of fayalite, silica, Si + K-rich glass, phosphate, and alkali feldspar are dispersed throughout. On the basis of plagioclase morphology and dimensions 12051, 12047, and 12054 are judged to have had more rapid cooling histories than 12063 and 12056.

12016. This is an equigranular, medium-grained microgabbro (Fig. 2a). In its mineralogy, textures, grain size, and sequence of crystallization it is essentially similar to 12056. However, the non-evolved bulk chemistry of 12016 is matched by modal differences (less ilmenite, more olivine) and textural relationships differ in such a way as to suggest a slower cooling history.

Olivine in 12016 occurs as resorbed grains within pyroxenes and as subhedral to euhedral grains enclosed in plagioclase. One olivine grain is substantially larger (1.2 mm) than others in the rock or in the porphyritic basalts. Pyroxenes tend to be elongate but more equant and less seriate than in 12056 (0.5–1.0 mm). The pyroxenes are complex intergrowths of augite and low-Ca pyroxene although augite dominates the cores. Unlike the chemically evolved ISBs, equant plagioclase dominates over lath-shaped grains. Ophitic texture is poorly developed. Rather, plagioclase occurs interstitially to segregations of pyroxene grains and partly resorbed olivine. Where plagioclase occurs as large branching grains, a subpoikilitic texture is developed. Ilmenite morphology is variable, including both highly elongate and equant grains (0.5 mm). However, compared to 12056, those in 12016 are more equant, less skeletal and tend also to be subpoikilitic.

The mesostasis textures in 12016 differ substantially from those in 12056. Figure 2c,d illustrates an anomalously large mesostasis segregation containing the same phases present in 12056, 12051, etc. The symplectoid texture is notably absent. Instead there are grains of fayalite, silica, and phosphate in a cryptocrystalline matrix which appears to consist of devitrified glasses of contrasting composition similar to those expected from immiscibility. Textural relations indicative of immiscibility are absent.

12005. This rock is heterogeneous, consisting of two distinct textural-mineralogical domains (Fig. 3). Large pyroxene oikocrysts (2–6 mm) enclose an early crystallizing assemblage of rounded and embayed olivine (Fig. 4c,d), single grains, and glomerophyric aggregates of spinel and minor associated metal. Pyroxene oikocrysts are bounded by narrow bands of an interstitial assemblage dominated by euhedral to subhedral olivine grains poikilitically enclosed in plagioclase. Scattered throughout the interstitial areas are irregular branching ilmenites that poikilitically enclose olivine and pyroxene. Additional phases are augite and low-Ca pyroxene, ulvöspinel, smaller euhedral grains of ilmenite, and minor holocrystalline mesostasis segregations consisting of plagioclase, K-feldspar, phosphate, ilmenite, troilite, and metal.

The duality of olivine textures present in 12056 and 12016 is rampant in 12005. The contrast between the textures and modal amount of olivine within pyroxene oikocrysts and those in the interstitial area is striking (Figs. 3, 4a,c). Extensive resorption of olivine in the early stages of crystallization is indicated. Several of the isolated anhedral olivine grains within the oikocrysts are in optical

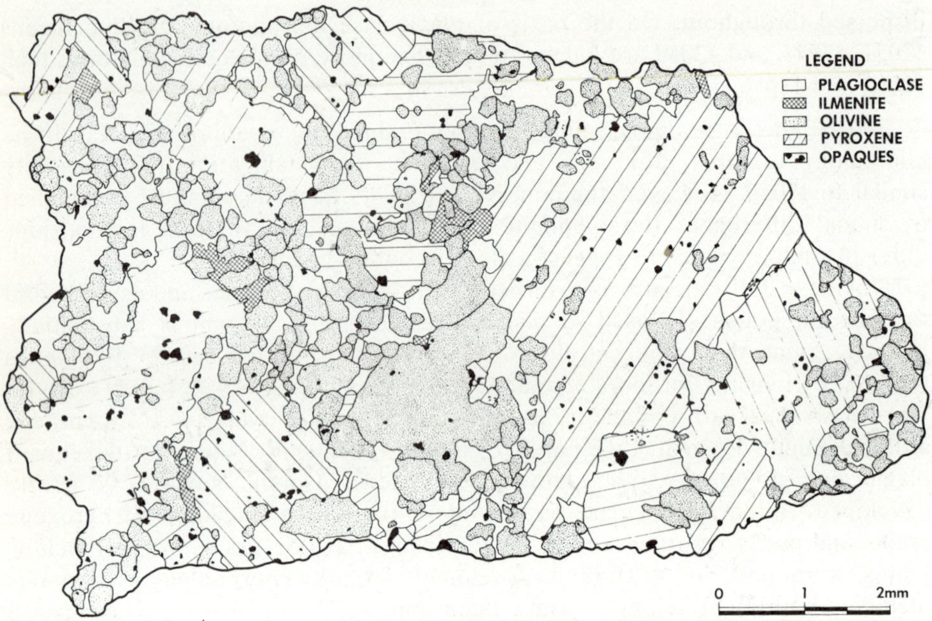

Fig. 3. *Camera Lucida* tracing of 12005,13. Individual large pyroxene grains are indicated by parallelism of hachures. "Opaques" includes spinel, metal, troilite, and small ilmenite grains.

continuity with neighboring grains suggesting 80% resorption of >1 mm olivines. There is a single very large olivine interstitial to several pyroxene oikocrysts which is embayed and irregular in outline but is still 3 mm in diameter (Figs. 3, 4d). It is in optical continuity with another isolated olivine implying an original diameter of >5 mm. The embayed grains within pyroxene are more magnesian (Fo_{66-60}) than the euhedral to subhedral grains intergrown with plagioclase (Fo_{61-57}).

Pyroxene oikocrysts are composite, with cores consisting primarily of augite and rims dominated by low-Ca pyroxene. However, intergrowths of low-Ca pyroxene are present in the cores of some oikocrysts, particularly as partial rims on rounded and embayed olivine. This relationship suggests that low-Ca pyroxene was formed initially in the peritectic reaction-relationship responsible for olivine resorption and the growth of the large oikocrysts. Subsequently augite became the dominant phase. The core-rim transition to low-Ca pyroxene probably reflects the incoming of plagioclase and ilmenite as pyroxenes intimately intergrown with plagioclase are compositionallly similar to the rims. Rim augite occurs only in the outermost margins of the oikocrysts. Both cores and rims contain microscopic exsolution lamellae that were not resolvable with the microprobe. A mineral orientation fabric is imparted to 12005 by the parallelism of the long dimensions of the pyroxene oikocrysts.

Lath-shaped plagioclase is very rare, the dominant morphologies being

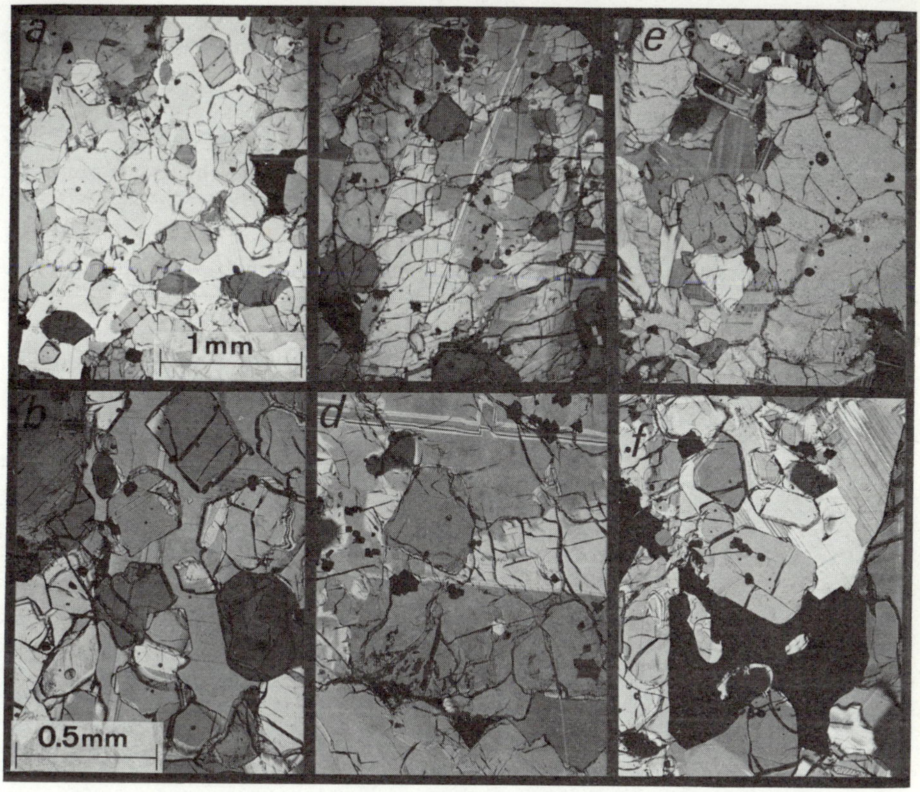

Fig. 4. Photomicrographs of 12005. All with partially crossed nicols: (a–c–d) same scale, (b–d–f) same scale (a). Abundant euhedral to subhedral olivines poikilitically enclosed in plagioclase. (b) Detail of same area. (c) Twinned pyroxene oikocryst. Numerous grains of anhedral olivine and euhedral spinel are included. (d) Detail of *upper left* portion of (c). Twin planes run horizontally. Augite core is at the *top*, rim and grain boundary with mesostasis segregations (triangular shape) are at the base. (e) *Right*—an anomalously large olivine grain containing abundant melt inclusions. *Upper left*—elongate olivine grain in optical continuity with large grain. (f) Ilmenite oikocryst enclosing olivine and pyroxene.

equant (0.5 mm) and large poikilitic grains (Fig. 4a,b). Plagioclase is intergrown with pyroxene oikocrysts only at their outermost margins. Similarly, ilmenite is present within the pyroxene oikocrysts only as a phase in crystallized melt inclusions and in the rims. Within the interstitial assemblage, ilmenite is present in two distinct habits. Prominent, irregular, branching ilmenites (up to 1 mm) poikilitically enclose olivine and pyroxene (Fig. 4f). Smaller grains lacking any included phases are generally idiomorphic and are equant to slightly elongate.

The modal amount, mineralogy, and textures of mesostasis segregations in 12005 contrast with those in 12056 and 12016. Mesostasis phases constitute less than 0.5% of the mode in 12005. Both a silica phase and strongly Fe-enriched ferromagnesian minerals are lacking. Instead the holocrystalline intergrowths consist of relatively sodic plagioclase (An_{73}), K-feldspar and phosphate plus ilmenite, metal, and troilite (Fig. 5). Thus, the only phases restricted to the mesostasis (K-feldspar and phosphate) are those minerals whose composition includes as an essential component an element that cannot be incorporated into the major constituent phases. Where present, mesostasis segregations are typically trapped between pyroxene oikocrysts or in embayments in the large poikilitic ilmenite grains.

Subsolidus reduction in ilmenite and ulvöspinel is widespread in 12005. Small veinlets of metal, typically associated with troilite, cut mafic silicates, and oxide phases.

12036. Sample 12036 exhibits the same type of textural-mineralogical heterogeneity as 12005. Pyroxene oikocrysts (5 mm) tend to be more equant than

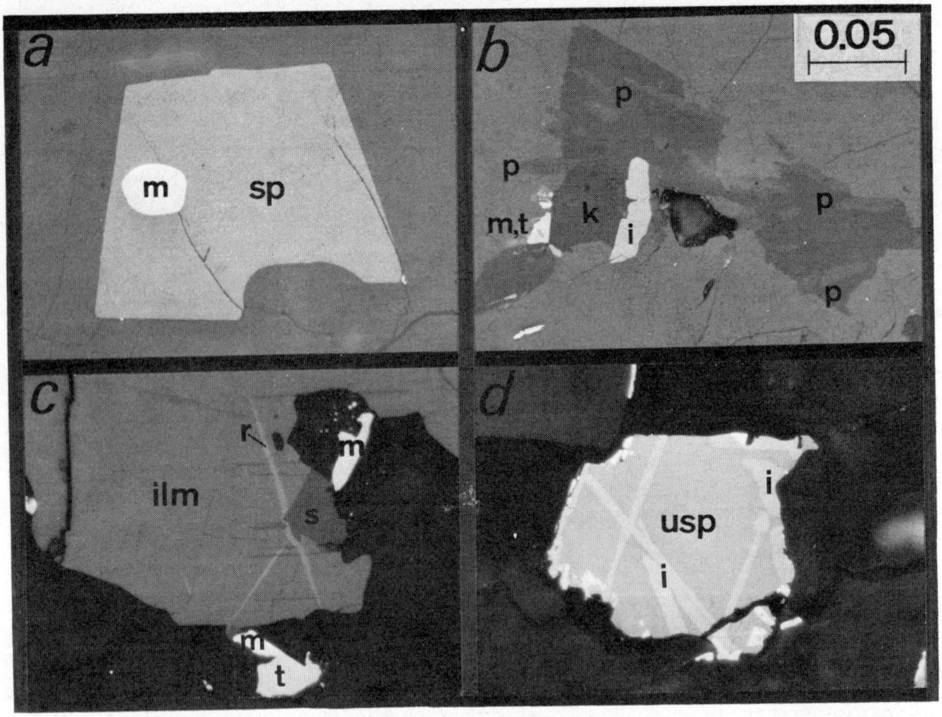

Fig. 5. Reflected light photomicrographs of oxide and mesostasis assemblages in 12005. All at same scale. (a) Metal droplet (m) inclusion in chromian spinel. (b) Holocrystalline segregation of mesostasis surrounded by pyroxene (see 4d). K-feldspar (k), ilmenite (i), metal and troilite (m,t), and acicular phosphate (p) included in plagioclase. (c) Spinel (s) and rutile (r) formed by subsolidus reduction of ilmenite (ilm). Reduction is best developed adjacent to small segregations of mesostasis which include metal and troilite. (d) Ilmenite (i) reduction lamellae in ulvöspinel (usp) with fine-grained intergrown metal and troilite.

in 12005 and contain less olivine as chadacrysts. Textural relations in the interstitial assemblage are similar to those in 12005. Keil *et al.* (1971) and Busche *et al.* (1971) provide additional petrographic details and mineral chemistry data.

IV. Fractionation—Crystallization Model for the Ilmenite Suite Basalts

Integration of petrographic, modal (Tables 1 and 2) and phase chemical data in this study with major, minor, and trace element analyses of Rhodes *et al.* (1977) and Nyquist *et al.* (1977) provides a basis for modeling of petrologic interrelationships among the demonstrably cogenetic ISBs. The 11 ISB samples have been ranked (from top to bottom in Table 1) in order of decreasing cooling rate on the basis of textural criteria; particularly plagioclase dimensions (cf., Walker *et al.*, 1976b). This study is primarily concerned with relating the petrography and phase chemistry of the individual samples to the dependent variables cooling history, fractionation history, and bulk chemistry. The basic petrologic framework for the discussion of mineral chemistry data which follows is summarized below.

(1) Papike *et al.* (1976) and others have noted the paucity of mare basalts with Fe/(Fe + Mg) = 0.49–0.50 and suggested that this is a fundamental characteristic of unfractionated partial melts; the bimodal distribution of compositions at higher and lower Fe/(Fe + Mg) reflecting complementary derivatives produced by near-surface crystal fractionation of the rarely sampled parental liquids. Olivine-porphyritic basalts with Fe/(Fe + Mg) = 0.51–0.49 are relatively abundant at the Apollo 12 site and serve well as parental magmas for fractionation modeling of the evolved basalts and partial olivine cumulates.

Table 1. Cooling rate sequence of Apollo 12 ilmenite suite basalts.

	Mg/(Mg + Fe)*
Porphyritic basalts (most rapid cooling)	
12008	0.500
12045	0.484
12022	—
Medium-grained subophitic basalts	
12047	0.364
12051	—
12054	0.382
12063	—
12056	0.434
Medium-grained equigranular basalt	
12016	0.499
Coarse-grained cumulates (slowest cooling)	
12036	0.577
12005	0.615

*Mg/(Mg + Fe) data taken from Rhodes *et al.* (1977).

Table 2. Modal analyses of medium- and coarse-grained ilmenite suite basalts.

	12047	12051	12063	12056	12016	12036	12005
Olivine	—	—	6.0	10.8	12.0	24	30.0
Pyroxene undif.	—	58.8	60.0	—	—	58	—
Low-Ca pyroxene	15.0	—	—	20.5	24.1	—	35.5
High-Ca pyroxene	33.4	—	—	28.7	28.0	—	21.0
Feldspar	38.0	26.5	25.0	28.8	29.1	12	11.0
Ilmenite	5.3	—	—	6.8	4.8	—	1.9
Spinel	1.0	—	2.0	1.6	1.6	—	0.5
Opaques undif.	—	9.3	6.0	—	—	5	—
Silica phase	5.3	2.8	1.6	0.8	Tr	—	—
Fayalite	—	—	—	1.5	Tr	—	—
Mesostasis undif.	2.0	2.6	5.0	1.0	0.6	—	0.1
Norm. oliv.	—	—	8.8	6.9	18.2	28.3	42.0
Norm. quartz	2.2	1.1	—	—	—	—	—

12047, 12056, 12016, 12005—This study.
12063—Average of two values—McGee *et al.* (1977); Papike *et al.* (1976).
12051—Average of two values—Brown *et al.* (1971); Papike *et al.* (1976).
12036—Prinz (1977) pers. comm.
Normative data calculated from analysis by Rhodes *et al.* (1977); Willis *et al.* (1971), and Compston *et al.* (1971).
Opaques undif. = spinel + troilite + metal + ilmenite.
Mesostasis undif. = K-feldspar + silica and K-rich glass ± fayalite ± silica phase + phosphate.

The best available candidates for basalts of a composition that could be parental to derivatives in the ilmenite and olivine basalt suites are the vitrophyres 12008 and 12009–12015, respectively, as the extremely rapidly chilled groundmasses of these rocks suggest minimal redistribution of phenocrysts subsequent to eruption. They are distinguished chemically by higher TiO_2 in 12008 and different trace element signatures, but in terms of major element characteristics (e.g., Fe/(Mg + Fe): 12008 = 0.50, 12015 = 0.49) they are fundamentally alike. Similar crystallization histories are also indicated by the fact that all three vitrophyres contain ≈ 15% equant olivine phenocrysts in a quenched groundmass that also includes some skeletal olivine. Experimental calibration of cooling rate dependent olivine morphologies in a synthetic 12009 composition led Donaldson *et al.* (1975) to conclude that the early equant olivine phenocrysts grew during relatively slow, pre-eruption cooling at rates of 1–9°C/hr in contrast to the much more rapidly chilled groundmass (40–100°C/hr). The compositional and textural similarities of 12008 and 12009 indicate to us that their crystallization histories must have been very similar and the major results (if not absolute cooling rate values) of the cooling rate experiments on 12009 can be extrapolated to 12008. The recognition of a two-stage cooling history is a very important step in interpreting the details of olivine fractionation in the ISBs. The major implication is that the magma parental to the fractionated ISBs was erupted at 80–100°C below its liquidus containing ≈15% olivine phenocrysts which grew at comparatively slow cooling rates over a wide temperature range.

The availability of olivine phenocrysts for fractionation is important in light of calculations by Rhodes *et al.* (1977) which demonstrate that chemical variations within the two groups can be largely accounted for by addition or subtraction of olivine equivalent to the average composition of phenocrysts in the vitrophyres. This implies that in coarse-grained rocks, olivine fractionation took place very early in the post-eruption cooling history.

Only very rapidly cooled portions of the eruptive unit, represented by the vitrophyres, did not fractionate abundant olivine. Groundmass textures in all three basalts are indicative of formation at substantial undercooling with the decreasing cooling rate sequence being 12008 → 12045 → 12022. The somewhat more slowly cooled porphyritic basalts 12022 and 12045 are slightly more evolved in composition (7 and 2% olivine removal, respectively) than 12008. Thus, the settling velocity of olivine was great enough that phenocrysts were lost from the cooling regime that gave rise to a variolitic groundmass texture. Olivine accumulation in rapidly cooled olivine basalts 12075, 12076, and 12004 (Walker *et al.*, 1976b; Rhodes *et al.*, 1977) is likewise indicative of the mobility of olivine phenocrysts in these low-viscosity magmas (Weill *et al.*, 1971; Table 2). As Walker *et al.* (1976b) have shown, the modal and normative olivine contents in these rocks agree quite closely (particularly for rapidly cooled samples). This provides an additional petrographic check on the fractionation model (see Table 2).

(2) The medium-grained, equigranular basalts are presumed to have formed under the chilled carapace of porphyritic basalts. They are characterized by non-porphyritic textures indicative of cooling rate histories intermediate between the porphyritic basalts and cumulates. There is an inverse correlation between cooling rate and percentage of modal olivine within this group. Samples 12051, 12054, and 12047, which are the most chemically evolved of the suite, lack normative and modal olivine (except for fayalite in the mesostasis) and exhibit textural relations suggestive of cotectic plagioclase and pyroxene crystallization. These textural-chemical relations indicate that olivine phenocrysts were removed completely from this section of the flow before pyroxene crystallization commenced. Modal and normative data agree quite well for each rock and with the proposition that the bulk compositions of 12051, 12054, and 12047 represent removal of about 15% olivine phenocrysts from the parental basalt. The less evolved members of this subgroup, 12063 and 12056, contain modal and normative olivine and are characterized by textural relations indicative of less rapid cooling histories. Stratigraphically these are placed below 12054, 12047, and 12051, near the lower level of the zone from which there was a net loss of olivine phenocrysts.

The medium-grained basalt 12016 has a major element bulk composition quite similar to the porphyritic basalts. Based on textural criteria, its position in the cooling rate sequence must be somewhere between the evolved sample 12056 and the cumulates 12036 and 12005. Thus, the stratigraphic position of 12016 is presumed to be at the level in the flow where the gain of olivine from above compensated for the loss by settling.

(3) Samples 12036 (Keil *et al.*, 1971; Busche *et al.*, 1972) and 12005 are the

complementary cumulus derivatives of the evolved medium-grained basalts. Consequently their stratigraphic positions must have been below the evolved samples. The normative olivine content of 12005 (42%) is about 2.5 times the amount in the parental basalt, in agreement with modeling calculations that suggest approximately 30% olivine accumulation relative to 12008. Rhodes *et al.* (1977) have shown, however, that the differentiation is more complex than simple olivine fractionation. Calculation of the fractionation on the basis of major elements does not agree in detail with models derived from application of partition coefficient data and magmaphile element concentrations (Ti, Ba, K, P, and REE). The cumulates are depleted in these elements relative to the values expected from simple olivine accumulation. Selective loss of the magmaphile elements by isothermal diffusion into the overlying liquid is supported by (1) the paucity of mesostasis phases enriched in the magmaphile elements, (2) the notable lack of Fe enrichment in the ferromagnesian minerals, and (3) the large pyroxene oikocrysts, formed by adcumulus growth, that modally dominate the two samples.

V. MINERAL CHEMISTRY

Olivine

Olivine is clearly the liquidus phase in the porphyritic ISBs. This was shown experimentally by Green *et al.* (1971) in 1 atm phase equilibria studies of 12022. Olivine crystallizes alone to 100°C below the liquidus where it is joined by spinel at 1200°C and then pyroxene, plagioclase, and ilmenite in the interval 1150–1125°C. The equilibrium liquidus olivine composition in these experiments is Fo_{77}. Weill *et al.* (1971) report olivine compositions in the range Fo_{76-58} in 12022. Butler (1972) found phenocryst cores in 12022 of Fo_{70-74} and Fo_{73} in 12008. The maximum Fo-contents of olivines in 12045 and 12008 analyzed in this study are Fo_{74} and Fo_{72}, respectively. Most of the olivine core compositions measured in this study and by Butler fall in the range $Fo_{71.5-69.5}$. Thus, the most magnesian core composition analyzed among the three rocks occurs in the most evolved sample, 12022. The sample with the lowest maximum Fo-content is 12008, and 12045 is intermediate in terms of its bulk composition and the maximum Fo-content of its phenocrystic olivine. The cores of the olivine phenocrysts are broad and weakly zoned. Sharply zoned rims, in which spinel octahedra are included, range in composition (in all three porphyritic basalts) from Fo_{67-65} to Fo_{60-54}. We concur with Butler (1972) that reequilibration between early formed phenocryst cores and residual liquid has not substantially modified them and is, therefore, not responsible for olivine less magnesian than the experimental liquidus composition. These more Fe-rich olivines are probably the result of crystallization over the 100°C temperature interval prior to the incoming of spinel (Green *et al.*, 1971).

The rims of olivine phenocrysts and skeletal microphenocrysts are essentially equivalent in composition to each other within and among the three porphyritic basalts (Fo_{67-54}). The termination of olivine crystallization occurs approximately at the incoming of the first pyroxene, as demonstrated by epitaxial overgrowths of augite on olivine phenocrysts in 12045 and 12022. Weill *et al.* (1971) inferred that this textural relationship indicates a peritectic reaction-relationship. An alternative mechanism suggested by a phenomenon observed in experimentally produced feldspar intergrowths (Lofgren and Gooley, 1977), that on rapid post-eruption cooling a substantial boundary diffusion layer formed around the olivine phenocrysts resulting in enhanced pyroxene nucleation.

The rounded and embayed morphology of olivine included in large pyroxene grains in 12056, 12016, and 12005 is evidence for a peritectic reaction-relationship. However, there is a duality of olivine textures in these rocks in that olivine surrounded by plagioclase tends to be euhedral to subhedral and exhibit no evidence of reaction. Moreover, the range of olivine compositions in 12056 and 12016 extends from Fo_{73} to Fo_5 and Fo_{67} to Fo_7, respectively. Among the three samples 12056, 12016, and 12005 there is an inverse relationship between Mg/(Mg + Fe) in the bulk rock and the most magnesian olivine analyzed (Fig. 6). However, there is a positive correlation between maximum Fo-content and the cooling rate sequence (Table 1) determined on the basis of textural criteria. This trend apparently reflects the cooling rate dependence of reequilibration (Moore and Evans, 1967). Similar compositional and textural patterns are present in the Apollo 12 olivine basalts (Walker *et al.*, 1976). The olivines in 12016 and 12005 have much lower Cr_2O_3 at a given Fo-content than those in the porphyritic basalts or even 12056 (Fig. 6). The lower Cr in olivine of the more slowly cooled rocks is also ascribed to reequilibration. The cores of the largest grains and those completely enclosed in pyroxene are the most magnesian in a given rock. Zoning

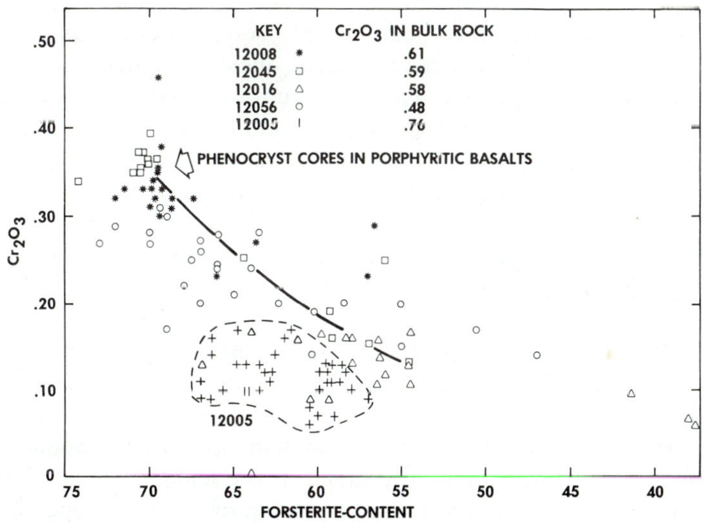

Fig. 6. Microprobe analyses of olivines. Fo-content vs. Cr_2O_3 for olivine more magnesian than Fo_{37}.

in these mantled grains is limited to ≈5% Fo whereas olivines which remained in contact with residual liquid exhibit much larger compositional differences between core and rim as well as asymmetric zoning in those grains partially enclosed by olivine.

On the basis of contrasting Cr and Ti concentrations in olivine from 12009–12004 and 12008–12022, Butler (1972) inferred two distinct Apollo 12 magma groups (for essentially texturally identical samples) that have now been proven valid (Rhodes *et al.*, 1977). Akella *et al.* (1976) and Huebner *et al.* (1976) measured Cr partitioning between olivine and liquid in synthetic systems under conditions compatible with lunar magmatism. Calculations of D_{Cr} for 12008 and 12009 with published data plus those accrued in this study are in agreement with these experimental studies. Akella *et al.* (1976) found that olivine and liquid did not equilibrate with respect to Cr in cooling rate runs (70 and 3.5°C/hr). This is consistent with the retention of high Cr-concentrations in the cores of olivine phenocrysts in 12008–12045–12022. It is emphasized that significant reequilibration has occurred in 12016 and 12005. There are no significant Ti variations in olivines among the five analyzed rocks. However, in the coarse-grained samples there is substantial reequilibration between adjacent ilmenite and olivine resulting in higher TiO_2 (up to 0.50 vs. ≈0.10) and lower Mg/(Mg + Fe) in the olivine. One example of this from 12005 is illustrated in Table 3, analyses 7 and 8. The details of olivine-spinel equilibration are discussed below in the spinel section.

Discussion of olivine chemistry

Rhodes *et al.* (1977) have modeled near-surface crystal fractionation in the ISBs using 12008 as the parental composition and an olivine of Fo_{72} as the fractionating phase that accounts for most of the compositional variation within the suite. This hypothesis has been tested from a petrographic and phase chemical standpoint in this paper. A related question that is more difficult to evaluate is: Does 12008 represent a primitive, unfractionated magma? One of the criteria which serves to identify basalts that could have crystallized *senso stricto* from a liquid is the composition of the most magnesian olivine present in the rock. Using K_D values of 0.33 and 0.30 for Fe–Mg partitioning between liquid and olivine (Roeder and Emslie, 1970), liquidus olivines of Fo_{75} and Fo_{77} are calculated for 12008. The absence of olivine phenocrysts as magnesian as these in 12008 suggests that its present composition is the result of a small net gain of olivine at some point. However, the decreasing cooling rate sequence of 12008 → 12045 → 12022 is inferred to represent a stratigraphically descending section at the top of a flow, in that decreasing cooling rate correlates positively with degree of olivine subtraction (Rhodes *et al.*, 1977; cf., Walker *et al.*, 1976b). The inverse correlation between the maximum Fo-content of olivine phenocrysts and Mg/(Mg + Fe) of the lava apparently arises from selective loss of early magnesian liquidus olivines from the topmost portion of the flow. One consequence to the compositional variation in phenocrystic olivine produced by monomineralic crystallization with falling temperature is that there will also be a grain size distribution in which phenocryst dimensions and Fo-content correlate

Table 3. Microprobe analyses of olivine.

	(1)	(2)	(3)	(4)	(5)	(6)	(7)	(8)	(9)
SiO_2	37.25	35.50	36.62	37.37	28.89	36.75	35.96	35.52	35.78
TiO_2	0.05	0.18	0.05	0.06	0.13	0.04	0.49	0.10	0.09
Al_2O_3	0.00	0.00	0.00	0.00	0.00	0.00	0.03	0.03	0.02
Cr_2O_3	0.33	0.23	0.13	0.29	0.01	0.16	0.07	0.09	0.11
FeO	25.27	35.90	29.02	24.93	64.00	29.64	34.67	36.66	34.75
MnO	0.25	0.30	0.34	0.26	0.63	0.28	0.33	0.36	0.34
MgO	35.76	26.75	32.92	35.94	3.13	33.13	29.11	27.59	28.58
CaO	0.29	0.43	0.25	0.30	0.61	0.28	0.18	0.28	0.26
Total	99.20	99.29	99.33	99.15	98.40	100.27	100.84	100.62	99.92
Si	0.944	0.999	0.994	0.996	1.002	0.990	0.987	0.989	0.993
Ti	0.001	0.004	0.001	0.001	0.003	0.001	0.010	0.002	0.002
Al	0.000	0.000	0.000	0.000	0.000	0.000	0.001	0.001	0.000
Cr	0.007	0.005	0.003	0.006	0.000	0.003	0.002	0.002	0.003
Fe	0.564	0.845	0.659	0.555	1.794	0.668	0.796	0.853	0.807
Mn	0.006	0.007	0.008	0.006	0.018	0.006	0.008	0.009	0.008
Mg	1.422	1.122	1.332	1.427	0.156	1.330	1.192	1.144	1.183
Ca	0.008	0.013	0.007	0.009	0.022	0.008	0.005	0.008	0.008
Total	3.002	2.995	3.004	3.000	2.995	3.007	3.001	3.008	3.003
Fo	71.6	57.0	66.9	72.0	8.0	66.6	60.0	57.3	59.4
Fa	28.4	43.0	33.1	28.0	92.0	33.4	40.0	42.7	40.6

(1) 12008—Most magnesian core composition.
(2) 12008—Most Fe-rich rim composition.
(3) 12016—Most magnesian composition analyzed. Core of anomalously large grain.
(4) 12056—Most magnesian composition. Core of large grain.
(5) 12056—Fayalite crystallized from Fe-rich immiscible fraction of residual mesostasis.
(6) 12005—Core of very large grain.
(7) 12005—Rim of euhedral grain partly enclosed in a poikilitic ilmenite grain within "interstitial" assemblage.
(8) 12005—Core of the same grain.
(9) 12005—Rim of euhedral grain adjacent to plagioclase in the "interstitial" assemblage.

positively. The relatively large and magnesian phenocrysts will settle selectively compared to the smaller, lower temperature phenocrysts (Walker *et al.*, 1976b). A second consequence of crystal settling within a lava flow, is that at a given level within the unit (except right at the flow boundaries) there will be partial compensation of phenocrysts from above for those lost downward (e.g., sample 12016 is apparently an example of this process in which a near balance was struck between addition and removal). The most magnesian olivine phenocryst core composition analyzed among the three rocks occurs in 12022, whereas the sample with the lowest maximum Fo-content is 12008. Significantly, the most magnesian olivine phenocryst in 12045 is also larger than any other in the thin section (0.7 mm vs. 0.2–0.5). There are also olivines in 12016 and 12005 that are much larger (1–5 mm) than those in the porphyritic basalts. Although some growth clearly occurred during settling, these large olivines are also embayed, raising the possibility that they are the remnants of a larger population of such grains now represented by smaller relics. Recognizing that the Fo_{72} olivine composition necessitated by the modeling calculations is probably the average of zoned phenocrysts (e.g., Fo_{76-69}), our findings are compatible with those of Rhodes *et al.* (1977). Similarly, small amounts of somewhat more Fe-rich olivine are required in fractionation modeling of the most evolved ISB samples (i.e., Fo_{69-60}). The range of zoning measured in natural olivine phenocrysts and the presence of Fo_{65} olivine at the olivine-pyroxene-plagioclase cotectic in Green *et al.*'s experiments are compatible with this premise.

Spinel

Microprobe analyses of spinels are presented in Table 4 and Figs. 7, 8, and 9. The paragenetic sequence of early titanian chromian spinel followed by chromian ulvöspinel recognized in the ISB samples is characteristic of low-Ti mare basalts (Apollo 12 and 15) and has been discussed in detail by El Goresy *et al.* (1976).

Table 4. Microprobe analyses of spinel.

	(1)	(2)	(3)	(4)	(5)	(6)	(7)	(8)	(9)
TiO_2	9.49	6.02	8.44	24.77	7.80	27.40	9.95	27.06	7.27
Al_2O_3	11.41	12.27	10.51	4.51	10.11	3.30	10.66	3.68	11.36
Cr_2O_3	38.40	45.78	41.19	13.30	39.49	11.83	37.57	12.59	43.11
FeO	34.54	28.13	32.93	53.80	39.75	55.51	35.27	53.43	32.31
MnO	0.72	0.68	0.80	0.64	0.54	0.44	0.54	0.40	0.58
MgO	4.66	7.20	5.26	1.74	0.81	0.74	4.21	1.69	4.90
CaO	0.09	0.05	0.11	0.29	0.00	0.06	0.08	0.07	0.04
Total	99.31	100.13	99.23	99.06	98.50	99.27	98.27	98.92	99.57
Ti	1.961	1.209	1.746	5.447	1.687	6.048	2.087	5.934	1.495
Al	3.693	3.860	3.409	1.553	3.424	1.140	3.507	1.264	3.661
Cr	8.337	9.657	8.961	3.075	8.972	2.744	8.289	2.901	9.323
Fe	7.933	6.278	7.579	13.156	9.555	13.626	8.232	13.028	7.391
Mn	0.168	0.153	0.186	0.160	0.132	0.108	0.127	0.098	0.134
Mg	1.906	2.864	2.158	0.757	0.345	0.325	1.749	0.736	1.998
Ca	0.025	0.014	0.031	0.090	0.000	0.018	0.022	0.023	0.020
Total	24.024	24.033	24.069	24.239	24.115	24.010	24.015	23.984	24.014
Cr/(Cr + Al)	0.693	0.714	0.724	0.664	0.724	0.706	0.703	0.697	0.718
Fe/(Fe + Mg)	0.806	0.687	0.778	0.946	0.965	0.977	0.825	0.947	0.787

n.d.—Not analyzed for.
(1) 12008—Relatively Ti-rich rim of euhedral spinel in groundmass.
(2) 12008—Most refractory core composition analyzed.
(3) 12045—Typical core of spinel enclosed within olivine phenocryst.
(4) 12045—Chromian ulvöspinel rim in groundmass.
(5) 12056—Low-Mg core of mantled grain.
(6) 12056—Low-Mg ulvöspinel adjacent to residual mesostasis.
(7) 12056—High-Mg core of a relatively unzoned spinel in the core of a pyroxene grain.
(8) 12056—Chromian ulvöspinel rim on chromian spinel, all enclosed in a pyroxene grain.
(9) 12056—High-Mg chromian spinel.

12008. Spinels in 12008 are optically homogeneous, a result confirmed by microprobe analyses that indicate limited chemical zoning entirely restricted to low-Ti compositions. The trend from core to rim is defined by an increase in TiO_2 (6.0→9.9 wt.%) accompanied by a slight decrease in Cr/Al (2.5→2.2) and an increase in Fe/(Fe + Mg) from 0.69 to 0.81. Chromian spinel in 12009 is similar in general character to that in 12008 but is 30–50% lower in TiO_2, apparently as a function of bulk rock chemistry. Donaldson *et al.* (1976) found that early spinel compositions (12009 experiments) were not a function of cooling rate (70–7°C/hr) and that at these rapid rates ulvöspinel did not appear.

12045. Early chromian spinels in 12045 partly overlap in composition with those in 12008 and extend the zoning trend of slightly increasing Ti, increasing Fe/(Fe + Mg) and decreasing Cr/(Cr + Al). This composite trend is well displayed in the Cr:Fe:Mg ternary plot in Fig. 7. The combined range of zoning in the two samples is as follows: Ti/(Ti + Cr + Al) = 0.08–0.20, Fe/(Fe + Mg) = 0.68–0.86, and Cr/(Cr + Al) = 0.72–0.70. This composite trend is equivalent to zoning trend #1 of El Goresy *et al.* (1976). Spinel grains in 12045 that are not protected by

Table 4. (*Continued*).

	(10)	(11)	(12)	(13)	(14)	(15)	(16)	(17)	(18)
TiO_2	25.87	8.68	28.23	8.39	6.64	21.36	6.78	24.47	1.78
Al_2O_3	4.52	11.29	3.35	10.72	11.75	5.32	12.03	5.39	45.98
Cr_2O_3	14.38	40.26	8.90	45.17	42.88	24.47	44.21	19.03	17.81
FeO	50.91	35.46	57.33	32.55	32.41	44.26	28.69	44.69	23.67
MnO	0.53	0.52	0.52	0.31	0.32	0.36	0.27	0.34	0.14
MgO	2.92	3.63	1.15	3.62	4.06	3.71	6.45	5.14	10.07
CaO	0.04	0.05	0.01	n.d.	n.d.	n.d.	n.d.	n.d.	n.d.
Total	99.17	99.88	99.49	100.75	98.05	99.48	98.43	99.06	99.45
Ti	5.590	1.797	6.230	1.712	1.389	4.534	1.388	5.156	0.306
Al	1.530	3.660	1.159	3.424	3.853	1.768	3.853	1.778	12.385
Cr	3.268	8.757	2.065	9.684	9.432	5.459	9.500	4.214	3.218
Fe	12.233	8.158	14.069	7.382	7.539	10.444	6.522	10.471	4.525
Mn	0.128	0.121	0.129	0.070	0.074	0.086	0.062	0.081	0.026
Mg	1.250	1.487	0.504	1.462	1.683	1.560	2.611	2.148	3.432
Ca	0.014	0.014	0.002	—	—	—	—	—	—
Total	24.012	23.995	24.158	23.734	23.969	23.852	23.936	23.848	23.892
Cr/(Cr + Al)	0.681	0.705	0.641	0.739	0.710	0.755	0.711	0.703	0.206
Fe/(Fe + Mg)	0.907	0.846	0.965	0.835	0.818	0.870	0.714	0.830	0.569

(10) 12016—High-Mg chromian ulvöspinel.
(11) 12016—Low-Mg chromian spinel.
(12) 12016—Low-Mg chromian ulvöspinel.
(13) 12005—Chromian spinel intergrown with reduced ilmenite.
(14) 12005—Typical chromian spinel enclosed in olivine.
(15) 12005—Chromian ulvöspinel also included in olivine.
(16) 12005—Typical chromian spinel enclosed in the core of a pyroxene.
(17) 12005—Chromian ulvöspinel enclosed in the rim of a pyroxene.
(18) 12005—Single grain of Cr–Al pleonaste in pyroxene.

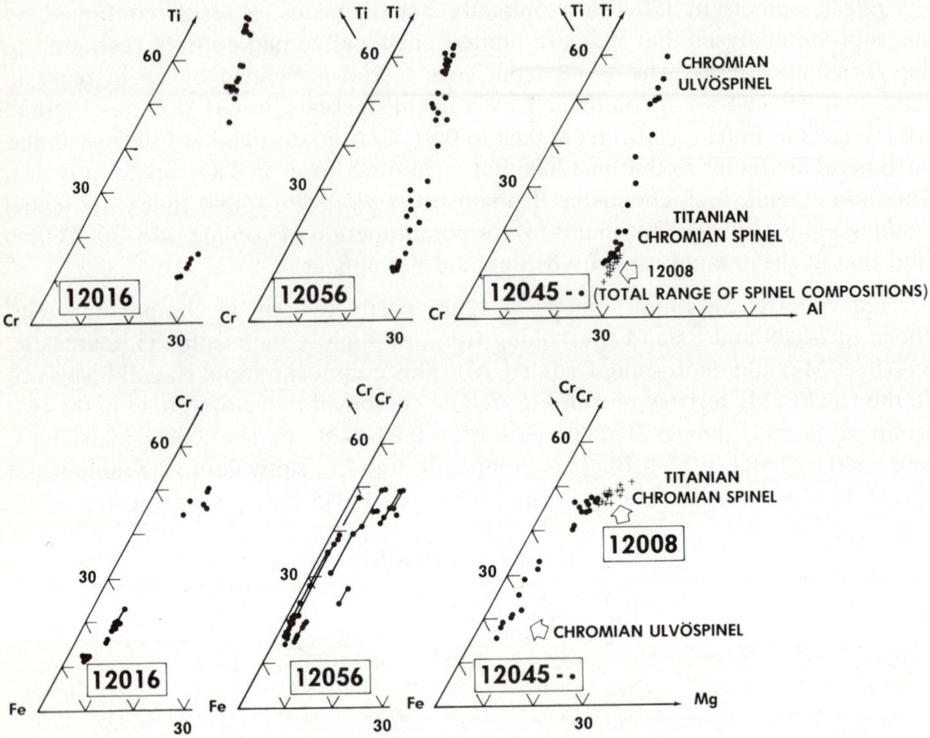

Fig. 7. Microprobe analyses of spinel in 12008, 12045, 12056, and 12016. Normalized compositions are plotted on normalized Ti:Cr:Al (top) and Cr:Fe:Mg (bottom) ternaries. Analyses from 12008 (+) and 12045 (○) are plotted together.

enclosure in olivine phenocrysts are zoned to ulvöspinel rims. Skeletal grains of ulvöspinel also occur separately in the groundmass. Optically the boundary between chromian spinel cores and ulvöspinel rims varies from sharp to gradational. The relative degree of idiomorphism of the cores correlates with the sharpness of the break between core and rim compositions. Although these textures are suggestive of minor resorption prior to precipitation of ulvöspinel rims, this paragenesis is not consistent with the rapid cooling history of this sample. The variations in apparent idiomorphism of the chromian spinel cores may be the result of varying geometry between core-rim boundaries and the thin section surface. The rims are compositionally zoned in the same sense as the cores but the relative increase in Ti/(Ti + Cr + Al) is far more abrupt. Ti/(Ti + Cr + Al) increases from 0.2 to 0.6, Fe/(Fe + Mg) increases sharply from 0.87 to 0.91 and thereafter is extended to 0.95, and Cr/(Cr + Al) decreases slightly from 0.70 to 0.66.

12056. The cooling rate difference between 12045 and 12056 is reflected in a marked change in the spinel parageneses. Relatively small idiomorphic spinel octahedra included within large pyroxene grains exhibit gradational zoning from chromian spinel cores to Ti-rich rims. The zoning trends in these grains are equivalent to these in early 12045 spinels and again are in agreement with trend #1 of El Goresy *et al.* (1976). Spinel grains not protected by pyroxene are either weakly zoned ulvöspinel with high Fe/(Fe + Mg) = 0.94–0.98 or composite grains of chromian spinel rimmed by ulvöspinel. In the latter case, resorption of earlier chromian spinel occurred prior to the precipitation of later Ti-rich ulvöspinel. In all grains the Cr/(Cr + Al) increases from core to rim. A characteristic feature of zoning in these spinels is that individual grains zoned from low to high Ti/(Ti + Cr + Al) have variable Fe/(Fe + Mg). Analyses 5, 6, 7, and 8 in Table 4 are examples of core-rim pairs in high-Mg and low-Mg grains. The high Fe/(Fe + Mg) grains are those which are not mantled by pyroxene implying that they are the result of Fe–Mg reequilibration between spinel and the Fe-rich residual liquid. Comparison to the 12008–12045 compositions demonstrates that Fe/Mg variations at a given TiO_2-content are not the result of equilibration to lower Fe/(Fe + Mg) between spinel and surrounding pyroxene, as the variation extends to higher, rather than lower, Fe/(Fe + Mg). Busche *et al.* (1972) recognized similar zoning patterns in ilmenite suite basalt 12051 including a large Fe/(Fe + Mg) spread in chromian spinel. However, we have not analyzed any of the nearly pure end-member ulvöspinel (>30 wt.% TiO_2) which they found.

12016. Spinel parageneses in 12016 are quite similar to those in 12056 with the exception that spinel compositions intermediate between early Cr spinel and Ti-rich ulvöspinel were not analyzed. This may be due to sampling, as relatively few analyses were obtained from this sample. The zoning data available indicate a sharp break between core and rim, a pattern corroborated by petrographic observations which indicate advanced resorption of cores prior to growth of the rim.

12005. Spinel parageneses in 12005 are unique among the samples we have studied in that (1) individual grains exhibit limited zoning and none is a composite of chromian spinel and ulvöspinel and (2) there is an extremely high correlation between spinel composition and the relative order in the crystallization sequence of the enclosing phase (Fig. 8a). Spinel enclosed in the cores of early anhedral olivine grains and in the cores of the pyroxene oikocrysts are restricted in composition to chromian spinel with less than 10 wt.% TiO_2. Chromian ulvöspinel is likewise restricted in distribution to the rims of large pyroxenes and the interstitial plagioclase-olivine-pyroxene assemblage. Spinel of a composition intermediate between the two is predictably located near the core-rim boundaries and in the cores of euhedral olivines within the interstitial assemblage. Linear arrays of spinel grains oriented normal to pyroxene and olivine grain boundaries exhibit compositional changes that mimic the composite trends derived from all the analyses.

Superimposed on these compositional variations are additional effects of subsolidus equilibration. The most pronounced of these is the Fe–Mg exchange between spinel and the enclosing silicates, olivine, and pyroxene. Those included in olivine generally have higher Fe/(Fe + Mg) than grains enclosed in pyroxene

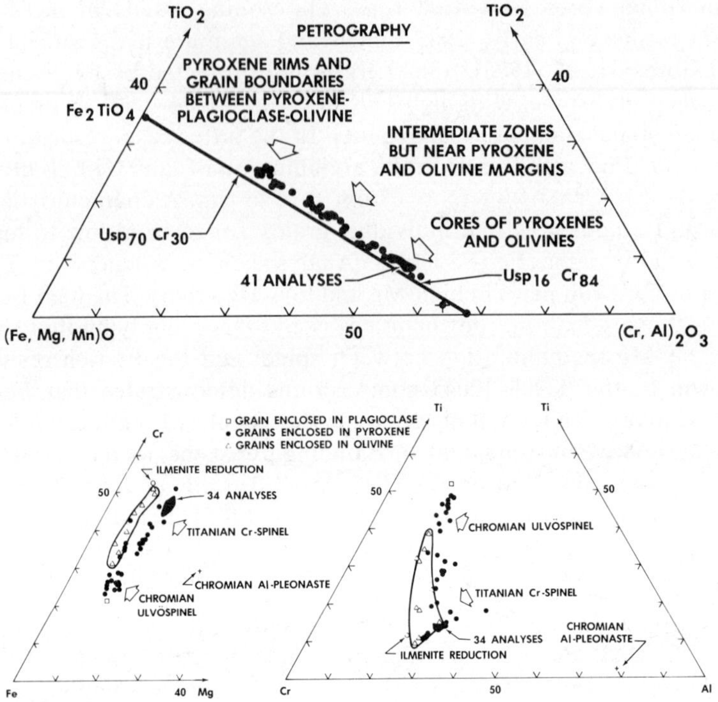

Fig. 8. Microprobe analyses of spinel in 12005. (a) Analyses plotted along the Usp–Cr join and (b) on the normalized Cr:Fe:Mg and Ti:Cr:Al ternaries.

(Fig. 8), indicative of an Fe–Mg exchange in which partitioning of Mg into olivine relative to spinel increases with falling temperature (Irvine, 1965). There is also a substantial difference between Cr/(Cr + Al) in spinels enclosed in olivine vs. pyroxene that implies a decrease in Cr/Al in spinels coexisting with pyroxene. Zoning in the larger grains of chromian spinel included in pyroxene probably records an igneous trend with reequilibration superimposed. The rim to rim microprobe traverse shown in Fig. 9 illustrates the combination. The sharp drop in Cr_2O_3 coupled with the slight increase in Al_2O_3 at the rim is inferred to be at least in part the result of reequilibration as the igneous trend is characterized by decreasing Al_2O_3. Note the substantial spread in Cr/Al in the Cr–Ti–Al ternary plot (Fig. 8) compared to the other samples. The data of Busche *et al.* (1972) record similar scatter in Cr/Al for spinels in cumulate 12036. Note also that it is the early spinel which exhibits the largest variation in Cr/Al suggesting that the extent of reequilibration decreased with falling temperature.

One of the spinel grains analyzed in 12005 is an anomalous Ti-poor chromian pleonaste. Three grains with similar compositions were analyzed in 12036 by Busche *et al.* (1971). The three grains in 12036 have higher Fe/(Fe + Mg) and Al_2O_3 and lower Cr_2O_3 than the 12005 analysis. Taken together, these four analyses define trends that are not continuous with the other spinels in these

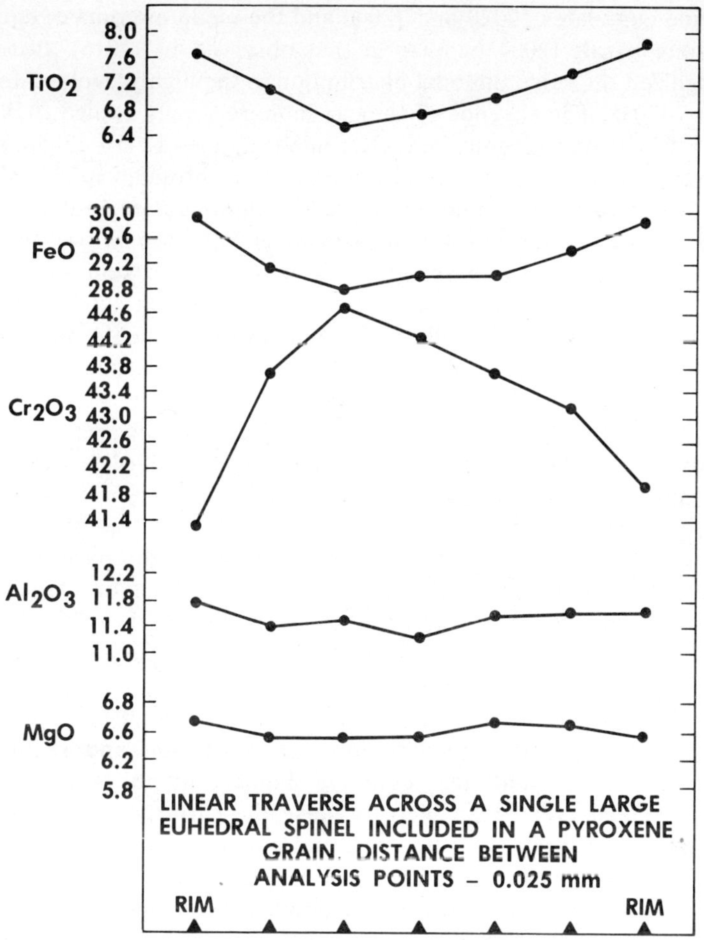

Fig. 9. Microprobe traverse illustrating zoning in a single grain of chromian spinel in 12005.

rocks in that Cr/Al decreases with increasing Fe/(Fe + Mg) and at the same Fe/(Fe + Mg) they contain far less TiO_2 than the early Ti-chromian spinels.

Discussion of spinel chemistry

The fundamental differences in the crystallization history and zoning patterns between spinels in slowly cooled and quickly cooled rocks outlined by El Goresy *et al.* (1976) are present in the ISB spinels. The porphyritic basalts exhibit zoning from early chromian spinel cores to chromian ulvöspinel rims with little or no resorption of the first phase prior to precipitation of the rims. Resorption textures are well developed in the medium-grained basalts resulting in composite trends. Haggerty and Meyer (1970) found a compositional gap in spinels from the partial olivine cumulates 12020 and 12040 and the medium-grained, equigranular pigeonite suite basalt 12064 (similar to that observed in 12016). Busche *et al.* (1972) recognized the same bimodal distribution in the highly evolved feldspathic microgabbro 12038. The absence of the gap in more slowly cooled rocks such as 12005 and 12036 as well as more quickly cooled samples 12051, 12056, and 12045 suggests that the combination of resorption of early chromian spinel followed by plating-on of chromian ulvöspinel is favored by intermediate cooling rates.

A third distinct pattern of textural-chemical variation is present in 12005 spinels. Idiomorphic chromian spinel in the cores of pyroxene oikocrysts that does not show Ti enrichment beyond the range present in 12008 is inferred to have accumulated (along with olivine phenocrysts) during the initial phase of crystal settling. The absence of ulvöspinel rims is probably due to the buffering effect of the olivine-liquid peritectic reaction that led to the growth of the large pyroxene oikocrysts which enveloped the low-Ti spinels. The presence of the first substantially Ti-enriched spinel at the core-rim boundary of the pyroxene oikocrysts suggests that it formed in response to a change in residual liquid composition although falling temperature may also be a factor in controlling spinel compositions. The Ti–Al substitution relations in the pyroxenes indicate that the core-rim boundary coincides with the incoming of plagioclase (see below). One possible interpretation is that the spinels with $TiO_2 > 10$ wt.% are free-grown grains that formed with something like their present composition in response to the gradually changing liquid composition. A second is that spinel more enriched in the ulvöspinel component than the initial composition is the result of complete reaction (combination of resorption and reequilibration) between early spinel and the evolving liquid composition. The second mechanism is considered most likely because the early cumulate spinels were probably distributed throughout the rock and not segregated in the areas which became the cores of pyroxene oikocrysts. In either case, the limited range of zoning in these spinels and the correspondence between composition and crystallization sequence are consistent with very slow cooling through the temperature range in which spinel formed. Note also that the maximum Ti-content of ulvöspinel in 12005 is less than in 12045, 12016, or 12056, consistent with its less evolved residual liquid and the extensive subsolidus reduction (El Goresy and Ramdohr, 1975).

One unlikely possibility for the origin of the pleonaste spinel is that it represents the true low-pressure liquidus phase or a high-pressure phase which, with falling temperature, tends to react out. None of the experimental studies of low-Ti mare basalts record such spinels but the ratio of spinel to melt would be limited by the initial Cr-contents of the magmas suggesting that they could be overlooked in the rocks and in experimental charges. A more probable alternative is that they are refractory xenocrystal material plucked from anorthositic or KREEPy basement rocks or their regolith prior to eruption. Although these anomalous spinels are the only petrographic evidence for assimilation in the Apollo 12 basalts, the aberrant trace element characteristics (Rhodes *et al.*, 1977) and $^{87}Sr/^{86}Sr$ initial ratio (Nyquist *et al.*, 1977) of sample 12036 provide some support for the presence of a very small amount of exotic material.

Pyroxene: porphyritic basalts

Augite phenocrysts in the porphyritic ISBs vary in morphology from dendrites (12008) to highly elongate (12045) to less elongate (12022) within the cooling rate sequence in which there is also an increase in the grain size and relative proportions of the phenocrysts. There are also marked variations in pyroxene composition among these nearly isochemical basalts (Table 5; Figs. 10, 11, and 12). Dendrites in 12008 and phenocrysts in 12045 and 12022 all plot near the diopside-hedenbergite join as do pyroxenes from quickly cooled Apollo 11 and 17 high-Ti basalts. They also have similarly high Ti- and Al-contents (e.g., Brown *et al.*, 1975) and Ti:Al = 0.35–0.52. Dendrites in 12008 have higher Cr_2O_3 than the other samples (Fig. 12). The vertical trend on the quadrilateral (Fig. 10a) is apparently the result of Ca enrichment with progressive crystallization as the low-Ca compositions are highest in Cr. Zoning trends in these extremely fine-grained dendrites have not been documented in detail but similar trends were found by Donaldson *et al.* (1975) in cooling rate runs on a 12009 analog composition. In 12045 and 12022 there are two distinct generations of pyroxene, phenocrysts, and groundmass, which have markedly different compositions. The pyroxene overgrowths on olivine phenocrysts and discreet pyroxene phenocrysts in 12045 are equivalent in composition and exhibit decreasing Ti, Cr, Al, and Ca with Fe enrichment in core to rim zoning. The compositional hiatus between phenocrysts and groundmass pyroxene is most dramatically expressed in lower Ti, Al, and Cr concentrations (Figs. 11 and 12). Individual grains are also strongly zoned with respect to Fe/Mg from core to rim with the latter approaching compositions near the ferrosilite-hedenbergite join (Fig. 10a). The groundmass pyroxenes in 12045 and 12022 are essentially identical in composition, apparently as the result of rapid co-crystallization with plagioclase and ilmenite at substantial undercooling.

Pyroxene: medium- and coarse-grained basalts

Microprobe analyses of pyroxene in 12056, 12016, and 12005 were collected primarily along linear traverses from core to rim or from rim to rim (10–25 points). Beginning and end points were pre-selected with the locations of intervening analysis points determined automatically by equal-increment spacing along the traverse. Although small grains intergrown with plagioclase were analyzed, the majority of analyses were made on large grains. The traverses were oriented normal to optical zoning boundaries that are generally parallel to the long dimension of the grains. This mode of data collection results in an unbiased survey reflecting the entire range of pyroxene compositions in the sample. However, where zoning trends are complex (e.g., intergrowths of high-Ca and low-Ca pyroxene) fine details may remain unresolved. In this study the distance between points (20–100 μm) was, in some cases, greater than the scale of two-pyroxene intergrowths which are present in these rocks.

Pyroxene analyses of 12056, 12016, and 12005 have been divided into cores and rims in plotting them on Figs. 10, 11, and 12. These distinctions have been made on the basis of location within the individual grain with respect to discrete

Table 5. Microprobe analyses of pyroxene.

	(1)	(2)	(3)	(4)	(5)	(6)	(7)	(8)	(9)
SiO_2	43.14	41.68	46.50	46.00	49.72	50.15	50.45	50.65	48.18
TiO_2	5.17	7.44	2.86	0.96	1.66	1.16	1.50	0.70	0.79
Al_2O_3	7.67	9.02	4.73	1.44	3.38	2.30	2.83	0.78	1.09
Cr_2O_3	1.03	0.55	0.20	0.06	0.89	0.72	0.80	0.20	0.17
FeO	16.79	13.26	18.98	40.09	13.18	17.76	15.30	25.89	33.50
MnO	0.29	0.20	0.38	0.66	0.26	0.31	0.22	0.38	0.45
MgO	10.13	8.29	8.89	2.47	16.27	19.80	16.62	15.78	8.85
CaO	15.08	18.64	16.73	7.81	13.84	6.89	11.92	5.00	6.24
Na_2O	0.08	0.15	0.03	0.01	0.06	0.01	0.00	0.00	0.01
Total	99.38	99.23	99.30	99.51	99.26	99.10	99.64	99.39	99.27
Si	1.676	1.615	1.820	1.941	1.872	1.893	1.896	1.963	1.954
Ti	0.151	0.217	0.084	0.031	0.047	0.033	0.043	0.002	0.024
Al^{IV}	0.324	0.385	0.180	0.059	0.128	0.102	0.104	0.036	0.046
Al^{VI}	0.027	0.027	0.038	0.013	0.022	0.000	0.021	0.000	0.006
Cr	0.032	0.017	0.006	0.002	0.027	0.022	0.024	0.006	0.006
Fe	0.545	0.429	0.621	0.414	0.415	0.560	0.481	0.839	1.136
Mn	0.010	0.007	0.013	0.023	0.008	0.010	0.007	0.012	0.015
Mg	0.587	0.479	0.519	0.155	0.913	1.114	0.931	0.912	0.535
Ca	0.628	0.774	0.701	0.353	0.558	0.279	0.480	0.208	0.271
Na	0.006	0.011	0.003	0.001	0.004	0.000	0.000	0.000	0.001
Total	3.985	3.960	3.985	3.992	3.995	4.013	3.987	3.996	3.993
Wo	35.7	46.0	38.1	18.4	29.6	14.3	25.4	10.6	14.0
En	33.3	28.5	28.2	88.1	48.4	57.0	49.2	46.5	27.5
Fs	31.0	25.5	33.7	73.6	22.0	28.7	25.4	42.8	58.5

(1) 12008—Dendritic microphenocryst.
(2) 12045—Core of brown phenocryst.
(3) 12045—Rim of same phenocryst.
(4) 12045—Rim of pale green groundmass pyroxene.
(5) 12016—Augite core of large grain.
(6) 12016—Low-Ca pyroxene intergrown with augite in the core of same large grain.
(7) 12016—Subcalcic augite in the core of another large grain.
(8) 12016—Low-Ca rim.
(9) 12016—Fe-rich, low-Ca grain margin adjacent to mesostasis.

zoning boundaries and certain compositional parameters (Fe/Fe + Mg, Cr_2O_3, and Ti:Al) which tend to vary discontinuously across these zoning boundaries. However, in some traverses the compositional variations were continuous, or different parameters were conflicting. Thus, the graphical distinction may be arbitrary for some points in the compositional ranges close to the core-rim boundaries. The internal zoning patterns of individual pyroxenes in the three rocks exhibit significant variation among the three rocks and are discussed separately below.

12056. The cores of large grains are dominated by augite although small amounts of low-Ca pyroxene are present. Hollister *et al.* (1971) found a similar pattern in the cores of 12063 pyroxenes, a rock close to 12056 in bulk com-

Table 5. (*Continued*).

	(10)	(11)	(12)	(13)	(14)	(15)	(16)	(17)	(18)
SiO_2	49.56	51.74	51.10	49.82	45.87	50.53	52.33	51.21	49.90
TiO_2	1.97	1.16	1.41	1.18	1.00	1.61	0.90	0.83	1.77
Al_2O_3	3.40	2.27	2.54	2.17	1.66	3.22	1.64	1.04	2.52
Cr_2O_3	0.94	0.60	0.68	0.25	0.21	0.86	0.71	0.32	0.60
FeO	12.05	17.68	15.33	24.27	35.49	10.72	15.71	20.10	13.37
MnO	0.24	0.29	0.26	0.36	0.43	0.23	0.31	0.38	0.22
MgO	15.09	19.05	17.42	13.23	3.44	16.26	21.80	20.35	16.12
CaO	16.23	7.15	10.46	7.95	11.13	16.47	6.81	5.32	15.20
Na_2O	0.04	0.06	0.02	0.04	0.08	0.07	0.00	0.03	0.07
Total	99.52	100.00	99.22	99.26	99.31	99.97	100.20	99.58	99.78
Si	1.866	1.927	1.918	1.936	1.916	1.879	1.927	1.931	1.878
Ti	0.056	0.033	0.040	0.035	0.032	0.045	0.025	0.024	0.050
Al^{IV}	0.134	0.073	0.082	0.064	0.082	0.121	0.071	0.046	0.112
Al^{VI}	0.016	0.027	0.030	0.035	0.000	0.020	0.000	0.000	0.000
Cr	0.028	0.018	0.020	0.008	0.007	0.025	0.021	0.010	0.018
Fe	0.379	0.551	0.481	0.789	1.240	0.333	0.484	0.634	0.421
Mn	0.008	0.009	0.008	0.012	0.015	0.007	0.010	0.012	0.007
Mg	0.846	1.057	0.973	0.766	0.214	0.902	1.196	1.144	0.904
Ca	0.654	0.285	0.421	0.331	0.498	0.656	0.269	0.215	0.613
Na	0.003	0.004	0.002	0.003	0.007	0.005	0.000	0.002	0.005
Total	3.991	3.984	3.977	3.977	4.011	3.995	4.002	4.018	4.009
Wo	34.8	15.1	22.4	17.5	25.5	35.0	14.0	10.9	32.0
En	45.0	55.8	51.9	40.6	11.0	48.5	61.2	57.1	47.0
Fs	20.2	29.1	25.7	41.9	63.5	17.5	24.8	32.0	22.0

(10) 12056—Augite in core of large grain.
(11) 12056—Relatively rare low-Ca pyroxene in core of large grain.
(12) 12056—Subcalcic augite in core of large grain.
(13) 12056—Rim adjacent to core of same grain—marks an abrupt shift in composition.
(14) 12056—Ferroaugite margin adjacent to mesostasis.
(15) 12005—Augite core of large grain.
(16) 12005—Low-Ca pyroxene in core of large grain.
(17) 12005—Relatively Fe-rich, low-Ca margin.
(18) 12005—Relatively Fe-rich augite margin.

position and cooling history. They performed zoning traverses with very closely spaced points across a few grains and found that the low-Ca phase apparently represents the innermost core, implying that the augite followed the low-Ca pyroxene in the crystallization sequence and nucleated epitaxially on it. If this inference is correct, it may reflect initial precipitation of a low-Ca pyroxene as the result of the olivine-liquid peritectic. Both 12063 and 12056 exhibit a discontinuity on the pyroxene quadrilateral between cores and rims. Individual zoning trends in rims are illustrated by arrows on the quadrilateral plot (Fig. 10). Ti and Al zoning in cores and rims is oscillatory and is generally positively correlated with Ca-content.

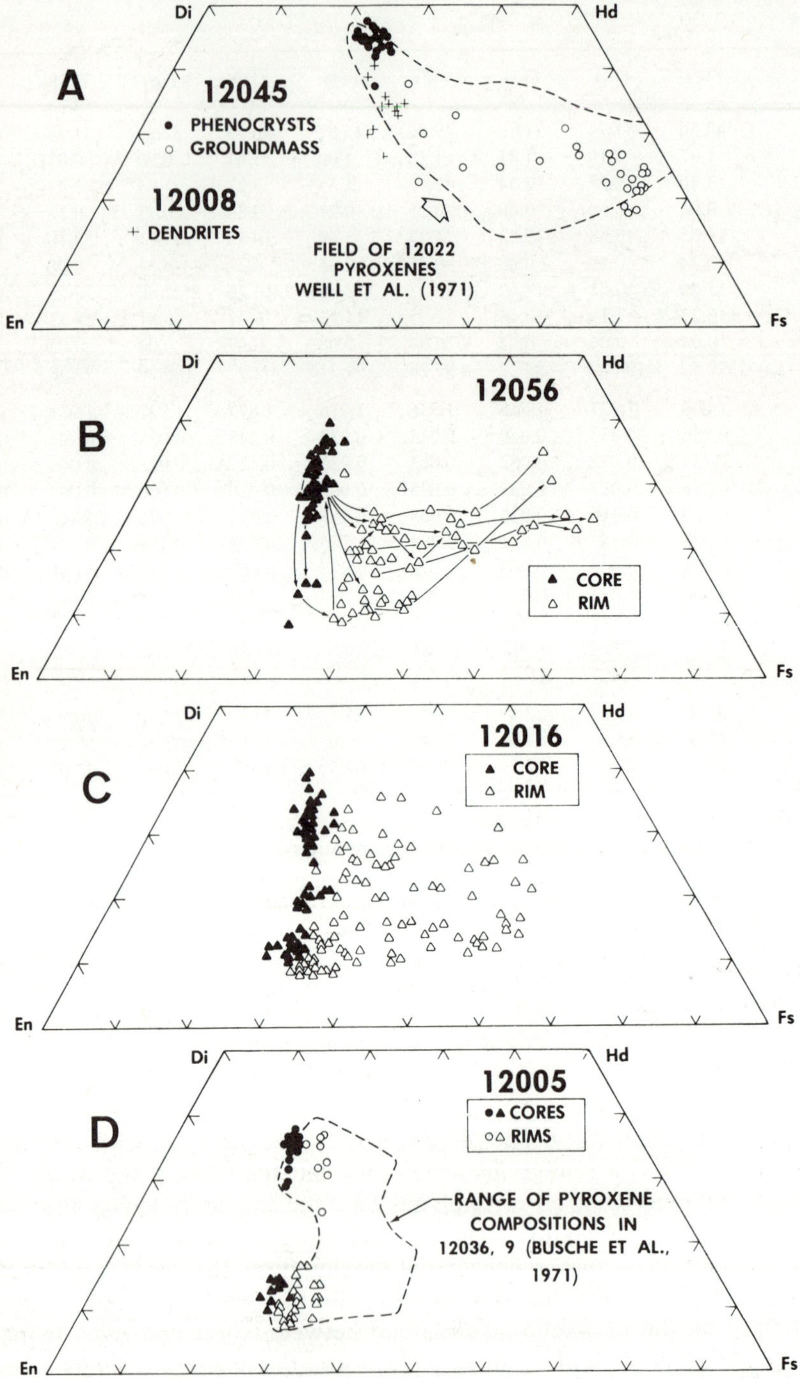

Fig. 10. Microprobe analyses of pyroxenes plotted in terms of quadrilateral components.

12016. Pyroxene zoning trends in 12016 are the most complex of the medium- to coarse-grained rocks, primarily due to abundant intergrowths of high-Ca and low-Ca pyroxene in both cores and rims. Arrows tracing the individual zoning trends are not shown on the quadrilateral plot in order to avoid clutter as they tend to be quite irregular. Like 12056, most symmetrically zoned grains in 12016 have augite cores intergrown with minor low-Ca pyroxene or subcalcic augite. These relationships have not been entirely resolved in genetic terms. A compositional hiatus between core and rim is not evident in the composite plot of analyses on the quadrilateral but it does occur in some of the individual grains.

12005. The majority of large grains are composed of augite cores rimmed by low-Ca pyroxene. Rim augite is most typically present in the outermost margin intergrown with the low-Ca phase. However, several of the largest grains include augite plus low-Ca pyroxene intergrowths in their cores. Locally, resorbed olivine grains are partly mantled by low-Ca pyroxene, again suggesting the two-pyroxene intergrowths are the result of a peritectic reaction. The lack of any truly subcalcic compositions contrasts with all the other samples studied (including 12036—Busche *et al.*, 1971) and may indicate an approach to an equilibrium solvus. The complete separation of augite and low-Ca pyroxene on the quadrilateral is matched by non-overlap between the two phases on the Ti–Al and Cr_2O_3 vs. Fe/(Fe + Mg) plots.

Discussion of pyroxene chemistry

Compositional differences between pyroxenes in the porphyritic basalts and the medium- to coarse-grained basalts are ascribed to the distinctly different cooling histories inferred for the two groups. The presence of only augite with high Ti and Al as well as high Ti:Al in 12008, 12045, 12022, and the rapidly cooled Apollo 11 and 17 high-Ti basalts is a characteristic expression of the relatively high Ti-contents of the rocks and the crystallization of pyroxene at substantial undercooling. The cooling rate dependence of augite phenocrysts in the porphyritic ISBs is demonstrated by its absence (pyroxene phase = pigeonite) in isothermal melting experiments on 12022 (Green *et al.*, 1971). Higher bulk Ti appears to favor augite over low-Ca pyroxene in rapidly cooled rocks as the latter phase is the dominant pyroxene in porphyritic olivine suite basalts such as 12002 (Walker *et al.*, 1976a) which appears to have had a cooling rate history similar to 12022.

Within the three porphyritic basalts there are also major differences among the three rocks that are inferred to be controlled by cooling rate. Although the augite phenocrysts in 12022 and 12045 plot in similar positions in terms of quadrilateral components they have quite different characteristics in terms of their Ti–Al relations. In contrast to the Ti and Al decrease with progressive crystallization in 12045 phenocrysts, those in 12022 (Weill *et al.*, 1971; Bence and Papike, 1972) show an increase in Ti and Al from core to rim and have lower

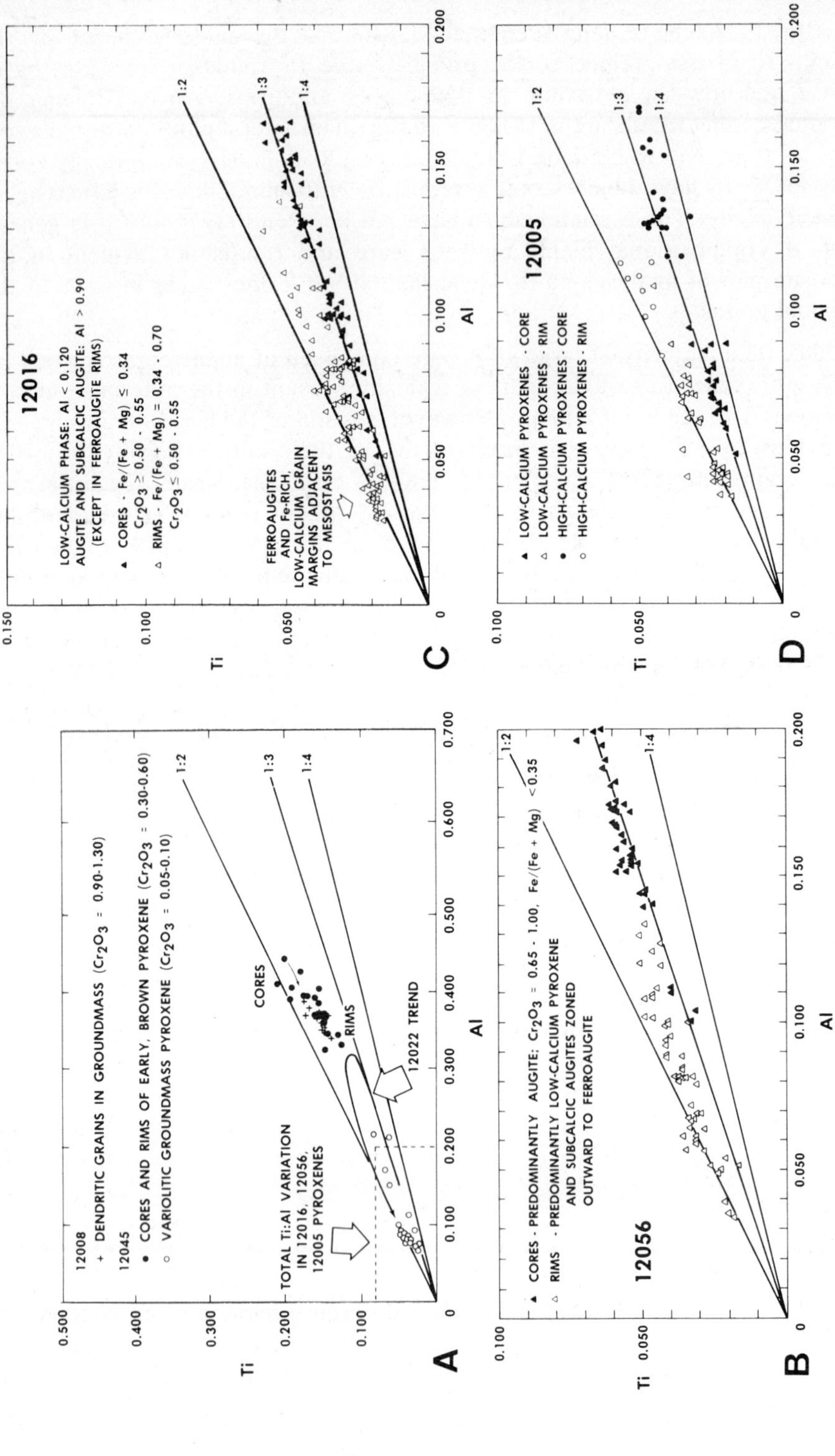

Fig. 11. Microprobe analyses of pyroxenes illustrating variations in Ti–Al relations. 12022 trend after Bence and Papike (1972).

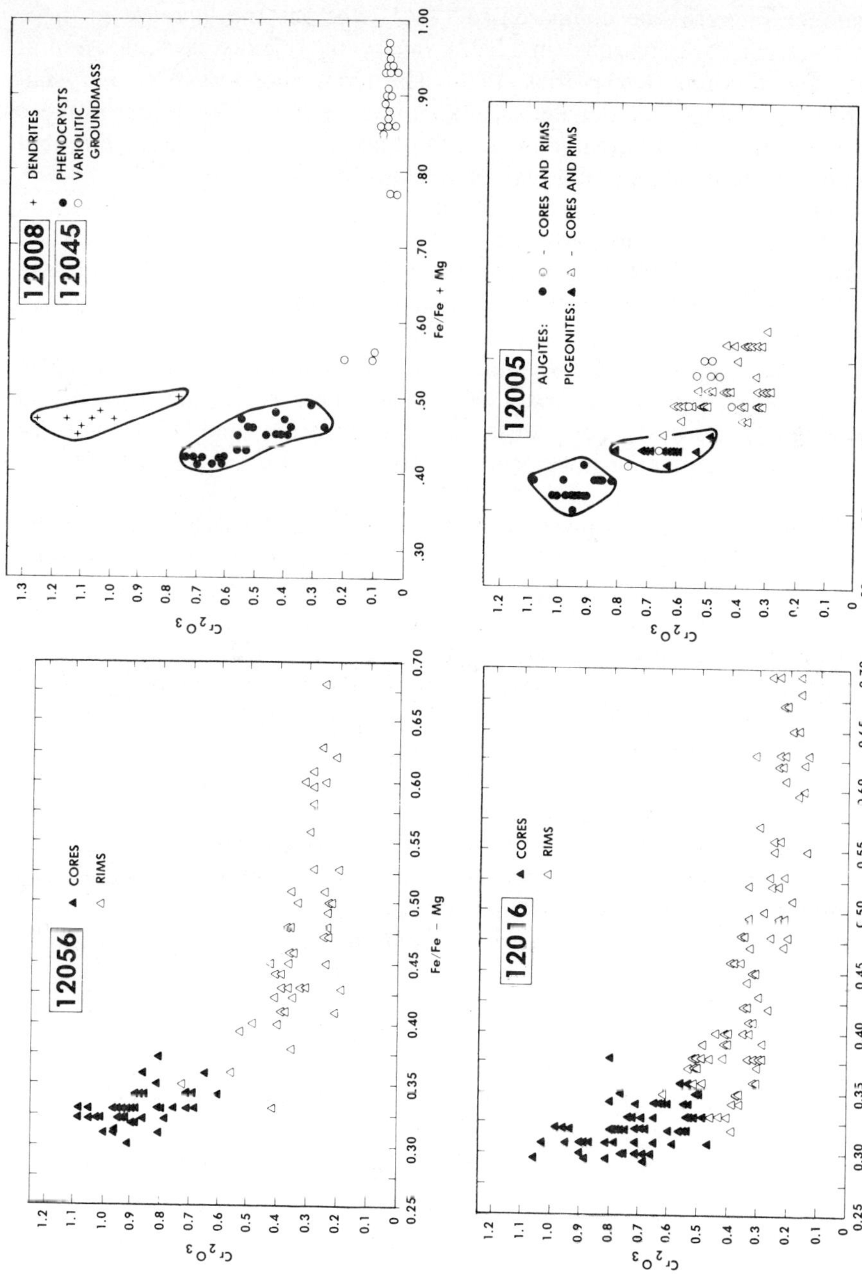

Fig. 12. Microprobe analyses of pyroxenes. Cr_2O_3 vs. Fe/(Fe + Mg). Note different scales for Fe/(Fe + Mg).

Ti:Al ratios (0.33) as opposed to 12045 (0.35–0.52). The much lower Ti and Al in the cores of the 12022 phenocrysts is due to less supression of the temperature of appearance of pyroxene compared to 12045, whereas the increase in these components from cores to rims (in 12022) apparently reflects crystallization at increasing undercooling (Kirkpatrick, 1976). The 12045 phenocrysts also crystallized with progressively greater undercooling, but the initial Ti-content in these pyroxenes is substantially greater than in the bulk composition suggesting that Ti-depletion of the residual liquid may be responsible for the trend.

In contrast to the porphyritic basalts, low-Ca pyroxene comprises a significant proportion of the pyroxene in the three medium- to coarse-grained rocks. The ratio of low-Ca pyroxene to augite increases in the sequence 12056→12016→12005 (Table 2). Within the same sequence there is a progressive change in the Ti:Al relations that appears to correspond to cooling rate differences (Fig. 11b,c,d). The maximum Ti and Al concentrations in the 12056 augites exceed those in 12016 and 12005. In all three rocks the low-Ca and high-Ca compositions in the cores of large grains have the same Ti:Al with the low-Ca pyroxene containing lower concentrations of Ti and Al as well as Cr (Figs. 11 and 12).

In all three rocks there are optical zoning boundaries between core and rim across which there is a marked shift to lower Ti, Al, and Cr accompanied by an increase in Ti:Al to ≈1:2 and a decrease of octahedral Al to zero. The ratio of 1:2 is typical of lunar pyroxenes crystallizing with plagioclase and the decrease in Ti and Al concentrations apparently marks the effect of ilmenite and plagioclase co-precipitation, depleting the residuum in these components. However, for a given rock the composite of many analyses indicates that there tends to be a gradation between core and rim compositions with respect to these elements. This pattern of gradual Ti–Al decrease contrasts with the phenocryst-groundmass relations in the porphyritic basalts. This is apparently due to the very rapid growth of plagioclase and ilmenite at substantial undercooling in more quickly cooled rocks as compared to the more slowly cooled samples in which there is a closer approach to equilibrium crystallization.

The degree of Fe enrichment in the ISB pyroxenes also varies substantially due to the cooling rate dependence of residual liquid fractionation. As expected, the most fractionated liquids are found in the porphyritic basalts in which the strongly zoned groundmass pyroxenes approach the hedenbergite-ferrosilite join. The Mg/(Mg + Fe) ranges of the pyroxenes in the three medium- and coarse-grained basalts are 12056—0.70–0.15, 12016—0.71–0.27, and 12005—0.75–0.63. The marked contrast between the 12005 pyroxenes and those in 12056–12016 is the result of the extensive adcumulus growth in the former.

Feldspar

A survey of the range of plagioclase compositions in 12056, 12016, and 12005 (Table 6; Fig. 13) was obtained by analyzing cores and rims of several grains in each sample. The plagioclase composition at various stages of crystallization has

Table 6. Microprobe analyses of feldspar.

	(1)	(2)	(3)	(4)	(5)	(6)	(7)	(8)	(9)
SiO_2	46.75	49.28	50.18	47.68	48.34	47.95	45.51	50.33	66.40
TiO_2	0.09	0.04	0.10	0.11	0.05	0.04	0.08	0.10	0.23
Al_2O_3	33.59	31.06	30.89	32.70	32.40	32.24	35.16	31.41	18.74
FeO	0.49	1.48	1.11	0.80	1.25	1.84	0.33	0.56	0.45
MgO	0.30	0.07	0.08	0.37	0.06	0.01	0.19	0.10	0.00
CaO	17.39	15.90	15.15	17.02	16.64	16.91	18.42	14.41	0.27
Na_2O	0.97	1.23	1.65	1.24	1.11	0.85	0.76	2.27	0.59
K_2O	0.06	0.58	0.61	0.07	0.24	0.54	0.06	1.02	15.19
BaO	0.00	0.13	0.07	0.00	0.00	0.10	n.d.	n.d.	n.d.
Total	99.60	99.75	99.85	100.00	100.08	100.49	100.49	100.20	101.87
Si	2.155	2.271	2.301	2.192	2.220	2.206	2.088	2.297	2.997
Ti	0.003	0.001	0.004	0.004	0.002	0.001	0.003	0.004	0.008
Al	1.827	1.687	1.670	1.772	1.754	1.748	1.901	1.689	0.997
Fe	0.019	0.057	0.043	0.031	0.048	0.071	0.013	0.021	0.017
Mg	0.021	0.005	0.006	0.025	0.004	0.001	0.013	0.007	0.000
Ca	0.860	0.785	0.744	0.838	0.819	0.834	0.905	0.705	0.013
Na	0.086	0.110	0.147	0.111	0.098	0.076	0.067	0.201	0.051
K	0.003	0.034	0.036	0.004	0.014	0.032	0.003	0.059	0.874
Ba	0.000	0.002	0.001	0.000	0.000	0.002	—	—	—
Total	4.973	4.953	4.951	4.976	4.958	4.970	4.994	4.985	4.959
An	90.6	84.5	80.3	87.9	88.0	88.5	92.8	73.1	1.4
Ab	9.1	11.8	15.9	11.6	10.5	8.1	6.9	20.8	5.4
Or	0.3	3.7	3.9	0.5	1.5	3.4	0.3	6.1	93.2

n.d.—Not analyzed for.
(1) 12016—Core of large equant grain with typical high An, low Fe/Fe + Mg.
(2) 12016—Rim adjacent to mesostasis.
(3) 12016—Rim adjacent to pyroxene.
(4) 12056—Core of large equant grain.
(5) 12056—Rim near mesostasis patch.
(6) 12056—Rim near mesostasis patch.
(7) 12005—Core of large equant grain with high An and low Fe/Fe + Mg.
(8) 12005—Sodic bytownite intergrown with K-feldspar.
(9) 12005—K-feldspar in holocrystalline patch of mesostasis.

been monitored by analyzing grains of various sizes and morphologies ranging from large and equant (early) to small lath-shaped grains (late). The Fe–Mg zoning in mare basalt plagioclase parallels the increase in Fe/Mg in coexisting liquid during progressive crystallization and can, therefore, be used as an index of fractionation against which An–Ab–Or variations in the plagioclase can be compared (e.g., Longhi *et al.*, 1976).

The cores of large equant grains are the most calcic compositions analyzed in all three rocks and also have the lowest Fe/(Fe + Mg). The cores of smaller lath-shaped plagioclase, that apparently formed later in the crystallization sequence (in any given rock) than the large blocky grains, have intermediate Fe/(Fe + Mg) which varies with proximity to mesostasis segregations. Within the

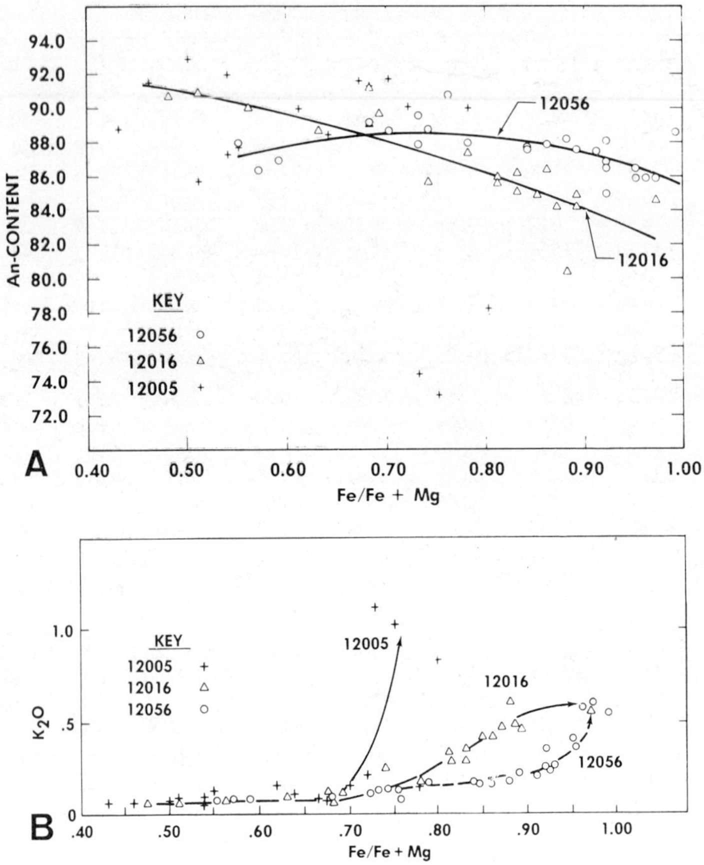

Fig. 13. Microprobe analyses of plagioclase. An–content (a) and K_2O (b) vs. Fe/(Fe + Mg).

fractionation-cooling rate sequence 12056→12016→12005 the most calcic plagioclases analyzed have successively lower Fe/(Fe + Mg) and higher An-contents (12056 = 0.55–88; 12016 = 0.47–91; 12005 = 0.43–93). This progression is consistent with predictions of the effect of plagioclase nucleation at decreasing increments of undercooling. The trend to higher An-content of the initial plagioclase may signify progressively less depression of the plagioclase liquidus from 12056→12016→12005 with decreasing cooling rate. This is corroborated by an increase in An-content in the 12056 plagioclases at Fe/(Fe + Mg) of 0.65–0.75. Lofgren (1973) showed that suppression of plagioclase crystallization at large undercoolings can result in reverse zoning of this type. Similar zoning relationships have been recognized in 12021 (Crawford, 1973) and 12002 (Longhi *et al.*, 1976).

The late-stage enrichment of Fe over Mg in plagioclase follows the pattern observed for olivine and pyroxene: the enrichment approaches an Fe/(Fe +

Mg) = 1.0 in both 12016 and 12056 and is less extreme in 12005 (0.78). The late-stage enrichment in K at a given Fe/(Fe + Mg) varies substantially among the three rocks.

Rim compositions in 12005 adjacent to and intergrown with K-feldspar and phosphate (i.e., holocrystalline mesostasis segregation) are sodic bytownite ($An_{78-73}Ab_{16-21}Or_{5-7}$). These are among the most sodic plagioclases analyzed in mare basalts (cf., Papike *et al.*, 1976). The most sodic plagioclase in 12036 is also bytownite with a high-K content (Or_5) indicating a similar zoning pattern (Keil *et al.*, 1971). Aided by a slow crystallization rate and abundant trapped liquid, plagioclase zoning in 12005 was suppressed, both in terms of An–Ab–Or and Fe/(Fe + Mg). Plagioclase rims, which grew adjacent to rare segregations of mesostasis, record the strong alkali enrichment in a vanishingly small amount of residual liquid.

The late-stage enrichment of alkalis in 12056 and 12016 can be directly contrasted due to similar degrees of Fe enrichment and the presence of discrete alkali-rich phases in their mesostasis segregations. Relative to 12016, the Na + K enrichment in 12056 plagioclases at Fe/(Fe + Mg) > 0.70 is suppressed (Fig. 13b). The symplectoid intergrowths in mesostasis segregations of 12056 are considered evidence of immiscibility in the residual liquid of 12056 at a less advanced stage of crystallization than in 12016, which lacks these intergrowths. It is suggested that plagioclase crystallization from the Fe-rich, alkali-depleted immiscible fraction produces plagioclase with similar characteristics. Powell *et al.* (1976) proposed this mechanism as an explanation for the high An-contents of acicular plagioclase in the mesostasis of highlands melt rocks analogous to those described by Brown and Peckett (1972). As the ratio of the Fe-rich to silica- and K-rich immiscible fractions is generally high, plagioclase adjacent to mesostasis segregations would most likely be in contact with the Fe-rich phase rather than the silica-rich fraction which is dispersed as small blebs. Non-equilibrium crystallization from the Fe-rich liquid could account for the observed alkali enrichment trends in the 12056 and 12016 plagioclases.

K-feldspar occurs in the holocrystalline mesostasis of 12005 (analysis 9, Table 6) and 12016. The silica- and K-rich phase in the mesostasis segregations of 12016 and 12056 appears to be primarily composed of a glass although Trzcienski and Kulick (1972) identified a Ba-rich K-feldspar in the mesostasis of 12063. The composition of the alkali-rich glass in 12016 is as follows: $SiO_2 = 61.98$, $TiO_2 = 0.21$, $Al_2O_3 = 22.0$, $FeO = 1.07$, $MgO = 0.02$, $CaO = 1.00$, $Na_2O = 0.32$, $K_2O = 11.61$, $BaO = 0.12$. A Ba phase was sought in the mesostasis areas but was not located. Ba-contents in most analyzed plagioclase grains were below detection limits but some rims adjacent to mesostasis contained resolvable amounts (0.01–0.10 wt.% BaO).

Ilmenite

A wide variety of ilmenite morphologies is present in the ISBs. The change from skeletal plates to more equant forms correlates with cooling rate (Usselman *et al.*, 1975). The wispy filamental ilmenite (<1 μm) in 12008 is too fine grained to be analyzed with the microprobe but the thin skeletal plates in 12045 (1–10 μm) and 12022 (5–25 μm) and the more coarse-grained laths and blocky grains in the other samples in the suite have been analyzed in this study (Fig. 14) or by other workers.

Ilmenite in 12045 is nearly stoichiometric $FeTiO_3$ with MgO < 0.1 in most grains (one at 0.4 wt.% MgO) and Cr_2O_3 < 0.2 wt.%. Similar compositions with slightly higher MgO (0.4–0.6 wt.%) were found for ilmenite in 12022 (Weill *et al.*, 1971; Cameron, 1971). The MgO-contents of ilmenite in the medium-grained basalts are more variable and higher. Grain to grain variations in ilmenite composition in 12016 and 12056 are much greater than within individual grains. Ilmenite within or adjacent to mesostasis segregations is low in MgO and Cr_2O_3 relative to those intergrown with plagioclase and ferromagnesian minerals. Ilmenite with a similar compositional range (0.11–1.68 wt.% MgO) is present in the evolved, medium-grained ISB 12051 (Keil *et al.*, 1971). Ilmenite in 12005 is characterized by a narrow compositional range and high MgO and Cr_2O_3 relative to 12016 and 12056. Keil *et al.* (1971) report MgO-contents of 3.0–4.6 wt.% in 12036, consistent with its intermediate chemistry and lack of Fe enrichment in its late-stage mesostasis relative to 12016 and 12056. Similar patterns of ilmenite

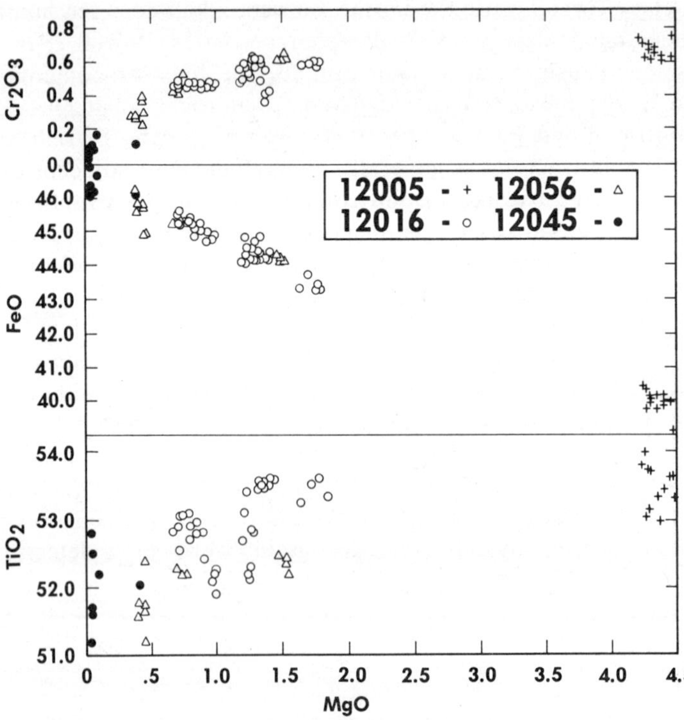

Fig. 14. Microprobe analyses of ilmenite in 12045, 12016, 12056, and 12005. Cr_2O_3, FeO, and TiO_2 vs. MgO.

compositional variation are present in fine-grained and coarse-grained members of the Apollo 12 olivine suite (e.g., El Goresy *et al.*, 1971). These observations are consistent with the conclusion of Usselman (1975) that ilmenite compositions are controlled by reequilibration with residual liquid and coexisting ferromagnesian phases.

Metal

Metallic Fe appears to have been a stable phase throughout most, if not all, of the crystallization history of the ISBs. It occurs in the rims of olivine phenocrysts in porphyritic basalts and is commonly intergrown with early spinel in all the samples studied. Metal inclusions have not been observed in the *cores* of olivine phenocrysts, except as the crystallization products of melt inclusions. Hewins and Goldstein (1974) recognized the same pattern in a detailed study of metal-olivine relationships in 12004 and 12022. Consequently, it has not been demonstrated that the parental liquid of the ISBs was saturated with metallic Fe under the P–T–f_{O_2} conditions at which liquidus olivine precipitated. The textural relations are compatible with the presence of a metal phase *on eruption* and do not preclude earlier metal saturation since the small number and grain size of the metal grains may account for the negative observation. However, modeling of Ni–Co fractionation in 12022 by Hewins and Goldstein (1974) suggests that olivine crystallized alone prior to the onset of metal saturation.

Brett *et al.* (1971) and Reid *et al.* (1970) analyzed metal in several Apollo 12 basalts including 12008 and 12022. Early metal enclosed in olivine phenocrysts tends to be strongly enriched in Ni (30–56% Ni) relative to the slightly later and more abundant population intergrown with spinel. Metallic Fe abundances reported in Gibson *et al.* (1977) indicate an accumulation of this phase in 12016, 12036, and 12005 relative to all the other ISB samples, consistent with the presence of metal inclusions in cumulus olivine and spinel. 12008 contains much less metal than 12045 (220 vs. 520 ppm) apparently as the result of quenching, which acted to suppress the growth of groundmass metal. In the medium- to coarse-grained basalts metal tends to be concentrated in the mesostasis segregations where it is typically intergrown with troilite. It also occurs associated with reduction assemblages in ilmenite and ulvöspinel where it forms veins crosscutting silicates and oxide phases and as micron-sized disseminated blebs. As noted above, the intensity of subsolidus reduction increases from 12056 → 12016 → 12005.

Three element analyses (Ni–Co–Fe) of metal in 12056, 12016, and 12005 are reported in Fig. 15. Metal grains intergrown with late-stage phases, such as troilite, in 12056 and 12016 are characterized by very low Ni and Co consistent with concentration of these elements in early crystallizing metal. Metal within olivine in 12056 has the highest Ni of any analyzed. Keil *et al.* (1971) report Ni–Co relations in metal of 12051 very similar to those in 12016 and 12056. Metal in 12005 (this study) and 12036 (Keil *et al.*, 1971) is characterized by relatively constant Co (1–2.5 wt.%) and Ni-contents which are intermediate between the

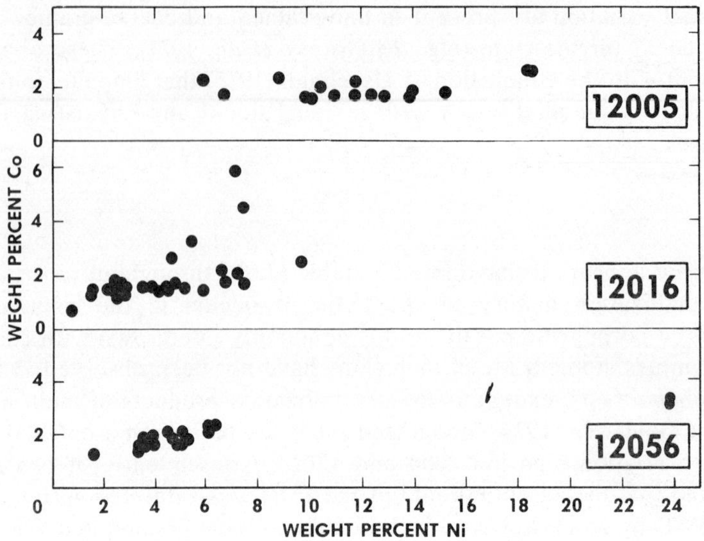

Fig. 15. Microprobe analyses of metal grains in 12005, 12016, and 12056 (Ni vs. CO).

strongly Ni-enriched early metal in olivine phenocrysts (>20 Ni wt.%) and the low-Ni late-stage metal associated with mesostasis segregations (<3–6 Ni wt.%). A similar pattern of Ni–Co variation was found by Dowty *et al.* (1974) in a range of textural types in Apollo 15 mare basalts (see their Fig. 8). The higher average Ni-content of the metal grains in the cumulates (12036–12005) probably reflects the presence of partly equilibrated Ni-rich metal added during olivine accumulation. This compositional pattern is similar to that recorded in ISB ilmenites.

VI. SUMMARY AND CONCLUSIONS

As vertical cross sections of lunar lava flows have been sampled by random impact excavation rather than systematic collection from continuous outcrop, reconstruction of petrogenetic relationships among cogenetic rocks must proceed by the interpretation of data gathered from the individual samples. Our ability to make such a reconstruction hinges on the assumptions that (1) the degree of differentiation indicated by the bulk chemistry of individual samples is a function of the cooling history at that particular location in the flow and (2) that each sample will have recorded in its mineralogy and textures a unique crystallization history. An internally consistent conceptual framework within which all the ilmenite suite basalts can be related has emerged in the integration of combined mineral chemistry data and petrographic interpretations with a fractionation model determined independently on the basis of bulk chemistry (Rhodes *et al.*, 1977). The major conclusion we have derived from this petrologic synthesis is that the bulk chemical variations in the ilmenite suite can be reconciled with differentiation within a single cooling unit. The critical observation is that included within the suite there are (1) olivine-porphyritic basalts, (2) medium-grained basalts with evolved compositions that can be related to the porphyritic basalts by subtraction of olivine and (3) their complementary derivatives, partial olivine cumulates.

The three olivine-porphyritic basalts 12008 (vitrophyre), 12045, and 12022 (variolitic groundmasses) comprise a *decreasing cooling rate sequence* corresponding to *increasing olivine removal* (up to 7% in 12022). The progressive coarsening of the groundmass phases and variations in the morphologies of various minerals that are produced by crystallization at different degrees of undercooling are clear indications of the cooling rate sequence. The Ti–Al relations among the phenocrystic and groundmass pyroxenes vary in a way entirely consistent with this sequence. On the basis of these relationships, the porphyritic basalts are inferred to represent the chilled carapace of the flow and the cooling rate hierarchy (fast→slow) is presumed to reflect a descending stratigraphic sequence. Walker *et al.* (1976b) reached an analogous conclusion for the Apollo 12 olivine basalt suite using a somewhat different approach. On the basis of a monotonic covariance between decreasing cooling rate and increasing olivine accumulation, they inferred that the olivine basalts (varying from vitrophyres to partial cumulates) represent samples from the base of a single flow.

The medium-grained, ophitic to subophitic, evolved basalts (12047, 12054, 12051, 12063, and 12056) are the best represented textural-chemical type among the ilmenite suite basalts. These are quite distinct texturally from the porphyritic basalts and have had much slower cooling histories. Chemically these range from quartz-normative basalts from which olivine removal has been 100% efficient (12047, 12054, and 12051) to less fractionated variants that have retained some of the original phenocrystic olivine (12063 and 12056). The inferred descending stratigraphic sequence corresponds to *decreasing cooling rate* and *decreasing olivine removal.* These trends exhibit the inverse correlation found among the porphyritic basalts. As a group they must have formed in the upper portion of the flow, below the porphyritic basalts and above the cumulates.

Sample 12016 is also an equigranular, medium-grained basalt. Its major element bulk chemistry is very much like the porphyritic basalts but its textures are indicative of a cooling history intermediate between the evolved basalts and the more slowly cooled cumulates. Consequently, its stratigraphic position also must have been intermediate between the zones of olivine removal and accumulation so that the amount lost by settling was just compensated for from above.

Mineral chemistry variations measured in 12056 and 12016 that are dependent on cooling rate (e.g., pyroxene compositions, spinel parageneses, and mineral reequilibration) are entirely consistent with radically lower cooling rates for the medium-grained basalts compared to the porphyritic basalts and with the detailed stratigraphic inferences. Most of the chemical variation among the ISBs can be accounted for by fractional crystallization of olivine with the composition of Fo_{72} plus minor chromian spinel and metal. This is consistent with the compositions of olivine phenocrysts in the porphyritic basalts. However, in the most evolved ISBs additional small amounts of more Fe-rich olivine (Fo_{69-60}) are required in the modeling calculations (Rhodes *et al.*, 1977). All the medium-grained basalts contain strongly zoned pyroxene and minor fayalite + silica + alkali-rich glass in their residual mesostases indicating substantial internal fractionation.

The fractionation modeling calculations of Rhodes *et al.* (1977) require approximately 30% olivine accumulation from a parent of 12008 composition to produce 12005. This sample is consequently the most extreme example of a mare basalt cumulate thus far recognized in the Apollo collection. Sample 12036 is tentatively included in the ISBs despite some anomalous chemical characteristics (Rhodes *et al.*, 1977). In virtually all respects (mineral chemistry, modal mineralogy, and bulk chemistry) it is intermediate between 12016 and 12005. However, pyroxene must also be involved as a fractionating phase in the cumulates. Petrographic support for this comes from the presence of large pyroxene oikocrysts in 12005 (and 12036) that include abundant embayed and rounded olivine. These features are interpreted as evidence for extensive adcumulus growth of pyroxene and an olivine-liquid peritectic reaction. Thus, the pyroxene is not a cumulus phase but a product of diffusion in intercumulus liquid. A second allied deviation from simple olivine fractionation is necessitated by the fact that the medium-grained, evolved basalts and cumulates are enriched and depleted respectively in magmaphile elements (e.g., Fe, Ti, K, P, and REE) relative to calculated values derived from modeling the fractionation on major elements and trace elements that are preferentially incorporated into early crystallizing phases (e.g., Co, Ni, and Cr). Sample 12016 is particularly enriched in magmaphile elements as befits its inferred stratigraphic position immediately above the cumulate horizon. This complexity is entirely consistent with the paucity of residual mesostasis and the unusually limited Fe enrichment recorded in the ferromagnesian minerals of 12005 and 12036. Additional features that are indicative of slow cooling in the presence of abundant intercumulus liquid abound in 12005. The plagioclase exhibits restricted Fe/Mg but greater than normal alkali enrichment. Spinel parageneses are unique in that none of the individual grains is zoned from titanian chromium spinel to chromium ulvöspinel even though the composite of spinel analyses shows an unbroken range between the most Ti-poor and Ti-rich compositions. The pyroxenes record evidence for slow cooling in their relatively coarse exsolution lamellae, low Ti and Al and low Ti:Al plus an absence of compositions with Ca-contents intermediate between augite and low-Ca pyroxene. The latter may be indicative of an approach to an equilibrium solvus.

Acknowledgments—We are indebted to Bob Myers and Ken Edloe (LEC) for software development in conjunction with automation of the MAC microprobe. Gary Lofgren and J. M. Rhodes contributed numerous invaluable insights concerning mare basalt petrology and geochemistry. An exceptionally thorough review by S. A. Morse contributed to improvement of the original manuscript. Elinor Stockton, Linda Dungan, and Lina Romero assisted in preparation of the manuscript. This paper constitutes Contribution No. 4 of the Basaltic Volcanism Study Project. The Project is organized and administered by the Lunar Science Institute/Universities Space Research Association under contract NSR 09–051–001 with the National Aeronautics and Space Administration.

References

Akella J., Williams R. J., and Mullins O. (1976) Solubility of Cr, Ti, and Al in co-existing olivine, spinel, and liquid at 1 atm. *Proc. Lunar Sci. Conf. 7th*, p. 1179–1194.

Brett R., Butler P., Meyer C., Reid A. M., Takeda H., and Williams R. (1971) Apollo 12 igneous rocks 12004, 12008, 12009, and 12022: A mineralogical and petrological study. *Proc. Lunar Sci. Conf. 2nd*, p. 301–317.

Brown G. M. and Peckett A. (1971) Selective volatilization on the lunar surface: Evidence from Apollo 14 feldspar phyric basalts. *Nature* **234**, 262–266.

Brown G. M., Emeleus C. H., Holland J. G., Peckett A., and Phillips R. (1971) Picrite basalts, ferrobasalts, feldspathic norites, and rhyolites in a strongly fractionated lunar crust. *Proc. Lunar Sci. Conf. 2nd*, p. 583–600.

Brown G. M., Peckett A., Emeleus C. H., Phillips R., and Pinsent R. H. (1975) Petrology and mineralogy of Apollo 17 mare basalts. *Proc. Lunar Sci. Conf. 6th*, p. 1–13.

Busche F. D., Conrad G. H., Keil K., Prinz M., Bunch T. E., Erlichman J. E., and Quaide W. L. (1971) Electron microprobe analyses of minerals from Apollo 12 lunar samples. Univ. New Mexico, Inst. Meteoritics, Spec. Pub. 3, 61 pp.

Busche F. D., Prinz M., Keil K., and Bunch T. E. (1972) Spinel and the petrogenesis of some Apollo 12 igneous rocks. *Amer. Mineral.* **57**, 1729–1747.

Butler P., Jr. (1972) Compositional characteristics of olivines from Apollo 12 samples. *Geochim. Cosmochim. Acta* **36**, 773–785.

Cameron E. N. (1971) Opaque minerals in certain lunar rocks from Apollo 12. *Proc. Lunar Sci. Conf. 2nd*, p. 193–206.

Compston W., Berry H., Vernon M. J., Chappell B. W., and Kaye M. J. (1971) Rb–Sr chronology and chemistry of lunar material from the Ocean of Storms. *Proc. Lunar Sci. Conf. 2nd*, p. 1471–1485.

Crawford M. L. (1973) Crystallization of plagioclase in mare basalts. *Proc. Lunar Sci. Conf. 4th*, p. 705–717.

Donaldson C. H., (1976) An experimental investigation of olivine morphology. *Contrib. Mineral. Petrol.* **57**, 187–213.

Donaldson C. H., Usselman T. M., Williams R. J., and Lofgren G. E. (1975) Experimental modeling of the cooling history of Apollo 12 olivine basalts. *Proc. Lunar Sci. Conf. 6th*, p. 843–869.

Dowty E., Prinz M., and Keil K. (1973) Composition, mineralogy and petrology of 28 mare basalts from Apollo 15 rake samples. *Proc. Lunar Sci. Conf. 4th*, p. 423–444.

El Goresy A., Prinz M., and Ramdohr P. (1976) Zoning in spinels as an indicator of the crystallization histories of mare basalts. *Proc. Lunar Sci. Conf. 7th*, p. 1261–1279.

El Goresy A. and Ramdohr P. (1975) Subsolidus reduction of lunar opaque oxides: Textures, assemblages, geochemistry, and evidence for a late stage endogenic gaseous mixture. *Proc. Lunar Sci. Conf. 6th*, p. 729–745.

El Goresy A., Ramdohr P., and Taylor L. A. (1971) The opaque minerals in the lunar rocks from Oceanus Procellarum. *Proc. Lunar Sci. Conf. 2nd*, p. 219–235.

Engel A. E. J., Engel C. G., Sutton A. L., and Myers A. T. (1971) Composition of five Apollo 11 and Apollo 12 rocks and one Apollo 11 soil and some petrogenetic considerations. *Proc. Lunar Sci. Conf. 2nd*, p. 439–448.

Gibson E. K., Jr., Brett R., and Andrawes F. (1977) Sulfur in lunar mare basalts as a function of bulk composition. *Proc. Lunar Sci. Conf. 8th*. This volume.

Green D. H., Ringwood A. E., Ware N. G., Hibberson W. O., Major A., and Kiss E. (1971) Experimental petrology and petrogenesis of Apollo 12 basalts. *Proc. Lunar Sci. Conf. 2nd*, p. 601–615.

Haggerty S. E. and Meyer H. O. A. (1970) Apollo 12: Opaque oxides. *Earth Planet. Sci. Lett.* **9**, 379–387.

Hewins R. H. and Goldstein J. I. (1974) Metal-olivine associations and Ni–Co contents in two Apollo 12 mare basalts. *Earth Planet. Sci. Lett.* **24**, 59–70.

Hollister L. S., Trzcienski W. E., Jr., Hargraves R. B., and Kulick C. G. (1971) Petrogenetic significance of pyroxenes in two Apollo 12 samples. *Proc. Lunar Sci. Conf. 2nd*, p. 529–557.

Huebner J. S., Lipin B. R., and Wiggins L. B. (1976) Partitioning of chromium between silicate crystals and melts. *Proc. Lunar Sci. Conf. 7th*, p. 1195–1220.

Irvine T. N. (1965) Chromian spinel as a petrogenetic indicator. Part I: Theory. *Can. J. Earth Sci.* **2**, 648–672.

James O. B. and Wright T. L. (1972) Apollo 11 and 12 mare basalts and gabbros: Classification, compositional variations, and possible petrogenetic relations. *Bull. Geol. Soc. Amer.* **83**, 2357–2382.

Keil K., Prinz M., and Bunch T. E. (1971) Mineralogy, petrology, and chemistry of some Apollo 12 samples. *Proc. Lunar Sci. Conf. 2nd*, p. 319–341.

Kirkpatrick R. J. (1976) Towards a kinetic model for the crystallization of magma bodies. *J. Geophys. Res.* **81**, 2565–2571.

Kushiro I. and Haramura H. (1971) Major element variation and possible source materials of Apollo 12 crystalline rocks. *Science* **171**, 1235–1237.

Lofgren G. E. (1973) Temperature induced zoning in synthetic plagioclase feldspar. In *The Feldspars* (W. S. Mackenzie and J. Zussman, eds.), p. 362–375. Manchester University Press.

Lofgren G. E. and Gooley R. (1977) Simultaneous crystallization of feldspar intergrowths from the melt. *Amer. Mineral.* **62**, 217–228.

Longhi J., Walker D., and Hays J. F. (1976) Fe and Mg in plagioclase. *Proc. Lunar Sci. Conf. 7th*, p. 1281–1300.

Moore J. G. and Evans B. W. (1967) The role of olivine in the crystallization of the prehistoric Makaopuhi tholeiitic lava lake. *Contrib. Mineral. Petrol.* **15**, 202–223.

Nyquist L. E., Bansal B. M., Wooden J. L., and Wiesmann H. (1977) Sr-isotopic constraints on the petrogenesis of Apollo 12 mare basalts. *Proc. Lunar Sci. Conf. 8th.* This volume.

Papike J. J., Hodges F. N., Bence A. E., Cameron M., and Rhodes J. M. (1976) Mare basalts: Crystal chemistry, mineralogy, and petrology. *Rev. Geophys. Space Phys.* **14**, 475–540.

Powell B. N., Dungan M. A., and Weiblen P. W. (1975) Apollo 16 feldspathic melt rocks: Clues to the magmatic history of the lunar crust. *Proc. Lunar Sci. Conf. 6th*, p. 415–433.

Reid A. M., Meyer C., Jr., Harmon R. S., and Brett R. (1970) Metal grains in Apollo 12 igneous rocks. *Earth Planet. Sci. Lett.* **9**, 1–5.

Rhodes J. M., Blanchard D. P., Dungan M. A., Brannon J. C., and Rodgers K. V. (1977) Chemistry of Apollo 12 mare basalts: Magma types and fractionation processes. *Proc. Lunar Sci. Conf. 8th.* This volume.

Roeder P. L. and Emslie R. F. (1970) Olivine-liquid equilibrium. *Contrib. Mineral. Petrol.* **29**, 275–289.

Schaal R. B. and Hörz F. (1977) Shock metamorphism of lunar and terrestrial basalts. *Proc. Lunar Sci. Conf. 8th.* This volume.

Trzcienski W. E., Jr. and Kulick C. G. (1972) Plagioclase and Ba–K phases from Apollo samples 12063 and 14310. *Proc. Lunar Sci. Conf. 3rd*, p. 591–602.

Usselman T. M. (1975) Ilmenite chemistry in mare basalts, an experimental study (abstract). In *Origins of Mare Basalts and their Implications for Lunar Evolution*, p. 164–168. The Lunar Science Institute, Houston.

Usselman T. M., Lofgren G. E., Donaldson C. H., and Williams R. J. (1975) Experimentally reproduced textures and mineral chemistries of high-titanium mare basalts. *Proc. Lunar Sci. Conf. 6th*, p. 997–1020.

Walker D., Kirkpatrick R. J., Longhi J., and Hays J. F. (1976a) Crystallization history of lunar picritic basalt sample 12002: Phase equilibria and cooling rate studies. *Bull. Geol. Soc. Amer.* **87**, 646–656.

Walker D., Longhi J., Kirkpatrick R. J., and Hays J. F. (1976b) Differentiation of an Apollo 12 picrite magma. *Proc. Lunar Sci. Conf. 7th*, p. 1365–1389.

Warner J. L. (1971) Lunar crystalline rocks: Petrology and geology. *Proc. Lunar Sci. Conf. 2nd*, p. 469–480.

Weill D. F., Grieve R. A., McCallum I. S., and Bottinga Y. (1971) Mineralogy-petrology of lunar samples. Microprobe studies of samples 12021 and 12022; viscosity of melts of selected lunar compositions. *Proc. Lunar Sci. Conf. 2nd*, p. 413–430.

Willis J. P., Ahrens L. H., Dauchin R. V., Erlank A. J., Gurney J. J., Hofmeyr P. K., McCarthy T. S., and Ovren M. J. (1971) Some interelement relationships between lunar rocks and fines and stony meteorites. *Proc. Lunar Sci. Conf. 2nd*, p. 1123–1138.

Proc. Lunar Sci. Conf. 8th (1977), p. 1383–1415.
Printed in the United States of America

Sr-isotopic constraints on the petrogenesis of Apollo 12 mare basalts

L. E. NYQUIST

NASA Johnson Space Center, Houston, Texas 77058

B. M. BANSAL, J. L. WOODEN, and H. WIESMANN

Lockheed Electronics Co., Inc., Houston, Texas 77058

Abstract—The Sr-isotopic systematics of a number of Apollo 12 basalts have been determined as part of a larger effort including petrographic, major, minor, and trace element studies. Crystallization ages (T) and initial $^{87}Sr/^{86}Sr$ (I) were determined for pigeonite basalts 12039 and 12055 and ilmenite basalt 12051. The results are: 12039: $T = 3.19 \pm .06$ AE, $I = 0.69965 \mp 5$, 12055: $T = 3.19 \pm .06$ AE, $I = 0.69962 \mp 4$; 12051: $T = 3.16 \pm .04$ AE, $I = 0.69938 \mp 2$. (2σ errors of the York (1966) program, 1 AE = 10^9 yr). These ages are in good agreement with values for Apollo 12 basalts obtained by other workers.

$^{87}Sr/^{86}Sr$ measurements for a larger suite of whole rock and plagioclase samples permit calculation of I-values for these samples. I-values for olivine and pigeonite basalts are indistinguishable, whereas I-values for ilmenite basalts are distinctly lower. This difference corresponds to other compositional differences, in particular TiO_2 and REE abundances. Initial $^{87}Sr/^{86}Sr$ is negatively correlated with Sm/Ce ratios. Alkali, alkaline earth, and REE abundances were also measured in this investigation for a sub-set of the total sample suite. A self-consistent model for the evolution of the Sr-isotopic composition and REE abundances can be formulated for basalt formation by small (2–10%) degrees of melting of cumulate sources composed of varying proportions of olivine, orthopyroxene, and clinopyroxene. An important feature of the model is the necessity to assume non-chondritic relative abundances of the REE in the parental liquid at the time of formation of the cululate sources. The model also provides a satisfactory description of the evolution of the Sr-isotopic composition and REE abundances of Apollo 17 high-Ti basalts and Apollo 15 low-Ti basalts. Two important features of the major element composition of the basalts, the "mg"-values and TiO_2 contents, also appear to be adequately described. Finally, a lunar "scenario" involving open system crystallization of a magma ocean from the crust downward is suggested. This scenario seems compatible with the concepts developed in this paper and other important observational constraints.

INTRODUCTION

THE SR-ISOTOPIC and trace element compositions of a number of Apollo 12 mare basalts have been measured in order to place constraints on their petrogenesis. This study is part of a cooperative effort including petrographic and major, minor, and trace element studies of a larger suite of samples (Dungan and Brown, 1977; Rhodes *et al.*, 1977). The delineation of Apollo 12 basalts into chemical sub-groups has been refined as a result of this combined study (see Rhodes *et al.*, 1977). Our ultimate goal is the determination of complete mineral isochrons for at least one member of each chemical subgroup and for any "exceptional" samples. In addition, whole rock $^{87}Sr/^{86}Sr$ measurements were made for the larger sub-set of samples for which alkali, alkaline earth, and rare earth elements (REE) were also measured. Finally, because of the usefulness of the initial $^{87}Sr/^{86}Sr$ ratios (I-values) for classification and petrogenetic interpretations, Rb–Sr analyses were made on plagioclase separates of several samples for which complete mineral isochrons were not determined. Emphasis was placed on "ilmenite" basalts with TiO_2 contents of 4–5% because: (i) they are intermediate in composition between Apollo 12 and 15 low-Ti and Apollo 11 and 17 high-Ti basalts; (ii) the Apollo 12 landing site is near a mare region of intermediate TiO_2 content (Pieters and McCord, 1976); and (iii) hitherto these basalts had been poorly represented in the lunar data base.

Our joint studies, combined with earlier work by a number of authors, have resulted in a significant clarification of the characteristics of the Apollo 12 ilmenite basalts. This chemical sub-group is found to be abundant at the Apollo 12 site. Whereas it is possible to relate the two other major subgroups, the olivine and pigeonite basalts, to each other by near surface fractionation processes; neither of these subgroups can be related by this process to the ilmenite basalts (Rhodes *et al.*, 1977). Two key differences between the ilmenite and olivine-pigeonite basalts are their differing Sr-isotopic and REE compositions. It is suggested that these differences are inherited from distinct basalt source regions. A self-consistent model for the evolution of the Sr-isotopic composition and the REE abundances of each major group can be formulated for basalt formation by small (2-10%) degrees of partial melting of cumulate sources composed of olivine (ol), orthopyroxene (opx), and clinopyroxene (cpx). An important feature of this model is the necessity to assume non-chondritic relative abundances of the REE in the parental liquid at the time of formation of the cumulate sources. Confidence in the calculational technique is increased by the ability to obtain calculated "mg" values (100 MgO/(MgO + FeO) in mole %) in agreement with those directly found by the methods of experimental petrology. Although the calculations presented here may change with the acquisition of new data and insights they hold promise for the eventual unification of the chemical and isotopic data for lunar mare basalts.

Finally a scenario of lunar differentiation is suggested which seems most compatible with the concepts developed in this paper and other important observational constraints. Whereas this scenario is similar in many respects to earlier ones suggested by other authors, it has some unique features and interesting implications for further lunar studies. In particular, the postulated sequence of mafic cumulates is inverted with respect to that of several current models of lunar structure (e.g., Taylor and Jakeš, 1974).

Analytical Procedures

All samples were crushed and ground in a boron carbide mortar. Samples for mineral separation were sieved into <44 μm, 44–74 μm, 74–149 μm, and >149 μm size fractions. The sieve was thoroughly cleaned and the nylon mesh replaced between samples. The mineral separates were primarily obtained from the 74–149 μm fractions.

12039,19

The allocated 3.02 g sample was crushed and ground. A 51 mg whole-rock sample was taken from this material. The mineral separate material was alternately sieved and crushed until the major portion of the sample was in the 74–149 μm size range. This size fraction was then separated into plagioclase, pyroxene, and ilmenite concentrates on a Frantz magnetic separator. Essentially pure mineral separates for plagioclase, brown pyroxene, yellow pyroxene, and ilmenite were prepared from these concentrates by handpicking under a binocular microscope.

12051,135

Re-allocation of the unused portion of 12051,40 previously studied by Hubbard and Gast (1971) was requested from the Lunar Sample Curator. 533 mg of this material were used for mineral separates and processed as described under 12039,19. Plagioclase, brown pyroxene, and ilmenite were the purified products.

12055,28

An original 3.23 g sample was crushed and ground. 50 mg of this material were taken for a whole rock sample and 977 mg for mineral separations. Purified mineral separates of plagioclase, brown pyroxene, and ilmenite were produced as described under 12039,19.

12022,168

Re-allocation of the unused portion of 12022,38 previously studied by Hubbard and Gast (1971) was requested from the Lunar Sample Curator. A 403 mg coarsely ground sample was finely ground. 20 mg were taken for whole-rock analyses. Because of the fine grain size of the ground mass of this rock, the sample was ground until the major portion of the sample was in the 44–74 μm size range. A plagioclase separate was prepared from this size fraction by taking the float from a 2.9 g/cm^3 heavy liquid (half bromoform and half tetrabromoethane). The plagioclase was cleaned by washing with spectrophotometric grade acetone before dissolution.

12014,18; 12016,9; 12017,36; 12031,25; 12054,62; 12056,13

Plagioclase separates were obtained from these samples as described for 12039,19. Some of these samples were only coarsely ground in order to preserve suitable material for future mineral separations. Unfortunately, this increases the probability that the 50–70 mg samples used for whole rock analyses were unrepresentative. This problem is particularly acute for 12031,25 which is very coarse grained. A 1.94 g sample of 12031,25 was coarsely ground and 72 mg used for whole rock analysis. Trace element abundances measured for this sample were ~20% lower than for 12031,23 analyzed by Rhodes *et al.* (1977). Samples of 12017,36 and 12056,13 are expected to be more representative because of the finer grain size of these basalts. Original sample weights before coarse grinding were 2.05 g for 12056,13 and 3.12 g for 12056,13. The original 2.75 g sample of 12014,18 was first coarsely ground and then one-half of it further ground to a fine grain size. A 50 mg whole rock sample was then taken from this fine material. Samples 12016,9 and 12054,62 were originally ~1.3 g and had been finely ground by J. M. Rhodes prior to XRF analysis. Thus, whole rock samples of 12017, 12016, and 12054 should be representative unless heterogeneities exist among ~1 g samples.

12005,7; 12008,48; 12015,9 + 11; 12036,24; 12045,12; 12047,3

The whole-rock samples for these rocks were taken from approximately 1.3 g samples allocated to J. M. Rhodes. These samples had been prepared by J. M. Rhodes by crushing and fine grinding in a boron carbide mortar. The ~50 mg samples used for analysis are expected to be representative of the larger 1.3 g samples.

Chemical and Mass Spectrometric Procedures

Chemical procedures for the whole-rock multielement analyses are essentially those of Gast *et al.* (1970). The major portion of each blank reported in Table 3 is derived from cross-contamination from the numerous spikes used. The ion exchange procedures for the mineral separates are similar to those used for the multielement analyses but with only a Rb and Sr mixed spike being used. Blanks are 0.045 ng for Rb and 0.5 ng for Sr with an uncertainty of ±50%. Mass spectrometric procedures are those by Nyquist *et al.* (1973, 1974). All data reported here were obtained by magnetic field scanning.

RESULTS

The Rb–Sr analytical data are given in Table 1. Single stage model ages and calculated initial $^{87}Sr/^{86}Sr$ (I) are also given for the whole rock analyses. The I-values are calculated assuming an average Apollo 12 basalt age of 3.25 AE from the data of Papanastassiou and Wasserburg (1970, 1971a,b). Calculated I-values are given for plagioclase separates when these are the only mineral separates analyzed for a given basalt.

Internal isochrons were determined for the mineral separate data given in Table 1 using the York (1966) regressions. Ages (T) and initial $^{87}Sr/^{86}Sr$ (I) found were 12039: $T = 3.19 \pm .06$ AE, $I = 0.69965 \mp 5$; 12051: $T = 3.16 \pm .06$ AE, $I = 0.69938 \mp 2$; 12055: $T = 3.19 \pm .06$ AE, $I = 0.69962 \mp 4$. The uncertainties are the 2σ values from the York program.

The mineral separate data are given in Fig. 1 in terms of deviations from the best fit isochron. As can be seen from Fig. 1, the 2σ error from the York (1966) program may be fortuitously small for 12051. We visually estimate the uncertainties in T and I as $\leqslant 0.14$ AE, and $\leqslant 0.00005$, respectively. Even with the smaller error limits our data for 12051 are in satisfactory agreement with $T = 3.26 \pm .10$ AE and $I = 0.69932 \mp 6$ as found by Papanastassiou and Wasserburg (1971a). The value of 3.58 ± 0.30 AE found by Compston *et al.* (1971) seemed to suggest that this ilmenite basalt might have an older age closer to that of the high-Ti basalts of Apollo 11 and 17. This, however, is not the case.

Table 2 summarizes Rb–Sr and ^{40}Ar–^{39}Ar ages of Apollo 12 basalts. The age data reported here are in general agreement with those of other workers. However, our data do not reflect the considerable range of published values. Some of the dispersion in the ages summarized in Table 2 is undoubtedly due to interlaboratory bias and/or systematic errors. We do not attempt to evaluate these factors here; a critical review of Rb–Sr data is given elsewhere (Nyquist, 1977).

Nine Rb–Sr ages for Apollo 12 basalts have been determined by Papanastassiou and Wasserburg (1970, 1971a,b) and provide a good data set for comparison to our own work. The ages reported here are slightly younger than six of the ages reported by Papanastassiou and Wasserburg (1970, 1971a,b) but are in nearly exact agreement with three others. Thus, there is no evidence of an interlaboratory bias in ages. Consequently, an average age of 3.25 AE, reflecting primarily the larger data base of Cal Tech values, was used in calculating I-values for the data in Table 1. Had we used the average age found in this

Table 1. Rb–Sr analytical data for Apollo 12 mare basalts.

Sample	Wt. (mg)	Rb (ppm)	Sr (ppm)	$^{87}Rb^{a}/^{86}Sr$	$^{87}Sr^{b}/^{86}Sr$	T_B[c]	T_L[d]	$I(3.25)$[e]	Type[f]
I. Whole rock and plagioclase data									
12005,7 WR[g]	38	0.501	78.2	0.0185 ± 6	0.70021 ± 10	4.19 ± .52	4.45 ± .52	0.69935 ± 10	I
12008,48 WR	38	0.681	133.4	0.0148 ± 4	0.70007 ± 7	4.55 ± .45	4.89 ± .45	0.69939 ± 7	I
12014,18 P1	9.3	0.599	323	0.00537 ± 5	0.69989 ± 9	—	—	0.69964 ± 9	O
12014,18 WR	50	0.926	90.4	0.0296 ± 4	0.70096 ± 7	4.38 ± .22	4.54 ± .22	0.69959 ± 7	
12015,9–11 WR	41	1.05	98.4	0.0309 ± 5	0.70093 ± 5	4.14 ± .18	4.29 ± .18	0.69950 ± 5	O
12016,9 P1	9.8	0.191	475	0.00116 ± 2	0.69943 ± 8	—	—	0.69938 ± 8	I
12017,36 P1	17	0.539	332	0.00470 ± 4	0.69976 ± 7	—	—	0.69954 ± 7	P
12017,36 WR	52	1.263	118	0.0310 ± 4	0.70100 ± 7	4.28 ± .21	4.44 ± .21	0.69957 ± 7	
12022,168 P1	2.1	0.959	343	0.0081 ± 1	0.69958 ± 14	—	—	0.69921 ± 14	I
12022,168 WR	22.1	0.708	134.7	0.0152 ± 1	0.69995 ± 7	3.91 ± .35	4.23 ± .34	0.69925 ± 7	
12031,25 P1	12	0.135	270	0.00145 ± 3	0.69963 ± 7	—	—	0.69956 ± 7	F?
12031,25 WR	73	0.797	153	0.0151 ± 1	0.70022 ± 6	5.15 ± .30	5.46 ± .30	0.69952 ± 6	
12036,24 WR	44	1.08	94.5	0.0331 ± 3	0.70099 ± 5	3.99 ± .15	4.14 ± .15	0.69946 ± 5	I?
12045,12 WR	49	0.717	143.3	0.0145 ± 1	0.69994 ± 5	4.05 ± .26	4.38 ± .26	0.69927 ± 5	I
12047,3 WR	44	0.990	164	0.0175 ± 2	0.70011 ± 6	4.04 ± .28	4.31 ± .28	0.69930 ± 6	I
12054,62 P1	3.8	0.408	388	0.00304 ± 5	0.69955 ± 10	—	—	0.69941 ± 10	I
12056,13 P1	6.9	0.299	433	0.00200 ± 3	0.69953 ± 7	—	—	0.69944 ± 7	I
12056,13 WR	50	0.785	161.5	0.0141 ± 1	0.69997 ± 4	4.31 ± .22	4.64 ± .22	0.69932 ± 4	
12076,18 WR	56	1.022	93.6	0.0316 ± 4	0.70112 ± 4	4.46 ± .14	4.61 ± .14	0.69966 ± 4	O
II. Whole rock and mineral separates of 12039,19								0.69965 ± 5	P
Br. Px	6.8	0.807	29.4	0.0794 ± 7	0.70319 ± 12	—	—	—	
W.R. 2	27.8	1.26	136.7	0.0266 ± 2	0.70083 ± 7	4.53 ± .22	4.71 ± .22	—	
Y. Px	6.65	1.501	38.6	0.1125 ± 10	0.70461 ± 19	—	—	—	
Ilm	7.9	2.163	36.6	0.1710 ± 13	0.70741 ± 23	—	—	—	
Plag 2	4.6	0.200	320.2	0.00181 ± 5	0.69969 ± 11	—	—	—	
W.R. 1	50.8	1.29	138.4	0.0270 ± 3	0.70091 ± 6	4.67 ± .20	4.84 ± .20	—	

Table 1. (*Continued*).

Sample	Wt. (mg)	Rb (ppm)	Sr (ppm)	$^{87}Rb^{a}/^{86}Sr$	$^{87}Sr^{b}/^{86}Sr$	T_B[c]	T_L[d]	$I(3.25)$[e]	Type[f]
III. Whole rock and mineral separates of 12051,135								0.69938 ± 2	I
Plag 1	6.7	0.360	384.4	0.00271 ± 3	0.69953 ± 10	—	—	—	
Brown Px	11.4	0.667	49.0	0.0394 ± 3	0.70116 ± 14	—	—	—	
Ilmenite	10.6	1.509	73.6	0.0593 ± 5	0.70204 ± 8	—	—	—	
Plag 2	12.8	0.404	391.5	0.00299 ± 3	0.69951 ± 5	—	—	—	
IV. Whole rock and mineral separates of 12055,28								0.69962 ± 5	P
W.R.	50	1.14	120.4	0.0274 ± 4	0.70087 ± 5	4.50 ± .19	4.68 ± .19	—	
Brown Px	12.8	0.660	41.97	0.0455 ± 5	0.70172 ± 9	—	—	—	
Y. Px	14.0	0.0987	7.78	—	—	—	—	—	
Plag 1	9.1	0.634	332.0	0.0055 ± 1	0.69980 ± 11	—	—	—	
Plag 2	15.7	0.602	310.5	0.0056 ± 1	0.69987 ± 7	—	—	—	
Ilm	6.2	3.53	95.5	0.1069 ± 8	0.70445 ± 8	—	—	—	
V. Concurrent analyses of NBS 987									
5/26/76					0.71021 ± 4				
7/14/76					0.71027 ± 4				
9/8/77					0.71019 ± 6				
10/19/77					0.71025 ± 5				
1/17/77					0.71021 ± 7				
4/7/77					0.71014 ± 6				

[a]Uncertainties correspond to last figures.
[b]Uncertainties correspond to last figures and are $2\sigma_m$. Normalized to $^{88}Sr/^{86}Sr = 8.37521$.
[c]Model age assuming $I = 0.69910$ (BABI plus our bias).
[d]Model age assuming $I = 0.69903$ (Apollo 16 anorthosites for $T = 4.6$ AE).
[e]Initial $^{87}Sr/^{86}Sr$ calculated assuming $T = 3.25$ AE except for 12039, 12051, and 12055 for which directly measured values are given.
[f]Basalt type from Rhodes *et al.* (1977): O = olivine, P = Pigeonite, I = ilmenite, F = feldspathic.
[g]WR = whole rock, Pl = plag, Br. Px = brown pyroxene, Y. Px = yellow pyroxene, Ilm = ilmenite.

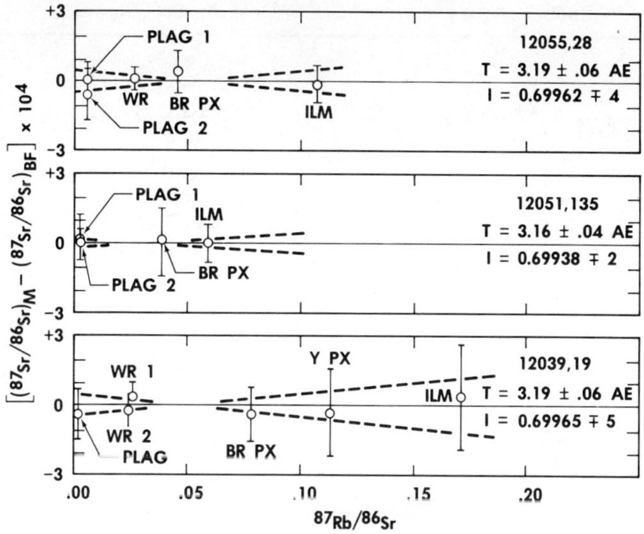

Fig. 1. Deviations of measured $^{87}Sr/^{86}Sr$ for mineral separates of 12039, 12051, and 12055 from the best fit isochron (York, 1966). Error limits (2σ) are shown by the dotted lines.

work, i.e., 3.18 AE, the difference in calculated I-values would have been at most 3 in 70,000.

Pieters and McCord (1976) have suggested that the ilmenite basalts may be derived from a small tongue of moderately high-Ti material a few kilometers to the east of the Apollo 12 landing site. These authors further suggest that the higher-Ti unit corresponds closely to a unit designated by Soderblom and Lebofsky (1972) as younger than the unit upon which the landing site is located. There is no evidence in the data summarized in Table 2 that the ilmenite basalts are detectably younger than the olivine and pigeonite basalts, however. On the other hand, the data do not exclude the hypothesis that the ilmenite basalts are stratigraphically younger than the olivine and pigeonite basalts as is suggested by Rhodes *et al.* (1977). The absolute age difference must be less than $\sim 10^8$ yr, if this is the case, however.

(T, I) parameters of Apollo 12 basalts are shown in Fig. 2. Interlaboratory bias is a severe problem for initial $^{87}Sr/^{86}Sr$ as is evident from Table 2. Thus, we have plotted only our own data and that from the Cal Tech laboratory. Since 1972 we have assumed an interlaboratory bias of 0.00012 when comparing data from the two laboratories (Nyquist *et al.*, 1972). Although initial $^{87}Sr/^{86}Sr$ determined for 12051 by the two groups differs by only 0.00006, the analytical uncertainty is great enough to be consistent with this previously established bias. Moreover, a bias correction of 0.00012 provides excellent agreement for other Apollo 12 data. Bias corrections for the data of Compston *et al.* (1971) and Murthy *et al.* (1971) are, in our opinion, less well known and I-values from these authors are not used in following discussions.

Table 2. Ages and initial $^{87}Sr/^{86}Sr(I)$ of Apollo 12 basalts

Sample	Type[a]	I	Rb–Sr (AE)	$^{40}Ar–^{39}Ar$ (AE)
12002	O	.69949 ± 5	3.36 ± .10 (1)	3.24 ± .05 (2)
				3.21 ± .04 (3)
12004	O	.69943 ± 5	3.29 ± .07 (4)	
		.69988 ± 7	2.94 ± .11 (6)	
		.69963 ± 6	3.01 ± .11 (7)	
		.69952 ± 4	3.03 ± .09 (7)	
12008	I	0.69923 ± 4 (5)[c]		3.18 ± .07 (8)[b]
				3.09 ± .07 (8)[b]
12009	O	0.69951 ± 5 (5)[c]		3.29 ± .07 (8)[b]
				3.17 ± .07 (8)[b]
12020	O			3.12 ± .04 (3)
12021	P	.69941 ± 5	3.33 ± .06 (5)	
12022	I			3.11 ± .04 (3)
12035	O	.69949	3.20 (5)	
12038	F	.69938 ± 8	3.28 ± .21 (6)	
12039	P	.69965 ± 5	3.19 ± .06 (9)	
12040	O	.69935 ± 3	3.30 ± .04 (5)	
		.69961 ± 8	3.21 ± .10 (6)	
12051	I	.69932 ± 6	3.26 ± .10 (1)	3.27 ± .05 (2)
		.69955 ± 16	3.58 ± .30 (6)	3.24 ± .05 (3)
		.69938 ± 2	3.16 ± .04 (9)	3.16 ± .05 (8)
				3.17 ± .07 (8)
12052	P	.69973 ± 10	3.28 ± .19 (6)	
		.69945 ± 9	2.92 ± .18 (7)	
12055	P	.69962 ± 4	3.19 ± .06 (9)	
12063	I	.69918 ± 6	3.30 ± .13 (5)	
		.69914 ± 3	3.34 ± .10 (7)	
12064	P	.69943 ± 6	3.18 ± .09 (5)	3.18 ± .01 (10)
				3.20 ± .04 (10)
12065	P	.69957 ± 5	3.16 ± .09 (5)	3.24 ± .05 (2)
				3.16 ± .04 (3)

[a]Basalt type from Rhodes *et al.* (1977): O = olivine, P = pigeonite, I = ilmenite, F = feldspathic.
[b]Total $^{40}Ar–^{39}Ar$ ages.
[c]Calculated assuming age = 3.25 AE.

References:

1. Papanastassiou and Wasserburg (1970).
2. Turner (1971).
3. Alexander *et al.* (1972), Alexander and Davis (1974).
4. Papanastassiou and Wasserburg (1971b).
5. Papanastassiou and Wasserburg (1971a).
6. Compston *et al.* (1971).
7. Murthy *et al.* (1971).
8. Stettler *et al.* (1973).
9. Present work.
10. Horn *et al.* (1975).

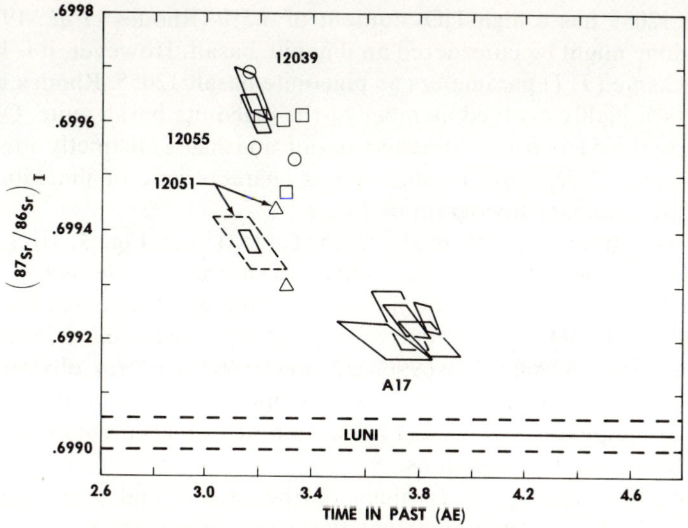

Fig. 2. (T, I) for Apollo 12 basalts as determined in this investigation. (T, I) for Apollo 17 basalts from Nyquist *et al.* (1974, 1975, 1976) are shown for comparison. Error polygons are calculated as described in Nyquist *et al.* (1976). The smaller polygon for 12051 is from the York program error limits; the larger polygon shows our visually estimated maximum error limits. Data for Apollo 12 basalts from Papanastassiou and Wasserburg (1970, 1971a,b) are shown as squares, circles, and triangles for olivine, pigeonite, and ilmenite basalts. These data have been adjusted for an interlaboratory bias of 0.00012. Our estimate of the lunar initial $^{87}Sr/^{86}Sr$ ratio, LUNI = 0.69903 ± 3, determined from Apollo 16 anorthosites is also shown.

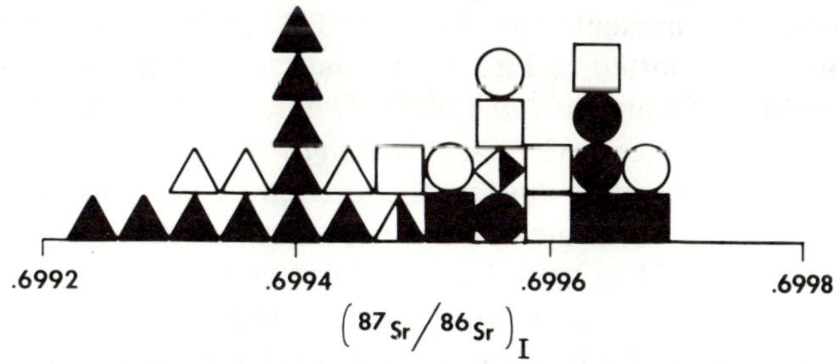

Fig. 3. Histogram of I-values from this investigation (filled symbols) and from the Cal Tech laboratory (open symbols, Papanastassiou and Wasserburg 1970, 1971a,b). Olivine, pigeonite, and ilmenite basalts are shown as squares, circles, and triangles, respectively. I-values calculated from whole rock and plagioclase measurements are included as well as directly determined ones. The quantization unit for I is 0.00004. I-values of 0.69938–0.69941 are shown plotted at 0.6994, etc. This figure contains more I-values than Figs. 2 and 5 because either the ages or the Sm/Ce ratio is undetermined for some samples.

Sample 12039 has a high-TiO_2 content of 4.5% (Rhodes *et al.*, 1977) and on this basis alone might be considered an ilmenite basalt. However, it is here shown to have the same (T, I) parameters as pigeonite basalt 12055. Rhodes *et al.* (1977) show it to be a highly evolved member of the pigeonite basalt suite. On the other hand, sample 12051 is a true ilmenite basalt and has a distinctly lower I-value than 12039 and 12055. Low I-values are a characteristic of ilmenite basalts as shown by the summary histogram of Fig. 3.

Two observations can be made from Table 1 and Fig. 3: (i) I-values for olivine and pigeonite basalts are indistinguishable and lie within the range 0.69947–0.69969; (ii) I-values for ilmenite basalts are lower and lie within the range 0.69924–0.69944. If only our own data are considered, average I-values from Table 1 are 0.69960 ± 7; 0.69960 ± 5; and 0.69934 ± 7 for olivine, pigeonite, and ilmenite basalts, respectively. (Quoted uncertainties are the standard deviation of the population of values.) Addition of Cal Tech data (open symbols) does not change these conclusions.

The complete overlap of I-values of the olivine and pigeonite basalts is consistent with the interpretation that these two basalt groups are comagmatic (Rhodes *et al.*, 1977). However, the range in I is ~50% greater than would be expected for repeat analyses of a standard. Similarly, the range in I for ilmenite basalts is also 50% larger than expected for sampling of a single magma. Thus, although the major and trace element abundances of most Apollo 12 basalts can be explained on the basis of only two magma types, the ilmenite and olivine-pigeonite magmas (Rhodes *et al.*, 1977), the Sr-isotopic data suggests the possibility that more than one batch of magma may contribute to each magma type. In this connection, it should be noted that Papanastassiou and Wasserburg (1970, 1971a) distinguish numerous lava flows on the basis of (T, I) parameters. These individual flows might be interpreted as due to the addition of new batches of magma into a few magma chambers.

Trace element analyses by mass spectrometric isotope dilution for nine Apollo 12 basalts are presented in Table 3. The chondrite-normalized Ba and REE abundances are plotted in Fig. 4. The data are separated according to the chemical sub-groups of Rhodes *et al.* (1977). The Ba and REE abundance patterns of the individual ilmenite basalts are strikingly parallel, although there is a two-fold dispersion in absolute abundances caused by near surface fractionation effects (Rhodes *et al.*, 1977). The abundance patterns of the olivine and pigeonite basalts are also parallel. Pigeonite basalts 12055 and 12039 have slightly higher trace element abundances than olivine basalts 12015 and 12014. (Basalt 12036 is a cumulate of uncertain affinity (Dungan and Brown, 1977; Rhodes *et al.*, 1977).) Thus, the similarity of the olivine and pigeonite basalts and the difference between them and the ilmenite basalts exists for relative Ba and REE abundances as well as for the initial $^{87}Sr/^{86}Sr$ values. This is illustrated explicitly in Fig. 5 where I-values are plotted versus Sm/Ce ratios, a measure of the slope of the light REE abundance pattern.

From Fig. 5 it is seen that a distinct negative correlation exists between I-values and Sm/Ce ratios for Apollo 12 basalts. A similar but somewhat poorer

Table 3. Alkali, alkaline earth, and REE in Apollo 12 mare basalts.

Sample No. Weight	12005,7 38 mg	12008,48 38 mg	12014,18 50 mg	12015,9–11 41 mg	12039,19 51 mg	12055,28 50 mg	12036,24 44 mg	12047,3 45 mg	12031,25 73 mg	Blanks ng
Li ppm	6.1	—	—	—	9.5	—	6.4	—	10.4	1.4
K ppm	275	449	397	450	576	512	508	602	414	120
Rb ppm	0.501	0.681	0.926	1.05	1.29	1.14	1.08	0.990	0.797	0.5
Sr ppm	78.2	133	90.4	98.4	138	120	94.5	164	153	1.1
Ba ppm	34.5	49.6	53.2	60.1	74.4	68.8	56.3	69.2	49.6	20
La ppm	3.62	—	—	—	7.25	—	5.03	—	5.01	1.4
Ce ppm	9.87	15.4	14.4	16.1	19.5	18.4	14.0	20.1	13.5	0.7
Nd ppm	8.38	13.9	10.9	12.2	15.0	14.0	11.1	17.5	10.5	0.3
Sm ppm	3.07	5.14	3.70	4.14	5.10	4.80	3.89	6.39	3.62	0.03
Eu ppm	0.687	1.17	0.795	0.869	1.17	1.05	0.861	1.45	1.14	0.02
Gd ppm	4.23	7.39	4.97	5.60	7.01	6.44	5.50	8.95	4.96	0.02
Dy ppm	5.25	8.88	5.92	6.70	8.17	7.80	6.36	10.9	6.22	0.02
Er ppm	3.10	5.27	3.56	4.07	4.92	4.63	3.76	6.40	3.77	0.02
Yb ppm	2.69	4.55	3.10	3.52	4.31	3.98	3.22	5.54	3.35	0.04
Lu ppm	0.390	0.657	0.433	0.486	0.604	0.562	0.469	0.798	0.486	0.01

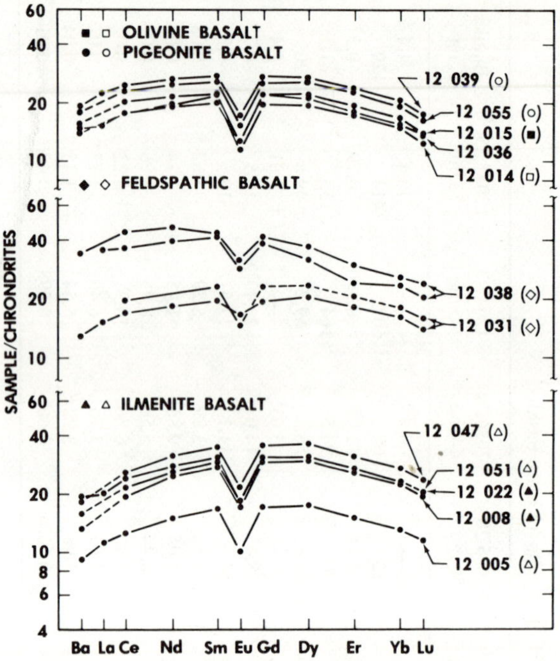

Fig. 4. Chondrite normalized Ba and REE abundances of Apollo 12 basalts. The data are separated according to the chemical sub-groups of Rhodes *et al.* (1977). The symbols used by these authors are given behind the sample numbers; filled symbols represent fine grained rocks, open symbols represent coarse grained rocks. The classification of 12036 is uncertain. Data from this investigation except for 12022 and 12051 (Hubbard and Gast, 1971), 12038 (Haskin *et al.*, 1971 and Schnetzler and Philpotts, 1971) and the "upper" sample of 12031 (Rhodes *et al.*, 1977). The dashed line for this latter sample shows interpolated values not directly measured.

correlation exists for I vs. TiO_2 (Nyquist *et al.*, 1977; Rhodes *et al.*, 1977). In the latter case the effect of near surface fractionation has obscured the original differences in TiO_2 of the parental magmas. Thus, for example, basalt 12039 has a high-TiO_2 content of 4.5%, equal to that of typical ilmenite basalts, but its age, initial $^{87}Sr/^{86}Sr$, and REE abundance pattern are shown in this investigation to be identical to that of the typical pigeonite basalt 12055. 12039 is thus a highly evolved pigeonite basalt, not an ilmenite basalt (see also Rhodes *et al.*, 1977).

Because I-values are unaffected by near surface fractionation and Sm/Ce ratios are insensitive to it, we interpret the characteristics displayed in Fig. 5 as inherited from the basalt sources. In the following discussion we will emphasize the ilmenite and olivine-pigeonite magmas. We shall ignore the possibility of magma subtypes and concentrate on the main (or possibly end-member) magma types. $Sm/Ce \approx 0.25$, $^{87}Sr/^{86}Sr \approx 0.69962$ will be taken as characteristic of olivine-pigeonite basalts and $Sm/Ce \approx 0.31$, $^{87}Sr/^{86}Sr \approx 0.69933$ as characteristic of the

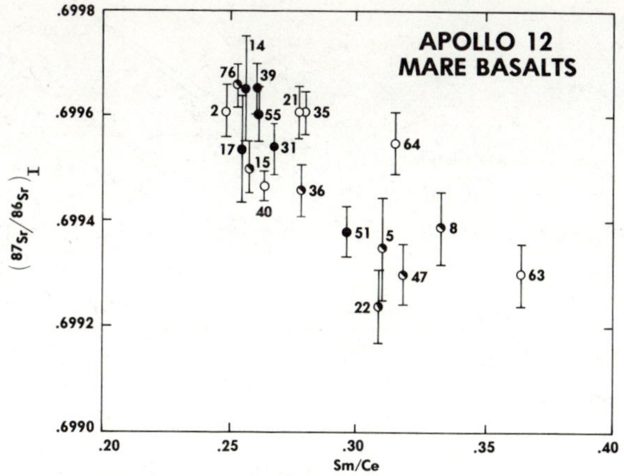

Fig. 5. Initial $^{87}Sr/^{86}Sr(I)$ vs. Sm/Ce for Apollo 12 basalts. Data shown as filled symbols are from this investigation for I-values determined from complete mineral isochrons or from plagioclase-whole rock pairs. Data shown as half-filled symbols are from this investigation for I-values which have been calculated from whole rock $^{87}Sr/^{86}Sr$ analyses assuming an age of 3.25 AE from the work of Papanastassiou and Wasserburg (1970, 1971a,b). I-values from these authors are shown as open symbols corrected for an interlaboratory bias of +0.00012. The corresponding Sm/Ce ratios are from this laboratory (Hubbard and Gast, 1971; Wiesmann and Hubbard, 1975), or from Schnetzler and Philpotts (1971) or Haskin *et al.* (1971).

ilmenite basalts. We shall not discuss the apparently feldspathic basalt 12031 or the unusual basalt 12036.

Discussion

Initial $^{87}Sr/^{86}Sr$ ratios are shown in Fig. 6 versus $^{87}Rb/^{86}Sr$ ratios for Apollo 12 basalts, Apollo 17 basalts, and Apollo 16 anorthosites. $^{87}Rb/^{86}Sr$ ratios for the Apollo 12 and Apollo 17 basalts have been corrected for radioactive decay during the last 3.25 and 3.8 AE, respectively. Because the following arguments are extremely sensitive to uncertainties in $^{87}Sr/^{86}Sr$ and because other correlative chemical data are necessary to the interpretations, only data from our laboratory are used. The evolution of $^{87}Sr/^{86}Sr$ with time is shown at the right of the figure for an undifferentiated reservoir with $^{87}Rb/^{86}Sr = 0.05$ and with an initial $^{87}Sr/^{86}Sr = .69903$, i.e., as found for Apollo 16 anorthosites corrected for radioactive decay during the last 4.6 AE (LUNI, Nyquist *et al.*, 1973, 1975). Single-stage isochrons through LUNI give $^{87}Sr/^{86}Sr$ in reservoirs of varying $^{87}Rb/^{86}Sr$ at 3.8 and 3.25 AE ago, respectively. Data for representatives of the two main Apollo 12 magma types plot within error limits of the single-stage isochron for evolution from 4.6 to 3.25 AE ago. The customary and most straightforward interpretation is that source regions of differing Rb/Sr were established ~4.6 AE ago and that basaltic magmas were produced from them 3.25 AE ago with little or no

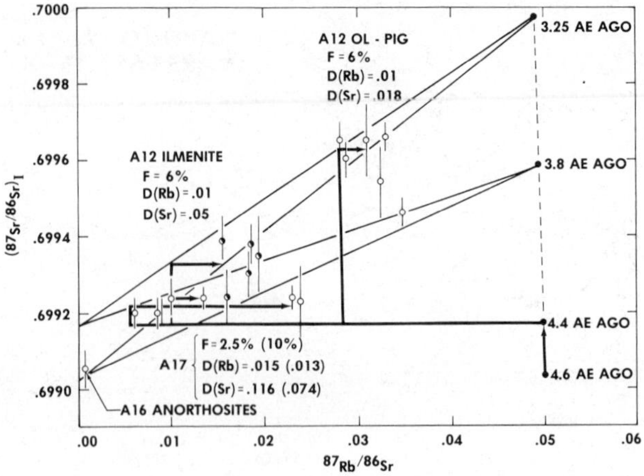

Fig. 6. Initial $^{87}Sr/^{86}Sr(I)$ vs. $^{87}Rb/^{86}Sr$ for Apollo 12 and 17 basalts. $^{87}Rb/^{86}Sr$ values are corrected for radioactive decay of ^{87}Rb during the crystallization age of the rocks. All data from our laboratory. Half-filled symbols are Apollo 12 ilmenite basalts. $^{87}Sr/^{86}Sr$ for Apollo 16 anorthosites is average measured value for low-Rb/Sr anorthosites (Nyquist *et al.*, 1973, 1975). The evolution of $^{87}Sr/^{86}Sr$ with time is shown at the right of the figure for an undifferentiated reservoir with $^{87}Rb/^{86}Sr = 0.05$ and initial $^{87}Sr/^{86}Sr = 0.69903$. Single stage isochrons giving $^{87}Sr/^{86}Sr$ as a function of $^{87}Rb/^{86}Sr$ are shown for reservoirs established at 4.6 AE ago with $^{87}Sr/^{86}Sr = 0.69903$ and at 4.4 AE ago with $^{87}Sr/^{86}Sr = 0.69917$, respectively. Hypothetical evolutionary paths are shown for the three basalt types. Values of distribution coefficients and degree of melting required for each path are given (see text). Values in parentheses for the Apollo 17 basalts correspond to the evolutionary path with the shorter "return leg."

fractionation of the Rb/Sr ratio. This is certainly an acceptable explanation for the Rb–Sr data alone. However, formation of mare basalts with *precisely zero* fractionation of Rb/Sr has quite different implications than their formation with a *small* Rb–Sr fraction. Moreover, we have argued previously that the Apollo 17 data provides clear evidence for fractionation of Rb/Sr during genesis of the Apollo 17 lavas (Nyquist *et al.*, 1976). The evolutionary path suggested for the Apollo 17 basalts is shown in Fig. 6: (i) evolution from $T = 4.6$ AE, $I =$ LUNI to $T = 4.4$ AE, $I = 0.69917$ with $^{87}Rb/^{86}Sr = 0.05$, (ii) a decrease in $^{87}Rb/^{86}Sr$ from 0.05 to 0.006 at formation of the source region of Apollo 17 basalts, (iii) evolution from $T = 4.4$ AE, $I = 0.69917$, to $T = 3.8$ AE, $I = 0.69922$ with this $^{87}Rb/^{86}Sr$ ratio, and (iv) an increase in Rb/Sr upon partial melting of the source to form the basalts. The reader is referred to the previous work for detailed justification of this hypothetical evolutionary path. The key arguments for it hinge on: (i) although there is a four-fold variation in Rb/Sr for Apollo 17 basalts, inital $^{87}Sr/^{86}Sr$ values are constant; (ii) backward extrapolation in time with reasonable fractionation of Rb/Sr at 3.8 AE ago leads to $^{87}Sr/^{86}Sr = 0.69917$ at 4.4 AE ago; and (iii) evolution from $T = 4.6$ AE, $I =$ LUNI to $T = 4.4$ AE, $I = 0.69917$ requires $^{87}Rb/^{86}Sr = 0.05$.

If the above is correct, then the Apollo 12 basalts also show the effects of fractionation of Rb/Sr at their genesis. That is, the Apollo 12 data in Fig. 6 lie to the right of a single stage isochron between 4.4 AE ago and 3.25 AE ago with $I = 0.69917$. One is thus led to consider the consequences for the Apollo 12 basalts of an evolution qualitatively similar to that suggested for the Apollo 17 basalts. In so doing it is recognized that much of the argument is within the error limits of the Rb–Sr data. Thus the ensuing discussion illustrates a possible but nonunique explanation of the data. The analytical uncertainties are such that the details of the conclusions could be considerably in error even if the approach is in general correct. It is thus necessary to search for self-consistency with other compositional data in order to evaluate the validity of the conclusions.

The hypothetical evolutionary paths to be evaluated for the Apollo 12 basalts are shown in Fig. 6. In order to evaluate the implications for basalt composition a unifying model is required. The model chosen is simple and limited in scope. That is, the evolution of each basalt type is considered separately. No attempt is made to account for basalt formation as the natural consequence of a grander scheme for the geochemical evolution of the moon. Rather, we will attempt to obtain "compositional snapshots" of individual episodes in lunar geochemical evolution. A broader understanding will then be sought by piecing together the individual episodes.

The assumptions used in the model calculations are: (i) During the earliest stages of lunar crust and mantle formation cumulates formed from an early liquid. These cumulates were later to be (partially) remelted to form basaltic magmas. The early liquid may, but need not, have been the result of formation of a lunar magma ocean. (ii) The equations of bulk equilibrium and mass balance are assumed adequate to describe element partitioning. Let C_0, C_c, and C_m represent element concentrations in the early liquid, the cumulate, and the later melt, respectively. For fractional crystallization one has

$$\frac{C_c}{C_0} = \frac{1-(1-X)^D}{X}, \tag{1}$$

for the average concentration of an element in the cumulate compared to the average concentration in the liquid over an interval during which X wt.% of the system crystallizes (cf. Haskin *et al.*, 1970). Expanding in Maclaurin's series yields $C_c/C_0 = D + (X/2)D(1-D) + \cdots$. We wish to follow the crystallization process only over the interval for which the source of a single basalt type is formed. This interval is unknown but is unlikely to be more than ~20% crystallization of the total system. Thus the error in neglecting the second term of the expression is small, at most ~10% for $D < 1$. Consequently, the approximation

$$\frac{C_c}{C_0} \approx D, \tag{2}$$

will be used. This is, of course, the same as the equilibrium relationship between the liquid and the last-formed crystals.

Following Weill *et al.* (1974) the equations of bulk equilibrium and mass balance give

$$\frac{C_m}{C_c} = \frac{1}{F + D(1 - F)}, \tag{3}$$

for partial melting, where F is the fraction of melt and C_m is the concentration in the melt. (iii) It is assumed that the degree of remelting is small so that no cumulus phases are exhausted and the bulk distribution coefficients are the same for melting as for accumulation, i.e., $D_{\text{melting}} = D_{\text{xtln}}$. Then, if E_1 and E_2 are the concentrations of elements "1" and "2", respectively, one has from Eqs. (2) and (3)

$$\left(\frac{E_1}{E_2}\right)_c \Big/ \left(\frac{E_1}{E_2}\right)_0 \approx \frac{D_1}{D_2}, \tag{4}$$

and

$$\left(\frac{E_1}{E_2}\right)_m \Big/ \left(\frac{E_1}{E_2}\right)_0 \approx \frac{D_1}{D_2}\frac{F + D_2(1 - F)}{F + D_1(1 - F)}, \tag{5}$$

where D_1 and D_2 are the bulk distribution coefficients and F is the percent melt. (iv) The basalt source is assumed to be composed primarily of olivine and pyroxene (Green *et al.*, 1971a,b; Green *et al.*, 1975).

Equations (4) and (5) form the basis of the model calculations of this paper. In outline the procedure is to: (i) choose $(^{87}Rb/^{86}Sr)_0 = 0.05$ to conform to the Apollo 17 basalt study (Nyquist *et al.*, 1976) and the hypothetical evolution paths shown in Fig. 6. (ii) Choose $\bar{D}(Rb) \approx 0.01$ where $\bar{D}$ is the composite distribution coefficient (cf. Weill *et al.*, 1974). The validity of this step rests on the similarity of D(Rb) for olivine (ol), orthopyroxene (opx), and clinopyroxene (cpx) (see, for example, Philpotts and Schnetzler, 1970a). (iii) $\bar{D}(Sr)$ is then calculated from Eq. (4) and $(^{87}Rb/^{86}Sr)_c$ read from Fig. 6. For example, in Fig. 6 the Apollo 12 ilmenite basalts imply $(^{87}Rb/^{86}Sr)_c = 0.01$. (iv) The degree of melting, F, is calculated from Eq. (5) and $(^{87}Rb/^{86}Sr)_m$ read from Fig. 6. For Apollo 12 ilmenite basalts we take $(^{87}Rb/^{86}Sr)_m = 0.016$ corresponding to 12008 and 12022 which according to Rhodes *et al.* (1977) most nearly represent the parental magma.

The values of $\bar{D}(Rb)$, $\bar{D}(Sr)$ and F required to model the suggested evolution of the Apollo 12 and 17 mare basalts are shown on Fig. 6 adjacent to the hypothetical evolutionary paths. In all cases the percent melting is small (2–10%) while $\bar{D}(Sr)$ varies from ~0.02 for Apollo 12 olivine-pigeonite basalts to ~0.1 for Apollo 17 basalts. Interestingly, F is calculated to be the same (6%) for both the ilmenite and olivine-pigeonite basalts of Apollo 12. A low value of $F = 2.5\%$ is required to explain the high $^{87}Rb/^{86}Sr \approx 0.02$ of Station 4 Apollo 17 basalts (cf. Nyquist *et al.*, 1976). As will be seen later in the discussion such a low value of F is at odds with the REE abundance pattern. An F of 10% corresponds to the alternate path shown for the Apollo 17 basalts.

The above conclusions are reasonable but quite model-dependent. It is thus highly desirable to check them for self-consistency against abundances and distribution coefficients of other LIL elements. To do this requires at least some knowledge of crystal-liquid distribution coefficients. Fortunately, the quantity and quality of data for distribution coefficients is steadily increasing and there now exist several experimental determinations of distribution coefficients as well as values determined from natural systems. Distribution coefficients chosen for this work are given in Table 4.

Self-consistency between Rb/Sr in Fig. 6 and Sm/Ce in Fig. 7 can be imposed by using the distribution coefficients for ol, opx, and cpx from Table 4. First, the percent cpx in a polymineralic source is calculated by letting $D(\text{ol}, \text{Sr}) = D(\text{opx}, \text{Sr}) = .01$. Next $\bar{D}(\text{Ce})$ is calculated since the olivine and opx D's are equal. Finally, $D(\text{Sm})$ is calculated from Eq. (5) using F from Fig. 6 and an assumed initial $(\text{Sm/Ce})_0$. An obvious choice is to set $(\text{Sm/Ce})_0 = 0.23$, i.e., the chondritic value. However, when this is done the resulting Sm distribution coefficients are less than expected for sources of the Apollo 12 basalts containing the calculated amounts of cpx and any ol:opx ratio. Distribution coefficients calculated in this way are shown in parentheses in Fig. 7.

Consideration of the total REE abundance pattern shows that the assumption of chondritic relative abundances of the REE in the early liquid is probably not a good one. The heavy REE give the best evidence of this. Distribution coefficients for ol and opx increase progressively from Gd to Yb and are nearly constant for cpx (Table 4). Polymineralic cumulates of these minerals formed from a liquid with chondritic relative REE abundances will almost certainly have a positive slope of the chondrite-normalized heavy REE abundance pattern. This is illustrated in Fig. 8 for two source mineralogies which could produce the Rb and Sr distribution coefficients indicated in Fig. 6 for Apollo 12 basalts. A cumulate consisting entirely of cpx would be expected to have a *slight* negative heavy REE slope (~10% from Gd to Yb). It is not possible to produce the pronounced negative slope of the chondrite-normalized heavy REE abundance pattern as observed for all mare basalts (and for almost all lunar rocks) by remelting such cumulates. Progressively smaller degrees of melting have the effect of causing the abundance pattern to monotonically approach that in the original liquid and will not cause a change in sign of the slope of the abundance pattern. These general considerations apply independently of whether the melting is modal as assumed throughout this discussion, or non-modal as suggested by Drake and Consolmagno (1976) for high-Ti basalts.

These considerations strongly imply that the relative REE abundances were non-chondritic in the liquid parental to the source cumulates. More specifically, the slope of the REE abundance pattern from Gd to Yb must have been negative. In their model Shih and Schonfeld (1976) accomplished this via prior crystallization of garnet, an approach which has been criticized by Drake and Consolmagno (1976). In keeping with the spirit of our approach to mare basalt genesis we need not at this juncture address the cause of the heavy REE depletion but simply note the necessity of its presence. However, it is worth noting that nearly all lunar rocks show a depletion of the heavy REE. Perhaps more significantly, the calculated *parental magmas* for the primitive dunite

Table 4. Distribution coefficients.

	ol[a]	opx[b]	cpx[c]	Cl(12)[d]	C3(12)	C2(15)	C2(17)
K	.0081	.0091	.040	.02	.010	.015	.027
Rb	.0084	.0091	.015	.01	.009	.010	.0125
Sr	.0094	.0104	.116	.05	.017	.032	.074
Ba	.0086	.011	.022	.015	.0096	.0115	.017
La	—	—	—	—	—	—	—
Ce	.009	.009	.077	.035	.014	.023	.05
Nd	.010	—	.174	—	.022	.044	—
Sm	.011	.022	.260	.11	.028	.063	.16
Eu[e]	.0098	.022	.151	.067	.020	.040	.10
Gd	.012	.037	.325	.14	.034	.078	.20
Dy	.014	—	.351	—	.038	.085	—
Er	.017	—	.330	—	.029	.083	—
Yb	.023	.17	.294	.17	.042	.080	.22

[a]Sample HHP-66-19 of Philpotts and Schnetzler (1970a) and Schnetzler and Philpotts (1970) following the choice of Shih *et al.* (1975). Values for Ce, Sm, Eu, Yb, are in good agreement with 1200°C experimental values of McKay and Weill (1977).

[b]Values for K and Sr from GSFC-184 of Philpotts and Schnetzler (1970a) and Schnetzler and Philpotts (1970). *D*(Rb) for GSFC-184 is three-fold higher than *D*(K) and *D*(Sr); it is here chosen = *D*(K). *D*(Ba) and REE are from Weill and McKay (1975) and are in good agreement with values from GSFC-184 except for Ce and Sm. The Gd value is obtained by interpolation between Sm and Yb. Like Ce and Sm it is two-to-three fold lower than the corresponding values for GSFC-184.

[c]Core augite HHP-66-19 of Philpotts and Schnetzler (1970) and Schnetzler and Philpotts (1970). Values are in good agreement with diopside composition A of Grutzeck *et al.* (1974).

[d]Cl(12), etc. refer to calculated cumulate sources shown in Figs. 9, 10, 11.

[e]Following Shih *et al.* (1975), *D*(Eu) for ol and cpx is taken as $D(\text{Eu}) = 0.8\ D(\text{Sr}) + 0.1\ (D(\text{Sm}) + D(\text{Gd}))$.

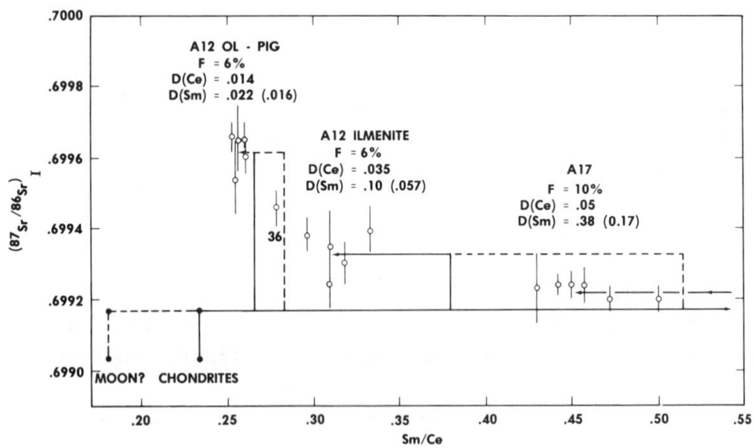

Fig. 7. Initial $^{87}Sr/^{86}Sr(I)$ vs. Sm/Ce ratios for Apollo 12 and 17 basalts. All data from our laboratory. Evolutionary paths which correspond to those of Fig. 6 are shown. The solid lines and parenthetical values of *D*(Sm) assume an initial chondritic Sm/Ce ratio. Dotted lines and *D*(Sm) values without pararentheses assume a lower "lunar" initial Sm/Ce ratio of 0.18 (see text).

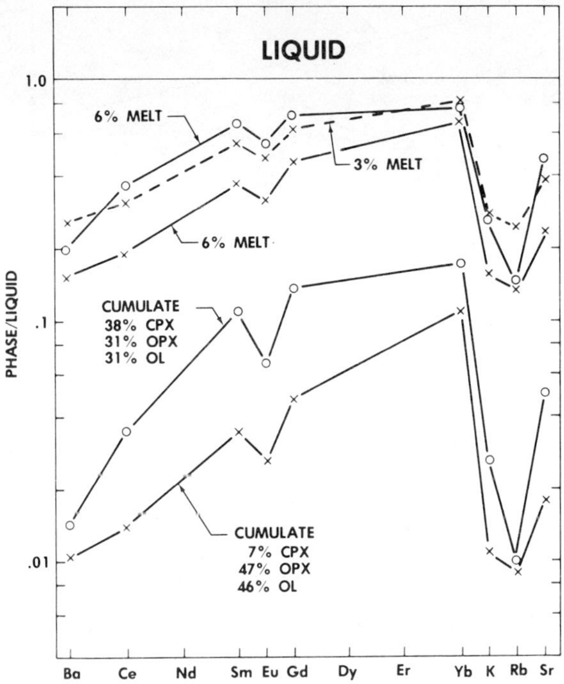

Fig. 8. Effect of accumulation and subsequent remelting for Ba and REE abundances and two hypothetical mafic cumulate mineralogies. Trace element abundances in the melt from the cumulates approach those in the liquid from which the cumulates formed at small degrees of melting. The slope of the abundance patterns of the melt must approach that in the original liquid in a monotonic manner without reversals in sign.

72417, troctolite 76535, and anorthosite 15415 are also heavy REE depleted (Laul and Schmitt, 1975; Haskin *et al.*, 1974; Hubbard *et al.*, 1971). If these primitive rocks formed from a relatively unevolved lunar magma ocean, then the heavy REE depletion is a primary lunar feature. We suggest this to be the case.

To continue our modeling of isotopic and trace element evolution during mare basalt genesis we adopt the practical expediency of incorporating the required heavy REE slope into the assumed parental liquid as shown in Fig. 9. The slope is determined from the chondrite normalized Dy/Yb ratio of Apollo 12 basalts (see Fig. 4). Most other lunar basalts have essentially the same Dy/Yb ratio. The same slope is extended to the light REE elements as well. This procedure is *ad hoc* and its validity must be judged from whether it is useful in explaining the isotopic and trace element systematics.

The assumption of non-chondritic relative abundances for the parental magma for the cumulates does allow the construction of a self-consistent picture of mare basalt petrogenesis as shown in Figs. 9, 10, and 11. The abundance patterns for the cumulates in these figures are calculated from Eq. (2), i.e.,

$$C_c = \bar{D}C_0, \tag{6}$$

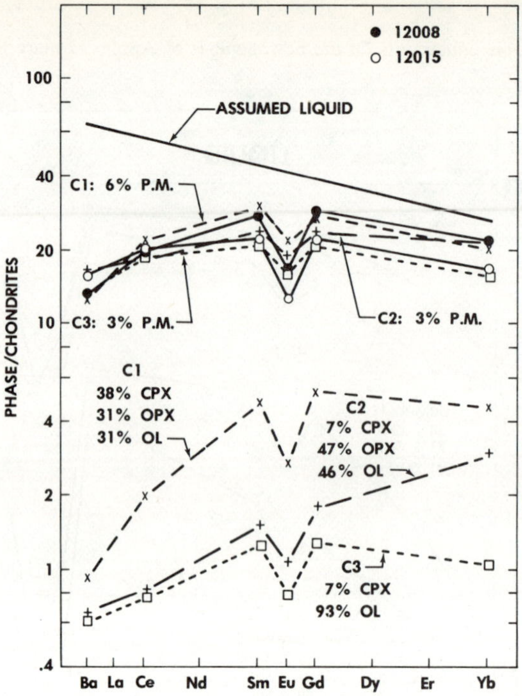

Fig. 9. Calculated chondrite-normalized Ba and REE abundance patterns for cumulates and melts from an assumed liquid of the composition shown. Good fits are obtained to the measured abundances of ilmenite basalt 12008 for 6% melting of cumulate C1. The light REE pattern of olivine basalt 12015 is satisfactorily fit by 3% melting of either C2 or C3. Proper fit to the heavy REE of 12015 is obtained only with C3.

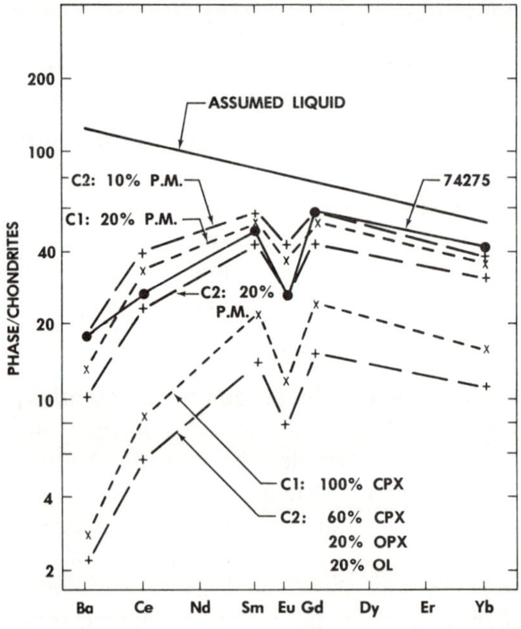

Fig. 10. Calculated chondrite-normalized Ba and REE abundance patterns for cumulates and melts from an assumed liquid of the composition shown. Moderately good fits to the abundance pattern of 74275 can be obtained for 10–20% melting of C2 or ~20% melting of C1.

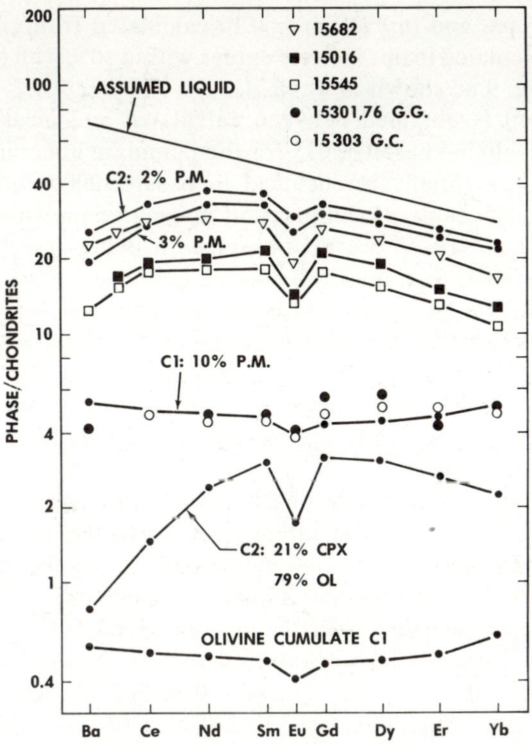

Fig. 11. Calculated chondrite-normalized Ba and REE abundance pattern for cumulates and melts from an assumed liquid of the composition shown. Moderately good fits to the abundance pattern of Apollo 15 quartz-normative (15682) and olivine-normative basalts (15016, 15545) can be obtained for 2–3% melting of cumulate C2. A moderately good fit to the Apollo 15 green glass data (Ridley *et al.*, 1973) is obtained for 10% melting of olivine cumulate C1. Data for 15016 from Helmke *et al.* (1973); all other data from this laboratory.

where D is the composite distribution coefficient. The patterns for the melts are calculated by combining Eqs. (3) and (6):

$$C_m = C_0 \frac{\bar{D}}{F + \bar{D}(1 - F)}, \tag{7}$$

C_0, C_m, and C_c are as defined for Eqs. (2) and (3). The abundance pattern of ilmenite basalt 12008, suggested by Rhodes *et al.* (1977) as representative of the parental magmas of Apollo 12 ilmenite basalts, is well matched for 6% melting of a cumulate containing 38% cpx and 31% of both opx and ol. The percentage cpx is obtained from $\bar{D}(\mathrm{Sr}) = 0.05$ (Fig. 6) as described above. The ol : opx ratio is not determined by the model calculations, it is thus assumed = 1. The 6% partial melting of the calculated source required to match the measured REE pattern is the same as that deduced from the Rb–Sr data. In Fig. 7 the Sm/Ce ratios of the ilmenite basalts are matched for (i) $(\mathrm{Sm/Ce})_0 = 0.18$ from Fig. 9, (ii) $\bar{D}(\mathrm{Ce}) = .035$

from the percent cpx, and (iii) $\bar{D}(Sm) = 0.10$ calculated from (i) and (ii) and 6% melting. $\bar{D}(Sm)$ calculated in this manner agrees within 10% with that calculated for cumulate C1 in Fig. 9 as shown in Table 4.

Good agreement is obtained between calculated and measured REE abundance patterns for olivine basalt 12015 for the cumulate mineralogy of C3 in Fig. 9. Basalt 12015 is essentially isochemical to basalt 12009 and its composition should give a good estimate of the parental magma composition for the olivine basalts (Rhodes *et al.*, 1977). 3% melting gives a better match to the absolute abundances of the REE than does 6% melting deduced from Rb–Sr if the initial REE abundance in the "early" liquid is assumed the same as for the ilmenite basalts. This discrepancy can easily be accounted for in the uncertainties of the Rb–Sr calculation. On the other hand 6% melting and $(Sm/Ce)_0 = 0.18$ in Fig. 8 gives $\bar{D}(Ce) = .014$ and $\bar{D}(Sm) = .022$ the same as for cumulate C3. It is thus concluded that there is good self-consistency for the Apollo 12 olivine-pigeonite basalts also.

It is more difficult to obtain self-consistency for the high-Ti Apollo 17 basalts. Whereas it is possible to obtain a rather good fit to the REE from cumulates containing predominantly cpx, 10–20% melting appears to be required. The best fit is for 20% melting of a cumulate containing 60% cpx (opx : ol = 1 is again assumed). This is inconsistent with the maximum of 10% melting previously inferred for at least the Apollo 17 Station 4 basalts (Nyquist *et al.*, 1976). 10% melting and $\bar{D}(Ce) = 0.05$ requires $\bar{D}(Sm) = 0.38$ for $(Sm/Ce)_0 = .18$ in Fig. 7. $\bar{D}(Sm) = 0.38$ seems too high compared to $\bar{D}(Ce) = .05$ (cf. Table 3); in this case the $\bar{D}(Sm) = 0.17$ calculated from an assumed chondritic $(Sm/Ce)_0$ is a better fit. Alternatively, 20% melting and $(Sm/Ce)_0 = .18$ gives $\bar{D}(Sm) = .18$ for $\bar{D}(Ce) = .05$ in good agreement with that of C3 in Table 3. Nevertheless, 10% melting seems the best compromise between the inferences drawn from the Rb–Sr data on the one hand and the REE data on the other.

It should be emphasized that although the self-consistency obtained for the Apollo 17 basalts is not quantitatively as excellent as for the Apollo 12 basalts; it is nevertheless qualitatively quite good. Rb–Sr and REE abundances are explained by comparatively small degrees of melting. Moreover, the absolute REE abundances in the early liquid are higher by a factor of 2 at the time of formation of the Apollo 12 sources. This is qualitatively consistent with the concept of an early liquid which evolves in composition. (See, for example, Philpotts and Schnetzler, 1970b; Taylor and Jakeš, 1974; Walker *et al.*, 1975a; Shih *et al.*, 1975; Shih and Schonfeld, 1976; Drake, 1976; Drake and Consolmagno, 1976).

The discussion to this point has been limited to Apollo 12 ilmenite and olivine-pigeonite basalts and Apollo 17 basalts because our data are most complete for them and because they represent examples of low, intermediate, and high-Ti lunar basalts. However, it is clear that the models presented here have implications for Sm–Nd systematics and vice-versa. Apollo 15 basalts are the only low-Ti basalts which have been studied by the Sm–Nd method. Lugmair and Marti (1977) conclude from Sm–Nd studies of 15555 that Sm/Nd must have been greater than chondritic in the source region. Sm/Nd was then decreased to approximately the chondritic value at the time of basalt genesis 3.3 AE ago. By inference, if the lunar Sm/Nd was chondritic (or less as suggested here) then the Sm/Nd systematics are also reflecting at least two episodes in which the Sm/Nd ratio has changed, qualitatively consistent with formation of mafic cumulate sources followed by subsequent remelting as considered here.

If the model presented here is applied to Apollo 15 basalts with initial $^{87}Sr/^{86}Sr \approx 0.69940$ (on our scale) and $^{87}Rb/^{86}Sr \approx 0.025$, then (i) $\bar{D}(Sr) = .032$ is found to correspond to $\bar{D}(Rb) = .01$, (ii) this in turn implies a source with ~21% cpx, and (iii) the percentage of partial melting is calculated to be 3%. The REE pattern for 3% melting of a cumulate of 21% cpx and 79% ol and other conditions as for the model calculations for Apollo 12 basalts is shown in Fig. 11. The match between the calculated pattern and that of quartz-normative basalt 15682 is good. The REE of olivine-normative basalts 15016 and 15545 are 30–45% lower in absolute abundance. Basalt 15016 is fine grained and suggested by Rhodes and Hubbard (1973) as a possible parental composition to the olivine-normative basalts. The average major element composition of the coarse grained basalt 15555 is close to that of 15016 and is also suggested as a possible parental composition by Rhodes and Hubbard (1973). However, the REE abundances reported by Schnetzler et al. (1972) are at least 50% lower than those of other Apollo 15 basalts, and must be considered suspect until sampling bias can be ruled out. Indeed, lugmair (1975) reports Sm and Nd abundances of 21 × chondritic for 15555, which differs by <4% from those of 15016 reported by Helmke *et al.* (1973). Rhodes and Hubbard (1973) pointed out a comparatively wide range of published major element analyses for this basalt. Thus, we consider the low REE abundances of 15555 reported by Schnetzler *et al.* (1972) as unrepresentative and the agreement between the model calculation and the measured REE abundances in Apollo 15 basalts to be satisfactory. Better agreement could have been obtained by lowering the abundance of REE in the early liquid by 30–40%.

Lugmair and Marti (1977) also report Nd-isotopic studies of the emerald green glass and conclude that they reflect isotopic evolution in a chondritic Sm–Nd reservoir prior to volcanic eruption of the glass. Following Ridley *et al.* (1973) and Green and Ringwood (1973) they infer that the "chondritic" reservoir was also "primordial". This need not have been the case, however. As shown in Fig. 11, the green glass could be the product of ~10% melting of a source in which the REE abundances were dominated by olivine. According to Green and Ringwood (1973) olivine would have been the liquidus phase to 260 km depth in the moon, i.e., down to the inferred depths of origin of most mare basalts.

The source mineralogies calculated here are intended to be used only in a qualitative or semi-quantitative manner. Distribution coefficients of the various minerals are not precisely known. Moreover, pyroxene distribution coefficients probably form more or less of a continuum corresponding to the continuum of pyroxene compositions. Thus it is the distribution coefficients used to fit the data which are more meaningful than the mineralogies assumed as an aid to arrive at them. It is partly for this reason that the distribution coefficients used to obtain the best fits are given explicitly in Table 4.

One key remaining question is whether the model presented here can account for the correlation between initial $^{87}Sr/^{86}Sr$, Sm/Ce, and TiO_2 for the Apollo 12 basalts (Nyquist *et al.*, 1977; Rhodes *et al.*, 1977). It has been possible in the above to account for the differing trace element and Sr-isotopic characteristics of the two main Apollo 12 magma types essentially by varying the "mix" in the basalt source of crystals produced from the same early liquid. The constraints provided by the isotopic and trace element data are probably not strong enough to imply *exactly* the same early liquid in both cases. However, the similar absolute abundances of the REE in the two magma types strongly imply a very similar starting liquid. Thus for TiO_2 the test of our model appears to be whether it predicts similar TiO_2 contents in the early liquid for the two Apollo 12 magma types.

To the best of our knowledge the relevant distribution coefficients are not available for a truly quantitative test of the model. Green *et al.* (1975) give $(TiO_2)_{liq}/(TiO_2)_{xtl} \approx 5$, 6.5–7, and 8 for high-Ca cpx, low-Ca cpx, and orthopyroxene, respectively, at high pressures. Corresponding olivine distribution coefficients are not given, but presumably are considerably higher. Green *et al.* (1971b) give $TiO_2 < 0.03\%$ in liquidus olivine at 1 bar for 12009. Liquidus pyroxenes at 1 bar contained ~0.8–1.0% TiO_2 and at 15 kbar from 0.34% (opx) to 0.84%. Thus for a rough calculation we assume the olivine (crystal/liquid) distribution coefficient to be 0.1 that of ortho-pyroxene. The resulting values used for calculation are: $D(\text{cpx}) = 0.2$; $D(\text{opx}) = 0.125$; $D(\text{ol}) = 0.01$. These values truly make Ti an "incompatible" element.

The composite distribution coefficients corresponding to C1 and C3 of Fig. 9 are then $\bar{D}(\text{Ti}) = 0.024$ and $\bar{D}(\text{Ti}) = 0.12$, respectively. The TiO_2 concentration calculated for the early liquid from Eq. (7) is in both cases $C_0(TiO_2) = 6.4\%$ for 3% melting in the case of the olivine pigeonite basalts and 6% melting for the ilmenite basalts. This exact agreement in $C_0(TiO_2)$ in the two cases is almost certainly fortuitous, but indicates the potential capability of the model of satisfying the TiO_2 test for the Apollo 12 basalts. TiO_2 contents in the ilmenite and olivine-pigeonite basalt sources are calculated to be 0.76% and 0.15%, respectively. The TiO_2 values calculated for the early liquid and the cumulate source from the distribution coefficients corresponding to the ilmenite basalts is relatively insensitive to uncertainty in $D(\text{ol})$.

For Apollo 15 basalts the cumulate mineralogy of C2 (Fig. 11) and 3% melting gives $\bar{D}(\text{Ti}) = 0.05$ and $C_0(TiO_2) = 3.7\%$. This is slightly lower than for the Apollo 12 basalts; analogous to the slightly lower REE abundances implied above for the early liquid at the time of formation of the Apollo 15 sources. The TiO_2 content of the source is calculated to be 0.18%.

For the high-Ti Apollo 17 basalts the cumulate mineralogy of C2 of Fig. 10 corresponds to a composite distribution coefficient $\bar{D}(\text{Ti}) = .15$. The formally calculated TiO_2 of the early liquid and the cumulate sources are 20% and 2.9%, respectively, for 10% melting. It is probable that TiO_2 in the liquid will not reach such a high level before the onset of precipitation of a Ti-mineral such as ilmenite (cf. Ringwood, 1976a). However, Green *et al.* (1975) report TiO_2 contents of up to 2.4% in high-Ca cpx on the liquidus of 74275 at 12 kbar, and Kesson (1975) reports TiO_2 contents of 3.7% in liquidus pyroxenes at 8 kbar for 70215. Thus for small degrees of melting as are suggested here, it may not be necessary to postulate a Ti-mineral in the source of high-Ti basalts. However, if ilmenite does exist in the source of the high-Ti basalts, its presence is unlikely to change the major conclusions presented above. This is because trace element distribution coefficients, at least of the REE, are similar for ilmenite, olivine, and orthopyroxene (McKay and Weill, 1976). The calculational procedure we have used will not be valid for TiO_2 in this case, however.

One can inquire whether the remaining major element composition of mare basalts is compatible with the ideas presented above. Insofar as pertinent data are available, this does appear to be the case. Thus Eqs. (4) and (5) above can be applied to the Mg/Fe ratio, i.e.,

$$\left(\frac{\mathrm{Mg}}{\mathrm{Fe}}\right)_m \Big/ \left(\frac{\mathrm{Mg}}{\mathrm{Fe}}\right)_0 = \frac{D(\mathrm{Mg})}{D(\mathrm{Fe})} \frac{F+(1-F)D(\mathrm{Fe})}{F+(1-F)D(\mathrm{Mg})}. \tag{8}$$

Weill and McKay (1975) have found $D(\mathrm{Mg}) \approx 3.3$ and $D(\mathrm{Fe}) \approx 1.0$ for olivine and orthopyroxene; we assume a similar value for clinopyroxene (see also Shih and Schonfeld, 1976, and references therein). With these parameters Eq. (8) can be used to calculate mg-values (mg = [100 MgO/MgO + FeO] in mole %) in the early liquid from the mg-values of the basalts and the melt percentages calculated from the isotopic and trace element arguments. Furthermore, the mg-values in the source regions of the mare basalts can be calculated from Eq. (4); i.e., $(\mathrm{Mg})_c = D(\mathrm{Mg})(\mathrm{Mg})_0$, etc. The results of these calculations are shown in Table 5.

Table 5. Calculated mg-values.

	F(%)[a]	$\left(\frac{\mathrm{Mg}}{\mathrm{Fe}}\right)_m \Big/ \left(\frac{\mathrm{Mg}}{\mathrm{Fe}}\right)_0$ [b]	$(\mathrm{mg})_m$[c]	$(\mathrm{mg})_0$[d]	$(\mathrm{mg})_s$[e]
12015	3	1.02	51	51	77
12008	6	1.04	50	49	76
74275	10	1.07	50	49	76
70215	10	1.07	44	42	71
Green Glass	10	1.07	62	60	83
15386	50?	1.5?	58	40?	75?
	75?	2.1?		40?	69?

[a]Percent partial melting.

[b]Mg and Fe in units of weight.

[c]mg = 100 MgO/MgO + FeO in molar units. Subscript "*m*" means melt, i.e., assumed equal to that in basalt.

[d]Subscript "0" means original early liquid.

[e]Subscript "*s*" means cumulate source regions.

The principal conclusion from Table 5 is that mg-values calculated for the source regions of mare basalts by this manner are in excellent agreement with the values determined directly by the methods of experimental petrology (Green *et al.*, 1971a,b; 1975). It is also worth noting that because the degree of remelting is small, the mg-value in the basalts is close to that calculated for the early liquid. The effect of larger values of F is illustrated for KREEP basalt 15386. It this basalt was produced by partial melting 3.9 AE ago then the degree of melting must have been comparatively large to explain the Rb-Sr model age of 4.3 AE. If 15386 represents a melt of 50–75% of its source region, and if the above equations are still applicable (i.e., no phases exhausted, etc.) then the mg-value for the 15386 source is comparable to or lower than that in the sources of mare basalts. Perhaps more significantly, the mg-value of the early liquid is lower than at the formation of the cumulate sources for mare basalts, as it would be for a progressively evolving composition. Perhaps the petrogenesis of KREEP and mare basalts should be considered together, rather than separately as has often been the tendency.

Summary and Conclusions

It has been possible to quantitatively account for the main features of the Sr-isotopic and trace element data of mare basalts in general and Apollo 12 basalts in particular. This has been done by considering each magma type to have evolved in turn via formation of cumulates from an early liquid followed by generation of basaltic magmas by small amounts of remelting. At least two features of the major element data, i.e., the TiO_2 contents and mg-values, appear to be adequately described in the scenario which has emerged. The methods used have been largely descriptive rather than predictive. Thus we prefer to call the picture of lunar evolution which has evolved a "scenario" rather than a model. It falls within the scope of so-called "cumulate melting" models, however, and we have benefited from the contributions of earlier workers (e.g., Philpotts and Schnetzler, 1970b; Taylor and Jakeš, 1974; Walker *et al.*, 1975a; Shih *et al.*, 1975; Shih and Schonfeld, 1976; Drake and Consolmagno, 1976). The considerations presented by the proponents of the hybridization and assimilation models (Hubbard and Minear, 1975, 1976; Ringwood and Kesson, 1976) have also proved enlightening. In general we have found the main theme of these latter models to be less useful to a quantitative understanding of the data, however. This is due in part to a lack of specificity in these models. The data presented here provide at least one specific test of them.

One might postulate that the Apollo 12 ilmenite basalts provide evidence of partial assimilation of ilmenite-clinopyroxene cumulate "pods" by primitive lunar material according to the model of Kesson and Ringwood (1976) and Ringwood and Kesson (1976). The Apollo 12 olivine-pigeonite basalts would represent melts directly from the interior or at most only slightly contaminated with assimilated material; Apollo 17 high-Ti basalts represent melts with a large complement of the ilmenite-cpx pods. It is then reasonable to expect that on a plot of initial $^{87}Sr/^{86}Sr$ vs. Sm/Ce (Fig. 7) that the Apollo 12 ilmenite basalts might lie on a mixing curve. This is not the case. The mixing curve, which is nearly a straight line and consequently not shown on the figure, completely misses at least four out of five data points of the ilmenite basalts. If data for Apollo 15 basalts were plotted on the figure, a mixing curve through their data field and that of the Apollo 17 basalts would intersect the Apollo 12 ilmenite basalts but would leave the Apollo 12 olivine-pigeonite basalts unexplained.

Another criticism of assimilation models can be made if the whole moon REE abundances are as low as ~2× chondritic as suggested, for example, by Ringwood (1976a,b). According to assimilation models the hi-Ti basalts inherited their REE from late stage ilmenite-clinopyroxene cumulates. The REE abundances in hi-Ti basalts are ~50× chondritic. Thus the liquid from which the ilm-cpx cumulates formed must have had REE abundances of ~200× chondritic. If, REE in the whole moon are only 2× chondritic, the liquid parental to the ilm-cpx cumulates must represent the last ~1% of solidification of a presumed magma ocean. Such a liquid seems an unlikely parent for the postulated cumulates. Firstly, the volume of cumulates which would be produced from it is extremely small. Secondly, the formation of ilmenite-bearing cumulates would have begun much earlier, when ~10% of the system remained (Ringwood, 1976a). Both problems, however, are alleviated if a higher whole moon REE abundance is assumed. (Similar considerations have been presented by Drake and Consolmagno (1976)).

Thus, without denying the possibility that some magma-mixing or assimilation may occur, we find the cumulate melting hypothesis far more useful in describing the main features of the isotopic and trace element systematics. As shown above, the working hypothesis which we have presented also describes important features of the major element composition of each magma type. We now briefly attempt to construct a more general lunar scenario which accommodates our interpretations for the individual mare magma types. To this end we adopt the widely-held concept of a lunar magma ocean. The interpretations prior to this point in the discussion do not rely on this concept although they imply it.

One important criticism of cumulate melting models has been their failure to account for the similar "mg" and Cr values of high- and low-Ti basalts)Ringwood, 1976a). This criticism derives from the belief that the cumulates are produced during closed system crystallization. If crystallization proceeded with the system open to the primordial lunar material, the system could be buffered against loss of Mg and Cr. Such would be the case if crystallization of the lunar magma ocean proceeded from the outside towards the inside. This is a feature of many lunar thermal models (e.g., Hubbard and Minear, 1975, 1976). On the other hand, Ringwood (1976a) and Walker *et al.* (1975b) argue that crystallization starts from the base of the magma ocean and proceeds outward leaving olivine, the first crystallizing phase, at the base of the magma ocean. In contrast, Nakamura *et al.* (1976) interpret the seismic data in terms of an olivine-rich upper mantle with mg-value decreasing with depth. This is also a feature of the recent lunar density model of Bills and Ferrari (1977). Most models of closed system magma ocean solidification produce the plagioclase-rich crust only relatively late in the crystallization sequence, after formation of the mafic cumulates which later participate in mare basalt petrogenesis (cf. Drake and Consolmagno, 1976). Yet crustal anorthosites, dunites, and troctolites have old ages, low initial $^{87}Sr/^{86}Sr$ and they apparently formed from parental magmas with REE abundances of only ~10–15× chondrites. If the crust and mantle sources for mare basalts formed from the same magma ocean, the arguments presented in the preceding discussion dictate that the solidification of the crust occurred prior to that of the mantle. We thus reject the concept that crust-mantle solidification occurred primarily upwards from the base of the magma ocean. We consider the observational evidence to be against this concept and the arguments for it inconclusive inasmuch as they rely on predicting the behavior of an immense

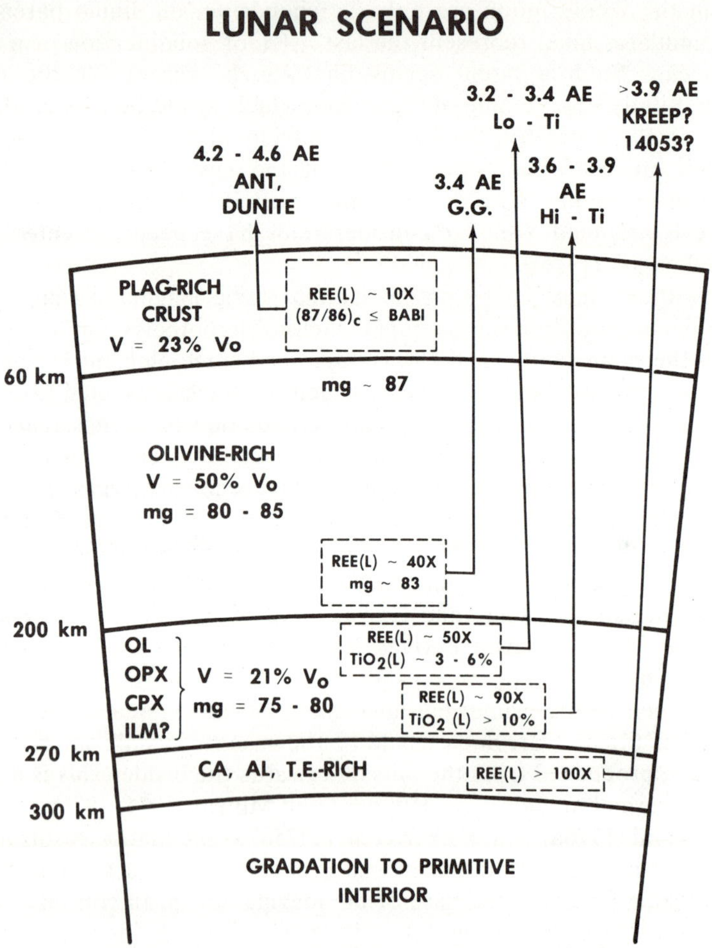

Fig. 12. Suggested lunar scenario based on the conclusions of this work and other key observational data (see text).

system in which the role and scale of major processes such as convection are not well understood.

The lunar scenario which we prefer is depicted in Fig. 12. Again, it is largely descriptive in nature. The intent is to rely either on direct observations or direct interpretations of observations insofar as possible. The main features and their rationale follow.

The depth of crust and upper mantle is taken from the interpretation of the seismic data by Nakamura *et al.* (1976). The region betwen 200–270 km is partly defined from the depths corresponding to conditions of multiple saturation of phases on the liquidi of primitive mare basalts, exclusive of the green and orange glass (see summary in Ringwood, 1976a). Mass balance for REE abundances were also considered in the definition of this region whose boundaries are somewhat arbitrary.

The plagioclase-rich crust is assumed to have formed first with a low initial $^{87}Sr/^{86}Sr$ ($\leq$BABI) as measured for Apollo 16 anorthosites. REE abundance in the parental liquid and thus presumably in the initial magma ocean are assumed to be ~10× chondritic from the calculated parental magmas for anorthosite 15415, troctolite 76535, and dunite 72417. Some process such as flotation or elutriation is presumed operative to form the plagioclase-rich crust coincident with early olivine crystallization. Thereafter crystallization is presumed to have occurred downward in the same sequence as given by models of closed system crystallization. The mg-value of the olivine is assumed to change from ~87 close to the base of the crust (dunite 72417 and troctolite 76535) to ~83 close to the base of the olivine portion of the upper mantle (green glass).

The source regions of mare basalts are assumed to lie beneath the olivine portion of the mantle and to consist of olivine, orthopyroxene, and clinopyroxene. This region is envisioned as heterogeneous on a comparatively small scale due to gravitational settling, convective stirring during crystallization, etc. This heterogeneity seems required to explain the production of at least two quite different Apollo 12 magmas from different sources at essentially the same time and place. Mg values are approximately 75–80 in the mare basalt sources, making the total range in mg in the upper mantle as deduced from sample studies (~77–87) close to that postulated by Nakamura *et al.* (1976) from the seismic data (80–85).

REE abundances in the magma ocean are assumed to vary from ~10× chondritic at the beginning of solifification to >100× chondritic near the end. The REE abundances calculated for the "early liquid" in the above discussion (REE(L) in Fig. 12) are shown for each magma type. Values of REE(L) of 40–50× chondritic are consistent with solidification of ~75% of the total system prior to formation of the cumulate sources of the low-Ti basalts. Progressive crystallization resulting in an approximately two-fold increase in REE(L) prior to crystallization of the sources of high-Ti basalts is assumed to have taken place. TiO_2 abundances in the early liquid at the time of formation of the low-Ti sources of 3–6% as calculated here would imply initial TiO_2 contents of the magma ocean of ~1%. This is two-to-three fold greater than given for the bulk moon by Ganapathy and Anders (1974) or Taylor and Jakeš (1974), but is in approximate agreement with ~10× chondritic for the REE abundances.

In this scenario Ca, Al, and trace elements are progressively enriched deep in the magma ocean as these elements are excluded from phases crystallizing above. It is further envisioned that the liquid is continually buffered against element loss by the "primordial" matter beneath. After solidification, the trace element enriched zone should also be the first to remelt, suggesting the interesting possibility that KREEP and/or some early feldspathic "mare" basalts like 14053 are produced by rather extensive melting (>50%) of this region. During remelting, isotherms are envisioned to move outward with time causing a small degree of melting in the source of hi-Ti basalts from 3.6–3.9 AE ago and in the source of low-Ti basalts from 3.2–3.4 AE ago. These melts become the parental magmas for the high- and low-Ti basalts, respectively.

The depth of origin of the basaltic magmas is envisioned to be very similar for both high- and low-Ti basalts. This is in agreement with the interpretations of Green *et al.* (1975), but not with those of Walker *et al.* (1975a). Our scenario places the source region of the high-Ti basalts immediately below that of the low-Ti basalts. The experimental studies do not seem to be definitive concerning the relative depths of origin of the two magma types (cf. Ringwood, 1976a).

Acknowledgments—Continual exchange of ideas and data with J. M. Rhodes, M. Dungan, and D. Blanchard during the course of parallel investigations has been very fruitful. We benefited from reviews by D. A. Papanastassiou and M. Tatsumoto.

REFERENCES

Alexander E. C., Jr. and Davis P. K. (1974) ^{40}Ar–^{39}Ar ages and trace element contents of Apollo 14 breccias; an interlaboratory cross-calibration of ^{40}Ar–^{39}Ar standards. *Geochim. Cosmochim. Acta* **38**, 911–928.

Alexander E. C., Jr., Davis P. K., and Reynolds J. H. (1972) Rare-gas analyses of neutron irradiated Apollo 12 samples. *Proc. Lunar Sci. Conf. 3rd*, p. 1787–1795.

Bills B. G. and Ferrari A. J. (1977) A lunar density model consistent with topographic, gravitational, librational, and seismic data. *J. Geophys. Res.* **82**, 1306–1314.

Compston W., Berry H., Vernon M. J., Chappell B. W., and Kage M. J. (1971) Rubidium-strontium chronology and chemistry of lunar material from the Ocean of Storms. *Proc. Lunar Sci. Conf. 2nd*, p. 1471–1485.

Drake M. J. (1976) Evolution of major mineral compositions and trace element abundances during fractional crystallization of a model lunar composition. *Geochim. Cosmochim. Acta* **40**, 401–411.

Drake M. J. and Consolmagno G. J. (1976) Critical review of models for the evolution of high-Ti mare basalts. *Proc. Lunar Sci. Conf. 7th*, p. 1633–1657.

Dungan M. A. and Brown R. W. (1977) The petrology of the Apollo 12 ilmenite basalt suite. *Proc. Lunar Sci. Conf. 8th*. This volume.

Ganapathy R. and Anders E. (1974) Bulk compositions of the moon and earth, estimated from meteorites. *Proc. Lunar Sci. Conf. 5th*, p. 1181–1206.

Gast P. W., Hubbard N. J., and Wiesmann H. (1970) Chemical composition and petrogenesis of basalts from Tranquility Base. *Apollo 11 Lunar Sci. Conf.* p. 1143–1163.

Green D. H. and Ringwood A. E. (1973) Significance of a primitive lunar basaltic composition present in Apollo 15 soils and breccias. *Earth Planet. Sci. Lett.* **19**, 1–8.

Green D. H., Ringwood A. E., Ware N. G., Hibberson W. O., Major A., and Kiss E. (1971b) Experimental petrology and petrogenesis of Apollo 12 basalts. *Proc. Lunar Sci. Conf. 2nd*, p. 601–615.

Green D. H., Ringwood A. E., Hibberson W. O., and Ware N. G. (1975) Experimental petrology of Apollo 17 mare basalts. *Proc. Lunar Sci. Conf. 6th*, p. 871–893.

Green D. H., Ware N. G., Hibberson W. O., and Major A. (1971a) Experimental petrology of Apollo 12 basalts: Part 1, Sample 12009. *Earth Planet. Sci. Lett.* **13**, 85–96.

Grutzeck M., Kridelbaugh S., and Weill D. (1974) The distribution of Sr and REE between diopside and silicate liquid. *Geophys. Res. Lett.* **1**, 273–275.

Haskin L. A., Allen R. O., Helmke P. A., Paster T. P., Anderson M. R., Korotev R. L., and Zweifel K. A. (1970) Rare earths and other trace elements in Apollo 11 lunar samples. *Proc. Apollo 11 Lunar Sci. Conf.*, p. 1213–1231.

Haskin L. A., Helmke P. A., Allen R. O., Anderson M. R., Korotev R. L., and Zweifel K. A. (1971) Rare-earth elements in Apollo 12 lunar materials. *Proc. Lunar Sci. Conf. 2nd*, p. 1307–1317.

Haskin L. A., Shih C.-Y., Bansal B. M., Rhodes J. M., Wiesmann H., and Nyquist L. E. (1974) Chemical evidence for the origin of 76535 as a cumulate. *Proc. Lunar Sci. Conf. 5th*, p. 1213–1225.

Helmke P. A., Blanchard D. P., Haskin L. A., Telander K., Weiss C., and Jacobs J. W. (1973) Major and trace elements in igneous rocks from Apollo 15. *The Moon* **8**, 129–148.

Horn P., Jessberger E. K., Kirsten T., and Richter H. (1975) ^{39}Ar–^{40}Ar dating of lunar rocks: Effects of grain size and neutron irradiation. *Proc. Lunar Sci. Conf. 6th*, p. 1563–1591.

Hubbard N. J. and Gast P. W. (1971) Chemical composition and origin of nonmare lunar basalts. *Proc. Lunar Sci. Conf. 2nd*, p. 999–1020.

Hubbard N. J., Gast P. W., Meyer C., Nyquist L. E., and Shih C. (1971) Chemical composition of lunar anorthosites and their parent liquids. *Earth Planet. Sci. Lett.* **13**, 71–75.

Hubbard N. J. and Minear J. W. (1975) A physical and chemical model of early lunar history. *Proc. Lunar Sci. Conf. 6th*, p. 1057–1085.

Hubbard N. J. and Minear J. W. (1976) Petrogenesis in a modestly endowed moon. *Proc. Lunar Sci. Conf. 7th*, p. 3421–3455.

Kesson S. E. (1975) Mare basalts: Melting experiments and petrogenetic interpretations. *Proc. Lunar Sci. Conf. 6th*, p. 921–944.

Kesson S. E. and Ringwood A. E. (1976) Mare basalt petrogenesis in a dynamic moon. *Earth Planet. Sci. Lett.* **30**, 155–163.

Laul J. C. and Schmitt R. A. (1975) Dunite 72417: A chemical study and interpretation. *Proc. Lunar Sci. Conf. 6th*, p. 1231–1254.

Lugmair G. W. (1975) Sm–Nd systematics of some Apollo 17 basalts (abstract) In *Papers Presented to the Conference on Origins of Mare Basalts and Their Implications for Lunar Evolution*, p. 107–109. The Lunar Science Institute, Houston.

Lugmair G. W. and Marti K. (1977) Evolution of the lunar interior. Sm–Nd systematics of Apollo 15 green glass and the question of the lunar initial $^{143}Nd/^{144}Nd$ (abstract). In *Lunar Science VII*, p. 597–599. The Lunar Science Institute, Houston.

McKay G. A. and Weill D. F. (1976) Petrogenesis of KREEP. *Proc. Lunar Sci. Conf. 7th*, p. 2427–2447.

Murthy V. R., Evensen N. M., Jahn B. M., and Coscio M. R., Jr. (1971) Rb–Sr ages and elemental abundances of K, Rb, Sr and Ba in samples from the Ocean of Storms. *Geochim. Cosmochim. Acta* **35**, 1139–1153.

Nakamura Y., Latham G. V., Dorman H. J., Duennbier F. K. (1976) Seismic structure of the moon: A summary of current status. *Proc. Lunar Sci. Conf. 7th*, p. 3113–3121.

Nyquist L. E. (1977) Lunar Rb–Sr chronology. *Phys. Chem. Earth* **10**, p. 103–142.

Nyquist L. E., Bansal B. M., and Wiesmann H. (1975) Rb–Sr ages and initial $^{87}Sr/^{86}Sr$ for Apollo 17 basalts and KREEP basalt 15386. *Proc. Lunar Sci. Conf. 6th*, p. 1445–1465.

Nyquist L. E., Bansal B. M., and Wiesmann H. (1976) Sr-isotopic constraints on the petrogenesis of Apollo 17 mare-basalts. *Proc. Lunar Sci. Conf. 7th*, p. 1507–1528.

Nyquist L. E., Bansal B. M., Wiesmann H., and Jahn B. M. (1974) Taurus-Littrow chronology: Some constraints on early lunar crustal development. *Proc. Lunar Sci. Conf. 5th*, p. 1515–1539.

Nyquist L. E., Bansal B. M., Wooden J. L., and Wiesmann H. (1977) Sr-isotopic constraints on the petrogenesis of Apollo 12 ilmenite basalts (abstract). In *Lunar Science VIII*, p. 735–737. The Lunar Science Institute, Houston.

Nyquist L. E., Hubbard N. J., Gast P. W., Bansal B. M., Wiesmann H., and Jahn B. (1973) Rb–Sr systematics for chemically defined Apollo 15 and 16 materials. *Proc. Lunar Sci. Conf. 4th*, p. 1823–1846.

Nyquist L. E., Hubbard N. J., Gast P. W., Church S. E., Bansal B. M., and Wiesmann H. (1972) Rb–Sr systematics for chemically defined Apollo 14 breccias. *Proc. Lunar Sci. Conf. 3rd*, p. 1515–1530.

Papanastassiou D. A. and Wasserburg G. J. (1970) Rb–Sr ages from the Ocean of Storms. *Earth Planet. Sci. Lett.* **8**, 269–278.

Papanastassiou D. A. and Wasserburg G. J. (1971a) Lunar chronology and evolution from Rb–Sr studies of Apollo 11 and 12 samples. *Earth Planet. Sci. Lett.* **11**, 37–62.

Papanastassiou D. A. and Wasserburg G. J. (1971b) Rb–Sr ages of igneous rocks from the Apollo 14 mission and the age of the Fra Mauro formation. *Earth Planet. Sci. Lett.* **12**, 36–48.

Pieters C. and McCord T. (1976) Characterization of lunar mare basalt types: I. A remote sensing study using reflection spectroscopy of surface soils. *Proc. Lunar Sci. Conf. 7th*, p. 2677–2690.

Philpotts J. A. and Schnetzler C. C. (1970a) Phenocryst-matrix partition coefficients for K, Rb, Sr, and Ba with application to anorthosite and basalt genesis. *Geochim. Cosmochim. Acta* **34**, 1549–1578.

Philpotts J. A. and Schnetzler C. C. (1970b) Apollo 11 lunar samples: K, Rb, Sr, Ba and rare-earth concentrations in some rocks and separated phases. *Proc. Apollo 11 Lunar Sci. Conf.*, p. 1471–1486.

Rhodes J. M. and Hubbard N. J. (1973) Chemistry, classification, and petrogenesis of Apollo 15 mare basalts. *Proc. Lunar Sci. Conf. 4th*, p. 1127–1148.

Rhodes J. M., Blanchard D. P., Brannon J. C., Rodgers K. V., and Dungan M. A. (1977) Chemistry of Apollo 12 mare basalts: Magma types and fractionation processes. *Proc. Lunar Sci. Conf. 8th*. This volume.

Ridley W. I., Reid A. M., Warner J. L., and Brown R. W. (1973) Apollo 15 green glasses. *Phys. Earth Planet. Interiors* **7**, 133–136.

Ringwood A. E. (1976a) Limits on the bulk composition of the moon. *Icarus* **28**, 325–349.

Ringwood A. E. (1976b) *Basaltic magmatism and the composition of the moon, part I: Major and heat-producing elements.* Publication 121, Research School of Earth Sciences, Australian National University, Canberra. 54 pp.

Ringwood A. E. and Kesson S. E. (1976) A dynamic model for mare basalt petrogenesis. *Proc. Lunar Sci. Conf. 7th*, p. 1697–1722.

Schnetzler C. C. and Philpotts J. A. (1970) Partition coefficients of rare earth elements between igneous matrix material and rock forming mineral phenocrysta-II. *Geochim. Cosmochim. Acta* **34**, 331–340.

Schnetzler C. C. and Philpotts J. A. (1971) Alkali, alkaline earth, and rare-earth element concentrations in some Apollo 12 soils, rocks, and separated phases. *Proc. Lunar Sci. Conf. 2nd*, p. 1101–1122.

Schnetzler C. C., Philpotts J. A., Nava D. F., Schumann S., and Thomas H. H. (1972) Geochemistry of Apollo 15 basalt 15555 and Soil 15531. *Science* **175**, 426–428.

Shih C.-Y., Haskin L. A., Wiesmann H., Bansal B. M., and Brannon J. C. (1975) On the origin of high-Ti mare basalts. *Proc. Lunar Sci. Conf. 6th*, p. 1255–1285.

Shih C.-Y. and Schonfeld E. (1976) Mare basalt genesis: A cumulate-remelting model. *Proc. Lunar Sci. Conf. 7th*, p. 1757–1792.

Soderblom L. A. and Lebofsky L. A. (1972) Techniques for rapid determination of relative ages of lunar areas from orbital photography. *J. Geophys. Res.* **77**, 279–296.

Stettler A., Eberhardt P., Geiss J., Grögler N., and Maurer P. (1973) Ar^{39}–Ar^{40} ages and Ar^{37}–Ar^{38} exposure ages of lunar rocks. *Proc. Lunar Sci. Conf. 4th*, p. 1865–1888.

Taylor S. R. and Jakeš P. (1974) The geochemical evolution of the moon. *Proc. Lunar Sci. Conf. 5th*, p. 1287–1305.

Turner G. (1971) ^{40}Ar–^{39}Ar ages from the lunar maria. *Earth Planet. Sci. Lett.* **11**, 169–191.

Walker D., Longhi J., Stopler E. M., Grove T. L., and Hays J. F. (1975a) Origin of titaniferous lunar basalts. *Geochim. Cosmochim. Acta* **39**, 1219–1235.

Walker D., Longhi J., and Hays J. F. (1975b) Differentiation of a very thick magma body and implications for the source regions of mare basalts. *Proc. Lunar Sci. Conf. 6th*, p. 1103–1120.

Weill D. F., McKay G. A., Kridelbaugh S. J., and Grutzeck M. (1974) Modeling the evolution of Sm and Eu abundances during lunar igneous differentiation. *Proc. Lunar Sci. Conf. 5th*, p. 1337–1352.

Wiesmann H. and Hubbard N. J. (1975) A compilation of the lunar samples data generated by the Gast, Nyquist, and Hubbard lunar sample PI-ships, Lyndon B. Johnson Space Center, Houston. Unpublished report.

York D. (1966) Least-squares fitting of a straight line. *Can. J. Phys.* **44**, 1079–1086.

Proc. Lunar Sci. Conf. 8th (1977), p. 1417–1428.
Printed in the United States of America

Sulfur in lunar mare basalts as a function of bulk composition

EVERETT K. GIBSON, JR.[1], ROBIN BRETT[2], and FIKRY ANDRAWES[3]

[1]Code SN7/Geochemistry Branch, NASA Johnson Space Center, Houston, Texas 77058
[2]U.S. Geological Survey, Reston, Virginia 22092
[3]Lockheed Electronics Company, Houston, Texas 77058

Abstract—Sulfur abundances in 18 Apollo 12 mare basalts range from 560 to 1520 ppm, with a median value of 1110 ppm, and metallic iron abundances range from 220 to 970 ppm, with a mean value of 660 ppm. No correlation between sulfur abundance and metallic iron content was observed for the Apollo 12 basalts and metallic iron abundances are not primarily caused by S loss. Sulfur contents of the mare basalts are directly related to the bulk composition of the rocks and especially to the TiO_2 content. Sulfur abundances in these mare basalts increase with increasing degrees of fractionation, and appear to be a consequence of concentration of S in the melts during fractionation. Unlike the Apollo 17 melts, the Apollo 12 melts were unsaturated with respect to sulfide. Changes in sulfur abundance due to T in saturated melts such as oceanic basalts are minimal because of concomitant changes in f_{O_2} and f_{S_2}. Composition also appears to control S content for Apollo 17 basalts. Cumulus processes may cause Fe–FeS enrichment.

INTRODUCTION

THE GEOCHEMISTRY of sulfur in lunar mare basalts has been the subject of considerable scrutiny. Lunar basalts typically contain troilite associated with metallic iron, an assemblage that Skinner (1970) attributed to an initially homogeneous sulfide liquid that separated immiscibly from the parent magma. Sato and Helz (1971) suggested that the loss by volatilization of S dissolved in a lunar basalt melt as S^{2-} could precipitate metallic Fe. Nash and Hansel (1973) pointed out that the equilibrium fugacity of S_2 for the reaction $Fe + 1/2\,S_2 = FeS$ at 1200°C is eight orders of magnitude above the confining pressure of the lunar atmosphere, so that S loss might be anticipated during eruption. Gibson and coworkers (Gibson and Moore, 1974; Gibson *et al.*, 1975, 1976) noted an inverse correlation between metallic Fe and total S abundances for the Apollo 15 and 17 mare basalts and suggested that the relationship resulted from either desulfuration of troilite or variations in the activities of metallic Fe and S in the melts from which the mare basalts crystallized. However, sulfur isotopic measurements on mare basalts have failed to find any major sulfur fractionation, which might be expected if substantial S loss had occurred (Gibson *et al.*, 1975). Brett (1976) recently reviewed the proposed processes for reduction of mare basalts by S loss. He suggested that S loss was the dominant mechanism in producing metallic Fe in mare basalts. Both Brett (1976) and Gibson *et al.* (1976) reported that the Apollo 11 and 17 mare basalt melts appeared to be saturated with sulfide, i.e., Fe–S–O liquid, and that the Apollo 12 and 15 mare basalt melts were not saturated with sulfide at the time of their extrusion.

For the present study we examined 18 previously unanalyzed Apollo 12 mare basalts for their total sulfur and metallic iron concentrations and the relationships of these two variables with the bulk composition of the samples. This study is a part of the consortium effort on the Apollo 12 mare basalts. Rhodes *et al.* (1977) have studied the major- and minor-element chemistry and have classified the rocks into four compositionally defined groups. Their classification is used in the present study. Nyquist *et al.* (1977) examined the isotopic systematics and geochronology of the samples, and Dungan and Brown (1977) completed a petrographic and a petrologic study of the samples. In our study we have considered the question: Are the sulfur and metallic iron relationships observed in lunar mare basalts a function of the initial composition of the melt, or has extensive sulfur loss taken place which resulted in reduction?

EXPERIMENTAL

Sulfur abundances were determined using a LECO IR-32 total sulfur analyzer. Samples ranging from 20 to 50 mg were heated in oxygen at 1650°C; the resulting SO_2 was measured using an infrared Luft-cell detector. The detection limit of the analyzer is 1 μg S. The accuracy of the sulfur analyses were checked against NBS standard steel 55e ($S = 110 \pm 10$ ppm) and USGS standard reference basalt BCR-1 ($S = 464 \pm 10$ ppm). Experimentally obtained values were within $\pm 5\%$ of the accepted values. Sample powders were prepared from 1- to 2-g rock chips by Rhodes *et al.* (1977) using a boron carbide crusher. Prior to sulfur analysis, metallic iron abundances were measured magnetically on the sample powders by R. V. Morris using the procedures of Pearce (1973). Total sulfur and metallic iron abundances for the 18 Apollo 12 basalts are given in Table 1.

Table 1. Sulfur abundances for Apollo 12 basalts.

Sample	Basalt type	Total sulfur (ppm)	Metallic Fe (ppm)
12005	cumulate	560 ± 20	760
12006	olivine	740 ± 20	970
12007	pigeonite	1200 ± 50	550
12008	ilmenite	1180 ± 40	220
12011	pigeonite	1050 ± 30	570
12012	olivine	890 ± 30	820
12014	olivine	850 ± 80	840
12015	olivine	830 ± 40	460
12016	ilmenite	1220 ± 20	790
12031	feldspathic	890 ± 20	630
12036	cumulate	1105 ± 30	970
12039	pigeonite	1410 ± 20	760
12043	pigeonite	1110 ± 60	730
12045	ilmenite	1240 ± 40	420
12047	ilmenite	1520 ± 30	460
12054	ilmenite	1410 ± 50	470
12055	pigeonite	1060 ± 20	760
12056	ilmenite	1510 ± 30	580

Discussion

(1) *Total sulfur measurements on Apollo 12 mare basalts*

Sulfur abundances for the 18 Apollo 12 mare basalts studied ranged from a low of 560 ± 20 ppm for cumulate 12005 to a high of 1520 ± 30 ppm for ilmenite basalt 12047. The median value for the 18 new analyses of Apollo 12 basalts is 1110 ppm, and the mean value is 1098 ppm. The previously reported 21 sulfur analyses of 13 Apollo 12 mare basalts had produced a range of 350–910 ppm with a mean of 679 ppm (Gibson, 1977). The present range in determined sulfur abundances for 31 Apollo 12 basalts is therefore 350–1520 ppm with a mean of 874 ppm. Dungan and Brown. (1977) and Rhodes *et al.* (1977) identified coarse-grained basalts 12005 and 12036 as cumulates, and we note that they are lower in total sulfur than other basalts of the same chemical groups. Gibson *et al.* (1976) had noted previously that the Apollo 17 mare basalts that are cumulates contain low sulfur.

(2) *Metallic iron abundances*

Metallic Fe abundances range from 220 to 970 ppm and have a mean value of 660 ppm. The vitrophyre 12008 contains the lowest amount of metallic iron (220 ppm) measured. No significant differences in metallic iron contents were noted between the fine-grained and coarse-grained Apollo 12 mare basalts. In addition, no groupings of metallic iron abundances were noted between the four chemical groups of Rhodes *et al.* (1977). Gibson *et al.* (1976) noted variations in metallic Fe abundances in Apollo 17 basalts both between fine-grained and coarse-grained basalts and between the three types of Apollo 17 mare basalts.

(3) *Total sulfur-metallic iron relationships*

Previous studies of Apollo 15 and 17 mare basalts (Gibson *et al.*, 1975, 1976; Brett, 1976) have shown that a negative correlation exists between the amount of metallic iron and the total sulfur content of the basalts. Previous workers (Gibson and Moore, 1974; Gibson *et al.*, 1975, 1976; Brett, 1976) suggested that the negative correlation between metallic iron and sulfur resulted from degassing of S and consequent reduction of the basalts prior to their final crystallization. In Fig. 1, the metallic iron versus sulfur relationships for the 18 Apollo 12 basalts are presented. Note that the strong negative correlation between Fe° and sulfur seen previously for the Apollo 15 and 17 basalts does not exist. The calculated correlation coefficient between Fe° and S for the 18 basalts is -0.5 and is not significant for the number of samples analyzed.

The data plotted in Fig. 1 show that the ilmenite basalts are grouped. The pigeonite basalts are clustered together in the center of the figure, and the olivine basalts are all on the left side of the plot. The observed distribution is a function of the sulfur content of the samples and not of the amount of metallic iron or of

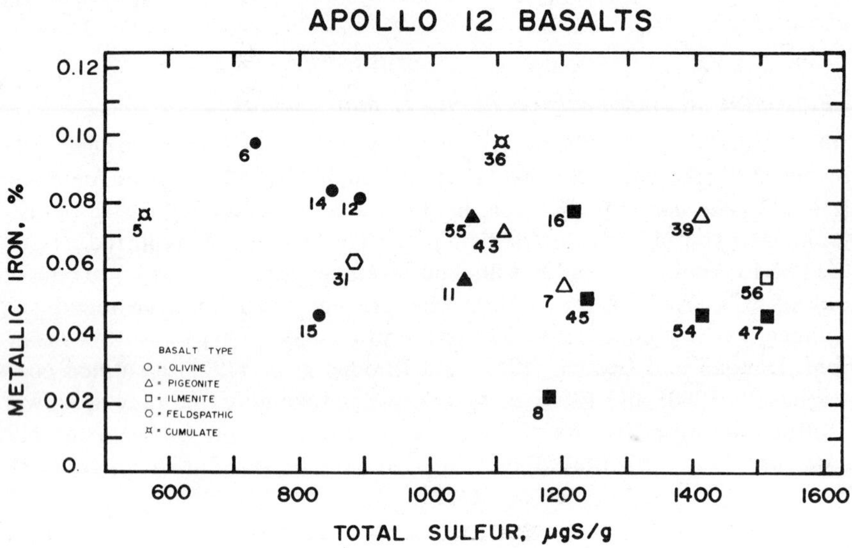

Fig. 1. Total sulfur and metallic iron abundance relationships for the Apollo 12 mare basalts. The inverse relationship between sulfur and metallic iron seen previously for the Apollo 15 and 17 mare basalts does not exist for the Apollo 12 basalts. The correlation coefficient for the 18 basalts is −0.54, which is not significant for the number of samples. Light symbols represent coarse-grained basalts; dark symbols represent fine-grained basalts. Metallic iron data are from R. V. Morris (unpublished).

the grain size of the basalts, which is a reflection of their relative cooling rates. Therefore, it appears that sulfur abundances measured in these mare basalts are not greatly dependent on post-eruptive outgassing processes, if at all, and may reflect their source regions and magma compositions.

One might argue that if the f_{O_2} of the basalts upon eruption was well above Fe° saturation, and was variable from melt to melt, then sulfur outgassing could have been important even though a correlation between Fe° and S would not be observed. However, there is good evidence that the bulk of Apollo 12 basalts were close to Fe° saturation on eruption, as shown by the presence of metallic iron in some of the minerals that were earliest to form (e.g., Brett *et al.*, 1971).

(4) *Relationship between sulfur abundances and rock chemistry*

Rhodes *et al.* (1977) have measured the major- and trace-element chemistry for the 18 Apollo 12 mare basalts and have classified the basalts into chemical groups. Their data confirm the earlier classifications of these basalts into four compositionally defined basalt types: olivine, pigeonite, ilmenite, and feldspathic basalts (James and Wright, 1972; Rhodes *et al.*, 1975; Papike *et al.*, 1976). The analytical data on the Apollo 12 mare basalts, as noted by Rhodes *et al.* (1977), are consistent with the derivation of the olivine and pigeonite basalts from a parental liquid by moderate amounts of olivine fractionation. The compositional hiatus between the pigeonite basalts and the olivine basalts, coupled with wide textural diversity of essentially isochemical samples, suggest that they have evolved from the olivine basalt composition by shallow subsurface fractionation prior to eruption rather than by fractionation within a single cooling unit after eruption. The compositions of the ilmenite basalts are also attributed by Rhodes *et al.* to processes of near-surface crystal fractionation. Rhodes *et al.* (1977) reported that no more than three magma types (olivine-pigeonite, ilmenite, and feldspathic) are required to produce the suite and that these three are derived from distinctly different source materials.

Gibson *et al.* (1976) noted that the total S and TiO_2 abundances were directly related for Apollo 17 mare basalts. They found that high-Ti basalts from Apollo 11 and 17 were enriched in S with respect to those from Apollo 12 and 15, which are lower in TiO_2 and S. The TiO_2 and S relationship for the Apollo 12 mare basalts analyzed in this study is given in Fig. 2. The correlation between S and TiO_2 for the 18 basalts is very high (correlation coefficient = +0.92). Olivine basalts and pigeonite basalts have distinct populations in Fig. 2. The ilmenite basalts are spread over a wider region and could possibly be separated into two subgroups. No apparent systematic differences are noted between fine- and coarse-grained basalts. Generally, the coarse-grained pigeonite basalt and ilmenite basalt lie toward the high sulfur end of their respective subgroups, but insufficient coarse-grained samples have been studied to verify that a trend exists.

The relationship between MgO/FeO and S is given in Fig. 3. Note that the olivine, pigeonite, and ilmenite basalts plot as distinct groups in Fig. 3. A similar relationship was found for the Apollo 17 basalts (Gibson *et al.*, 1976). Differences within each group may represent varying degrees of near-surface fractionation. The most S-rich samples are also the most fractionated basalts found in the Apollo 12 suite (Rhodes *et al.*, 1977).

The Behavior of Sulfur in Lunar Mare Basalts

Major-element compositions of the Apollo 12 and 17 basalts for which Gibson *et al.* (1976; this paper) had determined sulfur were recalculated in terms of mole fractions of the oxides, using the data of Rhodes *et al.* (1977 for Apollo 12 basalts; 1976 for Apollo 17 basalts).

$$\text{Log } C_s = a_0 + a_1 N_{FeO} + a_2 N_{CaO} + a_3 N_{MgO} + a_4 N_{TiO_2} + a_5 (N_{Na_2O} + N_{K_2O})$$

was then calculated for each rock for a temperature of 1200°C, where C_s = sulfide capacity[1], N = mole fraction, and $a_0 \ldots . a_5$ are regression coefficients.

[1]As Haughton *et al.* (1974) state: "The term sulfide capacity is not entirely satisfactory as the words might suggest to the unwary that it connotes the amount of sulfur in a silicate melt when equilibrated with a sulfide melt. This, of course, is not the case. The term is a pseudo-equilibrium constant and does not in any way, connote the saturation sulfur capacity."

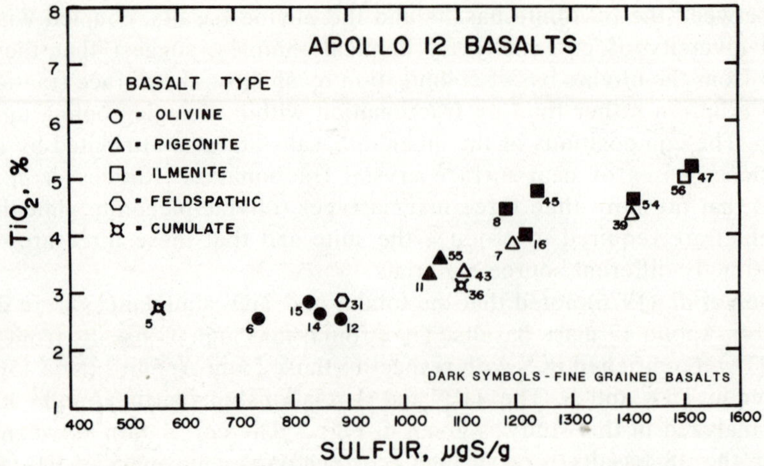

Fig. 2. Total sulfur and TiO_2 relationship for the Apollo 12 mare basalts. The correlation coefficient = +0.92. The TiO_2 data are from Rhodes *et al.* (1977). Light symbols represent coarse-grained basalts; dark symbols represent fine-grained basalts.

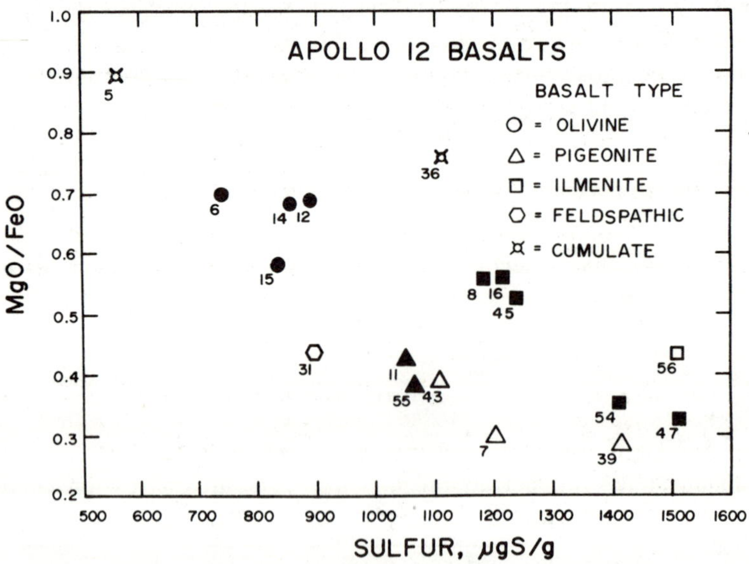

Fig. 3. Total sulfur and MgO/FeO relationship for the Apollo 12 mare basalts. Correlation coefficient = −0.73. The MgO and FeO data are from Rhodes *et al.* (1977). Light symbols represent coarse-grained basalts; dark symbols represent fine-grained basalts.

The formula is an empirical one from Haughton *et al.* (1974), whose use they state should be restricted to mafic melts. The choice of 1200°C is arbitrary, choice of another temperature would merely change all log C_s values linearly, according to the temperature correction suggested by Haughton *et al.*

$$C_s = S_m \sqrt{\frac{f_{O_2}}{f_{S_2}}}$$

where S_m = wt.% sulfur dissolved as sulfide in the silicate melt (Fincham and Richardson, 1954).

A plot of $-\log C_s$ versus measured S is shown for the Apollo 12 basalts in Fig. 4 and for the Apollo 17 basalts in Fig. 5.

Note the good correlation (correlation coefficient = −0.95) between log C_s and S for the 18 Apollo 12 basalts, especially if the cumulate 12036 is not considered. The groups reported by Rhodes *et al.* (1977) appear as entities on the diagram.

It is noteworthy that the correlation of log C_s and S is truly a reflection of the correlation of bulk composition with S rather than of FeO with S. Haughton *et al.* (1974) suggest that there is often a good correlation between FeO and S, but in the present case, no such correlation exists, the correlation coefficient being 0.04. The correlation between TiO_2 and S (Fig. 2) is strong however (+0.92).

It seems reasonable that the retentive power of a magma for S is directly related to bulk composition, and that one could therefore observe a correlation between S and composition as a result of outgassing. However, in the present case, since the correlation between S and metallic Fe abundance is extremely poor, such an explanation is unlikely. Brett (1976) suggested that iron abundance was related to S loss but pointed out that the poor correlation coefficient (0.59) between Fe° and S suggested that the melts had a range of initial S contents. Sulfur abundance appears to be related to the composition of the melt as has been observed by Mathez (1976) and Czamanske and Moore (1977) for oceanic basalts. The correlation suggests that the sulfur abundance in the Apollo 12 suite is related to differentiation conditions. the correlation between S content and bulk composition does not address the question of whether or not the Apollo 12 melts were saturated with respect to sulfide (see footnote 1). However, the much lower abundance of S in Apollo 12 mare basalts compared to Apollo 17 basalts, the presence of metallic iron inclusions with no troilite in early-formed minerals in Apollo 12 basalts (e.g., Brett *et al.*, 1971), and preliminary experimental studies by S. E. Kesson (pers. comm., 1977) on S solubility in Apollo 12 compositions all suggest that the Apollo 12 melts were not saturated with respect to sulfide, in agreement with Brett (1976) and Gibson (1976).

If f_{O_2} of the melts is taken from Sato *et al.* (1973) and Sato (1976), who find little variation in f_{O_2} in mare basalts at any given temperature, then f_{S_2} can be calculated for each rock at a given temperature from the relationship above. The range in calculated f_{S_2} at 1200°C for the Apollo 12 rocks studied is small ($-\log f_{S_2} = 4.5$–5) and may not be real if the small variations in f_{O_2} for mare basalts determined by Sato and the error in regression coefficients discussed by

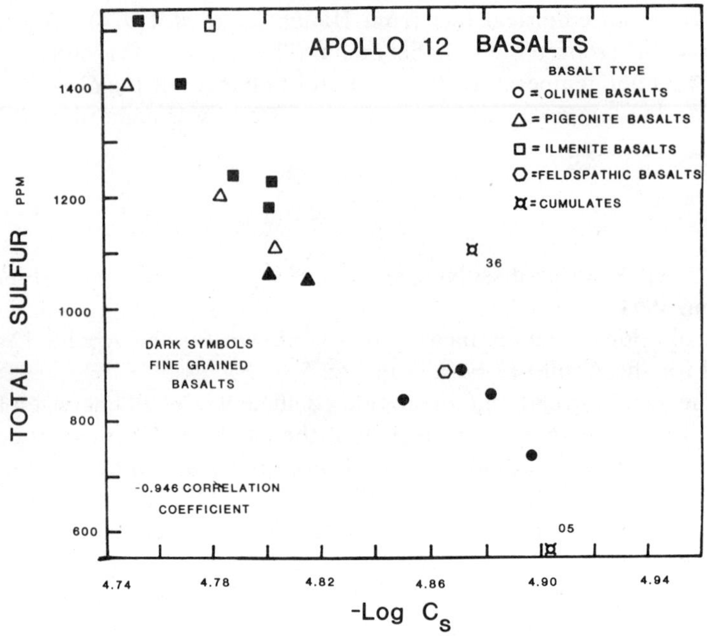

Fig. 4. Total sulfur data versus calculated sulfide capacity for the Apollo 12 mare basalts. Correlation coefficient = −0.946. The sulfide capacity, C_s at 1200°C, was calculated using the procedures of Haughton *et al.* (1974), the chemical data of Rhodes *et al.* (1977), oxygen and sulfur fugacity data from Sato and Helz (1971). The strong correlation between sulfur content and sulfide capacity for the Apollo 12 mare basalts is controlled by the composition of the melts and not by desulfuration processes. Light symbols represent coarse-grained basalts; dark symbols represent fine-grained basalts.

Haughton *et al.* (1974) are taken into account. Alternatively, if one assumes an average value for $\log f_{S_2}$, and uses the values of f_{O_2} and C_s noted above, one may calculate S contents of Apollo 12 basalts to within 100 ppm of their true values, in most cases. The fact that calculated results are so close to measured results is strong evidence that sulfur loss by volatilization was negligible.

The compositions of the rocks under discussion represent a range of liquidus temperatures of at least 100°C (e.g., Green *et al.*, 1971). If f_{O_2}/f_{S_2} is constant with temperature, Haughton *et al.* (1974) suggest that such a temperature difference should be reflected in a variation of about a factor of 6 in sulfur solubility based largely on some simple systems. Such extreme sulfur variations are clearly not seen in the present data. Czamanske and Moore (1977) have noted the same paradox with respect to S contents in suites of oceanic basalts. They state that changes in f_{O_2} and f_{S_2} during differentiation seem to have offset the effect of temperature and that crystallization and changes in FeO content have controlled S concentration in the melts. This conclusion agrees with the present calculations and data. It would be tempting to ascribe the lack of a temperature effect to near-isochemical, isothermal melts but contain differing amounts of

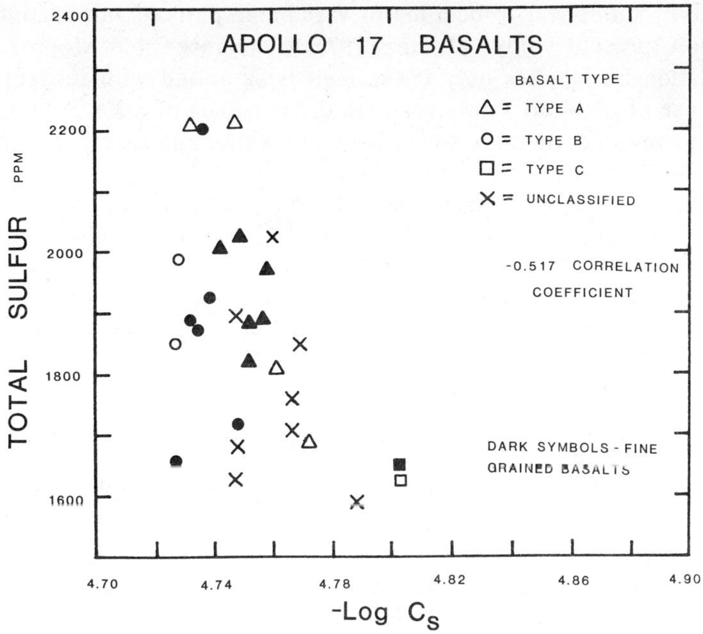

Fig. 5. Total sulfur data and calculated sulfide capacity for the Apollo 17 mare basalts. Correlation coefficient = − 0.517. The sulfide capacity, C_s at 1200°C, was calculated using the procedures of Haughton *et al.* (1974), the chemical data of Sato (1976). Light symbols represent coarse-grained basalts; dark symbols represent fine-grained basalts.

phenocrysts. However, the data of Czamanske and Moore are for glasses that represent a range of equilibrium temperatures. Also, in the case of the Apollo 12 basalts, not all the rocks can be made from 12009, the vitrophyre, or any theoretical composition by gravitational differentiation processes.

Recalling that $C_s = S_m(f_{O_2}/f_{S_2})^{1/2}$ (Fincham and Richardson, 1954), therefore $\log S_m = \log C_s + \frac{1}{2}\log f_{S_2} - \frac{1}{2}\log f_{O_2}$. But $\log C_s = a \times (10^4/T) + b$ where $a = -1.2$ (Haughton *et al.*, 1974), and $\log f_{O_2}$ may also be represented by an equation of the form $a \times (10^4/T) + b$. Log f_{O_2} vs T curves of both mare basalts and terrestrial basalts are parallel to the buffer curves involving iron, in which $a = -2.9 \pm 0.2$ (e.g., Sato *et al.*, 1973).

Log f_{S_2} is also of the form $a \times (10^4/T) + b$. Naldrett (1969, Appendix 2) summarized experimental data for Fe-S-O melts as a function of composition, f_{O_2}, f_{S_2}, and T. If "a" is calculated for isochemical polythermal systems (using more than 40 data points) from these data, $a = 1.4 \pm 0.3$. Therefore $\log S_m = a' \times (10^4/T) + b'$.

Summing the above, $a' = -0.5$. Therefore, considering the effects of T, f_{S_2}, and f_{O_2}, $\log S_m = -0.5 \times (10^4/T) + b$. The apparent effect of temperature on sulfur solubility is therefore fairly small at elevated temperatures if the effects of f_{O_2} and f_{S_2} are taken into account and provided that no appreciable amounts of crystals form, thus changing the bulk composition. This explains why temperature does not appear to be a major variable in plots of sulfur content versus composition (present study; Mathez, 1976; Czamanske and Moore, 1977). The above relationship applies only if the melt is saturated with respect to sulfide, since the use of Naldrett's data requires the presence of a Fe–S–O liquid. If the melt is not saturated, then the sulfur content will not change upon cooling and no temperature effect on S content will be observed.

The cumulate 12036 is considerably higher in S than would be anticipated by its composition (Fig. 4) and is also high in metallic Fe, which suggests that accumulation of small amounts of Fe–FeS liquid may have taken place. It is also noteworthy that the very Mg-rich cumulate 12005 is relatively rich in metallic Fe, suggesting that the process precipitating the metal was taking place during this early period of accumulation.

The Apollo 17 mare basalts show no correlation between S content and composition if the data are considered as a whole (Fig. 5). If they are divided up into type, following the classification of Rhodes *et al.* (1976), then a correlation is seen in the Type A rocks, and the same conclusions can be made about these as those made for the Apollo 12 basalts. The steepness in slope of the plot of Type A rocks is doubtlessly due to the high FeO and TiO_2 contents compared with those of the Apollo 12 basalts. Haughton *et al.* (1974) have shown that a plot of S vs FeO in basaltic melts containing immiscible Fe–S–O liquid becomes near vertical at about 25 wt.% FeO. The Type A rocks contain about 19 wt.% FeO and 12% TiO_2. Haughton *et al.* pointed out that the positive influence of TiO_2 content on C_s and sulfur content of a magma is nearly as strong as that of FeO. The Type B rocks may show a correlation between S content and composition if it is considered that at such high FeO and TiO_2 contents, the slope of the S vs log C_s curve is near vertical.

Data are insufficient for any conclusions to be drawn about sulfur content and composition for Type C rocks. The unclassified group probably represents several types and so contributes to the scatter in Fig. 5.

Clearly, the factors determining S content in Apollo 17 mare basalts are not totally understood. It is also clear that the excellent correlation between S and metallic iron contents in Apollo 17 mare basalts observed by Gibson *et al.* (1976) and Brett (1976) still needs to be explained. The mechanism of reduction of mare basalts still remains almost as much a mystery as it was in 1969—the mechanism of reduction by C and outgassing of CO and CO_2, discussed by Sato (1976), should be investigated in more detail.

CONCLUSIONS

The examination of sulfur abundances, metallic iron abundances, and rock chemistry for the Apollo 12 mare basalts in this study has resulted in the following conclusions:

(1) Total sulfur abundances found in this study for 18 new Apollo 12 basalts have extended the previously observed sulfur abundance ranges for the four types of mare basalts at the Apollo 12 site.

(2) No apparent correlation exists between the amount of total sulfur in the Apollo 12 basalts and their metallic iron contents, in contrast to the correlations seen previously for the Apollo 15 and 17 basalts.
(3) Sulfur abundances in the Apollo 12 mare basalts are directly related to the composition of the basalts, especially to the TiO_2 contents of the basalts and the degree of near-surface fractionation that the basalt has undergone.
(4) The bulk of the metallic iron in the Apollo 12 mare basalts does not appear to be related to sulfur loss from the melt, as proposed previously by Brett (1976) and by Gibson *et al.* (1976).
(5) The melts from which the Apollo 12 mare basalts crystallized were apparently unsaturated with sulfide, in agreement with Brett (1976) and Gibson (1976).
(6) Cumulus processes may cause Fe-FeS enrichments (e.g., 12036).
(7) Sulfur abundances in Apollo 17 Type A and Type B basalts may be compositionally controlled like the basalts from Apollo 12, but at higher contents of $(FeO + TiO_2)$.
(8) At present, reduction processes and processes of metallic iron formation in mare basalts are not completely understood.

Acknowledgments—The authors thank Richard V. Morris for allowing the use of his unpublished data on metallic iron abundances and acknowledge fruitful discussions with Thomas Cobleigh; very thoughtful reviews by Edward Stolper and P. L. Roeder improved the paper considerably. One of us (R. B.) acknowledges NASA Contract T1168A.

References

Brett R., Butler P., Jr., Meyer C., Jr., Reid A. M., Takeda H., and Williams R. (1971) Apollo 12 igneous rocks 12004, 12008, 12009, and 12022: A mineralogical and petrological study. *Proc. Lunar Sci. Conf. 2nd*, p. 301–317.

Brett R. (1976) Reduction of mare basalts by sulfur loss. *Geochim. Cosmochim. Acta* **40**, 997–1004.

Czamanske G. K. and Moore J. G. (1977) Composition and phase chemistry of sulfide globules in basalt from the Mid-Atlantic Ridge rift valley near 37°N latitude. *Bull. Geol. Soc. Amer.* In press.

Dungan M. A. and Brown R. W. (1977) The petrology of the Apollo 12 ilmenite basalt suite. *Proc. Lunar Sci. Conf. 8th*. This volume.

Fincham C. J. B. and Richardson F. D. (1954) The behaviour of sulfur in silicate and aluminate melts. *Phil. Trans. Roy. Soc. Lond.* **223**, 40–62.

Gibson E. K., Jr. (1977) Volatile elements, carbon, nitrogen, sulfur, sodium, potassium, and rubidium in the lunar regolith. *Phys. Chem. Earth* **10**, 57–62.

Gibson E. K., Jr. and Moore G. W. (1974) Sulfur abundances and distributions in the valley of Taurus-Littrow. *Proc. Lunar Sci. Conf. 5th*, p. 1823–1837.

Gibson E. K., Jr., Chang S., Lennon K., Moore G. W., and Pearce G. W. (1975) Sulfur abundances and distributions in mare basalts and their source magmas. *Proc. Lunar Sci. Conf. 6th*, p. 1287–1301.

Gibson E. K., Jr., Ussleman T. M., and Morris R. V. (1976) Sulfur in Apollo 17 basalts and their source regions. *Proc. Lunar Sci. Conf. 7th*, p. 1491–1505.

Green D. H., Ringwood A. E., Ware N. G., Hibberson W. O., Major A., and Kiss E. (1971) Experimental petrology and petrogenesis of Apollo 12 basalts. *Proc. Lunar Sci. Conf. 2nd*, p. 601–615.

Haughton D. R., Roeder P. L., and Skinner B. J. (1974) Solubility of sulfur in mafic magmas. *Econ. Geol.* **69**, 451–462.

James O. B. and Wright T. L. (1972) Apollo 11 and 12 mare basalts and gabbros: Classification, compositional variations, and possible petrogenetic relations. *Bull. Geol. Soc. Amer.* **83**, 2357–2382.

Mathez E. A. (1976) Sulfur solubility and magmatic sulfides in submarine basalt glass. *J. Geophys. Res.* **81**, 4269–4276.

Naldrett A. J. (1969) A portion of the system Fe-S-O between 900 and 1080°C and its application to sulfide ore magmas. *J. Petrol.* **10**, 174–201.

Nash W. P. and Hansel W. D. (1973) Partial pressures of oxygen, phosphorus and fluorine in some lunar lavas. *Earth Planet. Sci. Lett.* **20**, 13–27.

Nyquist L. E., Bansal B. M., Wooden J. L., and Wiesmann H. (1977) Sr-isotopic constraints on the petrogenesis of Apollo 12 mare basalts. *Proc. Lunar Sci. Conf. 8th.* This volume.

Papike J. J., Hodges F. N., Bence A. E., Cameron M., and Rhodes J. M. (1976) Mare basalts: Crystal chemistry, mineralogy, and petrology. *Rev. Geophys. Space Phys.* **14**, 475–540.

Pearce G. W. (1973 "Magnetism and lunar surface samples." Ph.D. dissertation, University of Toronto.

Rhodes J. M., Hodges F. N., and Papike J. J. (1975) Mare basalts: Major element composition and classification (abstract). In *Papers Presented to the Conference on Origins of Mare Basalts and Their Implications for Lunar Evolution*, p. 135–139. The Lunar Science Institute, Houston.

Rhodes J. M., Hubbard N. J., Wiesmann H., Rodgers K. V., Brannon J. C., and Bansal B. M. (1976) Chemistry, classification and petrogenesis of Apollo 17 mare basalts. *Proc. Lunar Sci. Conf. 7th*, p. 1467–1489.

Rhodes J. M., Blanchard D. P., Dungan M. A., Rodgers K. V., and Brannon J. C. (1977) Chemistry of Apollo 12 mare basalts: Magma types and fractionation processes. *Proc. Lunar Sci. Conf. 8th.* This volume.

Sato M. (1976) Oxygen fugacity and other thermochemical parameters of Apollo 17 high-Ti basalts and their implications on the reduction mechanism. *Proc. Lunar Sci. Conf. 7th*, p. 1323–1344.

Sato M. and Helz R. T. (1971) Oxygen fugacity values of Apollo 12 basaltic rocks. Second Lunar Sci. Conf., unpublished abstract.

Sato M., Hickling N., and McLane J. W. (1973) Oxygen fugacity values of Apollo 12, 14, and 15 lunar samples and reduced state of lunar magmas. *Proc. Lunar Sci. Conf. 4th*, p. 1061–1079.

Skinner B. J. (1970) High crystallization temperatures indicated for igneous rocks from Tranquillity Base. *Proc. Apollo 11 Lunar Sci. Conf.*, p. 891–895.

Proc. Lunar Sci. Conf. 8th (1977), p. 1429–1442.
Printed in the United States of America

Coarse-grained basalt 71597: A product of partial olivine accumulation

RICHARD D. WARNER, KLAUS KEIL, and G. JEFFREY TAYLOR

Department of Geology and Institute of Meteoritics, University of New Mexico, Albuquerque, New Mexico 87131

Abstract—Rock 71597 is a coarse-grained, plagioclase-poikilitic basalt that contains 15.8 wt.% MgO (Murali *et al.*, 1977) and 19.3 vol.% modal olivine. Olivine and ilmenite crystals occur in two size generations. Large (2–5 + mm) subequant olivine crystals (~18%, modally) have skeletal morphology and Fo_{73-75} cores, while small (0.1–0.5 mm) "matrix" olivines have prismatic, non-skeletal habit and Fo_{63-69} compositions. Large (up to 5 mm long) ilmenite crystals (~2%, modally) also exhibit skeletal morphology.

The presence of large olivine and ilmenite crystals with skeletal habits suggestive of crystallization under relatively rapid (~5°C/hr?) cooling in a rock otherwise of texture consonant with slower (<1°C/hr) cooling is best explained if the large skeletal crystals formed toward the more rapidly cooled margin of a flow and accumulated in 71597 by gravitational crystal settling toward the flow's interior. In support of this hypothesis, calculations show that the observed olivine compositions (Fo_{73-75}) are approximately 10 mol.% too Fe-rich to have crystallized from a melt of composition corresponding to that of 71597.

INTRODUCTION

ROCK 71597 IS A 12.35 g coarse-grained basalt collected with the Apollo 17 Station 1A rake samples. Hand-specimen inspection indicates that the rock is unusually olivine rich (Keil *et al.*, 1974). Instrumental neutron activation analysis of an 0.61 g chip of this rock (Murali *et al.*, 1977) shows it to be a high-Ti mare basalt of unique composition (Fig. 1). Compared to other Apollo 17 very high-Ti mare basalts, 71597 has a much higher MgO content and lower CaO, Na_2O, and K_2O abundances (Table 1). Its TiO_2 content (8.4 wt.%) is at the low end of the Apollo 17 range. As shown in Fig. 1, the composition of 71597 more closely resembles that of the Apollo 17 orange glass. However, even here there are compositional differences (Table 1): 71597 contains 2 wt.% more Al_2O_3, nearly 1 wt.% more MgO and CaO, and ~2 wt.% less FeO than does Apollo 17 orange glass.

The uniqueness of its chemistry has prompted us to undertake a detailed petrographic and mineralogic examination of 71597, with the purpose of determining whether (1) it crystallized from a melt of composition equivalent to that of the whole rock and, thus, represents a new magma type at the Apollo 17 site, or (2) it crystallized from a melt of composition similar to other Apollo 17 very high-Ti mare basalts, and owes its unique chemistry to olivine accumulation.

Table 1. Comparison of major element composition of 71597 with other Apollo 17 high-Ti mare basalts and Apollo 17 orange glass.

	71597[1]	Range of 71 Apollo 17 high-Ti mare basalts[2]	Apollo 17 orange glass[3]
TiO_2 (%)	8.4	7.8–14.5	8.87
Al_2O_3	7.9	7.0–11.7	5.85
Cr_2O_3	0.47	0.162–0.647	0.55
FeO	19.8	16.1–21.8	21.96
MnO	0.237	0.216–0.270	—
MgO	15.8	5.2–11.2	14.99
CaO	7.9	8.7–13.6	7.16
Na_2O	0.29	0.314–0.50	0.33
K_2O	0.027	0.04–0.11	—

[1]Analysis from Murali *et al.* (1977).
[2]Analyses from Laul *et al.* (1974), Warner *et al.* (1975a,b), and Murali *et al.* (1977).
[3]Analysis from Reid *et al.* (1973).

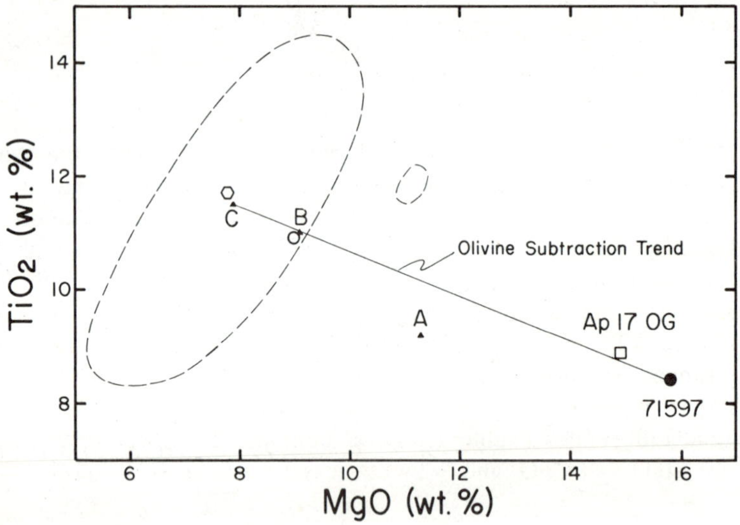

Fig. 1. Plot of TiO_2 vs. MgO (wt.%) showing composition of 71597 in relation to that of main group of Apollo 17 very high-Ti basalts (large dashed field), Station 4 MgO-rich basalts (small dashed field) and Apollo 17 orange glass. Open circle indicates composition of fine-grained basalt 71547 (Table 4, col. 2); open hexagon represents average composition of 22 fine-grained Apollo 17 basalts (Table 4, col. 4). Points A, B and C represent derivative compositions obtained by subtracting 18% olivine +2% ilmenite, 24% olivine, and 27% olivine (respectively) from 71597 composition (see discussion in text).

Method of Study

Two polished thin sections were made from the neutron-irradiated 0.61 g chip analyzed by INAA by Murali *et al.* (1977). To provide as much surface material as possible, five or six individual slices through the chip were mounted per thin section (giving a total surface area of $\sim 240\,mm^2$). The texture of the rock was studied with a petrographic microscope. A mode (6015 counts) was obtained by point counting both thin sections in reflected light; olivine was distinguished from pyroxene by changing to transmitted light. Quantitative mineral analyses were obtained by electron microprobe using methods described in Keil (1967). Complete mineral analyses were corrected for differential matrix effects following the procedure of Bence and Albee (1968) and Albee and Ray (1970).

Petrography

Rock 71597 is a coarse-grained basalt and in general has a plagioclase-poikilitic texture (Fig. 2) similar to other coarse-grained Apollo 17 very high-Ti basalts (Papike *et al.*, 1974; Warner *et al.*, 1975a). By volume, it contains 19.3% olivine, 39.4% pyroxene, 28.3% plagioclase, 11.8% ilmenite, 0.1% armalcolite, 0.2% spinel, 0.2% metal and troilite, 0.4% cristobalite, and 0.2% glassy mesostasis. The 19.3% modal olivine is more than twice as much as we have found in any of the other Apollo 17 rake samples (maximum, 8.3%—Warner *et al.*, 1975a)

Fig. 2. Transmitted light photomicrograph (X nicols) showing plagioclase-poikilitic texture of the rock. Most of field of view is occupied by a single plagioclase crystal which poikilitically encloses crystals of olivine (high relief), pyroxene, armalcolite (large opaque crystal at lower left) and ilmenite. Horizontal dimension equals 2.5 mm.

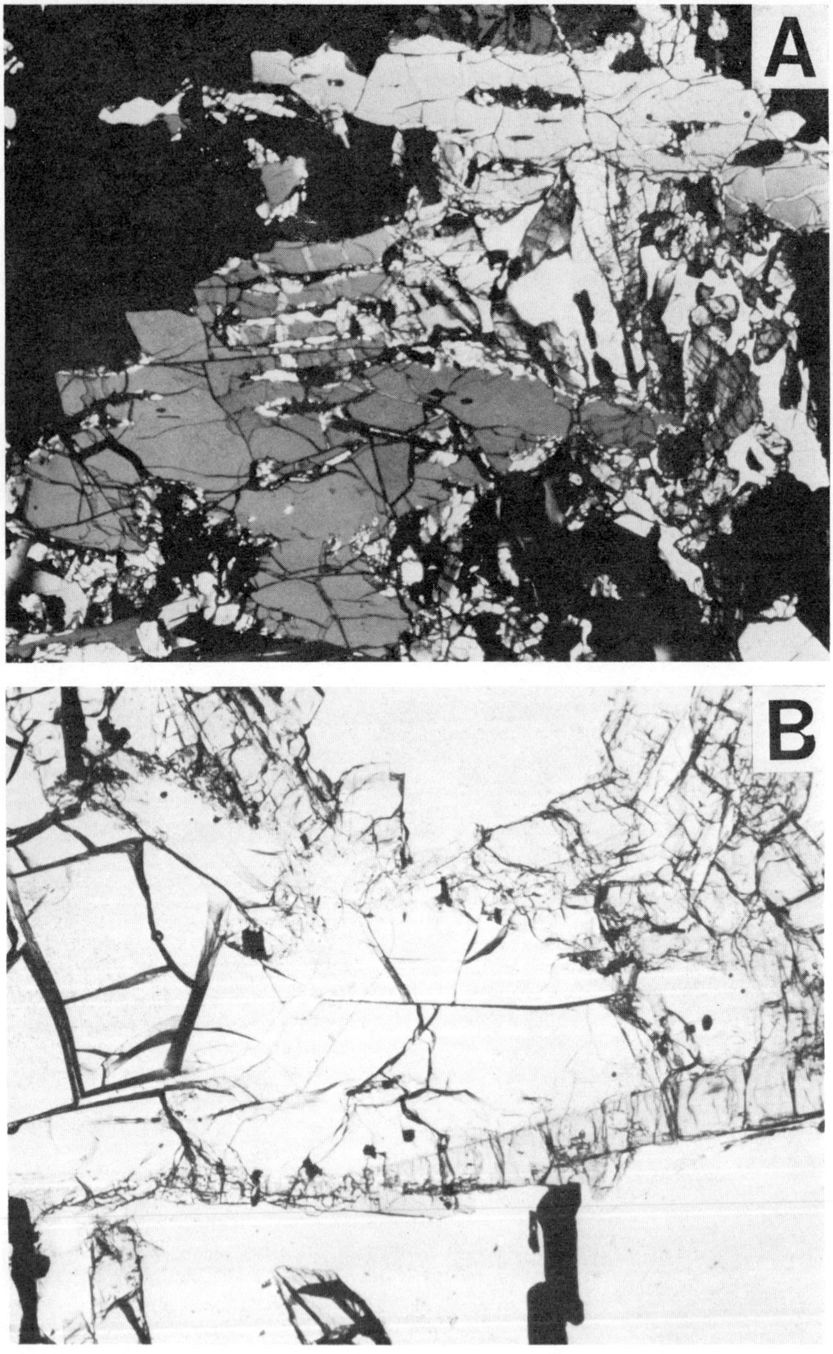

Fig. 3. (A) Transmitted light photomicrograph (partly X nicols) showing skeletal habit of large olivine crystal. Olivine at left center and at top of field of view is all part of a single crystal. Horizontal dimension equals 5.5 mm. (B) Transmitted light photomicrograph showing pyroxene mantling portion of large olivine crystal. Note small euhedral spinel crystals (opaque) near edge of the olivine. Horizontal dimension equals 1.8 mm.

and nearly twice the amount we have found for olivine-rich Station 4 basalts such as 74245 and 74275 (10.5 and 10.6%, respectively—unpublished data from our laboratory). Compared to other coarse-grained Apollo 17 very high-Ti basalts, 71597 has greater than 15 times as much olivine as the average modal abundance of olivine reported by Warner *et al.* (1975a).

All but 1–2% of the modal olivine is comprised of a small number of very large (2–5 + mm) crystals, most of which have irregular skeletal shapes (Fig. 3A). The large skeletal olivine crystals contain sparse melt inclusions as well as inclusions of euhedral Cr_2O_3-rich spinels. Margins of the olivine crystals are often corroded and are mantled by pyroxene (Fig. 3B). Although not pronounced, there is some tendency for these large olivines to cluster. For example, three such olivine crystals occur about 1 mm apart, but similar-sized areas elsewhere in the rock are devoid of large olivine grains. Compositionally, the large skeletal olivines have Fo_{73-75} cores (Table 2, col. 1). The crystals are zoned outwards to more Fe-rich margins adjacent to mantling pyroxene (Table 2, col. 2).

The remainder of the olivine in 71597 consists of numerous small crystals 0.1–0.5 mm long. The majority occur as euhedral, prismatic crystals poikilitically enclosed by plagioclase; a few occur as rounded grains surrounded by pyroxene. The small olivine crystals are unzoned and have compositions ranging from Fo_{63} to Fo_{69} (Table 2, col. 3). There is thus a small gap in composition between that of

Table 2. Electron microprobe analyses (wt.%) of olivine and pyroxene.

	1	2	3	4	5	6
SiO_2	37.9	36.6	36.8	51.3	47.4	53.3
TiO_2	.08	.09	.08	2.19	4.5	.73
Al_2O_3	*	*	*	2.66	6.4	.81
Cr_2O_3	.21	.15	.11	.43	.77	.23
FeO	22.5	29.2	29.1	10.7	8.3	19.0
MnO	.21	.35	.36	.21	.18	.35
MgO	38.2	33.1	32.8	16.1	13.6	22.2
CaO	.22	.20	.31	17.3	20.0	4.1
Total	99.3	99.7	99.6	100.9	101.2	100.7
Fo	75.2	66.9	66.8			
Fa	24.8	33.1	33.2			
En				46.7	41.7	62.0
Fs				17.3	14.3	29.8
Wo				36.0	44.1	8.2

Columns (1–2): Analyses of large skeletal olivine crystal. Analysis 1 from core, 2 from edge of crystal. (3): Analysis of small euhedral olivine crystal. (4). Analysis of pyroxene adjacent to large skeletal olivine crystal. (5): Analysis of titanaugite crystal. (6): Analysis of pigeonite crystal.

*Less than 0.01 wt.%.

the cores of the large skeletal crystals (Fo_{73-75}) and that of the small olivine grains (Fo_{63-69}—Fig. 4).

Pyroxene grains exhibit habits typical of those of other coarse-grained Apollo 17 very high-Ti basalts. Sector-zoned titanaugite crystals range up to ~2 mm across, but are most commonly 0.5–1 mm wide. The majority of the pyroxene occurs as anhedral, subequant to bladed crystals 0.2–0.3 mm wide. Compositions range from titanaugite (maximum 4.5 wt.% TiO_2; 6.5 wt.% Al_2O_3) to pigeonite (Fig. 4; Table 2); the extreme Fe enrichment typical of late-stage pyroxene crystallization in other Apollo 17 basalts (Papike *et al.*, 1974; Warner *et al.*, 1975a) is not present in 71597.

Plagioclase occurs as large platy crystals (up to 6 mm long and 2.5 mm wide) that poikilitically enclose crystals of olivine, pyroxene, and ilmenite (Fig. 2); compositions are An_{85-90}. A few smaller plagioclase crystals are present, associated with late-stage cristobalite and are of composition ~An_{80} (Fig. 4).

Ilmenite is the most abundant opaque oxide mineral. It exhibits a twofold grain-size distribution similar to (but less pronounced than) olivine in that there is a small number of very large (up to 5 mm long) crystals which are an order of magnitude bigger than the majority of ilmenite crystals. The large ilmenites (~2%, modally) are elongate (equant in cross section) skeletal crystals with sawtooth bladed margins (Fig. 5A). They contain 4.8–5.1 wt.% MgO (Table 3, col. 3). The remainder of the ilmenite occurs as subequant crystals 0.1–0.2 mm

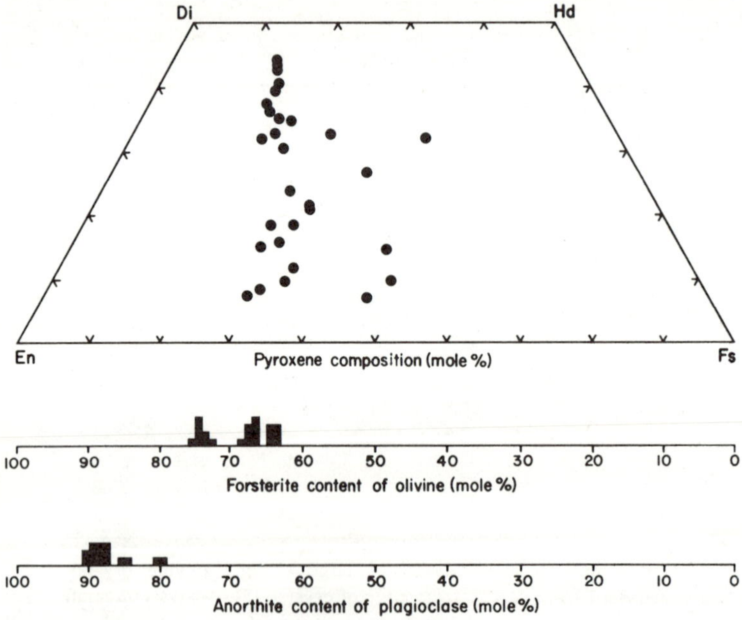

Fig. 4. Composition of pyroxenes (projected onto pyroxene quadrilateral), olivines and plagioclases in 71597. Olivine compositions $>Fo_{70}$ are from cores of large skeletal olivine grains; those $<Fo_{70}$ are from small "matrix" olivine crystals.

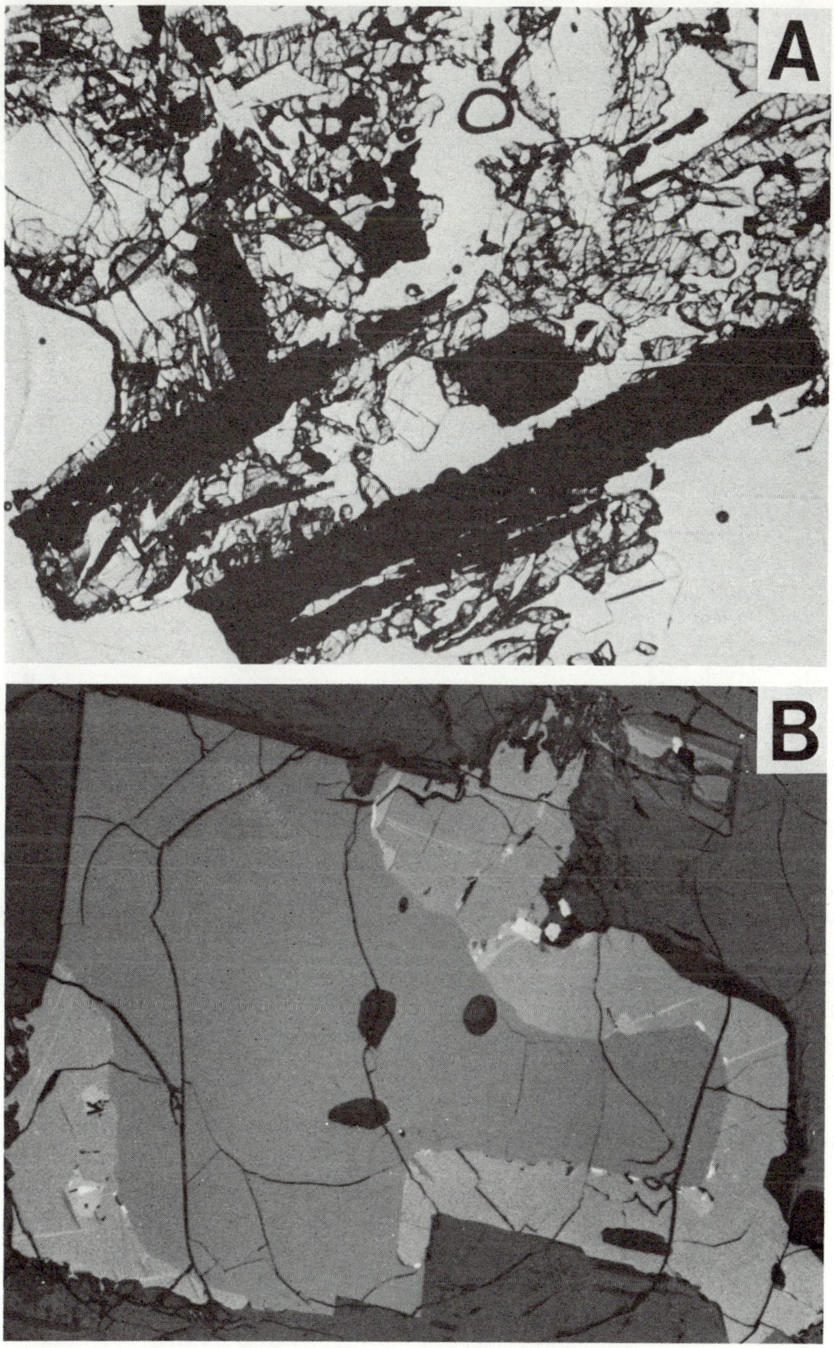

Fig. 5. (A) Transmitted light photomicrograph showing skeletal habit of large ilmenite crystals. Horizontal dimension equals 5.5 mm. (B) Reflected light photomicrograph showing armalcolite crystal mantled by ilmenite. Horizontal dimension equals 0.9 mm.

Table 3. Electron microprobe analyses (wt.%) of armalcolite, ilmenite, and spinel.

	1	2	3	4	5	6	7
SiO_2	.10	.04	.03	.08	n.a.	.07	.06
TiO_2	73.0	54.3	54.6	53.2	13.9	19.9	23.1
Al_2O_3	1.97	.03	*	.02	8.8	6.9	5.9
Cr_2O_3	1.48	.73	1.01	.81	32.3	25.2	20.5
FeO	17.7	39.6	40.8	41.5	37.7	41.7	47.6
MnO	.04	.54	n.a.	.40	.49	.52	.64
MgO	6.1	5.8	4.9	3.0	5.6	5.1	3.2
Total	100.4	101.0	101.3	99.0	98.8	99.4	101.0
Fe/(Fe + Mg)	.62	.79	.82	.89	.79	.82	.89

Columns (1–2): Analyses of ilmenite-mantled armalcolite crystal. Analysis 1 from armalcolite core, 2 from ilmenite mantle. (3): Analysis of large skeletal ilmenite crystal. (4): Analysis of small ilmenite lath. (5): Analysis of spinel grain inside large skeletal olivine crystal. (6): Analysis of spinel grain inside small euhedral olivine crystal. (7): Analysis of greater than 0.1 mm spinel crystal in "matrix".

*Less than 0.01 wt.%.

n.a. not analyzed.

wide or as laths generally <0.5 mm long. They show considerable range in MgO (1–8 wt.%) depending on their textural setting. For example, ilmenite crystals inside olivine melt inclusions or inside titanaugite crystals often contain >7 wt.% MgO, whereas ilmenite crystals associated with plagioclase and/or cristobalite often contain <2 wt.% MgO. Overall, ilmenite ranges in Fe/(Fe + Mg) from 0.71–0.96 (Fig. 6). Blebs and lamellae of rutile and aluminian chromite are commonly present in ilmenite.

Armalcolite is rare in 71597 (0.1%, modally), but where present it is usually partly or wholly mantled by ilmenite (Fig. 5B). The composite grains are usually large (≥0.5 mm across). Cores of armalcolite grains vary in Fe/(Fe + Mg) from 0.55 to 0.69 (Fig. 6); the most magnesian grains are zoned, with Fe/(Fe + Mg) increasing outwards, and Cr_2O_3 decreasing. Ilmenite mantling armalcolite contains nearly 6 wt.% MgO (Table 3, col. 2).

Spinel of intermediate chromian ulvöspinel composition (Fig. 6) is widely disseminated throughout the rock. A few crystals occur at the outer margins of some of the large skeletal olivines (Fig. 3B): the crystals are generally euhedral, 15–20 μm wide, and contain ≥30 wt.% Cr_2O_3 (Table 3, col. 5). Euhedral, 15–20 μm wide crystals of spinel are common as inclusions in the small (0.1–0.5 mm) olivine grains; they contain less Cr_2O_3 (usually ~25 wt.%—Table 3, col. 6). Euhedral to subhedral spinel crystals occur inside pyroxene grains and also are abundantly scattered throughout the rest of the rock where they overlap pyroxene-plagioclase grain boundaries. A number of the latter crystals exceed 0.1 mm in width. These "matrix" spinels are the least Cr_2O_3-rich found in the rock (Table 3, col. 7). Late stage ulvöspinel was not found (Fig. 6).

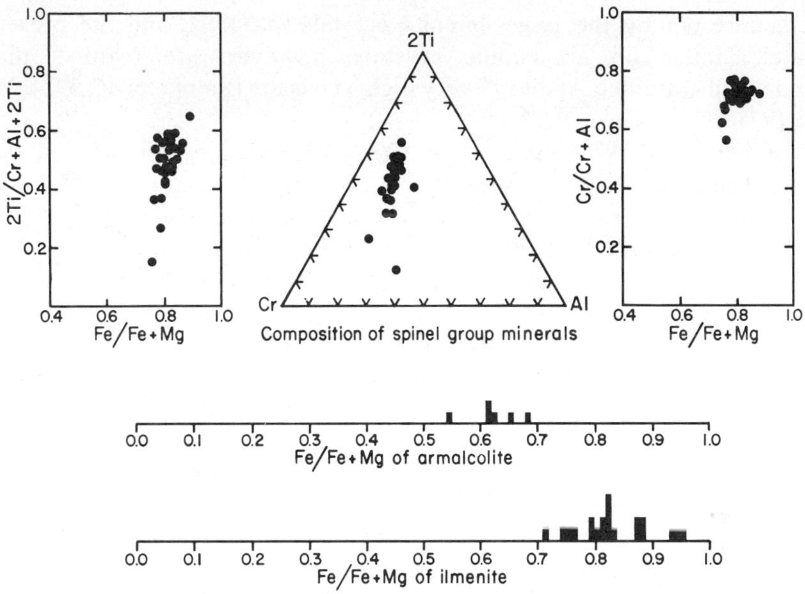

Fig. 6. Compositions of spinels (projected onto lunar spinel prism), armalcolites and ilmenites in 71597.

Cristobalite is less abundant (0.4%) in 71597 than in other coarse-grained Apollo 17 very high-Ti basalts (average, 2.4%—Warner *et al.*, 1975a) and is inhomogeneously distributed. It is most abundant in areas with least olivine. Metal and troilite (often intergrown) are widely disseminated in cristobalite-rich mesostasis and in ilmenite. A few tiny crystals of zirconolite also are present in 71597.

Summary of Petrologic Evidence

The presence of skeletal crystals is unusual for Apollo 17 coarse-grained very high-Ti basalts. Whereas olivines of size and crystal habit similar to the small olivine grains present in 71597 are typical for such basalts (Warner *et al.*, 1975a), the skeletal habit exhibited by the large (>2 mm) olivine crystals in 71597 is usually manifested only in fine-grained Apollo 17 very high-Ti basalts. The composition of the large skeletal olivine crystals (Fo_{73-75}) is also atypical for coarse-grained Apollo 17 basalts, as the most magnesian olivines present are usually about Fo_{70} (Warner *et al.*, 1976). However, they are typical of the most primitive olivine compositions ($\sim Fo_{75}$) encountered in fine-grained Apollo 17 very high-Ti basalts (Warner *et al.*, 1976). The occurrence of skeletal ilmenite crystals and armalcolite crystals that are mantled by ilmenite is also an unusual feature: normally, ilmenite crystals in coarse-grained Apollo 17 basalts have a rounded, amoeboidal shape (Papike *et al.*, 1974), and armalcolite occurs as discrete grains surrounded by titanaugite (El Goresy *et al.*, 1974). The skeletal habit manifested by the large ilmenite crystals in 71597, and the presence of ilmenite mantles on armalcolite crystals, however, are features that are common in fine-grained Apollo 17 very high-Ti basalts (Papike *et al.*, 1974; Warner *et al.*, 1975a).

In summary, 71597 contains two size generations of olivine and ilmenite crystals. The large olivine crystals have a skeletal habit and compositions which are usually found only in the more rapidly cooled, fine-grained Apollo 17 basalts. The habit of the large ilmenite crystals, together with the presence of ilmenite-mantled armalcolite crystals, are textural features also typical of fine-grained Apollo 17 basalts. From the presence of the above in a rock otherwise of a coarse-grained texture consonant with relatively slow cooling, it appears reasonable to conclude that the large skeletal crystals formed elsewhere and accumulated in 71597 through gravitational crystal settling.

Relation of 71597 to other Apollo 17 very High-ti Basalts

Textural evidence summarized in the preceding section favors the interpretation that 71597 owes its unique chemistry to accumulation of the large skeletal olivine crystals possibly coupled with minor ilmenite (and armalcolite) accumulation. A second line of evidence also indicates that 71597 cannot simply be the result of crystallization from a melt of composition corresponding to that of the whole rock: the compositions of the observed skeletal olivine crystals are not appropriate for crystallization from a parent bulk composition corresponding to that of 71597. Green *et al.* (1975) have measured the partitioning of iron and magnesium between olivine and high-Ti mare basalt liquids and have obtained a $K_D = 0.28$. Using this K_D value, and knowing the Fe/Mg ratio in 71597 (from the analysis given in Table 1), we can compute the liquidus olivine composition expected for a melt of 71597 composition. The result shows that the olivine should be of composition $\sim Fo_{84}$. Instead, the most magnesian olivine in 71597 is of composition Fo_{75}. Consequently, it is reasonable to conclude that the large skeletal olivine crystals did not precipitate from a melt of 71597 composition.

The Fo_{73-75} compositions of the large skeletal olivines in 71597 argue against derivation from the same flow that produced Station 4 MgO-rich basalts such as 74245 and 74275, as these basalts contain olivine phenocrysts of distinctly more magnesian ($\sim Fo_{80}$) composition (Brown *et al.*, 1975; Green *et al.*, 1975). However, olivines of composition Fo_{73-75} are precisely those that are found in the main group (Fig. 1) of Apollo 17 very high-Ti basalts, both as liquidus compositions in equilibrium melting experiments (e.g., Kesson, 1975; Green *et al.*, 1975) and as microphenocryst compositions in the natural rocks (Warner *et al.*, 1975a). We therefore conclude that 71597 is an olivine-enriched sample from a "normal" Apollo 17 basalt flow and attempt to explore further the nature of this relation below.

We have made calculations trying to relate the composition of 71597 with that of other Apollo 17 very high-Ti basalts via the observed "cumulate" assemblage. Such attempts can lead only to very crude approximations, for a number of reasons. First, the extremely large size of the olivine crystals, coupled with their tendency to cluster, make the representativeness of the observed mineral mode (also the bulk composition determined by INAA) questionable. Second, margins of the skeletal olivine crystals appear to have been resorbed, so that the original volume of olivine present was greater by some unknown amount. This is true for armalcolite, too, but the relative volume concerned in that case is probably trivial. Third, the skeletal olivine crystals are zoned, so that an appropriate average olivine composition is difficult to estimate. Since much of the zoning occurs adjacent to the crystal margins, and may have resulted from reaction after the crystals had settled, an estimated average composition close to that shown by the cores of the large olivine grains would probably be representative (say, Fo_{73}).

Using observed abundances of large skeletal crystals (~18% olivine and 2% ilmenite—and neglecting 0.1% armalcolite as trivial) and appropriate compositions (Fo_{73} for olivine—see above—and analysis given in Table 3, col. 3 for ilmenite), we have subtracted this assemblage from 71597 bulk composition (Table 1). The resultant composition is given in Table 4 (col. 1) and shown on Fig. 1 as A. It is apparent that this composition falls well outside the main Apollo 17 basalt field. Compared to an average composition for Apollo 17 very high-Ti basalt (Table 4, col. 4), the resultant composition is 2.5 wt.% too low in TiO_2 and 3.5 wt.% too high in MgO. Furthermore, the liquidus olivine computed to be in equilibrium with this composition (using $K_D = 0.28$) is Fo_{80}, which is 5 mol.%

Table 4. Results of fractionation calculations on 71597 composition.

	1	2	3	4	5
TiO_2 (%)	9.2	10.9	11.0	11.7	11.5
Al_2O_3	9.9	9.4	10.4	9.2	10.8
Cr_2O_3	.51	.44	.55	.38	.55
FeO	18.2	19.7	18.3	19.7	18.1
MnO	.23	.26	.24	.25	.23
MgO	11.3	9.0	9.1	7.8	7.9
CaO	9.8	10.2	10.3	10.7	10.7
Na_2O	.36	.36	.38	.38	.40
K_2O	.03	.07	.04	.05	.04

Columns (1): Composition obtained by subtracting 18% olivine and 2% ilmenite from 71597 bulk composition. (2): Composition of fine-grained Apollo 17 very high-Ti basalt 71547 (from Murali *et al.*, 1977). (3): Composition obtained by subtracting 24% olivine from 71597 bulk composition. (4): Average composition of 22 fine-grained Apollo 17 very high-Ti basalts analyzed by R. A. Schmitt and coworkers. (5): Composition obtained by subtracting 27% olivine from 71597 bulk composition.

more magnesian than the most magnesian olivine found in 71597. To relate 71597 more closely to other Apollo 17 basalt compositions, and to obtain a bulk composition that will yield an equilibrium liquidus olivine composition near Fo_{75}, thus require subtraction of more olivine than is observed in 71597. As noted above, petrographic evidence (marginal resorption of skeletal olivine grains) is strong that 71597 did contain more olivine than is now present. On the other hand, the deficiency in TiO_2 (Table 4, col. 1 vs. col. 4) suggests subtraction of less ilmenite (or at least a smaller proportion of ilmenite relative to olivine) if 71597 is derived from a magma of average Apollo 17 basalt composition.

In an attempt to elucidate more clearly the relation between 71597 and other Apollo 17 very high-Ti basalts, we have made two additional calculations. First, we have chosen the analysis (Table 4, col. 2) of a fine-grained basalt that has the highest ratio of MgO relative to TiO_2 (excluding Station 4 basalts, which have inappropriate olivine compositions—see above) and calculated the amount of olivine and/or ilmenite accumulation required to derive 71597 from this rock. The results (Table 4, col. 3; B in Fig. 1) show that a reasonable fit is obtained by subtracting 24% olivine (but no ilmenite) from 71597; subtraction of a small amount of chromian ulvöspinel would correct the difference in Cr_2O_3 and leave only Al_2O_3 (1 wt.% high) and FeO (1.4 wt.% low) as problems. In the second calculation we have attempted to relate 71597 to the average Apollo 17 basalt composition (Table 4, col. 4; Fig. 1) through olivine and/or ilmenite fractionation. The results show that subtraction of 27% olivine (again no ilmenite) yields a reasonably close fit (compare cols. 4 and 5 in Table 4 and C with average Apollo 17 basalt composition in Fig. 1). Not only do these two calculations provide close matches in bulk composition, but the computed equilibrium liquidus olivine compositions (Fo_{76} and Fo_{74}, respectively) also provide close matches to the observed olivine compositions in 71597 (Fo_{73-75}). Given the inherent uncertainties in the above calculations (such as uncertainty in representativeness of observed bulk composition and modal abundances; uncertainty in amount of olivine that has been resorbed; and uncertainty in choice of appropriate average composition for the zoned skeletal olivine crystals), the derivation of 71597 by major (24–27%) olivine, and possibly minor (0–2%) ilmenite crystal settling within a "normal" Apollo 17 basalt flow would certainly appear to be feasible.

Crystallization of 71597

Our model for the crystallization of 71597 is as follows. The overall coarse-grained texture suggests that the bulk of the rock's cooling occurred relatively slowly, probably in the interior of a flow. Comparison of the plagioclase-poikilitic texture in 71597 with textures experimentally reproduced by Usselmann *et al.* (1975) for high-Ti mare basalt compositions, indicates a cooling rate $<1°C/hr$. However, the skeletal habit of the large olivine and ilmenite crystals suggests that they crystallized at a faster cooling rate, possibly toward the edge of the flow. The habit of the large olivine crystals in 71597 is very similar to that developed in a 2°C/hr cooling experiment by Usselmann *et al.* (1975—Fig. 3e).

The large ilmenite crystals have habits that seem appropriate for cooling at ~7°C/hr, based on morphological characteristics summarized in Table 4 of Usselmann *et al.* (1975). The large skeletal crystals now present in 71597 therefore probably crystallized in a regime where the cooling rate was on the order of several degrees per hour. Of course, the olivine and ilmenite crystals need not have originated at the exact same levels of the flow—in fact, they probably did not. Subsequent to their initial crystallization, the skeletal crystals settled into the flow's interior. The very large size of the crystals, particularly of the olivines (which are more than twice the dimensions of any olivine crystals we have seen in other Apollo 17 very high-Ti basalts), may have resulted from additional crystallization as they settled. During subsequent slow (<1°C/hr) cooling in the flow's interior, some reaction took place between the large skeletal olivines and the surrounding melt, thereby enriching the residual melt in MgO and thus probably accounting for the absence of late stage ulvöspinel and lack of pronounced Fe enrichment in late crystallizing pyroxene.

Acknowledgments—We thank D. Lange for preparing polished probe mounts of 71597. This work was supported in part by the National Aeronautics and Space Administration, Grant NGL 32-004-063 (Klaus Keil, Principal Investigator). This paper has benefited from constructive comments provided by W. B. Bryan and D. T. Vaniman.

References

Albee A. L. and Ray L. (1970) Correction factors for electron probe microanalyses of silicates, oxides, carbonates, phosphates and sulfates. *Anal. Chem.* **42**, 1408–1414.

Bence A. E. and Albee A. L. (1968) Empirical correction factors for the electron microanalysis of silicates and oxides. *J. Geol.* **76**, 382–403.

Brown G. M., Peckett A., Emeleus C. H., Phillips R., and Pinsent R. H. (1975) Petrology and mineralogy of Apollo 17 mare basalts. *Proc. Lunar Sci. Conf. 6th*, p. 1–13.

El Goresy A., Ramdohr P., Medenbach O., and Bernhardt H. (1974) Taurus-Littrow TiO_2-rich basalts: Opaque mineralogy and geochemistry. *Proc. Lunar Sci. Conf. 5th*, p. 627–652.

Green D. H., Ringwood A. E., Hibberson W. O., and Ware N. G. (1975) Experimental petrology of Apollo 17 mare basalts. *Proc. Lunar Sci. Conf. 6th*, p. 871–893.

Keil K. (1967) The electron microprobe X-ray analyzer and its application in mineralogy. *Fortschr. Miner.* **44**, 4–66.

Keil K., Dowty E., and Prinz M. (1974) *Description, classification and inventory of 113 Apollo 17 rake samples from stations 1A, 2, 7 and 8.* 149 pp. NASA Johnson Space Center, Houston.

Kesson (1975) Mare basalts: Melting experiments and petrogenetic interpretations. *Proc. Lunar Sci. Conf. 6th*, p. 921–944.

Laul J. C., Hill D. W., and Schmitt R. A. (1974) Chemical studies of Apollo 16 and 17 samples. *Proc. Lunar Sci. Conf. 5th*, p. 1047–1066.

Murali A. V., Ma M.-S., Schmitt R. A., Warner R. D., Keil K., and Taylor G. J. (1977) Chemistry of 30 Apollo 17 rake basalts; 71597 a product of partial olivine accumulation (abstract). In *Lunar Science VIII*, p. 703–705. The Lunar Science Institute, Houston.

Papike J. J., Bence A. E., and Lindsley D. H. (1974) Mare basalts from the Taurus-Littrow region of the moon. *Proc. Lunar Sci. Conf. 5th*, p. 471–504.

Reid A. M., Lofgren G. E., Heiken G. H., Brown R. W., and Moreland G. (1973) Apollo 17 orange glass, Apollo 15 green glass and Hawaiian lava fountain glass (abstract). *EOS* (*Trans. Amer. Geophys. Union*) **54**, 607–608.

Usselmann T. M., Lofgren G. E., Donaldson C. H., and Williams R. J. (1975) Experimentally reproduced textures and mineral chemistries of high-titanium mare basalts. *Proc. Lunar Sci. Conf. 6th*, p. 997–1020.

Warner R. D., Keil K., Prinz M., Laul J. C., Murali A. V., and Schmitt R. A. (1975a) Mineralogy, petrology, and chemistry of mare basalts from Apollo 17 rake samples. *Proc. Lunar Sci. Conf. 6th*, p. 193–220.

Warner R., Keil K., Murali A. V., and Schmitt R. A. (1975b) Petrogenetic relationships among Apollo 17 basalts (abstract). In *Papers Presented to the Conference on Origins of Mare Basalts and Their Implications for Lunar Evolution*, p. 179–183. The Lunar Science Institute, Houston.

Warner R. D., Warren R. G., Mansker W. L., Berkley J. L., and Keil K. (1976) Electron microprobe analyses of olivine, pyroxene and plagioclase from Apollo 17 rake sample mare basalts. Special Publication No. 15, Univ. New Mexico, Institute of Meteoritics, 158 pp.

Proc. Lunar Sci. Conf. 8th (1977), p. 1443–1471.
Printed in the United States of America

Very low Ti (VLT) basalts: A new mare rock type from the Apollo 17 drill core

D. T. VANIMAN and J. J. PAPIKE

Department of Earth and Space Sciences, State University of New York, Stony Brook, New York 11794

Abstract—The Apollo 17 drill core contains phaneritic fragments, vitrophyres, and glass beads of a new very low Ti (VLT) mare basalt. VLT lithic fragments are characterized by TiO_2 content of ~0.5%, Mg/(Mg + Fe) ~ 0.52, CaO/Al_2O_3 ~ 0.9 and low alkali content (Na_2O ~ 0.1%, K_2O ~ 0.0%). The mineral systematics and modal composition of VLT basalt are similar to Apollo 12 and 15 low Ti basalts, but VLT basalts cannot be related to these mare basalts by crystal fractionation. VLT basalt is isochemical with some of the less mafic green glasses; fractionation of VLT magma from a liquid of green-glass composition is a possibility. Spectral reflectance studies indicate that VLT-type basalts may be relatively common in mare basins which have not been directly sampled. The discovery of this new basalt type emphasizes the importance of lunar soils as a depository of yet unrecognized lunar rock types.

INTRODUCTION

THE DARK BASALTIC MARE PLAINS cover approximately 17% of the lunar surface but comprise only about 1% of the lunar crust (Head, 1976). However, the importance of mare basalts far exceeds their volumetric abundance. The most abundant mare basalts flooded impact basins during a stage of intense lunar volcanism at 3.9–3.1 b.y. These rocks provide important insights into the geochemical nature and thermal history of the lunar interior. Recent reviews by Head (1976) on mare volcanism in space and time with emphasis on photogeologic and remote sensing studies, and by Papike *et al.* (1976) emphasizing sample studies, provide the reader with background information on these important rock types.

Post Apollo studies of mare basalts are extending the landing site data base. Earth-based spectral mapping and examination of soil samples for new basalt types that may have been carried large distances by meteorite impacts are two of the ways this is being accomplished. The small lithic fragments in soils are providing important clues to the geochemical nature of mare that were not sampled directly by the Apollo and Luna missions. These fragments are our only current source of new mare basalt types other than the ongoing Soviet sampling program.

In this paper we describe a new mare basalt type from the >0.02 mm soil fraction of the Apollo 17 drill core. Unlike most mare basalts, which have TiO_2 contents of ~1.5 wt.% to >13 wt.%, this new mare basalt has <0.5 wt.% TiO_2 (Vaniman and Papike, 1977; Taylor *et al.*, 1977). Green *et al.* (1972) described a Luna 16 mare basalt with 0.4 wt.% TiO_2, a rock similar to the new basalt in the

Apollo 17 core but higher in alkali content. The Apollo 17 core contains the first occurrence of many fragments with varied texture from a "very low titanium" (VLT) mare basalt.

The chemistry and mineralogy of VLT basalt place it in the mare basalt clan. Three chemical parameters effectively separate VLT basalt from the highland feldspathic basalts: VLT basalt has low alkali content (<0.2% $Na_2O + K_2O$, vs ~0.9%) high CaO/Al_2O_3 ratio (~0.9 vs 0.5), and low Mg/(Mg + Fe) ratio (~0.52 vs 0.7). Ryder *et al.* (1975) described olivine-normative pigeonite basalts from Boulder 1, Apollo 17 that are similar to VLT basalt and yet by their geologic setting must predate the major mare volcanism. Ryder and Taylor (1976) suggested that these basalts are evidence of mare-type volcanism predating the formation of large multiringed impact basins at 3.9 b.y. In comparison to the ancient basalts in Boulder 1, the VLT basalt fragments of the drill core are less siliceous and contain olivine richer in Mg(Fo_{85}). Ryder *et al.* (1977) showed that most basalt fragments within Boulder 1 are KREEP-rich (e.g., 0.25–1.05% K_2O); in this respect especially they are unlike the VLT basalts within the Apollo 17 drill core ($\leq$ 0.04% K_2O). The VLT basalts can probably be related to large mare areas which spectral reflectance studies show to be covered by previously unsampled basalts of very low Ti content.

The Apollo 17 drill core contains a significant population of phaneritic and vitrophyric VLT basalts and related glasses. The samples described here occur in a string of polished thin sections which provide a complete section of the upper 100 cm of soil at the drill site (core segments 70007, 70008, 70009). During our study we identified 3 phaneritic and 11 vitrophyric fragments of VLT basalt and 21 VLT glass beads in these thin sections (Fig. 1). However, since we examined only half of the polished thin sections for all possible VLT glass beads, we estimate that the population of VLT glass beads is ~40 in our sample. Taylor *et al.* (1977) also worked with polished thin sections of the core and described VLT phaneritic basalts in core sections 70007 and 70009. Thus, the recognition of VLT basalt by both groups points out the importance of this new rock type at the Apollo 17 site.

Interest in very low Ti mare basalts has increased with the return of Luna 24 samples from Mare Crisium. Preliminary data (Barsukov *et al.*, 1977) indicated that basalts from Crisium have <1% TiO_2. Current studies (Vaniman and Papike, 1977; Taylor *et al.*, 1977) show that the Luna 24 mare basalts contain 0.6–1.3% TiO_2. They are poorer in MgO (5.2–7.7%) and richer in Al_2O_3 (12.8–15.6%) and CaO (13.1–13.6%) than are the Apollo 17 VLT basalts (compare Table 2). The Apollo 17 and Luna 24 "very low titanium" basalts are separated by 700 km; the origins of these two widely separated basalts types are still obscure, and it remains to be demonstrated whether they can be related to a common magma type.

The spectral reflectance data of Pieters and McCord (1976) suggest that very low Ti basalts may be abundant in Mare Imbrium and Mare Frigoris (TiO_2 < 1.5 wt%). Pieters (pers. comm.) suggests that the original TiO_2 estimates for low Ti basalts may be too high; calibration of the reflectance intensities is still highly qualitative in soils of low TiO_2 content. Revision downward may extend the very low Ti basalt cover to the basins of Serenitatis, Crisium, Tranquillitatis, Nec-

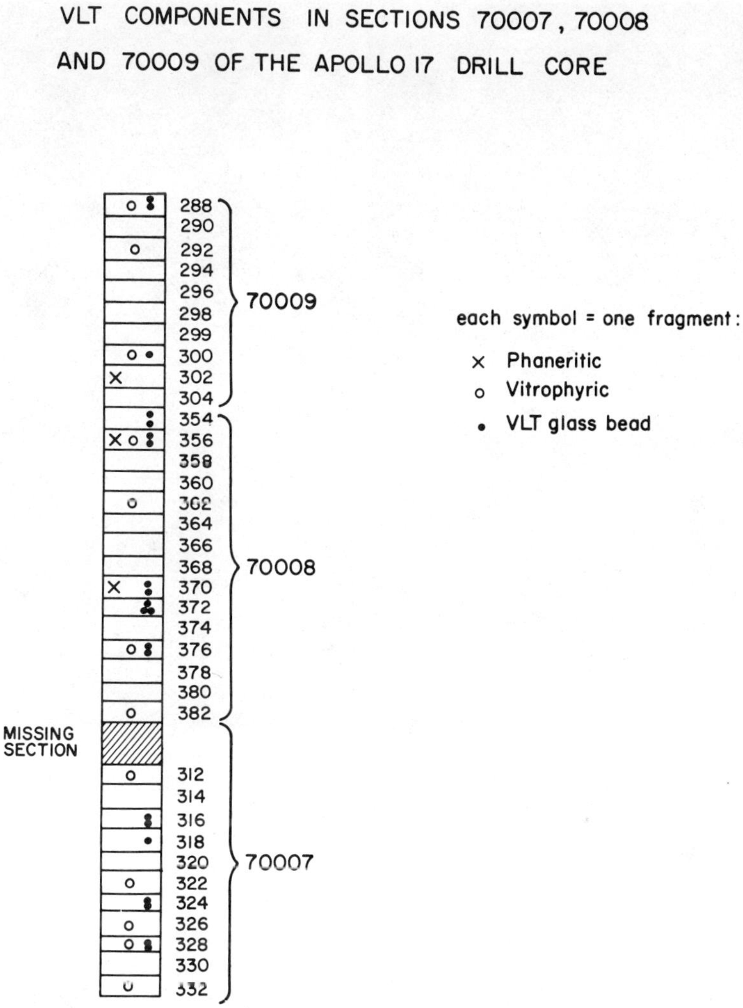

Fig. 1. Stratigraphic location of VLT lithic and glass fragments in the upper three segments of the Apollo 17 drill core.

taris, Humorum, and eastern Procellarum. There is good reason to believe that VLT-type basalts are a significant mare component.

Petrography of VLT Phaneritic Basalt

Phaneritic VLT basalts from the Apollo 17 drill core are medium-grained (0.2–0.4 mm) subophitic rocks (Fig. 2a,b). Subhedral pyroxene rims are intergrown with euhedral laths of later-crystallized plagioclase. The earliest phases to precipitate are chromian spinel and olivine. Euhedral spinels up to 0.05 mm are characteristic of VLT basalt. Zoned and rounded olivine grains have

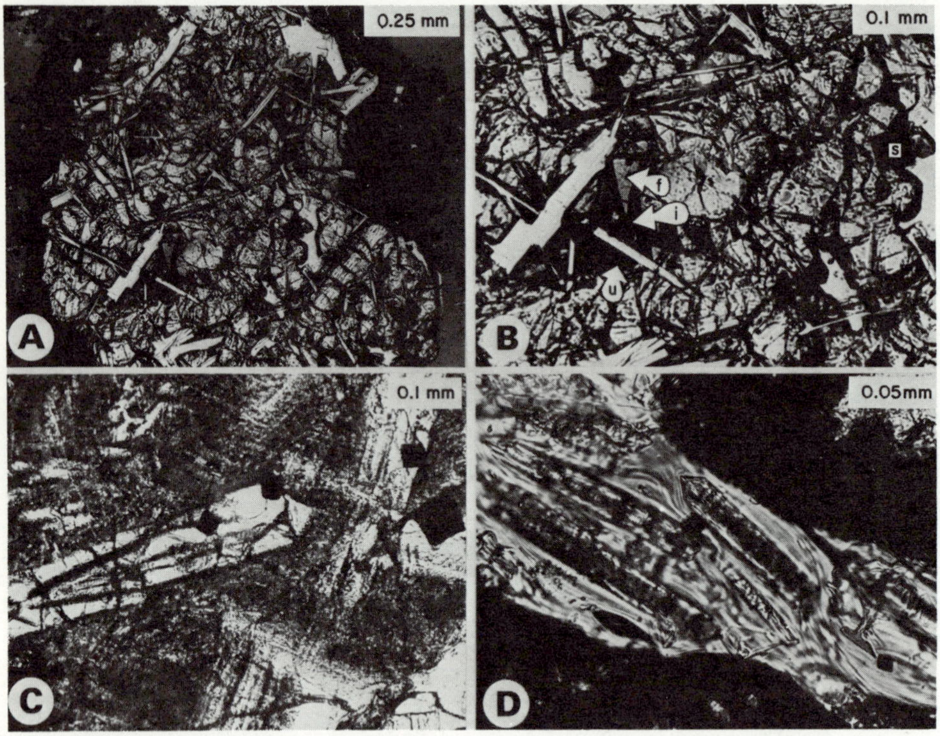

Fig. 2. Photomicrographs of VLT lithic fragments; (a): phaneritic fragment, PTS 70008,356, field of view = 2.0 mm; (b) phaneritic fragment, PTS 70008,370, field of view = 2.0 mm; (c) coarse crystalline spinel-olivine-pyroxene vitrophyre, PTS 70009,288, field of view = 1.0 mm; (d) glass-rich spinel-olivine vitrophyre, PTS 70008,362, field of view = 0.3 mm.

resorbed boundaries against pyroxene. Intergrain areas contain glass and anhedral late-stage crystals of CaFe pyroxene, ulvöspinel, ilmenite, silica, Fe-sulfide, and metallic iron. Modal data for two large phaneritic fragments in PTS* 70008,356 (1.75 mm fragment) and PTS 70008,370 (2.2 mm fragment) are given in Table 1.

PETROGRAPHY OF VLT VITROPHYRES

VLT vitrophyres contain chromian spinel and olivine ± pyroxene in a matrix of pale yellow/green to clear glass. Figure 2c,d shows glassy and variolitic textures in VLT vitrophyres. Olivine and pyroxene form skeletal crystals with grain sizes ranging from 0.001 to 0.1 mm. Spinel is always present and occurs as small (0.005 mm) equant octahedral or cruciform crystals which are deep crimson or opaque. No other minerals have been found in the vitrophyres.

---

*Polished thin section.

Table 1. VLT basalt modes.

	70008,356	70008,370
Px.	60.0	61.1
Plag.	31.1	27.7
Oliv.	5.3	4.3
Spinel	0.7	0.3
Ilm.	0.2	0.2
Fe	0.1	0.2
SiO_2	1.6	0.6
Mesostasis	—	4.2
Void space	1.0	1.4
	100.0	100.0

VLT Glasses

Yellow/green glass spheres with compositions close to VLT basalt but slightly richer in Fe and Mg are found throughout the drill core (Fig. 1). The spheres, which range from 0.02 to ~0.2 mm in diameter, form a uniform population tightly clustered in composition (ref. Vaniman and Papike, 1977b). Unfortunately, these glasses cannot be distinguished by color from the other pastel yellow/green glasses of different composition in the drill core. Each VLT glass bead is internally homogeneous, but there may be a variation in silica between beads. The chemistries of VLT glasses and lithics are contrasted in the concluding section of this paper.

Analytical Methods

Modal analyses

Modal analyses of two phaneritic and three vitrophyric fragments were obtained in reflected light. A counting interval of 0.02 mm was used. The counting grid was shifted and counting was repeated on each fragment until 1000 points were identified.

Microprobe analyses

Microprobe analyses were made on an ARL EMX-SM microprobe with three automated positionable spectrometers and one semi-fixed spectrometer (for Si). Counting intervals were standardized by beam integration; counting times for each element were ~20 sec. The accelerating potential was fixed at 15 kV and sample current was maintained at 0.015 μA (brass sample). Data were reduced on-line using the empirical matrix corrections methods of Bence and Albee (1968) and Albee and Ray (1970). Apollo 17 orange-glass mount 74220,39, studied by Rhodes *et al.* (1974), was used as a working standard. Broad-beam analyses using a 50 μ spot were made only on the vitrophyres with high glass contents (<33% crystals) and with crystals <20 μ in diameter. Broad-beam analyses are compared to analyses obtained by modal recombination (Table 2) and are plotted in Figs. 13 and 14.

Whole rock chemical analyses

Whole rock chemical analyses of vitrophyric and phaneritic VLT fragments were obtained by modal recombination of the mineral analyses (phaneritic rocks) or recombination of glass and mineral analyses (vitrophyres). This method is most reliable for vitrophyres, where the glass is homogeneous and the microphenocrysts are unzoned. Mineral zoning complicates the method when it is attempted in coarse-grained rocks. In VLT basalts the most severe problem is with pyroxene zonation. To account for this zonation, we used the average of 215 pyroxene analyses in the VLT basalt fragment of PTS 70008,356 and 135 pyroxene analyses in the VLT basalt fragment of PTS 70008,370. The pyroxene analyses were taken on grid locations and on evenly spaced single-grain traverses. The greatest variable in VLT pyroxene zonation is Mg: Fe content, and for this reason the constraints on MgO and FeO composition are weakest in the phaneritic analyses (Table 2). However, we believe that the method of modal recombination is justified because:

(1) The modal recalculations of phaneritic rocks are similar to the modal recalculations of the vitrophyres.
(2) Modal recalculations of vitrophyres agree well with broad-beam analyses of the same fragments.
(3) The empirical calculations of Lipin (1976) can be applied to the olivine microphenocrysts of the VLT vitrophyres to yield Fe/(Fe + Mg) ratios of the coexisting liquid. For these glassy rocks the Fe/(Fe + Mg) ratios are similar to those determined by modal recalculation.
(4) An automated but equivalent method (Gancarz and Albee, 1973) is used routinely and successfully by California Institute of Technology (for example, Dymek *et al.*, 1975).

COMPARISON OF PHANERITIC VLT BASALT AND OTHER MARE BASALTS

The average modal and chemical analyses of two VLT phanerites are compared to other mare basalts in Fig. 3. VLT basalt is pyroxene-rich and feldspar-poor, similar to the other mare basalts.

The data plotted in Fig. 3 show that modal percent opaques varies directly with weight percent TiO_2 in mare basalts. In all mare basalts other than VLT basalt the amount of modal ilmenite is much greater than spinel, but in VLT basalt spinel is more abundant than ilmenite (Table 1).

The "very high titanium" and "very low titanium" basalts fall at opposite ends of the mare compositional trends in Fig. 3. Although both extremes of TiO_2 concentration occur in the mare basalts of Taurus-Littrow Valley, these two basalt types cannot be related by any likely method of fractional crystallization or fractional fusion. Of all mare rock types the Apollo 12 and 15 low Ti basalts are most similar to VLT basalt, but VLT basalt is too Mg and Si-rich to be related to these low Ti basalts by olivine fractionation. Fractionation of Fe–Ti oxides cannot be assumed, for these are not liquidus phases in VLT basalt. Green glass is the only mare composition with high enough Mg: Fe to be parental to VLT basalt.

MINERALOGY OF VLT LITHICS

A schematic crystallization sequence for VLT lithics is shown in Fig. 4. Olivine (Fo_{74-75}) and chromian spinel are liquidus or near-liquidus phases; all vitrophyres have at least these two minerals. Pigeonite coexists with olivine in

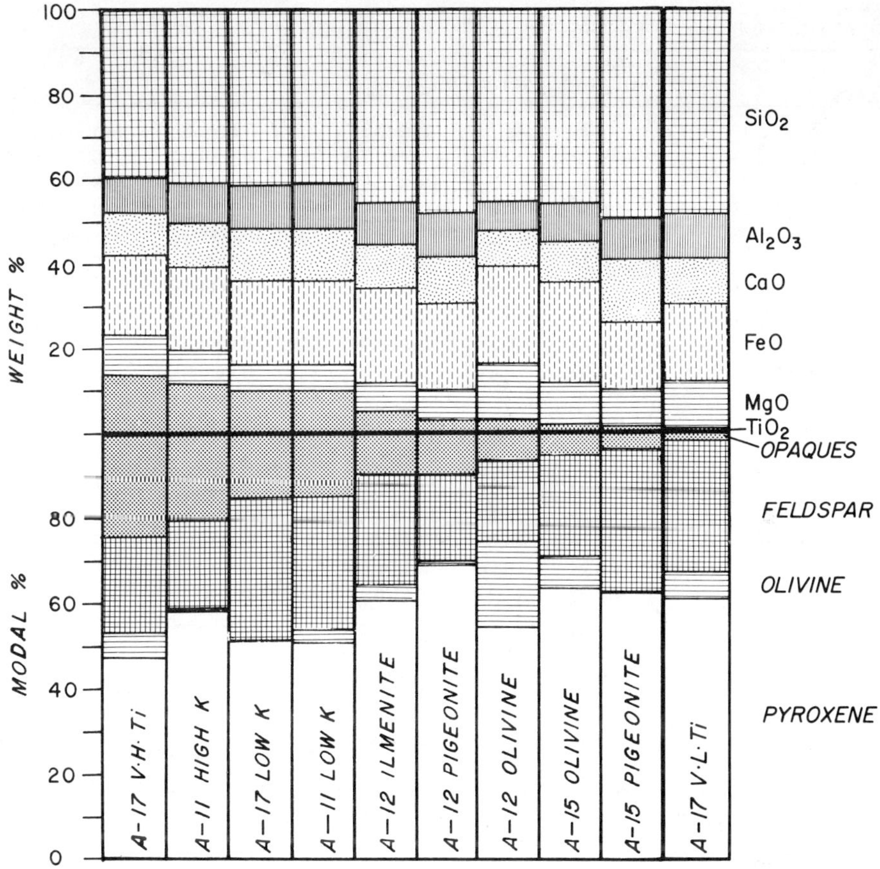

Fig. 3. Mare basalt chemical compositions (upper diagram) and modal compositions (lower diagram): Data summarized in Papike *et al.* (1976) and presented in this paper (Apollo 17 VLT basalt).

many vitrophyres, but in phaneritic rocks the olivine is zoned to Fo_{33}, and the rims are rounded and resorbed against adjacent pyroxene. No plagioclase was observed in the vitrophyres. Phaneritic VLT basalts have plagioclase laths which are intergrown only with the pyroxene rims. These observations suggest that the onset of plagioclase crystallization occurred late in the pyroxene growth history. Oxide zonations have a two-stage history: Fe-enrichment of chromian spinel, with slight increase in Ti, followed by cessation of spinel crystallization and finally a late-stage growth of ulvöspinel along with Fe-sulfide and Fe°.

Pyroxene

The first pyroxene to crystallize in VLT vitrophyres and phanerites is a pigeonite, $Ca_6Mg_{68}Fe_{26}$. Pyroxenes in the vitrophyres develop strong sectoral differences in Ca abundance, but these sectoral differences are suppressed in

Table 2. Bulk composition of VLT phanerites, vitrophyres, and glass.

	1	2	3	4	5
SiO_2	48.7	48.1	48.8(47.3)	47.0(46.6)	45.97
TiO_2	0.50	0.36	0.69(0.70)	0.67(0.62)	0.70
Al_2O_3	11.4	11.2	10.0(10.6)	10.4(10.8)	9.97
Cr_2O_3	0.60	0.60	0.69(0.71)	0.87(0.76)	0.48
FeO	19.0	18.2	17.9(18.1)	18.9(18.8)	18.66
MnO	0.26	0.26	0.30(0.27)	0.30(0.27)	0.24
MgO	9.4	11.0	11.8(11.6)	13.9(13.0)	14.19
CaO	10.2	10.2	9.4(10.0)	9.3(9.2)	9.53
Na_2O	0.15	0.15	0.06(0.08)	0.09(0.07)	0.19
K_2O	0.04	0.01	0.02(0.01)	0.02(0.01)	0.01
Σ	100.2	100.1	99.7(99.4)	101.1(100.1)	99.94
Fo	—	1.90	—(2.46)	7.42(5.95)	9.99
Fa	—	2.51	—(2.99)	7.90(6.75)	10.33
Di	16.79	16.99	16.03(17.38)	15.25(13.91)	17.30
Hy	49.10	45.82	53.17(44.80)	39.66(41.58)	32.30
Or	0.24	0.06	0.12(0.06)	0.12(0.06)	0.06
Ab	1.27	1.27	0.51(0.68)	0.76(0.59)	1.61
An	30.31	29.99	27.07(28.67)	27.83(28.99)	26.32
Ilm	0.95	0.68	1.31(1.33)	1.27(1.18)	1.33
Chrom	0.88	0.88	1.02(1.05)	1.28(1.12)	0.71
Qz	0.73	—	0.42(—)	—(—)	—
Mg/(Mg + Fe)	0.47	0.52	0.54(0.53)	0.57(0.55)	0.58
CaO/Al_2O_3	0.89	0.90	0.93(0.94)	0.90(0.86)	0.96

Column (1): Modal recombination of VLT phanerite, PTS 70008,370; Column (2): Modal recombination of VLT phanerite, PTS 70008,356; Column (3): Modal recombination of VLT vitrophyre, PTS 70007,328 (broadbeam analysis in parentheses); Column (4): Modal recombination of VLT vitrophyre, PTS 70008,362 (broadbeam analysis in parentheses); Column (5); Analysis of representative VLT glass bead, PTS 70007,328.

phaneritic rocks where an Fe-enrichment trend predominates in all sectors (Fig. 5). The same pattern of pyroxene sector zoning occurs in Apollo 12 basalts (Bence and Papike, 1972). The Ca-enrichment trend in phaneritic VLT is reversed where plagioclase began to precipitate and remove Ca from the melt. However, the Ca-enrichment trend is not interrupted in VLT vitrophyres, where plagioclase nucleation was suppressed.

High-Ca and low-Ca pyroxene zoning trends in a medium-grained (0.2 mm) VLT vitrophyre (PTS 70009,288; Fig. 2c) are listed in Table 3, and high-Ca/low-Ca trends of a VLT phanerite (PTS 70008,356; Fig. 2a,b) are listed in Table 4. The sectoral differences in Ca content cannot be distinguished optically and are revealed only by microprobe analysis. Zoning trends in two VLT pyroxene grains are shown in Fig. 6. Both grains have been sectioned approximately

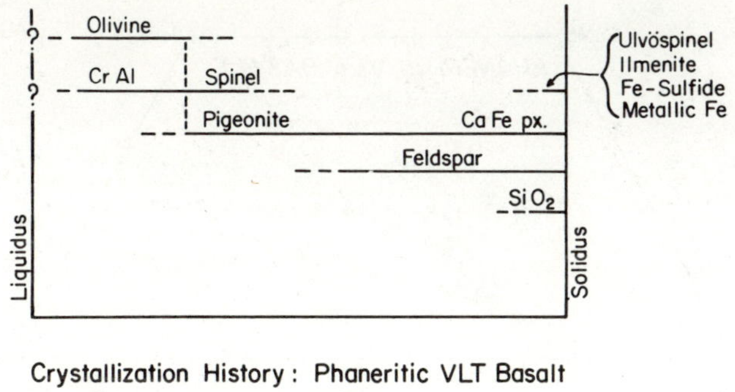

Fig. 4. Schematic crystallization history of VLT lithics.

perpendicular to *c*, and both have low-Ca pigeonite* cores. One of these crystals has only (110) faces parallel to *c*, and traverses across these faces show high-Ca zonation. The other crystal has (100), (110) and (010) faces parallel to *c*, and in this grain only the (110) sector is high in Ca. Sectoral chemistry of this second crystal is summarized in Table 5.

Minor element concentrations are also affected by sectoral control. The zonation of Cr, Al^{VI}, and Ti in phaneritic and vitrophyric VLT basalt are shown in Fig. 7.

Figure 7a shows an initial enrichment of Al^{VI} in the low-Ca pyroxene cores of phaneritic VLT basalt. Where sector zoning begins in the pyroxene rims, the high-Ca zoning trend (ref. Fig. 5a) is accompanied by high Al^{VI} concentration and the low-Ca zoning trend is poor in Al^{VI}. There is a strong tendency for $R^{2+}Al(SiAl)O_6$ to concentrate with Ca in the (110) sector. With precipitation of plagioclase the accumulation of Ca and Al^{VI} in (110) ceases and all sectors zone outward to a common Ti, Fe-rich pyroxene rim.

Figure 7b shows two quite different minor element zoning trends in pyroxene of a VLT vitrophyre. Both trends are dominated by Al^{VI}. In the high-Ca zoning trend (ref. Fig. 5b) there is an increase of $R^{2+}Al(SiAl)O_6$ with increase in Ca, but in the low-Ca zoning trend there is a decrease of $R^{2+}Al(SiAl)O_6$ and an increase in $R^{2+}TiAl_2O_6$ along with Fe enrichment. These opposed trends of Al^{VI} enrichment in the high-Ca trend and Al^{VI} depletion in the low-Ca trend are shown by the opposed arrows in Fig. 8b. The two trends might not be simultaneous.

In VLT vitrophyres the component $R^{2+}Al(SiAl)O_6$ is the most prominent substitution for Ca–Mg–Fe "quadrilateral" compositions. The concentration of this component in the vitrophyre pyroxenes is a result of suppressed plagioclase precipitation (see, for example, Bence and Papike, 1972); the extreme enrichment of Al in vitrophyre pyroxenes is shown in the plot of Ti vs Al (Fig. 8b). In VLT phanerites the pyroxene Ti : Al ratio initially follows the same high-Al trend as the vitrophyres, but departs from this trend when plagioclase begins to

*The term "pigeonite" is used for pyroxenes with an optic axial plane containing the *b* crystallographic axis.

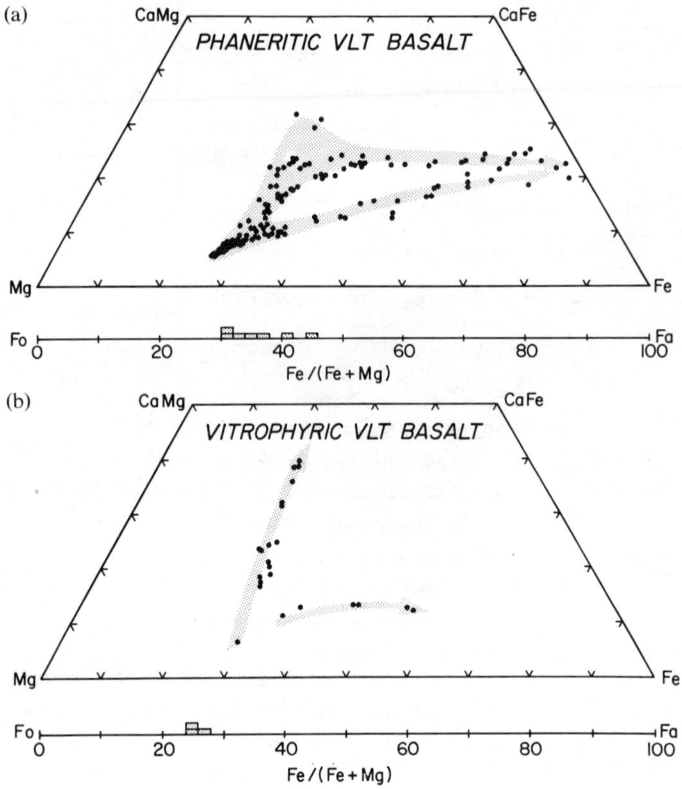

Fig. 5. Ca:Mg:Fe composition of pyroxenes and Mg:Fe composition of olivine; (a) phaneritic VLT basalt; (b) vitrophyric VLT basalt. Stippled arrows show zonation trends in pyroxenes.

precipitate. With subsequent crystallization the phanerite pyroxenes incorporate $R^{2+}TiAl_2O_6$ and approach a Ti:Al ratio of 1:2. The late-formed residual CaFe pyroxenes of the phaneritic VLT basalts have Ti:Al ratios that exceed 1:2. This may reflect some reduction of Ti^{4+} to Ti^{3+}, yielding the component $R^{2+}Ti^{3+}(SiAl)O_6$. The high-Ti CaFe pyroxenes are dark red, and are intergrown with a late-stage reduced assemblage of metallic Fe and Fe-sulfide.

Olivine

VLT vitrophyres have olivine of Fo_{74-75} with little zonation. Olivines of phaneritic VLT basalt are zoned from Fo_{71} to Fo_{55}. The magnesian olivines of phanerites are rich in Cr_2O_3 (0.39 wt.%) and low in MnO (0.23 wt.%) and TiO_2 (~0.00 wt.%) relative to Fe-enriched olivines (Cr_2O_3 = 0.15 wt.%, MnO = 0.42 wt.%, and TiO_2 = 0.09 wt.%). These concentrations of minor elements are within the range of mare basalt olivine compositions reported by Steele and Smith (1975).

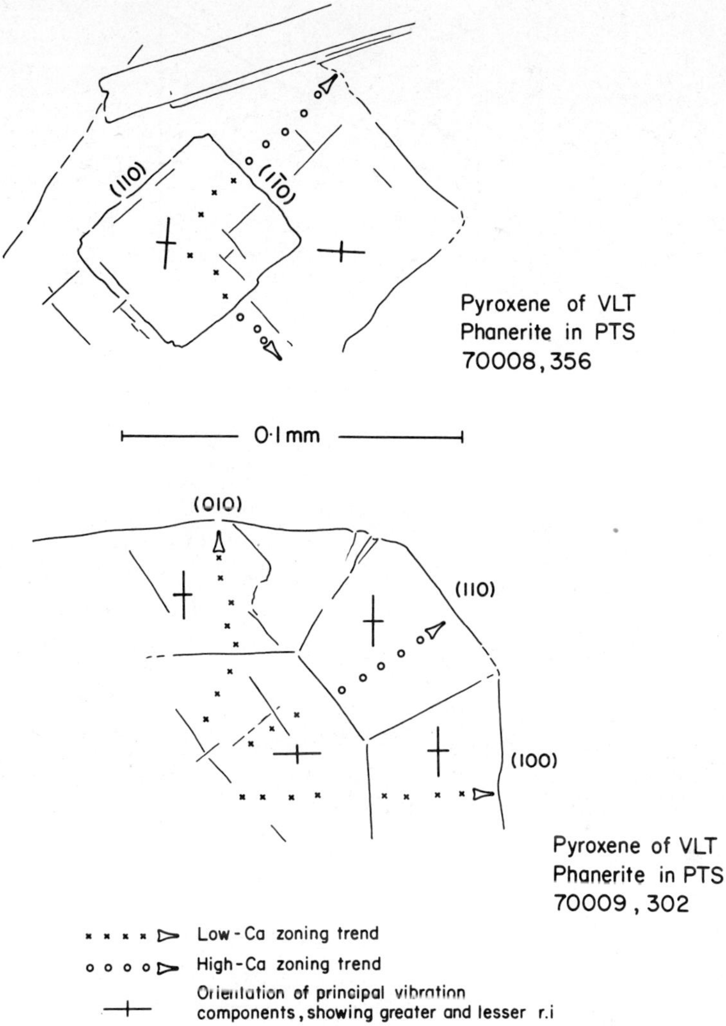

Fig. 6. Ca zoning trends in two pyroxene growths sectioned $\sim\perp C$. Note low-Ca "pigeonite" cores.

The unzoned, skeletal olivine crystals analyzed in the vitrophyres have a Mg:Fe composition of Fo_{75}. Olivine of this composition would be in equilibrium with a melt having a calculated Mg/(Mg + Fe) ratio of 0.50 (assuming $K_D = 0.33$; Lipin, 1976). This calculated Mg/(Mg + Fe) ratio is within the range of whole rock compositions determined from microprobe data (Fig. 11; Mg/(Mg + Fe) of VLT phanerites = 0.47–0.52, vitrophyres = 0.50–0.60). One vitrophyre is much richer than the others in normative olivine (Fig. 14); the olivine microphenocrysts of this rock may be more Mg-rich than Fo_{75}, but the crystals were too small to analyze.

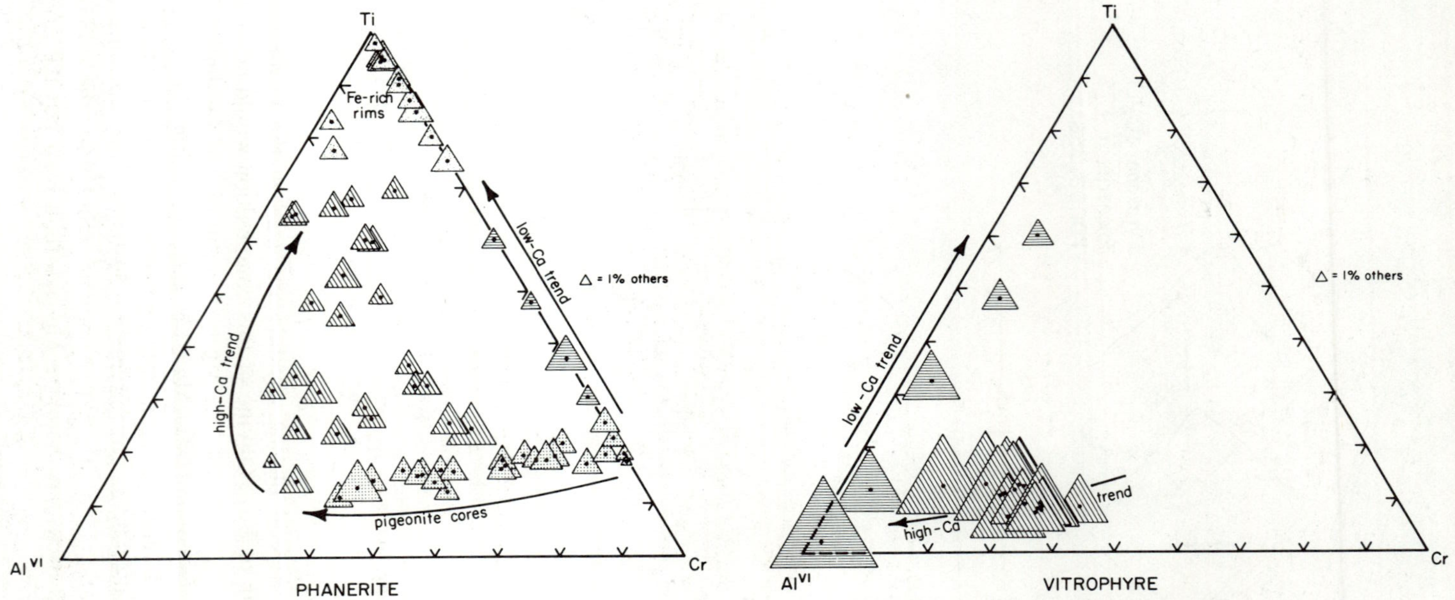

Fig. 7. Cr:Al^{VI}:Ti zonation; (a) phaneritic VLT basalt, PTS 70008,356; (b) vitrophyric VLT basalt, PTS 70009,288. Note that the edge dimension of the triangular symbol is proportional to the percentage of "others" components.

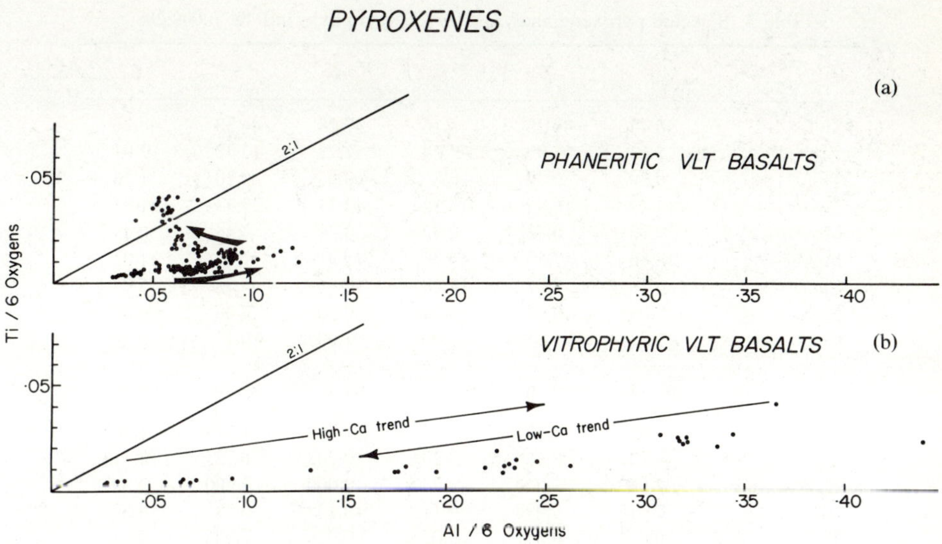

Fig. 8. Al:Ti zonation; (a) VLT phanerites; (b) vitrophyres.

Feldspar

Feldspars of VLT phanerites occur as euhedral laths ophitically intergrown in pyroxene rims, or as anhedral crystals between pyroxene grains. There is no consistent difference in composition between euhedral and anhedral feldspar. The range of feldspar compositions is restricted: An_{93-97} with 0.7 mole% Or (Table 7).

In Fig. 9 the Ab–An range of feldspar compositions in VLT basalt is compared to feldspar compositions of the common mare basalts. Feldspars of VLT basalt are at the Ca-rich end of the compositional range for mare basalt feldspars. The suppressed zonation and Ca-rich compositions of feldspar reflect the lower alkali content of VLT basalt relative to other mare basalts. The other mare basalts (ref. Papike *et al.*, 1976) have Na_2O contents of 0.17–0.54 wt.% (VLT = 0.01–0.15 wt.%) and K_2O contents of 0.04–0.32 wt.% (VLT = 0.01–0.04 wt.%). The low alkali content of VLT basalt is comparable to green glass.

Taylor *et al.* (1977) noted that Fe:Mg rises with increased albite content in VLT basalt feldspars. Our data show an increase of both Fe and K with Na enrichment (Table 7). Wenk and Wilde (1973) and Longhi *et al.* (1976a,b) showed that tetrahedral site deficiencies in Al + Si are common in mare feldspars; both groups conclude that this deficiency is due to coupled substitution of (Fe,Mg) + Si for 2Al in tetrahedral coordination:

$$CaAl_2Si_2O_8 \rightarrow Ca(Fe,Mg)Si_3O_8.$$

Longhi *et al.* (1976a,b) pointed out that Fe and Mg could be incorporated in sites of higher coordination as $(Fe,Mg)Al_2Si_2O_8$, but this substitution is uncommon in lunar feldspars. However, in VLT basalt there is a tendency for the more

Table 3. Selected pyroxene analyses*: VLT vitrophyre in PTS 70009,288.

	1	2	3	4	5	6
SiO_2	50.45	49.20	48.16	47.28	44.38	48.28
Al_2O_3	5.13	5.99	7.24	7.73	12.32	2.91
TiO_2	0.39	0.39	0.87	0.93	0.20	1.26
FeO	17.42	16.45	13.52	12.71	23.47	31.81
MnO	0.30	0.33	0.32	0.23	0.44	0.43
MgO	19.86	17.50	14.26	11.91	12.12	11.01
CaO	6.10	8.67	14.28	17.64	5.39	5.93
Na_2O	0.00	0.00	0.00	0.00	0.00	0.00
Cr_2O_3	1.53	1.57	2.06	1.00	0.17	0.17
Total	101.18	100.10	100.71	99.43	98.49	101.80
Si	1.853	1.835	1.795	1.792	1.719	1.887
Al^{IV}	0.147	0.165	0.205	0.208	0.281	0.113
$\Sigma_{tet.}$	2.000	2.000	2.000	2.000	2.000	2.000
Al^{VI}	0.075	0.098	0.113	0.138	0.281	0.021
Ti	0.011	0.011	0.024	0.026	0.006	0.037
Fe	0.535	0.513	0.421	0.403	0.760	1.040
Mn	0.009	0.010	0.010	0.007	0.015	0.014
Mg	1.087	0.973	0.792	0.673	0.700	0.641
Ca	0.240	0.347	0.570	0.716	0.224	0.248
Na	0.000	0.000	0.000	0.000	0.000	0.000
Cr	0.044	0.046	0.061	0.030	0.005	0.005
$\Sigma_{oct.}$ and VIII	2.001	1.998	1.991	1.993	1.991	2.006
$\Sigma_{cations}$	4.001	3.998	3.991	3.993	3.991	4.006
Wo	12.8	18.8	31.8	39.8	13.2	12.8
En	58.1	52.8	44.2	37.4	41.2	33.0
Fs	29.1	28.4	24.0	22.8	45.6	54.2

*All analyses normalized to six oxygens; Column (1): Analysis of pyroxene core; Columns (2–4): Analyses of high-Ca zoning trend; Columns (5–6): Analyses of low-Ca zoning trend.

Na-rich feldspars to have Fe and Mg in tetrahedral coordination and for the more Ca-rich feldspars to have Fe + Mg in both tetrahedral and higher coordination (Table 7).

Opaque minerals

The near-liquidus opaque phase of VLT basalt is chromian spinel with an Fe/(Fe + Mg) ratio of ~0.65 (Table 8, Fig. 10a). In phanerites the spinels are zoned to high Fe content with slight Ti enrichment (Ti ~ 0.03 atoms per four oxygens); there is a hiatus in the spinel crystallization history before discrete grains of late anhedral ulvöspinel appear. Ulvöspinel, ilmenite, metallic Fe,

Table 4. Selected pyroxene analyses*: phaneritic VLT basalt in PTS 70008,356.

	1	2	3	4	5	6	7	8	9	10	11	12	13	14
SiO_2	53.09	53.24	52.10	51.42	51.70	49.56	49.75	49.38	47.24	51.34	49.00	48.51	46.44	45.76
Al_2O_3	1.45	1.46	2.09	2.06	2.50	1.84	1.67	1.43	1.24	1.71	1.89	1.26	1.20	1.04
TiO_2	0.20	0.18	0.32	0.43	0.50	0.55	0.55	0.57	1.14	0.29	0.47	0.85	1.08	1.19
FeO	17.45	16.99	17.51	18.08	15.92	22.61	23.92	27.42	34.49	21.39	26.59	31.94	41.28	42.54
MnO	0.27	0.32	0.32	0.29	0.31	0.51	0.43	0.42	0.50	0.38	0.50	0.46	0.58	0.58
MgO	23.60	23.72	19.56	17.13	14.50	13.80	11.38	10.15	4.73	18.81	12.46	8.46	0.44	0.82
CaO	3.63	3.48	7.88	10.76	15.20	9.21	11.39	10.11	10.57	5.20	7.13	8.20	10.06	8.58
Na_2O	0.00	0.00	0.00	0.00	0.00	0.00	0.00	0.00	0.00	0.00	0.00	0.00	0.00	0.00
Cr_2O_3	0.73	0.78	1.01	0.94	1.01	0.43	0.64	0.64	0.29	0.71	0.48	0.29	0.02	0.06
Total	100.42	100.17	100.79	101.11	101.64	98.51	99.73	100.12	100.20	99.83	98.52	99.97	101.10	100.57
Si	1.945	1.950	1.929	1.920	1.924	1.937	1.944	1.947	1.936	1.940	1.941	1.949	1.947	1.938
Al^{IV}	0.055	0.050	0.071	0.080	0.076	0.063	0.056	0.053	0.060	0.060	0.059	0.051	0.053	0.052
$\Sigma_{tet.}$	2.000	2.000	2.000	2.000	2.000	2.000	2.000	2.000	1.996	2.000	2.000	2.000	2.000	1.989
Al^{VI}	0.007	0.013	0.020	0.011	0.034	0.022	0.021	0.014	0.000	0.016	0.029	0.008	0.006	0.000
Ti	0.006	0.005	0.009	0.012	0.014	0.016	0.016	0.017	0.035	0.008	0.014	0.026	0.034	0.038
Fe	0.535	0.520	0.542	0.565	0.496	0.739	0.782	0.904	1.182	0.676	0.881	1.073	1.448	1.506
Mn	0.008	0.010	0.010	0.009	0.010	0.017	0.014	0.014	0.017	0.012	0.017	0.016	0.021	0.021
Mg	1.288	1.295	1.079	0.953	0.804	0.804	0.663	0.597	0.289	1.059	0.736	0.506	0.027	0.052
Ca	0.142	0.137	0.313	0.431	0.606	0.386	0.477	0.427	0.464	0.211	0.303	0.353	0.452	0.389
Na	0.000	0.000	0.000	0.000	0.000	0.000	0.000	0.000	0.000	0.000	0.000	0.000	0.000	0.000
Cr	0.021	0.023	0.030	0.028	0.030	0.013	0.020	0.020	0.009	0.021	0.015	0.009	0.001	0.002
$\Sigma_{oct.}$ and VIII	2.007	2.003	2.003	2.009	1.994	1.997	1.993	1.993	1.996	2.003	1.991	1.991	1.989	2.008
$\Sigma_{cations}$	4.007	4.003	4.003	4.009	3.994	3.997	3.993	3.993	3.996	4.003	3.991	3.991	3.989	3.997
Wo	7.2	7.0	16.2	22.1	31.6	20.0	24.8	22.2	24.0	10.8	15.6	18.3	23.5	20.0
En	65.6	66.3	55.8	48.9	42.0	41.7	34.5	30.9	14.9	54.1	38.0	26.2	1.4	2.7
Fs	27.2	26.7	28.0	29.0	26.4	38.3	40.7	46.9	61.1	35.1	46.4	55.5	75.1	77.4

*All analyses normalized to six oxygens; Columns (1-2): Analyses of pigeonite cores; Columns (3–9): Analyses of high-Ca zonation trend; Columns (10–12): Analyses of low-Ca zonation trend; Columns (13–14): Analyses of Fe-rich pyroxene rims.

Table 5. Analyses from (100), (010), and (110) sectors of a single pyroxene*: phaneritic VLT basalt, PTS 70009,302.

	1	2	3	4	5	6	7	8	9	10	11	12	13	14
SiO_2	50.18	48.74	46.50	45.95	45.39	55.54	51.18	48.54	47.14	53.49	52.89	50.64	50.23	49.48
Al_2O_3	1.46	1.05	1.35	1.17	1.20	1.08	1.05	1.51	0.79	1.04	2.14	2.52	2.37	1.98
TiO_2	0.45	0.56	0.96	1.12	1.31	0.16	0.23	0.72	0.69	0.12	0.20	0.58	0.59	0.70
FeO	25.46	33.79	40.45	42.04	43.36	20.42	25.31	34.56	42.27	17.84	16.99	20.38	22.43	26.81
MnO	0.53	0.60	0.60	0.66	0.66	0.40	0.50	0.55	0.61	0.32	0.32	0.40	0.48	0.51
MgO	12.30	7.26	2.73	2.07	0.96	19.23	14.03	5.65	2.16	22.67	20.43	12.07	10.98	7.97
CaO	9.19	7.44	6.78	6.52	6.88	6.35	7.84	9.55	7.03	4.09	7.44	13.39	13.17	13.20
Na_2O	0.00	0.00	0.00	0.00	0.00	0.00	0.00	0.00	0.00	0.00	0.00	0.00	0.00	0.00
Cr_2O_3	0.36	0.19	0.12	0.08	0.09	0.54	0.38	0.19	0.09	0.61	0.97	0.67	0.55	0.35
Total	99.93	99.63	99.49	99.61	99.85	100.72	100.52	101.27	100.78	100.15	101.38	100.65	100.80	101.00
Si	1.956	1.976	1.956	1.948	1.937	1.961	1.969	1.953	1.972	1.969	1.936	1.934	1.934	1.943
Al^{IV}	0.044	0.024	0.044	0.052	0.060	0.039	0.031	0.047	0.027	0.031	0.064	0.066	0.066	0.057
$\Sigma_{tet.}$	2.000	2.000	2.000	2.000	1.997	2.000	2.000	2.000	2.000	2.000	2.000	2.000	2.000	2.000
Al^{VI}	0.023	0.026	0.023	0.007	0.000	0.009	0.017	0.025	0.012	0.014	0.028	0.048	0.042	0.035
Ti	0.013	0.017	0.030	0.036	0.042	0.004	0.007	0.022	0.022	0.003	0.006	0.017	0.017	0.021
Fe	0.830	1.146	1.423	1.491	1.548	0.638	0.815	1.163	1.479	0.548	0.520	0.651	0.723	0.881
Mn	0.017	0.021	0.022	0.024	0.024	0.013	0.016	0.019	0.022	0.010	0.010	0.013	0.016	0.017
Mg	0.714	0.439	0.171	0.131	0.061	1.070	0.805	0.339	0.135	1.243	1.115	0.687	0.630	0.467
Ca	0.384	0.323	0.306	0.296	0.315	0.254	0.323	0.412	0.315	0.161	0.292	0.548	0.544	0.555
Na	0.000	0.000	0.000	0.000	0.000	0.000	0.000	0.000	0.000	0.000	0.000	0.000	0.000	0.000
Cr	0.011	0.006	0.004	0.003	0.003	0.016	0.012	0.006	0.003	0.018	0.028	0.021	0.017	0.011
$\Sigma_{oct.}$ and VIII	1.992	1.978	1.979	1.988	1.993	2.004	1.995	1.986	1.988	1.997	1.999	1.985	1.989	1.987
$\Sigma_{cations}$	3.992	3.978	3.979	3.988	3.990	4.004	3.995	3.986	3.988	3.997	3.999	3.985	3.989	3.987
Wo	19.7	16.8	15.9	15.3	16.2	12.9	16.5	21.3	16.2	8.2	15.1	28.8	28.4	28.9
En	36.7	22.7	8.9	6.7	3.1	54.2	41.4	17.5	6.9	63.4	57.6	36.2	33.0	24.3
Fs	43.6	60.5	75.2	78.0	80.7	32.9	42.4	61.2	76.9	28.4	27.3	35.0	38.6	46.8

*All analyses normalized to six oxygens; Columns (1–5): Analyses of low-Ca zoning trend in sector (100); Columns (6–9): Analyses of low-Ca zoning trend in sector (010); Columns (10–14): Analyses of high-Ca zoning trend in sector (110).

Table 6. Selected olivine analyses*: phaneritic VLT basalts and VLT vitrophyres.

	1	2	3	4	5	6	7	8
SiO_2	37.24	36.57	36.43	36.99	36.34	34.99	37.98	37.71
Al_2O_3	0.17	0.22	0.27	0.13	0.17	0.20	0.18	0.21
TiO_2	0.01	0.03	0.06	0.03	0.05	0.09	0.07	0.06
FeO	28.25	32.53	35.11	29.85	33.76	39.35	23.06	23.43
MnO	0.28	0.39	0.37	0.34	0.35	0.42	0.29	0.26
MgO	35.23	30.31	29.07	33.85	30.82	26.42	37.94	38.30
CaO	0.28	0.28	0.30	0.33	0.35	0.29	0.41	0.39
Cr_2O_3	0.39	0.30	0.20	0.27	0.12	0.15	0.55	0.67
Total	101.85	100.75	101.81	101.79	101.96	101.91	100.48	101.03
Si	0.981	0.994	0.991	0.982	0.982	0.975	0.989	0.980
Al	0.005	0.007	0.009	0.004	0.005	0.006	0.006	0.006
Ti	0.000	0.001	0.001	0.001	0.001	0.002	0.001	0.001
Fe	0.622	0.740	0.799	0.663	0.763	0.917	0.502	0.509
Mn	0.006	0.009	0.008	0.008	0.008	0.010	0.006	0.006
Mg	1.382	1.228	1.179	1.340	1.241	1.097	1.473	1.483
Ca	0.008	0.008	0.009	0.010	0.010	0.009	0.012	0.011
Cr	0.008	0.006	0.004	0.006	0.002	0.003	0.011	0.014
$\Sigma_{cations}$	3.012	2.993	3.000	3.014	3.012	3.019	3.000	3.010
Fo	68.8	62.1	59.3	66.6	61.7	54.2	74.3	74.2
Fa	31.2	37.9	40.7	33.4	38.3	45.8	25.7	25.8

*All analyses normalized to four oxygens; Phanerites in columns (1–6) and vitrophyres in columns (7–8); Columns (1–3): PTS 70008,356; Columns (4–6): PTS 70008,370; Column (7): PTS 70007,328; Column (8): PTS 70008,382.

Fe-sulfide, and CaFe pyroxene occur together in late-stage assemblages. Vitrophyric VLT basalts lack these late-stage crystallization products, but the chromian spinels of vitrophyres are zoned to higher Ti content (Ti ~ 0.45 atoms per four oxygens) than the spinels in phaneritic VLT basalt.

Low Ti basalts, like VLT basalt, have chromian spinel as a liquidus phase at low pressure (Papike *et al.*, 1976). The crystallization path of spinel in Apollo 12 basalts is discontinuous between chromian spinel and ulvöspinel, similar to the discontinuity that is assumed to occur in VLT basalts (Fig. 10). Weill *et al.* (1971) related this spinel discontinuity in Apollo 12 basalt to changing conditions of crystallization that accompany eruption onto the lunar surface, but Busch *et al.* (1972) suggested that the spinel discontinuity in Apollo 12 basalts may be due to a loss of Cr to the first precipitated pyroxenes. Either explanation may be valid for the VLT basalts.

By comparison with the common mare basalts, spinels of the VLT basalt are most similar in composition to the spinels of Apollo 12, 15, and Luna 16 basalts (Fig. 10b; ref. Papike *et al.*, 1976). Spinels in all of these rock types have a common zoning trend beginning on the Fe–Mg–Cr–Al base of the spinel "prism"

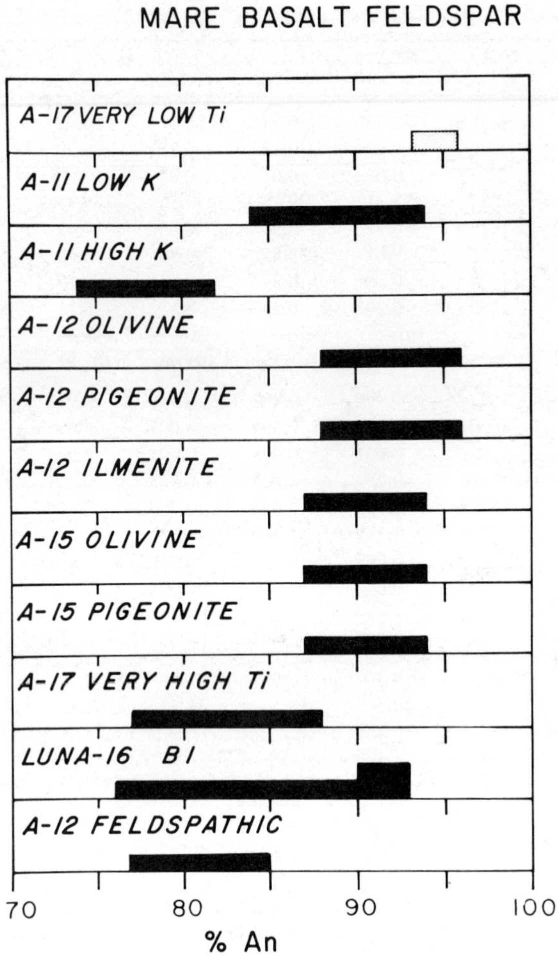

Fig. 9. Mole percent An in mare basalt feldspars: Data summarized in Papike *et al.* (1976) and in this paper (Apollo 17 very low Ti basalt).

(ref. Haggerty, 1972) and trending toward ulvöspinel with increase in Fe. The coupling of Fe and Ti variation in zoning trends of pyroxene and spinel occurs in both the low Ti and the VLT mare basalts.

Silica

Crystals of a silica polymorph comprise 0.6–1.6% of the VLT phanerites. The crystals occur as individual subhedral grains up to 0.05 mm long. Normative compositions of VLT basalt are SiO_2 deficient or only slightly silica saturated (Table 1); therefore, the large residual silica grains must have developed as a result of small scale fractionation resulting from metastable crystallization within VLT basalts.

Table 7. Selected feldspar analyses*: phaneritic VLT basalts in PTS 70008,356 and PTS 70008,370.

	1	2	3	4	5	6
SiO_2	45.03	45.06	46.17	44.47	45.40	47.01
Al_2O_3	34.93	34.65	33.72	35.17	34.79	32.67
FeO	0.90	0.98	0.98	0.65	0.64	1.34
MgO	0.26	0.13	0.18	0.19	0.26	0.15
CaO	19.42	19.21	19.08	18.97	18.53	18.48
Na_2O	0.46	0.50	0.54	0.34	0.56	0.71
K_2O	0.03	0.04	0.04	0.03	0.04	0.12
Total	101.03	100.57	100.71	99.82	100.19	100.48
Si	2.067	2.078	2.122	2.061	2.092	2.165
Al	1.891	1.884	1.828	1.922	1.889	1.774
Σ	3.958	3.962	3.950	3.983	3.981	3.939
Fe	0.034	0.038	0.038	0.025	0.025	0.052
Mg	0.018	0.009	0.012	0.013	0.018	0.010
Ca	0.955	0.949	0.940	0.942	0.915	0.912
Na	0.041	0.045	0.048	0.030	0.050	0.063
K	0.002	0.003	0.002	0.002	0.002	0.007
Σ	1.050	1.044	1.040	1.012	1.010	1.044
$\Sigma_{\text{total cations}}$	5.008	5.006	4.990	4.995	4.991	4.983
Or	0.2	0.3	0.2	0.2	0.2	0.7
Ab	4.1	4.5	4.8	3.1	5.2	6.4
An	95.7	95.3	95.0	96.7	94.6	92.9
Fe/(Fe + Mg)	0.66	0.81	0.75	0.66	0.58	0.83

*All analyses normalized to eight oxygens; Columns (1–3): PTS 70008,356; Columns (4–6): PTS 70008,370.

Chemical Characterization of VLT Basalt

The chemical composition of VLT basalt is similar to the common low Ti basalts, though lower in TiO_2 and alkalis (Fig. 3). The chemical systematics of pyroxene, olivine, and spinel in VLT basalt and common low Ti basalts reflect this similarity in major-element composition and suggest similar emplacement and cooling histories. Texture and major-element chemistry indicate that the VLT basalt is a mare rock type. Yet there are details of composition in which the VLT basalt is distinctive from the other mare basalts (e.g., the low alkali content which is reflected in Ca-rich plagioclase with restricted Ca–Na zonation). These distinctive chemical differences are summarized in Figs. 11–13.

Figure 11 is a plot of Mg/(Mg + Fe) ratios in common mare basalts and VLT basalt. The most Mg-rich mare basalts have Mg/(Mg + Fe) ratios of 0.52 (Papike *et al.*, 1976). Apollo 12 olivine basalts with higher ratios of Mg–Fe are probably olivine cumulate rocks (Walker *et al.*, 1972). Green glasses have an Mg/(Mg + Fe) range of about 0.56–0.65 (data base of Stolper, 1974). The VLT basalts span the gap between those mare basalts which are highest in Mg/Fe and the green glasses that are lowest in Mg/Fe.

Figure 12 is a three-axis plot comparing atomic Mg/(Mg + Fe), CaO/Al_2O_3 (wt.%) and TiO_2 (wt.%); these are effective chemical parameters for separating mare from highlands rocks. The data plotted include average basalt types from Apollo sites 11, 12, 15, and 17 (Papike *et al.*, 1976), average Apollo 15 green glass (Ridley *et al.*, 1972), and average Apollo 17 orange glass (Rhodes *et al.*, 1974). The plot also includes average VLT basalt and glass from the Apollo 17 drill core and average compositions of six highland melt-rock groups plus aluminous basalt. The seven groups of highland melt rock plus aluminous basalt were selected by cluster analysis of 148 analyses gathered from the literature on aluminous basalt, feldspathic basalt and poikilitic rocks (Naney *et al.*, 1977). The highland melt-rock groups are distinctively lower in CaO/Al_2O_3 (0.53–0.61) than mare basalts and higher in Mg/(Mg + Fe) (0.63–0.72). The low CaO/Al_2O_3 ratio of the highland melt rocks is distinctive. This ratio reflects the dominant feldspar component in highland melt rocks, for highland plagioclase has a CaO/Al_2O_3 ratio of ~0.55. The higher Mg/Fe of impact melts is due to mixing with an Mg-rich component in the highland crust (Wänke *et al.*, 1976). Aluminous basalts plot between mare and highland basalt groups; these basalts are believed to be extrusive (ref. Ridley, 1975; Irving, 1975). Within this array of melt rocks and mare basalts the VLT basalt plots closest to mare basalts and green glass in CaO/Al_2O_3, Mg/(Mg + Fe) and TiO_2 content.

Figure 13 is a diagram of wt.% TiO_2 vs wt.% MgO. The common high Ti and low Ti basalts plot as two distinct groups. The VLT basalt and VLT glasses plot together with green glasses as a third, equally distinctive group. The range of green glass compositions is from the data summary of Stolper (1974) representing Apollo 11, 14, and 15 and Luna 16. The VLT compositions can be plotted along with green glass on any number of bivariate oxide plots, but the result is always as seen in Fig. 13: The compositional ranges of VLT basalt and green glass overlap.

CONCLUSIONS

VLT basalt is chemically similar to low Ti basalts (e.g., Apollo 12) and very similar to green glasses. The crystal-chemical systematics of VLT basalt reflect these similarities, and speculations about the petrogenesis and possible mare sources of VLT basalt point to some interesting problems for further study.

A possible common origin for VLT basalt and green glass

Stolper (1974) emphasized the diversity of green glasses. He used ternary projections of An–Fo–Fa and SiO_2–Ol–An to present the green glass data along with his experimental results. Figure 14 shows VLT basalts and glasses plotted in these two ternary projections.

Table 8. Selected oxide analyses*: phaneritic VLT basalts and VLT vitrophyres.

	1	2	3	4	5	6	7	8
Al_2O_3	18.95	18.28	17.23	1.78	0.32	17.95	9.06	6.09
TiO_2	1.30	1.24	1.12	34.07	51.14	1.06	6.08	16.91
FeO	25.95	27.39	29.77	63.30	46.65	23.33	37.64	47.55
MnO	0.32	0.35	0.29	0.35	0.38	0.34	0.34	0.32
MgO	7.36	5.36	4.30	0.00	0.22	8.21	1.42	1.48
Cr_2O_3	46.61	46.58	47.08	0.39	0.08	48.57	46.46	28.41
Total	100.49	99.20	99.79	99.89	98.79	99.46	101.00	100.76
Ti	0.031	0.031	0.028	0.948	0.984	0.026	0.160	0.451
Fe	0.702	0.760	0.833	1.959	0.998	0.635	1.103	1.410
Mn	0.009	0.010	0.008	0.011	0.008	0.009	0.010	0.010
Mg	0.354	0.265	0.214	0.000	0.008	0.398	0.074	0.078
Cr	1.192	1.221	1.245	0.011	0.001	1.249	1.287	0.797
Al	0.722	0.714	0.680	0.078	0.010	0.688	0.371	0.255
$\Sigma_{cations}$	3.010	3.001	3.008	3.007	2.009	3.005	3.008	3.001
Fe/(Fe + Mg)	0.66	0.74	0.80	1.00	0.99	0.61	0.94	0.95
(Al/2):(Cr/2):Ti	37:60:03	36:61:03	34:63:03	04:01:95	—	35:63:02	19:65:16	13:41:46

*All analyses normalized to four oxygens; Columns (1–2): CrAl spinels in phanerite, PTS 70008,370; Column (3): CrAl spinel in phanerite, PTS 70008,356; Column (4): Ulvöspinel in phanerite, PTS 70008,356; Column (5): Ilmenite in phanerite, PTS 70008,370; Columns (6–8): CrAl spinels in vitrophyre, PTS 70007,316.

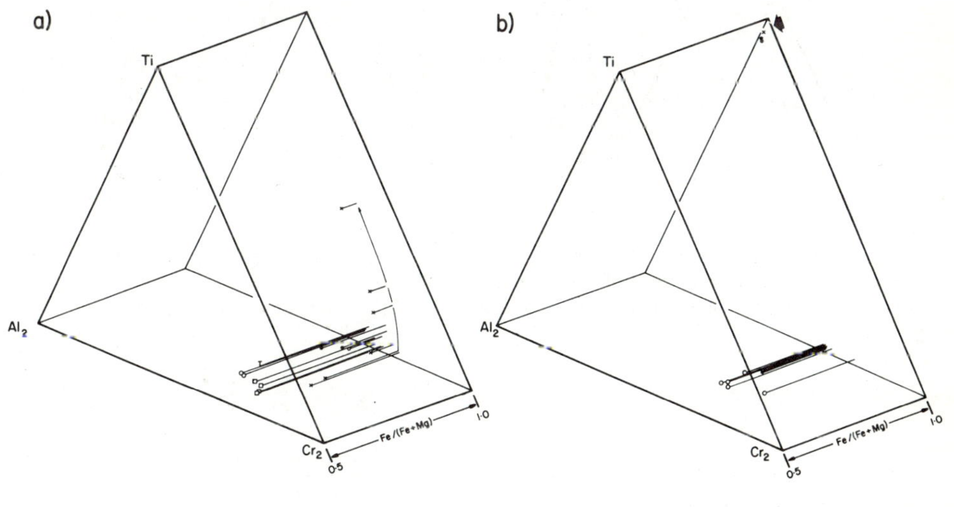

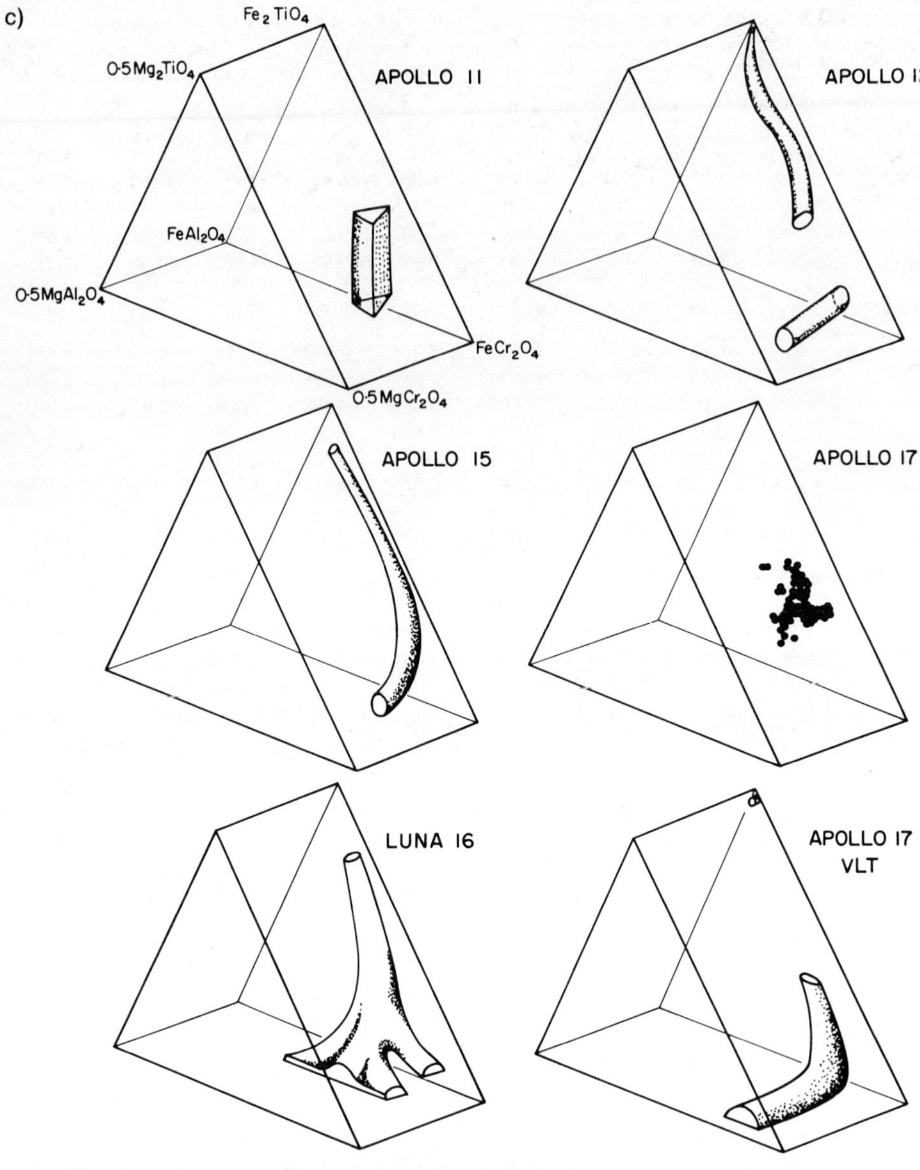

Fig. 10. Spinel compositions; (a) Apollo 17 VLT vitrophyres; (b) Apollo 17 VLT phanerites; (c) Comparison of zoning trends in spinels of VLT basalt and other mare basalts: data of Haggerty (1972) and VLT basalt data of this paper.

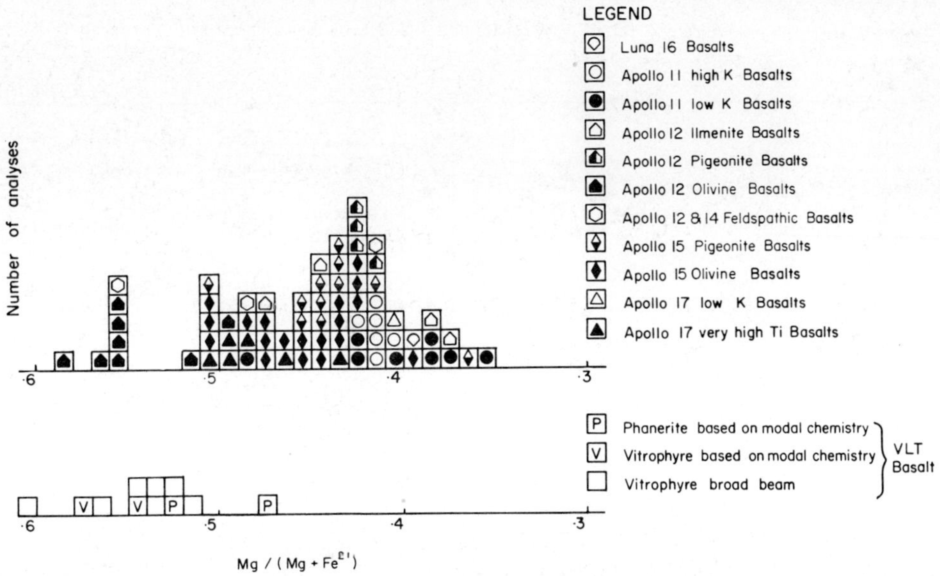

Fig. 11. Atom ratio Mg/(Mg + Fe) in VLT basalt compared to other mare basalts (data summary of Papike *et al.*, 1976).

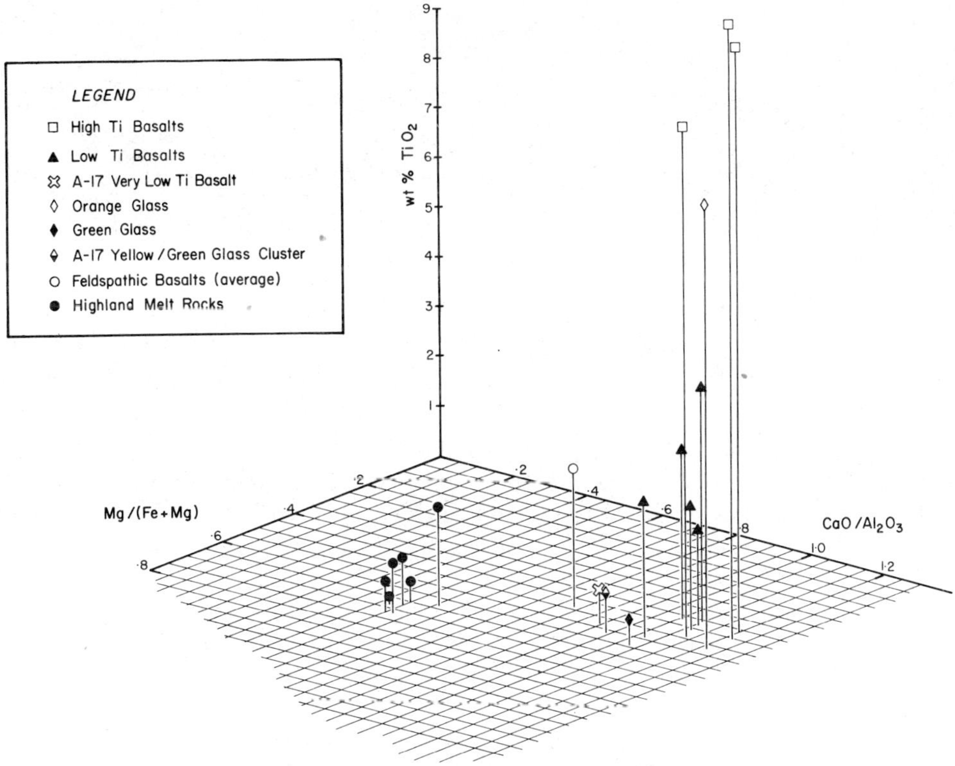

Fig. 12. Atom ratio Mg/(Mg + Fe), weight percent ratio CaO/Al_2O_3, and weight percent TiO_2 in mare basalts and highland melt rocks. Averages of other mare basalts (from Papike *et al.*, 1976) and the averages of six highland melt-rock groups and feldspathic basalts (cluster analysis of Naney and Papike, 1977) are compared to the average VLT basalt and VLT glass (data of this paper). The average Apollo 17 orange glass of Rhodes *et al.* (1974) and average green glass of Ridley (1972) are also plotted.

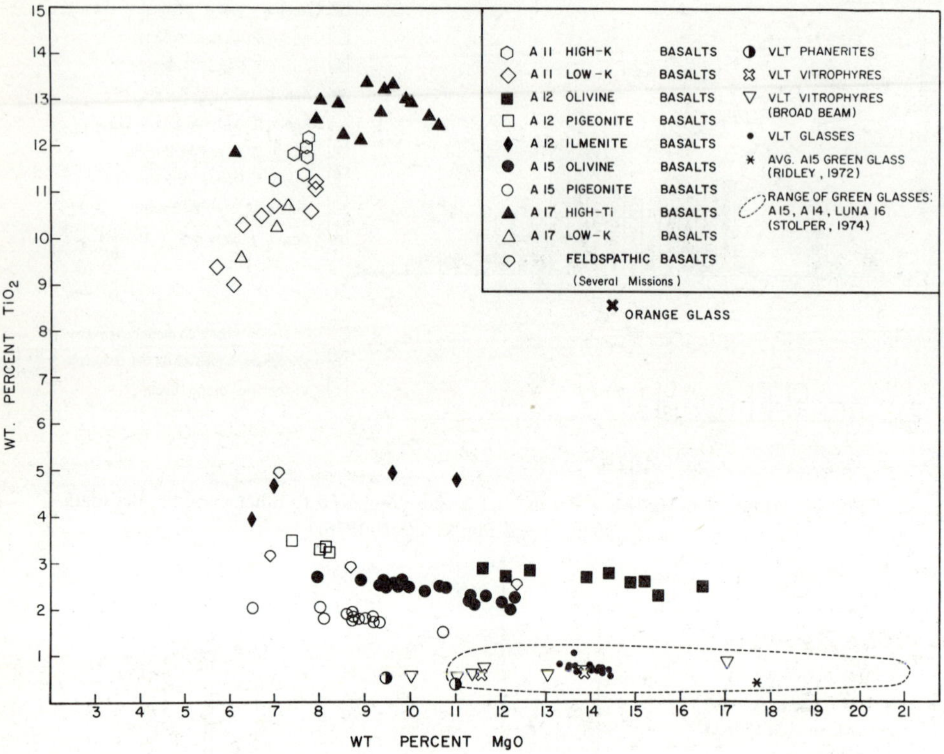

Fig. 13. MgO vs TiO_2 in high Ti, low Ti, and feldspathic mare basalts (data summary of Papike *et al.*, 1976) and in the Apollo 17 VLT basalts and glasses (data of this paper). Plotted points are individual analyses except for the averages of orange and green glass. An average of Apollo 15 green glass and the range of green glass compositions are plotted for comparison to VLT basalt.

In Fig. 14a the VLT glasses plot at the olivine-poor edge of Stolper's grouping for the majority of Apollo 15 and 14 green glasses ("group I"). Stolper *et al.* (1974) described a trend of Fo_{69} control in the compositional variation of the group I glasses. Figure 14a shows this trend of olivine control in the major axis of the ellipse that encloses the group I glass compositions. Figure 14a also shows that no olivine control is evident in the VLT glasses from the Apollo 17 drill core. The only variation observed is between lithic fragments: all VLT lithics, except for one vitrophyre, plot outside group I and along a trend toward the An–Fa join. The arrow in Fig. 14a is Stolper's (1974) experimental fractionation path for liquids derived from Apollo 15 green glass (sample SLS8). The VLT basalt compositions lie along this fractionation trend. It is important to note that this fractionation trend involves liquidus olivine and is not the fixed trend of non-liquidus Fo_{69} control which Stolper *et al.* (1974) showed for group I green glass. The VLT basalts and glasses do not show the anomalous olivine variation that affects the mafic green glasses.

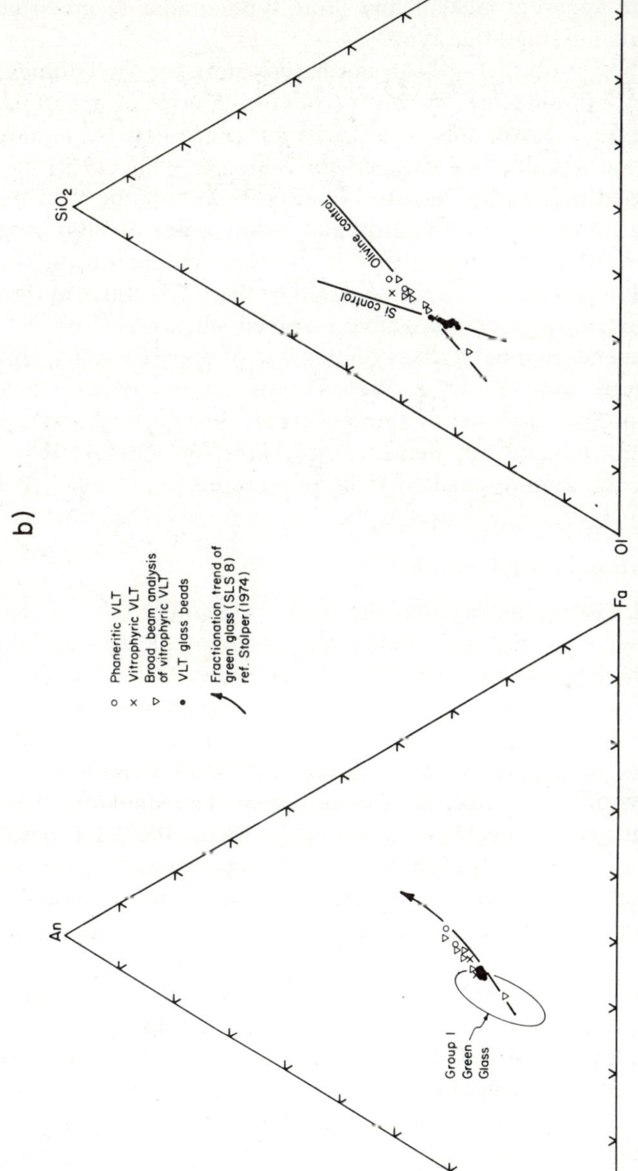

Fig. 14. Normative composition of VLT basalts and glasses. The same projections and norm calculation methods of Stolper (1974) are used. (a): An–Fo–Fa projection, including the prominent Apollo 15 "group I" green glass cluster of Stolper (1974). Stolper's experimental green glass fractionation trend is also shown. (b): SiO_2–Ol–An projection with trends of olivine and Si variation.

Arguments for a genetic relationship between VLT basalt and green glass are permissive at best, if based only on major-element data. However, the strong evidence of a lava-fountaining origin for green glass based on surface correlated volatiles (ref. Chou *et al.*, 1974, 1975; Meyer *et al.*, 1975) has always been weakened by the apparent lack of any lithic type similar to green glass. VLT basalt may be this missing lithic type.

Figure 14b shows the SiO_2–Ol–An normative plots for VLT lithics and VLT glasses of the Apollo 17 core. A line of olivine control is apparent for VLT lithics; as discussed above, this is a variation controlled by liquidus olivine rather than the non-liquidus Fo_{69} control which Stolper *et al.* (1974) showed to be a general feature of green glass. Figure 14b also shows that the VLT glasses may deviate from the trend of olivine control and follow a line of SiO_2 variation. The SiO_2 variation in VLT glasses appears in all plots comparing SiO_2 and other oxides. If the VLT glasses are directly related to the VLT lithics in the drill core, then the glass-forming process may have involved silica loss.

However, the end-member glasses on this line of apparent silica control differ in SiO_2 content by only 1.1 wt.%. We estimate our analytical uncertainty at ±0.4 wt.% in replicate analyses of these glasses; therefore, this range of silica variation is barely beyond our limits of precision. We suggest that this silica variation is real, but a more detailed study is required.

Low-P equilibration of VLT basalt

Walker *et al.* (1976) showed that the depth of equilibration of mare basalts may be estimated from the total FeO + MgO content of the basalt. Using the most magnesium-rich, silica-poor VLT vitrophyre the estimated depth of equilibration is <100 km. This estimate rests on the assumption that the liquid was saturated with both olivine and pyroxene at this shallow depth. Although this shallow depth is indicated by the multiple saturation hypothesis, we do not suggest a shallow olivine pyroxenite source region. The similarity between VLT basalt and green glass indicates a more likely origin: the VLT basalt may be obtained from a green glass-type magma of deep origin which was tapped repeatedly at higher levels or fractionated in a near-surface magma chamber. The ultimate source of VLT basalt may, therefore, be the source area of the refractory green glasses, at ~200–400 km depth (Taylor and Jakeš, 1974).

If VLT basalt is related to a green glass-type magma, that magma must fall near the magnesium-poor, silica-rich end of the group I green glass trend described by Stolper (1974; ref. Fig. 14a). The origins of the green glass liquids are obscured by the non-liquidus Fo_{69} variation of the group I trend. However, a specific member of the group I trend (i.e., the VLT glasses) might be chosen as a liquid from which a near-surface differentiation may produce the observed fractionation trend of phaneritic and vitrophyric VLT basalts. If olivine + pyroxene co-saturation did not occur during the fractionation, then the depth of last equilibration before eruption is not constrained.

The association of very high Ti and very low Ti basalts at the edge of

Serenitatis is intriguing. Longhi *et al.* (1974) showed that the suite of Apollo 17 very high Ti basalts can be related to each other by low-P fractionation of a high Ti liquid held at shallow depths. The substructure of the Serenitatis Basin may, therefore, contain holding chambers where such differentiation took place. VLT magma rising from greater depths may also have been held by such chambers and allowed to fractionate at low pressures. This process would permit olivine fractionation and low-pressure equilibration of VLT magma derived from a green glass-type melt.

Possible age of VLT basalt

Green glass has been dated at 3.38 ± 0.6 b.y. (^{40}Ar–^{39}Ar; Podosek and Huneke, 1973). Low Ti mare basalts have $^{40}Ar/^{39}Ar$ and Rb–Sr dates of 3.3 b.y. The VLT basalt type is most likely related to one of these two compositions, and, therefore, may also be a young mare basalt about 3.3 b.y. old. However, it must be noted that speculations about the age of VLT basalt are very tentative.

Possible mare sources of VLT basalt

A young very low Ti mare source may be nearby: crater counts show that central Serenitatis is covered by a young mare formation (Boyce and Dial, 1973), and the spectral reflectance data of Pieters and McCord (1976) indicate that central Serenitatis is floored by basalts with only 2–3% TiO_2. Pieters (pers. comm.) suggests that this TiO_2 estimate may be too high. Therefore, the Ti content of central Serenitatis flows may be as low as VLT basalt. The light soils of central Serenitatis approach to within 100 km of the Apollo site, and Head (pers. comm.) suggests that impacts on the light areas of Serenitatis may have delivered fragments of interior Serenitatis flows to the Apollo 17 site.

The Apollo 17 drill core studies demonstrate, once again, the importance of soil studies in the recognition of new rock types. Soil samples have the potential of providing rock fragments which have been delivered by distant impacts. Delano (1975) demonstrated this in his study of the mare component in the Apollo 16 drill core. Could soil samples from sites other than Apollo 17, Luna 16 and Luna 24 also contain very low Ti mare basalts? Using the data of Pieters and McCord (1976) and allowing for the possibility of overestimates of TiO_2 for low Ti basalts, it is possible that VLT mare basalts occur in Mare Imbrium (Apollo 15), Mare Tranquillitatis (Apollo 11), and eastern Procellarum (Apollo 12 and 14). It is not unreasonable to expect that several new mare rock types may be found in the soil collections from some of these landing sites.

Acknowledgments—We acknowledge Dr. N. Z. Boctor and Dr. B. R. Lipin for helpful reviews and suggestions. Discussions with Dr. T. L. Grove have been especially valuable. This research was funded by NASA grant #NSG-9044, which we gratefully acknowledge.

REFERENCES

Albee A. L. and Ray L. (1970) Correction factors for electron probe microanalysis of silicates, oxides, carbonates, phosphates and sulfates. *Anal. Chem.* **42**, 1408–1414.

Barsukov V. L. and Florensky V. I. (1977) The lunar soil from Mare Crisium: Preliminary data (abstract). In *Lunar Science VIII*, p. 61–63. The Lunar Science Institute, Houston.

Bence A. E. and Albee A. L. (1968) Empirical correction factors for electron microanalysis of silicates and oxides. *J. Geol.* **76**, 382–403.

Bence A. E. and Papike J. J. (1972) Pyroxenes as recorders of lunar basalt petrogenesis: Chemical trends due to crystal-liquid interaction. *Proc. Lunar Sci. Conf. 3rd*, p. 431–469.

Boyce J. M. and Dial A. L. (1973) Relative ages of some near-side mare units based on Apollo 17 metric photographs. In *Apollo 17 Prelim. Sci. Rep.*, NASA SP-330, p. 29-26 to 29-28.

Busche F. D., Prinz M., Keil K., and Bunch T. E. (1972) Spinels and the petrogenesis of some Apollo 12 igneous rocks. *Amer. Mineral.* **57**, 1729–1747.

Chou C.-L., Baedecker P. A., Bild R. W., and Wasson J. T. (1974) Volatile-element systematics and green glass in Apollo 15 lunar soils. *Proc. Lunar Sci. Conf. 5th*, p. 1645–1657.

Chou C.-L., Boynton W. V., Sundberg L. L., and Wasson J. T. (1975) Volatiles on the surface of Apollo 15 green glass and trace-element distributions among Apollo 15 soils. *Proc. Lunar Sci. Conf. 6th*, p. 1701–1727.

Delano J. W. (1975) Petrology of the Apollo 16 mare component: Mare Nectaris. *Proc. Lunar Sci. Conf. 6th*, p. 15–47.

Dymek R. F., Albee A. L., and Chodos A. A. (1975) Comparative mineralogy and petrology of Apollo 17 mare basalts: Samples 70215, 71055, 74255, and 75055. *Proc. Lunar Sci. Conf. 6th*, p. 49–77.

Gancarz A. J. and Albee A. L. (1973) Microprobe analysis of the bulk composition of phase aggregates. Proc. 8th Natl. Conf. on Electron Probe Analysis, New Orleans.

Green J. A., Kurat G., Keil K., and Prinz M. (1972) Electron microprobe analyses of lithic fragments, glasses, chondrites and minerals from Luna 16 fines. Spec. Pub. 5, p. 1–16, Univ. of New Mexico, Inst. of Meteorites.

Haggerty S. E. (1972) Luna 16: An opaque mineral study and a systematic examination of compositional variations of spinels from Mare Fecunditatis. *Earth Planet. Sci. Lett.* **13**, 328–352.

Head J. W., III (1976) Lunar volcanism in space and time. *Rev. Geophys. Space Phys.* **14**, 265–300.

Irving A. J. (1975) Chemical, mineralogical and textural systematics of non-mare melt rocks: Implications for lunar impact and volcanic processes. *Proc. Lunar Sci. Conf. 6th*, p. 363–394.

Lipin B. R. (1976) The origin of Fra Mauro basalts (abstract). In *Lunar Science VII*, p. 495–497. The Lunar Science Institute, Houston.

Longhi J., Walker D., Grove T. L., Stolper E. M., and Hays J. F. (1974) The petrology of the Apollo 17 mare basalts. *Proc. Lunar Sci. Conf. 5th*, p. 447–469.

Longhi J., Walker D., and Hays J. F. (1976a) Fe, Mg and silica in lunar plagioclase (abstract). In *Lunar Science VII*, p. 501–503. The Lunar Science Institute, Houston.

Longhi J., Walker D., and Hays J. F. (1976b) Fe and Mg in plagioclase. *Proc. Lunar Sci. Conf. 7th*, p. 1281–1300.

Meyer C., McKay D. S., Anderson D. H., and Butler P. (1975) The source of sublimates on the Apollo 15 green and Apollo 17 orange glass samples. *Proc. Lunar Sci. Conf. 6th*, p. 1673–1699.

Naney M. T., Papike J. J., and Vaniman D. T. (1977) Lunar highland melt rocks: Major element chemistry and comparisons with lunar highland glass groups (abstract). In *Lunar Science VIII*, p. 717–719. The Lunar Science Institute, Houston.

Papike J. J., Hodges F. N., Bence A. E., Cameron M., and Rhodes J. M. (1976) Mare basalts: Crystal chemistry, mineralogy and petrology. *Rev. Geophys. Space Phys.* **4**, 475–540.

Papike J. J., Lellis S. F., Becker R., and Vaniman D. T. (1977) The Apollo 17 drill core: Modal data (Sections 70007, 70008, 70009) (abstract). In *Lunar Science VIII*, p. 753–755. The Lunar Science Institute, Houston.

Pieters C. and McCord T. B. (1976) Characterization of lunar mare basalt types: I. A remote sensing study using reflection spectroscopy of surface soils. *Proc. Lunar Sci. Conf. 7th*, p. 2677–2690.

Podosek F. A. and Huneke J. C. (1973) Argon in Apollo 15 green glass spherules (15426): ^{40}Ar–^{39}Ar age and trapped argon. *Earth Planet. Sci. Lett.* **19**, 413–421.

Rhodes J. M., Rodgers K. V., Shih C., Bansal B. M., Nyquist L. E., Wiesmann H., and Hubbard N. J. (1974) The relationships between geology and soil chemistry of the Apollo 17 landing site. *Proc. Lunar Sci. Conf. 5th*, p. 1097–1117.

Ridley W. I. (1975) On high-alumina mare basalts. *Proc. Lunar Sci. Conf. 6th*, p. 131–145.

Ridley W. I., Reid A. M., Warner J. L., and Brown R. W. (1972) Apollo 15 green glass. *Phys. Earth Planet. Interiors* **7**, 133–136.

Ryder G., Stoeser D. B., Marvin U. B., Bower J. F., and Wood J. A. (1975) Boulder 1, Station 2, Apollo 17: Petrology and petrogenesis. *The Moon* **14**, 327–357.

Ryder G. and Taylor G. J. (1976) Did mare-type volcanism commence early in lunar history? *Proc. Lunar Sci. Conf. 7th*, p. 1741–1755.

Ryder G., Stoeser D. B., and Wood J. A. (1977) Apollo 17 KREEPy basalt: A rock type intermediate between mare and KREEP basalts. *Earth Planet. Sci. Lett.* **35**, 1–13.

Steele I. M. and Smith J. V. (1975) Minor elements in lunar olivine as a petrologic indicator. *Proc. Lunar Sci. Conf. 6th*, p. 451–467.

Stolper E. M. (1974) Lunar ultramafic glasses. B.A. thesis. Harvard University, Cambridge, Massachusetts. 178 pp.

Stolper E. M., Walker D., Longhi J., and Hays J. F. (1974) Compositional variation in lunar ultramafic glasses (abstract). In *Lunar Science V*, p. 749–751. The Lunar Science Institute, Houston.

Taylor G. J., Keil K., and Warner D. (1977) Very low-Ti mare basalts. *Geophys. Res. Lett.* **4**, 207–210.

Taylor S. R. and Jakeš P. (1974) The geochemical evolution of the moon. *Proc. Lunar Sci. Conf. 5th*, p. 1287–1305.

Vaniman D. T. and Papike J. J. (1977a) The Apollo 17 drill core: Chemistry and stratigraphy of monomineralic fragments and the discovery of a new very low Ti (VLT) mare basalt (abstract). In *Lunar Science VIII*, p. 952–954. The Lunar Science Institute, Houston.

Vaniman D. T. and Papike J. J. (1977b) The Apollo 17 drill core: Glass chemistry and intracore distributions (sections 70007, 70008, 70009) (abstract). In *Lunar Science VIII*, p. 955–957. The Lunar Science Institute, Houston.

Vaniman D. T. and Papike J. J. (1977c) Ferrobasalts from Luna 24. Unpublished manuscript.

Walker D., Longhi J., and Hays J. F. (1976) Heterogeneity in titaniferous lunar basalts. *Earth Planet. Sci. Lett.* **30**, 27–36.

Wänke H., Palme H., Kruse H., Baddenhausen H., Cendales M., Dreibus H., Hofmeister H., Jagoutz E., Palme C., Spettel B., and Thacker R. (1976) Chemistry of lunar highland rocks: A refined evaluation of the composition of the primary matter. *Proc. Lunar Sci. Conf. 7th*, p. 3479–3499.

Weill D. F., Grieve R. A., McCallum I. S., and Bottinga Y. (1971) Mineralogy-petrology of lunar samples: Microprobe studies of samples 12021 and 12022; Viscosity of melts of selected lunar compositions. *Proc. Lunar Sci. Conf. 2nd*, p. 413–430.

Wenk H. R. and Wilde W. R. (1973) Chemical anomalies of lunar plagioclase, described by substitution vectors and their relation to optical and structural properties. *Contrib. Mineral. Petrol.* **41**, 89–104.

Wood J. A. and Ryder G. (1977) The Apollo 15 green clods and the green glass enigma (abstract). In *Lunar Science VIII*, p. 1026–1028. The Lunar Science Institute, Houston.

Proc. Lunar Sci. Conf. 8th (1977), p. 1473–1487.
Printed in the United States of America

Lead isotope studies of mare basalt 70017

J. M. MATTINSON, G. R. TILTON, W. TODT[1], and J. H. CHEN

Department of Geological Sciences, University of California, Santa Barbara, California 93106

Abstract—Uranium, thorium, and isotopic lead data have been obtained on bulk samples and pyroxene, ilmenite, and plagioclase separated from basalt 70017. In a concordia diagram the whole rock, pyroxene, and ilmenite points plot along a chord intersecting the curve at 3.7 and 4.33 AE. The plagioclase data, obtained from 1 ng quantities of lead, do not plot on the chord, but are of uncertain accuracy because of the small amounts of lead involved. 70017 appears to plot along a distinctly different chord in a concordia diagram from 75055 (Tera and Wasserburg, 1976) and 75035 (Nunes *et al.*, 1974), two other Apollo 17 mare basalts for which U–Pb data on mineral separates are known (although maximum error limits in the 70017 analyses just allow for the possibility that all three basalts have a common chord). The lead data resemble Sm–Nd results (Lugmair, 1975) which show that in an isochron diagram the 70017 whole rock does not plot on the isochron that fits the whole rock data for basalts 75055, 75075, and 15555. Models to account for the Sm–Nd and U–Pb data are briefly discussed.

A 3.7 AE crystallization age for 70017 yields an initial radiogenic $^{207}Pb/^{206}Pb$ ratio of 1.32, when meteorite primordial lead is used to correct for non-radiogenic lead. This is the same kind of parentless lead suggested by previous studies of soils and soil breccias from Stations 0, 1, 6, 8, and 9 at Taurus-Littrow. Basalts like 70017 are thus potential sources of such lead. Since the initial radiogenic lead ratios in 75055 and 75035 are both 1.41, however, it is not yet apparent that lead from Taurus-Littrow mare basalts can entirely account for the soil and soil breccia parentless leads.

Uranium and thorium concentrations in ilmenite are approximately twice those in the pyroxene. The Th/U ratio in ilmenite is 2.2, a ratio much lower than those in either plagioclase or pyroxene. Ilmenite could therefore play an important role in fractionating U and Th in the source regions of basalts like 70017 that have low Th/U ratios.

INTRODUCTION

IN PREVIOUS STUDIES Church and Tilton (1974, 1975) showed that soils and soil breccias from Stations 0, 1, 6, 8, and 9 at Taurus-Littrow appear to contain large fractions of parentless radiogenic lead which, when corrected for non-radiogenic lead with meteoritic primordial lead, yield a ratio of 1.32 ± 0.01 for $^{207}Pb/^{206}Pb$. The lead from these stations has a distinctly different isotopic composition from that in the orange and black soils at Station 4 (Nunes *et al.*, 1974; Silver, 1974; Tera and Wasserburg, 1976). These relationships are depicted in Fig. 1.

The 1.32 initial ratio is problematic because it is not clear whether it is a result of the complicated mixing processes that must occur in the production of lunar regolith rendering it therefore of no fundamental significance, or whether the soils are dominated by some source such as crystalline rocks having such an

[1]Present address: Laboratorium für Geochronologie, Universität Heidelberg, 69 Heidelberg, Germany.

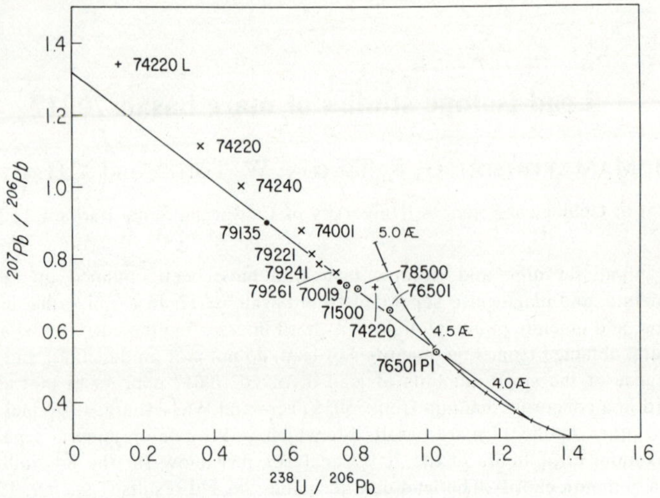

Fig. 1. Modified concordia diagram (Tera and Wasserburg, 1972) showing U–Pb isotope relationships in Apollo 17 soils and soil breccias. Open circles: soils from Church and Tilton (1974, 1975); filled circles: soil breccias from Church and Tilton (1975); × = soils from Nunes *et al.* (1974); + = soils from Tera and Wasserburg (1976). 74220 L is the acid-leached fraction from orange soil. 76501 Pl = separated plagioclase fraction. With the exception of station four soils the data define an isochron intersecting the concordia curve at 3.7 and 4.3 AE corresponding to an initial radiogenic $^{207}Pb/^{206}Pb$ ratio of 1.32 ± 0.01.

initial ratio. From the lead isotope systematics, Church and Tilton (1975) showed that the Apollo 17 mare basalts could be the source of the lead. Rare-earth and other trace-element studies on soil 71500,7 also support that hypothesis (Church *et al.*, 1976). These observations are consistent with a mare basalt source for the initial lead, but convincing proof requires confirmation through information from internal lead isochrons from possible basalt parents.

So far very few isotopic lead data are available from mineral separates of Apollo 17 mare basalts. Nunes *et al.* (1974) reported data for total rock, pyroxene, and plagioclase of 75035. From these data they derived an age of 3.54 ± 0.40 AE for the basalt (new decay constants) and 1.31 for the initial radiogenic $^{207}Pb/^{206}Pb$ ratio. These relationships are shown in Fig. 2, chord A. The large uncertainty in age is due to the large error limits in the plagioclase data. Newer Rb–Sr age-determination work by Murthy and Coscio (1976) yields an age of 3.81 ± 0.14 AE (2σ)—or 3.73 AE if $1.42 \times 10^{-11}\ yr^{-1}$ is used for the ^{87}Rb decay constant. Turner and Cadogan (1975) found an $^{40}Ar/^{39}Ar$ plateau age of 3.77 ± 0.04 AE. These data indicate that the mean U–Pb age derived by Nunes *et al.* (1974) may be too low.

Tera and Wasserburg (1974, 1976) reported isotopic lead data for whole rock and plagioclase from 75055, obtaining results shown in Fig. 2. The data plot along chord B, which corresponds to an age of 3.80 AE for the basalt, and an

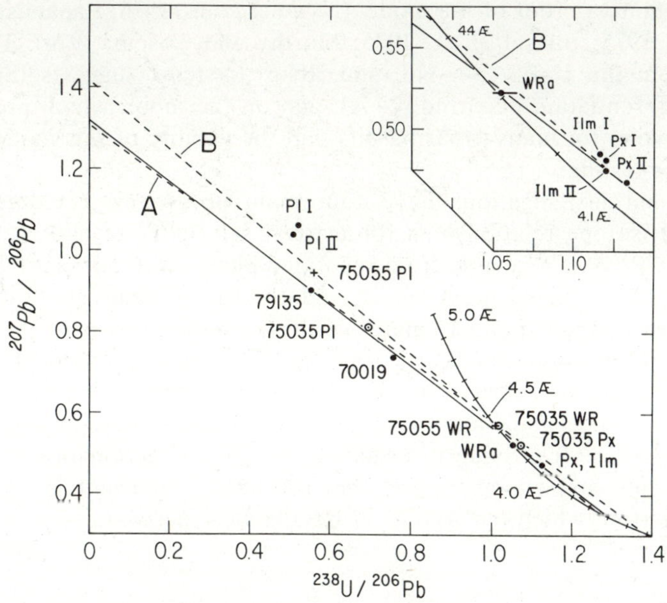

Fig. 2. Modified concordia diagram (Tera and Wasserburg, 1972) with U–Pb data for 70017,13 total rock (WRa), plagioclase (Pl), pyroxene (Px), and ilmenite (Ilm). Also shown are 75035 plagioclase, pyroxene and whole rock from Nunes *et al.* (1974) and 75055 whole rock and plagioclase from Tera and Wasserburg (1974, 1976). 70019 and 79135 are soil breccias from Church and Tilton (1975). Inset shows an enlarged portion of the upper concordia intercept area. Bar on whole rock indicates range of blank corrections. Chord A is the best fit to the 75035 data as given by Nunes *et al.* (1974); chord B is the best fit to the 75055 data as given by Tera and Wasserburg (1974, 1976).

initial radiogenic $^{207}Pb/^{206}Pb$ ratio of 1.41. The 3.80 AE crystallization age agrees closely with other age-determination work (Tera *et al.*, 1974). The initial ratio corresponds to evolution of radiogenic lead in a reservoir from 4.42 to 3.80 AE ago, starting from meteoritic primordial lead (Tera *et al.*, 1974). The 4.4 AE age for the reservoir is a value that occurs in other lunar U–Pb studies—e.g., basalts 15555 (Tera and Wasserburg, 1974) and 12064 (Papanastassiou *et al.*, 1977). Whole rock Sm–Nd data from 15555, 75055, and 75075 (Lugmair *et al.*, 1976; Lugmair, 1975) also plot along an isochron corresponding to an age of 4.40 ± 0.06 AE in a $^{143}Nd/^{144}Nd$–$^{147}Sm/^{144}Nd$ diagram. In this case the $^{143}Nd/^{144}Nd$ ratio at 4.4 AE is slightly higher than the meteoritic primordial value (as given by the Juvinas achondrite) 4.56 AE ago. On the other hand the lead data are calculated by assuming that lead in the basalt reservoir had an isotopic composition 4.4 AE ago exactly equal to that of primordial meteoritic lead. This assumption is probably not rigorously correct, although the consistency of the U–Pb and Sm–Nd data suggests that the resulting errors are not large. The most extensive data are available for the Rb–Sr system. Although the Rb–Sr results are consistent with the establishment of basalt reservoirs 4.4 AE ago, they do not provide a definitive proof of the model (Nyquist *et al.*, 1976; Papanastassiou and Wasserburg, 1975; Bansal *et al.*, 1975; Murthy and Coscio, 1976). The present evidence from the U–Pb, Sm–Nd, and Rb–Sr systems suggests that a major chemical differentiation occurred 4.4 AE ago on the moon which produced the source reservoirs for many lunar basalts, but the validity of this model certainly needs further testing.

The present investigations have four main objectives: (1) determine the internal lead isotope relationships for mare basalt 70017 to add to the limited amount of U–Pb data available from separated phases of lunar crystalline rocks; 70017 is a coarse-grained mare basalt, suitable for mineral separation; (2) test whether at least some Apollo 17 mare basalts might be sources of the parentless radiogenic lead with an apparent $^{207}Pb/^{206}Pb = 1.32$ that is found in many soils and soil breccias at Taurus-Littrow; (3) obtain further information on the possible significance and validity of the 4.4 AE differentiation age for basalt source reservoirs; (4) provide information on Th–U fractionation for minerals such as ilmenite and clinopyroxene that are likely important in the cumulate source rocks from which the Apollo 17 basalts were derived.

EXPERIMENTAL PROCEDURES

Sample handling and mineral separations

All work was done under a laminar flow of ultraclean air. The basalt sample, weighing about 5 g, was fragmented with a carbide-tipped chisel in an enclosed stainless steel container. The chisel operates through a viton "o"-ring in a plastic cover. The fragments were crushed in a small carbide diamond mortar and pestle that was specially made with close tolerances (*ca.* 0.001 inch) to minimize sample losses during crushing. About 1 g of the crushed sample was reserved for whole rock analyses and the balance used for mineral separations.

Mineral separates were prepared by sieving with nylon screens in stainless steel holders, followed by magnetic separation in an acetone medium in a Franz separator equipped with a closed quartz and teflon ramp. The latter process facilitates separation of fine-grained material and minimizes handling losses. Ilmenite and plagioclase, concentrated in the most magnetic and least magnetic fractions, respectively, were cleaned of impurities using a teflon vacuum device. Clinopyroxene grains were handpicked from the moderately magnetic fractions. Clinopyroxene grains showed well-developed zoning with pale yellowish-tan cores grading to dark brown rims. Some of the pyroxene grains may have cores of olivine (Longhi *et al.*, 1975). Three pyroxene fractions were prepared by handpicking light, medium, and dark colored pyroxene grains. Sample descriptions are in Appendix 1.

Chemical processing

Chemical procedures were designed to handle 10–50 mg samples. Sample powders were split into aliquots for concentration (ID) and isotopic composition (IC) determinations, and were acid washed in cold 1 N HCl for 10 min prior to digestion. This procedure has the disadvantages of possibly producing sample heterogeneity and greater blank variability between ID and IC runs compared to the solution aliquoting technique. Nevertheless, we feel that the major advantage of the method—excellent equilibration of sample and tracer—outweighs the disadvantages.

Pb, U, and Th were separated by the usual anion exchange methods (for example, see Church and Tilton, 1974) except that small samples (10–20 mg) were processed using only a 100 μl teflon column for Pb separation. During the early stages of the work, samples were digested in covered teflon beakers on a hot plate. Blanks for Pb were typically 0.5 ng. Subsequently, small teflon "bombs" were used at 150°C. Use of the bombs and other improvements in technique reduced the Pb blank to <0.2–0.3 ng. U blanks during the early work were *ca.* 0.02 ng, and were subsequently reduced to <0.01 ng. Blanks for Th were usually similar to the U blanks.

Even though the whole rock sample and mineral separates are very low in U and Th, the blank corrections for these elements are minor. The lead blank represents a substantial proportion of the total lead in many of the samples, however. Small variations in the size of the blank produce large variations in the corrected $^{206}Pb/^{204}Pb$ ratios of samples that contain small amounts of lead, highly radiogenic lead, or both. That our blank is somewhat variable and that we have not always successfully corrected for all the variations may be seen in the calculated blanks and corrected $^{206}Pb/^{204}Pb$ data for separate IC and ID digestions of the same sample (Table 2). This blank variability has a rather small effect on the corrected $^{207}Pb/^{206}Pb$ ratios, the concentrations of ^{206}Pb and ^{207}Pb, and the model ages calculated from them, however. We are continuing our efforts to reduce the levels of our blanks and to improve their reproductibility.

Mass spectrometry

The isotopic measurements were made on a 35 cm, 90° sector AVCO mass spectrometer with an electron multiplier and a remote-ranging Cary 401 vibrating reed electrometer. An on-line mini-computer controls magnetic peak switching and data acquisition and reduction. Pb was analyzed using the usual silica gel-H_3PO_4 technique. Our samples typically contained 0.6–5 ng of total Pb per analysis and were all run using the electron multiplier. Results were corrected for mass fractionation and discrimination via replicate analyses of National Bureau of Standards SRM-981. Observed $^{208}Pb/^{206}Pb$ and $^{207}Pb/^{206}Pb$ ratios are believed to be accurate to better than a few tenths percent and the $^{204}Pb/^{206}Pb$ ratios are believed to be accurate to better than 1% in most cases. The Pb concentrations before blank corrections are known to better than 1%.

U and Th were analyzed using the carbon emitter technique (Arden and Gale, 1974) and the electron multiplier. U typically ranged from 0.25–5 ng per analysis, and Th typically ranged from 0.75–15 ng per analysis. Replicate determinations on National Bureau of Standards samples SRM U-050 and SRM U-500 indicated that a simple square root of the mass correction sufficed to remove

Table 1. Element concentrations 70017,13.

	Weight (mg)	Concentration (ppm)[1]			$^{238}U/^{204}Pb$ (μ)	$^{232}Th/^{238}U$ (k)
		Pb	U	Th		
Whole rock A	60.7	0.1514	0.0730	0.2204	234	3.12
Acid wash[2]	(4060)	0.01113	0.00279	0.02906	227	10.75
WR plus AW[3]	—		0.0758	0.2495		3.40
Pyroxene I	26.5	0.0391	0.0732	0.2200	379	3.10
Pyroxene II	18.5	0.1553	0.0855	0.2584	399	3.12
Pyroxene III	10.2	0.0599	0.0275	0.0857	171	3.22
Ilmenite I	16.8	0.2189	0.1252	0.2601	283	2.09
Ilmenite II	20.8	0.3036	0.1749	0.4008	357	2.37
Plagioclase I	47.5	0.0284	0.00516	0.01590	69.1	3.10
Plagioclase II	48.2	0.0282	0.00467	0.01300	42.6	2.87

[1] Errors in concentrations are discussed in the text.

[2] Leach of large whole rock sample: 15 min in cold 2 N HCl. Concentrations in μg of element removed per gram of sample.

[3] U and Th values obtained by combining whole rock A and acid wash.

Table 2. Isotopic composition of lead, 70017,13.

Sample		Weight (mg)	Observed ratios $^{208}Pb/^{206}Pb$	Observed ratios $^{207}Pb/^{206}Pb$	Observed ratios $^{204}Pb/^{206}Pb$	Corrected ratios $^{208}Pb/^{206}Pb$	Corrected ratios $^{207}Pb/^{206}Pb$	Corrected ratios $^{204}Pb/^{206}Pb$	Blank, ng[1] calculated	Blank, ng measured
Whole rock A	IC	60.7	0.9499	0.5587	0.006530	0.8907	0.5443	0.004018	0.8	0.5
	ID	54.6	7.369	0.5502	0.005710	—	0.5467	0.004346	0.4	0.5
Whole rock B[2]	IC	29.3	0.8546	0.5372	0.003310	0.8281	0.5291	0.002206	0.17	0.17
Acid wash	IC	(1,841)	1.959	0.6483	0.005354	1.9561	0.6412	0.003434	0.9	1.0(est.)
	ID	(2,219)	5.917	0.6441	0.005207	—	0.6405	0.003456	1.0	1.0(est.)
Pyroxene I	IC	29.2	0.9692	0.5175	0.005933	0.9048	0.4990	0.003111	0.4	0.7
	ID	26.5	6.696	0.4994	0.005206	—	0.4860	0.002874	0.3	0.7
Pyroxene II	IC	19.2	0.8969	0.5030	0.00547	0.8288	0.4838	0.002642	0.3	0.29
	ID	18.5	8.082	0.5097	0.00564	—	0.4940	0.002768	0.3	0.29
Pyroxene III	IC	11.5	1.2052	0.5980	0.01671	0.9729	0.5348	0.006637	0.3	0.34
	ID	10.2	26.437	0.5976	0.01880	—	0.5275	0.005799	0.4	0.34
Ilmenite I	IC	17.8	0.7717	0.5287	0.00710	0.6808	0.5075	0.00382	0.5	0.34
	ID	16.8	6.070	0.5076	0.00522	—	0.4959	0.00308	0.3	0.34
Ilmenite II	IC	22.8	0.7150	0.4994	0.00383	0.6936	0.4941	0.00304	0.2	0.20
	ID	20.8	3.865	0.4979	0.00386	—	0.4939	0.00305	0.19	0.20
Plagioclase I	IC	43.5	1.147	1.0368	0.01531	0.9779	1.0769	0.00819	0.3	0.30
	ID	47.5	18.70	0.953	0.0139	—	1.0050	0.00764	0.3	0.30
Plagioclase II	IC	48.3	1.346	0.9956	0.02324	1.1010	1.0557	0.01279	0.6	0.34
	ID	48.2	19.26	0.9736	0.01636	—	1.0315	0.01054	0.3	0.34

[1]Isotopic composition of blank is: ^{204}Pb: ^{206}Pb: ^{207}Pb:^{208}Pb: 1.00 : 18.90 : 15.60 : 38.60. Calculated blanks are derived from reiterative calculations that minimize the differences in isotopic ratios of IC and ID determinations.

[2] ID determination failed.

Table 3. 70017,13 Model ages in 10^9 yr.[1]

Sample		$\frac{^{206}Pb}{^{238}U}$	$\frac{^{207}Pb}{^{235}U}$	$\frac{^{207}Pb}{^{206}Pb}$	$\frac{^{208}Pb}{^{232}Th}$
Whole rock A	IC	4.297	4.303	4.306	4.408
	ID	—	—	4.308	—
Whole rock B	IC	—	—	4.290	—
	ID	—	—	—	—
Acid wash	IC	5.198	4.754	4.569	4.034
	ID	—	—	4.567	—
Pyroxene I	IC	4.112	4.160	4.184	4.361
	IC	—	—	4.147	—
Pyroxene II	IC	4.080	4.123	4.193	3.980
	ID	—	—	4.223	—
Pyroxene III	IC	4.284	4.249	4.233	4.396
	ID	—	—	4.225	—
Ilmenite I	IC	4.110	4.169	4.198	4.425
	ID	—	—	4.175	—
Ilmenite II	IC	4.112	4.151	4.170	4.255
	ID	—	—	4.169	—
Plagioclase I	IC	6.917	5.743	5.339	7.948
	ID	—	—	5.234	—
Plagioclase II	IC	7.000	5.739	5.304	9.006
	ID	—	—	5.270	—

[1] Decay constants $^{238}U = 0.155125 \times 10^{-9}\ yr^{-1}$,
$^{235}U = 0.98485 \times 10^{-9}\ yr^{-1}$,
$^{232}Th = 0.049475 \times 10^{-9}\ yr^{-1}$.

mass discrimination introduced by the electron multiplier. Such corrections were applied to both the U and Th ratios. The measured concentrations of U and Th are thought to be accurate to better than 1% in most cases. The results of our analyses are given in Tables 1–3 and are shown graphically in Figs. 2–4.

Discussion

Pb, U, and Th concentrations

Pb, U, and Th concentrations of 70017 whole rock and its constituent minerals are among the lowest reported for lunar basalts. The U content of the total rock (0.0730 ppm, Table 1) falls at the bottom of the range indicated for a variety of crystalline rocks from the Apollo 11–17 sites (Fig. 1 of Laul and Fruchter, 1976). The lead concentration is likewise about half that found by Nunes *et al.* (1974) in seven other Apollo 17 mare basalts.

Th/U data

Most lunar basalts have a rather limited range of Th/U ratios (3.8 ± .2, see Silver, 1976; Laul and Fruchter, 1976). The Apollo 17 basalts are exceptions to this general pattern, however, with Th/U ratios ranging from 3.5 to 2.1 (Laul and Fruchter, 1976). Basalt 70017 falls within the range of the other Apollo 17 basalts with a total rock Th/U ratio of 3.12 or 3.4 (calculated by recombining the whole rock and acid wash U and Th values, see Table 1). In terms of the major minerals of 70017 (about 50% clinopyroxene, 25% ilmenite, and 25% plagioclase) pyroxene and ilmenite dominate the Th, U budget (see concentration data in Table 1). Plagioclase contributes only a minor porportion of the Th and U in the total rock. Pyroxene has a Th/U ratio slightly lower than that of the total rock (recombined with the acid wash) whereas ilmenite has a significantly lower one (Table 1). The only other significant contributor to the Th, U budget of 70017 is a labile component that is leached from the rock with cold HCl ("acid wash", Table 1). This labile component is highly enriched in Th (Th/U—10.75) and evidently is associated with a significant amount of initial radiogenic ^{206}Pb and ^{207}Pb as indicated by the old Pb–U model ages (Table 3) and the high ^{207}Pb/^{206}Pb ratio (Table 2). The labile component might be in the form of loosely bound intergranular material. The late-stage fluids of the basalt would concentrate U, Th, and Pb as the major minerals crystallized. Th, especially, would be enriched as ilmenite crystallized. Alternatively, the labile component might consist of a Th-rich minor phase such as whitlockite plus an initial lead-rich phase such as troilite. Acid washes of mineral separates also show higher than average Th/U ratios (Appendix 2) and high radiogenic ^{207}Pb/^{206}Pb ratios (not detectable in plagioclase—the acid wash of this mineral is dominated by modern terrestrial lead). Chemical studies of the acid leach should help resolve the origin of the labile component.

The Th/U ratios of pyroxene and ilmenite are significant insofar as they might be representative, at least qualitatively, of the Th/U ratios of the same minerals in the source regions of the basalts. Laul and Fruchter (1976) have attempted to explain the low and variable Th/U ratios of the Apollo 17 basalts by using a two-stage model of basalt evolution in which clinopyroxene controls U and Th distributions. These authors assume that clinopyroxene discriminates against Th relative to U (distribution coefficients used in their models: $D_{Th} =$ 0.050–0.030, $D_U = 0.10$–0.050), a relationship that is qualitatively supported by Th/U measurements for clinopyroxene in rock 12064 by Silver (1976). Our Th/U ratios for clinopyroxene and the total rock plus acid wash (Table 1) are compatible with these Kd values, especially considering that early crystallization of ilmenite at low pressure in 70017 (Longhi *et al.*, 1974) would have increased the Th/U ratio in the melt by the time pyroxene began to crystallize. Recent experimental work suggests that clinopyroxene may enrich Th relative to U at 20 kbar, however (Benjamin *et al.*, 1977), just the opposite of Laul and Fruchter's values and those observed by Silver (1976) and us (this report).

We should note that this high pressure experimental work may be more relevant to the problem of fractionation in the source regions of mare basalts than are Th/U ratios measured on clinopyroxenes crystallized at or near the lunar surface. If the distribution coefficients of Benjamin *et al.* (1977) are applicable, then the clinopyroxene model of Laul and Fruchter (1976) is invalid. We suggest that ilmenite might be an alternative to clinopyroxene as a source cumulate phase that could, through a two-stage evolution (Laul and Fruchter, 1976) produce low Th/U ratios in derivative basalts. Ilmenite has a Th/U ratio much lower than does the total rock or clinopyroxene, and has high Th and U concentrations as well. Our ilmenite data are, of course, for low pressure crystallization, and are subject to an additional uncertainty—questionable sample purity. Experimental work of the type reported by Benjamin *et al.* (1977) that included ilmenite would be most useful in resolving this problem.

The 70017 plagioclase also has slightly low Th/U ratios (3.10 and 2.87, Table 1). Silver (1976) has commented on generally low Th/U ratios in many lunar (and terrestrial) plagioclases and the role plagioclase and other minerals might play in determining Th/U ratios in basaltic rocks. However, the very low concentrations of Th and U in the lunar plagioclases suggest that their effect on the Th/U ratios of the source rocks or the basalts is negligible, as assumed by Laul and Fruchter (1976).

Isotopic data

Previous age-determination work on 70017: Nyquist *et al.* (1975) reported a Rb–Sr age of 3.67 ± 0.18 AE (2σ) for 70017 (λ for $^{87}Rb = 1.39 \times 10^{-11}\ yr^{-1}$), or 3.59 AE if $1.42 \times 10^{-11}\ yr^{-1}$, which we prefer, is used for the decay constant. Phinney *et al.* (1975) and Schaeffer *et al.* (1977) have published ^{40}Ar–^{39}Ar age data. Phinney *et al.* found no pronounced plateau in their release data, but considered the highest apparent age of 3.80 AE at 900°C as a possible maximum age, and the total rock age, 3.63 AE, as a minimum age. Schaeffer *et al.* (1977) listed ^{39}Ar–^{40}Ar ages on separated minerals from 70017. The maximum values obtained were 3.67 ± 0.12 AE from a plagioclase crystal and 3.62 ± 0.18 AE for cristobalite. We conclude that, in spite of considerable investigation, the crystallization age of 70017 is not precisely known. The data seem to indicate an age of around 3.70 AE, although a value as high as 3.80 AE cannot be excluded. In the latter case 70017 would be similar in age to several other Apollo 17 mare basalts whose ages are well determined (e.g., Nyquist *et al.*, 1975, 1976; Murthy and Coscio, 1967; Tera *et al.*, 1974).

Isotopic lead relationships: The U–Pb data are plotted in Fig. 2 in the modified concordia diagram (Tera and Wasserburg, 1972) that is appropriate for samples containing large fractions of inherited radiogenic lead (parentless lead). The lead ratios are corrected only for primordial lead from meteorites (Tatsumoto *et al.*, 1973; Tilton, 1973) since we have no exact knowledge of the isotopic composition of initial lead for the basalt.

From an analytical standpoint we place the greatest reliance on the ilmenite and pyroxene data. Both of these minerals dissolved readily in the chemical processing, and were analyzed after our chemical blanks had been reduced over the earlier work. Plagioclase dissolved with some difficulty, and the analyzed samples moreover contained only 1 ng of Pb and a few tenths nanograms of U and Th. The blank corrections are therefore more critical for the plagioclase analyses, and we also have no experience to test how accurately we can perform isotope dilution analyses on samples at such low levels for U, Th, and Pb. We followed the usual total spiking procedures, but feel further verification of the validity of the methods is needed. The total rock analysis is satisfactory from the standpoint of chemical processing, but was done in the earlier work where blanks were higher.

In Fig. 2 the pyroxene, ilmenite, and whole rock data yield a lower intersection on the concordia curve at *ca.* 3.7 AE, as shown by the solid chord in Fig. 2. Within error limits the data also fit a lower intersection at 3.8 AE, i.e., chord B in Fig. 2. The uncertainty in crystallization age from these U–Pb data is comparable to those from the Rb–Sr and K–Ar data.

We initially expected, by analogy with Rb–Sr data, to find a more radiogenic lead in ilmenite that would help to determine the lower intersection more accurately. Unfortunately the ilmenite and pyroxene data are almost indistinguishable. Since all of these analyses were made on samples of similar size and with similar blanks, it seems unlikely that intrinsic differences have been disguised by the analytical blanks. The small U enrichments found in ilmenite compared to the pyroxene values, have been compensated by comparable increases in lead concentrations. The U enrichment in the ilmenite agrees closely with the factor of two enrichment of K reported for 70017 ilmenite compared to 70017 pyroxene by Schaeffer *et al.* (1977). Nakamura *et al.* (1977), who also studied ilmenite from 15065, did find a substantially more radiogenic lead compared to other minerals, but the ilmenite does not plot along the chord determined by the other minerals in a concordia diagram.

If the crystallization age of 3.7 AE is correct, the 70017 data are compatible with an initial radiogenic $^{207}Pb/^{206}Pb$ ratio of 1.32, as shown in Fig. 2. In this case 70017 whole rock does not plot on the chord that contains 75055 and 75035 whole rocks. The upper intersection on concordia is 4.33 AE compared to 4.42 AE for 75055 and 75035. Lugmair (1975) reports similar differences from Sm–Nd data. In a $^{143}Nd/^{144}Nd$–$^{147}Sm/^{144}Nd$ diagram, Lugmair found that 70017 plots distinctly below the isochron that includes basalts 75075, 75055, and 15555. Several mechanisms have been cited to explain the observations: (1) the 70017 parent separated from a "chondritic" reservoir 4.28 ± 0.05 AE rather than 4.42 AE ago (Lugmair, 1975); (2) the "reservoir age" for 70017 is 4.42 AE, but its Sm/Nd ratio was lowered by $\sim 2\%$ at the time of crystallization compared to the other basalts (Lugmair, 1975); (3) the "reservoir age" for 70017 is 4.42 AE, but the Sm/Nd ratio in the reservoir was lower over the interval from 4.42 to 3.7 AE for 70017 than for the other basalts (Nyquist *et al.*, 1976). The 70017 U–Pb data, as shown in Fig. 2 are consistent with mechanism (1) and do not seem to plot along a 4.42–3.7 AE chord as mechanism (2) and (3) would require. More detailed discussion of these observations does not seem profitable until more precise U–Pb whole rock data have been obtained on 70017.

The 70017 data are consistent with an initial radiogenic $^{207}Pb/^{206}Pb$ ratio of 1.32, suggesting that basalts similar to 70017 could provide a source of the radiogenic lead observed in the soils and breccias shown in Fig. 1. However, even if the 70017 chord in Fig. 2 is correct, it is still not clear how much of a role 70017-type basalts play as a source of parentless lead at Taurus-Littrow because the Station 5 basalts, 75055 and 75035, clearly do not plot on the 70017 chord. We plan to measure internal U–Pb isochrons on basalts 75075 and 71055 to obtain further information in this question.

The lead isotopic composition relationships are shown in Figs. 3 and 4. Since the leads are highly radiogenic, the Tera *et al.* (1974) lead correlation diagram is the most appropriate one to use. The diagrams illustrate the close similarity of lead from pyroxene and ilmenite discussed above. It is not possible to add or subtract blank lead from the analyses so as to generate grossly different isotopic compositions, assuming that the true values for both minerals plot along a common regression line through 3.7 AE (or any other reasonable crystallization age) on the ordinate.

The lead isotopic composition data do not determine a $^{207}Pb/^{206}Pb$ age with sufficient accuracy to be useful. The whole rock, ilmenite, and pyroxene points provide very little leverage on the slope of a regression line through the data. The plagioclase points could provide suitable leverage, but the blank corrections critically influence those results. A regression through all of the data in Figs. 3 and 4 yields an age of 3.4 AE.

Using a crystallization age of 3.7 AE and an initial radiogenic $^{207}Pb/^{206}Pb$ ratio of 1.32, the isotopic composition of the initial lead in 70017 at the time of crystallization is given by the intersection of the regression lines through the data points in Fig. 4 with the line connecting primordial Pb and 1.32 on the ordinate. The approximate ranges of values are $^{206}Pb/^{204}Pb = 53–143$; $^{207}Pb/^{204}Pb - 68–186$.

CONCLUSIONS

(1) U–Pb data for 70017 total rock and separated pyroxene and ilmenite plot along a chord in a concordia diagram that intersects the curve at 3.7 and 4.33 AE, and indicates an initial radiogenic $^{207}Pb/^{206}Pb$ ratio of 1.32 at the time of crystallization of the basalt.

(2) The U–Pb data from 70017 appear to plot on a different chord from 75055 and 75035. Sm–Nd data for 70017 also do not plot on the isochron that fits the total rock data for 75055, 75075, and 15555, thus supporting the U–Pb data. We should point out, however, that applying the maximum error limits to the total rock U–Pb data just allows for the possibility that all the rocks plot on a common chord corresponding to a crystallization age of 3.8 AE and an initial radiogenic $^{207}Pb/^{206}Pb$—1.41.

(3) In terms of two-stage evolution models, assuming a crystallization age of 3.7 AE, both the U–Pb and Sm–Nd results indicate that the source for 70017 separated from a "chondritic reservoir" later than the sources for

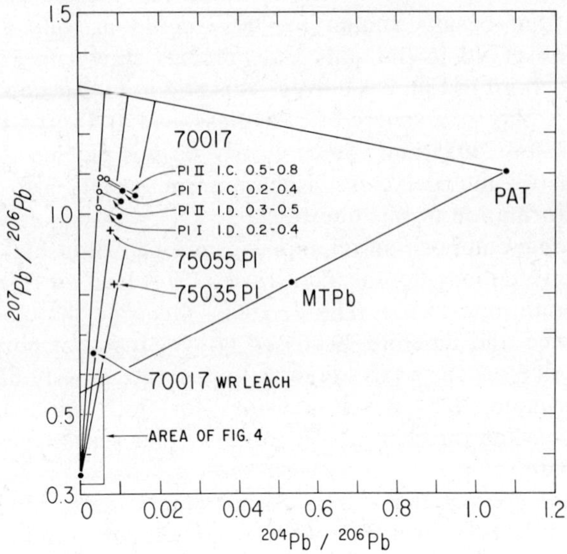

Fig. 3. $^{207}Pb/^{206}Pb$–$^{204}Pb/^{206}Pb$ diagram. PAT = meteoritic primordial lead, MTPb = modern terrestrial (lab contamination) lead. Plagioclase isotopic composition data for 70017 is shown for both IC runs and recalculated ID runs. Numbers give extremes of blank corrections (nanograms) indicated by solid circles (minimum blank) and open circles (maximum blank). Data for 75055 plagioclase (Tera and Wasserburg, 1974, 1976) and 75035 plagioclase (Nunes *et al.*, 1974) are included for reference. Middle line is a best fit to all the data; other two lines are drawn to include the extremes in data points. See text for discussion.

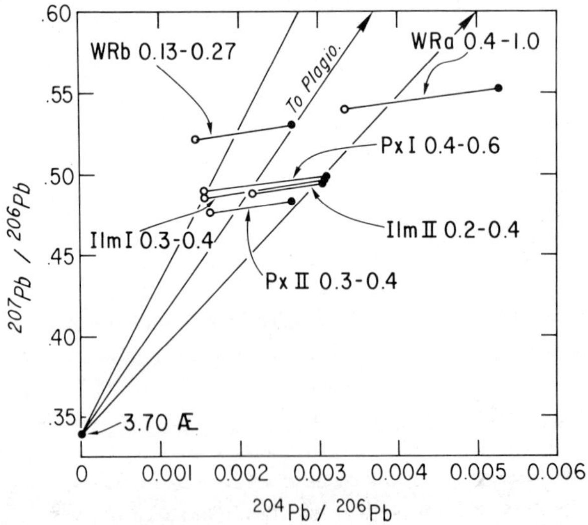

Fig. 4. Enlarged portion of Fig. 3 with data for 70017 minerals and total rock. Range of extreme blank corrections is shown as in Fig. 3. Abbreviations are the same as in Fig. 2.

the remaining basalts—4.33 vs 4.42 AE, approximately. Alternative models involving more than two stages can permit simultaneous separation of all basalt sources from a "chondritic reservoir".

(4) The Th/U ratio of 70017 is 3.12 (3.4 when recombined with the acid wash), well within the range of other Apollo 17 basalts, and considerably lower than the $3.8 \pm .2$ value that typifies most other lunar materials. The major minerals, especially ilmenite, tend to discriminate against Th relative to U. The low Th/U ratio (*ca.* 2.2) of ilmenite, and relatively high Th and U concentrations (*ca.* twice those of clinopyroxene) suggest that ilmenite might be an important factor in the fractionation of U and Th in the source regions of the Apollo 17 mare basalts.

Acknowledgments—This work was supported by NASA grant NGR-05-010-081. We thank M. Stein for valuable assistance with many aspects of the mass spectrometry. We are grateful to Evelyn Gordon, Diane Mondragon, David Doerner, and David Crouch for assistance in the preparation of the manuscript and figures.

REFERENCES

Arden J. W. and Gale N. H. (1974) Separation of trace amounts of uranium and thorium and their determination by mass spectrometric isotope dilution. *Anal. Chem.* **46**, 687–691.

Bansal B. M., Wiesmann H., and Nyquist L. E. (1975) Rb–Sr ages and initial $^{87}Sr/^{86}Sr$ ratios for Apollo 17 mare basalts. In *Papers Presented to the Conference on Origins of Mare Basalts and Their Implications for Lunar Evolution*, p. 1–5. The Lunar Science Institute, Houston.

Benjamin T. M., Burnett D. S., Ng A., and Seitz M. G. (1977) Laboratory studies of Pu–U–Th fractionation (abstract). In *Lunar Science VIII*, p. 94–96. The Lunar Science Institute, Houston.

Brown G. M., Peckett A., Emeleus C. H., Phillips R., and Pinsent R. H. (1975) Petrology and mineralogy of Apollo 17 mare basalts. *Proc. Lunar Sci. Conf. 6th*, p. 1–14.

Church S. E. and Tilton G. R. (1974) Lead isotope systematics of some Apollo 17 soils and some separated components from 76501. *Proc. Lunar Sci. Conf. 5th*, p. 1389–1400.

Church S. E. and Tilton G. R. (1975) Lead isotope systematics of soils and soil breccias from Taurus-Littrow (abstract). In *Lunar Science V*, p. 143–145. The Lunar Science Institute, Houston.

Church S. E., Tilton G. R., Wright J. E., and Lee-Hu C.-N. (1976) Volatile element depletion and $^{39}K/^{41}K$ fractionation in lunar soils. *Proc. Lunar Sci. Conf. 7th*, p. 423–439.

Laul J. C. and Fruchter J. S. (1976) Thorium and uranium variations in Apollo 17 basalts, and K–U systematics. *Proc. Lunar Sci. Conf. 7th*, p. 1545–1559.

Longhi J., Walker D., Grove T. L., Stolper E. M., and Hays J. F. (1974) The petrology of the Apollo 17 mare basalts. *Proc. Lunar Sci. Conf. 5th*, p. 447–469.

Lugmair G. W. (1975) Sm–Nd systematics of some Apollo 17 basalts. In *Papers Presented to the Conference on Origins of Mare Basalts and Their Implications for Lunar Evolution*, p. 107–110. The Lunar Science Institute, Houston.

Lugmair G. W., Scheinin N. B., and Marti K. (1975) Sm–Nd age and history of Apollo 17 basalt 75075: Evidence for early differentiation of lunar exterior. *Proc. Lunar Sci. Conf. 6th*, p. 1419–1429.

Murthy V. R. and Coscio M. R., Jr. (1976) Rb–Sr ages and isotopic systematics of some Serenitatis mare basalts. *Proc. Lunar Sci. Conf. 7th*, p. 1529–1544.

Nakamura N., Unruh D. H., Gensho R., and Tatsumoto M. (1977) Evolution history of lunar mare basalts: Apollo 15 samples revisited (abstract). In *Lunar Science VIII*, p. 712–714. The Lunar Science Institute, Houston.

Nunes P. D., Tatsumoto M., and Unruh D. M. (1974) U–Th–Pb systematics of some Apollo 17 lunar samples and implications for a lunar basin excavation chronology. *Proc. Lunar Sci. Conf. 5th*, p. 1487–1514.

Nyquist L. E., Bansal B. M., and Wiesmann H. (1975) Rb–Sr ages and initial $^{87}Sr/^{86}Sr$ for Apollo 17 basalts and KREEP basalt 15386. *Proc. Lunar Sci. Conf. 6th*, p. 1445–1465.

Nyquist L. E., Bansal B. M., and Wiesmann H. (1976) Sr isotopic constraints on the petrogenesis of Apollo 17 mare basalts (abstract). In *Lunar Science VII*, p. 636–638. The Lunar Science Institute, Houston.

Papanastassiou D. A., DePaolo D. J., Tera F., and Wasserburg G. J. (1977) An isotopic triptych on mare basalts: Rb–Sr, Sm–Nd, U–Pb (abstract). In *Lunar Science VIII*, p. 750–752. The Lunar Science Institute, Houston.

Papanastassiou D. A. and Wasserburg G. J. (1975) Rb–Sr study of a lunar dunite and evidence for early lunar differentiates. *Proc. Lunar Sci. Conf. 6th*, p. 1467–1489.

Phinney D., Kahl S. B., and Reynolds J. H. (1975) ^{40}Ar–^{39}Ar dating of Apollo 16 and 17 rocks. *Proc. Lunar Sci. Conf. 6th*, p. 1593–1608.

Schaeffer O. A., Müller H. W., and Grove T. L. (1977) Laser ^{39}Ar–^{40}Ar study of Apollo 17 basalts (abstract). In *Lunar Science VIII*, p. 837–839. The Lunar Science Institute, Houston.

Silver L. T. (1974) Patterns of U–Th–Pb distributions and isotope relations in Apollo 17 soils (abstract). In *Lunar Science V*, p. 706–708. The Lunar Science Institute, Houston.

Silver L. T. (1976) Implications of major mineral fractionation of thorium and uranium in some lunar materials (abstract). In *Lunar Science VII*, p. 809–811. The Lunar Science Institute, Houston.

Tatsumoto M., Knight R. J., and Allegre C. J. (1973) Time differences in the formation of meteorites as determined by $^{207}Pb/^{206}Pb$. *Science* **180**, 1279–1283.

Tera F., Papanastassiou D. A., and Wasserburg G. J. (1974) Isotopic evidence for a terminal lunar cataclysm. *Earth Planet. Sci. Lett.* **22**, 1–21.

Tera F. and Wasserburg G. J. (1972) U–Th–Pb systematics in three Apollo 14 basalts and the problem of initial Pb in lunar rocks. *Earth Planet. Sci. Lett.* **14**, 281–304.

Tera F. and Wasserburg G. J. (1974) U–Th–Pb systematics of lunar rocks and inferences about lunar evolution and the age of the Moon. *Proc. Lunar Sci. Conf. 5th*, p. 1571–1599.

Tera F. and Wasserburg G. J. (1976) Lunar ballgames and other sports (abstract). In *Lunar Science VII*, p. 858–860. The Lunar Science Institute, Houston.

Tilton G. R. (1973) Isotopic lead ages of chondritic meteorites. *Earth Planet. Sci. Lett.* **19**, 321–329.

Turner G. and Cadogan P. H. (1975) The history of lunar bombardment inferred from ^{40}Ar-^{39}Ar dating of highland rocks. *Proc. Lunar Sci. Conf. 6th*, p. 1509–1538.

APPENDIX 1: DESCRIPTION OF SAMPLES

Whole rock:	Coned from crushed 70017,13 total rock sample.
Acid wash:	Obtained by leaching a large, coarsely crushed whole rock sample in cold 2 N HCl for 15 min.
Pyroxene I:	Medium brown colored clinopyroxene.
Pyroxene II:	Dark brown colored clinopyroxene. Appears to represent the outer rims of large clinopyroxene grains.
Pyroxene III:	Light yellowish-tan clinopyroxene. Appears to represent early-formed cores of large clinopyroxene grains (may include some olivine).
Ilmenite I:	100–200 mesh fraction. Ilmenite with some intergrown pyroxene.
Ilmenite II:	Plus 100 mesh fraction, some intergrown pyroxene.
Plagioclase I and II:	Least magnetic fraction; virtually 100% clear, colorless grains.

APPENDIX II: 70017 MINERAL ACID WASH DATA

Sample	Concentrations (ppm)[1]			$\frac{^{238}U}{^{204}Pb}$	$\frac{^{232}Th}{^{238}U}$	Isotopic comp.[2]	
	Pb_{tot}	^{238}U	Th	(μ)	(k)	$\frac{^{207}Pb}{^{206}Pb}$	$\frac{^{204}Pb}{^{206}Pb}$
Pyroxene I	0.0181	0.00661	0.0360	64.4	5.6	0.697	0.02101
Ilmenite I	0.0656	0.01037	0.0994	19.4	9.9	0.731	0.02889
Plagioclase I	0.0610	0.00231	0.0842	34.8	3.78	0.803	0.0461

[1] Based on spiked run only. $^{208}Pb/^{206}Pb$ ratio for sample assumed. Concentrations in μg of element removed per gram of sample.

[2] Recalculated from spiked run. Assumed blank of 0.2 ng Pb subtracted.

Proc. Lunar Sci. Conf. 8th (1977), p. 1489–1499.
Printed in the United States of America

Laser ^{39}Ar–^{40}Ar study of Apollo 17 basalts

O. A. SCHAEFFER, H. W. MÜLLER, and T. L. GROVE

Department of Earth and Space Sciences, State University of New York at Stony Brook, Stony Brook, New York 11794

Abstract—The laser ^{39}Ar–^{40}Ar method has been used to study three Apollo 17 basalts: 70215, a basalt with a normal well-behaved ^{39}Ar–^{40}Ar release pattern; 70017, a basalt with a disturbed ^{39}Ar–^{40}Ar release pattern, showing a limited intermediate maximum age followed by a broad low age region; and 75035, a basalt with a pattern similar to 70017, followed by a high temperature maximum age. The laser study showed that all the mineral systems in 70215 had small and uniform temperature losses while for basalts 70017 and 75035 the laser ^{39}Ar–^{40}Ar ages were apparently mainly controlled by the minerals containing mesostasis inclusions, with the suggestion that the drop in ages observed by the conventional ^{39}Ar–^{40}Ar method was not only due to recoil of the ^{39}Ar during neutron irradiation but also due to gas loss from some minerals. From our measurements it appears that the plagioclases—at least for the intermediate temperature range between about 600°C and 1000°C—are the best minerals to use for a reliable age.

INTRODUCTION

IN A PREVIOUS STUDY (Plieninger and Schaeffer, 1976), it was shown that for the Apollo 15 basalt, 15607, certain mineral phases appeared to be retentive for argon while others were not. Basalt 15607 had previously been studied by the conventional ^{39}Ar–^{40}Ar method (Husain, 1974) and was found to have suffered extensive gas loss. In order to investigate whether different argon retentivity of mineral phases is a general characteristic of lunar basalts and to investigate whether the release of argon from selected mineral sites can lead to an understanding of peculiarities in the ^{39}Ar–^{40}Ar release patterns, the present study was undertaken.

The ages obtained by the laser release of the argon from neutron irradiated samples are total K–Ar ages. Compared to the ^{39}Ar–^{40}Ar stepwise heating experiments, these ages essentially integrate the complete ^{39}Ar–^{40}Ar temperature releases and are therefore directly equivalent to the total ^{39}Ar–^{40}Ar ages calculated from all temperature steps of the ^{39}Ar–^{40}Ar stepwise heating data.

By heating the samples at different temperatures after neutron irradiation and before the laser release, it is possible to obtain integrated ^{39}Ar–^{40}Ar release patterns in contrast to the normal differential patterns. This task can be executed repeatedly by increasing the preheating temperatures because of the tiny amount of material needed for a ^{39}Ar–^{40}Ar age determination with the laser technique. The gas released from the preheated sample then essentially integrates the temperature releases from the preheating temperature to the very end of the usual ^{39}Ar–^{40}Ar temperature releases.

The basic model assumption for this "preheating prior to lasering" technique is exactly the same as for the ^{39}Ar–^{40}Ar stepwise heating technique, i.e., to deplete unretentive phases to be able to measure separately the more chronologically meaningful retentive sites. This new technique is described in some detail by Müller *et al.* (1977).

For this study, we chose Apollo 17 basalts which had been previously studied by the conventional ^{39}Ar–^{40}Ar stepwise heating method. Basalt 70215, a porphyritic, spherulitic, high-titanium basalt (Kirsten and Horn, 1974) has a well-defined ^{39}Ar–^{40}Ar plateau with only minor argon loss and was chosen as an example of a well-defined ^{39}Ar–^{40}Ar plateau. Basalt 70017, a medium-grained, high-titanium basalt, has a ^{39}Ar–^{40}Ar release pattern which shows a distinct high temperature drop off (Phinney *et al.*, 1975). The high temperature drop off precluded the assignment of a reliable plateau age to this sample. One of our objectives is to investigate the processes responsible for the high temperature drop off, which are presently not fully understood. However, the ingenious experiment of Huneke and Smith (1976) provides an explanation for the existence of low ages in high temperature release patterns. Basalt 75035 is a medium-grained, high-titanium basalt which shows a complicated ^{39}Ar–^{40}Ar pattern at the high temperatures (Turner and Cadogan, 1975). There is a drop in age, not as extreme as for 70017, followed by an increase in age. This basalt was chosen to investigate the cause of the intermediate and high temperature ages and hopefully to decide whether the ~4.0 G.y. ages for the highest temperature fractions have chronological significance.

EXPERIMENTAL

The samples were polished and then irradiated in the core of the Brookhaven High Flux Beam Reactor receiving a total fast neutron dose of $\sim 10^{19}$. Gases were released by a laser beam, in a way similar to that described by Plieninger and Schaeffer (1976). The samples were then heated to various temperatures between 500°C and 1000°C, and selected minerals of the samples were again outgassed by a laser beam after each heating. As the surfaces studied were relatively small, it was often necessary to repolish the samples in order to expose fresh material for study.

The ages were calculated from the $^{40}Ar/^{39}Ar$ ratios (after corrections for ^{40}Ar solar wind and Ca-derived ^{39}Ar) using the equation:

$$t = 4.341 \log[1 + J(^{40}Ar/^{39}Ar)] \text{ G.y.}$$

The mean life of ^{40}K was taken as 1.885 G.y. A thorough and detailed discussion of the calculation procedure is presented elsewhere (Schaeffer and Schaeffer, 1977). The J value for the irradiation of basalt 70215 (Lab. No. 26 to 29 in Table 1) was 0.09558, and for all other samples it was 0.09435.

Unfortunately the masses 38 and 36 are usually completely masked by background. No correction for atmospheric ^{40}Ar could, therefore, be executed. However, the contribution from atmospheric ^{40}Ar is completely negligible for the laser technique as is evident from these and other (Plieninger and Schaeffer, 1976) measurements where reliable ^{39}Ar–^{40}Ar ages are obtained independent of the level of the ^{36}Ar background. Furthermore, the background (40/36) ratio is always <1, and the 40 level in the background was always much smaller than the intensity of mass 40 in the measurement.

One of the problems in the laser release of gases is to make certain that the gas released comes only from the minerals melted by the laser. It was observed that, for the case where a K-rich mesostasis is adjacent to a K-poor pyroxene, the laser site could be located on the pyroxene mineral within less than a pit diameter from the mesostasis without releasing any argon from the mesostasis. A previous experiment to investigate the release of argon from material adjacent to the laser pit (Plieninger and Schaeffer, 1976) also indicated clearly that only the material melted by the laser is outgassed. We feel, therefore, confident that the spatial resolution of the laser probe technique, i.e., the possibility to make an independent age determination of a mineral grain without interference from adjacent grains, is of the order of 10 μ.

Results and Discussion

70215,182

The high-titanium basalt 70215 consists of 7% olivine, 42% clinopyroxene, 29% plagioclase, 18% opaques (mostly ilmenite), and 4% silica phase (Longhi *et al.*, 1974; Dymek *et al.*, 1975). The spherulitic groundmass of olivine, ilmenite, plagioclase, and clinopyroxene is intergrown on a scale of a few micrometers which makes it impossible to get a separate analysis for the plagioclase with the laser probe. Analyses have been carried out on phenocrysts of olivine, pyroxene, and ilmenite-armalcolite and on the groundmass plagioclase-pyroxene aggregates.

We find for all constituents within the error bars the same total K–Ar age of $3.68 \pm .1$ G.y., see Table 1. This value is in fairly good agreement with the total K–Ar age from the stepwise heating experiments of Kirsten and Horn (1974) of $3.77 \pm .1$ G.y. For the gas released from mineral phases after heating the sample for 2 hr at 600°C, the ages are 3.82 ± 0.05 G.y. for all the mineral phases, in agreement with the plateau age of 3.84 ± 0.04 G.y. (Kirsten and Horn, 1974). Due to the fine-grained nature of basalt 70215, the ages given in Table 1 are composite ages of from 10 to 40 individual mineral grains. As a result, any variations in the ages of different grains of the same mineral will not be observed. The remarkable observation is that the different minerals analyzed behave in the same way as to loss of argon at the low temperatures. This observation is in contrast to that observed for basalt 15607 (Plieninger and Schaeffer, 1976) where the plagioclase showed little argon loss while the pyroxenes showed substantial gas loss.

From the comparison of the laser study to the normal ^{39}Ar–^{40}Ar study for basalt 70215, it is apparent that the sample behaves in a way which might be called classical for the ^{39}Ar–^{40}Ar method, i.e., the argon loss seems to be diffusive related. There appear to be sites in all the minerals which show the loss of gas. This behavior can tentatively be understood by the rapid cooling rates (25°C to 50°C/hour, Usselman *et al.*, 1975) during crystallization and by the fact that 70215 has experienced no significant reheating since crystallization. The rapid cooling event precluded efficient fractional crystallization (i.e., separation of a K-rich mesostasis) and distributed potassium evenly throughout the fine-grained microcristalline groundmass.

70017,25

Sample 70017 is a medium-grained, high-titanium basalt with textural relations suggesting slow cooling (Longhi *et al.*, 1974; Brown *et al.*, 1975). Individual

Table 1. Laser released Ar isotopes from various minerals of basalt 70215,182 in 10^{-12} cm^3.‡

Mineral	Lab. No.	Temp.†	^{40}Ar	^{39}Ar*	^{38}Ar*	^{37}Ar	^{36}Ar*	$^{40}Ar_K/^{39}Ar_K$	Age (G.y.)
Olivine	26	—	473.0 ± 16.5	7.70 ± .49	8.10 ± 3.62	648.3 ± 186.2	6.45 ± 6.45	61.30 ± 4.47	3.63 ± .10
Pyroxene	27	—	830.2 ± 15.4	12.74 ± .57	12.39 ± 4.60	1710.8 ± 222.1	10.62 ± 7.86	65.01 ± 3.15	3.72 ± .07
Plagioclase-px aggregates	28	—	1145.9 ± 28.9	18.46 ± .74	12.37 ± 5.58	1028.5 ± 281.6	10.59 ± 8.05	62.00 ± 2.94	3.65 ± .07
Armalcolite-ilmenite intergrowth	29	—	714.9 ± 25.6	11.02 ± .66	10.81 ± 6.26	593.2 ± 362.2	11.07 ± 7.86	64.64 ± 4.53	3.72 ± .10
Pyroxene	65	600°C	370.2 ± 4.3	5.19 ± .33	2.94 ± 2.94	655.3 ± 5.0	4.79 ± 4.79	71.07 ± 4.62	3.85 ± .09
Pyroxene	75	600°C	516.5 ± 5.1	7.56 ± .14	2.32 ± 1.47	854.9 ± 4.3	3.63 ± 2.87	68.16 ± 1.46	3.78 ± .04
Plagioclase-px aggregates	74	600°C	444.0 ± 5.5	6.47 ± .17	2.83 ± 1.47	624.9 ± 3.7	10.39 ± 2.40	67.86 ± 2.13	3.78 ± .05
Plagioclase-px aggregates	63	600°C	799.0 ± 8.3	11.35 ± .20	2.94 ± 2.94	1004.6 ± 8.2	2.87 ± 2.87	70.34 ± 1.45	3.83 ± .04
Armalcolite-ilmenite intergrowth	76	600°C	738.7 ± 8.1	10.56 ± .31	3.36 ± 1.57	866.2 ± 8.4	4.79 ± 4.79	69.81 ± 2.19	3.82 ± .05

*Corrected for neutron induced contributions.

†Preheated prior to lasering.

‡The exact amount of material melted by the laser pulses cannot be determined. The data can, however, within a factor of ~2 also be understood in units of 10^{-9} cm^3 STP/g.

minerals could easily be analyzed with the laser probe because of their large (1–2 mm) size. The proportions of minerals are 50% pyroxene, 26% plagioclase, 22% ilmenite, 1% olivine, 1% silica, and some native iron, troilite and mesostasis (Longhi *et al.*, 1974). The K–Ar ages (see Table 2) of unpreheated samples for the large plagioclase 70017,25,30 and for the cristobalite 70017,25,41 of 3.67 ± .12 G.y. and 3.62 ± .18 G.y., respectively, agree with the total K–Ar age of 3.63 ± .03 G.y. (Phinney *et al.*, 1975). These authors report for 70017 a ^{39}Ar–^{40}Ar release pattern that does not yield a well-defined high temperature plateau. Their release pattern is characterized by a maximum of 3.80 ± .03 G.y. at 900°C, followed by a high temperature decrease and a very slight high temperature increase at 1400°C. Basalt 70017 is sufficiently coarse grained so that each age represents the laser outgassing of from one to three individual grains. As a result, it is possible to determine whether the different plagioclase or pyroxene grains show similar ages. From Table 2, it can be seen that ages for the individual grains of the same mineral are different. The age differences for the pyroxenes are most likely due to the trapped melt inclusions (see Fig. 1a). There are no features observable under the microscope which would explain the different ages for the plagioclases. It is worthwhile to note that the cristobalite in 70017 is relatively clean and is associated with melt segregated mesostasis inclusions and, therefore, contains only minor amounts of potassium (Fig. 1a), (Dymek *et al.*, 1975). This is also verified by the measurement of 70017,25,41 which reveals a rather small amount of ^{39}Ar.

On preheating the sample to 600°C after neutron irradiation and before laser release of the argon, one sample of plagioclase and ilmenite shows an increase in age to an average value of 3.86 and 3.81 G.y., respectively. The average of these values is very close to the age for the other Apollo 17 basalts and to the maximum of 3.80 ± 0.03 G.y. at 900°C by the gas release curve of Phinney *et al.* (1975).

Although no preheated pyroxene analysis was done, we may speculate that the high temperature drop off in ages was caused by the recoil of ^{39}Ar during neutron irradiation from the potassium in abundant microscopic inclusions in the pyroxene. As the laser melts volumes of ~50 μm in diameter, well in excess of the recoil range of ^{39}Ar of ~1 μ, there would be no noticeable effect of recoil on the laser ^{39}Ar–^{40}Ar ages. On the other hand, if the inclusions degas their argon before the pyroxene itself, there will be a decrease in ages at the higher temperatures (see Huneke and Smith, 1976). There is also the possibility that some of the plagioclases lose appreciable Ar as one plagioclase sample gave an age of 3.61±.05 after preheating to 600°C.

75035,52

The modal analysis for the medium-grained, subophitic, high-titanium basalt 75035 is 44% clinopyroxene, 33% plagioclase, 15% ilmenite, 5% cristobalite, 2% pyroxferroite, and 1% native Fe, troilite and mesostasis (Longhi *et al.*, 1974). Our measurements in Table 3 show that in contrast to 70017 the potassium in

Table 2. Laser released Ar isotopes from various minerals of basalt 70017,25 in 10^{-12} cm^3.‡

Mineral	Lab. No.	Temp.†	^{40}Ar	^{39}Ar*	^{38}Ar*	^{37}Ar	^{36}Ar*	$^{40}Ar_K/^{39}Ar_K$	Age (G.y.)
Plagioclase	30	—	395.5 ± 17.4	6.18 ± .46	7.02 ± 5.09	986.9 ± 50.8	10.50 ± 9.10	63.45 ± 5.60	3.67 ± .12
Plagioclase	36	—	175.0 ± 7.5	3.41 ± .26	2.94 ± 2.94	234.4 ± 39.6	5.75 ± 5.75	50.67 ± 4.61	3.31 ± .13
Pyroxene	31	—	186.2 ± 7.7	3.28 ± .33	13.71 ± 5.48	2533.1 ± 122.4	8.42 ± 8.05	56.73 ± 6.24	3.49 ± .15
Pyroxene	32	—	149.3 ± 8.6	2.77 ± .39	14.27 ± 5.67	4593.5 ± 100.4	2.59 ± 2.59	53.82 ± 8.26	3.40 ± .21
Pyroxene	37	—	98.6 ± 7.5	2.36 ± .34	4.21 ± 3.03	1241.8 ± 45.9	5.36 ± 4.79	41.05 ± 6.90	2.99 ± .23
Ilmenite	33	—	665.1 ± 20.3	11.95 ± .88	14.51 ± 5.48	810.6 ± 61.6	7.67 ± 6.99	55.66 ± 4.44	3.46 ± .11
Ilmenite	34	—	299.3 ± 11.6	5.17 ± .38	4.89 ± 4.89	764.5 ± 44.9	3.75 ± 3.75	57.87 ± 4.82	3.52 ± .12
Cristobalite	41	—	104.8 ± 4.4	1.66 ± .19	2.94 ± 2.94	30.9 ± 30.9	6.03 ± 5.75	61.64 ± 8.33	3.62 ± .18
Plagioclase	62	600°C	390.9 ± 6.3	6.28 ± .12	4.09 ± .98	824.1 ± 5.3	12.26 ± 2.39	61.34 ± 1.80	3.61 ± .05
Plagioclase	67	600°C	171.7 ± 1.2	2.36 ± .08	5.82 ± 1.96	471.5 ± 3.4	9.85 ± 6.13	71.36 ± 3.24	3.86 ± .07
Ilmenite	61	600°C	76.3 ± 2.5	1.09 ± .09	6.05 ± 3.91	141.8 ± 4.0	4.79 ± 4.79	69.11 ± 6.64	3.81 ± .13

*Corrected for neutron induced contributions.

†Preheated prior to lasering.

‡The exact amount of material melted by the laser pulses cannot be determined. The data can, however, within a factor of ~2 also be understood in units of 10^{-9} cm^3 STP/g.

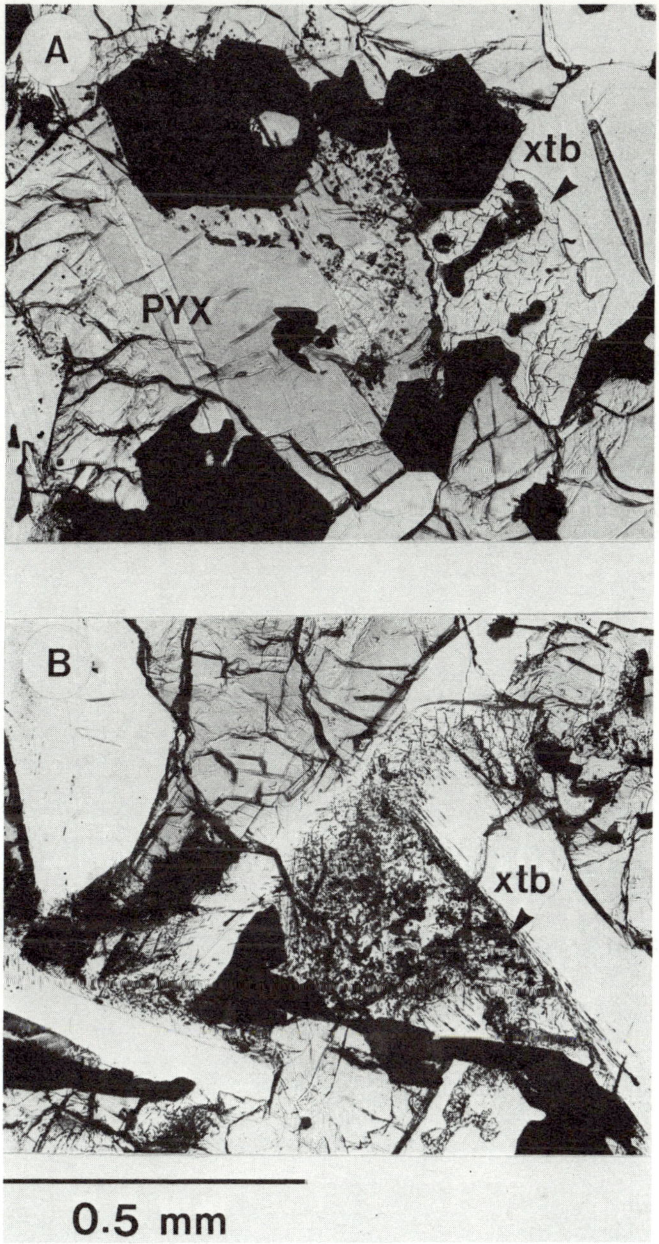

Fig. 1. (a) 70017,110. Large pyroxene crystal (pyx) contains micron-sized blebs of trapped silicate liquid near its rim. Cristobalite (xtb) contains large blebs of trapped mesostasis. (b) 75035,84. Cristobalite (xtb) is surrounded on two sides by plagioclase laths and on the bottom an ilmenite crystal. Note the abundant micron-sized blebs of trapped mesostasis that are peppered through the crystal. Plane polarized light. Same scale for both photographs.

Table 3. Laser released Ar isotopes from minerals of basalt 75035,52 in 10^{-12} cm^3.‡

Mineral	Lab. No.	Temp.†	^{40}Ar	^{39}Ar*	^{38}Ar*	^{37}Ar	^{36}Ar*	$^{40}Ar_K/^{39}Ar_K$	Age (G.y.)
Plagioclase[a]	40	—	117.2 ± 11.2	2.53 ± .15	2.94 ± 2.94	545.4 ± 54.5	9.64 ± 8.91	67.99 ± 6.63	3.78 ± .14
Plagioclase[a]	43	—	325.9 ± 18.4	4.79 ± .41	2.85 ± 1.96	583.9 ± 50.6	8.49 ± 4.31	67.29 ± 7.01	3.76 ± .15
Cristobalite[a]	39	—	2381.2 ± 50.3	39.15 ± 2.18	5.87 ± 5.87	28.6 ± 25.7	5.74 ± 5.74	60.79 ± 3.62	3.60 ± .09
Cristobalite[a]	44	—	3559.0 ± 80.7	55.72 ± 2.08	3.36 ± 3.13	29.6 ± 18.6	3.35 ± 3.35	63.86 ± 2.79	3.68 ± .07
Pyroxene[a]	38	—	106.4 ± 5.4	1.84 ± .36	6.30 ± 5.09	2245.1 ± 63.8	6.71 ± 6.71	57.10 ± 12.09	3.50 ± .29
Pyroxene[a]	42	—	426.6 ± 18.2	11.36 ± .56	3.36 ± 3.13	2315.2 ± 134.1	5.75 ± 5.75	37.39 ± 2.46	2.85 ± .09
Plagioclase[a]	52	650°C	281.7 ± 4.2	4.16 ± .18	2.87 ± 1.96	331.3 ± 37.4	4.79 ± 4.79	67.34 ± 3.19	3.76 ± .07
Plagioclase[a]	56	650°C	178.1 ± 2.7	2.51 ± .11	2.21 ± .78	269.8 ± 35.7	8.27 ± 2.59	69.50 ± 3.72	3.81 ± .08
Cristobalite[a]	53	650°C	336.4 ± 4.3	5.16 ± .15	2.78 ± .88	175.0 ± 31.2	7.62 ± 2.87	64.56 ± 2.19	3.69 ± .05
Cristobalite[a]	55	650°C	574.9 ± 6.4	8.81 ± .38	2.83 ± .88	321.9 ± 25.4	7.97 ± 2.01	64.83 ± 2.91	3.70 ± .07
Pyroxene[a]	54	650°C	132.0 ± 2.1	1.38 ± .18	10.87 ± .88	2818.4 ± 44.6	12.66 ± 3.07	93.43 ± 12.94	4.31 ± .19
Plagioclase[a]	57	900°C	246.2 ± 4.6	3.44 ± .21	2.39 ± .97	387.6 ± 43.1	9.59 ± 2.20	70.25 ± 4.77	3.83 ± .10
Cristobalite[a]	59	900°C	817.2 ± 10.5	11.68 ± .22	1.93 ± 1.08	73.2 ± 29.3	9.75 ± 2.20	69.53 ± 1.64	3.81 ± .04
Pyroxene[a]	58	900°C	5.2 ± 1.5	.32 ± .10	2.94 ± 1.27	1500.8 ± 21.9	2.49 ± 2.49	15.17 ± 8.26	1.68 ± .75
Pyroxene[a]	60	900°C	37.0 ± 1.5	.39 ± .16	2.35 ± 1.27	1493.6 ± 54.2	6.71 ± 6.71	86.88 ± 40.22	4.18 ± .64
Plagioclase[b]	70	800°C	229.0 ± 2.0	3.16 ± .08	3.86 ± 1.96	439.9 ± 3.3	4.79 ± 4.79	72.08 ± 2.15	3.87 ± .05
Plagioclase[b]	71	800°C	162.3 ± 1.3	2.12 ± .12	4.45 ± 1.47	477.2 ± 1.9	11.77 ± 2.20	73.96 ± 4.99	3.92 ± .09
Cristobalite[b]	78	800°C	57.7 ± .7	1.00 ± .08	.98 ± .98	39.2 ± 2.4	4.79 ± 4.79	55.15 ± 5.71	3.44 ± .14
Cristobalite[b]	79	800°C	267.0 ± 2.3	4.43 ± .09	2.88 ± .98	121.2 ± 3.1	7.64 ± 2.87	59.49 ± 1.55	3.56 ± .04
Pyroxene[b]	69	800°C	126.0 ± 2.2	1.51 ± .10	3.33 ± 3.33	910.7 ± 7.8	2.87 ± 2.87	83.17 ± 5.99	4.11 ± .10
Plagioclase[b]	80	1050°C	93.7 ± 3.6	0.73 ± .09	2.94 ± 2.94	201.2 ± 2.8	2.87 ± 2.87	127.41 ± 16.61	4.84 ± .18
Plagioclase[b]	85	1050°C	36.0 ± 1.5	.48 ± .07	2.44 ± 2.44	73.4 ± 4.4	6.71 ± 6.71	69.33 ± 14.92	3.81 ± .29
Cristobalite[b]	81	1050°C	19.2 ± 1.6	.37 ± .18	2.94 ± 2.94	8.62 ± 8.62	9.58 ± 9.58	40.00 ± 31.49	2.95 ± 1.1

*Corrected for neutron induced contributions.

†Preheated prior to lasering.

‡The exact amount of material melted by the laser pulses cannot be determined. The data can, however, within a factor of ~2 also be understood in units of 10^{-9} cm^3 STP/g.

[a,b]Analyses made on two different chips.

75035 is overwhelmingly found in the cristobalite and mesostasis that is often included as micron-sized blebs (Fig. 1b). The cristobalite contains at least a factor of 10 more argon than the average plagioclase. Although it is very difficult with the laser beam to distinguish between the pure mesostasis and cristobalite containing trapped mesostasis inclusions, our measurements seem to indicate that mesostasis can be distinguished by its high K/Ca ratio from the "pure" cristobalite. (Compare representative analyses of mesostasis and cristobalite in Apollo 17 high-titanium basalts: Roedder and Weiblen, 1975 and Dymek *et al.*, 1975.)

In order to investigate the loss of argon which is released at low temperatures, we have heated the neutron irradiated sample 75035 to 650°C, 900°C, and 1050°C for about 2 hr, polishing after each heating and investigating with the laser probe. The pyroxenes show a somewhat erratic behavior with a large spread in apparent ages from 1.68 to 4.31 G.y. (see Table 3). The two low ages of 3.50 and 2.85 G.y. for the unpreheated samples, however, indicate that the pyroxenes suffered strong argon losses. Some pyroxene (75035,52,58) preheated to 900°C seems to be almost completely outgassed and shows a low apparent age of 1.68 ± 0.75 G.y., while others (75035,52,60 and 75035,52,69) show high ages. This fact imposes the question of whether a high temperature drop off as seen for the pyroxenes (Turner and Cadogan, 1975) is a common feature and whether this drop off can be attributed *a priori* to the pyroxene.

Our combined mineralogical observations and laser released gas analyses suggest that there is a high temperature drop off in the cristobalite which is peppered with micrometer sized mesostasis but empty of large mesostasis blebs. However, we cannot exclude within our measurements that there is also a drop off in the pyroxene. In fact, the pyroxene in 75035 also contains tiny inclusions which may be responsible for such an effects. For the cristobalites the spread in apparent ages is considerably smaller than for the pyroxenes, and the relatively high ages for the unpreheated samples of 3.60 and 3.68 G.y. show that this mineral phase experienced only minor gas losses. The ages for the cristobalite high K-mesostasis intergrowth increase from ~3.60 to ~3.82 G.y. (chip a, Table 3), while the ages for the cristobalite that only contain micron-sized mesostasis inclusions (chip b, Table 3) are markedly lower. There is, therefore, a suggestion that the age decreases with increasing temperature of outgassing. The high temperature age decrease could be caused by recoil of ^{39}Ar similar as appeared to be the case for the pyroxenes in 70017. Some argon loss may have also occurred, as the measurements on chip b show a considerably lower Ar concentration than chip a.

It appears likely that the 600–1000°C plagioclase plateau seen in the ^{39}Ar–^{40}Ar gas release experiment (Turner and Cadogan, 1975) may be biased to lower ages by gas retained in the tiny micrometer sized mesostasis inclusions trapped in the cristobalite as well as by larger mesostasis blebs (i.e., it is likely that the mineral separation technique concentrated both cristobalite and plagioclase).

From Table 3 it is evident that the apparent age for the plagioclase remains essentially the same for the temperature range up to about 1000°C. Indeed, even the results for the unpreheated samples 75035,52,40 and 75035,52,43 of 3.78 and 3.76 G.y., respectively, are very close to the intermediate temperature ^{39}Ar–^{40}Ar plateau by Turner and Cadogan (1976) suggesting only minor Ar loss of the plagioclase. We cannot evaluate the erratic behavior of the plagioclase after preheating to 1050°C. The two ages of 3.81 and 4.84 G.y. obtained for this temperature step indicate, however, that the plagioclase crystal after such a preheating does not yield chronologically meaningful ages for 75035.

The argon data for 75035 allow us to suggest a sequence for the reliability of apparent ages from different mineral phases in the intermediate temperature range (~600°C to ~1000°C). The apparent ages are more reliable as we proceed from pyroxene to cristobalite and finally to plagioclase.

CONCLUSIONS

Our observations do not yield a simple picture, and it seems that we are still far away from understanding in detail every single stepwise heating gas release pattern. Even where we have similar gas release patterns, as in basalt 70017 and basalt 75035, the same mineral components in the different basalts do not show a similar behavior. Our measurements, however, underline the ^{39}Ar–^{40}Ar doctrine, that the plagioclases give the most reliable ages at least for the intermediate temperature release. However, the different results for the 1050°C plagioclase ages (75035,52,80 and 75035,52,85) set a clear-cut caution to the interpretation of high temperature plagioclase plateaus. On the other side, the high temperature increase observed in the stepwise heating experiment on 75035 seems to be solely attributable to the plagioclases as mentioned also by Turner and Cadogan (1975) and does not have any chronological significance. That is, there is no phase which shows a high age without extensive prior outgassing.

From our measurements, it appears that the cause of the high temperature drop off in age is not only due to recoil effects but also due to minerals which have lost appreciable argon. Other not well understood behavior in mineral systems seem to be producing low ages. For example, some plagioclase (70017,25,62) and some pyroxenes (75035) give lower ages than expected and in some cases unexplainably high ages.

It is possible that by combining the laser technique with the ^{39}Ar–^{40}Ar temperature release studies more precise ages can be determined, as we suggest in the case of 75035.

Acknowledgments—We would like to thank A. E. Bence for helpful discussion, T. Ludkewycz for operation of the mass spectrometer, and R. Schlott for skillful glass blowing. The work was supported by NASA grants NGL-33-015-174 and NGL-33-015-130.

REFERENCES

Brown G. M., Peckett A., Emeleus C. H., Phillips R., and Pinsent R. H. (1975) Petrology and mineralogy of Apollo 17 mare basalts. *Proc. Lunar Sci. Conf. 6th*, p. 1–13.

Dymek R. F., Albee A. L., and Chodos A. A. (1975) Comparative mineralogy and petrology of Apollo 17 mare basalts: Samples 70215, 71055, and 75055. *Proc. Lunar Sci. Conf. 6th*, p. 49–77.

Huneke J. C. and Smith S. P. (1976) The realities of recoil: ^{39}Ar recoil out of small grains and anomalous age patterns in ^{39}Ar–^{40}Ar dating. *Proc. Lunar Sci. Conf. 7th*, p. 1987–2008.

Husain L. (1974) ^{40}Ar–^{39}Ar chronology and cosmic ray exposure ages of the Apollo 15 samples. *J. Geophys. Res.* **79**, 2588–2606.

Kirsten T. and Horn P. (1974) Chronology of the Taurus-Littrow region III: Ages of mare basalts and highland breccias and some remarks about the interpretation of lunar highland rock ages. *Proc. Lunar Sci. Conf. 5th*, p. 1451–1475.

Longhi J., Walker D., Grove T. L., Stolper E. M., and Hays J. F. (1974) The petrology of the Apollo 17 mare basalts. *Proc. Lunar Sci. Conf. 5th*, p. 447–469.

Müller H. W., Plieninger T., James O., and Schaeffer O. A. (1977) Laser probe ^{39}Ar–^{40}Ar dating of materials from consortium breccia 73215. *Proc. Lunar Sci. Conf. 8th.* This volume.

Phinney D., Kahl S. B., and Reynolds J. H. (1975) ^{40}Ar–^{39}Ar dating of Apollo 16 and 17 rocks. *Proc. Lunar Sci. Conf. 6th*, p. 1593–1608.

Plieninger T. and Schaeffer O. A. (1976) Laser probe ^{39}Ar–^{40}Ar ages of individual mineral grains in lunar basalt 15607 and lunar breccia 15465. *Proc. Lunar Sci. Conf. 7th*, p. 2055–2066.

Roedder E. and Weiblen P. W. (1975) Anomalous low-K silicate melt inclusions in ilmenite from Apollo 17 basalts. *Proc. Lunar Sci. Conf. 6th*, p. 147–164.

Schaeffer G. A. and Schaeffer O. A. (1977) ^{39}Ar–^{40}Ar ages of lunar rocks. *Proc. Lunar Sci. Conf. 8th.* This volume.

Turner G. and Cadogan P. H. (1975) The history of lunar bombardment inferred from ^{40}Ar–^{39}Ar dating of highland rocks. *Proc. Lunar Sci. Conf. 6th*, p. 1509–1538.

Usselman T. M., Lofgren G. E., Donaldson C. H., and Williams R. J. (1975) Experimentally reproduced textures and mineral densities of high-titanium mare basalts. *Proc. Lunar Sci. Conf. 6th*, p. 997–1020.

Proc. Lunar Sci. Conf. 8th (1977), p. 1501–1520.
Printed in the United States of America

Cooling histories of Apollo 15 quartz-normative basalts

TIMOTHY L. GROVE

Department of Earth and Space Sciences, State University of New York, Stony Brook, New York 11794

DAVID WALKER

Hoffman Laboratory, Harvard University, Cambridge, Massachusetts 02138

Abstract—Experimental controlled cooling rate studies of vitrophyre 15597 are used to calculate the physical properties of a quartz-normative magma and cooling histories of Apollo 15 quartz-normative basalts (QNBs). The physical properties of a QNB magma are such that pyroxene phenocrysts do not settle appreciably during the cooling of the flow unit, consistent with the small degree of chemical differentiation that occurred between members of the suite. Two types of cooling histories characterize the QNBs: (1) The vitrophyres experienced initial slow cooling followed by rapid cooling; and (2) the coarse-grained microgabbros underwent slow, nearly linear cooling histories. Textural evidence in the lunar vitrophyres is consistent with the crystallization of these samples near the top of the flow. The initial slow followed by rapid cooling history may have resulted from the chilling of liquids that inherited early crystals in the interior of the flow. Cooling history estimates for the coarse-grained microgabbros suggest that some samples cooled at least 10 m from a flow contact.

INTRODUCTION

THE RESULTS OF COOLING RATE EXPERIMENTS conducted on a synthetic analog of vitrophyre 15597 have been used to estimate the thermal histories of Apollo 15 quartz-normative basalts (QNBs) and to characterize the physical properties of the quartz-normative liquid during crystallization. The possibility of QNB magma fractionation by gravitative settling of pyroxene phenocrysts in the local environment of solidification is evaluated from the physical properties of experimentally produced residual liquids. Bulk chemical and textural criteria of the QNB hand specimens are used to reconstruct the positions of the samples within a hypothetical lava flow and to estimate the size of that flow. The nearly isochemical nature of the texturally diverse QNB suite is related to the lack of differentiation by pyroxene settling.

NATURE OF THE QNB FLOW

Number of flows sampled

There is no convincing evidence that the QNBs were sampled from a single flow. The restricted chemistry of the suite argues for close genetic relation between the QNBs, but hand specimens may represent samples of several flows that were derived from the same near-surface magma chamber. On the other hand, they may represent samples of one flow. Helmke *et al.* (1973) use small differences in trace-element abundance to suggest that the quartz-normative vitrophyres 15597 and 15486 represent chilled samples of different magmas. These differences may be the result of heterogeneous contamination of a single magma upon eruption, but there are also systematic differences in mineral chemistries between these two vitrophyres (Grove and Bence, 1977) which may suggest different magma batches. Outcrops of flow units up to 20 m in thickness are visible on the other side of Hadley Rille (Lofgren *et al.*, 1975; Swann *et al.*, 1972) that may give an indication of the thickness of the QNB flow. If there is more than one flow unit represented in the samples, our approach for positioning a hand specimen is unaffected; and the processes discussed here should occur in each of the flow units.

Do the vitrophyres represent preserved liquid compositions?

The Apollo 15 QNB liquids are nearly cosaturated with olivine (Fo_{68}–Fo_{70}) and pyroxene (Wo_5En_{68}) at lunar surface conditions. Humphries *et al.* (1972), Longhi *et al.* (1972), Rhodes and Hubbard (1973), and Chappell and Green (1973) point out the difficulties with deriving the QNBs from Apollo 15 olivine-normative basalts by any type of equilibrium crystal-liquid fractionation. Unlike the Apollo 12 quartz-normative basalts, which can be related to the Apollo 12 picrites by olivine fractionation (Kushiro and Haramura, 1971) the Apollo 15 QNBs constitute a flow unit distinct from the olivine-normative basalts. Various possibilities can be envisioned for the origin of the QNB flow: (1) The QNB samples represent liquids produced directly by partial melting of the lunar interior, or (2) the QNB samples represent liquids which were produced by fractionation in a shallow magma chamber near the lunar surface and/or during extrusion and flow across the surface. Hypothesis (1) is rather difficult to reconcile with current moon models. A QNB liquid is multiply saturated at pressures less than 5 kbar. Approximately 2.5 kbar (or 50 km depth) is obtained using the curve of Walker *et al.* (1976b) for low-Ti basalt cosaturation, which places a proposed source region of olivine pyroxenite within the lunar crust. Accepting current models of a plagioclase rich crust to 60 km and an olivine pyroxenite mantle to 300–400 km (Toksöz *et al.*, 1974), requires that possibility (1) be rejected.

The vitrophyres represent preserved melt samples (Weigand and Hollister, 1973) and contain phases that crystallized from liquid with the sample's bulk composition (Grove and Bence, 1977). Differentiation of a more primitive liquid by olivine fractionation could have occurred in near-surface magma chambers which occasionally erupted QNB magma onto the lunar surface. The complementary crystal cumulates are stored at depth and as we shall see, little further fractionation of QNB magma occurs during solidification.

It is also possible that the QNBs sampled do not represent the liquid extruded and that differentiation of a more primitive liquid may have occurred during the course of its extrusion and flow across the lunar surface. Perhaps the entire flow was not sampled. The average soil at Hadley Rille is predominantly olivine normative (Rhodes and Hubbard, 1973; Duncan *et al.*, 1975) suggesting that the olivine normative basalts crop out at the surface and the underlying QNBs were exposed by meteorite impact. Thus, the QNB samples may represent the fractionated top of a flow unit. Schaber (1973) has observed lava flows in southwestern Imbrium that extend for up to 1000 km and estimates that it required from 10 to 20 days for the flow to cover that distance. Considering the settling velocities of olivine in a picritic flow (Walker *et al.*, 1976c) a surface residence time of 10 days is sufficient to allow a more olivine-normative precursor to undergo considerable olivine fractionation. Thus, a variant of hypothesis (2) which involves fractionation near the lunar surface probably explains the origin of the QNB liquid.

Determination of Cooling Histories

Experiments

The QNBs show quite a variation in texture (Lofgren *et al.*, 1975; Papike *et al.*, 1976), characterized by covariation in both pyroxene phenocryst size and groundmass plagioclase size (Fig. 6). Some controversy has been generated over the interpretation of porphyritic textures in Apollo 12 quartz-normative basalts (Hollister *et al.*, 1971; Bence and Papike, 1972; Boyd and Smith, 1971) and Apollo 15 QNBs (Dowty *et al.*, 1974). The basis of the controversy was whether or not a porphyritic texture can be generated by a simple cooling history. Lofgren *et al.* (1974) have shown that porphyritic textures can be produced in linear cooling rate experiments. Thus, a complex cooling history (e.g., early crystallization of phenocrysts at a slower cooling rate, followed by eruption, rapid cooling, and crystallization of the groundmass) is not necessary to produce the crystallization history recorded in porphyritic textures. We will attempt to reconstruct the cooling history which produced the members of the QNB suite.

Linear controlled cooling rates have been carried out from 0.5° to 150°C/hr in iron metal capsules sealed in silica glass tubes. Experiments were all initiated above the equilibrium liquidus temperature (Grove and Bence, 1977), cooled to a desired temperature at a desired rate, and then quenched. The technique allows one to determine the effect of cooling rate on the temperature of phase appearance, and to analyze residual liquid compositions during crystallization. All experiments contained pyroxene phenocrysts (with minor olivine and Cr spinel) and groundmass of plagioclase, ilmenite, and cristobalite. Porphyritic textures were produced in all linear cooling rate experiments.

As a function of increasing cooling rate the number of clinopyroxene phenocrysts increases and their size decreases. We have measured these two quantities in our experiments (nucleation density and growth size) and used them to estimate the cooling rate sustained during the early portion of the crystallization history. Pyroxene nucleation density was determined in experiments quenched over a range of temperatures, and it was found that most nucleii form in the 50°C temperature interval after pyroxene appearance. Growth then dominates in producing crystalline pyroxene down to the temperature of groundmass appearance. The size of groundmass plagioclase increases at slower rates (Walker *et al.*, 1976a), and is used to determine cooling history during the late stages of crystallization.

Early stage cooling rate

Nucleation density was determined by counting the number of pyroxenes in a thin section and dividing by the area. The change in nucleation density as a function of cooling rate in the experiments is plotted in Fig. 1 along with several values determined for lunar samples. Estimates of cooling rate for lunar samples are tabulated in Table 1. As a check on the nucleation density estimate, a measure of growth rate was also determined. It should be noted that sizes measured to give growth rate may in turn be dependent on nucleation density, because a large number of crystals will be smaller for a given amount of pyroxene crystallization. Two methods were employed. The first used the distance measured perpendicular to (110) from the center of the core pigeonite to the augite rim in phenocrysts with **c** vertical. The nucleation density and core growth estimates of cooling rate are internally consistent with one another, suggesting that growth and nucleation were similar in the experiments and in nature. This is true for 15499 and 15058 whose cooling rates during the early stages of crystallization can be deduced from the experiments (Figs. 1 and 2). This internal consistency is lacking for vitrophyre 15597, the complexities of which are described below. The second method measures the phenocryst (core plus rim) width perpendicular to (110) in crystals with **c**-axis vertical. The experiments and lunar samples are compared in Fig. 3 and rate estimates based on size are the ones tabulated in Table 1 for lunar samples. The first method,

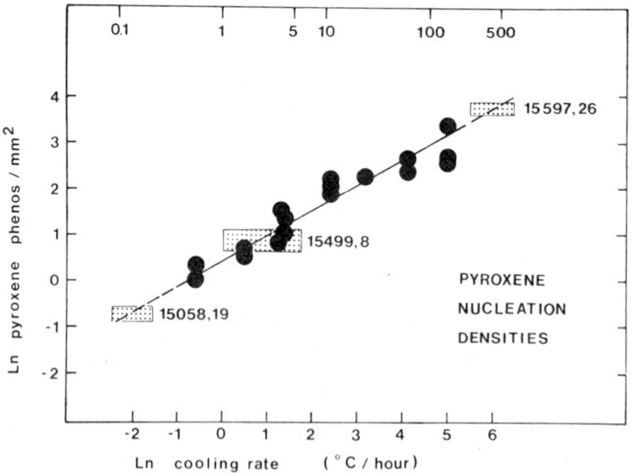

Fig. 1. Pyroxene phenocryst nucleation density in cooling rate experiments and lunar samples. Natural log of the number of pyroxenes/mm^2 measured in thin sections is plotted as a function of cooling rate. The nucleation densities measured in thin sections of lunar samples are used to estimate early cooling rate (Table 1).

Table 1. Cooling rate estimates for Apollo 15 QNBs.

Sample	Thin sections examined	Early-stage cooling rate: Pyroxene phenocryst nucleation density (pyx/mm^2)	Early-stage cooling rate: Rate estimate (°C/hr)	Integrated cooling rate: Total pyroxene phenocryst size (mm)*	Integrated cooling rate: Rate estimate (°C/hr)	Late-stage groundmass cooling rate: Plagioclase size (mm)†	Late-stage groundmass cooling rate: Rate estimate (°C/hour)	Final position from conductive boundary (cm)
Dune Crater vitrophyres								
15485	,3	2.8	3.75	0.23–0.26	3.75–10	0.004	85–250	6–9
15486	,22	2.5	1.75–3.75	0.38–0.47	3.75	0.004	85–250	6–9
15499	,95,8	2.0–3.4	1.75–3.75	0.51–0.70	1.75	0.011	30	15
Hadley Rille vitrophyres								
15597	,26,24,18,17	41–50	300–500	0.10–0.18 (0.44 megacryst)	10–150 (3.75)[a]	<0.001	300–500	4
15595	,33,35	5.1–6.1	10	0.30–0.37	3.75	0.008	60	12
15596	,16	2.2	3.75	0.45–0.51	1.75	0.021	13	24
Coarse-grained samples								
15058	,19,21,128	0.3–0.6	~0.04–0.2	1.00–1.18	b	0.32–0.36	0.1	300
15475	,16	0.3	~0.05	1.30–1.85	b	0.25–0.36	0.1	263
15495	,11	0.2	~0.05	1.11	b	0.11–0.17	0.5	133
15076	,17	0.5	0.1	1.37–1.78	b	0.30–0.45	0.2	263
15065	,91,89,93	0.1–0.4	0.03–0.1	1.48–1.96	b	1.10–3.0	0.01	1032
15085	,25	—	—	1.67–1.85	b	1.48–1.78	0.01	1032
Rake samples								
15125	,7	10.7	~30	0.15–0.23	~10	0.004	85–250	6–9
15666	6	1.1	0.5	0.83	0.5	0.012	34	15
15682	6	0.5	0.1	0.75–0.83	0.5	0.075	1.5	70

*Width measured perpendicular to (110) faces.

†Width measured perpendicular to (010) faces.

[a]Pyroxene megacryst in 15597,17.

[b]Pyroxene width is larger than the 0.5°C/hr experiments. No extrapolation of growth measurement was made to lower rates.

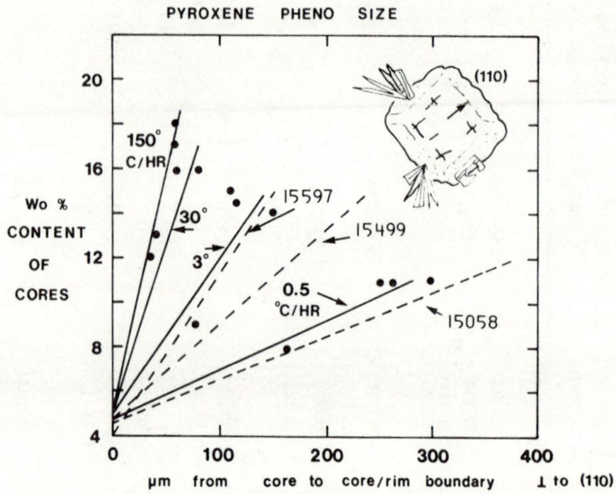

Fig. 2. The size (abscissa) and composition (ordinate) of the pigeonite cores of pyroxene phenocrysts in linear cooling rate experiments are shown by the solid lines. Rim compositions and phenocryst size in several cooling rate experiments are from phenocrysts that had **c** perpendicular to the thin section. The maximum extent of the line marks the distance from the core to pigeonite-augite rim boundary, and indicates the composition of the pigeonite at the boundary. A similar measurement made in lunar samples (dashed lines) is used to estimate the early cooling rate.

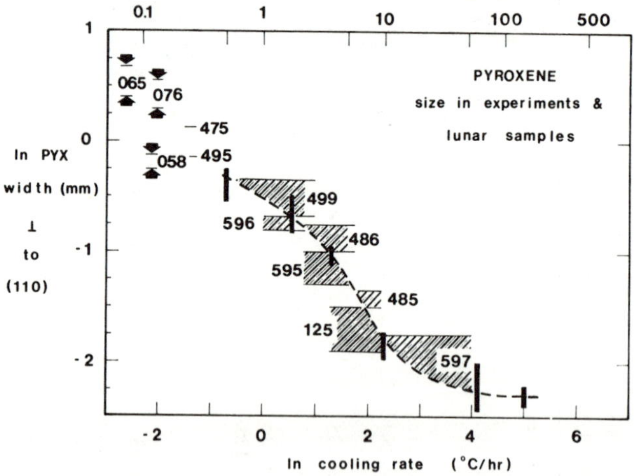

Fig. 3. Pyroxene phenocryst size measured in linear cooling rate experiments (solid bars) and lunar samples (hatched regions). Pyroxene size is measured perpendicular to (110) in phenocrysts oriented with the **c** axis vertical. Cooling rate estimates for lunar samples are tabulated in Table 1. The cooling rate estimate was not extrapolated to lunar rocks containing pyroxenes larger than those found in 0.5°C/hr experiments. Thus, the positioning of samples to the left of 0.5°C/hr has no significance.

core growth, gives an estimate of the rate dependency of growth during the early stages of crystallization and is comparable to nucleation density. The second method, total pyroxene phenocryst size, integrates both the early and later cooling and crystallization history experienced by the basalt and might be different than the nucleation estimate.

Late-stage cooling rate

Plagioclase displays a distinctive hollow, box-like morphology (Bryan, 1972; Crawford, 1973) in all cooling rate experiments, the lunar vitrophyres and in the coarse-grained microgabbros. In order to estimate late-stage cooling history, the width of groundmass plagioclase perpendicular to (010) was measured in the cooling rate experiments, and the change in spacing of (010) faces is plotted as a function of cooling rate in Fig. 4. Walker *et al.* (1976c) suggest that a linear relation on a log log plot such as Fig. 4 connects plagioclase size to cooling rate. We find that the expression

$$\log \frac{\partial T}{\partial t} = -1.71 \log(\text{width}) - 1.75, \qquad (1)$$

fits the data. Late-stage cooling rates were determined for the QNBs from this expression (Table 1). The late-stage rate estimate for the coarse-grained samples from plagioclase size has been compared to other cooling rate indicators by Walker *et al.* (1977). Zr partitioning between ilmenite and ulvöspinel (Taylor *et al.*, 1973) and the homogenization of compositional zonation in olivine (Walker *et al.*, 1977; Taylor *et al.*, 1977) give rate estimates that are similar to the plagioclase grain size estimate (0.3 – 3°C/day).

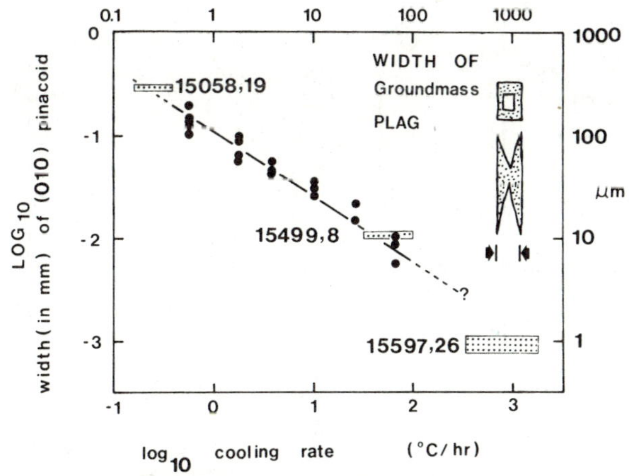

Fig. 4. Width of plagioclase groundmass measured in linear cooling rate experiments. Log_{10} of the perpendicular distance (in mm) is plotted against cooling rate. Late-stage cooling rate estimates are obtained for lunar samples by extrapolating this curve (see text and Table 1).

Cooling histories of Apollo 15 QNBs

The Dune Crater vitrophyres 15485, 15486, and 15499 were collected from a single boulder representing a 10–20 cm section of the QNB flow. Their cooling histories are characterized by an initial slow cooling rate followed by a rapid late-stage cooling rate (Table 1). The pyroxene nucleation densities preserved in the samples are identical, but there is an increase in pyroxene and plagioclase size from 15485 to 15486 to 15499. Thus, if nucleation density is used, the samples appear to have similar early cooling rates. Total pyroxene size, on the other hand, integrates cooling rate over a larger temperature interval and suggests that the samples experienced an increase in rate at lower temperatures which was larger for 15485 and 15486 than for 15499. The late-stage cooling rates for all three samples are more rapid than the initial rates. The lack of consistency between the nucleation density and total pyroxene growth estimates could result from a non-linear cooling history. The initial slow cooling followed by rapid late-stage crystallization experienced by the samples is not characteristic of the cooling history that exists near the contact of a flow (Jaeger, 1957). One should find a nearly uniform rapid cooling rate if the entire crystallization of the liquid occurred as close to a chilled contact as suggested by the groundmass cooling rates. If any variation of cooling rate occurs with temperature, the rates

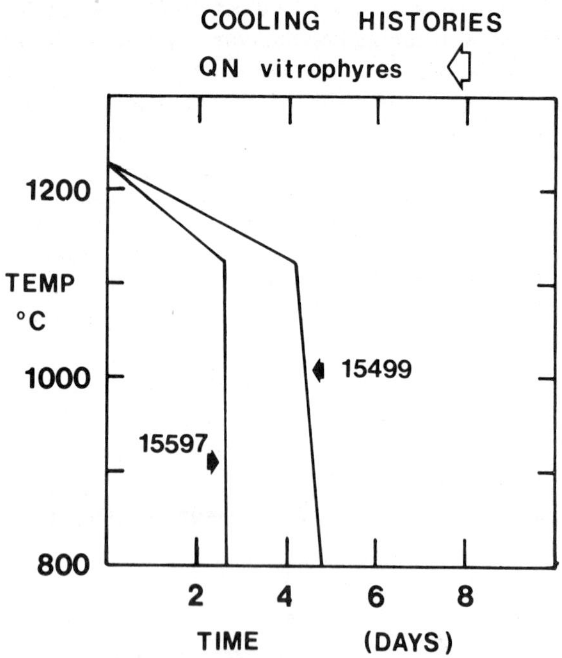

Fig. 5. Temperature-time cooling history for quartz-normative vitrophyres 15597 and 15499. Note that the cooling history is the reverse of what one would expect for crystallization by heat loss near the edge of a lava flow.

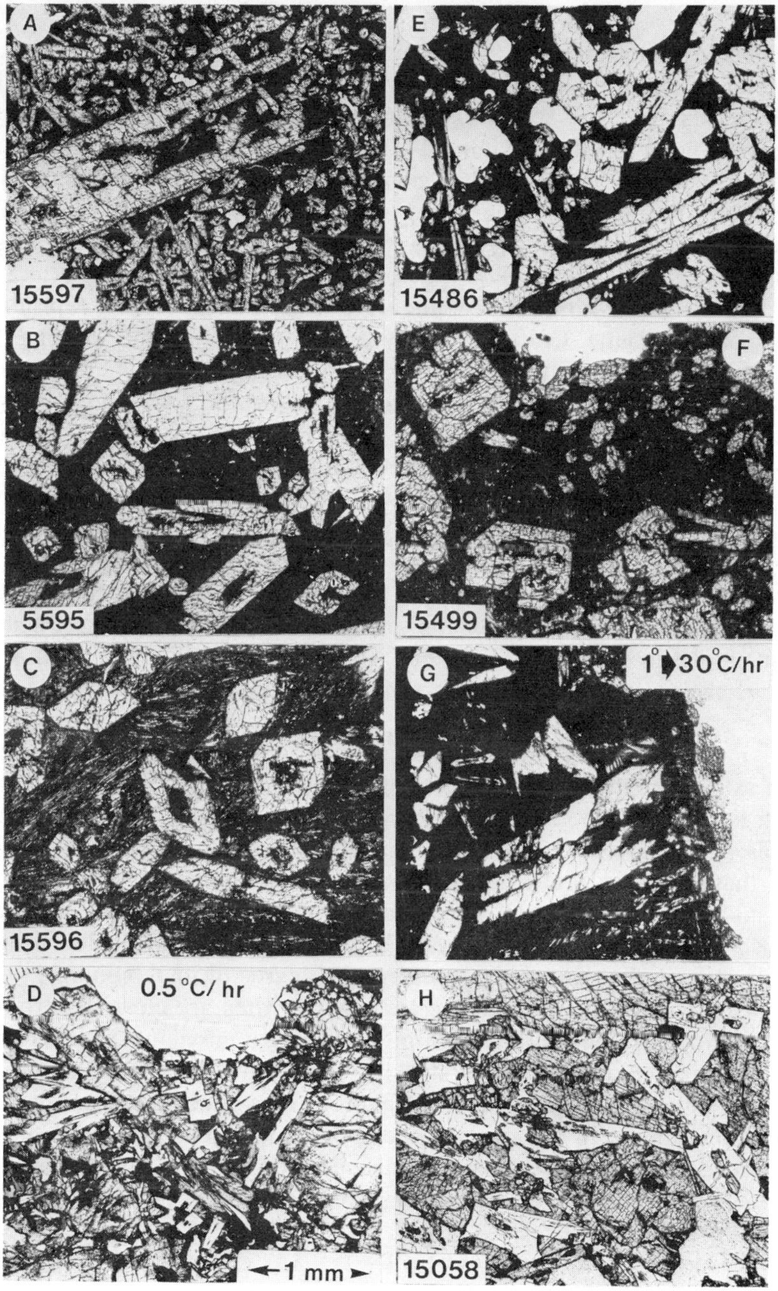

Fig. 6. Lunar and synthetic quartz-normative basalts. Hadley Rille vitrophyres; (a) 15597,17 (b) 15595,33 and (c) 15596,16. (d) Thin section of linear controlled cooling rate experiment (0.5°C/hr, 1225–975°C). Dune Crater vitrophyres; (e) 15486,22 and (f) 15499,8. (g) Controlled cooling rate experiment cooled from 1225°C at 1°C/hr for 4 days, then 30°C/hr for 12 hr to simulate 15499's cooling history. (h) Coarse-grained microgabbro 15058,21. Scale for all photos is the same.

should *decrease*, not *increase*. However, if the liquid grew its first pyroxenes in the interior of the flow and then was extruded and chilled, the cooling history could be explained. The actual displacement of the liquid mass relative to the chill margin need not be great (< 1 m) and might easily have occurred as a natural consequence of the emplacement process or as a consequence of minor convective transport of material within the top of a stationary solidifying flow. 15485 and 15486 would have been closest to the chill margin and 15499 slightly further away.

To see if our controlled cooling rate experiments could reproduce the textures of the Dune Crater vitrophyres by means of the cooling history suggested, we carried out a two-stage crystallization experiment designed to match the one proposed for 15499 (Fig. 5, 1°C/hr for 4.17 days, followed by 30°C/hr for 11.67 hr). The result (Fig. 6g) resembles the texture and chemistry of 15499 (Fig. 6f) rather closely. (Measured total pyroxene size in the experiment was 0.63 mm giving a rate estimate of ~1.75°C/hr. Plagioclase width equals 0.010 mm giving a late-stage rate estimate of 30°C/hr).

In fact, there is additional textural information that lends support to the proposed slow followed by fast cooling history. Skeletal pyroxenes smaller than the phenocrysts, yet larger than the groundmass pyroxene are present in the experiment and in 15499. The pyroxene is a ferroaugite with composition similar to that of the (110) epitaxial augite rims on phenocrysts, and nucleated in the liquid before plagioclase and after the pyroxene phenocrysts. Pyroxenes with this distinctive morphology and composition were not found in any linear cooling rate experiments. Linear cooling rate experiments contain only discrete phenocrysts in a groundmass of pyroxene and plagioclase. Pyroxenes similar to the ones grown in the 15499 analog experiment are found in the other lunar basalts that show textural evidence of a late-stage crystallization rate that was more rapid than the early rate. They can be seen in 15486 Fig. 6e) and have been described (Weigand and Hollister, 1973) in 15597. The new nucleii result from increasing the cooling rate and, therefore, the degree of undercooling in the later stages of crystallization. Nucleation and growth rates are close to zero at small degrees of undercooling with growth rate reaching a maximum at a lower degree of undercooling than nucleation density (Kirkpatrick, 1976). Thus, for large degrees of undercooling, nucleation is favored. This effect is present in the linear cooling experiments in the case of the groundmass pyroxene. Evidently, perturbing the degree of undercooling by discontinuously increasing the cooling rate has the effect of causing an extra nucleation event between the phenocryst and groundmass nucleation events.

The Hadley Rille vitrophyres [15597, 15595, 15596 (Fig. 6a,b,c,)] also have slow initial followed by rapid late-stage cooling histories. 15597 is especially intriguing because of the variety of early stage cooling rates that are estimated, depending on which pyroxene phenocrysts are measured or which method of rate estimation is chosen. Early cooling rates are: (1) 3.75°C/hr if the megacryst size is used (2) 10–150°C/hr if the size of the smaller, more abundant pyroxene phenocrysts are used and (3) 300–500°C/hour for nucleation density of the smaller pyroxenes. Chemical rate indicators for 15597 (Grove and Bence, 1977), give 3.75°C/hr or less. We suggest that the megacrysts in 15597 grew in a more slowly cooled environment for possibly a day or two (Fig. 5). The nucleation density recorded by the smaller pyroxenes may have been biased by concentration of nucleii during the event that increased the cooling rate.

The coarse-grained porphyritic samples (15058, 15475, 15495, 15076, 15065, 15085) appear to have experienced slow, nearly linear cooling rates throughout their entire cooling history (Table 1). We have not attempted to extrapolate the growth size curve (Fig. 4) to rates slower than 0.5°C/hr and have used only nucleation density to estimate early cooling history.

The textural relations in one of our slowly cooled experiments (0.5°C/hr) can be compared to porphyritic-subophitic basalt 15058 (Fig. 6d,h). The pyroxene phenocryst and groundmass plagioclase morphologies are quite similar, but our slowest cooling rate was not slow enough to reproduce the plagioclase and pyroxene sizes observed in the microgabbros. The relationships between early and late-stage rate estimates do not show a consistent trend for the coarse-grained samples. Our experiments do not allow us to attach any real significance to the differences in early- and late-stage cooling rates estimated for the coarser samples. It is possible that pyroxene nucleii in the interior portion of the flow were affected by movements in the magma after nucleation. Convective or flow processes might operate in the upper portion of a flow and the same process of pyroxene transport that is preserved in the vitrophyres may have operated in the flow interior. In the following section we will see that gravitative settling is not as effective a differentiation mechanism for pyroxene as it is for olivine in picritic basalts (Walker *et al.*, 1976c). Thus, any convective or flow process would more effectively redistribute the suspended pyroxene crystals than would settling.

Modeling the Differentiation by Crystal Settling of a QNB Magma at the Site of Solidification

Physical properties of the QNB magma

The controlled cooling rate studies on a synthetic analog of vitrophyre 15597 (Grove and Bence, 1977) have been used to characterize the changes in magma properties during crystallization. From our experiments, we know the compositions of residual liquids, the degree of crystallinity and the compositions and sizes of mineral phases as a function of temperature. This information is used to calculate liquid viscosity, density, and crystal settling velocities and to model the extent of differentiation by pyroxene phenocryst settling that might occur during the cooling of a QNB flow. The model of Walker *et al.* (1976c) computes the relative concentration of a phenocryst phase as a function of position during the cooling of a flow. QNB magma physical properties calculated from 1.75 and 0.5°C/hr rate experiments were used for the crystal settling calculations (Fig. 7). Residual liquid densities and viscosities were calculated after the method of

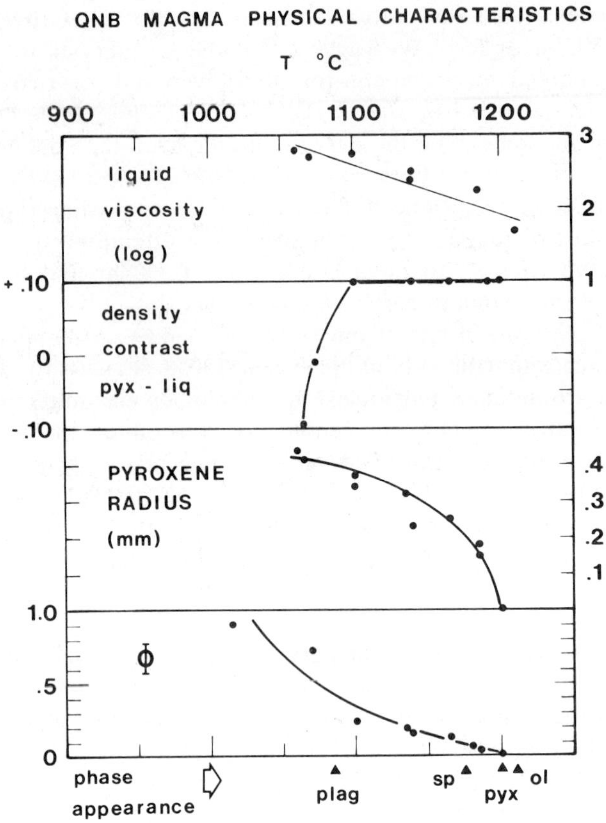

Fig. 7. Variation in physical properties of a quartz-normative liquid with temperature determined from 0.5 and 1.75°C/hr cooling rate experiments on the synthetic 15597 composition. Phase appearance sequence is shown at the bottom. Φ denotes degree of crystallinity. Pyroxene radius estimated by reducing volume of the lath to that of a sphere. Density contrast is ρ pyroxene $-\rho$ liquid. Liquid density and viscosity are calculated after Bottinga and Weill (1970, 1972).

Bottinga and Weill (1970, 1972) using the liquids from 1.75 and 0.5°C/hr cooling rate experiments. Pyroxene density was estimated by extrapolating the high-temperature cell dimension data for diopside (Cameron *et al.*, 1973) and clinoenstatite (Smyth, 1974) to an average pyroxene composition for the cores of clinopyroxene phenocrysts ($Wo_{10}En_{60}Fs_{30}$). The density difference between pyroxene and liquid is quite small and pyroxene should actually float in late-stage residual liquids (after plagioclase appearance). Pyroxene radius was calculated by estimating the volume of a pyroxene phenocryst and recalculating it to a sphere. Crystal settling velocities were calculated using a modified Stoke's law expression that accounts for both shape effects and the fact that the crystals are settling through a suspension (Walker *et al.*, 1976c, their Eq. (3) from Lewis *et al.*, 1949). An expression for the degree of crystallinity (ϕ) as a function of temperature required for the settling calculation was estimated from thin sections of experiments.

Differentiation by pyroxene settling

The physical properties of the quartz-normative magma (Fig. 7) were substituted into the computer program of Walker *et al.* (1976c) and pyroxene settling was calculated for 1, 10, and 50 m flow thicknesses. The solution of Jaeger (1968) of the heat flow problem relating temperature, time, and position in a flow with surface temperature of 0°C was used. The cooling calculation began at 1210°C and pyroxene growth and settling were initiated at 1185°C. Thermal diffusivity was taken as 0.01 cm^2/sec and the crystal settling calculation was stopped at 1075°C, the temperature of plagioclase appearance. This temperature provides a maximum value for crystal settling, because slower cooling rates in the interior

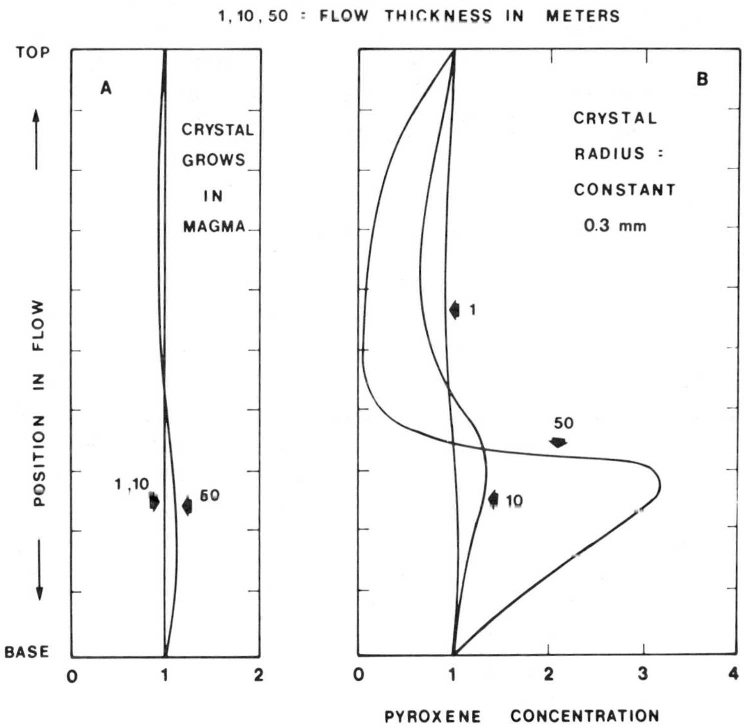

Fig. 8. Profiles of pyroxene settling for flows of 1, 10, and 50 m thickness computed for a QNB magma. The ordinate is scaled to total thickness. The abscissa is a relative scale that indicates concentration factor. One is the initial concentration of pyroxene in the flow at the time of nucleation or emplacement. In (a) the crystals nucleate and grow in a completely liquid flow. The settling calculation assumes nucleation density does not vary through the flow. (b) shows the pyroxene settling that would occur in a flow that initially contained large phenocrysts.

of a large flow will raise the temperature of plagioclase appearance and halt crystal settling at higher temperatures. The calculation does not account for nucleation differences as a function of initial position or release of latent heat of fusion during solidification.

Two sorts of settling profiles are calculated and shown in Fig. 8: One sort for a liquid magma that nucleates and grows crystals and one sort for a magma that initially contained crystals. The abscissa is the position in the magma and the ordinate is relative pyroxene phenocryst concentration. When pyroxene phenocrysts nucleate and grow in the magma very little crystal settling takes place, even in the 50 m flow. However, substantial pyroxene settling occurs if the flow is emplaced containing large crystals. Chemical and textural evidence suggests that no pyroxene of substantial size (except for the few isolated phenocrysts in 15597) was initially present in the QNB magma. The cooling rate estimates suggest that some nucleii inherited from the more slowly cooled interior were transported to rapidly chilled outer portions of the flow. Thus, a rather thick flow (up to 50 m) of quartz-normative basalt could crystallize at the lunar surface and undergo only slight fractionation due to crystal settling. (The maximum concentration factor for the 50 m flow is 1.15.) The limited extent of chemical variation that exists between the QNBs results from the inability of pyroxene to settle appreciably during the cooling of the flow.

RECONSTRUCTION OF THE QNB FLOW

Sample position in the QNB flow: relative distance from top or bottom

Walker *et al.* (1976c) have used the fact that grain size and normative mineral content show specific correlations as a function of position in a cooling unit to determine the position of a sample in a magma body which differentiates by crystal settling in the environment of solidification. Variation in the QNBs might result from limited amounts of pyroxene settling. Unfortunately, normative mineral content differences between QNB hand specimens are slight; the variation between the vitrophyres (e.g., 15597 and 15499, Fig. 9), which presumably cool too rapidly to be involved in the differentiation, is nearly equal to the differences between the coarse-grained samples. However, correlations in texture alone may be useful in establishing relative distances of the samples from a chill margin. Such a correlation exists, for example, in the Hadley Rille vitrophyres: 15597 (Fig. 6a) appears to have many small phenocrysts, 15595 (Fig. 6b) fewer larger phenocrysts, and 15596 (Fig. 6c) even fewer larger phenocrysts.

The modal amount of phenocrystic pyroxene decreases, as does nucleation density, with increasing grain size. We have seen that these differences are unlikely to result from crystal settling. Sketches of terrestrial examples of flow differentiation (Drever and Johnston, 1959, 1967) show increasing grain size away from the edge of the flow, but it is not apparent whether or not this is coupled with an increase or decrease in the modal and/or normative abundance of phenocrysts. In any case, since little actual differentiation has occurred in

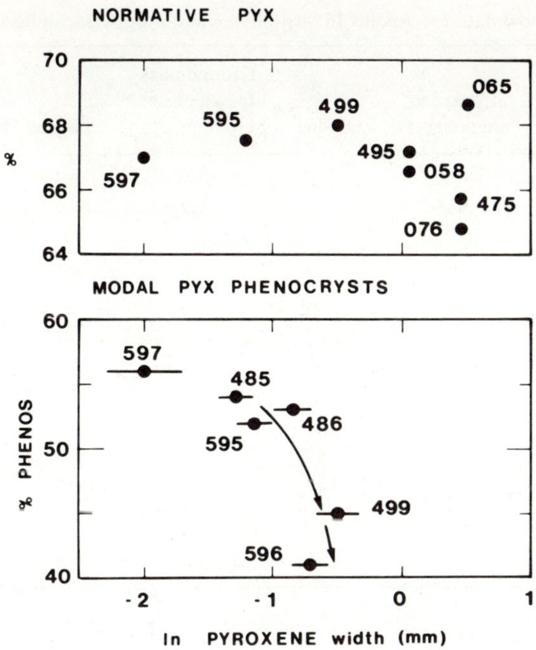

Fig. 9. Normative pyroxene calculated for chemically analyzed quartz-normative basalts is plotted against pyroxene size. Modal % of pyroxene phenocrysts (Table 2) for vitrophyres is plotted against pyroxene size. Note the correlation of increasing pyroxene size and decreasing modal pyroxene in the vitrophyres.

normative chemistry of the QNB suite, the need to ascribe the observed features to flowage differentiation is doubtful. The textural sequence probably results from differences in cooling rate which varies as a function of distance from a flow margin. At slower rates (lower nucleation density) the temperature interval between phenocryst pyroxene and groundmass pyroxene entry in the crystallization sequence is smaller than at faster cooling rates. In the narrower temperature interval less total phenocrystic pyroxene grows relative to groundmass pyroxene, even though the phenocrysts are larger. At faster cooling rates (higher nucleation density) the temperature interval for phenocryst growth is greater, groundmass pyroxene appearance is suppressed, and a larger proportion of phenocrystic pyroxene is produced. Cooling rates decrease away from a chill margin so the decrease in modal phenocrystic pyroxene with an increase in size is interpreted as a progression away from a chill margin. Abundant vesicles in the lunar outcrop (Swann *et al.*, 1972, AS15-87-11768) as well as in the hand specimens (e.g., 15485, 15486, 15499) suggest that this chill margin was the top of the flow.

Distance from chill margin

Although the pyroxene vitrophyres appear to have crystallized near the top of a flow where a fixed temperature or radiative boundary condition might be

Table 2. Modal data for Apollo 15 vitrophyres and porphyritic-spherulitic basalts.

		Pyroxene phenocrysts	Olivine	Groundmass (plagioclase, pyroxene, etc.)	Opaques	# of points	Grid size (mm)
Hadley	15597,26	55.8		43.9	0.3	1500	0.1
Rille	*	50.1		49.9	<0.1	—	—
	15595,33	52		48	—	1500	0.2
	*	48.1		51.6	0.3		
	15596,16	40.9		58.9	0.2	1600	0.2
Dune	15485,3	54.2	0.5	44.8	0.5	1500	0.2
Crater	15486,22	52.8	0.8	46.0	0.4	2800	0.2
	15499,95	44.9	1.0	53.8	0.3	3200	0.2
		45.5	1.3	53.1	0.2	1400	0.4
	†	41.8	0.8	57.3			
	15125,7	45.3	3.1	51.4	0.2	3100	0.1
Rake	a	44	4	51.6	0.4		

*Papike *et al.* (1976).
†Rhodes and Hubbard (1973).
[a]Dowty *et al.* (1973).

more appropriate to their cooling, the treatment of Jaeger (1957) for heat loss from a conductive boundary allows us to get a crude estimate of the distance of a sample from such a surface of heat loss and a minimum size for the QNB flow. The treatment of Jaeger (1957) refines Winkler's (1949) equation relating the position of a sample from a conductive boundary to the cooling rate recorded by a mineral phase.

$$x = \sqrt{\frac{k}{(\partial T/\partial T)}}. \tag{2}$$

The distance from the conductive boundary is x, k is a constant calculated from the temperature range of growth of the phase used to estimate cooling rate, the total range of solidification (1210–975°C), latent heat of fusion (80 cal/g), diffusivity of magma and wall rock (0.01 cm^2/sec) and specific heat (0.29 cal/g deg). The initial conductive boundary and country rock temperature was taken as 0°C. Plagioclase size is used to estimate the late-stage cooling rate in the vitrophyres and coarse-grained samples and is appropriate to estimates of the position of a sample at the end of crystallization.

The calculation of k is extremely sensitive to the temperature interval over which the cooling rate is recorded. Since plagioclase crystallization is systematically suppressed from its equilibrium temperature (1140°C) to lower temperatures (1050°C at 30°C/hr) separate k's are calculated for the slowly cooled coarse-grained samples and the rapidly cooled vitrophyres. The temperature intervals of plagioclase crystallization for the slowly cooled samples were taken as 1140–975°C or 1100–975°C ($k = 2.96$ and 2.66, respectively). For the rapidly cooled samples where plagioclase appearance is suppressed (1050–975°C) k was taken as 2.03. Combining the cooling rate estimate from plagioclase size (1) with (2) gives the positions of samples from a conductive heat loss boundary. The distance estimates indicate that some coarse-grained samples (Table 1) came

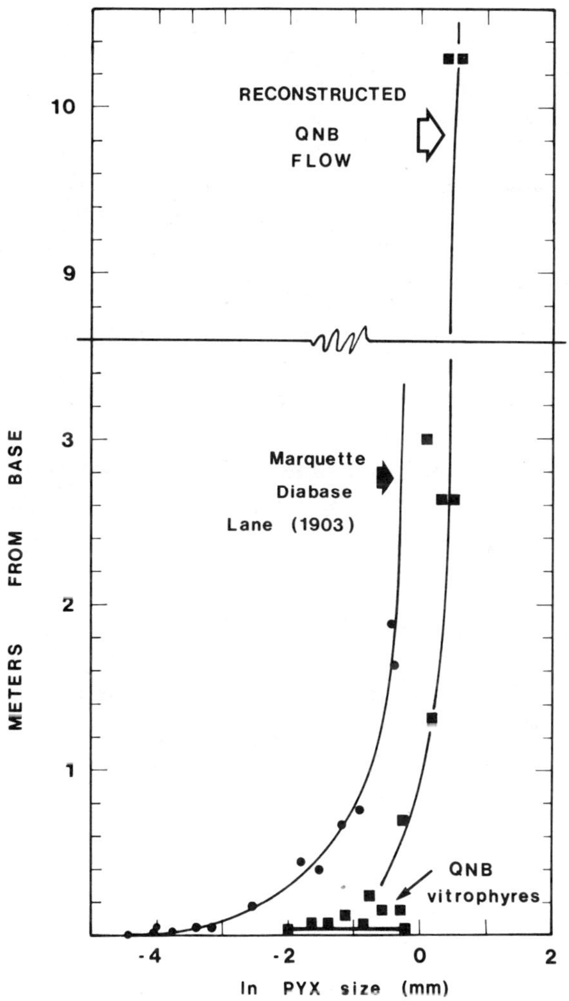

Fig. 10. The distance of QNB samples from a conductive heat loss boundary is plotted against pyroxene size. See Table 1 for lunar sample numbers. Note the wide range of pyroxene sizes preserved in the outer 20 cm of the flow margin. For comparison the measured size of augite crystals and their distance from the base of the Marquette diabase (Lane, 1903a,b) is also shown.

from 10 m within the flow. This means that the minimum flow thickness is at least 20 m and probably greater, since the calculation assumes only one nearby surface for conductive heat loss. If the vitrophyres represent the top of the flow, a constant temperature or radiative heat loss solution is more appropriate to their cooling history. Unfortunately, the mechanism of heat transfer to the lunar atmosphere is not fully characterized. Walker *et al.* (1976a) suggest that comparable cooling rates for a constant temperature boundary condition would exist about three times further from the flow surface. Conversely, radiative heat loss from the surface (Provost and Bottinga, 1972) requires that the samples be closer to the flow surface. No "cook book" method for relating cooling rate to position exists for either constant temperature or radiative heat loss.

The positions of hand samples from a conductive boundary are plotted for the QNB flow (Fig. 10) along with pyroxene size. For comparison the measured augite sizes and positions from the base of the Marquette diabase (Lane, 1903a,b) are also shown. The Marquette diabase shows a grain size versus distance distribution not unlike the one deduced from our calculated repositioning of hand specimens returned from Hadley Rille. Both show an approach to constant size 2 or 3 m from the flow contact.

CONCLUSIONS

Calculation of pyroxene phenocryst settling in a quartz-normative magma indicates that little differentiation occurs in flows up to 50 m in thickness. This property is reflected in the QNB suite by a wide variation in textures coupled with an absence of significant bulk rock compositional variation. The presence of vesicles in hand specimens of QNB vitrophyres suggests that they represent samples of the top of the QNB flow. The complex cooling histories deduced for these samples (initial slow cooling followed by rapid cooling) could result from the introduction of small intratelluric phenocryst-charged magma batches from more slowly-cooled portions of the flow into the flow's chill margin. Our simplified assumption of a single flow with simple boundary conditions does not account for the many complex cooling histories that could have occurred (e.g., extrusion of multiple flows on other pre-existing partially crystallized flows). However, the minimum thickness estimate for the flows of 20 m is not unlike the flow thicknesses that are observed in the outcrops on the far side of Hadley Rille.

Acknowledgments—This work was supported by NASA grant NGL-33-015-130 (TLG) and NGL-22-007-247 (J. F. Hays, principal investigator) and the Committee on Experimental Geology and Geophysics of Harvard University (DW). We thank D. H. Lindsley for assistance with experiments and A. E. Bence, J. Delano, J. J. Papike, and G. Lofgren for helpful discussion.

REFERENCES

Bence A. E. and Papike J. J. (1972) Pyroxenes as recorders of lunar basalt petrogenesis: Chemical trends due to crystal-liquid interaction. *Proc. Lunar Sci. Conf. 3rd*, p. 431–469.

Bottinga Y. and Weill D. F. (1970) Densities of liquid silicate systems calculated from partial molar volumes of oxide components. *Amer. J. Sci.* **269**, 169–182.

Bottinga Y. and Weill D. F. (1972) The viscosity of magmatic silicate liquids: A model for calculation. *Amer. J. Sci.* **272**, 438–475.

Boyd F. R. and Smith D. (1971) Compositional zoning in pyroxenes from lunar rock 12021, Oceanus Procellarum. *J. Petrol.* **12**, 439–464.

Bryan W. B. (1972) Morphology of quench crystals in submarine basalts. *J. Geophys. Res.* **77**, 5812–5819.

Cameron M., Sueno S., Prewitt C. T., and Papike J. J. (1973) High-temperature crystal chemistry of acmite, diopside, hedenbergite, jadeite, spodumene and ureyite. *Amer. Mineral.* **58**, 594–618.

Chappell B. W. and Green D. H. (1973) Chemical composition and petrogenetic relationships in Apollo 15 mare basalts. *Earth Planet. Sci. Lett.* **18**, 237–246.

Crawford M. L. (1973) Crystallization of plagioclase in mare basalts. *Proc. Lunar Sci. Conf. 4th.* p. 705–717.

Dowty E., Prinz M., and Keil K. (1973) Composition, mineralogy and petrology of 28 mare basalts from Apollo 15 rake samples. *Proc. Lunar Sci. Conf. 4th*, p. 423–444.

Dowty E., Keil K., and Prinz M. (1974) Lunar pyroxene-phyric basalts: Crystallization under supercooled conditions *J. Petrol.* **15**, 419–453.

Drever H. I. and Johnston R. (1959) The petrology of picritic rocks in minor intrusions a Hebridean group. *Trans. Roy. Soc. Edin.* **63**, 459–499.

Drever H. I. and Johnston R. (1967) Picritic minor intrusions. In *Ultramafic and Related Rocks* (P. J. Wyllie, ed.), p. 71–82. Wiley, New York.

Duncan A. R., Sher M. K., Abraham Y. C., Erlank A. J., Willis J. P., and Ahrens L. H. (1975) Compositional variability of the Apollo 15 regolith (abstract). In *Lunar Science VI*, p. 220–222. The Lunar Science Institute, Houston.

Grove T. L. and Bence A. E. (1977) Pyroxenes as recorders of lunar basalt petrogenesis revisited: An experimental study of pyroxene-liquid interaction (abstract). In *Lunar Science VIII*, p. 383–385. The Lunar Science Institute, Houston.

Helmke P. A., Blanchard D. P., Haskin L. A., Telander K., Weiss C., and Jacobs J. W. (1973) Major and trace elements in igneous rocks from Apollo 15. *The Moon* **8**, 129–148.

Hollister L. S., Trzcienski W. E., Jr., Hargraves R. B., and Kulick C. G. (1971) Petrogenetic significance of pyroxenes in two Apollo 12 samples. *Proc. Lunar Sci. Conf. 2nd*, p. 529–557.

Humphries D. J., Biggar G. M., and O'Hara M. J. (1972) Phase equilibria and origin of Apollo 15 basalts, etc. In *The Apollo 15 Lunar Samples* (J. W. Chamberlain and C. Watkins, eds.), p. 103–106. The Lunar Science Institute, Houston.

Jaeger J. C. (1957) The temperature in the neighborhood of a cooling intrusive sheet. *Amer. J. Sci.* **255**, 306–318.

Jaeger J. C. (1968) Cooling and solidification of igneous rocks. In *Basalts, the Poldevaart Treatise on Rocks of Basaltic Composition* (H. H. Hess and A. Poldevaart, eds.), p. 503–536. Wiley, New York.

Kirkpatrick R. J. (1976) Towards a kinetic model for the crystallization of magma bodies. *J. Geophys. Res.* **81**, 2565–2571.

Kushiro I. and Haramura H. (1971) Major element variation and possible source materials of Apollo 12 crystalline rocks. *Science* **171**, 1235–1237.

Lane A. C. (1903a) Studies of the grain of igneous intrusives. *Bull. Geol. Soc. Amer.* **14**, 369–384.

Lane A. C. (1903b) Porphyritic appearance of rocks. *Bull. Geol. Soc. Amer.* **14**, 385–406.

Lewis W. K., Gilliland E. R., and Bauer W. C. (1949) Characteristics of fluidized particles. *Ind. Eng. Chem.* **41**, 1104–1117.

Lofgren G. E., Donaldson C. H., and Usselman T. M. (1975) Geology, petrology and crystallization of Apollo 15 quartz-normative basalts. *Proc. Lunar Sci. Conf. 6th*, p. 79–99.

Lofgren G., Donaldson C. H., Williams R. J., Mullins O., Jr., and Usselman T. M. (1974) Experimentally reproduced textures and mineral chemistry of Apollo 15 quartz-normative basalts. *Proc. Lunar Sci. Conf. 5th*, p. 549–567.

Longhi J., Walker D., Stolper E. M., Grove T. L., and Hays J. F. (1972) Petrology of Mare/Rille basalts 15555 and 15065. In *The Apollo 15 Lunar Samples* (J. W. Chamberlain and C. Watkins, eds.), p. 131–134. The Lunar Science Institute, Houston.

Papike J. J., Hodges F. N., Bence A. E., Cameron M., and Rhodes J. M. (1976) Mare basalts: Crystal chemistry, mineralogy and petrology. *Rev. Geophys. Space Phys.* **14**, 475–540.

Provost A. and Bottinga Y. (1972) Rates of solidification of Apollo 11 basalt and Hawaiian tholeiite. *Earth Planet. Sci. Lett.* **15**, 325–337.

Rhodes J. M. and Hubbard N. J. (1973) Chemistry, classification and petrogenesis of Apollo 15 mare basalts. *Proc. Lunar Sci. Conf. 4th*, p. 1127–1148.

Schaber G. G. (1973) Lava flows in Mare Imbrium: Geologic evaluation from Apollo orbital photography. *Proc. Lunar Sci. Conf. 4th*, p. 73–92.

Swann G. A., Bailey N. G., Batson R. M., Freeman V. L., Hait M. H., Head J. W., Holt H. E., Howard K. A., Irwin J. B., Larson K. B., Muehlberger W. R., Reed V. S., Rennilson J. J., Schaber G. G., Scott D. R., Silver L. T., Sutton R. L., Ulrich G. E., Wilshire H. G., and Wolfe E. W. (1972) Preliminary geologic investigations of the Apollo 15 landing site. In *Apollo 15 Prelim. Sci. Rep.* NASA SP-289, p. 5-1 to 5-112.

Taylor L. A., McCallister R. H., and Sardi O. (1973) Cooling histories of lunar rocks based on opaque mineral geothermometers. *Proc. Lunar Sci. Conf. 4th*, p. 819–828.

Taylor L. A., Walker B. M., Onorato P. I. K., and Uhlmann D. R. (1977) Fe–Mg diffusion in olivine: A cooling rate indicator (abstract). In *Lunar Science VIII*, p. 928–930. The Lunar Science Institute, Houston.

Toksöz M. N., Dainty A. M., Solomon S. C., and Anderson K. R. (1974) Structure of the moon. *Rev. Geophys. Space Phys.* **12**, 539–567.

Walker D., Kirkpatrick R. J., Longhi J., and Hays J. F. (1976a) Crystallization history of lunar picritic basalt sample 12002: Phase equilibria and cooling rate studies. *Bull. Geol. Soc. Amer.* **87**. 646–656.

Walker D., Longhi J., and Hays J. F. (1976b) Heterogeneity in titaniferous lunar basalts. *Earth Planet. Sci. Lett.* **30**, 27–36.

Walker D., Longhi J., Kirkpatrick R. J., and Hays J. F. (1976c) Differentiation of an Apollo 12 picrite magma. *Proc. Lunar Sci. Conf. 7th*, p. 1365–1389.

Walker D., Longhi J., Stolper E. M., Grove T. L., and Hays J. F. (1977) Slowly cooled microgabbros 15065 and 15555 (abstract). In *Lunar Science VIII*, p. 964–966. The Lunar Science Institute, Houston.

Weigand P. W. and Hollister L. S. (1973) Basaltic vitrophyre 15597: An undifferentiated melt sample. *Earth Planet. Sci. Lett.* **19**, 61–74.

Winkler H. G. F. (1949) Crystallization of basaltic magma as recorded by variations of crystal size in dikes. *Mineral Mag.* **28**, 557–574.

Proc. Lunar Sci. Conf. 8th (1977), p. 1521–1547.
Printed in the United States of America

Slowly cooled microgabbros 15555 and 15065

DAVID WALKER,[1] JOHN LONGHI,[2] ANTONIO C. LASAGA,[1] EDWARD M. STOLPER,[1] TIMOTHY L. GROVE,[3] and JAMES F. HAYS[1]

[1]Hoffman Laboratory, Harvard University, Cambridge, Massachusetts 02138
[2]Department of Earth and Planetary Science, MIT, Cambridge, Massachusetts 02139
[3]Department of Earth and Space Sciences, SUNY Stony Brook, Stony Brook, New York 11794

Abstract—Experimental equilibration of 15555 and 15065 in high-purity iron capsules in a closed system causes little change in the sample's bulk composition. Comparison of mineral zoning trends with the experimental sequence suggests that the early crystallizing phases in the rock maintained surface equilibrium with the liquid. Olivine and pyroxene chemistry indicates that 15555 cooled slightly more rapidly than 15065 which cooled at ~0.2–0.3°C/day. Even though the cooling rates are slow, significant amounts of accumulated olivine or pyroxene are not found in these rocks—they represent liquid compositions. These liquids appear to have been fractionated during ascent through the crust (15065) or to have been derived from an olivine-pyroxene source (15555) of Fe/Fe + Mg ~ 0.73 at a depth between 150–200 km.

INTRODUCTION

THE APOLLO 15 MISSION landed near the shore of Palus Putredinis on the southeast margin of Mare Imbrium at the foot of the Apennine Mountains. Iron-rich vitrophyres, basalts, and gabbros of an olivine-normative suite (including 15555) and a quartz-normative suite (including 15065) were returned from the mare floor and the edge of Hadley Rille (LSPET, 1972). Members of these two suites were shown to have a wide range of textures, but relatively simple chemical variations which could be described principally in terms of olivine and pigeonite fractionation, respectively (Rhodes and Hubbard, 1973; Chappell and Green, 1973). The wide range of textures has been shown to be related to the cooling rate experienced by the sample during crystallization (Lofgren *et al.*, 1975; Grove, 1977; Bianco and Taylor, 1977). This paper gives an experimental demonstration of the chemical variations expected in residual liquids and accumulated crystals during crystallization of representative compositions. The experiments help to evaluate the relationship between rock compositions and liquid compositions, and the extent to which equilibrium was maintained during crystallization of the rocks studied. Furthermore, indirect estimation of the cooling rates during solidification of the relatively coarse grained rocks studied (15555 and 15065) allows estimation of magma body dimensions and of the length of time available for accomplishing differentiation in the local environment where the rocks solidified.

EXPERIMENTAL ($P \leqslant 1$ bar)

Iron-bearing basaltic compositions present special difficulties in experimental work because interaction with a container may produce a change in bulk composition of the charge. The possible effects and the dependence on P_{O_2} have been reviewed by Kesson and Lindsley (1976). We have investigated various procedures in the search for a technique which would maintain the bulk composition of a charge constant for a long enough period of time to establish equilibrium.

We report here the results of experiments on 15555 and 15065 by four techniques at pressures $\leqslant 1$ bar. Natural rock powder of ~20 μm grain size was used as starting material unless otherwise noted.

Method 1

Drilled Mo rod was used as a capsule in evacuated silica tubes. Experiments were typically of a duration from 1–3 days and iron loss ranged up to 20% of the amount present near the liquidus temperatures. The P_{O_2} of these experiments is unknown in detail but can be estimated to be effectively around 1–2 orders of magnitude below the Fe/FeO equilibrium (Walker *et al.*, 1973). For runs on the order of a few hours, the iron loss is less than 5% of the amount present. The phase equilibrium results for this set of experimental conditions (and the preliminary results in the 0–20 kbar pressure range) were presented by Longhi *et al.* (1972).

Method 2

In an attempt to investigate the effects of P_{O_2} on the crystallization sequence, a second series of experiments was performed in Mo foil envelopes in a gas-mixing furnace with $H_2/CO_2 = 5$ (about 1.16 log units below Fe/FeO) for durations of 6–24 hr. This experimental configuration resulted in iron losses of as much as 25% of the amount present.

Method 3

To reduce the iron losses, a third series of experiments was performed with $H_2/CO_2 = 4$ (about 0.95 log units below Fe/FeO) in Mo foil envelopes for durations of 1 hr. The results of these experiments were consistent with an essentially constant bulk composition, but the short duration of the runs leaves legitimate doubt about the establishment of equilibrium.

Method 4

Drilled Fe rod of high purity (~10 ppm impurities; Johnson-Matthey Grade 1) was used as a capsule material inside sealed silica tubes. Runs were typically of duration 1–3 days and iron loss or gain was ~ ±2% of the amount present.

Temperature calibration and control systems for these experiments are described in Walker *et al.* (1973, 1975) and Longhi *et al.* (1974). Typical charge masses are about 5–10 mg. Results are listed in Table 1. The Fe/Fe + Mg of the product phases, which is a useful index of Fe gain or loss, is also given. Only techniques (3) and (4) have products which do not show significant changes in charge bulk composition. Techniques (1) and (2) show significant Fe loss to the Mo containers. The loss is noticeably worse in technique (2) where the surface of the container/mass of charge ratio is greater than in (1). Technique (3) produced more heterogeneous pyroxenes than the other techniques and failed to produce pyroxene at 1200°C in 15555 even though pyroxene was present at 1224°C and 1175°C. These features indicate that 1 hr was not a sufficient length of time to establish an equilibrium. All four techniques produced somewhat different appearance fields for olivine and pyroxene. This is particularly obvious for 15065 olivine, which is present in technique (1) only at 1250°C, but present from 1273°C to 1124°C (with exceptions to be noted below) in technique (4).

Technique (4), Fe capsules in sealed silica tubes, gives the most satisfactory results for the present compositions in terms of maintaining bulk composition constant and allowing runs of long enough duration to approach equilibrium. Plagioclase appearance was reversed with a wide bracket, but dissolution temperatures are thought to be more representative of the equilibrium relations than precipitation temperatures because of the difficulty of plagioclase nucleation. Details of the phase chemistry produced by technique (4) are listed in Tables 2 and 3.

Use of Fe capsules in sealed silica tubes requires particular care in preparation. We have followed the procedures below which we believe to be important:

(a) Purity of the iron is paramount. Severe iron loss in the silicates can result from use of impure iron (Walker *et al.*, 1972; Grove *et al.*, 1973). Biggar *et al.* (1974) have suggested that trace amounts of carbon and phosphorus in iron in sealed system experiments may be capable of reducing a portion of the ferrous silicates in the charge.

(b) Machining of capsules and lids from rod stock should be done under heavy oil which should not be removed until the capsule is ready for immediate use. This procedure prevents oxidation. Oil can be removed with acetone and compressed air.

(c) Cleaning of the iron at high temperature in a hydrogen-bearing atmosphere to remove oxidation is to be avoided, because subsequent charges run in such capsules tend to reduce the silicates. This is presumably the result of hydrogen absorption by the iron.

(d) Baking of empty capsules within evacuated silica tubes ~12 hr at ~1200°C before use with a silicate charge is desirable to anneal the capsule and remove residual surface films. It is also desirable to equilibrate a charge in its melting range in the capsule in an evacuated silica tube as a further preliminary conditioning procedure. This method serves to "get" most remaining impurities, and although this charge may suffer iron gain or loss, subsequent charges run in such a prepared capsule have only limited (~ ±2% of Fe present) iron gain or loss.

(e) It is desirable to reuse a single capsule for the whole series of experiments on a given composition. This can be accomplished by placing only 5–10 mg of sample in a capsule with a capacity of several mm^3. The charge can be pried out afterwards. Unfortunately, this procedure promotes partial alkali extraction from the charge into the void space. Since the lids are not vapor seals, but only mechanical plugs, the alkalies may escape and combine with the silica tube but their actual disposition is unknown.

(f) Silica tubes with ~1.5 mm walls are sufficiently thick to prevent inward oxygen diffusion and are durable enough to sustain the amount of devitrification encountered in our run times.

(g) Starting materials should have an intrinsic P_{O_2} close to being consistent with iron saturation. Material with excess ferric iron will oxidize some of the iron capsule in the process of reaching iron-saturation equilibrium, leading to a change in bulk composition of the charge. This effect probably is not severe for materials which depart slightly from iron saturation, but examples of metal-bearing lunar basalts which gain iron when equilibrated in properly prepared iron capsules are known (Walker *et al.*, 1975). The P_{O_2} of the charges in the present experiments is not known in detail, but is thought to be relevant to the crystallization of the natural material because bulk composition is maintained and mineral chemistry of the experimental and natural silicates is comparable.

Our preferred results from technique (4) for 15555 may be compared to those reported by Kesson (1975) for a synthetic analog of 15555–15016 also run in high-purity iron in evacuated silica tubes. We confirm her conclusion that pyroxene entry by technique (1) was at too high a temperature. We also confirm the other phase appearance temperature she reported. There is minor disagreement about the composition of the liquidus olivine composition she reports of Fo_{75} (vs. Fo_{73} here). We prefer Fo_{73} because it conforms to the equilibrium olivine-liquid distribution coefficient ($K_D = 0.33$) observed in low-Ti mare basalts (Longhi *et al.*, 1975) for the bulk compositions run in both laboratories.

Table 1. Apollo 15 basalt

Mo rod—sealed tube							Mo foil—$5/6H_2/1/6CO_2$						
Run #	T°C Duration (hours)	Liq	Ol	Sp	Px	Plag	Run #	T°C Duration (hours)	Liq	Ol	Sp	Px	Plag
555-22	$1300_{(18.7)}$	45					555-85	$1299_{(9)}$	43				
555-18	$1275_{(12.5)}$	48	24				555-84	$1276_{(24)}$	37	ol			
555-17	$1250_{(12.3)}$	51	26				555-83	$1250_{(17)}$	38	21			
555-16	$1225_{(22.3)}$	52	27	62			555-82	$1225_{(24)}$	44	23		Wo_5En_{75}	
555-47	$1250/1225_{(6.5/38.3)}$	43	20	sp	Wo_4En_{77}								
555-11	$1200_{(18.3)}$	52	28	57	Wo_5En_{72}		555-81	$1203_{(6)}$	51	29	sp	Wo_6En_{71}	
555-12	$1175_{(58)}$	59	33	70	cpx		555-80	$1176_{(14)}$	50	25	sp	Wo_6En_{67}	
555-14	$1150_{(47)}$	58	36	sp	$Wo_{10}En_{59}$		555-79	$1151_{(10)}$	56	35	sp	Wo_9En_{54}	
555-15	$1125_{(76)}$	68	41	sp	$Wo_{11}En_{55}An_{91}$ $Wo_{31}En_{46}$		555-78	$1132_{(15)}$	63	ol	sp	$Wo_{15}En_{50}$	plag
555-46	$1150/1125_{(1.5/66.8)}$	67	38	sp	$Wo_{11}En_{58}$ An_{93}								
555-25	$1100_{(58)}$	75?	ol	sp	cpx	plag							
555-13	$1050_{(36)}$	absent	ol	sp	cpx	plag							
065-11	$1275_{(27.4)}$	43					065-76	$1276_{(24)}$	30				
065-18	$1263_{(43.1)}$	25			cpx								
065-10	$1250_{(15.4)}$	43	20				065-75	$1250_{(17)}$	43	22	54		
065-20	$1250_{(21)}$	37			Wo_4En_{80}								
065-19	$1240_{(46.1)}$	19			Wo_4En_{84}								
065-6	$1225_{(42)}$	42		sp	Wo_4En_{75}		065-74	$1225_{(24)}$	42		55	Wo_5En_{76}	
065-5	$1200_{(46.4)}$	52		sp	Wo_5En_{69}		065-73	$1203_{(6)}$	49	24	sp	Wo_6En_{64}	
065-4	$1178_{(47)}$	49		sp	Wo_7En_{65}		065-72	$1176_{(14)}$	51		sp	Wo_8En_{61}	
065-55	$1300/1160_{(0.5/44.5)}$	50		61	Wo_8En_{66}								
065-2	$1150_{(141.5)}$	60		70	Wo_9En_{64} An_{94}		065-71	$1151_{(10)}$	56		sp	$Wo_{10}En_{60}$	plag
065-15	$1170/1150_{(5.4/44.5)}$	60		sp	Wo_9En_{60}	plag							
065-3	$1125_{(228.3)}$	65		sp	Wo_8En_{62} $Wo_{15}En_{54}$	plag	065-70	$1132_{(15)}$	63		sp	$Wo_{14}En_{46}$ $Wo_{32}En_{45}$	plag

When two values of run temperature and duration are listed separated by a slash, the first values refer to an initial equilibration and the second values to the final equilibration before quenching. There was no intermediate quenching. The same convention is followed in Tables 2 and 3. Numbers listed for liquid (Liq), olivine (Ol), and spinel (Sp) are values of $Fe/Fe+Mg$ for those phases determined by electron microprobe. If a phase was present but not analysed, the abbreviation for the phase is listed.

EXPERIMENTAL (P 5–30 kilobars)

Runs in the range 5–30 kbar were carried out in a Boyd–England (1960) solid-medium, piston-cylinder apparatus using the operating procedures described in Walker *et al.* (1972). Pressure corrections of −8% for friction were used to estimate actual pressure of piston-in runs. No pressure correction was applied to the nominal e.m.f. of Pt/Pt10Rh thermocouples used to monitor and control temperature. Molybdenum capsules were used to contain the charges in general. However, at high pressures and at the high temperatures prevailing at the liquidus, iron loss by diffusion into the molybdenum proved to be significant. Some runs were performed in graphite capsules to try to circumvent the iron loss problem. This did not prove to be an especially satisfactory alternative since it can be demonstrated that graphite capsules leak at least half the time. This casts some doubt on their integrity the rest of the time.

The results of experiments at pressures from 5–30 kbar are listed in Tables 4 and 5. All high pressure experiments on 15065 were done in Mo capsules, a procedure which has probably enhanced experimental results ($P \leq 1$ bar).

Mo foil—4/5H_2/1/5CO_2—short runs						High—purity iron—sealed tube						
Run #	T°C Duration	Liq	Ol	Sp	Px	Run #	T°C Duration (hours)	Liq	Ol	Sp	Px	Plag
	All for one hour					555-96	$1304_{(16)}$	53				
555-86	1274	53	29			555-97	$1277_{(18.7)}$	52	27			
555-87	1249	57	31	69		555-98	$1233_{(25)}$	57	32			
555-88	1224	57	40	sp	(Wo_4En_{71})							
						555-99	$1201_{(20.7)}$	62	36	72		
555-89	1200	61	40	sp		555-100	$1175_{(30.2)}$	64	38	73		
555-90	1175	63	42	sp	$Wo_{11}En_{58}$	555-103	$1157_{(46.5)}$	66	40	sp	Wo_9En_{60}	
						555-102	$1148_{(38.5)}$	66	41	75	$Wo_{12}En_{53}$	plag
						555-101	$1124_{(50.1)}$	74	46	83	$Wo_{14}En_{51}$	An_{98}
											$Wo_{31}En_{40}$	
						555-105	$1176/1124_{(1/117)}$	73	48	81	$Wo_{34}En_{42}$	An_{99}
065-84	1274	50		69		065-104	$1273_{(22.8)}$	49	ol			
065-85	1249	51	26	sp		065-101	$1249_{(24.3)}$	53	28			
065-86	1224	54	30	77		065-100	$1225_{(20.5)}$	53	28	63	Wo_6En_{68}	
065-89	1200	57	32	sp	Wo_5En_{68}	065-99	$1205_{(22)}$	56	30	sp	Wo_5En_{68}	
065-90	1175	59	37	sp	Wo_5En_{68}	065-98	$1172_{(23.7)}$	61	35	69	Wo_8En_{62}	
						065-105	$1148_{(38.5)}$	67		sp	$Wo_{11}En_{59}$	An_{95}
						065-108	$1190/1134_{(5.3/43)}$	70	44	78	$Wo_{12}En_{54}$	An_{99}
						065-106	$1129_{(72.5)}$	71		sp	pig	plag
						068-102	$1128_{(70)}$	74		sp	$Wo_{12}En_{53}$	An_{96}
						065-96	$1125_{(47.2)}$	74	50	81	$Wo_{14}En_{49}$	An_{95}
						065-107	$1124_{(67)}$	75	49	83	$Wo_{12}En_{53}$	An_{98}
						068-103	$1101_{(76.7)}$	liq		sp	$Wo_{14}En_{45}$	plag

spinel stability. This conclusion is reached by analogy with the runs on 15555 at high pressures done in Mo, graphite, and high purity iron. Spinel shows enhanced stability in Mo and graphite relative to iron capsules, especially in the 5–10 kbar range. Neglecting spinel then, clinopyroxene is the liquidus phase for 15065 at elevated pressures. The observed clinopyroxene liquidus pressure-temperature gradient may be somewhat steeper than the real gradient as a result of iron loss to the Mo at the high temperatures reached at higher pressures. An exact determination of the magnitude of the possible effect was not pursued, since it was felt that the answer would add little to the understanding of the petrogenesis of 15065. The products recovered from experiments near the solidus at high pressure probably do not represent equilibrium assemblages since there appears to be a problem with reproducibility. The same is true for the experiments on 15555.

High pressure experiments in Mo capsules on 15555 reported by Longhi *et al.* (1972) indicated that this composition was cosaturated with olivine and clinopyroxene ± spinel near 10 kbar. Phase relations near this pressure at the liquidus were reinvestigated in high-purity iron. The cosaturation pressure is at about 8.5 kbar and spinel is not present at the liquidus. Phase compositions are listed in Table 3.

Kesson (1975) has also reported high pressure experimental results in iron on a 15016–15555 synthetic analog. She found the olivine-pyroxene cosaturation pressure to be about 12 kbar. The difference between the reported cosaturation pressures is thought to result partly from different pressure calibration procedures used in the two laboratories (see Johannes *et al.*, 1971).

Table 2. 15065 experiments in high-purity iron capsules—phase compositions.

		Bulk composition	065-104	065-101	065-100	065-99	065-98	065-105	065-108	065-106	065-102	065-96	065-107	065-103	065-112*
Liquid	SiO_2	47.1	47.4	47.5	48.0	47.8	47.2	46.1	46.0	47.2	45.6	45.8	45.9	present	41.6
	TiO_2	1.63	1.70	1.63	1.72	1.84	2.02	2.53	2.08	3.29	3.44	2.40	3.01		2.27
	Al_2O_3	9.17	9.23	9.39	10.2	10.7	11.7	11.1	11.4	10.2	10.2	10.3	10.4		9.36
	Cr_2O_3	.81	.73	.83	.66	.55	.40	.31		.34	.24	.27	.26		
	FeO	19.2	18.1	19.3	17.8	18.4	18.9	20.1	21.6	21.7	22.5	23.4	23.7		32.4
	MgO	10.4	10.6	9.63	8.96	7.96	6.85	5.55	5.27	4.92	4.49	4.65	4.32		2.94
	MnO	.30						.32		.30			.34		
	CaO	9.43	9.59	9.72	10.5	10.8	11.3	11.4	11.5	11.7	11.2	11.1	11.1		12.0
	Na_2O	.32	.01	.05	.09	.24	.40	.17		.04	.16	.33	.05		
	Σ	98.36	97.36	98.05	97.93	98.29	98.77	98.58	97.85	99.69	97.83	98.25	99.08		100.57
Olivine	SiO_2		present	36.7	35.6	36.3	35.6	[absent]	34.5	[absent]	[absent]	33.6	35.8	[absent]	34.1
	TiO_2			.08	.10	.11	.12		.11			.14	.14		.17
	Cr_2O_3			.69	.67	.57	.45					.37	.45		
	FeO			24.1	24.4	26.6	30.5		36.5			40.8	39.8		51.1
	MgO			35.6	35.9	34.4	32.1		26.2			23.1	22.9		15.2
	MnO												.41		
	CaO			.38	.43	.49	.47		.51			.62	.68		.86
	Σ			97.55	97.10	98.47	99.24		97.82			98.63	100.18		101.43
Spinel	TiO_2				3.11	present	3.41	present	2.97	present	present	3.21	7.45	present	present
	Al_2O_3				13.7		14.1		14.3			13.3	12.4		
	Cr_2O_3				48.2		46.6		45.0			46.4	38.8		
	FeO				24.4		27.0		30.3			31.0	35.2		
	MgO				8.08		6.81		4.92			4.02	4.06		
	MnO								.38				.37		
	CaO				.20		.45		.30			.23	.25		
	Σ				97.69		98.37		98.17			98.16	98.53		

Pyroxene	SiO_2			53.7	53.6	53.0	51.3	49.9	present	51.7	50.8	52.4	49.3	
	TiO_2			.28	.34	.38	.63	.56		.57	.66	.49	1.02	
	Al_2O_3			.80	.98	1.27	2.32	2.10		1.40	1.46	1.40	2.30	
	Cr_2O_3			.89	.87	.75	1.04	1.01		.66	.62	.69	.73	
	FeO			16.9	17.0	18.8	18.4	20.8		21.5	22.4	21.1	24.0	
	MgO			24.1	24.0	22.0	20.0	18.3		18.5	16.6	18.4	15.0	
	MnO			.34	.32	.37	.34	.34		.39	.40	.38	.42	
	CaO			2.91	2.68	3.78	5.08	5.58		5.73	6.46	5.82	6.57	
	Na_2O			.02	.04	.02	.03	.03		.03	.03	.03	.03	
	Σ			99.94	99.83	100.37	99.14	98.62		100.48	99.43	100.71	99.37	
Plagioclase	SiO_2						45.2	43.8	present	44.3	44.1	46.8	present	45.4
	Al_2O_3						33.6	33.5		33.9	34.0	35.5		34.4
	FeO						.93	1.23		1.01	1.22	.91		1.52
	MgO						.28	.43		.26	.20	.29		.15
	CaO						18.6	19.5		18.8	18.7	18.7		19.6
	Na_2O						.33	.09		.39	.53	.26		
	Σ						98.94	98.55		98.66	98.75	102.46		101.07
	T°C	1273	1249	1225	1205	1172	1148	1190/1134	1129	1128	1125	1124	1101	1124
	Hours	22.8	24.3	20.5	22	23.7	38.5	5.3/43	72.5	70	47.2	67	76.7	46.5

*Starting material was 21% FeO + 79% 15065.

Electron microprobe analyses are based on the matrix correction procedure of Bence and Albee (1968). Standards used are quartz, rutile, kyanite, uvarovite, hematite, enstatite, rhodonite, diopside, and labradorite. Same applies for Table 3.

Table 3. 15555 experiments in high-purity iron capsules—phase compositions.

		Bulk composition	555-97	555-98	555-99	555-100	555-103	555-102	555-101	555-105	555-115
		2, 3, 96									
Liquid	SiO_2	44.8	44.3	45.7	45.5	46.1	46.4	45.8	44.1	43.2	44.2
	TiO_2	2.49	2.68	2.78	2.84	2.82	3.09	3.26	4.71	3.60	2.65
	Al_2O_3	8.60	8.54	9.23	9.52	10.1	10.1	10.8	9.85	9.68	8.73
	Cr_2O_3	.69				.39	.36	.29	.23		
	FeO	22.2	20.6	20.7	21.1	21.0	21.8	20.0	23.4	23.5	20.8
	MgO	11.1	10.6	8.64	7.17	6.61	6.40	5.74	4.61	4.93	10.3
	MnO					.33	.30	.31	.35		
	CaO	9.21	9.74	10.1	10.8	11.1	11.7	11.7	11.0	11.6	9.71
	Na_2O	.34				.10	.03	.14	.11		
	Σ	99.43	96.46		96.93	98.55	100.18	98.04	98.36	96.51	96.39
Olivine	SiO_2		35.9	36.0	35.8	34.9	36.2	35.3	34.7	33.0	36.6
	TiO_2		.09	.11	.13	.15	.18	.16	.21	.17	.13
	Cr_2O_3					.37	.44	.34	.31		
	FeO		23.8	27.7	30.9	32.5	34.2	34.6	38.3	39.7	26.5
	MgO		36.2	33.3	31.0	29.9	29.3	27.5	24.8	24.3	34.7
	MnO					.38	.39	.43	.43		
	CaO		.37	.39	.44	.42	.53	.44	.44	.51	.50
	Σ		96.36	97.50	98.27	98.62	101.24	98.77	99.19	97.68	98.43

Spinel	TiO_2			4.87	4.67	present	6.59	12.2		5.11	
	Al_2O_3			13.2	12.4		12.4	9.44		11.6	
	Cr_2O_3			43.1	44.2		38.2	34.1		46.0	
	FeO			28.8	29.3		31.1	37.5		32.3	
	MgO			6.22	5.98		5.74	4.24		4.28	
	MnO				.39		.37	.44		.40	
	CaO			.32	.25		.60	.25		.28	
	Σ			96.51	97.19		95.00	98.17		99.97	
Pyroxene	SiO_2					52.8	51.5	51.5	49.07	48.9	52.5
	TiO_2					.52	.72	.81	1.79	1.06	.40
	Al_2O_3					1.09	1.48	1.59	4.07	2.82	1.52
	Cr_2O_3					.70	.61	.72	.88	1.04	
	FeO					19.9	21.5	20.8	15.8	14.7	16.3
	MgO					21.3	18.1	17.6	12.5	14.0	22.2
	MnO					.33	.40	.37	.34	.25	
	CaO					4.36	5.54	6.38	13.6	15.9	4.65
	Na_2O					.01	.02	.03	.03	.04	
	Σ					101.01	99.87	100.30	98.08	98.71	97.57
Plagioclase	SiO_2						present	45.3		43.0	
	Al_2O_3							33.0		33.5	
	FeO							2.56		.94	
	MgO							.56		.38	
	CaO							18.3		19.5	
	Na_2O							.24		.09	
	Σ							99.96		97.41	
	T°C	1277	1233	1201	1175	1159	1148	1124		1176/1124	1300
	Hours	18.7	25	20.7	30.2	46.5	38.5	50.1		1/117	2 (8.5 kb)

Table 4. 15065 high pressure experiments.

	Run #	T°C	Capsule	Duration	Products					
5 kbar	065-7	1275	Mo	2	glass	sp				
	065-9	1250	Mo	2.2	glass	sp				
	065-58	1240	Mo	3	glass	sp	cpx			
	065-21	1160	Mo	9	glass	sp	cpx	plag		
	065-41	1130	Mo	18.3	glass	sp	cpx	plag		
7.5 kbar	065-67	1280	Mo	2	glass		cpx			
10 kbar	065-22	1290	Mo	2	glass					
	065-12	1275	Mo	2	glass		cpx			
	065-13	1250	Mo	2.5	glass	sp	cpx			
	065-47	1200	Mo	2	glass	sp	cpx			
	065-8	1175	Mo	18.7	glass	sp	cpx	plag		
12 kbar	065-38	1300	Mo	2	glass		cpx			
15 kbar	065-39	1405	Mo	1.7	glass					
	065-36	1390	Mo	1.8	glass		cpx			
	065-34	1375	Mo	2.3	glass		cpx			
	065-29	1365	Mo	1.5	glass		cpx			
	065-28	1360	Mo	2	glass		cpx(q?)			
	065-26	1340	Mo	2	glass		cpx			
	065-25	1325	Mo	1.8	glass	(sp?)	cpx			
	065-23	1300	Mo	2	glass	(sp?)	cpx			
	065-43	1230	Mo	3.1	glass	sp	cpx	plag	silica	
	065-14	1200	Mo	9	glass	sp	cpx	plag	(silica?)	
	065-54	1200	Mo	41.8	glass	sp	cpx	plag	silica	
20 kbar	065-53	1475	Mo	3	glass		cpx			
	065-42	1450	Mo	1.8	glass		cpx		(silica?)	
	065-30	1400	Mo	2	glass		cpx			
	065-27	1350	Mo	2	glass		cpx			
	065-24	1200	Mo	10		sp	cpx	plag	silica	ilmenite
	065-57	1200	Mo	41.8		sp	cpx	plag	silica	garnet (ilm?)
25 kbar	065-37	1300	Mo	7.8	glass	(sp?)	cpx	(plag?)	(silica?)	
	065-35	1250	Mo	19	glass	sp	cpx	plag	silica	
	065-52	1250	Mo	24.3		sp	cpx	plag	(silica?)	garnet
	065-31	1204	Mo	10.1	glass	sp	cpx			
30 kbar	065-48	1250	Mo	26	glass	sp	cpx	plag	silica	garnet
	065-44	1200	Mo	17.2	glass	sp	cpx	plag	silica	ilm (gar?)

Crystallization of 15555 and 15065

The crystallization sequence of a rock may often be deduced from its texture and mineral zoning. A comparison between the petrographically-determined and experimental crystallization sequences allows an evaluation of the extent to which equilibrium was maintained during the rock's crystallization. To the extent that the natural crystallization is not strongly super-cooled, the experiments may be used to indicate a temperature scale for the natural event. The compositions of early crystallizing phases which have solid solutions, such as olivine and pyroxene, will be dependent upon the liquid composition from which they precipitate. If cores of natural minerals have the same composition as liquidus phases in experiments on the bulk composition of the rock, then the bulk composition of the rock is consistent with that of a liquid. If, however, the rock

Table 5. 15555 high pressure experiments.

	Run #	T°C	Capsule	Duration	Products						
5 kbar	555-4	1300	Mo	1.7	glass						
	555-110	1295	Fe	2	glass	ol					
	555-65	1287	Mo	2.1	glass						
	555-108	1285	Fe	2.2	glass	ol					
	555-5, 73	1250	Mo	2.2	glass	ol	sp				
	555-45	1125	Mo	24.2	glass	ol	sp	cpx	plag		
7.5 kbar	555-72	1290	Mo	2.1	glass	ol					
	555-70	1280	Mo	2	glass	ol	(sp)	(cpx)			
8.5 kbar	555-115	1300	Fe	2	glass	ol(Fo70)		cpx ($Wo_{10}En_{64}$)			
10 kbar	555-67	1325	graphite	2	glass						
	555-111	1320	Fe	2	glass						
	555-66	1312	graphite	2.1	glass			cpx			
	555-62	1300	graphite	2	glass			cpx			
	555-8	1300	Mo	2.2	glass						
	555-107	1300	Fe	2	glass	ol		cpx			
	555-58	1287	graphite	2.3	glass			cpx			
	555-61	1287	Mo	2.1	glass		sp	cpx			
	555-7	1275	Mo	2.7	glass	ol	sp	cpx			
	555-6	1250	Mo	2.1	glass	ol	sp	cpx			
	555-21	1200	Mo	13.3	glass	ol	sp	cpx			
12.5 kbar	555-68	1325	graphite	2.2	glass						
	555-63	1312	graphite	2.4	glass			cpx			
	555-64	1312	Mo	2	glass						
	555-60	1300	graphite	2	glass			cpx			
	555-71	1300	Mo	2	glass			cpx			
	555-44	1275	Mo	2.8	glass		sp	cpx			
15 kbar	555-19	1350	Mo	2	glass						
	555-38	1325	Mo	2.7	glass						
	555-9, 39	1300	Mo	2.3	glass			cpx			
	555-10	1275	Mo	2.2	glass			cpx			
	555-27	1200	Mo	14.2	glass	(ol?)	(sp?)	cpx			
20 kbar	555-59	1425	graphite	2.1	glass			cpx			
	555-50	1400	graphite	2.2	glass			cpx			
	555-57	1375	graphite	2.2	glass			cpx			
	555-52	1350	graphite	2.2	glass						
	555-53	1325	graphite	2.2	glass			cpx	(?opx)		
	555-29	1200	Mo	11.5	glass	ol	sp	cpx	garnet		
25 kbar	555-56	1450	graphite	2.2	glass			cpx			
	555-55	1400	graphite	2.2	glass			cpx	(gar?)		
	555-54	1360	graphite	2.1	glass			cpx	garnet		
	555-30	1200	Mo	18		(ilmenite?)					
							(sp?)	cpx	garnet	plag	silica

is not a liquid composition but contains a significant admixture of early phenocrysts, then experiments on the rock's bulk composition will produce liquidus phases with a content of the high-temperature component (e.g. forsterite or enstatite) higher than that observed in the petrographic phenocrysts. This effect results from the extra temperature increment required to dissolve the admixed phenocrysts. In principle, then, a comparison of liquidus phases and early phenocryst cores provides a test of whether a rock has admixed phenocrysts or

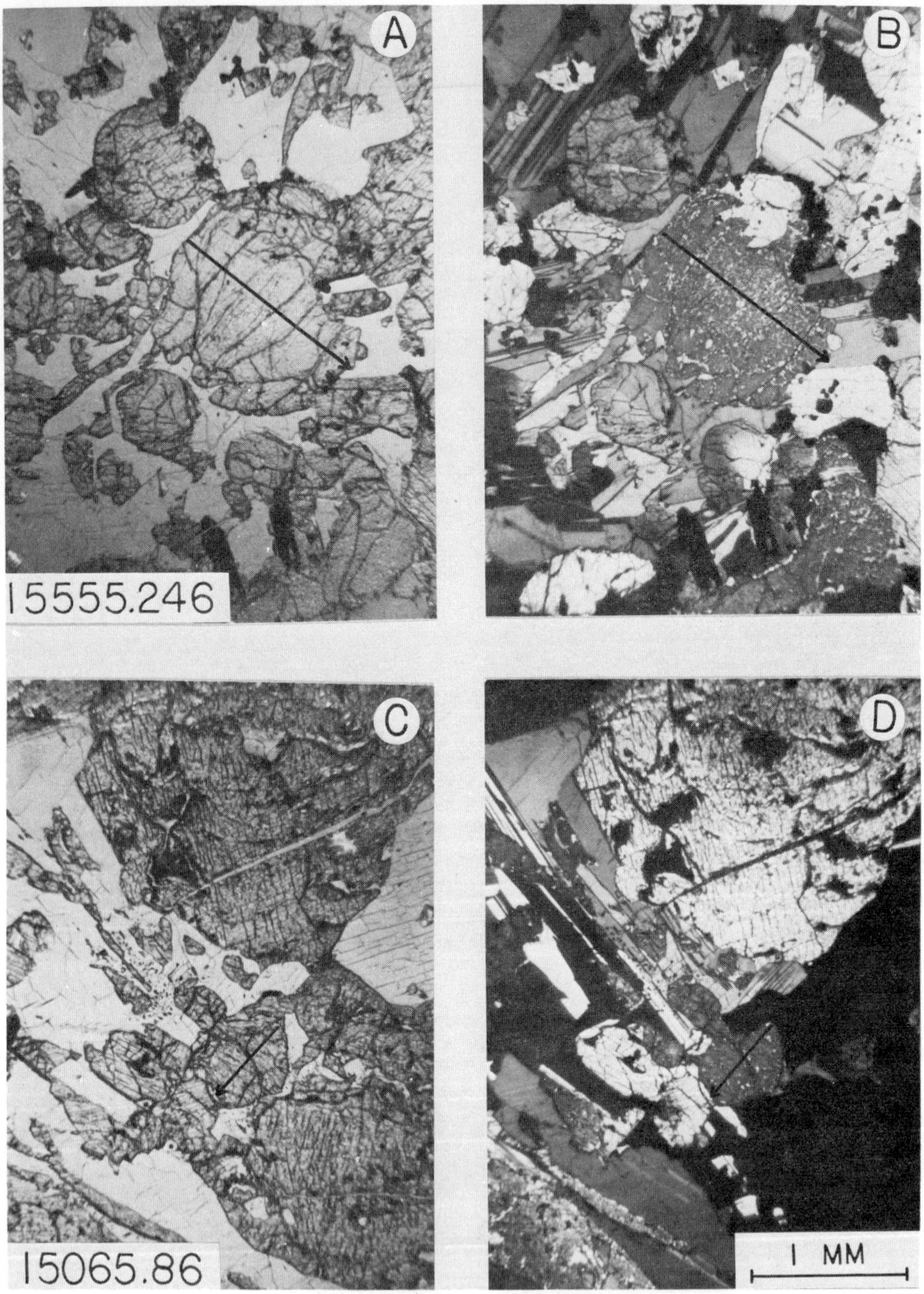

Fig. 1. Photomicrographs in transmitted light of thin sections of Apollo 15 microgabbros. B and D are the same subject as A and C respectively, but with crossed nicols. The scale is the same for all photos. Lines with arrow tips indicate the path of compositional profiles across olivine grains shown in Fig. 4.

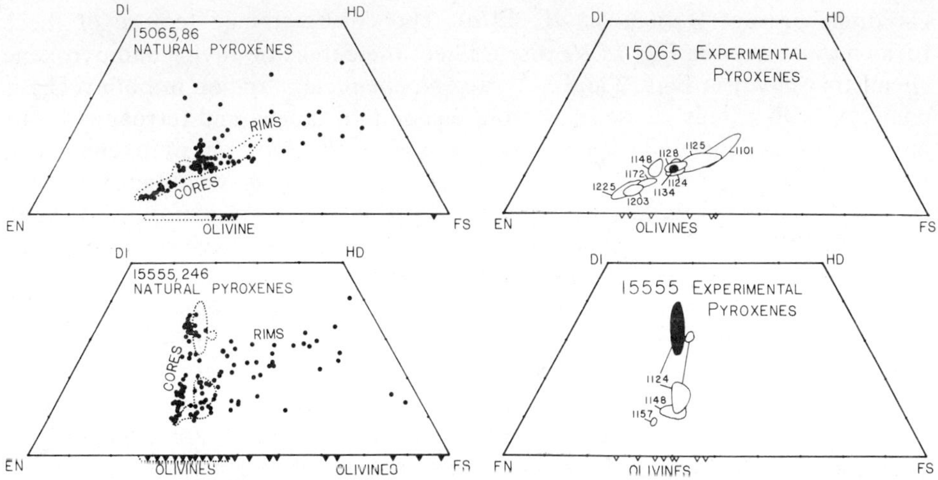

Fig. 2. Pyroxene compositions and olivine Fe/Fe + Mg ratios. Dotted envelopes in natural composition subfigures indicate the range of experimental phase compositions. Numbers on envelopes in experimental composition subfigures indicate equilibration temperatures. Filled envelopes are compositions produced in experiments where starting materials were first equilibrated at a temperature above the final temperature.

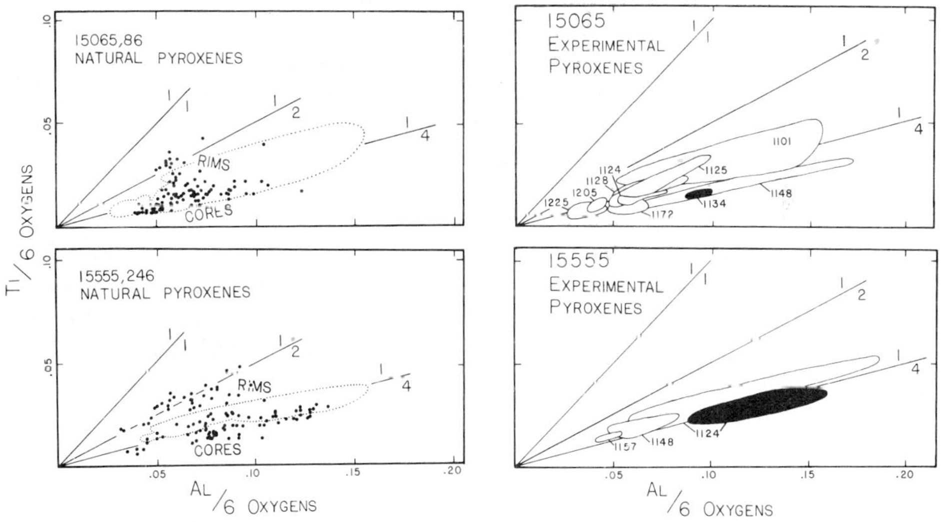

Fig. 3. Ti and Al in pyroxenes. Envelopes and numbers have the same significance as in Fig. 2. Note experimental Al–Ti heterogeneity.

is consistent with being a liquid composition. Examples of the use of this test and criticisms thereof may be found in Green *et al.* (1971), Grove *et al.* (1973), Longhi *et al.* (1974), and O'Hara *et al.* (1975).

We have previously described the petrography and given modes for 15555 and 15065 in Longhi *et al.* (1972) and discussed the plagioclase compositional variations present (Longhi *et al.*, 1976). The coarse-grained texture of these rocks may be seen in Fig. 1. We discuss here the details of olivine and pyroxene chemistry shown in Figs. 2 and 3. Pyroxene chemical variation in both rocks is complex, with extensive zonation from pigeonite to augite and ferroaugite. The most magnesian pigeonite compositions are generally found in pyroxene cores, but are also found occasionally in an annulus which is surrounded by and surrounds more evolved pyroxene compositions. Sector zoning is present, especially in 15555. These features and their interpretation have been described in more detail by Bence and Papike (1972) and Boyd (1972).

Also shown in Fig. 2 are the experimental pyroxene compositions, which show a more limited range than the natural ones. The experimental pyroxenes conform closely to the compositions of the natural pyroxene cores. Most importantly, the most magnesian pigeonite core compositions are duplicated by the most magnesian experimental pyroxenes produced nearest the liquidus. This would not be the case if the rock contained a significant admixture of early phenocrysts and suggests that the rock compositions are those of liquids, not crystal-liquid admixtures.

This coincidence also implies rather little super-cooling at pyroxene entry, which is consistent with the implied slow cooling rates recorded by the 1–5 mm grain size and the lack of fine-grained groundmass. Furthermore, the conformance of the core zoning sequences to the equilibrium crystallization trends suggests that the pyroxenes maintained surface equilibrium with the bulk of the sample during the early stages of crystallization when the sample was largely liquid. Equilibration could not have been more comprehensive than this or else the original core composition would have been erased. Zoning trends in the pyroxene *rims*, however, depart markedly from the equilibrium trends suggesting that if the pyroxenes maintained even surface equilibrium, it was not with the bulk of the largely crystalline sample. The difference in pyroxene trends in (olivine-normative) 15555 and (quartz-normative) 15065 is partly a result of differing bulk compositions. The augite branch on 15555 core trends is reflected in the equilibrium trend for 15555 whereas it is not in 15065. But the rim compositions of pyroxenes have differences between the two rocks which are probably amplified by differences in cooling rate, since the rim pyroxenes do not represent even surface equilibrium with the differing bulk compositions. The greater enrichment of Ti and Al in the 15555 rim pyroxenes than in 15065 is consistent with a more rapid cooling rate for 15555 (Donaldson *et al.*, 1975; Walker *et al.*, 1976a). We note that the Ti/Al in the experimental pyroxenes remains close to $\frac{1}{4}$ even after plagioclase appearance. The change to the $\frac{1}{2}$ trend in the rims of the natural pyroxenes is interpreted to mean that supercooling delayed plagioclase nucleation and that growth was rapid when plagioclase appeared, producing an Al depletion in the residual melt recorded by the pyroxenes. The experimental pyroxenes have a larger range of Al abundances than the natural pyroxene starting materials. This suggests that some of the pyroxene produced experimentally actually grew during the runs. The range of Al abundances is proportionally larger than the range observed for Fe, Mg, and

Ca abundances in a particular charge. Evidently, equilibration of Al is more difficult to produce than equilibration of the major pyroxene constituents.

Olivines in 15555 zone continuously over a wide range in composition: two profiles are shown in Fig. 4. A comparison of the experimental liquidus olivine (Fo_{73}) with the most magnesian core compositions (Fo_{71}) allows little (<4%) admixture of olivine phenocrysts to be present in the rock. (We actually interpret the small difference between Fo_{73} and Fo_{71} as a reequilibration phenomenon in our cooling models below.) The olivines indicate little supercooling or departure from surface equilibrium during the initial crystallization of 15555.

In contrast, olivines in 15065 (which are unstable in the bulk composition below 1100°C) show a very limited range of composition. Figure 4 shows two traverses across olivine grains. Each grain is homogeneous, but different grains have slightly different compositions. The olivine grains discussed here are relatively large (up to ~600 μm) anhedra, and are distinguishable from the late euhedral fayalite (Fa_{96}) intergrown with tridymite in the mesostasis. The liquidus olivine (Fo_{74}) is significantly more magnesian than the olivines found in 15065. This is not interpreted to imply the presence of admixed olivine phenocrysts in the rock because there is little olivine actually present (~1%). Almost 40% of admixed olivine of ~Fo_{55} would have to be present to shift the liquidus olivine to Fo_{74}. Furthermore, the pyroxenes do not show the anomaly expected if there were much admixed olivine present. Rather, it is thought that the cooling rate of

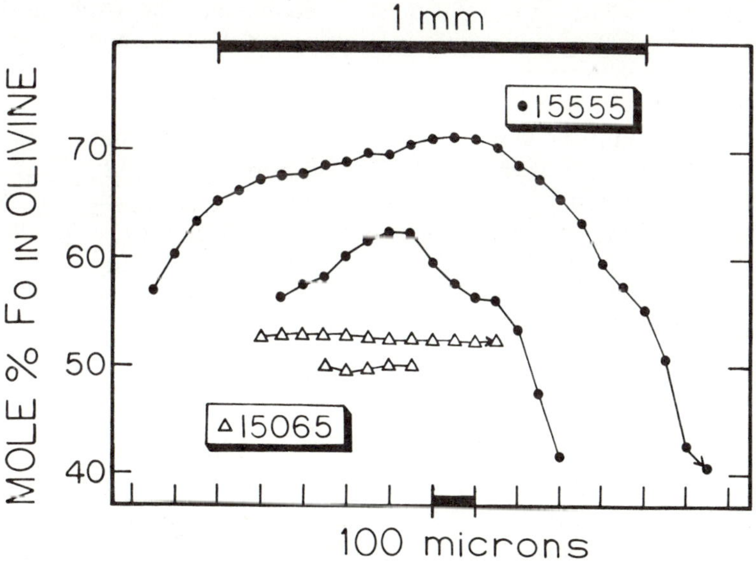

Fig. 4. Olivine composition profiles. Upper traverse for each rock is shown in Fig. 1. (Arrow tips correspond to arrow tips in Fig. 1.) Note extensive continuous normal zoning in 15555 and homogeneity in 15065 within grains but heterogeneity between grains.

15065 was slow enough for zoned olivines originally present to homogenize by diffusion. The cooling rates required to accomplish this are estimated below.

As a parenthetical remark, experimental complications seem to have been introduced by the extreme homogeneity of the olivines in the rock-powder starting material (total range only Fo_{55-50} with most at Fo_{52}). Runs at 1124 and 1125°C produced olivines Fo_{51-50}; the starting materials were recovered. However, at 1128, 1129, and 1148°C, *no* olivine was recovered in an especially diligent search of the products. There seems to be a "gap" in olivine recovery in this range below 1172°C where the product contains Fo_{65}. Electron probe analysis of the products does not suggest gain or loss of FeO from the bulk composition. It was clear that the iron-rich olivines in the starting material were dissolving in this olivine "gap" where a more magnesian olivine should be stable. For some reason, iron was not being exchanged for magnesium in the olivines. The olivines from the liquidus to 1172°C, on the other hand, are thought to have grown from experimentally produced nuclei. The high-temperature side of the "gap" is then closed by the ability of the charge to produce its own fresh nuclei. The low-temperature side of the "gap" is terminated at the temperature when the olivines in the starting material become stable. This explanation of the olivine "gap" was tested by equilibrating a charge on the high-temperature side of the gap at 1190°C where Fo_{68} should have been present and then lowering the temperature into the "gap." This experiment (065-108) had Fo_{56} present with pigeonite ($Wo_{12}En_{54}$), plagioclase, and spinel. On this basis, olivine should indeed be stable throughout the "gap." A curious implication of these experiments is that an experimentally grown magnesian olivine may be exchanged to an iron-rich olivine more easily than the starting material olivine may be exchanged to a magnesian composition.

This implication is necessary to explain the exchange $Fo_{68} \rightarrow Fo_{56}$ in experiment 065-108 as compared with the inability of the Fo_{52} in the starting material to become more magnesian as they should in the "gap." It was not known whether there was some peculiarity of the olivines in the starting material that was hindering the Fe–Mg exchange. To test this, ~20% wüstite was added to a charge 065-112 of 15065 and equilibrated at 1124°C. The olivines in the product were Fo_{35} (see Table 2), suggesting that the starting material olivines were capable of exchanging Fe for Mg easily but not *vice versa*. The details of this phenomenon are not understood.

Petrogenesis

We have argued on the basis of internal evidence that 15555 and 15065 represent the compositions of liquids. These samples are part of an olivine-normative and quartz-normative suite of samples, respectively, present in the Apollo 15 collection. The compositional and petrographic variations in these suites have been reviewed recently by Papike *et al.* (1976). The conclusion that 15555 and 15065 represent liquid compositions is reinforced by the existence of fine-grained to vitrophyric samples of similar composition. The chemical variations within the quartz-normative suite can be described by a magnesian pigeonite component and within the olivine-normative suite by an olivine component of composition $\sim Fo_{70-67}$ (Hubbard and Rhodes, 1973; Chappell and Green, 1973; Humphries *et al.*, 1972; Papike *et al.*, 1976). These phases occur near the liquidi of the appropriate compositions, which is consistent with the suggestion that fractionation of magnesian pigeonite and olivine is the cause of the chemical variations within the suites. Furthermore, we support Kesson's (1975) conclusion that the experimental residual liquids of 15555 show the chemical variation of the more evolved members of the olivine-normative suite. Both suites produce plagioclase-saturated residual liquids of similar composition, but differences in detail between these liquids, such as TiO_2 content, suggest that the two suites are not simply related to one another by fractionation near the lunar surface even with a plagioclase-saturated residual liquid as an intermediary.

O'Hara *et al.* (1975) have summarized and defended an alternate interpretation; namely that admixture of phenocrysts (with and without quench overgrowths) in a plagioclase saturated residual liquid is responsible for the compositional variation observed in the two suites. However, experiments on these rock compositions show that the olivine compositions in equilibrium with plagioclase-saturated liquids ($< \sim Fo_{60}$) are less magnesian than the olivines actually found in the rocks (Fo_{71} in 15555). O'Hara *et al.* (1975) account for this by proposing that the plagioclase-saturated liquid into which olivine was admixed was magnesian enough to grow the olivines observed, but that the admixture includes quench overgrowths of significantly more iron-rich olivine. In this case, experiments on the rock composition do not reproduce the original plagioclase-saturated liquid. However, according to this proposal, it is coincidental that the liquidus olivines should match the most magnesian phenocryst cores observed. The match will be observed only when the admixed material has the same Fe/Fe + Mg as the host liquid. It is unlikely that this admixture could occur in such a way that both olivine and pyroxene would show the observed similarities between natural and experimental Fe/(Fe + Mg) ratios. Accordingly, we accept the interpretation that some of the rocks represent liquid compositions.

The experimental results at elevated pressure may be used to constrain the petrogenesis of such liquids. The results for 15065 do not provide evidence of generation of magma at great depth in the moon. Rather, the olivine-pyroxene cosaturation at less than 5 kbar suggests that these compositions were last fractionated less than 100 km deep in the moon, possibly within the lunar crust during ascent. On the other hand, the cosaturation of 15555 with olivine and pyroxene at ~8.5 kbar (see Fig. 5) is consistent with segregation of this olivine-normative liquid from a source region at ~170 km depth. The compositions of the residual olivine and pyroxene are given in Table 3. The inferred source region for the Apollo 15 olivine-normative basalts is at a shallower level and has slightly higher Fe/(Fe + Mg) than the source region of the Apollo 12 olivine-normative basalts (Green *et al.*, 1971; Grove *et al.*, 1973). These source region relations are

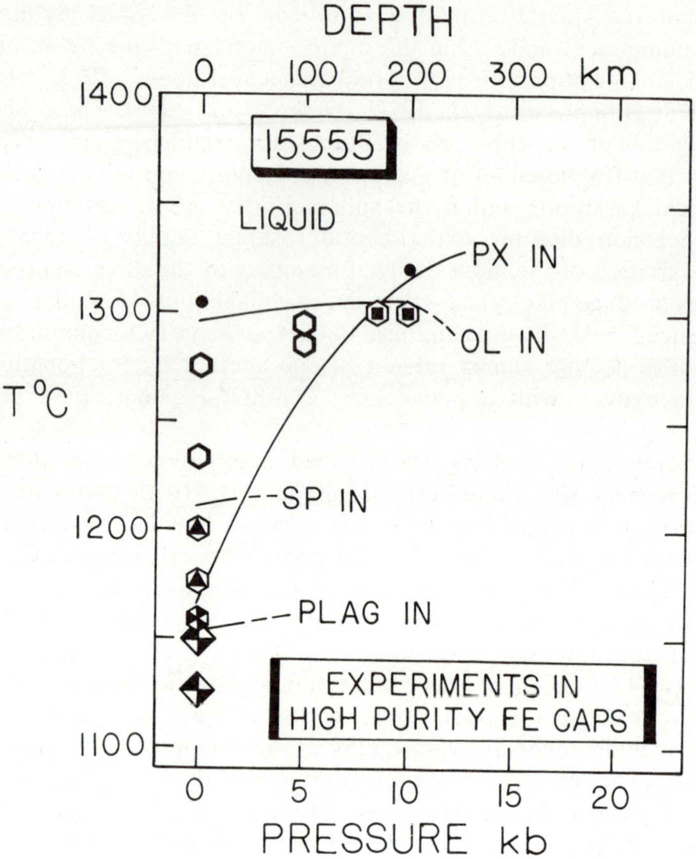

Fig. 5. Experiments on 15555 in high-purity iron capsules. Note cosaturation with olivine and pyroxene at pressure corresponding to lunar depths of 150–200 km.

consistent with the zonation expected in a large cumulate sequence produced during the early differentiation of the outer portions of the moon.

Cooling Rate

Laboratory experiments in which molten basaltic compositions are cooled at controlled rates have been successful in reproducing textural and chemical characteristics observed in mare basalts and in estimating cooling rates by direct comparison of rock and experiment (Lofgren *et al.*, 1975; Donaldson *et al.*, 1975; Grove, 1977; Bianco and Taylor, 1977; Walker *et al.*, 1976a). This approach is limited to relatively fine-grained textures because of experimental constraints. Although sufficiently long cooling times and large enough containers are not currently available for producing textures coarse enough to be directly comparable to gabbros, there are several indirect methods available for estimation of the cooling rate in coarse-grained rocks such as 15065. These methods are independent of one another and allow a certain amount of cross-comparison.

(a) *Zr partitioning*

Taylor *et al.* (1975) have used the temperature dependence of Zr distribution between ilmenite and ulvöspinel as the basis for a cooling rate indicator. Zr distribution continually readjusts in response to temperature change during cooling; the slower the cooling, the more comprehensive the readjustment and the lower the apparent temperature of equilibration (T_0). Taylor *et al.* (1975) showed that these considerations could be used to estimate the cooling rate (R) if the diffusion rates and form of the cooling curve were known. Diffusion coefficients (D) of the form:

$$D = D_0\, e^{-A/T}, \tag{1}$$

were assumed and the cooling rate was assumed to be linear:

$$T = T_0 - Rt, \tag{2}$$

where T and t are temperature and time, and A and D_0 are constants. A characteristic distance (X_c) over which Zr exchange occurs was replaced by the integral over the cooling history of the diffusion coefficient yielding the following relation:

$$R \simeq \frac{D_0 A}{X_c^2} \frac{(1 - 2T_0/A)}{(A/T_0)^2}\, e^{-A/T_0}. \tag{3}$$

The observed Zr distribution in 15065 was used by Taylor *et al.* (1975) to find an apparent temperature of equilibration which implied a cooling rate of ~3°C/day.

(b) *Olivine homogenization*

The observation that olivines in 15065 have homogenized can be coupled with known diffusion coefficients for Fe–Mg exchange to give another estimate of cooling rate. Buening and Buseck (1973) have determined Fe–Mg interdiffusion coefficients for olivine as a function of temperature, composition, P_{O_2}, and crystallographic direction. The slowest crystallographic direction will limit the homogenization rate, so we have used the data for the b direction. Since the formulation used does not consider compositional dependence of the diffusion coefficient, we have chosen ~Fo_{65} as representative. During the cooling and crystallization, P_{O_2} does not remain constant but changes in response to silicate-oxide interactions, so we have used a P_{O_2}, an order of magnitude below Fe/FeO, as suggested by the experimental work. These values can be used to interpolate and extrapolate A of 3.2×10^4 °K and D_0 of 3.0×10^{-2} cm²/sec from Buening and Buseck's data. Other information required to estimate cooling rate is the crystal size which is homogenized. The maximum diameter of olivines in 15065 is

~600 μm, so we have taken 300 μm as the characteristic radius for Fe–Mg exchange.

A model in which olivine grains (of different size with various proportions of iron-rich rims in the original zoning profile) are homogenized by internal diffusion after communication with the rock is lost appears to be appropriate. We noted above that olivine is unstable in the bulk rock composition below ~1125°C. Its presence is an indication that olivine was not in effective communication with the rest of the rock below ~1125°C. Furthermore, the observation that each olivine grain, although homogeneous, has a different composition also suggests that late equilibration was not with the bulk of the rock. We have approached this problem by assuming that spherical crystals of radius (a) with initial concentration profiles of forsterite content that are a function of radius (r) homogenize by diffusion as a closed system below a temperature of closure (T_c). The diffusion coefficient was assumed to be a function of temperature only, and dependence on composition and crystallographic direction were ignored. The equation for diffusion inside an isotropic sphere is

$$\frac{\partial c}{\partial t} = D\left(\frac{\partial^2 c}{\partial r^2} + \frac{2}{r}\frac{\partial c}{\partial r}\right), \tag{4}$$

or, if $u \equiv cr$

$$\frac{\partial u}{\partial t} = D(t)\frac{\partial^2 u}{\partial r^2}. \tag{5}$$

D depends on time (t) since temperature is a function of time during cooling. At this point we can eliminate the time dependence of D by introducing a dimensionless variable

$$t' = \int_0^t \frac{D(\tau)}{a^2}\, d\tau, \tag{6}$$

where τ is an integration parameter.

Since $dt'/dt = D(t)/a^2$ we can rewrite our differential equation

$$\frac{\partial u}{\partial t'} = a^2 \frac{\partial^2 u}{\partial r^2}. \tag{7}$$

We have now reduced the problem to a standard diffusion equation (constant coefficients). The boundary condition in our case is such that *no* material is exchanged between the mineral grain and the surroundings during the cooling. Hence

$$\left.\frac{\partial c}{\partial r}\right|_{r=a} = 0, \tag{8}$$

or

$$\left.\frac{\partial u}{\partial r}\right|_{r=a} = 0. \tag{9}$$

The initial condition to our problem is given by a distribution function

$$c(r, 0) = F(r) = g(r'), \tag{10}$$

where we have found it useful to introduce another dimensionless variable $r' \equiv r/a$. The solution to the differential equation with these conditions is given by

$$c(r', t') = 3 \int_0^1 g(r'')r''^2\, dr'' + \frac{2}{r'} \sum_n e^{-\beta_n^2 t'} \frac{\sin(\beta_n r')}{\sin^2 \beta_n} \int_0^1 g(r'') \sin(\beta_n r'')r''\, dr'', \tag{11}$$

where $\beta_n \cot \beta_n - 1 = 0$ and r'' is an integration parameter (see p. 97, Crank, 1975). If we let t become arbitrarily large, t' will approach a well-defined value

$$t' = \int_0^\infty \frac{D(\tau)d\tau}{a^2}. \tag{12}$$

The concentration profile $c(r', t')$ at this value of t' will, of course, be the observed "frozen" profile today. Thus we can relate t' to the cooling rate. If we use eqs. (1) and (2)

$$D(t) = D_0\, e^{-A/(T_c - Rt)} = D_0\, e^{-(A/T_c)/(1-Rt/T_c)}, \tag{13}$$

or expanding

$$\frac{1}{1 - \dfrac{Rt}{T_c}} = 1 + \frac{Rt}{T_c} + \cdots, \tag{14}$$

$$D(t) \sim D_0\, e^{-(A/T_c^2)Rt}\, e^{-(A/T_c)}. \tag{15}$$

As long as $Rt \ll T_c$; which is legitimate in our case, since after a reasonable drop in temperature, $\Delta T = Rt \approx 300°\text{C}$, there is very little diffusion. Thus the limiting value of t' will be

$$t' = \int_0^\infty \frac{D_0}{a^2} e^{-(A/T_c)}\, e^{-(A/T_c^2)R\tau}\, d\tau, \tag{16}$$

$$t' = \frac{D_0}{a^2} \frac{T_c^2}{RA} e^{-(A/T_c)}, \tag{17}$$

or

$$R = \frac{D_0}{a^2} \frac{T_c^2}{t'A} e^{-(A/T_c)}. \tag{18}$$

The last equation is the desired relation between R and t'.

Figure 6 shows the concentration profile as a function of scaled radius in the crystal for various values of t'. For $t' = 0$, the initial concentration profile is

$$g(r/a) = g(r') = 70\, e^{(-.714(r')^3)}, \tag{19}$$

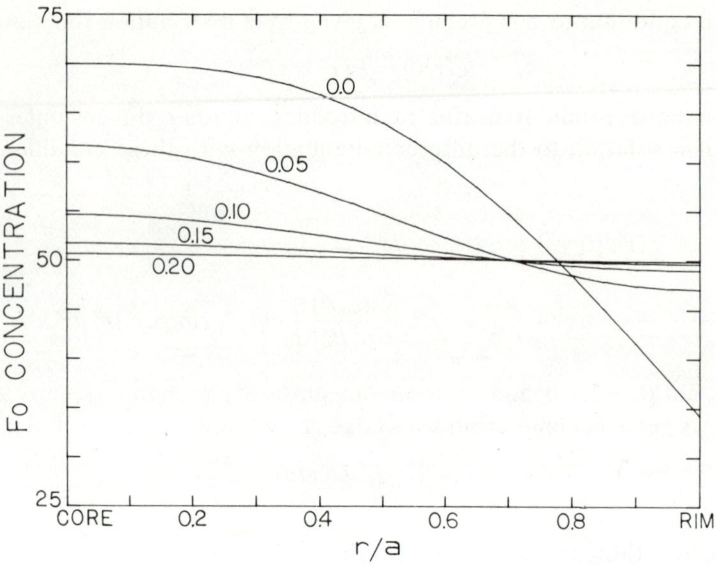

Fig. 6. Calculated core to rim compositional profiles in olivine for various values of t'. See text for discussion.

which gives an initial core composition similar to the equilibrium liquidus olivine, an initial profile shape similar to that in zoned phenocrysts which have not homogenized, and an integrated fayalite content such that the crystal will homogenize to the compositions observed. For larger values of the dimensionless parameter t', the profile decays as the crystal homogenizes. Evidently, values of t' of ~ 0.2 are sufficient to sensibly homogenize the crystals.

Equation (6) defining t' is of the same form as the expression for characteristic distance used by Taylor *et al.* (1975) if the constant a^2 is transposed to the left. Equation (18) is directly analogous to Taylor *et al*'s result above (3) except the characteristic distance (X_c) is replaced by the actual radius of the crystal (a) and T_c, the closure temperature at which the profile begins to decay, replaces T_0, the apparent temperature of equilibration. Also, t' is a dimensionless multiplier which serves the purpose of including the initial concentration profile and specific geometry of the problem which are not considered in Taylor *et al.*'s formulation. The factor $(1-2T_0/A)$ in Taylor's approximate formula is replaced by 1 in Eq. (18). This is a small difference, since $2T_0/A < 1/10$. T_c is taken as 1100°C since homogenization must have started by this temperature below which olivine would be consumed if communication with the rock were maintained. Then, using an (a) of 300 μm and the A and D_0 noted above, the cooling rate is calculated to be ~ 0.06°C/day. It is notable that this rate is compatible with the rate of ~ 0.02°C/day which could be formally calculated from Eq. (3), even though the assumptions involved are quite different and unrealistic in the case of Eq. (3). This compatibility might be anticipated from the similarity of Eqns. (3)

and (18). This suggests that the solution is more dependent on the diffusion constants (D_0 and A) and the diffusion length (a) than on the assumed boundary conditions or considerations of initial concentration profiles and grain geometry which only enter through the multiplier t'.

It may be unrealistic to expect that all the homogenization occurs below a certain closure temperature T_c. At these slow cooling rates, homogenization undoubtedly is in progress during crystal growth. This effect can be approximately accounted for by raising T_c to a value above the closure temperature. It has been shown that much of the growth of olivine in a dynamic cooling situation occurs within the first ~50°C of the olivine appearance temperature or by the appearance temperature of low-calcium pyroxene (Walker *et al.*, 1976; Grove, 1977). Such temperatures might be taken as more useful estimates of the temperature required to start homogenization, even though the initial profile is not fully established. In the present case, use of 1200°C for T_c instead of 1100°C raises the estimate of cooling rate from 0.06 to ~0.3°C/day. We adopt the higher T_c approximating completion of olivine growth and its cooling rate estimate of 0.3°C/day as more reasonable than the lower T_c approximating temperature of closure because homogenization certainly proceeds between the two temperatures. Although this treatment upsets the assumptions about the initial concentration profile, we saw above that the results are not particularly sensitive to this initial condition. In any case, the increase of T_c by as much as 100°C, which is a considerable fraction of the solidification interval, only changes the estimated cooling rate by a factor of ~5. This is to say that the theoretical shortcomings of this correction for homogenization during growth are unlikely to introduce order of magnitude errors, and hence the approach probably represents an improvement in estimating the cooling rate.

Cooling rates calculated on the basis of olivine homogenization are maximum possible rates because slower rates would also produce homogeneous olivines. Larger values of t' than 0.2, corresponding to slower cooling rates, all give flat profiles whereas smaller values of t' give a profile of non-degenerate shape that can be used to obtain an absolute cooling rate rather than just a maximum. This approach might be applicable to 15555 olivines which are not homogeneous. But the profiles shown in Fig. 4 compare rather poorly with those reported by Bell and Mao (1972) and Taylor *et al.* (1977), indicating a certain amount of complexity in 15555 olivine zoning which raises doubt as to whether a meaningful original profile can be estimated in a particular case. In contrast to the situation in 15065 where the original profile was reduced to flatness, the olivines in 15555 have undergone relatively little homogenization, as noted above in the comparison of experimental and natural core olivines. Consequently, the values of t' will be rather small and exert a large correction on the cooling rate in Eq. (18). A comparison of zoning profiles in Fig. 4 and Bianco and Taylor (1977) for 15555 experiments suggests that t' is probably less than 0.01 which gives a cooling rate of ~4°C/day when the size and temperature interval for 15555 olivine is considered. Although we do not claim to make this estimate with much confidence as a result of uncertainty about the initial profile and the sensitivity of the result to t', the solution is compatible with that given for 15555 by Bianco and Taylor (1977) and Taylor *et al.* (1977) of ~10°C/day.

Further limits on the 15065 cooling rate can be made by considering the preservation of pyroxene zoning. Since the characteristic diffusion distance is proportional to $\sqrt{Dt}$, the observation that compositional heterogeneities are preserved in 15065 pyroxenes on a ~10 μm scale, while they are erased on at least a ~300 μm scale in olivine requires that the diffusion coefficients for pyroxenes must be more than ~10^3 times smaller than those for olivine. Huebner *et al.* (1975) concluded that they were unlikely to be more than 10^4 times smaller on the basis of a compilation of available data. Therefore, the cooling rate estimated from olivine homogenization must not be far below the actual value or else the pyroxenes would homogenize on a scale greater than the ~10 μm observed.

(c) *Plagioclase size*

Grove (1977) observed a strong inverse correlation between cooling rate and plagioclase size in controlled cooling experiments on a quartz-normative Apollo 15 basalt composition similar to 15065. Figure 7 reproduces Grove's data and shows the extrapolation of cooling rate from the observed plagioclase size in 15065. The estimate of ~0.2°C/day is strikingly similar to the preferred value suggested above from olivine homogenization of ~0.3°C/day. These two estimates are preferred over the other method because the distances and diffusion

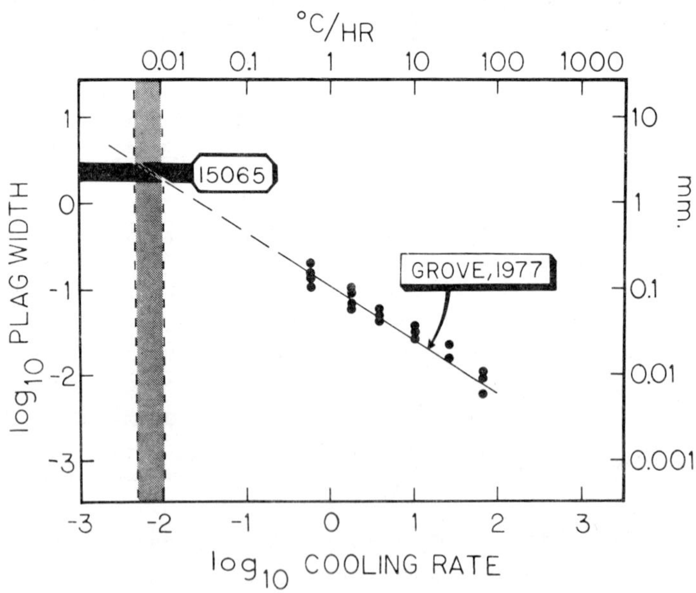

Fig. 7. Plagioclase width as a function of cooling rate. Extra-polation of Grove's data indicates a cooling rate of ~0.2°C/day for 15065.

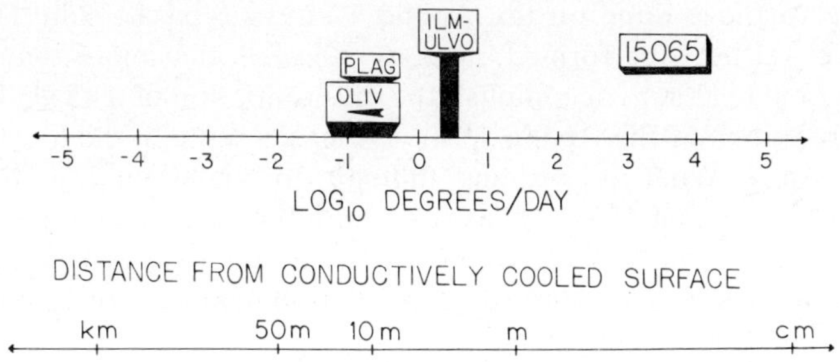

Fig. 8. Summary of results for various mineralogical cooling speedometers in 15065. Scale for distance from a chill margin is calculated for the plagioclase crystallization interval by the method of Jaeger (1957). Values used for crystallization intervals are from Grove's (1977) dynamic cooling experiments: solidification interval from 1210 to 975°C and plagioclase interval from 1140 to 975°C. Other constants used are: latent heat of 80 cal/g, heat capacity of 0.3 cal/g-deg, and thermal diffusivity of 0.01 cm^2/sec.

constants are better for olivine than for ulvöspinel and ilmenite. Nonetheless, the other method gives an estimate of 3°C/day which is no more than an order of magnitude from the preferred values. These various estimates are summarized in Fig. 8. The mutual consistency of estimates based on such various independent methods is encouraging.

Magma Size and Differentiation

Also shown in Fig. 8 is a scale axis which relates cooling rate to distance within a magma body. The calculation is based on the solution of Jaeger (1957) for the variation of cooling rate in a magma body as a function of distance from a surface of conductive heat loss. The values given in Fig. 8 were used to make the calculation for the case where magma intrudes country rock of the same material at 0°C. The preferred cooling rates for 15065 imply that 15065 was ~10 m from its nearest chill margin and that the magma body was "a few" tens of meters in minimum dimension. This size estimate is larger than that deduced by Brett (1975) because the cooling rates estimated here are slower than any available to Brett.

The magma body size and cooling time available for differentiation in the local environment of solidification can be evaluated in terms of the variability observed in the rock compositions. For the case of the 15065 magma body which takes at least a year or two to traverse the solidification interval in its interior, sufficient time and space are available to accomplish the limited amount of pigeonite fractionation observed by crystal settling (Grove and Walker, 1977). However, some of the variations observed in the vitrophyre samples which did not participate in prolonged cooling and differentiation may require additional differentiation by flowage or settling in a shallow pre-eruption magma chamber.

The cooling time for 15555 of a few months also implies a magma body more than a few meters thick. Although simple differentiation calculations of detail comparable to those done on the Apollo 12 olivine basalt suite (Walker *et al.*, 1976b) have not been performed, analogy suggests that more than enough time and room are available to accomplish the differentiation of a 15555 liquid into the other compositions of the Apollo 15 olivine basalt suite in the local environment of solidification. What is perhaps difficult to understand is how rocks as dissimilar as 15016 and 15555 in texture and inferred cooling history could have preserved the same bulk composition against gravitative olivine differentiation. This problem awaits a resolution—a resolution which may require discarding simple differentiation mechanisms.

Acknowledgments—This work was supported by NASA grant NGL-22-007-247 and the Committee on Experimental Geology and Geophysics of Harvard University. We appreciate review comments by J. S. Huebner, L. A. Taylor, and R. J. Williams.

REFERENCES

Bell P. M. and Mao H. K. (1972) Zoned olivine crystals in an Apollo 15 lunar rock. In *The Apollo 15 Lunar Samples* (J. W. Chamberlain and C. Watkins, eds.) p. 26–28. The Lunar Science Institute, Houston.

Bence A. E. and Albee A. L. (1968) Empirical correction factors for the electron microanalysis of silicates and oxides. *J. Geol.* **76**, 382–403.

Bence A. E. and Papike J. J. (1972) Pyroxenes as recorders of lunar basalt petrogenesis: Chemical trends due to crystal-liquid interaction. *Proc. Lunar Sci. Conf. 3rd*, p. 431–469.

Bianco A. S. and Taylor L. A. (1977) Lunar olivine normative basalts: Cooling rate experimentation (abstract). In *Lunar Science VIII*, p. 103–105. The Lunar Science Institute, Houston.

Biggar G. M., Humphries D. J., and O'Hara M. J. (1974) Experimental crystallization and geochemical interpretation of low titanium basalts (abstract). In *Lunar Science V*, p. 60–26. The Lunar Science Institute, Houston.

Boyd F. R. (1972) Zoned proxenes in lunar rock 15555. *Carnegie Inst. Wash. Yearb.* **71**, 459–463.

Boyd F. R. and England J. L. (1960) Apparatus for phase equilibrium measurements at pressures up to 50 kilobars and temperatures up to 1750°C. *J. Geophys. Res.* **65**, 741–748.

Brett R. (1975) Thickness of some lunar mare basalt flows and ejecta blankets based on chemical kinetic data. *Geochim. Cosmochim. Acta* **29**, 1135–1141.

Buening D. K. and Buseck P. R. (1973) Fe–Mg lattice diffusion in olivine. *J. Geophys. Res.* **78**, 6852–6862.

Chappell B. W. and Green D. H. (1973) Chemical compositions and petrogenetic relationships in Apollo 15 mare basalts. *Earth Planet. Sci. Lett.* **18**, 237–246.

Crank J. (1975) *The Mathematics of Diffusion.* 2nd edition. Clarendon Press, Oxford. 414 pp.

Donaldson C. H., Usselman T. M., Williams R. J., and Lofgren G. E. (1975) Experimental modeling of the cooling history of Apollo 12 olivine basalts. *Proc. Lunar Sci. Conf. 6th*, p. 843–869.

Green D. H., Ringwood A. E., Ware N. G., Hibberson W. O., Major A., and Kiss E. (1971) Experimental petrology and petrogenesis of Apollo 12 basalts. *Proc. Lunar. Sci. Conf. 2nd*, p. 601–615.

Grove T. L. (1977) Cooling histories of Apollo 15 quartz normative basalts (abstract). In *Lunar Science VIII*, p. 380–382. The Lunar Science Institute, Houston.

Grove T. L. and Walker D. (1977) Cooling histories of Apollo 15 quartz-normative basalts. *Proc. Lunar Sci. Conf. 8th.* This volume.

Grove T. L., Walker D., Longhi J., Stolper E., and Hays J. F. (1973) Petrology of rock 12002 and origin of picritic basalts at Oceanus Procellarum. *Proc. Lunar Sci. Conf. 4th*, p. 995–1011.

Huebner J. S., Ross M., and Hickling N. (1975) Significance of exsolved pyroxenes from lunar breccia 77215. *Proc. Lunar Sci. Conf. 6th*, p. 529–546.

Humphries D. J., Biggar G. M., and O'Hara M. J. (1972) Phase equilibria and origin of Apollo 15 basalts etc. In *The Apollo 15 Lunar Samples* (J. W. Chamberlain and C. Watkins, eds.) p. 103–107. The Lunar Science Institute, Houston.

Jaeger J. C. (1957) The temperature on the neighborhood of a cooling intrusive sheet. *Amer. J. Sci.* **255**, 306–318.

Johannes W., Bell P. M., Mao H. K., Boettcher A. L., Chipman D. W., Hays J. F., Newton R. S., and Seifert F. (1971) An interlaboratory comparison of piston-cylinder pressure calibration using the albite breakdown reaction. *Contrib. Mineral. Petrol.* **32**, 24–38.

Kesson S. E. (1975) Mare basalts: Melting experiments and petrogenetic interpretations. *Proc. Lunar Sci. Conf. 6th*, p. 921–944.

Kesson S. E. and Lindsley D. H. (1976) Mare basalt petrogenesis—a review of experimental studies. *Rev. Geophys. Space Phys.* **14**, 361–373.

Lofgren G. E., Donaldson C. H., and Usselman T. M. (1975) Geology, petrology, and crystallization of Apollo 15 quartz-normative basalts. *Proc. Lunar Sci. Conf. 6th*, p. 79–99.

Longhi J., Walker D., Stolper E. M., Grove T. L., and Hays J. F. (1972) Petrology of mare/rille basalts 15555 and 15065. In *The Apollo 15 Lunar Samples* (J. W. Chamberlain and C. Watkins, eds.) p. 131–134. The Lunar Science Institute, Houston.

Longhi J., Walker D., Grove T. L., Stolper E. M., and Hays J. F. (1974) The petrology of the Apollo 17 mare basalts. *Proc. Lunar Sci. Conf. 5th*, p. 447–469.

Longhi J., Walker D., and Hays J. F. (1975) Fe–Mg distribution between olivine and lunar basaltic liquids (abstract). *EOS* (*Trans. Amer. Geophys. Union*) **56**, 471.

Longhi J., Walker D., and Hays J. F. (1976) Fe and Mg in plagioclase. *Proc. Lunar Sci. Conf. 7th*, p. 1281–1300.

LSPET (1972) The Apollo 15 lunar samples: a preliminary description. *Science* **175**, 363–375.

O'Hara M. J., Humphries D. J., and Waterston S. (1975) Petrogenesis of mare basalts: Implications for chemical, mineralogical, and thermal models for the moon. *Proc. Lunar Sci. Conf. 6th*, p. 1043–1055.

Papike J. J., Hodges F. N., Bence A. E., Cameron M., and Rhodes J. M. (1976) Mare basalts: Crystal chemistry, mineralogy, and petrology. *Rev. Geophys. Space Phys.* **14**, 475–540.

Rhodes J. M. and Hubbard N. J. (1973) Chemistry, classification, and petrogenesis of Apollo 15 mare basalts. *Proc. Lunar Sci. Conf. 4th*, p. 1127–1148.

Taylor L. A., Ulhmann D. R., Hopper R. W., and Misra K. C. (1975) Absolute cooling rates of lunar rocks: Theory and application. *Proc. Lunar Sci. Conf. 6th*, p. 181–191.

Taylor L. A., Walker B. M., Onorato P. I. K., and Uhlmann D. R. (1977) Fe–Mg diffusion in olivine: A cooling rate indicator (abstract). In *Lunar Science VIII*, p. 928–930. The Lunar Science Institute, Houston.

Walker D., Longhi J., and Hays J. F. (1972) Experimental petrology and origin of Fra Mauro rocks and soil. *Proc. Lunar Sci. Conf. 3rd*, p. 797–817.

Walker D., Longhi J., Grove T. L., Stolper E., and Hays J. F. (1973) Experimental petrology and origin of rocks from the Descartes Highlands. *Proc. Lunar Sci. Conf. 4th*, p. 1013–1032.

Walker D., Longhi J., Stolper E. M., Grove T. L., and Hays J. F. (1975) Origin of titaniferous lunar basalts. *Geochim. Cosmochim. Acta* **39**, 1219–1235.

Walker D., Kirkpatrick R. J., Longhi J., and Hays J. F. (1976a) Crystallization history of lunar picritic basalt sample 12002: Phase equilibria and cooling rate studies. *Bull. Geol. Soc. Amer.* **87**, 646–656.

Walker D., Longhi J., Kirkpatrick R. J., and Hays J. F. (1976b) Differentiation of an Apollo 12 picrite magna. *Proc. Lunar Sci. Conf. 7th*, p. 1365–1389.

Proc. Lunar Sci. Conf. 8th (1977), p. 1549–1579.
Printed in the United States of America

Experimental study of pyroxene-liquid interaction in quartz-normative basalt 15597

TIMOTHY L. GROVE and A. E. BENCE

Department of Earth and Space Sciences, State University of New York,
Stony Brook, New York 11794

Abstract—Equilibrium and controlled cooling rate experiments on a synthetic analog of Apollo 15 quartz normative basalt 15597 document the effects of crystallization history on pyroxene chemistry. The partitioning of the major elements Ca, Mg, and Fe between pyroxene and liquid is rate independent, and the $K_D^{Fe/Mg} = X_{pyx}^{FeO} X_{liq}^{MgO} / X_{pyx}^{MgO} X_{liq}^{FeO}$ calculated for low-Ca pyroxene is temperature dependent ($K_D^{Fe/Mg} = 0.28$ at 1200°C and 0.17 at 975°C). From these cooling rate studies it appears that chemical trends observed in lunar pyroxenes are a consequence of three rate processes. These are: (1) rate dependent suppression of crystallization of other phases until lower temperature alters liquid composition and, therefore, the composition of the pyroxene that crystallizes; (2) the efficiency of fractional crystallization and the extent of crystal zoning are rate dependent; and (3) the partitioning of Al_2O_3, TiO_2, and Cr_2O_3 between pyroxene and liquid depends upon cooling rate. The concentrations of minor elements in early pyroxene record its early cooling rate.

INTRODUCTION

THE APOLLO 15 QUARTZ-NORMATIVE BASALTS (QNBs) comprise a texturally diverse group ranging from vitrophyres to gabbros, all with essentially identical chemical composition. The textures and the mineralogies of this isochemical suite reflect variations in the thermal histories of the basalts. The interpretation of the crystallization process recorded in the complex chemical trends observed in pyroxene in these and other basalts have been the center of much controversy (Bence *et al.*, 1971; Boyd and Smith, 1971, Hollister *et al.*, 1971; Bence and Papike, 1972; Dowty *et al.*, 1974). This study attempts to understand the chemical trends observed in the pyroxenes of Apollo 15 lunar QNBs through controlled cooling rate studies.

The Apollo 15 QNBs are an ideal suite to evaluate experimentally because the pyroxene in them displays a diversity of morphologies and chemical trends. The similarities in bulk chemistry require that only limited crystal-liquid fractionation has occurred between members of the quartz-normative suite. Rhodes and Hubbard (1973) suggest that the addition or subtraction of minor amounts of the liquidus phases olivine, pyroxene, and Cr spinel (up to 7%) are sufficient to explain the chemical variations. The limited degree of differentiation may have resulted because pyroxene, the dominant phenocryst phase, along with minor olivine and Cr spinel is unable to settle appreciably during the cooling of a QNB flow unit (Grove, 1977).

EXPERIMENTAL PROCEDURES

Selection of bulk composition

Starting material with the bulk composition of pyroxene vitrophyre 15597 was chosen for the equilibrium and controlled-cooling rate studies. The analysis of Chappell and Green (1973) is compared with microprobe analyses of glasses quenched from above the liquidus in both equilibrium (#5) and controlled-cooling experiments (#84) in Table 1. These data show that the synthetic glass is a good match for Chappell and Green's composition. The vitrophyric texture of 15597 indicates that it was rapidly chilled at the lunar surface, thus it probably represents a chilled melt composition (or melt plus crystal composition) that underwent little near-surface fractionation. Its composition is intermediate between the coarse-grained, more quartz-normative members of the suite (15076, 15058, 15475, 15495) which are possible differentiates and the coarse-grained more olivine-normative members of the suite (15065, 15118) which are possible cumulates. Thus, 15597 could be the bulk magma composition and the coarser-grained samples may have been derived from it by relatively small amounts of addition or subtraction of the liquidus phases (olivine, Cr-spinel, and pigeonite).

Experimental conditions

Equilibrium and controlled cooling-rate experiments were performed in high-purity iron capsules sealed in evacuated silica glass tubes (see Appendix). The iron capsule technique is appropriate for the crystallization of the Apollo 15 quartz-normative basalts, because the vitrophyres (15597, Weigand and Hollister, 1973) and the coarse-grained samples (15065, Longhi *et al.*, 1972; Brown *et al.*, 1972) contain iron blebs within early formed phases. Thus, the experiments impose an unknown but presumably close to realistic f_{O_2} on the crystallization (Kesson and Lindsley, 1976). The bulk composition of the glass starting material was unaffected by the iron container (Table 1) and no net iron loss or gain occurred during the early stages of crystallization.

Experimental results

The equilibrium phase relations are shown on the right side of Fig. 1 and summarized in Table 2. Olivine (Fo_{68}–Fo_{70}) appears as the liquidus phase at 1215°C and is reacted out by 1130°C (Fo_{56}). Low-Ca pyroxene ($Wo_5En_{68}Fs_{27}$) follows at 1197°C and becomes increasingly more Ca- and Fe-rich in lower temperature runs (Fig. 2, Table 1). Spinels that are too small to analyze with the electron microprobe appear at 1185°C and plagioclase crystallizes at 1143°C. Olivine and pyroxene are unzoned at higher temperatures but the pyroxene spans a range of compositions in the lower temperature experiments. The experimental liquid-olivine (Fo_{70}) pair yields a K_D ($K_D = X^{ol}_{FeO}X^{liq}_{MgO}/X^{ol}_{MgO}X^{liq}_{FeO}$) of 0.33 in close agreement with the experimental K_D reported for other low-Ti lunar basalts (Longhi *et al.*, 1975; Walker *et al.*, 1976). Lofgren *et al.* (1974) report a slightly more iron-rich liquidus olivine (Fo_{68}) and a more magnesian liquidus pyroxene ($Wo_4En_{75}Fs_{21}$) in their experiments on an average QNB composition. Their synthetic starting composition is more magnesian ($Fe/(Fe+Mg) = 0.52$ as compared with 0.57 for this study), so one would expect their liquidus olivine to be more magnesian (Fo_{73} for a K_D of 0.33). The experimental $K_D^{Fe/Mg}$ of 0.28 ± 0.04 for pyroxene and liquid as defined by Grove and Bence (1977) is similar to the calculated $K_D = 0.26$ for the liquidus pyroxene of Lofgren's experiments.

The low-pressure equilibrium phase relations obtained from the 15597 synthetic glass are similar to those of other workers. Lofgren *et al.*'s more magnesian starting material has both olivine and pyroxene on the liquidus at 1204°C. Humphries *et al.* (1972) studied several Apollo 15 lunar samples in Mo capsules held in a gas mixture at the iron–wüsite buffer. Their experiments on the bulk composition 15499 are closest to the 15597 glass, showing spinel and olivine as the liquidus phases (at 1222°C) followed by pyroxene at 1210°C. The higher f_{O_2} imposed by Humphries *et al.* probably explains the appearance of early spinel. Longhi *et al.* (1972) experimented with lunar sample 15065 (a

Table 1. Electron microprobe analyses of run products in equilibrium experiments and a representative starting liquid (#84) for cooling rate experiments.

Liquids (in wt.%) (Run No.) # of analyses	15597* (Chappell and Green, 1973)	#5 8	#84 3	#6 4	#8 3	#12 4	#14 5	#15 2
SiO_2	48.09	48.5(.5)†	47.9(.3)[a]	47.9(.1)	48.1(.4)	47.3(.1)	46.7(.2)	46.6(.1)
Al_2O_3	9.46	9.70(.44)	9.41(.03)	9.37(.22)	10.1(.2)	10.7(.2)	11.0(.1)	10.5(.1)
TiO_2	1.80	1.79(.04)	1.82(.03)	1.81(.11)	1.83(.05)	1.92(.04)	2.24(.07)	2.63(.09)
FeO	20.27	20.3(.3)	20.7(.3)	20.2(.2)	20.8(.4)	20.9(.6)	22.2(1.0)	23.0(.1)
MnO	0.30	0.25(.05)	0.34(.03)	0.27(.03)	0.30(.06)	0.29(.02)	0.29(.04)	0.30
MgO	8.76	8.70(.27)	8.56(.06)	8.26(.16)	7.10(.90)	6.06(.33)	4.59(.69)	4.09(.59)
CaO	10.45	10.3(.16)	10.4(.1)	10.5(.1)	10.7(.4)	11.6(.2)	11.6(.1)	10.8(.5)
Na_2O	0.32	0.10(.08)	0.18(.04)	0.21(.03)	0.16(.05)	0.22(.04)	0.21(.03)	0.26(.04)
K_2O	0.06	0.07(.01)	0.06(.01)	0.08(.01)	0.04(.04)	0.05(.03)	0.07(.04)	0.10(.01)
Cr_2O_3	0.48	0.47(.03)	0.51(.04)	0.47(.03)	0.43(.08)	0.33(.03)	0.19(.09)	0.20(.04)
Σ	100.00	100.12	99.9	99.1	99.6	99.4	99.1	98.5
Olivine				Fo_{70}	Fo_{67}	Fo_{61}	Fo_{56}	none
Pyroxene								
# of analyeses					3	7	6	8
SiO_2					52.9(.2)	52.1(.3)	51.4(.5)	50.8(.4)
Al_2O_3					0.83(.04)	1.07(.30)	1.31(.29)	1.61(.38)
TiO_2					0.28(.07)	0.29(.05)	0.44(.05)	0.49(.07)
FeO					17.9(.3)	19.3(.5)	20.6(.3)	22.5(.5)
MnO					0.28(.03)	0.33(.03)	0.32(.02)	0.37(.03)
MgO					24.5(.4)	21.6(.3)	19.5(.4)	17.2(.5)
CaO					2.68(.12)	3.80(.20)	5.65(.37)	6.46(.75)
Na_2O					0.0	0.0	0.0	0.0
K_2O					0.0	0.0	0.0	0.0
Cr_2O_3					0.82(.09)	0.71(.11)	0.70(.06)	0.60(.08)
Σ					100.2	99.2	99.9	100.0
Other phases		none	none			Cr spinel	Cr spinel plagioclase	Cr spinel plagioclase
Temperatures (°C)		1225	cooling rate expt.	1212	1194	1161	1138	1123

*Their analysis normalized to 100 wt.%.

†Parenthesized units represent one standard deviation of the replicate analyses in terms of the least units cited. Therefore, 48.5(0.3) should read 48.5 ± 0.3. Run duration = 8.3 hr.

[a]Cooling rate for this run was ~ 3.85°C/hr. Initial temperature = 1225°C, final temperature = 1208°C and total time was 4.4 hr.

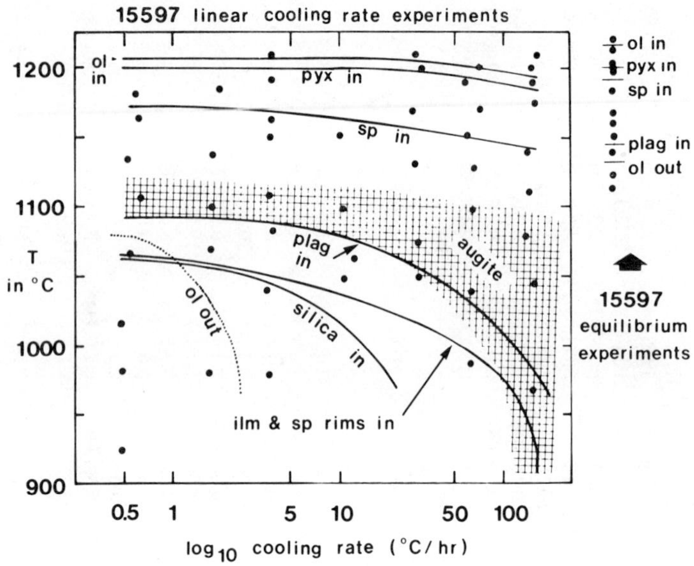

Fig. 1. Results of linear controlled cooling rate and equilibrium experiments on synthetic glass with the bulk composition of 15597. All cooling rate experiments were initiated at 1225°C ± 5°C, cooled at the rate desired to a temperature indicated by the black dot and quenched into water. Lines show the temperature below which a phase appears or disappears. The cross-hatched region indicates the presence of epitaxial (110) augite rims on pyroxene phenocrysts. See Table 2 for reversals of phase appearance in equilibrium runs.

more olivine-normative member of the suite) and found liquidus olivine at a higher temperature (1273°C) followed by spinel and pyroxene at 1225°C. These results are consistent with 15065 being an olivine plus pyroxene cumulate. The 15597 experiments are most applicable to the crystallization of Apollo 15 QNBs. Humphries *et al.* do not report detailed phase chemistry for their experiments or show that their liquids stayed on composition during the course of the experiment.

Controlled cooling rate experiments

Controlled cooling rate experiments were also performed on the 15597 synthetic composition. The experiments were carried out at linear rates of 0.5°, 1.75°, 3.75°, 10°, 30°, 60°, and 150°C/hr. The experiments were initiated approximately 10°C above the equilibrium liquidus, cooled at the linear rate to a desired temperature, then quenched into H_2O (see Appendix). The technique is similar to that employed by Walker *et al.* (1975) and allows the direct determination of the appearance temperature of phases as a function of cooling rate. Therefore, the effects of cooling can be monitored throughout the crystallization interval. The abscissa of Fig. 1 indicates the cooling rate for each experiment and the ordinate the temperature at which the experiment was quenched. Each point shows the temperature of quenching for an experiment carried out at a particular cooling rate.

Several aspects of the equilibrium crystallization history are changed by carrying out dynamic crystallization on 15597. The appearance temperature and chemistry of the phases olivine, pyroxene, and spinel are not significantly affected by cooling rate. Olivine appearance is suppressed by about 15°C at 150°C/hr, but only 5°C at 3.75°C/hr. The change in the appearance temperatures for pyroxene and spinel is of the same magnitude. The later appearing phases, on the other hand, are significantly affected by cooling rate. At the slowest rate (0.5°C/hr) plagioclase appearance is suppressed by 60°C.

Table 2. Equilibrium experiments on synthetic analog of 15597 in iron capsules sealed in evacuated SiO_2 glass tubes.

Run #	T°C	Duration (hr)	Starting* material	Products
5	1225	8.3	gl	gl†
6	1212	18	gl	gl, Fo_{70}†
7	1204	47.5	gl	gl, Fo_{68},
38	1200	25	#13	gl, ol
8	1194	44.8	gl	gl, Fo_{67}, cpx†
76	1178/1188	91.2/46.2	gl	gl, ol, cpx
9	1182	54.3	gl	gl, Fo_{66}, cpx, sp
11	1168	71.2	gl	gl, Fo_{64}, cpx, sp
12	1161	42.5	gl	gl, Fo_{61}, cpx, sp†
13	1150	135.5	gl	gl, Fo_{60}, cpx, sp
33	1145	68	#14	gl, ol, cpx, sp
14	1138	73	gl	gl, Fo_{56}, cpx, sp, plag†
74	1125/1139	88.2/119.6	gl	gl, ol, cpx, sp, plag
15	1123	119	gl	gl, cpx, sp, plag†

*Number indicates that the product from this experiment was used as starting material.

†Compositions listed in Table 1.

Note: subscripts to Fo indicate compositions of olivine as (Mg/[Mg + Fe]) × 100 in mole%. gl = glass, ol = olivine, cpx = clinopyroxene, sp = spinel, plag = plagioclase.

In the 150°C/hr run plagioclase is suppressed by ~ 160°C and is present as micron-sized microlites in the groundmass. The growth of an augite rim (hatched region in Fig. 1) on early pigeonite phenocrysts accompanies the suppression of plagioclase appearance. Upon plagioclase nucleation and growth, a groundmass of low calcium pyroxene and plagioclase is formed.

Olivine reacts out only in the very slowly cooled experiments (0.5° and 1.75°C/hr) and persists to lower temperatures in the cooling rate experiments (1080°C in the 1.75°C/hr runs) than in the equilibrium experiments (1130°C). Even then, an olivine crystal is occasionally trapped within a large pyroxene phenocryst and is preserved at lower temperatures. Generally olivine in contact with liquid dissolves. At cooling rates more rapid than 1.75°C/hr olivine is found surrounded by groundmass plagioclase and pyroxene. It is mantled by these phases and unable to react with the liquid. Cristobalite appears as a late-stage mineral in all experiments except the 150°C/hr runs. Ilmenite also appears as a late crystallizing phase slightly before cristobalite.

The solidus could not be determined in the equilibrium experiments because of the difficulty in distinguishing the abundant tiny crystals from liquid. The synthetic basalts are essentially completely crystallized by 975°C at 0.5°C/hr. A 60°C/hr run quenched at 890°C still contains small amounts of glass, but is > 99% crystalline. The 150°C/hr sample contains significant glass along with a quenched pyroxene-plagioclase groundmass at 750°C. At the rapid cooling rate the sample never really becomes completely crystalline.

All linear cooling rates produce a porphyritic texture as observed by Lofgren *et al.* (1974). The phenocrysts are clinopyroxene and the groundmass consists predominantly of a clinopyroxene-plagioclase aggregate. A discussion of the textures and their relation to crystallization history is given by Lofgren *et al.* (1975) and Grove and Walker (this volume).

DISCUSSION

Applicability of experiments to lunar basalt crystallization

A comparison of the mineral chemistries of the lunar samples to those of the equilibrium and cooling rate experiments tests the applicability of the experiments. Because little chemical variation exists between QNB textural types, the rapidly cooled and more slowly cooled basalts should have mineral chemistries that reflect their cooling histories. Lofgren *et al.* (1975) divide the Apollo 15 QNBs into vitrophyric, porphyritic-spherulitic, and porphyritic-subophitic varieties. An examination of the vitrophyric and porphyritic-spherulitic varieties shows that there are two types that differ in their phase chemistry but which have essentially the same bulk composition. One type collected near the Hadley Rille site includes samples 15597, 15595, and 15596. It is distinguished by the presence of magnesian clinopyroxene ($Wo_4En_{70}Fs_{26}$) and chromian spinel as phenocryst phases and the absence of olivine (Table 3). The second type which was collected near Dune Crater includes members that are slightly more olivine-normative (15485, 15486, 15499) and contains olivine (Fo_{68-70}) phenocrysts. Clinopyroxene phenocrysts are less magnesian ($Wo_5En_{67}Fs_{28}$).

The similar compositions of the near-liquidus phases in equilibrium experiments and the cores of minerals in the controlled-cooling rate experiments show that the iron capsule experiments are applicable to the crystallization of the Dune Crater vitrophyres. The phase chemistries of the Hadley Rille vitrophyres suggest crystallization under lower f_{O_2} conditions than their Dune Crater compositional equivalents. An experimental crystallization study of the Juvinas eucrite (Stolper, 1976) is pertinent to the difference in phase chemistries. With decreasing f_{O_2} iron blebs become more abundant in the Juvinas experimental charges, liquids become more magnesian and the liquidus mineral olivine is replaced by clinopyroxene. If the QNB vitrophyres are from the same cooling unit, these differences in f_{O_2} of crystallization are explainable. The Hadley Rille vitrophyres may be samples of the flow top (exposed to lunar vacuum) and the Dune Crater vitrophyres are from the flow bottom (higher f_{O_2} imposed by the magma).

The similar mineral chemistries in the experiments and the lunar samples are consistent with the crystallization of the Apollo 15 QNB suite near the lunar surface. The experiments indicate that the olivine and pyroxene in the lunar samples crystallized from a magma with bulk composition of 15597. The experimentally produced spinels and those found in the vitrophyres are slightly different in composition (Table 3). The lunar spinels in 15597 and 15486 have Cr/(Cr + Al) and Mg/(Mg + Fe) higher than their experimental counterparts. We suggest above that the two vitrophyres crystallized under slightly different f_{O_2} conditions, that are fairly close to those of the experiments. The higher Cr content of the lunar spinels suggests higher f_{O_2} (Akella *et al.*, 1976; Huebner *et al.*, 1976; Schreiber and Haskin, 1976; Lipin, 1977) than the experiments, but Mg/(Mg + Fe) of the liquid also controls Cr content. With higher Mg number more Cr is included in spinel. The difference between the lunar and experimental spinels may be related to differences in appearance temperature imposed by f_{O_2}. If spinel appeared in the lunar samples at higher temperature, the higher Cr and Mg content could be explained. Petrographic observations of the natural basalts (Weigand and Hollister, 1973; Bence and Papike, 1972) suggest the appearance of spinel with or before olivine and pyroxene.

Pertinence of the cooling rate experiments to nucleation and growth of phases in lunar QNBs

In the following discussion pyroxene chemistry will be explored further in an attempt to determine crystallization rate. Two experimental techniques have been applied to the dynamic crystallization of lunar basalts; the iron capsule technique (Walker *et al.*, 1975; this study) and the wire loop technique (Lofgren *et al.*, 1974; Donaldson *et al.*, 1975). The crystallization behavior of the system depends on the technique used. Walker *et al.* (1975) and this study (Fig. 1) find that liquidus phase appearance is not dependent on cooling rate while the temperature of appearance of the later phases (plagioclase, ilmenite, spinel) is systematically depressed with more rapid cooling rates. Furthermore, the chemistry of the liquidus phases is independent of cooling rate, while the later appearing minerals are affected. The many small iron blebs abundant in the liquid starting material probably provide nucleation sites for the formation of the first olivines and pyroxenes in iron capsule experiments. In the wire loop experiments the compositions of the first phases to crystallize vary with cooling rate (Donaldson *et al.*, 1975; Lofgren *et al.*, 1974; Usselman and Lofgren, 1976). We might expect that nucleation and growth behavior differs depending on technique, but a detailed comparison of the two techniques is not presently available. The question as to which technique is more applicable to the crystallization of lunar basalts should be rephrased to which technique is most successful for a specific basalt type? We suggest that the iron capsule technique is more applicable than the loop method in the case of the Apollo 12 picrites and the Apollo 15 QNBs. These basalts contained Fe blebs in their early liquids and have phenocryst core compositions that are identical in both the rapidly and slowly cooled samples. Comparisons of nucleation densities in the experimental and lunar samples (Grove, 1977) reveal that some QNB's (for example, 15597) contain more nucleii than any of the experiments performed in iron capsules. This is *not* to say that nucleation behavior in iron capsule experiments is applicable to all lunar basalt crystallization, but that in the case of the Apollo 15 QNBs, the evidence from the natural samples indicates that abundant nucleation sites were present. Hence, the Fe capsules may be an appropriate experimental technique.

Relationship of Pyroxene Chemistry to Crystallization History

The experiments show that pyroxene responds to crystallization history in the following manner: (1) The partitioning of major elements (Ca, Mg, and Fe) between pyroxene and liquid is not strongly dependent on cooling rate, so the effects of suppression of appearance temperature of other phases (e.g., plagioclase) and the efficiency of fractional crystallization control pyroxene chemistry; (2) the partitioning of minor elements (especially Al_2O_3 and TiO_2) are strongly rate dependent and may be useful as indicators of cooling rate. The following discussion treats two aspects of pyroxene chemistry. The first deals with the chemical variations in pyroxene as a function of cooling history, and the second discusses pyroxene-liquid partitioning behavior for major and minor elements in the cooling rate and equilibrium experiments.

Equilibrium experiments

The compositions of pyroxene in the equilibrium experiments are depicted in terms of Wo–En–Fs, Ti vs. Al per six oxygens and a plot of normalized Al^{VI}, Ti, and Cr (Figs. 2a, b, c). The major element compositions of equilibrium pyroxenes become more Ca and Fe-enriched at lower temperatures, but augite rims do not form. Titanium and aluminum (Fig. 2b) are present in small quantities (Table 1) and although the Ti-Al plot appears to follow a 1/4 trend, there is within the errors of the probe analyses no octahedrally coordinated Al. Aluminum fills the tetrahedral site, balancing the charge excess from Cr^{3+}, Ti^{4+}, and possibly Ti^{3+} that are present in octahedral coordination. Although the 1123°C equilibrium experiment appears to contain octahedrally coordinated Al (Fig. 2c), the amount of Al^{VI} is less than the errors associated with replicate probe analyses. Thus, equilibrium pyroxenes contain no appreciable Al^{VI}, either before or after the appearance of plagioclase.

Cooling rate experiments

The effect of cooling rate on pyroxene chemistry is profound, and at a glance it may give a rough idea of the crystallization history of a lunar quartz normative basalt. A number of groups (Boyd and Smith, 1971; Hollister *et al.*, 1971; Bence and Papike, 1972; Weigand and Hollister, 1973) noted the wide variation in texture and pyroxene chemistry in lunar quartz-normative basalts from the Apollo 12 and 15 sites, and attempted to estimate cooling histories and paragenetic sequences based on chemical trends. Can these trends be used to estimate the cooling history of lunar basalts? Our approach will be to understand the chemical variations as a function of cooling rate and the processes that cause the trends, then choose criteria that may serve as cooling rate indicators. Although the major element composition of the first pyroxene to crystallize does not change as a function of cooling history, the minor element contents of the first pyroxene to crystallize increase with more rapid cooling rate (Table 3). Suppression of later appearing phases (especially plagioclase) as a function of cooling rate affects both the major and minor element chemistry of pyroxene.

Pyroxene chemistries for the 150°, 60°, 3.75°, 1.75°, and 0.5°C/hr cooling rate experiments are shown in Figs. 3–7. Complete zoning profiles were measured for

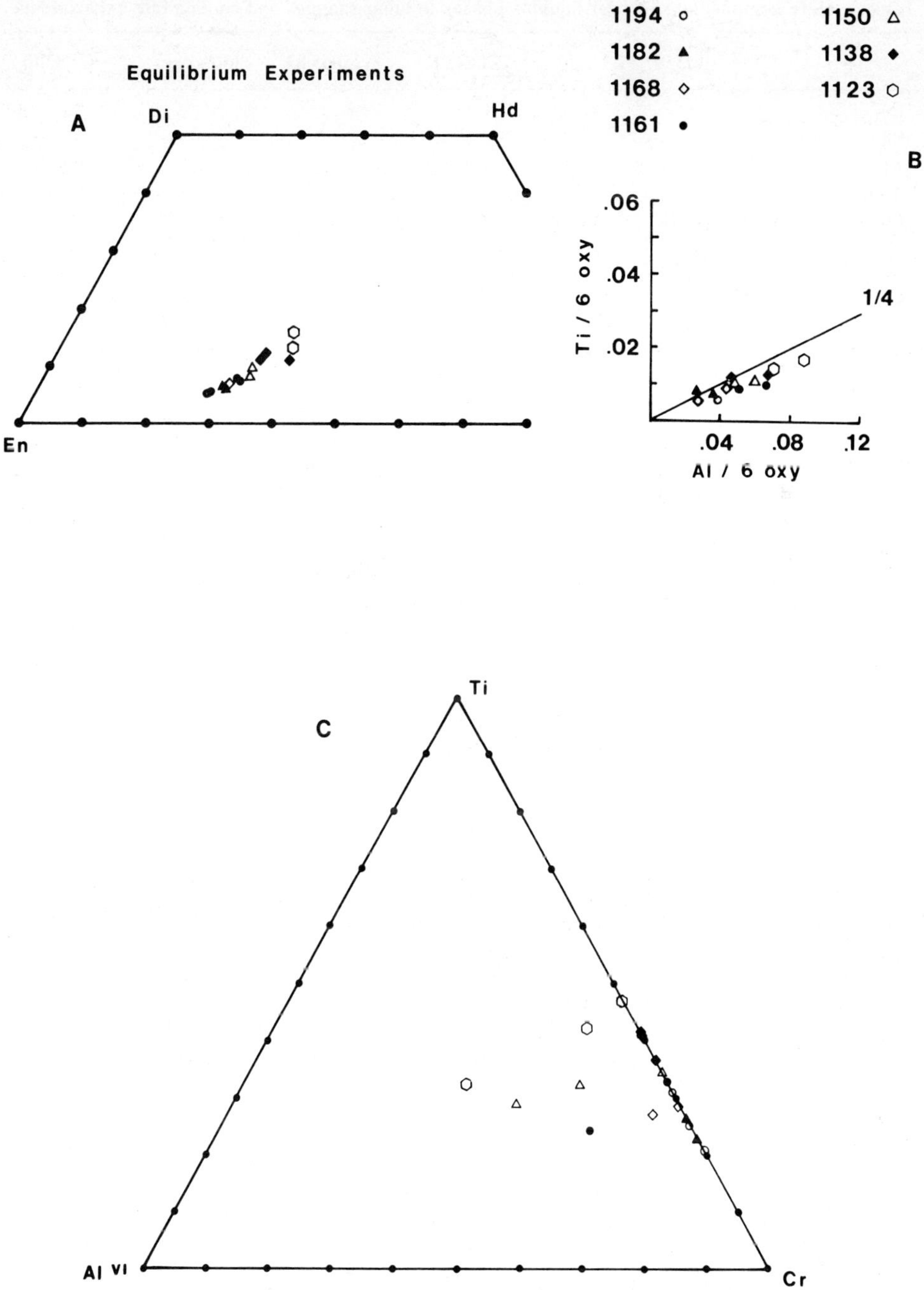

Fig. 2. Pyroxene compositions in equilibrium experiments. Symbols indicate the temperature at which the pyroxene was equilibrated. (a) Ca–Fe–Mg variations. (b) Ti and Al normalized to 6 oxygens. (c) Normalized Al^{VI}–Ti–Cr.

Table 3. Core compositions of near-liquidus phases in lunar samples and cooling rate experiments.

	15597,17	15486,22	15065,83	150°C/hr	0.5°C/hr
Pyroxene					
SiO_2	53.5	52.8	53.2	52.0	54.2
Al_2O_3	1.29	2.02	1.25	1.78	0.83
TiO_2	0.37	0.49	0.39	0.39	0.24
FeO	16.2	16.6	16.0	17.2	17.1
MnO	0.35	0.40	0.42	0.23	0.26
MgO	25.7	24.4	26.1	24.0	25.1
CaO	2.15	2.83	2.20	2.78	2.56
Cr_2O_3	0.97	1.27	1.00	1.21	0.88
Σ	100.6	100.8	100.6	99.6	101.2
Normalized to six oxygens					
Si	1.939	1.921	1.930	1.919	1.959
Al	0.055	0.086	0.054	0.077	0.035
Ti	0.010	0.014	0.011	0.011	0.006
Fe	0.491	0.503	0.485	0.532	0.518
Mn	0.011	0.012	0.013	0.007	0.008
Mg	1.380	1.320	1.411	1.322	1.353
Ca	0.084	0.110	0.085	0.110	0.099
Cr	0.028	0.037	0.029	0.036	0.025
Σ	4.008	4.003	4.018	4.014	4.003
Olivine					
SiO_2		37.0		37.3	37.9
Al_2O_3		0.55		0.00	0.00
TiO_2		0.14		0.11	0.03
FeO		28.4		27.6	29.0
MnO		0.38		0.23	0.30
MgO		34.7		34.8	33.8
CaO		0.35		0.48	0.33
Cr_2O_3		0.59		0.49	0.44
Σ		102.1		101.0	101.8
Fo		68		69	67
Fa		32		31	33
Spinel					
Al_2O_3	11.6	11.8		12.2	12.9
TiO_2	2.39	2.62		4.32	3.33
FeO	25.0	24.8		29.1	27.7
MnO	0.61	0.53		0.46	0.44
MgO	7.70	7.90		6.02	5.72
Cr_2O_3	55.3	55.3		50.0	50.1
Σ	102.6	103.0		102.1	100.2
Cr/(Cr + Al)	0.762	0.759		0.733	0.722
Mg/Mg + Fe	0.355	0.362		0.269	0.269
Iron metal spheres	In pyx cores with chromite in groundmass	In pyx with chromite in groundmass	In pyx cores	Liquids saturated with iron metal	

each experiment, but the data are summarized to give some idea of the pyroxene composition as a function of temperature. Thus, an experiment quenched at a particular temperature is shown with a specific symbol. The chemical changes that have occurred in the temperature interval between that experiment and the one quenched at a higher temperature are represented by the portion of the trend containing the symbol for the experiment under consideration. The most fractionated pyroxene (i.e., the composition at the crystal-liquid interface) is at the low-temperature end of each trend.

The entire crystallization interval in the 150°C/hr experiments is covered in approximately two hours. The initial pyroxene has major element chemistry similar to that in the equilibrium experiments (Fig. 3a) but the initial Ti and Al contents are significantly higher. The cores of pyroxene phenocrysts follow a pigeonite trend down to about 1079°C, where an epitaxial augite rim forms on (110) while the pigeonite trend continues on the (010) sector. The augite rim persists for the remainder of the crystallization history and zones to a ferroaugite ($Wo_{32}En_{25}Fs_{43}$ at 968°C). At 968°C a groundmass of submicron-sized pyroxene and plagioclase appears, but a substantial part of the charge is glass. Titanium and aluminum contents of the pyroxenes increase continuously during crystallization as a result of the suppression of plagioclase and ilmenite (Fig. 3b) while the Cr content steadily decreases. A substantial proportion of the Al is in octahedral coordination (approximately 38% of the total 8.7 wt.% Al_2O_3 in the ferroaugite rims).

At 60°C/hr the pigeonite cores follow essentially the same major element trend (Fig. 4a) but contain slightly less total Ti and Al (Fig. 4b). Again a (110) epitaxial augite rim grows on the pigeonite core and the (010) pigeonite sector persists. At the slower crystallization rate, however, a groundmass of pyroxene, plagioclase, and ilmenite forms. Nucleation of groundmass plagioclase, ilmenite, and pyroxene occurs between 1040° and 992°C. The effect on pyroxene chemistry is to stop the ferroaugite trend at $Wo_{33}En_{31}Fs_{36}$ and form groundmass low-Ca pyroxene (Wo_{20}–Wo_{25}). Comparison of the Ti vs. Al plot for 150° and 60°C/hr runs (Figs. 3b, 4b) shows that less total Ti and Al are incorporated in the pigeonite cores and epitaxial augite of the 60°C/hr runs. Octahedral Al is incorporated into the pyroxene (Fig. 4c) until plagioclase and groundmass pyroxene nucleate. The proportion of Al^{VI} decreases sharply in the groundmass pyroxene.

The pyroxene phenocrysts in the 3.75°C/hr cooling rate experiments (Fig. 5) are characterized by lower total Al and Ti and by a less pronounced (110) augite rim. Both phenomena are a consequence of the appearance of plagioclase at a higher temperature (1085°C). A distinct phenocryst-groundmass texture forms, but the groundmass pyroxenes are somewhat larger than in the 60°C/hr experiments and are zoned to almost completely Mg-free compositions. The variations of Al^{VI}, Ti, and Cr and Ti–Al show that less total Al_2O_3, Al^{VI}, and Ti are incorporated in the (110) augite rims before plagioclase and groundmass pyroxene appear and the incorporation of octahedral Al ceases. The groundmass pyroxene ranges from Wo_{10} to Wo_{20} (between 1080° and 1040°C) increasing to Wo_{20}

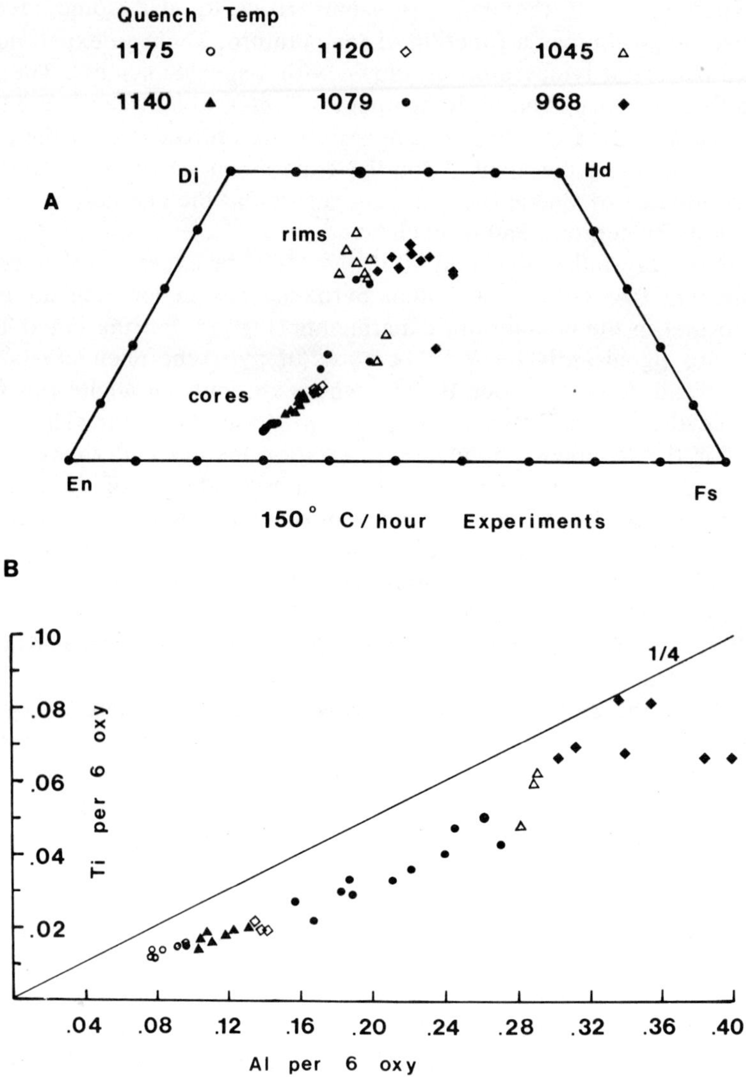

to Wo_{30} (1040° to 981°C) thus skirting the forbidden zone. Cr is exhausted from the liquid and Ti is incorporated as Ti^{3+} and Ti^{4+} as indicated by the 981°C Ti–Al data that lie between a 1/2 and 1/1 line in Fig. 5b.

At the slowest cooling rates (1.75°C/hr, Fig. 6 and 0.5°C/hr, Fig. 7) conditions closer to those of the equilibrium experiments are approached. The pigeonite phenocryst cores contain less total Ti and Al and no Al^{VI} (Figs. 6c, 7c). As in the equilibrium experiments the Ti–Al plots show a 1/4 trend. The appearance of plagioclase is suppressed from its equilibrium temperature of 1145 to 1085°C.

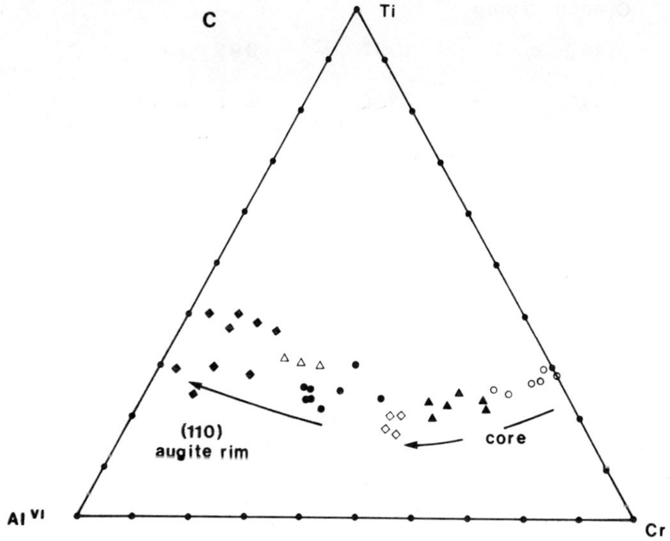

Fig. 3. Pyroxene compositional relations in 150°C/hr cooling rate experiments. The zoning trend is shown for the entire crystallization intervals, but to avoid overlapping analyses, only the zoning profile produced in the temperature interval between quenched runs is shown. Thus, each symbol represents the compositional variation produced between the quench temperature of the previous run and the quench temperature indicated by the symbol. The initial pyroxene to crystallize in all experiments is $Wo_5En_{68}Fs_{27}$. (a) Ca–Fe–Mg relations. Groundmass pyroxene was too small to analyze. (b) Ti and Al plot normalized to six oxygens. Note enrichment of aluminium in epitaxial (110) augite rims. (c) Normalized Ti–Cr–Al^{VI}.

Between approximately 1135°C and the appearance of plagioclase, some octahedral Al is incorporated into the pyroxene phenocrysts and an augite rim ($Wo_{33}En_{40}Fs_{27}$) forms on the pigeonite cores. At this point plagioclase nucleates at phenocryst-liquid interfaces and in the liquid. The Wo content of the rim-plus-groundmass pyroxene varies between Wo_{12} and Wo_{25}, increasing to Wo_{30} to avoid the "forbidden" zone near the Fs end of the quadrilateral (Figs. 6a, 7a). The experimental pyroxene trend then bifurcates, one trend approaches hedenbergite ($Wo_{42}Fs_{68}$) and the pyroxene is always found with fayalite. The other trend approaches $Wo_{14}Fs_{86}$ and "pyroxene" compositionally identical to pyroxferroite is found with plagioclase, ilmenite, cristobalite, and liquid, but never with fayalite.

The pyroxferroite-like mineral was found to persist in a 0.5°C/hr experiment cooled to 925°C. The presence of this phase and its avoidance of the hedenbergite-plus fayalite assemblage suggests that pyroxferroite may persist in rather slowly cooled basalts. Lindsley and Burnham (1970) show that the equilibrium assemblage was fayalite and silica at 980°C, and suggest that pyroxferroite would react out at cooling rates less than 3°C/hr. However, its existence in the 0.5°C/hr experiments and in rather slowly cooled Apollo 15 QNBs (e.g., 15058, 15065) suggest that it may persist metastably during the cooling of a lava flow.

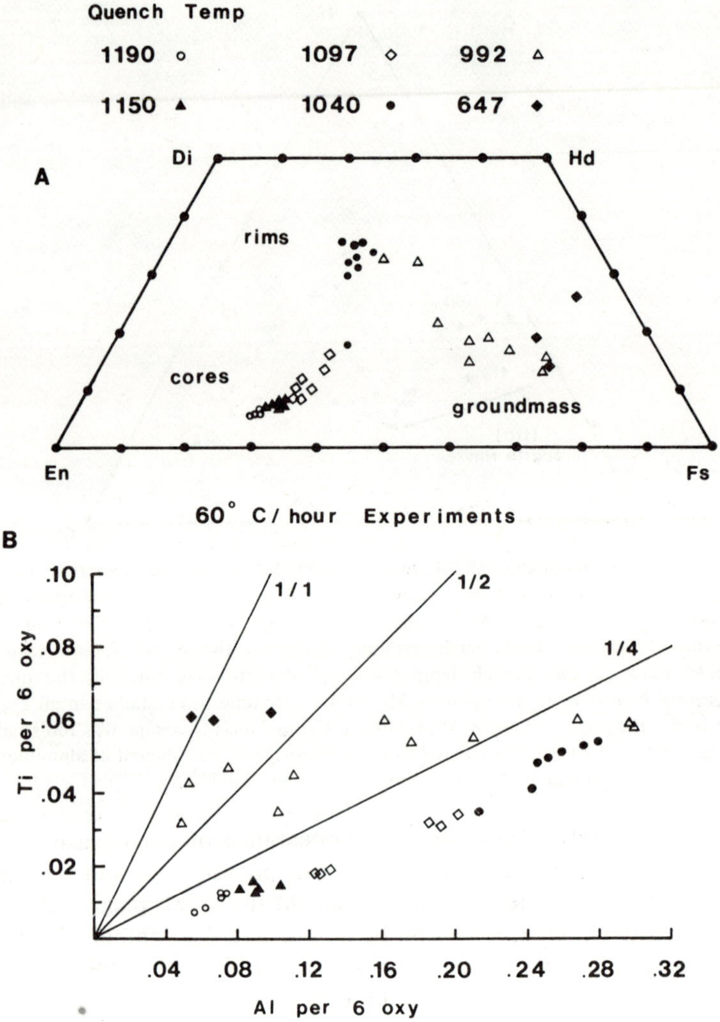

Partitioning of elements between pyroxene and liquid

We have just characterized the differences in pyroxene compositional variation that occur as a function of cooling rate. However, we would like to know the relationship between pyroxene and liquid through the crystallization sequence. Is there a rate dependency on pyroxene-liquid partitioning? If not, what combined effects operate to produce the zoning trends and is there any chemical parameter that we can measure in a lunar pyroxene that will give a clue as to its cooling history?

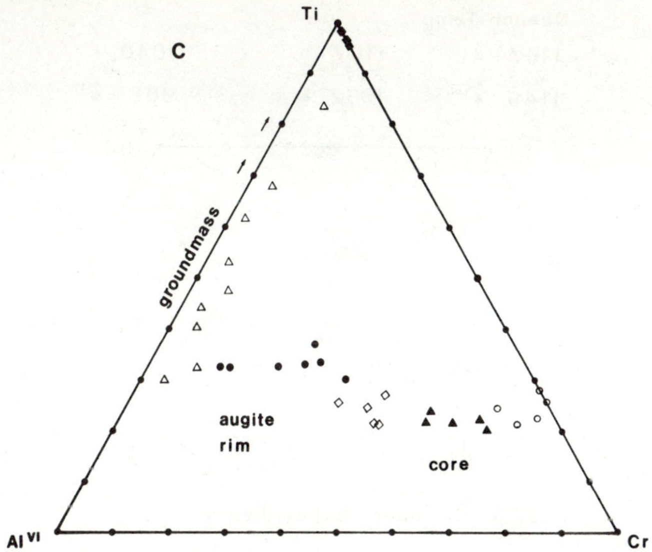

Fig. 4. Pyroxene compositional relations in 60°C/hr experiments. Symbols are as described in Fig. 3. (a) Ca–Fe–Mg relations. Note drop in Ca content of groundmass pyroxene (992°C experiment) upon nucleation of plagioclase. (b) Ti–Al relations. Note drop in total Al content as a result of plagioclase nucleation. (c) Al^{VI}–Ti–Cr plot. Note drop in octahedral aluminum content of groundmass pyroxene.

At least two important kinetic effects control pyroxene chemistry. Pyroxene appearance temperature is not significantly suppressed as a function of cooling rate (Fig. 1). This is reflected in a similarity in major element chemistry of phenocryst cores from all cooling rate experiments and in the lunar samples. Cooling rate affects the appearance temperature of later phases (e.g., plagioclase and ilmenite) thus altering the liquid line of descent which results in changes in the major and minor element chemistry of the pyroxene. To investigate the possible rate dependency on element partitioning between pyroxene and liquid, pyroxene-liquid pairs were analyzed in the equilibrium and cooling rate experiments. By rate dependency we mean the manner in which cooling rate changes the nucleation and growth behavior of the system.

A representative number of pyroxene rim and quenched liquid analyses were obtained for each experiment (see Appendix). Glass analyses used to calculate partitioning were made between 10 and 30 microns from the pyroxene-liquid interface. There is some variability in the liquid chemistry of the more rapidly cooled experiments, but those crystallized at rates slower than 3.75°C/hr do not contain appreciable compositional gradients in the liquid. The existence of compositional variations in the rapidly cooled experiments (30°C/hr and faster) as a result of differences in diffusion rates of elements away from the crystal-liquid interface is presently under study. Both the (010) pigeonite sector and the (110) epitaxial augite (when present) were analyzed and treated separately. Groundmass pyroxene that crystallizes with plagioclase invariably shows a wide

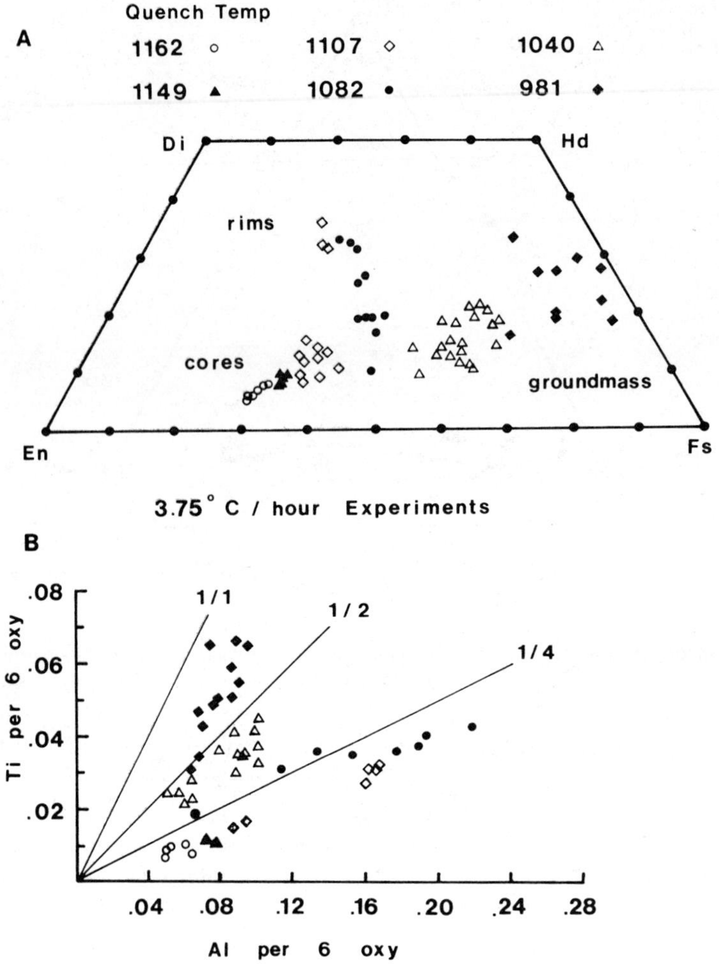

range of Ca contents (Wo_{12} to Wo_{30}). Ross *et al.* (1973) show that groundmass pyroxene in an Apollo 12 quartz-normative basalt contains two phases (a low-Ca and a high-Ca phase) when examined by single-crystal x-ray methods. Thus, the electron microprobe analyses may be averages of two-phase intergrowths.

Bulk distribution coefficients

Bulk distribution coefficients, D (where $D = X_{\text{oxide}}^{\text{pyroxene}}/X_{\text{oxide}}^{\text{liquid}}$ in mole%), have been determined for FeO, MgO, CaO, Al_2O_3, TiO_2, and Cr_2O_3 in the equilibrium and controlled cooling rate experiments. The D's are classified as either low-Ca clinopyroxene or augite. Low-Ca clinopyroxene includes the phenocryst cores,

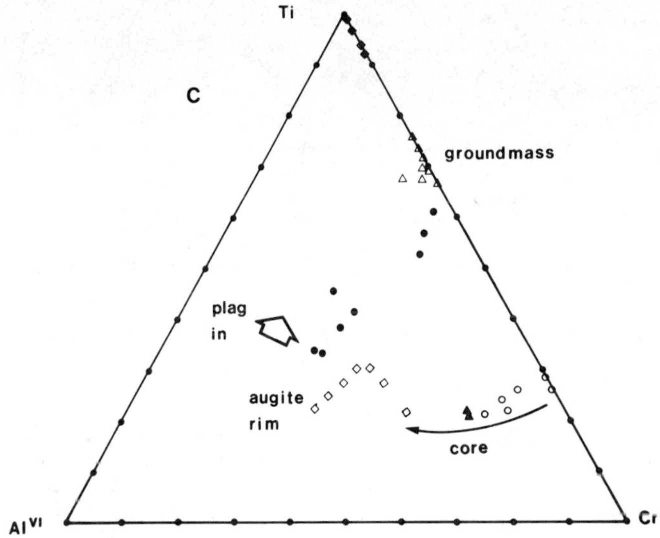

Fig. 5. Pyroxene zoning trends in 3.75°C/hr experiments. Symbols are as described in Fig. 3. (a) Pyroxene quadrilateral. (b) Ti–Al relations. Note less total Al included in augite rims than in more rapidly cooled experiments. (c) Al^{VI}–Ti–Cr.

(010) sectors and the groundmass pyroxene, and varies from Wo_{12} to Wo_{25}. The augite *D*'s are analyses of (110) epitaxial augite quenched in the cooling rate experiments for which plagioclase appearance is suppressed.

The bulk *D*'s for FeO, MgO, and CaO are shown for both low-Ca pyroxene and augite in log *D* vs. $1/T$ plots in Fig. 8. The symbols denote *D*'s for both linear cooling rate experiments and equilibrium experiments. The most striking feature of the bulk partition coefficients for major elements is that the partitioning is nearly independent of cooling rate. This means that the differences between pyroxene zoning trends observed in the linear cooling experiments are the result of other factors that are rate dependent. The efficiency of fractional crystallization through crystal zoning and the changes in liquid composition imposed by the suppression of appearance temperature of later phases control pyroxene chemistry.

There does appear to be a small, yet systematic, difference in the bulk *D* for MgO as a function of cooling rate, and MgO appears to be preferentially partitioned into the liquid at faster cooling rates. Lofgren's (1974) experiments in the system Ab–An show that supercooling reduces plagioclase appearance temperature and changes the composition of the first crystalline phases to more Na-rich compositions. In the case of pyroxene a slightly more Fe-rich phase may result at more rapid cooling rates. This effect is consistent with the observation in the linear cooling rate experiments (Fig. 1) which show a slight depression in pyroxene appearance temperature. Donaldson *et al.* (1975) have shown that the effect of supercooling a picrite liquid changes the first olivine to more Fe-rich

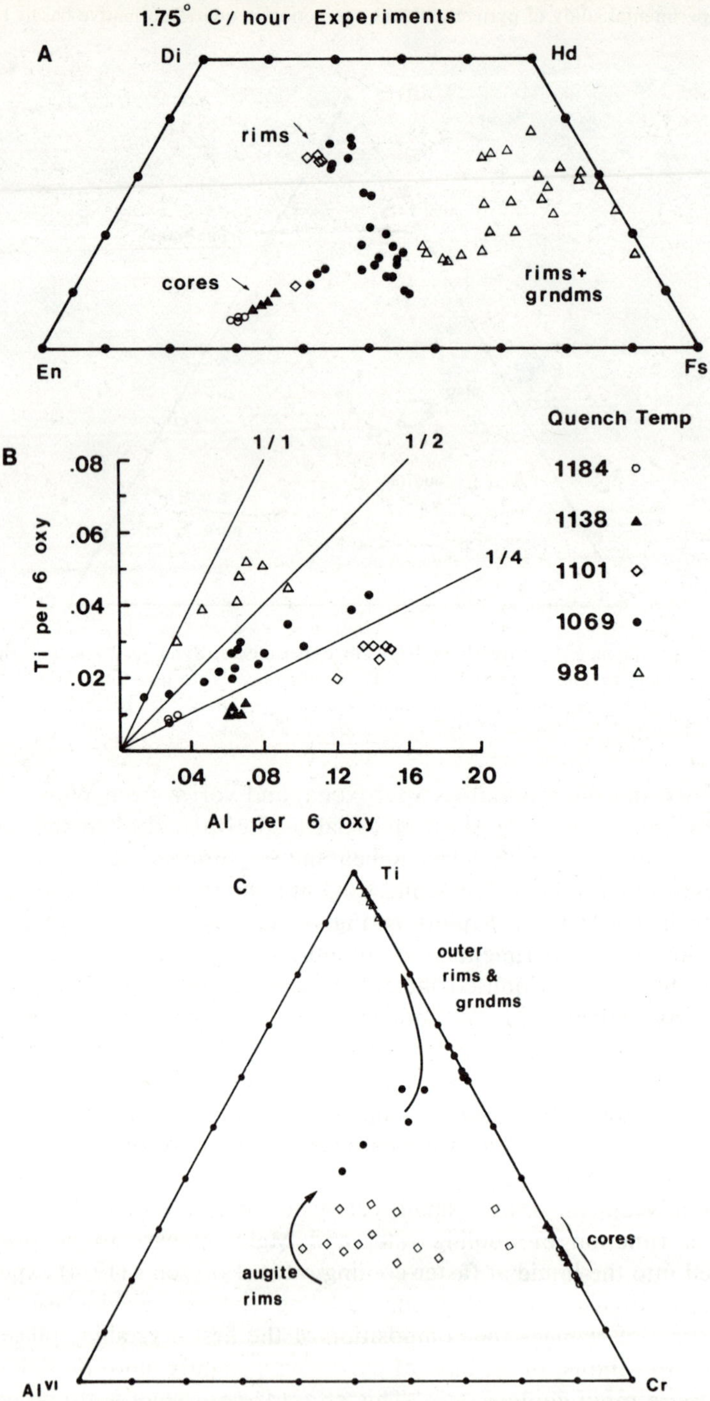

Fig. 6. Pyroxene compositional relations in 1.75°C/hr experiments. Symbols as in Fig. 3. (a) Pyroxene quadrilateral. (b) Ti–Al relations. Note total Al included in pyroxene decreases. (c) Al^{VI}–Ti–Cr. Note initial pyroxenes contain only tetrahedrally coordinated Al.

compositions. There does not seem to be a systematic shift in the bulk D's for CaO and FeO (see Appendix).

Linear least squares fits were obtained for the bulk distribution coefficients from both the equilibrium and linear cooling rate experiments. Equations were calculated for MgO and FeO using the 33 pyroxene-liquid pairs plotted in Fig. 8. The resulting Eqs. (1) for MgO and (2) for FeO give correlation coefficients of 0.90 and 0.92, respectively. By subtracting Eq. (2) from (1) the temperature dependence of the distribution coefficient $K_D^{Fe/Mg}$ for low-Ca clinopyroxene (3) is found. $K_D^{Fe/Mg}$ varies from 0.28 at the first appearance of pyroxene (1200°C) to 0.17 at 975°C.

$$\log D_{MgO} = 3.40\left(\frac{10^3}{T, K}\right) - 1.89 \quad (1)$$

$$\log D_{FeO} = 1.31\left(\frac{10^3}{T, K}\right) - 1.00 \quad (2)$$

$$\log X_{FeO}^{pyx}X_{MgO}^{liq}/X_{MgO}^{pyx}X_{FeO}^{liq} = \log K_D^{Fe/Mg} = -2.08\left(\frac{10^3}{T, K}\right) + 0.89 \quad (3)$$

Roeder and Emslie (1970) and Longhi *et al.* (1975) find that the K_D for olivine is essentially temperature independent in the range 1400–1000°C. The $K_D^{Fe/Mg}$ for low-Ca clinopyroxene shows a marked temperature dependence. Roeder and Emslie (1970) also point out that the data of Medaris (1969) and Williams (1971) for partitioning of iron and magnesium between olivine and calcium-free pyroxene are temperature independent, implying that these two phases can be used to determine the Fe/(Fe + Mg) ratio of the liquid. In the case of the present experiments, olivine cocrystallizing with low-Ca clinopyroxene follows a temperature independent $K_D^{Fe/Mg}$, but the pyroxene does not.

Pyroxene in Apollo 16 breccias and melt rocks (Bence *et al.*, 1973; Walker *et al.*, 1973) shows a temperature dependency of Fe–Mg partitioning with other ferromagnesian phases. Noritic breccias (equilibrated to low temperatures) and feldspathic basalts (rapidly cooled) with essentially the same bulk Fe/(Fe + Mg) contain olivine of similar composition, but the coexisting pyroxene in the noritic breccias are more magnesian than the pyroxene in the rapidly cooled feldspathic basalts.

The temperature dependent $K_D^{Fe/Mg}$ may be useful as an indication of the appearance temperature of the first low-Ca clinopyroxene. If pyroxene is not the liquidus phase, it may be difficult to reconstruct the composition of the liquid from which the first pyroxene crystallized. The CaO bulk D appears to be quite sensitive to temperature and may also be a useful indicator of the crystallization temperature of early low Ca clinopyroxene.

Bulk D's for CaO, FeO, and MgO were calculated for the few (110) epitaxial augite rim and liquid pairs that were present in the quenched cooling rate experiments (Fig. 8). The augite coexists with olivine, the low-Ca clinopyroxene, and spinel. The bulk D for CaO shows little temperature dependence and is nearly equal to 1. The $K_D^{Fe/Mg}$ calculated for ten augite-liquid pairs ranges from 0.23 at 1120°C to 0.21 at 975°C and appears to be temperature independent.

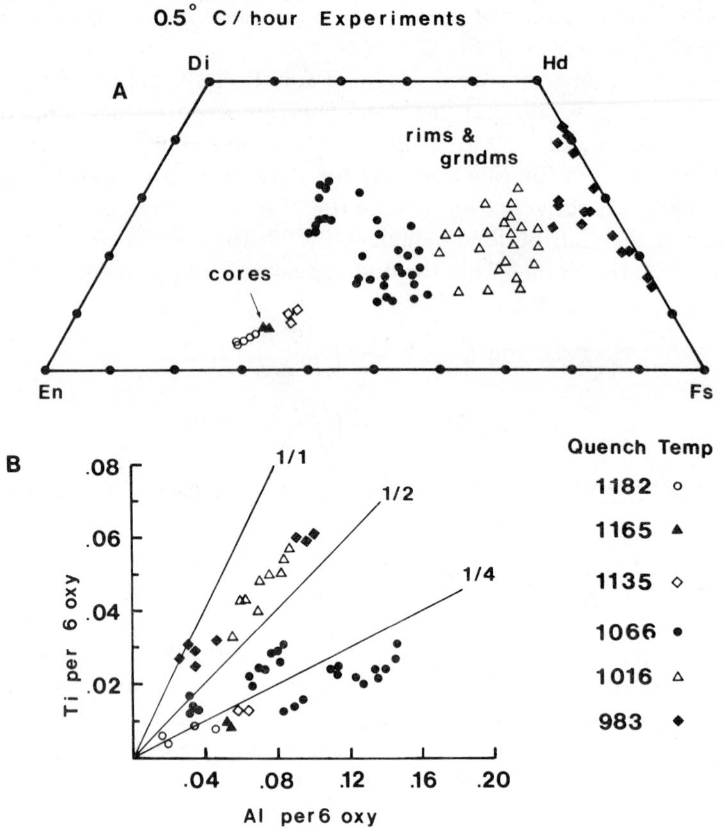

Applicability of QNB data to the crystallization of other lunar basalts

Extrapolation of the bulk *D*'s for quartz-normative basalts to other lunar basalt compositions must be carried out with caution. One would expect pyroxene-liquid partitioning to be affected by changes in liquid composition and external conditions (f_{O_2}, pressure, etc.). Longhi *et al.* (1975) found that the $K_D^{Fe/Mg}$ for olivine-liquid pairs in experiments on high titanium basalts was quite different (0.26–0.29) from the $K_D^{Fe/Mg}$ for low Ti and terrestrial basalts (0.3–0.33). The change is ascribed to differences in the fayalite activity of the melt (Longhi *et al.*, 1975) between the two systems. To evaluate the applicability of our partition coefficients to other lunar basalt systems, the experiments of Walker *et al.* (1976) on picrite 12002 and Longhi *et al.* (1974) on the Apollo 17 high-Ti basalts have been compared. The bulk *D*'s for the QNB's were used to estimate the equilibration temperatures of Walker's experiments. CaO, the most sensitive to temperature gives estimates that are consistently 10°–15°C low. The partitioning of Fe and Mg is not sensitive to the small temperature range, and gives estimates that are within ±25°C of the actual experiment.

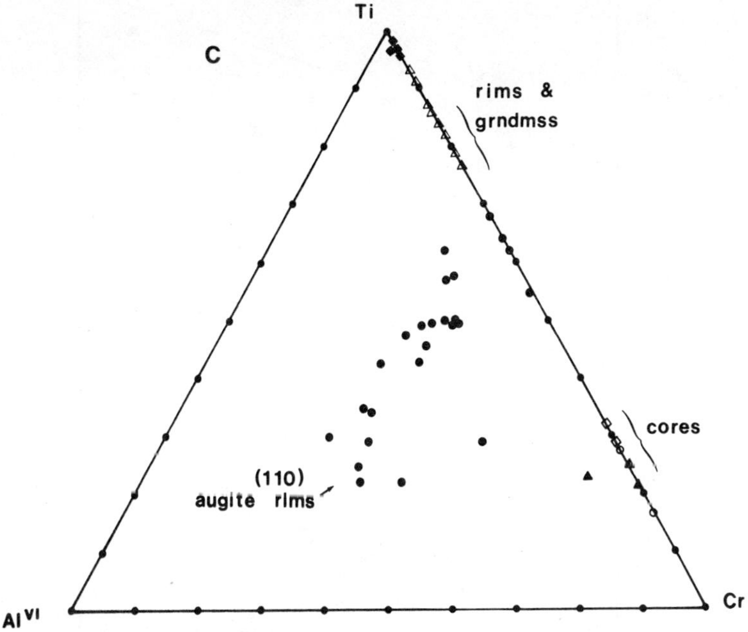

Fig. 7. Pyroxene compositional relations in 0.5°C/hr experiments. Symbols are as described in Fig. 3. (a) Pyroxene quadrilateral. Note late-stage pyroxene becomes Ca-rich and is found with fayalite or follows a low Ca trend to a composition that is similar to lunar pyroxferroite. (b) Ti–Al relations. Note less total Al is included in pyroxene cores. (c) Al^{VI}–Ti–Cr. Note octahedrally coordinated Al is present only in experiment where plagioclase appearance has been suppressed.

It is not possible to estimate the crystallization temperature of the Apollo 17 augites. CaO, MgO, and FeO fail to give consistent or even reasonable temperature estimates. This failure probably results from differences between the two basalt systems both in bulk composition and coexisting phases. The high-Ti basalt augite is in equilibrium with plagioclase, ilmenite, olivine, ± pigeonite while the QNB augites coexist with spinel, low-Ca clinopyroxene and olivine. Surprisingly, the pigeonite analysis reported by Longhi *et al.* (1974) in their 70017 experiments yields a temperature estimate when compared to the low-Ca clinopyroxene data that is about 20°C too high.

Bulk partitioning behavior of the minor elements

Bulk partition coefficients were determined for Al_2O_3, TiO_2, and Cr_2O_3 (Fig. 9) for low-Ca pyroxene and augite. In contrast to the major elements a rate dependency is observed for the partitioning of Al_2O_3 and TiO_2 between pyroxene and liquid. At rapid cooling rates both are partitioned preferentially in the pyroxene. The data for Cr_2O_3 are somewhat ambiguous. The 150°C/hr and 60°C/hr data give higher D's than the more slowly cooled experiments, but the

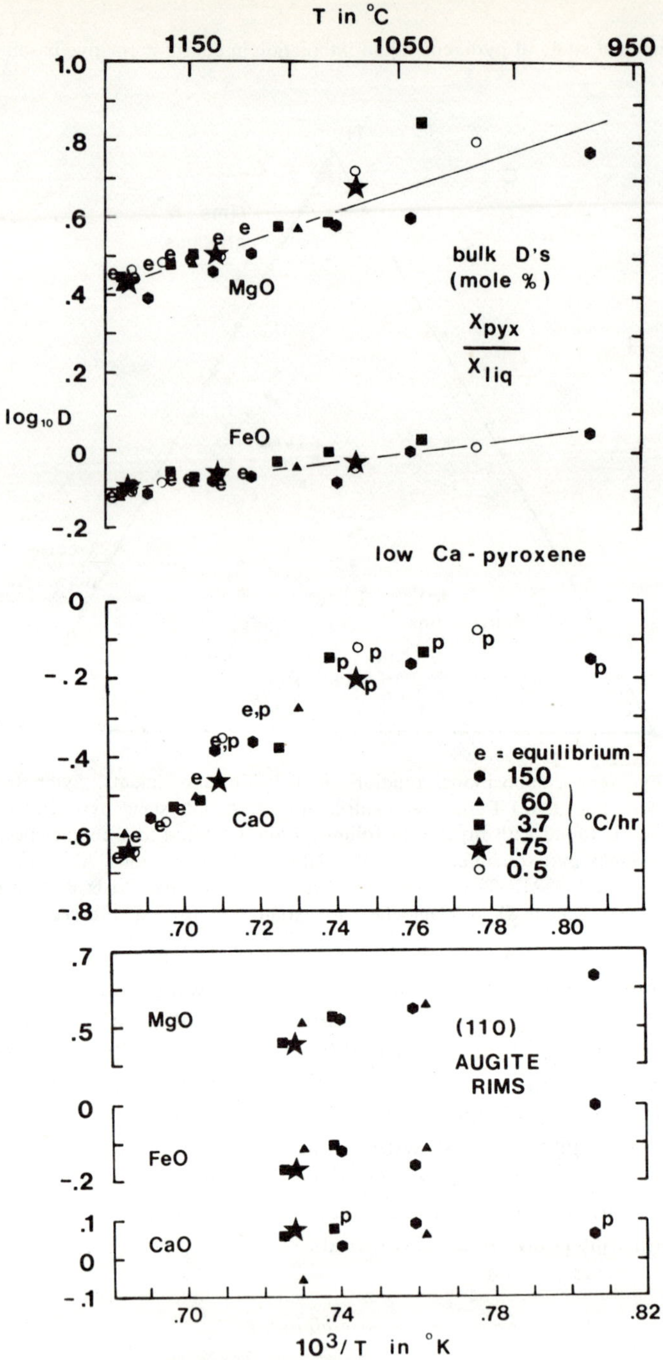

Fig. 8. Experimentally determined bulk D's for FeO, MgO, and CaO in low-Ca pyroxene and augite. The results of both equilibrium (e) and controlled cooling rate experiments are presented on $\log_{10} D$ vs. $1/T$ (°K) plots. Different symbols indicate the cooling rate of the experiment. D's for the cooling rate experiments were calculated from multiple analyses of pyroxene rims and coexisting liquid. p indicates the presence of plagioclase. Lines fitted to MgO and FeO data are used to calculate a $K_D^{Fe/Mg}$ (see text).

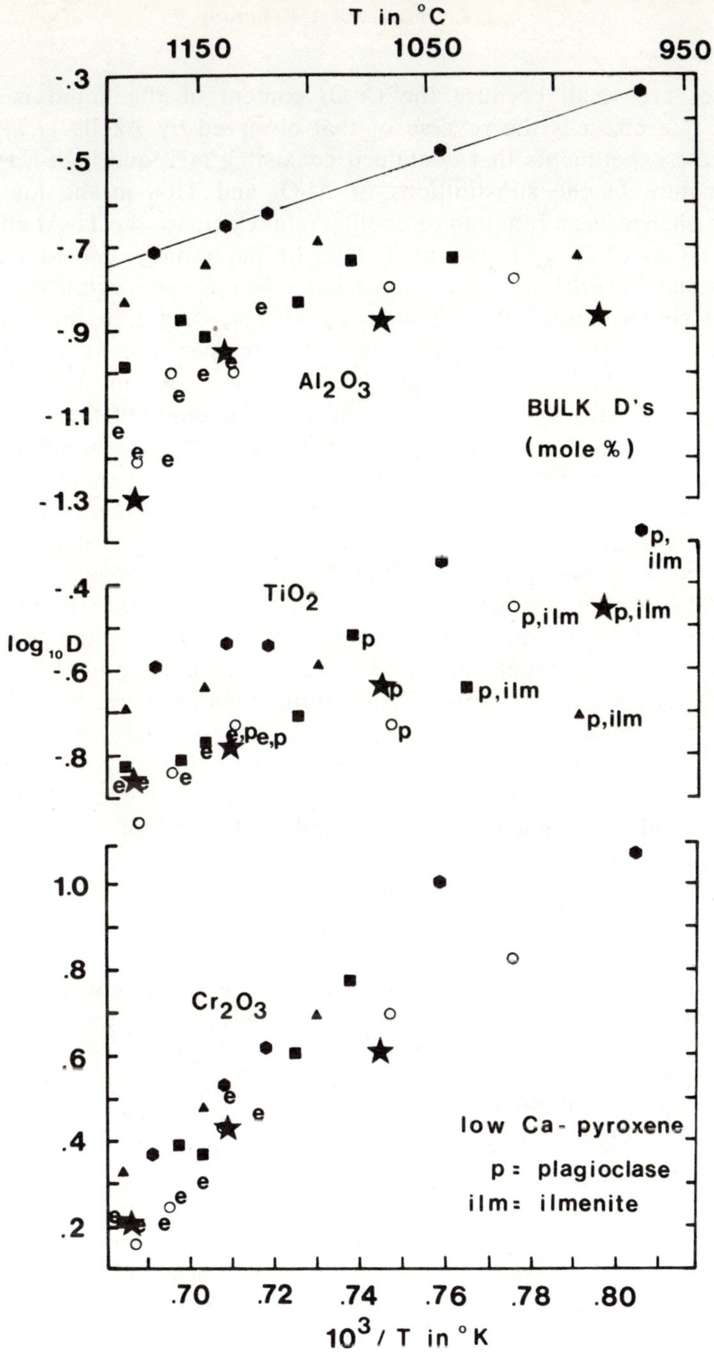

Fig. 9. Experimentally determined bulk *D*'s for Al_2O_3, TiO_2, and Cr_2O_3 in low-Ca pyroxene. Symbols as in Fig. 8. The presence of plagioclase and/or ilmenite is indicated by *p* and ilm. The temperature dependence of minor element partitioning is best displayed by the Al_2O_3 data. A line is arbitrarily drawn through the Al_2O_3 bulk *D*'s for 150°C/hr experiments to illustrate the increased partitioning of Al_2O_3 into pyroxene.

differences are small because the Cr_2O_3 content of the liquid is low. This apparent rate effect is the reverse of that observed by Akella *et al.* (1976) in cooling rate experiments that contained coexisting melt and high-Ca pyroxene.

The nature of the substitutions of Al_2O_3 and TiO_2 in the low-Ca clinopyroxene changes as a function of cooling rate. Compare the Ti–Al and Al^{VI}–Ti–Cr plots (Figs. 2–7) with the bulk *D*'s. In the rapidly cooled experiments (150°C/hr and 60°C/hr) the earliest pyroxene contains some octahedral Al, but in the more slowly cooled experiments no Al^{VI} is present in the early-formed pyroxene, although it does appear at lower temperature as the equilibrium appearance of plagioclase is suppressed. Three important substitutions are recognized: $Al^{VI}Al^{IV}$, $Ti^{VI}Al_2^{IV}$ and $Cr^{VI}Al^{IV}$. The amounts of these substitutions increase with increasing cooling rate, as indicated by the greater amounts of total Cr, Al, and Ti in the first pyroxenes to crystallize from the rapidly cooled runs. With crystallization the $Cr^{VI}Al^{IV}$ component continually decreases, the $Al^{VI}Al^{IV}$ component increases as long as plagioclase appearance is suppressed, disappearing after plagioclase nucleates, and the $Ti^{VI}Al_2^{IV}$ component increases continually. This is observed for all the cooling rate experiments. The bulk partition coefficients for augite show the same rate dependence as the low-Ca clinopyroxene, and the amounts of Cr, Al, and Ti are greater in the more calcic pyroxenes. The amount of "others" substitution in a pyroxene increases with its Ca content.

The rate dependence of minor element partitioning may serve as a useful indicator of the cooling rate experienced by early pyroxene. In the following section it will be applied to lunar samples. The minor elements that are incorporated during the late stages of cooling are related to rate in a complex manner. The suppression of the appearance temperature of late phases (e.g., plagioclase and ilmenite) affects the liquid line of descent and the partitioning of minor and major elements into the pyroxene. It is difficult to estimate late-stage cooling rates using minor element chemistry, because it is not known when the rate was initiated.

Cooling rates of lunar samples

Several criteria have been applied to estimate the thermal histories of Apollo 15 quartz-normative basalts. Lofgren *et al.* (1975) have compared phenocryst and groundmass textures of linear cooling rate experiments to the lunar samples and estimated the cooling history during the early and late stages of crystallization. Grove (1977) and Grove and Walker (this volume) measure nucleation densities of phenocryst pyroxene and sizes of groundmass plagioclase in linear cooling rate experiments to determine early and late-stage cooling rates. Rates determined by these two techniques differ slightly but are in general agreement. Both groups find that the cooling history of the vitrophyric and porphyritic-spherulitic QNBs are characterized by a slow initial cooling rate followed by a rapid cooling rate.

Do minor and major element chemistry support or further complicate the rate estimates obtained from textures? We will use the minor element chemistry of the cores of pyroxene phenocrysts to estimate the early cooling history of two vitrophyres (15597 and 15486) and two porphyritic-subophitic basalts (15058 and 15065). The degree of Al-enrichment found in the epitaxial augite rims, which can be conveniently read from the end of the 1/4 trend on a Ti vs. Al per 6 oxygen plot, is used to estimate the late-stage cooling rate. The Al enrichment value does not give a unique temperature at which late-stage cooling can be calculated, since plagioclase appearance and thus the temperature at which the switch from a 1/4 to a 1/2 trend occurs is dependent on cooling rate.

The major and minor element trends in 15597, 15486, 15065, and 15058 give a general idea of their respective cooling rates. Mineral chemistry for 15597 can be found in Weigand and Hollister (1973) and Fig. 10; for 15486, Albee *et al.* (1972) and Fig. 10; for 15058, Bence and Papike (1972); and for 15065, Walker *et al.* (1977) and Fig. 10. Comparison of the linear cooling rate experiments to the lunar samples shows that the general zoning trends of 15597 and 15486 (Fig. 10) are closest to the 150°C/hr experiments. 15058 has zoning trends that appear to have formed at a cooling rate slower than 0.5°C/hr (Fig. 7), and 15065 chemical variations nearly approach those of the equilibrium experiments (Fig. 2). Close inspection of representative pyroxene core compositions (Table 3 and Fig. 11) shows that the minor element chemistries of 15597, 15058, and 15065 pyroxenes are essentially identical and most similar to the 3.75°C/hr experiments. This indicates that pyroxene cores from these rocks formed under similar conditions, probably fairly close to equilibrium. Since the conditions of our experiments may not have duplicated those operating at or near the lunar surface, it is probably too much to expect the lunar pyroxenes precisely to match the experimental ones. The assignment of an absolute rate estimate of 3.75°C/hr on core pyroxenes from these three samples is probably incorrect. It is interesting, however, that three samples with such a variation in late-stage cooling histories have pyroxene cores with similar minor element chemistries. It is proposed that the pyroxene cores crystallized under essentially equilibrium conditions near the lunar surface. Pyroxene cores in 15486 have minor element chemistries that match the 150°C/hr experiments.

The rate estimates from minor elements are qualitatively in agreement with those that use textural criteria, and the complex cooling history of vitrophyre 15597 (Grove, 1976) is confirmed by the minor elements. The early cooling rate appears to have been significantly slower than the late-stage cooling history. The chemical data for 15486 indicate that this sample records nucleation and growth in a very rapid cooling environment for its entire crystallization history (at least 150°C/hr or greater) and conflict with the rate estimates based on textural criteria (Grove, 1977). The conflict may result from inability to analyze the core of an early pyroxene. If we assume that the QNBs represent one flow unit, the crystallization must have been rather complex. Some samples (15597) contained small pyroxene nuclei that formed within the interior of the flow, and some (15486) were apparently totally liquid, both being brought in close contact to lunar regolith and very rapidly chilled. The distance of the vitrophyres from the lava-regolith contact must have been on the order of 4–6 cm, for cooling rates as rapid at 150°C/hr to be operative (Grove, 1977).

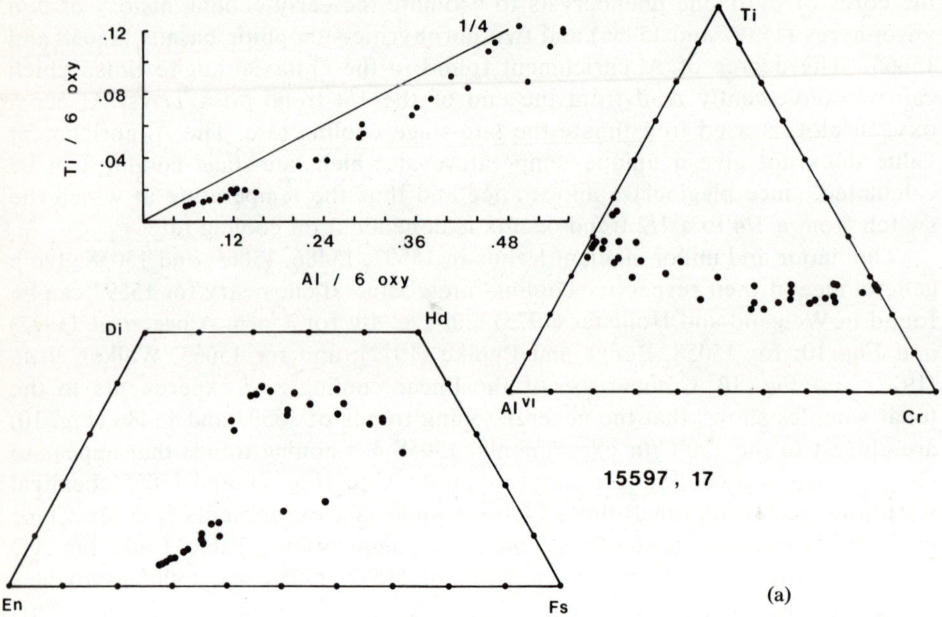
.12
.08
.04
Ti / 6 oxy
.12
.24
.36
.48
Al / 6 oxy
1/4
Ti
Al VI
Cr
Di
Hd
En
Fs
15597, 17

(a)

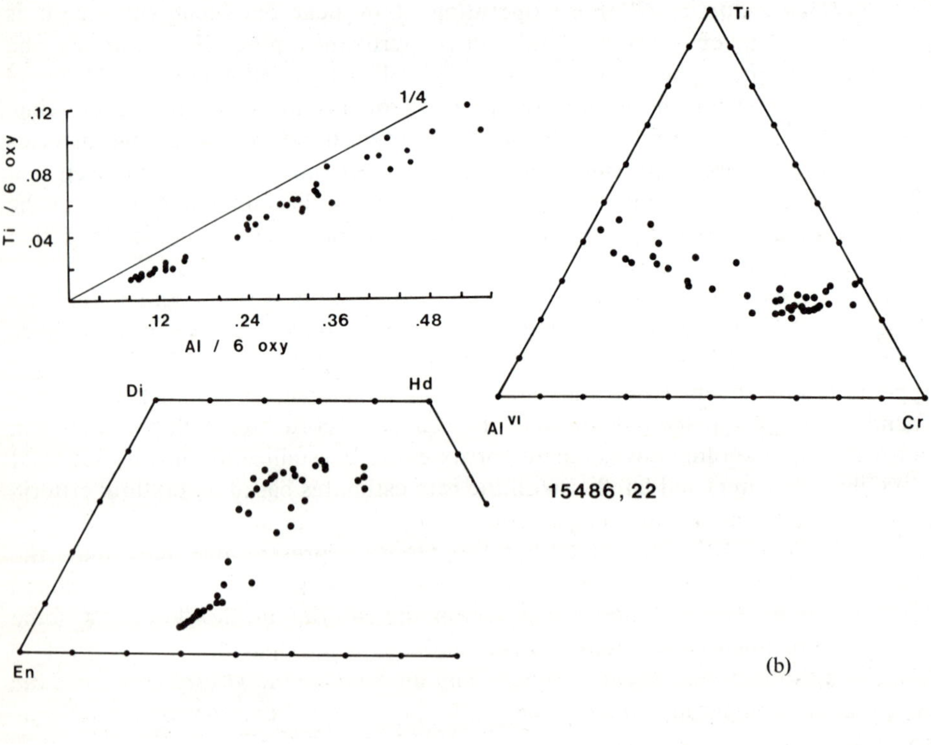
.12
.08
.04
Ti / 6 oxy
.12
.24
.36
.48
Al / 6 oxy
1/4
Ti
Al VI
Cr
Di
Hd
En
15486, 22

(b)

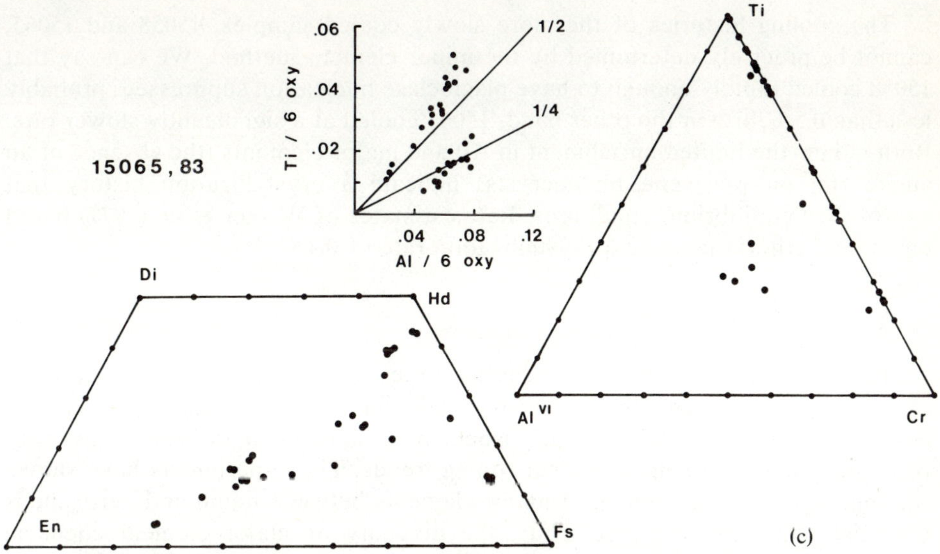

Fig. 10. Pyroxene compositional variations in Apollo 15 quartz-normative basalts. (a) 15597 and (b) 15485 are vitrophyres. (c) 15065 is a slowly cooled porphyritic-subophitic basalt.

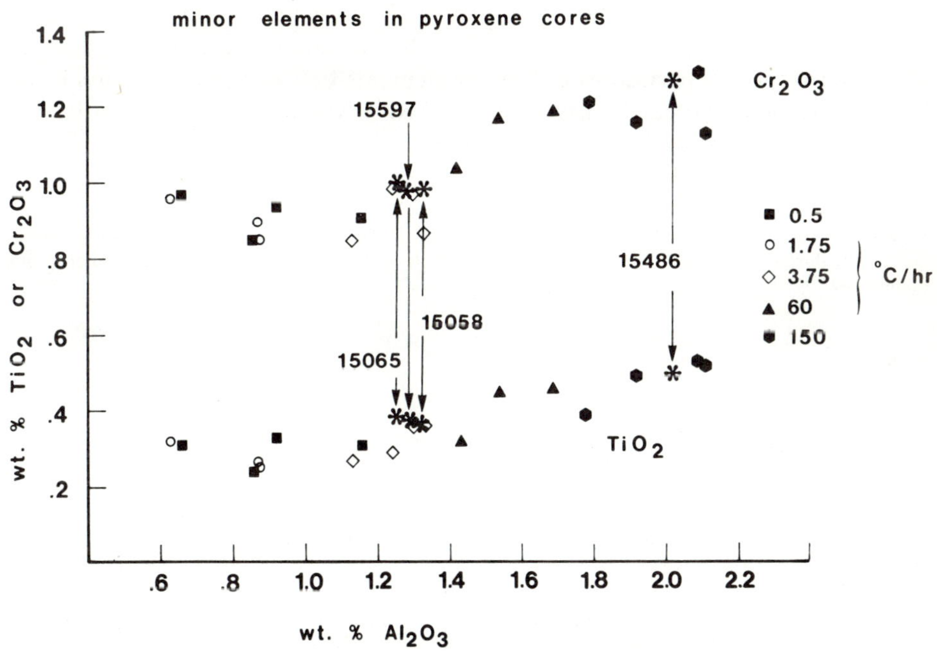

Fig. 11. Comparison of minor element contents of pyroxene cores in Apollo 15 quartz-normative basalts with the pyroxene cores of the controlled cooling rate experiments. Weight percent Al_2O_3 is plotted against wt.% TiO_2 and Cr_2O_3. Asterisks indicate lunar samples and other symbols indicate the cooling rate experiments.

The cooling histories of the more slowly cooled samples, 15058 and 15065, cannot be precisely determined by the minor element method. We can say that 15058 cooled rapidly enough to have plagioclase nucleation suppressed, probably less than 0.5°C/hr. On the other hand, 15065 cooled at a significantly slower rate. Both minor (the limited enrichment in Al) and major elements (the absence of an augite rim on pyroxene phenocrysts) indicate a crystallization history that approached equilibrium conditions. Rate estimates of Walker *et al.* (1977) based on several criteria indicate a crystallization rate of 0.1°C/day.

SUMMARY

These experiments were conducted to evaluate the effects of dynamic crystallization on pyroxene chemistry. The lunar QNB system was chosen because it contains a suite of nearly isochemical samples that display a diversity of textures and pyroxene chemical zoning trends. The experiments have shown the following: (1) partitioning of major elements between liquid and pyroxene is essentially rate independent. Thus, the diversity of major element chemical trends observed in the QNB pyroxenes result from the efficiency of fractional crystallization through crystal zoning and the relative diffusion rates of the pyroxene constituents through the melt; (2) the lack of suppression of pyroxene nucleation is reflected in the compositional similarity of the early pyroxenes in all cooling rate experiments; (3) the suppression of appearance temperature of later phases alters the liquid line-of-descent and the composition of the pyroxene; (4) partitioning of minor elements between pyroxene and liquid is rate dependent and can be used in a qualitative way to estimate the early- and late-stage cooling histories of lunar samples.

Acknowledgments—This work was supported by NASA grant NGL-33-015-130. We thank D. H. Lindsley for assistance with experiments and J. Delano, D. Walker, G. Lofgren, and J. J. Papike for discussion during the course of the work. Walter Holzwarth maintained the electron microprobe facility.

REFERENCES

Akella J., Williams R. J., and Mullins O. (1976) Solubility of Cr, Ti and Al in co-existing olivine, spinel and liquid at 1 atm. *Proc. Lunar Sci. Conf. 7th*, p. 1179–1194.

Albee A. L. and Ray L. (1970) Correction factors for electron probe microanalysis of silicates, oxides, carbonates, phosphates and sulfates. *Anal. Chem.* **42**, 1408–1414.

Albee A. L., Chodos A. A., and Gancarz A. J. (1972) Petrology of Apollo 15 sample 15486. In *The Apollo 15 Lunar Samples* (J. W. Chamberlain and C. Watkins, eds.), p. 20–25. The Lunar Science Institute, Houston.

Bence A. E. and Albee A. L. (1968) Empirical correction factors for electron microanalysis of silicates and oxides. *J. Geol.* **76**, 382–403.

Bence A. E. and Papike J. J. (1972) Pyroxenes as recorders of lunar basalt petrogenesis: Chemical trends due to crystal-liquid interaction. *Proc. Lunar Sci. Conf. 3rd*, p. 431–469.

Bence A. E., Papike J. J., and Lindsley D. H. (1971) Crystallization history of clinopyroxenes in two porphyritic rocks from Oceanus Procellarum. *Proc. Lunar Sci. Conf. 2nd*, p. 559–574.

Bence A. E., Papike J. J., Sueno S., and Delano J. W. (1973) Pyroxene poikiloblastic rocks from the lunar highlands. *Proc. Lunar Sci. Conf. 4th*, p. 597–611.

Boyd F. R. and Smith D. (1971) Compositional zoning in pyroxenes from lunar rock 12021, Oceanus Procellarum. *J. of Petrol.* **12**, 439–464.

Brown G. M., Emeleus C. H., Holland J. G., Peckett A., and Phillips R. (1972) Petrology, mineralogy and classification of Apollo 15 mare basalts. In *The Apollo 15 Lunar Samples* (J. W. Chamberlain and C. Watkins, eds.), p. 40–44. The Lunar Science Institute, Houston.

Chappell B. W. and Green D. H. (1973) Chemical composition and petrogenetic relationships in Apollo 15 mare basalts. *Earth Planet. Sci. Lett.* **18**, 237–246.

Donaldson C. H., Usselman T. M., Williams R. J., and Lofgren G. E. (1975) Experimental modeling of the cooling history of Apollo 12 olivine basalts. *Proc. Lunar Sci. Conf. 6th*, p. 843–869.

Dowty E., Keil K., and Prinz M. (1974) Lunar pyroxene-phyric basalts: Crystallization under supercooled conditions. *J. of Petrol.* **15**, 419–453.

Grove T. L. (1976) Crystallization histories of Apollo 15 quartz-normative basalts (abstract). *Geol. Soc. Amer.* **8**, 894.

Grove T. L. (1977) Cooling histories of Apollo 15 quartz-normative basalts (abstract). In *Lunar Science VIII*, p. 380–382. The Lunar Science Institute, Houston.

Grove T. L. and Bence A. E. (1977) Pyroxenes as recorders of lunar basalt petrogenesis revisited: An experimental study of pyroxene-liquid interaction (abstract). In *Lunar Science VIII*, p. 383–385. The Lunar Science Institute, Houston.

Hollister L. S., Trzcienski W. E., Jr., Hargraves R. B., and Kulick C. G. (1971) Petrogenetic significance of pyroxenes in two Apollo 12 samples. *Proc. Lunar Sci. Conf. 2nd*, p. 529–557.

Huebner J. S., Lipin B. R., and Wiggins L. B. (1976) Partitioning of chromium between silicate crystals and melts. *Proc. Lunar Sci. Conf. 7th*, p. 1195–1220.

Humphries D. J., Biggar G. M., and O'Hara M. J. (1972) Phase equilibria and origin of Apollo 15 basalts, etc. In *The Apollo 15 Lunar Samples* (J. W. Chamberlain and C. Watkins, eds.), p. 103–106. The Lunar Science Institute, Houston.

Kesson S. E. (1975) Mare basalts: Melting experiments and petrogenetic interpretations. *Proc. Lunar Sci. Conf. 6th*, p. 921–944.

Kesson S. E. and Lindsley D. H. (1976) Mare basalt petrogenesis—A review of experimental studies. *Rev. Geophys. Space Phys.* **14**, 361–373.

Lindsley D. H. and Burnham C. W. (1970) Pyroxferroite: Stability and x-ray crystallography of synthetic $Ca_{0.15}Fe_{0.85}SiO_3$ pyroxenoid. *Science* **168**, 364–367.

Lindsley D. H., Kesson S. E., Hartzman M. J., and Cushman M. K. (1974) The stability of armalcolite: Experimental studies in the system MgO–Fe–Ti–O. *Proc. Lunar Sci. Conf. 5th*, p. 521–534.

Lipin B. R. (1977) The effect of chromium on model lunar highland compositions (abstract). In *Lunar Science VIII*, p. 586–588. The Lunar Science Institute, Houston.

Lofgren G. (1974) An experimental study of plagioclase crystal morphology: Isothermal crystallization. *Amer. J. Sci.* **274**, 243–273.

Lofgren G., Donaldson C. H., Williams R. J., Mullins O., Jr., and Usselman T. M. (1974) Experimentally reproduced textures and mineral chemistry of Apollo 15 quartz-normative basalts. *Proc. Lunar Sci. Conf. 5th*, p. 549–567.

Lofgren G. E., Donaldson C. H., and Usselman T. M. (1975) Geology, petrology and crystallization of Apollo 15 quartz normative basalts. *Proc. Lunar Sci. Conf. 6th*, p. 79–99.

Longhi J., Walker D., Stolper E. M., Grove T. L., and Hays J. F. (1972) Petrology of mare/Rille basalts 15555 and 15065. In *The Apollo 15 Lunar Samples* (J. W. Chamberlain and C. Watkins, eds.), p. 131–134. The Lunar Science Institute, Houston.

Longhi J., Walker D., Grove T. L., Stolper E. M., and Hays J. F. (1974) The petrology of the Apollo 17 mare basalts. *Proc. Lunar Sci. Conf. 5th*, p. 447–469.

Longhi J., Walker D., and Hays J. F. (1975) Fe–Mg distribution between olivine and lunar basaltic liquids (abstract). *EOS* (*Trans. Amer. Geophys. Union*) **56**, 471.

Medaris L. G. (1969) Partitioning of Fe^{++} and Mg^{++} between coexisting synthetic olivine and orthopyroxene. *Amer. J. Sci.* **267**, 945–968.

Rhodes J. M. and Hubbard N. J. (1973) Chemistry, classification and petrogenesis of Apollo 15 mare basalts. *Proc. Lunar Sci. Conf. 4th*, p. 1127–1148.

Roeder P. L. and Emslie R. F. (1970) Olivine-liquid equilibrium. *Contrib. Mineral. Petrol.* **29**, 275–289.

Ross M., Huebner J. S., and Dowty E. (1973) Delineation of the one atmosphere augite-pigeonite miscibility gap for pyroxenes from lunar basalt 12021. *Amer. Mineral.* **58**, 619–635.

Schreiber H. D. and Haskin L. A. (1976) Chromium in basalts: Experimental determination of redox states and partitioning among synthetic silicate phases. *Proc. Lunar Sci. Conf. 7th*, p. 1221–1259.

Stolper E. (1976) Experimental melting of the Juvinas eucrite. In *Progress in Experimental Petrology* (G. M. Biggar, ed.), p. 223–227. National Environmental Research Council, Edinburgh.

Usselman T. M. and Lofgren G. E. (1976) The phase relations, textures and mineral chemistries of high-titanium mare basalts as a function of oxygen fugacity and cooling rate. *Proc. Lunar Sci. Conf. 7th*, p. 1345–1363.

Walker D., Longhi J., Grove T. L., Stolper E., and Hays J. F. (1973) Experimental petrology and origin of rocks from Descartes Highlands. *Proc. Lunar Sci. Conf. 4th*, p. 1013–1032.

Walker D., Kirkpatrick R. J., Longhi J., and Hays J. F. (1975) 12002 revisited (abstract). In *Lunar Science VI*, p. 841–843. The Lunar Science Institute, Houston.

Walker D., Kirkpatrick R. J., Longhi J., and Hays J. F. (1976) Crystallization history of lunar picritic basalt sample 12002: Phase equilibria and cooling rate studies. *Bull. Geol. Soc. Amer.* **87**, 646–656.

Walker D., Longhi J., Stolper E. M., Grove T. L., and Hays J. F. (1977) Slowly cooled microgabbros 15065 and 15555 (abstract). In *Lunar Science VIII*, p. 964–966. The Lunar Science Institute, Houston.

Weigand P. W. and Hollister L. S. (1973) Basaltic vitrophyre 15597: An undifferentiated melt sample. *Earth Planet. Sci. Lett.* **19**, 61–74.

Williams R. J. (1971) Reaction constants in the system Fe–MgO–SiO_2–O_2 at 1 atm between 900 and 1300°C: Experimental results. *Amer. J. Sci.* **270**, 354–360.

Williams R. J. and Mullins O. (1976) A system using solid ceramic oxygen electrolyte cells to measure oxygen fugacities in gas-mixing systems. NASA TM X-58167, p. 4.

APPENDIX

Experimental techniques

A silicate glass with the bulk composition reported for 15597 by Chappell and Green (1973) was used as starting material for the equilibrium and cooling rate experiments (Table 1). Oxide mixes were prepared from Johnson Matthey high-purity Fe sponge, Fe_2O_3, SiO_2, and other oxide materials after the technique of Lindsley *et al.* (1974). Glass starting materials were obtained by fusing 1 g quantities of the oxide mix in Fe° containers sealed in evacuated SiO_2 glass tubes. The resulting glass starting materials were ground under alcohol in an agate mortar for 1 hr. As a consequence of the preparation of glasses in iron containers, the starting materials contained tiny blebs of metallic iron.

Equilibrium experiments were carried out in Johnson Matthey high-purity (99.9998% Fe) iron capsules with friction fitted lids and samples were prepared in a manner similar to that of Kesson (1975). Cooling rate experiments were performed in high-purity Johnson Matthey (99.99% Fe) capsules. 0.15–0.20 g of glass were packed into iron capsules (0.445 cm inner diameter by 0.600 cm high) with friction-fitted lids. The sample was dried for 30 min under vacuum at 800°C with an "oxygen getter" (Fe sponge in a cooler portion of the tube) to prevent oxidation of the Fe capsule and the silicate glass.

Capsules in evacuated silica glass tubes were suspended in vertical Pt–Rh wound resistance furnaces. A Pt–10Rh thermocouple placed near the furnace winding supplied the input to a proportional controller. A thermocouple paced next to the sample was used to monitor temperature. The monitor thermocouple was calibrated against the melting point of gold (1064°C). Controlled cooling rate experiments were performed using two methods. (1) A stepping motor was connected to the set point dial of a Leeds and Northrup Electromax controller and was programmed to decrease the temperature in 1/3°C steps over a range of rates and temperature spans. (2) A Data Trak programmer (Research, Inc.) was used to add or subtract a millivoltage signal to the control thermocouple after the method developed by Williams and Mullins (1976).

Phase appearance reversals in equilibrium experiments were accomplished by taking a previous run that contained the phase in question and holding it at a temperature above the phase appearance to see if it would dissolve. Also, glass starting material was held at a temperature in which the phase was stable for a sufficient time to nucleate the phase, then the temperature was raised. The run was allowed to re-equilibrate and then examined (Table 2). All controlled cooling-rate experiments were initiated at 1225°C ± 5°C, approximately 10°C above the equilibrium liquidus. Samples were held at 1225°C for 15 to 30 min, then cooled to a desired temperature and quenched into water. It was found in long-term cooling rate experiments that nearly all the magnesium in the residual liquid could be partitioned into the silicates. The resulting late-stage pyroxenes are slightly more iron-enriched than those observed in the lunar samples. It is impossible in these long-term experiments to evaluate whether iron gain from the capsule caused the extreme differentiation to Mg-free phases, or whether there is a difference in the conditions imposed on crystallization by the experiment and those experienced by the very slowly cooled lunar samples. Whatever the effect might be, we do observe the breakdown of ulvöspinel in our experiments to ilmenite and iron. This late-stage reaction occurs in the lunar basalts, and indicates that both the experiments and the basalts experienced very low f_{O_2} conditions during final solidification.

Analytical procedures

All pyroxene-liquid data were obtained on an automated 4-spectrometer ARL-EMX SM microprobe using real-time data reduction. The matrix correction procedures of Bence and Albee (1968) and modifications of Albee and Ray (1970) were used. For the bulk D calculations between 6 and 12 representative liquid analyses were averaged for each experiment. Similarly, the rim compositions of several pyroxene grains (5 to 10 rim analyses from different grains) were obtained and averaged.

Proc. Lunar Sci. Conf. 8th (1977), p. 1581–1592.
Printed in the United States of America

Cooling rate estimations based on kinetic modelling of Fe–Mg diffusion in olivine

L. A. TAYLOR

The University of Tennessee, Knoxville, Tennessee 37916

P. I. K. ONORATO and D. R. UHLMANN

Massachusetts Institute of Technology, Cambridge, Massachusetts 02139

Abstract—Olivine in lunar and terrestrial rocks, as well as in meteorites, commonly is compositionally zoned. This zonation can be used to estimate cooling rates and thermal history. The Fe^{++} and Mg^{++} diffusion in olivine is a function of: (1) temperature, (2) composition, (3) f_{O_2}, and (4) crystallographic direction. An experimental check of published interdiffusion data has verified the accuracy of the study of Buening and Buseck (1973); therefore, diffusion data from their thorough study were used in the present investigation. A finite, one-dimensional kinetic model was developed which considers all of the above factors. Because the "as-solidified" profile of a given olivine is not known, a step-function, with infinite gradient, is assumed; the position of this step is based on mass balance considerations of the measured compositional profile. This assumption directs attention to the *minimum cooling rate* associated with the preservation of a given gradient.

The linear cooling rates of lunar rocks 12002 and 15555 were estimated using this olivine cooling-rate indicator as 10°C/day and 5°C/day, respectively. These values agree well but are slightly lower than those obtained by dynamic crystallization studies (10–20°C/day). The olivine-based estimates are minima, and a more detailed knowledge of the starting, "as-solidified", compositional profile would allow for calculations of absolute cooling rates.

INTRODUCTION

VARIOUS EXPERIMENTAL and theoretical approaches have been employed to estimate the relative and absolute cooling rates of lunar rocks. Dynamic crystallization studies, e.g., Lofgren *et al.* (1974) and Walker *et al.* (1976), have shown that variations in cooling rate can account for much of the textural and mineralogical variations in mare basalts. However, this type of approach has several uncertainties built into it, as discussed, for example, by Bianco and Taylor (1977a,b), Lofgren (1977), Lofgren and Usselman (1977), and Walker *et al.* (1976). The kinetics of the re-equilibration of elemental partitionings, e.g., Zr between ilmenite and ulvöspinel determined by Taylor *et al.* (1975), have been modelled so that absolute cooling rates can be estimated for certain lunar rocks. However, concentrations of the pertinent minor or trace element must be sufficiently high for precise analyses to be performed. In addition, theoretical applications of kinetics to lunar studies have proven invaluable (e.g., Onorato *et al.*, 1976; Simonds, 1975). The present study was conducted to illustrate the use of diffusion kinetics and modelling of a mineral phase to obtain cooling-rate estimates.

Of the major igneous rock-forming minerals (e.g., pyroxene, plagioclase, olivine), the complications of the phase equilibria and kinetics within the pyroxene and plagioclase systems are well known. However, the relatively simpler chemistry of olivines should lend itself to geothermometry. Nevertheless, studies to date along these lines have not proven fruitful.

The present study was directed at the olivine system, in particular the diffusion of Fe^{++} and Mg^{++} within compositionally zoned olivines. Previous studies (Chao *et al.*, 1975; Huebner, 1976) have considered the effects of cooling on the zonation of olivines (in 77115); however, no kinetic models were presented. Partitioning of Fe and Mg between olivine and orthopyroxene has been shown (Medaris, 1969) to not be significantly temperature dependent over the temperature range 900–1300°C. Hence, diffusion across olivine-pyroxene grain boundaries can be assumed to be minimal, and the re-equilibration of compositionally zoned olivines is largely one of response to prolonged cooling. Thus, a given primary compositional gradient will tend to decrease (i.e., "smooth out") with time at slow cooling rates until the olivine is chemically homogeneous. Some of the preliminary results of this study were presented by Taylor *et al.* (1977).

INTERDIFFUSION COEFFICIENTS

Experiments have been conducted by Buening and Buseck (1973) and Misener (1973) to determine interdiffusion coefficient data for Fe and Mg in olivine. Buening and Buseck (1973) used a crystal of one composition and a powder of another to make their diffusion couple; Misener (1973) used two crystals of different compositions placed together. In such experiments, the question of the quality of the grain to grain contacts is always present.

The diffusion data are derived from experimentation wherein two different compositions with the same crystallographic orientation are placed in contact to form a diffusion couple. After annealing at a given temperature and f_{O_2} for a predetermined time, the compositional profile across the boundary of the two phases is measured by electron microprobe. This profile is subjected to the Boltzmann–Matano solution (Crank, 1956) from which the interdiffusion coefficient, $\tilde{D}$, can be determined.

Isothermal heating experiments were conducted during the present study to test the validity of interdiffusion coefficient determinations of Buening and Buseck (1973) and Misener (1973) for Fe^{++} and Mg^{++} diffusion between natural specimens of forsterite (Fo_{92}) and fayalite (Fa_{100}). Chips from these phases were polished (1 μm diamond compound in the last step), sized by sawing into 3 mm cubes, and joined together (with *c* axes parallel) across the cleaned, polished surfaces by placing them in $Ag_{70}Pd_{30}$ foil, in turn wrapped with Pt wire. The reactants, Fo_{92} and Fa_{100}, therefore, defined a diffusion couple for this experiment.

Two sets of experiments were designed in order to check for possible grain boundary effects upon diffusion of cation species between phases in the couple: (1) a sample, prepared as above, was sealed in an evacuated silica tube, duplicating the technique used by Misener (1973); and (2) a free-floating silica rod, exerting a load of about 5 lb/in^2 on the sample in the vertical position, was sealed in an evacuated silica tube. The loaded capsule was designed to minimize grain boundary effects upon cation diffusion rates.

Both charges were then heated at 1000°C for 311 hours, quenched, and examined with a binocular microscope to check the nature of the grain boundary contact surface. No apparent displacement along this surface could be discerned. Each couple was mounted in epoxy and a polished surface prepared such that the compositional profiles, measured with an automated, MAC 400S electron microprobe, were those across the center of the contact surface between forsterite and fayalite.

Optical observations of the fayalite in some experiments revealed evidence for noticeable oxidation of the fayalite to magnetite + silica. It would appear that some oxidation occurred late during the annealing process. The experiments were consequently performed in triplicate. One set of charges was, in fact, subjected to heating at 400°C for 3 hours under a dynamic vacuum prior to sealing of the silica tube. Compositional profiles away from the grain boundary into Fo_{92} were obtained for the charges without evidence of oxidation. The profiles are somewhat flatter than those obtained by Misener (1973) and about the same as calculated for $f_{O_2} = 10^{-12}$ atm from the data of Buening and Buseck (1973). This indicates that the interdiffusion data by the latter authors is probably closer to correct. Also, no differences between profiles in the loaded and unloaded samples were detected. Therefore, the interdiffusion coefficient data for Fe^{++} and Mg^{++} as determined by the very thorough investigation of Buening and Buseck (1973) were used in the present study.

Buening and Buseck (1973) showed that the interdiffusion of Fe^{++} and Mg^{++} in olivine is a function of (1) temperature, (2) composition, (3) f_{O_2}, and (4) crystallographic direction. In the forsterite-fayalite system at an f_{O_2} of 10^{-12} atm and in the c direction, the interdiffusion coefficient, $\tilde{D}_c$, is

$$\tilde{D}_c = \exp(-.0501\, X_{Fe_2SiO_4} - 14.03) \exp - \left\{ \frac{(1.373ev - .0095\, X_{Fe_2SiO_4})}{kT} \right\}, \quad (1)$$

for $T < 1125°C$. They showed that the diffusion coefficient varies with oxygen fugacity (i.e., partial pressure) as

$$D = K \cdot f_{O_2}^{(.172)}, \quad (2)$$

with an accuracy in the exponent of ± 0.022. This is consistent with a simple vacancy process which exhibits a pressure dependence of $f_{O_2}^{1/6}$. This factor was used to determine the diffusion over a range of temperatures and fugacities of oxygen.

$$\tilde{D}_c = 10^2 f_{O_2}^{1/6} \exp(-.0501\, X_{Fe_2SiO_4} - 14.03) \exp - \left\{ \frac{(1.373ev - .0095\, X_{Fe_2SiO_4})}{kT} \right\}. \quad (3)$$

The range of oxygen fugacities which corresponds to lunar conditions is about 0.5 log atm below the iron-wüstite buffer curve (Sato *et al.*, 1973). The temperature dependence of the log of the oxygen fugacity is approximately linear.

$$\log_{10} f_{O_2} = .015\, T(°K) - 34.6 \quad (4)$$

Buening and Buseck (1973) show a significant anisotropy in the diffusion coefficient. No analytical expression was given for this difference, but the tabulated data indicate that $\tilde{D}_c$ is greater than $\tilde{D}_a$ by a factor of about 4 at 1050°C. As an approximation,

$$\tilde{D}_a = \frac{\tilde{D}_c}{4}, \tag{5}$$

for all temperatures. The anisotropy increases as the temperature decreases so the approximation overestimates the extent of diffusion below 1050°C and to a lesser degree underestimates it above 1050°C.

Kinetic Model

The minimum cooling rate necessary for the preservation of compositional zoning in an olivine grain has been determined. At equilibrium, there are no concentration gradients in the forsterite-fayalite system. The extent to which a grain approaches equilibrium during cooling depends upon the "*as-solidified*" concentration profile, as well as upon the thermal history (cooling rate). Lacking information about the as-solidified concentration profile, it has been assumed that at the solidus, there was a step-function in the forsterite concentration within the grain. That is, initially there are no compositional gradients within the grain other than the infinite gradient at the boundary between the Fo-rich core and a Fa-rich rim.

This assumption directs attention to *the minimum cooling rates associated with the preservation of a given gradient* since a system with an initial gradient would equilibrate more rapidly than one with a step-function profile. Likewise, the largest compositional gradients found in the olivine crystals of a particular sample were used for this analysis, and the crystallographic direction of the profile is noted. It has also been assumed that there is no flux either across the center of the grain or across the grain boundary. In order to determine the relative initial sizes of the core and the rim, x_1 and x_2 for the step-function of the model, it was assumed that the grain was effectively a closed system and that mass balance was maintained, and the initial Fe concentrations in the core, C_1, and the rim, C_2, as determined by the microprobe analyses, were varied slightly.

Attention was specifically directed to a finite system, rather than an infinite system as described by Lasaga and Richardson (1975). The equation to be solved is the one-dimensional diffusion equation in which the diffusion coefficient is a function of composition and therefore position. The diffusion coefficient is also a function of temperature which is, itself, a function of time.

$$\frac{\partial C}{\partial t} = \frac{\partial}{\partial X}\left\{\tilde{D}_a(C(X),\, T(t))\,\frac{\partial C}{\partial X}\right\}, \tag{6}$$

where C is the mole fraction of Fe and $\tilde{D}_a$ is the interdiffusion coefficient in the a crystallographic direction as calculated from Eqs. 1–5 which take f_{O_2} into account. The initial conditions are:

$$\text{for:}\quad 0 \leqslant x < x_1 \qquad \text{for:}\quad x_1 < x \leqslant x_1 + x_2$$
$$C(x,0) = C_1 \qquad C(x,0) = C_2. \tag{7}$$

These conditions can be changed to include segregation during solidification if the distribution coefficient is known as a function of temperature. Such modified conditions can readily be handled by the program used to solve the diffusion problem.

Equation (6) has been solved by replacing it with its finite difference analog, applying the boundary conditions, and solving the resultant set of simultaneous equations by the Thomas tri-diagonal method. Due to the non-linearity of the equation, it was necessary to iterate the procedure to approach the correct solution.

Concentration profiles were calculated for several cooling rates and sets of initial conditions. The values of x_1, x_2, C_1, and C_2 indicate possible solidus conditions in the lunar samples investigated. These parameters could not be changed independently because mass balance and the size of the grain had to be maintained.

For cooling rates larger than about 5°C/day, there was little change in the forsterite profile at temperatures below about 750°C. Cooling rates smaller than about 1°C/day allowed diffusion to be effective at lower temperatures. *Since the diffusion coefficient depends on concentration, each case (i.e., olivine grain) must be considered separately.* The absolute concentration, as well as the concentration gradient, are important parameters for specifying the minimum cooling rates.

Application

Two lunar samples were chosen (i.e., 12002 and 15555) for application of the olivine kinetic model presented above. Cooling rates based on dynamic crystallization studies (Bianco and Taylor, 1977a,b; Walker *et al.*, 1976) have been estimated for these samples. Therefore, these samples provide a means of comparison between our theoretical approach and the empirical studies. Inasmuch as these lunar specimens are compositionally and texturally different, different cooling rate estimates should be obtained from compositional gradients in their olivines.

The applicability of the kinetic model is dependent upon a well-defined choice of olivine from which a diffusion couple can be described. Realizing that a thin section represents but a random cut through the olivine grains, it is imperative that one search the section for the most suitable grains. In the case of samples 12002 and 15555, this entailed finding and measuring olivines which had compositionally different cores and rims. Profiles across these grains should flatten over cores and rims, thereby suggesting that the plane of the section cuts near or through the center of the crystal. Initially, several large, equant grains (Fig. 1) were chosen which exhibited differences in birefringence between rim

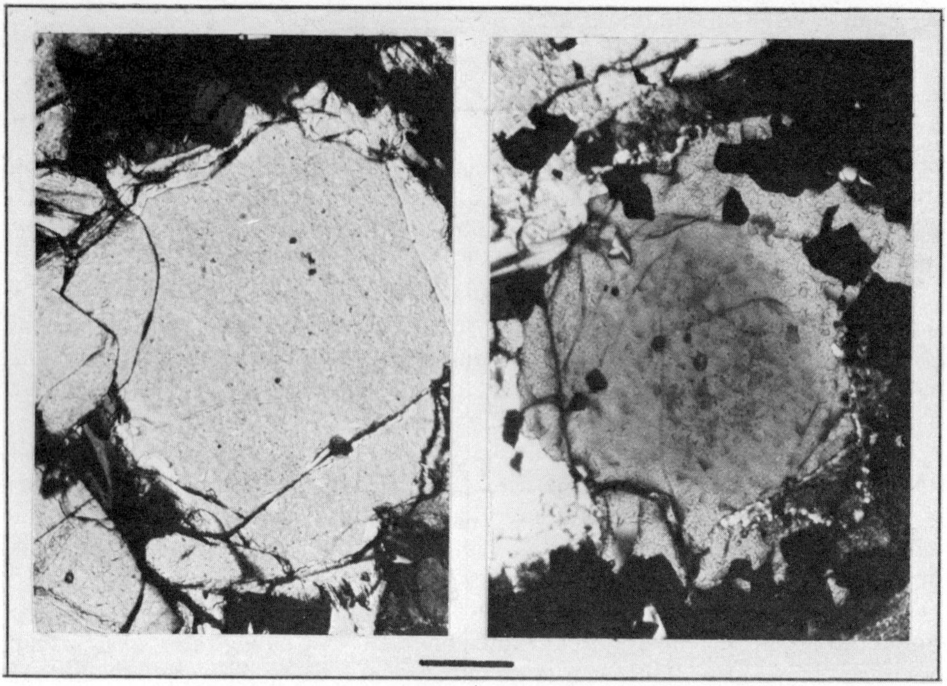

Fig. 1. Representative olivine grains from 15555 (left photo) and 12002 (right). Scale bar = 125 μm.

and core (transmitted light, crossed polars). Several crystals were subjected to electron microprobe analyses in order to determine a grain with the largest compositional gradient. The regular geometries of the crystal cores and rims provided a means for choosing grains suspected of having symmetrical compositional profiles. Those grains containing melt or crystal inclusions were not chosen, since compositional gradients are perturbed in the vicinity of these inclusions. Core and rim geometries and compositional profiles were defined by traversing each olivine grain such that compositional variations were known at most points on the section of the grain. No measurable oscillatory or reversed zoning was present in any of the analyzed grains.

Compositional profiles of the olivines in 12002 and 15555 selected for modelling are shown in Figs. 2 and 3, where two 12002 olivines (Fig. 2) are profiled along the *a* direction and the 15555 grain (Fig. 3) is profiled along *a* (N–S) and *c* (E–W) directions. The upper profile in Fig. 2 and the N–S profile in Fig. 3, both in the *a* crystallographic direction, were used for the modelling discussed below. Those traverses exhibiting a maximum compositional difference between rim and core (15555—50 mol% Fo; 12002—15 mol% Fo) were chosen.

Based on the compositional profiles of the olivines in these two rocks (Figs. 2 and 3), and assuming that the system was closed and mass balance was

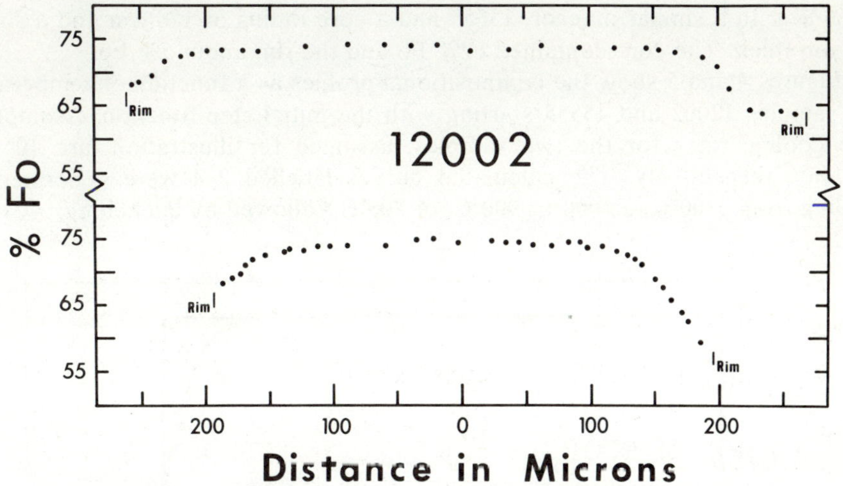

Fig. 2. Compositional profiles across two olivine grains in 12002.

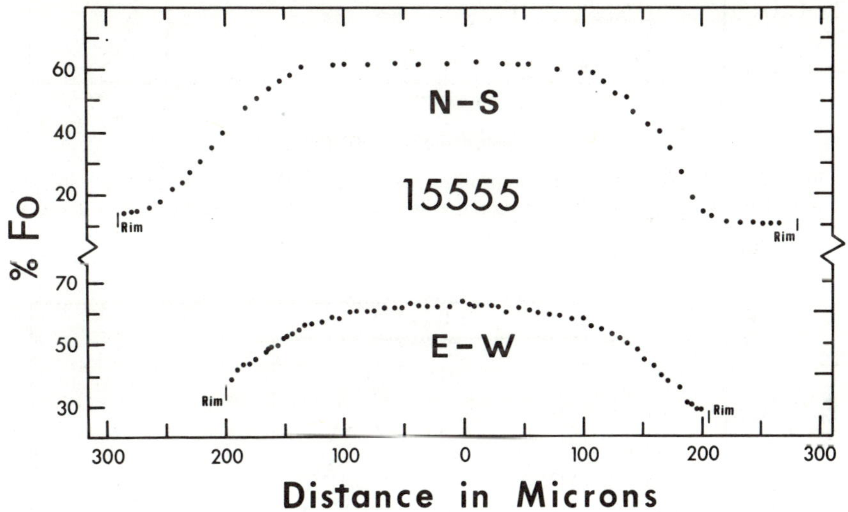

Fig. 3. Compositional profiles across an olivine grain in 15555 taken in two perpendicular directions.

maintained, the dimensions of initial step-function were determined. This was done by integrating the measured forsterite content over the radius of the grain, assuming initial values of the concentration in the core and rim and determining the relative size of the core and rim such that the forsterite content is constant. According to the best fit of calculated and measured forsterite distributions, sample 12002 olivine originally had a core radius of 234 μm and a rim thickness of 60 μm. In a similar manner, 15555 had a core radius of 188 μm and a rim of 106 μm thick. The core contained 62% Fo and the rim about 5% Fo.

Figures 4 and 5 show the compositional profiles as a function of temperature for samples 12002 and 15555 starting with the initial step-function assumption. The cooling rates for the two samples, assumed for illustration, are 10° and 5°C/day, respectively. The calculated curves labelled 2–4 were generated by cooling from 1100°C to 1000°C, 900°C, or 800°C, followed by quenching. As to be

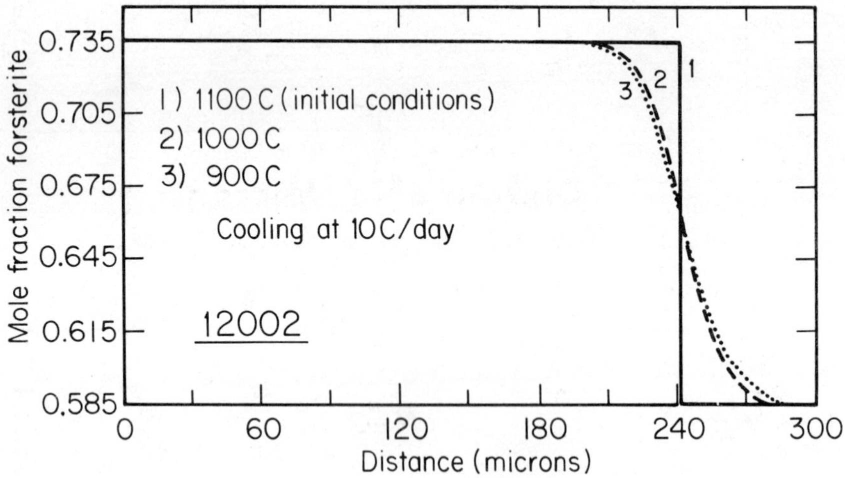

Fig. 4. Calculated compositional profiles for a 12002 olivine at a cooling rate of 10°C/day. Curve 1 is the step-function profile used as the starting point. Curves 2 and 3 are for cooling to the respective temperatures.

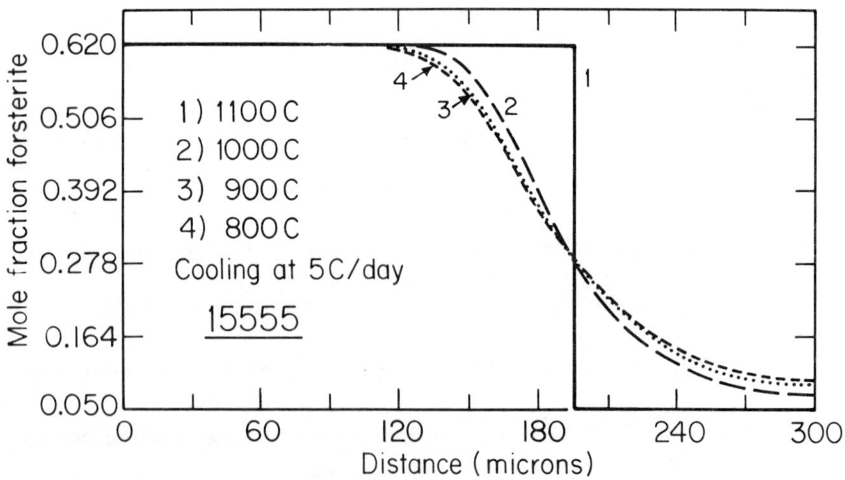

Fig. 5. Calculated compositional profiles for a 15555 olivine at a cooling rate of 5°C/day. See Fig. 4 for additional explanation.

expected, the temperature to which effective diffusion is still operative is lower for the slower cooling rate. Below 900°C in the case of 12002 at 10°C/day (Fig. 4) and 800°C for 15555 at 5°C/day (Fig. 5), there is little change in the profiles—i.e., they converge.

Figures 6 and 7 show the calculated diffusion profiles for 12002 and 15555 as a function of cooling rate. These gradients were calculated using the kinetic model discussed above. The open circles represent compositions of the actual

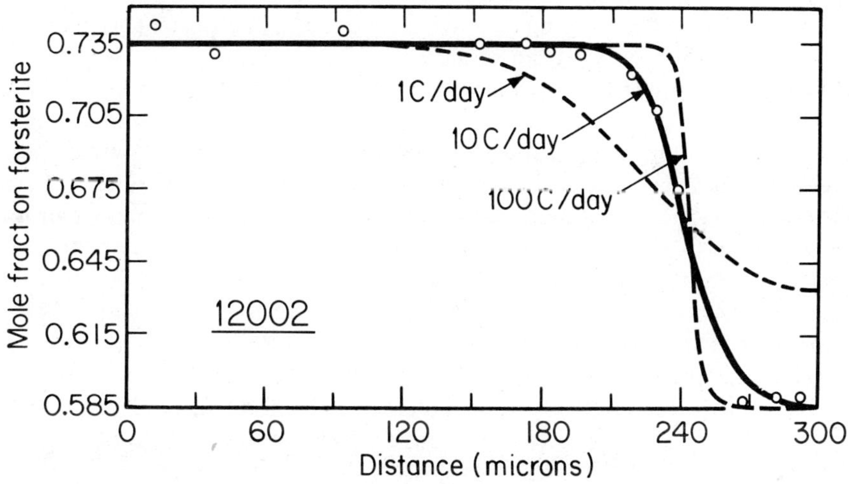

Fig. 6. Calculated compositional profiles for a 12002 olivine at different cooling rates. Open circles are measured compositions from Fig. 2.

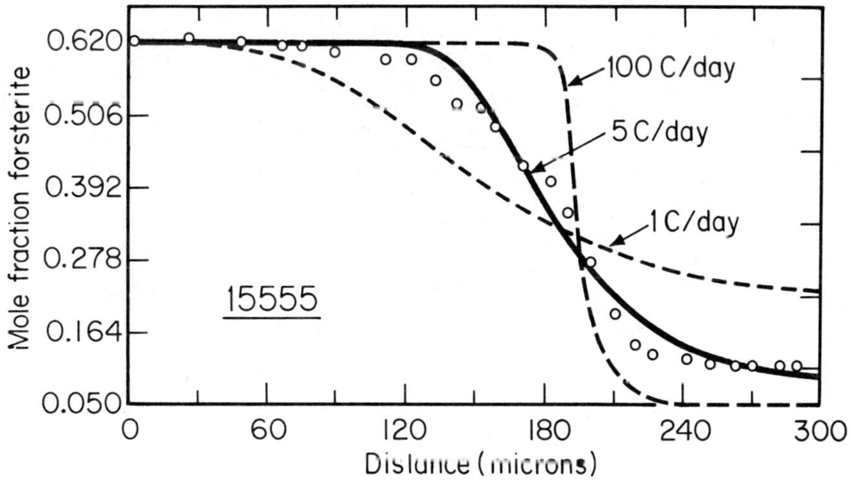

Fig. 7. Calculated compositional profiles for a 15555 olivine at different cooling rates. Open circles are measured compositions from Fig. 3.

crystals as determined by electron microprobe and are also given in Figs. 2 and 3. The agreement between the calculated and observed profiles is good for *10°C/day and 5°C/day for 12002 and 15555,* respectively. It is noteworthy that Walker *et al.* (1977) using a less rigorous kinetic treatment, arrived at about this same value for 15555.

Error evaluation. A change in the assumed initial core and rim concentrations has a small effect on the calculated cooling rate. Due to the assumption that the olivine grain is effectively isolated from its surroundings, the location of the concentration gradient within the grain and the final concentration gradient would be changed. The cooling rate which results in the best fit to the experimental data would vary less than a factor of 2.

If the sample cooled from a temperature higher than 1100°C, a higher cooling rate would be required to maintain the observed profile. It was found that increasing the initial temperature to 1200°C increased the cooling rate to 50°C/day for the calculations based on sample 12002. At high temperatures substantial diffusion takes place, resulting in a decrease in the concentration gradient at 1100°C. However, a cooling rate of 50°C/day for sample 12002 is not in accord with previous estimates.

The cooling rate varies linearly with the interdiffusion coefficient, and therefore, an error in the diffusivity has a linear effect on the calculated cooling rate. For example, if the diffusivity is a factor of 10 larger than the actual diffusivity for all conditions, the cooling rate calculated for sample 12002 is a factor of 10 faster than the actual cooling rate. According to Buening and Buseck (1973), the values of $\tilde{D}_c$ at $p_{O_2} = 10^{-12}$ atm as calculated from Eq. (1) are within $\pm 22\%$ of the measured values for $T < 1125°C$. Although this error is relatively large, its effect on the calculated cooling rate is small compared to that of an initial concentration gradient or a higher initial temperature.

Another source of possible error is the three-dimensional character of the grain as opposed to the one-dimensional diffusion equation. When Eq. (6) was replaced by the diffusion equation in spherical coordinates, ignoring the anisotropy of diffusion, there was no difference in the final concentration gradient and hence in the calculated cooling rate. In a forthcoming paper we will present several refinements to the kinetic model, with a detailed analysis for a finite plate and sphere geometries. However, it is usually not feasible to determine the complete compositional character of an olivine crystal necessary for the detailed three-dimensional modelling. Therefore, although less rigorous in approach, the one-dimensional model outlined above lends itself to application more readily.

DISCUSSION

The cooling rates of 10°C/day for 12002 and 5°C/day for 15555, based on the above kinetic model, compare favorably with those obtained from controlled crystallization studies by Walker *et al.* (1976) and Bianco and Taylor (1977a,b). These investigators estimated rates of 10–20°C/day for both of these rocks. In general, the results from the present study are a bit lower by a factor of two to three. However, it should be emphasized that our *olivine-based cooling-rate estimates are minima*, since we used the step function as the starting point rather than an "as-solidified" profile. For the "as-solidified" profile to change to the observed profile would require less time at a given temperature—i.e., a faster cooling rate. With more refined knowledge of the starting profile, a more detailed statement of the absolute cooling rate is possible.

The applications to petrology of this diffusion model for olivine have many important ramifications for the study of lunar and terrestrial rocks, as well as meteorites. Differences in cooling rates between otherwise similar rocks can be discerned allowing for reconstruction of the thermal history of the rocks. Use of the cooling-rate indicator is not limited to igneous rocks; in fact, it will find wide application in metamorphic studies, for example, in the investigations of lunar clasts in breccias (Chao *et al.*, 1975; Huebner, 1976), as well as in the effects of metamorphism in certain meteorite groups.

Acknowledgments—We would like to thank Drs. P. Buseck, J. Goldstein, R. Housley, S. Huebner, T. Usselman, D. Walker, and R. J. Williams for their helpful discussions and suggestions. Assistance with the electron microprobe analyses and experimentation by Dr. B. M. Walker is also acknowledged. We also thank the University of Tennessee for providing substantial support for the purchase of the electron microprobe. The research presented in this paper was supported by NASA grants NGL-43-001-140 (L.A.T.) and NGR 22-009-646 (D.R.U.).

References

Bianco A. S. and Taylor L. A. (1977a) Lunar olivine normative basalts: Cooling rate experimentation (abstract). In *Lunar Science VIII*, p. 103–105. The Lunar Science Institute, Houston.

Bianco A. S. and Taylor L. A. (1977b) Applications of dynamic crystallization studies: Lunar olivine normative basalts. *Proc. Lunar Sci. Conf. 8th*. This volume.

Buening D. K. and Buseck P. R. (1973) Fe–Mg lattice diffusion in olivine. *J. Geophys. Res.* **78**, 6852–6862.

Chao E. C. T., Minkin L. T., Thompson C. L., and Huebner J. S. (1975) The petrogenesis of 77115 and its xenocrysts: Description and preliminary interpretation. *Proc. Lunar Sci. Conf. 6th*, p. 493–515.

Crank J. (1956) *The Mathematics of Diffusion*, p. 236–245. Clarendon Press, Oxford.

Huebner J. S. (1976) Diffusively rimmed xenocrysts in 77115 (abstract). In *Lunar Science VII*, p. 396–398. The Lunar Science Institute, Houston.

Lasaga A. C. and Richardson S. M. (1975) Mathematical model of diffusion in retrograde metamorphism (abstract). *EOS* (*Trans. Amer. Geophys. Union*) **56**, 459.

Lofgren G. (1977) Dynamic crystallization experiments on the origin of textures in impact generated liquids. *Proc. Lunar Sci. Conf. 8th*. This volume.

Lofgren G., Donaldson C. H., Williams R. J., Mullins O., Jr., and Usselman T. M. (1974) Experimentally reproduced textures and mineral chemistry of Apollo 15 quartz normative basalts. *Proc. Lunar Sci. Conf. 5th*, p. 549–567.

Lofgren G. and Usselman T. M. (1977) Dynamic crystallization experiments on the origin of textures in impact generated liquids (abstract). In *Lunar Science VIII*, p. 589–591. The Lunar Science Institute, Houston.

Medaris L. G., Jr. (1969) Partitioning of Fe^{++} and Mg^{++} between coexisting synthetic olivine and orthopyroxene. *Amer. J. Sci.* **267**, 945–968.

Misener J. D. (1973) Ph.D. thesis, University of British Columbia, Vancouver.

Onorato P. I. K. and Uhlmann D. R. (1976) Heat flow in impact melts: Apollo 17 Station 6 Boulder and some applications to other breccias and xenolith laden melts. *Proc. Lunar Sci. Conf. 6th,* p. 181–191.

Sato M., Hickling N. L., and McLane J. E. (1973) Oxygen fugacity values of Apollo 12, 14, and 15 lunar samples and reduced state of lunar magmas. *Proc. Lunar Sci. Conf. 4th,* p. 1061–1079.

Simonds C. H. (1975) Thermal regimes in impact melts and the petrology of the Apollo 17 Station 6 Boulder. *Proc. Lunar Sci. Conf. 6th,* p. 641–672.

Taylor L. A., Uhlmann D. R., Hopper R. W., and Misra K. C. (1975) Absolute cooling rates of lunar rocks: Theory and application. *Proc. Lunar Sci. Conf. 6th,* p. 181–191.

Taylor L. A., Walker B. M., Onorato P. I. K., and Uhlmann D. R. (1977) Fe–Mg diffusion in olivine: A cooling rate indicator (abstract). In *Lunar Science VIII,* p. 928–930. The Lunar Science Institute, Houston.

Walker D., Kirkpatrick R. J., Longhi J., and Hays J. F. (1976) Crystallization of lunar picritic basalt sample 12002: Phase equilibria and cooling rate studies. *Bull Geol. Soc. Amer.* **87**, 646–656.

Walker D., Longhi J., Stolper E. M., Grove T. L., and Hays J. F. (1977) Slowly cooled microgabbros 15065 and 15555 (abstract). In *Lunar Science VIII,* p. 964–966. The Lunar Science Institute, Houston.

Proc. Lunar Sci. Conf. 8th (1977), p. 1593–1610.
Printed in the United States of America

Applications of dynamic crystallization studies: Lunar olivine-normative basalts

ALDO S. BIANCO and LAWRENCE A. TAYLOR

Department of Geological Sciences, The University of Tennessee, Knoxville, Tennessee 37916

Abstract—Dynamic crystallization studies were performed on two synthetic glasses of olivine-normative basalt composition. The glasses are similar in composition to the mare olivine-normative basalt samples 12009 and 15555. Experiments were designed to study the effects of viscosity (primarily a function of FeO differences between the two melts) and the initial temperature of cooling on mineral chemistry, texture, and temperature of appearance of phases.

Olivine compositions seem to indicate that there are no significant differences in the degree of undercooling at which olivine nucleates, between the two types of melt cooled at the same rate. Olivine-nucleation densities, however, differ. There are more nuclei in the low-FeO melt for equivalent cooling rates. Comparison of the olivine-nucleation density (grains/mm^2) in sample 15555 with the nucleation densities in the high-FeO melt indicate that 15555 was cooled at a rate of approximately 1–0.5°C/hr at the time of olivine nucleation and growth.

Texture, mineral chemistry, and temperature of appearance of phases are greatly affected by the temperature at which cooling is initiated. The crystallization behavior of a melt is probably affected just as much by the temperature at which cooling initiates as it is by the cooling rate, and a wide variety of textures and mineral chemistries may be produced by varying these parameters. Extreme caution must therefore be exercised in applying experimental data to natural rock systems.

INTRODUCTION

EXPERIMENTAL CRYSTALLIZATION STUDIES at controlled cooling rates have successfully duplicated the mineral chemistries and textures of several mare basalt rocks (e.g., Donaldson *et al.*, 1975a; Grove, 1977; Grove and Bence, 1977; Lofgren *et al.*, 1974, 1975; Usselman *et al.*, 1975, 1976; Walker *et al.*, 1975, 1976a). Systematic changes in mineral chemistry, in the temperature of appearance of phases, in texture, and in crystal morphologies occur as a function of cooling rate and degree of undercooling. The results of these studies have been used to estimate cooling rates of certain mare basalts. However, it has become apparent recently that the number of parameters to be considered in such applications is large, and a certain caution must be exercised in applying these experimental data to natural rock systems. Factors which, to greater or lesser extents, influence the crystallization behavior of melts include: experimental technique, largely a function of sample container; sample mass and its relationship to container mass; surface area of contact between sample and container; the shape of the melt; the composition of the gas phase in contact with the charge (Lofgren and Donaldson, 1975); oxygen fugacity conditions during cooling (Usselman *et al.*, 1975), and the temperature relative to the liquidus at which cooling initiates (Usselman and Lofgren, 1976; Bianco and Taylor, 1977; Lofgren, 1977).

The viscosity of a melt is another important factor which affects the crystallization process, since diffusion rates are related to viscosity. Diffusion rates increase as viscosity decreases, and therefore, the viscosity of a melt will have an affect on the degree of supercooling achieved before nucleation occurs in a cooling experiment, as well as having an affect on growth rates. Therefore, dynamic crystallization experiments were performed in order to study the effects of viscosity on the mineral chemistry and texture of mare olivine-normative basalts.

It is known that the precooling history of melting and nucleation can affect parameters such as nucleation densities, crystal morphology, and gross textures (Donaldson, 1976; Lofgren, 1977; Usselman and Lofgren, 1976). In order to further investigate the effects that the precooling history of the charge has on the crystallization behavior of the melt, subliquidus cooling experiments were performed. In this portion of the present study, we investigate the effects of cooling initiated at subliquidus temperatures on the degree of undercooling of plagioclase.

Selection of Starting Material

The FeO content of a melt will significantly affect its viscosity (Bottinga and Weill, 1972). It was, therefore, considered desirable to study a rock type that had a large variation in this component, hence, the olivine-normative basalt group. Two glasses of olivine-normative basalt composition were prepared and are similar in composition to Apollo 12 and 15 mare basalt samples 12009 and 15555. The viscosities of the two glasses as calculated from the method of Bottinga and Weill (1972) are 230 poise (high FeO) and 300 poise (low FeO) at 1300°C, that is, the low-FeO melt is about 30% more viscous at this temperature.

The criteria for the selection of the glass compositions (Table 1) were: (1) a large difference in FeO content was necessary, (2) the MgO/MgO + FeO ratio had to be within the lunar range, and (3) the synthetic should be similar in composition to specific natural samples, thereby allowing for comparisons of the synthetic analog to the natural rock. The compositions chosen based on these criteria are realistic analogs of samples 12009 and 15555. However, because of the above limitations, the CaO and Al_2O_3 contents of the two synthetic materials are not identical.

Experimental Procedure

Silicate glasses were prepared from appropriate oxide and carbonate starting material. For each glass, 8 g of the initial mixture was fine ground under toluene. The mixture was placed in a Pt crucible which was suspended in a glow-bar furnace through which pure argon flowed. The mixture was initially heated at 800°C. After 2 hr, the temperature was slowly raised to 1300°C and maintained at this temperature for 20 min, at which time the Pt crucible was quenched in cold water. Two additional cycles of regrinding and melting were performed. The glasses were then analyzed by the electron microprobe and checked for homogeneity.

Crystallization experiments were performed in CO–CO_2 gas-mixing furnaces using the platinum wire loop technique (Donaldson *et al.*, 1975b) to support the samples. Two charges (a low FeO and a

Table 1. Compositions of lunar analog starting materials.

	High Fe[1]	15555*	Low Fe[1]	12009†	Lunar range
SiO_2	43.32	44.29	44.29	45.03	43–47
TiO_2	2.88	2.34	3.70	2.90	2–5
Al_2O_3	8.78	8.23	12.16	8.59	8–13
Cr_2O_3	0.65	0.58	0.61	0.55	0.1–0.7
FeO	23.70	23.13	17.65	21.03	17–24
MgO	11.88	11.18	9.43	11.55	6–14
CaO	6.96	9.42	10.58	9.42	7–12
MnO	0.39	0.31	0.42	0.23	0.2–0.4
Na_2O	0.35	0.28	0.39	0.28	0.1–0.4
K_2O	0.17	0.04	0.09	n.a.	0.0–0.4
Totals	99.08	99.80	99.20	99.58	
$\frac{MgO}{MgO + FeO}$	0.33	0.33	0.35	0.35	0.30–0.45

[1]Electron-microprobe analyses of batch starting material.
*Analysis by Chappell and Green (1973).
†Analysis by Compston *et al.* (1971).

high FeO) were placed in the furnace simultaneously, separated by about $\frac{1}{8}$ of an inch such that all variables, except bulk composition, could be kept constant. Oxygen fugacities were monitored by a solid-electrolyte cell and controlled at approximately 0.5 log atm below the iron-wüstite buffer curve.

For the isothermal quench experiments, the starting material was placed in the furnace as a finely ground glass that had been pressed into pellets. The charges were placed directly into the furnace at the desired temperature. The duration of the isothermal experiments ranged from 6 to 24 hr. For the constant cooling-rate experiments, the charges were melted for 2 hr at 1290°C (representing a temperature which is about 10°C and 30°C above the liquidus for the high-FeO melt and low-FeO melt, respectively). The furnace was then cooled at a constant rate to approximately the equilibrium solidus temperature (1075°C) and then quenched. Some cooling experiments were performed such that the samples were placed in the furnace at an initial temperature of 1200°C. The charges were maintained at this subliquidus temperature for 2 hr before controlled cooling was initiated; termination of these experiments occurred at 1100°C, slightly above the equilibrium solidus. These experiments will be referred to as the subliquidus experiments.

All of the quenched products were mounted in epoxy, and both thin and polished sections were prepared for optical examination and electron-microprobe analyses. Compositional data were obtained with an automated MAC 400 S electron microprobe, using appropriate synthetic and natural standards. Analyses of the glasses were performed with a defocussed beam with a spot size of about 30 μm in diameter; crystalline phases were analyzed with a 1–2 μm diameter beam. All the analyses were conducted at 15 kV with a beam current of 0.035–0.050 μA. Correction for matrix effects were applied according to procedures described by Bence and Albee (1968) and Albee and Ray (1970).

Electron-microprobe analyses of a charge melted at 1290°C for 6 hr (and then quenched) showed no significant compositional changes from the batch-melted material. Iron loss was detected only within 10 μm of the platinum wire. There was a tendency for chromite and metal to nucleate adjacent to the wire. In some cooling experiments, there was a tendency for the silicate phases to nucleate adjacent to the wire, however, most run products have their silicate phases uniformly distributed.

For all the cooling-rate experiments and for some selected isothermal experiments, the original polished sections were reground and repolished several times. This was necessary in order to obtain reliable statistics for parameters such as nucleation density and mean grain size, as well as obtaining a three-dimensional view of the texture and crystal morphology. It was also observed that variations in modal percentage of phases occurred as sections were ground to different levels. In a few instances, for example, olivine or pyroxene were not observed in the first cut, but appeared upon further grinding.

CRYSTALLIZATION SEQUENCE FROM ISOTHERMAL EXPERIMENTS

The crystallization sequence of the two synthetic melts, as determined from the isothermal experiments, is compared in Fig. 1 with those determined for 15555 (Longhi *et al.*, 1972) and 12009 (Green *et al.*, 1971). The high-FeO melt is almost identical in composition to 15555 (Table 1); therefore, it is not surprising that they have similar crystallization temperatures and sequences for the various phases. The low-FeO melt is slightly different in composition from 12009, and this is reflected in slight differences in crystallization temperatures for the various phases. In fact, the low-FeO glass is depleted in FeO and enriched in CaO and Al_2O_3 relative to 12009; these differences result in the earlier appearance of plagioclase in the synthetic system and also expand the spinel stability range. Thus, differences observed between the phase relations of the synthetic melts and the natural rocks are consistent with minor bulk compositional differences between the materials.

MINERAL COMPOSITIONS

Olivine

The compositions of the olivines grown in the isothermal experiments (Fig. 2) define the olivine solidus of the melts. The liquidus temperatures are 1282 ± 7°C (high-FeO melt, Fo = 75 mole%) and 1257 ± 7°C (low-FeO melt, Fo = 79 mole%).

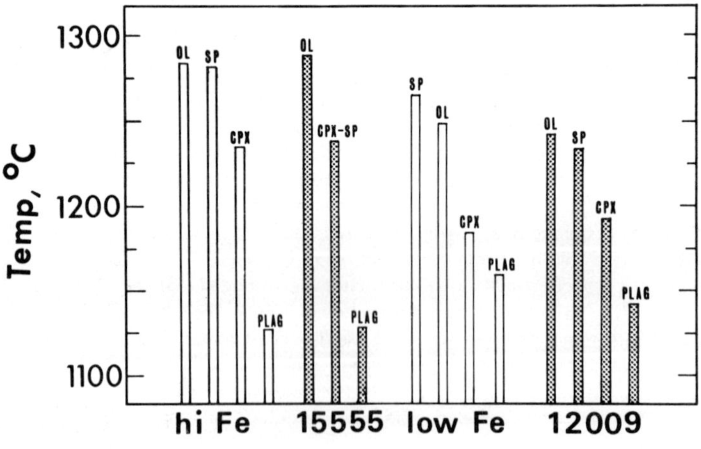

Fig. 1. Crystallization sequence of high-FeO and low FeO compositions as determined from isothermal experiments. The crystallization sequence of 15555 (Longhi *et al.*, 1972) and 12009 (Green *et al.*, 1971) are shown for comparison.

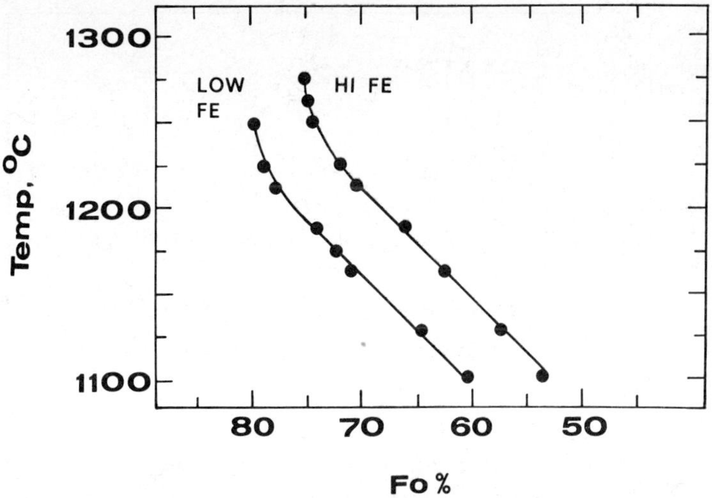

Fig. 2. Composition of olivines (Fo mole%) as a function of temperature as determined from the isothermal experiments.

The quenched products in the isothermal experiments show homogeneous olivine grains whose chemistry did not vary within the uncertainties of the microprobe analyses either within or among various grains. Olivine-liquid partitioning of Fe and Mg was determined for several isothermal temperatures in both high-FeO and low-FeO melts. $K_D^{Fe/Mg}$ values between 0.31 and 0.34 were obtained. These values agree well with those of Roeder and Emslie (1970).

The core to rim compositions of the olivines as a function of cooling rate are shown in Fig. 3. Note that as the cooling rate increases, the core composition decreases in forsterite content. The composition of the cores can be used to estimate the temperature of olivine nucleation (compare core composition from Fig. 3 with equilibrium composition from Fig. 2). Olivine would appear to nucleate and grow from both melts at approximately 45°C undercooling for a cooling rate of 48°C/hr, for example. Similar comparisons for the other cooling rates show that for any given cooling rate, both types of melt are undercooled by almost equal amounts.

The use of olivine core compositions as indication of degree of undercooling assumes that the olivines nucleating in the cooling-rate experiments at a particular temperature have identical compositions to those grown in isothermal experiments at that same temperature, as discussed by Donaldson *et al.* (1975a). Hopper and Uhlmann (1974), however, have theoretically shown that this need not be true. Upon due consideration it is felt that the inferred nucleation temperatures are probably within ±10°C of the actual values. Thus, it appears that there is no significant difference in the degree of undercooling before olivine nucleation occurs, for the two types of melt as a function of cooling rate.

Detailed electron microprobe traverses were performed on some selected

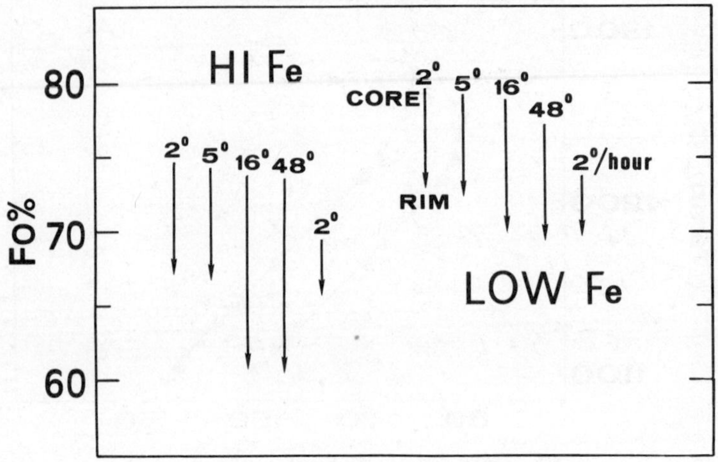

Fig. 3. Core to rim composition of olivines as a function of cooling rate. Notice that the subliquidus experiments (2°C/hr) have the lowest core composition for each type of melt.

olivine grains from the various cooling-rate experiments (Fig. 4). Most crystals possess shallow compositional gradients in the core regions and steep gradients in the core to rim regions. These synthetic olivines lack the shallow compositional gradients in the rim regions that are found in natural olivines (Taylor *et al.*, 1977).

Pyroxene

Pyroxenes in the isothermal experiments are pigeonites and sub-calcic augites which range in composition from Wo_5En_{79} to $Wo_{32}En_{48}$ (low-FeO melt) and Wo_2En_{70} and Wo_6En_{65} (high-FeO melt); the Wo content increases as a function of decreasing temperature, in effect, a reflection of the changing composition of the pyroxene solidus as a function of temperature.

All the pyroxenes in the cooling-rate experiments (Fig. 5) are normally zoned from a less calcic core to a more calcic rim. The FeO/FeO + MgO ratios of the core increase as cooling rate increases. Also, the FeO/FeO + MgO ratio systematically increases for a given Wo content, as cooling rate increases. These observations suggest that the temperature of pyroxene nucleation decreases as the cooling rate increases. The greater enrichment in Wo content for the pyroxenes which grew from the low-FeO melt may be attributed to the higher Cao content in the starting material.

In the subliquidus experiments (Fig. 6), the pyroxenes in the low-FeO charge show an initial trend of Ca enrichment followed by Ca depletion upon nucleation of plagioclase. This charge contains about 20% by volume plagioclase. The high-FeO charge does not follow this trend of Ca depletion, but instead, the

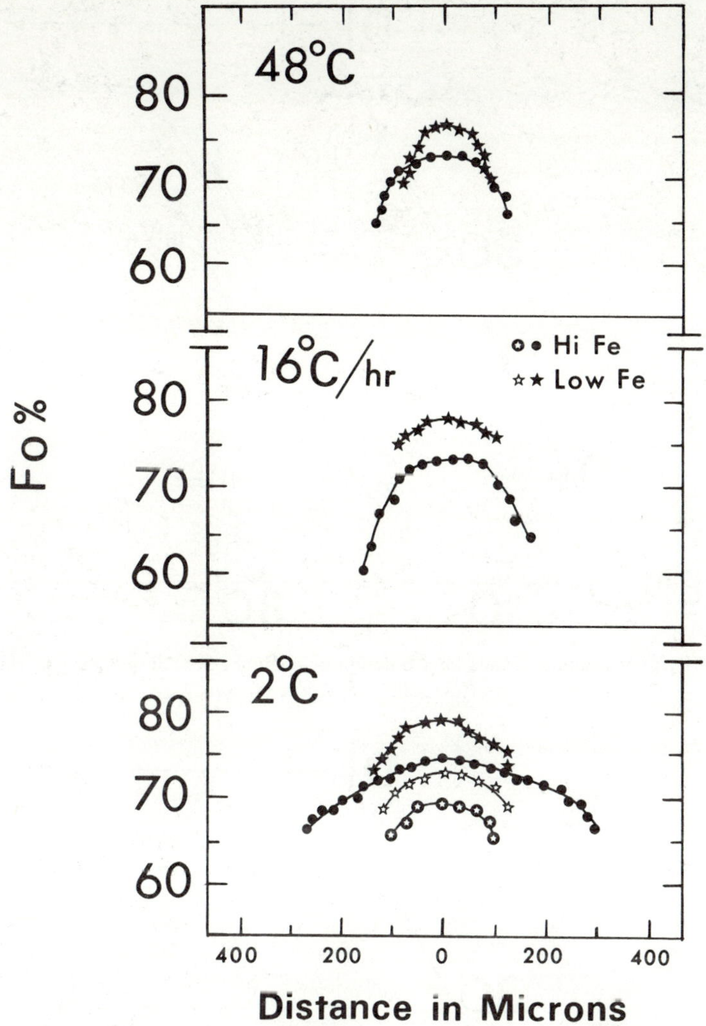

Fig. 4. Detailed electron microprobe traverses across selected olivine grains. Notice the shallow compositional gradients in the core regions and the steep gradients in the core to rim regions. The subliquidus experiments (2°C/hr) are represented by the open symbols.

pyroxenes continuously zone from a less calcic core to a more calcic rim. This charge only contains about 1–3% plagioclase.

All the pyroxenes in the isothermal and cooling-rate experiments (with the exception of the low-FeO, 2°C/hr, subliquidus experiment) have Ti:Al ratios (based on 6 oxygen atoms) of approximately 1:4. As cooling rate and degree of undercooling increase, the Al and Ti contents systematically increase (the ratio remains 1:4). Individual crystals in the cooling-rate experiments, zone from Al- and Ti-deficient cores to Al- and Ti-rich rims. These data reflect the general

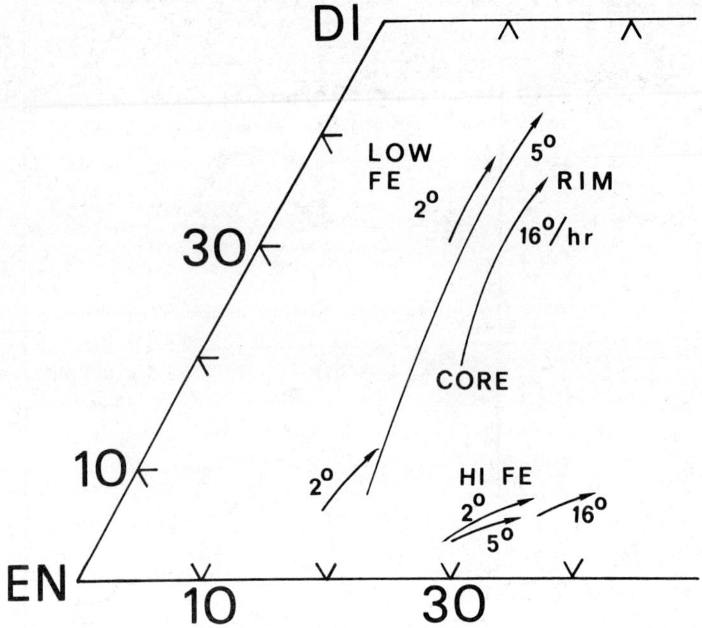

Fig. 5. Pyroxene zoning trends as a function of cooling rate. All cooling experiments initiated cooling above the liquidus.

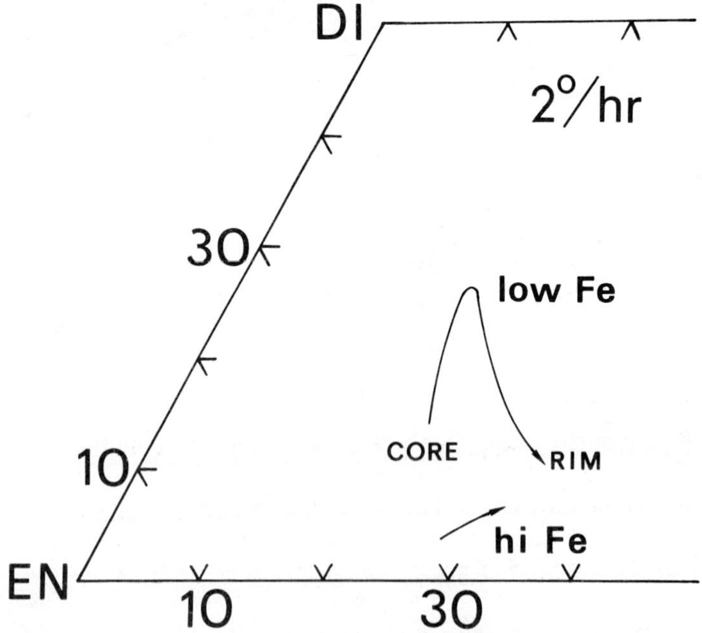

Fig. 6. Pyroxene zoning trends in the subliquidus experiments (2°C/hr).

tendency for increasing incorporation of Ti and Al with decreasing temperatures, as well as with increasing cooling rates. For any one cooling rate, the pyroxenes in the low-FeO melt have higher contents of Al and Ti and exhibit a greater extent in zonation than the pyroxenes in the high-FeO charge. These observations probably reflect the differences in bulk chemistry between the two melts.

The pyroxenes in the subliquidus experiments tend to have Ti:Al ratios that are greater than 1:4. In fact, pyroxenes in the low-FeO melt are zoned from ratios of 1:4 in the cores to ratios of 1:2 in the rims, reflecting the sensitivity of pyroxene composition to plagioclase nucleation and growth.

It is difficult to estimate the temperature of pyroxene nucleation as a function of cooling rate based solely on compositional data. This is a result of the complex multicomponent solid solution in the pyroxenes as well as of fractionation effects. The composition of the liquid from which pyroxene nucleates and grows is affected by the previous nucleation and growth of olivine and spinel. This degree of modification to the liquid composition (i.e., fractionation) is a direct function of the cooling rate. An additional complication arises from the lack of plagioclase nucleation in the cooling-rate experiments (with the exception of the subliquidus experiments). Therefore, a determination of the pyroxene solidus as a function of cooling rate is not feasible at the present time. Several complications mentioned above (e.g., changing liquid composition) do not allow for a direct comparison with the pyroxene isothermal solidus.

Textural Variations

Systematic variations in crystal morphology with cooling rate were observed for melts of both compositions. All the run products are hypocrystalline. At all cooling rates, the textures are porphyritic, olivine crystals are larger than pyroxene crystals, and only one generation of olivine and pyroxene formed. Plagioclase nucleated only in the subliquidus experiments; the low-FeO charge cooled at 2°C/hr contained approximately 20% plagioclase crystals, and plagioclase formed as dendritic intergrowths with pyroxene and as isolated hollow laths (Fig. 7a). Many of the plagioclase crystals appear to be attached to pyroxenes and are enclosed by glass; they probably nucleated heterogeneously on the pyroxene-melt interfaces. Hollow skeletons of plagioclase which seem to be entirely surrounded by glass could have nucleated in a homogeneous mode; the cores of these crystals are also filled by glass.

Olivine morphology changes considerably as a function of cooling rate. At the slowest cooling rates investigated, the olivine crystals occur as non-skeletal, subhedral, equant to subequant crystals (Fig 7b). At faster cooling rates, the shapes progress to externally euhedral but internally skeletal (Fig. 7c), through hollow crystals, to elongate, internally and externally skeletal crystals (Fig. 7d,e), to chain-like forms. This progression of morphology as a function of cooling rate is similar to that observed by Donaldson *et al.* (1975a) in their study of the olivine-normative basalt sample 12009. At any one cooling rate, more than one type of morphology may occur. For example, the low-FeO charge cooled at

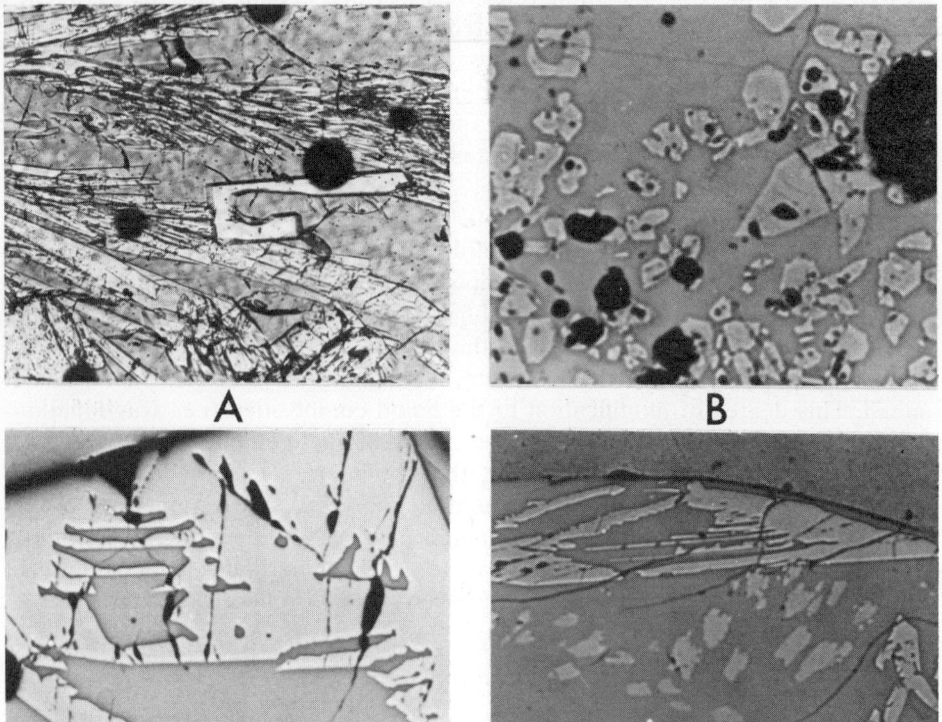
A
B
C
D
E
F

5°C/hr contains elongate, internally and externally skeletal crystals as well as subhedral, subequant grains (Fig. 7f).

Pyroxene morphology also changes as a function of cooling rate, but to a lesser extent than olivine. At the slowest cooling rates, pyroxenes occur as subhedral, equant crystals and as elongate externally euhedral but internally skeletal crystals (Fig. 8a,b). As the rate of cooling increases, the shapes progress to hollow skeletal and elongate chain-like skeletal. As with the olivines, more than one type of morphology may form at any one cooling rate.

Chromium spinels are present at all cooling rates; however, they do not exhibit a large range in morphology. Most are euhedral to subhedral grains, but a few occur as dendrites (Fig. 8c). The spinels are either surrounded by glass or are found as inclusions in olivine (Fig. 8a). They are located primarily at the base of the charge as a result of crystal settling and near the rims as a result of heterogeneous nucleation at the melt surface. The spinels are also commonly associated with native Fe grains.

Textural and morphological variations were observed as a result of initiating cooling at different temperatures. When the low-FeO melt was cooled at 2°C/hr from above the liquidus temperature (Fig. 7b), olivines and pyroxenes occur as small euhedral, equant grains. However, in an experiment at the same cooling rate, where cooling was initiated below the liquidus, the low-FeO melt crystallized to dendritic intergrowths of pyroxene and plagioclase, and externally euhedral but internally skeletal olivines (Fig. 7a). This is contrasted with the high-FeO melt (2°C/hr), where large olivines (up to 1.5 mm in length) and large pyroxenes of equant, subhedral to externally euhedral but internally skeletal crystals occurred (Fig. 8b,d). Again, for the high-FeO melt, but cooling from subliquidus temperatures (2°C/hr), pyroxene and olivine crystals were found as small euhedral to subhedral, equant grains (Fig. 8e). The nucleation density of olivines and pyroxenes (grains/mm^2) in this charge is significantly larger than in the charge where cooling was initiated at above the liquidus.

Viscosity Effects on Mineral Composition and Texture

Before discussing the effects that viscosity may have on the mineral chemistry and texture, other effects must be taken into account. As mentioned in the

Fig. 7. Reflected light photomicrographs of run products. (a) Low-FeO melt, 2°C/hr subliquidus experiment. Dendritic intergrowths of pyroxene and plagioclase. Notice the hollow plagioclase crystal in center. Externally euhedral, but internally skeletal olivine crystals are shown along bottom. Width of field = 600 μm. (b) Low-FeO melt, 2°C/hr above liquidus experiment. Euhedral to subhedral pyroxenes and olivines. Width of field = 1200 μm. (c) High-FeO melt, 5°C/hr. Externally euhedral but internally skeletal olivine grains. Width of field = 1200 μm. (d) Low-FeO melt, 5°C/hr. Skeletal olivine grains at the top and elongate skeletal olivine in SE corner. Small pyroxene grains in middle. Width of field = 600 μm. (e) Low-FeO melt, 16°C/hr. Elongate, internally and externally skeletal olivine grains. Width of grains = 600 μm. (f) Low-FeO melt, 5°C/hr. Notice two distinct olivine morphologies; elongate, internally and externally skeletal crystals, as well as subhedral, subequant grains. Width of field = 600 μm.

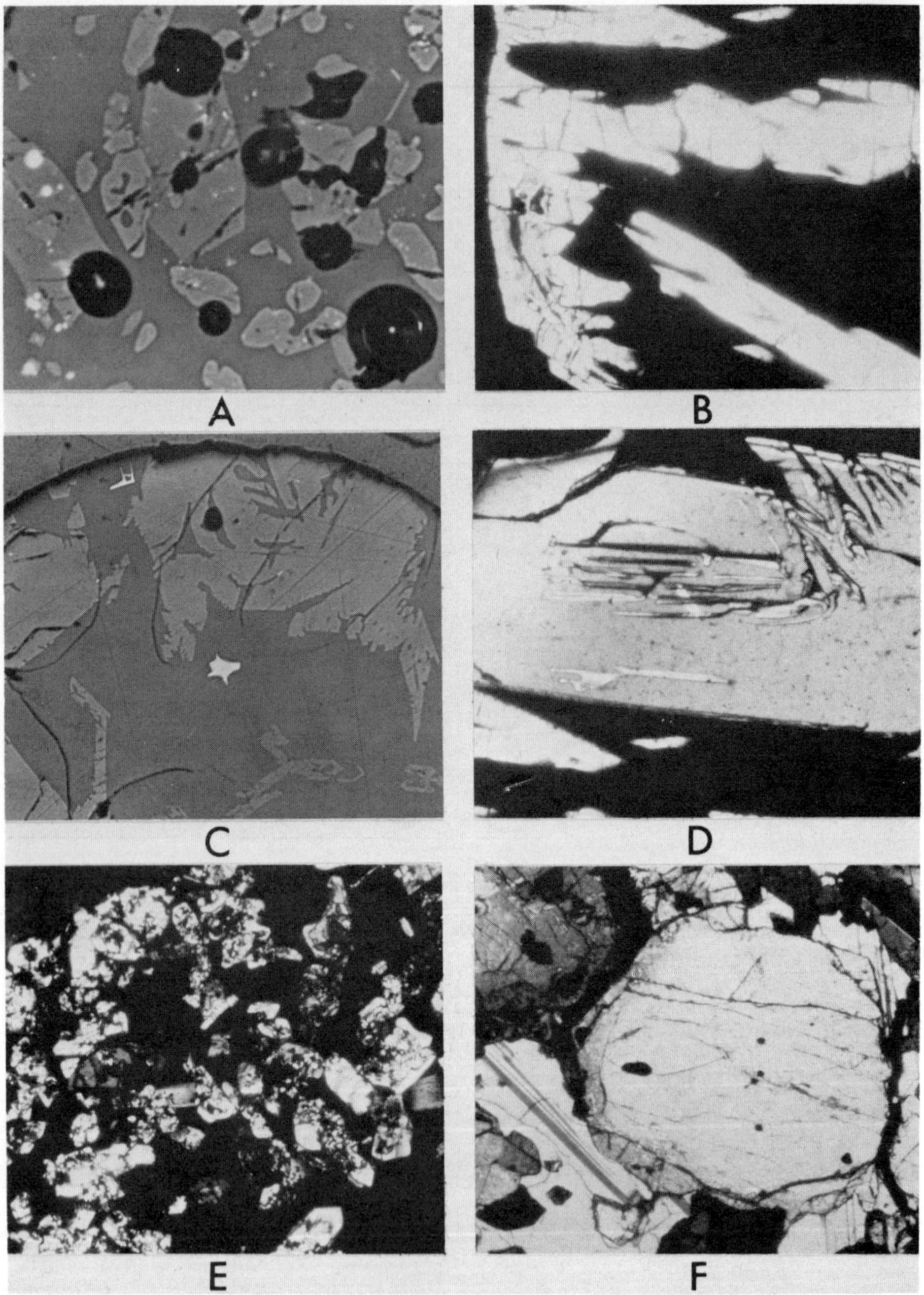
A
B
C
D
E
F

experimental procedure section, for each cooling experiment, two charges (a low FeO and a high FeO) were run simultaneously so that both melts could experience identical cooling and f_{O_2} conditions. However, because the two melts have different liquidus temperatures, a superheating effect existed which may have partially masked any viscosity effects. The low-FeO melt initiated cooling at a ΔT_s value of 30°C, while the high-FeO melt initiated cooling at a ΔT_s value of 10°C. It is reasoned that superheat will tend to delay nucleation by possibly altering the melt structure, as well as by destroying any Angstrom sized particles in the melt. However, Donaldson (1976) observed that olivine-nucleation rate and growth rate cannot be altered by large superheating (up to 100°C). Thus, it is assumed that for the values of ΔT_s in these experiments, superheat does not affect the olivine-nucleation temperature.

Olivine chemistry indicates that there are no significant differences in the degree of undercooling at which olivine nucleation occurred between the two types of melt, for the same cooling rate. It therefore appears that viscosity differences between the two melts does not measurably affect the crystallization behavior of olivine.

For any given cooling rate, olivines in the low-FeO charge are smaller in width and have a larger nucleation density than olivines in the high-FeO charge (Table 2). Assuming that olivines nucleated from both melts at approximately the same degree of undercooling, for any given cooling rate, then the olivines in the high-FeO melt would be growing at higher temperatures (approximately 25°C higher) and for a longer time than the olivines in the low-FeO melt. This could account for crystal size differences between the two melts. However, the growth rate is also affected by the viscosity of the melt (diffusion rates increase as viscosity decreases) and olivines from the high-FeO melt would be expected to have faster growth rates than olivines from the low-FeO melt. Therefore, both viscosity and the temperature at which individual phases grow will affect the olivine growth rate in the same sense.

Fig. 8. Reflected and transmitted photomicrographs. (a) High-FeO melt, 2°C/hr subliquidus experiment. Euhedral to subhedral olivines and pyroxenes. Abundant native iron and spinel inclusions in both the olivines and pyroxenes. Width of field = 1200 μm. (b) High-FeO melt, 2°C/hr above liquidus experiment. Elongate, externally subhedral pyroxene grains, many of which are internally skeletal. Width of field = 1200 μm. (c) High-FeO melt, 16°C/hr. All grains are olivines and chromium spinels. Notice skeletal spinel (upper part of photo) with dendritic extensions. Also, notice the wide variety in olivine morphology. (d) High-FeO melt, 2°C/hr above liquidus experiment. Extremely large externally euhedral but internally skeletal olivine grain. Width of field = 1200 μm. (e) High-FeO melt, 2°C/hr subliquidus experiment. Small, euhedral to subhedral olivine and pyroxene grains. Note the contrast in nucleation densities between this charge and charge in photograph (d). Width of field = 1200 μm. (f) Natural sample 15555 showing rounded subhedral olivine grain surrounded by smaller olivines and pyroxenes and plagioclase. Compare the size and shape of natural olivine with synthetic olivine grown from high-FeO melt at 2°C/hr above liquidus experiment (d). Width of field = 1200 μm.

Table 2. Nucleation densities (grains/mm^2), mean width (μm), and morphological types of olivine as a function of cooling rate and bulk composition.

	High-FeO melt			Low-FeO melt		
	5°C/hr	16°C/hr	48°C/hr	5°C/hr	16°C/hr	48°C/hr
Nucleation density	1.2	1.5	5.2	4.3	6.8	12.7
Mean width	170	130	80	110	85	15
Morphological type*	A,B	A,B,C	D,E	B,C,D	D,E	E,F

*A = Tabular; B = Internal subhedral skeletal; C = Hollow; D = Elongate internal and external skeletal; E = Chain-like; F = Dendritic.

Again, if it is assumed that olivines nucleated from both melts at approximately the same degree of undercooling, for any cooling rate, then the olivine-nucleation density data indicate that the number of nucleation events are not solely a function of the degree of undercooling. Instead, each bulk composition possesses a unique relationship between the degree of undercooling and the number of nucleation events.

The nucleation rate equation and the growth rate equation contain two exponential terms (one related to the bulk free energy of crystallization and the other related to the activation energy necessary to move an atom, ion, or molecule across the interface between the crystal and the melt) which affect the forms of the growth rate and the nucleation rate versus degree of undercooling curves. The bulk free energy term increases with degree of undercooling, whereas the activation energy term decreases with degree of undercooling, being a function of atomic mobility. As a result, both growth rate and nucleation rate have maximum values at specific degrees of undercooling, where one term balances the other. These two exponential terms are a function of various parameters (e.g., temperature, viscosity of the melt, latent heat of fusion, structure of the melt, and structure of the phase precipitating from the melt). Growth rates, for example, will be controlled by diffusion in the melt, flow of latent heat away from the growing crystal surface, and by reactions at the crystal-melt interface (Kirkpatrick, 1975; Donaldson, 1975). Therefore, any viscosity effects on olivine-nucleation density or grain size (Table 2) will be masked by these other factors mentioned above. However, based on the observation that olivines nucleated from both melts at approximately the same degree of undercooling, for any cooling rate and taking into consideration all the variables which affect nucleation and growth, then it can be inferred that viscosity differences between the two melts has little affect on the crystallization process.

COMPARISON OF SYNTHETIC ROCKS WITH NATURAL SAMPLES

Although the cooling experiments were not completely successful in duplicating all the textures found in the natural rocks, crystal morphologies occurring in the natural rocks were reproduced in the experiments. Olivines in the 2°C/hr charge (high-FeO melt), for example, are similar in size and morphology to olivines in 15555 (Fig. 8f). However, it is difficult to estimate cooling rate based on morphology alone, because at any one cooling rate, a range in morphologies may occur.

The olivine-nucleation density in the high-FeO melt decreases as cooling rate decreases. By using the same method that Walker *et al.* (1976a) employed in their study of lunar picrite melt 12002, a cooling rate of approximately 0.5–1.0°C/hr was estimated for 15555 (compare the olivine-nucleation density of 15555 with the olivine-nucleation density data of the high-FeO cooling-rate experiments). This cooling rate is applicable for the beginning of olivine crystallization. This rate agrees favorably with that determined by Taylor *et al.* (1977) in their kinetic-model study based on Fe–Mg diffusion in olivine.

Precooling History of the Melt

A silicate melt possesses an "internal structure" which is dependent on its temperature. For each temperature an "equilibrium" arrangement of the Si–O–M (M = metals) bonds exist, in terms of bond lengths and strengths, and probably, bond angles (Bottinga and Weill, 1972; Hess, 1970). According to our reasoning, before a melt can nucleate a crystalline phase, it must possess a certain "internal structure" in order to facilitate this nucleation. When cooling is initiated at any value of ΔT_S or when cooling occurs at any rate, the nucleation of a phase is delayed until the melt has rearranged its *initial* "internal structure" to the structure which is in near-equilibrium with the crystalline phase. This results in a lag time for the rearrangement of the melt, and hence for the nucleation of the phase.

By initiating cooling at subliquidus temperatures, the melt is given time to attain its"equilibrium internal structure" for that temperature. Therefore, when upon cooling, the melt approaches the equilibrium temperature of a phase (e.g., 1140°C for plagioclase in 15555), its liquid "structure" is closer to the "equilibrium structure" than for a melt which initiated cooling at a higher temperature. As a consequence, a lower degree of undercooling is necessary to nucleate that particular phase. This is exactly what was observed during the present study.

The experimental crystallization studies of Donaldson *et al.* (1975a), Usselman and Lofgren (1976), and Walker *et al.* (1976a), as well as the present study, demonstrate the rate of cooling effects on the temperature of appearance of phases. Plagioclase is especially suppressed. This is demonstrated by the failure of plagioclase to nucleate in any of our experiments where cooling was initiated at above liquidus temperatures and quenched at 1075°C. Walker *et al.* (1976a) have observed that even at extremely slow cooling rates the amount of plagioclase suppression is significant (e.g., at a cooling rate of 0.4°C/hr, plagioclase nucleation is suppressed by 120°C).

Plagioclase, however, did nucleate in our 2°C/hr, *subliquidus* experiments, where the melt initiated cooling at 1200°C and was quenched at 1100°C. The equilibrium temperature of appearance for plagioclase in the high-FeO melt is at 1140°C. Therefore, the *maximum* degree of undercooling for plagioclase in this experiment was 40°C. Preliminary experimentation indicates that the same melt cooled from above the liquidus at 2°C/hr undercools over 200°C before plagioclase nucleation occurs. Thus, as reported by Bianco and Taylor (1977), the present study indicates that the *temperature at which plagioclase nucleation appears is not greatly depressed for a melt which initiates cooling at subliquidus temperatures.* Usselman and Lofgren (1976), in their study of high-titanium mare basalts, also observed that plagioclase nucleation is enhanced by initiating cooling at subliquidus temperatures.

Lofgren and Donaldson (1975), Donaldson (1976), Usselman and Lofgren (1976), and Lofgren (1977) have demonstrated that by initiating cooling at subliquidus temperatures, parameters such as crystal morphology, incubation period, nucleation sites and nucleation densities, and gross textures are greatly affected. Their observations, in conjunction with the present study, suggest that by initiating cooling at different temperatures below the liquidus, the subsequent crystallization history of the melt will be affected. A magma which erupts free of crystals, for example, will have a different crystallization history and will produce a rock with minerals of different physical and chemical characteristics than the same magma which erupts with some crystals, i.e., at subliquidus temperatures.

Summary

Conclusions which can be drawn from this study are:

(1) Comparison of mineral composition and textures of the two melts (a high-FeO and a low FeO melt) as a function of cooling rate indicates that the viscosity difference between the two melts does not measurably affect the crystallization process. Since the two melts bracket the upper and lower limit in FeO for olivine-normative basalts, it may be safe to assume that viscosity differences have played no significant role in development of textures and mineral chemistry found in the natural olivine-normative basalts.

(2) Lunar sample 15555 is estimated to have a cooling rate of approximately 1–0.5°C/hr for the beginning of olivine crystallization, based on olivine-nucleation density data. This is a valid statement assuming that the melt erupted free of crystals.

(3) Texture, crystal morphology, mineral composition, and temperature of appearance of phases are greatly affected by initiating cooling at subliquidus temperatures. The present study demonstrates that the temperature at which cooling initiates is probably as important as the subsequent cooling rate, and a wide variation in textures and mineral chemistries can be produced by varying these parameters.

Acknowledgments—We wish to express our appreciation to Drs. T. Grove, G. Lofgren, T. Usselman, B. Walker, and R. J. Williams for helpful suggestions and many stimulating discussions. Drs. C. H. Donaldson and T. L. Grove are acknowledged for the review of the manuscript. Many thanks are due Dr. G. Lofgren for use of some of his lab facilities at J.S.C. and Messrs. O. Mullins and D. Smith for assistance in the lab. We also thank the University of Tennessee for providing substantial support for the purchase of the electron microprobe. The research presented in this paper was supported by NASA grant NGL-43-001-140.

REFERENCES

Albee A. and Ray L. (1970) Correction factors for electron probe microanalysis of silicates, oxides, carbonates, phosphates, and sulfates. *Anal. Chem.* **42**, 1408–1414.

Bence A. E. and Albee A. (1968) Empirical correction factors for the electron microanalysis of silicates and oxides. *J. Geol.* **76**, 382–403.

Bianco A. S. and Taylor L. A. (1976) Cooling rate experimentation on olivine normative mare basalts: Effects of FeO content on mineral texture and chemistry. *Geol. Soc. Amer. Abstracts with Programs* **8**, 778.

Bianco A. S. and Taylor L. A. (1977) Lunar olivine normative basalts: Cooling rate experimentation (abstract). In *Lunar Science VIII*, p. 103–105. The Lunar Science Institute, Houston.

Bottinga Y. and Weill D. F. (1972) The viscosity of magmatic silicate liquids: A model for calculation. *Amer. J. Sci.* **272**, 438–475.

Chappell B. W. and Green D. H. (1973) Chemical compositions and petrogenetic relationships in Apollo 15 mare basalts. *Earth Planet. Sci. Lett.* **18**, 237–246.

Compston W., Berry H., Vernon M. J., Chappell B. W., and Kaye M. J. (1971) Rubidium-strontium chronology and chemistry of lunar material from the Ocean of Storms. *Proc. Lunar Sci. Conf. 2nd*, p. 1471–1485.

Donaldson C. H. (1975) A petrogenetic study of harrisite in the Isle of Rhum pluton, Scotland. Ph.D. thesis, St. Andrews University, Scotland. 138 pp.

Donaldson C. H. (1976) An experimental investigation of olivine morphology. *Contrib. Mineral. Petrol.* **57**, 187–213.

Donaldson C. H., Usselman T. M., Williams R. J., and Lofgren G. E. (1975a) Experimental modeling of the cooling history of Apollo 12 olivine basalts. *Proc. Lunar Sci. Conf. 6th*, p. 843–869.

Donaldson C. H., Usselman T. M., Williams R. J., and Lofgren G. E. (1975b) A sampling holding technique for study of crystal growth in silicate melts. *Amer. Mineral.* **60**, 324–326.

Green D. H., Ringwood A. E., Ware N. G., Hibberson W. D., Major A., and Kesson S. (1971) Experimental petrology and petrogenesis of Apollo 12 basalts. *Proc. Lunar Sci. Conf. 2nd*, p. 601–615.

Grove T. L. (1977) Cooling histories of Apollo 15 quartz normative basalts (abstract). In *Lunar Science VIII*, p. 380–382. The Lunar Science Institute, Houston.

Grove T. L. and Bence A. E. (1977) Pyroxenes as recorders of lunar basalt petrogenesis revisited: An experimental study of pyroxene-liquid interaction (abstract). In *Lunar Science VIII*, p. 383–385. The Lunar Science Institute, Houston.

Hess P. C. (1970) Polymer model of silicate melts. *Geochim. Cosmochim. Acta* **35**, 289–306.

Hopper R. W. and Uhlmann D. R. (1974) Solute redistribution during crystallization at constant velocity and constant temperature. *J. Crys. Growth* **21**, 203–213.

Kirkpatrick J. R. (1975) Crystal growth from the melt. *Amer. Mineral.* **60**, 798–814.

Lofgren G. E. (1977) Dynamic crystallization experiments bearing on the origin of textures in impact generated liquids. *Proc. Lunar Sci. Conf. 8th.* This volume.

Lofgren G. E. and Donaldson C. H. (1975) Phase relations and non-equilibrium crystallization of ocean ridge tholeiite from the Nazca Plate (abstract). EOS (*Trans. Amer. Geophys. Union*) **56**, 468.

Lofgren G. E., Donaldson C. H., and Usselman T. M. (1975) Geology, petrology and crystallization of Apollo 15 quartz-normative basalts. *Proc. Lunar Sci. Conf. 6th*, p. 79–100.

Lofgren G. E., Donaldson C. H., Williams R. J., Mullins O., Jr., and Usselman T. M. (1974) Experimentally reproduced textures and mineral chemistry of Apollo 15 quartz-normative basalts. *Proc. Lunar Sci. Conf. 5th*, p. 549–567.

Longhi J., Walker D., Stopler E. H., Grove T. L., and Hays J. F. (1972) Petrology of Mare/Rille basalts 15555 and 15065. In *The Apollo 15 Lunar Samples* (J. W. Chamberlain and C. Watkins, eds.), p. 131–134. The Lunar Science Institute, Houston.

Roeder P. L. and Emslie R. F. (1970) Olivine-liquid equilibrium: *Contrib. Mineral. Petrol.* **29**, 275–289.

Taylor L. A., Walker B. M., Onorato P. I. K., and Uhlmann D. R. (1977) Fe–Mg diffusion in olivine: A cooling rate indicator (abstract). In *Lunar Science VIII*, p. 928–930. The Lunar Science Institute, Houston.

Usselman T. M. and Lofgren G. E. (1976) The phase relations, textures, and mineral chemistries of high titanium mare basalts as a function of oxygen fugacity and cooling rate. *Proc. Lunar Sci. Conf. 7th*, p. 1345–1363.

Usselman T. M., Lofgren G. E., Donaldson C. H., and Williams R. J. (1975) Experimentally reproduced textures and mineral chemistries of high-titanium mare basalts. *Proc. Lunar Sci. Conf. 6th*, p. 997–1020.

Walker D., Longhi J., Stolper E. M., Grove T. L., and Hays J. F., (1975) Origin of titaniferous lunar basalts. *Geochim. Cosmochim. Acta* **39**, 1219–1235.

Walker D., Kirkpatrick R. J., Longhi J., and Hays J. F. (1976a) Crystallization history of lunar picritic basalt sample 12002: Phase equilibria and cooling rate studies. *Bull. Geol. Soc. Amer.* **87**, 646–656.

Walker D., Kirkpatrick R. J., Longhi J., and Hays J. F. (1976b) Olivine nucleation in lunar basaltic melts in iron capsules (abstract). *EOS* (*Trans. Amer. Geophys. Union*) **57**, 356.

Proc. Lunar Sci. Conf. 8th (1977), p. 1611–1624.
Printed in the United States of America

Apollo 17 TiO_2-rich basalts: Reverse spinel zoning as evidence for subsolidus equilibration of the spinel-ilmenite assemblage

A. EL GORESY and P. RAMDOHR

Max-Planck-Institut für Kernphysik, P.O. Box 103980, 6900 Heidelberg/Germany (F.R.)

Abstract—A detailed electron microprobe study of chromian ulvöspinels occuring in TiO_2-rich basalts indicates a bimodal distribution in the initial $FeO/(FeO + MgO)$ (FFM) and $TiO_2/(TiO_2 + Cr_2O_3 + Al_2O_3)$ ratios of primary chromian ulvöspinels. Initial FFM and $TiO_2/(TiO_2 + Cr_2O_3 + Al_2O_3)$ ratios of chromian ulvöspinels in plagioclase poikilitic basalts vary from 0.79 to 0.90 and 0.30 to 0.45 respectively. The initial values for chromian ulvöspinels in olivine porphyritic basalts vary from 0.89 to 0.98 and 0.40 to 0.57 respectively. These differences are suggestive of crystallization under different f_{O_2} conditions whereby the conditions during crystallization of plagioclase poikilitic basalts were relatively more reducing than for olivine porphyritic basalts.

Reverse zoning in chromian ulvöspinels was encountered in numerous TiO_2-rich basalts. This unique zoning is characterized by sharp decrease in $TiO_2/(TiO_2 + Cr_2O_3 + Al_2O_3)$ ratio from core to rim but only when ilmenite occurs at the spinel boundary. This reverse zoning is indicative of subsolidus equilibration between chromian ulvöspinel and ilmenite and not of subsolidus reduction. With decreasing temperature Cr_2O_3, Al_2O_3, and V_2O_3 partition between coexisting ilmenite and chromian ulvöspinel in favor of the spinel. This mechanism was also probably active during and after subsolidus reduction of lunar chromian ulvöspinel and hence a major part of the reverse chemical variation of those spinels may have formed due to subsolidus equilibration rather than subsolidus reduction alone.

INTRODUCTION

PLAGIOCLASE POIKILITIC ILMENITE BASALTS and olivine porphyritic ilmenite basalts which were recovered from the Taurus-Littrow site (Apollo 17) are characterized by their high content in opaque oxide minerals. In both basalt types ilmenite is the dominant opaque mineral with variable amounts of armalcolite and chromian ulvöspinel (El Goresy *et al.*, 1973, 1974; Haggerty, 1973). The difference in textures of silicates and opaque minerals in the two basalt types was interpreted by El Goresy *et al.* (1974) as an evidence for different crystallization sequences. El Goresy *et al.* (1974) also emphasize that these two sequences are not the result of different cooling rates. The two crystallization sequences observed were duplicated experimentally using the same starting material but at different oxygen fugacities (Usselman and Lofgren, 1976). This indicates that f_{O_2} was the major controlling factor for generation of the two crystallization sequences.

In the majority of the Apollo 17 TiO_2-rich basalts, chromian ulvöspinel crystallized along with olivine and armalcolite and extended to lower temperatures thus covering a great deal of the cooling history. Hence, it is expected that cooling of lunar basaltic liquids at different oxygen fugacities should reveal variations in the spinel compositions since $FeO/(FeO + MgO)$ ratios of these melts are dependent on f_{O_2}. Subsolidus reduction reactions of lunar opaque minerals and the partitioning of minor and major elements among members of the reduced assemblages was extensively discussed by El Goresy and Ramdohr (1975). Although chromian ulvöspinel and ilmenite frequently occur together without any sign of subsolidus reduction, the possibility of subsolidus equilibration and its influence on the compositional variations of spinels in TiO_2-rich basalts was not explored. Experimental evidence for subsolidus equilibration between ilmenite and coexisting chromian ulvöspinel was first presented by Usselman (1975). In his experiments, Usselman noted that the spinel compositions with falling temperatures in the ilmenite$_{ss}$–spinel$_{ss}$–rutile–iron system show the identical chemical trends as spinels which are believed to have undergone subsolidus reduction. It is the intention of the present report to (1) document spinel compositions in both plagioclase poikilitic basalts and olivine porphyritic basalts and (2) genetically evaluate zoning trends of chromian ulvöspinel which resulted from primary crystallization and subsolidus equilibration. Chemical zoning trends were traced with an automated electron microprobe on numerous individual spinel grains in plagioclase poikilitic ilmenite basalts and olivine porphyritic ilmenite basalts (samples 74242,19, 70017,109, 70035,16, 70215,159, and 71155,36). Before analyses spinels were carefully scanned in reflected light with special interest given to the assemblage chromian ulvöspinel–ilmenite, especially the contact and zoning relationships at their mutual boundaries. Experimental investigations in the system FeO–TiO_2–Cr_2O_3 in equilibrium with metallic Fe° were also carried out, the results of which are published in this volume (Knecht *et al.*, 1977).

Results

Spinel initial compositions in Apollo 17 basalt types

Microscopic investigations of zoning features in chromian ulvöspinels in TiO_2-rich basalts are a difficult task. This is mainly due to the fact that these spinels though chemically zoned, fail in many cases to display this zoning optically. It is therefore necessary to trace the zoning trends with the electron microprobe on a large number of spinels to ascertain their characteristics. Single point analyses of spinels from both basalt types were plotted in the spinel prism to search for any relationship between spinel chemistry and rock type. Since the analyzed spinels displayed various zoning relationships depending on the coexisting assemblage only spinels without subsolidus reaction rims were selected for this plot. The FFM and $TiO_2/(TiO_2 + Cr_2O_3 + Al_2O_3)$ ratios in the spinel cores are assumed to represent the initial ratios fixed during growth from the basaltic liquid.

Despite considerable overlap, evidence for bimodal distribution of both initial FFM and $TiO_2/(TiO + Cr_2O_3 + Al_2O_3)$ ratios of spinels was found (Fig. 1). Both initial FFM and $TiO_2/(TiO_2 + Cr_2O_3 + Al_2O_3)$ ratios of spinels in plagioclase poikilitic basalts are generally lower than the initial ratios of spinels in olivine porphyritic basalts. Initial FFM ratios of spinels in plagioclase basalts vary between 0.79 and 0.90 whereas the initial ratios for spinels in olivine porphyritic

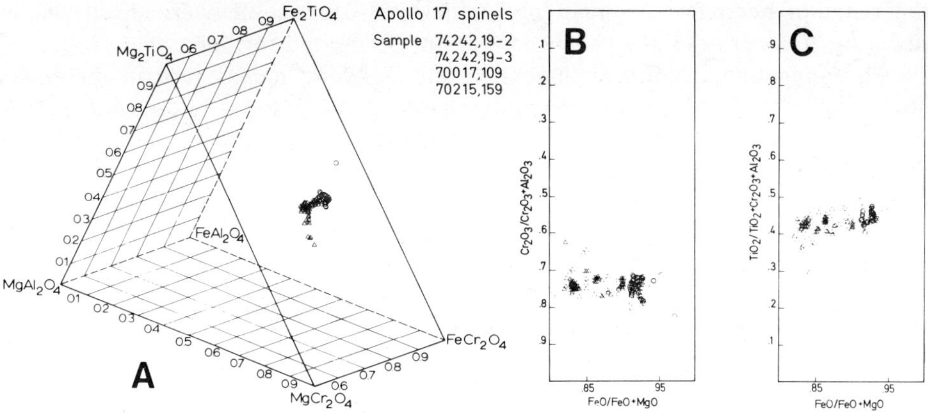

Fig. 1. Composition of cores of primary chromian ulvöspinel as plotted in the modified spinel prism. (A) Bimodal distribution of chromian ulvöspinels in plagioclase poikilitic basalts (triangles) and in olivine porphyritic basalts (circles). (B) Base of the prism demonstrating the well developed bimodal distribution. (C) Front face of the prism showing the lower initial FFM and $TiO_2/TiO_2 + Cr_2O_3 + Al_2O_3$ in spinels in plagioclase poikilitic basalts (triangles) and higher initial ratios in spinels in olivine porphyritic basalts (circles).

basalts vary between 0.89 and 0.98 (Fig. 1). $TiO_2/(TiO_2 + Cr_2O_3 + Al_2O_3)$ ratios for spinels in plagioclase poikilitic basalts range from 0.30 to 0.45 and for spinels in olivine porphyritic basalts from 0.40 to 0.57. However, initial ratios of chromian ulvöspinels occuring as inclusion in olivines in some olivine porphyritic basalts do fall within the range of the spinels in plagioclase poikilitic basalts. The bimodal distribution of the initial ratios displayed in Fig. 1 strongly suggests that chromian ulvöspinels in plagioclase poikilitic basalts did crystallize from basaltic liquids with relatively lower FFM ratios perhaps due to crystallization under relatively reducing conditions as compared to olivine porphyritic basalts. This conclusion is in good agreement with experimental studies of Usselman and Lofgren (1976) and Sato (1976).

Subsolidus equilibration of chromian ulvöspinel and ilmenite

Apart from spinels displaying subsolidus reduction reactions or ilmenite exsolution features (Haggerty, 1973; El Goresy *et al.*, 1974, 1975) chromian ulvöspinels occur in three distinct assemblages in TiO_2-rich basalts:

(1) Chromian ulvöspinel with armalcolite reacted to ilmenite.
(2) Chromian ulvöspinel with ilmenite rims. Both 1 and 2 are enclosed in olivine, pyroxene, plagioclase, or in the interstices.
(3) Idiomorphic chromian ulvöspinel without ilmenite or armalcolite as inclusions in olivine, pyroxene, or in the interstices between various silicates.

Textures of the first assemblage and the reverse spinel zoning resulted from the reaction between chromian ulvöspinel and armalcolite were discussed in detail by El Goresy *et al.* (1974) and will not be treated further.

Chromian ulvöspinel with ilmenite rims is by no means rare in the rocks studied. The assemblage was encountered in both rock types but more frequent in coarser grained basalts. Size of the ilmenite rims varies considerably in the same samples. Early crystallized chromian ulvöspinel present as inclusions in olivine displays the ilmenite rims as narrow separate patches, however, in many cases they occur along all sides of the spinel grains and almost completely surround them (Fig. 2). The ilmenite rims show both idiomorphic and curvate surface contacts to the ulvöspinel even in the same grain. Two possible mechanisms may be responsible for the formation of the ilmenite rims. (1) Reaction of the cooling basaltic liquid with the chromian ulvöspinel thus depositing ilmenite

$$Fe_2TiO_4 + \underset{\text{(from melt)}}{TiO_2} \longrightarrow 2\ FeTiO_3.$$

This reaction would lead to a partial depletion of the liquid in TiO_2.

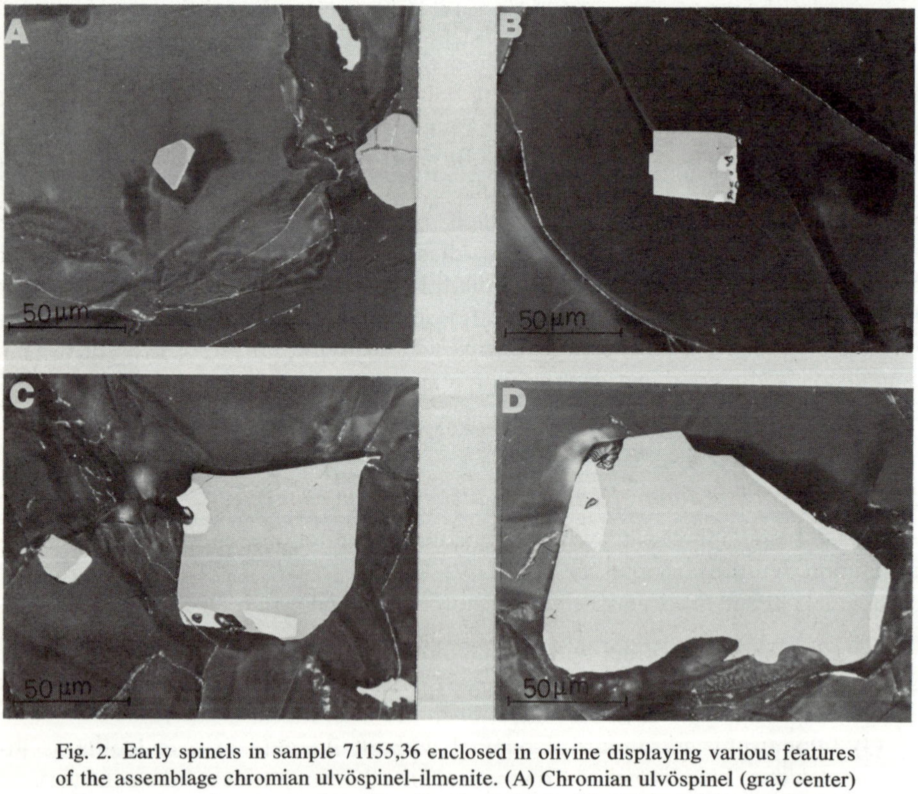

Fig. 2. Early spinels in sample 71155,36 enclosed in olivine displaying various features of the assemblage chromian ulvöspinel–ilmenite. (A) Chromian ulvöspinel (gray center) with a narrow almost complete ilmenite rim (white). (B) Idiomorphic chromian ulvöspinels with rims. (C) Chromian ulvöspinels with partial idiomorphic ilmenite rims. (D) Chromian ulvöspinel with various ilmenite rims occurring on different sides of the grain.

(2) Simultaneous precipitation of chromian ulvöspinel and minor ilmenite. The textures (Fig. 2) are strongly suggestive that both mechanisms were responsible for the formation of this assemblage. Subsolidus reduction is not the responsible mechanism since: (1) Ilmenite formed by subsolidus reduction is always oriented along (111) planes of the chromian ulvöspinel and (2) metallic Fe was never encountered with the ilmenite rims described here, regardless of the amount of ilmenite. Exsolution of ilmenite from a nonstoichiometric ulvöspinel is also a very unlikely mechanism because the exsolution textures are distinctly different from the intergrowth reported. Furthermore, none of the microprobe analyses of chromian ulvöspinel in the studied assemblage give any evidence for non-stoichiometry. Chromian ulvöspinels crystallized later in the sequence and occuring in the interstices usually display large ilmenite rims. Both mechanisms described above are also responsible for this intergrowth although nucleation site of ilmenite around preexisting chromian ulvöspinel was probably the dominating mechanism. Silicates were often found to be trapped between ilmenite and ulvöspinel (Fig. 3). All chromian ulvöspinel with ilmenite rims were found to display reverse zoning features. The reflection color changes gradually from tan in the core to blue or bluish-gray towards the ulvöspinel–ilmenite boundaries (Fig. 3) indicating an enrichment in the chromite molecule. This reverse zoning is restricted to the ilmenite–ulvöspinel boundaries and was never observed at ulvöspinel–silicate contacts. These features strongly suggest that the reverse zoning was formed due to a subsolidus reaction involving both chromian ulvöspinel and ilmenite but not the silicates. Careful inspection of chromian ulvöspinel grains without ilmenite rims revealed no evidence for such reverse zoning. Electron microprobe tracing of zoning trends in numerous spinel grains in various basalts of the two major TiO_2-rich basalt types indicate the presence of two distinct zoning trends characterized by certain coupled oxide variations from core to rim:

(1) Sharp increase in $TiO_2/(TiO_2 + Cr_2O_3 + Al_2O_3)$; slight increase in $FeO/(FeO + MgO)$ (FFM) ratio, and decrease in V_2O_3 content.
(2) Sharp decrease in $TiO_2/(TiO_2 + Cr_2O_3 + Al_2O_3)$.

The first trend is characteristic of primary zoning and is encountered in idiomorphic spinels which are not in any reaction relationship with either armalcolite or ilmenite (Fig. 5). The path of the trend is the same in both plagioclase poikilitic and olivine porphyritic basalts regardless if the spinel crystallized early (enclosed in olivine) or later in the sequence. The most striking feature is that a trend characteristic of sharp decrease in FFM ratio is not found. Such a trend may be deduced from plotting all single analyses points together as shown in Fig. 1. The second trend was encountered in spinels mantled by ilmenite. It is also identical to the zoning trend found in spinels reacted with armalcolite (El Goresy *et al.*, 1974). The major oxide variations of this trend are the same in spinels occurring with ilmenite regardless if they are early or late in the crystallization sequence. However, certain rocks are characterized by specific oxide variations (Figs. 4 and 5). In olivine porphyritic basalt 70215,159 the

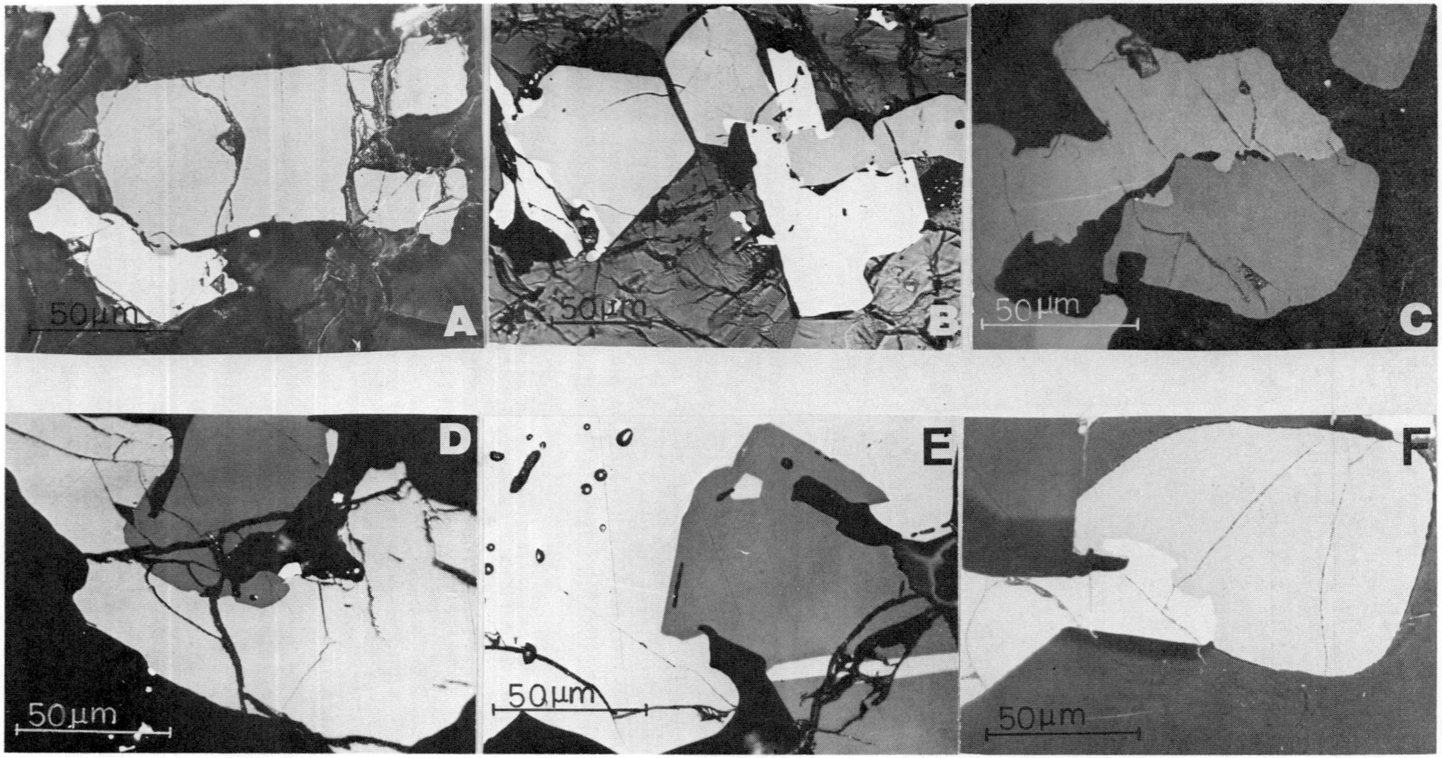

Fig. 3. Younger generations of chromian ulvöspinels in sample 71155,36 displaying various degrees of subsolidus equilibration. (A) Incipient reaction between chromian ulvöspinel (gray, center) and ilmenite (white). Note silicates trapped between ilmenite and ulvöspinel. Several grains of chromian ulvöspinel with various degrees of reaction. (B) Chromian ulvöspinel (gray) with curved surface to ilmenite. Note dark rim at the contact indicating enrichment in chromite. (C) Chromian ulvöspinel with extensive reaction features with ilmenite (white) and dark chromite-rich rim only at the contact to ilmenite. (D) Chromian ulvöspinel with a dark chromite-rich boundary only at the ilmenite (white) contact. (E) and (F) Chromian ulvöspinel (gray) with incipient reaction with ilmenite (white). Notice that all textures shown here, are distinctly different from subsolidus reduction reactions.

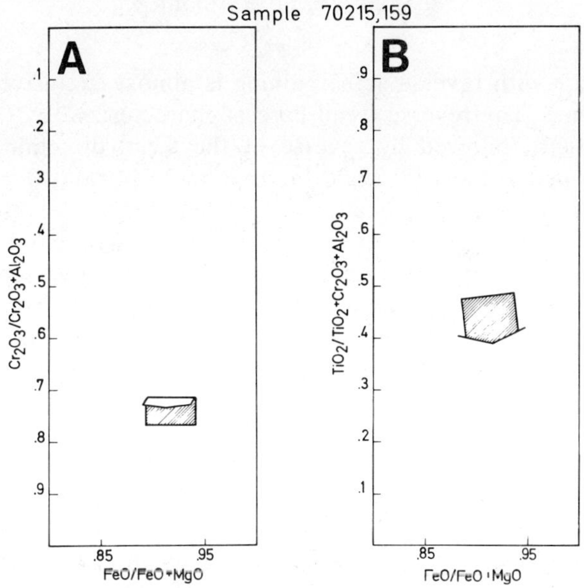

Fig. 4. Zoning trends of equilibrated spinels in sample 70215,159. (A) Base of the prism with the trend indicative of increase in $Cr_2O_3/(Cr_2O_3 + Al_2O_3)$ followed by reverse of the trend. (B) Front face of the prism showing the sharp decrease in $TiO_2/(TiO_2 + Cr_2O_3 + Al_2O_3)$ and slight increase in FFM ratio from core to rim.

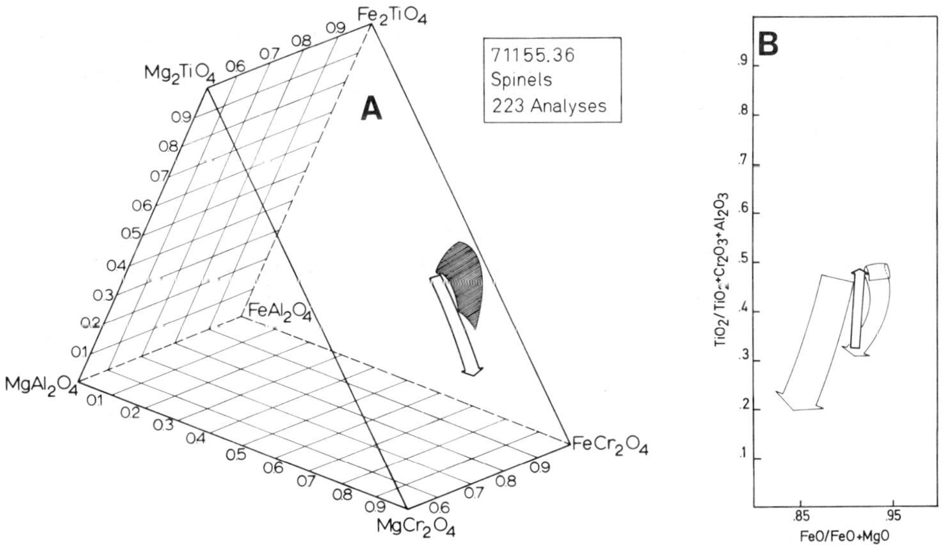

Fig. 5. Zoning relationships in individual zoned spinels in sample 71155,36. (A) Inverted zoning in early spinels due to subsolidus equilibration. Hatched area encloses primary zoning trend and equilibration trend in later spinel generations. (B) Zoning relationships as seen in the front face of the prism. Sharp decrease in $TiO_2/(TiO_2 + Cr_2O_3 + Al_2O_3)$ and decrease in FFM ratios of all equilibrated spinels. Primary zoning trend indicative of sharp increase in $TiO_2/(TiO_2 + Cr_2O_3 + Al_2O_3)$ lies between the two trends of equilibrated spinels.

ilmenite-spinel pair with reverse spinel zoning is almost exclusively present as inclusions in olivine. The reverse trend here is characterized by (1) increase in $Cr_2O_3/(Cr_2O_3 + Al_2O_3)$ followed by reverse of the trend direction close to the spinel-ilmenite boundary, and (2) slight increase in FFM ratio (Fig. 4). Detailed analyses of the host olivine indicate no evidence for spinel-olivine reequilibration. In sample 71155,36 a distinct difference in the coupled oxide substitutions between reversely zoned early spinels and the reversely zoned later spinel generations was found. Early spinels show a drop in FFM ratio from 0.86 to 0.84 along with sharp decrease in $TiO_2/(TiO_2 + Cr_2O_3 + Al_2O_3)$ ratio from 0.46 down to 0.20 (Fig. 5). These variations are coupled with an increase in V_2O_3 content. Later reversely zoned spinels in the same samples show first a slight increase (primary) in $TiO_2/(TiO_2 + Cr_2O_3 + Al_2O_3)$ from 0.47 to 0.50 followed by a sharp decrease in the ratio to 0.30. This variation is coupled with a decrease in FFM ratio and V_2O_3 content. These features with all its varieties are indicative of a subsolidus equilibration process *not induced by subsolidus reduction processes.* Part of the ilmenite may have formed by reaction between the basaltic liquid and the spinel. Hence some of the reverse features could have resulted from liquid solid reactions. However, the majority of the reverse zoning, specifically in later spinel generations in sample 71155,36, is almost certainly the product of subsolidus equilibration.

DISCUSSION

Two important conclusions could be drawn from the present investigations: (1) Subsolidus equilibration between ilmenite and chromian ulvöspinel may partially or even completely erase the primary zoning relationship of spinels established during crystallization. (2) Subsolidus reduction is not the only process responsible for reverse zoning in chromian ulvöspinel. For this reason, subsolidus reduction is not the only mechanism responsible for *reverse zoning trends in reduced spinels.* By the first appearance of ilmenite formed by (1) subsolidus reduction, (2) reaction between spinel and basaltic liquid, or (3) precipitation from the liquid around spinel, equilibration between spinel and ilmenite will start. Enrichment of the host spinel in chromite will take place and the coexisting ilmenite will be depleted in Cr_2O_3. After termination of the reduction process subsolidus equilibration between ilmenite and the spinel host will continue. Hence the reverse zoning may be primarily the result of subsolidus equilibration rather than the reduction processes alone. In a recent study of Apollo 14 sample 14072 Haggerty (1977) concluded that the lower limit of chromian ulvöspinel reduction is ~47 mole% Fe_2TiO_4. This conclusion appears to be in conflict with the findings of El Goresy *et al.* (1972) that the lower limit for spinel composition formed by subsolidus reduction in sample 14072 (Fig. 6d of El Goresy *et al.*, 1972) is Chr_{68}–Usp_{32}. The present findings, however, strongly suggest that such a lower limit is mainly the result of subsolidus equilibration between ilmenite and the host spinel. The lower limit of chromian ulvöspinel only due to the reduction processes at the prevailing f_{O_2} and temperature may have a higher Fe_2TiO_4 content than given by Haggerty (1977).

The present findings have recently gained ample support through comprehensive experimental studies in the system FeO–TiO_2–Cr_2O_3 in coexistence with metallic Fe° (Knecht *et al.*, 1977). Knecht *et al.* (1977) report that the slope of the spinel-ilmenite boundary of the spinel + ilmenite + ferropseudobrookite (or rutile) triangle is continuously tilted in a counterclockwise direction with decreasing temperature (Fig. 9 of Knecht *et al.*, 1977). The spinel becomes enriched in chromite molecule whereas ilmenite is simultaneously depleted in Cr_2O_3 as a result of subsolidus equilibration. This equilibration process extends even after disappearance of chromian pseudobrookite (<1038°C) leading to further enrichment of the spinel in the chromite molecule and depletion of ilmenite in Cr_2O_3. Hence, we conclude that subsolidus equilibration between chromian ulvöspinel and ilmenite best explains the reverse zoning features observed in lunar chromian ulvöspinels. We also propose that this process plays a major role during and after subsolidus reduction reactions.

Cationic relationships and substitutions

Tracing cation substitutional trends in zoned individual primary spinel is a powerful tool in delineating the crystallization sequence of mare basalts (El Goresy *et al.*, 1976). In TiO_2-rich basalts analyses of the cation substitutional trends of spinels displaying the two zoning trends discussed above should reveal: (1) Information about the chemical changes which took place in the cooling basaltic liquid during growth of the spinels, and (2) cation substitutional changes which took place during subsolidus equilibration between chromian ulvöspinel and coexisting ilmenite. The substitutional relationship among the trivalent cations (Cr, Al, V) is of special importance in view of the recent experimental studies. Investigations in the system FeO–TiO_2–Cr_2O_3 in equilibrium with metallic Fe° (Knecht *et al.*, 1977) indicate that the partitioning of Cr between spinel and ilmenite is temperature-dependent and the chemistry of the coexisting spinel-ilmenite pair is a potential lunar thermometer. Application of the experimental results on lunar rocks to deduce equilibration temperatures is only possible if the partitioning behavior of Al between spinel and ilmenite in the system FeO–TiO_2–Al_2O_3 is similar to the partitioning of Cr between the same phases in the system FeO–TiO_2–Cr_2O_3. According to Usselman (1975) the effects of MgO and Al_2O_3 on the spinel trend in FeO–TiO_2–Cr_2O_3–Fe system appear to cancel one another. However, the relationship will become more complex due to changes in the stability of the phase assemblages. The application of chromian ulvöspinel and ilmenite as lunar thermometer will be a subject of a later report (Woermann and El Goresy, 1977).

Cr–Al substitutional trends

The two chemical zoning trends encountered in Apollo 17 chromian ulvöspinels are also well developed in the Cr–Al substitutional trends. Subsolidus equilibration between chromian ulvöspinel and ilmenite may only partially erase the primary zoning features in some rocks, whereas the primary zoning patterns may be completely erased in others. Substitutional relationships in equilibrated early chromian ulvöspinels in sample 70215,159 indicate preservation of the primary substitutional trend in the spinel cores only (Fig. 6). The trend indicates a descent to lower Cr and Al concentration along the 2 : 1 Cr/Al ratio line. It then turns back toward higher concentrations with a slight shift to higher Cr/Al ratios. The deflection in the substitutional trend is interpreted as due to subsolidus equilibration with the ilmenite rim. Evidently, the partitioning behavior of Cr and Al are not quite similar in this particular assemblage in 70215. In 71155,36 three distinct substitutional relationships were encountered: (1) A substitutional trend indicative of increase in Cr and Al concentrations from core to rim accompanied by continuous decrease of the Cr/Al ratio from a ratio close to 2 : 1 to a ratio lying between 2:1 and 1:1 (Fig. 7a). This trend is characteristic of the early spinels equilibrated with ilmenite. Surprisingly, the Cr and Al behavior in early spinels is different from its behavior in 70215. Perhaps this reflects different cooling rates. (2) The second substitutional trend is restricted to spinels with primary zoning features. It is indicative of decrease in Cr and Al concentrations with an increase of the Cr/Al ratio from 2:1 in the cores to 3:1 in the rims (Fig. 7b). This trend is indicative of simultaneous crystallization of pyroxene and plagioclase. (3) The third trend was encountered in later chromian ulvöspinel generations which equilibrated with ilmenite. It starts at a Cr/Al ratio of 3:1 and takes a slight descent along this line, before it takes a complete deflection of the trend towards higher Cr and Al concentrations, with no evidence for substantial change in the Cr/Al ratio (Fig. 7c). In those spinels subsolidus equilibration did

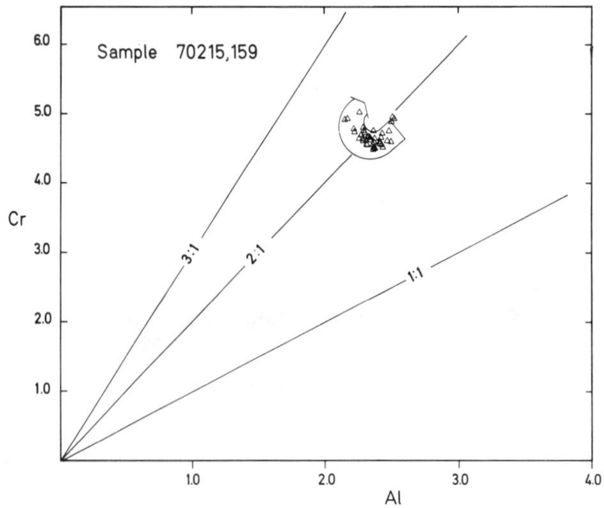

Fig. 6. Cr–Al substitutional relationships of equilibrated spinels in sample 70215,159. Note reverse of the trend due to subsolidus equilibration.

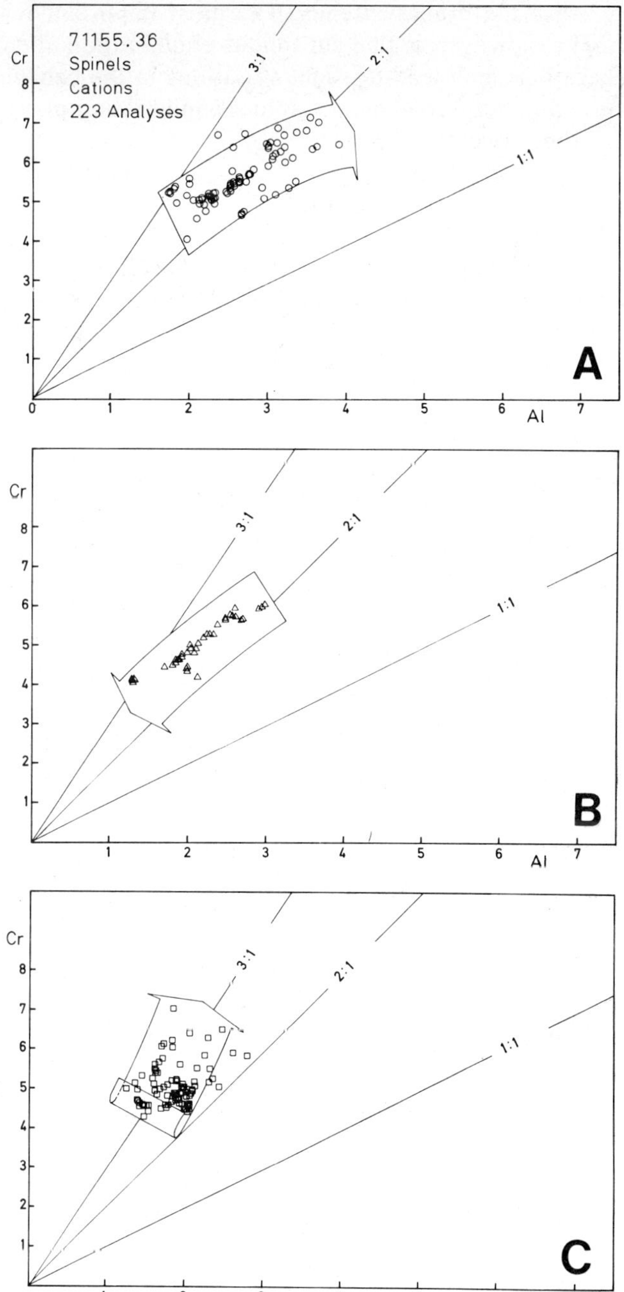

Fig. 7. Cr–Al substitutional relationship of equilibrated and primary spinels in sample 71155,36. (A) Substitutional trend for early equilibrated spinels. (B) Substitutional trend for primary zoning. (C) Substitutional trend for equilibrated spinels of younger generations.

not completely erase the primary trends. The most important feature deduced from Figs. 7a and c, however, is that subsolidus equilibration of early spinel and later spinel generations may lead to slight variations in the behaviors of Cr and Al. Perhaps the temperature of ilmenite-spinel equilibration plays a key role in the behaviors of these two elements.

V–Cr and V–Al substitutional trends

The three substitutional trends encountered in the Cr–Al substitutional plots were also well developed in the V–Cr and V–Al relationships (Fig. 8). Vanadium behaves like Cr and Al during subsolidus equilibration beween chromian ul-

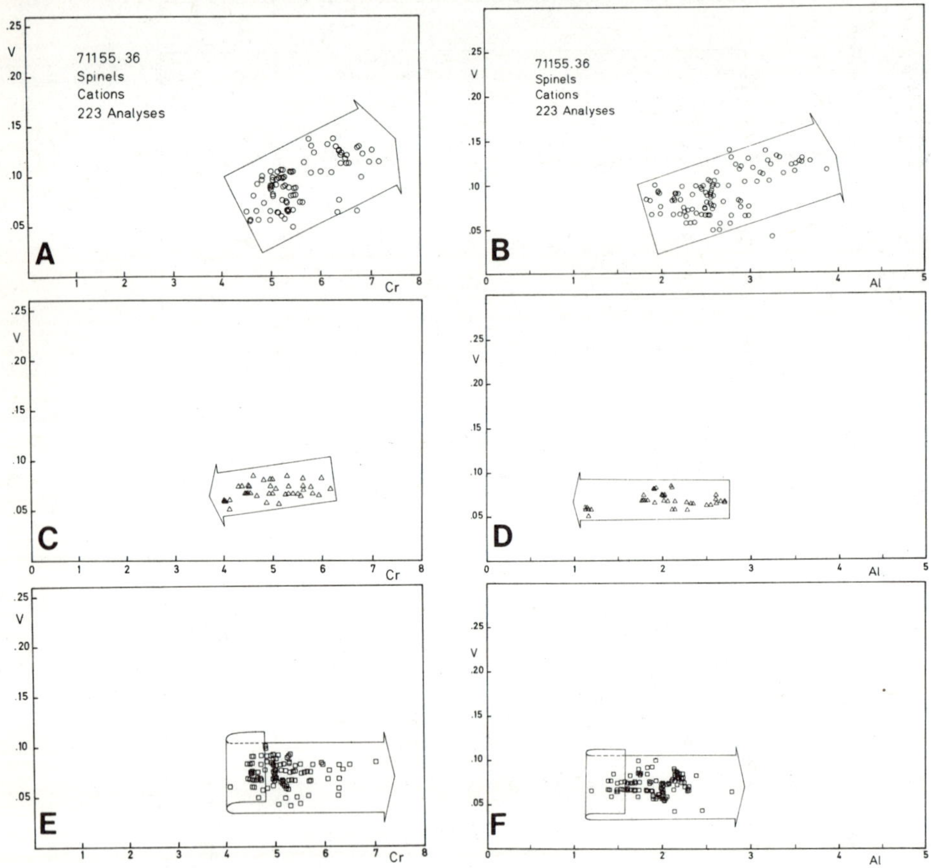

Fig. 8. V–Cr and V–Al substitutional trends of spinels in sample 71155, 36. (A) and (B) V–Cr and V–Al substitutional trends for early equilibrated spinels indicative of increase of the trivalent cations from core to rim. (C) and (D) Substitutional relationship for primary spinels showing decrease of all trivalent cations from core to rim. (E) and (F) Substitutional relationships for equilibrated later spinels indicating reversal of the trend thus suggesting that the primary zoning trend was not completely erased.

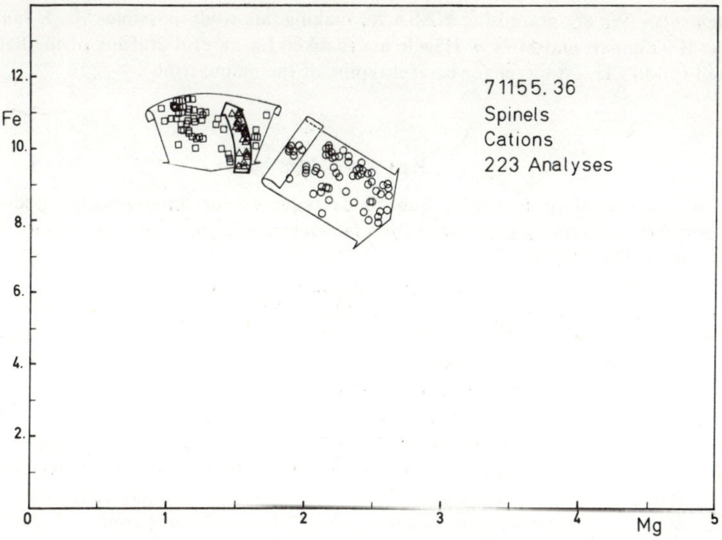

Fig. 9. Fe–Mg substitutional relationship for equilibrated early spinels (circles), primary spinels (triangles), and later equilibrated spinels (squares) in sample 71155,36.

vöspinel and ilmenite. This behavior is quite evident in the early equilibrated spinels (Figs. 8a and b) where V increases upon increase of Cr and Al from core to rim. Due to the low V content in the later equilibrated spinel generation, the V–Cr and V–Al relationships are not quite clear. No evidence for increase of the V content upon increase of Cr and Al contents was found (Figs. 8e and f). During growth of spinel from the basaltic liquid, V, Cr, and Al sympathetically decrease from core to rim (Figs. 8c and d). This suggests that all three elements are substituting for Ti^{4+}. The general substitutional behavior of the trivalent cations is in agreement with the predictions of Knecht *et al.* (1977) thus probably allowing the application of the equilibrated oxide pair as a lunar thermometer.

Fe–Mg substitutional trends

The Fe–Mg substitutional relationships of: (1) early equilibrated chromian ulvöspinels, (2) late equilibrated chromian ulvöspinels, and (3) primary zoning in chromian ulvöspinels in sample 71155,36 are shown in Fig. 9. The substitutional trend of the early equilibrated spinels is indicative of an antipathetic Fe–Mg relationship characterized by increase of Fe with decrease of Mg, followed by reversal of the trend towards higher Mg and lower Fe values from core to rim. A similar trend was also found for later equilibrated spinels although the slope of the trend is much steeper. Primary Fe–Mg substitutional trend developed during growth of spinels from the liquid is indicative of an antipathetic Fe–Mg relationship towards higher Fe and lower Mg values from core to rim.

Acknowledgments—We are grateful to NASA for making this study possible. Mr. J. Janicke, Mr. H. Weber, Mrs. H Träumer, and Miss S. Hägele are thanked for careful drafting of the figures. We are also indebted to Miss H. Urmitzer for careful typing of the manuscript.

REFERENCES

El Goresy A. and Ramdohr P. (1975) Subsolidus reduction of lunar opaque oxides: Textures, assemblages, geochemistry, and evidence for a late-stage endogenic gaseous mixture. *Proc. Lunar Sci. Conf. 6th*, p. 729–745.

El Goresy A. and Ramdohr P. (1976) Zoning in spinels as an indicator of the crystallization histories of mare basalts. *Proc. Lunar Sci. Conf. 7th*, p. 1261–1279.

El Gorsey A., Ramdohr P., and Medenbach O. (1973) The opaque minerals in Apollo 17 samples (abstract). *EOS* (*Trans. Amer. Geophys. Union*) **54**, 591–592.

El Goresy A., Ramdohr P., Medenbach O., and Bernhardt H.-J. (1974) Taurus-Littrow TiO_2-rich basalts: Opaque mineralogy and geochemistry. *Proc. Lunar Sci. Conf. 5th*, p. 627–652.

El Goresy A., Taylor L. A., and Ramdohr P. (1972) Fra Mauro crystalline rocks: Mineralogy, geochemistry and subsolidus reduction of opaque minerals. *Proc. Lunar Sci. Conf. 3rd*, p. 333–349.

Haggerty S. (1973) Apollo 17: Armalcolite paragenesis and subsolidus reduction of chromian-ulvöspinel and chromian picroilmenite (abstract). *EOS* (*Trans. Amer. Geophys. Union*) **54**, 593–594.

Haggerty S. (1977) Apollo 14: Oxide, metal and olivine mineral chemistries in 14072 with a bearing in the temporal relationships of subsolidus reduction (abstract). In *Lunar Science VIII*, p. 392–394. The Lunar Science Institute, Houston.

Knecht B., Simons B., Woerman E., and El Goresy A. (1977) Phase relations in the system Fe–Cr–Ti–O and their application in lunar thermometry. *Proc. Lunar Sci. Conf. 8th*. This volume.

Sato M. (1976) Oxygen fugacity and other thermochemical parameters of Apollo 17 high-Ti basalts and their implications on the reduction mechanism. *Proc. Lunar Sci. Conf. 7th*, p. 1323–1344.

Usselman T. M. (1975) Ilmenite chemistry in mare basalts, an experimental study. In *Papers Presented to the Conference on Origins of Mare Basalts and Their Implications for Lunar Evolution*, p. 164–168. The Lunar Science Institute, Houston.

Usselman T. M. and Lofgren G. E. (1976) The phase relations, textures, and mineral chemistries of high-titanium mare basalts as a function of oxygen fugacity and cooling rate. *Proc. Lunar Sci. Conf. 7th*, p. 1345–1363.

Woermann E. and El Goresy A. (1977) The assemblage chromian ulvöspinel-ilmenite: a new lunar geothermometer. EPSL. In preparation.

Proc. Lunar Sci. Conf. 8th (1977), p. 1625–1638.
Printed in the United States of America

Cooling rates for lunar samples determined with a diffusion model for phosphide exsolution

R. H. HEWINS

Department of Geological Sciences, Rutgers University, New Brunswick, New Jersey 08903

J. I. GOLDSTEIN

Department of Metallurgy and Materials Science, Lehigh University, Bethlehem, Pennsylvania 18015

Abstract—A numerical model for diffusion-controlled phase growth has been applied to the exsolution of phosphide lamellae in lunar metal grains. Computer simulations reproduce observed composition profiles, and reveal the influence of cooling rate on dimensional and compositional parameters of phosphide and metal. At lower cooling rates, phosphide lamellae are larger and the concentration of P in the metal host close to the interface is lower. Cooling rates inferred for Apollo 16 samples, based on compositions and dimensions of the phosphide-metal grains, are mostly in the range 1–100°C/day. These rates correspond to burial depths of 5–0.5 m for melt rocks and 3–0.3 m for breccias. This is in good agreement with thicknesses of lunar cooling units determined by other techniques.

INTRODUCTION

COOLING RATE IS A FACTOR which strongly influences the textures and mineralogical details of lunar rocks. Attempts have been made to determine cooling rates from crystal morphology (Lofgren *et al.*, 1975) and opaque oxide compositions (Taylor *et al.*, 1975). Some lunar rocks contain metallic grains consisting of lamellae of phosphide (schreibersite) exsolved from kamacite (alpha Fe-Ni). Since the Fe–Ni–P system is well known (Doan and Goldstein, 1970; Heyward and Goldstein, 1973; Norkiewicz and Goldstein, 1975), a mathematical model for diffusion-controlled phase growth can be used to simulate this exsolution. Calculations show that compositions and dimensions of phosphide lamellae and their host metal grains are strongly influenced by cooling rate. In this paper, cooling rates for lunar samples are obtained by matching measured phosphide data with calculated values.

COMPUTER SIMULATION OF PHOSPHIDE EXSOLUTION

A numerical model for diffusion-controlled phosphide growth in kamacite in the ternary system Fe–Ni–P has been developed by Randich (1975a). The assumptions of this model and its application are further discussed by Randich and Goldstein (1975) and Goldstein and Randich (1977). The temperature variation of two- and three-phase boundaries of the Fe–Ni–P phase diagram (Doan and Goldstein, 1970) and of diffusion coefficients (Heyward and Goldstein, 1973; Norkiewicz and Goldstein, 1975) is incorporated in a computer simulation program. Exsolution of a phosphide lamella may be simulated by providing the computer program with data such as bulk composition, initial and final transformation temperature, and cooling rate. The program calculates a composition profile in a discontinuous fashion, as temperature decreases and time increases. Final interface compositions and lamella size are the important parameters for interpreting natural phosphides cooled at unknown rates.

The basic assumptions in the computer program HEX (Randich, 1975b) are that interface equilibrium is maintained at all times and that phase growth is controlled by diffusion. Plane front, one-dimensional growth is considered. The program was designed for an investigation of phosphide exsolution in hexahedrites (Randich, 1975a; Randich and Goldstein, 1977). These meteorites apparently formed at the very low cooling rates (~0.5–25°C/m.y.) associated with the interiors of asteroidal bodies. HEX therefore specifically assumes that bulk equilibrium is maintained in the metal-phosphide assemblage down to moderately low temperatures (~400°C). Clearly this assumption may not be valid for the higher cooling rates experienced near or on planetary surfaces. Therefore, for lunar samples, a modified version of HEX was written without the assumption of bulk equilibrium. In the modified program, a phosphide nucleus (halfwidth 0.01 μ) is assumed to form after a small amount of undercooling (~50°C). The fundamental assumptions of interface equilibrium and diffusion-controlled phase growth are maintained in the modified program.

The general characteristics of the exsolution process under lunar conditions can be illustrated using a bulk composition of 5.5 wt.% Ni, 0.5 wt.% P, and 94.0 wt.% Fe in the program. Figure 1 shows composition profiles for Ni and P for phosphide growing in kamacite as a function of cooling rate after cooling to 400°C. The effect of lowering the cooling rate is to increase the size (halfwidth) of the lamella. This lamella size is, potentially at least, a measure of cooling rate. Note that Ni and P are partitioned into the phosphide lamella and the metal phase is depleted in these elements close to the interface. As the cooling rate is lowered, the metal phase as a whole is more strongly depleted in P which goes to form the phosphide. In addition, at lower cooling rates the composition gradients become less steep, indicating lesser departure from equilibrium.

The influence of final growth temperature on composition profiles is shown in Fig. 2. At 500°C the composition gradients are very small, indicating very slight departures from equilibrium. Note that on cooling to lower temperatures, the additional growth of the lamella is negligible. However, the interface depletion of Ni and P in the metal is marked, as required by the phase diagram. On cooling from 400°C to 250°C, the diffusion rates are so low that the composition profile is changed only in the immediate vicinity of the interface.

Interface compositions are commonly used to indicate final temperatures of equilibration. The spatial resolution of the conventional electron probe is such that, in the Fe–Ni system, analyses characteristic of a single phase can be obtained to within about 1 μ of a grain boundary (Miyake and Goldstein, 1974). When data are collected on a traverse with 1 μ steps, the analyses closest to the interface but still representative of only one grain are, in general, about 1.5 μ

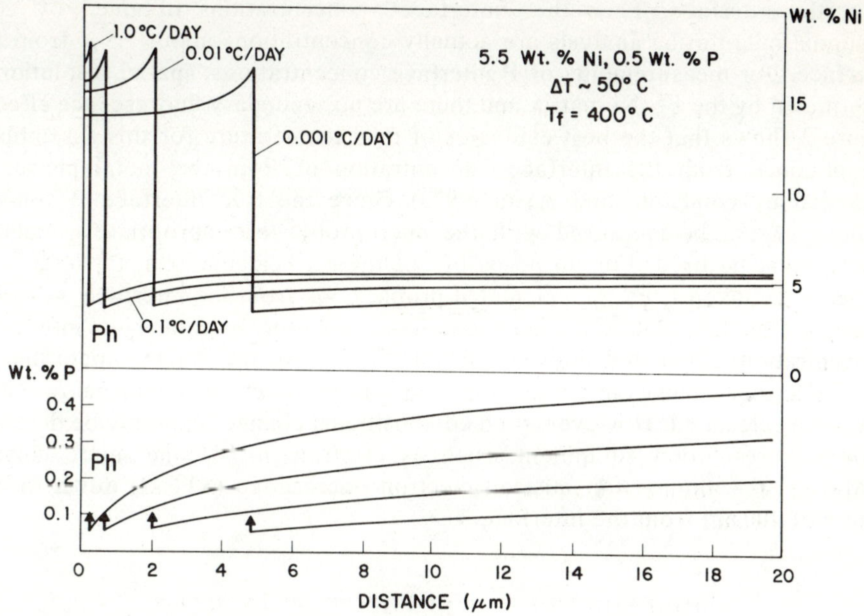

Fig. 1. Concentration profiles for Ni and P across the interface between a phosphide lamella and host metal grain, as a function of cooling rate.

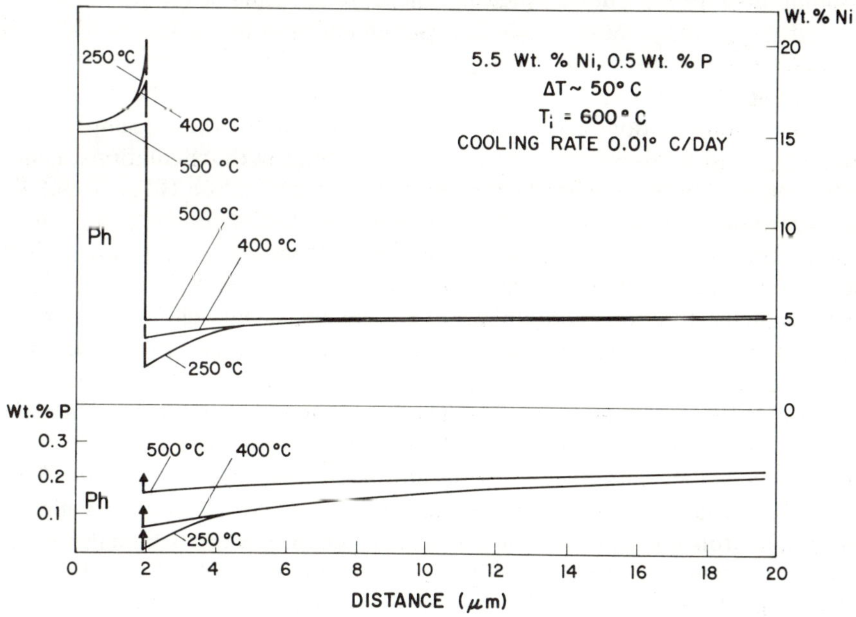

Fig. 2. Concentration profiles for Ni and P across the interface between a phosphide lamella and host metal grain, as a function of final temperature.

from the interface. Thus the "interface" concentrations obtained by conventional microprobe analysis are actually concentrations about 1.5 μ from the interface. For measurements of P interface concentrations, spatial resolution is determined by the Fe–Ni matrix and there are no secondary fluorescence effects. Figure 2 shows that the best estimates of final temperature for this assemblage are obtained from the interface concentration of P in the metal phase, as indicated by Goldstein and Axon (1973). Since the true interface P concentrations cannot be measured with the microprobe, the appropriate calculated values must be used. For purposes of cooling rate calculations, P "interface" values were taken from the calculated profile 1.5 μ from the interface, in metal. These are the data plotted for interpretation of microprobe data. The "interface" concentrations calculated show significant changes down to final temperatures of 250°C. Further changes in composition close to the interface occur below 250°C and were calculated. However, such compositional changes can only be detected with high resolution equipment, such as electron microscope microanalyzer (EMMA) or scanning transmission electron microscope (STEM) down to distances of 100 nm from the interface.

SIMULATION OF LUNAR PHOSPHIDE LAMELLAE

The bulk composition of lunar metal-phosphide grains is variable. Published analyses of such two-phase grains concentrated from lunar soil provide a good indication of the range in bulk compositions. Bulk compositions for metal phosphide in soil 63501 determined by microprobe in the scanning mode (Goldstein and Axon, 1973) show a typical range and are plotted in Fig. 3. The sample is bimodal in terms of P. A large group of grains has less than 1% P and a smaller group contains several percent P. Most grains contain between 4% and 7% Ni. Two compositions have been chosen to represent these groups, 5.5% Ni with 0.5% P and 5.5% Ni with 2.0% P. Phosphide growth calculations using the low P composition have been discussed in the earlier section (Figs. 1 and 2).

In addition to bulk composition, cooling rate, initial temperature and final temperature, the diffusion field length is an important variable in the simulation program. The diffusion field length is the distance from which P is drawn into the growing lamella. In grains containing several lamellae, the field length is half the distance between growing phosphides. Figure 4 shows several photomicrographs of lamellae in metal-phosphide particles. In all of these, the diffusion field length is half the distance between lamellae. The field length will determine the amount of P available for growth, and will influence the size of the lamella that can be grown. Under some circumstances the diffusion field length will influence the composition gradients and therefore the "interface" compositions as measured with the electron microprobe. Values of field length (i.e., half lamellar spacing) observed in lunar samples range from 10 to 100 μ, while the full width of observed lamellae ranges from much less than 1 μ to at most 10 μ. These values may be confirmed from recorded observations (Fig. 4; Figs. 5 and 7 in Goldstein and Axon, 1973; Fig. 6 in Gooley *et al.*, 1973; Fig. 4 in Misra and Taylor, 1975).

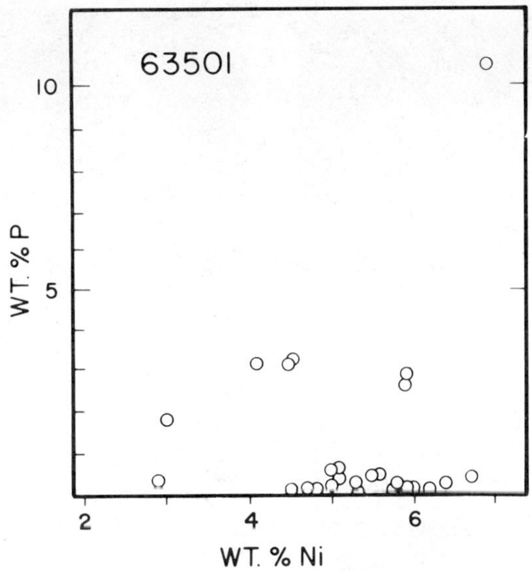

Fig. 3. Bulk compositions of all two-phase metal-phosphide particles separated from soil 63501 (Goldstein and Axon, 1973).

The influence of diffusion field length on lamella size (halfwidth) is illustrated in Fig. 5. The computer runs were made for a bulk composition of 0.5% P, a nucleation temperature of 600°C and field lengths of 10, 25, 50, 100 μ, and higher values. Although the sizes of lamellae are essentially fixed at high temperatures, as shown in the previous section, the halfwidths plotted were obtained from runs with a final temperature 250°C.

Figure 5 shows that lower diffusion field lengths give smaller lamellae, since the amount of P available influences the size to which the phosphide can grow. It also confirms the earlier observation that lower cooling rates give larger lamellae. Note that the curve for a cooling rate of 0.001°C/day falls quite close to the calculated line for infinitely low cooling rate. The case of an infinitely low cooling rate is attained when all the possible P has diffused into the phosphide. This condition yields the maximum possible lamella halfwidth.

Several important conclusions can be drawn from Fig. 5. First, for grains of metal phosphide containing an average of 0.5% P, the lamella halfwidth and diffusion field length uniquely define the cooling rate. Second, if the lamellae attained a total width in excess of 1 μ, cooling rates of 0.1°C/day or lower were necessary. Smaller phosphides cooled more quickly and these phosphides cannot be investigated with the microprobe.

Figure 5 can also be used to illustrate the lamella sizes of metal-phosphide grains with 2% P. In the case of P-rich grains, exsolution begins at very high temperatures (Doan and Goldstein, 1970) and a close approach to equilibrium is maintained down to moderate temperatures for all, even high, cooling rates. Runs were made with an equilibrium version of the computer program to obtain

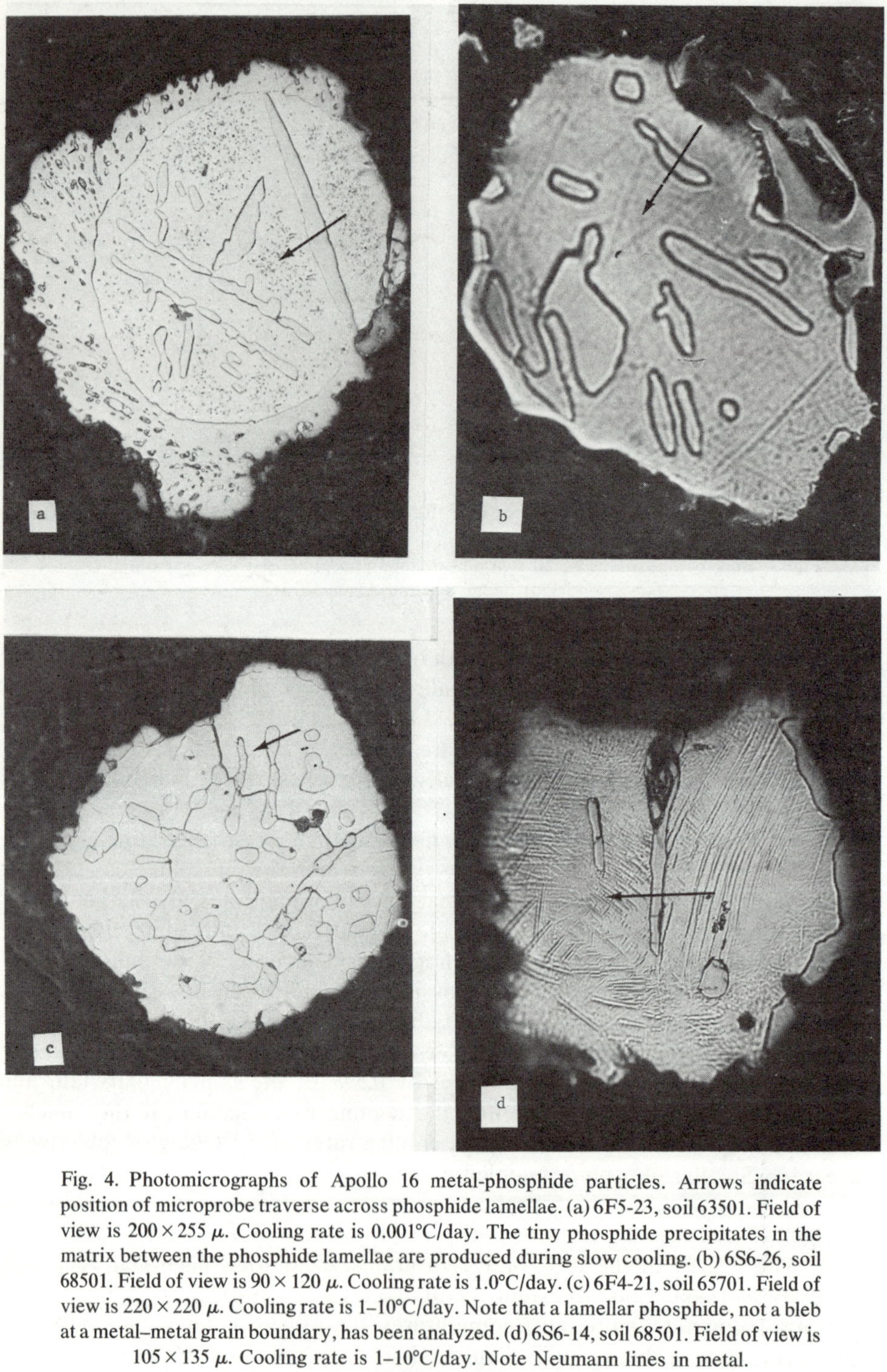

Fig. 4. Photomicrographs of Apollo 16 metal-phosphide particles. Arrows indicate position of microprobe traverse across phosphide lamellae. (a) 6F5-23, soil 63501. Field of view is 200 × 255 μ. Cooling rate is 0.001°C/day. The tiny phosphide precipitates in the matrix between the phosphide lamellae are produced during slow cooling. (b) 6S6-26, soil 68501. Field of view is 90 × 120 μ. Cooling rate is 1.0°C/day. (c) 6F4-21, soil 65701. Field of view is 220 × 220 μ. Cooling rate is 1–10°C/day. Note that a lamellar phosphide, not a bleb at a metal–metal grain boundary, has been analyzed. (d) 6S6-14, soil 68501. Field of view is 105 × 135 μ. Cooling rate is 1–10°C/day. Note Neumann lines in metal.

Table 1. Compositional and dimensional data for phosphide lamellae and host metal.

Sample*	Par-ticle #	Half-width microns	Field length microns	Bulk P wt.%	"Interface" P	Cooling rate °C/day
6F4	21	2.1	9	2.0	0.18	>1.0, <10.0
6F5	23	3.0	23	2.0	0.05	~0.001
6S6	14	1.5	12	2.0	0.20	>1.0, <10.0
6S6	26	1.6	10	2.0	0.16	~1.0
6F2	1	3.2	18	2.0	0.60	>100.0
6F2	2	2.5	40	2.0	0.36	~100.0
6F2	18	3.2	32	1.0	0.18	~1.0

*6F4, soil 65701; 6S6, soil 68501; 6F2, soil 63501; 6F5, soil 63501 (as described in Goldstein and Axon, 1973).

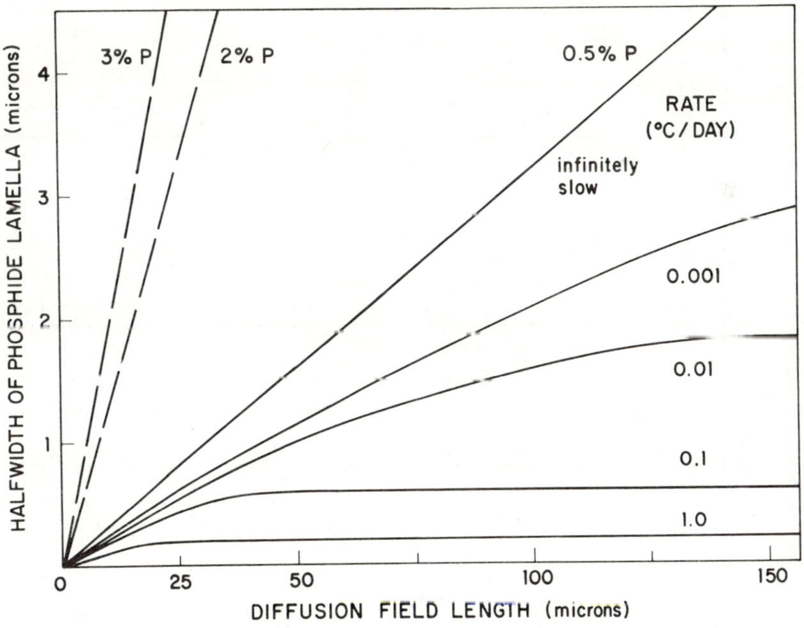

Fig. 5. Size (halfwidth) of phosphide lamellae grown in metal as a function of cooling rate and diffusion field length. These computer simulations were run for a bulk composition of 0.5 wt.% P and 5.5 wt.% Ni and a temperature of final equilibration of 250°C. For P contents of 2–3 wt.%, growth at high temperature is so extensive that lamella size for any cooling rate is given by the infinitely slow case.

composition information. These runs confirm that, even for high cooling rates of 1.0°C/day, the lamella sizes plot essentially on the curve for infinitely low cooling rate. A curve showing lamella size for grains with 3% P at an infinitely low cooling rate is also plotted in Fig. 5. Note that for P-rich grains, no information about cooling rate can be gathered from the lamella size.

The calculated "interface" P concentrations have been plotted in Fig. 6 and following figures. Note that the P number used is the calculated P concentration at 1.5 μ from metal-phosphide interface. This value is that P concentration which can be measured with the microprobe. In Fig. 6, the "interface" composition is shown for a final temperature of 250°C, as a function of cooling rate and diffusion field length. The true interface P concentration is 0.005 wt.% P at 250°C and is independent of the cooling rate and field length. The calculated "interface" compositions plotted in Fig. 6 are seen to be strongly dependent on cooling rate. At lower cooling rates, there is a close approach to equilibrium and consequently the "interface" P concentration is lower. Calculations with 5.5 wt.% Ni, 2.0 wt.% P alloys show that the "interface" composition is essentially the same as for 0.5 wt.% P for the same cooling rate. As shown in Fig. 6, diffusion field length has a small influence on "interface" composition at field

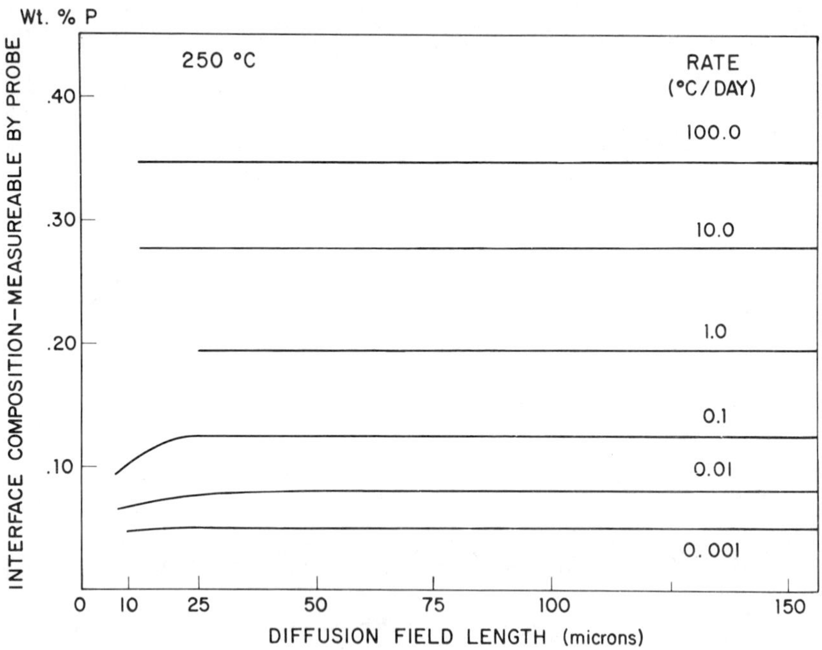

Fig. 6. P concentration at 1.5 μ from the interface (the "interface" composition measurable by electron microprobe) at a final temperature of 250°C as a function of cooling rate. These curves were calculated for a bulk composition of 5.5 wt.% Ni and 0.5 wt.% P. The curves for a composition of 5.5 wt.% Ni and 2.0 wt.% P fall very close to these.

lengths greater than 25 μ. Figure 6 shows that for all metal-phosphide grains (0.5–2.0 wt.% P) that had a singular cooling history to low temperature, the cooling rate can be obtained using measured "interface" compositions and diffusion field lengths. This cooling rate should also match, at the 0.5 wt.% P composition, the estimate of cooling rate obtained from lamella size (Fig. 5). If no match occurs, this may mean that the cooling pattern was interrupted by excavation. Ejection of material at moderate temperature by meteorite impact would have caused rapid quenching.

"Interface" P contents of the metal phase obtained at 400°C and 500°C are shown in Figs. 7 and 8 respectively. On comparing P concentrations for 250°C and 400°C (Figs. 6 and 7) two points become apparent. For low cooling rates, "interface" P concentrations are lower at the lower temperature because of greater approach to equilibrium. The true equilibrium interface composition at 400°C is 0.06 wt.% P. However, the "interface" P concentrations at the higher cooling rates ($\geqslant$0.1°C/day) do not change on going from 400°C to 250°C. This may be readily understood by reference to Fig. 2. At low temperatures, diffusion can modify compositions only immediately adjacent to the true interface. For a high cooling rate, little low temperature modification will occur. Thus for high cooling rates, the point at which the microprobe is measuring "interface" composition is too far from the true interface to detect the changes going on.

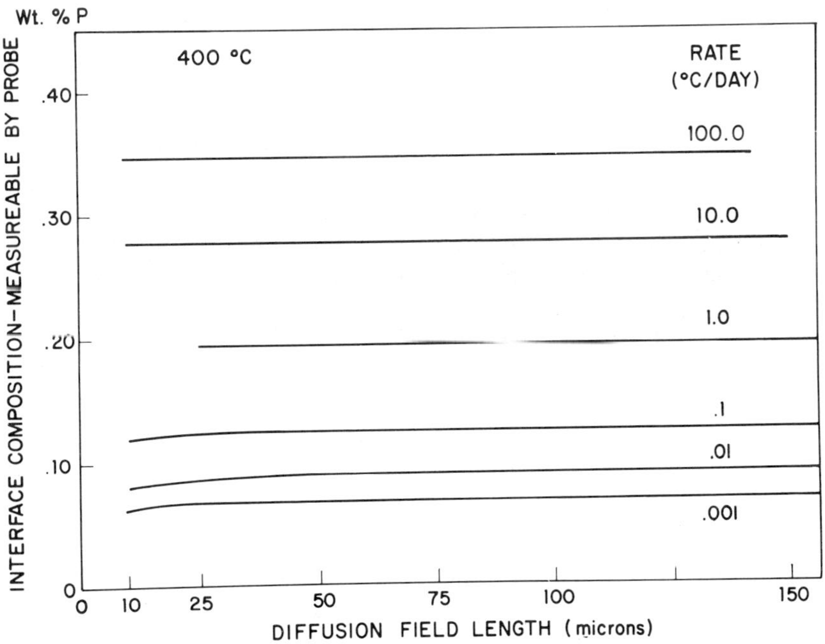

Fig. 7. "Interface" P concentration of the metal phase at a final temperature of 400°C, as a function of cooling rate. Curves for bulk compositions of 0.5 wt.% P, 5.5 wt.% Ni (shown) and 2.0 wt.% P, 5.5 wt.% Ni are almost identical.

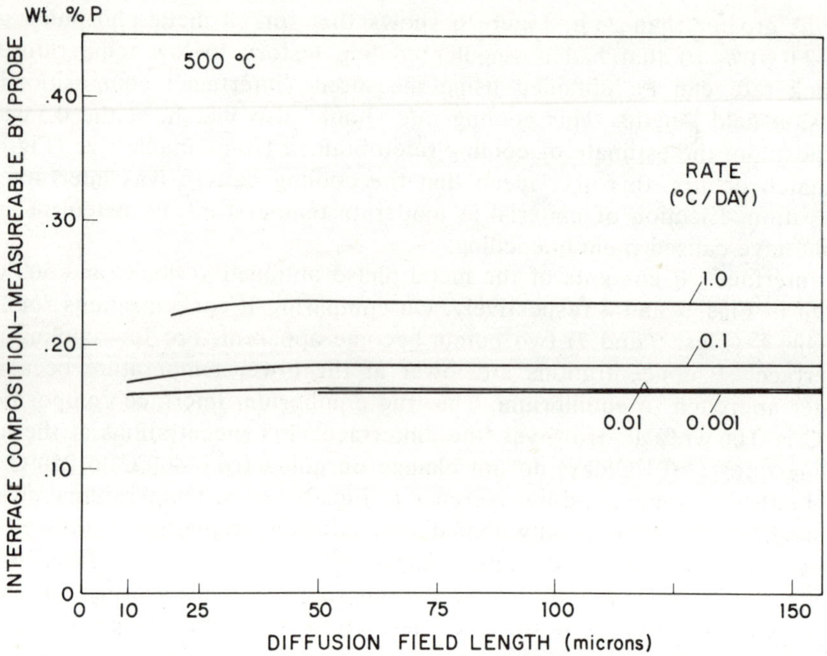

Fig. 8. "Interface" P concentration for a bulk composition of 0.5 wt.% P, 5.5 wt.% Ni, and a final temperature of 500°C, as a function of cooling rate.

"Interface" compositions at 500°C are shown in Fig. 8. All the curves are significantly higher than the lower temperature curves, which is consistent with Fig. 2. Note that at 500°C, the compositions for rates of 0.01°C/day and 0.001°C/day virtually coincide. This indicates a very close approach to equilibrium is maintained down to 500°C, which again is consistent with the nearly flat profiles of Fig. 2. The true interface concentration is approximately 0.16 wt.% P at 500°C.

For a metal-phosphide grain containing 0.5 wt.% P, the cooling rate may be obtained from the lamella size (Fig. 5) and then the temperature of excavation or effective quenching may be estimated from "interface" composition (Figs. 6, 7, 8). For P-rich grains (>0.5 wt.% P) such unique solutions may not be possible. However, in most cases, constraints can be placed on the cooling rate and effective final temperature from the "interface" composition.

Cooling History of Lunar Phosphides

Metal-phosphide grains are found in a number of lunar rocks, particularly impact melts and breccias. In many cases, however, the phosphides occur as equant blebs at grain boundaries rather than as the lamellae for which this diffusion model was developed. Suitable lamellar phosphides are observed in

Apollo 16 melt rocks and breccias (e.g., Misra and Taylor, 1975). Similar phosphides in large metal grains, presumably derived from similar rocks, have been magnetically concentrated from Apollo soils 63501, 65701 and 68501 (Goldstein and Axon, 1973). Compositions typical of these grains (0.5 wt.% P, 5.5 wt.% Ni and 2.0 wt.% P, 5.5 wt.% Ni) have been used in these simulations. Therefore, compositional and dimensional measurements can be taken from these grains to obtain cooling rate information, as long as composition profiles indicate growth during cooling. Grains with reversed P gradients indicating dissolution on reheating were rejected. The data for accepted grains are listed in Table 1, along with the calculated cooling rates based on the appropriate figures. Figure 4 illustrates four of the analyzed particles. Arrows indicate the position of the electron probe trace by which "interface" P was measured. All of the phosphides chosen are approximately lamellar in shape.

All of the Apollo 16 phosphide-metal grains for which analyses were made are P-rich ($\gtrsim$1 wt.% P). None of the large group containing about 0.5 wt.% P developed sufficiently wide phosphide lamellae to warrant microprobe analysis. This implies cooling at a rate of 0.1°C/day or higher, for all of the low P metal grains. Absolute cooling rates for the P-rich particles fall mainly in the range 1–100°C/day (Table 1). Cooling rates based on oxide compositions and crystal morphology have been applied only to basaltic rocks. Mare basalts and gabbros have inferred cooling rates of 5–24°C/day (Lofgren *et al.*, 1975) and 3–7°C/day (Taylor *et al.*, 1975). These cooling rates for the high temperature range are considered equivalent to a flow thickness of 2–3 m (Lofgren *et al.*, 1975). Previous estimates of cooling rates of non-mare rocks, also using metallic minerals, were based on less precise cooling rate models than the present one. Brett (1975) has used available cooling rates as a basis for the important inference that breccia cooling units are very thin sheets (1–5 m). The determination of absolute cooling rates from phosphide exsolution is therefore a useful development for examining models of breccia formation.

Smith and Goldstein (1977) presented a rapid graphical method for converting cooling rates into burial depths. The phosphide data of Table 1 are consistent with depths of burial between 10 and 1 m, according to this method. The conversion technique is based on the calculations of Taylor and Heymann (1971) for the cooling of a chondritic composition cooling unit by radiation from both sides. However, as shown by Taylor and Heymann, this model deviates significantly at low temperatures from the actual case of a layer on the surface of a planet cooling by radiation from the top of the layer and conduction from the base. The cooling rates determined for phosphide exsolution are those effective at 600–500°C, where the maximum phosphide growth took place. Because this is in the low temperature range, the conversion method of Smith and Goldstein (1977) is not the most appropriate.

Equations for the cooling of a surface flow by radiation from the top and conduction from the base have been presented by Jaeger (1968) and have been used by Brett (1975) for lunar breccia cooling units. Cooling rates determined in this study have been converted to burial depths using the Jaeger model. The equation used is

$$T/T_0 = \phi(\xi, \tau) - \phi(2 - \xi, \tau), \tag{1}$$

(Jaeger, 1968) where T_0 is the initial temperature of the flow (°C) and T is the temperature at any point in the flow (°C). The terms ξ and τ are dimensionless parameters and

$$\phi(\xi, \tau) = \frac{1}{2}\left(\operatorname{erf}\frac{\xi+1}{2\tau^{1/2}} - \operatorname{erf}\frac{\xi-1}{2\tau^{1/2}}\right). \tag{2}$$

The flow has a thickness $2a$ and is assumed to have a surface temperature of 0°C. The term ξ is the ratio x/a where x is the distance from the center of the flow. In this problem we calculate the cooling rate at the center of the cooling unit, $x = 0$, where $\xi = 0$. The term τ is given by:

$$\tau = \frac{kt}{a^2}, \tag{3}$$

where t is the time after deposition (sec) and k is the thermal diffusivity (cm^2/sec). The thermal diffusivity is assumed not to vary with temperature and is equal to $K/\rho C$, where K is the thermal conductivity of the cooling unit and basement (cal/cm sec°K), ρ is the density of the cooling unit and basement (g/cm^3), and C is the specific heat of cooling unit and basement (cal/g°K). The conversion of absolute cooling rates to burial depths is complicated by the fact that thermal conductivity is strongly influenced by the porosity of the rocks (Langseth *et al.*, 1976). Thus a certain cooling rate would imply a greater thickness for a solid igneous rock than for breccia or regolith, because the latter have lower thermal conductivities. The exact nature of the host in which the analyzed Apollo 16 metal-phosphide particles cooled is actually unknown. Three

Table 2. Calculated depths for cooling rates 1 and 100°C/day in the range 600–500°C.

	Model I (solid igneous rock)		Model II (breccia)		Model III (regolith)		Brett (1975) (ejecta blanket)	
Conductivity K (cal/cm sec°K)	4.8×10^{-3}*		1.4×10^{-3}*		2.4×10^{-5}*		9×10^{-4}†	
T_0, deposition temperature (°C)	1250		1000		750		1250	
ρ(g/cm^3)	2.9†		2.9†		2.9†		2.9†	
C(cal/g deg)	0.4†		0.4†		0.4†		0.4†	
τ_1, τ_2	$\tau_1 = 0.4$	$\tau_2 = 1.0$	$\tau_1 = 0.2$	$\tau_2 = 1.0$	$\tau_1 = 0.2$	$\tau_2 = 0.4$	$\tau_1 = 0.4$	$\tau_2 = 1.0$
T/T_0	0.605	0.298	0.885	0.298	0.885	0.605	0.605	0.298
T_1, T_2(°C)	$T_1 = 756$	$T_2 = 372$	$T_1 = 885$	$T_2 = 298$	$T_1 = 664$	$T_2 = 454$	$T_1 = 756$	$T_2 = 372$
Rate (°C/day)	1	100	1	100	1	100	1	100
a (m)	4.8	0.48	2.8	0.28	0.43	0.04	2.1	0.21

*Langseth *et al.* (1976).
†Brett (1975).

alternative conversions of rate to depth were therefore made, assuming the thermal conductivity and temperature of deposition of igneous rock, breccia and regolith, respectively. The adopted initial temperatures T_0 were chosen to represent temperature above the solidus, temperature between solidus and sintering point, and temperature below that at which sintering can occur. Details of these calculations, which follow the Jaeger method, are given briefly below.

Equation (1) was solved for T, using values of τ for which values of the function ϕ were tabulated by Jaeger (1968). The time t (sec) to reach a temperature can be calculated, assuming various values for a, using Eq. (3). Cooling rates may be calculated by considering the difference in τ values, τ_1 and τ_2, between successive T values, T_1 and T_2, in the temperature range of interest, 600–500°C. The depth a was calculated for the observed phosphide cooling rates (C.R.), 1–100°C/day, using the expression

$$a = \sqrt{\frac{k(T_1 - T_2)}{(\tau_2 - \tau_1)(\text{C.R.})}}. \tag{4}$$

Results are given in Table 2. A calculation using the data of Brett (1975) is also given in Table 2. Note that the highest calculated depth of burial is about 5 m for an impact melt sheet and about 3 m for porous impact breccia for the 1°C/day cooling rate. For the one slow cooled soil particle (~0.001°C/day, Table 1) the depth of burial is ~100 m. The above calculations support the conclusion of Brett (1975) that in general lunar impact rocks are derived from local craters rather than from basin-forming events.

General Applicability of Technique

Cooling rates for lunar samples can thus be obtained from simple composition and dimension data for metal-phosphide particles. There are several limitations to the application of this method, not the least of which is that not all lunar rocks contain metal grains with phosphides. Furthermore, the phosphides must be lamellae, not blebs formed at metal–metal grain boundaries. The composition profiles must be consistent with growth during cooling. With dissolution or reheating, the P gradients are reversed and then the method is not applicable. Given a suitable phosphide, the interface P concentration, the size of the lamella (halfwidth) and the diffusion field length must be determined accurately for determination of cooling rate. The determinative curves used here were drawn so that the apparent "interface" compositions resolved by electron microprobe could be used with confidence. However, halfwidth and field length may not be easy to determine, because the phosphide lamellae may be distributed irregularly, oriented at angles to one another and inclined to the plane of the section. These problems must be overcome by judicious choice of the traverses across the lamellae.

One important finding is that bulk composition is not a significant variable, so long as the Ni content is not so high that taenite is present. Thus the figures presented here can be used for cooling rate determination for the majority of lunar phosphides. The agreement between the burial depths obtained here and by other techniques indicates that further application of the phosphide method for cooling rate determination of lunar rocks is warranted.

Acknowledgments—E. Randich and A. Moren assisted with the development of computer programs. J. Friel, A. Romig, and D. Bush helped with the operation of an automated microprobe. L. Dunkin assisted with data processing and plotting. The paper was critically and helpfully reviewed by R. Brett and R. Gooley. The work was aided by NASA grant NGR-39-007-056 (J. I. Goldstein, Principal Investigator) and support from Rutgers Research Council.

REFERENCES

Brett R. (1975) Thicknesses of some lunar mare basalt flows and ejecta blankets based on chemical kinetic data. *Geochim. Cosmochim. Acta* **39**, 1135–1141.

Doan A. S. and Goldstein J. I. (1970) The ternary phase diagram, Fe–Ni–P. *Met. Trans.* **1**, 1759–1767.

Goldstein J. I. and Axon H. J. (1973) Composition, structure and thermal history of metallic particles from 3 Apollo 16 soils, 65701, 68501 and 63501. *Proc. Lunar Sci. Conf. 4th*, p. 751–775.

Goldstein J. I. and Randich E. (1977) Variation of interface compositions during diffusion-controlled precipitate growth in ternary systems. *Met. Trans A* **8A**, 105–109.

Gooley R. C., Brett R., and Warner J. L. (1973) Crystallization history of metal particles in Apollo 16 rake samples. *Proc. Lunar Sci. Conf. 4th*, p. 799–810.

Heyward T. R. and Goldstein J. I. (1973) Ternary diffusion in the α and γ phases of the Fe–Ni–P system. *Met. Trans.* **4**, 2335–2342.

Jaeger M. C. (1968) Cooling and solidification of igneous rocks in basalt. In *The Poldervaart Treatise on Rocks of Basaltic Composition* (H. H. Hess and A. Poldervaart, eds.), p. 503–536. Wiley-Interscience, New York.

Langseth M. A., Keihm S. J., and Peters K. (1976) Revised lunar heat-flow values. *Proc. Lunar Sci. Conf. 7th*, p. 3143–3171.

Lofgren G. E., Donaldson C. H., and Usselman T. M. (1975) Geology, petrology and crystallization of Apollo 15 quartz-normative basalts. *Proc. Lunar Sci. Conf. 6th*, p. 79–99.

Misra K. C. and Taylor L. A. (1975) Characteristics of metal particles in Apollo 16 rocks. *Proc. Lunar Sci. Conf. 6th*, p. 615–639.

Miyake G. I. and Goldstein J. I. (1974) The Tucson meteorite. *Geochim. Cosmochim. Acta* **38**, 1201–1212.

Norkiewicz A. S. and Goldstein J. I. (1975) Ternary dissolution kinetics in the Fe–Ni–P system. *Met. Trans. A* **6A**, 891–900.

Randich E. (1975a) Non-isothermal diffusion-controlled phase growth in ternary systems and its application to iron meteorites. Ph.D. thesis, Lehigh University, Bethlehem, Pennsylvania.

Randich E. (1975b) HEX. A computer program for simulating the growth of schreibersite in hexahedrites. Lehigh University Computer Center, Bethlehem, Pennsylvania. Unpublished report.

Randich E. and Goldstein J. I. (1975) Non-isothermal diffusion-controlled growth in ternary systems. *Met. Trans. A* **6A**, 1553–1560.

Randich E. and Goldstein J. I. (1977) Cooling rates of the hexahedrites. Submitted to *Geochim. Cosmochim. Acta.*

Smith B. A. and Goldstein J. I. (1977) The metallic microstructures and thermal histories of severely reheated chondrites. *Geochim. Cosmochim. Acta* **41**, 1061–1072.

Taylor G. J. and Heymann D. (1971) Postshock thermal histories of reheated chondrites. *J. Geophys. Res.* **76**, 1879–1893.

Taylor L. A., Uhlmann D. R., Hopper R. W., and Misra K. C. (1975) Absolute cooling rates of lunar rocks: Theory and application. *Proc. Lunar Sci. Conf. 6th*, p. 181–191.

Proc. Lunar Sci. Conf. 8th (1977), p. 1639–1672.
Printed in the United States of America

Rb–Sr and Sm–Nd chronology and genealogy of mare basalts from the Sea of Tranquility

D. A. PAPANASTASSIOU, D. J. DEPAOLO, and G. J. WASSERBURG

The Lunatic Asylum of the Charles Arms Laboratory, Division of Geological and Planetary Sciences, California Institute of Technology, Pasadena, California 91125

Abstract—Rb–Sr and Sm–Nd ages of two Apollo 11 mare basalts are 3.64 ± 0.05 and 3.57 ± 0.03 AE for high-K basalt 10072, and 4.01 ± 0.11 and 3.88 ± 0.06 AE for low-K basalt 10062. Rb–Sr, Sm–Nd, and ^{40}Ar–^{39}Ar ages are in good agreement and show an extensive time interval for filling of the Sea of Tranquility, presumably by thin lava flows, in agreement with similar observations for the Ocean of Storms. The older ages indicate that mare filling may have started prior to the Imbrium event. Initial Sr and Nd isotopic compositions on Apollo 11 basalts identify at least two parent sources producing basalts. The low-K basalts may have been produced from a single reservoir over an extended time interval. The Sm–Nd isotopic data demonstrate that low-K and high-Ti basalts from Apollo 11 and 17 have been derived from totally distinct reservoirs while low-Ti Apollo 15 mare basalt sources have Sm/Nd similar to the sources of Apollo 11 basalts. It appears that groupings of mare basalt based on Ti content and on isotopic data do not coincide. Apollo 11 high-K basalts show no significant Sm/Nd fractionation near the time of crystallization in contrast to characteristically large Rb/Sr fractionation for these basalts.

INTRODUCTION

IN THIS REPORT we provide a review of the chronology of Apollo 11 mare basalts and present new Rb–Sr and the first Sm–Nd internal isochrons on two Apollo 11 mare basalts. The Sm–Nd work is an extension of the original development and application of Sm–Nd dating to problems of lunar chronology and petrogenesis by Lugmair and his colleagues (Lugmair, 1975; Lugmair, *et al.*, 1975). The aims of the present study are as follows:

(a) Determination of more refined Rb–Sr ages and of the first Sm–Nd ages on A-11 samples and comparison of the chronologies defined by Rb–Sr, Sm–Nd, and ^{40}Ar–^{39}Ar.

(b) Determination based on age, initial Sr, and initial Nd compositions of how many distinct magma sources are required for the A-11 basalts.

(c) Comparison of Rb–Sr and Sm–Nd chemical fractionation especially for the high-K A-11 basalts which are characterized by large Rb–Sr fractionation near the time of crystallization.

(d) Comparison of Sm–Nd fractionation of A-11 and A-17 mare basalts. These basalts have similar major element compositions, including high-Ti contents. The A-17 basalts show the effects of large Sm–Nd fractionation (Lugmair *et al.*, 1975) near the time of formation of the moon. In contrast the data to be presented show that A-11 basalts underwent considerably less early Sm–Nd fractionation.

(e) Review of the time constraints for REE and Rb–Sr fractionation on the moon and of the constraints on petrogenetic models based on Rb–Sr, Sm–Nd, and on Eu fractionation.

We present data for a low-K and a high-K Apollo 11 basalt by both Rb–Sr and Sm–Nd. The choice of the samples was predicated on the need for crystallization ages on more low-K basalts to compare with the ^{40}Ar–^{39}Ar gas retention ages, and on the importance of comparing Rb–Sr and Sm–Nd fractionation for A-11 high-K rocks which indicate significant Rb–Sr fractionation near the time of crystallization of these rocks.

Review

The great majority of reported Rb–Sr data on Apollo 11 rocks were based on the first sample allocation and examination seven years ago. Although some workers presented Rb–Sr whole-rock data on Apollo 11 samples, we shall restrict the discussion to the determinations of internal isochrons, since our purpose is to establish a precise chronology of crystallization at this site. Crystallization ages were reported by Papanastassiou *et al.* (1970), Gopalan *et al.* (1970), and Compston *et al.* (1970). Gopalan *et al.* (1970) reported ages on two high-K basalts which had relatively large uncertainties. Their report of an old age for 10024 was not confirmed (Papanastassiou and Wasserburg, 1971a). Compston *et al.* (1970) provided well-defined internal isochrons on two high-K basalts; the reported ages have been revised (De Laeter *et al.*, 1973) and are now in good agreement with the data presented by Papanastassiou *et al.* (1970). Papanastassiou *et al.* (1970) determined internal isochrons on four A-11 high-K rocks and on two low-K rocks. We have listed in Table 1 all precise determinations of Rb–Sr ages and initial $^{87}Sr/^{86}Sr$ on A-11 basalts. The first Rb–Sr data on Apollo 11 showed: (a) a uniform crystallization age for all samples, (b) two distinct rock groups based on initial $^{87}Sr/^{86}Sr$ indicating that the rocks were produced from at least two distinct magma reservoirs, (c) characteristically young whole-rock model ages for the high-K rocks indicating that, for these rocks, Rb/Sr was significantly increased at or very near the time of crystallization, and (d) low initial $^{87}Sr/^{86}Sr$ for all samples indicating that the source regions and presumably the moon formed with low Rb/Sr and with initial $^{87}Sr/^{86}Sr$ near BABI.

Extensive work on Apollo 11 samples by the ^{40}Ar–^{39}Ar technique was reported by Turner (1970). This work established: (a) the existence of measured ages for low-K rocks as old as 3.92 AE, (b) that the high-K rocks had suffered significant ^{40}Ar loss so that an ^{40}Ar–^{39}Ar apparent age plateau was not defined by whole rock data, and (c) that in a specific case of high-K basalt 10072 an apparently well-defined younger age was obtained. These data indicated that there existed differences in the ages of Apollo 11 basalts possibly as large as 4×10^8 yr. A report by the Lunatic Asylum (1970) showed that plagioclase was significantly more retentive for ^{40}Ar than whole rocks. Turner *et al.* (1971, 1972)

Table 1. Apollo 11 mare basalts.

Sample	K Content	$(^{87}Sr/^{86}Sr)_I$[a]	Rb–Sr[a] (AE)	$^{40}Ar–^{39}Ar$[b] (AE)	
10003	L	0.69909 ∓ 4	3.84 ± 0.08 (1)	3.92 ± 0.07	TR;S (6)
				3.91 ± 0.03	P1;B (7)
10017	H	0.69932 ∓ 5	3.59 ± 0.05 (2)	> 3.23 ± 0.06	TR;S (6)
		0.69930 ∓ 10	3.71 ± 0.11 (3)		
10020	L	—	—	(3.77 ± 0.03)	TR;B (8)
10022	H	—	—	3.59 ± 0.06	TR;S (6)
				(3.58 ± 0.04)	TR;B (8)
10024	H	0.69935 ∓ 8	3.61 ± 0.07 (4)	> 3.48 ± 0.05	TR;S (6)
10029	L	—	—	(3.89 ± 0.05)	TR;B (10)
10032	L	—	—	3.58 ± 0.05	TR;B (10)
10044	L	0.69909 ∓ 7	3.71 ± 0.11 (2)	3.73 ± 0.05	TR;S (6)
				3.71 ± 0.03	TR;B (8)
10045	L	—	—	(3.75 ± 0.03)	TR;B (8)
10047	L	—	—	3.74 ± 0.03	TR;B (7)
10049	H	—	—	(3.45 ± 0.04)	TR;B (8)
10050	L	—	—	3.75 ± 0.03	P1;B (8)
10057	H	0.69939 ∓ 5	3.63 ± 0.04 (2)	> 3.39 ± 0.04	TR;S (6)
10058	L	0.69906 ∓ 5	3.63 ± 0.20 (2)	(3.71 ± 0.03)	TR;B (8)
10062	L	0.69901 ∓ 4	4.01 ± 0.11 (5)	3.82 ± 0.06	TR;S (6)
10069	H	0.69926 ∓ 8	3.68 ± 0.06 (2)	> 3.48 ± 0.04	TR;B (8)
10071	H	0.69923 ∓ 8	3.68 ± 0.02 (2)	3.51 ± 0.06	P1;B (9)
10072	H	0.69926 ∓ 5	3.64 ± 0.05 (5)	3.52 ± 0.05	TR;S (6)
		0.69940 ∓ 12	3.71 ± 0.11 (3)	3.57 ± 0.04	TR;B (8)
				3.62 ± 0.04	P1;B (8)

[a]Uncertainties are 2σ; initial $^{87}Sr/^{86}Sr$ from De Laeter *et al.* (1973) may be subject to interlaboratory bias.

[b]TR = Total rock sample; Pl – Plagioclase; ages in parentheses by the Berne group indicate intermediate temperature plateau only, and are considered by this group less reliable (Geiss *et al.*, 1977). S = Sheffield, B = Berne.

References: (1) Papanastassiou and Wasserburg (1975a). (2) Papanastassiou *et al.* (1970). (3) DeLaeter *et al.* (1973). (4) Papanastassiou and Wasserburg (1971a). (5) Present work. (6) Turner (1970). (7) Stettler *et al.* (1974). (8) Geiss *et al.* (1977). (9) Stettler *et al.* (1973). (10) Guggisberg *et al.* (1977).

and Podosek *et al.* (1972) have shown that plagioclase is the most retentive mineral. Podosek *et al.* (1972) have demonstrated differences in apparent $^{40}Ar/^{39}Ar$ between plagioclase and whole-rock plateau ages for a mare basalt even though both plateaus appeared well defined. There is good agreement between $^{40}Ar–^{39}Ar$ plateau ages defined by plagioclase and Rb–Sr internal isochron ages. Exceptions to the good agreement between $^{40}Ar–^{39}Ar$ plagioclase plateau ages and Rb–Sr ages exist for highland rocks with a more complex history (cf. Jessberger *et al.*, 1974). Some exceptions, e.g., for the troctolite 76535, are not well understood. We consider that the $^{40}Ar–^{39}Ar$ technique yields reliable ages for plagioclase from igneous rocks which exhibit little gas loss and high temperature plateau ages defined by a significant fraction of the ^{40}Ar

released in several steps. Subsequent to the work by Turner, the Berne group reexamined rock 10003 (Stettler *et al.*, 1974) and confirmed its old age. Furthermore this group has provided a large number of $^{40}Ar-^{39}Ar$ analyses on Apollo 11 basalts (see Table 1 for references). The $^{40}Ar-^{39}Ar$ ages on Apollo 11 mare basalts are shown in Table 1. The ages shown include: (a) whole-rock analyses and (b) analyses of plagioclase mineral separates. Some of the ages are based on "intermediate" temperature plateaus which are considered less reliable than ages defined by plagioclase high temperature plateaus (Geiss *et al.*, 1977). These details have been noted in Table 1. The data show a range in gas retention ages from 3.91 to 3.45 AE. Geiss *et al.* (1977) have proposed that the rocks may be subdivided in three to four age groups which coincide with petrologic types defined by Warner (1971). We note from Table 1 that there is in fact only a small number of $^{40}Ar-^{39}Ar$ high temperature plateau ages obtained on plagioclase separates.

The chronology of A-11 basalts has so far been based only on Rb–Sr and $^{40}Ar-^{39}Ar$ data. No crystallization ages have been reported yet by the U–Pb or Sm–Nd techniques for rocks of this mission. The data summarized above indicate the following major problems:

(a) Rb–Sr data on only a limited number of A-11 low-K basalts exist and these data are not of sufficient precision to resolve some of the small age differences indicated by $^{40}Ar-^{39}Ar$ ages. This problem is due to the rarity and difficulty of isolating high-Rb/Sr phases in low-K mare basalts.

(b) The Rb–Sr data did not exhibit the wide spread in ages observed by $^{40}Ar-^{39}Ar$. This may be due to the limited Rb–Sr data on A-11 low-K rocks which have yielded old $^{40}Ar-^{39}Ar$ ages. In addition some of the young $^{40}Ar-^{39}Ar$ ages contributing to the wide spread were obtained on high-K basalts which exhibit significant gas loss and may not be reliable.

SAMPLING

High-K basalt 10072

This sample is the coarsest of the A-11 high-K rocks which are as a group fine-grained. Despite the coarser size of the pyroxene, interstitial, late crystallizing phases appear as small inclusions in all major phases and their exclusion from mineral separates of the major phases has required extreme care. A 4.2 g sample of the rock which included lunar exterior surfaces was received from the curator. These lunar exterior surfaces were removed using a tungsten chisel, and a sample of the total rock from totally interior material was obtained for Rb–Sr and Sm–Nd. Our aim was to obtain: (a) high purity plagioclase separates free of interstitial phases for both Rb–Sr and Sm–Nd, (b) high purity pyroxene especially for the Sm–Nd work, and (c) separates highly depleted in plagioclase and enriched in ilmenite which acts as a Sr-free carrier phase for the higher Rb/Sr interstitial phases.

Mineral separates were obtained from the debris produced during chiselling off of the exterior surfaces and separately from totally interior material. Two mineral separates were obtained from the chipping debris and are labelled CD in Table 2. Mineral separations involved: (a) only mechanical means (crushing, sieving) and magnetic separations (such separates are labelled M in the tables and figures), and (b) additional density separations using organic liquids ($CH_2I_2 + CH_3COCH_3$) followed by magnetic separations (labelled L).

In this study we found no evidence of differential leaching by the organic liquids in agreement with all our earlier work on mare basalts and in contrast to our work on the dunite 72417 (Papanastassiou and Wasserburg, 1975b). For Sm–Nd, improvement in the range in Sm/Nd for the plagioclase and pyroxene end members was obtained by first using density and magnetic separations for material crushed to less than 75 μm and then reprocessing samples highly enriched in plagioclase and in pyroxene after crushing to less than 44 μm. For Rb–Sr the small amount needed of highest purity plagioclase were obtained by handpicking.

Low-K basalt 10062

We received from the curator an 8.1 g sample which had been previously allocated for mineralogical investigations; the sample had been handled and exposed to terrestrial contamination. The sample also included exterior lunar surfaces. For this work most exterior surfaces were removed to produce essentially interior material. Complete removal of all exterior surfaces was not possible because of the concomitant decrease in the useful sample weight. Mineral separations were obtained by the techniques described above. The rock is fine-grained so that the production of mineral separates required extensive work. High purity mineral separates of pyroxene and plagioclase were obtained (better than 99% pure as shown by grain counts using the electron microprobe). These separates provided a sufficient spread for Sm/Nd. However, the mineral separations did not result in the identification of high-Rb/Sr materials. A survey by the electron microprobe of grain mounts of density separates, including separates enriched in cristobalite and in ilmenite, revealed no significant high-K concentration spots. Similarly, chemical analyses of small aliquots (0.1–1 mg) of mineral separates for K, Rb, and Sr showed no high Rb/Sr. The low abundance of mesostasis and the fine grain size (Carter and MacGregor, 1970) have prevented the separation of higher Rb/Sr phases. An ilmenite separate produced by magnetic separation showed a small enrichment of $^{87}Sr/^{86}Sr$ over the total rock. A second ilmenite separate produced by density separations did not show a significant improvement due to the presence of small amounts of plagioclase in binary grains. Rb–Sr concentrations in a small sample obtained by handpicking of clean ilmenite grains of size 75–44 μm under a binocular microscope showed that some improvement of Rb/Sr was possible. Therefore, two man-weeks were invested in handpicking clean ilmenite, free of plagioclase. The final separate shows an enrichment of $^{87}Sr/^{86}Sr$ over the plagioclase of 0.3%. It is clear from these results that it may be impossible to obtain sufficiently high-Rb/Sr phases in extremely low-K and fine-grained samples to provide Rb–Sr ages with an uncertainty lower than 0.1 AE. However, in these cases reasonable enrichments in Sm–Nd have been obtained which result in a precise age determination.

Analytical Procedures

All samples were dissolved following standard procedures using HF and $HClO_4$. Small (<1%) aliquots of each sample were used for establishing the concentrations of K, Rb, Sr, Ba, Sm, and Nd in a single mass spectrometer run. The major sample aliquots were then spiked in an optimal fashion. We have checked the cross-contamination between K, Rb, Sr, Sm, and Nd tracers. In all cases the cross-contamination is negligible except for Nd in the ^{41}K–^{40}K spike solution. This effect has been reduced to negligible levels by adding less of the K spike, or eliminated for some samples by not spiking the Sm–Nd aliquot for K.

Rb–Sr

Rb and Sr blanks were monitored regularly and are 0.01–0.003 ng for Rb and 0.1–0.05 ng for Sr and are negligible for the data presented here. The concentrations in the tracer solutions were checked with four Rb and four Sr gravimetric solutions which were independently prepared using Johnson–Matthey spectrographically pure RbCl and $SrCO_3$ and NBS SRM 987 $SrCO_3$. The stoichiometry of the Johnson-Matthey salts has been confirmed in the past (Wasserburg *et al.*, 1962).

Using the four sets of Rb and Sr gravimetric solutions, the concentrations of the single ^{87}Rb tracer solution and *two* ^{84}Sr tracer solutions, in use, each agreed to 0.1%. A fifth gravimetric normal of SRM 987 gave results for *both* Sr tracer solutions 0.4% higher than the rest. We attribute this highly atypical result to unknown problems during the preparation of this gravimetric solution and have discarded this datum in view of the close agreement of the results of four other determinations using independently prepared gravimetric solutions. Sr mass spectrometric analyses were performed with new zone-refined Ta ribbon which shows no detectable ^{85}Rb signal during the Sr run (Wasserburg *et al.*, 1977). This has resulted in increased ease of analyzing Sr, as data collection is not restricted by waiting for the Rb signal to die.

Sm–Nd

Sm and Nd were isolated using chemical techniques described by Eugster *et al.* (1970) and DePaolo and Wasserburg (1976a). Rare earth elements were separated from major elements on a 1 cm diameter cation exchange column using HCl as elutriant. Sm and Nd were then separated on a second cation column using as elutriant 0.2M 2-methyllactic acid with pH adjusted to approximately 4.7.

Samples were spiked with tracers enriched in ^{150}Nd and ^{147}Sm. Tracer concentrations were calibrated using gravimetric standard solutions which were made by dissolving ultra pure metal chunks of Sm and Nd obtained from Ames Laboratory, Iowa. The chunks of metal were analyzed for impurities and weighed at Ames prior to shipment in evacuated containers. Upon arrival at Caltech, the metal chunks were weighed and the weights agreed with those measured at Ames to better than 0.01%. Sm concentrations were calculated using the Sm isotopic composition reported by Russ *et al.* (1971) and Russ (1974). Nd concentrations were calculated using the isotopic composition of normal Nd reported by DePaolo and Wasserburg (1976a,c). $^{147}Sm/^{144}Nd$ ratios are precise to better than 0.1%, where this error represents the sum of errors resulting from weighing of spikes and uncertainties of isotope ratios of the spikes measured in the mass spectrometer.

Nd was measured in the mass spectrometer as NdO^+ and Sm was measured as SmO^+. In this work, all Nd isotopic measurements were made on spiked samples. The ^{150}Nd tracer is approximately 96% ^{150}Nd. The small amounts of the other Nd isotopes in the tracer significantly change the measured isotope ratios and must be corrected. These corrections are made using the isotopic composition of the tracer measured by us. This measurement is quite precise, but additional uncertainty must be added to the measured ratios due to the inability to measure precisely the instrumental mass discrimination for the tracer. The uncertainty in mass discrimination is believed to be about 0.05% per mass unit. The net correction to $^{143}Nd/^{144}Nd$ under our spiking conditions (^{150}Nd-tracer added equal to twice ^{150}Nd in sample) is approximately $0.1500 \pm 0.0003\%$ where the uncertainty is due to uncertainty in the tracer composition. The added uncertainty to $^{143}Nd/^{144}Nd$ from this effect is negligible. Nd isotopic ratios were normalized to $^{146}Nd/^{142}Nd = 0.636155$ to remove the effects of instrumental isotope fractionation. This is equivalent to normalizing to $^{150}Nd/^{142}Nd = 0.2096$ for the unspiked samples (DePaolo and Wasserburg, 1976b) and is essentially the average value that is measured in the Lunatic I mass spectrometer (Wasserburg *et al.*, 1969) if no correction is made for mass fractionation.

In unspiked samples, $^{150}Nd/^{142}Nd$ is used for the determination of mass fractionation because the difference of 8 mass units for these isotopes minimizes the error in determining the fractionation factor per mass unit. Since we use a ^{150}Nd tracer, for spiked samples the $^{146}Nd/^{142}Nd$ ratio is used for the determination of fractionation. This ratio still provides a difference of 4 mass units, involves abundant Nd isotopes, and the contributions to ^{142}Nd and ^{146}Nd from the ^{150}Nd tracer are small.

In all Nd mass spectrometer runs made in our laboratory, all Nd isotope ratios are measured. Data for the non radiogenic isotopes for five samples from 10072 and five recently measured terrestrial samples are shown in Fig. 1. Measured values of $^{142}Nd/^{144}Nd$ and $^{145}Nd/^{144}Nd$, which can be measured with precision similar to $^{143}Nd/^{144}Nd$, agreed from run to run to ±0.005% or better. The isotopic composition of lunar Nd appears to be identical to that of terrestrial Nd to within the present limits of resolution of ~2 parts in 10^5. Lugmair *et al.* (1975) first reported precise measurements of $^{143}Nd/^{144}Nd$ and $^{142}Nd/^{144}Nd$ in the Juvinas achondrite and in reagent Nd. However, these numbers were revised (Lugmair, pers. comm.). The revised value for $^{143}Nd/^{144}Nd$ measured in Juvinas by Lugmair was subsequently used by DePaolo and Wasserburg (1976a) and reported by Lugmair *et al.* (1976). High precision isotopic abundances of Nd for all isotopes were first reported by DePaolo and Wasserburg (1976a,c). The means of data shown in Fig. 1 agree with the means of measurements made by DePaolo and Wasserburg (1976a,b) to ±2 parts in 10^5 for all nonradiogenic isotopes. Measurements made by Lugmair *et al.* (1976) on the troctolite 76535 and their reagent Nd standard agree reasonably well with our values (after renormalization) except for $^{145}Nd/^{144}Nd$, which they measure ~2 parts in 10^4 higher, and $^{142}Nd/^{144}Nd$, which shows substantial variation in their data (2 parts in 10^4). This difference is far outside analytical uncertainty and is not understood. Measurements made by Nakamura *et al.* (1976) also differ from ours by substantial amounts. Their measured $^{142}Nd/^{144}Nd$ is greater than ours by 2 parts in 10^4 and $^{145}Nd/^{144}Nd$ is less than ours by about 2 parts in 10^4. Nakamura *et al.* (1976) made their measurements on Nd^+ rather than on NdO^+ as is done in this laboratory. We correct our data for the oxygen isotopic composition using the composition reported by Nier (1950). However, we measure the oxygen isotopic composition of NdO^+ to be substantially different from Nier's values (see discussion below). If we were to correct our data using our measured oxygen composition, it would increase our calculated $^{142}Nd/^{144}Nd$ by 0.8 parts in 10^4 and decrease our $^{145}Nd/^{144}Nd$ by 0.4 parts in 10^4. This change would bring our measured ratios closer to those measured by Nakamura *et al.* (1976) but would still leave a substantial discrepancy between the data from the two laboratories. The data on the nonradiogenic Nd isotopes provide an important indicator of data precision and a check on the absence of data artifacts which is not available for either Sr or Pb. In view of the high precision necessary for Nd isotopic work, collecting and reporting these data are necessary for an evaluation of the Nd data.

Total chemistry blank for Nd is ~70–100 pg and for Sm, although not measured, is presumed to be in the range 20–30 pg. No blank corrections were necessary to either Nd isotope ratios or Sm/Nd ratios. No corrections for neutron capture effects were applied to Nd isotope ratios. During a run, possible interfering species are monitored at masses 156($^{140}Ce^{16}O$), 157($^{141}Pr^{16}O$), and 170($^{154}Sm^{16}O$). We generally measure the ratios $156/160 \approx 10^{-5}$; $157/160 = 10^{-4} - 10^{-3}$; $170/160 \approx 5 \times 10^{-6}$ (mass 160 is $^{144}Nd^{16}O$) during a run, and no corrections to Nd isotope ratios are necessary. Only in one sample (10072, pyroxene-B) a correction of 0.006% was made to $^{143}Nd/^{144}Nd$ due to the contribution of $^{141}Pr^{18}O$.

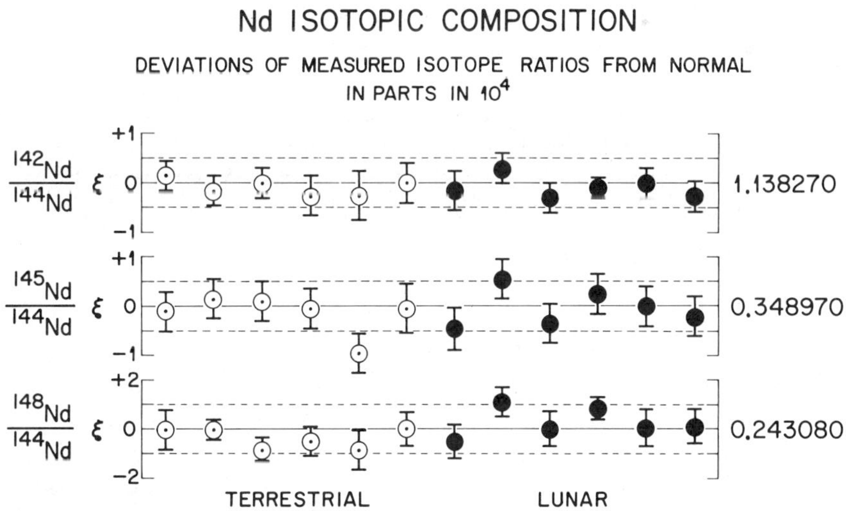

Fig. 1. Measured values of non-radiogenic Nd isotope ratios plotted as deviations (ξ) in parts in 10^4 from the mean values for these ratios as measured over the past year given at right. $\xi = ((R_i - R_i^{GM})/R_i^{GM}) \times 10^4$ (GM = grand mean). Data points represent terrestrial and lunar samples measured during this work; error bars are $2\sigma_{mean}$.

Oxygen isotopic abundance

Oxygen isotopic corrections to Nd isotope ratios were made assuming that the ratios $^{18}O/^{16}O = 0.002045$ and $^{17}O/^{16}O = 0.0003708$. These ratios are those measured by Nier (1950). Since a large variability of oxygen isotope ratios is found in nature, we have made preliminary measurements of $^{18}O/^{16}O$ and $^{17}O/^{16}O$ to check if the ratios given above are the same as those found in the Nd oxide ion beam. This was done by measuring the ratios $^{150}Nd^{18}O/^{150}Nd^{16}O$ and $^{150}Nd^{17}O/^{150}Nd^{16}O$ on a sample of our enriched ^{150}Nd tracer and on a spiked sample Nd run. The results obtained so far indicate $^{18}O/^{16}O = 0.00211–0.00213$ and $^{17}O/^{16}O = .000385–0.000392$. There is some variability in these ratios between runs as well as variation within a run. The measured $^{18}O/^{16}O$ is higher than the value given by Nier by about 4%. This difference could be the result of mass fractionation occurring: (a) in the mass spectrometer during evaporation of Nd_2O_3 from the filament as NdO^+ with concomitant loss of O atoms, (b) during oxidation of the Nd sample by heating in air, or (c) in Nier's spectrometer. It appears, however, that relative to the Nier values, our measured $^{17}O/^{16}O$ is higher than it should be relative to $^{18}O/^{16}O$ if the differences were only due to mass dependent fractionation. At present, the uncertainty in our $^{17}O/^{16}O$ is too large to evaluate critically this possibility. A more detailed description of this experiment will be given elsewhere. For our purposes here, it is only necessary to note that if our measured values were used to correct our Nd isotope ratios instead of the Nier values, the $^{143}Nd/^{144}Nd$ ratios reported here would be increased by 2 parts in 10^5, which is approximately half the analytical uncertainty. Variation of O isotope composition between and during runs causes effects about one fourth this large, which are negligible. However the effects on Nd isotopic composition from correcting for the O isotope composition depend upon the type of sample being analyzed (e.g., spiked or unspiked) and on which pair of isotopes is used for the mass discrimination correction. The measured $^{143}Nd/^{144}Nd$ of Lugmair (1975) and Lugmair *et al.* (1975, 1976) would also be changed by +2 parts in 10^5 if our measured O ratios were the same as those found in the NdO^+ beam in their mass spectrometer. On our unspiked runs (DePaolo and Wasserburg, 1976a,b) a change of +4 parts in 10^5 would result.

Mass fractionation correction

Our normalization of the Nd isotopic data for instrumental mass fractionation is not identical to that used by Lugmair and his coworkers. A change in the normalization constant for any element changes the calculated relative proportions of the isotopes of that element and thus changes the atomic weight of the element. Relative to the normalization used by Lugmair and coworkers, our normalization causes us to calculate a 0.05% lower concentration of ^{144}Nd in a sample and 0.15% lower $^{143}Nd/^{144}Nd$. These differences cause us to calculate an isochron age which is lower by 0.2% than we would calculate if we used the other normalization. The shift in $^{143}Nd/^{144}Nd$ is large but uniform and can be easily calculated. Our data can be made comparable to those of Lugmair by multiplying our $^{143}Nd/^{144}Nd$ by 1.001563 and our $^{147}Sm/^{144}Nd$ by 1.0005.

ANALYTICAL RESULTS

Rb–Sr

The Rb–Sr data are shown in Table 2 and in Figs. 2 and 3. Data have been obtained on seven samples of 10072. The data show a range in $^{87}Sr/^{86}Sr$ of 1.7% and permit a precise age determination. All data lie on a straight line with a best fit slope corresponding to an age of 3.64 ± 0.05 AE (2σ uncertainties; $\lambda(^{87}Rb) = 0.0139\ AE^{-1}$). The initial $^{87}Sr/^{86}Sr$ is 0.69928 ∓ 0.00006; samples of plagioclase with $^{87}Sr/^{86}Sr$ very near the initial value have been analyzed. The data show no evidence of differential leaching of Rb or Sr for the samples exposed to organic liquids. Similarly, samples obtained from the chipping debris (CD) which included lunar exterior surfaces are in agreement with data obtained from interior samples. The total rock has radiogenic $^{87}Sr/^{86}Sr$ and yields a well-defined model age $T_{BABI} = 3.83 \pm 0.03$ AE. This model age is characteristic of all previously analyzed high-K A-11 samples. The age, initial Sr and model age are in total agreement with determinations on all other A-11 high-K basalts (Papanastassiou *et al.*, 1970).

Table 2. Rb–Sr analytical results.

Sample	Weight[a] (mg)	K (ppm)	Rb[b] (10^{-8} mole/g)	^{88}Sr[b] (10^{-8} mole/g)	$^{87}Rb/^{86}Sr$[c] ($\times 10^2$)	$^{87}Sr/^{86}Sr$[d]
10072						
Plag-M	2.0	3320	1.075	434.1	0.578	0.69958 ± 4
Plag-M (CD)	5.9	2544	4.021	400.9	2.340	0.70036 ± 10
Plag + Quint-L	4.2	3584	7.07	419.1	3.935	0.70138 ± 6
Pyroxene-L	42*	380	1.023	32.46	7.35	0.70307 ± 5
Total-M	116*	2752	7.33	165.6	10.32	0.70463 ± 5
Ilmenite M (CD)	21	3046	8.66	139.3	14.50	0.70678 ± 4
Ilmenite-L	9.8	2204	6.80	68.98	23.00	0.71129 ± 9
10062						
Plag-M	2.4	—	0.175	325.9	0.126	0.69908 ± 6
Plag-L	19	445	0.238	374.8	0.148	0.69910 ± 5
Total-M	117*	607	0.950	181.2	1.223	0.69970 ± 6
Pyroxene-L	35*	142	0.2201	38.23	1.343	0.69976 ± 6
Ilmenite-M	19*	311	0.603	55.01	2.555	0.70048 ± 5
Ilmenite-L	23	352	0.731	43.60	3.910	0.70116 ± 9
				43.56	3.914	0.70128 ± 5

[a]Samples marked by an asterisk were analyzed also for Sm–Nd.
[b]Concentrations calculated using $^{85}Rb/^{87}Rb = 2.591$; $^{86}Sr/^{88}Sr = 0.1194$; and $^{84}Sr/^{88}Sr = 0.006748$.
[c]Uncertainties ±0.4%.
[d]Uncertainties are $\pm 2\sigma$, in the last significant figures.

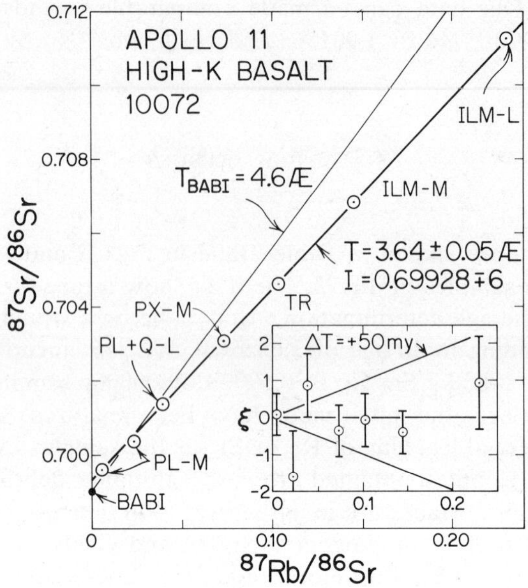

Fig. 2. Rb–Sr evolution diagram for high-K basalt 10072. Inset shows deviations (ξ) of the $^{87}Sr/^{86}Sr$ data off the best fit line in parts in 10^4. T, I uncertainties are 2σ; the range in measured $^{87}Sr/^{86}Sr$ is 1.7%.

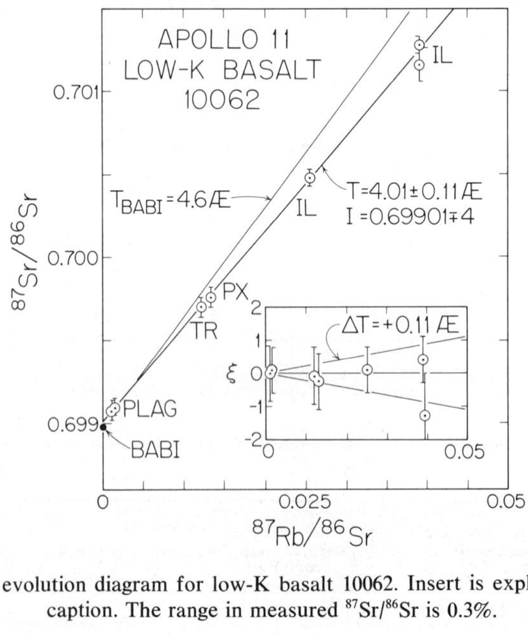

Fig. 3. Rb–Sr evolution diagram for low-K basalt 10062. Insert is explained in Fig. 2 caption. The range in measured $^{87}Sr/^{86}Sr$ is 0.3%.

Data have been obtained on six samples from 10062. The data show a range in $^{87}Sr/^{86}Sr$ of 0.3%. All data lie on a straight line on the Sr evolution diagram corresponding to $T = 4.01 \pm 0.11$ AE and $(^{87}Sr/^{86}Sr)_I = 0.69901 \mp 0.00004$ (2σ). All data lie on the best fit line well within experimental uncertainties except for one duplicate, less precise analysis of an ilmenite sample. The $^{87}Sr/^{86}Sr$ measurements are of high quality and the lower precision in the age as compared to 10072 is primarily due to the low enrichment in $^{87}Sr/^{86}Sr$ for the mineral separates obtained. The initial $^{87}Sr/^{86}Sr$ is very close to the value of BABI; the total rock has a model age $T_{BABI} = 4.1 \pm 0.3$ AE.

Sm–Nd

The Sm–Nd data for six samples from high-K basalt 10072 and five samples from low-K basalt 10062 are given in Table 3. The range of $^{143}Nd/^{144}Nd$ obtained for both samples is about 50 parts in 10^4 and enables a precise age determination for both rocks. As seen in Fig. 4, the data points from 10072 all lie within 2 parts in 10^5 of a straight line and determine an excellent isochron corresponding to an age of 3.57 ± 0.03 AE (2σ uncertainties; $\lambda(^{147}Sm) = 0.00654\ AE^{-1}$). The 10062 data (Fig. 5) also all lie well within uncertainties (4 parts in 10^5) of the best fit line and yield an age of 3.88 ± 0.06 AE. The Sm–Nd age of 10072 is slightly lower than the Rb–Sr age of 3.64 ± 0.05 AE but is just within analytical errors. For 10062, the Sm–Nd age is in reasonable agreement with the less precise Rb–Sr age of 4.01 ± 0.11 AE. Comparison of the Sm–Nd and the Rb–Sr results shows that the age of 10072 is well resolved from that of 10062 for both techniques. This

Table 3. Sm–Nd analytical results.

Sample	Weight[a] (mg)	Sm[b] (10^{-8} mole/g)	^{144}Nd[b] (10^{-8} mole/g)	$^{147}Sm/^{144}Nd$[c]	$^{143}Nd/^{144}Nd$[d]
10072					
Quint-L	3.5	34.975	26.716	0.1963	0.511998 ± 16
Plag-L	26	3.098	2.515	0.1847	0.511721 ± 18
Ilmenite-L	21	14.027	10.623	0.1980	0.512035 ± 21
Total-M	116*	14.681	10.680	0.2061	0.512238 ± 17
Pyroxene-A-L	42*	5.306	2.930	0.2715	0.513788 ± 15
Pyroxene-B-L	15	4.591	2.391	0.2879	0.514154 ± 17
10062					
Pyroxene-L	35*	4.705	2.499	0.2823	0.514258 ± 20
Pyroxene-M	20	4.985	2.791	0.2679	0.513847 ± 23
Total-M	117*	8.412	5.848	0.2157	0.512524 ± 19
Ilmenite-M	19*	3.852	2.693	0.2145	0.512487 ± 53
Plag-L	29	1.1518	0.9155	0.1887	0.511839 ± 20

[a]Samples marked by an asterisk were analyzed for Rb–Sr also.
[b]Concentrations calculated using $^{147}Sm/^{148}Sm = 1.33386$, $^{150}Nd/^{144}Nd = 0.238585$.
[c]Uncertainty ±0.1%.
[d]Measured on spiked samples; normalized to $^{146}Nd/^{142}Nd = 0.636155$.

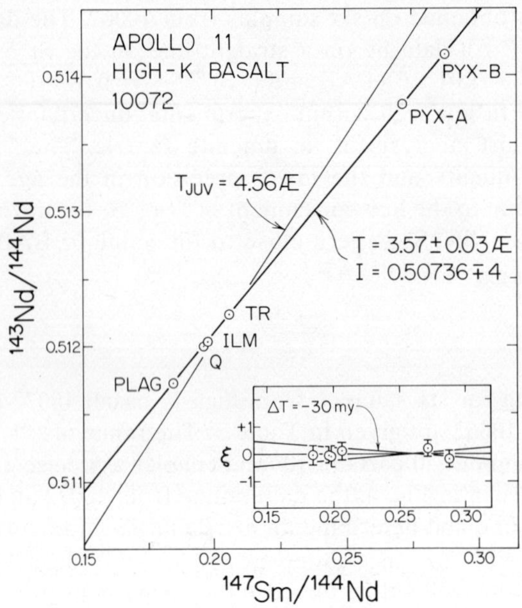

Fig. 4. Sm–Nd evolution diagram for high-K basalt 10072. The precise age and initial $^{143}Nd/^{144}Nd$ result from the large range of $^{143}Nd/^{144}Nd$ of 50 parts in 10^4 and the fact that all data lie within 0.2 parts in 10^4 of the isochron. Inset shows deviations of $^{143}Nd/^{144}Nd$ from the best fit line in parts in 10^4. The Juvinas isochron is shown for comparison.

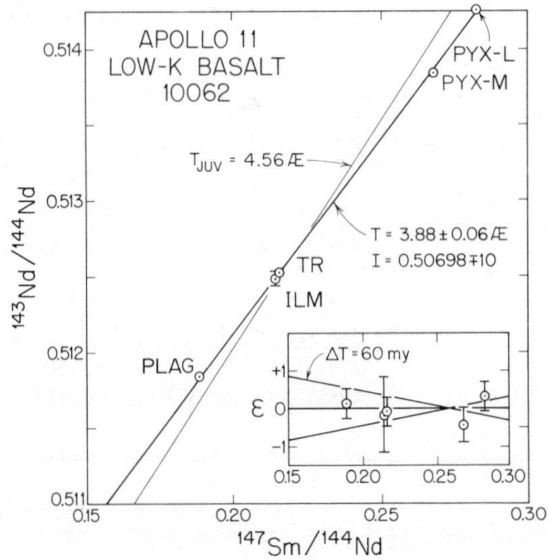

Fig. 5. Sm–Nd evolution diagram for low-K basalt 10062. The range in $^{145}Nd/^{144}Nd$ is 50 parts in 10^4. The Juvinas isochron is shown for comparison.

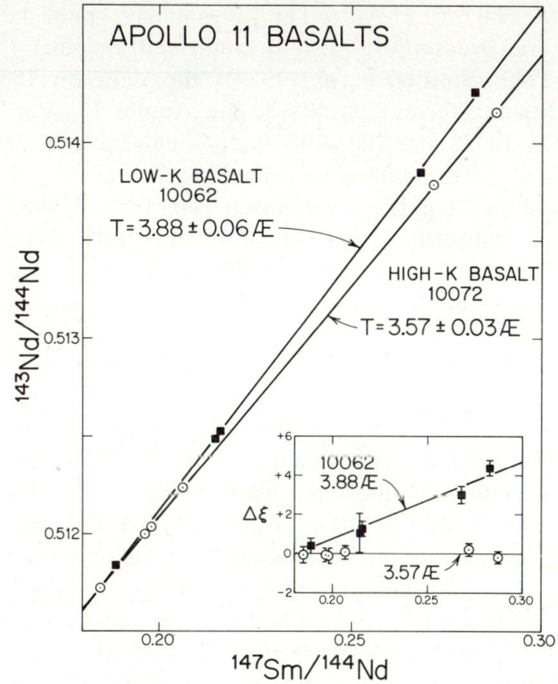

Fig. 6. Sm–Nd evolution diagram showing data for both low-K basalt 10062 and high-K basalt 10072. The data define two well-resolved linear arrays and indicate an unambiguous difference in ages for the two basalts. The inset shows the deviations in parts in 10^4 of the 10062 data points from the isochron defined by the 10072 data.

can be seen most clearly for the Sm–Nd results by inspection of Fig. 6 in which the age difference from the distinct slopes of the isochrons is readily apparent. Since the extremum points on the Sm–Nd isochrons are plagioclase and pyroxene, which are ubiquitous in lunar basalts, the Sm–Nd system could provide precise ages for most lunar basalts. In particular, precise Sm–Nd ages are obtainable on low-alkali basalts, which are difficult to date by Rb–Sr due to the rarity of high-Rb/Sr phases.

Discussion

The Rb–Sr ages of all measured A-11 mare basalts are listed in Table 1; the ages for 10072 and 10062 for all dating schemes are given in Table 4. The Rb–Sr ages for all six high-K basalts are tightly clustered within 0.09 AE from 3.59 to 3.68 AE; sample 10072 falls well within this range. The A-11 high-K basalt data are consistent with crystallization at a single time or within a very narrow time interval. The Rb–Sr ages on four low-K mare basalts have larger uncertainties of ~±0.1 AE (except 10058, Table 1). The data yield a range in ages from

3.63 ± 0.20 AE to 4.01 ± 0.11 AE. The previously published Rb–Sr data (Papanastassiou and Wasserburg, 1975a) confirmed an older ^{40}Ar–^{39}Ar age for 10003 (Turner, 1970; Stettler *et al.*, 1974); the data on 10062 confirm the existence of additional ancient basalts at the Apollo 11 site. The Rb–Sr data indicate that rocks 10058 and 10044 are indistinguishable in age, rock 10062 is distinctly older and rock 10003 is intermediate in age. Rb–Sr ages for these low-K basalts indicate the existence of three essentially distinct ages.

The Sm–Nd crystallization ages for 10072 and 10062 (Table 4) are clearly distinct and are in good agreement with the Rb–Sr ages. The precise Sm–Nd data on 10062 demonstrate that the Sm–Nd system can result in higher resolution measurements for these rocks than Rb–Sr. The Sm–Nd data clearly permit the identification of a 0.3 AE interval for volcanism at the A-11 site. Additional Sm–Nd data are required to resolve possible small time differences for some A-11 basalts.

The data presented here, which agree with ^{40}Ar–^{39}Ar data (Turner, 1970; Geiss *et al.*, 1977), demonstrate that mare filling at the Sea of Tranquility occurred over an extended time interval of ~0.4 AE. This observation is consistent with the observation of extended intervals of mare filling for the Ocean of Storms of ~0.3 AE. Similarly for Taurus-Littrow, the younger age of the orange glass determined by U–Pb (3.48 AE; Tera and Wasserburg, 1976) and by ^{40}Ar–^{39}Ar (3.54 ± 0.05; cf. Huneke *et al.*, 1973, also note discussion of trapped ^{40}Ar problems) provides strong evidence for extended volcanism at this valley. The sampling of rocks with distinct ages at locations on the moon which do not have significant contributions from large distant impacts, indicates that the filling of a single mare basin consisted of thin flows over relatively long time intervals ($2–4 \times 10^8$ yr). These flows must comprise a veneer over the older debris blanket contained in the basin. Both the flows and the older debris must contribute significantly to the local soil chemistry.

The observation of crystallization ages for mare basalts approaching 4.0 AE is of great importance in connection with the time of excavation of the large basins. Similarly the observation of mare basalt ages older than 4.0 AE would provide important constraints for the thermal evolution of the moon in the interval 4.4–4.0 AE. From the combined Rb–Sr and Sm–Nd data on 10062, an age for 10062 not less than 3.90 AE is obtained. Independent of uncertainties associated with the determination of absolute decay constants (see below), these data show that the age of a mare basalt is identical to or older than the inferred

Table 4. Age summary (AE).

	Rb–Sr	Sm–Nd	^{40}Ar–^{39}Ar	I_{Sr}	I_{Nd}
10062	4.01 ± 0.11	3.88 ± 0.06	3.82 ± 0.06[a]	0.69901 ∓ 4	0.50698 ∓ 10
10072	3.64 ± 0.05	3.57 ± 0.03	3.62 ± 0.04[b]	0.69928 ∓ 6	0.50736 ∓ 4

[a]Turner (1970) whole rock data.
[b]Geiss *et al.* (1977) plagioclase data.

age of the Imbrium event (cf. Papanastassiou and Wasserburg, 1971b). If mare volcanism was triggered by large impacts, e.g., by weakening of the lunar crust and outer mantle, the data on 10062 indicate that an event slightly prior to the Imbrium impact may be responsible for the initial volcanism at the Sea of Tranquility. Alternatively the observation of mare basalt crystallization ages older than the terminal lunar cataclysm would indicate that mare volcanism occurred largely independently of the major impacts on the moon. Based on data on 10003 Geiss *et al.* (1977) have discussed the possibility that mare basalt lavas had begun to invade the Tranquillitatis Basin when the Imbrium impact occurred, so that the Tranquillitatis Basin may be filled with alternating layers of basalt and debris. The identification of mare volcanism at 4.0 AE and at possibly earlier times remains an important problem in lunar chronology and evolution. The data show a distinct overlap between "early" mare volcanism and major impact melting and metamorphism between 4.0 and 3.8 AE. An incisive analysis of the nature and sequence of events in this period has yet to be presented.

Age comparison and ^{87}Rb half-life

Rb–Sr and Sm–Nd crystallization ages have been reported on four mare basalts: 10062 and 10072 (this work); 75075 (Lugmair *et al.*, 1975; Murthy and Coscio, 1976); and 15065 (Papanastassiou and Wasserburg, 1973; Nakamura *et al.*, 1977). The data of 10072 are of sufficient precision to attempt to check on the ^{87}Rb half-life. The half-life of ^{147}Sm has been determined most recently by Gupta and MacFarlane (1970) as 106 ± 2 AE. This value is also essentially the mean of the most precise determinations (cf., Donhoffer, 1963). Much of the quoted uncertainty in the reported half-lives was the result of uncertainty in the isotopic composition of Sm (cf., Wright *et al.*, 1961), which has since been determined more precisely (Russ *et al.*, 1971). It thus appears that the best value for the ^{147}Sm half-life is 106 AE ($\lambda = 0.00654$ AE^{-1}) and the uncertainty in this number is about $\pm 1\%$. The Rb–Sr and Sm–Nd data on 10072 show a difference in age of 0.07 AE. If this difference is due to the uncertainty in the half-life of ^{87}Rb, a downward revision in the ^{87}Rb half-life by 2% is indicated corresponding to a revised $\lambda(^{87}\text{Rb}) = 0.0142$ AE^{-1}. This would be in close agreement to the revision of $\lambda(^{87}\text{Rb})$ required for concordance with the U–Pb system using the revised U decay constants (Jaffey *et al.*, 1971). We note that use of the less precise Sm–Nd data on 15065 and on 75075 would result in different conclusions about $\lambda(^{87}\text{Rb})$. Although arguments about half-lives based on high quality age determinations of good igneous rocks are of merit, we believe that Rb–Sr ages should still be reported using $\lambda = 0.0139$ AE^{-1}, until the ^{87}Rb decay constant is in fact accurately determined by mass spectrometric methods or until more high quality age comparisons are made. In cases where detailed comparison is necessary for ages obtained by different techniques, it may be advisable to use the revised half-life for establishing agreement or disagreement between the different techniques.

From a comparison of reliable ^{40}Ar–^{39}Ar age determinations for samples subject to small gas loss and no serious trapped ^{40}Ar problems (cf. review by

Turner, 1977; Tera *et al.*, 1974; Stettler *et al.*, 1974; Kirsten and Horn, 1974) it appears that there is good agreement between Rb–Sr crystallization ages and ^{40}Ar–^{39}Ar gas retention ages. Therefore a revision of $\lambda(^{87}Rb)$ by +2% would require an increase in $\lambda(^{40}K)$ of 3% (for constant λ_e/λ) in order to avoid having ^{40}Ar–^{39}Ar ages systematically older than Rb–Sr ages. Additional absolute calibrations of the ^{40}Ar–^{39}Ar technique are desirable for a detailed comparison of ages.

For 10072, the results of all three methods are in close agreement when the plagioclase data for ^{40}Ar–^{39}Ar are considered (Geiss *et al.*, 1977). For 10062, the Rb–Sr and ^{40}Ar–^{39}Ar are barely in agreement; the Sm–Nd age has an intermediate value and is in agreement within uncertainties with both the Rb–Sr and ^{40}Ar–^{39}Ar ages. However, it is possible that the ^{40}Ar–^{39}Ar apparent age obtained on a total rock sample is slightly depressed due to ^{40}Ar loss. There is good agreement for all other Rb–Sr and ^{40}Ar–^{39}Ar ages for A-11 except for rock 10071 as pointed out by Geiss *et al.* (1977). The Rb–Sr data on 10071 should possibly be reexamined; similarly Rb–Sr data on 10049 would be useful for comparing with the much younger ^{40}Ar–^{39}Ar for this sample, although most probably the intermediate temperature ^{40}Ar–^{39}Ar plateau age on a total rock cannot be considered reliable (Geiss *et al.*, 1977). We note from the data in Table 1 that there is still only a limited number of ^{40}Ar–^{39}Ar high temperature plateaus on plagioclase.

Artifacts from mixing lines

In our work on Rb–Sr crystallization ages we have pointed out the possibility that the linear arrays obtained on Rb–Sr evolution diagrams may be the result of mixing of only two end members, plagioclase and late crystallizing interstitial phases. In such a case it is in principle possible to obtain an erroneous age if one of the end members has been an open system subsequent to the crystallization of the rock. We have expended a considerable effort in analyzing phases with distinct major element chemistry and K, Rb, Sr abundances (cristobalite, ilmenite, quintessence) for the purpose of establishing the presence in individual samples of more than two phases which are carriers of Rb and Sr. In many instances it has been possible to demonstrate by element correlations (Papanastassiou and Wasserburg, 1971a; Wasserburg and Papanastassiou, 1971b) that more than two phases were sampled and that therefore the data defined true isochrons. In contrast to the Rb–Sr system, where incompatible elements in the late crystallizing phases may be easily mobilized, the Sm–Nd age determinations depend primarily on Sm and Nd in major minerals (pyroxene and plagioclase) in which the REE would be expected to be much less mobile. The agreement of Sm–Nd and Rb–Sr crystallization ages demonstrates that for mare basalts the phases used for dating have not been disturbed since the time of crystallization. This conclusion is contrary to the discussion by Birck *et al.* (1975), who have again reexamined this problem and who believe that the possibility that the isochrons are the result of mixing lines may not permit reliable Rb–Sr age determinations.

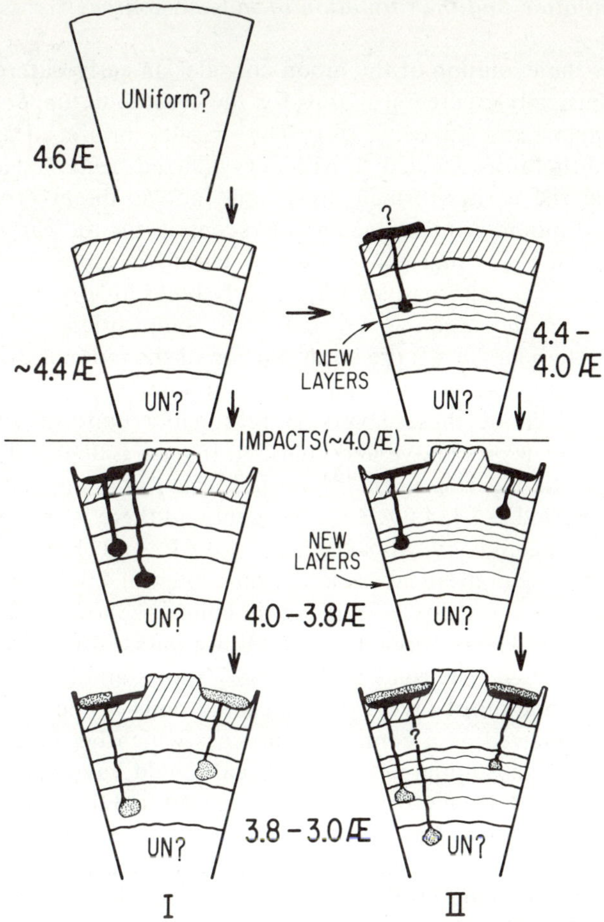

Fig. 7. Idealized representation of two possible versions for the formation of parent reservoirs for mare basalts. The coarse layering shown represents the chemically distinct reservoirs formed during the early lunar differentiation near 4.4 AE while the finer-scale layering represents reservoirs formed during the period 4.4–3.0 AE. The ruled layer represents the lunar crust. The solid and stippled blobs and layers represent mare basalt magmas generated by melting in the different layers and then erupted at the surface as lava flows filling the great lunar basins which were excavated by major impacts at ~4.0 AE. The solid blobs represent magmas formed before or near the time of impacts while the stippled blobs and flows represent the later stages of mare volcanism. UN represents possible undifferentiated material in the lunar interior from which some lunar rocks may be derived. In I, the differentiation of the lunar crust and mantle into a layered structure takes place during a single event at ~4.4 AE and volcanism is restricted to times more recent than the time of major impacts at ~4.0 AE. II depicts the possibility that volcanism and the production of new layers was an ongoing process which continued without interruption from 4.4 to 3.0 AE essentially independently of impacts at 4.0 AE.

Lunar differentiation and the formation of magma sources

Models for the evolution of the moon consider an early differentiation which resulted in a basically layered structure for the crust and "upper mantle" of the moon (cf. Taylor and Jakeš, 1974). The basalts produced in the interval 4.0–3.0 AE are therefore viewed as being crystallized from magmas which were produced from the earlier-formed layers and not as direct products from an undifferentiated moon. A highly schematic scenario for the early history of the moon is shown in Fig. 7 and will serve as a basis for subsequent discussion of the isotopic data. Two alternatives for an early lunar history are shown. In both cases, we assume the moon is formed as a compositionally uniform sphere ~4.6 AE ago. At about 4.4 AE the outer portion of the moon is differentiated and a sequence of distinct layers is formed corresponding to the crust and a layered upper mantle. Each of these layers is reasonably uniform within itself but distinct from other layers above and below it. In case I, depicted on the left half of the diagram, the layering in the moon is formed in a single distinct event at ~4.4 AE. The period ~4.4–4.0 AE is quiescent and there is only rare volcanism. Volcanism is triggered by impacts at ~4.0 AE and basaltic magmas are generated by melting in these layers in the time interval 4.0–3.0 AE. The magmas are envisioned volumetrically insignificant in comparison to the layers they were melted from, so the layers formed at 4.4 AE are undisturbed by the removal of magmas to the surface. The right half of Fig. 7 shows another possibility (II). In this case, the formation of layers or segregated pods in the older layers in the lunar interior is viewed as a continuing process which began at 4.4 AE. Under these conditions a substantial amount of basalt could have been extruded prior to 4.0 AE. For basalt magma formation between 4.4 to 3.0 AE, magmas are being melted from layers formed at 4.4 AE, and, in addition, from layers that were formed at times distinctly younger than 4.4 AE. During the time that the more ancient basalts were being produced, even more new layers were being formed in the lunar mantle. Basaltic magmas formed later are then derived from layers of a variety of ages. Some magmas are melted from 4.4 AE old layers, others are melted from layers formed 4.4–4.0 AE ago, and still others are derived from layers which were not formed until much more recently than 4.0 AE. New layers formed during the era of basaltic volcanism could have formed as a result of magma removal from pre-existing layers. For instance, a large amount of magma removed from one part of a layer could cause that part of the layer to be left with a different composition, thereby creating in effect a new sublayer. By very limited partial melting, it is also possible to obtain basaltic magmas from a single layer at successively younger times without significantly affecting the composition of the layer. In this case the same layer can provide similar basaltic melts over a prolonged interval. For our purposes different layers may have distinct major and trace element abundances. To the extent that different layers have a different Rb/Sr or Sm/Nd, they will be evolving different $^{87}Sr/^{86}Sr$ and $^{143}Nd/^{144}Nd$ from 4.4 AE onward. At any given time after 4.4 AE, each layer will have its own characteristic $^{87}Sr/^{86}Sr$ and $^{143}Nd/^{144}Nd$. If two batches of basaltic magma were formed at ~3.9 AE in two different layers and were then erupted at the surface, we would observe a difference in initial Sr and Nd in the basalts which would indicate that they were derived from different "layers" or "parent reservoirs." The major issues are whether substantial differentiation of the old layered mantle took place after ~4.4 AE and whether there exists a significant volume of the moon which remained undifferentiated. In the ensuing discussion we will use the terms "layer" and "reservoir" or "parent reservoir" interchangeably. In each case these terms are used to describe regions of the lunar interior analogous to the layers described above from which basaltic magmas have been melted and erupted at the surface.

Figure 8 shows a schematic evolution of a radiogenic isotope in two reservoirs A and B. Basalts (A1 and B1) produced at the same time with different I values must have been melted from different reservoirs ("layers") with distinct (*P*/*D*) ratios. Basalts A1 and A2 produced at different ages were extracted from a single reservoir A without altering the *P*/*D* of the parent reservoir A. This simple evolution model, where basalt parent reservoirs are formed at one time T_0 (~4.4 AE), and the basalts are formed at later times is analogous to the lunar history scheme I of Fig. 7. For model II of Fig. 7, where parent reservoirs are being formed after T_0, more complicated trajectories are possible in Fig. 8. For instance, it is possible to produce basalt A2 from a reservoir B′ which at some time prior to T_2 (e.g., at T_1) was produced from B with a *P*/*D* sufficiently large to evolve from point B1 to point A2 in the available time $T_1 - T_2$. The extent to which a reservoir can reach A2 from B1 is only constrained by the size of fractionation in *P*/*D*.

We consider the evolution of the Sr and Nd isotopic composition prior to the time of crystallization of a basalt. The ^{87}Sr and ^{143}Nd evolution is constrained to

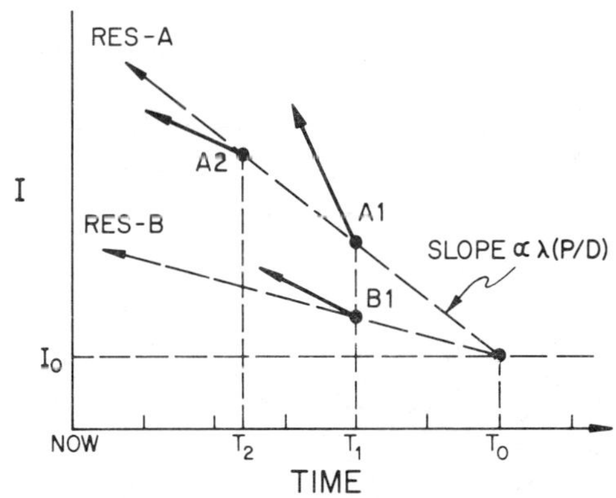

Fig. 8. Evolution of a radiogenic isotope in two distinct reservoirs as a function of time; the difference in slopes is proportional to the parent/daughter ratio (*P*/*D*) in the rock, the magma and the parent reservoir. Basalts A1 and B1 produced at the same time from distinct reservoirs will have distinct initial isotopic composition as determined from their isochrons. Basalt A2 is compatible with production from reservoir A; however, alternate, more complex or independent evolution paths are also possible.

yield the initial $^{87}Sr/^{86}Sr$ (I_{Sr}) and $^{143}Nd/^{144}Nd$ (I_{Nd}) determined by the isochrons at the time of crystallization. For Rb–Sr (Papanastassiou and Wasserburg, 1971a) and Sm–Nd we get similar equations:

$$I_{Sr} - I_{Sr}^{\circ} = \lambda_{Rb}(^{87}Rb/^{86}Sr)_1 \cdot \Delta T_1 + \lambda_{Rb}(^{87}Rb/^{86}Sr)_2 \cdot \Delta T_2 + \cdots$$
$$I_{Nd} - I_{Nd}^{\circ} = \lambda_{Sm}(^{147}Sm/^{144}Nd)_1 \cdot \Delta T_1 + \lambda_{Sm}(^{147}Sm/^{144}Nd)_2 \cdot \Delta T_2 + \cdots.$$

If T_0 is the time of formation of the moon and T_x is the time of crystallization of a particular basalt then:

$$\Delta T_1 + \Delta T_2 + \Delta T_3 + \cdots = T_0 - T_x$$

The ratios $(^{87}Rb/^{86}Sr)_1/(^{87}Rb/^{86}Sr)_2$ and $(^{147}Sm/^{144}Nd)_1/(^{147}Sm/^{144}Nd)_2$ are the fractionation factors for Rb–Sr and for Sm–Nd between stages 1 and 2. I_{Sr}° and I_{Nd}° are the initial ratios with which the moon formed. For two basalts with identical ages, a difference in the I values requires differences in the P/D ratios in parent sources lasting for some finite time intervals ΔT_i. The size of the differences in I_{Sr} and I_{Nd} depends on the degree of fractionation of P/D elements and on the length of the intervals ΔT_i for which particular $(P/D)_i$ values remain constant. Differences in I_{Sr} and in I_{Nd} provide distinct information about the parent reservoirs. One of the basic distinctions between Rb–Sr and Sm–Nd systematics is that the maximum magnitude of Sm–Nd fractionations is expected to be much smaller than possible Rb–Sr fractionations. Therefore, production of large differences in I_{Nd} *requires* relatively large time intervals ΔT_i.

Initial $^{87}Sr/^{86}Sr$ values of lunar rocks are presented and discussed in terms of a (T, I) diagram identical to Fig. 8. However, because Sm/Nd varies by a much smaller amount relative to the average Sm/Nd of most reservoirs than does Rb/Sr, we have represented the Nd isotopic data in a different way, as discussed at some length by DePaolo and Wasserburg (1976b). In this representation, the initial ratios are compared to the evolution of $^{143}Nd/^{144}Nd$ in a reference reservoir (CHUR) which has a chondritic Sm/Nd and initial Nd equal to that measured in Juvinas (Lugmair *et al.*, 1975; Lugmair *et al.*, 1976). Initial Nd for a rock which crystallized at T is given as the fractional deviation in parts in 10^4 of the initial $^{143}Nd/^{144}Nd$ from the $^{143}Nd/^{144}Nd$ in CHUR ($I_{CHUR}(T)$). This quantity is termed ϵ_I^{CHUR} and defined

$$\epsilon_I^{CHUR} = \frac{I_{ROCK} - I_{CHUR}(T)}{I_{CHUR}(T)} \times 10^4.$$

The initial $^{87}Sr/^{86}Sr$ data for Apollo 11 basalts are shown in Fig. 9, and the initial $^{143}Nd/^{144}Nd$ data for all mare basalts measured are shown in Fig. 10. For Sr a single straight line evolution curve can be drawn through all of the A-11 low-K basalts, which indicates that these basalts could all have been derived from the same reservoir at distinct times. Because of the very low Rb/Sr in the parent reservoirs of these basalts, departures from a single reservoir are difficult to detect. In contrast to this result, Beaty *et al.* (1977) have presented arguments for distinct groups of A-11 low-K basalts based on major element chemistry, and stated that the observed major element differences are neither the result of

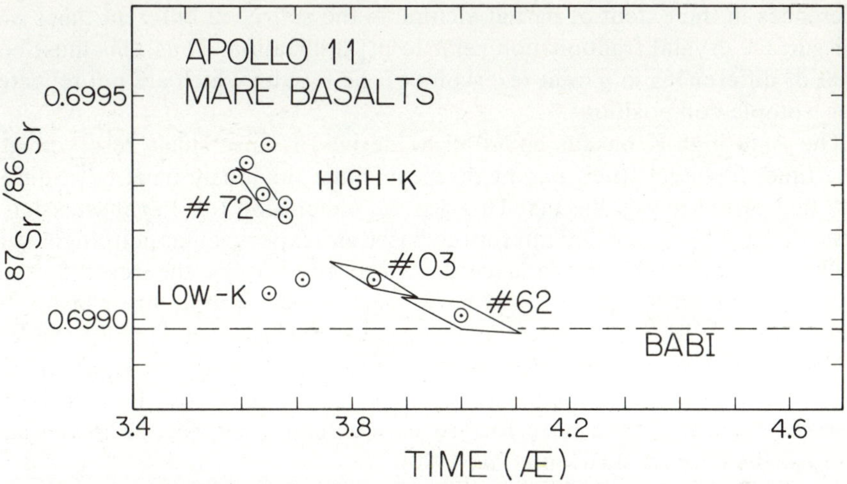

Fig. 9. Age (T), Initial $^{87}Sr/^{86}Sr$ (I) for Apollo 11 mare basalts. All high-K rocks (including 10072) are tightly grouped in both T and I. Low-K rocks show a distinct range in T and relatively lower I values. Unlabelled low-K rocks are 10044 and 10058.

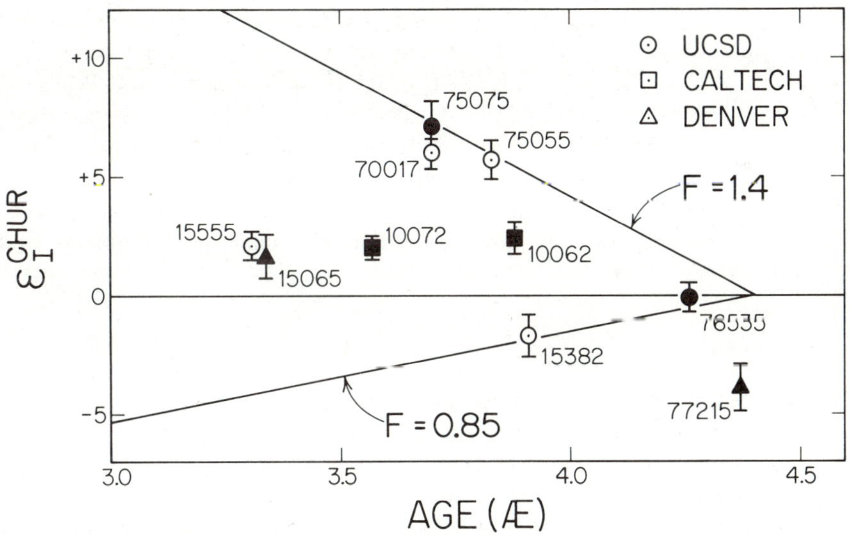

Fig. 10. Fractional deviations in parts in 10^4 of the initial $^{143}Nd/^{144}Nd$ of lunar rocks from the evolution of $^{143}Nd/^{144}Nd$ in a chondritic Sm/Nd reservoir (CHUR) plotted versus time. Solid lines diverging from CHUR at 4.4 AE represent Nd evolution in reservoirs fractionated from CHUR at that time with Sm/Nd 1.4 and 0.85 times chondritic. Filled points are samples for which Sm–Nd internal isochrons have been determined. Open points are calculated from whole rock Sm–Nd data of Lugmair (1975) and Rb–Sr internal isochron ages. CHUR line is calculated using modern $^{143}Nd/^{144}Nd = 0.511836$ and $^{147}Sm/^{144}Nd = 0.1936$ (De Paolo and Wasserburg, 1976a).

differences in the extent of partial melting in the source at different times nor of near surface crystal fractionation prior to crystallization. Thus, this must be the result of differences in parent reservoirs of the basalts which are not reflected in their isotopic compositions.

The A-11 high-K basalts could all be derived from a single reservoir at the same time; however, their parent reservoir must obviously have been different from that of the low-K basalts. This general conclusion (cf. Papanastassiou and Wasserburg, 1971a) and the inferences based on experimental petrologic studies (Walker *et al.*, 1975) are in accord. Based on Rb–Sr, the reservoir which produced the high-K basalts could have been formed early in lunar history with a relatively low Rb/Sr or as late as the time of extrusion of some of the low-K basalts ($\sim$3.9 AE) with a considerably higher Rb/Sr. If the high Rb/Sr of the high-K basalts was a characteristic of their *parent* reservoir, then that parent reservoir could not have been formed earlier than $\sim$3.9 AE, the model age of these basalts (Papanastassiou *et al.*, 1970).

For Sm–Nd, rocks 10072 and 10062 were clearly produced at distinct times. From Fig. 10, ϵ_I^{CHUR} for 10072 is slightly below that for 10062, even though 10072 is the younger basalt. Therefore, it is not possible to derive 10072 from the same reservoir as 10062. This conclusion is in agreement with the Sr data. It can also be seen in Fig. 10 that Apollo 15 basalts 15555 and 15065 (Lugmair, 1975; Nakamura *et al.*, 1977) could be derived from a reservoir similar to that of 10072, a somewhat surprising result considering the striking chemical differences in major elements between Apollo 11 and 15 basalts (Table 5). This conclusion on A-15 basalts is in agreement with Rb–Sr data (Papanastassiou and Wasserburg, 1973). It can also be seen that Apollo 17 basalts must be derived from a different reservoir than the Apollo 11 basalts as discussed more fully below. We note in Fig. 10 that sample 77215 falls in a nearly inaccessible part of the diagram and this may be the result of experimental artifacts, or due to partial isotopic equilibrium achieved for this sample, which is a clast in a highland breccia (Nakamura *et al.*, 1976).

Sources of high-Ti mare basalts

Models of the early history of the moon depict the entire outer moon as being extensively melted very soon after the moon accreted (Taylor and Jakeš, 1974). In a model such as this, the layering of the outer moon represents crystal cumulates which formed as different minerals crystallized from this magma (Philpotts and Schnetzler, 1970). The isotopic data can be related to the mineralogy of these layers and thus are directly applicable to the problem of the characterization of magma sources and lunar interior structure. The relationship of Rb/Sr and Sm/Nd to mineralogy is summarized below. The major minerals which crystallized from the magma are believed to be olivine, orthopyroxene, clinopyroxene, plagioclase, and ilmenite or spinel (Hodges and Kushiro, 1974). Of these, olivine, orthopyroxene, and ilmenite would crystallize with very low concentrations of REE, Rb, and Sr, so crystallization of these minerals from the magma will merely cause the concentrations of the trace elements in the residual liquid to be increased with *no* significant change of Rb/Sr or Sm/Nd. Clinopyroxene will crystallize with a higher Sm/Nd and lower Rb/Sr than the magma from which it crystallized (Shih *et al.*, 1975). Thus layers rich in clinopyroxene will have high Sm/Nd and low Rb/Sr, and the liquid left after clinopyroxene has crystallized will have a low Sm/Nd and slightly higher Rb/Sr. Plagioclase will crystallize with a low Sm/Nd and extremely low Rb/Sr (Drake and Weill, 1975), so layers rich in plagioclase will reflect these traits as well as have a high-Eu concentration. Residual liquids left after most of the magma is crystallized will have high concentrations of REE and Rb, high Rb/Sr and somewhat low Sm/Nd relative to the original magma. The consequences of layering in the lunar mantle for Rb–Sr and Sm–Nd systematics in second generation magmas would be directly dependent on the different proportions of clinopyroxene and plagioclase (or their high, P, T equivalents) and the presence of interstitial, trace-element rich material. In most models, plagioclase and residual liquids are expected to be more abundant in the upper part of the magma body; i.e., at relatively shallow depths in the lunar mantle. In addition, Ma *et al.* (1976) have emphasized the role of individual trace phases such as apatite in determining the trace element pattern in the layers. However, the following discussion will be limited to major phases.

It has been observed that the low-K, high-Ti basalts found on the Apollo 17 and 11 missions are similar in both major and trace element abundances (cf. Table 5) and that their compositions can be related by near surface removal of olivine (Walker *et al.*, 1975; Dymek *et al.*, 1975). It might be expected, then, that these magmas were formed by melting of the same mineralogical "layer" in the moon. The ages and initial $^{87}Sr/^{86}Sr$ of these basalts are very similar, so from Sr isotopes it could be concluded that these magmas originated in the same layer of the moon. Inspection of Fig. 10, however, shows that ϵ_I^{CHUR} of Apollo 17 basalts is +6 to +7, while for Apollo 11 low-K basalt it is about +2. This indicates that the A-17 basalts were derived from a layer which had a much higher Sm/Nd than did the source of the low-K A-11 basalts. If it is assumed that the moon formed 4.55 AE ago with $^{143}Nd/^{144}Nd$ equal to the initial ratio of the Juvinas achondrite (Lugmair *et al.*, 1975), then the layer from which the Apollo 17 basalts were derived must have had Sm/Nd at least 30% higher than chondritic. This represents a very high Sm/Nd and indicates that this layer must have been dominantly composed of clinopyroxene or possibly small amounts of garnet, the only major minerals which can have such high Sm/Nd. The layers from which the Apollo 11 basalts were derived, however, must have included higher proportions of plagioclase or residual liquid which have lower Sm/Nd. The initial $^{87}Sr/^{86}Sr$ of the low-K basalts indicate that the source layer of the A-11 basalts had Rb/Sr the same or lower than did the parent layer of the A-17 basalts. It is difficult for the A-11 *source* layer to be enough enriched in residual liquid to have a lower Sm/Nd and still have a low Rb/Sr. However, an increased amount of plagioclase in the A-11 (low-K) source layer would provide a low Rb/Sr and Sm/Nd. This conclusion is not in disagreement with experimental results on

Table 5. Chemical composition of mare basalts.

	10072[a]	10062[b]	75075[c]	15555[d]
SiO_2	40.53	38.8	38.51	44.57
TiO_2	11.74	10.3	13.33	2.10
Al_2O_3	8.52	12.1	8.29	8.69
FeO	19.76	18.3	18.85	22.53
MnO	0.24	0.27	0.25	0.29
MgO	7.68	7.21	9.68	11.36
CaO	10.42	12.0	10.17	9.40
Na_2O	0.54	0.69	0.37	0.27
K_2O	0.27	0.07	0.11	0.04
P_2O_5	0.14	<0.02	0.12	0.06
S	0.24	—	—	0.06
Cr_2O_3	0.35	0.25	0.55	0.61
K	2752	607	435	230
Rb	6.26	0.81	0.46	0.54
Sr	176	192	165	74.1
Sm	22.1	12.6	7.22	3.87
Nd	64.8	35.5	17.0	11.8
Sm/Nd[e]	0.341	0.355	0.425	0.328
Sm/Eu[e]	8.7	6.6	4.7	4.4

[a]Maxwell *et al.* (1970), trace elements from our data and Haskin *et al.* (1970).

[b]Rose *et al.* (1970), trace elements from our data and Gast *et al.* (1970).

[c]Rose *et al.* (1974), trace elements from Shih *et al.* (1975), Lugmair *et al.* (1975).

[d]Papike *et al.* (1976), trace elements from Schnetzler *et al.* (1972), Lugmair *et al.* (1975).

[e]Sm/Eu in chondrites = 2.6; Sm/Nd in chondrites = 0.32.

Apollo 11 compositions (Walker *et al.*, 1975) which indicate that plagioclase was not a *residual* mineral in the source, since *all* of the plagioclase present in the source layer could have entered the melt. The higher initial $^{87}Sr/^{86}Sr$ and relatively low ϵ_I^{CHUR} of the Apollo 11 high-K basalts indicate that their source layer was slightly enriched in residual liquid component (and possibly also depleted in plagioclase) compared to the sources of the low-K basalts. The amounts of olivine, orthopyroxene, and ilmenite are not indicated by the isotopic data, since these minerals are essentially devoid of the trace elements considered here.

In summary, it appears that high-Ti mare basalts have been derived from a multiplicity of mineralogically distinct "layers" in the lunar mantle. The Apollo 17 basalts must be derived from a layer comprised dominantly of clinopyroxene or perhaps some garnet and devoid of trapped late-stage residual liquid. The Apollo 11 low-K basalts must be derived from a similar clinopyroxene-rich layer, but containing significantly greater amounts of plagioclase. Apollo 11 high-K basalts parent sources could correspond to a clinopyroxene-rich layer which contains a small amount (1%) of trapped K, REE and Rb-rich liquid. In the simplest models (Taylor and Jakeš, 1974) this would place the source layer of the Apollo 17 basalts at the greatest depth, with the source layers of the Apollo 11 low-K basalts and Apollo 11 high-K basalts at successively shallower levels.

Sm–Nd fractionation during basalt formation and the time of the early lunar differentiation

We have reviewed recently the evidence for early lunar differentiation based on Rb–Sr and U–Pb (Papanastassiou and Wasserburg, 1976). We view as the strongest current evidence for early lunar differentiation the existence of two ancient rocks (a troctolite and a dunite) and of a polymict breccia (12013) highly enriched in KREEPUTh elements and with Rb–Sr systematics strongly indicating production of this material at ~ 4.5 AE. While the Rb–Sr and Sm–Nd data on troctolite 76535 indicate an ancient age, there exists a significant discrepancy. The distinct behavior of Sm–Nd and of Rb–Sr systematics for the troctolite (Lugmair *et al.*, 1976; Papanastassiou and Wasserburg, 1976) requires further investigation, although possible mechanisms for making the Rb–Sr and Sm–Nd data compatible have been presented which consider the Rb–Sr ancient age as the primary age. We have also reviewed extensively in the past the Rb–Sr model ages of soils and most breccias and mare basalts which indicate in a general way an early interval for lunar differentiation. More recently Nyquist *et al.* (1976) have presented more complex two-stage models to interpret the Apollo 17 mare basalt Rb–Sr data. These models require precursors to the parent sources of mare basalts from 4.6–4.4 AE with a significant range in Rb/Sr. Furthermore, the Rb/Sr in these early reservoirs are much higher than observed in basalts today. Whereas these models are constructed to assure that the Rb/Sr in the basalts are higher than in the parent reservoirs, these models do not provide a better definition of the chronology of early lunar differentiation processes. This is basically due to the small initial $^{87}Sr/^{86}Sr$ enrichments found for mare basalts which makes these calculations relatively insensitive. By contrast, U–Pb data on mare basalts and on highland samples yield a well-defined upper intersection with concordia at 4.42 AE (Tera and Wasserburg, 1974, 1976). Data on concordant whole rocks as well as isochrons for mare basalts of distinct ages define a unique intersection at 4.42 AE. Althouth this time is model dependent, it implies a precise time at which mare basalt reservoirs were produced. We review now the evidence for early lunar differentiation based on Sm–Nd.

The first Sm–Nd data bearing on early lunar differentiation were obtained by Lugmair *et al.* (1975) and Lugmair (1975). Their data on mare basalt 75075 showed that the fractionated rare earth element pattern relative to chondrites of Apollo 17 basalts was a characteristic of the parent reservoir of these rocks. To estimate the time at which the parent reservoirs acquired their fractionated Sm/Nd, Lugmair *et al.* (1975) calculated two types of model ages, which are illustrated in Fig. 12. T_{CHUR} ages (also called T_{ICE} by Lugmair *et al.*, 1976) are based on the intersection of a basalt evolution line with the chondritic evolution line. The chondritic evolution line is calculated from data on the Juvinas achondrite. Sm–Nd *model* ages can also be defined which are analogous to T_{BABI} Rb–Sr model ages by extrapolation of the evolution of mare basalts to the initial $^{143}Nd/^{144}Nd$ of Juvinas. Lugmair *et al.* (1975) found that T_{CHUR} of 75075 was close to 4.4 AE and noted that this number was similar to the 4.42 AE event indicated by U–Pb. They noted that this T_{CHUR} age could be explained if the parent reservoir for this basalt had formed at 4.4 AE with the Sm/Nd measured in 75075 from a pre-existing chondritic Sm/Nd reservoir. If T_{CHUR} ages of additional mare basalts were to cluster very near 4.4 AE, then this would provide substantial evidence for the time of early differentiation on the moon. In addition, this would imply that the Sm/Nd of each mare basalt was the same as the Sm/Nd of its parent reservoir and would therefore represent a strong constraint on the petrogenesis of mare basalts. To further investigate this possibility Lugmair (1975) obtained Sm–Nd data on three additional mare basalts. He found that 75075, 75055, and 15555 whole rocks defined an isochron indicating an age of 4.40 ± 0.06 AE and having initial $^{143}Nd/^{144}Nd$ equal to the value expected in a chondritic reservoir at 4.40 AE. This is equivalent to saying that the three basalts have T_{CHUR} ages which are all equal to 4.40 AE. These data thus provided support for the simple model for 75075 and were entirely consistent with a model for lunar differentiation in which the moon was formed with a chondritic Sm/Nd and then differentiated at 4.4 AE to form the different parent reservoirs of the mare basalts, and the basalts were formed from 4.0–3.0 AE with Sm/Nd unchanged from that of their respective parent reservoirs. One additional basalt, 70017, did not lie on the isochron defined by the other three rocks, but could be made consistent with the early differentiation at 4.4 AE if a change of Sm/Nd in the basalt of 2% relative to the parent reservoir had occurred during genesis of 70017 at ~3.7 AE. Thus, all the data suggested that less than about 2% change of Sm/Nd occurred during melting extrusion and crystallization of the basalts. However, the T_{CHUR} ages of Lugmair *et al.* (1975) and Lugmair (1975) were calculated relative to the $^{143}Nd/^{144}Nd$ of Juvinas prior to its revision. The revision of the Juvinas data causes the T_{CHUR} ages of those basalts to be increased by amounts greater than or equal to 100 m.y. depending upon the Sm/Nd of the basalt. This change makes the whole rock isochron of Lugmair (1975) more difficult to interpret. If, for instance, the 4.4 AE isochron age is taken as the age of the basalt parent reservoirs, then the initial $^{143}Nd/^{144}Nd$ from this isochron is not consistent with chondritic evolution prior to 4.4 AE. Assuming that the three point total rock isochron defines a true age, the distinct T_{CHUR} ages of the basalts indicate that the Sm/Nd of the basalts is substantially different (~4–6%) from their parent reservoirs. This situation represents a substantially different Nd evolution model than the simple two-stage model suggested by Lugmair (1975). Alternatively, it can be considered that the isochron is an accident and has no real time meaning. In any case, the revision of the Juvinas data has necessitated some major changes in interpretation of the Sm–Nd isotope, which have not been explored in the literature to date. However, it should be made clear that the conclusions from model age calculations discussed here and by Lugmair are very sensitive to the choice of parameters. As in the case of Sr evolution, it is not evident that a definite assignment for primary source ages is possible if multi-stage models must be invoked. Certainly, the Sm–Nd data show evidence of early lunar differentiation but do not precisely define the time of that event.

The revision of the Juvinas data and the acquisition of a substantial amount of new data necessitate reevaluation of the Sm–Nd evidence bearing on the time of early lunar differentiation and the behavior of rare earth elements at the time of formation of the basalts. The two are intimately related so we have chosen to discuss them in terms of a diagram (Fig. 11) which shows the relationship between the time of formation of the source reservoirs of mare basalts and the inferred Sm/Nd fractionation during genesis and crystallization of the basalts.

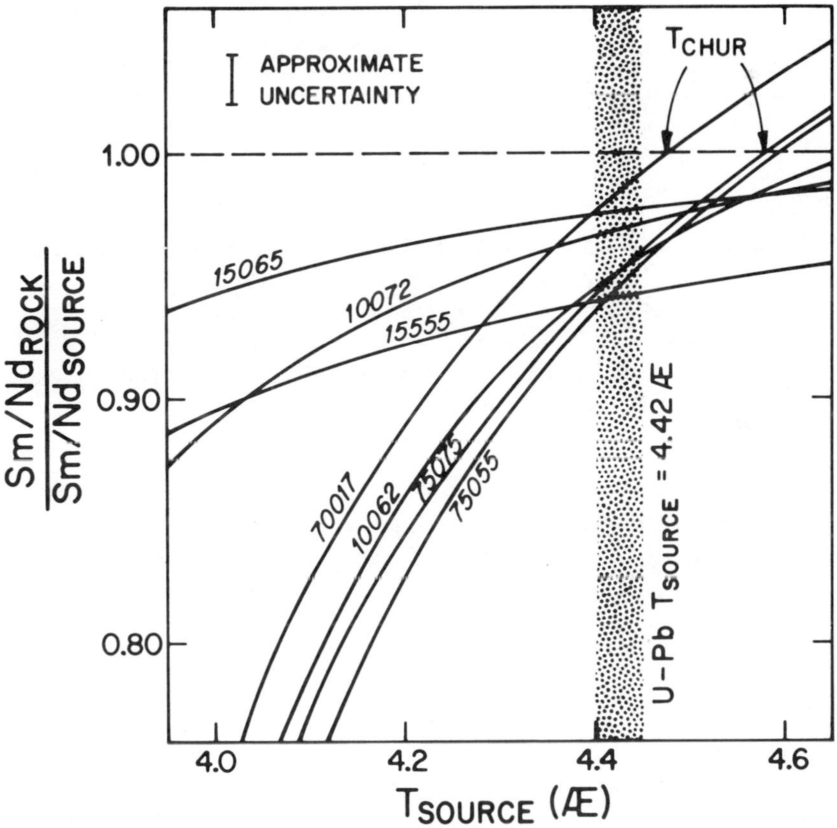

Fig. 11. Sm/Nd fractionation at the time of crystallization of mare basalts as a function of the time of formation of the source. We assume a single-stage evolution of the source and that the source was derived from a chondritic Sm/Nd reservoir. For a source age of 4.429 AE Sm/Nd fractionation range of 0.94–0.98 is obtained. For a fractionation of unity the T_{CHUR} ages are obtained (see Fig. 12). Note bunched intersections of curves in the interval 4.4–4.6 AE which, however, do not precisely define a unique T_{source}. Rock 70017 appears unique.

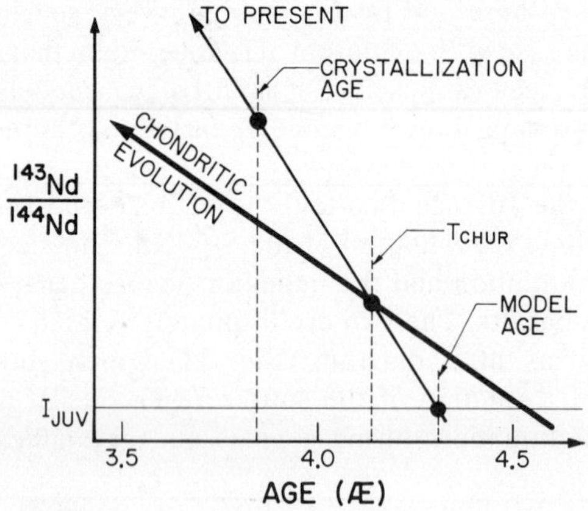

Fig. 12. Comparison of the evolution of $^{143}Nd/^{144}Nd$ in a sample and a reservoir with a chondritic Sm/Nd. The initial $^{143}Nd/^{144}Nd$ and the crystallization age are determined from an internal isochron. Extrapolation of the evolution line of $^{143}Nd/^{144}Nd$ in the sample to earlier times yields the T_{CHUR} age where it intersects the chondritic curve and the model age (equivalent to T_{BABI} for Rb–Sr) where it intersects the initial $^{143}Nd/^{144}Nd$ of Juvinas (I_{JUV}).

This diagram is calculated assuming a homogeneous moon originated with Sm/Nd and $^{143}Nd/^{144}Nd$ equal to that in the Juvinas achondrite (revised data, Lugmair *et al.*, 1976) and that at some later time, T_{source}, the source reservoir for each basalt formed with a fixed Sm/Nd ratio such that it could evolve to the initial $^{143}Nd/^{144}Nd$ of that basalt at the time of crystallization. The curves shown on this diagram are calculated from the following equation:

$$f = \frac{(\mathrm{Sm/Nd})_{rock}}{(\mathrm{Sm/Nd})_{source}} = \frac{\lambda(^{147}\mathrm{Sm}/^{144}\mathrm{Nd})_{rock}(T_s - T_x)}{[\epsilon_I^{CHUR}(T_x)\cdot I^{CHUR}(T_x)]\cdot 10^{-4} + \lambda(^{147}\mathrm{Sm}/^{144}\mathrm{Nd})_{CHUR}(T_s - T_x)}$$

where $T = 0$ refers to the present; $I^{CHUR}(0) = 0.511836$; $\lambda = 0.00654\ \mathrm{AE}^{-1}$; $(^{147}Sm/^{144}Nd)_{CHUR} = 0.1936$; T_x is the crystallization age of the basalt; and $T_s = T_{source}$, the time at which the parent reservoir of the basalt was fractionated from a chondritic reservoir. In addition, $I^{CHUR}(T) = I^{CHUR}(0) - \lambda T(^{147}Sm/^{144}Nd)$. For convenience we have chosen a chondritic Sm/Nd evolution prior to T_{source}. However, these calculations can also be performed with Sm/Nd values distinct from the chondritic value for the stage prior to T_{source}. At this time it is not in fact known whether the moon formed with chondritic Sm/Nd.

Referring to Fig. 11, if the source reservoir of 70017 was formed at 4.4 AE ago, then the Nd isotopic data indicate that the Sm/Nd of 70017 is 0.97 of Sm/Nd in its source reservoir, or that a 3% Sm/Nd fractionation occurred during the genesis of 70017 near the time of its crystallization at 3.7 AE. However, if the source reservoir of 70017 was formed at 4.2 AE, the data indicate that approximately 11% Sm/Nd fractionation occurred at 3.7 AE. The curves on this diagram intersect the dashed line corresponding to $(Sm/Nd)_{rock}/(Sm/Nd)_{source} = 1.0$, at a time T_{source} which is equal to the T_{CHUR} age of the basalt. Thus $T_{CHUR} = 4.47$ AE for 70017. If the curves for two basalts intersect on this diagram, then T_{source} of the intersection is equal to the age of an isochron drawn for the total rock data for the two basalts. As can be seen in this figure, if the source regions for the mare basalts were formed from an original chondritic reservoir at about 4.4 AE ago, then the inferred Sm/Nd fractionation factors for the basalts show a narrow range of about 0.94–0.98. If the source regions were formed at younger times, for instance 4.3 or 4.2 AE ago, then, especially for Apollo 17 basalts, the inferred Sm/Nd fractionation at the time of crystallization would change to values of 0.9 or less. This would constitute quite significant Sm/Nd fractionation. It should be noted (Fig. 11) that essentially regardless of the age of the source of the mare basalts, some Sm–Nd fractionation, about 2–6%, must be inferred at the time of crystallization of the basalts. Also, it can be seen that T_{CHUR} ages show quite a large spread, some being much greater than 4.6 AE, and therefore do not provide a precise estimate of the time of the early lunar differentiation. Nevertheless, the data in Fig. 10 show that large fractionations of Sm/Nd occurred in the formation of the parent reservoirs of mare basalts at times significantly older than the time of mare volcanism (Lugmair *et al.*, 1975). The data base is still small, but the range of ϵ_I^{CHUR} values seen in Fig. 10 indicates that, if the moon started out with chondritic Sm/Nd, then Sm/Nd fractionations which occurred *early* in lunar history range from 0.85 to a maximum of about 1.4. In contrast to this large Sm–Nd fractionation, the Sm–Nd fractionation which occurred at the time of mare basalt crystallization is probably on the order of 5%. Thus it can be concluded that the Sm–Nd data clearly indicate the existence of an early lunar differentiation, but presently do not precisely define the age of that event.

Lunar Sm/Nd

Interpretation of much of the Nd isotopic data is dependent on the Sm/Nd and initial $^{143}Nd/^{144}Nd$ of the moon. To evaluate the problem of lunar Sm/Nd it is instructive to compare lunar and terrestrial isotopic data. DePaolo and Wasserburg (1976a,b) have determined ϵ_I^{CHUR} for a variety of terrestrial igneous rocks of ages ranging from 0 to 3.6 AE. Most of these samples have ϵ_I^{CHUR} approximately equal to zero. These results indicate the existence of major magma sources with chondritic Sm/Nd over at least 4/5 of the earth's history. These autohrs consider this to be strong evidence that the earth has chondritic Sm/Nd to within a few percent. The most aberrant samples are mid-ocean ridge basalts, but even these samples are derived from a reservoir which has had an average Sm/Nd only about 10% higher than chondritic. Thus, a reasonable case can be made for a chondritic Sm/Nd for the earth.

In contrast to the apparent uniformity of Sm/Nd in the earth, lunar samples (Fig. 10) have ϵ_I^{CHUR} which deviate greatly from 0 and have a range comparable to that observed in young terrestrial basalts. This range exists despite the fact that the lunar samples are very old. Thus, large variations of ϵ_I^{CHUR} were able to evolve in the moon over a time interval of only 1 AE, whereas similar variations on the earth took 4.5 AE to develop. The lunar data indicate a range of Sm/Nd of some 40–50% in magma sources on the moon compared to the ~10% on the earth. Of the ten data shown in Fig. 10, only one has $\epsilon_I^{CHUR} = 0$. Lugmair and Marti (1977) have interpreted their data on green glass balls as indicating that some magmas were differentiated from a deep chondritic lunar source. However, there is a general lack of evidence for the existence of a long-lived chondritic Sm/Nd reservoir in the moon as found in the earth. The variability of Sm/Nd in the moon makes it impossible to deduce with confidence the bulk lunar Sm/Nd.

The contrast between the lunar and terrestrial Nd isotope data may imply a grossly different early differentiation history for these two planets. The lunar data indicate that reservoirs with greatly different Sm/Nd were formed very early in lunar history and preserved over at least 1.5 AE until the end of mare volcanism. Rare earth element abundance and Rb–Sr data also suggest early lunar differentiation in order to account for the Eu anomaly. The Sm/Nd and Eu fractionations probably resulted from the same lunar differentiation event, although it is possible that they are the result of different processes. The earth seems to have remained a fairly uniform Sm/Nd reservoir over most of its early history. A possible explanation for this difference between the earth and the moon is that a profound differentiation event, such as affected the moon, never occurred in the earth. However, this would be surprising since the gravitational energy possibly available for the moon is much smaller than for the earth. Alternatively, the differentiation began in both planets early but it was "frozen in" on the moon at an early stage due to the limited energy available. In this case, the lunar reservoirs which formed early with variable Sm/Nd remained subsequently largely undisturbed and were able to evolve their widely varying $^{143}Nd/^{144}Nd$. On the earth, however, sufficient energy may have been available so that mixing processes, e.g., convection, competed with differentiation and prevented reservoirs from being isolated long enough to evolve isotopic differences. In any case, this contrast appears to represent a first-order difference in planetary evolution and must be closely related to the early thermal histories of the two planets.

Petrogenesis of Apollo 11 high-K basalts

Most lunar basalts have Rb–Sr model ages of approximately 4.5 AE (cf. Papanastassiou and Wasserburg, 1971a). This indicates that Rb–Sr fractionation at the time of basalt crystallization was small (less than a factor of two). The Apollo 11 high-K basalts are an exception to this rule. Their model ages are approximately 3.8–3.9 AE, very close to the age of their crystallization at 3.65 AE. These low model ages correspond to about a factor of five enrichment in Rb/Sr at the time of crystallization. Thus these basalts have apparently been affected by some process during their formation which makes them unique among lunar mare basalts. Nyquist *et al.* (1976) showed that some Apollo 17 basalts also show a smaller but significant Rb/Sr fractionation. The Sm–Nd isotopic investigation of two Apollo 11 basalts presented in this paper was undertaken to determine if the apparent uniqueness of these basalts based on

Rb–Sr fractionation also applies to REE fractionation. As can be seen in Fig. 11, if the age of the source of all the basalts is the *same* and is approximately 4.4–4.5 AE then the Sm–Nd fractionation factor for the Apollo 11 high-K basalts is essentially identical to the fractionation factors found in low-K Apollo 11 basalts and in most other mare basalts. Thus there is no evidence for a unique fractionation history of the Apollo 11 basalts based on Sm–Nd isotopic systematics.

Any hypothesis to explain the formation of high-K basalts must account for the following: (1) Rb/Sr fractionation factor of about five, (2) lack of large Sm/Nd fractionation, (3) major element compositional similarity to low-K and high-Ti basalts and similarity in Sr concentrations, (4) high concentrations of KREEPUTh elements, (5) large negative Eu anomaly, and (6) absence of plagioclase as a near-liquidus phase at any pressure. Shih *et al.* (1975) have suggested that the high REE concentrations, relatively flat REE pattern, and fractionated Rb/Sr could be explained if the high-K basalts were derived by a very small degree (~ 1%) of partial melting of a source rock with clinopyroxene as the dominant residual phase. The Nd isotopic data indicate that if the source reservoir of high-K basalt 10072 was formed at 4.4 to 4.6 AE, then its Sm/Nd is only 2–5% lower than that of its source (Fig. 11). In contrast, the model of Shih *et al.* (1975) predicts that high-K magmas have Sm/Nd ~20–30% lower than their source. As can be deduced from Fig. 11, such large Sm/Nd fractionation during generation of high-K basalts could only be the case if the parent reservoir for these basalts was formed at ~3.8 AE rather than in an early lunar differentiation at 4.4–4.5 AE.

The basic contrast of large Rb/Sr and small Sm/Nd fractionation at the time of crystallization of the Apollo 11 high-K rocks significantly complicates petrogenetic models for these rocks. Significant change in Rb/Sr with moderate fractionation in Sm–Nd can be obtained by removal of plagioclase, but for these high-K basalts, plagioclase is neither a residual phase at the time of magma removal nor a liquidus phase at any time during transit of the magma to the surface. It only begins to crystallize during the later stages of crystallization after extrusion. Thus if plagioclase crystallization is responsible for the unique trace element characteristics of high-K basalts, then these characteristics must have been generated during the crystallization of the magma after extrusion. A conceivable way in which this could happen is by redistribution within a single lava flow of the last remaining 1–2% of liquid after the flow is mostly crystalline and a large amount of plagioclase has crystallized. Simplified calculations indicate that these residual liquids could have extremely high Rb/Sr but Sm/Nd only slightly different from the original magma. Preferential concentration of this late liquid in certain parts of a lava flow could result in segments of the flow being enriched in trace elements and having high Rb/Sr and thus resembling the Apollo 11 high-K basalts. In this manner these basalts could be derived from lavas which originally had lower Rb/Sr and lower trace element concentrations more typical of other mare basalts. However, the existence of a mechanism by which the late-stage liquid could be extracted from a major volume of a mostly crystalline lava flow and concentrated in a different portion of the flow is not clear, especially in thin lava flows. A petrogenetic model for the origin of high-K basalts remains as a perplexing problem.

Acknowledgment—Through the offices of The Curator of Lunar Samples, well documented materials from all the Apollo missions have been made available for use. The continuing accessibility of lunar samples as well as their careful preservation and preparation is gratefully acknowledged. This research was supported by NASA grant No. NGL 05-002-188. This work is contribution No. 2909 of the Division of Geological and Planetary Sciences, California Institute of Technology.

REFERENCES

Beaty D. W., Dymek R. F., and Albee A. L. (1977) Petrographic investigation of sample 10003, the oldest mare basalt (abstract). In *Lunar Science VIII*, p. 79–81. The Lunar Science Institute, Houston.

Birck J.-L., Fourcade S., and Allegre C. J. (1975) $^{87}Rb/^{86}Sr$ age of rocks from the Apollo 15 landing site and significance of internal isochrons. *Earth Planet. Sci. Lett.* **26**, 29–35.

Carter J. L. and MacGregor I. D. (1970) Mineralogy, petrology and surface features of some Apollo 11 samples. *Proc. Apollo 11 Lunar Sci. Conf.*, p. 247–265.

Compston W., Chappell B. W., Arriens P. A., and Vernon M. J. (1970) The chemistry and age of Apollo 11 lunar material. *Proc. Apollo 11 Lunar Sci. Conf.*, p. 1007–1028.

DeLaeter J. R., Vernon M. J., and Compston W. (1973) Revision of lunar Rb–Sr ages. *Geochim. Cosmochim. Acta* **37**, 700–702.

DePaolo D. J. and Wasserburg G. J. (1976a) Nd isotopic variations and petrogenetic models. *Geophys. Res. Lett.* **3**, 249–252.

DePaolo D. J. and Wasserburg G. J. (1976b) Inferences about magma sources and mantle structure from variations of $^{143}Nd/^{144}Nd$. *Geophys. Res. Lett.* **3**, 743–746.

DePaolo D. J. and Wasserburg G. J. (1976c) Nd isotopic variations and petrogenetic models (abstract). *EOS (Trans. Amer. Geophys. Union)* **57**, 351.

Donhoffer D. (1963) Determination of the half-lives of radioactive nuclides ^{147}Sm and ^{176}Lu occurring in nature by means of liquid scintillators. *Nucl. Phys.* **50**, 489–496.

Drake M. J. and Weill D. F. (1975) The partition of Sr, Ba, Ca, Y, Eu, and other REE between plagioclase feldspar and magmatic liquid: An experimental study. *Geochim. Cosmochim. Acta* **39**, 689–712.

Dymek R. F., Albee A. L., and Chodos A. A. (1975) Comparative mineralogy and petrology of Apollo 17 mare basalts: Samples 70215, 71055, 74255, and 75055. *Proc. Lunar Sci. Conf. 6th*, p. 49–77.

Eugster O., Tera F., Burnett D. S., and Wasserburg G. J. (1970) The isotopic composition of Gadolinium and neutron capture effects in some meteorites. *J. Geophys. Res.* **75**, 2753–2768.

Gast P. W., Hubbard N. J., and Wiesmann H. (1970) Chemical composition and petrogenesis of basalts from Tranquility Base. *Proc. Apollo 11 Lunar Sci. Conf.*, p. 1143–1163.

Geiss J., Eberhardt P., Grögler N., Guggisberg S., Maurer P., and Stettler A. (1977) Absolute time scale of lunar mare formation and filling. *Phil. Trans. Roy. Soc. Lond.* **285A**, 151–158.

Gopalan K., Kaushal S., Lee-Hu C., and Wetherill G. W. (1970) Rb–Sr and U–Th–Pb ages of lunar materials. *Proc. Apollo 11 Lunar Sci. Conf.*, p. 1195–1206.

Guggisberg S., Eberhardt P., Geiss J., Grögler N., and Stettler A. (1977) Youngest and oldest mare basalts: The temporal extent of mare filling (abstract). In *Lunar Science VIII*, p. 386–388. The Lunar Science Institute, Houston.

Gupta M. C. and MacFarlane R. D. (1970) The natural alpha radioactivity of Samarium. *J. Inorg. Nucl. Chem.* **32**, 3425–3432.

Haskin L. A., Allen R. O., Helmke P. A., Paster T. P., Anderson M. R., Korotev R. L., and Zweifel K. A. (1970) Rare earths and other trace elements in Apollo 11 lunar samples. *Proc. Apollo 11 Lunar Sci. Conf.*, p. 1213–1231.

Hodges F. M. and Kushiro I. (1974) Apollo 17 petrology and experimental determination of differentiation sequences in model moon compositions. *Proc. Lunar Sci. Conf. 5th*, p. 505–520.

Huneke J. C., Jessberger E. K., Podosek F. A., and Wasserburg G. J. (1973) ^{40}Ar–^{39}Ar measurements in Apollo 16 and 17 samples and the chronology of metamorphic and volcanic activity in the Taurus-Littrow region. *Proc. Lunar Sci. Conf. 4th*, p. 1725–1756.

Jaffey A. H., Flynn K. F., Glendenin L. E., Bentley W. C., and Essling A. M. (1971) Precision measurements of half-lives and specific activities of ^{235}U and ^{238}U. *Phys. Rev.* **C4**, 1889–1906.

Jessberger E. K., Huneke J. C., and Wasserburg G. J. (1974) Evidence for a ~4.5 AE age of plagioclase clasts in a lunar highland breccia. *Nature* **248**, 199–202.

Kirsten T. and Horn P. (1974) Chronology of the Taurus-Littrow region III: Ages of mare basalts and highland breccias and some remarks about the interpretation of lunar highland rock ages. *Proc. Lunar Sci. Conf. 5th*, p. 1451–1475.

Lugmair G. W. (1975) Sm–Nd systematics of some Apollo 17 basalts (abstract). In *Papers presented to the Conference on Origins of Mare Basalts and their Implications for Lunar Evolution*, p. 107–110. The Lunar Science Institute, Houston.

Lugmair G. W. and Marti K. (1977) Evolution of the lunar interior: Sm–Nd systematics of A15 green glass and the question of the lunar initial $^{143}Nd/^{144}Nd$ (abstract). In *Lunar Science VIII*, p. 597–599. The Lunar Science Institute, Houston.

Lugmair G. W., Marti K., Kurtz J. P., and Scheinin N. B. (1976) History and genesis of lunar troctolite 76535 or: How old is old? *Proc. Lunar Sci. Conf. 7th*, p. 2009–2033.

Lugmair G. W., Scheinin N. B., and Marti K. (1975) Sm–Nd age and history of Apollo 17 basalt 75075: Evidence for early differentiation of the lunar exterior. *Proc. Lunar Sci. Conf. 6th*, p. 1419–1429.

Lunatic Asylum (1970) Ages, irradiation history, and chemical composition of lunar rocks from the Sea of Tranquility. *Science* **167**, 463–466.

Ma M.-S., Murali A. V., and Schmitt R. A. (1976) Chemical constraints for mare basalt genesis. *Proc. Lunar Sci. Conf. 7th*, p. 1673–1695.

Maxwell J. A., Peck L. C., and Wiik H. B. (1970) Chemical composition of Apollo 11 lunar samples 10017, 10020, 10072 and 10084. *Proc. Apollo 11 Lunar Sci. Conf.* p. 1369–1374.

Murthy V. R. and Coscio M. R., Jr. (1976) Rb–Sr ages and isotopic systematics of some Serenitatis mare basalts. *Proc. Lunar Sci. Conf. 7th*, p. 1529–1544.

Nakamura N., Tatsumoto M., Nunes P. D., Unruh D. M., Schwab A. P., and Wildeman T. R. (1976) 4.4 b.y. old clast in Boulder 7, Apollo 17: A comprehensive chronological study by U–Pb, Rb–Sr and Sm–Nd methods. *Proc. Lunar Sci. Conf. 7th*, p. 2309–2333.

Nakamura N., Unruh D. M., Gensho R., and Tatsumoto M. (1977) Evolution history of lunar mare basalts: Apollo 15 samples revisited (abstract). In *Lunar Science VIII*, p. 712–714. The Lunar Science Institute, Houston.

Nier A. O. (1950) A redetermination of the relative abundances of the isotopes of carbon, nitrogen, oxygen, argon and potassium. *Phys. Rev.* **77**, 789–793.

Nyquist L. E., Bansal B. M., and Wiesmann H. (1976) Sr isotopic constraints on the petrogenesis of Apollo 17 mare basalts. *Proc. Lunar Sci. Conf. 7th*, p. 1507–1528.

Papanastassiou D. A. and Wasserburg G. J. (1971a) Lunar chronology and evolution from Rb–Sr studies of Apollo 11 and 12 samples. *Earth Planet. Sci. Lett.* **11**, 37–62.

Papanastassiou D. A. and Wasserburg G. J. (1971b) Rb–Sr ages of igneous rocks from the Apollo 14 mission and the age of the Fra Mauro formation. *Earth Planet. Sci. Lett.* **12**, 36–48.

Papanastassiou D. A. and Wasserburg G. J. (1973) Rb–Sr ages and initial Sr in basalts from Apollo 15. *Earth Planet. Sci. Lett.* **17**, 324–337.

Papanastassiou D. A. and Wasserburg G. J. (1975a) A Rb–Sr study of Apollo 17 Boulder 3: Dunite clast, microclasts, and matrix (abstract). In *Lunar Science VI*, p. 631–633. The Lunar Science Institute, Houston.

Papanastassiou D. A. and Wasserburg G. J. (1975b) Rb–Sr study of a lunar dunite and evidence for early lunar differentiates. *Proc. Lunar Sci. Conf. 6th*, p. 1467–1489.

Papanastassiou D. A. and Wasserburg G. J. (1976) Rb–Sr age of troctolite 76535. *Proc. Lunar Sci. Conf. 7th*, p. 2035–2054.

Papanastassiou D. A., Wasserburg G. J., and Burnett D. S. (1970) Rb–Sr ages of lunar rocks from the Sea of Tranquility. *Earth Planet. Sci. Lett.* **8**, 1–19.

Papike J. J., Hodges F. N., Bence A. E., Cameron M., and Rhodes J. M. (1976) Mare basalts: crystal chemistry, mineralogy and petrology. *Rev. Geophys. Space Sci.* **14**, 475–540.

Philpotts J. A. and Schnetzler C. C. (1970) Apollo 11 lunar samples: K, Rb, Sr, Ba and rare-earth concentrations in some rocks and separated phases. *Proc. Apollo 11 Lunar Sci. Conf.*, p. 1471–1486.

Podosek F. A., Huneke J. C., and Wasserburg G. J. (1972) Gas retention and cosmic-ray exposure ages of lunar rock 15555. *Science* **175**, 423–425.

Rose H. J., Cuttitta F., Dwornik E. J., Carron M. K., Christian R. P., Lindsay J. R., Ligon D. T., and Larson R. R. (1970) Semimicro X-ray fluorescence analysis of lunar samples. *Proc. Apollo 11 Lunar Sci. Conf.*, p. 1493–1497.

Rose H. J., Jr., Cuttitta F., Berman S., Brown F. W., Carron M. K., Christian R. P., Dwornik E. J., and Greenland L. P. (1974) Chemical composition of rocks and soils at Taurus-Littrow. *Proc. Lunar Sci. Conf. 5th*, p. 1119–1133.

Russ G. P., III (1974) Neutron stratigraphy in the lunar regolith. PhD. thesis, California Institute of Technology, Pasadena.

Russ G. P., Burnett D. S., Lingenfelter R. E., and Wasserburg G. J. (1971) Neutron capture on ^{149}Sm in lunar samples. *Earth Planet. Sci. Lett.* **13**, 53–60.

Schnetzler C. C., Philpotts J. A., Nava D. F., Schuhmann S., and Thomas H. H. (1972) Geochemistry of Apollo 15 basalt 15555 and soil 15531. *Science* **175**, 426–428.

Shih C., Haskin L. A., Wiesmann H., Bansal B. M., and Brannon J. C. (1975) On the origin of high-Ti mare basalts. *Proc. Lunar Sci. Conf. 6th*, p. 1255–1285.

Stettler A., Eberhardt P., Geiss J., Grögler N., and Maurer P. (1973) ^{39}Ar–^{40}Ar ages and ^{37}Ar–^{38}Ar exposure ages of lunar rocks. *Proc. Lunar Sci. Conf. 4th*, p. 1865–1888.

Stettler A., Eberhardt P., Geiss J., Grögler N., and Maurer P. (1974) On the duration of lava flow activity in Mare Tranquillitatis. *Proc. Lunar Sci. Conf. 5th*, p. 1557–1570.

Taylor S. R. and Jakeš P. (1974) The geochemical evolution of the moon. *Proc. Lunar Sci. Conf. 5th*, p. 1287–1305.

Tera F., Papanastassiou D. A. and Wasserburg G. J. (1974) Isotopic evidence for a terminal lunar cataclysm. *Earth Planet. Sci. Lett.* **22**, 1–21.

Tera F. and Wasserburg G. J. (1974) U–Th–Pb systematics on lunar rocks and inferences about lunar evolution and the age of the moon. *Proc. Lunar Sci. Conf. 5th*, p. 1571–1599.

Tera F. and Wasserburg G. J. (1976) Lunar ball games and other sports (abstract). In *Lunar Science VII*, p. 858–860. The Lunar Science Institute, Houston.

Turner G. (1970) Argon-40/Argon-39 dating of lunar rock samples. *Proc. Apollo 11 Lunar Sci. Conf.*, p. 1665–1684.

Turner G. (1977) Potassium–argon chronology of the moon. *Phys. Chem. Earth* **10**, 145–195.

Turner G., Huneke J. C., Podosek F. A., and Wasserburg G. J. (1971) ^{40}Ar–^{39}Ar ages and cosmic ray exposure ages of Apollo 14 samples. *Earth Planet. Sci. Lett.* **12**, 19–35.

Turner G., Huneke J. C., Podosek F. A., and Wasserburg G. J. (1972) ^{40}Ar–^{39}Ar systematics in rocks and separated minerals from Apollo 14. *Proc. Lunar Sci. Conf. 2nd*, p. 1589–1612.

Walker D., Longhi J., Stolper E. M., Grove T. L., and Hays J. F. (1975) Origin of titaniferous lunar basalts. *Geochim. Cosmochim. Acta* **39**, 1219–1236.

Warner J. L. (1971) Lunar crystalline rocks: Petrology and geology. *Proc. Lunar Sci. Conf. 2nd*, p. 469–480.

Wasserburg G. J. and Papanastassiou D. A. (1971) Age of an Apollo 15 mare basalt; lunar crust and mantle evolution. *Earth Planet. Sci. Lett.* **13**, 97–104.

Wasserburg G. J., Papanastassiou D. A., Nenow E. V., and Bauman C. A. (1969) A programmable magnetic field mass spectrometer with on-line data processing. *Rev. Sci. Inst.* **40**, 288–295.

Wasserburg G. J., Tera F., Papanastassiou D. A., and Huneke J. C. (1977) Isotopic and chemical investigations on Angra dos Reis. *Earth Planet. Sci. Lett.* **35**, 294–316.

Wasserburg G. J., Wetherill G. W., Silver L. T., and Flawn P. T. (1962) A study of the ages of the precambrian of Texas. *J. Geophys. Res.* **67**, 4021–4047.

Wright P. M., Steinberg E. P., and Glendenin L. E. (1961) Half-life of Samarium-147. *Phys. Rev.* **123**, 205–208.

Proc. Lunar Sci. Conf. 8th (1977), p. 1673–1696.
Printed in the United States of America

Evolution of mare basalts: The complexity of the U–Th–Pb system

DANIEL M. UNRUH and MITSUNOBU TATSUMOTO

U.S. Geological Survey, MS 963, Box 25046, Denver, Colorado 80225

Abstract—We have attempted to gain more insight into mare basalt evolution by performing a very detailed leaching and mineral separation U–Th–Pb systematics study on mare basalt 15085. We have found that ~20–50% of the U, Th, and Pb resided on the grain boundaries or in the mesostasis and that the $^{207}Pb/^{206}Pb$ ratios of the grain boundaries and crystal interiors are distinctly different. These distinct trends appear to represent either continuous or episodic post-crystallizational disturbances to the U–Th–Pb system of this rock. Using U and Pb partition coefficients, we have concluded that existing two- and three-stage U–Pb evolution models do not accurately describe mare basalt genesis. We have proposed an alternative two-stage + KREEP mixing model as a simple approximation to U–Pb evolution in lunar rocks. Most Rb–Sr and Sm–Nd data are compatible with this model.

INTRODUCTION

LUNAR MARE BASALTS are the only "young" (~3–4 b.y.) volcanic rocks yet observed on the moon and it is generally believed that they represent partial melting of cumulates which formed during the latter stages of lunar primary differentiation. However, as Ringwood and Kesson (1976) pointed out, simple evolution such as varying degrees of partial melting of a common source, or "two-stage" cumulate remelting will not satisfactorily account for all of the features observed in mare basalts such as their Mg/Fe ratios and Cr_2O_3 contents. Therefore, these authors proposed that mare basalts are the products of multi-stage differentiation and magma mixing. Alternatively mare basalts may have originated from late-stage "layered cumulates" (Shih and Schonfeld, 1976) or from cumulates containing both early- and late-crystallizing material (Drake and Consolmagno, 1976).

U–Pb data (Tera and Wasserburg, 1974, 1976) suggested that evolution of most lunar rocks (both mare and non-mare) could be approximated by two stages and that primary lunar differentiation occurred at ~4.4 b.y. (10^9 yr) ago. Sm–Nd data (Lugmair *et al.*, 1975), based on "chondritic" model age calculations, initially supported a two-stage model and the 4.4 b.y. primary differentiation age. However, the "chondritic" initial $^{143}Nd/^{144}Nd$ ratio has recently been redetermined (Lugmair, 1976; Nakamura *et al.*, 1976, 1977) and the recalculated "chondritic" model ages now range from ~3.3 b.y. for mare basalt 15426 (Lugmair *et al.*, 1977) to ~6.2 b.y. for mare basalt 15555 (Lugmair *et al.*, 1976). Therefore, based on the revised Sm–Nd data it seems reasonable to conclude that at least some mare basalts have had a more complex evolutionary history. Nunes *et al.* (1975) have proposed that more complex models should be used to interpret U–Pb data, although the U–Pb data alone do not yield a unique interpretation (e.g., compare Nunes *et al.*, 1975 with Tera and Wasserburg, 1974). Nunes *et al.* (1975) have also suggested that the apparent 4.4 b.y. U–Pb upper concordia intercept could also be interpreted as a KREEP-mixing line, thus our group has referred to this discordia line as the *KREEP line*, and we have been skeptical about the significance of the 4.42 b.y. age.

One should note, however, that all of the previously mentioned age interpretations are based on certain assumptions and that the apparent older model ages will vary with the assumptions used in constructing the model (e.g., Tera and Wasserburg, 1972a, b, 1974; Tatsumoto and Unruh, 1976).

Another problem encountered with a two-stage evolution model for mare basalts is that the Pb–Pb and U–Pb (parent daughter) whole-rock "isochrons" (e.g., Tatsumoto, 1970; Tatsumoto *et al.*, 1971, 1972) are not in agreement with either the Rb–Sr internal or whole-rock isochron ages (e.g., Papanastassiou and Wasserburg, 1970, 1971a, 1973; Papanastassiou *et al.*, 1970; Nyquist *et al.*, 1973; Compston *et al.*, 1972) or the 4.4 b.y. "primary differentiation" age. In order to more closely examine the discrepancies between the Pb–Pb and Rb–Sr and Sm–Nd ages, as well as to determine more about U–Th–Pb evolution in mare basalts, we have performed a rather detailed U–Th–Pb systematics study on low-Ti mare basalt 15085 collected at Elbow Crater near the Appenine Front.

SAMPLES AND SAMPLE PREPARATION

Rock 15085 is a coarse-grained porphyritic clinopyroxene (pigeonite) basalt containing 66% zoned clinopyroxene, 22% plagioclase feldspar, 4% tridymite, 1.5% ilmenite, 1.5% ulvospinel, and a trace of troilite and Cr-spinel (Wilshire and Brett, 1971). This sample of 15085 was a re-allocation that had previously been partially crushed and treated with heavy liquids. Even though the sample had been washed with acetone, there is still the possibility that some Pb contamination was induced by treatment with the heavy liquids (as we discuss in the Appendix, the contamination does not cause a serious problem in the data interpretation). Nevertheless, this sample was chosen for the present study because it was the only large sample of coarse-grained mare basalt in our possession. Furthermore the "clean" nature (lack of large inclusions) of most of the crystals, especially plagioclase, enabled us to obtain good mineral separations.

The partially crushed sample was sieved into +50, −50–+100, and −100 mesh sizes. Mineral separations were done by handpicking from the +50 mesh material. Additional opaque (OP in Figs. and Tables, =US + ILM + TROIL + Cr-SPIN) material was obtained from the −50–+100 mesh fraction. The zoned pyroxene was separated into two fractions, green pigeonite cores (PXL in Tables and Figs.) and brown augite rims (PXD) (Wilshire and Brett, 1971). The plagioclase separate (PL) was more than 99% free of mafic material but probably contained some tridymite.

All whole-rock samples and mineral separates were leached with ~0.5 ml of cold 2 N HCl for 10 min prior to decomposition. The leaching serves two purposes. First, it removes any surficial Pb contamination that may have been picked up during the mineral separation processes and previous treatment with heavy liquids. The second purpose of the acid leach is to remove any Pb indigenous to the sample that resides in micro-fractures or on grain boundaries. The HCl leaching should not remove any Pb, U, or Th residing within the crystal lattice (Shimizu, 1974; although his leaching experiments were performed on unfractured terrestrial mineral grains). Thus, by comparing the data of the acid leach with that of the residue we may be able to determine if such processes as Pb diffusion (or intermediate daughter diffusion) or recent shock metamorphism have played a role in the U–Th–Pb evolution of this rock.

Analytical Procedures

Wet chemical procedures have been previously described by Tatsumoto and Unruh (1976). The diameter of the HBr anion exchange column used for the Pb separation has been increased from 0.08 to 0.15 cm to provide more consistent Pb recovery for larger samples.

In this study all whole-rock (WR) and mineral fractions were divided into two portions. One portion was analyzed for its Pb isotopic composition (designated by a "P" in Tables and Figs.) and the other portion was "spiked" with a ^{208}Pb, ^{235}U, ^{230}Th-enriched solution for U, Th, and Pb concentration determinations ("C" in Tables and Figs.). Prior to spiking and analysis all samples were first leached with cold 2 N HCl for about 10 min. The leachate from the Pb isotopic composition (P) analysis was aliquoted into two portions, one of which was spiked for U, Th, and Pb concentration determinations (P–LC in Tables and Figs.). The Pb isotopic composition of the leachate was determined from the other portion of solution (P–LP in Tables and Figs.). The entire leachate from the powder split used for the U, Th, and Pb concentration determinations (C) was spiked (C–LC).

Samples were run on an NBS-type tandem mass spectrometer equipped with a pulse counter on the second stage (Tatsumoto and Unruh, 1976). Whenever the signal was sufficiently strong, the sample was measured by both the pulse counter and Faraday cage collector on the first stage. The data obtained by pulse counting were almost always within uncertainty the same as that obtained by the Faraday cage, and no systematic biases were observed, so that the data obtained by pulse counting could be directly compared to the data obtained by the Faraday cage.

Lead blanks for the residues ranged from 0.3 ± 0.1 ng(10^{-9}g) for 50–100 mg sized samples to 0.20 ± 0.05 ng for 10-30 mg sized samples and leaches. The ^{206}Pb blank/sample ratios (see Tatsumoto and Unruh, 1976, for details) ranged from ~ 0.01 in the opaque and -100 mesh residue samples to ~ 0.08 in the pyroxene-core residue fraction (and as high as 0.15 in the leach fraction of this sample). The ^{204}Pb blank/sample ratios ranged from ~ 0.04 to ~ 1.0 (average ~ 0.3) and thus there are relatively large uncertainties associated with the blank-corrected ratios involving ^{204}Pb. U and Th blanks were less than 0.005 ng and ^{238}U blank/sample ratios generally ranged from 0.005 to 0.0005. Exceptions were the plagioclase and pyroxene-core separates where the ^{238}U blank/sample ratio ranged from 0.011 to 0.067.

Experimental Results

The relative amounts of U, Th, and Pb that were extracted from WR samples and mineral separates of 15085 by leaching with cold 2 N HCl are shown in Table 1 and the absolute U, Th, and Pb concentrations in the leaches (–LC) and residues (undesignated) are shown in Table 2. From Table 1 it is evident that significant portions of U, Th, and Pb are easily leachable, and thus we infer that these portions of U, Th, and Pb originally resided on grain boundaries, in micro-fractures, or in the intersticies or residual quench phases. This observation is not especially surprising since Pb, U, and Th are all incompatible elements in most basic minerals ($D^{s/l} \ll 1$ for most rock-forming silicates) and we might expect crystals to be progressively enriched in incompatible elements outward from the center during crystal growth. Therefore, the U, Th, and Pb in the residue fractions is probably not uniformly distributed. Tera *et al.* (1972) and Tera and Wasserburg (1974, 1975) have also observed large amounts of easily leachable Pb, U, and Th in anorthosite 15415 and in mare basalts 15555 and 75055.

Another interesting result of the leaching is the apparent depletion of ^{207}Pb relative to ^{206}Pb in the leached fractions (Table 1). The Pb isotopic compositions

Table 1. Percentages of U, Th, and Pb leached from mineral separates of 15085 by cold 2 N HCL.

	% leached						
Sample[1]	U	Th	Pb	^{204}Pb	^{206}Pb	^{207}Pb	^{208}Pb
WR1P*	42.6	41.0	45.3	52.8	46.0	41.8	45.8
WR1C	40.7	38.2	40.8	28.8	42.1	35.9	41.9
PXDP*	23.8	25.0	30.4	52.2	29.8	27.2	31.8
PXDC	31.4	30.1	36.1	63.9	35.4	34.3	37.4
< 100 P*	38.8	38.5	71.1	95.9	64.6	73.2	74.5
OPP*	33.3	34.4	38.2	53.6	37.3	32.7	42.1
PLP*	42.2	—	68.7	70.8	66.7	63.8	68.8
PLC	50.0	—	49.2	53.1	47.1	43.8	49.6
PXLC	14.1	—	33.3	35.1	32.4	33.8	33.7

[1] WR = whole rock, PXD = brown pyroxene rims (augite), < 100 = −100 mesh whole sample, OP = opaques, PL = plagioclase, PXL = green pyroxene cores (pigeonite).

* Percentages were calculated by assuming that the corresponding "P" and "C" residue analyses have the same ^{206}Pb, U, and Th concentrations.

of the leach and residue fractions are shown in Table 3 and these data are plotted on a $^{206}Pb/^{204}Pb$ vs. $^{207}Pb/^{204}Pb$ diagram in Fig. 1. Unleached whole-rock analyses from 15085 and other Elbow Crater basalts (Tatsumoto *et al.*, 1972) are shown for comparison. The leaches and residues form distinctly separate trends in Fig. 1 and the apparent ^{207}Pb depletion relative to ^{206}Pb in the leaches is quite obvious. Notice also that the unleached whole-rock analyses (open circles) lie between the two trends.

^{208}Pb, on the other hand, appears to be enriched relative to ^{206}Pb in the leach fractions (Table 3). However, as previously mentioned, it is possible that the leach fractions contain a significant amount of modern terrestrial Pb contamination induced by prior sample handling. Addition of Pb contamination would increase both the $^{207}Pb/^{206}Pb$ and $^{208}Pb/^{206}Pb$ ratios (and decrease the $^{206}Pb/^{204}Pb$ ratio) of most lunar samples. Since in most cases the $^{208}Pb/^{206}Pb$ ratios of the leach fractions are only slightly higher than those of the residues, the apparent ^{208}Pb enrichment could easily be due to contamination. The $^{207}Pb/^{206}Pb$ ratios, however are distinctly lower than those of the residues, and *this trend cannot be explained by pre-analysis Pb contamination in the leach fractions.*

The U–Pb data from Tables 2 and 3 are plotted on Wetherill's (1956) U–Pb evolution diagram in Fig. 2. The distinct trends defined by the leach and residue fractions are again quite evident. The data are corrected for "initial" Pb in two ways: (1) nominal blank corrections with the remaining ^{204}Pb removed as

Table 2. U, Th, and Pb concentrations of mineral separates and mineral separate leaches from 15085.

Sample[1]	Weight (mg)	ppm U[2]	ppm Th[2]	ppm Pb[3]	$\frac{^{232}Th}{^{238}U}$	$\frac{^{238}U^{2,3}}{^{204}Pb}$
WR1C	94.70	0.0630 ± 0.0006	0.250 ± 0.002	0.116 ± 0.003	4.09	227 ± 30
WR1C–LC	(94.70)	0.0433 ± 0.0004	0.155 ± 0.001	0.0800 ± 0.0005	3.70	281 ± 13
WR1P–LC	(35.33)	0.0467 ± 0.0004	0.174 ± 0.001	0.096 ± 0.001	3.85	127 ± 10
WR2C	21.19	0.0639 ± 0.0009	0.253 ± 0.005	0.111 ± 0.005	4.10	297 ± 50
PXDC	84.25	0.110 ± 0.001	0.439 ± 0.002	0.195 ± 0.003	4.13	364 ± 45
PXDC–LC	(84.25)	0.0504 ± 0.0004	0.189 ± 0.004	0.110 ± 0.001	3.88	94 ± 2
PXDP–LC	(35.33)	0.0345 ± 0.0003	0.146 ± 0.001	0.085 ± 0.001	4.38	115 ± 7
< 100C	69.06	0.150 ± 0.001	0.541 ± 0.005	0.267 ± 0.002	3.71	456 ± 50
< 100P–LC	(22.22)	0.095 ± 0.001	0.339 ± 0.002	0.657 ± 0.005	3.65	11.7 ± 0.2
OPC	9.94	0.291 ± 0.003	0.978 ± 0.008	0.415 ± 0.010	3.47	924 ± 100
OPP–LC	(9.95)	0.145 ± 0.001	0.513 ± 0.004	0.257 ± 0.004	3.66	405 ± 60
PLC	100.98	0.0041 ± 0.0002	—	0.042 ± 0.003	—	10.0 ± 1
PLC–LC	100.98	0.0041 ± 0.0001	0.0189 ± 0.0001	0.0406 ± 0.0006	4.76	8.0 ± 0.2
PLP–LC	(27.07)	0.0030 ± 0.0002	0.0137 ± 0.0001	0.092 ± 0.002	4.77	2.5 ± 0.3
PXLC	57.91	0.0073 ± 0.0003	—	0.045 ± 0.006	—	11.6 ± 0.9
PXLC–LC	57.91	0.0012 ± 0.0001	0.0056 ± 0.0001	0.020 ± 0.002	4.68	4.2 ± 0.6

[1] Leaches are designated by "–LC". Residues are undesignated.

[2] Uncertainties primarily correspond to 2 σ from the mass spectrometer run and ±50% of the blank.

[3] Uncertainties primarily correspond to ±0.05–0.10 ng Pb blank, 2 σ from the mass spectrometer and ±0.1% of the spike concentration.

[4] Sample weights in parentheses (leaches) were calculated based on the total weight prior to leaching and aliquoting.

meteorite primordial Pb (open symbols; Tatsumoto *et al.*, 1973) and (2) removal of all ^{204}Pb and commensurate amounts of the other three Pb isotopes as terrestrial contamination (closed symbols). These "initial Pb" corrections represent the reasonable maximum extremes in terms of model ages only, and since no attempt has been made to correct the data for Pb derived from U decay prior to the first observable event, the upper intercept ages have no strict age significance. However, both the leach and residue trends (except PL) suggest a lower concordia intercept age of about 1 b.y. regardless of the type of correction made (broken lines in Figs. 2a and b).

A tie-line connecting PL and the lower cluster of points (nominal blank correction, Fig. 2a, solid line) intersects concordia at ~4.3 and ~2.5 b.y. A similar tie-line for the data corrected for maximum blank (solid line, Fig. 2b) intersects concordia at ~4.7 and ~3.0 b.y.

The apparent ~1.0, 2.5, and 3.0 b.y. lower concordia intercepts (Figs. 2a and b) indicate that some sort of post-crystallizational disturbance has affected the U–Pb systematics of this rock. The disturbance might, only in part, be artificially

Table 3. Pb isotopic compositions of mineral separates and mineral separate leaches from 15085.

Sample[2]	Weight[3] (mg)	Blank (ng)	Observed $^{206}Pb/^{204}Pb$	Corrected for blank[1] $^{206}Pb/^{204}Pb$	$^{207}Pb/^{206}Pb$	$^{208}Pb/^{206}Pb$	$^{207}Pb/^{204}Pb$	$^{208}Pb/^{204}Pb$
WR1P	123.15	0.40	141 ± 1	158 ± 30	0.4917 ± 0.0045	1.151 ± 0.011	77.8 ± 15	182.1 ± 35
WR1P–LP	(79.99)	0.20	107 ± 1	117 ± 7	0.4126 ± 0.0021	1.143 ± 0.0053	48.3 ± 3.1	134.0 ± 8.5
WR1P–LC	(35.33)	0.15	101 ± 1	121 ± 7	0.4134 ± 0.0029	—	49.9 ± 3.4	—
WR1C	94.70	0.41	154 ± 3	183 ± 41	0.4768 ± 0.0050	—	86.8 ± 20	—
WR1C–LC	(94.70)	0.20	196 ± 1	239 ± 10	0.3678 ± 0.0018	—	87.8 ± 6	—
WR2C	21.19	0.25	140 ± 5	232 ± 32	0.4764 ± 0.0053	—	110 ± 15	—
PXDP	99.86	0.40	305 ± 6	317 ± 15	0.4466 ± 0.0013	1.079 ± 0.003	142 ± 10	342 ± 25
PXDP–LP	(64.53)	0.20	100 ± 1	124 ± 6	0.3930 ± 0.0021	1.187 ± 0.008	48.6 ± 2.5	146 ± 8
PXDP–LC	(35.33)	0.20	104 ± 1	127 ± 7	0.3890 ± 0.0025	—	49.3 ± 2.7	—
PXDC	84.25	0.41	286 ± 9	296 ± 20	0.4434 ± 0.0016	—	131 ± 9	—
PXDC–LC	(84.25)	0.20	86 ± 1	91.5 ± 1.5	0.4219 ± 0.0015	—	38.6 ± 0.7	—
<100P	66.11	0.41	275 ± 5	346 ± 35	0.4660 ± 0.0022	1.084 ± 0.006	161 ± 10	376 ± 20
<100P–LP	(43.89)	0.20	26.66 ± 0.07	26.81 ± 0.08	0.6939 ± 0.0004	1.730 ± 0.002	18.61 ± 0.06	46.40 ± 0.15
<100P–LC	(22.22)	0.20	26.62 ± 0.09	26.77 ± 0.10	0.6954 ± 0.0010	—	18.62 ± 0.08	—
<100C	69.06	0.41	285 ± 3	367 ± 30	0.4593 ± 0.0013	—	169 ± 17	—
OPP	25.15	0.15	488 ± 2	650 ± 55	0.4277 ± 0.0010	0.9675 ± 0.0030	278 ± 24	629 ± 55
OPP–LP	(15.20)	0.15	230 ± 1	334 ± 60	0.3482 ± 0.0025	1.181 ± 0.009	117 ± 20	396 ± 70
OPP–LC	(9.95)	0.20	198 ± 5	329 ± 57	0.3458 ± 0.0027	—	114 ± 20	—
OPC	9.94	0.20	249 ± 5	628 ± 80	0.4624 ± 0.0028	—	290 ± 40	—
PLP	82.10	0.40	30.0 ± 0.2	30.7 ± 0.3	0.8686 ± 0.0011	1.597 ± 0.008	26.7 ± 0.3	49.1 ± 0.6
PLP–LP	(55.03)	0.25	24.83 ± 0.03	25.3 ± 0.1	0.7654 ± 0.0011	1.766 ± 0.004	19.4 ± 0.1	44.7 ± 0.2
PLC	100.98	0.20	31.4 ± 0.2	32.7 ± 0.4	0.8530 ± 0.0020	—	27.8 ± 0.4	—
PLC–LC	(100.98)	0.20	25.1 ± 0.3	25.6 ± 0.3	0.7458 ± 0.0016	—	19.1 ± 0.2	—
PXL–P	40.57	0.25	25.0 ± 0.2	26.1 ± 0.3	0.7286 ± 0.0034	1.750 ± 0.009	19.0 ± 0.2	45.7 ± 1.2
PXLP–LP	(28.41)	0.20	21.5 ± 0.2	22.8 ± 0.3	0.7649 ± 0.0035	1.858 ± 0.012	17.4 ± 0.3	42.3 ± 0.7
PXLC	57.91	0.25	25.3 ± 0.1	26.3 ± 0.2	0.7136 ± 0.0035	—	18.8 ± 0.2	—
PXLC–LC	(57.91)	0.20	20.26 ± 0.02	21.2 ± 0.2	0.7985 ± 0.0053	—	16.9 ± 0.2	—

[1] Uncertainties primarily correspond to ±0.05–0.10 ng Pb blank, 2σ from the mass spectrometer and ±0.1% of the spike concentration. Since "C" samples were spiked with ^{208}Pb, no ^{208}Pb isotopic compositional data was obtained for these samples.

[2] Leaches are designated by "–LC" or "–LP". Residues are undesignated. Pb isotopic compositional analyses are designated by a "P" (e.g., WR1P). Pb isotopic compositions obtained from the concentration run are designated by a "C" (e.g., WR1C).

[3] Sample weights in parentheses (leaches) were calculated based on the total weight prior to leaching and aliquoting.

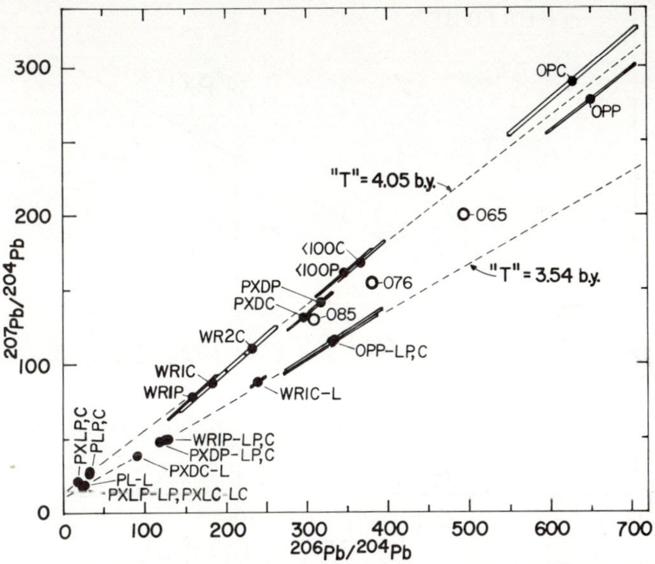

Fig. 1. $^{206}Pb/^{204}Pb$ vs. $^{207}Pb/^{204}Pb$ plot of mineral separate residues and leaches from 15085. The leaches and residues define distinctly different trends yielding apparent ages of 3.54 ± 0.06 and 4.05 ± 0.08 b.y., respectively. The ages probably have no significance, and are much older than Rb–Sr and Sm–Nd ages of ~3.3–3.4 b.y. for this rock. The two trends suggest that post-crystallizational disturbances have affected the Pb–Pb system. The "old" ages suggest ^{207}Pb enrichment from an outside source or lack of isotopic homogenization prior to crystallization.

induced (i.e., contamination, leaching, or prior treatment with heavy liquids) but is most likely due to natural processes such as meteorite impact or trace-element diffusion. Each of these possibilities will be discussed in detail in the next section of this paper.

In order to more closely examine the individual U–Pb and Th–Pb systems, the U–Pb data from Tables 2 and 3 are plotted on $^{238}U/^{204}Pb$ vs. $^{206}Pb/^{204}Pb$, $^{235}U/^{204}Pb$ vs. $^{207}Pb/^{204}Pb$, and $^{232}Th/^{204}Pb$ vs. $^{208}Pb/^{204}Pb$; parent vs. daughter diagrams in Fig. 3. These types of diagrams are similar to the commonly used Rb–Sr and Sm–Nd evolution diagrams. The $^{238}U/^{204}Pb$–$^{206}Pb/^{204}Pb$ residue data (Fig. 3a, open circles) define a line corresponding to an "age" of 3.44 ± 0.22 b.y. (2σ; York, 1969; lower broken line). The large uncertainty in the "age" is a result of large errors in the ^{204}Pb concentrations (see "Analytical Procedures" section). The apparent age is in excellent agreement with the Rb–Sr age of 3.4 b.y. (Papanastassiou and Wasserburg, 1973; two points only) and the Sm–Nd age of 3.34 ± 0.09 b.y. (Nakamura *et al.*, 1977) for rock 15065, another Elbow Crater basalt. The apparent agreement between the U–Pb and Rb–Sr ages may be fortuitous since the apparent U–Pb age is controlled by the opaque (OP) fraction (i.e., deletion of this point yields a 3.68 ± 0.23 b.y. age). The OP fraction is the only non-silicate fraction, and it is quite possible that we have preferentially leached Pb relative to U from this fraction. Tera and Wasserburg

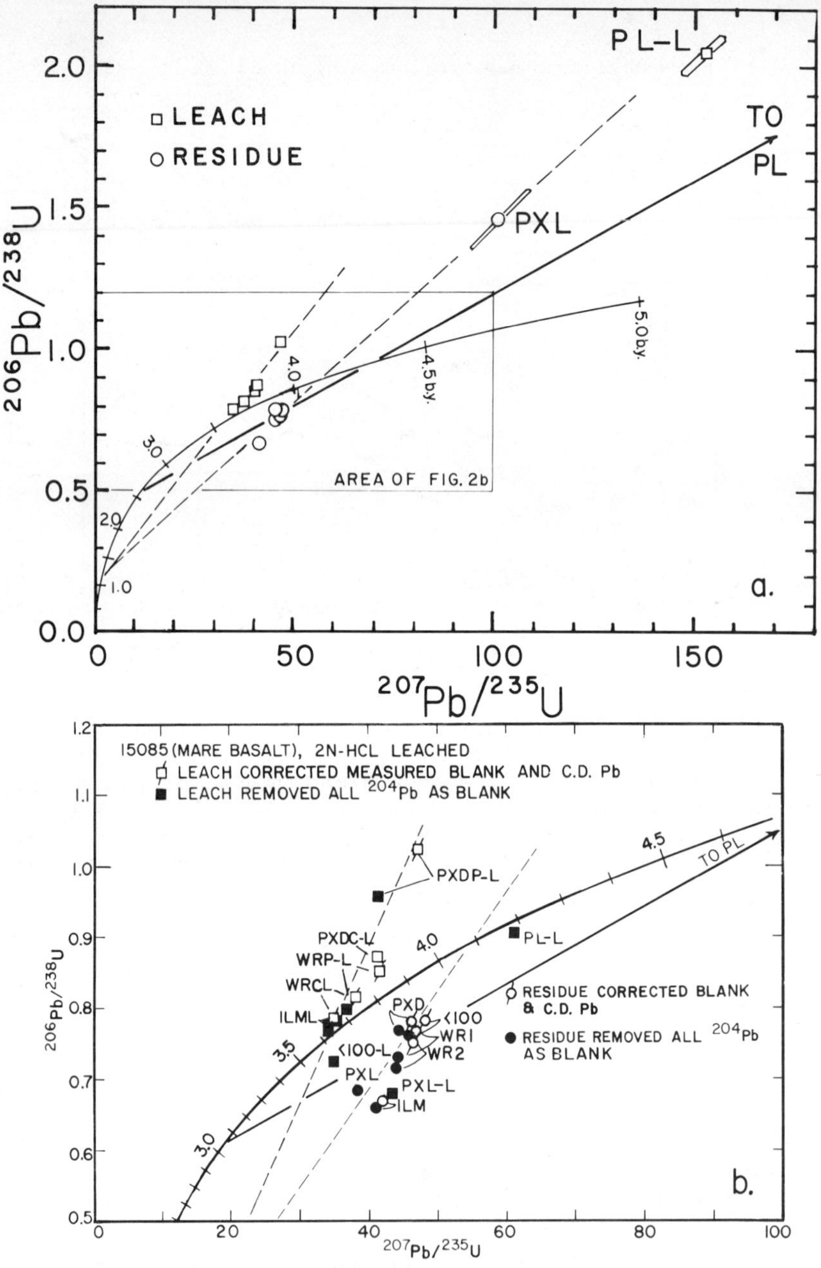

Fig. 2. U–Pb evolution diagram for 15085. Data are corrected for Canyon Diablo troilite Pb (open symbols) and all ^{204}Pb as modern terrestrial Pb (blank; closed symbols). (a) The leach (squares) and residue (circles) data define two distinct trends yielding upper concordia intercepts of ~3.6 and 4.1 b.y., respectively (broken lines). These upper intercepts probably have no strict age significance due to improper initial Pb correction. An ~1.0 b.y. "metamorphic" event is also suggested by the apparent lower concordia intercept. The solid line is a tie-line connecting PL and the lower cluster of points and intersects concordia at ~4.3 and 3.0 b.y. (b) The expanded portion of Fig. 2a shows the relations of individual data in more detail. The broken lines are those shown in Fig. 2a. The solid line connecting PL and the lower cluster intersects concordia at ~4.7 and 2.5 b.y.

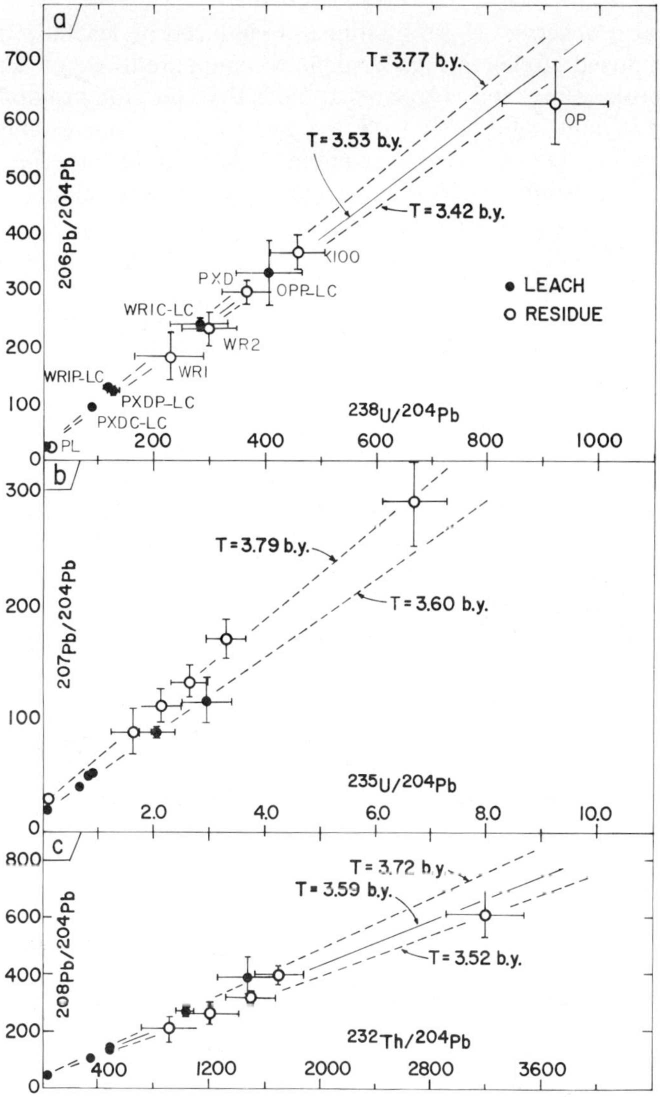

Fig. 3. U–Pb and Th–Pb parent-daughter evolution diagrams for 15085. (a) $^{238}U/^{204}Pb$ vs. $^{206}Pb/^{204}Pb$ plot. The residues (open circles) define an "age" (lower broken line) of 3.42 ± 0.22 b.y. (95% confidence) which is in agreement with Rb–Sr and Sm–Nd ages. The leaches (closed circles) define a 3.77 ± 0.26 b.y. "age" (upper broken line). Combination of the data (solid line) yields a 3.53 ± 0.09 b.y. "age". (b) $^{235}U/^{204}Pb$ vs. $^{207}Pb/^{204}Pb$ plot. The residues (solid) line define a 3.79 ± 0.10 b.y. apparent age. The leach data (broken line) yield a 3.60 ± 0.12 b.y. "age". Both ages are older than the Rb–Sr and Sm–Nd ages which suggests ^{207}Pb enrichment relative to U (addition of "old" Pb). (c) $^{232}Th/^{204}Pb$ vs. $^{208}Pb/^{204}Pb$ plot. The residue trend (lower broken line) defines a 3.5 ± 0.5 b.y. apparent age, whereas the leaches (upper broken line) define a 3.72 ± 0.12 b.y. "age". However this very old age is mostly controlled by OP–L, which appears to have suffered Th–Pb fractionation during leaching. The combined data (solid line) yield an apparent age of 3.59 ± 0.25 b.y.

(1974) have also observed U–Pb fractionation induced by leaching. The trend of the leaches (closed circles in Fig. 3a) defines an apparent age of 3.77 ± 0.16 b.y. (2 σ, upper broken line) and is somewhat older than the trend established by the residues, but is indistinguishable from the 3.68 ± 0.23 b.y. age established by the residue (excluding OP) trend. The combined data (solid line, Fig. 3a) yield a 3.53 ± 0.22 b.y. apparent age. This age is in reasonable agreement with the Rb–Sr and Sm–Nd ages for Elbow Crater basalts.

The ^{207}Pb–^{235}U data (Fig. 3b), however, show trends opposite those exhibited by the ^{238}U–^{206}Pb data. The leaches define a line corresponding to an age of 3.60 ± 0.12 b.y. (lower line) and lie below the 3.79 ± 0.10 b.y. trend of the residues (upper line), thus indicating ^{207}Pb depletion in the grain boundaries relative to the crystals. The anomalously old ages obtained suggest overall enrichment of ^{207}Pb relative to uranium. These observations suggest that we may be dealing with a combination of some form of Pb diffusion and either addition of Pb from an outside source, or lack of isotopic homogenization prior to crystallization.

The ^{232}Th–^{208}Pb data (Fig. 3c) show trends opposite those exhibited by the ^{207}Pb–^{235}U data previously discussed. The residue data (lower line, Fig. 3c) correspond to an "age" of 3.5 ± 0.5 b.y., which is higher but not statistically different from the ~3.4 b.y. crystallization age. The leach data, on the other hand, define an "age" of 3.78 ± 0.12 b.y. (upper line). These observations indicate ^{208}Pb enrichment on the grain boundaries relative to Th, and overall enrichment of ^{208}Pb (or depletion of Th) is suggested by the old "ages". The combined data (solid line, Fig. 3c) yield a 3.59 ± 0.25 b.y. apparent age. This age is in reasonable agreement with the ~3.4 b.y. crystallization age.

Preferential leaching of Pb relative to U and Th would account for the trends observed in Figs. 3a and c, but preferential leaching of U relative to Pb is indicated in Fig. 3b. Thus, preferential leaching alone cannot explain the trends observed in Fig. 3.

DISCUSSION

Episodic disturbances to the U–Pb system

The young lower concordia intercepts observed in Fig. 2, the old apparent ages observed in Figs. 1 and 3, and the different trends observed for the leaches and residues all suggest that some type of post-crystallizational disturbance to the U–Pb system has occurred. The disturbance may have been of a continuous nature (see next section) or it may have been an episodic event. This "single event" could possibly be a result of sample handling (i.e., contamination or preferential leaching of U, Th, and Pb by prior treatment with heavy liquids), however we deem this quite unlikely, based on our discussion in the Appendix to this paper. The lower concordia intercept age of ~1.0 b.y. (Figs. 2a and b) corresponds reasonably well to the age assigned to Elbow Crater (early Copernican; Swann *et al.*, 1972) from which the Elbow Crater basalts 15065, 15076, and 15085 may be ejecta (Swann *et al.*, 1972). Minor cratering might be expected to have a slight effect on the U–Pb systematics of a rock, since evidence for this type of event has been well documented for breccias and soils (e.g., Tatsumoto, 1970; Tatsumoto *et al.*, 1971; Silver, 1971). We have no thermodynamic calculations to prove that U or Pb can be moved during minor cratering, and thus the mechanism that physically moves the U or Pb remains somewhat obscure (at least to us). However, as we previously mentioned up to 50% of the Pb, U, and Th resides on the grain boundaries (Table 1). Thus it may be possible for minor shock accompanied by slight heating to cause partial reequilibration of the U–Th–Pb system or even U, Th, or Pb loss from the whole-rock system, and the possibility of an episodic disturbance should be considered.

Careful reexamination of the whole-rock data of Tatsumoto *et al.* (1972) from Elbow Crater basalts and Papanastassiou and Wasserburg's (1973) Rb–Sr internal isochrons from Elbow Crater basalts suggest that these data are consistent with the possibility of a recent disturbance to the U–Pb and Rb–Sr systematics of minor-impact ejecta. Three Elbow Crater whole-rock samples, 15065, 15076, and 15085, were analyzed by Tatsumoto *et al.* (1972), and Rb–Sr internal isochrons for these three samples were measured by Papanastassiou and Wasserburg (1973). The Rb–Sr ages are 3.28 ± 0.04, 3.33 ± 0.08, and 3.40 ± 0.04 b.y. (two points only) for 15065, 15076, and 15085, respectively. There is a slight correlation between the Rb–Sr ages and the Pb–Pb model ages, which increase from 3.89 b.y. (15065) to 3.92 b.y. (15085). A Pb–Pb whole rock plot (see Figs. 1 and 4) yields an ~ 3.8 b.y. apparent age. The increasing ages can also be correlated to a decrease in Pb radiogenicity and an increase in the degree of discordancy on a U–Pb evolution diagram (see Tatsumoto *et al.*, 1972). The data may indicate varying degrees of radiogenic Pb loss during a minor impact (Elbow Crater?) event. Pb redistribution within the rocks would also be reflected in the apparent old ages of the 15085 mineral separates (Figs. 1–4). The apparent correlation between the Rb–Sr ages and Pb radiogenicity may indicate slight Rb loss or redistribution as well. Defining a trend with only three samples is quite speculative however, and other mechanisms may also account for the data.

If the apparent differences in the Rb–Sr ages are real differences in formation ages, thus indicating magma extrusion over a prolonged time interval, then the Pb–Pb and U–Pb whole-rock "isochrons" would have no strict age significance since the samples would not truly be coeval. The old ages suggested by the U–Pb and Pb–Pb mineral separate data, however, would not be explained in this manner unless Pb isotopes and/or U–Pb ratios were not homogenized prior to crystallization. This seems somewhat unreasonable in view of the strong evidence for Sr isotopic homogenization in the source of these three samples (Papanastassiou and Wasserburg, 1973).

Episodic disturbances to meteorites have been well documented (e.g., Chen and Tilton, 1976; Tatsumoto *et al.*, 1976; Unruh *et al.*, 1977). These severe disturbances generally produce a scatter in the U–Pb parent daughter relationships, although very recent disturbances have little effect on the Pb–Pb data. Because the 15085 U–Pb data show rather well-defined trends instead of a scatter, we cannot conclude that the disturbance must have been episodic, although we cannot completely rule out this possibility.

Trace-element diffusion in 15085

Generally, Pb diffusion models have been involved when the U–Pb data appeared to define a "metamorphic event" when there was no geologic evidence or other isotopic evidence to support the observed "event". This may be the case with 15085. Several models for Pb diffusion have been proposed to explain various discordant patterns observed in terrestrial zircons (e.g., Tilton, 1960; Wetherill, 1963; Wasserburg, 1963). In Tilton's model the rate of diffusion of the three radiogenic Pb isotopes is proportional to the rate at which they are produced (inversely proportional to the half-lives). Pb diffusion would produce apparent depletion of ^{207}Pb relative to ^{206}Pb in the host phase (crystals) and assuming the "whole rock" was a closed system we would expect to see enrichment of ^{207}Pb relative to ^{206}Pb on the grain boundaries or in the interstitial material. From Fig. 1 it is evident that we observe just the opposite trend. The grain boundaries (leaches) are depleted in ^{207}Pb relative to the crystals (residue). Tie-lines in Fig. 2 connecting the various leach and residue fractions have a negative slope, whereas Pb volume diffusion would predict a positive slope (cf. Tilton, 1960). U and Pb combined diffusion (Wasserburg, 1963) would yield trends similar to those of Tilton's volume diffusion model. Continuous U diffusion (with insignificant Pb diffusion) would account for the apparent ^{207}Pb enrichment in the residues but we would also expect to see enrichment of U relative to Pb in the leach fraction (Figs. 2 and 3a) and we do not.

Differences in α-recoil energies would also favor migration of ^{207}Pb over that of ^{206}Pb (Russel and Ahrens, 1957). Tilton (1960), based on calculations by Bøggild (1941), concluded that α-recoil would produce only enough energy to move a ^{238}U nucleus a maximum of about 10^{-5} cm, and thus would not have a substantial effect on Pb diffusion out of a fairly large crystal, unless the crystal was highly fractured. If the crystal is small or highly fractured, then α-recoil might provide enough energy to move the Pb to the grain boundary or fracture. If this were true then grain boundary diffusion would control Pb migration and we would again expect to see ^{207}Pb enrichment in the interstices.

There are further complications with 15085 as opposed to zircons, however, that do not allow us to rule out the possibility that Pb diffusion has played a role in the U–Th–Pb evolution of this rock. If we assume that the U in the leaches is actually indigenous to the grain boundaries (no leaching-induced U–Pb fractionation) then we can conclude, based on the U content of the leaches, that a significant amount (up to 50%; Table 1) of the radiogenic Pb has actually been produced on the grain boundaries or in the interstices. Pb diffusion along grain boundaries would be two dimensional (Tilton, 1960). Thus Pb isotopes (and U–Pb ratios) might more easily be reequilibrated along grain boundaries (and even lost from the system as a whole) than within a crystal if Pb diffuses by a random-walk mechanism. Thus, our method of recombining the data discussed in the Appendix (Table 4, Fig. 4) may not strictly be valid. If we look only at the residue data (solid lines, Fig. 2), then we can interpret these apparent residue trend-lines as diffusion trajectories. Since the data consist of essentially only two points, PL and the lower cluster, these data could define either a curve or a line.

In either case, the upper concordia intercept would have no age significance due to improper initial Pb correction (the "true" diffusion trajectory has been rotated clockwise).

Intermediate daughter diffusion (Rn) may also account for the 15085 data. Frisen and Heymann (1972) have calculated the probabilities of Rn escape from lunar regolith under varying temperature and grain-size conditions. ^{222}Rn is an intermediate daughter in the ^{238}U–^{206}Pb decay series and has a half life of 3.825 days. ^{219}Rn is an intermediate daughter in the ^{235}U–^{207}Pb decay series and ^{220}Rn is an intermediate daughter in the ^{232}Th–^{208}Pb decay series. These two isotopes have half-lives of 3.9 and 54.5 sec, respectively. The curves calculated by Frisen and Heymann (1972) showed a direct correlation between probability of escape and the half-life of the Rn isotope (probability of ^{222}Rn escape is *much* greater than that of ^{220}Rn and ^{219}Rn). The probability of escape of all three isotopes decreased as the grain size of the regolith decreased, and increased as temperature increased.

Silver's (1973) work on lunar regolith indicates that Pb would be lost preferentially to U (presumably by diffusion processes) just from exposure to "sunlight" (higher temperature). Surface soil samples 63501,43 was in full exposure to the sunlight (Silver, pers. comm., 1977) and contains 25% less Pb (yet 30% more U) than does surface sample 63321,22 which was collected only 15 m away and from a location in "the shade" (Silver, 1973). Furthermore, a surface sample collected in a sun-exposed area showed Pb depletion relative to the subsurface sample, and the anomalously old yet "normal" discordant (Pb loss relative to U) patterns exhibited by the surface samples indicates preferential loss of ^{206}Pb relative to ^{207}Pb and thus is in support of Frisen and Heymann's (1972) predictions. Diffusion trajectories predicted by this type of model would not be greatly different than those predicted by volume diffusion mechanisms (Tilton, 1960; Wetherill, 1963; Wasserburg, 1963). The trajectories would be "steeper" (the apparent lower concordia intercept would be less than 0 if proper initial Pb corrections were made), but tie-lines connecting the leach and residue data in Fig. 2 should still have a positive slope. We would expect the probabilities of Rn escape to be much smaller in a rock than in unconsolidated regolith, but significant Rn diffusion would produce ^{206}Pb depletion relative to ^{207}Pb in the host phase. If Rn diffuses through a crystal but is retained on the grain boundaries we can explain the apparent ^{206}Pb enrichment (^{207}Pb depletion) on the grain boundaries relative to the crystal observed in Fig. 1. However, we would predict that Rn produced on the grain boundaries would have a much higher probability of escape and of reequilibration than that of Rn produced in the crystal lattice (again using a two-dimensional random-walk argument). Thus, the suggestion that Rn diffusion has played a role in the U–Pb evolution of this rock is somewhat speculative.

The 15085 data clearly indicate that the U–Pb system has suffered some type of post-crystallizational disturbance. However, we are unable to find a unique interpretation to explain the data. The disturbance may have been episodic, continuous diffusion, or a combination of both.

U–Pb implications to mare basalt evolution

Mare basalts are generally considered to have originated from some type of cumulate remelting (e.g., Ringwood and Essene, 1970; Schnetzler and Philpotts, 1971). These early cumulate remelting theories, however, do not completely satisfy all of the major-, minor-, and trace-element characteristics of mare basalts (e.g., Ringwood, 1976; Shih *et al.*, 1975) and more complex petrogenic models involving hybrid magmas (Ringwood and Kesson, 1976) or layered cumulates (Shih and Schonfeld, 1976) have been proposed. Comprehensive reviews of the various cumulate remelting theories may also be obtained from Ringwood (1976) and Drake and Consolmagno (1976).

(a) $^{238}U/^{204}Pb$ *(μ) in the mare basalt source-cumulates.* The solid/liquid distribution coefficients (*D*) of U and Th are much less than 1 for most basic minerals (e.g. Seitz, 1973; Onuma *et al.*, 1968; Nagasawa and Wakita, 1968). Not much is known about *D*-values for Pb however U–Pb data from terrestrial basalts suggest that these values are about 20 times higher than those of U, but are still less than 1 (e.g., Trueil, 1973; Tatsumoto, 1977). Therefore, basaltic cumulates (mare basalt source) should have lower U, Th, and Pb abundances and lower $^{238}U/^{204}Pb$ (μ) ratios than either the residual or parental liquids, provided little or no intercumulus liquid was incorporated in the cumulate. Partial remelting of these cumulates would produce even lower U and Pb abundances and μ values in the residual phase, whereas the liquid (mare basalts) would contain higher U and Pb abundances than the parental cumulate but the μ value of the liquid would be either slightly higher than that of the parental cumulate (small degrees of partial melting) or virtually unchanged (large degrees of partial melting).

If we follow the crystallization sequence of Drake and Consolmagno (1976, Fig. 1) then the mare basalt source-cumulates would not have formed until 50% of the original liquid as olivine, 10% as orthopyroxene, and 20% as anorthosite had fractionally crystallized.

If we can approximate partition coefficients as being approximately equal to the true distribution coefficients then we can at least semi-quantitatively predict the effects of previous fractional crystallization on the liquid from which mare basalt source-cumulates formed. Reported U partition coefficients for clinopyroxene range from ~0.001 (diopside; Seitz, 1973) to ~0.01–0.1 (glass-primary pyroxene pairs; Kleeman *et al.*, 1969). We will use an intermediate value of $D_U = 0.05$ which is within the range of 0.02–0.08 (augite) reported by Nagasawa and Wakita (1968) and in reasonable agreement with the value of 0.02 (augite) reported by Onuma *et al.* (1968). Using this value and based on the ratios of U-abundances in coexisting CPX and OL, and CPX and OPX (Kleeman *et al.*, 1969), we calculate $D_U^{OL} \simeq 10^{-4}$ and $D_U^{OPX} \simeq 0.0005$. Even if we use an unreasonably high value for $D_{Pb} = 0.1$, the μ value of the residual liquid (μ_1), after 50% crystallization of olivine and 10% orthopyroxene (Drake and Consolmagno, 1976), will only be about 10% higher than the μ value (μ_0) of the initial liquid (even using $D_{Pb} = 1.0$ will only cause the μ value of the liquid to increase by a factor of ~2.5). An ~10% increase in the initial μ value would not be detectable within present analytical uncertainties.

We have calculated a value of $D_U \simeq 0.005$ for lunar anorthosite (PL) using a K/U ratio of $\sim 10^5$ for anorthosite (e.g., Tatsumoto *et al.*, 1972), $D_K^{PL} = 0.2$ and the K/U of the "whole moon" = 2400 (e.g., Laul and Fruchter, 1976). Using a U concentration of 43 ppb for the whole moon (Langseth *et al.*, 1976), we calculate the U concentration in the liquid after OL and OPX crystallization to be about 105 ppb. Using this value and a 2 ppb U concentration for anorthosites, we obtain $D_U \simeq 0.02$. Again we will compromise and use $D_U^{PL} = 0.01$. U–Pb two-stage model calculations for anorthosite 60015 (Nunes *et al.*, 1973) yield an initial $^{238}U/^{204}Pb$ ratio (μ) of ~200 as compared with the observed μ value of ~1.5. Based on these values and $D_U \simeq 0.01$ we arrive at $D_{Pb} \simeq 1.3$. However, Tera *et al.* (1972) concluded that up to 75% of the Pb found in anorthosite 15415 may not be indigenous to the anorthosite. Thus, our calculated range for D_{Pb} is 0.3–1.3 and we will arbitrarily use a value of 1.0.

If 50% of the remaining liquid fractionally crystallized as anorthosite (20% of the total original liquid, Drake and Consolmagno, 1976), we calculate the μ value of this second residual liquid (μ_2) to be $\sim 2 \times \mu_1$ or $2.2 \times \mu_0$.

We will next assume that the low-Ti mare basalt source-cumulates consisted of OL, OPX, and CPX (Drake and Consolmagno, 1976; Shih and Schonfeld, 1976). Although CPX and PL were probably crystallizing simultaneously we will assume that crystallization of the CPX in the cumulate did not commence until after anorthositic crust formation since this assumption will have the most adverse affect on our conclusions reached at the end of this section.

As previously stated, we will use $D_U^{CPX} = 0.05$ and we will assume $D_{Pb}^{CPX} = 0.5$ and that 50% of the remaining liquid crystallizes as CPX in cumulates. As previously discussed, OL and OPX contain very little U or Pb and thus we approximate $\mu_{CPX} = \mu_{cumulate}$ (μ_3). The calculated value for μ_3 is $\sim 0.12 \times \mu_2$ or $0.25 \times \mu_0$. If we consider the *maximum* of about 4% liquid (Drake and Consolmagno, 1976) to be incorporated into the cumulate, we obtain $\mu_3 \simeq 0.17 \times \mu_2$ or $\sim 0.37 \times \mu_0$. The "uncertainty" in the value for μ_3 is probably on the order of a factor of 5 based on all of the assumptions required in obtaining this number. However, the point we wish to stress is that the mare basalt source-cumulates would not have had a μ value drastically larger than the original μ value of the outer several hundred kilometers of the moon (unless 99% of the Pb was volatilized out of the magma ocean).

(*b*) *U–Pb evolution models for mare basalts.* If lunar mare basalts are indeed products of cumulate remelting then U–Pb evolution in mare basalts could be expressed as follows:

$$\left(\frac{^{206}Pb}{^{204}Pb}\right)_{observed} = \left(\frac{^{206}Pb}{^{204}Pb}\right)_0 + \mu_0(e^{\lambda T_0} - e^{\lambda T_1}) + \mu_1(e^{\lambda T_1} - e^{\lambda T_2}) + \mu_2(e^{\lambda T_2} - 1).$$

Similar equations exist for the $^{207}Pb/^{204}Pb$ and $^{208}Pb/^{204}Pb$ data. In this equation T_0 = the age of lunar accretion and/or remelting of the outer several hundred kilometers (e.g., Ringwood, 1976); T_1 = cumulate formation age; T_2 = cumulate remelting and crystallization; $\mu_0 = {}^{238}U/{}^{204}Pb$ ratio from T_0 to T_1; $\mu_1 = {}^{238}U/{}^{204}Pb$ ratio from T_1 to T_2; μ_2 = observed ${}^{238}U/{}^{204}Pb$ ratio; $\lambda = {}^{238}U$ decay constant; and $({}^{206}Pb/{}^{204}Pb)_0$ = original Pb isotope ratio of the moon, presumably Canyon Diablo troilite Pb (Tatsumoto *et al.*, 1973). U–Pb fractionation induced by segregation of the anorthosite crust prior to cumulate formation is ignored for the present discussion. Based on our previous calculations $\mu_1 < \mu_0$ and $\mu_2 \geqslant \mu_1$ (μ_2 may be 2–3 times greater than μ_1 if a small degree of partial melting is considered).

Tera and Wasserburg (e.g., 1974, 1976) have elected to approximate mare basalt evolution as two stages, by assuming that μ_0 was sufficiently small such that no significant U–Pb evolution occurred during the time interval T_0 to T_1, yet their calculated μ_1 value for low-Ti basalt 15555 (Tera and Wasserburg, 1974) was $\simeq 265$ thus contradicting our calculated constraints. The calculated T_1 value for 15555 using this model is $\simeq 4.4$ b.y., and Tera and Wasserburg (1974) argued that this age supported their 4.4 b.y. primary differentiation age ("the time of formation of the magma reservoirs from which mare basalts were later derived as secondary differentiates"; Papanastassiou *et al.*, 1977): However, if a more reasonable value for μ_0 is chosen then the calculated value for T_1 will decrease (i.e., correction of U–Pb data for more evolved Pb will shift the upper concordia intercept on a U–Pb evolution diagram to a younger apparent age; see Tera and Wasserburg, 1974; Tatsumoto and Unruh, 1976). For example, Nunes *et al.* (1975) assumed $\mu_0 = 170$ (arbitrary). The calculated value for T_1 is 3.93 b.y. ($T_0 = 4.57$ b.y., $T_2 = 3.3$ b.y.). The calculated value for μ_1 is ~ 280 and is still too large relative to μ_0, however, their μ_1/μ_0 ratio of ~ 1.65 is still within the realm of possibility considering the uncertainty in our calculation. If we assume a higher value for μ_0, T_1 is further reduced however μ_1 increases. If a lower μ_0 value is used, then T_1 becomes older and μ_1 approaches the two-stage value of 265 reported by Tera and Wasserburg (1974) and in either case the μ_1/μ_2 ratio increases. The calculations of Nunes *et al.* (1975) yield approximately the minimum μ_1/μ_0 ratio and are therefore most consistent with our predictions. Three-stage calculations for the Elbow Crater basalts (Tatsumoto *et al.*, 1972) yield similar values for T_1 (3.8–4.0 b.y.) at the minimum μ_1/μ_0 ratios, although these minimum ratios are somewhat higher than for 15555. We are therefore unable to rigorously explain Apollo 15 mare basalt data by a three-stage U–Pb evolution model under our imposed constraint, $\mu_0 > \mu_1$.

Reasons for the apparent high μ_1/μ_0 ratios may be, in part, the function of Pb contamination (see Appendix), however, even increasing the measured blanks of both the data of Tatsumoto *et al.* (1972) and Tera and Wasserburg (1974) by an order of magnitude will not appreciably decrease the μ_1/μ_0 ratio, although both μ_1 and μ_0 increase (i.e., μ_1 approaches μ_0 only when all ${}^{204}Pb$ is contamination and as T_1 approaches T_0). A more likely explanation for the calculated high μ_1/μ_0 ratios is that our assumed model is incorrect.

We previously suggested the possibility that $\mu_1 \simeq \mu_2$ in our three-stage model. In this case, the WR data could be approximated by two-stage evolution (this practice is commonly employed when Pb–Pb whole-rock isochrons are used to define the source age). However, the constraint $\mu_0 > \mu_1 \simeq \mu_2$ would still be necessary. Using this model, the calculated μ_1/μ_0 ratios are even higher than those calculated using Nunes *et al.'s* (1975) three-stage model so we consider this model unlikely. Furthermore, Rb–Sr whole-rock data (Compston *et al.*, 1972; Nyquist *et al.*, 1974) for Apollo 15 mare basalts indicate that Rb–Sr fractionation did occur during crystallization (Shih and Schonfeld, 1976). In this regard, Apollo 15 mare basalts may be somewhat unique.

If lunar primary differentiation was completed very early in the history of the moon ($T_0 = T_1$ in our three-stage model; Schonfeld, 1976), then the value of μ_0 becomes irrelevant. However, we are then forced to conclude that either the 4.4 b.y. upper concordia intercepts defined by some mare basalts are in error (possibly as a result of post-crystallizational disturbances to the internal U–Pb systematics similar to those observed for 15085), or that the moon is only 4.4 b.y. old [how then do we explain the 4.56 b.y. Rb–Sr age of troctolite 76535 and the 4.49 b.y. Rb–Sr age of dunite 72417 (Papanastassiou and Wasserburg, 1975, 1976)?]. Tatsumoto *et al.* (1972) found that calculated two-stage model ages for mare basalts were consistent with an ~4.57 b.y. upper concordia intercept. Single-stage Sm–Nd calculations (Lugmair *et al.*, 1975, 1976; N. Nakamura, pers. comm.) for many mare basalts would be consistent with either a 4.4 or 4.56 b.y. apparent source age, as would Rb–Sr whole-rock isochrons for mare basalts, except Apollo 15 (for a summary, see Shih and Schonfeld, 1976).

Rb–Sr whole-rock data for KREEP-rich material indicate 4.1–4.3 b.y. ages (e.g., O'Nions and Pankhurst, 1972; Nyquist *et al.*, 1973). Assuming our two-stage model, these ages suggest either that primary differentiation "slowed down" after mare basalt source-cumulate formation, or more likely that the KREEP source is a secondary differentiate. The 3.9–4.4 b.y. U–Pb "cataclysm" line (Tera *et al.*, 1974a,b) obtained from non-mare rocks would therefore represent nothing more than a KREEP-mixing line as Nunes *et al.* (e.g., 1975) proposed. The 4.4 b.y. upper intercept would reflect an upper limit to the age of the KREEP source, both because of improper initial Pb correction, and because KREEP is mixed with older material. For example, if we assume that the KREEP source is 4.3 b.y. old (the upper concordia intercept in a U–Pb concordia diagram is ~4.3 b.y.) then the calculated μ value from 4.57 to 4.3 b.y. is ~250. This value is quite reasonable based on our previous discussion and based on calculated first stage μ values for anorthosites (200, e.g., Nunes *et al.*, 1973) and mare basalts (100–300, Tatsumoto, 1970; Nunes *et al.*, 1973, 1974; Tera and Wasserburg, 1975) although it is low compared to first-stage μ's calculated from KREEP basalts (~500, e.g., Nunes *et al.*, 1973). These data suggest that 4.3 b.y. may still represent an upper limit to the KREEP-source age.

The apparent 4.4 b.y. upper concordia intercepts for U–Pb internal isochrons from mare basalts (Tera and Wasserburg, 1974, 1976) may also reflect KREEP contamination. However, one should note, that within the uncertainties of the U–Pb data, the uncertainty in the Rb–Sr crystallization age (lower concordia intercept), and considering the possibilities of U–Pb internal migration (as we see in 15085), Tera and Wasserburg's data would not be inconsistent with either an ~4.3 or ~4.55 b.y. model upper intercept.

Support for this two-stage + KREEP-mixing model exists in the literature. Schonfeld (1976) has also used KREEP contamination to explain the Rb–Sr data of anorthositic rocks (although he used 4.4 b.y. as the mare basalt source-cumulate age). He found that after removal of calculated KREEP contamination Rb–Sr whole-rock data for these samples yielded an age of 4.63 ± 0.10 (using $\lambda_{^{87}Rb} = 1.39 \times 10^{-11}\ yr^{-1}$). Papanastassiou and Wasserburg (1976) have also concluded that lunar primary differentiation could have occurred very early in the history of the moon based on their 4.61 ± 0.07 b.y. Rb–Sr internal isochron age for troctolite 76535. However, the Sm–Nd (Lugmair *et al.*, 1976) and Ar–Ar (Bogard *et al.*, 1975; Huneke and Wasserburg, 1975) ages of 76535 are about 4.2 b.y.

This two-step KREEP-mixing model would not predict the intermediate (~3.8–4.1 b.y.) Pb–Pb whole-rock isochron ages for mare basalts (e.g., Tatsumoto, 1970; Tatsumoto *et al.*, 1971, 1972). Therefore, more complex U–Pb evolution models (four or more stages) may be required to define U–Pb evolution in mare basalts. Possible sources for this third U–Pb fractionation event (T') are as follows: (1) Based on our previous discussion for 15085, U–Pb systems of mare basalt whole rocks may have suffered a post-crystallizational disturbance, either natural or analytically induced ($T' < T_2$). (2) The mare basalt sources may be a result of "secondary" differentiation (cumulates formed from a liquid which was previously fractionated by segregation of the anorthositic crust at a time distinctly younger than the "primary" differentiation ($T_1 < T' < T_0$). (3) Prior to or during partial melting the mare basalt parent-cumulates may have mixed or hybridized with earlier cumulates, undifferentiated primitive material (Ringwood and Kesson, 1976) or with KREEP (e.g., Nunes *et al.*, 1975) at some time significantly prior to crystallization ($T_2 < T' < T_1$). If a four or more stage U–Pb model is required to explain these data then assignment of ages to the various events would be quite speculative since there are not enough constraints to yield a unique age interpretation. Therefore, we prefer to interpret lunar U–Pb data by our two-stage + KREEP-mixing model, primarily due to its simplicity.

SUMMARY

(1) Our detailed leaching study on 15085 indicates that some type of post-crystallizational disturbance has affected the internal U–Pb systematics of this rock. Whether this disturbance was episodic (meteorite impact-induced) or of a continuous nature (diffusion) is not readily apparent.

(2) Our 15085 study also indicates that up to 50% of the U, Th, and Pb are easily leachable from all minerals in this rock. These data suggest that the bulk of the Pb, U, and Th may reside on grain boundaries or in the interstices or mesostasis. The different U–Pb and Pb–Pb trends exhibited by the leaches and residues suggest that either the disturbance to the U–Pb system may have had different effects on the U and Pb on the grain

boundaries as opposed to that in the crystals, or that Pb isotopes were not homogenized among minerals prior to crystallization.

(3) A two-stage KREEP-mixing model (Nunes *et al.*, 1975; Schonfeld, 1976) has been proposed to explain U–Pb evolution in mare basalts. This model has been presented solely to explain the U–Pb data, although Rb–Sr and Sm–Nd data are also compatible with many of our arguments.

(4) If lunar primary differentiation occurred at a time distinctly different from lunar accretion then more complex evolution models are required to interpret the U–Pb data. A three-stage model has been shown to be unsatisfactory in rigorously explaining the U–Pb data under our calculated constraints. A four or more stage U–Pb evolution model would not yield a unique interpretation, so we have elected to interpret lunar U–Pb data by the two-stage + KREEP-mixing model until future constraints can be imposed.

(5) Based on reported and calculated U and Pb partition coefficients we conclude that the *initial* $^{238}U/^{204}Pb$ ratio of the moon was quite high, ~100–300. Therefore, in all of the previously discussed U–Pb evolution models, the 4.4 b.y. upper concordia intercept (Tera *et al.*, 1974b; Tera and Wasserburg, 1974, 1976) has no physical significance other than as an upper limit to either the cumulate formation age (three-stage model) or the age of the KREEP source (two-stage model). Using the two-stage model, our preferred age for the KREEP source is 4.2 ± 0.2 b.y. (see also Nunes *et al.*, 1975; Nyquist *et al.*, 1973; O'Nions and Pankhurst, 1972). However, many of our arguments are based on our calculated values for D_U and D_{Pb}. These values are not really proven and Pb solid-liquid partition coefficients should be determined before our arguments can be considered irrefutable.

Acknowledgments—This work was supported primarily by NASA Interagency transfer order T-2407A. Comments by R. E. Zartman and K. R. Ludwig are appreciated. We are also indebted to Sandra L. Blanchard for work above and beyond the call of duty in preparing this manuscript to meet the deadline. Technical reviews are responsible for some of the expanded discussions (i.e., the Appendix). Other comments by these reviewers were appreciated.

References

Bogard D. D., Nyquist L. E., Bansal B. M., Wiesmann H., and Shih C. Y. (1975) 76535: An old lunar rock. *Earth Planet. Sci. Lett.* **26**, 69–80.

Bøggild J. K. (1941) Range-velocity relation for fission fragments in helium. *Phys. Rev.* **60**, 827–830.

Chen J. H. and Tilton G. R. (1976) Isotopic lead investigations of the Allende carbonaceous chondrite. *Geochim. Cosmochim. Acta* **40**, 635–643.

Compston W., de Laeter J. R., and Vernon M. J. (1972) Strontium isotope geochemistry of Apollo 15 basalts. In *The Apollo 15 Lunar Samples* (J. W. Chamberlain and C. Watkins, eds.), p. 347–351. The Lunar Science Institute, Houston.

Church S. E. and Tatsumoto M. (1975) Lead isotopic relations in oceanic ridge basalts from the Juan de Fuca-Gorda Ridge area, N. E. Pacific Ocean. *Contrib. Mineral. Petrol.* **53**, 253–279.

Drake M. J. and Consolmagno G. J. (1976) Critical review of models for the evolution of high-Ti mare basalts. *Proc. Lunar Sci. Conf. 7th*, p. 1633–1657.

Frisen L. J. and Heymann D. (1972) Model for Radon diffusion through the lunar regolith. *The Moon* **3**, 461–471.

Huneke J. C. and Wasserburg G. J. (1975) Trapped ^{40}Ar in troctolite 76535 and evidence for enhanced ^{40}Ar–^{39}Ar age plateaus (abstract). In *Lunar Science VI*, p. 417–419. The Lunar Science Institute, Houston.

Jaffey A. H., Flynn K. F., Glendenin L. E., Bentley W. C., and Essling A. M. (1971) Precision measurement of half-lives and specific activities of ^{235}U and ^{238}U. *Phys. Rev.* **C4**, 1889–1906.

Kleeman J. D., Green D. H., and Lovering J. F. (1969) Uranium distribution in ultramafic inclusions from Victorian basalts. *Earth Planet. Sci. Lett.* **5**, 449–458.

Langseth M. G., Keihm S. J., and Peters K. (1976) The revised lunar heat flow values (abstract). In *Lunar Science VII*, p. 474–475. The Lunar Science Institute, Houston.

Laul J. C. and Fruchter J. S. (1976) Thorium and uranium variations in Apollo 17 basalts, and K–U systematics. *Proc. Lunar Sci. Conf. 7th*, p. 1545–1559.

Lugmair G. W., Scheinin N. B., and Marti K. (1975) Sm–Nd age and history of Apollo 17 basalt 75075. Evidence for early differentiation of the lunar exterior. *Proc. Lunar Sci. Conf. 6th*, p. 1419–1429.

Lugmair G. W., Marti K., Kurtz J. P., and Scheinin N. B. (1976) History and genesis of lunar troctolite 76535 or: How old is old? *Proc. Lunar Sci. Conf. 7th*, p. 2009–2033.

Nagasawa H. and Wakita H. (1968) Partition of U and Th between augite and host lavas. *Geochim. Cosmochim. Acta* **32**, 917–921.

Nakamura N., Tatsumoto M., Nunes P. D., Unruh D. M., Schwab A. P., and Wilderman T. R. (1976) 4.4 b.y.-old clast in Boulder 7, Apollo 17: A comprehensive chronological study by U–Pb, Rb–Sr and Sm–Nd method. *Proc. Lunar Sci. Conf. 7th*, p. 2309–2333.

Nakamura N., Unruh D. M., Gensho R., and Tatsumoto M. (1977) Evolution history of lunar mare basalts: Apollo 15 samples revisited. In *Lunar Science VIII*, p. 712–713. The Lunar Science Institute, Houston.

Nunes P. D., Tatsumoto M., Knight R. J., Unruh D. M., and Doe B. R. (1973) U–Th–Pb systematics of some Apollo 16 lunar samples. *Proc. Lunar Sci. Conf. 4th*, p. 1797–1822.

Nunes P. D., Tatsumoto M., and Unruh D. M. (1974) U–Th–Pb systematics of some Apollo 17 lunar samples and implications for a lunar basin excavation chronology. *Proc. Lunar Sci. Conf. 5th*, p. 1487–1514.

Nunes P. D., Tatsumoto M., and Unruh D. M. (1975) U–Th–Pb systematics of anorthositic gabbros 78155 and 77017—implications for early lunar evolution. *Proc. Lunar Sci. Conf. 6th*, p. 1431–1444.

Nyquist L. E., Hubbard N. J., Gast P. W., Bansal B. M., Wiesmann H., and Jahn B. (1973) Rb–Sr systematics for chemically defined Apollo 15 and 16 materials. *Proc. Lunar Sci. Conf. 4th*, p. 1823–1846.

Nyquist L. E., Bansal B. M., Wiesmann H., and Jahn B. M. (1974) Taurus-Littrow chronology: Some constraints on early lunar crustal development. *Proc. Lunar Sci. Conf. 5th*, p. 1515–1539.

Nyquist L. E., Bansal B. M., and Wiesmann H. (1975) Rb–Sr ages and initial ^{87}Sr^{86}Sr for Apollo 17 basalts and KREEP basalt 15386. *Proc. Lunar Sci. Conf. 6th*, p. 1445–1466.

O'Nions R. K. and Pankhurst R. J. (1972) A 4.3 aeon pre-Imbrium event. *Nature* **237**, 446–447.

Onuma N., Higuchi H., Wakita H., and Nagasawa H. (1968) Trace element partition between two pyroxenes and the host lava. *Earth Planet. Sci. Lett.* **5**, 47–51.

Papanastassiou D. A. and Wasserburg G. J. (1970) Rb–Sr ages from the Ocean of Storms. *Earth Planet. Sci. Lett.* **8**, 269–278.

Papanastassiou D. A. and Wasserburg G. J. (1971) Lunar chronology and evolution from Rb–Sr studies of Apollo 11 and 12 samples. *Earth Planet. Sci. Lett.* **11**, 37–63.

Papanastassiou D. A. and Wasserburg G. J. (1973) Rb–Sr ages and the initial strontium in basalts from Apollo 15. *Earth Planet. Sci. Lett.* **17**, 324–337.

Papanastassiou D. A. and Wasserburg G. J. (1975) Rb–Sr study of a lunar dunite and evidence for early lunar differentiates. *Proc. Lunar Sci. Conf. 6th*, p. 1467–1489.

Papanastassiou D. A. and Wasserburg G. J. (1976) Rb–Sr age of troctolite 76535. *Proc. Lunar Sci. Conf. 7th*, p. 2035–2054.

Papanastassiou D. A., Wasserburg G. J., and Burnett D. S. (1970) Rb–Sr ages of lunar rocks from the Sea of Tranquillity. *Earth Planet. Sci. Lett.* **8**, 1–19.

Papanastassiou D. A., DePaolo D. J., Tera F., and Wasserburg G. J. (1977) An isotopic triptych on mare basalts: Rb–Sr, Sm–Nd, U–Pb (abstract). In *Lunar Science VIII*, p. 750–752. The Lunar Science Institute, Houston.

Patterson C. C. (1955) The $^{207}Pb^{206}Pb$ ages of some stone meteorites. *Geochim. Cosmochim. Acta* **7**, 151–153.

Ringwood A. E. (1976) Limits on the bulk composition of the moon. *Icarus* **28**, 325–349.

Ringwood A. E. and Essene E. (1970) Petrogenesis of Apollo 11 basalts, internal constitution and origin of the moon. *Proc. Apollo 11 Lunar Sci. Conf.*, p. 769–799.

Ringwood A. E. and Kesson S. E. (1976) A dynamic model for mare basalt petrogenesis. *Proc. Lunar Sci. Conf. 7th*, p. 1697–1722.

Russel R. D. and Ahrens L. H. (1957) Additional regularities among discordant lead-uranium ages. *Geochim. Cosmochim. Acta* **11**, 213–218.

Schnetzler C. C. and Philpotts J. A. (1971) Alkali, alkaline earth and rare earth element concentrations in some Apollo 12 soils, rocks and separated phases. *Proc. Lunar Sci. Conf. 2nd*, p. 1101–1122.

Schonfeld E. (1976) Chronology of the early lunar crust. *Proc. Lunar Sci. Conf. 7th*, p. 2093–2105.

Seitz M. G. (1973) Uranium and thorium partitioning in diopside-melt and whitlockite-melt systems. *Carnegie Inst. Wash. Yearb.* **72**, 581–586.

Shih C. and Schonfeld E. (1976) Mare basalt genesis: A cumulate remelting model. *Proc. Lunar Sci. Conf. 7th*, p. 1757–1792.

Shih C., Haskin L. A., Wiesmann H., Bansal B. M., and Brannon J. C. (1975) On the origin of high-Ti mare basalts. *Proc. Lunar Sci. Conf. 6th*, p. 1255–1285.

Shimizu N. (1974) An experimental study of the partitioning of K, Rb, Cs, Sr, and Ba between clinopyroxene and liquid at high pressures. *Geochim. Cosmochim. Acta* **38**, 1789–1798.

Silver L. T. (1970) Uranium-thorium-lead isotopes in some Tranquillity Base samples and their implications for lunar history. *Proc. Apollo 11 Lunar Sci. Conf.*, p. 1533–1574.

Silver L. T. (1971) U–Th–Pb isotope systems in Apollo 11 and 12 regolithic materials and a possible age for the Copernicus impact event (abstract). *EOS* (*Trans. Amer. Geophys. Union*) **52**, 534.

Silver L. T. (1973) Uranium-thorium-lead isotopic characteristics in some regolith materials from the Descartes Region (abstract). In *Lunar Science IV*, p. 672–674. The Lunar Science Institute, Houston.

Swann G. A., Bailey N. G., Batson R. M., Freeman V. L., Hait M. H., Head J. W., Holt H. E., Howard K. A., Irwin J. B., Larson K. B., Muehlberger W. R., Reed V. S., Rennilson J. H., Schaber G. G., Scott D. R., Silver L. T., Sutton R. L., Ulrich G. E., Wilshire H. G., and Wolfe E. W. (1972) Preliminary geologic investigation of the Apollo 15 landing site. In *Apollo 15 Prelim. Sci. Rep.*, NASA SP-289, p. 5-1 to 5-112.

Tatsumoto M. (1970) Age of the moon: An isotopic study of U–Th–Pb systematics of Apollo 11 lunar samples—II. *Proc. Apollo 11 Lunar Sci. Conf.*, p. 1595–1612.

Tatsumoto M. (1977) Isotopic composition of Pb in oceanic basalt and its implication to mantle evolution. In *Trace Elements and their Role in Igneous Petrology* (C. J. Allegre and S. R. Hart, eds.). Elsevier, Amsterdam. In press.

Tatsumoto M., Hedge C. E., Knight R. J., Unruh D. M., and Doe B. R. (1972) U–Th–Pb, Rb–Sr and K measurements on some Apollo 15 and 16 samples. In *The Apollo 15 Lunar Samples* (J. W. Chamberlain and C. Watkins, eds.), p. 391–395. The Lunar Science Institute, Houston.

Tatsumoto M., Knight R. J., and Allegre C. J. (1973) Time differences in the formation of meteorites as determined from the ratio of lead-207 to lead-206. *Science* **180**, 1279-1283.

Tatsumoto M., Knight R. J., and Doe B. R. (1971) U–Th–Pb systematics of Apollo 12 lunar samples. *Proc. Lunar Sci. Conf. 2nd*, p. 1521–1546.

Tatsumoto M. and Unruh D. M. (1976) KREEP basalt age: Grain by grain U–Th–Pb systematics study of the quartz monzodiorite clast 15405,88. *Proc. Lunar Sci. Conf. 7th*, p. 2107–2129.

Tatsumoto M., Unruh D. M., and Desborough G. A. (1976) U–Th–Pb and Rb–Sr systematics of Allende and U–Th–Pb systematics of Orgueil. *Geochim. Cosmochim. Acta* **40**, 617–634.

Tera F., Papanastassiou D. A., and Wasserburg G. J. (1974a) The lunar time-scale and a summary of isotopic evidence for a terminal lunar cataclysm (abstract). In *Lunar Science V*, p. 792–794. The Lunar Science Institute, Houston.

Tera F., Papanastassiou D. A., and Wasserburg G. J. (1974b) Isotopic evidence for a terminal lunar cataclysm. *Earth Planet. Sci. Lett.* **22**, 1–21.

Tera F., Ray L. A., and Wasserburg G. J. (1972) Distribution of Pb–U–Th in lunar anorthosite 15415 and inferences about its age. In *The Apollo 15 Lunar Samples* (J. W. Chamberlain and C. Watkins, eds.), p. 396–401. The Lunar Science Institute, Houston.

Tera F. and Wasserburg G. J. (1972a) U–Th–Pb systematics in three Apollo 14 samples. *Earth Planet. Sci. Lett.* **14**, 281–304.

Tera F. and Wasserburg G. J. (1972b) U–Th–Pb systematics in lunar highland samples from the Luna 20 and Apollo 16 missions. *Earth Planet. Sci. Lett.* **17**, 36–51.

Tera F. and Wasserburg G. J. (1974) U–Th–Pb systematics on lunar rocks and inferences about lunar evolution and the age of the moon. *Proc. Lunar Sci. Conf. 5th*, p. 1571–1599.

Tera F. and Wasserburg G. J. (1975) The evolution and history of mare basalts as inferred from U–Th–Pb systematics (abstract). In *Lunar Science VI*, p. 807–809. The Lunar Science Institute, Houston.

Tera F. and Wasserburg G. J. (1976) Lunar ballgames and other sports (abstract). In *Lunar Science VII*, p. 858–860. The Lunar Science Institute, Houston.

Trueil M. (1973) Criteres petrologiques geochimiques et structuraux de la genese et de la differenciation des magmas basaltiques: Exemple de Lafar. Dr.S. thesis, Universite de Paris-Sud, Center Dorsay.

Tilton G. R. (1960) Volume diffusion as a mechanism for discordant lead ages. *J. Geophys. Res.* **65**, 2933–2945.

Unruh D. M., Nakamura N., and Tatsumoto M. (1977) History of the Pasamonte achondrite: Relative susceptibility of the Sm–Nd, Rb–Sr, and U–Pb systems to metamorphic events. *Earth Planet. Sci. Lett.* In press.

Wasserburg G. J. (1963) Diffusion processes in lead-uranium systems. *J. Geophys. Res.* **68**, 4823–4846.

Wasserburg G. J. and Papanastassiou D. A. (1971) Age of an Apollo 15 mare basalt: Lunar crust and mantle evolution. *Earth Planet. Sci. Lett.* **13**, 97–104.

Wetherill G. W. (1956) Discordant uranium-lead ages. *Trans. Amer. Geophys. Union* **37**, 320–326.

Wetherill G. W. (1963) Discordant uranium-lead ages, part 2: Discordant age resulting from diffusion of lead and uranium. *J. Geophys. Res.* **68**, 2957–2965.

Wilshire H. G. and Brett P. R. (1971) Thin section description of 15085. In *The Apollo 15 Lunar Sample Information Catalog*, p. 84. MSC-03209, NASA/Johnson Space Center, Houston.

York D. K. (1969) Least square fitting of a straight line with correlated errors. *Earth Planet. Sci. Lett.* **5**, 320–324.

APPENDIX

Modern Pb contamination and heavy-liquid leaching

The effects of sample handling and exposure to heavy liquids on the U–Pb system is not known. We will first consider the possibility that modern terrestrial Pb contamination has been induced via these processes. If such is the case, then this contamination should affect only the leach data. As an extreme correction we have removed all of the ^{204}Pb in the leach fractions (and corresponding amounts of the other three isotopes) as contamination. We have then added the remaining ^{206}Pb, ^{207}Pb, and ^{208}Pb and the U and Th in the leach fraction to those in the residue fractions. The ^{207}Pb–^{206}Pb data combined in this manner are shown in Fig. 4. Data from previous whole-rock analyses (Tatsumoto *et al.*, 1972; open circles) are also plotted. These data fall, within uncertainty, on the line established by the 15085 separation. These data yield an apparent age of 3.86 ± 0.08 b.y. (2σ), which is still significantly older than the 3.46 b.y. Rb–Sr and Sm–Nd ages. Rb–Sr whole-rock data for Apollo 15 mare basalts also define the same apparent age as the Rb–Sr internal isochrons (e.g., Shih and Schonfeld, 1976; Papanastassiou and Wasserburg, 1973). The ^{238}U–^{206}Pb, ^{235}U–^{207}Pb, and ^{232}Th–

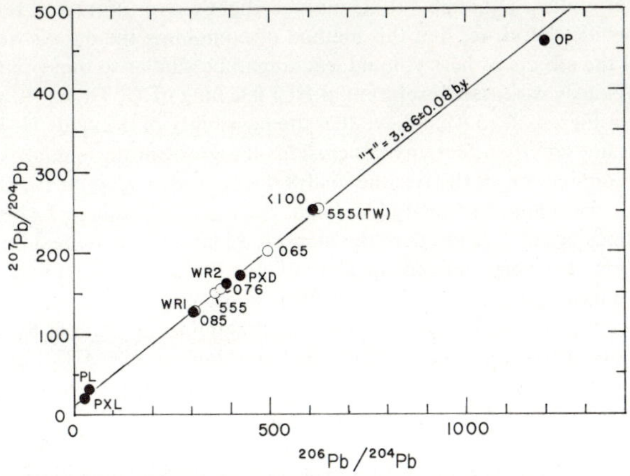

Fig. 4. $^{206}Pb/^{204}Pb$ vs. $^{207}Pb/^{204}Pb$ plot for "combined" leach and residue data from 15085. The data are "combined" by assuming that all ^{204}Pb (and commensurate ^{206}Pb, ^{207}Pb, and ^{208}Pb) in the leach fractions represents modern terrestrial contamination. The data yield an apparent age of 3.86 ± 0.08 b.y. (2σ). Also shown are Apollo 15 whole-rock data from Tatsumoto *et al.* (1972) and Tera and Wasserburg's (1974) data for 15555 (TW).

^{208}Pb combined data (not shown) yield apparent ages of 3.63 ± 0.17, 3.77 ± 0.14, and 3.5 ± 0.5 b.y., respectively. Thus, even this rather extreme correction is not sufficient to reduce the apparent ages to the expected value of ~ 3.4 b.y. The validity of combining data in the manner previously discussed may be tested by comparing the combined data for WR1 and WR2 to those obtained by Tatsumoto *et al.* (1972) from an unleached, untreated sample. This has been done in Table 4. The U and Th concentrations of our combined data (WR1C and WR2C) are about 6–10% lower than those of the untreated sample (WR). These data suggest that small portions of U and Th could have been leached during the heavy-liquid treatment. However, the differences in the data could just as well be due to sample in homogeniety, since 15085 is rather coarse grained and the variations between WR1C and WR2C are approximately the same as those between Tatsumoto *et al.*'s WR and WR2C. Furthermore, the variations are even less than those observed in many fine-grained terrestrial basalts (e.g., Church and Tatsumoto, 1975) and in other mare basalts (cf., 15555, Tatsumoto *et al.*, 1972).

The Pb concentration of WR1C and WR2C are also lower than that of Tatsumoto *et al.*'s whole-rock but the Pb isotopic compositions and the $^{238}U/^{204}Pb$ (μ) ratios are quite comparable within

Table 4. Comparison of "combined" 15085 whole-rock data with previous whole-rock data.

085 sample	ppm U	ppm Th	ppm Pb	$\frac{^{232}Th}{^{238}U}$	$\frac{^{238}U}{^{204}Pb}$	$\frac{^{206}Pb}{^{204}Pb}$	$\frac{^{207}Pb}{^{206}Pb}$	$\frac{^{208}Pb}{^{206}Pb}$
WR[1]	0.118	0.459	0.208	4.01	382	309	0.4231	1.087
WR1C[2]	0.106	0.405	0.186	3.93	383	305	0.4172	1.117
WR2C[2]	0.111	0.426	0.174	3.97	514	390	0.4152	0.9790

[1] Tatsumoto *et al.* (1972).

[2] This work. Leach and residue data are combined with all ^{204}Pb in the leach fraction removed as modern terrestrial Pb.

analytical blank uncertainties, although WR2C may be slightly over-corrected relative to Tatsumoto *et al.*'s WR. We conclude therefore, that this method of combining the data is quite reasonable.

We would expect the effects of heavy-liquid leaching to be similar to those that we observe in our HCl leaches. As previously discussed preferential HCl leaching of U, Th, or Pb will not account for the trends observed in Figs. 1–4, so we believe that the possibility of previous U, Th, or Pb extraction by heavy liquids will not greatly affect any conclusions drawn from these data.

Analytical Pb contamination in the residue analysis would also account for the anomalously old ages in Figs. 1–4, but the amount of analytical blank required to obtain ~3.4 b.y. ages in Figs. 1–4 would be about 10 times larger (~2 ng) than the measured blank for some residue samples (e.g., PL, WR1, PXD). We deem this very unlikely in view of our blank reproducibility over the past few years. Furthermore, unless the U–Pb ratios and Pb isotopic compositions of all mineral separates and whole-rocks (except possibly PL) were identical, "contaminated" data in Figs. 1–4 would yield a scatter and not the discreet trends we observe (i.e., the data in Fig. 4 would not define a single blank-mixing line, but a different mixing line would be observed for each whole-rock and mineral separate).

We conclude, therefore, that while significant Pb contamination may exist in the leach fractions, the anomalously old ages (Figs. 1–4) cannot be explained by Pb contamination.

Proc. Lunar Sci. Conf. 8th (1977), p. 1697–1729.
Printed in the United States of America

Shock metamorphism of lunar and terrestrial basalts

RAND BRIAN SCHAAL

Lockheed Electronics Company, Inc., Houston, Texas 77058

FRIEDRICH HÖRZ

NASA Johnson Space Center, Houston, Texas 77058

Abstract—Lonar Crater (India) basalt and lunar basalt 75035 were shock loaded under controlled laboratory conditions up to 1000 kbar, the highest peak pressure obtained to date in successful shock recovery; most of the experiments were performed in a CO/CO_2 (1:1) environment evacuated to 10^{-7} torr. The classification scheme of progressive shock metamorphism devised by Kieffer *et al.* (1976a) for basalt from Lonar Crater also applies to lunar basalts. The major shock features are as follows: Class 1 (0 to ~250 kbar): brittle and plastic deformation; shock lamellae in plagioclase; Class 2 (~250 to ~450 kbar): complete phase transition of plagioclase to maskelynite; Class 3 (~450 to ~600 kbar): onset of feldspar fusion, predominantly along grain boundaries; Class 4 (~600 to ~800 kbar): vesicular and flowed feldspar melts with loss of original rock texture; Class 5 (>800 kbar): melting of pyroxene and ilmenite, in addition to plagioclase, generating "whole rock" melts. The lunar plagioclase ($\approx An_{87}$) in rock 75035 converts to maskelynite at approximately the same pressure as terrestrial plagioclase; however, lunar feldspar tends to be slightly more resistant to melting at higher pressures. Feldspar melting occurs in a wide pressure range, ~450–800 kbar, and basaltic "whole rock" melting requires shock pressures in excess of 800 kbar.

Out of a total of 152 basalt specimens in the curatorial thin section collection, only three rocks show shock effects as severe as Class 2 features. The natural features are identical to their experimental analogs. These three rocks, 15684, 12054, and 79155, are coincidentally draped with a thick glass of basaltic composition suggesting a genetic relationship between the impacts which produced the shock features and the glasses.

The scarcity of shocked basalt hand samples in contrast to the abundance of shock-produced agglutinates and homogeneous glass spheres in the lunar regolith attest to the dominant role of micrometeorite impact in the evolution of the lunar regolith. While agglutinates are produced by impacts into fine-grained soils, homogeneous glasses are generated by impacts onto solid rocks. Thus, statistical, geochemical surveys of such glass populations may help identify their parent lithologies. We also speculate that the overall glass content is lower in asteroidal regoliths and larger in Mercurian regoliths as compared to that in lunar regoliths.

INTRODUCTION

AGGLUTINATES, HOMOGENEOUS GLASS SPHERULES, and diaplectic feldspars are abundant in the lunar regolith and indicate a prolonged, repetitive nature of the lunar meteorite bombardment. To account for these abundant shock-produced particles, any given surface point must have suffered numerous impacts, even on relatively young maria. In order to improve our understanding of these cumulative shock effects it is essential to determine the effects of single impacts, i.e., the response of material during the passage of a single shock wave. This can be accomplished by examining materials shock loaded in the laboratory or lunar and terrestrial materials involved only in a single impact event. This study is a petrographic investigation not only of experimentally shocked terrestrial and lunar basalts but also of naturally shocked lunar basalt samples which exhibit shock effects more severe than fracturing on hand specimen scales.

Previous investigations of experimentally and naturally shocked basalts serve as a framework for this study (James, 1969; Short, 1969; Chao *et al.*, 1970; Dence *et al.*, 1970; Schaal *et al.*, 1976; Kieffer *et al.*, 1976a). In particular, Kieffer *et al.* (1976a) distinguish five grades of progressive shock damage ranging from brittle failure to whole rock melting in naturally shocked basalt from Lonar Crater, India; absolute pressure calibration of the shock intensities was accomplished via controlled laboratory shock recovery experiments which duplicated natural effects from Class 1 through Class 4 in shots up to 642 kbar.

The first part of this report extends results of recovery experiments on Lonar basalt to approximately 1000 kbar, the highest shock pressures to date at which recovery was successful. The major part, however, describes shock features in a suite of analogous shock recovery experiments up to 1000 kbar on lunar basalt 75035 and in naturally shocked lunar basalts. The main objective is to document artificial and natural shock effects in lunar basalts. Moreover, the obvious parallelism between the Lonar and lunar studies makes it possible to contrast and compare terrestrial and lunar analogs in order to assess the fundamental effects of shock metamorphism on basaltic rocks. Discussion also focuses on the origin of variegated glasses which coat the shocked lunar basalts. Finally, we speculate on the role of shock in the evolution of asteroidal and Mercurian regoliths.

SHOCK RECOVERY EXPERIMENTS

A 20 mm flat plate accelerator is used in the shock-loading experiments. Details of this apparatus and the experimental procedures, including recovery of the shocked materials, are described by Hörz (1970), Gibbons *et al.* (1975), and Kieffer *et al.* (1976a). Lonar basalt target discs were 7 mm in diameter, 0.5 mm thick, and were shock loaded in an atmospheric vacuum of $2–4 \times 10^{-2}$ torr. Shots at 614 and 742 kbar, however, were only 6 mm in diameter, with the same thickness, and, more significantly, were shocked in a newly developed auxiliary vacuum chamber housed inside the normal, large impact chamber which is normally evacuated to 10^{-2} torr. The new auxiliary chamber may be pumped down to 10^{-7} torr and can be repeatedly flushed with any gas or gas mixture(s). The objective of this improvement is to better approximate ambient lunar impact conditions, especially oxygen fugacity, by repeatedly flushing with a CO/CO_2 (1:1) mixture and evacuating to as low as 10^{-7} torr during the actual experiment. The 614 and 742 kbar Lonar experiments served as test shots for this new vacuum system. They were performed at 2×10^{-6} torr of a CO/CO_2 (1:1) gas mixture.

Lunar basalt 75035 was selected for the shock experiments because of its unfractured and holocrystalline texture and its medium grain size (Longhi *et al.*, 1974), making each target disc of 5 mm diameter and 0.5 mm thickness fairly representative of the whole rock. Basalt 75035 in turn is fairly representative of the Apollo 17 low-K high-Ti basalts (Rhodes *et al.*, 1977; Longhi *et al.*, 1974; Papike *et al.*, 1976) and has geochemical and petrological affinities to some Apollo 11 and 12 basalts; thus, 75035 reflects a common lunar basalt type. Chip 75035,23 was sliced into thin wafers which in turn were cored with a diamond coring device to produce discs 5 mm in diameter which were ground and polished to 0.5 mm in thickness. Cutting, coring, and polishing were accomplished using high purity mineral oils and great care was taken to prevent contamination with H_2O. All ten recovery experiments utilized the new high vacuum system and CO/CO_2 (1:1) gas mixtures at pressures ranging from 10^{-5} to 10^{-8} torr. Additional details of the experiments are given in Tables 1 and 2. Chemical analyses were made with an ARL-EMX electron microprobe using the analytical techniques discussed by Schaal *et al.* (1976).

Table 1. Experimental shock effects in basalt from Lonar Crater, India.

	—600 kbar—	Class 4		—800 kbar— Class 5	
Shock pressure (± 3%)	614 kbar	696 kbar	742 kbar	974 kbar	996 kbar
Texture	Microporphyritic texture destroyed; vesiculated	Ophitic texture destroyed; vesiculated	Mostly frothy; vesicular	Frothy; fragmented	Frothy; fragmented
Plagioclase	Flowed colorless glass; some vesicles; amorphous outlines	Flowed, colorless vesicular glass	Flowed, colorless, vesicular glass	Flowed colorless glass mixed with colored glasses; highly vesiculated	Flowed glass mixed in colored glasses; highly vesiculated
Pyroxene	Granulated; displaced along microfaults; reduced birefringence	Crushed; granulated; mosaicism	Granulated; fragmented; mosaicism	Flowed greenish glass mixed with colorless glass; relict grains granulated	Flowed, mixed, greenish glass; relict grains granulated
Ulvöspinel	Fractured	Fractured	Fractured	Granulated	Fragmented; mixed
Palagonite	Isotropic; flowed; reddish brown color preserved	Not recognizable	Isotropic; flowed; dark brown color	Isotropic; flowed; dark brown color	Not recognizable
Sample recovery	~75%	~30%	~75%	~40%—2 sections	~15%
Flyer plate	W*	W	W	W	W
Sample assembly	SS304†	SS304	FS77‡	FS77	FS77
Target cylinder	SS304	SS304	SS304	SS304	SS304
Projectile velocity (km/sec, ± 1%)	1.87	2.08	1.66	2.07	2.11

*W = 99.9% W.
†SS304 = Stainless steel 304 (68% Fe, 21% Cr, 9% Ni).
‡FS77 = Fansteel 77(89% W, 7% Ni, 4% Cu).

Table 2. Shock effects in basalt 75035.

Shock pressure (±3%)	Class 1	—250 kbar—	Class 2	—450 kbar—	Class 3
	93 kbar	204 kbar	302 kbar	400 kbar	494 kbar
Texture	Ophitic texture preserved	Ophitic texture preserved	Ophitic texture preserved	Ophitic texture preserved	Ophitic texture preserved
Plagioclase	Grains fractured; some granulation; undulatory and patchy extinction	Intense fracturing; abundant micro faults; undulatory extinction	Partial to total conversion to maskelynite; grain outlines preserved; few fractures	All grains converted to maskelynite; grains fractured and deformed	All grains converted to maskelynite or plagioclase glass with incipient flow; intensely fractured and granulated; mosaicism
Pyroxene	Irregular and cleavage fractures; some granulation	Closed spaced irregular fractures; some granulation; mosaicism	Fractured and granulated; mosaicism	Intensely fractured and granulated; mosaicism	Intensely fractured and granulated; mosaicism
Ilmenite	Fractured; shock twins	Fractured; shock twins	Fractured; well developed shock twins	Fractured; disjointed; shock twins	Fractured; disjointed; shock twins
Cristobalite	Polygonal fractures; mosaicism	Irregular and polygonal fractures; mosaicism	Abundant irregular fractures; relict polygonal fractures; isotropic, diaplectic	Intensely fractured; relict polygonal fractures; isotropic; diaplectic	Intensely fractured; relict polygonal fractures; isotropic; diaplectic (~15% of mode)
Sample recovery	~80%	~85%	~90%	~30%	~65%
Flyer plate	Al2024*	SS304†	W‡	W	W
Sample assembly	Al2024	SS304	SS304	SS304	SS304
Target cylinder	Al2024	SS304	SS304	SS304	SS304
Projectile velocity (km/sec, ±1%)	1.10	0.97	1.03	1.31	1.56
Vacuum pressure (torr)	5.3×10^{-7}	1.2×10^{-6}	1.4×10^{-5}	3.1×10^{-8}	1.2×10^{-6}

*Al2024 = Aluminum 2024 (93.4% Al, 4.5% Cu, 1.5% Mg, 0.6% Mn).
†SS304 = Stainless steel 304 (68% Fe, 21% Cr, 9% Ni).
‡W = 99.9% W

Table 2. (*Continued*).

Shock pressure (± 3%)	—600 kbar—	Class 4		—800 kbar— Class 5	
	606 kbar	668 kbar	743 kbar	853 kbar	999 kbar
Texture	Modified by flow, but recognizable	Modified by flow, but recognizable	Slightly frothy; texture destroyed	Frothy; ophitic texture destroyed	Frothy; glassy
Plagioclase	Slightly flowed colorless glass; sparse vesicles	Slightly flowed colorless glass; sparse vesicles	Flowed, vesicular, colorless glass	Frothy, flowed glass; dark color	Frothy, highly vesiculated flowed glass; mixed colored glass
Pyroxene	Granulated; mosaicism; reduced birefringence	Intensely fractured; mosaicism	Granulated; fragments disjointed	Melting on edges; granulated; mosaicism	Flowed, mixed greenish glass; relict grains fractured
Ilmenite	Fractured; disjointed; shock twins	Fractured; fragmented; shock twins	Fractured, granulated; shock twins	Fragmented; shock twins	Fragmented; shock twins
Cristobalite	Not recognizable	Fractured; partially fused; some relict polygonalism; isotropic	Fused to colorless silica glass; few fractures	Fused to colorless silica glass; amorphous outline	Colorless silica glass intermixed with other glasses
Sample recovery	~ 10%	~ 85%	~ 80%	~ 70%—3 sections	~ 15%—2 sections
Flyer plate	W*	W	W	W	W
Sample assembly	SS304†	SS304	FS77°	FS77	FS77
Target cylinder	SS304	SS304	SS304	SS304	SS304
Projectile velocity (km/sec, ± 1%)	1.86	2.01	1.67	1.80	2.12
Vacuum pressure (torr)	3.6×10^{-6}	3.0×10^{-6}	2.3×10^{-6}	2.3×10^{-5}	2.7×10^{-5}

*W = 99.9% W.
†SS304 = Stainless steel 304 (68% Fe, 21% Cr, 9% Ni).
‡FS77 = Fansteel 77 (89% W, 7% Ni, 4% Cu).

EXPERIMENTALLY SHOCKED LONAR BASALTS

Kieffer *et al.* (1976a) reported shock recovery experiments of Lonar basalt up to 642 kbar. We now describe additional experiments at 614, 696, 742, 974, and 996 kbar. Experimental conditions for shots 696, 974, and 996 kbar were identical to those in earlier shots (Kieffer *et al.*, 1976a). The main petrographic results are described in Table 1 and illustrated in Fig. 1. The observed progression of shock metamorphism is consistent with our previous studies. The boundary between Classes 4 and 5 may be placed at approximately 800 kbar. No whole rock melts are observed in the 742 kbar experiment, though vesicular plagioclase glass is abundant. In the 974 and 996 kbar experiments the total amount of material converted to whole rock melt is ≈80%; detrital pyroxene and

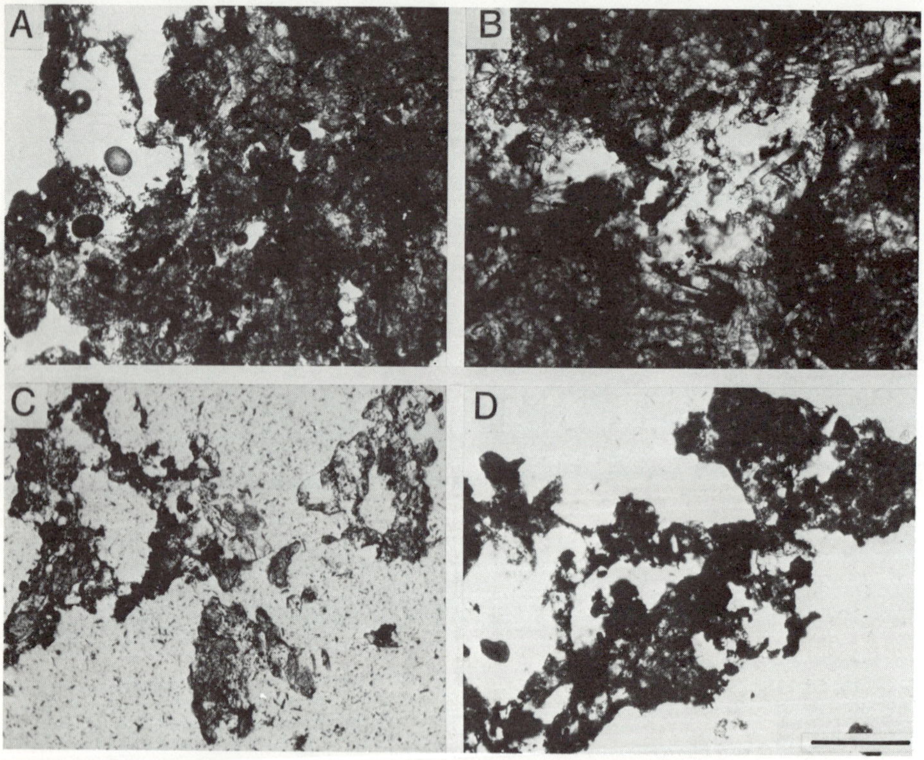

Fig. 1. Experimentally shocked basalt from Lonar Crater (India) representing shock Classes 4 and 5. Scale bar is 0.2 mm. (1A) Plane polarized transmitted light. Sample shocked to 696 kbar. Note vesiculation and flow of colorless plagioclase glass (left) and loss of porphyritic texture. Class 4. (1B) Sample shocked to 742 kbar. Note again vesiculation and drastic modification of original texture. All white grains are isotropic plagioclase glasses, either diaplectic or fused. Class 4. (1C) Sample shocked to 974 kbar showing frothy glass with schlieren and granulated inclusions of highly disjointed pyroxene. Most of the glass is "whole rock" melt, but some schlieren have strong affinities to monomineralic compositions. Class 5. (1D) Sample shocked to 999 kbar. Vesiculation and frothy texture are similar to those in 1C. Class 5.

ulvöspinel fragments are the only crystalline minerals remaining. Judging from these experiments, pressures in excess of 1 Mbar are necessary to cause complete melting of the Lonar basalt. These experiments also demonstrate that a variety of shock melting phenomena (e.g., selective melting of feldspars, vesiculation of feldspar melt, and onset of "whole rock" melting) develop in confined, seemingly random areas; the volumetric proportions of these glass types increase over a wide range of pressures, from ~400 to >1000 kbar. The shock experiments at 614 and 742 kbar performed in the new auxiliary vacuum system at 10^{-7} CO/CO_2 pressure fit perfectly well into the above scheme of progressive shock metamorphism. Varying oxygen fugacities compared to the Gibbons *et al.* (1975) and Kieffer *et al.* (1976a) as well as our 700, 974, and 996 kbar experiments at atmospheric vacuum of 10^{-2} torr did not lead to significant changes in the recovered products on our scales of observation. No shock reduction was observed, e.g., Fe°, and dissemination of sample assembly materials (Gibbons *et al.*, 1975) was largely eliminated by using a tungsten alloy (FS 77; Table 1) to contain the sample. Better controlled experiments with variable f_{O_2} are planned for the future, however, and constitute the main rationale for the development of the auxiliary vacuum system.

Experimentally Shocked Lunar Basalt

Unshocked basalt 75035 is a low-K, high-Ti, medium-grained, subophitic basalt described by Longhi *et al.* (1974). Their modal analysis gives 44% clinopyroxene, 33% plagioclase, 15% ilmenite, 5% cristobalite, 2% pyroxferroite, 1% native iron plus traces of troilite, and some mesostasis. Plagioclase and pyroxene are relatively unfractured (Fig. 2) and ilmenite is untwinned. Basalt 75035 samples were successfully recovered from ten experiments in which shock pressures ranged from 93 to 999 kbar.

Following the classification of Kieffer *et al.* (1976) the progressive damage and phase transitions of feldspar also serve as the best indicators of shock intensity in the lunar basalt. The most severe metamorphic effect in each thin section was used for classification of shock intensity and associated pressure calibration. The major shock effects observed in our experiments are summarized in Table 2 and illustrated in Figs. 3, 4, and 5.

Class 1 (<250 kbar): Shock features in feldspar and pyroxene grains are characterized by increasing fracturing up to approximately 200–250 kbar (Fig. 3A–D). Feldspar grains in rare cases display shock lamellae (e.g., Stöffler, 1972). Shock damage in ilmenite is characterized by abundant twinning even at 93 kbar, in agreement with Sclar (1971) and Minkin and Chao (1971). These features are typical of Class 1 features in the Lonar basalt (Kieffer *et al.*, 1976a).

Class 2 (~250–450 kbar): The characteristic Class 2 features are based on the conversion of plagioclase to diaplectic feldspar with reduced birefringence and, especially, to diaplectic feldspar glass, maskelynite, in the 302 kbar shot (Fig. 3E–H). Some diaplectic feldspar grains display shock lamellae. Pyroxene grains

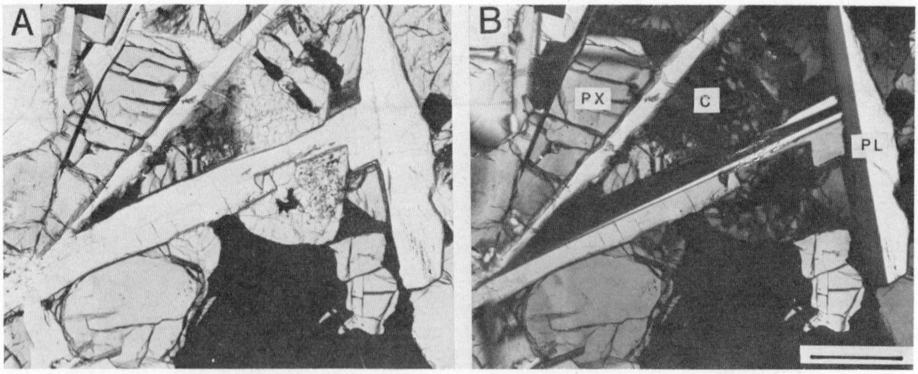

Fig. 2. Unshocked basalt 75035,76. Scale bar is 0.2 mm. (2A) Note medium-grained, subophitic texture; sparse fractures in pyroxene (PX); tabular shapes of plagioclase laths (PL); polygonal fractures in cristobalite (C); and interstitial ilmenite. Plane polarized light. (2B) Crossed polarizers. Note even birefringence in pyroxenes, undeformed twins in plagioclase, and nearly isotropic cristobalite.

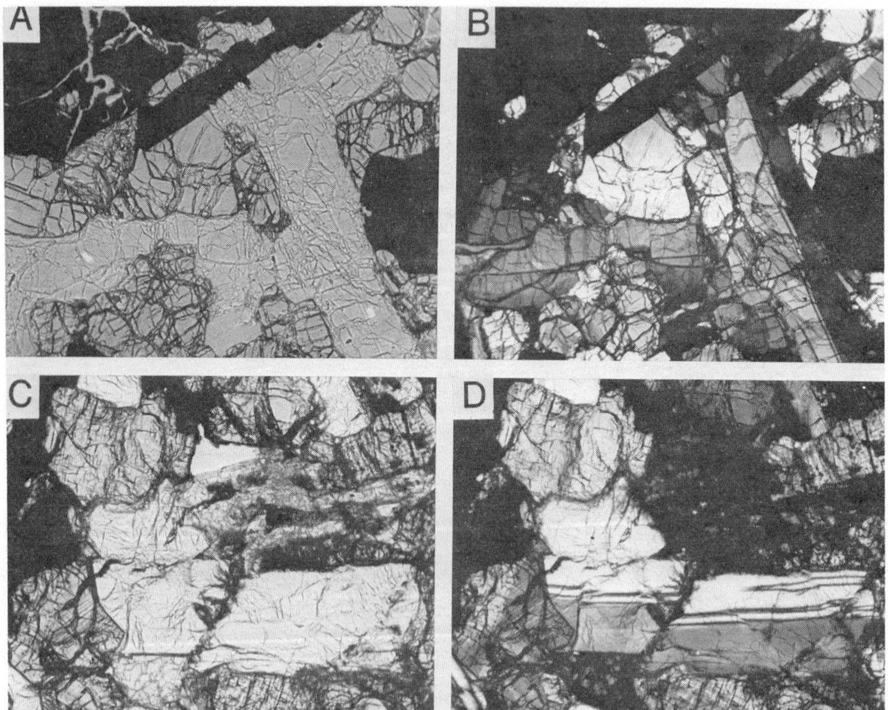

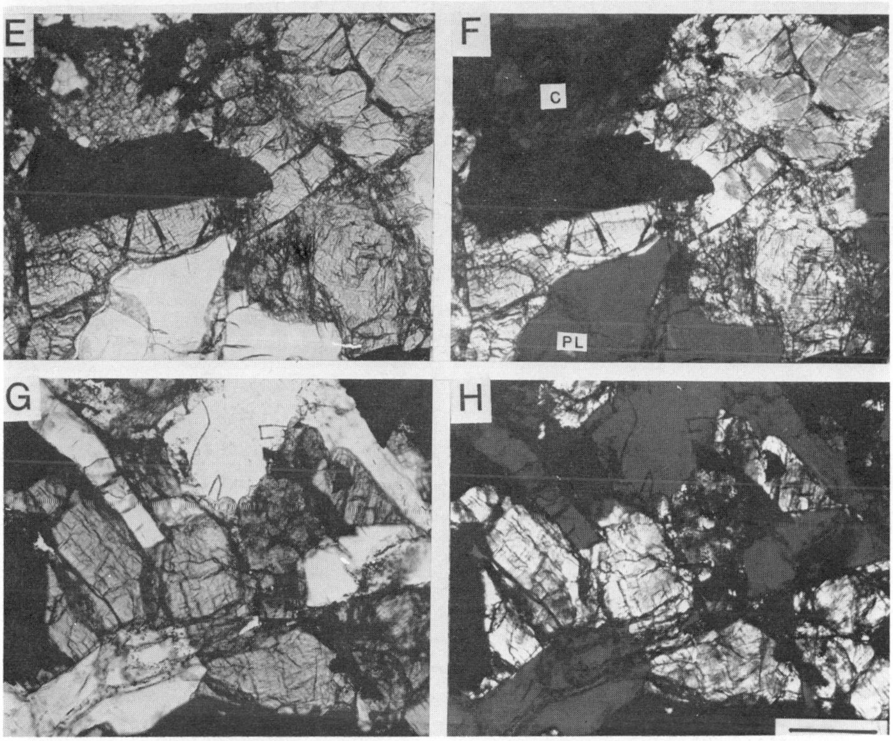

Fig. 3. Experimentally shocked lunar basalt 75035 representing shock metamorphic Classes 1 and 2. Scale bar for all photomicrographs is 0.2 mm. (3A) Class 1. Sample shocked to 93 kbar showing subophitic texture and abundant fractures in all component minerals. Plane polarized light. (3B) Same view as (3A), but with crossed polarizers. Note patchy extinction in plagioclase and pyroxene grains. (3C) Class 1. Sample shocked to 204 kbar. Note increased fine scale fracturing of all minerals. Plane light. (3D) Same view as (3C), but with crossed polarizers. Note distinct microfault offsets in twinned plagioclase, uneven extinction, and areas of decreased birefringence, i.e., almost isotropic patches. Deformation lamellae are visible in pyroxene grains, e.g., lower left. The cristobalite is also nearly isotropic. (3E) Class 2. Sample shocked to 302 kbar. Plane light. (3F) Same view as in (3E), but with polarizers crossed at 80°. Plagioclase (PL) is transformed into isotropic maskelynite containing sparse, conchoidal fractures. Note patchy extinction and shock lamellae in the highly fractured pyroxene grain to the upper right. Cristobalite (C) is also converted to diaplectic glass. (3G) Class 2. Sample shocked to 400 kbar. Plane light. (3H) Same view as (3G), but with polarizers crossed at 80°. All plagioclase and cristobalite are converted to diaplectic glasses. Ophitic texture is preserved. Mosaicism and patchy extinction are prominent in pyroxenes.

are progressively more fractured and granulated although grain boundaries are generally preserved. Ilmenite grains are twinned and in some cases bent.

Class 3 (~450 to ~600 kbar): The distinction between Class 2 and 3 features is subtle, but important. Class 3 feldspar is fused to a glass which shows signs of incipient flow (Fig. 4A–D). This selective melting of feldspar starts on grain

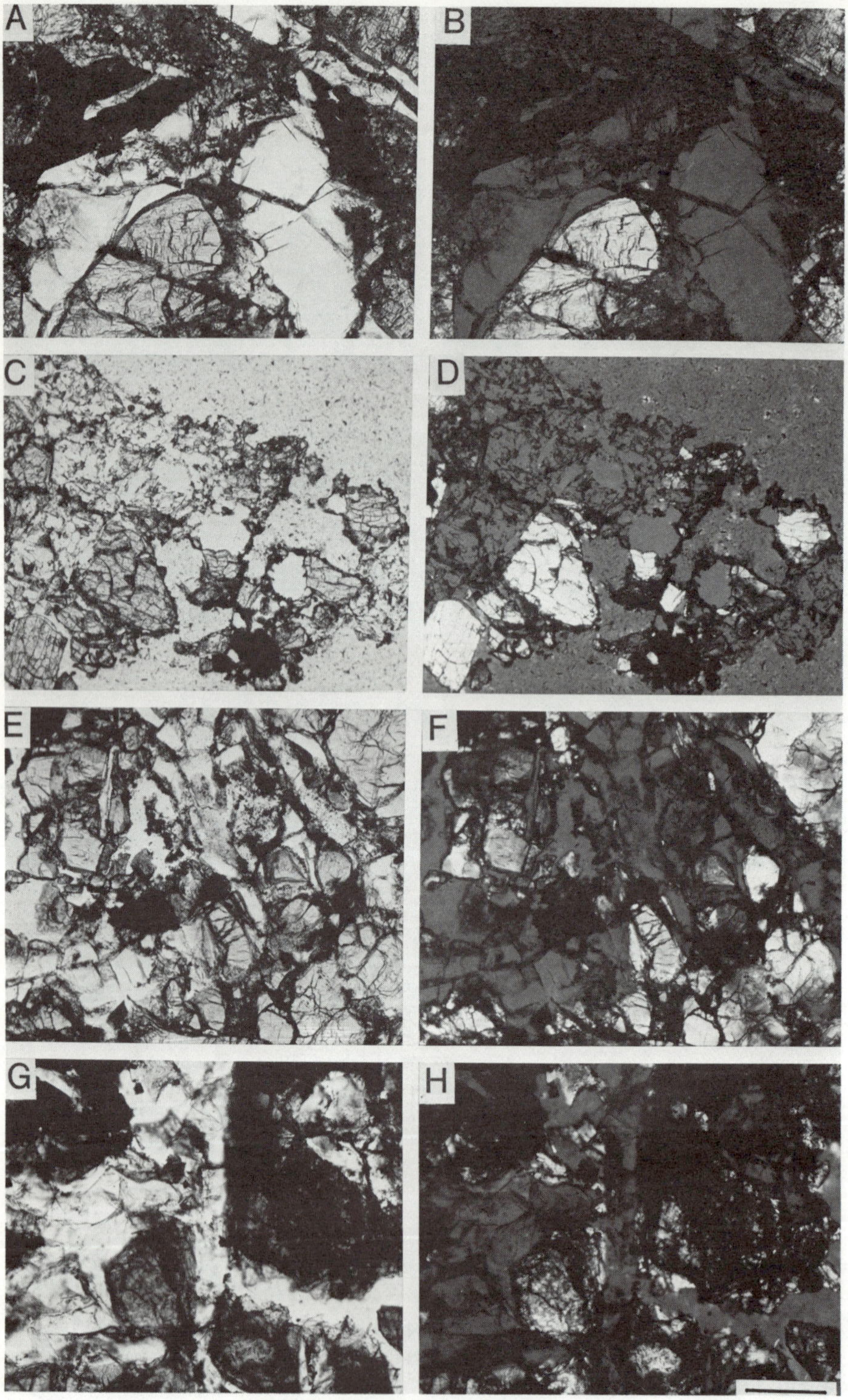
A
B
C
D
E
F
G
H

boundaries so that the glass conforms to the shape of fractured contiguous pyroxene grains, thereby modifying the initial rock texture. The total amount of feldspar glass increases with progressive pressure but does not amount to more than 10% of all feldspars present in the 494 kbar shot.

Class 4 (~600–800 kbar): The index for Class 4 features is the presence of vesicles in flowed, colorless, feldspar glass. Flow is prominent and may engulf several neighboring grains and penetrate fractures. Pyroxenes are more intensely granulated and fragments may be disjointed from their parent grains; some may be dragged along by the feldspar melt. Ilmenite is also highly granulated and disjointed. The overall rock texture has dramatically changed to an aggregate of intensely flowed feldspar melts and relatively large, detrital pyroxene, and ilmenite fragments.

Class 5 (>800 kbar): Between 800 and 1000 kbar all component mineral phases including pyroxenes and ilmenite show signs of melting (Fig. 5A–D). Thus, a large variety of glasses is observed, ranging from almost pure mineral compositions in individual schlieren to two or three phase mixtures and, finally, to whole rock melts. The 999 kbar recovery is dominated by whole rock melts; however, some 20% relict pyroxene and ilmenite fragments are dispersed in the flowed glass.

Throughout the range of progressive shock damage produced in these basalts composition variations of major mineral components was not detectable in microprobe analyses (see also Schaal *et al.*, 1976). Analyses showed that plagioclase glasses in samples shocked to pressures between 450 and 1000 kbar have compositions which are identical to those in maskelynites in shots between 250 and 450 kbar, in diaplectic plagioclase, <200 kbar, as well as in unshocked plagioclase (Table 3). Compositions of clinopyroxenes in the entire suite of shocked materials also demonstrate a coincidence with those in unshocked

---

Fig. 4. Experimentally shocked lunar basalt 75035 representing shock-metamorphic Classes 3 and 4. Scale bar is 0.2 mm for all photomicrographs. (4A) Class 3. Sample shocked to 494 kbar. Plane polarized light. (4B) Same view as (4A), but with polarizers crossed at 80°. Note complete conversion of plagioclase to isotropic glass with only slight modification of initial ophitic textures. Incipient melting is observed along some plagioclase boundaries. The only birefringent phase remaining is pyroxene. (4C) Class 4. Sample shocked to 606 kbar. Plane light. (4D) Same view as (4C), but with polarizers crossed at 80°. Plagioclase is totally converted to a colorless glass with slight vesiculation and incipient flow. Original ophitic texture is modified by flow. Large pyroxene grains are fragmented but retain normal birefringence. (4E) Class 4. Sample shocked to 668 kbar. Plane light. (4F) Same view as (4E), but with polarizers crossed at 80°. About 10% of plagioclase is fused into a vesicular, slightly flowed glass; the remaining plagioclase is maskelynite. Pyroxene and ilmenite grains are fragmented and disjointed. Initial rock textures are destroyed. (4G) Class 4. Sample shocked to 743 kbar. Plane light. (4H) Same view as (4G), but with polarizers crossed at 80°. Highly fluid plagioclase glass comprises a "matrix" enclosing pyroxene and ilmenite fragments. Pyroxenes are intensely fractured and show extreme patchy extinction (and on rare occasions an apparent decrease in birefringence). The glass is moderately vesiculated creating a slightly frothy texture.

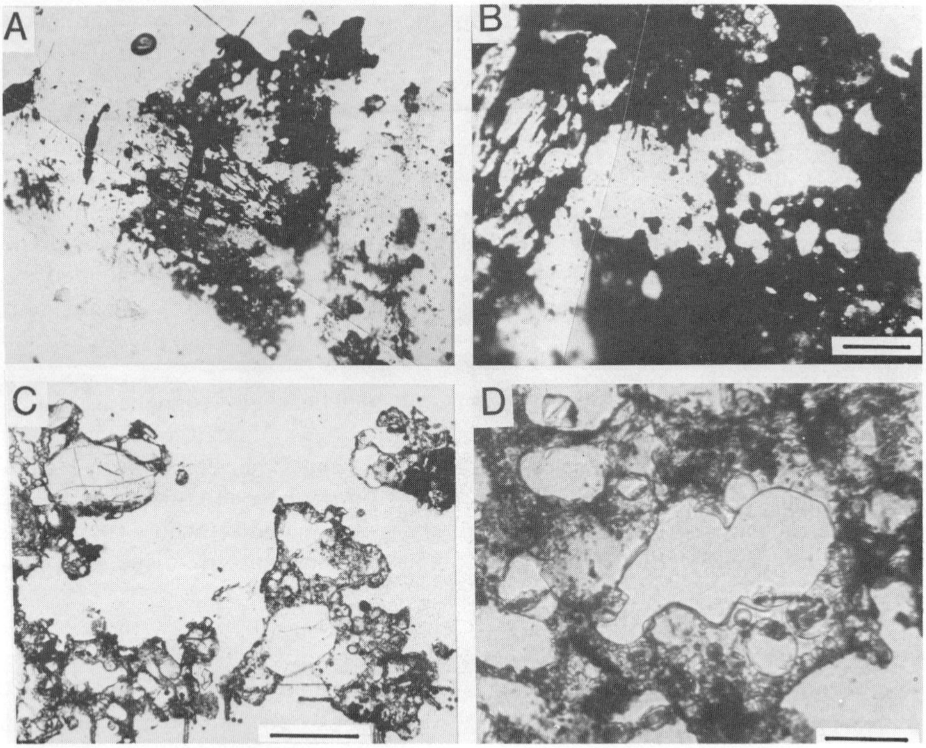

Fig. 5. Experimentally shocked lunar basalt 75035 representing shock-metamorphic Class 5. Transmitted plane polarized light in each photomicrograph. (5A) Sample shocked to 853 kbar. Note abundance of highly vesiculated, frothy glass, much of which is made up of "whole rock" melt according to microprobe analyses, though individual schlieren may be made up of monomineralic compositions. All feldspar is molten; however, pyroxene and ilmenite fragments are abundant. Scale bar for (5A) and (5C) is 0.2 mm. (5B) Enlargement of (5A) showing details of vesicles and flow features in glass. Scale bar is 50 μm, (5C) Sample shocked to 999 kbar. Note prominent, frothy texture of glass and detrital fragments of pyroxene and ilmenite grains. Relict grains may be aligned in flow features. Pyroxenes display reduced birefringence and edge melting. (5D) Enlargement of (5C) showing details of vesicles and flow features in glass. Scale bar is 50 μm.

samples (Fig. 6). In addition, the same is true for cristobalite. Nearly pure silica glasses (>96% SiO_2) are present as transparent schlieren within Class 5 glasses in samples shocked to 853 and 999 kbar, representing monomineralic remnants of cristobalite. Quantitative microprobe analyses in Class 5 samples were seriously impaired by several factors: (1) small quantities of sample recovered and extreme difficulties in preparing high quality polished thin sections, (2) intense, small-scale fracturing in relict crystalline grains, (3) small exposure area of frothy glass between vesicles, and (4) small-scale heterogeneity in mixed glasses prohibiting precise relocation of probe beam on successive runs. However, based on qualitative analyses we conclude that chemical fractionation caused by

Table 3. Compositions of experimentally shocked plagioclase, 75035.

Pressure	Average composition Or	Ab	An	No. of grains	Range An
Unshocked	.5	16.6	82.9	15	82.9–88.0
Unshocked	1.7	17.6	80.7	1	—
93 kbar	.3	11.8	87.9	10	85.1–89.0
204 kbar	.5	13.7	85.8	10	80.5–88.3
302 kbar	.5	12.6	86.9	10*	79.7–89.0
400 kbar	.5	12.7	86.8	10*	81.4–89.1
494 kbar	.6	14.2	85.2	10*	80.5–89.3
668 kbar	.4	12.9	86.7	10*	82.3–88.8
853 kbar	.9	15.0	84.1	1†	—
999 kbar	.6	13.5	85.9	3†	84.3–89.2

*Maskelynite.
†Plagioclase glass.

selective alkali volatilization, by preferential assimilation, by crystal differentiation, or by other means induced by a single impact event is not appreciable in concentrations sensitive to probe analyses, in agreement with Schaal *et al.* (1976).

In summary, the shock features and associated peak pressures in lunar basalt 75035 are compatible with those observed in the Lonar basalt. The scheme of progressive shock metamorphism of Lonar basalt proposed by Kieffer *et al.* (1976) also applies to lunar basalt 75035. One subtle difference, however, is the

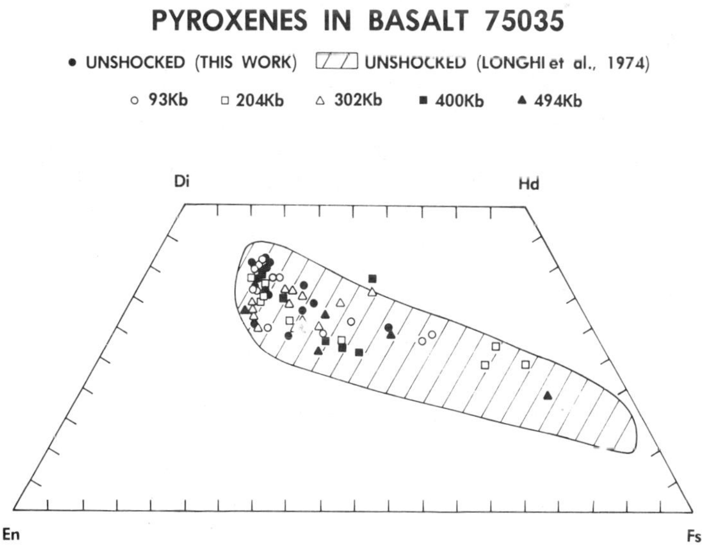

Fig. 6. Pyroxene quadrilateral showing range in compositions of shocked and unshocked pyroxenes in lunar basalt 75035.

somewhat increased stability of the lunar plagioclase which vitrifies at slightly higher pressures, i.e., at 250 kbar, and melts less profusely between 500 and 600 kbar than the terrestrial analog; some maskelynite grains are still present at 853 kbar. Thus, while the pressure brackets for the various shock grades are more or less identical, the amounts of feldspar and whole rock melts tend to be somewhat less in the lunar basalt at any given pressure. This may be understood in view of the higher melting point of the lunar anorthitic plagioclase (An_{87}; $\approx$1526°C) versus the Lonar plagioclase (An_{60}; $\approx$1474°C/1 bar, 0% H_2O; Bowen, 1913). Another possible factor could be the grain size; the Lonar basalt is significantly finer grained. Furthermore, Ahrens *et al.* (1977) report an equation of state for lunar basalt 70215 which has a lower shock impedance than corresponding terrestrial analogs and the "mixed phase" regime may possibly extend up to 800 kbar. In addition, Lonar basalts contain hydrous palagonite which may lower the overall melting temperature of the rock and/or the contiguous feldspar grains.

NATURALLY SHOCKED LUNAR BASALTS

Though it is commonly accepted that all lunar basalt specimens had experienced at least one meteorite impact that brought them to the surface for collection by the astronauts, it is also commonly agreed that the number of hand specimens displaying genuine shock features is extremely rare in the Apollo basalt collection (Chao *et al.*, 1970; Dence *et al.*, 1970; Engelhardt *et al.*, 1970; Quaide and Bunch, 1970). The mare basalt collection differs dramatically in this aspect from the suite of highland rocks which generally tend to be intensely shocked or even composed of impact melts. Classification schemes of progressive shock metamorphism for mare basalts in the past were based on particles recovered from the regolith and on inferences from terrestrial rocks, mostly granitic (Chao *et al.*, 1970; Dence *et al.*, 1970; Engelhardt *et al.*, 1970). We carried out a systematic search through the entire curatorial thin section collection which disclosed that out of 152 specimens available only three lunar basalts are shocked throughout with shock damage more severe than fracturing. These samples are 15685, 12054, and 79155 (we thank T. Usselmann for bringing sample 79155 to our attention), none of which were previously described in terms of shock metamorphic features.

Basalt 15684

Basalt 15684 is a pigeonite basalt chip approximately 8 × 6 mm in size. This sample was retrieved from an aggregate of rock chips cemented by a black glass matrix. Although a porphyritic subophitic texture is well preserved within the chip, 80–90% of the tabular to acicular plagioclase laths are isotropic maskelynite, accounting for 28% of the mode (Fig. 7A,B). Clinopyroxenes comprise 66% of the mode; pigeonite predominates (Fig. 8). Ilmenite accounts for ~8%, and trolite and kamacite are present in trace amounts; compositions are listed in

Table 4. Electron microprobe analyses of minerals and glasses in basalt 15684.

	(1)	(2)	(3)	(4)	(5)	(6)	(7)	(8)	(9)
SiO_2	48.03	51.47	53.84	48.61	—	48.17	50.68	39.64	47.20
TiO_2	.06	.65	.29	.96	52.33	.11	1.14	9.52	2.16
Al_2O_3	32.74	2.51	1.25	1.05	.04	32.01	4.67	6.26	9.56
FeO	.78	15.20	17.04	32.02	45.95	1.53	19.92	25.54	20.22
MnO	.00	.27	.28	.43	.29	.04	.26	.32	.27
MgO	.21	15.63	23.43	6.73	.13	.45	15.26	7.38	9.29
CaO	17.01	13.21	3.06	10.05	—	17.14	7.75	10.83	10.02
Na_2O	1.16	—	—	—	—	1.13	.28	.25	.41
K_2O	.06	—	—	—	—	.05	—	.08	.11
Cr_2O_3	—	.76	.66	.14	.07	.01	.68	.34	.49
Total	100.05	99.70	99.85	100.00	98.81	100.64	100.64	100.16	99.73

(1) Maskelynite, average of 10 grains.
(2) Augite, average of 4 grains.
(3) Pigeonite, average of 4 grains.
(4) Ferroaugite, average of 8 grains.
(5) Ilmenite, average of 7 grains.
(6) Colorless glass, average of 4 spots, 2 analyses for K_2O, and 2 for Cr_2O_3.
(7) Yellow green glass, single spot.
(8) Reddish brown glass, average of 7 spots, 2 analyses for K_2O, and 5 for Cr_2O_3.
(9) Greenish yellow to yellowish brown glass, average of 16 spots, 10 analyses for K_2O, and 6 for Cr_2O_3.

Table 5. Electron microprobe analyses of minerals and glasses in basalt 12054.

	(1)	(2)	(3)	(4)	(5)	(6)	(7)	(8)	(9)
SiO_2	46.44	46.97	49.58	50.15	51.36	—	49.12	40.71	43.27
TiO_2	.00	.00	1.40	1.28	.89	52.33	1.42	7.78	4.72
Al_2O_3	33.04	32.55	2.36	2.13	1.40	—	2.65	5.71	9.51
FeO	.54	.61	13.74	16.00	18.75	45.45	17.74	29.37	19.57
MnO	.00	.00	.24	.26	.26	—	.26	.32	.24
MgO	.28	.20	15.68	18.08	20.75	.17	13.92	1.92	9.22
CaO	18.42	18.75	14.15	9.61	4.85	—	12.77	10.06	11.45
Na_2O	.95	1.01	.02	.02	.01	—	.07	.19	.23
K_2O	.00	.01	—	—	—	—	—	—	—
Cr_2O_3	—	—	.79	.73	.57	—	.52	.13	.36
Total	99.67	100.10	97.96	98.26	98.84	97.95	98.47	96.19	98.57

(1) Plagioclase, average of 5 grains.
(2) Maskelynite, average of 5 grains.
(3) Augite, average of 4 grains.
(4) Subcalcic augite, average of 3 grains.
(5) Pigeonite, average of 3 grains.
(6) Ilmenite, average of 6 grains.
(7) Pale green glass, average of 4 spots.
(8) Reddish brown glass, single spot.
(9) Yellowish orange to light brown glass, average of 25 spots.

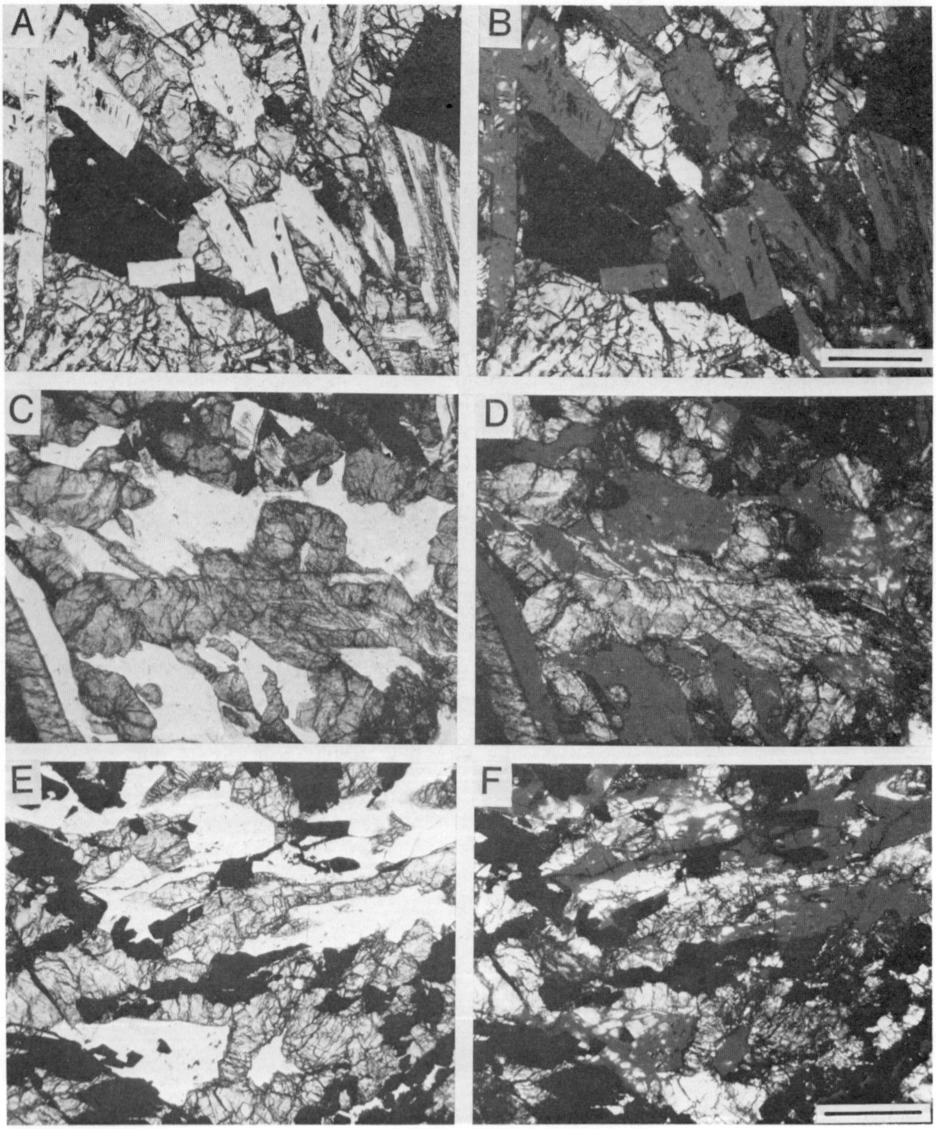
A
B
C
D
E
F

Table 4. Moderate shock effects are pervasive in thin section 15684,3. Prevalence of diaplectic feldspar glass throughout the section indicates a Class 2 shock history, with peak pressures less than 450 kbar. Pyroxene grains are granulated and mosaic; some contain closely spaced fractures and shock lamellae. The pressure distribution appears to have been uniform throughout this sample.

Basalt 12054

Ilmenite basalt 12054 (thin section 12054,79) contains 29% plagioclase, 62% clinopyroxene, 11% ilmenite, and traces of troilite and kamacite. Mode is based on a 3000-spot point count. The range of compositions of 40 plagioclase grains determined by microprobe analysis is from $An_{89}Ab_{11}$ to $An_{93}Ab_7$ with an average of $An_{91}Ab_9$ (Table 5). Analyses of 60 randomly selected pyroxene grains demonstrate a wide range of compositions between augite, subcalcic augite, pigeonite, ferroaugite, and subcalcic ferroaugite (Fig. 8). Augite and pigeonite are most abundant, consistent with trends reported for most Apollo 12 ilmenite basalts (Papike *et al.*, 1976). Individual pyroxene grains may be zoned with wide compositional variations between augite and subcalcic ferroaugite; exsolution lamellae are common. Ilmenite grains contain less than 1% MgO. Amorphous blebs of troilite and kamacite are generally closely associated or intergrown with metallic iron, perhaps representing a product of desulfurization of troilite (Gibson *et al.*, 1976; Brett, 1976) or a late-stage product of liquid immiscibility (Frondel, 1975, p. 20).

Weak to moderate shock effects representing Class 1 and Class 2 pressures are dominant petrographic features throughout the section (Fig. 7C,D). Diaplectic feldspars are highly fractured and exhibit pronounced undulatory extinction and decreased birefringence. Some contain up to three independently oriented systems of planar shock lamellae whose crystallographic planes are not measured at present. Maskelynite occurs throughout the section but is most abundant near one end where the sample is coated with glass. Analyses of 19

Fig. 7. Naturally shocked lunar basalts. (7A) Basalt 15684,3. The fine- to medium-grained subophitic texture shown in (7A) consists of tabular plagioclase maskelynite laths, fractured pyroxene grains, and interstitial ilmenite grains. Plane polarized transmitted light. (7B) Same view as (7A) but with polarizers crossed at 80°. Note that plagioclase is partially to totally converted to maskelynite, typical of Class 2 shock intensity. Pyroxene grains display patchy extinction and are moderately fractured. Scale bar is 0.2 mm. (7C) Basalt 12054,79. Unfractured maskelynite grains are interspersed with intensely fractured pyroxene and ilmenite grains in medium-grained subophitic texture. Plane light. Scale bar is 0.5 mm for (6C–F). (7D) Same view as (7C), but with polarizers crossed at 80°. Some maskelynite grains contain relict domains of plagioclase crystal structure (white patches). Pyroxenes grains show patchy extinction. These features represent Class 2 shock intensity. (7E) Basalt 79155,60. This medium-grained ophitic basalt contains slightly fractured plagioclase-maskelynite grains and intensely fractured pyroxene and ilmenite grains. Plane light. (7F) Same view as (7E), but with polarizers crossed at 80°. Note relict crystalline domains in maskelynite grains and mosaicism in pyroxene grains, characteristic of Class 2 shock intensity.

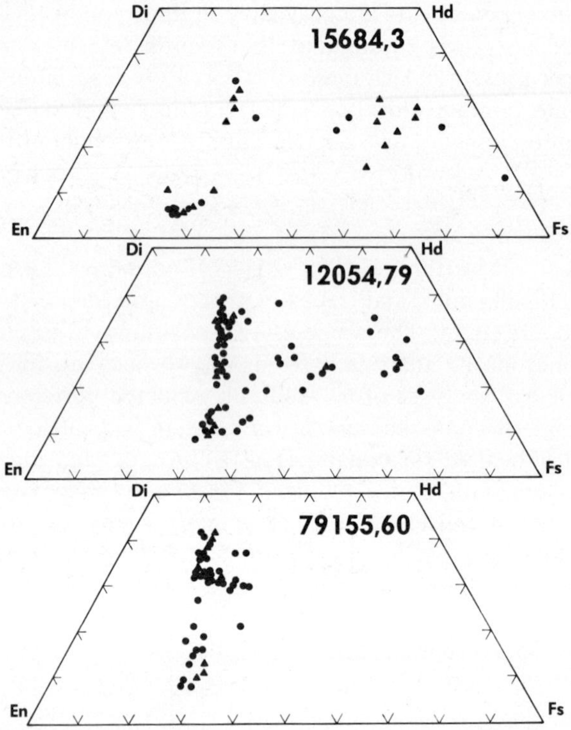

Fig. 8. Pyroxene quadrilaterals showing the variation of compositions of pyroxene grains in 15684,3, 12054,79, and 79155,60. Circles represent uncorrected compositions based on three element analyses only (Ca, Fe, Mg), while triangles represent corrected compositions based on nine element analyses presented in Table 4, 5, and 6, respectively.

maskelynite grains range from $An_{87}Ab_{12}Or_1$ to $An_{93}Ab_7$ with an average of $An_{91}Ab_9$, nearly identical with the unshocked plagioclase compositions (Table 5). Pyroxenes are intensely fractured or granulated and have suffered marked decrease in birefringence. Shocked ilmenite grains are also fractured and many display bent twin lamellae less than 15 μm wide. Thus, we estimate that peak shock pressures did not exceed 450 kbar within 12054.

Basalt 79155

Basalt 79155 is a high-Ti basalt. Thin section 79155,60 contains 19% plagioclase, 52% clinopyroxene, 24% ilmenite, and traces of olivine, troilite, and kamacite. Compositions of 28 plagioclase grains range from $An_{83}Ab_{17}$ to $An_{88}Ab_{12}$ with an average of $An_{87}Ab_{13}$ (Table 6). Augite and pigeonite are predominant pyroxenes and are strongly zoned; olivine (Fo_{69}) forms the core of some augite grains. Subcalcic augite is minor (Fig. 8) ilmenite grains contain less than 2% MgO (Table 6). Troilite and kamacite are usually intergrown.

Table 6. Electron microprobe analyses of minerals and glasses in basalt 79155.

	(1)	(2)	(3)	(4)	(5)	(6)	(7)	(8)	(9)
SiO_2	47.48	47.46	49.40	52.41	—	47.86	43.33	37.65	41.21
TiO_2	.06	.07	2.57	1.46	53.01	.08	7.26	16.35	9.38
Al_2O_3	33.30	33.26	3.62	1.96	.42	33.08	8.93	9.27	11.41
FeO	.35	.33	10.41	17.05	43.16	.55	16.40	20.44	17.05
MnO	.02	.02	.21	.28	.41	.02	.25	.34	.26
MgO	.28	.34	15.82	21.13	1.70	.27	10.57	7.95	7.92
CaO	16.83	16.91	16.91	5.60	—	16.62	11.20	9.18	10.78
Na_2O	1.39	1.37	.06	.00	—	1.45	.46	.38	.54
K_2O	.04	.03	—	—	—	.05	.06	.06	.10
Cr_2O_3	—	—	.69	.41	.90	—	.35	—	.32
Total	99.75	99.79	99.69	100.30	99.60	99.98	98.81	101.62	98.97

(1) Plagioclase, average of 5 grains.
(2) Maskelynite, average of 5 grains.
(3) Augite, average of 6 grains.
(4) Pigeonite, average of 3 grains.
(5) Ilmenite, average of 5 grains.
(6) Colorless glass, single spot.
(7) Yellowish orange glass, average of 5 spots, 2 analyses for K_2O, and 3 for Cr_2O_3.
(8) Reddish brown glass, single spot.
(9) Yellowish brown glass, average of 18 spots, 11 analyses for K_2O, and 7 for Cr_2O_3.

Rock 79155 displays an irregular spatial distribution of weak to moderate shock features and contains intergranular "pools" of vesicular glass randomly dispersed within the section. However, igneous textures are well preserved (Fig. 7E,F). More than half of the section contains fractured feldspar, pyroxene, and ilmenite grains with typical Class 1 cataclastic features. Some diaplectic feldspar grains display microfaults and planar shock lamellae. Interspersed among the fractured plagioclase crystals, accounting for the remainder of the section, are plagioclase grains which are partially to totally converted to maskelynite; one end of an individual grain may be crystalline (birefringent) and the other vitreous (isotopic) while retaining the initial crystal outline. Maskelynite grains signifying a Class 2 pressure regime are most abundant near the glassy zones. Most pyroxene grains are intensely fractured or granulated; undulatory extinction and mosaicism are widespread. Some pyroxenes display finely spaced shock lamellae, different from fractures, cleavages, or exsolution lamellae. Thus, unlike samples 15684 and 12054, which are fairly homogeneously shocked throughout the entire thin section, basalt 79155 displays general, Class 1 features and irregular pockets, a few millimeters across, of Class 2 features concentrated predominantly, though not exclusively, in areas where whole rock melts occur.

The similarity of these shock effects and those produced experimentally in rock 75035 indicate that the natural shock effects in lunar basalt can be simulated and successfully calibrated in the laboratory. The congruence of shock features in these two sets of samples is striking. We estimate that samples 15684, 12054,

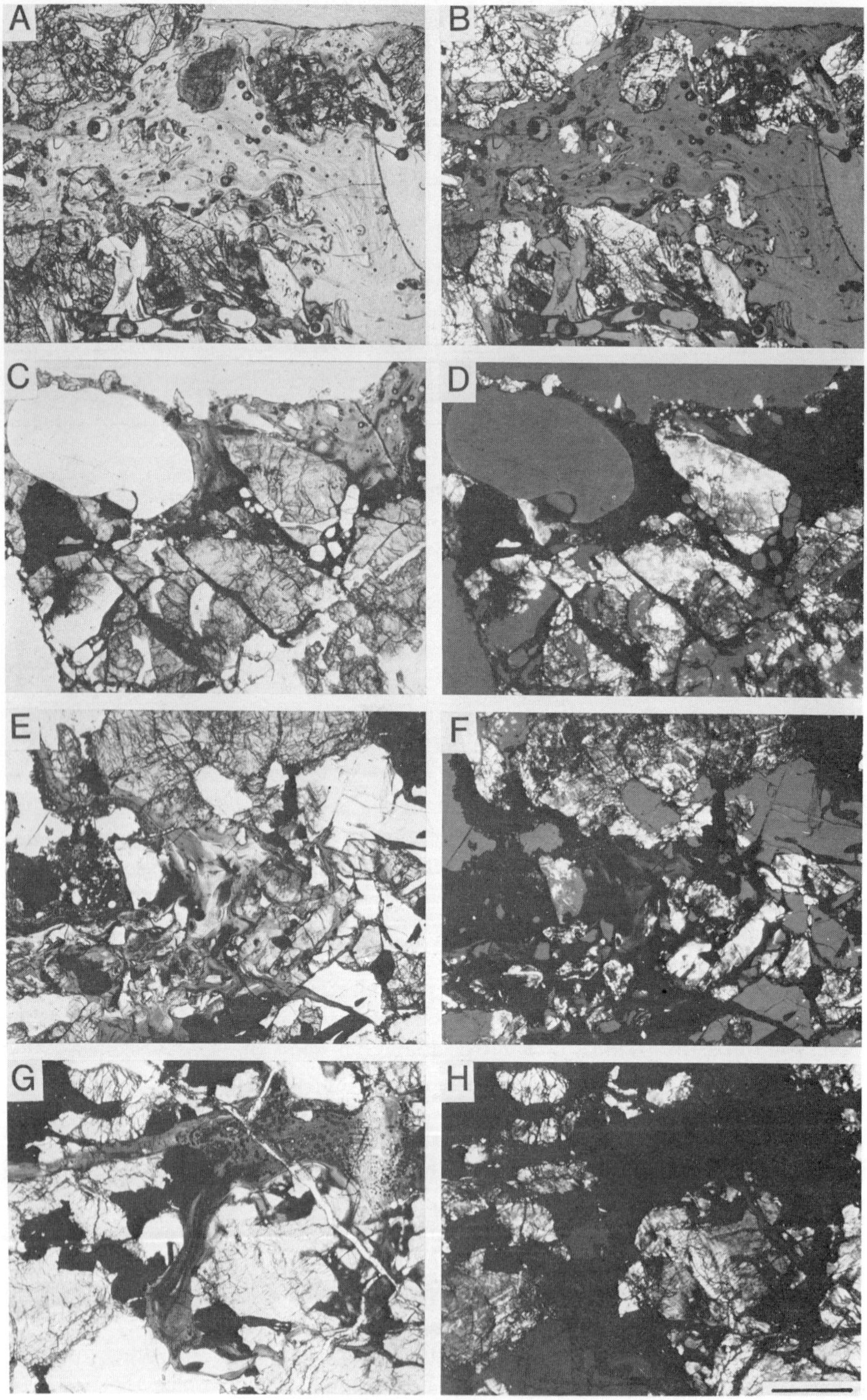
A
B
C
D
E
F
G
H

and 79155 have experienced maximum shock pressures on the order of 300–450 kbar.

Basaltic Glass Coatings

Each of the three naturally shocked basalts discovered in our survey is draped by a relatively thick, vesicular, variegated glass coating. The converse is not true, i.e., other basalts draped with glass are not necessarily shocked, e.g., rock 12017. We consider this correlation more than fortuitous and present textural and geochemical evidence that may illuminate the origin of these glass coatings. Shock melting by a single cratering event results in minor chemical modifications of the target material, as illustrated in experimentally produced whole rock melts, in the glass liner of a microcrater on rock 10085 (Schaal *et al.*, 1976), and in many other impact melts (e.g., Engelhardt, 1967; Simonds *et al.*, 1976). Therefore, it should be possible to establish the source area of these large glass coatings, especially to discriminate between a regolith or basaltic parent material.

15684 glass coating

A flowed vesicular glass coats nearly the entire perimeter of 15684 in thin section. The glass coating formed a thermal aureole in many of the grains it coats: edge melting is visible in pyroxene grains (Fig. 9A,B). The predominant colors in this glass are moderate yellow green (5GY 7/4; GSA rock-color chart). Flow features are accentuated with colorless and moderate reddish brown (10R 4/6) schlieren and with flow bands consisting of aligned kamacite, troilite, and schreibersite spherules up to 6 μm in diameter. An additional light brown (5YR 5/6) glass occupies a round patch within the basalt and a vesicular dark yellowish

Fig. 9. Glass coatings on naturally shocked lunar basalts. Scale bar is 0.5 mm in each photomicrograph. (9A) Basalt 15684,3. A flowed glass layer coats most of rock 15684 and contains vesicles, colored schlieren, flow bands of aligned metal spherules and shocked basaltic inclusions. Plane polarized light. (9B) Same view as (9A), but with polarizers crossed at 80°. Birefringence is visible in crystalline inclusions in isotropic glass. (9C) Basalt 12054,79. This highly vesiculated glass coats one side of rock 12054. The glass contains colored schlieren, flow bands of aligned metal spherules, and crystalline basalt inclusions. (9D) Same view as (9C), but with polarizers crossed at 80°. Birefringent pyroxene and basalt inclusions are recognizable in isotropic glass. (9E) Rock 79155,60. Colorless and colored glasses are interflowed in fractures within rock 79155. Colorless schlieren formed from plagioclase melts, and dark colored schlieren formed from ilmenite and pyroxenes. Plane light. Scale bar is 0.5 mm. (9F) Same view as (9E), but with polarizers crossed at 80°. Shocked basaltic fragments and vesicles are identifiable in glass. Note abundance of maskelynite. (9G) Rock 79155,60. This highly vesiculated, colored glass is injected into fractures in the rock interior through both maskelynite (lower center) and pyroxene grains (upper left). Multicolored schlieren enhance flow features. Plane light. Scale bar is 0.5 mm. (9H) Same view as (9G), but with polarizers crossed at 80°. The abundance of glass is made evident by the extent of the black area.

brown (10YR 5/2) glass fills fractures within the basalt. Glass colors are correlated to their compositions (Table 4). The majority of glass compositions cluster near typical Apollo 15 pigeonite basalt compositions on the ternary diagram (Fig. 10), though individual glass areas may constitute only a two or three phase mixture. However, most of the glass draping 15684 more closely resembles Apollo 15 basalt compositions than Apollo 15 soils, and, therefore, we conclude that it is derived from a (local?) basaltic substrate. Direct geochemical comparison with basalt 15684 is not possible because whole rock analyses for this specimen are not available.

12054 glass coating

A layer of yellowish orange to reddish brown glass up to 0.5 mm thick coats sample 12054 on the edge where shock effects are most severe. The glass did not form *in situ*, however. A thermal aureole less than 1 mm deep is visible in grains which the glass coats. Feldspar grains are partially to totally melted and show signs of incipient flow. Edges of most pyroxene grains coated with glass are rounded by edge-melting and form a veneer of grayish yellow green (5GY 7/2) to pale green (5G 7/2) glasses less than 0.1 mm thick with compositions nearly identical to those of underlying pyroxene grains. Flow schlieren in this "pyroxene glass" are traceable to distances of up to 0.4 mm. The major glass coating contains oblong vesicles up to 1.2 mm long (Fig. 9C,D). Pale yellowish orange (10YR 8/6) glasses are mixtures of plagioclase plus pyroxene com-

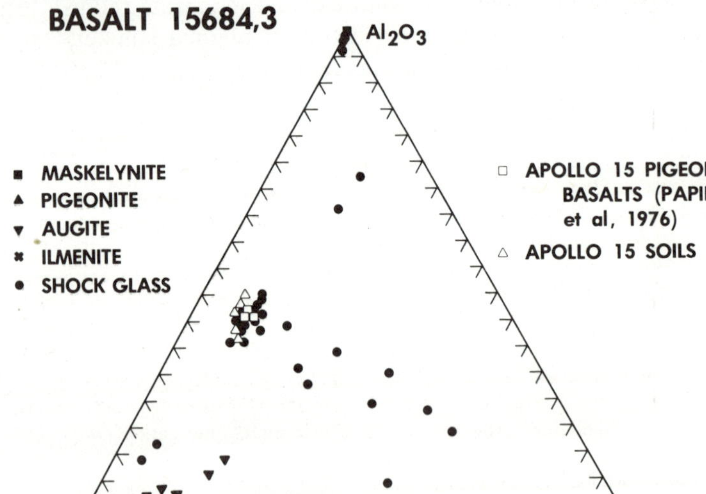

Fig. 10. Ternary diagram showing composition variation of shock glasses coating basalt 15684 with respect to the major rock-forming minerals. Glass compositions show closer affinity to other Apollo 15 pigeonite basalts than to Apollo 15 soils.

ponents; moderate reddish brown (10R 4/6) glasses are enriched in ilmenite components; and dark yellowish orange (10YR 6/6) to light brown (5YR 5/6) glasses are mixtures of all the basaltic components (Table 5). These prominant orange and brown glasses are most similar to the 12054 whole-rock sample (Rhodes *et al.*, 1977) or to a typical Apollo 12 ilmenite basalt in composition (Fig. 11); they show much less resemblance to Apollo 12 soil compositions (Frondel *et al.*, 1971). Thus, the 12054 glass is a flowed basaltic melt which thermally modified sample 12054 along its contact. The glass was deposited on 12054 as an impact melt splash derived from a basalt similar to 12054.

Both the thermally modified pyroxene grains and the green pyroxene glasses contain minute dendritic metal particles less than 1 μm across. These dendrites have medium reflectance. Furthermore, the brown glass coating contains two types of metal particles: one is the dendritic crystallites similar to those found in the green glass and in the pyroxenes. The other type consists of spherules with high reflectance and larger sizes than the dendritic particles. The dendrites are more abundant than the spherules. However, both particle types are too small to be accurately analyzed by conventional electron microprobe techniques.

79155 glass coating

A black glass coats the 79155 hand specimen but most of this crust is broken off the sample in thinsection 79155,60. Nevertheless, glass zones possibly related

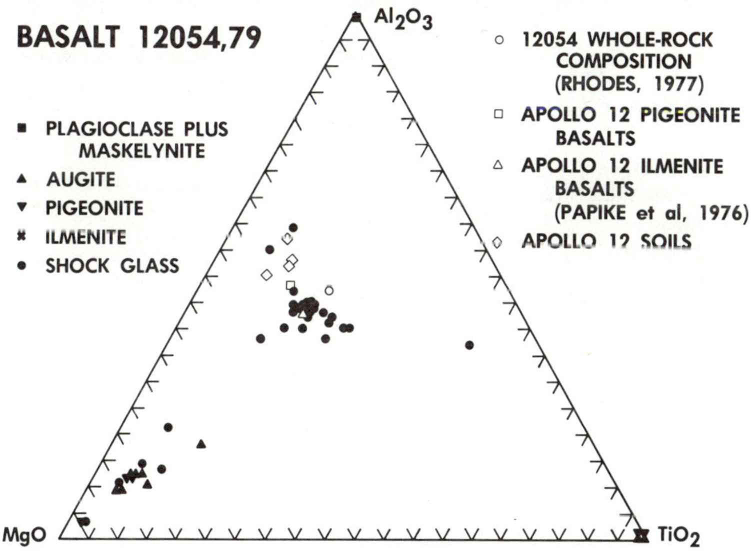

Fig. 11. Ternary diagram showing composition variation of shock glasses coating basalt 12054 with respect to the major rock-forming minerals. Glass compositions and the 12054 whole rock composition more closely resemble compositions of Apollo 12 pigeonite and ilmenite basalts than the local soils.

to the coating are preserved in intergranular spaced and in fractures covering approximately 6% of the interior of the section. The glass contains vesicles up to 0.5 mm in diameter and exhibits prominant flow features and schlieren accentuated by a wide range of colors and by flow bands of metal spherules less than 1 μm in diameter (Fig. 9E–H). Microprobe analyses again demonstrate the correlation of color with composition in the glasses: colorless glasses are composed predominantly of plagioclase components (relatively Al-rich): pale to dark yellowish or orange glasses (10YR 8/6 and 6/6) are enriched in pyroxene components (Mg-rich); moderate reddish brown glasses (10R 4/6) are enriched in ilmenite components (Ti-rich); and moderate yellowish brown glasses (10YR 5/4) are mixtures of feldspar, pyroxene, and ilmenite components (Table 6) in proportions the same as in the host basalt. Plots of these compositions on the ternary diagram show a closer affinity to Apollo 17 basalts than to Apollo 17 soils (Fig. 12). The slight K_2O enrichment in the brown glass (Table 6) is probably derived from K-rich mesostasis. XRF whole rock analysis gives a K_2O abundance of 0.06% similar to that in these brown glasses (Rhodes, pers. comm.). These findings are consistent with high alkali and silica contents reported in melt inclusions and in mesostasis (Weiblen and Roedder, 1976). Thus, we conclude that the glass within 79155 is a basaltic glass which was mobile but incompletely mixed; it certainly is not as homogeneous as glasses coating 15684 and 12054.

The origin of this basaltic glass is difficult to interpret because of a duality in the evidence. One line of evidence suggests an allochthonous origin; glass coats the surface of rock 79155 and fills fractures like thin feeder dikes to internal pools (Fig. 8). This suggests that a melt splashed onto the rock surface and

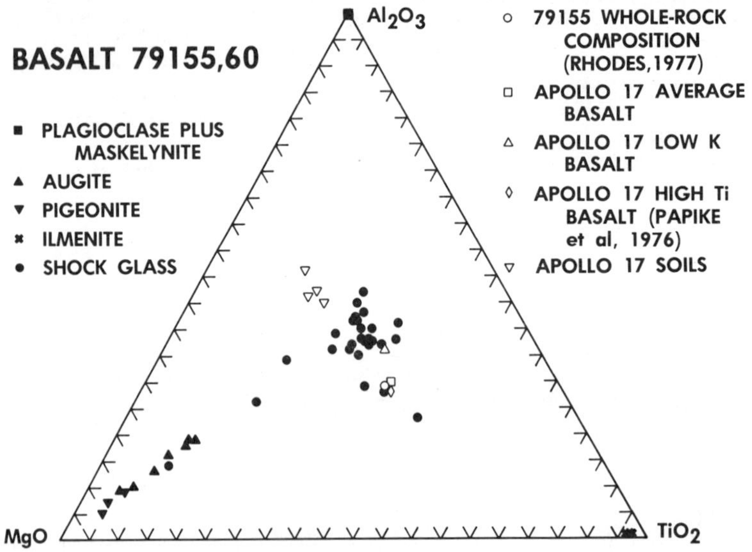

Fig. 12. Ternary diagram showing composition variation of shock glasses coating basalt 79155 with respect to the major rock-forming minerals. Glass compositions and the 79155 whole rock composition are more similar to Apollo 17 basalts than to local soils.

injected into fractures. Other evidence suggests that shock-induced melting may have formed the glass *in situ*. Nearly monomoneralic and bimineralic glasses are incompletely mixed and are located in patches where shock effects (i.e., fracturing and presence of maskelynite; see above) are most abundant. This suggests shock pressures and temperatures were focused along grain boundaries and formed melt in highly localized zones of pressure concentrations. The question whether the glass "ponds" formed *in situ* by shock melting or whether they are spatially and, thus, genetically related to the exterior glass splash remains unresolved.

We conclude that the glasses associated with each of these naturally shocked basalts was derived from basaltic targets most likely from the local substrate and not from their respective local regoliths.

Discussion

The following discussion purposely exceeds the scope of the specific experiments and observations reported above in order to summarize some of our previous shock experiments on basaltic targets and to discuss implications of these studies on the evolution of basaltic regoliths on other planetary surfaces.

Shock metamorphism of basalts

Our studies of the shock metamorphism of basaltic targets pursued over the last three years (Gibbons *et al.*, 1975; Schaal *et al.*, 1976; Kieffer *et al.*, 1976a) reveal a fairly close congruence of shock features between terrestrial and lunar basalts, despite some differences in phase chemistry and grain size. Within the limitations of our analytical techniques, the major diagnostic shock features occur at similar if not identical peak pressures and the various stages of shock metamorphism in solid basalts span similar pressure ranges. Furthermore, the congruence of shock effects in naturally and experimentally shocked specimens is striking; most naturally occurring petrographic features may be duplicated in the laboratory.

Feldspar remains the most diagnostic component phase for determining shock intensity. It displays a variety of deformation features and phase transitions which are highly sensitive to specific pressure histories. This is consistent with the stages of progressive shock metamorphism observed in terrestrial granitic rocks in which all framework silicates (quartz, K-feldspar, and plagioclase) may undergo basically similar phase changes (e.g., Stöffler, 1972). Lunar Ca-rich plagioclase is somewhat more resistant to melting than terrestrial plagioclase. Also, the mafic and oxide minerals in both rock types appear to display less dramatic and certainly less systematic deformation features and little or no phase transitions until shock temperatures are reached which are sufficient to produce "whole rock melts". The twinning of ilmenite under shock conditions constitutes a valid shock indicator at very low pressures in agreement with Sclar (1971) and Minkin and Chao (1971). The onset of "shock lamellae" in pyroxenes is ill defined at the moment in our materials, but occurs at pressures of ~200 kbar. Their usefulness as pressure indicators, however, is limited because the onset of their formation approximately coincides with the more readily diagnosed formation of diaplectic feldspar glass. Furthermore, pyroxene lamellae are created through all degrees of shock up to the melting stage. Thus, the framework silicates appear to be best suited for pressure calibration of specific shock histories.

Examination of basalts shocked between 100 and 1000 kbar made it possible to characterize several stages in progressive shock melting. Shock-induced selective melting of plagioclase occurs over a very wide pressure range, starting with grain-boundary melting at ~450 kbar, resulting in pronounced flow and vesiculation at ~650 kbar, and finally contributes to mixed mineral melts and whole rock melts at >800 kbar. The amount of feldspar glass increases from a few localized occurrences at ~450–500 kbar to >90% at > 900 kbar; however, variations in the exact amount of melt present at intermediate pressures may vary, as exemplified by experiments on both the Lonar and lunar basalts. Thus, calibration of peak shock pressures in naturally shocked rocks between ~500 and 800 kbar is exceedingly difficult, because the occurrence of vesiculated feldspar glass at >650 kbar may be influenced by the amount of volatiles, e.g., H_2O, rather than by the specific P/T history. The production of feldspar melts mixed with diaplectic feldspar at shock pressures between 500 and 800 kbar demonstrates that the shock pressure distribution is heterogeneous on millimeter scales. Even the melts generated in the 974, 996, and 999 kbar shots contain relatively large amounts of pyroxene fragments and are characterized by numerous schlieren intermixed with "whole rock melts". Therefore, we conclude that the production of relatively homogeneous basaltic melts, such as the large glass coatings or the glass spheres in the Apollo collection, require ≫1000 kbar for their formation. Unfortunately, recovery attempts of Lonar basalt targets at ~1200 kbar were not successful, despite considerable efforts to improve the target container design.

Interpretation of the observed features with respect to equation of state (EOS) measurements on dense "basaltic" rocks (e.g., Vacaville basalt, Jones *et al.*, 1965 and Centreville diabase, McQueen *et al.*, 1967) is difficult because in-shock temperatures were not calculated and release adiabats were not measured. Release adiabats are available for lunar basalt 70215 only up to 200 and >1100 kbar pressure (Ahrens *et al.*, 1977). The EOS data available are, thus, not suitable to interpret specific features between 400 and ~1000 kbar in both the Lonar and the lunar basalts. These pressures are within the so-called "mixed phase" regime where normal and high density phases coexist; upon release high pressure phases convert to phases such as maskelynite or feldspar glasses, with densities lower than their initial densities. Plagioclase is the only mineral that demonstrably undergoes shock-induced density changes which are retained in recovery products. Pyroxene is relatively unaffected, though Ahrens and Gaffney (1971) interpret the onset of a "mixed phase" regime, i.e., the partial conversion to a high pressure phase, at 135 kbar in enstatite targets. Adiabatic release of such a mixed phase regime should lead to significantly decreased densities of pyroxenes at pressures between 200 and 600 kbar, however, slightly decreased density as evidence by decreased birefringence is observed in our experiments only at pressures >600 kbar. Ilmenite also does not display any evidence of significantly reduced density at pressures as high as 900 kbar, although King and Ahrens (1976) observe a phase change at 320 kbar. Thus, plagioclase alone may bear the brunt of the shock compression and, as a consequence, may constitute the major variable in EOS measurements on basaltic targets.

Gibbons *et al.* (1975) report shock recovery experiments on particulate plagioclase and pyroxene mixtures as a simplified terrestrial analog to simulate the shock behavior of a lunar "basaltic" regolith. The resulting stages of progressive shock metamorphism differ significantly from those reported above on dense basalt targets. In accordance with theoretical considerations presented by Kieffer *et al.* (1976b) and Ahrens and Cole (1974), these porous aggregates yielded significantly more feldspar melt as well as whole rock melt at relatively lower pressures, e.g., at 514 kbar all feldspar is molten and ~50% of the entire sample was converted to an intergranular "whole rock" melt. We emphasize again the pronounced difference in the shock behavior between dense and porous targets, especially at 200–500 kbar pressures. Shock experiments on powdered basalt 75035 are planned to better characterize these differences.

Basaltic glass coatings

Our chemical analyses of the glass coatings on 15684, 12054, and 79155 compared with the bulk compositions of proximal basalt and regolith samples indicate that the glasses are derived from a basaltic substrate rather than from the local regolith. This interpretation strongly suggests that each glass coating is the product of a single impact. Estimates of the size of the parent crater are speculative, although the total penetration must have been deeper than the respective regolith thickness. Unfortunately, it is presently not possible to discern whether the glass coating and the solid-state shock effects were produced by the same impact or whether two separate impacts were responsible. However, the fact that the only shocked specimens recognized to date are also draped with a basaltic glass leads us to believe that the solid-state shock effects and the glass production are genetically related, and that they are the result of a single impact. Dating of the shock event(s) and the glass formation via $^{40}Ar/^{39}Ar$ laser probe methods (e.g., Plieninger and Schaeffer, 1976) may resolve the possibility of genetic relationship(s).

Furthermore, samples 15684, 12054, and 79155 may be valuable materials to study a variety of shock-related questions in lunar sample research; they are well suited to assess the possible disturbance of isotope systems used for absolute age dating, $^{40}Ar/^{39}Ar$ stepwise heating techniques, in particular, because the specimens are shocked to ~300–400 kbar during a single event. Also, a survey of the erasure of solar particle tracks and fission tracks by the passage of a shock wave may be of interest. Microchemical analysis may evaluate the origin of the metal particles in the glass coating, by shock reduction in the absence of solar-wind derived hydrogen and/or by dissemination of meteoritic or indigenous metal. The value of these samples rests in the fact that they are shock products of a single impact.

Shocked lunar basalt and mare regoliths

The distribution of shock effects in the Apollo "basalt" collection is markedly bimodal in a comparison of hand specimens with regolith constituents. Out of approximately 152 hand specimens (>10 g) studies in our survey, only three samples displayed genuine shock features more severe than fracturing on hand-specimen scales. Small-scale shock effects attributed to microcraters on exposed rock surfaces (e.g., Schaal *et al.*, 1976) were not considered. In contrast, mare regolith constituents contain widespread shock features and abundant shock glasses which may commonly exceed 50% by volume (e.g., Heiken, 1974). Thus, the cumulative impact energy deposited into the regolith by micrometeroids is indeed dramatically higher than the energy expended into the substrate by larger meteorites (Gault *et al.*, 1974). The scarcity of "shocked" basalt hand specimens is in fair agreement with the amount of material shock metamorphosed during single cratering events (e.g., Hörz, 1969; Stöffler *et al.*, 1975; Pohl *et al.*, 1977).

Quaide and Oberbeck (1975) demonstrated that the overall growth of the regolith is accomplished by adding freshly excavated material on top of an existing debris layer; according to Stöffler *et al.* (1975) this material is in large part "unshocked". Therefore, it is highly probably that most regolith glasses, including agglutinates, were subsequently formed by micrometeoroid impact into fine-grained regolith (Mendell and McKay, 1975; Gibbons *et al.*, 1976; Hu and Taylor, 1977). Such conclusions are probably also applicable to the more homogeneous, often spherical, glass particles characterized chemically by Reid *et al.* (1972) and many others. They also formed from micrometeorite impact, like the "agglutinates", except that their target sites were solid rock surfaces, i.e., rock fragments scattered on the lunar surface. The shock experiments described may explain the marked absence of mafic glasses of monomineralic compositions in the statistical glass surveys (e.g., Reid *et al.*, 1972): pyroxene and ilmenite (possibly also olivine, which, unfortunately, was not present in 75035) start to melt under shock conditions only after most of the plagioclase is completely molten. The pyroxene and ilmenite melts, thus, are immediately in contact with feldspar melts and mechanical mixing is accomplished rapidly upon pressure release, forming "whole rock" melts. Such whole rock melts will dominate the glass population and slight chemical variations are expected, clustering around the whole rock composition. These considerations lend additional support to the conclusion that discrete clusters of glass compositions are related to the bounds of parent-rock compositions.

Asteroidal regoliths

We will now apply similar considerations to speculate on the lithologic properties of basaltic regoliths on planetary bodies other than the moon. In doing so, we must clearly separate the effects of a single impact onto a solid bedrock versus the effects of repetitive bombardment which inevitably creates progressively more comminuted and finer-grained particulate material.

First, we consider a single impact on a fresh basaltic surface in the asteroid belt where mean collision velocities are a few kilometers per second (Hartman, 1972). With Fig. 13 we estimate the intensity of shock effects created by impact either by a chondritic meteorite (basalt: density ≈ 3 g/cc) or by an iron meteorite (≈7 g/cc) on a basaltic target using a mean impact velocity of 5 km/sec. Peak pressures achieved with the basaltic projectile are on the order of 500 kbar, barely sufficient to selectively melt some feldspar components and to preserve pyroxene and oxide components. Solid-state shock effects are dominant features in the impact ejecta. However, impact with an iron meteorite at 5 km/sec creates peak pressures in excess of 800 kbar and produces significant amounts of feldspar melts with minor amounts of "whole rock" melts.

Next, what if subsequent impacts occur on progressively more comminuted deposits? Gibbons *et al.* (1975) reported that significant amounts of "whole rock" melt are generated at pressures as low as 400 kbar in fine-grained pyroxene-plagioclase mixtures. Thus, given sufficiently repetitive bombardment, significant amounts of fine-grained regolith would be generated, which in turn would lead to increased glass production, i.e., "agglutinates" in a textural and chemical sense (Mendell and McKay, 1975; Hu and Taylor, 1977). The absolute amount, of course, will depend on the cumulative impact history.

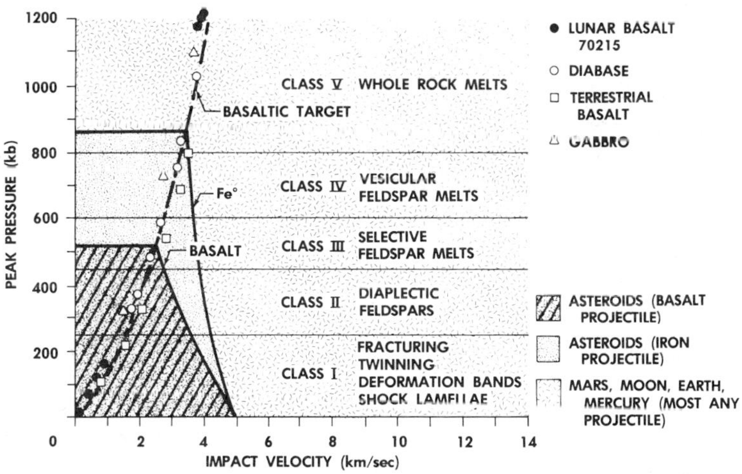

Fig. 13. Equation-of-state curves for "basaltic" materials in the pressure-particle velocity plane and projected shock damage in solid basalt targets using iron and basalt projectiles at 5 km/sec impact velocity typical for asteroidal collisions. (Lunar basalt: Ahrens *et al.*, 1977. All other equation-of-state data compiled by v. Thiel, 1967).

Recent spectral reflectance measurements of asteroidal surfaces, specifically the basaltic surface of Vesta (McCord *et al.*, 1970; Matson *et al.*, 1977), reveal a distinct lack of glassy surface materials. "Agglutinates" are also conspicuously absent in meteoritic "regolith" breccias (Kerridge and Kieffer, 1977). This lack of glass is consistent with our model of impact of a basaltic projectile on a solid basalt target at 5 km/sec (Fig. 13). Impact into fine-grained particulate targets is necessary to produce glass at that velocity (Gibbons *et al.*, 1975). Matson *et al.* (1977) discuss a variety of possible factors to account for the Vesta observations including total bombardment history, mean impact velocity and compositional differences. Our experimentally shocked lunar and Lonar basalts demonstrate that differences in modal compositions in basalts do not affect glass production significantly.

Matson *et al.* (1977) argue that the cumulative bombardment history on asteroids is rather similar to that of the inner planets. While this view may be correct for the early, large-scale bombardment history, it cannot be extended to micrometeoroids (the major lunar glass producer) because "large"-scale projectiles and micrometeoroids are derived from different sources. The micrometeoroid population is thought to originate predominantly, if not exclusively, from comets (e.g., Whipple, 1967; Zook, 1977). Micrometeoroid detectors on board Pioneer 10 and 11 measured a factor of 3 decrease in micrometeoroid activity going from 1.0 to 3.0 a.u. (Dohnanyi, 1975); conversely, recent HELIOS 1 measurements reveal a factor of approximately 20 more micrometeoroids at 0.3 a.u. than at 1 a.u. (Grün *et al.*, 1977). These measurements attest to an abundance of cometary particles in the inner solar system. In addition, the mean impact velocities vary systematically as a function of heliocentric distance (Dohnanyi, 1975; Grün *et al.*, 1977). Micrometeorite velocities in the asteroidal belt are uncertain at present, but Dohnanyi (1975) estimates that ~ 14 km/sec is a reasonable mean velocity. Thus, micrometeoroids are capable of generating impact melts in the asteroid belt, although the absolute micrometeorite flux and mean velocities there are lower than those on the lunar surface. These factors account for a relative lower glass production in the asteroid belt but do not totally prevent it.

Therefore, some other factor must be responsible for the apparent absence (=paucity?) of glass in the IR investigations of Vesta. We postulate that Vesta has a relatively coarse-grained surface deposit which prevents the formation of soil agglutinates. Micrometeoroid impacts on "pristine" rock surfaces generate relatively small amounts of glass and will not form agglutinates, the dominant lunar regolith glass. The grain size of such a deposit must be larger than the dimensions of typical microcraters (0.1–3 mm in diameter); therefore, we suggest that grain sizes on Vesta are greater than 1 cm.

Speculations concerning the possible origin of a regolith on Mercury must consider that both the spatial density and the impact velocity of micrometeoroids are highest on Mercury than on all other planetary surfaces. Such impact conditions should create extremely glass-rich and finely comminuted surface deposits. Whether the total amount of glass on Mercury will exceed the amount in the lunar regolith remains uncertain because the kinetics of glass formation strongly depend on the rate of replenishing the entire deposit with freshly excavated, pristine material and, thus, depends on the flux of relatively massive projectiles that penetrate bedrock (Mendell and McKay, 1975). If large-scale crater production rates are similar on Mercury and on the Moon (e.g., Gault *et al.*, 1975; Wetherill, 1975), then there will be approximately the same volume of pristine substrate material excavated per unit time; however such material exposed on the Mercurian surface will be more thoroughly reworked due to the higher micrometeoroid flux. In addition, large-scale secondary craters will be less widespread on Mercury than on the Moon (Gault *et al.*, 1975; Oberbeck and Aggarwal, 1977), allowing a relatively larger intercrater surface area to remain undisturbed per unit time; this in turn will lead to increasingly more comminution and reworking by micrometeoroids. Therefore, the Mercurian regolith may be finer grained and contain more agglutinitic glass than the lunar regolith.

Acknowledgments—We appreciate discussions with S. W. Kieffer and R. V. Gibbons and the skilled technical support from T. D. Thompson, Lockheed Electronics Co., Inc., Houston.

Portions of this study were supported by NASA grant NSG-7052 to UCLA and by the Lunar Science Institute, where RBS was a Visiting Graduate Fellow during the summer of 1976. The Lunar Science Institute is operated by the Universities Space Research Association under contract number NSR-09-051-001 with the National Aeronautics and Space Administration. This paper constitutes the Lunar Science Institute Contribution No. 285.

References

Ahrens T. J. and Cole D. M. (1974) Shock compression and adiabatic release of lunar fines from Apollo 17. *Proc. Lunar Sci. Conf. 5th*, p. 2333–2345.

Ahrens T. J. and Gaffney E. S. (1971) Dynamic compression of enstatite. *J. Geophys. Res.* **76**, 5504–5513.

Ahrens T. J., Jackson I., and Jeanloz R. (1977) Dynamic properties of ilmenite rich mare basalt and the relative ages of lunar cratered surfaces (abstract). In *Lunar Science VIII*, p. 1–3. The Lunar Science Institute, Houston.

Bowen N. L. (1913) The melting phenomena of plagioclase feldspar. *Amer. J. Sci.* **35**, 583.

Brett R. (1976) Reduction of mare basalts by sulfur loss. *Geochim. Cosmochim. Acta* **40**, 997–1004.

Chao E. C. T., James O. B., Minkin J. A., Boreman J. A., Jackson E. D., and Raleigh C. B. (1970) Petrology of unshocked crystalline rocks and evidence of impact metamorphism in Apollo 11 returned lunar sample. *Proc. Apollo 11 Lunar Sci. Conf.*, p. 287–314.

Dence M. R., Douglas J. A. V., Plant A. G., and Traill R. J. (1970) Petrology, mineralogy and deformation of Apollo 11 samples. *Proc. Apollo 11 Lunar Sci. Conf.*, p. 315–340.

Dohnanji J. S. (1975) Sources of interplanetary dust: Asteroids. In *Interplanetary Dust and Zodiacal Light* (H. Elsaesser and H. Fechtig, eds.), p. 187–205. Springer Verlag, New York.

Engelhardt W. v. (1967) Chemical composition of Ries glass bombs. *Geochim. Cosmochim. Acta* **31**, 1677–1689.

Engelhardt W. v., Arndt J., Müller W. F., and Stöffler D. (1970) Shock metamorphism of lunar rocks and origin of the regolith at the Apollo 11 landing site. *Proc. Apollo 11 Lunar Sci. Conf.*, p. 363–384.

Frondel C., Klein C., and Ito J. (1971) Mineralogical and chemical data on Apollo 12 lunar fines. *Proc. Lunar Sci. Conf. 2nd*, p. 719–726.

Frondel J. W. (1975) *Lunar Mineralogy*. John Wiley and Sons, New York. 323 pp.

Gault D. E., Guest J. G., Murray B. J., Dzurisin D., and Malin M. C. (1975) Some comparisons of impact craters on Mercury and the Moon. *J. Geophys. Res.* **80**, 2444–2460.

Gault D. E., Hörz F., Brownlee D. G., and Hartung J. B. (1974) Mixing of the lunar regolith. *Proc. Lunar Sci. Conf. 5th*, p. 2365–2385.

Gibbons R. V., Hörz F., and Schaal R. B. (1976) The chemistry of some individual lunar soil agglutinates. *Proc. Lunar Sci. Conf. 7th*, p. 405–422.

Gibbons R. V., Morris R. V., Hörz F., and Thompson T. D. (1975) Petrographic and ferromagnetic resonance studies of experimentally shocked regolith analogs. *Proc. Lunar Sci. Conf. 6th*, p. 3143–3171.

Gibson E. K., Usselman T. M., and Morris R. V. (1976) Sulfur in the Apollo 17 basalts and their source regions. *Proc. Lunar Sci. Conf. 7th*, p. 1491–1505.

Grün E., Fechtig H., Kissel J., and Gammelin P. (1977) Micrometeoroid data from the first two orbits of Helios I. *J. Geophys. Res.* In press.

Hartman W. K. (1972) *Moons and Planets.* Bogden and Quigley, New York. 404 pp.

Heiken G. (1974) A catalog of lunar soils. NASA Johnson Space Center, Curator's Office. 221 pp.

Hörz F. (1969) Structural and mineralogical evaluation of an experimentally produced impact crater in granite. *Contrib. Mineral. Petrol.* **21**, 365–377.

Hörz F. (1970) A small ballistic range for impact metamorphism studies. NASA-TN-D5787. 17 pp.

Hu H. N. and Taylor L. A. (1977) Agglutinate formation: Lack of chemical fractionation (abstract). In *Lunar Science VIII*, p. 463–465. The Lunar Science Institute, Houston.

James O. B. (1969) Shock and thermal metamorphism of basalt by nuclear explosion, Nevada Test Site. *Science* **166**, 1615–1620.

Jones A. H., Isbell W. M., and Maiden C. J. (1965) Measurement of the very high pressure properties of materials using a light gas gun. G. M. Defense Research Laboratories, Technical Report # TR65-84.

Kerridge J. F. and Kieffer S. W. (1977) A constraint on impact theories of chondrule formation. *Earth Planet. Sci. Lett.* **35**, 35–42.

Kieffer S. W., Phakey P. P., and Christie J. M. (1976b) Shock processes in porous quartzite: Transmission electron microscope observations and theory. *Contrib. Mineral. Petrol.* **59**, 41–93.

Kieffer S. W., Schaal R. B., Gibbons R. V., Hörz F., Milton D., and Duba A. (1976a) Shocked basalt from Lonar Impact Crater (India) and experimental analogues. *Proc. Lunar Sci. Conf. 7th*, p. 1391–1412.

King D. A. and Ahrens T. J. (1976) Shock compression of ilmenite. *J. Geophys. Res.* **81**, 931–935.

Longhi J., Walker D., Grove T. L., Stolper E. M., and Hays J. F. (1974) The petrology of the Apollo 17 mare basalts. *Proc. Lunar Sci. Conf. 5th*, p. 447–469.

Matson D. L., Johnson T. V., and Veeder G. J. (1977) Soil maturity and planetary regoliths: The Moon, Mercury and asteroids (abstract). In *Lunar Science VIII*, p. 625–627. The Lunar Science Institute, Houston.

McCord T. B., Adams J. B., and Johnson T. V. (1970) Asteroid Vesta: Spectral reflectivity and compositional implications. *Science* **168**, 1445–1447.

McQueen R. G., March S. P., and Fritz J. N. (1967) Hugoniot equation of state of twelve rocks. *J. Geophys. Res.* **72**, 4999–5036.

Mendell W. W. and McKay D. S. (1975) A lunar soil evolution model. *The Moon* **13**, 285–292.

Minkin J. A. and Chao E. C. T. (1971) Single crystal X-ray investigation of deformation in terrestrial and lunar ilmenite. *Proc. Lunar Sci. Conf. 2nd*, p. 237–246.

Oberbeck V. R. and Aggarwal H. R. (1977) Topographic analysis of lunar secondary craters of Copernicus and implications. *Proc. Lunar Sci. Conf. 8th.* Vol. 3.

Papike J. J., Hodges F. N., Bence A. E., Cameron M., and Rhodes J. M. (1976) Mare basalts: Crystal chemistry, mineralogy and petrology. *Rev. Geophys. Space Phys.* **14**, 475–540.

Plieninger T. and Schaeffer O. A. (1976) Laser probe ^{39}Ar–^{40}Ar ages of individual mineral grains in lunar basalt 15607 and lunar breccia 15465. *Proc. Lunar Sci. Conf. 7th*, p. 2055–2066.

Pohl J., Stöffler D., Gall H., and Ernstson K. (1977) The Ries impact crater. In *Impact and Explosion Cratering* (D. J. Roddy, R. O. Pepin and R. B. Merrill, eds.). Pergamon Press, New York. In press.

Quaide W. L. and Bunch T. E. (1970) Impact metamorphism of lunar surface materials. *Proc. Apollo 11 Lunar Sci. Conf.*, p. 711–729.

Quaide W. L. and Oberbeck V. R. (1975) Development of the mare regolith: Some model considerations. *The Moon* **13**, 27–55.

Reid A. M., Warner J., Ridley W. I., Johnston D. A., Harmon R. S., Jakeš P., and Brown R. W. (1972) The major element compositions of lunar rocks as inferred from glass compositions in the lunar soils. *Proc. Lunar Sci. Conf. 3rd*, p. 363–378.

Rhodes J. M., Blanchard D. P., Brannon J. C., Rodgers K. V., and Dungan M. A. (1977) Chemistry of Apollo 12 mare basalts: Magma types and fractionation processes. *Proc. Lunar Sci. Conf. 8th.* This volume.

Schaal R. B., Hörz F., and Gibbons R. V. (1976) Shock metamorphic effects in lunar microcraters. *Proc. Lunar Sci. Conf. 7th*, p. 1039–1054.

Sclar C. B. (1971) Shock-induced features of Apollo 12 microbreccias. *Proc. Lunar Sci. Conf. 2nd*, p. 817–832.

Short N. M. (1969) Shock metamorphism of basalt. *Modern Geology* **1**, 81–95.

Simonds C. H., Warner J. L., Phinney W. C., and McGee P. E. (1976) Thermal model for impact breccia lithification: Manicouagan and the moon. *Proc. Lunar Sci. Conf. 7th*, p. 2509–2528.

Stöffler D. (1972) Deformation and transformation of rock-forming minerals by natural and experimental shock processes I: Behavior of minerals under shock compression. *Fortschr. Miner.* **49**, 50–113.

Stöffler D., Gault D. E., Wedekind J., and Polkowski G. (1975) Experimental hypervelocity impact into quartz sand: Distribution and shock metamorphism of ejecta. *J. Geophys. Res.* **80**, 4062–4077.

Thiel v. (1966) Compendium of shock wave data. Lawrence Radiation Laboratory, Livermore, Report; U.C.RL-50108.

Weiblen P. W. and Roedder E. (1976) Compositional interrelationships of mare basalts from bulk chemical and melt inclusion studies. *Proc. Lunar Sci. Conf. 7th*, p. 1449–1466.

Wetherill G. (1975) Late heavy bombardment of the moon and terrestrial planets. *Proc. Lunar Sci. Conf. 6th*, p. 1539–1561.

Whipple F. L. (1967) On maintaining the meteoritic complex. *Smithson. Astrophys. Observ. Spec. Rep.* No. 239, p. 1–46.

Zook H. A. (1977) Temporal and spatial variations of the interplanetary dust flux. In *Papers Presented to the XXth Plenary Meeting of COSPAR and Associated Activities.* Tel-Aviv, Israel, 7–18 June 1977.

Proc. Lunar Sci. Conf. 8th (1977), p. 1731–1750.
Printed in the United States of America

Compositional dispersions in lunar and terrestrial basalts

L. A. HASKIN[1], J. W. JACOBS[1], J. C. BRANNON[2], and M. A. HASKIN[1]

[1]Department of Earth and Planetary Sciences and McDonnell Center for the Space Sciences, Washington University, St. Louis, Missouri 63130
[2]Lockheed Electronics Company, Inc., Houston, Texas 77058

Abstract—Variations in concentrations of major and trace elements (La, Sm, Eu, Yb, Sc, Co, SiO_2, TiO_2, Al_2O_3, FeO, MgO, CaO) among suites of lunar mare basalts have been compared with those for suites of terrestrial basalt samples. The terrestrial samples have known field and, in some cases, genetic relationships. Lunar basalt suites Apollo 11 type A (high K), Apollo 17 type A fine grained, Apollo 17 type B fine grained, and Apollo 12 pigeonite basalts show only slightly more variability in composition than found for vertical sections of single flows of homogeneous terrestrial basalts. The Apollo 15 quartz and olivine normative basalt suites and Apollo 12 ilmenite and pigeonite basalt suites have compositions slightly more variable than found in successive lavas that are not regarded as chemically differentiated relative to each other (i.e., not showing compositional changes attributable to magmatic fractionation). These Apollo 12 and 15 basalt suites are not more variable, however, than horizontal sections of two terrestrial flows. Most of the dispersion in composition of the coarser grained lunar rocks stems from the sampling of the rocks for analysis. Only the suite of Apollo 11 type B (low K) basalts is as variable in composition as suites of terrestrial basalts that are clearly differentiated relative to each other (i.e., show compositional differences attributable to magma fractionation). Based on analogy with terrestrial basalts, there is no reason to presume that the Apollo collections from any given landing site represent many or even several lava-forming events. We cannot yet explain quantitatively the causes of the observed compositional variations within relatively undifferentiated suites of terrestrial or lunar basalts.

INTRODUCTION

MARE BASALT COMPOSITIONS are important to our inferences of overall lunar composition, e.g., relative depletion in volatile and siderophile elements and possible enrichment in refractory lithophile elements relative to earth and chondritic meteorites, nature of interior regions (as interpreted by experimental petrologists), compositions of broad regions of mare terrain (as inferred from spectral analysis). The validity of these inferences depends in part on some rather mundane characteristics of the lunar basalt samples that affect assumptions about what those samples represent. To what extent do the samples collected at a given Apollo site represent an entire mare or some other large region as opposed to a few local flows? Does any analyzed sample represent a primitive liquid from partial melting, or have all undergone substantial differentiation or assimilation? Do the basalts adequately represent their magma types, or even their flows? Because sampling of lunar basalts is limited in areal extent and without normal geological control, these characteristics cannot be established directly. One approach to understanding their probable importance is comparison with terrestrial analogs, with proper awareness of known and unsuspected differences in situation between earth and moon. To provide adequate sets of data for such comparisons, we have begun extensive analyses of suites of terrestrial basalts.

As an example of how our impressions of mare basalt nature can vary, consider how many different basalts are represented by the Apollo 17 samples and the processes responsible for their chemical and petrological differences. Experimental petrologists selected very few of those basalts as representing primitive compositions and argued for derivation of the rest by near surface crystal fractionation (e.g., Longhi *et al.*, 1974). Petrologists found no reason to require numerous separate basalt groups (see Papike *et al.*, 1976 for a summary). Rhodes *et al.* (1976), mainly on the basis of major-element composition, defined only a few groups of basalts and suggested that variations in composition within the groups resulted from near surface fractionation.

On the other hand, Shih *et al.* (1975) in a study of trace elements, mainly rare earths (REE), concluded that near surface fractional crystallization could not account for the different compositions from a single starting composition. They did not find clear trends of trace-element covariation expected for several groups of basalts each of which might independently have undergone such fractionation. Absence of such trends does not rule out that process, it merely fails to support it. They concluded that partial melting to different extents of a homogeneous source material could not provide such differences as those in Eu/Sm among different possible starting compositions for possible groups that might have then undergone fractional crystallization. They argued that the minimum requirement for production of the basalts was partial melting of a slightly chemically inhomogeneous but isotopically homogeneous source. This leaves the impression that many basalts and many events were sampled at the Apollo 17 site (and by analogy, at the Apollo 11, 12, and 15 sites).

Such arguments as those of Shih *et al.* (1975) are based on current mathematical models for trace-element partitioning. The strength of those arguments depends on the accuracy with which the models used actually describe the behavior of trace elements during the natural processes that produced the basalt specimens analyzed. These models, based on Nernst distribution coefficients, are on firm chemical grounds (e.g., McIntire, 1963). It is less clear that the conditions of complete equilibrium or surface equilibrium during fractional crystallization or of complete equilibrium or incremental partial melting are apt descriptions of the petrological processes of basalt formation. There may be other, unrecognized, processes that affect trace-element distributions in a more drastic manner than do the above processes. If so, those processes must be identified and partitioning models modified to include them. Certainly the geochemically coherent REE group should be amenable to quantitative modelling if the petrological processes that fractionate it can be determined. Quantitative description of REE behavior will be a much stronger constraint on knowledge of basalt origin and history than qualitative comparisons, provided that the processes primarily responsible for REE fractionation are also the principal processes that produce basalts.

The variations in REE concentrations among samples of Apollo 17 mare basalts are not great. The range for Ce is 11.3–29 ppm; that for Ce/Sm is 2.0–2.6; for Sm/Eu 3.8–5.7. These ranges are substantial in comparison with uncertainties in analytical technique ($\sim \pm 1-2\%$). Variations in the ratio ppmCe/ppmSm (a measure of fractionation of light REE from medium REE) versus ppm Ce (a measure of the extent of fractional crystallization or partial melting) (Haskin *et al.*, 1970) for Apollo 17 basalt samples analyzed in the JSC mass spectrometry laboratory (Shih *et al.*, 1975; Rhodes *et al.*, 1976) are shown in Fig. 1. Derivation of the samples by any simple process of partial melting to different extents or of fractional crystallization of a single liquid, according to current models, would give a straight or slightly curved line. If different minerals entered or left the process, a discontinuity in slope might occur. If two or three sources or parents were involved, more or less parallel lines might result. No such simple trends appear in Fig. 1. The extent of fractional crystallization of a single liquid needed to account for the range in Ce concentration would be ~60%. That to account for the range for Ce/Sm would be ~75%. These values assume the unreasonably favorable case of crystallizing only clinopyroxene. For a precipitate that is only one-third clinopyroxene, the ranges correspond to > 95% crystallization, which are totally unreasonable. The difficulties in trying to model quantitatively details of

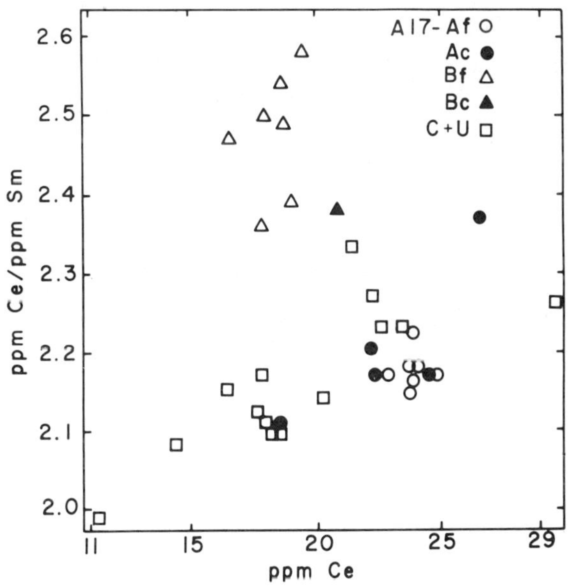

Fig. 1. Concentration ratios of Ce to Sm have been plotted against concentrations of Ce for Apollo 17 basalts analyzed by Shih *et al.* (1975) and Rhodes *et al.* (1976). No clearcut trends are observed, either for the group as a whole, or for the major suites A and B as defined by Rhodes *et al.*, even when coarser-grained members (Ac and Bc) are omitted. Symbols C and U refer to tentative small subgroup C and basalts of undetermined kinship.

trace-element distributions have been considered by many investigators (e.g., Drake and Consolmagno, 1976) and are not further discussed here.

Instead, this paper compares the dispersions in composition observed among samples of terrestrial basalts with known field relationships to those of various subgroups of lunar mare basalts. This comparison indicates that there are not substantial reasons based on trace-element data to require that large numbers of different basalts were sampled at each Apollo site. The results do indicate effects of processes not included in current models for trace-element partitioning.

BASALT SUITES

The following suites of basalt, were considered.

I. Samples from a single basalt flow.
   - A. Vertical section, 11 m thick tholeiite flow, Iceland; ~50 samples from 31 locations analyzed by Watkins *et al.* (1970) for major elements; 11 samples analyzed for REE by Haskin *et al.* (1971a).
   - B. Vertical section, 6.7 m thick flow, Cascades mesa tholeiitic basalt, central Oregon; 5 samples analyzed for major and trace elements (Brannon, Haskin, and McBirney, to be published).
   - C. Horizontal section, flow similar to and located near that of B; 5 samples analyzed for major and trace elements (Brannon, Haskin, and McBirney, to be published).
   - D-1. Vertical section 11.7 km from vent, McCartys basalt (tholeiitic), New Mexico; 6 samples analyzed for major elements by Carden and Laughlin (1974).
   - D-2. Vertical section, 26.1 km from vent, McCartys basalt; 3 samples.
   - D-3. Vertical section, 41.5 km from vent, McCartys basalt; 3 samples.
   - D-4. Type I basalt (<4 km from vent) McCartys basalt; 9 samples.
   - D-5. Type II basalt, McCartys basalt; 28 samples.
   - D-6. All samples, McCartys basalt.
   - E. Vertical section, 69 m thick tholeiitic basalt, Makaopui lava lake, Hawaii; 7 samples, major elements, Evans and Moore, (1967).

II. Chemically undifferentiated, successive flows (i.e., lacking compositional differences attributable to magmatic fractionation).
   - A. Basalt group E, high-alumina tholeiite, Steens Mt., Oregon; 9 samples, major elements, Gunn and Watkins (1970), REE, Helmke and Haskin (1973).
   - B. Columbia River basalt, upper Picture Gorge series; 9 samples, trace elements, Osawa and Goles (1969).
   - C. Columbia River basalt, lower Picture Gorge series; 11 samples, trace elements, Osawa and Goles (1969).
   - D. Columbia River basalt, upper Yakima series; 11 samples, trace elements, Osawa and Goles (1969).
   - E. Columbia River basalt, upper Yakima series; 7 samples, trace elements, Osawa and Goles (1969).

III. Related basalts with clear pattern of relative differentiation (i.e., compositional differences attributable to magmatic fractionation).
   - A. Keweenawan olivine tholeiites, north shore, Lake Superior; 15 samples from separate flows, major and trace elements, Haskin, Brannon, and Green, to be published.
   - B. Steens Mt., Oregon; 63 flows, major elements, Gunn and Watkins (1970) and 52 flows, REE, Helmke and Haskin (1973).

IV. Basalts with similar petrology and geological setting from a single broad area.
   - A. Cascades mesa basalts, SE quadrant of Oregon; 59 samples, major and trace elements, Brannon, Haskin, and McBirney, to be published.

The following suites of basalt of different extents of unknown relationships were compared with the above.

1. DSDP Leg 37, hole 332B, "Plagioclase ± olivine" tholeiites, mid-Atlantic ridge; 15 samples major and trace elements, Blanchard *et al.* (1976).
2. DSDP Leg 37, hole 332B, picrites; 4 samples, Blanchard *et al.* (1976).
3. DSDP Leg 37, hole 332B, "olivine" tholeiites; 4 samples, Blanchard *et al.* (1976).
4. Apollo 11 soil 10084; major elements by 6 analysts, trace elements by 7 analysts, used as estimate of dispersion caused by analytical procedures on suites of Apollo 11 and 12 basalts. Analytical data for all Apollo 11 and 12 samples used in this paper are from the following references: trace elements, Gast *et al.* (1970), Goles *et al.* (1970a,b), Haskin *et al.* (1970), Philpotts and Schnetzler (1970), Wänke *et al.* (1970), Brunfelt *et al.* (1971), Goles *et al.* (1971), Haskin *et al.* (1971b), Hubbard and Gast (1971), Schnetzler and Philpotts (1971), Smales *et al.* (1971), Taylor *et al.* (1971), Wakita *et al.* (1971), and Wänke *et al.* (1971); major elements, Agrell *et al.* (1970), Compston *et al.* (1970), Engel and Engel (1970), Maxwell *et al.* (1970), Rose *et al.* (1970), Cuttitta *et al.* (1971), Compston *et al.* (1971), Engel *et al.* (1971) Kushiro and Haramura (1971), Maxwell and Wiik (1971), Scoon (1971), and Willis *et al.* (1971, 1972).
5. Apollo 11, type A basalt (see Papike *et al.*, 1976 for discussion of mare basalt classification); 8 separate collected specimens, 19 samples by 7 different analysts for trace elements, 9 samples by 5 analysts for major elements. The average of the results of all analysts for each specimen was used to represent that specimen in the suite. This method of averaging was used for all Apollo 11 and 12 basalt suites.
6. Apollo 11, type B basalt; 8 specimens, 20 samples by 6 analysts for trace elements, 10 samples by 5 analysts for major elements.
7. Apollo 12, olivine basalt; 8 specimens, 22 samples by 8 analysts for trace elements, 15 samples by 6 analysts for major elements.
8. Apollo 12, ilmenite basalt; 4 specimens, 10 samples by 5 analysts for trace elements, 7 samples by 6 analysts for major elements.
9. Apollo 12, pigeonite basalt; 4 specimens, 9 samples by 6 analysts for trace elements, 9 samples by 6 analysts for major elements.
10. Apollo 15, quartz-normative basalt; 7 specimens analyzed for major and trace elements by Helmke *et al.* (1973).
11. Apollo 15, olivine-normative basalt; 13 specimens analyzed for major and trace elements by Helmke *et al.* (1973).
12. Apollo 17, type A, fine-grained basalt; 7 specimens analyzed for major and trace elements by Rhodes *et al.* (1976) and Shih *et al.* (1975).
13. Apollo 17, type B, fine-grained basalt; 8 specimens analyzed for major and trace elements by Rhodes *et al.* (1976) and Shih *et al.* (1975).
14. Apollo 17, type A, fine-grained basalt (olivine microporphyritic ilmenite basalt, higher Sm group); 6 specimens analyzed by Laul, Murali, and Schmitt (Warner *et al.*, 1975).
15. Apollo 17, type B, fine-grained basalt (olivine microporphyritic ilmenite basalt, lower Sm group); 9 specimens analyzed by Laul, Murali, and Schmitt (Warner *et al.*, 1975).

Results

Results of the dispersion analyses for these suites of basalts are given in Tables 1–3. Dispersions were determined for trace elements La, Sm, Eu, Yb, Sc, and Co. La, Sm, and Yb are representative incompatible elements that concentrate to different extents in early liquids during partial melting or in residual liquids during fractional crystallization. Eu has an affinity for plagioclase, especially in lunar environments, which are strongly reducing so that a high fraction of that element is in the 2+ oxidation state. Sc is a more or less

Table 1. Mean values, standard deviations, and ranges for trace elements in basalt suites.

	La			Sm			Eu			Yb			Sc			Co		
	Mean	%S.D.	Range	Mean	%S.D.	Range	Mean	%S.D.	Range	Mean	%S.D.	Range	Mean	%S.D.	Range	Mean	%S.D.	Range
I-A Ice.	10.2	10	1.4	4.4	9.5	1.4	1.58	6.6	1.2	2.1	7.5	1.3	—	—	—	—	—	—
I-B Cas.V.	10.7	2.7	1.08	4.8	3.1	1.09	1.59	2.8	1.07	3.3	3.0	1.06	32	1.8	1.05	44	4.4	1.1
I-C Cas.H	10.0	36	2.0	4.7	26	1.7	1.56	18	1.5	3.1	25	1.6	30	2.8	1.07	43	9.3	1.2
II-A St.E.	15.8	12	1.4	6.2	5.3	1.2	1.95	4.6	1.1	3.1	13	1.5	—	—	—	—	—	—
II-B UPG	11.3	12	1.4	3.9	7.8	1.3	—	—	—	3.0	8.7	1.3	40	9.5	1.3	40	5.3	1.2
II-C LPG	7.6	8.9	1.3	3.1	8.5	1.4	—	—	—	2.2	20	2.0	42	3.5	1.1	42	5.3	1.2
II-D LY	24	5.9	1.2	6.4	7.1	1.2	—	—	—	2.9	8.6	1.3	34	1.3	1.06	39	2.4	1.08
II-E Y	19.1	5.1	1.2	4.2	8.0	1.3	—	—	—	2.2	13	1.5	33	2.1	1.06	38	5.2	1.2
III-A Kew.	7.9	54	3.8	3.0	44	3.8	1.08	31	2.6	1.77	35	3.8	27	8.9	1.5	48	10	1.4
III-B St.	21	30	3.5	7.3	24	3.6	2.2	18	2.5	3.4	20	3.5	—	—	—	—	—	—
IV-A Cas.	10.2	68	15	4.0	51	8.6	1.39	45	3.5	2.6	37	5.1	32	14	2.4	48	15	1.8
1 DS PL.	5.0	13	1.3	2.4	9.5	1.3	0.91	8.0	1.3	2.3	10	1.3	40	4.6	1.2	—	—	—
2 DS PI	1.74	4.1	1.1	1.12	2.2	1.1	0.49	6.9	1.2	1.44	4.8	1.1	33	3.8	1.1	—	—	—
3 DS O1	2.6	23	1.8	1.63	26	1.8	0.64	13	1.4	1.81	8.7	1.2	36	3.8	1.1	—	—	—
4 10084	16.0	11	1.3	13.3	5.4	1.2	1.83	9.7	1.3	10.8	11	1.5	61	5.0	1.1	29	10	1.2
5 A11-A	25	8.6	1.3	20	7.7	1.3	2.2	5.0	1.2	18.9	7.7	1.3	78	6.7	1.2	28	8.6	1.3
6 A11-B	10.5	26	2.1	13.7	25	2.0	2.1	21	1.8	13.0	23	1.7	83	10	1.3	14.9	17	1.7
7 A12 OL	5.3	15	1.6	3.9	9.2	1.4	0.84	6.9	1.3	3.4	7.2	1.2	42	6.1	1.2	60	6.8	1.2
8 A12 IL	6.2	6.2	1.2	5.8	6.1	1.1	1.30	7.4	1.2	5.3	4.8	1.1	60	4.3	1.1	39	26	1.9
9 A12 PG	6.8	4.5	1.1	5.0	6.8	1.1	1.07	2.4	1.1	4.3	7.1	1.1	50	6.6	1.2	37	9.5	1.3
10 A15 Q	5.9	14	1.5	4.0	12	1.5	0.98	12	1.3	2.7	10	1.3	44	10	1.3	46	17	1.5
11 A15 OL	4.9	14	1.7	3.5	15	1.7	0.89	12	1.6	2.3	12	1.6	44	6.3	1.2	52	5.6	1.1
12 A17-Af	6.7	3.8	1.1	10.9	3.6	1.1	2.2	4.3	1.2	10.5	5.0	1.2	80	2.6	1.1	17.8	7.1	1.3
13 A17-Bf	5.8	6.0	1.2	7.6	8.0	1.3	1.53	7.1	1.3	7.6	5.1	1.2	84	2.3	1.1	18.5	12	1.3
14 A17-AL	6.5	6.4	1.2	10.2	5.5	1.2	2.0	5.8	1.2	9.0	8.0	1.3	77	5.1	1.2	19.1	6.9	1.2
15 A17-BL	5.3	8.5	1.3	7.0	9.3	1.3	1.43	8.7	1.3	6.7	6.7	1.3	79	3.2	1.1	21	11	1.5

Table 2. Mean values, standard deviations, and ranges for major elements in basalt suites.

	SiO_2			TiO_2			Al_2O_3			FeO			MgO			CaO		
	Mean	%S.D.	Range	Mean	%S.D.	Range	Mean	%S.D.	Range	Mean	%S.D.	Range	Mean	%S.D.	Range	Mean	%S.D.	Range
I-A Ice.	47.9	0.63	1.03	1.95	4.9	1.23	16.3	2.7	1.13	10.8	3.3	1.18	6.4	4.3	1.18	12.5	2.4	1.12
I-B Cas.V.	49.1	1.10	1.03	1.40	5.9	1.18	16.6	3.7	1.10	11.1	1.8	1.05	6.6	5.0	1.11	9.3	2.0	1.04
I-C Cas.H.	49.5	2.7	1.06	1.37	9.5	1.28	16.8	4.2	1.11	10.5	1.7	1.04	7.1	5.6	1.15	9.2	7.5	1.17
I-D1 Mc.V.	50.3	0.74	1.02	1.47	3.5	1.07	14.9	0.5	1.01	10.4	1.6	1.04	7.9	2.1	1.06	9.2	0.97	1.02
I-D2 Mc.V.	50.2	0.69	1.01	1.53	3.8	1.07	14.8	1.7	1.03	10.5	1.0	1.02	8.0	2.4	1.05	9.0	2.0	1.04
I-D3 Mc.V.	50.7	1.31	1.03	1.53	3.8	1.07	15.0	0.4	1.01	10.5	1.3	1.03	7.8	4.0	1.08	9.2	0.51	1.01
I-D4 Mc.1.	52.4	0.96	1.03	1.48	3.0	1.07	15.4	1.7	1.05	9.9	1.1	1.03	6.2	8.3	1.22	9.8	1.9	1.06
I-D5 Mc.2.	50.4	1.14	1.04	1.49	2.3	1.07	14.9	0.8	1.03	10.4	1.3	1.05	8.0	3.1	1.11	9.2	1.6	1.06
I-D6 Mc.	50.9	2.0	1.07	1.50	3.1	1.14	15.0	1.9	1.09	10.3	2.5	1.09	7.5	10.7	1.4	9.3	5.5	1.48
I-E Mak.	49.7	1.8	1.05	2.4	10.1	1.32	13.0	5.4	1.19	11.2	1.2	1.04	9.7	22	2.0	10.5	2.9	1.10
II-A St.E.	47.8	1.25	1.05	2.2	4.8	1.18	16.5	2.8	1.11	12.5	3.2	1.14	5.9	6.8	1.30	9.3	5.5	1.27
III-A Kew.	48.5	2.1	1.08	1.21	41	3.6	17.8	6.6	1.24	10.4	16.3	1.7	7.8	20	2.1	10.9	8.8	1.5
III-B St.	49.0	3.2	1.15	2.3	19.8	3.2	16.5	10.7	1.7	12.0	13.0	2.4	4.9	23	2.9	8.9	13.0	1.8
IV-A Cas.	48.5	3.9	1.20	1.37	34	5.1	17.1	6.1	1.39	10.2	11.8	1.8	8.0	17	2.6	10.0	11.1	1.7
1 DS PL	49.1	1.09	1.03	1.00	7.9	1.27	15.3	6.9	1.22	9.2	7.9	1.28	7.5	8.8	1.34	12.1	5.0	1.18
2 DS PI	45.9	3.1	1.06	0.53	4.7	1.12	13.9	3.7	1.09	8.9	1.5	1.03	13.4	12.2	1.34	10.9	7.2	1.19
3 DS OL	46.8	1.36	1.03	0.73	14.7	1.4	15.6	4.0	1.10	9.2	4.3	1.11	9.0	18	1.5	11.6	5.5	1.12
4 10084	42.0	0.7	1.02	7.5	1.6	1.05	13.9	2.3	1.06	15.7	1.7	1.04	7.9	1.7	1.04	12.0	1.00	1.03
5 A11-A	40.4	1.03	1.03	11.8	2.9	1.08	8.8	13	1.36	19.3	2.0	1.05	7.6	3.5	1.10	10.6	2.9	1.09
6 A11-B	40.7	2.9	1.07	10.6	10.4	1.4	10.2	7.4	1.26	18.7	4.9	1.15	7.0	13.1	1.4	11.5	3.7	1.10
7 A12 Ol	44.3	1.6	1.04	2.6	7.3	1.27	8.0	5.0	1.18	21.2	2.5	1.09	14.3	11	1.4	8.5	5.7	1.18
8 A12 IL	44.5	3.7	1.08	4.6	9.6	1.25	9.8	7.7	1.18	20.8	4.2	1.09	8.5	25	1.7	10.8	9.5	1.24
9 A12 PG	46.5	0.5	1.01	3.4	4.0	1.09	10.4	3.3	1.07	19.7	1.8	1.04	7.9	4.9	1.12	11.0	2.2	1.05
10 A15 Q	48.9	2.5	1.07	2.0	11.2	1.32	9.6	3.7	1.09	20.4	1.6	1.04	8.6	11.5	1.38	9.9	9.9	1.32
11 A15 Ol	45.9	2.3	1.08	2.7	13.6	1.4	8.4	6.7	1.27	22.9	4.3	1.15	10.3	4.5	1.17	9.7	4.4	1.17
12 A17-Af	39.2	1.7	1.04	12.1	3.8	1.11	8.8	2.2	1.06	19.0	1.1	1.03	8.2	4.7	1.13	10.9	3.0	1.08
13 A17-Bf	33.6	2.5	1.08	12.3	7.4	1.22	9.1	7.0	1.18	19.3	2.6	1.07	7.8	5.8	1.21	11.0	4.0	1.13
14 A17-AL	—	—	—	12.5	4.6	1.13	8.8	3.9	1.10	19.0	6.2	1.16	8.1	7.4	1.23	10.3	5.7	1.16
15 A17-BL	—	—	—	12.2	6.8	1.25	8.9	7.0	1.28	19.4	3.6	1.12	7.6	10	1.34	10.4	4.7	1.19

Table 3. Mean values, standard deviations, and ranges for element ratios in basalt suites.

	La/Sm			Sm/Eu			La/Yb			Fe/(Fe + Mg)		
	Mean	%S.D.	Range	Mean	%S.D.	Range	Mean	%S.D.	Range	Mean	%S.D.	Range
I-A Ice.	2.4	5.2	1.16	2.8	4.4	1.11	5.0	3.1	1.11	0.49	2.6	1.14
I-B Cas.V	2.2	3.5	1.07	3.0	1.5	1.03	3.2	4.3	1.11	0.49	2.8	1.08
I-C Cas.H.	2.1	9.8	1.26	3.0	8.2	1.21	3.2	11	1.28	0.46	2.8	1.08
II-A St.E.	2.6	13	1.35	3.2	4.7	1.17	5.1	12	1.4	0.54	3.0	1.12
II-B UPG	2.9	7.1	1.26	—	—	—	3.8	15	1.6	—	—	—
II-C LPG	2.5	8.9	1.4	—	—	—	3.5	16	1.7	—	—	—
II-D LY	3.8	3.8	1.12	—	—	—	8.4	5.6	1.17	—	—	—
II-E Y	4.5	9.2	1.32	—	—	—	8.6	9.8	1.28	—	—	—
III-A Kew.	2.6	14	1.6	2.6	14	1.5	4.1	27	2.5	0.43	16	1.6
III-B St.	2.9	17	1.5	3.3	11	1.9	6.3	20	1.8	0.58	8.9	1.5
IV-A Cas.	2.4	28	3.3	2.9	8.7	1.4	4.0	60	11	0.42	14	1.9
1 DS PL	2.1	8.7	1.5	2.6	4.7	1.17	2.2	15	2.0	0.41	5.6	1.19
2 DS PI	1.56	3.3	1.07	2.3	6.1	1.15	1.21	4.6	1.11	0.27	9.5	1.27
3 DS OL	1.62	19	1.5	2.5	17	1.5	1.42	20	1.7	0.37	11	1.28
4 10084	1.22	11	1.25	7.3	6.4	1.21	1.48	13	1.37	0.53	1.2	1.04
5 A11-A	1.27	4.5	1.16	9.4	5.9	1.20	1.35	8.2	1.31	0.59	1.2	1.04
6 A11-B	0.78	24	1.8	6.5	8.8	1.29	0.82	22	1.8	0.60	5.4	1.20
7 A12 OL	1.42	16	1.6	4.7	4.8	1.16	1.55	9.2	1.29	0.46	6.7	1.18
8 A12 IL	1.07	9.0	1.22	4.5	2.1	1.05	1.19	9.1	1.24	0.58	8.6	1.20
9 A12 PG	1.36	5.0	1.13	4.7	5.5	1.14	1.60	6.0	1.16	0.58	1.4	1.03
10 A15 Q	1.47	6.8	1.24	4.1	11	1.34	2.2	9.6	1.36	0.57	4.8	1.14
11 A15 OL	1.37	4.5	1.17	4.0	4.9	1.19	2.1	3.9	1.13	0.56	2.2	1.07
12 A17-Af	0.62	2.3	1.07	5.0	2.2	1.08	0.65	3.5	1.13	0.57	4.6	1.19
13 A17-Bf	0.76	5.3	1.14	4.9	1.5	1.04	0.76	3.1	1.08	0.59	3.4	1.12
14 A17-AL	0.64	8.4	1.27	5.1	6.6	1.22	0.73	9.4	1.34	0.57	4.4	1.13
15 A17-BL	0.76	10.3	1.4	4.9	5.9	1.17	0.79	8.8	1.4	0.59	3.5	1.12

compatible element, having distribution coefficients greater than unity for pyroxene, as is Co, with distribution coefficients greater than unity for olivine and pyroxene. Thus, somewhat contrasting behaviors during partial melting or fractional crystallization are represented by these six elements.

Dispersions were determined for major element oxides SiO_2, TiO_2, Al_2O_3, total Fe as FeO, MgO, and CaO. The extent of variation of major elements is restricted somewhat by the consideration only of basalts. Elements of contrasting behavior within that context are considered. In most situations, ferromagnesian minerals crystallize early or melt late relative to feldspars.

Use only of element concentrations provides an indication of compositional change but limited information about differentiation among a suite of basalts. For example, local accumulation of olivine could cause substantial variation in concentrations of incompatible elements such as REE without causing any significant change in relative REE abundances. Similarly, loss or accumulation of feldspar should significantly change the abundance of Eu relative to that of Sm. In order to monitor to some degree changes caused by simple dilution as opposed to those corresponding to general differentiation, dispersions were determined for La/Sm, Sm/Eu, La/Yb, and the atomic ratio Fe/(Fe + Mg).

In order to provide a convenient measure of variability in concentration, we calculated standard deviations for each element measured and each ratio in each basalt suite. For some suites, the distribution of results is more or less normal and the standard deviation can be regarded in the usual way as indicating the dispersion to be expected for sampling and analyzing a random specimen from that suite. For others, there is a clear trend of differentiation or a multimodal (non-normal) distribution, leading to rather large values for the standard deviation. In these cases, the standard deviation is merely the result of some analytical data being processed through a mathematical formula and serves as an indicator of large spreads in concentration. By using the standard deviation we minimize effects from any single peculiar sample that might be present in an otherwise reasonably homogeneous suite. However, ranges in variation are also important observations. Thus, we have tabulated ranges in terms of the ratio of the maximum observed value to the minimum observed value.

The purpose of this work is to provide a comparison of compositional dispersion between suites of basalts whose member specimens are in some known relationship to each other (from field, petrographic, and compositional evidence) and that of suites of basalts whose member specimens are in unknown relationship to each other. Such a comparison can be made directly from the data in Tables 1–3. To provide a more convenient means for comparison, we introduce an "index of dispersion," defined as follows: The observed value for any parameter (e.g., standard deviation for ppm La) was assigned the value of 4 if it was smaller than the larger of the corresponding dispersions for the Iceland single flow or the Cascades vertical section from a single flow (set I). It was assigned the value 3 if it was less than the largest value observed for any suite of

Table 4. Indices of dispersion for samples of known relationship.

	Trace		Major		Ratio		Ave.
	S.D.	Range	S.D.	Range	S.D.	Range	
I-A Ice.	4	4	4	4	4	4	4
I-B Cas.V.	4	4	4	4	4	4	4
I-D1 Mc.V.	—	—	4.0	4.0	—	—	4.0
I-D2 Mc.V.	—	—	4.0	4.0	—	—	4.0
I-D3 Mc.V.	—	—	3.7	4.0	—	—	3.9
I-D5 Mc.2	—	—	3.8	3.8	—	—	3.8
I-D4 Mc.1	—	—	3.7	3.8	—	—	3.8
II-A St.E.	3.5	3.6	3.5	3.5	3.0	3.3	3.4
II-E Y	3.4	3.4	—	—	3.0	3.5	3.3
II-D LY	3.8	3.4	—	—	3.0	3.0	3.3
II-B UPG	3.2	3.2	—	—	3.0	3.0	3.2
II-C LPG	3.4	3.4	—	—	3.0	3.0	3.2
I-D6 Mc.	—		3.2	3.0	—	—	3.1
I-C Cas.H.	2.2	2.2	2.7	3.2	3.0	3.0	2.7
I-E Mak.	—	—	2.5	2.8	—	—	2.7
III B St.	2.3	2.0	2.0	2.0	2.0	2.0	2.1
III A Kew.	2.2	2.0	2.0	2.0	2.0	2.0	2.0
IV A Cas.	1.0	1.0	1.8	1.5	1.5	1.3	1.4

chemically undifferentiated, successive flows (set II) but too large to merit a value of 4. It was assigned a value of 2 if it was less than the larger value observed for the related, differentiated flows (set III) but too large to merit a value of 3. It was assigned a value of 1 if it was less than the value for the Cascades mesa basalts (set IV) but too large to merit a value of 2. It was assigned a value of zero if it exceeded the value for the Cascades mesa basalts.

Averages were computed for ranges and for standard deviations for each class of data (i.e., trace element, major element, ratio). These averages are given in Table 4, along with a grand average for the six averages just described. This index is analogous to the grade-point index of a student and like that analog is very useful but by no means a substitute for considering carefully the characteristics of each individual suite to determine whether the index accurately reflects the properties of that suite.

Discussion

Before examining the results in detail, it is necessary to discuss briefly some observed effects of sampling and analysis. In defining boundaries for index values, we relied as much as possible on sample suites for whose analyses we were confident we knew the precision was much smaller than the observed dispersion of results. The effects of analysis on the dispersion are well documented for our own measurements by neutron activation (Jacobs *et al.*, 1977); they are shown as standard deviations and ranges in Fig. 2 for multiple analyses of the Knippa basalt, presumed homogeneous (as powdered material from the same bottle) for this test of precision. Except for the least variable of basalt suites, such analytical errors are small or negligible.

For Apollo 11 and 12 basalts, effects of interlaboratory bias must be considered because no single laboratory analyzed a large enough number of samples to serve as a suite. Most of the analysts whose data were used for the Apollo 11 and 12 suites also analyzed Apollo 11 lunar soil 10084. We have assumed 10084 to be homogeneous and have attributed variations in values reported for it to be a combination of random analytical uncertainty and interlaboratory bias. The magnitudes of the variations are given in Tables 1–3. Some are shown in Fig. 2. Indices of dispersion are included in Table 4 and, to the extent that 10084 is homogeneous, indicate the minimum range of dispersion that can be expected for Apollo 11 and 12 basalts on analytical grounds. For basalt suites from later missions, data from only one laboratory were used for any given element, thus eliminating effects of interlaboratory differences.

Effects of sampling can be overwhelming. Three separate chips of Apollo 17 medium-grained basalt 70135 weighing 0.25 to 3.3 g were analyzed in different laboratories (see Haskin and Korotev, 1977 for discussion). The ranges observed are shown in Fig. 2. Those for incompatible elements greatly exceed those observed for the terrestrial basalt suites with related members and exceed those for the lunar basalt suites compiled. The principal reasons for these wild variations in concentrations of incompatible elements are differing quantities of

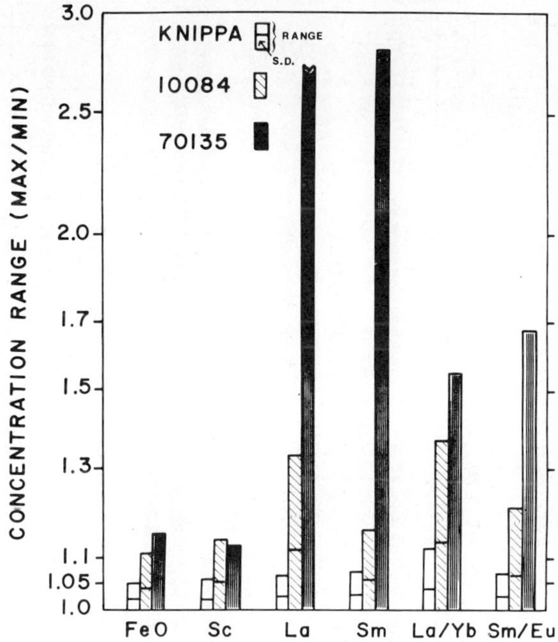

Fig. 2. The histogram shows contributions of analytical uncertainty to observed trace-element dispersions. Lefthand bars are standard deviations and ranges observed for repeated neutron activation analysis of Knippa basalt (Jacobs *et al.*, 1977). Center bars show observed dispersions for seven analysts for Apollo 11 lunar soil 10084. Righthand bars show possible effects of sampling on dispersion; they represent the range observed for three samples of basalt 70135. Ordinate is (1 + standard deviation) for standard deviation or range as maximum observed value divided by minimum observed value.

final liquid (mesostasis) in the different chips analyzed (Haskin and Korotev, 1977). The variations stand as stark reminders of what problems inadequate sampling can cause. In fact, the ranges for REE concentrations among the three chips exceed those for all other reported analyses of Apollo 17 basalts combined.

Rhodes *et al.* (1974) indicate that variations are most pronounced in the coarser-grained lunar basalts. Thus, the finer-grained Apollo 11 basalts and the selected fine-grained Apollo 17 basalts should be relatively free of these effects. The magnitude of the sampling problem in this work is not known, but is clearly much smaller than for the three chips of 70135, as indicated by the much lower dispersions. The lower dispersions probably cannot be accounted for on the basis of sample size, as most samples allocated for chemical analysis were in the range of 1–2 g and samples of 70135 weighing 1.6 g and 3.3 g yielded values for REE approximately a factor of 2 apart. The dispersions discussed in this work are small enough to provide useful comparisons.

Typical results are illustrated in Fig. 3 for terrestrial basalts. Values plotted

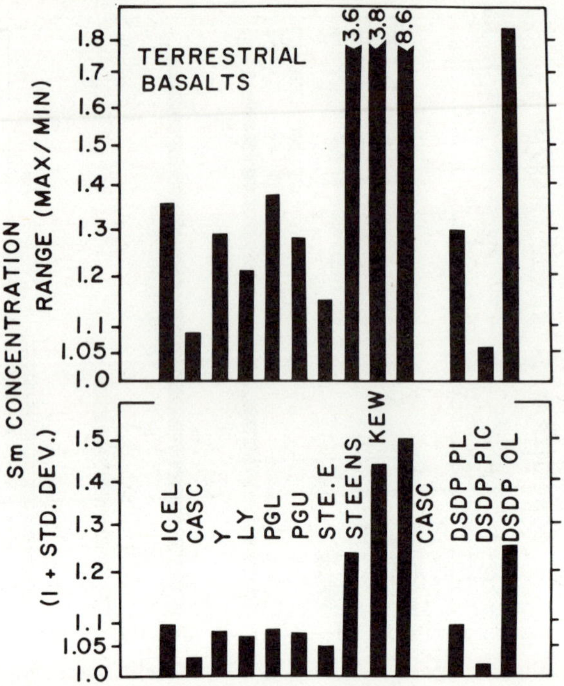

Fig. 3. Dispersions in Sm concentration for suites of terrestrial samples are shown as (1 + standard deviation) and range. Refer to section on basalt suites to understand symbols. Lefthand CASC refers to Cascades vertical section, single flow and righthand CASC refers to all Cascades basalt.

are one plus standard deviation and range, both on a logarithmic ordinate. The Iceland single flow and the Cascades vertical flow are both fine-grained, homogeneous basalts in thin section, not showing evidence for *in situ* differentiation. Wood *et al.* (1976) attribute part of the compositional variability of the Icelandic samples to zeolite facies metamorphism, however. Samples from these flows are used as a measure of minimum dispersion to be expected for reasonable sampling of homogeneous, undifferentiated terrestrial basalts. Thus, they are given a value of 4 in all categories in Table 4. Standard deviations for trace elements for the Cascades vertical section are comparable to our analytical precision and the samples must be regarded as identical with respect to those trace elements (and others not reported here). Standard deviations for most major elements significantly exceed the expected precision for their measurement and are larger than those for some of the other sample suites. These values were not determined in our laboratory, but values for FeO were obtained in our neutron activation analyses; the mean was 11.0 to .27 (±2.4%), comparable to our analytical uncertainty and in good agreement with the value of 11.0 (±1.8%) based on the analyses supplied by A. McBirney (pers. comm.). It seems unlikely that the flow is less homogeneous with respect to major elements than with respect to trace elements, but that is what these data indicate. Major-element data for the Iceland single flow (vertical section) and the three vertical sections of McCartys basalt show less dispersion. Trace element (REE) data for the Iceland basalt are more dispersed than for the Cascades vertical flow, but were obtained by earlier, less precise procedures, so there is probably some contribution from analytical procedures to their dispersion.

The vertical sections of the Iceland, Cascades, and McCartys flows are all quite homogeneous. The horizontal section of the Cascades basalt is clearly inhomogeneous in trace elements but less clearly so in major elements. The trace-element distribution is more or less bimodal. The explanation for this dispersion is not obvious or known. A similar trend of vertical homogeneity but horizontal inhomogeneity occurs in the McCartys basalt (Carden and Laughlin, 1974); the inhomogeneity extends to Sr isotopes (Brookings *et al.*, 1975). Neither of these flows shows any clear pattern of differentiation or assimilation of crustal matter. By contrast, the thick vertical section of the Makaopuki lava lake does show a clear trend of differentiation during cooling (Evans and Moore, 1968) with consequent greater dispersion for most major elements than found in the vertical sections of the thinner flows.

The samples of related, undifferentiated flows (group II) have dispersions of trace elements comparable to or slightly larger than those observed for the single flows. The Steens Mt. E basalts, the only group II representatives for which major-element data are available, have dispersions for those elements slightly greater than those of group I (vertical).

The basalt suites of groups I (vertical) and II are set well apart from those of groups III and IV in dispersions of both major and trace elements as measured by both standard deviations and ranges. The group I horizontal samples, Cascades and McCartys (D6), have standard deviations intermediate for most elements between those of group I (vertical) and II and those of groups III and IV. The ranges for major elements in the Makaopuki samples are substantially less than those for groups III and IV.

The dispersions in the trace-element ratios for group I (vertical) samples are lower than most of those of group II, both in standard deviation and range. The standard deviations for the trace-element ratios of the group III samples exceed those of group II substantially in most cases, but the ranges exceed those of group II less markedly; absence of data for Eu and Mg for most group II samples make these trends less certain. The dispersions in ratios for the Cascades mesa basalts (IV-A) significantly exceed those for group III for La/Sm and La/Yb, but not for Sm/Eu or Fe/(Fe + Mg). The general increase in dispersion in going from group I to group IV suites is summarized in the indices of Table 4 as well as in the more detailed data of Tables 1–3.

The dispersions observed for the DSDP plagioclase basalts (1 DS PL) exceed those for single flow vertical sections or related, undifferentiated basalts (Tables 1–5, Fig. 3). This is reasonable in that the basalts in this suite were not contiguous in the core taken from hole 332B and it is not known how many flows they represent. The dispersions for the DSDP picritic basalts are comparable to

Table 5. Indices of dispersion for DSDP and lunar basalts.

	Trace		Major		Ratios		Ave.
	S.D.	Range	S.D.	Range	S.D.	Range	
1 DS PL	2.8	3.4	2.7	2.7	2.8	2.3	2.8
2 DS PI	3.4	3.8	3.0	3.2	2.8	3.3	3.3
3 DS OL	2.4	2.6	1.8	3.3	1.3	2.3	2.3
4 10084	2.7	3.2	4.0	4.0	3.0	3.0	3.3
9 A12 PG	3.5	3.5	3.8	4.0	3.3	3.5	3.6
12 A17-Af	3.5	3.5	3.5	3.8	3.5	3.3	3.5
5 A11-A	3.3	3.5	3.2	3.7	3.3	3.3	3.4
13 A17-Bf	3.0	3.2	2.7	3.0	3.3	4.0	3.2
14 A17-AL	3.3	3.7	2.6	3.6	2.5	3.0	3.1
15 A17-BL	3.0	3.0	2.4	2.6	2.5	3.3	2.8
11 A15 OL	2.3	2.5	2.7	2.8	3.5	3.0	2.8
8 A12 IL	2.8	3.2	1.7	2.5	3.0	3.0	2.7
7 A12 OL	2.8	2.5	2.3	2.7	2.3	2.3	2.5
10 A15 Q	1.7	2.2	2.7	2.7	2.5	3.0	2.5
6 A11-B	1.5	2.0	2.3	2.7	1.8	1.8	2.0

those for related, undifferentiated flows. Only two of the four picritic samples were adjacent in the core, and this degree of homogeneity is surprising in olivine cumulative rocks. It is perhaps an accident of so few (4) samples in the suite. The DSDP olivine basalts show substantial dispersions, comparable to those of the suites of related, differentiated basalts. The olivine basalts also were not all contiguous in the core. Overall, the DSDP basalts show compositional dispersions consistent with both our limited knowledge of their geological relationships and the scale derived from basalts of better known relationships.

Data for the suites of lunar mare basalts are given in Tables 1–3 and 5. Typical results are shown in Fig. 4. Most of the lunar basalt suites show dispersions comparable to those of related, chemically undifferentiated suites of terrestrial basalts or smaller. The low dispersions for the Apollo 12 pigeonite and Apollo 11 type A (high K) basalts are especially interesting in that analytical uncertainty, as judged by the results for 10084, is sufficient to account for essentially all of them. The low dispersions of the Apollo 17 group A and group B basalts are less surprising in that these groupings were based on composition and only finer-grained basalts were considered in order to reduce the dispersions. Figure 4 shows how broad the dispersion becomes if all Apollo 17 basalts are included (in this case, those reported in Warner *et al.*, 1975). The dispersions for groups A and B fine-grained basalts based on data of Rhodes *et al.* (1976) are somewhat lower than those of Laul, Murali, and Schmitt (Warner *et al.*, 1975). This arises from larger standard deviations, but not ranges, for trace elements and larger standard deviations and ranges for major elements in the case of Laul *et al.* This is probably in part a result of their somewhat less precise methods of analysis for major elements, but mainly a consequence of their use of smaller samples than those used by Rhodes *et al.* The average compositions

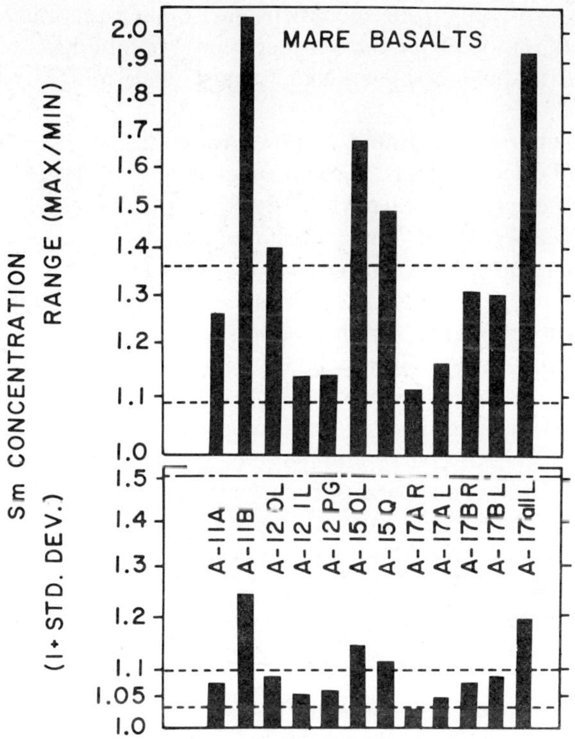

Fig. 4. Dispersions in Sm concentration for suites of lunar mare basalts. The lower pair of horizontal lines shows the Sm dispersions for the Iceland and Cascades (vertical) single flows, used to define dispersion index 4. The upper pair of horizontal lines shows the ranges for those same samples.

determined for the two types of Apollo 17 basalt by the two groups of analysts are in good agreement (Tables 1–3). This is of particular interest since those analyzed by Laul *et al.* are portions of small rocks from rake samples mostly from Stations 1 and 8 whereas those analyzed by Rhodes *et al.* are fragments of large rocks from all mare sites.

The Apollo 11 type B basalt suite show the greatest dispersions. The type B rocks are coarser grained than the type A specimens, but are nevertheless rather fine grained. Effects of coarser-grain size, if any, should be offset somewhat because several samples of most Apollo 11 type B rocks were analyzed. The Apollo 17 mare basalts, which show overall nearly as large a dispersion, clearly encompass at least two compositional groups. It is tempting to consider that the Apollo 11 B basalts also constitute two or more groups. On the basis of major- and trace-element compositions, no consistent groups stand out (e.g., 10044, 10050, and 10058 are relatively low in FeO, 10020, 10050, and 10045 are relatively low in La, etc.). However, Stettler *et al.* (1977) have argued that two age groups are represented among Apollo 11-B basalts. Warner (1971) has divided the Apollo 11-B basalts into two separate types on petrographic grounds. The dispersions for the two types (four specimens each) taken separately are smaller for most parameters, more so for the Warner type 2 ophitic basalts than for the type 1 ophitic basalts, but not by enough to raise their indices much above a value of 2.

Substantial dispersions are found for the Apollo 12 olivine and Apollo 12 ilmenite basalt suites. These suites show clear evidence of crystal fractionation, pointed out by numerous investigators at the time the Apollo 12 basalts were first studied (e.g., Warner, 1971). Thus, substantial dispersion might be expected, and values of dispersion indices are similar to those for the terrestrial, single-flow horizontal sections and the Makaopuki lava lake. They are not as low as the values for the terrestrial related, differentiated flows.

There is substantial dispersion in values for Apollo 15 olivine-normative and quartz-normative basalts. Apollo 15 basalts have been analyzed by Rhodes and Hubbard (1973) who discuss their dispersions for major elements, particularly the problems of sampling of coarse-grained lunar rocks. Most of the dispersions for major elements for seven olivine-normative basalts analyzed by Rhodes and Hubbard are smaller than those analyzed by Helmke *et al.* (1973) but most of those for six quartz-normative basalts are greater than found by Helmke *et al.* The differences are insufficient to change the indices of dispersion for the major elements as given in Table 5. The average concentrations obtained by Rhodes and Hubbard agree with those obtained by Helmke *et al.* to within the dispersion observed by one group of analysts or the other for every element for both the olivine-normative and the quartz-normative basalts. Rhodes and Hubbard reported REE data for only six specimens, insufficient for a meaningful comparison, but with only one sample lying outside the range observed by Helmke *et al.* (15682,4, quartz-normative). The Apollo 15 basalt suites of Helmke *et al.* (1973) were used in the dispersion analyses to eliminate any effects of interlaboratory bias. It should be noted, however, that a sample of olivine-normative 15555 analyzed by Schnetzler *et al.* (1972) has trace-element concentrations substantially lower than those observed for other Apollo 15 basalts. Rhodes and Hubbard (1973) have tabulated major elements for 7 samples of 15555 and shown substantial dispersions (index of 2.7, same as for the entire group of samples, Table 5). Thus, sampling problems appear to be the predominant cause of the observed dispersions for the Apollo 15 basalt suites.

The purpose of this paper has been to compare dispersions in composition for suites of lunar basalts with those observed for suites of terrestrial basalts of known field or genetic relationship. This was done because of the implications of variations in trace-element concentration, especially REE, as interpreted by current models, with respect to the number of different basalts sampled, the types of processes that produced them, the homogeneity of composition of their source regions, etc. The comparison indicates that several of the Apollo basalt suites (Apollo 11-A, Apollo 17 A and B fine-grained basalts, and Apollo 12 pigeonite basalts), as sampled for analysis are scarcely more variable in composition than vertical sections of single flows of several terrestrial basalts. These groups are a little less variable in composition than series of successive terrestrial flows that can be regarded as chemically undifferentiated relative to each other. The Apollo 15 olivine and quartz normative basalts and the Apollo 12 ilmenite and olivine basalts show compositional variations no more pronounced than can be found within single terrestrial flows, whose inhomogeneity in horizontal section has been demonstrated. Much of the reason for the dispersion among suites of Apollos 12 and 15 basalts, however, is in the sampling of the parent rocks for analysis. In fact, the inhomogeneities of these parent rocks on a scale of centimeters clearly overshadow any differences that might exist among different general regions of the lava flows from which these samples might have derived. That is, it probably would not help to grind up the entire parent rocks of the collection because the individual parent rocks are themselves inadequate samples of their flows on a scale of tens of meters or kilometers. Except for brief mention of their broad dispersions, the coarser-grained Apollo 17 rocks (e.g., Papike *et al.*, 1976; Rhodes *et al.*, 1976) have been ignored. Only one basalt suite, Apollo 11-B (low K), shows variability that exceeds that for all Apollo 17 basalts taken together and comparable to those for terrestrial basalts that are clearly differentiated with respect to each other. Independent evidence (age measurements, Stettler *et al.*, 1974) indicates that at least two different flows are represented by the Apollo 11-B basalts. To the extent that analogy between the lunar and terrestrial basalts holds, there is no firm reason on the basis of trace element data to presume that the Apollo basalts at any site represent many, or even several, different lava flows. Nor is there reason to believe that the samples represent differentiates by fractional crystallization of one or a few flows, except on a distance scale of centimeters as indicated by differences among analyzed samples of single specimens (Rhodes and Hubbard, 1973; Haskin and Korotev, 1977). Data for major elements, included in the above analysis, support those conclusions. Studies of crystallization ages and isotope initial ratios are consistent with them (see Papike *et al.*, 1976, Table 1a for data sources searched; also, Nyquist *et al.*, 1977, and L. Nyquist, pers. comm.). As mentioned above, petrographic evidence does not require that many flows were sampled. At least some lunar basalt flows are very large, so that multiple impacts into the surface in the vicinity of a landing site need not sample numerous flows. Nor are the flows so thick that substantial, 10 m scale differentiation during crystallization would have to be expected (e.g., an Imbrium flow, 400 km in extent, 7–20 m thick across a flow lobe 190 km from source vent, Moore and Schaber, 1975).

However, evidence for the number of flows sampled is circumstantial, not observational. Furthermore, the compositional and isotopic heterogeneities within single terrestrial flows tens of kilometers in extent are a signal for caution in interpreting compositional clusters of basalts of the same age from a single Apollo site as necessarily indicating different flows. Perhaps a single eruption can bring more than one composition or even type of basaltic lava to the surface.

It is beyond the scope of this paper to discuss possible reasons for the extent of variability of composition within single terrestrial flows or among related flows. Nevertheless, careful measurements of major- and trace-element compositions document that variability exists well outside of analytical and centimeter scale sampling uncertainties. The present models for trace element distribution as applied to supposed rock-forming processes do not account for this variability. Until that variability can be properly modelled, interpretations from present models about dctailed as opposed to general characteristics of basalts, their source regions, and the processes that produced them should be regarded with reservation.

Acknowledgments—We thank M. J. Drake and F. A. Frey for helpful comments on part of this work. We thank M. Bewig for assistance in preparing the manuscript and D. Shuey and L. Ross for assistance in preparing the figures. This paper constitutes contribution No. 2 of the Basaltic Volcanism Study Project. The Project is organized and administered by the Lunar Science Institute/Universities Space Research Association under Contract NSR 09-051-001 with the National Aeronautics and Space Administration. This work was supported in part by NASA grant NSG-9055.

REFERENCES

Agrell S. O., Scoon J. H., Muir I. D., Long J. V. P., McConnell J. D. C., and Peckett A. (1970) Observations on the chemistry, mineralogy and petrology of some Apollo 11 lunar samples. *Proc. Apollo 11 Lunar Sci. Conf.*, p. 93–128.

Blanchard D. P., Rhodes J. M., Dungan M. A., Rodgers K. V., Donaldson C. H., Brannon J. C., Jacobs J. W., and Gibson E. K., Jr. (1976) The chemistry and petrology of basalts from Leg 37 of the Deep-Sea Drilling Project. *J. Geophys. Res.* **81**, 4231–4246.

Brookins D. G., Carden J. R., and Laughlin A. W. (1975) Additional note on the isotopic composition of strontium in McCartys flow, Valencia county, New Mexico. *Earth Planet Sci. Lett.* **25**, 327–330.

Brunfelt A. O., Heier K. S., and Steinnes E. (1971) Determination of 40 elements in Apollo 12 materials by neutron activation analysis. *Proc. Lunar Sci. Conf. 2nd*, p. 1281–1290.

Carden J. R. and Laughlin A. W. (1974) Petrochemical variations within the McCartys basalt flow, Valencia county, New Mexico. *Bull. Geol. Soc. Amer.* **85**, 1479–1484.

Compston W., Berry H., Vernon M. J., Chappell, B. W., and Kaye M. J. (1971) Rubidium-strontium chronology and chemistry of lunar material from the Ocean of Storms. *Proc. Lunar Sci. Conf. 2nd*, p. 1471–1485.

Compston W., Chappell B. W., Arriens P. A., and Vernon M. J. (1970) The chemistry and age of Apollo 11 lunar material. *Proc. Apollo 11 Lunar Sci. Conf.*, p. 1007–1027.

Cuttitta F., Rose H. J., Jr., Annell C. S., Carson M. K., Christian R. P., Dwornik E. J., Greenland L. P., Helz A. W., and Ligon D. T., Jr. (1971) Elemental composition of some Apollo 12 lunar rocks and soils. *Proc. Lunar Sci. Conf. 2nd*, p. 1217–1229.

Drake M. J. and Consolmagno G. J. (1976) Critical review of models for the evolution of high-Ti mare basalts. *Proc. Lunar Sci. Conf. 7th*, p. 1633–1657.

Engel A. E. J. and Engel C. G. (1970) Lunar rock compositions and some interpretations. *Proc. Apollo 11 Lunar Sci. Conf.*, p. 1081–1084.

Engel A. E. J., Engel C. G., Sutton A. L., and Myers A. T. (1971) Composition of five Apollo 11 and Apollo 12 rocks and one Apollo 11 soil and some petrogenic considerations. *Proc. Lunar Sci. Conf. 2nd*, p. 439–448.

Evans B. W. and Moore J. G. (1968) Mineralogy as a function of depth in the prehistoric Makaopuki tholeiitic lava lake, Hawaii. *Contrib. Mineral. Petrol.* **17**, 85–115.

Gast P. W., Hubbard N. J., and Wiesmann H. (1970) Chemical composition and petrogenesis of basalts from Tranquillity Base. *Proc. Apollo 11 Lunar Sci. Conf.*, p. 1143–1163.

Goles G. G., Duncan A. R., Lindstrom D. J., Martin M. R., Beyer R. L., Osawa M., Randle K., Meek L. T., Steinborn T. L., and McKay S. M. (1971) Analyses of Apollo 12 specimens: Compositional variations, differentiation processes, and lunar soil mixing models. *Proc. Lunar Sci. Conf. 2nd*, p. 1063–1081.

Goles G. G., Randle K., Osawa M., Lindstrom, D. J., Jerome D. Y., Steinborn T. L., Beyer R. L., Martin M. R., and McKay S. M. (1970) Interpretations and speculations on elemental abundances in lunar samples. *Proc. Apollo 11 Lunar Sci. Conf.*, p. 1177–1194.

Goles G. G., Randle K., Osawa M., Schmitt R. A., Wakita H., Ehmann W. D., and Morgan J. W. (1970) Elemental abundances by instrumental activation analyses in chips from 27 lunar rocks. *Proc. Apollo 11 Lunar Sci. Conf.*, p. 1165–1176.

Gunn B. M. and Watkins N. D. (1970) Geochemistry of the Steens Mountain basalts, Oregon. *Bull. Geol. Soc. Amer.* **81**, 1497–1516.

Haskin L. A., Allen R. O., Helmke P. A., Paster T. P., Anderson M. R., Korotev R. L., and Zweifel K. A. (1970) Rare earths and other trace elements in Apollo 11 lunar samples. *Proc. Apollo 11 Lunar Sci. Conf.*, p. 1213–1231.

Haskin L. A., Helmke P. A., Allen R. O., Anderson M. R., Korotev R. L., and Zweifel K. A. (1971b) Rare earth elements in Apollo 12 lunar materials. *Proc. Lunar Sci. Conf. 2nd*, p. 1307–1317.

Haskin L. A., Helmke P. A., Paster T. P., and Allen R. O. (1971a) Rare earths in meteoritic, terrestrial, and lunar matter. In *Activation Analysis in Geochemistry and Cosmochemistry*, (A. Brunfelt and E. Steinnes, eds.) p. 201–218. University of Oslo Press, Oslo.

Haskin L. A. and Korotev R. L. (1977) Test of a model for trace element partition during closed-system solidification of a silicate liquid. *Geochim. Cosmochim. Acta.* **41**, 921–939.

Helmke P. A., Blanchard D. P., Haskin L. A., Telander K., Weiss C., and Jacobs J. W. (1973) Major and trace elements in igneous rocks from Apollo 15. *The Moon* **8**, 129–148.

Helmke P. A. and Haskin L. A. (1973) Rare-earth elements, Co, Sc, and Hf in the Steens Mountain basalts. *Geochim. Cosmochim. Acta* **37**, 1513–1529.

Hubbard N. J. and Gast P. W. (1971) Chemical composition and origin of nonmare lunar basalts. *Proc. Lunar Sci. Conf. 2nd*, p. 999–1020.

Jacobs J. W., Korotev R. L., Blanchard D. P., and Haskin L. A. (1976) A well-tested procedure for instrumental neutron activation analysis of silicate rocks and minerals. *J. Radioanal. Chem.* In press.

Kushiro I. and Haramura H. (1971) Major element variation and possible source materials of Apollo 12 crystalline rocks. *Science* **171**, 1235–1237.

Longhi J., Walker D., Grove T. L., Stolper E. M., and Hays J. F. (1974) The petrology of the Apollo 17 mare basalts. *Proc. Lunar Sci. Conf. 5th*, p. 447–469.

Maxwell J. A., Peck L. C., and Wiik H. B. (1970) Chemical composition of Apollo 11 lunar samples 10017, 10020, 10072 and 10084. *Proc. Apollo 11 Lunar Sci. Conf.*, p. 1369–1374.

Maxwell J. A. and Wiik H. R. (1971) Chemical composition of Apollo 12 lunar samples 12004, 12033, 12051, 12052 and 12065. *Earth Planet. Sci. Lett.* **10**, 285–288.

McIntire W. L. (1963) Trace element partition coefficients—a review of theory and applications to geology. *Geochim. Cosmochim. Acta* **27**, 1209–1264.

Moore H. J. and Schaber G. G. (1975) An estimate of the yield strength of the Imbrium flows. *Proc Lunar Sci. Conf 6th*, p. 101–118.

Nyquist L. E., Bansal B. M., Wooden J. L., and Wiesmann H. (1977) Sr-isotopic constraints on the petrogenesis of Apollo 12 ilmenite basalts (abstract). In *Lunar Science VIII*, p. 735–737. The Lunar Science Institute, Houston.

Osawa M. and Goles G. G. (1969) Trace element abundances in Columbia River basalts. *Proc. 2nd Columbia River Basalt Symposium*, (E. H. Gilmore and D. Stradling, eds.) p. 55–71. Eastern Washington State College Press, Cheney, Wash.

Papike J. J., Hodges F. N., Bence A. E., Cameron M., and Rhodes J. M. (1976) Mare basalts: Crystal chemistry, mineralogy, and petrology. *Rev. Geophys. Space Phys.* **14**, 475–540.

Philpotts J. A. and Schnetzler C. C. (1970) Apollo lunar samples: K, Rb, Sr, Ba, and rare-earth concentrations in some rocks and separated phases. *Proc. Apollo 11 Lunar Sci. Conf.*, p. 1471–1486.

Rhodes J. M. and Hubbard N. J. (1973) Chemistry, classification, and petrogenesis of Apollo 15 mare basalts. *Proc. Lunar Sci. Conf. 4th*, p. 1127–1148.

Rhodes J. M., Hubbard N. J., Wiesmann H., Rodgers K. V., Brannon J. C., and Bansal B. M. (1976)

Chemistry, classification, and petrogenesis of Apollo 17 mare basalts. *Proc. Lunar Sci. Conf. 7th*, p. 1467–1489.

Rose H. J., Jr., Cuttitta F., Dwornik J., Carron M. K., Christian R. P., Lindsay J. R., Ligon D. F., and Larson R. R. (1970) Semimicro X-ray fluorescence analysis of lunar samples. *Proc. Apollo 11 Lunar Sci. Conf.*, p. 1493–1497.

Schnetzler C. C. and Philpotts J. A. (1971) Alkali, alkaline earth, and rare-earth element concentrations in some Apollo 12 soils, rocks and separated phases. *Proc. Lunar Sci. Conf. 2nd*, p. 1101–1122.

Schnetzler C. C., Philpotts J. A., Nava D. F., Schuhmann S., and Thomas H. H. (1972) Geochemistry of Apollo 15 basalt 15555 and soil 15531. *Science* **175**, 426–428.

Scoon J. H. (1971) Chemical analyses of lunar samples 12040 and 12064. *Proc. Lunar Sci. Conf. 2nd*, p. 1259–1260.

Shih C. Y., Haskin L. A., Wiesmann H., Bansal B., and Brannon J. C. (1975) On the origin of high-Ti mare basalts. *Proc. Lunar Sci. Conf. 6th*, p. 1255–1285.

Smales A. A., Mapper D., Webb M. S. W., Webster R. K., Wilson J. D., and Hislop J. S. (1971) Elemental composition of lunar surface material (part 2). *Proc. Lunar Sci. Conf. 2nd*, p. 1253–1258.

Stettler A., Eberhardt P., Geiss J., Grogler N., and Maurer P. (1974) On the duration of lava flow activity in Mare Tranquillitatis, *Proc. Lunar Sci. Conf. 5th*, p. 1557–1570.

Taylor S. R., Rudowski R., Muir P., Graham A., and Kaye M. (1971) Trace element chemistry of lunar samples from the Ocean of Storms. *Proc. Lunar Sci. Conf. 2nd*, p. 1083–1099.

Wakita H., Rey P., and Schmitt R. A. (1971) Abundances of the 14 rare-earth elements and 12 other trace elements in Apollo 12 samples: Five igneous and one breccia rocks and four soils. *Proc. Lunar Sci. Conf. 2nd*, p. 1319–1329.

Wänke H., Rieder R., Baddenhausen H., Spettel B., Teschke F., Quijano-Rico M., and Balacescu A. (1970) Major and trace elements in lunar material. *Proc. Apollo 11 Lunar Sci. Conf.*, p. 1719–1727.

Wänke H., Wlotzka F., Baddenhausen H., Balacescu A., Spettel B., Teschke F., Jagoutz E., Kruse H., Quijano-Rico M., and Rieder R. (1971) Apollo 12 samples: Chemical composition and its relation to sample locations and exposure ages, the two component origin of the various soil samples and studies on lunar metallic particles. *Proc. Lunar Sci. Conf. 2nd*, p. 1187–1208.

Warner J. L. (1971) Lunar crystalline rocks: Petrology and geology. *Proc. Lunar Sci. Conf. 2nd*, p. 469–480.

Warner R. D., Keil K., Prinz M., Laul J. C., Murali A. V., and Schmitt R. A. (1975) Mineralogy, petrology, and chemistry of mare basalts from Apollo 17 rake samples. *Proc. Lunar Sci. Conf. 6th*, p. 193–220.

Watkins N. D., Gunn B. N., and Coy-yll R. (1970) Major and trace element variations during the initial cooling of an Icelandic lava. *Amer. J. Sci.* **268**, 24–49.

Willis J. P., Ahrens L. H., Danchin R. V., Erlank A. J., Gurney J. J., Hofmeyr P. K., McCarthy T. S., and Orren M. J. (1971) Some inter-element relationships between lunar rocks and fines, and stony meteorites. *Proc. Lunar Sci. Conf. 2nd*, p. 1123–1138.

Willis J. P., Erlank A. J., Gurney J. J., Theil R. H., and Ahrens L. H. (1972) Major, minor, and trace element data for some Apollo 11, 12, 14, and 15 samples. *Proc. Lunar Sci. Conf. 3rd*, p. 1269–1273.

Wood D. A., Gibson I. L., and Thompson R. N. (1976) Elemental mobility during zeolite facies metamorphism of the tertiary basalts of eastern Iceland. *Contrib. Mineral. Petrol.* **55**, 241–254.

Proc. Lunar Sci. Conf. 8th (1977), p. 1751–1765.
Printed in the United States of America

Examination of the liquid line of descent of mare basalts in light of data from melt inclusions in olivine

PAUL W. WEIBLEN

Department of Geology and Geophysics, University of Minnesota, Minneapolis, Minnesota 55455

Abstract—Data on 226 melt inclusions and their host olivines (173 new analyses) are presented and interpreted in light of general liquid lines of descent and olivine fractionation. In contrast to the older conceptions of the liquid line of descent of basaltic liquids on an FeO/(FeO + MgO) vs. SiO_2 variation diagram, the melt inclusions in olivine along with those in ilmenite and late-stage mesostasis in mare basalts define a complex trend. It is complicated by the effects of ilmenite, plagioclase, and pyroxene fractionation within the melt inclusions in olivine, the fact that most of the basalts were multiply saturated during olivine fractionation and the phenomenon of late-stage silicate immiscibility. The first two factors can account for the observed marked divergence of the melt inclusion data from simple early olivine fractionation trends. The scatter in the data eliminates resolution of the different mare basalt types on the basis of the silica and FeO/(FeO + MgO) variation in the melt inclusion data from olivine.

Homogenization experiments on the melt inclusions in olivine would eliminate the effects of daughter mineral fractionation and could provide useful data on the actual liquid line of descent. These data might make it possible to resolve differences in trends for the different mare basalts, which in turn would make it possible to use melt inclusions in soil fragments to relate the fragments to specific mare basalt types.

INTRODUCTION

THE CRYSTALLIZATION SEQUENCES of mare basalts have been relatively firmly established on the basis of: (1) experimental studies (see ref. in Kesson and Lindsley, 1976), (2) textural studies of the rocks (Papike *et al.*, 1976), and (3) data from silicate melt inclusions (Weiblen and Roedder, 1976). In contrast, the nature of the actual liquid line of descent of the different mare basalt types (Papike *et al.*, 1976) is not well defined except for limited analyses of the melt products from experimental studies of the crystallization sequences (e.g., Longhi *et al.*, 1975) or late-stage silicate immiscibility (Roedder and Weiblen, 1971; Rutherford *et al.*, 1974). The recognition of possible early silicate immiscibility (20% melt) by Hollister and Crawford (1977) in Apollo 15 sample 15382 raises further questions about the actual liquid line of descent in the mare basalts. In this paper the general trend of the liquid line of descent of mare basalts as defined by melt inclusions is discussed in terms of the variation in silica and the Fe/(FeO + MgO) ratio. The data are contrasted with bulk rock compositional variations and previous conceptions of the evolution of basaltic liquids. In addition, one aspect of the early part of the liquid line of descent of mare basalts, the effect of olivine fractionation, is evaluated and contrasted with data from melt inclusions in olivine and with bulk rock compositions. In a companion paper (Roedder and Weiblen, 1977) the late part of the liquid line of descent represented by the mesostasis is characterized and the origin of variations in compositions examined in light of melt inclusion studies.

Current Status of the Data Base for Melt Inclusions in Olivine

The data base for melt inclusions in olivine now consists of 226 analyses of melt inclusions and hosts in Apollo 11, 12, 15, 16, and 17 as well as Luna 16 and 20 samples (Table 1).[1] Of these, 173 are new analyses. The reasons for analytical scatter in the analyses of melt inclusions are discussed in a companion paper (Roedder and Weiblen, 1977) and in Roedder and Weiblen (1971).

General Trend of the Liquid Line of Descent of Lunar Rocks and Melt Inclusions in Olivine on FeO/(FeO + MgO) vs. SiO_2 Variation Diagrams

The experimental work of Osborn (1959) in the FeO, Fe_2O_3, MgO, SiO_2 system represents a landmark in characterization of the liquid line of descent of basaltic melts. As a result of Osborn's (1959) work the variation diagram of FeO/(FeO + MgO) ratio vs. SiO_2 has been widely used in the literature (e.g., Carmichael, 1964). The diagram illustrates differences between distinctive terrestrial magma differentiation trends which are due to both bulk composition and controls on oxygen fugacity, as shown in Fig. 1. Until the recent recognition of silicate immiscibility (McBirney, 1975) the *calculated* liquid line of descent of the Skaergaard magma was considered to represent the general trend of tholeiitic basalts. Gaps in trends such as that shown by the 1965 Makaopuhi lava lake data on Hawaiian tholeiite were considered to represent sample bias rather than a fundamental break in the trend. The variation in bulk rock composition of mare lavas and a variety of Apollo 14 materials are shown in Fig. 2. This figure illustrates two relatively well-established characteristics of mare lavas: (1) the similarity of Apollo 11 and 17 mare lava compositions with regard to silica content and a small but systematic variation in FeO/FeO + MgO ratio; (2) the similarity of Apollo 12 and 15 basalts with both higher silica content and a wider range of FeO/FeO + MgO variation than the Apollo 11 and 17 rocks.

[1]The 226 analyses, along with our 518 previously-published and 484 new analyses on melt inclusions in other phases and ~544 not directly related to melt inclusion analyses are listed in Table 1. Each of the 1772 entries has a sequential tape number, a reference number (the unpublished analyses bear reference numbers 261–744), and are grouped into categories based on the nature of the sample analyzed, reference where published, or both. In order to save space, Table 1 has been archived. See NAPS document no. 03095 for 113 pages of supplementary material. Order from ASIS/NAPS c/o Microfiche Publications, P.O. Box 3513, Grand Central Station, New York, NY 10017. Remit in advance for each NAPS accession number. Institutions and organizations may use purchase orders when ordering, however, there is a billing charge for this service. Make checks payable to Microfiche Publications. Photocopies are $28.25. Microfiche are $3.00 each. Outside the United States and Canada, postage is $3.00 for a photocopy or $1.00 for a fiche. The data can also be obtained as a magnetic tape or on punched cards from the authors.

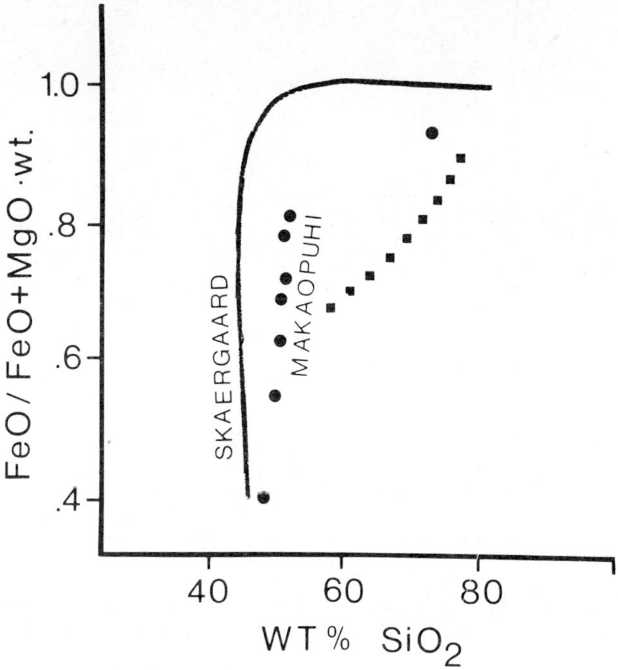

Fig. 1. FeO/(FeO + MgO) vs. SiO_2 variation diagram for terrestrial basalts. The solid line is the general trend of *calculated* liquid compositions of the Skaergaard and the dashed line the trend of calc-alkaline lavas. The former is presumed to represent a general basaltic trend formed during crystallization under conditions of constant bulk composition and the latter under externally controlled oxygen fugacity (modified from Osborn, 1959). The solid dots are the electron microprobe analyses of glasses in the 1965 Makaopuhi lava lake samples (Wright and Weiblen, 1968). No intermediate compositions between 55 and 75% silica were found in the Makaopuhi samples.

The variation in Apollo 14 KREEP rocks and associated materials is shown (only) to illustrate that as complete a differentiation trend as the Makaopuhi trend (but also with a break in the SiO_2 content at high FeO/FeO + MgO ratios) can be found in the lunar samples. The significance of this is beyond the scope of this paper and will be presented elsewhere (Weiblen and Roedder, in prep.).

In Fig. 3 the record of the melt compositions in early crystallizing olivine is contrasted with the late-stage differentiates. Despite the scatter which will be discussed below, the olivine melt inclusion data define a slightly positive trend which lies to the right of the bulk compositions of the rocks. This is consistent with olivine fractionation as shown in Fig. 4. The diagram gives the *illusion* that olivine fractionation alone could give rise to compositions that would split into high-silica and high-iron melts. Because silicate immiscibility does not appear to have occurred until over 90% of the melt was crystallized, it is clear that ilmenite, pyroxene, and plagioclase crystallization was involved. Pyroxene and

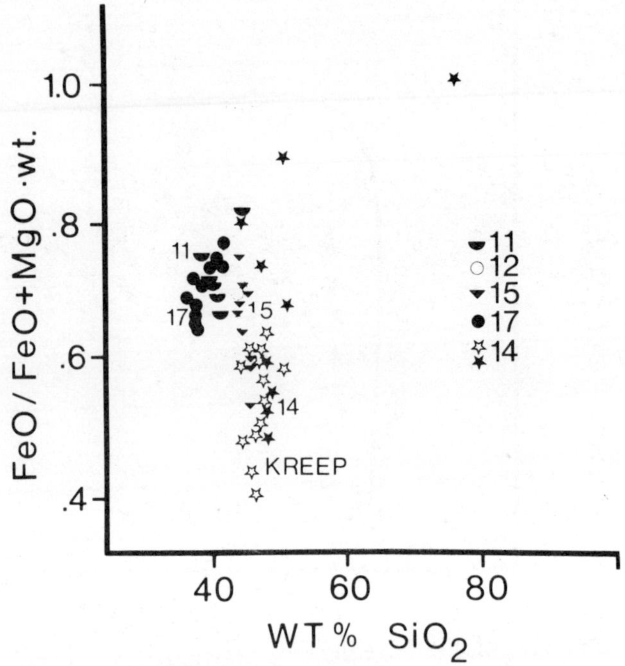

Fig. 2. FeO/(FeO + MgO) vs. SiO_2 variation diagram for Apollo rocks. The limited trends of Apollo 11 (hemispheres) and 17 (solid dots) are distinct from the Apollo 12 (open circles) and 15 (triangles) rocks. Apollo 14 materials include miscellaneous breccia and glass compositions (solid stars) and KREEP basalts (open stars). The Apollo 14 trend is similar to the Makaopuhi trend (Fig. 1). The data and their references for this figure are from Papike *et al.* (1976).

plagioclase crystallization however would produce only a small negative slope in the trend and only fractionation of pyroxene would increase the FeO/FeO + MgO ratio. Ilmenite on the other hand would have a different effect on compositional parameters, sharply increasing SiO_2 and sharply decreasing the FeO/FeO + MgO ratio. (The actual amount of MgO would also be decreasing during ilmenite crystallization due to pyroxene crystallization.)

The wide scatter of melt inclusion compositions in ilmenite is shown in Fig. 5. During ilmenite, pyroxene, and plagioclase crystallization the possibility for inhomogeneity to develop in local volumes of melt gradually increases as the residual melt becomes more viscuous and diffusion rates decrease. However, as found, the scatter should lie between the two immiscible melts.

The presumed high-iron immiscible melt composition of Hollister and Crawford (1977) plots at the lower end of the FeO/(FeO + MgO) trend of the high-iron melts (Fig. 3). Possibly immiscibility occurs for certain bulk compositions only after olivine and minor ilmenite, pyroxene, and plagioclase fractionation.

In summary the probable liquid line of descent of mare basalts is far more

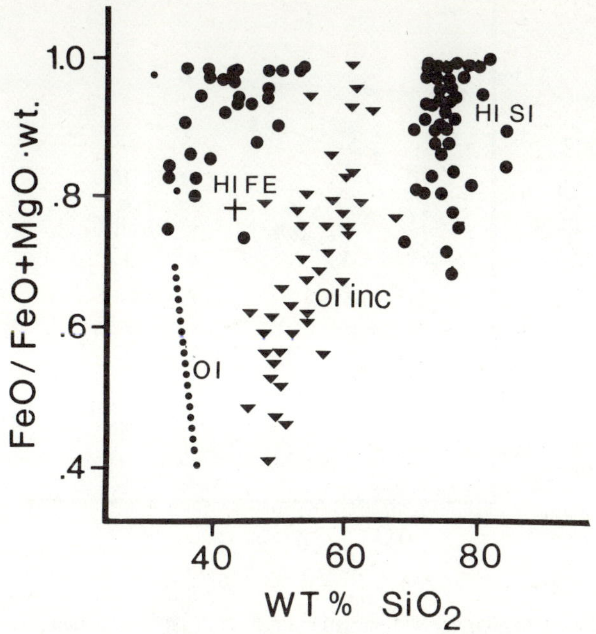

Fig. 3. FeO/(FeO + MgO) vs. SiO_2 variation diagram for melt inclusions in olivine (solid triangles), their host olivines (small dots), high-iron immiscible silicate glasses (large dots, $SiO_2 < 60\%$) high-silica immiscible silicate glasses (large dots, $SiO_2 > 60\%$). The melt inclusion data on melt inclusions in olivine are shown in greater detail in Figs. 7–11. Presumably the two immiscible glasses could be derived from some of the melt inclusion compositions which plot between the two conjugate glasses. See text for further discussion. The early high-iron glass of Hollister and Crawford (1977) (Cross) might have been produced by a composition similar to the melt inclusions with FeO/(FeO + MgO) ratios of about 0.7. The melt inclusion data with FeO/(FeO + MgO) ratios <0.5 are from Apollo 16 high-alumina basalt 62295 and are shown only for comparison. Data are listed in Table 1. See Footnote 1.

complex than is suggested by the older versions of terrestrial data and their interpretations. Based on the data in Fig. 5, the early trend controlled by olivine is marked by a slight positive slope. Variations in ilmenite, pyroxene, and plagioclase crystallization would produce widely varying compositions but silicate immiscibility eventually occurs. At what stage it is not now possible to predict, except for the restricted compositions investigated experimentally (Roedder and Weiblen, 1970; Rutherford *et al.*, 1974). The high-silica immiscible melt compositions difine a slightly negative trend which might reflect the effect of ilmenite crystallization prior to the onset of immiscibility or a unique crystal fractionation within the immiscible high-silica melt itself (Roedder and Weiblen, 1977).

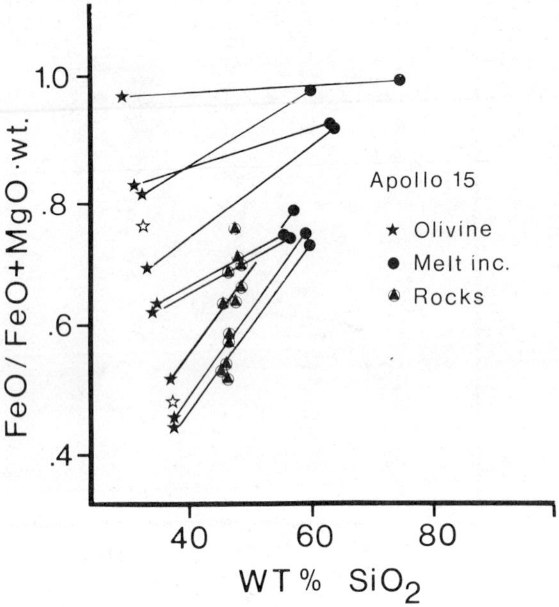

Fig. 4. FeO/(FeO + MgO) vs. SiO_2 variation diagram of Apollo 15 melt inclusions, host olivines, and rocks. The tie lines are drawn between inclusion and host olivine. Such straight lines on a plot of ratios versus an element do not intersect the bulk composition. As can be seen, the olivine data shows a slight curvature (this is more obvious on Figs. 6–10). Thus a straight line connecting melt and olivine in the pure olivine system will not intersect the bulk composition. In the basalt system the bulk rock must, however, lie between the olivine and its equilibrium melt composition as the data shows. Fractionation of equilibrium olivine (surface equilibrium) will produce a series of olivine and melt pairs which form a regular ruled surface on this plot. The Apollo 15 data approach this. Fractionation of disequilibrium composition olivines could account for the scatter, but the reasons for dispersion of the data are more complex as discussed in the text.

Olivine Fractionation and Observed Trends of Individual Mare and Hawaiian Basalts

The general trend of the melt inclusion data shown in Fig. 3 raises two questions. One, how does this trend compare with calculated trends of different basalt types and two, are there specific trends distinguishable for each of the mare basalt types?

The calculated liquid lines of descent for equilibrium, incremental olivine fractionation of the average compositions of the different mare lava compositions defined by Papike *et al.* (1976) are shown in Fig. 6. The calculated trends in Fig. 6 are based on fractionation of 2.5 wt.% increments of equilibrium composition olivine calculated on the basis of the experimentally determined melt-crystal partition coefficient (K_D) of 0.3 for iron and magnesium:

$$K_D = \frac{X_{Fe} \text{ in olivine}}{X_{Fe} \text{ in glass}} \times \frac{X_{Mg} \text{ in glass}}{X_{Mg} \text{ in olivine}},$$

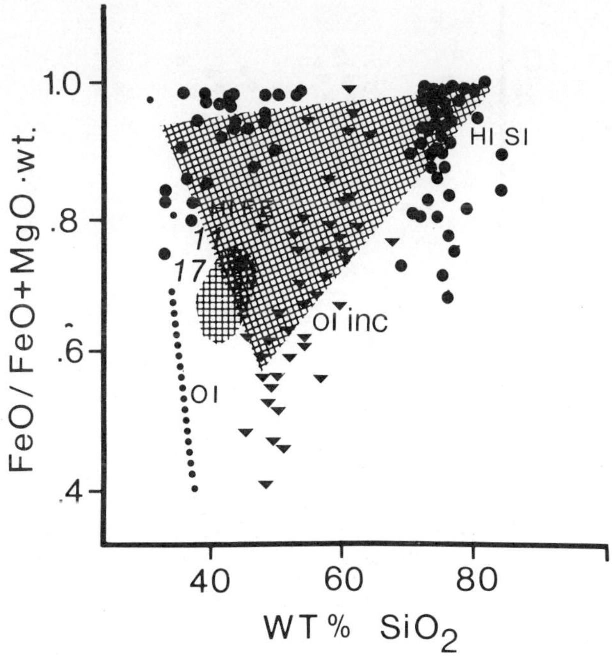

Fig. 5. FeO/(FeO + MgO) vs. SiO_2 variation diagram of all melt inclusions and glass compositions in mare basalts. Same symbols as Fig. 3. Apollo 11 and 17 rock compositions are outlined by elliptically shaped cross-hatched pattern. Wide scatter of melt inclusion compositions in ilmenite are confined to the triangular cross-hatched pattern. Note that the ilmenite data do not extend to as low FeO/(FeO + MgO) ratios as the olivine data, which is consistent with olivine preceding ilmenite in the crystallization sequence (Weiblen and Roedder, 1976). See text for further discussion. All data listed in Table 1 for the melt inclusions and their hosts. Bulk compositions from Papike *et al.* (1976).

(Roeder and Emslie, 1970; Mysen, 1975; Walker *et al.*, 1976). If the range in reported silica contents of these different basalts in not due to analytical error then it is clear from this figure that the different mare basalt types cannot be related to each other by simple olivine fractionation. Furthermore, if the variations in composition of the olivine melt inclusions were *dominated* by olivine control they would define separate trends on the FeO/(FeO + MgO) vs. SiO_2 variation diagram.

The melt inclusion data from olivine in Apollo 11 low K, Apollo 12, 15, and 17, and the 1965 Makaopuhi lava lake basalts are shown in Figs. 7–11, respectively. The data for these figures are listed in Table 1 and the ranges, averages, and standard deviations are listed in Table 2. One general characteristic of these data is that it all plots to the right of the calculated equilibrium fractionation trends. Different increments of fractionation of *equilibrium* composition olivine will only move data points along the calculated trends not to the right. Fractionation of disequilibrium composition olivine will move compositions to the

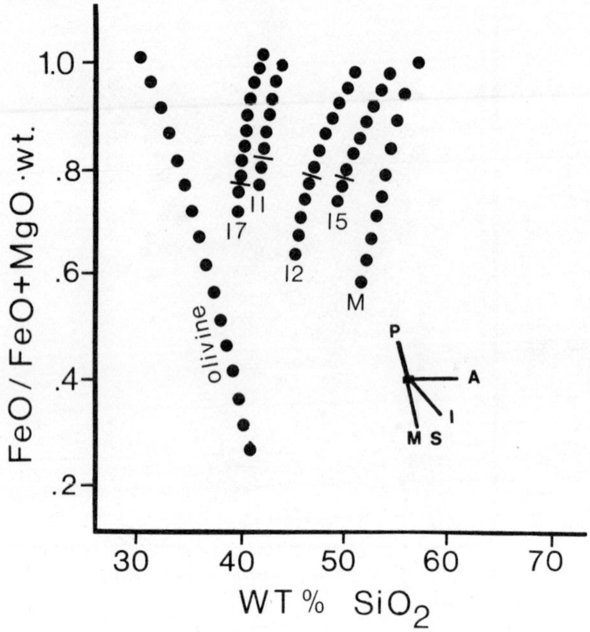

Fig. 6. FeO/(FeO + MgO) vs. SiO_2 variation diagram for olivine fractionation trends of Apollo and Hawaiian lavas. Trends were calculated using a value of 0.3 for the partition coefficient for iron and magnesium between melt and crystals (see text). Data points represent 2.5 wt.% repetitive subtraction of equilibrium composition olivine from the initial compositions. The different trends from left to right show the effect of increases in SiO_2 and are calculated from averages given by Papike *et al.* (1976). Apollo 11, 12, 15, 17, and Hawaiian (M) (Makaopuhi lava lake, Wright and Weiblen, 1968). All the trends are slightly convex away from the ordinate. Tie lines (not shown) define ruled surfaces. The data points represent the maximum amount of olivine that can be fractionated from each of the compositions. The bars normal to the trends indicate the composition of the lavas after subtraction of the amount of modal olivine reported by Papike *et al.* (1976) for each of the basalts. The general displacement of the olivine fractionation trends by the fractionation of pyroxene (p), ilmenite (i), armalcolite (a), metallic iron (m), and iron sulfide (s) are indicated for a hypothetical melt composition. On this and Figs. 7–11, the same symbols have been used for both olivine and melt compositions as a matter of convenience.

upper right as shown in Fig. 4 if the olivines are more magnesian and the lower right if they are less magnesian than the equilibrium composition. Mass balance calculations show, however, that ilmenite, pyroxene, and plagioclase are all required to produce the observed melt inclusion compositions. These phases, along with early armalcolite and chrome-spinel, are all involved to various degrees in the evolution of the lava compositions during olivine crystallization for all the mare basalt types.

In addition to changes in lava compositions prior to trapping, melt inclusion compositions are of course affected by crystallization of daughter minerals. Metallic iron, iron sulfide, ilmenite, and plagioclase have all been identified as

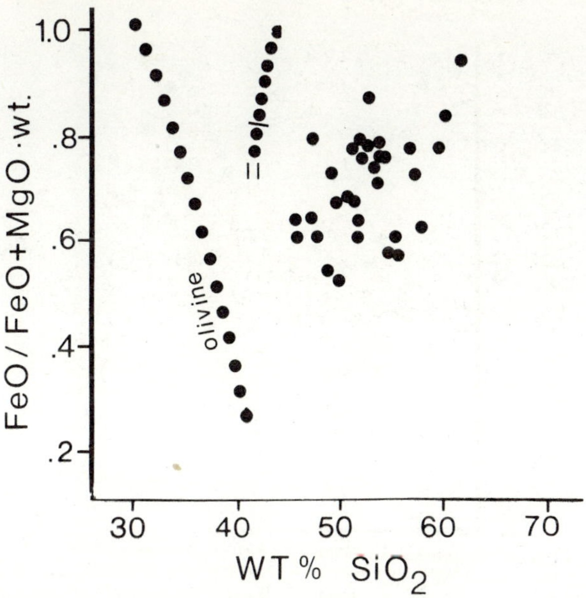

Fig. 7. FeO/(FeO + MgO) vs. SiO_2 diagram for calculated olivine fractionation trends and olivine melt inclusion data of Apollo 11 low K basalts (Data from Table 1; samples listed in Table 2). Note the data are significantly displaced from the calculated olivine control trend. As discussed in the text, this must reflect significant ilmenite fractionation (see Fig. 6).

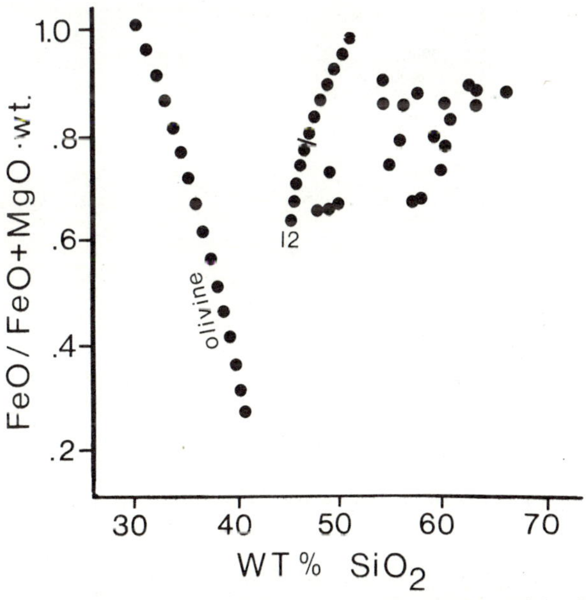

Fig. 8. FeO/(FeO + MgO) vs. SiO_2 variation diagram for calculated olivine fractionation trends and data from melt inclusions in olivine in Apollo 12 basalts (See Table 2). The melt inclusion data are displaced to the right of the olivine fractionation trend but do not show as strong an effect of ilmenite fractionation (not displaced to the *lower* right) as the Apollo 11 samples. This reflects the lower TiO_2 content of the Apollo 12 basalts.

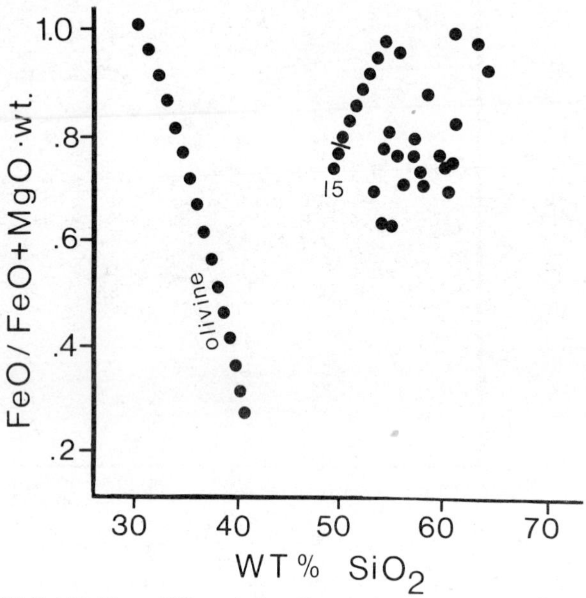

Fig. 9. FeO/(FeO + MgO) vs. SiO_2 variation diagram for calculated olivine fractionation trends and melt inclusion data from Apollo 15 basalts (See Table 2). The effect of ilmenite fractionation is apparent but less pronounced than for the Apollo 11 melt inclusions. The scatter could be due to disequilibrium olivine fractionation. (Fig. 4).

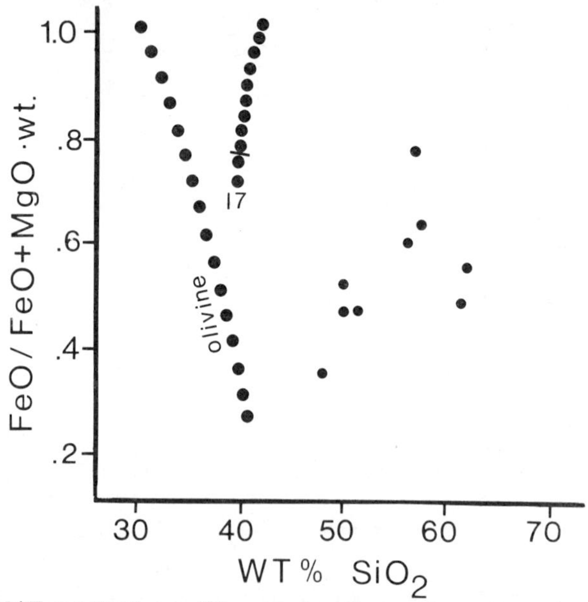

Fig. 10. FeO/(FeO + MgO) vs. SiO_2 variation diagram for calculated olivine fractionation trends and melt inclusion data from Apollo 17 basalts (See Table 2). These data show the greatest effect attributable to ilmenite fractionation of all Apollo basalts. Compare Figs. 7–9.

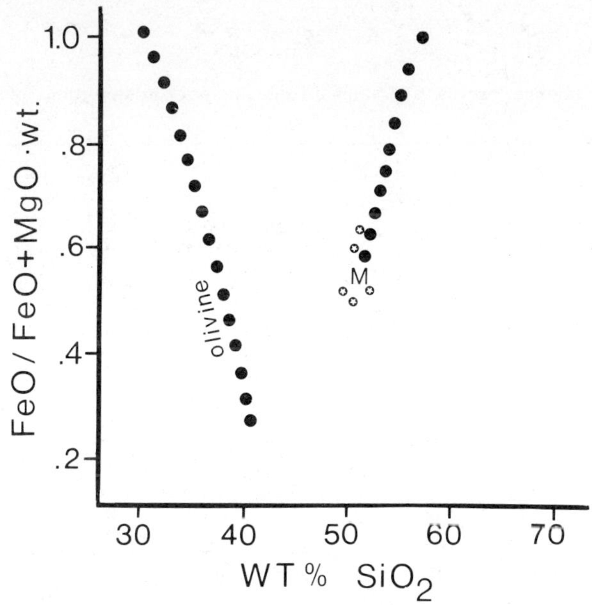

Fig. 11. FeO/(FeO + MgO) vs. SiO_2 variation diagram for calculated olivine fractionation trends and melt inclusion compositions of the Hawaiian Makaopuhi lava lake samples. The melt inclusion data (open circles) coincide more closely with the calculated trend than is the case with the Apollo samples. (Compare Figs. 7–10.) The lower FeO/(FeO + MgO) ratios may reflect compositions of melts trapped prior to eruption and at an earlier stage of olivine fractionation than the observed bulk composition.

daughter minerals in melt inclusions in olivine (Roedder and Weiblen, 1971). Similarly fine to coarse pyroxene is found in most melt inclusions in olivine (Roedder and Weiblen, 1977). Analyses of melt inclusions in olivine in which daughter minerals are fortuitously absent in the volume analyzed could account for the dispersion of the data shown in Figs. 7–10.

The effects of daughter mineral fractionation are illustrated in Fig. 6. It is evident that metallic iron, iron sulfide, armalcolite, chrome-spinel, and ilmenite could all be involved in the observed dispersion of the melt inclusion data. The relative abundance of these phases indicates that ilmenite is the principal control. If in any of the mare basalt types, olivine dominated in the evolution of the liquid over any appreciable interval of crystallization, one would expect the melt inclusion data to extrapolate toward the olivine fractionation trends. Significantly only the Apollo 12 and to a lesser degree the Apollo 15 data show a hint of this behavior (Figs. 8 and 9). This is consistent with the interpretation that these basalts have been affected by olivine fractionation (Kesson and Lindsley, 1976, p. 363). It is interesting to note that the terrestrial Makaopuhi lava lake melt inclusion data (Fig. 11) do cluster around the olivine control trend. In this case, crystallization occurred at sufficiently high oxygen fugacity to

Table 2. Maxima, minima, ranges, and standard deviations of compositions of melt inclusions in olivine.

Number	(1)	(2)	(3)	(4)	(5)	(6)	(7)	(8)	(9)	(10)
SiO_2	61.90	39.30	51.88	5.33	56.00	46.00	52.04	3.91	75.00	43.30
Al_2O_3	26.00	7.20	14.50	4.75	23.00	10.80	17.70	4.61	21.80	3.51
FeO	20.70	1.77	9.06	3.89	12.90	.56	6.08	4.44	27.60	2.93
MgO	15.20	.14	4.31	3.07	8.74	1.46	4.17	2.86	16.60	.80
CaO	21.90	9.57	16.38	2.82	19.40	10.30	15.62	4.33	15.90	3.46
Na_2O	.96	0	.29	.24	.54	.12	.29	.17	.89	.14
K_2O	1.01	0	.24	.26	.54	.12	.26	.17	6.40	0
TiO_2	9.80	.35	3.01	1.66	3.19	1.13	2.01	.75	5.60	.29
P_2O_5	0	0	0	0	0	0	0	0	.23	0
MnO	1.14	.07	.32	.26	.21	.05	.12	0	.53	0
Cr_2O_2	1.37	0	.10	.24	0	0	0	0	.15	0
WFEMG	.58	.93	.68		.60	.28	.59		.62	.79
Total	101.55	97.81	100.09	.91	100.73	96.28	98.28	1.90	102.08	97.52

Number	(11)	(12)	(13)	(14)	(15)	(16)	(17)	(18)	(19)	(20)
SiO_2	56.95	6.66	76.70	40.40	57.88	7.76	61.90	40.70	55.40	6.96
Al_2O_3	12.62	3.68	21.50	6.03	13.71	2.46	16.40	5.63	11.63	2.77
FeO	9.61	5.29	29.40	1.75	9.20	4.95	24.50	5.70	9.38	5.89
MgO	4.90	3.82	14.00	0	2.85	2.76	21.70	4.68	8.17	5.73
CaO	12.56	2.94	18.30	1.53	12.32	3.80	17.90	3.02	13.47	4.21
Na_2O	.40	.20	1.31	0	.33	.33	.77	0	.32	.28
K_2O	.58	1.47	6.88	0	.63	1.60	.09	0	.04	0
TiO_2	1.77	1.20	6.71	.60	2.16	1.26	2.78	1.17	1.95	.65
P_2O_5	.02	0	.25	0	.09	0	.05	0	.01	0
MnO	.20	.14	.44	0	.13	.10	.36	.08	.19	.10
Cr_2O_2	.01	0	1.30	0	.10	.22	.53	0	.13	.20
WFEMG	.66		.68	1.00	.76		.53	.55	.53	
Total	99.62	1.02	102.29	94.85	99.41	1.41	101.46	99.63	100.68	.52

(1) Data are listed in the order: maxima, minima, mean, and standard deviations for each mare basalt group.

(1–4) Data for 33 analyses of Apollo 11 low K basalt 10020,41,51,82,107.

(5–8) Data for 5 analyses of Apollo 11 low K basalt 10062,39.

(9–12) Data for 22 analyses of Apollo 12 basalt 12035,23; 2 analyses of 12018,83, and 1 analysis of 12036,8.

(13–16) Data for 4 analyses of Apollo 15 basalt 15532,11; 11 analyses of 15545,64 (olivine basalt); 9 analyses of 15555,34 (olivine basalt); 2 analyses of 15602,30; and 10 analyses of 15598,11.

(17–20) Data for 3 analyses of Apollo 17 basalt 71135,28 and 6 analyses of 78505,65.

(2) Table 1 contains 3 additional analyses of Luna 16 and 20; 4 analyses of 62295; and 2 analyses of 14321,25.

(3) WFEMG refers to FeO/(FeO + MgO) in weight percent.

(4) Total refers to totals of the original data.

eliminate metallic iron as a phase. Furthermore, ilmenite follows plagioclase and pyroxene in the crystallization sequence (Wright and Weiblen, 1968).

Displacement of the Apollo 11 and 17 melt inclusion data to lower Fe/(Fe + Mg) ratios and higher silica than the olivine control trends is quite marked. Armalcolite is an important phase in these basalts, but because of the similarity in Fe and Mg compositions of this phase and the bulk rock (El Goresy *et al.*, 1974, p. 641; Papike *et al.*, 1976, p. 480) fractionation of armalcolite cannot cause the observed scatter. Again ilmenite is the most likely dominant control. If the scatter is related to daughter mineral fractionation, the fact that the data do not extrapolate toward the olivine control trends supports the experimental and textural observations that ilmenite crystallized in significant amounts before or during olivine crystallization (Papike *et al.*, 1976) in these basalts.

In summary, the answer to the first question posed above is that the data on melt inclusions in olivine are highly divergent from simple trends predicted by olivine fractionation. The answer to the second question is that the dispersion of the data from individual lava types eliminates resolution of individual trends of the liquid line of descent.

Conclusions

It is clear that the actual liquid line of descent of the different mare basalts based on current experimental data and data from melt inclusions is far more complex than that of terrestrial basalts based on pre-Apollo data.

As discussed above, the data on melt inclusions in olivine are disappointing in terms of providing straightforward data and simple interpretations of the early liquid line of descent of the different mare lava types. The observed textures and compositions, however, are characteristics of the lavas which reflect unique cooling histories and these data are useful in themselves. There is, however, still a way in which the melt inclusions can be used to further our knowledge of the liquid line of descent and the details of the origin of the petrologically and geochemically important late-stage mesostasis. Melt inclusions can be homogenized in heating experiments to provide an analyzable sample of the *true* composition of the melt when it was trapped (Roedder and Weiblen, 1970). Compositional data from such experiments on melt inclusions in olivine will still reflect the effects of ilmenite, pyroxene, and plagioclase fractionation, but of the *bulk* melt rather than the *trapped* melt.

Homogenization of melt inclusions in heating experiments is laborious, but it provides a means of obtaining equilibrium melt compositions as well as data on crystallization temperatures and rates (Roedder and Weiblen, 1970). Only one analysis of an homogenized melt inclusion in olivine has been published (Roedder and Weiblen, 1970). This analysis (from Apollo 11 low K basalt 10020) corresponds essentially to the bulk rock composition, which suggests that the scatter in the 10020 melt inclusion data could be due solely to daughter mineral fractionation. More analyses of this type are obviously needed if we wish to sort out the effects of daughter mineral fractionation and ascertain if melt inclusions in olivine record significant and reliable information on the liquid line of descent of the different mare basalt types. If this is the case, data from melt inclusions in olivine would make it possible to define the magma types of soil fragments, relate them to known magma types, and aid in a systematic search for undetected magma types in the Apollo and Luna soil and core samples (Vaniman and Papike, 1977). Analyses of homogenized melt inclusions are also needed to relate data from olivines in a meaningful way to the problem of the relationship between early olivine fractionation and the eventual compositional variations in the late-stage differentiates and the problem of the stage at which silicate immiscibility occurs in the different mare basalt types.

Acknowledgments—I am indebted to Michael Cohen for able assistance in the computer graphics and to Elizabeth Ihrig for assistance in editing and typing the original manuscript. The final manuscript benefited from reviews by Edwin Roedder, Maria Crawford, and Michael Rhodes. Support for this study from National Aeronautics and Space Administration grant NSG-9050 is gratefully acknowledged.

References

Carmichael I. S. E. (1964) The petrology of Thingmuli, a tertiary volcano in eastern Iceland. *J. Petrol.* **5**, 435–460.

El Goresy A., Ramdohr P., Medenbach O., and Bernhardt H. J. (1974) Taurus-Littrow TiO_2-rich basalts: Opaque mineralogy and geochemistry. *Proc. Lunar Sci. Conf. 5th*, p. 627–652.

Hollister L. S. and Crawford M. L. (1977) Evidence for a direct genetic connection between Apollo 15 KREEP and Apollo 12 and 15 basalts (abstract). In *Lunar Science VIII*, p. 452–454. The Lunar Science Institute, Houston.

Kesson S. E. and Lindsley D. H. (1976) Mare basalt petrogenesis—A review of experimental studies. *Rev. Geophys. Space Phys.* **14**, 361–373.

Longhi J., Walker D., Grove T. L., Stolper E. M., and Hays J. F. (1974) The petrology of the Apollo 17 mare basalts. *Proc. Lunar Sci. Conf. 5th*, p. 447–469.

McBirney A. R. (1975) Differentiation of the Skaergaard intrusion. *Nature* **253**, 691–694.

Mysen B. (1975) Partitioning of iron and magnesium between crystals and partial melts in peridotite upper mantle. *Contrib. Mineral. Petrol.* **52**, 69–76.

Osborn E. F. (1959) Role of oxygen pressure in the crystallization and differentiation of basaltic magma. *Amer. J. Sci.* **257**, 609–647.

Papanastassiou D. A., Wasserburg G. J., and Burnett D. S. (1970) Rb–Sr ages of lunar rocks from the Sea of Tranquillity. *Earth Planet. Sci. Lett.* **8**, 1–19.

Papike J. J., Hodges F. N., and Bence A. E. (1976) Mare basalts: Crystal chemistry, mineralogy, and petrology. *Rev. Geophys. Space Phys.* **14**, 475–540.

Roedder E. and Weiblen P. W. (1970) Lunar petrology of silicate melt inclusions, Apollo 11 rocks. *Proc. Apollo 11 Lunar Sci. Conf.*, p. 801–837.

Roedder E. and Weiblen P. W. (1971) Petrology of silicate melt inclusions, Apollo 11 and Apollo 12 and terrestrial equivalents. *Proc. Lunar Sci. Conf. 2nd*, p. 507–528.

Roedder E. and Weiblen P. W. (1977) Compositional variation in late-stage differentiates in mare lavas, as indicated by silicate melt inclusions. *Proc. Lunar Sci. Conf. 8th*. This volume.

Roeder P. L. and Emslie R. F. (1970) Olivine-liquid equilibrium. *Contrib. Mineral. Petrol.* **29**, 275–289.

Rutherford M. J., Hess P. C., and Daniel G. H. (1974) Experimental liquid line of descent and liquid immiscibility of basalt 70017. *Proc. Lunar Sci. Conf. 5th*, p. 569–583.

Walker D., Kirkpatrick R. J., Longhi J., and Hays J. F. (1976) Crystallization history of lunar picritic basalt sample 12002: Phase-equilibria and cooling-rate studies. *Bull. Geol. Soc. Amer.* **87**, 646–656.

Weiblen P. W. and Roedder E. (1976) Compositional interrelationships of mare basalts from bulk chemical and melt inclusion studies. *Proc. Lunar Sci. Conf. 7th*, 1449–1466.

Wright T. L. and Weiblen P. W. (1968) Mineral composition and paragenesis in tholeiitic basalt from Makaopuhi lava lake, Hawaii (abstract). *Geol. Soc. Amer. Spec. Paper* **115**, 242–243.

Vaniman D. R. and Papike J. J. (1977) The Apollo 17 drill core: Chemistry and stratigraphy of monomineralic fragments and the discovery of a new very low Ti (VLT) mare basalt (abstract). In *Lunar Science VIII*, p. 952–954. The Lunar Science Institute, Houston.

Proc. Lunar Sci. Conf. 8th (1977), p. 1767–1783.
Printed in the United States of America

Compositional variation in late-stage differentiates in mare lavas, as indicated by silicate melt inclusions

EDWIN ROEDDER

959 National Center, U.S. Geological Survey, Reston, Virginia 22092

PAUL W. WEIBLEN

Department of Geology and Geophysics, University of Minnesota, Minneapolis, Minnesota 55455

Abstract—Approximately 1000 electron microprobe analyses (484 previously unpublished) of silicate melt inclusions and of their host and associated phases in lunar mare basalts are presented (in archival form). Only about 100 analyses are of inclusions in olivine—the rest are mainly from late stages of crystallization, and hence provide some insight into the late differentiation trends for these magmas. There are three compositional types of late-stage melts: immiscible "high-Fe" and "high-K" melts, and a possibly contemporaneous but enigmatic "low-K" melt. The factors affecting the interpretation of these data are reviewed, particularly the fragmental nature of the record and crystallization of the trapped melt. The compositional data differ from sample to sample, thus permitting a grouping and recognition of some rock types; they verify but do not explain the previously reported strongly bimodal distribution of K_2O values for inclusions in ilmenite; and they delineate some differentiation trends (e.g., increasing SiO_2, K_2O, and FeO/FeO + MgO), but some unanswered questions remain. Thus, silicate liquid immiscibility seems to occur at ~2% K_2O, but the stage cannot be specified exactly, and the origin of the low-K, high-Si melt inclusions in ilmenite reported earlier remains obscure.

INTRODUCTION

WE HAVE MADE 977 electron microprobe analyses of melt inclusions and of their host and associated phases in lunar mare basalts, of which 484 are new; the rest have been published (Roedder and Weiblen, 1970, 1971, 1972a,b, 1974, 1975; Powell and Weiblen, 1972; Weiblen and Roedder, 1973). As a result of the sample allotments, this work covered a wide variety of mare samples; the main goal was to investigate the liquid line of descent for lunar mare magmas through analyses of silicate melt inclusions.

Although volumetrically insignificant on the moon, the late-stage differentiates are extremely important in problems of lunar chronology. The "magic" component in lunar soils has been discussed by several authors (e.g., Papanastassiou *et al.*, 1970; Evensen *et al.*, 1974). Evensen (pers. comm., 1977) has reviewed the data on this "magic" component and shows that: (1) it is found in all soils sampled on the moon, (2) it consists of varying proportions of two fractions, one derived from the local rocks, and the other derived from rocks older than the mare lavas, (3) it was transported and mixed in local soils as *fragmental* material, and (4) it occurs in increasing amounts in the smaller size fractions of the soils. Thus beyond the considerations of petrologic curiosity, it is important to characterize and understand the origin of the mesostasis in the mare lavas. For the latter problem, we must know the details of the liquid line of descent in the mare lavas which leads to these compositions. The recognition of possible early silicate immiscibility (>20% melt) by Hollister and Crawford (1977) in an Apollo 15 sample gives additional impetus to define the liquid line of descent in mare lavas.

In this paper, we review the factors that must be considered in interpreting data from silicate melt inclusions, present and discuss the new analytical data, and attempt a synthesis of the old and new data. In particular, we characterize the late-stage mesostasis and discuss problems of its origin. In a companion paper (Weiblen, 1977), the early part of the liquid line of descent is examined in light of data from melt inclusions in olivine.

INTERPRETATION OF DATA FROM SILICATE MELT INCLUSIONS

Although silicate melt inclusions provide some types of data not obtainable from any other source, they are not a panacea for problems in petrogenesis. A series of factors that can cause difficulty or ambiguity in interpretation has been discussed by Roedder (1976) and, for the most part, need not be repeated here. However, three are particularly pertinent to the lunar data:

1. *The fragmental nature of the record*

In addition to the problem inherent in the nature of the lunar rock sampling, it is necessary to remember that inclusions represent defects in the otherwise orderly process of growth of host crystals. As such, they can form by many different mechanisms, but in the lunar samples, most melt inclusions have been trapped by three processes: *First*, early skeletal or tubular growth may be enclosed by later solid growth, trapping melt. This may result in a single, relatively large inclusion, as in some "cored phenocrysts" of plagioclase and pyroxene, where the composition, nucleation, and growth rate were favorable (e.g., see Donaldson *et al.*, 1975), but such large inclusions are relatively uncommon. Lunar olivines, however, commonly contain one or more inclusions that have an almost random spatial distribution, and have no noticeable bias toward occurrence in the cores. We presume that these inclusions are also the result of skeletal growth on crystal edges at various stages, as is sometimes noted in terrestrial olivine phenocrysts. Some terrestrial olivines commonly have axially arrayed inclusions (Roedder and Weiblen, 1971, their Fig. 1). We do not know what variables cause this difference. The only other lunar mineral that has consistently trapped melt inclusions by skeletal growth is ilmenite, which sometimes contains as much as 50% extraneous material (e.g., 10047). Most such inclusions consist of single crystals, optically continuous with nearby crystals and hence presumably connected with them in the third dimension, but inclusions representing the trapping of a homogeneous melt are also common in ilmenite (Roedder and Weiblen, 1970, their Figs. 2 and 4). *Second*, the enclosure of another crystal by the growing host crystal may trap some melt as well. The most common examples are melt inclusions attached to spinel euhedra enclosed within early olivine. *Third*, immiscible liquid droplets present within the liquid from which the host crystal is growing may adhere to the crystal surface and become trapped. By our interpretation, this last process may explain the very common occurrence of a row of tiny inclusions of glass of potash granite composition near the margin of many of the crystals of plagioclase, pyroxene, and other late phases in most of the mare basalts (e.g., Roedder and Weiblen, 1970, their Figs. 14, 15, and 17; 1971, Fig. 21). Earlier crystal growth is free of such inclusions, presumably because the liquid was a single phase at that time. Several features of these inclusions seem to preclude an origin for them based on a sudden change in conditions causing skeletal growth and hence a row of inclusions (e.g., there is no preferential occurrence of such rows on the relatively fastest growing faces), and the chemical compositions of some of the host phases, such as ilmenite, which could hardly grow rapidly from a melt of potash granite composition.

We have previously shown (Weiblen and Roedder, 1976) that the crystallization sequence in mare basalts is spinel, olivine, ilmenite, and then plagioclase plus pyroxene, leading to a late-stage separation of two immiscible melts. We have found a few melt inclusions in spinel, and many in olivine, giving a reasonable sampling of melts through the early stages of crystallization. However, the bulk of the crystals formed from a mare magma are pyroxene and plagioclase, and except for the last stages of growth of these two, when two liquids were present, they seldom have trapped analyzable inclusions. Ilmenite crystallizes over a wide range, and contains abundant inclusions, but most of these are in the part that crystallized late, when the residual melt contained 60% or more of SiO_2. Therefore, melt inclusions present us with an incomplete record of the sequence of melts at best, since most are incorporated in either very early or very late phases.

A dilemma concerning the origin of the two grossly different inclusion compositions found in ilmenite (Roedder and Weiblen, 1975) is still unresolved. Both contain approximately 75% SiO_2, but whereas one group contains ~6% K_2O, the other averages near to 0% K_2O.

2. *The possible effects of cooling rate on late-stage melt composition*

J. E. Quick (pers. comm., 1977) has suggested that the composition of late-stage, high-silica melts might be a function, in part, of the cooling rate. Grove and Bence (1977) showed by experimental runs on a synthetic analog of the Apollo 15 quartz-normative basalt 15597 that the sequence of phases, and their compositions (particularly pyroxene) changed significantly with cooling rate, from 0.5 to 150°C/hr. These differences affected the distribution of both major and minor elements between crystals and liquid, and hence could be expected to affect the composition of the late-stage liquids.

The optimum test of this effect on late-stage melts would be to compare the late-stage melt compositions in pairs of rocks of identical bulk composition but differing only in cooling rate. There is a range of grain sizes among the samples we have examined, presumably covering a rather broad range of cooling rates (Usselman *et al.*, 1975), but we cannot know whether the cooling rate has been a significant factor in the late-stage melt composition since the bulk compositions are far from identical. A high cooling rate would also favor an increased compositional range for the residual melts due to local crystal-liquid equilibria.

3. *Crystallization of the trapped melt*

Except for those with very high-silica content, most lunar inclusions have crystallized to some degree. If this crystallization involves only deposition of the host mineral on the walls, or epitaxial growth of a single daughter crystal (e.g., an ilmenite plate on the olivine host (Roedder and Weiblen, 1970), the residual melt represents the original melt minus the amounts of the crystallized phases. The amount of crystallization of the host phase on the walls can be large (~30%; Anderson, 1974). As many crystalline inclusions are essentially opaque, some may have no *visible* daughter ilmenite or other coarse phases, i.e., none intersecting the polished surface, simply because of the plane of the random cut. As a result of this crystallization, on any type of variation diagram, the composition of the liquid at the time of trapping should fall along the control line between the present composition of the inclusion and the appropriate crystallized phase(s).

If, as is common, the residual melt around the daughter crystal has also crystallized to a feathery intergrowth of pyroxene, etc. (Figs. 1–4), analyses of the mixture will show scatter reflecting this heterogeneity, adding to the scatter from loss due to crystallization of host or daughter phases and from systematic analytical errors (Albee *et al.*, 1977). Sometimes most of the crystals are fine enough ($\leq 2\ \mu m$; Figs. 1 and 3) for a defocused beam (10–20 μm) to give a reasonably good average analysis, but if the crystals are coarse (Figs. 2 and 4), and/or the inclusion is small, no analytical method on a single random section can yield a valid analysis. Unfortunately, in most inclusions analyzed, at least some of the crystals exceeded 2 μm.

New data obtained

To fill gaps in what must remain at best a fragmental record because of the nature of the original sampling on the moon, and to minimize the difficulties of interpretation inherent in the above, we selected additional slides that were optimum for the purpose, and made 484 new analyses. Most of these were of inclusions in olivine and ilmenite, in Apollo 11 low-K basalts 10020,51,−82, and −107; 10044,41; 10047,11; and 10062,39; Apollo 12 low-Ti olivine basalt 12035,23 and unclassified basalt 12012,7; Apollo 15 low-Ti olivine basalt 15545,64 and unclassified basalt 15598,11; and unclassified Apollo 17 basalts 71135,28 and 78505,65 (classifications from Papike *et al.*, 1976). Some of the results from these new analyses were presented in graphical form earlier (Weiblen and Roedder,

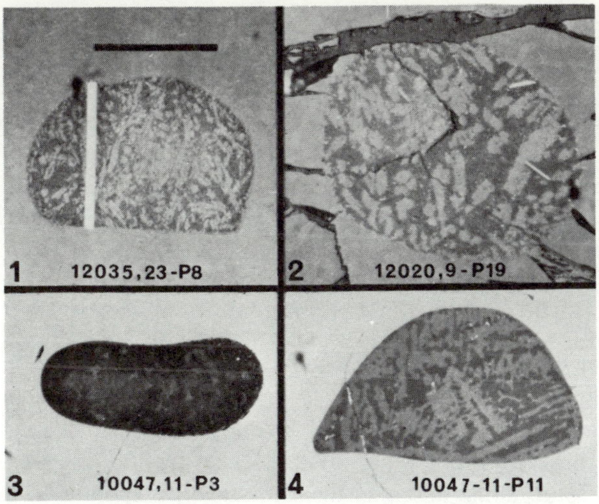

Figs. 1–4. Fine- and coarse-grained crystalline silicate melt inclusions in olivine and ilmenite from the samples indicated, in reflected plain light, all at the same magnification. (Scale bar = 20 μm.)

Fig. 1. Typical crystallized inclusion in olivine of low-Ti olivine basalt 12035. An epitaxial daughter crystal of ilmenite (vertical bar) is embedded in a granular matrix consisting of feathery pyroxene crystals in glass(?). Analysis of this matrix (Table 1, tape no. 77) shows 53.9% SiO_2 and 0.15% K_2O.

Fig. 2. Coarsely crystallized inclusion in olivine. This might well also have a hidden epitaxial daughter crystal like that in Fig. 1. The coarseness of the crystals would make defocused-beam analysis rather inaccurate without corrections (Albee *et al.*, 1977), even if this particular section were truly representative.

Figs. 3–4. Melt inclusions in ilmenite from low-K basalt slide 10047,11. The inclusion in Fig. 3 apparently consists of essentially all glass, and a few feathery crystals, whereas that in Fig. 4 apparently nucleated pyroxene earlier in the cooling cycle and hence is now rather coarse.

1976), but the full analyses are given for the first time as Table 1[1]. The methods of analysis and data reduction are the same as described earlier (Roedder and Weiblen, 1973, p. 1032) and are given in the notes to Table 1. Only about 100 of the analyses in Table 1 are of inclusions in early phases such as olivine; most of the other inclusions analyzed (in ilmenite or interstitial) are high in SiO_2 (>70%) and hence presumably trapped during late-stage crystallization.

---

[1]The 484 new analyses, along with our 1288 previously published analyses of lunar samples, (and including ~770 not directly pertinent to melt inclusions), are listed in Table 1. Each of the 1772 entries has a sequential tape number, a reference number (the unpublished analyses bear reference numbers 261–744), and are grouped into categories based on the nature of the sample analyzed, where published, or both. In order to save space, the table, consisting of 75 pages, has been archived. See NAPS document no. 03095 for 113 pages of supplementary material. Order from ASIS/NAPS c/o Microfiche Publications, P.O. Box 3513, Grand Central Station, New York, NY 10017. Remit in advance for each NAPS accession number. Institutions and organizations may use purchase orders when ordering, however, there is a billing charge for this service. Make checks payable to Microfiche Publications. Photocopies are $28.25. Microfiche are $3.00 each. Outside the United States and Canada, postage is $3.00 for a photocopy or $1.00 for a fiche. The data can also be obtained as a magnetic tape or punched cards from the authors.

OVERALL CHARACTERIZATION OF LATE-STAGE DIFFERENTIATES

It is easy to infer from the references to late-stage "mesostasis" in the literature on the mare basalts that there is essentially only one composition. In the following brief review, we want to emphasize that not only are there three quite distinct major compositional types, which we have termed high-Fe, high-K, and low-K, but that these three types are found in specific textural occurrences, and also reveal compositional variations within each for individual rocks or groups of rocks.

Textural and mineralogical features

The late-stage differentiates occur: (1) as silicate melt inclusions within crystals, (2) as interstitial mesostasis, in the mare basalts, and (3) as clasts in breccias. All three of these types of materials may contribute to the "exotic" component in lunar soils, and it is the late-stage differentiates that have yielded the minor but geochemically important lunar phases such as zirconolite, pyroxferroite, tranquillityite, apatite, whitlockite, and celsian, as well as the less exotic cristobalite, fayalite, and iron-rich pyroxenes. As a result of the absence of volatiles and hence high viscosity of these melts (other than the high-Fe), and the relatively high cooling rates, an appreciable fraction of these late-stage differentiates has remained as glass. In view of the possible significance, the partitioning of geochemically important elements between these various phases, and the melt, should be pursued experimentally.

The three major compositional types are not all found in each of the three major textural occurrences. Only the following have been found:

	High-Fe	High-K	Low-K
Melt inclusions in:			
Ilmenite	×	×	×
Pyroxene		×	
Plagioclase	×	×	
Pyroxferroite		×	
Cristobalite	×	×	
Mesostasis (interstitial inclusions)	×	×	
Clasts in breccias		×	

The absence of recognizable high-Fe melt as inclusions in pyroxene and pyroxferroite is expectable, since high-Fe melt crystallizes readily when in contact with such phases. Similarly, the post-shock metamorphism that has affected many of the breccias would also cause any clasts of high-Fe glass to crystallize and hence lose their identity. The occurrence of the low-K melts solely as inclusions in ilmenite is another perplexing, but possibly significant aspect to the problem of the nature and origin of this peculiar melt. Rare intergrowths of crystalline plagioclase and silica have been described in some

Apollo 17 breccias (Ryder *et al.*, 1975) that might represent the crystallized equivalents of low-K melts. The most notable occurrence of the high-K melt composition as clasts is the Apollo 14 breccias (Roedder and Weiblen, 1972, their Table 6), but they are also common in some Apollo 17 breccias, particularly 72255 (Stoeser *et al.*, 1974; Ryder *et al.*, 1975).

Bulk chemical composition

In Table 2, we list the average chemical composition of the three major chemical types of late-stage differentiates from Apollo 15 and 17 samples. These specific samples were used, rather than a grand average of all analyses, as there are significant differences between the data from various missions (e.g., Roedder and Weiblen, 1972b, p. 281). The averages in Table 2 are based on our previously published analyses (also included in Table 1). Analyses from individual rock groups, and particularly from individual inclusions, will differ from these averages, as detailed below, but most differences in averages are not large.

Grouping and Recognition of Rock Types on the Basis of Late-Stage Inclusion Analyses

Many classifications of mare basalts have been made on the basis of bulk chemistry, the most thoroughly documented being that of Papike *et al.* (1976). The differences in composition between these types, though analytically significant, are small. The late-stage melt inclusions, however, represent a sample of the residue after crystallization of ~95% of the original magma (Roedder and Weiblen, 1970). For those elements that are concentrated in the late-stage melts, absolute differences are enhanced. Thus inclusions from different magmas should show large relative differences, even though the differences in the original magma compositions were small. We have found that this is indeed true. Differences in the composition of the melt inclusions in the various samples and rock types are far larger than the expected analytical errors and are sufficient to permit recognition of some rock groups, at least, even as small fragments in breccias. This has the added advantage of requiring only nondestructive electron microprobe analyses, whereas bulk chemistry for either major or minor elements on a crystalline rock requires the destruction of an appreciable amount of sample, and hence would not be suitable for small fragments in breccias.

In order to recognize relationships between the many variables represented in Table 1 (sample number, rock type, mission, host mineral, stage of differentiation, and eight major elements), a variety of computer plots was made. These are far too numerous and large to be reproduced here, so we present only a few of them. The relationships are much more easily visualized by the reader, and can be presented much more briefly, when described in terms of how the points fall on a diagram, so we will use this artifice for presenting the salient points from other diagrams not reproduced here. Copies of these various diagrams are available from the authors.

Table 2. Average chemical composition of the three major types of late-stage differentiates. Data from Table 1. Number of analyses averaged given in parentheses.

	Apollo 17 high K (32)				Apollo 15 high Fe (8)				Apollo 17 low K (57)			
	Max.	Min.	Mean	S.D.	Max.	Min.	Mean	S.D.	Max.	Min.	Mean	S.D.
SiO_2	80.3	67.8	76.31	2.69	46.7	36.9	42.44	3.25	88.5	64.0	76.79	5.86
Al_2O_3	17.1	8.34	12.10	1.76	11.9	2.50	5.63	2.99	16.7	5.83	11.37	2.25
FeO	8.46	0.77	2.20	1.33	42.6	19.5	30.19	8.23	11.8	0.15	2.96	2.34
MgO	0.66	<0.05	0.23	0.20	5.30	0.34	1.77	1.72	2.19	<0.05	0.67	0.55
CaO	5.80	<0.05	1.80	1.40	20.9	9.90	13.00	3.69	10.6	2.57	6.66	2.14
Na_2O	1.43	<0.05	0.41	0.32	0.34	<0.05	0.05	0.10	1.91	<0.05	0.39	0.37
K_2O	10.30	3.98	5.95	1.47	1.52	0.01	0.33	0.52	0.36	<0.05	0.04	<0.10
TiO_2	2.19	0.05	0.68	0.42	7.10	3.50	5.13	1.18	1.76	0.14	0.84	0.30
P_2O_5	0.65	<0.05	0.05	0.14	2.05	0.13	0.73	0.71	0.11	<0.05	0.01	<0.10
MnO	0.11	<0.05	0.01	<0.10	0.47	0.28	0.40	<0.10	0.24	<0.05	0.02	<0.10
BaO	2.11	<0.05	0.13	0.40	—	—	—	—	—	—	—	—
Cr_2O_3	0.05	<0.05	<0.01	<0.10	0.18	<0.05	0.10	<0.10	0.07	<0.05	<0.01	<0.10
Totals	101.75	97.68	99.86	1.06	101.92	98.00	99.76	1.40	101.69	97.79	99.76	0.96

In Figs. 5 and 6, we illustrate a few of the differences between groups. In Fig. 5, for example, note that the points for 10062 are offset from those for 10044 and 10047. This difference is important, as 10044, 10047, and 10062 are *all* classified as Apollo 11 low-K basalts (Papike *et al.*, 1976). Although the grain size of 10062 is smaller than that of the other two, so the cooling rates were probably not the same, there are also differences in the bulk chemistry of these rocks, particularly in Si/Al and Fe/Mg. In Fig. 6 note that the average for Apollo 14 granitic clasts is distinct from the "granitic" differentiates found in the Apollo VHTi basalts. Similar distinctions were found between the "fields" of 10047 and all Apollo 15 high-K inclusions on plots of K_2O vs. SiO_2 and $FeO/(FeO + MgO)$ vs. SiO_2. Very few points have been excluded from these plots on the basis of the factors discussed above, but this exclusion results in a tighter clustering of points in each group.

A strong clustering of points is found on many of the other plots. Thus, on a plot of $FeO/(FeO + MgO)$ vs. SiO_2 for all high-K inclusions from all missions (304 analyses, including Luna 16 and 20), more than two-thirds of the points fell in an area that constituted only about 5% of the area necessary to include all points. The points representing averages for individual rock samples or rock groups show a much smaller compositional range than the individual analyses, of course, but even these averages reveal analytically significant differences. As an example, on Fig. 7 we plot K_2O vs. SiO_2 for the high-K inclusion averages (18) for various rock groups; these range from 5–9% K_2O and from 73 to 78% SiO_2. Similarly, in the various groups of low-K inclusions, although the variations in

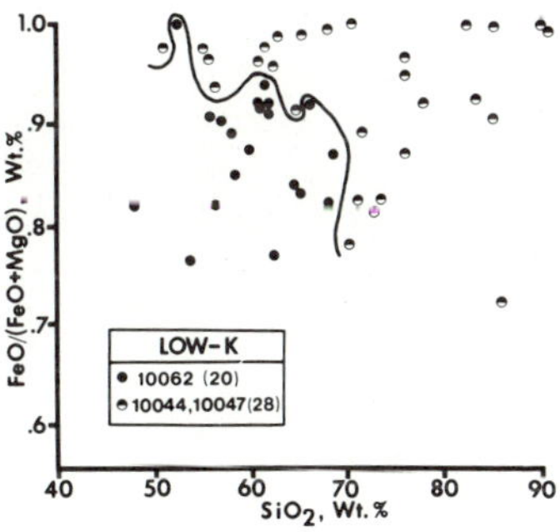

Fig. 5

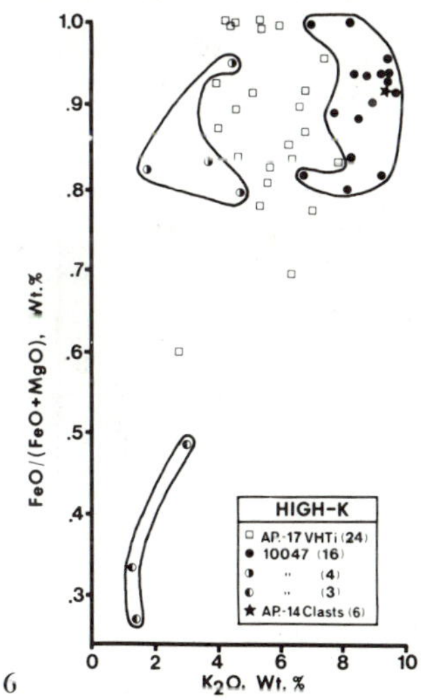

Fig. 6

Figs. 5–6. Plots of the composition of silicate melt inclusions in ilmenite from various samples, grouped as indicated for comparison purposes. The number of points plotted is given in parentheses. Data from Table 1. The lines outlining the "fields" are drawn solely as a visual aid.

K_2O and FeO/(FeO + MgO) are small, the CaO ranges from 5 to 8% and the SiO_2 from 74 to 81%. Yet each of the averages represents very sizeable numbers of samples.

We have no ready explanation for many features in such plots. Thus, the inclusions in some *individual* rock samples show significant subgroups. Twenty-three high-K melt inclusions were analyzed from slide 10047,11. Of these, 16 fall in a relatively tight cluster on all plots (e.g., Fig. 6), and the remaining seven, although essentially identical with the 16 in terms of occurrence and Si, Ca, and Na, are markedly different in other respects and fall in two separate subgroups. Such groups are too tightly clustered to have the differences assigned to simple random sampling and analytical errors; *ad hoc* explanations of these differences can usually be found, such as local equilibrium reflecting the crystallization of another phase nearby, but these explanations are not very satisfactory.

One characteristic trend was found—a small but significant *negative* correlation between K_2O and SiO_2 within various groups of samples. This is apparent in a plot of all high-K averages (Fig. 7), where the 18 points delineate a rough line with a negative slope. We believe this is only in part a result of the requirement of a negative correlation between any two major constituents, if all others are held constant. The points outlining such trends, which are based on data on the residual melts in a series of independent, different magmas, do not necessarily imply derivation of one inclusion composition from another, as might be assumed for a similar trend observed in a suite from a single sample.

BIMODALITY OF POTASSIUM VALUES

Roedder and Weiblen (1975) showed that 87 melt inclusions either in or interstitial to late ilmenite all had ~76% SiO_2, but they had a very strongly

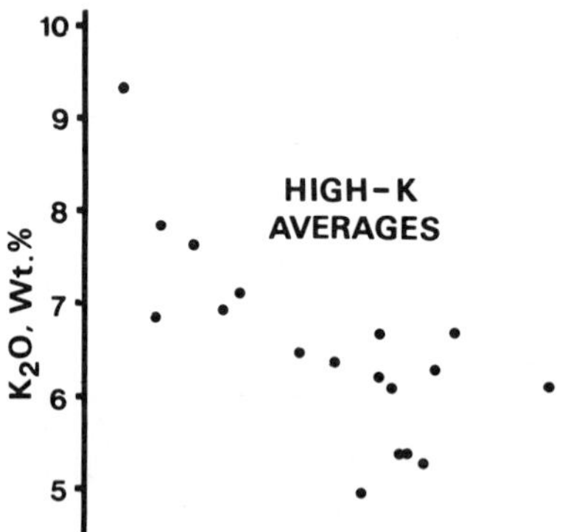

Fig. 7. Plot of the average compositions of high-K silicate melt inclusions (both in ilmenite and interstitial) from various individual samples or missions (data from Table 1). Individual points represent averages of 3–35 analyses, with an average of 11 analyses.

bimodal K_2O distribution: one group averaged 0.03% K_2O, whereas the other averaged 6.4%, and only three were in the range 0.15–4.05% K_2O. A new plot of K_2O for the 309 such inclusions in Table 1 (Fig. 8) shows that although there are now analyses in almost every K_2O concentration class interval, the bimodality is statistically secure, one group averaging ~6% and the other near 0% K_2O. Note that the average of all 309 analyses falls at ~2.8% K_2O, at the center of the low point on the histogram. We still do not have a satisfactory explanation for the origin of the extremely low-K melts. We have previously listed 13 suggested mechanisms (Roedder and Weiblen, 1975), but found them all unsatisfactory to some degree.

Delineation of Differentiation Trends

We have found a plot of K_2O vs. SiO_2 to be particularly illustrative of the compositional range of late-stage differentiates. In Fig. 9, we show K_2O vs. SiO_2 for our data on all types of inclusions, and in Fig. 10 we show an expansion of the low-K part of Fig. 9. Also plotted are the residual melts found by Rutherford *et al.* (1974) in laboratory fractionation studies on a synthetic composition

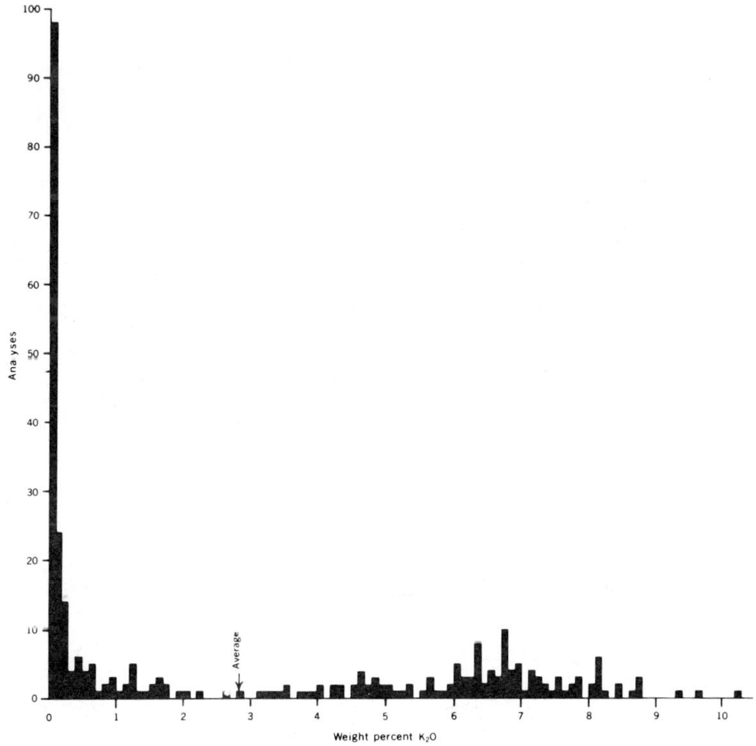

Fig. 8. Frequency diagram for K_2O contents of 309 inclusions (interstitial and in ilmenite) in Apollo 11, 12, 15, and 17 basalts. The data from individual missions differ significantly (Table 1), but all yield similar bimodal K_2O distributions.

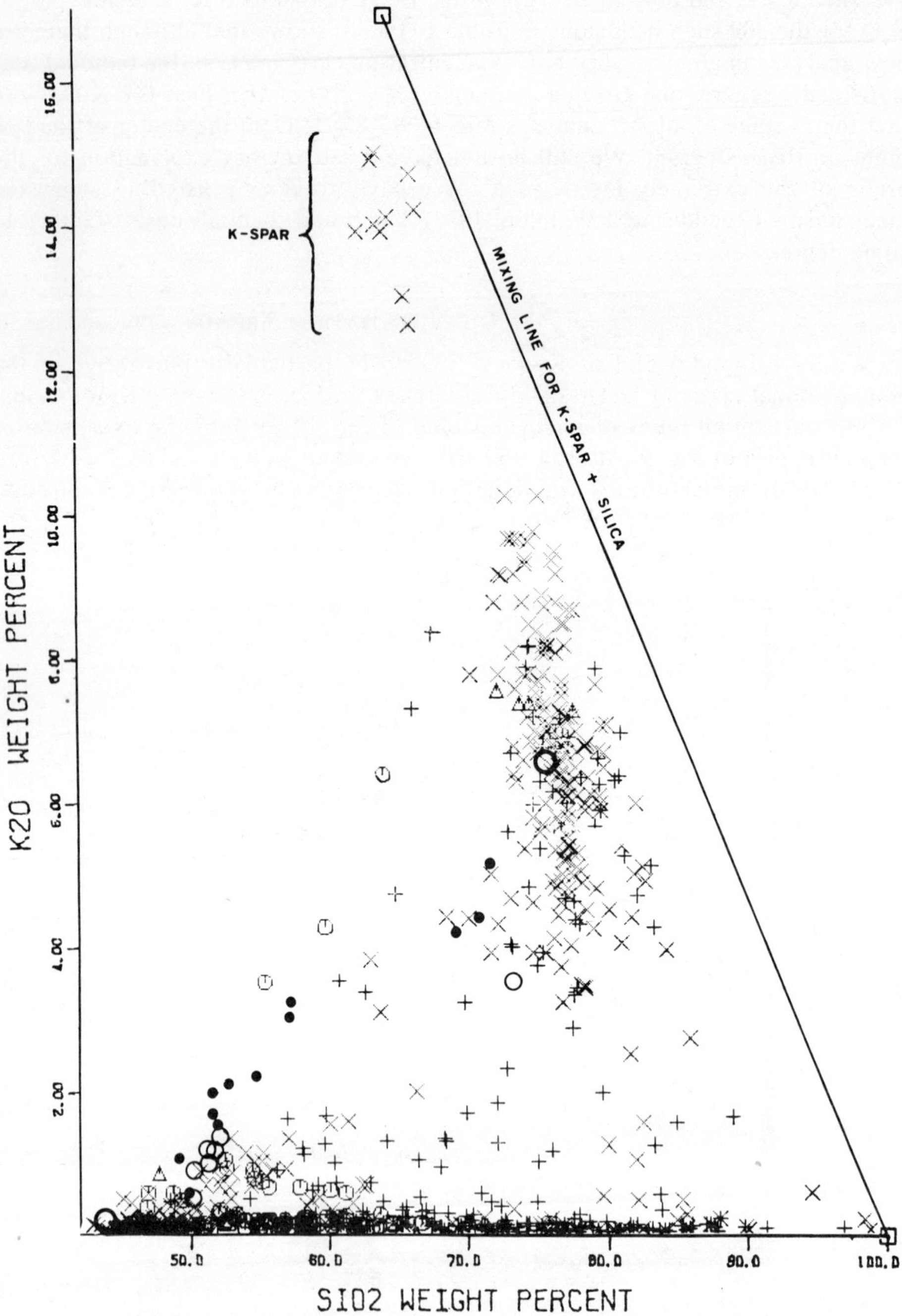
K2O WEIGHT PERCENT
SIO2 WEIGHT PERCENT
K-SPAR
MIXING LINE FOR K-SPAR + SILICA
2.00
4.00
6.00
8.00
10.00
12.00
14.00
16.00
50.0
60.0
70.0
80.0
90.0
100.0

duplicating 70017, and those found in some Hawaiian samples (taken from Weiblen and Roedder, 1976, p. 1455; Wright and Weiblen, 1968; Wright *et al.*, 1976).

Several features are evident on these two plots. *First*, although some ilmenite trapped early melts that were just as low in SiO_2 and K_2O as those trapped in olivine, and hence some of the ilmenite crystallized early, no inclusions with very high SiO_2 and only three with high K_2O were trapped in olivine. This suggests that crystallization of olivine generally ceased before such melts were formed. A more detailed discussion of the early differentiation trends shown on the lower left side of Figs. 9 and 10 is given in a companion paper by Weiblen (1977). It is also evident that there is no real difference in the compositional range for inclusions in ilmenite vs. those that are interstitial, except in the very high-K range. *Second*, a significant gap at 2–3% K_2O may possibly indicate liquid immiscibility, although this gap occurs over a range of SiO_2 values. All points from Rutherford *et al.* (1974) above 1.9% K_2O represent analyses of fine-grained *mixtures* of two immiscible melts (P. C. Hess, pers. comm., 1977). *Third*, if immiscibility involves simply the splitting of a liquid to form high-Fe and high-Si melts without simultaneous crystallization, then the parent liquid just before splitting has to lie along a line connecting the two compositions just after the liquid has split. As there are no analyses with 2–3% K_2O near the line connecting these two, either we have fortuitously failed to sample the liquid just before splitting, which seems unlikely in view of the number and SiO_2 range of our analyses, or some other mechanism must be involved. Since the melt probably arrived at the field of immiscibility as a consequence of crystallization, it is probable that additional separation of the immiscible melts also entails formation of a third phase, such as SiO_2, so that the trends would not be explicit on Figs. 9 and 10 without knowledge of the relative amounts of phases. *Fourth*, FeO/(FeO + MgO) generally increases with degree of crystallization. We have examined this ratio for the analyses plotted on Figs. 9 and 10 and find that it increases with SiO_2 and that on a plot vs. K_2O, low values for the ratio occur only with low values of K_2O. At high-silica values the ratio is variable; this is probably due to the fact that since most of the magnesia has been used up before the melt reaches even 60% SiO_2, formation of relatively minor amounts of crystals may effect large changes in the ratio. *Fifth*, we had

Fig. 9. K_2O vs. SiO_2 for all 826 selected melt inclusion analyses: octagons with vertical line = inclusions in olivine; triangles = inclusions in spinel, plagioclase, and pyroxene; (+) = inclusions in ilmenite; (×) = re-entrant and interstitial inclusions, plus ten K-spars (bracketed; these contain significant barium). For comparison, we include data on the residual melts obtained after partial crystallization in experimental runs on composition 70017 (solid circles; Rutherford *et al.*, 1974); a differentiated series of samples from Hawaii, ending in a granophyre (small open circles; Weiblen and Roedder, 1976, p. 1455; Wright and Weiblen, 1968; Wright *et al.*, 1976); and average high-Fe and high-K inclusions (Apollo 11; large circles). Only the top edges of the symbols for points along the abscissa show in this plot; the portion from 0–2% K_2O is expanded in Fig. 10. A mixing line between stoichiometric K-spar and SiO_2 (squares) is indicated.

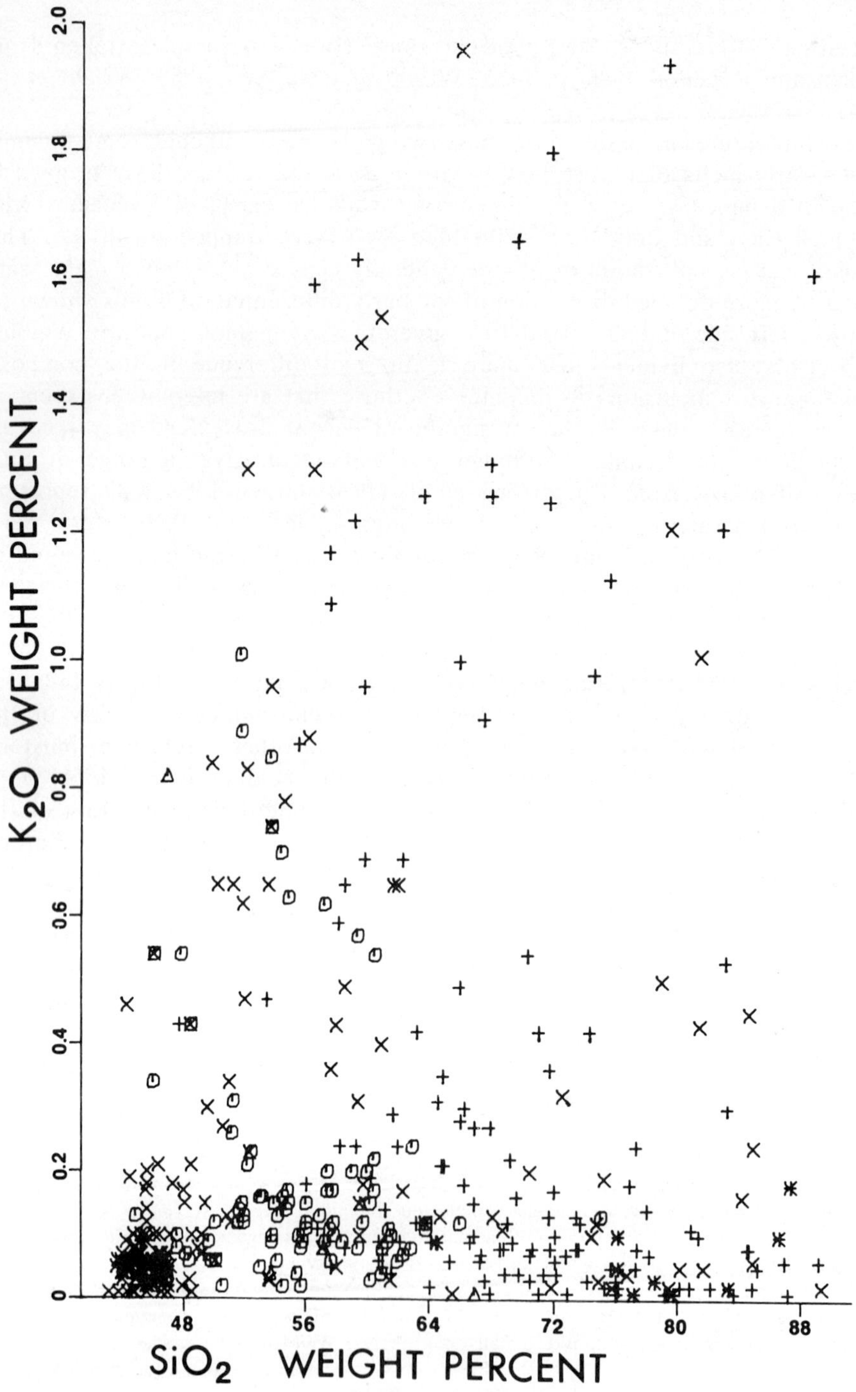

Fig. 10. K_2O vs. SiO_2 for all inclusion analyses with less than 2% K_2O (expanded from bottom edge of Fig. 9). Symbols as given in Fig. 9.

hoped that such plots would show a relatively linear differentiation trend, as would be required if all the inclusion samples were glassy and cogenetic. The spread of the data on Figs. 9 and 10 and similar plots presumably reflects the combination of different trends for the various original magma compositions, liquid immiscibility, local variations in crystal-liquid equilibria at the time of trapping, sample heterogeneity from crystallization, and analytical errors. The broad scatter in the high-K range and the extremely high SiO_2 values are probably due mainly to inhomogeneity from K-spar and silica minerals in the inclusions; these minerals cannot be easily distinguished from high-Si glass in reflected light or by cathodoluminescence in this kind of occurrence. As the high-Si, high-K inclusions contain only 70–90% normative K-spar plus silica, they do not fall on the K-spar/silica mixing line (Fig. 9) but are dispersed more or less parallel to it. *Last*, the trend from Rutherford *et al.* (1974) and from the Hawaiian samples fits a curved line between the bulk rock composition of mare basalts (which would plot in the lower left corner of Fig. 9) and our Apollo 11 high-K melt average. In theory, these high-K compositions can be derived from the bulk rock by crystallizing olivine, ilmenite, and pyroxene. Most inclusion compositions fall on the higher SiO_2, lower K_2O side of this line, however. As the parent magmas have very low K_2O, the crystals must have even lower K_2O to yield a net K_2O enrichment in the melt. Thus, 97% crystallization with 0% K_2O will leave 3% of melt enriched 33-fold in K_2O (e.g., from 0.04 to 1.32%). Rutherford *et al.* (1974) assumed 98.3% crystallization of K_2O-free phases (including plagioclase) to obtain the 3% K_2O melts. Many lunar plagioclases have enough K_2O to result in a negative slope on Fig. 9. We found ~20% modal plagioclase in 70017; this would have to average 0.25% K_2O for the calculated residual liquids to show 0% K_2O enrichment. Although analyses of plagioclase from a similar basalt, 75083 (Papike *et al.*, 1976) show 0.02–0.28% K_2O, it seems highly unlikely that the crystallization of plagioclase could keep the residual liquid depleted in K_2O to the extent shown by many points on Figs. 9 and 10. In fact, in 70035, we found that even plagioclase crystallizing from a *high*-K melt contained only 0.35% K_2O in the outermost zone. Thus the paradox remains: all available data point to K_2O enrichment in the residual melts, yet many glassy inclusions in ilmenite have low K_2O and high SiO_2, and no mechanism seems adequate to explain this fact (Roedder and Weiblen, 1975).

Conclusions

Many of the problems in interpretation and much of the scatter found in the composition of silicate melt inclusions in lunar samples can be attributed to the incomplete nature of the record and to crystallization after trapping. These sources of difficulty were minimized by the selection of appropriate sample material. The resulting data bank: (1) makes possible the grouping and recognition of some rock types, even in breccia fragments, on the basis of analysis of melt inclusions; (2) verifies the still-inexplicable bimodal distribution of K_2O values for high-Si melt inclusions in ilmenite; and (3) permits delineation of some differentiation trends, and, in particular, suggests the intervention of silicate liquid immiscibility when these residual melts have reached about 2% K_2O.

Acknowledgments—We are indebted to Michael Cohen for assistance in the computer graphics. The manuscript benefited from provocative reviews by Rosalind Helz, Bruce Lipin, L. S. Hollister, and J. E. Quick. Support for this study from National Aeronautics and Space Administration grants T-1170-A and NSG-9050 is gratefully acknowledged.

REFERENCES

Albee A. L., Quick J. E., and Chodos A. A. (1977) Source and magnitude of errors in "broad-beam analysis" (DBA) with the electron probe (abstract). In *Lunar Science VIII*, p. 7–9. The Lunar Science Institute, Houston.

Anderson A. T., Jr. (1974) Evidence for a picritic, volatile-rich magma beneath Mt. Shasta, California. *J. Petrol.* **15**, 243–267.

Donaldson C. H., Usselman T. M., Williams R. J., and Lofgren G. E. (1975) Experimental modeling of the cooling history of Apollo 12 olivine basalts. *Proc. Lunar Sci. Conf. 6th*, p. 843–869.

Evensen N. M., Murthy V. R., and Coscio M. R., Jr. (1974) Provenance of KREEP and the exotic component: Elemental and isotopic studies of grain size fractions in lunar soils. *Proc. Lunar Sci. Conf. 5th*, p. 1401–1417.

Grove T. L. and Bence A. E. (1977) Pyroxenes as recorders of lunar basalt petrogenesis revisited: An experimental study of pyroxene-liquid interaction (abstract). In *Lunar Science VIII*, p. 383–385. The Lunar Science Institute, Houston.

Hollister L. S. and Crawford M. L. (1977) Evidence for a direct genetic connection between Apollo 15 KREEP and Apollo 12 and 15 basalts (abstract). In *Lunar Science VIII*, p. 452–454. The Lunar Science Institute, Houston.

Papanastassiou D. A., Wasserburg G. J., and Burnett D. S. (1970) Rb–Sr ages of lunar rocks from the Sea of Tranquillity. *Earth Planet. Sci. Lett.* **8**, 1–19.

Papike J. J., Hodges F. N., Bence A. E., Cameron M., and Rhodes J. M. (1976) Mare basalts: Crystal chemistry, mineralogy, and petrology. *Rev. Geophys. Space Phys.* **14**, 475–540.

Powell B. N. and Weiblen P. W. (1972) Petrology and origin of lithic fragments in the Apollo 14 regolith. *Proc. Lunar Sci. Conf. 3rd*, p. 837–852.

Roedder E. (1976) Petrologic data from experimental studies on crystallized silicate melt and other inclusions in lunar and Hawaiian olivine. *Amer. Mineral.* **61**, 684–690.

Roedder E. and Weiblen P. W. (1970) Lunar petrology of silicate melt inclusions, Apollo 11 rocks. *Proc. Apollo 11 Lunar Sci. Conf.*, p. 801–837.

Roedder E. and Weiblen P. W. (1971) Petrology of silicate melt inclusions, Apollo 11 and Apollo 12 and terrestrial equivalents. *Proc. Lunar Sci. Conf. 2nd*, p. 507–528.

Roedder E. and Weiblen P. W. (1972a) Petrographic features and petrologic significance of melt inclusions in Apollo 14 and 15 rocks. *Proc. Lunar Sci. Conf. 3rd*, p. 251–279.

Roedder E. and Weiblen P. W. (1972b) Silicate melt inclusions and glasses in lunar soil fragments from the Luna 16 core sample. *Earth Planet. Sci. Lett.* **13**, 272–285.

Roedder E. and Weiblen P. W. (1973) Petrology of some lithic fragments from Luna 20. *Geochim. Cosmochim. Acta* **37**, 1031–1052.

Roedder E. and Weiblen P. W. (1974) Petrology of clasts in lunar breccia 67915. *Proc. Lunar Sci. Conf. 5th*, p. 303–318.

Roedder E. and Weiblen P. W. (1975) Anomalous low-K silicate melt inclusions in ilmenite from Apollo 17 basalts. *Proc. Lunar Sci. Conf 6th*, p. 147–164.

Rutherford M., Hess P. C., and Daniel G. H. (1974) Experimental liquid line of descent and liquid immiscibility for basalt 70017. *Proc. Lunar Sci. Conf. 5th*, p. 569–583.

Ryder G., Stoeser D. B., Marvin U. B., and Bower J. F. (1975) Lunar granites with unique ternary feldspars. *Proc. Lunar Sci. Conf. 6th*, p. 435–449.

Stoeser D. B., Marvin U. B., Wood J. A., Wolfe R. W., and Bower J. F. (1974) Petrology of a stratified boulder from South Massif Taurus-Littrow. *Proc. Lunar Sci. Conf. 5th*, p. 355–377.

Usselman T. M., Lofgren G. E., Donaldson C. H., and Williams R. J. (1975) Experimentally reproduced textures and mineral chemistries of high-titanium mare basalts. *Proc. Lunar Sci. Conf. 6th*, p. 997–1020.

Weiblen P. W. (1977) Examination of the liquid line of descent of mare basalts in light of data from melt inclusions in olivine. *Proc. Lunar Sci. Conf. 8th*. This volume.

Weiblen P. W. and Roedder E. (1973) Petrology of melt inclusions in Apollo samples 15598 and 62295, and of clasts in 67915 and several lunar soils. *Proc. Lunar Sci. Conf. 4th*, p. 681–703.

Weiblen P. W. and Roedder E. (1976) Compositional interrelationships of mare basalts from chemical and melt inclusion studies. *Proc. Lunar Sci. Conf. 7th*, p. 1449–1466.

Wright T. L., Peck D. L., and Shaw H. R. (1976) Kilauea lava lakes; natural laboratories for study of cooling, crystallization, and differentiation of basaltic magma. *Amer. Geophys. Union. Monograph No. 19*, p. 375–390.

Wright T. L. and Weiblen P. W. (1968) Mineral composition and paragenesis in tholeiitic basalt from Makaopuhi lava lake, Hawaii (abstract). *Geol. Soc. Amer. Spec. Paper 115*, p. 242–243.

Proc. Lunar Sci. Conf. 8th (1977), p. 1785–1807.
Printed in the United States of America

Redox states of Ti, Zr, Hf, Cr, and Eu in basaltic magmas: An experimental study

HENRY D. SCHREIBER

Department of Chemistry, Virginia Military Institute, Lexington, Virginia 24450

Abstract—Synthetic silicate melt analogs of basaltic magmas were used to study the dependence of the redox states of various multivalent elements (Ti, Zr, Hf, Cr, Eu) on the bulk melt composition, imposed oxygen fugacity and melt temperature. The synthetic iron-free melt compositions were represented by compositions in the forsterite-anorthite-silica and forsterite-anorthite-diopside systems. Experimental results allowed the assignment of a scale of relative reduction potentials of redox couples for these elements in the two synthetic silicate melts; in order of ease of reduction, Cr(III) → Cr(II) > Eu(III) → Eu(II) >>> Ti(IV) → Ti(III). Neither Zr(IV) nor Hf(IV) could be reduced to their trivalent states, even under very reducing conditions which simulated lunar magmatic conditions.

Terrestrial magmas contain Ti, Zr, Hf, Cr, and Eu almost exclusively in their corresponding oxidized redox states: tetravalent Ti, Zr, and Hf, and trivalent Cr and Eu. The presence of excess quantities of the oxidizing agent Fe(III) precludes the existence of appreciable amounts of reduced redox states such as Ti(III), Cr(II) and Eu(II) in terrestrial magmas. The experimental studies predict that lunar magmas possess essentially all their Hf and Zr as Hf(IV) and Zr(IV), 1–7% of their Ti as Ti(III), 40–80% of their Eu as Eu(II), and 70–95% of their Cr as Cr(II). Since the observed mineralogies and the elemental distribution patterns of these important geochemical elements in lunar basalts are partially dependent upon the redox state of the elements in the magmas from which the basalts crystallized, these estimates provide semi-quantitative constraints for models of the geochemical evolution of the moon.

INTRODUCTION

STUDIES INTO THE REDOX STATE of planetary magmas provide important constraints on the origin and development of planetary bodies (Sato, 1976). Differences in the redox state of lunar versus terrestrial magmas have been mainly ascribed to the much lower oxygen fugacity that prevailed during mare basalt crystallization compared to that of terrestrial basalt crystallization. One effect of the low lunar oxygen fugacity was the stabilization of reduced redox states of some multivalent elements in the lunar environment. Whereas important geochemical elements such as Fe, Ti, Cr, and Eu have existed almost exclusively as Fe(III)–Fe(II), Ti(IV), Cr(III), and Eu(III) ions in terrestrial magmas, lunar magmas contained Fe as Fe(II)–Fe(0) with possible small amounts of Fe(III) and significant concentrations of reduced ions such as Ti(III), Cr(II), and Eu(II) (e.g., Taylor, 1975). Exotic reduced redox states of Zr (e.g., Garg and Ehmann, 1976) and Nb (e.g., Erlank *et al.*, 1972) have also been postulated to have occurred in the lunar magmas under the extremely reducing conditions.

The occurrence of reduced redox states of important geochemical elements such as Fe, Ti, Cr, and Eu affected both the stability and chemistry of mare basalt minerals (Papike *et al.*, 1976). A schematic diagram which illustrates the relationship of the initial redox state of a typical crystallizing lunar magma and its subsequent mineralogy and elemental distribution is shown in Fig. 1. Since Fe(II), Ti(IV), and Cr(III) can be major constituents of important rock-forming minerals, the activities of each of these species in the initial magma individually regulate the solubility as well as the chemistry of their respective mineral phases. For example, the melt's Cr(III) concentration, not its total Cr concentration, may control the precipitation and composition of the spinel phase which crystallizes from the liquid (Schreiber and Haskin, 1976). While Cr(III) is a lattice-forming constituent of spinel and may be critical for the crystallization of this mineral, Cr(II) is not a building-block of spinel and will not dictate any constraints on spinel's crystallization. Ions of a particular redox couple also fractionate by trace-element substitution of different amounts of each ion into various minerals, since the crystal/liquid partition coefficients depend upon each ion's unique size, charge, and crystal field strength with respect to available crystal lattice sites. Eu(III)–Eu(II) fractionation by the preferential incorporation of Eu(II) into plagioclase is a classic illustration of this process. Whereas Eu(II) is readily accepted into plagioclase lattice sites, Eu(III) and the other trivalent rare earths are not. Since Eu is very sensitive to the crystallization of plagioclase, the Eu content with respect to the other rare earths in a particular rock provides valuable insight into the conditions which accompanied its crystal-

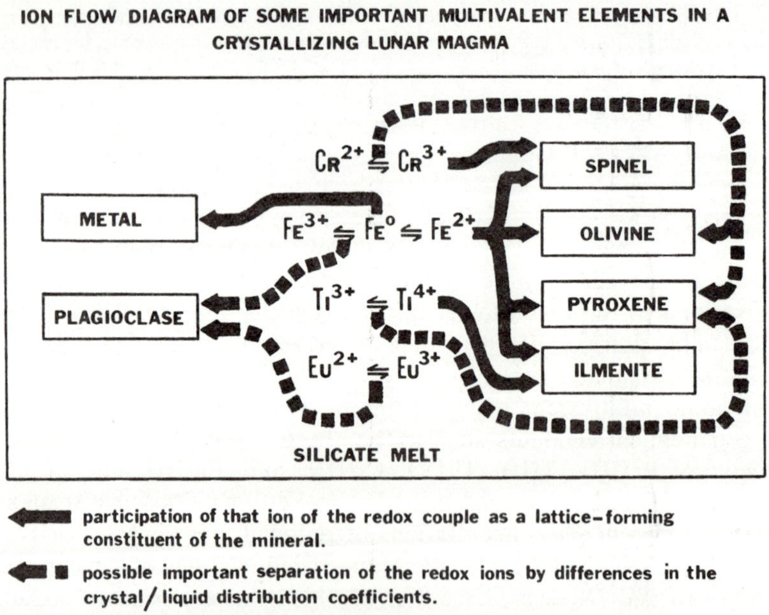

Fig. 1. A schematic illustration of the role of different redox states of Cr, Fe, Ti, and Eu in the crystallization of a typical lunar magma. Some important minerals as well as modes of ion fractionation were not included in the diagram for the sake of simplicity.

lization. Thus, the observed mineralogy and elemental distribution patterns in lunar basalts reflect the proportions of reduced and oxidized ions of Fe, Ti, Cr, and Eu which were present in the initial magma.

For a particular lunar basalt sample the different relative concentrations of Ti(III) with respect to Ti(IV), of Cr(II) to Cr(III) and of Eu(II) to Eu(III) in the various minerals can be viewed as possible records of the initial redox conditions of the magma from whence the rock evolved. However, complexities in the interpretation of such relic indicators are introduced by various crystallization models and by post-crystallization processes, such as subsolidus reductions, which have probably altered the original redox state (e.g., Haggerty, 1977). As an example of the latter situation, although Fe(0) is a ubiquitous component of lunar basalts, Fe(0) may not have been present in the initial magma, but may have been produced in the emplaced magma or resulting rock by many plausible mechanisms (e.g., Sato, 1976; Tsay and Bauman, 1977). Nevertheless, a determination of the conditions that existed at the time of lunar magma generation must await the experimental calibration of redox state ratios of multivalent elements in the initial magma and in early crystallizing phases such as olivine, pyroxene, and Fe–Ti oxides (Papike *et al.*, 1976). The purpose of this study is to quantify these redox state ratios of Ti, Zr, Cr, and Eu in synthetic silicate melt analogs of lunar basaltic magmas in order to decipher as unambiguously as possible this history of lunar basalts from their observed mineralogies and elemental distribution patterns.

Experimental Procedures and Results

The experimental approach

The nature of the Ti(IV)–Ti(III), Zr(IV)–Zr(III), Hf(IV)–Hf(III), and Eu(III)–Eu(II) redox equilibria in silicate melts was investigated in synthetic samples under controlled conditions. These experiments complement the previously studied Cr(VI)–Cr(III)–Cr(II) redox system under analogous conditions. Silicate melt samples containing a trace amount of an individual multivalent element were synthesized as a function of bulk melt composition, of melt temperature and of imposed oxygen fugacity.

Bulk compositions FAS and FAD (Table 1), both of which melt at about 1500°C, served as silicate melt models for basaltic magmas (Schreiber and Haskin, 1976). FAS and FAD are acronyms which represent compositions in the *f*orsterite-*a*northite-*s*ilica and *f*orsterite-*a*northite-*d*iopside systems respectively. Composition FAS simulates magmas important in lunar petrogenesis, while composition FAD is a diopside-rich silicate. These synthetic systems differ from actual magmas mainly by the replacement of the more realistic (Mg, Fe)-olivine component with the pure forsterite end member.

Melt temperatures of 1500°C and 1550°C were used in conjunction with experimentally imposed oxygen fugacities which spanned the range of lunar and terrestrial magmatic environments. In this paper the term oxygen fugacity will be used interchangeably with the symbol P_{O_2} in order to indicate the approximate equivalence of oxygen fugacity and partial pressure imposed on the samples.

Table 1. Bulk melt compositions of the synthetic systems with comparisons to the compositions of terrestrial and lunar basalts.

	Synthetic melts[1]		Natural basalts[5]	
Wt.% oxide	FAS[2]	FAD[3]	Earth	Moon
SiO_2	53.9	51.1	44–53	37–49
Al_2O_3	10.7	2.9	13–19	7–14
FeO	—	—	7–14	18–23
MnO	—	—	0.09–0.3	0.21–0.29
MgO	29.3	26.1	4–10	6–17
CaO	5.7	19.1	8–12	8–12
Na_2O	—	—	1.8–3.8	0.1–0.5
K_2O	—	—	0.3–1.9	0.2–0.3
TiO_2	(1.3)[4]	(1.3)[4]	0.9–3.3	0.3–13
Cr_2O_3	(0.1–0.6)	(0.1–0.8)	0.005–0.04	0.12–0.70
Eu(ppm)	(9300)	(10600)	1.9[6]	0.21–2.6

[1]Each synthetic melt contained only a single redox oxide, either Ti, Cr, or Eu oxide.

[2]53% forsterite, 27% anorthite, 20% silica by weight.

[3]25% forsterite, 6% anorthite, 69% diopside by weight.

[4]Approximate concentration of ZrO_2 or HfO_2 when used in place of TiO_2.

[5]10–90% frequency range of terrestrial continental basalts and range of lunar maria, from Taylor (1975).

[6]Haskin *et al.* (1968).

Sample preparation

Silicate melt samples were prepared from powdered glasses of bulk composition FAS or FAD which contained 0.5–1.2 wt.% of either TiO_2, ZrO_2, HfO_2, Eu_2O_3, Gd_2O_3 or Cr_2O_3. The procedure for the preparation is described in detail by Schreiber and Haskin (1976). That redox equilibrium was established in the silicate liquids was determined by time studies. After redox equilibrium in the melt was attained at the desired temperature and oxygen fugacity for a particular sample, the sample was rapidly quenched to a glass for subsequent analysis.

Inherent in this sample preparation procedure is the assumption that the redox equilibrium established in the melt is not perturbed by the quench. Although others (e.g., Ulmer *et al.*, 1976) have discussed the difficulties in preserving the actual redox state of a melt during a quench, circumstantial evidence of the validity of the quench assumption in this study is provided by: (1) duplicate samples which were quenched at different rates yielded identical redox states, and (2) the agreement of the results with theoretical models. In most cases the samples were quenched in less than a second by electrically dropping the samples from the furnace into a container of water.

Sample analysis

Similar procedures to those established by Schreiber and Haskin (1976) for Cr redox state determinations in glasses were employed for the determination of Ti, Zr, Hf, and Eu redox state concentrations in the quenched glasses.

Total trace element contents of the synthetic glasses were measured by standard electron microprobe techniques on polished chips of the glasses. Although the total trace element content of every sample was determined on the electron microprobe, the analyzed concentration always agreed within the accuracy of the analyses to its original concentration determined by the weight of the trace element oxide added to the bulk glass powder. Because of the mobility of Cr at high temperatures in the synthetic systems, the total concentrations of Cr in Cr-doped samples varied from sample-to-sample in the previous study (Schreiber and Haskin, 1976). As a check on sample homogeneity, major element concentrations (Ca, Al, Mg, and Si) were also determined for each glass sample.

Chemical redox titrations were used to measure the concentration of Cr(II) or Cr(VI) in Cr-doped glasses, of Ti(III) in Ti-doped glasses, of Zr(III) in Zr-doped glasses, of Hf(III) in Hf-doped glasses, and of Eu(II) in Eu-doped glasses. The method for the redox titrations followed the basic design of Close *et al.* (1958), Johnston (1965), and Close and Tillman (1969); specific procedures are described in Schreiber and Haskin (1976) and in more detail in Schreiber (1976).

In order to complement and confirm the redox concentrations obtained from chemical titrations, a portion of each glass sample was also analyzed by electron paramagnetic resonance (EPR) methods. Instrumental parameters and procedures are discussed in Schreiber and Haskin (1976); for quantitative comparison of samples, every analysis was normalized to a common set of EPR reference conditions. Typical X-band EPR spectra of Ti, Eu, and Cr in the synthetic glasses are compared in Fig. 2. The EPR method monitored the Cr(III) concentration in the sample through the intensity of the Cr(III) signal without interference from Cr(II) or Cr(VI), since these two Cr ions do not produce an EPR signal at ambient temperature. In Ti-doped glasses, Ti(III) gave a characteristic intense EPR band (e.g., Bell *et al.*, 1974) while Ti(IV) did not show any EPR signal. Since Zr and Hf are in the same periodic group as Ti, both Zr(III) and Hf(III) in the silicate glasses may display an EPR resonance analogous to Ti(III). No EPR signal was observed in any sample containing Zr or Hf in this study. However, although Zr(III) and Hf(III) may give EPR resonances, the lack of the resonances is not strong evidence for the absence of Zr(III) and Hf(III) in the melts because of matrix effects. The X-band EPR spectral pattern for Eu in the synthetic glasses was complex and was characteristic of Eu(II) in the samples; Eu(III) shows no EPR absorption under the experimental conditions. The Gd(III) EPR spectrum was similar to that of Eu(II) in pattern but the relative intensities of the corresponding peaks were different. Since the peak heights of the EPR signals only provided relative concentrations of the individual ions, the EPR measurements in this study were calibrated to absolute concentrations by the complementary redox titration results.

The sensitivity and accuracy of the calibration scheme for determining trace redox concentrations of Cr, Ti, and Eu in glasses are illustrated in Fig. 3, which correlates the EPR signal intensities with the concentrations of the corresponding redox states inferred from the chemical redox titrations. The well-defined calibrations of Fig. 3 attest to the reliability of the results, despite the small sample sizes and trace concentrations. The extremely low concentrations of Ti(III) in the samples necessitate a curve instead of a straight line for calibration in this system; the curve is probably due to small systematic errors at the detectability limit of the titration technique.

For the results reported in this study, concentrations of particular ions were defined by the EPR intensity measurements calibrated to absolute weight % units by the redox titrations as per Fig. 3. Use of the EPR measurements instead of the direct application of the redox titration determinations reduced the random scatter in the experimental points.

Experimental results

Chemical equations for the pertinent Cr, Ti, and Eu redox equilibria can be written according to the convention established by Johnston (1965) for trace elements in silicate melts. These equations,

$$4Cr_{(m)}^{3+} + 2O_{(m)}^{2-} = 4Cr_{(m)}^{2+} + O_{2(v)}, \quad (1)$$

$$4Ti_{(m)}^{4+} + 2O_{(m)}^{2-} = 4Ti_{(m)}^{3+} + O_{2(v)}, \quad (2)$$

$$4Eu_{(m)}^{3+} + 2O_{(m)}^{2-} = 4Eu_{(m)}^{2+} + O_{2(v)}, \quad (3)$$

explicitly demonstrate the dependence of the redox ion concentrations on the oxygen fugacity and the melt composition, which controls the oxide ion activity of the melt.

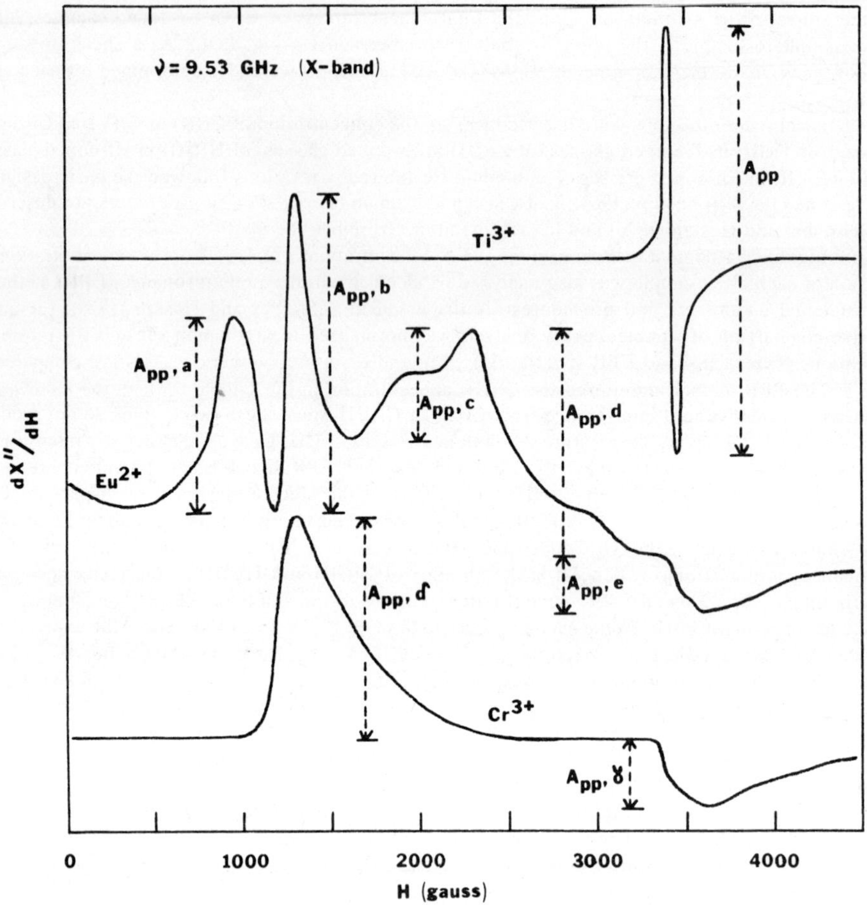

Fig. 2. Representative EPR spectra of Ti-, Eu-, and Cr-doped glasses of synthetic composition. The general features of the three representative spectra were the same for both synthetic compositions. The spectra are shown as the derivative of the absorption versus the magnetic field strength. A_{pp} is the intensity of the resonance signal. The intensity of each EPR signal reflected the concentration of the particular resonance species present in the glass; this in turn made the intensity, but not the spectral shapes, dependent upon the synthesis conditions of the sample. Only a single resonance is characteristic of Ti(III) in the Ti-doped samples. All "peak-to-peak" intensities measured in the complex Eu(II) spectrum (a through e) are roughly proportional to the Eu(II) concentration in the glass. $A_{pp}(\delta)$ is the Cr(III) resonance observed in Cr-doped samples; $A_{pp}(\gamma)$ as well as other potential complications in the EPR spectra in the Cr system are discussed in Schreiber and Haskin (1976).

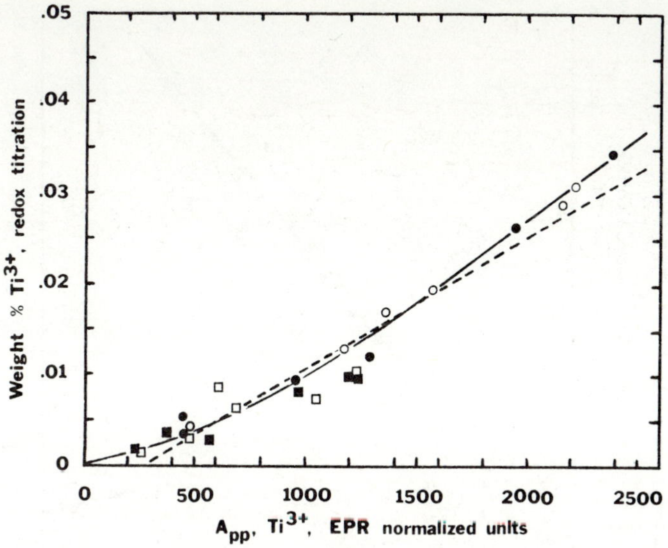

(A)

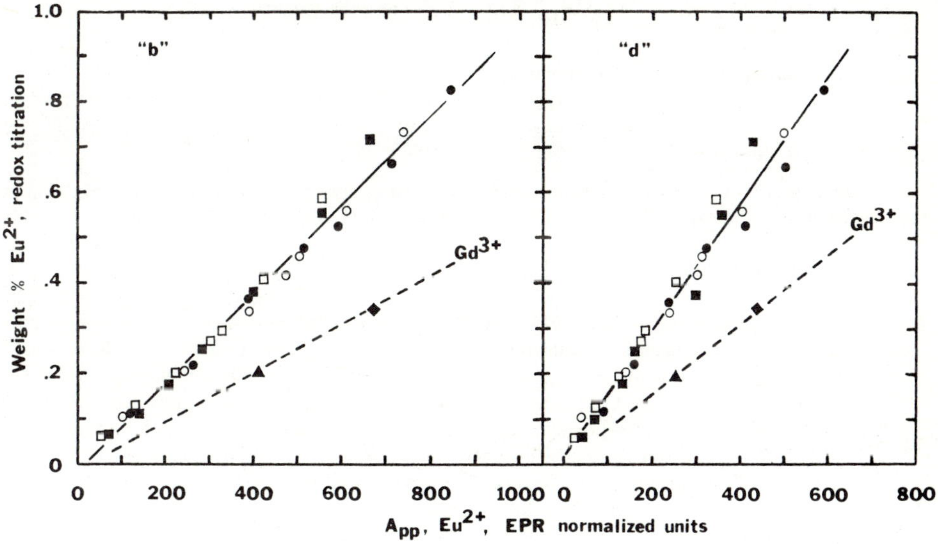

(B)

Fig. 3. Calibration of the Ti(III), Eu(II), and Cr(III) EPR resonance intensities to the wt.% of each ion determined by chemical redox titrations for the synthetic glasses. The different symbols represent the synthesis conditions for the samples: solid symbols = 1500°C, open symbols = 1550°C, circles = composition FAS, and squares = composition FAD. (A) Ti(III) calibration curve is the solid line while the least-squares straight line is the dashed line. (B) Eu(II) calibration is shown by the solid least-squares line for the two most intense features in the Eu(II) EPR spectra (A_{pp}-"b" and A_{pp}-"d" from Fig. 2). For comparison a calibration of Gd(III) for the same spectral features is shown by the dashed line, where the triangle = composition FAS and the diamond = composition FAD.

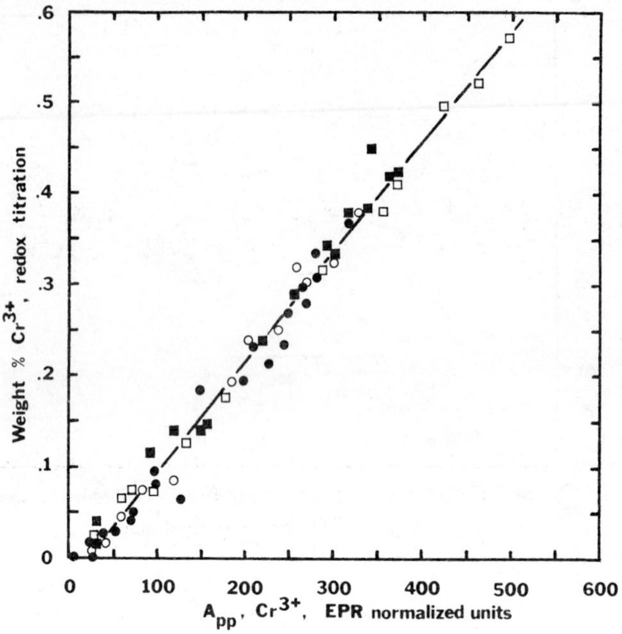

(C)

Fig. 3(*cont.*). (C) Cr(III) calibration is shown by the solid least-squares line for A_{pp}-"δ" in the EPR spectra. Weight % Cr(III) determined by chemical redox titrations equals total Cr less analyzed Cr(II) or Cr(VI).

The dependence of the redox equilibria on the imposed oxygen fugacity allows an examination of the consistency of the experimental measurements with Eqs. (1)–(3). As a consequence of the redox equilibria expressions, a plot of $-\log P_{O_2}$ vs $\log R$, where R is equal to the ratio of the reduced to oxidized ion concentrations (e.g., $[Cr^{2+}]/[Cr^{3+}]$), should be linear with a slope of 4 for the Cr(III)–Cr(II), Ti(IV)–Ti(III), and Eu(III)–Eu(II) systems or of 4/3 for the Cr(VI)–Cr(III) system (e.g., Johnson, 1965; Schreiber and Haskin, 1976). Indeed, such a plot for each redox couple in each bulk melt composition and at each synthesis temperature produced well-defined straight lines whose slopes agreed within experimental error to those predicted from the equilibrium expressions. These plots are shown in Fig. 4 for the Cr, Ti, and Eu redox equilibria and are also a convenient graphical means for displaying all experimental data. The trivalent states of Zr and Hf were never found in any of the silicate melts, even those synthesized under the most reducing conditions. This seems to be consistent with the chemistry of the Ti, Zr, and Hf systems, since only low concentrations of Ti(III) existed in the synthetic melts under reducing conditions and periodic properties dictate Zr(IV) and Hf(IV) to be even harder to reduce to the trivalent state than Ti(IV).

Since the equilibria for reduction as written in Eqs. (1)–(3) are typically endothermic, higher temperatures of synthesis should produce a greater proportion of the reduced redox state with respect to the oxidized redox state for a particular element in the melt. Figure 4 illustrates that the temperature shifts the redox equilibria in the expected direction. The temperature dependence of the redox equilibria was also used to calculate the thermodynamic standard enthalpy of reduction ($\Delta H°$) through a plot of $\log R$ vs T^{-1} where T is the temperature in °K (e.g., Morris *et al.*, 1974; Schreiber and Haskin, 1976). The experimental $\Delta H°$ for the reduction of Cr(III)$\rightarrow$Cr(II), of Ti(IV)$\rightarrow$Ti(III), and of Eu(III)$\rightarrow$Eu(II) in the synthetic silicate melts are listed in Table 2.

The bulk melt composition in which the trace element redox equilibrium is established is a major factor in the degree of reduction of the multivalent element in that liquid (e.g., Morris and Haskin, 1974; Lauer and Morris, 1977). Relative to melt composition FAS at the same temperature and

Table 2. Standard enthalpy changes ($\Delta H°$) for the reduction equilibria of Ti, Eu, and Cr in the synthetic silicate melts.

	$\Delta H°$(kcal/mole of O_2)		
Redox couple	Melt FAS	Melt FAD	Average
Ti(IV) $\rightarrow$ Ti(III)	70 ± 28	71 ± 19	70 ± 23
Eu(III) $\rightarrow$ Eu(II)	44 ± 35	36 ± 16	40 ± 25
Cr(III) $\rightarrow$ Cr(II)	35 ± 12	38 ± 3	36 ± 8

oxygen fugacity, melt composition FAD consistently stabilized a greater portion of the Cr, Ti, or Eu in their respective higher redox states—that composition FAD is a more "oxidizing" melt is shown by the results in Fig. 4. The relative compositional dependence observed for the trace element redox equilibria in silicate melts for this study is consistent with the comprehensive experimental/theoretical model of Lauer and Morris (1977) and is believed to be mainly a consequence of the variation in Al and Ca contents between compositions FAS and FAD.

Geochemical Applications

Reduction potentials of redox couples in silicate melts

The geochemical significance of this experimental study is summarized in Fig. 5, which compares the redox equilibria of Ti, Eu, and Cr in the synthetic Fe-free melts. By use of a common reference temperature, oxygen fugacity, and bulk melt composition in Fig. 5, the position of the equilibrium lines establishes a primitive scale of relative reduction potentials (E^*) of the redox couples in silicate melt solutions. Thus, the electromotive-force series of Cr(III) $\rightarrow$ Cr(II) > Eu(III) $\rightarrow$ Eu(II) >>> Ti(IV) $\rightarrow$ Ti(III) indicates the relative ease of reduction of these ions in identical silicate melts. The trend of the experimental enthalpies of reduction (Table 2) effectively substantiates this order in terms of the energy required for the reduction reaction. However, within experimental uncertainty the reduction scale is much better defined by the dependence of the equilibria on the oxygen fugacity (Fig. 5) than by the enthalpies of reduction (Table 2). According to standard reduction potentials ($E°$) of these redox couples in water (Latimer, 1952; Dellein *et al.*, 1976; Morss, 1976), the electromotive-force series observed in water is directly opposite, namely Ti(IV) $\rightarrow$ Ti(III) $\gg$ Eu(III) $\rightarrow$ Eu(II) $\simeq$ Cr(III) $\rightarrow$ Cr(II) in order of ease of reduction. A direct correlation between aqueous $E°$ and silicate melt E^* was experimentally observed to a first approximation by Johnston (1965).

The results of this study for Ti, Eu, and Cr redox couples were combined in Fig. 6 with those of the systematic study of Johnston (1965) for one-electron reductions of other metal ions in sodium disilicate melts. Since both this study and Johnston (1965) measured the redox equilibrium of Ti(IV)–Ti(III), albeit at different temperatures in different silicate melts, the Ti value served as a

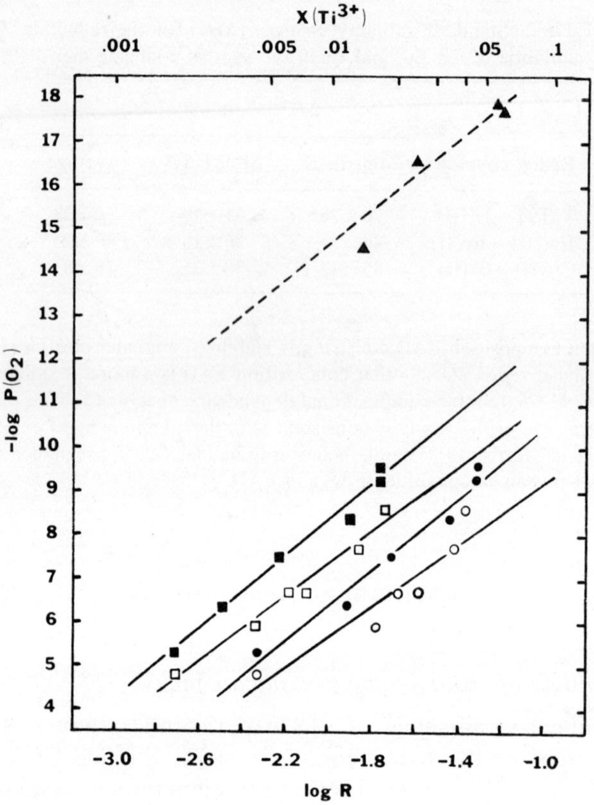
X(Ti^{3+})
.001
.005
.01
.05
.1
18
17
16
15
14
13
12
11
10
9
8
7
6
5
4
-log P(O$_2$)
-3.0
-2.6
-2.2
-1.8
-1.4
-1.0
log R

(A)

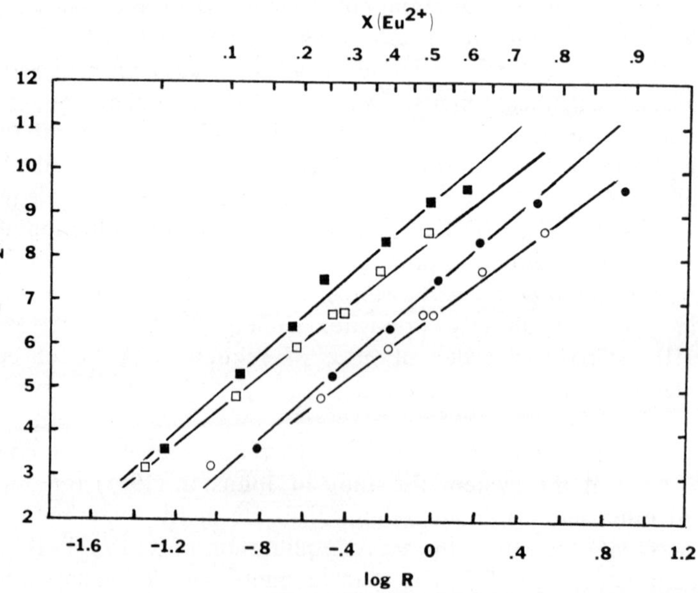
X(Eu^{2+})
.1
.2
.3
.4
.5
.6
.7
.8
.9
12
11
10
9
8
7
6
5
4
3
2
-log P(O$_2$)
-1.6
-1.2
-.8
-.4
0
.4
.8
1.2
log R

(B)

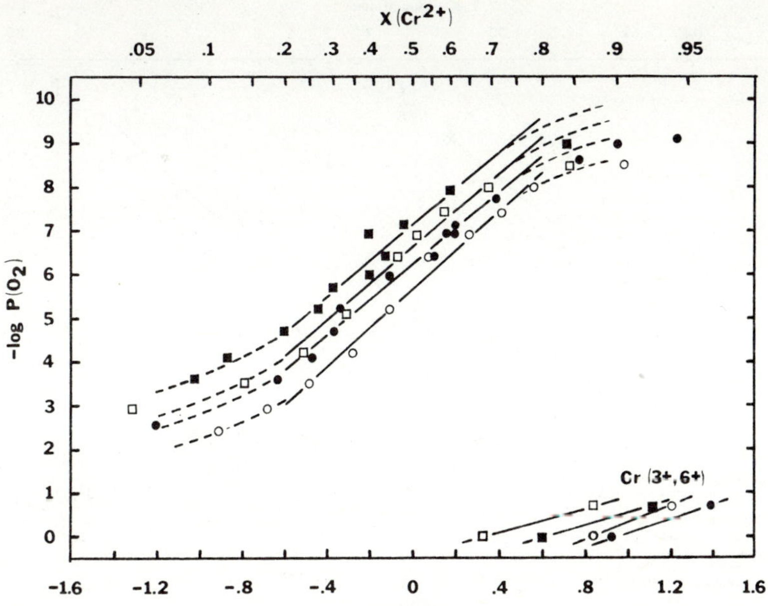

(C)

Fig. 4. Experimental results for the redox equilibria of Ti, Eu, and Cr in the synthetic silicate melts with respect to the oxygen fugacity, temperature and melt composition. The x-axis of the graph represents the degree of reduction in the particular equilibrium, the y-axis is an indication of the oxygen fugacity, solid/open symbols illustrate the synthesis temperature (solid = 1500°C, open = 1550°C), and symbol shape represents the bulk melt compositions (circle = FAS, square = FAD). All straight lines have been determined by least-squares analyses and have slopes which conform to theoretical expectations within experimental error. (A) The Ti(IV)–Ti(III) system with $R = [Ti^{3+}]/[Ti^{4+}]$ and $X(Ti^{3+}) = [Ti^{3+}]/\Sigma[Ti^{n+}]$. The solid triangles and the dashed line are the experimental results of Johnston (1965) for the Ti(IV)–Ti(III) equilibrium at 1085°C in sodium disilicate melts for comparison. The large difference between the results of this study with those of Johnston (1965) in this graph is due to the correspondingly large differences in melt temperature and composition. (B) The Eu(III)–Eu(II) system with $R = [Eu^{2+}]/[Eu^{3+}]$ and $X(Eu^{2+}) = [Eu^{2+}]/\Sigma[Eu^{n+}]$. (C) The Cr(VI)–Cr(III) and Cr(III)–Cr(II) systems with $R = [Cr^{3+}]/[Cr^{6+}]$ or $[Cr^{2+}]/[Cr^{3+}]$ and $X(Cr^{2+}) = [Cr^{2+}]/\Sigma[Cr^{n+}]$. Departure from linearity indicated by the dashed lines for the Cr(III)–Cr(II) equilibrium may be due to analytical limitations or to competing equilibria with other Cr redox states.

reference point for relating the two studies. With the tenuous oversimplication that solvation and thermal effects for the other redox couples were similar to those for Ti(IV)–Ti(III), an appropriate scale factor could be calculated to relate the two systems semi-quantitatively. The internal consistency of Fig. 6, which plots silicate melt E^* versus aqueous $E°$, provides confidence that the assumptions made in combining the two sets of results must be at least approximately valid. The significance of Fig. 6 is that it further expands the electromotive-force series in silicate melt solutions to include other geochemically important redox

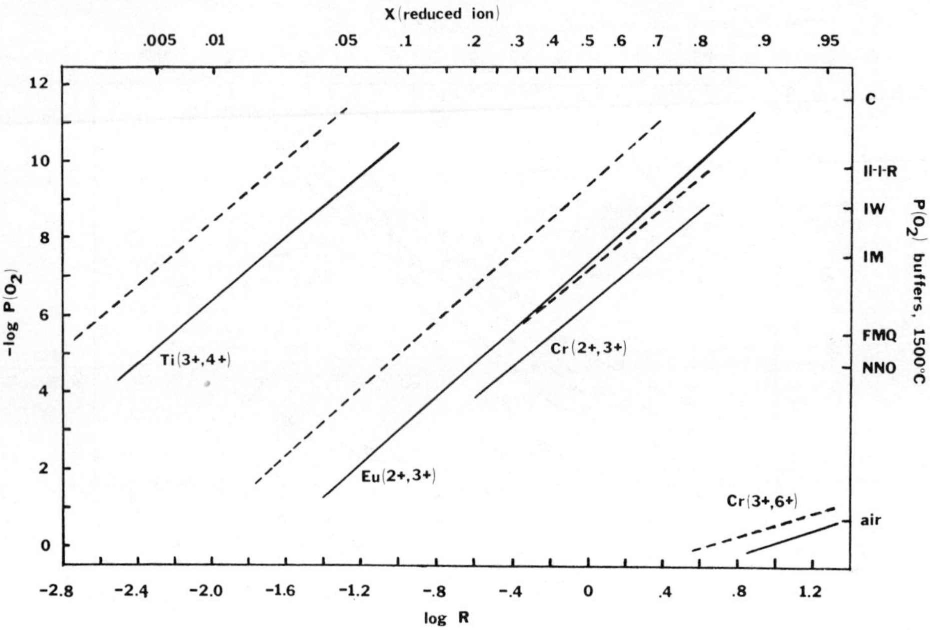

Fig. 5. A summary of the Ti(IV)–Ti(III), Eu(III)–Eu(II), Cr(III)–Cr(II), and Cr(IV)–Cr(III) redox equilibria in synthetic Fe-free silicate melts at 1500°C with respect to imposed oxygen fugacity and bulk melt composition. Symbolism key: dashed line = melt composition FAD; solid line = melt composition FAS; $P(O_2)$ = oxygen fugacity; R = ratio of reduced to oxidized redox state concentrations of the particular redox couple, as defined in Fig. 4; X (reduced ion) = fraction of the total element concentration present in the reduced redox state, as defined in Fig. 4; the mineral assemblages for the oxygen fugacity buffers are NNO = nickel-nickel oxide, FMQ = fayalite-magnetite-quartz, IM = iron-magnetite, IW = iron-wustite, Il-I-R = ilmenite-iron-rutile, and C = graphite at 1 bar in CO_2–CO gas. The oxygen fugacities of the buffers are based upon calculated values (Huebner, 1971) extrapolated to a temperature of 1500°C.

couples; in terms of ease of reduction, Co(III) → Co(II) > Mn(III) → Mn(II) > Ce(IV) → Ce(III) ≫ Fe(III) → Fe(II) > V(V) → V(IV) > {Cr(III) → Cr(II) ≃ V(IV) → V(III) ≃ Eu(III) → Eu(II)} ≫ Ti(IV) → Ti(III). Geochemical consequences of this order of reduction potentials in silicate melts will be discussed in later sections. The weakness in the overall trend for a correlation between aqueous $E°$ and silicate melt E^* is illustrated by the dashed line in Fig. 6. An interesting observation from Fig. 6 is that all M(III)–M(II) redox couples lie above the dashed line while all M(IV)–M(III) and M(V)–M(IV) redox couples lie below the dashed line. The chemical reasons for this apparent behavior are being investigated.

Redox states of Ti, Zr, Hf, Eu, and Cr in terrestrial magmas

The intrinsic oxygen fugacities of most terrestrial magmas have been buffered near those defined by fayalite-magnetite-quartz (e.g., Nash and Hausel,

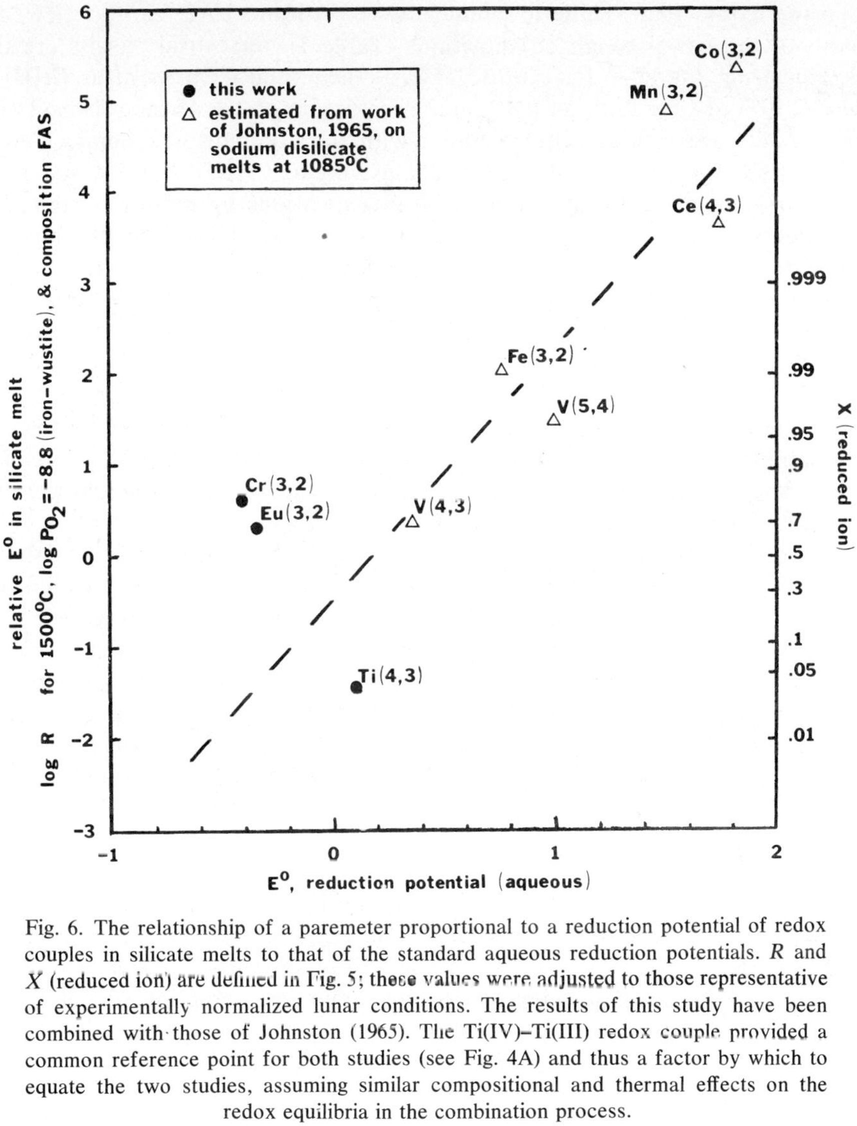

Fig. 6. The relationship of a paremeter proportional to a reduction potential of redox couples in silicate melts to that of the standard aqueous reduction potentials. R and X (reduced ion) are defined in Fig. 5; these values were adjusted to those representative of experimentally normalized lunar conditions. The results of this study have been combined with those of Johnston (1965). The Ti(IV)–Ti(III) redox couple provided a common reference point for both studies (see Fig. 4A) and thus a factor by which to equate the two studies, assuming similar compositional and thermal effects on the redox equilibria in the combination process.

1973). Although magmas on the earth have evolved under a wide range of oxygen fugacities, temperatures, and liquid compositions, a reasonable set of typical conditions for terrestrial basaltic magmas comprise an oxygen fugacity of $10^{-7} - 10^{-8}$ atm, a temperature of 1200 ± 150°C, and a melt composition shown in Table 1. In order to compare directly the experimental results of this study to the environment of terrestrial magmas, an extrapolation of temperature-oxygen fugacity relations from 1200°C to 1500°C indicates that at 1500°C the corresponding oxygen fugacity of the magmas would be about 10^{-3}–10^{-6} atm (Schreiber and Haskin, 1976).

To the extent that synthetic compositions FAS and FAD can be viewed as analogs of terrestrial basalt compositions (Table 1), terrestrial basalts are then estimated from Fig. 5 to have up to 55% of their total Cr present as Cr(II), as much as 30% of their Eu as Eu(II), and a maximum of 1% of their Ti as Ti(III). These results are in clear contradiction with the common geochemical knowledge that Cr, Eu, and Ti exist almost exclusively as Cr(III), Eu(III), and Ti(IV) in terrestrial magmas. In agreement with investigations on natural systems, the experiments predict Zr and Hf to occur in the tetravalent state in the magmas.

The reason for the apparent contradiction which develops when the experimental results of Fig. 5 are directly applied to terrestrial systems is the absence of iron oxides in synthetic compositions FAS and FAD. All terrestrial magmas have iron oxide as a major compositional component but, more importantly, have an appreciable portion of this Fe present as Fe(III) in addition to Fe(II). The existence of significant amounts of Fe(III), in great excess over the minor amounts of total Cr, Eu, and Ti, precludes the stabilization of appreciable amounts of reduced redox states of Cr, Eu, and Ti in the terrestrial magmas. The higher reduction potential of the Fe(III)–Fe(II) redox couple in silicate melts in comparison to the reduction potentials of Cr(III)–Cr(II), Eu(III)–Eu(II), and Ti(IV)–Ti(III) in the same solvent (see Fig. 6) indicates that Fe(III) would have spontaneously oxidized any Cr(II), Eu(II), or Ti(II) present. The large excess of Fe(III) over these ions would make this reduction near complete. Thus, with the buffering capacity of the large concentrations of the Fe(III)–Fe(II) couple in terrestrial magmas, it is difficult to imagine stabilization of large amounts of reduced redox states such as Cr(II), Eu(II), and Ti(III) in these magmas. However, these reduced ions could occur under special conditions, as the above argument is valid only in very general terms. V would also be expected to exist mainly as V(V) in terrestrial magmas from the same argument when applied to the results of Fig. 6. By use of analogous arguments for the redox couples of higher reduction potentials than the Fe(III)–Fe(II) couple in silicate melts, the oxidized redox states of Ce, Co, and Mn in terrestrial magmas should be relatively rare.

The previous paragraph viewed the direct mutual interaction of Fe(III) with Ti(III), Cr(II), and Eu(II) in terrestrial magmas as a means for the rationalization of the experimental results. This is perhaps an oversimplified representation but allows a straightforward understanding of silicate melt processes in terms of aqueous solution concepts. An alternate and probably equivalent approach is to change the redox states of Ti, Cr, and Eu which are present in a magma by the variation in their activity coefficients due to the presence of the Fe(III)–Fe(II) couple in the melt.

Redox states of Ti, Zr, Hf, Eu, and Cr in lunar magmas

Whereas terrestrial magmas evolved under oxygen fugacities buffered near those of fayalite-magnetite-quartz, the oxygen fugacity of lunar magmas was limited near that of iron-wustite or ilmenite-iron-rutile (Nash and Hausel, 1973).

As such, the intrinsic oxygen fugacity of lunar basalts was about five orders of magnitude less than their terrestrial analogs, or about 10^{-12}–10^{-13} atm at a melt temperature of 1200°C (e.g., Taylor, 1975; Papike *et al.*, 1976). Normalization of this oxygen fugacity to a melt temperature of 1500°C yields a corresponding oxygen fugacity of about 10^{-9}–10^{-10} atm for direct comparison to the experimental results of Fig. 5 (Schreiber and Haskin, 1976).

Insofar as the Fe-free synthetic compositions reflect lunar basalt compositions (Table 1), the synthetic silicate melts at a simulated lunar atmosphere predict lunar magmas to contain about 1–7% of their Ti as Ti(III), 40–80% of their Eu as Eu(II), and 70–95% of their Cr as Cr(II). Even under these very reducing conditions of the lunar atmosphere, all Zr and Hf are predicted to exist in the tetravalent redox state. Unlike the terrestrial magmas where the experimental results could not be directly applied because of the absence of Fe in the synthetic melt compositions, lunar magmas will be shown to be compatible with the results from the experimental systems.

That 70% (±20%) of the Eu in the lunar maria exists as Eu(II) is a generally accepted average (Philpotts, 1970; Drake, 1975; Taylor, 1975). If this value is representative of the magmas which were the precursors of the maria, 70% of the Eu as Eu(II) is quite consistant with the estimated range based on the experiments. Upon crude normalization of the Cr, Ti, and Zr redox ratios to the composition, temperature and oxygen fugacity which defines 70% Eu(II) on Fig. 5, lunar magmas would furthermore be predicted to possess initially about 90% of their Cr as Cr(II), about 4% of their Ti as Ti(III), and essentially 0% of their Zr as Zr(III) on the average. Lunar environments may have been locally more oxidizing or more reducing and consequently change these absolute percentages, but the relative Cr–Eu–Ti redox calibration defined by Fig. 5 should still remain valid.

The presence of the reduced redox states of Cr, Eu, and Ti in lunar basalts is generally accepted. Inferences from lunar chemistry and mineralogy as well as direct observations by spectroscopic means indicate the stabilization of Cr(II), Eu(II), and Ti(III) in lunar materials under the reducing atmosphere of the moon. However, the exact amounts of the Cr(II), Eu(II), and Ti(III) ions present in lunar magmas have generally remained unknown. Comparisons of the experimental predictions of the fraction of Cr, Eu, Ti, and Zr in their corresponding reduced redox states with the quantitative estimates of other workers on lunar or simulated-lunar materials are made in Table 3.

Cr: Schreiber and Haskin (1976) review most of the previous evidence pertaining to the redox state of Cr on the moon. Much of this earlier work attempted to ascertain whether all the Cr in lunar basalts was Cr(III) or whether small amounts of Cr(II) might exist under the lunar environment. The results of this study confirm the conclusions of Schreiber and Haskin (1976) that most of the Cr in lunar basalts is Cr(II). Furthermore, the recent works of Akella *et al.* (1976), Huebner *et al.* (1976), McKay and Weill (1976), and Lipin (1977) on synthetic analogs of lunar basalts substantiate this claim that Cr(II) is the predominant Cr species in lunar magmatic processes. Since Cr is a critical

Table 3. Comparison of the fraction of a particular element in its reduced redox state {X (reduced ion) defined in Fig. 4} as estimated in synthetic silicate melts in this study with that based upon lunar or simulated-lunar materials in other studies.

Redox couple	This study X(reduced ion)	X(reduced ion)	Previous studies material	Reference
Cr(III)–Cr(II)	0.7–0.95	0.65–1.0	Synthetic glass	Schreiber and Haskin (1976)
		0.9–0.95	Synthetic glass[1]	Muan (1976)
Eu(III)–Eu(II)	0.4–0.8	0.7(±.2)	Lunar basalts	Philpotts (1970), Drake (1975)
Ti(IV)–Ti(III)	0.01–0.07	0.01–0.10	Synthetic glass	Wells and Hapke (1977)
		0.03(±.03)	Plagioclase	Weeks (1973)
		0.04–0.13	Armalcolite	Range consistent with many works[2]
		0.08–0.25	Cr–Ti–spinel	Agrell *et al.* (1970)
		0.3–0.4(±.3)	Pyroxene	Sung *et al.* (1974)
Zr(IV)–Zr(III)	0	0.15–0.25	Lunar basalts	Postulated by Garg and Ehmann (1976)

[1]Composition of silicate melt was not representative of lunar basalts, and the imposed conditions were somewhat more reducing than lunar.

[2]Lindsley *et al.* (1974), Smyth (1974), Green *et al.* (1975), Kesson and Lindsley (1975), Usselman and Lofgren (1976), Wechsler *et al.* (1976), Friel *et al.* (1977).

element for the understanding of lunar petrogenesis, models for the geochemical evolution of the moon (Taylor, 1976; Taylor and Jakeš, 1977) have rapidly assimilated the existence of Cr as Cr(II) into their scenarios of lunar history.

Eu: The importance of the Eu(II)–Eu(III) redox equilibrium for models of lunar petrogenesis is well established through the fractionation of Eu(II) from the trivalent rare earth elements in the crystallization of common minerals (e.g., Weill and Drake, 1973; Taylor, 1975). Since the fraction of Eu(II) in lunar magmas provides a crucial constraint upon the reliability of this experimental study, it is very comforting to see the agreement between the results of this study with those of independent measurements on lunar rocks (Table 3).

Ti: Besides the quantitative estimates of the fraction of Ti present as Ti(III) in various minerals listed in Table 3, Ti(III) has been qualitatively identified as a viable species in lunar basalts by EPR studies (e.g., Weeks, 1973; Bell *et al.*, 1974, 1976), by optical absorption crystal-field spectra (e.g., Burns *et al.*, 1972, 1976; Mao *et al.*, 1973; Bell *et al.*, 1974, 1976; Loeffler *et al.*, 1975), by optical reflection spectroscopy (e.g., Hua *et al.*, 1976), by electron microprobe band shifts (e.g., Pavicevic *et al.*, 1972), and by implications from chemical analyses (e.g., Boyd and Smith, 1971; Bence and Papike, 1972; Friel *et al.*, 1977). Ti(III) has also been postulated to participate in a variety of subsolidus reduction mechanisms in lunar basalts, mainly in the production of metallic Fe (e.g., Brett *et al.*, 1971; El Goresy *et al.*, 1974; Haselton and Nash, 1975; Sato, 1976). The comparisons of Table 3 appear to indicate that a plausible range for the Ti(III) content of the initial lunar magmas is from 1 to 10% of the total Ti. Most lunar minerals seem to contain the same approximate Ti(III) fraction as in the initial magma; thus, with respect to Ti(IV), Ti(III) was not strongly rejected or preferentially incorporated in any common lunar mineral with the possible exception of pyroxene. Highly reduced lunar pyroxenes have up to 40% of their Ti as Ti(III) (Sung *et al.*, 1974), and an even more reduced meteoritic pyroxene from Allende has up to 70% of its Ti as Ti(III) (Dowty and Clark, 1973). The preferential incorporation of Ti(III) in pyroxenes is compatible with its favorable crystal field stabilization energy. Since Ti is a major component of the lunar maria and since there is widespread agreement that Ti(III) is present in the maria, the experimental calibration of the amount of Ti(III) actually present will provide important constraints on models for the formation of the lunar mare basalts.

Zr, Hf: Ehmann and associates (e.g., Chyi and Ehmann, 1973; Garg and Ehmann, 1976; Garg *et al.*, 1977) have argued that lunar Zr/Hf fractionation may be partly due to the selective reduction of some Zr to Zr(III) while Hf remains as Hf(IV) in lunar magmas. Analyses of Zr/Hf ratios for pyroxenes, olivines, plagioclases, and ilmenites from lunar rocks provide credence to the hypothesis that the Zr(III) with respect to Zr(IV) and Hf(IV) was preferentially excluded from the earliest formed mineral phases. However, within the detection limits of the analytical procedure, Zr-doped samples never contained any Zr(III) in the synthetic silicate melts, even under very reducing conditions. The fractions of reduced species in Zr-, Hf-, and un-doped samples were identical within experimental error. That essentially no Zr(III) ions are stabilized in the synthetic analogs of lunar magmas is consistent with the expectation that Zr(IV) is much harder to reduce than Ti(IV) and with the relatively small amounts of Ti(III) stabilized under lunar conditions. Thus, Zr/Hf fractionations observed in lunar basalts cannot be caused by the selective reduction of Zr(IV) and the subsequent exclusion of Zr(III) from the earliest crystallized phases: Instead the fractionation is probably a result of small differences in the Zr(IV) and Hf(IV) crystal/liquid partition coefficients (Laul *et al.*, 1974).

V, Nb: Figure 6 has been graphed in order to allow direct expression of the percentage of a particular redox couple in its reduced redox state under lunar-simulated conditions, using 70% of the Eu as Eu(II) as a reference point. Upon recognition that extrapolations in this manner are only very approximate, the prediction that V might exist as a complex equilibrium of V(V), V(IV), V(III), and possibly even V(II) redox states in lunar magmas is evident. A rough estimate of the proportion of V in each redox state is 5% V(V), 25% V(IV), and 70% V(III) in lunar magmas. A systematic investigation into the redox state of V in lunar basalts and its incorporation into various minerals may be very fruitful in terms of the understanding of lunar basalt petrogenesis. Since Nb occurs in the same periodic group as V, similar reduced redox states of Nb may also be expected for this element in lunar magmas. Whereas the terrestrial chemistry of Nb is probably governed by the properties of Nb(V), some speculation has focused on the possibility of reduced redox states such as Nb(IV) or Nb(III) being stabilized in lunar and meteoritic minerals (El Goresy, 1971; Erlank *et al.*, 1972; Duncan *et al.*, 1973; Marvin, 1975). Nevertheless, this experimental study provides support for the potential importance of reduced redox states of Nb and V in lunar and meteoritic processes.

Fe: The results from Fig. 6 predict that less than 1% of the total Fe in lunar basalts is present as Fe(III); the remainder of the Fe probably occurs as Fe(II), although the Fe(II)–Fe(0) equilibrium was not investigated. Whereas the direct application of the results of this experimental study to terrestrial basalts was limited by the presence of appreciable amounts of Fe(III) in the natural systems, this constraint does not pertain to lunar situations. The reduced redox states of Ti, Cr, and Eu can exist in lunar magmas, since there is no large excess of Fe(III) to oxidize them. Weeks (1973) and Burns *et al.* (1976) have also reasoned that the reducing conditions required for the production of Ti(III) appear to exclude Fe(III) on the lunar surface. Although Mössbauer and EPR spectroscopy are quite sensitive to the measurement of Fe(III) contents of natural minerals, a wide range of estimates for the actual Fe(III) content of lunar materials has been reported. Most of the lunar Fe(III) was probably incorporated into plagioclase which may have from 1 to 10% of its total Fe as Fe(III) (e.g., Hafner *et al.*, 1971a; Schurmann and Hafner, 1972; Niebuhr *et al.*, 1973). However, Fe(III) is probably very rare on the moon (e.g., Sato *et al.*, 1973; Frondel, 1975) and most likely accounts for less than 1% of the total Fe present (e.g., Hafner *et al.*, 1971b; Burns and Vaughan, 1975; Tsay and Baumann, 1977; Wells and Hapke, 1977) in agreement with the experimental extrapolations. Even though Fe(0) is ubiquitous in lunar basalts, Fe(0) may not have been present in the initial lunar magma but was produced in post-emplacement events (e.g., Sato, 1976, and Haggerty, 1977, review the variety of mechanisms proposed for the production of Fe(0) in the lunar basalts). Thus, the Fe(II)–Fe(0) redox equilibrium would probably not interfere with predictions made on the redox states of other elements in the initial lunar magmas. Since essentially all Fe in lunar magmas exists as Fe(II), its effect upon the redox ratios of Ti, Cr, and Eu in lunar magmas would be entirely compositional, analogous to changing the Ca(II) or Mg(II) content of the melts. For example, the $[Ti^{3+}]/[Ti^{4+}]$ ratio varied from about 0.1 for Fe-free glasses to less than 0.01 for high-Fe glasses under a simulated lunar atmosphere according to Wells and Hapke (1977); this is interpreted as a purely compositional effect similar to the difference in the Ti(IV)–Ti(III) redox equilibrium between synthetic compositions FAS and FAD (e.g., predictable as per the model of Lauer and Morris, 1977).

In fact, it may be the presence of Fe in lunar magmas which allows the direct application of these experimantal results obtained for a single multivalent element in synthetic silicate melts to the lunar magmas. Because of differences in relative reduction potentials of the Ti(IV)–Ti(III), Cr(III)–Cr(II), and Eu(III)–Eu(II) couples in silicate melts as determined in this study, a complex mutual interaction by electron exchange of these redox couples might be expected to result for a melt (such as a lunar magma) which contained all multivalent elements at their natural concentrations. If these interactions disturbed the individual redox equilibria appreciably (or, similarly, if drastic changes in the activity coefficients of the redox ions occurred), the application of the results of this experimental study to natural systems would be severely limited. Clearly, there cannot be appreciable redox interaction among the multivalent components in lunar magmas, or the agreement between the results reported in this experimental study with the previous lunar estimates would be surprisingly fortuitous. However, the Fe as a major component in lunar magmas is postulated to define internally (redox "buffer") the oxide ion activity of the melt and thus individual redox equilibria such as eqs. (1)–(3). Similar to an acid-base pH buffer with respect to addition of small amounts of H^+ or OH^- in aqueous systems, the large excess of Fe(II) as a potential redox ion in silicate melts will effectively resist changes in oxide ion activity with minor additions of multivalent elements such as Ti, Cr, and Eu. Therefore, the results of the individual redox equilibria reported in this study are directly applicable to similar equilibria in lunar magmas.

Conclusions

Typical terrestrial basaltic magmas contain Cr, Eu, Ti, Zr, Hf, V, and Nb mainly in their corresponding oxidized redox states; that is, trivalent Cr and Eu, tetravalent Ti, Zr, and Hf, and probably pentavalent V and Nb. The presence of a large excess of the oxidizing agent Fe(III) in terrestrial magmas precludes the occurrence of appreciable amounts of the reduced redox states of these multivalent elements.

Because of their evolution at low oxygen fugacities and their virtual lack of Fe(III), lunar magmas typically contain reduced redox states of Cr, Eu, Ti, and possibly V and Nb, but not Zr and Hf. In a crystallizing lunar magma the relative percentages of these reduced redox states with respect to the total elemental concentrations are estimated to be about 70–95% Cr(II), 40–80% Eu(II), 1–7% Ti(III), 0% Zr(III), 0% Hf(III), and possibly significant amounts of V(IV), V(III), Nb(IV), and Nb(III). Less than 1% of the total Fe is present as Fe(III). Since many of these multivalent elements impose critical constraints on models of lunar petrogenesis, the experimental calibration of redox state ratios in lunar magmas reported in this study provides a foundation for semi-quantitative refinements of these models.

Acknowledgments—Much of the experimental work in this study was performed during the author's tenure at NASA Johnson Space Center while he was affiliated with the University of Wisconsin, Madison. This manuscript benefited greatly from the constructive reviews of J. S. Huebner, D. H. Lindsley, and G. C. Ulmer, and from discussions of the author with H. V. Lauer on the redox behavior of trace ions in silicate melts. I also thank C. W. Hartless for her helpful assistance in the preparation of this manuscript.

REFERENCES

Agrell S. O., Scoon J. H., Muir I. D., Long J. V. P., McConnell J. D. C., and Pecket A. (1970) Observations on the chemistry, mineralogy and petrology of some Apollo 11 lunar samples. *Proc. Apollo 11 Lunar Sci. Conf.*, p. 93–128.

Akella J., Williams R. J., and Mullins O. (1976) Solubility of Cr, Ti, and Al in coexisting olivine, spinel, and liquid at 1 atm. *Proc. Lunar Sci. Conf. 7th*, p. 1179–1194.

Bell P. M., Mao H. K., and Weeks R. A. (1974) Crystal-field transition and electron paramagnetic resonance of trivalent titanium in iron-free magnesium-aluminum-calcium-silicate glass synthesized under controlled oxygen fugacity. *Carnegie Inst. Wash. Yearb.* **69**, 228–229.

Bell P. M., Mao H. K., and Weeks R. A. (1976) Optical spectra and electron paramagnetic resonance of lunar and synthetic glasses: A study of the effects of controlled atmosphere, composition, and temperature. *Proc. Lunar Sci. Conf. 7th*, p. 2543–2559.

Bence A. E. and Papike J. J. (1972) Pyroxenes as recorders of lunar basalt petrogenesis: Chemical trends due to crystal-liquid interaction. *Proc. Lunar Sci. Conf. 3rd*, p. 431–469.

Boyd F. R. and Smith D. (1971) Compositional zoning in pyroxenes from lunar rock 12021, Oceanus Procellarum. *J. Petrol.* **12**, 439–464.

Brett R., Butler P., Meyer C., Reid A. M., Takeda H., and Williams R. (1971) Apollo 12 igneous rocks 12004, 12008, 12009, and 12022; A mineralogical and petrological study. *Proc. Lunar Sci. Conf. 2nd*, p. 301–317.

Burns R. G., Huggins F. E., and Abu-Eid R. M. (1972) Polarized absorption spectra of single crystals of lunar pyroxenes and olivines. *The Moon* **4**, 93–102.

Burns R. G., Parkin K. M., Loeffler B. M., Leung I. S., and Abu-Eid R. M. (1976) Further characterization of spectral features attributable to titanium on the moon. *Proc. Lunar Sci. Conf. 7th*, p. 2561–2578.

Burns R.G. and Vaughan D. J. (1975) Polarized electronic spectra. In *Infrared and Raman Spectroscopy of Lunar and Terrestrial Minerals* (C. Karr, ed.), p. 39–72. Academic Press, New York.

Chyi L. L. and Ehmann W. D. (1973) Zirconium and hafnium abundances in some lunar materials and implications of their ratios. *Proc. Lunar Sci. Conf. 4th*, p. 1219–1226.

Close W. P., Shepherd H. M., and Drummond C. H. (1958) Determination of several valences of iron, arsenic and antimony, and selenium in glass. *J. Amer. Ceram. Soc.* **41**, 455–460.
Close W. P. and Tillman J. F. (1969) Chemical analysis of some elements in oxidation-reduction equilibria in silicate glasses. *Glass Tech.* **10**, 134–146.
Dellien I., Hall F. M., and Hepler L. G. (1976) Chromium, molybdenum, and tungsten: Thermodynamic properties, chemical equilibria, and standard potentials. *Chem. Rev.* **76**, 283–310.
Dowty E. and Clark J. R. (1973) Crystal structure refinement and optical properties of a Ti^{3+} fassaite from the Allende meteorite. *Amer. Mineral.* **58**, 230–242.
Drake M. J. (1975) The oxidation state of europium as an indicator of oxygen fugacity. *Geochim. Cosmochim. Acta* **39**, 55–64.
Duncan A. R., Erlank A. J., Willis J. P., and Ahrens L. H. (1973) Composition and inter-relationships of some Apollo 16 samples. *Proc. Lunar Sci. Conf. 4th*, p. 1097–1113.
El Goresy A. (1971) Meteoritic rutile: A niobium-bearing mineral. *Earth Planet. Sci. Lett.* **11**, 359–361.
El Goresy A., Ramdohr P., Medenbach O., and Bernhardt H. (1974) Taurus-Littrow TiO_2-rich basalts: Opaque mineralogy and geochemistry. *Proc. Lunar Sci. Conf. 5th*, p. 627–652.
Erlank A. J., Willis J. P., Ahrens L. H., Gurney J. J., and McCarthy T. S. (1972) Inter-element relationships between the moon and stony meteorites with particular reference to some refractory elements (abstract). In *Lunar Science III*, p. 239–241. The Lunar Science Institute, Houston.
Friel J. J., Harker R. I., and Ulmer G. C. (1977) Armalcolite stability as a function of pressure and oxygen fugacity. *Geochim. Cosmochim. Acta* **41**, 403–410.
Frondel J. W. (1975) *Lunar Mineralogy.* John Wiley, New York. 323 pp.
Garg A. N. and Ehmann W. D. (1976) Zr–Hf fractionation in chemically defined lunar rock groups. *Proc. Lunar Sci. Conf. 7th*, p. 3397–3410.
Garg A. N., Ali M. Z., and Ehmann W. D. (1977) The distribution of Zr and Hf in lunar minerals and rock 12013 (abstract). In *Lunar Science VIII*, p. 346–348. The Lunar Science Institute, Houston.
Green D. H., Ringwood A. E., Hibberson W. O., and Ware N. G. (1975) Experimental petrology of Apollo 17 mare basalts. *Proc. Lunar Sci. Conf. 6th*, p. 871–893.
Hafner S. S., Virgo D., and Warburton D. (1971a) Oxidation state of iron in plagioclase from lunar basalts. *Earth Planet. Sci. Lett.* **12**, 159–166.
Hafner S. S., Virgo D., and Warburton D. (1971b) Cation distributions and cooling history of clinopyroxenes from Oceanus Procellarum. *Proc. Lunar Sci. Conf. 2nd*, p. 91–108.
Haggerty S. E. (1977) Apollo 14: Oxide, metal and olivine mineral chemistries in 14072 with a bearing on the temporal relationships of subsolidus reduction (abstract). In *Lunar Science VIII*, p. 392–394. The Lunar Science Institute, Houston.
Haselton J. D. and Nash W. P. (1975) A model for the evolution of opaques in mare lavas. *Proc. Lunar Sci. Conf. 6th*, p. 747–755.
Haskin L. A., Haskin M. A., Frey F. A., and Wildeman T. R. (1968) Relative and absolute terrestrial abundances of the rare earths. In *Origin and Distribution of the Elements* (L. H. Ahrens, ed.), p. 889–912. Pergamon, New York.
Hua C. T., Dollfus A., and Mandeville J. C. (1976) Ultraviolet diffuse reflectance spectroscopy for lunar, meteoritic, and terrestrial samples. *Proc. Lunar Sci. Conf. 7th*, p. 2605–2622.
Huebner J. S. (1971) Buffering techniques for hydrostatic systems at elevated pressures. In *Research Techniques for High Pressure and High Temperature* (G. C. Ulmer, ed.), p. 123–177. Springer-Verlag, New York.
Huebner J. S., Lipin B. R., and Wiggins L. B. (1976) Partitioning of chromium between silicate crystals and melts. *Proc. Lunar Sci. Conf. 7th*, p. 1195–1220.
Johnston W. D. (1965) Oxidation-reduction equilibria in molten $Na_2O{\cdot}2SiO_2$ glass. *J. Amer. Ceram. Soc.* **48**, 184–190.
Kesson S. E. and Lindsley D. H. (1975) The effects of Al^{3+}, Cr^{3+}, and Ti^{3+} on the stability of armalcolite. *Proc. Lunar Sci. Conf. 6th*, p. 911–920.
Latimer W. M. (1952) *The Oxidation States of the Elements and their Potentials in Aqueous Solutions*, 2nd ed. Prentice-Hall, New York. 392 pp.
Lauer H. V. and Morris R. V. (1977) Redox equilibria of multivalent ions in silicate glasses. *J. Amer. Ceram. Soc.* In press.

Laul J. C., Hill D. W., and Schmitt R. A. (1974) Chemical studies of Apollo 16 and 17 samples. *Proc. Lunar Sci. Conf. 5th*, p. 1047–1066.

Lindsley D. H., Kesson S. E., Hartzman M. J., and Cushman M. K. (1974) The stability of armalcolite: Experimental studies in the system MgO–Fe–Ti–O. *Proc. Lunar Sci. Conf. 5th*, p. 521–534.

Lipin B. R. (1977) The effect of chromium on model lunar highland compositions (abstract). In *Lunar Science VIII*, p. 586–588. The Lunar Science Institute, Houston.

Loeffler B. M., Burns R. G., and Tossell J. A. (1975) Metal→metal charge transfer transitions: Interpretation of visible-region spectra of the moon and lunar materials. *Proc. Lunar Sci. Conf. 6th*, p. 2663–2776.

Mao H. K., Virgo D., and Bell P. M. (1973) Analytical and experimental study of iron and titanium in orange glass from Apollo 17 soil sample 74220. *Proc. Lunar Sci. Conf. 4th*, p. 397–412.

Marvin U. B. (1975) The perplexing behavior of niobium in meteorites and lunar samples (abstract). *Meteoritics* **10**, 452–454.

McKay G. A. and Weill D. F. (1976) Petrogenesis of KREEP. *Proc. Lunar Sci. Conf. 7th*, p. 2427–2447.

Morris R. V. and Haskin L. A. (1974) EPR measurement of the effect of glass composition on the oxidation states of europium. *Geochim. Cosmochim. Acta* **38**, 1435–1445.

Morris R. V., Haskin L. A., Biggar G. M., and O'Hara M. J. (1974) Measurement of the effects of temperature and partial pressure of oxygen on the oxidation states of europium in silicate glasses. *Geochim. Cosmochim. Acta* **38**, 1447–1459.

Morss L. R. (1976) Thermochemical properties of yttrium, lanthanium, and the lanthanide elements and ions. *Chem. Rev.* **76**, 827–841.

Muan A. (1976) Mineral equilibria in an interdisciplinary perspective. *Amer. Mineral.* **61**, 355–365.

Nash W. P. and Hausel W. D. (1973) Partial pressures of oxygen, phosphorous, and fluorine in some lunar lavas. *Earth Planet. Sci. Lett.* **20**, 13–27.

Niebuhr H. H., Zeira S., and Hafner S. S. (1973) Ferric iron in plagioclase crystals from anorthosite 15415. *Proc. Lunar Sci. Conf. 4th*, p. 971–982.

Papike J., Hodges F., Bence A., Cameron M., and Rhodes J. (1976) Mare basalts: Crystal chemistry, mineralogy and petrology. *Rev. Geophys. Space Phys.* **14**, 475–540.

Pavicevic M., Ramdohr P., and El Goresy A. (1972) Electron microprobe investigations of the oxidation states of Fe and Ti in ilmenite in Apollo 11, Apollo 12, and Apollo 14 crystalline rocks. *Proc. Lunar Sci. Conf. 3rd*, p. 295–303.

Philpotts J. A. (1970) Redox estimation from a calculation of Eu^{2+} and Eu^{3+} concentrations in natural phases. *Earth Planet. Sci. Lett.* **9**, 257–268.

Sato M. (1976) Oxygen fugacity and other thermochemical parameters of Apollo 17 high-Ti basalts and their implications on the reduction mechanism. *Proc. Lunar Sci. Conf. 7th*, p. 1323–1344.

Sato M., Hickling N. L., and McLane J. E. (1973) Oxygen fugacity values of Apollo 12, 14, and 15 lunar samples and reduced state of lunar magmas. *Proc. Lunar Sci. Conf. 4th*, p. 1061–1079.

Schreiber H. D. (1976) The Experimental Determination of Redox States, Properties, and Distribution of Chromium in Synthetic Silicate Phases and Application to Basalt Petrogenesis. Ph.D. thesis, University of Wisconsin. 222 pp.

Schreiber H. D. and Haskin L. A. (1976) Chromium in basalts: Experimental determination of redox states and partitioning among synthetic silicate phases. *Proc. Lunar Sci. Conf. 7th*, p. 1221–1259.

Schurmann K. and Hafner S. S. (1972) On the amount of ferric iron in plagioclase from lunar igneous rocks. *Proc. Lunar Sci. Conf. 3rd*, p. 615–621.

Smyth J. R. (1974) The crystal chemistry of armalcolites from Apollo 17. *Earth Planet. Sci. Lett.* **24**, 262–270.

Sung C. M., Abu-Eid R. M., and Burns R. G. (1974) Ti^{3+}/Ti^{4+} ratios in lunar pyroxenes: Implications to depth of origin of mare basalt magma. *Proc. Lunar Sci. Conf. 5th*, p. 717–726.

Taylor S. R. (1975) *Lunar Science: A Post Apollo View.* Pergamon, Elmsford, New York. 372 pp.

Taylor S. R. (1976) Geochemical constraints on the composition of the moon. *Proc. Lunar Sci. Conf. 7th*, p. 3461–3477.

Taylor S. R. and Jakeš P. (1977) Geochemical evolution of the moon: Taylor-Jakes-Bence models revisited (abstract). In *Lunar Science VIII*, p. 934–936. The Lunar Science Institute, Houston.

Tsay F. D. and Bauman A. J. (1977) Implications of the occurrence of Fe^{3+} and Fe^0 in lunar samples (abstract). In *Lunar Science VIII*, p. 943–945. The Lunar Science Institute, Houston.

Ulmer G. C., Rosenhauer M., Woermann E., Ginder J., Drory-Wolff A., and Wasilewski P. (1976) Applicability of electrochemical oxygen fugacity measurements to geothermometry. *Amer. Mineral.* **61**, 653–660.

Usselman T. M. and Lofgren G. E. (1976) The phase relations, textures, and mineral chemistries of high-titanium mare basalts as a function of oxygen fugacity and cooling rate. *Proc. Lunar Sci. Conf. 7th*, p. 1345–1363.

Wechsler B. A., Prewitt C. T., and Papike J. J. (1976) Chemistry and structure of lunar and synthetic armalcolite. *Earth Planet. Sci. Lett.* **29**, 91–103.

Weeks R. A. (1973) Paramagnetic resonance spectra of Ti^{3+}, Fe^{3+}, and Mn^{2+} in lunar plagioclases. *J. Geophys. Res.* **78**, 2393–2401.

Weill D. F. and Drake M. J. (1973) Europium anomaly in plagioclase feldspar: Experimental results and semiquantitative model. *Science* **180**, 1059–1060.

Wells E. and Hapke B. (1977) Lunar soil: Iron and titanium bands in the glass fraction. *Science* **195**, 977–979.

Proc. Lunar Sci. Conf. 8th (1977), p. 1809–1829.
Printed in the United States of America

Apollo 14: Oxide, metal, and olivine mineral chemistries in 14072 with a bearing on the temporal relationships of subsolidus reduction

STEPHEN E. HAGGERTY

Department of Geology, University of Massachusetts, Amherst, Massachusetts 01003

Abstract—An optical and electron microprobe study of basalt 14072, which is an intensely reduced sample, has shown that the degree of decomposition of chrome-bearing ulvöspinels can be related to crystallization paragenesis, to the compositions of metallic iron, and to troilite precipitation. These spinels have three modes of occurrence: Early Mg, Al, and Cr-rich varieties are present in olivine, pyroxene, and plagioclase; in cores with mantles of chromian-ulvöspinel; and as discrete crystals interstitial to groundmass silicates. Spinels present as inclusions in either silicates or as cores to zoned crystals are non-reduced, whereas those occurring as mantles are partially reduced and contain minor ilmenite and ± metallic iron. Interstitial spinels are late stage and are Fe + Ti enriched, are intensely reduced, and are characterized by abundant ilmenite and by large concentrations of metallic iron + troilite. Initial crystallization and reduction reequilibration yield compositions which are virtually identical and are, on average, Usp_{47}. A meaningful correlation has been established for the Co contents of metallic iron and these demonstrate that the early metals are enriched in Co (1.2–1.6 wt.%), that metals associated with reduction have intermediate values (0.6–1.0 wt.% Co), and that metals in mesostases, which are typical in tridymite eutectoids, have extremely low concentrations (<0.1 wt.% Co). The presence of troilite as a major constituent of the reduced assemblage in Fe + Ti-rich spinels suggests that these spinels were undergoing reduction at a stage in the crystallization sequence of high sulfur activity. Troilite is also present in eutectic intergrowths with metallic iron, as discrete immiscible globules along silicate grain boundaries and in the mesostases; FeS is absent, however, in silicate melt inclusions, and in mildly reduced mantle spinels. Complexly zoned phenocrystic olivine (Fo_{63-79}) displays Cr-rich metal-bearing symplectites along decorated dislocations and these are inferred to have developed as a result of an early episode of subsolidus olivine reduction. No direct evidence for late-stage olivine (Fo_{35}) reduction is noted; however, Fe + SiO_2 eutectoids in mesostases could have developed from fayalite decomposition. A magmatic-deuteric model is proposed for reduction and this model requires, in part, that a continuous sequence of reduction ensued among liquid, solid, and gas in a "closed system" of the type which may be anticipated in regions of extremely thick lava flows, or in regions of magma crystallization at hypabyssal depths. Based on available experimental and thermodynamic data final equilibration is estimated to have been in the range 700–1000°C and $f_{O_2} = 10^{-16}$–10^{-23} atm.

INTRODUCTION

THE NATURE OF THE REDUCING AGENT and the possible mechanisms proposed for subsolidus reduction of lunar basalts are summarized in two excellent reviews by Brett (1976), and by El Goresy (1976) as follows: (1) the initial presence of Ti^{3+} and Cr^{2+} in the melt, (2) vacuum pumping of oxygen on extrusion, (3) sulfur loss from lunar magmas, (4) alkali volatilization, (5) reduction by carbon or by CO_2 at shallow depths, (6) extrusion of basalts over a solar-wind hydrogen saturated regolith, and (7) reduction by gaseous mixtures of CO and carbonyl iron compounds, $Fe(CO)_5$. Although it is now widely recognized that lunar basalts are in a highly reduced state, and that the presence of metallic Fe is ubiquitous, the number of samples in which intense reduction is observed is small, and this has led to a further proposition that reduction was most likely sporadic rather than one associated with the normal course of crystallization events, or one directly linked to mineral or bulk rock chemistry (Haggerty, 1972).

Minerals displaying subsolidus reduction are chromian-ulvöspinels which decompose to a Cr-enriched spinel + Mg-rich ilmenite + Fe, ilmenites which decompose to chromite + rutile + Fe, and fayalite which yields SiO_2 + Fe. This mineral reduction sequence corresponds to the order of decomposition susceptibility at high temperatures and is representative also of the abundance in which these assemblages are encountered in the lunar samples.

Among the problems which have hindered a unified interpretation for the reduction of lunar basalts are the following: (1) in each of the reactions noted above troilite is also commonly present, and although processes of subsequent sulfidation have been invoked, none of the proposed mechanisms adequately account for the presence of FeS in association with Fe; (2) both calculated and experimental data for the regions of oxide stability as a function of T°C and f_{O_2} atm are currently only available for end-member compositions, and hence their application to the more complex mineral chemistries of lunar basalts is limited; and (3) this limitation also effectively precludes any rational estimate of whether a proposed reduction agent is at all plausible, i.e., whether the finite reduction capacity* of the gas species can indeed induce reduction of minerals of specific compositions, within the constraint that reduction ensues in the subsolidus.

This study was initiated with three prime objectives: (1) to define the distribution of troilite and Fe, and to evaluate the role of FeS in subsolidus reduction assemblages; (2) to define the lower limit of stabilizing elements such as Mg, Al, and Cr in ulvöspinel to determine the terminal compositions at which reduction is inhibited; and (3) to compare the mineral chemistries of decomposed olivine with unaltered olivine in an attempt to bracket T°C and f_{O_2}. The sample chosen for detailed study (4 PTS of 14072) was selected because it is one of the most intensely reduced samples returned from the moon (El Goresy *et al.*, 1972), and because it forms an excellent comparison with a companion sample (14053) which is also intensely reduced and for which an extensive mineralogical and petrological data base already exists (El Goresy *et al.*, 1972; Longhi *et al.*, 1972; Haggerty, 1972).

SPINEL MINERAL CHEMISTRY

Spinels in sample 14072 have three modes of occurrence: (1) as inclusions in silicates, (2) spinels which are strongly zoned with core-mantle relationships, and

*Finite reduction capacity is defined here as an equilibrium buffering constant for a specific gas or gaseous mixture and the ability (capacity) of these gas species to induce decomposition. The term is dependent on thermodynamic and kinetic factors in the context of mineral stability but is independent of whether the system is open or closed.

(3) interstitial spinels which are unzoned but are intensely reduced. Examples of each of these three types are illustrated in Fig. 1.

The silicate included spinels are present in olivine, clinopyroxene, and rarely in plagioclase. These spinels exhibit mild zoning with Cr, Al, and Mg preferentially concentrated in the cores and Fe + Ti at the margins. Zoning is not optically discernable and there is no evidence for subsolidus reduction. Spinels within this category may be broadly classified as Al–Mg–Ti chromites although there are systematic compositional differences which are dependent on the host silicate phase (Table 1). Those included in olivine are aluminian-magnesian chromites with MgO = 5 wt.% and Al_2O_3 = 19–21 wt.%; where clinopyroxene is the host phase the range in compositions is much greater and these spinels are either Al–Mg chromites or Al–Mg–Ti chromites. Rare occurrences of chromite in plagioclase exhibit compositions equivalent to Ti–Al–Mg chromites with TiO_2 = 16 wt.%, Al_2O_3 = 8 wt.%, and MgO = 3 wt.%.

The second suite of spinels have typical core-mantle relationships, where the core is an aluminous magnesian chromite and the mantle is chromian ulvöspinel. These spinels are typical of many low-Ti mare basalts and are characteristic of samples from the Apollo 12 and 15 sites. The spinels are either interstitial to the silicate matrix but the more common disposition is one in which the chromite constituent is partially armored in either olivine or pyroxene, and where the Fe–Ti-rich mantle protrudes into the groundmass. Chromite cores exhibit no evidence of subsolidus reduction whereas the chromian-ulvöspinel mantles on these cores are commonly (~80%) decomposed to an assemblage of oriented ilmenite lamellae in association with an enriched Cr–Al spinel ± Fe ± FeS. The systematics of this reduction pattern are similar to those previously reported for example by El Goresy *et al.* (1972) and by Haggerty (1972) and in an *idealized* form for a member of the solid solution series, chromite-ulvöspinel, the reaction may be expressed as follows:

$$\chi(\underset{\text{solid solution}}{FeCr_2O_4 + Fe_2TiO_4}) \rightarrow Y(\underset{\text{solid solution}}{FeCr_2O_4 + Fe_2TiO_4}) + FeTiO_3 + \underset{\text{metal}}{Fe} + \tfrac{1}{2}O_2,$$

with χ having Ti > Cr and Y having Ti < Cr.

The zonation trends for mantles which have undergone subsolidus reduction are not as clearly defined as they are for those composite crystals which remain pristine, and this is due to the enrichment in Cr and the liberation of Fe + Ti to form ilmenite + metallic iron in the former. However, there is a limit to which ilmenite lamellae are present in these reduced mantles and the compositions of the spinel host adjacent to these lamellae are equivalent, on average, to 47 mole% ulvöspinel. This limit is also approximately that of the most Ti-rich spinels analyzed at the outermost margins of core-mantle type crystals which have not undergone decomposition, so that the extent of subsolidus reduction, based on the redistribution of Fe, Ti, and Cr, is very close to those of spinel compositions which have resulted from initial crystallization.

Spinels in the third category are unzoned, are present as interstitial constituents to the silicate matrix, are partially to intensely reduced with variable to

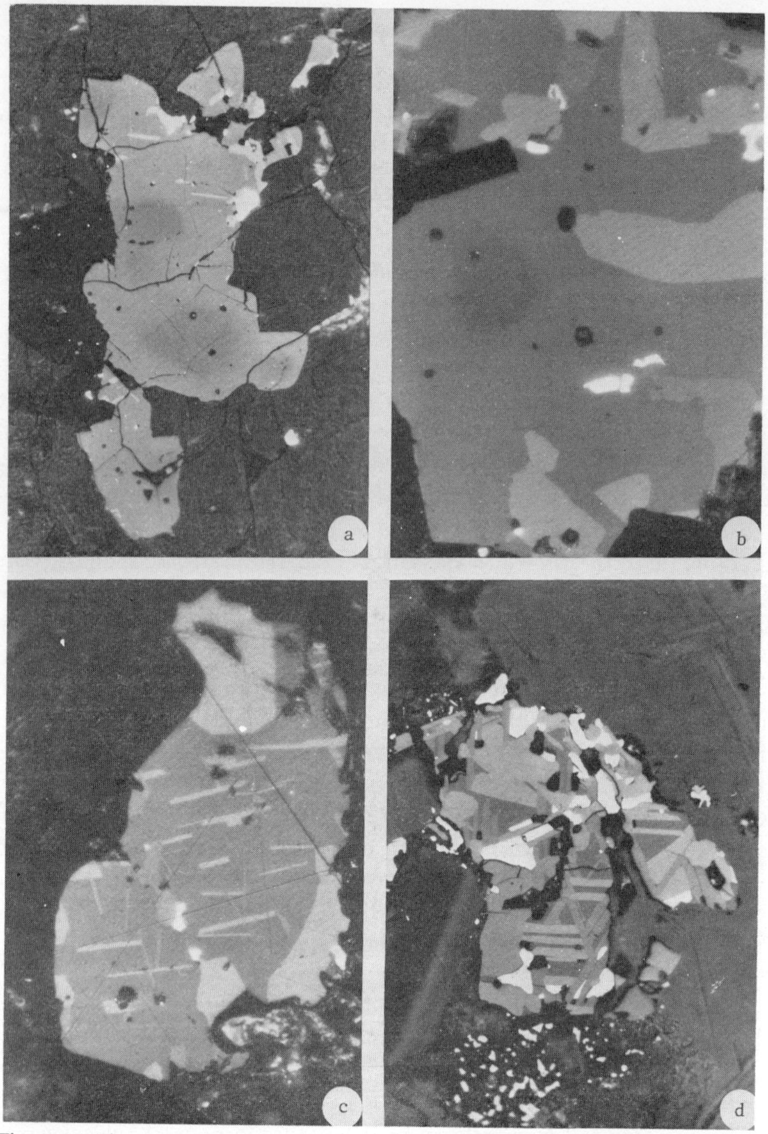

Fig. 1. (a) Five irregularly shaped chromite cores (dark gray) mantled by a continuous overgrowth of chromian-ulvöspinel (gray). Note the diffuse contact between the mantle and the respective cores and note also that the fine ilmenite lamellae (light gray) are restricted to the mantle. The bright flecks are metallic iron and the host is pyroxene + plagioclase. (b) A single chromite nucleus (dark gray) mantled by chromian-ulvöspinel and containing thick internal and external composite crystals of ilmenite with dispersed metallic iron (white). (c) Well-developed ilmenite lamellae along {111} spinel planes and in external composite association in a host of chromian-ulvöspinel. The white blebs are metallic iron and the surrounding black area is the silicate host. (d) An original Fe + Ti-rich chromian-ulvöspinel which is intensely reduced to a Cr-rich spinel (dark gray triangular and rectangular areas) + ilmenite (light gray) + metallic iron (white). This reduced complex is interstitial to clinopyroxene (right), to plagioclase (left), and abuts against a metallic iron-rich mesostasis. The black holes within the spinel represent either plucked regions of metallic iron or alternatively are the result of volume shrinkage during reduction.

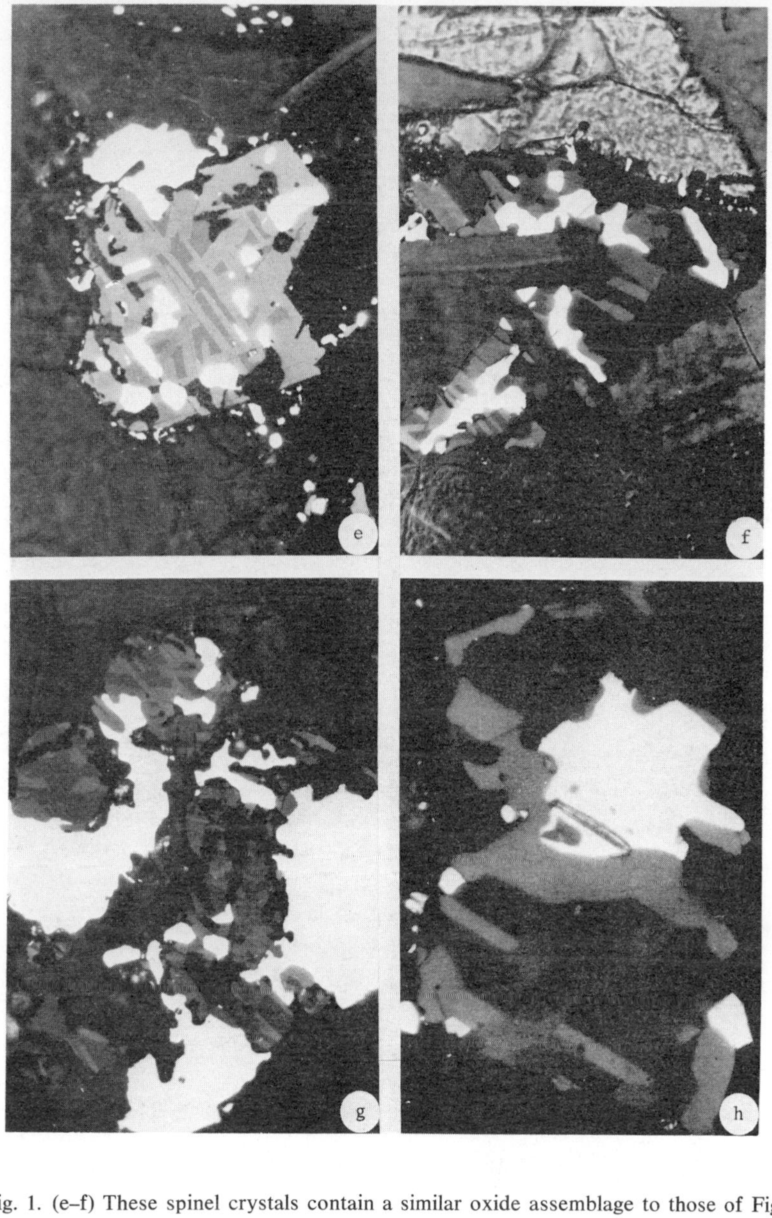

Fig. 1. (e–f) These spinel crystals contain a similar oxide assemblage to those of Fig. 1(d) but the bright irregular to blocky crystals are a mixture of metallic iron + troilite. Note the clearly defined nature of these crystals as interstitial to the surrounding ophitic assemblage of clinopyroxene + plagioclase. (g–h) In these two crystals the extremely large white blebs are troilite, the metallic iron is also white but finer grained and blocky. The dark gray host is chromian-ulvöspinel and the lighter gray lamellae are ilmenite. Scale: photomicrographic widths for Figs. 1(a), (d), (e), (f), and (g) = 0.15 mm; and for Figs. 1(b), (c), and (h) = 400 μm. Oil immersion objectives with mixed reflected and transmitted light.

Table 1. Representative spinel compositions—sample 14072.

	Spinels enclosed in silicates							
	Olivine	Olivine	Olivine	Cpx	Cpx	Cpx	Cpx	Plag
MgO	5.08	6.53	4.76	4.01	3.56	3.88	2.91	2.73
Al_2O_3	21.40	19.88	19.29	17.90	14.20	13.92	9.34	8.26
SiO_2	—	—	—	—	—	—	—	—
ZrO_2	0.07	—	0.06	—	0.16	0.10	0.12	0.08
CaO	—	—	0.03	0.08	0.04	0.08	0.07	0.13
TiO_2	2.90	2.54	3.72	4.20	5.96	7.78	16.37	16.78
Cr_2O_3	39.12	41.56	40.36	40.07	41.20	38.86	28.05	28.19
MnO	0.40	0.33	0.28	0.35	0.40	0.36	0.29	0.46
FeO	30.58	28.97	31.99	33.26	35.18	35.12	42.81	43.74
Total	99.55	99.81	100.49	99.87	100.70	100.10	99.96	100.37
	Core-mantle relationships							
	#21 Core	Int	Margin	#18 Core	Int	Margin	Core	Margin
MgO	4.29	3.07	2.83	4.40	3.35	3.16	4.34	3.12
Al_2O_3	21.31	13.71	8.74	19.19	11.96	8.19	20.58	8.92
SiO_2	—	—	—	—	—	—	—	—
ZrO_2	0.07	0.11	0.09	—	0.02	—	—	0.03
CaO	0.02	0.04	0.06	0.02	0.06	0.04	—	0.05
TiO_2	2.74	11.06	16.31	3.59	12.29	17.30	2.69	16.08
Cr_2O_3	39.21	33.17	30.26	40.08	31.00	26.19	40.76	27.80
MnO	0.28	0.19	0.44	0.35	0.35	0.49	0.41	0.36
FeO	32.28	39.00	42.65	31.90	40.85	44.59	32.03	44.31
Total	100.20	100.35	101.38	99.53	99.88	99.96	100.81	100.67
	Reduced chromian-ulvöspinels							
	Fe + FeS-free #6		Minor Fe #13		Abundant Fe + FeS #11		Abundant Fe + FeS #10	
MgO	3.91	3.86	2.66	2.72	1.40	1.49	1.03	1.00
Al_2O_3	10.58	8.34	9.16	7.89	8.82	10.05	5.90	5.88
SiO_2	—	—	—	—	—	—	—	—
ZrO_2	0.15	0.22	0.01	0.23	0.12	0.15	0.19	0.10
CaO	0.03	0.07	0.02	0.02	0.02	—	0.03	0.03
TiO_2	13.32	16.52	12.78	17.21	12.65	11.19	11.96	11.94
Cr_2O_3	31.43	27.99	34.88	27.52	33.17	37.25	40.93	40.42
MnO	0.32	0.33	0.45	0.43	0.43	0.43	0.51	0.47
FeO	39.34	42.03	39.81	43.27	42.52	39.83	39.15	39.68
Total	99.08	99.36	99.77	99.29	99.13	100.39	99.70	99.52

Table 1. (*Continued*).

	Representative ilmenite compositions							
	#6	#13	#11	#10	Discrete		Mesostasis	
MgO	4.40	3.07	1.65	1.50	1.42	4.21	1.66	1.69
Al_2O_3	0.07	—	0.04	0.07	0.10	0.03	0.09	0.12
SiO_2	—	—	—	—	—	—	—	—
ZrO_2	0.03	0.22	0.14	0.12	0.09	0.04	0.10	0.11
CaO	0.05	0.07	0.04	0.03	—	0.11	0.17	0.25
TiO_2	52.42	52.24	51.67	50.55	51.35	53.95	50.65	50.83
Cr_2O_3	1.13	0.90	0.52	0.56	0.44	0.16	0.45	0.61
MnO	0.52	0.30	0.46	0.43	0.43	0.44	0.51	0.45
FeO	42.29	42.44	45.97	46.05	46.07	41.93	46.29	45.97
Total	100.91	99.24	100.49	99.31	99.90	100.87	99.92	100.03

Abbreviations: Cpx is clinopyroxene; Plag is plagioclase. Core, intermediate, and margin refer to representative analyses of spinels exhibiting core-mantle relationships. For reduced chromian-ulvöspinels, assemblages with minor Fe or assemblages in which Fe + FeS is absent are for spinels with low concentrations of ilmenite. These spinels are interstitial (#6 and 13) to the silicate matrix and are referred to as being partially reduced. Spinels with abundant Fe + FeS (#10 and 11) are also interstitial but contain high concentrations of ilmenite and are intensely reduced.

abundant ilmenite ± Fe ± FeS, and are inferred to have had initial crystallization compositions which were substantially more enriched in ulvöspinel than spinels which are now present in the rock. As a group, these spinels may be classified as Ti–Al–Mg chromites, although in all cases there are substantial variations among TiO_2, Al_2O_3, and MgO (Table 1). A comparison of these reduced spinels with those that are present in some clinopyroxene crystals, with those present in plagioclase, and with those present in reduced mantles on unreacted cores, shows that there is a striking similarity among compositions (Table 1). Here once again, therefore, is the suggestion that the limit to which these initial TiO_2-enriched spinels have equilibrated is approximately equivalent to the maximum enrichment exhibited by spinels which nucleated and grew prior to the onset of either pyroxene or plagioclase crystallization.

Electron microprobe data for these three spinel groups are illustrated in Figs. 2 and 3 in the lunar multicomponent spinel prism and in base and rear rectangular projections of the prism. Important facets to note are the clearly defined and aluminous-rich nature of spinels incorporated as early crystallization products in olivine; that the core-mantle trends overlap in part with spinels in clinopyroxene; and that the reduced spinels cluster both at the terminal limits of the primary spinels and extend also to higher values of Fe/Fe + Mg.

The emphasis here, and elsewhere in this paper, is that the mantles on early formed chromites predate the crystallization of interstitial spinels and the former also clearly predate the precipitation of plagioclase. These relationships are reflected in the trends of reduction reequilibration (Fig. 2) for the interstitial spinels which are toward $FeCr_2O_4$ and are, therefore, indicative of aluminian

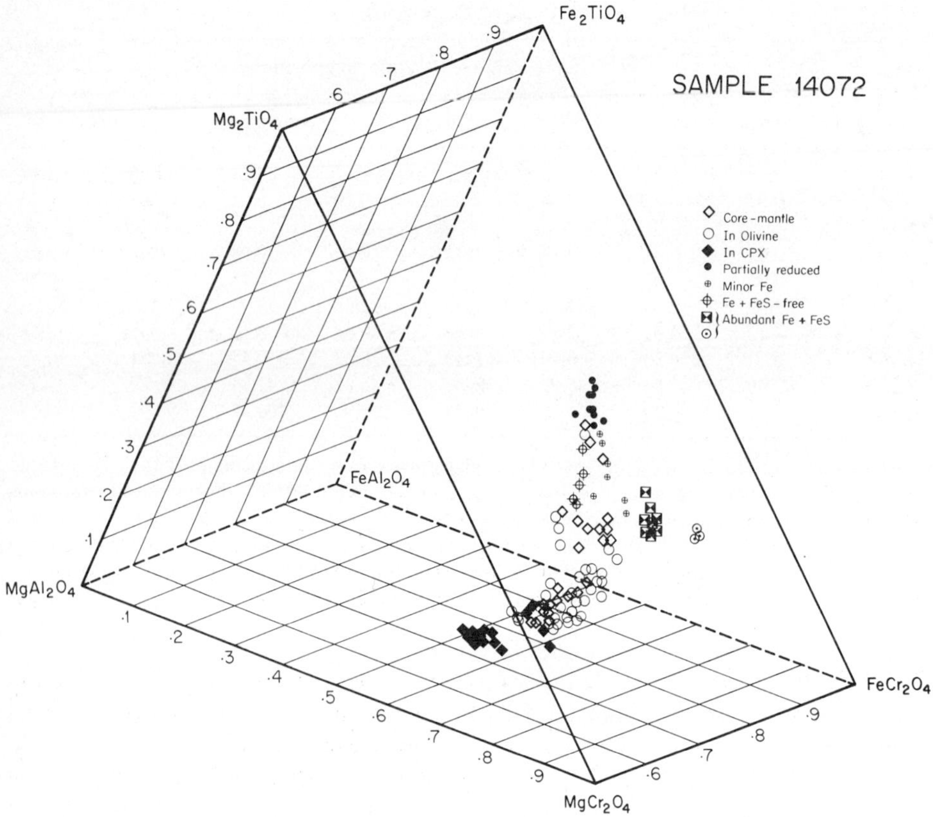

Fig. 2. Electron microprobe analyses for spinel compositions from a variety of associations in sample 14072. Spinels in olivine or pyroxene are confined largely to the base of the prism; those with core-mantle relationships (see Figs. 1a–b) originate on the prism base and trend toward ulvöspinel (Fe_2TiO_4) at the apex; partially reduced interstitial spinels and those forming mantles on chromite, cluster at the high-Ti end of the spectra; those that are intensely reduced (see Figs. 1d–h) are interstitial, contain abundant Fe + FeS and trend toward chromite ($FeCr_2O_4$). The initial crystallization trend, therefore, differs rather markedly from the reduction trend and this difference is attributed to aluminian enrichment in early formed spinels and Fe + Cr enrichment in later formed spinels.

depletion in the melt induced by the formation of plagioclase; thus the critical and terminal composition of Usp_{47} refers specifically to the limits of crystallization which were reached, and are still preserved, in both primary and reduced assemblages, and does not refer to the intensely reduced interstitial spinels whose initial compositions, although unknown, are now at the stability limit defined by the more stable and aluminous-rich varieties.

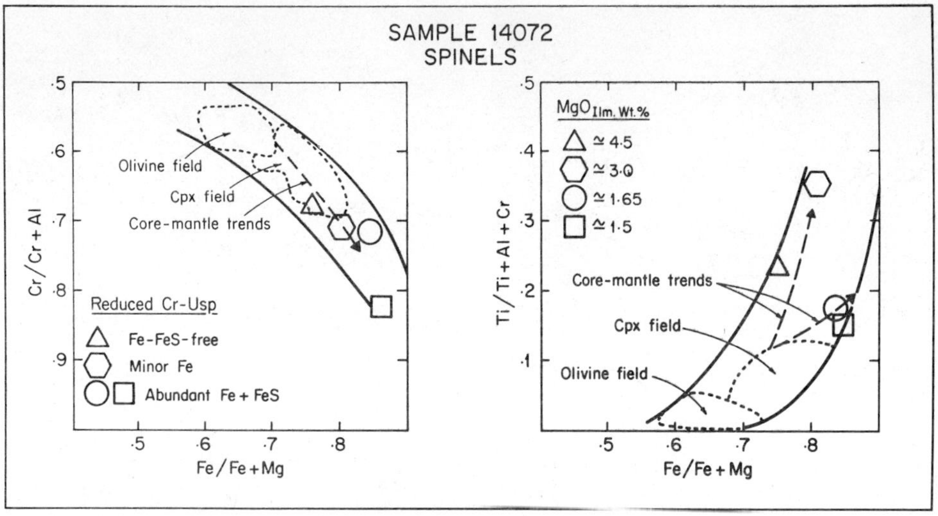

Fig. 3. (a–b) Base and rear rectangular projections of the spinel data shown within the prism in Fig. 2. The olivine and Cpx fields refer to spinels enclosed in olivine and clinopyroxene, respectively. The core-mantle trends are for electron microprobe traverses across crystals similar to those illustrated in Figs. 1a–b. The distribution of Fe + FeS is shown in Fig. 3a and the average values of MgO in ilmenites associated with spinel reduction are summarized in Fig. 3b. Fe–FeS free and minor Fe are associated with interstitial type spinels; spinels containing abundant Fe + FeS are late-stage interstitial spinels with high concentrations of ilmenite.

Ilmenite Mineral Chemistry

Ilmenites in sample 14072 are also present in three distinct modes of occurrence: (1) as large (150–200 μm) euhedral laths interstitial to the silicate matrix, (2) as smaller (~20 μm) subhedral crystals in late residual mesostases, and (3) as products of subsolidus reduction.

The large discrete ilmenites exhibit mild normal zoning with slightly enriched Mg cores and Mg-depleted crystal margins, which rarely differ by more than 1 wt.% MgO. The maximum MgO content recorded in this study is 4.21 wt.% MgO, although the range in compositions vary to a minimum of <1 wt.% with no preferred population apparent. A number of ilmenites contain silicate melt inclusions and immiscible blebs of Fe + FeS, some are traversed by thin metallic iron veinlets (Fig. 4), but none were observed to exhibit any evidence of the reduction reaction: ilmenite → rutile + chromite + Fe which is so typical of Apollo 17 high-Ti mare basalt ilmenites.

The second group of mesostasis ilmenites are relatively uncommon and these ilmenites are uniformly low in MgO with an average composition of 1.7 wt.%. The associated mineralogy of these mesostases is SiO_2 (tridymite), clinopyroxene, olivine, Fe + FeS, whitlockite, and glass.

Ilmenites in the third group are those present in spinels which have undergone subsolidus reduction (Fig. 1). For the most part these ilmenites exhibit a

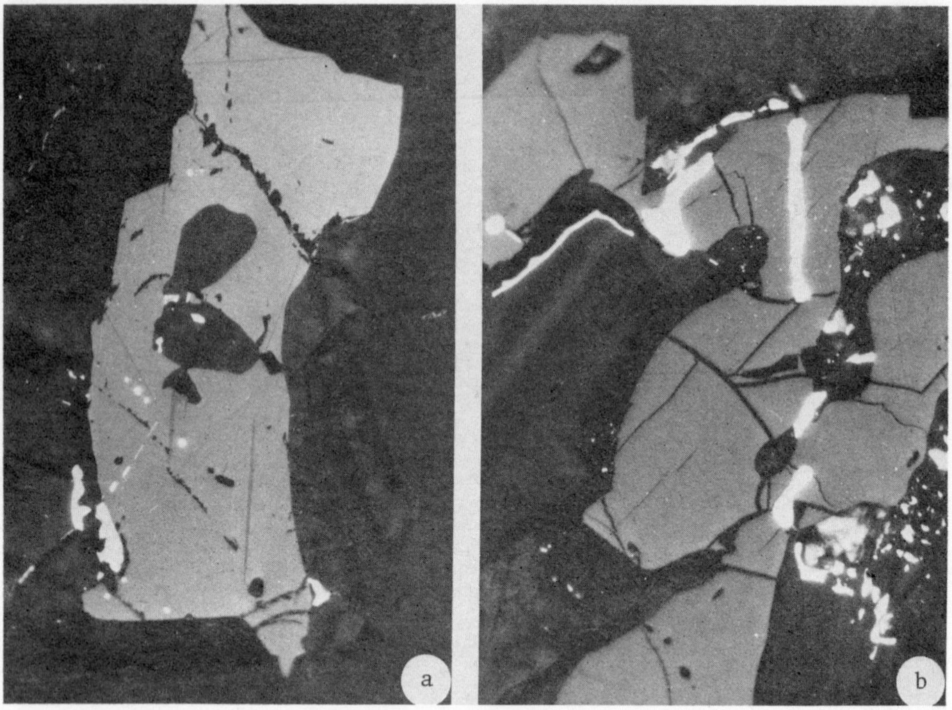

Fig. 4. (a–b) Two large, primary anhedral ilmenite crystals enclosing ameboid silicate melt inclusions which consist of glass + metallic iron + troilite. Veinlets of iron are present in the ilmenite and in the silicate groundmass. The vertical gray lamellae in (a) are chromite. Scale: photomicrographic widths are = 0.15 mm. Oil immersion objectives with mixed reflected and transmitted light.

well-controlled crystallographic orientation along {111} spinel planes although in many instances the textures are best described as internal or external composite. Metallic iron in association with troilite may or may not be present but a clearly defined relationship exists on the modal abundance of ilmenite and the abundance of Fe + FeS. Where ilmenite is present in small concentrations, minute flecks of Fe are visible at the spinel-ilmenite contact but in most instances neither Fe nor FeS are present. For large percentages of ilmenite there are also large concentrations of blocky metallic iron and/or troilite. The compositions of these ilmenite types vary in MgO contents and the range is typically from 1–4.5 wt.%. An outstanding feature is that many of these ilmenites have Cr_2O_3 contents which are in excess of 1 wt.%, a factor which distinguishes this class of ilmenites from the remaining two groups of primary ilmenites.

Compositional data for these three ilmenite types are listed in Table 1 and are illustrated in Fig. 5, for variations of Fe vs. Mg. Although there are several gaps in this overall trend a relatively well-defined relationship exists with three dominantly developed populations. Significant aspects of the relationship are the uniformly low-Mg but high-Fe concentrations in the mesostasis ilmenites, and

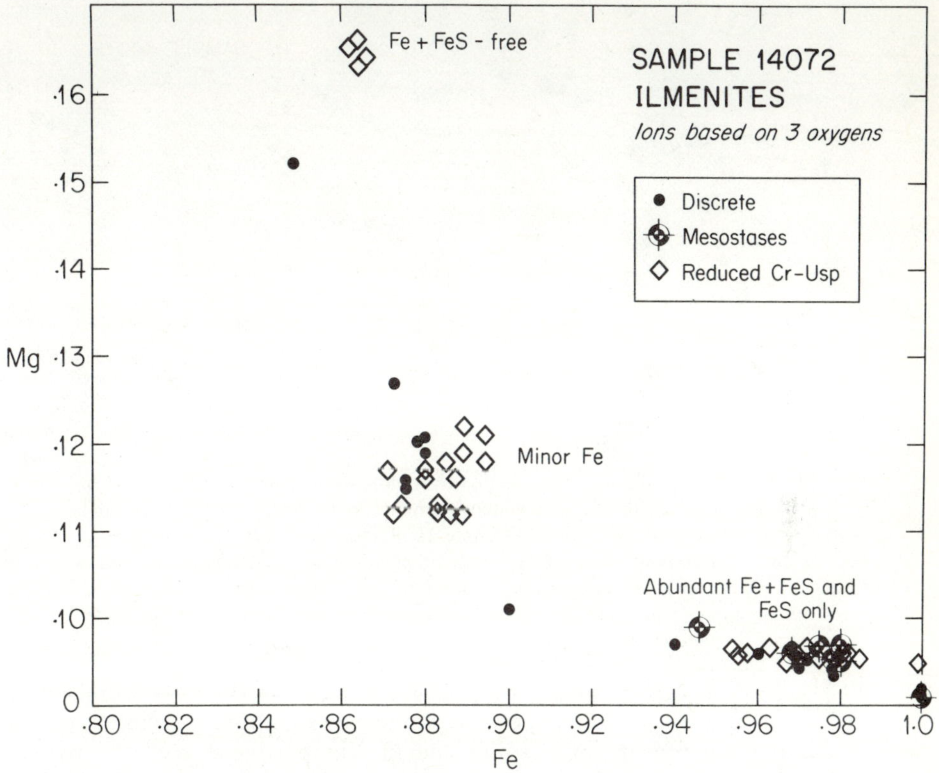

Fig. 5. Ilmenite compositions determined by electron microprobe plotted as a function of Fe vs. Mg. Discrete and mesostases ilmenites refer to primary crystalline phases; reduced Cr-Usp and the distribution of Fe + FeS refer to assemblages associated with the decomposition of chromian ulvöspinel.

that both the large discrete ilmenites as well as those present in subsolidus reduced associations span the entire observed range.

Metal Chemistry

Metallic iron is present in at least six modes of occurrence: (1) in silicate melt inclusions in ilmenite, (2) as veinlets transecting ilmenite crystals and clinopyroxene, (3) in discrete or in eutectic intergrowths with troilite, (4) in mesostases as discrete grains, (5) in pseudo-eutectoid intergrowths with tridymite in mesostases, and (6) in association with ilmenite in spinels which have undergone subsolidus reduction.

The chemistry of these metal types is summarized in Fig. 6 as a function of Ni vs. Co. The meteoritic metal field as proposed by Goldstein and Yakowitz (1971) is included for the purposes of comparison but note that the field has been extrapolated here to values of less than 4 wt.% Ni. The metal chemistry of 14072 is broadly related to paragenesis and the following significant patterns

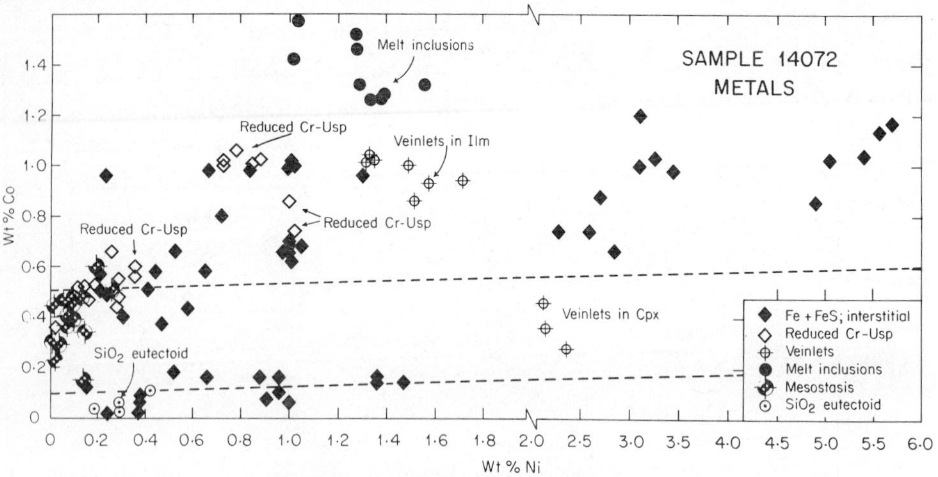

Fig. 6. Electron microprobe data for metallic iron plotted as a function of Ni vs. Co. The dashed lines enclose the field of meteoritic metal as proposed by Goldstein and Yakowitz (1971), but extrapolated to Ni contents of less than 4 wt.%. An important aspect of this diagram is the location of metal compositions associated with chromian-ulvöspinel reduction. These are located at intermediate values of wt.% Co and lie between those metals which are present in silicate melt inclusions in ilmenite (Fig. 4), and those in mesostases SiO_2 eutectoids (Fig. 7).

emerge: (1) there is a wide range in Ni and Co contents for interstitital metal which is either associated with FeS or is sulfide-free, (2) metal veinlets are of two forms, and are present in ilmenite and in clinopyroxene with the former having higher Co contents, (3) the highest Co contents are associated with metals in early and trapped melt inclusions in ilmenite, (4) the lowest Co contents are in metal-tridymite eutectics, (5) the most surprising result is that metals associated with chromian-ulvöspinel decomposition have Co and Ni contents which are intermediate between the crystallization extremes of melt inclusions and late-stage mesostases, and (6) a large proportion of the analyses fall within the lunar metal field, although a significant number also occupy the *extrapolated* meteoritic field, and these include the veinlets in clinopyroxene. If the Co contents of these metals are taken to be indicative of the stages of metal precipitation (Hewins and Goldstein, 1974; Goldstein *et al.*, 1974), then the following points emerge: (1) interstitial metals precipitated over a wide range of crystallization temperatures, (2) because the metals in the reduced spinel assemblages have intermediate Co contents to those in melt and mesostasis associations, decomposition is most likely to have resulted before the final stage of crystallization.

Oxide–Metal–Sulfide Relationships

The distribution of metallic iron and troilite as noted above is widely variable and this is so also for Fe + FeS in those spinels which have undergone sub-solidus reduction. Three classes of distribution emerge: (1) those that are Fe + FeS free, (2) those that contain minor concentrations of Fe, and (3) those that contain abundant Fe + FeS.

Examples typical of the distribution of Fe + FeS as a function of reduced spinel mineral chemistry (Fig. 3a) show that Fe + FeS free (i.e., ilmenite only) assemblages are more magnesian than either of the assemblages in which either minor Fe is present, or in which both Fe + FeS are abundant; for the last group, although Cr/Cr + Al ratios differ for approximately equivalent Fe/Fe + Mg ratios a reasonably well-defined progressive decay is apparent. With respect to coexisting ilmenite compositions (Fig. 3b) the more magnesian spinels yield more magnesian ilmenites, but reduced spinels with only minor metallic iron have the highest TAC ratios (Ti/Ti + Al + Cr) for intermediate values of Fe/Fe + Mg. The significance of the TAC ratio (Haggerty, 1972) is that it is both a measure of the degree of initial crystallization (i.e., olivine-included spinels have low TAC values, whereas mantles on chromite cores have higher values), and is also a measure of the degree of reequilibration (i.e., the lower the TAC ratio the more intensely reduced is the spinel) given the fact that the presence of ilmenite is indicative of an initially higher ulvöspinel component, where $TAC_{Usp} = 1$. Thus the interpretation of Fig. 3b is that the least equilibrated spinels contain minor metallic iron (TAC = 0.37), those with no metallic iron are magnesian-enriched spinels and have TAC = 0.23, and the most intensely reduced group have abundant metallic iron + troilite, with TAC = 0.15–0.18. Although the compositions of the ilmenites which are associated with these spinels show that the more magnesian spinels yield more magnesian ilmenite and no metallic iron, a more explicit division of ilmenite mineral chemistry (Fig. 5) demonstrates that at least three distinct fields are present: a metal and sulfide-free field (high Mg); an intermediate, minor Fe field (intermediate Mg); and an abundant Fe + FeS field (low Mg). Discrete primary ilmenites are present in two of the three fields and are also present between the major fields. However, a notable point is that late-stage mesostasis ilmenite is confined exclusively to the low-Mg and abundant Fe + FeS region. That metal-free and metal-bearing assemblages are related to the Fe/Fe + Mg ratios of spinels is at first sight a clear and simple function of initial mineral chemistry and this is reflected also in the compositions of coexisting ilmenites where Mg partitioning into ilmenite, on reduction, results in $Mg_{ilm} > Mg_{sp}$. However, the distribution of FeS cannot be adequately explained by this simple viewpoint and interpretation.

The wide dispersion of Co contents at the two crystallization extremes of melt inclusions (1.4 wt.% Co) and tridymite eutectics (0.1 wt.% Co) and the observation that metallic iron which is associated with spinel reduction lies between these limits is a clear endorsement to the suggestion made above that final crystallization had not yet been reached by the time that the onset of reduction had begun. If the metal paragenetic model is now integrated with the compositions of spinels and of ilmenite and the distribution of Fe + FeS is also considered, we are forced to the conclusion that the onset of reduction was prior to final consolidation and that Mg-rich spinels (Fe + FeS free) and spinels with intermediate Fe/Fe + Mg ratios (minor Fe) were reduced at temperatures greater than the upper limit of troilite precipitation. The more stable compositions (Mg, Cr, Al) were thus reduced at elevated temperatures and the less stable compositions (Fe, Ti) in the field of FeS stability; thus *initial* reduction probably preceded even that of interstitial chromian-ulvöspinel crystallization.

OLIVINE MINERAL CHEMISTRY

Olivines in sample 14072 are representative of a range in the stages of crystallization and a variety of compositions are present. Large corroded phenocrysts contain Al–Mg chromite + metal-free silicate melt inclusions. These olivines are complexly zoned with core compositions of Fo_{79} to margins of Fo_{63} (Table 2) and are characterized by symplectic Cr-rich metal-bearing decorated dislocations (Fig. 7). Olivines are also present in clinopyroxene and in plagioclase and these crystals, which are unzoned, have compositions in the range Fo_{46-52}. Minor amounts of olivine are present in the mesostases and their compositions are, on average, Fo_{35}. Neither of the latter Fe-rich species exhibit signs of decomposition and nor do these contain decorated dislocations. Decomposition of olivine to Fe + SiO_2 has, however, been reported by Taylor *et al.* (1972) and although this assemblage was not observed in the present study

Table 2. Representative olivine compositions—sample 14072.

	1 Margin	2	3	4	5 Core	6 Melt	7 Sym	8 Chr	9 Plag	10 Mesos
Mole%										
Fo	62.9	66.4	71.5	73.8	78.6	69.5	77.5	69.4	46.5	35.0
Wt.%										
Cr_2O_3	0.02	0.01	0.08	0.06	0.16	0.04	0.14	0.39	0.08	0.02

Analyses 1–8 olivine megacrysts; 1–5 zoning trend; 6 adjacent to melt inclusions; 7 adjacent to Cr-rich symplectite; 8 adjacent to chromite; Analysis 9 olivine in plagioclase; Analysis 10 olivine in mesostasis.

	Symplectites in olivine						
Sample #	14072	14072	76535	72415	72417	12018	15557
SiO_2	37.91	40.76	37.6	38.4	39.4	35.9	37.4
Al_2O_3	0.03	1.61	6.1	4.58	4.07	2.34	0.96
FeO	19.54	16.47	12.1	9.50	9.71	23.6	21.2
MgO	38.70	23.63	24.0	15.30	18.60	19.1	19.5
CaO	1.58	2.26	2.9	16.5	10.6	2.06	1.78
TiO_2	—	0.11	0.6	0.20	0.26	0.53	0.31
MnO	0.01	0.18	0.2	0.40	0.74	0.56	0.40
Cr_2O_3	3.02	14.71	16.3	16.1	16.8	15.5	18.4
NiO	—	—	—	<0.05	<0.05	—	—
ZrO_2	0.05	0.15	—	—	—	—	—
Total	100.84	99.88	100.0	100.98	100.18	99.59	99.95

Analyses for 14072 are from this study and the remainder are from Bell *et al.* (1975).

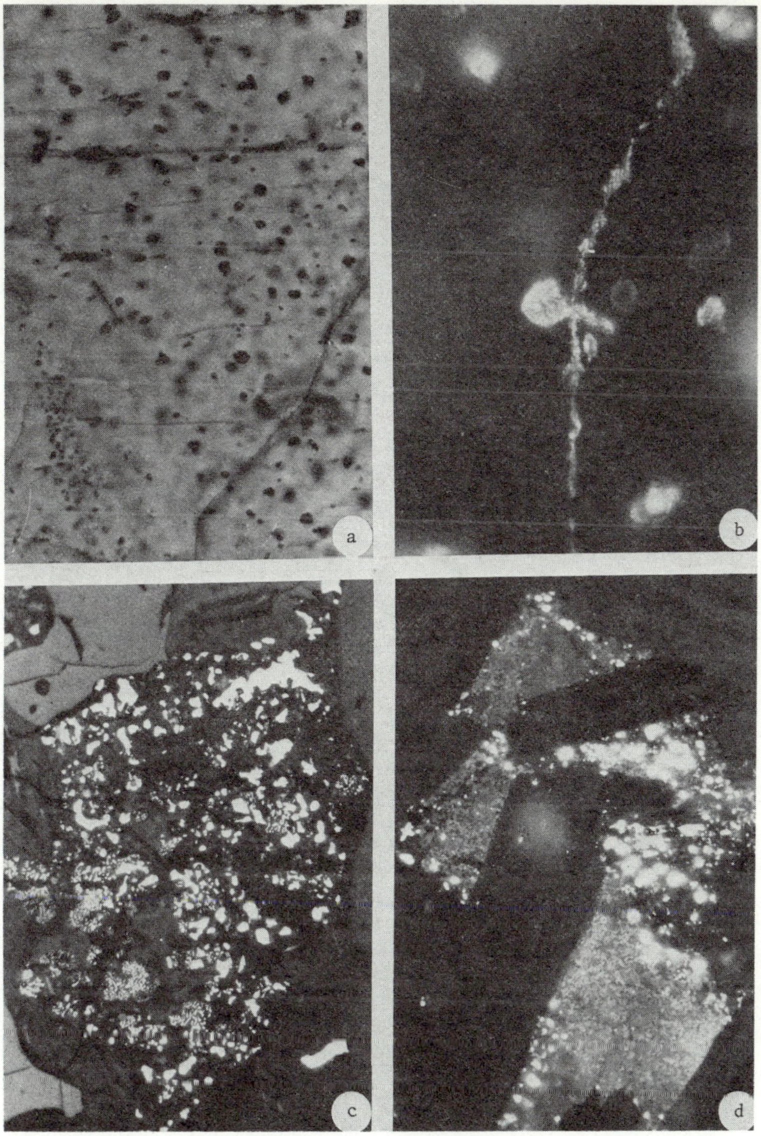

Fig. 7. (a–b) Metal-bearing Cr-rich decorated dislocations in olivine (see Table 2 for electron microprobe data), with (a) illustrating the abundance of these inclusions in transmitted light, and (b) the fine textured nature of the symplectic assemblage. These decorated dislocations are restricted to phenocrystic olivine in 14072. (c–d) Metallic iron-bearing mesostases of two distinct types: (c) contains well-developed symplectites in association with tridymite in a eutectoid-like intergrowth, and although olivine is not present, this assemblage could have resulted by fayalite decomposition; (d) is a glass + whitlockite-rich mesostasis, the bright blebs are metallic iron + troilite and interstitial SiO_2 is also present. The high relief gray phases at the upper and lower left of (c) are ilmenite, and the surrounding silicates in both photomicrographs are plagioclase + clinopyroxene. Scale: photomicrographic widths: (a) = 400 μm; (b) = 150 μm; (c) and (d) = 0.15 mm. Oil immersion objectives with (a) in transmitted light and (b–d) in mixed reflected and transmitted light.

the possibility does exist that the metal-tridymite eutectics are the result of fayalite reduction.

The phenocrystic olivine is of particular interest in the framework of the model being proposed for subsolidus reduction, insofar as reduction is considered to have started before final consolidation. These olivines are also of interest because of the inclusions which are present and because there is the suggestion of decomposition. Olivine compositions adjacent to melt inclusions and adjacent to Al–Mg chromite are virtually indistinguishable (Fo_{70}), which is in contrast to olivine compositions adjacent to symplectic dislocations which are typically Fo_{78} and are, therefore, equivalent to olivine compositions at the cores of zoned crystals.

The symplectites within these olivines are of two forms: (1) clots vary in diameter from <1 μm to a maximum of 20 μm, and (2) meandering stringers (1–5 μm in width) of the symplectic assemblage are pervasive throughout the olivine and commonly radiate from the larger clots. In transmitted light the clots are virtually opaque but in extremely thin portions of the section a deep red coloration is apparent. In reflected light these ovoid blebs are in high relief and a symplectic texture is apparent with at least three phases present: a white high reflectivity constituent, which is interpreted to be metallic Fe; a subdued gray component, which is considered to be chromite; and the host, which approximates that of the large olivine and is most likely a silicate. Electron microprobe beam-location techniques utilizing back-scattered electron images of these blebs were still insufficient to determine the compositions of individual phases, and oxide or metal analytical totals were erratic. Enlarging the electron beam to a 10 μm spot size, however, resulted in good analytical totals and examples of two extreme compositions, representative of the range determined, are listed in Table 2. For these analyses SiO_2 is relatively constant (~38.5 wt.%) and there is little variation in Al_2O_3 (.03–1.6 wt.%), CaO (1.6–2.3 wt.%), MnO, TiO_2, and ZrO_2. The major elements which do vary are FeO (16.5–19.5 wt.%), MgO (23.6–38.7 wt.%), and Cr_2O_3 (3.0–14.7 wt.%). Although the stringers are optically similar to the clots, analyses of the former were unsuccessful.

Similar symplectic intergrowths have been reported by Roedder and Weiblen (1971), and by Bell *et al.* (1975) in olivines from mare basalts; symplectic assemblages are also reported from a lunar dunite (Bell *et al.*, 1975) and a lunar troctolite (Bell *et al.*, 1975; Dymek *et al.*, 1975; Gooley *et al.*, 1974; Haggerty, 1975). For the troctolite there is abundant evidence for the reaction of olivine + plagioclase → clinopyroxene + orthopyroxene + spinel, but for the dunite and mare basalts the two contrasting viewpoints currently held are: (1) decomposition of a preexisting Cr-rich garnet, and (2) diffusion of minor elements out of olivine (Bell *et al.*, 1975). Compositions of the symplectites in the troctolite (76535), the dunite (72415–72417) and in two mare basalts (12018, 15557) are listed in Table 2 together with those obtained for 14072. There are relatively wide variations for all elements except SiO_2 (35–40 wt.%). However, if only the basalts are compared (14072, 12018, and 15557) a much closer agreement results, with 14072 showing lower FeO and Cr_2O_3 contents but with higher MgO contents. Although a metal-bearing symplectite has not previously been reported in samples other than 14072, there are clear indications that at least chromite is present. The Cr_2O_3 contents of the large olivines vary from .02 to 0.16 wt.% (Table 2). A maximum value of 0.39 wt.% was obtained adjacent to an included chromite but this may have resulted by edge effect electron excitation. Groundmass olivines or those included in other silicates have values of approximately 0.05 wt.%. Data reported by Weiblen in Bell *et al.* (1975) show that other symplectite-bearing basaltic olivines contain between 0.13 to 0.34 wt.% Cr_2O_3 which is a range comparable to, but slightly higher, on average, than those for 14072.

Although no reliable analytical data were obtained on the symplectic stringers there is little doubt that these are mineralogically similar to the symplectite blebs. A large body of literature is currently available on decorated dislocations in terrestrial olivine (e.g., Haggerty and Baker, 1967; Champness, 1970; Kohlstedt and Vander Sande, 1975; Kohlstedt *et al.*, 1976; Zuench and Green, 1976) and these examples would tend to indicate that the lunar symplectites are texturally identical, although the former are clearly the result of oxidation. Bell, in Haggerty *et al.* (1970), demonstrated that chromium in an Apollo 11 basaltic olivine is present as Cr^{2+} and although the state of Cr is not known for the 14072 symplectites, the association with metallic Fe as a derived product of olivine decomposition would tend to suggest that the symplectites were formed by subsolidus reduction. We have demonstrated experimentally that the reduction of terrestrial olivine (Fa_{80}) does yield metallic iron and one lunar example is known, sample 14053 (El Goresy *et al.*, 1972; Haggerty, 1972; Longhi *et al.*, 1972). Although the presence of a Cr-rich constituent has not been demonstrated, the textural similarity between the 14072 decorated dislocations and those in terrestrial olivines is indisputable. The proposal by Roedder and Weiblen (in Bell *et al.*, 1975) therefore, is favored for the migration of minor elements and the suggestion, furthermore, is made here that this migration is along favorable site dislocations in the olivine. This proposal is supported by the present study which has demonstrated that zones depleted in FeO are particularly characteristic of olivine immediately adjacent to the symplectite assemblage. In summary, the symplectites are considered to have formed by early subsolidus reduction of phenocrystic olivine, and this may have taken place simultaneously with the decomposition of chromian-ulvöspinel mantles on chromite.

Temporal, T°C and f_{O_2} Relationships for Subsolidus Reduction

To evaluate the timing of subsolidus reduction in sample 14072, and to estimate T°C and f_{O_2}, the following data summary is pertinent: (1) early spinels, which are Al, Mg, and Cr-rich exhibit no evidence of subsolidus reduction, (2) lower aluminian–magnesian–chromian spinels which form mantles on early chromite do exhibit reduction and the assemblage is ilmenite with minor concentrations of metallic iron; troilite is absent, (3) spinels which precipitated at a late stage in the crystallization sequence are intensely reduced and these contain abundant Fe + FeS, (4) in some instances only metallic iron or troilite is present in the most reduced spinels, (5) troilite is absent in silicate melt inclusions in olivine, but is present in a small number of eutectic intergrowths with metallic iron and is also present as a minor constituent in the residual mesostases assemblages, (6) metals associated with spinel decomposition have Co contents which are intermediate between those metals which are present in silicate melt inclusions and those present in late-stage mesostases, (7) the terminal compositions for spinels crystallizing from the melt or for spinels which have reequilibrated by reduction is approximately 47 mole% Fe_2TiO_4, (8) the ilmenite mineral chemistry of reduced spinels is related to spinel composition, to the modal abundances of Fe + FeS, and to the proportion of ilmenite present in these reduction assemblages, (9) primary ilmenite compositions span the entire range of compositions exhibited by ilmenites in subsolidus spinel associations, and (10) the earliest formed magnesian olivines display metal-bearing Cr-rich decorated dislocations, olivines of intermediate composition are non-reduced, and metallic iron-tridymite eutectics in mesostases may have resulted from fayalite decomposition.

The distribution of troilite suggests that sulfur was an active component during the reduction of late formed spinels and that the activity of sulfur was extremely low during the early stages of crystallization. A subsequent sulfidation event could not have taken place, and if it did the selective transformation of Fe → FeS would have to be accounted for, noting that many spinels have both Fe + FeS present in the decomposition assemblage. The suggestion, therefore, is that the reduction of these spinels occurred in the field of troilite stability. Because the more Al, Mg, and Cr-rich spinels also exhibit reduction and because only ilmenite ± Fe is present, this decomposition must have been active at temperatures higher than those at which troilite precipitated. These latter spinels are more stable than the later Ti-rich varieties with respect to reduction and that applies also to olivine. Given the fact that the more stable constituents (i.e., Mg-rich phenocrystic olivine and Al–Mg–Cr-rich mantles on chromite) display decomposition, that the more Fe-rich varieties of olivine are stable, and that troilite is absent, it follows that a high temperature episode of reduction was active prior to the crystallization of these olivines and prior also to the crystallization of chromian-ulvöspinel. Since troilite is only present in those spinels which were originally enriched in Fe_2TiO_4, and since the metals associated with this phase of reduction are enriched in Ni and Co it implies that a substantial fraction of residual melt must still have been present and that it was in a highly reduced state. This suggestion is corroborated by the analytical data for primary spinel compositions and for spinel compositions which have undergone reduction which show that the terminal compositions for both are 47 mole% Fe_2TiO_4. Therefore reduction and crystallization were very nearly that of an equilibrium condition, a situation which seems otherwise implausible in the absence of an actively crystallizing liquid. In summary, therefore, it is proposed that reduction was continuous with crystallization and that the more stable constituents were reduced at elevated temperatures and the less stable components at correspondingly lower temperatures in the field of active troilite precipitation.

An apparent anomaly in this model is that the more Fe-rich olivines are not decomposed as is the case in 14053, where the initial compositions must have approached Fa_{100} (El Goresy *et al.*, 1972; Haggerty, 1972). A possible explanation for the lack of decomposition may simply be that these olivines (Fo_{35-46}) are stable even under the condition at which late-stage chromian-ulvöspinel is unstable. The relative degrees of reduction susceptibility may also arise as a consequence of kinetics and this aspect has been addressed previously in the context of olivine and of ilmenite decomposition by Taylor *et al.* (1972) and by El Goresy *et al.* (1972). A major drawback currently in any attempt to evaluate relative mineral stabilities, in terms of T°C and f_{O_2}, is the absence of experimental data for members of intermediate compositions. Thermodynamic data are available for ulvöspinel and chromite (Haggerty, 1971) and for the olivine series (Nitsan, 1974). In addition, experimental data do exist for ulvöspinel, for ilmenite (Taylor *et al.*, 1972) and also for fayalite (Williams, 1971). Taylor *et al.* (1972) have noted that the fayalite → quartz + iron curve intersects the ulvöspinel → ilmenite + iron curve at approximately 950°C and at an $f_{O_2} = 10^{-18}$ atm; lower values of f_{O_2} are required for fayalite reduction below this temperature, whereas a lower f_{O_2} is required for ulvöspinel reduction above 950°C. The dilution of Mg, Al, and Cr in ulvöspinel will increase its stability to lower values of f_{O_2} (Haggerty, 1972), and this applies also to higher concentrations of MgO in fayalite. At 950°C, chromite is stable to approximately 10^{-19} atm (Haggerty, 1971) and the range for the olivine series at this temperature is $10^{-16.5}$–$10^{-18.5}$ as reported by Nitsan (1974). Although the range in possible f_{O_2} can only be approximated, Fo_{50} and Usp_{50} are likely to decompose at the following values of T°C and f_{O_2} atm, respectively: at 1000°C, $10^{-16.3}$ and $10^{-16.8}$; at 900°C, $10^{-17.9}$ and $10^{-18.6}$; at 800°C, $10^{-19.9}$ and $10^{-21.0}$; and at 700°C, $10^{-22.3}$ and $10^{-23.5}$ atm. For both reactions metallic iron will be a decomposition constituent. It is important to emphasize that the estimates for olivine decomposition are closer to reality than those of the spinel mineral group because in the latter the influence of either MgO or Al_2O_3 has not been considered. In all cases, however, these two minerals, at the 50 mole% level, are likely to be no more than an order of magnitude apart in terms of f_{O_2} stability for the temperature range 700–1000°C. Because olivines that are more Fe-rich than Fo_{50} (for example those that are present in plagioclase and in the mesostases) have not decomposed, and because there is no evidence for the decomposition of ilmenite, the final equilibration range for subsolidus reduction is most likely to have been between 700 and 1000°C and between 10^{-16} and 10^{-23} atm. Most lunar basalts are considered to have crystallized at values of f_{O_2} in the range 10^{-13}–10^{-14} atm at ~1125°C (e.g., Sato and Helz, 1971; Sato *et al.*, 1973) and since intense subsolidus reduction is relatively rare in mare basalts one may confidently assume that samples such as 14053 and 14072 are extreme examples of reduction reequilibration. These samples are furthermore the only known examples that exhibit the decomposition of olivine to $Fe + SiO_2$. Although the nature of the reducing agent remains unknown the metal chemistry and the distribution of troilite suggests that reduction was active during the entire period of crystallization and deuteric cooling. Such a condition could have resulted in an extremely thick lava flow (i.e., 10–20 m), in a flow which was buried soon after eruption, or alternatively in a unit which cooled at hypabyssal levels. In this magmatic-deuteric model the reducing agent would be confined and a continuous interaction among liquid, solid, and gaseous constituents would result.

SUMMARY AND CONCLUSIONS

Sample 14072 is an intensely reduced basalt which exhibits a wide range in spinel reduction and in olivine decomposition. The more stable Mg, Al, and Cr spinels are free of reduction, those slightly enriched in Fe + Ti exhibit mild reduction and metallic iron is either a minor constituent or is absent. Advanced reduction is only observed in late-stage chromian-ulvöspinels and in these the abundance of metallic iron and of troilite is proportional to the abundance of ilmenite which results from decomposition. For olivine, it is only the more magnesian and early formed varieties which have decomposed and in these, metal-bearing Cr-rich symplectites which decorate dislocations in corroded phenocrysts are considered to have developed in an early episode of reduction. The Co contents of metallic iron are interpreted here as being indicative of stages of crystallization, with high-Co values appearing in trapped silicate melt inclusions and low-Co values present in tridymite eutectoids in mesostases which also contain a complex phosphate assemblage and glass. Cobalt contents of metals associated with spinel reduction have intermediate values between the extremes of early and late-stage precipitates and because these are also associated with troilite, reduction is considered to have taken place during the precipitation of FeS in the presence of an actively crystallizing liquid. The model invoked for reduction is a magmatic-deuteric process which requires an environment approximating that of a closed system in which the volatile constituents are contained and in which reduction proceeds continuously with crystallization. The respective responses to reduction and to crystallization were close to that of equilibrium based on the terminal compositions of both primary and reequilibrated spinels with ulvöspinel = 47 mole%.

Acknowledgments—This study was supported by NASA under grant NGR-22-010-089 which is gratefully acknowledged. The Department of Geology's electron microprobe facility was supported by the National Science Foundation and by the University of Massachusetts under grant GA-74206. Maintenance, instrumental design and soft ware modifications to the electron microprobe were skillfully effected by Dr. G. Yagunoff. Fruitful but at times heated discussions have resulted from views aired with R. Brett, A. El Goresy, E. Gibson, D. Lindsley, M. Sato, and L. Taylor on the problems and mechanisms of reduction but their stimulating insights have been invaluable. The manuscript was substantially improved by the reviewers K. Misra, J. Friel, and R. Williams.

References

Bell P. M., Mao H. K., Roedder E., and Weiblen P. W. (1975) The problem of the origin of symplectites in olivine-bearing lunar rocks. *Proc. Lunar Sci. Conf. 6th*, p. 231–248.

Brett R. (1976) Reduction of mare basalts by sulfur loss. *Geochim. Cosmochim. Acta* **40**, 997–1004.

Champness P. E. (1970) Nucleation and growth of iron oxides in olivines, $(MgFe)_2SiO_4$. *Mineral. Mag.* **37**, 790–800.

Dymek R. F., Albee A. L., and Chodos A. A. (1975) Comparative petrology of lunar cumulative rocks of possible primary origin: Dunite 72415, troctolite 76535, norite 78235, and anorthosite 62237. *Proc. Lunar Sci. Conf. 6th*, p. 301–341.

El Goresy A. (1976) Oxide minerals in lunar rocks. In *Oxide Minerals. Min. Soc. Amer. Short Course Notes* (D. Rumble III, ed.), Vol. 3, p. EG 1–46.

El Goresy A., Taylor L. A., and Ramdohr P. (1972) Fra Mauro crystalline rocks: Mineralogy, geochemistry and subsolidus reduction of the opaque minerals. *Proc. Lunar Sci. Conf. 3rd*, p. 339–349.

Goldstein J. I., Hewins R. H., and Axon H. J. (1974) Metal silicate relationships in Apollo 17 soils. *Proc. Lunar Sci. Conf. 5th*, p. 653–671.

Goldstein J. I. and Yakowitz H. (1971) Metallic inclusions and metal particles in the Apollo 12 soil. *Proc. Lunar Sci. Conf. 2nd*, p. 177–191.

Haggerty S. E. (1971) Subsolidus reduction of lunar spinels. *Nature Phys. Sci.* **234**, 113–117.

Haggerty S. E. (1972) Apollo 14: Subsolidus reduction and compositional variation of spinels. *Proc. Lunar Sci. Conf. 3rd*, p. 305–332.

Haggerty S. E. (1975) Geochemistry of opaque oxides in troctolites and basalts from Taurus Littrow (abstract). In *Lunar Science VI*, p. 321–323. The Lunar Science Institute, Houston.

Haggerty S. E. and Baker I. (1967) The alteration of olivine in basaltic and associated lavas. Part 1. High temperature alteration. *Contrib. Mineral. Petrol.* **16**, 233–257.

Haggerty S. E., Boyd F. R., Bell P. M., Finger L. W., and Bryan W. B. (1970) Opaque minerals and olivine in lavas and breccias from Mare Tranquillitatis. *Proc. Apollo II Lunar Sci. Conf.*, p. 513–518.

Hewins R. H. and Goldstein J. I. (1974) Metal-olivine associations and Ni–Co contents in two Apollo 12 mare basalts. *Earth Planet. Sci. Lett.* **24**, 59–70.

Kohlstedt D. L. and Vander Sande J. B. (1975) An electron microscope study of naturally occurring oxidation produced precipitates in iron-bearing olivines. *Contrib. Mineral. Petrol.* **53**, 13–24.

Kohlstedt D. L., Goetze C., Durham W. B., and Vander Sande J. B. (1976) New technique for decorating dislocations in olivine. *Science* **191**, 1045–1047.

Longhi J., Walker D., and Hays J. F. (1972) Petrography and crystallization history of basalts 14310 and 14072. *Proc. Lunar Sci. Conf. 3rd*, p. 131–139.

Nitsan U. (1974) Stability field of olivine with respect to oxidation and reduction. *J. Geophys. Res.* **79**, 706–711.

Roedder E. and Weiblen P. W. (1972) Petrographic features and petrologic significance of melt inclusions in Apollo 14 and 15 rocks. *Proc. Lunar Sci. Conf. 3rd*, 251–279.

Sato M. and Helz R. T. (1971) Oxygen fugacity values of Apollo 12 basaltic rocks. In Unpublished Papers from Second Lunar Science Conference.

Sato M., Hickling N., and Mclane J. E. (1973) Oxygen fugacity values of Apollo 12, 14 and 15 lunar samples and reduced state of lunar magmas. *Proc. Lunar Sci. Conf. 4th*, p. 1061–1079.

Taylor L. A., Williams R. J., and McCallister R. H. (1972) Stability relations of ilmenite and ulvöspinel in the Fe–Ti–O system and application of these data to lunar mineral assemblages. *Earth Planet. Sci. Lett.* **16**, 282–288.

Williams R. J. (1971) Reaction constants in the system Fe–MgO–SiO_2–O_2 at 1 atm, between 900° and 1300°C: experimental results. *Amer. J. Sci.* **270**, 334–347.

Zeunch D. H. and Green H. W. (1976) Naturally decorated dislocations in olivine from periodotite xenoliths. *Geology* preprint.

Proc. Lunar Sci. Conf. 8th (1977), p. 1831–1838.
Printed in the United States of America

Structural studies on tranquillityite and related synthetic phases

B. M. GATEHOUSE

Chemistry Department, Monash University, Clayton, Victoria, 3168, Australia

IAN E. GREY

CSIRO Division of Mineral Chemistry, P.O. Box 124, Port Melbourne, Victoria, 3207, Australia

J. F. LOVERING and D. A. WARK

Department of Geology, University of Melbourne, Parkville, Victoria, 3052, Australia

Abstract—Tranquillityite, $Fe_8^{2+}Zr_2Ti_3Si_3O_{24}$, which also contains minor amounts of Ca, Al, Mn, Cr, Nb, REE, Hf, and U, occurs in association with late-stage interstitial phases in both mare and non-mare basalts from all Apollo sites. In its natural state it is metamict and does not give an X-ray diffraction pattern. Reconstitution of the structure was carried out in both air and vacuum and the reconstituted mineral was studied by single-crystal X-ray diffraction methods. It was found to have a fluorite-related structure. Attempts to synthesize tranquillityite in the laboratory are described. A structure determination for a related MnZrTi-silicate synthetic phase was carried out and its possible relationship to the tranquillityite structure is discussed.

INTRODUCTION

TRANQUILLITYITE $[Fe_8^{2+}(Y+Zr)_2Ti_3Si_3O_{24}]$ a completely new silicate mineral, was originally described by Lovering *et al.* (1971) and Lovering and Wark (1971) as a late-stage interstitial phase in basaltic rocks from the Apollo 11 and 12 sites. More recent work by Lovering and Wark (1975) has established that tranquillityite occurs in both mare and non-mare basalts from all Apollo sites so that it is a common product of the late-stage crystallization of a wide variety of lunar basaltic magmas.

In the original description by Lovering *et al.* (1971) it was shown that lunar tranquillityite contained minor amounts of Ca, Al, Mn, Cr, Nb, REE, and Hf, along with trace amounts (i.e., ~50 ppm to a few hundred ppm) of uranium. Although it was previously thought that lunar tranquillityite had a crystalline structure, more recent work by us has shown that it is metamict, presumably as a result of radiation damage from the uranium in the mineral structure undergoing α-decay over periods in excess of 3×10^9 yr. The present study was instituted to determine the detailed structure of tranquillityite both by means of structural studies on synthetic tranquillityite-like phases and by attempts to reconstitute the structure of natural tranquillityites extracted from lunar basalts.

X-RAY DIFFRACTION STUDIES ON NATURAL TRANQUILLITYITE

Presumably because of radiation damage from the contained uranium, all tranquillityite grains extracted from lunar basalts are found to be metamict. Long-exposure oscillation photographs failed to produce any observable diffraction effects; however, single-crystal diffraction patterns could be obtained for tranquillityite grains which had been heated in air to 800°C for 30 min. A precession photograph of the resulting reconstituted mineral is shown in Fig. 1(i). It is seen that the diffraction spots are quite broad and diffuse, indicating domain formation within the crystal. From measurements of the spot broadening the average domain diameter is calculated to be 40 Å. Longer heating time did not improve the quality of the diffraction patterns from the crystals. When heated at temperatures above 900°C, the tranquillityite structure began to break down, as evidenced by the change in the diffraction patterns.

Precession photographs such as the one shown in Fig. 1(i) could be indexed on a face-centred cubic cell, $a = 4.85(3)$ Å. The intensity distribution of reflections is consistent with a fluorite-related structure. The anion to metal ratio in tranquillityite is 1.5 whereas in fluorite it is 2 and so anion vacancies exist in the former. However, for the mineral reconstituted in air, the unit cell shows no detectable deviation from cubic symmetry and no evidence of superlattice reflections, and so it is concluded that the oxygen vacancies are disordered.

In more recent experiments we have carried out reconstitution experiments in vacuo. Grains of tranquillityite were heated to 800°C in sealed silica tubes evacuated to 10^{-4} torr. Following this treatment, the mineral gave diffraction patterns comprising very sharp diffraction spots, in contrast to the air-reconstituted samples. Precession photographs for the mineral reconstituted in vacuo are shown in Fig. 1(ii) and (iii).

Furthermore, for the mineral reconstituted in vacuo the diffraction patterns showed a small rhombohedral distortion from cubic symmetry. Unit cell parameters were calculated from a least-squares fit of 20 values of strong reflections, obtained on a PW1100 four-circle diffractometer. They are $a_{rh} = 4.743(5)$ Å; $\alpha_{rh} = 88.57(6)$ Å. A number of weaker reflections were also recorded which could be indexed on a "double" rhombohedral cell with $a_{rh} = 2 \times 4.743 = 9.486$ Å. However, the assignment of a superlattice on the basis of our measured data is somewhat tenuous because the crystals which we have studied have been very small (maximum dimension ~35 μm) and the measured intensities of all but the subcell reflections would be considered as "unobserved" in terms of the counting statistics.

ATTEMPTED LABORATORY SYNTHESES OF TRANQUILLITYITE

In addition to the diffraction studies on the natural mineral, we have attempted to prepare samples of synthetic tranquillityite. Mixtures of powdered oxides were weighed out according to the composition for the mineral and were pelleted and reacted for various times in both sealed tubes and controlled

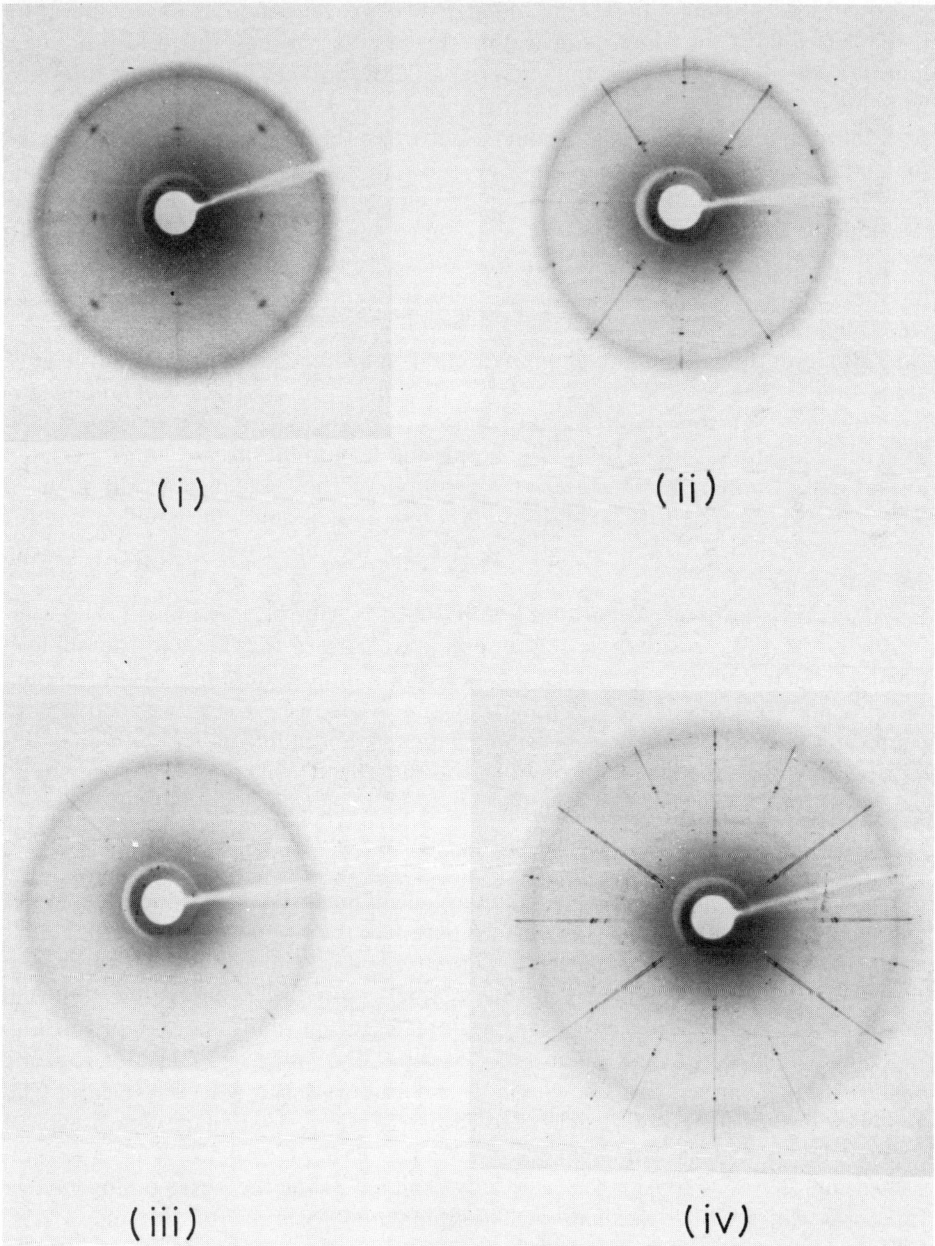

Fig. 1. Precession photographs for (i) tranquillityite reconstituted in air at 800°C; (ii) and (iii) tranquillityite reconstituted in vacuo at 800°C; (iv) synthetic MnTiZr-silicate. (Doubling of relections towards the centres of photographs is an artifact caused by slight buckling of the film in the cassette).

gaseous buffers using CO_2/H_2 mixtures. However, reactions in the temperature range 800–1500°C (at which temperature the sample melted), failed to produce a fluorite-type phase. In a subsequent series of experiments the synthetic mixtures were heated in air at various temperatures. At 1350°C, zircon, $ZrTiO_4$, and pseudobrookite were the only products, but when the temperature was increased to 1500°C, the X-ray diffractograms on the reacted samples showed additional lines due to a cubic fluorite-related phase with $a = 4.74$ Å. It was observed that when about 10 at.% of the iron was replaced by manganese, the product obtained from 1500°C heat in air had the fluorite phase as the major one in the reaction mixture and at 50 at.% replacement, only lines due to fluorite-related phases were observed.

SEM and reflected light examination of polished sections of this 50 at.% preparation revealed the presence of a polyphase intergrowth containing zirconolite-like and tranquillityite-like phases, together with ZrY-oxide, MnFe-titanate, and Mn-silicates. Both zirconolite and tranquillityite are fluorite-related phases with similar XRD patterns. Fortuitously the Mn-silicate, although of different structure, had its main diffraction line coincident with the main line for the fluorite-related structures, giving rise to the apparently "pure" fluorite diffractogram.

A further synthetic preparation, with 100 at.% Mn, was examined with substantially the same results, i.e., zirconolite-like MnZrY-titanate and tranquillityite-like MnTiZr-silicate as major phases, with minor amounts of ZrY-oxide, Mn-titanate and Mn-silicate forming the matrix. However, in this second preparation there was better separation of the "tranquillityite"/"zirconolite" intergrowth, with some clean 50 μm euhedral rhombs of MnTiZr-silicate able to be removed for X-ray crystallographic study. Euhedral or subhedral cores of Mn-tranquillityite tended to be surrounded by an irregular rosette of Mn-zirconolite. The minor Mn-titanate occurred in a characteristic acicular habit reminiscent of skeletal ilmenite in lunar basalt mesostases. In fact, the entire assemblage of synthetic phases closely parallels the assemblage of zirconolite-tranquillityite-baddeleyite-ilmenite-fayalite typical of lunar basalt mesostases. The phase assemblage obtained in the synthetic mixture is compared with that observed for a typical natural lunar assemblage in the photomicrographs in Fig. 2. Other illustrations are given in Lovering and Wark (1971). Microprobe analyses were carried out on a number of grains of the MnTiZr-silicate. The compositions of different grains were found to be very similar, with a typical analysis giving 45.66% Mn, 4.35% Si, 4.25% Y, 7.85% Zr, and 5.27% Ti. A crystal measuring $35 \times 35 \times 10$ μm was carefully isolated from the surrounding matrix for X-ray diffraction examination. The calculated formula for this phase normalized to 24 oxygens (*vide infra*) is:

$$Mn_{10.82}Ti_{1.43}(Y,Zr)_{1.74}Si_{2.01}O_{24}.$$

By comparison with the composition of tranquillityite, $Fe_8Ti_3(Y,Zr)_2Si_3O_{24}$, it is apparent that although atomic proportions vary from one to the other, both

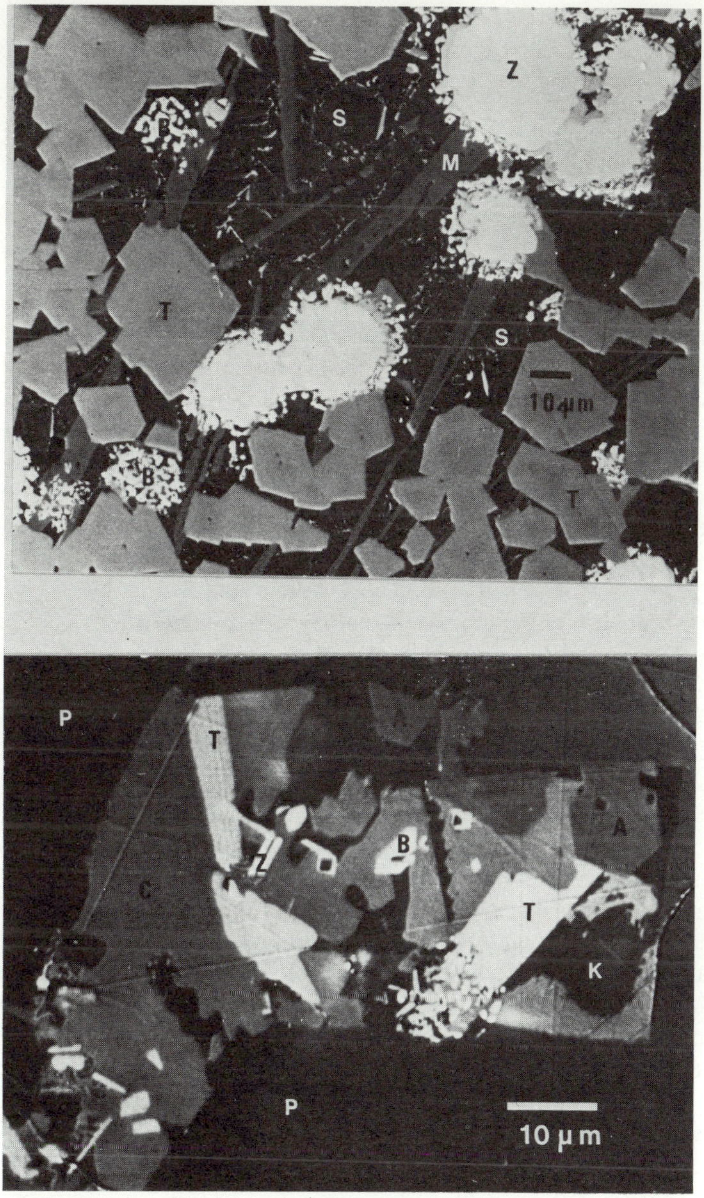

Fig. 2(i). Backscattered electron photograph of phase assemblage in a synthetic tranquillityite preparation in which Mn replaces all Fe. (ii) Oil immersion reflected light photograph of natural tranquillityite-zirconolite-baddeleyite assemblage in mesostasis of mare basalt 10047, 20. A = apatite; B = baddeleyite or YZr-oxide; K = potassium feldspar; M = Mn-titanate; P = plagioclase; S = Mn-silicate; T = tranquillityite or MnTiZr-silicate; Z = zirconolite or MnYZr-titanate; C = clinopyroxene.

compounds do contain the same cation types and they both have the same metal: oxygen ratio of 2:3. Furthermore, the X-ray diffraction data for the two are very similar, as will be reported below, and so it was decided to proceed with a crystal structure analysis of the synthetic phase in the expectation that it would assist in the interpretation of the natural material.

STRUCTURAL STUDIES IN MnTiZr-SILICATE

A precession photograph of the MnTiZr-silicate is shown in Fig. 1(iv). The close similarity between the diffraction pattern of the synthetic phase and that of the natural tranquillityite, Fig. 1(iii), is obvious. However, the pattern for the synthetic phase has, in addition to the main fluorite sub-cell spots, a number of weaker superlattice reflections which could all be indexed on a tetragonal cell with $a = 9.501(4)$ Å and $c = 19.049(8)$ Å. The systematic extinctions uniquely defined space group $I4_1/acd$. Similar unit cell dimensions have been reported for the mineral braunite, Mn_7SiO_{12} (J.P.R. de Villiers, 1975), which also has the same space group.

A set of X-ray intensity data was collected for the MnZrTi-silicate on a PW1100 automatic diffractometer, using MoKα radiation. The braunite-type structure was found to be consistent with the Patterson map and so the atomic coordinates for braunite were used to initiate refinement of the structure. Because of the very small size and poor quality of the crystal [weak reflections from a second phase appear in the precession photograph, Fig. 1(iv)], a satisfactory convergence of the structure refinement was not achieved. However, the final R factor of 0.10 for 82 reflections with $I > 2\sigma I$ and the reasonable atomic distances obtained support the general correctness of the braunite-type structure. It may be described as a fluorite-related structure in which one quarter of the oxygens have been removed in an ordered way to give three main types of metal-oxygen coordination polyhedra, i.e., tetrahedron, octahedron, and cube. We find that in the MnTiZr-silicate, silicon orders into the tetrahedral site, zirconium and yttrium into the cubic site, and trivalent manganese plus titanium into two octahedral sites, and divalent manganese into a third octahedral site. From the structure solution the partitioning of manganese between the divalent and trivalent oxidation states has been determined, and combined with the microprobe analysis, has been used to calculate the unit cell composition:

$$Mn^{2+}_{9.14}Mn^{3+}_{12.49}Ti_{2.86}Zr_{2.24}Y_{1.24}Si_{4.03}O_{48}.$$

The corresponding ideal end member composition is

$$Mn^{2+}_{8}Mn^{3+}_{16}Zr_4Si_4O_{48} = Mn_6ZrSiO_{12},$$

i.e., it is derived from the braunite composition by the replacement of 2 Mn^{3+} by $Mn^{2+} + Zr^{4+}$.

Application to the Structure of Tranquillityite

The determination of the structure of the synthetic MnTiZr-silicate has some relevance to tranquillityite because it affords an explanation of the apparent contradictory facts that the lunar mineral has a fluorite-related structure, yet contains a significant amount (~ 13–16 wt.%) of silica. Prior to the determination of the structure of braunite all reported fluorite-related structures had metal atoms in six-fold coordination or higher (e.g., pyrochlore, C-rare earth, M_7O_{12} structures). The braunite structure, however, is an illustration of the high degree of adaptability of fluorite-type structures, in incorporating four-, six-, and eight-coordinated metal atoms. In the MnZrTi-silicate the silicon, titanium and manganese, and yttrium and zirconium are ordered onto a cation sublattice exactly as in fluorite, and the range of different coordinations is achieved solely by the ordered removal of some oxygens from the fluorite anion sublattice together with small movements of the remaining anions.

A similar situation may be envisaged for tranquillityite, which has the same types of cation i.e., Si, Zr, Ti but with Fe in place of Mn, and also the same metal:oxygen ratio of 2:3. Although the density of natural tranquillityite has not been measured a calculation of it, based on the "double" rhombohedral cell, i.e., $a = 9.486$ Å, $\alpha = 88.57°$ and two formula units of $Fe_8Zr_2Ti_3Si_3O_{24}$ per unit cell, gives $\rho = 4.83 \text{ g cm}^{-3}$, which is very reasonable for a fluorite-related structure. For comparison, the density of braunite is 4.77 g cm^{-3}. By analogy with the structure for the MnZrTi-silicate, it is reasonable to assume that the 32 metal ions per unit cell in tranquillityite will also be ordered on a fluorite cation sublattice, and that 16 of the fluorite anion positions per unit cell are unoccupied and the remaining 48 anion sites are occupied in an ordered way by oxygen, to give tetrahedral coordination around the silicon, and higher coordination spheres about the other metals.

Unfortunately, the single crystals of tranquillityite which have been extracted were too small to carry out a complete structure analysis; further structural studies of this interesting lunar mineral are now proceeding on larger, more suitable samples.

Acknowledgments—We wish to thank the National Aeronautics and Space Administration, the Lunar Sample Analysis Planning Team and, in particular, the Lunar Sample Curator and his assistants for their continuing and generous help in selecting appropriate thin sections from the collection and for making up special thick sections for this work. We also thank the Department of Science, Commonwealth Scientific and Industrial Research Organization, the Australian Institute of Nuclear Science and Engineering, and the Australian Research Grants Committee for support of this work.

References

De Villiers J. P. R. (1975) The crystal structure of braunite with reference to its solid-solution behaviour. *Amer. Mineral.* **60**, 1098–1104.

Lovering J. F. and Wark D. A. (1971) Uranium-enriched phases in Apollo 11 and Apollo 12 basaltic rocks. *Proc. Lunar Sci. Conf. 2nd*, p. 151–158.

Lovering J. F. and Wark D. A. (1975) The lunar crust: Chemically defined rock groups and their K–U fractionation. *Proc. Lunar Sci. Conf. 6th*, p. 1203–1217.

Lovering J. F., Wark D. A., Reid A. F., Ware N. G., Keil K., Prinz M., Bunch T. E., El Goresy A., Ramdohr P., Brown G. M., Peckett A., Phillips R., Cameron E. N., Douglas J. A. V., and Plant A. G. (1971) Tranquillityite: A new silicate mineral from Apollo 11 and Apollo 12 basaltic rocks. *Proc. Lunar Sci. Conf. 2nd*, p. 39–45.

Proc. Lunar Sci. Conf. 8th (1977), p. 1839–1847.
Printed in the United States of America

Applications of a new logical search on the lunar data base for use in pattern recognition classification of Apollo 15 mare basalts

D. D. PRATT, C. B. MOORE, M. L. PARSONS, and D. L. ANDERSON

Arizona State University, Tempe, Arizona 85281

Abstract—Pattern recognition techniques were used to study 51 sets of chemical analyses on 30 Apollo 15 mare basalts. Two chemical and petrological basaltic groups were found; olivine-normative and quartz-normative with no chemical subgroups present in either. Coarse-grained samples 15555 and 15065 presented a problem due to large variations in individual analyses. A computer data base tape containing about 190,000 geochemical, chemical, age dating, and petrological analyses compiled by Warner and coworkers at the NASA Johnson Space Center in Houston was obtained, and a program incorporating Boolean Search capabilities was adapted to search this file for additional analyses in order to replace single analyses with averages. Twenty additional sets of data on Apollo 15 basaltic samples were found with the search and their classification determined. Two samples, 15388 and 15385, were found by pattern recognition to have chemical analyses quite distinct from either of the two major Apollo 15 basaltic groups.

INTRODUCTION

THE APOLLO MISSIONS RETURNED a variety of samples from various parts of the lunar surface. Chemical analyses of these samples have produced large quantities of major and trace element compositional data. It is imperative to compile the data in some useful form so that analyses of interest can be found. This has been done by Jeffery Warner and coworkers at the NASA Johnson Space Center in Houston.

A computer data base containing about 190,000 geochemical, chemical, age dating, petrological, and mineral analyses has been collected. The data on the tape include that taken from the Preliminary Examining Team's results in *Science* plus all lunar science articles in *Lunar Science* abstracts, the *Proceedings* of the lunar science conferences, the *Apollo 15 Lunar Samples*, *Earth and Planetary Science Letters*, *Geochimica et Cosmochimica Acta*, *Nature*, *Science*, and *Journal of Geophysical Research.* A periodic update on the tape is made.

A major problem in interpretation of the data obtained from this tape is the magnitude of information on each sample. Data have been collected in a variety of laboratories and procedures have varied. Published correlations and comparisons usually include those found useful in classical petrochemical studies and are limited to one or more binary scatter diagrams or plots. It is desirable to study a complete segment of the data base simultaneously to find natural classifications based on the total data.

Apollo 15 mare basalts are reported as an example for this study. The initial data base included that compiled by Butler (1976). Computer implemented pattern recognition techniques were used to test the classification scheme. A search of the lunar data base was then made to find additional samples to add to the study, and check analyses on some coarse-grained samples that presented classification problems.

APOLLO 15 MARE BASALTS

It is generally accepted that two major chemical groups of mare basalts were returned during the Apollo 15 mission (Butler, 1976). Each of these classes have petrologically been further divided into two textural groupings according to thin section examinations. The first major group has been labeled olivine-normative basalt, characterized by higher FeO and TiO_2 values and lower SiO_2 composition; the second group contains quartz-normative basalts with lower FeO and TiO_2 and higher SiO_2 values (Butler, 1976). The quartz-normative basalt samples were collected from widely distributed regions of the Apollo 15 landing site while the olivine-normative were mostly confined to Station 9A at Hadley Rille (LSPET, 1972). In this paper pattern recognition techniques were used to investigate further subdivisions of the two major classes based solely on chemical data.

PATTERN RECOGNITION

In pattern recognition studies, each variable (in this case, the measured value for each element) is treated as a separate dimension. This is normally the case when one element percentage is plotted against that of a second element. Pattern recognition extends this idea to *n*-dimensions so all *n* variables may be considered simultaneously. Visual plots in *n*-dimension are not possible, but statistical manipulations and distances can be defined in *n*-dimensions, and various types of dimension reduction into two dimensions can be made.

The pattern recognition program used was obtained from B. R. Kowalski, University of Washington (Duewer *et al.*). The program, referred to as ARTHUR, was developed primarily for chemical and spectroscopical data classifications. Data are usually first autoscaled by subroutine SCALE*. This process gives the measured values for each element (called a feature) a mean of zero and a standard deviation of one to give equal weight to all variables. This subroutine also calculates means, standard deviations, and ranges (Kowalski *et al.*, 1972).

After data are autoscaled, many possibilities exist. If data without previous knowledge of categories are to be examined, a process called unsupervised learning is used. In unsupervised learning, natural classifications are sought by comparing similarities (usually based on a distance measure) between samples. Two choices are present in this package: TREE (minimal spanning tree) and HIER (hierarchical clustering). Similar types of clustering have been used in geochemical literature (Howarth, 1973).

*All subroutines in the pattern recognition program will be identified by their subroutine names and are given in capital letters.

After categories are found by unsupervised learning, supervised learning methods are available to check how well data separation has been achieved and to define the separating surfaces. Supervised learning includes subroutine KNN which checks category classifications on the basis of the K nearest neighbors where K is 1, 3, 4, 5, 6, 7, 8, 9, and 10 (Cover, 1967). A pattern belongs to the category which is represented most often among its K nearest neighbors.

The data can be plotted into two dimensions, or reduced in the number of features and then plotted into two dimensions. Reduction is accomplished by finding which features give the most useful information for group separation. This is done by subroutine SELECT which produces weighted features which are linearly independent and ordered according to the various system correlations (Kowalski *et al.*, 1972). The best features are chosen to reduce the data to m-dimensions. KARLOV (Karhunen–Loeve) transformation plots can be made (Kowalski *et al.*, 1973). The transformation (also known as principal component analysis) is an orthogonal rotation of the m-dimensional coordinate system, such that the first new coordinate is in the direction of largest variance and the second is orthogonal to the first and contains the second largest amount. If the first two coordinates are now plotted, the mjaority of the variance or information will be retained.

A second type of plotting is nonlinear mapping, NLM, which maps from the original n-dimensions (or m-dimensions if SELECT has been used first) to two or three dimensional space (Kowalski *et al.*, 1973), while minimizing the interpattern distance errors.

The dividing surfaces between categories can be found. MULTI iteratively develops $(n - 1)$ dimensional hyperplanes separating given categories of patterns represented in n-dimensional space. PLANE is similar to MULTI except that it is a binary classifier (Nillsson, 1965). It develops separating hyperplanes between two given categories at a time. Least squares multilinear regressions (LEAST) and stepwise multilinear regressions (STEP) are also available. Correlations can be found (CORREL).

Pattern Recognition Philosophy

Each of the pattern recognition techniques has its limitations. Interpretation of results can be difficult due to distortions in the summary plots and interpreter's bias (Butler, 1976). This is the reason for use of multiple techniques. Pattern recognition gains its power through operator interactive comparisons of many statistically unrelated techniques, and it is only through similar results from all of the techniques that truly significant classifications can be defined.

The Search Routine

The search routine is a major adaptation/remodeling of a program written to search a bibliographic data base of crystallographic references (Motherwell). The program is based on decoding of structured questions using Boolean logic.

Multiple questions can be "asked" per search, each being "asked" simultaneously, thus allowing the data file to be read only once. (In computer searching input/output costs can be enormous. Since this file contains about 190,000 individual analyses, the one pass method is advantageous.)

Each analysis on the tape is comprised of nine fields: Mission number, sample number, phase of analysis (all, part, etc.) element (element, oxide, or dating technique), value*, unit of value*, tag (numbers assigned to replicate analysis for individuality), method of analysis, and accession number (reference number for complementary bibliographic data base).

Even though the program uses a single pass method, reading 190,000 card images is expensive. To save on computer time, the original file has been restructured in two ways. First, the card images are converted to binary and second, the data are written in 1792 word blocks†. This reduces the number of reads to about 950, each analysis being stripped out of the block. Compared to the original form, the blocked-binary data can be read and decoded in 1/100 the time.

Questions are input as data strings connected by one of three logical operators (AND, OR, or NOT). Multiple search fields can be specified per question with multiple strings per field. This allows the structuring of complex questions which can save on the printing of extraneous analyses. Each string is stored previous to the search along with its field and logical importance.

For each analysis, *all* the necessary fields are searched for matches to *all* the strings of *all* the questions. If all strings to a certain question are found, then the logical outcome is calculated. If the answer is true, the analysis is stored, sorted for that particular question. Once the entire file has been searched, the questions are printed separately with all their matched analyses.

RESULTS OF CLASSIFICATION

Unsupervised pattern recognition techniques were initially used. One sample, 15557 was found to be peculiar in its analyses. The published data for SiO_2 on this mare basalt according to Cuttitta *et al.*, (1973) was 25.74% well outside of the range of normal Apollo 15 basalts. A search of the lunar data base obtained 179 pieces of data on this rock including four individual SiO_2 analyses; 45.74, 45.01, 45.06, and 25.74%. This search confirmed the belief that the 25.74% analysis was probably a typographical error and should have been 45.74%. Initially unscaled data were used for obtaining a hierarchical clustering dendrogram similar to the procedure of Butler (1976). Similarity of samples was defined by a distance measurement. Three categories were obtained similar to those described by Butler (1976).

The major problem of Butler's work is that the use of unscaled raw data produces large biases in the results obtained. Raw data are difficult to analyze

*Not searchable in current program version.

†This represents the most efficient blocking on UNIVAC 1100/42 computer system.

because of the differences in the relative magnitude of measurements. When hand drawn, two-dimensional plots are made, the scientist automatically scales the data. In order to do the same in a computer, some type of scaling must be used. The utility of pattern recognition techniques such as hierarchical clustering lies in the ability to study all variables simultaneously without previous bias. Failure to scale the data before use gives prejudices towards those whose ranges are the largest in the entire data base regardless of intercategory ranges. Therefore, only FeO, MgO, Al_2O_3, and SiO_2 are being given approximately equal weight in Butler's work. The other six major elements have already been biased against, being less than 50% as important as the first four. A better approach, it is felt, is to scale the data initially to remove these biases.

When this is done, a slight problem with separating the two major petrological categories is found. On further investigation, it is seen that the only problems occur with the two Apollo 15 mare coarse-grained basalts 15065 and 15555. Previous investigators have discussed these two samples at great length, emphasizing the relationship between subsample size and average grain size (Rhodes *et al.*, 1973). Nava (1974) states that different small-sized fragments of these two could classify each within both olivine-normative and quartz-normative. This is exactly what was found in our study. By searching the data base tape 1635 analyses on 15555 were found including 14 sets of major elemental analyses plus major analyses on different fractions of the rock. With this type of data at hand, an average value was obtained and used in place of individual analyses. Sample 15065 has 445 analyses including 12 sets of major elemental data. An average was calculated and used.

Using scaled data for the remaining samples plus these averages the dendrogram given in Fig. 1 was obtained. The numbers listed are the last three digits of the NASA assigned sample number. A clear distinction between the two petrological groups can now be seen breaking at a similarity of 0.42. The average values for 15555 and 15065 are shown with a star. Using this unbiased method the olivine-normative samples no longer show two categories, but simply fall into one large one. To fully utilize the power of pattern recognition techniques, other methods were also investigated using the scaled data base. A KARLOV plot and a nonlinear mapping study produced similar results. This strongly suggests the fallacy of the olivine-normative subdivision.

Additional Samples

The 20 Apollo 15 mare basalt samples in Table 1 were added to the data to determine their classification. Results of five pattern recognition techniques on the autoscaled data are shown with the symbol 1 for quartz-normative and 2 for olivine-normative. Identical classification assignments by all pattern recognition techniques are shown for 19 of these mare basalts. Sample 15388, a coarse basalt, gave classification difficulty, this is also seen in the sample's analyses. The values for Al_2O_3 (12.8%), FeO (17.6%), MgO (7.7%), and TiO_2 (5.5%) are outside the ranges of normal Apollo 15 basalts. Three hundred analyses on 15388

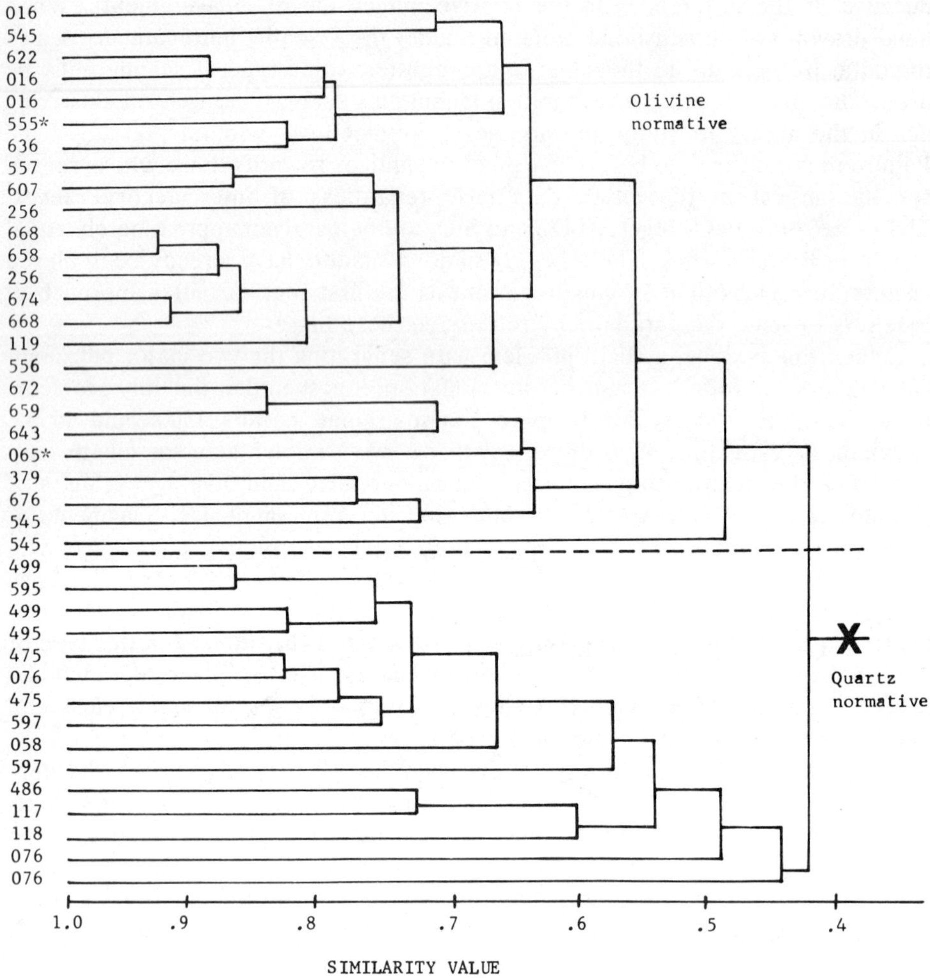

Fig. 1. Hierarchical clustering of 40 Apollo 15 mare basalts. Numbers refer to the last three digits of the NASA assigned sample number.

were found from a search including three sets of major elemental data. The importance of this lies in the fact that triplicate analyses agreed very closely. This confirms the idea that this sample is indeed a unique basalt chemically dissimilar to either of the Apollo 15 groups found, agreeing with Laul *et al.* (1972).

Sample 15385 also was shown to be unique by pattern recognition techniques. A search on this sample gave 176 analyses including three sets of major elemental data that agreed with the values used in this study. The unusual characteristics of these samples can be distinctively seen as well by a KARLOV plot (Fig. 2). This shows the plot of the total set with category zero representing

Table 1. Unknown classification results.

	KNN	KARLOV	MULTI	HIER	TREE
15016	2	2	2	2	2
15058	1	1	1	1	1
15065	2	2	2	2	2
15076	1	1	1	1	1
15105	2	2	2	2	2
15118	1	1	1	1	1
15256	2	2	2	2	2
15385	2	2	2	2	2
15388	2	1	2	2	1
15499	1	1	1	1	1
15545	2	2	2	2	2
15555	2	2	2	2	2
15556	2	2	2	2	2
15557	2	2	2	2	2
15607	2	2	2	2	2
15620	2	2	2	2	2
15633	2	2	2	2	2
15666	1	1	1	1	1
15672	2	2	2	2	2
15676	2	2	2	2	2

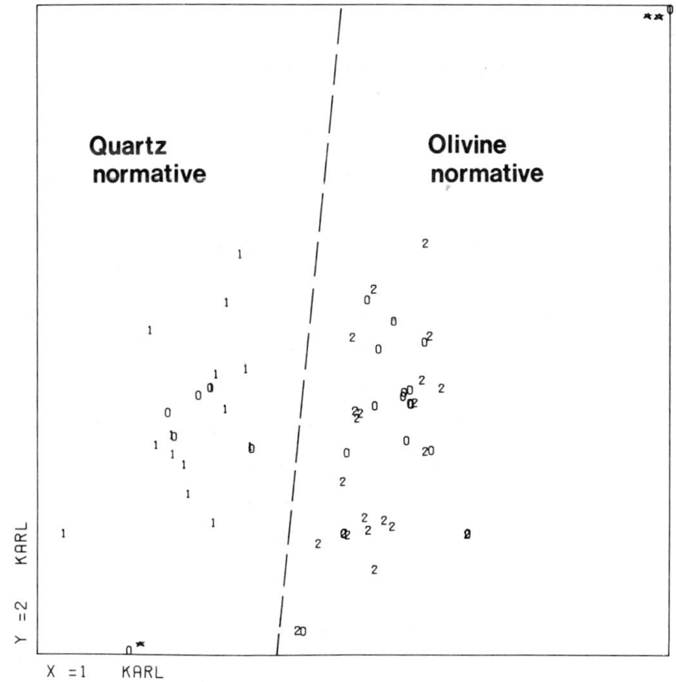

Fig. 2. Karhunen–Loeve plot of knowns and unknowns. 1 = quartz-normative; 2 = olivine-normative; 0 = unknowns; ★ = 15388; ★★ = 15385.

these 20 unknowns. The zero in the figure with one star represents sample 15388; with two stars sample 15385. The remainder of the added unknown samples can be easily categorized into one of the two major groups using Fig. 4 since all are similar to other samples within the groups. The results are summarized in Table 1.

CONCLUSIONS ON THE CLASSIFICATION

(1) It is imperative to use some type of scaling on divergent data before doing multivariate analyses to remove initial bias.
(2) Two major Apollo 15 mare basalt classes are present; olivine-normative and quartz-normative. No subcategories have been found in this study.
(3) Pattern recognition has proven itself as a viable method for studying large data bases. Through comparisons of the individual techniques natural classification results can be obtained.
(4) Apollo 15 samples 15388 and 15385 do not resemble either of the major classes of Apollo 15 mare basalts.
(5) Apollo 15 samples 15555 and 15065 are coarse-grained and multiple analyses on different fragments do not agree. Averages must be used in classification procedures.

It is important to evaluate trace elemental data on these samples. We have previously shown that rare earths are very important in the classification of Apollo 17 mare basalts (Pratt *et al.*, 1977). The compilation of trace element data on Apollo 15 basalts is presently being undertaken.

CONCLUSION ON THE SEARCH

The results demonstrate just one possible use of the search program. Taking into account the total information available and the size of the file, the utility of such a computerized search cannot be overemphasized.

The program is available from the authors, along with a program to restructure the file. The data are available either from the authors or Jeff Warner at NASA.

Acknowledgments—This study was supported in part by NASA grant NGL-03-001-001. We sincerely appreciate the efforts of Jeffery Warner at NASA Johnson Space Center and his coworkers for the lunar data base, B. Kowalski for the pattern recognition program, Sam Motherwell for the original version of the search program, and the Arizona State University computer services for computer time.

REFERENCES

Butler J. C. (1976) Variations in chemical composition of Apollo 15 mare basalts. *Proc. Lunar Sci. Conf. 7th*, p. 1429–1447.

Cover T. M. and Hart P. E. (1967) Nearest neighbor pattern classification. *IEEE Trans. on Info. Theory* **IT-13**, 21.

Cuttitta F., Rose H. J., Jr., Annell C. S., Carron M. K., Christian R. P., Ligon D. T., Jr., Dwornik E. J., Wright T. L., and Greenland L. P. (1973) Chemistry of twenty-one igneous rocks and soils returned by the Apollo 15 mission. *Proc. Lunar Sci. Conf. 4th*, p. 1081–1096.

Duewer D. L., Koskinen J. R., and Kowalski B. R. "ARTHUR", available from B. R. Kowalski, Laboratory for Chemometrics, Department of Chemistry, BG-10, University of Washington, Seattle, Washington, 98195.

Howarth R. J. (1973) The pattern recognition problem in applied geochemistry. *Proc. 4th Int. Geochem. Exploration Symposium*, London, 1972 (M. J. Jones, ed.), p. 259–273. Institute of Mining and Metallurgy, London.

Kowalski B. R. and Bender C. F. (1972) Pattern recognition. A powerful approach to interpreting chemical data. *J. Amer. Chem. Soc.* **94**, 5632–5639.

Kowalski B. R. and Bender C. F. (1973) Pattern recognition II. Linear and nonlinear methods for displaying chemical data. *J. Amer. Chem. Soc.* **95**, 686–693.

Laul J. C. and Schmitt R. A. (1972) Bulk and REE abundances in three Apollo 15 igneous rocks and six basaltic rake samples. In *The Apollo 15 Lunar Samples* (J. W. Chamberlain and C. Watkins, eds.), p. 225–227. The Lunar Science Institute, Houston.

LSPET (1972) The Apollo 15 lunar samples: A preliminary description. *Science* **175**, 363–375.

Motherwell S. *Cambridge Crystallographic Files Users Manual*, University Chemical Laboratory, Lensfield Rd., Cambridge, England.

Nava D. F. (1974) Chemical composition of some soils and rock types from the Apollo 15, 16, and 17 lunar sites. *Proc. Lunar Sci. Conf. 5th*, p. 1087–1096.

Nillson N. J. (1965) *Learning Machines*. McGraw-Hill, New York. 137 pp.

Pratt D. D., Moore C. B., and Parsons M. L. (1977) Pattern recognition techniques applied to chemical classifications of Apollo 17 mare basalts (abstract). In *Lunar Science VIII*, p. 790–792. The Lunar Science Institute, Houston.

Rhodes J. M. and Hubbard N. J. (1973) Chemistry, classification, and petrogenesis of Apollo 15 mare basalts. *Proc. Lunar Sci. Conf. 4th*, p. 1127–1148.

Proc. Lunar Sci. Conf. 8th (1977), p. 1849–1867.
Printed in the United States of America

Composition and origin of plagioclase, pyroxene, and olivine clasts of lunar breccias 14006, 14063, 14066, 14311, 14320, and 14321

D. STÖFFLER and H.-D. KNÖLL
Institute of Mineralogy, University of Münster, D-44 Münster, Germany

Abstract—From the correlation of chemical components (An vs. FeO in plagioclase; En–Fs–Wo and CaO vs. TiO_2 in pyroxene; Fo vs. CaO in olivine) the petrological provenance of the mineral clast population of crystalline and detrital Fra Mauro breccias was deduced by a model calculation. 6072 analyses of corresponding minerals taken from the literature were used to calculate the proportions of the three major groupings of lunar igneous rocks: mare basalts (MB), anorthositic-noritic-troctolitic rocks (ANT), and KREEP- and high-alumina basalts (HAB). Mineral clasts from such rocks and from ultramafic rocks (dunites) occur in variable proportions in all breccias. Olivine is predominantly derived from dunitic-troctolitic rocks, pyroxene, and plagioclase mainly from HAB- and ANT-lithologies. A significant portion of the latter minerals fits an origin from pre-Imbrian mare basalt like rocks. These characteristics favor previous interpretations concerning an origin of the crystalline Fra Mauro breccias from high temperature impact formations produced in large local pre-Imbrian cratering events rather than in the Imbrian event itself.

1. INTRODUCTION

THE FRA MAURO FORMATION which was sampled by the Apollo 14 landing mission is generally believed to have been deposited as a result of the impact formation of the Imbrian basin. Since this important geological unit bears valuable information about the pre-Imbrian history of the lunar crust and the processes accompanying the formation of the Imbrium Basin as well, the study of Fra Mauro breccias has been intensified recently by a number of laboratories (Duncan *et al.*, 1975; Wood, 1976, 1977; Phinney *et al.*, 1976; Stöffler *et al.*, 1976; Warner *et al.*, 1977).

We have demonstrated on the basis of grain size statistics that the clast population of Fra Mauro (Apollo 14) breccias with crystalline or detrital matrices originates from an immature impactoclastic detritus developed by modest reworking of primordial igneous rocks and impact breccias derived from these rocks (Stöffler *et al.*, 1976). From the mineral chemistry of plagioclase clasts it was concluded that different types of comminuted "igneous" source rocks had contributed to the mineral clast population of the breccias. The present paper adds new data on the mineral chemistry of plagioclase, pyroxene, and olivine obtained in the same and in additional breccia samples. These and the previous analyses are based on the well established fact that the major and minor element composition of the rock-forming minerals is diagnostic of the type of igneous rock in which they crystallized. Relevant data for lunar igneous rocks used in this study have been summarized by Smith (1974), Smith and Steele (1976), and Papike *et al.* (1976).

2. Samples and Methods of Investigation

Thin sections from five breccia samples with crystalline matrix (14006, 14066, 14311, 14320, 14321) and from one breccia with detrital matrix (14063) were used for microprobe analysis of randomly selected mineral clasts. In the case of breccias with crystalline matrix only mineral fragments from the dark matrix (microbreccia 2 and 3 of Grieve *et al.*, 1975) of 14321 and from the dark matrix portions of 14066, 14311, and 14320 were analyzed. The lighter portions of the matrix were avoided because the distinction between clastic and matrix minerals is much less reliable compared to the dark matrix. The analyzing electron beam (spot size: 10–15 μm) was restricted to the center of the mineral clasts in order to avoid reaction rims. Chemical zoning was not considered. No clasts smaller than 20 μm were analyzed. Si, Al, Fe, Mg, Ca, Ti, Cr, and Mn were determined in pyroxene and olivine. Ca, Na, and Fe were measured in plagioclase.* Reduction of the microprobe data was done by the correction procedures of Sweatman and Long (1972) and of Goldstein and Comella (1969). The resulting data were reduced by computer plotting of binary and ternary diagrams. A computer program was written to convert such raw data plots into plots in which the frequency of data points in small compositional ranges (in percent of the total number of points) or the actual number of data points in these ranges was printed (Fig. 1). The same procedure was applied for the literature data of plagioclase, pyroxene, and olivine from lunar rocks of igneous or "meta-igneous" origin (Fig. 2; see Section 3).

3. Composition of Plagioclase, Pyroxene, and Olivine in Various Lunar Igneous Rocks

The mineral chemistry of the main rock-forming minerals of lunar igneous rocks which could be potential contributors to the clast population of Fra Mauro breccias has been systematically investigated in the past. Data of many authors were reviewed recently by Smith (1974), Frondel (1975), Smith and Steele (1976), and Papike *et al.* (1976). Chemical data for plagioclase, pyroxene, and olivine were extracted from the literature and arranged into four major groups of igneous rocks:

(1) mare basalt (MB),
(2) anorthosite-norite-troctolite (ANT),
(3) KREEP and high alumina basalt (HAB),
(4) ultramafic rocks.

In the MB-group 743, 2073, and 1194 analyses were obtained for plagioclase, pyroxene, and olivine, in the ANT- and HAB-groups, the corresponding figures are 594, 552, 402, and 192, 178, 144, respectively. The data were taken from the following publications: plagioclase and olivine data from Dowty *et al.*, (1973), Hlava *et al.* (1973), Conrad *et al.* (1973), Dowty *et al.* (1976), and Warner *et al.* (1976); pyroxene data from Dowty *et al.* (1973), Hlava *et al.* (1973), Conrad *et al.* (1973), Dowty *et al.* (1976), Warner *et al.* (1976), McCallum *et al.* (1974), Powell and Weiblen (1972), Meyer and McCallister (1973), Tarasov *et al.* (1973), Ashwal (1975), Dixon and Papike (1975), Bence and Papike (1972), Kushiro *et al.* (1972), Ridley *et al.* (1972), Longhi (1972), Green *et al.* (1972), Brown *et al.* (1972), Papike *et al.* (1974), Dymek *et al.* (1975b), Taylor and Misra (1975), and Drake and Klein (1973). Correlation analyses of all elements (oxides) were made for each mineral within the three main types of igneous lithologies (MB, HAB, ANT; see above) in order to find element correlations most suitable for differentiating between the three types of rocks. The number of available data from ultramafic rocks was insufficient for this kind of statistical analysis.

---

*The original (raw) data are not completely presented here. They can be obtained from the authors upon request.

MOL PCT. AN

		91.0	92.0	93.0	94.0	95.0	96.0	97.0	98.0	99.0	100.0
W	0.05	0	0	0	0	0	0	0	0	0	0
T	0.10	0	0	0	0	0	0	0	0	0	0
.	0.15	0	0	0	0	0	0	0	0	0	0
	0.20	0	0	0	0	0	0	0	0	0	0
P	0.25	0	0	0	0	0	0	0	0	0	0
C	0.30	0	0	0	0	0	0	0	0	0	0
T	0.35	1	0	0	1	0	0	0	0	0	0
.	0.40	1	0	1	0	0	0	0	0	0	0
	0.45	0	0	0	1	0	0	0	0	0	0
F	0.50	1	0	1	2	0	0	0	0	0	0
E	0.55	0	1	1	1	0	2	0	0	0	0
O	0.60	0	0	0	1	1	0	0	0	0	0
	0.65	0	1	1	0	1	0	0	0	0	0
	0.70	1	0	0	0	0	0	0	0	0	0
	0.75	1	1	1	1	0	0	0	0	0	0
	0.80	0	1	0	1	0	0	0	0	0	0
	0.85	0	0	0	1	0	0	0	0	0	0
	0.90	0	0	0	0	0	0	0	0	0	0
	0.95	0	0	0	0	0	0	0	0	0	0
	1.00	0	0	0	0	0	0	0	0	0	0

Fig. 1. Section of a computer plot of the An (mol.%) and FeO content (wt.%) of plagioclase clasts from breccia 14066 (thin sections 7, 8, and 46). The number of data points is printed for small ranges in composition (1% An and 0.05% FeO). Example: two analyses fall into the range from 93–94% An and 0.45–0.50% FeO. The section of the plot shown contains 36 data points, the total number of data points is 71.

PLAGIOCLASE ANT
NUMBER OF DATA POINTS: 594

MOL PCT. AN

		91.0	92.0	93.0	94.0	95.0	96.0	97.0	98.0	99.0	100.0
W	0.05	0	17	0	17	17	0	34	101	34	67
T	0.10	0	0	0	17	0	34	84	135	101	84
.	0.15	0	51	17	101	118	34	286	404	286	185
	0.20	34	152	135	185	320	135	539	741	236	168
P	0.25	0	101	168	236	303	185	589	606	387	168
C	0.30	51	67	34	67	219	51	168	67	67	67
T	0.35	17	84	84	51	135	51	118	51	101	17
.	0.40	0	51	17	17	51	17	17	0	0	0
	0.45	0	17	0	0	0	17	34	0	0	0
F	0.50	0	17	0	0	0	0	17	0	0	0
E	0.55	0	0	0	0	0	0	0	0	0	0
O	0.60	0	0	0	17	17	0	0	0	0	0
	0.65	0	0	0	0	0	0	0	0	0	0
	0.70	0	0	17	0	0	0	0	0	0	0
	0.75	0	0	0	0	0	0	0	0	0	0
	0.80	17	0	0	0	0	0	0	0	0	0
	0.85	0	0	0	0	0	0	0	0	0	0
	0.90	0	0	0	0	0	0	0	0	0	0
	0.95	0	0	0	17	0	0	0	0	0	0
	1.00	0	0	0	0	0	0	0	0	0	0

Fig. 2. Section of a computer plot of the An- (mol.%) and FeO-content (wt.%) of plagioclase from lunar anorthositic-noritic-troctolitic rocks (ANT). Data taken from the literature as referenced in the text (Section 3). The percentage (×100) of data points is printed for small ranges in composition. For further explanation see Fig. 1.

Computer print-outs of the type described in Section 2 were made for the following correlations: FeO–$CaAl_2Si_2O_8$(An) for plagioclase; $MgSiO_3$(en)–$FeSiO_3$(fs)–$CaSiO_3$(wo), CaO–TiO_2 for pyroxene; and $MgSiO_4$(fo)–CaO for olivine. Two examples of such plots for plagioclase (Figs. 1 and 2) are shown. A complete example of the compositional variation of plagioclase for MB-, HAB-, and ANT-lithologies is shown in an additional plot (Fig. 3) which was made for the convenience of the reader because the print-outs are difficult to read. In using such element correlations to characterize the mineral chemistry of the various groups of lunar igneous rocks it should be kept in mind that these groups are not all equally well defined. The assignment of any rock sample to the group of mare basalts or ultramafic rocks seems to be most unequivocal. However, in

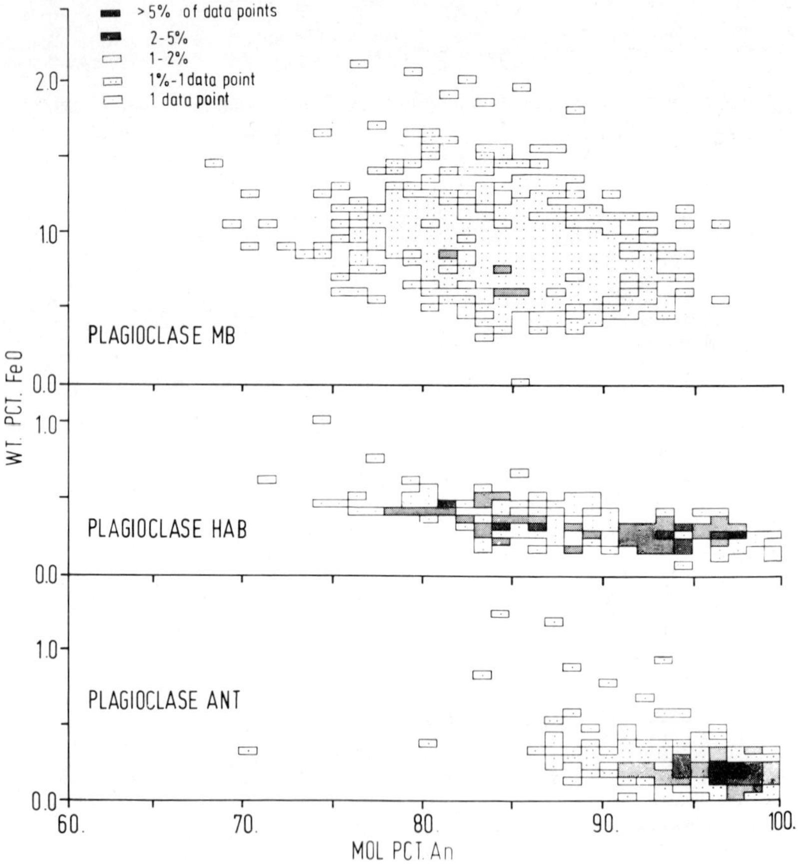

Fig. 3. An-content (mol.%) versus FeO (wt.%) of plagioclase of the main groups of lunar igneous rocks; transformation of a computer print-out of the type shown in Fig. 2. Number of analyses for MB, ANT, and HAB are 743, 594, and 192, respectively. Data taken from the literature referenced in the text (Section 3). MB = mare basalts, ANT = anorthositic-noritic-troctolitic, and HAB = KREEP- and high-alumina basalt.

the case of ANT-rocks and KREEP- and high-alumina basalts which both underwent at least part of the complex bombardment history of the lunar highlands, the primary igneous origin—plutonic or volcanic—is not always certain because of the effect of impact metamorphism. Consequently, it cannot be ruled out that impact melt rocks, either chemically unchanged or hybridized by mixing of different primary igneous and impactoclastic rocks, are contained in these groups of lithologies.

Since impact metamorphism does not produce a major fractionation of primary igneous rocks by partial melting or vaporization on a grand scale we do believe that most of the chemical characteristics of the primary igneous rocks have not been changed drastically by subsequent impact metamorphism although their texture may have been changed in the course of impact melting and subsequent crystallization. Moreover, observations from terrestrial craters and impact formations clearly indicate that the proportion of the total of displaced rocks in any impact event which changes texture and possibly chemistry due to strong impact metamorphism is relatively small, probably less than 5% (Grieve *et al.*, 1977; Pohl *et al.*, 1977; Stöffler, 1975, 1977). Mineral fragments with essentially unchanged major element chemistry derived from the primordial target rock can even be found in impact melt rocks which very often bear a large fraction of unshocked rock and mineral debris (Dence, 1971; Engelhardt, 1972). Depending on the thermal history of the melt-clast mixture a variable portion of clasts, however, may be affected by chemical reaction with the melt (see Simonds *et al.*, 1976).

4. Results

Results of microprobe analyses of generally 30–80 grains of each of the three main types of mineral clasts were reduced to the same set of element correlation plots for each measured breccia as it was done for the literature data described in Section 3. Computer print-outs from which the frequency of data point for small compositional ranges can be read (e.g., Fig. 1) were then compared with the same type of print-outs (e.g., Fig. 2) which were obtained for the three main igneous lithologies, namely, mare basalts (MB), ANT-rocks (ANT), and KREEP- and high-alumina basalts (HAB). Previous data for plagioclase of breccias 14006, 14311, 14320, and 14321 (Stöffler *et al.*, 1976) were included in this procedure. Theoretically, each measured data point for any mineral clast could fall into four types of areas in the literature data plots which either belong to one, two, three or none of the fields covered by data points of three classes of igneous lithologies (MB, ANT, HAB; compare Fig. 1 with Fig. 3, for example). The assignment of measured data points to any of the three source rock lithologies was made according to the following procedure: Data points falling in none of the three fields were counted separately and their percentage of the total number of data points was recorded (termed component X in Fig. 4). Data points falling in only one of the fields, that means in areas where the three lithologies do not overlap, were fully counted. Points falling into two or three fields at a time were counted after being corrected by a factor which accounts for the probability of a

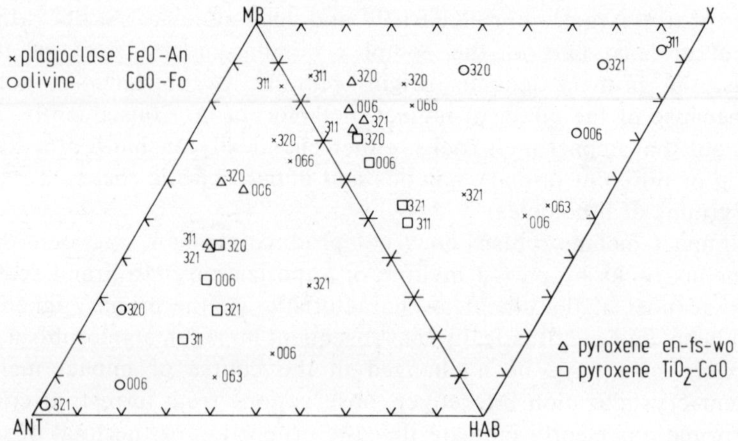

Fig. 4. Proportions of ANT-, HAB-, and MB-lithologies in the mineral clast population of Fra Mauro breccias as calculated from the chemical composition of the mineral clasts. The data were obtained by comparing computer print-outs of the measured An–FeO, en–fs–wo, CaO–TiO_2, Fo–CaO correlations of plagioclase, pyroxene, and olivine clasts with corresponding plots of the literature data from ANT-, HAB-, and MB-rocks. (For detailed explanation of the procedure see text, Section 4.) *X* gives the percentage of data points not fitting the composition of ANT-, HAB-, and MB-rocks. Numbers refer to the Apollo 14 breccias ANT, HAB, and MB explained in Fig. 3.

mineral clast belonging to the two or three potential source rock lithologies. This factor was calculated from the relative frequency with which literature data points in each of the overlapping fields of MB-, ANT-, and HAB-lithologies occurred at the position of the measured data point in the compositional plot in question. The relative proportions of MB-, ANT-, and HAB-lithologies from which the measured mineral clast population can be derived was then calculated for various element correlations and plotted in Fig. 4. Since a good statistical base for the compositional fields of the mineral phases of ultramafic rocks is not available at present, the proportion of such rock material within the total mineral clast population of the breccias could not be obtained by this model calculation. Therefore, only the presence of "dunitic" olivine was checked visually by a comparison of our raw data plots (e.g., Fig. 11) with the literature data for dunites (e.g., Smith and Steele, 1976).

For pyroxene the relative proportions of orthopyroxene (<5% wo), pigeonite (5–30% wo), and augite (>30% wo) was calculated from the pyroxene quadrilaterals for each sample (Fig. 5) using the raw data plots of Fig. 10. In addition, linear regression analyses of the FeO–$CaAl_2Si_2O_8$-correlation in plagioclase clasts, and of the CaO–Mg_2SiO_4-correlation in olivine clasts were made. Resulting parameters are plotted in Figs. 6–8.

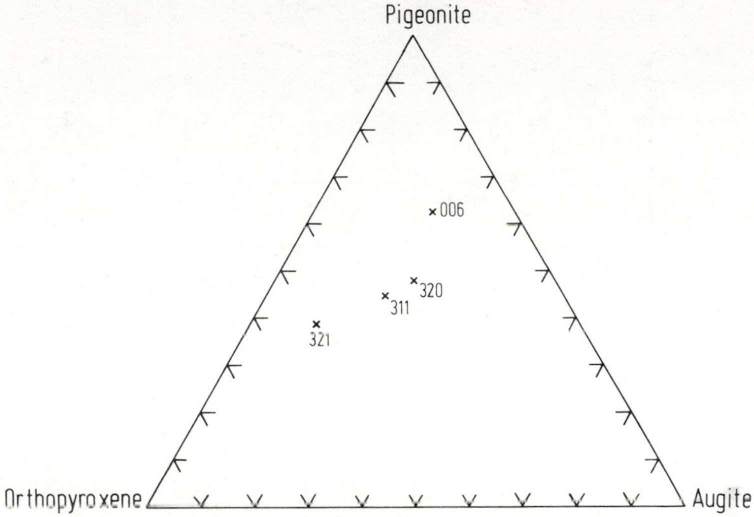

Fig. 5. Proportions of augite (wo > 30%), pigeonite (wo 5–30%), and orthopyroxene (wo < 5%) in Apollo 14 breccias (sample numbers indicated) calculated from the compositional data of Figs. 10a, b, c, d.

5. Discussion

The modal composition of the mineral clast population in various Apollo 14 breccias has been analyzed by a number of authors (e.g., Chao *et al.*, 1972; Engelhardt *et al.*, 1972; Lindsay, 1972; Wilshire and Jackson, 1972). Modal data of the breccias discussed in this paper are summarized in Table 1 (see also Stöffler *et al.*, 1976). It has been suggested from the relative abundances of plagioclase, pyroxene, olivine, and ilmenite and from grain size statistics that the mineral clast population of both breccias with crystalline and detrital (breccia 14063) matrix was produced by impact comminution and reworking of a variety of pre-Imbrian source rocks (Stöffler *et al.*, 1976). The mineral chemistry of the major clastic silicate phases as shown in Figs. 9–11 allows a more quantitative estimate of the proportion of these source rocks contributing to the clast population of the analyzed breccias. The characteristics of the three major silicates shall be discussed here in some detail.

Plagioclase

The multi-rock provenance of plagioclase in all samples becomes quite obvious from Figs. 4, 6, and 8 (see also Stöffler *et al.*, 1976). However, the compositional range for the plagioclase clasts differs distinctly among the measured breccias. Plagioclase of breccias 14066, 14311, and 14320 is characterized by a relatively high Fe-content, and a small standard deviation of the An-content in 14311 and 14320, whereas in 14006, 14321, and 14063 the Fe-content is lower and the standard deviation of the An-content is markedly larger

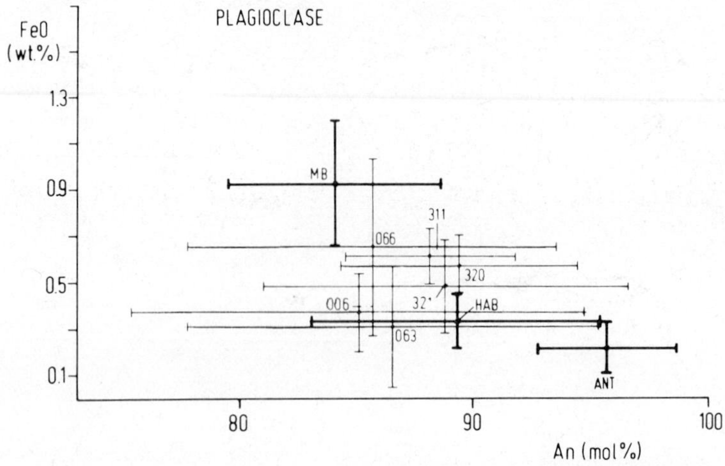

Fig. 6. Averages and standard deviations (1σ) of the An- (mol.%) and FeO- (wt.%) content of plagioclase from the main lunar igneous lithologies (heavy lines) and from the clast population of Apollo 14 breccias (weak lines). MB = mare basalts (743 data points), HAB = KREEP- and high-alumina basalts (192 data points), and ANT = anorthositic-noritic-troctolitic rocks.

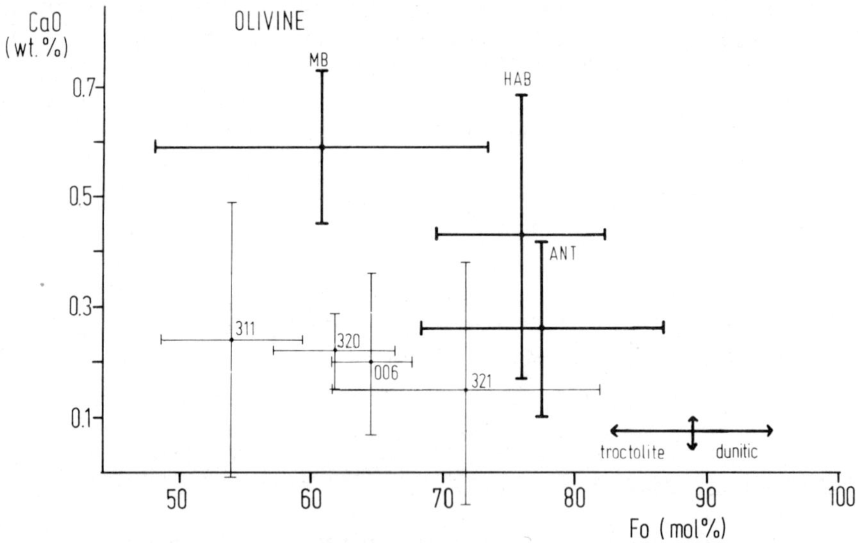

Fig. 7. Averages and standard deviations (1σ) of the Fo (mol.%) and CaO (wt.%) content of olivine from the main lunar igneous lithologies (heavy lines) and from the clast population of Apollo 14 breccias (weak lines). Symbols as in Fig. 6. MB, HAB, ANT, 14006, 14311, 14320, and 14321 include 1194, 144, 402, 38, 15, 30, and 67 analyses, respectively. Arrows indicate the compositional range of olivine found in dunitic and troctolitic rocks according to Steele and Smith (1975).

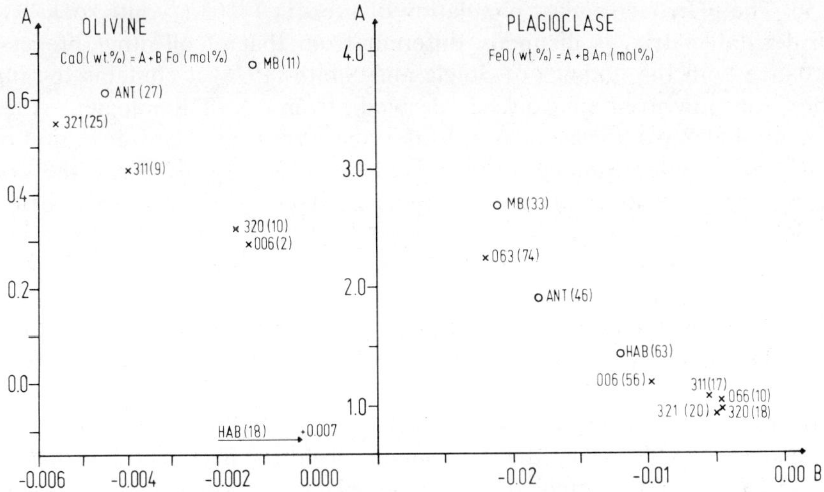

Fig. 8. Parameters of linear regression analyses of the FeO–An and the CaO–Fo correlations of plagioclase and olivine, respectively, as defined by the equations given at the top of the diagrams. Crosses: mineral clasts in breccias. Circles: minerals in lunar igneous rocks. The data base and the symbols are the same as for Figs. 6 and 7. Numbers in parentheses refer to the regression coefficients (in %). Note the strong deviation of the olivine parameters from either of the three source rock lithologies (ANT, HAB, MB), and the marked differences of the plagioclase clast population between breccias with crystalline (006, 066, 311, 320, 321) and detrital (063) matrices.

Table 1. Modal composition of mineral clasts (>0.02 mm) in Apollo 14 breccias. The remaining clasts are recrystallized glass (diaplectic glass and quenched rock melt).

	Plagio-clase	Pyroxene + olivine	Ratio plag:pyr	Opaques	Lithic clasts	Total
Breccias with crystalline matrix						
14006,8	46.7	28.3	1.65	2.1	21.5	98.6
14066,50	25.6	45.3	0.56*	1.2	15.1	87.2
14311,89	47.2	26.5	1.78	2.6	22.8	99.1
14320,8	39.4	17.6	2.24	3.6	35.6	96.2
14321,215	53.3	26.3	2.03	1.1	18.3	99.0
Breccia with detrital matrix						
14063,56	35.0	17.0	2.06	2.7	7.9	62.6

*This seems to be an extreme value; ratios of 1.0 to 1.5 occur in other sections of the same rock (compare Wilshire and Jackson, 1972 and Lindsay, 1972).

(Fig. 9). The plagioclase clast population of breccia 14063 ("white rock") which has a detrital matrix, is distinctly different from that of all other breccias. In accordance with the findings of Steele and Smith (1976) it contains the highest fraction of low-iron plagioclase derived from ANT-lithologies (Fig. 4). Moreover, 14063 plagioclase clasts display the largest standard deviation of the An- and FeO-content among all measured breccias (Fig. 6). Since the crystallization temperature of the matrix of breccias 14066, 14311, and 14320 (belonging to grades 7, 5, and 6, respectively, of Warner's 1972 classification) was relatively high and the bulk matrix is rich in Fe, the high Fe-content of the plagioclase clasts in these breccias could be due to secondary equilibrium diffusion. Then one should expect high Fe-plagioclase clasts also in breccia 14006 which has grade 6. This is not the case (Fig. 6). Moreover, observed matrix-clast reactions are restricted to the clast rims (Warner, 1972) and recrystallization effects are absent in the cores of the clasts (Warner *et al.*, 1977) from which our analyses were taken. Provided that the Fe-content of lunar plagioclase is indeed an indicator of the melt composition from which it crystallized (Smith, 1974) the following conclusions concerning the provenance of plagioclase clasts in the polymict Fra Mauro breccias can be drawn from Figs. 4, 6, and 8: (1) the measured breccias contain variable proportions of ANT-, HAB-, and MB-lithologies in their plagioclase population; (2) a group of breccias (14066, 14311, and 14320) with a large content of high-iron plagioclase clasts possibly derived from MB-lithologies, can be distinguished from a group of breccias (14063, 14006, and 14321) with a low content of high-iron plagioclase clasts; and (3) a variable amount of plagioclase occurs in these breccias whose composition does not fit the known compositional range of lunar igneous plagioclase. This will be further discussed at the end of this chapter.

Pyroxene

As in the case of plagioclase the compositional range of pyroxene clasts in breccias 14006, 14311, 14320, and 14321 can only be matched by a multi-rock source as is seen from the enstatite (en)–ferrosilite (fs)–wollastonite (wo) plot (Fig. 10a–d) and from the CaO–TiO_2 correlation, both of which are evaluated in Fig. 4. Again the breccias have different pyroxene populations although the differences are less distinct than in the case of plagioclase. In all breccias the pyroxene population can be considered as a mixture of augite (wo $> 30\%$), pigeonite (wo 5–30%), and orthopyroxene (wo $< 5\%$) which does not appear to represent an equilibrium assemblage (Fig. 5). It contrasts strongly to the composition of the *in situ* crystallized matrix pyroxenes in breccias 14006, 14066, and 14311 which have a rather constant orthopyroxene composition as measured by Warner (1972). The composition of pyroxene clasts in breccia 14063 as measured by Steele and Smith (1976) indicates that this breccia contains a distinctly smaller fraction of pigeonite than the crystalline matrix breccias of Fig. 5. This confirms again that ANT-lithologies prevail in the mineral clast population of 14063 (Fig. 4, see also Steele and Smith, 1976). Applying the

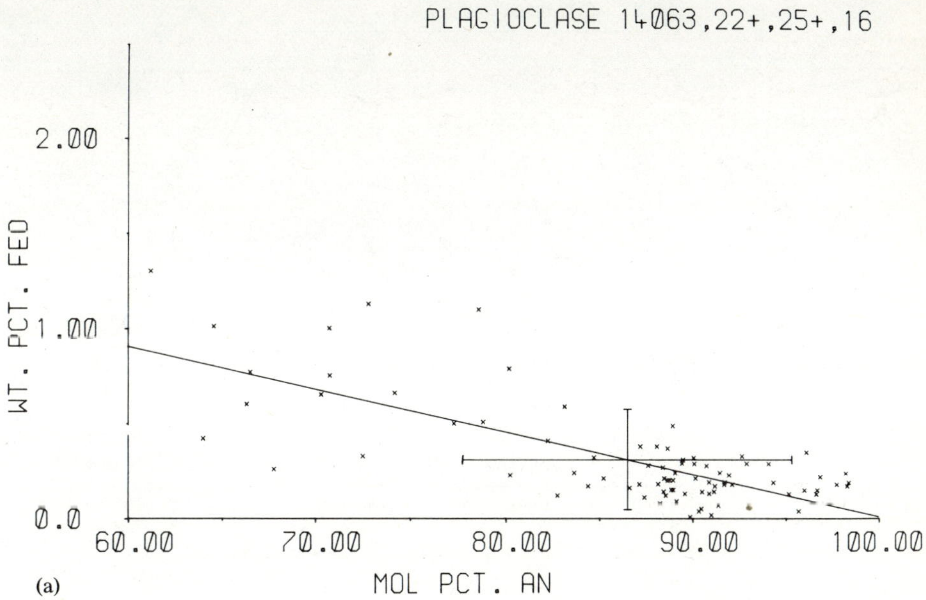

(a)

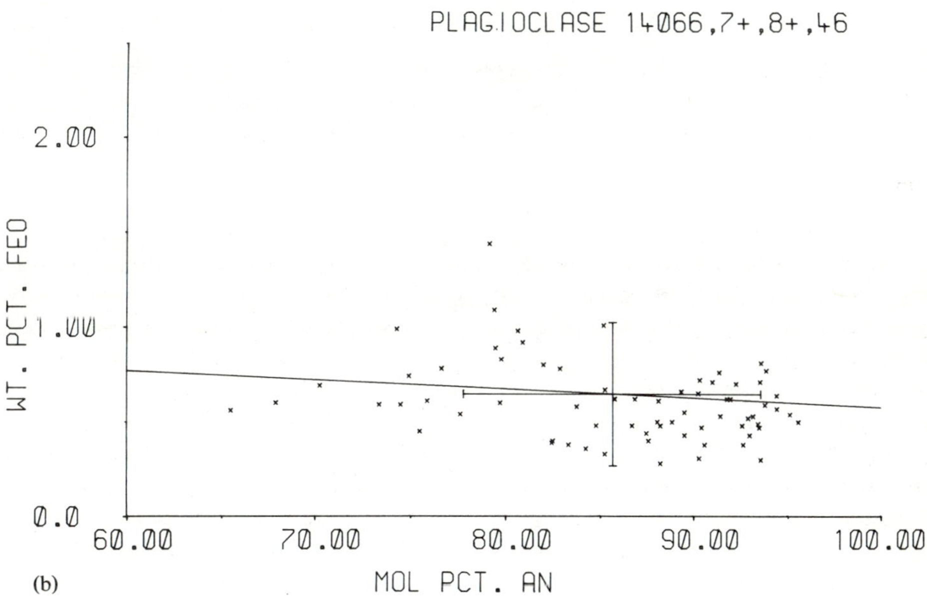

(b)

Fig. 9a,b. Composition of plagioclase clasts (>20 μm) from breccias 14063 and 14066 (82 and 71 analyses, respectively); average, standard deviation (1σ), and linear regression are indicated.

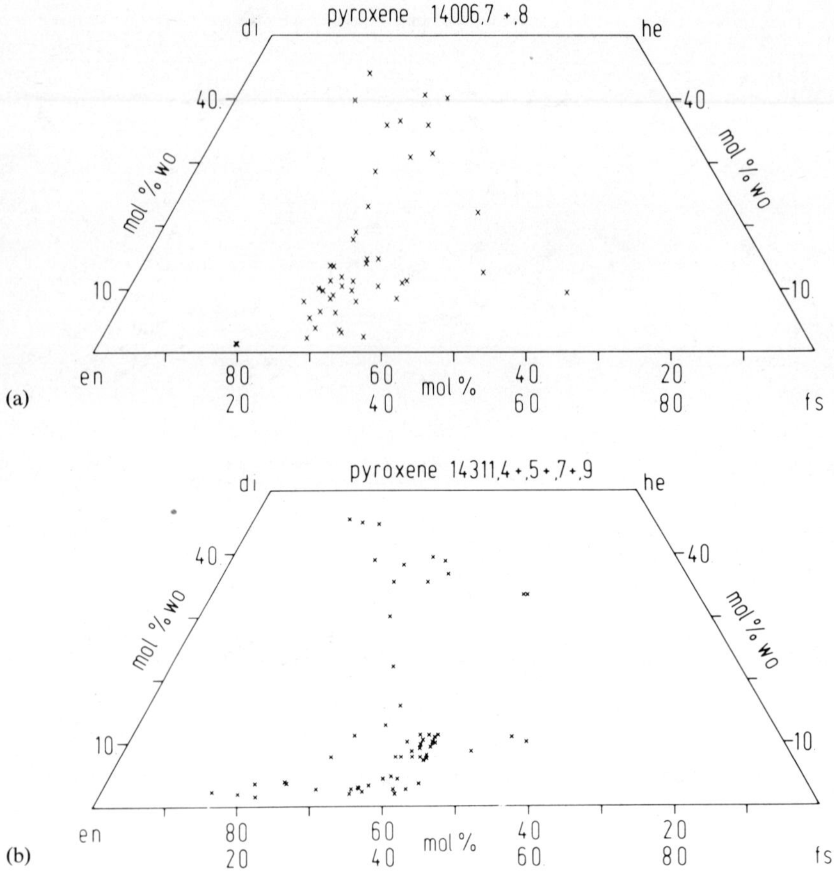

$CaO–TiO_2$ correlation for the calculation of the ANT-, HAB-, and MB-proportions the amount of MB-lithology contributing to the clast population is consistently decreased by about 20% as compared to the corresponding proportions which were calculated from the en–fs–wo plots (Fig. 4). Because of the greater overlap of the compositional fields for ANT, HAB, and MB in the latter plots, our calculation method yields an actual over-estimate of pyroxene from mare-like rocks. The proportion of pyroxene which does not fit the compositional range of pyroxene from known lunar rocks is relatively small in all breccias (ranging from 8–14%).

Olivine

As seen from Figs. 7, 8, and 11a–d, especially from the relatively large proportion of olivine clasts which fall outside the compositional range of known ANT-, HAB-, and MB-lithologies (Fig. 4), types of source rocks other than those inferred for plagioclase and pyroxene are required for the olivine clasts of

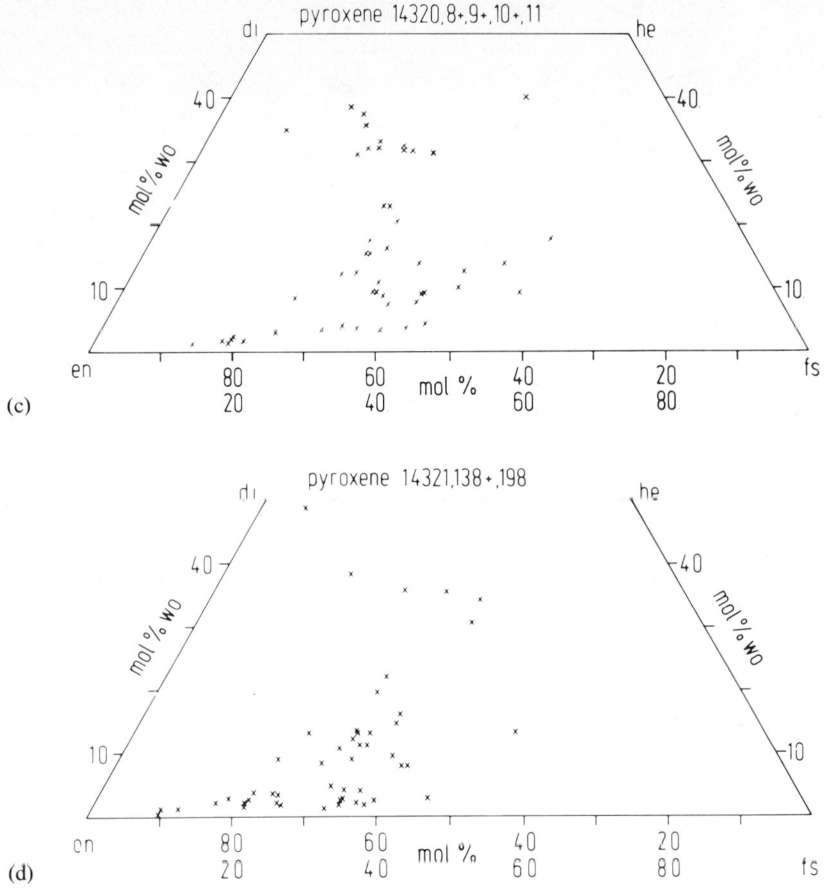

Fig. 10a,b,c,d. Composition of pyroxene clasts (>20 μm) from breccias 14006, 14311, 14320, and 14321 (53, 63, 50, and 43 analyses, respectively).

breccias 14006, 14311, 14320, and 14321. The low Ca-content and the rather narrow compositional range in 14006 and 14320 (small standard deviations, Fig. 7) suggest that the olivine clasts of these breccias may have a monolithic source which is presumably troctolitic to dunitic (compare data of Steele and Smith, 1975 and Dymek *et al.*, 1975a), although the Mg-content is relatively low (Fig. 7). The composition of olivine in 14311 is unusually low in Mg and differs from all Apollo 14 breccias so far measured as noted by Cameron and Fisher (1975). Breccia 14321 carries olivine clasts of rather variable composition.

What can be concluded from the data discussed above in terms of the petrological history and of the type and abundance of pre-Imbrian lithologies represented in the mineral clast population of the breccias? Before we make an attempt at interpreting the data, the method of data reduction and subsequent model calculations has to be reviewed critically. We believe that there are three problems which should be considered in order to prevent misinterpretations.

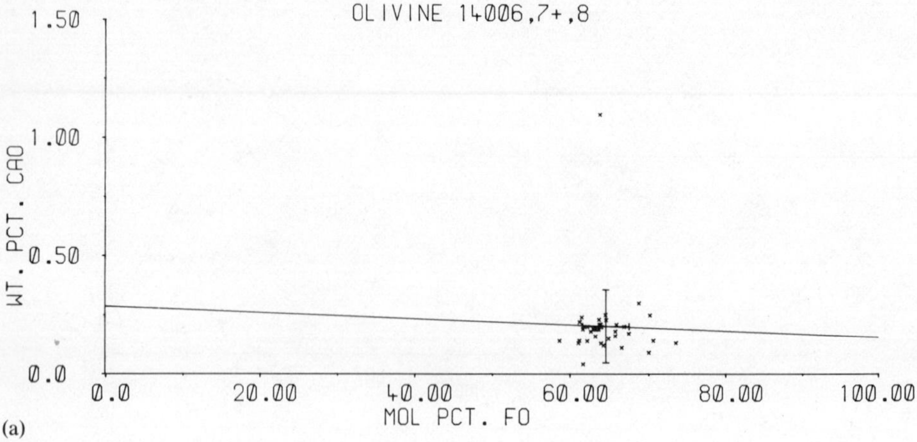

(a)

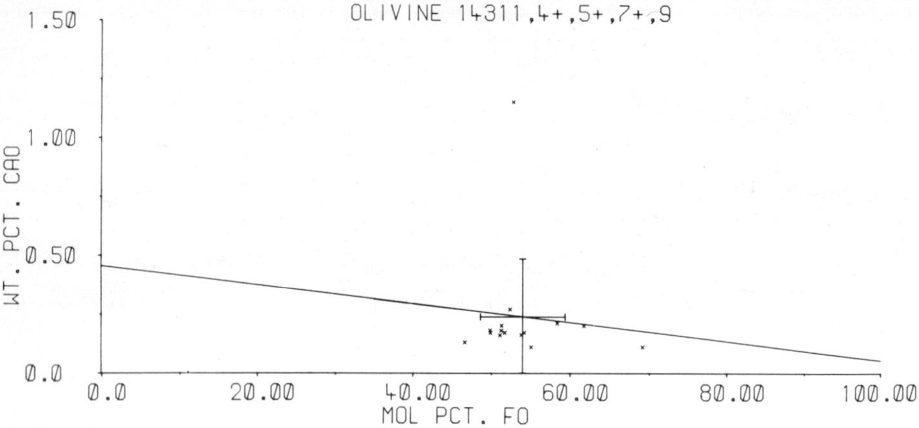

(b)

First, the model is applicable and meaningful only if any secondary change of the mineral clast chemistry by post-brecciation matrix-clast diffusion processes can be considered to be minor or absent. We think that this assumption is justified for the following reasons. Mineral and lithic clasts do not show any noticeable sign of annealing and recrystallization except for rim reactions in some breccias (see Warner, 1972; Cameron and Fisher, 1975; Warner *et al.*, 1977). Moreover, the composition of plagioclase and pyroxene clasts is distinctly different from that of their matrix counterparts (Warner, 1972; Stöffler *et al.*, 1976). This holds also for the Fe-content of plagioclase, which is markedly higher in the matrix (Stöffler *et al.*, 1976).

Second, our method of calculating the proportions of source rock lithologies (ANT, HAB, MB) implies that a certain number of mineral clasts could be accommodated with more than one type of source rock. Therefore the percentages of ANT, HAB, and MB in Fig. 4 are based on probabilities with which particular clasts can be assigned to the potential source rocks. The actual

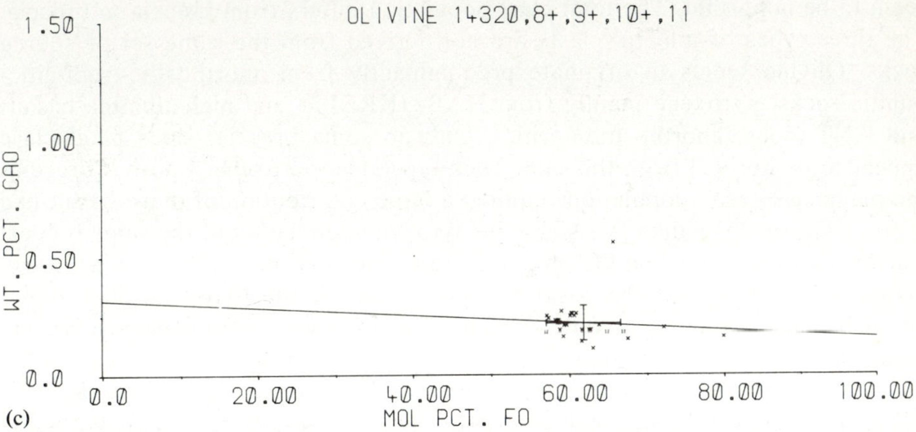

(c)

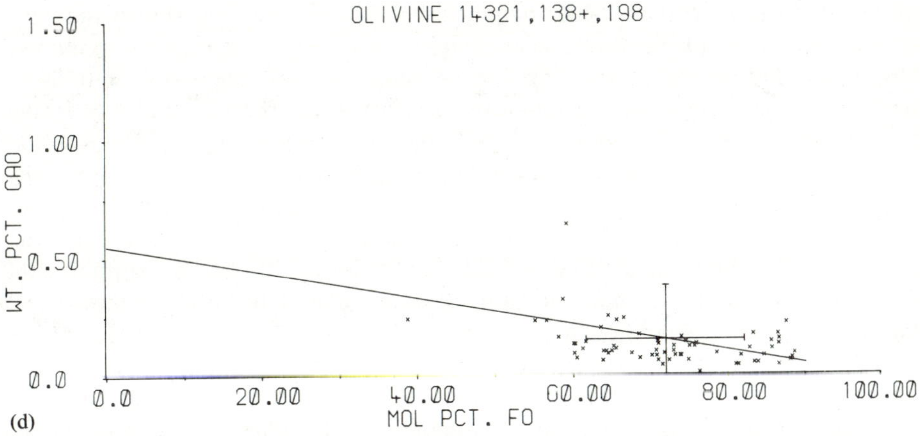

(d)

Fig. 11a,b,c,d. Composition of olivine clasts (>20 μm) from breccias 14006, 14311, 14320, and 14321 (38, 15, 30, and 67 analyses, respectively; solid lines as in Fig. 9).

percentage may be higher or lower by ~10%. An additional although minor uncertainty to these figures is introduced by the fact that ultramafic rocks such as dunites were not considered in our model calculations (see Section 4).

Third, the evaluated literature data may not be statistically representative for the actual variation of the mineral chemistry of lunar igneous rocks. In addition, all data for mare basalt minerals were taken from rocks which are younger than the breccias. When our model calculation yields a certain abundance of MB-rocks, this only means that the clast population of a breccia contains minerals which crystallized from silicate liquids similar to post-Imbrian mare basalts.

Concerning the petrological provenance of the silicate mineral clasts of the Fra Mauro breccias studied in this paper the following general characteristics seem to be important. The total clast population differs from breccia to breccia. The three types of mineral clasts are not derived from the same set of source rocks. Olivine seems to originate predominantly from anorthositic-troctolitic-dunitic rocks, pyroxene mainly from HAB- (KREEP- and high-alumina basalt) and ANT-rocks (anorthositic-noritic). Only in some breccias does plagioclase appear to be derived from the same rock type(s) as pyroxene. In other breccias the plagioclase clast population requires a large contribution of mare basalt-like rocks. This creates a dilemma since the pyroxene population of the same breccia can be extremely low in MB-lithology (e.g., breccia 14311, Fig. 4). One explanation could be that the basaltic source rocks of the pyroxene clast population had a very low pyroxene content. The only known pre-Imbrian basalts which are clasts in Fra Mauro breccias, do carry too much pyroxene (10–60 vol.% according to Wilshire and Jackson, 1972), notwithstanding the fact that the FeO content of plagioclase in these basalts is too low (mostly below 0.6 wt.%, e.g., Ridley, 1975, see also data reviewed in Stöffler *et al.*, 1976) to account for the mare basalt-like plagioclase clasts of the Fra Mauro breccias. Alternatively, the plagioclase clasts may be derived in part from unknown pre-Imbrian lithologies or they may have acquired their high iron content by some secondary process prior to their incorporation into the Fra Mauro breccias. In any case, there is a definite contribution of pre-Imbrian mare basalt-like lithologies to all measured breccias showing up in the pyroxene and plagioclase clast population (see also Stöffler *et al.*, 1976; our previous estimates of the percentage of such rocks are somewhat lower than the present estimates because of the different methods of data reduction). This finding supports the presence of pre-Imbrian mare volcanism as postulated by other authors (e.g., Ryder and Taylor, 1976; Hawke and Head, 1977). Part of the clasts which do not fit the known igneous lithologies may be derived from pre-Imbrian impact melt rocks of hybrid chemistry which was caused by shock fusion of heterogenous target lithologies. In general, the provenance of the mineral clasts discussed above agrees reasonably well with the provenance of the lithic clasts analyzed in the same and other Fra Mauro breccias (e.g., Steele and Smith, 1976; Wood, 1976, 1977). The implications of these results for the geological provenance of the Fra Mauro breccias and the origin of the Fra Mauro Formation shall not be discussed here in detail. The reader is referred to our previous paper (Stöffler *et al.*, 1976) in this respect. We only like to add that the variability of the petrological provenance of mineral clasts in Apollo 14 breccia samples found in this study confirms the previously stated view (Stöffler *et al.*, 1974, 1976) that these breccias represent clasts within the Fra Mauro Formation which were probably derived from KREEP- and high-alumina basalt-rich impactites formed in local pre-Imbrian craters (Head and Hawke, 1975; Hawke and Head, 1977) and incorporated in the Fra Mauro Formation by secondary mass transport (Oberbeck, 1975). Results of the Imbrium Consortium (Wood, 1976, 1977), in particular of Ryder and Bower (1976, 1977) concerning the mineral clast chemistry of breccias 14064, 14082, and 14312 can be interpreted in the same way. Also Warner *et al.* (1977) do concur with this view and argue in favor of a fragment-laden impact melt origin of the Fra Mauro breccias displaying a crystalline matrix. As discussed in Stöffler *et al.* (1976) our data are not in contradiction to this interpretation.

Acknowledgments—We thank Mrs. G. Grant, Miss F. Möllers, Miss U. Lenzen, H. Rettmar, W. U. Reimold, and R. Ostertag, Institute of Mineralogy, University of Münster, and D. Mangliers, University of Tübingen, for their very valuable technical assistance.

Financial support by the Deutsche Forschungsgemeinschaft (German Science Foundation) is gratefully acknowledged.

References

Ashwal L. D. (1975) Petrologic evidence for a plutonic igneous origin of anorthositic norite clasts in 67955 and 77017. *Proc. Lunar Sci. Conf. 6th*, p. 221–230.

Bence A. E. and Papike J. J. (1972) Pyroxenes as recorders of lunar basalt petrogenesis. *Proc. Lunar Sci. Conf. 3rd*, p. 431–469.

Brown G. M., Emeleus C. H., Holland J. G., Pecket A., and Phillips R. (1972) Mineral chemical variations in Apollo 14 and Apollo 15 basalts and granitic fractions. *Proc. Lunar Sci. Conf. 3rd*, p. 141–157.

Cameron K. L. and Fisher G. W. (1975) Olivine matrix-reactions in thermally metamorphosed Apollo 14 breccias. *Earth Planet. Sci. Lett.* **25**, 197–207.

Chao E. C. T., Minkin J. A., and Best J. B. (1972) Apollo 14 breccias: General characteristics and classification. *Proc. Lunar Sci. Conf. 3rd*, p. 645–659.

Conrad G. H., Hlava P. F., Green J. H., Moore R. B., Moreland G., Dowty E., Prinz M., Keil K., Nehru C. E., and Bunch T. E. (1973) Electron microprobe analyses of lithic fragments and their minerals from Luna 20 fines. Spec. Publ. No. 12, Univ. New Mexico. 62 pp.

Dence M. R. (1971) Impact melts. *J. Geophys. Res.* **76**, 5552–5565.

Dence M. R. and Plant A. G. (1972) Analysis of Fra Mauro samples and the origin of the Imbrium Basin. *Proc. Lunar Sci. Conf. 3rd*, p. 379–399.

Dixon J. R. and Papike J. J. (1975) Petrology of anorthosites from the Descartes region of the moon: Apollo 16. *Proc. Lunar Sci. Conf. 6th*, p. 263–291.

Dowty E., Green J. A., Hlava P. F., Keil K., Moore R. B., Nehru C. E., Prinz M., and Warner R. D. (1976) Electron microprobe analyses of minerals from Apollo 16 rake samples. Institute of Meteoritics, Spec. Publ. No. 14, Univ. New Mexico, Albuquerque. 141 pp.

Dowty E., Prinz M., Nehru C. E., Moreland G., Moore R. B., Keil K., Hlava P. F., and Green J. A. (1973) Electron microprobe analyses of minerals from Apollo 15 mare basalt rake samples. Institute of Meteoritics, Spec. Publ. No. 9, Univ. New Mexico, Albuquerque. 113 pp.

Drake J. C. and Klein C., Jr. (1973) Lithic fragments and glasses in microbreccia 15086: Their chemistry and occurrence. *Proc. Lunar Sci. Conf. 4th*, p. 467–479.

Duncan A. R., Grieve R. A. F., and Weill D. F. (1975) The life and times of Big Bertha: Lunar breccia 14321. *Geochim. Cosmochim. Acta* **39**, 265–273.

Dymek R. F., Albee A. L., and Chodos A. A. (1975a) Comparative petrology of lunar cumulate rocks of possible primary origin: Dunite 72415, troctolite 76535, norite 78235, and anorthosite 62237. *Proc. Lunar Sci. Conf. 6th*, p. 301–341.

Dymek R. F., Albee A. L., and Chodos A. A. (1975b) Comparative mineralogy and petrology of Apollo 17 mare basalts: Samples 70215, 71055, 74255, and 75055. *Proc. Lunar Sci. Conf. 6th*, p. 49–77.

Engelhardt W. v. (1972) Shock produced rock glasses from the Ries crater. *Contrib. Mineral. Petrol.* **36**, 265–292.

Engelhardt W. v., Arndt J. A., Stöffler D., and Schneider H. (1972) Apollo 14 regolith and fragmental rocks, their compositions and origin by impacts. *Proc. Lunar Sci. Conf. 3rd*, p. 753–770.

Frondel J. W. (1975) *Lunar Mineralogy.* Wiley-Interscience, New York. 323 pp.

Goldstein J. I. and Comella P. A. (1969) A computer program for electron probe microanalysis in the fields of metallurgy and geology. Goddard Space Flight Center, Greenbelt, Maryland.

Green D. H., Ringwood A. E., Ware N. G., and Hibberson W. O. (1972) Experimental petrology and petrogenesis of Apollo 14 basalts. *Proc. Lunar Sci. Conf. 3rd,* p. 197–206.

Grieve R. A. F., Dence M. R., and Robertson P. B. (1976) The generation and distribution of impact melts: Implications for cratering processes (abstract). In *Papers Presented to the Symposium on Planetary Cratering Mechanics,* p. 40–42. The Lunar Science Institute, Houston.

Grieve R. A. F., McKay G. A., Smith H. D., and Weill D. F. (1975) Lunar polymict breccia 14321: A petrographic study. *Geochim. Cosmochim. Acta* **39**, 229–245.

Hawke B. R. and Head J. W. (1977) Pre-Imbrian geologic history of the Apollo 14 region: Relation to Fra Mauro formation sample provenance. In *Studies by the Imbrium Consortium,* Vol. 2 (J. A. Wood, ed.). LSI Contr. No. 268D, p. 97–112. Center for Astrophysics, Cambridge.

Head J. W. and Hawke B. R. (1975) Geology of the Apollo 14 region (Fra Mauro): Stratigraphic history and sample provenance. *Proc. Lunar Sci. Conf. 6th,* p. 2483–2501.

Hlava P. F., Green J. A., Prinz M., Nehru C. E., Keil K., Dowty E., and Bunch T. E. (1973) Apollo 15 rake sample microbreccias and non-mare rocks: Bulk rock, mineral and glass electron microprobe analyses. Institute of Meteoritics, Spec. Publ. No. 11, Univ. New Mexico, Albuquerque. 78 pp.

Kushiro I., Ikeda Y., and Nakamura Y. (1972) Petrology of Apollo 14 high-alumina basalt. *Proc. Lunar Sci. Conf. 3rd,* p. 115–129.

Lindsay J. F. (1972) Sedimentology of clastic rocks from the Fra Mauro region of the moon. *J. Sediment. Petrol.* **42**, 19–32.

Longhi J., Walker D., and Hays J. F. (1972) Petrography and crystallisation history of basalts 14310 and 14072. *Proc. Lunar Sci. Conf. 3rd,* p. 131–139.

McCallum I. S., Mathez E. A., Okamura F. P., and Ghose S. (1974) Petrology and crystal chemistry of poikilitic anorthositic gabbro 77017. *Proc. Lunar Sci. Conf. 5th,* p. 287–302.

Meyer H. O. A. and McCallister R. H. (1973) Mineralogy and petrology of Apollo 16: Rock 60215, 13. *Proc. Lunar Sci. Conf. 4th.* p. 661–665.

Oberbeck V. R. (1975) The role of ballistic erosion and sedimentation on lunar stratigraphy. *Rev. Geophys. Space Phys.* **13**, 337–362.

Papike J. J., Bence A. E., and Lindsley D. H. (1974) Mare basalts from the Taurus-Littrow region of the moon. *Proc. Lunar Sci. Conf. 5th,* p. 471–504.

Papike J. J., Hodges F. N., Bence A. E., Cameron M., and Rhodes J. M. (1976) Mare basalts: crystal chemistry, mineralogy, and petrology. *Rev. Geophys. Space Phys.* **14**, 475–540.

Phinney E. C., McKay D. S., Simonds C. H., and Warner J. L. (1976) Lithification of vitric- and clastic-matrix breccias: SEM petrography. *Proc. Lunar Sci. Conf. 7th,* p. 2469–2492.

Pohl J., Stöffler D., Gall H., and Ernstson K. (1976) The Ries impact crater. In *Impact and Explosion Cratering* (D. J. Roddy, R. O. Pepin, and R. B. Merrill, eds.). Pergamon Press Inc., New York. In press.

Powell B. N. and Weiblen P. W. (1972) Petrology and origin of lithic fragments in the Apollo 14 regolith. *Proc. Lunar Sci. Conf. 3rd,* p. 837–852.

Ridley W. I. (1975) On high-alumina mare basalts. *Proc. Lunar Sci. Conf. 6th,* p. 131–145.

Ridley W. I., Brett R., Williams R. J., Takeda H., and Brown R. W. (1972) Petrology of Fra Mauro basalt 14310. *Proc. Lunar Sci. Conf. 3rd,* p. 159–170.

Ryder G. and Bower J. F. (1976) Petrology. In *Interdisciplinary Studies by the Imbrium Consortium,* Vol. 1 (J. A. Wood, ed.). LSI Contr. 267D, p. 22–37, p. 41–50, p. 55–66. Center for Astrophysics, Cambridge.

Ryder G. and Bower J. F. (1977) Petrology. In *Interdisciplinary Studies by the Imbrium Consortium,* Vol. 2 (J. A. Wood, ed.). LSI Contr. 268D, p. 1–7, p. 13–15. Center for Astrophysics, Cambridge.

Ryder G. and Taylor G. J. (1976) Did mare-type volcanism commence early in lunar history? *Proc. Lunar Sci. Conf. 7th,* p. 1741–1755.

Simonds C. H., Warner J. L., Phinney W. C., and McGee P. E. (1976) Thermal model for impact breccia lithification: Manicouagan and the moon. *Proc. Lunar Sci. Conf. 7th,* p. 2509–2528.

Stöffler D. (1975) Ries crater breccias and planetary impact formations. *Fortschr. Miner.* Spec. Issue to Vol. 52: IMA-papers 9th Meeting Berlin-Regensburg 1974, p. 385–387.

Stöffler D. (1977) Research drilling Nördlingen 1973: polymict breccias, crater basement and cratering model of the Ries impact structure. *Geol. Bavarica* **75**, 443–458.

Stöffler D., Dence M. R., Graup G., and Abadian M. (1974) Interpretation of ejecta formations at the Apollo 14 and 16 sites by a comparative analysis of experimental, terrestrial, and lunar craters. *Proc. Lunar Sci. Conf. 5th*, p. 137–150.

Stöffler D., Knöll H.-D., Reimold W.-U., and Schulien S. (1976) Grain size statistics, composition and provenance of fragmental particles in some Apollo 14 breccias. *Proc. Lunar Sci. Conf. 7th*, p. 1965–1985.

Smith J. V. (1974) Lunar mineralogy: A heavenly detective story. Presidential Address, Part 1. *Amer. Mineral.* **59**, 231–243.

Smith J. V. and Steele I. M. (1976) Lunar mineralogy: A heavenly detective story. Part II. *Amer. Mineral.* **61**, 1059–1116.

Steele I. M. and Smith J. V. (1975) Minor elements in lunar olivine as a petrologic indicator. *Proc. Lunar Sci. Conf. 6th*, p. 451–467.

Steele I. M. and Smith J. V. (1976) Mineralogy and petrology of complex breccia 14063,14. *Proc. Lunar Sci. Conf. 7th*, p. 1949–1964.

Sweatman T. R. and Long J. V. P. (1969) Quantitative electron-probe microanalyses of rock-forming minerals. *J. Petrol.* **10**, 332–379.

Tarasov L. S., Nasarov M. A., Shevaleevsky I. D., Makarov E. S., and Ivanov V. I. (1973) Mineralogy of anorthositic rocks from the region of the crater Apollonius C (Luna 20). *Proc. Lunar Sci. Conf. 4th*, p. 333–349.

Taylor L. A. and Misra K. C. (1975) Pyroxene-phyric basalt 15075: Petrography and petrogenesis. *Proc. Lunar Sci. Conf. 6th*, p. 165–179.

Warner J. L. (1972) Metamorphism of Apollo 14 breccias. *Proc. Lunar Sci. Conf. 3rd*, p. 623–643.

Warner J. L., Simonds C. H., Phinney W. C., Geeslin J. H., and Bickel C. E. (1977) Petrology and crystallization history of Apollo 14 impactites (abstract). In *Lunar Science VIII*, p. 982–984. The Lunar Science Institute, Houston.

Warner R. D., Warren R. G., Mansker W. L., Berkley J. L., and Keil K. (1976) Electron microprobe analyses of olivine, pyroxene, and plagioclase from Apollo 17 rake sample mare basalts. Institute of Meteoritics, Spec. Publ. No. 15, Univ. New Mexico, Albuquerque. 158 pp.

Wilshire H. G. and Jackson E. D. (1972) Petrology and stratigraphy of the Fra Mauro formation at the Apollo 14 site. U.S. Geol. Survey Prof. Paper 785, 1–26.

Wood J. A. (ed.) (1976) Interdisciplinary Studies by the Imbrium Consortium. Vol. 1. LSI Contr. No. 267D. Center for Astrophysics, Cambridge. 155 pp.

Wood J. A. (ed.) (1977) Interdisciplinary Studies by the Imbrium Consortium. Vol. 2. LSI Contr. No. 268D. Center for Astrophysics, Cambridge. 125 pp.

Proc. Lunar Sci. Conf. 8th (1977), p. 1869–1893.
Printed in the United States of America

Apollo 14 revisited, or breccias aren't so bad after all

CHARLES H. SIMONDS

The Lunar Science Institute, 3303 NASA Road 1, Houston, Texas 77058

WILLIAM C. PHINNEY and JEFFREY L. WARNER

NASA Johnson Space Center, Houston, Texas 77058

PATRICIA E. MCGEE, JILL GEESLIN, ROY W. BROWN, and J. MICHAEL RHODES

Lockheed Electronics Co., Inc., Houston, Texas 77058

Abstract—Restudy of the large Apollo 14 rocks supplemented by petrographic and chemical studies of the previously undescribed comprehensive sample suggests that the Apollo 14 samples were formed by the same processes which operated to form the range of impactities found at Apollo 16 and 17 and in terrestrial craters, especially those in crystalline targets. The abundant crystalline matrix breccias of the Apollo 14 samples generally have higher clast contents and less refractory clast populations than the clast-laden impact melts that are abundant at the other Apollo non-mare landing sites. These differences can be explained by mixing a greater amount of cold clastic debris into the superheated melt formed during the impacts that produced the Apollo 14 breccias. This could result in thermal equilibration of the clast-melt mixture to temperatures near or below the solidus rather than near or below the liquidus as is the case for impact melts. The other major lithology at Apollo 14, the vitric matrix breccias, contains agglutinates and has bulk compositions matching the local soil. The vitric matrix breccias cannot be the protolith of crystalline matrix breccias, because they contain too little coarse debris, too few mineral clasts, and are compositionally distinctive from the crystalline breccias. Textural and compositional variations among the breccias indicate that several impact events are responsible for the breccia population.

INTRODUCTION

SINCE THE LAST SERIES of petrologic examinations of the Apollo 14 collection by Chao *et al.* (1972), Christie *et al.* (1973), von Engelhardt *et al.* (1972), Lally *et al.* (1972), Quaide and Wrigley (1972), Warner (1972), Wilshire and Jackson (1972), and Williams (1972) a great deal has been learned about impact formed rocks, the geology of the moon, and cratering. Data presented in this paper include a compilation of available bulk chemical analyses including twenty new major element analyses from the comprehensive sample, petrographic modes of clast types, petrographically determined grain sizes, mineral chemistry for plagioclase and mafic mineral clasts, and matrix petrography using both the optical and scanning electron microscopes. 54 Apollo 14 rocks for which thin sections are available, have been examined petrographically. These rocks include 24 from the comprehensive sample which previously had not been available. The samples are classified into groups which can be related in terms of petrogenesis to lithologies from the Apollo 16 and 17 landing sites and from

Table 1. Classification.

Sample, thin section	Mass	Petrographic category	Porosity	Fractures SEM	Fractures PET	Warner (1972)	Chao (1972)	von Englehardt (1972)	Quaide and Wrigley (1972)	Wilshire and Jackson	Clasts >1 mm
14006,5	12.13	CMB(FM)	Low	Some	Few	6	2cU	Cry.	Ann.	C	
14041,5 and 6	166.27	VMB	Low		Few					F_1	2
14042,7	103.19	VMB	High			1	1a			F_1	10
14045,7,8	65.24	VMB	High							F_1	1
14047,53	242.01	VMB	High		Few	1	1a			F_1	3
14049,38 and 39 and 41	200.13	VMB	Int.		V.few	2	1a	Reg.		F_1	1
14053,6	251.32	Mare B.							Basalt	B	
14055,7 and 11	110.99	VMB	Int.		Many	1	1a	Reg.	Reg.	F_1	1
14063,51	135.45	LMB	High		Many	3	2a	W.R.	W.R.	F_3	23
14064,21	107.53	CMB(FM)	Low		V.few					F_3	40
14066,47 and 48 and 50	509.80	CMB(FM)	Many	Some	Few	7	2cS	Cry.	Ann.	F_4	25
14068,8	35.47	CMB(CL)	Int.		V.few	8				C	
14072,12	45.06	Mare B.								B	
14073,7	10.35	CMB(CL)								B	
14082,10 and 40	62.63	LMB	High		Few?	3	2a	W.R.		F_2	10
14169,4 and 5	part of 14303-4 [78.66]	CMB clast			Few					F_4	25
14171,7 and 11	37.79	CMB(FM) clast		Some	Few	4	2cS			F_3	30
14172,3 and 4	32.10	CMB(FM) clast								F_4	6
14173,3	17.59	Gabbro clast								F_4	20
14179,3	3.03	Ann. feld. gabbro clast								F_4	20
14180,3 and 4	4.75	Crushed troc. clast								F_4	20
14181,4 and 5	2.48	CMB(CF) clast								F_4	1
14187,2 and 3	1.09	CMB(CL) clast								F_4	50
14250,34	4.06	VMB	Low		V.few					F_1	10
14253,4 and 3		CMB(FM)	V.low		V.few						

14255,5 and 6	22.15	VMB	V.low		Cooling cracks?					F_2	2
14264,5 and 6 and 7	117.87	VMB	Low		V.few					F_4	10
14265,8	65.79	VMB	Low		Glass filled					F_2	10
14266,4	6.95	CMB(FM)	Low		Cooling cracks					F_4	25
14269,4 and 5	17.19	VMB	Low		Cooling cracks						
14270,9	25.59	CMB(FM)	Int.		Few		2cU			F_2	1
14271,10,11,12,13,14	97.41	VMB	Low		V.few						
14274,4,5	15.18	CMB	Low		V.few					C	
14275,45	12.46	VMB	Low		None					F_2	10
14276	12.75	CMB(CL)								B	
14278,4,5	7.60	VMB	V.low		None					F_2	20
14279,4,5	5.67	CMB(FM)	V.low		Few					F_4	1
14283,5,6	1.25	VMB	Low		None						
14301,13,78,81	1360.60	VMB	High		Few	1	1a			F_2	18
14303,46 and 52	898.40	CMB(FM)	Low int?	Some	V.many	6	2cS			F_4	28
14304,13	2997.9 part of 14303	CMB(FM)	Low	Some	Mod.	6	2cS			F_4	25
14305,98 and 95	2497.50	CMB(FM)	Low	Some	Many	6	2cS			F_4	28
14306,4 and 69 and 65	534.50	CMB(FM)	V.low	Many	Mod. many	4	2cS		Ann. brec.	F_4	15
14307,40 and 31	155.	VMB	V.low		Cooling cracks?	1	1c	Reg. brec.		F_2	20
14310,173	3439.	CMB(CF)	V.low		Few				B	B	
14311,84 and 93 and 94	3204.40	CMB(FM)	Many int.	None	None	5	2cU		Ann. brec.	F_4	20
14312,5 and 13	299.00	CMB(FM)	Low	V.many	Many	7	2cS			F_4	45
14313,41 and 48	144.0	VMB	Low		None	1	1b		Reg. B.	F_2	5
14314,10 and 11	115.70	CMB(FM)	Low	Many	Many	7	2cS	Reg. brec.		F_4	25
14315,7 and 10	115.0	VMB	Low	Int.	Many	3	3	Reg. brec.		F_2	20
14318,5 and 11 and 45	38.20	VMB	Mod.-	Int.	Many	3	3		Reg. brec.	F_2	30
14319,4 and 5 and 19	211.60	CMB(FM)	Low	Some	Some	7	2cS			F_4	39
14320,6 and 4 and 7	64.90	CMB(FM)	Low	Some	Many	6	2cS		Ann. brec.	F_4	13
14321,197 and 210	8998.0	CMB(FM)	Low		Mod.	4	2b		Ann. brec.	F_4	30

Footnotes to Table 1.

Classification of Chao *et al.* (1972).
(1) Regolith microbreccias.
  (a) unshocked porous.
  (b) compact, nonporous.
  (c) shocked.
(2) Fra Mauro breccias.
  (a) unannealed or slightly annealed, feldspathic breccias.
  (b) moderately annealed breccias.
  (c) strongly annealed (thermally metamorphosed) breccias: U—Unshocked, S—Shocked.
(3) Spherule-rich transported microbreccias.

Classification scheme of von Engelhardt *et al.*, (1972).
  Glass-rich breccias.
  Glass-poor breccias with fragmental matrices.
  Glass-poor breccias with crystalline matrices.

Classification scheme of Quaide and Wrigley (1972).
  Regolith breccias.
  White rock breccias.
  Annealed breccias.

Classification scheme of Wilshire and Jackson (1972).
  B—Homogeneous crystalline rocks, basaltic.
  C—Homogeneous crystalline rocks, metaclastic.
  F—Fragmental rocks, light clasts dominant.
  F_1—Fragmental rocks, light clasts dominant, matrix friable.
  F_2—Fragmental rocks, light clasts dominant, matrix coherent to moderately coherent.
  F_3—Fragmental rocks, dark clasts dominant, matrix friable.
  F_4—Fragmental rocks, dark clasts dominant, matrix coherent to moderately coherent.

terrestrial impact structures. The number and types of petrogenetically significant impacts at the Apollo 14 landing site are evaluated.

Particular emphasis is placed on study of the two main breccia types of KREEP-like composition, the vitric and the crystalline breccias, in order to test first the suggestion of Warner (1972) that those two groups represent a sequence of progressive metamorphism and second the other models for breccia lithification. We also seek to understand the petrogenetic relation between the bulk of the crystalline matrix breccias at Apollo 14 and the macroscopically and compositionally similar clast-laden impact melts such as the Apollo 17 Station 6 boulder.

Observations

Table 1 presents a sample by sample classification of the Apollo 14 rocks based on the specific thin sections examined. The groups are related to criteria previously proposed by Christie *et al.* (1972), von Engelhardt *et al.* (1972), Lally *et al.* (1973), Quaide and Wrigley (1972), and Simonds *et al.* (1976). Also given are the specific designations by previous workers. The definitions of the groups are as follows:

Crystalline matrix breccias. Impactites having tough holocrystalline matrices with at least some vestiges of clasts and siderophile contents (when analyses are available) suggesting some meteorite contamination. Rocks in this group show recrystallization effects that fall in Group A of Christie *et al.* (1973). All Apollo 14 rocks in this group have a KREEP-like composition, although some small clasts of crystalline breccias-in-breccias are more feldspathic than KREEP. The crystalline matrix breccias may be subdivided into three groups on the basis of correlated variations in clast content, biasing of clast population to refractory materials, and the size and degree of crystal morphology development in matrix feldspar.

(1) Clast-free impact melts: Subophitic igneous textured rocks with isolated, fine-grained regions of groundmass (suggestive of initial thermal, nucleation, or compositional heterogeneities) and high siderophile (meteorite contamination) contents compatible with an impact rather than volcanic origin.

(2) Clast-bearing impact melt rocks: With 1–15% clasts, mostly refractory mineral clasts, very few if any glass-bearing clasts, and matrices with subhedral plagioclase.

(3) Fra Mauro breccias: With over 15% clasts, with numerous lithic and relatively low melting-point mineral clasts, few if any glass clasts, and sub-10 μm subhedral to anhedral plagioclase in the matrix. This group grades into the clast-bearing melts; the term Fra Mauro breccias is used following the suggestion of Chao *et al.* (1972) because of the abundance of that lithology at the Apollo 14 (Fra Mauro) landing site.

Vitric matrix breccias. Fine-grained impactites, of variable coherence, with a fragmental texture, abundant glass and glass-bearing clasts, and abundant low melting-point mineral and lithic clasts including some agglutinates. Rocks in this group tend to lack recrystallization effects and, thus, fall in Group B of Christie *et al.* (1973). Varying amounts of glass between grains bind the rocks together (Lally *et al.*, 1973; Phinney *et al.*, 1976). These samples texturally resemble vitric breccias from Apollos 11, 12, 15, 16, and 17 and Luna 16, 20, and 24. The bulk composition of this group is KREEP-like, and closely matches the composition of local soil, including its characteristic meteoritic contamination.

Light matrix breccias. Friable, porous, fragmental breccias, with a range of clast types (only a few percent of which are glass clasts). Rocks in this group lack recrystallization effects and thus fall in Christie *et al.*'s (1973) Group B. These rocks are more feldspathic than the Apollo 14 crystalline or vitric matrix breccias. The rocks have meteorite contamination indicated by high siderophile abundances.

Mare basalts. Igneous-textured rocks with high iron contents chracteristic of mare basalts from Apollo's 11, 12, 15, and 17, but compositionally distinct from the major groups found at those landing sites. The rocks have low siderophile abundances suggesting little if any meteoritic contamination. The Apollo 14 basalts typically date at 3.9 AE or older (Compston *et al.*, 1972), that is they are older than the mare basalts from the other Apollo sites.

It is obvious from the above scheme that there has been some retrenchment from the classification scheme suggested by Warner (1972). The three grades of vitric matrix breccias (1–3) are combined into a single category because our SEM investigations of these rocks suggest that matrix textures of these grades are heterogeneous and intermingled on a scale of millimeters. It is our observation that porosity and glass content, suggested to form a continuous sequence by Christie *et al.* (1973), are also heterogeneous within individual samples. All of the rocks in grades 4–7 (Warner, 1972) fall in the category of Fra Mauro breccias, and it may prove possible to further subdivide this group on the basis of matrix studies, however two features make further subdivision difficult: (1) the abundant fracturing that is characteristic of many of the rocks at the Apollo 14 site obscures many aspects of their textures and (2) as discussed below, matrix texture varies on a millimeter scale. The single sample that represented Warner's Group 8 (14068) still falls in a separate category on the basis of its low clast content and well formed igneous texture. The rock is one of the clast-laden, impact melts.

The classification presented does not consider either fracturing or porosity to be significant variables. Petrographic and SEM examination of crystalline breccias reveals numerous fractures on a wide range of scales. It is on the basis of these fractures that Chao *et al.* (1972) distinguished between shocked and unshocked samples. These fractures cut the rock's matrix thus postdate breccia lithification, and may well be due to reworking of pre-existing crystalline breccias by Imbrium secondaries as suggested by Chao *et al.* (1972) and

Oberbeck *et al.* (1975). In our experience, these fractures are unusually abundant in Apollo 14 rocks relative to Apollo 16 or 17 highland lithologies. Measurements were made of petrographic modes and grain sizes in order to test whether (1) the vitric and crystalline breccias could be cogenetic and (2) if there were consistent correlations of clast type, matrix texture and the fraction of matrix.

Modal and Grain Size Observations

Models for the formation of clast-laden, impact-melt rocks as mixtures of superheated silicate melt and cold clastic debris (Simonds, 1975; Simonds *et al.*, 1976) suggest that there should be a correlation between clast abundance and the biasing of the clast population to refractory compositions.

The length of intercept and lithology of all clasts greater than 39 μm were recorded by a line intercept technique in which thin sections were traversed in a series of equally spaced parallel lines. This lowest measured size was chosen for convenience on the microscope used. The modal percentages of the different lithologies are equivalent to that which would be obtained if a point counting technique were used. The sizes, however, are not directly comparable with data obtained by some other procedures. For example, the data of Stöffler *et al.*

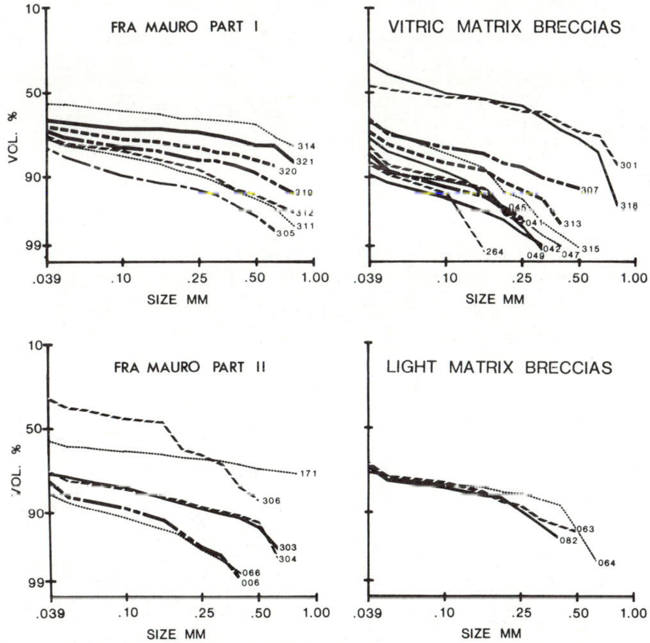

Fig. 1. Cumulative volume percent less than a given size for Fra Mauro vitric and light matrix breccias. Only last three digits of generic sample numbers as listed in Table 2 are given. All material less than 39 μm was defined as matrix and its total intercept length measured but not the individual grain sizes.

Table 2. Modes (vol.%).

		Light matrix breccias (LMB)			Vitric matrix breccias (VMB)										
Sample number		14063	14064	14082	14041	14042	14047	14049	14055	14264	14301	14307	14313	14315	14318
Comma numbers		,51	,21	,10	,5	,12	,53	,38 ,39 ,41	,7	,7	,78 ,13	,40	,48	,10 ,7	,5 ,11
Matrix (<39 μ)		73	74	71	79½	88	85½	78	72	83	46	65	72	66	42½
Clasts (>39 μ):	Plagioclase	9	5½	10½	3½	1	½	2	1	3	2	1½	4	3½	2
	Mafic	5½	2	2	2½	2	2	1	½	—	2½	3	1	½	1½
	Ilmenite	TR	—	TR	—	—	TR	TR	—	—	TR	—	TR	TR	—
	Hetergenous glass	TR	½	TR	TR	2	4	1½	5½	2	1	½	2	3½	2
	Homogenous glass	—	—	—	2	TR	½	½	1	2	TR	½	TR	1½	½
	Agglutinates	—	—	—	TR	—	TR	3	—	2	—	1	—	—	—
	VMB	7½	—	3½	4½	1	6½	6	7	1	5	5½	6½	7	8½
	CMB	1	12	7	3	2	1	2½	2	4	28½	6½	5	3½	27
	Granulitic	1	—	1½	4½	4	—	4½	11	3	10	9	7	9½	11
	Other metamorphic	2	—	—	TR	—	—	—	—	—	1½	—	½	1	—
	Mare basalt	TR	—	TR	—	—	TR	½	—	—	3½	6½	½	3½	4
	Feldspathic basalt	½	5	3	TR	TR	—	½	—	TR	TR	TR	½	½	1
	Crushed	½	—	1½	½	—	TR	—	—	—	TR	½	1	TR	—
	Pore	—	½	—	—	TR	—	TR	—	—	TR	½	TR	TR	—
	Other	TR	½	TR	—	—	—	—	—	—	—	—	TR	TR	—

		Crystalline matrix breccias (CMB)												
Sample number		14006	14066	14171	14303	14304	14305	14306	14311	14312	14314	14319	14320	14321
Comma numbers		,5	,48 ,47 ,50	,8	,46	,2	,98	,69 ,4	,94	,15	,11 ,10	,4	,6 ,4	,97
Matrix (<39 μ)		77	$83\frac{1}{2}$	57	$75\frac{1}{2}$	74	80	36	75	75	$55\frac{1}{2}$	$71\frac{1}{2}$	$69\frac{1}{2}$	66
Clasts (>39μ):	Plagioclase	11	$8\frac{1}{2}$	3	10	3	7	$5\frac{1}{2}$	$6\frac{1}{2}$	10	$9\frac{1}{2}$	10	6	$9\frac{1}{2}$
	Mafic	7	$3\frac{1}{2}$	4	$4\frac{1}{2}$	5	7	$1\frac{1}{2}$	3	3	4	$4\frac{1}{2}$	$4\frac{1}{2}$	4
	Ilmenite	TR	TR	—	$\frac{1}{2}$	—	TR	TR	—	TR	TR	—	TR	—
	Hetergenous glass	—	—	—	—	—	—	TR	—	—	—	—	—	$\frac{1}{2}$
	Homogenous glass	—	—	—	—	—	—	—	—	—	—	—	—	—
	Agglutinates	—	—	—	—	—	—	—	—	—	—	—	TR	TR
	VMB	—	—	—	$\frac{1}{2}$	—	—	TR	—	1	TR	$\frac{1}{2}$	1	$2\frac{1}{2}$
	CMB	$\frac{1}{2}$	—	34	7	6	1	52	$2\frac{1}{2}$	2	23	2	$2\frac{1}{2}$	$5\frac{1}{2}$
	Granulitic	1	$1\frac{1}{2}$	—	$\frac{1}{2}$	2	$1\frac{1}{2}$	$2\frac{1}{2}$	$4\frac{1}{2}$	3	3	$\frac{1}{2}$	$3\frac{1}{2}$	$2\frac{1}{2}$
	Other metamorphic	—	$\frac{1}{2}$	TR	TR	—	—	—	—	$\frac{1}{2}$	4	$6\frac{1}{2}$	3	$4\frac{1}{2}$
	Mare basalt	$\frac{1}{2}$	$\frac{1}{2}$	—	TR	—	—	—	—	TR	—	—	4	—
	Feldspathic	1	$\frac{1}{2}$	2	1	10	3	2	TR	5	—	3	—	$3\frac{1}{2}$
	Crushed	—	TR	—	—	—	—	$\frac{1}{2}$	—	$\frac{1}{2}$	—	$\frac{1}{2}$	$2\frac{1}{2}$	1
	Pore	2	1	TR	$\frac{1}{2}$	—	$\frac{1}{2}$	TR	$8\frac{1}{2}$	$\frac{1}{2}$	TR	$\frac{1}{2}$	TR	$\frac{1}{2}$
	Other	TR	$\frac{1}{2}$	—	—	—	TR	TR	TR	—	—	$\frac{1}{2}$	$3\frac{1}{2}$	—

(1976) were collected by determining the equivalent radius of a particle with the same area as the object in question: in effect, an average radius. Because many of the traverses using our procedure do not cross a clast at its maximum dimension, the average size presented in Fig. 1 is smaller than that determined by the procedure of Stöffler *et al.* (1976).

The distinction between clasts and matrix is arbitrary in most vitric and some Fra Mauro breccias since the grain size of clastic fragments is seriate. Therefore, the very small cold fragments that have not been digested, as was the case in the clast-laden impact melts, are still present (Simonds, 1975; Onorato *et al.*, 1977). The modes so determined are given in Table 2, and the grain size information is in Table 2 and Fig. 1.

The *plagioclase*, *mafic* minerals and *ilmenite* clast categories used in Table 2 are self-explanatory. The *heterogeneous glass* clasts range from schlieren-filled fragments to glasses with spherulitic devitrification. Recognition of *agglutinates* was done using reflected light, where such particles are readily observed to contain round pores $\frac{1}{10}$ to $\frac{1}{4}$ the diameter of the particle. *Vitric matrix breccia* (VMB) clasts are similar to agglutinates in transmitted light; both have a fine-grained brown matrix. *Vitric matrix breccias*, however, have relatively tiny pores if any. *Granulitic* clasts are similar to sample 79215 (Bickel *et al.*, 1976) and 76235 (Simonds, 1975; Warner *et al.*, 1977); they are typically feldspathic norites or troctolites containing feldspar with polygonal outlines and mafic minerals that are generally confined to corners or grain boundaries between feldspars. *Crystalline matrix breccias* (CMB) run the full range from Fra Mauro breccias to those with a clast-laden poikilitic texture. The distinction between the *mare basalts* and the *feldspathic basalts* is based largely on the pyroxene color. If a clast has pink pyroxene, the fragment is considered a *mare basalt*; pale yellow-green pyroxene suggests a *feldspathic basalt.* Many of the *feldspathic basalts* are similar to small areas of 14310, inferred to be an impact melt. Others, however, have the needle-like, curving feldspar characteristic of some of the Apollo 15 KREEP basalts, such as 15382. The small size of the clasts, however, does not allow consistent subdivision of the different textures. *Crushed* lithologies include a variety of intensely cataclastic, feldspar-rich, plutonic lithologies. In sample 14320 the bulk of the $3\frac{1}{2}\%$ *others* consists of light matrix breccia.

CLAST MICROPROBE ANALYSES

The compositions of mineral clasts in the breccias can reveal two types of information: First, by examining the fraction of feldspar clasts more refractory than the matrix feldspar one can determine the extent of digestion of the less shocked material incorporated in the breccia; and second, by comparing mineral and lithic clast populations, one can subdivide groups of breccias whose bulk compositions form clusters. Microprobe data on clast compositions are presented in Figs. 2 and 3. The technique of analyses is the same as was used for Apollo 16 (Warner *et al.*, 1973; Simonds *et al.*, 1973) and Apollo 17 samples

(Simonds, 1975; Warner *et al.*, 1976). For each sample three elements were analyzed in 100 feldspar clasts and all mafic clasts from one or two thin sections. For these experiments only the KREEP composition breccias were analyzed; the light matrix breccias were excluded. Figure 2 suggests that samples 14305, 14311, and 14314 are distinct in mineral clast population, from the other crystalline breccias, and from each other. Data on the mafic mineral clast are few and therefore, conclusions about their range in composition are quite uncertain.

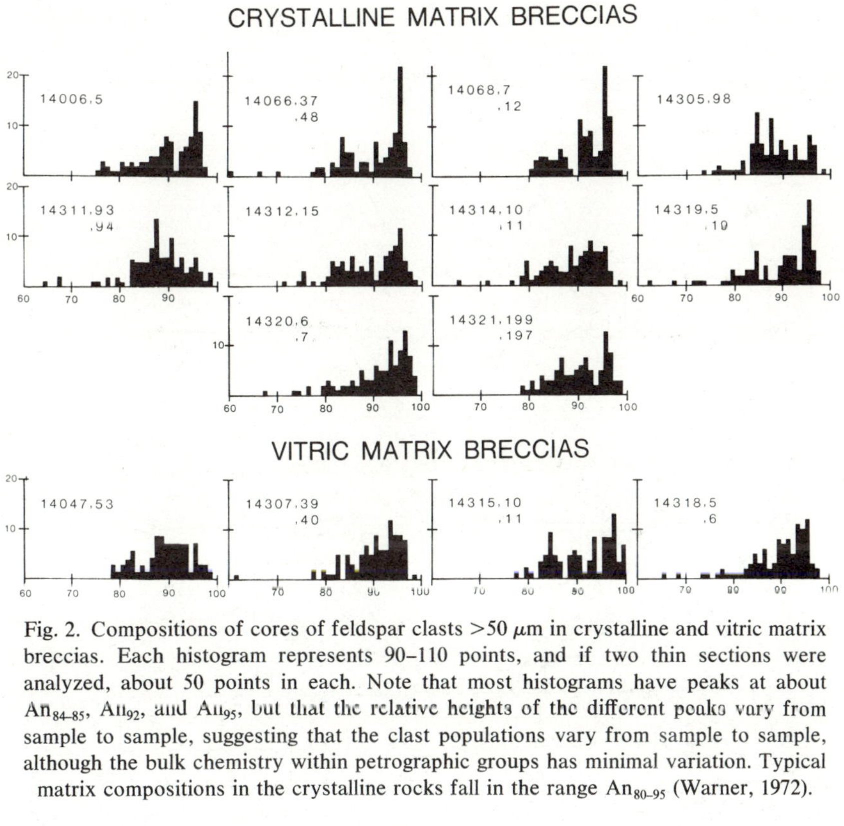

Fig. 2. Compositions of cores of feldspar clasts >50 μm in crystalline and vitric matrix breccias. Each histogram represents 90–110 points, and if two thin sections were analyzed, about 50 points in each. Note that most histograms have peaks at about An_{84-85}, An_{92}, and An_{95}, but that the relative heights of the different peaks vary from sample to sample, suggesting that the clast populations vary from sample to sample, although the bulk chemistry within petrographic groups has minimal variation. Typical matrix compositions in the crystalline rocks fall in the range An_{80-95} (Warner, 1972).

Bulk Composition

Twenty new bulk chemical analyses of clasts, matrices, and glass coatings are summarized in Table 3. Analyses done by X-ray fluorescence utilized 100–280 mg of material fused in a lanthanum-bearing lithium borate flux following procedures outlined in Norish and Hutton (1969). These procedures are those used customarily in this laboratory for lunar and terrestrial materials. Na was analyzed by instrumental neutron activation. The electron microprobe analyses were carried out on replicate 10 mg samples ground, fused, and analyzed according to procedures described by Brown (1977).

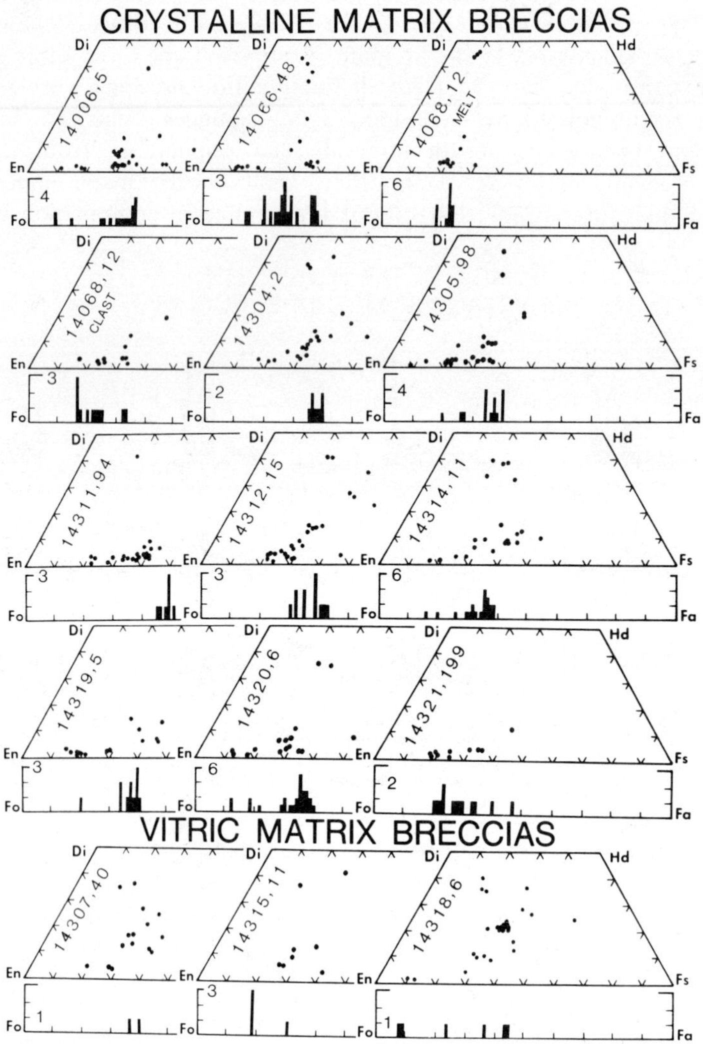

Fig. 3. Compositions of the cores of mafic mineral clasts greater than 50 μm in crystalline and vitric matrix breccias. All mafic clasts in the sections indicated were analyzed. Sample 14068, the only clast-laden melt analyzed, had a unique, very magnesian composition and a new set of matrix analyses were made.

DISCUSSION

Petrogenesis of vitric matrix breccias

Vitric matrix breccias have been sampled by all of the Apollo and Luna sample-return missions. Several specific questions may be asked about the origin of this rock type, specifically:

(1) Are they made from lithified soil?
(2) Are they co-genetic with the Fra Mauro breccias?
(3) What is the process of their lithification?
(4) What is their age?
(5) What type of crater is responsible for their formation?

One line of evidence suggesting that these breccias contain at least some lunar soil is the incorporation of agglutinates, which comprise up to 58% of a mature regolith at Apollo 14 (Heiken, 1974). However, the very small fraction of agglutinates (Table 2) suggests that the soil which made up the breccias was very immature or that only a small fraction of the recognizable protolith was agglutinate-rich soil. A second line of evidence is the similarity in chemical composition between the soil and the vitric breccias. Other features common to both lunar soil and the breccias include the presence of solar flare particle tracks (Dran *et al.*, 1973); excess iron, much of which is in fine grained, superparamagnetic sizes (Gose *et al.*, 1973); and trapped solar wind gases (Bogard and Nyquist, 1973) all of which indicate direct exposure at the lunar surface.

The bulk composition, grain size, and clast modes of the vitric breccias suggest that they are not co-genetic with the crystalline Fra Mauro breccias. Figure 4 presents MgO and Al_2O_3 data for the bulk samples; i.e., matrix plus small clasts, for the full range of Apollo 14 materials. Note that the vitric breccias are systematically depleted in MgO and generally more enriched in Al_2O_3 than the crystalline breccias. The crystalline breccias therefore, are not made by simply metamorphosing the Apollo 14 vitric breccias. The grain size data given in Fig. 1 shows that the vitric breccias typically have more matrix and smaller clasts than the crystalline breccias. Although the processes involved in the formation of the crystalline breccias probably increases grain sizes, this increase should occur only for the materials whose sizes were initially less than about 10 μm and should not produce any recognizable effect on the plots shown. Similar arguments were presented by Stöffler *et al.* (1976). The final and most conclusive argument that the vitric breccias are not the protoliths to the Fra Mauro breccias is the higher mineral clast content of the Fra Mauro breccias. In order to make the crystalline breccias from the soils, it would be necessary to create clasts.

The process of soil or vitric breccia lithification is a subject of controversy. Models for the lithification of soil, *in situ*, by shock waves with less than 100 kbar peak pressure have been proposed by Chao *et al.* (1972), Christie *et al.* (1973), and Kieffer (1975). They argue that the complex interactions of shock waves with boundaries between grains in porous aggregates result in glass formation at the contacts between grains. Christie *et al.* demonstrate the existence of the predicted glass at the grain boundaries, and they further suggest, on the basis of examination of two Apollo 14 and one Apollo 15 breccias, that there is a greater amount of glass in the breccias with lower porosities. They indicate that such a relation would correspond to increasing shock pressure for lithification. The alternate model (McKay and Morrison, 1971; Waters *et al.*,

Table 3. Major element analyses.

Sample number	14169,2	14173,2	14181,1	14253,2	14255,2	14255,3	14264,2	14264,3	14265,3	14265,4
Type	Clast CMB(FM)	Gabbro clast	Clast CMB(CF)	Matrix CMB(FM)	Glass coat	Matrix VMB	Matrix VMB	Clast CMB(CL)	Glass coat	Matrix VMB
Method	EMP	EMP	EMP	EMP	EMP	XRF	XRF	EMP	EMP	XRF
SiO_2	47.35	41.78	47.59	48.15	47.055	47.35	49.28	46.19	47.53	47.36
TiO_2	1.52	0.89	2.14	1.15	1.72	1.75	2.05	1.80	1.84	1.76
Al_2O_3	17.00	7.12	15.90	17.65	17.87	17.45	16.29	11.81	17.43	17.51
Cr_2O_3	0.21	0.53	0.20	0.21	0.21			0.54	0.22	
FeO	10.05	21.21	9.54	8.65	9.85	10.46	10.23	16.43	10.06	10.62
MnO			.11			0.16	0.13			0.14
MgO	10.47	22.82	11.97	10.75	9.40	9.48	9.08	11.36	9.25	9.26
CaO	10.53	5.24	9.60	10.49	11.16	11.09	10.12	10.26	11.25	11.17
Na_2O	0.75	0.20	1.02	0.68	0.62	0.66	0.93	0.37	0.60	0.68
K_2O	0.43	0.13	.90	0.57	0.46	0.47	0.82	0.61	0.46	0.47
P_2O_5			.76			0.45	0.66			0.50
S						0.09	0.09			0.09
Total	98.29	99.91	99.72	98.31	98.33	99.41	99.68	99.39	98.66	99.50

1971; Simonds, 1973; Uhlmann *et al.*, 1975; Phinney *et al.*, 1976) for the formation of the vitric breccias involves thermal welding or sintering by small amounts of glass. In this case the glass forms from the small volume near the point of impact and becomes turbulently mixed with less-shocked debris either in the crater or in material deposited on its rim. Proponents of both models agree that small amounts of glass in the finest-grained matrix materials act as the binder in the rocks. Furthermore, both the sintering and shock lithification models predict that greater amounts of glass will increase the coherence of the rocks due to greater amounts of binder glass and be associated with lower porosities. Lower porosities result from the shrinkage accompanying sintering, or result from slightly higher shock pressures which result in the formation of more glass. Neither model suggests that the coarser mineral and lithic fragments be subjected to high shock pressures; i.e., pressures high enough to produce distinctive petrographic effects. Thus, at present, understanding of the mechanism of breccia lithification is at an impasse.

The age of the vitric breccias cannot be directly determined by radiometric techniques. Petrographically similar breccias from other landing sites, however, contain fragments of mare basalt similar to local rocks dated at 3.9–3.2 AE. Therefore, the vitric breccias at the other landing sites formed after major mare flooding, that is after the conclusion of the most intense bombardment of the moon. The mare-covered areas near the other Apollo sites have very few craters even a kilometer across. Therefore, the soil breccias at the other landing sites were formed by small cratering events. The vitric breccias at Apollo 14 contain numerous fragments of the other lithologies found at the landing site and

Table 3. (*Continued*).

Sample number	14266,2	14269,2	14271,6	14271,7	14271,8	14274,2	14275,2	14278,2	14279,2	14289,3
Type	Matrix CMB(FM)	Matrix VMB	Clast I CMB(FM)	Clast III CMB(CL)	Matrix VMB	Matrix CMB(CL)	Matrix VMB	Matrix VMB	Matrix CMB(FM)	Matrix VMB
Method	EMP	XRF	XRF	EMP	XRF	XRF	XRF	EMP	EMP	EMP
SiO_2	47.89	47.78	48.39	47.38	48.30	46.40	46.80	46.93	48.21	47.39
TiO_2	1.72	1.72	1.59	1.68	2.06	1.75	2.00	2.17	1.66	1.96
Al_2O_3	15.74	17.59	15.95	16.29	16.48	16.09	17.24	17.24	16.18	17.85
Cr_2O_3	0.24			0.15				0.22	0.22	.21
FeO	10.77	10.36	10.41	9.33	10.51	10.95	10.16	9.92	10.21	9.51
MnO		0.15	0.14		0.15	0.16	0.15			
MgO	10.37	9.23	12.62	11.66	9.54	11.47	10.38	9.77	9.83	9.23
CaO	10.30	11.19	9.85	10.00	10.07	10.11	10.55	10.75	10.23	10.91
Na_2O	0.73	0.68	0.83	0.90	0.89	0.76	0.83	0.75	0.76	0.74
K_2O	0.59	0.49	0.27	0.66	0.80	0.34	0.52	0.61	0.78	0.60
P_2O_5		0.46	0.65		0.65	0.62	0.54			
S		0.08	0.06		0.09	0.08	0.07			
Total	98.35	99.73	100.76	98.06	99.54	98.73	99.24	98.36	98.08	98.41

probably are simply derived by events unrelated to any of the major local craters or the Imbrium event.

Petrogenesis of the Fra Mauro breccias

It was argued by Simonds *et al.* (1976) that the Fra Mauro breccias were formed simply as mixtures of superheated impact melt and cold fragmental debris in a manner similar to impact melts. The only distinction between the clast-laden impact melts and the Fra Mauro breccias is that the Fra Mauro breccias had a larger fraction of fragmental debris which equilibrated to relatively low, probably subsolidus temperatures. Scanning electron micrography of ion etched thin sections, using procedures described in Phinney *et al.* (1976) illustrate a continuous petrographic sequence from the impact-melt rocks with low clast contents and euhedral matrix minerals (e.g., the Apollo 17 Station 6 boulder of Simonds, 1975; Simonds *et al.*, 1976; Warner *et al.*, 1976) to the Fra Mauro breccias with high clast contents and anhedral matrix minerals.

The boulder samples display a continuous textural gradation from relatively coarse grained and diabasic (Fig. 5A) through finer grained and poikilitic (Fig. 5B) to very fine grained and subophitic (Fig. 5C). The boulder sample (76255) with the greatest volume of clastic fragments and the least volume of crystalline matrix reveals a possible genetic relationship between the clastic fragments and the crystalline material. The margins of many clastic fragments show vesiculation and the rounded to irregular shapes of plastic flow (Fig. 5D). Some of this plastic material has separated into discrete fragments that are a few

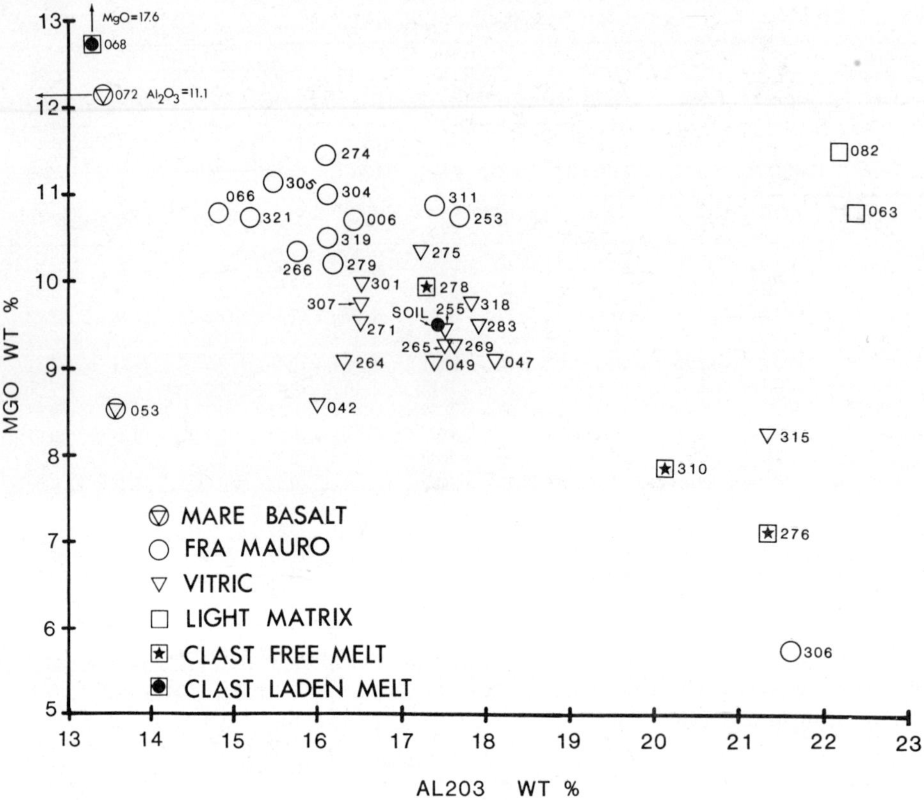

Fig. 4. MgO–Al_2O_3 plot for analyses of bulk samples of matrix plus small clasts. Note that each matrix type defines a resolvable compositional field with the average Apollo 14 surface soil falling in the center of the range of the vitric matrix breccias.

micrometers across and have the composition of the parent minerals (Fig. 5E). Within these zones containing discrete fragments are 20–100 μm patches of fine-grained crystalline matrix material forming irregular connections between the original clasts (Fig. 5F). Apparently some of the discrete fragments coalesce into small patches and crystallize. The textures in these patches are similar to the more massive very fine-grained types shown in Fig. 5C.

Many Apollo 14 Fra Mauro matrices (Fig. 5G) are essentially identical to the finer-grained textures of the boulder samples (Fig. 5C). These contain tabular plagioclase crystals that interlock with pyroxene crystals. Euhedral to subhedral ilmenite grains are also present. The Apollo 14 Fra Mauro matrices also display textures that show a mosaic of equant, anhedral crystals of plagioclase, pyroxene, and ilmenite (Fig. 5H). A gradation of textures intermediate to these is also present (Fig. 5I). In some Fra Mauro matrix breccias, such as 14319, this entire range of textures is found to be gradational over the distance of a few millimeters. The tabular feldspar illustrated in Fig. 5K occurs within a few millimeters of the region illustrated in Fig. 5L.

Comparisons between clast-laden melt rocks from the Apollo 17 Station 6 boulder and the Fra Mauro breccias indicate a series of correlated variations in matrix texture, the relative abundances of different clast types, relative abundances of different composition plagioclase clasts, and the total volume fraction of matrix. These correlations are:

(1) The increasing presence of euhedral plagioclase in the matrix with increasing fraction of matrix as illustrated in Fig. 5.
(2) The increasing proportion of refractory mineral clasts relative to less refractory lithic and glass-bearing clasts with increasing fraction of matrix as illustrated in Fig. 6.
(3) The increasing proportion of the more refractory, high An content, plagioclase clasts with increasing matrix content as illustrated in Fig. 7.

The fact that these correlations exist and that they can be seen in suites of samples from different landing sites suggests that first, the processes of formation of the two groups, Fra Mauro breccias and clast-laden melts, are related and second, samples which must have been lithified by different impact events indicate a systematic behavior which carries over from impact to impact. The observed correlations are compatible with the previously published model for impactites as mixtures of superheated melt from near the point of impact and less heated fragmental debris from nearer the periphery of the excavation (Simonds, 1975 and the papers referenced therein). Thermal calculations by Simonds *et al.* (1976) and Onorato *et al.* (1977) based on this model, with realistic values of grain size and thermal physical properties suggest that (1) thermal equilibration on a scale of less than a centimeter between clasts and matrix takes on the order of a hundred seconds to smooth out thermal variations greater than a few degrees and (2) the fraction of clast digestion that results in biasing of observed clast population to refractory materials is solely a function of the temperature to which the clast-melt mixture equilibrates and the relative melting points of clasts and matrix.

The correlation between matrix texture and clast content is probably due to a combination of two effects. First, as is generally true for the crystallization of silicate melts (Kingery *et al.*, 1976), and as illustrated in the experiments on mare basalt crystallization by Lofgren *et al.* (1974), crystallization at higher temperatures or slower cooling rates generally produces more euhedral minerals. Second, the fact that melts with initially higher clast contents may have contained more small clasts to serve as nuclei (Ryder and Bower, 1976; Lofgren, 1977). Since crystal morphology is largely a function of growth rather than nucleation processes, plagioclase morphology may provide a crude geothermometer to determine the temperature to which the clast-melt mixture equilibrated. Grain size on the other hand is probably more a function of the abundance of the relict plagioclase grains that serve as nuclei, although the occurrence of relict cores such as illustrated in Fig. 5J does not obviously

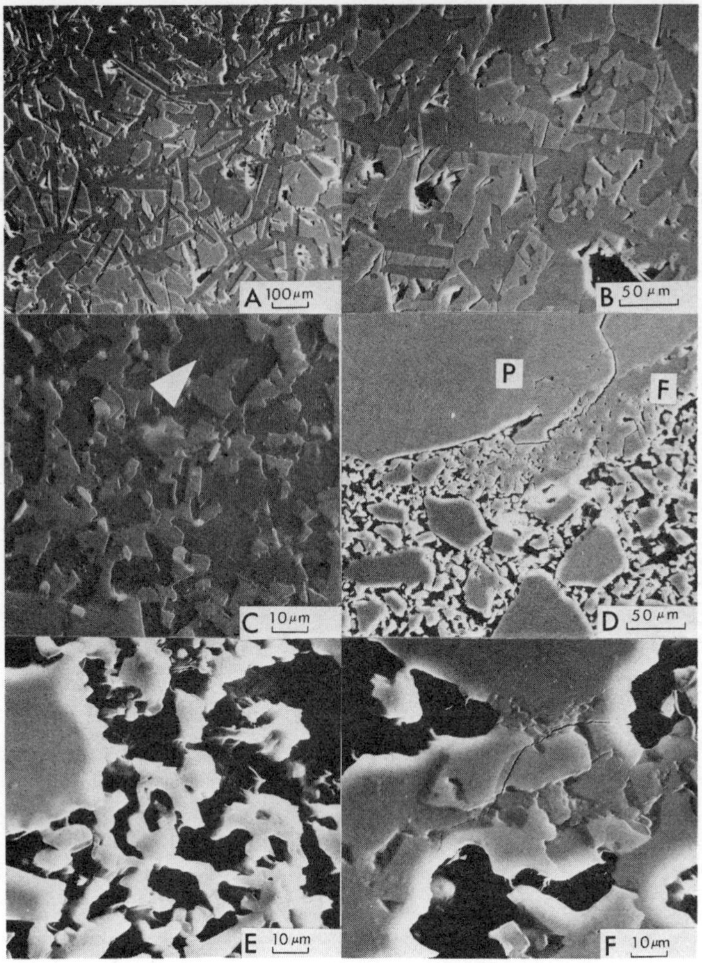

Fig. 5. SEM photomicrographs showing the progression of matrix textures from clast-laden impact melts (76215, 76295) through melt invading friable clasts (76255) to a spectrum of Fra Mauro breccias (14305, 14311, 14303, and 14321). The 76××× group are from the 10 m thick melt sheet sampled at the Apollo 17 Station 6 boulder. Note that there is a general correlation of the degree of plagioclase grain shape development with clast content, except for 14321. (A) 76215,57. Diabasic texture (98% matrix). Note tabular euhedral plagioclase (dark gray) and low-Ca pyroxene (medium gray). (B) 76215,56. Finer grained poikilitic texture (91% matrix). The euhedral to subhedral plagioclase (darker gray) occurs within oikocrysts of low-Ca pyroxene (medium gray). (C) 76295,79. Very fine-grained, subophitic texture (83% matrix). Plagioclase (dark gray) is euhedral to subhedral, low-Ca pyroxene (medium gray) occurs in subophitic patches and ilmenite (white) is euhedral and tabular. Arrow points to small relict core in plagioclase, possibly a core serving as a nucleus for the grain. (D) 76255,74. The margins of clastic fragments (only about 60% matrix) show vesiculation, rounding, and irregular shapes. This clast contains a contact between plagioclase (F) and pyroxene (P) both of which show signs of melting and plastic flow. (E) 76255,76. Fragments displaying signs of melting and plastic flow. These fragments have the same composition as the mineral clasts of Fig. 5D which is from less than 1 cm away. (F) 76255,76. One of many 20–100 μm patches of crystalline material that forms irregular connections between clasts in the same part of sample as Figs. 5E and 5D. Darker gray material is plagioclase; medium gray is low-Ca pyroxene; one needle of ilmenite (white) occurs at lower right.

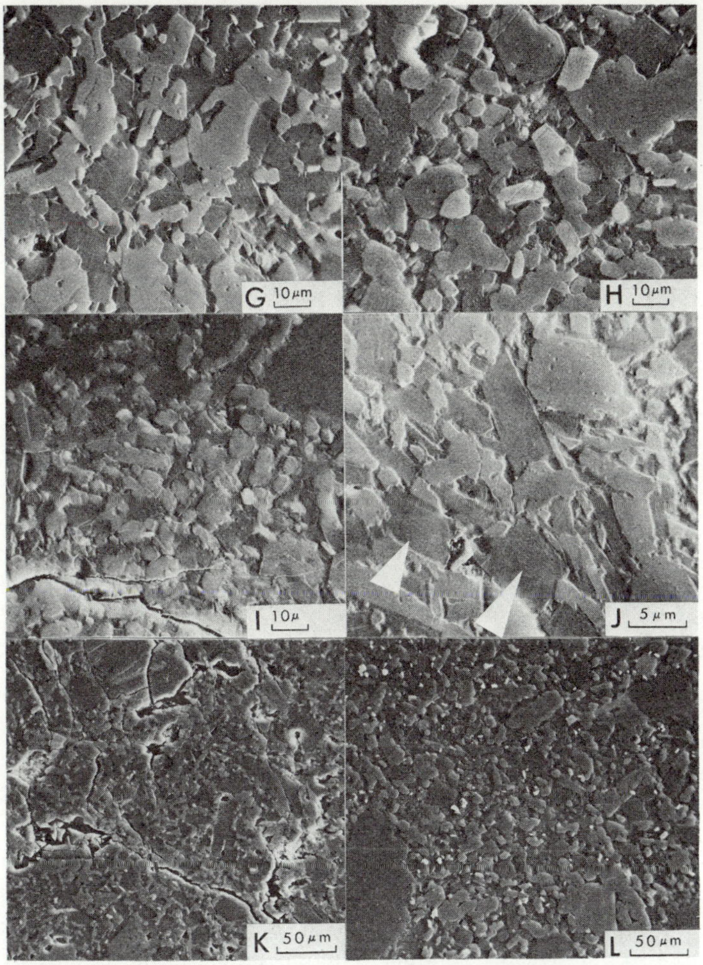

Fig. 5. (G) 14305,100. Very fine-grained, subophitic crystalline matrix similar to (C) (80% matrix). Plagioclase (dark gray) is euhedral to mostly subhedral, low-Ca pyroxene (medium gray) occurs as subhedral patches, and ilmenite (white) is euhedral and tabular. (H) 14311,13. Mosaic of equant, anhedral crystals of plagioclase (dark gray), pyroxene (medium gray), and ilmenite (light gray) (75% matrix). One ilmenite crystal at left center is euhedral. (I) 14303,50. Texture that appears to be intermediate between (G) and (H) ($74\frac{1}{2}$% matrix). The same minerals occur in each. Note plagioclase mineral clasts at upper right and left center. (J) 14321,29. Crystalline matrix texture in which many of the plagioclase (dark gray) and pyroxene (medium gray) grains consist of irregularly shaped cores (arrows) of pre-existing minerals with overgrowths of the same mineral. The local area on which these observations were made has about 90% matrix, in contrast to the other sections for which detailed petrography was done (Table 1). (K) 14319,4. Tabular feldspar encased in granular pyroxene and ilmenite. (L) 14319,4. Mosaic of equant anhedral feldspar in a more clast-rich zone a few millimeters from that illustrated in (K).

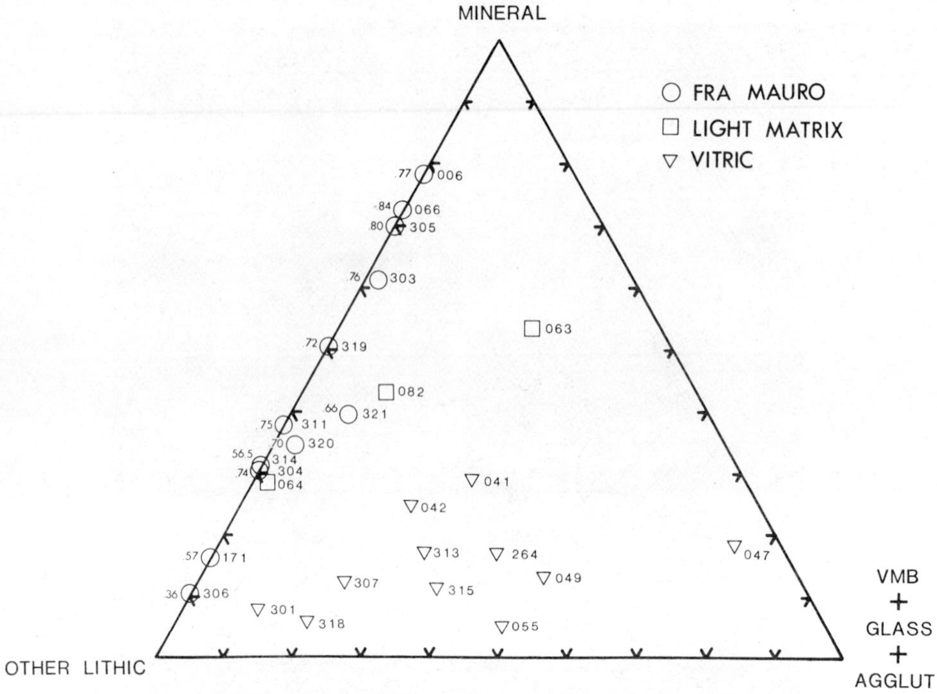

Fig. 6. Ratio of clast abundances in Apollo 14 breccias. The component defining the lower right corner is all glass bearing clasts, vitric matrix breccias, glass, and agglutinates. Only the last three digits of sample numbers are given. Numbers to left of triangle are volume fraction of matrix in the Fra Mauro breccias. That fraction correlates directly with mineral clast abundance.

correlate to total clast content. To produce plagioclase with euhedral to subhedral forms (Figs. 5A, B, C, G, J, and K) crystallization must take place at temperatures well above the solidus. Whereas anhedral morphologies (Figs. 5H, I, and L) are comparable to some textures produced by subsolidus crystallization (Kingery *et al.*, 1976).

The observation of different textures within small regions that should thermally equilibrate within a few minutes suggests that the crystal morphology is determined at least in part by extremely local thermal conditions. Or else, as implied by the studies of Lofgren (1977), grain shape is somehow determined by nucleation. It is interesting to note that the spherulitic to dendritic crystal morphologies, characteristic of crystallization at near-solidus temperatures, is not observed. Their lack may be due to incomplete sampling or, possibly, to abundant nuclei suggested to be present in impact melts (Lofgren, 1977; Ryder and Bower, 1976). The nuclei may, in some way, promote crystallization of the more equant morphologies, rather than the spherulites and dendrites which crystallize in clast-free materials at similar temperatures.

The observed correlation of variations in matrix abundance and extent of

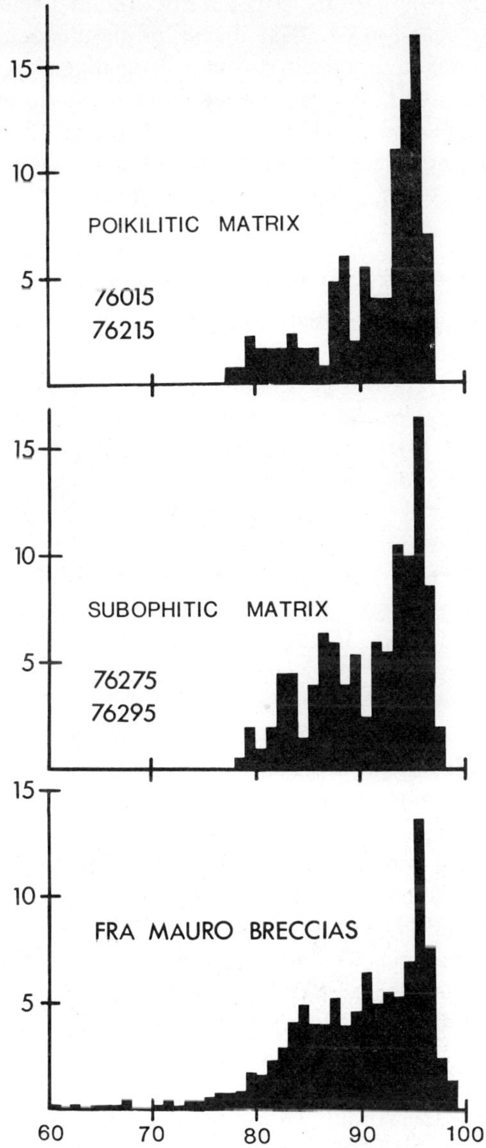

Fig. 7. Comparison of compositions of cores of feldspar clasts in poikilitic matrix samples from Apollo 17 Station 6 boulder (91% matrix), subophitic matrix samples from that boulder (about 83% matrix) and the Apollo 14 Fra Mauro breccias (average 69% matrix). Note that with increasing matrix content the fraction of clasts with calcium content less than An_{85} decreases. Such a comparison of clasts appears valid since the bulk composition of all rocks is KREEPy.

clast digestion compares favorably with that predicted by the calculations referred to above. The calculations, particularly the most refined set of Onorato *et al.* (1977), suggest that 5–15% clast digestion should occur in samples that equilibrate to subsolidus temperatures and ~70% digestion in samples which equilibrate to temperatures halfway between solidus and liquidus. The latter will digest most lithic clasts and plagioclase less refractory than the average melt plagioclase. Virtually no clast digestion should occur in melt-clast mixtures equilibrating to temperatures >200° below the solidus.

CRYSTALLINE BRECCIAS AND HISTORY OF APOLLO 14 LANDING SITE

Several impact events are probably responsible for the formation of Apollo 14 breccias. These events may be similar to the large pre-Imbrium craters discussed by Head and Hawke (1975) and Hawke (1977). Large gaps in composition separate (1) most of the crystalline breccias, (2) clast-free impact melts 14276 and 14310, and (3) crystalline matrix breccia 14306 with its unique composition and clast population. Because studies of terrestrial, impact melts of both the clast-laden, and clast-free types suggest that each impact event produces a very thorough mixing of the materials in such rocks (Simonds *et al.*, 1976), a minimum of three separate events is implied for production of the Apollo 14 crystalline breccias. The variation in abundance of different compositions of plagioclase clasts in these crystalline rocks suggests that possibly more than one event may be necessary to form the Fra Mauro breccias whose compositions display a single cluster on Fig. 4. The very complete mixing is not characteristic of less shocked and melted breccias such as the light matrix breccias. These breccias could represent a less shocked part of the debris from the same crater that produced any of the other three compositional groups, or they could be derived from a totally separate impact.

The large amount of melt found in the crystalline breccias is incompatible with ejection from a basin as far away as Imbrium, a point previously made by von Engelhardt *et al.* (1972) and Chao *et al.* (1972). We prefer to interpret the Apollo 14 Fra Mauro breccias as rocks formed on the continuous ejecta of large craters, in positions similar to suevites in the Ries (Pohl *et al.*, 1977) or the tagamites (clast-laden melts) of Popagai (Masaitis *et al.*, 1975). A photogeologic survey by Hawke (1977) suggests that the Apollo 14 site lies on the rims of several local craters, a position likely to have produced the fragment-rich breccias that comprise the Apollo 14 sample collection. This contrasts with the Apollo 16 landing site which lies within a series of overlapping craters (Head, 1975), a position more likely to have produced thick masses of the impact melts that comprise much of the Apollo 16 sample collection (Warner *et al.*, 1973).

Still another impact event may be responsible for the abundant post-lithification fracturing observed by petrographic microscope and, at a sub-micrometer scale, by SEM. As indicated above, presumably a number of younger events are responsible for the production of the vitric breccias.

Conclusions

(1) The vitric matrix breccias are made in part from agglutinate-bearing soil.
(2) Because of their finer grain size, lower clast content, lower abundance of mineral clasts, and different major element composition than the Fra Mauro breccias, the vitric breccias are not the protoliths of the Fra Mauro breccias.
(3) The Fra Mauro breccias, crystalline breccias typically with over 15% clasts, form from mixtures of impact melt and cold clastic debris in a manner similar to the crystalline rocks with fewer clasts.
(4) Those Fra Mauro breccias with relatively high matrix contents (75–80%), euhedral matrix plagioclase, and evidence of extensive clast digestion are inferred to have crystallized and thermally equilibrated at temperatures well above the solidus. Other Fra Mauro breccias have a wide range in plagioclase morphologies, intermediate matrix contents (65–75%), and less extensive clast digestion. Those with the lowest fraction of matrix have the most anhedral plagioclase and least evidence for clast digestion and are inferred to have thermally equilibrated to temperatures well below the solidus.

Acknowledgments—The manuscript was improved by reviews by Gary Lofgren, Fred Hörz, Charles Meyer Jr., Susan W. Kieffer, and Richard A. F. Grieve. The participation of C.H.S. was supported by the Lunar Science Institute which is operated by the Universities Space Research Association under Contract No. NSR-09-051-001 with the National Aeronautics and Space Administration. This paper constitutes the Lunar Science Institute Contribution No. 286.

References

Bickel C. E., Warner J. L., and Phinney W. C. (1976) Petrology of 79215: Brecciation of a lunar cumulate. *Proc. Lunar Sci. Conf. 7th*, p. 1793–1819.

Bogard D. D. and Nyquist L. E. (1972) Noble gas studies on regolith materials from Apollo 14 and 15. *Proc. Lunar Sci. Conf. 3rd*, p. 1797–1819.

Brown R. W. (1977) A simple fusion technique for whole rock analysis with the electron microprobe. *Geochim. Cosmochim. Acta* **41**, 435–438.

Chao E. C. T., Minkin J. A., and Best J. B. (1972) Apollo 14 breccias: General characteristics and classification. *Proc. Lunar Sci. Conf. 3rd*, p. 645–660.

Christie J. M., Griggs D. T., Heuer A. H., Nord G. L., Jr., Radcliffe S. V., Lally J. S., and Fisher R. M. (1973) Electron petrography of Apollo 14 and 15 breccias and shock produced analogues. *Proc. Lunar Sci. Conf. 4th*, p. 365–382.

Compston W., Vernon M. S., Berry H., Rudowski R., Gray C. M., Ware N., Chappell B. W., and Kaye M. (1972) Apollo 14 mineral ages and the thermal history of the Fra Mauro Formation. *Proc. Lunar Sci. Conf. 3rd*, p. 1487–1501.

Dran J. C., Duraud J. P., Maurette M., Durrieu L., Jouret C., and Legressus C. (1972) Track metamorphism in extraterrestrial breccias. *Proc. Lunar Sci. Conf. 3rd*, p. 2883–2903.

Engelhardt W. v., Arndt J., Stöffler D., and Schneider H. (1972) Apollo 14 regolith and fragmental rocks, their compositions and origin by impacts. *Proc. Lunar Sci. Conf. 3rd*, p. 753–770.

Gose W., Pearce G. W., Strangway D. W., and Larson E. E. (1972) Magnetic properties of Apollo 14 breccias and their correlation with metamorphism. *Proc. Lunar Sci. Conf. 3rd*, p. 2387–2395.

Hawke B. R. (1977) Pre-Imbrian geology of the Apollo 14 region: Evidence for local origin of KREEP (abstract). *EOS Trans. Amer. Geophys. Union* **58**, 425.

Head J. W. (1975) Stratigraphy of the Descartes Region (Apollo 16): Implications for the origin of samples. *The Moon* **11**, 77–99.

Head J. W. and Hawke B. R. (1975) Geology of the Apollo 14 region (Fra Mauro): Stratigraphic history and sample provenance. *Proc. Lunar Sci. Conf. 6th*, p. 2483–2501.

Heiken G. H. (1974) A Catalogue of Lunar Soils. NASA Johnson Space Center, Curators Office. 221 pp.

Kieffer S. W. (1975) From regolith to rock by shock. *The Moon* **13**, 301–320.

Kingery W. D., Bowen H. K., and Uhlmann D. R. (1976) *Introduction to Ceramics*. Wiley, New York. 1032 pp.

Lally J. S., Fisher R. M., Christie J. M., Griggs D. T., Heuer A. H., Nord G. L., Jr., and Radcliffe S. V. (1972) Electron petrography of Apollo 14 and 15 rocks. *Proc. Lunar Sci. Conf. 3rd*, p. 401–422.

Lofgren G. E. (1977) Dynamic crystallization experiments bearing on the origins of textures in impact generated liquids. *Proc. Lunar Sci. Conf. 8th*. This volume.

Lofgren G., Donaldson C. H., Williams R. J., Mullins O., Jr., and Usselman T. M. (1974) Experimentally produced textures and mineral chemistry of Apollo 15 quartz normative basalts. *Proc. Lunar Sci. Conf. 5th*, p. 549–567.

Masaitis V. L., Mikhaylov M. V., and Selivanouskaya T. V. (1975) *Papagai Meteorite Crater*. Pub. of All Union Sci. Res. Geol. Inst. Nauk, Moscow. 124 pp. (in Russian).

McKay D. S. and Morrison D. A. (1971) Lunar breccias. *J. Geophys. Res.* **76**, 5658–5669.

Norrish K. and Hutton J. T. (1969) An accurate X-ray spectrographic method for the analysis of a wide range of geological samples. *Geochim. Cosmochim. Acta* **33**, 431–454.

Oberbeck V. R., Morrison R. H., and Hörz F. (1975a) Transport and emplacement of crater and basin deposits. *The Moon* **13**, 9–26.

Onorato P. I. K., Uhlmann D. R., and Simonds C. H. (1977) Numerical calculations of thermal history of Manicouagan impact melt sheet. Submitted to *J. Geophys. Res.*

Phinney W. C., McKay D. S., Simonds C. H., and Warner J. L. (1976) Lithification of vitric- and clastic-matrix breccias: SEM petrography. *Proc. Lunar Sci. Conf. 7th*, p. 2469–2492.

Pohl J., Stöffler D., Gall H., and Evanstson K. (1977) The Ries impact crater. In *Impact and Explosion Cratering* (D. J. Roddy, R. O. Pepin, and R. B. Merrill, eds.) Pergamon Press, New York. In press.

Quaide W. and Wrigley R. (1972) Apollo 14 regolith and fragmental rocks, their compositions and origin by impacts. *Proc. Lunar Sci. Conf. 3rd*, p. 771–784.

Ryder G. and Bower J. F. (1976) Poikilitic KREEP impact melts in the Apollo 14 white rocks. *Proc. Lunar Sci. Conf. 7th*, p. 1925–1948.

Simonds C. H. (1973) Sintering and hot pressing of Fra Mauro composition glass and the lithification of lunar breccias. *Amer. J. Sci.* **273**, 428–439.

Simonds C. H. (1975) Thermal regimes in impact melts and the petrology of the Apollo 17 Station 6 boulder. *Proc. Lunar Sci. Conf. 6th*, p. 641–672.

Simonds C. H., Warner J. L., and Phinney W. C. (1973) Petrology of Apollo 16 poikilitic rocks. *Proc. Lunar Sci. Conf. 4th*, p. 613–632.

Simonds C. H., Warner J. L., Phinney W. C., and McGee P. E. (1976) Thermal model for impact breccia lithification: Manicouagan and the moon. *Proc. Lunar Sci. Conf. 7th*, p. 2509–2528.

Stöffler D., Knöll H. D., Reinold W.-U., and Schulien S. (1976) Grain size statistics, composition, and provenance of fragmental particles in some Apollo 14 breccias. *Proc. Lunar Sci. Conf. 7th*, p. 1965–1985.

Uhlmann D. R., Klein L., Onorato P. I. K., and Hopper R. W. (1975) The formation of lunar breccias; sintering and crystallization kinetics. *Proc. Lunar Sci. Conf. 6th*, p. 693–705.

Warner J. L. (1972) Metamorphism of Apollo breccias. *Proc. Lunar Sci. Conf. 3rd*, p. 623–643.

Warner J. L., Simonds C. H., and Phinney W. C. (1973) Apollo 16 rocks classification and petrogenetic model. *Proc. Lunar Sci. Conf. 4th*, p. 481–504.

Warner J. L., Simonds C. H., and Phinney W. C. (1976) Apollo 17 Station 6 boulder sample 76255; absolute petrography of breccia matrix and igneous clasts. *Proc. Lunar Sci. Conf. 7th*, p. 2233–2250.

Warner J. L., Phinney W. C., Bickel C. E., and Simonds C. H. (1977) Feldspathic, granulitic impactites and pre-final bombardment lunar evolution. *Proc. Lunar Sci. Conf. 8th.* This volume.

Waters A. C., Fisher R. V., Garrison R. E., and Wax D. (1971) Matrix characteristics and origin of lunar breccia samples 12034 and 12073. *Proc. Lunar Sci. Conf. 2nd*, p. 893–907.

Williams R. J. (1972) The lithification and metamorphism of lunar breccias. *Earth Planet. Sci. Lett.* **16**, 250–256.

Wilshire H. G. and Jackson E. D. (1972) Petrology and stratigraphy of Fra Mauro formation at Apollo 14 site. *U.S. Geol. Survey Prof. Paper 785.* 26 pp.

Proc. Lunar Sci. Conf. 8th (1977), p. 1895–1923.
Printed in the United States of America

Petrology of Apollo 15 black-and-white rocks 15445 and 15455—Fragments of the Imbrium impact melt sheet?

GRAHAM RYDER and J. F. BOWER

Center for Astrophysics (Smithsonian Astrophysical Observatory and Harvard College Observatory), 60 Garden Street, Cambridge, Massachusetts 02138

Abstract—Two macroscopically similar black-and-white rocks, 15445 and 15455, were collected from the rim of Spur Crater on the Apennine Front. 15445 contains a lithic clast assemblage of plutonic/metamorphic spinel troctolite, troctolite, norite, and anorthosite. Its fine-grained, vesicular, black, coherent matrix consists of a melt-bonded aggregate of small mineral clasts, which are mainly olivine, plagioclase, and pink spinel. Rare augite occurs as clasts, but low-Ca pyroxene is absent. The melt has crystallized to olivine, plagioclase, and opaque minerals, with some pyroxene. The matrix which has a magnesian low-K Fra Mauro composition, is interpreted as an impact melt.

15455 is very similar in chemistry and clast population to 15445, except that the matrix we have studied (a vein cutting a large norite) is finer grained than that of 15445. Spinel troctolite clasts have not been reported, but pink spinel is conspicuous in the matrix. The norite, the largest white clast, is mineralogically similar to those in 15445, except that it has a greater abundance and variety of minor phases.

15445 and 15455 are not only similar to each other in all respects, including meteoritic trace elements, but they are distinct from any other large (> 25 g) samples from the Apollo 15 site. Their location and characteristics suggest that they are samples of an impact melt sheet which forms a bedrock unit of the Apennine Front. We believe that such a melt sheet was not formed in a local small-scale event, and that it was produced during the Imbrium impact event. While broadly similiar in petrology and chemistry to the Apollo 17 boulders (interpreted by others as fragments of the Serenitatis melt sheet) some significant differences, which have implications for lunar crustal structure, exist.

INTRODUCTION

THE APOLLO 15 MISSION to Hadley Delta collected a number of samples from the lower slope of the Apennine Front, which is an outer ring of the Imbrium Basin. Among these samples are the distinctive and similar-appearing black-and-white rocks 15445 and 15455, which were collected from the rim of Spur Crater (Station 7). 15445 was lying near a 1.5 × 0.6 m boulder that contained white clasts, and it was collected because the astronauts believed it was likely to have once been part of the boulder (Bailey and Ulrich, 1975). 15445 and the boulder are probably representative of the subregolithic material characteristic of the Apennine Front at the elevation sampled, and their position on the rim crest of Spur Crater suggests that they were derived from the maximum sample depth (~20 m) from within the crater (Swann *et al.*, 1972). 15455, lying 10 m from the same boulder on the Spur Crater rim, was the largest of numerous rock fragments in an area of ~100 m^2 that are characterized by the presence of high-albedo clasts, associated with high-albedo soils (Swann *et al.*, 1972). Because the locations of 15445 and 15455 suggest that they might be samples of the

Apennine Front bedrock, they take on a special significance because of the paucity of other possible samples of this bedrock, which, given the proximity of the Imbrium Basin, is likely to consist of Imbrium ejecta. (Virtually all of the other large (> 25 g) samples collected from the Apennine Front are brown glass matrix soil breccias, containing green glass, mare basalt, and KREEP basalt fragments. They clearly postdate the Imbrium impact event and the resulting Front, and are not a part of the bedrock.)

15445 was chosen for study by the Imbrium Consortium (leader J. A. Wood) because, although earlier studies had been made of the rock (Ridley *et al.*, 1973; Anderson, 1973) it was felt that a more thorough study could possibly provide additional information about the pre-Imbrium crust. The macroscopic similarities of 15445 and 15455 clearly made a comparison of the two rocks desirable. However, published work on 15455 did not include a petrographic description, other than a preliminarly outline in the Apollo 15 Lunar Science Information Catalog (1971). Therefore we obtained the only two thin sections available to us from the SSPL to make a preliminary study of 15455.

This paper describes the petrology of 15445 and 15455, using both our own data and previously published observations. Our microprobe analyses were made using a modified version of the techniques and reduction procedures of Wood *et al.* (1971). The thin sections studied are listed in Ryder and Bower (1976, 1977) along with additional tabulated data. Most of our thin sections of 15445 were cut specifically for the Imbrium Consortium, and in general we have not studied thin sections made previously for other investigators.

We have done less work on 15455 than 15445, but enough to show that the two black-and-white rocks are very similar to each other, a conclusion that is supported by other published data, particularly trace-element chemistry. 15445 and 15455 are probably part of the same bedrock unit at the Apollo 15 site. Our observations suggest that this bedrock unit is an impact melt sheet, and our conclusion is that it is a part of the melt sheet produced by Imbrium impact event.

PETROLOGY OF SAMPLE 15445

Introduction

Sample 15445 (287.2 g) consists of white clasts set in a dark-colored, coherent, fine-grained crystalline matrix (Figs. 1 and 2). Some of the matrix surfaces show an abundance of tiny vesicles which, in a few areas, are elongated and roughly aligned (Marvin, 1976). Globules of metal or sulfide occur in a few of the vesicles. The white clasts make up about 25% of the rock (Marvin, 1976). The original cutting photographs of 15445 designated the clasts as A, B, . . . etc.; the Imbrium Consortium uses a different identification scheme, and for clarity and for easier comparison with other Imbrium Consortium publications, both designations will be given here e.g., Clast A/Lithology 45E. Designations such as

Fig. 1. (A) Sample 15445. (B) Sample 15455.

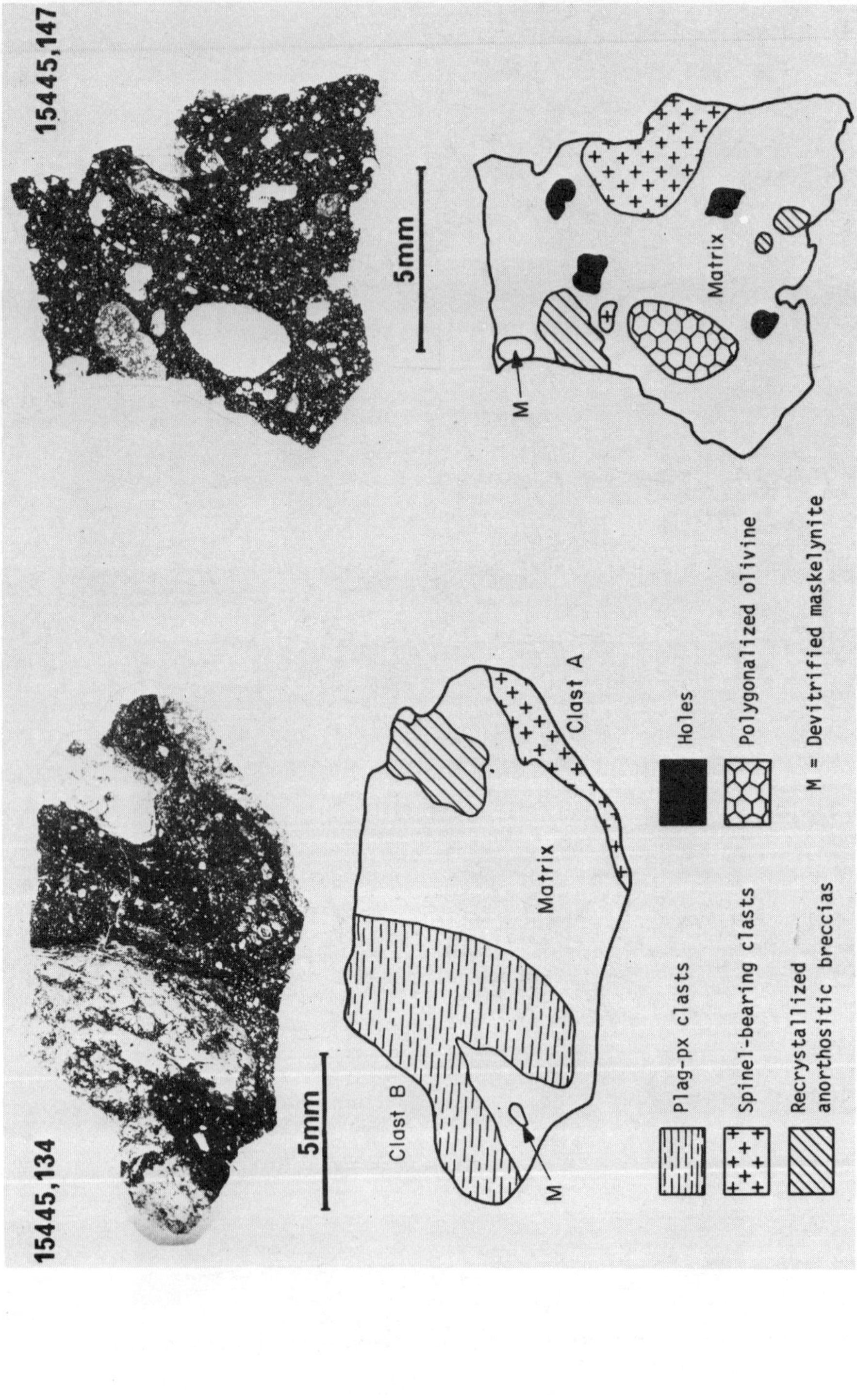

Fig. 2. Plane light photomicrographs and maps of thin sections 15445,134 (left) and 15445,147 (right).

Lithology 45E refer to specific clasts, not to clast types; the only exception is Lithology 45A which refers to all matrix samples. (Note that *Type A* of Ridley *et al.* (1973) does not refer to *Clast A*, but to norite clasts exemplified by *Clast B*.)

Lithic clasts

Several types of lithic clasts occur in 15445. Some distinction between them can be made macroscopically (Marvin, 1976), but some macroscopically similar lithologies are quite different microscopically. Spinel troctolite, norite, and anorthosite are the predominant clast types. A macroscopically identified gabbro lithology has not been observed in the thin sections which we have studied. Other lithic clast types are present, but, significantly, do not include polymict breccia fragments or other surface-derived fragments such as basaltic-textured clasts.

Spinel troctolite. Several of the clasts in 15445 consist of a cataclastic assemblage of olivine, plagioclase, and an aluminous spinel e.g., Clast A/Lithology 45E (Figs. 2 and 3). Some authors have referred to Clast A/Lithology 45E as a peridotite, but our macroscopic observations and thin section studies indicate that it has too much plagioclase (30–40%) to be termed a peridotite; spinel troctolite is preferable. A summary of data from this clast is given in Table 1 and spinel compositions are shown in Fig. 4. Clearly, some variability is apparent in the clast. It is free of meteoritic contamination (Gros *et al.*, 1976) and is a crushed, friable, once coarse-grained (>2 mm), plutonic or metamorphic rock, notable for its Mg-rich olivine and its Mg–Al spinel (pleonaste). Steele and Smith (1975) remarked on the extremely low-Ca content of the olivine, which is the lowest found in any analyzed lunar olivine and makes this olivine distinct from those in the 72415-7 dunite and the 76535 troctolite. The low-Ca content is interpreted as resulting from a slow-cooling and/or deep-seated origin.

We have not found pyroxene in our thin sections of Clast A/Lithology 45E (cf., Ridley *et al.*, 1973; Anderson, 1973), but we have found small amounts of pyroxene in a spinel troctolite clast in thin sections 15445,145-,147 (Fig. 3C). In this clast low-Ca pyroxene is more abundant than high-Ca pyroxene, and is more Fe-rich (En_{87} vs. En_{92}) and much less aluminous (Ryder and Bower, 1976) than the pyroxene Ridley *et al.* (1973) and Anderson (1973) report for Clast A/Lithology 45E. Analyzed olivine and spinel are compositionally similar to those in Clast A/Lithology 45E.

Ridley *et al.* (1973) reported REE data for Clast A/Lithology 45E, and Blanchard *et al.* (1977) reported REE data for Clast F/Lithology 45C, which is also a spinel troctolite. Both patterns show low abundances of light REEs and exceptionally low Eu, but that of Clast A/Lithology 45E is enriched in heavy REEs, while that of Clast F/Lithology 45C has low abundances of heavy REEs, producing a "normal" looking pattern. Ridley *et al.* (1973) believe that a history involving garnet accumulation accounts for the enrichment in heavy REEs found in Clast A/Lithology 45E, but such a history would have produced a pattern with

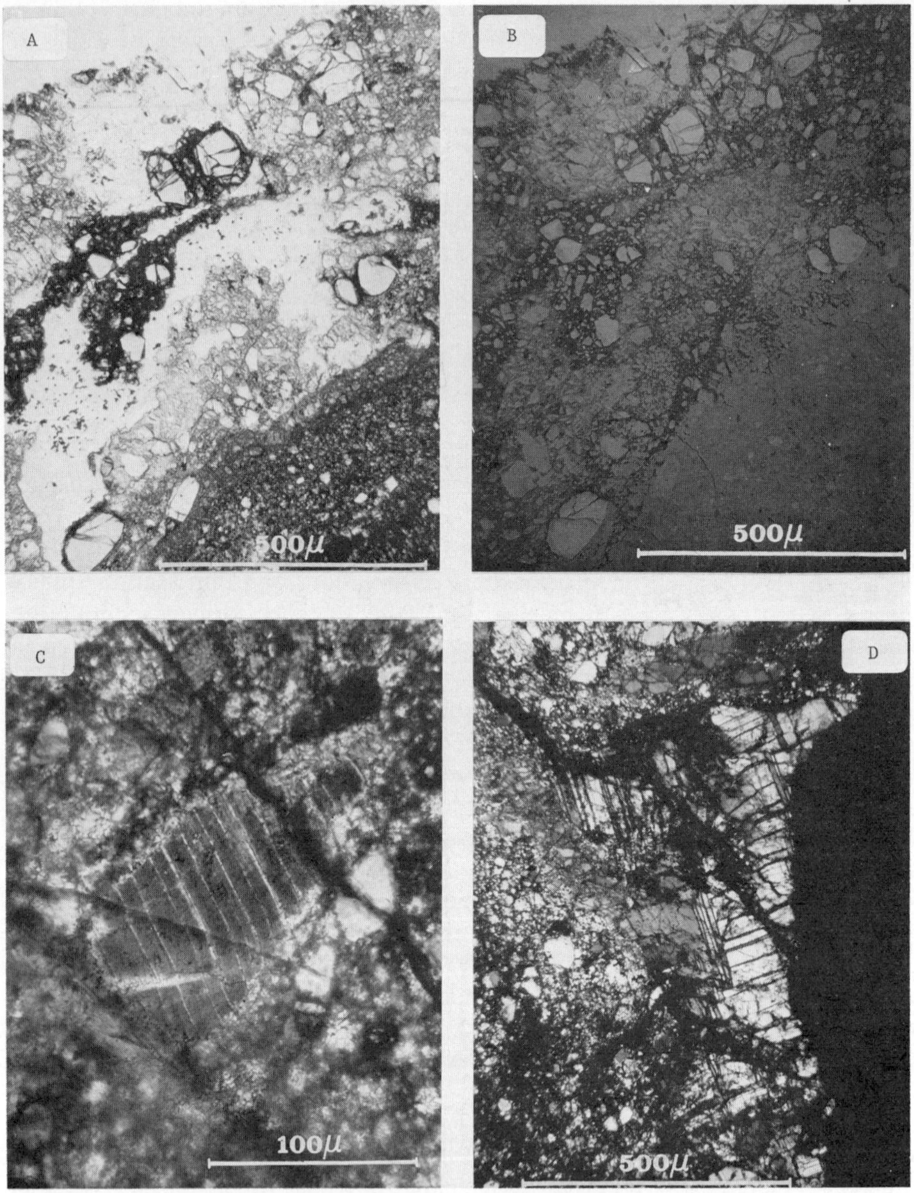

Fig. 3. Photomicrographs of spinel troctolite clasts. (A) Clast A in 15445,134. Pink spinel occurs in crushed dark schlieren, plagioclase in white zones, and olivine in gray zones. Lower right is 15445 matrix (plane light). (B) Same field as Fig. 3A. Clast A is crushed and friable compared to 15445 matrix (lower right) (reflected light). (C) Augite in crushed clast similar to Clast A in 15445,146. Note exsolved lamellae of orthopyroxene. Dark lines crossing field are fractures in the thin section (crossed polarizers). (D) Plagioclase in crushed clast similar to Clast A in 15445,146. Some plagioclase aggregates reach almost 2 mm across (crossed polarizers).

Table 1. Summary of Clast A/Lithology 45E spinel troctolite data for 15445.

Mineral	Proportion	Composition
Olivine	(a) present (b) >40% (c) ~50%	(a) Fo_{91-88} (b) Fo_{91} (c) Fo_{90-92} (d) Fo_{92}
Plagioclase	(a) minor (b) 5% (c) 30–40%	(a) An_{98} (b) An_{90-95}(?) (c) An_{89-95}
Pyroxene	(a) present (b) <40% (c) not found	(a) En_{92}, Al_2O_3 2–3% (b) En_{91}, Al_2O_3 5%
Pleonaste	(a) present (b) 15% (c) 10–20%	(a) 13% Cr_2O_3, 9.5% FeO (b) 14% Cr_2O_3, 9.4% FeO (c) 10.2–15.3% Cr_2O_3, 8.0–12.6% FeO
Others	(b) traces (c) traces	(b) rutile(?), armalcolite (c) unidentified opaques

(a) Ridley *et al.* (1973).
(b) Anderson (1973).
(c) This study.
(d) Steele and Smith (1975).

a steep slope showing a considerable enrichment of Yb over Tb, as is typical of garnet (e.g., Arth and Hanson, 1975). Instead, the analysis of Clast A/Lithology 45E shows an enrichment *en masse* in heavy REEs. If this pattern represents REE abundances accurately, Clast A/Lithology 45E is substantially different from Clast F/Lithology 45C, and the reason for its enrichment in heavy REEs remains elusive.

The spinel troctolite clasts in 15445 are modally and mineralogically similar to a small fragment (commonly known as PST) from breccia 67435, but PST has preserved its cumulate texture (Prinz *et al.*, 1973). Spinel troctolite fragments in the Apollo 17 soils are mineralogically similar but contain more plagioclase, about 60–70% (Bence *et al.*, 1974). Dymek *et al.* (1976) report plagioclase-rich spinel cataclasites with more iron-rich olivine (Fo_{72}) from an Apollo 17 boulder. The restricted compositional range of olivine within a sample is an outstanding feature of all these samples. Unfortunately, none of these spinel troctolite clasts has been analyzed for rare earth elements. (It should be noted that this group of coarse-grained spinel troctolites is distinct from the basaltic-textured "(spinel-) troctolites" found among highland rocks, in particular at the Apollo 16 site.)

The spinel troctolite clasts are probably from cumulates that underwent slow cooling and possibly some post-crystallization equilibration; alternatively, Bence and McGee (1976) have proposed that they are the residuum of partial melting

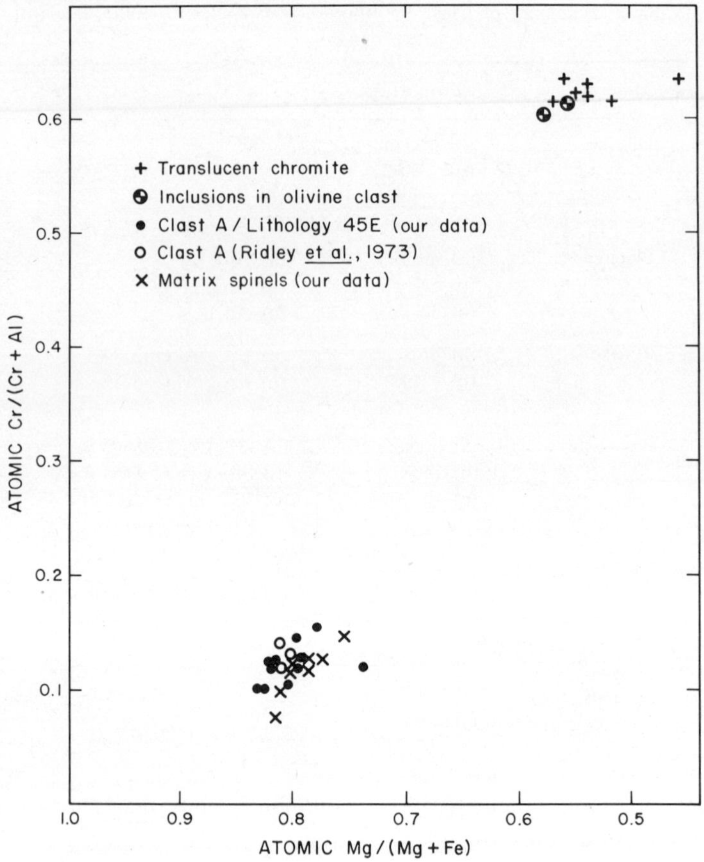

Fig. 4. Compositions of spinel in clasts and matrix of 15445.

deep in the crust. If they are direct cumulates, the parent magma must have been very aluminous (~20% Al_2O_3) and magnesian for spinel to have crystallized. For spinel to have been preserved, it must have been very efficiently separated from the liquid, or the liquid itself must have been spinel-normative (i.e., it lay outside of the Ol–Si–An pseudoternary system).

Norite. The largest clast in 15445, Clast B/Lithology 45D, is a norite, consisting of 60–65% plagioclase (An_{94-95}) and 35–40% low-Ca pyroxene (En_{77-81}, cf., Ridley *et al.*, 1973, En_{80-85}). The pyroxenes contain ~1.5% Al_2O_3, ~0.4% TiO_2, and ~0.7% Cr_2O_3. As shown in Figs. 2, 5A, and 5B, the clast is cataclasized; nonetheless it is free of meteoritic contamination (Gros *et al.*, 1976). We have observed silica and an opaque mineral (ilmenite?) as minor phases and Ridley *et al.* (1973) report accessory armalcolite and ilmenite in this clast. A minor vein system of Fe-metal (possibly secondary, resulting from the formation of the matrix) cuts the pyroxene locally; pyroxene adjacent to Fe-metal is not sig-

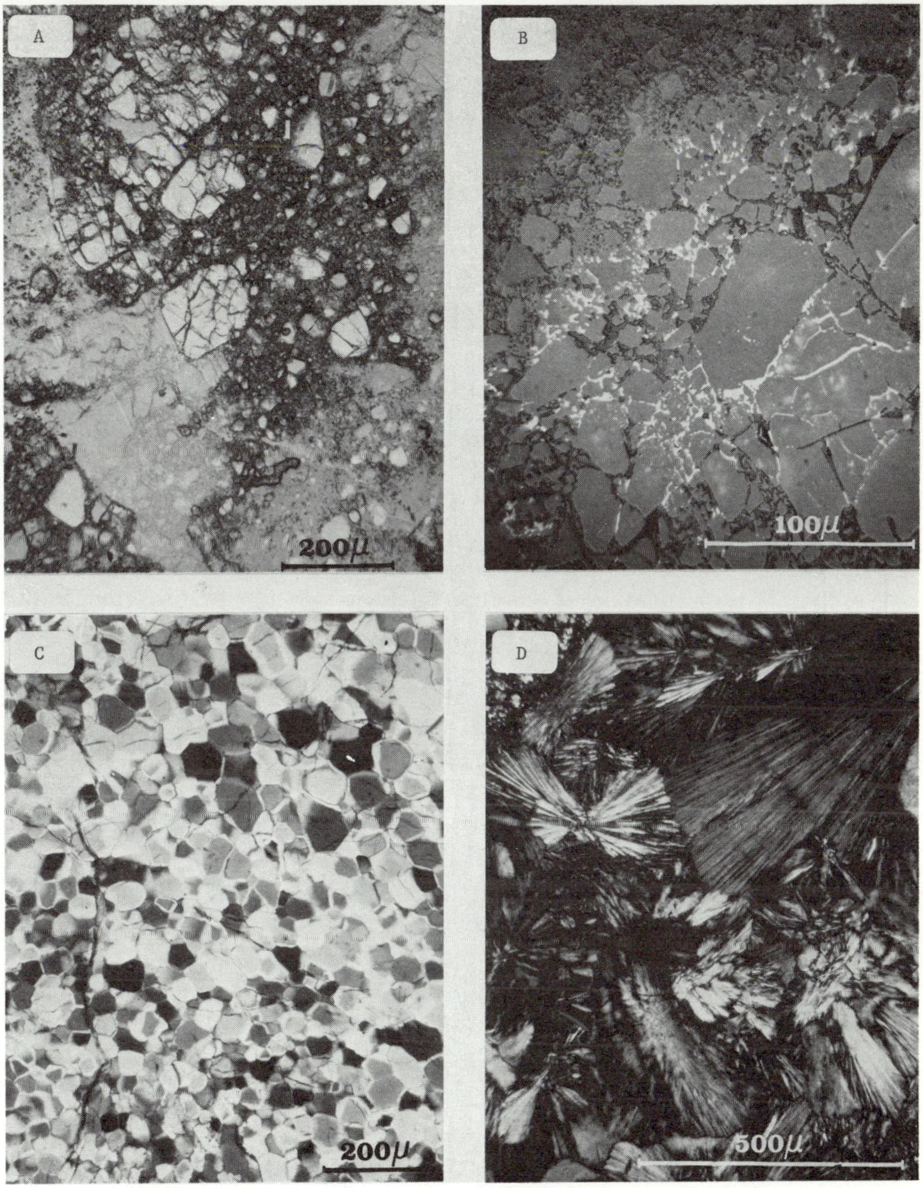

Fig. 5. Photomicrographs of clasts in 15445. (A) Clast B (norite) in 15445,134. Pyroxene (white grains in darker zone) and plagioclase (light color) (plane light). (B) Fe-metal (white) veining pyroxene (gray) in Clast B in 15445,135. Metal may not be primary (reflected light). (C) Polygonal granoblastic olivine (Fo_{87}) in 15445,147 (crossed polarizers). (D) Spherulitic plagioclase clast, probably devitrified maskelynite, in 15445,92 (crossed polarizers).

nificantly different in composition from other pyroxene (Fig. 6). Relict textures suggest that Clast B/Lithology 45D was originally coarse grained (>1 mm) and may have been poikilitic, compatible with an origin involving accumulation from a silicate melt. The mineralogy of Clast B/Lithology 45D is broadly similar to the 78235 norite (McCallum and Mathez, 1975) and to the large norite clast in 15455 (this paper) except that these latter norites have a greater variety of minor phases.

Whole-rock REE data for the Clast B/Lithology 45D was reported by Ridley *et al.* (1973) and by Blanchard *et al.* (1977). The REE abundance levels and patterns are very similar to each other and demonstrate that the norite has somewhat higher absolute REE abundances than the spinel troctolite. The positive Eu anomaly strongly suggests plagioclase accumulation. The pattern and abundance levels are similar to those in the 78235 norite (Winzer *et al.*, 1975) but differ significantly from those of the 72255 Civet Cat norite (Blanchard *et al.*, 1975) and the 72215 norite (Winzer *et al.*, 1974), which have higher total REE abundances and negative Eu anomalies. Possibly the gross similarity in mineralogy and trace-element chemistry between the 15445 norite and the 78235 norite allow the arguments of Winzer *et al.* (1975) concerning the origin of the 78235 norite to be applied to the 15445 norite, although mineral separates were not analyzed in the case of 15445. Winzer *et al.* suggest that the 72215 norite was in equilibrium with a liquid of approximately low-K Fra Mauro composition, which is compatible with constraints on the liquid derived from the norite mineralogy alone (see Ridley *et al.*, 1973).

Other fragments of similar norite occur in 15445: a part of a thin section ostensibly from Clast E/Lithology 45B (see below) which displays a cumulate texture (Figs. 7A, 7B), and an undesignated clast (Fig. 7C) have a mineralogy similar to that of Clast B/Lithology 45D. However, low-Ca pyroxene is not a part of the monomineralic clast population of the black matrix of 15445.

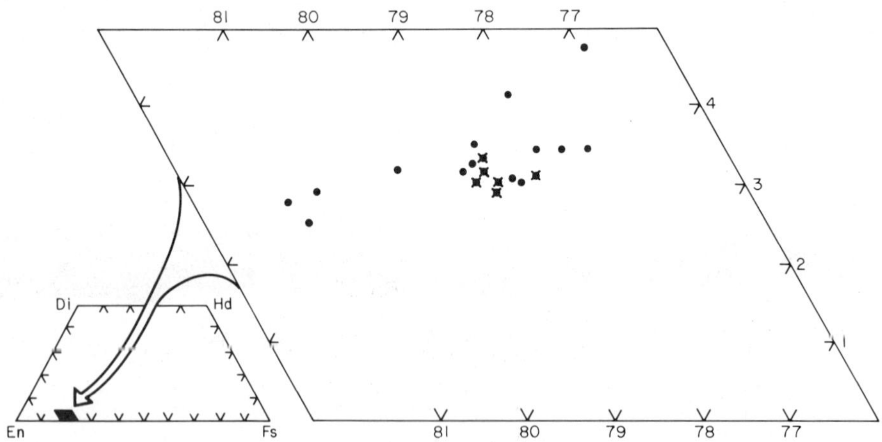

Fig. 6. Compositions of pyroxene in norite (Clast B/Lithology 45D); points with crosses are pyroxenes adjacent to Fe-metal veins.

Anorthosite. A number of small clasts in the thin sections consist entirely, or almost entirely, of brecciated and subsequently annealed plagioclase grains (Fig. 2). The only one analyzed had plagioclase compositions of An_{96-97}. The relatively large size of some of these clasts (several millimeters in diameter) suggests that they are from anorthosites *per se* or very coarse-grained (>5 mm) multiphase rocks. Several of the clasts studied macroscopically by Marvin (1976), including Clast E/Lithology 45B, appeared to be virtually all plagioclase. Thin sections of this clast provide conflicting evidence, however, and therefore will be discussed separately below (Subsection *Clast E/Lithology 45B*).

"*Gabbroic*" *clasts.* Marvin (1976) observed rare "gabbroic" clasts that contain yellow-brown pyroxene, in contrast to the green pyroxene present in the norite (Clast B/Lithology 45D). These lithic clasts are not present in the thin sections, so their character remains unknown. Rare monomineralic fragments of augite in the matrix suggest the presence of a gabbroic lithic type in 15445.

Clast E/Lithology 45B. Macroscopically Clast E/Lithology 45B is white, and free of pink spinel and green mafic minerals (Marvin, 1976). A small (2 × 3 mm) thin section of Clast E/Lithology 45B (15445,139) consists mainly of plagioclase (>95%), with minor amounts of a crushed unidentified mafic phase and very rare small grains of ilmenite and Fe-metal. The plagioclase has been intensely cataclasized and granulated, but was subsequently sintered, so little pore-space remains. On the evidence of 15445,139, Clast E/Lithology 45B is an anorthosite; this conclusion is supported by the REE pattern and the low abundance of FeO reported for Clast E/Lithology 45B by Blanchard *et al.* (1977).

However, according to the documentation, thin section 15445,221 also comes from this clast, and this section is entirely different (Fig. 7A). It is heterogeneous, and consists of zones of crystalline plagioclase + olivine, and crystalline plagioclase + low-Ca pyroxene. The whole is injected by a brownish glass (most of which has devitrified into a fine-grained mass) that contains inclusions of plagioclase, olivine, and pyroxene, as well as a vein-like mass of ilmenite and ulvöspinel. The composition of the glass (Table 2) is troctolitic, and distinct from the matrix of 15445. In the areas of crystalline clasts, both olivine and pyroxene show crystalline grain boundaries with plagioclase and appear to have a cumulate texture (e.g., Fig. 7B), but there is no evidence that pyroxene and olivine coexist in the same original lithology; rather, it appears as if two distinct lithic types, a norite and a troctolite, were mixed and injected by the glass veins. The olivine in the section is Fo_{82-84}, and the pyroxene is $En_{80-82}Wo_{3-4}$, with minor elements similar to the pyroxene in Clast B/Lithology 45D. Plagioclase is more abundant than either pyroxene or olivine, and compositions range from An_{93} to An_{97}; no distinction can be made between the compositions of plagioclase associated with pyroxene and plagioclase associated with olivine. On the evidence from section 15445,221, therefore, Clast E/Lithology 45B is a complex clast, not a simple anorthosite.

Whether or not this section is from Clast E/Lithology 45B, (there is always the possibility of an error in documentation), it has some interesting features.

Table 2. Analyses of materials in 15445 and 15455.

Wt.%	15445,221 glass veins			15445,92 matrix	15445,147 matrix		15445,25 matrix[a]	15455,14 matrix[b]
SiO_2	40.2	42.2	43.4	45.7	45.3	44.6		47.3
TiO_2	0.18	0.17	0.21	1.56	1.70	1.47	1.49	1.35
Al_2O_3	25.4	25.4	24.5	17.3	17.5	16.66	16.2	17.1
Cr_2O_3	0.15	0.16	0.14	0.18	0.15			0.26
FeO	4.8	4.7	4.9	7.9	9.5	9.83	10.2	8.79
MnO	0.08	0.09	0.09	0.11	0.16	0.14		
MgO	13.7	13.6	13.7	13.4	15.7	16.0	15.5	13.3
CaO	13.7	13.8	13.6	11.5	9.7	10.04	9.6	10.6
Na_2O	0.45	0.51	0.52	0.62	0.81		0.55	0.58
K_2O	0.07	0.07	0.07	0.22	0.18			0.17
BaO	0.03	0.04	0.02	0.06	0.11			
P_2O_5	0.01	0.03	0.02	0.18	0.27	0.21		
Ni				0.04				
S				0.04		0.06		
Sum	98.77	100.76	101.24	99.01	101.12			99.5
Method	DBA(25 μ)	DBA(25 μ)	DBA(25 μ)	DBA(100 μ)	DBA(100 μ)		XRF	—
No. of points	9	20	10	10	10			
MgO/MgO + FeO	.74	.74	.74	.63	.62	.62	.60	
CIPW norm								
Fo	24.2	23.6	23.7	10.8	17.3		10.6	6.5
Fa	6.7	6.4	6.6	4.3	7.3		5.4	3.0
En	0.0	0.0	0.0	18.4	13.9		19.2	24.1
Fs	0.0	0.0	0.0	6.6	5.3		8.9	9.9
Wo	8.3	5.1	2.2	5.2	1.2		2.5	3.8
Or	0.4	0.4	0.4	1.3	1.1		1.0	1.0
Ab	3.8	4.3	4.3	5.3	6.8		4.8	4.9
An	49.1	55.7	61.3	44.3	43.2		43.8	43.9
Ilm	0.3	0.3	0.4	3.0	3.2		2.9	2.6
Cr	0.2	0.2	0.2	0.3	0.2		0.3	0.4
Ap	0.0	0.1	0.0	0.4	0.6		0.5	—
Tr	0.0	0.1	0.0	0.1	—		0.2	—
Cor	6.9	3.9	0.9	—	—		—	—

[a]Ridley *et al.* (1973).
[b]Taylor *et al.* (1973).

(1) The crystalline troctolite in 15445,221 is more Fe-rich than the 76535 troctolite. It is not known if the troctolite in 15445,221 is free of meteoritic contamination. (2) Troctolitic glass, whose normative mineralogy is compatible with derivation by total shock-melting of the troctolite lithology, is present. The normative corundum (Table 2) may indicate that spinel was present in the

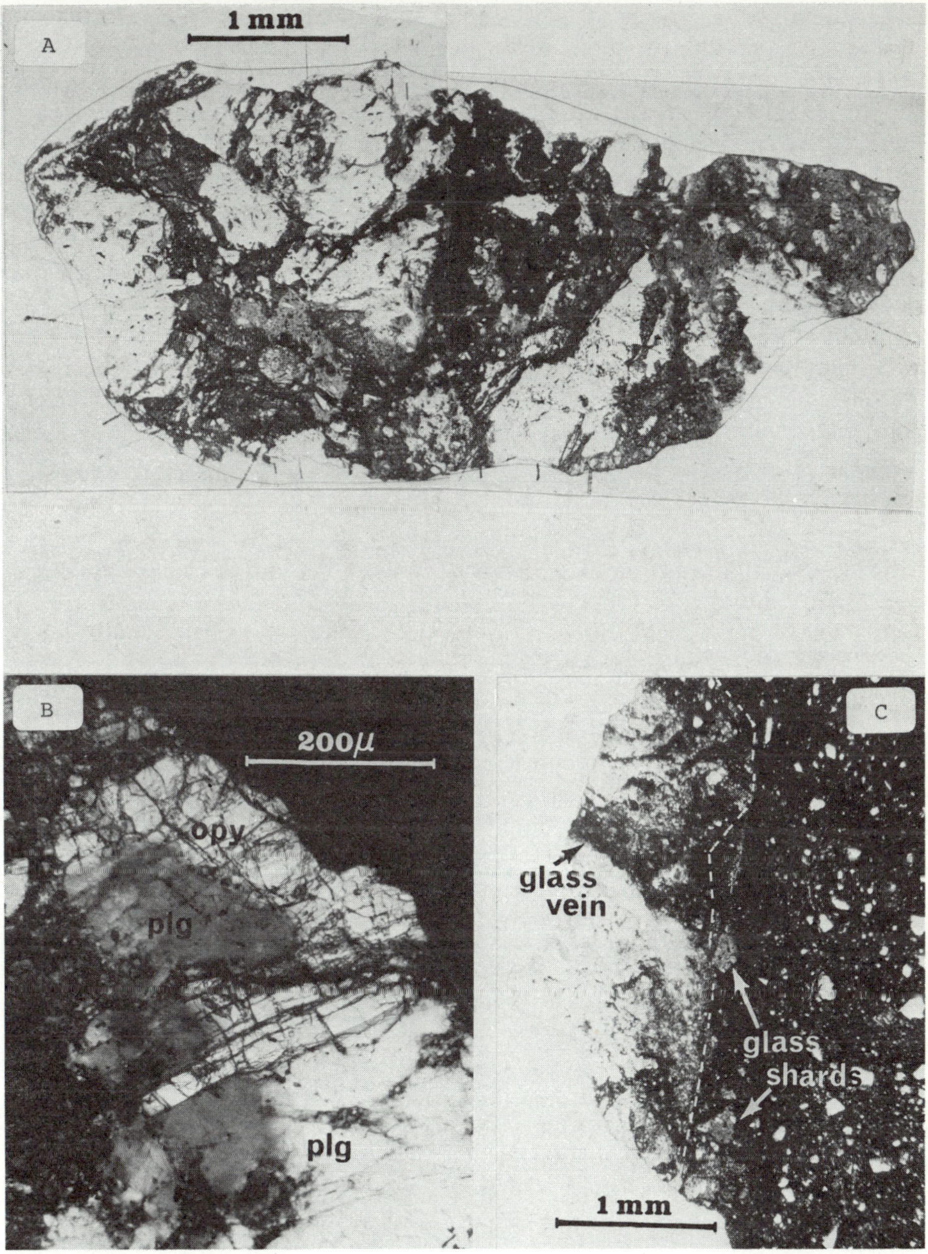

Fig. 7. Photomicrographs of clasts in 15445. (A) Photomicrograph of thin section 15445,221 (plane light). White areas are plagioclase-rich norite and troctolite. Gray areas are glassy veins. Black area in center is ilmenite and ulvöspinel in vein. (B) Detail of thin section 15445,221, showing igneous texture in norite (crossed polarizers). opy, pyroxene; plg, plagioclase. (C) Detail of thin section 15445,220 (plane light). Glass vein cuts plagioclase-rich clast. Dark area to right is matrix.

parent. If this glass was produced by total shock-melting, as seems likely, then the troctolite contained about 70% plagioclase and 30% olivine. If it was produced by the event that created the matrix of 15445, then the troctolite glass was prevented from mixing with the matrix melt for some reason, and was also prevented from crystallizing to the same degree as the 15445 matrix. Another thin section, 15445,220, whose location within 15445 is unknown, contains identical-appearing brown glass that forms a vein through a plagioclase-rich norite and partially surrounds this clast (Fig. 7C). Part of the glass surrounding the clast is mixed with matrix; part occurs as well-defined shards or clasts within the matrix. It is possible that the glass was formed by shock-melting of a troctolite lithology during the same event that formed the matrix of 15445, but was chilled (perhaps during a ballistic phase, or merely because of contact with cold clasts) prior to incorporation in the matrix, and was therefore prevented from mixing with the matrix.

Other lithic clasts. Most other small clasts are monomineralic aggregates. *Polygonalized olivines* are clasts with a polygonal granoblastic texture, consisting solely of olivine (Fig. 5C). The largest such clast in the thin sections has a diameter of 3 mm, and the clasts invariably have smooth, curving boundaries (Fig. 2). In the two clasts analyzed, olivine has a fairly constant composition of Fo_{87-88}. The protolith was either a dunite or a coarse-grained (>3 mm) granoblastic troctolite. The metamorphism clearly *preceded* the formation of 15445. *Spherulitic plagioclase* masses (Fig. 5D), up to ~1 mm in diameter, are most typically rounded and smooth, but some are irregular. At least some clearly crystallized *in situ* from the circumference inward. The teardrop- or lapilli-shape of several of these masses suggests that at least some of the aggregates crystallized from a plagioclase liquid or devitrified from a high temperature, plastic, form of maskelynite. Rare *granulitic ANT* clasts are small fragments with a triple-junction texture and a plagioclase-mafic mineralogy (Fig. 8A); an analytical study of these fragments has not been made.

15445 Matrix

The matrix of 15445 (Lithology 45A) is dark colored, coherent, locally stratified, fine grained, and vesicular (Figs. 2, 8B, 8C). The matrix is an aggregate of mineral clasts and minerals that crystallized at least partly from a silicate melt, as demonstrated in particular by the presence of euhedral, skeletal olivines (Fig. 8D). It has siderophile abundances within the normal range for meteoritic-contaminated highlands breccias and similar to that of the 15455 matrix (Gros *et al.*, 1976). We conclude therefore that it is probably a fragment-laden impact melt. Except for the larger mineral grains, which are obviously not the products of melt crystallization, distinction between clastic and crystallized melt phases is not an objective exercise; the finest clastic material and the coarsest melt material overlap in grain size. For this reason, the distinction between melt groundmass and bulk matrix (= groundmass + clasts), which is theoretically possible (and made by James (1976) for a similar matrix texture in rock 73215), is

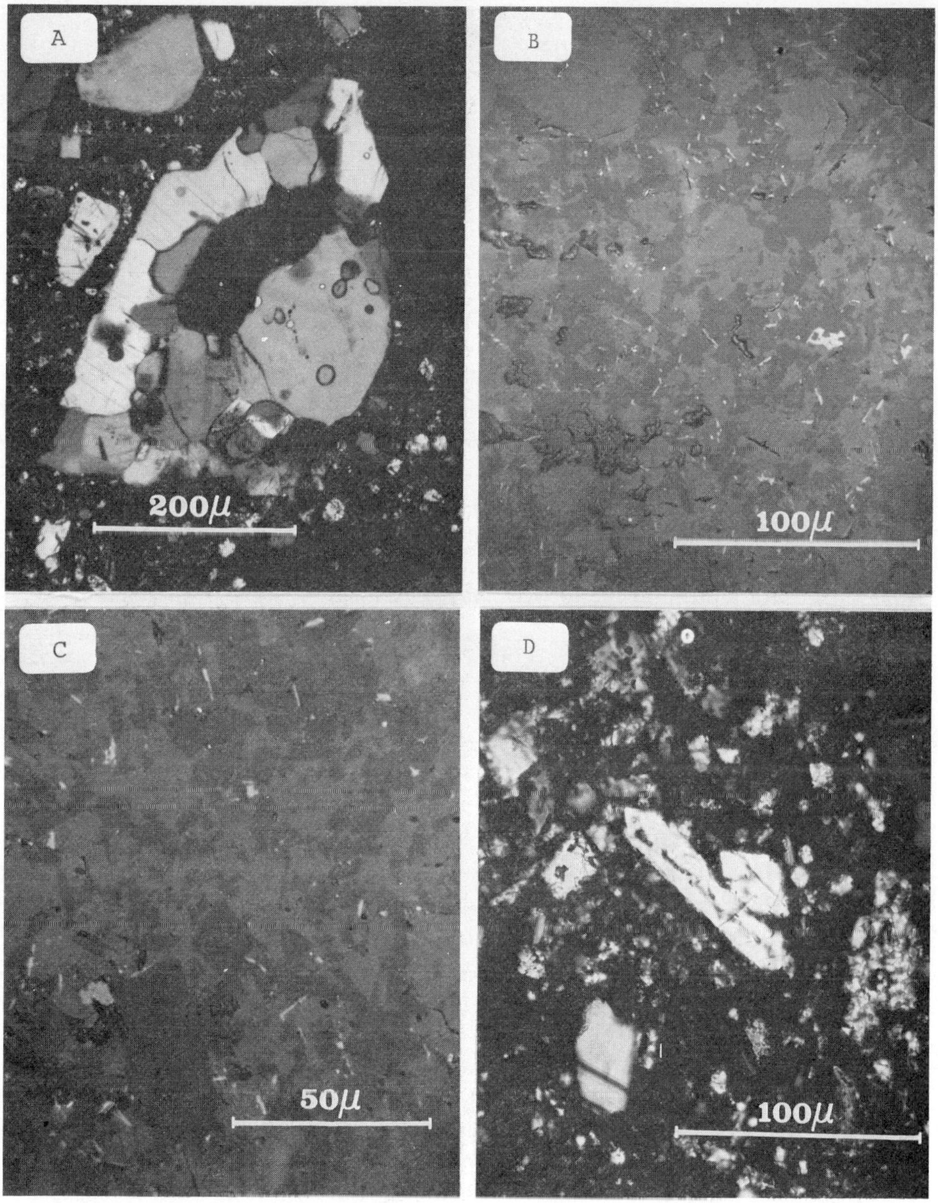

Fig. 8. Photomicrographs of clast and matrix in 15445. (A) Granulitic ANT clast in 15445,147 (crossed polarizers). (B, C) Typical matrix textures in 15445,147 (reflected light). Dark gray, plagioclase; light gray, mafics; white laths, mainly ilmenite. (D) Skeletal olivines in matrix of 15445,92, indicative of silicate melt in the matrix (crossed polarizers).

no more than an approximation in the present case. Obvious clastic material includes plagioclase, olivine, pleonaste spinels, very minor augite, and minor opaques including chromite. The melt phase crystallized plagioclase, olivine, at least *some* low-Ca pyroxene, and opaque phases. Most of the latter are needles of ilmenite, but may include armalcolite and chromite. Metal and sulfide grains are scattered throughout the matrix.

Textural characteristics of the matrix. While mineral and lithic clasts are abundant, perhaps as much as 70% of the matrix is a finer-grained mass of interlocking anhedra of olivine and plagioclase, mainly 5–20 μ in diameter, with intergranular needles of ilmenite (Figs. 8B, 8C). Some of the scattered skeletal olivines, which are an integral part of the mass, are more than 100 μ long. Bulk analyses of the matrix (Table 2) suggest that pyroxene is present in the matrix, but mineral analyses (below) record only rare low-Ca pyroxene in grains large enough for analysis. However, reflectivity differences demonstrate that pyroxene as well as olivine is present in the finest-grain sizes of the matrix. Some of the matrix tends to a micropoikilitic texture, with stubby plagioclase chadacrysts enclosed in olivine oikocrysts. Interstitial glass is present in the matrix, but rare. Although the matrix texture is fairly homogeneous, local variations exist, particularly near the borders of the large lithic clasts. The eutaxitic structure results from differences in grain size, vesicularity, and mineralogy. Flow around the larger clasts is indicated.

Mineral fragments and clast-matrix interactions. Many analyses of matrix mineral fragments have been made, but as noted above, distinction between clastic material and crystallized melt material is not always possible. Most of the grains identified as clastic in Fig. 9 are $>50\ \mu$ in diameter. The other grains, most of which are $<50\ \mu$ in diameter (except for euhedral, skeletal grains) probably include some clastic material, but are interpreted primarily as grains that crystallized from the melt.

The mineral clasts, usually less than 200 μ and rarely larger than 1 mm in diameter, are predominantly plagioclase and olivine, with less abundant pleonaste, rare augite, and rare chromite.

Plagioclase: Analyzed fragments have a compositional range from An_{97} to An_{65}, but most are An_{97-93} (Fig. 9B). No appreciable primary zoning occurs in any of these grains, but in a few cases detectable zoning produced by reaction with the matrix is present in the outer few microns. Several fragments contain inclusions; a few grains contain abundant inclusions of silica. The compositional distribution is not particularly diagnostic of the source rocks of these fragments, since these calcic plagioclases fall into the range for the 15445 lithic clasts described above, and indeed into the range for many other lunar lithologies. The rare presence of more sodic plagioclases, however, indicates a source not sampled among the lithic clasts.

Olivine: A considerable range is apparent in the histograms of olivine compositions (Fig. 9A). As in the plagioclase, no primary zoning has been detected, but in some cases reaction with the matrix melt produced $\leqslant 20\ \mu$ wide

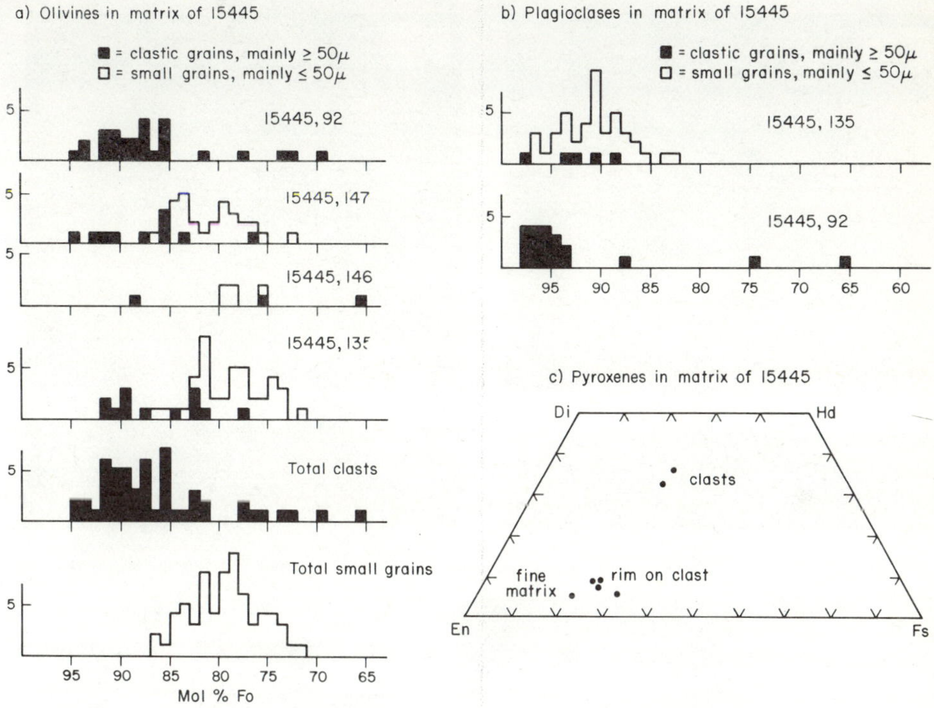

Fig. 9. Compositions of matrix mineral grains in 15445.

borders that are slightly more Fe-rich. In a few cases, an olivine grain has a corona of a different olivine that trapped blebs of feldspar and Fe-metal (Fig. 10A) during its growth. The single analyzed grain of this type was Fe-rich ($\sim Fo_{65}$) with a more magnesian corona ($\sim Fo_{75}$). Some grains contain inclusions of other phases, such as chromite. One such analyzed grain has an Mg–Al chromite phase (Fig. 4) in a host that is Fo_{88}. Many of the clastic olivine grains have compositions that are clearly compatible with derivation from the spinel troctolite lithic type, and others are compatible with derivation from the observed troctolitic lithic type. The more Fe-rich olivines are not represented in lithic clasts analyzed in 15445, but have been observed in other lithic clasts from the lunar highlands (e.g., Ryder *et al.*, 1975; Fig. 8). They may be from the rare granulitic ANT clasts observed in 15445.

Pyroxene: Augite is a rare clastic phase, and is always overgrown with pigeonite (Fig. 10B). This overgrowth zone, a few microns wide, contains trapped inclusions of ilmenite and plagioclase. Augite has not been observed in any of the lithic clasts in 15445 thin sections. None of the augite clasts is greater than about 50 μ in diameter. Low-Ca pyroxene is conspicuous by its absence: we have not observed a definite clast. Ridley *et al.* (1973) reported rare orthopyroxene clasts, similar in composition to those which we believe crystal-

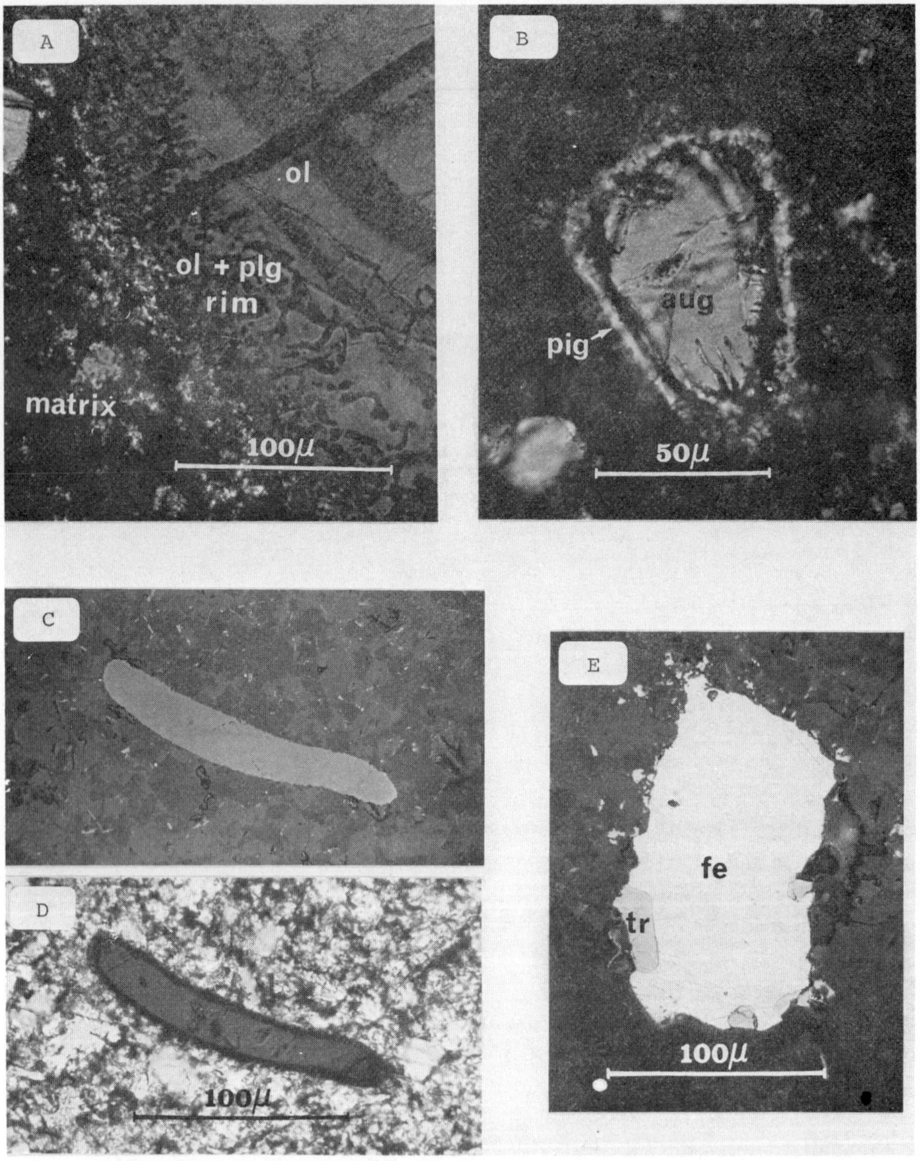

Fig. 10. Matrix phases in 15445. (A) Clast-matrix rimming relation in 15445,146 (crossed polarizers). Clast is olivine (Fo_{65}), rim is more magnesian olivine and plagioclase. (B) Clast-matrix rimming relation in 15445,135 (crossed polarizers). Opaque area between pigeonite rim and augite clast is mainly ilmenite and plagioclase. (C) Chromite fragment in 15445,147 (reflected light). Shape is probably due to reaction with the matrix melt. (D) Plane light photomicrograph of chromite fragment in Fig. 10C, illustrating its translucency. (E) Typical Fe-metal/troilite particle in 15445,147 (reflected light). fe, Fe-metal with 5–9% Ni; tr, troilite.

lized from the melt, and Ridley *et al.* (1973) did not distinguish a melt phase. The absence of clasts of low-Ca pyroxene is particularly interesting, since the largest lithic clast in 15445 is a friable norite, and we might therefore expect low-Ca pyroxene to be an important contributor to the mineral clast population. That it is not is significant.

Pleonaste: Although pleonaste (pink spinel) is not abundant, it is ubiquitous, and is locally enriched, forming schlieren. Single grains are rarely larger than ~50 μ in diameter. Compositions of these clastic spinels are given in Fig. 4; they have a range similar to those in the spinel troctolites, but appear to be slightly richer in Fe, probably because of reaction with the matrix. All the pleonaste grains in the matrix have a very narrow (1–2 μ) corona consisting of an opaque phase that has the optical characteristics of ulvöspinel, and minor plagioclase.

Chromite: Most of the chromite occurs as rare small opaque grains, which we have not analyzed. There is one large 150 μ long by 30 μ wide grain which is translucent through the 30 μ thickness of the thin section (Fig. 10C, 10D). Its red-brown color is uniform across the grain. Analyses (Fig. 4 and Ryder and Bower, 1977) show that this chromite is also compositionally distinctive, having high Mg and low Ti contents similar to some terrestrial (non-opaque) chromites. The chromite is compositionally similar to those included in the olivine clast (Fo_{88}) noted above, except that the latter chromites contain more Ti. The only other lunar chromites reported to have similar compositions are less magnesian chromite from 76535 troctolite (e.g., Dymek *et al.*, 1975) and chromite from 72415-7 dunite (Dymek *et al.*, 1976); none of these are reported as translucent. The shape of the chromite grain (Fig. 10C) suggests that it has reacted with the melt, and its composition is slightly more Fe-rich toward the edges, presumably from such reaction. A corona of an opaque phase (ulvöspinel?) exists, but is even narrower than those on the clastic pleonastes. The chromite contains a small, unidentified, anisotropic opaque inclusion, but there is no sign of any type of symplectite intergrowth like that found in 72415-7 dunite and several troctolites (e.g., Dymek *et al.*, 1975). It seems likely that the chromite is derived from a magnesian, olivine-rich lithic type.

Melt phases. Most of the melt phases large enough to analyze are olivine and plagioclase, in approximately equal amounts. Other phases include laths of ilmenite, at least some pyroxene, and rare interstitial glass. Fe-metal and troilite blebs, associated with vesicles in the matrix, appear to have crystallized from a vapor phase.

Plagioclase: Plagioclase grains less than 50 μ in diameter have compositions ranging from An_{96} to An_{82} with a peak in the distribution at about An_{90}. They are more sodic overall than the large, obviously clastic plagioclase (Fig. 9B) i.e., the clastic assemblage is more refractory than the melt assemblage.

Olivine: The small olivine grains have compositions ranging from Fo_{86} to Fo_{71}, with a normal distribution peaked at Fo_{78-79} (Fig. 9A). The skeletal olivines are zoned in the range of ~Fo_{85-80} compatible with their early crystallization from a melt phase. Overall, these small olivines are more Fe-rich (less refractory) than the obviously clastic material.

Pyroxene: Only one small discrete grain of pyroxene was analyzed, but pigeonite, which clearly crystallized from the melt, overgrows the augite clasts. Compositions are given in Figure 9C. It seems likely that much of the finest-grained mafic material is low-Ca pyroxene, since the norm of the bulk matrix analyses contains abundant pyroxene.

Fe-metal and troilite: Particles of Fe-metal and troilite are scattered through the matrix, and are often associated with vesicles, as noted in the macroscopic observations (Marvin, 1976). Several of the grains are more than 100 μ in diameter. Most frequently the bulk of any given grain is Fe-metal with troilite blebs occurring around the edges of the metal (Fig. 10E). Occasionally troilite occurs as tiny discrete particles in the matrix. Microprobe analyses show that the Ni content of the metal ranges mainly from 5–9%, with the complete range occurring within the larger single grains; one analysis had 16% Ni. Presumably the metal is internally structured, but etching experiments have not been done. The Ni contents are similar to those of metals in the 15455 black matrix (Hewins and Goldstein, 1975). The association of Fe-metal and troilite with vesicles suggests that they crystallized from a vapor phase late in the formation of the 15445 matrix; definite clastic metal has not been recognized.

Brief summary of 15445

15445 is an assemblage of magnesian plutonic and metamorphic lithic clasts of deep-crustal origin, embedded in a matrix of mineral fragments bonded by a melt phase. Clasts of surficial origin are absent. The melt phase has a Mg-rich low-K Fra Mauro composition and has crystallized into plagioclase, prominent olivine, and a pyroxene component, as well as opaques.

PETROLOGY OF 15455

Introduction

Sample 15455 (937.2 g) is macroscopically similar to 15445 (Fig. 1), but few results of studies of it have yet been published. For our purposes the most significant are chemical analyses of a white clast and of a matrix sample by Taylor *et al.* (1973), and analyses for siderophile and volatile trace elements in clasts and matrix by Ganapathy *et al.* (1973). Both the white clasts analyzed, one reportedly an anorthosite, the other a leucogabbro, were free of meteoritic contamination. However, the matrix has absolute and relative abundances of siderophiles similar to the matrix of 15445 (Gros *et al.*, 1976), which *is* contaminated with meteoritic material.

Because no petrographic study of 15455 had been published other than a preliminary description in the Apollo 15 Lunar Sample Information Catalog (1971), we obtained thin sections 15455,26 and 15455,28 for study. The thin sections are two in a group of serial sections from a chip of the large white clast (P. Butler Jr., pers. comm.) and include a vein of black matrix (Fig. 11A). A few small lithic clasts occur in the black vein (Fig. 11B, 11C).

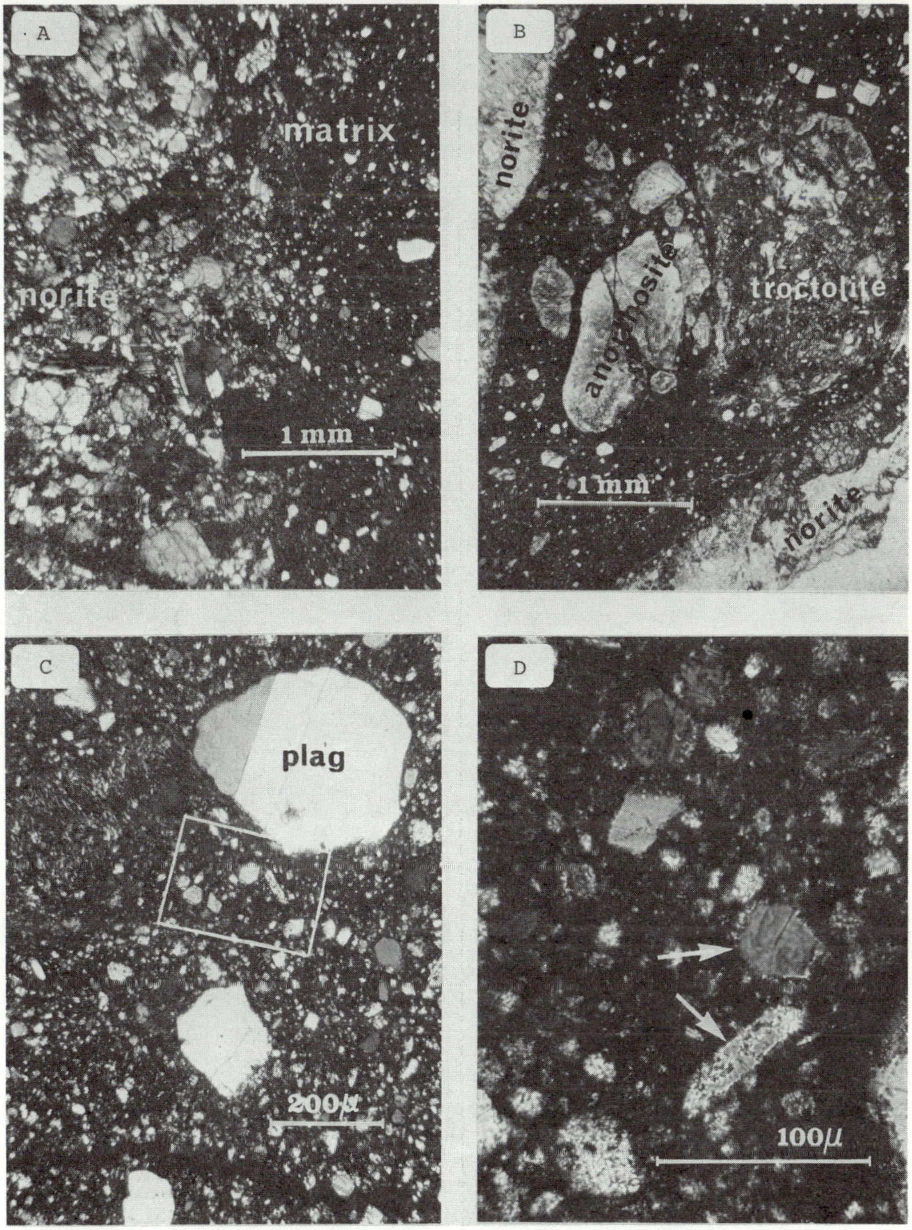

Fig. 11. Photomicrographs of 15455,28. (A) General view of 15455, showing cataclastic norite with twinned plagioclase grains, and fragment-laden matrix (crossed polarizers). (B) General view of matrix vein (black) cutting the norite, and containing rounded lithic clasts (plane light). (C) Matrix texture, showing abundant small clasts and common large clasts (crossed polarizers). Inset is Fig. 11D. (D) Detail of 11C (crossed polarizers). Arrows point to euhedral, melt-crystallized olivines.

Lithic clasts

Norite. The large white clast which forms the bulk of the two thin sections is a norite (Figs. 11A, 12A, 12B, 12C) consisting mainly of plagioclase (~70%) and orthopyroxene (~30%). The compositions of these two major phases (Fig. 13) are similar to those in the norite in 15445. A variety of accessory phases is also present. The norite is cataclasized and friable (Figs. 11A, 12A, 12B), in sharp

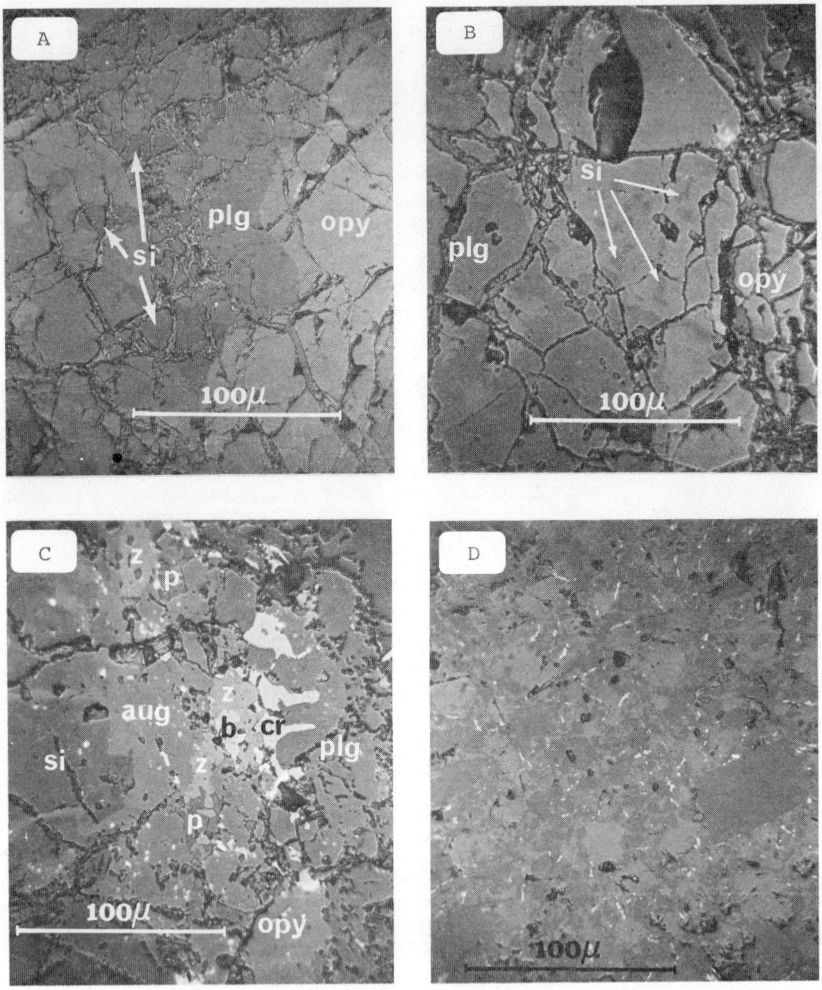

Fig. 12. Photomicrographs of 15455,28. (A–C) Details of norite illustrating mineralogy and igneous textures (reflected light). si, silica; plg, plagioclase; opy, orthopyroxene; z, zircon; p, phosphate; b, baddeleyite; cr, chromite; aug, augitic diopside. (D) Typical matrix texture, showing similarity to 15445 matrix, except for finer grain size (reflected light). Dark gray, plagioclase; light gray, mafics; white laths, mainly ilmenite; very dark patches are vesicles and plucked areas.

contrast to the coherent black matrix (Fig. 12D). In spite of this cataclasis, primary textures have been preserved in the norite (Figs. 12A, 12B, 12C).

The orthopyroxene is $En_{80-83}Wo_{1-3}$, contains ~1% Al_2O_3, ~0.4% TiO_2, and ~0.5% Cr_2O_3, and is thus very similar to that in the 15445 norite. Augitic diopside, a minor phase, contains lower abundances of these minor elements (~0.6% Al_2O_3, ~0.3% TiO_2, ~0.3% Cr_2O_3) than the orthopyroxene, implying that clinopyroxene and orthopyroxene are not in equilibrium in as much as under equilibrium conditions the clinopyroxene would contain greater abundances of these elements. In addition the augitic diopside is a more magnesian pyroxene than would be expected in equilibrium with the orthopyroxene. The minor element distribution between pyroxenes, as well as the interstitial nature of the clinopyroxene (Fig. 12C), suggests that the norite is igneous, and that the augitic diopside crystallized much later than the orthopyroxene, following depletion of the magma in Al_2O_3, TiO_2, and Cr_2O_3. Plagioclase, $An_{91.5}$–An_{95}, has a slightly wider range of composition than that in the 15445 norite.

In contrast to the 15445 norite, but like 78235 (McCallum and Malthez, 1975) the 15455 norite contains a variety of accessory phases. *Silica*, *armalcolite* (75% TiO_2; 16% FeO; 8% MgO), *chromite* (6.7% MgO; 9–10% Al_2O_3; 55% Cr_2O_3; 25–30% FeO), *ilmenite* (3% MgO) with areas of *rutile* exsolution, a *phosphate*,

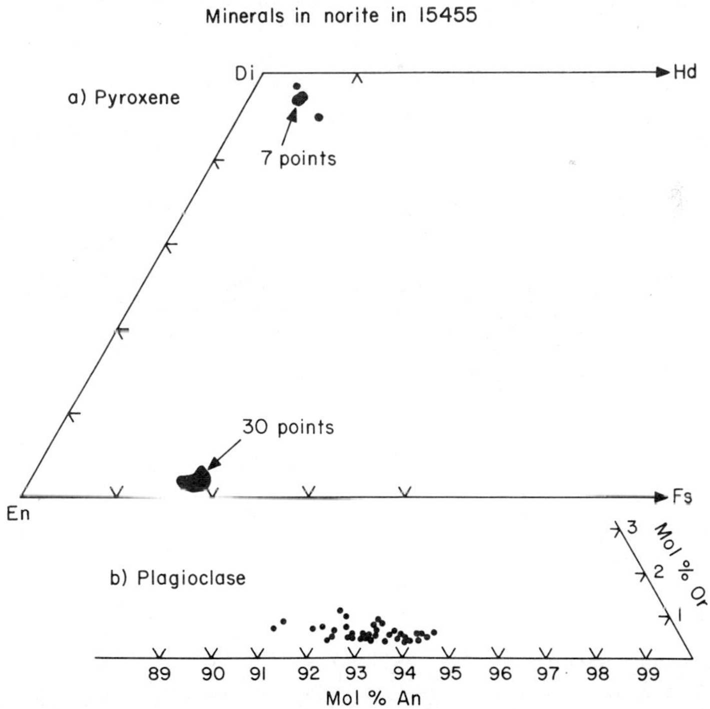

Fig. 13. Compositions of pyroxenes and plagioclase in the large white norite clast in 15455.

zircon, *baddeleyite*, *Fe-metal*, and *troilite* have been identified. In rare cases these accessory phases occur together in interstitial areas with clinopyroxene (Fig. 12C) and probably crystallized from a residual liquid; these areas are mineralogically more complex than other diopside-bearing assemblages in lunar norites (e.g., Irving *et al.*, 1974; Sclar and Bauer, 1975) and appear to be primary rather than shock induced or from extraneous fluids. Armalcolite and some of the chromites are not interstitial and occur as tiny discrete grains embedded in plagioclase and pyroxenes. The presence of these minor phases suggests an igneous origin for the norite, a suggestion supported by the relict textures observed in the norite which are not (metamorphic) triple junctions. It seems likely that the norite is a cumulate, and that the accessory phases (with the exception of the Fe-metal, armalcolite, and chromite grains, which probably crystallized at an earlier stage) crystallized from a trapped residual liquid. Fe-metal in the norite (referred to as the anorthositic facies of 15455) was analyzed by Hewins and Goldstein (1975) and was found to contain up to 10% Co. Hewins and Goldstein (1975) believed that a correlation between Ni and Co in the metal was suggestive of an igneous fractionation trend for the 15455 norite.

Taylor *et al.* (1973) analyzed a sample which is probably the large white clast of norite (P. Butler, pers. comm.). However, a CIPW norm of the analysis contains 20% olivine and only 7% pyroxene; furthermore the low REE abundances and the REE pattern are those normally associated with plagioclase alone. Because the analyzed material was handpicked separates, the analysis may be unrepresentative of the norite as a whole (S. R. Taylor, pers. comm.).

Other lithic clasts

A few small cataclastic lithic clasts within the black matrix vein include anorthosite and troctolite (Fig. 11B). The troctolite clast, which is 2 mm long, consists of olivine, plagioclase, and minor augite, in a brecciated mass whose grains are $<100\,\mu$ in diameter. The composition of the olivine is Fo_{85}, while plagioclase ranges from An_{94-96}. An unidentified opaque mineral occurs in this clast. The anorthosite clast is also a fine-grained cataclasite, and is cut by a matrix stringer (Fig. 11B). Plagioclase compositions range from An_{91-95}, with a peak at An_{93}, and the clast is therefore more sodic than anorthosites like the large sample 15415. Although mafic minerals do not occur in this clast, no conclusion can be made as to whether the parent rock was anorthosite or some coarse-grained mafic-bearing rock. Spherulitic plagioclase aggregates, similar to those observed in 15445, occur as clasts in 15455.

15455 matrix

The black competent matrix intrudes the norite and other clasts (Figs. 1, 11B) and clearly behaved as a fluid during the formation of the 15455 parent unit. The overall characteristics of the matrix (Figs. 11C, 12D) are similar to those of the matrix of 15445, and the presence of skeletal olivines (Fig. 11D) confirms this similarity and emphasizes the presence of a silicate-melt component in the matrix. Presumably the matrices of both 15445 and 15455 were produced by impact melting.

The portion of the matrix studied in the thin sections has a texture similar to that of the matrix of 15445, but has a finer grain size (Fig. 12D). The more rapid cooling that caused the finer grain size was probably a result of the injection of the melt as a narrow vein, and it seems likely that elsewhere the matrix of 15455 has a coarser grain size, perhaps similar to that of 15445.

Analyses of olivine and plagioclase mineral grains in the matrix of 15455,28 are shown in Fig. 14. A comparison with the data for the matrix of 15445 (Fig. 9) shows that there is no significant compositional difference between the minerals of the two matrices. Pink spinel, ilmenite, Fe-metal, troilite, and other minor phases occur in the 15455 matrix. Apart from one large clast of cataclastic orthopyroxene, (mineralogically identical to pyroxenes in the nearby norite clast) pyroxene is apparently absent from the matrix clast population, as it virtually is from the 15445 matrix. Fe-metal grains in the matrix of 15455 are texturally identical to those of 15445, and our analyses, and those of Hewins and Goldstein (1975) show a similar range of Ni (5–9%).

An analysis of the matrix of 15455 (Table 2), reported by Taylor *et al.* (1973) shows that it is a magnesian low-K Fra Mauro composition chemically similar to the matrix of 15445 (Table 2). A matrix sample was analyzed for meteoritic trace elements by Ganapathy *et al.* (1973) and both the abundances and inter-element ratios are similar to those of the 15445 matrix (Gros *et al.*, 1976; Higuchi and Morgan, 1976).

Summary of 15455

Like 15445, 15455 is an assemblage of lithic clasts of deep-crustal origin, embedded in a matrix of mineral fragments bonded by a melt phase. In its major- and trace-element chemistry, (including meteoritic siderophiles) the nature of

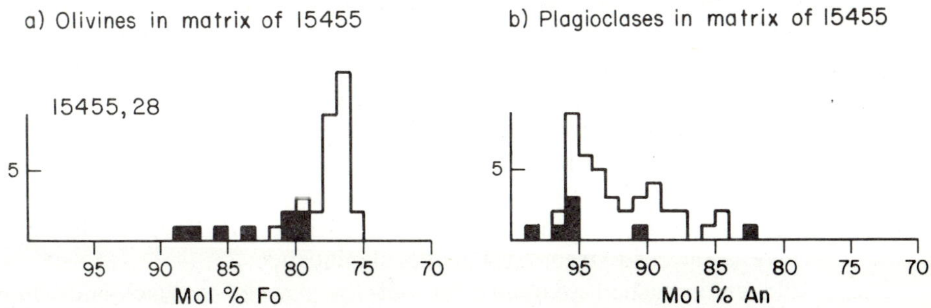

Fig. 14. Compositions of minerals in matrix of 15455. Shaded are large (> 50 μ, clastic) grains, unshaded are small (< 50 μ, mainly melt) grains, as in Fig. 9.

15455 is a close match to that of 15445, as it is petrographically. The matrix sample we studied, a narrow vein ~4 mm wide cutting the large norite clast, is finer grained than the matrix of 15445. We suspect that the matrix of 15455 away from the clast may be of a grain size similar to that of 15445.

Origin of 15445 and 15455

15445 and 15455 are distinct from any other of the large (>25 g) rocks (virtually all of which are regolith breccias) collected from the Apennine Front. They are, however, very similar to each other in overall appearance, lithic clast population, and mineral fragment population, and in the texture, mineralogy, and chemistry of their matrices. These similarities suggest they were once part of the same bedrock unit, and were excavated by the Spur Crater impact. The 1.5 m boulder and other nearby unsampled fragments that contain light-colored clasts are probably derived from the same bedrock unit.

From the characteristics of 15445 and 15455 we conclude that this bedrock unit is an impact melt sheet which has a composition equivalent to a magnesian low-K Mauro basalt. The indicated deep-seated origin for the clast population of this bedrock, and the fact that low-K Fra Mauro is not the composition of the upper part of the lunar crust (Taylor, 1975) suggests that the inferred impact was not a small-scale local event. Instead, the characteristics of the bedrock are those to be expected of a large-scale impact, and we suggest that it was produced by the Imbrium event, which is responsible for the Apennine Front itself. Beneath this impact melt bedrock unit there is likely to be an abundant amount of fragmental material ejected by the same event.

The Apollo 15 black-and-white rocks have some similarities to the several boulders sampled at the Apollo 17 site, which are generally believed to be fragments of the Serenitatis melt sheet (e.g., Winzer *et al.*, 1977). Significantly however, the meteoritic trace-element signatures of the Apollo 15 black-and-white rocks, though similar to each other, are distinct from those of the Apollo 17 boulders (Higuchi and Morgan, 1975). Thus 15445 and 15455 are unlikely to be remnants of the Serenitatis melt sheet. In addition, the black-and-white rocks are slightly, but significantly more magnesian, more mafic, and less KREEP-rich than Apollo 17 boulders (which are chemically and texturally more similar to the Apollo 16 poikilitic rocks). In all these rocks, the matrix composition cannot be derived from the clast population, which is significantly more plagioclase-rich and magnesian, and less KREEP-rich, that is, the impact melt (matrix) is not derived from the same locality as the clast population. This is also indicated by the paucity of shock features in clasts with respect to highly shocked (melted) matrix (R.A.F. Grieve, pers. comm.). The petrographic nature of the KREEP source remains unknown. No primary clasts of low-K Fra Mauro basalt composition survive, a fact of extreme, if elusive, significance.

One of the distinguished characteristics of the Apollo 15 black-and-white rocks is the virtual lack of monomineralic pyroxene fragments in the clast population, despite the fact that norite is conspicuous among the large lithic clasts. This feature cannot be explained by preferential digestion of pyroxene (which is less refractory than olivine and plagioclase) because the matrices are so fine grained that they must have solidified before appreciable digestion of any phases had taken place. Therefore, pyroxene was not available for addition to the clast population during the time that the melt was superheated enough to intimately mix in small clasts, although a pyroxene component must have been present in the source of the melt itself. By the time the melt had overtaken and swamped a noritic lithology it may have merely become too viscous to intimately incorporate small fragments, a hypothesis supported by the flow structure observed around the large clasts (Fig. 2). For this reason we suspect that noritic rocks occur at a higher level in the crust than do troctolites, at least in the Imbrium region. This influences our belief that the Apollo 15 melt sheet (15445 and 15455) is derived from a greater depth than the Apollo 17 melt sheet (Apollo 17 boulders) and subsequently influences the crustal model outlined in our companion paper (Ryder and Wood, 1977).

In summary, the characteristics of 15445 and 15455 suggest that they are fragments of a bedrock unit that was created as an impact melt sheet (or pond) by the Imbrium impact event. None of the other larger (>25 g) Apollo 15 rocks fulfill the expected characteristics of such a sheet. We conclude, therefore, that these two rocks are our best samples of deep crustal material excavated by that event.

Acknowledgments—P. Butler, Jr. kindly retraced the documentation of 15455 samples. Detailed constructive reviews by A. L. Albee and R. A. F. Grieve considerably improved the manuscript. This research was supported by NASA grant NGL-09-015-150 to J. A. Wood.

References

Anderson A. T. (1973) The texture and mineralogy of lunar peridotite, 15445,10. *J. Geol.* **81**, 219–226.

Apollo 15 Lunar Sample Information Catalog (1971) MSC 03209, NASA Johnson Space Center, Houston.

Arth J. G. and Hanson G. N. (1975) Geochemistry and origin of the early Precambrian crust of northeastern Minnesota. *Geochim. Cosmochim. Acta* **39**, 325–362.

Bailey N. G. and Ulrich G. E. (1975) *Apollo 15 Voice Transcript.* U.S.G.S. Report No. USGS-GD-74-029.

Bence A. E., Delano J. W., Papike J. J., and Cameron K. L. (1974) Petrology of the highlands massifs at Taurus-Littrow: An analysis of the 2–4 mm soil fraction. *Proc. Lunar Sci. Conf. 5th*, p. 785-827.

Bence A. E. and McGee J. (1976) Significance of the assemblage anorthite–aluminous enstatite–forsterite–aluminous spinel in the lunar highlands. *Geol. Soc. Amer. Abstracts with Programs* **8**, 772.

Blanchard D. P., Haskin L. A., Jacobs J. W., Brannon J. C., and Korotev R. L. (1975) Major and trace element chemistry of Boulder 1 at Station 2, Apollo 17. *The Moon* **14**, 359–371.

Blanchard D. P., Brannon J. C., Haskin L. A., and Jacobs J. W. (1977) Chemistry of sample 15445. In *Interdisciplinary Studies by the Imbrium Consortium*, Vol. 2 (J. A. Wood, ed). LSI Contr. No 268D, p. 60. Center for Astrophysics, Cambridge.

Dymek R. F., Albee A. L., and Chodos A. A. (1975) Comparative petrology of lunar cumulate rocks of possible primary origin: Dunite 72415, troctolite 76535, norite 78235, and anorthosite 62237. *Proc. Lunar Sci. Conf. 6th*, p. 301–341.

Dymek R. F., Albee A. L., and Chodos A. A. (1976) Petrology and origin of Boulders #2 and #3, Apollo 17 Station 2. *Proc. Lunar Sci. Conf. 7th*, p. 2335–2378.
Ganapathy R., Morgan J. W., Krähenbühl U., and Anders E. (1973) Ancient meteoritic components in lunar highlands rocks: Clues from trace elements in Apollo 15 and Apollo 16 samples. *Proc. Lunar Sci. Conf. 4th*, p. 1239–1261.
Gros J., Takahashi H., Morgan J. W., and Anders E. (1976) Siderophile and volatile elements in Apollo 14 and 15 rocks of the Imbrium Consortium. In *Interdisciplinary Studies by the Imbrium Consortium*, Vol. 1 (J. A. Wood, ed.). LSI Contr. No. 267D, p. 125–131. Center for Astrophysics, Cambridge.
Hewins R. H. and Goldstein J. I. (1975) The provenance of metal in anorthositic rocks. *Proc. Lunar Sci. Conf. 6th*, p. 343–362.
Higuchi H. and Morgan J. W. (1975) Ancient meteoritic component in Apollo 17 boulders. *Proc. Lunar Sci. Conf. 6th*, p. 1625–1651.
Irving A. J., Steele I. M., and Smith J. V. (1974) Lunar norite fragments and associated diopside veins. *Amer. Mineral.* **59**, 1062–1068.
James O. B. (1976) Petrology of aphanitic lithologies in consortium breccia 73215. *Proc. Lunar Sci. Conf. 7th*, p. 2145–2178.
Marvin U. B. (1976) Sample 15445 hand specimen descriptions and processing. In *Interdisciplinary Studies by the Imbrium Consortium*, Vol. 1 (J. A. Wood, ed.). LSI Contr. No. 267D, p. 108–110. Center for Astrophysics, Cambridge.
McCallum I. S. and Mathez E. A. (1975) Petrology of noritic cumulates and a partial melting model for the genesis of Fra Mauro basalts. *Proc. Lunar Sci. Conf. 6th*, p. 395–414.
Prinz M., Dowty E., Keil K., and Bunch T. E. (1973) Spinel troctolite and anorthosite in Apollo 16 samples. *Science* **179**, 74–76.
Ridley W. I., Hubbard N. J., Rhodes J. M., Weismann H., and Bansal B. (1973) The petrology of lunar breccia 15445 and petrogenetic implications. *J. Geol.* **81**, 621–631.
Ryder G. and Bower J. F. (1976) Sample 15445 petrology. In *Interdisciplinary Studies by the Imbrium Consortium*, Vol. 1 (J. A. Wood, ed.). LSI Contr. No. 267D, p. 111–115. Center for Astrophysics, Cambridge.
Ryder G. and Bower J. F. (1977) Sample 15445 petrology and sample 15455 petrology. In *Interdisciplinary Studies by the Imbrium Consortium*, Vol. 2 (J. A. Wood, ed.). LSI Contr. No. 268D, p. 42–60, 63–70. Center for Astrophysics, Cambridge.
Ryder G. and Wood J. A. (1977) Serenitatis and Imbrium impact melts: Implications for large-scale layering in the lunar crust. *Proc. Lunar Sci. Conf. 8th*. Vol. 1.
Ryder G., Stoeser D. B., Marvin U. B., Bower J. F., and Wood J. A. (1975) Boulder 1, Station 2, Apollo 17: Petrology and petrogenesis. *The Moon* **14**, 327–357.
Sclar C. M. and Bauer J. F. (1975) Shock-induced subsolidus reduction-decomposition or orthopyroxene and shock-induced melting in norite 78235. *Proc. Lunar Sci. Conf. 6th*, p. 799–820.
Steele I. M. and Smith J. V. (1975) Minor elements in lunar olivine as a petrologic indicator. *Proc. Lunar Sci. Conf. 6th*, p. 451–467.
Swann G. A., Bailey N. G., Batson R. M., Freeman V. L., Hait M. H., Head J. W., Holt H. E., Howard K. A., Irwin J. B., Larson K. B., Muehlberger W. R., Reed V. S., Rennilson J. J., Schaber G. G., Scott D. R., Silver L. T., Sutton R. L., Ulrich G. E., Wilshire H. G., and Wolfe E. W. (1972) Preliminary Geologic Investigation of The Apollo 15 Landing Site. *NASA SP-289*. p. 5-1 to 5-112.
Taylor S. R. (1975) *Lunar Science—A Post-Apollo View*. Pergamon Press, New York. 372 pp.
Taylor S. R., Gorton M. P., Muir P., Nance W., Rudowski R., and Ware N. (1973) Lunar highlands composition: Apennine Front. *Proc. Lunar Sci. Conf. 4th*, p. 1445–1459.
Winzer S. R., Nava D. F., Lum R. K. L., Schuhmann S., Schuhmann P., and Philpotts J. A. (1975) Origin of 78235, a lunar norite cumulate. *Proc. Lunar Sci. Conf. 6th*, p. 1219–1229.
Winzer S. R., Nava D. F., Schuhmann S., Kouns C. W., Lum R. K. L., and Philpotts J. A. (1974) Major, minor and trace element abundances in samples from the Apollo 17 Station 7 boulder: Implications for the origin of early lunar crust rocks. *Earth Planet. Sci. Lett.* **23**, 439–444.

Winzer S. R., Nava D. F., Schuhmann P. J., Lum R. K. L., Schuhmann S., Lindstrom M. M., Lindstrom D. J., and Philpotts J. A. (1977) The Apollo 17 "melt sheet:" Chemistry, age, and Rb/Sr systematics. *Earth Planet. Sci. Lett.* **33**, 389–400.

Wood J. A., Marvin U. B., Reid J. B., Jr., Taylor G. J., Bower J. F., Powell B. N., and Dickey J. S., Jr. (1971) Mineralogy and petrology of the Apollo 12 lunar sample. *Smithsonian Astrophysical Observatory Special Report 333.* 272 pp.

Proc. Lunar Sci. Conf. 8th (1977), p. 1925–1941.
Printed in the United States of America

Apollo 15 breccia rake samples—mineralogy of lithic and mineral clasts

IAN M. STEELE[1], ANTHONY J. IRVING[2], and JOSEPH V. SMITH[1]

[1]Department of the Geophysical Sciences, University of Chicago, Chicago, Illinois 60637
[2]The Lunar Science Institute, 3303 NASA Road 1, Houston, Texas 77058

Abstract—The major features of 22 Apollo 15 breccia rake samples are described, and 30 lithic clasts in these samples are classified as KREEP basalts (9), mare basalts (11), or "ANT" (10) based on chemical data for plagioclase, pyroxene, and olivine. Notable clasts include:

(1) a coarse-grained (4 mm) norite in 15689, with two chemically distinct diopsides, chromite, whitlockite, K–Na–Si–Al phase, Fe-metal, troilite, and Ca–Cr–Zr armalcolite occurring in veins similar to those in large norite 78235;
(2) an olivine-plagioclase-pyroxene clast in 15364 with mare-basalt affinity and unusually well-developed euhedral plagioclase;
(3) an ultrabasic clast in 15360 with olivine (Fo_{91}) and minor low- and high-Ca pyroxene, plagioclase, and chromite. The olivine has an unusually low CaO content (0.03 wt.%), and other chemical features suggest that this sample is unlike ultrabasic samples described to date;
(4) pyroxene-plagioclase-olivine clast in 15146 which is mineralogically similar to a mare basalt cumulate but has low-Fe plagioclase and very restricted pyroxene compositions typical of "ANT". The bulk composition closely matches that of Apollo 15 "green glass" to which it may be related.

Numerous isolated exsolved pyroxene grains were analyzed (bulk and individual lamellae) and the range of compositions indicates a slowly cooled mare basalt source. In terms of exsolution features and compositions of tiny exsolved opaque phases these pyroxenes closely resemble those in large anorthosite 67075. A unique intergrowth of Fe-metal (Ni 7.0 wt.%) and $\sim Cr_{0.1}Fe_{0.9}S$ of possible meteoritic origin is present in 15316. The pyroxene-plagioclase-olivine clast (15146) as well as other conflicting chemical data may result from metamorphism of a mare-type basalt or from plutonic crystallization of magma with mare basalt affinity.

INTRODUCTION

WITH THE EXCEPTION of mare basalts, the Apollo 15 samples are dominated by dark matrix- and glass-rich breccias. Notable exceptions include such samples as 15415 and 15362 (anorthosites), 15455 and 15418 (anorthositic norites), and 15386 and 15382 (KREEP basalts). Little mineralogic data can be obtained from the fine-grained matrix of breccias and thus we concentrate on the chemistry of mineral and lithic clasts. The combination of texture (for lithic clasts) and chemistry can be used for interpretation by comparison with the larger lunar samples. We report here data for 22 breccia rake samples with particular emphasis on chemistry of mineral and lithic clasts; preliminary data have been previously reported for these samples (Steele *et al.*, 1972a; Steele and Smith, 1977).

ANALYTICAL TECHNIQUES

Data were obtained by a variety of techniques: (1) Rapid surveys of whole breccias as well as individual lithic clasts were made by obtaining count rates for Mg, Fe, and Ca on random mafic grains using an electron probe. Approximate matrix corrections were made by assuming the balance from 100% to be SiO_2, and compositions were then plotted on a pyroxene quadrilateral; (2) Electron-probe analyses were obtained for plagioclase for minor (Mg, Fe, K) and major (Ca, Na) elements. The anorthite content was obtained by reference to calibration curves based on synthetic glasses, while the minor element contents were derived by reference to mineral standards; (3) Total analyses were made using an energy dispersive spectrometer with an on-line computer. Data were processed according to procedures described by Reed and Ware (1975). Detection levels are about 0.2 oxide wt.% and accuracy as estimated from standards is $\sim\pm5\%$ of the amount of present at the 1.0% level and $\pm2\%$ at levels $>\sim5\%$. (4) Total analyses were made using computer controlled wavelength dispersive spectrometers with on-line data reduction using standard ZAF correction procedures. These analyses were used to obtain accurate values of minor elements with detection levels below 100 ppmw.

GENERAL SAMPLE DESCRIPTION

Table 1 presents a list of the samples studied, a summary of relative amounts (visual estimation) of the major components (glass, lithic clasts, mineral clasts, fine-grained matrix, porosity), and specific comments for each sample. The majority of breccia samples are from Station 7 (Spur Crater) and Station 2 (St. George Crater) at the base of Hadley Delta and presumably represent material mostly derived from this slope. Samples from Station 9A (Rille area) are dominated by mare basalts, but two breccias are present in our samples. We use the term mare basalt in reference to the commonly observed mare samples indicating rapid cooling of surface flows; mare-type basalt is used in a wider sense to indicate any rock showing chemical affinity to mare basalts. Only two samples (15357, 15364) contain no glass whereas a variety of brown, yellow, red, and green glass is present in all others as veins, coatings or spheres; only rarely is devitrified glass present. The abundance of glass and fine-grained matrix clearly indicates that most samples have not been annealed, at least in the most recent stage. Examples of the range of textures, including holocrystalline, vesicular, and glassy, are shown in Figs. 1a–f (see caption for details).

GENERAL MAFIC MINERALOGY—UNIQUE Fe–Cr–S INTERGROWTH

Optical study, random mafic analyses, and detailed analyses of selected clasts showed nothing particularly unusual about the general chemistry of these samples, except for several features described later. Figures 2a–d show the general distribution of pyroxene compositions in 15146, 15322, 15347, and 15364, respectively (other similar plots for 15316, 15335, and 15357 were given in Steele *et al.*, 1972a). These illustrate several points: (1) 15146 shows a restricted range of compositions for pyroxene, but with a rather high Ca content and absence of optically visible exsolution lamellae. These features are typical of early-crystallizing pyroxenes in mare basalt rocks (but see later for conflicting data); (2) 15322 shows a wide range of compositions which can be interpreted in terms of

Table 1. Sample numbers, percentage of major components and brief comments for breccia rake samples studied.

	% glass	% lithic clasts	Clast type†	% mineral clasts	% fine matrix	% porosity	Comments
Station 2							
15135,6-7	(45)*	5	M	20	(45)*	30	Dark vesicular agglutinate. Largest mineral clasts are shocked plagioclase.
15146,1	tr.	15	M?	75	10	0	Near-monomict breccia. One coarse igneous clast. Abundant shock features.
Station 7							
15316,2	10	5	A	40	40	5	Shocked anorthosite clast. Reaction between matrix and mineral clasts.
15322,1	30	5	K,A,B	30	35	0	Many glass spheres. Meta-breccia clasts.
15326,1	50	0	—	5	45	0	Mostly glass spheres or devitrified spheres.
15329,6	50	5	K,A	30	15	0	Fine glass and matrix are difficult to distinguish. Mineral clasts mostly plagioclase.
15331,2	35	10	M,A,B	25	30	0	Several anorthositic clasts.
15335,2	40	10	M,A,B	15	25	10	Glass and matrix are difficult to distinguish. Glass veins present.
15341,1	20	5	A	15	60	0	Very dark breccia.
15344,2	15	35	K,B	35	15	0	One large KREEP basalt clast. Glass coated.
15347,2	15	5	M,A,B	40	35	5	Several igneous-textured clasts.
15350,2	(65)	5	K,M,B	30	(65)	0	Fine glass and matrix are difficult to distinguish.
15351,2	15	5	B	15	65	0	Very dark breccia.
15353,2	35	15	M,B	20	30	0	One large igneous clast.
15357,2	0	5	A	95	0	0	Holocrystalline breccia. Large plagioclase grains.
15360,2	20	5	K,M,O	30	45	0	Much of glass is devitrified.
15364,1	0	20	O	80	0	0	Holocrystalline. Mostly plagioclase. One clast has euhedral plagioclase.
15366,1	95	0	—	0	5	0	Green glass breccia.
15378,2	20	5	K	30	45	0	Dark breccia. Several igneous-textured clasts.
15389,1	(75)	5	B	0	(75)	20	Agglutinate
Station 9A							
15687,7-8	(60)	15	M	15	(60)	10	Agglutinate. Mare basalt + clast.
15689,6	5	60	O	15	20	0	Dominated by large pyroxene-maskelynite clast.

*Parentheses indicate that distinction could not be made between glass and fine matrix.
†M = mare basalt; A = anorthosite; K = KREEP basalt; B = breccia; O = other.

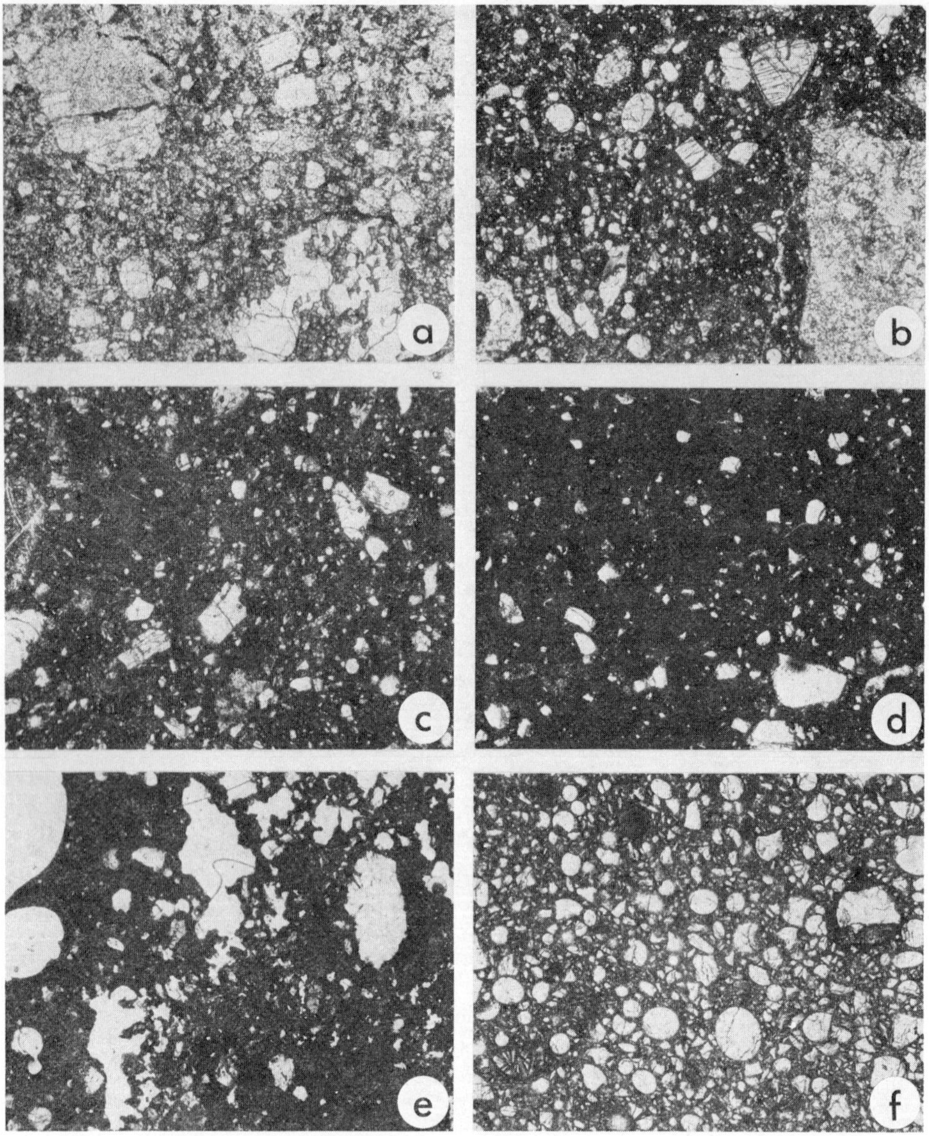

Fig. 1. Photomicrographs of Apollo 15 breccias illustrating range of textures and components: (a) 15364—holocrystalline breccia with large clast composed of plagioclase and olivine at lower-right; (b) 15331—note dominant dark matrix, presence of glass spheres, mineral clasts and one large fine-grained holocrystalline lithic clast; (c) 15316—note few glass spheres and predominance of dark matrix and mineral clasts; (d) 15351—nearly totally composed of dark matrix with scattered mineral and rare glass clasts; (e) 15135—breccia with numerous vesicles and rare mineral clasts; (f) 15366—breccia composed almost entirely of green glass spheres or fragments. The finest material also has the green glass composition. All photos: plain-light; max dimension = 2.7 mm.

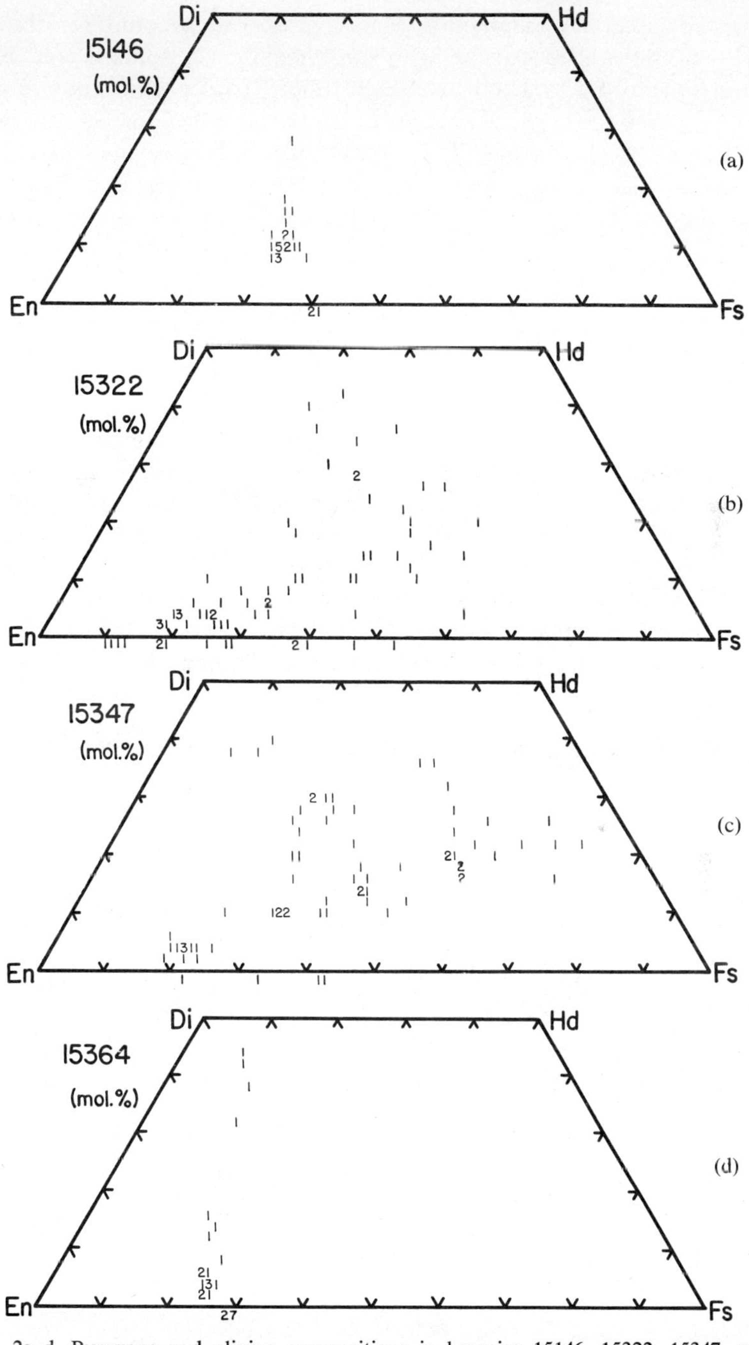

Fig. 2a–d. Pyroxene and olivine compositions in breccias 15146, 15322, 15347, and 15364, respectively. Data obtained by three element analyses as described in text. Numbers on figures indicate number of analyses which plot at that composition; olivine is plotted on base of quadrilateral. See text for discussion.

at least three general components. Mare basalt can account for the Fe-rich olivine (Fo < ~82) and pyroxene with intermediate Ca content (see Fig. 1 of Steele *et al.* (1972b) for typical mare basalt trend). The concentration of analyses at $En_{80-70}Fs_{15-25}Wo_5$ is typical of KREEP basalt (see Fig. 5 of Steele *et al.* (1972b) or Irving (1975) for KREEP basalt trend). The olivine cluster at $\sim Fo_{89}$ suggests an ultrabasic component as described by Smith and Steele (1976). Although other minor components may be present, the major features in 15322 can be explained by the above components; (3) 15347 shows a mixture of mare-type material and KREEP. The mare component is indicated by the olivine which is only very rarely present in KREEP (e.g., Steele *et al.*, 1972b; Dowty *et al.*, 1976); (4) 15364 (Figs. 1a, 2d) differs from 15146 (Fig. 2a) by having widely separated Mg-rich high- and low-Ca pyroxene compositions and olivine. These compositions are distinctive as discussed later.

In all breccia rake samples except 15689 the largest mineral clasts are plagioclase, suggestive of an anorthosite component which is present in the larger rocks at the Apollo 15 site (e.g., 15415) as well as in clasts. The grain size of pyroxene (except for 15689), olivine and minor phases is generally about $\frac{1}{10}$ as large as that of the largest plagioclase (~1 mm).

Glass fragments and spheres are common in nearly all samples, but were not studied in detail. Reconnaissance analyses showed two distinct groupings previously recognized (e.g., Warner *et al.*, 1972): (1) "green glass" (Al_2O_3 7.5, FeO 20.0, MgO 18.0 wt.%) and (2) "aluminous basalt glass" (Al_2O_3 18.3, FeO 9.0, MgO 9.3 wt.%). Of particular interest is the source of the green glass and whether any of the crystalline lithic clasts have comparable compositions.

An extensive search for unusual mineral compositions was made using qualitative energy dispersive displays. Compared with similar surveys of Apollo 14 samples, Zr-bearing phases and phosphates are rare. A very unusual 70 μm metal intergrowth was found in 15316. The host was Fe-metal with ~7 wt.% Ni and <0.3% Co. Within this metal are numerous 5–10 μm, subround inclusions of a phase close to $(Fe_{0.9}Cr_{0.1})S$ in composition. Scanning X-ray images for SK_α and CrK_α indicate, as do reflected light optics, a homogeneous sulfide phase. Nichiporuk and Chodos (1959) recorded 5.6 wt.% Cr in graphite-free troilite nodules from the Coahuila hexahedrite, and smaller values for other iron meteorites. A meteoritic origin for the lunar fragment seems plausible. A two-phase assemblage of submicroscopic daubreelite-troilite cannot be ruled out, but no further speculation is warranted for this unique occurrence. This breccia is otherwise not unusual except for a large 0.2 mm grain of $MgAl_2O_4$-rich pink spinel.

Breccia clasts—general

Table 2 enumerates 30 clasts which have been studied, and provides brief descriptions of each. Only clasts large enough relative to grain size were chosen so textural and chemical data could be representative of a larger parental sample. Clasts which were themselves breccia were not included. Clast size

Table 2. Description of clasts in Apollo 15 rake samples listed in Table 1.

	Pyx type*	Pl type†	Brief description‡
15146	O?	LFe,LCa	Pyx–pl–ol–il–cm–Fe. Coarse grained (1 mm) equigranular minerals not zoned.
15316	O	LFe,HCa	Mostly shocked pl (.3 mm) with minor (<5%) very fine pyx.
15322	K	LFe,LCa	Pl (70%), pyx (25%), and residual areas with il and phosphate. Very fine grained.
15329A	K	LFe,HCa	Lath-shaped pl with interstitial pyx; minor il. Small clast.
15329B	K	LFe,HCa	Similar to 15329A.
15331A	O?	LFe,HCa	Pl (80%)–pyx (20%) with fine grain size (<.2 mm).
15331B	M	HFe	Variolitic texture. Pyx–pl–il. Small clast.
15335A	O	LFe,HCa	Equant pl (~0.05 mm, 80%) and ol (~0.05 mm, 20%). Equilibrium texture suggests metamorphic effect.
15335B	M	HFe	Fine grained (<.1 mm). Pl–pyx–ol with igneous texture. Reaction with matrix.
15344	K	LFe,LCa	Large clast of pl–pyx–mesostasis (glass–il–phos). Igneous texture.
15347A	M	HFe	Variolitic texture but shocked. Pl up to 0.5 mm with interstitial pyx.
15347B	M	HFe	Similar to 15347A but finer grained and not shocked.
15347C	M	HFe	Similar to 15347B but il present.
15347D	M	HFe	Similar to 15347C.
15350A	K	LFe,LCa	Large pyx (50%, <0.25 mm) in finer pyx–pl matrix. Lath-shaped pl.
15350B	M	HFe	Coarse pyx (0.3 mm) with minor pl. Small clast.
15350C	K	LFe,LCa	Lath-shaped pl (50%, 0.2 mm), equant pyx, minor il.
15353A	M	HFe	Equigranular pl (50%, 0.1 mm)–pyx (45%)–ol–il with igneous texture.
15353B	O	LFe,HCa	Small pl–pyx clast.
15360A	O	LFe,HCa	Small clast of pl–pyx–ol with ol dominant.
15360B	M	HFe	Small pyx–pl clast.
15360C	K	LFe,LCa	Small pl–pyx–il clast.
15364A	O	LFe,HCa	Euhedral and anhedral pl (0.7 mm) poikilitically enclosed by ol; minor pyx.
15364B	O	LFe,HCa	Similar to 15364A except for no euhedral pl.
15378A	K	LFe,LCa	Igneous texture with large (0.3 mm) pl, pyx, and residuum of il–glass–pl.
15378B	K	LFe,LCa	Similar to 15378A.
15687A	M	HFe	Igneous texture; equal amounts of pl and pyx (0.2 mm) and minor spinel.
15687B	M	HFe	Similar to 15687B.
15689A	O	LFe,HCa	Small shocked clast with large pl (0.4 mm) and minor pyx. Subophitic texture.
15689B	O	LFe,HCa	Large clast of shocked norite. Pyx is dominant with 4 mm grain size. Pl is maskelynite. Minor Fe and cm.

*Pyroxene type—K = KREEP, M = mare basalt, O = other.
†Plagioclase type—LFe = low Fe, HFe = high Fe, LCa = low Ca, HCa = high Ca.
‡Pyx = pyroxene, pl = plagioclase, il = ilmenite, cm = chromite, Fe = metal, ol = olivine.

ranged from 0.1 to 10 mm^2 with most near 0.2 mm^2. Compositions of mafic minerals and plagioclase have been used previously to classify lunar samples and for each clast these data are summarized in Table 2 and ranges given in detail on Figs. 3a–c and 4. Figure 3a shows a common pattern typical of KREEP basalt similar to large sample 15386 (Steele *et al.*, 1972b). Olivine is not present and pyroxenes are Mg-rich (up to $mg = 0.85$) [=atomic Mg/(Mg + Fe)] but extend to Fe-rich compositions. Figure 3b shows a second group with pyroxenes having

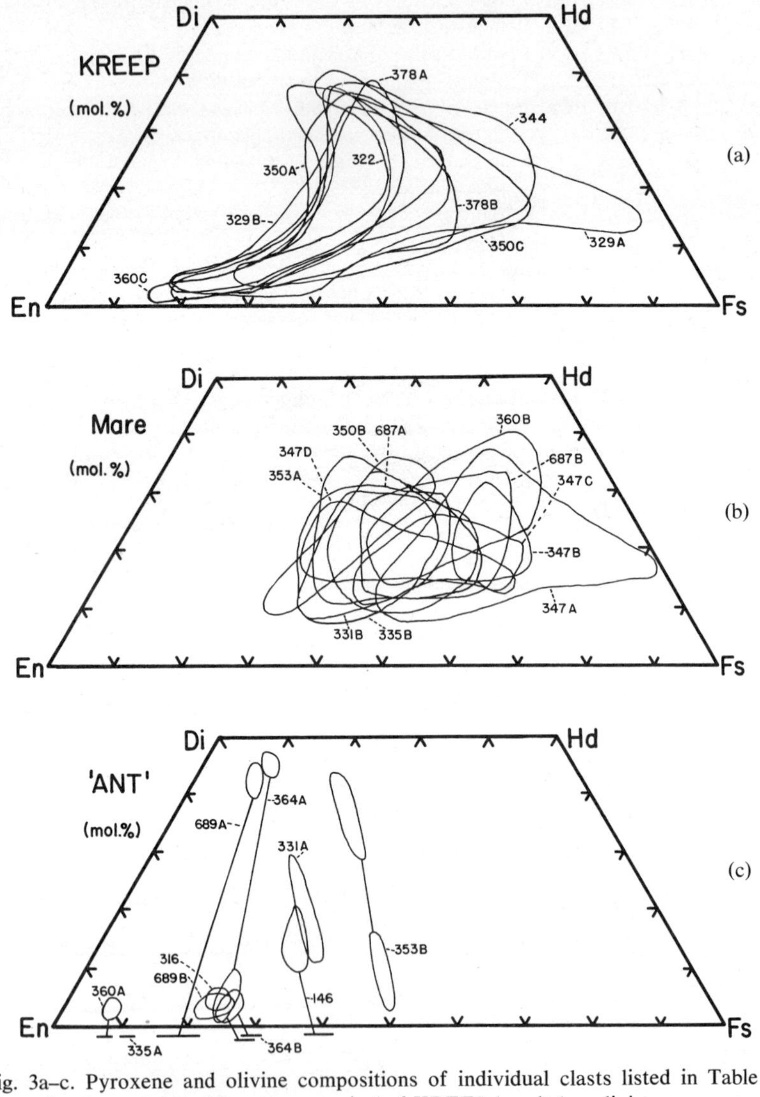

Fig. 3a–c. Pyroxene and olivine compositions of individual clasts listed in Table 2. Figure 3a shows those with patterns typical of KREEP basalt (no olivine was present); Fig. 3b shows those with mare basalt patterns; Fig. 3c shows those with "ANT" suite patterns. Numbers refer to the last three digits of the sample number.

mg < 0.65 and extending to more Fe-rich compositions; olivine is generally present. This is the typical trend for mare basalt (Steele *et al.*, 1972b). Figure 3c shows a third type of trend characterized by generally Mg-rich compositions, restricted range for any one clast, and except for 15331A and 15146 equilibrium compositions. These features are typical of the general "ANT" group of lunar material (see Smith and Steele (1976) for comments on "ANT").

Minor elements, including K and Fe (especially Fe because of ease in determination using an electron probe), in combination with the Ca content (expressed as anorthite content) of plagioclase have also proven useful for this three-fold division. Figure 4 shows data for iron for the 30 lithic clasts (3–4 analyses for each clast) with symbols based on the classification using mafic mineral data (Fig. 3a–c). It is quite apparent that plagioclase compositions also reflect these groupings: (1) KREEP basalt plagioclase has low Fe (<~0.7 wt.%) and low Ca (< ~ An_{90}); (2) mare basalt plagioclase has high Fe (Fe > ~ 0.5 wt.%) and high Ca content (> ~ An_{88}); (3) ANT plagioclase has low Fe (< 0.2 wt.%) and high Ca content (> An_{90}).

These data are summarized in Table 2 under pyroxene and plagioclase type and show a 1:1 correspondence (with one exception) between the clast type based on either pyroxene or plagioclase chemistry. This is reassuring because recognition of parental rock type can be based only on the composition of individual mineral grains, especially plagioclase. This fact is significant for samples which are dominated by single mineral grains, such as many soils or fine-grained separates. Table 2 does have one exception in that sample 15146 appears to have an "ANT"-like pyroxene composition and KREEP-like plagioclase composition; this sample will be discussed in detail in the next section.

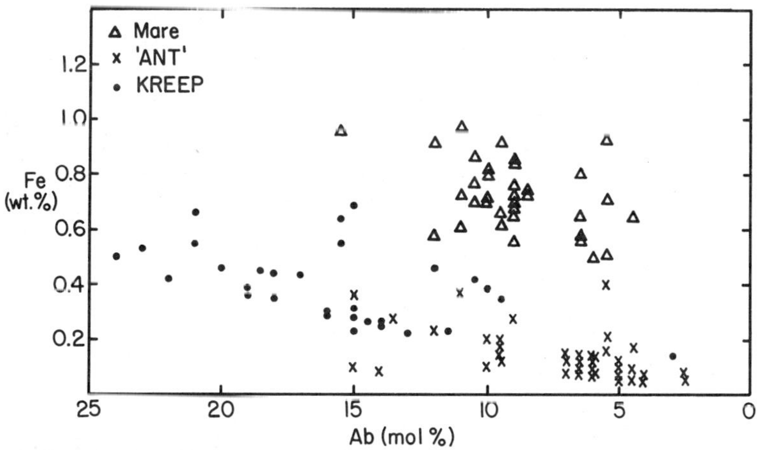

Fig. 4. Plot of Fe (wt.%) content vs. Ab content of plagioclase for clasts listed in Table 2. Symbols are assigned according to clast type based on mafic mineral compositions (Fig. 3a–c, Table 2). Note that with only several exceptions, the assignment based on Fe in plagioclase matches that based on mafic compositions.

Selected clasts—details

(a) *KREEP basalt clasts* (Fig. 5a). Apollo 15 samples contained the most abundant samples of KREEP material with crystalline melt textures (e.g., Steele *et al.*, 1972b; Dowty *et al.*, 1976). Many small KREEP basalt clasts have been recognized in the soils (e.g., Basu and Bower, 1976) and breccias (e.g., Reid *et al.*, 1977) from Apollo 15 indicating that, whatever the source, it is common. Of the 30 clasts in Table 2, nine are classed as KREEP basalt. Typical features of these clasts include plagioclase laths with interstitial pyroxene and conspicuous dark mesostasis areas composed of silica-rich glass, ilmenite, and phosphates. Typical analyses for 15344 are given in Table 3, analyses (1)–(5).

(b) *Mare basalt clasts.* Eleven clasts have chemical features consistent with mare basalt origin (Table 2). None of these clasts is large enough to permit classification based only on textural evidence which emphasizes the importance of a chemical classification based on individual phase composition. The presence of mare clasts in these breccias is not surprising based on the proximity to mare source areas. No detailed study was made of these clasts.

(c) *Norite clast* (Fig. 5b). Noritic breccia is a common material on the moon but few examples of a coarse-grained parent have been reported; an important one is 78235 (McCallum and Mathez, 1975; Steele, 1975). The norite clast in 15689 is remarkably similar to large sample 78235. This clast is dominated by shocked low-Ca pyroxene ($En_{74}Fs_{23}Wo_3$) and isotropic plagioclase (maskelynite, An_{91}) with minor equigranular diopside ($Al_2O_3 \sim 1$ wt.%). Several veins, intermittent in two dimensions, occur in this clast and their mineralogy again is remarkably similar to those in other norite samples (e.g., Irving *et al.*, 1974): chromite, whitlockite, Ca–Cr–Zr armalcolite, K–Na–Si–Al phase, diopside (Al_2O_3 $\sim$4 wt.%), troilite, and Fe-metal ($Co_{2.7}$, $Ni_{5.7}$) (analyses 6–12, Table 3). Differences between this vein assemblage and the assemblage in 78235 include the absence of rutile, baddeleyite, and zircon and presence of an armalcolite and Na–K–Al–Si phase in 15689. Haggerty (1973) noted that complex armalcolite commonly is associated with the other oxide phases mainly rutile + baddeleyite suggesting decomposition; possibly, this relation holds for 78235 but in 15689 decomposition has not occurred. Ryder and Bower (1977) describe an interstitial assemblage within a norite clast in 15455 which contains armalcolite, zircon, baddeleyite, and rutile, among other phases; the relation, if any, between armalcolite and the other phases could possibly be deduced from this clast in 15455. The Na–K–Si–Al phase is too silica-rich for a feldspar, but a mixture of SiO_2 and Na–K-feldspar is possible as is glass: three analyses were very similar (Table 3, analysis 10) indicating no variation in composition within this clast.

McCallum and Mathez (1975) discussed the possible relation between KREEP and norites through partial melting of the latter, but ruled out derivation of KREEP by partial melting because of a large discrepancy between observed and calculated Eu anomalies. The calculated values were based only on whitlockite/liquid REE relations. Because the complex armalcolite phase carries appreciable REE and is at least as abundant as phosphates (in 15689), the

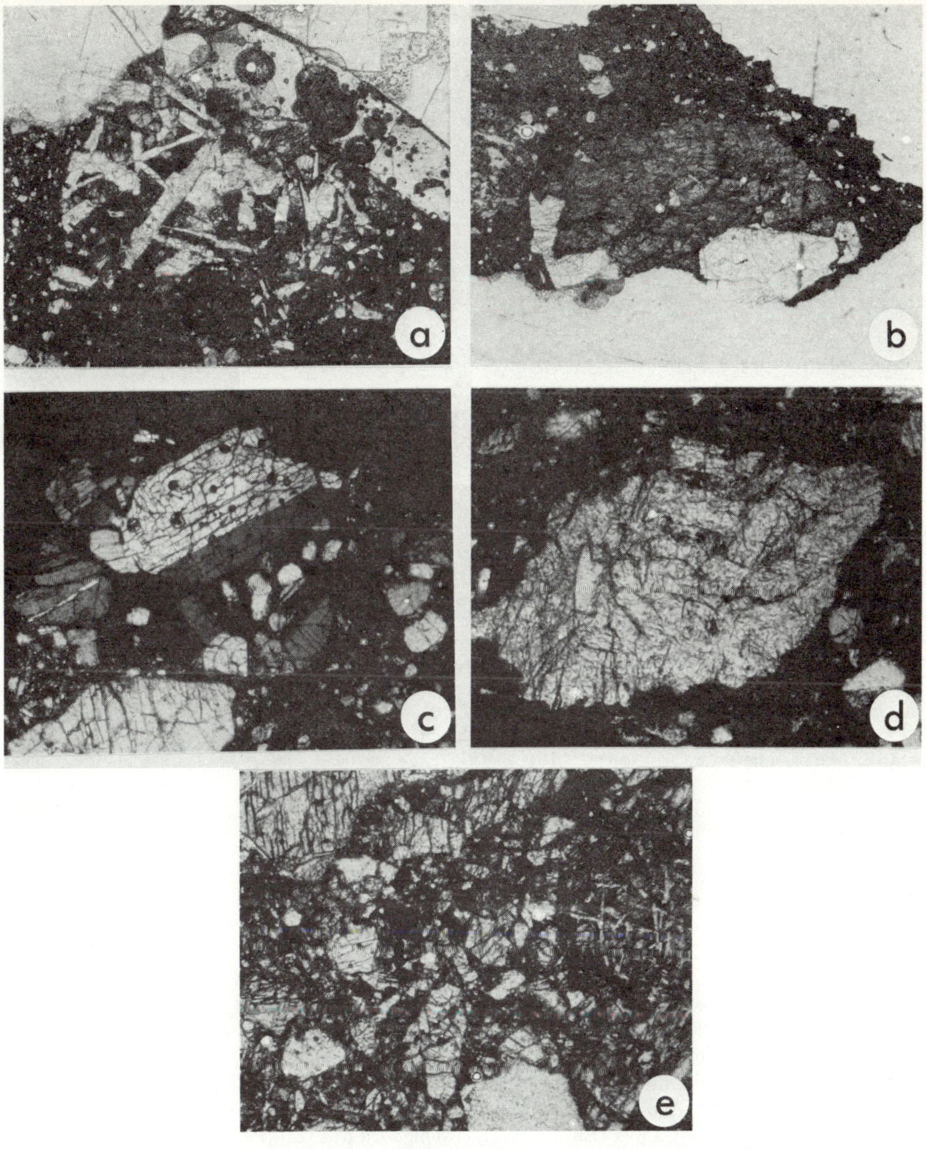

Fig. 5a–e. Photomicrographs of clasts described in text: (a) 15344 KREEP basalt clast—light laths are plagioclase with interstitial darker pyroxene and mesostasis. Visicular glass rind is at top and dark matrix at bottom (max dimension = 6 mm); (b) Norite clast 15689—light colored plagioclase with large single grain of darker pyroxene in dark matrix (max dimension = 6 mm); (c) Large euhedral plagioclase in 15364 (top center) surrounded by olivine (at extinction) with numerous smaller plagioclase inclusions. Top of photo is epoxy, bottom is breccia matrix (max dimension = 1.2 mm); (d) Ultrabasic clast in 15360—nearly all olivine with one grain of plagioclase (light) (max dimension = 0.7 mm); (e) pyroxenite(?) clast in 15164 composed mostly of pyroxene and olivine but large clast at top-left contains plagioclase (not shown). Compositions of mineral clasts within breccia matrix are the same as minerals in clast. Note igneous textured clast at center-right suggesting that breccia is actually polymict (max dimension = 1.2 mm). All photos in plain light except (c) which is with polars crossed.

Table 3. Selected analyses of phases in Apollo 15 clasts 15344, 15689B, 15364A, 15360A, and 15146.

	15344 KREEP basalt					15689B norite						
	(1)	(2)	(3)	(4)	(5)	(6)	(7)	(8)	(9)	(10)	(11)	(12)
Na_2O	0.04	1.86	0.01	0.39	0.01	0.03	0.20	0.29	1.01	2.6	na	0.02
MgO	11.1	na	27.9	0.05	1.40	26.1	15.4	14.5	0.04	nd	1.33	5.92
Al_2O_3	1.24	33.0	2.40	11.7	4.57	1.30	1.92	4.23	34.8	12.9	1.44	21.2
SiO_2	49.2	48.5	53.5	79.0	0.08	55.4	53.0	51.5	46.6	77.2	0.23	0.10
K_2O	0.04	0.14	0.01	6.26	nd	nd	nd	0.03	0.07	3.6	na	0.01
CaO	10.5	16.3	1.55	0.81	0.10	1.75	19.9	22.1	19.2	0.5	3.82	0.07
TiO_2	1.03	na	0.73	0.59	22.0	0.76	1.52	2.22	0.03	nd	60.4	1.51
Cr_2O_3	0.32	na	0.90	nd	19.5	0.81	0.80	1.36	na	nd	12.9	45.6
MnO	0.40	na	0.20	0.02	0.38	0.29	0.14	0.13	na	nd	0.07	0.22
FeO	25.0	0.39	12.6	1.61	49.1	14.3	7.62	5.45	0.05	nd	8.72	26.8
NiO	0.01	nd	nd	0.02	0.05	0.03	nd	0.06	na	nd	na	nd
Σ	98.88	100.19	99.80	100.45	97.19	100.77	100.50	101.87	101.80	97.5	93.66	101.45

	15364A				15360A				15146				
	(13)	(14)	(15)	(16)	(17)	(18)	(19)	(20)	(21)	(22)	(23)	(24)	(25)
Na_2O	na	nd	0.57	nd	0.01	0.10	na	nd	nd	nd	nd	nd	nd
MgO	36.7	26.5	0.06	2.5	33.7	18.0	49.6	8.9	3.5	18.9	3.6	16.4	28.2
Al_2O_3	0.01	1.27	36.6	12.0	0.90	1.72	0.03	13.0	nd	0.7	5.0	1.9	nd
SiO_2	37.6	54.0	44.2	0.2	57.8	53.6	41.3	0.3	nd	52.5	0.2	51.6	35.5
K_2O	na	nd	0.05	nd	0.01	0.01	na	nd	nd	nd	nd	nd	nd
CaO	0.10	1.85	19.5	0.5	2.20	23.0	0.03	0.5	nd	5.4	0.2	15.0	0.2
TiO_2	0.04	0.64	0.04	5.1	0.35	0.79	0.04	0.5	56.4	0.8	26.0	1.1	nd
Cr_2O_3	0.01	0.37	na	45.7	0.40	0.70	0.03	55.9	0.5	0.1	17.2	0.5	nd
MnO	0.27	0.24	na	nd	0.09	0.09	0.06	0.4	nd	0.1	0.3	nd	0.2
FeO	25.4	14.9	0.17	32.9	4.65	1.83	7.10	20.3	41.3	21.1	49.0	13.3	34.6
NiO	nd	0.02	na	nd	nd	0.01	0.04	nd	nd	nd	nd	nd	nd
Σ	100.13	99.79	101.19	98.9	100.11	99.85	98.23	99.8	101.7	99.6	101.5	99.8	98.7

(1) High-Ca pyx ($En_{34}Fs_{43}Wo_{23}$); (2) plagioclase (An_{79}); (3) low-Ca pyx core ($En_{77}Fs_{20}Wo_3$); (4) silica-rich glass in mesostasis; (5) Cr-ulvöspinel; (6) low-Ca pyx ($En_{74}Fs_{23}Wo_3$); (7) high-Ca pyx (matrix) ($En_{45}Fs_{13}Wo_{42}$); (8) high-Ca pyx (vein) ($En_{44}Fs_9Wo_{47}$); (9) maskelynite (An_{91}); (10) K–Na–Al–Si phase in vein, BaO = 0.7 (EDS); (11) "armalcolite" phase in vein. $ZrO_2 = 4.75$, REE present but not analyzed; (12) chromite in vein; (13) olivine (Fo_{72}) poikilitically enclosing euhedral plagioclase; (14) low-Ca pyx ($En_{73}Fs_{23}Wo_4$); (15) euhedral plagioclase (An_{97}); (16) chromite with euhedral plagioclase (EDS); (17) low-Ca pyx ($En_{89}Fs_7Wo_4$); (18) high-Ca pyx ($En_{51}Fs_3Wo_{46}$); (19) olivine (Fo_{91}); (20) chromite (EDS); (21) ilmenite (EDS); (22) low-Ca pyx ($En_{55}Fs_{34}Wo_{11}$) (EDS); (23) Cr-ulvöspinel (EDS); (24) high-Ca pyx ($En_{47}Fs_{22}Wo_{31}$) (EDS); (25) olivine (Fo_{59}) (EDS). nd = not detected; na = not analyzed; EDS = energy dispersive system; pyx = pyroxene.

calculated values may be in error depending on the relative abundance of REE-bearing armacolite and phosphate. Unfortunately, relevant data for armalcolite are not known to extend such calculations.

(d) *Clast 15364A with euhedral plagioclase.* Figure 5c illustrates the unique euhedral, chemically unzoned, form of plagioclase enclosed poikilitically by olivine; we have not previously observed this feature in lunar rocks on such a relatively large scale. Free growth of the plagioclase before the olivine is indicated. The composition of the plagioclase (low Fe, high Ca) indicates affinities to the ANT group of rocks while the rather high Ca content of the pyroxene is common in mare basalts. The remaining portion of 15364 is similar mineralogically to the clast shown in Fig. 5c and thus appears to be a monomict breccia, but this portion does not have euhedral plagioclase grains. Ti-bearing chromite (Table 3, analysis 16) is present both in the clast and in the breccia matrix indicating a complication with the ANT group assignment based on plagioclase because Ti-chromite is characteristic of mare basalts. Sample 67075 has similar chemical features (Ti-chromite, low-Fe plagioclase) possibly indicating genetic similarities.

The above chemical data are ambiguous in that 15364A has features of both the ANT and mare basalt rock types. The possibility that this sample might result from annealing of mare-type basalt raises the difficult problem of recognition of such rocks. Although most samples of mare basalt must derive from the late lava flows on the surfaces of mare basins, there is no reason why some mare-type basalts should not have annealed to gabbroic rocks either in magma chambers in the crust, or perhaps even in the mare basins themselves. Ryder and Taylor (1976) discussed the possibility of early mare-type volcanism, but did not discuss the possible effects of associated plutonism or metamorphic recrystallization.

(e) *Clast 15360A with ultrabasic affinity.* We have previously described many examples of "dunitic" or "troctolitic" fragments which are characterized by Mg-rich olivine (e.g. Steele and Smith, 1975). Clast 15360A (Fig. 5d) is an additional example but with somewhat different chemical features. The high Mg content of the olivine (Fo_{91}) and the very low Ca content (~ 0.03 wt.%) matches only with the olivine from spinel cataclasite 15445 (Steele and Smith, 1975) and 73263,1,11 (Bence *et al.*, 1974); the common "dunitic" olivine (Fo_{88}) has CaO $\sim$0.075 wt.%. Unlike 15445 and 73263,1,11, a Mg–Al spinel is not present although chromite is, and the Al_2O_3 content of the low-Ca pyroxene is much lower; these suggest that 15360A is unique. Plagioclase (An_{93}) is present in 15360A. The clast is too small to deduce textural relations, but all phases show shock effects. One cannot rule out meteoritic contamination, but there is nothing "non-lunar" about the fragment; it appears to be an additional bit of evidence for a diverse ultrabasic suite on the moon.

(f) *Clast 15146—pyroxenite* (?) (Fig. 5e). This clast is composed of five phases—pyroxene ($En_{55}Fs_{34}Wo_{11}$), plagioclase (An_{85-90}), olivine (Fo_{59}), ilmenite, and Cr-ulvöspinel (Table 3—analyses 21–25)—which could represent a typical mare-basalt assemblage. It was previously noted that the plagioclase was low in Fe, unlike mare basalt samples, and that the pyroxenes, like in "ANT" samples, did not show a wide range of composition; however, the pigeonitic composition does not indicate low temperature equilibration as is found in most ANT samples. These chemical features suggest that 15146 is apparently unique and may represent a deviant from the normal mare basalt crystallization products. It is also notable that the bulk chemistry would closely correspond to "green glass" composition (e.g., Warner *et al.*, 1972) if the mode is low-Ca pyroxene ≈55%, plagioclase ≈20%, and olivine ≈25%—close to that visually estimated in the clast (~65% pyx, ~15% pl, ~15% ol; accuracy poor due to coarse grain size). However, the TiO_2 content of "green glass" (~0.4 wt.%) is too low to allow the ~1% of ilmenite observed in the clast which together with the silicates would give ~1 wt.% bulk TiO_2. Further data are needed (e.g., REE abundances) to test the origin of this clast, but it may turn out to be a crystalline relative to "green glass". The low TiO_2 content of this clast might indicate a relationship to the VLT (very low Ti) mare basalts described by Vaniman and Papike (1977) and Taylor *et al.* (1977), but details such as the Fe content of the plagioclase and mineral zoning are different. This clast as well as other rare samples of mare-type basalts are an indication of a diverse suite of mare-type basalts in contrast to those collected on the mare surfaces.

(g) *Exsolved pyroxene clasts.* Twenty-five small, exsolved pyroxene grains were observed in these breccias, especially those which contained some dark matrix. Special care was made to analyze (by EDS) the lamellae in each as well as to obtain a bulk composition by using a large, but stationary, beam and applying matrix corrections assuming a homogeneous material. Individual lamellae ranged from about 15 μm to the limit of optical resolution. X-ray scanning images showed only one set of lamellae in all grains examined.

The major element variations are shown on a pyroxene quadrilateral (Fig. 6). The maximum *mg* of the bulk composition is ~0.74 and the range extends to very Fe-rich compositions. Based on just bulk pyroxene compositions the trend is similar to that observed in other Apollo 15 mare samples which show no coarse exsolution (e.g., Dowty *et al.*, 1974) and a possible interpretation is that these grains represent a more slowly cooled mare basalt. Another feature on Fig. 6 is the suggestion of a gap in the bulk trend at an *mg* of ~0.60 possibly indicating the beginning of plagioclase crystallization. Dowty *et al.* (1974) noted abrupt changes in pyroxene trends at a similar composition and attributed this to plagioclase crystallization. A similar gap was noted by McCallum *et al.* (1975) in

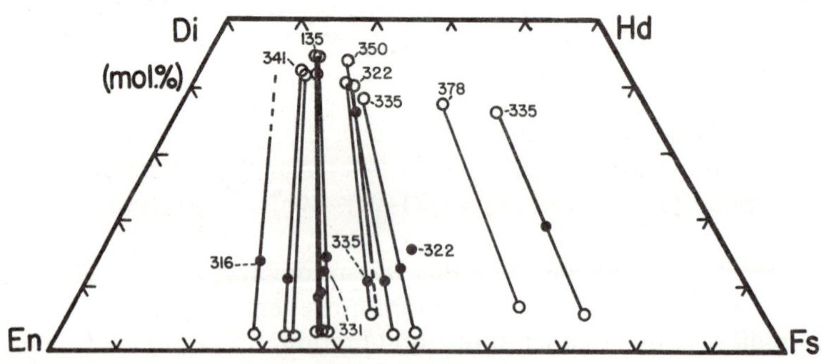

Fig. 6. Pyroxene quadrilateral showing compositions of bulk (solid circles) and high- and low-Ca lamellae (open circles) of exsolved pyroxene grains. Numbers give last three digits of breccia listed in Table 1.

67075 pyroxene compositions but at a more Fe-rich composition. These chemical similarities among the pyroxenes might suggest similar crystallization histories except for cooling rates.

Very commonly, tiny opaque phases (ilmenite and Ti-bearing chromite) are oriented at high angles to the pyroxene lamellae (Fig. 7a,b) and always centered on the high-Ca lamellae. Apparently, after pyroxene exsolution began, minor elements such as Ti, Cr, and Al which are always enriched in high-Ca relative to low-Ca pyroxene, migrated to form the oriented opaques. Okamura *et al.* (1976) described similar features but on a coarser scale in 67075 and commented on the mechanism of formation. Again this suggests a similarity between the thermal history of these exsolved pyroxene grains and those in 67075.

Conclusions

The clasts recognized in these Apollo 15 breccias provide additional evidence for the common occurrence of KREEP basalts at the Apollo 15 site and contamination from the nearby mare regions. The unusual clasts such as norite and the ultrabasic sample provide additional samples of a poorly defined group of rocks from the moon which could represent early crystallizing products. Some evidence, mostly from conflicting chemical features and exsolved

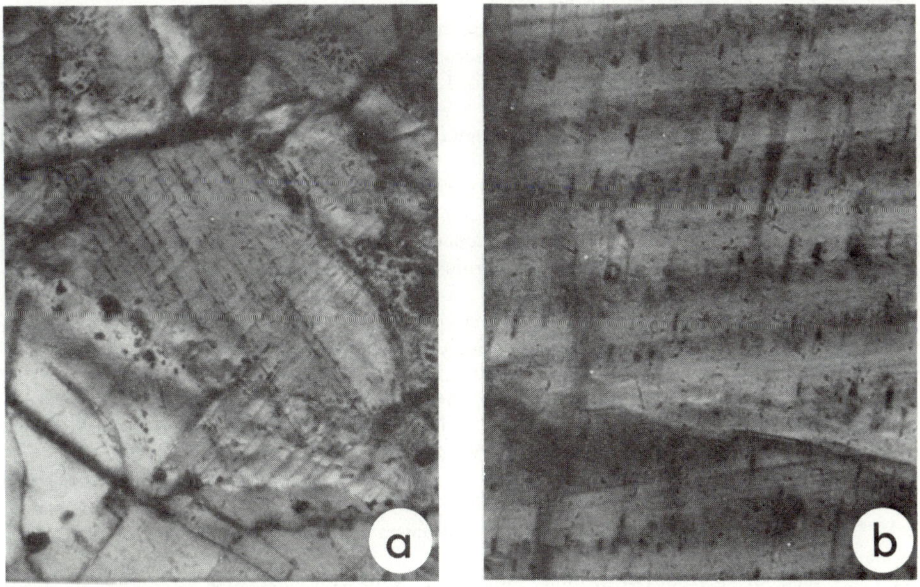

Fig. 7a–b. Photomicrographs of exsolved pyroxene grains showing details of opaque exsolution: (a) Grain in 15331. Trace of pyroxene lamellae is NE–SW while trace of opaques is NW–SE. Note that opaques are not continuous but are centered on high-Ca lamellae (max dimension = 0.23 mm); (b) Grain in 15378. Trace of pyroxene is E–W while that of opaques is N–S. Again note concentration of opaques on high lamellae (max dimension = 0.1 mm, oil immersion).

pyroxene grains indicates that some mare-type samples are represented by metamorphosed or plutonic samples in contrast to the abundant near-surface lava flows. One sample having mare basalt affinities also might provide a link between the abundant "green glass" and a crystalline rock.

Acknowledgments—We thank I. Baltuska, O. Draughn, and T. Solberg for technical help, I. McCallum for review, and NASA for grant 14-001-171 Res. This work is sponsored in part by the Lunar Science Institute, operated by the Universities Space Research Association under contract No. NSR-09-051-001 with the National Aeronautics and Space Administration. This paper constitutes the Lunar Science Institute Contribution No. 287.

References

Basu A. and Bower J. F. (1976) Petrography of KREEP basalt fragments from Apollo 15 soils. *Proc. Lunar Sci. Conf. 7th*, p. 659–678.

Bence A. E., Delano J. W., Papike J. J., and Cameron K. L. (1974) Petrology of the highlands massifs at Taurus-Littrow: An analysis of the 2–4 mm soil fraction. *Proc. Lunar Sci. Conf. 5th*, p. 785–827.

Dowty E., Keil K., and Prinz M. (1974) Lunar pyroxene-phyric basalts: Crystallization under supercooled conditions. *J. Petrol.* **15**, 419–453.

Dowty E., Keil K., Prinz M., Gros J., and Takahashi H. (1976) Meteorite-free Apollo 15 crystalline KREEP. *Proc. Lunar Sci. Conf 7th*, p. 1833–1844.

Haggerty S. E. (1973) Armalcolite and genetically associated opaque minerals in the lunar samples. *Proc. Lunar Sci. Conf. 4th*, p. 777–797.

Irving A. J. (1975) Chemical, mineralogical and textural systematics of non-mare melt rocks: Implications for lunar impact and volcanic processes. *Proc. Lunar Sci. Conf. 6th*, p. 363–394.

Irving A. J., Steele I. M., and Smith J. V. (1974) Lunar noritic fragments and associated diopside veins. *Amer. Mineral.* **59**, 1062–1068.

McCallum I. S. and Mathez E. A. (1975) Petrology of noritic cumulates and a partial melting model for the genesis of Fra Mauro basalts. *Proc. Lunar Sci. Conf. 6th*, p. 395–414.

McCallum I. S., Okamura F. P., and Ghose S. (1975) Mineralogy and petrology of sample 67075 and the origin of lunar anorthosites. *Earth Planet. Sci. Lett.* **26**, 36–53.

Nichiporuk W., and Chodos A. A. (1959) The concentration of vanadium, chromium, iron, cobalt, nickel, copper, zinc, and arsenic in the meteoritic iron sulfide nodules. *J. Geophys. Res.* **64**, 2451, 2463.

Okamura F. P., McCallum I. S., Stroh J. M., and Ghose S. (1976) Pyroxene-spinel intergrowths in lunar and terrestrial pyroxenes. *Proc. Lunar Sci. Conf. 7th*, p. 1889–1899.

Reed S. J. B. and Ware N. G. (1975) Quantitative electron microprobe analysis of silicates using energy-dispersive X-ray spectrometry. *J. Petrol.* **16**, 499–519.

Reid A. M., Richardson S. H., and Duncan A. R. (1977) Non-mare "igneous" clasts in Apollo 15 breccias (abstract). In *Lunar Science VIII*, p. 801–803. The Lunar Science Institute, Houston.

Ryder G. and Bower J. F. (1977) Petrology of Apollo 15 black-and-white rocks 15445 and 15455 — fragments of the Imbrium impact melt sheet? *Proc. Lunar Sci. Conf. 8th*. This volume.

Ryder G. and Taylor G. J. (1976) Did mare-type volcanism commence early in lunar history? *Proc. Lunar Sci. Conf. 7th*, p. 1741–1755.

Smith J. V. and Steele I. M. (1976) Lunar mineralogy: A heavenly detective story. Part II. *Amer. Mineral.* **61**, 1059–1116.

Steele I. M. (1975) Mineralogy of lunar norite 78235; second lunar occurrence of $P2_1ca$ pyroxene from Apollo 17 soils. *Amer. Mineral.* **60**, 1086–1091.

Steele I. M. and Smith J. V. (1975) Minor elements in lunar olivine as a petrologic indicator. *Proc. Lunar Sci. Conf. 6th*, p. 451–467.

Steele I. M. and Smith J. V. (1977) Apollo 15 breccia rake samples—description and mineralogy of lithic clasts and exsolved pyroxene grains (abstract). In *Lunar Science VIII*, p. 899–901. The Lunar Science Institute, Houston.

Steele I. M., Smith J. V., and Grossman L. (1972a) Mineralogy and petrology of Apollo 15 rake samples: II. Breccias (abstract). In *The Apollo 15 Lunar Samples* (J. W. Chamberlain and C. Watkins, eds.), p. 161–164. The Lunar Science Institute, Houston.

Steele I. M., Smith J. V., and Grossman L. (1972b) Mineralogy and petrology of Apollo 15 rake samples: I. Basalts (abstract). In *The Apollo 15 Lunar Samples* (J. W. Chamberlain and C. Watkins, eds.) p. 158–160. The Lunar Science Institute, Houston.

Taylor G. J., Keil K., and Warner R. D. (1977) Very low Ti mare basalts. *Geophys. Res. Lett.* **4**, 207–210.

Vaniman D. T. and Papike J. J. (1977) The Apollo 17 drill core: Chemistry and stratigraphy of monomineralic fragments and the discovery of a new *very low* Ti (VLT) mare basalt (abstract). In *Lunar Science VIII*, p. 952–954. The Lunar Science Institute, Houston.

Warner J., Ridley W. I., Reid A. M., and Brown R. W. (1972) Apollo 15 glasses and the distribution of non-mare crustal rock types (abstract). In *The Apollo 15 Lunar Samples* (J. W. Chamberlain and C. Watkins, eds.) p. 179–181. The Lunar Science Institute, Houston.

Proc. Lunar Sci. Conf. 8th (1977), p. 1943–1966.
Printed in the United States of America

The petrology and geochemistry of impact melts, granulites, and hornfelses from consortium breccia 61175

STEPHEN R. WINZER[1], D. F. NAVA[2], M. MEYERHOFF[1], D. J. LINDSTROM[3], R. K. L. LUM[3], M. M. LINDSTROM[3], P. SCHUHMANN[3], S. SCHUHMANN[2], AND J. A. PHILPOTTS[2]

[1]Martin Marietta Laboratories, 1450 S. Rolling Road, Baltimore, Maryland 21227
[2]Astrochemistry Branch Laboratory for Extraterrestrial Physics, Goddard Space Flight Center, Greenbelt, Maryland 20771
[3]Department of Geology, University of Maryland, College Park, Maryland 20742

Abstract—Clasts and matrix from consortium breccia 61175 have been analyzed for major, minor, and trace elements. Detailed petrographic analyses of 58 clasts have been completed. The matrix of 61175 is anorthositic in composition, with lithophile trace element abundances from 20 to 40× chondrites. Clasts comprise impact melt rocks, xenocryst and xenolith-free VHA and anorthositic basalts, anorthosite, and ANT granulites and hornfelses. The VHA and anorthositic basalts are interpreted as impact melts, and all are probably formed by impact into anorthositic or anorthositic gabbro target rocks and their metamorphic equivalents. Hornfelses were probably formed by incorporation of breccias or pre-existing melt rocks into a melt sheet prior to cooling. Melt-rock lithophile trace element abundances range from very low, with patterns similar to the anorthosites, to higher abundances with patterns resembling Apollo 16 KREEP. This range may be indicative of more than one melt sheet.

INTRODUCTION

LUNAR SAMPLE 61175 is a light matrix polymict breccia collected at Station 1 near the rim of Plum Crater, Apollo 16 landing site. As described and classified by LSPET (1972) it is one of a small number of samples containing the widest variety of clasts of any breccia type returned from the site (Fig. 1). Because of the variety of rock types present as clasts, this rock was designated for consortium study, and, early in 1976, such a consortium effort was organized. The Principal Investigators of the consortium are E. Anders, J. Geiss, D. F. Nava, O. A. Schaeffer, and M. Tatsumoto. The consortium study is being directed by the senior author.

61175 is of more than usual interest because of the variety of rock types present as clasts, and because at least three major processes are indicated by the clasts. Major rock types include extrusive and plutonic igneous rocks, impact melts, and fine- and coarse-grained metamorphic rocks (hornfelses and granulites). Almost all rock types described as single specimens or in breccias and rake samples from other stations at the Apollo 16 site are represented in the clast population of 61175. This paper describes the major rock types found in 61175, and discusses their petrologic implications. In addition, major, minor, and lithophile trace element data are presented for the bulk rock and two major clast types separated and designated for detailed consortium study (age, isotopic systematics, and volatile trace element analysis).

MATRIX

The friable matrix of 61175 consists chiefly of comminuted mineral fragments, from several millimeters down to several microns in size. The mineral fragments are held together by a network of glass, which probably comprises less than 5% of the total mass of the boulder (see Fig. 2). The glass is compositionally heterogeneous with matrix network glasses occasionally showing elevations in Na and K contents relative to Na and K contents in mineral fragments or lithic clasts.

Table 1 presents modal data for five polished electron microprobe mounts made from large (several grams each) chips resulting from further cutting of 61175,19. Modes from the five mounts reflect the variability in distribution and size of lithic clasts of the major rock types present. Grouped under breccias are a range of rock types including cataclastic ANT breccias, melt rocks, and the fine-grained metamorphic rocks (hornfelses). The latter are included among the breccias because, apart from their recrystallized, equilibrated matrices, they closely resemble the melt rocks optically, and cannot always be identified with certainty in the absence of chemical information. Likewise, granulites, which comprise the bulk

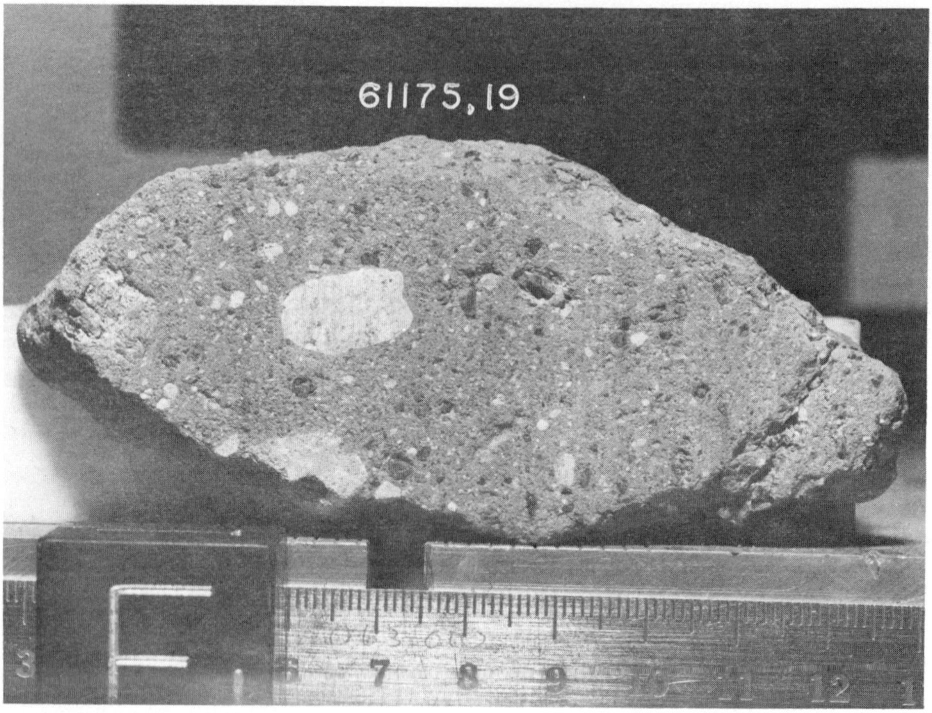

Fig. 1. Photograph of 61175,19. This is the original slab, taken from the center of 61175. The large white clast is a moderately shocked anorthosite. Smaller crystalline clasts are ANT and granulites, while the small dark gray clasts are impact melts or hornfelses (NASA/JSC photo 5-73-22080).

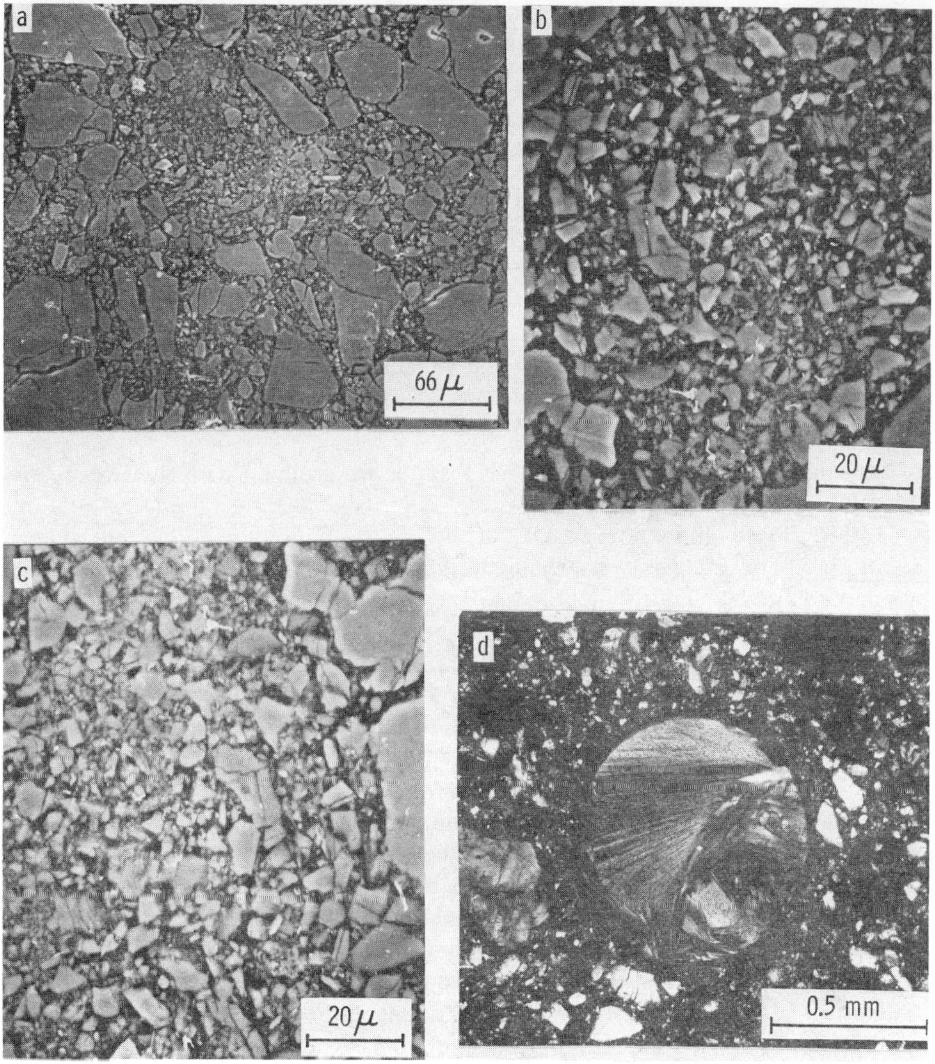

Fig. 2. SEM photomicrographs of polished thin sections of the matrix of 61175. (a) Mineral fragments (mainly plagioclase) and small lithic fragments (lower right corner) in fine-grained groundmass. (b) Higher magnification photo of relatively glass-free portion of the fine-grained matrix. Again, many of the fragments are plagioclase, but a small proportion in this size range are glass. (c) Higher magnification photomicrograph of a more glass-rich portion of the matrix. These glassy areas interlock together to form a network which holds the rock together. (d) Droplet chondrule in 61175 matrix. The chondrule is filled with plagioclase.

Table 1. Modal analysis of thin sections from 61175.

Sample #	61175,9	61175,101	61175,104	61175,95	61175,98	(Average all sections)
01	4.6	3.9	6.1	2.1	4.5	4.4
Plag	66.8	77.7	68.2	52.8	72.5	67.7
Opx	0.2	0.6	0.06	—	0.5	0.2
Cpx	0.04	—	0.9	0.05	0.05	0.2
Ilm	0.04	0.03	0.3	0.11	0.05	0.15
Clasts						
Basalt	7.0	9.1	9.1	4.5	7.2	7.5
Breccia	14.3	5.1	11.2	37.3	12.2	14.8
Anorthosite	4.6	2.8	2.0	1.4	2.4	3.0
Norite	1.8	0.03	0.9	0.13	0.1	0.7
Troctolite	0.7	0.03	1.4	1.6	0.6	1.0
Number of points	6324	3877	5315	3785	1952	21253

of the coarse-grained rocks of the ANT suite, are grouped with their respective unmetamorphosed ANT suite rocks.

Lithic clasts comprise up to 45% of the breccia boulder sections, but average about 20%. These clasts were defined somewhat arbitrarily on the basis of the number of mineral grains included in the clast, thus coarse-grained fragments with only three or four crystals were not classified as rock types, whereas fine-grained lithic fragments 0.5–1 mm in size were often considered classifiable. This arbitrary classification tends to bias the mode toward finer-grained rock types, as the coarse-grained ANT rock types would have to be several millimeters in largest dimension to be classifiable. Another caveat to be considered when using these modes is the distinction between basalts and impact melts. All rocks classified as impact melts contain mineral and/or lithic fragments, some of which have been variably shocked. There are clasts, however, which are texturally and mineralogically similar to the impact melts found in 61175, but which contain no lithic or mineral fragments. As the textures of the fragment free rocks also resemble those of some lunar basalts, they cannot be classified as impact melts with certainty and hence are grouped with the basalts.

Optical, SEM and SEM/EDS examination of the matrix of 61175 suggests several conclusions. First, the matrix is composed of comminuted mineral and lithic fragments of several different rock types, including at least one apparently not represented in the clast populations. Coarse spinel fragments, up to several millimeters in longest dimension, occur scattered throughout the matrix of 61175. Such large spinels are not found in the coarse-grained clast suite. Coarse-grained spinel troctolites are described from other Apollo 16 rocks (Prinz *et al.*, 1973), thus it is possible that this rock type may be the source of these spinels. Second, the lithification temperature of the breccia was not high. Evidence against high temperature lithification includes lack of sintering of mineral and glass fragments, high porosity, lack of divitrification of included glasses and absence of reaction rims around included clasts and mineral fragments. Third, the matrix of 61175 probably contains a significant regolith component, as is suggested by the presence of glass shards and droplets and one or two well-formed chondrules, Fig. 2d. Fourth, the bulk matrix has lower lithophile trace element abundances than Apollo 16 KREEP and higher abundances than the anorthosites (Fig. 3).

Clasts

Lithic clasts range in size from less than 1 mm up to about 2 cm. The majority of the clasts are shocked and brecciated ANT rocks or melt rocks, with basalts, metamorphic rocks and igneous plutonic rocks present in decreasing order of abundance.

Because of the difficulties encountered in identifying basalts and melt rocks, the two are grouped together under the superheading "Rapidly cooled igneous rocks". Slowly cooled (plutonic) igneous rocks and their metamorphic equivalents and the hornfelses are discussed separately.

Rapidly Cooled Igneous Rocks

Basalts

The difficulty in differentiating basalts and melt rocks on the basis of texture and mineralogy has already been noted. Many clasts which texturally resemble lunar basalts have bulk and mineral compositions similar to the impact melts, but lack lithic clasts and mineral fragments commonly found in those rocks classified as impact melts. Where field relations are unknown, the distinction between

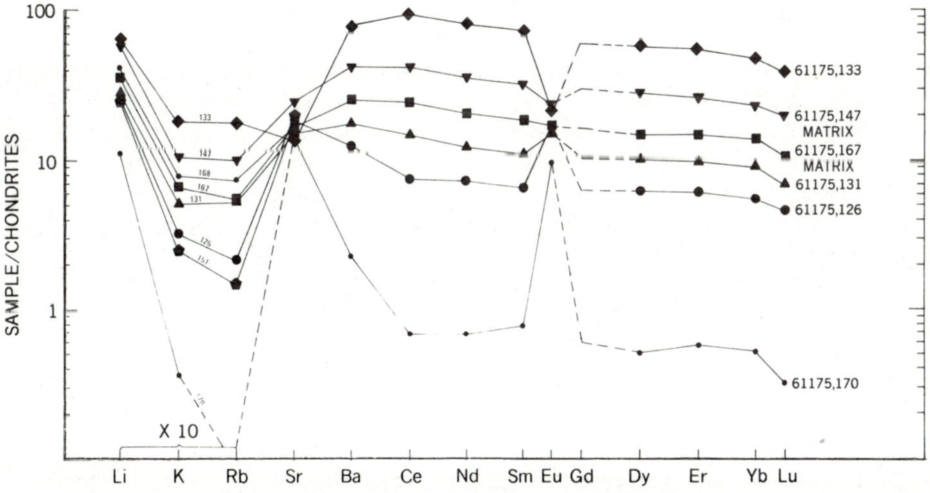

Fig. 3. Lithophile trace element abundances from clasts and matrix of 61175. 61175,133 and ,126 are dark clasts and are probably melt rocks. 61175,131 and ,170 are white clasts, and are moderately shocked anorthosites.

clast-free impact melts and internally derived extrusive rocks is probably impossible without extensive mineralogical, chemical, and isotopic data. The following guidelines have been used when classifying clasts from 61175: (1) Rocks which contained Fe metal with 4% or more Ni were classified as impact melts if they also contained xenocrysts and/or xenoliths. (2) Fine-grained rocks which contained magnesian spinel and magnesian olivine (Fo_{70} or greater) were classified as impact melts as this is an unlikely assemblage in basalts. It should be realized that all of these rocks may be impact melts and that application of the term basalt to these rocks is a convenience justified by their texture, but without genetic implications.

Figure 4 shows the range of textures and mineralogy found in basalt clasts from 61175. Clasts fitting the criteria outlined above show a considerable range in texture and mineralogy. Hyalocrystalline rocks consisting of euhedral plagioclase laths in glass of pyroxene composition (Fig. 4a) form one extreme, while holocrystalline varieties with a microdiabasic texture (Fig. 4b) form the opposite extreme. Most basalts are holocrystalline rocks containing plagioclase, clinopyroxene, orthopyroxene, olivine, and ilmenite (Fig. 4c). Most also exhibit a complex mesostasis which contains one or two pyroxenes, ilmenite, metallic iron, zircon or a mineral with a high zirconium content, sodic plagioclase, apatite, and occasionally K-feldspar or glass. All larger clasts with basaltic textures and mineralogy are high alumina, low titania basalts.

Pyroxene compositions from basalt clasts are presented in Fig. 5. Most pyroxenes are zoned, generally from low Ca, En_{55-60} to higher Ca, En_{45-50}. The most extreme zoning observed in a single crystal is from Wo_7En_{72} to $Wo_{14}En_{57}$, and the pyroxene compositions in that rock range from $Wo_7En_{72}Fs_{21}$ to $Wo_{28}En_{27}Fs_{45}$. Pigeonites are most commonly zoned while augites occur as smaller grains with restricted compositional ranges. Olivines are not significantly zoned. Typical zoning ranges from 2 to 4 mol.% (Fig. 5). Plagioclase compositions in all basalts generally range from An_{90-97}, with larger phenocrysts zoned up to 5 mol.% ($An_{97\rightarrow 92}$).

The chief differences among the basalt clasts analyzed appear to be in degree of crystallinity, texture, crystallization history, and the nature of the liquidus phase. Plagioclase is the principal liquidus mineral; however, in a few cases, olivine has clearly crystallized first on the basis of textural evidence. Plagioclase is most often followed by pyroxene or ilmenite but in some cases both crystallized together. The different crystallization histories create a textural series ranging from hyalocrystalline through intersertal to intergranular and, more rarely, subophitic. A few of the more coarsely crystalline basalts exhibit a diabasic texture, with microphenocrysts of zoned plagioclase (An_{97-92}) in a groundmass of euhedral plagioclase laths, subhedral pyroxene, olivine and ilmenite (Fig. 4c). The more coarsely crystalline rocks lack the complex mesostasis regions found in the fine-grained intersertal basalts.

Several conclusions can be made regarding the basalt component of 61175: (1) All the basalts appear to be the high Al_2O_3, low TiO_2 variety (when large enough to be classified reliably). Most fall into the field of anorthositic basalts as defined by Irving (1975). (2) They form two groups on the basis of crystallization history, those which crystallized plagioclase first and those which crystallized olivine first. (3) Their pyroxene and olivine compositions are either similar to or slightly more magnesian than the high Al_2O_3, low TiO_2 basalts described in other Apollo 16 samples (e.g., Delano, 1975). (4) The wide range of textures either indicates differing cooling histories or nucleation rates for the basalts present.

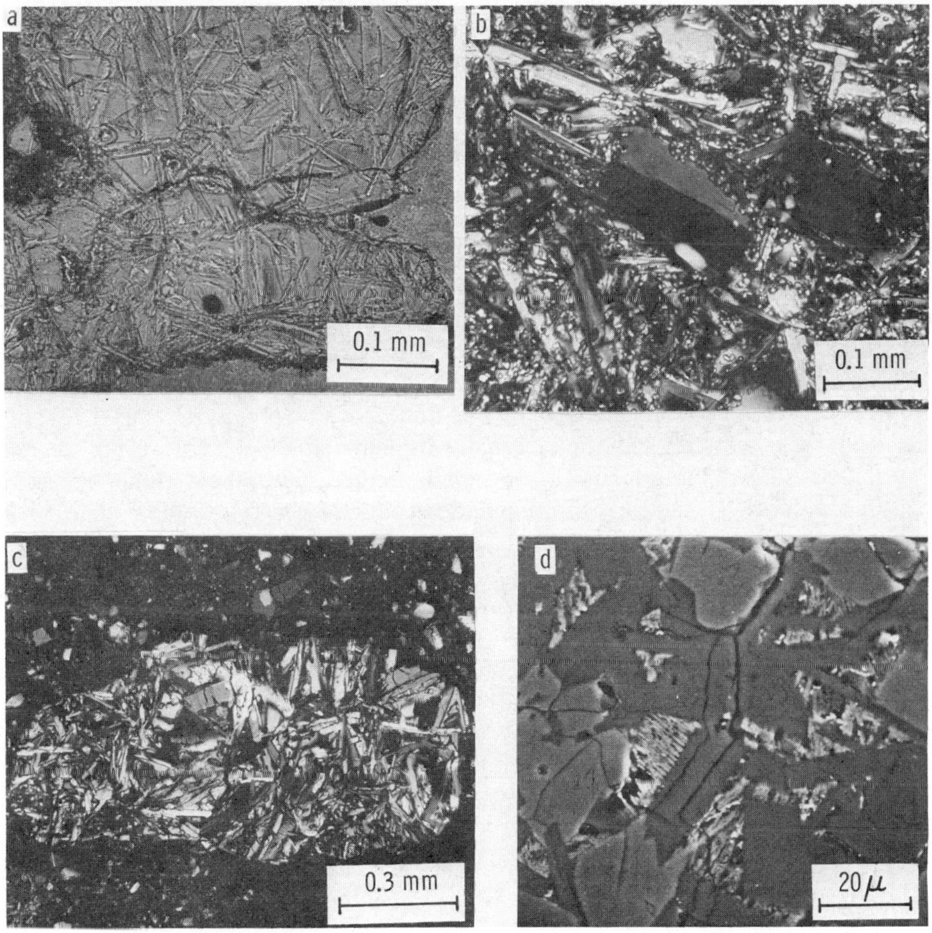

Fig. 4. Textures of "basalt" clasts. (a) Hyalocrystalline. Euhedral plagioclase laths (An_{95-97}) in random orientation in a glass matrix. The glass has a composition close to that of augite. (b) Microdiabase porphyry. Phenocrysts of plagioclase in a groundmass of euhedral to subhedral plagioclase and subhedral pyroxene. (c) Typical basalt texture. Euhedral, tabular plagioclase phenocrysts (zoned $An_{97\rightarrow 90}$) in a matrix of euhedral to subhedral plagioclase laths ($An_{90\rightarrow 95}$), low calcium pyroxene ($Wo_{5-8}En_{71-73}Fs_{20-24}$), and olivine ($Fo_{77-78}$), glass and minor ilmenite. (d) Scanning electron photomicrograph of a polished thin section of olivine basalt. This photomicrograph shows a typical mesostasis texture, containing pyroxene (darker gray), glass (lighter gray), ilmenite, and Fe metal (bright spots).

Impact melt rocks

By far the most common, and most variable, suite of clasts found in 61175 are the impact melt rocks. Impact melts range in texture and composition from glasses to holocrystalline rocks in which the groundmass texturally and mineralogically resembles the basalts (Fig. 6). Glass shards, or fragments of glass and maskelynite up to 5 mm in longest dimension are present (Fig. 6a). These glasses have compositions similar to those of plagioclase found in the clasts (Fig. 7). Other glasses occur as devitrified melt with few mineral fragments to devitrified fragment-laden melts (the fine-grained hyalocrystalline melt rocks) (Fig. 6b). The crystalline or partly crystalline melt rocks range in texture from wispy fine-grained hyalocrystalline melts with acicular to subhedral plagioclase, often with glass cores and familiar "swallow tail" configuration (Fig. 6c) to fragment-laden microdiabases with euhedral feldspar phenocrysts and a groundmass of euhedral to subhedral plagioclase, low calcium pyroxene, pigeonite, and olivine. The plagioclases occur as either subradial growths or as subaligned to aligned laths suggestive of flowage (Fig. 6b). Many melt rocks contain opaque phases which are concentrated in patches (Fig. 6d) of anhedral grains of Fe (up to 9% Ni), troilite, ilmenite, and occasionally a phosphide (schreibersite?).

Xenocrysts include plagioclase, olivine and, more rarely, pyroxene. A few of the melt rocks contain xenoliths, usually as very small polycrystalline aggregates. One such xenolith has an annealed texture, but others retain igneous textures. Xenocryst and xenolith populations range from 0% to more than 50% of the total clast.

Pyroxene and olivine compositions from melt-rock clasts are presented in Fig. 8. Pyroxene compositions generally overlap those of the basalts found as clasts in 61175, as do the olivines. In most melt rocks, groundmass pyroxenes and olivines are quite magnesian [Fo_{70} or higher (to Fo_{95}), En_{70-80} for low Ca

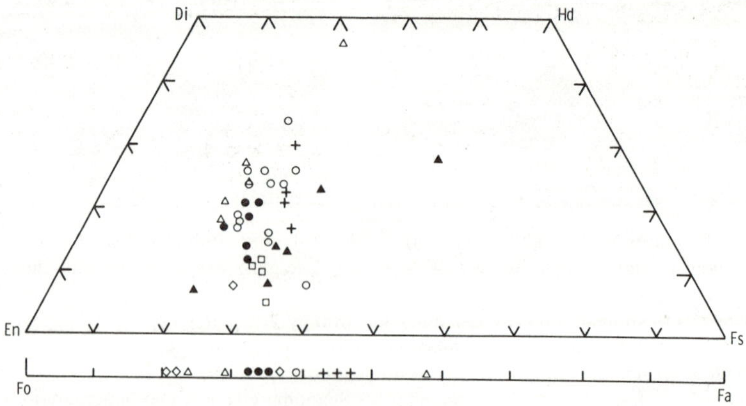

Fig. 5. SEM/EDS analyses of pyroxenes and olivine from basalt clasts. Different symbols represent analyses of minerals from different clasts. Most points represent several analyses.

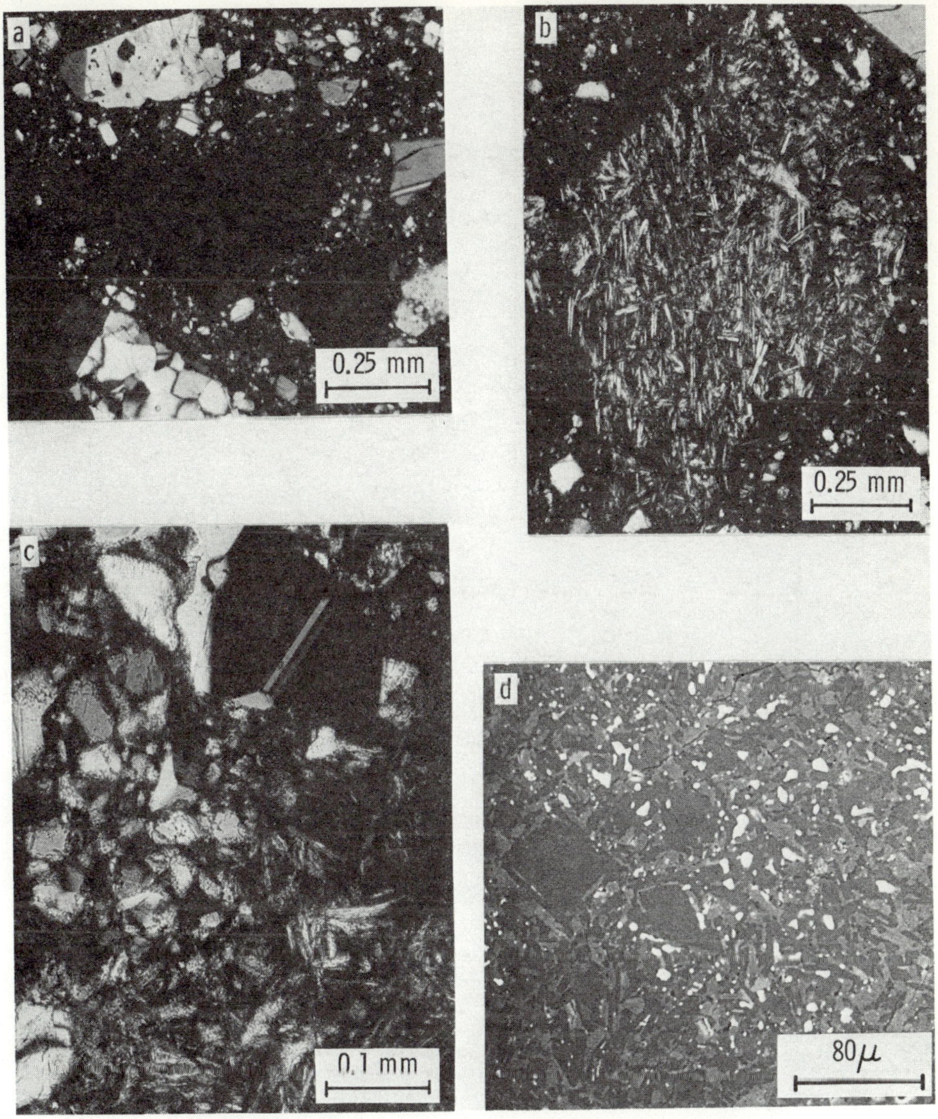

Fig. 6. Textures of melt rocks from 61175. (a) Maskelynite grain in matrix of 61175. (b) Coarse-grained melt with intersertal texture. Plagioclase laths are aligned, indicative of flowage in the melt. Xenocrysts (longer subhedral to anhedral grains) of plagioclase ($\sim An_{95}$) and one small portion of an annealed xenolith (near the top of the clast) mark this as a melt rock. The melt matrix consists of plagioclase ($An_{83\rightarrow95}$, both zoned and unzoned) pyroxene ($Wo_5En_{59}Fs_{36}$), olivine (Fo_{72-75}) and glass. (c) Fine-grained, clast-laden melt rock. An annealed ANT clast (upper right) with a developing hornfels texture is enclosed in a fine-grained matrix of plagioclase (An_{88-95}), pyroxene ($Wo_{17-31}En_{63-47}Fs_{18-20}$), and olivine ($Fo_{40-49}$). Pyroxenes with En contents as low as 33 also occur in this rock. These and the iron-rich olivines may be xenocrysts. (d) Scanning electron photomicrograph of a metal-rich portion of a melt-rock clast. Dark gray grains are phenocryst and groundmass plagioclase, lighter grays are olivine, pyroxene, and glass. Bright anhedral blebs are Fe metal with Ni contents of up to 9 wt.%, troilite and minor phosphide. Subhedral to euhedral lath-shaped grains are ilmenite.

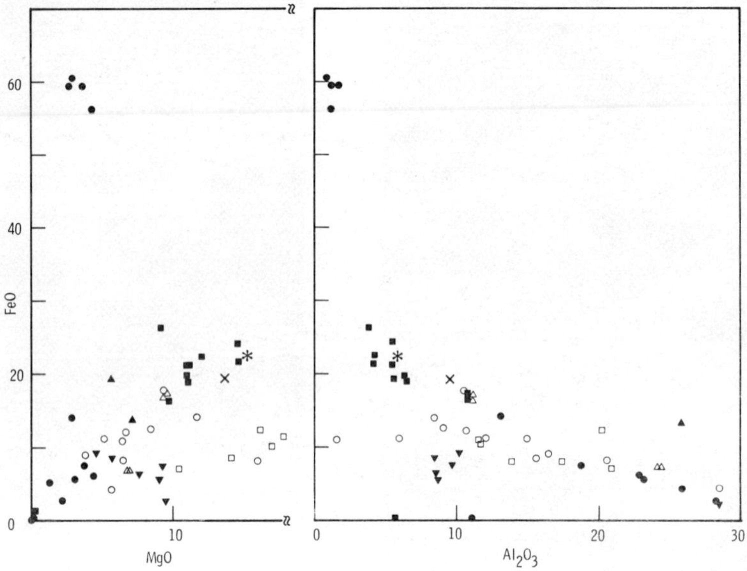

Fig. 7. SEM/EDS analyses of glasses from melt rocks and matrix of 61175. Different symbols indicated different clasts. *Composition of orange glass (74220) (Nava, 1974, *Proc. Lunar Sci. Conf. 5th*, p. 1087–1096) × Average Composition of Apollo 16 glasses, from Glass, 1976, *Proc. Lunar Sci. Conf. 7th*, p. 679–693. Filled circles represent glass from a breccia of mare basalt composition. The range of glass compositions is a result of differences in degree of crystallization, liquidus temperatures and crystallization histories of melt rocks found as clasts in 61175.

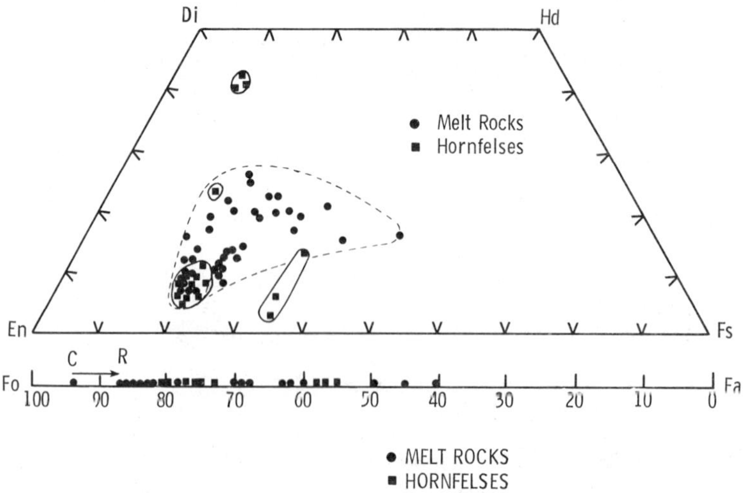

Fig. 8. SEM/EDS analyses of pyroxenes and olivines in melt rocks and hornfelses from 61175.

pyroxene]. Plagioclases generally range from An_{88} to An_{95}, but sodic plagioclase occurs occasionally. K-feldspars, apatite, and whitlockite are rare or absent.

Accessory phases include magnesian spinel, chromite, troilite, Fe(Ni) metal, schreibersite, ilmenite, armalcolite, a zirconium mineral, and rutile. Accessory phases are not always found in the mesostasis, but may occur as individual grains scattered throughout the rock, or in aggregates concentrated in certain areas of a clast. The texture of some of these aggregate areas suggests that they result from breakdown and melting of xenoliths, thus some of the accessory phases may not have crystallized from the melt formed by the impact.

Plutonic Rocks and Metamorphic Equivalents

Few pristine plutonic igneous clasts are found in 61175. Most rocks of the ANT suite are recrystallized or brecciated, thus obscuring original textures and making rock classification difficult. Both granulites and igneous rocks are grouped together in this section for convenience. A few of the coarse-grained diabases or microtroctolites are included with the plutonic rocks (Fig. 9a), but the most common plutonic rock is anorthosite, which occurs as larger clasts up to 2 cm long (Fig. 9b). The anorthosites have been subjected to shock pressures up to the plagioclase/maskelynite transition, and some have been shocked to the point where some melting has occurred. Plagioclase in anorthosites is coarse grained (to several millimeters), and has compositions between An_{95} and An_{100}. Minor and accessory minerals include magnesian olivine, orthopyroxene, and occasionally ilmenite or spinel.

More commonly found are coarse-grained granulitic anorthosites, norites, or troctolites. These clasts exhibit a range of metamorphic textures, including granular anhedral groundmass plagioclase exhibiting smooth boundaries and triple junctions indicative of an equilibration during subsolidus recrystallization (Fig. 9c). Included in this groundmass are anhedral, irregular orthopyroxene poikiloblasts which enclose anhedral plagioclase, and, rarely, ilmenite (Fig. 9d). In several clasts, these orthopyroxene poikiloblasts have exsolved a high calcium pyroxene as lamellae up to 0.1 mm wide (Fig. 9d). The composition of the major minerals in any one granulite clast is usually quite constant. Generally, olivine and pyroxene are unzoned, plagioclase is often unzoned but euhedral to subhedral grains are occasionally found to have more calcic cores and more sodic rims. Anhedral plagioclase grains are occasionally found enclosing small rounded olivine grains, or euhedral ilmenites. Larger anhedral magnesian ilmenites are found in some clasts.

Pyroxene and olivine compositions from plutonic igneous and granulitic ANT rocks (Fig. 10) exhibit a very small range in composition. Olivines range from $Fo_{67\rightarrow72}$, pyroxenes ranges from $Wo_5En_{67}Fs_{28} \rightarrow Wo_5En_{71}Fs_{24}$ and Wo_{35-40} En_{48-46} Fs_{17-14} in one clast, to $Wo_{4-10}En_{67-60}Fs_{28-33}$ and $Wo_{25-45}En_{56-38}Fs_{13-19}$ (as lamellae in the low Ca pyroxene).

The texture and restricted range of compositions of the minerals from the recrystallized ANT rocks are indicative of high temperature metamorphism. The

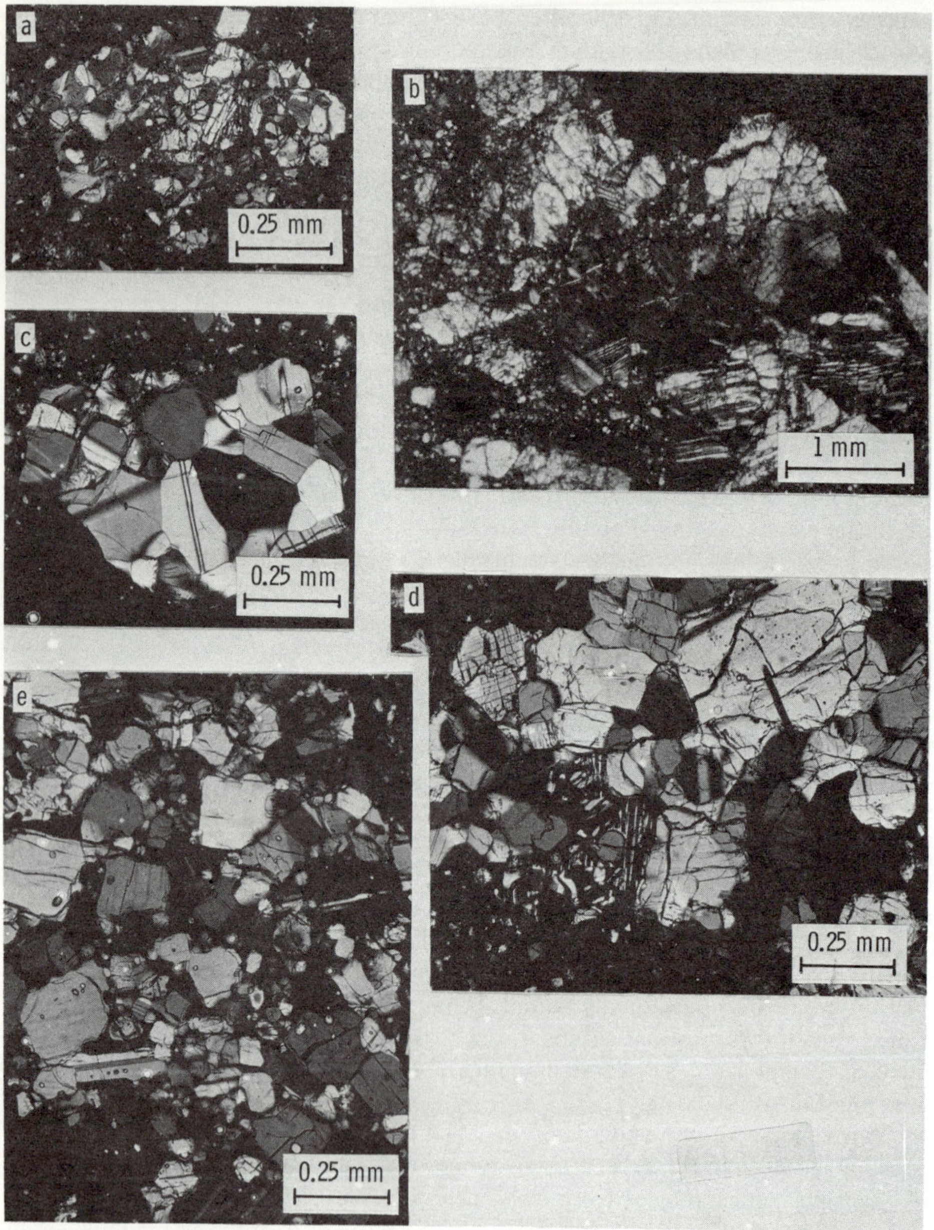
a
0.25 mm
b
1 mm
c
0.25 mm
d
0.25 mm
e
0.25 mm

textures and grain sizes are comparable to terrestrial granulites, and are, by inference, the result of a rather long residence time at high temperatures. Metamorphism has been sufficiently thorough to reduce chemical gradients between minerals and within minerals. The presence of such thoroughly metamorphosed rocks raises several interesting questions about the lunar crust at the Apollo 16 site. These will be discussed in the conclusions.

HORNFELSES

The hornfelses are a group of clasts characterized by textural variety and chemical homogeneity. Superficially, some of the hornfelses resemble the melt rocks, however certain textural differences may be discerned. The typical hornfels texture is one of anhedral to subhedral plagioclase porphyroblasts or xenocrysts surrounded by anhedral olivine or pyroxene (or both) and plagioclase. These anhedral grains form an interlocking mass which, when viewed at moderate magnification with a scanning electron microscope, exhibit smooth curving boundaries and triple junctions (Fig. 11). This basic texture has several varieties, one of the more important of which is the development of porphyroblastic pyroxenes in the groundmass surrounding the xenocrysts (or xenoliths). Where this texture is best displayed, large subhedral to euhedral pyroxene porphyroblasts enclose plagioclase xenocrysts (sometimes subhedral to euhedral) and groundmass plagioclase (anhedral) (Fig. 11c,d). The interstitial areas between the pyroxene porphyroblasts consist of granular olivine and plagioclase with smooth boundaries and triple junctions. This textural type has been found in only one clast so far. Anhedral pyroxene porphyroblasts are more commonly found, but most common are small, anhedral pyroxenes with anhedral olivine or plagioclase inclusions which make up the groundmass along with single anhedral olivines, plagioclases, and pyroxene. In the coarse-grained porphyroblastic varieties, two pyroxenes have been found. The main porphyroblast is a low Ca pyroxene (often orthopyroxene) hosting small high Ca pyroxenes of irregular shape and size.

Fig. 9. Textures of ANT suite clasts from 61175. (a) Micro troctolite-granulite containing olivine (Fo_{77}), plagioclase (An_{95}) and minor low-Ca pyroxene (En_{70}). (b) Shocked anorthosite; this is part of a large (~1 cm) brecciated clast. Some plagioclase has been partially transformed to maskelynite; other plagioclases show deformation lamellae (lower left). Very little glass is present. Plagioclase composition is An_{95-100}. (c) Granulite. Anhedral plagioclase, with curving boundaries and triple junctions is indicative of re-equilibration by subsolidus recrystallization. Subhedral orthopyroxene grain is located in the upper left corner. (d) Norite granulite containing anhedral to subhedral plagioclase, some slightly zoned (An_{97-95}). Anhedral orthopyroxene host ($Wo_4En_{63-67}Fs_{28-33}$) (lower left) encloses anhedral olivine. The bat shaped opaque mineral is a magnesian ilmenite. (e) Poikiloblastic norite-granulite. A single large, anhedral orthopyroxene grain (at extinction) encloses anhedral to subhedral plagioclase (some with thin rims), anhedral olivine and anhedral to subhedral ilmenite. The groundmass (upper left) is granoblastic, and contains plagioclase, olivine, and pyroxene.

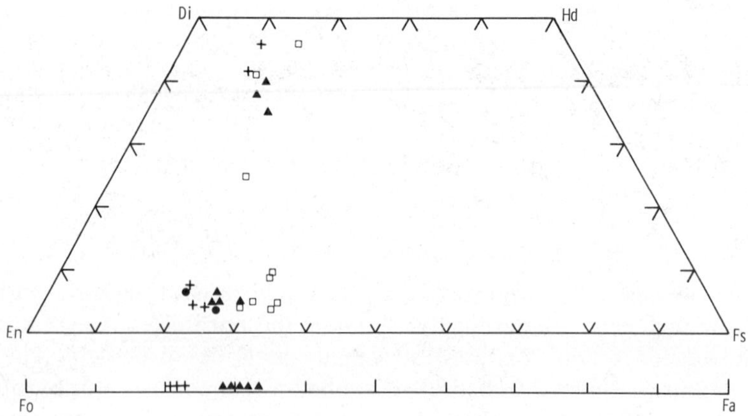

Fig. 10. SEM/EDS analyses of pyroxenes from ANT-suite rocks. Squares are mineral compositions from the clast shown in Fig. 6d, triangles from the clast shown in Fig. 6e. Most points represent several analyses.

All hornfelses are characterized by an absence of glass and by chemical homogeneity among the minerals present. Olivines and pyroxenes are unzoned and of constant composition throughout the clast (Fig. 8), in contrast to olivine and pyroxene compositions from the melt rocks in which considerable chemical heterogeneity is exhibited throughout a single clast. Plagioclase is also generally unzoned and of constant composition, but in the larger crystals and in the xenocrysts, zoning does occur in the form of calcic cores (usually most of the

Fig. 11. Textures of hornfelses from 61175. (a) Granoblastic-hornfels consisting of olivine ($\sim Fo_{77}$), plagioclase ($\sim An_{95}$), and minor low Ca pyroxene. (b) Unusual hornfelsed gabbro (?). This clast has mineral compositions distinct from all other hornfelses analyzed. Olivines (Fo_{55-58}) are more iron-rich than those from the other hornfelses (typically Fo_{74-80}) and the pyroxenes are also more iron-rich (En_{56-60}) than those of all other hornfelses (typically $\sim En_{75}$). (c) Scanning electron photomicrograph of the texture of the hornfels shown in (a). Anhedral plagioclase (dark gray) is surrounded by, and occasionally found as tiny inclusions in olivine (light gray). Curved boundaries show good triple junctions. Minerals are unzoned, and compositions vary only slightly throughout the clast. (d) Poikiloblastic hornfels. Relatively large (.5–.7 mm) subhedral orthopyroxene poikiloblasts ($Wo_{4-8}En_{72-77}Fs_{18-20}$) enclose anhedral to subhedral plagioclase ($\sim An_{88-90}$), subhedral plagioclase, xenocrysts and anhedral olivine (Fo_{78-81}). Interstitial material surrounding the pyroxene crystals is a granoblastic matrix of olivine and plagioclase (same composition as enclosed grains) with curved boundaries and triple junctions. Larger, annealed xenocrysts are seen in the lower half of the clast. (d) Scanning electron photomicrograph of the boundary between orthopyroxene poikiloblast and groundmass. The area of the photomicrograph is a portion of the lower boundary of the large orthopyroxene grain at the top of the clast in (d). Note the anhedral to subhedral plagioclases and iron metal grains in the pyroxene (right half of the photograph) and the anhedral-granoblastic texture of the surrounding matrix (left half of photograph). The large orthopyroxene poikiloblasts contain irregular patches(?) of high calcium pyroxene ($Wo_{40-42}En_{48-50}Fs_{10-11}$). This texture, with the exception of grain size, is nearly the same as that of the granulite shown in Fig. 9e.

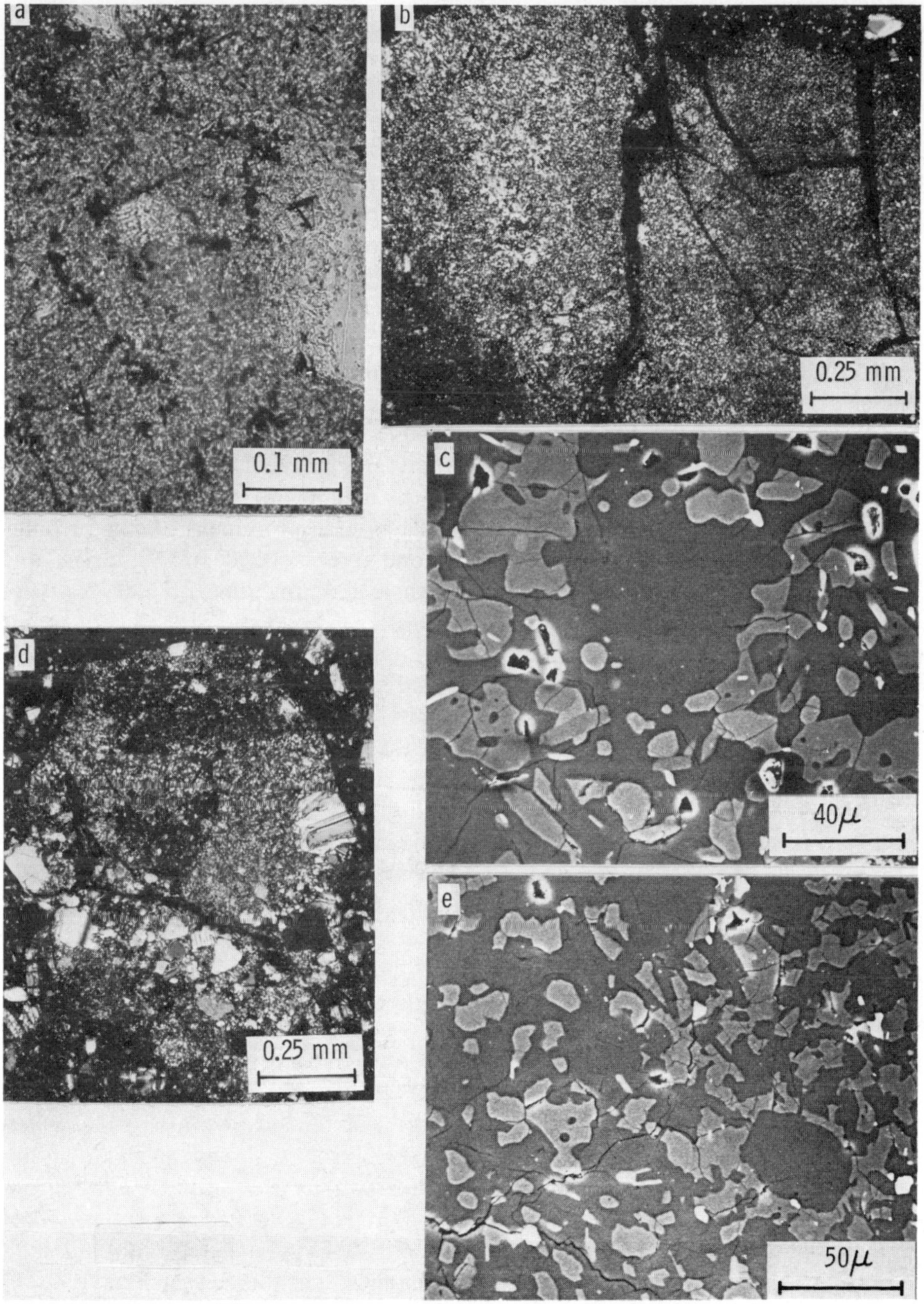
a
0.1 mm
b
0.25 mm
c
40μ
d
0.25 mm
e
50μ

grain) and thin, more sodic, rims. The rim composition is usually the same as the composition of the ground mass plagioclase.

Accessory minerals include Fe metal, with the same range of Ni content as the melt rocks (4.5–9%), troilite, ilmenite, occasional apatite, and phosphide. Sodic feldspar, K-feldspar, and whitlockite are generally absent.

In summary, the hornfelses exhibit an equilibrated texture, lack of zoning and chemical homogeneity among mineral phases, which is indicative of a close approach to chemical equilibrium. In this manner the hornfelses are similar to the coarse-grained granulites. In fact, with the exception of grain size, the textures and mineral compositions of these two groups are remarkably similar.

BULK COMPOSITION

In order to relate the textural types found in 61175 to one another, most of the clasts were analyzed using a defocussed beam technique. These analyses, 58 in all, were done as follows: As with the mineral analyses, a scanning electron microscope with energy dispersive spectrometer was used. The microscope was set in scanning mode, and, where possible, the magnification adjusted to include the entire clast. Where irregular clasts or clasts with prominent cracks or holes were to be analyzed, the analysis was done over several smaller areas and averaged. Three repeat analyses, each with a counting time of 120 sec, were done. This is sufficient to keep the counting error below 1% for all but minor elements. The data were corrected using the Bence/Albee correction procedures and then recorrected following the method outlined by Albee *et al.* (1977). Comparison of normative mineral compositions with the actual compositions obtained by SEM/EDS analysis shows that the differences between the two are small (1–2 mol.% An, Fo, En, or Wo) for the major mineral components.

The bulk chemical analyses of ANT and granulitic ANT, basalt, melt rocks, and hornfelses are plotted on a ternary diagram (Fig. 12) after Walker *et al*, (1973a). No single group of rocks, classified on the basis of texture and mineralogy, is identifiable on the basis of bulk chemistry. All groups overlap one another, and most lie in the plagioclase primary crystallization field. The majority of rocks analyzed fit the criteria for use of this diagram, including the "basalts". Three clasts, all breccias containing significant glass, have compositions far from the compositional plane of the diagram and are not plotted. The position of rocks in the plagioclase field deserves comment. As indicated earlier, basalts can be divided into two groups based on the liquidus phase (plagioclase or olivine). The same is true for the impact melts, and may be extended to those rocks which apparently crystallized magnesian spinel first. All rocks plotted on this diagram would be expected to have plagioclase as the liquidus phase, under equilibrium conditions. The fact that textural examination reveals the presence of other liquidus phases suggests that other factors, such as undercooling, may be controlling the liquidus phase. Another possibility, considered less likely, is that these phases could be relicts, or were nucleated around relicts as the melts cooled.

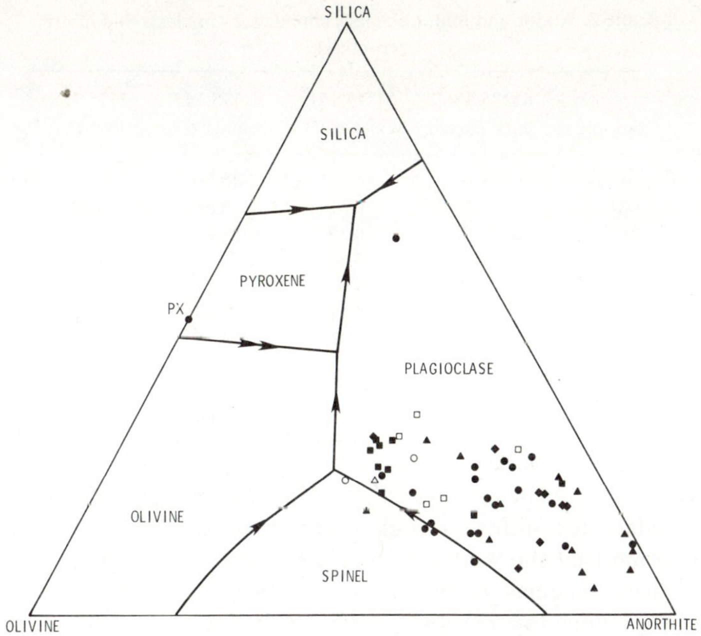

Fig. 12. Ternary diagram (after Walker, 1973b) of bulk compositions of clasts from 61175. ▲—ANT, ■—hornfels, ●—impact melts, and ◆—basalts. Unfilled symbols mark clasts whose classification is uncertain.

Other trends are evident from this plot. First, the basalts are clustered toward the An apex of the triangle, along with the more feldspathic granulites. Many hornfelses cluster near the spinel-plagioclase-olivine piercing point, but examples of both basalts and hornfelses can be found throughout the entire range. The melt rocks (excluding breccias and the odd glassy compositions) span the entire range defined by the feldspathic granulites and the most ferromagnesian hornfelses.

A few clasts and matrix samples from 61175 have been separated and analyzed for major elements by semi-micro atomic absorption spectrophotometry (Nava and Philpotts, 1973) (Table 2). Large-ion-lithophile trace element abundances were also obtained on these samples, following the method of Schnetzler *et al.* (1967). Differences in major element abundances in the separated samples of matrix and clasts analyzed by atomic absorption spectrophotometry mainly reflect different amounts of mafic and feldspathic components, with the matrix being somewhat more mafic than the clasts. Bulk matrix samples analyzed are identical within analytical error (for major elements), and their counterparts are to be found in other Apollo 16 boulders, for instance 67455 fragment-laden melt (Lindstrom *et al.*, 1977) and 67435 glass coating (Warner *et al.*, 1976). Counterparts to this composition can be found both at the North Ray and South Ray Crater sites. This is not surprising in view of the generally anorthositic nature of the Apollo

Table 2. Major and minor element abundances in clasts and matrix from 61175.

Sample	61175,126 dark clast	61175,151 white clast	61175,147 matrix	61175,168 matrix
SiO_2	45.5	44.6	45.5	45.6
TiO_2	0.47	0.47	0.53	0.52
Al_2O_3	28.78	30.29	27.88	27.87
FeO	3.38	2.80	4.33	4.25
MnO	0.05	0.04	0.06	0.06
MgO	4.46	3.14	4.88	4.76
CaO	16.75	16.16	16.14	16.19
Na_2O	0.58	0.56	0.51	0.50
Cr_2O_3	0.05	0.04	0.06	0.06
Total	100.02	98.10	99.89	99.81

16 breccia boulders. The clasts differ from the matrix, and their chemical differences reflect the different rock types involved. Sample 61175,151 is an anorthosite, somewhat shocked and brecciated, and comparable in composition to other anorthosite breccias from the Apollo 16 site; 61175,126 is a melt rock compositionally similar to those analyzed by SEM/EDS. It is also similar to the bulk matrix of 61175.

Large-ion-lithophile trace element abundances for separated clasts and matrix are presented in Table 3 and Fig. 3. Compared to the matrix compositions for the Apollo 17 boulders, the matrix trace element abundances in 61175 are more variable, and are considerably depleted in KREEP (as discussed earlier). The matrix trace element abundances are higher than those of the more feldspathic breccias such as 67455 (Lindstrom *et al.*, 1977), but are lower in abundance than Apollo 16 KREEP (Hubbard *et al.*, 1973). The matrix LIL trace element abundances are consistent with the modal data, and SEM/EDS data which suggest that the majority of the matrix is anorthositic. Glass, which binds the matrix together, is of variable composition, and is irregularly distributed. As the most potassic compositions come from the glass, it is reasonable to assume that the glass is the major contributor to the KREEP component of the matrix.

An insufficient number of clasts have been analyzed for trace elements to make firm conclusions about the relationship of clasts to one another, however some general remarks can be made. First, the anorthosite breccias (subjected to shock pressures above the maskelynite phase transition, but below those needed for significant melting) show typical anorthosite lithophile trace element patterns with a range of from 1 to 10× chondrites. More surprising, however, are the low REE abundances in one of the dark melt-rock clasts analyzed (61175,126). This pattern is typical of a highlands anorthosite. The rare-earth element abundances in the other dark clast analyzed (61175,133) are higher than the other clasts or matrix analyzed, and approach Apollo 16 KREEP in abundances (Hubbard *et al.*, 1973). The spread in lithophile trace element abundances between the melt rocks analyzed differs from that of single terrestrial impact melt sheets. In

Table 3. Lithophile trace element abundances in matrix and clasts from 61175.

Sample	61175,126 dark clast	61175,131 white clast	61175,133 dark clast	61175,147 matrix	61175,151 white clast	61175,167 matrix	61175,168	61175,170 white clast	BCR-1
Li	4.80	4.98	11.5	10.1	4.48	6.50	7.37	2.00	12.4
K	318	504	1800	1050	244	666	794	36.1	13600
Rb	0.636	1.57	5.41	2.97	0.437	1.64	2.19		49.0
Sr	200	174	149	246	219	177	182	163	332
Ba	45.7	65.3	280	153		91.8		8.34	663
Ce	5.87	11.4	76.0	33.0		18.9		0.541	53.2
Nd	4.18	7.05	48.2	21.1		11.8		0.392	28.7
Sm	1.17	2.04	13.2	5.91		3.37		0.144	6.40
Eu	1.19	1.03	1.50	1.66		1.20		0.695	1.85
Gd									
Dy	1.85	2.97	16.6	8.41		4.40		0.155	
Er	1.09	1.75	9.94	4.69		2.59		0.104	3.60
Yb	1.01	1.67	9.06	4.29		2.51		0.095	3.37
Lu	0.152	0.226	1.27	0.654		0.347		0.108	0.446
Zr						194			
Hf									
Weight	0.02347g	0.03894	0.01709	0.13265	0.0702	0.07355	0.11954	0.01343	0.11339

terrestrial impact melts (e.g., Winzer 1975; Winzer and Lum, 1975; Floran *et al.*, 1977) the trace element abundances and patterns are usually nearly identical. The same observation can be made for the Apollo 17 melts (e.g., Winzer *et al.*, 1977). The spread in abundances and differences in the patterns for the two melt-rock clasts analyzed may indicate that more than one melt sheet, or more than one impact event is represented in the clast population of 61175. As some of the clasts analyzed by SEM/EDS techniques are low in K, they may also be low in REEs and may represent an impact melt with a very low KREEP component. Another possible explanation which cannot be discounted, is that of sample heterogeneity. Both melt-rock clasts are quite small, and, although they are individual clasts from different parts of 61175, they may not be representative of any given melt-sheet composition.

DISCUSSION

The clasts in 61175 exhibit a remarkable textural diversity yet, when bulk composition and mineral compositions are compared between groups of samples, this diversity tends to disappear. Unlike other major lunar highlands samples, whose clasts differ significantly in chemical composition (e.g., Winzer *et al.*, 1977; James, 1977) no group of clasts in 61175 appears distinct from another except on the basis of texture. All mineral compositions from the various clast types overlap one another, and only those clasts with a metamorphic texture form a distinct group on the basis of their restricted mineral compositions. This restricted group is a sub-group of all other clasts in terms of mineral composition.

The simplest explanation for these similarities is that all igneous textured rocks except the ANT-suite rocks are impact melts whereas the hornfelses are their metamorphosed equivalents. There is no doubt that the melt rocks, as described in the earlier portion of this paper, are of impact origin. The more difficult assertion to prove is that those rocks with basaltic textures, free of xenoliths or xenocrysts, are also impact generated melt rocks.

Evidence in favor of such an interpretation is indirect, but suggestive. First, all igneous textured, xenolith and xenocryst-free rocks plot in either the anorthositic basalt or VHA basalt fields dilineated by Irving (1975). They are similar or identical in composition to coarse-grained igneous rocks of the highlands ANT suite. Experimental studies on anorthositic basalts and VHA basalts (such as 68415 Walker *et al.*, 1973a,b) indicate that these compositions would have plagioclase, spinel, and corundum on the liquidus at 15 kbar, and this assemblage contradicts that derived from basalt parent materials. It was judged more likely that this type of rock represented shallow melting of plagioclase cumulates. Second, those clasts which do contain iron metal show a nickel abundance which is considerably higher than that of iron metal in mare basalts. Such elevations (generally up to 9% Ni in these rocks) are usually considered to be indicative of meteoritic origin (e.g., Hewins and Goldstein, 1975). Siderophile trace element abundances for these clasts should help in their interpretation. Third, the bulk compositions of the melt rocks and "basalts" and the compositional ranges of their major mineral phases are identical. Detailed petrological studies of terrestrial impact melts (e.g., Grieve, 1975; Floran *et al.*, 1977) indicate that rocks with highly variable textures can be formed by the crystallization of melts with relatively constant composition. Winzer (1975), Winzer *et al.* (1976, 1977), and Grieve (1975) have established that impact melts have major and trace element compositions explainable in terms of total melting and mixing of target rocks. The compositional similarities of impact melts and "basalts" to ANT-suite rocks found as clasts in 61175 suggest that both are impact generated total melts derived from such rocks. The presence of xenoliths and xenocrysts of ANT rocks in the impact melts supports this hypothesis, as do the trace element abundances.

On the basis of the evidence cited above, we favor the hypothesis that all non-plutonic, igneous-textured rocks are impact melts. The possibility still exists, however, that the basaltic-textured rocks originated by internal processes, such as shallow partial melting; however, such processes seem unlikely. Age dating and lithophile trace element data for these clasts should help resolve the problem of their origin.

The origin of the metamorphic rocks remains to be explained. Considerable controversy has arisen over the origin of the fine-grained rocks containing sieve-textured pyroxenes. Depending on interpretation, these rocks have been variously labelled poikilitic (e.g., Simonds *et al.*, 1973) or poikiloblastic (e.g., Albee *et al.*, 1973). More recently, poikilitic pyroxenes have been reported from terrestrial impact melts (Grieve, 1975; Floran *et al.*, 1977), but poikilitic plagioclase is more common. Experimental studies (Lofgren, 1977) suggest that poikilitic textures can be formed under restricted conditions of nucleation and under-cooling. These poikilitic melts have several textural features in common with one another, the most important of which are the dominance of euhedral to subhedral feldspars, as found in poikilitic pyroxenes, and the presence of euhedral to subhedral crystalline material in the groundmass surrounding the poikilitic minerals. The groundmass may be glassy or completely crystalline.

The 61175 hornfelses, unlike the poikilitic rocks from terrestrial melts, and Apollo 17 poikilitic rocks lack the dominant euhedral to subhedral plagioclases found as inclusions in the pyroxenes. Those hornfelses not containing sieved pyroxenes have a granular texture, with anhedral to subhedral plagioclase, pyroxene, and olivine. The texture ranges through rocks containing pyroxenes or olivines with inclusions of plagioclase and/or pyroxene or olivine through those with larger, sieve-like pyroxenes including anhedral (usually round) olivine and anhedral to subhedral plagioclase. The interstitial material in *all* hornfelses is characterized by anhedral to subhedral plagioclase and pyroxene and anhedral olivine. These minerals form the granular base, with 120° intersections indicative of textural equilibration by subsolidus recrystallization. The base texture is *not* reported from any of the poikilitic impact melts, although recrystallization is evident in partially digested clasts in the Manicouagan impact melt (Floran *et al.*, 1977). On the basis of the dominant crystal form of included minerals and interstitial groundmass in the 61175 hornfelses these rocks must have undergone considerable subsolidus recrystallization, and thus the textures observed are considered to be dominantly metamorphic. The textures are thus termed poikiloblastic where larger pyroxene grains have formed.

Euhedral to subhedral feldspars do occur in the poikiloblastic rocks, as well as in the granoblastic hornfelses. Most of these are clearly xenocrysts, but occasional laths occur in the groundmass. These plagioclases generally have the same composition as those which are anhedral, thus they appear to be chemically equilibrated with the rest of the rock. It is uncertain whether these are xenocrysts, or whether their form is controlled by the subhedral mafic phases. Euhedral to subhedral plagioclase and other minerals are not uncommon in metamorphic rocks (see Spry, 1969 plates 15, 18, 21, and 23 for examples), thus their presence does not exclude a metamorphic origin.

The presence of xenocrysts in the hornfelses is strong evidence that the original material could have been an impact generated melt rock. Another possibility, which has been raised by Bence *et al.* (1973) and Albee *et al.* (1973), is that the original rock was a glassy breccia or agglutinate and that temperatures were sufficiently high to cause partial melting of the original rock. Evidence for this hypothesis includes the presence of euhedral crystals, and the absence of less refractory phases such as pyroxene among the xenocrysts. This explanation cannot be excluded with the evidence presently available.

If partial melting is an important process in the origin of the poikiloblastic rocks, it could be an important factor in producing both the coarse-grained granulitic rocks and the fine-grained hornfelsic rocks. The poikiloblastic granulites differ only in grain size from the hornfelses, the textures are similar or identical, as are the mineral compositions. There is considerable evidence from large terrestrial melt sheets that digestion or partial digestion of even very large lithic clasts can occur. This process is similar to anatexis in high-grade metamorphic terranes, and is a combination of metamorphic and igneous processes, producing textures indicative of both subsolidus recrystallization and melting. The chief drawback to the clast digestion process is that, more often than not, the melt which is formed in the clast remains as a glass and does not recrystallize. In addition, no coarse-grained rocks equivalent in texture to the granulites have been found in included clasts in terrestrial melt sheets. Evidence from regional metamorphic terrains suggests that development of such textures requires long time periods at high temperatures. Insufficient data exist to quantify either the cooling times for large melt sheets, or the actual time needed to form the coarser-grained granulitic textures, thus the plausibility of this hypothesis cannot be quantitatively evaluated.

CONCLUSIONS

On the basis of the available evidence, the different lithotypes found in 61175 can be explained by a multi-stage process. The first stage would be formation of a relatively large melt sheet(s) by impact(s) into highland ANT-suite rocks. These impacted rocks may include granulites which are metamorphic equivalents of the original ANT rocks formed by processes akin to terrestrial regional metamorphism prior to impact (see Winzer, 1976). As the melt expanded radially outward from the point of impact it would have swept up and included clasts of original country rock, and breccias formed during the impact. The melt would solidify, forming a variety of textures according to its proximity to cold country rock and the size and number of included xenocrysts and xenoliths. Included breccia or possibly pre-existing melt-rock clasts would undergo high temperature thermal metamorphism possibly including extensive partial melting following their inclusion in the melt. Following solidification of the melt, and final cooling, the resulting melt sheet was again the target of an impact, which possibly formed the 61175 breccia from melt sheet and regolith components. More than one melt sheet may have been involved, as suggested by the limited trace element data available.

Acknowledgment—This research was partially supported by NASA/GSFC contract NAS-5-23544 to Martin Marietta Laboratories and grant NSG-5121 to the University of Maryland. The manuscript benefitted from helpful reviews by R. J. Floran and U. B. Marvin.

References

Albee A. L., Gancarz A. J., and Chodos A. A. (1973) Metamorphism of Apollo 16 and 17 and Luna 20 metaclastic rocks at about 3.95 AE: Samples 61156, 64423,14-2, 65015, 67483,15-2, 76055, 22006, and 22007. *Proc. Lunar Sci. Conf. 4th*, p. 569–595.

Albee A. L., Quick J. E., and Chodos A. A. (1977) Source and magnitude of errors in "broad-beam analysis" (DBA) with the electron probe (abstract). In *Lunar Science VIII*, p. 7–9. The Lunar Science Institute, Houston.

Bence A. D., Papike J. J., Sueno S., and Delano J. W. (1973) Pyroxene poikiloblastic rocks from the lunar highlands. *Proc Lunar Sci. Conf. 4th*, p. 597–611.

Delano J. W. (1975) Petrology of the Apollo 16 mare component: Mare Nectaris. *Proc. Lunar Sci. Conf. 6th*, p. 15–47.

Floran R. J., Grieve R. A. F., Phinney W. C., Warner J. L., Simonds C. H., and Blanchard D. P. (1977) Manicouagan impact melt, Quebec. Part I: Stratigraphy, petrology and chemistry. *J. Geophys. Res.* In press.

Grieve R. A. F. (1975) Petrology and chemistry of the impact melt at Mistastin Lake crater, Labrador. *Bull. Geol. Soc. Amer.* **86**, 1617–1629.

Hewins R. H. and Goldstein J. I. (1975) The provenance of metal in anorthositic rocks (abstract). In *Lunar Science VI*, p. 358–360. The Lunar Science Institute, Houston.

Hubbard N. J., Rhodes J. M., Gast P. W., Bansal B. M., Shih C.-Y., Wiesmann H. and Nyquist L. E. (1973) Lunar rock types: The role of plagioclase in non-mare and highland rock types. *Proc. Lunar Sci. Conf. 4th*, p. 1297–1312.

Irving A. J. (1975) Chemical, mineralogical and textural systematics of non-mare melt rocks: Implications for lunar impact and volcanic processes, *Proc. Lunar Sci. Conf. 6th*, p. 363–394.

James O. B. (1976) Petrology of aphanitic lithologies in consortium breccia 73215. *Proc. Lunar Sci. Conf. 7th*, p. 2145–2178.

LSPET (1972) Preliminary examination of lunar samples. In *Apollo 16 Prelim. Science Rep.* NASA SP-315, p. 7–1 to 7–58.

Lindstrom M. M., Nava D. F., Lindstrom D. J., Winzer S. R., Lum R. K. L., Schuhmann P. J., Schuhmann S., and Philpotts J. A. (1977) Geochemical studies of the white breccia boulders at

North Ray Crater, Descartes region of the Lunar Highlands. *Proc. Lunar Sci. Conf. 8th.* This volume.

Lofgren G. E. (1977) Dynamic crystallization experiments bearing on the origin of textures in impact generated liquids (abstract). In *Lunar Science VIII*, p. 589–590. The Lunar Science Institute, Houston.

Nava D. F. and Philpotts J. A. (1973) A lunar differentiation model in light of new chemical data on Luna 20 and Apollo 16 soils. *Geochim. Cosmochim. Acta* **37**, 963–973.

Prinz M., Dowty E., Keil K., and Bunch T. E. (1973) Spinel troctolite and anorthosite in Apollo 16 samples. *Science* **179**, 74–76.

Schnetzler C. C., Thomas H. H., and Philpotts J. A. (1967) Determination of rare-earth elements in rocks and minerals by mass spectrometric stable isotope dilution technique. *Anal. Chem.* **39**, 1888–1890.

Simonds C. H., Warner J. L., and Phinney W. C. (1973) Petrology of Apollo 16 poikilitic rocks. *Proc. Lunar Sci. Conf. 4th*, p. 613–632.

Spry A. (1969) *Metamorphic Textures.* Pergamon Press, Oxford. 350 pp.

Walker D., Longhi J., Grove T. L., Stolper E. M., and Hays J. F. (1973a) Experimental petrology and origin of rocks from the Descartes Highlands. *Proc. Lunar Sci. Conf. 4th*, p. 1013–1032.

Walker D., Longhi J., Grove T. L., Stolper E. M., and Hays J. F. (1973b) Origin of lunar feldspathic rocks. *Earth Planet. Sci. Lett.* **20**, 325–336.

Warner R. D., Planner H. N., Keil K., Murali A. V., Ma M.-S., Schmitt R. A., Ehmann W. D., James W. D., Jr., Clayton R. N., and Mayeda T. K. (1976) Consortium investigation of breccia 67435. *Proc. Lunar Sci. Conf. 7th*, p. 2379–2402.

Winzer S. R. (1975) Does impact produce chemical fractionation? (abstract). *EOS* (*Trans. Amer. Geophys. Union*) **56**, 389–390.

Winzer S. R. (1976) Studies of consortium breccia 61175 (abstract). *Meteoritics* **11**, 390–393.

Winzer S. R., Nava D. F., Schuhmann S., Lum R. K. L., and Philpotts J. A. (1975) Origin of the Station 7 boulder : A note. *Proc. Lunar Sci. Conf. 6th*, p. 707–710.

Winzer S. R., Nava D. F., Schuhmann P. J., Lum R. K. L., Schuhmann S., Lindstrom M. M., Lindstrom D. J., and Philpotts J. A. (1977) The Apollo 17 "Melt Sheet": Chemistry, age and Rb/Sr systematics. *Earth Planet. Sci. Lett.* **33**, 389–400.

Proc. Lunar Sci. Conf. 8th (1977), p. 1967–1986.
Printed in the United States of America

Apollo 16 white boulder consortium samples 67455 and 67475: Petrologic investigation

JEAN A. MINKIN, CAROLYN L. THOMPSON, and E. C. T. CHAO

U.S. Geological Survey, National Center, Reston, Virginia 22092

Abstract—Samples 67455 and 67475, from a large white boulder on the rim of North Ray Crater, were selected for systematic study by the International Consortium because the boulder appears to be an ejecta block which represents a major lithology of the ejecta stratigraphic section penetrated by the North Ray cratering event. 67455 is a polymict white feldspathic impact ejecta breccia. Distinguishing characteristics of this and other samples from near the rim of North Ray Crater suggest the possibility that these breccias come from a different source area or at least a different part of the same source area as compared with other Apollo 16 light matrix breccias. Sample 67475 is an impact melt which we believe, on the basis of present evidence, was produced by the same event that produced 67455. This event resulted in the transport of this assemblage to the site in the foothills of the Smoky Mountains, where the boulder was subsequently excavated by the North Ray cratering event. Final interpretation of whether the white boulder represents ejecta from the Nectaris event must await age determinations on selected glass clasts from 67455 and 67475.

INTRODUCTION

LUNAR SAMPLES 67455 AND 67475 were dislodged from a white breccia boulder situated on the southeast rim of North Ray Crater, located in the foothills of the Smoky Mountains of Descartes (Fig. 1). The boulder, approximately 5 m long and 2 m high (Hörz *et al.*, 1972), appears to be an ejecta block which represents a major lithology of the ejecta stratigraphic section excavated by the North Ray cratering event some 50 m.y. ago (Drozd *et al.*, 1974). The pre-impact stratigraphic sequence has been interpreted by Hodges *et al.* (1973) as Cayley Formation (possibly Orientale ejecta) overlying Imbrium ejecta, which in turn overlie a layer of different morphology which may perhaps also be Imbrium or possibly Nectaris ejecta. Sample 67455, totalling 942 g, is the largest sample of white matrix breccia collected at Apollo 16 Station 11. 67475, which weighed 175 g, was an isolated melt clast in the white boulder. Parts of both samples were distributed to investigators in an early allocation, but only fragmentary data were obtained in this initial effort.

In the interest of determining the possible source area(s) and the probable age(s) of the event(s) which produced and deposited 67455 and 67475, a systematic study by the International Consortium was organized with E. C. T. Chao as consortium leader and with responsibility for mineralogy-petrology. Other investigators in this consortium are E. Anders (siderophile and volatile trace elements), M. D. Fuller (magnetic properties), J. Geiss ($^{40}Ar/^{39}Ar$ ages and light rare-gas studies), D. F. Nava (major, minor, trace and rare-earth element chemistry), P. B. Price (fission track studies), G. W. Reed, Jr. (labile and non-labile elements), and M. Tatsumoto (Rb/Sr and U–Th–Pb systematics).

Preliminary results of our min-pet study of 67455 have been briefly reported earlier (Chao *et al.*, 1977). A detailed description of our investigations on both samples is presented below.

SAMPLE DESCRIPTIONS

67455 is a friable white polymict feldspathic breccia which by the time it arrived at the Lunar Receiving Laboratory had broken into several lumps weighing more than 50 g each, plus a large amount of small fragments and powdery material (Fig. 2). The material is so friable that the larger pieces tend to break apart with the slightest amount of handling, and the surfaces of the fragments are heavily coated with powder. It was therefore not possible to accomplish a preliminary classification of fragment population in this sample by the usual examination of sample surfaces to identify clast types and relations. Instead, loose material in the sample tray was sifted through a 2 mm mesh sieve, and more than 1700 fragments remaining on the sieve surface were then

Fig. 1. Traverse map of the Apollo 16 landing site. (NASA Pan Photo 4618; map overlay from U.S.G.S. Interagency Report: Astrogeology 54, Fig. 4, p. 17.)

examined for preliminary classification and selection of subsamples for allocation to consortium members (Fig. 3).

In contrast to sample 67455, 67475 is a very coherent rock, with dimensions approximately $6 \times 6 \times 4.5$ cm (Fig. 4). It is principally a brownish-gray fragment-laden melt, in which fragments are much more abundant than the interstitial glass. Adhered to the surface of the coherent part is white, friable, fragmental material similar to that of 67455. Subsamples of material typical of 67475 were chosen and distributed to members of the consortium.

Petrography and Mineral Chemistry

Sample 67455

Our study of 67455 encompassed microscopic examination of two large thin sections 67455,50 and ,57 (Fig. 5) of adjacent areas which contain many clasts of the breccia assemblage; and 11 other thin sections of selected smaller fragments: 67455,94; ,96; ,97; ,99; ,102; ,106; ,112; ,155; ,160; ,217; and ,219. On the basis of petrographic study, areas were chosen for electron microprobe analysis and scanning electron microscope examination.

Fragment population of the breccia (Table 1a) was estimated from counts of fragments greater than 40 μm in size in the two large thin sections. About 70% of the clast population is interpreted as anorthositic material. Other clasts include gabbroic anorthosite, feldspathic microbreccia, and melt glass. Most of the groundmass of the breccia consists of angular to subangular plagioclase grains.

(a) *Clast types*

(1) *Anorthosites.* The anorthosites, which account for 11–14% of the estimated breccia fragment population (Table 1a), are moderately coherent, weakly shocked, and nonporous, and range from approximately 0.5 to 1 mm in size. The fragments have a cataclastic texture and usually consist of 3 or 4 large plagioclase grains and many crushed smaller grains. The larger grains are angular and unzoned, but exhibit polysynthetic twinning and undulatory extinction. The crushed and broken grains show random extinction in polarized light, but they retain some optical continuity with the unbroken grains. There is no evidence of annealing in these fragments, as the large anorthite grains are not polyhedral.

Three types of anorthosites are observed (although they may be parts of the same parent body): (1) anorthite with no interstitial minerals, (2) anorthite with interstitial iron-rich olivine and minor orthopyroxene (Fig. 6), and (3) anorthite with interstitial pigeonite (some with exsolved augite lamellae) and augite. In all three types of anorthosite, plagioclase grains contain stringers of rod-like inclusions of olivine, as shown in Fig. 6. The interstitial minerals occur as broken and randomly oriented grains at anorthite grain boundaries.

The plagioclase grains in all three types have a range of composition An_{94-98}, with most grains in the range An_{96-98}. The rod-like inclusions in the anorthite

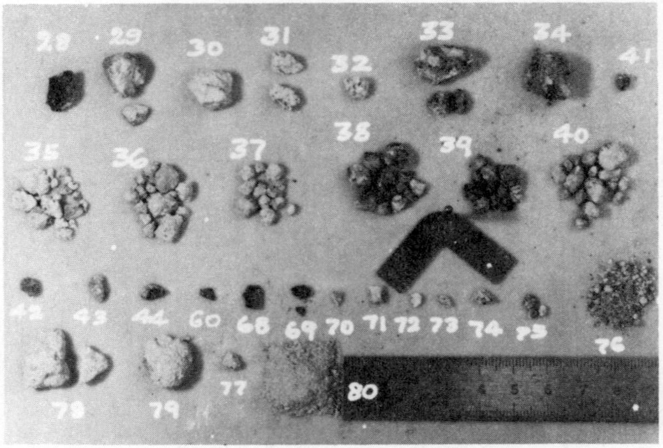

Fig. 2. Photograph showing sample 67455 as received at the Lunar Receiving Laboratory; largest fragment is 7 × 5 × 5 cm (NASA photo S-72-38194).

Fig. 3. Documentation photo of the 67455 materials after sieving and preliminary classification for allocation for consortium study (NASA photo S-74-34504).

Fig. 4. Photograph of the very coherent rock 67475 with adhered white friable, fragmental material similar to 67455 (NASA photo S-72-37956).

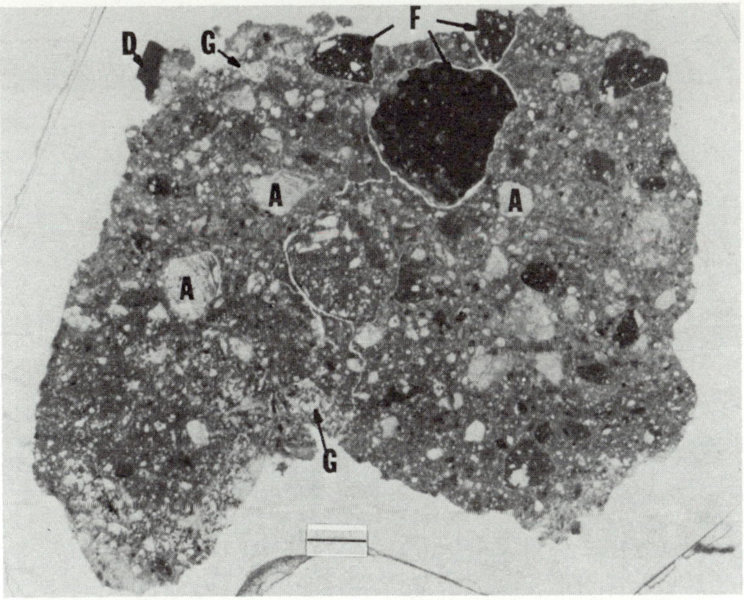

Fig. 5. Whole thin section photograph of 67455,57. Transmitted light. Some major clasts include: (A) cataclastic anorthosites with mafic minerals, (D) devitrified glass, (F) fragment-laden melt, and (G) recrystallized gabbroic anorthosites. Bar scale 1 mm.

grains are very narrow and about 3–5 μm long at most. Thus they could not be analyzed quantitatively with the electron microprobe, but were identified as olivine by semi-quantitative analysis with an energy-dispersive analyzer on the scanning electron microscope. Interstitial olivines are iron-rich and fall in the range Fo_{49-54}. The accompanying minor orthopyroxene has approximate molar composition $Wo_2En_{60}Fs_{38}$. Interstitial iron-rich pigeonite has a range of composition $Wo_{5-7}En_{40-43}Fs_{55-50}$, and the composition of the augite is $Wo_{40-45}En_{35-30}Fs_{25}$.

(2) *Gabbroic anorthosite.* 8–9% of the clasts (Table 1a) are gabbroic anorthosites. These range in size from less than 0.5 to about 2 mm, are fairly coherent, and exhibit textures which range from those showing little recrystallization to those with rounded, equigranular grains (Fig. 7) and triple boundary junctions. Some have poikiloblastic textures where the pyroxenes appear to surround the plagioclase and olivine. The major minerals of the gabbroic anorthosites are plagioclase An_{92-97}, olivine Fo_{46-55}, pigeonite $Wo_{7-18}En_{54-48}Fs_{39-34}$, augite $Wo_{20-40}En_{50-40}Fs_{30-20}$, orthopyroxene $Wo_{2-3}En_{63-56}Fs_{35-41}$, and interstitial ilmenite and iron metal.

(3) *The feldspathic microbreccias.* 3–8% of the clast population (Table 1a) are coherent feldspathic clasts about 200 μm in size at most. They are observed both as separate entities and within dark glassy clasts. These microbreccias consist entirely of small, randomly oriented plagioclase grains packed closely together. The grains are almost uniformly less than 30 μm in size, in contrast to the size of

Table 1.

(a) Fragment population (> 40 μm) of 67455

Fragment type	67455,50 (1554 counts)	67455,57 (825 counts)
Gabbroic anorthosite	7.9	9.1
Feldspathic microbreccia	2.7	6.8
Cataclastic anorthosite with olivine	0.7	3.4
Cataclastic anorthosite with pyroxene	5.0	3.4
Cataclastic anorthosite	8.3	4.4
Plagioclase	51.0	46.2
Olivine	3.7	6.2
Pyroxene	8.2	3.2
Glass:		
devitrified and brown with schlieren	0.5	0.9
annealed	0.8	0
coated grains	1.4	5.1
Fragment-laden melt:		
with plagioclase laths	3.4	2.5
without plagioclase laths	4.2	6.5
Opaque	2.2	2.3
Total	100.0	100.0

(b) Fragment population (> 40 μm) of thin section 67475,82 (903 counts)

Fragment type	Fragment-laden melt	Metal- and silica-rich breccia clast	Light-colored olivine-bearing breccia clast
Gabbroic-troctolitic anorthosite	4.5	0.4	15.3
Cataclastic anorthosite w/mafic minerals	0.3	—	—
Cataclastic anorthosite	4.2	—	0.5
Anorthosite w/polygonal texture	1.0	—	1.6
Crushed anorthite stringers	0.5	—	—
Plagioclase	71.7	51.5	44.2
Olivine	2.9	—	13.7
Clinopyroxene	5.3	16.5	8.4
Orthopyroxene	1.0	—	5.8
Silica	2.0	11.8	1.1
Feldspar w/silica lamellae	1.4	6.3	—
NiFe	0.04	—	0.5
Fe-metal	0.9	7.2	—
Ilmenite	2.1	3.4	4.2
Troilite	0.4	—	4.7
P-rich grains	0.3	2.5	—
Glass: devitrified	1.5	0.4	—
Total	100.04	100.0	100.0

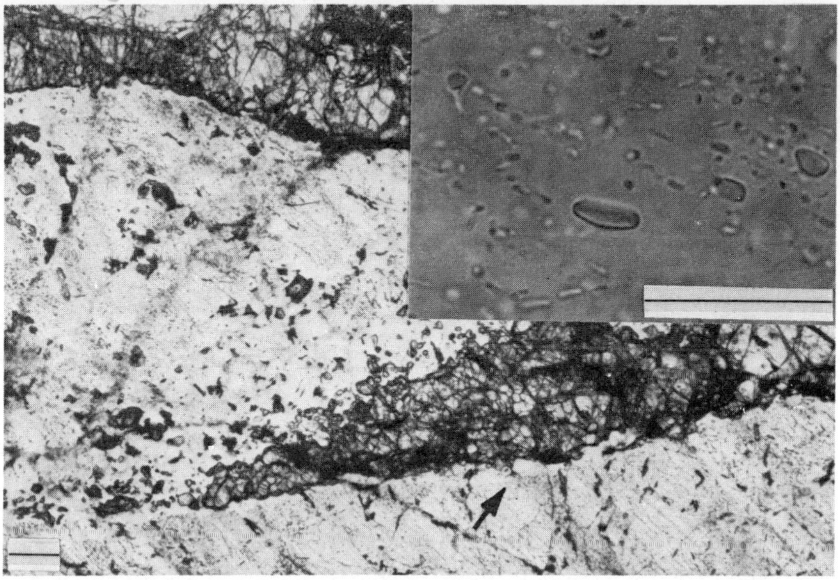

Fig. 6. Photomicrograph of cataclastic anorthosite with interstitial olivine and minor orthopyroxene (darker areas). Inset shows the rod-like high-index olivine inclusions in anorthite in the area indicated by the arrow. Transmitted light. Bar scale for large photo 100 μm, inset 25 μm.

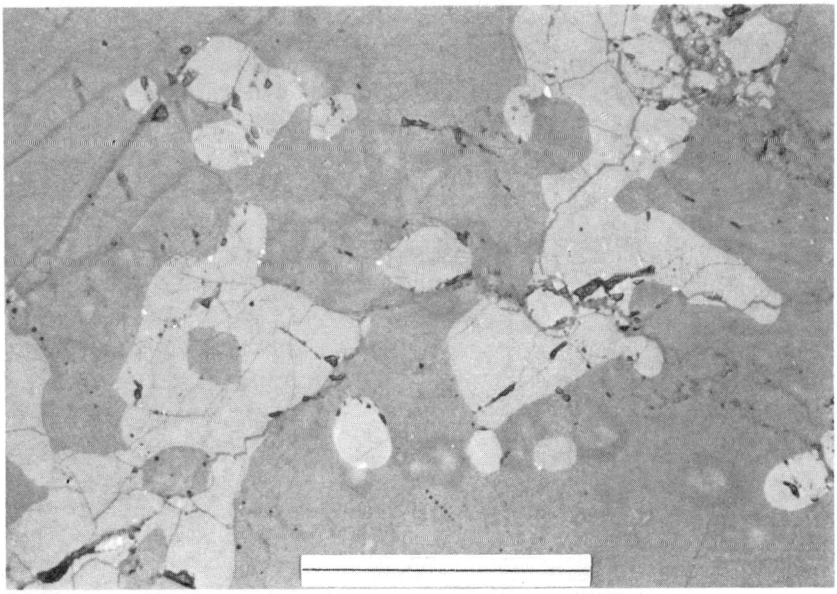

Fig. 7. Photomicrograph showing textural features of a recrystallized gabbroic anorthosite clast in 67455,50. Shown are rounded grains of pyroxene and olivine (light gray) and plagioclase (darker gray). Pyroxene often surrounds plagioclase. Reflected light. Bar scale 100 μm.

the grains in the cataclastic anorthosites. The plagioclase in the microbreccias is commonly twinned and shows some wavy extinction. No other minerals are present. The bulk composition of such a microbreccia in a dark glassy clast is given in Table 2.

(4) *Melt-glass clasts.* 11–15% of the fragments are melt-glass clasts, some containing slightly more glass than others.

(a) Fragment-laden melt clasts make up 8–9% of the fragment population of 67455, as seen from Table 1a. These clasts, up to 600 × 900 μm in size, consist mainly of closely packed small plagioclase xenocrysts (15–20 μm) with a small amount of interstitial melt. Some contain skeletal-hemihedral, hollow-core lath plagioclase grains (about 30 μm long and only a few microns wide) which crystallized from the melt in the matrix, as shown in Fig. 8. Molar compositions of the plagioclase xenocrysts range from An_{87} to An_{97} and the compositions of the laths are An_{94-96}. Some of the plagioclase laths have scalloped edges and form interconnecting networks around the pyroxenes. Xenocrysts of mafic minerals observed include olivine Fo_{46-70}, orthopyroxene $Wo_{2-3}En_{58-77}Fs_{40-20}$, and augite $Wo_{31-43}En_{41-38}Fs_{27-19}$.

(b) 3–6% of the clasts (Table 1a) are classified as glassy. Some unannealed glasses are found which are xenocryst-free and exhibit flow structures (schlieren). These clasts are usually about 65 μm across. There are also occasional occurrences of strained glass (about 40 μm in size) which display random wavy extinction. Other unannealed glasses, up to 0.9 mm across, have very fine devitrified textures displaying many small laths.

Annealed glass clasts are rare in 67455. They range in size from 150 to 800 μm, and are very coherent, hard and nonporous, with discontinuous tension fractures. Such clasts are usually surrounded by more porous areas in the breccia. Even within the same thin section the degree of annealing is varied as gauged by the variation in grain size of pyroxenes, from incipient nucleation to the more advanced stages where the pyroxene network grows and surrounds the plagioclase, developing into fine-grained, optically continuous porphyroblasts, and with still more annealing taking on a coarse-grained hornfelsic texture.

Ranges of bulk compositions of various glass and melt clasts, as determined by defocused beam electron microprobe analyses, are given in Table 2. Molar compositions of minerals in the various clast types are plotted in Fig. 9, and selected mineral analyses are presented in Table 3.

Sample 67475:

Detailed examination of thin sections 67475,81; ,82 (Fig. 10); and ,90 identified this rock as a fragment-laden melt containing two major clast types (Table 1b): a dark clast rich in metallic iron and silica, and a light-colored olivine-bearing clast. Other fragments present include anorthosites, olivine-bearing gabbroic anorthosite, and devitrified glass. The brown fragment-laden

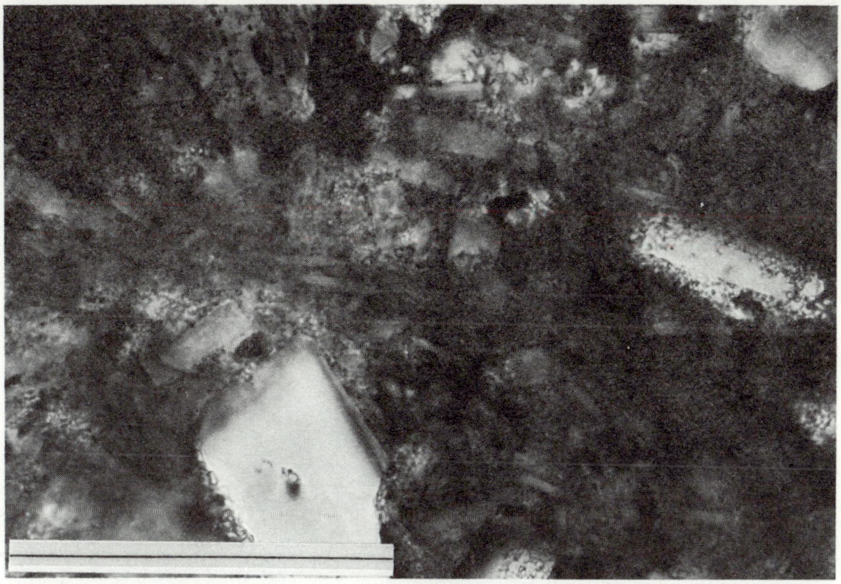

(a)

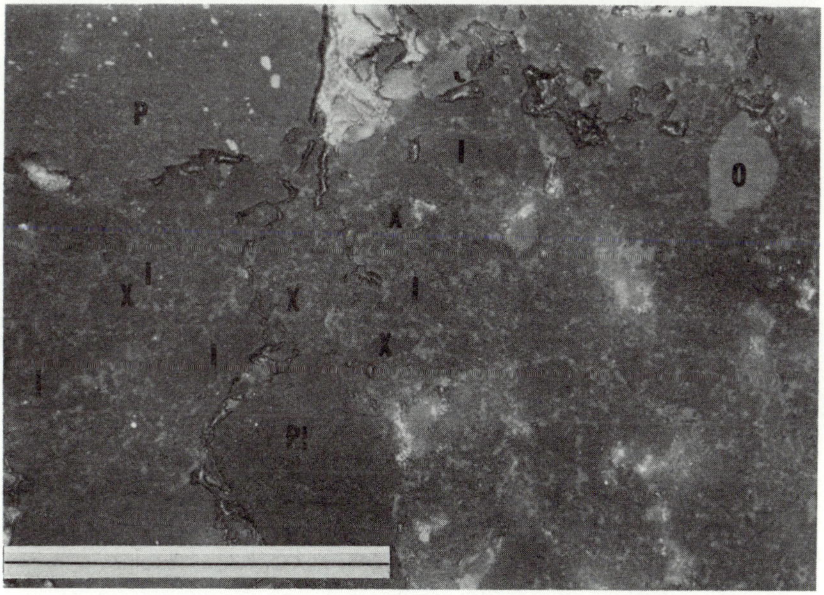

(b)

Fig. 8. (a) Photomicrograph showing plagioclase laths in fragment-laden melt in 67455,99. The laths are aligned lengthwise in a flow pattern. Transmitted light. Bar scale 100 μm. (b) Same area as Fig. 8a shown in reflected light. The light-reflecting interstitial melt occurs between small laths (1) and xenocrysts (X) of plagioclase. Larger xenocrysts of pyroxene (P), olivine (O), and plagioclase (Pl) are found randomly throughout the melt. Bar scale 100 μm.

Table 2. Bulk compositions of glasses, fragment-laden melts, and feldspathic microbreccias in 67455 and 67475 (oxide wt.%).*

	Glasses							
	Strained		Unannealed Brown with schlieren		Devitrified		Annealed	
	Average (2 analyses)	Range	Average (5 analyses)	Range	Average (9 analyses)	Range	Average (20 analyses)	Range
Subsample No.	67455,99-A2		67455,57-1; ,94-A		67475,81-A2; ,81-B		67455,94-B; ,96; ,112; ,217; ,219	
SiO_2	44.8	44.3–45.3	44.0	42.7–44.8	44.2	42.0–45.9	42.7	41.4–44.2
TiO_2	0.17	0.17–0.17	0.18	0.12–0.32	0.34	0.14–0.51	0.19	0–0.36
Al_2O_3	31.4	30.6–32.3	30.4	30.2–31.2	33.0	31.1–34.7	32.2	26.3–37.1
FeO	3.05	2.90–3.20	4.8	3.07–6.9	2.58	1.03–3.60	4.6	0.52–8.5
MgO	1.61	1.45–1.77	3.32	2.05–4.6	1.76	0.62–2.76	2.88	0.30–4.8
CaO	17.9	17.8–17.9	16.9	16.6–17.6	17.8	16.5–18.8	16.8	14.4–19.6
Na_2O	0.43	0.42–0.44	0.29	0.27–0.35	0.51	0.48–0.64	0.41	0.28–0.58
K_2O	0.16	0.13–0.19	0.10	0.07–0.20	0.14	0.11–0.20	0.14	0–0.28
MnO	0.04	0.03–0.05	0.10	0.06–0.13	—	0–0.04	0.06	0–0.20
Cr_2O_3	0.07	0.07–0.07	0.04	0–0.15	—	0–0.07	0.07	0–0.29
Total	99.6		100.1		100.3		100.0	
Mg/Mg + Fe	0.42		0.49		0.48		0.46	

	Fragment-laden melts						Feldspathic microbreccias	
	No skeletal plagioclase		With skeletal plagioclase		Annealed			
	Average (7 analyses)	Range	Average (10 analyses)	Range	Average (4 analyses)	Range	Average (2 analyses)	Range
Subsample No.	67455,57-,11; ,97-2C		67455,99-A1; ,102; ,155; 67475,81-A1		67455,57-4; ,57-10		67455,96	
SiO_2	43.6	41.8–45.5	44.2	42.5–45.4	42.2	41.7–42.9	42.6	41.5–43.6
TiO_2	0.16	0–0.39	0.30	0.11–0.73	0.12	0–0.20	0.14	0.09–0.19
Al_2O_3	31.8	28.6–36.3	30.0	26.9–32.9	33.6	29.4–36.8	36.4	33.5–39.4
FeO	3.68	0–6.7	4.9	3.36–7.3	4.9	3.11–8.4	0.82	0.80–0.85
MgO	2.43	0.29–4.2	3.87	1.74–4.0	3.16	1.99–5.4	0.74	0.57–0.90
CaO	17.3	15.4–19.7	16.6	15.2–18.0	15.8	13.2–17.6	18.0	17.6–18.3
Na_2O	0.46	0.29–0.65	0.53	0.40–0.66	0.52	0.46–0.57	0.49	0.46–0.55
K_2O	0.08	0–0.21	0.18	0.06–0.30	0.12	0.05–0.16	0.11	0–0.22
MnO	0.08	0.06–0.13	0.09	0–0.17	0.06	0–0.11	—	0
Cr_2O_3	0.05	0–0.08	0.03	0–0.12	—	0–0.03	0.04	0.03–0.04
Total	99.6		100.2		100.5		99.3	
Mg/Mg + Fe	0.48		0.52		0.47		0.56	

*Defocused beam electron microprobe analyses.
—Below detectable limit.

Table 3. Electron microprobe analyses of selected minerals in 67455 and 67475 (oxide wt.%).

	67455									67455 and 67475			
	Cataclastic anorthosite				Gabbroic anorthosite					Plagioclase		Fragment-laden melt	
Subsample No.	Anorthite ,106	Interstitial olivine ,106	Interstitial pigeonite ,219	Interstitial augite ,219	Anorthite ,57-1	Olivine ,57-8	Orthopyroxene ,57-8	Pigeonite ,57-3	Augite ,57-3	Xenocryst ,57-14	"Matrix" lath ,57-14	Olivine ,219-B	Orthopyroxene ,57-4
SiO_2	43.3	33.7	49.6	51.3	44.6	33.8	53.2	51.9	51.1	45.2	44.4	35.1	54.4
TiO_2	n.d.*	0.03	0.08	0.17	n.d.	—	0.20	0.14	0.75	n.d.	n.d.	0.07	0.55
Al_2O_3	35.3	0.05	0.87	0.39	34.5	0.21	1.29	0.55	1.46	34.3	35.3	0.35	0.97
FeO	0.35	42.4	30.6	14.4	0.22	41.4	23.2	25.0	15.7	0.19	0.26	37.2	13.6
MgO	n.d.	23.7	13.5	11.1	n.d.	22.9	21.8	19.3	13.7	n.d.	n.d.	25.8	28.9
CaO	19.8	—	2.98	21.3	20.0	0.39	1.34	3.80	16.6	18.5	18.9	0.45	1.50
Na_2O	0.24	—	—	0.03	0.48	n.d.	n.d.	n.d.	n.d.	0.89	0.60	n.d.	n.d.
K_2O	0.18	n.d.	n.d.	n.d.	0.19	n.d.	n.d.	n.d.	n.d.	0.12	0.08	n.d.	n.d.
MnO	n.d.	0.35	0.46	0.23	n.d.	0.48	0.31	0.40	0.30	n.d.	n.d.	0.43	0.26
Cr_2O_3	n.d.	n.d.	0.15	0.14	n.d.	—	0.19	0.06	0.73	n.d.	n.d.	0.05	0.25
BaO	n.d.	n.d.	n.d.	n.d.	n.d.	n.d.	n.d.	n.d.	n.d.	n.d.	n.d.	n.d.	n.d.
Total	99.3	100.3	98.2	99.1	100.0	99.2	101.5	101.2	100.3	99.3	99.6	99.3	100.3
Mg/Mg + Fe		0.44	0.38	0.52		0.43	0.58	0.52	0.55			0.49	0.75
An	96.8				94.8					91.3	94.1		
Ab	2.1				4.1					8.0	5.4		
Or	1.1				1.1					0.7	0.5		
Wo			6.5	44.4			2.7	7.5	34.7				2.9
En			41.2	32.2			61.0	53.5	39.8				76.8
Fs			52.3	23.4			36.3	39.0	25.5				20.3
Fo		49.9				49.6						55.3	
Fa		50.1				50.4						44.7	

Table 3. (*Continued*).

	67475									
	Olivine-bearing gabbroic anorthosite clast		Metal- and silica-rich breccia					Light-colored olivine-bearing breccia		
	Olivine	Augite	K-feldspar	Labradorite	Olivine	Augite	"Matrix" pyroxene	Orthopyroxene	Augite	Olivine
Subsample No.	,81-A4	,81-A4	,90-A	,90-B	,81-D	,90-B	,90-A	,81-C	,81-E	,81-C
SiO_2	36.7	52.4	62.1	56.9	31.8	47.1	50.8	54.1	49.3	36.7
TiO_2	0.05	1.45	n.d.	n.d.	0.14	0.79	1.19	0.37	0.70	0.04
Al_2O_3	0.30	2.00	20.6	27.6	0.14	0.89	1.28	1.17	0.52	0.20
FeO	22.1	7.5	0.24	0.46	57.4	25.3	30.8	14.4	26.5	28.2
MgO	38.8	17.2	n.d.	n.d.	10.1	5.0	11.2	27.5	3.31	32.9
CaO	0.27	17.6	2.56	11.9	0.25	19.5	5.8	1.89	17.5	0.29
Na_2O	n.d.	0.14	0.29	3.96	n.d.	n.d.	n.d.	—	n.d.	n.d.
K_2O	n.d.	n.d.	13.6	0.68	n.d.	n.d.	n.d.	n.d.	n.d.	n.d.
MnO	0.21	0.15	n.d.	n.d.	0.91	0.54	0.77	0.13	0.37	0.36
Cr_2O_3	0.05	0.63	n.d.	n.d.	—	0.12	0.09	0.41	0.05	—
BaO	n.d.	n.d.	2.40	0.16	n.d.	n.d.	n.d.	n.d.	n.d.	n.d.
Total	98.4	99.0	101.8	101.7	100.3	99.2	101.9	100.0	98.3	98.8
Mg/Mg + Fe	0.71	0.76			0.19	0.25	0.37	0.74	0.15	0.62
An			13.3	59.3						
Ab			2.7	36.1						
Or			84.0	4.1						
Wo		37.1				42.2	12.7	3.7	41.0	
En		50.6				15.0	34.4	74.4	10.8	
Fs		12.3				42.8	52.9	21.9	48.2	
Fo	75.8				23.1					67.5
Fa	24.2				76.9					32.5

*n.d. = not determined.
— = below detectable limit.

melt is very similar in texture and composition to that observed in 67455 (Fig. 8, Table 2, and Lindstrom *et al.*, 1977). As in 67455 this melt is dominated by closely packed small (less than 30 μm) plagioclase xenocrysts of subhedral to rounded shapes, with the melt itself occurring only interstitial to the plagioclase grains and containing occasional euhedral to hemihedral-skeletal laths of plagioclase. The plagioclase compositions are similar to those in the fragment-laden melt of 67455. Interspersed through the melt are large (up to 200 μm) xenocrysts

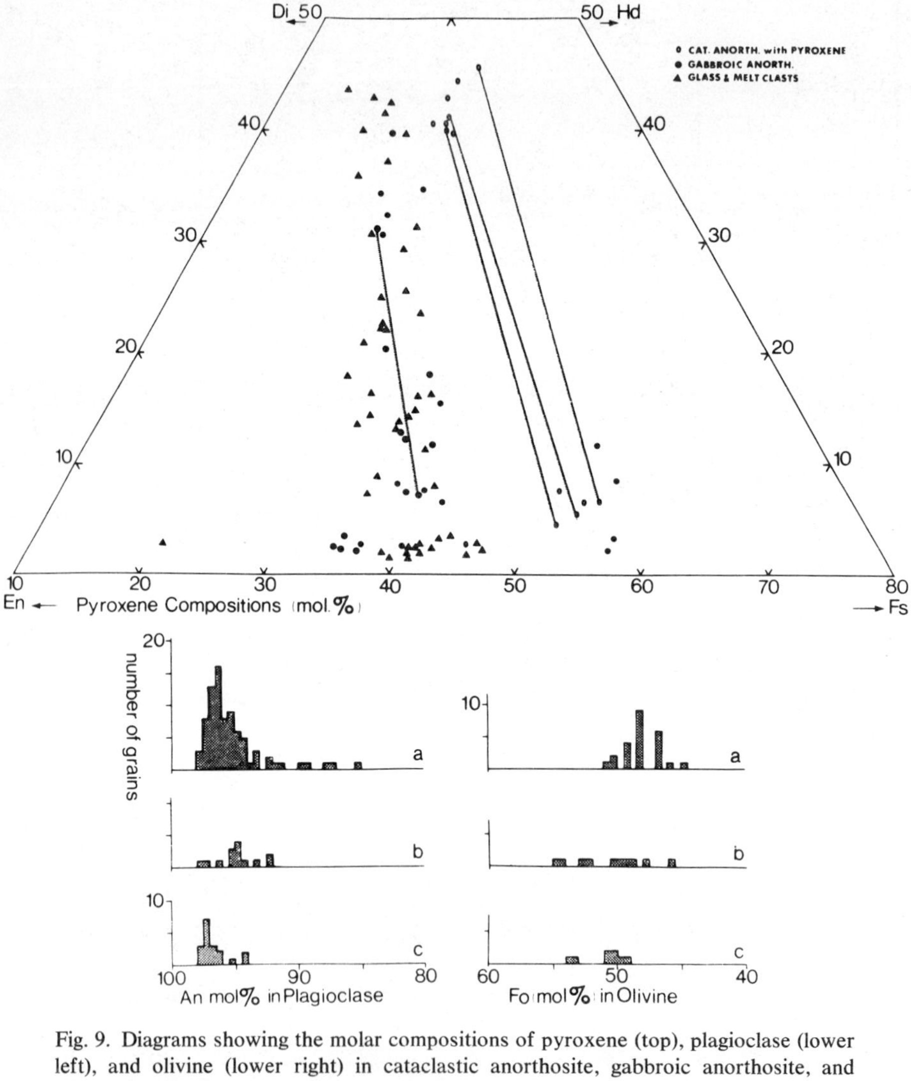

Fig. 9. Diagrams showing the molar compositions of pyroxene (top), plagioclase (lower left), and olivine (lower right) in cataclastic anorthosite, gabbroic anorthosite, and melt-glass clasts in sample 67455. Symbols used in the pyroxene diagram are identified above. For the plagioclase and olivine diagrams, (a) shows compositions in melt-glass clasts, (b) in gabbroic anorthosite, and (c) in cataclastic anorthosite.

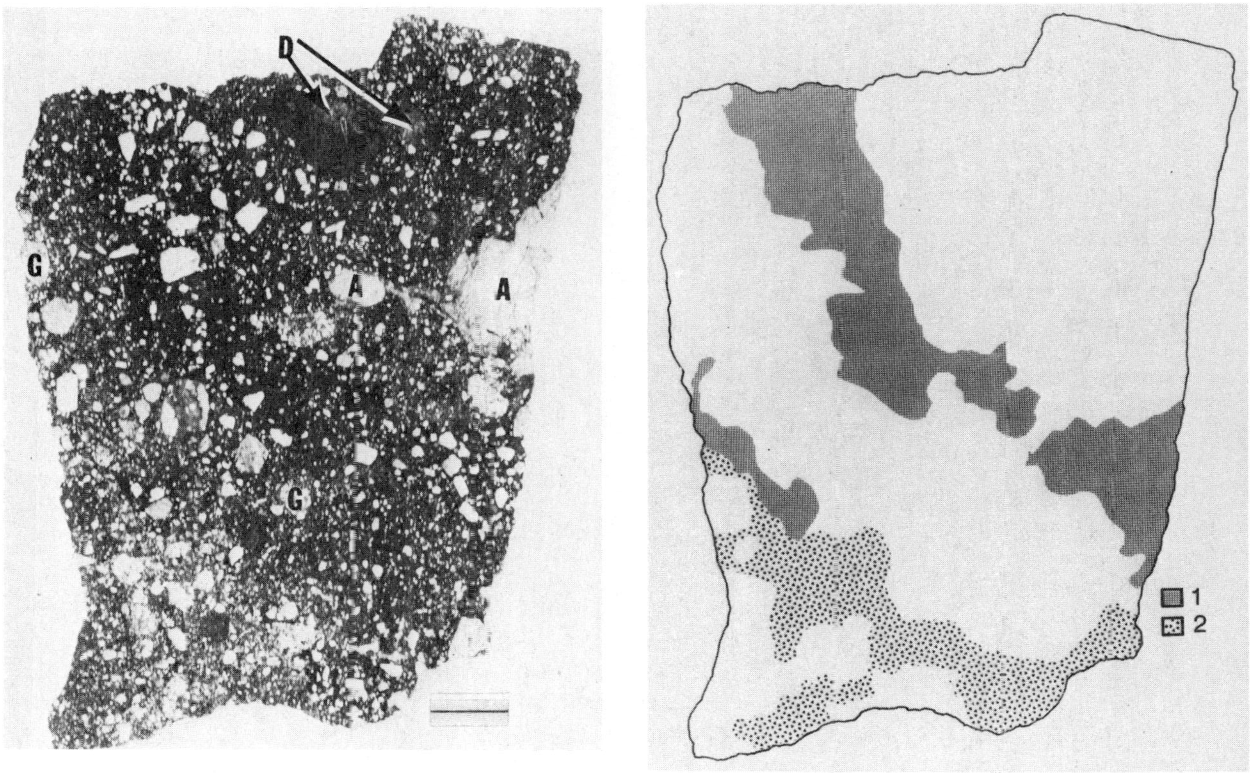

Fig. 10. (a) Whole thin section photograph of 67475,82. Transmitted light. Bar scale 1 mm. Labeled clasts are: (A) cataclastic anorthosites, (D) devitrified glass, and (G) recrystallized olivine-bearing gabbroic anorthosites. (b) Map of thin section outlining two major clasts. (1) metal- and silica-rich dark breccia; (2) light-colored olivine-bearing breccia. White area represents the fragment-laden melt with clasts as identified in Fig. 10a.

of plagioclase (An_{88-98}), olivine (Fo_{56-66}), orthopyroxene $Wo_{3-4}En_{71-64}Fs_{26-32}$, augite $Wo_{42}En_{41-47}Fs_{17-11}$, and silica. Small rounded blebs of metallic iron, troilite and Fe–Ni–Co (Fe 95%, Co 0.2%, Ni 4%) are distributed throughout the melt.

Clasts

(1) *Metal- and silica-rich breccia.* This breccia, which constitutes about 12% of the area of thin section 67475,82 (Fig. 10) is rich in grains and globules of metallic iron and contains large grains (up to 160 × 400 μm) of silica "peppered" with holes, ferroaugite $Wo_{32-42}En_{22-15}Fs_{46-43}$, apatite, whitlockite, and ilmenite (Fig. 11). The feldspars present include K-feldspar $Or_{82-84}Ab_{2-3}An_{16-13}$, usually in bands alternating with lamellae of silica, and grains of labradorite $An_{53-63}Ab_{39-32}Or_{8-5}$. Olivine, observed in only one area, is very iron-rich, Fa_{76-78}. Troilite is also rare. The "matrix" of this breccia consists of fine-grained pyroxenes, $Wo_{4-13}En_{45-34}Fs_{51-53}$. The metallic iron occurs as an irregular blebby intergrowth with silica, and as separate grains. There appears to be a distinct boundary between the clast and the fragment-laden melt.

(2) *The light-colored olivine-bearing breccia* accounts for about 8% of the area of 67475,82 (Fig. 10). It contains silica, plagioclase An_{92-96} and more rarely $An_{60}Ab_{35}Or_5$, orthopyroxene $Wo_4En_{72-80}Fs_{24-15}$, ferroaugite $Wo_{35-42}En_{18-10}Fs_{47-48}$, olivine Fo_{63-68} and ilmenite, with abundant troilite interspersed among the grains in some areas (Fig. 12). Clasts of the recrystallized olivine-bearing gabbroic anorthosite found elsewhere in the fragment-laden melt (see below) are also

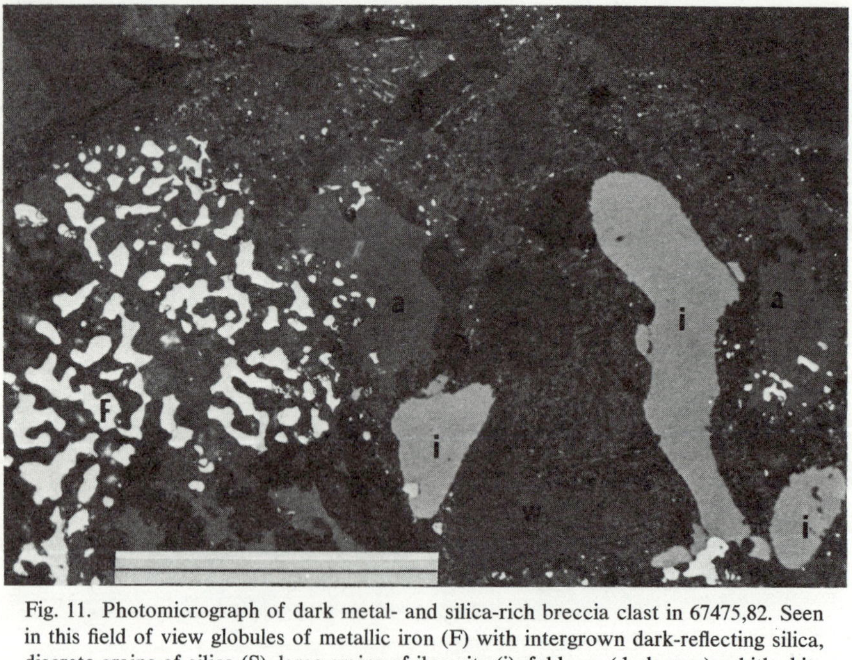

Fig. 11. Photomicrograph of dark metal- and silica-rich breccia clast in 67475,82. Seen in this field of view globules of metallic iron (F) with intergrown dark-reflecting silica, discrete grains of silica (S), large grains of ilmenite (i), feldspar (dark gray), whitlockite (w) and ferroaugite (a). Reflected light, bar scale 100 μm.

abundant in this clast; also present are a few recrystallized anorthosites with polygonal textures. This breccia appears more porous than the metal- and silica-rich breccia.

(3) *Anorthosites.* Cataclastic anorthosites (An_{96}) found in 67475 differ from those in 67455 in that no rod-like olivine inclusions are found in the plagioclase, and mafic minerals associated with the anorthite were observed in only one cataclastic anorthosite fragment in this sample. Such clasts in 67475 are generally smaller than those in 67455, with a maximum size about 0.5 mm. Two other types of anorthosite clasts observed in 67475 include a very small (about 40 μm) recrystallized anorthosite with polygonal texture, and crushed anorthosites which form stringers of anorthite fragments in close proximity to the cataclastic anorthosites.

(4) *Recrystallized olivine-bearing gabbroic anorthosite.* In this type of lithic clast (average size about 0.6 mm) triple-boundary junctions are observed and often the pyroxene forms porphyroblasts surrounding rounded plagioclase grains, similar in textural appearance to the gabbroic anorthosites of 67455 (Fig. 7). The major mineral constituents of this lithic type include plagioclase An_{93-96}, olivine Fo_{74-77}, diopsidic augite $Wo_{34-38}En_{52-50}Fs_{14-12}$, and ilmenite. The mafic minerals here are considerably more magnesian than those of the gabbroic anorthosites of 67455.

(5) *Devitrified glass.* These fragments are up to 1.6 mm long and 1.0 mm wide,

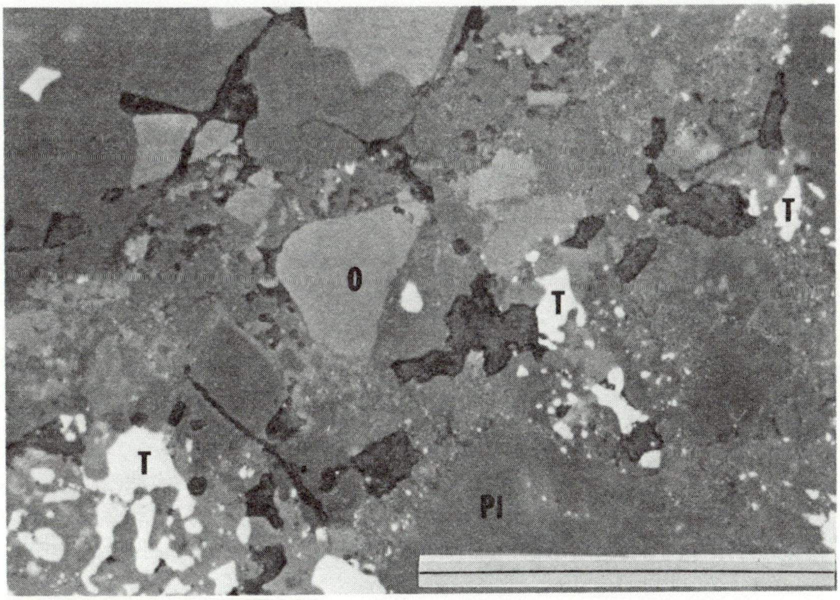

Fig. 12. Photomicrograph of light-colored olivine-rich breccia clast in 67475,82. At the upper left is a recrystallized olivine-bearing gabbroic anorthosite clast. The dominant metal of this clast is troilite (T). Also labeled are a large olivine (O) and plagioclase grain (Pl). Reflected light. Bar scale 100 μm.

and are similar to the devitrified glass in sample 67455. However, in 67475 devitrification in such clasts is more developed, clearly showing plagioclase and pyroxene crystals closely intergrown to produce a "feathery" texture.

Bulk compositions of devitrified glass and fragment-laden melt from 67475 are included in Table 2. Compositions of some of the distinctive minerals found in the clasts in this sample are tabulated in Table 3.

DISCUSSION AND CONCLUSIONS

The small amounts of glass clasts and the absence of glass spherules in the white feldspathic breccia 67455 indicate clearly that this is not a regolith, but rather a polymict impact ejecta breccia. On the basis of present indications we believe that the clast assemblage of 67455 was derived from a different source area, or perhaps from a different part of the source area, from which most other Apollo 16 light matrix breccias originated. Three factors supporting this suggestion are as follows:

(1) Most other Apollo 16 anorthosites do not contain olivines or pyroxenes as iron-rich as those associated with the cataclastic anorthosite in 67455 (Table 3 and Fig. 9), nor have investigators of other Apollo 16 anorthosites reported stringers of olivine inclusions like those in the plagioclase grains of 67455 anorthosite (Fig. 6). The only other occurrences of olivines and pyroxenes as iron-rich as those in the 67455 anorthosites are those reported in anorthositic breccia 67075 (also collected from the rim of North Ray Crater) by McCallum *et al.* (1975), and the iron-rich pyroxenes in anorthosite described by Dixon and Papike (1975) for a 2–4 mm coarse fines sample 63503,13,21 from Station 13, which is also situated close to North Ray Crater (Fig. 1). A unique source area is implied if the olivines and pyroxenes associated with some anorthosites found near the rim of North Ray Crater are more iron-rich than those of other Apollo 16 anorthositic rocks.

(2) 67455 has a stronger positive Eu anomaly and a lower abundance of other rare-earth elements than do most other Apollo 16 breccias (Lindstrom *et al.*, 1977). A similar pattern was found for 67075 by Haskin *et al.* (1973). The REE abundance probably reflects the high plagioclase content of these breccias (Lindstrom *et al.* (1977) report 86–96% normative plagioclase contents for their analyses of 67455 subsamples). Thus it appears that 67455 and other white matrix breccias from the rim of North Ray Crater are more plagioclase-rich than most other Apollo 16 breccias, again implying a distinctive source area.

(3) Several samples from the North Ray Crater site contain a unique meteoritic component. Hertogen *et al.* (1977) found a very high Ir/Au ratio (their group 7) in 67455 and 67075 and four other North Ray Crater samples. This component is, however, not related to the North Ray impact

event, since it is found in old breccia clasts. Hertogen *et al.* (1977) also found this meteoritic component in two Apollo 14 samples (14063 and 14064) and Apollo 15 sample 15418, ejecta which are possibly associated with the Imbrium structure. This unique meteoritic component, however, is not found in Apollo 16 samples from localities other than the North Ray Crater site. Since meteoritic components can be derived from any impact, large or small, it would be difficult if not impossible to attribute or restrict any meteoritic component to a single large basin event. The presence of a unique meteoritic component however, does distinguish some ejecta excavated by the North Ray Crater as coming from a different source area compared to other Apollo 16 samples.

A fourth possible distinction may emerge when age data are obtained on the glassy clasts in 67455. Because of the apparently polymict nature of the subsamples analyzed, attempts to obtain a $^{39}Ar/^{40}Ar$ age for 67455 failed to give a well-defined plateau (Schaeffer and Schaeffer, 1977). However, Turner *et al.* (1973) did obtain a well-defined $^{40}Ar/^{39}Ar$ plateau for 67075, yielding an age of $4.04 \pm .05$ AE. If all the glassy clasts in 67455 turn out to be as old or older than this age this will imply for these North Ray Crater anorthosites a source older and distinct from that of some other Apollo 16 anorthositic rocks, which have reported ages as low as $3.89 \pm .01$ AE (Schaeffer and Schaeffer, 1977).

The presence of many small meteoritic metallic particles confirms that 67475 is an impact melt. The two major clast types it contains, the dark iron- and silica-rich breccia and the light-colored olivine-bearing breccia, have not been observed among the clasts of 67455. In fact, ferroaugites as calcium- and iron-rich as those in the clasts of 67475 (Table 3) have rarely been reported in other Apollo 16 samples. Warner *et al.* (1973) found a few augites in this range of composition in 64817, a porphyritic basalt from Station 4 on the slope of Stone Mountain. Sparse occurrences of this composition were also reported by Delano *et al.* (1973) in light matrix breccia fragments in Apollo 16 coarse fines. The station(s) from which these samples came were not identified. The metal- and silica-rich clasts represent a highly differentiated residuum in an environment of highly calcic anorthosites, an association similar to that of the iron-rich olivines and pyroxenes with the 67455 anorthosites. The mineral assemblages in both 67455 and 67475 are difficult to interpret petrogenetically, demonstrating the complex nature of this white boulder. Hertogen *et al.* (1977) determined that 67475 has a slightly lower Ir/Au ratio (their meteoritic group 6) than that of 67455. This determination may have been affected by the abundant metallic iron and troilite in 67475, the presence of which may tend to obscure somewhat the level of the noble metals present. Although there are some differences between 67455 and 67475, we believe that there is a strong possibility that they came from the same general source, since the glasses and fragment-laden melts in the two samples are compositionally very similar (Table 2 and Lindstrom *et al.*, 1977).

We suggest on the basis of the data presently available, that 67455 and 67475 may contain a significant portion of ejecta from Nectaris. The results of age determinations on the glassy clasts of 67455 and 67475 should help to confirm or reject the Nectaris hypothesis.

Acknowledgment—This work was supported by NASA contract T-2357A. We are grateful to M. M. Lindstrom and M. R. Dence for their constructive reviews of this manuscript.

REFERENCES

Chao E. C. T., Minkin J. A., and Thompson C. L. (1977) Petrology of consortium sample 67455, from a white-matrix breccia boulder near the rim of North Ray Crater, Descartes (abstract). In *Lunar Science VIII*, p. 166–168. The Lunar Science Institute, Houston.

Delano J. W., Bence A. E., Papike J. J., and Cameron K. L. (1973) Petrology of the 2–4 mm soil fraction from the Descartes region of the moon and stratigraphic implications. *Proc. Lunar Sci. Conf. 4th*, p. 537–551.

Dixon J. R. and Papike J. J. (1975) Petrology of anorthosites from the Descartes region of the moon: Apollo 16. *Proc. Lunar Sci. Conf. 6th*, p. 263–291.

Drozd R. J., Hohenberg C. M., Morgan C. J., and Ralston C. E. (1974) Cosmic-ray exposure history at the Apollo 16 and other lunar sites: lunar surface dynamics. *Geochim. Cosmochim. Acta* **38**, 1625–1642.

Haskin L. A., Helmke P. A., Blanchard D. P., Jacobs J. W., and Telander K. (1973) Major and trace element abundances in samples from the lunar highlands. *Proc. Lunar Sci. Conf. 4th*, p. 1275–1296.

Hertogen J., Janssens M.-J., Takahashi H., Palme H., and Anders E. (1977) Lunar basins and craters: Evidence for systematic compositional changes of bombarding population. *Proc. Lunar Sci. Conf. 8th*. Vol. 1.

Hodges C. A., Muehlberger W. R., and Ulrich G. E. (1973) Geologic setting of Apollo 16. *Proc. Lunar Sci. Conf. 4th*, p. 1–25.

Hörz F., Carrier W. D., III, Young J. W., Duke C. M., Nagle J. S., and Fryxell R. (1972) Apollo 16 special samples. In *Apollo 16 Prelim. Sci. Rep.* NASA SP-315, p. 7–24 to 7–54.

Lindstrom M. M., Nava D. F., Lindstrom D. J., Winzer S. R., Lum R. K. L., Schuhmann P. J., Schuhmann S., and Philpotts J. A. (1977) Geochemical studies of the white breccia boulders at North Ray Crater, Descartes region of the lunar highlands. *Proc. Lunar Sci. Conf. 8th*. This volume.

McCallum I. S., Okamura F. P., and Ghose S. (1975) Mineralogy and petrology of sample 67075 and the origin of lunar anorthosites. *Earth Planet. Sci. Lett.* **26**, 36–53.

Schaeffer O. A. and Schaeffer G. A. (1977) ^{39}Ar–^{40}Ar ages of lunar rocks (abstract). In *Lunar Science VIII*, p. 840–842. The Lunar Science Institute, Houston.

Turner G., Cadogan P. H., and Yonge C. J. (1973) Argon selenochronology. *Proc. Lunar Sci. Conf. 4th*, p. 1889–1914.

Warner J. L., Simonds C. H., and Phinney W. C. (1973) Apollo 16 rocks: Classification and petrogenetic model. *Proc. Lunar Sci. Conf. 4th*, p. 481–504.

Proc. Lunar Sci. Conf. 8th (1977), p. 1987–2006.
Printed in the United States of America

Petrology of crystalline matrix breccias from Apollo 17 rake samples

RICHARD D. WARNER, G. JEFFREY TAYLOR, and KLAUS KEIL

Department of Geology and Institute of Meteoritics, University of New Mexico,
Albuquerque, New Mexico 87131

Abstract—The petrology, mineralogy, and composition of 13 crystalline matrix breccias from the Apollo 17 rake samples are described in detail. The breccias have matrices of uniform modal mineralogy (plagioclase, 50–54%; olivine and pyroxene, 41–46%) but diverse textures (micro-subophitic; poikilitic; microgranular-micropoikilitic). Matrix compositions show limited major-element variations (SiO_2, 46.4–50.2 wt.%; Al_2O_3, 17.4–20.3 wt.%; FeO, 7.4–8.8 wt.%; MgO, 8.1–11.6 wt.%; CaO, 11.1–12.1 wt.%). Two of the breccias have matrices of high-K (>0.5 wt.% K_2O) KREEP composition; the remainder are of medium-K KREEP affinity.

Clast assemblages contain a higher proportion of plagioclase than observed in the matrices and also are more calcic. Mafic mineral clasts, particularly olivine, are more magnesian in average composition than matrix minerals. The most common lithic clast type encountered has ANT mineralogy and usually a fine-grained, recrystallized texture; devitrified clasts of anorthositic composition and fine-grained breccia clasts are also abundant. Two troctolite clasts in breccia 72735 preserve an igneous, possibly primary, texture: mineral compositions are nearly identical with those reported for coarse-grained troctolite 76535 (Dymek *et al.*, 1975). A coarse-grained clast of granitic mineralogy in breccia 72539 has a graphic texture and contains feldspar of ternary composition similar to those reported by Ryder *et al.* (1975).

The majority of bulk matrix compositions plot on or very near the plagioclase-olivine cotectic in the system olivine-anorthite-silica. If the matrix compositions represent impact total melts (Grieve *et al.*, 1974; Dence *et al.*, 1976), the inferred cotectic control requires that the source material was itself an igneous differentiate with compositions lying along the plagioclase-olivine cotectic. Alternatively, the proximity of the matrix compositions to the plagioclase-olivine cotectic could be accounted for if the breccia matrices represent a suite of impact-generated partial melts.

INTRODUCTION

A SIGNIFICANT FRACTION of non-mare rocks returned by the Apollo 17 mission consists of abundant mineral and lithic clasts set in a fine-grained crystalline matrix. Because of the crystalline character of their matrices and because of their clastic nature, Simonds *et al.* (1974) termed such rocks "crystalline matrix breccias". With the exception of the work of Simonds *et al.* (1974), most petrologic studies of these breccias have been directed at one or a few selected samples and, thus, have had limited applicability insofar as the overall range of matrix lithologies represented is concerned. In this paper we present the results of a detailed petrological study of 13 crystalline breccias of variable matrix textures from the Apollo 17 rake sample collection. Our examination of the 13 breccias has been directed primarily toward detailed characterization of textures, bulk compositions, and mineralogy of the breccia matrices, in order to provide insight into their petrogenesis. At the same time we have surveyed, described, and analyzed mineral and lithic clast populations, with emphasis being placed on the search for material of possibly primary crustal origin.

MATERIALS and METHODS

Samples were received in the form of small chips taken from the parent rocks; the chips ranged in weight from 0.23 to 0.77 g. The samples were sliced with a wire saw and polished thin sections prepared, several slices being mounted per thin section. In several instances two polished thin sections were prepared from a single sample.

Textures were determined with a petrographic microscope equipped for both reflected and transmitted light. Modes (Table 1) were obtained in reflected light at 1000× magnification (mineral and lithic clast types were identified by switching to lower magnifications). An average of ~3000 points was counted per breccia. Olivine and pyroxene mineral clasts were fairly readily distinguished from coarse matrix grains by differences in texture: mineral clasts have regular shapes whereas matrix grains are usually irregular and enclose and are penetrated by numerous plagioclase crystals. On the other hand, it was generally difficult to distinguish small plagioclase mineral clasts from coarse matrix grains. In most of the breccias the size of the finest recognizable clasts was approximately 50 μm: consequently, a 50 μm grain size cutoff was adopted between clasts ($\geq$50 μm) and matrix (<50 μm—with the exception of >50 μm poikilitic mafic (and opaque) crystals in poikilitic-textured matrices). In the finest-grained breccias this may have resulted in a small overestimation of the matrix component.

Mineral analyses were obtained by electron microprobe using methods described in Keil (1967). Complete mineral analyses were corrected for differential matrix effects following the procedure of Bence and Albee (1968) and Albee and Ray (1970). Bulk analyses of selected lithic clasts were obtained by the broad beam technique (Prinz *et al.*, 1971). Matrix bulk analyses (Table 2) were also obtained via the broad beam technique, in the following manner. A 200 μm diameter beam was placed on areas specially selected for absence of clasts; areas with abundant cracks or voids were also selectively avoided. A total of 15 or 20 such 200 μm diameter areas were analyzed per given thin section slice and the results averaged. This process was repeated for three or four different slices, so that a total of 60 individual 200 μm diameter areas were analyzed per sample. Reproducibility between individual slices in a given sample was customarily to better than 1% for SiO_2, 2% for Al_2O_3 and CaO, and 3% for FeO, MgO and Na_2O. The remaining oxides, all of which (except TiO_2) are present at concentration levels $\leq$1 wt.%, were of poorer reproducibility; K_2O and P_2O_5 showed the greatest intrasample variation. The bulk analyses were not corrected for differential matrix effects. However, the standards used were minerals selected for their similarity in composition with the matrix minerals (i.e., synthetic anorthite for CaO, Al_2O_3, and SiO_2; Marjalahti olivine for FeO, MnO, and MgO; A209 pyroxene for TiO_2; synthetic spinel for Cr_2O_3; Or-1 K-feldspar for Na_2O and K_2O and apatite for P_2O_5). Hence, matrix effects on the bulk analyses are probably small.

PETROGRAPHY and MINERALOGY of BRECCIA MATRICES

The modal and bulk composition data presented in Tables 1 and 2 indicate that there is very little chemical or mineralogical variation among the crystalline matrix breccias studied. Rocks 72735 and 72558 are set apart from the remaining breccias by having higher K_2O contents (0.89 and 0.57 wt.%, respectively) and lower MgO/MgO + FeO ratios. REE abundances reported for 72735 are 1.5–1.8× higher than those of other crystalline matrix breccias (Murali *et al.*, 1977; Laul *et al.*, 1975), and the overall REE pattern is similar to that for high-K KREEP (Murali *et al.*, 1977). The remaining 11 crystalline matrix breccias are generally of medium-K KREEP affinity (77518 tending toward high-K, and 72535 toward low-K KREEP type compositions). They may be subdivided into three groups on the basis of matrix textures. Four breccias (all from Station 2) have a microsubophitic matrix texture. Three breccias (all from Station 7) have a poikilitic texture. The remaining four rocks exhibit a microgranular to micro-

Table 1. Modes (vol.%) of Apollo 17 crystalline matrix breccias.

	Microsubophitic				Poikilitic			Microgranular-micropoikilitic				High-K KREEP	
	72535	72536	72539	72738	77515	77539	77545	72548	72549	72736	77518	72558	72735
Points counted	3222	3496	4072	2561	3427	3132	1487	3595	2464	3097	4398	1098	2827
Matrix	84.8	82.9	87.7	86.9	87.0	29.1	74.6	80.4	84.3	72.2	84.7	51.9	72.9
Mineral clasts	11.0	12.9	7.6	10.1	10.9	3.8	9.4	16.4	14.2	12.0	12.6	9.0	12.7
Lithic clasts	4.2	4.2	4.7	3.0	2.1	67.1	16.0	3.2	1.5	15.8	2.7	39.1	14.3
Mineral clasts													
Plagioclase	7.1	8.9	5.5	7.1	7.7	2.1	6.7	10.9	10.1	8.1	7.5	4.8	8.0
Olivine/pyroxene	3.7	3.9	1.9	2.9	3.0	1.7	2.7	5.4	4.1	3.8	4.8	4.1	4.7
Opaque oxide	tr	tr	0.1	tr	—	—	—	tr	—	tr	0.1	—	tr
Metal/troilite	0.2	0.1	0.1	0.1	0.2	—	—	0.1	tr	0.1	0.2	0.1	—
Other	—	—	tr	tr	—	—	—	—	—	—	—	—	—
Total	11.0	12.9	7.6	10.1	10.9	3.8	9.4	16.4	14.2	12.0	12.6	9.0	12.7
Lithic clasts													
ANT	1.9	2.6	1.2	1.6	0.4	66.9	14.9*	2.3	1.0	0.9	0.5	38.8	10.6
Devitrified anorthosite	0.6	1.0	0.9	0.4	0.8	0.1	0.9	0.3	0.2	7.8	0.4	0.1	2.4
Breccia	1.4	0.6	2.2	0.5	0.4	—	—	0.4	0.3	7.1	1.6	—	1.3
Other	0.3	tr	0.4	0.5	0.5	0.1	0.2	0.2	—	tr	0.2	0.2	tr
Total	4.2	4.2	4.7	3.0	2.1	67.1	16.0	3.2	1.5	15.8	2.7	39.1	14.3
Percent of matrix (normalized to 100)													
Plagioclase	52.9	52.3	51.6	50.8	52.7	50.8	53.2	54.4	53.7	50.3	51.9	53.9	52.7
Olivine/pyroxene	43.8	44.3	45.0	43.9	44.5	45.2	44.0	43.3	43.7	46.2	44.1	40.9	42.2
Opaque oxide	2.9	3.0	2.5	2.4	2.0	2.2	1.6	1.7	1.4	2.1	1.7	1.7	1.2
Metal/troilite	0.1	0.3	0.4	0.2	0.2	0.2	0.2	0.1	0.4	0.6	0.3	0.5	0.3
Other	0.2	tr	0.6	2.7	0.7	1.5	1.0	0.5	0.8	0.8	2.0	3.0	3.6

*Includes polygonal dunite (14.5%).

Table 2. Broad beam bulk analyses of crystalline breccia matrices.

	Microsubophitic				Poikilitic			Microgranular-micropoikilitic				High-K KREEP	
	72535	72536	72539	72738	77515	77539	77545	72548	72549	72736	77518	72558	72735
SiO_2	47.9	46.7	47.1	46.4	48.3	48.1	49.6	48.1	48.8	47.5	47.1	50.2	50.1
TiO_2	1.68	1.63	1.77	1.69	1.51	.84	.77	1.47	.95	.67	1.25	.76	.67
Al_2O_3	18.1	18.2	17.4	18.7	18.2	17.7	17.7	20.3	19.1	19.3	19.7	19.4	20.2
Cr_2O_3	.17	.16	.15	.16	.19	.16	.19	.15	.17	.16	.15	.16	.18
FeO	8.7	8.8	8.4	8.0	8.1	7.8	7.4	7.4	7.8	7.7	7.8	8.5	7.9
MnO	.13	.13	.12	.11	.09	.11	.11	.11	.11	.13	.12	.16	.12
MgO	10.6	11.1	11.1	10.1	11.0	11.3	11.5	9.3	11.2	11.6	10.5	8.7	8.1
CaO	11.9	11.5	11.3	11.8	11.4	11.1	11.4	12.1	12.0	11.9	11.7	11.3	11.5
Na_2O	.54	.57	.59	.82	.70	.73	.72	.60	.58	.72	.72	.85	.68
K_2O	.07	.21	.16	.23	.28	.27	.21	.27	.27	.26	.42	.57	.89
P_2O_5	.27	.30	.28	.27	.27	.26	.23	.15	.37	.27	.31	.25	.32
Total	100.0*	99.3	98.4	98.3	100.0*	98.4	99.8	100.0*	101.4	100.2	99.8	100.9	100.7
MgO/MgO + FeO	.55	.56	.57	.56	.58	.59	.61	.56	.59	.60	.57	.51	.51

*Normalized to 100.0 wt.%.

poikilitic matrix texture. Matrix textures and mineralogy in each of the above types as well as in high-K KREEP breccias 72735 and 72558 are described in the following paragraphs.

Microsubophitic matrix breccias (72535, 72536, 72539, 72738)

Rocks belonging to this group are extremely fine-grained, matrix-rich breccias. Macroscopically, they are described in the "Apollo 17 Lunar Sample Information Catalog" as blue-gray breccias; in thin section they appear as dark matrix breccias. The samples are all characterized by a porous (12–20% void space) basaltic-textured matrix consisting of euhedral 2–15 μm wide and 10–

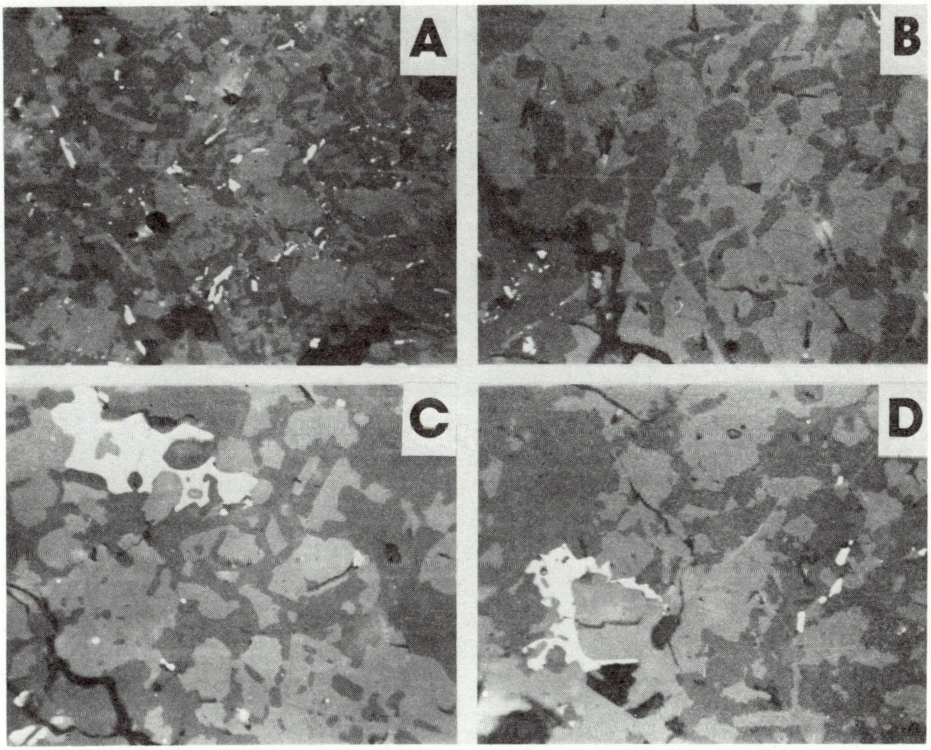

Fig. 1. Reflected light photomicrographs illustrating various matrix textures encountered. Horizontal dimension equals ~0.2 mm in all photomicrographs. (A) Microsubophitic matrix texture in breccia 72536. Note euhedral, lath-like habit of plagioclase (dark gray) crystals. (B) Poikilitic matrix texture in breccia 77545. Note laths to stubby crystals of plagioclase (dark gray) enclosed by single pyroxene oikocryst (light gray). (C) Microgranular-micropoikilitic matrix texture in breccia 72736. (D) Matrix texture in high-K KREEP breccia 72735. Note lack of lath-like plagioclase crystals, extreme irregularity of crystal outlines on a fine scale, and presence of extremely fine-grained interstitial phase of K-feldspar composition (darkest gray).

30μm long plagioclase laths subophitically enclosed by irregular mafic crystals (Fig. 1A). In some samples (e.g., 72535) mafic grains locally are ophitic to micropoikilitic in habit. Opaque minerals (mostly ilmenite) occur as irregular discrete rods 1–5 μm wide and up to 20 μm long, except in 72539, where they are present as extremely minute (<1 μm) grains which tend to occur in clusters at the edge of mafic grains. Tiny grains of metal and troilite are widely disseminated in the matrix. Several breccias (e.g., 72536) contain matrix heterogeneities, with patches of finer-grained material locally present. The finer-grained areas are generally also of microsubophitic texture but with grain sizes reduced by a factor of two or three. Contact with the "normal" matrix areas is often fairly sharp.

Relative proportions of matrix phases in the microsubophitic-textured breccias are extremely uniform, averaging about 52% plagioclase, 44% olivine + pyroxene, and nearly 3% ilmenite (Table 1). Matrix plagioclase compositions generally range An_{84-97}, but an occasional less calcic composition is also encoun-

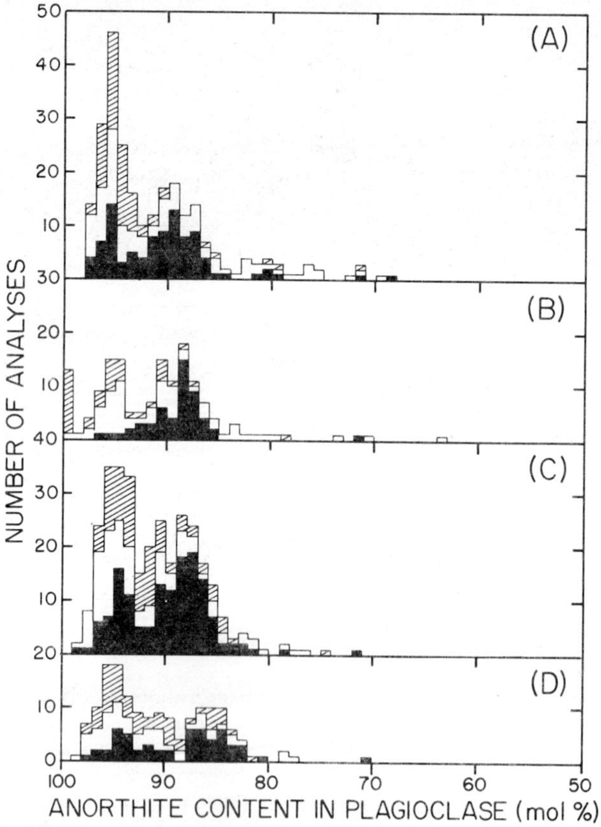

Fig. 2. Histograms of anorthite content in plagioclase from (A) microsubophitic, (B) poikilitic, (C) microgranular-micropoikilitic and (D) high-K KREEP breccias. Symbols: filled, matrix grains; blank, mineral clasts; diagonal lines, grains in lithic clasts.

tered (Fig. 2A). There is a bimodality in matrix plagioclase compositions, with maxima at approximately An_{89} and An_{95}. Olivine and low-Ca pyroxene are the predominant mafic phases, with minor high-Ca pyroxene also present. Matrix olivine compositions range Fo_{67-78} (Fig. 3A); within individual samples the compositional variation is very limited (e.g., Fo_{69-71} in 72535 and Fo_{70-74} in 72539). Representative matrix pyroxene compositions are shown in Fig. 4A. Low-Ca pyroxenes exhibit considerable variation in Ca content; they are mostly pigeonitic in composition. Opaque phases encountered are predominantly ilmenite (Fe/(Fe + Mg), 0.71–0.83) with subordinate chromite (Fe/(Fe + Mg), 0.70–0.82).

Poikilitic matrix breccias (77515, 77539, 77545)

Matrices of these rocks are considerably coarser grained than the microsubophitic breccias and the rocks are therefore much lighter in color. Their matrices are characterized by oikocrysts of low-Ca pyroxene that enclose abundant euhedral tablets or laths of plagioclase and minor rounded olivine

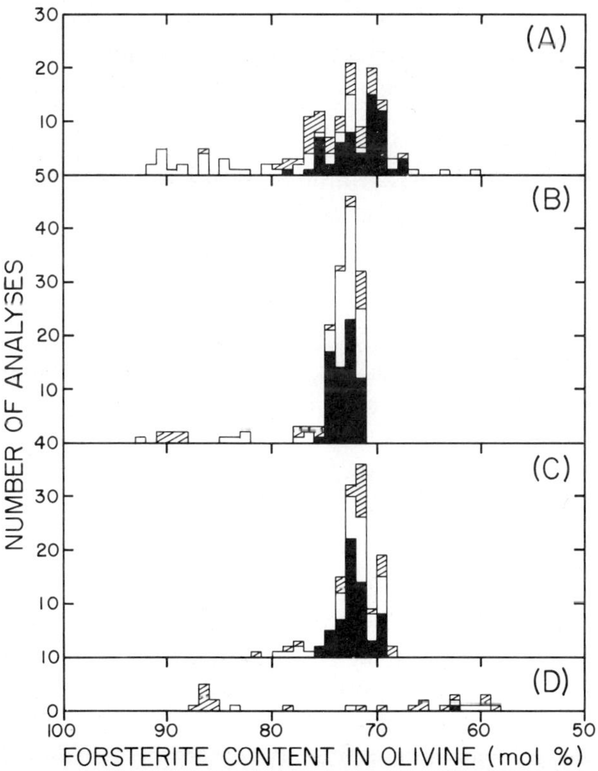

Fig. 3. Histograms of forsterite content in olivine from (A) microsubophitic, (B) poikilitic, (C) microgranular-micropoikilitic and (D) high-K KREEP breccias. Symbols: filled, matrix grains; blank, mineral clasts; diagonal lines, grains in lithic clasts.

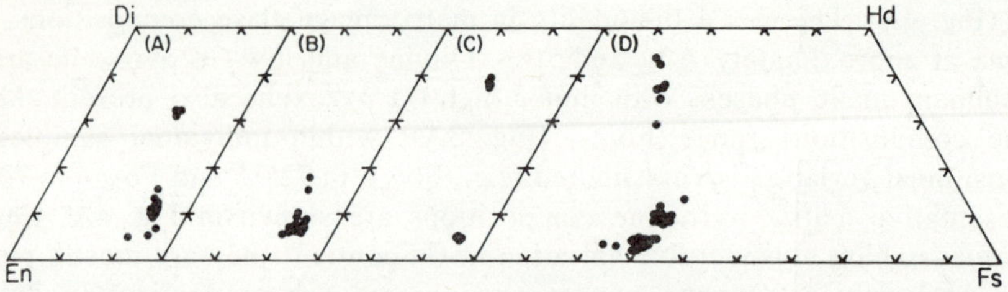

Fig. 4. Matrix pyroxene compositions from representative breccias projected onto pyroxene quadrilateral. Analyses from (A) microsubophitic matrix breccia 72535, (B) poikilitic matrix breccia 77545, (C) microgranular matrix breccia 72548, and (D) high-K KREEP breccia 72735.

grains (Fig. 1B). The oikocrysts are 0.2–0.4 mm long except in 77545 where some exceed 1 mm in length. Plagioclase chadacrysts are mostly 5–20 μm wide and 10–40 μm long; many exhibit a slight rounding at the corners suggestive of some matrix recrystallization. The olivine grains enclosed by the pyroxene oikocrysts are generally equant crystals 20–30 μm and larger in diameter. Ilmenite (with abundant chromite and rutile blebs and/or lamellae) is the predominant opaque mineral and occurs as irregular, often poikilitic grains up to 0.2 mm long. Armalcolite of similar habit is present as sparse grains in 77539. Metal and troilite are also present in the matrix and occur almost exclusively as small disseminated crystals. Minor interoikocryst areas are scattered throughout the matrices; these areas contain mostly feldspar of habit and grain size similar to the plagioclase chadacrysts. Vesicles (up to 1 mm wide in 77545) are abundant in all poikilitic matrix breccias.

As in the microsubophitic matrix breccias, modal matrix plagioclase contents range from 51–53% (Table 1). Plagioclase compositions range from An_{85} to An_{96} (Fig. 2B). Pyroxene oikocrysts are almost exclusively low-Ca pyroxene: only one high-Ca pyroxene grain ($En_{47}Fs_{12}Wo_{41}$) was encountered (in 77515). The oikocrysts are dominantly pigeonitic in composition (Fig. 4B). Matrix olivine compositions show very limited compositional variation (Fo_{71-74}—Fig. 3B). Ilmenite also shows restricted compositional variation (Fe/(Fe + Mg), 0.77–0.80 in 77515 and 77545, and 0.72–0.77 in 77539). Armalcolite in 77539 ranges 0.52–0.55 in Fe/(Fe + Mg).

Microgranular-micropoikilitic matrix breccias (72548, 72549, 72736, 77518)

These breccias are also coarser grained than the Station 2 microsubophitic breccias and appear as light matrix breccias in thin section. Matrix textures within a single specimen generally vary from microgranular to micropoikilitic, depending on size of mafic grains (Fig. 1C). Matrix plagioclase occurs as laths (3–15 μm wide, 10–40 μm long) or stubby grains (20–30 μm wide); most crystals have rounded corners, suggestive of fairly extensive matrix recrystallization. Olivine and pyroxene are present as subequant grains 15–25 μm across or as irregular micropoikilitic crystals on the order of 50–100 μm in length. Ilmenite (often with chromite/rutile blebs and lamellae) is the most common opaque mineral present and varies in habit from predominantly equant (2–20 μm wide) grains (e.g., 72548) to irregular micropoikilitic crystals up to 0.2 mm long. Armalcolite is present in some rocks (e.g., 72549); baddeleyite has also been identified. Small grains of metal and troilite (mostly $< 10\ \mu$m) are widely scattered throughout the matrices. Small amounts of an interstitial K-rich phase (mostly K-feldspar?) are present at plagioclase-plagioclase or plagioclase-mafic grain contacts: this phase is generally 2–3 μm or less in size.

Breccia 77518 contains an area of K–Si–Al-rich glass (79–81 wt.% SiO_2; 12–13 wt.% Al_2O_3; 7–8 wt.% K_2O) at one edge of one of the thin section slices (since the material we received is a chip from the parent rock, we have no way of knowing whether of not this glassy area was surficially exposed on the parent rodk). The glass contains sparse tiny euhedral mafic crystals. The glassy area is separated from the breccia matrix by a transition zone in which the proportion of glass regularly decreases with increasing distance from the edge of the section.

Matrix plagioclase exhibits a range in composition from An_{71} to An_{98} (Fig. 2C); the majority of grains are An_{85-96} in composition. As in the microsubophitic matrix breccias, there is a distinct bimodality, with maxima at An_{87-88} and An_{94}. Olivine compositions are restricted in range and vary from Fo_{69-70} in 72548 to Fo_{71-75} in 72549, 72736, and 77518 (Fig. 3C). Matrix pyroxenes include both low-Ca (orthopyroxene) and high-Ca phases except in 77518, where only low-Ca pyroxene is apparently present. Compositional variation is extremely limited, both within and between samples; orthopyroxene analyses cluster near $En_{72}Fs_{23}Wo_5$ while high-Ca pyroxenes cluster near $En_{48}Fs_{14}Wo_{38}$. Width of the two-pyroxene gap is larger than in the microsubophitic breccias (Fig. 4C). Ilmenite ranges in Fe/(Fe + Mg) from 0.63–0.82; armalcolite in 72549 has Fe/(Fe + Mg) between 0.41 and 0.51. Analysis of the interstitial K-rich phase is difficult because of its small grain size. Results of several measurements show that it is mostly K-feldspar of composition $An_3Ab_{1-2}Or_{95-96}$.

High-K KREEP matrix breccias (72558, 72735)

As noted earlier, these two Station 2 breccias are distinguished from the other crystalline matrix breccias studied by having higher K_2O contents and lower MgO/MgO + FeO values (Table 2). They also tend to have a higher proportion of clastic material (Table 1). In general, the matrices are characterized by irregular to subequant mafic grains surrounded by anhedral plagioclase (Fig. 1D). Matrix grain boundaries are extremely irregular in outline, with numerous tiny inflections. In 72735 most of the plagioclase present in the matrix occurs in fine-grained patches rather than as whole grains. Tiny ($\leq 2\ \mu$m) patches of a K-rich interstitial silicate phase are common, particularly within the fine-grained plagioclase areas. Analyses of this phase in 72735 show it to be K-feldspar. Ilmenite is the chief opaque mineral and is present as rods or equant grains several micrometers wide or as irregular micropoikilitic crystals. Chromite is occasionally present, both as discrete grains and as lamellae in ilmenite (associated with rutile). Baddeleyite is also present in the matrix. Small grains of metal and troilite are sparsely disseminated throughout the matrices of both 72558 and 72735.

Analyses of matrix plagioclase indicate a range in composition from An_{81-97} (plus one analysis of An_{71}) with a bimodal grouping containing maxima at An_{84-87} and An_{93-94} (Fig. 2D). Interstitial K-feldspar is mostly too small for electron microprobe analysis, but results of several good analyses show it to be of composition $An_{3-7}Ab_{3-6}Or_{90-94}$. The mafic phase in the matrix is almost wholly pyroxene; only one olivine grain (Fo_{63}) was encountered in matrix surveys of both 72558 and 72735. Compositionally, matrix pyroxenes are more Fe-rich than in other crystalline matrix breccias (Fig. 4D), reflecting the lower bulk matrix MgO/MgO + FeO values of these particular breccias (Table 2). As shown by Fig. 4D, pyroxenes in the high-K KREEP breccias are also more variable in composition. Ilmenite ranges in Fe/(Fe + Mg) from 0.80 to 0.83; chromite in 72735 ranges from 0.79–0.85 in Fe/(Fe + Mg).

CLAST POPULATIONS

The total volume of clastic material present in thin sections of the breccias we have studied ranges from 12% to about 70% (Table 1). For the most part, samples with an apparent high clast content are biased by the presence of one or two large lithic clasts. For example, 66% of our thin section of 77539 consists of a single granoblastic anorthosite clast, and 14.5% of 77545 is covered by a single dunite clast. Allowing for sampling problems such as these, the proportion of clastic material present is relatively uniform.

Mineral clasts

Abundance and relative proportions of mineral clasts are quite similar for the 13 breccias studied. Plagioclase is most abundant, followed by olivine and pyroxene. Small amounts of ilmenite, Fe–Co–Ni metal and troilite are present in many of the rocks; rare clasts of pink spinel, chromite, cristobalite, and K-feldspar have also been observed.

By volume, plagioclase constitutes about two-thirds of the mineral clasts in nearly all of the rocks studied. They vary in form from subangular to rounded: in a general way, the degree of angularity decreases between the finest-grained (microsubophitic) breccias and those with coarser-grained, partially recrystallized matrices (microgranular-micropoikilitic). Size of plagioclase clasts is generally seriate, ranging from a maximum of 0.4 to >1 mm in length down to matrix size grains. Marginal overgrowths are occasionally present, most commonly in breccias that have undergone matrix recrystallization. A small number of clasts display shock effects (e.g., undulose extinction; fractures with recrystallized veinlets). Compositionally, plagioclase mineral clasts exhibit an overall range from An_{63-99} (Fig. 2). The majority (>75%) of clasts analyzed coincide in composition with the characteristic matrix compositional range ($\sim An_{84-96}$); approximately equal numbers of clasts are more calcic and less calcic than this range. Zoning of plagioclase clasts is uncommon, but marginal overgrowths of different composition than crystal cores are fairly commonly observed, especially in the microgranular-micropoikilitic matrix breccias. Marginal overgrowths are generally An_{85-90} in composition, irrespective of whether the associated clast is more or less calcic.

Olivine mineral clasts exhibit a similar range in size as do plagioclase, but with proportionately fewer clasts at the maximum end of the range. The majority of olivine clasts are subrounded or rounded, but a few are subangular. Clast-matrix reaction is more pronounced among olivine clasts than plagioclase. Shock features are present in some grains. The overall range in compositions encountered is Fo_{59-92} (Fig. 3). For the most part, mineral clasts of olivine have cores at least as magnesian as matrix olivine. In the microsubophitic matrix breccias, many olivine clasts are significantly more magnesian than matrix olivine, whereas in the other crystalline matrix breccias most clasts are identical in composition with the matrix and only a few are more magnesian (Fig. 3). The more Mg-rich olivine clasts often are surrounded by reaction rims with compositions approaching that of the matrix olivine.

Pyroxene mineral clasts are fairly abundant in most crystalline matrix breccias, but are absent in poikilitic breccias 77515 and 77545. The overall size range encountered is similar to that for olivine and plagioclase clasts. Like olivine, pyroxene clasts tend to be subrounded or rounded rather than subangular. A number of pyroxene grains have been shocked, as evidenced by the presence of undulose extinction, strain fractures and, in one crystal, kink banding. Pyroxene mineral clasts are predominantly orthopyroxene. Many orthopyroxene grains contain exsolutions of high-Ca pyroxene. In most cases the high-Ca pyroxene is present as very narrow (1–2 μm and less up to a maximum of $\sim$5 μm thick) parallel lamellae. Oriented submicron-sized exsolutions of an opaque phase (probably ilmenite) are occasionally present as well. A few pyroxene grains consist of approximately equal amounts of orthopyroxene and high-Ca pyroxene, present as alternating subparallel lamellae and blebs 10–20 μm or more thick. Discrete mineral clasts of pigeonite and high-Ca pyroxene are also found. Representative analyses of pyroxene clasts are shown in Fig. 5. Orthopyroxene clasts vary in composition from $En_{84}Fs_{12}Wo_4$ to $En_{64}Fs_{33}Wo_3$; many are more Mg-rich than average matrix pyroxenes. Pigeonite clasts trend to more Fe-rich compositions, with the majority being 15–25 mol.% richer in $FeSiO_3$ than typical matrix pyroxenes. Most high-Ca pyroxene mineral clasts analyzed also tend to be more Fe-rich than corresponding matrix pyroxenes. Zoning in pyroxene mineral clasts is rare; Fig. 5 shows a traverse across one unusual crystal continuously zoned from a ferropigeonite core to an augite rim. Pyroxenes with extreme compositions (either at Mg-rich or Fe-rich end) sometimes have marginal overgrowths with compositions close to that of matrix grains.

Opaque oxide clasts observed are mostly ilmenite and are most abundant in

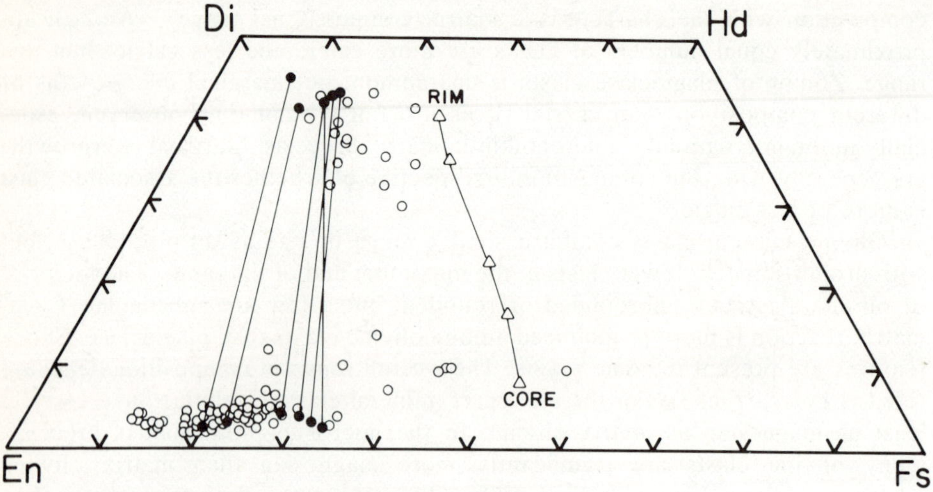

Fig. 5. Compositions of pyroxene mineral clasts projected onto pyroxene quadrilateral. Connected solid circles represent coexisting host-lamellae pairs. Connected triangles represent core to rim traverse across single zoned mineral clast.

the microsubophitic matrix breccias. Ilmenite clasts are small (<0.2 mm), generally subequant in shape and rounded or subrounded in form, and have compositions (Fe/(Fe + Mg), 0.75–0.83) similar to matrix ilmenite. Subangular to subrounded clasts (up to 0.4 mm long) of pink spinel are fairly common in microsubophitic breccia 72539 and occasionally present in other crystalline matrix breccias. Spinel clasts in microgranular-micropoikilitic matrix breccias contain plagioclase reaction coronas similar to those described by Dymek *et al.* (1976). The spinels contain (in wt.%) 57–63% Al_2O_3, 5–8% Cr_2O_3, 9–18% FeO, and 16–23% MgO. Rare small (<0.1 mm) mineral clasts of chromite-rich (41–47 wt.% Cr_2O_3) spinel are present in a few (microsubophitic) breccias.

Clasts of metal and troilite, or of both together, are present in most of the rocks studied (Table 1). They generally occur as irregular globules up to 0.3 mm in diameter. The most common metal contains 6–8 wt.% Ni and 0.7–0.9 wt.% Co; a few grains have edges that are richer in Ni (10–15 wt.%) but poorer in Co (0.5–0.7 wt.%). A second group of metal analyses cluster near 20 wt.% Ni and 1.6 wt.% Co.

Lithic clasts

The most common lithic clasts encountered are of ANT material; devitrified clasts of anorthositic composition and fine-grained breccia clasts are also common (Table 1). The majority of lithic clasts fall within the size range exhibited by plagioclase and mafic mineral clasts (0.1–1 mm). A number of clasts in the 1–2 mm range are exposed in the thin sections studied, but only one clast >2 mm

Table 3. Broad beam bulk analyses of selected lithic clasts.

	(1)	(2)	(3)	(4)	(5)
SiO_2	44.6	44.3	41.8	43.2	48.1
TiO_2	.06	.58	.10	.34	1.65
Al_2O_3	35.6	34.7	17.1	14.2	19.9
Cr_2O_3	n.a.	.01	.09	.15	.16
FeO	.29	.54	8.2	8.8	7.3
MnO	.01	*	.10	.12	.14
MgO	.30	.35	23.0	24.2	11.0
CaO	19.3	18.1	9.1	8.2	12.5
Na_20	.55	.80	.44	.42	.56
K_2O	.09	.22	.12	.14	.18
P_2O_5	n.a.	.09	.04	.10	.28
Total	100.8	99.7	100.1	99.9	101.8
MgO/MgO + FeO	.51	.39	.74	.73	.60

n.a. = not analyzed. * – less than 0.01 wt.%.

Columns: (1) Granoblastic anorthosite clast (~6 mm) in 77539.
(2) Devitrified anorthosite clast (~2 mm) in 72736.
(3) Troctolite clast (1 × 1.3 mm) in 72735.
(4) Troctolite clast (~1.5 mm) in 72735.
(5) Microgranular breccia clast (~2 mm) in 72736.

(a ~6 mm granoblastic anorthosite clast in 77539). Broad beam bulk analyses of selected lithic clasts are given in Table 3.

Most ANT clasts consist of fine-grained (<0.1 mm) material that has been moderately to extensively recrystallized. The more highly annealed clasts have a granoblastic texture with 120° triple junctions. Several coarser-grained clasts (chiefly anorthositic) also have a granoblastic texture. A few coarse-grained clasts have a poikilitic texture and are characterized by large (up to 1 mm) oikocrysts of low-Ca (in some cases high-Ca) pyroxene which enclose abundant plagioclase chadacrysts and minor rounded olivine grains (Fig. 6A). The plagioclase grains typically have rounded outlines, suggesting a metamorphic origin for the poikilitic texture, or at least prolonged metamorphic annealing.

Only two ANT clasts have been observed in which an original igneous texture has been preserved. Both are trocolites and both occur in high-K KREEP breccia 72735. One clast (~1 × 1.3 mm) has a cumulate texture, with 0.1–0.2 mm anhedral olivine (Fo_{86-87}) grains poikilitically enclosed by a few large (>1 mm long) plagioclase (An_{92}) crystals (Fig. 6B). The clast also contains one ~20 μm diameter crystal of pink spinel and trace amounts of ilmenite plus chromite. A bulk analysis of the clast is given in Table 3, column 3. The other troctolite clast is ~1.5 mm in diameter. It has a basaltic (intergranular) texture but has subsequently suffered minor shock deformation. Olivine grains are of

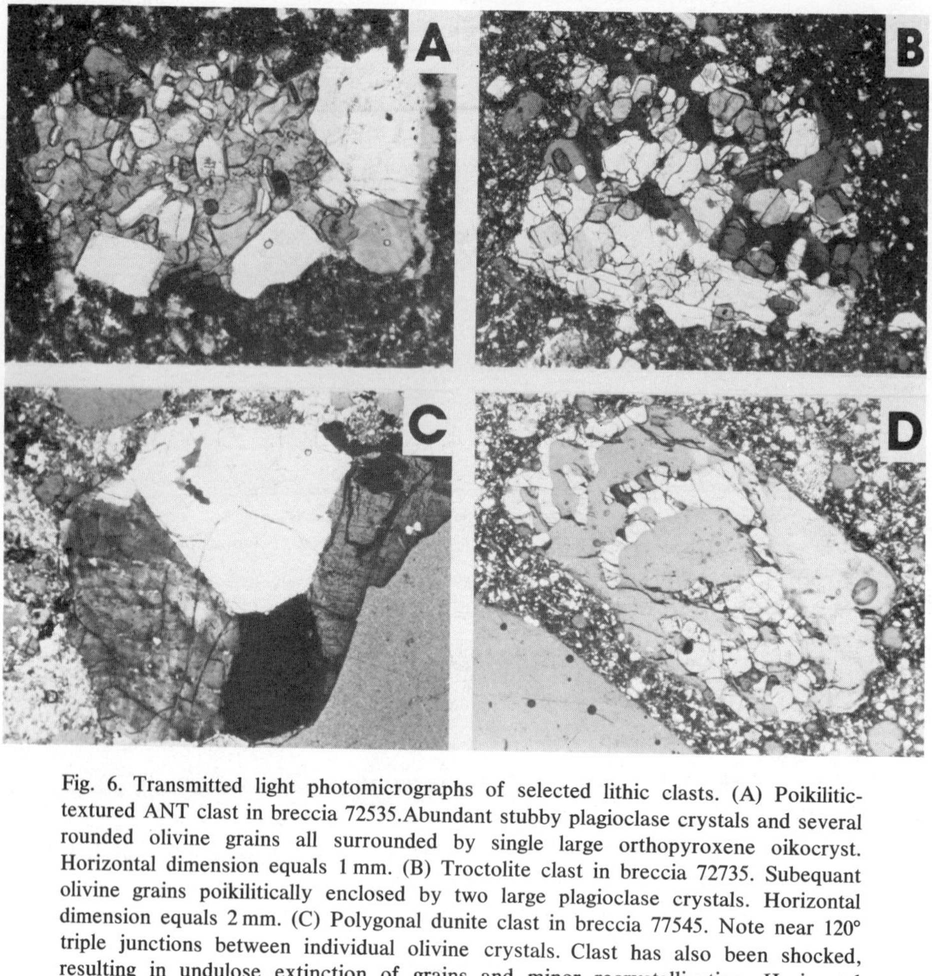

Fig. 6. Transmitted light photomicrographs of selected lithic clasts. (A) Poikilitic-textured ANT clast in breccia 72535. Abundant stubby plagioclase crystals and several rounded olivine grains all surrounded by single large orthopyroxene oikocryst. Horizontal dimension equals 1 mm. (B) Troctolite clast in breccia 72735. Subequant olivine grains poikilitically enclosed by two large plagioclase crystals. Horizontal dimension equals 2 mm. (C) Polygonal dunite clast in breccia 77545. Note near 120° triple junctions between individual olivine crystals. Clast has also been shocked, resulting in undulose extinction of grains and minor recrystallization. Horizontal dimension equals 2 mm. (D) Graphic granite clast in breccia 72539. Clast consists of single large crystal of K-feldspar, zoned crystal of ternary feldspar composition (right edge), and single wormy silica grain. Horizontal dimension equals 1 mm.

similar size, shape and composition (Fo_{86-87}) as in the first clast. Plagioclase grains tend to be smaller and occur as irregular crystals (some of which are poikilitic) and laths (up to 0.8 mm long); they range from An_{91-96}. Orthopyroxene ($En_{85-86}Fs_{11-12}Wo_{2-3}$) and high-Ca pyroxene ($En_{49}Fs_5Wo_{46}$) are present between olivine crystals. Scattered crystals of pink spinel (14–20 wt.% MgO; 11–20 % FeO; 51–59% Al_2O_3; 10–16% Cr_2O_3), ZrO_2-righ (~ 5 wt.%) armalcolite (Fe/(Fe + Mg), 0.53–0.58) and high-MgO (13–14 wt.%) ilmenite are also present. The bulk composition of this clast is slightly less feldspathic and more mafic than the first clast (Table 3, column 4).

A large polygonal dunite clast is present in 77545. It is part of a 6×6 mm angular clast exposed on the breccia surface (Keil *et al.*, 1974). The clast has a coarse granoblastic texture, with 0.5–1 mm size olivine grains intersecting at near 120° triple junctions (Fig. 6C). The clast has been shocked, resulting in undulose extinction of the olivine grains and minor recrystallization along fractures. It is composed almost entirely of olivine of composition $\sim Fo_{89}$; minor amounts of chromite are present, confined to narrow zones along olivine-olivine grain boundaries.

Devitrified clasts of anorthositic composition are a distinctive type that is present in all crystalline matrix breccias studied. Although many probably represent single crystal plagioclase that has been maskelynitized and devitrified, there are several large (~ 2 mm) clasts of this type which are considerably bigger than any plagioclase mineral clasts observed. Many devitrified anorthosite clasts, especially the larger ones, contain numerous small crystals of olivine, pyroxene, ilmenite, etc. A bulk composition of one of the larger clasts is given in Table 3 (column 2): its relatively high TiO_2 precludes derivation from single crystal plagioclase.

Fine-grained breccia clasts with a variety of matrix textures also constitute an abundant lithic clast type. The most common types encountered have an extremely fine-grained microgranular texture. Many breccia clasts are of similar mineralogy and bulk composition as the host breccias (e.g., Table 3, column 5); more feldspathic breccia clasts are also frequently present.

A few lithic clasts are present which cannot conveniently be grouped under the above categories. For the most part they include basaltic-textured clasts of uncertain origin. Unique among clast types observed is a coarse-grained clast with granitic mineralogy in 72539. It has a graphic texture (Fig. 6D) and consists of a single large crystal of K-feldspar ($An_3Ab_3Or_{94}$), a zoned feldspar grain of ternary composition ($\sim An_{49}Ab_{16}Or_{35}$), a large wormy silica grain, and a small crystal of ilmenite (Fe/(Fe + Mg), 0.91). The ternary feldspar composition reported here is similar to those that have been reported by Ryder *et al.* (1975—Table 2, columns 8–10) from Boulder #1, Apollo 17, Station 2.

Discussion

Nature of breccia matrices

Textural aspects. The 13 crystalline matrix breccias studied exhibit a rather wide range of matrix textures. Four of the breccias have an extremely fine-grained microsubophitic texture indicative of rapid crystallization from a melt (presumably of impact origin). Another three breccias have a poikilitic texture, for which an origin by crystallization from an impact melt also seems likely (e.g., Simonds *et al.*, 1973). Breccias with a microgranular to micropoikilitic texture have had their original matrix textures somewhat obscured by solid state recrystallization: this recrystallization is manifested by a greater degree of homogenization of pyroxene compositions than is present in other crystalline matrix breccias (Fig. 4C). However, the fine grain size of the breccias, together with the lath-shaped habit of many plagioclase crystals, suggest that the matrices of these breccias also originated through crystallization from a melt. The origin of the matrix texture in the two breccias with high-K KREEP composition is ambiguous. Although fine grained, they lack lath-like plagioclase crystals that would be expected if the matrices crystallized from a melt. On the other hand, the extreme irregularity of matrix grain shapes on a fine scale, coupled with the relatively large spread in matrix mineral compositions (Fig. 4D), precludes the possibility of extensive solid state recrystallization.

Compositional aspects. A striking feature of the breccias studied is that, despite the variety of matrix textures displayed, matrix compositions show limited variability (Table 2). For example, the overall range in SiO_2 is from 46.4 to 50.2 wt.%; similarly, Al_2O_3 ranges from 17.4 to 20.3 wt.%, FeO from 7.4 to 8.8 wt.%, MgO from 8.1 to 11.6 wt.%, and CaO from 11.1 to 12.1 wt.%. This relative constancy in composition is reflected by the modal data as well: the

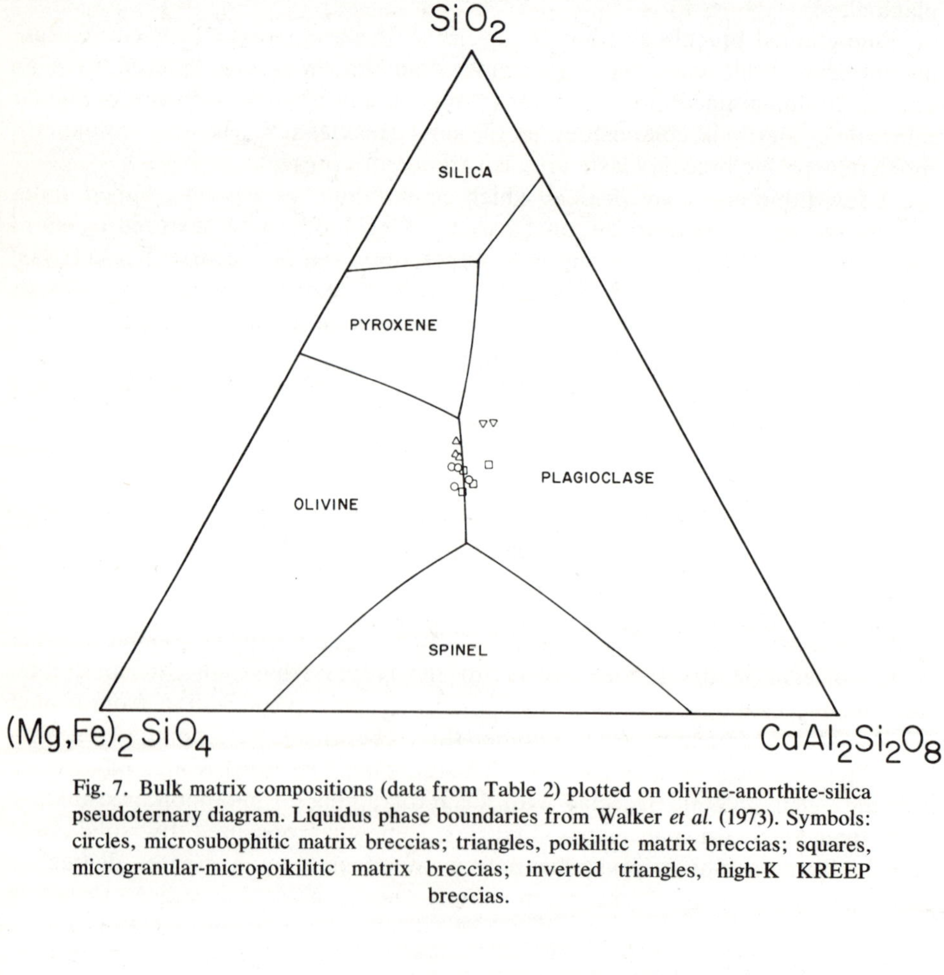

Fig. 7. Bulk matrix compositions (data from Table 2) plotted on olivine-anorthite-silica pseudoternary diagram. Liquidus phase boundaries from Walker *et al.* (1973). Symbols: circles, microsubophitic matrix breccias; triangles, poikilitic matrix breccias; squares, microgranular-micropoikilitic matrix breccias; inverted triangles, high-K KREEP breccias.

amount of matrix plagioclase varies only from 50 to 54 vol.%, while that of olivine plus pyroxene varies only from 41 to 46 vol.% (Table 1).

The significance of the matrix compositions observed is best understood with reference to the pseudoternary liquidus diagram for the system olivine-anorthite-silica (Walker *et al.*, 1973). Figure 7 is a plot of the bulk matrix compositions (from Table 2) on this diagram. It shows that all but three of the crystalline matrix breccias have matrix compositions that lie along or very near the plagioclase-olivine cotectic. These include all four microsubophitic matrix breccias, all three poikilitic-textured breccias, and three of the microgranular-micropoikilitic matrix breccias. The three points that lie away from the plagioclase-olivine cotectic (all in the plagioclase field) correspond to the two high-K KREEP breccias and microgranular breccia 72548.

In light of current models proposing that the matrices of many highland breccias represent impact total melts (Grieve *et al.*, 1974; Dence *et al.*, 1976), the observed cotectic control over matrix compositions (Fig. 7) is somewhat difficult to explain. Such models would require that the target material itself had compositions lying along the cotectic. Thus, a two-stage evolution would be involved, in which a prior episode of partial melting produced compositions along the plagioclase-olivine cotectic, and then in a subsequent impact event (or events) these compositions were totally remelted and mixed with foreign clasts to produce the crystalline matrix breccias at the Apollo 17 site. As pointed out by James (1976), a major difficulty with this hypothesis is that lithic clasts of suitable source rocks are lacking (nearly all returned highland rocks are more feldspathic than the matrix compositions).

Alternatively, the cotectic control inferred for most Apollo 17 crystalline matrix breccias could be explained if these breccias represent impact generated partial melts, as previously suggested by Warner *et al.* (1974). In light of this alternative, it is interesting to note that of the crystalline matrix breccias studied in this investigation, those with textures most indicative of crystallization from a melt (i.e., microsubophitic and poikilitic matrix breccias) all have matrix compositions lying along the plagioclase-olivine cotectic, whereas the two breccias with textures least indicative of melt crystallization (high-K KREEP breccias 72558 and 72735) have matrix compositions that lie away from the cotectic (Fig. 7).

Nature of clast assemblages

Comparison of the relative proportions of plagioclase and mafic mineral clasts to their abundances in the breccia matrices reveals that the clastic component is significantly richer in plagioclase than the matrix component (Table 1). By volume, the ratio of plagioclase mineral clasts to those of olivine and pyroxene is close to or greater than 2 : 1 in nearly all of the breccias examined. In contrast, the relative matrix proportions of plagioclase to olivine and pyroxene is $1.2 \pm 0.1 : 1$. The lithic clast component is predominantly ANT plus devitrified anorthosite, and thus also tends to be more plagioclase-rich than the matrix.

Average mineral compositions in the clast component are also different from those of matrix minerals. For example, the average composition of plagioclase in mineral and lithic clasts is more calcic than the average matrix plagioclase composition (An_{91} vs. An_{89}). This is true not only for the 13 breccias studied as a whole, but for each individual textural category as well. Olivine in mineral and lithic clasts is more magnesian than matrix olivine (average, Fo_{74} vs. Fo_{72}). In this case, there is a more pronounced difference in the microsubophitic matrix breccias (average in clasts, Fo_{76} vs. average in matrix, Fo_{71}), less pronounced in poikilitic matrix breccias (Fo_{74} vs. Fo_{72}) and no difference in composition in microgranular-micropoikilitic matrix breccias (both Fo_{72}). Presumably this pattern does not reflect a different clast assemblage in the different breccia types but rather selective digestion of the most magnesian clasts in the more recrystallized microgranular-micropoikilitic breccia matrices. Pyroxene in mineral and lithic clasts also averages somewhat higher MgO/MgO + FeO than does matrix pyroxene.

Lithic clasts encountered consist predominantly of reworked material (i.e., recrystallized ANT clasts, devitrified clasts of anorthositic composition, fine-grained breccia clasts). The two troctolite clasts in 72735 are an exception in that they preserve an original, possibly primary, igneous texture. Compared to other samples considered as likely candidates for primary igneous lunar highland rocks, such as spinel troctolites 67435 (Prinz *et al.*, 1973) and 65785 (Dowty *et al.*, 1974) and troctolite 76535 (Gooley *et al.*, 1974; Dymek *et al.*, 1975), the two troctolite clasts in 72735 are considerably finer grained. However, they are roughly similar in bulk composition (e.g., MgO/MgO + FeO) and mineralogy. In particular, mineral compositions in the troctolites are nearly identical to those reported by Dymek *et al.* (1975) for troctolite 76535 (e.g., Fo_{86-87} vs. Fo_{86-88}; $En_{85-86}\ Fs_{11-12}\ Wo_{2-3}$ vs. $En_{86}\ Fs_{12}\ Wo_{2}$; $En_{49}\ Fs_{5}\ Wo_{46}$ vs. $En_{48.5}\ Fs_{4.4}\ Wo_{43.6}$; An_{91-96} vs. An_{95-97}), so that a close genetic association between the clasts in 72735 and coarse-grained troctolite 76535 appears probable.

The coarse-grained clast of granitic mineralogy in 72539 is further evidence for the presence of a granitic source material within the lunar crust. Its graphic texture and the presence of feldspar with ternary composition suggest derivation from a similar source as the granite clasts described by Ryder *et al.* (1975) from Boulder #1 (also from Station 2).

Relation to other Apollo 17 breccias

The type and variety of matrix textures exhibited by the 13 breccias studied in this investigation are very similar to those reported for other Apollo 17 breccias. For example, the microsubophitic and microgranular-micropoikilitic matrix textures observed in our Station 2 rake samples closely resemble matrix textures encountered in Boulders #2 and #3 from Apollo 17, Station 2 (Dymek *et al.*, 1976; Simonds *et al.*, 1974), Station 3 breccias such as 73215 (James *et al.*, 1975; James, 1976) and 73235 (Dence *et al.*, 1976), Unit C (76255, 76275, 76295) of the Station 6 boulder (Simonds, 1975; Warner *et al.*, 1976), and a few Station 7 breccias (such as 77075 and 77115—Simonds *et al.*, 1974). Similarly, the poikilitic matrix texture exhibited by Station 7 rake samples (77515, 77539, 77545) closely resembles matrix textures described in Unit B (76015, 76215) of the Station 6 boulder (Simonds, 1975) and Station 7 breccia 77135 (Simonds *et al.*, 1974). Where bulk matrix compositions have been determined in these breccias, they are, for the most part, very similar to those we have analyzed (e.g., Simonds, 1975; James, 1976; Dymek *et al.*, 1976). That is, they tend to fall within a restricted range that plots close to the olivine-plagioclase cotectic in the olivine-anorthite-silica pseudoternary liquidus diagram (James, 1976).

We have found no apparent analogue among other Apollo 17 non-mare rocks for high-K KREEP breccias 72735 and 72558. For example, Simonds *et al.* (1974 — Table 1 and Fig. 4) report no Apollo 17 breccias with >0.5 wt.% K_2O. However, Bence *et al.* (1974) describe a high-K KREEP fragment from a 2–4 mm soil (72703): like 72735 and 72558, it is significantly more silica-rich than other crystalline matrix breccia types (Fig. 15—Bence *et al.*, 1974) and contains mafic phases more Fe-rich (e.g., olivine of $\sim Fo_{60}$).

Clast types and populations observed in other Apollo 17 breccias are similar to those described in this report. Our observations that clast assemblages are more plagioclase-rich and that clast plagioclase is more calcic, olivine more forsteritic, and pyroxene more enstatitic than average matrix compositions have been noted by numerous investigators (e.g., Simonds, 1975; James, 1976; Dymek *et al.*, 1976). This relation has been observed in Apollo 14 breccias (Ryder and Bower, 1976a), Apollo 15 black and white breccias (e.g., Ryder and Bower, 1976b), and Apollo 16 poikilitic rocks (e.g., Simonds *et al.*, 1973), and so appears to be a general characteristic of lunar crystalline breccias.

Acknowledgments—We thank D. Lange for preparing polished probe mounts of the breccias; W. Mansker, H. Planner, and R. Warren for assistance in electron microprobe analysis; and G. Conrad for maintenance of electron microprobe facilities. This work was supported in part by NASA, grant NGL-32-004-063 (Klaus Keil, Principal Investigator).

References

Albee A. L. and Ray L. (1970) Correction factors for electron probe microanalyses of silicates, oxides, carbonates, phosphates and sulfates. *Anal. Chem.* **42**, 1408–1414.

Bence A. E. and Albee A. L. (1968) Empirical correction factors for the electron microanalysis of silicates and oxides. *J. Geol.* **76**, 382–403.

Bence A. E., Delano J. W., Papike J. J., and Cameron K. L. (1974) Petrology of the highlands massifs at Taurus-Littrow: An analysis of the 2–4 mm soil fraction. *Proc. Lunar Sci. Conf. 5th*, p. 785–827.

Dence M. R., Grieve R. A. F., and Plant A. G. (1976) Apollo 17 grey breccias and crustal composition in the Serenitatis Basin region. *Proc Lunar Sci. Conf. 7th*, p. 1821–1832.

Dowty E., Keil K. and Prinz M. (1974) Igneous rocks from Apollo 16 rake samples. *Proc. Lunar Sci. Conf. 5th*, p. 431–445.

Dymek R. F., Albee A. L., and Chodos A. A. (1975) Comparative petrology of lunar cumulate rocks of possible primary origin: Dunite 72415, troctolite 76535, norite 78235, and anorthosite 62237. *Proc. Lunar Sci. Conf. 6th*, p. 301–341.

Dymek R. F., Albee A. L., and Chodos A. A. (1976) Petrology and origin of Boulders #2 and #3, Apollo 17 Station 2. *Proc. Lunar Sci. Conf. 7th*, p. 2335–2378.

Gooley R., Brett R., Warner J., and Smyth J. R. (1974) A lunar rock of deep crustal origin: Sample 76535. *Geochim. Cosmochim. Acta* **38**, 1329–1340.

Grieve R. A. F., Plant A. G., and Dence M. R. (1974) Lunar impact melts and terrestrial analogs: Their characteristics, formation and implications for lunar crustal evolution. *Proc. Lunar Sci. Conf. 5th*, p. 261–273.

James O. B. (1976) Petrology of aphanitic lithologies in consortium breccia 73215. *Proc. Lunar Sci. Conf. 7th*, p. 2145–2178.

James O. B., Brecher A., Blanchard D. P., Jacobs J. W., Brannon J. C., Korotev R. L., Haskin L. A., Higuchi H., Morgan J. W., Anders E., Silver L. T., Marti K., Braddy D., Hutcheon I. D., Kirsten T., Kerridge J. F., Kaplan I. R., Pillinger C. T., and Gardiner L. R. (1975) Consortium studies of matrix of light gray breccia 73215. *Proc. Lunar Sci. Conf. 6th*, p. 547–577.

Keil K. (1967) The electron microprobe X-ray analyzer and its application in mineralogy. *Fortschr. Mineral.* **44**, 4–66.

Keil K., Dowty E., and Prinz M. (1974) *Description, Classification and Inventory of 113 Apollo 17 Rake Samples from Stations 1A, 2, 7 and 8.* NASA-Lyndon B. Johnson Space Center, Houston, Texas. 149 pp.

Laul J. C. and Schmitt R. A. (1975) Chemical composition of Apollo 17 samples: Boulder breccias (2), rake breccias (8), and others (abstract). In *Lunar Science VI*, p. 489–491. The Lunar Science Institute, Houston.

Murali A. V., Ma M.-S., Laul J. C., and Schmitt R. A. (1977) Chemical composition of breccias, feldspathic basalt and anorthosites from Apollo 15 (15308, 15359, 15382 and 15362), Apollo 16 (60618 and 65785), Apollo 17 (72435, 72536, 72559, 72735, 72738, 78526 and 78527) and Luna 20 (22012 and 22013) (abstract). In *Lunar Science VIII*, p. 700–702. The Lunar Science Institute, Houston.

Prinz M., Bunch T. E., and Keil K. (1971) Compositions and origin of lithic fragments and glasses in Apollo 11 samples. *Contrib. Mineral. Petrol.* **32**, 211–230.

Prinz M., Dowty E., Keil K., and Bunch T. E. (1973) Spinel troctolite and anorthosite in Apollo 16 samples. *Science* **179**, 74–76.

Ryder G. and Bower J. F. (1976a) Poikilitic KREEP impact melts in the Apollo 14 white rocks. *Proc. Lunar Sci. Conf. 7th*, p. 1925–1948.

Ryder G. and Bower J. F. (1976b) Sample 15445: Petrology. In *Interdisciplinary Studies by the Imbrium Consortium*, Vol. 2 (J. A. Wood, ed.). LSI Contr. No. 268D, Center for Astrophysics, Cambridge.

Ryder G., Stoeser D. B., Marvin U. B., and Bower J. F. (1975) Lunar granites with unique ternary feldspars. *Proc. Lunar Sci. Conf. 6th*, p. 435–449.

Simonds C. H. (1975) Thermal regimes in impact melts and the petrology of the Apollo 17 Station 6 boulder. *Proc. Lunar Sci. Conf. 6th*, p. 641–672

Simonds C. H., Warner J. L., and Phinney W. C. (1973) Petrology of Apollo 16 poikilitic rocks. *Proc. Lunar Sci. Conf. 4th*, p. 613–632.

Simonds C. H., Phinney W. C., and Warner J. L. (1974) Petrography and classification of Apollo 17 non-mare rocks with emphasis on samples from the Station 6 boulder. *Proc. Lunar Sci. Conf. 5th*, p. 337–353.

Walker D., Longhi J., Grove T. L., Stolper E., and Hays J. F. (1973) Experimental petrology and origin of rocks from the Descartes Highlands. *Proc. Lunar Sci. Conf. 4th*, p. 1013–1032.

Warner J. L., Simonds C. H., and Phinney W. C. (1974) Impact-induced fractionation in the lunar highlands. *Proc. Lunar Sci. Conf. 5th*, p. 379–397.

Warner J. L., Simonds C. H., and Phinney W. C. (1976) Apollo 17, Station 6 boulder sample 76255: Absolute petrology of breccia matrix and igneous clasts. *Proc. Lunar Sci. Conf. 7th*, p. 2233–2250.

Proc. Lunar Sci. Conf. 8th (1977), p. 2007–2027.
Printed in the United States of America

Petrology of 78155: An early, thermally metamorphosed polymict breccia

C. E. BICKEL

NASA Johnson Space Center, Houston, Texas 77058 and Department of Geology, San Francisco State University, San Francisco, California 94132

Abstract—Polymict breccia 78155 has a bulk composition of anorthositic norite and belongs to a small group of rocks that were brecciated and thermally metamorphosed before the final lunar bombardment of 4.0–3.9 AE. Much of the rock is granoblastic matrix that was produced by the thermal metamorphism of brecciated precursors. The matrix consists of anorthite and pigeonite with minor augite, olivine, Cr-spinel, ilmenite, troilite, and Fe–Ni–Co metal. The mafic silicates of the matrix have restricted compositional ranges. Minor mineral clasts (anorthite, pyroxene, and olivine) and rare lithic clasts are relics of the precursors. Some of the lithic clasts are fine-grained, igneous-textured anorthosite with very restricted mafic silicate compositions. These clasts formed by impact melting of anorthosite. A second group of lithic clasts consists of coarser-grained rocks with both igneous and metamorphic textures. Bulk and mafic silicate compositions of these clasts vary greatly. The clasts represent the lunar crust prior to 4.2 AE, the age of the granoblastic matrix. The persistence of uninverted pigeonite in matrix and clasts indicates that metamorphism was at a very high temperature and may have been terminated catastrophically. About 20% of the rock is angular, very fine-grained crushed material that is derived from the granoblastic matrix and clasts. This explains the friable nature of the rock and records that it was further brecciated after thermal metamorphism.

INTRODUCTION

LUNAR SAMPLE 78155 has an ^{40}Ar–^{39}Ar age of 4.2 AE (Turner and Cadogan, 1975) and thus is among the most ancient rocks returned from the moon. Early surveys of the lunar samples indicated that the rock is crushed and has a relatively coarse-grained, metamorphic texture (LSPET, 1973b; Wilshire, 1974). The rock has appeared in petrogenetic schemes (e.g., Hubbard *et al.*, 1974; Hollister, 1975; Taylor and Bence, 1975). Because it has not previously been studied in detail and the original lunar sample investigations yielded contradictory opinions as to whether the rock is monomict or polymict (LSPET, 1973a,b) there has been uncertainty about its mixed character (e.g., Hewins and Goldstein, 1975). This petrographic study establishes that the rock is indeed polymict and belongs to a small group of rocks (e.g., 67955, 76230, 77017, 79215) that appear to have been brecciated and thermally metamorphosed before the final lunar bombardment of 4.0–3.9 AE (Warner *et al.*, 1977a). Sample 78155 is important because its clast population reveals the nature of rocks that resided at or near the lunar surface before 4.2 AE.

Sample Description

Sample 78155 is a composite consisting of over 50 angular, friable, light-gray fragments, the largest measuring 6.5 × 4.5 × 3 cm, that together weigh 401 g (LSPET, 1973a). They were collected from a small, subdued crater located in distinctly darker-colored regolith on the wall of a 15 m diameter crater at Station 8 in the Taurus-Littrow Valley. Presumably these fragments represent the projectile responsible for the small crater.

This study of polished thin sections 78155, 7, 8, 42, 43, 45, 48, and ,104 cut from fragments ,1 ,22 and ,25 indicates that these fragments are essentially identical. Hereafter in this paper the samples will be treated as one rock.

Sample Composition and Terminology used

Sample 78155 is a holocrystalline, weakly coherent polymict breccia. Table 1 lists the major constituents of the rock together with their abundances, grain sizes, and sizes of clasts and fragments. Most of the rock is granoblastic matrix of anorthite, pigeonite, augite, olivine, and opaque minerals. The granoblastic texture is interpreted to have formed by extensive annealing of brecciated

Table 1. Lithology of 78155.

Lithology	Approximate % of rock*	Maximum size of fragment or clast (mm) #	Grain size (μm)
Granoblastic matrix	65	4.6 × 3.0	Anorthite: 20–100, mafic silicates: 2–40, oxides: 0.5–75
Mineral clasts			
Anorthite	10	3 × 2	—
Pyroxene†	3	0.8 × 0.6	—
Olivine	<1	0.6 × 0.2	—
Polymineralic lithic clasts			
Fine grained, felty textured anorthosite (Type I)	1–2	2.8 × 1.0	Finer grained: Anorthite: 8 × 40 to 16 × 80; mafic silicates: 8 × 12. Coarser grained: Anorthite: 20 × 160 to 30 × 200; mafic silicates: 15 × 25, wedge shaped up to 60 μm long
Annealed Type I	1–2	1.2 × 0.8	Same as Type I, but with greater ranges in sizes of mafic silicates
Medium and coarse grained (Type II)	<1	1.5 × 1.0	See Table 3
Crushed material	20	0.1	Angular fragments ≤ 100 μm across

*Mode based on visual estimates and limited point counting.
#All the constituents are seriate in size.
†Pigeonite is much more abundant than augite.

precursors. Much coarser-grained mineral clasts of the same silicates are set in the granoblastic matrix. There are two distinct varieties of polymineralic lithic clasts. The more abundant variety (Type I) is fine-grained anorthosite with a felty texture. The less abundant variety (Type II) consists of coarser-grained clasts of diverse composition and texture. Finally, there are small, angular rock and mineral fragments that compose veins and irregularly shaped regions. This incoherent crushed material appears to be locally derived from disaggregated granoblastic matrix and clasts. Clearly the sample was brecciated before and after it was thermally metamorphosed.

In this paper "grain size" refers to crystal size of coherent rock or minerals: granoblastic matrix or clasts in the granoblastic matrix. Grain size of the granoblastic matrix and annealed clasts is an effect of thermal metamorphism of the entire sample; grain sizes of clasts with relict textures reflect igneous and metamorphic processes that preceded the first brecciation. The term "matrix" always refers to granoblastic matrix. The terms "crushed material" and "fragments" refer to the incoherent products of the later brecciation. For purposes of tabulation "crushed material" is here defined as angular mineral and lithic fragments under 100 μm across. Because all types of mineral and lithic clasts found as fragments are also set in granoblastic matrix, the term "clast" can always be taken to denote inclusions in the granoblastic matrix. But it should be understood that some of the clasts became detached from the granoblastic matrix during the later brecciation.

Bulk composition

Table 2 presents an X-ray fluorescence analysis and CIPW norm of 78155 that essentially represents the granoblastic matrix and mineral clasts. The normative mineralogy of the bulk rock is that of an anorthositic norite.

PETROGRAPHY

Granoblastic matrix

Granoblastic matrix makes up most of 78155 and occurs as fragments that range up to millimeters across. The largest fragments are subrounded to subangular and the smallest are subangular to angular (Fig. 1A). They are separated by crushed material and by mineral clasts. The granoblastic matrix is composed of anorthite, pigeonite, augite, olivine, Cr-spinel, ilmenite, troilite, and Fe–Ni–Co metal. The mode varies within and between fragments, ranging from anorthositic norite to noritic anorthosite (Table 2). In a few regions the matrix is enriched in a particular mafic silicate near clasts of that mineral (Table 2, column 6).

Anorthite occurs as irregularly shaped polygons that display infrequent polysynthetic twinning and sharp to weakly undulose extinction. Shock-induced features such as patchy extinction and offset twin lamellae are not present.

Table 2. Composition of 78155.

Chemical analysis 78155,2*		CIPW norm		
SiO_2	45.57	Orthoclase	0.47	Normative plagioclase (mole%)
TiO_2	0.27	Albite	2.79	An 95.3
Al_2O_3	25.94	Anorthite	69.06	Ab 4.1
Cr_2O_3	0.14	Diopside	4.88	Or 0.6
FeO	5.82	Hypersthene	16.71	
MnO	0.10	Olivine	5.03	Normative pyroxene (mole%)
MgO	6.33	Chromite	0.21	Wo 11.1
CaO	15.18	Ilmenite	0.51	En 59.5
Na_2O	0.33	Apatite	0.09	Fs 29.4
K_2O	0.08	Pyrite	0.07	
P_2O_5	0.04	Total	99.82	(Fe + Mn)/(Fe + Mn + Mg) in normative
S	0.04			silicates = 33.0 atom%
Total	99.84			

*Analysis by J. M. Rhodes (LSPET, 1973b).

	Modes based on visual point counts		
	(1)	(2)	(3)
Anorthite	73.5	85.3	90–95
Mafic silicates	25.7	14.4	5–10
Cr-spinel and ilmenite	0.4	0.2	0.05 (Cr-spinel and ilmenite, troilite and metal)
Troilite and metal	0.4	0.1	
Points counted	1122	2871	7140

(1) Representative granoblastic matrix.
(2) Atypically coarse-grained, anorthositic granoblastic matrix.
(3) Compositional range of fine-grained, felty-textured anorthositic lithic clasts (Type I).

	Proportions of the mafic silicates (%), based on point counts with the electron microprobe								
	(1)	(2)	(3)	(4)	(5)	(6)	(7)	(8)	(9)
Pigeonite	74	86	79	74	64	29	55	69	61
Augite	12	4	6	24	29	12	0	20	9
Olivine	14	10	15	2	7	59	45	11	30
Points counted	184	74	33	46	14	17	22	18	40

(1) Average of granoblastic matrix.
(2)–(5) Granoblastic matrix of individual fragments.
(6) Granoblastic matrix near a clast of olivine.
(7) Two Type I lithic clasts.
(8) Another Type I lithic clast.
(9) Average of (7) and (8).

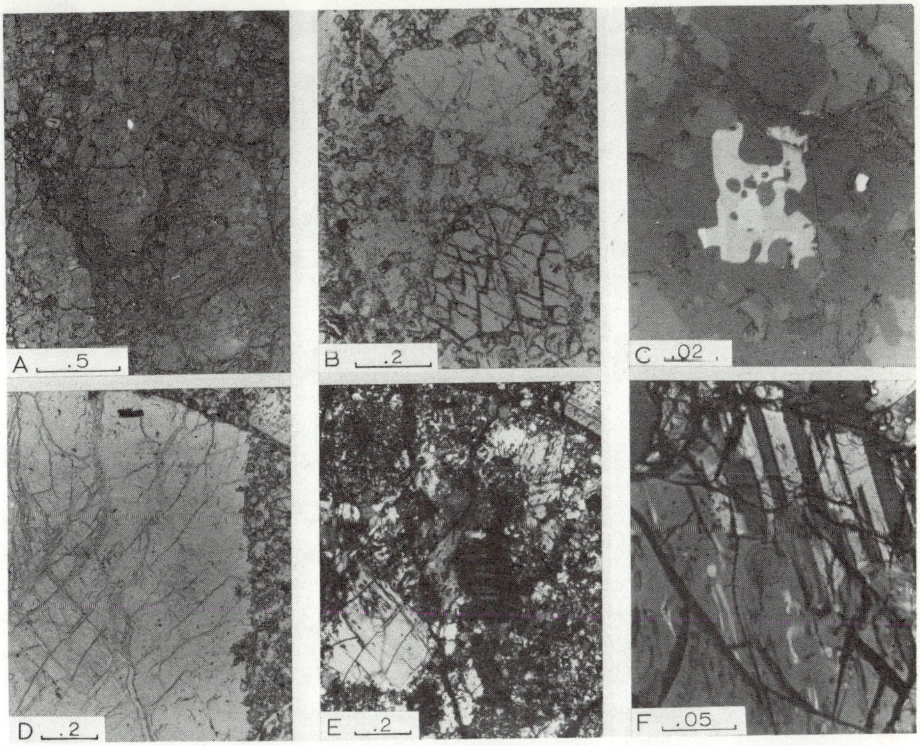

Fig. 1. Textural features of 78155. Scales in millimeters. (A) Illustration of the rock's incoherence. Fragments of granoblastic matrix are set in crushed material. Reflected light. (B) Granoblastic matrix enclosing a small clast of anorthite and of pyroxene. Plane polarized light. (C) Poikiloblastic Cr-spinel with exsolved ilmenite and minor Fe metal. The metal occupies the tip of the promontory at the lower left corner of the spinel. The spinel is surrounded by anorthite and mafic silicates of the granoblastic matrix. Reflected light. (D) Contact between a large anorthite clast and granoblastic matrix (,48). Plane polarized light. (E) Same view as Fig. 1D with crossed polarizers, illustrating shock induced features (felty texture) of the clast. (F) Low-Ca pyroxene clast of 78155,48. The polysynthetic twinned pyroxene is pigeonite (Wo_{10}) and the rest of the clast (gray) is very low-Ca pyroxene (Wo_{4-5}) with small, irregular patches of exsolved augite (white wisps). Crossed polarizers. The Wo content of the grain from lower left to upper right across the photo is plotted in Fig. 6.

Smaller grains of mafic silicates and opaque minerals are located primarily along the boundaries and triple junctions between anorthite polyhedra (Fig. 1B). Grains of mafic silicates larger than about 10–20 μm are irregular in shape, with scalloped boundaries. Smaller grains are bead shaped with smooth boundaries. Many of the smallest grains are enclosed within anorthite.

The most abundant oxide is Cr-spinel; most occurs as subhedral grains, from 0.5 to 75 μm across, many of which are poikiloblastic with rounded inclusions of anorthite. The Cr-spinel also contains exsolved ilmenite and Fe metal (Fig. 1C). Rarely the oxides occur in anorthite-mantles complexes (up to 200 μm in total diameter) similar to those of sample 79215 (Bickel *et al.*, 1976). In these complexes a central cluster of spinel and ilmenite, with or without metal, is surrounded by anorthite crystals with or without blebs of oxide and metal. Rare ilmenite occurs as individual, lath-shaped grains not associated with spinel.

Fe–Ni–Co metal and troilite occur individually and as composite grains that are mostly about 2–6 μm and generally range up to 50 μm across. The metal and troilite are irregularly shaped with cuspate boundaries or are rounded. Metal is about twice as abundant as troilite.

Monomineralic clasts

Monomineralic clasts of anorthite, pyroxene, and olivine occur within the granoblastic matrix and as fragments in the crushed material. They display continuous ranges in size, and the smallest clasts are difficult to distinguish from the matrix grains.

Anorthite. Over 95% of the anorthite clasts are irregularly shaped, monocrystalline fragments that have irrational boundaries and display polysynthetic twinning and sharp or weakly undulose extinction. Near their edges some of these clasts contain blebs of mafic silicates, generally about 2–8 μm across, which form "necklaces". Others contain isolated blebs or trains of mafic silicates. Very few anorthite clasts contain blebs of Mg–Al spinel. In rare monocrystalline anorthite clasts microfaults offset twin planes and crystal boundaries, presumably in response to shock. The extinction of other rare clasts is irregular and patchy. Four of the sectioned clasts consist entirely of felty-textured intergrowths of weakly elongated but generally anhedral grains that average about 50 μm in length. These clasts were observed only within one region that measures about 2 × 2 mm (78155,45). A particularly large clast contains similar fine-grained felty anorthite irregularly intergrown with patches (hundreds of micrometers across) of strongly undulose anorthite (Figs. 1D and 1E). This suggests that these two textures may have been produced by a single episode of shock followed by annealing; the felted areas were the more severely shocked and may have been vitrified. Rare polycrystalline clasts consist of mosaics of coarse anorthite. Different mosaic-textured clasts have slightly different grain sizes, generally ranging from 80 to 160 μm. The mosaic texture presumably results from thorough annealing of strongly shocked or vitrified anorthite. All the types of anorthite clasts are found within identical granoblastic matrix.

Mafic silicates. The most abundant mafic clasts are pigeonite, but there are numerous clasts of augite and olivine and very rare clasts of very low-Ca pyroxene with probable orthorhombic symmetry. The clasts have irregular or blockly shapes with scalloped boundaries and exhibit sharp or weakly undulose extinction. Rare pigeonite and augite clasts are polysynthetically twinned (Fig. 1F). The planar twin lamellae range from 1 to 35 μm across and are discontinuous. Most pyroxene clasts do not display microscopic exsolution features; a few have exsolution lamellae that generally are under 1 μm across. Rare clasts consist of mosaics of pyroxene or olivine and thus are monomineralic, but polycrystalline. Their origin is probably the same as that of the mosaic-textured anorthite clasts—i.e., annealing of shocked minerals.

Polymineralic lithic clasts

Polymineralic lithic clasts are set in crushed material and in the granoblastic matrix. Thus they must have experienced the inferred thermal metamorphism responsible for the development of the granoblastic matrix.

Fine-grained, felty-textured anorthosite clasts (*Type I*). Fine-grained clasts with the composition of anorthosite are distinguished from the granoblastic matrix by the following criteria (Figs. 2A, 2B):

(1) The clasts are exceptionally rich in anorthite (Table 2).
(2) The anorthite occurs as tabular crystals with abundant carlsbad twinning.
(3) Wedge-shaped or nearly equant grains of pigeonite and olivine occupy the interstices between the anorthite laths. This relationship is the same as in intergranular volcanic rocks, but the texture is best termed "felty" to emphasize the very high feldspar content of the clasts. The texture strongly suggests that these clasts crystallized from an anorthosite melt.
(4) Augite is exceptionally rare in these clasts.
(5) Within an individual clast the mafic silicates have a much smaller range of grain size and shape than in the granoblastic matrix. Rarely several adjacent mafic silicates are in optical continuity.
(6) The average grain size is smaller than in the granoblastic matrix.

Grains of Cr-spinel with exsolved ilmenite and Fe metal, platy ilmenite, and composite grains of Fe metal and troilite are very rare in these clasts. Some of the clasts contain inclusions of blocky, relatively coarse anorthite with or without "necklaces" of mafic silicates (Figs. 2A and 2B). These inclusions, which average about 5% of the Type I clasts, are probably xenocrysts. Inclusions of irregular-shaped xenocrystic (?) pigeonite are very rare.

There is a continuum between clasts with igneous textures, which have sharp boundaries with the granoblastic matrix, and rare, generally granoblastic-textured regions of the matrix in which a few lath-shaped anorthite grains are present. As part of the continuum there are indistinctly bounded clasts in which most of the anorthite is somewhat elongated, but is untwinned and irregular in shape. The mafic silicates in these clasts are not wedge shaped and have a greater range of size than in the distinctly igneous-textured clasts. The textural variation of the Type I lithic clasts is interpreted to be a function of their pre-annealing strain. Those that were most strained have undergone the most profound solid-state recrystallization, and modification of their original igneous textures.

Medium- and coarse-grained lithic clasts (*Type II*). One coarse-grained and 12 medium-grained lithic clasts were distinguished from the granoblastic matrix

(Fig. 2; Table 3). Their individual bulk compositions have limited significance because they are small relative to their grain sizes. Most range mineralogically from anorthositic norite to noritic anorthosite.

The medium-grained clasts belong to four textural-compositional groups (Types IIA–IID, Table 3). Seven clasts have igneous textures (Type IIA; Figs. 2C and 2D) and are distinguished by the presence of tabular plagioclase. Type IIA clasts with under 80% plagioclase have subophitic textures in which optically continuous pyroxenes partly enclose euhedral laths of plagioclase; the pyroxenes are separated by coarser, anhedral plagioclase. In three Type IIA clasts irregular-shaped or rounded olivines are enclosed by pyroxene or separate

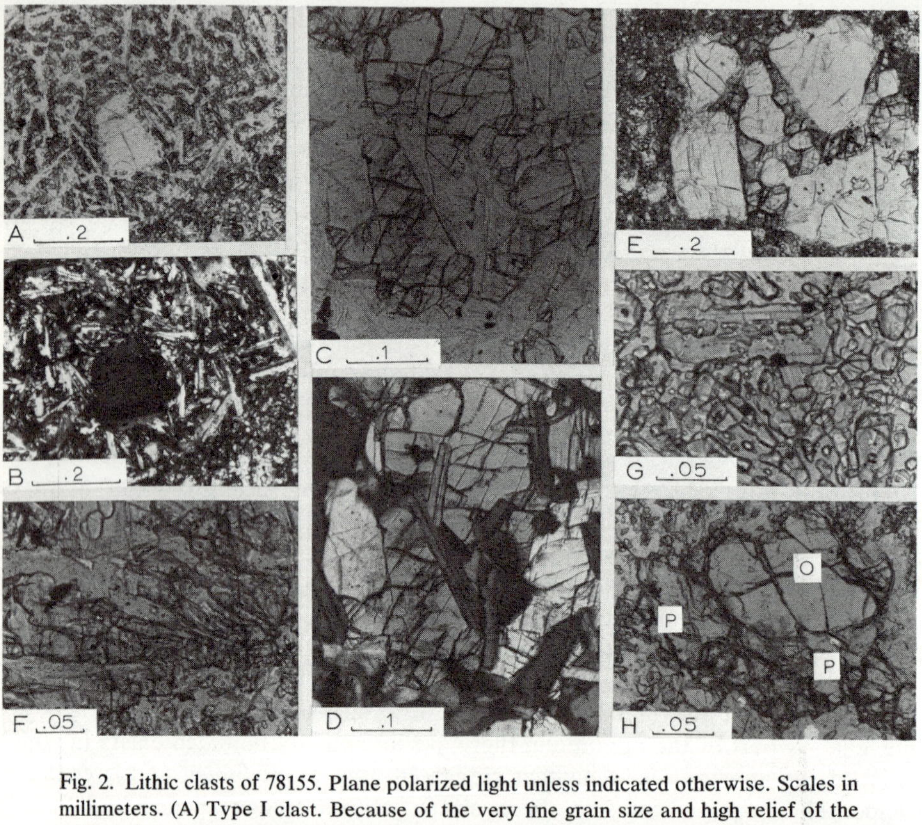

Fig. 2. Lithic clasts of 78155. Plane polarized light unless indicated otherwise. Scales in millimeters. (A) Type I clast. Because of the very fine grain size and high relief of the mafic minerals their apparent abundance (viewed with transmitted light) is much greater than their true abundance (viewed with reflected light). The large crystal of anorthite with a necklace of mafic minerals probably is a xenocryst. (B) Same view as Fig. 2A, with crossed polarizers. (C) Type IIA clast (,48). (D) Same view as Fig. 2C, with crossed polarizers, illustrating subophitic texture: pigeonite partly encloses tabular plagioclase. (E) Type IIB clast (,42). The pyroxene is optically continuous. (F) Type IIC clast. Pyroxene has recrystallized as sub-radiating columnar crystals. (G) Type IID clast. The clast is composed almost entirely of olivine and anorthite. (H) Type IIE clast. Rounded olivine (O) is partly mantled by pigeonite (P).

the anhedral plagioclase grains. In the Type IIA clast with over 80% plagioclase optically continuous, wedge-shaped pyroxene grains occupy the interstices between laths of plagioclase.

The textures of Types IIB, C, and D lithic clasts are interpreted as metamorphic textures that predate the thermal metamorphism that produced the granoblastic matrix. Three Type IIB clasts consist of coarse, polygonal grains of anorthite that are separated and enclosed by coarse, optically continuous pyroxene. (Fig. 2E). Their textures are poikiloblastic (or possibly poikilitic), and are similar to that of 77017. A single Type IIC clast is distinguished by subradiating, skeletal, columnar pyroxene grains enclosed or separated by subhedral plagioclase (Fig. 2F). Wedge- and irregular-shaped pyroxene grains are also present. A single Type IID troctolitic clast is distinct in composition as well as texture (Fig. 2G). The olivines are seriate in size. Olivine grains longer than 40 μm are rod shaped ($\leq 240 \times 50\ \mu$m) and grains less than 40 μm long are slightly elongated or equant ($\geq 10\ \mu$m). Some of the larger grains contain rounded inclusions of anorthite and opaque minerals. Pigeonite is very rare. Euhedral or poikiloblastic spinel ranges from 2 to 40 μm across. Poikiloblastic anorthite encloses the other minerals.

The coarse-grained precursors that contributed the large mineral clasts of 78155 appear to be represented by a single coarse-grained, Type IIE lithic clast (Fig. 2H). The olivine of this clast is rounded and mantled by pigeonite.

Crushed material

78155 has been fractured and crushed into fragments that range from 4.6×3 mm to under a micrometer across. Angular mineral and lithic fragments under 100 μm across are referred to as crushed material. Together with slightly larger fragments, crushed material forms anastomosed networks of veinlets and irregularly shaped regions (Fig. 1A). Veinlets of crushed material generally are under 200 μm across and regions of crushed material devoid of larger fragments measure, at most, about 500 μm across. The crushed material includes up to about 40% pore space. In many regions the fragments can be fitted together. Apparently crushing generally involved differential movements of, at most, hundreds of micrometers. Open fractures from 1 to 20 μm across cut the crushed material and coherent fragments.

Mineral Chemistry

Representative chemical analyses of silicates and oxides are presented in Table 4. No phosphates were observed in 78155. The composition of the metal is discussed by Hewins and Goldstein (1975).

Anorthite

The average composition of the plagioclase in 78155 is about $An_{94.5}Ab_{4.7}Or_{0.8}$. Figure 3 demonstrates that the anorthite in the Type I lithic clasts displays the least compositional range. Anorthite of the monomineralic anorthite clasts, Type II lithic clasts, and the granoblastic matrix displays about the same

Table 3. Characteristics of representative

Textural type	Thin section	Size of clast (mm)	Approx. modal plag. (%)	Ca/Ca + Na of plag. (atom%)	Mafic mineralogy
II-A	,42	1.5 × 1.0	75	94.5 zoned	Pig + oliv(?)
Igneous	,47	0.4 × 0.25	60	94.5	Oliv ⩾ pig
	,48	0.9 × 0.6	70	96	Pig > aug
	,48	0.3 × 0.15	60	—	Pyx > oliv
	,104	0.6 × 0.3	85	97	Pig ~ aug
	,104	1.0 × 0.25	40	95	Pig ⩾ oliv ~ aug
II-B	,42	0.6 × 0.5	85	94	Augite
Poikiloblastic	,48	0.35 × 0.15	80	—	Pyx intergrowths
	,104	0.7 × 0.3	80	96	Pig + pyx intergrowths
II-C radiating, columnar pyroxene	,48	1.0 × 0.4	74	92.5	Pig > aug
II-D granoblastic troctolite	,48	1.5 × 1.2	50	95	Oliv ⩾ pig
II-E coarse-grained igneous	,48	0.7 × 0.4	25	—	Oliv ~ pig

compositional range with respect to Ab and An, but the Or content of the coarse anorthite clasts is slightly less than elsewhere in the rock. Anorthite of at least one of the Type IIA lithic clasts is compositionally zoned (Fig. 3D). The FeO content is about 0.3 wt.% in the monomineralic clasts, 0.3–0.5% in the granoblastic matrix, 0.5% in the Type I lithic clasts, and ranges from 0.4 to 0.9% in the Type II lithic clasts.

Olivine

Most of the olivine in 78155 has a limited range of composition (Fig. 4). Olivines from monomineralic clasts and the granoblastic matrix have identical compositional ranges (Fa_{35-38}). Olivines from Type I lithic clasts are very slightly enriched in iron (Fa_{36-40}), and olivines from Type II lithic clasts display a relatively great compositional range (Fa_{26-38}). The Ca content of the olivine is low (average 0.14–0.19 wt.% CaO) except in olivine from Type I lithic clasts (average 0.4 wt% CaO). The high CaO content of the olivine in these fragments

Type II lithic clasts in 78155.

Fe/Fe + Mg(atom%)			Average grain size (μm)	
Pigeonite	Augite	Olivine	Plagioclase	Mafic silicates
17	—	—	150 × 60	Oliv: 50–100; pig: poikilitic up to 600 across
25	—	28.5	160 × 20 and 80 × 80	Oliv: 120 × 80; pig: 240
33.5	29	—	150 × 25 and 200 × 100	500
—	—	—	30 × 40	Pyx: 250 × 100; oliv: 30
33	29.5	—	250 × 60 and 160 × 160	200 × 50
33	29.5	38	160 × 60	Oliv: 60 × 40; pyx: 240 × 160 and 80 × 80
—	25.5	—	200 × 150	450 (one grain)
—	—	—	100 × 50	200 × 50 (one grain)
33	30.5	—	240 × 200	320 × 200 (one grain)
33	29.5	—	250 × 100	150 × 30 radiating to 400 × 100 blocky
34	—	38	80	See text
32.5	—	36.5	100 × 100 recrystallized	Oliv: 300 × 250 enclosed by pig: 600 × 150

supports textural evidence that they cooled in an extrusive environment (Steele and Smith, 1975) and chemical evidence that there was limited re-setting of mineral compositions during thermal metamorphism.

Pyroxene

Except in two Type II lithic clasts there is a "preferred" pyroxene composition of $Wo_{10}En_{61}Fs_{29}$ (Fig. 4). The dominant pyroxene is pigeonite with a positive sign and very low 2 V. The proximity of the pigeonite composition to that of the rock's normative pyroxene (Table 2) explains the scarcity of augite. Despite the narrow compositional range of pigeonite in the granoblastic matrix (Fe/(Fe + Mg) = 30.5–34%) there is a detectable compositional difference between fragments (Fig. 5). Large pigeonite clasts display a greater range of Fe/Mg than do pigeonites of small clasts and the granoblastic matrix.

There is little difference in the Wo content of the low-Ca pyroxene throughout the rock. Low-Ca pyroxene with less than 5 mole% Wo is exceedingly rare. It is

Table 4. Mineral compositions of 78155.

	1	2	3	4		5	6	7	8
SiO_2	44.8	36.7	53.0	49.6		—	—	—	—
TiO_2	—	0.04	0.4	0.6		53.8	14.4	0.8	0.1
Al_2O_3	35.6	0.00	1.0	4.1		0.00	8.4	18.1	67.0
Cr_2O_3	—	0.04	0.6	0.9		0.6	31.2	47.4	1.8
FeO	0.5	31.4	19.0	13.8		36.6	37.6	25.4	9.6
MnO	—	0.3	0.3	0.3		0.3	0.3	0.3	0.00
MgO	—	31.6	21.7	16.3		6.5	5.4	6.3	21.3
CaO	18.4	0.16	4.5	14.1		—	—	—	—
Na_2O	0.6	0.00	0.02	0.05		—	—	—	—
K_2O	0.2	—	—	—		—	—	—	—
Total	100.1	100.2	100.5	99.7		97.8	97.3	98.3	99.8
Ionic proportions									
O	32.0	24.0	24.0	24.0		6.0	32.0	32.0	32.0
Si	8.264	5.979	7.831	7.456		—	—	—	—
Ti	—	0.005	0.048	0.066		1.987	3.059	0.168	0.009
Al	7.748	0.000	0.175	0.731		0.000	2.795	5.677	15.752
Cr	—	0.005	0.072	0.103		0.024	6.966	9.960	0.281
Fe	0.083	4.281	2.348	1.736		1.501	8.871	5.638	1.597
Mn	—	0.041	0.039	0.033		0.013	0.081	0.065	0.000
Mg	—	7.675	4.776	3.656		0.476	2.288	2.505	6.336
Ca	3.645	0.028	0.706	2.275		—	—	—	—
Na	0.207	0.000	0.006	0.015		—	—	—	—
K	0.035	—	—	—		—	—	—	—
Total cations	19.982	18.014	16.001	16.071		4.001	24.060	24.013	23.975
Or/Fo/Wo	0.9	64.2	9.0	29.7	Ulvospinel		38.5	2.1	0.1
Ab/Fa/En	5.3	35.8	61.0	47.7	Chromite		35.7	62.4	1.8
An/ /Fs	93.8		30.0	22.6	Picotite		8.2	—	—
					Hercynite		—	4.8	18.1
					Spinel		17.6	30.7	80.0

(1) Anorthite of granoblastic matrix. (2) Olivine of granoblastic matrix. (3) Pigeonite of composite pyroxene clast. (4) Augite that contacts pigeonite of analysis 3. (5) Ilmenite intergrown with Cr-spinel in anorthite-mantled complex of opaque minerals (granoblastic matrix). (6) Cr-spinel intergrown with ilmenite of analysis 5. (7) Low-Ti Cr-spinel from a relatively large, rounded, compositionally zoned spinel of the granoblastic matrix (see Fig. 7). Not visibly asociated with ilmenite. (8) Mg–Al spinel bleb in anorthite clast.

restricted to rare pyroxene clasts, xenocrysts in Type I lithic clasts, and three Type II lithic clasts, two of which are anomalously magnesian. One large clast composed of very-low-Ca pyroxene and pigeonite was traversed with the electron microprobe (Figs. 1F and 6). Most of the clast is untwinned pyroxene with Wo_{4-5}, a high 2 V, and presumed orthorhombic symmetry. It has exsolved irregular patches of augite that average about 5×30 μm (white wisps in Fig. 1F, represented by one point on Fig. 6). The polysynthetic twinned pyroxene of Fig. 1F is pigeonite with traces of exsolved augite. Figure 6 illustrates that the transition from Wo_5 to Wo_{10} occurs over about 20 μm or less, depending on the

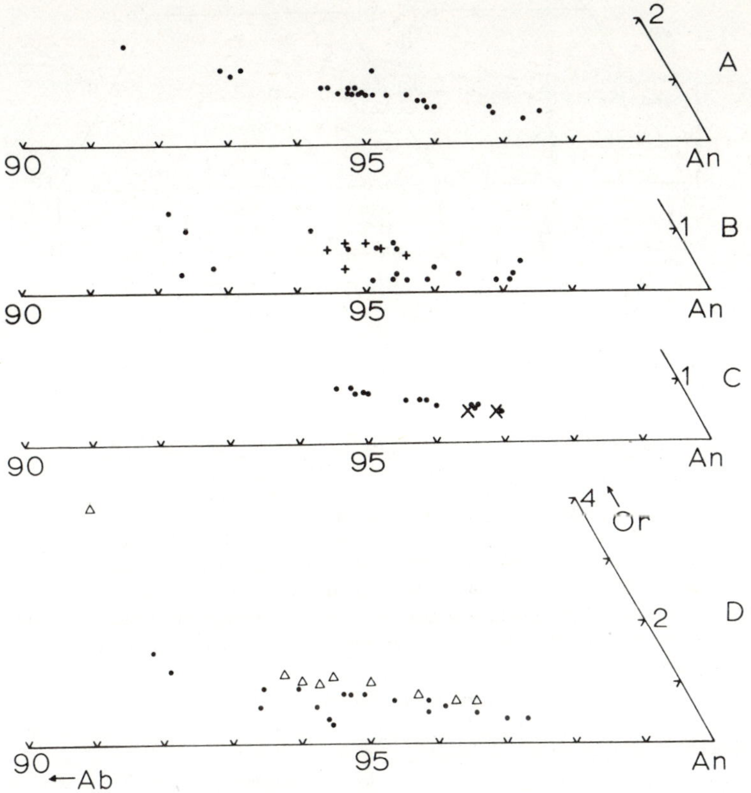

Fig. 3. Compositions of plagioclase from 78155 (mole%). (A) Granoblastic matrix. (B) Monomineralic clasts of anorthite. (+) indicates small clasts transitional with the granoblastic matrix. (C) Type I lithic clasts. (×) indicates xenocrysts. (D) Type II lithic clasts. (△) indicates plagioclase from a single large Type IIA clast in 78155,42.

inclination of the contact. Because pigeonite and very-low-Ca pyroxene are each restricted to a discrete portion of the clast and the very-low Ca pyroxene is enriched in Mg/(Mg + Fe) by 2 atom%, the two pyroxenes are taken to have crystallized together from a silicate melt, and thus both are inferred relics of the rock's precursor. The exsolution of augite from each is taken to have occurred before the annealing of 78155.

Oxides

The spinels of 78155 define two distinct compositional groups (Fig. 7) between which there was never chemical equilibrium (McCallum *et al.*, 1974). Rare Mg–Al spinels are transparent and colorless or faintly pink. They occur only as blebs in monocrystalline clasts of anorthite and are not associated with ilmenite. Opaque Cr-spinels with exsolved ilmenite and metal are abundant in the granoblastic matrix (Fig. 1C). Their large variation of Ti results from the

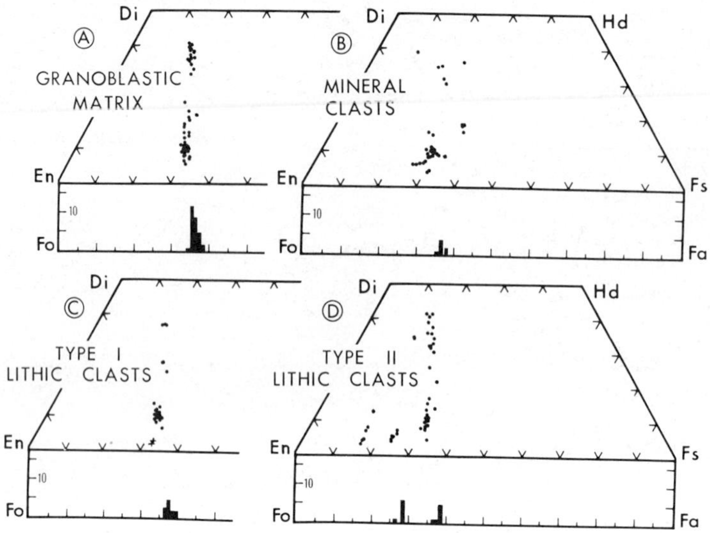

Fig. 4. Pyroxene and olivine compositions of 78155. (A) Granoblastic matrix. (B) Mineral clasts. (C) Type I lithic clasts. (+) indicates two xenocrysts with very low Wo contents. Compositions of other xenocrysts lie within the pigeonitic cluster. (D) Type II lithic clasts. Analyses with low Fs and low Fa come from only two clasts.

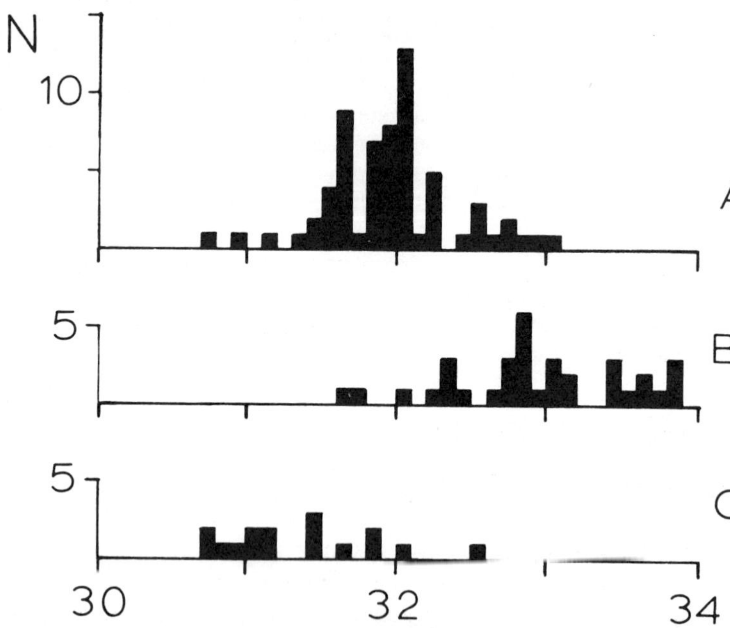

Fig. 5. Fe/(Fe + Mg) of pigeonite in different fragments of the granoblastic matrix (atom%). (A) 78155,7, two fragments. (B) 78155,42, two fragments. (C) 78155,7, one fragment and 78155,42, one fragment.

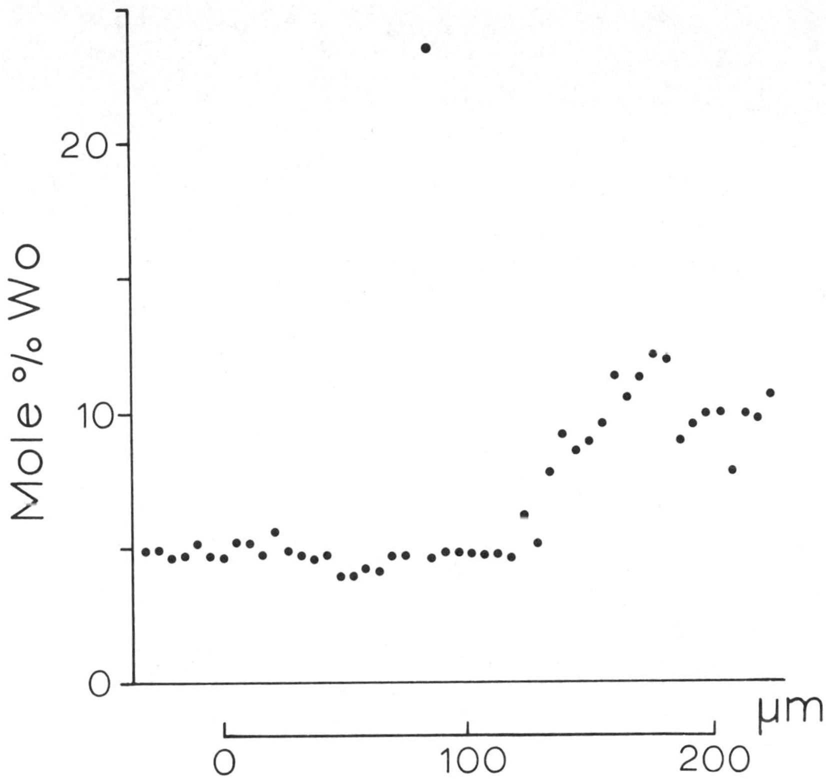

Fig. 6. Profile showing the Ca content of low Ca pyroxene clast in 78155,48. The traverse runs from lower left to upper right across Fig. 1F.

incomplete reduction of an ulvöspinel component to ilmenite and Fe metal. The ilmenite of 78155 has a high MgO content (3.5–6.5 wt.%).

Isotopic Ages

Isotopic studies have established the unusual antiquity of 78155. Turner and Cadogan (1975, p. 1526) report an "extremely well defined" ^{40}Ar–^{39}Ar plateau age of 4.22 ± 0.04 AE, "identical to its total argon age". This dates the last event that has affected the chemistry of the rock. The ^{40}Ar–^{39}Ar ages of most lunar polymict breccias from the highlands date the basin-forming impacts of 4.0–3.9 AE. 78155 lies above concordia on a $^{206}Pb/^{238}U$–$^{207}Pb/^{235}U$ diagram. Nunes *et al.* (1974) interpret this to mean that the rock may have been injected with excess Pb, perhaps in a soil component, during impacts between 4.2 and 4.5 b.y. ago. With further study of the U–Th–Pb systematics, Nunes *et al.* (1975) recognized that 78155 contains a minimum of two different initial leads, but proposed that lead has been isolated in the sample since about 4.23 AE. Nyquist

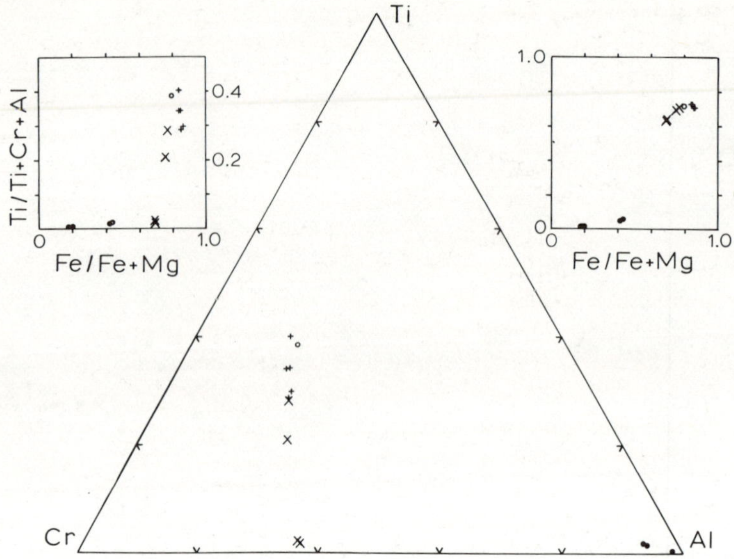

Fig. 7. Compositions of spinels of 78155 represented in three projections (molecular proportions). Symbols identify the textural varieties as follows: (●) Mg–Al spinel blebs in anorthite clasts; (+) blocky, euhedral or subhedral Cr-spinels with exsolved ilmenite and Fe metal (granoblastic matrix); (×) analyses of a single, relatively large, rounded Cr-spinel not associated with ilmenite (granoblastic matrix); (○) Cr-spinel from anorthite-mantled complex of spinel, ilmenite, and Fe metal.

et al. (1974) report the Rb–Sr whole rock model age of 78155 as about 4.4–4.5 AE. Turner and Cadogan (1975) report an exposure age of 30 m.y.

THE PROTOLITH: A RESTRICTED SET OF EARLY LUNAR ROCKS

Most of 78155 is granoblastic matrix and crushed material in which there is no vestige of a precursor. Nevertheless, three distinct types of precursors are established on the basis of the clasts (Table 1). Mineral clasts are by far the most abundant and their coarse grain sizes suggest that they are derived from plutonic rocks (primitive cumulates?). The compositional and proportional correspondence of silicates of the granoblastic matrix and mineral clasts indicates that most of 78155 is derived from crushed and recrystallized coarse-grained rock represented by these clasts. The plutonic precursors had a limited range of mineral compositional and their average bulk composition is inferred to be that of the present sample. The blebs of colorless spinel in some anorthite clasts indicates that Mg–Al spinel crystallized early in at least some of the plutonic precursors.

The Type I lithic clasts crystallized from anorthosite melt. Because melting of the average lunar crust (anorthositic norite) would not produce such a magma, it must have originated by total impact melting of pre-existing anorthosites. The precursors belonged to the ferroan anorthosite suite of Dowty *et al.* (1974). Anorthositic impact-melt rocks have been found as clasts in other rocks, but they are largely restricted to light matrix breccias, which, like 78155, crystallized before 4.0 AE (Warner and Bickel, 1976). Apparently relatively large target areas of anorthosite were present only before the planet-wide mixing that coincided with the final intense lunar bombardment.

The medium-grained Type II lithic clasts have neither the coarse grain size nor the texture of lunar cumulates. They probably originated by remelting and by metamorphism of crustal rocks very early in lunar history. Some have mineral compositions similar to the granoblastic matrix, but the distinctly different mineral compositions in others demonstrates that thermal metamorphism did not efficiently equilibrate Fe and Mg between clasts and matrix (Fig. 4).

The rare-earth elemental abundance pattern of 78155 shows weak enrichment relative to chondrites and a small, positive europium anamoly (Hubbard *et al.*, 1974). This contrasts strongly with the REE pattern of rocks containing appreciable KREEP. Although Nunes *et al.* (1975) determined that the initial isotopic composition of lead in 78155 was distinctly different from that of KREEP, they proposed on the basis of the rock's U content that it may contain as much as 5 or 6% of the KREEP component. Clearly the rock's chief precursors belong to the ANT suite. If any KREEP component is present it probably belongs to rare, medium-grained Type II lithic fragments and the small portion of the matrix that is derived from them. Warner and Bickel (1976) have proposed that rocks rich in KREEP were not present at the lunar surface when the precursors of 78155 were assembled to form the present breccia.

Nature of the Metamorphism

The texture of the granoblastic matrix in 78155 is essentially the same as in 79215 (Bickel *et al.*, 1976; Warner *et al.*, 1977a). The matrices of these rocks are distinctly finer grained than such Apollonian metamorphic rocks as 15415 and 76535, and their mineral compositions are not homogeneous over distances of 1 cm. They cannot have been annealed over the time period requisite for deep-seated, Apollonian metamorphism ($\sim 10^8$ yr; Stewart, 1975). Furthermore, the brecciated character and high siderophile trace elemental abundances of these rocks indicate that they had early surface histories (Morgan *et al.*, 1974; Higuchi and Morgan, 1975). James (1974) has established that none of the breccias produced by the major, basin-forming impacts has been intensely annealed, and that rocks with granoblastic textures such as 78155 occur only as clasts in those breccias. This implies that the thermal energy necessary to produce these metamorphic rocks was available near the lunar surface only before the final bombardment which formed the youngest multi-ringed basins (Warner *et al.*, 1977a). Based on pyroxene compositions, thermal metamorphism of 78155 occurred at temperatures between 1100 and 1200°C (Ross and Huebner, 1975).

PYROXENE COMPOSITIONS AND THE EQUILIBRATION OF 78155

The pyroxenes of 78155 are noteworthy in two important respects: (1) based on their Ca contents, the pyroxenes of the granoblastic matrix appear to have last equilibrated at the unusually high temperature of 1100–1200°C and (2) the Ca contents of most pyroxenes in mineral and lithic clasts are the same as in the granoblastic matrix (Fig. 4). The first observation indicates that 78155 was thermally metamorphosed at very high temperature. Because many low-Ca pyroxenes of equivalent Fe/Mg, from lunar plutonic and metamorphic rocks contain under 5 mole% Wo (Bickel and Warner, 1977, Fig. 2), the second observation suggests that during thermal metamorphism the pyroxenes of most clasts in 78155 reequilibrated with respect to their Wo contents. This is remarkable because magnesian pyroxenes of some Type II lithic clasts have not exchanged Mg and Fe with the granoblastic matrix and a structural change is involved in converting orthopyroxene to pigeonite.

Of greater interest is the fact that the pigeonite of 78155 did not invert when the rock cooled. Because there is no evidence that any pigeonite of the granoblastic matrix has inverted, the very-low-Ca pyroxene of rare clasts such as that illustrated in Fig. 1F is taken to be relict from the rock's precursor. Uninverted magnesian pigeonite is not found in terrestrial mafic plutonic and metamorphic rocks which cooled slowly. There is textural evidence that pigeonite has inverted in many lunar plutonic rocks (e.g., 62237, Dymek *et al.*, 1975; norite and gabbro clasts of 76255, Warner *et al.*, 1976) and in 79215, a granoblastic metamorphic rock in which the final equilibration temperature of the pyroxenes was about 1000°C (Bickel *et al.*, 1976).

But uninverted magnesian pigeonite is found in 77017, which has been interpreted both as a plutonic (Ashwal, 1975) and a metamorphic (McCallum *et al.*, 1974) rock. Ashwal has proposed that in 77017, pigeonite (nearly identical with that of 78155) has persisted because its inversion may be promoted by an intergranular fluid phase that was not present in all lunar plutonic rocks. However, there is no independent evidence for such a fluid phase in the lunar plutonic or metamorphic rocks with inverted pigeonite. McCallum *et al.* (1974) argue that the poikiloblastic texture of 77017 was produced in a thick, hot ejecta blanket by devitrification of glass in which numerous plagioclase nuclei formed early and rare pyroxene nuclei formed late. They contend that devitrification provided an internal heat source and that pigeonite remained metastable because the cooling rate of the rock increased substantially when devitrification was completed. Such an explanation may apply to the metastability of pigeonite in 78155. Rocks 77017 and 78155 have nearly identical compositions (LSPET, 1973b) and both contain mineral and lithic clasts, but 77017 is coarser grained than the granoblastic matrix of 78155, which suggests that 77017 may have cooled more slowly. If both samples were vitrified before their thermal metamorphism, nucleation of pyroxene may have been inhibited in slower-cooled (?) 77017 to produce a poikiloblastic texture, but may have been promoted in 78155 to produce a granoblastic texture. However, there is no direct evidence that 78155 ever contained glass.

Because the solidus temperature of 78155 is about 1200°C (Walker *et al.*, 1973) and the pyroxenes appear to have last equilibrated at 1100–1200°C, the annealing of the granoblastic matrix must have occurred over a temperature range of a few tens of degrees. Huebner *et al.* (1975) have demonstrated that pigeonite probably remained metastable in 77215 during a period of at most 30,000 years when the rock cooled from about 1160 to 930°C; augite lamellae up to 30 μm across exsolved from the metastable pigeonite. Sample 78155 must have cooled rapidly at least through this temperature range in which metastable pigeonite exsolves augite and then inverts to orthopyroxene. At lower temperatures the unmixing of pyroxenes is inhibited by low diffusivities of migrating species. The exsolution features and extreme variation of Ti in the Cr-spinels of 78155 suggest that reduction of their ulvöspinel component to ilmenite and Fe metal occurred sluggishly in response to decreasing temperature and fugacity of oxygen. Because no rutile was observed in 78155, it appears that as the rock cooled, subsolidus reduction, as well as re-equilibration of pyroxene, did not proceed to the extent that it did in 79215 (Bickel *et al.*, 1976). Sample 78155 appears to have cooled more rapidly than 79215.

Quenching explains the persistence of magnesian pigeonite in terrestrial and lunar volcanic rocks. Hewins and Goldstein (1977) have established that metallic grains in Apollo 16 soil 63501 cooled slowly to 500°C and then were quenched. They suggest that excavation was responsible for the quenching. On the moon large impacts could have quenched certain plutonic and metamorphic rocks from temperatures near the solidus; some deeply buried rocks may have been catastrophically uncovered or ejected into debris blankets. Although meteorite impacts generally heat target rocks, impacts may catastrophically increase cooling rates of rocks that were deeply buried. The annealing of 78155 may have been catastrophically terminated by such an event, presumably dated by the rock's 4.2 AE ^{40}Ar–^{39}Ar age.

Inferred Sample History

Based on the compositional and proportional correspondence between the mineral clasts and the granoblastic matrix the major precursor of 78155 was plutonic rocks that may have been primitive cumulates. The lithic clasts represent rocks produced between the formation of the initial lunar crust and the 4.2 b.y. ^{40}Ar–^{39}Ar age of the rock. Type I clasts are derived from impact-melted ferroan anorthosite, and the medium-grained Type IIA igneous clasts crystallized from magma of unknown origin, but which probably postdates the inferred primitive moon-wide magma ocean. Types II B, C, and D metamorphic clasts are derived from pre-existing rocks that may have been brecciated before they were thermally metamorphosed. Their metamorphic textures appear to pre-date the thermal metamorphism of the rock's granoblastic matrix. Most of the protolith of 78155 was severely crushed and possibly vitrified before its annealing to produce the granoblastic matrix. Thermal metamorphism of the rock probably terminated at 4.2 AE; termination may have been abrupt. There is textural evidence of only two brecciations of 78155. The first preceeded its thermal metamorphism, and the second produced the crushed material. If the rock were severely fragmented about 4.2 b.y. ago it would not have remained coherent when much later ejected to Station 8. Final brecciation to produce incoherent fragments must have coincided with its ejection to the regolith at Station 8, where it produced a shallow crater. This event is probably dated by the rock's 30 m.y. exposure age.

SUMMARY AND CONCLUSIONS

Although 78155 clearly is polymict, it is distinguished from most lunar polymict breccias by three important features which are related to the rock's unusually great age. The rock has a relatively coarse-grained, granoblastic matrix (a product of thermal metamorphism), restricted mineral compositions, and a restricted and unusual set of mineral and lithic clasts. The latest events that have affected the chemistry of this rock appear to pre-date the final intense lunar bombardment of 4.0–3.9 AE. The survey of James (1974) and re-investigations of Apollo 14 breccias by Warner *et al.* (1977b) indicate that breccias produced during and after the final intense lunar bombardment were not subjected to significant thermal metamorphism. The rock's inferred precursors and relict clast population are important because they represent the lunar crust before the final intense bombardment.

Acknowledgments—I thank J. Warner for helpful discussions and advice throughout the course of this study and C. Hardy for typing a draft of the paper. The paper has benefitted from thoughtful reviews by R. H. Hewins and G. Heiken. My work was supported by the 1976 NASA–American Society for Engineering Education Summer Faculty Fellowship Program administered by the University of Houston.

REFERENCES

Ashwal L. D. (1975) Petrologic evidence for a plutonic igneous origin of anorthositic norite clasts in 67955 and 77017. *Proc. Lunar Sci. Conf. 6th*, p. 221–230.

Bickel C. E. and Warner J. L. (1977) Survey of lunar plutonic and granulitic lithic fragments (abstract). In *Lunar Science VIII*, p. 112–114. The Lunar Science Institute, Houston.

Bickel C. E., Warner J. L., and Phinney W. C. (1976) Petrology of 79215: Brecciation of a lunar cumulate. *Proc. Lunar Sci. Conf. 7th*, p. 1793–1819.

Dowty E., Prinz M., and Keil K. (1974) Ferroan anorthosite: A widespread and distinctive lunar rock type. *Earth Planet. Sci. Lett.* **24**, 15–25.

Dymek R. F., Albee A. L., and Chodos A. A. (1975) Comparative petrology of lunar cumulate rocks of possible primary origin: Dunite 72415, troctolite 76535, norite 78235, and anorthosite 62237. *Proc. Lunar Sci. Conf. 6th*, p. 301–341.

Hewins R. H. and Goldstein J. I. (1975) The provenance of metal in anorthositic rocks. *Proc. Lunar Sci. Conf. 6th*, p. 343–362.

Hewins R. H. and Goldstein J. I. (1977) Cooling history of Apollo 16 phosphide-metal assemblage (abstract). In *Lunar Science VIII*, p. 436–438. The Lunar Science Institute, Houston.

Higuchi H. and Morgan J. W. (1975) Ancient meteoritic component in Apollo 17 boulders. *Proc. Lunar Sci. Conf. 6th*, p. 1625–1651.

Hollister L. S. (1975) Evolution of the moon between 4.6 and 3.3 AE. *Proc. Lunar Sci. Conf. 6th*, p. 1159–1178.

Hubbard N. J., Rhodes J. M., Wiesmann H., Shih C.-Y., and Bansal B. M. (1974) The chemical definition and interpretation of rock types returned from non-mare regions of the moon. *Proc. Lunar Sci. Conf. 5th*, p. 1227–1246.

Huebner J. S., Ross M., and Hickling N. (1975) Significance of exsolved pyroxenes from lunar breccia 77215. *Proc. Lunar Sci. Conf. 6th*, p. 529–546.

James O. B. (1974) Lunar highlands breccias generated by major impacts. In *Proc. Sov.-Amer. Conf. on Cosmochemistry of the Moon and Planets (Moscow)*. June 4–8, 1974. Nauka, Moscow, 1975. Also NASA SP-370, in press.

LSPET (1973a) *Apollo 17 Sample Information Catalogue.* MSC 03211. Johnson Space Center, Houston. 447 pp.

LSPET (1973b) *Apollo 17 Prelim. Sci. Rep.*, NASA SP-330.

McCallum I. S., Mathez E. A., Okamura F. P., and Ghose S. (1974) Petrology and crystal chemistry of poikilitic anorthositic gabbro 77017. *Proc. Lunar Sci. Conf. 5th*, p. 287–302.

Morgan J. W., Ganapathy R., Higuchi H., Krähenbühl U., and Anders E. (1974) Lunar basins: Tentative characterization of projectiles, from meteoritic elements in Apollo 17 boulders. *Proc. Lunar Sci. Conf. 5th*, p. 1703–1736.

Nunes P. D., Tatsumoto M., and Unruh D. M. (1974) U–Th–Pb systematics of some Apollo 17 lunar samples and implications for a lunar basin excavation chronology. *Proc. Lunar Sci. Conf. 5th*, p. 1487–1514.

Nunes P. D., Tatsumoto M., and Unruh D. M. (1975) U–Th–Pb systematics of anorthositic gabbros 78155 and 77017—implications for early lunar evolution. *Proc. Lunar Sci. Conf. 6th*, p. 1431–1444.

Nyquist L. E., Bansal B. M., Wiesmann H., and Jahn B.-M. (1974) Taurus-Littrow chronology: Some constraints on early lunar crustal development. *Proc. Lunar Sci. Conf. 5th*, p. 1515–1539.

Ross M. and Huebner J. S. (1975) A pyroxene geothermometer based on composition-temperature relationships of naturally occurring orthopyroxene, pigeonite, and augite (abstract). *Int. Conf. Geothermometry Geobarometry*, p. 173–176. Pennsylvania State University.

Steele I. M. and Smith J. V. (1975) Minor elements in lunar olivine as a petrologic indicator. *Proc. Lunar Sci. Conf. 6th*, p. 451–467.

Stewart D. P. (1975) Apollonian metamorphic rocks (abstract). In *Lunar Science VI*, p. 774–776. The Lunar Science Institute, Houston.

Taylor S. R. and Bence A. E. (1975) Evolution of the lunar highland crust. *Proc. Lunar Sci. Conf. 6th*, p. 1121–1141.

Turner G. and Cadogan P. H. (1975) The history of lunar bombardment inferred from ^{40}Ar–^{39}Ar dating of highland rocks. *Proc. Lunar Sci. Conf. 6th*, p. 1509–1538.

Walker D., Grove T. L., Longhi J., Stolper E. M., and Hays J. F. (1973) Origin of lunar feldspathic rocks. *Earth Planet. Sci. Lett.* **20**, 325–336.

Warner J. L. and Bickel C. E. (1976) Petrologic constraints on the age of KREEP (abstract). *EOS (Trans. Amer. Geophys. Union)* **57**, 1019.

Warner J. L., Bickel C. E., Phinney W. C., and Simonds C. H. (1977a) Feldspathic, granulitic impactites that pre-date the final lunar bombardment (abstract). In *Lunar Science VIII*, p. 979–981. The Lunar Science Institute, Houston.

Warner J. L., Simonds C. H., Phinney W. C., Geeslin J. H., and Bickel C. E. (1977b) Petrology and crystallization history of Apollo 14 impactites (abstract). In *Lunar Science VIII*, p. 982–984. The Lunar Science Institute, Houston.

Warner J. L., Simonds C. H., and Phinney W. C. (1976) Apollo 17, Station 6 boulder sample 76255: Absolute petrology of breccia matrix and igneous clasts. *Proc. Lunar Sci. Conf. 7th*, 2233–2250.

Wilshire H. G. (1974) Provenance of terra breccias (abstract). In *Lunar Science V*, p. 846–847. The Lunar Science Institute, Houston.

Proc. Lunar Sci. Conf. 8th (1977), p. 2029–2050.
Printed in the United States of America

4.2–4.3 AE anorthositic soil fragments: equilibrated or unequilibrated polycomponent systems?

J. W. DELANO and A. E. BENCE

Department of Earth and Space Sciences, State University of New York, Stony Brook, New York 11794

Abstract—Mineralogical studies of five 2–4 mm anorthositic soil fragments whose argon-argon spectra contain high-temperature components with ages of 4.2–4.3 AE indicate that four of the five are polymict breccias. One of the breccia fragments (67483,13,8) is monomict and its $^{40}Ar/^{39}Ar$ plateau age of 4.26 ± 0.05 AE may record the time of excavation. The four polymict samples contain both young (~4.0 AE) low-temperature and old (>4.2 AE) high-temperature argon components. However, the proportions of these gas components do not correlate with the clast/matrix ratio or with the degree of annealing of matrix. Two possibilities must be considered: (1) The "ages" are composite resulting from the mixing of gas from ancient clasts and younger matrix, or (2) the high-temperature "ages" represent the time of formation or annealing of the breccia. At present, it is not possible to resolve the two. If the ages are true formational ages, these samples provide a window through the 4.0 AE "cataclysm" and may supply information on the lithology of the ancient lunar crust.

INTRODUCTION

SEVERAL CRYSTALLINE, anorthositic 2–4 mm soil fragments, from the Apollo 16 and 17 sites have $^{40}Ar/^{39}Ar$ plateau "ages" of 4.2–4.3 AE (Schaeffer and Husain, 1973; Schaeffer and Schaeffer, 1977). Comparably high argon-argon "ages" are reported for a few large highland rocks from both sites (Kirsten and Horn, 1974; Husain and Schaeffer, 1975; Turner and Cadogan, 1975). Two contrasting interpretations of the 4.2–4.3 AE ages have been presented in the literature: (1) the "ages" represent true excavation ages (Schaeffer and Husain, 1973) and (2) the "ages" are composites with contributions from old (≳4.4 AE) clasts and young (~3.9 AE) matrix (Huneke *et al.*, 1973; Jessberger *et al.*, 1974a,b; Tera *et al.*, 1974). Since these interpretations bear directly on the controversy surrounding the "terminal cataclysm" (Tera *et al.*, 1974; Schaeffer and Husain, 1974), considerable effort has gone into attempts to resolve the problem.

The detailed $^{40}Ar/^{39}Ar$ analyses of mineral separates from crystalline highland and mare rocks reveals plagioclase with high-temperature ages that are older than the whole rock ages for some, but not all, of the rocks studied (Turner *et al.*, 1972; Huneke *et al.*, 1973; Jessberger *et al.*, 1974a,b; Horn *et al.*, 1975; Turner and Cadogan, 1975). In the case of the highland breccias, the plagioclase can have a relatively clear high-temperature signature giving an "age" significantly older than the whole rock plateau "age". In one case Jessberger *et al.* (1974b) suggest that ~4.5 AE plagioclase relicts are responsible for an old, high-temperature component in the argon spectrum. The bulk of the argon released from plagioclase separates in this highlands breccia, 65015, defines a plateau age of ~3.98 AE which is thought to be a metamorphic age. Jessberger *et al.* (1974b) suggest that the breccia is composed of at least two components having distinctly different ages. The 3.98 AE metamorphic event did not result in complete equilibration of the argon among the various components. The recognition of these components in breccias whose ages fall between 4.0 and 4.6 AE is crucial to arguments bearing on the terminal cataclysm or cataclysms. If the various components were equilibrated under different conditions at different times, it may be possible to use mineral chemistry to distinguish the components in the breccia and to reconstruct its petrogenetic history.

The petrographies and phase chemistries of five crystalline anorthositic fragments with $^{40}Ar/^{39}Ar$ plateau "ages" ranging from 4.1 to 4.3 AE are reported. Our objectives are to establish: (a) whether the individual fragments are monomict or polymict breccias; and (b) if polymict, whether the clasts and matrix have equilibrated chemically. If equilibration of a polymict breccia has occurred, the 4.2–4.3 AE ages may have real significance.

METHOD OF ANALYSIS AND SAMPLES STUDIES

All of the samples reported here are from the Apollo 16 and 17 2–4 mm soil fractions. In each case a preliminary macroscopic description was made of the fragment which was then split in two by chiselling. One fragment was transferred to O. A. Schaeffer for dating and a polished thin section was prepared from the other half for which detailed petrographic descriptions were obtained. Finally, an exhaustive electron microprobe study was performed to document inter- and intragrain chemical variations. For this study an automated, 4-spectrometer ARL-EMX-SM electron microprobe was used. Normal operating conditions were: 15 kV accelerating potential and 0.015 μA specimen current (normalized on brass). For specific minor-element analysis operating conditions were varied to give optimum signals. Data reduction procedures are those of Bence and Albee (1968) using a matrix modified from that reported by Albee and Ray (1970).

The samples studied in this manner are given in Table 1.

Table 1.

Sample	Station	$^{40}Ar/^{39}Ar$ age (AE)	$K_{D\,Opx\text{-}Cpx}^{Fe/Mg}$*	$K_{D\,Ol\text{-}Opx}^{Fe/Mg}$†	Matrix equilibration T (°C)
67483,13,6	11	≥ 4.2[1]	—	—	(> solidus T)
67483,13,8	11	4.26 ± 0.05	0.71	1.44	$900 > T \geq 810$
67483,14,2	11	4.24 ± 0.05	—	—	(> solidus T)
73263,1,1	3	4.23 ± 0.05[1]	0.86	1.18	$1070 > T \geq 925$
73263,1,6	3	4.18 ± 0.2[1]	0.86	1.15	$1070 > T \geq 925$

$$*K_{D\,Opx\text{-}Cpx}^{Fe/Mg} = \left(\frac{Fe}{Mg}\right)_{Cpx} \Big/ \left(\frac{Fe}{Mg}\right)_{Opx}.$$

$$\dagger K_{D\,Ol\text{-}Opx}^{Fe/Mg} = \left(\frac{Fe}{Mg}\right)_{Ol} \Big/ \left(\frac{Fe}{Mg}\right)_{Opx}.$$

[1]High-temperature fraction.

RESULTS

67483,14,2 (Station 11, rim of North Ray Crater)

This fragment is a polymict clast-laden, fine-grained melt rock with three distinguishable components (Fig. 1a): (1) gray intersertal matrix (~50%), (2) dark, angular, lithic clasts (~30%), and (3) mono-mineralic clasts dispersed in both 1 and 2 (~20%). The gray intersertal matrix has a foliated texture in which plagioclase clasts and laths exhibit a preferred orientation. Rare olivine and pyroxene clasts are scattered throughout. The matrix consists of fine-grained ($\leqslant 20\,\mu$) plagioclase and ferromagnesian silicates. It is characterized by Fe/(Fe + Mg) ratios of ~0.45, Al_2O_3 concentrations of ~30 wt.% (by broad-beam microprobe analysis) and $K_2O \sim 0.02$ wt.%. The compositions of the matrix phases are quite variable, reflecting the lack of attainment of chemical equilibrium during rapid cooling from a melt. The dark, lithic clasts are finer grained than the matrix, contain fewer plagioclase clasts and are not foliated. However, they are compositionally similar to the matrix.

Plagioclase clasts ($\geqslant 0.02$ mm), exhibit a variety of thermal and shock histories and constitute two compositional groups (Fig. 2a); one group at An_{94} and a second group at An_{96}. A shattered, polycrystalline, anorthositic fragment (0.05 mm) has the composition An_{96}. Plagioclase minor-element (Fe and Mg) variations are distinguishable between individual clasts and between the two compositional groups (Table 2).

Plagioclase clasts with core CaO contents in excess of An_{95} tend to have FeO and MgO concentrations below detectability limits. The more sodic plagioclases have 0.01–0.02 wt.% MgO and FeO in excess of ~0.07 wt.% in their cores. Microprobe profiles for Fe, Ca, and Mg from cores to rims of the clasts indicate partial equilibration of the rims with the matrix.

The olivine clasts are compositionally variable with a range from Fo_{51} to Fo_{65} and a maximum in the vicinity of Fo_{61} (Fig. 2b). Intragrain zoning is minimal. Thus, the clasts have not equilibrated with the matrix assemblage. The range of olivine clast compositions is strong evidence for a number of chemically distinct sources for these components.

Rare pyroxenes occur as fine-grained clasts and appear not to have equilibrated with the matrix. One clast contains both high- and low-Ca pyroxenes apparently in an exsolution relationship (Fig. 2c). Low temperatures of equilibration are indicated by their compositions, and it is concluded that the pyroxene exsolution relationships were established prior to incorporation in the melt. The variation in pyroxene compositions (Fig. 2c) also indicates a variety of petrologically distinct sources.

The $^{40}Ar/^{39}Ar$ release pattern for this fragment contains a moderately well-defined plateau (Fig. 2d) and a high-temperature (1400°C and 1600°C fractions) plateau age of 4.24 ± 0.05 AE was assigned by Schaeffer and Husain (1973). The similar behaviors of $^{38}Ar_{SP}/^{37}Ar$, $^{36}Ar_{SP}/^{37}Ar$, and $^{40}Ar/^{39}Ar$ (age) were suggested by these workers as evidence that the breccia has been a closed system since its formation (at 4.24 AE). The detailed study reported here, however, indicates that

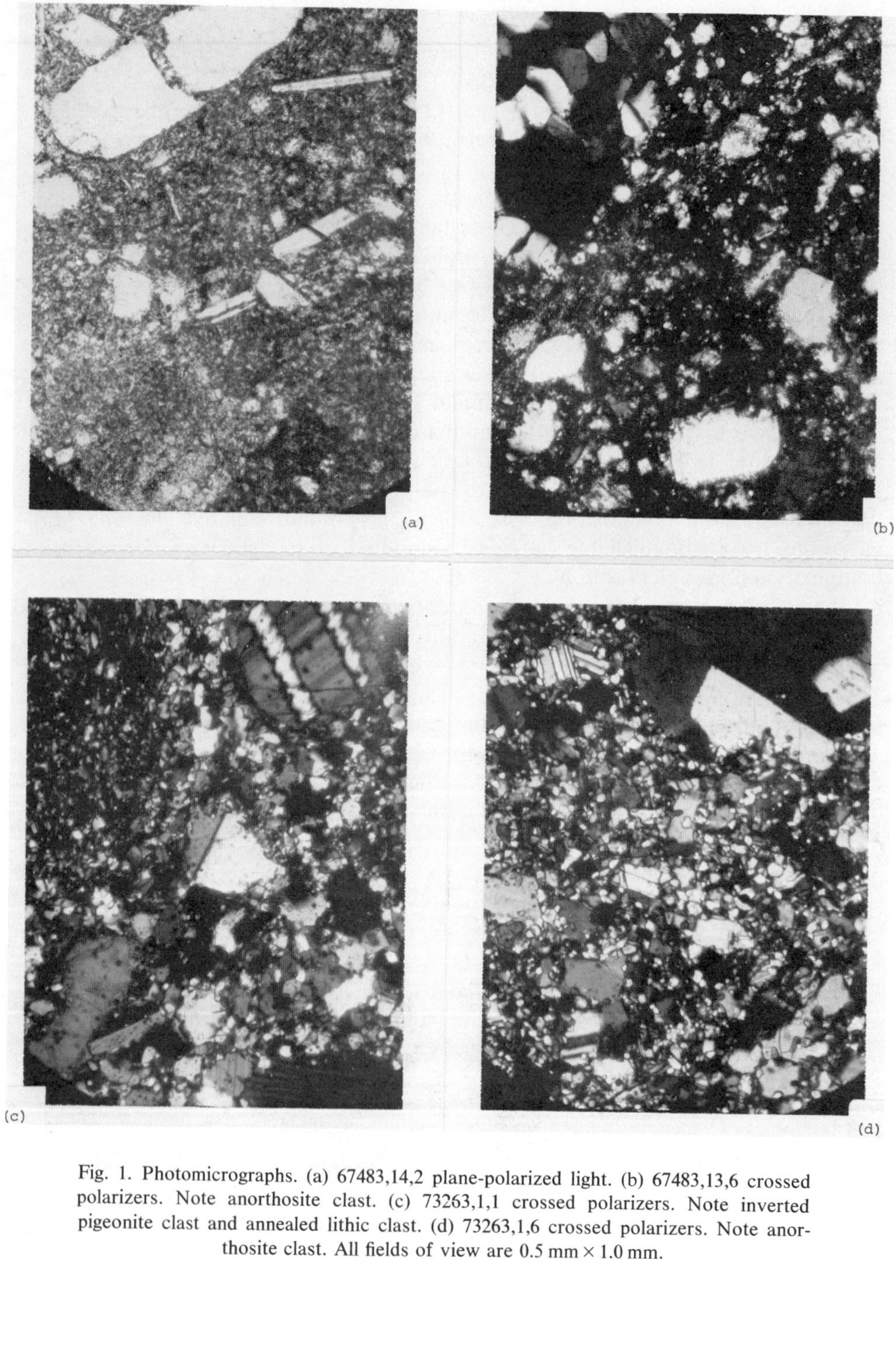

Fig. 1. Photomicrographs. (a) 67483,14,2 plane-polarized light. (b) 67483,13,6 crossed polarizers. Note anorthosite clast. (c) 73263,1,1 crossed polarizers. Note inverted pigeonite clast and annealed lithic clast. (d) 73263,1,6 crossed polarizers. Note anorthosite clast. All fields of view are 0.5 mm × 1.0 mm.

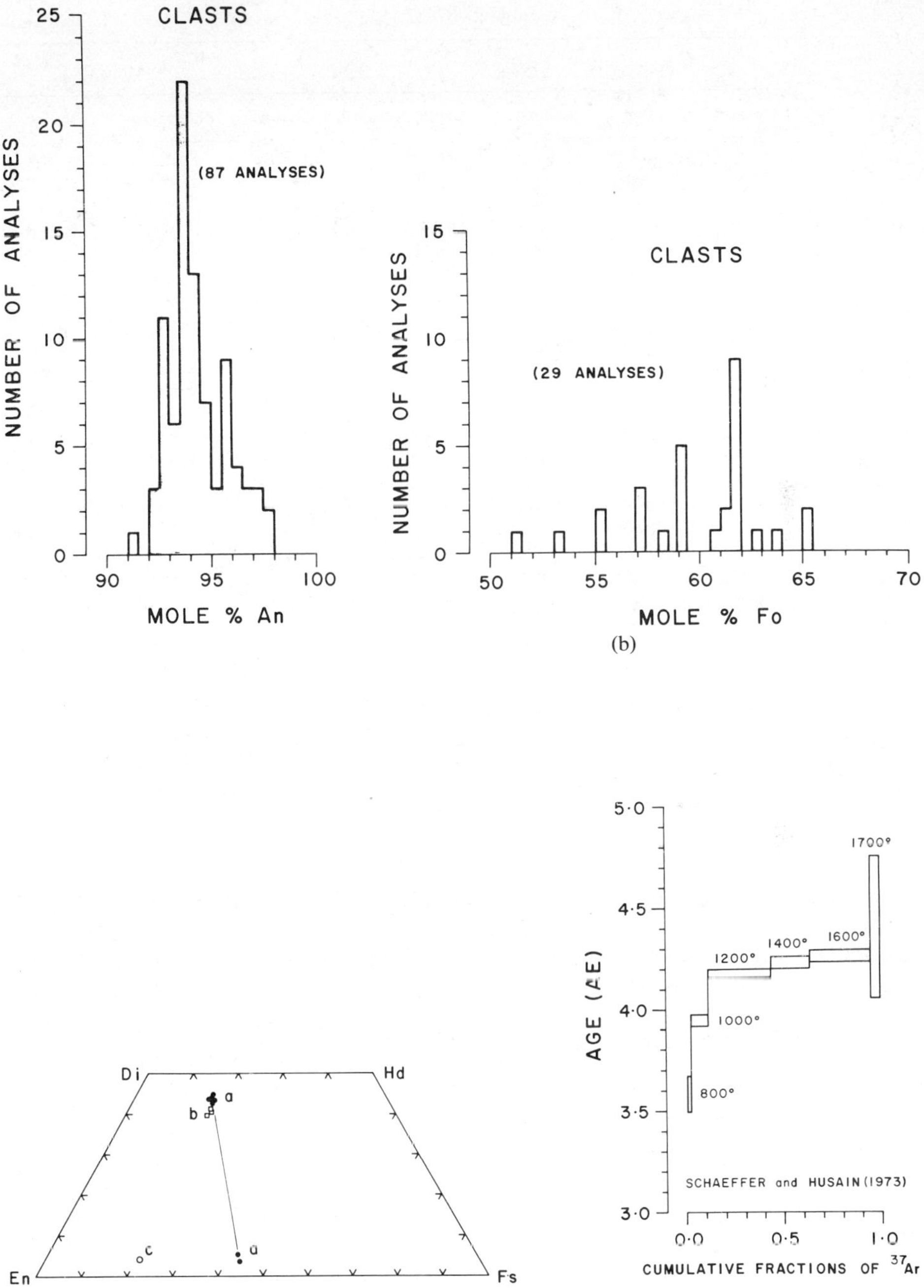

Fig. 2. 67483,14,2. (a) Plagioclase clast compositions. (b) Olivine clast compositions. (c) Pyroxene clasts: *a*—exsolved pyroxene clast. *b,c*—individual clasts. (d) $^{40}Ar/^{39}Ar$ temperature release spectrum.

Table 2. 67483,14,2.

	Clasts					Matrix	
	Plagioclase	Olivine	Olivine	Augite	Orthopyroxene	Plagioclase	Olivine
SiO_2	43.1	36.1	35.6	50.1	53.4	44.2	34.9
Al_2O_3	35.4	0.18	0.17	2.39	1.41	36.0	0.22
TiO_2	—	—	—	1.61	1.14	—	—
FeO	0.07	32.2	35.1	11.2	13.4	0.19	39.0
MnO	—	0.36	0.42	0.21	0.24	—	0.36
MgO	0.06	31.0	28.6	14.1	27.5	0.12	25.0
CaO	19.4	0.28	0.28	19.5	2.06	19.5	0.36
Na_2O	0.62	—	—	0.01	0.00	0.46	—
K_2O	0.03	—	—	—	—	0.01	—
Cr_2O_3	—	0.04	0.02	0.54	0.56	—	0.00
Σ_{oxides}	98.7	100.2	100.2	99.7	99.7	100.5	99.8
Si	2.071	0.987	0.987	1.890	1.928	2.035	0.992
Al	1.914	0.006	0.006	0.107	0.060	1.953	0.007
Ti	—	—	—	0.046	0.031	—	—
Fe	0.003	0.736	0.814	0.354	0.403	0.007	0.927
Mn	—	0.008	0.010	0.007	0.007	—	0.009
Mg	0.004	1.263	1.184	0.795	1.480	0.008	1.059
Ca	0.955	0.008	0.008	0.789	0.080	0.964	0.011
Na	0.055	—	—	0.001	0.000	0.041	—
K	0.002	—	—	—	—	0.001	—
Cr	—	0.001	0.000	0.016	0.016	—	0.000
	An 94.38	Fo 62.94	58.95	Wo 40.57	4.05	An 95.86	Fo 53.09
	Ab 5.44	Fay 37.06	41.05	En 40.90	75.11	Ab 4.09	Fay 46.91
	Or 0.18			Fs 18.54	20.84	Or 0.05	

the story is more complicated. The breccia is a clast-laden melt rock in which plagioclase clasts dominate. The compositions of all the mineral clasts indicate a variety of provenances and a lack of equilibration of the clasts with the groundmass at the time of or subsequent to the formation of the breccia. Since it has already been shown that the argon in ancient plagioclase clasts may not isotopically re-equilibrate even at melt temperatures (Jessberger *et al.*, 1974b), there is no reason to believe that the argon has re-equilibrated between anorthositic clasts and melt matrix in sample 67483,14,2. Some support for this interpretation (lack of equilibration) is seen when the $^{40}Ar/^{39}Ar$ and $^{39}Ar_{SP}/^{37}Ar$ (K/Ca) releases (Schaeffer and Husain, 1973) are compared. The low-temperature fractions have low ages and high K/Ca ratios (0.0035 at 800°C). This is consistent with an origin in the quenched groundmass. The high-temperature fractions give the high ages and lower K/Ca ratios (0.0020 at 1600°C)—consistent with argon release from old, refractory plagioclase clasts. The major difficulty with this suggestion is that the proportion of clasts to matrix (~20:80) in the fragment studied petrographically appears to be too low to account for the large volume of old argon. It is possible that the segment studied petrographically and the segment dated differ in their clast/matrix proportions.

67483,13,6 (Station 11, rim of North Ray Crater)

Sample 67483,13,6 is a polymict, anorthositic breccia containing ~20% plagioclase clasts (>25μ) set in a fine-grained matrix (Fig. 1b). Single-crystal clasts dominate. A few of the larger (250–450 μ) clasts are polycrystalline with variable grain sizes and textures suggesting a wide range of thermal and shock histories. Clast compositions range from $An_{98.5}$ to An_{86} with distinct interclast differences (Fig. 3a and Table 3). Core-to-rim microprobe profiles of the clasts reveal some zoning with respect to Fe and Mg (especially within 20 μ of the clast rims). Both elements are concentrated near the rims, thereby suggesting at least partial equilibration of these elements with the matrix. A single euhedral 20 μ spinel clast (Table 3) in contact with the matrix appears unreacted.

The matrix (~80% of the rock) consists of fine-grained, granular olivine, orthopyroxene, and plagioclase. The ferromagnesians are arranged in radiating feathery aggregates. Their small grain size (<2–3 μ) precludes quantitative analysis by electron microprobe. Matrix plagioclases (Fig. 3b) are more restricted in composition ($An_{93.5-96}$) than the clasts. It would appear that annealing of the breccia resulted in equilibration of the groundmass plagioclase and the clast rims. We suggest that the breccia was a rapidly cooled, clast-laden melt rock prior to annealing.

The variable compositions of the plagioclase clasts suggest that a number of anorthositic sources have contributed to the clast population. The unreacted spinel, the Fe–Mg zoning at the rims of the plagioclase clasts, and the interclast compositional and textural differences are consistent with this rock being a polymict breccia that has equilibrated only to a limited degree.

The published $^{40}Ar/^{39}Ar$ release pattern of Schaeffer and Husain (1973) is shown in Fig. 3c. These workers do not assign a specific age due to the absence of a plateau and the paucity of data points but suggest that the sample has an age in excess of 4.2 AE. We believe that this sample contains two components (clasts and matrix) having significantly different ages. The data are consistent with a complex petrogenesis involving incorporation of ancient, ~4.5 AE, plagioclase fragments in a melt probably formed by impact. The quenched, clast-laden, melt rock was annealed at about 4.0 AE. This event was of sufficient magnitude to recrystallize the groundmass and partially re-equilibrate the rims of the plagioclase clasts. The cores of the plagioclase clasts probably were not degassed either during incorporation in the melt or subsequently during annealing.

67483,13,8 (Station 11, rim of North Ray Crater)

This is a shocked, coarse-grained, monomict, anorthositic gabbro containing 60% plagioclase ($\geqslant 0.05$ mm) and ~40% olivine and pyroxene ($\geqslant 0.02$ mm). The plagioclase is An_{97} with no intragrain zoning of either major or minor elements (Fig. 4a). The olivines (Fo_{59}) and the pyroxenes (Opx; $Wo_3En_{65.5}Fs_{31.5}$ and augite; $Wo_{42.5}En_{43}Fs_{14.5}$) are also uniform in composition (Figs. 4b,c). Representative analyses are reported in Table 4.

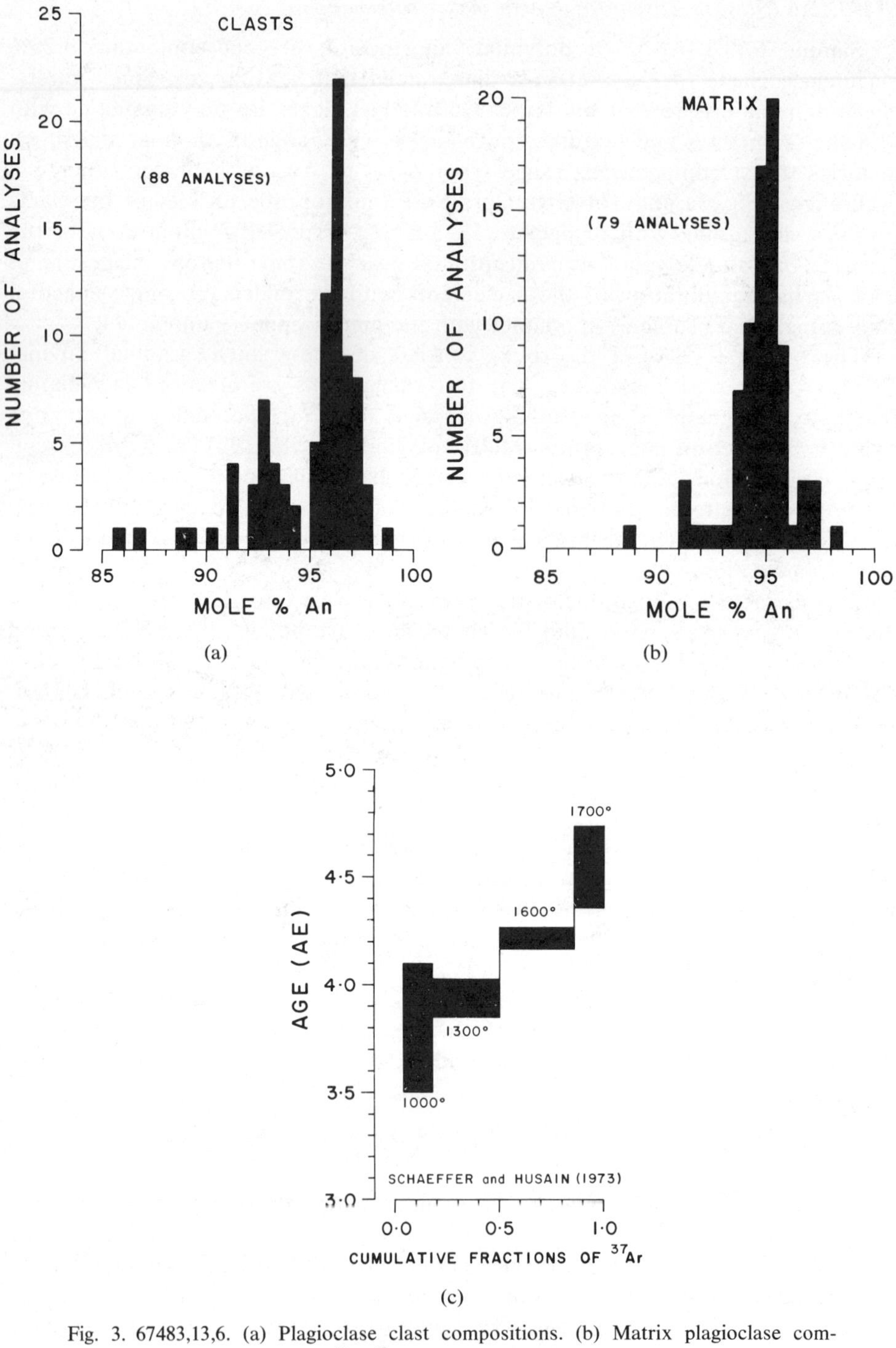

Fig. 3. 67483,13,6. (a) Plagioclase clast compositions. (b) Matrix plagioclase compositions. (c) $^{40}Ar/^{39}Ar$ temperature release spectrum.

Table 3. 67483,13,6.

	Clasts			Groundmass
	Plagioclase (core)	Plagioclase (rim)	Spinel	Plagioclase
SiO_2	43.3	45.6	0.00	44.9
Al_2O_3	36.7	35.4	58.1	35.9
TiO_2	—	—	0.39	—
FeO	0.08	0.16	20.8	0.18
MnO	—	—	0.13	—
MgO	0.00	0.08	13.8	0.09
CaO	19.8	18.8	0.13	19.3
Na_2O	0.15	0.78	—	0.65
K_2O	0.01	0.04	—	0.04
Cr_2O_3	—	—	8.31	—
Σ_{oxides}	100.0	100.9	101.7	101.1
Si	2.002	2.083	0.000	2.055
Al	2.003	1.911	1.808	1.938
Ti	—	—	0.008	—
Fe	0.003	0.006	0.460	0.007
Mn	—	—	0.003	—
Mg	0.000	0.006	0.545	0.007
Ca	0.981	0.919	0.004	0.945
Na	0.014	0.069	—	0.058
K	0.000	0.003	—	0.002
Cr	—	—	0.173	—
An	98.55	92.77		94.06
Ab	1.40	6.97		5.73
Or	0.05	0.25		0.21

The distribution of Fe and Mg between coexisting orthopyroxene and augite ($K_D = 0.71$) indicates an equilibration temperature of $\geqslant 810°C$ (Lindsley *et al.*, 1974). The distribution of Fe and Mg between coexisting olivine and orthopyroxene ($K_D = 1.44$) is consistent with an equilibration at $\sim 900°C$ (Williams, 1971). These data suggest complete Fe–Mg equilibration of the coexisting phases (Table 1).

The argon release data of Schaeffer and Husain (1973) indicate that about 80% of the argon has an age of 4.2–4.3 AE (Fig. 4d). Since all phases appear to have equilibrated with respect to their major and minor components prior to the shock event it seems reasonable to assume that the argon also equilibrated at that time. The low ages of the low-temperature fraction may be a reflection of the shock event. Thus, it appears that the high-temperature plateau age of 4.26 ± 0.05 AE obtained for 67483,13,8 represents a true event—probably the time of excavation.

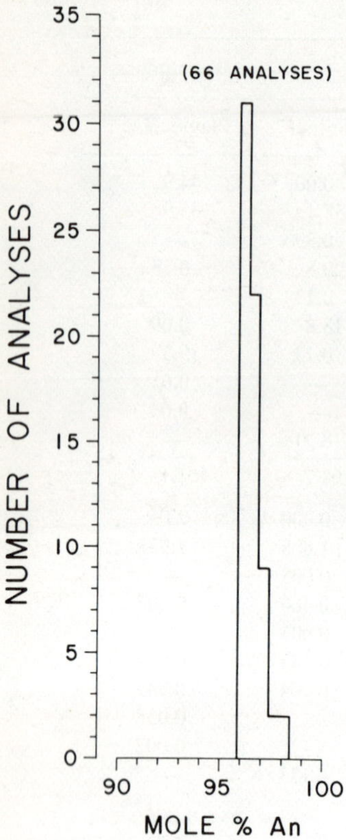

(a)

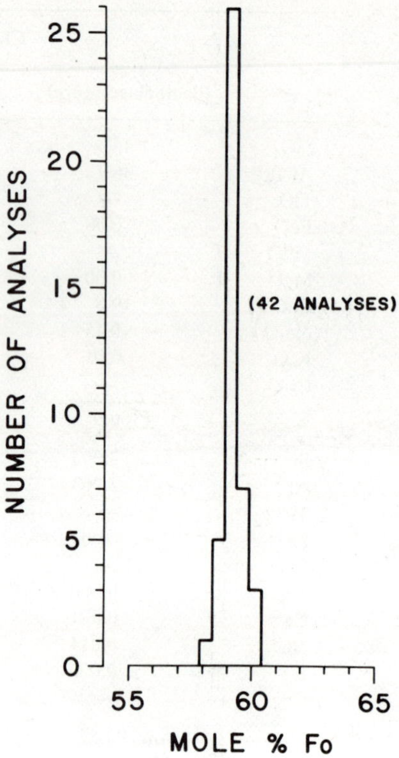

(b)

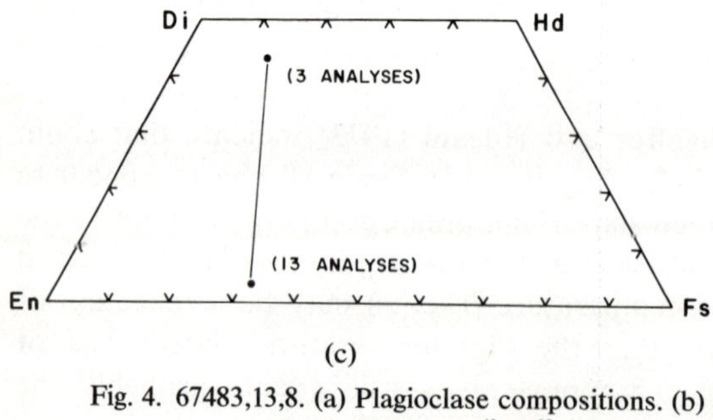

(c)

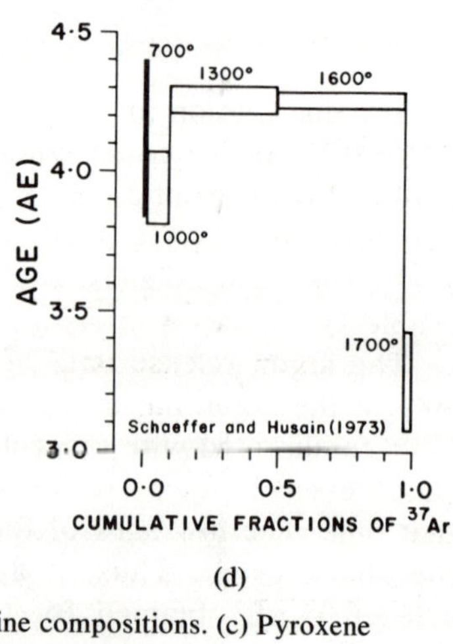

(d)

Fig. 4. 67483,13,8. (a) Plagioclase compositions. (b) Olivine compositions. (c) Pyroxene compositions. (d) $^{40}Ar/^{39}Ar$ temperature release spectrum.

Table 4. 67483,13,8.

	Olivine	Orthopyroxene	Augite	Plagioclase
SiO_2	35.8	52.8	52.7	44.5
Al_2O_3	0.00	0.66	1.68	36.2
TiO_2	—	0.44	0.90	—
FeO	35.2	20.1	8.66	0.12
MnO	0.37	0.29	0.15	—
MgO	28.9	23.5	15.4	0.06
CaO	0.08	1.65	21.4	19.3
Na_2O	—	0.00	0.02	0.36
K_2O	—		—	0.03
Cr_2O_3	0.06	0.21	0.47	—
Σ_{oxides}	100.4	99.7	101.4	100.6
Si	0.989	1.962	1.932	2.046
Al	0.000	0.029	0.072	1.958
Ti	—	0.012	0.025	—
Fe	0.816	0.625	0.265	0.005
Mn	0.009	0.009	0.005	—
Mg	1.193	1.301	0.846	0.004
Cr	0.002	0.066	0.841	0.951
Na	—	0.000	0.002	0.033
K	—	—	—	0.002
Cr	0.001	0.062	0.014	—
	Fo 59.14	Wo 3.27	42.97	An 96.52
	Fay 40.86	En 65.03	43.23	Ab 3.30
		Fs 31.70	13.80	Or 0.18

73263,1,1 (Station 3, landslide from South Massif)

This is a polymict, anorthositic breccia consisting of ~25% clasts (≥25 μ) in a fine-grained matrix (Fig. 1c). The clasts consist of single crystals of plagioclase (An_{95-97}), olivine (Fo_{76-78}), orthopyroxene ($Wo_{2-4}En_{73-77}Fs_{19-24}$), and spinel in decreasing order of abundance. One inverted pigeonite clast containing lamellae of calcic pyroxene ($Wo_{44}En_{46}Fs_{10}$) in an orthopyroxene host ($Wo_3En_{74}Fs_{23}$) was observed (Fig. 5e). Polycrystalline clasts include: (a) anorthosite fragments, whose textures indicate a variety of thermal and shock histories; and (b) a fine-grained, recrystallized, noritic fragment containing plagioclase (An_{96}), olivine (Fo_{77}), orthopyroxene ($Wo_4En_{76}Fs_{20}$), and clinopyroxene ($Wo_{38}En_{51}Fs_{11}$).

Chromian pleonaste spinel (Table 5) occurs in a reaction complex with ilmenite and is surrounded by a plagioclase aureole devoid of any ferromagnesian phases. This complex is similar to those observed in 79215 (Bickel *et al.*, 1976; McGee and Bence, 1977). These are thought to have equilibrated at temperatures of the order of 800–900°C.

The fine-grained matrix (2–25 μ) consists of plagioclase (An_{94-97}), olivine (Fo_{75-77}), poikilitic orthopyroxene ($Wo_{3-7}En_{72-77}Fs_{19-21}$), and calcic clinopyroxene

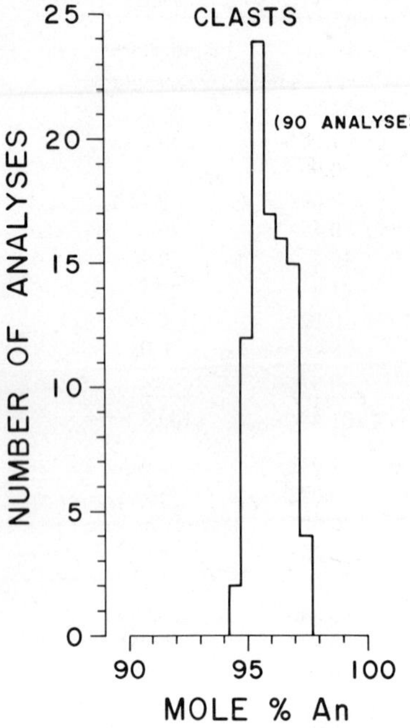
CLASTS
(90 ANALYSES)
NUMBER OF ANALYSES
0
5
10
15
20
25
90
95
100
MOLE % An

(a)

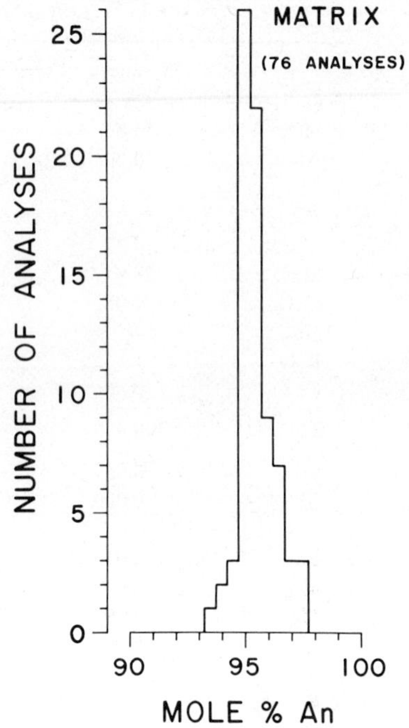
MATRIX
(76 ANALYSES)
NUMBER OF ANALYSES
0
5
10
15
20
25
90
95
100
MOLE % An

(b)

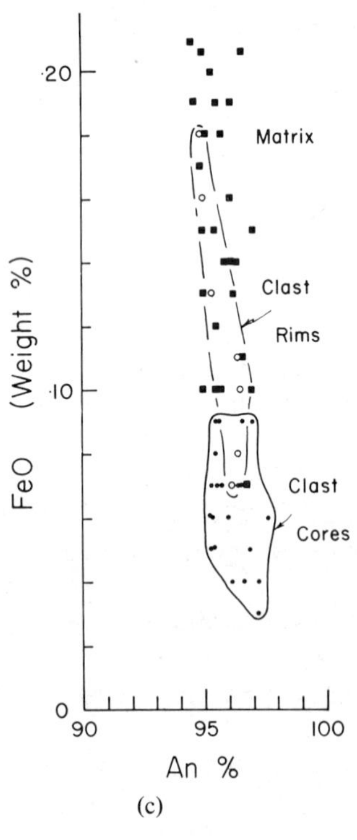
Matrix
Clast
Rims
Clast
Cores
FeO (Weight %)
0
·10
·20
90
95
100
An %

(c)

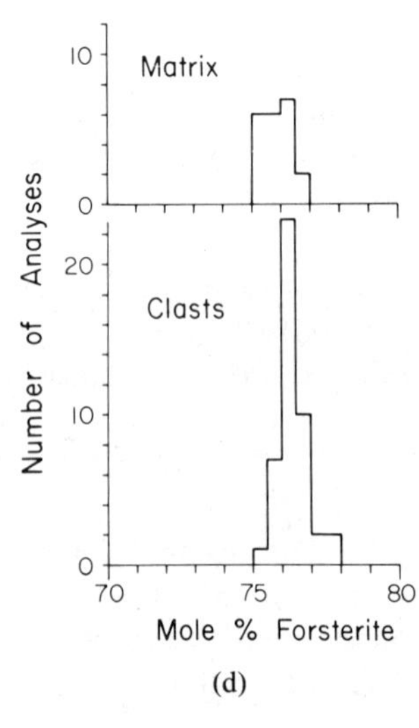
Matrix
Clasts
Number of Analyses
0
10
0
10
20
70
75
80
Mole % Forsterite

(d)

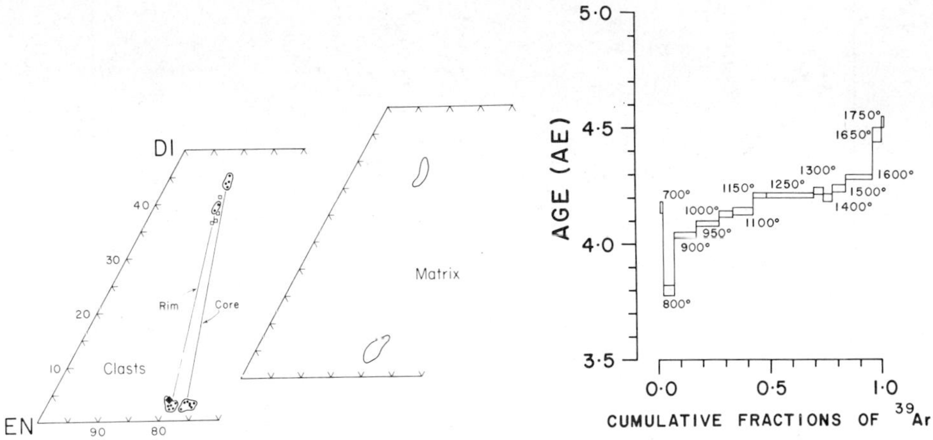

Fig. 5. 73263,1,1. (a) Plagioclase clast compositions. (b) Matrix plagioclase compositions. (c) FeO concentrations in plagioclase. (d) Clast and matrix olivine compositions. (e) Pyroxene compositions. Core and rim pyroxenes from inverted pigeonite clasts are joined by tie-lines. Solid diamond and open squares are from separate clasts. Fields of matrix compositions are outlined. (f) $^{40}Ar/^{39}Ar$ temperature release spectrum from Schaeffer and Schaeffer (1977).

($Wo_{36-40}En_{49-53}Fs_{11-12}$) in decreasing order of abundance. Representative analyses of the clast and matrix phases are given in Table 5. Slight but significant compositional differences can be seen between the matrix and the clast phases (Figs. 5a–e).

Groundmass and clast plagioclases have identical, tightly restricted, major-element compositional ranges (Figs. 5a,b). However, the concentrations of Fe and Mg in the clast cores differ systematically from their concentrations in the clast rims and groundmass plagioclases (Fig. 5c). These differences suggest that only the rims of the plagioclase clasts have equilibrated with the groundmass assemblage.

Groundmass and clast olivines differ slightly in their fayalite contents. Although some compositional overlap is observed the groundmass olivines have systematically higher fayalite contents (Fo_{75-77}) than the clasts olivines (Fo_{76-78}) (Fig. 5). The narrow compositional range of the groundmass olivines suggest that they have equilibrated. The narrow compositional range of olivine clasts in this polymict breccia suggests a common source.

The width of the miscibility gap in the core of the inverted pigeonite clast (Wo_3–Wo_{44}) is greater than the phase separation of coexisting clinopyroxene-orthopyroxene in the groundmass (Wo_4–Wo_{39}). The rim of the inverted pigeonite clast contains pyroxenes with compositions similar to the groundmass pyroxenes (Fig. 5e and Table 5). These observations suggest that inversion of the pigeonite occurred prior to its incorporation in the breccia and equilibration of the clast rim occurred during annealing of the breccia. The distribution of Fe and Mg between the augite lamellae and orthopyroxene in the clast ($K_D \sim 0.68$) is

Table 5. 73263,1,1.

	Clasts						Groundmass			
				Inverted Pigeonite						
	Plagioclase	Olivine	Orthopyroxene	Augite	Orthopyroxene	Spinel	Orthopyroxene	Augite	Olivine	Plagioclase
SiO_2	43.9	38.9	54.8	52.3	54.3	0.00	54.2	51.5	38.4	44.9
Al_2O_3	35.3	0.00	0.87	1.96	0.98	30.8	1.36	2.58	0.00	35.5
TiO_2	—	—	0.29	0.49	0.13	3.42	0.94	1.88	—	—
FeO	0.03	21.4	13.2	6.29	16.0	22.8	13.0	7.09	21.8	0.18
MnO	—	0.20	—	—	—	0.39	—	—	0.24	—
MgO	0.00	39.7	28.7	16.9	27.3	10.7	28.5	17.8	39.5	0.05
CaO	19.5	0.14	1.93	22.1	1.27	—	2.21	19.1	0.33	19.6
Na_2O	0.28	—	—	—	—	—	—	—	—	0.40
K_2O	0.06	—	—	—	—	—	—	—	—	0.13
Cr_2O_3	—	0.01	0.49	0.53	0.32	31.4	0.49	0.71	0.03	—
Σ_{oxides}	99.1	100.4	100.3	100.6	100.3	99.5	100.7	100.7	100.3	100.8
Si	2.047	1.002	1.957	1.919	1.958	—	1.931	1.882	0.994	2.061
Al	1.943	0.000	0.037	0.085	0.042	1.098	0.057	0.111	0.000	1.921
Ti	—	—	0.008	0.014	0.004	0.078	0.025	0.052	—	—
Fe	0.001	0.461	0.394	0.193	0.482	0.577	0.388	0.217	0.471	0.007
Mn	—	0.004	—	—	—	0.010	—	—	0.005	—
Mg	0.000	1.526	1.527	0.923	1.469	0.484	1.510	0.970	1.525	0.003
Ca	0.976	0.004	0.074	0.869	0.049	—	0.084	0.748	0.009	0.964
Na	0.025	—	—	—	—	—	—	—	—	0.035
K	0.003	—	—	—	—	—	—	—	—	0.008
Cr	—	0.000	0.014	0.015	0.009	0.750	0.014	0.021	0.001	—
	An 97.18	Fo 76.65	Wo 3.70	43.77	2.45		Wo 4.26	38.66	Fo 76.2	An 95.7
	Ab 2.49	Fa 23.35	En 76.57	46.51	73.46		En 76.19	50.14	Fa 23.8	Ab 3.5
	Or 0.34		Fs 19.73	9.72	24.09		Fs 19.55	11.20		Or 0.8

consistent with equilibration at a temperature of ⩾810°C (Lindsley *et al.*, 1974). A $K_D \simeq 0.86$ for the matrix pyroxenes indicates equilibration at >810°C. A $K_D \simeq 1.18$ for olivines and orthopyroxenes is consistent with equilibration at 925–1070°C (Williams, 1971) (Table 1).

The higher temperature of matrix equilibration suggests relatively rapid cooling of the breccia after the incorporation of the clasts. The occurrence of the inverted pigeonite (⩾810°C) clast with only a small reaction rim (>810°C) in contact with the matrix is consistent with that inference. The preservation of anorthositic fragments with a variety of shock textures indicates that the relatively high-temperature annealing event was not prolonged.

The argon-argon temperature release pattern for fragment 73263,1,1 (Schaeffer and Schaeffer, 1977) (Fig. 5f) contains a high-temperature plateau with an average age of 4.23 ± 0.05 AE and a less well-defined low-temperature plateau for the 1000°C and 1100°C fractions of 4.14 ± 0.07 AE. The argon and petrographic data are consistent with, at least, a two-component system comprised of old, unequilibrated, anorthositic, and plagioclase single-crystal clasts (>4.2 AE) and younger, recrystallized matrix (⩽4.1 AE).

We suggest that sample 73263,1,1 was a breccia (possibly a clast-laden melt) containing old clasts, incorporated in an ejecta blanket that was annealed at about 4.1 AE in a transient high-temperature event.

73263,1,6 (Station 3, landslide from South Massif)

This sample is an annealed anorthositic breccia containing ~20% clasts dominated by single-crystal plagioclase fragments (An_{93-98}) (Fig. 6a) and one annealed polycrystalline anorthosite (An_{95}) (Table 6) (Fig. 1d). Rare olivine clasts have an extremely restricted compositional range centered at Fo_{77-78} (Fig. 6c and Table 6). Single-crystal clasts of orthopyroxene ($Wo_3En_{78}Fs_{19}$) and clinopyroxene ($Wo_{45}En_{47}Fs_8$) are also rare. The anorthosite clast contains <5% ferromagnesians consisting of olivine (Fo_{78}), orthopyroxene ($Wo_4En_{77}Fs_{19}$) and clinopyroxene ($Wo_{39}En_{50}Fs_{11}$). With the exception of the single anorthosite clast, the tightly restricted compositional ranges of the other clasts suggests a common source for each phase.

The matrix of 73263,1,6 consists of plagioclase (An_{91-98}), orthopyroxene laths ($Wo_4En_{77}Fs_{19}$), granular clinopyroxene ($Wo_{39}En_{50}Fs_{11}$), and granular olivine (Fo_{77-78}) (Figs. 6b,c,d). On the basis of major and minor elements, the clast and matrix plagioclase and olivine are indistinguishable. The pyroxenes, however, have small but distinct differences. Clast pyroxenes are more magnesian and contain lower concentrations of aluminum, titanium, and chromium than the matrix pyroxenes. Clast orthopyroxenes are less calcic than their matrix counterparts (Wo_{2-3} vs. Wo_{3-4}) and the clast augites are more calcic but less aluminous than the matrix augites (Wo_{42-44} vs. Wo_{36-41}) (Figs. 6d,e). Pyroxenes in the anorthositic clast are most similar to the matrix pyroxenes.

Iron-magnesium partitioning between the pyroxene pairs ($K_D \sim 0.87 \pm 0.01$) both within the matrix and within the anorthositic clast indicate that subsolidus

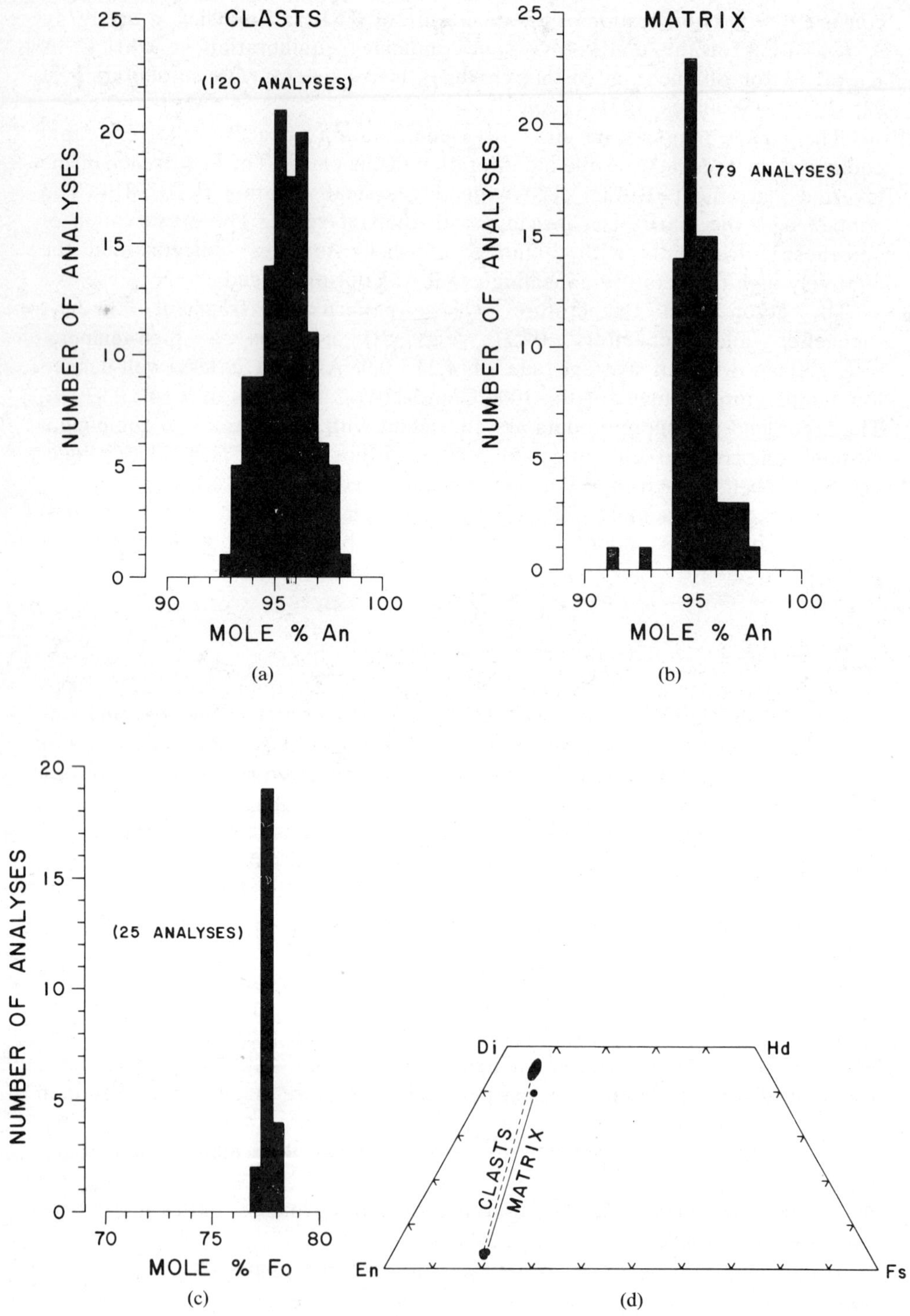
CLASTS
(120 ANALYSES)
NUMBER OF ANALYSES
MOLE % An
MATRIX
(79 ANALYSES)
NUMBER OF ANALYSES
MOLE % An
(25 ANALYSES)
NUMBER OF ANALYSES
MOLE % Fo
Di
Hd
En
Fs
CLASTS
MATRIX

(a) (b) (c) (d)

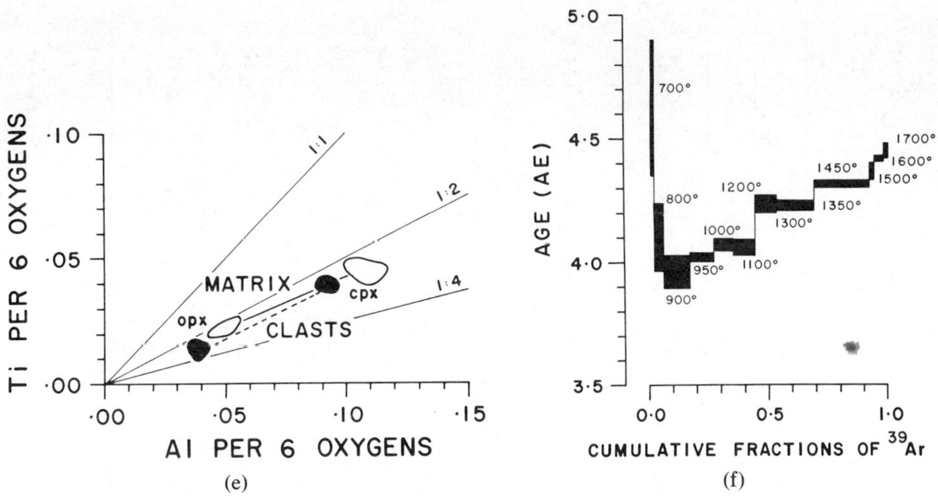

Fig. 6. 73263,1,6. (a) and (b) Clast and matrix plagioclases. (c) Clast and matrix olivine compositions. (d) Clast and matrix pyroxenes. (e) Ti vs. Al for matrix and clast pyroxenes. (f) $^{40}Ar/^{39}Ar$ temperature release spectrum from Schaeffer and Schaeffer (1977).

equilibration occurred at temperatures >810°C (Lindsley *et al.*, 1974). A K_D of ~1.15 for the olivines and orthopyroxenes indicates that their equilibration occurred at 925–1070°C (Williams, 1971) (Table 1).

The argon-argon temperature release pattern for 73263,1,6 (Schaeffer and Schaeffer, 1977) is shown in Fig. 6c. The four low-temperature fractions (900–1100°C) give a moderately defined plateau consistent with an age of ~4.0 AE. The 1200–1450°C fractions (~50% of the ^{39}Ar) give a less well-defined plateau at 4.2–4.3 AE. These data are consistent with a two-component system containing old (>4.2 AE) plagioclase clasts in a young (~4.0 AE) annealed, matrix. The annealing event at ~4.0 AE was insufficient to equilibrate the argon.

Discussion and Conclusions

Four of the breccias having ages ≳4.2 AE (67483,14,2; 67483,13,6; 73263,1,1; and 73263,1,6) are polymict. We suggest that the portions of the argon-argon spectrum having ages greater than 4.0 AE originate in unequilibrated relicts of greater age—possibly as old as the original crust (4.4–4.6 AE). The activation energies required to initiate diffusion of major and minor elements between silicate phases comprising the clasts and matrix are higher than the activation energies required for argon diffusion (Hofmann and Hart, 1975). Consequently, equilibration of major and minor elements between matrix and clast phases in a polymict breccia may be considered strong evidence favoring argon equilibration. However, lack of minor-element equilibration does not preclude the possibility that the argon did equilibrate. Diffusion is a rate process, therefore,

Table 6. 73263,1,6.

	Clasts				Anorthosite		Matrix			
	Orthopyroxene	Augite	Olivine	Plagioclase	Plagioclase	Orthopyroxene	Orthopyroxene	Augite	Olivine	Plagioclase
SiO_2	54.8	52.1	38.2	43.8	44.5	53.4	53.7	51.4	38.1	44.7
Al_2O_3	0.92	2.18	0.00	35.7	34.6	1.23	1.21	2.47	0.05	35.4
TiO_2	0.52	1.43	—	—	—	0.87	0.78	1.62	—	—
FeO	12.3	4.91	20.7	0.08	0.20	12.6	12.9	6.45	21.9	0.26
MnO	—	—	0.24	—	—	—	—	—	0.23	—
MgO	28.8	16.8	41.0	0.11	0.12	27.9	29.3	17.5	39.5	0.11
CaO	1.54	21.5	0.12	19.7	19.8	2.52	1.93	18.9	0.19	19.3
Na_2O	—	—	—	0.34	0.47	—	—	—	—	0.46
K_2O	—	—	—	0.08	0.19	—	—	—	—	0.10
Cr_2O_3	0.46	0.73	0.13	—	—	0.48	0.55	0.81	0.01	—
Σ_{oxides}	99.3	99.7	100.4	99.8	99.9	99.0	100.4	99.2	100.0	100.3
Si	1.964	1.914	0.984	2.032	2.065	1.935	1.919	1.901	0.990	2.061
Al	0.039	0.095	0.000	1.954	1.896	0.053	0.051	0.108	0.001	1.923
Ti	0.014	0.040	—	—	—	0.024	0.021	0.045	—	—
Fe	0.368	0.151	0.446	0.003	0.008	0.381	0.385	0.200	0.477	0.010
Mn	—	—	0.005	—	—	—	—	—	0.005	—
Mg	1.539	0.921	1.574	0.008	0.008	1.505	1.561	0.963	1.531	0.007
Ca	0.059	0.846	0.003	0.977	0.984	0.098	0.074	0.749	0.005	0.954
Na	—	—	—	0.030	0.043	—	—	—	—	0.041
K	—	—	—	0.005	0.011	—	—	—	—	0.006
Cr	0.013	0.021	0.003	—	—	0.014	0.016	0.024	0.000	—
	Wo 3.01	44.10	Fo 77.71	An 96.49	94.83	Wo 4.94	Wo 3.66	39.17	Fo 76.06	An 95.29
	En 78.29	48.03	Fay 22.29	Ab 3.02	4.10	En 75.85	En 77.30	50.38	Fay 23.94	Ab 4.10
	Fs 18.71	7.88		Or 0.48	1.07	Fs 19.21	Fs 19.04	10.45		Or 0.61

the duration of a heating event, as well as its temperature, determines the extent to which components are equilibrated.

The texture of sample 67483,14,2 indicates that it is a clast-laden, melt rock and its $^{40}Ar/^{39}Ar$ release spectrum (Fig. 2d) contains a young, low-temperature ($\leq$1000°C) component and an old ($\geq$4.2 AE), high-temperature component. A logical interpretation of this spectrum consistent with the petrography is that the sample contains ancient (>4.2 AE) clasts that were incorporated into a younger (probably 4.0 AE) melt formed in an impact event. As noted above, however, the proportions of clast to matrix and their relative concentrations of potassium are not consistent with this idea. Furthermore, it is unlikely that all of the clasts would be very old. This may be a simple problem of sample homogeneity. An alternative, but less likely, explanation is that the high-temperature plateau age of 4.2 AE dates the formation of the melt rock itself. Subsequent reheating (possibly at ~4.0 AE) resulted in partial argon loss from the least retentive sites in the glassy matrix. It is not, however, possible to distinguish between these two models with the information presently available.

The $^{40}Ar/^{39}Ar$ spectrum obtained on sample 67483,13,6 (Fig. 3c) is generally similar to that obtained on 67483,14,2. In this case, however, there is evidence for annealing possibly through reheating or in a slow cooling history. One of the clasts appears to be a slightly annealed, shocked anorthosite fragment and the matrix appears to be a recrystallized melt. Although the equilibration of major and minor elements between clasts and matrix has not occurred, some equilibration of the argon might be expected during annealing. We suggest that 67483,13,6 is a slightly recrystallized clast-laden melt rock. The lithologic and isotopic data are consistent with the incorporation of plagioclase clasts and anorthosite fragments in a shock melt formed and annealed at about 4.0 AE.

The two Apollo 17 fragments (73262,1,1 and 73263,1,6) are the most severely recrystallized samples studied. The partitioning of Fe and Mg between the matrix phases suggests that this recrystallization occurred at temperatures in the range 925–1070°C. This temperature range is near the low-pressure solidus for their composition and should seemingly have been sufficient to thoroughly degas argon from the matrix. However, nearly 60% of the radiogenic argon is older than 4.2 AE (Fig. 5f). This is not similar to the relative proportions of clasts and matrix. Consequently, some old argon that did not re-equilibrate during annealing must be retained in matrix phases. The variation of K/Ca with ^{39}Ar fraction released (Schaeffer and Schaeffer, 1977) suggests two-component mixing. High K/Ca ratios (~0.007) are associated with the low-temperature release of argon, presumably in some matrix phases. High-temperature gases are characterized by lower K/Ca ratios (~0.005), a relationship to be expected if this argon is a characteristic of the plagioclase clasts.

Similar arguments apply to sample 73263,1,6. It too was annealed at very high temperatures. However, equilibration of the argon in that annealing event was not complete. Nearly 60% of the argon (Fig. 6f) gives an age between 4.0 and 4.1 AE. High K/Ca ratios (~0.008) are associated with the young argon, whereas the majority of the old argon can be attributed to plagioclase clasts with low

K/Ca ratios (~0.006). A lack of correlation of clast/matrix ratios with the proportions of old and younger gas again suggests an old component within the matrix of the specimen.

Of the five samples studied, only anorthositic gabbro 67483,13,8 appears to be truly monomict. Thus, an argon-argon plateau age of 4.24 ± 0.05 AE (Fig. 4d) might represent an excavation age. This age is similar to the 4.26 ± 0.02 AE argon-argon age obtained for troctolite 76535 (Husain and Schaeffer, 1975) and the 4.28 ± 0.05 AE argon-argon age obtained for poikilitic gabbro 78503,13A (Kirsten and Horn, 1974). Husain and Schaeffer suggest that the 4.26 AE age is the time of the Serenitatis Basin forming event, and the predominance of 3.90–4.0 AE ages for highlands samples reflects the dominance of younger, Imbrium ejecta at the highland landing sites.

The most striking feature of all of the samples studied is the presence of a significant amount of 4.2–4.3 AE argon. The proportion of this old, high-temperature, component to the younger (~4.0 AE) component correlates poorly with the relative proportions of clasts and groundmass. Also, the degree of matrix recrystallization and/or groundmass-clast-equilibration inferred from major- and minor-element partitioning correlates poorly with the proportion of the old gas component. Even where it appears that the matrix and clasts are largely chemically equilibrated, two component "ages" persist. A lack of argon equilibration in these highly annealed samples is an unsatisfying explanation because the proportions of gas components and matrix/clast ratios do not agree. Additional studies of argon diffusion mechanisms appear mandated. In particular, *in situ* dating of the matrix and clasts of these old breccias through the laser technique (Plieninger and Schaeffer, 1976; Müller *et al.*, 1977) may resolve the problem.

An alternative explanation—that the old, high-temperature argon plateaus date the time of formation or annealing of the breccia must be entertained. Although the majority of highlands samples was affected by the event at ~4.0 AE, a few samples apparently excavated by earlier events may have retained the signature of their earlier history. Argon-argon "ages" of ~4.25 AE, whether they constitute the majority of the temperature spectrum or only the high-temperature fraction appear to have some real significance in lunar crustal evolution. These samples may provide a "window" through the ~4.0 AE cataclysm and supply information concerning lithology of the ancient lunar crust.

Neither 67483,14,2 nor 67483,13,8 are highly magnesian samples. Their Mg/(Mg + Fe) ratios of ~0.6 are consisent with anorthositic highland crust. The separation of the crustal components from the differentiating magma ocean must, for most lunar bulk compositions, have occurred after significant evolution of the magma had occurred. As a consequence, ferromagnesian phases within the primitive plagioclase-rich crustal rocks should be enriched in their iron components. This relationship is observed in the anorthosites (e.g., McCallum *et al.*, 1975; Dixon and Papike, 1975). The chemistries, mineralogies and ages of 67483,14,2 and 67483,13,8 are consistent with their formation in the primitive lunar anorthositic crust and subsequent excavation at about 4.25 AE.

The old components of 67483,13,6; 73263,1,1; and 73263,1,6 almost certainly formed in the primitive highlands crust. However, without localized *in situ* dating of the breccias, discrimination of these old components is at best uncertain. Consequently, the correlation of age components in a total rock argon-argon spectrum with the lithologic components must be made with caution.

Acknowledgments—We wish to thank O. A. Schaeffer for providing the argon-argon data for 73263,1,1 and 73263,1,6 prior to publication. This research was supported by NASA grant #NGL-33-015-130.

References

Albee A. L. and Ray L. (1970) Correction factors for electron probe microanalysis of silicates, oxides, carbonates, phosphates, and sulfates. *Anal. Chem.* **42**, 1408–1414.

Bence A. E. and Albee A. R. (1968) Empirical correction factors for electron microanalysis of silicates and oxides. *J. Geol.* **76**, 382–403.

Bickel C. E., Warner J. L., and Phinney W. C. (1976) Petrology of 79215: Brecciation of a lunar cumulate. *Proc. Lunar Sci. Conf. 7th*, p. 1793–1819.

Dixon J. R. and Papike J. J. (1975) Petrology of anorthosites from the Descartes region of the moon: Apollo 16. *Proc. Lunar Sci. Conf. 6th*, p. 263–291.

Hofmann A. W. and Hart S. R. (1975) An assessment of local and regional isotopic equilibrium in a partially molten mantle. *Carnegie Inst. Wash. Yearb.* **74**, 195–210.

Horn P., Jessberger E. K., Kirsten T., and Richter H. (1975) ^{39}Ar–^{40}Ar dating of lunar rocks: Effects of grain size and neutron irradiation. *Proc. Lunar Sci. Conf. 6th*, p. 1563–1591.

Huneke J. C., Jessberger E. K., Podosek F. A., and Wasserburg G. J. (1973) ^{40}Ar/^{39}Ar measurements in Apollo 16 and 17 samples and the chronology of metamorphic and volcanic activity in the Taurus-Littrow region. *Proc. Lunar Sci. Conf. 4th*, p. 1725–1756.

Husain L. and Schaeffer O. A. (1975) Lunar evolution: The first 600 million years. *Geophys. Res. Lett.* **2**, 29–32.

Jessberger E. K., Huneke J. C., Podosek F. A., and Wasserburg G. J. (1974a) High resolution argon analysis of neutron-irradiated Apollo-16 rocks and separated minerals. *Proc. Lunar Sci. Conf. 5th*, p. 1419–1449.

Jessberger E. K., Huneke J. C., and Wasserburg G. J. (1974b) Evidence for a ~4.5 AE age of plagioclase clasts in a lunar highland breccia. *Nature* **248**, 199–201.

Kirsten T. and Horn P. (1974) ^{39}Ar–^{40}Ar chronology of the Taurus-Littrow region II: A 4.28 b.y. old troctolite and ages of basalts and highland breccia. *Proc. Lunar Sci. Conf. 5th*, p. 1451–1475.

Lindsley D. H., King H. E., Jr., and Turnock A. C. (1974) Compositions of synthetic augite and hypersthene coexisting at 810°C: Application to pyroxenes from lunar highlands rocks. *Geophys. Res. Lett.* **1**, 134–136.

McCallum I. S., Okamura F. P., and Ghose S. (1975) Mineralogy and petrology of sample 67075 and the origin of lunar anorthosites. *Earth Planet. Sci. Lett.* **26**, 36–53.

McGee J. and Bence A. E. (1977) Oxide phase assemblages in lunar breccia 79215: Limitations on T-f_{O_2} conditions of metamorphism (abstract). *EOS* (*Trans. Amer. Geophys. Union*) **58**, 430.

Müller H. W., Plieninger T., James O. B., and Schaeffer O. A. (1977) Laser probe ^{40}Ar–^{39}Ar dating of materials from consortium breccia 73215 (abstract). In *Lunar Science VIII*, p. 697–699. The Lunar Science Institute, Houston.

Plieninger T. and Schaeffer O. A. (1976) Laser probe ^{39}Ar–^{40}Ar ages of individual mineral grains in lunar basalt 15607 and lunar breccia 15465. *Proc. Lunar Sci. Conf. 7th*, p. 2055–2066.

Schaeffer O. A. and Husain L. (1973) Early lunar history: Ages of 2 to 4 mm soil fragments from the lunar highlands. *Proc. Lunar Sci. Conf. 4th*, p. 1847–1863.

Schaeffer O. A. and Husain L. (1974) Early lunar history: Ages 2 to 4 mm soil fragments from the lunar highlands. *Proc. Lunar Sci. Conf. 4th*, p. 1847–1863.

Schaeffer G. A. and Schaeffer O. A. (1977) ^{39}Ar–^{40}Ar ages of lunar rocks. *Proc. Lunar Sci. Conf. 8th.* This volume.

Tera F., Papanastassiou D. A., and Wasserburg G. J. (1974) Isotopic evidence for a terminal lunar cataclysm. *Earth Planet. Sci. Lett.* **22**, 1–21.

Turner G. and Cadogan P. H. (1975) The history of lunar bombardment inferred from ^{40}Ar–^{39}Ar dating of highland rocks. *Proc. Lunar Sci. Conf. 6th*, p. 1509–1538.

Turner G., Huneke J. C., Podosek F. A., and Wasserburg G. J. (1972) Ar^{40}–Ar^{39} systematics in rocks and separated minerals from Apollo 14. *Proc. Lunar Sci. Conf. 3rd*, p. 1589–1612.

Williams R. J. (1971) Reaction constants in the system Fe–MgO–SiO_2–O_2: I. Experimental Results. *Amer. J. Sci.* **270**, 334–360.

Proc. Lunar Sci. Conf. 8th (1977), p. 2051–2066.
Printed in the United States of America

Feldspathic granulitic impactites and pre-final bombardment lunar evolution

JEFFREY L. WARNER and WILLIAM C. PHINNEY

NASA Johnson Space Center, Houston, Texas 77058

CHARLES E. BICKEL

San Francisco State University, San Francisco, California 94122

CHARLES H. SIMONDS

The Lunar Science Institute, Houston, Texas 77058

Abstract—Attention is called to a group of rocks called feldspathic granulitic impactites. These samples are characterized by a high modal plagioclase content of between 70 and 80%. The samples are impactites in that they display a clast-matrix structure and are enriched in trace siderophile elements that are indicative of meteorite components. The matrices of these samples display granulitic textures indicative of rather complete solid-state annealing. Although the ages of these samples are not conclusively determined, we argue that they all formed in the period after consolidation of the lunar crust and before the final bombardment.

The granulitic impactites contain essentially no KREEP component, indicating that KREEP appeared on the lunar surface mostly after the formation of the granulitic impactites, at about the start of the final bombardment. This relation constrains theories as to the time of KREEP formation.

The granulite metamorphism indicated by the matrix textures of these samples requires 1000°C temperatures for prolonged periods of time. Such conditions may have been achieved early in lunar history when still hot impactite sheets could have been buried by layers of younger ejecta that were themselves hot.

We observe that there is a temporal relationship between the types of lunar impactites: granulitic impactites formed before the final bombardment, crystalline-matrix breccias formed during the final bombardment, and vitric-matrix breccias formed after the final bombardment. This sequence is consistent with our thermal model of breccia lithification and our understanding of the decay of the meteorite flux rate.

INTRODUCTION

ESTABLISHING THE HISTORY of the moon during its first 600 m.y. of evolution continues to be a major goal of the lunar sample analysis program, yet it has been elusive because almost all lunar samples from the highlands have yielded ages in the range 3.84 – 4.05 AE (Wetherill, 1975), and samples from the lunar mare are even younger. This tight cluster of highland ages has been interpreted to date a final lunar bombardment (Tera *et al.*, 1974): a period of intense meteorite impact during which most of the returned samples formed. But the final lunar bombardment did not obliterate all evidence of older samples. Anorthosites and Mg-rich plutonic rocks probably represent materials that formed in the original lunar magma ocean (Dymek *et al.*, 1975; Warner, 1975). Some of those plutonic rocks have subsequently undergone Apollonian metamorphism (Stewart, 1975).

In this paper we define a suite of impactites that is petrographically and chemically distinct and apparently formed before the final lunar bombardment. The group was first discussed as a coherent set of lunar materials by Simonds *et al.* (1974) who referred to "rocks with 70–80% feldspar" from Apollo 17. For reasons presented later we believe that these rocks are the oldest impactites returned from the moon. They are common in the Apollo collections. Members of this group from Apollo 11 were described by Wood *et al.* (1970) at the First Lunar Science Conference, and other members from every landing site have been described at subsequent Lunar Science Conferences. This suite includes five large discrete samples, but most members occur as clasts in breccias that formed during or after the final lunar bombardment, or as lithic fragments in the regolith.

The chief thrust of this contribution is to point out the common chemical and textural characteristics of these impactites, and to speculate on some of the implications of such characteristics. Their chemical composition is that of anorthositic norite or troctolite and thus they approximate the lunar crustal composition. The textures all display granulitic matrices suggestive of a high temperature metamorphic history. Based on these characteristics we call members of this suite of impactites *feldspathic granulitic impactites*, or simply *granulitic impactites*. Note that we use the term "impactite" to mean any impact produced rock. We are not following the AGI glossary definition that restricts "impactite" to glassy impact produced rocks.

Very similar rocks belonging to this suite have been referred to as hornfelses (e.g., James, 1977b; Roedder and Weiblen, 1974) and granulites (e.g., Bickel *et al.*, 1976). The term "hornfels" emphasizes the rather fine grain size of these rocks in comparison to many regionally metamorphosed terrestrial rocks. Unfortunately, the term "hornfels" generally connotes a magmatic source for the heat of thermal metamorphism, and there is no reason to believe that such was the case for the granoblastic and poikiloblastic textured rocks described in this paper. However, these rocks appear to have been thermally metamorphosed at shallow depths in the lunar crust. The term "granulite" generally connotes a rock from a very widespread terrestrial metamorphic terrain that was metamorphosed at considerable depth, in some cases near the base of the crust. The mineral assemblages of lunar granoblastic and poikiloblastic rocks are compatible with the use of either term.

The existence of feldspathic granulitic impactites implies certain facts about the chemistry and the thermal environment of the lunar surface before the final bombardment. The absence of KREEP from all these rocks implies that KREEP was not present on the lunar surface when they formed. The high temperature metamorphic history of these samples represents more prolonged annealing at higher, but subsolidus, temperature conditions than have apparently affected any younger rocks. Such conditions may be explained by a higher heat flow and/or a more intense flux of meteorite bombardment during the time when the granulitic impactites formed.

Description of Feldspathic Granulitic Impactites

Criteria for sample identification

Three criteria are used to distinguish feldspathic granulitic impactites.

First, the samples must be impactites. We have previously established the criteria to recognize the lithified deposits of impact events known as breccias or impactites (Simonds *et al.*, 1975) as (1) their clast-matrix structure and (2) their enrichment in trace siderophile elements of meteoritic origin by at least an order-of-magnitude over the indigenous lunar abundances as observed in igneous rocks such as mare basalts.

Second, the samples must display granulitic (metamorphic) matrix textures. Crystal forms are anhedral and equant and boundaries form smooth curves that approach 120° triple junctions.

Third, the samples have modal mineral compositions with between 70 and 80% plagioclase. This mode translates into a major-element composition that approximates the model lunar crust with about 25 wt.% Al_2O_3.

Surely a complete sampling of the moon will reveal rocks that satisfy criteria 1 and 2 but not 3, because older rocks have diverse compositions. However, such samples have not yet been identified within the Apollo collections suggesting that impactites with granulitic textures *and* a feldspathic composition are the most abundant type of granulitic impactite on the moon.

Occurrence

Feldspathic granulitic impactites occur as clasts in breccias and as fragments in regolith samples. Five are large discrete samples: 67955 (163 g), 76230/76235 (77 g), 77017 (1730 g), 78155 (401 g), and 79215 (554 g). Sample 76230/76235 is a clast in the Apollo 17, Station 6 boulder—a portion of an impact-melt breccia sheet (Simonds, 1975). Although the other four were collected as loose rocks from the regolith, they surely were formerly clasts within breccias.

There are several sets of feldspathic granulitic impactite clasts in breccias and fragments in regolith that have been described in the literature. These include:

(1) Many of the gabbroic anorthosite fragments described by Wood *et al.* (1970) in the Apollo 11 collections (e.g., Wood's Figs. 3a,b,c).
(2) The granulitic (mosaic) textured and poikiloblastic textured fragments and clasts reported by Bickel and Warner (1977).
(3) The three anorthositic gabbro clasts in 73215 (29.9, 46.25, and 46.33) reported by James (1977a) and Blanchard *et al.* (1977).

(4) The clast upon which the original data for Apollo 17, Station 2 boulder sample 72335 was obtained (Laul and Schmitt, 1974; Dymek *et al.*, 1976).
(5) Granulitic and poikiloblastic "ANT" clasts in Apollo 17, Station 2 boulder samples 72215 and 72235 (Ryder *et al.*, 1975).
(6) Several lithic clasts in 67915 described by Roedder and Weiblen (1974).
(7) "White rind" and "white clast" in Apollo 17, Station 6 boulder sample 76315 (Simonds, 1975).

Impact origins

In order to show that these samples originated as lithified impact debris one must demonstrate that each displays a clast-matrix structure, that each is polymict, and that each has an enriched trace siderophile element spectrum. Parts of these tasks are difficult because the granulitic impactites have undergone post crystallization thermal metamorphism. That process, and perhaps the original lithification process, yielded a restricted and highly refractory clast assemblage.

Sample 78155 contains the clearest evidence for an impactite origin. The report by Bickel (1977) describes two types of lithic clasts. The first are fine-grained anorthosites with a felty texture in which the interstices between tabular plagioclase are occupied by crystals of pigeonite and olivine. The second type of lithic clasts are coarse grained and display a range in composition (40–80 vol.% plagioclase; the major mafic mineral is olivine in some, low-Ca pyroxene in others, and augite in one) and texture (subophitic, poikiloblastic, and granoblastic). This sample has an enriched trace siderophile element spectrum (Morgan *et al.*, 1974) that fits into one of the more poorly defined siderophile groups (Higuchi and Morgan, 1975).

Sample 77017 has been interpreted by some workers as igneous (Ashwal, 1975); however, McCallum *et al.* (1974), who have done the most detailed petrography of 77017, describe the rock as a breccia that contains "... lithic clasts of anorthosite and troctolite and single crystals of rounded olivine and rounded to subhedral plagioclase". The sample has an enriched trace siderophile element spectrum (Morgan *et al.*, 1974) that matches one of the groups identified with the final bombardment (Higuchi and Morgan, 1975).

Sample 79215 was described by Bickel *et al.* (1976). They demonstrate that 79215 has a clast-matrix structure. Evidence that 79215 is polymict is not compelling, probably because the annealing of 79215 has homogenized compositional differences of over a few mole percent. Bickel *et al.* describe millimeter to centimeter sized regions of the rock with different textures and modes (olivine-rich clusters and a region with lath-shaped ilmenite) that are here interpreted as lithic clasts. According to Morgan *et al.* (1975) the trace siderophile element spectrum is enriched and very different from all known groups.

Sample 76230/76235 has a clast-matrix structure similar to 77017. Some megacrysts may be mineral clasts. Rounded regions about 1 mm across are interpreted as anorthosite lithic clasts. The anorthosites are the only evidence that the rock may be polymict since all mineral compositions have been homogenized (Simonds, 1975). Isolated olivine oikocrysts may mark the location of former clasts. Siderophile elements have not been determined.

As illustrated in Fig. 7 of Nord *et al.* (1975) and described by Hollister (1973), sample 67955 has a clast-matrix structure similar to 79215. Hollister's description suggests that the rock contains lithic clasts of various types. The trace siderophile element spectrum has been determined by Ganapathy *et al.* (1974); it is enriched and poorly matches one of the groups typical of the final bombardment (Higuchi and Morgan, 1975).

Some lithic clasts in crystalline-matrix breccias display matrix textures and modal compositions similar to granulitic impactites. Each displays a clast-matrix structure defined mostly by mineral clasts. Trace siderophile elements are enriched in those clasts for which data exist (Laul and Schmitt, 1974).

Granulitic matrix textures

The matrices of these impactites all display granulitic (metamorphic) textures. Crystal forms are anhedral and equant with boundaries that form smooth curves and meet at roughly 120° triple junctions. There are two textural types. One type displays granoblastic or mosaic textures (Fig. 1A)—networks of 10–50 μm plagioclase crystals define the mosaic patterns, and smaller mafic minerals occur along boundaries and at triple junctions (see descriptions of

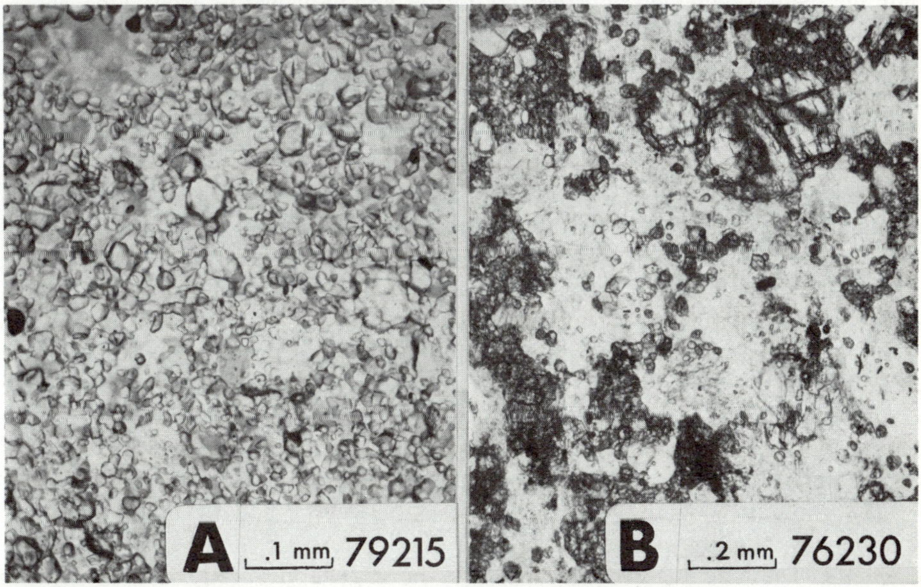

Fig. 1. Photomicrographs in transmitted light illustrating the matrix texture of the granulitic impactites. (a) An example of the granoblastic or mosaic texture in sample 79215. (b) An example of the poikiloblastic texture in sample 76230.

78155 and 79215 in Bickel *et al.*, 1976 and Bickel, 1977). A second type of impactite displays poikiloblastic textures (Fig. 1B)—anhedral plagioclase crystals with smooth boundaries (up to 100 μm across) enclose rounded olivine blebs and the plagioclase itself is partly enclosed by pyroxene oikocrysts (see descriptions of 77017 and 76235 in McCallum *et al.*, 1974 and Simonds, 1975).

Mineral chemistry in the granulitic impactites is generally homogeneous. Crystals are not zoned and the composition of most mineral clasts matches the composition of the same mineral in adjacent matrix (see rock descriptions referenced above). However, the scale length of the homogeneity is on the order of millimeters. For example, Bickel *et al.* (1976) have demonstrated that there are systematic differences in the Fe/Mg ratio of coexisting olivine and low-Ca pyroxene in different regions of 79215.

Mineral clasts have been annealed. Feldspar clasts commonly display olivine necklaces but melt inclusions are not observed.

Modes and chemical compositions

All granulitic impactites from the lunar highlands contain between 70 and 80 modal percent calcic plagioclase along with low-Ca pyroxene, high-Ca pyroxene, and olivine. Differences in the ratio of the mafic minerals occur both from sample to sample and within each sample (e.g., see Bickel *et al.*, 1976 for details concerning the modal range within 79215). Fe–Ni metal, ilmenite, spinel, and phosphates total about 1% of each sample. Rocks of this modal composition, if plutonic, would be called anorthositic norites and anorthositic troctolites.

The major-element compositions of feldspathic granulitic impactites, as expected from the modal composition, approximate proposed model lunar crustal compositions (e.g., Taylor and Bence, 1975) and the composition of "highland basalt" as defined by the Apollo Soil Survey (Reid *et al.*, 1972). Major-element data is set out in Table 1.

Trace-element data for granulitic impactites is shown in relation to chondritic abundances in Fig. 2. We note that although the texture and mode of the clast in 72335 is appropriate for this discussion, the chemistry is not, because it is contaminated with KREEP-type material: the analyzed fragment is described as a surface chip with an exterior patinated surface (Laul and Schmitt, 1974). In addition, we note the large samples all contain less La than 14 times chondrites and have a positive Eu anomaly. By contrast the three samples with a negative Eu anomaly and La about 22 times chondrites are clasts removed from breccias. These relations suggest that these clasts are partly contaminated with KREEP-type material from the matrices of the host breccias.

The granulitic impactites have La values between 7 and 14 times chondrites and display slightly positive slopes in the pattern of the heavy rare-earth elements. They define a compositionally homogeneous group that is distinct from KREEP and also distinct from Taylor and Bence's (1975) postulated average lunar crust which displays a characteristic KREEP pattern. The granulitic impactites' pattern is indicative of much less fractionation from chondritic

Table 1. Major-element composition of some feldspathic, granulitic impactites.

	67955[1]	76230[2]	77017[2]	78155[2]	79215[3]
SiO_2	45.0	44.5	44.1	45.6	43.8
TiO_2	.3	.2	.4	.3	.3
Al_2O_3	27.7	27.0	26.6	25.9	27.7
Cr_2O_3	—	.1	.1	.1	.2
FeO	3.8	5.1	6.2	5.8	4.6
MnO	.05	.06	.08	.1	.06
MgO	7.1	7.6	6.1	6.3	6.3
CaO	15.5	15.2	15.4	15.2	15.9
Na_2O	.4	.4	.3	.3	.5
K_2O	.03	.06	.06	.08	.1
P_2O	—	.05	.03	.04	.4
Total	99.9	100.3	99.4	99.7	99.9

[1]LSPET (1973a).
[2]LSPET (1973b).
[3]Bickel *et al.* (1976).

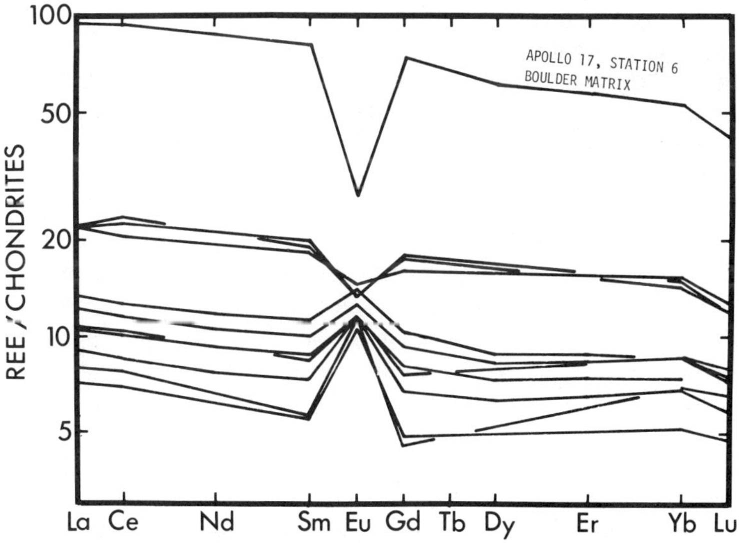

Fig. 2. Plot of rare-earth element spectra normalized to chondritic abundances. The ten patterns with flat or slightly positive curves for the heavy rare-earth elements are the granulitic impactites. The average pattern for the matrix of the Apollo 17, Station 6 boulder, a typical Apollo 17 type KREEP sample, is shown for comparison. It is suggested that the three patterns with La about 20 times chondrites are contaminated. Thus only the seven patterns with La between 7 and 14 times chondrites represent granulitic impactites (see text for discussion). Data from Blanchard *et al.* (1975, 1977) and Hubbard *et al.* (1974).

abundances, and in fact, may be indicative of the true rare-earth element pattern for the lunar crust. The implications of granulitic impactite chemistry is discussed at some length by Blanchard *et al.* (1977).

Ages of granulitic impactites

Only fragmentary data exist on the ages of these samples. One large sample and several granulitic impactite clasts extracted from breccias yield ages between 4.2 and 4.3 AE. Two other large samples yield ages between 3.9 and 4.0 AE—ages that correspond with the final lunar bombardment. We argue from the geologic occurrences that the ages of the two younger samples are reset, and conclude that the feldspathic granulitic impactites were produced and metamorphosed before the final lunar bombardment.

Sample 78155 has a superior Ar-plateau age of $4.22 \pm .04$ AE (Turner and Cadogan, 1975) that has been confirmed by a Pb-isotopic study (Nunes *et al.*, 1975).

The three granulitic impactite clasts from 73215 described by James (1977a) have Ar-plateaus obtained by Jessberger *et al.* (1976). Each clast yielded a double plateau with a low temperature age of 4.06–4.07 AE and a high temperature age of 4.22–4.26 AE. The young plateau corresponds to the age of the host breccia and the older plateau is considered by Jessberger *et al.* to represent the age of the clast. The ages of these small clasts have been partially reset.

Two other large granulitic impactites have yielded Ar-plateau ages. Cadogan and Turner (1976) have determined a 3.94 AE plateau for 76230/76235, and the weighted mean of one plateau age by Kirsten and Horn (1974) and one plateau age by Phinney *et al.* (1975) for 77017 is 3.98 AE. These plateau ages are considered reset by events that post-date granulite metamorphism of the rocks. Sample 76230/76235 is from a 1 m clast in the Apollo 17, Station 6 boulder described by Simonds (1975). Cadogan and Turner (1976) have obtained acceptable plateaus from 13 samples from the boulder including six matrix samples and two samples of 76235; all plateau ages fall between 3.92 and 4.02. Surely 76230/76235, a clast in the boulder, must be older than the matrix. Since the measured plateau ages are the same, the age of the clast must have been reset when the matrix formed. Unfortunately no evidence exists as to the original age of 76230/76235.

It is less certain that the plateau age of 77017 represents a post-metamorphic event. Although 77017 was not chipped from a boulder, the only similar material at the Apollo 17 site occurs as clasts in boulders, and all of the Apollo 17 boulders display radiometric ages that fall within the time interval of the final lunar bombardment. This suggests that 77017 is analogous to 76230/76235 and that its Ar-plateau age may be reset.

Although we are confident that the granulitic impactites formed before the final lunar bombardment, we are puzzled by which samples display reset ages and which samples retain their formational age. For example, at Apollo 17 the two samples that are completely reset are large (tens of centimeters to 1 m), whereas the centimeter sized clasts recovered from 73215 retain evidence of their original age. Perhaps the tendency, or capacity, for ages of clasts to be reset is determined in part by the nature of the host breccia. The small samples that are not reset come from the South Massif and the large samples that are reset come from the North Massif. Other workers have suggested that the South Massif represents a higher stratigraphic level compared to the North Massif (ALGIT, 1973). And a higher stratigraphic level may suggest a more rapid cooling history for the host breccias, an inference confirmed by each sample's petrography (James, 1977a; Simonds, 1975).

In the discussion of sample 79215, Bickel *et al.* (1976) suggested that the unusual trace siderophile element spectrum of that sample was a criterion for a pre-final bombardment age. Although that proposed age criterion appears not to be generally applicable, they were apparently correct in their inferred age based on their serendipity and not on their adroitness (i.e., they were right for reasons that are not generally true).

Discussion

This suite of impactites that may pre-date the final lunar bombardment provides our best petrologic record (thus far) of the development of the lunar surface after the solidification of the hypothesized lunar magma ocean at about 4.4 AE, and before the start of the final lunar bombardment at about 4.1 AE. In this section we explore several chemical and textural characteristics that are shared by all members of the suite of granulitic impactites, and then speculate on the significance of those common characteristics in terms of lunar evolution assuming that the granulitic impactites formed in the 4.4–4.1 AE interval.

Composition of the lunar crust and the age and origin of KREEP

Feldspathic granulitic impactites have compositions similar to those of lunar plutonic rocks. Their major-element compositions are close to the composition calculated for the lunar crust. The compositions plot far from any point of multiple saturation on Walker's pseudoternary diagram (Walker *et al.*, 1973) suggesting that cumulate processes were involved in the original generation of the chemical systems. Lithophile trace elements are low compared to most breccias that formed during or after the final bombardment, and the rare-earth element patterns are only slightly fractionated with respect to chondrites. These chemical features indicate that the precursors for the granulitic impactites were dominated by plutonic, perhaps cumulate, igneous rocks.

The rare-earth element patterns for granulitic impactites are different from those of impactites that formed during or after the final bombardment—the former materials display a slope for the heavy rare earths that is flat or slightly positive, whereas all impactite materials that formed later in lunar history display a uniform negative slope that is characteristic of KREEP. Mixing models by Schonfeld (1975; pers. comm., 1977) indicate that granulitic impactites contain only 3–4% KREEP. By contrast, later-formed breccias contain up to 100% KREEP. All mixing models yield results that are a function of the input components. A small and reasonable change in Schonfeld's feldspathic component will yield models for the granulitic impactites that contain no KREEP.

The chemical composition of the lunar surface, as recorded by impactites, appears to have undergone a major change at about the time of the final bombardment. Impacts produce violent and extreme mixing and it is inferred that impactites record the average composition of the target material of any given impact (Simonds *et al.*, 1976a). The observed composition of granulitic impactites (assuming they are not allochthonous blocks that formed in the farside highlands or some other trace element-poor region of the lunar surface) suggests that the pre-bombardment crust was essentially without KREEP. In contrast, KREEP composition material dominates the chemistry of rocks that formed during the final bombardment.

The hypothesized arrival of KREEP material at the lunar surface must have followed the formation of the granulitic impactites and approximately coincided with the start of the final bombardment. This age for KREEP agrees with the crystallization age for both igneous and brecciated KREEP (Meyer, 1977). If KREEP originated as a volcanic rock, its interval of extrusion corresponded with the interval of the final bombardment, perhaps because the large impacts of the final bombardment induced KREEP volcanism in a manner similar to that proposed by French (1970) for the triggering of volcanism by the Sudbury event. On the other hand, if KREEP formed as a late differentiate within the lunar crust, its arrival on the lunar surface corresponded with the interval of the final bombardment because the multi-ringed basins formed by the final bombardment were required to excavate it. In either case important speculations may be made about the moon's history.

If KREEP is volcanic then there are three major types of volcanic rocks returned from the moon: KREEP (the oldest), high-Ti, high-Fe mare basalts (intermediate in age), and low-Ti, high-Fe mare basalts (youngest). These relations are illustrated in the top of Fig. 3. The young high-Ti basalts predicted for Western Procellarum based on crater morphology and reflectance spectra (Boyce, 1974; Pieters and McCord, 1976) were not identified among the Apollo 12 coarse fines (Lellis and Papike, 1975) and may not exist. This picture of lunar volcanic activity is remarkably simple and is not much more complex than the pattern inferred in 1971 when Warner (1971) first suggested that the depth to the source region of lunar volcanics increases with time. Such a "zone refinement" for the moon is illustrated in Fig. 4. A similar "zone refinement" has been proposed for Mars (Schonfeld, 1977), and is probably to be expected in the future history of the earth.

If KREEP was a late differentiate within the lunar crust then it must have existed as a significant still-molten unit within the crust at the time of final bombardment. Such a concept is barely within the bounds of lunar thermal history calculations. Furthermore, this residual melt must have been within the range of excavation of the major basins—a few tens of kilometers. The surface

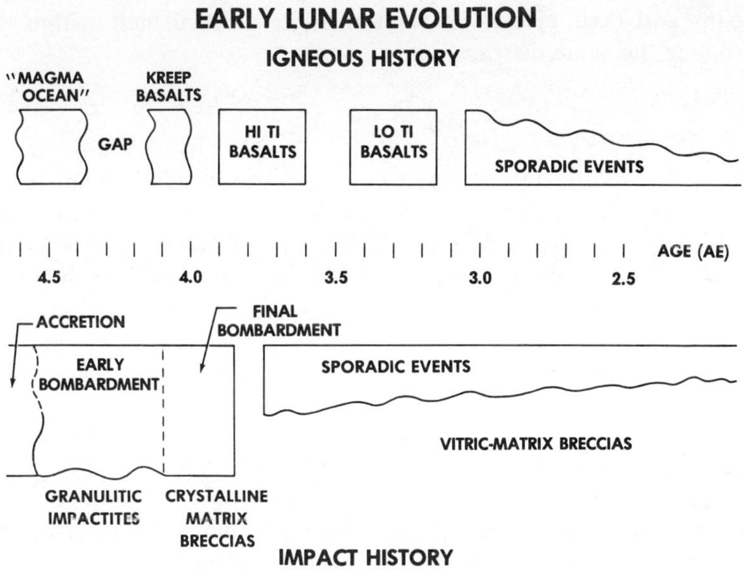

Fig. 3. Schematic illustration of the chief events in lunar evolution that are recorded in the lunar samples.

distribution of KREEP components as determined from the orbital gamma-ray detectors indicates a lack of this component on the farside and an irregular distribution on the nearside. In fact, the distribution on the nearside is so closely associated with the Imbrium and Procellarum basins and their ejecta that Schonfeld and Meyer (1973) were led to postulate volcanic KREEP units within

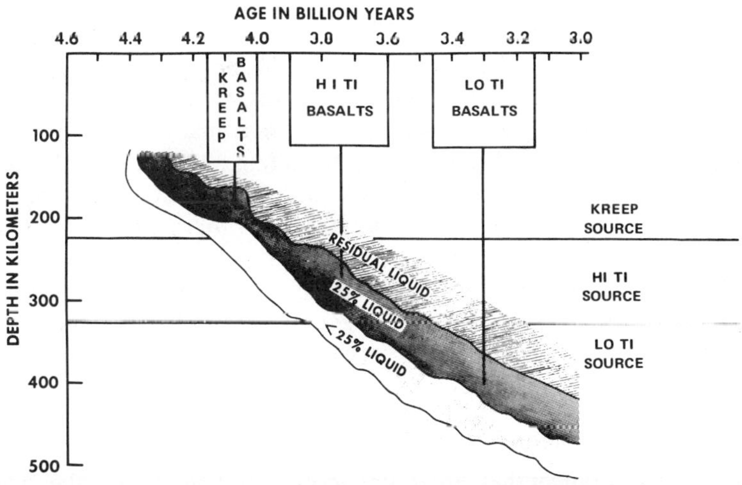

Fig. 4. Schematic illustration of some aspects of lunar structure based on a simple zone refinement model for lunar volcanic rocks and assuming that KREEP is one of the volcanics. Thermal model taken from McConnell and Gast (1972).

these basins and their ejecta. Excavation of a residual melt within the crust would produce the same distribution.

A lunar granulite metamorphism

The granulitic matrix textures and homogeneous mineral chemistry of these samples indicate high-grade metamorphism that is characterized by rather complete solid-state annealing or recrystallization. This homogenization with a scale length of about a millimeter contrasts with the Apollonian metamorphism of Stewart (1975) with a scale length of about a centimeter. The Apollonian metamorphism affected samples such as 15415 and 76535 that are primitive cumulates that cooled slowly under appreciable cover in the original lunar crust. Metamorphism of the granulitic impactites occurred later, nearer the lunar surface, and more quickly. The granulitic metamorphism may have been promoted by an intense meteorite bombardment that preceded the final lunar bombardment.

Granulitic texture does not characterize matrices of crystalline-matrix (impact-melt) breccias that formed during the final bombardment. Granulitic impactites are found only as clasts in those breccias (James, 1977b). This suggests that the high temperatures and prolonged heating required to develop granulitic textures was achieved only before the final bombardment. If such conditions were achieved more recently the granulites produced were not sampled by the Apollo program, perhaps because basin-sized impacts which are required to excavate granulites did not occur more recently.

Formation of granulitic textures under dry conditions where there is no evidence for a vapor or a melt requires extensive volume and grain boundary diffusion of Si and Al in plagioclase. Such diffusion processes are inferred to be ineffective at less than 1000°C. Evidence for temperatures in excess of 1100°C are inferred from coexisting uninverted pigeonite and low-Ca augite in 78155 (Bickel, 1977), and temperatures of between 1050 and 1100°C are inferred from mineral chemistry in 77017 (McCallum *et al.*, 1974).

Apparently such high temperatures were not maintained for prolonged periods in the larger terrestrial impact structures (e.g., Manicouagan) or typically in the lunar impact-melt rocks. We suggest that these conditions can be achieved by burial of several kilometers where the 1000°C isotherm is relatively shallow. These conditions could have been achieved early in lunar history when there was a higher impact flux and a higher heat flow, and still hot impactites could have been buried by layers of hot younger ejecta. Following this model the observed Ar-plateau ages are likely to represent the times of excavation.

A temporal sequence of impactite textures

This paper has argued that lunar impactites formed before the final bombardment have undergone a thermal metamorphism and display matrices with granulitic textures. Investigations by us (Simonds *et al.*, 1976a, 1977) and others (Englehardt *et al.*, 1972; and Quaide and Wrigley, 1974) have demonstrated that vitric-matrix breccias are younger than the final lunar bombardment. The "brown glass matrix breccias" of Apollo 15 are vitric matrix and they contain lithic clasts of Apollo 15 type mare basalt. This demonstrates that they are younger than 3.2 AE. The vitric-matrix breccias from Station 9 at Apollo 17 contain spheres of the orange glass which post-date the final bombardment. The vitric-matrix breccias at Apollo 14 appear to be younger than the crystalline-matrix breccias because the compositions of the former, but not the latter, match the composition of local soil. And finally, essentially all breccias found at mare sites are vitric and contain mare basalt material.

Apparently crystalline-matrix and/or melt-rock breccias were the chief breccia type that formed during the final bombardment. This is demonstrated by the 3.85–4.1 AE ages obtained on many crystalline-matrix breccias.

This temporal sequence of impactites is illustrated on the bottom half of Fig. 3. One wonders if such relationships will hold for the samples to be collected in future lunar exploration missions. We note that during the late 19th and early 20th centuries, at a stage in the development of terrestrial geology that is somewhat similar to the current status of lunar geology, all high-grade metamorphosed rocks were considered Precambrian. There will surely be exceptions to these very simple relations depicted in Fig. 3. For instance we know of one sample (15256) that has the composition of a mare basalt and yet is a crystalline-matrix breccia, albeit with an unusual texture.

We can suggest an explanation for the observed temporal sequence of breccia types. Each type represents a different thermal history. All impactite types form from the violent mixing of cold fragmental material and superheated melt. During first stage cooling the mixture reaches a rapid thermal equilibrium in a few tens of seconds, and during second stage cooling the deposit loses heat to its surroundings in a manner similar to a normal volcanic flow. The breccia type achieved by each impactite sample is a function of that sample's first and second stage cooling history. According to this model (that we have elaborated upon in some of our recent works, e.g., Simonds *et al.*, 1967b) the vitric-matrix breccias originated as fragment-rich mixtures that equilibrated to subsolidus temperatures during the first stage of cooling, and had a cool environment during the second stage of cooling. The crystalline-matrix breccias originated as fragment-poor mixtures that reached supersolidus temperatures during the first stage of cooling, and had a cool environment during the second state of cooling. The granulitic impactites originated as fragment-poor mixtures that equilibrated to supersolidus temperatures during the first stage of cooling, and lost heat to a hot environment during the second stage of cooling.

The thermal environments required to produce granulitic impactites, crystalline-matrix breccias, and vitric-matrix breccias can respectively be achieved by numerous large impacts; fewer, smaller impacts; and sporadic, small impacts. This sequence of events is similar to the cratering history of the moon. Based on this model we predict that on other cratered planets such as Mercury and Mars the relative ages of impactites may be established on the basis of their petrography.

Acknowledgments—Reviews by Richard Williams, Jeffrey Taylor, Martin Prinz, and Steve Winzer have greatly improved the manuscript. The speculative discussion is totally our responsibility. One of the authors (C.H.S.) is a staff scientist at the Lunar Science Institute. The Lunar Science Institute is operated by the Universities Space Research Association under contract No. NSR-09-051-001 with the National Aeronautics and Space Administration. This paper constitutes the Lunar Science Contribution No. 288.

REFERENCES

AFGIT (Apollo Field Geology Investigation Team) (1973) Geologic exploration of Taurus–Littrow: Apollo 17 landing site. *Science* **182**, 672–680.

Ashwal L. D. (1975) Petrologic evidence for a plutonic igneous origin of anorthositic norite in 67955 and 77017. *Proc. Lunar Sci. Conf. 6th*, p. 221–230.

Bickel C. E. (1977) Petrology of 78155: An early, thermally metamorphosed polymict breccia. *Proc. Lunar Sci. Conf. 8th.* This volume.

Bickel C. E. and Warner J. L. (1977) Survey of lunar plutonic and granulitic lithic fragments (abstract). In *Lunar Science VIII*, p. 112–114. The Lunar Science Institute, Houston.

Bickel C. E., Warner J. L., and Phinney W. C. (1976) Petrology of 79215: Brecciation of a lunar cumulate. *Proc. Lunar Sci. Conf. 7th*, p. 1793–1819.

Blanchard D. P., Haskin L. A., Jacobs J. W., Brannon J. C., and Korotev R. L. (1975) Major and trace element chemistry of Boulder 1 at Station 2, Apollo 17. *The Moon* **14**, 359–371.

Blanchard D. P., Jacobs J. W., and Brannon J. C. (1977) Chemistry of ANT clasts and a felsite from consortium breccia 73215 and of gabbroic anorthosite 79215. *Proc. Lunar Sci. Conf. 8th.* This volume.

Boyce J. M., Dial A. L., and Soderblom L. A. (1974) Ages of the lunar nearside light plans and maria. *Proc. Lunar Sci. Conf. 5th*, p. 11–23.

Cadogan P. H. and Turner G. (1976) The chronology of the Apollo 17 Station 6 Boulder. *Proc. Lunar Sci. Conf. 7th*, p. 2267–2285.

Dymek R. F., Albee A. L., and Chodos A. A. (1975) Comparative petrology of lunar cumulate rocks of possible primary origin: Dunite 72415, troctolite 76535, norite 78235, and anorthosite 62237. *Proc. Lunar Sci. Conf. 6th*, p. 301–341.

Dymek R. F., Albee A. L., and Chodos A. A. (1976) Petrology and origin of Boulders #2 and #3, Apollo 17 Station 2. *Proc. Lunar Sci. Conf. 7th*, p. 2335–2378.

Engelhardt W. von, Arndt J., Stöffler D., and Schneider H. (1972) Apollo 14 regolith and fragmental rocks, their compositions and origin by impacts. *Proc. Lunar Sci. Conf. 3rd*, p. 753–770.

French B. M. (1970) Possible relations between meteorite impact and igneous petrogenesis, as indicated by the Sudbury structure, Ontario, Canada. *Bull. Volcanol.* **34**, 466–517.

Ganapathy R., Morgan J. W., Higuchi H., Anders E., and Anderson A. T. (1974) Meteoritic and volatile elements in Apollo 16 rocks and in separated phases from 14306. *Proc. Lunar Sci. Conf. 5th*, p. 1659–1683.

Gros J., Takahashi H., Hertogen J., Morgan J. W., and Anders E. (1976) Composition of the projectiles that bombarded the lunar highlands. *Proc. Lunar Sci. Conf. 7th*, p. 2403–2425.

Higuchi H. and Morgan J. W. (1975) Ancient meteoritic component in Apollo 17 boulders. *Proc. Lunar Sci. Conf. 6th*, p. 1625–1651.

Hollister L. S. (1973) Sample 67955: A description and a problem. *Proc. Lunar Sci. Conf. 4th*, p. 633–641.

Hubbard N. J., Rhodes J. M., Wiesmann H., Shih C.-Y., and Bansal B. M. (1974) The chemical definition and interpretation of rock types returned from the non-mare regions of the moon. *Proc. Lunar Sci. Conf. 5th*, p. 1227–1246.

James O. B. (1977a) Petrology of four clasts from consortium breccia 73215 (abstract). In *Lunar Science VIII*, p. 502–504. The Lunar Science Institute, Houston.

James O. B. (1977b) Lunar highlands breccias generated by major impacts. In *Proc. Soviet-Amer. Conf. on the Cosmochemistry of the Moon and Planets* (Moscow). June 4–8, 1974, p. 624-629. Also NASA SP-370, in press.

Jessberger E. K., Kirsten T., and Staudacher Th. (1976) Argon-argon ages of consortium breccia 73215. *Proc. Lunar Sci. Conf. 7th*, p. 2201–2215.

Kristen T. and Horn P. (1974) Chronology of the Taurus–Littrow region III: Ages of mare basalts and highland breccias and some remarks about the interpretations of lunar highland rock ages. *Proc. Lunar Sci. Conf. 5th*, p. 1451–1475.

Laul J. C. and Schmitt R. A. (1974) Chemical composition of boulder 2 rocks and soils, Apollo 17, Station 2. *Earth Planet. Sci. Lett.* **23**, 206–219.

Lellis S. and Papike J. J. (1975) Apollo 12 basalts: Petrography and interstation distributions (abstract). In *Papers Presented to the Conference On Origins of Mare Basalts and their Implications for Lunar Evolution*, p. 94–98. The Lunar Science Institute, Houston.

LSPET (Lunar Sample Preliminary Examination Team) (1973a) The Apollo 16 lunar samples: A petrographic and chemical description of samples from the lunar highlands. *Science* **179**, 23–34.

LSPET (Lunar Sample Preliminary Examination Team) (1973b) Apollo 17 lunar samples: Chemical and petrographic description. *Science* **182**, 659–672.

McCallum I. S., Mathez E. A., Okamura F. P., and Ghose S. (1974) Petrology and crystal chemistry of poikilitic anorthositic gabbro 77017. *Proc. Lunar Sci. Conf. 5th*, p. 287–302.

McConnell R. K. and Gast P. W. (1972) Lunar thermal history revisited. *The Moon* **5**, 41–51.

Meyer C., Jr. (1977) Petrology, mineralogy, and chemistry of KREEP basalt. *Phys. Chem. Earth*. In press.

Morgan J. W., Ganapathy R., Higuchi H., Krähenbühl U., and Anders E. (1974) Lunar basins: Tentative characterization of projectiles, from meteoritic elements in Apollo 17 boulders. *Proc. Lunar Sci. Conf. 5th*, p. 1703–1736.

Morgan J. W., Higuchi H., Ganapathy R., and Anders E. (1975) Meteoritic material in four terrestrial meteorite craters. *Proc. Lunar Sci. Conf. 6th*, p. 1609–1623.

Nord G. L., Jr., Christie J. M., Hever A. H., and Lally J. S. (1975) North Ray Crater breccias: An electron petrographic study. *Proc. Lunar Sci. Conf. 6th*, p. 779–797.

Nunes P. D., Tatsumoto M., and Unruh D. M. (1975) U–Th–Pb systematics of anorthositic gabbros 78155 and 77017—implications for early lunar evolution. *Proc. Lunar Sci. Conf. 6th*, p. 1431–1444.

Phinney D., Kahl S. B., and Reynolds J. H. (1975) ^{40}Ar–^{39}Ar dating of Apollo 16 and 17 rocks. *Proc. Lunar Sci. Conf. 6th*, p. 1593–1608.

Pieters C. and McCord T. B. (1976) Characterization of lunar mare basalt types: I. A remote sensing study using reflection spectroscopy of surface soils. *Proc. Lunar Sci. Conf. 7th*, p. 2677–2690.

Quaide W. and Wrigley R. (1972) Mineralogy and origin of Fra Mauro fines and breccias. *Proc. Lunar Sci. Conf. 3rd*, p. 771–784.

Reid A. M., Ridley W. I., Harmon R. S., Warner J. L., Brett R., Jakeš P., and Brown R. W. (1972) Highly aluminous glasses in lunar soils and the nature of the lunar highlands. *Geochim. Cosmochim. Acta* **36**, 903–912.

Roedder E. and Weiblen P. W. (1974) Petrology of clasts in lunar breccia 69715. *Proc. Lunar Sci. Conf. 5th*, p. 303–318.

Ryder G., Stoeser D. B., Marvin U. B., Bower J. F., and Wood J. A. (1975) Boulder 1, Station 2, Apollo 17: Petrology and petrogenesis. *The Moon* **14**, 327–357.

Schonfeld E. (1975) A model for the lunar anorthositic gabbro. *Proc. Lunar Sci. Conf. 6th*, p. 1375–1384.

Schonfeld E. (1977) Martian volcanism (abstract). In *Lunar Science VIII*, p. 843–845. The Lunar Science Institute, Houston.

Schonfeld E. and Meyer C., Jr. (1973) The old Imbrium hypothesis. *Proc. Lunar Sci. Conf. 4th*, p. 125–138.

Simonds C. H. (1975) Thermal regimes in impact melts and the petrology of the Apollo 17 Station 6 boulder. *Proc. Lunar Sci. Conf. 6th*, p. 641–672.

Simonds C. H., Phinney W. C., and Warner J. L. (1974) Petrography and classification of Apollo 17 non-mare rocks with emphasis on samples from the Station 6 boulder. *Proc. Lunar Sci. Conf. 5th*, p. 337–353.

Simonds C. H., Warner J. L., and Phinney W. C. (1975) Criteria for the impact origin of lunar fragmental-matrix and melt-rock matrix polymict breccias (abstract). EOS (*Trans. Amer. Geophys. Union*) **56**, 1071.

Simonds C. H., Warner J. L., and Phinney W. C. (1976a) Thermal regimes in cratered terrain with emphasis on the role of impact melt. *Amer. Mineral.* **61**, 569–577.

Simonds C. H., Warner J. L., Phinney W. C., and McGee P. (1976b) Thermal model for breccia lithification: Manicouagan and the moon. *Proc. Lunar Sci. Conf. 7th*, p. 2509–2528.

Simonds C. H., Phinney W. C., Warner J. L., McGee P. E., Geeslin J., Brown R., and Rhodes J. M. (1977) Apollo 14 revisited, or breccias aren't so bad after all. *Proc. Lunar Sci. Conf. 8th.* This volume.

Stewart D. B. (1975) Apollonian metamorphic rocks (abstract). In *Lunar Science VI*, p. 774–776. The Lunar Science Institute, Houston.

Taylor S. R. and Bence A. E. (1975) Evolution of the lunar highland crust. *Proc. Lunar Sci. Conf. 6th*, p. 1121–1141.

Tera F., Papanastassiou D. A., and Wasserburg G. J. (1974) Isotopic evidence for a terminal lunar cataclysm. *Earth Planet. Sci. Lett.* **22**, 1–21.

Turner G. and Cadogan P. H. (1975) The history of lunar bombardment inferred from ^{40}Ar–^{30}Ar dating of highland rocks. *Proc. Lunar Sci. Conf. 6th*, p. 1509–1538.

Walker D., Grove T. L., Longhi J., Stolper E. M., and Hays J. F. (1973) Origin of lunar feldspathic rocks. *Earth Planet. Sci. Lett.* **20**, 325–336.

Warner J. L. (1971) Lunar crystalline rocks: Petrology and geology. *Proc. Lunar Sci. Conf. 2nd*, p. 469–480.

Warner J. L. (1975) Mineralogy, petrology, and geochemistry of the lunar samples. *Rev. Geophys. Space Phys.* **13**, 107–113 and 163–168.

Warner J. L., Simonds C. H., and Phinney W. C. (1976) Apollo 17, Station 6 boulder sample 76255: Absolute petrology of breccia matrix and igneous clasts. *Proc. Lunar Sci. Conf. 7th*, p. 2233–2250.

Wetherill G. W. (1975) Late heavy bombardment of the moon and terrestrial planets. *Proc. Lunar Sci. Conf. 6th*, p. 1539–1561.

Wood J. A., Dickey J. S., Jr., Marvin U. B., and Powell B. N. (1970) Lunar anorthosites and a geophysical model of the moon. *Proc. Apollo 11 Lunar Sci. Conf.*, p. 965–988.

Proc. Lunar Sci. Conf. 8th (1977), p. 2067–2078.
Printed in the United States of America

Crystallization kinetics, viscous flow, and thermal history of lunar breccia 67975

D. R. UHLMANN, L. C. KLEIN, and C. A. HANDWERKER

Department of Materials Science and Engineering, Massachusetts Institute of Technology, Cambridge, Massachusetts 02139

Abstract—The viscous flow and crystallization behavior of the matrix composition in lunar breccia 67975 have been determined over wide ranges of temperature. The variation of viscosity with temperature is closely described by the model of Bottinga and Weill (1972), and even more closely by the model of Shaw, 1972. The maximum crystal growth rate for the 67975 matrix composition, about 7.3×10^{-4} cm min^{-1}, is observed at a temperature of about 990°C ($\Delta T = 220$°C). This maximum growth rate is slower than that of any other lunar composition studied to date.

Time-temperature-transformation (TTT) and continuous cooling (CT) curves were constructed from the measured crystal growth rates and viscosities, and used to estimate the coolng rate required to form bodies with the degree of crystallinity observed in the matrix of the breccia. The detailed analysis of crystallization statistics has been used to describe the state of crystallinity in a cooled body of the 67975 composition, and the results are compared with the predictions of the CT treatment. It is estimated from the kinetic analysis that the 67975 breccia cooled at a rate of about 0.06°C min^{-1} over the temperature range below the liquidus. From an analysis of viscous sintering under nonisothermal conditions, it is concluded that significant sintering will be achieved for particles which come in contact at any temperature above or somewhat below the liquidus.

Heat flow analyses have been carried out for the cooling of a melt-clast mixture by heat flow from hot melt to cold clasts, and also for cooling by heat flow to the surroundings for various boundary conditions of physical interest. The results indicate that the 67975 breccia did not cool in the size of the lunar sample, but as part of a much larger body. Annealing of the matrix glass in the 67975 breccia increases its density, which suggests that it formed by cooling from the molten state rather than by direct shock-induced transition from crystal to glass.

INTRODUCTION

PREVIOUS WORK in our laboratory (e.g., Uhlmann *et al.*, 1975) has analyzed the process of breccia formation by viscous sintering in a stress-free environment. This process had been advanced by Simonds (1973) as the principal mechanism of formation of clastic or fragmental breccias. The kinetic analysis by Uhlmann *et al.* uses a combined treatment of crystallization and viscous sintering to derive information about the thermal histories of clastic breccias.

Application of the analysis to a number of lunar breccias (e.g., Klein *et al.* 1975; Uhlmann and Klein, 1976) has indicated that it is capable of providing a close estimate of the conditions for forming sintered bodies having various degrees of partial crystallinity. In particular, the approach seems well-suited for evaluating the thermal histories of highly vitric breccias.

The elucidation of thermal histories using the kinetic treatment of breccia formation requires detailed information about the viscous flow and crystallization behavior of the matrix material. The kinetic treatment of crystallization is based on the construction of time-temperature-transformation (TTT) and continuous cooling (CT) curves corresponding to the degree of crystallinity observed in the matrix (~1%). The TTT curves are constructed, using the formal theory of transformation kinetics from measured crystal growth rates and calculated nucleation frequencies.

The nucleation frequencies are calculated using the classical theory of homogeneous nucleation with measured viscosities over a wide range of temperatures. In these calculations, a nucleation barrier of 60 kT at a relative undercooling, ΔT_r, of 0.2 has been assumed. Here $\Delta T_r = \Delta T/T_E$, where ΔT is the undercooling and T_E is the liquidus temperatures. The use of classical nucleation theory and the range of the assumed nucleation barrier are in accord with results obtained by Klein and Uhlmann (1976) in Lunar Composition 70019.

Continuous cooling (CT) curves for the two cooling histories of interest, constant rate cooling and logarithmic cooling, are obtained from the TTT curve using the procedure described previously (Onorato and Uhlmann, 1976). From the CT curves corresponding to the degree of crystallinity observed in the matrix of the breccia, it is possible to estimate the rate at which the lunar sample cooled.

In pursuing the model of breccia formation by viscous sintering, the viscosity data were also used to estimate the times required at various temperatures to achieve a substantial degree of sintering, such as that observed in the matrix of the breccia. A significant degree of sintering corresponds to the development of necks between the particles which are comparable to the particle radius in extent. Using the method described previously (Uhlmann *et al.*, 1975) for viscous sintering under conditions of continuous cooling, it has been possible to estimate the minimum temperature at which the particles of matrix came into contact.

The information obtained in this way can be combined with detailed heat flow calculations for geometries of interest to obtain further insight into the conditions of formation of the sample. A model which includes the heat sinking effects of entrapped clasts on the initial cooling of a molten ejecta blanket (Simonds, 1975) has been applied to the 67975 sample. On this basis, the physical situation during cooling has been divided into two separate heat flow problems, one appropriate at short times when the melt cools very rapidly by heat flow to the clasts, and the second appropriate at long times when the blanket cools more slowly by conduction and radiation to the surroundings.

As a complement to the information obtained using the analysis described above, a more detailed kinetic description of the development of partial crystallinity in lunar bodies has been carried out. By using a crystal distribution function (Hopper *et al.*, 1974a), essentially complete statistical information on the state of crystallinity of a body can be obtained.

In the present paper, the analysis is applied to lunar breccia 67975. Measurements of viscosity and crystal growth rate are reported, and are used with the kinetic analysis to estimate the rate at which the sample cooled. The predictions of the analysis based on continuous cooling curves are compared for the first time with those based on the analysis of crystallization statistics; and the results are combined with heat flow calculations to estimate the minimum size of the cooling body.

Sample 67975 is a friable, highly vitric breccia weighing 447 g which was returned from the southeast rim of North Ray Crater. The lunar sample is approximately 11.5 × 8 × 7 cm in size, and consists of about 25% inhomogeneously distributed clasts and about 75% dark glass matrix. Most of the clasts are finely granular fragments of light matrix breccia; the average clast size is less than about 0.1 mm.

The composition of the matrix glass, which contains about 1% crystals, was determined by electron beam microprobe analysis of representative areas of sample 67975,98. The average composition is shown in Table 1, together with the matrix compositions of other Apollo 16 samples previously examined in our laboratory.

Experimental Procedure

The materials studied in this investigation were synthesized from reagent grade raw material powders. The melting was carried out in multikilogram batches in a large-volume Mo-wound furnace under conditions of low oxygen activity (in Mo crucibles in a 5% N_2–95% N_2 forming gas atmosphere). The melting procedure and sample preparation for the viscous flow study have been previously described in greater detail (Cukierman *et al.*, 1972).

Viscosities in the molten range were determined using a rotating-cylinder viscosimeter, in which the torque required to rotate a bob, suspended in a cylinder containing the specimen, at a given angular velocity was measured. The instrument is a modification of one described previously (Cukierman *et al.*, 1972). Viscosities greater than about 10^9 poise were determined using a bending-beam viscosimeter in which the rate of viscous deformation of a simple beam was measured (Cukierman *et al.*, 1972). Both viscosimeters were provided with an atmosphere at 5% H_2–95% N_2 forming gas; both were calibrated using NBS reference materials.

The measurements of crystal growth rates were carried out on small glass specimens (typically 3 mm × 3 mm × 3 mm) cut from the material used in the viscous flow study. The growth rates were determined from the measured thicknesses of the external crystal layers on specimens held at a given crystallization temperature for various periods of time. See Scherer *et al.* (1972) for details. The crystallization furnace was provided with a gas-mixing system for controlling the oxygen fugacity to the conditions suggested by Sato *et al.* (1973). A solid ceramic electrolyte cell was used to monitor the oxygen fugacity continuously in the furnace, as described by Williams (1972).

Table 1. Compositions of Apollo 16 samples (wt.%).

	67975	68502	65016	60095
SiO_2	44.7	45.8	44.2	46.4
TiO_2	0.7	0.0	0.6	0.0
Al_2O_3	30.4	27.3	26.5	23.5
FeO	2.5	5.0	5.5	6.9
MgO	1.2	5.3	7.3	10.5
CaO	18.3	15.3	15.3	12.1
Na_2O	0.8	0.6	0.4	0.0
K_2O	0.3	0.1	0.1	0.8

VISCOUS FLOW AND CRYSTALLIZATION BEHAVIOR

The variation of viscosity with temperature for the 67975 composition is shown in Fig. 1. These data are compared in Fig. 2 with previously-reported results for samples 60095, 65016, and 68502. The dashed interpolations of each curve in Fig. 1 and Fig. 2 represent regions of temperature in which reliable flow data could not be obtained because of crystallization. (The samples were held at each temperature for 30–60 min to assure thermal equilibrium.) Data in the high-temperature range below the liquidus were obtained by superheating the specimens well above their liquidus to melt out retained heterogeneous nuclei.

The glass transition temperature (T_g) of the 67975 composition is about 680°C. This is the same as the T_g of the 60095 composition, and lower than the glass transition temperatures of 65016 (700°C) and 68502 (740°C). The apparent activation energies for viscous flow in the high-temperature (fluid) range and the low-temperature (highly viscous) range around T_g are tabulated in Table 2. Also

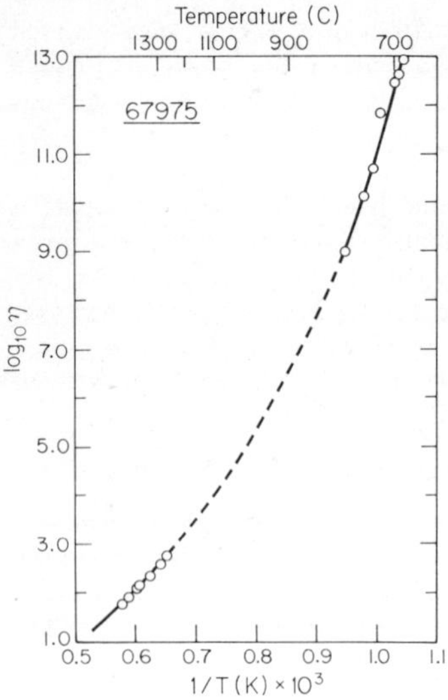

Fig. 1. Viscosity versus temperature relation for lunar composition 67975.

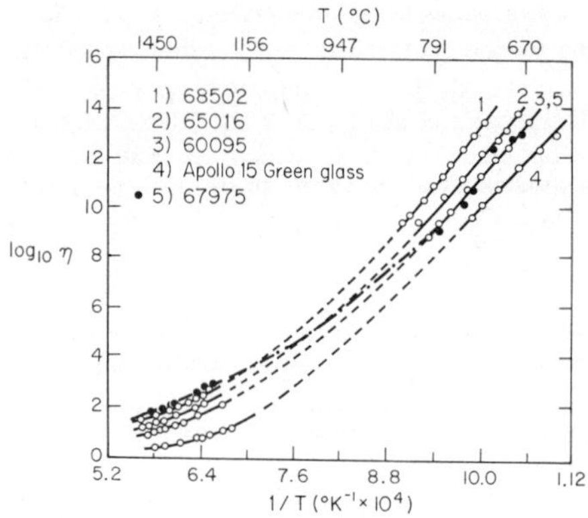

Fig. 2. Viscosity versus temperature relations for Apollo 16 samples 60095, 65016, 68502, and 67975.

Table 2. Activation energies for viscous flow for Apollo 16 compositions.

	Activation energy, ΔE (kcal mole^{-1})			
	67975	60095	65016	68502
ΔE, high T	60	55	60	60
ΔE, low T	180	155	145	200

shown for comparison are the apparent activation energies (ΔE) for viscous flow of the other Apollo 16 compositions. It is seen that the ΔE values for all these materials in the high temperature region are about 55–60 kcal mole^{-1}; while in the low-temperature region, a much greater range in ΔE is found (145–200 kcal mole^{-1}).

At all temperatures, the crystallization of the 67975 composition proceeded uniformly from the external surfaces of the samples; and the thickness of the crystal layer increased linearly with time at all temperatures. Growth rates were determined from the slopes of least-square lines drawn through the thickness versus time-at-temperature plots. The growth rates of the 67975 composition and two previously-studied Apollo 16 compositions vary with temperature as shown in Fig. 3. The maximum growth rate of the 67975 composition is about 7.3×10^{-4} cm min^{-1}; it occurs at an undercooling of about 220°C (the liquidus temperature was determined as 1210 ± 10°C).

Discussion of Viscous Flow and Crystallization Behavior

The melting conditions used in preparing the present samples are somewhat more reducing than those generally estimated for the lunar environment (Sato *et al.*, 1973). Differences in the state of reduction over the range of concern for the moon have been shown to have an insignificant effect on liquid viscosity in the high-temperature region for lunar composition 15555, which contains 22.5 wt.% FeO (Cukierman and Uhlmann, 1974). Since the present breccia composition contain only 2.5 wt.% FeO, the effects of the state of reduction on melt viscosity should be minimal.

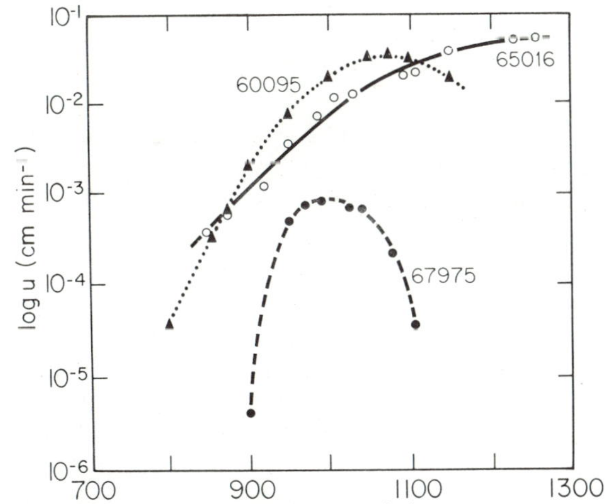

Fig. 3. Crystal growth rate versus temperature relations for lunar compositions 67975, 60095, and 65016.

For the 67975 liquid, the high-temperature flow data are predicted within 0.2 in $\log_{10} \eta$ by the semi-empirical model of Bottinga and Weill (1972). The semi-empirical model of Shaw (1972) provides even better agreement with the experimental data.

The overall variation of viscosity with temperature for 67975 places it well within the range observed previously for other Apollo 16 compositions (Fig. 2). In the high-temperature, fluid region, the viscosity of 67975 is higher than any other Apollo 16 sample investigated.

Of particular importance in considering the crystallization behavior, the liquidus temperature of 67975 is quite low (1210°C); and the viscosity at the liquidus is higher than that of any other lunar composition investigated in our laboratory. This high viscosity at the liquidus is responsible for the much lower growth rate observed for 67975 than for the other Apollo 16 compositions (see Fig. 3).

The 67975 composition has the smallest FeO content (2.5 wt.%) and the largest $SiO_2 + Al_2O_3$ content (75.1 wt.%) of any lunar composition studied to date in our laboratory. The mole fraction of $SiO_2 + Al_2O_3$ (71.4 mole%) places this composition in the range of good commercial oxide glass-formers. The high $SiO_2 + Al_2O_3$ content is undoubtedly responsible for the relatively high viscosity of the material.

BRECCIA FORMATION

The kinetic model for breccia formation discussed in the Introduction can be used to estimate the conditions under which breccias of various types can be formed. These conditions are estimated by constructing continuous cooling (CT) curves corresponding to the degree of crystallinity observed in the breccias. First, time-temperature-transformation (TTT) curves are constructed from the expression for the volume fraction, V_c/V, crystallized in a time t:

$$\frac{V_c}{V} \approx 1 - \exp\left(-\frac{\pi}{3} I_v u^3 t^4\right), \tag{1}$$

where I_v is the nucleation rate per unit volume and u is the growth rate.

The TTT curves are constructed by calculating the time corresponding to a given value of V_c/V for each of a series of temperatures. Measured values of the crystal growth rate (Fig. 3 above) have been employed; and the nucleation rate has been calculated by assuming a barrier for homogeneous nucleation of 60 kT at a relative undercooling ($\Delta T/T_E$) of 0.2. Here ΔT is the undercooling and T_E is the liquidus temperature. This assumption is in accord with observations on the glass-forming tendencies of other lunar compositions as well as experimental results on the 700019 composition. The nucleating effect of entrapped clasts has been neglected because work on the other lunar compositions and experience with the 67975 composition in our laboratory indicate good agreement between predictions of the kinetic treatment assuming only homogeneous nucleation and the experimental results. From the results of Onorato and Uhlmann (1976), this would imply that the clasts are not potent nucleating agents for the melts considered—at least in the critical temperature range of rapid overall crystallization kinetics.

The TTT curve for $V_c/V = 10^{-2}$ (the observed degree of crystallinity in the matrix) for lunar composition 67975 is shown in Fig. 4. Also shown are continuous cool (CT) curves corresponding to this degree of crystallinity, calculated on the basis of constant (linear) cooling rates as well as logarithmic cooling rates.

The CT curves, are constructed from corresponding isothermal TTT curves following the approach of Grange and Kiefer (1941) as elaborated and applied to the problem of glass formation by Onorato and Uhlmann (1976).

From the results shown in Fig. 4, it is concluded that by cooling at a constant rate, the degree of crystallinity observed in the matrix of the 67975 breccia would have formed at a cooling rate of about 0.08°C min^{-1}. This is a much slower cooling rate than those estimated for other Apollo 16 samples studied in our laboratory (60095, 9×10^{2}°C min^{-1}; 65016, 2×10^{3}°C min^{-1}).

This difference in estimated critical cooling rates correlates well with experience in forming the three compositions as glasses in the laboratory. The 67975 material is by far the best glass-former of any lunar composition investigated to date. This ease of glass formation is not surprising, since the 67975 composition is close to some commercial alkaline earth aluminosilicates. Indeed, this composition has a usefully wide working range, and would make a fine, rather refractory commercial glass for applications where the optical absorption due to the 2.5% FeO would not present a limitation.

The conditions required to form a glass of the 67975 composition have also been evaluated using the analysis of crystallization statistics (Hopper *et al.*, 1974b). This analysis uses a crystal distribution function, $\psi(r, t, R)$, which provides an essentially complete description of the state of crystallinity in a body. This function is defined such that the number of crystals in a volume

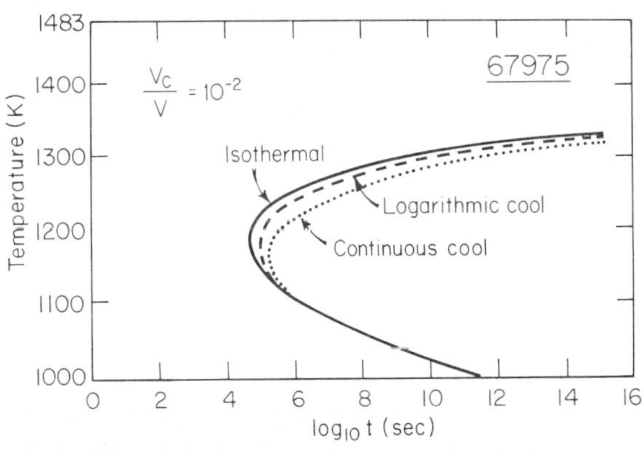

Fig. 4. Time-temperature-transformation (TTT) and continuous cooling (CT) curves for lunar composition 67975. Volume fraction crystallized = 10^{-2}.

element dv at r having radii between R and $R + dR$ is

$$N = \psi(r, t, R)\, dv\, dR. \tag{2}$$

With this form of kinetic analysis, it is possible to describe crystallization which takes place at temperatures below the nose of the CT curve (which is usually neglected in estimating cooling rates from CT curves). Using the cooling rate estimated for 67975 from the CT analysis (0.08°C min^{-1}), the distributions of crystals were calculated for various temperatures during cooling. The results indicate that the approximations used in constructing CT curves from TTT curves provide a useful description of the volume fraction crystallized at temperatures between the melting point and the nose of the CT curve, but in detail that the CT analysis overestimates somewhat the degree of crystallinity in this range. As an example, the analysis of crystallization statistics (CS) indicates a fraction crystallized of about 1.2×10^{-3} for 67975 at the temperature of the nose of the CT curve (893°C), while the CT analysis indicates a fraction crystallized of 10^{-2}.

The CS calculations also indicate that significant crystallization takes place at temperatures between the nose of the CT curve (T_N) and the glass transition (T_g). For the 67975 composition cooled at 0.08°C min^{-1}, for example, the volume fraction crystallized at the glass transition is calculated as about 3×10^{-3}. Such an increase by about a factor of 3 in the fraction crystallized between T_N and T_g is found over a wide range of cooling rates.

It is seen, therefore, that the CT analysis overestimates the degree of crystallinity at T_N, and by neglecting crystallization at temperatures below T_N, overestimates by a smaller factor the fraction crystallized in a cooled body. Overall, the CT analysis overestimates the critical cooling rate for the 67975 composition by about a factor of 1.3. The cooling rate estimated by the analysis of crystallization statistics is about 0.06°C min^{-1}; and this value will be taken as the cooling rate of the 67975 breccia since the CS analysis avoids the assumptions used in constructing CT curves.

As indicated previously (Uhlmann *et al.*, 1975), the condition for significant welding of the matrix particles in a breccia can be expressed:

$$\int_{t_{\text{con}}}^{t_{\text{sin}}} \frac{dt'}{\eta[T(t')]\gamma} \approx \frac{R}{\gamma}, \tag{3}$$

where R is the radius of the particles, γ is the surface energy, η is the viscosity, t_{con} is the time at which the particles came in contact, and t_{sin} is the time at which significant sintering has taken place.

For typical silicate liquids, γ is the range of 300 ergs cm^{-2} and R is ~15–20 μ for the matrix particles in the 67975 breccia. Using these values, the relation of Eq. (3) can be used to estimate the time and temperature at which the matrix particles came in contact for any given thermal history. This evaluation can be carried out numerically by taking t_{sin} as the time at which the cooling curve intersects the CT curve—or, for cooling histories which do not involve intersection with the CT curve (which pass beyond the nose of the CT curve), by taking t_{sin} as the time at which the temperature reaches the glass transition temperature. For the case of cooling at a constant rate for a material whose viscosity over the temperature range of interest can be approximated by an Arrhenius relation, one can evaluate the minimum temperature at which the matrix particles came in contact (Uhlmann *et al.*, 1975).

For lunar composition 67975, the minimum contact temperatures for any cooling rates which intersect the CT curve are only slightly above the temperatures of the curve. It is concluded that contact of the matrix particles at any temperature above or somewhat below the liquidus is sufficient to produce substantial sintering of the bodies for cooling rates of interest.

Heat Flow Analyses

The thickness of a body of the 67975 composition which is obtainable with the observed degree of crystallinity of the 67975 matrix is estimated to order-of-magnitude accuracy as about 25 cm. This estimate is obtained from the relation:

$$y \approx (D_{th}\tau_N)^{1/2} \tag{4}$$

where the thermal diffusivity, D_{th}, has been taken as 3×10^{-3} cm^2 sec^{-1}, a value typical for iron-containing slags (J. F. Elliott, MIT, pers. comm.), and τ_N is the time at the nose of the TTT curve. The estimated thickness is larger by about a factor of two or so of the dimensions of the lunar sample ($11.5 \times 8 \times 7$ cm).

Heat flow analyses for various cases of interest can provide more precise estimates of the size of body which can be formed with the degree of crystallinity of the 67975 matrix. The first of these analyses is based on the two-stage cooling model suggested by Simonds (1975) and applied to the Apollo 17, Station 6 boulder by Onorato *et al.* (1976). This model treats the cooling behavior of a clast-laden melt in terms of two distinct stages, the first involving heat flow from hot melt to cold clasts and the second involving heat flow from the melt to its surroundings.

The first stage cooling, involving heat flow to localized sinks, is very rapid. For the 67975 composition, for example, the temperature would drop from 1300°C to 1089°C in only 10 sec for a melt containing 25 vol.% clasts with a mean size of 90 μm. It is found, however, that the temperature of thermal equilibration of the melt-clast system is above the nose of the CT curve for this composition for any initial clast content and initial temperature which gives the final volume fraction of clasts observed in the lunar sample.

This result leaves the requirement of cooling by heat flow to the surroundings in order to form the breccia (cooling to sufficiently low temperatures that the rate of crystallization is insignificant). Three models have been employed in treating this cooling [see Onorato *et al.* (1976) for details of calculations]: (1) cooling of the melt sheet by conduction to a cold basement below and by radiation from the top surface; (2) cooling by conduction to a cold basement below and to a similar cold layer on top (corresponding to the melt sheet being covered by a cold blanket); and (3) cooling by radiation as a single body, modelled as a sphere. The last problem has been treated by Hopper *et al.* (1974a).

For a melt sheet cooled by conduction below and radiation above, it is found that the critical cooling rate (0.06°C min^{-1}) would be achieved at the slowest-cooling part of a body about 75 cm thick. For a melt sheet cooled by conduction at both surfaces, the critical cooling rate would be reached at the slowest-cooling part (the center) of a body about 60 cm thick; and for a spherical body cooling by radiation, the critical rate would be achieved at the center of a body about 1.3 m in radius.

From these results, it is concluded that the 67975 breccia could have formed with the degree of crystallinity observed in the matrix even at the center of a much larger body than the lunar sample. Further, a body having the size of the lunar sample would have cooled much more rapidly than is consistent with the observed degree of crystallinity. On this basis, it is concluded that the 67975 breccia did not cool in its present form, but as part of a larger body.

ANNEALING OF GLASS AND CONCLUDING DISCUSSION

The discussion of the previous sections has been based on the view that the 67975 breccia formed by the coalescence of clast-laden melt particles, rather than by shock lithification. This view is consistent with the sizable porosity, the friability, and the absence of extensive fracturing in the matrix of this breccia.

To differentiate between viscous coalescence and shock lithification, a small section of matrix glass (about 0.01 g) was extracted from the sample and subjected to an annealing schedule (one hour at the glass transition temperature followed by an overnight furnace cool to ambient). The index of refraction was determined by the Central Illumination Method (Wahlstrom, 1969) both prior to and after the annealing. The index was observed to increase upon annealing, by about 0.004, indicating relaxation of the glass to a more dense state during the anneal. This behavior is characteristic of glasses initially cooled through the glass transition more rapidly than during the annealing treatment. The behavior is contrary, however, to the characteristics of densified glasses produced directly by shock, and provides strong evidence that the matrix glass in the 67975 breccia formed by cooling of molten material, rather than by a shock-induced direct crystal-glass transition.

It is concluded, therefore, that the kinetic analysis of breccia formation by viscous coalescence can well be applied to the 67975 breccia. This analysis indicates that the breccia cooled at a rate of about 0.06°C min^{-1}, a cooling rate which is slower than that of other highly vitric clastic lunar breccias whose thermal histories have been determined. The 67975 composition is, in fact, the best glass-former of any lunar composition studied to date, and compares favorably in this regard with several commercial glass compositions. The origin of this ease of glass formation is the relatively high $SiO_2 + Al_2O_3$ content and the relatively low liquidus temperature, which combine to produce a high viscosity at the liquidus.

The cooling rate of about 0.06°C min^{-1} estimated for the 67975 composition is inconsistent with the sample having cooled in its present form, save in an environment of very restricted heat flow. Rather, the cooling rate suggests that the breccia cooled as part of a much larger body than the lunar sample.

Finally, the results of the present study indicate how a combined treatment of crystallization kinetics, viscous sintering and heat flow can be used to provide insight into the thermal histories of lunar breccias. The technique works most effectively for breccias which contain a significant amount of glass in the matrix. For breccias which are completely crystallized and exhibit varying degrees of metamorphism, the present analysis can provide useful limits on the cooling history such as an upper limit to the rate at which they could have cooled and a lower limit to the size of the cooling body. In this case, the results provided by the analysis should be complemented by microstructural studies of the type reported by Lofgren *et al.* (1974) and by studies of kinetic equilibration of the type reported by Taylor *et al.* (1975), Huebner (1976), and Taylor *et al.* (1977).

Acknowledgments—Financial support for the present work was provided by the National Aeronautics and Space Administration. This support is gratefully acknowledged, as are stimulating discussions with P. I. K. Onorato of M.I.T. and C. H. Simonds of the Lunar Science Institute.

References

Bottinga Y. and Weill D. F. (1972) The viscosity of magmatic silicate liquids: A model for calculation. *Amer. J. Sci.* **272**, 438–475.

Cukierman M., Tutts P. M., and Uhlmann D. R. (1972) Viscous flow behavior of Lunar Compositions 14259 and 14310. *Proc. Lunar Sci. Conf. 3rd*, p. 2619–2625.

Cukierman M. and Uhlmann D. R. (1974) Effects of iron oxidation state on viscosity, Lunar Composition 15555. *J. Geophys. Res.* **79**, 1594–1598.

Grange R. A. and Kiefer J. M. (1941) Transformation of austenite on continuous cooling and its relation to transformation at constant temperature. *Trans. A.S.M.* **29**, 85–115.

Hopper R. W., Onorato P. I. K., and Uhlmann D. R. (1974a) Thermal histories and crystal distributions in partly devitrified lunar glasses cooled by radiation. *Proc. Lunar Sci. Conf. 5th*, p. 2257–2273.

Hopper R. W., Scherer G., and Uhlmann D. R. (1974b) Crystallization statistics, thermal history and glass formation. *J. Non-Cryst. Solids* **15**, 45–62.

Huebner J. S. (1976) Diffusively rimmed xenocrysts in 77115 (abstract). In *Lunar Science VII*, p. 396–398. The Lunar Science Institute, Houston.

Klein L. C., Onorato P. I. K., Uhlmann D. R., and Hopper R. W. (1975) Viscous flow, crystallization behavior, and thermal histories of lunar breccias 70019 and 79115. *Proc. Lunar Sci. Conf. 6th*, p. 579–593.

Klein L. C. and Uhlmann D. R. (1976) The kinetics of lunar glass formation, revisited. *Proc. Lunar Sci. Conf. 7th*, p. 1113–1121.

Lofgren G., Donaldson C. H., Williams R. J., Mullins O., Jr., and Usselman T. M. (1974) Experimentally reproduced textures and mineral chemistry of Apollo 15 quartz normative basalts. *Proc. Lunar Sci. Conf. 5th*, p. 549–567.

Onorato P. I. K. and Uhlmann D. R. (1976) Nucleating heterogeneities and glass formation. *J. Non-Cryst. Solids* **22**, p. 367–378.

Onorato P. I. K., Uhlmann D. R., and Simonds C. H. (1976) Heat flow in impact melts: Apollo 17 Station 6 Boulder and some applications to other breccias and xenolith laden melts. *Proc. Lunar Sci. Conf. 7th*, p. 2449–2467.

Sato M., Hickling N. L., and McLane J. E. (1973) Oxygen fugacity values of Apollo 12, 14, and 15 lunar samples and reduced state of lunar magmas. *Proc. Lunar Sci. Conf. 4th*, p. 1061–1079.

Scherer G., Hopper R. W., and Uhlmann D. R. (1972) Crystallization behavior and glass formation of selected lunar compositions. *Proc. Lunar Sci. Conf. 3rd*, p. 2627–2637.

Shaw H. R. (1972) Viscosities of magmatic silicate liquids: An empirical method of prediction. *Amer. J. Sci.* **272**, 870–893.

Simonds C. H. (1973) Sintering and hot pressing of Fra Mauro composition glass and the lithification of lunar breccias. *Amer. J. Sci.* **273**, 428–439.

Simonds C. H. (1975) Thermal regimes in impact melts and the petrology of the Apollo 17 Station 6 boulder. *Proc. Lunar Sci. Conf. 6th*, p. 641–672.

Taylor L. A., Uhlmann D. R., Hopper R. W., and Misra K. C. (1975) Absolute cooling rates of lunar rocks based on the kinetics of Zr diffusion in opaque oxides: Applications to Apollo 15 rocks from Elbow Crater. Companion paper. *Proc. Lunar Sci. Conf. 6th*, p. 181–191.

Taylor L. A., Onorato P. I. K. and Uhlmann D. R. (1977) Cooling rate estimations based on kinetic modeling of Fe–Mg diffusion in olivine. *Proc. Lunar Sci. Conf. 8th.* This volume.

Uhlmann D. R. and Klein L. C. (1976) Crystallization kinetics, viscous flow, and thermal histories of lunar breccias 15286 and 15498. *Proc. Lunar Sci. Conf. 7th*, p. 2529–2541.

Uhlmann D. R., Klein L. C., Onorato P. I. K., and Hopper R. W. (1975) The formation of lunar breccias: Sintering and crystallization kinetics. *Proc. Lunar Sci. Conf. 6th*, p. 693–705.

Wahlstrom E. E. (1969) *Optical Crystallography*, p. 112–114. John Wiley and Sons, New York.

Williams R. J. (1972) A solid ceramic electrolyte system for measuring redox conditions in high temperature gas mixing studies. NASA TM X-58105. Johnson Space Center, Houston.

Proc. Lunar Sci. Conf. 8th (1977), p. 2079–2095.
Printed in the United States of America

Dynamic crystallization experiments bearing on the origin of textures in impact-generated liquids

GARY E. LOFGREN

NASA Johnson Space Center, Houston, Texas 77058

Abstract—Dynamic crystallization experiments on a synthetic glass of 14310 composition have shown that the density and distribution of crystals present at the initiation of cooling can drastically affect the texture of the crystallized product. Experiments were conducted in one atmosphere gas-mixing furnaces at an oxygen fugacity slightly below iron-wüstite. Cooling histories were either linear or nonlinear following a temperature versus either square root of time or log time relationship. The precooling crystallization and/or melt history was systematically varied to produce different densities and distributions of crystals at the initiation of cooling. The nucleation density, unfortunately, was not consistent for equivalent precooling histories because of its random nature thus adding some measure of uncertainty to the conclusions. When cooled from a complete liquid, the experimental charges were porphyritic with 1–2 mm phenocrysts set in a very fine-grained matrix (< 0.1 mm). This result suggests that 14310 did not cool from a complete melt. When cooled from a crystal-liquid mixture, the experimental charges had widely varying textures depending on the density and distribution of crystals that act as nucleation sites once cooling is initiated. If the nuclei were dense and evenly distributed, poikilitic textures usually developed with olivine and/or possibly orthopyroxene enclosing euhedral plagioclase microlites. If the nuclei were irregularly distributed, widely divergent textures developed side by side within 1–2 mm of each other. In one charge the texture changed from medium-grained subophitic to fine-grained intersertal to very fine-grained poikilitic over a space of 3 mm. These results suggest that the close spatial association of widely divergent textures in the matrices of the poikilitic breccias and in rocks with basaltic textures that display similar closely spaced textural variation can be the result of the density and distribution of potential nucleation sites. The nucleation sites could be micron to sub-micron crystals residual to the impact melting process or could be cold-injected clasts. Whatever the source of the nucleation sites, it is clear that local variations in cooling rate in the matrices of the breccias (although possible) are not necessary.

INTRODUCTION

FELDSPATHIC BASALT, 14310, has a basaltic texture even though it is most likely one of the few examples of crystallization from a clast-free, impact-generated liquid returned from the moon (James, 1973). Breccias of the same approximate bulk composition have been described by numerous workers. They must have formed as some combination of impact-generated liquid and either partial-melt residual or cold-injected clasts. The variety of matrix textures is usually attributed to variations in the cooling history of the mixture (Warner *et al.*, 1973). One of the more controversial matrix textures characterized by euhedral (usually tabular) plagioclase enclosed in poikilitic low-Ca pyroxene has been most recently discussed by Ryder and Bower (1976). Controversies initially centered around whether this texture developed from the initial crystallization of a liquid, the devitrification of a glass, or the solid-state metamorphism of an initially clastic breccia. Ryder and Bower support the first case and suggest that nucleation is an important factor in producing the poikilitic texture.

An even more perplexing problem is the close spatial association of such widely divergent textures as intersertal and poikilitic occurring within a few millimeters of each other (Fig. 1) and described by Simonds (1975). 14310 is characterized by textural variations from ophitic to subophitic to fine-grained intersertal on a centimeter scale (Fig. 6b). These variations have been attributed to local variations in volatile content or cooling rate (Ridley *et al.*, 1972; Crawford and Hollister, 1974).

These hypotheses are amenable to experimental testing and this paper reports the results of dynamic crystallization experiments on a synthetic glass that resembles 14310 in composition. There is an infinite variety of possible cooling histories that could be employed. An attempt was made to use cooling histories that could give some insight into how the same bulk composition material could produce either ophitic basaltic textures or poikilitic breccia matrix textures, often in the same rock. Nucleation appears to be a major variable so attempts were made to systematically vary the density and distribution nuclei at the initiation of cooling. Unfortunately it is very difficult to duplicate experimentally the widely held idea that the clasts are injected as cold fragments in the hot,

Fig. 1. Photomicrograph of 76215,59 showing the close spatial association of subophitic, intersertal, and poikilitic textures. The poikilitic crystals are not easily seen; they enclosed the fine plagioclase laths on the right side of the figure.

impact-generated liquid. The experimental conditions come closer to duplicating a partially melted crystalline precursor in which the remaining clasts are hot and genetically related to the liquid. In addition the crystal size is uniformly small. This last problem does not appear to be a major one because nucleation and growth on large (> 2–3 mm) pre-existing clasts is not common.

Impact craters are neither common nor, more important, well preserved on earth and thus most of our understanding of lunar impact processes must come from examination of randomly collected samples of the products of that process. The experimental duplication of some of the features in these samples can hopefully reduce some of the ambiguity.

Experimental Techniques

A synthetic homogeneous glass that is similar in composition to 14310 (Table 1) was prepared by repeated fusion of an oxide/carbonate mixture. The finely powdered glass was pressed into pellets which were attached to 8 mil platinum wire (Donaldson *et al.*, 1975). The pellets were suspended in a gas-mixing furnace (Williams and Mullins, 1976). An appropriate mixture of CO and CO_2 provided the atmosphere in the furnace which was held slightly below the iron-wüstite buffer curve.

Constant temperature crystallization, melting, and dynamic cooling experiments were conducted in the manner described by Lofgren *et al.* (1974), Usselman *et al.* (1975), and Usselman and Lofgren (1976). One additional important variable in these experiments is the systematic variation of the precooling crystallization and/or melting history of the charge (Table 2). Three basic precooling histories were used (a) melting—where the pellet becomes a sphere of liquid, (b) crystallization—where the pellet is partially crystallized at a subliquidus temperature of 1280°C; there are approximately 15% plagioclase crystals and 85% liquid, and (c) crystallization and melting—where the pellet

Table 1. Chemical compositions of 14310, the synthetic starting material, and lunar breccias with poikilitic textures.

	(1)	(2)	(3)	(4)
SiO_2	47.20	48.82	45.6–51.4	44.65–47.04
TiO_2	1.24	1.16	0.53–3.60	0.72–1.72
Al_2O_3	20.10	20.50	15.8–24.9	17.18–22.94
Cr_2O_3	0.18	0.07	0.02–0.10	not reported
FeO	8.38	7.69	3.9–9.3	7.08–10.53
MgO	7.87	7.78	4.5–7.9	9.9–13.15
MnO	0.11	n.d.	0.03–0.15	0.07–0.13
CaO	12.30	12.51	9.7–12.5	10.41–13.34
Na_2O	0.63	0.60	0.69–1.13	0.3–0.56
K_2O	0.49	0.39	0.55–1.74	0.11–0.35
	99.84	99.51		

(1) Published analysis of 14310.
(2) Synthetic starting material.
(3) Ranges of compositions (defocused beam analysis) for Apollo 14 KREEP-rich rocks show poikilitic textures (Ryder and Bower, 1976).
(4) Compositions of Apollo 16 and 17 poikilitic rocks (Simonds *et al.*, 1973).

Table 2. Run data for dynamic cooling experiments on synthetic powdered glass starting material of approximate 14310 composition. The precooling histories are shown in columns 2 and 3. The completely melted runs are entered in the melt column with melt temperatures ≥ 1300°C. The partially crystallized charges are entered in the crystallize glass column. The crystallized, then partially melted runs have entries in both columns. The sequence of events is read from left to right.

Run No.	Temp/hr crystallize glass	Temp/hr melt	Cooling history	Final temp	$-\text{Log} f_{O_2}$ Start cooling	$-\text{Log} f_{O_2}$ Finish cooling
44	1280/6	—	Quench	1280	11.3	—
41	1280/24	—	Quench	1280	11.2	—
43	1280/88	—	Quench	1280	11.2	—
40	1150/53	1280/6	Quench	1280	11.1	—
42	1150/49	1280/24	Quench	1280	11.2	—
51	1150/51	1280/88	Quench	1280	11.2	—
31	—	1300/19	2°C/hr	1002	11.2	15.2
47	—	1320/6	2°C/hr	1002	10.9	14.6
6	—	1320/3	5°C/hr	990	10.9	15.0
29	1280/6	—	2°C/hr	1005	11.3	14.5
45	1280/24	—	2°C/hr	1000	11.3	15.1
32	1280/88	—	2.3°C/hr	1000	11.2	15.4
10	1280/24	—	4.5°C/hr	1058	11.3	14.1
18	1280/19	—	Temp vs. log time	984	11.2	16.0
17	1280/20	—	Temp vs. log time	1010	11.2	15.2
16	1280/23	—	33.6°C/hr	1021	11.2	14.0
37	1150/88	1280/6	2°C/hr	1003	11.3	15.6
33	1150/50	1280/24	2°C/hr	1000	11.2	15.2
38	1152/54	1280/88	2°C/hr	999	11.2	15.1
36	1145/68	1280/7	Temp vs. $\sqrt{\text{time}}$	1000	11.3	15.4
46	1150/52	1281/24	Temp vs. $\sqrt{\text{time}}$	1000	11.3	15.3
48	1150/52	1280/96	Temp vs. $\sqrt{\text{time}}$	1000	11.2	15.2

is crystallized below 1200°C and then partially melted at 1280°C. The first precooling history requires nucleation before crystallization, the other two have plagioclase crystals present before cooling is initiated.

None of these precooling crystallization and/or melting histories is meant to duplicate any kind of natural process. They are used only to produce a variety of nucleation site densities and distributions. The cooling curves used are shown in Fig. 2. There are two basic types, linear or constant cooling rate curves and nonlinear curves where temperature varies as a function of the square root of time.

EXPERIMENTAL RESULTS

The textural variations observed in the experimentally crystallized charges appear to be more a function of the precooling melting, crystallization, and nucleation history than of the subsequent cooling history. The most distinctive

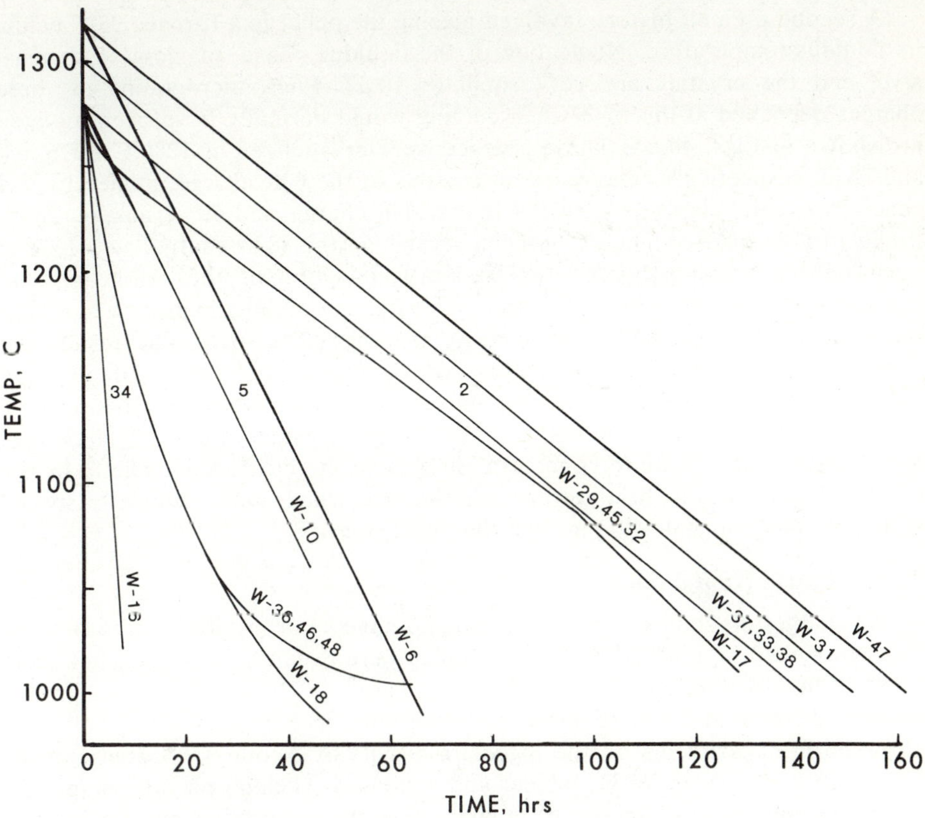

Fig. 2. Cooling curves for experiments listed in Table 2. The precooling histories are not shown. The numbers 34, 5, and 2 are the approximate cooling rates for the adjacent curves.

difference in texture is between those runs in which a nucleation event was required after cooling was initiated and those in which nuclei were already present.

Precooling melt history

Three basically different kinds of precooling melt histories were used in an attempt to control and vary the presence, density, and distribution of potential nucleation sites. The starting material is a finely powdered glass pressed into pellets as described earlier. If this pellet is placed in the furnace above its liquidus temperature (~ 1310°C), a uniform melt is produced (Fig. 3a). This melt may contain angstrom-sized, refractory, heterogeneous nucleation sites, or there may be certain degrees of polymerization in the melt that may act as homogeneous nucleation sites, but there are no cognate crystalline phases present to act as nucleation sites.

A second premelt history involved placing the pellet in a furnace 25°C below its liquidus temperature. Nucleation of the liquidus phase, plagioclase, is massive, and the crystals are very small ($< 10\,\mu$). Photomicrographs of three charges quenched at the time when cooling would normally have been initiated are shown in Figs. 4d,e,f. These charges were crystallized at 1280°C for 6, 24, and 88 hr, respectively. There are no crystals in the 6 hr charge, numerous and generally evenly distributed crystals in the 24 hr charge, and numerous, but more irregularily distributed, coarser-grained crystals in the 88 hr charge.

A third precooling history involved crystallization of the charge at temperatures near 1150°C where the ratio of crystals to liquid is higher, then raising the temperature to 1280°C and partially melting the charge. The results are shown in Figs. 5d,e,f. The 6 hr charge has numerous nuclei in a definite swirl pattern that creates areas that have a high crystal density and areas that have no crystals at all. The 24 hr charge has fewer, larger crystals in a less well-defined swirl pattern and a definite concentration around the outer edge. The 88 hr run has far fewer and much larger crystals that are widely and irregularly spaced with a definite concentration around the outer edges.

Crystallization from a liquid

All charges crystallized from a total liquid have a porphyritic texture (Fig. 3). The run melted at 1300°C (W-31) is included here because runs quenched from 1300°C showed that no feldspar had nucleated in 24 hr. These porphyritic textures develop at linear (constant) cooling rates in a manner similar to those observed in experiments on the pigeonite basalts of Apollo 15 (Lofgren *et al.*, 1974; Grove, 1977). In W-31, tabular and somewhat skeletal plagioclase phenocrysts are enclosed in a matrix of tabular plagioclase microlites and fan spherulitic pyroxene (Fig. 3c); there is some acicular to fibrous plagioclase intergrown with the fibrous pyroxene. The plagioclase phenocrysts are commonly enclosed in an envelope of finely fibrous pyroxene that coarsens outward and the matrix plagioclase is usually in the coarser-grained areas between the pyroxene fan spherulites (Fig. 6d). A few larger tabular pigeonite and orthopyroxene crystals are also located in these coarser-grained areas. Some of these pyroxenes appear to comprise pigeonite mantles on orthopyroxene cores.

There are textural differences between W-31 and W-47 that must be related to melting temperature because they were cooled at the same rate. The phenocrysts in W-47 are slightly more skeletal and the matrix is finer grained (Fig. 3b). The matrix plagioclase microlites are much smaller, the pyroxene fibers in the fan spherulites have significantly smaller diameters, and pyroxenes larger than $10\,\mu$ are rare (Fig. 6c).

The more rapidly cooled W-6 (Figs. 2, 3d) has significantly more elongate and skeletal plagioclase phenocrysts and a much finer matrix. Plagioclase and pyroxene fibers intergrown in fan spherulites comprise most of the matrix. Elongate, highly skeletal crystals with parallel extinction may be olivine.

The plagioclase phenocrysts show very slight normal zoning, less than 2 mole % An. In W-31 the phenocrysts average $Or_{0.8}Ab_{6.5}An_{92.7}$ and the matrix

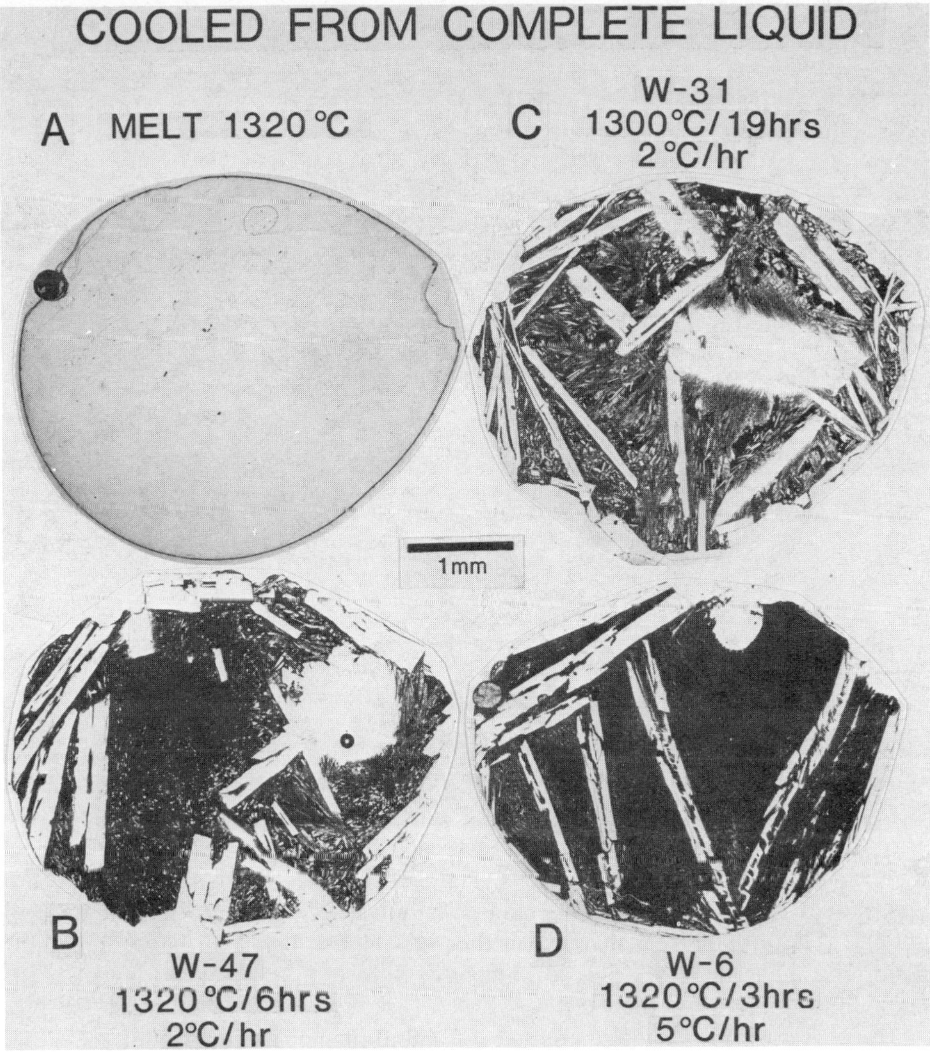

Fig. 3. Photomicrographs of charges crystallized from a completely melted charge. (a) Completely melted charge that represents the state of the liquid at the initiation of cooling. (b) W-47. (c) W-31. (d) W-6.

plagioclase is only slightly more sodic, $An_{91.5}$, but is significantly more Or rich, $Or_{1.6}$. The phenocrysts in W-47 are compositionally indistinguishable from those in W-31 averaging $Or_{0.9}Ab_{5.3}An_{93.8}$. The matrix plagioclase is very similar but also slightly more calcic than in W-31 (Table 3). In the more rapidly cooled W-6, the plagioclase is significantly more sodic and Or rich, averaging $Or_{1.1}Ab_{10.5}An_{88.5}$ reflecting the more rapid cooling rate for this charge. The matrix plagioclase is too small to analyze.

Table 3. Plagioclase compositions in dynamic cooling experiments.

	Or	Ab	An
Cooled from complete melt at 1320°C			
W-47 at 2°C/hr	0.9	5.3	93.8
W-6 at 5°C/hr	1.1	10.5	88.5
Cooled from complete melt at 1300°C			
W-31	0.8	6.5	92.7
Cooled from melt plus crystals at 1280°C			
W-29 (all at 2°C/hr)	0.7	5.9	93.4
W-32 (all at 2°C/hr)	0.7	3.5	95.8
W-37 (all at 2°C/hr)	1.0	7.0	92.0
W-33 (all at 2°C/hr)	0.9	5.7	93.4
W-10 at 5°C/hr	0.7	7.7	91.6
W-36 at average 18°C/hr	1.0	10.0	89.0
W-16 at 34°C/hr	1.0	9.6	89.4
Groundmass plagioclase			
W-47	1.6	5.4	92.9
W-31	1.6	6.9	91.5
W-32	1.3	3.9	94.8

Crystallization from a partially crystallized charge

Charges that were crystallized at 1280°C for 6, 24, and 88 hours and then cooled showed the most erratic variations in texture (Fig. 4). The three charges on the left side of Fig. 4 vary only in the precooling history; they were all cooled from 1280°C at 2°C/hr. The charges in the central vertical row (Figs. 4d,e,f) have an identical precooling history to the crystallized charges on their immediate left and should represent the density and distribution of nuclei at the initiation of cooling. The crystallized charges on the right (Figs. 4g,h,i) all have precooling histories similar to W-41 (Fig. 4e), but have different cooling histories from the crystallized charges on their left.

W-29 (Fig. 4a) is very fine grained. Euhedral plagioclase microlites are either enclosed in large, poikilitic olivine crystals (Fig. 6f), or in mesostasis of a glass and fine-grained pyroxene. This texture contrasts markedly with the porphyritic textures of W-45 and W-32 even though all three have the same cooling history. W-29 and W-18 have very similar poikilitic textures yet have very different cooling histories (Fig. 2). W-10 (Fig. 4g) has a seriate, intersertal texture compared to W-45 (Fig. 4b). W-16 (Fig. 4i) and W-45 (Fig. 4b), which have the same precooling history, are the only two that can be simply related by cooling history; the finer-grained, porphyritic texture of W-16 compared to W-45 being the natural consequence of its faster cooling rate. This lack of systematic behavior is not asthetically pleasing, but, as will be explained more fully later, is the key to understanding the close spatial association of the widely different textures in the matrices of breccias.

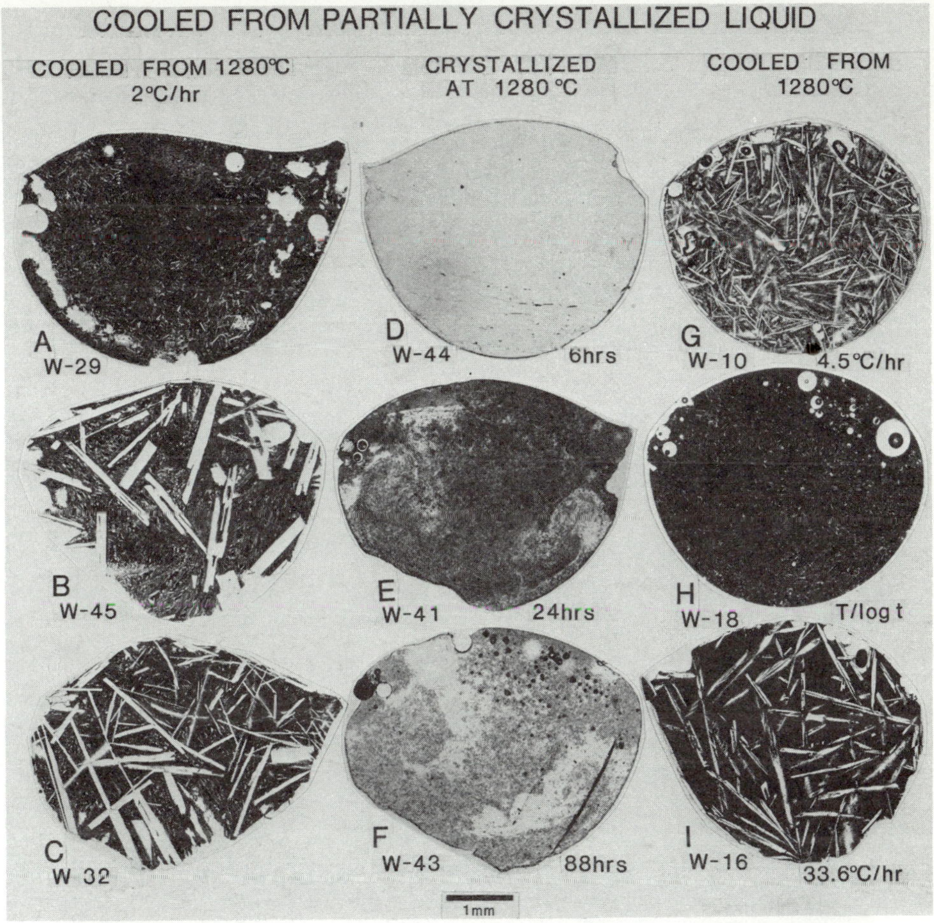

Fig. 4. Photomicrographs of partially crystallized charges cooled from 1280°C. The charges were crystallized at 1280°C for various times (see below) and those quenched from that temperature are shown in the center vertical column. The charges in the left-hand column were cooled from 1280°C at 2°C/hr; those in the right-hand column had different cooling histories (see below and text for details). (a) W-29. (b) W-45. (c) W-32. (d) W-44, crystallized six hours. (e) W-41, crystallized 24 hours. (f) W-43, crystallized 88 hours. (g) W-10, cooled at 4.5°C/hr. (h) W-18, temperature versus log time cooling curve. (i) W-16, cooled at 33.6°C/hr.

The compositions of the feldspars do appear to be systematically related to cooling rate becoming systematically more sodic with faster cooling rate (Table 3, compare W-29 and 32 with W-10 and 16). Even though many crystals in W-29 resemble the groundmass plagioclase in W-47 and 31 in shape and size, they are not nearly as Or rich. The crystals are rarely zoned more than 2% An.

Crystallization from a partially melted charge

Crystallization from a previously crystallized, then partially melted charge is more systematic than the previous case. Two different cooling histories were systematically studied. Linear cooling from 1280°C at 2°C/hr. (Figs. 5a,b,c) and nonlinear cooling that approximates a temperature versus square root of time function (Fig. 2: 36, 46, 48 and Figs. 5g,h,i). The nonlinear rates are initially faster, approximately 10–20°C/hr between 1280°C and about 1050°C, and then

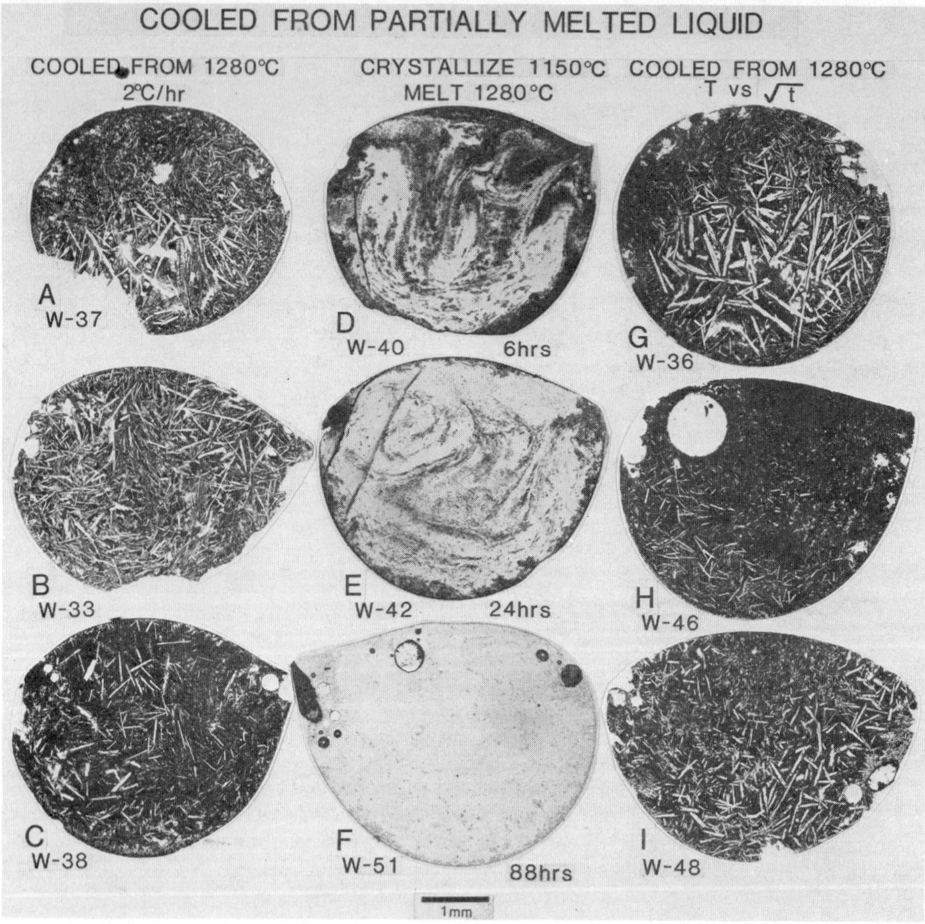

Fig. 5. Photomicrographs of partially melted charges cooled from 1280°C. The charges were first crystallized at 1150°C then partially melted at 1280°C before cooling. Runs quenched after partial melting for various times (see below), but before cooling, are shown in the central vertical column. The charges in the left-hand column were cooled at 2°C/hr and those in the right-hand column with a temperature versus square root of time curve (see Fig. 2 and text for details). (a) W-37. (b) W-33. (c) W-38. (d) W-40, melted for 6 hours. (e) W-42, melted for 24 hours. (f) W-51, melted for 88 hours. (g) W-36. (h) W-46. (i) W-48.

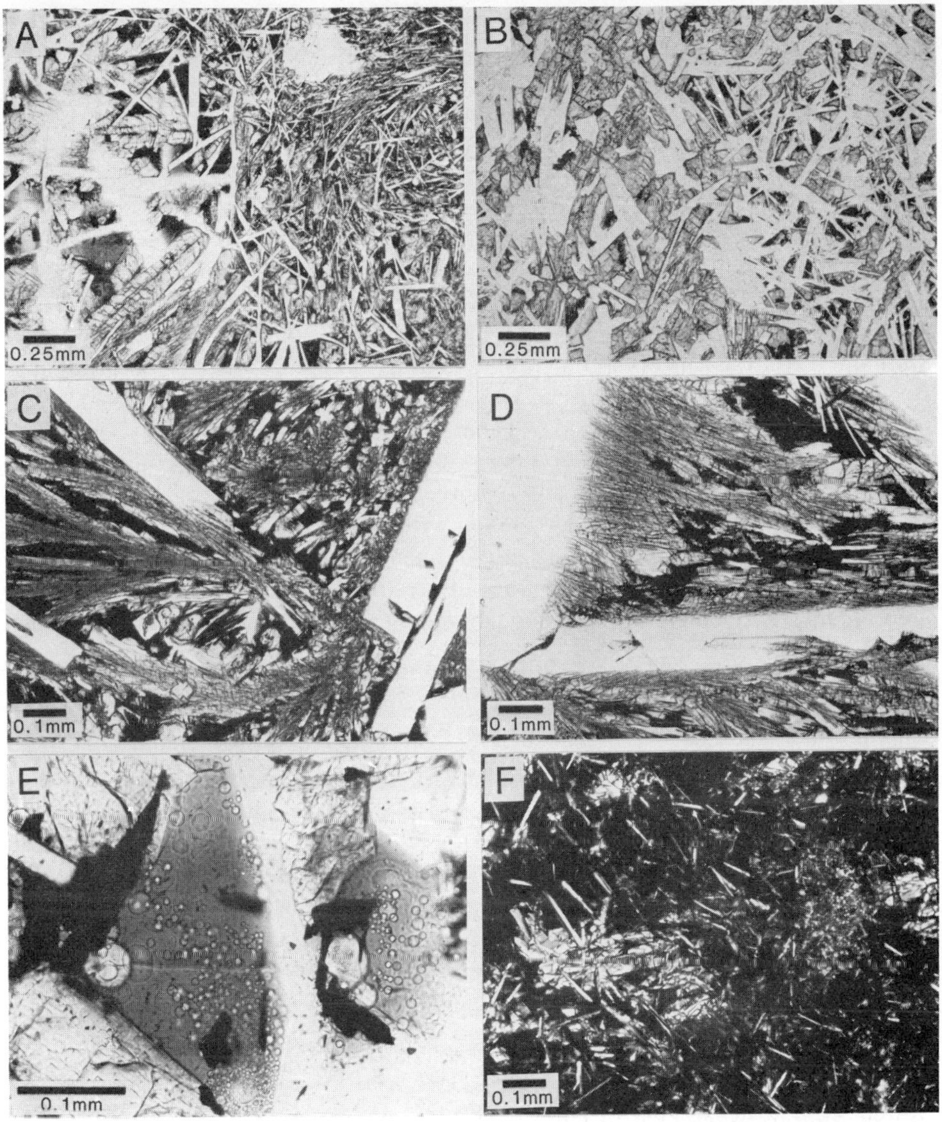

Fig. 6. Photomicrographs of run products and lunar samples showing textural details and comparisons. (a) W-37 showing the striking range of textures in the approximately 2 mm width of the photograph. (b) 14310,186 showing a similar but not so drastic change in texture in the same distance. (c) Detail in the matrix of W-47 showing the fine-grained pyroxene fan spherulites and coarser-grained interspherulitic areas. (d) Detail of W-31 in contrast to W-47 showing the much coarser-grained matrix. (e) Liquid immiscibility in W-37; clear droplets are enclosed in a reddish-brown glass. (f) Poikilitic olivine enclosing euhedral plagioclase microlites in W-29.

become much slower than 2°C/hr during the final 50°C of cooling. The textures are generally finer grained, reflecting the initially more rapid cooling (portions of W-36, however, are as coarse grained as any of the 2°C/hr linear runs) and have slightly better developed poikilitic texture that most likely developed at the slowest cooling rates.

The most striking feature of the crystallized charges shown in Fig. 5 is the close spatial association of very different textures. This feature is most obvious in W-37 and W-36 (Figs. 5a,g). Within 2–3 mm the textures in W-37 change from subophitic and coarse-grained to fine-grained intersertal, to very fine-grained poikilitic (Fig. 6a). A similar change in W-36 is seen from porphyritic to poikilitic. W-33 is dominated by subtle and closely spaced changes from coarse-grained to fine-grained intersertal. The remaining crystallized charges show less dramatic variations in crystal size and texture randomly dispersed throughout the charge.

The plagioclase compositions (Table 3), as with the previously described runs, vary systematically with cooling rate. The plagioclase in the initially more rapidly cooled W-36 is significantly more sodic than the more slowly cooled W-37 and W-33. The increased cooling rate, however, does not significantly affect the Or content. Only the groundmass plagioclase which presumably crystallized from a more potassium-rich residual melt, shows significant Or enrichment. As with the previous charges the zoning is rarely more than 2–3 mole% An.

Liquid immiscibility

Liquid immiscibility is present in nearly all the crystallized charges. It is interstitial to the major phases and appears to occur when the charge is between 80 and 90% crystallized. The nearly colorless presumably SiO_2-rich blebs are set in a deep brown glass matrix (Fig. 6e). There is a small degree of coalescence of the silica-rich melt into larger masses. The immiscibility occurs so late in the crystallization sequence it does not appear to affect the texture in the manner observed by Hollister and Crawford (1977).

DISCUSSION

Application of experiment to lunar samples

The application of experimental studies to understanding features observed in rocks always requires an assumption that natural processes can be simulated in generally shorter term and volumetrically restricted laboratory experiments. In this case, the sizes of the crystals relative to the volumes used and the general success in replicating the natural textures suggests that the volume and time considerations are not serious deficiencies.

A more serious potential problem is the duplication of the nucleation process. Some impact-generated melts no doubt carry unmelted crystals at least in the sub-micron size ranges. There is also the hypothesis that most clasts in breccias are injected cold into superheated melts (von Engelhardt, 1971; Simonds *et al.*, 1973). The smaller of these clasts would be digested but many larger clasts would have been reduced to micron to sub-micron sizes and remain. Any of these clasts could act as nucleation centers for crystallization of the enclosing liquid. Since renewed growth on large clasts greater than a few millimeters is rare, it is most likely that only the smaller, presumably crystal fragments, act as nucleation sites.

The fine-grained crystals, whether they be residual or injected, are generally distributed in a very uneven pattern in obvious impact-generated glasses. They are usually concentrated in patches enclosing areas free of crystals. It is this distribution of fine crystals that needs to be duplicated at the beginning of a cooling experiment to simulate the crystallization of an impact-generated liquid. This irregular distribution appears to be most easily obtained by partially melting a crystalline charge (Figs. 5d,e,f) even though this most likely is not the process in nature. It is impossible to duplicate the injection of cold clasts during cooling with the present experimental apparatus. Because of the sample size, it is also impossible to have large clasts present in the melt. Cooling from 1280°C in the experiments with some irregular distribution of clasts may well simulate the natural case after the cold clasts have come to some thermal steady state with the melt.

The temperature chosen to start all subliquidus cooling is 1280°C. There is nothing magic about this temperature. Almost any temperature between 1230 and 1290°C could have been used. The criteria for choosing a temperature are primarily the percent crystals versus melt and what phase(s) should be present when cooling is initiated. As with any cooling experiments, the number of potential histories is infinite and the attempt here was to show how nucleation density and distribution could, in principle, affect texture for a limited range of cooling conditions.

The effect of nucleation versus cooling history

It is readily evident from comparison of the textural variations produced in the experimental charges that cooling history alone cannot explain the differences between charges. It is even more difficult to explain the differences in texture within a single charge by differences in cooling history. The other obvious variables are nucleation rate, density, and distribution. The nucleation rate is very difficult to determine so only the density and distribution will be treated here.

An attempt was made in this study to systematically compare the precooling density and distribution of nuclei to the final texture. To do this it is necessary to assume that in two different charges with the same precooling history—one crystallized with a known cooling history and the other quenched at the point when cooling was initiated in the crystallized charge—the density and distribution of nuclei would be the same. It is, of course, impossible to verify this point by actual comparison. We can, however, make some logical assumptions about the nucleation density and distribution and the resulting texture and then check for consistency of pattern.

If we compare W-37 and W-40, for example, the crystal-free glassy areas in W-40 would be the most favorable location for the coarse-grained, subophitic parts of W-37, and the crystal-choked areas in W-40 would be more favorable for the fine-grained poikilitic parts of W-37. The variations in crystal density in W-40 would account for the gradations in grain size in W-37. If these analogies hold, it is apparent that the density and distribution of crystals in W-37 must have been somewhat, but not significantly different than in W-40. The crystal-free area in W-37 must have been in the lower central part of the charge rather than in the U-shaped area in W-40.

By this reasoning the textures in W-36, W-33, W-46, W-10, W-31, W-37, and W-6 fit the precooling crystal density and distribution pattern of their early-quenched equivalent. The textures in several other charges do not fit at all. If W-29 cooled from a condition as shown in W-44, it should be porphyritic and not have the fine-grained poikilitic texture it displays. W-44 is a more likely precursor for W-45 and W-41, a more likely precursor for W-29 than is the observed result. It would be more logical for W-51 to be the precursor of W-32 and not W-43.

The logical conclusion of this dilemma is that the nucleation process, at least in this case, is not systematic. The most erratic results were obtained in the case where nucleation was required at 1280°C, not the partial melting of previously existing crystalline material. This may be largely the effect of nucleation rate which is not capable of being treated in this study. It is clear that the details of the nucleation need further study.

Application to 14310

The original purpose of this study was to experimentally duplicate the crystallization histories of feldspathic lunar basalts in general and 14310 as a representative of that group of basalts in particular. The textural variations in 14310 have been described by many authors, but the most definitive work is by James (1973). There is a close spatial association of different grain sizes and textures on a millimeter to centimeter scale (Fig. 6b). These variations are too large to be duplicated accurately in a 4–6 mm diameter experimental charge. Similar variations, however, were produced in experimental charges, albeit on a smaller scale, that do give insight as to how the textural variations could occur. Runs W-37 and W-33 (Figs. 5b and 6a) come closest to duplicating the ophitic to subophitic to intersertal textures and the grain size difference in close spatial association as observed in 14310. Thus it seems highly probable that the irregular distribution of crystals that acted as nucleation centers for later growth caused the variations.

The experimental results strongly suggest that 14310 did not crystallize from a liquid totally free of at least sub-micron crystals capable of acting as nucleation sites. In all experimental runs cooled from a complete liquid the result is a porphyritic texture as seen in Fig. 3. In addition, these nucleation sites must have been irregularily distributed to account for the textural variation.

Although some textural features of 14310 were reproduced reasonably well, the mineralogy and phase compositions were not. Orthopyroxene is rare. This may result from the lack of complete crystallization and the absence of a slow subsolidus cooling history. There was not time for the olivine to react with the melt to form orthopyroxene. The plagioclase in 14310 is highly and erratically zoned (Ridley *et al.*, 1972; Longhi *et al.*, 1972; Crawford and Hollister, 1974). Similar zoning was not observed in the experimental charges. Both of these facts suggest that 14310 cooled more slowly between about 1225 and 1050°C than the experimentally crystallized charges.

The origin of 14310 has generated considerable controversy about whether it is an impact melt of surficial materials as the direct result of a meteorite impact or whether it is the result of partial melting within the lunar crust or mantle. If the origin of the crystals acting as nucleation centers could be determined, it would be possible to say something definitive about the origin of the 14310 liquid. Even though no clasts are found in 14310, James (1973) believes it to be the product of total impact melting on the lunar surface. The crystals necessary to produce the textural variations in this case could be the sub-micron sized relicts of the impact melting process. They would be nearly impossible to observe in thin section, but their distribution in the melt could have dominated the subsequent textural development during crystallization. If this melt was subjected to injection of cold clasts, then they were nearly completely digested so that only micron sized particles remained when the melt started to cool. This seems unlikely, however, because it is difficult to envision the broad spectrum of fragment sizes that would have been injected being all digested so that only sub-micron size crystals remained.

It is possible that, if 14310 was derived as a partial melt (Crawford and Hollister, 1974), it could have carried with it sufficient crystals or have been sufficiently cooled before eruption to produce enough crystals to act as nucleation centers as described above. Textural heterogeneities similar to those observed in 14310 have been documented in high titanium lunar basalts by Walker *et al.* (1976) and reproduced experimentally by Usselman and Lofgren (1976). The irregular distribution appears to derive from the flow characteristics of the lava.

Thus, the results of these experiments cannot uniquely distinguish between an igneous or an impact melt origin. It does seem reasonable that the impact processes would be more adept at producing irregular distribution of nuclei than would igneous flow processes.

Application to matrix textures in breccias

Ryder and Bower (1976) suggested that the presence in impact melts of numerous plagioclase nuclei at the initiation of cooling is an important precursor to the development of poikilitic matrix textures. In spite of the lack of a detailed understanding of nucleation the results of this study confirm the validity of their hypothesis. The poikilitic textures were developed most readily in the parts of the experimental charges where the plagioclase crystals are the most densely concentrated. The nature of the poikilitic phase in the experimental charges is difficult to determine because the densely spaced, fine-grained plagioclase leave few areas of the enclosing crystal large enough to analyze. The few poikilitic crystals examined to date are olivine. Olivine is in reaction relationship with the 14310 liquid and should react with the liquid to form orthopyroxene. The experimental charges were quenched while liquid still remained and many have cooling rates too rapid to allow this reaction to take place.

The importance of the distribution and density of nucleation sites to the textural development of breccia matrices has not been discussed previously. It is readily apparent from the experiments reported here that it may be the key factor to the development of the close spatial association of such widely divergent textures as the ophitic and poikilitic varieties. There now exists direct experimental evidence that these quite different textures can develop quite close together without requiring major differences in cooling histories over the millimeter to centimeter distances involved. Onorato *et al.* (1976) have suggested, based on thermal calculations, that local thermal gradients do exist when the melt and cold clast are initially mixed. The thermal equilibration time, however, is only 100 sec and it is doubtful if sufficient crystallization could have taken place in response to these thermal gradients in this very short time.

Acknowledgments—This study could not have been completed without the participation of several individuals. R. J. Williams prepared the starting material. Dennis Smith and T. M. Usselman made the experimental runs. O. Mullins performed his usual magic in keeping the equipment operating flawlessly. Discussion with R. J. Williams and T. M. Usselman were of great assistance. C. H. Simonds kindly reviewed the manuscript.

REFERENCES

Crawford M. L. and Hollister L. S. (1974) KREEP basalt: A possible partial melt from the lunar interior. *Proc. Lunar Sci. Conf. 5th*, p. 399–419.

Donaldson C. H., Williams R. J., and Lofgren G. E. (1975) A sample holding technique for study of crystal growth in silicate melts. *Amer. Mineral.* **60**, 324–326.

Grove T. L. (1977) Cooling histories of Apollo 15 quartz normative basalts (abstract). In *Lunar Science VIII*, p. 380–382. The Lunar Science Institute, Houston.

Hollister L. S. and Crawford M. L. (1977) Evidence for direct genetic connection between Apollo 15 KREEP and Apollo 12 and 15 basalts (abstract). In *Lunar Science VIII*, p. 452–454. The Lunar Science Institute, Houston.

James O. B. (1973) Crystallization history of lunar feldspathic basalt 14310. *U.S. Geol. Survey Prof. Paper 841*, 29 pp.

Lofgren G. E., Donaldson C. H., Williams R. J., Mullins O., Jr., and Usselman T. M. (1974) Experimentally reproduced textures and mineral chemistry of Apollo 15 quartz normative basalts. *Proc. Lunar Sci. Conf. 5th*, p. 549–567.

Longhi J., Walker D., and Hays J. F. (1972) Petrography and crystallization history of basalts 14310 and 14072. *Proc. Lunar Sci. Conf. 3rd*, p. 131–139.

Onorato P. I. K., Uhlmann D. R., and Simonds C. H. (1976) Heat flow in impact melts: Apollo 17 Station 6 boulder and some applications to other breccias and xenolith laden melts. *Proc. Lunar Sci. Conf. 7th*, p. 2449–2467.

Ridley W. I., Brett R., Williams R. J., Takeda H., and Brown R. W. (1972) Petrology of Fra Mauro basalt 14310. *Proc. Lunar Sci. Conf. 3rd*, p. 159–170.

Ryder G. and Bower J. F. (1976) Poikilitic KREEP impact melts in the Apollo 14 white rocks. *Proc. Lunar Sci. Conf. 7th*, p. 1925–1948.

Simonds C. H. (1975) Thermal regimes in impact melts and the petrology of the Apollo 17 Station 6 boulder. *Proc. Lunar Sci. Conf. 6th*, p. 641–672.

Simonds C. H., Warner J. L., and Phinney W. C. (1973) Petrology of Apollo 16 poikilitic rocks. *Proc. Lunar Sci. Conf. 4th*, p. 613–632.

Usselman T. M. and Lofgren G. E. (1976) The phase relations, textures, and mineral chemistries of high-titanium mare basalts as a function of oxygen fugacity and cooling rate. *Proc. Lunar Sci. Conf. 7th*, p. 1345–1363.

Usselman T. M., Lofgren G. E., Donaldson C. H., and Williams R. J. (1975) Experimentally reproduced textures and mineral chemistries of high titanium mare basalts. *Proc. Lunar Sci. Conf. 6th*, p. 997–1020.

von Engelhardt W. (1971) Detrital impact formations. *J. Geophys. Res.* **76**, 5566–5574.

Walker D., Longhi J., and Hays J. F. (1976) Heterogeneity in titaniferous mare basalts. *Earth Planet. Sci. Lett.* **30**, 27–36.

Warner J. L., Simonds C. H., and Phinney W. C. (1973) Apollo 16 rocks: Classification and petrogenetic model. *Proc. Lunar Sci. Conf. 4th*, p. 481–504.

Williams R. J. and Mullins O. (1976) A system using solid ceramic oxygen electrolyte cells to measure oxygen fugacities in gas-mixing systems. NASA TM X-58167, 34 pp.

Proc. Lunar Sci. Conf. 8th (1977), p. 2097–2123.
Printed in the United States of America

Experimental melting relations of 63545, 76015, and 76055

JOHN W. DELANO*

Department of Earth and Space Sciences, State University of New York,
Stony Brook, New York 11794

Abstract—The liquidus phase-relations of highlands rocks 63545, 76055, and 76015 have been determined. These experiments constrain the origin of noritic breccias and "very high alumina" (VHA) basalts.

63545 (VHA basalt) has pleonaste spinel as its liquidus phase in the pressure interval from 0 to 30 kbar. At 30 kbar, spinel + garnet occur on the liquidus. Garnet is the liquidus phase at pressures greater than 30 kbar.

76055 (noritic breccia) has olivine as its liquidus phase in the pressure interval from 0 to 23 kbar. Olivine and orthopyroxene are simultaneously saturated at the liquidus at 23 kbar. At pressures greater than 23 kbar, orthopyroxene is the liquidus phase.

76015 (noritic breccia) has olivine as its liquidus phase at 0 kbar. Olivine + spinel coexist on the liquidus in the pressure interval from about 5 to 12 kbar. Olivine + spinel + orthopyroxene are simultaneously on the liquidus at 12 kbar. Orthopyroxene + spinel are the liquidus phases at pressures greater than 12 kbar.

Experimental results suggest that these compositions do not represent magmas derived by partial melting of either cosmic or differentiated source regions at any pressure in the moon.

INTRODUCTION

THE HIGHLANDS NORITIC ROCKS consist of KREEP "basalts" (Meyer *et al.*, 1971) (e.g., 15386), noritic breccias (Rhodes *et al.*, 1974) (e.g., 76015), "spinel" troctolites (Prinz *et al.*, 1973) (e.g., 62295), and "very high alumina (VHA) basalts" (Hubbard *et al.*, 1973a) (e.g., 63545). These rocks are all characterized by noritic major-element chemistries (Walker *et al.*, 1972, 1973a, b; Hubbard *et al.*, 1972, 1973b, 1974; Nava, 1974) with sub-chondritic (Ca/Al) ratios of ~0.8 (chondritic value is ~1.1), negative Eu and Sr anomalies, high (La/Yb) ratios of 3 to 4 (chondritic value is 1.5), and rare earth element (REE) abundances of ⩾50 over chondrites (e.g., Hubbard and Gast, 1971; Hubbard *et al.*, 1972, 1973b, 1974; Nava, 1974; Shih, 1976; Wakita *et al.*, 1975). They have impact metamorphic/crystallization ages of ~3.9 AE, as well as Rb/Sr and U/Pb model ages of 4.4 ± 0.1 AE (e.g., Jessberger *et al.*, 1974; Nyquist *et al.*, 1973, 1974; Kirsten and Horn, 1974; Papanastassiou and Wasserburg, 1971, 1972, 1976; Schaeffer and Husain, 1973; Tera and Wasserburg, 1974; Tera *et al.*, 1973, 1974).

While it appears that the trace element and isotopic systems in the noritic rocks can generally be regarded as having been closed on the whole-rock scale

*Present address: Research School of Earth Sciences, The Australian National University, P.O. Box 4, Canberra, A.C.T. 2600, Australia.

for ~4.4 AE (Nyquist *et al.*, 1974; Tera *et al.*, 1974), uncertainty presently exists as to whether this is also true of the major-element compositions. The uncertainty arises due to the presence of complex brecciated textures with abundant rock and mineral clasts in virtually all samples of the highlands noritic suite (e.g., Bence *et al.*, 1973; Chao *et al.*, 1974; Irving, 1975; Simonds, 1975). Of petrologic importance is whether these clasts are cognate inclusions (Hubbard *et al.*, 1974) or xenocrysts and xenoliths (Dowty *et al.*, 1974; Meyer *et al.*, 1974).

The mineral and rock clasts are generally devoid of shock deformation or polymict textures, thereby suggesting that they did not have a significant regolith history prior to the 3.9 events (Chao *et al.*, 1974; James *et al.*, 1975). Furthermore, the clasts have chemical compositions, which while being too restricted to have been derived from a typically heterogeneous regolith (Chao *et al.*, 1974) are too diverse to have been derived by impact comminution of a pre-existing homogeneous plutonic or volcanic sequence (Albee *et al.*, 1973; Chao *et al.*, 1975; Irving, 1975; James, 1976; Simonds, 1975). Petrologists who have examined these lithologies generally agree that the chemical compositions of the matrix (e.g., melt) and clast population are similar (Chao, 1973; Simonds, 1975). However, the various petrologists seem to disagree as to the precise aspects of the small chemical differences between the matrix and clast population.

PETROGENETIC HYPOTHESES

Conceptually, there are three categories of plausible hypotheses for the petrogenesis of the highlands noritic compositions: (1) internally generated liquids, (2) regolith-contaminated, internally generated liquids, and (3) impact-produced mixtures of ancient highlands components.

(Hypothesis 1) Liquids generated from the lunar interior, when emplaced in shallow pressure regimes, will have chemistries that reflect the residual mineralogy of the source, the extent of partial melting, and the extent of fractionation en route to the emplacement site. Walker *et al.* (1973a, b) have suggested that low pressure partial melting of differentiated crustal compositions may have produced KREEP and "spinel" troctolites. However, as yet, no crustal candidates have been recognized that could satisfy both the major- and trace-element requirements of this model (McCallum and Mathez, 1975). For example, Shih (1976) has contended that KREEP liquids could be derived by fractional crystallization of an initially undifferentiated (i.e., cosmic) reservoir during the global differentiation event at 4.6–4.4 AE, while, on the other hand, Weill and McKay (1975) and Hubbard and Minear (1975) have suggested that KREEP could form by equilibrium crystallization of an initially cosmic reservoir.

(Hypothesis 2) Bence *et al.* (1974) and Chao *et al.* (1974) have speculated that the clast-laden noritic rocks could have been derived when melts, that were produced by partial melting of some unspecified source region, incorporated clasts during passage through the thick highlands regolith.

(Hypothesis 3) The mixing of ancient highlands components (e.g., KREEP and ANT) by impact processes to yield "spinel" troctolites, noritic breccias, and VHA "basalts" has been advocated by Dowty *et al.* (1974), Schonfeld (1974), Simonds *et al.* (1975), Wänke *et al.* (1975), Keil *et al.* (1975), and Meyer *et al.* (1974).

Clearly, no consensus presently exists among lunar scientists as to the petrogenesis of the highlands noritic rocks. The experiments to be discussed in this paper were initiated for the purpose of placing additional constraints on the origin of these noritic compositions. Specifically, the hypothesis (group 1) dealing with partial melting of both cosmic and differentiated source regions will be evaluated using the experimental results.

Due to the presence of negative Eu and Sr anomalies in highlands noritic rocks, the hypothesis dealing with partial melting (or equilibrium crystallization) of a cosmic source region requires that plagioclase be a residual phase in the source region. That is, if this model is applicable, plagioclase must be a liquidus phase at some pressure in these samples. The absence of this characteristic would effectively rule out this hypothesis for the origin of these highlands chemistries.

A hypothetical origin by partial melting of a differentiated source region (i.e., "highland basalt") can also be evaluated. Since plagioclase persists as a residual phase throughout the entire low-pressure melting sequence of "highland basalt" (Walker *et al.*, 1973a, b), plagioclase would be required to occur on the liquidus of noritic rocks at pressures of <5 kbar if this hypothesis were applicable.

Experimental candidates

Chemical and isotopic parameters

Three compositions (76015, 76055, 63545; Table 1) have been experimentally investigated. These samples were selected for the following reasons:

(1) These samples are among the few highlands noritic rocks for which a comprehensive array of chemical and isotopic data have been reported.
(2) 76015, 76055, and 63545 span a substantial portion of the chemical range between KREEP and VHA basaltic end-members (Fig. 2).
(3) 76015 and 63545 are members of abundant chemical suites (i.e., noritic breccias and VHA "basalt", respectively) in the lunar highlands. These two major suites have never before been experimentally investigated. 76055 is chemically distinct from all highlands compositions but may be related to the noritic breccias by addition of olivine.
(4) The model ages (Rb/Sr) of these samples indicate a possible age of formation at ~4.4 AE (i.e., near to the time of the global differentiation event).

Table 1. Electron microprobe analyses of starting materials used in 63545, 76055, and 76015 experiments.

	Sample Compositions			Starting Materials (±1σ)		
	63545[a]	76055[b,d]	76015[c]	63545[e]	76055[f]	76015[g]
SiO_2	43.5	45.4	46.4	43.8 ± 0.2	45.4 ± 0.1	46.2 ± 0.2
TiO_2	0.96	1.32	1.53	1.00 ± 0.03	1.33 ± 0.04	1.53 ± 0.07
Al_2O_3	22.5	16.2	17.7	23.1 ± 0.1	16.6 ± 0.1	18.0 ± 0.1
FeO	7.23	9.24	9.41	6.5 ± 0.3	9.5 ± 0.1	9.5 ± 0.1
MnO	0.07	0.11	0.13	0.03 ± 0.01	0.11 ± 0.02	0.16 ± 0.03
MgO	12.3	17.2	12.6	12.3 ± 0.1	16.7 ± 0.2	12.4 ± 0.1
CaO	12.8	9.58	11.0	13.0 ± 0.2	9.4 ± 0.1	10.7 ± 0.1
Na_2O	0.38	0.52	0.72	0.21 ± 0.01	0.68 ± 0.03	0.94 ± 0.02
K_2O	0.12	0.21	0.27	0.09 ± 0.01	0.39 ± 0.02	0.50 ± 0.04
Cr_2O_3	0.11	0.19	0.19	0.10 ± 0.02	0.17 ± 0.02	0.17 ± 0.03
Total:	100.0	100.0	100.0	100.1	100.3	100.1
Fe/(Fe + Mg)	0.25	0.23	0.29	0.23	0.24	0.30
AB	3.22	4.48	6.09	1.78	5.75	7.95
AN	59.33	41.20	44.27	61.82	41.09	43.42
OR	0.71	1.24	1.60	0.53	2.30	2.95
OL	24.17	25.71	14.94	20.57	27.92	19.35
FA	7.17	7.10	5.12	5.54	8.06	6.77
FO	17.00	18.61	9.82	15.02	19.86	12.58
PYR	10.56	24.56	29.87	13.39	20.44	23.27
DI	3.35	5.07	8.33	2.14	4.45	7.82
HY	7.21	19.49	21.54	11.25	15.99	15.45
ILM	1.82	2.51	2.91	1.90	2.53	2.91
CR	0.16	0.28	0.28	0.15	0.25	0.25

[a]Hubbard *et al.* (1973).
[b]Nava (1974).
[c]Phinney *et al.* (1974).
[d]Apollo 17 Preliminary Examination Team (1973).
[e]Experimental runs #32 and 42.
[f]Experimental runs #19 and 24.
[g]Experimental runs #6 and 7.

(5) The fractionated light rare-earth element (REE) abundance patterns suggest that clinopyroxene was involved in their petrogenesis (Hubbard and Gast, 1971). However, Walker *et al.* (1973a, b) had demonstrated that pyroxene was not likely to be a liquidus phase in these compositions at *low*-pressure. Hence, these experiments were designed to determine whether clinopyroxene might occur on the *high*-pressure liquidi of these chemistries.

63545 is a VHA basalt (Irving, 1975) with a ($^{87}Rb/^{87}Sr$) T_{BABI} model age of 4.5 ± 0.2 AE (Nyquist *et al.*, 1973), a (CaO/Al_2O_3) ratio of 0.57 (Hubbard *et al.*,

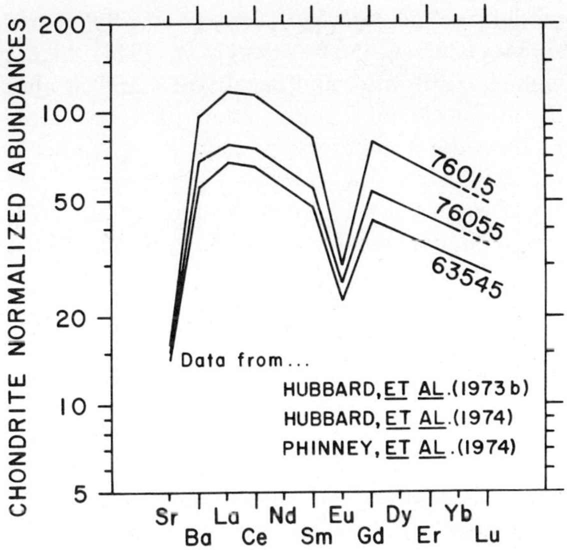

Fig. 1. Chondrite-normalized abundance patterns in 76015, 76055, and 63545.

1973b), high rare-earth abundances relative to chondrites (Fig. 1), a high (La/Yb) ratio relative to chondrites, and negative Eu and Sr anomalies (Hubbard *et al.*, 1973b). The 63545 major-element composition reported by Hubbard *et al.* (1973b) plots within the spinel liquidus field (Fig. 2) on the olivine-anorthite-silica pseudoternary of Walker *et al.* (1973b).

76055 is an olivine-enriched micronorite hornfels (Chao, 1973) with a

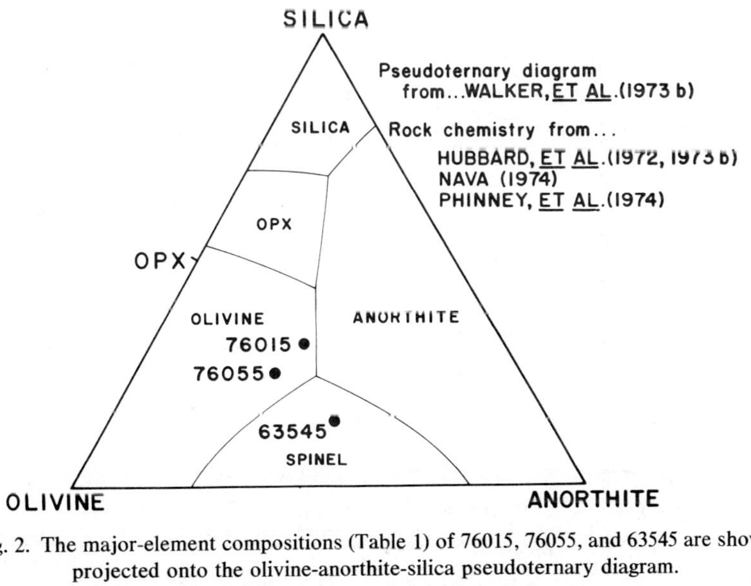

Fig. 2. The major-element compositions (Table 1) of 76015, 76055, and 63545 are shown projected onto the olivine-anorthite-silica pseudoternary diagram.

($^{87}Rb/^{87}Sr$)T_{BABI} model age of 4.45 AE (Tera *et al.*, 1974) and 4.39 ± 0.10 AE (Nyquist *et al.*, 1974), (CaO/Al_2O_3) = 0.59 (Nava, 1974), high rare-earth abundances (Fig. 1), high (La/Yb) ratio, and negative Eu and Sr anomalies (Hubbard *et al.*, 1974). The major-element composition of 76055 plots in the olivine liquidus field (Fig. 2) on the pseudoternary diagram of Walker *et al.* (1973b).

76015 is a poikilitic noritic breccia (Simonds, 1975) possessing a ($^{87}Rb/^{87}Sr$)T_{BABI} model age of 4.4 ± 0.1 AE (Nyquist *et al.*, 1974), (CaO/Al_2O_3) = 0.63, high rare-earth abundances (Fig. 1), high (La/Yb) ratio, and negative Eu and Sr anomalies (Phinney *et al.*, 1974). This composition projects into the olivine liquidus field (Fig. 2) of the pseudoternary diagram (Walker *et al.*, 1973b).

PETROGRAPHIC DESCRIPTIONS

63545

No petrographic descriptions have been published for 63545. This brief description is based upon microscopic examination (no electron microprobe data) of petrographic thin section 63545,6. This sample is a fine-grained, crystalline melt rock (Fig. 3a) consisting of pink spinel, plagioclase, olivine, pyroxene (?), interstitial glass, minor oxides, and metal. Plagioclase clasts are present, some of which display possible shock deformational features. One recrystallized anorthositic clast (Fig. 3b) was observed in 63545,6.

The texturally derived crystallization sequence indicates that spinel was the liquidus phase followed by plagioclase and then by olivine. This sequence is entirely consistent with its bulk composition (Fig. 2).

76055

According to Chao (1973), 76055 is "... an olivine-bearing micronorite hornfels with abundant xenocrysts and xenoliths. It consists of two parts: a large clast of olivine micronorite hornfels with well-developed porphyroblastic texture, surrounded by highly vesicular olivine micronorite hornfels with incipient development of the porphyroblastic textures. This rock contains xenoliths of granulated dunite, xenocryst-laden olivine-plagioclase rocks, granulated and recrystallized anorthosite, annealed shock-vitrified plagioclase glass, and well-polygonalized anorthosite".

Additional information on the mineralogy and petrology of 76055 has been reported by Albee *et al.* (1973), Chao (1973), and Warner *et al.* (1973).

76015

According to Simonds (1975), 76015 is a vesicular, poikilitic impact melt with a homogeneous matrix. The lithic clast population is extremely diverse consisting of dunite, anorthosites, annealed anorthositic troctolites, and a shocked troctolite.

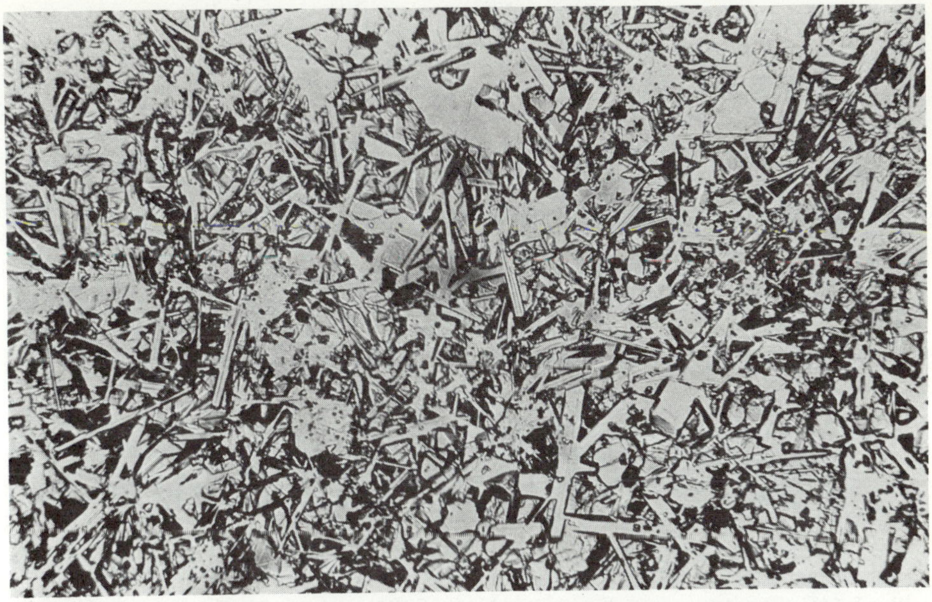

(a)

(b)

Fig. 3. (a) Plane-light photomicrograph of subophitic texture in 63545. The white, rectangular-shaped mineral is plagioclase with minute, crystalline inclusions of pleonaste spinel. The gray, high-relief phase is olivine. The black interstitial regions are glass. Dimension along the lower edge is ~0.5 mm. (b) Plane-light photomicrograph of fine-grained, recrystallized anorthositic clast in 63545. The granular, gray, high-relief phases in the clast are olivine and low-Ca pyroxene. Dimensions along the lower edge of view are ~0.5 mm.

For additional information on the mineralogy and petrology of 76015, see Simonds (1975).

EXPERIMENTAL TECHNIQUES

The starting materials used in these experiments were prepared from oxide-silicate mixes. The oxide-silicate components (without Fe sponge) were homogenized with an automatic mortar by grinding under alcohol for ~5 hr. Following homogenization, the Fe sponge was added to the mix and ground thoroughly for ~3 hr. To avoid oxidation, care was exercised to minimize the time duration that the Fe-sponge was in contact with the alcohol. The mixture was next dried in a vacuum oven at ~110°C for 1 hr.

The starting material was prepared in a large volume (1.4 cm^3) 99.9% Fe capsule in which the oxide-silicate mixture had been loosely packed. The capsule was loosely fitted with an Fe-lid. This was then placed inside a silica-glass tube (closed off at one end) and a capillary tube was drawn in the glass just above the top of the Fe-capsule using an oxygen + natural gas torch. Care was exercised to avoid any oxidation of the Fe-capsule and its contents during this heating. A pellet of Fe-sponge was placed inside the silica glass tube at the top of the capillary tube before the entire assembly was attached to a vacuum pump. The sample was evacuated at room temperature for ~20 min before the assembly was carefully lowered into a vertical furnace for drying under vacuum at 800°C. This drying stage lasted for 10–15 min. Next, the assembly was withdrawn from the furnace so that the capillary could be severed and fused by the torch.

A silica-glass hook was fused to the top of the evacuated silica-glass tube containing the dried sample. This assembly was then hung on a thin Pt-wire next to the thermocouple and slowly (over a time interval of ~1 hr) lowered into the hotspot of a vertical, Pt-wound furnace at 1260°C. After 1 hr in the hotspot, the thin Pt-wire on which the charge was suspended in the furnace was severed by applying an electric current to the wire. The charge free-fell through the furnace into a container of cold water. After the silica-glass tubing was opened, the starting material was extracted from the Fe capsule and ground under spec-pure acetone to a fine ($\leq 10\ \mu$) power to ensure homogeneity.

Most experiments were conducted in high-purity (Johnson Matthey Grade 1, 99.998%) Fe capsules; the remainder were run in Mo-capsules. All charges run at $P = 0$ kbar were tightly loaded into the capsules with loosely fitted lids, placed inside silica-glass tubes, evacuated and dried at 800°C for 1 hr with an Fe-sponge "oxygen getter" at the top of the glass capillary. The use of high purity Fe-capsules imposes a realistic oxygen fugacity. All experiments at $P = 0$ kbar were hung on thin Pt-wire in a vertical Pt-wound furnace and free-fell into cold water for a quick (<5 sec) quench.

Charges run at $P = 5$–30 kbar were tightly loaded into Fe-capsules but lids were not put on at this stage. Each capsule was placed inside a silica-glass tube and dried under vacuum as described previously. When the charge was ready to be used, the evacuated silica-glass tube containing the charge was cracked open and the Fe-container was tightly fitted with a Fe-lid, and immediately assembled in the high-pressure cell. During this procedure, the sample was exposed to air for less than 15 sec.

All experiments at $P = 0$ kbar were performed in a vertical Pt-wound furnace with temperature controlled by means of a stepless proportional controller, with input from a Pt–$Pt_{10}Rh$ thermocouple in an Al_2O_3 tube adjacent to the windings. Temperatures were monitored by a similar thermocouple whose exposed junction lay adjacent to the evacuated silica-glass tube in the hotspot. This recording thermocouple was regularly calibrated against the melting point of Au (1063°C).

High-pressure (5–30 kbar) experiments were performed in a single-stage piston-cylinder apparatus, similar to that described by Boyd and England (1963) but with the charge instead of the Pt–$Pt_{10}Rh$ or $W_{97}Re_3$–$W_{75}Re_{25}$ thermocouple tip centered in the hot spot. The thermocouple and capsule were separated from one another by a thin ceramic disk so as to prevent contamination of the thermocouple. The thermal profile inside the furnace is known, and all reported temperatures have been adjusted accordingly; no correction has been made for the pressure effect on thermocouple e.m.f. Experimental initial pressures were held 1 kbar above the desired nominal pressure for ~5–10 min, then adjusted to nominal pressure cold, brought to temperature, and again adjusted over intervals of $\leqslant$10 min to nominal pressure. This procedure is believed to approximate the true pressure more closely than either instroke or outstroke techniques. The precision of pressure measurements is believed to be ±1 kbar. During experiments, temperatures were controlled by a stepless proportional controller, and the total uncertainty in temperature measurements is estimated to be ±10°C. All furnace assemblies were pre-dried in a nitrogen atmosphere at ~1100°C for 25 min.

Phase-entry reversals were achieved in the following manner: The starting assemblage, which contained the phase in question, was held at a temperature just above the phase-entry temperature long enough for the phase to disappear. This procedure was reversed in a subsequent run where a starting assemblage without the phase in questions was held at this same temperature, and then the temperature was lowered 5–10°C to allow the phase to nucleate just inside its stability field. This 5–10°C temperature interval brackets the phase entry.

Experimental Results

63545

The results of the crystallization experiments conducted on the 63545 composition (Table 1) are summarized in Figs. 4a, b. At 0 kbar, spinel is the liquidus phase. The liquid becomes saturated with plagioclase in the interval 1289°–1299°C. Olivine (Table 2) begins to precipitate at 0 kbar in the interval 1265°–1275°C; spinel disappears by reaction with the liquid between 1225° and 1236°C; and pyroxene begins crystallizing in the interval 1214°–1225°C.

At pressures up to ~30 kbar, pleonaste spinel is the liquidus phase (Table 2;

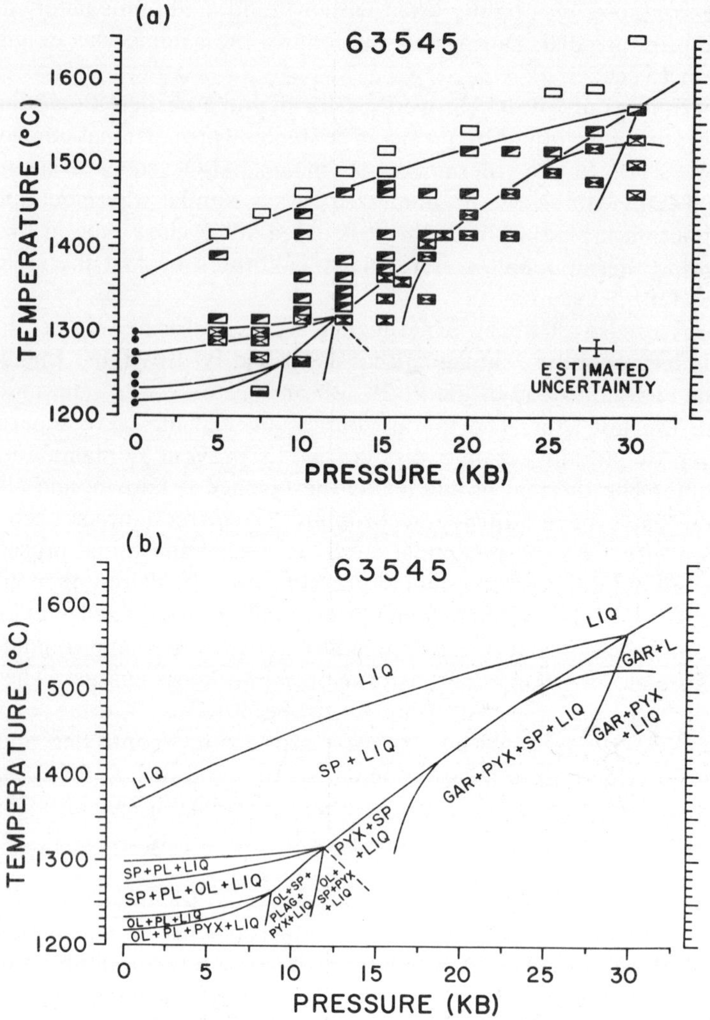

Fig. 4. (a) The melting relations of 63545 at iron saturation are shown as a function of pressure and temperature. Experimental data with identical symbols have identical phase assemblages. The estimated uncertainties in pressure and temperature (for all charges run at $P \geq 5$ kbar) are ±1 kbar and ±10°C. (b) Identification of phase fields illustrated in (a).

Figs. 4a, b) at which point it is replaced by garnet (Table 2). Pyroxene (Table 2) is stable near, but not at, the liquidus in the pressure range of 25–30 kbar. The olivine (Table 2) stability field never approaches the 63545 liquidus at any of the pressures investigated. Spinel and garnet appear simultaneously on the liquidus at ~30 kbar and ~1550°–1600°C.

Chemical analyses of liquids in equilibrium with these solid phases are provided in Table 3.

Table 2. Selected electron microprobe analyses of minerals in 63545 experiments.

	Olivine				Pyroxene				Garnet				Spinel
	(1)	(2)	(3)	(4)	(5)	(6)	(7)	(8)	(9)	(10)	(11)	(12)	(13)
SiO_2	39.5	39.7	40.7	38.6	48.3 ± 0.5	51.5	47.3	46.7	41.6	41.3	40.9	41.0	0.78
TiO_2	0.00	0.00	0.00	0.00	1.3 ± 0.2	0.40	0.71	0.74	1.03	0.89	1.25	0.92	0.22
Al_2O_3	0.00	0.00	0.00	0.00	9.2 ± 0.6	7.5	14.1	14.7	23.2	23.7	23.1	23.8	65.6
FeO	14.5	10.6	10.3	14.8	10.6 ± 0.5	4.8	5.2	4.0	9.3	7.8	9.9	8.1	14.0
MnO	n.a.	n.a.	n.a	n.a.	0.13 ± 0.01	n.a.	n.a.	n.a.	0.1	n.a.	n.a.	n.a.	n.a.
MgO	46.4	50.2	48.5	46.0	19.0 ± 0.2	22.5	16.5	16.5	16.0	18.6	16.7	17.7	18.6
CaO	0.44	0.47	0.54	0.52	11.4 ± 0.6	14.2	16.2	17.7	10.2	7.7	8.8	8.6	n.a.
Na_2O	n.a.	n.a.	n.a.	n.a.	0.05 ± 0.02	n.a.	n.a.	n.a.	0.00	n.a.	n.a.	n.a.	n.a.
K_2O	n.a.	n.a.	n.a.	n.a.	n.a.	n.a.	n.a.	n.a.	0.00	n.a.	n.a.	n.a.	n.a.
Cr_2O_3	n.a.	n.a.	n.a.	n.a.	0.08 ± 0.02	n.a.	0.05	n.a.	0.12	0.09	n.a.	n.a.	0.82
Total:	100.8	101.0	100.0	99.9	100.1	100.9	100.1	100.3	101.6	100.1	100.7	100.1	100.0
Fe/(Fe + Mg)	0.15	0.11	0.11	0.15	0.24	0.11	0.15	0.12	0.25	0.19	0.25	0.20	0.30
NA	0.000	0.000	0.000	0.000	0.004	0.000	0.000	0.000	0.000	0.000	0.000	0.000	0.000
MG	1.721	1.830	1.776	1.728	1.033	1.187	0.884	0.882	1.698	1.976	1.788	1.887	0.701
AL	0.000	0.000	0.000	0.000	0.396	0.313	0.597	0.621	1.947	1.990	1.955	2.006	1.954
SI	0.983	0.971	0.999	0.973	1.762	1.822	1.700	1.674	2.962	2.942	2.937	2.931	0.020
K	0.000	0.000	0.000	0.000	0.000	0.000	0.000	0.000	0.000	0.000	0.000	0.000	0.000
CA	0.012	0.012	0.014	0.014	0.446	0.538	0.624	0.630	0.778	0.588	0.677	0.659	0.000
TI	0.000	0.000	0.000	0.000	0.036	0.011	0.019	0.020	0.055	0.048	0.067	0.049	0.004
CR	0.000	0.000	0.000	0.000	0.002	0.000	0.001	0.000	0.007	0.005	0.000	0.000	0.016
MN	0.000	0.000	0.000	0.000	0.004	0.000	0.000	0.000	0.006	0.000	0.000	0.000	0.000
Fe^{2+}	0.302	0.217	0.212	0.312	0.323	0.142	0.156	0.120	0.554	0.465	0.594	0.484	0.295
Σ	3.017	3.029	3.001	3.027	4.005	4.011	3.982	3.996	8.006	8.013	8.018	8.016	2.991

(1) Run #12 @ $P = 0.0$ kbar, $T = 1248°C$.
(2) Run #74 @ $P = 5.0$ kbar, $T = 1300°C$.
(3) Run #26 @ $P = 7.5$ kbar, $T = 1290°C$.
(4) Run #80 @ $P = 7.5$ kbar, $T = 1270°C$.
(5) Run #34 @ $P = 10.0$ kbar, $T = 1265°C$.
(6) Run #37 @ $P = 12.5$ kbar, $T = 1340°C$.
(7) Run #3 @ $P = 15.0$ kbar, $T = 1365°C$.
(8) Run #79 @ $P = 17.5$ kbar, $T = 1405°C$.
(9) Run #31 @ $P = 17.5$ kbar, $T = 1340°C$.
(10) Run #7 @ $P = 20.0$ kbar, $T = 1415°C$.
(11) Run #54 @ $P = 22.5$ kbar, $T = 1415°C$.
(12) Run #50 @ $P = 25.0$ kbar, $T = 1490°C$.
(13) Run #34 @ $P = 10.0$ kbar, $T = 1265°C$.
n.a. = not analyzed.

Table 3. Selected electron microprobe analyses of liquids in 63545 experiments.

	(1)	(2)	(3)	(4)	(5)	(6)	(7)	(8)	(9)
SiO_2	44.7±0.5	45.0±0.2	45.2±0.3	45.7±0.2	45.4±0.2	43.8±0.1	45.9±0.3	43.8±0.3	47.3
TiO_2	0.96±0.04	1.03±0.05	1.05±0.07	1.44±0.02	2.19±0.14	1.01±0.01	1.07±0.04	0.99±0.05	1.05
Al_2O_3	20.7±0.4	20.1±0.4	19.5±0.7	18.0±0.7	18.3±0.5	23.2±0.1	17.6±0.4	22.9±0.1	19.3
FeO	8.4±0.6	7.9±0.1	8.4±0.2	10.2±0.2	12.1±0.3	6.4±0.1	8.7±0.2	6.5±0.5	6.7
MnO	0.04±0.02	0.04±0.02	0.05±0.02	0.08±0.01	0.12±0.03	0.03±0.01	0.05±0.03	0.03±0.01	0.05
MgO	11.9±0.1	12.4±0.1	13.0±0.1	12.5±0.3	11.5±0.7	12.2±0.1	12.6±0.4	12.4±0.1	12.0
CaO	13.2±0.3	13.0±0.1	12.5±0.2	11.8±0.2	10.6±0.3	13.0±0.2	13.9±0.3	12.9±0.2	13.5
Na_2O	0.13±0.01	0.17±0.03	0.17±0.02	0.15±0.04	0.09±0.03	0.20±0.01	0.16±0.03	0.22±0.01	0.19
K_2O	0.09±0.01	0.07±0.04	0.09±0.02	0.15±0.01	0.11±0.02	0.09±0.01	0.13±0.01	0.09±0.01	0.12
Cr_2O_3	0.00	0.00	0.00	0.05±0.01	0.15±0.04	0.09±0.02	0.05±0.01	0.10±0.02	0.0
Total:	100.1	99.7	100.0	100.1	100.6	100.0	100.2	99.9	100.2
AB	1.10	1.44	1.44	1.27	0.76	1.69	1.35	1.86	1.61
AN	55.63	53.88	52.18	48.00	49.20	62.14	46.92	61.23	51.45
OR	0.53	0.41	0.53	0.89	0.65	0.53	0.77	0.53	0.71
OL	17.37	16.01	16.76	12.91	8.56	19.98	15.22	20.51	6.63
FA	5.92	5.04	5.35	4.75	3.58	5.34	5.08	5.50	1.87
FO	11.44	10.97	11.41	8.15	4.98	14.63	10.14	15.01	4.76
PYR	23.67	26.02	27.06	34.20	37.01	13.63	33.79	13.77	37.81
DI	7.96	8.56	7.93	8.55	2.76	1.89	17.80	2.22	12.46
HY	15.70	17.46	19.13	25.65	34.25	11.74	15.99	11.56	25.36
ILM	1.82	1.96	1.99	2.73	4.16	1.92	2.03	1.88	1.99
CR	0.00	0.00	0.00	0.07	0.24	0.13	0.07	0.15	0.00

(1) Run #53 @ $P = 0.0$ kbar, $T = 1299$°C.
(2) Run #52 @ $P = 0.0$ kbar, $T = 1289$°C.
(3) Run #51 @ $P = 0.0$ kbar, $T = 1275$°C.
(4) Run #5 @ $P = 0.0$ kbar, $T = 1265$°C.
(5) Run #12 @ $P = 0.0$ kbar, $T = 1248$°C.
(6) Run #42 @ $P = 5.0$ kbar, $T = 1415$°C.
(7) Run #74 @ $P = 5.0$ kbar, $T = 1285$°C.
(8) Run #32 @ $P = 7.5$ kbar, $T = 1440$°C.
(9) Run #77 @ $P = 7.5$ kbar, $T = 1300$°C.

The pressure, temperature, duration, and products of all 63545 experiments used in constructing Fig. 4 are contained in Table 4.

76055

The results of the crystallization experiments conducted on the 76055 composition (Table 1) are summarized in Figs. 5a, b. At 0 kbar, olivine (Table 5) is the liquidus phase. Plagioclase saturation occurs in the interval 1244°–1225°C.

At pressures up to ~23 kbar (Fig. 5), olivine is the liquidus phase (Table 5) at which point it is replaced by orthopyroxene (Table 5). Garnet and spinel are stable near, but not at, the liquidus in the pressure range of 20–25 kbar. Olivine and orthopyroxene appear simultaneously on the liquidus at ~23 kbar and ~1520°C.

Chemical analyses of liquids in equilibrium with these solid phases are provided in Table 6.

The pressure, temperature, duration, and products of all 76055 experiments used in constructing Fig. 5 are contained in Table 7.

Table 3. (*Continued*).

(10)	(11)	(12)	(13)	(14)	(15)	(16)	(17)
46.8 ± 0.6	46.6 ± 0.3	47.3 ± 0.2	46.4 ± 0.7	46.4	44.6 ± 0.9	46.4 ± 0.2	45.8 ± 0.4
1.19 ± 0.08	1.01 ± 0.04	1.02 ± 0.04	1.04 ± 0.06	1.06	0.93 ± 0.02	1.01 ± 0.09	1.09 ± 0.04
18.4 ± 0.5	18.6 ± 0.5	18.6 ± 0.4	20.8 ± 0.8	21.0	22.2 ± 0.5	19.7 ± 0.6	21.0 ± 0.5
7.5 ± 0.1	7.6 ± 0.1	6.9 ± 0.1	5.2 ± 0.1	6.66	6.99 ± 0.02	7.7 ± 0.2	8.1 ± 0.3
0.09 ± 0.01	0.07 ± 0.03	0.07 ± 0.02	0.00	0.07	0.10 ± 0.01	0.07 ± 0.01	0.04 ± 0.02
12.0 ± 0.3	11.81 ± 0.05	11.64 ± 0.08	12.1 ± 0.1	11.4	12.6 ± 0.3	11.9 ± 0.1	9.8 ± 0.4
13.1 ± 0.3	13.3 ± 0.1	14.1 ± 0.4	13.8 ± 0.3	13.0	12.4 ± 0.2	13.0 ± 0.4	14.2 ± 0.5
0.18 ± 0.01	0.20 ± 0.02	0.19 ± 0.01	0.22 ± 0.01	0.21	0.18 ± 0.03	0.16 ± 0.02	0.29 ± 0.02
0.14 ± 0.01	0.11 ± 0.02	0.09 ± 0.01	0.10 ± 0.02	0.12	0.17 ± 0.02	0.11 ± 0.02	0.16 ± 0.02
0.0	0.05 ± 0.03	0.05 ± 0.03	0.04 ± 0.02	0.0	0.10 ± 0.01	0.06 ± 0.03	0.04 ± 0.01
99.4	99.3	99.9	99.7	99.9	100.3	100.1	100.5
1.52	1.69	1.61	1.86	1.78	1.52	1.35	2.45
48.98	49.53	49.63	55.47	56.00	59.27	52.71	55.53
0.83	0.65	0.53	0.59	0.71	1.00	0.65	0.95
7.08	8.19	5.96	9.09	8.05	18.38	9.66	11.34
2.17	2.60	1.77	2.02	2.35	5.26	3.07	4.22
4.91	5.60	4.19	7.07	5.70	13.13	6.58	7.12
38.72	37.28	40.18	30.65	31.36	18.18	33.73	28.12
12.88	13.26	16.33	10.35	6.82	1.81	9.50	12.10
25.84	24.02	23.85	20.30	24.54	16.37	24.24	16.02
2.26	1.92	1.94	1.98	2.01	1.77	1.92	2.07
0.00	0.07	0.07	0.06	0.00	0.15	0.09	0.06

(10) Run #26 @ $P = 7.5$ kbar, $T = 1290$°C.
(11) Run #20 @ $P = 12.5$ kbar, $T = 1365$°C.
(12) Run #37 @ $P = 12.5$ kbar, $T = 1340$°C.
(13) Run #83 @ $P = 12.5$ kbar, $T = 1330$°C.
(14) Run #6 @ $P = 15.0$ kbar, $T = 1390$°C.
(15) Run #71 @ $P = 20.0$ kbar, $T = 1540$°C.
(16) Run #62 @ $P = 20.0$ kbar, $T = 1465$°C.
(17) Run #50 @ $P = 25.0$ kbar, $T = 1490$°C.

76015

The results of the crystallization experiments conducted on the 76015 composition (Table 1) are summarized in Figs. 6a, b. At 0 kbar olivine is the liquidus phase. In the pressure interval from $\leqslant 5$ to ~ 12 kbar, olivine and spinel (Table 8) are cosaturated at the liquidus. Orthopyroxene (Table 8) is the liquidus phase, accompanied by spinel, at $\geqslant 12$ kbar. Olivine + spinel + orthopyroxene occur simultaneously on the liquidus at ~ 12 kbar and ~ 1350°C.

Chemical analyses of liquids in equilibrium with these solid phases are provided in Table 9.

The pressure, temperature, duration, and products of all 76015 experiments used in constructing Fig. 6 are contained in Table 10.

Discussion

Partial melting (or equilibrium crystallization) of cosmic source regions

Since none of these three highlands noritic compositions has plagioclase as a liquidus phase at any pressure, the negative Eu and Sr anomalies (Fig. 1) cannot

Table 4. 63545 experiments.

P (kbar)	Run #	T(°C)	Capsule	Hours	Products	Probe analyses	S.A.*
0.0	53	1299	Fe	3.0	sp + liq	liq	c
0.0	2	1289	Fe	7.8	sp + plag + liq		b
0.0	52	1289	Fe	1.5	sp + plag + liq	liq	c
0.0	51	1275	Fe	2.5	sp + plag + liq	liq	c
0.0	5	1265	Fe	5.3	ol + sp + plag + liq	liq	b
0.0	12	1248	Fe	15.3	ol + sp + plag + liq	ol, liq	b
0.0	15	1236	Fe	23.5	ol + sp + plag + liq		c
0.0	19	1225	Fe	51.3	ol + plag + liq		b
0.0	9	1214	Fe	51.3	ol + plag + pyx + liq		b
5.0	42	1415	Fe	1.3	liq	liq	a
5.0	44	1390	Fe	2.0	sp + liq		a
5.0	14	1315	Fe	5.8	sp + liq		b
5.0	74	1300	Fe	2.5	sp + plag + liq	ol, liq	b
5.0	16	1290	Fe	3.0	ol + sp + plag + liq		b
7.5	32	1440	Fe	1.0	liq	liq	b
7.5	25	1315	Fe	2.8	sp + liq		c
7.5	77	1300	Fe	2.8	sp + plag + liq	liq	c
7.5	26	1290	Fe	4.8	ol + sp + plag + liq	ol, liq	b
7.5	80	1270	Fe	4.8	ol + sp + plag + liq	ol	b
7.5	81	1230	Fe	6.8	ol + plag + pyx + liq		c
10.0	28	1465	Fe	1.0	liq		b
10.0	40	1440	Fe	1.0	sp + liq		a
10.0	36	1415	Fe	2.0	sp + liq		b
10.0	13	1365	Fe	1.8	sp + liq		c
10.0	10	1340	Fe	3.0	sp + liq		b
10.0	75	1325	Fe	2.3	sp + liq		c
10.0	11	1315	Fe	3.3	sp + plag + liq		a
10.0	34	1265	Fe	3.3	ol + sp + plag + pyx + liq	sp, pyx	b
12.5	63	1490	Fe	0.8	liq		b
12.5	41	1465	Fe	1.3	sp + liq		a
12.5	22	1385	Fe	2.0	sp + liq		a
12.5	20	1365	Fe	2.5	sp + liq	liq	a
12.5	76	1350	Fe	2.0	sp + liq		c
12.5	37	1340	Fe	2.5	sp + liq ± pyx	pyx, liq	a
12.5	83	1330	Fe	3.3	sp + liq	liq	d
12.5	21	1315	Fe	3.0	sp + pyx + liq		c
15.0	30	1515	Fe	0.8	liq		b
15.0	35	1490	Fe	1.0	liq		a
15.0	39	1475	Fe	1.0	sp + liq		a
15.0	38	1465	Fe	1.0	sp + liq		d
15.0	1	1415	Fe	1.3	sp + liq		b
15.0	6	1390	Fe	2.0	sp + liq	liq	b
15.0	78	1375	Fe	1.5	sp + liq (+ pyx?)		d
15.0	3	1365	Fe	2.8	sp + pyx + liq	pyx	b
15.0	4	1340	Fe	3.5	sp + pyx + liq		b
15.0	33	1315	Fe	3.5	sp + pyx + liq		b
16.0	82	1360	Fe	2.0	sp + pyx + liq		a
17.5	27	1465	Fe	1.3	sp + liq		a

Table 4. (*Continued*).

P (kbar)	Run #	T(°C)	Capsule	Hours	Products	Probe analyses	S.A.*
17.5	18	1415	Fe	1.5	sp + liq		b
17.5	79	1405	Fe	2.0	sp + pyx + liq	pyx	b
17.5	17	1390	Fe	2.3	sp + pyx + gar + liq		c
17.5	31	1340	Fe	2.8	sp + pyx + gar + liq	gar	b
18.5	84	1415	Fe	1.7	sp + pyx + liq		b
20.0	46	1540	Fe	1.0	liq		e
20.0	71	1540	Mo	0.7	liq	liq	a
20.0	48	1515	Fe	1.3	sp + liq		a
20.0	23	1465	Fe	2.0	sp + liq		b
20.0	62	1465	Fe	2.0	sp + liq	liq	a
20.0	8	1440	Fe	1.5	sp + pyx + gar + liq		d
20.0	7	1415	Fe	1.8	sp + pyx + gar + liq	gar	b
22.5	45	1515	Fe	1.8	sp + liq		d
22.5	43	1465	Fe	2.3	sp + pyx + gar + liq		d
22.5	54	1415	Fe	1.8	sp + pyx + gar + liq	gar	c
25.0	64	1585	Mo	0.3	liq		b
25.0	49	1515	Fe	2.0	sp + liq (+pyx?)		a
25.0	50	1490	Fe	2.0	sp + pyx + gar + liq	gar, liq	e
27.5	66	1590	Mo	0.2	liq		e
27.5	56	1540	Fe	0.8	sp + pyx + liq		a
27.5	68	1520	Fe	0.8	sp + pyx + gar + liq		b
27.5	59	1480	Fe	0.3	sp + pyx + gar + liq		d
30.0	69	1650	Mo	0.1	liq		e
30.0	67	1565	Mo	0.3	sp + gar + liq		d
30.0	86	1530	Fe	2.0	gar + liq		b
30.0	85	1500	Fe	2.3	pyx + gar + liq		e
30.0	73	1465	Fe	0.5	pyx + gar + liq		b

Abbreviations: ol = olivine, sp = pleonaste spinel, plag = plagioclase, pyx = pyroxene, gar = garnet, liq = liquid.

*Starting assemblages (S.A.): a, liq; b, sp + liq; c, ol + plag + liq; d, sp + pyx + liq; e, sp + pyx + gar + liq.

be accounted for by partial melting of any cosmic source region at any pressure. Furthermore, clinopyroxene, which is required to cause most of the light rare-earth enrichment (Hubbard and Gast, 1971; Delano, 1977), is also not on the liquidus. Therefore, the partial melting hypothesis is concluded to be untenable as regards the petrogenesis of 63545, 76055, and 76015.

Partial melting of differentiated source regions

Crust. The absence of a liquidus multiple saturation point at low pressures (<5 kbar) involving plagioclase ± other phases (e.g., spinel, olivine) is evidence that 63545, 76055, and 76015 are not partial melts of any anorthositic source (e.g., highland basalt).

Mantle. Any partial melt of the differentiated lunar mantle should possess a

Table 5. Selected electron microprobe analyses of minerals in 76055 experiments.

	Olivine									
	(1)	(2)	(3)	(4)	(5)	(6)	(7)	(8)	(9)	(10)
SiO_2	40.5 ± 0.3	40.6	39.8 ± 0.4	39.4	38.7 ± 0.5	40.3 ± 0.6	39.7 ± 0.2	39.4 ± 0.3	40.7 ± 0.1	40.6
TiO_2	n.a.	0.00	0.06 ± 0.05	0.08	0.12 ± 0.04	n.a.	0.08 ± 0.02	n.a.	n.a.	n.a.
Al_2O_3	n.a.	0.00	0.00	0.00	0.00	n.a.	0.00	0.04 ± 0.02	n.a.	n.a.
FeO	10.9 ± 0.3	11.3	12.0 ± 0.3	14.7	16.6 ± 0.2	10.3 ± 0.1	10.5 ± 0.2	11.7 ± 0.1	9.5 ± 0.3	10.0
MnO	n.a.	0.11	0.13 ± 0.02	n.a.	n.a.	n.a.	n.a.	0.07 ± 0.01	n.a.	n.a.
MgO	48.8 ± 0.6	47.5	47.0 ± 0.4	44.6	44.5 ± 0.9	49.7 ± 0.7	48.7 ± 0.4	48.4 ± 0.4	49.9 ± 0.1	50.0
CaO	0.34 ± 0.02	0.38	0.38 ± 0.03	0.35	0.36 ± 0.02	0.37 ± 0.03	0.40 ± 0.04	0.39 ± 0.04	0.38 ± 0.04	0.37
Na_2O	n.a.	0.00	0.00	n.a.	n.a.	n.a.	n.a.	n.a.	n.a.	n.a.
K_2O	n.a.	0.00	0.00	n.a.	n.a.	n.a.	n.a.	n.a.	n.a.	n.a.
Cr_2O_3	n.a.	0.15	0.17 ± 0.03	n.a.	n.a.	n.a.	n.a.	0.11 ± 0.02	n.a.	n.a.
Total:	100.5	100.0	99.5	99.1	100.3	100.7	99.4	100.1	100.5	101.0
Fe/(Fe + Mg)	0.11	0.12	0.12	0.16	0.17	0.10	0.11	0.12	0.10	0.10
Na	0.000	0.000	0.000	0.000	0.000	0.000	0.000	0.000	0.000	0.000
Mg	1.783	1.747	1.746	1.683	1.678	1.810	1.800	1.788	1.813	1.813
Al	0.000	0.000	0.000	0.000	0.000	0.000	0.000	0.001	0.000	0.000
Si	0.992	1.002	0.992	0.997	0.979	0.985	0.984	0.976	0.992	0.987
K	0.000	0.000	0.000	0.000	0.000	0.000	0.000	0.000	0.000	0.000
Ca	0.009	0.010	0.010	0.009	0.010	0.010	0.011	0.010	0.010	0.010
Ti	0.000	0.000	0.001	0.002	0.002	0.000	0.001	0.000	0.000	0.000
Cr	0.000	0.003	0.003	0.000	0.000	0.000	0.000	0.002	0.000	0.000
Mn	0.000	0.002	0.003	0.000	0.000	0.000	0.000	0.001	0.000	0.000
Fe^{2+}	0.223	0.233	0.250	0.311	0.351	0.210	0.218	0.242	0.194	0.203
Σ	3.008	2.997	3.005	3.002	3.019	3.015	3.014	3.022	3.008	3.013

(1) Run #20 @ $P = 0.0$ kbar, $T = 1300$°C.
(2) Run #2 @ $P = 0.0$ kbar, $T = 1275$°C.
(3) Run #4 @ $P = 0.0$ kbar, $T = 1244$°C.
(4) Run #7 @ $P = 0.0$ kbar, $T = 1225$°C.
(5) Run #10 @ $P = 0.0$ kbar, $T = 1198$°C.
(6) Run #25 @ $P = 7.5$ kbar, $T = 1385$°C.
(7) Run #16 @ $P = 15.0$ kbar, $T = 1405$°C.
(8) Run #22 @ $P = 15.0$ kbar, $T = 1380$°C.
(9) Run #6 @ $P = 17.5$ kbar, $T = 1465$°C.
(10) Run #11 @ $P = 17.5$ kbar, $T = 1440$°C.

multiple saturation point on its liquidus within the pressure interval from 3 to 48 kbar. The 3 kbar value corresponds to the pressure at the top of the lunar mantle (e.g., Latham *et al.*, 1973). The 48 kbar value is the pressure at the center of the moon. This large interval of pressure results from the present uncertainty as to the maximum depth to which the moon differentiated at 4.6–4.4 AE (e.g., Taylor and Jakeš, 1974; Kesson and Ringwood, 1976; Solomon and Chaiken, 1976).

63545 does not possess a significant multiple saturation point. The coexistence of spinel and garnet on the liquidus at ~30 kbar is less a petrologically significant, multiple saturation than a pressure transition of the liquidus phase. Hence, 63545 is not a partial melt from the differentiated lunar mantle.

76055 has olivine and orthopyroxene coexisting on the liquidus at ~23 kbar. While this is permissive evidence for 76055 being a partial melt from the differentiated mantle at a depth of ~450–500 km, severe problems arise from that inference. The chemical structure of the differentiated lunar mantle inferred from 76055 is not readily amenable to the scenario deduced from mare basalt data. For example, the Apollo 15 green glass, possessing only a small negative

Table 5. (*Continued*).

Olivine		Pyroxene					Garnet		Spinel
(11)	(12)	(13)	(14)	(15)	(16)	(17)	(18)	(19)	(20)
39.5 ± 0.3	40.5	51.6 ± 0.5	50.1 ± 0.6	51.5 ± 0.6	51.7 ± 0.7	49.5 ± 0.4	40.7 ± 0.5	40.8 ± 0.4	0.04
n.a.	n.a.	0.47 ± 0.02	0.45 ± 0.03	0.42 ± 0.04	0.35 ± 0.04	0.38 ± 0.03	1.2 ± 0.1	0.80 ± 0.12	0.00
0.08 ± 0.05	n.a.	9.0 ± 1.5	10.2 ± 0.5	9.9 ± 0.7	7.9 ± 0.8	9.4 ± 0.8	24.0 ± 0.7	23.4 ± 0.7	67.8
11.0 ± 0.3	9.7	7.1 ± 0.1	6.7 ± 0.1	6.7 ± 0.2	6.9 ± 0.2	7.2 ± 0.1	9.6 ± 0.2	9.6 ± 0.3	7.9
0.06 ± 0.02	n.a.	0.09 ± 0.01	0.10 ± 0.01	0.08 ± 0.02	0.09 ± 0.02	0.03 ± 0.01	0.15 ± 0.01	0.10 ± 0.01	n.a.
49.7 ± 0.4	49.9	30.0 ± 0.4	29.8 ± 0.3	29.1 ± 0.2	30.5 ± 0.4	19.0 ± 0.5	19.3 ± 0.5	18.6 ± 0.6	22.8
0.39 ± 0.02	0.26	2.4 ± 0.4	2.5 ± 0.1	2.5 ± 0.2	2.6 ± 0.1	13.7	6.4 ± 0.3	7.2 ± 0.5	n.a.
n.a.	n.a.	0.03 ± 0.03	0.02 ± 0.02	0.05 ± 0.02	0.06 ± 0.01	n.a.	n.a.	n.a.	n.a.
n.a.	n.a.	n.a.	n.a.	n.a.	n.a.	n.a.	n.a.	n.a.	n.a.
0.09 ± 0.02	n.a.	0.30 ± 0.03	0.27 ± 0.04	0.25 ± 0.02	0.22 ± 0.03	0.15 ± 0.04	0.29 ± 0.06	0.26 ± 0.03	1.43
100.8	100.5	101.0	100.1	100.5	100.3	99.4	101.5	100.8	100.0
0.11	0.10	0.12	0.11	0.11	0.11	0.17	0.22	0.22	0.16
0.000	0.000	0.002	0.001	0.003	0.004	0.000	0.000	0.000	0.000
1.818	1.816	1.547	1.549	1.503	1.583	1.027	2.034	1.980	0.838
0.002	0.000	0.367	0.419	0.404	0.324	0.402	1.999	1.969	1.970
0.969	0.988	1.784	1.747	1.784	1.800	1.795	2.877	2.914	0.001
0.000	0.000	0.000	0.000	0.000	0.000	0.000	0.000	0.000	0.000
0.010	0.009	0.089	0.093	0.093	0.097	0.532	0.485	0.551	0.000
0.000	0.000	0.012	0.012	0.011	0.009	0.010	0.064	0.043	0.000
0.002	0.000	0.008	0.007	0.007	0.000	0.004	0.016	0.015	0.028
0.001	0.000	0.003	0.003	0.002	0.003	0.001	0.009	0.006	0.000
0.226	0.199	0.205	0.195	0.194	0.201	0.218	0.567	0.573	0.163
3.028	3.012	4.017	4.028	4.001	4.028	3.991	8.051	8.051	3.000

(11) Run #12 @ $P = 17.5$ kbar, $T = 1415°C$.
(12) Run #15 @ $P = 20.0$ kbar, $T = 1470°C$.
(13) Run #22 @ $P = 15.0$ kbar, $T = 1380°C$.
(14) Run #12 @ $P = 17.5$ kbar, $T = 1415°C$.
(15) Run #1 @ $P = 20.0$ kbar, $T = 1440°C$.
(16) Run #9 @ $P = 22.5$ kbar, $T = 1465°C$.
(17) Run #28 @ $P = 25.0$ kbar, $T = 1475°C$.
(18) Run #18 @ $P = 20.0$ kbar, $T = 1400°C$.
(19) Run #28 @ $P = 25.0$ kbar, $T = 1475°C$.
(20) Run #1 @ $P = 20.0$ kbar, $T = 1440°C$.
n.a. = not analyzed.

Eu anomaly (Taylor *et al.*, 1973), seems to have originated from a mantle source at a depth of ~400 km (Green and Ringwood, 1972). However, a literal interpretation of the 76055 results would indicate that the mantle between 450–500 km has a very large negative Eu anomaly. This relationship is not tenable and must be interpreted as indicating that 76055 cannot be a melt derived from that depth in the moon.

76015 has olivine + spinel + orthopyroxene as liquidus phases at ~12 kbar. While this data is permissive evidence for 76015 being a partial melt from the differentiated lunar mantle at a depth of ~250 km, severe difficulties would seem to arise as to reconciling this with the information indicating that low- and high-TiO_2 mare basalts were derived from similar regions of the mantle (e.g., Grove *et al.*, 1973; Green *et al.*, 1975; Kesson, 1975). For example, the Fe/(Fe + Mg) ratio of the hypothetical residue that yielded 76015 is ~0.1, whereas the value determined from mare basalts is 0.2–0.3.

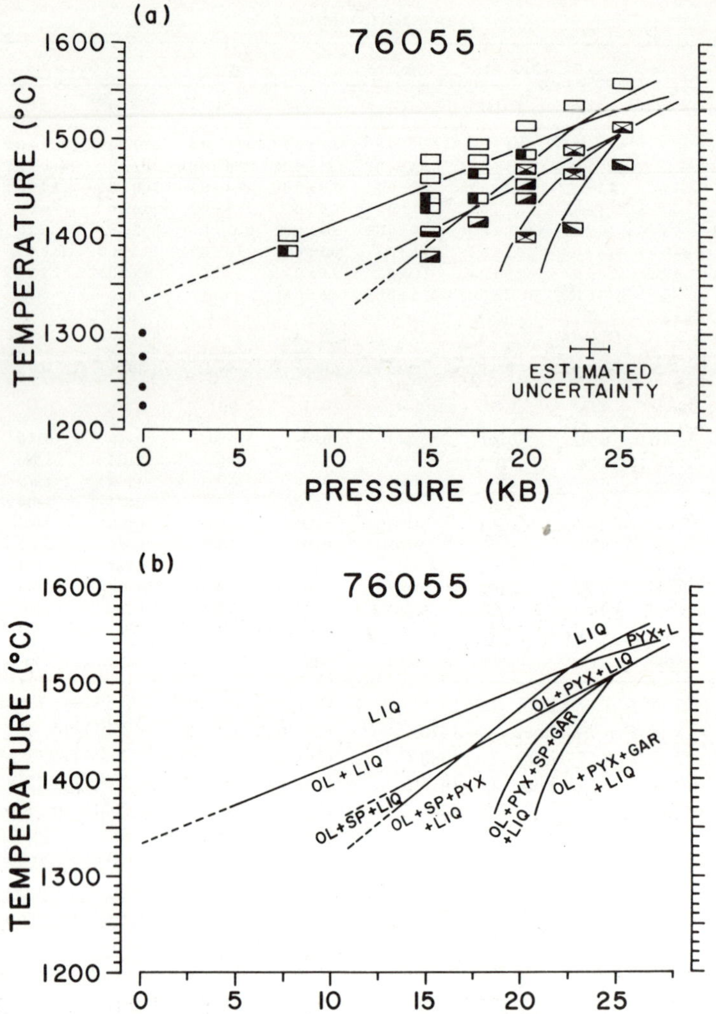

Fig. 5. (a) The melting relations of 76055 at iron saturation are shown as a function of pressure and temperature. Experimental data with identical symbols have identical phase assemblages. The estimated uncertainties in pressure and temperature (for all charges run at $P \geqslant 5$ kbar) are ± 1 kbar and ± 10°C. (b) Identification of phase fields illustrated in (a).

Other petrogenetic hypotheses

The experimental melting relations of 63545, 76055, and 76015, when examined in conjunction with experimental petrology and trace element chemistry of mare basalts, would seem to make an origin of these noritic rocks by partial melting of either cosmic or fractionated source regions at any pressure in

Table 6. Selected electron microprobe analyses of liquids in 76055 experiments.

	(1)	(2)	(3)	(4)	(5)	(6)
SiO_2	45.7 ± 0.2	46.5 ± 0.2	46.5 ± 0.3	45.4 ± 0.1	45.3 ± 0.1	45.4 ± 0.2
TiO_2	1.40 ± 0.05	1.51 ± 0.05	1.54 ± 0.02	1.32 ± 0.04	1.33 ± 0.04	1.42 ± 0.02
Al_2O_3	18.2 ± 0.1	18.3 ± 1.1	18.8 ± 0.4	16.6 ± 0.1	16.5 ± 0.1	17.4 ± 0.1
FeO	9.5 ± 0.2	9.4 ± 0.1	9.3 ± 0.1	9.5 ± 0.1	9.5 ± 0.1	9.9 ± 0.1
MnO	0.07 ± 0.02	0.13 ± 0.02	0.11 ± 0.03	0.10 ± 0.01	0.11 ± 0.02	0.08 ± 0.01
MgO	13.3 ± 0.3	12.1 ± 0.1	10.8 ± 0.2	16.5 ± 0.1	16.8 ± 0.2	13.6 ± 0.2
CaO	10.5 ± 0.1	10.7 ± 0.1	11.4 ± 0.1	9.43 ± 0.07	9.3 ± 0.1	10.7 ± 0.1
Na_2O	0.73 ± 0.04	1.1 ± 0.1	1.00 ± 0.07	0.68 ± 0.04	0.67 ± 0.01	0.80 ± 0.05
K_2O	0.44 ± 0.01	0.49 ± 0.02	0.40 ± 0.07	0.39 ± 0.02	0.38 ± 0.02	0.45 ± 0.02
Cr_2O_3	0.18 ± 0.03	0.21 ± 0.03	0.17 ± 0.01	0.16 ± 0.02	0.18 ± 0.02	0.15 ± 0.01
Total:	100.0	100.4	100.0	100.1	100.1	99.9
AB	6.18	9.31	8.46	5.75	5.67	6.77
AN	45.08	43.55	45.63	41.09	40.89	42.56
OR	2.60	2.90	2.36	2.30	2.25	2.66
OL	20.69	19.53	14.99	27.26	27.90	23.88
FA	6.92	6.87	5.61	7.94	8.01	8.13
FO	13.77	12.66	9.38	19.33	19.89	15.75
PYR	22.54	21.98	25.40	20.93	20.57	21.12
DI	5.66	7.72	8.87	4.57	4.21	8.51
HY	16.88	14.27	16.54	16.35	16.36	12.61
ILM	2.66	2.87	2.92	2.51	2.53	2.70
CR	0.27	0.31	0.25	0.24	0.27	0.22

(1) Run #20 @ $P = 0.0$ kbar, $T = 1300°C$.
(2) Run #2 @ $P = 0.0$ kbar, $T = 1275°C$.
(3) Run #4 @ $P = 0.0$ kbar, $T = 1244°C$.
(4) Run #24 @ $P = 7.5$ kbar, $T = 1400°C$.
(5) Run #19 @ $P = 15.0$ kbar, $T = 1460°C$.
(6) Run #22 @ $P = 15.0$ kbar, $T = 1380°C$.

the moon most unlikely. Of the hypotheses outlined in the beginning of this paper, the following still remain as possible mechanisms for the origin of these highlands noritic rocks.

(a) Fractional crystallization during the early lunar differentiation event.
(b) Regolith-contaminated melts produced by internal processes.
(c) Impact-fused mixtures of ancient crustal components.

Since the experimental data on these highlands noritic chemistries is not easily reconcilable with petrogenetic schemes for mare basalts, presently available information on textures and chemistry (e.g., Simonds, 1975) would tend to favor hypothesis (c). This suggestion, however, does not necessarily apply to some KREEP "basalts" (e.g., 15382, 15386, 15434) which are neither clast-laden (Nyquist *et al.*, 1974) nor meteorite-contaminated (e.g., Morgan *et al.*, 1973) and hence share few chemical and textural similarities with 63545, 76055, and 76015.

Table 7. 76055 experiments.

P (kbar)	Run #	*T* (°C)	Capsule	Hours	Products	Probe analyses	S.A.*
0.0	20	1300	Fe	0.8	ol + liq	ol, liq	b
0.0	2	1275	Fe	3.8	ol + liq	ol, liq	b
0.0	4	1244	Fe	13.0	ol + liq	ol, liq	c
0.0	7	1225	Fe	63.8	ol + plag + liq	ol	b
0.0	10	1198	Fe	58.0	ol + plag + liq	ol	b
7.5	24	1400	Fe	1.3	liq	liq	b
7.5	25	1385	Fe	1.5	ol + liq	ol	a
15.0	14	1480	Fe	1.0	liq		b
15.0	19	1460	Fe	1.3	liq	liq	b
15.0	5	1440	Fe	1.3	ol + liq		a
15.0	13	1430	Fe	2.8	ol + liq		b
15.0	16	1405	Fe	2.0	ol + sp + liq	ol	b
15.0	22	1380	Fe	3.2	ol + sp + pyx + liq	ol, pyx, liq	b
17.5	17	1495	Mo	0.5	liq		b
17.5	23	1480	Fe	1.0	liq		b
17.5	6	1465	Fe	1.3	ol + liq	ol	a
17.5	11	1440	Fe	2.0	ol + liq	ol	b
17.5	12	1415	Fe	2.0	ol + sp + pyx + liq	ol, pyx	b
20.0	21	1515	Mo	0.5	liq		a
20.0	31	1485	Fe	0.8	ol + liq		e
20.0	15	1470	Fe	1.0	ol + pyx + liq	ol	b
20.0	34	1455	Fe	2.0	ol + sp + pyx + liq		b
20.0	1	1440	Fe	2.0	ol + sp + pyx + liq	pyx, sp	b
20.0	18	1400	Fe	2.5	ol + sp + pyx + gar + liq	gar	d
22.5	33	1535	Mo	0.2	liq		b
22.5	30	1490	Fe	0.8	ol + pyx + liq		e
22.5	9	1465	Fe	1.5	ol + sp + pyx + gar + liq	pyx	d
22.5	37	1410	Fe	1.7	ol + pyx + gar + liq		e
25.0	35	1560	Mo	0.1	liq		d
25.0	36	1515	Mo	0.2	ol + pyx + liq		e
25.0	28	1475	Fe	2.0	ol + pyx + gar + liq	pyx, gar	b

*Starting assemblages (S.A.): a, liq; b, ol + liq; c, ol + plag + liq; d, ol + pyx + liq; e, ol + gar + liq.

In addition, the KREEP composition has been experimentally demonstrated as having liquidus phase relations consistent with a true igneous origin (Walker *et al.*, 1973a, b; Lipin, 1976).

Mixing models should be refined to account for these highlands noritic rocks by using ancient highlands components. Wänke *et al.* (1975) and Taylor and Bence (1975) seem to have embarked on fruitful calculations in this quest for the ancient highlands components.

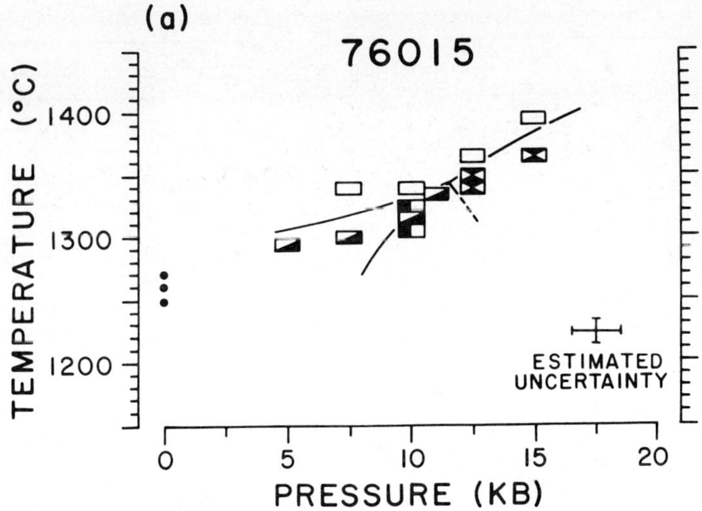

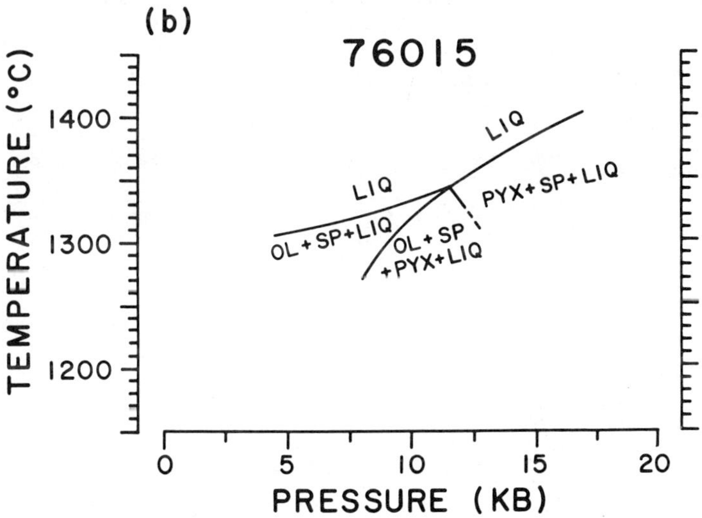

Fig. 6. (a) The melting relations of 76015 at iron saturation are shown as a function of pressure and temperature. Experimental data with identical symbols have identical phase assemblages. The estimated uncertainties in pressure and temperature (for all charges run at $P \geq 5$ kbar) are ±1 kbar and ±10°C. (b) Identification of phase fields illustrated in (a).

Table 8. Selected electron microprobe analyses of minerals in 76015 experiments.

	Olivine					Pyroxene		
	(1)	(2)	(3)	(4)	(5)	(6)	(7)	(8)
SiO_2	39.2	39.6	39.8	39.0 ± 0.2	39.4 ± 0.2	51.5	53.4 ± 0.2	50.9
TiO_2	0.00	0.00	0.06	0.07 ± 0.03	0.09 ± 0.02	0.60	0.44 ± 0.05	0.48
Al_2O_3	0.12	0.00	0.00	0.00	0.00	7.56	5.0 ± 0.3	8.7
FeO	11.4	12.6	12.6	12.5 ± 0.1	12.9 ± 0.2	8.10	7.8 ± 0.2	7.8
MnO	n.a.	n.a.	0.07	0.13 ± 0.01	0.11 ± 0.01	0.12	0.11 ± 0.02	0.09
MgO	48.7	48.3	46.9	47.1 ± 0.5	46.6 ± 0.2	28.8	30.6 ± 0.5	28.9
CaO	0.40	0.38	0.37	0.44 ± 0.06	0.43 ± 0.08	2.78	2.34 ± 0.04	3.0
Na_2O	n.a.	n.a.	0.00	0.00	0.00	0.00	0.00	0.00
K_2O	n.a.	n.a.	0.00	0.00	0.00	0.00	0.00	0.00
Cr_2O_3	n.a.	n.a.	0.14	0.14 ± 0.05	0.13 ± 0.02	0.35	0.31 ± 0.05	0.30
Total:	99.8	100.9	99.9	99.4	99.7	99.8	100.0	100.2
Fe/(Fe + Mg)	0.12	0.13	0.13	0.13	0.13	0.14	0.12	0.13
Na	0.000	0.000	0.000	0.000	0.000	0.000	0.000	0.000
Mg	1.802	1.776	1.740	1.761	1.738	1.511	1.597	1.510
Al	0.004	0.000	0.000	0.000	0.000	0.314	0.206	0.359
Si	0.973	0.977	0.990	0.978	0.986	1.813	1.869	1.784
K	0.000	0.000	0.000	0.000	0.000	0.000	0.000	0.000
Ca	0.011	0.010	0.010	0.012	0.012	0.105	0.088	0.113
Ti	0.000	0.000	0.001	0.001	0.002	0.016	0.012	0.013
Cr	0.000	0.000	0.003	0.003	0.003	0.010	0.009	0.008
Mn	0.000	0.000	0.001	0.003	0.002	0.004	0.003	0.003
Fe^{2+}	0.237	0.260	0.262	0.262	0.270	0.238	0.228	0.229
Σ	3.025	3.023	3.007	3.019	3.011	4.010	4.012	4.019

(1) Run #1 @ $P = 0.0$ kbar, $T = 1269$°C.
(2) Run #2 @ $P = 0.0$ kbar, $T = 1260$°C.
(3) Run #3 @ $P = 0.0$ kbar, $T = 1248$°C.
(4) Run #11 @ $P = 7.5$ kbar, $T = 1300$°C.
(5) Run #10 @ $P = 10.0$ kbar, $T = 1325$°C.
(6) Run #10 @ $P = 10.0$ kbar, $T = 1325$°C.
(7) Run #12 @ $P = 12.5$ kbar, $T = 1350$°C.
(8) Run #4 @ $P = 15.0$ kbar, $T = 1365$°C.
n.a. = not analyzed.

Table 9. Selected electron microprobe analyses of liquids in 76015 experiments.

	(1)	(2)	(3)	(4)	(5)	(6)	(7)	(8)	(9)
SiO_2	46.3 ± 0.3	46.1 ± 0.3	47.1 ± 0.1	46.2 ± 0.2	46.2 ± 0.3	46.2 ± 0.1	46.3 ± 0.2	46.1 ± 0.1	45.7 ± 0.3
TiO_2	1.59 ± 0.09	1.59 ± 0.05	1.60 ± 0.06	1.54 ± 0.10	1.61 ± 0.05	1.52 ± 0.04	1.63 ± 0.04	1.58 ± 0.04	1.59 ± 0.02
Al_2O_3	18.1 ± 0.1	18.4 ± 0.1	18.3 ± 0.2	18.0 ± 0.1	18.4 ± 0.2	17.9 ± 0.1	18.5 ± 0.1	18.0 ± 0.1	17.8 ± 0.2
FeO	9.7 ± 0.1	9.5 ± 0.1	9.7 ± 0.2	9.4 ± 0.1	9.54 ± 0.09	9.5 ± 0.1	9.4 ± 0.2	9.5 ± 0.3	9.6 ± 0.2
MnO	0.13 ± 0.04	0.12 ± 0.03	0.10 ± 0.01	0.17 ± 0.02	0.14 ± 0.02	0.14 ± 0.03	0.16 ± 0.02	0.14 ± 0.03	0.12 ± 0.02
MgO	11.9 ± 0.1	11.4 ± 0.1	11.0 ± 0.1	12.4 ± 0.1	11.1 ± 0.2	12.4 ± 0.1	10.8 ± 0.2	11.4 ± 0.3	11.8 ± 0.2
CaO	10.8 ± 0.2	11.1 ± 0.2	11.0 ± 0.4	10.7 ± 0.1	11.0 ± 0.1	10.7 ± 0.1	11.5 ± 0.2	11.1 ± 0.2	11.3 ± 0.4
Na_2O	0.78 ± 0.06	0.93 ± 0.05	0.66 ± 0.02	0.95 ± 0.02	1.17 ± 0.07	0.91 ± 0.02	1.00 ± 0.03	1.20 ± 0.05	1.00 ± 0.05
K_2O	0.38 ± 0.04	0.50 ± 0.03	0.41 ± 0.02	0.50 ± 0.03	0.52 ± 0.05	0.50 ± 0.04	0.56 ± 0.03	0.50 ± 0.02	0.51 ± 0.01
Cr_2O_3	0.16 ± 0.03	0.18 ± 0.03	0.16 ± 0.02	0.19 ± 0.02	0.18 ± 0.04	0.15 ± 0.03	0.14 ± 0.04	0.14 ± 0.04	0.15 ± 0.03
Total:	99.8	99.8	100.0	100.1	99.9	99.9	100.0	99.7	99.6
AB	6.60	7.87	5.58	8.12	9.90	7.70	8.46	10.15	8.46
AN	44.76	44.56	45.76	43.33	43.42	43.28	44.34	42.25	42.57
OR	2.25	2.95	2.42	2.95	3.07	2.95	3.31	2.95	3.01
OL	14.85	17.19	8.76	19.34	18.85	18.81	16.89	20.14	20.66
FA	5.38	6.30	3.33	6.71	7.04	6.58	6.36	7.41	7.47
FO	9.47	10.90	5.44	12.64	11.81	12.23	10.53	12.74	13.19
PYR	23.12	23.96	34.22	23.11	21.30	24.07	23.70	20.95	21.62
DI	7.14	8.52	7.16	7.89	9.05	7.93	10.32	10.39	10.93
HY	20.98	15.44	27.07	15.22	12.25	16.14	13.37	10.56	10.70
ILM	3.02	3.02	3.04	2.92	3.06	2.89	3.10	3.00	3.02
CR	0.24	0.27	0.24	0.28	0.27	0.22	0.21	0.21	0.22

(1) Run #1 @ $P = 0.0$ kbar, $T = 1269°C$.
(2) Run #2 @ $P = 0.0$ kbar, $T = 1260°C$.
(3) Run #3 @ $P = 0.0$ kbar, $T = 1248°C$.
(4) Run #7 @ $P = 7.5$ kbar, $T = 1340°C$.
(5) Run #11 @ $P = 7.5$ kbar, $T = 1300°C$.
(6) Run #6 @ $P = 10.0$ kbar, $T = 1340°C$.
(7) Run #10 @ $P = 10.0$ kbar, $T = 1325°C$.
(8) Run #12 @ $P = 12.5$ kbar, $T = 1350°C$.
(9) Run #4 @ $P = 15.0$ kbar, $T = 1365°C$.

Table 10. 76015 experiments.

P (kbar)	Run #	T (°C)	Capsule	Hours	Products	Probe analyses	S.A.*
0.0	1	1269	Fe	5.0	ol + liq	ol, liq	b
0.0	2	1260	Fe	11.0	ol + liq	ol, liq	b
0.0	3	1248	Fe	21.5	ol + liq	ol, liq	b
5.0	16	1295	Fe	1.5	ol + sp + liq		a
7.5	7	1340	Fe	1.5	liq	liq	b
7.5	11	1300	Fe	1.7	ol + sp + liq	ol, liq	d
10.0	6	1340	Fe	1.8	liq	liq	b
10.0	10	1325	Fe	1.5	ol + sp + pyx + liq	ol, pyx, liq	a
10.0	9	1315	Fe	1.7	ol + sp + liq		b
10.0	13	1305	Fe	1.5	ol + sp + pyx + liq		b
11.0	14	1335	Fe	1.3	ol + sp + liq		d
12.5	5	1365	Fe	0.7	liq		b
12.5	12	1350	Fe	1.0	sp + pyx + liq	pyx, liq	d
12.5	8	1340	Fe	1.5	sp + pyx + liq		b
15.0	15	1395	Fe	1.2	liq		d
15.0	4	1365	Fe	0.8	sp + pyx + liq	pyx, liq	a

*Starting assemblages (S.A.): a, liq; b, ol + liq; c, ol + sp + liq; d, pyx + sp + liq.

CONCLUSIONS

The 63545, 76055, and 76015 compositions do not represent magmas derived by partial melting of either cosmic or differentiated source regions at any pressure in the moon. While other hypotheses cannot be ruled out, presently available information would seem to favor an origin by impact-fusion of several highlands components.

Acknowledgments—Special thanks are due to Professor Donald H. Lindsley (State University of New York at Stony Brook) for his expert guidance throughout this study. This research was funded by NASA grant NGL-33-015-130.

REFERENCES

Albee A. L., Gancarz A. J., and Chodos A. A. (1973) Metamorphism of Apollo 16 and 17 and Luna 20 metaclastic rocks at about 3.95 AE: Samples 61156, 64423, 14-2, 65015, 67483, 15-2, 76055, 22006, and 22007. *Proc. Lunar Sci. Conf. 4th*, p. 569–595.

Bence A. E., Delano J. W., Papike J. J., and Cameron K. L. (1974) Petrology of the highlands massifs at Taurus-Littrow: An analysis of the 2–4 mm soil fraction. *Proc. Lunar Sci. Conf. 5th*, p. 785–827.

Boyd F. R. and England J. L. (1963) Effect of pressure on the melting of diopside $CaMgSi_2O_6$ and albite $NaAlSi_3O_8$ in the range up to 50 kilobars. *J. Geophys. Res.* **68**, 311–323.

Chao E. C. T. (1973) The petrology of 76055, 10, a thermally metamorphosed fragment-laden olivine micronorite hornfels. *Proc. Lunar Sci. Conf. 4th*, p. 719–732.

Chao E. C. T., Minkin J. A., and Thompson C. L. (1974) Preliminary petrographic description and geologic implications of the Apollo 17 station 7 boulder consortium samples. *Earth Planet. Sci. Lett.* **23**, 413–428.

Chao E. C. T., Minkin J. A., Thompson C. L., and Huebner J. S. (1975) The petrogenesis of 77115 and its xenocrysts: Description and preliminary interpretation. *Proc. Lunar Sci. Conf. 6th,* p. 493–515.

Delano J. W. (1977) Petrogenesis of highlands noritic rocks: Calculation of some trace element constraints. Manuscript in preparation.

Dowty E., Prinz M., and Keil K. (1974) "Very High Alumina Basalt": A mixture and not a magma type. *Science* **183**, 1214–1215.

Green D. H. and Ringwood A. E. (1972) Significance of Apollo 15 mare basalts and "primitive" green glasses in lunar petrogenesis. In *The Apollo 15 Lunar Samples* (J. W. Chamberlain and C. Watkins, eds.), p. 82–84. The Lunar Science Institute, Houston.

Green D. H., Ringwood A. E., Hibberson W. O., and Ware N. G. (1975) Experimental petrology of Apollo 17 mare basalts. *Proc. Lunar Sci. Conf. 6th,* p. 871–893.

Grove T. L., Walker D., Longhi J., Stolper E., and Hays J. F. (1973) Petrology of rock 12002 and origin of picritic basalts at Oceanus Procellarum. *Proc. Lunar Sci. Conf. 4th,* p. 995–1011.

Hubbard N. J. and Gast P. W. (1971) Chemical composition and origin of nonmare lunar basalts. *Proc. Lunar Sci. Conf. 2nd,* p. 999–1020.

Hubbard N. J., Gast P. W., Rhodes J. M., Bansal B. M., Wiesmann H., and Church S. E. (1972) Nonmare basalts: Part II. *Proc. Lunar Sci. Conf. 3rd,* p. 1161–1179.

Hubbard N. J. and Minear J. W. (1975) A physical and chemical model of early lunar history. *Proc. Lunar Sci. Conf. 6th,* p. 1057–1085.

Hubbard N. J., Rhodes J. M., and Gast P. W. (1973a) Chemistry of lunar basalts with very high alumina contents. *Science* **181**, 339–342.

Hubbard N. J., Rhodes J. M., Gast P. W., Bansal B. M., Shih C.-Y., Wiesmann H., and Nyquist L. E. (1973b) Lunar rock types: The role of plagioclase in non-mare and highland rock types. *Proc. Lunar Sci. Conf. 4th,* p. 1297–1312.

Hubbard N. J., Rhodes J. M., Wiesmann H., Shih C.-Y., and Bansal B. M. (1974) The chemical definition and interpretation of rock types returned from the non-mare regions of the moon. *Proc. Lunar Sci. Conf. 5th,* p. 1227–1246.

Irving A. J. (1975) Chemical, mineralogical and textural systematics of non-mare melt rocks: Implications for lunar impact and volcanic processes. *Proc. Lunar Sci. Conf. 6th,* p. 363–394.

James O. B. (1976) Petrology of the aphanitic lithologies in consortium breccia 73215 (abstract). In *Lunar Science VII,* p. 420–422. The Lunar Science Institute, Houston.

James O. B., Brecher A., Blanchard D. P., Jacobs J. W., Brannon J. C., Korotev R. L., Haskin L. A., Higuchi H., Morgan J. W., Anders E., Silver L. T., Marti K., Braddy D., Hutcheon I. D., Kirsten T., Kerridge J. F., Kaplan I. R., Pillinger C. T., and Gardiner L. R. (1975) Consortium studies of matrix of light gray breccia 73215. *Proc. Lunar Sci. Conf. 6th,* p. 547–577.

Jessberger E. K., Huneke J. C., and Wasserburg G. J. (1974) Evidence for a 4.5 aeon age of plagioclase clasts in a lunar highlands breccia. *Nature* **248**, 199–202.

Keil K., Warner R. D., Prinz M., and Dowty E. (1975) Rocks 60618 and 65785: Evidence for admixture of KREEP in lunar impact melts. *Geophys. Res. Lett.* **2**, 369–372.

Kesson S. E. (1975) Mare basalts: Melting experiments and petrogenetic interpretations. *Proc. Lunar Sci. Conf. 6th,* p. 921–944.

Kesson S. E. and Ringwood A. E. (1976) Mare basalt petrogenesis in a dynamic moon. *Earth Planet. Sci. Lett.* **30**, 155–163.

Kirsten T. and Horn P. (1974) Chronology of the Taurus-Littrow region III. Ages of mare basalts and highland breccias and some remarks about the interpretation of lunar highland rock ages. *Proc. Lunar Sci. Conf. 5th,* p. 1451–1475.

Latham G., Ewing M., Dorman J., Nakamura Y., Press F., Toksöz N., Sutton G., Duennebier F., and Lammlein D. (1973) Lunar structure and dynamics—Results from the Apollo passive seismic experiment. *The Moon* **7**, 396–420.

Lipin B. R. (1976) The origin of Fra Mauro basalts (abstract). In *Lunar Science VII,* p. 495–497. The Lunar Science Institute, Houston.

LSPET (1973) Apollo 17 lunar samples: Chemical and petrographic description. *Science* **182**, 659–672.

McCallum I. S. and Mathez E. A. (1975) Petrology of noritic cumulates and a partial melting model for the genesis of Fra Mauro basalts. *Proc. Lunar Sci. Conf. 6th,* p. 395–414.

Meyer C., Jr., Brett R., Hubbard N. J., Morrison D. A., McKay D. S., Aitken F. K., Takeda H., and Schonfeld E. (1971) Mineralogy, chemistry, and origin of the KREEP component in soil samples from the Ocean of Storms. *Proc. Lunar Sci. Conf. 2nd,* p. 393–411.

Meyer C., Jr., Anderson D. H., and Bradley J. G. (1974) Ion microprobe mass analysis of plagioclase from "non-mare" lunar samples. *Proc. Lunar Sci. Conf. 5th,* p. 685–706.

Morgan J. W., Krähenbühl U., Ganapathy R., Anders E., and Marvin U. B. (1973) Trace element abundances and petrology of separates from Apollo 15 soils. *Proc. Lunar Sci. Conf. 4th,* p. 1379–1398.

Nava D. F. (1974) Chemical compositions of some soils and rock types from the Apollo 15, 16, and 17 lunar sites. *Proc. Lunar Sci. Conf. 5th,* p. 1087–1096.

Nyquist L. E., Hubbard N. J., Gast P. W., Bansal B. M., Wiesmann H., and Jahn B. (1973) Rb-Sr systematics for chemically defined Apollo 15 and 16 materials. *Proc. Lunar Sci. Conf. 4th,* p. 1823–1846.

Nyquist L. E., Bansal B. M., Wiesmann H., and Jahn B.-M. (1974) Taurus-Littrow chronology: Some constraints on early lunar crustal development. *Proc. Lunar Sci. Conf. 5th,* p. 1515–1539.

Papanastassiou D. A. and Wasserburg G. J. (1971) Rb-Sr ages of igneous rocks from the Apollo 14 mission and the age of the Fra Mauro formation. *Earth Planet. Sci. Lett.* **12**, 36–48.

Papanastassiou D. A. and Wasserburg G. J. (1972) Rb-Sr systematics of Luna 20 and Apollo 16 samples. *Earth Planet. Sci. Lett.* **17**, 52–63.

Papanastassiou D. A. and Wasserburg G. J. (1976) Early lunar differentiates and lunar initial $^{87}Sr/^{86}Sr$ (abstract). In *Lunar Science VII,* p. 665–667. The Lunar Science Institute, Houston.

Phinney W. C., Anders E., Bogard D., Butler P., Gibson E., Gose W., Heiken G., Hohenberg C., Nyquist L., Pearce W., Rhodes M., Silver L., Simonds C., Strangeway D., Turner G., Walker R., Warner J., and Yukas D. (1974) Progress report: Apollo 17, station 6 boulder consortium. In *Lunar Science V, Suppl. A,* p. 7–13. The Lunar Science Institute, Houston.

Prinz M., Dowty E., Keil K., and Bunch T. E. (1973) Mineralogy, petrology and chemistry of lithic fragments from Luna 20 fines: Origin of the cumulate ANT suite and its relationship to high-alumina and mare basalts. *Geochim. Cosmochim. Acta* **37**, 979–1006.

Rhodes J. M., Rodgers K. V., Shih C., Bansal B. M., Nyquist L. E., Wiesmann H., and Hubbard N. J. (1974) The relationships between geology and soil chemistry at the Apollo 17 landing site. *Proc. Lunar Sci. Conf. 5th,* p. 1097–1117.

Schaeffer O. A. and Husain L. (1973) Early lunar history: Ages of 2 to 4 mm soil fragments from the lunar highlands. *Proc. Lunar Sci. Conf. 4th,* p. 1847–1863.

Schonfeld E. (1974) The contamination of lunar highlands rocks by KREEP: Interpretation by mixing models. *Proc. Lunar Sci. Conf. 5th,* p. 1269–1286.

Shih C.-Y. (1976) On the origin of KREEP basalt (abstract). In *Lunar Science VII,* p. 800–802. The Lunar Science Institute, Houston.

Simonds C. H. (1975) Thermal regimes in impact melts and the petrology of the Apollo 17 station 6 boulder. *Proc. Lunar Sci. Conf. 6th,* p. 641–672.

Simonds C. H., Warner J. L., and Phinney W. C. (1976) Clast-melt interactions in lunar and terrestrial impact melts (abstract). In *Lunar Science VII,* p. 812–814. The Lunar Science Institute, Houston.

Solomon S. C. and Chaiken J. (1976) Thermal expansion and thermal stress in the moon and terrestrial planets: Clues to early thermal history (abstract). In *Lunar Science VII,* p. 825–827. The Lunar Science Institute, Houston.

Taylor S. R. and Bence A. E. (1975) Evolution of the lunar highland crust. *Proc. Lunar Sci. Conf. 6th,* p. 1121–1141.

Taylor S. R., Gorton M. P., Muir P., Nance W., Rudowski R., and Ware N. (1973) Lunar highlands composition: Apennine Front. *Proc. Lunar Sci. Conf. 4th,* p. 1445–1459.

Taylor S. R. and Jakeš P. (1974) The geochemical evolution of the moon. *Proc. Lunar Sci. Conf. 5th,* p. 1287–1305.

Tera F., Papanastassiou D. A., and Wasserburg G. J. (1973) A lunar cataclysm at ~3.95 AE and the structure of the lunar crust (abstract). In *Lunar Science IV,* p. 723–725. The Lunar Science Institute, Houston.

Tera F., Papanastassiou D. A., and Wasserburg G. J. (1974) Isotopic evidence for a terminal lunar cataclysm. *Earth Planet. Sci. Lett.* **22**, 1–21.

Tera F. and Wasserburg G. J. (1974) U-Th-Pb systematics on lunar rocks and inferences about lunar evolution and the age of the moon. *Proc. Lunar Sci. Conf. 5th,* p. 1571–1599.

Wakita H., Laul J. C., and Schmitt R. A. (1975) Some thoughts on the origin of lunar ANT-KREEP and mare basalts. *Geochem. J.* **9**, 25–41.

Walker D., Longhi J., and Hays J. F. (1972) Experimental petrology and origin of Fra Mauro rocks and soil. *Proc. Lunar Sci. Conf. 3rd,* p. 797–817.

Walker D., Grove T. L., Longhi J., Stolper E. M., and Hays J. F. (1973a) The origin of lunar feldspathic rocks. *Earth Planet. Sci. Lett.* **20**, 325–336.

Walker D., Longhi J., Grove T. L., Stolper E., and Hays J. F. (1973b) Experimental petrology and origin of rocks from the Descartes Highlands. *Proc. Lunar Sci. Conf. 4th,* p. 1013–1032.

Wänke H., Palme H., Baddenhausen H., Dreibus G., Jagoutz E., Kruse H., Palme C., Spettel B., Teschke F., and Thacker R. (1975) New data on the chemistry of lunar samples: Primary matter in the lunar highlands and the bulk composition of the moon. *Proc. Lunar Sci. Conf. 6th,* p. 1313–1340.

Warner J. L., Simonds C. H., Phinney W. C., and Gooley R. (1973) Petrology and genesis of two "igneous" rocks from Apollo 17 (76055 and 77135) (abstract). *EOS* (*Trans. Amer. Geophys. Union*) **54**, 620–621.

Weill D. F. and McKay G. A. (1975) The partitioning of Mg, Fe, Sr, Ce, Sm, Eu, and Yb in lunar igneous systems and a possible origin of KREEP by equilibrium partial melting. *Proc. Lunar Sci. Conf. 6th,* p. 1143–1158.

Proc. Lunar Sci. Conf. 8th (1977), p. 2125–2135.
Printed in the United States of America

Phase relations in the system Fe–Cr–Ti–O and their application in lunar thermometry

BARBARA KNECHT, BRUNO SIMONS and EDUARD WOERMANN

Institut für Kristallographie, Rheinisch-Westfälische Technische Hochschule Aachen, Aachen, Germany

AHMED EL GORESY

Max Planck Institut für Kernphysik, Heidelberg, Germany

Abstract—Phase relationships in the system $FeO–Cr_2O_3–TiO_2$ in equilibrium with metallic iron have been investigated in the compositional range of interest to lunar petrology.

With decreasing temperature the spinel phase coexisting with ilmenite is depleted in its ulvite component and the ilmenite in its eskolaite component. This relationship is quantitatively evaluated for sp + ilm + fps three phase assemblages and qualitatively for sp + ilm two phase assemblages. It explains the phenomenon of "reverse zoning" in lunar spinels.

The temperature dependence of the following phase assemblages has been determined:

(1) Spinel + ilmenite + rutile assemblages are restricted to temperatures below 1038°C.
(2) Spinel + ilmenite + pseudobrookite assemblages are stable above 1038°C.
(3) Ilmenite + pseudobrookite + rutile assemblages are restricted to a temperature range between 1038 and 1068°C.
(4) Spinel and rutile coexist at temperatures below *ca.* 1180°C.
(5) Two pseudobrookite phases coexist between *ca.* 1180 and 1230°C.

INTRODUCTION

LUNAR SPINELS GENERALLY EXHIBIT zonal structures. Haggerty (1972), El Goresy *et al.* (1976) and others reported compositional changes in the spinel phase precipitated during the solidification of lunar basaltic magma, ranging from early crystallizing titanian chromites to late chromian ulvites. The simultaneous variations of other element abundances, e.g., Fe/(Fe + Mg) (FFM), do not seem to exhibit the same regularity of trend. In some spinel grains from Apollo 17 basalts, however, El Goresy and Ramdohr (1975, 1977) observed a reversal of the normal Ti/(Ti + Cr) trend. Since the reported reverse zoning is restricted to spinel-ilmenite contacts it appears to be due to subsolidus exchange reactions between these two phases. An enrichment of the $FeCr_2O_4$ component in the spinel phase at lower temperatures with a simultaneous decrease in Cr_2O_3 in coexisting ilmenite has also been reported by Usselman (1975) in a series of experiments.

In this paper an attempt is made to find a theoretical explanation for the reversal in spinel zoning and to present a quantitative experimental evaluation of its temperature dependence by the delineation of phase relationships in the system $FeO–Cr_2O_3–TiO_2$ in equilibrium with metallic iron in a series of subsolidus isotherms.

EXPERIMENTS

Experimental investigations in the system Fe–Cr–Ti–O were restricted to compositions which are of interest to lunar petrology.

Since metallic iron is a common accessory mineral in lunar basalts all experiments were conducted in equilibrium with a metal phase. Thus the compositions of all oxide phases involved are located on the ternary boundary surface of the metal stability volume in the quaternary system. In this surface some iron may be in the trivalent state, some titanium will occur in the reduced, trivalent state (MacChesney and Muan, 1961; Simons, 1974), and some chromium in the divalent state (Ulmer and White, 1966; Ulmer 1970). Thus the section investigated does not coincide with the system $FeO–Cr_2O_3–TiO_2$. This has to be kept in mind, although the components will be quoted as FeO, Cr_2O_3, and TiO_2 and diagrams will be constructed accordingly.

Furthermore, since neither wüstite nor eskolaite or the various Andersson-phase solid solutions (Flörke and Lee, 1970; Anderson *et al.*, 1959) were encountered in lunar basalts it is justified to limit the investigations to the section $FeCr_2O_4–Fe_2TiO_4–TiO_2$.

Starting mixtures were prepared by weighing and mixing appropriate amounts of TiO_2 (Merck No. 808), Cr_2O_3 (Riedel de Haen No. 12233) Fe_2O_3 (Merck No. 3924), and Fe (Merck No. 3819).

In order to ensure a true equilibrium condition with metallic iron an excess of 10% Fe was added to every batch. One part of the samples was pre-reacted in iron crucibles at 1200°C, another part was adopted for final equilibration as raw oxide mixture. Thus for a set of experiments below 1200°C the equilibrium state was approached from two directions. No differences were observed in the final results of both sets of runs.

For final equilibration approximately 0.4 g of each mixture was placed into an iron crucible with a tight fitting lid and sealed into an evacuated silica glass capsule. The samples were heated in conventional, vertical tube furnaces. Reaction times varied between 10 days at 1000°C to 24 hours at 1250°C. These limits were not to be exceeded in order to avoid recrystallization and consequent imbrittlement of the silica glass. The samples were quenched by rapid immersion into cold water. The recovery of unoxidized iron crucibles and of dry samples was accepted as an indication that the silica glass capsules were intact and gas tight throughout the whole heating period.

The reaction products were investigated microscopically using reflected light and oil immersion, and by X-ray diffractometry for identification of the oxide phases and to ascertain that metallic iron was a coexisting phase. The compositions of the oxide phases were determined by electron microprobe analysis. In order to ensure equilibrium conditions only oxide grains that were in mutual contact with coexisting phases were analyzed. Repetitions of the analyses on other oxide grains in the same sample and in other samples synthesized under similar conditions proved that reproducibility usually was much better than $\pm 1\%$ for each component analyzed. The analytical data are thus better than indicated by the triangular symbols in Fig. 1(a–f). The electron microprobe analyses provided data on the concentration of Fe, Cr, and Ti in the oxide phase. These data were recalculated as FeO, Cr_2O_3, and TiO_2, respectively. Since the majority of the results totalled more than 100% a stoichiometric, synthetic ferropseudobrookite was used as continuous monitor after each series of analyses. The analyses of the monitor summed close to 100%. Hence we believe that the excess in the totals is due to the presence of Cr^{2+} and Ti^{3+} in our charges.

Quantitative data on the Cr^{2+} concentration in the system Fe–Cr–O and on the Ti^{3+} concentration in the system Fe–Ti–O are given by Ulmer and White (1966) and Ulmer (1970) and by MacChesney and Muan (1961) and Simons (1974), respectively. Due to the fact that neither the electron microprobe nor wet chemical analyses can distinguish between ion assemblages $Ti^{4+} + Cr^{2+}$ and $Ti^{3+} + Cr^{3+}$ it is impossible to measure or calculate the Cr^{2+} and the Ti^{3+} concentrations of crystalline phases in the quaternary system, although their composition points in the phase diagram can be determined with high precision quoting the concentrations of the components Fe, Cr, Ti, and O.*

*Analytical data for 180 different phases have been tabulated. Copies can be obtained from the authors upon request.

RESULTS

The results are presented graphically for various isothermal sections in Fig. 1(a–f). From an inspection of these diagrams several unique properties of the system $FeO–Cr_2O_3–TiO_2$ will be evident.

It is conspicuous that in all sections the composition points of the phases determined by microprobe analysis do not coincide with the stoichiometric points of pure phases nor with lines of solid solutions in the $FeO–Cr_2O_3–TiO_2$ diagram. This deviation is due partly to defect concentrations in the phases concerned, mainly, however, to the formation of trivalent titanium and divalent chromium, leading to complex solid solutions, e.g., to spinels with the components $Fe_2TiO_4–FeTi_2O_4$ $CrCr_2O_4–FeCr_2O_4$, to ilmenites with the components $FeTiO_3–Cr_2O_3–Ti_2O_3$ and to pseudobrookite type phases with the components $FeTi_2O_5–TiCr_2O_5–Ti_3O_5$.

Projections from the true composition points of these phases in the Fe–Cr–Ti–O tetrahedron toward the oxygen apex onto the $FeO–Cr_2O_3–TiO_2$ plane will result in points displaced towards Ti- and Cr-rich compositions as compared with simple solid solutions in the projection plane.

Further striking features are the changing phase assemblages in the isothermal sections and the simultaneous temperature dependent shift of the spinel + ilmenite + ferropseudobrookite three phase triangle.

THE SPINEL COMPOSITION AS A LUNAR THERMOMETER

The stability relations of oxide phases in equilibrium with metallic iron in the system Fe–Ti–O as a function of temperature and oxygen fugacity attracted considerable interest in recent years. Experimental data have been presented by Taylor *et al.* (1972), Sato *et al.* (1973), Lindsley *et al.* (1974), Saha and Biggar (1974), Merritt and Turnbull (1974), Grey *et al.* (1974), and Simons and Woermann (1977). Figure 2 shows the stability relations of the iron-titanium oxide phases as determined by Simons and Woermann (1977).

From Fig. 2 it appears that the stability field of the ferropseudobrookite phase is narrowing toward lower temperatures. It finally pinches out at 1068°C (point L in Fig. 3). Likewise, the ilmenite field is narrowing toward higher temperatures. Lindsley *et al.* (1974) pointed out that a further projection of the monovariant lines spinel + ilmenite + iron and ilmenite + ferropseudobrookite + iron toward higher temperatures will ultimately lead to an intersection of these lines in an invariant point M (Fig. 3), from which another monovariant line spinel + ferrospeudobrookite + iron will originate. M thus constitutes the theoretical upper stability limit of ilmenite in the system Fe–Ti–O. However, from MacChesney and Muan (1961) as well as from various other investigations it is known that ilmenite melts congruently at 1400°C. M is thus actually located above the liquidus temperature in this system.

By addition of another component—e.g., Cr_2O_3—to the system Fe–Ti–O the invariant point M will be converted to a monovariant line $M–N$ (see Fig. 3). Since the phase assemblage spinel + ilmenite + ferropseudobrookite is encoun-

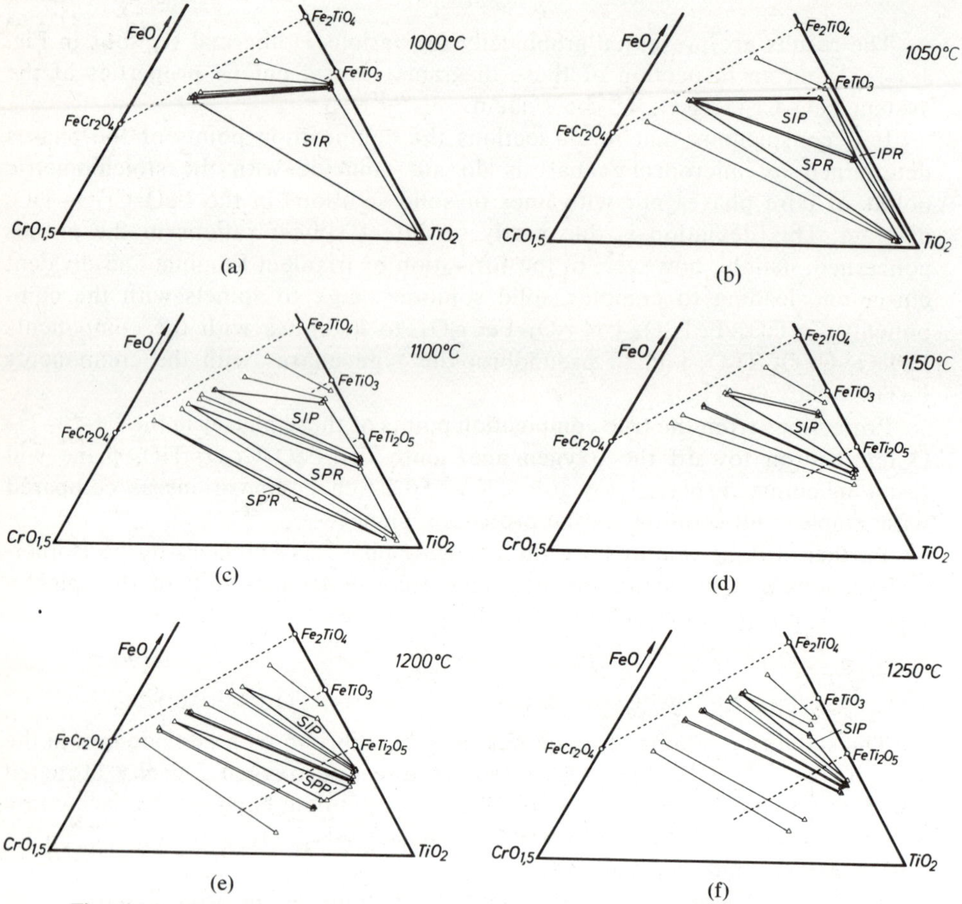

Fig. 1(a–f). Isothermal sections through the system Fe–Cr–Ti–O. Triangles designate electron microprobe data points. The oxide phases are located on the surface of the metal stability volume. They are projected along lines through the oxygen apex onto the FeO–Cr_2O_3–TiO_2 plane.

tered in isothermal sections of the system FeO–Cr_2O_3–TiO_2 at temperatures considerably below the hypothetical point M, e.g., at 1100°C in Fig. 1, it is evident that the line M–N must slope toward lower temperatures.

The addition of Cr_2O_3 results in more or less extensive ranges of spinel, ilmenite, or pseudobrookite solid solutions. Ilmenite is capable of incorporating only small amounts of eskolaite. Much greater solubilities exist among end members in the spinel series. According to Muan *et al.* (1972) the ulvite-chromite series is complete at least at temperature above 1000°C and probably even to much lower temperatures. Since the stability of chromite is considerably higher than that of ulvite, increasing chromium contents of the spinel phase will result in increasing stabilities of the solid solution, i.e., decreasing oxygen fugacities along the line M–N.

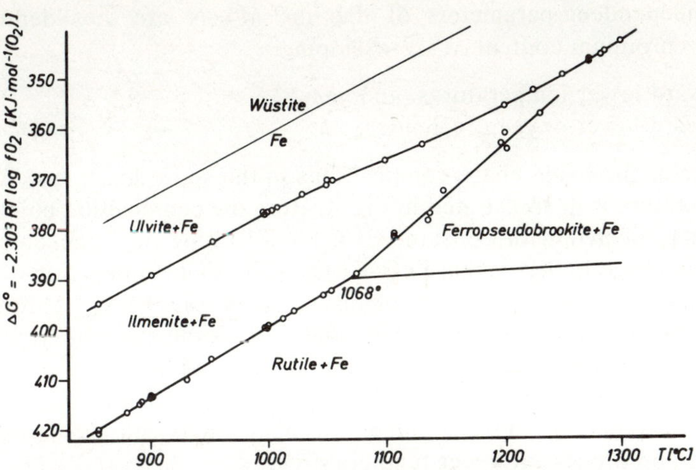

Fig. 2. Stability relations in the system Fe–Ti–O in equilibrium with metallic iron as determined by e.m.f. measurements (Simons and Woermann, 1977).

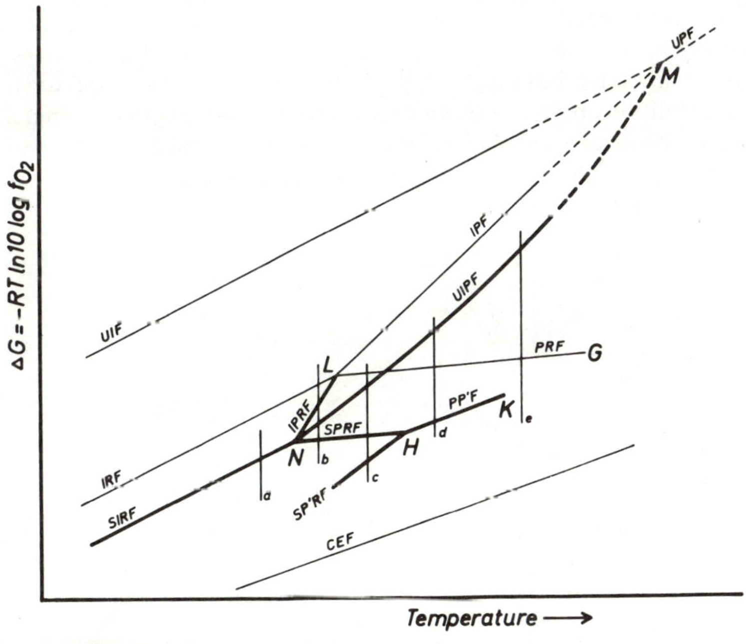

Fig. 3. Schematic $\Delta G/T$ diagram showing the stability relations in the systems Fe–Ti–O (thin lines) and Fe–Cr–Ti–O (thick lines) in equilibrium with metal—C = chromite, E = eskolaite, F = metallic iron, I = ilmenite, P, P′ = ferropseudobrookite, R = rutile, S = spinel, U = ulvite.

Two independent parameters of the line $M-N$ are thus deduced. With increasing chromium content M–N is sloping:

(a) toward lower temperatures, and
(b) toward lower oxygen fugacities.

The change of the oxide phase compositions in the $sp_{ss} + ilm_{ss} + fps_{ss}$ assemblage with temperature is demonstrated in Fig. 4. Here the composition points of Fig. 1 have been projected with constant Cr/(Cr + Ti) ratios onto the corresponding solid solution lines in the system FeO–Cr_2O_3–TiO_2. The composition of the spinel phase is sensitive to temperature changes, being enriched in chromium with decreasing temperature. At the same time, coexisting ilmenite is depleted in chromium, while the composition of the ferropseudobrookite phase remains fairly constant. The compositional shifts correspond to the "reverse zoning" sequence in spinel in TiO_2-rich lunar basalts, which can be explained as a re-equilibration process at lower temperatures.

The temperature dependence of the spinel composition in the Fe–Cr–Ti–O system may create a convenient lunar thermometer.

Quantitative relations between the Cr(Cr + Ti) ratio of the spinel phase, the temperature and the coexisting phases are given in Fig. 5. In accordance with phase rule requirements, exact temperatures can only be determined from spinels in four phase assemblages under isobaric conditions, e.g., from a $sp_{ss} + ilm_{ss} + fps_{ss}$ + metal assemblage.

In most practical cases it may safely be assumed that the activity of iron is close to unity in lunar basalts. It will, however, be problematic to ascertain equilibrium conditions between three oxide phases. The reverse zoning, e.g., has mostly been observed along spinel + ilmenite contacts with no ferropseudobrookite present. In this case the spinel composition alone cannot be employed to fix the equilibration temperature.

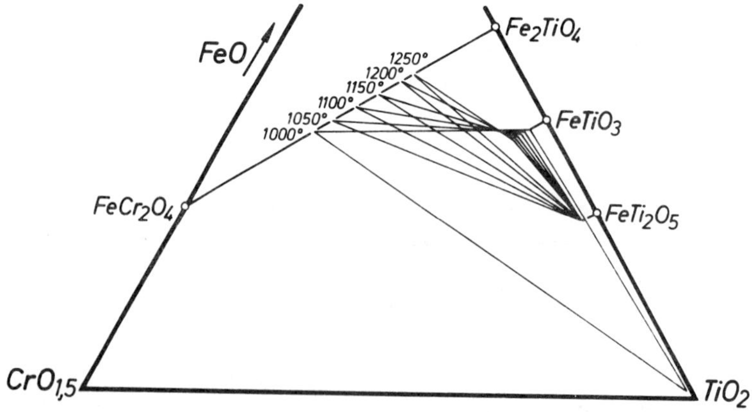

Fig. 4. Temperature dependence of the spinel + ilmenite + ferropseudobrookite phase assemblage. The triangles are constructed from the data points in Figs. 1(a–f).

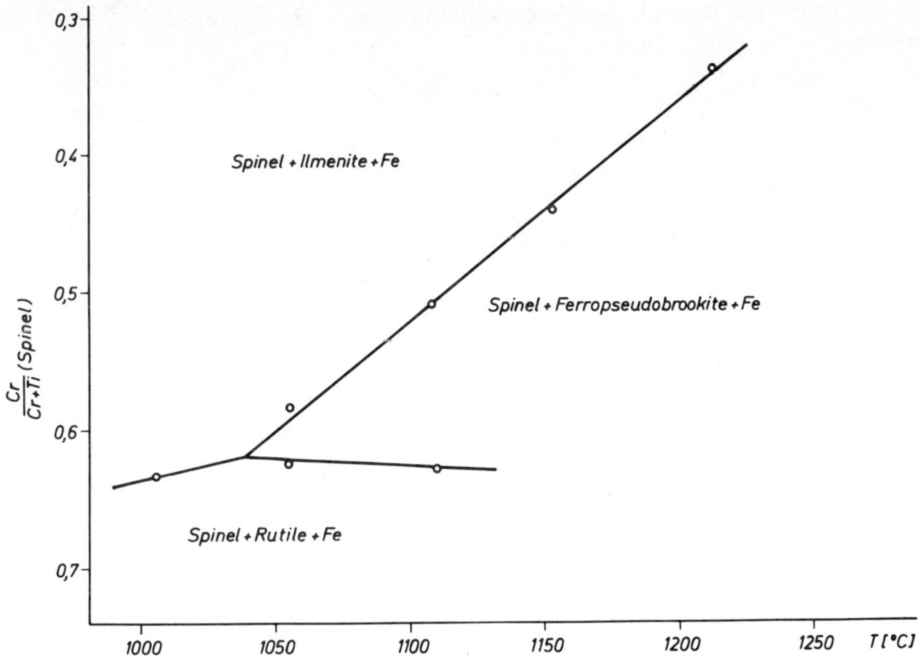

Fig. 5. The composition of the spinel phase as a function of temperature and coexisting phases.

In order to utilize the phase compositions even in cases of only two coexisting oxide phases as a lunar thermometer it is required to have another variable fixed. In case of a spinel-ilmenite paragenesis this may be achieved by the determination not only of the spinel but also of the ilmenite composition.

From Fig. 4 it is evident that the slope of the spinel-ilmenite boundary of the spinel + ilmenite + ferropseudobrookite triangle—and likewise the slope of all other spinel-ilmenite tie lines—is continuously tilted in a counterclockwise direction with decreasing temperature. This feature demonstrates an exchange reaction

$$\underset{\text{in spinel}}{Fe_2TiO_4} + \underset{\text{in ilmenite}}{Cr_2O_3} = \underset{\text{in spinel}}{Fe_2CrO_4} + \underset{\text{in ilmenite}}{FeTiO_3}$$

with a sensitively temperature dependent equilibrium constant.

This reaction shows a remarkable similarity to the temperature dependent exchange reaction

$$\underset{\text{in spinel}}{Fe_2TiO_4} + \underset{\text{in ilmenite}}{Fe_2O_3} = \underset{\text{in spinel}}{Fe_3O_4} + \underset{\text{in ilmenite}}{FeTiO_3}$$

which has been developed by Lindsley as a geothermometer (Lindsley, 1963; Buddington and Lindsley, 1964).

Experimental work is in progress to provide more quantitative data in order to determine the tie-line-temperature relationship in the system Fe–Cr–Ti–O for application as a lunar thermometer.

THE PHASE ASSEMBLAGES

Due to the low temperature instability of ferropseudobrookite with respect to ilmenite + rutile the line *M–N* in Fig. 3 will terminate in the invariant point *N* which is characterized by the phase assemblage $sp_{ss} + ilm_{ss} + fps_{ss} + rt_{ss}$ (+Fe). The temperature of *N* in the system Fe–Cr–Ti–O was determined experimentally to be 1038 ± 4°C.

Since both oxygen fugacity and temperature are intensive parameters, (see Lindsley *et al.*, 1968) it is possible to construct a log f_{O_2}/T or a G/T diagram by applying the method of Schreinemakers (Zen, 1966). The slope of the monovariant curve *M–N* is determined by the parameters defined above. The curve *N–L* is fixed by the known temperatures of the end points. The relative positions of the other monovariant curves around point *N* (Fig. 3) are provided by the Schreinemakers analysis.

At temperatures below *N* isothermal sections through the system FeO–Cr_2O_3–TiO_2, e.g., line "a" in Fig. 3, will intersect the monovariant line $sp_{ss} + ilm_{ss} + rt_{ss}$ (+Fe). A corresponding three phase triangle is outlined in Fig. 1a for the temperature of 1000°C. At temperatures between *N* and *L* (line "b") three monovariant lines will be intersected with corresponding three phase triangles $ilm_{ss} + fps_{ss} + rt_{ss}$ (+Fe), $sp_{ss} + ilm_{ss} + fps_{ss}$ (+Fe) and $sp_{ss} + fps_{ss} + rt_{ss}$ (+Fe) showing in the concentration diagram (Fig. 1b) for 1050°C. At a temperature above *L* line "c" intersects the monovariant curves $sp_{ss} + ilm_{ss} + fps_{ss}$ (+Fe) and $sp_{ss} + fps_{ss} + rt_{ss}$ (+Fe). In the concentration diagram for 1100°C (Fig. 1c), however, an additional three phase triangle $sp_{ss} + fps'_{ss} + rt_{ss}$ (+Fe) at higher chromium concentrations appears. Thus a new monovariant line must be constructed in Fig. 3. At 1200°C (Fig. 1e) only two triangles are observed with the introduction of again a new assemblage $sp_{ss} + fps_{ss} + fps'_{ss}$ (+Fe). In the temperature range between 1100°C and 1200°C the triangles $sp_{ss} + fps_{ss} + rt_{ss}$ (+Fe) and $sp_{ss} + fps'_{ss} + rt_{ss}$ (+Fe) must have been approaching each other until the respective sp + rt tie lines became identical. The ensuing phase assemblage $sp_{ss} + fps_{ss} + fps'_{ss} + rt_{ss}$ (+Fe) resulted in the constitution of a new invariant point *H*. Its temperature is estimated to be 1180 ± 20°C. At higher temperatures a degenerated monovariant line $fps_{ss} + fps'_{ss}$ (+Fe) originates from *H*, which is intersected by line "d" in Fig. 3. In the concentration diagram at 1250°C (Fig. 1f) and on the corresponding line "e" in Fig. 3 only one three phase triangle is intersected while the $sp_{ss} + fps_{ss} + fps'_{ss}$ (+Fe) triangle disappeared. Both ferropseudobrookite phases merged at *K* which is the critical temperature of a pseudobrookite solvus estimated to be around 1230 ± 20°C. More experimental work is needed to bracket the temperatures of *H* and *K*.

From Fig. 3 the stability ranges of the phase assemblages in the system Fe–Ti–Cr–O in equilibrium with metallic iron can be deduced.

(1) Spinel + ilmenite + rutile assemblages are restricted to temperatures below 1038°C (point N).
(2) Spinel + ilmenite + pseudobrookite assemblages are stable above 1038°C (point N).
(3) Ilmenite + pseudobrookite + rutile assemblages are restricted to a temperature range between 1038 and 1068°C (N–L).
(4) Spinel and rutile coexist at temperatures below *ca.* 1180°C (point H).
(5) Two pseudobrookite phases P and P′ coexist between *ca.* 1180 and 1230°C (H–K).
(6) Due to the low temperature instability of pseudobrookite structure phases the spinel + pseudobrookite′ + rutile assemblage must be terminated at low temperatures.
(7) Rutile is stable in equilibrium with metallic iron up to 1180°C. Above this temperature it will be replaced by a Magnéli phase Ti_nO_{2n-1} (Andersson *et al.*, 1957). G is a critical point in the solvus of the $FeTi_2O_5$–Ti_3O_5 solid solution series at 1340°C which intercepts the titanium dioxide-metallic iron tie lines (Ender and Woermann, 1977).

Discussion

Evidently the stability ranges of the various phase assemblages outlined in Fig. 3 as well as the temperature dependence of the spinel composition in equilibrium with ilmenite and ferropseudobrookite in Fig. 4 and the projected spinel-ilmenite tie line relationship represent potential lunar thermometers in the FeO–Cr_2O_3–TiO_2 system.

In lunar oxides, however, additional elements—mainly Al and Mg—are contained in considerable quantities. Their influence on the phase relationships should be known in order to gain a lunar thermometer for practical application. Preliminary data reported by Usselman (1975) show that the Cr/(Cr + Ti) ratio of the spinel phase decreases by incorporation of Mg and increases by addition of Al.

The system Fe–Al–Ti–O has been investigated by Muan *et al.* (1972) at the 1300°C isotherm. Here, again, the spinel + ilmenite + pseudobrookite three phase triangle is observed. Thus the theoretical deduction of phase relations in the Fe–Cr–Ti–O system holds with no exception also for the Fe–Al–Ti–O system at least up to the solvus in the spinel composition: with decreasing temperature the spinel phase will be depleted in its ulvite component. Since also the magnesium containing systems MgO–Al_2O_3–TiO_2 and MgO–Cr_2O_3–TiO_2 (Muan *et al.*, 1971) and even MgO–"Fe_2O_3"–TiO_2 (Woermann *et al.*, 1969) exhibit the same general pattern in their titanium-rich section it can be assumed that the same "reverse zoning" trend will be followed even by complex spinel and ilmenite compositions.

The stability ranges of the phase assemblages in Fig. 3, however, will be drastically influenced by addition of further components. The most conspicuous example is given by the stabilizing effect of magnesium on the ferropseudo-brookite phase as described by Lindsley *et al.* (1974).

Thus more detailed work is required to develop a reliable lunar thermometer.

Acknowledgments—We are indebted to D. H. Lindsley for valuable discussions and comments. Critical reviews by N. Z. Boctor and T. M. Usselman were helpful in improving the manuscript. We gratefully acknowledge the financial support of this project by the Deutsche Forschungsgemeinschaft, Bonn.

References

Andersson S., Collén B., Kuylenstierna U., and Magnéli A. (1957) Phase analysis studies on the titanium-oxygen system. *Acta Chem. Scand.* **11**, 1641–1652.

Andersson S., Sundholm A., and Magnéli A. (1959) A homologous series of mixed titanium chromium oxides $Ti_{n-2}Cr_2O_{2n-1}$ isomorphous with the series Ti_nO_{2n-1} and V_nO_{2n-1}. *Acta Chem. Scand.* **13**, 989–997.

Buddington A. F. and Lindsley D. H. (1964) Iron-titanium oxide minerals and synthetic equivalents. *J. Petrol.* **5**, 310–357.

El Goresy A., Prinz M., and Ramdohr P. (1976) Zoning in spinels as an indicator of the crystallization history of mare basalts. *Proc. Lunar Sci. Conf. 7th*, p. 1261–1279.

El Goresy A. and Ramdohr P. (1975) Subsolidus reduction of lunar opaque oxides: Textures, assemblages, geochemistry, and evidence for a late-stage endogenic gaseous mixture. *Proc. Lunar Sci. Conf. 6th*, p. 729–745.

El Goresy A. and Ramdohr P. (1977) Apollo 17 TiO_2-rich basalts: Reverse spinel zoning as evidence for subsolidus equilibration of the spinel-ilmenite assemblage. *Proc. Lunar Sci. Conf. 8th.* This volume.

Ender A. and Woermann E. (1977) The pseudobrookite phase in the system Fe–Ti–O in equilibrium with metallic iron. Manuscript in preparation.

Flörke O. W. and Lee Ch. W. (1970) ANDERSSON-Phasen, dichteste Packung und WADSLEY-defekte im System Ti–Cr–O. *J. Solid State Chem.* **1**, 445–453.

Grey I. E., Reid A. F., and Jones D. G. (1974) Reaction sequences in the reduction of ilmenite: 4.—Interpretation in terms of the Fe–Ti–O and Fe–Mn–Ti–O phase diagrams. *Trans. Inst. Min. Met.* **83**, C105–111.

Haggerty S. E. (1972) Luna 16: An opaque mineral study and a systematic examination of compositional variations of spinels from Mare Fecunditatis. *Earth Planet. Sci. Lett.* **13**, 328–352.

Lindsley D. H. (1963) Fe-Ti-oxides in rocks as thermometers and oxygen barometers. Equilibrium relations of coexisting pairs of Fe–Ti-oxides. *Carnegie Inst. Wash. Yearb.* **62**, 60–66.

Lindsley D. H., Speidel D. H., and Nafziger R. H. (1968) P–T–fO_2 relations for the system Fe–O–SiO_2. *Amer. J. Sci.* **266**, 342–360.

Lindsley D. H., Kesson S. E., Hartzman M. J., and Cushman M. K. (1974) The stability of armalcolite: Experimental studies in the system MgO–Fe–Ti–O. *Proc. Lunar Sci. Conf. 5th*, p. 521–534.

MacChesney J. B. and Muan A. (1961) Phase equilibria at liquidus temperatures in the system iron oxide-titanium oxide at low oxygen pressures. *Amer. Mineral.* **46**, 572–582.

Merrit R. R. and Turnbull A. G. (1974) A solid state cell study of oxygen activities in the Fe–Ti–O system. *J. Solid State Chem.* **10**, 252–259.

Muan A., Hauck J. and Löfall T. (1972) Equilibrium studies with a bearing on lunar rocks. *Proc. Lunar Sci. Conf. 3rd*, p. 185–196.

Muan A., Hauck J., Osborn E. F., and Schairer J. F. (1971) Equilibrium relations among phases occurring in lunar rocks. *Proc. Lunar Sci. Conf. 2nd*, p. 497–505.

Saha P. and Biggar G. M. (1974) Subsolidus reduction equilibria in the system Fe–Ti–O. *Indian J. Earth. Sci.* **1**, 43–49.

Sato M., Hickling N. L., and McLane J. E. (1973) Oxygen fugacity values of Apollo 12, 14 and 15 lunar samples and reduced state of lunar magmas. *Proc. Lunar Sci. Conf. 4th*, p. 1061–1079.

Simons B. (1974) *Zusammensetzung und Phasenbreiten der Fe–Ti–Oxide im Gleichgewicht mit metallischem Eisen.* Diplomarbeit, RWTH Aachen, West Germany. 104 pp.

Simons B. and Woermann E. (1977) Stability relations in the system Fe–Si–Ti–O in equilibrium with metal. Manuscript in preparation.

Taylor L. A., Williams R. W., and McCallister R. H. (1972) Stability relations of ilmenite and ulvöspinel in the Fe–Ti–O system and application of these data to lunar mineral assemblages. *Earth Planet. Sci. Lett.* **16**, 282–288.

Ulmer G. C. (1970) Chromite spinels. In *High Temperature Oxides.* (A. M. Alper, ed.), Vol. 1, p. 251–314. Academic Press, New York.

Ulmer G. C. and White W. B. (1966) Existence of chromous ion in the spinel solid solution series $FeCr_2O_4$–$MgCr_2O_4$. *J. Amer. Ceram. Soc.* **49**, 50–51.

Usselman T. M. (1975) Ilmenite chemistry in mare basalts, an experimental study (abstract). In *Papers Presented to the Conference on Origins of Mare Basalts and Their Implications for Lunar Evolution*, p. 164–168. The Lunar Science Institute, Houston.

Woermann E., Brezny B., and Muan A. (1969) Phase equilibria in the system MgO-iron oxide-TiO_2 in air. *Amer. J. Sci.* **267A**, 463–479.

Zen E. (1966) Construction of pressure-temperature diagrams for multicomponent systems after the method of Schreine-makers—a geometric approach. *U.S. Geol. Survey Bull.* **1225**, 1–56.

Proc. Lunar Sci. Conf. 8th (1977), p. 2137–2151.
Printed in the United States of America

Geochemical studies of the White Breccia Boulders at North Ray Crater, Descartes region of the lunar highlands

M. M. LINDSTROM,[1] D. F. NAVA,[2] D. J. LINDSTROM,[1]
S. R. WINZER,[3] R. K. L. LUM,[1] P. J. SCHUHMANN,[1]
S. SCHUHMANN,[2] and J. A. PHILPOTTS[2]

[1]Geology Department, University of Maryland, College Park, Maryland 20742
[2]Astrochemistry Branch, Laboratory for Extraterrestrial Physics, NASA Goddard Space Flight Center, Greenbelt, Maryland 20771
[3]Martin Marietta Laboratories, Baltimore, Maryland 21227

Abstract—Major and trace element compositions are presented for samples from the white boulders at the rim of North Ray Crater. The largest boulder, 67455, was studied as part of a consortium headed by E. C. T. Chao and we present data for nine clasts as well as the bulk breccia. The compositional variations among the clasts from 67455 imply that they are related and probably represent a mixing series. Samples of the other two breccias, 67415 and 67435, do not appear to be closely related to breccia 67455.

The bulk composition of breccia 67455 can be matched by a mixture of its analyzed clasts, but the proportions of clasts required by this compositional model do not agree with those of Chao's petrologic studies. An unanalyzed component, such as the gabbroic anorthosite, may provide a solution to the problem.

The components of 67455 are distinct from most other Apollo 16 breccias in their high Fe/Mg ratios and very low REE contents. The other North Ray Crater breccias are similar to anorthositic breccias from other Apollo 16 sampling localities and are distinct from the KREEP-enriched breccias which are abundant on the Cayley Plains. These two major rock types, anorthositic rocks and Apollo 16 KREEP, may represent ejecta from two basin-forming events, but if so it is unlikely that the KREEP rocks were derived from Orientale, which Hodges *et al.* (1973) suggested was the source of the Cayley Plains. While Apollo 16 KREEP may represent Imbrium ejecta, it seems as likely that the anorthositic rocks are of local origin than that they were transported to the site by a major impact.

INTRODUCTION

THE WHITE BRECCIA BOULDERS were sampled near the southwest rim of North Ray Crater at the Apollo 16 landing site. These dominantly light matrix breccias and the nearby dark matrix breccias of House, Outhouse, and Shadow Rocks are ejecta from North Ray Crater (AFGIT, 1973). The crater, which has been dated at 50 m.y. (Marti *et al.*, 1973), is 1 km in diameter and 230 m deep and excavated material from the deepest level at the Apollo 16 site. Ulrich (1973) suggested that the light matrix breccias are samples from moderate depth in the crater, while the dark matrix breccias are from the lowest stratigraphic level. Hodges *et al.* (1973) proposed a pre-impact stratigraphy for the North Ray Crater site: Cayley Formation, presumably Orientale ejecta, overlies Imbrium ejecta which in turn overlies a different morphology which may be either Imbrium or Nectaris ejecta. The study of North Ray Crater breccias and comparison to breccias from the

Cayley Plains provides insight into the possible source regions of the various units.

SAMPLE DESCRIPTIONS

The samples of the White Breccia Boulders included in this study are: 67415, an anorthositic breccia; 67435, a dark matrix breccia (Keil consortium); 67455, a light matrix breccia (Chao consortium); and 67475, a large clast of dark matrix breccia taken from the 67455 boulder. The following sample descriptions are based both on the authors' examination of thin sections and on studies in the literature.

67415 is described briefly in the Apollo 16 Lunar Sample Information Catalogue (1972). 67415 is dust-covered and very friable and broke into numerous fragments in transit. It is a crushed anorthositic breccia composed of 85% matrix (consisting of finely ground plagioclase and pyroxene), 8% plagioclase, 5% pyroxene, and 4% anorthositic fragments. It has been classified by Wilshire *et al.* (1973) as a B1 group breccia, having both light matrix and light clasts. Our sample, 67415,3, is representative of the bulk breccia.

67435 has been described by Warner *et al.* (1976) as a glass-coated, clast-rich breccia. It is of particular interest because it contains spinel troctolite clasts described by Prinz *et al.* (1973). This gray and white breccia consists of two major lithologies: gray, aphanitic-matrix clasts and white, sugary-textured matrix. The clasts consist of a dark poikilitic matrix enclosing xenocrysts of plagioclase, olivine, and metal, and xenoliths of recrystallized ANT rocks and feldspathic breccias. Warner *et al.* (1976) consider the poikilitic textured matrix to be an impact melt with composition similar to that of other Apollo 16 poikilitic rocks. The white matrix of 67435 is highly feldspathic, containing numerous feldspathic breccia clasts and xenocrysts with plagioclase dominant over orthopyroxene and olivine. Our sample, 67435,40, is a split of the bulk gray and white breccia.

67455 has been described by Chao *et al.* (1977) and Minkin *et al.* (1977). It is a very friable light matrix breccia which contains dark glassy clasts and is thus a B2 breccia (Wilshire *et al.*, 1973). It broke into numerous pieces on handling, so that interrelationships among clasts are generally unknown. Chao has classified the clasts and used two methods to do population studies: (1) megascopic inspection of sieved coarse (>2 mm) particles, and (2) petrographic examination of a large clast-rich thin section. Both population studies are given in Table 1. The two results are similar, the major difference being the recognition (in thin section) of a separate group of gabbroic anorthosites. Minkin *et al.* (1977) give a detailed modal analysis of 67455 thin sections which shows that only 11–15% of the fragments are clasts of cataclastic anorthosites. About 60% are single mineral fragments of plagioclase, olivine, or pyroxene. They attribute these to the breakdown of cataclastic anorthosites and conclude that 70% of the material is derived from cataclastic anorthosite. The clast types are described in more detail by Minkin *et al.* (1977). The clasts analyzed were selected by Chao to represent

Table 1. Clast populations in breccia 67455.

	Megascopic inspection of sieved particles[1]	Petrographic examination of thin section[2]
Cataclastic anorthosites	80	70
Gabbroic anorthosites	—	10
Microbreccias	11	7–8
Fragment laden melts	7	12–13
Recrystallized ANT, hornfels	2	1

[1]Chao-consortium report—pers. comm.
[2]Chao *et al.* (1977).

the various rock types. Sample 67455,151 was chosen to be representative of the bulk breccia.

The cataclastic anorthosites are medium grained, crushed, and broken, but only weakly shocked and unannealed. Three types of anorthosites are found: (1) anorthite with interstitial pigeonite (our samples 67455,32 and 67455,134), (2) anorthite with interstitial Fe-rich olivine (our sample 67455,127), and (3) anorthite with no interstitial minerals visible.

The gabbroic anorthosites are coherent fragments made up of equigranular anorthite and Fe-rich olivine, orthopyroxene, and clinopyroxene, with interstitial ilmenite and metal. Textures range from those which suggest little recrystallization to those with rounded grains and 120° triple junctions, occasionally with pyroxenes surrounding plagioclase and olivine in hornfelsic texture.

Feldspathic microbreccias occur both as separate clasts and within dark glassy clasts. These breccias consist of small plagioclase grains packed together in random orientation, with no other minerals present.

Fragment-laden melts are dark gray to brown melts containing varying amount of xenocrysts and xenoliths. Some are coherent clasts which consist dominantly of closely packed small plagioclase xenocrysts with only a small amount of interstitial melt. These include small amounts of mafic xenocrysts (olivine, orthopyroxene, augite) and a few small xenoliths. Sample 67455,113 is typical of these fragment-laden melts. Other melts are more friable and contain large clasts of feldspathic microbreccias. These melts, also designated gray glassy microbreccias, include our samples 67455,139 and 67455,140. Some melt clasts have annealed textures. The degree of annealing varies from incipient nucleation to coarse-grained hornfelsic texture. Sample 67455,120 is a hornfelsic melt rock which was separated into dark (noritic) hornfelsic melt and light anorthositic clasts.

A small number of clasts in 67455 are xenocryst-free glasses. Some exhibit flow structures, others are devitrified, and still others are annealed.

67475 is a large (175 g) clast from the 67455 boulder. It is a very coherent brownish-gray fragment-laden melt, with fragments dominant over glass. The major clasts in 67475 are described by Minkin *et al.* (1977). The anorthosite, gabbroic anorthosite, and breccia clasts in 67475 are distinct from corresponding clasts in 67455 in their mineral compositions and associations. An unusual metal- and Si-rich breccia contains phases not found in 67455 clasts (silica, apatite, whitlockite) and has feldspars rich in K and Na and very Fe-rich mafic phases. Our sample, 67475,101, represents the xenocryst-rich matrix of the melt rock.

CHEMICAL COMPOSITIONS OF THE WHITE BRECCIA BOULDERS

Chemical analyses of bulk samples of breccias 67415, 67435, 67455, and 67475 and clasts from 67455 are given in Table 2. Major and minor element analyses were done by semi-micro flame atomic absorption spectrophotometry combined with colorimetry for Ti and P (Nava and Philpotts, 1973). Trace element abundances were determined by stable isotope dilution mass spectrometry (Schnetzler *et al.*, 1967). Comparison to other analyses of these breccias shows that our bulk sample of 67455 is very similar to those analysed by Rose *et al.* (1973) and Wänke *et al.* (1973). Our sample of 67435 has slightly lower abundances of Fe, Mg, and REE (and correspondingly higher Al abundances) than the matrix sample analyzed by Warner *et al.* (1976).

All of the samples are highly feldspathic, ranging from 72% normative plagioclase in 67415 and 67435, through 86% in bulk samples of 67455 and 67475, to 96% in some of the anorthosites. The major compositional variations among these samples can be observed in variations of FeO or MgO, Al_2O_3, and lithophile trace element contents (K, Rb, REE). Figures 1a and b show variations of FeO and MgO with Al_2O_3 content. The ten samples of 67455 and 67475 define good linear correlations for both FeO vs Al_2O_3 and MgO vs Al_2O_3. Both 67415 and 67435 fall off the extrapolations of these lines, and are higher in MgO and lower in FeO relative to the 67455 lines. The linear correlations for 67455 and 67475 samples reflect the variations in normative proportions of plagioclase relative to pyroxene and olivine, and imply that all the 67455 samples may be related by mixing of two components while the 67415 and 67435 samples are not closely related to the 67455 breccia samples.

Figure 2 is a plot of chondrite-normalized lithophile trace element abundances for these breccia samples. All of the samples except 67435 have REE patterns dominated by anorthositic rocks. The REE pattern of 67435 reflects the importance of a KREEP-rich component in this breccia which has an anorthositic gabbro bulk composition. This is consistent with the interpretation of Warner *et al.* (1976) that 67435 is a mixture of dark, poikilitic clasts and light, feldspathic matrix.

Sm is plotted versus Al_2O_3 in Fig. 3.* The small clasts from 67455 again define a good linear correlation, but the 67455 bulk sample lies distinctly above this line and 67475 has an even higher Sm concentration. Breccias 67415 and 67435 also have higher Sm contents and fall off the 67455 clast line. These data

*Sm was selected as representative of the REE exclusive of Eu; it is possible with the low abundances and small sample sizes, that some of the Ce and Nd values are not of the same quality as the Sm values.

Table 2. Major and trace element composition of samples from the White Breccia Boulders.

	Anorthositic breccia 67415,3	Gray and white breccia 67435,40	Dark matrix breccia 67475,101	Light matrix breccia 67455,151	Fragment laden melt 67455,113	Noritic hornfels 67455,120	Anorthositic hornfels 67455,120	Cataclastic anorthosite 67455,127	Px-bearing anorthosite 67455,32	Px-bearing anorthosite 67455,134	Gray glassy microbreccia 67455,139	Gray glassy microbreccia 67455,140
wt.%												
SiO_2	44.6	45.2	44.4	44.5	44.4	44.8	44.4	44.0	44.9	45.3	44.1	44.6
TiO_2	0.32	0.66	0.36	0.21	0.23	0.23	<0.05	<0.05	0.13	0.05	0.18	0.13
Al_2O_3	26.04	25.71	30.87	30.75	28.69	28.96	34.50	34.21	32.28	33.03	30.17	30.59
FeO	4.64	5.03	2.91	3.58	4.99	5.04	0.61	1.46	2.62	1.94	4.11	3.90
MnO	0.06	0.07	0.04	0.05	0.07	0.07	0.01	0.02	0.04	0.03	0.05	0.05
MgO	7.77	6.75	2.57	2.83	3.47	3.87	0.51	0.94	1.31	1.12	2.81	2.69
CaO	15.05	15.16	17.21	17.13	17.01	17.32	19.83	18.73	18.10	18.45	17.68	17.75
Na_2O	0.52	0.52	0.60	0.44	0.47	0.31	0.33	0.32	0.42	0.42	0.39	0.37
K_2O	0.04	0.13	0.04	0.02	0.02	0.02	0.02	0.02	0.03	0.02	0.02	0.02
P_2O_5	0.03	0.11	<0.02	0.03	0.02	0.04	0.03	0.04	<0.02	0.04	0.04	0.04
Cr_2O_3	0.08	0.10	0.03	0.05	0.05	0.03	0.005	<0.005	0.02	0.01	0.04	0.03
Total	99.15	99.44	99.05	99.59	99.42	100.69	100.29	99.79	99.87	100.41	99.59	100.17
ppm												
Li	6.06	9.69	5.33	4.00	4.25	3.18	2.59	2.57	3.47	3.29	3.79	3.41
K	340	1173	339	183	188	160	174	144	235	195	182	163
Rb	0.964	4.50	0.672	0.378	0.169	0.634	0.706	0.492	0.870	0.751	0.559	0.529
Sr	177	170	216	174	152	144	161	155	116	163	150	148
Ba	61.2	163	48.5	23.6	19.4	13.2	11.5	9.97	13.6	11.3	12.7	23.6
Ce	9.62	35.7	6.66	3.32	2.96	2.47	0.513	0.660	0.924	0.676	1.66	3.29
Nd	6.23	23.8	4.22	2.08	1.92	1.88	0.330	0.283	0.673		1.14	1.54
Sm	1.84	6.61	1.21	0.641	0.60	0.571	0.112	0.0601	0.228	0.122	0.341	0.379
Eu	1.11	1.31	1.28	0.905	0.792	0.745	0.727	0.687	0.802	0.799	0.739	0.724
Gd		8.14	1.43						0.324			
Dy	2.75	8.85	1.68	1.11	0.777	1.01	0.249	0.080	0.394	0.199	0.620	0.595
Er	1.77	5.41	1.06	0.681	0.536	0.642	0.118	0.050	0.235	0.124	0.388	0.374
Yb	1.75	4.87	1.01	0.597	0.454	0.632	0.135		0.250	0.150	0.399	0.388
Lu	0.255	0.70	0.140	0.084	0.089	0.087	0.0245		0.030	0.0215	0.0566	0.0561

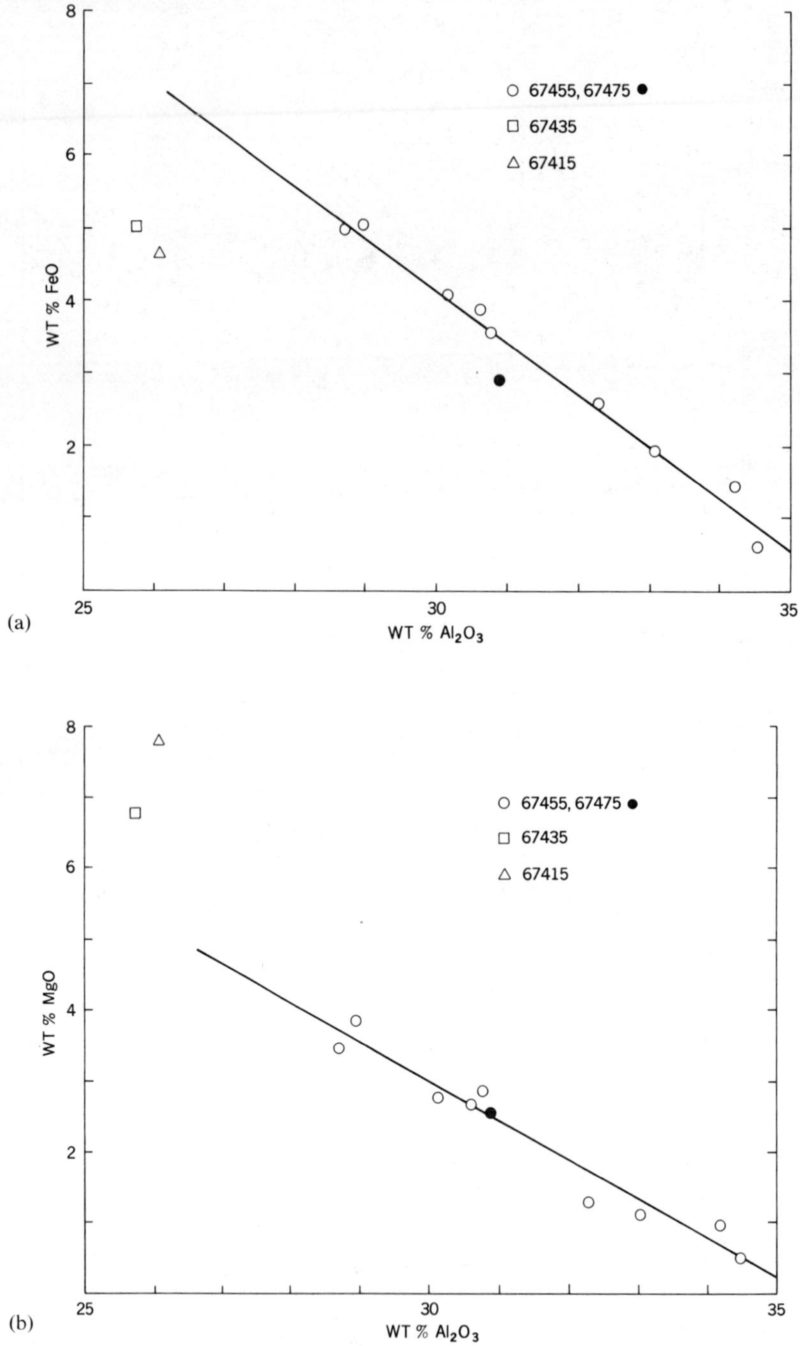

Fig. 1. Variation of (a) FeO and (b) MgO with Al_2O_3 content for samples from the White Breccia Boulders. The lines represent linear regressions on the 67455 and 67475 data.

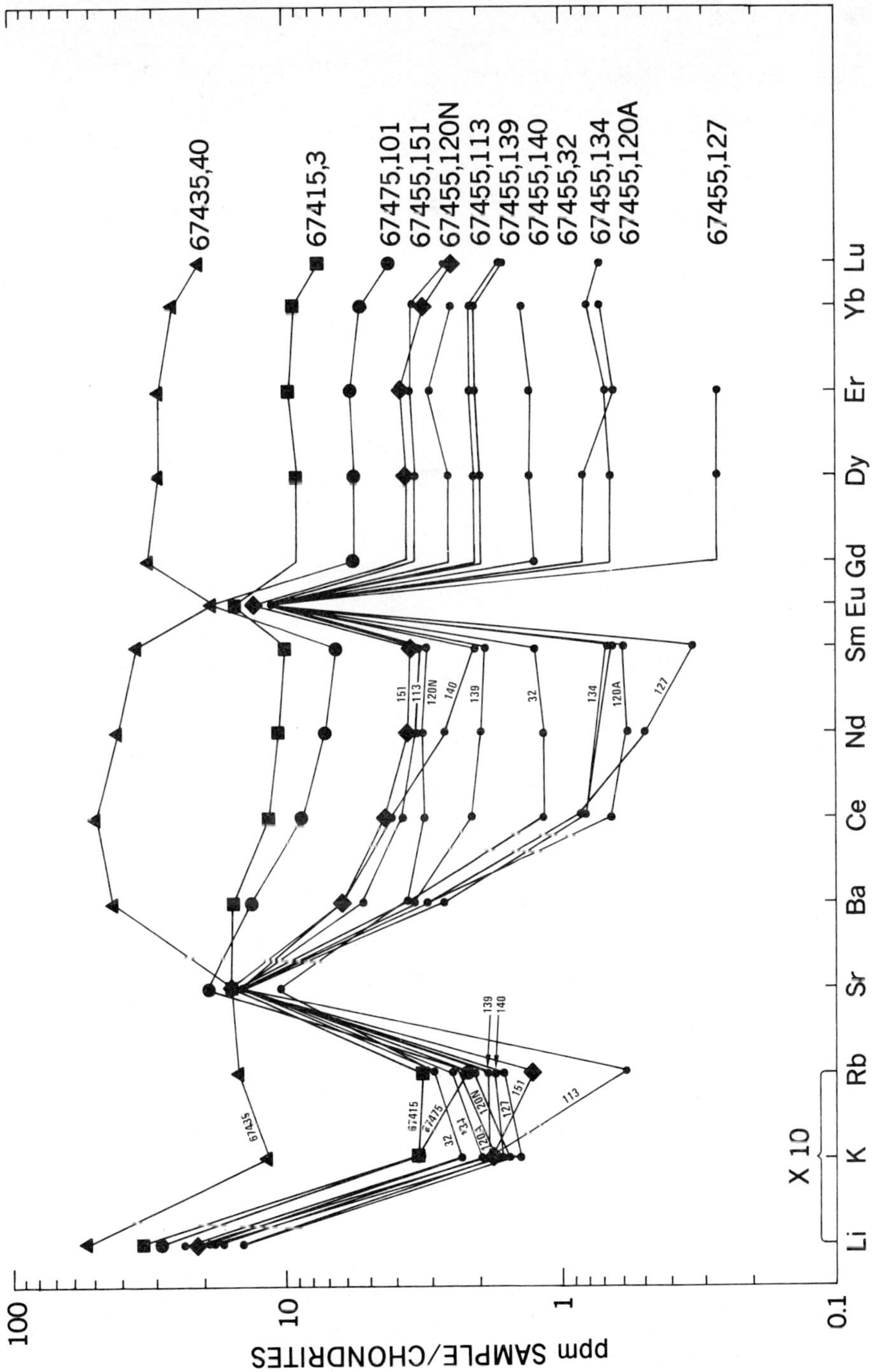

Fig. 2. Lithophile trace element abundances for samples from the White Breccia Boulders. The light REE analyses for 67455,140 and 67455,127 are not reliable because of large blank and spike corrections on very small anorthositic samples.

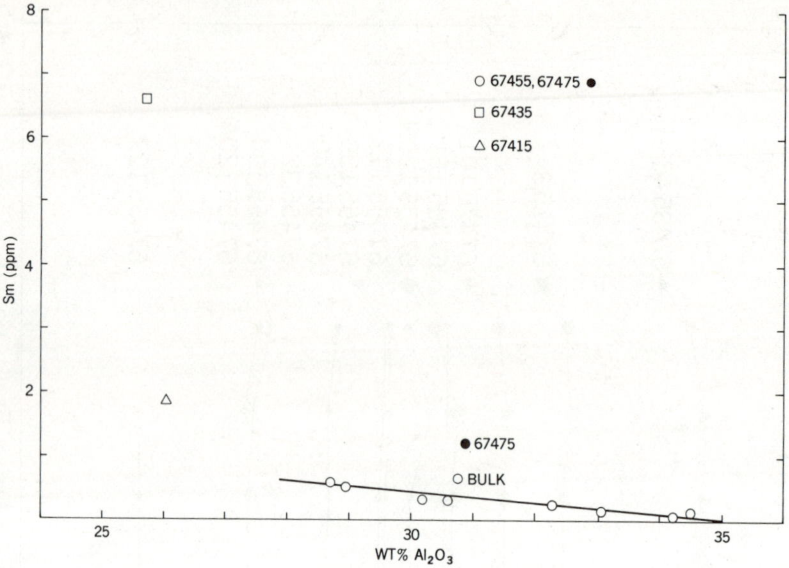

Fig. 3. Variation of Sm with Al_2O_3 content for samples from the White Breccia Boulders. The line is a linear regression for 67455 clast data.

illustrate that the REE are not generally correlated with major element composition. Whereas 67415 and 67435 are very similar in bulk composition, 67435 has nearly four times the REE contents.

The small clasts from 67455 appear to be related to each other by mixing of only two major components. The textural evidence supports this suggestion. The most Fe-, Mg- and REE-rich clasts are the fragment-laden melt and noritic hornfels, while the Al-rich rocks are relatively unshocked and unmelted anorthosites. The intermediate clasts are melt rocks containing large clasts of feldspathic breccias. The bulk sample of 67455 then represents incorporation of a more REE-enriched material to this series of REE-poor clasts.

If the 67455 bulk sample can be represented as a mixture of these analyzed clasts, then the compositional data may be used to evaluate the proportions of the various clasts in the bulk breccia. The three major compositional components are assumed to be: (1) fragment-laden melt and hornfelsic equivalents which are relatively rich in Fe and Mg, (2) anorthosites, the most common of which are those with interstitial pyroxene, and (3) 67475, also a fragment-laden melt and the only clast analyzed with REE enriched over the bulk sample. A mixture of approximately equal amounts of anorthosites and Fe, Mg-rich fragment-laden melts would fit the major element contents of the bulk sample. The addition of about 30% 67475 material is required to satisfy the REE data. Thus the compositional data suggest clast populations of 35% anorthosites and 65% dark clasts (35% Fe-enriched melt rocks, and 30% REE-enriched melt rocks). These proportions do not match those of Chao's population studies (Table 1), which indicate that 67455 contains 70–80% anorthositic clasts, or minerals derived from them. It is indeed unlikely that this light-colored breccia is made up of 65% dark clasts. Clast 67475 is a large clast in the 67455 boulder, yet similar material has not been found in the 67455 thin sections (Minkin *et al.*, 1977). Hence 67475 may not be a good sample of the REE-enriched component in 67455. The source of the REE may instead be in some as yet unanalyzed component. Two such components exist: the breccia matrix and the gabbroic anorthosites. The matrix is an unlikely source for REE-enriched material because what little material there is between the larger clasts in this loosely lithified breccia appears to be crushed clast material and is likely to be compositionally similar to the bulk breccia. The gabbroic anorthosites have a bulk composition that is more mafic than the anorthosites, and probably resembles the 67455 bulk sample compositionally. These gabbroic anorthosites are light-colored clasts which can provide much of the Fe and Mg in the bulk sample. However, more than the observed 10% of this component would be required to offset the low abundances of these elements in the anorthosites. If some of the mineral fragments are derived from gabbroic anorthosites, this component could represent more than the 10% observed as lithic clasts. The gabbroic anorthosites would also have to be somewhat richer in REE than the other clasts in 67455 because the bulk sample is richer in REE than any of the small analyzed clasts. A mixture of 30% gabbroic anorthosites (having $Al_2O_3 \sim 28$–30% and REE $\sim 10\times$ chondrites) with 20% fragment-laden melt (+hornfels + microbreccias) and 50% anorthosites will match the bulk composition of 67455 and make the amount of light, anorthositic rocks 80%, and dark, glassy rocks 20%. We cannot, however, exclude the possibility that a small amount of material richer still in Mg and Fe and in REE may contribute to the 67455 bulk sample even though no such material has as yet been observed.

Discussion

Comparison of these White Breccia Boulder samples with other breccias from the Apollo 16 site requires detailed studies of both the petrology and chemical composition of clasts from other breccias. Most breccia studies report only petrology or only bulk composition of the materials and the samples are not characterized sufficiently for comparison. Three studies report both the petrology and geochemistry of Apollo 16 breccias: Warner *et al.* (1976) report on 67435, to which we add an analysis of the bulk breccia; Fruchter *et al.* (1974) discuss 66055; and Winzer *et al.* (1977) describe 61175. As in 67455, the clasts in 67435, 66055, and 61175 show a variety of textures and include shocked igneous (cumulate) rocks, breccias, impact melts, and thermally metamorphosed rocks. Unlike 67455, in addition to KREEP-free anorthositic rocks, all three breccias contain clasts akin to Apollo 16 KREEP-rich rocks. These clasts have REE abundances $\sim 100\times$ chondrites and Al_2O_3 contents (when known) of 18–24%, and thus are similar to Apollo 16 KREEP and poikilitic rocks (Hubbard *et al.*, 1973; Drake *et al.*, 1974). While the origin of this Apollo 16 KREEP component is unknown since the samples are not igneous rocks, it does appear to be common

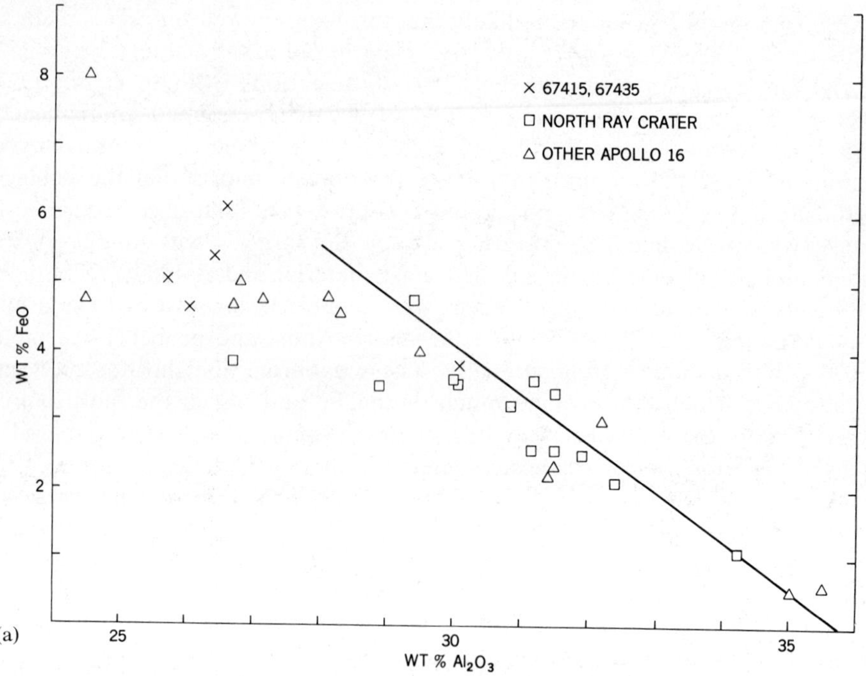
× 67415, 67435
□ NORTH RAY CRATER
△ OTHER APOLLO 16
WT % FeO
8
6
4
2
25
30
35
WT % Al_2O_3

(a)

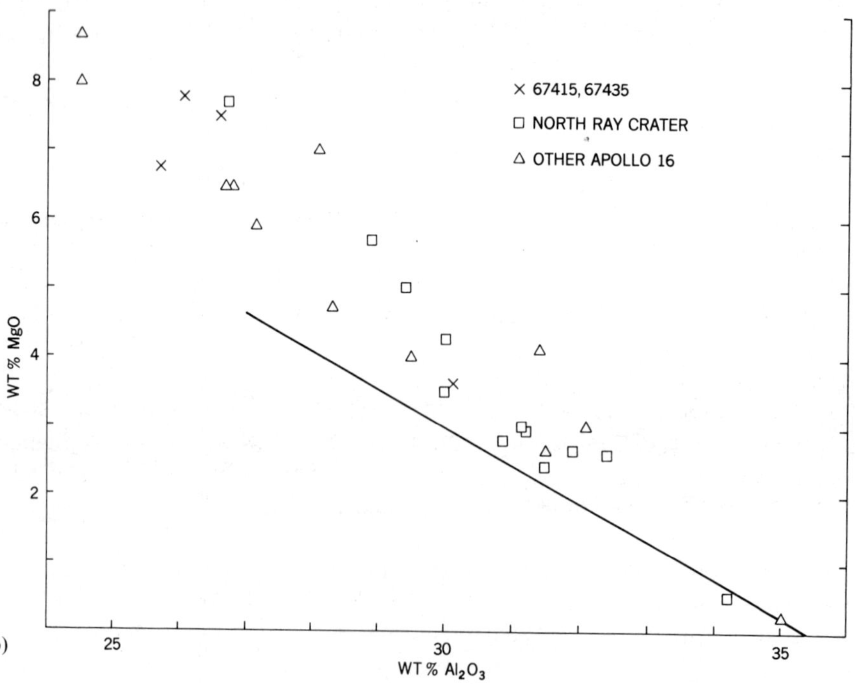
× 67415, 67435
□ NORTH RAY CRATER
△ OTHER APOLLO 16
WT % MgO
8
6
4
2
25
30
35
WT % Al_2O_3

(b)

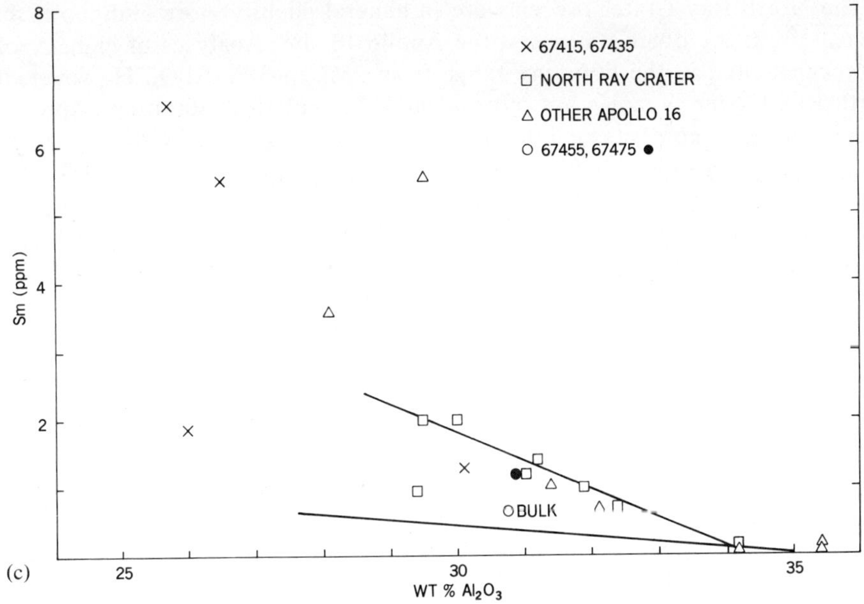

Fig. 4. Variation of (a) FeO, (b) MgO, and (c) Sm with Al_2O_3 contents for samples of other Apollo 16 breccias. For (a) and (b), the lines are the linear regressions on 67455 data for (c) both the 67455 regression line and one for the other anorthositic breccias are given. Sources: Brunfelt *et al.* (1973); Duncan *et al.* (1973); Fruchter *et al.* (1974); Haskin *et al.* (1973); Hubbard *et al.* (1973); Laul and Schmitt (1973); Laul *et al.* (1974); LSPET (1973); Rose *et al.* (1973); Taylor *et al.* (1973); Wänke *et al.* (1973); Wänke *et al.* (1974); Warner *et al.* (1976); Winzer *et al.* (1977).

to many Apollo 16 breccias and soils. The bulk rock or matrix samples of these breccias have REE contents of 25–50 × chondritic values compared to 4 × chondritic values for 67455. Since the anorthositic clasts have REE contents of 0.5–15 × chondrites, these breccias appear to contain a significant proportion of the KREEP-rich rocks.

Comparison of these breccias to others from the Apollo 16 site relies primarily on analyses of the compositions of bulk breccia samples. The White Breccia Boulders are similar in bulk composition to most of the other breccias at North Ray Crater. (Analyses of 60017, 63335, 63355, 67016, 67031, 67075, 67095, 67115, 67915, and 67955 are cited by various authors listed in the references.) These breccias are highly feldspathic, most containing about 29–32% Al_2O_3, although two of them (67095, 63355) have Al_2O_3 contents of 22%. Only three (67435, 67095, 63355) of the 14 analyzed samples have REE abundances indicating a significant amount of an Apollo 16 KREEP component. The eleven other breccias have REE abundances of about 1–10 × chondritic values and have positive Eu anomalies. These appear to be essentially KREEP-free anorthositic rocks. The 11 anorthositic breccias include both light and dark matrix breccias but no systematic compositional differences between the two types are evident.

The North Ray Crater breccias are in general slightly more aluminous than breccias from the other stations at the Apollo 16 site. Analyses of eight Apollo 16 breccias cited in the literature range from 25% to 35% Al_2O_3. However, the North Ray Crater breccias are more clearly distinct from the other Apollo 16 breccias in their contents of lithophile trace elements. Five of the eight other Apollo 16 breccias analyzed have REE abundances greater than $30\times$ chondritic values, most likely reflecting the importance of a KREEP component in these breccias. The remaining breccias are KREEP-free anorthositic breccias similar to those at North Ray Crater (64435 and monomict breccia 60025 have $Al_2O_3 \sim$ 32–35% and REE 1–$5\times$ chondrites).

More detailed comparisons can be made using Fig. 4 which shows the variation diagrams for FeO, MgO, and Sm versus Al_2O_3 for many Apollo 16 breccias. The lines are the linear regressions for the 67455 data alone (Figs. 1 and 3). The points represent literature data for other North Ray Crater breccias and breccias from other Apollo 16 sampling locations. While most of the points represent bulk breccias analyses, a few represent clasts, glass coatings, or separated light and dark lithologies. In Fig. 4a the points for other breccias cluster near the FeO/Al_2O_3 correlation for 67455 samples when Al_2O_3 is greater than 29% but fall below the line at lower Al_2O_3 contents. The MgO versus Al_2O_3 plot (Fig. 4b) shows that points for other breccias fall distinctly above the 67455 correlation line and define a separate correlation. The Fe/Mg ratios of 67455 samples are thus distinctly higher than those of most other Apollo 16 breccias. The Sm versus Al_2O_3 plot (Fig. 4c) resembles Fig. 4b in that all points at Al_2O_3 greater than 29% define a line distinct from that of the 67455 clasts. Large clast 67475 falls near this line and 67455 bulk breccia falls between the lines. The 67455 bulk sample may thus represent addition of an anorthositic component with slightly higher REE contents to its REE-poor clasts. Points at Al_2O_3 contents lower than 29% scatter and represent addition of varying amounts of Apollo 16 KREEP. The clasts in breccia 67455 are compositionally distinct from most other Apollo 16 samples in their high Fe/Mg ratio and very low REE contents. This distinction is most pronounced in the more mafic rocks (the fragment-laden melts) since the trends for 67455 clasts and other breccia samples intersect at nearly pure anorthite. A few samples resemble the 67455 materials, in particular breccia 67075 (Haskin *et al.*, 1973), but this material appears to be rare. Minkin *et al.* (1977) show that the mafic minerals in 67455 have higher Fe/Mg ratios than those of most other Apollo 16 materials. They resemble the phases in breccia 67075 (McCallum *et al.*, 1975), but are more Fe-rich than those of another North Ray Crater breccia (67955; Hollister, 1973) or in the ferroan anorthosites studied by Dowty *et al.* (1973). Thus 67455 represents a distinct material which is relatively rare at the Apollo 16 site and is not representative of the majority of North Ray Crater materials. The source of this rare material is unknown, but it seems unnecessary to attribute such rare material to a major basin, i.e., Nectaris as suggested by Chao *et al.* (1977). Indeed, the siderophile element studies of Hertogen *et al.* (1977) show that 67455, 67475, and 67075 all contain different meteorite components and thus are unlikely to represent ejecta from the same major impact. It seems more likely that they were deeply buried local material.

Most of the North Ray Crater breccias are compositionally similar to the other Apollo 16 anorthositic breccias. These are distinct from the KREEP-enriched breccias which are more abundant on the Cayley Plains than at North Ray Crater. The higher proportion of KREEP-rich material at the Cayley plains and higher abundance of anorthositic material at North Ray Crater were also observed in studies of soil particles by Delano *et al.* (1973), Taylor *et al.* (1973), and Drake *et al.* (1974). Hertogen *et al.* (1977) found that the Apollo 16 KREEP-rich breccias all contain the same meteorite component, while the anorthositic breccias exhibit several different meteoritic components, all distinct from that of the KREEP-rich breccias. If, as Hodges *et al.* (1973) suggest, the Cayley Plains and North Ray Crater materials represent ejecta from two major impact basins, then Apollo 16 KREEP represents ejecta from Orientale and the anorthositic materials represent Imbrium ejecta. However, as Drake *et al.* (1974) point out, Orientale is not a likely source for this KREEP material because the orbital gamma ray data show Orientale to have low contents of KREEP-related elements K and Th, while Imbrium material has very high concentrations of these elements (Metzger *et al.*, 1974). The contribution of Orientale ejecta to the Apollo 16 site is thus unknown. While the Apollo 16 KREEP rocks may represent Imbrium ejecta and the anorthositic rocks represent Nectaris ejecta, it is also possible, as suggested by Taylor *et al.* (1973), and supported by the existence of several meteorite components, that the anorthositic rocks are derived from local bedrock.

Acknowledgments—We thank Dr. E. C. T. Chao for providing carefully chosen samples of 67455 and 67475 and the petrologic background for interpretation of our results. Discussions of these samples with J. A. Minkin and C. L. Thompson were very valuable. The sample of 67435 from the consortium of Dr. K. Keil is also appreciated. A review by R. F. Dymek was very helpful. This work was supported in part by NASA/GSFC grant NSG 5121 to the University of Maryland.

References

AFGIT (Apollo Field Geology Investigation Team) (1973) Apollo 16 exploration of Descartes: A geologic summary. *Science* **179**, 62–69.

Apollo 16 Lunar Sample Information Catalogue (1972) (P. Butler, Jr., ed.). MSC-03210, Curators Office, NASA Johnson Space Center, Houston. 372 pp.

Brunfelt A. O., Heier K. S., Nilssen B., Sundovoll B., and Steinnes E. (1973) Geochemistry of Apollo 15 and 16 materials. *Proc. Lunar Sci. Conf. 4th*, p. 1209–1218.

Chao E. C. T., Minkin J. A., and Thompson C. L. (1977) Petrology of consortium sample 67455, from a white-matrix breccia boulder near the rim of North Ray Crater, Descartes (abstract). In *Lunar Science VIII*, p. 166–168. The Lunar Science Institute, Houston.

Delano J. W., Bence A. E., Papike J. J., and Cameron K. L. (1973) Petrology of the 2–4 mm soil fraction from the Descartes region of the moon and stratigraphic implications. *Proc. Lunar Sci. Conf. 4th*, p. 537–551.

Dowty E., Prinz M., and Keil K. (1974) Ferroan anorthosite: A widespread and distinctive lunar rock type. *Earth Planet. Sci. Lett.* **24**, 15–25.

Drake M. J., Taylor G. J., and Goles G. G. (1974) Descartes Mountains and Cayley Plains: Composition and provenance. *Proc. Lunar Sci. Conf. 5th*, p. 991–1008.

Duncan A. R., Erlank A. J., Willis J. P., and Ahrens L. H. (1973) Composition and interrelationships of some Apollo 16 samples. *Proc. Lunar Sci. Conf. 4th*, p. 1097–1113.

Fruchter J. S., Kridelbaugh S. J., Robyn M. A., and Goles G. G. (1974) Breccia 66055 and related clastic materials from the Descartes region, Apollo 16. *Proc. Lunar Sci. Conf. 5th*, p. 1035–1046.

Haskin L. A., Helmke P. A., Blanchard D. P., Jacobs J. W., and Telander K. (1973) Major and trace element abundances in samples from the lunar highlands. *Proc. Lunar Sci. Conf. 4th*, p. 1275–1296.

Hertogen J., Janssens M.-J., Takahashi H., Palme H., and Anders E. (1977) Lunar basins and craters: Evidence for systematic compositional changes of bombarding population. *Proc. Lunar Sci. Conf. 8th*. Vol. 1.

Hodges C. A., Muehlberger W. R., and Ulrich G. E. (1973) Geologic setting of Apollo 16. *Proc. Lunar Sci. Conf. 4th*, p. 1–25.

Hollister L. S. (1973) Sample 67955: A description and a problem. *Proc. Lunar Sci. Conf. 4th*, p. 633–641.

Hubbard N. J., Rhodes J. M., Gast P. W., Bansal B. M., Shih C. Y., Wisemann H., and Nyquist L. E. (1973) Lunar rock types: The role of plagioclase in non-mare and highland rock types. *Proc. Lunar Sci. Conf. 4th*, p. 1297–1312.

Laul J. C., Hill D. W., and Schmitt R. A. (1974) Chemical studies of Apollo 16 and 17 samples. *Proc. Lunar Sci. Conf. 5th*, p. 1047–1066.

Laul J. C. and Schmitt R. A. (1973) Chemical composition of Apollo 15, 16 and 17 samples. *Proc. Lunar Sci. Conf. 4th*, p. 1349–1367.

LSPET (1973) The Apollo 16 lunar samples: Petrographic and chemical description. *Science* **179**, 23–34.

McCallum I. S., Okamura F. P., and Ghose S. (1975) Mineralogy and petrology of sample 67075 and the origin of lunar anorthosites. *Earth. Planet. Sci. Lett.* **26**, 36–53.

Marti K., Lightner B. D., and Osborn T. W. (1973) Krypton and xenon in some lunar samples and the age of North Ray Crater. *Proc. Lunar Sci. Conf. 4th*, p. 2021–2036.

Metzger A. E., Trombka J. I., Reedy R. C., and Arnold J. R. (1974) Element concentrations from lunar orbital gamma-ray measurements. *Proc. Lunar Sci. Conf. 5th*, p. 1067–1078.

Minkin J. A., Thompson C. L., and Chao E. C. T. (1977) Apollo 16 white boulder consortium samples 67455 and 67475: Petrologic investigation. *Proc. Lunar Sci. Conf. 8th*. This volume.

Nava D. F. and Philpotts J. A. (1973) A lunar differentiation model in light of new chemical data on Luna 20 and Apollo 16 soils. *Geochim. Cosmochim. Acta* **37**, 963–973.

Prinz M., Dowty E., Keil K., and Bunch T. E. (1973) Spinel troctolite and anorthosite in Apollo 16 samples. *Science* **179**, 74–76.

Rose H. J., Jr., Cuttitta F., Berman S., Carron M. K., Christian R. P., Dwornik E. J., Greenland L. P., and Ligon D. T., Jr. (1973) Compositional data for twenty-two Apollo 16 samples. *Proc. Lunar Sci. Conf. 4th*, p. 1149-1158.

Schnetzler C. C., Thomas H. H., and Philpotts J. A. (1967) Determination of rare-earth elements in rocks and minerals by mass-spectrometric stable isotope dilution technique. *Anal. Chem.* **39**, 1888–1890.

Taylor G. J., Drake M. J., Hallam M. E., Marvin U. B., and Wood J. A. (1973) Apollo 16 stratigraphy: The ANT hills, the Cayley Plains, and a pre-Imbrium regolith. *Proc. Lunar Sci. Conf. 4th*, p. 553–568.

Taylor S. R., Gorton M. P., Muir P., Nance W. B., Rudowski R., and Ware N. (1973) Composition of the Descartes region, lunar highlands. *Geochim. Cosmochim. Acta* **37**, 2665–2683.

Ulrich G. E. (1973) A geologic model for North Ray Crater and stratigraphic implications for the Descartes region. *Proc. Lunar Sci. Conf. 4th*, p. 27–39.

Wänke H., Baddenhausen H., Dreibus G., Jagoutz E., Kruse H., Palme H., Spettel B., and Teschke F. (1973) Multielement analyses of Apollo 15, 16, and 17 samples and the bulk composition of the moon. *Proc. Lunar Sci. Conf. 4th*, p. 1461–1483.

Wänke H., Palme H., Baddenhausen H., Dreibus G., Jagoutz E., Kruse H., Spettel B., Teschke F., and Thacker R. (1974) Chemistry of Apollo 16 and 17 samples: Bulk composition, late stage accumulation and early differentiation of the moon. *Proc. Lunar Sci. Conf. 5th*, p. 1307–1336.

Warner R. D., Planner H. N., Keil K., Murali A. V., Ma M.-S., Schmitt R. A., Ehmann W. D., James W. D., Jr., Clayton R. N., and Mayeda T. K. (1976) Consortium investigation of breccia 67435. *Proc. Lunar Sci. Conf. 7th*, p. 2379–2402.

Wilshire H. G., Stuart-Alexander D. E., and Jackson E. D. (1973) Apollo 16 rocks: Petrology and classification. *J. Geophys. Res.* **78**, 2379–2392.

Winzer S. R., Nava D. F., Meyerhoff M., Lindstrom D. J., Lum R. K. L., Lindstrom M. M., Schuhmann P. J., Schuhmann S., and Philpotts J. A. (1977) The petrology and geochemistry of impact melts, granulites and hornfelses from consortium breccia 61175. *Proc. Lunar Sci. Conf. 8th*. This volume.

Proc. Lunar Sci. Conf. 8th (1977), p. 2153–2189.
Printed in the United States of America

Chemical compositions and possible immiscibility of two silicate melts in 12013

J. E. QUICK and A. L. ALBEE

Division of Geological and Planetary Sciences, California Institute of Technology, Pasadena, California 91125

M.-S. MA, A. V. MURALI, and R. A. SCHMITT

Department of Chemistry and the Radiation Center, Oregon State University, Corvallis, Oregon 97331

Abstract—12013 is a complex mixture of gray breccia and black breccia. The gray breccia is comprised of mostly basaltic and gabbroic lithic clasts, and plagioclase and pyroxene mineral clasts set in a matrix of smaller clasts and a felsite of granitic composition. The black breccia is comprised of mostly mineral clasts, some lithic clasts, and felsite blebs enclosed in an aphanitic groundmass of basaltic composition.

Textural relationships indicate that the felsite and the black breccia groundmass crystallized from melts which co-existed prior to the final lithification of 12013 but failed to mix. The major, minor, and trace element compositions of these two lithologies have been determined by a detailed instrumental neutron activation analysis-microprobe study.

Both the felsite and the black breccia groundmass have extreme enrichments of incompatible elements, and, therefore, present a unique opportunity to study two highly-evolved lunar lithologies which occur in close association in the same sample. The critical feature of their compositions is a "splitting up" of incompatible element associations so that the highest REE and P concentrations are in the black breccia groundmass, and the highest K, Ba, and Rb concentrations are in the felsite. This unusual distribution of incompatible elements may be due to silicate liquid immiscibility (SLI) involving genetically-related felsite and black-breccia-groundmass melts. The major element chemistries of these two lithologies appear compatible with a cogenetic relationship involving SLI, but details of the REE abundances in the felsite and the black breccia groundmass are at odds with experimentally determined REE partition coefficients for immiscible silicate melts (Watson, 1976). In view of the limited amount of experimental data on trace element partitioning during SLI, we conclude that final interpretation of a possible felsite-black breccia groundmass genetic relationship must await experiments on melts of more appropriate bulk compositions.

I. INTRODUCTION

THE ORIGIN of highly evolved lunar compositions (i.e., those enriched in "incompatible elements" such as Ba, U, K, REE, P, Th, Zr, Hf, etc.) is one of the fundamental problems of lunar petrogenesis. Were all such highly evolved compositions produced during a single, discrete episode early in lunar evolution and only slightly modified by subsequent impact processes (Albee and Gancarz, 1974), or were some produced during later events which may have occurred as recently as 3.9 b.y.? Could these highly evolved compositions be produced simply by conventional igneous processes such as partial melting of highland rocks (Walker *et al.*, 1973) or fractional crystallization of basaltic magmas

(Ryder *et al.*, 1975), or were more exotic processes such as silicate liquid immiscibility required (Rutherford *et al.*, 1976)? These questions have been difficult to address because of the rarity of large samples with evolved compositions. Evolved compositions were first recognized in the returned samples as a MAGIC component in regolith samples, and then as a petrologically-intangible, chemical component (KREEP) and tangible, but volumetrically insignificant, residual melts in interstices of mare basalts. Later, evolved compositions were recognized as glass or agglutinate fragments and crystalline clasts in complex breccias. Lunar rock 12013 is a breccia with numerous remarkable petrologic and chemical attributes which make it one of the most significant lunar samples to the problem of evolved lunar compositions.

An extreme enrichment in incompatible elements in 12013 was recognized soon after its return by the Apollo 12 mission, and 12013 became the subject of a consortium investigation involving chemical, isotopic, and petrologic studies (*Earth Planet. Sci. Lett.*, vol. 9, 1970). These studies were published without significant interchange of data between consortium members. Sample distribution for chemical and isotopic studies was based on visual discrimination between "black", "white", and "gray" lithologies in a thin slab, and only later were detailed petrographic investigations made on thin sections facing this slab (James, 1970; Albee *et al.*, 1970; Drake *et al.*, 1970). We now know that the samples distributed were, in fact, heterogeneous mixtures involving both clastic and melt components. Therefore, chemical and isotopic data published to date, for the most part, report the chemistry of complex mixtures. Nevertheless, these studies demonstrated the presence in 12013 of two distinct chemical components highly enriched in incompatible trace elements but having different REE patterns (Wakita and Schmitt, 1970; Schnetzler *et al.*, 1970; Hubbard *et al.*, 1970). Rb-Sr isotopic studies (Albee *et al.*, 1970) suggest that many of these characteristics were acquired at ~4.5 AE. Therefore, 12013 may be recording the oldest lunar event that led to the production of highly-evolved compositions.

12013 is a mixture of gray breccia and black breccia, each of which is comprised of both clastic and melt components (Quick and Albee, 1976). Although many of the same types of clasts occur in both breccias, there are significant differences in clast populations. Plagioclase mineral clasts are the most abundant clasts and lithic fragments are rare in the black breccia. In contrast, lithic clasts are more abundant than mineral clasts in the gray breccia. Basalt and microgabbro are the most common lithic clasts in the gray breccia, but ANT-suite rocks dominate the lithic clast suite in the black breccia. The melt components are now crystallized to a fine-grained intergrowth of K-feldspar and silica, which we term felsite, and an extremely fine-grained intergrowth of plagioclase and pyroxene which forms the groundmass of the black breccia. In terms of major elements, the felsite has a "granitic" composition and the black breccia groundmass has a "basaltic" composition. Both felsite and black breccia groundmass have highly evolved trace element chemistry, and it is the abundance of these two components which gives 12013 its high whole-rock enrichments in the incompatible elements. The gray breccia and black breccia correspond to the "gray" and "black" lithologies of the original studies of 12013, and the felsite corresponds to the "white" lithology.

12013 offers an opportunity to study two distinct, highly-evolved lithologies in one rock. We have undertaken a combined petrographic, microprobe, and instrumental neutron activation analysis study to document the chemistry of these two trace element-enriched lithologies. This should provide a basis for comparison of these two lithologies with each other and with other highly evolved lunar lithologies which have been recently characterized chemically (Nyquist *et al.*, 1977; Blanchard *et al.*, 1977).

II. Petrography

12013 is an extremely complex, heterogeneous rock and we will present only a brief petrographic outline to serve as a foundation for interpretation of the major, minor, and trace element data. More comprehensive petrographic descriptions have been presented by James (1970), Drake *et al.* (1970), and Albee *et al.* (1970).

12013 is a complex mixture on both macroscopic and microscopic scales. Two different breccias are readily identified on the macroscopic scale. A black breccia occurs as blebs and schlieren within a more abundant gray breccia. The boundaries between these two lithologies are generally sharp and well-defined by abrupt changes in color. Locally, the boundaries are smooth and cuspate (Fig. 1).

Gray breccia

The gray breccia is a mixture of lithic clasts, mineral clasts, and a fine-grained intergrowth, which we term felsite, comprised predominantly of Ba-rich K-feldspar and silica minerals. Large lithic and mineral clasts are set in a matrix of smaller lithic and mineral clasts and felsite. Clasts have a serital size distribution ranging from 4 mm to less than 10 μm in diameter. Felsite occurs as irregularly shaped, relatively clast-free, 20–500 μm diameter patches. A few patches of the felsite have sharply defined boundaries. Locally, felsite has extensively pervaded the interstices between clasts.

The suite of lithic clasts is dominated by feldspathic basalt clasts; these include aphanitic, variolitic, subophitic and microgabbroic textural varieties. The mineralogy and mineral chemistry of these clasts are similar. A typical basalt is 40–50% plagioclase (An_{64} to An_{94}) and 40–50% low-Ca pyroxene ($Wo_4En_{54}Fs_{42}$ to $Wo_7En_{48}Fs_{45}$), with high-Ca pyroxene ($Wo_{43}En_{37}Fs_{20}$ to $Wo_{35}En_{39}Fs_{26}$), olivine (Fo_{50} to Fo_{60}), ilmenite and chromite comprising less than 10% collectively. Olivine is more abundant in the microgabbros and may constitute as much as 30% of such clasts. The lithic clast suite also includes granulated norite and a single plagioclase-quartz clast. Norite is comprised of about 60–70% orthopyroxene ($Wo_2En_{80}Fs_{18}$ to $Wo_2En_{69}Fs_{29}$) and about 30–40% plagioclase ($>An_{90}$).

Plagioclase ($\sim An_{70}$ to $>An_{90}$) and low-Ca pyroxene ($Wo_4En_{60}Fs_{36}$ to

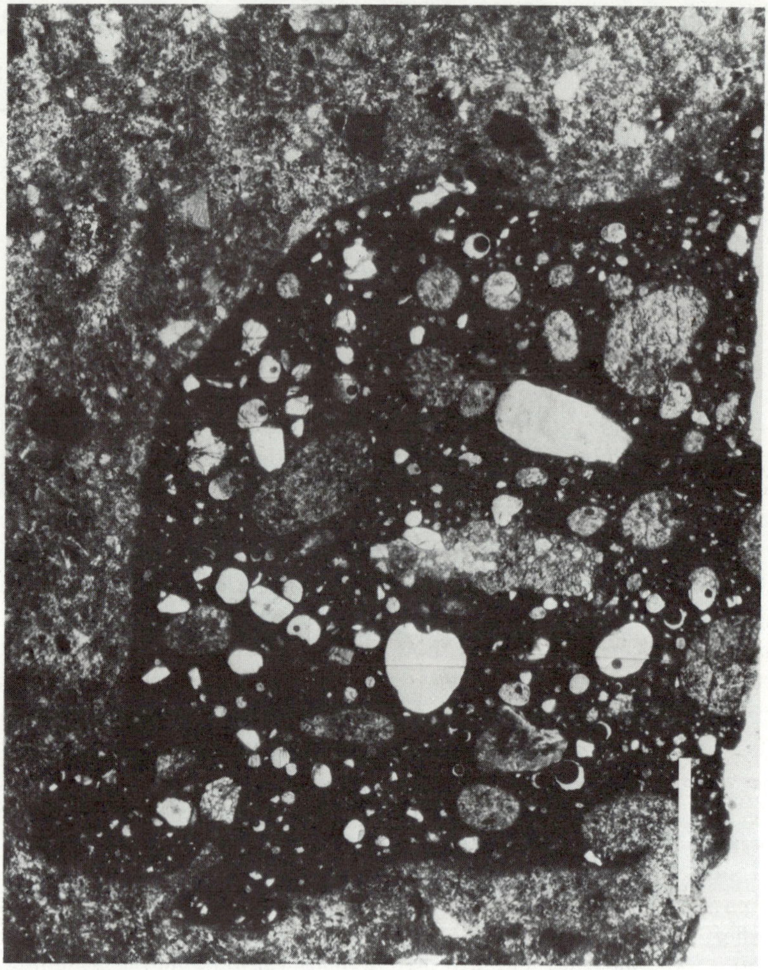

Fig. 1. Photomicrograph of a patch of black breccia in 12013,09,13. Note smooth, cuspate contact with surrounding gray breccia, and the occurrence of numerous, spheroidal, felsite "blebs" in the black breccia. A K X-ray scan of the same field of view is illustrated in Fig. 11. Bar scale is 1 mm.

$Wo_7En_{50}Fs_{43}$) are the most abundant mineral clasts. Relatively sodic plagioclase (An_{50} to An_{70}), high-Ca pyroxene ($Wo_{41}En_{38}Fs_{21}$ to $Wo_{36}En_{39}Fs_{25}$), olivine (Fo_{45} to Fo_{65}), ilmenite, and chromite are less abundant clasts.

Felsite is comprised of ~50% K-feldspar and ~40% tridymite and quartz inverted from tridymite. In most places, quartz is more abundant than tridymite. Minor phases include Nb-rich ilmenite (~1.5 wt.% Nb_2O_5), pyroxene ($Wo_9En_{41}Fs_{50}$ to $Wo_{37}En_{29}Fs_{34}$), and K-rich glass. Whitlockite, zircon, and zirconolite occur in trace amounts and have a nonuniform distribution throughout the felsite. The typical felsite texture is illustrated in Figs. 2A and B. Acicular

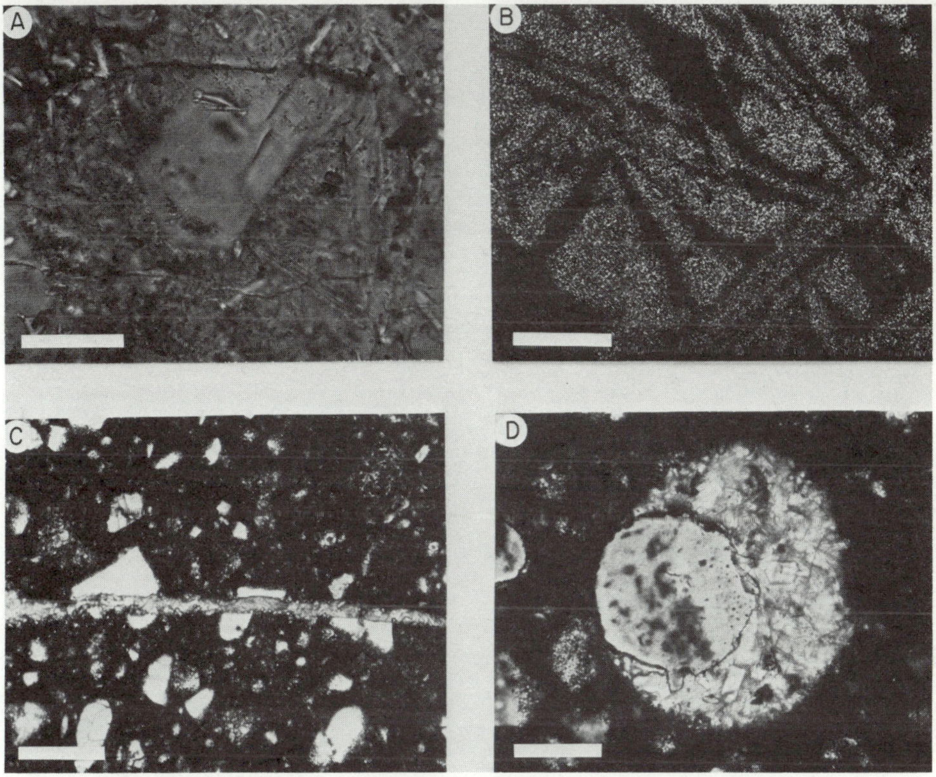

Fig. 2. Photomicrographs and a K X-ray scan illustrating felsite textures in 12013, and two modes of occurrence of felsite in the 12013 black breccia. (A) Typical felsite in the gray breccia of 12013,09,15—A large K-feldspar microphenocryst in an aphanitic matrix of quartz needles and K-feldspar. Bar scale is 500 μm. (B) K X-ray scan illustrating the typical felsite intergrowth of acicular quartz (inverted from tridymite) and K-feldspar (white areas). Bar scale is 25 μm. (C) 12013,09,14—Felsite vein cutting black breccia. Note that the vein cuts several plagioclase clasts. Bar scale is 200 μm. (D) 12013,09,13—Felsite "bleb" in black breccia. Note the occurrence of a vesicle within the "bleb", and K-feldspar microphenocrysts in the felsite. Bar scale is 100 μm.

tridymite and quartz inverted from tridymite occur in a fine-grained groundmass of anhedral K-feldspar, glass, ilmenite plates, and rare whitlockite, zircon, and zirconolite. Quartz or tridymite rods range in length from about 2 mm to 20 μm, but are generally less than 500 μm long. K-feldspar is optically continuous over areas up to 1 mm in diameter. Pyroxene forms extremely acicular crystals up to about 6 mm long. Locally, K-feldspar forms blocky microphenocrysts up to 600 μm wide; some of these microphenocrysts have labradorite cores.

Plagioclase clasts in contact with felsite are commonly rimmed by K-feldspar. Locally, plagioclase grains, which are isolated within K-feldspar, have simultaneous extinction, suggesting extensive replacement of plagioclase by K-feldspar. Felsite has reacted with olivine clasts to produce orthopyroxene

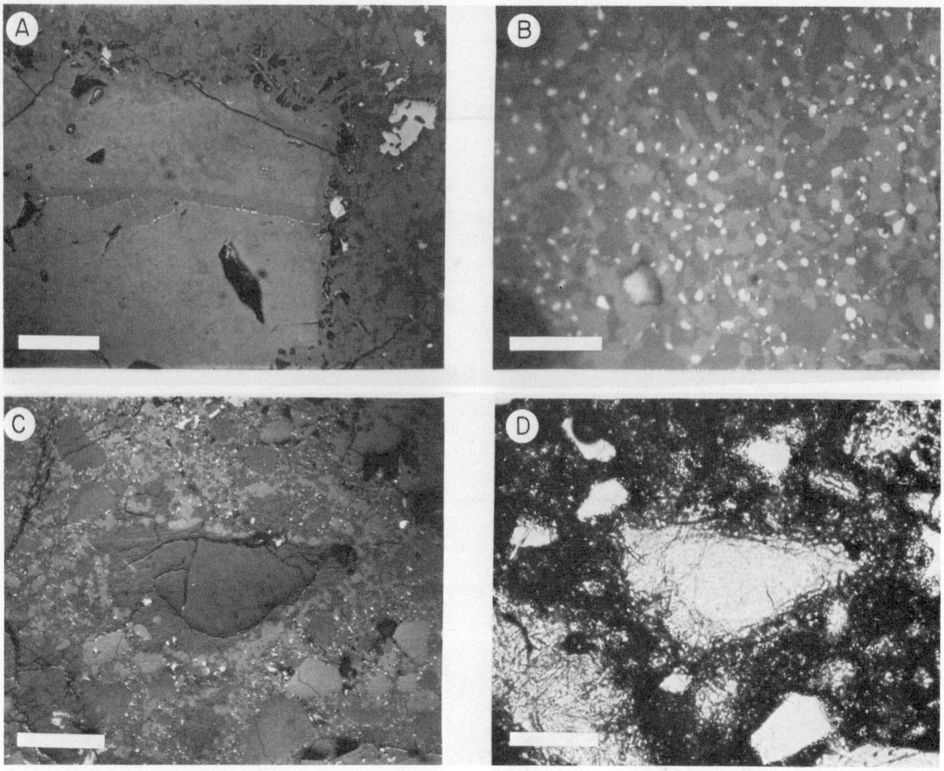

Fig. 3. Photomicrographs illustrating some melt-clast reactions and the typical texture of the black breccia groundmass. (A) 12013,09,15—Reflected light photomicrograph of an orthopyroxene lathwork (upper right) in the gray breccia produced by a reaction between felsite, which fills gray breccia interstices, and olivine (center). Interstices between laths are generally voids. Compare with Fig. 4. Bar scale is 100 μm. (B) Reflected light photomicrograph of the typical texture of the black breccia groundmass. Minerals, from highest to lowest reflectivity, are: ilmenite, low-Ca pyroxene, and plagioclase. Additional phases which are not resolvable in the photograph are apatite, whitlockite, sulfide, and Zr-bearing minerals. Bar scale is 25 μm. (C) 12013,09,15—Reflected light photomicrograph of a rounded tridymite clast in the breccia. Tridymite is completely surrounded by felsite rim which is, in turn, surrounded by a discontinuous rim of low-Ca pyroxene. Bar scale is 100 μm. (D) Same field of view as (C) but in transmitted light.

($\sim Wo_3En_{65}Fs_{32}$). This orthopyroxene forms thin ($<20\ \mu$m) continuous rims on olivine and, in many instances, grades outward, away from the olivine clast into an orthopyroxene lathwork (Fig. 3A). The interstices between the orthopyroxene laths are generally voids.

Black breccia

The black breccia in 12013 is comprised of lithic and mineral clasts and ovoid felsite blebs enclosed in an aphanitic groundmass. Lithic clasts are typically granulated and sheared-out over distances as great as 5 mm. Less deformed lithic clasts range in maximum dimension from about 2 mm down to about 100 μm, and mineral clasts range in diameter from about 2 mm down to about 10 μm. Vesicles are abundant in some parts of the black breccia and range in diameter from about 20 to about 300 μm.

Lithic fragments are much less abundant and generally smaller than in the gray breccia. Clast lithologies include norite, feldspathic basalt, and anorthosite. In contrast to the gray breccia, anorthosite and norite, which are granulated fragments of coarse-grained rocks, are the most abundant lithic clasts. Anorthosite is comprised of plagioclase (An_{50} to An_{70}). Most norite fragments are comprised of about 60% plagioclase ($>An_{90}$) and about 40% orthopyroxene ($Wo_2En_{80}Fs_{18}$ to $Wo_3En_{56}Fs_{41}$). One norite clast is comprised of low-Ca pyroxene ($Wo_3En_{58}Fs_{39}$ to $Wo_7En_{49}Fs_{44}$) which poikilitically encloses euhedral plagioclase (An_{88} to An_{96}). Feldspathic basalt clasts are texturally and mineralogically similar to basalt clasts in the gray breccia and are comprised of subequal amounts of plagioclase (An_{79} to An_{84}) and low-Ca pyroxene ($Wo_6En_{51}Fs_{43}$ to $Wo_{14}En_{46}Fs_{40}$), and minor amounts of high-Ca pyroxene ($Wo_{38}En_{39}Fs_{23}$), olivine ($\sim Fo_{45}$) and ilmenite.

One unusual type of clast is comprised of predominantly granulated plagioclase (An_{50} to An_{60}) and tridymite. Interstices between plagioclase and tridymite grains are filled with felsite. Granulated plagioclase is optically continuous over areas up to 5 mm in diameter. In some clasts, plagioclase is almost completely replaced by K-feldspar. Nb-rich ilmenite ($Nb_2O_5 > 1.0$ wt.%) and pyroxene ($Wo_{38}En_{36}Fs_{26}$) may or may not be present. Locally, relict allotriomorphic granular texture is preserved.

Plagioclase (An_{48} to An_{100}) and low-Ca pyroxene ($Wo_3En_{77}Fs_{20}$ to $Wo_8En_{46}Fs_{46}$) are the most abundant mineral clasts. Other, less abundant mineral clasts include maskelynite, tridymite, olivine (Fo_{55} to Fo_{65}), high-Ca pyroxene ($Wo_{42}En_{38}Fs_{20}$ to $Wo_{39}En_{83}Fs_{28}$), ilmenite, and chromite. Apatite, whitlockite, and zircon clasts occur in trace amounts.

The interstices between mineral and lithic clasts are filled by a fine-grained ($\lesssim 5$ μm) equigranular groundmass of 40–50% plagioclase (An_{90} to An_{67}), 40–50% low-Ca pyroxene ($Wo_3En_{56}Fs_{41}$ to $Wo_5En_{46}Fs_{49}$), ~5% ilmenite, ~5% phosphate minerals and trace amounts of Zr-bearing minerals. Figure 3B illustrates the typical groundmass texture.

The black breccia groundmass has reacted with olivine to produce orthopyroxene, with textural relations similar to the felsite-olivine reaction relationships. Orthopyroxene forms continuous rims up to 20 μm thick on olivine clasts. These rims grade outward, away from the olivine, into a corona of euhedral orthopyroxene laths, which is developed up to ~100 μm from olivine grains. The typical texture of this corona is illustrated in Fig. 4. The contact between the corona of orthopyroxene laths and the typical groundmass is gradational. In contrast to the felsite-olivine reaction zones, interstices between orthopyroxene laths are filled by plagioclase, ilmenite, and phosphate. The effect of this reaction on groundmass bulk chemistry has been determined by electron microprobe

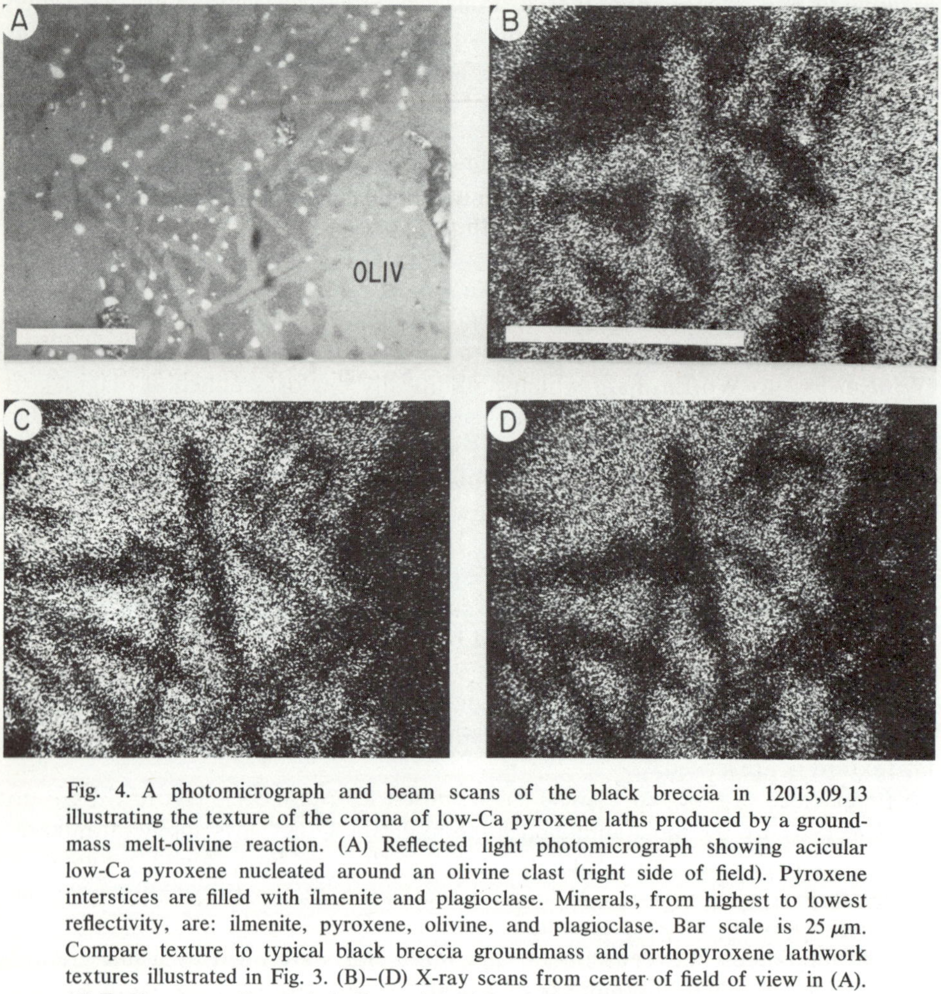

Fig. 4. A photomicrograph and beam scans of the black breccia in 12013,09,13 illustrating the texture of the corona of low-Ca pyroxene laths produced by a groundmass melt-olivine reaction. (A) Reflected light photomicrograph showing acicular low-Ca pyroxene nucleated around an olivine clast (right side of field). Pyroxene interstices are filled with ilmenite and plagioclase. Minerals, from highest to lowest reflectivity, are: ilmenite, pyroxene, olivine, and plagioclase. Bar scale is 25 μm. Compare texture to typical black breccia groundmass and orthopyroxene lathwork textures illustrated in Fig. 3. (B)–(D) X-ray scans from center of field of view in (A). (B) Fe X-ray scan. Bar scale is 50 μm. (C) Al X-ray scan. Same field of view as (B). (D) Ca X-ray scan. Same field of view as (B).

analysis of 25 μm diameter areas of the groundmass at various distances from a large (~500 μm diameter) olivine grain (Fo_{63}). Figure 5 illustrates the variation in FeO/(FeO + MgO) of the groundmass as a function of distance from the olivine clast. FeO/(FeO + MgO) is relatively constant at distances greater than 100 μm from the olivine clast where the groundmass is equigranular. Within the corona of orthopyroxene laths, FeO/(FeO + MgO) decreases abruptly as the olivine clast is approached. Effects of groundmass reactions with other clasts have not been determined, but the composition of the groundmass in areas more than 100 μm from olivine clasts appears to be relatively constant. This suggests that other clast-groundmass reactions have not greatly altered the groundmass composition. Textures do, however, suggest the black-breccia-groundmass melt

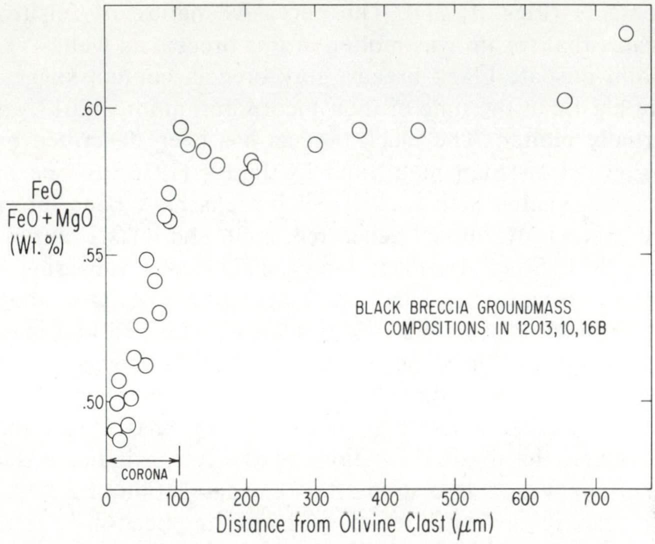

Fig. 5. Black breccia groundmass composition (determined by DBA) in black breccia sample 16B as a function of distance from the edge of the large olivine clast (Fo_{63}) illustrated in Figs. 6 and 7. The limit of the corona of low-Ca pyroxene laths is shown in the lower left corner, and the texture in this area is similar to that illustrated in Fig. 4. Areas further away from the olivine clast have textures similar to that illustrated in Fig. 3B.

reacted with plagioclase, tridymite, phosphate, and zircon clasts. Many plagioclase clasts appear corroded. Apatite, whitlockite, tridymite, and zircon clasts are well rounded suggesting resorption.

Felsite occurs in the black breccia in veins which cut both the groundmass and mineral clasts (Fig. 2C), as distinct, locally-vesicular ovoid blebs up to 1 mm in the longest dimension (Fig. 2D), as rims or partial rims on individual rounded clasts of tridymite or plagioclase (Fig. 3C,D), or as patches and interstitial material in granulated plagioclase-tridymite clasts. Felsite rims on tridymite clasts are partially rimmed by pyroxene coronas (Fig. 3C,D). Some felsite blebs also have partial coronas pyroxene.

Melt components

Textural evidence suggests that two melts—a felsite melt and the black-breccia-groundmass melt—coexisted in 12013 prior to its final lithification. Abundant vesicles in the black breccia indicate that its groundmass was a melt; the fine-grained, equigranular texture may have formed by devitrification of a glass or quenching of a melt. The occurence of felsite veinlets that cut the black breccia, and locally vesicular felsite blebs demonstrates that felsite in the black breccia was molten at one time. Furthermore, the sharp, elliptical contacts of the felsite blebs against the black breccia groundmass host are strongly suggestive of fluid-fluid contacts (Figs. 1, 2D). The pervasive nature of felsite in the gray breccia indicates that felsite was molten in this breccia as well.

The smooth, cuspate black breccia-gray breccia contact suggests that both breccias were plastic at the time of their incorporation into 12013, and therefore, probably partially molten. The black breccia has been described by Quick and Albee (1976) as "clast-laden melt" and by James (1976) as one member of a series of "fragment-laden melt rocks." Such rocks have been interpreted to be produced by mixing of impact-generated melt and clasts during an impact (Grieve *et al.*, 1974; Simonds, 1975; James, 1975a). We subscribe to this interpretation for the origin of the 12013 black breccia and its groundmass melt and suggest that the black breccia clast-laden melt was formed and juxtaposed with the gray breccia during the same impact event. Felsite was incorporated into both breccias at this time. Isolated clast-free felsite patches with sharp, polygonal boundaries suggest some, if not all, felsite may have been added to 12013 as clasts. However, the textural relations of the felsite indicate that some was added as a melt, or was melted soon after the aggregation of 12013.

Two petrographic observations have major implications for studying the trace element chemistry of 12013. First, almost all phases which concentrate incompatible elements have crystallized from the melt components—either the granitic-composition felsite melt or the basaltic-composition black-breccia-groundmass melt. These phases are: Minute phosphate minerals and Zr-bearing minerals in the black breccia groundmass; and whitlockite, apatite, zircon, zirconolite, and glass in the felsite. Clasts of zircon, apatite, and whitlockite in both breccias are insignificant ($<0.1\%$) in the modal abundance. Therefore, the incompatible element chemistry of a fragment of gray breccia will reflect felsite abundance and composition, and the incompatible element chemistry of a fragment of black breccia will reflect groundmass plus felsite abundance and composition. Second, it is virtually impossible to mechanically separate the felsite or the black breccia groundmass from clasts. Therefore, direct determination of the trace element chemistry of the felsite and the black breccia groundmass is not possible. For this reason, the composition of one of the melts can be determined only by analyzing bulk breccia samples and then subtracting out the contributions of the clasts and the other melt from the analyses (Ma *et al.*, 1977).

III. Sample Processing and Analytic Technique

During the preliminary studies on 12013, the rock was cut into slabs, and slab 12013,10 was sent to the California Institute of Technology. There it was broken into small fragments, many of which were distributed for chemical and isotopic studies (Anderson, 1970). From the remaining fragments of slab 12013,10, we handpicked two black breccia samples and two gray breccia samples taking care to minimize clast components and maximize melt components. Black breccia samples were handpicked from slab fragments 12013,10,09 and 12013,10,16B. Gray breccia samples were handpicked from slab fragments 12013,10,16B and 12013,10,28 and consisted of five and nine tiny fragments, respectively. These samples will be referred to herein as black breccia samples 09 and 16B, and gray breccia samples 16B and 28. However, these tiny samples must not be construed to be representative of the fragments from which they were handpicked. Additional larger fragments of black breccia (12013,10,01A) and of gray breccia (12013,10,12A) were also selected for the siderophile element studies of E. Anders, University of Chicago.

All samples were sent to the Oregon State University Radiation Center for determination of 29 major, minor, and trace element abundances by instrumental neutron activation analysis (INAA). Sample preparation and the detailed INAA procedures are essentially the same as reported by Wakita *et al.* (1970), and Laul and Schmitt (1973). Because of the minute sample sizes, small conical micro-polyvials ($\frac{1}{2}$ mil) were used as sample and standard containers during neutron activation and gamma-ray counting to minimize the geometric errors.

Following INAA, samples 01A and 12A were sent to E. Anders. The other four samples were used for detailed petrographic and electron microprobe study. Each fragment was mounted in epoxy and ground down. At evenly spaced intervals, surfaces were polished, photographed, mapped, and point counted. Two surfaces studied within black breccia sample 16B and one surface studied within one fragment of gray breccia sample 16B are illustrated in Figs. 6, 7, and 8. This phase of our project was designed to: (1) identify clasts with high trace element concentrations, (2) determine the mineralogy and mineral chemistry of the most abundant clasts, and (3) determine approximate modal abundance of clasts, felsite, and black breccia groundmass.

The microprobe study was performed with an automated MAC-5-SA3 electron microprobe equipped with three crystal spectrometers and an ORTEC Si(Li) solid state detector interfaced with an NS880 multichannel analyzer. Microprobe point counts were performed on gray breccia samples using the technique described by Chodos *et al.* (1977). The fine grain size of the black breccia groundmass, and the occurrence of plagioclase clasts and groundmass plagioclase with similar compositions prohibits microprobe point-counting as a means of discriminating groundmass in the black breccia samples. These samples were point counted by placing a grid over a mapped photomasaic of each layer and counting the phase under each grid point. Clasts were distinguished from groundmass on the basis of grain size and morphology. In general, there is a distinct gap in grain size between the groundmass ($\leqslant 5\ \mu m$) and clasts ($\geqslant 10\ \mu m$), and clasts are less equant and more angular than groundmass phases. Major element chemistry of the black breccia groundmass and felsite was determined by defocused beam analysis (DBA) of 25–200 μm diameter, clast-free areas using data reduction procedures described by Albee *et al.* (1977). For spot diameters $<50\ \mu$, DBA were performed by both energy dispersive analysis using a solid state detector (EDA) and wavelength dispersive analysis with conventional crystal spectrometers (WDA). EDA was used for spot diameters $>50\ \mu$. Defocused beam analyses run simultaneously with both EDA and WDA yield results that agree within the rather large uncertainties inherent in DBA. Areas of black breccia groundmass were carefully chosen for DBA on the basis of fine grain size and equant texture in order to avoid clast contamination and the effects of olivine-black breccia groundmass reactions. Mineral clasts were analyzed for 9 to 25 elements for each phase by WDA. Operating conditions were 15 kV accelerating voltage for both WDA and EDA, 0.05 μA sample current (on brass) for WDA, and 0.005 μA sample current for EDA. The system is interfaced to PDP-8/L and PDP 11 computers for control and on-line data processing (Chodos *et al.*, 1973; Chodos *et al.*, 1977).

Final estimates of felsite and black breccia groundmass abundance in each sample were based on weighted, least-squares, mass balance calculations (Reid *et al.*, 1973). The abundance of the two melt-derived lithologies and clasts in each sample were calculated by balancing the whole-sample composition, determined by INAA, with representative compositions of the melt-derived lithologies and clasts. The calculations were performed solely with elements routinely analyzed by microprobe. The least-squares calculations were checked by comparing calculated mineral modes with modes determined from point-count data. Trace element abundances of felsite and black breccia groundmass were calculated assuming the incompatible trace elements (REE, U, Th, etc.) are concentrated in these two components. The validity of this assumption, i.e., that the trace element contribution of the clasts is negligible, was confirmed by electron microprobe analyses of the mineral and lithic clasts, and by the mass balance calculations for K and Ba.

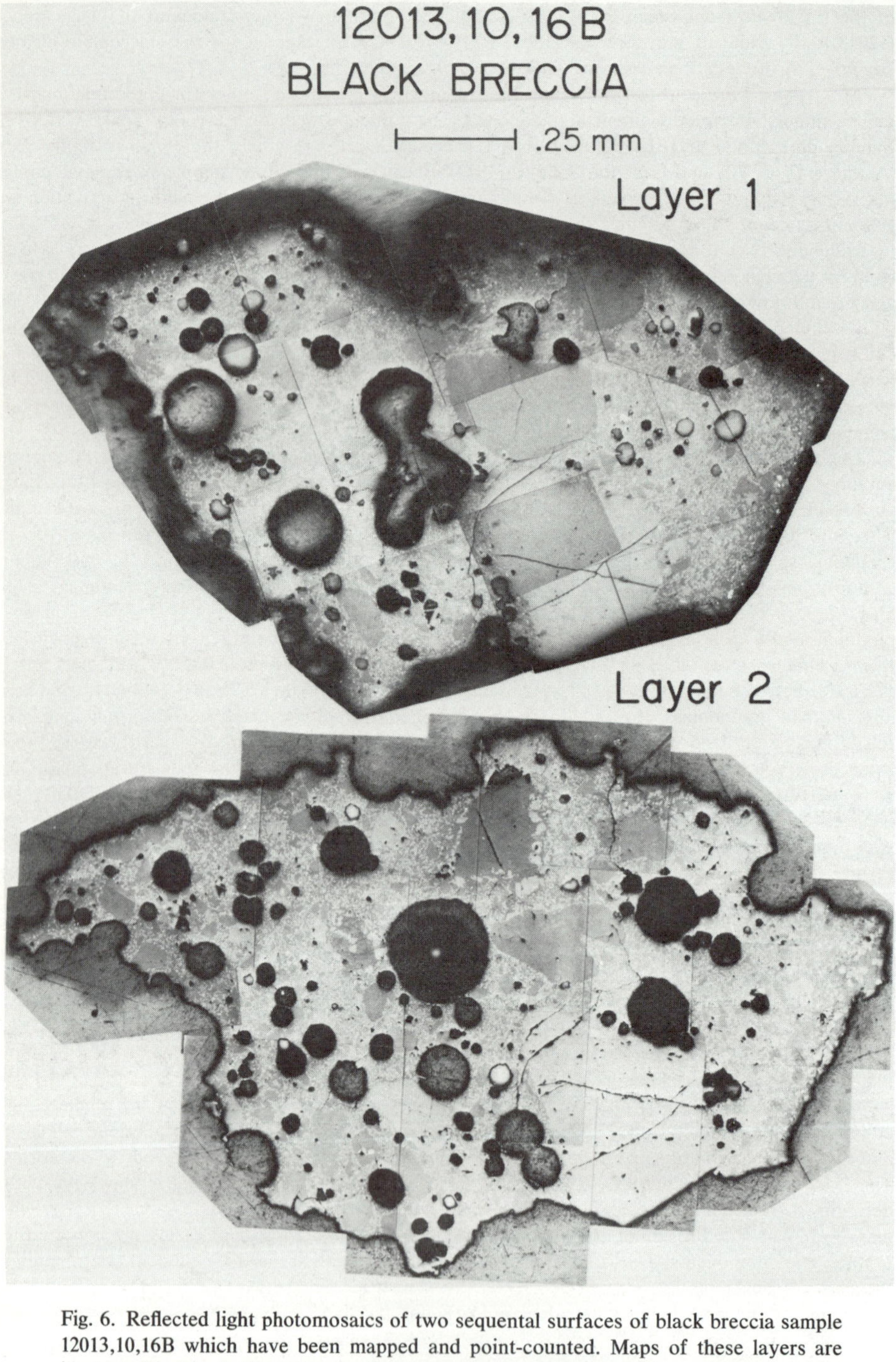

Fig. 6. Reflected light photomosaics of two sequental surfaces of black breccia sample 12013,10,16B which have been mapped and point-counted. Maps of these layers are illustrated in Fig. 6. Note abundant vesicles. A tridymite clast with a continuous felsite rim occurs near the top edge of Layer 2. The olivine clast referenced in Fig. 5 occurs near the lower, right edge of both layers.

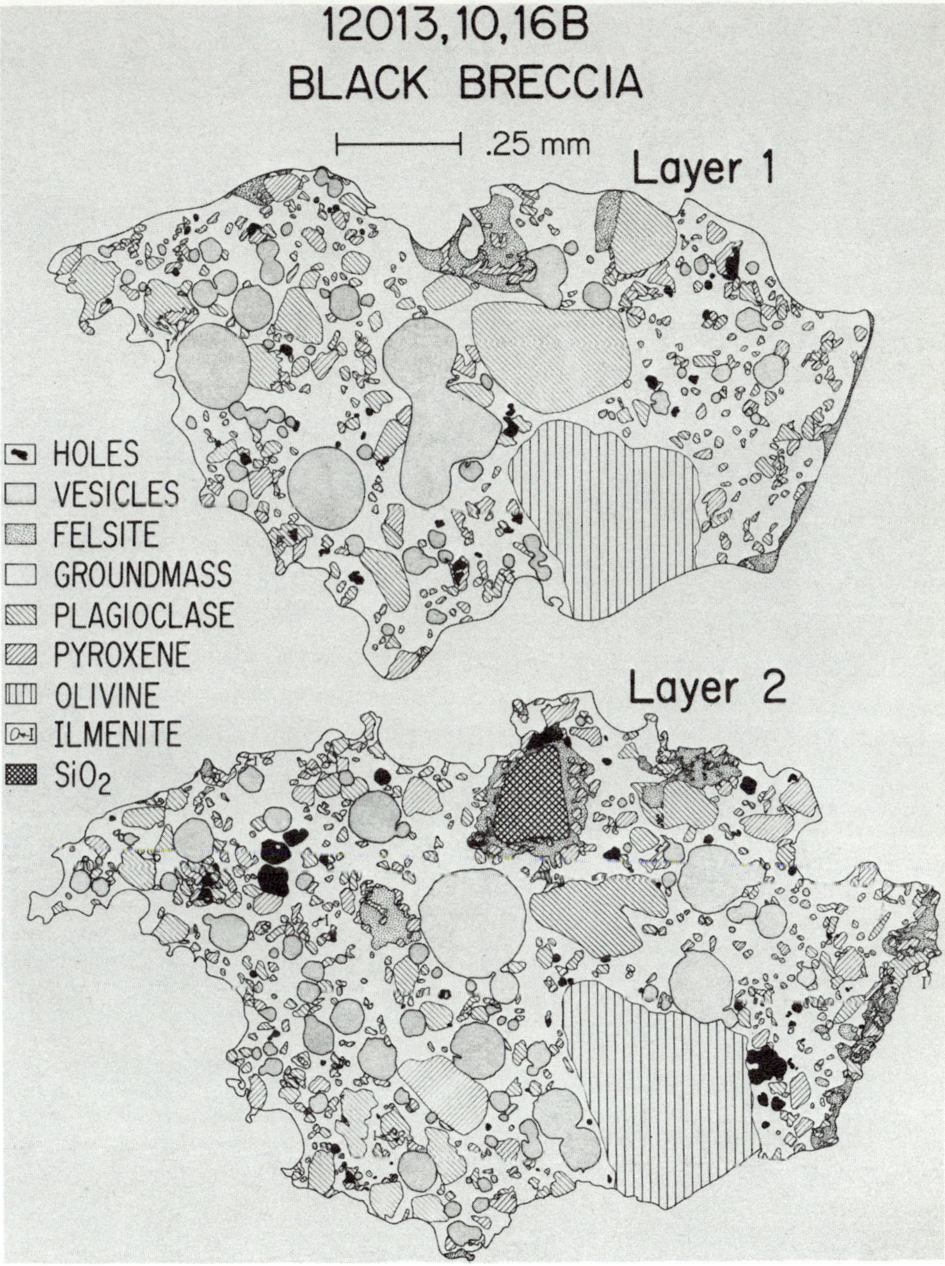

Fig. 7. Maps of the sequential surfaces of black breccia sample 12013,10,16B which are illustrated in Fig. 6. Holes are formed by plucking during polishing.

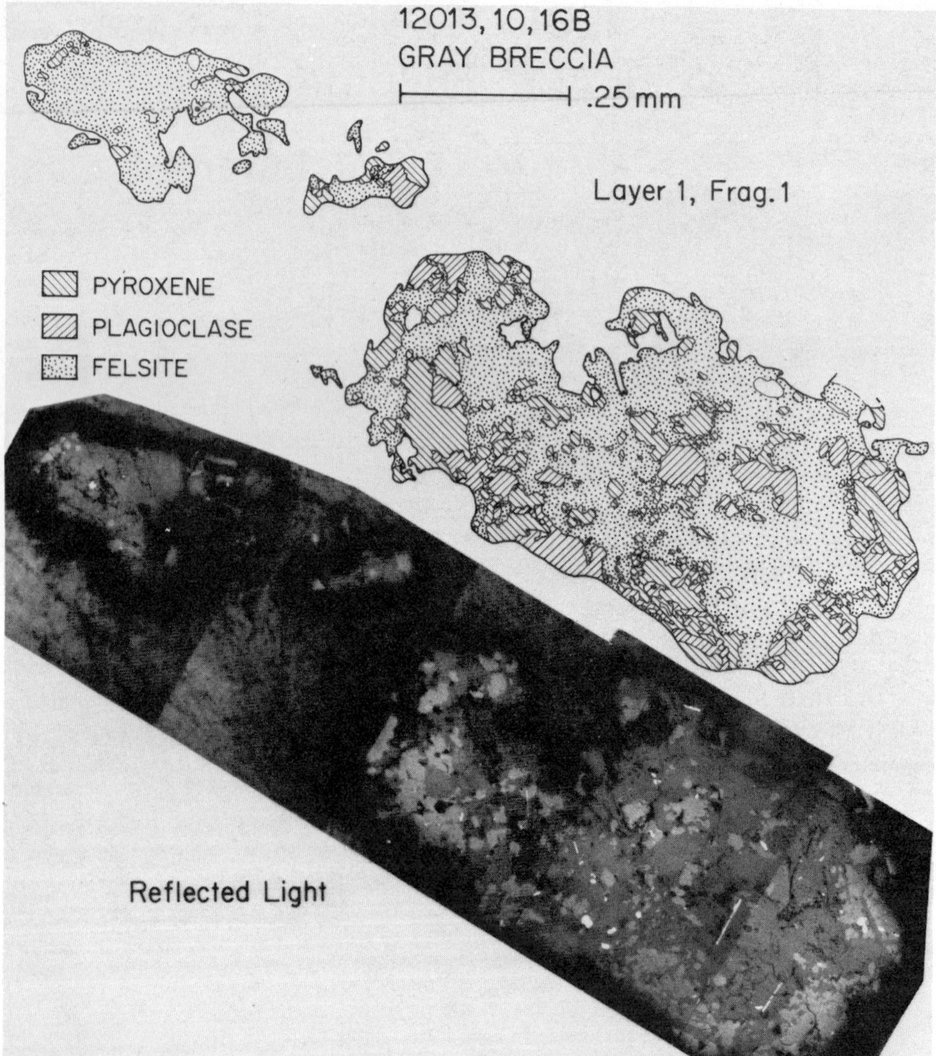

Fig. 8. Reflected light photomosaic and map of one point-counted surface of one fragment of gray breccia sample 12013,10,16B.

IV. Results

The "whole-rock" INAA data for our gray breccia and black breccia samples are listed in Table 1. The patterns of incompatible elements, normalized to chondritic abundances, of the six samples are illustrated in Fig. 9. Microprobe analyses of felsite and black breccia groundmass are tabulated in Table 2. The results of the constrained least-squares analyses applied to the trace elements

Table 1. Elemental abundances in 12013,10 rock fragments determined by INAA[a].

Element	12013,10,16 Black breccia (1.52 mg)	12013,10,09 Black breccia (2.04 mg)	12013,10,01A Black breccia (6.16 mg)	12013,10,16b Gray breccia (1.21 mg)	12013,10,28 Gray breccia (0.98 mg)	12013,10,12A Gray breccia (5.25 mg)
TiO_2(%)	1.9	4.6	3.3	0.75	0.86	0.3
Al_2O_3	12.5	11.9	14.6	11.1	9.8	10.1
FeO	14.2	13.3	13.7	9.6	6.0	14.0
MgO	10.7	8.6	9.0	6.3	5.5	8.5
CaO	7.3	8.9	9.7	4.1	3.8	4.6
Na_2O	1.45	1.47	1.45	1.28	1.17	1.12
K_2O	0.37	0.50	0.40	3.74	3.03	2.6
MnO	0.158	0.164	0.166	0.122	0.077	0.150
Cr_2O_3	0.130	0.144	0.162	0.151	0.086	0.252
Sc(ppm)	23	30	28	21	17	25
V	33	46	34	47	42	61
Co	29	24	27	12	8.6	24
Zr	940 ± 310	1140 ± 280	2070 ± 200	1130 ± 310	720 ± 310	690 ± 150
Hf	32	38	63	23	25	27
Th	17.2	17.0	24.5	44.8	35.6	47.5
U	5.3 ± 0.9	6.2 ± 0.3	10.4 ± 2.3	12.1 ± 1.4	10.4 ± 0.9	14.2 ± 1.0
Ba	660 ± 190	1430 ± 140	390 ± 150	4150 ± 200	4160 ± 290	3760 ± 160
La	135	130	135.7	64	39	56.5
Ce	333	344	347	170	100	151
Nd	242	219	215 ± 46	91	59	74 ± 23
Sm	53	53	59.0	20	13	19.0
Eu	2.52 ± 0.37	3.29 ± 0.30	3.89	1.85 ± 0.04	1.74 ± 0.11	2.14
Tb	10.6	11.4	12.5	4.9	3.7	5.2
Dy	67	70	75	38	23	33
Yb	36	37	39	35	25	30
Lu	4.8	4.5	5.2	5.2	3.0	4.2
Ta	3.9	9.1	5.3	6.6	5.0	7.1
Rb	16 ± 13	33 ± 17	10 ± 7	124 ± 21	101 ± 21	87 ± 11
Cs	1.8 ± 0.8	2.1 ± 0.6	1.7 ± 0.4	4.0 ± 0.7	4.3 ± 0.3	3.1 ± 0.6

[a]Estimated errors due to counting statistics are: Al_2O_3, Na_2O, MnO, Sc, La, and Sm, $\sim \pm 1$–3%; FeO, K_2O, Cr_2O_3, Co, Ce, Tb, Yb, Th, and Hf, $\sim \pm 5\%$; TiO_2, CaO, Dy, Lu, and Ta, $\sim \pm 5$–15%; MgO, V, Ba, Cs, Eu, and U, $\sim \pm 10$–30%; Zr, $\sim \pm 30$–40%; Nd and Rb, ~ 20–60%.

are presented in Table 3, and are illustrated in Fig. 10. Tables 4 and 5 summarize the compositional data input into the least-squares calculation for black breccia sample 16B and gray breccia sample 16B, along with the point-count determined modal mineralogy, and the modal mineralogy and bulk composition calculated by constrained least-squares.

The constrained least-squares calculations were performed using several constraints from the measured modal mineralogy of each sample to prevent negative mineral abundances from appearing in the solution. In the analysis of black breccia 16B, black breccia groundmass was constrained to be 50 ± 5 wt.% and augite was constrained to be 1 ± 1 wt.%. Similarly, tridymite was constrained to be $.5 \pm .5$ wt.% in the analysis of gray breccia 16B. Simultaneous solution for olivine, pyroxene, and tridymite abundances is possible because the average pyroxene compositions in the samples studied are not colinear with the olivine and tridymite compositions. Abundant tiny ilmenite inclusions occur in some

Table 2. Defocused beam microprobe analyses of black breccia groundmass and felsite in 12013,09,13.

	Black breccia groundmass		Felsite*			
	(1)		(2)		(3)	
	wt.%	1 σ	wt.%	1 σ	wt.%	1 σ
Na_2O	1.9	0.3	1.4	0.2	1.4	0.4
MgO	7.4	1.4	0.5	0.5	0.7	1.1
Al_2O_3	13.2	1.8	12.1	0.8	11.9	1.8
SiO_2	50.1	1.7	73.1	1.5	73.0	3.6
K_2O	0.06	0.03	6.7	0.6	6.8	1.1
CaO	9.5	1.1	1.5	0.6	1.4	1.1
TiO_2	2.2	0.5	n.a.		0.6	0.9
Cr_2O_3	0.13	0.04	<0.1		<0.1	
MnO	0.15	0.07	<0.1		<0.1	
FeO	12.3	1.9	1.3	0.9	1.4	1.8
BaO	0.03	0.03	n.a.		0.73	0.24
NiO	0.02	0.03	<0.1		<0.1	
P_2O_5	1.2	0.2	0.2	0.2	0.13	0.26
SO_3	0.1	0.1	<0.1		<0.1	
ZrO_2	0.3	0.6	n.a.		<0.1	
Sum	98.59		96.8		98.06	

(1) Average and standard deviation of 23 WDA analyses. Spot size 30 μm.

(2) Average and standard deviation of 15 EDA analyses. Spot size 200 μm.

(3) Average and standard deviation of 45 WDA analyses. Spot size 40 μm.

n.a. – not analyzed.

*Column (2) is our preferred estimate of the felsite composition for all oxides except BaO and TiO_2 for which we use the results in column (3).

pyroxene clasts in the black breccia sample, and, because of their small size, were not point-counted. As a result, the abundance of ilmenite is much lower in the point-count results than in the least-squares solution for this sample. In the gray breccia sample, pyroxene is higher and olivine is lower in the point-count results than in the least-square solution because the reaction, melt + olivine = orthopyroxene, has replaced olivine xenocrysts with orthopyroxene. Taking these observations into consideration, the calculated mineralogy and point-counted mineralogy are in excellent agreement. This agreement strongly supports the methodology used in this project, and we feel we have accurately determined the abundance of the phases and the compositions of the melt-derived lithologies in each sample.

The REE patterns in Fig. 9 form two distinct groups. Black breccia fragments have tightly-grouped REE patterns with the highest total REE abundances and a

Table 3. Calculated trace element abundances[a] in the felsite and the black breccia groundmass in fragments of 12013,10.

	Black breccia groundmass		Felsite	
Element	12013,10,16B	12013,10,09	12013,10,16B	12013,10,28
Sc	43	55	37	38
V	58	81	83	95
Co	56	45	21	19.4
Zr	1700	2010	1990	1620
Hf	60	71	41	56
Th	27.1	23.5	79.1	80.2
U	9.4	11.3	21.9	24.8
La	261	245	113	88
Ce	640	648	300	225
Nd	471	416	161	133
Sm	103	101	35	29
Eu	4.75	6.14	3.26	3.92
Tb	20.5	21.6	8.6	8.3
Dy	128	131	67	52
Yb	66	66	62	56
Lu	9.1	8.6	10.2	8.3
Ta	6.7	16.6	11.6	11.3
Rb	11	37	219	228
Cs	2.9	3.3	7.1	9.7

[a]All abundances in ppm. See text for calculation procedure.

negative Eu anomaly. Gray breccia fragments show a greater range in REE abundances than do the black breccia fragments, and possess a negative Eu anomaly superimposed on a distinctive, V-shaped REE pattern. These results are similar to those from the preliminary investigations of 12013. Our black breccia REE patterns are similar to those of the "dark-colored portions" (12013,10-#10, #15 and #8 matrix) analyzed by Schnetzler *et al.* (1970) and the "black material" (12013,10-05D) analyzed by Hubbard *et al.* (1970). Our gray breccia REE patterns are similar to those of the "light-colored portions" (12013,10-#37 + 24, #8, and #15) analyzed by Schnetzler *et al.* (1970) and "white material" (12013,10-05L) analyzed by Hubbard *et al.* (1970).

The REE patterns calculated for the felsite in gray breccia samples 16B and 28, and for the black breccia groundmass in black breccia samples 16B and 09 (Fig. 10) are basically similar to the "whole-rock" breccia patterns, but enriched in REE by a factor of ~2×. The close grouping of the REE patterns indicates that the black breccia groundmass is homogeneous on a scale of ~1.0 mg. These two patterns are similar to those of KREEP (Schmitt and Laul, 1973), but the absolute REE abundances are similar to those observed in 15405,85, a quartz monzodiorite sample (Nyquist *et al.*, 1977) and are the highest ever observed in any lunar lithology. The spread between the felsite patterns suggest that the felsite is not homogeneous on a scale as small as 1.0 mg. Furthermore, K and Ba

Table 4. Least-squares analysis of black breccia sample 16B: Compositional data used in the calculation, the modal mineralogy determined by visual point-counting, and the modal mineralogy calculated by least-squares. For discussion, see text.

	Calcic plag.	Sodic plag.	High-Ca pyx.	Low-Ca pyx.	Tridymite	Olivine	Ilmenite	Chromite	Felsite	Ground-mass	Bulk composition (INAA)	Bulk composition Least-squares[c]
Wt.% (point-count)[a]	15.7_3		15.4_6		1.6_2	12.6_1	0.1_4	0.0_2	4.4_1	49.7_9		
Wt.% (least-squares)[b]	10.1_0	6.2_4	0.8_4	13.3_7	1.6_2	11.9_6	1.3_5	0.0_4	4.7_2	49.7_6		
Data for least-squares calculation												
Na_2O	0.74	5.45	0.15	0.04	0.0	—	—	—	1.3_5	1.9_4	1.45 ± 0.03	1.4_5
MgO	0.02	0.08	12.34	18.77	0.02	31.83	3.22	2.68	0.5	7.4_5	10.7 ± 1.1	10.2_1
Al_2O_3	35.76	26.32	0.71	0.38	0.72	0.0	0.01	10.32	12.0_5	13.2_4	12.5 ± 0.1	12.4_8
SiO_2	44.06	56.75	51.81	51.48	98.22	37.96	0.04	0.0	73.1_4	50.1	—	49.8_8
K_2O	0.03	0.37	—	—	0.06	—	—	—	6.6_9	0.0_6	0.37 ± 0.02	0.3_7
CaO	19.31	10.43	19.79	2.74	0.04	0.09	—	—	1.5_4	9.5_2	7.3 ± 0.7	7.9_6
TiO_2	0.01	0.0	0.69	0.39	0.15	0.06	53.12	9.06	0.0	2.2_2	1.9 ± 0.1	1.8_8
Cr_2O_3	—	—	0.21	0.20	—	0.04	1.26	38.94	0.0_1	0.1_3	0.13 ± 0.01	0.1_8
MnO	—	—	0.25	0.44	—	0.12	0.28	0.57	0.0_6	0.1_5	0.158 ± 0.005	0.1_6
FeO	0.14	0.0	14.78	25.71	0.08	33.28	42.94	37.26	1.3	12.2_8	14.2 ± 0.4	14.3_4
BaO	0.0	0.0	—	—	0.0	—	—	—	0.73	0.0_3	0.07 ± 0.02	0.0_5
Sum	100.07	99.40	100.73	100.15	99.29	103.38	100.87	98.83	97.37	97.12		

[a]Calculated from point-count data (Vol.%; 23,909 points) using phase densities.
[b]Solution to least-squares calculation.
[c]Calculated from the least-squares-determined modal mineralogy using the compositional data for each phase.

Table 5. Least-squares analysis of gray breccia sample 16B: Compositional data used in the calculation, the modal mineralogy determined by microprobe point-counting, and the modal mineralogy calculated by least-squares. For discussion, see text.

	Calcic plag.	Sodic plag.	High-Ca pyx.	Low-Ca pyx.	Tridymite	Olivine	Ilmenite	Chromite	Felsite	Bulk composition (INAA)	Bulk composition Least-squares[c]
Wt.% (point-count)[a]	12.9_8		29.1_7		<0.1	0.1_3	0.7_1	0.0	57.0		
Wt.% (least-squares)[b]	5.2_4	8.3_5	0.8_2	21.7_4	0.4_0	4.5_8	1.2_2	0.1_9	57.4_6		
Data for least-squares calculation											
Na_2O	0.74	5.45	0.15	0.0	0.0	—	—	—	1.3_5	1.28 ± 0.04	1.2_7
MgO	0.02	0.08	13.10	15.69	0.02	31.83	3.22	2.68	0.5	6.3 ± 1.2	5.5_3
Al_2O_3	35.76	26.32	0.81	0.31	0.72	0.0	0.01	10.32	12.0_5	11.1 ± 0.1	11.0_9
SiO_2	44.06	56.75	51.64	51.38	98.22	37.96	0.04	0.0	73.1_4	—	62.8_0
K_2O	0.03	0.37	—	—	0.06	—	—	—	6.6_9	3.74 ± 0.19	3.8_8
CaO	19.31	10.43	19.30	2.21	0.04	0.09	—	—	1.5_4	4.1 ± 0.6	3.4_1
TiO_2	0.01	0.0	0.68	0.36	0.15	0.06	53.12	9.06	0.0	0.75 ± 0.11	0.7_5
Cr_2O_3	—	—	0.42	0.24	—	0.04	1.26	38.94	0.0_1	0.151 ± 0.005	0.1_5
MnO	—	—	0.05	0.35	—	0.12	0.28	0.57	0.0_6	0.122 ± 0.002	0.1_2
FeO	0.14	0.0	13.54	29.91	0.08	33.28	42.94	37.26	1.3	9.6 ± 0.3	9.5_0
BaO	0.0	0.0	—	—	0.0	—	—	—	0.73	0.46 ± 0.02	0.4_2
Sum	100.07	99.40	99.69	101.45	99.29	103.38	100.87	98.83	97.37		

[a]Calculated from point-count data (Vol.%; 4969 points) using phase densities.
[b]Solution to least-squares calculation.
[c]Calculated from the least-squares-determined modal mineralogy using the compositional data of each phase.

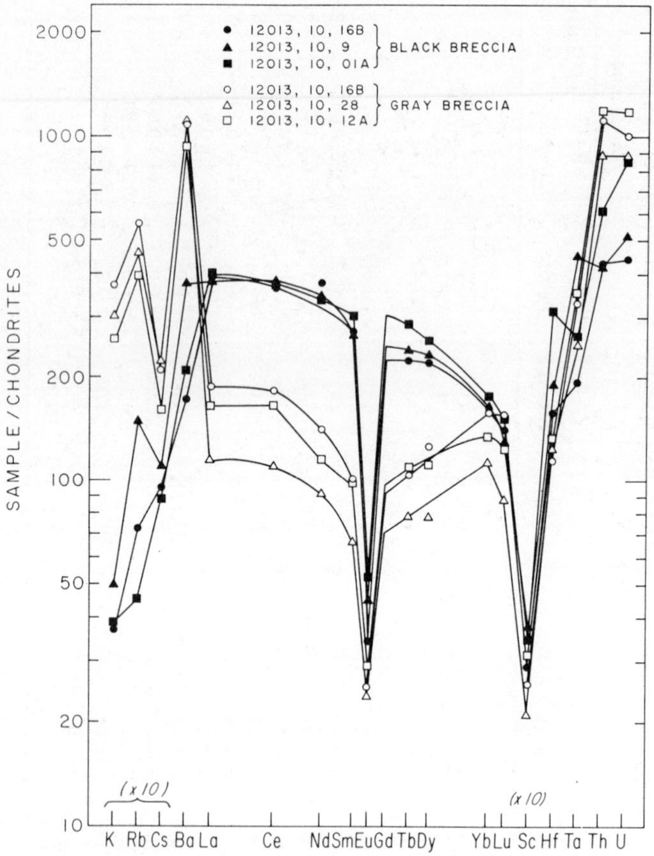

Fig. 9. Chondritic-normalized K, Rb, Cs, Ba, REE, Sc, Hf, Ta, and U abundances in six black breccia and gray breccia samples from 12013,10.

abundances do not correlate with REE abundance in the gray breccia. This may be explained by different amounts of Ba-rich K-feldspar in each of the gray breccia samples. The distinctive V-shaped REE pattern is similar to the pattern reported by Blanchard *et al.* (1977) for a felsite clast from 73215.

V. DISCUSSION

The main purpose of this discussion is to evaluate possible genetic relationships between the felsite and the black breccia groundmass. Two observations are critical to the problem. First, the association in 12013 of two petrographically and compositionally distinct lithologies in near equal abundance and both with extreme trace element enrichment is a singular occurrence among lunar samples. 12013 is a polymict, clastic rock, probably of impact origin, and, therefore, this association may be chance juxtaposition of totally unrelated lithologies. However, in view of the scarcity of highly evolved lunar lithologies, a genetic

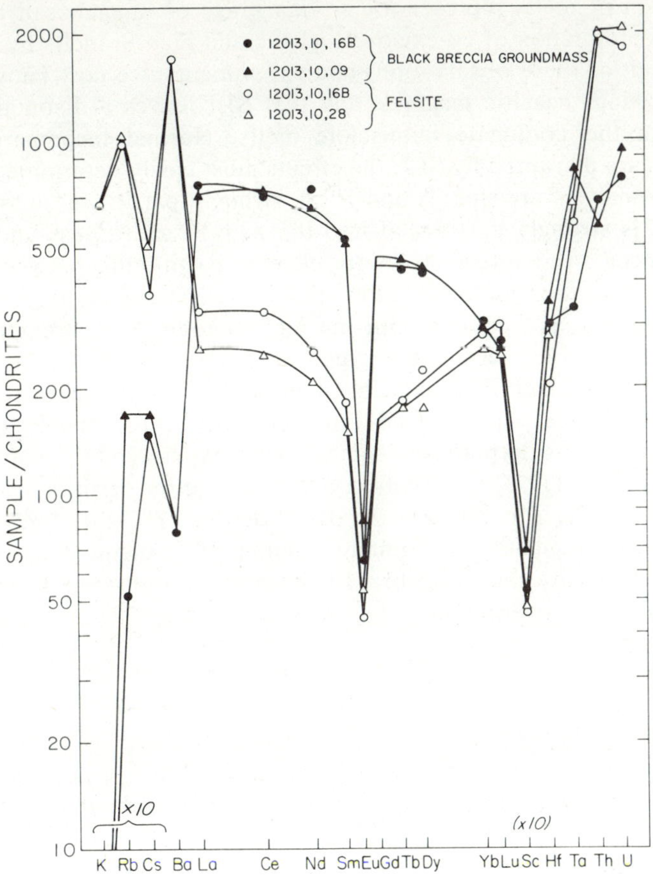

Fig. 10. Chondritic-normalized K, Rb, Cs, Ba, REE, Sc, Hf, Ta, and U abundances in the black breccia groundmass of black breccia samples 12013,10,16B and 12013,10,09, and the felsite of gray breccia samples 12013,10,16B and 12013,10,28.

relationship between the black breccia groundmass and the felsite seems plausible. Second, the probability of a genetic relationship is enhanced by the chemical compositions of these two lithologies, which almost demand an episode of silicate liquid immiscibility at some point in their evolutions.

Criteria for silicate liquid immiscibility (SLI)

Roedder and Weiblen (1970) first documented evidence for SLI in lunar basalts. The examples cited were late-stage, coexisting, high-Fe, and high-Si glasses in the interstices of Apollo 11 mare basalts. Numerous examples of SLI have since been recognized in lunar basalts returned by subsequent missions (Roedder and Weiblen, 1971, 1972; Powell *et al.*, 1975). Hess *et al.* (1975) and Rutherford *et al.* (1976) demonstrated experimentally that continued fractional crystallization of melts representing a wide range of lunar basalt compositions leads to the production of co-existing high-Fe and high-Si melts by SLI.

As a result of these observations and experiments, we now know that SLI is common in lunar basaltic magmas, and that SLI leaves a distinctive chemical signature on the conjugate immiscible melts. Normal incompatible element associations are disrupted by SLI; the effects most easily determined by electron microprobe analysis are that Ti and P are strongly partitioned into the high-Fe melt and K is strongly partitioned into the high-Si melt. Hess and Rutherford (1974) advanced a theoretical model for element partitioning between immiscible silicate melts based on ionic field strength (cation charge/cation radius). According to this hypothesis, cations with the highest ionic field strength and which cannot tetrahedrally coordinate oxygen (e.g., U, Th, REE, P, Zr, Ti) are partitioned into the high-Fe immiscible melt, and low-ionic field strength cations (e.g., K, Ba, Rb) are partitioned into the high-Si immiscible melt. Partial confirmation of this hypothesis comes from experimental investigations by Ryerson and Hess (1975) and Watson (1976) on cation partitioning (specifically Mg, P, Ca, Ti, Cr, Mn, Sr, Zr, Cs, Ba, La, Sm, Yb, Lu, and Ta) between immiscible silicate melts. As common incompatible element associations are strongly disrupted by the onset of SLI, it should be possible to recognize the effects of SLI in a given lithology by measuring certain key element ratios such as K/La and K/P.

Felsite as a product of SLI

Hess *et al.* (1975) suggested that the felsite in 12013 was a direct result of SLI, and that the felsite crystallized from a high-Si melt that had completely separated from the conjugate high-Fe melt. The critical point of their argument was that K/La, Rb/Yb, and Ba/La ratios were approximately an order of magnitude higher than the "average" lunar element ratios (Taylor and Jakeš, 1974) of K/La = 70, Rb/Yb = 0.8, and Ba/La = 13. Using published data on 12013 (Drake *et al.*, 1970; Hubbard *et al.*, 1970), Hess *et al.* (1975) inferred that these ratios in the felsite were K/La = 520, Rb/Yb = 3, and Ba/La = 90. Our data on the felsite composition yields ratios of K/La = 554, Rb/Yb = 3.8, and Ba/La = 65, which are in close agreement and clearly demonstrate that the normal associations of K, Rb, and Ba with REE do not exist in the felsite. This "splitting up" of incompatible elements is a characteristic feature of SLI. Hess *et al.* (1975) argued that crystal fractionation was an improbable mechanism for producing the felsite composition, since fractionation of >6 wt.% whitlockite from a parent granitic melt would be required to produce the observed felsite composition. This would constrain the parent melt to have at least 4 wt.% P_2O_5. In as much as no lunar granite has been reported with P_2O_5 contents this high, they rejected fractional crystallization for lack of a suitable parent. Furthermore, an origin of the 12013 felsite by SLI is supported by compositional similarities with experimentally produced high-Si melts (Hess *et al.*, 1975), and late-stage high-Si immiscible melts in lunar basalts.

Black breccia groundmass as a product of SLI

Compositionally, the 12013 black breccia groundmass is a low-K feldspathic basalt. However, it is distinguished from all other lunar feldspathic basalts by extremely high concentrations of REE, Th, U, and P. Our calculated K/La, Rb/Yb, and Ba/La ratios for the black breccia groundmass are 2.4, 0.2, and 1.2, respectively. These values are approximately an order of magnitude lower than the "average" lunar ratios (Taylor and Jakeš, 1974), suggesting that, as in the case of the felsite, normal incompatible element associations have been disrupted by SLI. The major element chemistry of the black breccia groundmass is not typical of late-stage, high-Fe immiscible melts commonly observed in lunar basalts (Roedder and Weiblen 1970, 1971, 1972; Powell *et al.*, 1975) but high-Fe immiscible melts with similar major element chemistries have been produced experimentally (Hess *et al.*, 1975).

James (1976) classified the 12013 black breccia as one member of a "gradational series of highlands breccias" which included: Apollo 17 light-gray breccias, blue-gray breccias, and green-gray breccias; Apollo 16 poikilitic rocks; Apollo 15 and 16 "black and white" rocks; and dark clasts in Apollo 14 thermally metamorphosed breccias. James (1976) noted that these were all "fragment-laden melt rocks" with similar textures, major element compositions, and REE element patterns. Using published data, we calculated the K/La, Rb/Yb, and Ba/La ratios for the following members of the series: 73215,161 and 73215,184 (James *et al.*, 1975b); 73215,46 (Blanchard *et al.*, 1976); 64015 (Bansal *et al.*, 1973); 15455 black (Taylor *et al.*, 1973); and 14306 dark (Taylor *et al.*, 1972). The range in K/La values for these samples is 44 to 133, the range in Rb/Yb values is 0.3–0.6, and the range in Ba/La values is 10 to 12. These values are close to the "average" lunar element ratios, and K/La and Ba/La are an order of magnitude higher than the respective ratios for the 12013 black breccia groundmass. While the 12013 black breccia may be classified as a member of this "gradational series of highlands breccias", we feel that the unusual composition of the black breccia groundmass strongly suggests an episode of SLI occurred at some time in its history.

Minor phases in 12013

The distribution of elements among the phases in 12013 must be understood before possible genetic relationships can be discussed. Microprobe analyses of gray breccia phases demonstrate that K and Ba are concentrated in K-feldspar microphenocrysts, interstitial K-feldspar, and trace amounts of glass, all of which formed from the felsite melt. Rb-Sr isotopic studies (Albee *et al.*, 1970) on heavy liquid separates of felsite demonstrated that the highest Rb concentrations were in separates rich in K-feldspar. Therefore, we conclude that K, Ba, and Rb concentrations in the gray breccia and felsite are controlled by the modal abundance of K-feldspar. The remaining incompatible elements are more difficult to assign to phases in the felsite. Candidates for concentrating REE, U, and Th are whitlockite, apatite, zircon, zirconolite, and possibly residual glass. The occurrences of these phases were discussed above. We emphasize that clastic phosphate and zircon grains are extremely rare in the gray breccia, and absent in gray breccia samples #16B and #28.

Therefore, the abundances of most trace elements in the gray breccia are exclusively controlled by phases which crystallized from the felsite melt, and understanding the distribution of these elements in the gray breccia is equivalent to understanding their distribution in the felsite. Given the compositions of these phases, we can estimate the modal abundances required to produce the observed whole-sample trace element concentration, or estimate the whole-sample concentration from measured modes. By comparing these estimates with observed modal abundances or element concentrations, it is possible to determine which phases control the trace element chemistry of the gray breccia and felsite. The compositions, including some trace element data, of zircon, apatite, whitlockite, and zirconolite (phase β) in 12013 have been documented by Haines *et al.* (1971). The highest REE concentrations occur in whitlockite ($REE_2O_3 \sim 4.9$ wt.%) and zirconolite ($REE_2O_3 \sim 2.3$ wt.%). By comparison, apatite and zircon have negligible REE abundance ($REE_2O_3 < 0.5$ wt.%). The highest U and Th concentrations occur in zirconolite, which contains ~3.4 wt.% UO_2 and ~4.1 wt.% ThO_2. Zircon, apatite, and whitlockite all contain less than 300 ppm U; Th concentrations for these phases are not known. Whitlockite is the most abundant phosphate mineral in the gray breccia, and the gray breccia REE concentrations may be explained primarily in terms of whitlockite abundance. Point-count results on gray breccia sample 28 indicate it contains 0.28 ± 0.07 wt.% phosphate-rich phases. Assuming these phosphates to be entirely whitlockite, we calculate whole-sample La, Ce, and Nd concentrations of 29 ± 7, 88 ± 22, 48 ± 12 ppm, respectively, which compare favorably with the whole-sample abundances determined by INAA (Table 1).

U and Th concentrations are controlled by the abundance of zirconolite. Based on point-count data, zircon, apatite, and whitlockite could contribute no more than 10% of the observed U abundance in gray breccia sample 28. The U and Th content of gray breccia sample 28 could be explained by 0.03 to 0.09 wt.% zirconolite. This is compatible with our point-count results which determined the abundance of Zr-rich phases (zircon + zirconolite) to be approximately 0.22 ± 0.08 wt.%. Analyses from Haines *et al.* (1971) indicate Hf is almost exclusively partitioned into zircon. Using this data, the amount of zircon required to explain the Hf abundance in gray breccia sample 28 is ~0.09. Therefore, our computed values for zircon + zirconolite are 0.12–0.18 wt.%, which is close to the measured amount.

A good estimate of the abundance of zirconolite in 12013 is important because of its high U and Th concentrations. Haines *et al.* (1971) noted a high Nb concentration in 12013 zirconolite (~8 wt.% Nb_2O_5) and suggested that, if most of the Nb were concentrated in zirconolite, up to 0.3 wt.% zirconolite would be required to account for the 12013 whole-rock Nb abundance of

170 ppm (LSPET, 1970). As a result of this study, we now know that Nb_2O_5 ranges from 0.5 wt.% in clastic ilmenite to >1.5 wt.% in ilmenite which crystallized from felsite. Using the 12013 whole-rock Ti abundance of 1.2 wt.% (LSPET, 1970), and assuming all Ti to be in ilmenite, we calculate ~4 wt.% ilmenite in 12013. Taking 0.5 wt.% Nb_2O_5 as a conservative estimate of Nb abundance in 12013 ilmenite, we calculate a contribution of 140 ppm Nb to the 12013 whole-rock composition. The remaining 30 ppm Nb can be accounted for by only 0.05 wt.% zirconolite, which is in agreement with estimates based on U and Th abundances.

We conclude that, in the felsite and gray breccia, K, Ba, and Rb concentrations reflect K-feldspar abundance, REE concentrations reflect mainly whitlockite abundance, and U and Th reflect mainly zirconolite abundance.

The minute grain size of the 12013 black breccia groundmass make a detailed analysis of incompatible element distributions difficult. The K in this lithology (0.06 wt.% K_2O) is probably concentrated in plagioclase which is 40–50 wt.% of the groundmass and contains 0.1–0.3 wt.% K_2O. No other K-bearing phases have been identified with the microprobe. We infer that Rb and Ba are also concentrated in this phase because their chemistries are similar to that of K. REE, U, and Th are concentrated in submicron to ~5 μm diameter phosphate-bearing phases and submicron diameter Zr-bearing phases. Both apatite and whitlockite have been identified on the basis of presence or absence of Cl, but the compositions of these phases have not been determined. The nature of the Zr-bearing phases has not been determined and they could be zirconolite and/or zircon. If the analyses of apatite, zircon, whitlockite, and zirconolite by Haines *et al.* (1971) were applicable, whitlockite would have to be the most abundant phosphate in order to account for the extremely high REE abundances in the groundmass, and zirconolite would have to be present to account for the high U and Th abundances. However, Burnett *et al.* (1971) demonstrated that U concentrations in 12013 phosphate minerals increase dramatically with decreasing grain size. Presumably, other incompatible elements also have the highest concentrations in the smallest accessory phases. Therefore, it is premature to speculate on the exact nature of the phosphate and Zr-bearing phases in the black breccia groundmass.

Felsite-black breccia groundmass relationship

As discussed above, many of the textural and chemical attributes of the black breccia groundmass and felsite in 12013 are compatible with a cogenetic relationship involving SLI. Abundant textural evidence suggests that these two lithologies crystallized from melts which co-existed but failed to mix. Diffusion zones should exist at the contacts of the felsite and black breccia groundmass if they crystallized from miscible melts. For example, in one experiment by Yoder (1973), diffusion zones 60 μm wide were produced between basaltic- and granitic-compositioned melts in contact for one hour at 1200°C. Coronas around olivine clasts in the black breccia indicate that diffusion occurred in the groundmass melt over distances on the order of 100 μm. However, compositional changes across the felsite-black breccia groundmass contacts are abrupt (Figs. 11 and 12), and similar to compositional changes across the interface of co-existing high-Fe and high-Si glasses in the interstices of lunar basalts (Fig. 13). The similarity of the felsite and black breccia groundmass to conjugate high-Si and high-Fe melts is even more striking on inspection of the major element chemistry. P is enriched in the black breccia groundmass ($P_2O_5 >$ 1 wt.%) and K is only present at trace levels (<0.1 wt.%), while the felsite is strongly enriched in K and depleted in P. Furthermore, the major element compositions of both felsite and black breccia groundmass are similar to those of experimentally produced conjugate high-Si and high-Fe immiscible melts (Hess *et al.*, 1975). We conclude that the textural relations and major element chemistries of the black breccia groundmass and felsite argue strongly for a cogenetic relationship involving SLI.

The trace element chemistry of the felsite and black breccia groundmass is more problematic. To test the trace elements for compatibility with SLI we have

Fig. 11. K X-ray scan of the patch of black breccia (12013,09,13) shown in Fig. 1. Felsite shows up as areas of high-K concentration (white). Note that K is concentrated in felsite "blebs" within the black breccia, and in the interstices between clasts of the surrounding gray breccia.

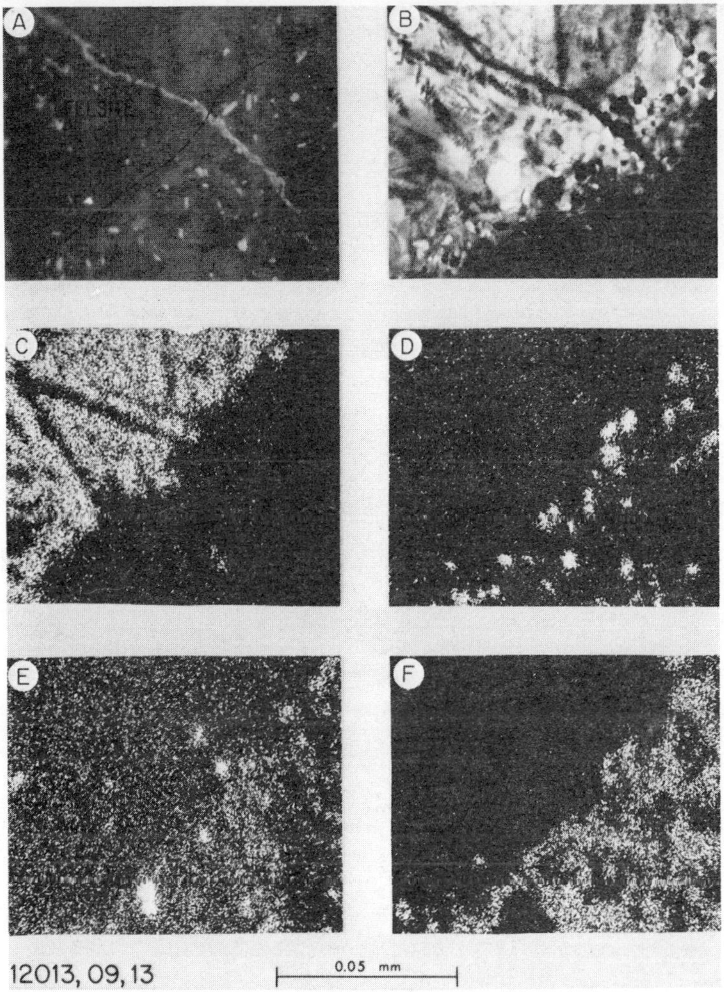

Fig. 12. 12013,09,13—Photomicrographs and X-ray scans of the boundary between a felsite "bleb" and surrounding black breccia groundmass illustrating the abrupt compositional transitions across the black breccia groundmass-felsite contact. Field of view is the same for all photos and scans. (A) Reflected light. (B) Transmitted light. The felsite-groundmass contact appears gradational because it dips gently to the lower right hand corner of the field of view. (C) K X-ray scan. (D) Ti X-ray scan. (E) P X-ray scan. (F) Fe X-ray scan.

compared the interelement ratios between the felsite and black breccia groundmass from sample 16B. The enrichment factors, i.e., ratios of the interelement felsic ratios to the interelement mafic ratios, e.g., $(K/La)_{felsic}/(K/La)_{mafic}$ (abbreviated hereafter as $(K/La)_{f/m}$) are inconsistent with the theoretical predictions (Hess and Rutherford, 1974) and the experimental data (Ryerson and Hess, 1975; Watson, 1976) for trace element partitioning between two immiscible melts. The

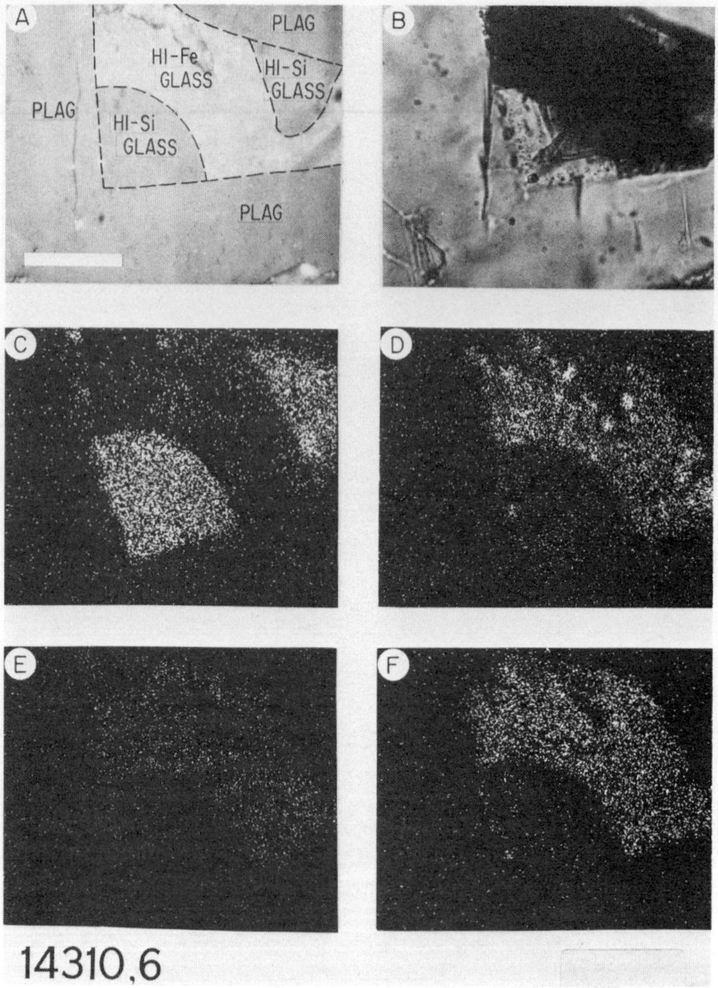

Fig. 13. 14310,6—Photomicrographs and X-ray scans of late-stage, coexisting, immiscible silicate glasses. The bar scale in (A) is 25 μm. All photos and scans are approximately the same scale. (A) Reflected light. (B) Transmitted light. (C) K X-ray scan. (D) Ti X-ray scan. (E) P X-ray scan. (F) Fe X-ray scan.

observed enrichment factors are: $(K/La)_{f/m} = 246$, $(K/Lu)_{f/m} = 94$, $(K/U)_{f/m} = 46$, $(K/Th)_{f/m} = 37$, $(K/Ta)_{f/m} = 63$, and $(La/Lu)_{f/m} = 0.4$. Based upon the theoretical predictions and the experimental results, the relative values of the enrichment factors should be $(K/Th)_{f/m} \sim (K/U)_{f/m} > (K/Lu)_{f/m} > (K/La)_{f/m}$ which is inconsistent with our observations on 12013. Furthermore, $(La/Lu)_{f/m}$ should be >1 for an immiscible melt pair since both La and Lu have the same charge (+3) but Lu has a smaller ionic radius and therefore greater ionic field strength. In addition to the enrichment factor argument, experimental evidence (Watson, 1976) indicates that the REE patterns for two equilibrium immiscible melts should be similar because only small differences exist between melt-melt partition coefficients for different REE.

For comparison, we have initiated a combined INAA-microprobe study of a terrestrial alkaline sill for which Philpotts (1976) has argued the importance of SLI (Dymek *et al.*, 1977). Our sample of this sill is comprised of white syenitic segregations (ocelli) in a black, fine-grained, lamprophyre matrix. Both the lamprophyre matrix and the ocelli are comprised of plagioclase, K-feldspar, kaersutitic amphibole, titanomagnetite, aegirine-augite, sphene, calcite, pyrite, analcime, and apatite (Dymek *et al.*, 1977). The mineralogy of these two lithologies differ primarily in modal abundance. Philpotts argued from petrographic and experimental evidence that the ocelli were formed by separation of an immiscible syenite melt from the lamprophyre melt. We will treat the ocelli and lamprophyre matrix as analogues of the felsite and black breccia groundmass, respectively.

The major, minor, and trace element concentrations of two samples of ocelli and two samples of lamprophyre matrix are given in Table 6, and the REE patterns for these samples are plotted in Fig. 14. J. A. Philpotts (pers. comm., 1977) has also determined REE abundances in ocelli and matrix samples from the same sill. Philpotts obtained REE abundances for the matrix that are in close agreement with our results, but REE abundances for the ocelli that are approximately a factor of 2 greater than in our ocelli samples. Also, the ocelli REE pattern determined by Philpotts more closely parallels our matrix REE patterns than our ocelli REE patterns.

Caution must be exercised in comparing this terrestrial sample with 12013 since the ocelli-lamprophyre system is silica-undersaturated and rich in alkalies. One effect of this bulk chemistry may be the absence of strong K-partitioning between ocelli and lamprophyre matrix. Nevertheless, several interesting observations may be made. First, contrary to theoretical predictions, U and Th are partitioned into the high-Si melt (ocelli) not the high-Fe melt (matrix). As a result of U and Th partitioning into the ocelli, the enrichment factors for the ocelli-lamprophyre system are $(K/Lu)_{f/m} \sim (K/La)_{f/m} > (K/Th)_{f/m} \sim (K/U)_{f/m}$ which is a similar order to 12013, but inverse to theoretical predictions. On the basis of these preliminary results, we suggest that the ionic field strength model for cation partitioning between immiscible melts may not be valid for all cations (e.g., U and Th), or may require modifications to accomodate all bulk compositions of melt systems which undergo SLI. Therefore, SLI in 12013 cannot be disputed simply on the basis of higher U and Th enrichments in the felsite compared to the black breccia groundmass. Second, as predicted by Watson's experimental investigation (Watson, 1976), the REE patterns of the syenite ocelli and lamprophyre matrix are approximately parallel. The partition coefficients, however, are smaller for La and Lu (~2) and much larger for Sm (~10) than Watson's (4–6). This discrepancy may reflect dependence on bulk chemistry or simply equilibration at different points on the two-liquid solvus. For any given bulk composition undergoing SLI, melt-melt partition coefficients will be a function of the temperature at which the two-liquid field is encountered. We

Table 6. Elemental abundances in ocelli and lamprophyre matrix of a terrestrial alkaline sill determined by INAA[a].

	OKA-4-1 matrix (13.87 mg)	OKA-4-2 matrix (58.10 mg)	OKA-4-3 ocelli (37.75 mg)	OKA-4-4 ocelli (23.06 mg)
TiO_2(%)	3.4	3.1	0.7	0.9
Al_2O_3	16.0	16.2	23.4	22.3
FeO	10.0	9.9	1.9	1.5
MgO	—	4.0[b]	0.5[b]	—
CaO	11.1	10.7	2.9	2.0
Na_2O	4.9	5.10	9.62	9.61
K_2O	2.5	2.2	2.3	3.2
MnO	0.368	0.279	0.092	0.085
Cr_2O_3(ppm)	5	9	—	—
Sc	6.3	6.6	0.16	0.12
V	281	266	44	33
Co	17.0	16.6	2.4	2.2
Zr	470	500	295	300
Hf	12.5	13.8	4.8	3.9
Th	14.2	15.8	16.5	19.6
U	3.5	3.2	4.5	4.4
Ba	640 ± 85	705 ± 90	945 ± 130	1560 ± 150
La	123.8	121.2	78.8	63.9
Ce	234	260	105	77
Nd	109 ± 55	118 ± 25	17 ± 9	24 ± 6
Sm	20.0	20.1	2.8	1.6
Eu	5.93	6.02	0.87	0.51
Dy	16.4	11.7	2.5	—
Tb	2.1	2.3	.47	.30
Yb	3.8	3.6	1.2	1.0
Lu	0.47	0.46	0.17	0.15
Ta	11.1	12.1	1.8	0.8
Cs	1.1	1.1	1.6	1.7
Rb	59	53	55	70
Au(ppb)	16	7	8	12

[a]Estimated errors due to counting statistics are: Al_2O_3, Na_2O, MnO, Sc, La, and Sm, ~ ±1–3%; FeO, K_2O, Cr_2O_3, Co, Ce, Tb, Yb, Th, and Hf, ~ ±5%; TiO_2, CaO, Dy, Lu, Cs, Rb, and Ta ~ ±5–15%; V, Ba, Au, Eu, Zr, and U, ~ ±10–30%; Nd, ~ ±50%.

[b]MgO determined by microprobe analysis of fused glass beads.

emphasize that Watson's REE partition coefficients are not rigorously applicable to all examples of SLI, and that REE partition coefficients could be a function of bulk chemistry or equilibration temperature.

Summarizing the preceding discussion, both 12013 felsite and black breccia groundmass have unusual chemical compositions which seem to require SLI to break up normal incompatible element associations. The major element chemsitries of these two lithologies are compatible with an origin as genetically-related conjugate immiscible melts, but some aspects of the trace element

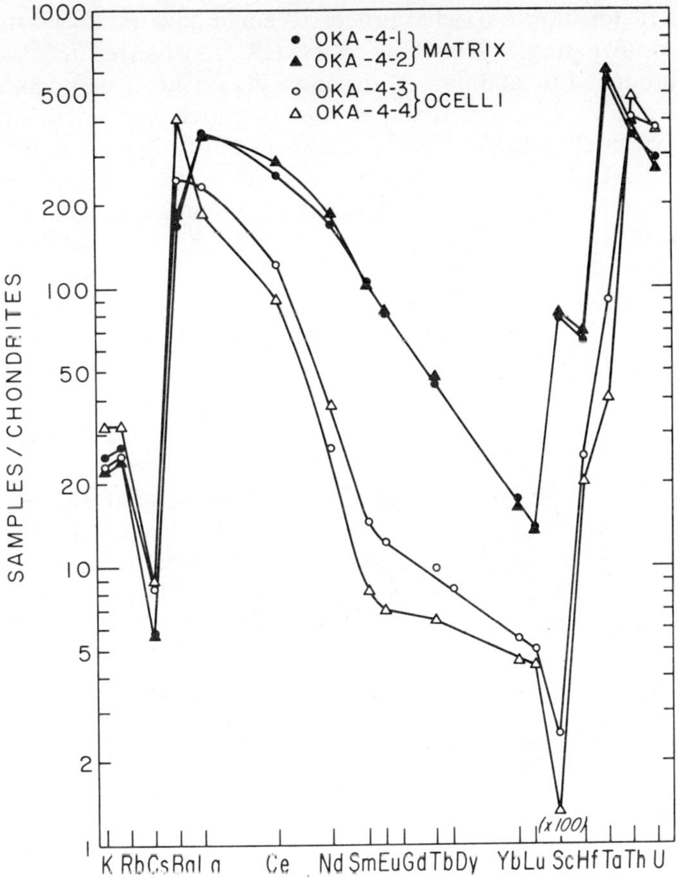

Fig. 14. Chondritic-normalized K, Rb, Cs, Ba, REE, Sc, Hf, Ta, and U abundances in two samples of syenite ocelli and two samples of lamprophyre matrix from a terrestrial lamprophyre dike. Sample occurrence and petrography have been described by Dymek *et al.* (1977). Philpotts (1976) postulated that the syenite ocelli and lamprophyre matrix crystallized from immiscible melts.

chemistries of the felsite and black breccia groundmass do not comply with theoretically and experimentally derived criteria for SLI. In particular, the distinctive, V-shaped REE pattern of the felsite is inconsistent with theoretical criteria, experimentally-derived criteria, and our terrestrial analogue. It is possible that the felsite REE pattern may have been altered by a post-SLI process, in which case, it should be possible to estimate the initial REE pattern of the felsite at the onset of SLI, using the experimental data of Watson (1976), and reconstruct the observed V-shaped pattern from this initial pattern by some petrographically-recognizable process.

If, at the onset of SLI, the felsite REE pattern was approximately parallel to the observed black breccia groundmass pattern but lower in absolute abundances by a factor of 4 to 6, what processes could have produced the observed V-shaped felsite pattern? Schnetzler *et al.* (1970) suggested the V-shape may have been produced by addition of zircon to the felsite. This hypothesis seems unlikely because we have not found zircon associated with the felsite except in trace amounts (≤.2%). Also, zircon occurs as tiny, submicron-sized, euhedral crystals in interstices between silica needles and K-feldspar; these were almost certainly crystallized from the felsite melt rather than added as xenocrysts.

Could intermediate REE be removed by fractionation of a phase or phases which concentrate intermediate REE after the felsite was incorporated into 12013? Apatite, whitlockite, zircon, and zirconolite occur in trace amounts (<0.01%) in the felsite. The REE patterns of these phases are not documented, but they do have non-uniform distributions in the felsite. Therefore, if any of these phases concentrated intermediate REE, the V-shaped pattern may simply reflect an incomplete sampling of the felsite, in that the phase rich in the intermediate REE was not representatively sampled. This seems unlikely since the felsite V-shape REE pattern is an intrinsic property of all 12013 gray breccia and felsite analyses (Schnetzler *et al.*, 1970; Hubbard *et al.*, 1970; Wakita and Schmitt, 1970; Ma *et al.*, 1977). These analyses were made on "large" (15–100 mg) samples from many different parts of slab 12013,10. Failure to obtain a representative sampling seems unlikely in view of the large amount of material analyzed.

Could intermediate REE have diffused out of the felsite melt and into the black breccia? Phosphate minerals are more abundant in the 12013 black breccia groundmass than in the felsite, and it is possible that the crystallization of these phases may have depleted nearby groundmass and felsite melts in the intermediate REE. If diffusion of REE into the black breccia were the mechanism, diffusion must have been extremely efficient over distances on the order of centimeters. However, black breccia groundmass-olivine reactions indicate that diffusion in 12013 was probably limited to a few hundred microns. In addition, the V-shaped felsite REE pattern crosses the black breccia groundmass REE pattern at Lu, indicating that some process such as plagioclase fractionation, in addition to intermediate REE removal, is required to produce an overall increase in REE abundances in the felsite.

Could multiple phase fractional crystallization have occurred after SLI but prior to the formation of 12013 and have been obscured by the breccia forming process? Using the REE abundances in the black breccia groundmass and felsite of 12013,10,16B we semi-quantitatively tested this possibility by: (1) assuming the black breccia groundmass is the high-Fe immiscible melt, (2) calculating the REE abundances in the felsite progenitor assuming that it was depleted in each REE by a factor of 4.5—the average of Watson's partition coefficients for REE between immiscible melts (Watson, 1976)—relative to the conjugate high-Fe melt, and (3) calculating the minimum fractional crystallization, *f*, and the bulk partition coefficients, $\bar{D}$, required to generate the observed felsite REE abundances from those of the hypothetical progenitor. For the last calculation, we used the equation of Shaw (1970):

$$\frac{C_i^1}{C_{i_0}} = (1-f)^{\bar{D}_i - 1}$$

where,

$C_i^1 \equiv$ concentration of element i in the residual liquid,

$C_{i_0} \equiv$ concentration of element i in the unfractionated liquid,

$f \equiv$ degree of fractional crystallization

$= \dfrac{\text{weight of crystals}}{\text{initial weight of melt}}$,

$\bar{D}_i \equiv$ bulk partition coefficient for element i

$= \dfrac{\text{concentration of } i \text{ in the bulk crystal residue}}{\text{concentration of } i \text{ in the residual melt}}$.

The equation assumes continuous separation of crystals from the melt. Values for C_i^1 correspond to the observed REE abundances in the felsite, and the values for C_{i_0} correspond to the inferred REE abundances in the unfractionated progenitor of the felsite melt. By assuming a value for f, we may calculate the bulk distribution coefficients for each REE.

Bulk distribution coefficients less than zero have no physical meaning. To insure $\bar{D}_{REE} > 0$ for all REE, the degree of fractional crystallization has to be >80%. The range of $\bar{D}_{REE}$ calculated for $x = 0.8 - 0.95$ are: La 0.59–0.78, Ce .54–0.76, Sm 0.74–0.86, Eu −.31–0.63, Tb 0.61–0.79, Dy 0.48–0.72, Yb 0.12–0.53, and Lu 0.05–0.49. The calculated $\bar{D}_{REE}$ patterns are similar to those for apatite in dacite and granite (Nagasawa, 1970) but the absolute values are smaller indicating that apatite would only be a small fraction of the crystal residue. Due to the uncertainties in the magnitudes of REE partition coefficients for apatite, whitlockite, and zirconolite, as well as in plagioclase and K-feldspar (Nagasawa, 1971), we have not performed a detailed mass balance calculation. This calculation suggests it may be possible to generate the felsite REE pattern in 12013 by a large degree of fractional crystallization (>80%) from a felsite progenitor which has a REE pattern parallel to that of the black breccia.

In any case, the required fractional crystallization would, unless the melt had a eutectic composition, so alter the major element chemistry that the felsite composition would no longer reflect SLI. Watson's experiments on trace element partitioning between immiscible melts suggest that the felsite is not generally related to the black breccia groundmass by SLI. However, considering the complexity of the SLI process and the paucity of experimental data for systems with different bulk compositions, it seems premature to rule out such a genetic relationship between the felsite and the black breccia groundmass.

VI. Conclusions

Any model for the origin of 12013 must explain the following observations. First, 12013 is comprised of two breccias—a gray breccia and a black breccia—

with somewhat different clast populations. Second, textural evidence suggests that two immiscible melts co-existed in 12013. One of these melts has crystallized to form the groundmass of the black breccia. The other melt has crystallized to a felsite that occurs in both the black breccia and the gray breccia. Third, the incompatible elements in both the felsite and black breccia groundmass have been "split up" in a manner suggestive of SLI. Fourth, some details of the trace element chemistries of the felsite and black breccia groundmass are, based on available experimentally and theoretically derived criteria, inconsistent with a cogenetic relationship for these two lithologies involving SLI.

Two very different models for the origin of 12013 may be considered. The first model assumes that the felsite melt and the black breccia groundmass melt were produced by complete melting of genetically-related rocks during an impact event and that these melts were mixed with clasts during the impact process to form the gray breccia and the black breccia. A possible scenario would be an impact into a highly-differentiated basaltic intrusion in which late-stage granophyres have separated. Such a model accounts for the evidence suggesting that these two melts are related by SLI and accounts for the somewhat different clast populations.

An alternative model is that the felsite melt and the black-breccia-groundmass melt formed by impact partial melting within two different, genetically unrelated progenitors. A possible scenario for this model would be an impact into two unrelated impact sheets—one basaltic in composition and the other containing abundant granitic material. Such a model explains much of the petrographic evidence, but it must be assumed that the two rocks fortuitously produced apparently immiscible melts, both of which were unusually rich in incompatible elements.

Many other models can be envisaged, but it is clear that the key is a final determination of whether the two melts are truly genetically related by SLI. The final solution to this problem requires experimental determination of trace element partition between melts with the exact compositions determined in this paper.

Acknowledgments—We wish to thank R. F. Dymek, D. A. Papanastassiou, and A. A. Chodos for assistance in this project. R. F. Dymek provided samples and participated in the study of the ocellar lamprophyre samples. D. Papanastassiou assisted in the selection of the 12013 samples and A. A. Chodos is responsible in large part for the quality of the electron microprobe data.

We particularly wish to thank O. B. James, with whom we have undertaken a detailed petrologic examination of 12013. Our understanding of this complex rock has been greatly enhanced by her petrographic observations and ideas on the origin of lunar breccias.

This manuscript has benefitted from reviews by R. F. Dymek, A. J. Gancarz, O. B. James, G. W. Reed, and E. Roedder.

Our research has been supported by NASA grants NGL-05-002-338 and NGL-38-002-039. This work constitutes Caltech Contribution No. 2906.

REFERENCES

Albee A. L., Burnett D. S., Chodos A. A., Haines E. L., Huneke J. C., Papanastassiou D. A., Podosek F. A., Russ G. P., and Wasserburg G. J. (1970) Mineralogic and isotopic investigations on lunar rock 12013. *Earth Planet. Sci. Lett.* **9**, 137–163.

Albee A. L. and Gancarz A. J. (1974) Petrogenesis of lunar rocks: Rb-Sr constraints and lack of H_2O. In *Proc. Sov.–Amer. Conf. Cosmochemistry of the Moon and Planets (Moscow)*, NASA SP-370. In press.

Albee A. L., Quick J. E., and Chodos A. A. (1977) Source and magnitude of errors in "broad-beam" analysis (DBA) with the electron probe (abstract). In *Lunar Science VIII*, p. 7–9. The Lunar Science Institute, Houston.

Anderson D. H. (1970) Introduction. The preliminary examination and preparation of lunar sample 12013. *Earth Planet. Sci. Lett.* **9**, 94–102.

Bansal B. M., Gast P. W., Hubbard N. J., Nyquist L. E., Rhodes J. M., Shih C.-Y., and Wiesmann H. (1973) Lunar rock types (abstract). In *Lunar Science IV*, p. 48–50. The Lunar Science Institute, Houston.

Blanchard D. P., Jacobs J. W., Brannon J. C., and Haskin L. A. (1976) Major and trace element compositions of matrix and aphanitic clasts from consortium breccia 73215. *Proc. Lunar Sci. Conf. 7th*, p. 2179–2187.

Blanchard D. P., Brannon J. C., Jacobs J. W., and Haskin L. A. (1977) Major and trace element abundances in anorthositic gabbro clasts and a clast of K-rich felsite from consortium breccia 73215 (abstract). In *Lunar Science VIII*, p. 124–126. The Lunar Science Institute, Houston.

Burnett D., Monnin M., Seitz M., Walker R., and Yuhas D. (1971) Lunar astrology—U-Th distributions and fission-track dating of lunar samples. *Proc. Lunar Sci. Conf. 2nd*, p. 1503–1519.

Chodos A. A., Albee A. L., Gancarz A. J., and Laird J. (1973) Optimization of computer-controlled quantitative analysis of minerals. *Proc. 8th Nat. Conf. on Electron Probe Analysis.* New Orleans, Louisiana.

Chodos A. A., Albee A. L., and Quick J. E. (1977) The use of energy dispersive analysis for the study of phase aggregates (abstract). *12th Nat. Conf. on Electron Probe Analysis.* Boston, Massachusetts.

Drake M. J., McCallum I. S., McKay G. A., and Weill D. F. (1970) Mineralogy and petrology of Apollo 12 sample no. 12013: A progress report. *Earth Planet. Sci. Lett.* **9**, 103–123.

Dymek R. F., Quick J. E., and Albee A. L. (1977) The role of silicate liquid immiscibility: Constraints from a terrestrial example (abstract). In *Lunar Science VIII*, p. 272–274. The Lunar Science Institute, Houston.

Grieve R. A. F., Plant A. G., and Dence M. R. (1974) Lunar impact melts and terrestrial analogs: Their characteristics, formation and implications for crustal evolution. *Proc. Lunar Sci. Conf. 5th*, p. 261–273.

Haines E. L., Albee A. L., Chodos A. A., and Wasserburg G. J. (1971) Uranium-bearing minerals of lunar rock 12013. *Earth Planet. Sci. Lett.* **12**, 145–154.

Hess P. C. and Rutherford M. J. (1974) Elemental fractionation between immiscible melts (abstract). In *Lunar Science V*, p. 328–330. The Lunar Science Institute, Houston.

Hess P. C., Rutherford M. J., Guillemette R. M., Ryerson F. J., and Tuchfeld H. A. (1975) Residual products of fractional crystallization of lunar magmas: An experimental study. *Proc. Lunar Sci. Conf. 6th*, p. 895–909.

Hubbard N. J., Gast P. W., and Wiesmann H. (1970) Rare earth, alkaline and alkali metal and 87/86 Sr data for subsamples of lunar sample 12013. *Earth Planet. Sci. Lett.* **9**, 181–184.

James O. B. (1970) Petrology of lunar microbreccia 12013,6. *U.S. Geol. Survey Interagency Report: Astrogeology* **23**.

James O. B. (1975a) Lunar highlands breccias generated by major impacts. In *Proc. Sov.–Amer. Conf. on Cosmochemistry of the Moon and Planets (Moscow)*, NASA SP-370. In press.

James O. B. (1975b) Petrography of the matrix of light gray (consortium) breccia 73215 (abstract). In *Lunar Science VI*, p. 438–440. The Lunar Science Institute, Houston.

James O. B. (1976) Petrology of aphanitic lithologies in consortium breccia 73215. *Proc. Lunar Sci. Conf. 7th*, p. 2145–2178.

Laul J. C. and Schmitt R. A. (1973) Chemical composition of Luna 20 rocks and soil and Apollo 16 soils. *Geochim. Cosmochim. Acta* **37**, 927–942.

LSPET (1970) Preliminary examination of lunar samples from Apollo 12. *Science* **167**, 1325–1339.

Ma M.-S., Murali A. V., Schmitt R. A., Quick J. E., and Albee A. L. (1977) Major, minor and trace elements compositions of two petrologically significant lithologies in 12013 (abstract). In *Lunar Science VIII*, p. 602–604. The Lunar Science Institute, Houston.

Nagasawa H. (1970) Rare earth concentrations in zircons and apatites and their host dacites and granites. *Earth Planet. Sci. Lett.* **9**, 359–364.

Nagasawa H. (1971) Partitioning of Eu and Sr between coexisting Plagioclase and K-feldspar. *Earth Planet. Sci. Lett.* **13**, 139–144.

Nyquist L. E., Wiesmann H., Shih C.-Y., and Bansal B. M. (1977) REE and Rb-Sr analysis of 15405 quartz-monzodiorite (super KREEP) (abstract). In *Lunar Science VIII*, p. 738–740. The Lunar Science Institute, Houston.

Philpotts A. R. (1976) Silicate liquid immiscibility: its probable extent and petrogenetic significance. *Amer. J. Sci.* **276**, 1147–1177.

Powell B. N., Dungan M. A., and Weiblen P. W. (1975) Apollo 16 feldspathic melt rocks: Clues to the magmatic history of the lunar crust. *Proc. Lunar Sci. Conf. 6th*, p. 415–433.

Quick J. E. and Albee A. L. (1976) 12013 revisited—Two clast-laden melts (abstract). In *Lunar Science VII*, p. 712–714. The Lunar Science Institute, Houston.

Reid M. J., Gancarz A. J., and Albee A. L. (1973) Constrained least-squares analysis of petrologic problems with an application to lunar sample 12040. *Earth Planet. Sci. Lett.* **17**, 433–445.

Roedder E. and Weiblen P. W. (1970) Lunar petrology of silicate melt inclusions, Apollo 11 rocks. *Proc. Apollo 11 Lunar Sci. Conf.*, p. 801–837.

Roedder E. and Weiblen P. W. (1971) Petrology of silicate melt inclusions. Apollo 11 and 12 and terrestrial equivalents. *Proc. Lunar Sci. Conf. 2nd*, p. 507–528.

Roedder E. and Weiblen P. W. (1972) Petrographic features and petrologic significance of melt inclusions in Apollo 14 and 15 rocks. *Proc. Lunar Sci. Conf. 3rd*, p. 251–279.

Rutherford M. J., Hess P. C., Ryerson F. J., Campbell H. W., and Dick P. A. (1976) The chemistry, origin and petrogenetic implications of lunar granite and monzonite. *Proc. Lunar Sci. Conf. 7th*, p. 1723–1740.

Ryder G., Stoeser D. B., Marvin U. B., and Bower J. F. (1975) Lunar granites with unique ternary feldspars. *Proc. Lunar Sci. Conf. 6th*, p. 435–449.

Ryerson F. J. and Hess P. C. (1975) The partitioning of trace elements between immiscible silicate melts (abstract). *EOS* (*Trans. Amer. Geophys. Union*) **56**, 470.

Schmitt R. A. and Laul J. C. (1973) A survey of the selenochemistry of major, minor and trace elements. *The Moon* **8**, 182–209.

Schnetzler C. C., Philpotts J. A., and Bottino M. L. (1970) Li, K, Rb, Sr, Ba and rare-earth concentrations, and Rb-Sr age of lunar rock 12013. *Earth Planet. Sci. Lett.* **9**, 185–192.

Shaw D. M. (1970) Trace element fractionation during anatexis, *Geochim. Cosmochim. Acta* **34**, 237–243.

Simonds C. H. (1975) Thermal regimes in impact melts and the petrology of the Apollo 17 Station 6 boulder. *Proc. Lunar Sci. Conf. 6th*, p. 641–672.

Taylor S. R., Kaye M., Muir P., Nance W., Rudowski R., and Ware N. (1972) Composition of the lunar uplands: Chemistry of Apollo 14 samples from Fra Mauro. *Proc. Lunar Sci. Conf. 3rd*, p. 1231–1249.

Taylor S. R., Gorton M. P., Muir P., Nance W., Rudowski R., and Ware N. (1973) Lunar highlands composition: Apennine Front. *Proc. Lunar Sci. Conf. 4th*, p. 1445–1459.

Taylor S. R. and Jakeš P. (1974) The geochemical evolution of the moon. *Proc. Lunar Sci. Conf. 5th*, p. 2387–1306.

Walker D., Grove T. L., Longhi J., Stolper E. M., and Hays J. F. (1973) Origin of lunar feldspathic rocks. *Earth Planet. Sci. Lett.* **20**, 325–336.

Wakita H. and Schmitt R. A. (1970) Elemental abundances in seven fragments from lunar rock 12013. *Earth Planet. Sci. Lett.* **9**, 169–176.

Wakita H., Schmitt R. A., and Rey P. (1970) Elemental abundances of major, minor and trace elements in Apollo 11 lunar rocks, soil and core samples. *Proc. Apollo 11 Lunar Sci. Conf.*, p. 1685–1717.

Watson E. B. (1976) Two-liquid partition coefficients: Experimental data and geochemical implication. *Contrib. Mineral. Petrol.* **56**, 119–134.

Yoder H. S. (1973) Contemporaneous basaltic and rhyolitic magmas. *Amer. Mineral.* **58**, 153–171.

Proc. Lunar Sci. Conf. 8th (1977), p. 2191–2213.
Printed in the United States of America

On the chemistry of lunar samples and achondrites. Primary matter in the lunar highlands: A re-evaluation.

H. Wänke, Hildegard Baddenhausen, K. Blum, M. Cendales, Gerlind Dreibus, H. Hofmeister, H. Kruse, E. Jagoutz, Christl Palme, B. Spettel, Ramaben Thacker, and Else Vilcsek

Max-Planck-Institut für Chemie (Otto-Hahn-Institut), Abteilung Kosmochemie, Saarstrasse 23, 6500 Mainz, Germany

Abstract—From our multielement analyses program we report on new data of nine Apollo 15 and one Apollo 17 breccias as well as six achondrites, including three new meteorites from the Antarctic. Beryllium and vanadium were added to our list of elements.

From the compositional data of three Antarctic achondrites, we classified Yamato 1 as a howardite, Yamato 74-450 as a eucrite and Yamato 74-013 as a diogenite.

The existence of primary matter in the lunar highlands, previously proposed by us (Wänke *et al.*, 1974a), is confirmed. However, it was found that the Mg-rich component which we originally attributed as pure primary matter, contains in fact about 41% of a material derived from residual melts of the early lunar differentiation to which the outer regions of the moon were subjected. According to our new calculations, the concentration of refractory elements in the primary matter, from which at least the outer regions of the moon accreted, is reduced by about a factor of 1.3, compared to our previous results. Compared to Cl abundances, the enrichment of refractories comes down to about a factor of 1.4.

Introduction

In order to obtain further insight in the Mg-rich component observed in the lunar highlands, we have analyzed nine Apollo 15 breccias and one Apollo 17 sample. We have included V and Be to the list of elements determined since these elements could be of importance to the understanding of the lunar fractionation processes. For both elements, the data in the literature are scarce and show considerable discrepancies. Altogether we have analyzed up to 57 elements including some for which practically no literature data are existing, like W and As. We have also measured V in a number of lunar samples for which we had put aside 50 mg aliquots of samples previously analyzed by us. We also report on the multielement analyses of six achondrites, including three new ones from the Antarctic.

Analytical Methods

The methods used for the analyses were similar to those described previously (Wänke *et al.*, 1970, 1973). The recent modifications and additions are briefly summarized as follows:

Samples between 0.2 and 1 g were pulverized and several aliquots were used according to the following scheme.

I. *Aliquot of about 300 mg*

Irradiation 1. Fast neutrons (14 MeV); $^3H(d, n)^4He$ (Philips neutron generator PW 5320), 5 × 10 sec. ^{16}N and ^{28}Al were counted with a 5″ × 5″ NaI crystal for the determination of O and Si. Pneumatic sample transfer between neutron generator and counter.

Irradiation 2. Fast neutrons (14 MeV), 3 × 5 min. ^{56}Mn, ^{28}Al, ^{27}Mg, and ^{24}Na were counted with a large volume Ge(Li) detector for the determination of Fe, Mg, and Al. Silicon was used as an internal standard. Pneumatic sample transfer.

Irradiation 3. Thermal neutrons (TRIGA with 1.8×10^{12} n/cm² sec), 3 × 5 min. ^{52}V, ^{28}Al, ^{27}Mg, ^{49}Ca, and ^{51}Ti were counted with a large volume Ge(Li) detector for the determination of V, Al, Mg, Ca, and Ti. Pneumatic sample transfer between reactor and counter.

Irradiation 4. Thermal neutrons (TRIGA, 6 hr with 7×10^{11} n/cm² sec). During a period of 4 weeks, the following activities were counted in five counting periods varying between 1 hr and 2 days for the determination of the respective elements: ^{24}Na, ^{42}K, ^{47}Ca, ^{46}Sc, ^{48}Sc (from Ti), ^{51}Cr, ^{56}Mn, ^{58}Co (from Ni), ^{59}Fe, ^{60}Co, ^{95}Zr, ^{131}Ba, ^{140}La, ^{141}Ce, ^{153}Sm, ^{152}Eu, ^{160}Tb, ^{165}Dy, ^{166}Ho, ^{175}Yb, ^{177}Lu, ^{181}Hf, ^{182}Ta, ^{192}Ir, ^{233}Pa (from Th), and ^{239}Np (from U).

Irradiation 5. Thermal neutrons (TRIGA, 6 hr with 7×10^{11} n/cm² sec). The samples were decomposed under vacuum with Na_2O and radiochemically processed to separate and determine the following elements: Ti (via ^{48}Sc), Cu, Zn, Ga, Ge, As, Se, Rb, Sr, Cs, Ba, La, Ce, Pr, Nd, Sm, Eu, Gd, Tb, Dy, Ho, Er, Yb, Lu, W, Hf, Ta, Re, Au, and U (via ^{135}Xe). For details of the chemical separation see Rieder and Wänke (1969) and Wänke *et al.* (1970).

II. *Aliquots of about 50 mg*

Irradiation with thermal neutrons (TRIGA, 2 hr with 7×10^{11} n/cm² sec). The samples were mixed with V_2O_5 and pyrohydrolysis was applied for the separation of the halogens. ^{38}Cl, ^{82}Br, and ^{128}I were counted for the determination of Cl, Br, and I. Lithium was determined via ^{18}F produced via the reactions $^6Li(n, \alpha)t$ and $^{16}O(t, n)^{18}F$.

After γ-counting the native F was determined with a specific fluoride electrode. For details of this step see Dreibus *et al.* (1976a).

III. *Aliquots of about 50 mg*

Beryllium was separated using ^7Be as tracer and determined by flameless atomic absorption. For details see Vilcsek (1977).

IV *Aliquots of about 100 mg*

Aliquots of about 100 mg were used for XRF analysis. Glass beads were formed using a flux of $Li_2B_4O_7$ and $NaNO_3$. The dilution by the flux was approximately 1 : 10. No heavy absorber was added and the interelement effects were corrected by the fundamental parameter method. The following major and trace elements were measured using the same glass beads: Mg, Al, Si, Ca, Ti, Fe, P, S, K, Cr, Mn, Co, Ni, Zn, Cu, Y, Zr, Nb, and Sr.

Results

The analytical results are listed in Tables 1–5.

All the Apollo 15 breccias analyzed in this work are mixtures of low-Ti mare basalts and KREEP-rich highland samples, as is clearly demonstrated in Figs. 1, 2, and 3. In all three plots (Fe vs Sc, V vs Sc, and Al vs Ga) the data points for the Apollo 15 breccias fall between those of the KREEP-rich highland samples and the low-Ti mare basalts. The content of Ni and other siderophile elements is low compared to most of the Apollo 16 breccias.

From the parallelism of Ni and Mg for most of the Apollo 16 samples, we have previously assumed Ni and the other siderophiles to be part of a Mg-rich primary component (Wänke *et al.*, 1976). Many, but not all, of the Apollo 16 breccias also show a positive correlation between Ni and La, indicating that KREEP and the Ni containing components were added in varying amounts, having a fixed Ni/La ratio which was established prior to the formation of the lunar anorthosites. The Ni/La ratios of the Apollo 15 breccias are considerably lower, but fall in the wide range of the KREEP containing Apollo 14 and 17 breccias (Fig. 4).

We have added V to the list of elements which we determine in our multielement analyses, because lunar sample data reported in the literature show large discrepancies. For example, for highland breccias Brunfelt *et al.* (1973) reported V/Sc weight ratios between 5 and 16, i.e., in the range of the C1 ratio of 9. However, Laul and Schmitt (1973) found, for similar samples, a V/Sc ratio of 2 to 3. Our results for V are reported in Table 5. Our measurements were carried out on homogenized aliquots of samples which we have analyzed previously in our multielement analyses program, hence, the data are directly comparable.

As can be seen from Fig. 2, all our analyses of lunar highland samples have V/Sc ratios between 2 and 3.6. The higher value corresponds to KREEP-poor samples. A similar V/Sc ratio is observed for low-Ti mare basalts, while high-Ti basalts have V/Sc ratios down to 0.4.

It is not possible to directly compare our data on Be with those of the literature as no data of this element exist for the samples we have analyzed. For KREEP-rich samples, the Be values of Rose *et al.* (1972, 1973) yield a La/Be ratio of about 8, while we find a La/Be ratio of about 6. However, the data are too scarce for any definite conclusion. Beryllium seems to be an element of high geochemical importance, as it correlates well with Li, the only element which seems to follow Mg during the condensation sequence of the solar nebula (Dreibus *et al.*, 1976b).

From the compositional data of three meteorites from the Antarctic, we classified Yamato 1 as a howardite, Yamato 74-450 as a eucrite and Yamato 74-013 as a diogenite. For additional discussion of the other analytical data and especially those of the achondrites, see the companion papers of Rammensee and Wänke (1977) and Dreibus *et al.* (1977).

Table 1. Major and minor elements in nine Apollo 15 and one Apollo 17 breccias.

Element	Method	15025-7	15027-2	15028-8	15288-11	15295-20	15325-2	15330-4	15498-52	15515-39	78503-14	Accuracy %
Major elements in %												
O	IFNAA	42.96	43.19	42.82	43.10	43.32	44.07	43.78	41.71	42.69	42.66	1
Mg	XRF	5.86	5.53	5.52	6.55	6.12	6.41	6.96	5.49	6.58	—	2
	IFNAA	6.12	5.60	5.59	6.56	6.22	6.58	7.11	5.55	6.49	5.75	2
	ITNAA[(2)]	5.63	5.34	5.42	6.67	6.26	6.64	7.18	5.38	—	—	4
	av.	5.94	5.54	5.54	6.57	6.18	6.51	7.05	5.50	6.53	—	
Al	XRF	6.85	7.10	7.24	7.90	8.52	8.56	8.39	5.24	6.20	—	2
	IFNAA	7.13	7.48	7.44	8.01	8.59	8.82	8.88	5.10	6.18	8.73	2
	ITNAA[(2)]	6.80	7.32	7.36	8.15	8.93	9.05	8.81	5.30	—	—	3
	av.	6.94	7.29	7.34	7.99	8.62	8.75	8.65	5.19	6.19		
Si	XRF	21.90	22.53	22.54	21.15	21.46	22.01	21.28	21.87	21.54		2
	IFNAA	22.66	23.20	22.93	21.67	21.92	22.80	21.69	22.05	21.53	19.43	1
	av.	22.50	23.06	22.85	21.56	21.82	22.63	21.61	22.01	21.53		
Ca	XRF	7.44	7.47	7.39	7.90	8.07	7.91	8.26	7.54	7.45		2
	ITNAA[(2)]	7.41	7.45	7.58	7.79	8.28	8.08	8.75	7.41	—		4
	ITNAA	7.60	7.45	7.20	7.86	7.96	7.69	8.06	7.30	7.22	7.25	7
	av.	7.44	7.46	7.41	7.88	8.10	7.93	8.33	7.50	7.43		
Ti	XRF	1.12	1.13	1.09	0.94	0.89	0.77	0.67	1.19	1.08	—	3
	ITNAA[(2)]	1.05	1.11	1.02	0.95	0.89	0.80	0.66	1.19	—	—	7
	RNAA	1.03	—	1.02	—	0.88	—	0.65	1.10	—	3.45	10
	av.	1.10	1.13	1.07	0.94	0.89	0.77	0.67	1.18			

Fe	XRF	11.60	10.69	10.92	10.01	9.09	8.67	9.78	14.90	13.32	—	3
	IFNAA	11.80	11.01	10.91	9.93	9.07	9.08	10.05	14.57	12.79	9.07	3
	ITNAA	11.90	11.21	11.03	10.51	9.38	8.88	9.74	15.02	13.28	9.70	2
	av.	11.80	11.03	10.98	10.24	9.23	8.87	9.82	14.88	13.17	9.49[1]	
Minor elements in ppm												
Na	ITNAA	3680	4450	4335	3300	3690	3780	2990	1890	2820	2830	2
P	XRF	1200	1720	1570	860	970	1080	590	380	870	—	10
S	XRF	580	1040	860	420	610	450	850	670	530	—	10
K	XRF	2300	3510	3410	1580	1890	2060	1020	320	1420	—	5
	ITNAA	2260	3530	3330	1550	1840	2040	1060	370	1360	640	5
	av.	2280	3520	3370	1565	1865	2050	1040	340[1]	1390		
Cr	XRF	2890	2550	2540	2800	2400	2310	2320	3990	3270	—	3
	ITNAA	2980	2660	2580	2770	2460	2250	2290	4030	3250	2270	2
	av.	2950	2620	2570	2780	2440	2270	2300	4020	3260		
Mn	XRF	1550	1470	1460	1410	1250	1220	1330	2040	1770	—	3
	ITNAA	1550	1510	1480	1380	1240	1230	1290	2080	1780	1365	2
	av.	1550	1500	1470	1390	1245	1225	1300	2070	1775		
Σ		99.90	100.2	99.43	99.31	99.10	100.6	100.8	98.91	99.68	101.5	

[1] Accuracy reduced by a factor of 2.
[2] 5 min irradiation.

Table 2. Trace elements in nine Apollo 15 and one Apollo 17 breccias.

Element	Method	15025-7	15027-2	15028-8	15288-11	15295-20	15325-2	15330-4	15498-52	15515-39	78503-14	Accuracy %
Li ppm	RNAA	16.6		26.2		14.3		9.8	7.7			5
Be ppm	AAS	5.26	—	7.59	—	5.47	—	1.91	1.12	—	—	5
F ppm	SIONE	94		84		59		51	31			5
Cl ppm	RNAA	28.3		29.7		20.4		16.4	4.2			5
Sc ppm	ITNAA	31.6	30.8	29.9	27.1	24.7	23.3	25.4	45.9	35.3	45.4	2
V ppm	ITNAA	108	97.9	95.6	95.1	76.9	77.4	86.8	182	—	—	5
Co ppm	ITNAA	41.5	38.9	39.1	44.6	39.4	35.7	46.6	44.9	49.4	28.3	2
Ni ppm	ITNAA	210	180	200	200	250	180	300	90[1]	230	190	10
Cu ppm	RNAA	19.8	—	5.29		4.72		11.2	7.73		—	5
Zn ppm	RNAA	10.0	—	8.0		18.0		42.4	15[1]		—	10
Ga ppm	RNAA	3.92	—	3.36		4.17		4.99	3.00		—	5
Ge ppm	RNAA	0.36[1]	—	0.30[1]		0.50		0.79	<0.25		—	15
As ppb	RNAA	39	—	63		23		24	1.5[1]		—	10
Se ppm	RNAA	0.23	—	0.35		0.15[1]		0.48	~0.03		—	20
Br ppm	RNAA	0.073	—	0.24		0.073		0.084	0.034		—	5
Rb ppm	RNAA	6.27	—	10.7		5.70		2.81	0.80[1]		—	10
Sr ppm	RNAA	139		139	125[1]	135	140[1]	110	75		—	5
	XRF	135	145	148	133	147	138	118	83	114	—	5
Y ppm	XRF	115	158	154	83	101	96	57	35	76	—	5
Zr ppm	XRF	472	662	666	324	394	405	196	94	306	220[1]	5
Nb ppm	XRF	35	47	48	23	28	29	12	10[1]	23	—	10

Cs ppm	RNAA	0.30		0.53		0.27		0.13	0.040(1)			10
Ba ppm	ITNAA	375	515	501	246	279	290	136	60	200	92	5
La ppm	ITNAA	33.6	47.3	46.9	23.4	27.7	29.5	13.5	5.93	20.5	6.80	2
Ce ppm	ITNAA	93.6	129	130	64.0	74.3	81.1	35.7	18.0	54.2	21.0	5
Pr ppm	RNAA	11.9	—	16.7		9.83		4.90	2.40		—	10
Nd ppm	RNAA	58	75	74	39	47		24	12		16	10
Sm ppm	ITNAA	14.9	20.4	19.7	11.1	12.1	12.6	6.36	3.58	8.80	5.47	2
Eu ppm	ITNAA	1.58	1.81	1.77	1.35	1.47	1.45	1.06	0.76	1.15	1.32	3
Gd ppm	RNAA	18.7	—	26.2		14.2		7.63	4.91		—	5
Tb ppm	ITNAA	3.28	4.54	4.53	2.30	2.58	2.71	1.35	0.79	2.14	1.40	5
Dy ppm	ITNAA	19.3	26.4	26.9	14.7	15.9	18.0	8.02	5.35	12.8	9.03	5
Ho ppm	RNAA	4.3	—	5.6		3.2		1.8	1.09		2.0	10
Er ppm	RNAA	12.4	—	17.4		9.75		5.45	3.19		—	7
Yb ppm	ITNAA	11.5	15.7	15.8	8.18	9.07	9.76	5.01	2.62	7.28	5.32	3
Lu ppm	ITNAA	1.54	2.17	2.18	1.09	1.26	1.32	0.68	0.37(1)	0.98	0.72	3
Hf ppm	ITNAA	11.8	17.0	17.0	8.22	9.65	10.0	4.72	2.49	7.50	4.94	3
Ta ppm	ITNAA	1.55	2.05	2.01	1.11	1.17	1.27	0.65	0.39	0.92	0.88	5
W ppm	RNAA	0.68	—	0.98		0.55		0.25	0.096(1)			5
Re ppb	RNAA	0.7	—	0.51		0.71		1.2	—			30
Ir ppb	ITNAA	5	3	—	—	—	—	—	—	—	7.5	30
Au ppb	RNAA	2.1	—	4.0		2.9		2.9	0.13(1)			5
Th ppm	ITNAA	4.9	7.45	7.49	3.70	3.89	4.68	2.2	1.08	3.2	0.71	10
U ppm	RNAA	1.43	2.3(1)	2.37		1.04		0.58	0.18			5

(1)Accuracy reduced by a factor of 2.

Table 3. Major and minor elements in achondrites and in BCR-1.

Element	Yamato 74-450 eucrite	Pasamonte eucrite	Yamato 1 howardite	Ellemeet diogenite	Yamato 74-013 diogenite	Johnstown diogenite	BCR-1	Accuracy %
Major elements in %								
O	—	42.7	—	44.07	—	43.9	45.40	1
Mg	4.57	3.62	12.89	15.57	14.87	15.36	2.11	4
Al	6.05	6.52	2.26	0.54[1]	0.38[1]	0.79	7.17	2
Si	22.57	23.30	23.76	24.24	24.09	24.54	25.52	2
Ca	7.11	7.37	2.74	0.79[1]	0.94[1]	1.31[1]	5.03	2
Ti	0.54	0.45	0.14	0.019[1]	0.034[1]	0.063[1]	1.36	3
Fe	14.43	14.54	12.99	13.34	13.07	12.37	9.39	2
Minor elements in ppm								
Na	3800	3690	930	260	103	150	24110	2
P	—	332	—	103	88	180	160	10
S	2360	1050	1105	870	4340	2200	360	10
K	450	440	70	22	4.6[1]	9.2[1]	14030	5
Cr	2710	2140	6970	11620	10050	5610	13[1]	2
Mn	4110	4260	4050	4210	4010	3750	1410	2
Σ		99.69		100.3		99.45	99.97	

[1] Accuracy reduced by a factor of 2.

The Mg-rich Component of the Lunar Highlands

(a) *Introduction*

We have considerable evidence that most of the lunar highland samples with their different chemical composition are not related by igneous processes but seem to represent mechanical mixtures of components of quite different origin. Exceptions are rocks of clearly cumulate origin, the lunar anorthosites, dunites and troctolites. Such an interpretation has been put forward by a number of authors on the basis of quite different arguments (Nava and Philpotts, 1973; Dowty *et al.*, 1974a; Warner *et al.*, 1974; Schonfeld, 1974; Wänke *et al.*, 1974a).

With some exceptions (ferroan anorthosites and samples with very high KREEP content) lunar highland samples have Mg/Fe ratios >1. In our preceding papers (Wänke *et al.*, 1974a, 1975, 1976) we have demonstrated that the composition of these highland samples can be explained as mechanical mixtures of three components:

(1) a trace element enriched component (KREEP) with Mg/Fe <1;
(2) a trace element poor anorthositic component;
(3) a Mg-rich component with Mg/Fe >1.

Multicomponent models have also been put forward by other authors. However, besides differences in the approach, disagreement exists about the nature of the Mg-rich component. In the mixing model of Schonfeld (1974) and Boynton *et al.* (1975), the high Mg/Fe ratios are explained by admixture of a dunitic component. However, lunar dunites (72417) and troctolites (76535) have Fe/Sc and Mg/Cr ratios differing from those of the average lunar highland breccias by a factor between 2 and 5 (Wänke *et al.*, 1976). Hence, these rocks cannot be the major source of the Mg-rich component.

We have previously advocated a primary origin of the Mg-rich component for two reasons (Wänke *et al.*, 1974a, 1975, 1976):

(1) nearly proportional relation of Mg with Ni in most lunar highland samples;
(2) nearly chondritic element ratios were found for a number of pairs of elements for the Mg-rich component (i.e., Al/Sc, Ca/Al, Mg/Si, and Mg/Cr).

According to our interpretation, this primary component never underwent large-scale magmatic fractionations and was added to the moon after a solid and differentiated (feldspathic) crust had already been formed.

However, it was pointed out by Ringwood (1975, pers. comm.) that in a crystallization process after separation of olivine and plagioclase, the remaining liquid would, for many elements, be compositionally similar to the primary matter (liquid). Especially since we treat the lunar anorthosites as an independent component, material from such residual liquids can be described as primary matter minus olivine.

Table 4. Trace elements in achondrites and in BCR-1.

Element	Yamato 74-450 eucrite	Pasamonte eucrite	Yamato 1 howardite	Ellemeet diogenite	Yamato 74-013 diogenite	Johnstown diogenite	BCR-1	Accuracy %
Li ppm	10.0	10.0	3.4	2.28	1.46	2.7(1)		5
Be ppm	0.709		0.097		—			5
F ppm		52	~6	—	1.5(1)	~2(1)	473	5
Cl ppm	—	26	7.5		2.6(1)	5.5	50.5	5
Sc ppm	30.0	33.9	18.1	14.5	11.7	15.8	30.7	2
V ppm	67.5	60.0	124	264	165	115	417	6
Co ppm	12.0	6.92	20.5	26.8	30.0	38.1	33.5	2
Ni ppm	—	—	—	23	—	150	—	20
Cu ppm	2.22	17.5	4.03	14.2	8.43	4.2	15.8	5
Zn ppm	1.44	12.8	<3	8.40	0.49(1)	0.65	120	10
Ga ppm	1.46	1.73	0.51	0.27	0.12	0.18	23.5	10
Ge ppm	~0.04	~0.06	—	<0.06	~0.04	~0.13	1.47	10
As ppb	5.6	120	5.6	13	32	32	710	10
Se ppm	0.37	0.091	0.55	0.22	1.2	—	—	20
Br ppm	0.047	0.149	0.074		0.017(1)	0.051	0.077	5
Rb ppm	0.45(1)	0.60(1)	~0.49	~0.17	0.045(1)		47.0	5
Sr ppm	85	76	22	0.86(1)	0.67(1)	—	338	10
Y ppm							39	10
Zr ppm	68	54	9	<1	<1	<1	188	10
Nb ppm		10(1)		<5	<5	<5	13.6	10

I ppm	—	0.056	—			0.017[1]	0.050	20
Cs ppm	0.019[1]	0.050	~0.007	~0.005	~0.002	—	1.00	10
Ba ppm	45.4	37.2	8.25	0.60[1]	~0.2	—	595	10
La ppm	4.79	3.50	0.68[1]	0.024	0.018[1]	0.063[1]	26.1	5
Ce ppm	12.0	9.8	1.67	0.13[1]		—	53.8	10
Pr ppm	1.9	1.3	0.26			0.021	7.2	15
Nd ppm	3.8	6.6	1.4			—	35	15
Sm ppm	2.78	1.98	0.42	0.020	0.016	0.059	6.40	5
Eu ppm	0.70	0.64	0.17	0.0045	0.0068	0.011	2.02	5
Gd ppm	3.4	2.73	0.54	—	—	0.24	6.2	10
Tb ppm	0.66	0.52	0.10	0.006[1]	0.0053	—	1.0	10
Dy ppm	4.2	3.50	0.66	0.57	0.048	0.21	6.4	5
Ho ppm	0.97	0.75	0.16	0.015	0.015	0.059	1.5	15
Er ppm	2.75	2.25	0.51	0.057	0.050	0.14	4.0	10
Yb ppm	2.61	2.20	0.49	0.068	0.073	0.17	3.2	5
Lu ppm	0.36	0.32	0.073[1]	0.011	0.013	0.027	0.51	5
Hf ppm	2.08	1.59	0.33[1]	—	—	—	4.8	5
Ta ppm	0.30	0.20	0.042[1]	~0.004	0.006	—	0.79	10
W ppm	0.085	0.093	0.016	0.010	0.0079	0.0035[1]	0.47	10
Re ppb	0.7	1.3	3.3	0.06	—	—	0.9	20
Ir ppb	—	—	—	—	—	6.4	—	30
Au ppb	0.27	3.7	0.45	1.1	0.35	1.7	0.60	10
Th ppm	—	—	—	—	—	—	6.21	5
U ppb	163		22	1.5[1]	<1.5	2.8[1]	—	10

[1]Accuracy reduced by a factor of 2.

Table 5. Vanadium in lunar samples and meteorites.

Lunar highland samples

Sample	V in ppm
14163.126	42.9 ± 1.0
15025.7	107.9 ± 3.5
15027.2	97.9 ± 2.4
15028.8	95.6 ± 3.7
15245.8	83.9 ± 2.9
15288.11	95.1 ± 6.2
15295.20	76.9 ± 3.3
15325.2	77.4 ± 3.7
15330.4	86.8 ± 4.3
15498.52	181.9 ± 5.9
60315.87	39.6 ± 2.3
61016.151	22.1 ± 3.1
62235.91	38.8 ± 3.0
62295.52	29.1 ± 1.7
64815.9	51.9 ± 1.6
65015.133	36.7 ± 1.5
66075.31	22.3 ± 3.5
67015.106	12.0 ± 3.3
67016.173	21.2 ± 4.2
67075.11	22.5 ± 2.4
67415.7	22.0 ± 3.2
67435.39	29.8 ± 2.1
67455.15	17.8 ± 2.3
68815.130	22.5 ± 1.0
73235.91	50.7 ± 1.2
77035.61	37.9 ± 1.4
78155.127	38.7 ± 0.9

Lunar basalts

Sample	V in ppm
10017.33	68.4 ± 1.6
10057.40	61.5 ± 1.2
12004.33	185.2 ± 6.2
12053.2	170.7 ± 5.2
12063.69	130.5 ± 2.9
15085.50	156.0 ± 5.7
15495.52	179.4 ± 5.3
72155.30	106.5 ± 2.8
75035.46	31.7 ± 2.2

Achondrites

Sample		V in ppm
Pasamonte	eu	60.0 ± 3.9
Jonzac	eu	73.3 ± 3.8
Béréba	eu	64.5 ± 2.9
Yamato 74-450	eu	67.5 ± 2.3
Juvinas	eu	73.4 ± 4.6
Serra de Magé	eu	110.9 ± 3.7
Malvern	ho	91.4 ± 3.6
Yamato 1	ho	123.6 ± 4.7
Johnstown	di	115.4 ± 4.4
Yamato 74-013	di	164.6 ± 6.8
Ellemeet	di	264.1 ± 15
Orgueil	Cl	55.4 ± 2.1

Taylor and Bence (1975) attribute the high-Mg content of the lunar highlands either to a frozen surface layer of whole-moon composition or to a continuing infall of material at a late stage in the accretionary process similar to Wänke *et al.* (1974a). Taylor and Bence (1977) discuss additional processes for the origin of the Mg-rich component and state: "Combination of these processes are quite possible."

We will demonstrate that the Mg-rich component is probably again a mixture of true undifferentiated primary matter and material originating from melts residual after olivine separation from an early lunar crystallization process.

(b) *Composition of the Mg-rich component as derived from mixing computations*

In our recent paper (Wänke *et al.*, 1976), we have shown that the composition of the Mg-rich component can be computed from a ternary mixing diagram with KREEP and lunar anorthosites as the other endmembers. In this computation, we used the KREEP-rich breccia 14321 as the KREEP endmember.

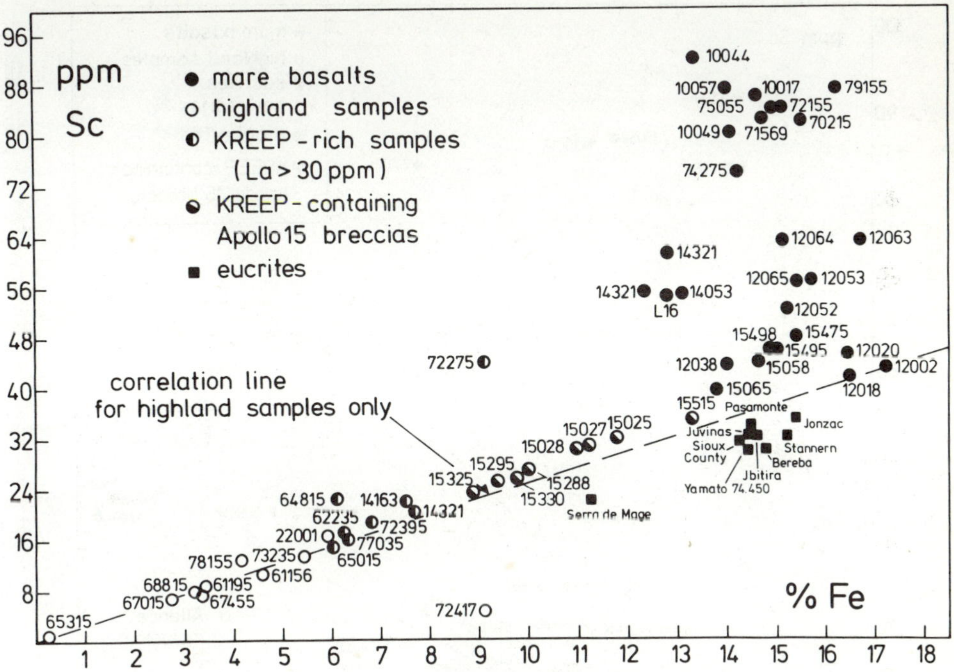

Fig. 1. Sc vs Fe. KREEP-rich and KREEP-poor highland samples plot along a well-defined correlation line. Hence, we have to conclude that the Fe/Sc ratio was not changed during the fractionations involved. Low-Ti mare basalts are also found close to this correlation line as are the eucrites. The KREEP containing Apollo 15 breccias fall between the highland samples and the low-Ti mare basalts. All data from the Mainz laboratory, this work and Wänke *et al.* (1970, 1971, 1972, 1973, 1974a, 1975, 1976).

We have previously underlined the fact that all highland samples have an identical pattern of incompatible elements, which is also identical to that of the KREEP-rich sample 14321 (Palme and Wänke, 1975). Hence, using La as one of these incompatible elements as a measure of the KREEP component, we can subtract this component and have reduced the number of components to two. The following compositional data of sample 14321 were used (Wänke *et al.*, 1976): 8.64% Al, 7.02% Ca, 7.00% Mg, 7.51% Fe, 22.75% Si, 0.13% Cr, and 20.9 ppm Sc and 91.9 ppm La.

Figure 5 shows the two-component mixing diagram obtained in this way. We restricted these computations to KREEP-poor samples (i.e., to samples with La < 25 ppm), and took only samples for which we have data from our own analyses. As before, for the computation we used the composition of the mixtures *without making assumptions as to the possible endmembers.*

There is little doubt about the nature of the left endmember, the lunar anorthosites. The composition of the right-hand side endmember, the Mg-rich component, was again calculated with the assumption that this component is of primary origin. Hence, we found its position in the mixing diagram at the point

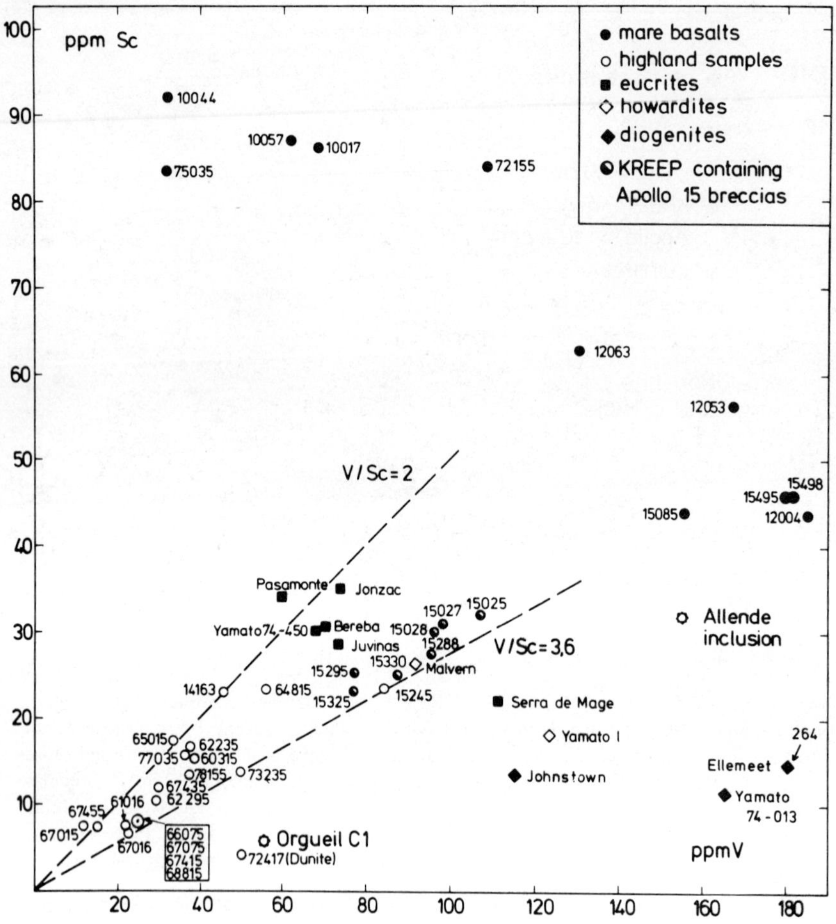

Fig. 2. Sc vs V. Contrary to the constant Sc/Fe ratio in Fig. 1, the V/Sc ratio differs for KREEP-rich and KREEP-poor highland samples. The V/Sc ratio of the latter is almost identical to that of the low-Ti mare basalts. The KREEP-containing Apollo 15 breccias are found in between the highland samples and the low-Ti basalts. Eucrites plot within the range of the lunar highland samples. Data for 72417 from Laul and Schmitt (1975), all other data from this work.

along the x-axis where the Al/Sc ratio approaches the chondritic value of 1370. The general philosophy behind this approach is the fact that both Al and Sc belong to the group of refractory elements and no fractionation of these elements from each other or from the other refractory elements has ever been observed in the solar nebula.

The resulting composition of the Mg-rich component is listed in the first column of Table 6. The results differ only slightly from our previous calculation (Wänke *et al.*, 1976). The values for the Mg/Cr and Ca/Al ratios are close to the Cl values, however, the deviation of the Mg/Si ratio seems significant.

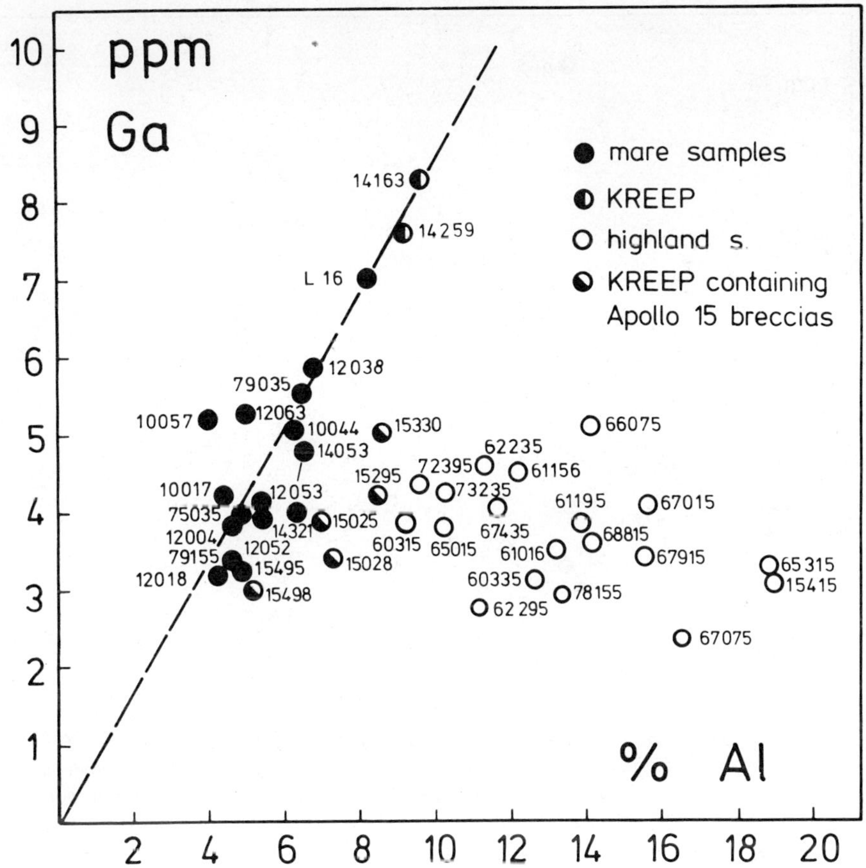

Fig. 3. Ga vs Al. The KREEP containing Apollo 15 samples are again found in between the mare basalts and anorthosite containing highland samples. This indicates that these samples are mixtures of at least three components: low-Ti mare basalts, KREEP, and anorthosites.

(c) *Origin of the Mg-rich component and its relation to unfractionated primary matter*

Von Michaelis *et al.* (1969) found for C1, C2, and C3 chondrites an almost identical Mg/Si ratio of 0.91, while Teschke and Wänke (1974) gave the following values for this ratio: Orgueil (C1) = 0.86, Murray (C2) = 0.89, and Vigarano (C3) – 0.93. For the following, we will assume a value of 0.91 to be the correct C1 Mg/Si ratio and will further assume that this value is also valid for the primary matter of which at least the outer regions of the moon accreted.

For the Mg-rich component we have calculated a considerably lower Mg/Si ratio of 0.71. Hence, it seems that the Mg-rich component cannot represent pure primary matter.

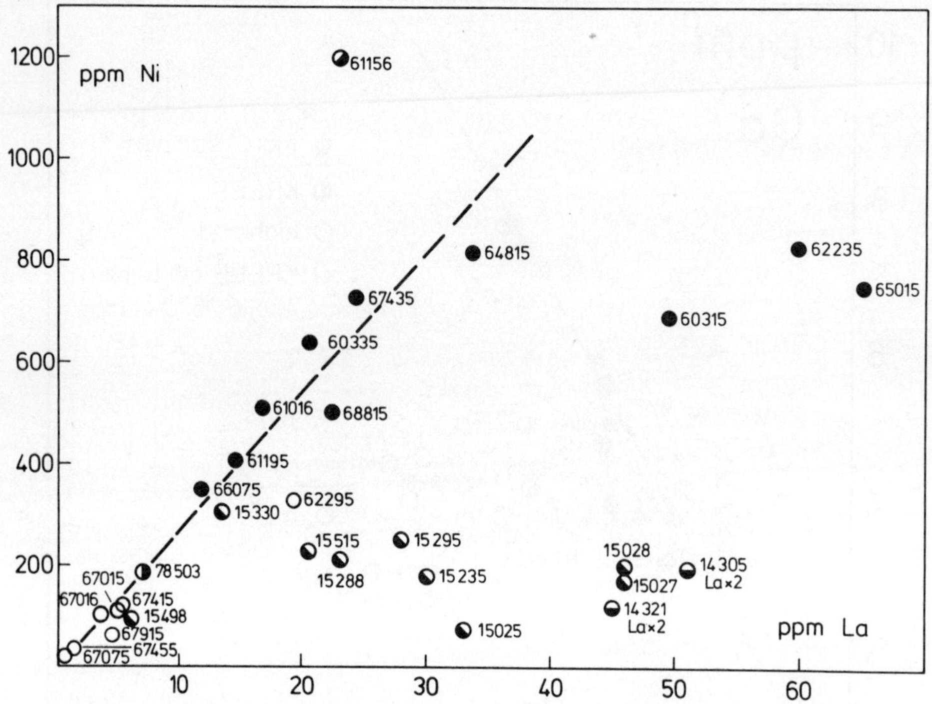

Fig. 4. Ni vs La. The dashed line is the correlation of all samples marked with full symbols, except samples 60315, 62235 and 65015. For all samples with full symbols a correlation between Ni and Mg is observed, too (Wänke *et al.*, 1976). The proportionality between Ni and La for many Apollo 16 samples indicates that the primary component which is the main source of both, Mg and Ni, was added to these samples together with the KREEP component in a fixed ratio. The Ni/La ratios of the Apollo 15 breccias are lower and fall in the wide range of the Apollo 14 and 17 breccias. All data from the Mainz laboratory.

The upper part of Fig. 6 outlines a simplified version of an evolutionary model of the outer regions of the moon proposed by Ryder and Wood (1977). Let us assume a totally molten outer layer of the moon down to a depth of about 50–100 km at an early stage. With separation of olivine the Mg/Si ratio of the residual liquid will decrease. After complete solidification we have three layers of quite different composition. In principle, large impacts could again mix these layers establishing the original composition of the primitive liquid. However, more likely the deep-seated olivine layer will remain almost untouched by such an impact-induced mixing process. Wood (1974) argues that the mafic minerals sank to the floor of the layer and were permanently removed from the system, while plagioclase floated to the surface and later on, due to impacts, mingled with the residual magma.

Assuming the lunar dunite 72417 to be representative of the olivine cumulates and the Mg-rich component from Table 6 to consist of true primary matter and

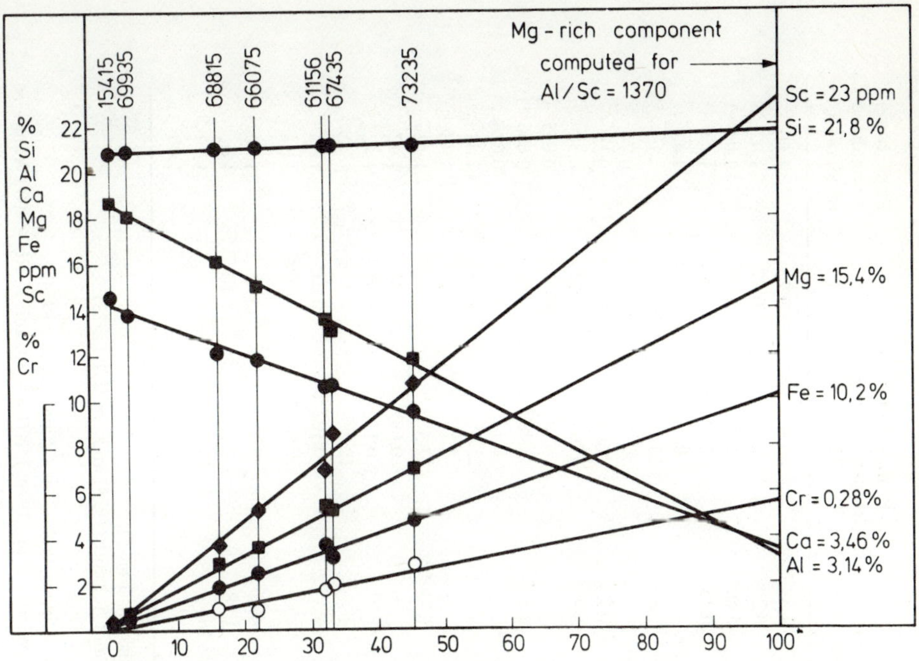

Fig. 5. After subtraction of KREEP (see text), the binary mixing diagram was computed using the composition of only those samples which were also used for the ternary diagram of Wänke *et al.* (1976), i.e., samples with La < 25 ppm and samples which are clear differentiates were excluded. All data from the Mainz laboratory.

material from residual liquids, we calculate that we have to add 31% olivine to the Mg-rich component derived from our mixing computation to raise the Mg/Si ratio from 0.71 to 0.91. Using the experimental olivine-liquid partition coefficients determined by Weill and McKay (1975), we obtain a similar result for the amount of olivine which we have to add to bring the Mg/Si ratio to the C1 value. Hence, to account for the missing olivine, the composition of the primary matter can be found as the sum of 69% of the Mg-rich component plus 31% of olivine. Since sample 72417, which was assumed to represent the olivine cumulate, has an Al/Sc ratio almost identical to the C1 value, its addition to the Mg-rich component derived from the mixing computation has no consequences for this computation.

The composition of the pure primary component after correction for the missing olivine is also given in Table 6. The resulting normative mineral composition is listed in Table 7. As we find for the primary matter an olivine content of 52%, we deduce that the Mg-rich component of the lunar highland samples consists in the average of 59% of true primary matter and 41% of material from melts remaining after olivine separation. Figure 6 shows a sketch of the proposed model.

For the weight ratio Mg/Fe of the pure primary component we now find a

Table 6. Calculation of the composition of the primary component in lunar highland samples.

	Mg-rich comp.	Mg-rich comp. (met. free)	Olivine[1] 72417	Mg-rich comp. × 0.69	Olivine × 0.31	Prim. comp. = Mg-rich comp. + 31% oli.	Cl	Prim. comp. normalized to Si and Cl
Mg%	15.4	15.9	27.4	10.97	8.49	19.5	9.25	1.00
Al	3.14	3.24	0.64	2.24	0.20	2.43	0.82	1.42
Si	21.8	22.5	18.70	15.53	5.80	21.30	10.20	1.00
Ca	3.46	3.57	0.82	2.46	0.25	2.71	0.90	1.44
Ti	—	—	—	—	—	0.12[2]	0.041	1.42
Fe (FeO)	10.2	10.5	9.1	7.25	2.82	10.1	17.8	—
V ppm						86[3]	55.4	0.74
Cr	2800	2900	2200	2000	682	2680	2300	0.56
Mn	1500	1600	850	1100	264	1360	1700	0.38
Na						450[4]	4800	0.045
K						48[4]	480	0.048
Rb						0.12[4]	2.2	0.026
Sr						25.5[2]	8.6	1.42
Ba						7.1[2]	2.4	1.42
Sc	23	23	4.3	16.3	1.3	17.5	5.9	1.42
La						0.68[2]	0.23	1.42
Eu						0.15[2]	0.05	1.42
Zr						12.2[2]	4.1	1.42
Th						0.102[2]	0.035	1.42
U						0.027[2]	0.009	1.42
Al/Sc	1370	1370	1490	1370	1490	1380	1370	1.00
Ca/Al	1.10	1.10	1.28	1.10	1.28	1.12	1.10	1.01
Mg/Cr	55	55	126	55	126	73	40	1.79
Mg/Si	0.71	0.71	1.47			0.91	0.91	1.00

Mg-rich component. Composition derived from the binary mixing diagram (Fig. 5).
[1]Lunar dunite 72417. Composition derived from the analyses of Laul and Schmitt (1975) and LSPET (1973).
[2]Calculated relative to Al assuming Cl abundance ratios.
[3]Calculated relative to Al assuming abundance ratios equal to those in Allende inclusions (Wänke *et al.*, 1974b).
[4]From element correlations observed in lunar samples: Na/Eu = 3000 (Wänke *et al.*, 1974a), K/La = 70, and K/Rb = 400 (Wänke *et al.*, 1973).

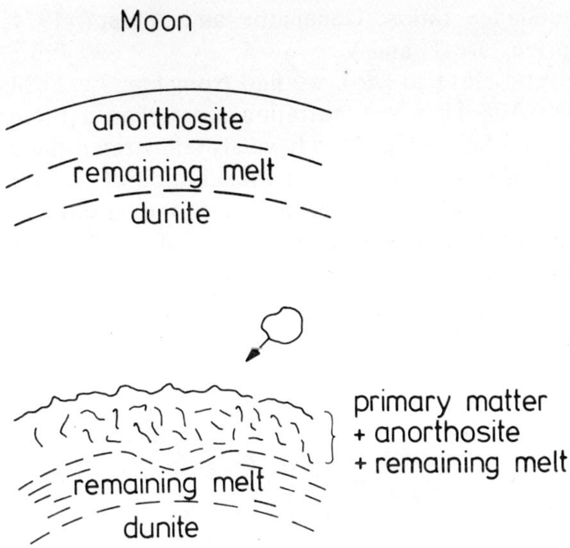

Fig. 6. Evolutionary model of the outer regions of the moon.

Table 7. Normative mineral composition of the primary component of the lunar highlands in wt.% (metal free).

	%		%
Ab	0.51	Fs	7.21
An	12.2	Fo	39.1
Wo	2.74	Fa	12.6
En	24.7	Cm	0.58

value of 1.93, compared to our previous estimate (Wänke *et al.*, 1976) of 1.71, while for the residual melts, which have supplied a considerable portion of the Mg-rich component, we calculate an average Mg/Fe ratio of about 1. The ferroan anorthosites recognized first as individual rock type by Dowty *et al.* (1974b) have Mg/Fe ratios <1. Thus, the mafic component of these rocks may come from such liquids.

Let us look for further evidence for the proposed presence of material from residual melts in the Mg-rich component. Certainly, the mafic elements Ti, Fe, and Sc would be enriched in these melts as would be the incompatible elements. A crude estimate leads to an enrichment factor of about 2 to 3. Most of the elements not listed in column one of Table 6 are highly incompatible elements and we cannot obtain meaningful results on their concentration in the Mg-rich component due to their abundance in KREEP. Among the refractory elements which as a group should be present in any primary material of the solar system in chondritic abundance ratios (Ganapathy and Anders, 1974; Wänke *et al.*, 1974a), the exceptions are Ti and V.

For a La/Mg ratio close to zero, we find from Fig. 7 a Ti/Mg ratio of 0.020, corresponding to 0.31% Ti, a concentration more than a factor of 2 too high relative to Al, Ca, and Sc. In Fig. 7 we have also illustrated the drastic change of the extrapolated Ti/Mg ratio, because of the addition of olivine. If we add only 20% olivine to the pure highland breccia 73235, the Ti content of the primary component, as calculated from the extrapolation of the Ti/Mg ratio (dotted line in Fig. 7), would be reduced from 0.31 to 0.14% Ti.

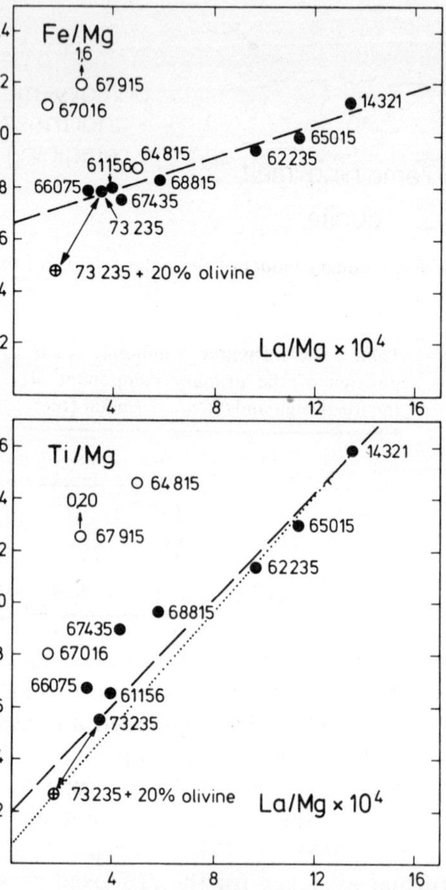

Fig. 7. In both diagrams the addition of KREEP to samples in which the primary matter is the main source of the mafic elements Fe, Mg, and Ti, is illustrated. Samples containing material which is probably supplied from residual melts in higher than average proportions (open symbols) show considerably higher Fe/Mg and Ti/Mg ratios. In the upper figure the dashed line corresponds to the correlation line of the samples marked with full symbols. In the lower figure the dashed line connects KREEP-rich samples with the sample with the lowest Ti/Mg ratio, the highland breccia 73235. For further details see text. All data from the Mainz laboratory.

If all non-feldspathic elements are supplied by the primary component, we would expect a constant V/Sc ratio for all highland samples. From Fig. 2, it is evident that the KREEP-rich samples (14163, 62235, 60315) have a V/Sc ratio almost a factor of 2 lower than most of the KREEP-poor samples (i.e., 61016, 67016, 66075, 67075, 67415, 68815, and 73235). The V/Sc ratio of the Allende inclusion is 4.8, i.e., even higher than the ratio observed in the KREEP-poor samples. The V/Sc ratio for Orgueil is a factor of 2 higher than the value in the Allende inclusion. This difference reflects the considerably lower enrichment factor of V in the Allende inclusions as compared to all the other refractory elements (Wänke *et al.*, 1974b).

Contrary to our previous suggestions (Wänke *et al.*, 1974a, 1975), we conclude that the mafic elements Fe, Mg, Ti, Sc, and V of the KREEP-rich samples cannot originate from the primary component or at most, the contribution of primary matter to these samples can only be small.

Hence, the KREEP-rich samples probably represent true lunar differentiates (KREEP basalts) of a highly advanced state to which varying amounts of anorthosites were added mechanically.

Summary

We conclude that the Mg-rich component observed in almost all lunar highland samples is, in fact, again a mixture of true primary matter and material from residual melts, probably resulting from fractional crystallization of the totally molten outer regions of the moon. We estimate a mean concentration of 61% unfractionated primary matter in the Mg-rich component. As can be seen from Table 6, we find for the primary matter an almost chondritic composition for the non-siderophile and non-volatile elements. For the refractory elements normalized to Si an enrichment factor of only 1.4 is obtained, while the moderately volatile elements Mn, Na, K, and Rb are depleted by factors of 2.6, 22, 21, and 38, respectively. Compared to our previous calculation (Wänke *et al.*, 1976), the abundance of the refractory elements in the primary matter (high temperature condensates) reduces from 18% to 14%.

In a companion paper of Dreibus *et al.* (1977) we present evidence in favor of the assumption that the composition of the primary matter of the outer region of the moon, in fact, does reflect the composition of the bulk moon. There we also demonstrate the perfect agreement of our new results on the lunar bulk composition with those of Taylor (1976) and Ringwood (1976), which were obtained by completely different methods.

Acknowledgments—We are grateful to NASA for making available the lunar material for this investigation. The samples were activated in the TRIGA-research reactor of the Institut für Anorganische Chemie und Kernchemie der Universität Mainz. We wish to thank the staff of the TRIGA-research reactor and the members of the staff of our Institute who are not listed as authors, in particular Mr. P. Deibele and Miss H. Prager. For supplying the samples of achondrites we are indebted to the following donors: Yamato 1, Yamato 74-450, Yamato 74-013—Prof. M. Shima,

Tokyo; Ellemeet—Dr. K. Fredriksson, U.S. National Museum, Smithsonian Institution, Washington; Jonzac—Dr. G. Kurat, Naturhistorisches Museum, Vienna; Malvern—Prof. L. H. Ahrens, University, Cape Town; Béréba and Serra de Magé—Prof. J. Fabries, Muséum National d'Histoire Naturelle, Paris. We wish to thank the Deutsche Forschungsgemeinschaft for financial support.

REFERENCES

Boynton W. V., Baedecker P. A., Chou C.-L., Robinson K. L., and Wasson J. T. (1975) Mixing and transport of lunar surface materials: Evidence obtained by the determination of lithophile, siderophile, and volatile elements. *Proc. Lunar Sci. Conf. 6th*, p. 2241–2259.

Brunfelt A. O., Heier K. S., Nilssen B., Sundvoll B., and Steinnes E. (1973) Geochemistry of Apollo 15 and 16 materials. *Proc. Lunar Sci. Conf. 4th*, p. 1209–1218.

Dowty E., Prinz M., and Keil K. (1974a) Very high alumina basalt: A mixture and not a magma type. *Science* **183**, 1214–1215.

Dowty E., Prinz M., and Keil K. (1974b) Ferroan anorthosite: A widespread and distinctive lunar rock type. *Earth Planet. Sci. Lett.* **24**, 15–25.

Dreibus G., Spettel B., and Wänke H. (1976a) Determination of lithium and halogens and the significance of lithium to the understanding of cosmochemical processes. *Proc.* 1976 *Intern. Conf. Modern Trends in Activation Analysis*, p. 1022–1031. GDCH, München.

Dreibus G., Kruse H., Palme H., and Wänke H. (1976b) Lithium as a correlated element and on the bulk composition of the Moon and the parent body of the eucrites (abstract). In *Lunar Science VII*, p. 208–210. The Lunar Science Institute, Houston.

Dreibus G., Kruse H., Spettel B., and Wänke H. (1977) The bulk composition of the moon and the eucrite parent body. *Proc. Lunar Sci. Conf. 8th.* Vol. 1.

Ganapathy R. and Anders E. (1974) Bulk compositions of the moon and earth, estimated from meteorites. *Proc. Lunar Sci. Conf. 5th*, p. 1181–1206.

Laul J. C. and Schmitt R. A. (1973) Chemical composition of Apollo 15, 16 and 17 samples. *Proc. Lunar Sci. Conf. 4th*, p. 1349–1367.

Laul J. C. and Schmitt R. A. (1975) Dunite 72417: A chemical study and interpretation. *Proc. Lunar Sci. Conf. 6th*, p. 1231–1254.

LSPET (1973) Apollo 17 samples: Chemical and petrographic description. *Science* **182**, 659–672.

Nava D. F. and Philpotts J. A. (1973) A lunar differentiation model in light of new chemical data on Luna 20 and Apollo 16 soils. *Geochim. Cosmochim. Acta* **37**, 963–973.

Palme H. and Wänke H. (1975) A unified trace-element model for the evolution of the lunar crust and mantle. *Proc. Lunar Sci. Conf. 6th*, p. 1179–1202.

Rammensee W. and Wänke H. (1977) On the partition coefficient of tungsten between metal and silicate and its bearing on the origin of the moon. *Proc. Lunar Sci. Conf. 8th.* Vol. 1.

Rieder R. and Wänke H. (1969) Study of trace element abundance in meteorites by neutron activation. In *Meteorite Research* (P. Millman, ed.), p. 75–86. D. Reidel, New York.

Ringwood A. E. (1976) *Basaltic Magmatism and the Bulk Composition of the Moon, Part I. Major and Heat Producing Elements.* Publication 1221, Research School of Earth Sciences, Australian National University, Canberra. 54 pp.

Rose H. J., Jr., Cuttitta F., Annell C. S., Carron M. K., Christian R. P., Dwornik E. J., Greenland L. P., and Ligon D. T., Jr. (1972). Compositional data for twenty-one Fra Mauro lunar materials. *Proc. Lunar Sci. Conf. 3rd*, p. 1215–1229.

Rose H. J., Jr., Cuttitta F., Berman S., Carron M. K., Christian R. P., Dwornik E. J., Greenland L. P., and Ligon D. T., Jr. (1973) Compositional data for twenty-two Apollo 16 samples. *Proc. Lunar Sci. Conf. 4th*, p. 1149–1158.

Ryder G. and Wood J. A. (1977) Serenitatis and imbrium impact melts: Implications for lunar crustal composition and stratigraphy (abstract). In *Lunar Science VIII*, p. 826–828. The Lunar Science Institute, Houston.

Schonfeld E. (1974) The contamination of lunar highland rocks by KREEP: Interpretation by mixing models. *Proc. Lunar Sci. Conf. 5th*, p. 1269–1286.

Taylor S. R. (1976) Geochemical constraints on the composition of the moon. *Proc. Lunar Sci. Conf. 7th*, p. 3461–3477.

Taylor S. R. and Bence A. E. (1975) Evolution of the lunar highland crust. *Proc. Lunar Sci. Conf. 6th*, p. 1121–1141.

Taylor S. R. and Bence A. E. (1977) Primary chemical components in the highland crust (abstract). In *Lunar Science VIII*, p. 931–933. The Lunar Science Institute, Houston.

Teschke F. and Wänke H. (1974) Major element analysis of extraterrestrial rock samples with 14 MeV neutrons. *Radiochem. Radioanal. Lett.* **18**, 341–348.

Vilcsek E. (1977) Determination of Be and Li by flameless atomic absorption spectroscopy. In preparation.

Von Michaelis H., Ahrens L. H., and Willis J. P. (1969) The composition of stony meteorites II. The analytical data and an assessment of their quality. *Earth Planet. Sci. Lett.* **5**, 387–394.

Wänke H., Rieder R., Baddenhausen H., Spettel B., Teschke F., Quijano-Rico M., and Balacescu A. (1970) Major and trace elements in lunar material. *Proc. Apollo 11 Lunar Sci. Conf.*, p. 1719–1727.

Wänke H., Wlotzka F., Baddenhausen H., Balacescu A., Spettel B., Teschke F., Jagoutz E., Kruse H., Quijano-Rico M., and Rieder R. (1971) Apollo 12 samples: Chemical composition and its relation to sample locations and exposure ages, the two component origin of the various soil samples and studies on lunar metallic particles. *Proc. Lunar Sci. Conf. 2nd*, p. 1187–1208.

Wänke H., Baddenhausen H., Balacescu A., Teschke F., Spettel B., Dreibus G., Palme H., Quijano-Rico M., Kruse H., Wlotzka F., and Begemann F. (1972) Multielement analyses of lunar samples and some implications of the results. *Proc. Lunar Sci. Conf. 3rd*, p. 1251–1268.

Wänke H., Baddenhausen H., Dreibus G., Jagoutz E., Kruse H., Palme H., Spettel B., and Teschke F. (1973) Multielement analyses of Apollo 15, 16 and 17 samples and the bulk composition of the moon. *Proc. Lunar Sci. Conf. 4th*, p. 1461–1481.

Wänke H., Palme H., Baddenhausen H., Dreibus G., Jagoutz E., Kruse H., Spettel B., Teschke F., and Thacker R. (1974a) Chemistry of Apollo 16 and 17 samples: Bulk composition, late stage accumulation and early differentiation of the moon. *Proc. Lunar Sci. Conf. 5th*, p. 1307–1335.

Wänke H., Baddenhausen H., Palme H., and Spettel B. (1974b) On the chemistry of the Allende inclusions and their origin as high temperature condensates. *Earth Planet. Sci. Lett.* **23**, 1–7.

Wänke H., Palme H., Baddenhausen H., Dreibus G., Jagoutz E., Kruse H., Palme C., Spettel B., Teschke F., and Thacker R. (1975) New data on the chemistry of lunar samples: Primary matter in the lunar highlands and the bulk composition of the moon. *Proc. Lunar Sci. Conf. 6th*, p. 1313–1340.

Wänke H., Palme H., Kruse H., Baddenhausen H., Cendales M., Dreibus G., Hofmeister H., Jagoutz E., Palme C., Spettel B., and Thacker R. (1976) Chemistry of lunar highland rocks: A refined evaluation of the composition of the primary matter. *Proc. Lunar Sci. Conf. 7th*, p. 3479–3499.

Warner J. L., Simonds C. H., and Phinney W. C. (1974) Impact-induced fractionation in the lunar highlands. *Proc. Lunar Sci. Conf. 5th*, p. 379–397.

Weill D. F. and McKay G. A. (1975) The partitioning of Mg, Fe, Sr, Ce, Sm, Eu, and Yb in lunar igneous systems and a possible origin of KREEP by equilibrium partial melting. *Proc. Lunar Sci. Conf. 6th*, p. 1143–1158.

Wood J. A. (1974) A survey of lunar rock types and comparison of the crust of earth and moon. In *Proc. Sov.-Amer. Conf. on Cosmochemistry of the Moon and Planets* (*Moscow*). NASA SP-370. In press.

Proc. Lunar Sci. Conf. 8th (1977), p. 2215–2235.
Printed in the United States of America

Pristine nonmare rocks and the nature of the lunar crust

PAUL H. WARREN and JOHN T. WASSON

Institute of Geophysics and Planetary Physics, Department of Earth and Space Sciences,
University of California, Los Angeles, California 90024

Abstract—Critical parameters for recognizing pristine nonmare rocks are reviewed, and a list of pristine rocks and clasts is given. These parameters include: low siderophile and non-KREEPlike incompatible element concentrations; coarse granularity, phase homogeneity, coarse exsolution lamellae and other textural characteristics; radiometric ages $\geqslant 4.2$ G.y.; and other isotopic evidence. Siderophile contents $\leqslant 3 \times 10^{-4}$ times those in CI chondrites provide the most reliable evidence for pristinity.

Relative proportions of pristine rock types are *not* in accord with abundances of primary rock types inferred from statistics amassed on soil fragments and glasses. In particular, much lower abundances of anorthositic norites and noritic anorthosites are found among the pristine rocks. Anorthosites and troctolites probably were the most important rocks in the primary nonmare crust; considerably later, norites (*sensu stricto*) were intruded. The reported "anomalous" negative correlation between Mg/(Fe + Mg) in mafics and An content in plagioclase among anorthosites disappears when the data set is confined to pristine samples. Some ferroan anorthosites have cumulus olivines, but so does 76335, a magnesian monomict rock with 88% plagioclase; between the ferroan and magnesian groups is a hiatus in Mg/(Fe + Mg) in mafics.

"Intermediate compositions, which may be mechanical mixtures, are suspect." (Wood, 1975)

INTRODUCTION

IT CAME AS A SURPRISE to many investigators that the returned samples from Apollo 16, the only Apollo mission to a site well within the lunar highlands, included no purely igneous rocks; almost all samples were heterogeneous breccias. There is now little dispute that the few rocks at first considered igneous (e.g., 65015, 68415, 68416) were so only in the sense that they had been made molten by impacting meteoroids. Although a few pristine, endogenously igneous nonmare samples have been found at the Apollo 14–17 sites, no adequate survey of these has been published. Detailed crustal formation models have been postulated on the basis of estimates about the abundances of primary rock components in the nonmare crust from major element compositional groupings of soil glasses (e.g., Walker *et al.*, 1973a), soil lithic fragments (e.g., Prinz *et al.*, 1973b), and full-sized rocks, few of which are pristine (e.g., Wood, 1975). It appears that interpretations of these groupings as predominant compositions among primary igneous rocks must be challenged. Compositions indistinguishable from mean regolith compositions are particularly likely to reflect recrystallization or melting of mixtures of unrelated rock types.

This paper will show that interdisciplinary study of the nonmare lunar rocks rooted in trace element, major element, and isotopic data plus petrographic evidence, can succeed in amassing a large suite of demonstrably pristine rocks, and that the relative numbers of these rocks are *not* in accord with statistics amassed on soil fragments and glasses. By "pristine" we mean rocks with primary compositions (albeit not necessarily textures) produced by lunar endogenous igneous processes; melt rocks and crystalline matrix breccias produced by impact processes are excluded. A petrographic synonym for pristine would be "unremelted, monomict".

EARLIER WORK

We began our search for pristine rocks by exhaustively compiling published analysis of Apollo rocks. This approach was similar to that of Wood (1975), who compiled analyses of 192 lunar (both mare and highlands) rocks, plus 661 "clasts and lithic fragments from soil samples." We plotted the most representative analysis available for each >5 g rock on a SiO_2–$Mg_{1.4}Fe_{0.6}SiO_4$–$CaAl_2Si_2O_8$ pseudoternary diagram (Walker *et al.*, 1973a). Figures 1 and 2 show separate diagrams for Apollo 14–17 missions. Although these diagrams are collectively similar to Wood's Fig. 4, Fig. 2 shows that troctolitic rocks are far less abundant at Apollo 16 then they seem to be at Luna 20 (Prinz *et al.*, 1973b). Note that there is a tight clustering of anorthosites, and a broad hiatus between this cluster and the majority of the Apollo 16 points. Such a pattern suggests that anorthosites might be surviving "end-member" rocks, whereas the majority trend could be produced by mixing of anorthosites with materials located near the plagioclase-olivine cotectic line or beyond.

PRISTINITY EVALUATION: A SUITE OF PRISTINE ROCKS

Our endeavor has been to generate diagrams similar to Figs. 1 and 2 but including only pristine rocks. To this end we have established a series of "attributes" common among pristine rocks that can serve to differentiate between them and nonpristine rocks. These are summarized in Table 1 together with the properties of 25 rocks that are almost certainly pristine.

Siderophile element data are crucial for recognizing pristine rocks and clasts. Low concentrations of siderophiles (such as Au, Ir, Ni, Os, or Re) can be taken as *prima facie* evidence for pristinity, since all nonmare samples having siderophile concentrations $\leq 2.5 \times 10^{-4}$ times those in CI chondrites appear to be endogenous igneous rocks or monomict (cataclastic) breccias. Many coarse-grained low-siderophile rocks have already been characterized by petrographers as cumulates (see columns 1, 3, and 7, Table 1).

Very low incompatible element concentrations also may indicate "pristinity" for adcumulate rocks with low contents of "trapped liquid", since early cumulate phases are poor hosts for incompatibles. Pristine rocks frequently have low siderophiles but not especially low incompatibles, it is true; but we know of no rocks having incompatible concentrations $<5 \times 10^{-3}$ times those in average high-K Fra Mauro rocks (Wasson *et al.*, 1977) that do not appear pristine. Incompatible contamination in nonpristine rocks presumably comes from rego-

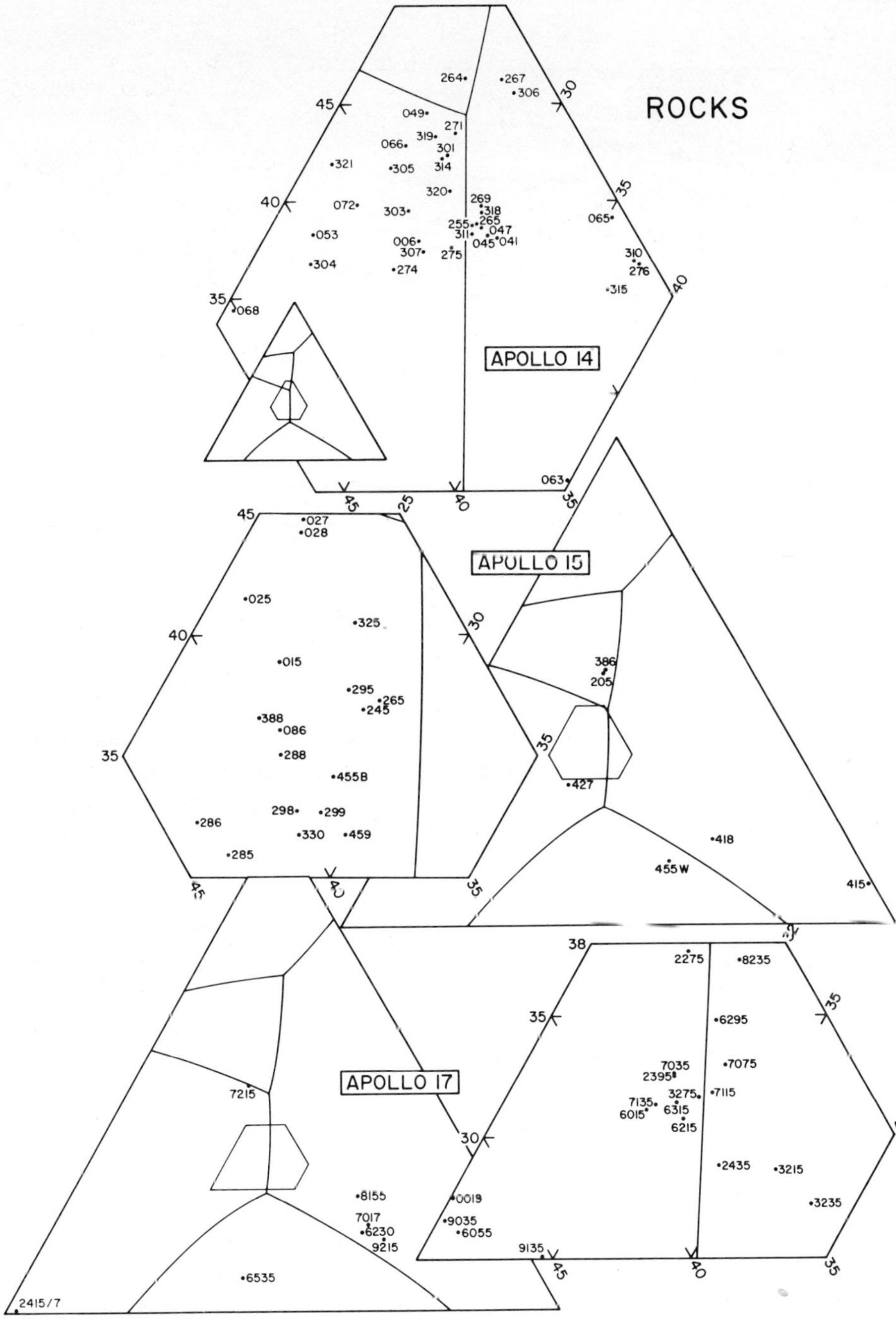

Fig. 1. Large (>5 g) nonmare rocks from the Apollo 14, 15, and 16 sites on the SiO_2–$(Mg,Fe)_2SiO_4$–$CaAl_2Si_2O_8$ pseudoternary diagram. Excepting pristine rocks, individual analyses—*not* averages—are plotted. A few extremely heterogeneous rocks were omitted to avoid unnecessary confusion. See Fig. 3 for labels for fields and components.

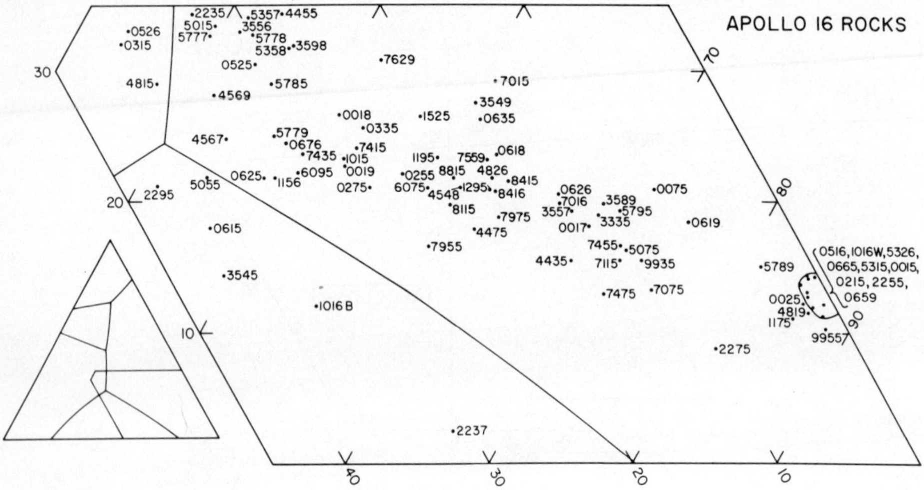

Fig. 2. Large (>5 g) Apollo 16 rocks on the pseudoternary diagram. Data were processed as described for Fig. 1.

lith materials (rich in KREEP disseminated by impacts) incorporated into breccias and impact-generated melts.

The third column in Table 1 relates to grain sizes. Textural pristinity is rare even among rocks which are compositionally pristine; most have been altered by severe shock events. However, most compositionally pristine, non-KREEPy rocks preserve vestiges of former coarse granularities, consistent with origins in slowly cooling plutonic environments.

Narrow mineral compositional ranges (column 5 in Table 1) are expected in monomict rocks, and are less likely in polymict rocks. Moreover, unzoned mineral grains can be expected in slowly cooled plutonic rocks, whereas zoned grains can be expected in rocks which crystallized in impact-melt pools too small to cool slowly. The final textural entry in Table 1, "cumulate" character (column 7) is given with the following caveat. Relict cumulate textural features have not been reported in any anorthosites or in the dunite. However, due to the low probability of obtaining pure molten plagioclase or olivine, monomineralic plagioclase or olivine rocks are sensibly considered to have cumulate origins.

The remaining two attributes of pristine rocks covered in Table 1, antiquity and low $^{87}Sr/^{86}Sr$, are telling because nonmare intrusive igneous activity has probably been rare since 4.2 G.y. ago. Anorthosites are too poor in incompatibles to permit Rb–Sr or Sm–Nd dating, but their extremely low $^{87}Sr/^{86}Sr$ ratios "are consistent with their derivation during the early lunar differentiation" (Dymek *et al.*, 1975).

Note in Table 1 that the "attributes" are roughly in order of increasing weight for pristinity characterization from right to left. The reader is cautioned that the Procrustean bed of Table 1 could never cover all the evidence weighed in judging some rocks pristine!

Table 1. Key attributes of pristine rocks and clasts*.

	1 low (<2 × 10⁻⁴ × chondrites) siderophiles	2 low (<5 × 10⁻³ × KREEP) incompatibles	3 coarse (>3 mm) grains	4 (>4.2 G.y.) antiquity	5 phase homo-geneity	6 low (<0.6992) $^{87}Sr/^{86}Sr$	7 "cumulate" character	Figs. 3 and 5 data sources	
Anorthosites									
15362		1,13	2		2			2	15362
15415	3,4†	3,5,6	7		9	11,12	8,16	6,7	15415
60015	4†	13,14	15,16		16	17	16	14,49	60015
60025	18,19	13,18,19,20	15,16,21,22		21	17,23	16,21	21,50	60025
60215	24		15,25		25		16	24	60215
61016c	19	19,26,27	15		16	11,59	16	51	61016c
62236		31	15						62236
62255		30	15					30	62255
62275		31			34			34	62275
64435c	4†	32							64435c
65315	33	33					16	33	65315
67075c	18	18						18	67075c
Troct. anorth.									
62237		28	28,29		28,29		28,29	28,29	62237
Anorthosite troct.									
76255c1	37	37			38			38	76255c1
Troctolites									
67435c			34		34		34	34	67435c
76535	36,40†		29,39,40	41	29,40	41	29,40	52	76535
15445c‡	37								15445c
Dunite									
72415/7	35,29†	35,42	29,39	43	29	43	29	29	72415/7
Anorth. norite									
15455c	4,†44		7,8				7,8	53	15455c

Table 1. (*Continued*).

	1 low ($<2\times10^{-4}\times$ chondrites) siderophiles	2 low ($<5\times10^{-3}\times$ KREEP) incompatibles	3 coarse (>3 mm) grains	4 (>4.2 G.y.) antiquity	5 phase homo geneity	6 low (<0.6992) $^{87}Sr/^{86}Sr$	7 "cumulate" character	Figs. 3 and 5 data sources	
Norites									
72255c	36		45	46	45	46		55	72255c
78235	47†		29,39,47		29,47		29,47	29,56	78235
Gabbro									
76255c2	37							38	76255c2
Qtz. monzodiorite									
15405c	37			48				54	15405c
KREEPy basalts									
15382	37							57,58	15382
72275c	36							55	72275c

*Designated by appendage of c onto sample number.
†Metal grains with Co/Ni too high to be of meteoritic origin.
‡Chemistry and petrography both poorly characterized.

References: (1) Murali *et al.* (1977); (2) Dowty *et al.* (1972); (3) Morgan *et al.* (1972); (4) Hewins and Goldstein (1975); (5) Hubbard *et al.* (1971); (6) Wänke *et al.* (1975); (7) LSPET (1971); (8) Ryder and Bower (1977); (9) James (1972); (10) Hargraves and Hollister (1972); (11) Nyquist *et al.* (1973); (12) Wasserburg and Papanastassiou (1971); (13) Laul and Schmitt (1973); (14) Taylor *et al.* (1973a); (15) LSPET (1972); (16) Dixon and Papike (1975); (17) Nunes *et al.* (1974); (18) Haskin *et al.* (1973); (19) Krähenbühl *et al.* (1973); (20) Nakamura *et al.* (1973); (21) Walker *et al.* (1973); (22) Hodges and Kushiro (1973); (23) Papanastassiou and Wasserburg (1972); (24) Rose *et al.* (1975); (25) Meyer and McCallister (1973); (26) Hubbard *et al.* (1974); (27) Philpotts *et al.* (1973); (28) This work; (29) Dymek *et al.* (1975); (30) Taylor *et al.* (1974); (31) Clark and Keith (1973); (32) Laul and Schmitt (1974); (33) Wänke *et al.* (1974); (34) Prinz *et al.* (1973); (35) Highuchi and Morgan (1975); (36) Morgan *et al.* (1974); (37) Gros *et al.* (1976); (38) Warner *et al.* (1976); (39) LSPET (1973); (40) Gooley *et al.* (1974); (41) Papanastassiou and Wasserburg (1976); (42) Laul and Schmitt (1975); (43) Papanastassiou and Wasserburg (1975); (44) Ganapathy *et al.* (1973); (45) Stoeser *et al.* (1974); (46) Gray *et al.* (1974); (47) McCallum and Mathez (1975); (48) Nyquist *et al.* (1977); (49) Juan *et al.* (1974); (50) Rose *et al.* (1973); (51) Nava (1974); (52) Rhodes *et al.* (1974); (53) Taylor *et al.* (1973b); (54) Taylor (1976); (55) Blanchard *et al.* (1974); (56) Winzer *et al.* (1975); (57) Hubbard *et al.* (1973); (58) Dowty *et al.* (1976); (59) Tera *et al.* (1973).

THE SURPRISING RARITY OF ANORTHOSITIC NORITES

Conspicuous by its absence among proven pristine rocks is noritic anorthosite (often given the more general name, gabbroic anorthosite). Anorthositic norite (≈anorthositic gabbro or "highland basalt") is represented by only one large clast, 15455c in our nomenclature. The petrographic evidence that this large clast is a pristine anorthositic norite is strong (Ryder and Bower, 1977), but there is a great deal of confusion associated with the identification of the several "white" clasts from 15455 which various groups have analyzed. Ganapathy *et al.* (1973) found two different clasts to be free of meteoritic siderophiles, but did not determine major elements; we assume that at least one of these was the large anorthositic norite. Taylor *et al.* (1973b) analyzed a "white" clast which they designated highland basalt, but which has a norm corresponding to a troctolite (Ryder and Bower, 1977). In all probability the clast analyzed by Taylor *et al.* was a troctolite, but it still appears to be pristine, since its Ni (12.0 ppm) and incompatible element (~0.02 times KREEP, see Fig. 4) concentrations are low, even for a troctolite. In this paper we designate Taylor's "white" clast 15455w, to distinguish it from 15455c.

In Fig. 3 we show pristine nonmare rocks (and clasts) on the pseudoternary. This is to be compared with Fig. 2. (Only Apollo 16 and Luna 20 went to anorthositic highlands areas. Our (unpublished) Luna 20 diagram is essentially the same as Fig. 15 in Prinz *et al.*, 1973b.) Essentially the same "sequence" appears in Fig. 2 as in Fig. 15 in Prinz *et al.* (1973b), Fig. 3 in Walker *et al.* (1973a), and Fig. 4 in Wood (1975). This trend, extending from along the olivine-plagioclase cotectic line towards the anorthosite cluster, and consisting of noritic anorthosites and anorthositic norites has been interpreted by the authors just mentioned as the principal differentiation trend of primary nonmare cumulates. Statistics amassed on soil lithic fragments (e.g., Delano *et al.*, 1973; Prinz *et al.*, 1973b) and glasses (e.g., Ridley *et al.*, 1973; Simkin *et al.*, 1973; Walker *et al.*, 1973a, Fig. 3), as well as rocks irrespective of pristinity (Fig. 2), all are similar, reinforcing the notion that this trend encompasses most of the rocks which formed the upper crust during the primary differentiation of the moon. Anorthositic norite was inferred to be the most abundant member of the sequence (Walker *et al.*, 1973a) at Apollo 16 as well as at Luna 20 (Prinz *et al.*, 1973b). Wood (1975) proposed that troctolites and "ferroan" anorthosites formed secondarily out of "main sequence" rocks.

The distribution of pristine rocks on Fig. 3 is radically different from distributions among soil lithic fragments, glasses, and nonpristine rocks. The absence of pristine rocks from the Apollo 16 trend militates against the hypothesis that this is a primary differentiation sequence. No major element data are available for 15455c, the only pristine anorthositic norite (see discussion of 15455 clasts above). Apparently, contrary to earlier beliefs, rocks which composed the primary nonmare crust were, in order of decreasing abundance among Apollo samples: Anorthosites, troctolites, and dunites (or, in the *nomenclature of Wager* (1968), plagioclase, plagioclase-olivine, olivine-plagioclase, and olivine cumulates). The most feldspathic pristine norite yet analyzed, 78235, is only 50% plagioclase and 50% low-Ca pyroxene (McCallum and Mathez, 1975).

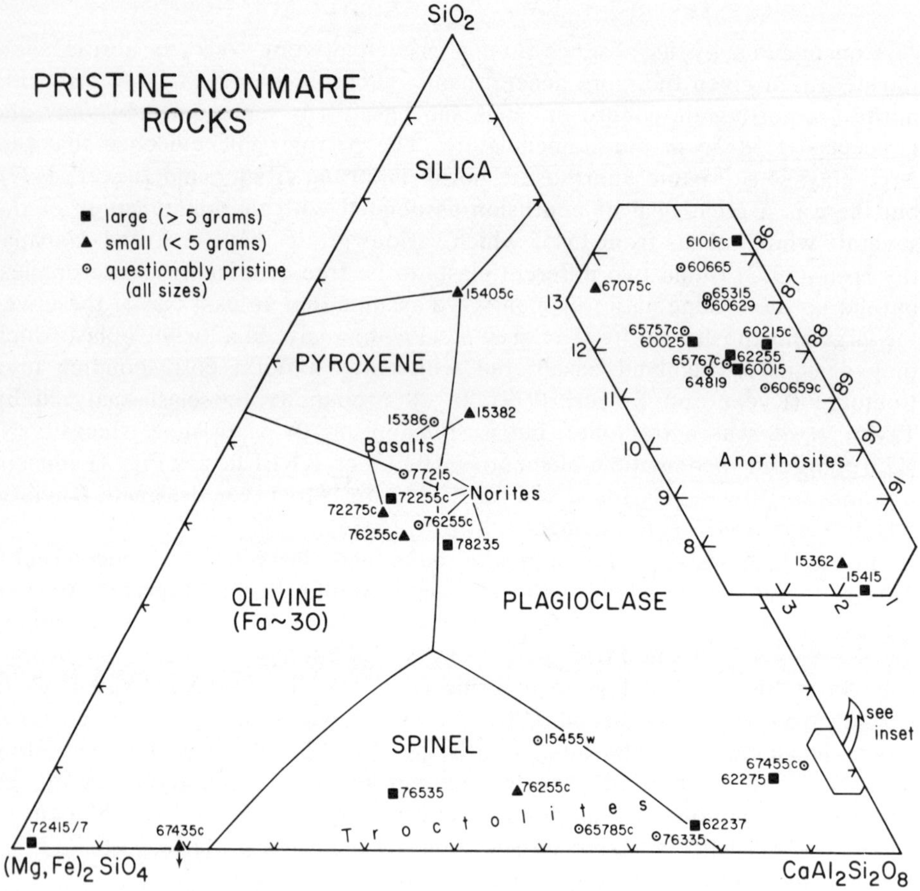

Fig. 3. Pristine nonmare rocks on the SiO_2–$(Mg,Fe)_2SiO_4$–$CaAl_2Si_2O_8$ pseudoternary diagram. Data sources for all filled symbols are as listed in Table 1. Lower case "c" after sample No. denotes a clast in a polymict breccia.

The relative abundances of pristine anorthosites, troctolites, and dunite are compatible with a stratified sample source: Plagioclase cumulates atop plagioclase, olivine cumulates (Wager, 1968, page 583, nomenclature: a comma is used in place of a hyphen between the cumulus mineral names when their relative abundances are unspecified) atop olivine cumulates. Only very large impacts plumbed below the anorthosites to dig up troctolites, and none but the largest excavated dunite. This model is supported by the dominance of anorthosites at Apollo 16, remote from all mare basins, whereas less feldspathic pristine rocks of deeper origin were common at Apollo 15 next to the Imbrium Basin and at Apollo 17 between the Serenitatis and Tranquillitatis basins. Furthermore, troctolites are relatively abundant at Luna 20, near the Fecunditatis Basin. Into this "layer cake", younger norites later intruded. 72255c has been dated by

Rb/Sr at 4.2 G.y. (Compston and Gray, 1974), and 77215 has been dated by Rb/Sr at 4.4 G.y. (Nunes *et al.*, 1976).

In addition to the aforementioned rock types, Fig. 3 includes one each of gabbro and quartz monzodiorite clasts, and three KREEPy basalts. All of these are relatively young, and relatively fine grained.

Evidence from Incompatible Elements

In Fig. 4 incompatible elements in pristine rocks are KREEP normalized and plotted in order of decreasing KREEP/CI-chondrite ratio. Figure 1 in our companion (Wasson *et al.*, 1977) paper (see text there for additional explanation of the figure) shows that Apollo 16 nonpristine rocks having KREEP concentrations as low as 10% still display flat, nondescript patterns when so plotted, cogent evidence that the incompatibles present in most highlands breccias, "melt" and "poikilitic" rocks, originated as an admixed KREEP-rich component. At less than 10% KREEP positive anomalies in K and Ba become visible.

Figure 4 shows the contrasting situation for some pristine nonmare rocks. Incompatible concentrations are low, and the patterns show anomalies at K and Ba and sometimes at Hf as well. Preliminary INAA data from our laboratory are plotted for two troctolitic anorthosites: 62237, which is almost certainly pristine, and 76335, which very probably is pristine. Patterns for 76335 and 15455w, especially the latter, also display a negative slope in their rare earth elements (superposed on the chondrite normalized negative slope of KREEP). The similarity in patterns between 76335 and 15455w supports the Ryder and Bower (1977) suggestion that 15455w is a troctolite, not a norite.

We have examined incompatible element patterns from many literature sources besides those shown in Fig. 4, and our conclusion remains the same. All rocks having such low incompatible concentrations and/or sloping, anomalous patterns appear to be pristine.

Evidence from Mg/Fe Ratios

Figure 5 shows plots of MgO/FeO vs. Al_2O_3. Like the pseudoternary diagrams, these diagrams illustrate how dissimilar pristine rocks are to the Apollo 16 and Apollo 17 nonpristine rocks. Averaged soil lithic fragments and glasses from the Apollo 16 and Apollo 17 sites collectively occupy the same ranges on this type of diagram as do the unsorted large rocks shown in Fig. 5a.

Figure 5 again illustrates that the make-up of the lunar crust inferred from pristine-sample evidence is distinctly different from that based on all rocks. In particular we receive a false impression regarding the compositional variance among primary rocks, if we fail to separate wheat from chaff.

The point already has been made that plagioclase-pyroxene cumulates (anorthositic norites, noritic anorthosites) appear to have been relatively rare compared to plagioclase cumulates and olivine, plagioclase cumulates. Note in Fig.

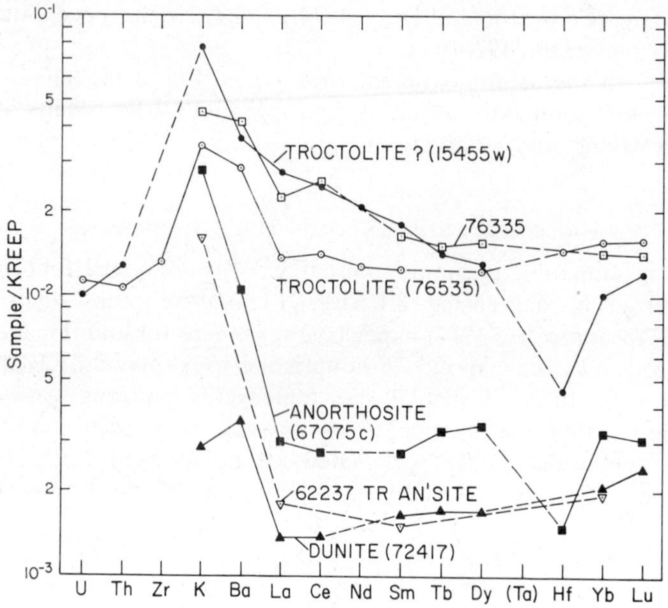

Fig. 4. Incompatible element concentrations of typical pristine nonmare rocks, normalized to best estimate of KREEP (Wasson *et al.*, 1977), in order (rightward) of decreasing enrichment in KREEPy basalt relative to CI chondrites. Data sources: 15455w, Taylor *et al.* (1973); 62237, our preliminary INAA data; 67075c, Haskin *et al.* (1973), except Ba, which is a 1σ lower limit estimated using Ba/La ratios averaged for 8 other anorthosites; 72417, Laul and Schmitt (1975); 76335, our preliminary INAA data; 76535, Haskin *et al.* (1974), except Th, Keith *et al.* (1974).

5b that all rocks having $20 < Al_2O_3 < 34\%$ are troctolites or troctolitic anorthosites.

Independent Evidence from Soils

Mixing model calculations have been used by many groups (e.g., Boynton *et al.*, 1975, 1976; Schonfeld, 1974) to estimate the relative concentrations of different components in Apollo 16 soils. There is a general consensus that the set of end components includes KREEPy basalt, an extralunar (meteoritic) component, and one or more local components rich in plagioclase. Figure 6 shows Apollo 16 soils on the pseudoternary diagram. This distribution is a microcosm of the Apollo 16 rock pattern (Fig. 2). Mature (i.e., well-mixed) soils occupy an oblong field near the center of the nonpristine rock field. Immature soils scatter into an area below and to the right of the mature samples, apparently reflecting recent contributions of anorthosites and to a lesser degree anorthositic troctolites. This is independent evidence that the latter rock types are dominant beneath the Apollo 16 regolith. The Apollo 16 soil fields in Fig. 5, derived analogously to the Fig. 6 fields, appear consistent with these conclusions. That

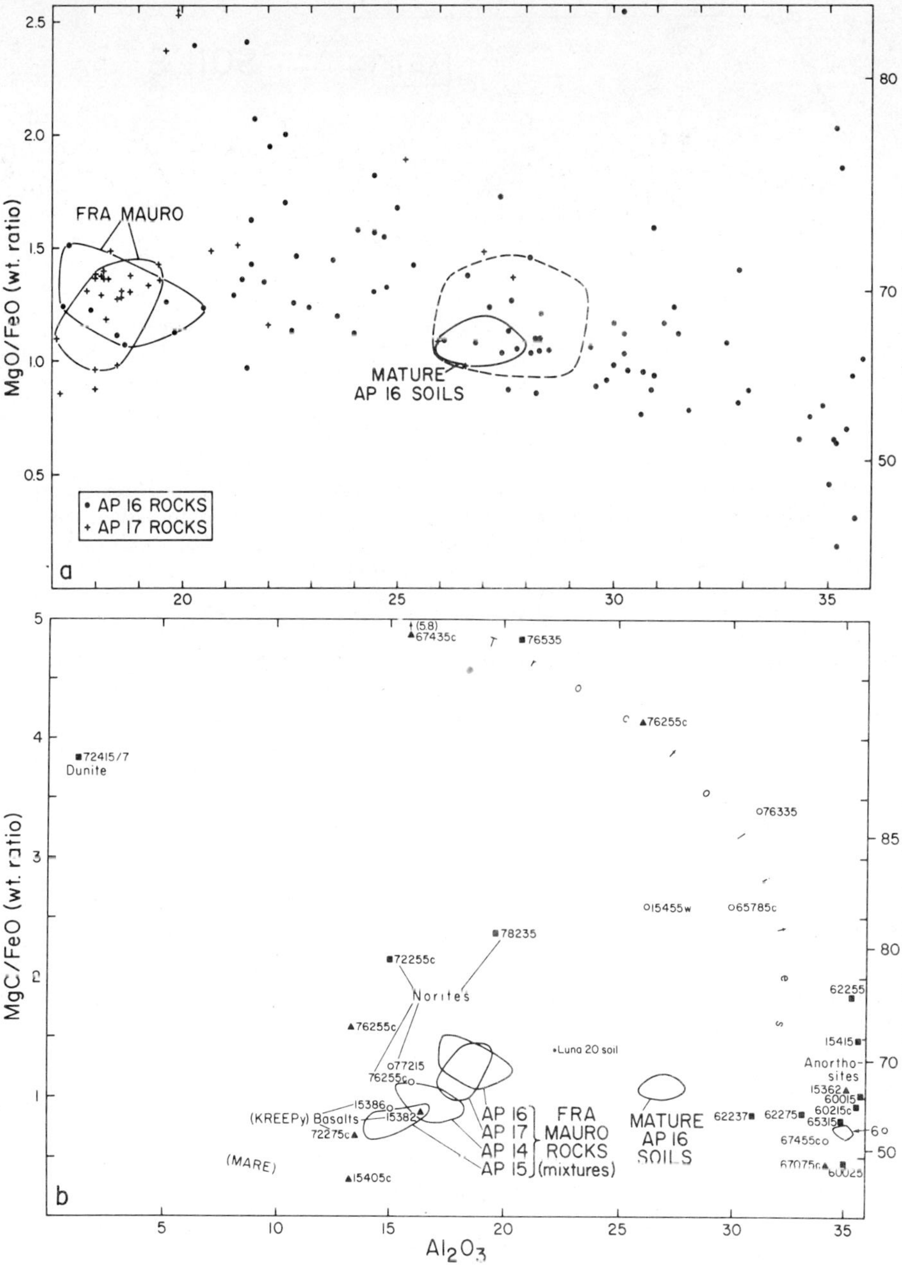

Fig. 5. (a) MgO/FeO vs. Al_2O_3 in large (>5 g) Apollo 16 and feldspathic (Al_2O_3 > 17%) Apollo 17 rocks. Data were selected as described for Fig. 1. (b) MgO/FeO vs. Al_2O_3 in pristine nonmare rocks. Note change in scale. Symbols, sample numbering and data sources as in Fig. 3.

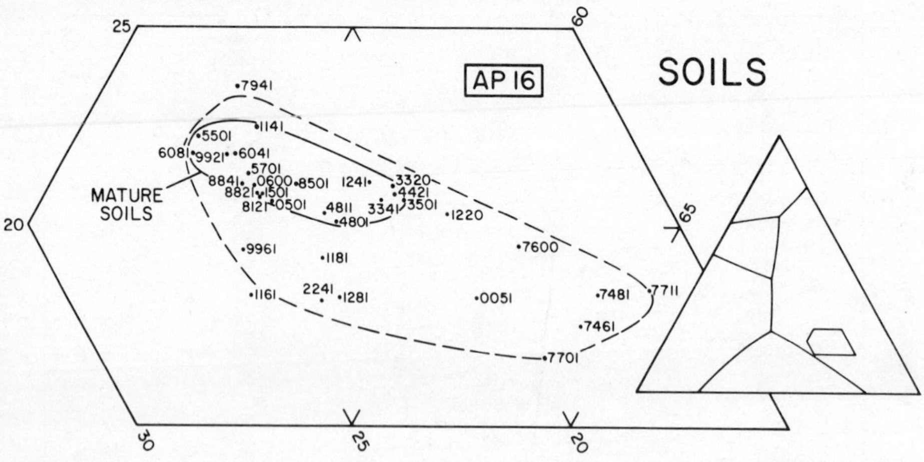

Fig. 6. Apollo 16 soils on the pseudoternary diagram. Individual analyses—*not* averages—are plotted. Soils 64501 and 69941 were omitted to avoid confusion; the available data for them are discrepant.

the compositions of immature soils do not scatter in the direction of the KREEPy basalts indicates either that no subregolith deposits of such materials are now present or, if present, that they were not penetrated by recent deep cratering events such as North and South Ray craters.

INDEPENDENT EVIDENCE FROM THE ORBITAL γ-RAY SPECTROMETRY EXPERIMENT

That the lunar crust does not largely consist of anorthositic norite and noritic anorthosite is also evidenced by the low Th concentrations observed in the farside upper regolith (Metzger *et al.*, 1974) by orbital γ-ray spectrometry. Whereas around the Apollo 16 site Th was found to be 2.1 μg/g in the upper few decimeters, concentrations found in six farside regions were 0.4–0.6 μg/g. A great deal of this Th undoubtedly was disseminated as impact ejecta from KREEP-rich regions such as Fra Mauro; 0.3 μg/g is probably a conservative upper limit for mean Th concentrations in pristine farside bedrock. This limit is far below concentrations in "feldspathic basalts" having anorthositic norite to noritic anorthosite compositions supposedly inherited directly from impact-remelted cumulates. Specifically, Walker *et al.* (1973b) proposed that 68415 was produced in this manner, but 68415 has a Th content of 1.3 μg/g (Tera *et al.*, 1973). Considering how anorthositic 68415 is, 79.3% plagioclase (Helz and Appleman, 1973), the hypothetical cumulate precursor is unrealistically Th-rich.

Since actual farside bedrock Th concentrations may be appreciably <0.3 μg/g, the most abundant rock type in the outermost 1–3 km of the farside crust may well be anorthosite.

Pristine Rocks are not Secondary

There is no doubt that major impacts sometimes generated melts, and that this is precisely the origin of "metaigneous" or "melt" rocks such as 14310 and 68415 that still show crystal relicts from the precursor rocks. A key question is whether some pristine rocks (Table 1) were also produced from these impact melts. For example, Wood (1975) postulated that most anorthosites and troctolites formed as secondary cumulates in great pools of impact-generated melts of "main sequence" primary cumulates.

We consider this mode of origin unlikely for any of the anorthosites and troctolites included in Table 1, with the unlikely exception of 67435c, which contains low-Co, near "meteoritic" metal (Prinz *et al.*, 1973a) (67435c is an exceptional case: Metal compositions in 76535 (Gooley *et al.*, 1974), 15455c, 78235/8, and several anorthosites (Hewins and Goldstein, 1975), are not in the meteoritic range). Adcumulus growth can only occur in magma chambers that cool very slowly. Constituents must have enough time to diffuse to the growing "cumulus" crystals, and diffusion rates are very dependent upon temperature. If the melt cools too rapidly, the proportion of material formed from trapped liquid increases (see Wager, 1968). It seems doubtful that impacts could form melt pools sufficiently slow in cooling to generate extreme *adcumulate* rocks (no zoning, no "trapped liquid") such as 62237 and the monomineralic anorthosites. Even if we assume that such rocks could form near the lunar surface, we cannot explain why neither the hypothetical precursor rocks nor the complementary rocks predicted for fractional crystallization of anorthositic, noritic liquids are abundant among pristine rocks.

Detailed textural evidence also supports a plutonic origin for many of the pristine rocks. The many Table 1 rocks with marked phase homogeneity (see column 5 in the table) apparently cooled very slowly, i.e., at depth. Extensive subsolidus equilibration of phases (particularly high-Ca and low-Ca pyroxenes) to very low temperatures, excellent evidence for slow cooling, is found in 62237 and 76535 (Dymek *et al.*, 1975); a whole suite of anorthosites (Dixon and Papike, 1975); 72255c (Stoeser *et al.*, 1974); 67075c (McCallum *et al.*, 1975); and 15415 (Hargraves and Hollister, 1972). The presence of symplectites has been interpreted as evidence for a deep origin for 76535 (Gooley *et al.*, 1974) and for 72415 (Bell and Mao, 1975). Although this interpretation has been called into question by Dymek *et al.* (1975), they nonetheless agree that 76535 probably originally accumulated at considerable depth. Stewart (1975) discussed the implications of the above-mentioned textural features for both 15415 and 76535, and concluded that "Apollonian metamorphism" deep within the lunar crust produced them.

New Data: 62237 and 76335

Troctolitic anorthosite 62237 was analyzed by electron microprobe and studied petrographically by Dymek *et al.* (1975), who observed that it has vestiges of a primary cumulate texture, despite extensive crushing. Our study of thin section 62237,8 confirms the observation of Dymek *et al.* that 62237 is a monomict brecciated plagioclase-olivine cumulate. Uncrushed plagioclase grains are up to 3 mm across, and olivines only slightly shattered are up to 2 mm. Plagioclase is remarkably uniform $An_{96.8\pm.4}Ab_{2.5\pm.4}Or_{0.3\pm.1}$ (17 analyses of 4 widely separated grains). Olivine composition appears equally uniform at about $Fo_{60.9}$ (13 analyses of 3 widely separated grains). Clark and Keith (1973) found 0.043 μg/g U (about 0.008 times average high-K Fra Mauro) in 62237. Our preliminary INAA data (see Fig. 4) on other incompatibles are even lower.

Dymek *et al.* reported 82.7% modal plagioclase and 16.1% model olivine. The visually estimated mode of our section is ~89% plagioclase and only ~10% olivine, and preliminary INAA data on our chip indicate a composition still more feldspathic. Furthermore, Stuart-Alexander and Reid (1972) reported only 10% "honey yellow" mafic (olivine) in a binocular description, not 20%, as Dymek *et al.* implied. The bulk composition used in Figs. 3 and 5 comes from the phase composition data of Dymek *et al.*, but *our mode.*

We have also begun analysis of brecciated troctolitic anorthosite thin section 76335,29, which shows a vestigial cumulate texture similar to 62237 (see Fig. 7, and compare with Fig. 10 in Dymek *et al.*, 1975). Shock effects are much more severe throughout 76335,29 than in portions of 62237,8, but sharp, angular outlines of original olivine grains remain, though more or less shattered into many optically continuous pieces. Intact plagioclase grains are up to 4 mm across, and original olivine grains were at least 2 mm across. The plagioclase is remarkably uniform $An_{95.6\pm.6}Ab_{3.7\pm.4}Or_{0.5\pm.1}$ (47 analyses of 11 widely separated grains) and so is the olivine, $Fo_{86.8\pm.2}$ (35 analyses of 9 widely separated grains). The visually estimated mode of 76335,29 is olivine ~12%, plagioclase ~88%; other phases (including pyroxenes) are at vanishingly low abundances. No metal grains could be found. The amount of olivine and its (original) granularity imply that, like 62237, 76335 originated as a plagioclase-olivine cumulate. If olivine had crystallized entirely from trapped intercumulus liquid, sundry other interstitial phases would surely be present.

No chemical analysis had hitherto been made of 76335. *Preliminary* INAA results on chip 76335,38 are as follows: 31% Al_2O_3, 17% CaO, 2.2% FeO, 7.5% MgO; and we calculate ~43% SiO_2 from our electron probe data. Preliminary data for incompatible elements are plotted in Fig. 4.

An–Fo Trends

Steele and Smith (1971) noted that for feldspathic fragments from Apollo 11 soils olivine Fo content correlated inversely with plagioclase An, a trend contrary to that predicted by the terrestrially inviolate Bowen reaction principle. Steele and Smith (1973) selected "only those fragments with igneous-looking textures and relatively coarse grain size" (almost all soil fragments or clasts from breccias) and found a similar correlation using data through Apollo 16. Warner *et al.* (1976a) constructed a plot for rocks and clasts "petrographically interpreted as being plutonic and potentially early crustal cumulates," which

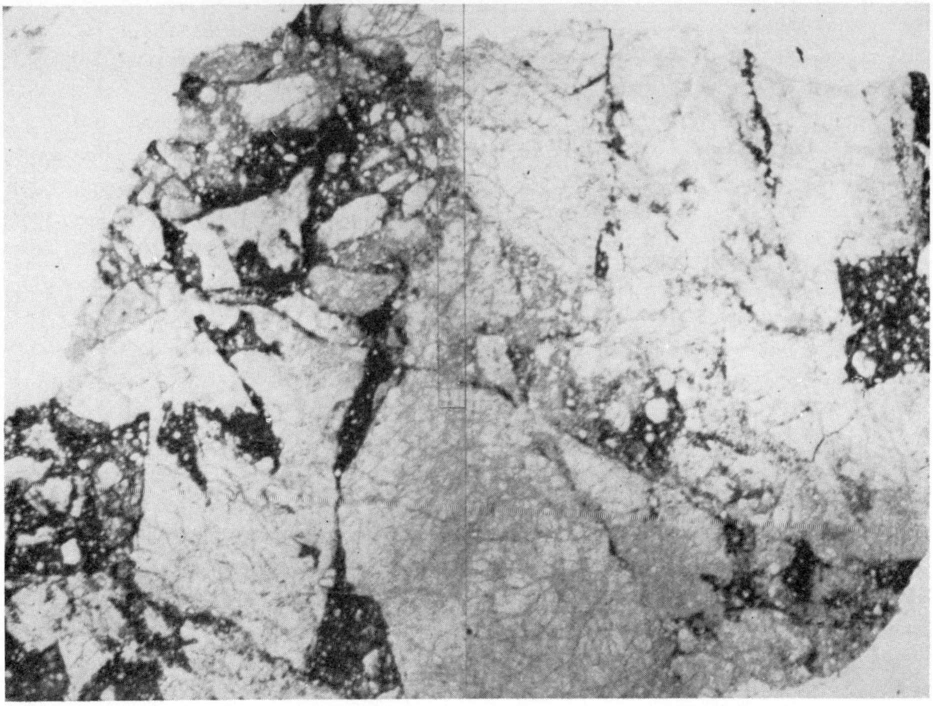

Fig. 7. Transmitted plane-polarized light photomicrograph of polished thin section 76335,29. Field of view represents 12 × 8 mm. The angular, dark gray regions are comprised of comminuted olivines. The area shown is approximately 95% of 76335,29.

they maintained confirmed the reality of the Steele and Smith trend, at least for true anorthosites. They also noted a separate "normal" trend in their data for "Mg-rich plutonic rocks."

Figure 8 is a Fo–An plot for pristine rocks only. The "Mg-rich" trend of Warner *et al.* is readily visible, although it depends upon very few pristine rocks. The negative Steele-Smith type trend has essentially disappeared, however. In fact, anorthosites show a general trend which is positive.

One of the most interesting things about Fig. 8 is the occurrence of rocks with *cumulus olivines* among the *anorthosite* cluster (in *lower* right). 62237 definitely contains cumulus olivine, and probably 60025 and 62275 both do also. Even if these rocks originally were heteradcumulates and not true adcumulates, it is most significant that the poikilitic crystals had to be in equilibrium with the magma just as the cumulus crystals had to. They would have been absent as cumulus crystals only by some quirk of the sedimentation. The olivine content of 62275 is 6% (Prinz *et al.*, 1973a), and 60025 according to the sample information catalog has areas of up to 20% olivine. Other plagioclase-olivine cumulates, epitomized by 76335, occur in the *upper* right corner of the diagram. At present there is a broad *hiatus* between the "magnesian" and "ferroan"

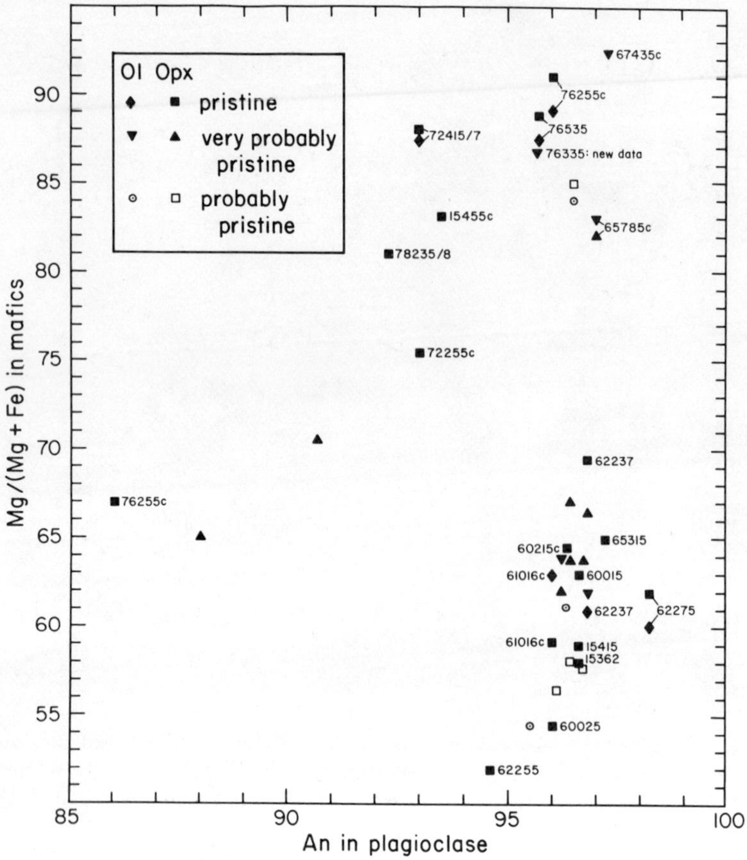

Fig. 8. Mg/(Mg + Fe) in mafics vs An in plagioclase among pristine nonmare rocks. In order of increasing An, unlabeled rocks and clasts are: 76255c, 72215, 67915c, 60516, 60665, 65757c, 65789 (olivine *and* pyroxene), 65767c, 64819, 60618c (olivine *and* pyroxene), 65326, 60659c, 60629.

clusters. Although study of additional rocks may fill this gap, it seems reasonable to attempt to predict it with petrogenetic models. The great range for Mg/(Mg + Fe) in samples with essentially identical An has seemingly been accommodated by recent models proposed for crystallization of a global magma ocean by Drake (1975) and Longhi (1977), but these models do not predict the hiatus. Warner *et al.* (1976a) included 76335 types in a "normal" (positively sloping) trend, and 62237 types in a Steele and Smith type trend, and argued that the two suites cannot have formed from "the same magma." The revision in the slope of the trend in the lower cluster to be "normal" (positive) might blunt their argument that the rocks of that trend formed by "different chemical processes," but their point that the two clusters are distinct seems well taken.

CONCLUSIONS

1. Anorthositic norites and noritic anorthosites were *rare* as primary nonmare rocks, not (as earlier believed) especially abundant.
2. In general, interpretations of compositional trends and loose groupings observed in soil glasses, soil lithic fragments, and rocks irrespective of petrography as *endogenous* should be reconsidered. As an example, low-K Fra Mauro "basalts" probably are preponderantly simple mechanical mixtures, not products of partial melting as proximity of soil glass groupings to the plagioclase-olivine cotectic line on the SiO_2–Ol–An pseudoternary diagram (latest misconstruction: Taylor and Bence, 1977) suggests.
3. The corollary to (1) is that *anorthosites* and *troctolites* were important constituents of the primary nonmare crust. Important volumes of norite (*sensu stricto*) were intruded sometime later.
4. The Steele-Smith negative Fo–An trend is not observed among pristine anorthosites; nevertheless, the Warner *et al.* (1976a) distinction between ferroan and magnesian trends for Mg/(Mg + Fe) vs. An appears to be solid.
5. Mechanical *mixing* appears to have been the dominant petrogenetic process on the highlands (i.e., the *top* of the nonmare crust). Although it caused pristine highlands rocks to be exceedingly rare, the meteoroid bombardment which caused the mixing was probably useful for excavating a few pristine samples from deep within the primary crust. Anorthosites far outnumber all other types of large pristine samples from Apollo 16, but pristine rocks from sites next to basins are more diverse, presumably as a result of basins having excavated deeper strata.

Acknowledgments—We are indebted to H. J. Chun for expert typing, to R. E. Jones for education in microprobe techniques, and to J. R. Arnold and W. V. Boynton for useful discussions. This research was mainly supported by NASA grant NGL 05-007-367.

REFERENCES

Bell P. M. and Mao H. K. (1975) Cataclastic plutonites: Possible keys to the evolutionary history of the early moon (abstract). In *Lunar Science VI*, p. 34–35. The Lunar Science Institute, Houston.

Blanchard D. P., Haskin L. A., Jacobs J. W., Brannon J. C., and Korotev R. L. (1974) Major and trace elements in rocks 72215, 72235, 72255, and 72275 from Boulder 1, Station 2, Apollo 17. In *Interdisciplinary Studies of Samples from Boulder 1, Station 2, Apollo 17* (J. A. Wood, ed.), Vol. 2, p. IV-1–IV-15. Smithsonian Astrophysical Observatory, Cambridge, Massachusetts.

Boynton W. V., Baedecker P. A., Chou C.-L., Robinson K. L., and Wasson J. T. (1975) Mixing and transport of lunar surface materials: Evidence obtained by the determination of lithophile, siderophile and volatile elements. *Proc. Lunar Sci. Conf. 6th*, p. 2241–2259.

Boynton W. V., Chou C.-L., Robinson K. L., Warren P. H., and Wasson J. T. (1976) Lithophiles, siderophiles and volatiles in Apollo 16 soils and rocks. *Proc. Lunar Sci. Conf. 7th*, p. 727–742.

Clark R. S. and Keith J. E. (1973) Determination of natural and cosmic ray induced radionuclides in Apollo 16 lunar samples. *Proc. Lunar Sci. Conf. 4th*, p. 2105–2113.

Compston W. and Gray C. M. (1974) Rb–Sr age of the civet cat clast, 72255,41. In *Interdisciplinary Studies of Samples from Boulder 1, Station 2, Apollo 17* (J. A. Wood, ed.), Vol. 1, p. 139–143. Smithsonian Astrophysical Observatory, Cambridge, Massachusetts.

Delano J. W., Bence A. E., Papike J. J., and Cameron K. L. (1973) Petrology of the 2–4 mm soil fraction from the Descartes region of the moon and stratigraphic implications. *Proc. Lunar Sci. Conf. 4th*, p. 537–551.

Dixon J. R. and Papike J. J. (1975) Petrology of anorthosites from the Descartes region of the moon: Apollo 16. *Proc. Lunar Sci. Conf. 6th*, p. 263–291.

Dowty E., Keil K., and Prinz M. (1972) Anorthosite in the Apollo 15 rake sample from Spur crater. In *The Apollo 15 Lunar Samples* (J. W. Chamberlain and C. Watkins, eds.), p. 62–65. The Lunar Science Institute, Houston.

Dowty E., Keil K., Prinz M., Gros J., and Takahashi H. (1976) Meteorite-free Apollo 15 crystalline KREEP. *Proc. Lunar Sci. Conf. 7th*, p. 1833–1844.

Drake M. J. (1975) Lunar anorthosite paradox: An alternative explanation. *Proc. Lunar Sci. Conf. 6th*, p. 293–299.

Dymek R. F., Albee A. L., and Chodos A. A. (1975) Comparative petrology of lunar cumulate rocks of possible primary origin: Dunite 72415, troctolite 76535, norite 78235, and anorthosite 62237. *Proc. Lunar Sci. Conf. 6th*, p. 301–341.

Ganapathy R., Morgan J. W., Krähenbühl U., and Anders E. (1973) Ancient meteoritic components in lunar highland rocks: Clues from trace elements in Apollo 15 and 16 samples. *Proc. Lunar Sci. Conf. 4th*, p. 1239–1261.

Gooley R., Brett R., and Warner J. (1974) A lunar rock of deep crustal origin: Sample 76535. *Geochim. Cosmochim. Acta* **38**, 1329–1339.

Gray C. M., Compston W., Foster J. J., and Rudowski R. (1974) Rb–Sr ages of clasts from within Boulder 1, Station 2, Apollo 17. In *Interdisciplinary Studies of Samples from Boulder 1, Station 2, Apollo 17* (J. A. Wood, ed.), Vol. 2, p. VII–1–VII-10. Smithsonian Astrophysical Observatory, Cambridge, Massachusetts.

Gros J., Takahashi H., Hertogen J., Morgan, J. W., and Anders E. (1976) Composition of the projectiles that bombarded the lunar highlands. *Proc. Lunar Sci. Conf. 7th*, p. 2403–2425.

Hargraves R. B. and Hollister L. S. (1972) Mineralogic and petrologic study of lunar anorthosite slide 15415,18. *Science* **175**, 430–432.

Haskin L. A., Helmke P. A., Blanchard D. P., Jacobs J. W., and Telander K. (1973) Major and trace element abundances in samples from the lunar highlands. *Proc. Lunar Sci. Conf. 4th*, p. 1275–1296.

Haskin L. A., Shih C.-Y., Bansal B. M., Rhodes J. M., Wiesmann H., and Nyquist L. E. (1974) Chemical evidence for the origin of 76535 as a cumulate. *Proc. Lunar Sci. Conf. 5th*, p. 1213–1225.

Helz R. T. and Appleman D. E. (1973) Mineralogy, petrology, and crystallization history of Apollo 16 rock 68415. *Proc. Lunar Sci. Conf. 4th*, p. 643–659.

Hewins R. H. and Goldstein J. I. (1975) The provenance of metal in anorthositic rocks. *Proc. Lunar Sci. Conf. 6th*, p. 343–362.

Higuchi H. and Morgan J. W. (1975) Ancient meteoritic component in Apollo 17 boulders. *Proc. Lunar Sci. Conf. 6th*, p. 1625–1651.

Hodges F. N. and Kushiro I. (1973) Petrology of Apollo 16 lunar highland rocks. *Proc. Lunar Sci. Conf. 4th*, p. 1033–1048.

Hubbard N. J., Gast P. W., Meyer C., Nyquist L. E., and Shih C. (1971) Chemical composition of lunar anorthosites and their parent liquids. *Earth Planet. Sci. Lett.* **13**, 71–75.

Hubbard N. J., Rhodes J. M., Gast P. W., Bansal B. M., Shih C.-Y., Wiesmann H., and Nyquist L. E. (1973) Lunar rock types: The role of plagioclase in non-mare and highland rock types. *Proc. Lunar Sci. Conf. 4th*, p. 1297–1312.

Hubbard N. J., Rhodes J. M., Wiesmann H., Shih C.-Y., and Bansal B. M. (1974) The chemical definition and interpretation of rock types returned from the non-mare regions of the moon. *Proc. Lunar Sci. Conf. 5th*, p. 1227–1246.

James O. B. (1972) Lunar anorthosite 15415: Texture, mineralogy, and metamorphic history. *Science* **175**, 432–435.

Juan V. C., Chen J. C., Huang C. K., Chen P. Y., and Wang Lee C. M. (1974) Petrology and chemistry of some Apollo 16 lunar samples (abstract). In *Lunar Science V*, p. 394–396. The Lunar Science Institute, Houston.

Keith J. E., Clark R. S., and Bennett L. J. (1974) Determination of natural and cosmic ray induced radionuclides in Apollo 17 lunar samples. *Proc. Lunar Sci. Conf. 5th*, p. 2121–2138.

Krähenbühl U., Ganapathy R., Morgan J. W., and Anders E. (1973) Volatile elements in Apollo 16 samples: Implications for highland volcanism and accretion history of the moon. *Proc. Lunar Sci. Conf. 4th*, p. 1325–1348.

Laul J. C. and Schmitt R. A. (1973) Chemical composition of Apollo 15, 16, and 17 samples. *Proc. Lunar Sci. Conf. 4th*, p. 1349–1367.

Laul J. C. and Schmitt R. A. (1974) Breccias 64435, 63335, and 63355 (abstract). In *Lunar Science V*, p. 435–437. The Lunar Science Institute, Houston.

Laul J.C. and Schmitt R. A. (1975) Dunite 72417: A chemical study and interpretation. *Proc. Lunar Sci. Conf. 6th*, p. 1231–1254.

Longhi J. (1977) Magma oceanography: 2. Chemical evolution (abstract). In *Lunar Science VIII*, p. 592–594. The Lunar Science Institute, Houston.

LSPET (1971) Lunar Sample Information Catalog—Apollo 15. NASA MSC-03209. 302 pp.

LSPET (1972) Lunar Sample Information Catalog—Apollo 16. NASA MSC-03210. 372 pp.

LSPET (1973) Lunar Sample Information Catalog—Apollo 17. NASA MSC-03211. 447 pp.

McCallum I. S. and Mathez E. A. (1975) Petrology of noritic cumulates and a partial melting model for the genesis of Fra Mauro basalts. *Proc. Lunar Sci. Conf. 6th*, p. 395–414.

McCallum I. S., Okamura F. P., and Ghose S. (1975) Mineralogy and petrology of sample 67075 and the origin of lunar anorthosites. *Earth Planet. Sci. Lett.* **26**, 36–53.

Metzger A. E., Trombka J. I., Reedy R. C., and Arnold J. R. (1974) Element concentrations from lunar orbital gamma-ray measurements. *Proc. Lunar Sci. Conf. 5th*, p. 1067–1078.

Meyer H. O. A. and McCallister R. H. (1973) Mineralogy and petrology of Apollo 16: Rock 60215,13. *Proc. Lunar Sci. Conf. 4th*, p. 661–665.

Morgan J. W., Krähenbühl U., Ganapathy R., and Anders E. (1972) Trace elements in Apollo 15 samples: Implications for meteorite influx and volatile depletion on the moon. *Proc. Lunar Sci. Conf. 3rd*, p. 1361–1376.

Morgan J. W., Ganapathy R., Higuchi H., Krähenbühl U., and Anders E. (1974) Lunar basins: Tentative characterizations of projectiles, from meteoritic elements in Apollo 17 boulders. *Proc Lunar Sci. Conf. 5th*, p. 1703–1736.

Murali A. V., Ma M.-S., Laul J. C., and Schmitt R. A. (1977) Chemical composition of breccias, feldspathic basalt and anorthosites from Apollo 15 (15308, 15359, 15382 and 15362), Apollo 16 (60618 and 65785), Apollo 17 (72435, 72536, 72559, 72735, 72738, 78526 and 78527) and Luna 20 (22012 and 22013) (abstract). In *Lunar Science VIII*, p. 700–702. The Lunar Science Institute, Houston.

Nakamura N., Masuda A., Tanaka T., and Kurasawa H. (1973) Chemical compositions and rare-earth features of four Apollo 16 samples. *Proc. Lunar Sci. Conf. 4th*, p. 1407–1414.

Nava D. F. (1974) Chemical compositions of some soils and rock types from the Apollo 15, 16, and 17 lunar sites. *Proc. Lunar Sci. Conf. 5th*, p. 1087–1096.

Nunes P. D., Knight R. J., Unruh D. M., and Tatsumoto M. (1974) The primitive nature of the lunar crust and the problem of initial Pb isotopic compositions of lunar rocks: A Rb–Sr and U–Th–Pb study of Apollo 16 samples (abstract). In *Lunar Science V*, p. 559–561. The Lunar Science Institute, Houston.

Nunes P. D., Nakamura N., and Tatsumoto M. (1976) 4.4 B.Y.-old clast in Boulder 7, Apollo 17 (abstract). In *Lunar Science VII*, p. 631–632. The Lunar Science Institute, Houston.

Nyquist L. E., Hubbard N. J., Gast P. W., Bansal B. M., Weismann H., and Jahn B. (1973) Rb–Sr systematics for chemically defined Apollo 15 and 16 materials. *Proc. Lunar Sci. Conf. 4th*, p. 1823–1846.

Nyquist L. E., Wiesmann H., Shih C.-Y., and Bansal B. M. (1977) REE and Rb–Sr analysis of 15405 quartz-monzodiorite (super-KREEP) (abstract). In *Lunar Science VIII*, p. 738–740. The Lunar Science Institute, Houston.

Papanastassiou D. A. and Wasserburg G. J. (1972) Rb–Sr systematics of Luna 20 and Apollo 16 samples. *Earth Planet. Sci. Lett.* **17**, 52–63.

Papanastassiou D. A. and Wasserburg G. J. (1975) Rb–Sr study of a lunar dunite and evidence for early lunar differentiates. *Proc. Lunar Sci. Conf. 6th*, p. 1467–1489.

Papanastassiou D. A. and Wasserburg G. J. (1976) Rb–Sr age of troctolite 76535. *Proc. Lunar Sci. Conf. 7th*, p. 2035–2054.

Philpotts J. A., Schuhmann S., Kouns C. W., Lum R. K. L., Bickel A. L., and Schnetzler C. C. (1973) Apollo 16 returned lunar samples: Lithophile trace-element abundances. *Proc. Lunar Sci. Conf. 4th*, p. 1427–1436.

Prinz M., Dowty E., Keil K., and Bunch T. E. (1973a) Spinel troctolite and anorthosite in Apollo 16 samples. *Science* **179**, 74–76.

Prinz M., Dowty E., Keil K., and Bunch T. E. (1973b) Mineralogy, petrology and chemistry of lithic fragments from Luna 20 fines: Origin of the cumulate ANT suite and its relationship to high-alumina and mare basalts. *Geochim. Cosmochim. Acta* **37**, 979–1006.

Rhodes J. M., Rodgers K. V., Shih C., Bansal B. M., Nyquist L. E., Wiesmann H., and Hubbard N. J. (1974) The relationships between geology and soil chemistry at the Apollo 17 landing site. *Proc. Lunar Sci. Conf. 5th*, p. 1097–1117.

Ridley W. I., Reid A. M., Warner J. L., Brown R. W., Gooley R., and Donaldson C. (1973) Glass compositions in Apollo 16 soils 60501 and 61221. *Proc. Lunar Sci. Conf. 4th*, p. 309–321.

Rose H. J., Cuttitta F., Berman S., Carron M. K., Christian R. P., Dwornik E. J., Greenland L. P., and Ligon D. T., Jr. (1973) Compositional data for twenty-two Apollo 16 samples. *Proc. Lunar Sci. Conf. 4th*, p. 1149–1158.

Rose H. J., Jr., Baedecker P. A., Berman S., Christian R. P., Dwornik E. J., Finkelman R. B., and Schnepfe M. M. (1975) Chemical composition of rocks and soils returned by the Apollo 15, 16, and 17 missions. *Proc. Lunar Sci. Conf. 6th*, p. 1363–1373.

Ryder G. and Bower J. F. (1977) Sample 15455: A petrological survey. In *Interdisciplinary Studies by the Imbrium Consortium*, Vol. 2 (J. A. Wood, ed.). LSI Contr. No. 268D, p. 63–70. Center for Astrophysics, Cambridge, Massachusetts.

Schonfeld E. (1974) The contamination of lunar highland rocks by KREEP: Interpretation by mixing models. *Proc. Lunar Sci. Conf. 5th*, p. 1269–1286.

Simkin T., Noonan A. F., Switzer G. S., Mason B., Nelen J. A., Melson W. G., and Thompson G. (1973) Composition of Apollo 16 fines 60051, 60052, 64811, 64812, 67711, 67712, 68821, and 68822. *Proc. Lunar Sci. Conf. 4th*, p. 279–289.

Steele I. M. and Smith J. V. (1973) Mineralogy and petrology of some Apollo 16 rocks and fines: General petrologic model of moon. *Proc. Lunar Sci. Conf. 4th*, p. 519–536.

Stewart D. B. (1975) Apollonian metamorphic rocks—The products of prolonged subsolidus equilibration (abstract). In *Lunar Science VI*, p. 774–776. The Lunar Science Institute, Houston.

Stoeser D. B., Marvin U. B., Wood J. A., Wolfe R. W., and Bower J. F. (1974) Petrology of a stratified boulder from South Massif, Taurus-Littrow. *Proc. Lunar Sci. Conf. 5th*, p. 355–377.

Stuart-Alexander D. and Reid A. M. (1972) Binocular description of 62237. In *Apollo 16 Lunar Sample Information Catalog*, p. 167. MSC-03210, NASA/Johnson Space Center, Houston.

Taylor G. J. (1976) Further petrological studies of the KREEP-rich quartz monzodiorite. In *Interdisciplinary Studies by the Imbrium Consortium*, Vol. 1 (J. A. Wood, ed.). LSI Contr. No. 268D, p. 94–95. Center for Astrophysics, Cambridge, Massachusetts.

Taylor S. R. and Bence A. E. (1977) Primary chemical components in the highland crust (abstract). In *Lunar Science VIII*, p. 931–933. The Lunar Science Institute, Houston.

Taylor S. R., Gorton M. P., Muir P., Nance W. B., Rudowski R., and Ware N. (1973a) Composition of the Descartes region, lunar highlands. *Geochim. Cosmochim. Acta* **37**, 2665–2683.

Taylor S. R., Gorton M. P., Muir P., Nance W., Rudowski R., and Ware N. (1973b) Lunar highlands composition: Apennine Front. *Proc. Lunar Sci. Conf. 4th*, p. 1445–1459.

Taylor S. R., Gorton M., Muir P., Nance W., Rudowski R., and Ware N. (1974) Lunar highland composition (abstract). In *Lunar Science V*, p. 789–791. The Lunar Science Institute, Houston.

Tera F., Papanastassiou D. A., and Wasserburg G. J. (1973) A lunar cataclysm at ~3.95 AE and the structure of the lunar crust (abstract). In *Lunar Science IV*, p. 723–725. The Lunar Science Institute, Houston.

Wänke H., Palme H., Baddenhausen H., Dreibus G., Jagoutz E., Kruse H., Spettel B., Teschke F., and Thacker R. (1974) Chemistry of Apollo 16 and 17 samples: Bulk composition, late stage accumulation and early differentiation of the moon. *Proc. Lunar Sci. Conf. 5th*, p. 1307–1335.

Wänke H., Palme H., Baddenhausen H., Dreibus G., Jagoutz E., Kruse H., Palme C., Spettel B., Teschke F., and Thacker R. (1975) New data on the chemistry of lunar samples: Primary matter in the lunar highlands and the bulk composition of the moon. *Proc. Lunar Sci. Conf. 6th*, p. 1313–1340.

Wänke H., Palme H., Kruse H., Baddenhausen H., Cendales M., Dreibus G., Hofmeister H., Jagoutz E., Palme C., Spettel B., and Thacker R. (1976) Chemistry of lunar highland rocks: A refined evaluation of the composition of the primary matter. *Proc. Lunar Sci. Conf. 7th*, p. 3479–3499.

Wänke H., Blum K., Dreibus G., Hofmeister H., Kruse H., and Spettel B. (1977) Primary matter in the lunar highlands: A reevaluation (abstract). In *Lunar Science VIII*, p. 973–975. The Lunar Science Institute, Houston.

Wager L. R. (1968) Rhythmic and cryptic layering in mafic and ultramafic plutons. In *Basalts: The Poldervaart Treatise on Rocks of Basaltic Composition, Vol. 2* (H. H. Hess and A. Poldervaart, eds.), p. 573–622. Wiley, New York.

Walker D., Grove T. L., Longhi J., Stolper E. M., and Hays J. F. (1973a) Origin of lunar feldspathic rocks. *Earth Planet. Sci. Lett.* **20**, 325–336.

Walker D., Longhi J., Grove T. L., Stolper E., and Hays J. F. (1973b) Experimental petrology and origin of rocks from the Descartes Highlands. *Proc. Lunar Sci. Conf. 4th*, p. 1013–1032.

Warner J. L., Simonds C. H., and Phinney W. C. (1976a) Genetic distinction between anorthosites and Mg-rich plutonic rocks: New data from 76255 (abstract). In *Lunar Science VII*, p. 915–917. The Lunar Science Institute, Houston.

Warner J. L., Simonds C. H., and Phinney W. C. (1976b) Apollo 17, Station 6 boulder sample 76255: Absolute petrology of breccia matrix and igneous clasts. *Proc. Lunar Sci. Conf. 7th*, p. 2233–2250.

Wasserburg G. J. and Papanastassiou D. A. (1971) Age of an Apollo 15 mare basalt; lunar crust and mantle evolution. *Earth Planet. Sci. Lett.* **13**, 97–104.

Wasson J. T., Warren P. H., Kallemeyn G. W., McEwing C. E., and Boynton W. V. (1977) SCCRV, a major component of highlands rocks. *Proc. Lunar Sci. Conf. 8th.* This volume.

Winzer S. R., Nava D. F., Lum R. K. L., Schuhmann S., Schuhmann P., and Philpotts J. A. (1975) Origin of 78235, a lunar norite cumulate. *Proc. Lunar Sci. Conf. 6th*, p. 1219–1229.

Wood J. A. (1975) Lunar petrogenesis in a well-stirred magma ocean. *Proc. Lunar Sci. Conf. 6th*, p. 1087–1102.

Proc. Lunar Sci. Conf. 8th (1977), p. 2237–2252.
Printed in the United States of America

SCCRV, a major component of highlands rocks

JOHN T. WASSON, PAUL H. WARREN, GREGORY W. KALLEMEYN, COLIN E. MCEWING, DAVID W. MITTLEFEHLDT, and WILLIAM V. BOYNTON

Institute of Geophysics and Planetary Physics, Department of Earth and Space Sciences, Department of Chemistry, University of California, Los Angeles, California 90024

Abstract—Concentrations of 36 elements are reported for 13 highlands rocks. Seven of these rocks have high concentrations of a mafic component characterized by high concentrations of Mg, Sc, Ti, V, Cr, Mn, and Fe and assigned the acronym SCCRV based on the chemical symbols of three of these elements. Concentrations of these seven elements in SCCRV were estimated by an iterative series of least-squares fittings of four components (chondritic matter, anorthosite, KREEP, and SCCRV) to the seven SCCRV-rich rocks. The concentrations are quite similar to those of the "primary matter" component of Wänke and coworkers with the exception that Ti is about 4× higher in SCCRV than in the primary matter component. The Ti/Sc ratio of SCCRV is 6–7× higher than observed in chondritic meteorites or in Ca–Al-rich clasts from CV chondrites, and seemingly inconsistent with an origin as a late primordial accretate. Insufficient evidence is available to choose between one-source or multiple-source lunar origins of SCCRV, but its early formation ≥4.25 G.y. ago is inconsistent with an origin by surficial mixing of mare basalts and an ultramafic such as dunite.

The bulk compositions of most of our samples place them in the anorthositic norite-noritic anorthosite compositional range. Incompatible element ratios are generally indistinguishable from those in KREEP, excepting small positive *K* anomalies in low-KREEP, high-anorthosite samples.

INTRODUCTION

MOST HIGHLANDS ROCKS have high siderophile contents indicating that they are not pristine (see summary in Warren and Wasson, 1977). As a result, several authors have adopted the working hypothesis that they are mixtures, and attempted to infer from their compositions the nature of their major components by means of least-squares fitting techniques (Schonfeld, 1974; Wänke *et al.*, 1974, 1975, 1976, 1977; Boynton *et al.*, 1976). There is general agreement that the set of end components includes at least one rich in plagioclase, one rich in KREEP, and an extralunar component (ELC).

Wänke *et al.* (1974, 1975, 1976, 1977) noted that Mg and several transition elements are enriched in highlands rocks, and proposed that these resulted from admixture of a single component in addition to ELC, anorthosite, and KREEP. They inferred that this component was primordial consisting of a mixture of CI-like chondritic matter and nebular high-temperature materials similar to the Ca–Al-rich clasts in the Allende chondrite.

It seemed of some importance to investigate in more detail the properties of this Mg-rich component in order (1) to determine whether a single component can account for the data on highlands rocks, and if so, (2) to better define its composition and (3) to examine models for its formation. To this end a suite of highlands samples rich in mafics was obtained and their composition studied.

Also included in our study were additional highlands samples selected on the basis of other properties.

EXPERIMENTAL

Samples (*ca.* 50–300 mg) were analyzed by instrumental neutron activation analysis (INAA) for 30 mainly lithophile elements, then by radiochemical neutron activation analysis (RNAA) for 9–10 mainly volatile and siderophile elements; Ni, Ir, Au, and sometimes U were determined by both techniques. Sample preparation was carried out in the UCLA clean room and handling was kept to a minimum. Samples which consisted of fragments ≤1 mm in size were transferred directly to polyethylene vials for irradiation. Larger fragments were gently crushed with a steel percussion mortar and pestle before transferral to the poly vials.

The INAA procedure, which generally takes about 4 weeks, was as described by Boynton *et al.* (1976). Afterwards, the samples were transferred into silica tubes and sealed for the high flux irradiation and subsequent RNAA determinations. The RNAA procedure was described in detail by Sundberg and Boynton (1977).

SAMPLE SELECTION AND DESCRIPTION

Most of the samples were Apollo 16 rocks selected on the basis of high mafic contents, and most of these were small (5–30 g) rake samples. Others were chosen on the basis of their potentially interesting siderophile compositions (67035, 67955) or unusual bulk compositions (77545).

60666. Microprobe data of Dowty *et al.* (1974) showed 60666 to have a spinel troctolite composition unique among Apollo 16 rocks; its texture is that of a fine-grained melt rock. Binocular inspection suggested that our chip was richer in large plagioclase relicts than that of Dowty *et al.*, and this was confirmed by our higher Al and Ca concentrations.

61569. Simonds *et al.* (1973) classified 61569 as a poikilitic melt rock and reported an unusually high modal olivine content of 22%. Our dark gray chip was too fine grained to be characterizable with binocular microscope.

63549. LSPET (1972) designated 63549 a fine-grained, crystalline rock; Hubbard *et al.* (1973) called it a gabbroic anorthosite. Our chips appeared homogeneous (although not equigranular). Our data agreed well with those of Hubbard *et al.*

63556. Simonds *et al.* (1973) indicated that 63556 was a fine-grained, poikilitic melt rock that plotted near the olivine-plagioclase-spinel point on the Walker *et al.* (1973) pseudoternary diagram. Our chip appeared to be inhomogeneous, but analyses of separate chips showed excellent agreement except for siderophiles. Our data agree well with unpublished data of Rhodes. 63556 is one of the KREEPiest Apollo 16 rocks.

64569. Simonds *et al.* (1973) reported 64569 to be a melt rock with an unusually high modal olivine content of 21%. Our chip was a variegated gray in color, and too fine grained to recognize minerals during binocular examination. Our Mg, Al, Ca, and Fe concentrations are about 10% higher than those of Floran *et al.* (1976).

64815. Another poikilitic melt rock, Simonds *et al.* (1973) found 64815 to contain 9% modal olivine and 34% modal orthopyroxene. In contrast, LSPET (1972) designated it a crushed ultramafic rock with mafic silicates accounting for 75–80% of the rock. Our chip appeared to be inhomogeneous. There is rather poor agreement between our Mg, Al, Ca, and Fe data and those of Hubbard *et al.* (1973) and Wänke *et al.* (1976), suggesting that samples of ~200 mg are not large enough to be representative.

65015. According to Simonds *et al.* (1973), 65015 is a melt rock containing 29% modal pigeonite; LSPET (1972) designated it a monomict gabbroic anorthosite breccia. Numerous groups have analyzed 65015; our data agree well with most previous analyses. Our earlier 65015 INAA data (Boynton *et al.*, 1975) were less precise than the current set and we have given extra weight to the latter in the derivation of the mean. 65015 is another of the KREEPiest Apollo 16 rocks.

65055. Warner *et al.* (1973) designated 65055 a porphyritic basalt, and suggested that these formed when breccias were heated to almost total melting. The high siderophile contents indicate that it is not pristine, and conflict with the classification of LSPET (1972) as a crystalline, possibly magmatic anorthosite. Our data show general agreement with those of Christian *et al.* (1976).

65779. Dowty *et al.* (1974) described 65779 as a fine-grained melt rock laden with plagioclase relicts; their microprobe bulk analysis showed a high MgO concentration of 10.1%. There is general agreement between our data and those of Dowty *et al.* except that our Fe concentration is 35% higher than theirs.

67035. The LSPET (1972) description indicated that 67035 was heterogeneous on a scale of centimeters. It was selected for analysis because we thought it might be North Ray Crater ejecta containing a record of the regolith siderophile composition in ancient times. Although binocular examination of the very fine-grained, friable rock suggested that it was homogeneous, separate "chips" showed significant differences in composition. Laul and Schmitt (1973) analyzed 67031, "a fragment of gray breccia 67035", and reported a composition intermediate between our replicates.

67559. Steele and Smith (1973) found 67559 to be "igneous-looking" and dominated by lath-shaped plagioclase (<1 mm long), which is also the grain size observed in our chips. Some inhomogeneity is indicated by the report of LSPET (1972) of plagioclase laths as large as 2 × 10 mm. The laths might be relicts from partial melting. Agreement between our data and those of Nava (1974) is satisfactory except for Al and Mg (our data are 9% higher and 15% lower, respectively).

67955. LSPET (1972) described 67955 as an extremely fine grained, shocked "monolithologic" rock. It was chosen because it was one of only two non-glass members of siderophile group 5 which Ganapathy *et al.* (1974) suggest originated as ejecta from the Nectaris Basin. Our data are in general agreement with LSPET (1972) XRF data, but our siderophile data are significantly different from those of Ganapathy *et al.*

77545. Described by LSPET (1973) as an unusually porous (25% cavities) "metaclastic" (i.e., annealed) breccia, it was studied by us because of the unusual composition reported by Laul and Schmitt (1975a). Our Al value is 2× higher than that of Laul and Schmitt. Our data show 77545 to be a rather typical Apollo 17 KREEPy noritic breccia.

RESULTS

Table 1 lists compositional data for 13 lunar rocks. Seven samples were analyzed in duplicate. The first replicates for three of these (63549, 65055, and 67955) were published by Boynton *et al.* (1976). Three previous replicates for RNAA-determined siderophiles and volatiles in 65015 were published by Wasson *et al.* (1975), and a single replicate of INAA-determined elements by Boynton *et al.* (1975). The older 65015 INAA data are not as precise as the current values, and have been given less weight in the determination of the mean.

Wasson *et al.* (1977) proposed that a KREEP normalization provides the most informative way to plot incompatible element data on lunar rocks. Following an example of Palme and Wänke (1975), elements are arranged in order of decreasing KREEP/CI ratio. The composition of KREEP is listed in Table 2 and discussed in a section below.

Figure 1 shows incompatible element ratios for the 13 rocks included in our current study. It is readily apparent that the incompatible patterns of samples having KREEP contents ⩾0.1 do not deviate significantly from a line with slope zero. The KREEP contents near 0.5 of 65015 and 63556 are among the highest known in Apollo 16 samples. It is also of interest that the nonpristine rocks 64815 and 77545 having essentially identical KREEP contents of about 32% have

Table 1. Concentrations of 36 elements in 13 highlands rocks. Concentrations in $\mu g/g$

	Na	Mg	Al	K	Ca	Sc	Ti	V	Cr	Mn	Fe	Co	Ni	Zn	Ga	Ge	Zr	Ru
60666,6	3.50	39	157	0.70	111	6.5	3.0	27*	0.82	540	43.2	53	800	≤5.7	3.6	530	204	46
61569,5	3.46	60	116	1.55	92	12.4	6.0	29*	1.11	660	57.6	54	1000	—	4.1	—	250	—
63549,9	3.60	26	158	0.56	121	7.8	2.2	22	0.64	460	34.9	19.5	220	≤4.1	2.6	138	—	8‡
mean‖	3.40	27	147	0.54	115	7.3	2.4	18	0.61	440	33.0	18.2	210	1.1	2.6	117	—	8‡
63556,6	4.41	62	104	2.82	86	15.7	7.0	35	1.25	820	63.4	26.7	346	≤2.3	4.7	850	800	43
	4.31	62	104	3.09	84	15.0	7.2	34	1.19	810	67.8	55	730	—	—	—	720	—
64569,6	3.74	75	119	1.63	84	13.3	6.2	40	1.32	760	66.2	59	930	≤3.5	4.6	2300	360	77
64815,10	3.92	76	106	2.20	84	21.7	9.0	51	1.62	890	70.3	37.3	460	≤5.8	5.1	1250	500	51
65015,63	3.88	60	109	2.48	84	16.1	7.3	40	1.32	850	64.8	49.1	—	—	—	—	920	—
mean¶	4.10	60	109	2.48	85	16.5	7.3	40	1.29	900	66.5	40	580	0.7	4.6	504	920	—
65055,19§	3.68	23	171	0.64	114	8.3	2.5	23†	0.68	520	36.9	22.6	170	≤7.1	2.7	100	—	—
mean‖	3.59	27	156	0.61	115	7.5	2.6	20†	0.63	480	34.6	27.0	318	0.6	2.8	170	—	26
65779,3	3.72	60	125	1.72	91	11.6	5.7	38	1.16	680	64.5	69	1080	≤9.1	4.3	2800	450	70
67035,31,	3.97	13	174	0.45	122	5.2	1.7	9†	0.34	290	20.5	5.3	23	≤2.9	4.1	28	57†	28
18	3.56	29	138	0.38*	111	6.5	1.9	13†	0.46	420	29.6	10.4	60*	—	4.7	—	32‡	—
67559,4	3.81	23	158	0.67	118	8.8	2.8	21*	0.69	450	33.5	20.6	257	≤5.6	3.8	320*	—	26
67955,35	3.60	50	151	0.53	107	7.1	1.9	25*	0.87	420	34.7	19.6	170	≤7.4	4.2	140	—	12
mean‖	3.55	46	145	0.50	107	7.3	1.8	25*	0.87	420	34.4	18.8	178	5.0	4.1	120	—	14
77545,3	5.25	78	99	2.00	79	17.0	9.0	—	1.40	—	69.1	13.5	60	1.9	5.5	50	560	<4
	5.20	83	105	2.44	85	17.6	10.9	49	1.44	930	75.5	17.5	62	≤5.3	5.3	120	590	—

*Uncertainty of 10–20%.
†Uncertainty of 20–40%.
‡Uncertainty of 40–80%.
§This replicate small, assigned ½ weight in the calculation of the mean.
‖Previous replicate published by Boynton *et al.* (1976).
¶Previous replicates published by Boynton *et al.* (1975).

very similar, high contents of the mafic component SCCRV discussed in the following section. The same amounts of rather similar ingredients were mixed at locations 1000 km apart.

The most striking secondary feature of the incompatible patterns is the positive K anomaly that is found in rocks with KREEP contents ≤6%. The K content in our most anorthositic sample, 67035A, is about 3–4 times greater than that obtained by interpolation between U, Th, and the light rare earths. Slight positive Zr and Ba anomalies are significant at about 95% confidence. An enrichment of K and Ba in lunar anorthosites was also observed by Palme and Wänke (1975).

Figure 2 shows the positions of our samples on the Walker *et al.* (1973) olivine-silica-anorthite pseudoternary. Silicon was not determined, rather inferred from the observed mineralogy. The SiO_2 concentrations (in %) used were 44 (60666 and 67035), 45 (63549, 65055, 67559, 67955), 46 (65779, 61569), and 47 for the remaining rocks. Literature data for other Apollo 16 rocks are plotted as a background field. The two KREEPy rocks 63556 and 65015 fall as expected near the center of the Apollo 16 Fra Mauro rock field. One rock plots in the spinel except Na, Mg, Al, K, Ca, Ti, Cr, and Fe, mg/g and Ge, Ru, Cd, In, Ir, and Au, ng/g.

Cd	In	Ba	La	Ce	Nd	Sm	Eu	Tb	Dy	Yb	Lu	Hf	Ta	Ir	Au	Th	U
≤17	≤7.6	132	11.4	28	19	4.9	1.20	1.02	7.4	3.6	0.49	3.8	0.39	28	9.0	1.69	0.48
—	5.5	180	18.1	45	25	8.0	1.33	1.65	8.8	5.8	0.82	5.9	0.67	20	18	2.8	0.76
—	6.7	83	6.8	16	11	2.8	1.08	0.60	3.8	2.2	0.31	2.2	0.24	9	3.6	1.03	0.37*
5.8	6.8	80	6.6	16	11	2.7	0.98	0.59	3.4	2.1	0.28	2.0	0.23	9	3.4	0.96	0.37*
≤9	0.12	550	55.1	128	80	23.1	2.01	4.6	30	16.4	2.28	17.4	1.8	13	5.5	8.3	2.2
—	—	500	51.6	128	77	22.3	1.86	4.5	27	15.5	2.18	17.2	1.8	20†	13	8.2	2.0
≤7	≤4.5	270	26.2	63	44	11.3	1.50	2.4	17	8.7	1.20	8.5	0.88	19	20	4.0	1.1
≤7	≤3.1	410	34.3	86	55	15.7	1.70	3.4	23	11.7	1.64	11.1	1.1	9†	8.4	5.6	1.6
—	—	560	55.6	132	80	24.8	2.10	5.2	33	17.1	2.44	19.2	2.1	—	—	9.4	2.6
16	0.3	540	55.0	132	80	25.0	2.19	5.2	33	18.0	2.50	20.5	2.2	13	9.9	9.4	2.6
≤6	6.4	82	7.4	19	11	3.3	1.20	0.75	—	2.6	0.35	2.4	0.30†	10	3.0	1.26	0.3†
1.9	6.4	79	6.5	17	10	2.8	1.07	0.62	—	2.3	0.31	2.2	0.27†	10	4.3	0.95	0.3†
≤53	≤50	300	29.5	70	50	12.7	1.60	2.5	16	9.1	1.25	9.2	1.0	26	24	4.3	1.3
≤12	0.46	35†	1.9	5	3‡	0.9	1.15	0.20	—	0.9	0.12	0.7*	0.08†	—	0.18	0.35	0.06†
—	0.48	49*	2.8	7	4*	1.2	1.06	0.28	2.0†	1.1	0.16	1.0	0.14†	—	1.62	0.50	0.14†
—	≤11	76	7.2	18	11	3.1	1.20	0.66	4.3	2.4	0.34	2.4	0.28	11	5.0	1.10	0.39*
3.5	0.79	78	5.3	13	8	2.1	1.06	0.51	2.6*	2.0	0.28	1.6	0.24	8	2.2	1.08	0.33
≤10	≤1.6	68	5.1	12	8	2.1	1.06	0.48	2.6*	2.0	0.28	1.6	0.22	8	2.1	1.02	0.30
2.8	0.35	340	32.2	82	51	15.4	2.00	3.1	23	11.0	1.52	11.8	1.4	1.0	0.8	5.4	1.4
<10	0.32	380	35.7	85	59	15.3	2.15	3.2	19	11.6	1.62	12.1	1.5	—	1.5	5.3	1.4

field, but in the absence of hard SiO_2 data its placement is not conclusive. SCCRV and KREEP-rich rocks 77545 and 64815 fall near the spinel-olivine-anorthite piercing point, but this is not petrogenetically significant since their high siderophile contents indicate that they are almost certainly multicomponent mixtures. Most of our samples fall in the anorthositic norite-noritic anorthosite band that extends between the Fra Mauro and the anorthosite fields, and has the mature soil field in its midsection. Our duplicate analyses of friable rock 67035 are both shown on the diagram to emphasize its heterogeneous nature.

Composition of SCCRV

Wänke *et al.* (1974, 1977) used mixing calculations to demonstrate the presence in highlands rocks of a mafic component having high concentrations of Mg, Sc, Ti, V, Cr, Mn, and Fe, and suggested that this component was "primary matter", interplanetary material accreted after formation of the lunar anorthositic crust. The presence of such a primitive component in the highlands crust would provide an important constraint for models of accretionary processes occurring during the first few hundred million years of solar system history. As a result we have attempted to evaluate the "primary matter" hypothesis using a somewhat different approach. Because we believed that the component might be of lunar provenance we have chosen to designate it by a neutral acronym, SCCRV (pronounced *scurv*), based on the chemical symbols of three of the seven

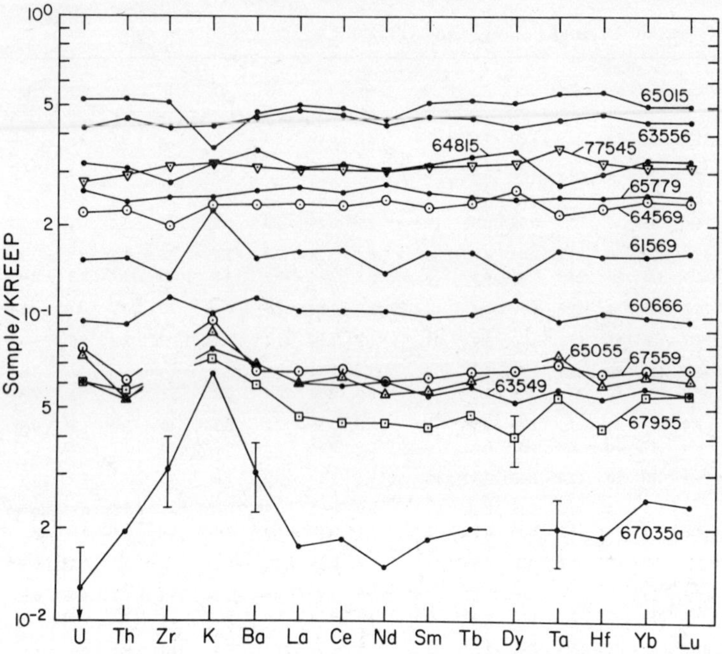

Fig. 1. KREEP-normalized incompatible element concentrations are flat in rocks containing >10% KREEP, indicating a single original source of KREEP. Positive K anomalies are observed in rocks having KREEP contents <10%, and slightly elevated Ba and Zr contents are found in high (91%) anorthosite rock 67035A.

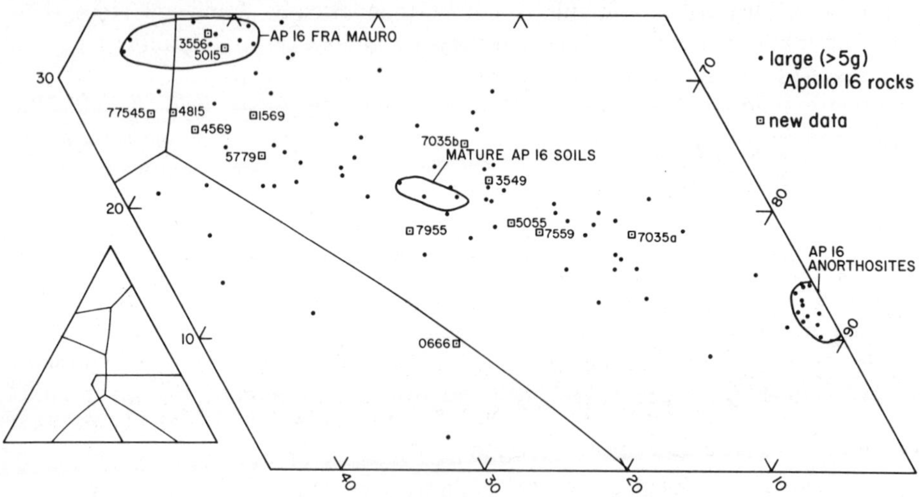

Fig. 2. Most of our current suite of rocks lie together with other Apollo 16 rocks in the anorthositic norite-noritic anorthosite band on an olivine-silica-anorthite pseudoternary diagram. All these rocks have high ($>2\times10^{-4}$ CI chondrites) siderophile contents and most are breccias or show petrographic evidence of being remelted. As a result they are all believed to be nonpristine mixtures of unrelated components.

main elements that it contributes to the lunar highlands. The reader should bear in mind that the remaining SCCRVy elements Mg, Ti, Mn, and Fe are just as important as those in the acronym.

Wänke *et al.* (1977) derived the composition of their "primary matter" component by least-squares calculations using a limited set of key elements: O, Si, Mg, Fe, Ca, and Al. (1) The recent extralunar contribution of Fe is subtracted on the basis of the assumption that this component contributes all the Ni, and that its Fe/Ni ratio is 15, slightly lower than chondritic; correction for extralunar contributions of other elements is not made. (2) The contribution of KREEP to elements other than incompatibles is made on the basis of the assumption that KREEP has the composition of rock 14321. Because this correction introduces high uncertainties in certain elements (e.g., Ca, Al), rocks having bulk La $\geq$ 25 μg/g were not used to define the composition of the primary matter component. (3) Following these corrections the key-element data are arranged into an abundance sequence extending from anorthosite towards the primary matter component. Only the relative amounts of the second component are estimated during this step. (4) The composition of the primary component is defined to have a Ca/Al ratio of 1.08 g/g, as observed in most classes of chondritic meteorites. An Al/Sc ratio of 1.36 was also used to determine this point. (5) Additional mixing calculations are used to determine the concentrations of several other elements in the primary matter component.

The composition of this "primary matter" given by Wänke *et al.* (1977) is listed in Table 3. They suggest that this composition resulted from accretion of material that was 85% CI chondrite and 15% high-temperature material resembling the Ca–Al-rich clasts in Allende and that prior to incorporation into highlands rocks this was modified by loss of most Fe as metal and 16.7% of the non-siderophile mass as an olivine cumulate.

Our approach is the following: Since corrections for other components are least when the concentration of the mafic component is greatest, we confined our estimation of SCCRV to highlands rocks 61569, 63556, 64569, 64815, 65015, 65779, and 77545 that have moderate to large contents of mafic elements (SCCRV contents $\geq$14%). Rock 67955 was not used despite its nominally high (16%) SCCRV content because its mafic element pattern was distinctly different from those in the other seven rocks. SCCRV was approximated and then refined by an iterative technique based on least-squares mixing calculations (Boynton *et al.*, 1975) fitting four components to these seven rocks: extralunar component (ELC), anorthosite, KREEP, and SCCRV.

Although the exact composition of the ELC is not known, it is probably chondritic; if so, it contributes only a minor amount of the non-siderophiles to lunar rocks, and the composition chosen little affects the fitting of the other components. We used the composition of Boynton *et al.* (1975) except that Mg was increased from 100 to 120 mg/g.

We used pure anorthosite rather than noritic anorthosite or gabbroic anorthosite because pure anorthosite occurs in pristine form, whereas pristine samples of the latter types have not been found (Warren and Wasson, 1977). The

Table 2. Composition of anorthosite, KREEP, and SCCRV used in least-squares mixing models. The anorthosite and KREEP composition have undergone significant revision relative to those listed in Boynton *et al.* (1975). The composition of SCCRV, except for seven (SCCRV) elements, is not precisely known (see text). Units μg/g except Na, Mg, Si, Al, Ca, Ti, and Fe, mg/g, and Ge, Ru, Cd, In, Ir, and Au, ng/g.

Element	Na	Mg	Si*	Al	K	Ca	Sc	Ti	V	Cr	Mn	Fe	Co	Ni	Zn	Ga	Ge	Zr*	Ru*	Cd
Anorth	3.0	1.6	207	186	0.058	137	0.6	0.1	5.0	30	67	2.7	0.7	1.0	0.3	4.0	5	3	0.1	1
KREEP	6.4	64	224	87	6.9	66	22	10.0	43	1250	1080	82	35	28	3.6	7.5	40	1740	2	15
SCCRV	0.5	182	200	15	0.03	20	37.1	17.2	99	3500	1560	135.4	38	120	0.0	4	30	5	3	1

Element	In	Ba	La	Ce	Nd*	Sm	Eu	Tb	Dy	Yb	Lu	Hf	Ta	Ir	Au	Th	U*
Anorth	0.2	4.8	0.12	0.3	0.14	0.04	1.1	0.01	0.06	0.03	0.004	0.014	0.004	0.006	0.01	0.013	0.004
KREEP	1.2	1150	110	270	180	49	3.0	10.0	65	36	5.0	37	4.0	0.10	0.20	18	5.0
SCCRV	0.2	5.0	0.20	0.54	0.36	0.12	0.09	0.03	0.18	0.13	0.02	0.08	0.01	0.40	0.40	0.15	0.04

*Si, Zr, Ru, Nd, and U not in our current mixing program but included for completeness.

anorthosite composition is listed in Table 2; it has undergone substantial revision relative to that of Boynton *et al.* (1975).

Pristine KREEP-rich rocks have been found at the Apollo 15 and Apollo 17 sites, but the concentrations of incompatible elements in these basalts are only 44–72% of the highest values found in Apollo 14 samples. The conclusion of Schonfeld (1974) seems inescapable; the fact that the KREEP pattern of incompatible elements remains the same independent of the concentration of incompatibiles is inconsistent with differing degrees of partial melting (Gast, 1972). Thus, the differing concentrations of incompatibles must result chiefly from dilution, presumably by wall rock prior to or during extrusion.

Since the samples having the highest contents of incompatibles are not pristine, and the concentrations of incompatibles are too low to form major mineral phases (and thus set an upper limit on their concentrations), there is no means to precisely define the composition of KREEP. The best choice seems to be to assume that the lunar samples having the highest KREEP contents are reasonable approximations of "original KREEP" (or borrowing a German prefix, urKREEP). The relative concentrations of the incompatibles must be correct, but urKREEP probably had higher absolute concentrations. The KREEP composition used is given in Table 2; it is largely based on our analyses of three 2–4 mm grains from Cone Crater soil 14143 and Wänke *et al.* (1972) analyses of 14305,81 and 14321,184,25. A detailed discussion will be published elsewhere (Wasson *et al.*, 1977). The composition differs significantly from that proposed by Laul and Schmitt (1973) used in our earlier mixing calculations (Boynton *et al.*, 1975, 1976), particularly in terms of its higher concentrations of Mg, Cr, Mn, and Fe, and lower concentrations of Na, Al, K, and Ca.

Our previous calculations (Boynton *et al.*, 1976) had shown that the "missing component" could be approximated by a mixture consisting of 4 parts dunite, 3 parts Apollo 15 pigeonite basalt, and 1 part Apollo 11 low-K basalt. After fitting each of our samples to ELC, anorthosite, KREEP, and this zeroth approximation of SCCRV, we calculated ELC-, anorthosite- and KREEP-free concentrations of the characteristic elements. Element/Cr ratios (Cr is the most precisely determined element) were calculated and means estimated visually from diagrams such as Fig. 3. Because of its exceptionally high SCCRV content (32%), 64815 was given extra weight in the visual determination of means. The high SCCRV rock 77545 was given less weight because SCCRV may show a geographical variation in composition, and we wanted the average from this work to correspond to Apollo 16 SCCRV.

If SCCRV consists of a single lunar rock type, it seems reasonable to expect it to be mafic to ultramafic. A composition of SCCRV was obtained by assuming ultramafic concentrations of Na_2O, SiO_2, Al_2O_3, K_2O, and CaO, subtracting the sum from one, and constraining the concentrations of the oxides of the seven SCCRV elements to sum to this remainder. The concentrations of all other elements are very rough approximations largely based on the Laul and Schmitt (1975b) analysis of dunite 72417; SCCRV contents of mafics or siderophiles were set about a factor of 2 lower and incompatibles about a factor of 2 higher than

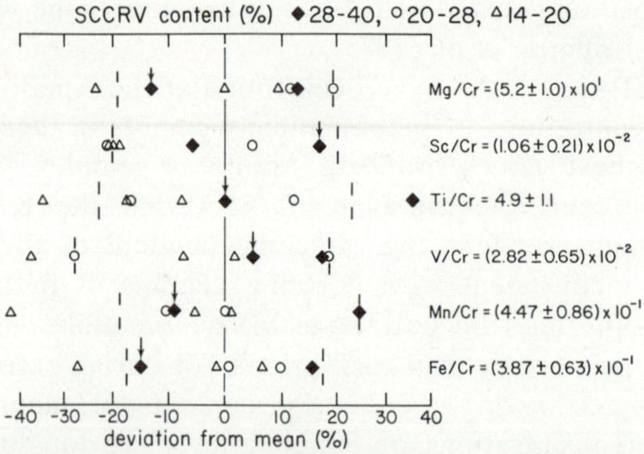

Fig. 3. The composition of SCCRV was estimated iteratively from a series of element/Cr diagrams of which this was the last; samples with high mafic contents were fitted by a mixing routine, ELC, anorthosite and KREEP were subtracted, and the remainder defined as SCCRV. Mean element/Cr ratios were estimated visually giving extra weight to SCCRV-rich Apollo 16 rock 68415, whose position is indicated by the vertical arrow. The indicated uncertainty limits on the mean were estimated visually to enclose at least five of the seven rocks.

the dunite concentrations. The assumed values are listed in Table 2.

The entire procedure was repeated twice in order to refine the results. The final ratios normalized to the mean ratio are shown in Fig. 3. Uncertainty limits corresponding approximately to standard deviations of an individual determination were chosen to enclose at least five of the seven points. With the exception of the V/Cr and Mn/Cr envelopes these were drawn symmetrically about the mean.

Concentrations and uncertainty limits at approximately 80% confidence are listed in Table 3 for the seven SCCRVy elements. The Cr uncertainty was estimated directly from the calculated SCCRV concentrations using Student's *t* distribution. Errors in the other six elements were obtained from this value and the limits on the ratios estimated from Fig. 3. Note that these are limits on precision; there is no reliable way to estimate the possible systematic bias resulting from errors in the compositions used for the other components (especially KREEP), or from uncertainties in the concentrations of other SCCRV oxides, particularly SiO_2. The low reduced χ^2 values discussed below suggest that uncertainties associated with the composition of the components are not a serious error source.

Also listed in Table 3 are the Wänke *et al.* (1977) estimates of their "primary component". The fourth column shows primary component/SCCRV ratios normalized at Cr. Six of the elements agree to within ±23%, but the Ti concentration of SCCRV is about a factor of 4 higher than that of the primary component. Since our estimates are based primarily on SCCRV-rich samples, they should be more accurate than those of Wänke *et al.* However, there is no doubt that they will evolve as additional SCCRV-rich samples are studied.

The relative amounts of the four components in our suite of highlands

Table 3. Derived concentrations of seven characteristic elements in SCCRV, a comparison of SCCRV with the "primary component" (P.C.) of Wänke *et al.* (1977), and the fraction of each element contributed by SCCRV to rock 64815.

	SCCRV (this work)	P.C. (Wänke *et al.*)	P.C./SCCRV (Cr norm'd)	64815 SCCRV contribution
Mg (mg/g)	182 ± 18	154	1.06	0.711
Sc (μg/g)	37.1 ± 3.8	23	0.77	0.614
Ti (mg/g)	17.2 ± 2.1	3.1	0.23	0.620
V (μg/g)	99 ± 12	~70	0.88	0.649
Cr (μg/g)	3500 ± 250	2800	1.00	0.702
Mn (μg/g)	1560 ± 170	1500	1.20	0.541
Fe (mg/g)	135 ± 12	102	0.94	0.574

Table 4. Least-squares estimates of component concentrations (in %) in highlands rocks. Rocks having high concentrations of mafic elements are well fit with these four components.

Rock	ELC	Anorthosite	KREEP	SCCRV	Sum	Reduced χ^2
60666	7.6	76.3	9.8	10.8	105	0.83
61569	5.3	54.5	15.9	22.1	98	0.31
63549	1.8	77.4	5.6	13.1	98	1.38
63556	2.6	35.2	45.8	15.1	99	0.25
64569	6.0	48.8	23.0	23.3	101	0.66
64815	2.0	40.3	32.0	31.8	106	0.61
65015	3.3	34.4	51.1	13.5	102	0.30
65055	2.4	80.1	5.8	13.1	101	1.41
65779	8.1	52.7	25.1	16.2	102	0.42
67035A	0.1	90.9	1.9	8.2	101	5.33
67035B	0.4	75.7	2.6	13.1	92	3.62
67559	2.2	81.8	6.4	12.7	103	3.41
67955	1.5	74.7	4.5	15.9	96	6.05
77545	0.2	39.5	31.3	30.6	102	0.72

samples are listed in Table 4, together with the calculated sums and reduced χ^2 values. With the exception of 67035B and 67955 all the sums are in the range 98–106%, and with the exception of the Station 11 anorthosite-rich samples 67035, 67559, and 67955, reduced χ^2 values are in the range 0.3–1.4. The relative uncertainties assigned each element in the fitting program were as follows: Al, Ca, Sc, Cr, Fe—5%; Mg—6%; Ti, Mn, La, Sm, Th—10%; Eu—15%; Ba, Ce, Tb, Dy, Yb, Lu, Hf—20%; Ta—30%; Ni, Ge, Ir, Au—40%; Co—80%; and Na, K, Zn, Ga, Cd, In—∞. These latter elements are highly volatile, and often result from regolith deposition of a volatile "fifth" component (Boynton *et al.*, 1976). By giving them zero weight we could estimate the amount expected from the four components without influencing the fit or the reduced χ^2 value.

Their high χ^2 values indicate that our four-component model is inadequate for Station 11 samples, many of which were excavated and/or mixed by the North Ray Crater impact. Poor fits to SCCRV elements contribute most to the high χ^2 values, but to date we cannot discover a pattern. For example, in 67955 the main problem is with Ti (experimental value 2× low) and to a lesser degree with Cr and Mg (values 1.3× high), whereas in 67559 the problems are Mg and Sc (1.3× low and high, respectively). Two equally plausible hypotheses are: (1) the components in the Station 11 rocks are the same as those from other Apollo 16 stations, but for accidental reasons at the latter stations the mafic components were homogenized to produce "SCCRV"; (2) there are components in the Station 11 rocks that are present in only negligible amounts in rocks from other stations.

ORIGIN OF SCCRV

We propose that the three most plausible hypotheses for the origin of SCCRV are the following:

(1) SCCRV is primordial (= interplanetary, chondritic) material (e.g., Wänke's primary matter).
(2) SCCRV consists entirely or mainly of a single type of lunar rock.
(3) SCCRV resulted from the mixing of two or more lunar materials.

As noted by Wänke *et al.* (1977), the chief compositional problem facing the "primary matter" model is the high ratio of Ti to other SCCRVy elements. For example, the Ti/Sc ratio in SCCRV is 464 g/g, whereas that in CI chondrites is about 64, and that in Allende Ca–Al-rich clasts about 76 (Wänke *et al.*, 1974). Even the "primary matter" Ti/Sc ratio of 135 estimated by Wänke *et al.* (1977) is a factor of 2 higher than these ratios. Their estimate is based on a trend observed on a plot of a Ti/Mg vs La/Mg for highlands samples from several landing sites. However, if one only considers the Apollo 16 samples on their diagram, a Ti/Mg (and through this, a Ti/Sc) ratio about 3× higher is inferred.

The very high Ti contents of Apollo 11 and 17 basalts are commonly attributed to the melting of ilmenite-rich materials at depth; since these basalts have Ti/Sc ratios (~830 in a typical Apollo 11 low-K basalt) significantly higher than those in SCCRV, the high ratio in SCCRV does not seem inconsistent with a lunar origin.

H. Wänke (pers. comm., 1977) suggests that the high Ti contents in SCCRV-rich rocks might be associated with KREEP rather than SCCRV. If this were so, we would expect Ti to be strongly correlated with highly incompatible elements such as La. Bulk Ti/La ratios in our samples, plotted on a line diagram in Fig. 4 (upper half), demonstrate that little covariance exists between Ti and REE. Whereas the range in Ti/La ratios is a factor of 2.5, the range in Ti/Cr values (Fig. 4, lower half) is only a factor of 1.5, and this reduces to a factor of 1.36 if the set is restricted to Apollo 16 samples. It is statistically unlikely that a KREEP-rich component with highly variable Ti/La ratio would have been mixed

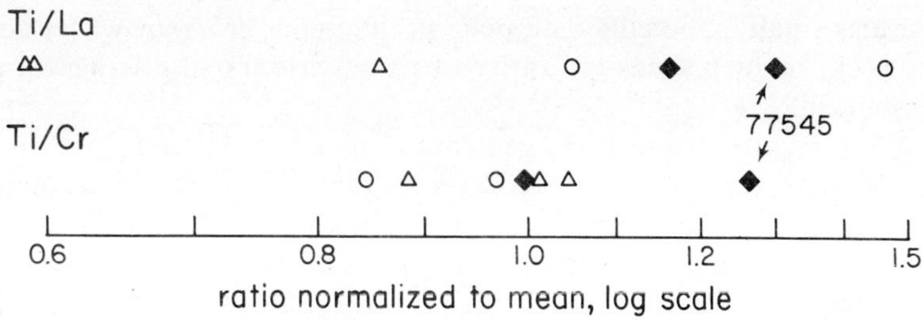

Fig. 4. Among our seven SCCRV-rich rocks Ti/La ratios vary by a factor of 2.5 whereas Ti/Cr ratios vary by a factor of 1.5, or 1.36 if 77545 is ignored. These data are inconsistent with the hypothesis that six of the seven SCCRV elements originated as a late primordial accretate, whereas the bulk of the Ti is of lunar origin and was brought to the lunar surface together with KREEP.

with an independent mafic component in just the right proportions to produce a constant Ti/Cr ratio.

Another problem with a primordial origin is the requirement that siderophiles be removed by loss of metal while molten. If, as Wänke *et al.* (1977) propose, olivine is to be removed as well the problems are compounded. It appears that melting must be total if siderophiles are to be lost. As noted by Warren and Wasson (1977), all impact-remelted rocks appear to be contaminated by siderophiles of extralunar derivation, indicating that incomplete removal of metal is the rule in impact melt pools. Further, as noted by various authors, only a small fraction of the ejecta from an impact is melted; generally the bulk remains unmelted, and the amount ejected as gas and dust may be larger than that melted.

If we make the unwarranted assumption that all of the late-accreted "primary matter" ends up in large, low-viscosity melts, a second problem is encountered. Such a melt would probably fractionally crystallize, and in fact Wänke *et al.* (1977) now suggest that olivine was lost by this mechanism. But during fractional crystallization the more refractory elements should fractionate from those less refractory, and materials having a wide range of Mg/Fe or Cr/Sc ratios be produced. Perhaps one can argue that later mixing could rehomogenize the ratios, but this mixing must not include the *lost* olivine and metal.

The arguments about the necessity for total melting and for rehomogenization following fractionation are more suggestive than conclusive. However, they do strengthen the conclusion based on the high Ti content that SCCRV is not of primordial origin.

The current evidence does not allow an unambiguous choice between the two lunar hypotheses. If SCCRV was a single magma or cumulate, a diligent search should turn up a few pure or nearly pure samples. No such rocks are known. We have not yet thoroughly examined the data on soil glasses and lithic fragments, but surveys such as those of Simkin *et al.* (1973) and Ridley *et al.* (1973) have not resolved any major groupings of particles with SCCRVy compositions. However, it may be that SCCRV was emplaced coevally with anorthosite as small mafic cumulate pods in the upper crust, or was a very early extrusive rock. In such cases the rarity of its occurrence in pristine form would be understandable.

If SCCRV resulted from homogenization of materials from diverse sources (e.g., one or more types of mare basalts with dunite or another ultramafic), one would expect to observe regional and probably also local variations in interelement ratios. In the previous section we pointed out that SCCRVy element concentrations in Apollo 16 Station 11 rocks are often not in the SCCRV proportions, but that there is inadequate data to assess the source of these variations. In Fig. 3 we find that, with the exception of Sc/Cr (and Mg/Cr and V/Cr ratios in 64569), 77545 has a higher element/Cr ratio than other Apollo 16 rocks. Although this may reflect a (modest) regional variation in SCCRV composition, it may also have resulted from slightly atypical Sc and Cr values in 77545 relative to other SCCRV-rich Apollo 17 rocks. A more thorough examination of the composition of Apollo 17 SCCRV is needed to answer this question.

A possible significant observation is the following. Among the seven samples with high ($\geq$13.5%) SCCRV contents and low ($\leq$0.7) χ^2 values, the SCCRV content exceeds the KREEP content only once (in 61569), and then only by a factor of 1.4. In the three rocks (64569, 64815, 77545) having the highest SCCRV contents the amounts of these two components are essentially identical. Why should high SCCRV contents be so closely associated with high KREEP contents? A possible explanation is provided in a model in which SCCRV is wall rock dissolved in a KREEPy magma. The great similarity of 77545 and 64815 demands that this magma have an extensive distribution on the lunar surface. During later mixing by impacts some of these materials were diluted with other SCCRV-free components. If this model were correct, the mafics present in other samples having high mafic/incompatible element ratios originated in another fashion (e.g., as mixtures of mare basalts and dunite).

AGE OF SCCRV

Many Apollo 17 highlands rocks are SCCRV-rich and similar to our 77545 in composition; consortium breccia 73215 is typical. Most clasts in 73215 are SCCRVy, and all have ^{39}Ar–^{40}Ar ages in the range 4.04–4.28 G.y. (Jessberger *et al.*, 1976). The most precise age is 4.24 ± 0.01 G.y. for clast 73215,46,10; its composition as given by Blanchard *et al.* (1976) is slightly higher in anorthosite and slightly lower in KREEP and SCCRV than 77545.

Clearly 4.24 G.y. is a minimum age for the onset of SCCRV formation. Such a high age is not inconsistent with an origin together with original KREEP *ca.* 4.3–4.4 G.y. ago (Schonfeld and Meyer, 1972; Palme, 1977), but militates against SCCRV consisting of mixtures of dunite and mare basalt since no mare basalts are known with ages $\geq$3.9 G.y. Although one might plausibly argue that some mare basalt generation began 4.3 G.y. ago, it seems unlikely that the area subtended by maria at that time was great enough to account for the large amounts of SCCRV present in rocks on the nearside of the moon.

In fact, the high age seems generally inconsistent with SCCRV models involving the "accidental" mixing of two or more components on the lunar surface. Although there is no problem accounting for Mg and possibly Cr and V with ancient ultramafic rocks, there are no known rocks having ages ≥4.3 G.y. that can serve as sources of the other four SCCRV elements. A mixing of rocks at depth, e.g., as wall rocks along different portions of a magmatic conduit, is an acceptable variation of the single lunar source hypothesis.

Acknowledgments—We are indebted to the JSC curatorial staff and to the LSAPT board for the provision of interesting lunar samples. We also thank H. J. Chun, M. Gjerde, and F. Henriquez for their supporting efforts, and R. Korotev for a constructive review. Sample irradiations at UCLA and University of Missouri Reactors were competently handled by C. Ashbaugh, S. Gunn, and their associates. This research was supported by NASA grant NGL-05-007-367.

References

Blanchard D. P., Jacobs J. W., Brannon J. C., and Haskin L. A. (1976) Major and trace element composition of matrix and aphanitic clasts from consortium breccia 73215. *Proc. Lunar Sci. Conf. 7th,* p. 2179–2187.

Boynton W. V., Baedecker P. A., Chou C.-L., Robinson K. L., and Wasson J. T. (1975) Mixing and transport of lunar surface materials: Evidence obtained by the determination of lithophile, siderophile and volatile elements. *Proc. Lunar Sci. Conf. 6th,* p. 2241–2259.

Boynton W. V., Chou C.-L., Robinson K. L., Warren P. H., and Wasson J. T. (1976) Lithophiles, siderophiles and volatiles in Apollo 16 soils and rocks. *Proc. Lunar Sci. Conf. 7th,* p. 727–742.

Christian R. P., Berman S., Dwornik E., Rose H. J., and Schnepfe M. M. (1976) Composition of some Apollo 14, 15, and 16 lunar breccias and two Apollo 15 fines (abstract). In *Lunar Science VII,* p. 138–140. The lunar Science Institute, Houston.

Dowty E., Keil K., and Prinz M. (1974) Igneous rocks from Apollo 16 rake samples. *Proc. Lunar Sci. Conf. 5th,* p. 431–445.

Floran R. J., Phinney W. C., Blanchard D. P., and Warner J. L. (1976) A comparison between the geochemistry and petrology of Apollo 16—terrestrial impact melt analogs (abstract). In *Lunar Science VII,* p. 263–265. The Lunar Science Institute, Houston.

Ganapathy R., Morgan J. W., Higuchi H., and Anders E. (1974) Meteoritic and volatile elements in Apollo 16 rocks and in separated phases from 14306. *Proc. Lunar Sci. Conf. 5th,* p. 1659–1683.

Gast P. W. (1972) The chemical composition and structure of the moon. *The Moon* **5**, 121–148.

Hubbard N. J., Rhodes J. M., Gast P. W., Bansal B. M., Shih C.-Y., Wiesmann H., and Nyquist L. E. (1973) Lunar rock types: The role of plagioclase in non-mare and highland rock types. *Proc. Lunar Sci. Conf. 4th,* p. 1297–1312.

Jessberger E. K., Kirsten T., and Staudacher T. (1976) Argon–argon ages of consortium breccia 73215. *Proc. Lunar Sci. Conf. 7th,* p. 2201–2215.

Laul J. C. and Schmitt R. A. (1973) Chemical composition of Apollo 15, 16, and 17 samples. *Proc. Lunar Sci. Conf. 4th,* p. 1349–1367.

Laul J. C. and Schmitt R. A. (1975a) Chemical composition of Apollo 17 samples: Boulder breccias (2), rake breccias (8), and others (abstract). In *Lunar Science VI,* p. 489–491. The Lunar Science Institute, Houston.

Laul J. C. and Schmitt R. A. (1975b) Dunite 72417: A chemical study and interpretation. *Proc. Lunar Sci. Conf. 6th,* p. 1231–1254.

LSPET (1972) Lunar Sample Information Catalog – Apollo 16. NASA MSC-03210. 372 pp.

LSPET (1973) Lunar Sample Information Catalog – Apollo 17. NASA MSC-03211. 447 pp.

Nava D. F. (1974) Chemical compositions of some soils and rock types from the Apollo 15, 16, and 17 lunar sites. *Proc. Lunar Sci. Conf. 5th,* p. 1087–1096.

Palme H. (1977) On the age of KREEP. Submitted to *Geochim. Cosmochim. Acta.*

Palme H. and Wänke H. (1975) A unified trace-element model for the evolution of the lunar crust and mantle. *Proc. Lunar Sci. Conf. 6th,* p. 1179–1202.

Ridley W. I., Reid A. M., Warner J. L., Brown R. W., Gooley R., and Donaldson C. (1973) Glass compositions in Apollo 16 Soils 60501 and 61221. *Proc. Lunar Sci. Conf. 4th,* p. 309–321.

Schonfeld E. (1974) The contamination of lunar highland rocks by KREEP: Interpretation by mixing models. *Proc. Lunar Sci. Conf. 5th,* p. 1269–1286.

Schonfeld E. and Meyer C. (1972) The abundances of components of the lunar soils by a least-squares mixing model and the formation age of KREEP. *Proc. Lunar Sci. Conf. 3rd,* p. 1397–1420.

Simkin T., Noonan A. F., Switzer G. S., Mason B., Nelen J. A., Melson W. G., and Thompson G. (1973) Composition of Apollo 16 fines 60051, 60052, 64811, 64812, 67711, 67712, 68821, and 68822. *Proc. Lunar Sci. Conf. 4th,* p. 279–289.

Simonds C. H., Warner J. L., and Phinney W. C. (1973) Petrology of Apollo 16 poikilitic rocks. *Proc. Lunar Sci. Conf. 4th,* p. 613–632.

Steele I. M. and Smith J. V. (1973) Mineralogy and petrology of some Apollo 16 rocks and fines: General petrologic model of moon. *Proc. Lunar Sci. Conf. 4th,* p. 519–536.

Sundberg L. L. and Boynton W. V. (1977) Determination of ten trace elements in meteorites and lunar materials by radiochemical neutron activation analysis. *Anal. Chim. Acta* **89**, 127–140.

Walker D., Longhi J., Grove T. L., Stolper E., and Hays J. F. (1973) Experimental petrology and origin of rocks from the Descartes Highlands. *Proc. Lunar Sci. Conf. 4th,* p. 1013–1032.

Wänke H., Baddenhausen H., Balacescu A., Teschke F., Spettel B., Dreibus G., Palme H., Quijano-Rico M., Kruse H., Wlotzka F., and Begemann F. (1972) Multielement analyses of lunar samples and some implications of the results. *Proc. Lunar Sci. Conf. 3rd,* p. 1251–1268.

Wänke H., Palme H., Baddenhausen H., Dreibus G., Jagoutz E., Kruse H., Spettel B., Teschke F., and Thacker R. (1974) Chemistry of Apollo 16 and 17 samples: Bulk composition, late stage accumulation and early differentiation of the moon. *Proc. Lunar Sci. Conf. 5th,* p. 1307–1335.

Wänke H., Palme H., Baddenhausen H., Dreibus G., Jagoutz E., Kruse H., Palme C., Spettel B., Teschke F., and Thacker R. (1975) New data on the chemistry of lunar samples: Primary matter in the lunar highlands and the bulk composition of the moon. *Proc. Lunar Sci. Conf. 6th,* p. 1313–1340.

Wänke H., Palme H., Kruse H., Baddenhausen H., Cendales M., Dreibus G., Hofmeister H., Jagoutz E., Palme C., Spettel B., and Thacker R. (1976) Chemistry of lunar highland rocks: A refined evaluation of the composition of the primary matter. *Proc. Lunar Sci. Conf. 7th,* p. 3479–3499.

Wänke H., Blum K., Dreibus G., Hofmeister H., Kruse H., and Spettel B. (1977) Primary matter in the lunar highlands: A reevaluation (abstract). In *Lunar Science VIII,* p. 973–975. The Lunar Science Institute, Houston.

Warner J. L., Simonds C. H., and Phinney W. C. (1973) Apollo 16 rocks: Classification and petrogenetic model. *Proc. Lunar Sci. Conf. 4th,* p. 481–504.

Warren P. H. and Wasson J. T. (1977) Pristine nonmare rocks and the nature of the lunar crust. *Proc. Lunar Sci. Conf. 8th.* This volume.

Wasson J. T., Chou C.-L., Robinson K. L., and Baedecker P. A. (1975) Siderophiles and volatiles in Apollo-16 rocks and soils. *Geochim. Cosmochim. Acta* **39**, 1475–1485.

Wasson J. T., Warren P. H., and Boynton W. V. (1977) KREEP: Composition and origin. In preparation.

Proc. Lunar Sci. Conf. 8th (1977), p. 2253–2300.
Printed in the United States of America

^{39}Ar–^{40}Ar ages of lunar rocks

GEORGE A. SCHAEFFER and OLIVER A. SCHAEFFER

Department of Earth and Space Sciences, State University of New York at Stony Brook, Stony Brook, New York 11794

Abstract—The ^{39}Ar–^{40}Ar ages for 21 lunar samples from Apollo 14, 15, 16, and 17 are presented. The cosmic-ray exposure ages, Ca, and K contents are also measured. Mare basalt 15075, a slowly cooled pyroxene-phyric basalt, has a ^{39}Ar–^{40}Ar plateau age of 3.45 ± 0.2 G.y. This is one of the oldest mare basalts found at Apollo 15. 65777, a recrystallized noritic breccia, has a plateau age of 3.72 ± 0.02 G.y., relatively young for a highland rock. This rock is probably related to a cratering event over 200 m.y. post Imbrium. Similarly breccia 68516 was formed 150 m.y. post Imbrium. Histograms for the ^{39}Ar–^{40}Ar plateau ages strongly suggest that the peak in ages at Apollo 14, 16, and 17 of about 4 G.y. is caused by the dominance of Imbrium ejecta at these landing sites.

INTRODUCTION

RADIOMETRIC AGES determined by the ^{39}Ar–^{40}Ar method have been of importance in elucidating lunar chronology, especially so for the anorthositic suite of lunar highland rocks for which other radiometric methods are very difficult. We have been engaged in a ^{39}Ar–^{40}Ar survey of coarse fine rock fragments and large rocks since the Apollo 14 sample return. This paper represents the final part of this survey. There are now a sufficient number of highland samples from Apollo 14, 16, and 17 to adequately characterize the age distribution of the Apollo rocks.

One of the major questions of lunar chronology is the first 600 m.y. The difficulty with elucidating this period in the evolution of the moon is the fact that all the Apollo landing sites are close to the Imbrium Basin, and as a result are likely to be dominated by ejecta from Imbrium. As a result, it is difficult to decide whether the preponderance of radiometric ages near 4.0 G.y. is due to an intense cratering event of rather short duration at 4.0 G.y. ago or whether rocks reset by the Imbrium event dominate the sample return.

Of prime concern in understanding the early moon is the question, do the ^{39}Ar–^{40}Ar ages which are clearly older than the Imbrium event ~4.0 G.y. ago represent distinct events on the lunar surface or are they composite ages of materials of 4.0 and 4.6 G.y. age, as might be expected if there was a short intense bombardment about 4.0 G.y. ago? In order to shed some light on this problem, a petrologic study by Delano and Bence (1977) has been made for several of the anorthositic coarse fine fragments which gave ^{39}Ar–^{40}Ar ages near 4.25 G.y. The present results will also be discussed in this context.

EXPERIMENTAL

The monitor

In determining an age by the ^{39}Ar–^{40}Ar method, it is necessary to determine the neutron fluence in order to know the production rate for ^{39}Ar from ^{39}K for the spectrum of neutrons which irradiate the sample. The most precise way to accomplish this is to irradiate a sample, the monitor, for which K and Ar contents have been accurately determined of about the same Ar/K ratio as the sample to be dated. As a result, the ^{39}Ar–^{40}Ar method determines an age relative to a monitor. It is thus necessary to know the $^{40}Ar/^{40}K$ ratio with good accuracy for the monitor. For the lunar ^{39}Ar–^{40}Ar ages, we have chosen a 2.7 G.y. old amphibolite from the Northern Lights Gneiss of northern Minnesota (Hanson *et al.*, 1971). A purified hornblende sample was carefully separated from the rock. The hornblende separate, NL-25, was carefully analyzed for K and ^{40}Ar in several laboratories. The results are presented in Table 1. The best K value is clearly that of Garner and Machlan of the Bureau of Standards. The ^{40}Ar value of Hanson was made by isotope dilution with a ^{38}Ar spike and as such is chosen for the ^{40}Ar value. It should be noted that practically all the values agree within the experimental uncertainties. As a result, the $^{40}Ar/^{40}K$ atom ratio, based on 1.167×10^{-4} for the $^{40}K/K$ ratio becomes 0.3530 ± 0.0022 for NL-25.

The homogeneity of the NL-25 hornblende was tested by irradiating samples from two different mineral separations. The results are presented in Table 2. It is clear that the $^{40}Ar^*/^{39}Ar^*$ ratio is the same for the two samples to within the experimental error of ±2%.

The monitor was compared to BS-1 both in our laboratory and at Menlo Park by M. Lanphere. The results are given in Table 3. Within the experimental errors, the two comparisons are in agreement. The ^{40}Ar in BS-1 was also measured by G. N. Hanson (pers. comm., 1972) and found to be 4915×10^{-8} cc/g compared to a Menlo Park value of 4890×10^{-8} cc/g (Dalrymple and Lanphere, 1971).

A comparison of monitors was made by Alexander and Davies (1974). The Berkeley, St. Severin; Cal Tech, hb 3 gr; Menlo Park, BS-1; Stony Brook, NL-25; and Toronto Gill

Table 1. Analysis of NL-25 monitor.

	Method	Analyst
Potassium, %		
0.3070 ± 0.0008	Isotope dilution	Garner and Machlan National Bureau of Standards
0.302 ± 0.006	Isotope dilution	G. N. Hanson State University of New York at Stony Brook
0.310 ± 0.006	Neutron activation	O. Müller Max-Planck-Institut für Kernphysik
0.304 ± 0.006	X-ray fluorescence	R. Stoenner Brookhaven National Laboratory
0.316 ± 0.001	Atomic absorption	O. Müller
0.296 ± 0.006	X-ray fluorescence	E. Norton Brookhaven National Laboratory
Argon, 10^{-8} cc/g		
7250 ± 40	Isotope dilution	G. N. Hanson
7160 ± 120	Mass spectrometry	T. Kirsten Max-Planck-Institut für Kernphysik
Calcium, %		
8.04 ± 0.01	Neutron activation	O. Müller
Sodium, %		
0.84 ± 0.01	Neutron activation	O. Müller

Table 2. Comparison of two separate preparations of NL-25.

Sample		$^{40}Ar^*/^{39}Ar^*$	
NL-25,1:	3616	33.38 ± 0.19	
	3716	33.56 ± 0.17	
	3816	34.26 ± 0.15	
	3916	34.76 ± 0.11	
	Average		33.99 ± 0.64
NL-25,2:	3216	33.32 ± 0.18	
	3316	33.59 ± 0.17	
	3416	33.50 ± 0.39	
	3516	34.05 ± 0.11	
	Average		33.62 ± 0.31
Over all average			33.80 ± 0.51

Table 3. Comparison of NL 25 and BS-1.

A. Stony Brook analysis of BS-1: Sample No.	$^{40}Ar^*/^{39}Ar^*$	$^{40}Ar/^{40}K$, atom ratio
4016	0.946 ± 0.013	0.00976 ± 0.0013
4116	0.995 ± 0.014	0.01027 ± 0.0014
4216	1.002 ± 0.027	0.01035 ± 0.0027
4316	1.056 ± 0.017	0.01090 ± 0.0017
	Average 1.000 ± 0.045	0.01032 ± 0.00047
	Menlo Park value(1) for BS-1	0.00985 ± 0.00004

B. Menlo Park analysis of NL-25: Sample No.	Age,(2) 10^6 yr	$^{40}Ar/^{40}K$, atom ratio
VIII-A-7	2443 ± 27	0.299 ± 0.006
VIII-B-7	2659 ± 22	0.349 ± 0.005
XII-A-5	2614 ± 23	0.338 ± 0.005
XII-B-2	2649 ± 22	0.346 ± 0.005
Average of VIII-B-7, XII-A-5, and XII-B-2		0.344 ± 0.006
	Stony Brook value for NL-25	0.353 ± 0.002

(1)Calculated from 8.84 ± .03% K_2O and $2.183 \pm 0.006 \times 10^{-9}$ mol ^{40}Ar/g from Dalrymple and Lanphere (1971).

(2)M. Lanphere, pers. comm. (1973).

Quarry Nepheline were compared in a single irradiation. Except for the Toronto monitor all agreed within the experimental errors of 1–2%.

Sample preparation and neutron irradiation

Prior to irradiation, the lunar samples were cleaned ultrasonically with reagent grade methanol. If the sample was a breccia, a small fragment <1 mg was analyzed for Ar to detect the presence of solar wind. All samples which showed excessive solar wind were not further analyzed. After the ultrasonic cleaning, the samples were weighed in thin Al foil containers and loaded in a quartz tube carefully noting the distance along the tube for correction of the neutron gradient in the reactor, usually on the order of 1%. Monitors of NL-25 were interspersed among the samples to measure the neutron fluence. The gradient was measured by interspersing 1.0 cm pieces of Ni wire, ~10 mg each.

The samples are irradiated in the core of the Brookhaven National Laboratory High Flux Beam Reactor. The usual flux associated with a 24 hr irradiation is on the order of 10^{19} neutrons/cm^2. This flux is sufficient for the usual potassium concentrations found in lunar material. However, for samples with low potassium contents, the time period has been doubled to reduce the error associated with the ^{39}Ar measurement.

The samples are usually allowed to stand at the reactor facilities for one to two months, thus decreasing the exposure of laboratory personnel to harmful radiation. The material is usually run within the year following irradiation in order that the ^{37}Ar is still well above the limit of detection.

Since November 1974, the procedure for analyzing the monitors and nickel wire data has been changed. The previous procedure is described by Husain (1974). The nickel wire, instead of being a continuous wire taped to the outside of the quartz irradiation tube, is now cut into six 1.0 cm sections and weighed to better than 1‰ accuracy before neutron irradiation. These six wires are wrapped in Al foil and spaced throughout the samples, neighboring the four hornblende monitors, with two extra wires at each end of the tube.

The first materials analyzed are the monitors. These are run with a total melt at 1600°C. The gas is expanded into the mass spectrometer in three steps. This is a check on the linearity of the amplifier or multiplier, for each step is in a different range of amplification. Any discrepancy in the ratios measured is a sign of a non-linear system.

During normal sample analysis, a change in gain of a factor of 1.5 is occasionally experienced. The monitor run with the three expansion method causes the gain to be changed over a range of 1.8. The measured non-linearity over this range is $<0.017\%$ change in ratio per 1% change in gain.

The nickel wires are counted on Ge–Li γ-ray detectors. The wires are mounted on aluminum cards for ease of handling and to insure similar counting geometries from sample to sample. The wires usually have 5–10 mV activity, and thus require shielding between them and the detector. Without this shielding, the dead time of the multi-channel analyzer increases to such a degree that the counting rate is only a function of the mass of the wire. The specific activity of ^{58}Co is measured relatively from sample to sample to determine the flux gradient along the samples irradiated. Figure 1 shows the results from irradiation #21 on 3/27/75 and for irradiation #22 on 5/13/75.

^{58}Co is produced from the $^{38}Ni(n, p)^{58}Co$ reaction. This reaction was selected to monitor the fast neutron fluence as it is produced by approximately the same energy neutron as produce ^{39}Ar from ^{39}K. If pure nickel is used as the target, there are very few other gamma emitting isotopes produced. The actual counting is performed by setting a narrow discriminating window around the 0.81 MeV peak, characteristic of ^{58}Co.

The monitor data are corrected for the gradient. The neutron fluence is then determined for each sample from the average monitor $^{40}Ar^*/^{39}Ar^*$ value, normalizing to the gradient obtained from the Ni wire. These are the $^{40}Ar^*/^{39}Ar^*$ values listed in Table 4.

The production rates for ^{39}Ar and ^{37}Ar from potassium and calcium used to calculate the absolute concentrations in the unknown samples are simple averages of the monitors irradiated. These production rates are not corrected for the gradient. The correction is, in the worst case, no larger than 2%. The standard gas pipettes on the mass spectrometers are about 10% accurate for absolute gas concentrations. As such, the pains to correct for a 2% variance are masked by the uncertainty in the mass spectrometer sensitivity.

Argon analysis

The measurements are made with a glass 20 cm radius 60° sector mass spectrometer. The sensitivity and discrimination at mass 40 is checked daily to be about 2×10^{-11} ccSTP/mV and .5% per mass, respectively. After each sample analysis, the furnace and Ti getter are degassed at 1750°C and 800°C, respectively, for 4 hr.

Two or three system blanks are run prior to each sample. If they are within 25% of previous blanks, a new sample is started. If not, the furnace is degassed 4 hr more, and the process continued until a suitable blank is achieved.

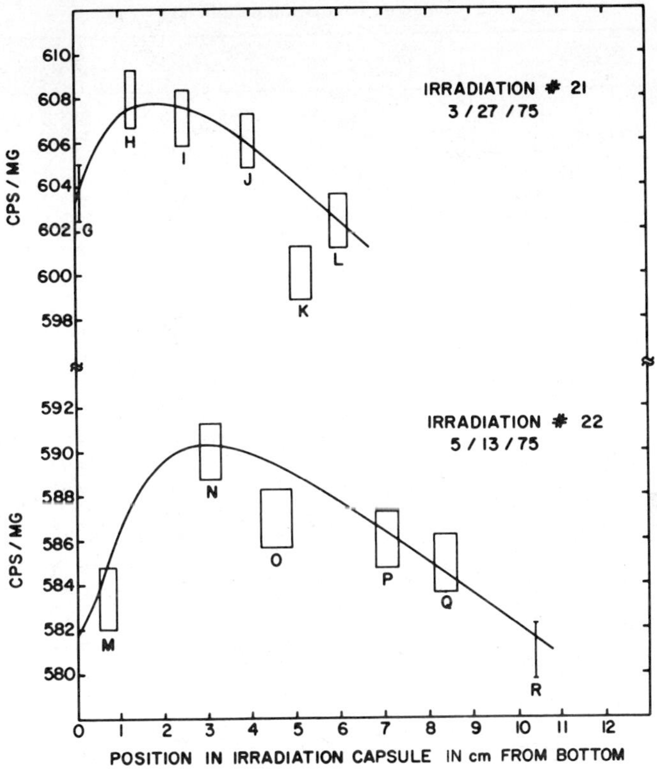

Fig. 1. Fast neutron gradient, as measured by ^{58}Co from Ni, during the neutron irradiations at the Brookhaven National Laboratory High Flux Beam Reactor.

A graph of system blanks of the furnace used since October 1974 is given in Fig. 2. The ^{40}Ar and ^{36}Ar at high temperatures are apparently gases diffusing from the hot glass walls. The background peak at mass 39 has two sources: (1) ^{39}K evaporating from the tungsten filament and entering the ion beam, and (2) impurities associated with traces of hydrocarbons. The ^{37}Ar blank is associated with ^{37}Cl which grows when more samples having high chloride concentrations are run. ^{38}Ar is usually at the limit of detection. At the highest temperatures, a slight amount is seen which is most likely air argon.

The ion source of the mass spectrometer has undergone two changes to achieve the low blanks for ^{39}Ar shown. First, an additional slit was placed between the filament and ionization chamber slit. This allows back biasing of the ^{39}K ions to reduce the probability of them entering the ion beam. Second, a platinum box was constructed to house the filament. The platinum box is electrically connected to one side of the filament. Since it is only millimeters from the filament, it becomes sufficiently hot to getter hydrogen. This minimizes the amount of hydrocarbons produced in the source which usually interfere with masses 39 and 38.

Data handling

While the peaks at masses 40, 39, 38, 37, and 36 are manually set, the mass spectral data collection and processing have been automated to facilitate isotopic intensity data reductions immediately after each temperature fraction.

The data are measured with a conventional Cary vibrating reed electrometer; the output of which

Table 4. Irradiation parameters.

Sample	$^{40}Ar^*/^{39}Ar^*$	$^{39}Ar^*/\%K$	$^{37}Ar/\%Ca$
14310,101:4021	31.79 ± .46	735.0 ± 10.0	385.0 ± 5.0
15075,24:3221	31.83 ± .46	735.0 ± 10.0	385.0 ± 5.0
15075,24:3121	31.97 ± .46	735.0 ± 10.0	385.0 ± 5.0
15505,72:4421	32.06 ± .46	735.0 ± 10.0	385.0 ± 5.0
60618,13:3321	31.86 ± .46	735.0 ± 10.0	385.0 ± 5.0
61135,10:0620,1	33.00 ± .10	711.0 ± 10.0	362.0 ± 5.0
61135,10:0620,2	33.00 ± .10	711.0 ± 10.0	362.0 ± 5.0
61135,10:0520	33.00 ± .10	711.0 ± 10.0	362.0 ± 5.0
62237,12:4621	32.17 ± .46	735.0 ± 10.0	385.0 ± 5.0
65777,2:0720	33.10 ± .20	700.0 ± 10.0	360.0 ± 5.0
65785,13:4721	32.13 ± .46	735.0 ± 10.0	385.0 ± 5.0
67035,20:0920	33.10 ± .20	700.0 ± 10.0	360.0 ± 5.0
67455,58:1120	33.18 ± .20	700.0 ± 10.0	360.0 ± 5.0
67618,1:1320	33.38 ± .20	700.0 ± 10.0	360.0 ± 5.0
68516,2:1520	33.81 ± .20	700.0 ± 10.0	360.0 ± 5.0
70255,36:4121	31.95 ± .46	735.0 ± 10.0	385.0 ± 5.0
14167,6,3	113.5 ± 0.5	231.47 ± 2.96	104.75 ± 2.9
14167,6,7	113.5 ± 0.5	231.47 ± 2.96	104.75 ± 2.9
14192,1,1	113.5 ± 0.5	231.47 ± 2.96	104.75 ± 2.9
14257,12,2	273.5 ± 3.0	85.5 ± 2.0	42.5 ± 1.0
61503,1,11:7213	15.20 ± .20	1500.0 ± 30.0	780.0 ± 10.0
72703,9,5:0219	32.20 ± .20	725.0 ± 10.0	385.0 ± 5.0
73263,1,1:1720	34.14 ± .20	700.0 ± 10.0	360.0 ± 5.0
73263,1,3:1820	34.25 ± .20	700.0 ± 10.0	360.0 ± 5.0
73263,1,6:1920	34.35 ± .20	700.0 ± 10.0	360.0 ± 5.0
73263,1,9:0419	32.20 ± .20	725.0 ± 10.0	385.0 ± 5.0
74243,2,1:1915	33.20 ± .20	706.2 ± 10.0	372.2 ± 5.0
76503,6,5:0319	32.20 ± .20	725.0 ± 10.0	385.0 ± 5.0
68815,141,B	33.10 ± 0.53	706.3 ± 10.0	371.5 ± 5.0
68815,133,B	33.16 ± 0.53	706.3 ± 10.0	371.5 ± 5.0
68815,133,C	32.96 ± 0.53	706.3 ± 10.0	371.5 ± 5.0
68815,67,D	33.17 ± 0.53	706.3 ± 10.0	371.5 ± 5.0

drives a voltage to frequency converter and a strip chart recorder for the operator's visual aid. The output of the voltage to frequency convertor is integrated for a selectable gate time period, usually 1 or 10 sec depending on the peak amplitude. The scalar is automatically cycled, selectable for 1 to 5 gate time periods. As such, an operator can select an integration period for a measurement from 1 to 50 sec.

After each cycle, the scalar is read by the on-line PDP11/10 processor. As the first cycle of the series is read, an internal clock is automatically read to calculate the seconds lapsed since the gas was introduced into the mass spectrometer tube. At the end of the last cycle, the mean and standard deviation of the data input is calculated. For each mass, the measurement of the peak top is followed by the measurement of the neighboring base line. The mean base line is subtracted from the mean peak top, and the result with the corresponding time and mass is outputed to the operator for verification before entering the data into the summary data file.

After the desired measuring sequence has been run, the data file for the fraction is scanned for each group of like mass values. If less than three measurements for a mass are found, a simple mean is calculated. Otherwise, a linear regression is performed in addition to the arithmetic mean. The

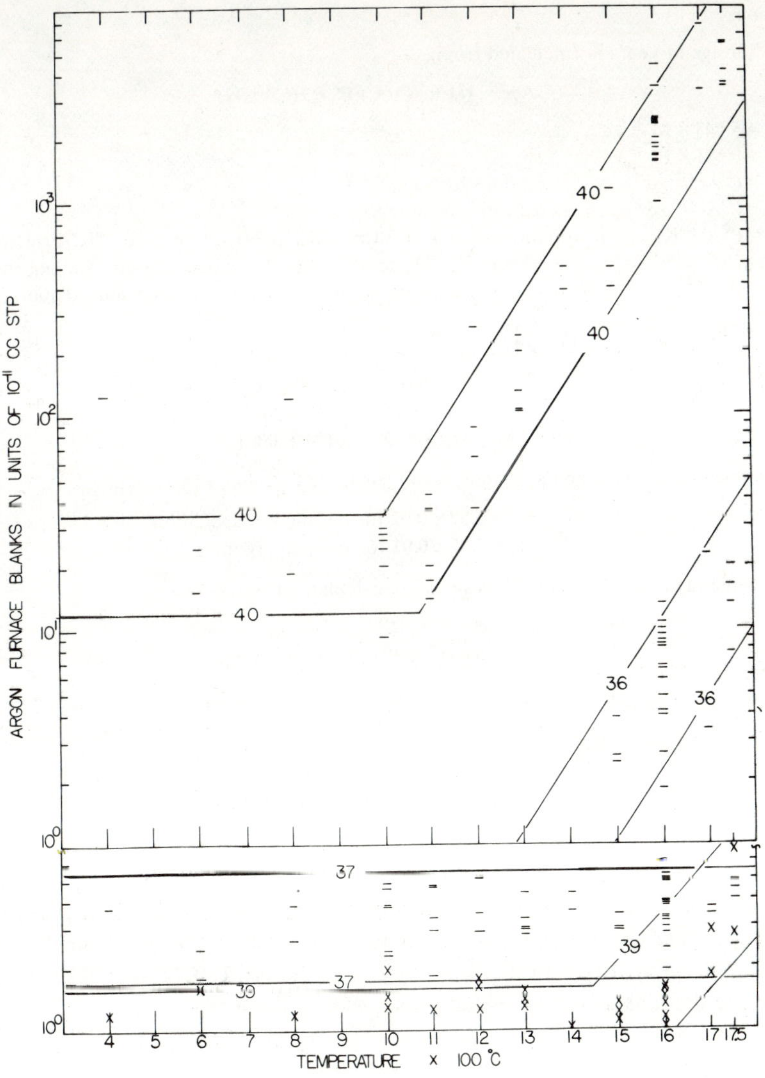

Fig. 2. Argon furnace blanks.

results are stored onto a mass storage device and can be recalled at any time after calculation.

Upon completion of a sample, all the isotopic data are tabulated, and the associated blank data are analyzed. This procedure has not been automated as it is the step which requires analysis of the data for validity and is the step necessary to monitor the proper functioning of the mass spectrometer. Without such an operator input, the data produced by an automated experiment can mask irregularities in instrument behavior and falsify the measured values.

The corrected data are then entered into the Center Computer facility at Stony Brook for final calculation. The turn around time is such that within 2 hr after a fraction has been measured the final age of that fraction can be obtained.

Calculations

^{39}Ar–^{40}Ar age in years is calculated from:

$$\text{Age} = 1/\lambda \ln [(\lambda/\lambda_e)(C_m R_s/R_m) + 1],$$

where $\lambda = 5.543 \times 10^{-10}$ yr^{-1},
$\lambda_e = 0.581 \times 10^{-10}$ yr^{-1},
$R_s = (^{40}Ar^*/^{39}Ar^*)_s$ corrected ratio for sample,
$R_m = (^{40}Ar^*/^{39}Ar^*)_m$ corrected ratio for monitor,
$C_m = (^{40}Ar/^{40}K)_m$ atom ratio for monitor (0.3530 ± 0.0022 for NL-25, based on $^{40}K/K = 1.167 \times 10^{-4}$).

The ratios $^{40}Ar^*/^{39}Ar^*$ are corrected for ^{39}Ar produced in the neutron irradiation and for trapped ^{40}Ar. The ^{39}Ar is corrected by using the following ratios obtained from neutron irradiation of Ca and K salts:

$$^{37}Ar/^{39}Ar \text{ from Ca} = 1{,}550 \pm 70,$$
$$^{37}Ar/^{38}Ar \text{ from Ca} = 19{,}300 \pm 3000,$$
$$^{37}Ar/^{36}Ar \text{ from Ca} = 4{,}320 \pm 20,$$
$$^{38}Ar/^{39}Ar \text{ from K} = .0100 \pm .0001.$$

The trapped argon, Ar_{tr}, and spallation argon, Ar_{sp}, are calculated by assuming:

$$(^{36}Ar/^{38}Ar)_{sp} = 0.62 \pm 0.02 \text{ for cosmic ray spallation},$$
$$(^{36}Ar/^{38}Ar)_{tr} = 5.35 \pm 0.01 \text{ for trapped argon}.$$

The cosmic ray exposure ages, T_e in m.y., are calculated from:

$$T_e = \frac{100}{1.69} \frac{(^{38}Ar_{sp}/^{39}Ar^*)_m}{(^{37}Ar/^{39}Ar^*_m)(\%Ca/^{37}Ar)_m}$$

using a cosmic ray production rate of 1.69×10^{-8} cc ^{38}Ar per gram Ca each million years.

RESULTS AND DISCUSSION

The argon isotopic results for all samples measured are presented in Table 5. Summary of the exposure and ^{39}Ar–^{40}Ar ages is found in Table 6. The data are usually also plotted either as a temperature release curve or, in the case where there is appreciable trapped ^{36}Ar, as a plot of $^{39}Ar^*/^{36}Ar_{sw}$ vs. $^{40}Ar^*/^{36}Ar_{sw}$. For samples with very disturbed argon systems, no plot is made. The individual samples are discussed in the order presented in Table 5.

14310,101 (Fig. 3)

Sample 14310,101 is a basaltic sample from Apollo 14. The temperature release pattern shown in Fig. 3 shows a plateau of 15 fractions from 675° to 1025°C with an average age of 3.92 ± 0.01 and contains 72% of the total ^{39}Ar released from potassium. The fractions before 675°C show a loss of ^{40}Ar, and the temperatures above 1025° show a high temperature drop off which might be due to gas released from low-K pyroxenes. The age agrees well with that obtained by Turner *et al.* (1971) of 3.89 ± 0.04 G.y. for a plagioclase separate.

The K/Ca ratio varies between 0.05 and 2.32 in an irregular manner over the temperatures analyzed. A cosmic ray exposure age could not be reported as there appears to be excess ^{38}Ar, probably due to the presence of Cl.

Table 5. Isotopic argon results for ^{39}Ar–^{40}Ar studies.

Temp. (°C)	$^{36}Ar/^{38}Ar$[(a)]	$^{38}Ar/^{37}Ar$	$^{39}Ar/^{37}Ar$	$^{40}Ar^*/^{39}Ar^*$[(b)]	$^{39}Ar \times 10^{-8}$ ccSTP/g	Exp. age ($\times 10^6$ yr)	Age ($\times 10^9$ yr)
14310,101:4021 (19.7 mg Basalt)							
500	1.896 ± 67.5	.00053 ± 86.7	.1294 ± 60.0	24.6 ± 1.9	18.35 ± 1.9	9. ± 114.9	2.313 ± 1.2
600	.506 ± 63.5	.00166 ± 63.2	.1623 ± 60.0	50.7 ± .9	23.01 ± .8	—	3.368 ± .4
650	.398 ± 28.4	.00333 ± 60.8	.1866 ± 60.0	65.5 ± 1.5	26.47 ± 1.4	—	3.780 ± .6
675	.666 ± 15.0	.00306 ± 61.0	.1694 ± 60.0	71.2 ± 1.6	24.01 ± 1.6	69. ± 61.3	3.918 ± .6
700	.812 ± 18.6	.00227 ± 61.7	.1107 ± 60.0	73.6 ± .8	15.70 ± .8	49. ± 62.4	3.971 ± .3
725	.464 ± 33.2	.00417 ± 100.9	.1836 ± 100.0	72.3 ± 1.1	15.62 ± 1.1	—	3.943 ± .4
750	.936 ± 17.1	.00490 ± 100.6	.1981 ± 100.0	71.0 ± .9	16.86 ± .9	103. ± 101.0	3.911 ± .3
775	.675 ± 12.0	.01252 ± 100.2	.2946 ± 100.0	71.6 ± .8	16.71 ± .7	280. ± 100.3	3.926 ± .3
800	.652 ± 7.2	.00944 ± 100.2	.1723 ± 100.0	72.6 ± .9	14.66 ± .9	212. ± 100.2	3.948 ± .3
825	.649 ± 5.1	.05395 ± 285.7	.8042 ± 285.7	72.7 ± .9	15.96 ± .9	1214. ± 285.8	3.952 ± .3
850	.606 ± 7.1	.02292 ± 100.1	.2753 ± 100.0	71.5 ± .9	15.61 ± .8	—	3.924 ± .3
875	.627 ± 6.9	.01936 ± 233.4	.1888 ± 233.3	71.3 ± 1.2	16.06 ± 1.2	438. ± 233.4	3.920 ± .4
900	.631 ± 2.5	.01491 ± 33.4	.0964 ± 33.3	71.8 ± .8	16.40 ± .7	337. ± 33.4	3.930 ± .3
925	.693 ± 1.8	.01304 ± 15.6	.0497 ± 15.5	72.4 ± .8	18.19 ± .8	291. ± 15.7	3.945 ± .3
950	.649 ± 3.3	.00919 ± 8.0	.0271 ± 7.9	73.5 ± 1.8	19.99 ± 1.8	207. ± 8.2	3.969 ± .6
975	.661 ± 3.2	.01157 ± 16.9	.0460 ± 16.7	72.7 ± 1.5	15.64 ± 1.4	260. ± 17.1	3.952 ± .5
1000	.718 ± 7.6	.03121 ± 100.2	.1797 ± 100.0	71.3 ± 1.8	10.19 ± 1.6	692. ± 100.3	3.919 ± .6
1025	.711 ± 7.4	.23022 ± 999.9	1.1895 ± 999.9	70.8 ± 3.9	6.75 ± 3.8	5113. ± 999.9	3.908 ± 1.4
1075	.678 ± 3.8	.13944 ± 500.0	.5553 ± 500.0	66.2 ± 1.0	6.30 ± .9	3119. ± 500.0	3.798 ± .3
1150	.723 ± 3.5	.00704 ± 23.4	.0236 ± 23.3	67.4 ± 1.4	5.75 ± 1.4	156. ± 23.5	3.827 ± .5
1250	.677 ± 4.8	.05676 ± 111.2	.0998 ± 111.1	64.8 ± 1.3	5.09 ± 1.2	1270. ± 111.2	3.763 ± .5
1350	.659 ± 6.5	.02223 ± 46.8	.0369 ± 46.6	62.4 ± 2.4	4.50 ± 2.2	499. ± 46.9	3.700 ± .9
1450	.803 ± 23.0	.00274 ± 26.2	.0023 ± 15.9	53.7 ± 12.8	.98 ± 8.2	60. ± 29.7	3.461 ± 5.1
1600	.722 ± 28.7	.00331 ± 68.7	.0040 ± 68.9	72.4 ± 21.6	.34 ± 17.6	73. ± 69.5	3.945 ± 7.5
1750	1.271 ± 29.7	.00237 ± 70.6	.0024 ± 72.9	144.5 ± 39.1	.20 ± 29.4	46. ± 73.5	5.127 ± 10.5

(a) All errors are relative errors in percent

(b) ^{40}Ar from the decay of ^{40}K only.

Table 5. (*Continued*).

Temp. (°C)	$^{36}Ar/^{38}Ar$[(a)]	$^{38}Ar/^{37}Ar$	$^{39}Ar/^{37}Ar$	$^{40}Ar^{*}/^{39}Ar^{*}$[(b)]	$^{39}Ar \times 10^{-8}$ ccSTP/g	Exp. age ($\times 10^{6}$ yr)	Age ($\times 10^{9}$ yr)
15075,24:3221 (31.2 mg Basalt)							
500	.634 ± 41.2	.00167 ± 74.4	.0255 ± 71.5	29.3 ± 3.4	.79 ± 3.2	38. ± 75.4	2.554 ± 1.8
600	.727 ± 34.2	.00397 ± 75.4	.0487 ± 71.5	43.2 ± 3.7	.76 ± 3.5	88. ± 76.8	3.121 ± 1.6
700	.668 ± 11.5	.00387 ± 36.8	.0282 ± 35.9	52.4 ± 3.2	1.76 ± 3.2	87. ± 37.2	3.421 ± 1.3
800	.671 ± 5.6	.01326 ± 60.2	.0401 ± 60.1	57.6 ± 3.7	1.78 ± 3.6	297. ± 60.2	3.569 ± 1.4
900	.677 ± 1.3	.01416 ± 14.3	.0269 ± 14.5	54.1 ± 2.3	2.93 ± 2.2	317. ± 14.4	3.469 ± .9
1000	.682 ± 2.4	.01040 ± 4.3	.0097 ± 4.4	53.3 ± 1.6	4.18 ± 1.6	232. ± 4.6	3.447 ± .6
1100	.615 ± 2.1	.01233 ± 2.7	.0116 ± 2.9	52.5 ± 1.5	5.07 ± 1.4	—	3.421 ± .6
1200	.642 ± 2.6	.01209 ± 7.1	.0040 ± 7.5	51.6 ± 2.9	2.58 ± 2.8	272. ± 7.3	3.395 ± 1.2
1300	.648 ± 2.6	.01315 ± 4.8	.0013 ± 6.6	57.1 ± 4.8	1.24 ± 4.8	296. ± 5.1	3.557 ± 1.9
1400	.639 ± 2.3	.01219 ± 2.5	.0007 ± 7.0	63.5 ± 6.9	.92 ± 6.7	275. ± 2.9	3.727 ± 2.6
1500	.674 ± 4.8	.01360 ± 5.8	.0010 ± 9.8	31.2 ± 19.9	.46 ± 8.5	304. ± 6.2	2.638 ± 10.4
1600	.714 ± 4.5	.00880 ± 19.8	.0004 ± 60.7	43.0 ± 93.9	.05 ± 57.5	195. ± 20.0	3.112 ± 41.4
1750	.582 ± 5.0	.01097 ± 10.6	.0005 ± 42.4	7.0 ± 999.9	.06 ± 41.1	—	.983 ± 999.9
15075,24:3121 (27.4 mg Basalt)							
500	2.180 ± 46.7	.00059 ± 92.6	.0114 ± 83.6	42.1 ± 6.3	.79 ± 6.2	9. ± 108.9	3.073 ± 2.8
600	.933 ± 32.5	.00100 ± 101.1	.0099 ± 100.3	45.2 ± 7.2	1.15 ± 7.1	21. ± 101.9	3.183 ± 3.1
700	.705 ± 11.9	.00637 ± 100.4	.0414 ± 100.0	45.4 ± 2.6	2.41 ± 2.5	142. ± 100.5	3.189 ± 1.1
800	.674 ± 4.0	.00968 ± 50.1	.0299 ± 50.0	53.6 ± 1.6	3.47 ± 1.4	217. ± 50.1	3.448 ± .6
900	.658 ± 1.8	.00728 ± 16.7	.0119 ± 16.8	54.2 ± 2.2	4.14 ± 2.2	164. ± 16.8	3.465 ± .9
1000	.625 ± 2.2	.01037 ± 7.2	.0094 ± 7.4	54.1 ± 1.9	7.68 ± 1.8	235. ± 7.4	3.462 ± .7
1050	.627 ± 5.3	.01248 ± 20.2	.0154 ± 20.1	52.4 ± 2.1	4.47 ± 2.0	282. ± 20.4	3.414 ± .8
1100	.611 ± 3.6	.01033 ± 25.0	.0103 ± 25.1	52.6 ± 2.6	2.39 ± 2.6	—	3.420 ± 1.0
1200	.613 ± 4.2	.01320 ± 14.1	.0031 ± 14.8	53.7 ± 4.4	2.58 ± 4.4	—	3.453 ± 1.8
1300	.586 ± 2.6	.01490 ± 9.1	.0016 ± 11.6	55.4 ± 7.5	1.05 ± 7.3	—	3.502 ± 2.9
1400	.618 ± 1.6	.01226 ± 22.3	.0011 ± 26.4	43.9 ± 15.5	.59 ± 14.2	—	3.137 ± 6.8
1500	.614 ± 3.7	.01389 ± 58.9	.0026 ± 60.7	34.4 ± 18.1	.51 ± 15.0	—	2.773 ± 9.0
1750	.597 ± 5.5	.00869 ± 37.2	.0006 ± 59.5	52.4 ± 138.3	.09 ± 46.5	—	3.411 ± 55.7

15505,72:4421 (10.8 mg Breccia with glass splash)							
500	5.352 ± 1.6	.04272 ± 100.0	.0176 ± 100.1	49.9 ± 4.2	16.22 ± 4.2	—	3.331 ± 1.7
600	5.492 ± .5	.10705 ± 100.0	.0105 ± 100.2	142.9 ± 6.4	15.53 ± 6.3	—	5.093 ± 1.7
650	5.348 ± 2.4	.12490 ± 100.0	.0041 ± 101.2	217.3 ± 43.4	6.11 ± 15.8	1. ± 999.9	5.839 ± 10.2
675	5.463 ± 1.1	.24286 ± 100.0	.0050 ± 101.0	343.7 ± 13.9	5.56 ± 13.9	—	6.672 ± 2.9
700	5.448 ± 1.0	.38234 ± 100.0	.0061 ± 100.7	362.9 ± 11.7	3.39 ± 11.7	—	6.771 ± 2.4
750	5.373 ± .7	.57688 ± 50.0	.0064 ± 50.6	375.3 ± 7.9	7.03 ± 7.9	—	6.833 ± 1.6
775	5.376 ± .6	.47680 ± 100.0	.0050 ± 100.9	314.4 ± 13.2	8.37 ± 13.2	—	6.509 ± 2.8
800	5.229 ± .5	1.67311 ± 100.0	.0226 ± 100.1	192.9 ± 100.0	8.35 ± 3.5	970. ± 102.6	5.625 ± 24.4
825	5.255 ± 1.2	.35608 ± 28.6	.0063 ± 28.8	148.5 ± 100.0	8.13 ± 3.6	162. ± 74.3	5.160 ± 26.6
850	5.198 ± .4	.40056 ± 100.0	.0100 ± 100.2	104.9 ± 100.0	7.36 ± 6.7	291. ± 101.5	4.556 ± 30.1
900	5.235 ± .3	.58014 ± 100.0	.0156 ± 100.1	97.8 ± 100.0	8.61 ± 4.8	318. ± 101.3	4.436 ± 31.0
950	5.242 ± 1.0	.93679 ± 181.8	.0219 ± 181.9	112.4 ± 100.0	4.45 ± 5.5	486. ± 189.2	4.674 ± 29.4
1000	5.082 ± .8	.52567 ± 100.0	.0109 ± 100.2	121.9 ± 100.0	4.04 ± 6.5	674. ± 101.1	4.815 ± 28.5
1050	5.128 ± 1.5	.48702 ± 100.0	.0077 ± 100.4	162.2 ± 100.0	2.86 ± 8.6	517. ± 106.5	5.316 ± 25.8
1150	5.027 ± .6	.69603 ± 66.7	.0103 ± 66.9	169.4 ± 100.0	5.70 ± 5.3	1076. ± 67.5	5.393 ± 25.5
1250	4.785 ± 1.3	1.40784 ± 200.0	.0204 ± 200.1	163.9 ± 100.0	3.77 ± 7.2	3807. ± 200.4	5.334 ± 25.7
1350	4.843 ± 2.0	.01772 ± 57.2	.0004 ± 115.2	221.4 ± 100.0	.48 ± 100.0	43. ± 60.6	5.873 ± 23.4
1500	4.374 ± 3.8	.00834 ± 100.0	.0006 ± 141.4	41.4 ± 100.0	.36 ± 100.0	39. ± 101.9	3.043 ± 45.1
1600	4.381 ± 4.5	.00590 ± 100.1	.0007 ± 141.4	144.1 ± 100.0	.24 ± 100.0	27. ± 102.8	5.108 ± 26.9
1750	3.970 ± 4.6	.00588 ± 100.1	.0006 ± 141.4	165.4 ± 100.0	.24 ± 100.0	39. ± 101.4	5.351 ± 25.7
60618,13:3321 (19.2 mg Crystalline)							
500	4.303 ± 63.5	.00393 ± 66.6	.3498 ± 20.1	22.2 ± 1.6	2.59 ± 1.4	20. ± 325.4	2.175 ± 1.0
600	19.229 ± 115.3	.00012 ± 180.3	.1910 ± 150.0	21.8 ± 3.4	2.82 ± 3.3	—	2.152 ± 2.2
700	2.786 ± 100.0	.00021 ± 100.0	.0479 ± 5.7	44.3 ± 5.7	4.25 ± 5.7	3. ± 208.7	3.157 ± 2.5
800	.074 ± 100.1	.00017 ± 7.6	.0143 ± 5.6	69.9 ± 1.7	3.53 ± 1.6	—	3.882 ± .6
900	.231 ± 72.8	.00027 ± 17.2	.0097 ± 5.6	75.1 ± 1.5	4.21 ± 1.4	—	4.000 ± .5
1000	.281 ± 40.1	.00026 ± 6.6	.0046 ± 5.0	76.5 ± 4.9	5.72 ± 4.9	—	4.032 ± 1.7
1100	.353 ± 8.3	.00023 ± 4.4	.0029 ± 3.2	72.8 ± 1.9	2.51 ± 1.9	—	3.949 ± .7
1200	.270 ± 101.0	.00026 ± 14.1	.0034 ± 3.4	68.6 ± 3.3	1.79 ± 3.2	—	3.851 ± 1.2
1300	.618 ± 4.0	.00031 ± 3.6	.0026 ± 3.7	45.4 ± 2.8	1.55 ± 2.7	—	3.195 ± 1.2
1400	6.738 ± 102.5	.00004 ± 22.9	.0028 ± 5.6	66.2 ± 26.1	.41 ± 4.7	—	3.792 ± 9.5
1500	.615 ± 56.1	.00021 ± 4.6	.0022 ± 4.4	80.3 ± 3.8	.55 ± 3.7	—	4.113 ± 1.3
1600	1.966 ± 141.4	.00026 ± 101.7	.0020 ± 23.8	115.7 ± 14.8	.14 ± 14.7	4. ± 169.5	4.735 ± 4.3
1700	4.695 ± 102.2	.00020 ± 100.2	.0009 ± 22.7	313.7 ± 22.0	.09 ± 21.8	1. ± 830.7	6.516 ± 4.6

(a) All errors are relative errors in percent.

(b) ^{40}Ar from the decay of ^{40}K only.

Table 5. (*Continued*).

Temp. (°C)	$^{36}Ar/^{38}Ar^{(a)}$	$^{38}Ar/^{37}Ar$	$^{39}Ar/^{37}Ar$	$^{40}Ar^*/^{39}Ar^{*(b)}$	$^{39}Ar \times 10^{-8}$ ccSTP/g	Exp. age ($\times 10^6$ yr)	Age ($\times 10^9$ yr)
61135,10:0620,1 (13.8 mg Anorthosite chip)							
300	3.573 ± 59.0	.13918 ± 71.8	.4322 ± 60.1	278.1 ± 33.5	.09 ± 33.4	1113. ± 173.0	6.233 ± 7.4
400	.991 ± 11.8	.09173 ± 9.2	.7934 ± 3.9	47.6 ± 1.1	1.62 ± 1.0	1800. ± 11.2	3.212 ± .5
500	2.412 ± 18.5	.01948 ± 18.2	1.0484 ± 1.8	64.2 ± .4	9.29 ± .3	258. ± 33.3	3.687 ± .1
550	5.136 ± 20.6	.01242 ± 9.6	1.1026 ± 1.1	83.9 ± .6	14.68 ± .5	12. ± 497.5	4.127 ± .2
600	4.952 ± 6.0	.01396 ± 6.4	.9845 ± 3.3	89.3 ± 1.1	20.00 ± 1.1	25. ± 80.5	4.232 ± .4
650	5.357 ± 6.7	.01483 ± 6.7	.8808 ± .4	86.9 ± .5	27.90 ± .3	—	4.186 ± .2
675	8.628 ± 15.6	.00780 ± 15.4	.8290 ± 1.2	84.7 ± 1.1	21.68 ± 1.1	—	4.144 ± .4
700	5.429 ± 6.2	.00667 ± 6.4	.6782 ± 2.8	79.6 ± 1.4	28.27 ± 1.2	—	4.039 ± .5
725	2.729 ± 17.6	.00676 ± 13.4	.5903 ± 3.2	74.7 ± 1.4	21.04 ± 1.4	80. ± 29.5	3.934 ± .5
775	4.887 ± 2.9	.00961 ± 2.8	.2809 ± .8	69.2 ± .6	33.10 ± .5	20. ± 33.5	3.808 ± .2
825	4.965 ± 1.0	.00837 ± 2.5	.1470 ± 2.4	70.1 ± .8	28.33 ± .7	14. ± 14.6	3.829 ± .3
860	4.750 ± 3.3	.00719 ± 3.2	.0896 ± .8	71.7 ± .8	18.42 ± .7	19. ± 29.0	3.868 ± .3
900	4.711 ± 1.3	.00610 ± 1.3	.0674 ± .9	73.4 ± .7	15.42 ± .6	18. ± 11.1	3.905 ± .2
950	4.948 ± 1.7	.00818 ± 1.4	.0633 ± .6	74.9 ± .8	18.19 ± .6	15. ± 22.3	3.938 ± .3
1000	4.692 ± 2.0	.00839 ± 2.1	.0575 ± 1.0	75.4 ± .9	22.20 ± .7	25. ± 16.5	3.949 ± .3
1050	4.516 ± 5.9	.00847 ± 5.4	.0798 ± 1.1	74.5 ± 1.0	13.79 ± .9	32. ± 37.1	3.931 ± .3
1100	4.718 ± 3.0	.00954 ± 2.9	.0750 ± .5	71.9 ± .6	8.50 ± .4	27. ± 25.0	3.872 ± .2
1150	5.186 ± 3.3	.00583 ± 3.8	.0584 ± 2.5	64.2 ± 1.3	5.08 ± 1.2	4. ± 107.5	3.686 ± .5
1200	3.663 ± 12.7	.00410 ± 10.6	.0500 ± 2.4	63.4 ± 1.2	5.54 ± 1.1	31. ± 36.7	3.667 ± .4
1250	3.884 ± 6.6	.00186 ± 6.4	.0432 ± 1.2	62.7 ± 1.1	5.73 ± 1.1	12. ± 23.6	3.647 ± .4
1300	1.140 ± 33.1	.00103 ± 31.8	.0326 ± 2.2	64.7 ± 2.0	4.69 ± 2.0	19. ± 40.5	3.699 ± .8
1400	.453 ± 88.4	.00068 ± 13.5	.0256 ± 1.7	64.2 ± .8	4.40 ± .7	—	3.686 ± .3
1500	.947 ± 14.3	.00063 ± 10.1	.0329 ± 2.1	68.6 ± 1.0	16.65 ± .9	12. ± 12.5	3.794 ± .4
1600	1.889 ± 45.2	.00386 ± 43.7	.0299 ± 2.9	78.2 ± 3.4	.55 ± 2.8	60. ± 67.9	4.011 ± 1.2
61135,10:0620,2 (3.8 mg Anorthosite fine powder)							
500	6.099 ± 9.7	.05014 ± 13.0	.0564 ± 9.9	242.1 ± 2.7	2.57 ± 2.7	—	5.982 ± .6
600	5.396 ± 2.8	.08172 ± 2.9	.0365 ± 4.2	491.9 ± 4.1	3.36 ± 4.1	—	7.278 ± .8
700	5.431 ± 1.1	.11083 ± 1.1	.0223 ± 3.5	567.0 ± 3.5	5.85 ± 3.5	—	7.540 ± .6
750	5.522 ± .8	.14305 ± 1.4	.0181 ± 6.4	541.0 ± 6.4	3.24 ± 6.3	—	7.453 ± 1.2
800	5.139 ± .8	.15668 ± 1.0	.0138 ± 3.5	396.6 ± 8.2	3.99 ± 3.4	149. ± 20.3	6.881 ± 1.6

850	5.337 ± .9	.10328 ± 1.0	.0131 ± 3.9	234.4 ± 10.0	5.28 ± 3.9	6. ± 367.0	5.923 ± 2.3
900	5.072 ± .9	.12038 ± 2.7	.0133 ± 3.9	225.5 ± 10.7	4.83 ± 2.8	150. ± 17.0	5.854 ± 2.5
950	5.133 ± 1.2	.09211 ± 1.1	.0109 ± 1.8	232.4 ± 9.5	4.07 ± 1.7	90. ± 30.7	5.908 ± 2.2
1000	5.157 ± 1.1	.10965 ± 1.7	.0091 ± 4.4	295.2 ± 11.5	3.32 ± 4.1	95. ± 29.4	6.342 ± 2.5
1050	5.150 ± .9	.09563 ± 1.8	.0074 ± 2.6	332.6 ± 10.3	3.46 ± 2.0	86. ± 25.1	6.559 ± 2.2
1100	5.231 ± .8	.07289 ± .9	.0070 ± 4.4	293.1 ± 10.4	3.14 ± 4.4	39. ± 39.0	6.329 ± 2.3
1150	5.190 ± 1.1	.05698 ± 1.3	.0071 ± 4.8	239.8 ± 10.0	2.95 ± 4.7	41. ± 37.5	5.965 ± 2.3
1200	5.246 ± 1.2	.05434 ± 1.3	.0071 ± 4.5	236.5 ± 9.8	3.08 ± 4.5	26. ± 61.8	5.939 ± 2.3
1250	5.264 ± 2.5	.03342 ± 2.6	.0065 ± 5.3	179.5 ± 9.4	2.62 ± 5.3	13. ± 160.0	5.445 ± 2.4
1350	5.250 ± 2.6	.02055 ± 2.8	.0068 ± 2.8	131.7 ± 6.7	2.62 ± 2.6	9. ± 139.1	4.899 ± 1.9
1450	4.109 ± 15.6	.00463 ± 15.4	.0051 ± 9.4	83.9 ± 10.3	1.46 ± 9.4	26. ± 66.8	4.126 ± 3.4
1650	2.439 ± 30.1	.00516 ± 29.6	.0108 ± 7.2	126.0 ± 7.6	.97 ± 7.1	68. ± 54.5	4.822 ± 2.2
61135,10:0520 (45.9 mg Anorthosite)							
600	5.308 ± .1	.07272 ± .2	.0435 ± .2	283.1 ± 1.6	7.20 ± .2	14. ± 30.2	6.265 ± .3
650	5.497 ± .2	.10106 ± .2	.0290 ± .2	370.1 ± .1	2.44 ± .1	—	6.754 ± .0
700	5.384 ± .2	.11050 ± .3	.0269 ± .2	329.3 ± .1	1.69 ± .1	—	6.540 ± .0
750	5.395 ± .2	.12325 ± .4	.0215 ± .6	330.5 ± .5	2.88 ± .5	—	6.548 ± .1
775	5.383 ± .2	.12925 ± .4	.0190 ± .7	280.7 ± .6	2.23 ± .6	—	6.250 ± .1
800	5.329 ± .2	.12523 ± .3	.0163 ± .7	248.0 ± 8.3	2.11 ± .6	12. ± 68.9	6.025 ± 1.9
825	5.231 ± .1	.12444 ± .2	.0143 ± .7	255.1 ± 8.9	2.02 ± .7	67. ± 10.9	6.076 ± 2.0
875	5.323 ± .4	.12451 ± .4	.0122 ± .3	266.4 ± 10.2	2.96 ± .2	15. ± 86.9	6.155 ± 2.3
900	5.233 ± .1	.10426 ± .2	.0112 ± .7	239.5 ± 10.2	2.23 ± .6	55. ± 9.4	5.963 ± 2.3
950	5.054 ± .3	.09698 ± .3	.0101 ± .6	237.2 ± 10.1	2.70 ± .5	129. ± 7.2	5.945 ± 2.3
1000	5.249 ± .3	.06143 ± .3	.0093 ± .5	191.7 ± 9.0	2.97 ± .5	28. ± 20.6	5.562 ± 2.2
1050	4.640 ± 2.2	.02838 ± .7	.0081 ± .6	148.9 ± 5.4	2.49 ± .6	91. ± 15.0	5.115 ± 1.4
1150	5.076 ± .1	.02838 ± .1	.0075 ± .5	156.6 ± 6.1	3.54 ± .5	35. ± 4.5	5.203 ± 1.6
1200	4.953 ± .2	.03831 ± .1	.0082 ± .7	168.3 ± 6.9	1.98 ± .7	68. ± 4.1	5.330 ± 1.8
1300	5.056 ± .3	.03919 ± .3	.0089 ± .4	149.9 ± 7.4	4.11 ± .4	52. ± 6.9	5.127 ± 2.0
1350	4.981 ± .3	.01859 ± .2	.0071 ± .7	119.1 ± 5.4	2.41 ± .7	31. ± 5.4	4.724 ± 1.6
1450	4.795 ± .5	.01118 ± .4	.0075 ± .7	107.2 ± 3.3	2.39 ± .7	28. ± 4.9	4.544 ± 1.0
1550	4.187 ± 1.5	.00434 ± 1.1	.0065 ± .9	94.9 ± 1.7	1.73 ± .9	23. ± 6.5	4.336 ± .5
1650	2.941 ± 5.9	.00148 ± 4.9	.0040 ± .8	97.7 ± .9	.70 ± .7	16. ± 11.7	4.384 ± .3

[a] All errors are relative errors in percent.
[b] ^{40}Ar from the decay of ^{40}K only.

Table 5. (*Continued*).

Temp. (°C)	$^{36}Ar/^{38}Ar^{(a)}$	$^{38}Ar/^{37}Ar$	$^{39}Ar/^{37}Ar$	$^{40}Ar^{*}/^{39}Ar^{*(b)}$	$^{39}Ar \times 10^{-8}$ ccSTP/g	Exp. age ($\times 10^6$ yr)	Age ($\times 10^9$ yr)
62237,12:4621 (14.1 mg White anorthositic breccia)							
500	6.252 ± 106.0	.00017 ± 160.1	.0008 ± 160.1	154.3 ± 100.0	.13 ± 100.0	—	5.222 ± 26.3
600	7.755 ± 71.8	.00024 ± 121.9	.0007 ± 141.4	41.3 ± 100.0	.13 ± 100.0	—	3.037 ± 45.2
700	4.259 ± 16.0	.00472 ± 200.5	.0064 ± 201.3	28.5 ± 24.7	.61 ± 22.8	25. ± 213.9	2.499 ± 13.6
800	2.629 ± 8.3	.00549 ± 63.0	.0032 ± 63.9	34.7 ± 15.3	.96 ± 13.2	71. ± 64.4	2.780 ± 7.5
900	1.314 ± 3.6	.00598 ± 25.0	.0016 ± 27.1	62.3 ± 11.0	1.23 ± 10.5	116. ± 25.1	3.679 ± 4.1
1000	.895 ± 3.0	.00648 ± 27.8	.0011 ± 33.2	72.1 ± 18.4	.77 ± 18.1	138. ± 27.9	3.917 ± 6.5
1100	1.172 ± 6.4	.11925 ± 999.9	.0344 ± 999.9	43.9 ± 21.9	.65 ± 21.2	2386. ± 999.9	3.129 ± 9.6
1200	.818 ± 7.7	.00344 ± 41.4	.0007 ± 58.7	86.8 ± 42.1	.42 ± 41.8	75. ± 41.5	4.227 ± 13.7
1300	.700 ± 10.5	.00327 ± 66.8	.0007 ± 91.1	77.1 ± 62.3	.56 ± 62.1	73. ± 66.8	4.028 ± 21.2
1400	.685 ± 6.6	.00404 ± 21.6	.0004 ± 102.3	80.0 ± 100.0	.10 ± 100.0	90. ± 21.8	4.091 ± 33.6
1500	.649 ± 13.3	.00538 ± 166.8	.0011 ± 194.4	199.9 ± 100.0	.13 ± 100.0	121. ± 166.8	5.683 ± 24.2
1600	.727 ± 29.3	.02139 ± 999.9	.0073 ± 999.9	154.0 ± 100.0	.14 ± 100.0	473. ± 999.9	5.219 ± 26.3
1750	.910 ± 9.8	.00232 ± 62.7	.0005 ± 117.9	134.3 ± 100.0	.14 ± 100.0	49. ± 62.8	4.978 ± 27.6
65777,2:0720 (56.8 mg Crystalline recrystallized noritic breccia)							
600	3.050 ± 1.5	.02045 ± 1.8	.8878 ± 1.1	12.6 ± .4	33.52 ± .1	211. ± 3.8	1.479 ± .4
650	4.724 ± 2.5	.01241 ± 2.6	.5477 ± 1.0	32.7 ± .4	19.48 ± .3	35. ± 21.1	2.649 ± .2
700	5.065 ± 2.4	.00750 ± 2.4	.3608 ± .7	54.4 ± .6	22.32 ± .4	10. ± 44.4	3.417 ± .2
750	4.878 ± 2.4	.00536 ± 2.4	.2792 ± .8	66.0 ± .4	24.90 ± .2	11. ± 27.3	3.727 ± .1
800	5.199 ± 1.6	.00471 ± 1.6	.1850 ± .3	68.6 ± .4	27.80 ± .3	3. ± 55.9	3.788 ± .1
850	5.021 ± 1.7	.00533 ± .9	.1029 ± .8	69.0 ± .8	24.68 ± .7	8. ± 26.6	3.798 ± .3
900	5.003 ± 1.0	.00511 ± .9	.0582 ± .7	70.8 ± .7	15.47 ± .5	8. ± 15.5	3.841 ± .2
950	5.103 ± 1.6	.00626 ± 1.7	.0378 ± .9	74.2 ± 1.0	12.17 ± .5	7. ± 35.2	3.918 ± .3
1000	4.965 ± .8	.00548 ± .7	.0292 ± .4	73.9 ± .7	8.58 ± .3	9. ± 11.0	3.912 ± .3
1050	4.984 ± 1.9	.00614 ± 1.5	.0296 ± .5	71.8 ± 1.0	6.87 ± .5	10. ± 27.0	3.864 ± .4
1100	4.917 ± 1.3	.00767 ± 1.1	.0345 ± .3	66.3 ± .9	4.20 ± .3	15. ± 16.2	3.734 ± .3
1200	4.819 ± .9	.00890 ± .8	.0301 ± 1.2	60.0 ± 1.6	4.66 ± 1.1	21. ± 9.2	3.572 ± .6
1300	4.874 ± 2.5	.00497 ± 1.7	.0143 ± 1.8	59.9 ± 2.2	1.90 ± 1.7	11. ± 26.7	3.570 ± .9
1400	9.012 ± 97.7	.00023 ± 97.7	.0142 ± .8	66.0 ± .9	9.23 ± .8	—	3.727 ± .3
1450	.469 ± 104.4	.00240 ± 100.0	.0121 ± 3.0	69.2 ± 2.9	.37 ± 2.9	—	3.804 ± 1.0
1550	6.100 ± 101.3	.00028 ± 100.0	.0108 ± 2.6	72.0 ± 2.7	.41 ± 2.6	—	3.868 ± 1.0
1650	1.771 ± 100.0	.00111 ± 100.0	.0082 ± 13.5	84.2 ± 14.0	.08 ± 13.5	18. ± 149.5	4.128 ± 4.7

65785,13:4721 (15.4 mg Crystalline)							
500	4.885 ± 102.0	—	—	11.2 ± .7	18.19 ± .5	—	1.384 ± .7
600	2.389 ± 62.3	.00444 ± 68.6	1.3447 ± 28.6	26.6 ± .6	23.11 ± .5	63. ± 116.2	2.407 ± .3
700	.731 ± 8.0	.01189 ± 7.6	.4818 ± 6.0	54.5 ± 1.3	40.22 ± 1.3	263. ± 8.3	3.466 ± .5
750	.703 ± 3.1	.01315 ± 3.1	.2578 ± 3.1	73.6 ± 1.9	25.96 ± 1.8	293. ± 3.5	3.954 ± .7
800	.635 ± 5.0	.01231 ± 1.4	.1258 ± 2.8	73.9 ± 2.5	22.55 ± 2.4	278. ± 2.1	3.961 ± .9
850	.667 ± 2.6	.01202 ± 2.9	.0604 ± 2.6	75.0 ± 1.3	19.42 ± 1.2	270. ± 3.4	3.985 ± .4
900	.653 ± 2.3	.01195 ± 2.5	.0306 ± 2.4	75.8 ± .7	12.86 ± .6	269. ± 2.9	4.003 ± .2
1000	.644 ± 1.7	.01358 ± 3.4	.0153 ± 4.4	76.6 ± 2.8	12.14 ± 2.8	306. ± 3.7	4.020 ± 1.0
1100	.677 ± 2.3	.01299 ± 1.3	.0112 ± 2.7	75.4 ± 2.6	7.36 ± 2.6	291. ± 2.0	3.995 ± .9
1200	.716 ± 2.7	.01478 ± 4.3	.0124 ± 5.4	63.2 ± 3.4	2.16 ± 3.3	328. ± 4.6	3.705 ± 1.2
1300	.693 ± 3.3	.01460 ± 2.2	.0060 ± 4.1	66.3 ± 3.9	1.90 ± 3.8	325. ± 2.7	3.783 ± 1.4
1400	1.109 ± 18.9	.01026 ± 25.6	.0069 ± 25.0	47.5 ± 11.5	.68 ± 10.8	208. ± 28.9	3.253 ± 4.9
1500	.687 ± 10.3	.01641 ± 11.0	.0098 ± 10.1	3.6 ± 100.0	.34 ± 7.1	366. ± 12.1	.565 ± 243.1
1700	.556 ± 6.4	.01459 ± 5.7	.0058 ± 8.4	29.9 ± 13.4	.70 ± 6.9	—	2.568 ± 7.2
67035,20:0920 (55.6 mg Anorthosite)							
600	3.291 ± .1	.00723 ± .4	.0114 ± .4	79.4 ± .3	.43 ± .3	67. ± 1.6	4.029 ± .1
700	4.329 ± .4	.00660 ± .2	.0097 ± .3	56.5 ± .3	.90 ± .3	30. ± 2.6	3.478 ± .1
800	5.056 ± 2.0	.00827 ± 2.0	.0096 ± .7	64.0 ± .7	1.86 ± .7	11. ± 37.1	3.676 ± .2
900	4.424 ± .3	.00846 ± .3	.0073 ± .5	68.4 ± .5	3.38 ± .5	35. ± 2.6	3.784 ± .2
1000	3.580 ± .4	.00469 ± .3	.0067 ± .5	74.0 ± .5	5.48 ± .5	35. ± 2.0	3.913 ± .2
1050	2.646 ± .8	.00282 ± .8	.0070 ± .5	76.3 ± .5	3.59 ± .5	34. ± 2.2	3.964 ± .2
1100	2.233 ± 1.3	.00246 ± 1.3	.0075 ± .4	75.5 ± .4	2.66 ± .4	34. ± 2.7	3.947 ± .1
1200	2.298 ± 1.7	.00286 ± 1.6	.0081 ± .5	72.9 ± .5	3.82 ± .5	39. ± 3.2	3.888 ± .2
1300	2.532 ± .7	.00305 ± .7	.0072 ± .5	74.2 ± .5	4.13 ± .5	38. ± 2.0	3.917 ± .2
1400	2.402 ± 5.6	.00210 ± 5.6	.0056 ± .7	78.4 ± .7	2.54 ± .7	28. ± 10.3	4.010 ± .2
1500	1.355 ± 4.5	.00187 ± 4.5	.0058 ± .5	82.7 ± .5	.75 ± .5	34. ± 6.1	4.098 ± .2
1600	.916 ± 8.9	.00177 ± 8.9	.0057 ± .6	81.7 ± .5	.39 ± .5	35. ± 10.9	4.079 ± .2

[a]All errors are relative errors in percent.
[b]^{40}Ar from the decay of ^{40}K only.

Table 5. (*Continued*).

Temp. (°C)	$^{36}Ar/^{38}Ar$[(a)]	$^{38}Ar/^{37}Ar$	$^{39}Ar/^{37}Ar$	$^{40}Ar^*/^{39}Ar^*$[(b)]	$^{39}Ar \times 10^{-8}$ ccSTP/g	Exp. age ($\times 10^6$ yr)	Age ($\times 10^9$ yr)
67455,58:1120 (48.6 mg Anorthosite)							
600	1.196 ± 10.1	.00232 ± 10.1	.0070 ± .4	107.8 ± .4	.61 ± .4	43. ± 13.1	4.543 ± .1
700	1.267 ± 7.8	.00150 ± 7.8	.0067 ± .5	52.5 ± .5	1.17 ± .4	27. ± 10.3	3.357 ± .2
800	1.144 ± 2.5	.00163 ± 2.5	.0052 ± .6	54.4 ± .6	2.68 ± .6	31. ± 3.5	3.412 ± .2
900	.859 ± 1.6	.00166 ± 1.4	.0039 ± .8	69.9 ± .8	3.62 ± .8	33. ± 2.2	3.816 ± .3
1000	.683 ± 1.9	.00166 ± 1.6	.0039 ± 1.2	74.5 ± 1.2	3.19 ± 1.2	35. ± 2.3	3.921 ± .4
1100	.669 ± 2.8	.00172 ± 2.2	.0046 ± .8	71.0 ± .8	2.50 ± .8	36. ± 3.0	3.842 ± .3
1200	.677 ± 2.8	.00183 ± 2.3	.0046 ± .8	66.9 ± .8	2.27 ± .8	38. ± 3.0	3.745 ± .3
1300	.608 ± 2.7	.00184 ± 2.4	.0034 ± 1.2	75.6 ± 1.2	2.38 ± 1.2	—	3.946 ± .4
1400	.594 ± 7.5	.00163 ± 5.7	.0032 ± 1.7	81.3 ± 1.7	.70 ± 1.7	—	4.066 ± .6
1500	.657 ± 20.6	.00149 ± 16.5	.0038 ± 3.7	76.5 ± 3.5	.31 ± 3.4	31. ± 18.9	3.965 ± 1.2
1600	.565 ± 5.5	.00167 ± 4.8	.0035 ± 1.1	80.4 ± 1.2	1.35 ± 1.1	—	4.047 ± .4
1700	.495 ± 141.4	.00281 ± 100.0	.0064 ± 1.9	84.1 ± 26.3	.05 ± .5	—	4.121 ± 8.8
67618,1:1320 (46.3 mg Plagioclase rich crystalline)							
600	2.552 ± 6.8	.00612 ± 4.1	.3595 ± 1.3	11.3 ± 1.2	41.09 ± 1.2	77. ± 9.4	1.355 ± 1.3
700	3.948 ± 2.0	.00624 ± 2.0	.0932 ± .5	33.1 ± .6	21.78 ± .4	39. ± 7.7	2.656 ± .3
750	4.241 ± 1.2	.00693 ± 1.2	.0486 ± .9	46.8 ± 1.1	9.08 ± .8	34. ± 5.9	3.169 ± .5
850	3.926 ± 1.2	.00705 ± 1.0	.0288 ± 1.7	56.2 ± 1.8	17.36 ± 1.6	45. ± 4.4	3.454 ± .7
900	3.180 ± 2.9	.00466 ± 1.9	.0232 ± 1.2	64.7 ± 1.2	7.72 ± 1.1	45. ± 5.3	3.680 ± .5
950	2.566 ± 2.0	.00373 ± 1.7	.0214 ± 1.1	64.5 ± 1.0	5.08 ± .8	46. ± 3.7	3.676 ± .4
1050	2.875 ± 1.6	.00575 ± 1.6	.0227 ± 1.0	56.0 ± 1.2	7.21 ± 1.0	64. ± 3.7	3.450 ± .5
1150	2.187 ± 3.7	.00450 ± 2.6	.0252 ± 1.2	45.6 ± .8	6.42 ± .7	64. ± 4.9	3.128 ± .4
1250	1.540 ± 5.4	.00246 ± 5.4	.0161 ± 1.3	42.2 ± 1.1	6.90 ± 1.0	42. ± 7.7	3.014 ± .5
1350	1.261 ± 6.3	.00222 ± 5.1	.0122 ± 2.3	50.8 ± 2.0	1.56 ± 1.8	41. ± 6.9	3.295 ± .8
1450	31.716 ± 3.4	.00224 ± 3.6	.0126 ± 2.2	55.6 ± 2.0	2.31 ± 1.8	—	3.438 ± .8
1550	.259 ± 103.4	.00377 ± 26.6	.0112 ± 10.1	27.4 ± 18.4	.16 ± 8.8	—	2.392 ± 10.6
1650	.495 ± 141.4	.00693 ± 101.1	.0168 ± 25.7	12.1 ± 88.7	.07 ± 20.8	—	1.427 ± 85.3

68516,2:1520 (48.6 mg Breccia, 80% shocked glass)							
600	2.007 ± 4.1	.06436 ± 19.7	.2095 ± 19.7	86.8 ± 3.5	1.15 ± 3.4	963. ± 20.4	4.144 ± 1.2
700	1.728 ± 5.0	.02554 ± 12.2	.2567 ± 11.2	29.6 ± 1.7	2.62 ± 1.5	414. ± 13.4	2.482 ± 1.0
800	3.354 ± 9.0	.01023 ± 9.2	.1487 ± 2.5	40.3 ± 1.0	4.30 ± .6	91. ± 24.0	2.923 ± .5
900	3.505 ± 1.8	.00903 ± 2.2	.0921 ± 1.9	54.5 ± 1.4	7.82 ± 1.3	75. ± 5.5	3.386 ± .6
950	2.559 ± 6.2	.00426 ± 6.4	.0572 ± 1.7	65.0 ± .9	5.66 ± .9	53. ± 12.1	3.668 ± .4
1000	2.229 ± 5.4	.00427 ± 4.8	.0522 ± 3.6	71.2 ± .8	4.92 ± .5	60. ± 7.5	3.815 ± .3
1050	2.158 ± 4.2	.00381 ± 3.9	.0323 ± 1.6	69.8 ± 1.5	2.93 ± 1.4	54. ± 6.7	3.783 ± .5
1100	1.923 ± 7.3	.00340 ± 6.6	.0286 ± 2.1	64.1 ± 1.9	1.70 ± 1.6	52. ± 10.5	3.645 ± .7
1200	.612 ± 3.6	.00630 ± 2.0	.0236 ± 2.1	61.1 ± 2.0	2.62 ± 2.0	—	3.567 ± .8
1300	.151 ± 6.0	.00502 ± 2.6	.0208 ± 1.6	70.5 ± 1.4	6.53 ± 1.4	—	3.799 ± .5
1350	.020 ± 100.0	.00492 ± 2.1	.0180 ± 2.3	69.9 ± 2.4	2.32 ± 2.3	—	3.785 ± .9
1400	.101 ± 8.2	.00427 ± 1.3	.0263 ± .4	66.5 ± .4	19.65 ± .4	—	3.704 ± .1
1450	.058 ± 23.5	.00357 ± 2.1	.0220 ± .5	70.8 ± .6	11.68 ± .5	—	3.807 ± .2
1500	.027 ± 33.6	.00417 ± 1.0	.0145 ± .7	70.8 ± .8	4.90 ± .6	—	3.805 ± .3
1550	.018 ± 50.6	.00417 ± 1.0	.0145 ± .7	70.4 ± .9	4.90 ± .6	—	3.797 ± .3
1600	.056 ± 10.9	.00395 ± 1.2	.0106 ± 1.2	72.8 ± 1.2	5.79 ± 1.2	—	3.853 ± .4
1650	.858 ± 34.6	.00253 ± 9.7	.0076 ± 6.6	76.0 ± 9.0	.41 ± 6.4	51. ± 13.3	3.922 ± 3.1
1750	.695 ± 106.0	.00384 ± 35.9	.0143 ± 12.2	155.7 ± 23.5	.14 ± 9.5	80. ± 43.7	5.150 ± 6.3
70255,36:4121 (41.8 mg Fine-grained basalt)							
500	1.517 ± 26.7	.01416 ± 102.5	1.4238 ± 100.0	45.0 ± 1.4	3.80 ± 1.3	260. ± 104.9	3.177 ± .6
600	.665 ± 36.2	.01521 ± 25.3	.9327 ± 22.9	49.8 ± 5.9	4.48 ± 5.7	341. ± 26.6	3.335 ± 2.4
700	.748 ± 8.9	.01222 ± 8.3	.2868 ± 6.3	61.6 ± .5	7.34 ± .5	269. ± 9.0	3.672 ± .2
750	.922 ± 4.8	.01039 ± 2.9	.0947 ± 3.1	66.3 ± 1.2	3.43 ± 1.0	220. ± 3.4	3.792 ± .4
850	.715 ± 2.8	.01340 ± 15.0	.0438 ± 14.9	68.1 ± 1.4	5.02 ± 1.4	297. ± 15.1	3.834 ± .5
900	.640 ± 2.9	.00998 ± 2.8	.0227 ± 3.2	70.4 ± 2.1	4.46 ± 2.1	225. ± 3.2	3.890 ± .8
1000	.601 ± 7.1	.01067 ± 5.7	.0214 ± 4.2	67.2 ± 1.1	4.72 ± .9	—	3.814 ± .4
1100	.600 ± 7.5	.01375 ± 12.3	.0273 ± 12.4	62.3 ± 3.1	2.79 ± 3.1	—	3.691 ± 1.2
1250	.583 ± 5.1	.01115 ± 2.9	.0033 ± 3.4	63.1 ± 2.1	5.11 ± 2.0	—	3.711 ± .8
1450	.637 ± 5.2	.01136 ± 5.3	.0020 ± 3.6	57.1 ± 6.3	.28 ± 3.4	256. ± 6.2	3.551 ± 2.4
1550	.640 ± 4.7	.01250 ± 4.8	.0023 ± 14.9	46.8 ± 33.9	.06 ± 14.3	282. ± 5.2	3.238 ± 14.4
1650	.977 ± 23.7	.03151 ± 68.7	.0053 ± 120.2	819.6 ± 100.0	.01 ± 100.0	660. ± 69.9	8.284 ± 16.6
1750	1.481 ± 103.1	.01571 ± 56.0	.0040 ± 111.8	543.0 ± 100.0	.01 ± 100.0	291. ± 72.0	7.520 ± 18.3

[a] All errors are relative errors in percent.
[b] ^{40}Ar from the decay of ^{40}K only.

Table 5. (*Continued*).

Temp. (°C)	$^{36}Ar/^{38}Ar^{(a)}$	$^{38}Ar/^{37}Ar$	$^{39}Ar/^{37}Ar$	$^{40}Ar^*/^{39}Ar^{*(b)}$	$^{39}Ar \times 10^{-8}$ ccSTP/g	Exp. age ($\times 10^6$ yr)	Age ($\times 10^9$ yr)
14167,6,3 (58.3 mg)							
500	1.094 ± 63.0	.02346 ± 47.0	.5082 ± 3.1	49.9 ± 9.1	1.50 ± 2.2	130. ± 60.0	1.628 ± 7.7
600	1.380 ± 9.7	.02189 ± 7.9	.2885 ± .6	142.7 ± 1.0	5.43 ± .6	113. ± 11.2	3.004 ± .5
700	1.924 ± 2.5	.02434 ± 2.2	.2484 ± .2	229.5 ± .3	14.76 ± .2	109. ± 4.6	3.749 ± .1
800	2.481 ± 2.0	.01711 ± 1.8	.1602 ± .2	247.3 ± .3	16.73 ± .2	64. ± 4.5	3.871 ± .1
900	.823 ± 7.2	.00906 ± 4.7	.1073 ± .4	238.1 ± .5	8.26 ± .4	53. ± 6.3	3.810 ± .2
1000	1.168 ± 8.8	.00843 ± 6.8	.0866 ± .7	188.0 ± 1.0	4.93 ± .7	46. ± 9.2	3.429 ± .4
1200	.773 ± 3.8	.00575 ± 2.3	.0249 ± .6	176.3 ± .8	6.00 ± .6	34. ± 4.0	3.329 ± .3
1500	.964 ± 13.3	.00407 ± 9.3	.0181 ± 2.1	224.1 ± 2.9	1.55 ± 2.1	23. ± 11.9	3.711 ± 1.1
14167,6,7 (60.1 mg)							
500	2.114 ± 110.2	.00526 ± 100.0	.3394 ± 1.7	30.3 ± 10.5	1.99 ± 1.6	22. ± 168.1	1.143 ± 12.7
600	2.225 ± 23.4	.00566 ± 21.4	.2509 ± .5	155.6 ± .8	6.38 ± .5	23. ± 37.3	3.135 ± .3
700	2.427 ± 6.3	.00862 ± 5.9	.2703 ± .2	237.7 ± .2	16.52 ± .2	33. ± 11.3	3.807 ± .1
800	2.146 ± 4.0	.00688 ± 3.7	.1832 ± .1	246.4 ± .2	22.49 ± .1	29. ± 6.8	3.866 ± .1
900	.758 ± 6.9	.00710 ± 4.3	.1231 ± .3	245.7 ± .3	12.37 ± .3	42. ± 5.9	3.861 ± .1
1000	.603 ± 13.3	.00923 ± 7.0	.0857 ± .8	238.5 ± 1.0	4.11 ± .8	—	3.812 ± .4
1500	.619 ± 5.0	.00648 ± 2.7	.0445 ± .4	151.8 ± .7	7.93 ± .4	—	3.098 ± .3
14192,1,1 (9.8 mg)							
500	2.335 ± 107.6	.41584 ± 107.6	7.9153 ± 42.0	36.1 ± 29.9	4.35 ± 5.3	1633. ± 183.8	1.299 ± 31.6
700	1.328 ± 7.4	.01912 ± 6.0	.0524 ± 2.2	60.1 ± 7.8	10.36 ± 2.2	100. ± 8.6	1.839 ± 5.8
800	1.759 ± 32.6	.00855 ± 28.5	.1606 ± 1.6	188.6 ± 2.2	14.96 ± 1.5	40. ± 43.2	3.434 ± .9
900	.937 ± 28.6	.00743 ± 19.7	.1603 ± .9	225.8 ± 1.2	24.76 ± .9	43. ± 24.5	3.723 ± .5
1000	19.047 ± 42.7	.00571 ± 42.7	.2026 ± 1.2	369.3 ± 1.4	18.85 ± 1.2	—	4.546 ± .4
1100	.535 ± 38.6	.02683 ± 18.5	.1845 ± 2.8	178.1 ± 4.1	8.44 ± 2.7	—	3.344 ± 1.7
1200	.340 ± 8.8	.06143 ± 8.8	.2004 ± 2.8	107.6 ± 5.8	8.44 ± 2.7	—	2.593 ± 3.1
1500	.251 ± 24.8	.11012 ± 6.2	.1931 ± 3.6	71.5 ± 10.6	6.48 ± 3.5	—	2.049 ± 7.1
14257,12,2 (14.7 mg)							
700	3.371 ± 1.0	.37457 ± 1.2	.4436 ± 1.1	86.0 ± 2.6	16.17 ± .8	392. ± 3.7	1.289 ± 2.7
800	2.405 ± .9	.31851 ± 1.0	.3216 ± 1.0	160.3 ± 1.5	15.41 ± .8	496. ± 2.9	1.960 ± 1.0
900	1.656 ± .9	.28947 ± .9	.2361 ± 1.1	325.9 ± 1.2	13.04 ± 1.0	565. ± 2.7	2.924 ± .6
1000	1.733 ± 1.5	.31847 ± 1.6	.1993 ± 2.3	391.9 ± 2.4	6.13 ± 2.1	609. ± 3.2	3.203 ± 1.0
1200	3.122 ± 2.3	.44781 ± 3.0	.1633 ± 6.4	150.1 ± 10.7	2.13 ± 6.1	527. ± 6.3	1.881 ± 7.8
1500	1.010 ± .7	.27364 ± .5	.0484 ± 2.8	476.2 ± 3.0	4.75 ± 2.7	628. ± 2.5	3.508 ± 1.2

61503,1,11:7213 (17.2 mg Medium-grained poikilitic breccia)							
600	33.064 ± 100.1	.00007 ± 102.3	.1871 ± 21.4	11.2 ± .2	67.06 ± .1	—	2.250 ± .1
700	44.477 ± 100.0	.00010 ± 102.5	2474 ± 22.6	29.8 ± .3	62.90 ± .2	—	3.700 ± .1
800	45.595 ± 100.0	.00009 ± 102.4	.2963 ± 22.2	36.5 ± .2	83.94 ± .1	—	4.033 ± .1
900	7.109 ± 14.5	.00046 ± 25.7	.1061 ± 21.2	35.6 ± .3	78.80 ± .2	—	3.990 ± .1
1000	2.408 ± 2.6	.00116 ± 20.2	.0275 ± 20.1	36.8 ± .5	46.94 ± .5	33. ± 20.5	4.044 ± .2
1100	2.331 ± 2.2	.00132 ± 20.2	.0165 ± 20.1	37.4 ± .8	31.01 ± .8	39. ± 20.4	4.073 ± .3
1200	3.276 ± 2.9	.00197 ± 20.9	.0147 ± 20.8	33.1 ± 1.1	7.45 ± 1.0	40. ± 21.9	3.871 ± .4
1300	2.065 ± 4.6	.00134 ± 20.8	.0090 ± 20.6	33.9 ± 2.1	5.23 ± 1.6	43. ± 21.4	3.909 ± .7
1400	.615 ± 10.0	.00093 ± 20.6	.0088 ± 20.4	33.0 ± 2.5	7.42 ± 1.6	—	3.868 ± .9
1500	.530 ± 16.4	.00093 ± 20.2	.0071 ± 20.2	34.0 ± 7.4	2.44 ± 2.1	—	3.916 ± 2.6
1650	.959 ± 12.1	.00102 ± 23.1	.0072 ± 21.9	45.2 ± 12.3	2.17 ± 2.3	43. ± 23.8	4.392 ± 3.8
1750	6.583 ± 35.6	.00462 ± 181.7	.0154 ± 178.7	237.6 ± 47.1	.22 ± 13.0	—	7.365 ± 8.8
72703,9,5:0219 (8.6 mg Plagioclase phenocrysts in dark matrix)							
600	4.041 ± 10.3	.00745 ± 57.9	.0049 ± 57.6	88.0 ± 7.7	.34 ± 7.5	47. ± 69.9	4.249 ± 2.5
700	5.371 ± 5.1	.02385 ± 49.1	.0191 ± 48.9	76.1 ± 1.9	.79 ± 1.7	—	4.006 ± .6
800	4.864 ± 1.6	.02137 ± 30.7	.0247 ± 30.7	72.4 ± .8	3.78 ± .8	50. ± 35.2	3.923 ± .3
900	3.486 ± 1.1	.01517 ± 13.8	.0322 ± 13.7	76.9 ± .5	10.79 ± .5	135. ± 14.1	4.024 ± .2
1000	1.780 ± 1.6	.00670 ± 7.3	.0270 ± 7.2	75.6 ± .2	17.67 ± .2	115. ± 7.5	3.994 ± .1
1100	2.049 ± .8	.00706 ± 4.1	.0166 ± 4.1	76.7 ± .3	19.08 ± .3	112. ± 4.4	4.019 ± .1
1200	2.700 ± .9	.00857 ± 7.6	.0163 ± 7.6	73.8 ± .7	7.55 ± .7	109. ± 7.8	3.954 ± .3
1300	1.326 ± 2.0	.00562 ± 5.7	.0161 ± 5.6	70.1 ± .6	12.29 ± .5	108. ± 6.0	3.871 ± .2
1400	.981 ± 4.9	.00424 ± 4.3	.0175 ± 2.7	72.0 ± .6	11.31 ± .5	89. ± 5.2	3.915 ± .2
1500	.763 ± 19.4	.00387 ± 22.1	.0128 ± 20.9	71.2 ± 1.9	2.23 ± 1.1	85. ± 22.7	3.897 ± .7
1600	.741 ± 32.6	.00416 ± 34.8	.0122 ± 16.0	78.9 ± 1.8	3.59 ± 1.6	92. ± 39.3	4.065 ± .6
1700	1.048 ± 8.2	.00406 ± 17.4	.0102 ± 17.5	93.7 ± 2.2	2.75 ± 1.1	84. ± 17.6	4.354 ± .7

[a] All errors are relative errors in percent.

[b] ^{40}Ar from the decay of ^{40}K only.

Table 5. (*Continued*).

Temp. (°C)	$^{36}Ar/^{38}Ar^{(a)}$	$^{38}Ar/^{37}Ar$	$^{39}Ar/^{37}Ar$	$^{40}Ar^*/^{39}Ar^{*(b)}$	$^{39}Ar \times 10^{-8}$ ccSTP/g	Exp. age ($\times 10^6$ yr)	Age ($\times 10^9$ yr)
73263,1,1:1720 (44.4 mg Recrystallized anorthosite)							
600	4.287 ± 11.8	.00608 ± 11.5	.0148 ± 4.8	280.8 ± 4.8	.29 ± 4.7	29. ± 58.9	6.189 ± 1.1
700	3.369 ± 4.7	.00657 ± 4.7	.0203 ± 2.0	89.4 ± 1.5	.92 ± 1.5	58. ± 12.6	4.176 ± .5
800	2.538 ± 1.4	.00661 ± 1.4	.0145 ± 1.3	71.7 ± 1.2	2.28 ± 1.2	83. ± 3.0	3.811 ± .4
900	1.673 ± .7	.00563 ± .6	.0135 ± .6	82.4 ± .6	5.82 ± .5	93. ± 1.7	4.041 ± .2
950	1.044 ± 2.1	.00476 ± 1.6	.0135 ± .7	85.4 ± .7	4.97 ± .6	92. ± 2.5	4.099 ± .2
1000	.982 ± 1.6	.00471 ± 1.2	.0134 ± .9	87.2 ± .8	3.44 ± .8	92. ± 2.1	4.136 ± .3
1100	1.139 ± 1.5	.00531 ± 1.5	.0132 ± 1.1	87.5 ± .9	4.88 ± .9	100. ± 2.4	4.141 ± .3
1150	.982 ± 1.7	.00486 ± 1.3	.0130 ± .7	91.2 ± .5	3.00 ± .5	95. ± 2.2	4.210 ± .2
1250	.771 ± .9	.00475 ± .5	.0121 ± .7	91.3 ± .6	11.11 ± .5	97. ± 1.6	4.213 ± .2
1300	.728 ± 4.1	.00468 ± 1.4	.0114 ± 1.2	92.6 ± 1.2	2.34 ± 1.2	97. ± 2.3	4.235 ± .4
1400	.775 ± 4.3	.00469 ± 1.7	.0110 ± .9	90.7 ± .8	2.03 ± .7	96. ± 2.5	4.202 ± .2
1500	.743 ± 1.9	.00457 ± 1.2	.0114 ± 1.0	92.8 ± .9	3.00 ± .9	94. ± 2.0	4.239 ± .3
1600	.812 ± .5	.00481 ± .7	.0114 ± .6	95.8 ± .4	6.65 ± .3	98. ± 1.7	4.294 ± .1
1650	.734 ± 5.8	.00484 ± 2.1	.0092 ± 2.3	106.2 ± 2.2	1.27 ± 2.2	100. ± 3.0	4.469 ± .7
1750	.940 ± 4.5	.01562 ± 3.2	.0355 ± 1.7	109.5 ± 1.5	.99 ± 1.4	308. ± 4.2	4.521 ± .5
1800	1.122 ± 62.2	.01128 ± 11.7	.0373 ± 4.1	153.2 ± 3.7	.39 ± 3.5	214. ± 22.0	5.104 ± 1.0
73263,1,3:1820 (31.2 mg Igneous anorthosite)							
600	5.657 ± 1.3	.05342 ± 1.4	.0067 ± 10.0	778.6 ± 10.0	.19 ± 10.0	—	8.059 ± 1.7
700	5.265 ± 1.7	.05111 ± 2.2	.0224 ± 3.9	133.4 ± 4.3	.55 ± 3.6	19. ± 104.8	4.856 ± 1.2
800	4.910 ± 1.0	.03474 ± 1.0	.0064 ± 2.7	67.9 ± 8.7	.74 ± 2.7	68. ± 12.8	3.718 ± 3.2
900	4.247 ± .7	.01855 ± .9	.0029 ± 2.4	58.8 ± 9.8	.95 ± 2.3	92. ± 3.7	3.486 ± 3.9
1000	3.171 ± .8	.01082 ± .7	.0018 ± 2.5	70.6 ± 5.9	1.05 ± 2.5	105. ± 2.3	3.780 ± 2.1
1100	3.763 ± 2.4	.01025 ± 2.4	.0014 ± 2.8	63.5 ± 9.2	1.01 ± 2.8	73. ± 8.1	3.608 ± 3.5
1200	1.783 ± .7	.00645 ± .7	.0013 ± 2.6	72.2 ± 3.3	1.53 ± 2.6	103. ± 1.8	3.817 ± 1.2
1300	1.315 ± 1.1	.00577 ± 1.4	.0014 ± 4.3	74.0 ± 4.4	.54 ± 4.1	104. ± 2.2	3.858 ± 1.6
1400	.740 ± 3.3	.00470 ± 2.4	.0012 ± 8.1	56.3 ± 8.5	.25 ± 8.0	97. ± 3.1	3.418 ± 3.4
1500	.688 ± 4.0	.00479 ± 3.9	.0010 ± 4.1	81.9 ± 4.1	.71 ± 4.0	100. ± 4.7	4.024 ± 1.4
1600	.695 ± 2.6	.00480 ± 1.5	.0011 ± 6.8	67.4 ± 8.3	.31 ± 6.8	100. ± 2.3	3.706 ± 3.1
1750	.728 ± 2.8	.00492 ± 1.7	.0013 ± 6.6	103.9 ± 7.2	.31 ± 6.6	102. ± 2.5	4.425 ± 2.2

73263,1,6:1920 (35.4 mg Recrystallized anorthosite)							
600	5.050 ± 2.5	.03588 ± 2.5	.0171 ± 5.2	218.3 ± 5.2	.32 ± 5.2	48. ± 44.6	5.723 ± 1.3
700	4.831 ± .9	.04525 ± 1.0	.0177 ± 16.3	120.4 ± 16.3	.81 ± 16.3	105. ± 9.1	4.674 ± 4.8
800	4.602 ± 1.0	.04986 ± 2.1	.0171 ± 2.8	89.8 ± 2.1	2.39 ± 2.1	167. ± 7.5	4.175 ± .7
900	3.438 ± 1.2	.02901 ± 3	.0154 ± .8	79.4 ± .8	6.40 ± .8	248. ± 2.6	3.968 ± .3
950	1.818 ± .5	.01557 ± 3	.0155 ± .6	79.8 ± .6	5.88 ± .6	246. ± 1.6	3.978 ± .2
1000	1.785 ± 2.0	.01487 ± 1.1	.0149 ± 1.1	82.3 ± 1.1	4.44 ± 1.1	237. ± 2.4	4.028 ± .4
1100	2.586 ± 1.7	.02012 ± 1.9	.0151 ± 1.0	82.7 ± .4	5.24 ± .4	249. ± 3.7	4.037 ± .1
1200	1.754 ± 1.1	.01575 ± 9	.0143 ± 1.9	91.1 ± 1.9	5.14 ± 1.9	254. ± 2.0	4.198 ± .6
1300	1.287 ± .6	.01419 ± 6	.0143 ± 1.0	90.0 ± 1.1	7.01 ± 1.0	258. ± 1.6	4.178 ± .4
1350	1.164 ± 1.5	.01468 ± 9	.0121 ± 1.1	90.0 ± 1.1	1.73 ± 1.0	275. ± 1.9	4.178 ± .4
1450	.699 ± .6	.01251 ± 6	.0129 ± .7	94.6 ± .7	13.76 ± .6	261. ± 1.6	4.262 ± .2
1500	.732 ± 2.7	.01231 ± 1.8	.0131 ± 1.7	97.5 ± 2.1	1.08 ± 1.5	255. ± 2.5	4.313 ± .7
1600	.629 ± 4.2	.01176 ± 1.0	.0130 ± .9	100.2 ± .9	2.10 ± .8	249. ± 1.9	4.359 ± .3
1700	.764 ± 2.4	.01207 ± 1.6	.0115 ± 1.7	103.0 ± 1.8	1.04 ± 1.6	248. ± 2.4	4.405 ± .6
1750	.580 ± 121.1	.01642 ± 100.7	.0159 ± 43.4	224.0 ± 41.9	.08 ± 41.6	—	5.770 ± 10.0
73263,1,9:0419 (11.2 mg Plagioclase and olivine phenocrysts in black matrix)							
600	1.333 ± 7.3	.01179 ± 52.9	.0920 ± 52.9	37.1 ± 1.3	2.16 ± 1.3	227. ± 52.9	2.874 ± .6
700	2.893 ± 3.5	.01865 ± 134.0	.2245 ± 134.0	64.6 ± 1.0	3.21 ± .9	219. ± 134.0	3.736 ± .4
800	2.970 ± 4.8	.02226 ± 30.1	.1911 ± 80.0	75.0 ± .6	5.16 ± .6	254. ± 80.7	3.981 ± .2
900	2.293 ± 2.8	.01324 ± 23.0	.0794 ± 22.9	81.0 ± .7	9.37 ± .6	194. ± 23.2	4.110 ± .2
1000	1.134 ± 1.4	.01215 ± 10.6	.0579 ± 10.5	78.2 ± .4	14.65 ± .4	245. ± 10.7	4.051 ± .1
1100	.799 ± .2	.00916 ± 5.3	.0398 ± 5.4	77.3 ± .4	14.09 ± .4	200. ± 5.5	4.033 ± .1
1200	.996 ± 2.2	.00989 ± 10.5	.0330 ± 10.4	74.2 ± .5	5.78 ± .5	206. ± 10.6	3.963 ± .2
1300	.649 ± 3.1	.01030 ± 4.0	.0301 ± 3.2	74.4 ± .2	24.07 ± .2	232. ± 4.4	3.967 ± .1
1350	.665 ± 1.6	.00984 ± 7.0	.0281 ± 6.9	73.5 ± .4	8.64 ± .4	221. ± 7.1	3.948 ± .1
1450	.700 ± 2.9	.00954 ± 12.5	.0262 ± 12.4	71.3 ± .8	4.57 ± .7	212. ± 12.6	3.899 ± .3
1550	.712 ± 3.2	.00794 ± 11.0	.0183 ± 11.0	75.8 ± 1.3	2.74 ± 1.1	176. ± 11.1	3.999 ± .4
1650	.899 ± 4.7	.00928 ± 22.4	.0198 ± 22.5	82.7 ± 1.4	1.36 ± .7	198. ± 22.5	4.145 ± .5

(a) All errors are relative errors in percent.
(b) ^{40}Ar from the decay of ^{40}K only.

Table 5. (*Continued*).

Temp. (°C)	$^{36}Ar/^{38}Ar$[(a)]	$^{38}Ar/^{37}Ar$	$^{39}Ar/^{37}Ar$	$^{40}Ar^*/^{39}Ar^*$[(b)]	$^{39}Ar \times 10^{-8}$ ccSTP/g	Exp. age ($\times 10^6$ yr)	Age ($\times 10^9$ yr)
74243,2,1:1915 (24.2 mg Ilmenite variolite)							
600	12.772 ± 101.0	.00036 ± 101.3	.3187 ± 16.5	65.5 ± 3.2	14.05 ± 3.1	—	3.708 ± 1.2
700	1.606 ± 15.4	.00172 ± 17.5	.1288 ± 11.9	75.1 ± 3.5	8.60 ± 3.4	30. ± 22.4	3.932 ± 1.2
800	1.391 ± 12.7	.00254 ± 18.8	.0615 ± 14.7	71.1 ± 3.5	8.64 ± 3.4	47. ± 22.0	3.844 ± 1.2
900	1.331 ± 11.8	.00276 ± 18.4	.0419 ± 15.2	71.7 ± 5.2	5.83 ± 5.1	51. ± 21.0	3.857 ± 1.8
1000	.615 ± 22.0	.00251 ± 11.9	.0273 ± 10.4	69.6 ± 3.4	4.85 ± 3.4	—	3.807 ± 1.2
1100	.589 ± 20.8	.00253 ± 16.8	.0252 ± 14.3	62.3 ± 6.0	2.73 ± 5.9	—	3.629 ± 2.3
1200	.383 ± 28.1	.00271 ± 7.9	.0066 ± 8.2	58.9 ± 3.1	3.58 ± 3.0	—	3.539 ± 1.2
1300	.621 ± 8.0	.00279 ± 7.4	.0031 ± 8.2	60.5 ± 3.9	4.30 ± 3.9	61. ± 7.6	3.582 ± 1.5
1450	.504 ± 20.8	.00276 ± 15.8	.0023 ± 15.3	16.9 ± 24.8	.38 ± 8.7	—	1.794 ± 19.0
1700	2.257 ± 32.2	.00426 ± 59.1	.0032 ± 65.9	479.4 ± 41.8	.07 ± 41.5	61. ± 72.9	7.219 ± 8.0
76503,6,5:0319 (28.8 mg Recrystallized breccia)							
600	2.329 ± 7.4	.02446 ± 72.7	.0321 ± 72.7	64.9 ± 1.6	.41 ± 1.5	354. ± 73.0	3.744 ± .6
700	2.573 ± 4.2	.01099 ± 12.6	.0169 ± 12.4	70.8 ± .6	.91 ± .5	146. ± 13.6	3.886 ± .2
800	2.099 ± 1.7	.00936 ± 5.6	.0147 ± 5.6	73.6 ± .4	2.34 ± .3	146. ± 5.9	3.949 ± .1
900	1.728 ± 1.0	.00861 ± 2.8	.0146 ± 2.7	75.2 ± .2	5.77 ± .2	149. ± 3.2	3.987 ± .1
1000	1.085 ± 1.3	.00736 ± 1.5	.0143 ± 1.2	75.4 ± .3	11.20 ± .2	150. ± 2.2	3.990 ± .1
1150	1.063 ± 1.5	.00744 ± 1.1	.0133 ± .9	75.6 ± .3	8.51 ± .3	153. ± 1.8	3.994 ± .1
1250	.910 ± 2.4	.00734 ± 2.3	.0126 ± 2.3	78.6 ± .3	4.33 ± .3	156. ± 2.7	4.060 ± .1
1350	.725 ± 2.1	.00728 ± 1.9	.0091 ± 1.9	79.3 ± .6	4.44 ± .5	161. ± 2.4	4.074 ± .2
1450	.644 ± 1.6	.00765 ± 1.6	.0116 ± 1.5	82.2 ± .5	10.63 ± .4	172. ± 2.1	4.134 ± .2
1500	.513 ± 60.1	.00880 ± 35.6	.0126 ± 35.7	84.4 ± 16.7	.25 ± 1.8	—	4.178 ± 5.5
1600	.638 ± 19.0	.00805 ± 11.8	.0099 ± 11.9	92.0 ± 7.0	.55 ± .8	182. ± 12.2	4.323 ± 2.2
1700	1.354 ± 26.8	.06080 ± 350.0	.0843 ± 350.0	180.5 ± 19.0	.21 ± 2.7	1163. ± 350.1	5.499 ± 4.7

68815,141,B:5422 (38.6 mg Black glass of breccia)							
500	1.622 ± 3.6	.12595 ± 26.7	.4237 ± 26.7	62.7 ± 2.1	1.96 ± 1.7	2169. ± 26.8	3.643 ± .8
600	.974 ± 5.6	.04486 ± 13.0	.1459 ± 12.7	23.5 ± 3.1	1.83 ± 3.0	907. ± 13.4	2.204 ± 1.9
700	.723 ± 3.4	.02310 ± 3.8	.0870 ± 3.6	16.1 ± 1.7	4.17 ± 1.6	494. ± 4.2	1.740 ± 1.3
800	.515 ± 1.1	.01140 ± 1.8	.0396 ± 2.0	14.0 ± .9	6.79 ± .8	—	1.582 ± .7
900	.616 ± 1.4	.00644 ± 1.5	.0241 ± 1.2	15.2 ± .9	7.67 ± .7	—	1.674 ± .8
1000	.801 ± 2.1	.00899 ± 2.8	.0313 ± 2.7	21.5 ± 1.2	5.95 ± 1.1	189. ± 3.2	2.085 ± .8
1100	.893 ± 3.7	.00928 ± 3.1	.0370 ± 2.6	23.0 ± 1.0	4.68 ± 1.0	191. ± 3.7	2.175 ± .6
1200	.715 ± 4.1	.00869 ± 5.3	.0374 ± 5.2	19.9 ± 1.0	3.41 ± 1.0	186. ± 5.6	1.990 ± .7
1300	.623 ± 2.4	.00761 ± 2.4	.0242 ± 1.7	33.5 ± 1.5	8.79 ± 1.1	166. ± 3.0	2.687 ± .7
1400	.622 ± 2.8	.00709 ± 1.6	.0215 ± 1.6	39.2 ± 1.5	4.57 ± 1.5	155. ± 2.3	2.914 ± .7
1500	.556 ± 2.6	.00708 ± 3.3	.0351 ± 2.9	29.7 ± .3	18.55 ± .3	—	2.518 ± .2
1600	.598 ± 1.1	.00892 ± .9	.0202 ± .9	52.8 ± .6	21.86 ± .5	—	3.371 ± .2
1700	.341 ± 2.3	.00998 ± 1.8	.0206 ± 1.8	65.6 ± .9	12.61 ± .9	—	3.717 ± .3
1750	4.964 ± 100.5	.02141 ± 36.4	.0353 ± 40.5	81.5 ± 20.4	.11 ± 20.4	38. ± 999.9	4.073 ± 6.9
68815,133,B:6522 (52.9 mg Black glass of breccia)							
500	.586 ± 8.7	.02585 ± 48.1	.1550 ± 47.6	51.5 ± 1.7	1.53 ± 1.7	—	3.328 ± .7
600	.041 ± 100.3	.02328 ± 53.3	.1893 ± 52.7	17.7 ± 1.6	1.69 ± 1.5	—	1.847 ± 1.2
700	.023 ± 100.1	.01476 ± 19.4	.1053 ± 18.9	11.1 ± 1.9	2.62 ± 1.6	—	1.344 ± 1.9
800	.042 ± 22.8	.01591 ± 8.5	.1104 ± 8.4	9.8 ± 1.2	6.21 ± 1.1	—	1.229 ± 1.3
900	.009 ± 100.0	.01060 ± 5.9	.0759 ± 5.6	12.4 ± .8	6.40 ± .7	—	1.457 ± .8
1000	.017 ± 67.2	.01451 ± 17.8	.0878 ± 17.8	18.1 ± 1.8	4.65 ± 1.6	—	1.874 ± 1.3
1100	.016 ± 100.1	.00996 ± 5.8	.0499 ± 4.4	24.5 ± .8	5.40 ± .8	—	2.252 ± .5
1200	.019 ± 14.3	.00921 ± 5.3	.0306 ± 5.0	34.3 ± .5	11.49 ± .4	—	2.716 ± .2
1250	.023 ± 13.6	.00905 ± .9	.0260 ± .9	54.1 ± .5	15.34 ± .5	—	3.404 ± .2
1300	.026 ± 6.6	.00723 ± 1.7	.0226 ± 1.6	61.8 ± 1.0	16.21 ± 1.0	—	3.616 ± .4
1350	.025 ± 13.4	.00813 ± 3.2	.0350 ± 3.0	47.2 ± .9	11.33 ± .8	—	3.194 ± .4
1400	.011 ± 100.0	.00606 ± 7.4	.0184 ± 7.4	63.2 ± 2.1	2.43 ± 2.1	—	3.654 ± .8
1500	.025 ± 8.0	.00869 ± 2.8	.0281 ± 2.7	60.6 ± .6	14.69 ± .5	—	3.586 ± .2
1550	.023 ± 15.9	.00939 ± 2.6	.0249 ± 2.3	67.1 ± 1.5	12.52 ± 1.4	—	3.750 ± .6
1600	.018 ± 40.0	.00742 ± 8.9	.0157 ± 8.5	69.9 ± 1.4	2.63 ± 1.4	—	3.818 ± .5
1650	.027 ± 100.0	.00563 ± 8.8	.0131 ± 8.4	69.4 ± 2.5	.74 ± 1.2	—	3.805 ± .9
1700	.117 ± 100.7	.00479 ± 33.4	.0155 ± 31.5	77.5 ± 10.2	.23 ± 3.9	—	3.987 ± 3.5
1750	.125 ± 100.8	.00412 ± 31.2	.0092 ± 29.2	42.5 ± 44.4	.15 ± 6.0	—	3.032 ± 20.1

[a] All errors are relative errors in percent.

[b] ^{40}Ar from the decay of ^{40}K only.

Table 5. (*Continued*).

Temp. (°C)	$^{36}Ar/^{38}Ar^{(a)}$	$^{38}Ar/^{37}Ar$	$^{39}Ar/^{37}Ar$	$^{40}Ar^*/^{39}Ar^{*(b)}$	$^{39}Ar \times 10^{-8}$ ccSTP/g	Exp. age ($\times 10^6$ yr)	Age ($\times 10^9$ yr)
68815,133,C:6722 (44.7 mg White clast of breccia)							
500	2.183 ± 72.1	.00213 ± 122.2	.0048 ± 111.1	396.9 ± 48.5	.07 ± 48.5	31. ± 155.5	6.885 ± 9.7
600	—	—	.0070 ± 316.2	5.7 ± 100.0	.03 ± 100.0	—	.808 ± 169.9
700	—	—	.0263 ± 301.2	5.1 ± 29.1	.12 ± 26.6	—	.745 ± 53.6
800	—	—	.0144 ± 101.3	6.7 ± 17.1	.20 ± 16.2	—	.925 ± 25.4
900	.314 ± 104.5	.00106 ± 52.5	.0096 ± 44.1	3.0 ± 13.0	.31 ± 10.4	—	.478 ± 37.3
1000	.163 ± 101.2	.00158 ± 36.9	.0095 ± 34.3	6.2 ± 8.8	.40 ± 8.2	—	.869 ± 13.8
1100	.082 ± 100.3	.00190 ± 21.5	.0070 ± 21.1	33.7 ± 6.8	.49 ± 6.7	—	2.699 ± 3.4
1200	.035 ± 83.6	.00213 ± 8.6	.0081 ± 8.5	57.4 ± 2.4	1.40 ± 2.4	—	3.507 ± .9
1300	.062 ± 16.3	.00184 ± 3.7	.0082 ± 2.7	68.0 ± 1.4	4.88 ± 1.4	—	3.783 ± .5
1350	.050 ± 100.1	.00206 ± 14.1	.0078 ± 13.8	67.4 ± 4.1	.83 ± 4.0	—	3.769 ± 1.5
1450	.069 ± 59.3	.00413 ± 22.3	.0137 ± 22.2	64.2 ± 4.2	.88 ± 3.8	—	3.688 ± 1.6
1550	.059 ± 33.8	.00259 ± 6.9	.0090 ± 7.1	68.5 ± 2.5	1.89 ± 2.3	—	3.793 ± .9
1650	.073 ± 16.2	.00248 ± 3.9	.0091 ± 4.0	69.9 ± 1.5	3.41 ± 1.3	—	3.826 ± .5
1750	.063 ± 7.1	.00245 ± 2.4	.0088 ± 2.5	69.2 ± .8	19.26 ± .8	—	3.810 ± .3
1800	.466 ± 18.7	.00408 ± 23.5	.0061 ± 24.7	64.7 ± 27.9	.37 ± 8.9	—	3.701 ± 10.3

68815,67,D:6222 (21.3 mg White clast of breccia)							
500	5.389 ± 7.9	.01194 ± 111.3	.0169 ± 112.1	1122.4 ± 14.5	.45 ± 14.5	—	8.801 ± 2.3
600	—	—	.0102 ± 157.4	5.8 ± 36.9	.20 ± 33.0	—	.823 ± 61.5
700	—	—	.0410 ± 286.1	4.6 ± 18.1	.43 ± 15.3	—	.678 ± 36.6
800	—	—	—	22.1 ± 28.1	.23 ± 27.9	—	2.120 ± 18.2
900	.917 ± 133.3	.00084 ± 137.3	.0056 ± 109.2	9.8 ± 29.6	.16 ± 28.9	17. ± 151.1	1.234 ± 33.0
1000	1.000 ± 138.8	.00036 ± 107.8	.0025 ± 57.3	24.9 ± 30.3	.15 ± 30.1	7. ± 130.0	2.275 ± 18.3
1100	.158 ± 101.1	.00149 ± 35.6	.0044 ± 36.0	31.5 ± 16.0	.41 ± 16.0	—	2.594 ± 8.5
1200	.076 ± 101.1	.00113 ± 13.7	.0032 ± 14.4	51.7 ± 8.4	.80 ± 8.2	—	3.335 ± 3.4
1250	.128 ± 30.5	.00067 ± 5.3	.0028 ± 4.6	62.2 ± 3.0	2.37 ± 2.9	—	3.628 ± 1.1
1300	.127 ± 30.5	.00079 ± 8.3	.0027 ± 4.6	69.4 ± 2.2	3.93 ± 2.2	—	3.805 ± .8
1350	.136 ± 103.0	.00135 ± 43.4	.0048 ± 50.2	48.9 ± 35.5	.60 ± 35.3	—	3.248 ± 15.0
1400	.520 ± 137.4	.00051 ± 107.4	.0027 ± 59.5	32.2 ± 40.5	.23 ± 29.4	—	2.627 ± 21.1
1450	1.039 ± 141.4	.00609 ± 875.3	.0640 ± 870.1	10.0 ± 100.0	.22 ± 29.9	121. ± 878.7	1.249 ± 109.9
1500	1.039 ± 141.4	.00035 ± 111.8	.0023 ± 69.2	17.0 ± 100.0	.14 ± 47.9	7. ± 136.0	1.799 ± 76.3
1550	.240 ± 102.6	.00096 ± 39.3	.0039 ± 36.6	45.1 ± 26.5	.36 ± 18.2	—	3.122 ± 11.7
1600	5.940 ± 101.5	.00022 ± 104.8	.0027 ± 40.2	70.3 ± 27.3	.26 ± 25.2	—	3.825 ± 9.8
1650	1.086 ± 35.3	.00149 ± 82.7	.0058 ± 81.7	94.0 ± 23.2	.33 ± 20.8	29. ± 85.1	4.309 ± 7.4
1700	.214 ± 24.0	.00118 ± 8.0	.0037 ± 6.6	69.6 ± 3.2	2.66 ± 2.1	—	3.811 ± 1.2
1750	.493 ± 26.3	.00163 ± 59.0	.0050 ± 39.3	52.6 ± 24.5	.59 ± 11.7	—	3.361 ± 10.0
1800	1.687 ± 119.7	.00123 ± 151.3	.0099 ± 144.0	57.1 ± 77.3	.59 ± 71.5	21. ± 180.6	3.490 ± 30.4

[a] All errors are relative errors in percent.
[b] ^{40}Ar from the decay of ^{40}K only.

Table 6. ^{39}Ar–^{40}Ar ages and ^{37}Ar–^{38}Ar exposure ages.

Sample No.	Type	Plateau Range, °C	Plateau Age, G.y.	%[1]	Total K–Ar Age, G.y.	E.A.[2]	%Ca[3]	ppm K[3]
			Large rocks					
Apollo 14								
14310,101	Bas	675–1025	3.92 ± 0.01	72	3.78 ± 0.01	×38	10.1	4481
Apollo 15								
15075,24	Bas	900–1200	3.45 ± 0.20	65	3.39 ± 0.08	274	12.0	307
15075,24	Bas	800–1300	3.45 ± 0.20	71	3.39 ± 0.01	258	11.4	426
15505,72	Brec	No plateau			5.18 ± 0.05	604	10.3	1554
Apollo 16								
60618,13	Melt	900–1100	4.00 ± 0.02	41	3.50 ± 0.01	×38	11.9	410
61135,10	Ant	725–1100	3.90 ± 0.10	52	3.94 ± 0.01	28	8.4	4855
61135,10	Ant	No plateau			6.27 ± 0.05	61	14.7	799
61135,10	Ant	No plateau			5.70 ± 0.05	44	12.2	714
62237,12	Brec	No plateau			3.59 ± 0.05	329	11.3	81
65777,2	Rnb	750–1050	3.72 ± 0.02	56	3.29 ± 0.10	8	7.9	3095
65785,13	Troc	750–1100	3.97 ± 0.02	50	3.40 ± 0.01	271	8.6	2552
67035,20	Ant	1000–1300	3.95 ± 0.05	65	3.89 ± 0.01	35	11.6	428
67455,58	Ant	No plateau			3.81 ± 0.01	35	13.8	297
67618,1	Cry	No plateau			2.59 ± 0.01	50	8.4	1810
68516,2	Brec	1300–1600	3.80 ± 0.05	54	3.65 ± 0.01	50	9.9	1286

Apollo 17								
70255,36	Bas	850–1000	3.84 ± 0.02	34	3.67 ± 0.01	310	6.3	565
				Coarse fines				
Apollo 14								
14167,6,3	Melt	800–900	3.87 ± 0.06	42	3.60 ± 0.01	71	6.2	2555
14167,6,7	Melt	800–900	3.86 ± 0.01	49	3.62 ± 0.01	130	5.2	3102
14192,1,1	Nor	No plateau			3.28 ± 0.02	40	6.3	4175
14257,12,2	Melt	No plateau			2.25 ± 0.01	536	6.6	6740
Apollo 16								
61503,1,11	Poik	800–1100	4.02 ± 0.02	61	3.67 ± 0.01	40	10.0*	2637
Apollo 17								
72703,9,5	Dmb	800–1100	4.01 ± 0.01	52	3.98 ± 0.01	110	13.0	1271
73263,1,1	R Ant	1150–1600	4.23 ± 0.05	53	4.19 ± 0.01	109	11.7	763
73263,1,3	I Ant	800–1300	3.90 ± 0.20	72	4.09 ± 0.02	87	13.3	116
73263,1,6	R Ant	800–1100	3.99 ± 0.03	38	4.16 ± 0.01	221	11.2	820
73263,1,9	Dmb	1000–1550	3.99 ± 0.05	70	3.97 ± 0.01	215	6.3	1322
74243,2,1	Il Var	700–1000	3.85 ± 0.05	53	3.76 ± 0.01	61	7.5*	751
76503,6,5	Brec	800–1150	3.99 ± 0.01	52	4.04 ± 0.01	150	10.0	684

(1) % of ^{39}Ar on plateau.
(2) Exposure age, m.y. ×38 means excess 38, exposure age not obtainable.
(3) 10% error assumed.

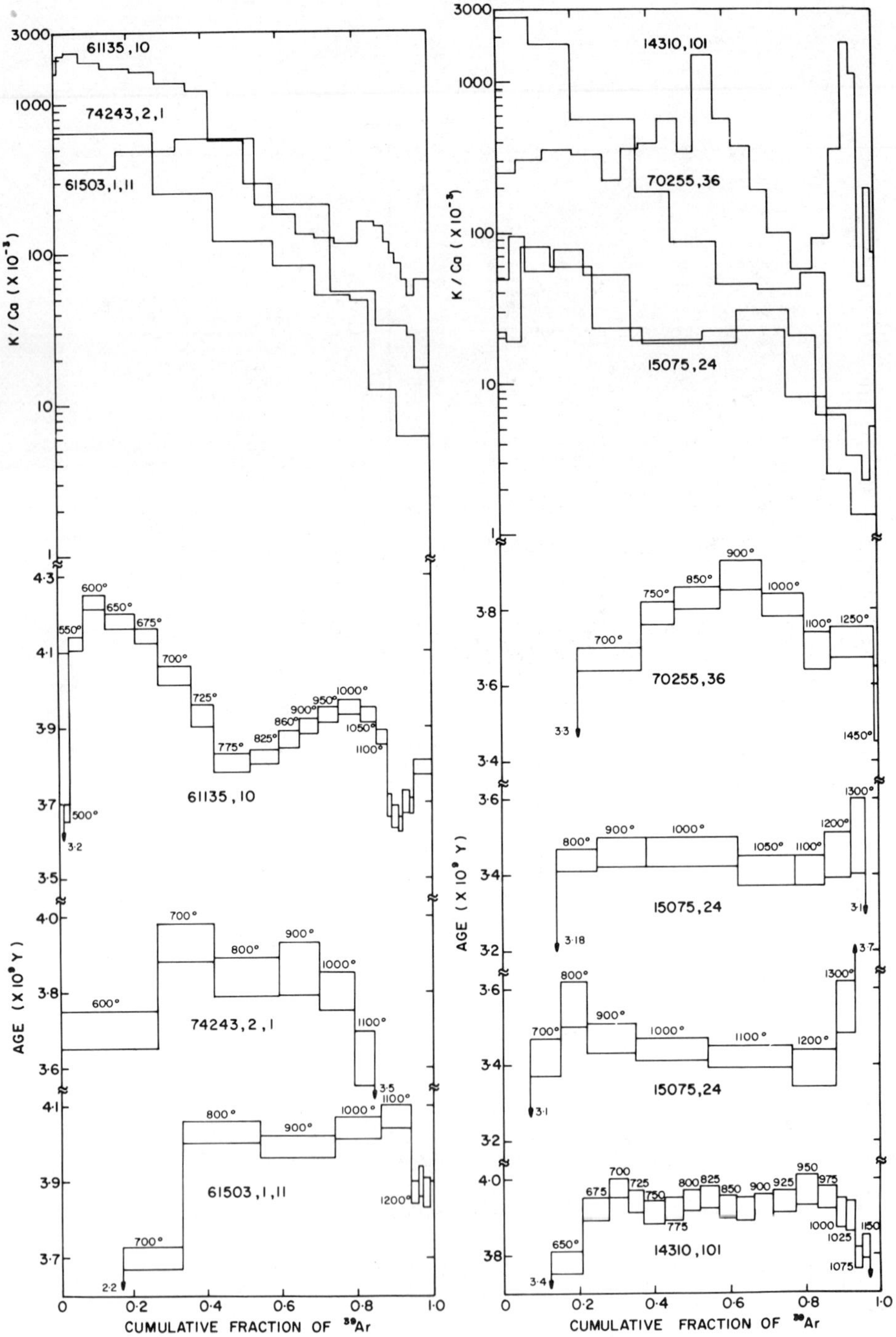

Fig. 3. ^{39}Ar–^{40}Ar age, 10^9 yr, and K/Ca versus the fractional ^{39}Ar releases for lunar rocks.

15075,24 (Fig. 3)

Sample 15075,24 is a mare basalt from Apollo 15. The data from the two analyses run are shown in Fig. 3. Both show a similar temperature pattern with some ^{40}Ar loss in the temperature fractions below 800°C and a plateau roughly between 800°C and 1300°C, containing 70% of the total ^{39}Ar released from potassium. The plateau age is 3.45 ± 0.20 G.y. This basalt is among the oldest mare basalts found at the Apollo 15 landing site.

The K/Ca ratio decreases continuously over the entire analysis from 0.09 at 600°C to about 0.001 at 1600°C. The ^{38}Ar–Ca exposure ages of 274 and 258 m.y. as reported in Table 6 are averages of the exposure ages over the plateau range.

15505,72 (not plotted)

Sample 15505,72 is a breccia with a large glass splash. The argon isotopic data presented in Table 5 indicate this sample has a large trapped component. The $^{36}Ar/^{38}Ar$ ratio is relatively constant at a value near 5.1. A ^{39}Ar–^{40}Ar age pattern gives little information. It increases from 3.33 G.y. at 500°C to a maximum of 6.83 G.y. at 750°C, only to decrease to 4.44 at the 900°C extraction; after which it increases to 5.3 G.y. at 1750°C. The $^{40}Ar/^{36}Ar_{sw}$ ratio is relatively constant between 800°C and 1350°C with a value near 0.8. With such a low ratio, any attempt to correct the trapped ^{40}Ar component is plagued with uncertainties in the $^{40}Ar/^{39}Ar$ ratio at, or near, 100%.

The K/Ca ratio varies with no pattern between 0.00073 and 0.044. The exposure age of 604 m.y. is the average exposure age over the entire experiment. The total range spans 1 m.y. to 3800 m.y., and consequently little meaning is given to the average age. This sample has its argon distributed in too complex a manner to give a meaningful ^{39}Ar–^{40}Ar age.

60618,13 (Fig. 4)

Sample 60618,13 is a large crystalline rock sample. The temperature release pattern shown in Fig. 4 shows a plateau over the three temperature fractions between 900°C and 1100°C with an average age of 4.00 ± 0.02 G.y. and which contains 41% of the total ^{39}Ar released from potassium. The temperature fractions before 900°C show large losses of ^{40}Ar with an age of 2.17 at 500°C. Also, a high temperature drop off is seen with a decrease in age to 3.19 G.y. at 1300°C.

The K/Ca ratio decreases continuously from 0.68 at 500°C to 0.002 at 1700°C. No cosmic-ray exposure age could be calculated as amounts of excess ^{38}Ar are seen throughout the entire analysis.

61135,10 (Fig. 3)

Sample 61135,10 is an anorthositic whole rock sample. Three different experiments were performed on this sample. Two yielded results with large

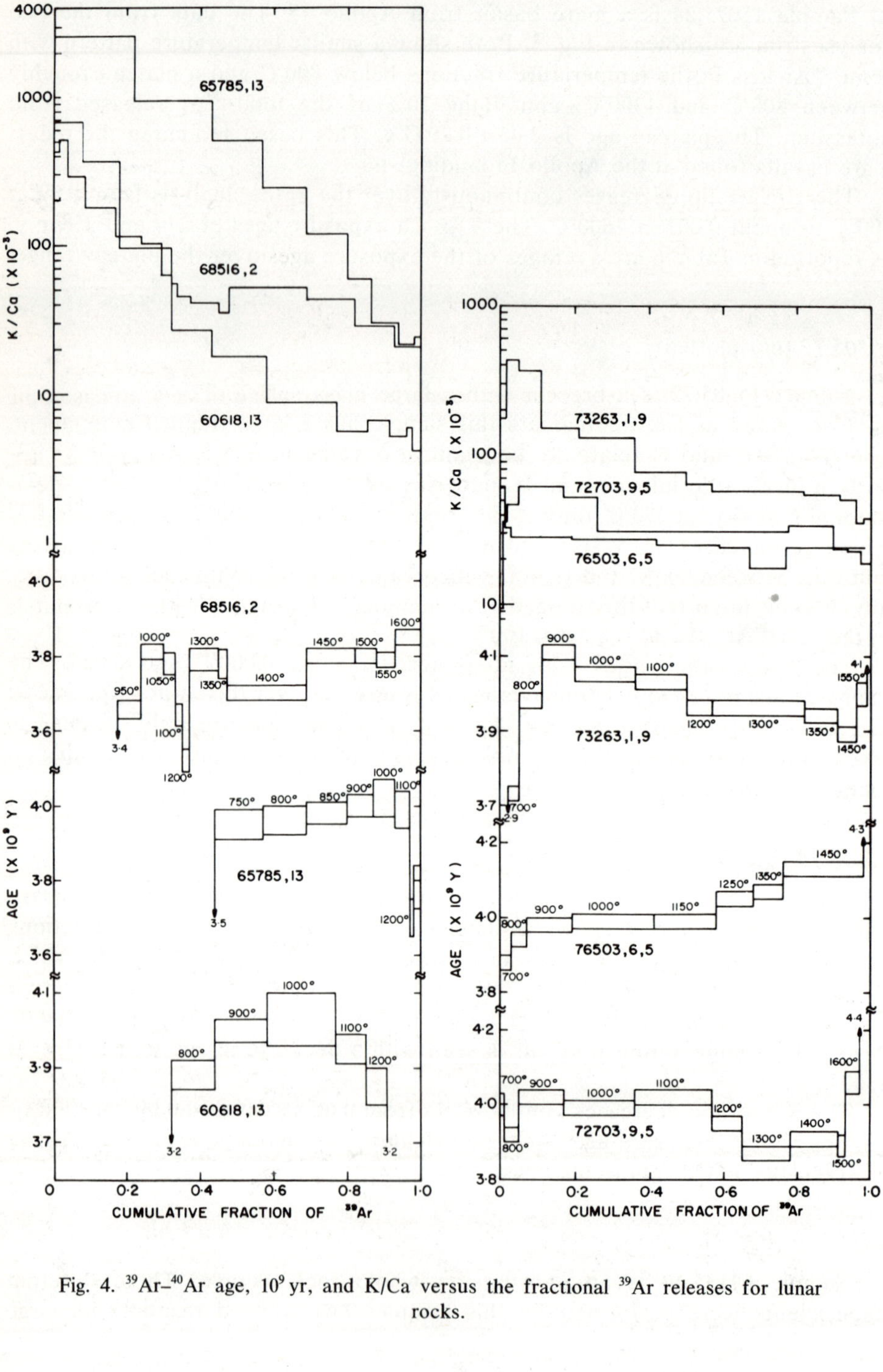

Fig. 4. ^{39}Ar–^{40}Ar age, 10^9 yr, and K/Ca versus the fractional ^{39}Ar releases for lunar rocks.

amounts of trapped gas, and as such no plateau is obtained. The third sample gave a reasonable plateau. The first sample run was a random picking of this sample, while the other two were splits of a second duplicate sample which was irradiated alongside the first sample. The sample which gave a plateau was a single large chunk picked and rewrapped after irradiation. The remainder of the sample consisted of a fine powder. The inescapable conclusion is that the fine powder of this sample is the location of these large amounts of trapped gas. It is puzzling that there is also a gross difference between the chemistry of this one chunk and the other samples as reported in Table 6. The single chip contained almost 0.5% potassium compared to an average of 750 ppm for the other two. The calcium deviated in the opposite sense. The high potassium piece contained 8.4% calcium compared to an average of 13% for the other two.

The data of the two samples with the large trapped components are not illustrated graphically as they provide little information. The plateau of the third is shown in Fig. 3. The low temperature fractions below 550°C show evidence of lost ^{40}Ar. The five fractions from 550°C to 700°C increase to an age of 4.23 G.y. at 600°C, and then decrease to an age of 3.93 G.y. to start the plateau at the 725°C fraction. The nine fractions on the plateau from 725°C to 1100°C give a weighted mean age of 3.90 ± 0.10 G.y. The range of ages over this plateau fluctuated from 3.81 to 3.95 G.y. After the 1100°C fraction, the age drops to 3.65 G.y. as typical with many lunar samples. The plateau includes 52% of the total ^{39}Ar released from potassium.

The ^{38}Ar–Ca cosmic-ray exposure age of 28 m.y. is the average age of the exposure ages over the plateau temperature range. In light of the variability of the exposure ages ranging from 1800 m.y. at 400°C to 4.0 at 1150°C, this average age might be viewed of little importance as it appears there is interference from a trapped ^{38}Ar component.

62237,12 (not plotted)

Sample 62237,12 is a white anorthositic breccia. The argon isotopic data presented in Table 5 do not show a plateau. The low temperature fractions from 500°C to 1200°C show signs of a trapped component with $^{40}Ar/^{36}Ar_{sw}$ ratios below 100. However, it is not a single component as these fractions do not fall on a line in a plot of $^{40}Ar/^{36}Ar_{sw}$ vs. $^{39}Ar/^{36}Ar_{sw}$. The ^{39}Ar–^{40}Ar age varies between 2.49 and 5.68 G.y. The ^{38}Ar–Ca exposure age varies between 24 and 2385 m.y. Also, the K/Ca ratio is highly variable with values between 0.00076 and 0.0673. This breccia has a very disturbed argon system.

65777,2 (Fig. 6)

Sample 65777,2 is a crystalline recrystallized nortic breccia. In light of the low $^{40}Ar/^{36}Ar_{sw}$ ratios, the data are represented by plotting $^{40}Ar/^{36}Ar_{sw}$ vs. $^{39}Ar/^{36}Ar_{sw}$ shown in Fig. 6. The seven temperature fractions from 750°C to 1050°C fall on a well defined line with an intercept of 10.7 ± 3.6 and a slope of

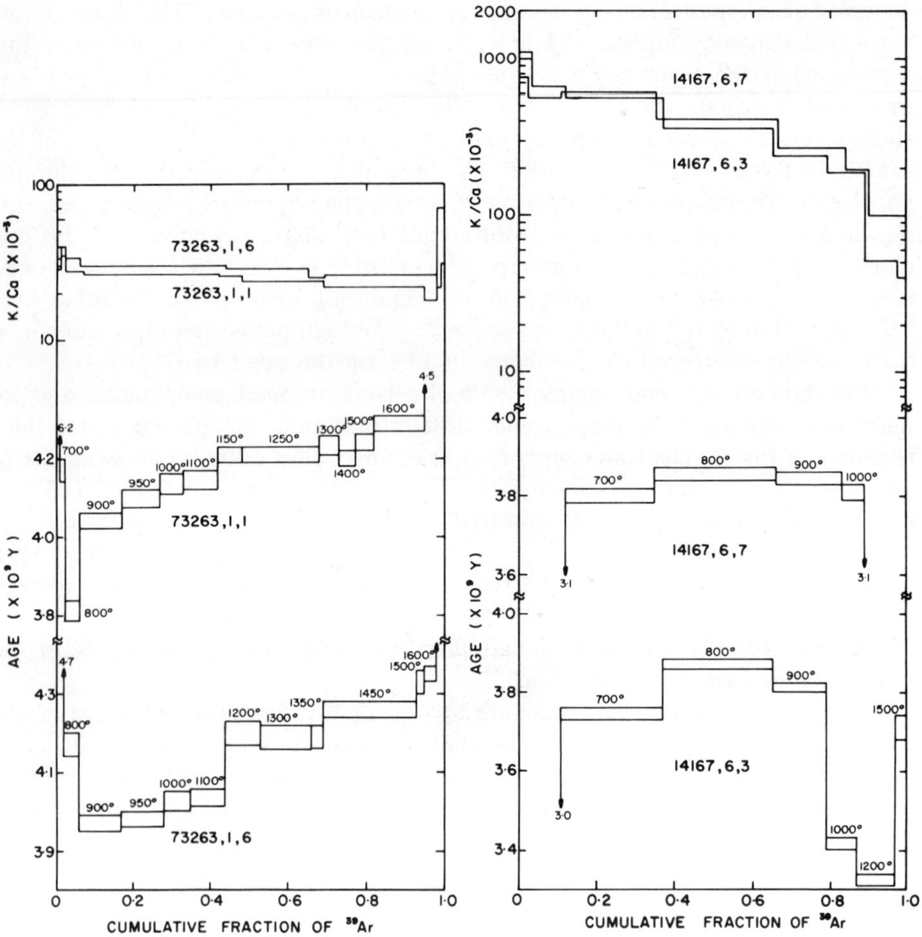

Fig. 5. ^{39}Ar–^{40}Ar age, 10^9 yr, and K/Ca versus the fractional ^{39}Ar releases for lunar rocks.

65.8 ± 0.7 which represents an age of 3.72 ± 0.02 G.y. as reported in Table 6. The low temperature fractions below 750°C show large amounts of ^{40}Ar have been lost through diffusive processes. Above 1100°C, the age drops off to an age of 3.57 G.y, a common occurrence in lunar samples. The ^{39}Ar–^{40}Ar plateau age of this sample of $3.72 \pm .02$ G.y. is among the youngest highland rock ages found; it is 200 m.y. post Imbrium. It was probably reset in a relatively large but not basin forming event.

The ^{38}Ar–Ca exposure age of 8 m.y. is the average exposure over the plateau temperature fractions. The K/Ca ratio decreases continuously over the entire temperature range from 1.7 at 600°C to 0.016 at 1650°C.

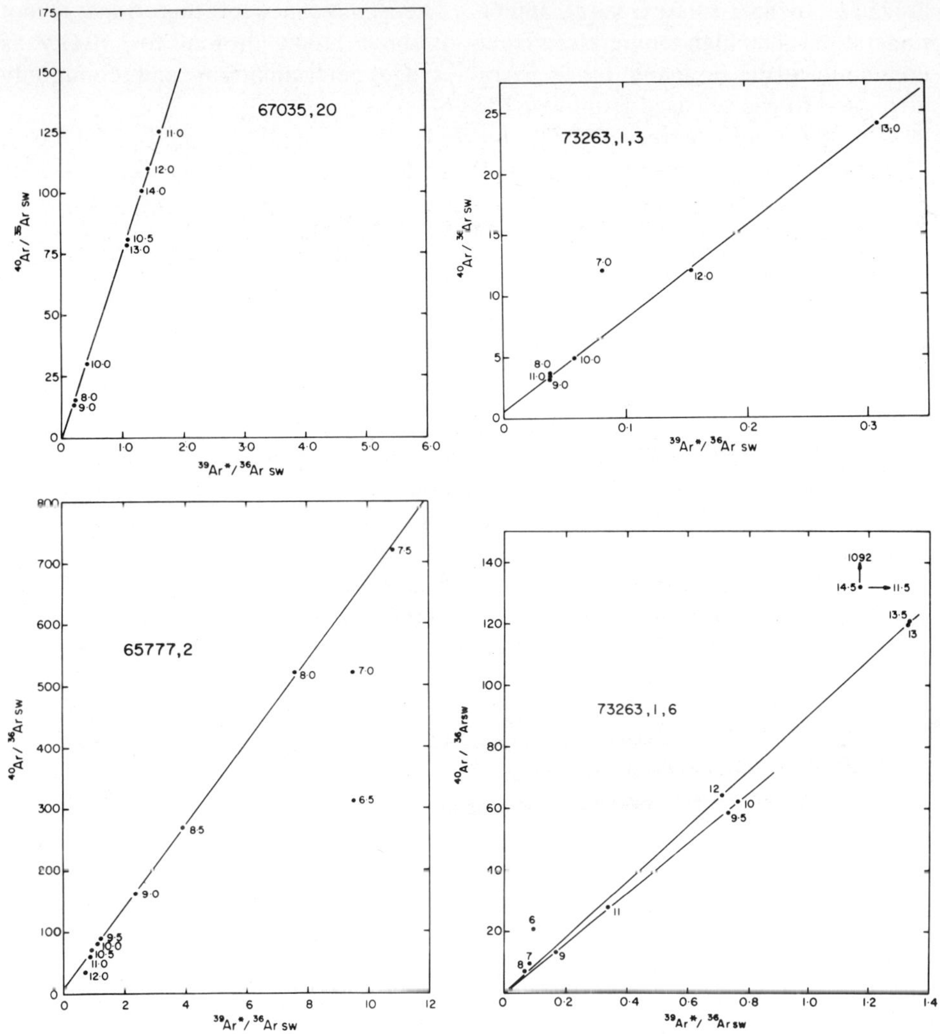

Fig. 6. $^{40}Ar/^{36}Ar_{sw}$ vs. $^{39}Ar/^{36}Ar_{sw}$ for lunar rocks. The numbers beside the points represent temperatures in hundreds of degrees Celsius.

65785,13 (Fig. 4)

Sample 65785,13 is a troctolitic rock sample from Apollo 16. The gas release pattern shown in Fig. 4 shows a plateau over six fractions between 750°C and 1100°C. The three fractions before the plateau show large amounts of gas loss with the age increasing from 1.38 G.y. at 500°C to 3.47 G.y. at 700°C. The plateau which follows contains 50% of the total ^{39}Ar released from potassium. The weighted mean plateau age was calculated to be 3.97 ± 0.02 G.y. However, there appears to be a definite slope to the plateau as it increases from 3.95 ± 0.04 G.y.

at 750°C to 4.02 ± 0.05 G.y. at 1000°C. The significance of this slope is not understood. The high temperature fractions above 1100°C drop off to 3.70 G.y. as commonly seen in many lunar ^{39}Ar–^{40}Ar age determinations and commonly attributed to gas released from low-K pyroxenes.

The K/Ca ratio decreases continuously from 2.63 at 600°C to 0.013 at 1700°C. The ^{38}Ar–Ca exposure age of 271 m.y. is the average exposure age over the entire experiment.

67035,20 (Fig. 6)

Sample 67035,20 is an anorthositic large rock sample. In light of the low $^{40}Ar/^{36}Ar_{sw}$ ratios, the argon isotopic data have been represented in Fig. 6, plotted as $^{39}Ar/^{36}Ar_{sw}$ vs. $^{40}Ar/^{36}Ar_{sw}$. The five fractions from 1000°C to 1300°C define a line with an intercept of 0.0 ± 0.2 and a slope of 74.6 corresponding to an age of 3.90 ± 0.05 G.y. 65% of the total ^{39}Ar released from potassium was released over this plateau region. The temperature fractions before 1000°C show diffusive gas loss. An intercept of 0.0 in the $^{39}Ar/^{36}Ar_{sw}$ vs. $^{40}Ar/^{36}Ar_{sw}$ is evidence of a trapped ^{36}Ar component without a correlated trapped ^{40}Ar component.

The K/Ca ratio varies slightly over the entire temperature range of this experiment from a high value of 0.023 at 600°C to a minimal value of 0.011 at 1400°C. The ^{38}Ar–Ca cosmic-ray exposure age varies directly with the K/Ca ratio, with an average value of 35 m.y.

67455,58 (not plotted)

Sample 67455,58 is an anorthositic large rock. The argon isotopic data presented in Table 5 show a scatter of ages from 3.35 to 4.54 G.y. No plateau was found in the $^{40}Ar/^{39}Ar$ ratios of the gas released. The low temperature fractions show ^{40}Ar has been lost through diffusive processes. The age increases continuously from 3.35 G.y. at 700°C to 3.92 G.y. at 1000°C, only to decrease to 3.74 G.y. at 1200°C. The four high temperature fractions, 1300°C–1600°C, fluctuate around an average age of 4.00 G.y.

The K/Ca ratio decreases from 0.014 at 600°C to 0.0078 at 900°C. The K/Ca ratios of the temperature extractions above 900°C range from 0.0064 to 0.0092 with no apparent pattern. There is evidence of excess ^{38}Ar above 1200°C. As such, the ^{38}Ar–Ca exposure age of 35 m.y. is the average age calculated from the gas released below the 1300°C fraction.

67618,1 (not plotted)

Sample 67618,1 is a plagioclase-rich crystalline rock. The argon isotopic data of sample 67618,1 presented in Table 5 show a similar age pattern as sample 67455,58. There is no plateau in the $^{40}Ar/^{39}Ar$ ratio over the entire temperature range of the analysis. The age starts at 1.35 G.y. at 600°C and climbs continuously to 3.68 G.y. at 900°C, only to turn around and decrease to 3.01 G.y. at

1250°C. The high temperature fractions above 1250°C contain only 3% of the total gas released from potassium and show a slight increase in the age of 3.43 G.y. There is evidence of a trapped ^{40}Ar component with the $^{40}Ar/^{36}Ar_{sw}$ ratios averaging around 100. However, in light of the irregular age pattern, no attempt was made to correct for the trapped component.

The K/Ca ratios start high at 0.71 for the 600°C and decrease quickly to 0.057 by the 850°C. After the 850°C fraction, the K/Ca ratio decreases slowly to a value of 0.28 at 1550°C. The ^{38}Ar–Ca cosmic-ray exposure age ranges from 34 to 77 m.y. with an average value of 50 m.y.

68516,2 (Fig. 4)

Sample 68516,2 is a breccia which is 80% shocked glass. The temperature release pattern of the argon isotopes is shown in Fig. 4. The low temperature fractions show evidence of diffusive loss of ^{40}Ar until the 1000°C fraction with the age increasing from 2.5 G.y. at 700°C to 3.81 G.y. at 1000°C. The three fractions after 1000°C, however, dip to an age of 3.57 G.y. at 1200°C as if the 1000°C fraction was the outgassing of a single mineral phase. The seven extractions between 1300°C and 1600°C, with the exception of two fractions, give a plateau age of 3.80 ± 0.05 G.y. The 1400°C fraction deviates from this average to a low of 3.70 G.y., a two sigma deviation, and the 1600°C fraction gives a higher age of 3.85, a one sigma deviation. Fifty-four per cent of the total ^{39}Ar released from potassium was released over this plateau region.

The K/Ca ratio shows a gradual decrease from 0.51 at 700°C to 0.21 at 1600°C. The ^{38}Ar–Ca cosmic-ray exposure age of 50 m.y. reported in Table 6 is the minimum exposure age, as this sample contains excess ^{38}Ar thought to be neutron induced from chlorine.

70255,36 (Fig. 3)

Sample 70255,36 is a fine-grained mare basalt. The temperature release pattern of the $^{40}Ar/^{39}Ar$ ratio shown in Fig. 3 shows a plateau similar to Apollo 17 basalts (Schaeffer *et al.*, 1976). The four low temperature fractions from 500°C to 750°C indicate ^{40}Ar was lost. The age increases from 3.18 G.y. at 500°C to 3.79 G.y. at 750°C. The next three fractions, from 850°C to 1000°C, yield a plateau age of 3.84 ± 0.02 and include 54% of the total ^{39}Ar released from potassium. The age decreases to 3.70 G.y. at 1100°C and 1250°C, at which temperature the sample is totally outgassed.

The K/Ca ratio decreases continuously over the entire experiment from 2.78 at 500°C to 0.004 at 1550°C. The ^{38}Ar–Ca exposure age of 310 m.y. is the average of the exposure ages over the entire analysis.

61503,1,11 (Fig. 3)

61503,1,11 is a medium-grained poikilitic coarse fine rock fragment. The

$^{40}Ar/^{39}Ar$ temperature release pattern shown in Fig. 3 indicates a plateau in the $^{39}Ar–^{40}Ar$ over the range of extraction temperatures from 800°C to 1100°C. The four fractions included in this range represent 61% of the total ^{39}Ar released from potassium and give a weighted mean age of 4.02 ± 0.02 G.y. The characteristics of the plateau of 61503,1,11 are not unusual for lunar samples. The two lower temperature fractions, 600°C and 700°C, both show diffusive gas loss while at higher temperatures the four fractions between 1200°C and 1500°C show a decrease in age to about 3.90 G.y., which might be attributed to gas released from low-K pyroxenes.

Since this sample was analyzed 746 days after neutron bombardment, an assumed calcium concentration of $10.0 \pm 2\%$ was used to correct the ^{39}Ar. The ^{40}Ar was not affected as this sample contains little trapped ^{40}Ar. The correction to the ^{39}Ar was also minimal due to the relatively high concentration of potassium in this sample of 2637 ppm. However, the K/Ca ratios in Table 5 and the exposure age of 40 m.y. reported in Table 6 should be viewed as approximate values as these numbers are directly tied to the %Ca in a sample. The K/Ca ratios show no dramatic changes; starting at approximately 0.5, they decrease continuously to a value of about 0.01 at the higher temperature fractions.

14167,6,3 (Fig. 5)

Sample 14167,6,3 is an Apollo 14 coarse fine sample. The temperature release pattern of the argon isotopes is shown in Fig. 5. There is evidence of a large amount of loss of ^{40}Ar in the temperature fractions below 800°C, with the 600°C fraction at 3.00 G.y. and the 700°C fraction at 3.75 G.y. The 800°C and 900°C fractions form a two point plateau which represents 42% of the total ^{39}Ar released from potassium and have a weight mean age of $3.87 \pm .06$ as reported in Table 6. The temperatures above 900°C show a sharp high temperature drop off to a value near 3.4 G.y. This is attributed to gas released from low potassium pyroxenes.

The K/Ca ratio decreases continuously from a high value of 1.14 at 500°C to a low of 0.04 at 1500°C. The ^{38}Ar–Ca exposure age of 71 m.y. is the average of all the exposure ages.

14167,6,7 (Fig. 5)

Sample 14167,6,7 is an Apollo 14 coarse fine sample. The temperature release pattern of the argon isotopes is shown in Fig. 5. There is evidence of a large amount of loss of ^{40}Ar in the temperature fractions below 800°C, with the 600°C fraction at 3.13 G.y. and the 700°C fraction at 3.81 G.y. The 800°C and 900°C fractions form a two point plateau which includes 49% of the total ^{39}Ar released from potassium and have a weight mean of $3.86 \pm .01$ G.y. The temperature fractions above 900°C show a high temperature drop off with ages decreasing to 3.1 G.y. This is attributed to gas released from potassium pyroxenes.

The K/Ca ratio decreases continuously from a high value of 0.77 at 500°C to a low value of .401 at 1500°C. The ^{38}Ar–Ca exposure age of 30 m.y. is the average of all the exposure ages.

14192,1,1 (not plotted)

Sample 14192,1,1 is an Apollo 14 coarse fine sample. The argon isotopic data reported in Table 5 show a release pattern with no plateau. The low temperature fractions increase from an age of 1.3 G.y. at 500°C to a maximum value of 4.5 G.y. at 1000°C, after which the age decreases to a value of 2.0 G.y. at 1500°C.

The K/Ca ratio is about constant at a value near 0.40 with the exception of the first two fractions which have values of 17.9 at 500°C and 0.12 at 700°C. The ^{38}Ar–Ca exposure age of 40 m.y. is the minimum exposure age as there is evidence of excess ^{38}Ar.

14257,12,2 (not plotted)

Sample 14257,12,2 is an Apollo 14 coarse fine sample. The argon isotopic data reported in Table 5 show a release pattern with no plateau. The age generally increases from 1.3 G.y. at 700°C to a value of 3.5 G.y. at 1500°C, with the exception of the 1200°C fraction which gives an age of 1.9 G.y.

The K/Ca ratio continuously decreases from 0.92 at 700°C to 0.09 at 1500°C. The ^{38}Ar–Ca exposure age of 536 m.y. is the average exposure age over the entire experiment.

72703,9,5 (Fig. 4)

Sample 72703,9,5 is a dark matrix breccia coarse fine with plagioclase phenocrysts in the dark matrix (Bence, pers. comm., 1977). The gas release pattern of this sample is shown in Fig. 4. The three fractions, 900°C, 1000°C, and 1100°C, include 52% of the total ^{39}Ar released from potassium. The weighted mean age for these fractions is 4.01 ± 0.01 G.y. There is evidence of a trapped ^{40}Ar component in the 600°C and 700°C fractions. However, the trapped ^{40}Ar component is outgassed by 800°C as this fraction shows diffusive gas loss. The temperature fractions above 1100°C decrease to an age of 3.87 G.y. which might be attributed to gas released from low-K pyroxenes.

This sample is relatively high in calcium concentration with a weight per cent of $13.0 \pm 0.4\%$. The K/Ca ratio is surprisingly constant over the high temperature fractions: 1100°C, 1200°C, and 1300°C with values of 0.03194, 0.03140, and 0.03106, respectively. The low temperature fractions show a marked increase in the ratio from 0.009 at 600°C to 0.062 at 900°C. It is curious that the fractions which define the age plateau show a marked decrease in the K/Ca ratio, as the K/Ca ratio is thought to reflect the nature of the mineral phases being outgassed. Intuitively, a constant K/Ca ratio would imply that a definite mineral phase is being outgassed. The opposite is indeed the case with this sample. The ^{38}Ar–Ca exposure age, however, conforms to the pattern seen in the K/Ca ratio. A relatively constant value of 110 m.y. has been measured for the 1100°C, 1200°C, and 1300°C fractions.

73263,1,1 (Fig. 5)

Sample 73263,1,1 is a recrystallized anorthositic coarse fine sample. The temperature release pattern shown in Fig. 5 shows a plateau between 1150°C and 1600°C with an average age of 4.23 ± 0.05 G.y. and including 53% of the total ^{39}Ar released from potassium. There appears to be a trapped ^{40}Ar component in the low temperature fractions evidenced by $^{40}Ar/^{36}Ar_{sw}$ ratios below 100 for the 700°C and 800°C fractions. However, this component is not present in the higher temperature fractions as this ratio is well over 1000 after 1250°C. There is also evidence of diffusive gas loss at 800°C as the age falls to 3.80 G.y. The age increased to 4.13 at 1000°C, and it might be suggested that a two fraction plateau exists for the 1000°C and 1100°C fractions with an age of 4.14 ± 0.07 G.y.

The K/Ca ratio remains relatively constant over all temperature fractions at a value of about 0.026. The ^{38}Ar–Ca exposure age of 109 m.y. is the average of the exposure ages over the total experiment.

73263,1,3 (Fig. 6)

Sample 73263,1,3 is an igneous anorthositic coarse fine sample. In light of the low $^{40}Ar/^{36}Ar_{sw}$ ratios, the argon isotopic data are represented in Fig. 6 as a plot of $^{40}Ar/^{36}Ar_{sw}$ vs. $^{39}Ar/^{36}Ar_{sw}$. The points for the six temperature fractions between 800°C and 1300°C define a line with an intercept of 0.7 and a slope of 75.2 corresponding to an age of 3.90 ± .2 G.y. The temperature fractions below 800°C show that a much larger ^{40}Ar trapped component is present, and the fractions above 1300°C vary erratically between 3.42 and 4.43 G.y.

The K/Ca ratio decreases from 0.04 at 700°C to 0.002 at 1500°C with a constant value of about 0.0028 for 1100°C to 1400°C. The ^{38}Ar–Ca exposure age of 87 m.y. is the average exposure age over the entire analysis.

73263,1,6 (Fig. 6)

Sample 73263,1,6 is a recrystallized anorthositic coarse fine. In light of the low $^{40}Ar/^{36}Ar_{sw}$ ratios, the argon isotopic data are represented in Fig. 6 as a plot of $^{40}Ar/^{36}Ar_{sw}$ vs. $^{39}Ar/^{36}Ar_{sw}$. The points for the five fractions between 800°C and 1100°C defines a line with an intercept of 0.0 and a slope of 80.62 corresponds to a plateau age of 3.99 ± 0.03 G.y. The 600°C and 700°C fractions contain a larger ^{40}Ar trapped component and consequently are found above the line. However, the three temperature fractions 1200°C, 1300°C, and 1350°C define a plateau age of 4.18 ± 0.2 G.y. as shown in Fig. 5. This may represent a relict age after recrystallization at 3.99 G.y.

The K/Ca ratio is relatively constant at a value of 0.027. The ^{38}Ar–Ca exposure age of 221 m.y. is the average exposure age of all the temperature fractions.

73263,1,9 (Fig. 4)

Sample 73263,1,9 is a dark matrix breccia which contains plagioclase and olivine phenocrysts (Bence, pers. comm., 1977). The gas release pattern of 73263,1,9 shown in Fig. 4 shows a pattern which might be considered two plateaus. The temperature fraction below 800°C shows ^{40}Ar losses while the 900°C fraction gives an indication of a trapped ^{40}Ar component. The trapped ^{40}Ar is suggested by a high $^{40}Ar/^{39}Ar$ ratio as well as a relatively high $^{36}Ar/^{38}Ar$ ratio for the fractions neighboring the 900°C fraction. However, no correction was made for this ^{40}Ar trapped component as the $^{36}Ar/^{38}Ar$ ratio drops quickly from the 2.97 value at 800°C to a near spallation ratio of 0.65 at 1300°C.

The plateau range from 1000°C to 1550°C as reported in Table 6 includes seven fractions and represents 70% of the total ^{39}Ar released from the potassium. However, this plateau is by no means flat. The 1000°C and 1100°C fractions with an average age of 4.04 G.y. might represent the age of a particular mineral phase while the latter three fractions, 1200°C, 1300°C, and 1350°C, with an average age of 3.96 G.y., are representative of a higher melting mineral phase. It is apparent that the 3.99 ± 0.05 G.y. average age reported in Table 6 is merely the weighted mean of these plateau areas.

The K/Ca ratios shed no immediate light on this discontinuity other than the fact that the high temperature plateau is a factor of three less rich in potassium with a K/Ca ratio near 0.05 compared to a ratio near 0.15 for the 1000°C fraction. The suggestion might be made that the high temperature plateau represents gas released from low-K pyroxenes.

The ^{38}Ar–Ca cosmic-ray exposure age varies from 176 to 254 m.y. with no apparent pattern over the entire temperature range. The age of 215 m.y. is the average of the ages over the entire range.

74243,2,1 (Fig. 3)

74243,2,1 is an ilmenite variolite coarse fine sample. The temperature release pattern shown in Fig. 3 indicates a plateau ranging from the 700°C to the 1000°C fractions. The gas release is typical of a basaltic sample. The plateau includes four temperature fractions including 53% of the total ^{39}Ar released from potassium. The 600°C shows diffusive loss of ^{40}Ar, and the temperature fractions above 1000°C show a high temperature decrease in the $^{40}Ar/^{39}Ar$ ratios, which can be attributed to gas released from low-K pyroxenes.

Since the sample was analyzed 966 days after the neutron bombardment, an assumed value of $7.5 \pm 0.5\%$ was used for the calcium concentration. This value is thought to be typical for basaltic samples. Half of the total ^{37}Ar was calculated as being released from the 1300°C fraction. This again is typical for basaltic samples. As such, the exposure age of 61 m.y. for this fraction was considered to be the best estimate of the ^{38}Ar–Ca exposure age. The K/Ca ratios decrease continuously over the entire temperature extractions.

76503,6,5 (Fig. 4)

76503,6,5 is a recrystallized breccia coarse fine sample. The gas release pattern for 76503,6,5 is shown in Fig. 4. A small amount of ^{40}Ar has been lost through diffusive processes in the 600°C and 700°C fractions. The following four fractions, from 800°C to 1150°C, which contain 52% of the total ^{39}Ar released from potassium, form a plateau with a weighted mean value of 3.99 ± 0.01 G.y. The next two fractions, 1250°C and 1350°C, seem to form a second plateau with an age of 4.06 G.y. These are followed by a single fraction at 1450°C with 20% of the ^{39}Ar released from potassium and an age of 4.13 G.y.

It might be suggested that these apparent steps in the plateau may indeed be different mineral phases or inclusions in the breccia. This notion is supported by the K/Ca ratios. A constant value of 0.028 is measured for the low temperature step, a value of 0.025 for the intermediate step, and a single value of 0.017 for the high temperature fractions. The ^{38}Ar–Ca exposure age also reflects older ages at the higher temperatures. A value of 150 m.y. has been measured by the low temperature fractions while at 1450°C the exposure age climbs to 172 m.y. All errors over this range are about ±4 m.y. Consequently, this difference of 22 m.y. might indeed be a remnant before the recrystallization age of 3.99 G.y.

Exposure ages of 68815

Five unirradiated chips 68815,141,B; 133,C; 133,B; 68; and 67,D were analyzed for ^{40}Ar, ^{38}Ar, and ^{36}Ar. The results are shown in Table 7. Four showed less than 37% of the ^{38}Ar is of solar-wind origin, while for 68815,141,B almost 96% of the ^{38}Ar is attributed to a solar-wind component. Such a large correction for the trapped component yields an error in the exposure age greater than 100% of the value. Consequently, the data for 68815,141,B are not used in calculating the exposure age for the breccia 68815. The other four samples give exposure ages ranging from 1.51 to 2.43 m.y. with an average of 1.83 ± 0.24 m.y. It should be noted, the age which was not included in this average for 68815,141,B of 3.89 ± 4.76 m.y. agrees well within the error.

Excess ^{38}Ar of 68815

The measured exposure age of 1.83 ± .24 m.y. for the breccia allows one to calculate the absolute amount of ^{38}Ar which was produced in the reactor from chlorine. The results are given in Table 8. The concentration of chlorine cannot be calculated from this data, as the production rate for chlorine has not been measured. However, relative comparisons between the samples are valid since, to within 2%, they were all exposed to the same neutron flux. The black glass samples have three to ten times as much chlorine as the clasts. This is also the

Table 7. 68815: ^{38}Ar–Ca exposure age and argon isotopic data from unirradiated samples.

Sample	Type	^{40}Ar $\times 10^{-8}$ cc/g	$^{40}Ar/^{36}Ar$	$^{36}Ar/^{38}Ar$	%SP	$^{38}Ar_{sp}$ $\times 10^{-8}$ cc/g	%Ca	Exp. age $\times 10^6$ yr
68815,141,B	GLS	4076.7 ± 410	49.8 ± 7	5.15 ± 0.73	4.2	0.67 ± 0.82	10.14 ± 0.2	3.89 ± 4.76
68815,133,C	CLT	2528.5 ± 250	2974.7 ± 420	1.93 ± 0.27	72.3	0.32 ± 0.05	10.65 ± 0.2	1.77 ± 0.28
68815,133,B	GLS	6731.6 ± 670	6799.6 ± 961	2.36 ± 0.33	63.3	0.26 ± 0.05	10.13 ± 0.2	1.51 ± 0.29
68815,68	GLS	6645.8 ± 660	19546.5 ± 2765	0.77 ± 0.11	96.8	0.42 ± 0.05	10.15 ± 0.3	2.43 ± 0.29
68815,67,D	CLT	858.5 ± 86	3577.1 ± 505	0.75 ± 0.11	97.3	0.31 ± 0.05	11.25 ± 0.5	1.62 ± 0.26

Table 8. 68815: ^{39}Ar–^{40}Ar plateau data.

Sample	Type	Plateau range	% ^{39}Ar	Plateau age $\times 10^9$ yr	K–Ar age $\times 10^9$ yr	%Ca	ppm K	$^{38}Ar_{Cl}$ $\times 10^{-8}$ cc/g
68815,141,B	GLS	No plateau			2.681 ± 0.003	10.14	1458	29.69 ± 2.9
68815,133,B	GLS	No plateau			3.015 ± 0.003	10.13	1646	32.39 ± 3.2
68815,133,C	CLT	1550–1750	72	3.811 ± 0.012	3.686 ± 0.007	10.65	489	8.89 ± 0.8
68815,67,D	CLT	No plateau			3.54 ± 0.02	11.25	214	3.26 ± 0.03

case if similar calculations are performed on the previously reported 68815 samples by Schaeffer *et al.* (1976).

68815 (Fig. 7)

The argon isotopic dates for the four samples 68815,141,B; 133,B; 133,C; and 67,D are given in Table 8. The temperature release patterns for these same samples are shown in Fig. 7. The results will be discussed by sample type, i.e., clast and black glass.

Clasts. The two white clast samples, 68815,133,C and 68815,67,D, are similar in many respects.

Clast 68815,133,C gives a three temperature plateau containing 72% of the ^{39}Ar released with a weighted age of $3.81 \pm 0.01 \times 10^9$ yr. The bulk of the gas is released at 1300°C and between 1650°C and 1750°C with a minimum value between 1350°C and 1550°C.

Clast 68815,67,D, due to a very low potassium concentration of 214 ppm, gives very poor results. All fractions are within a factor of four of the system blanks. However, due to the very consistent system blanks over the past year, a more precise correction could be made. Like 68815,133,C, the bulk of the ^{39}Ar was released at 1300°C and 1700°C. Both of these fractions gave ages identical, within the error, to the previous clast, their being $3.81 \pm 0.04 \times 10^9$ yr and $3.81 \pm .05 \times 10^9$ yr, respectively. The ^{39}Ar–^{40}Ar age of the two clasts, 68815,133,C and 68815,67,D, is $3.81 \pm 0.04 \times 10^9$ yr.

In both clasts, the temperature fractions which give the plateau age are the fractions which have the lowest K/Ca ratio. The ratio increases, however, by a factor of 2 for 68815,133,C and a factor of at least 10 for 68815,67,D for the temperature fractions between 1300°C and 1550°C.

It appears as though the low temperature and high temperature fractions which give plateau ages are of a very pure mineral phase. The release of gas seems to be associated with a limited temperature range during which a phase change occurs. This is evident by the fine resolution of the 68815,67,D temperature steps yielding only two fractions on the plateau, 1300°C and 1700°C fractions. The intermediate fractions, having relatively high potassium concentrations, might well be traces of black glass which have large amounts of ^{40}Ar loss.

Black glass. Samples 68815,141,B and 68815,133,B are black glass chips. The release patterns, shown in Fig. 7, show large amounts of ^{40}Ar have been lost.

The two samples reported by Schaeffer *et al.* (1976) show a similar loss, though a plateau age was reported for the higher temperatures. If all the black glass samples of 68815 are plotted on a ^{40}Ar vs. ^{39}Ar graph, all the points fall on or below an isochron corresponding to the age of 3.76×10^9 yr. As such, all of the black glass ages must be considered minimum ages.

Combining the results presented here and those reported by Schaeffer *et al.* (1976), the following chronology is suggested for breccia 68815. The time of brecciation is set by the two clasts reported here at 3.81×10^9 yr. The black glass,

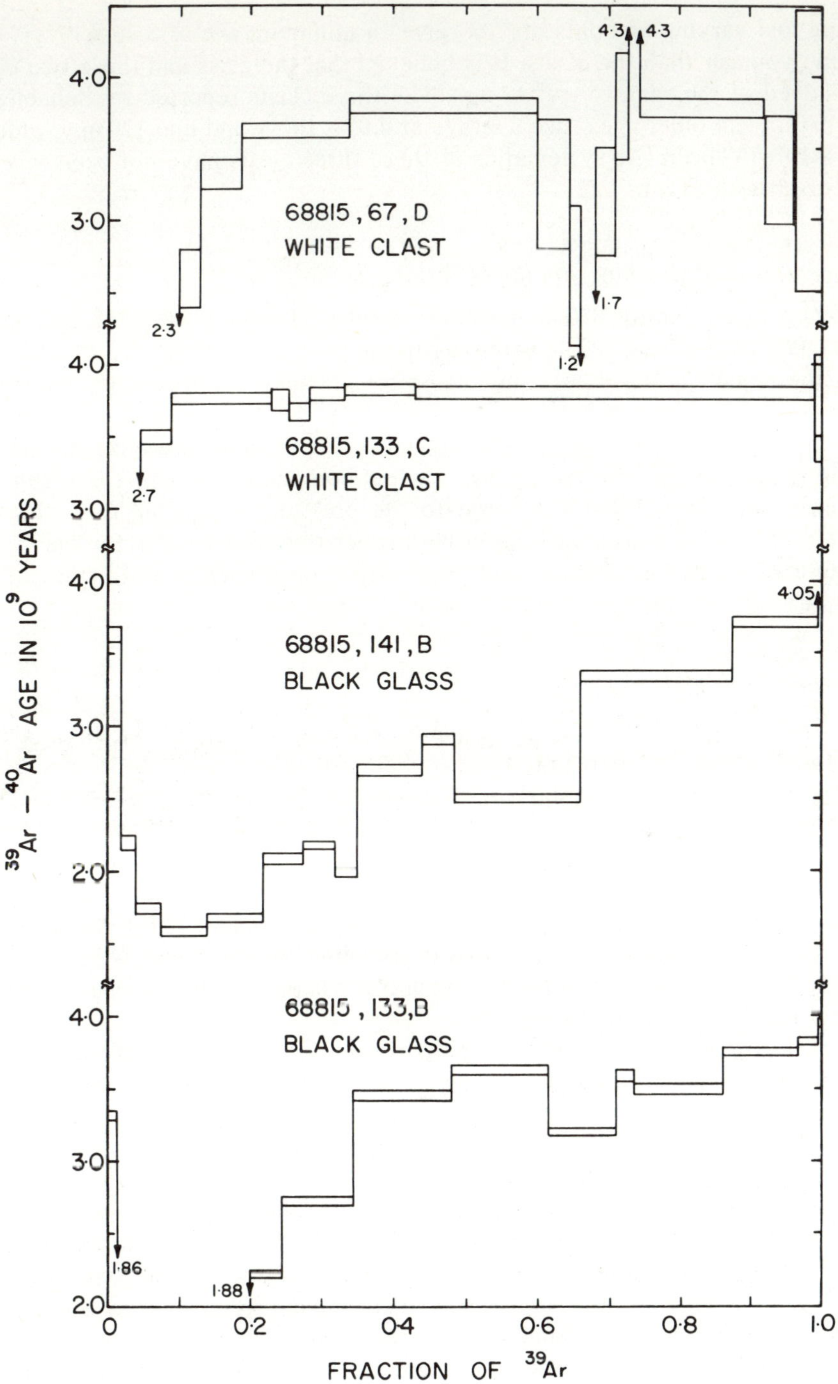

Fig. 7. ^{39}Ar–^{40}Ar age, 10^9 yr, for clasts and glass for lunar rock 68815. Some errors, for graphic purposes, are represented at larger values on the plots than in the tables.

having lost varying amounts of ^{40}Ar, gives a minimum age of 3.76×10^9 yr, only 40 m.y. younger than the clasts. It is believed that the glass and these two clasts are indeed of the same ^{39}Ar–^{40}Ar age. The three clasts reported by Schaeffer *et al.* (1976) yield older ages, two average at 4.05×10^9 yr and one 120 m.y. older at 4.12×10^9 yr. The K–Ar systematics of these three clasts have not been reset by the brecciation event.

Recalculation of previous Apollo 14 results

The previous Stony Brook Apollo 14 results (Husain *et al.*, 1971, and Sutter *et al.*, 1971) have been recalculated, as the previous results were not calculated with the same accuracy as is now possible. They were previously calculated manually with a somewhat less accurate $^{40}Ar/^{40}K$ atom ratio for the NL-25 monitor. For the recalculation, the original recorder charts were reread and the resulting peak heights put through the current computer program. The results are presented in Table 9 and compared to the previous results in Table 10. The present recalculated ages agree well with other radiometric ages for the Apollo 14 suite of rocks and are in agreement within the experimental errors of the previous values. The values are all about 150 m.y. older.

Distribution of highland ages

The distribution of ^{39}Ar–^{40}Ar highland rock ages for Apollo 14, 16, and 17 is presented in Fig. 8. The results prior to 1976 are summarized by Turner (1976). The present histograms contain, in addition, all the more recent results which come from various publications in the reference list. The noninclusion of Rb–Sr or U–Pb ages will not markedly change the histograms as there are only a few samples for which a Rb–Sr or U–Pb age is available and not a ^{39}Ar–^{40}Ar age. The number of samples—26 for Apollo 14, 49 for Apollo 16, and 42 for Apollo 17—is sufficient for a statistically meaningful sampling of the lunar highland rocks returned by the Apollo program. It is doubtful whether further measurement will greatly modify the distributions.

The histograms for the various sites have similarities and differences. The striking similarity is the peaking of ages near 4.0 G.y. This has been variously interpreted as a lunar cataclysm at 4.0 G.y. or as the preponderance of Imbrium ejecta in the highland rocks returned from the Apollo 14, 16, and 17 landing sites. The age peaking is most noticeable in the Apollo 14 and 17 results.

For Apollo 14, there is no evidence for ages older than 4.0 G.y., while for both other sites there is abundant evidence for rocks with ages approaching 4.3 G.y. The ages of the Apollo 16 samples do not show a marked peaking in age. The ages are more or less statistically distributed in time with a broad peak between 3.95 and 4.20 G.y. Such a spread in ages is compatible with the idea of Kirsten and Horn (1974) that the highland rock age histograms represent statistical lifetimes of rocks on the lunar surface. The marked peaking of the Apollo 14 and 17 histograms seems to indicate that these samples contain a large

Table 9. Recalculated Apollo 14 coarse fines.

Sample No.	Plateau Range, °C	Plateau Age, G.y.	%[(1)]	Total K–Ar age, G.y.	E.A.[(2)]	%Ca[(3)]	ppm K[(3)]
14053,34,1	700–900	3.97 ± 0.03	64	3.76 ± 0.01	×38	3.73	646
14152,1,1	300–1000	3.95 ± 0.01	58	3.87 ± 0.01	124	5.82	2988
14152,1,2	700–1000	3.89 ± 0.01	87	3.82 ± 0.01	120	5.49	3044
14167,6,1	700–1000	3.89 ± 0.01	77	3.68 ± 0.01	54	3.94	3628
14167,6,4	800–900	3.83 ± 0.02	47	3.64 ± 0.01	46	3.39	3057
14193,2,1	700–1000	3.94 ± 0.02	71	3.59 ± 0.01	60	5.00	2862
14257,12,1	No plateau			3.54 ± 0.01	57	4.88	7748
14257,12,3	No plateau			4.26 ± 0.01	731	7.17	4205
14310,100,1	800–1000	3.92 ± 0.01	61	3.80 ± 0.01	225	7.50	3745

[(1)]% of ^{39}Ar on plateau.
[(2)]Exposure age, m.y. ×38 means excess 38, exposure age not obtainable.
[(3)]10% error assumed.

Table 10. Comparison of previously published Apollo 14 coarse fine ages with the recalculated ages.

	Old plateau		Recalculated plateau	
Sample No.	Range, °C	Age, G.y.	Range, °C	Age, G.y.
14053,34,1	700–1500	3.83 ± 0.15	700–900	3.97 ± 0.03
14152,1,1	600–1500	3.77 ± 0.15	800–1000	3.95 ± 0.02
14152,1,2	700–1500	3.75 ± 0.05	700–1000	3.89 ± 0.01
14167,6,1	700–1500	3.76 ± 0.04	700–1000	3.89 ± 0.01
14167,6,4	700–1500	3.70 ± 0.03	800–900	3.83 ± 0.02
14193,2,1	700–1500	3.75 ± 0.10	700–1000	3.94 ± 0.02
14257,12,1	900–1500	3.78 ± 0.01	No plateau	
14257,12,3	900–1500	3.50 ± 0.12	No plateau	
14310,100,1	700–1500	3.74 ± 0.04	800–1000	3.92 ± 0.01

complement of rocks with ages near 4 G.y. and strongly suggests that these peaks in age are due to ejecta from major lunar basin events. The lack of ages older than 4.0 G.y. at the Fra Mauro landing site can be understood in view of the likely possibility that the Imbrium ejecta at this site are very thick and completely dominate the sample return. This is also born out in the fact that there are only a few non-brecciated rocks among the Apollo 14 sample return. The Apollo 14 and 17 landing sites are closer to the Imbrium Basin than Apollo 16, so the Imbrium ejecta should predominate more in the Apollo 14 and 17 samples than in the Apollo 16 samples.

It has been suggested that the old ages, 4.1–4.3 G.y., are not of chronological importance but are due to the mechanical mixing of minerals during a cratering event of 4.0 G.y., the late heavy bombardment age, and ~4.5 G.y., the age of the original lunar crust. Delano and Bence (1977) have made a petrologic study to see whether the minerals in certain of the rocks are chemically equilibrated or not. They have found evidence for 67483,13,8 that chemical equilibration has taken place, while for 73263,1,1 and 73263,1,6 there is a distinct lack of chemical equilibration. In light of these results, there is probably the need to reevaluate some of the old ages. However, the petrologic evidence clearly shows the existence of rocks with a chronological significance in the range 4.2–4.3 G.y. Moreover, the unequilibrated rocks are not necessarily of no chronological significance as it is possible that the mineral systems were open for loss of Ar in the lunar crust before being brought to the surface; hence, one could have a meaningful K–Ar age for a rock which is chemically unequilibrated. This question can likely be resolved by the study of individual mineral grains with the laser technique (e.g., Plieninger and Schaeffer, 1976).

There is a distinct difference of about 0.1 G.y. in the age peak for Apollo 14 samples and the Apollo 17 samples. According to Turner *et al.* (1973) the age difference represents the time interval between the Imbrium and the Serenitatis

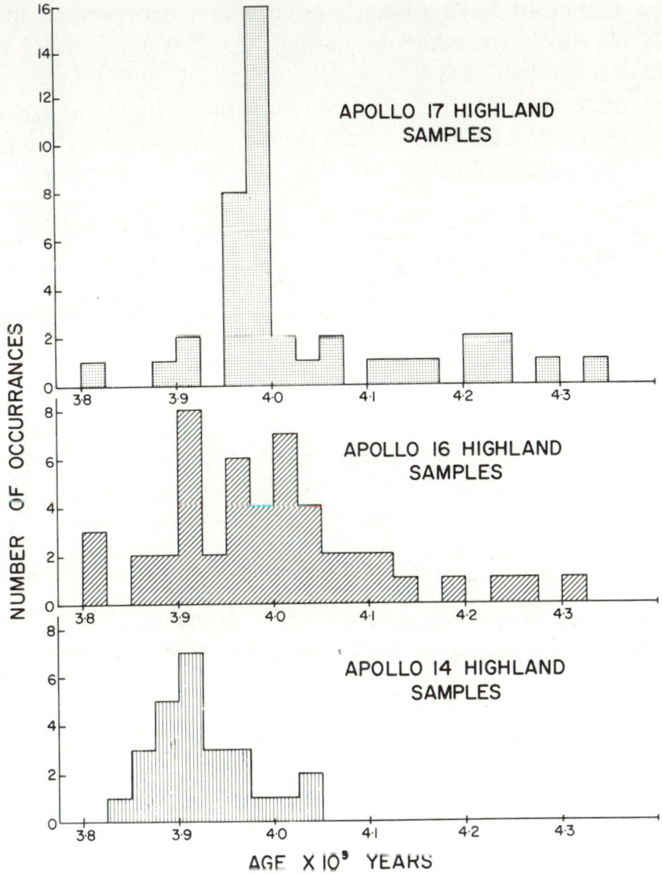

Fig. 8. Histograms of $^{39}Ar–^{40}Ar$ ages for highland rocks from Apollo 14, 16, and 17 samples.

events. The Apollo 14 breccias are from a thick ejecta blanket geologically related to the Imbrium event. A large number of the Apollo 17 breccias are from the massifs; in many cases derived from boulders which can be seen to have rolled down the massifs. The massifs were uplifted by the Serenitatis event and overlain by a thick layer of Serenitatis ejecta. It is not possible to rule out from the photogeology that the massifs are also mantled by a relatively substantial depth of Imbrium ejecta on top of the Serenitatis ejecta. However, in this case, one would expect no age difference.

In view of the recent results for the ages of the James consortium breccia 73215 (Jessberger *et al.*, 1977; Müller *et al.*, 1977), it seems possible that the age difference may be due to a similar effect for the entire suite of Apollo 17 breccias. For 73215 it was found that the matrix is not completely degassed and that the breccia forming event is determined by the ages of small felsic clasts which show evidence of having been molten when incorporated in the breccia (James, 1977). On this basis, while the normal ^{39}Ar–^{40}Ar age is 4.0 G.y., the actual breccia formation age is 3.92 G.y.

While the histogram of the ages of the Apollo highland rocks does not unequivocally prove that Imbrium ejecta dominate at all the landing sites, there is at least a strong suggestion that they do.

Acknowledgments—We would like to thank A. E. Bence for useful discussion concerning the petrology of the samples studied, T. Ludkewycz for operating the mass spectrometer, and R. Schlott for his skill in making the glass apparatus and especially the furnace used for the heating of lunar samples. We would like to acknowledge the helpful comments of the reviewers, J. H. Reynolds and R. S. Lewis. This work was supported by NASA grant NGL-33-015-174.

REFERENCES

Alexander E. C., Jr. and Davies P. K. (1974) ^{40}Ar–^{39}Ar ages and trace element contents of Apollo 14 breccias; an interlaboratory cross-calibration of ^{40}Ar–^{39}Ar standards. *Geochim. Cosmochim. Acta* **38**, 911–928.

Dalrymple G. B. and Lanphere M. A. (1971) $^{40}Ar/^{39}Ar$ technique of K–Ar dating, a comparison with the conventional technique. *Earth Planet. Sci. Lett.* **12**, 300–308.

Delano J. W. and Bence A. E. (1977) 4.2–4.3 AE Anorthositic soil fragments: Equilibrated or unequilibrated polycomponent systems. *Proc. Lunar Sci. Conf. 8th.* This volume.

Hanson G. N., Goldich S. S., Arth J. G., and Yardley D. H. (1971) Age of the early Precambrian rocks of the Saganaga Lake—Northern Light Lake Area, Ontario—Minnesota. *Can. J. Earth Sci.* **8**, 1110–1124.

Husain L. (1974) ^{40}Ar–^{39}Ar chronology and cosmic ray exposure ages of Apollo 15 samples. *J. Geophys. Res.* **79**, 2588–2606.

Husain L., Sutter J. F., and Schaeffer O. A. (1971) Ages of crystalline rocks from Fra Mauro. *Science* **173**, 1235–1236.

James O. B. (1977) Petrology of four clasts from consortium breccia 73215. *Proc. Lunar Sci. Conf. 8th.* This volume.

Jessberger E. K., Kirsten T., and Staudacher Th. (1977) One rock and many ages—further K–Ar data on consortium breccia 73215. *Proc. Lunar Sci. Conf. 8th.* This volume.

Kirsten T. and Horn P. (1974) Chronology of the Taurus–Littrow region III: Ages of mare basalts and highland breccias and some remarks about the interpretation of lunar highland rock ages. *Proc. Lunar Sci. Conf. 5th*, p. 1451–1475.

Müller H. W., Plieninger T., James O. B., and Schaeffer O. A. (1977) Laser probe ^{39}Ar–^{40}Ar dating of materials from consortium breccia 73215. *Proc. Lunar Sci. Conf. 8th.* This volume.

Plieninger T. and Schaeffer O. A. (1976) Laser probe ^{39}Ar–^{40}Ar ages of individual mineral grains in lunar basalt 15607 and lunar breccia 15465. *Proc. Lunar Sci. Conf. 7th*, p. 2055–2066.

Schaeffer O. A., Husain L., and Schaeffer G. A. (1976) The duration of volcanism in Mare Serenitatis. *Earth Planet. Sci. Lett.* **31**, 358–368.

Stettler A., Eberhardt P., Geiss J., Grögler N., and Maurer P. (1973) Ar^{39}–Ar^{40} ages and Ar^{37}–Ar^{38} exposure ages of lunar rocks. *Proc. Lunar Sci. Conf. 4th*, p. 1865–1888.

Sutter J. F., Husain L., and Schaeffer O. A. (1971) $^{40}Ar/^{39}Ar$ ages from Fra Mauro. *Earth Planet. Sci. Lett.* **11**, 249–253.

Turner G. (1976) Potassium–Argon chronology of the moon. To be published.

Turner G., Huneke J. C., Podosek F. A., and Wasserburg G. J. (1971) ^{40}Ar–^{39}Ar ages and cosmic ray exposure ages of Apollo 14 samples. *Earth Planet. Sci. Lett.* **12**, 19–35.

Turner G., Cadogan P. H., and Yonge C. J. (1973) Argon selenochronology. *Proc. Lunar Sci. Conf. 4th*, p. 1889–1914.

Proc. Lunar Sci. Conf. 8th (1977), p. 2301–2314.
Printed in the United States of America

The history of the Apollo 17 Station 7 boulder

N. NAKAMURA*

Colorado School of Mines, Department of
Chemistry and Geochemistry, Golden, Colorado 80401

M. TATSUMOTO

U.S. Geological Survey, Mail Stop 963,
Box 25046, Denver, Colorado 80225

Abstract—Age determination for dark dikelets 77075 which cut the 4.36 b.y. (10^9 year) old noritic breccia 77215 of Station 7 boulder of Apollo 17 was performed by the Rb–Sr and Sm–Nd methods. The Rb–Sr data (excluding a whole rock sample and an olivine fraction) define a regression line which corresponds to an age of 4.07 ± 0.09 b.y. (2σ) (using the new ^{87}Rb decay constant of 1.41×10^{-11} yr^{-1}) and an initial $^{87}Sr/^{86}Sr$ ratio of 0.69918 ± 0.00016 (2σ) providing that Sr isotopes in plagioclase xenocrysts equilibrated with that of matrix when 77075 formed. Due to small Sm/Nd fractionation among the density separates, Sm–Nd data do not define a precise isochron. We conclude that the Rb–Sr age (~4.07 b.y.) of 77075 approximates the intrusion age of the dikelets and 77075 is not an extension of 77115 which formed 3.75 b.y. ago. Thus, four events are recorded in the Station 7 boulder. Lunar history during the first ~0.6 b.y. is also discussed.

INTRODUCTION

WE WILL REPORT a chronological study conducted by the Rb–Sr and Sm–Nd methods on dark dikelets 77075 in the Station 7 boulder, Apollo 17.

Four specimens, 77135, 77115, 77075, and 77215, collected from a boulder at Taurus-Littrow Station 7 of Apollo 17 have been studied extensively by a consortium led by E. C. T. Chao. Field relations (Schmitt, 1973), photographic observations (Butler and Dealing, 1974), and a petrographic study (Chao *et al.*, 1974) indicated that the noritic breccia clast (represented by sample 77215) is the oldest material in the boulder. The clast was surrounded by "fragment-laden pigeonite feldspathic basalt" (represented by 77115), and cut by dark dikelets (represented by 77075). The dikelets were observed by Schmitt (1973) to be a continuation of 77115. The melt 77115 cooled, was fractured, and was then engulfed by a highly vesicular rock (represented by 77135) which is the youngest material in the boulder (Chao *et al.*, 1974).

Previous chronological studies of the Station 7 boulder components conducted by our group (Nunes *et al.*, 1974; Nakamura *et al.*, 1976) using Sm–Nd, Rb–Sr, and U–Pb methods and by the Bern group (Stettler *et al.*, 1974, 1975) using the ^{39}Ar–^{40}Ar method are generally in agreement with the geologic observations. The ages so far obtained form three clusters: ~4.36 b.y., ~3.95 b.y.,

*On Leave from Department of Earth Sciences, Kobe University, Kobe 657, Japan.

and ~3.75 b.y. The noritic breccia 77215 was dated to be ~4.36 b.y. old by Rb–Sr and Sm–Nd methods. The ^{39}Ar–^{40}Ar ages of 77215, 77075, and some clasts in 77115 and 77135 are ~3.95 b.y. old. The ^{39}Ar–^{40}Ar and Rb–Sr ages of the matrices 77115 and 77135 are 3.7–3.8 b.y. old.

Winzer *et al.* (1977), however, debated the interpretation of multi-stage Boulder 7 formation events. Based on the nearly identical major, minor, and trace element characteristics of Apollo 17 boulders, they suggested that Apollo 17 boulders originated from norites and possibly anorthositic gabbros and dunites about 4.4 b.y. old that suffered a major impact event 4.0 b.y. ago.

Our interpretation (Nakamura *et al.*, 1976) of the history of the Station 7 boulder was that either: (1) the 4.36 b.y. old ANT material was cut by material formed during the ~3.95 b.y. Serenitatis event and young (~3.75 b.y.) components from the Imbrium Basin excavation event were added to the boulder, or (2) the ~3.75 b.y. ages may represent partial resetting of the isotopic systems by intermediate-size post-basin excavation impact events. The existing data did not provide a unique interpretation.

No chronological data other than two whole-rock ^{39}Ar–^{40}Ar ages (3.90 b.y. and 3.93 b.y. existed for 77075 owing to a difficulty in separating minerals due to the extremely fine-grained texture of the rock. However, we believe that the age of 77075 may provide the key to the evolution of the Station 7 boulder as well as increase our understanding of lunar history prior to ~4.0 b.y. ago. Thus, we have undertaken a Rb–Sr and Sm–Nd study of 77075.

SAMPLES

The black dikelet 77075 is aphanitic and contains scattered specks (xenocrysts) of plagioclase, olivine, and pyroxene (Chao *et al.*, 1974). A sample of 77075,22 was crushed in a stainless steel mortar. Several chunks were separated for the whole-rock sample. Owing to the extremely fine-grained texture of 77075, the sample was then further crushed and sieved with a 200 mesh nylon sieve to obtain a better mineral separation. Plagioclase xenocrysts were, unfortunately, disintegrated during this process. Olivine (plus a small amount of pyroxene) xenocrysts and a matrix portion in which no observable xenocrysts existed were handpicked from the +200 mesh material using a binocular microscope under laminar flow. The rest of the material retained on a 200 mesh sieve was separated into density fractions of $\rho < 2.8$, $2.8 < \rho < 3.1$, $3.1 < \rho < 3.2$, and $\rho > 3.2\ g/cm^3$ using bromoform, methylene iodide, and acetone. The finely crushed material ($<74\ \mu m$) was also separated into density fractions of $\rho < 2.8$, $2.8 < \rho < 3.0$, $3.0 < \rho < 3.25$ (not analyzed), and $\rho > 3.25\ g/cm^3$. The $\rho < 2.8$ fractions, consisting mostly of plagioclase, from the two density separation procedures were combined for analysis.

ANALYTICAL PROCEDURES

The analytical techniques were similar to those previously reported (Nakamura *et al.*, 1976) in which a high-pressure liquid chromatography technique for elemental separation was employed. Samples were spiked with ^{41}K, ^{87}Rb, and ^{84}Sr, decomposed using an HF–$HClO_4$–HNO_3 mixture in a teflon bomb, converted to chlorides, and aliquoted from a clear HCl solution. The main aliquot was used for measurement of the Nd isotopic composition and K, Rb, and Sr analyses, and the small aliquot was spiked with ^{150}Sm and ^{146}Nd for Sm and Nd concentration determinations. K, Rb, and Sr were separated on a $\Phi\ 0.4 \times H\ 30$ cm column using AG50W, X-12, 200–400 mesh ion exchange resin by eluting with 2N-HCl under a pressure of 4000 g/cm^2. The REE fraction was eluted with 4N-HCl. Sm and Nd were then separated from other REE using a micro column (Φ 0.08 × H 30 cm with AG50W, X-12, ~20 μm particle size) and a 0.2M α-hydroxyisobutyric acid solution. Blank values were within the range reported previously (Nakamura *et al.*, 1976, Table 1).

The Sr isotopic composition was corrected for mass spectrometer fractionation by normalizing to $^{86}Sr/^{88}Sr = 0.1194$, then standardized to the NBS certified $^{87}Sr/^{86}Sr$ ratio of 0.71014 for SRM-987 for interlaboratory comparison. The Nd isotopic composition was normalized to $^{150}Nd/^{144}Nd = 0.236433$.

Results

Rb–Sr

The results of Rb–Sr analyses of the black dikelet 77075,22 are shown in Table 1 and depicted on a $^{87}Sr/^{86}Sr$ vs. $^{87}Rb/^{86}Sr$ diagram in Fig. 1.

As Chao *et al.* (1974) reported, 77075 contains scattered small xenocrysts of plagioclase, olivine, and pyroxene. We have tried to separate all of these xenocrysts from the matrix under a microscope in an attempt to date the matrix. However, some xenocrysts, of course, might have remained in the matrix. The handpicked yellowish colored xenocrysts are believed to be olivine (plus a small amount of pyroxene). The Rb–Sr data, except the handpicked olivine fraction, define a best fit line which yields a slope corresponding to an age (T) of 4.18 ± 0.08 b.y. (2σ) with an initial $^{87}Sr/^{86}Sr$ ratio (I) of 0.69913 ± 0.00013 (2σ) using $\lambda_{87_{rb}} = 0.0139 \times 10^{-9}\ yr^{-1}$. The slope and the initial ratio are calculated using York's program (1969) and the errors are two standard deviations.

This age corresponds to 4.09 ± 0.08 b.y. if we use the new ^{87}Rb decay constant of $1.42 \times 10^{-11}\ yr^{-1}$ which is recommended by the International Union of Geological Sciences. In our previous paper (Nakamura *et al.*, 1976), we used

Table 1. Rb–Sr data of 77075.

Sample	Weight (mg)	K (%)	Rb (ppm)	Sr (ppm)	$^{87}Rb/^{86}Sr$ [1]	$^{87}Sr/^{86}Sr$ [2]
Handpicked fraction						
Whole rock	22.75	0.1937	5.927	161.72	0.1060	0.70554 ± 4
Matrix	5.13	0.1996	5.978	153.49	0.1126	0.70583 ± 4
Olivine	0.62	0.0160	0.878	18.78	0.1352	0.70869 ± 8
Density separates of > 74 μm fraction						
$\rho < 2.8$ [3]	9.42	0.1132	2.604	211.24	0.0356	0.70123 ± 6
$2.8 < \rho < 3.1$	38.33	0.2090	6.274	165.35	0.1098	0.70571 ± 4
$3.1 < \rho < 3.2$	6.05	0.1834	6.079	135.35	0.1299	0.70690 ± 4
$\rho > 3.2$	2.81	0.0961	3.554	66.25	0.1552	0.70827 ± 10
Density separates of < 74 μm fraction						
$2.8 < \rho < 3.0$	0.9	0.1247	3.613	140.56	0.0743	0.70358 ± 6
$\rho > 3.25$	0.75	0.0505	1.764	42.92	0.1189	0.70630 ± 8

[1]Uncertainties are estimated to be ≲0.3%.

[2]Uncertainties correspond to last significant figure and are 2σ mean.

[3]Combined with the $\rho < 2.8$ separate of <74 μm fraction. Density is in g/cm^3.

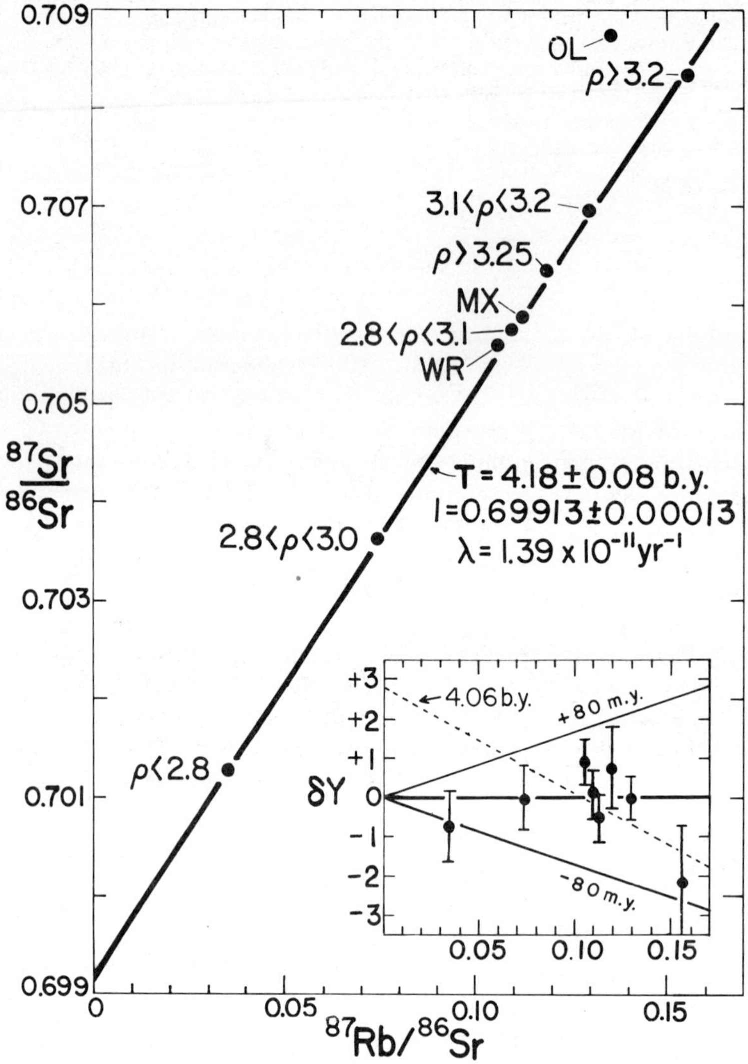

Fig. 1. Rb–Sr evolution diagram for the dark dikelet 77075 from the Apollo 17 Station 7 boulder. Uncertainties are smaller than the dots shown. Data of density fractions and handpicked matrix (MX) and whole rock (WR) samples form a line with a slope that corresponds to an age of 4.18 ± 0.08 b.y. (2σ) using $\lambda_{87} = 1.39 \times 10^{-11}\ yr^{-1}$. The handpicked olivine xenocrysts (OL) do not fall on the line established by the other separates. The insert shows the relative deviation of the measured $^{87}Sr/^{86}Sr$ ratios (δY in 10^4) from the best fit line. If we eliminate all fractions which possibly contained xenocrysts ($\rho < 2.8$, WR, and OL), the Rb–Sr data define an isochron (shown by dotted line) which corresponds to 4.06 ± 0.12 b.y.

$\lambda_{^{87}Rb} = 1.41 \times 10^{-11}\ yr^{-1}$ in order to bring Rb–Sr ages into agreement with U–Pb ages calculated using the new U decay constants of Jaffey *et al.* (1971). The isochron age of 77075 becomes 4.12 ± 0.08 b.y. using the latter value for $\lambda_{^{87}Rb}$. This is our preferred age since we believe that the IUGS value is too large (the IUGS value produces an ~90 m.y. reduction at ~4 b.y. in the age relative to the old decay constant, whereas our value produces an ~60 m.y. reduction as do the new U decay constants). In order to avoid confusion in comparison between ages obtained by this work and our previous paper, we will use $\lambda_{^{87}Rb} = 1.41 \times 10^{-11}\ yr^{-1}$ throughout the remainder of this paper. All ages should be reduced by ~30 m.y. for comparison with the IUGS value for $\lambda_{^{87}Rb}$.

The initial $^{87}Sr/^{86}Sr$ ratio of 77075 appears to be slightly higher than that of noritic breccia 77215 and similar to that of pigeonite basalts 77115 and 77135, however these differences are not distinct within the assigned error range. Although the K/Rb, Rb/Sr relations of the separates indicate that the dikelet sample is not a two-phase mixture, it does not definitely prove that the isochron is not a mixing line since the "mineral" separates are not very pure.

Sm–Nd

The Sm–Nd data for 77075 are shown in Table 2 and Fig. 2. The $^{147}Sm/^{144}Nd$ ratios in the density fractions measured do not show significant fractionation and range from 0.1645 in the $\rho < 2.8$ fraction to 0.1740 in the $\rho > 3.2$ fraction. Furthermore, data points for the whole rock, and $2.8 < \rho < 3.1$, and $\rho > 3.2$ fractions fall close together. Since the line consists essentially of two points, the Sm–Nd data do not define a precise isochron. The nominal value for the age is 4.13 ± 0.82 b.y. (2σ) and the initial $^{143}Nd/^{144}Nd$ ratio is 0.5073 ± 0.0010 (2σ). Although the age has large uncertainties, all data points lie well on the isochron; and the initial ratio appears to be reasonable and is slightly higher than that (0.5068 ± 1) of 77215 (Nakamura *et al.*, 1976) which is the oldest component of the Station 7 boulder. The nominal age is superficially in agreement with the Rb–Sr isochron age.

Table 2. Sm–Nd data of 77075.

Sample	Weight (mg)	Sm (ppm)	Nd (ppm)	$\frac{^{147}Sm}{^{144}Nd}$	$\left(\frac{^{143}Nd}{^{144}Nd}\right)^1$
$\rho < 2.8$[2]	9.42	10.23	37.59	0.1645 ± 2	0.511801 ± 37
$2.8 < \rho < 3.1$	38.33	15.18	53.04	0.1732 ± 1	0.512040 ± 19
Whole rock	22.75	14.10	49.00	0.1739 ± 1	0.512050 ± 19
$\rho > 3.2$	2.81	6.58	22.86	0.1740 ± 4	0.512094 ± 40

[1]Normalized to $^{150}Nd/^{144}Nd = 0.236433$. Errors correspond to last significant figures and are 2σ mean.

[2]Density is in g/cm^3.

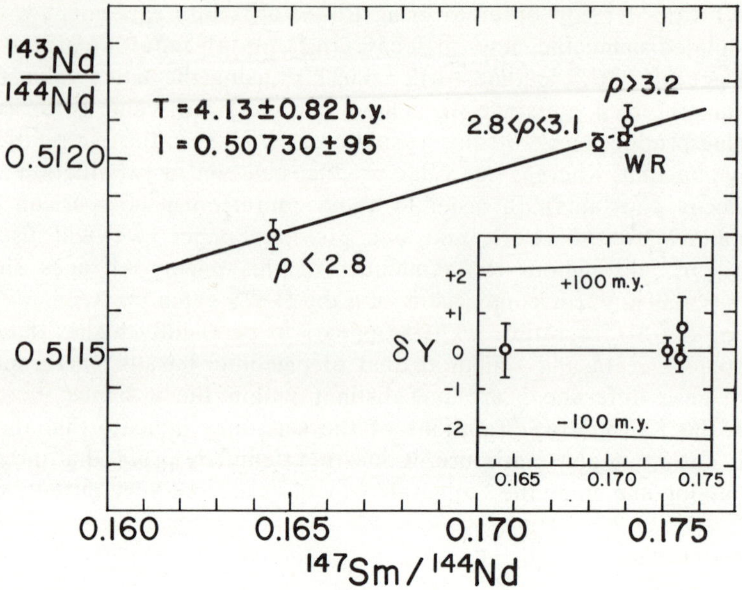

Fig. 2. Sm–Nd evolution diagram for the dark dikelet 77075. The Sm/Nd ratios show little fractionation among the density fractions and the data do not define a precise isochron. The nominal value for the age is 4.13 ± 0.82 b.y. (2σ) and the initial $^{143}Nd/^{144}Nd$ ratio is 0.5073 ± 0.0010 (2σ).

Discussion

The dating of the dark dikelets could solve quite important problems that exist in the interpretation of lunar evolution history as, for example, whether Apollo 17 site boulders all originated from a single ~3.95 b.y. impact event into 4.36 b.y. old ANT material (Winzer *et al.*, 1977). Furthermore, the age of 77075 may provide knowledge of the early history of the moon beyond 4 b.y. The Rb–Sr "isochron" age of 4.12 ± 0.08 b.y. appears to be distinctly older than the 3.95 b.y. age assigned for the Serenitatis event. However, since some data points deviate from the Rb–Sr "isochron", we think that some justifications are required before the Rb–Sr age is accepted as a true age.

The model age of the olivine relative to the observed initial $^{87}Sr/^{86}Sr$ ratio for the rock is 4.83 b.y. and that relative to BABI is 4.92 b.y. These model ages are too old and suggest that either the Sr isotopic composition is in error or that older olivine xenocrysts lost Rb relative to Sr during the ~4.12 b.y. event.

The handpicked olivine was, unfortunately, overspiked with ^{84}Sr and the $^{87}Sr/^{86}Sr$ ratio is somewhat uncertain due to a comparatively large correction for the ^{84}Sr spike contribution to ^{87}Sr and ^{86}Sr peaks. The spike correction, however, amounted to about 4% of the radiogenic $^{87}Sr/^{86}Sr$ ratio and, thus, it seems improbable that the displacement of the olivine fraction from the isochron can be attributed to experimental error. It is clear that the olivine xenocrysts did not reach isotopic equilibrium with the matrix.

The K, Rb, and Sr concentrations of the olivine xenocrysts of 77075 are more than ten times higher than those of olivine from troctolite 76535 (Papanastassiou and Wasserburg, 1976), indicating that the fraction may have contained either pyroxene xenocrysts or inclusions in the olivine. However, the K/Rb and Rb/Sr ratios of the olivine xenocrysts are exactly in the range reported for the olivine fraction of troctolite 76535. This means that these ratios alone may not be critical factors for defining whether the fractions are pure olivine or two component mixtures.

If the regression line corresponds to a mixing line, the age is significant only if the components are cogenetic and coeval. As stated above, the mineral separations were of low quality, due to the fine-grained texture of the rock. However, if the regression line represents two-component mixing between matrix (plagioclase) and xenocrysts (olivine), the olivine data should not deviate from the trend established by the other fractions.

Close examination of Fig. 1 shows that the whole-rock (WR) fraction also lies slightly above the best fit line. This deviation seems reasonable since the whole-rock fraction represents a mixture of both xenocrysts and matrix, and should be excluded from the isochron. Excluding WR and OL, the data define an isochron which corresponds to $T = 4.07 \pm 0.09$ b.y. (2σ) and $I = 0.69918 \pm 0.00016$ (2σ). The I is comparable to those of 77215 and 77115 (Nakamura *et al.*, 1976). The $\rho > 3.2$ fraction lies slightly below the isochron. The reasons for this deviation are not obvious. The deviation may be analytically induced or may be a result of the existence of a minor trace-element rich phase (phosphates?) whose isotopic systems are easily reset by post-crystallizational disturbances. On the other hand, this fraction lies above the trend established by the Sm–Nd data (Fig. 2), although this deviation is not distinct within the assigned uncertainties.

The $\rho < 2.8$ fraction which is mostly a mixture of plagioclase in the matrix and a small amount of plagioclase xenocrysts also needs close examination. If we reject the $\rho < 2.8$ fraction, under the assumption that it was not equilibrated when the rock formed, as well as the WR and OL fractions, the remaining six fractions define a line corresponding to $T = 4.00 \pm 0.12$ b.y. (2σ) and $I = 0.69932 \pm 0.00023$ (2σ). The obtained age appears to be smaller than the age including the $\rho < 2.8$ fraction because of strong influence by the $\rho > 3.2$ fraction. If rejection of the $\rho < 2.8$ fraction is valid, the obtained age is now slightly larger, but not distinguishable within experimental error, from the Serenitatis age(?) of 3.95 ± 0.05 b.y. assigned in our previous paper (Nakamura *et al.*, 1976). However, the initial $^{87}Sr/^{86}Sr$ obtained from the latter isochron appears too high when compared to Apollo 17 lunar rocks (see Nyquist *et al.*, 1976, considering -0.00011 for interlaboratory bias). Thus, we think the 4.0 b.y. age for 77075 is probably the lowest limit and we prefer the 4.07 b.y. age in which the plagioclase fraction is included in the isochron. In this case, we must assume that either Sr isotopes of plagioclase xenocrysts reached equilibrium with those of the matrix or the abundance of plagioclase xenocrysts is small enough such that their Rb–Sr system is not influencing the isochron.

A similar case can be found in the literature. Papanastassiou and Wasserburg (1976) observed that the 4.55 b.y. Rb–Sr isochron age of troctolite 76535 is distinctly older than the ages (about 4.25 b.y.) determined by Sm–Nd (Lugmair *et al.*, 1976) and ^{39}Ar–^{40}Ar methods (Husain and Schaeffer, 1975; Bogard *et al.*, 1975; Huneke and Wasserburg, 1975). In the Rb–Sr isochron of Papanastassiou and Wasserburg (1976), four data points of five olivine fractions, which contain more radiogenic ^{87}Sr than plagioclase and pyroxene, essentially controlled the slope of the 4.56 b.y. isochron, whereas only the olivine fraction deviated from the 4.26 b.y. Sm–Nd isochron (Lugmair *et al.*, 1976), which consisted of plagioclase, "symplectite", magnetic, and pyroxene separates. Papanastassiou and Wasserburg (1976, p. 2046) preferred to interpret that the 4.56 b.y. Rb–Sr age is the troctolite formation age and that the Sm–Nd and ^{39}Ar–^{40}Ar ages mark the time of excavation of the troctolite; providing that olivine is rigid and minor inclusions in the olivine did not equilibrate chemically and isotopically with other rock components, whereas the Sm–Nd systems of plagioclase, pyroxene, and the minor phases of whitlockite and chlorapatite which significantly control the Sm–Nd distribution among major minerals possibly equilibrated prior to the excavation of the troctolite. In a similar way, it may be possible to interpret that the olivine Rb–Sr system of the xenocrysts which contained inclusions enriched in Rb and Sr did not equilibrate with the matrix when they were engulfed, but that the Rb–Sr system of plagioclase reached equilibrium. Although Sr isotopic equilibration was not attained during brecciation (and consolidation) in the plagioclase of several breccias such as 65015 (Papanastassiou and Wasserburg, 1972) and 76055 (Tera *et al.*, 1974b), we believe Sr in the plagioclase xenocrysts of 77075 reached isotopic equilibrium with the matrix, because the xenocrysts are small ($<100\ \mu m$) and they were totally engulfed by an aphanitic matrix.

The partition coefficients of incompatible elements are small for olivine, and the high concentration of K, Rb, and Sr in our olivine xenocryst fraction appears mainly to be due to an impure separate containing pyroxene. We consider the pyroxene xenocrysts to have preferentially lost Rb and that the single-stage model age is older. On the other hand plagioclase xenocrysts were not rigid and did not remain closed systems when they were incorporated into 77075. In our previous study of 77215 (Nakamura *et al.*, 1976), two pyroxene fractions of clast 77215,145, obtained by handpicking and by a heavy liquid separation also plotted above the isochron, indicating Rb loss during a later event. Thus we prefer to conclude that the age (including $\rho < 2.8$ fraction) corresponds to the formation age of 77075 for the reasons discussed above.

Some 4.1 b.y. ages have been previously reported for Apollo 17 boulders and breccias. Compston *et al.* (1975) reported a 4.11 ± 0.05 b.y. age for the Civet Cat clast of shocked norite 72255 in Boulder 1 from the South Massif (Apollo 17, Station 2). They concluded that this age corresponds to the original cooling age as a plutonic igneous rock. This certainly agrees with the age of our present results and tends to support the theory that 4.1 b.y. igneous activity existed in the Apollo 17 site area. However, it should be noticed that the Civet Cat clast is noritic breccia while 77075 is a pigeonite feldspathic basalt. Compston *et al.*

(1975) reported a 3.96 ± 0.04 b.y. Rb–Sr age for pigeonitic basalt clasts in 72255. On the other hand, they also reported a 3.99 b.y. Rb–Sr model age for the granitic clast from the same boulder, suggesting that a variety of rock types existed in the area at ~4 b.y. ago.

Most $^{39}Ar/^{40}Ar$ ages of Apollo 17 boulders and breccias collected from North and South Massif are ~3.95 b.y. (Stettler *et al.*, 1974, 1975; Kirsten and Horn, 1974; Leich *et al.*, 1975; Turner and Cadogan, 1975; Husain and Schaeffer, 1975; Phinney *et al.*, 1975). Exceptions are a young age of 3.75 b.y. obtained for basalts of the Station 7 boulder and two high temperature ^{39}Ar–^{40}Ar ages of 4.17 and 4.04 b.y. for plagioclase and olivine, respectively, in a clast of 77115 (Station 7 boulder, Stettler *et al.*, 1975) and a 4.04 b.y. age for 76315 from the Station 6 boulder (Turner and Cadogan, 1975).

It has been interpreted (e.g., Turner *et al.*, 1973; Nakamura *et al.*, 1976) that the 3.95 b.y. ages correspond to the Serenitatis Basin excavation event (although Compston *et al.*, 1975, suggested the event is younger and Husain and Schaeffer, 1975, suggested it is older than 3.95 b.y.) and intermediate ^{39}Ar–^{40}Ar ages of 4.0–4.20 b.y. represent partial resetting of older material by a younger 3.95 b.y. event. Using a recently advanced Ar dating technique in which small clasts are individually melted using a laser beam, Schaeffer's group (Müller *et al.*, 1977) revealed that plagioclase and maskelynite clasts from breccia 73215 of Apollo 17 show older ages ranging from 4.04 to 4.11 b.y. Although they did not conclude whether the ages correspond to "true" ages of clasts or ages modified by partial outgassing during breccia aggregation, these ages are in agreement with the age of 77075 obtained in this study.

History of the Station 7 boulder

In our previous paper (Nakamura *et al.*, 1976), we summarized (Fig. 6) that the ages determined by K–Ar (Stettler *et al.*, 1974, 1975), Rb–Sr, Sm–Nd, and U–Pb methods for the Station 7 boulder form three age clusters; the oldest cluster contains the Rb–Sr and Nd–Sm ages of ~4.36 b.y. for noritic breccia 77215, the second cluster is ~3.95 b.y. and consists of almost all ^{39}Ar–^{40}Ar ages and the youngest one consists of 3.7–3.8 b.y. ages determined by ^{39}Ar–^{40}Ar, Rb–Sr, and U–Pb method for matrices of 77115 and 77135, although some ages are not of the best quality due to difficulties associated with age determinations for rocks which include clasts and xenocrysts. These ages probably indicate maximum ages. Now, one more age of ~4.07 b.y. between the first and second age clusters is added from our present study and the following history is given for the Station 7 boulder:

(1) We believe that the ~4.36 b.y. age of clast 77215 is a real age and it corresponds to the formation of the norite since the ages obtained by both the Rb–Sr and Sm–Nd methods are in excellent agreement (Nakamura *et al.*, 1976). We presume that Sm/Nd ratios in individual minerals did not change during the subsequent low-temperature breccia formation

and it is also unlikely to have been isotopically disturbed by the later processes.

(2) The norite was then brecciated, aggregated 4.36–4.07 b.y. ago (77215 was incorporated as a clast), and cut by veinlets 77075 at 4.07 b.y. ago. Alternatively, 77075 was formed by the Serenitatis event (3.95 b.y.) if we exclude the $\rho < 2.8$ fraction from the isochron. However, even in the latter case, the aggregation of 77215 should have occurred before the Serenitatis event (i.e., suggesting one event between 4.36 and 3.95 b.y.).

(3) The noritic breccia 77215 and dikelets 77075 were engulfed in pigeonite basalt (77115) 3.75 b.y. ago.

(4) 77115 cooled and fractured (Chao *et al.*, 1974) and was then engulfed by 77315 at 3.75 b.y. ago (the formation of 77115 and 77135 was during one 3.75 b.y. event or two events in close succession 3.75 b.y. ago).

Although the ~4.07 and 3.95 b.y. ages recorded in Boulder 7 are also found in other Apollo 17 boulders and breccias collected from North and South Massifs, the oldest (4.36 b.y.) and youngest (3.75 b.y.) ages observed in Boulder 7 are not common among Apollo 17 boulders.

The 4.25 b.y. age for the troctolite 76535 (Lugmair *et al.*, 1976; Husain and Schaeffer, 1975; Bogard *et al.*, 1975; Huneke and Wasserburg, 1975) is one of the oldest ages obtained by the ^{39}Ar–^{40}Ar and Sm–Nd methods from the Apollo 17 site. Kirsten and Horn (1974) also reported a 4.22 b.y. ^{39}Ar–^{40}Ar age for gabbro 78503 collected from the base of the Sculptured Hills. The oldest age of 4.36 b.y. measured in the Station 7 boulder is distinctly older than these oldest ^{39}Ar–^{40}Ar and Sm–Nd ages found for boulders from Apollo 17, and we believe that this age represents a lower limit to the initiation of the anorthositic lunar crust formation.

It is generally accepted that the deepest profiles of Serenitatis and Tranquilitatis ejecta are exposed at the Apollo 17 landing site at the rim of the Serenitatis basin. The Tranquilitatis event is older than the Serenitatis event. If we assign a 3.95 b.y. age to Serenitatis, as originally proposed by Turner *et al.* (1973) and if we assign the 4.36 b.y. age as the norite formation age, then the event that occurred during 4.07–4.36 b.y. may correspond to the Tranquilitatis event. Because 77075 is not a pseudotachylite produced by an *in situ* shock process (Chao *et al.*, 1974), we conclude that brecciation of 77215 occurred prior to 4.07 b.y. by an impact (Tranquilitatis?) event (i.e., the 4.07 b.y. age is the lower limit for brecciation) and penetration of dikelets 77075 into the breccia 77215 occurred 4.07 b.y. ago.

Papanastassiou and Wasserburg (1976, p. 4046) preferred to interpret that the 4.55 b.y. Rb–Sr age obtained for the troctolite 76535 corresponds to the formation, but the ~4.2 b.y. of Sm–Nd and ^{39}Ar–^{40}Ar ages correspond to an excavation time by an impact. This interpretation and the 77215 brecciation time, which occurred >4.07 b.y. ago, appear to contradict the hypothesis of a terminal lunar cataclysm (e.g., Tera *et al.*, 1974a,b).

The youngest age recorded in the matrices of 77115 and 77135 has been debated as to whether it corresponds to an impact event or an igneous event

(Chao *et al.*, 1974; Storey *et al.*, 1974; Winzer *et al.*, 1977). Although no 3.75 b.y. age has yet been reported for other Apollo 17 highland rocks (a number of 3.75 b.y. ages for Apollo 17 mare basalts are reported), including the results of the laser probe ^{39}Ar–^{40}Ar dating by Schaeffer's group (Müller *et al.*, 1977), we prefer to interpret that the youngest 3.75 b.y. age recorded in Station 7 boulder corresponds to the Imbrium event. Although most investigators concluded that the Imbrium event is ~3.82 b.y. old, Kirsten and Horn (1974) argued that the Imbrium impact might be as young as ≤3.72 b.y. Alternatively, the age is slightly younger than the true age of Imbrium due to later disturbances by one or more intermediate-size post-basin-excavation events.

Rocks Older than 4.0 b.y.

In the past few years, several rocks older than 4.0 b.y. have been found on the moon. These rocks should provide important clues to the first ~0.6 b.y. of lunar evolution. The troctolite 76535 has yielded the oldest Rb–Sr age yet found on the moon (4.55 ± 0.07 b.y.; Papanastassiou and Wasserburg, 1976), although the Sm–Nd age (Lugmair *et al.*, 1976) and Ar–Ar ages (Bogard *et al.*, 1974; Husain and Schaeffer, 1975; Huneke and Wasserburg, 1975) are all about 4.2 b.y. Although the younger ages have been attributed to the excavation event, we believe that further verification of the Rb–Sr age is needed before this age can be accepted as the true formation age and its origin as a product of lunar primary differentiation. Papanastassiou and Wasserburg (1975) have also obtained a 4.49 b.y. age for dunite 74217. These authors interpreted that the dunite may also be a product of lunar primary differentiation. Our group (Nakamura *et al.*, 1976 and this work) has added two more rocks (both from Boulder 7) to the small menagerie of old rocks. Rock 77215, a noritic breccia (Chao *et al.*, 1974), has yielded the oldest age (4.36 b.y.) confirmed by both Rb–Sr and Sm–Nd methods, and 77075, an aphanitic basalt has yielded the youngest age (4.0–4.07 b.y.) of the old group, although this age also requires further verifications. Anorthosites, which generally yield ≃4.2 b.y. Ar–Ar ages (excavation ages?) and 4.1 b.y. ages from other Apollo 17 boulders round out the menagerie.

Papanastassiou and Wasserburg (1975) concluded that the Rb–Sr age of the dunite 74217 (a similar argument would hold for 77215) was in reasonable agreement with the 4.42 b.y. "primary differentiation age" defined by U–Pb systematics using a two-stage model (Tera and Wasserburg, 1974, 1975). Because the ~4.4 b.y. ages were, in turn, in agreement with the time for lunar rare-earth element fractionation (4.36 b.y.; Lugmair *et al.*, 1975) using a single-stage evolution model from chondritic abundances for mare basalt 75075, these authors (Papanastassiou and Wasserburg, 1975; 1976; Papanastassiou *et al.*, 1977) concluded that the ~4.4 b.y. ages were closely related to the moon's primary differentiation.

We, however, have argued against the concept of the 4.4 b.y. primary differentiation age (Nunes *et al.*, 1975; Tatsumoto and Unruh, 1976; Unruh and

Tatsumoto, 1977) for the following reasons: (1) We believe that the apparent 4.42 b.y. common upper intercept on the U–Pb evolution diagram may be an illusion due to the near tangency of the ~3.9–4.4 b.y. chord with the concordia curve (i.e., multi-stage evolution, or "primary" differentiation over a prolonged period would not be distinguishable from apparent two-stage evolution; Nunes *et al.*, 1975). Alternatively, the apparent chord may be nothing more than a KREEP-mixing line, since a small amount of KREEP material should dominate the U–Pb systematics of a rock. (2) The 4.42 b.y. upper intercept age was obtained using Canyon Diablo troilite Pb as the initial Pb (two-stage model). If the moon was initially volatile-depleted (large initial $^{238}U/^{204}Pb$ ratio) at the time of accretion (e.g., Ganapathy and Anders, 1974) or differentiation of its outer several hundred kilometers ~4.56(?) b.y. ago, then the Pb isotopic composition may have evolved significantly during the early history of the moon and correction for more evolved Pb will lower the apparent older intercept (by as much as 200 m.y. by three-stage model; Nunes *et al.*, 1975). (3) The 4.36 b.y. age of rare-earth element differentiation from the chondritic abundance in basalt 75075 has been recalculated to be as old as 4.55 b.y. because of the revised $^{143}Nd/^{144}Nd$ value for the Juvinas achondrite (Lugmair *et al.*, 1976), which was also confirmed by Nakamura *et al.* (1976, 1977). The Sm–Nd model ages thus far measured for mare basalts range from 3.35 b.y. for 15426 (Lugmair *et al.*, 1977) to 6.18 b.y. for 15555 (Lugmair *et al.*, 1976). These results indicate that the Sm–Nd systematics of some lunar rocks cannot be explained by single-stage differentiation from the chondritic value. Thus, it seems at present, there is no way to unprejudicially prove that the 4.4 b.y. age is the time for the lunar primary differentiation and alternative explanations should be considered, although such an age is attractive considering gradual cooling of a 4.56 b.y. old lunar body from the surface.

The isotopic data thus far accrued do not provide a unique interpretation with regard to the first ~0.6 b.y. of lunar history. The U–Pb and Sm–Nd data do not require the data to be explained by two stages of evolution. Indeed, some Sm–Nd data (e.g., 15555) *demand* a more complex interpretation, as do the U–Pb systematics of some mare basalts (cf., Unruh and Tatsumoto, 1977). The ~4.07 b.y. ages of 77075 and other Apollo 17 boulders may indicate either a major meteorite impact or igneous activity ~4.07 b.y. ago in the Apollo 17 area. The ~4.36 b.y. age and deep-seated origin (Chao *et al.*, 1974) of 77215 suggest that this rock may also be a product of early lunar differentiation, although such an interpretation is quite speculative.

The fact that we see four events recorded in this boulder leads us to speculate that modifications to the isotopic systems of some rocks (or their sources?) may have occurred during the ~4.0–4.56 b.y. interval. Thus, the possibility of multi-stage evolution should at least be considered when discussing the evolution of lunar rocks.

Acknowledgments—This study is a part of the Station 7 boulder consortium study led by E. C. T. Chao. Comments given by G. R. Tilton, D. M. Unruh, G. W. Lugmair, and Herbert Palme on the manuscript are appreciated. We are also indebted to Sandra L. Blanchard for pounding her fingers to the bone to prepare this manuscript. This work was mianly supported by NASA Interagency transfer order T-2407A.

REFERENCES

Bogard D. D., Nyquist L. E., Bansal B. M., and Wiesmann H. (1974) 76535: An old lunar rock? (abstract). In *Lunar Science V*, p. 70–72. The Lunar Science Institute, Houston.

Bogard D. D., Nyquist L. E., Bansal B. M., Wiesmann H., and Shih C.-Y. (1975) 76535: An old lunar rock. *Earth Planet. Sci. Lett.* **26**, 69–80.

Butler P. and Dealint T. E. (1974) The dissection and consortium allocation of Apollo 17 lunar rock from the boulder at Station 7. *Earth Planet. Sci. Lett.* **23**, 429–434.

Chao E. C. T., Minkin J. A., and Thompson C. L. (1974) Preliminary petrographic description and geologic implications of the Apollo 17 Station 7 boulder consortium samples. *Earth Planet. Sci. Lett.* **23**, 413–428.

Compston W., Foster J. J., and Gray C. M. (1975) Rb–Sr ages of clasts from within Boulder 1, Station 2, Apollo 17. *The Moon* **14**, 445–462.

Ganapathy R. and Anders E. (1974) Bulk compositions of the moon and earth, estimated from meteorites. *Proc. Lunar Sci. Conf. 5th*, p. 1181–1206.

Husain L. and Schaeffer O. A. (1975) Lunar evolution: The first 600 million Years. *Geophys. Res. Lett.* **2**, 29–32.

Huneke J. C. and Wasserburg G. J. (1975) Trapped ^{40}Ar in troctolite 76535 and evidence for enhanced ^{40}Ar–^{39}Ar age plateaus (abstract). In *Lunar Science VI*, p. 417–419. The Lunar Science Institute, Houston.

Jaffey A. H., Flynn K. F., Glendenin L. E., Bentley W. C., and Essling A. M. (1971) Precision measurement of half-lives and specific activities of ^{235}U and ^{238}U. *Phys. Rev.* **C4**, 1889–1906.

Kirsten T. and Horn P. (1974) Chronology of the Taurus-Littrow region III: Ages of mare basalts and highland breccias and some remarks about the interpretation of lunar highland rock ages. *Proc. Lunar Sci. Conf. 5th*, 1451–1475.

Leich D. A., Kahl S. B., Kirschbaum A. R., Niemeyer S., and Phinney D. A. (1975) Rare gas studies on Boulder 1, Station 2, Apollo 17 (abstract). In *Lunar Science VI*, p. 501–503. The Lunar Science Institute, Houston.

Lugmair G. W. and Marti K. (1977) Evolution of the lunar interior: Sm–Nd systematics of A15 green glass and the question of the lunar initial ^{143}Nd/^{144}Nd (abstract). In *Lunar Science VIII*, p. 597–599. The Lunar Science Institute, Houston.

Lugmair G. W., Marti K., Kurtz J. P., and Scheinin N. B. (1976) History and genesis of lunar troctolite 76535 or: How old is old? *Proc. Lunar Sci. Conf. 7th*, p. 2009–2033.

Lugmair G. W., Scheinin N. B., and Marti K. (1975) Sm–Nd age and history of Apollo 17 basalt 75075. Evidence for early differentiation of the lunar exterior. *Proc. Lunar Sci. Conf. 6th*, p. 1419–1429.

Müller H. W., Plieninger T., James O. B., and Schaeffer O. A. (1977) Laser probe ^{40}Ar–^{39}Ar dating of materials from consortium breccia (abstract). In *Lunar Science VIII*, p. 697–699. The Lunar Science Institute, Houston.

Nakamura N., Tatsumoto M., Nunes P. D., Unruh D. M., Schwab A. P., and Wildeman T. R. (1976) 4.4 b.y.-old clast in Boulder 7, Apollo 17: A comprehensive chronological study by U–Pb, Rb–Sr and Sm–Nd methods. *Proc. Lunar Sci. Conf. 7th*, p. 2309–2333.

Nakamura N., Unruh D. M., Gensho R., and Tatsumoto M. (1977) Evolution history of lunar mare basalts: Apollo 15 samples revisited (abstract). In *Lunar Science VIII*, p. 712–713. The Lunar Science Institute, Houston.

Nunes P. D., Tatsumoto M., and Unruh D. M. (1974) U–Th–Pb and Rb–Sr systematics of Apollo 17 Boulder 7 from the north massif of the Taurus-Littrow Valley. *Earth Planet. Sci. Lett.* **23**, 445–452.

Nunes P. D., Tatsumoto M., and Unruh D. M. (1975) U–Th–Pb systematics of anorthositic gabbros 78155 and 77017—implications for early lunar evolution. *Proc. Lunar Sci. Conf. 6th*, p. 1431–1444.

Nyquist L. E., Bansal B. M., and Wiesmann H. (1976) Sr isotopic constraints on the petrogenesis of Apollo 17 mare basalts. *Proc. Lunar Sci. Conf. 7th*, p. 1507–1528.

Papanastassiou D. A. and Wasserburg G. J. (1972) Rb–Sr systematics of Luna 20 and Apollo 16 samples. *Earth Planet Sci. Lett.* **17**, 52–63.

Papanastassiou D. A. and Wasserburg G. J. (1975) Rb–Sr study of a lunar dunite and evidence for early lunar differentiates. *Proc. Lunar Sci. Conf. 6th*, p. 1467–1489.

Papanastassiou D. A. and Wasserburg G. J. (1976) Rb–Sr age of troctolite 76535. *Proc. Lunar Sci. Conf. 7th*, p. 2035–2054.

Papanastassiou D. A., DePaolo D. J., Tera F., and Wasserburg G. J. (1977) An isotopic triptych on mare basalts: Rb–Sr, Sm–Nd, U–Pb (abstract). In *Lunar Science VIII*, p. 750–752. The Lunar Science Institute, Houston.

Phinney D., Kahl S. B., and Reynolds J. H. (1975) ^{40}Ar–^{39}Ar dating of Apollo 16 and 17 rocks. *Proc. Lunar Sci. Conf. 6th*, p. 1593–1608.

Schmitt H. H. (1973) Apollo 17 report on the Valley of Taurus-Littrow. *Science* **182**, 681–690.

Stettler A., Eberhardt P., Geiss J., and Grögler N. (1974) ^{39}Ar–^{40}Ar ages of samples from the Apollo 17 Station 7 boulder and implication for its formation. *Earth Planet. Sci. Lett.* **23**, 453–461.

Stettler A., Eberhardt P., Geiss J., Grögler N., and Guggisberg S. (1975) Age sequence in the Apollo 17 Station 7 boulder (abstract). In *Lunar Science VI*, p. 771–773. The Lunar Science Institute, Houston.

Storey W. C., Humphries D. J., and O'Hara M. J. (1974) Experimental petrology of sample 77135. *Earth Planet. Sci. Lett.* **23**, 435–438.

Tatsumoto M. and Unruh D. M. (1976) KREEP basalt age: Grain by grain U–Th–Pb systematics study of the quartz monzodiorite clast 15405,88. *Proc. Lunar Sci. Conf. 7th*, p. 2107–2129.

Tera F., Papanastassiou D. A., and Wasserburg G. J. (1974a) The lunar time-scale and a summary of isotopic evidence for a terminal lunar cataclysm (abstract). In *Lunar Science V*, p. 792–794. The Lunar Science Institute, Houston.

Tera F., Papanastassiou D. A., and Wasserburg G. J. (1974b) Isotopic evidence for a terminal lunar cataclysm. *Earth Planet. Sci. Lett.* **22**, 1–21.

Tera F. and Wasserburg G. J. (1974) U–Th–Pb systematics on lunar rocks and inferences about lunar evolution and the age of the moon. *Proc. Lunar Sci. Conf. 5th*, p. 1571–1599.

Tera F. and Wasserburg G. J. (1975) The evolution and history of mare basalts as inferred from U–Th–Pb systematics (abstract). In *Lunar Science VI*, p. 807–809. The Lunar Science Institute, Houston.

Turner G. and Cadogan P. H. (1975) The history of lunar bombardment inferred from ^{40}Ar–^{39}Ar dating of highland rocks. *Proc. Lunar Sci. Conf. 6th*, p. 1509–1538.

Turner G., Cadogan P. H., and Yonge C. J. (1973) Argon selenschronology. *Proc. Lunar Sci. Conf. 4th*, p. 1889–1914.

Unruh D. M. and Tatsumoto M. (1977) Evolution of mare basalts: The complexity of the U–Th–Pb system. *Proc. Lunar Sci. Conf. 8th.* This volume.

Winzer S. R., Nava D. F., Schuhmann P. J., Lum R. K. L., Schuhmann S., Lindstrom M. M., Lindstrom D. J., and Philpotts J. A. (1977) The Apollo 17 "Melt Sheet", chemistry, age and Rb/Sr systematics. *Earth Planet. Sci. Lett.* **33**, 389–400.

York D. K. (1969) Least square fitting of a straight line with correlated errors. *Earth Planet. Sci. Lett.* **5**, 320–324.

Proc. Lunar Sci. Conf. 8th (1977), p. 2315–2319.
Printed in the United States of America

68815 revisited

SUSAN DUST and GHISLAINE CROZAZ

McDonnell Center for the Space Sciences, Washington University, St. Louis, Missouri 63130

Abstract—Remeasurement of the particle track density as a function of depth in lunar breccia 68815 has yielded galactic cosmic-ray track densities consistent with a 2 m.y. surface exposure, in good agreement with results by other authors. The empirical track production spectrum of Walker and Yuhas (1973) is confirmed. The solar-flare track profile is typical and shows no evidence of large-scale chipping. Radionuclide profiles (^{53}Mn) investigated by the Arnold group should thus be undisturbed and should provide a test of the Reedy–Arnold model and of the constancy of solar and galactic cosmic rays.

INTRODUCTION

LUNAR ROCK 68815 was collected from a meter-sized, highly angular boulder located on a ray of the young, bright-rayed South Ray Crater. Since the boulder shows little evidence of rounding or chipping, a short surface exposure time for 68815 is indicated. Previous work on this sample has confirmed its short exposure on the surface and simple irradiation history. Behrmann *et al.* (1973) obtained an exposure age of 2.0 ± 0.2 m.y. based on ^{81}Kr–Kr studies, and concluded that the absence of prominent neutron effects implied burial in excess of 7 m prior to ejection. Walker and Yuhas (1973) used the track data on 68815 and the known surface exposure age to derive an empirical track production spectrum. This spectrum was shown to be consistent with the contemporary VH differential energy spectrum averaged over solar cycle 19 (Yuhas and Walker, 1973). No evidence was seen for changes in spectral shape or for variations in absolute flux (to within 20%) when averaged over the 2 m.y. exposure age of 68815.

Because of its short exposure age and its simple irradiation history, 68815 makes possible a crucial test of the production rate of ^{53}Mn, both in the galactic (deep depth) and solar-flare (shallow depth) regions. The latter is particularly interesting because of suggestions that the solar flare production rate might have been higher in the past than previously assumed. For example, track data on rock 14310 have been interpreted as indicating a relatively short exposure (<3 m.y.) in its more recent orientation (Crozaz *et al.*, 1972). However, ^{53}Mn activities in the surface regions are high (Imamura *et al.*, 1974), indicating a longer exposure (>5 m.y.) to solar flares if the sun has been constant and if the ^{53}Mn production rates are taken from the Reedy-Arnold values (Reedy and Arnold, 1972). Recent measurements of ^{53}Mn in core tubes have also given high values (Nishiizumi *et al.*, 1976).

The present track measurements were undertaken as a complement to radionuclide work by the Arnold group on a selected sample of 68815. The two primary objectives were: (1) to assure that the rock surface had a normal complement of solar-flare tracks and had not been recently disturbed, and (2) to fill in a data gap between 5 mm and 2 cm in the earlier track studies on this rock.

SAMPLE PREPARATION, CUTTING, AND TRACK COUNTING

A column from a corner of 68815,234 was sent to us by the Arnold group at San Diego (the larger piece is being used for radionuclide measurements). The column extends from the exposed surface of the sample to a depth of ~21 mm. A polished thick section extending the width of the column from the top (exposed) surface to the 5 mm depth was prepared. After etching in boiling 6.25 N NaOH, the sample was examined under a scanning electron microscope (SEM) for solar flare particle tracks. Track counts were obtained for feldspar crystals only. Three track profiles were established, each beginning at a different location on the exposed surface, and extending to depths of $\sim 100\mu$. Further etching of the thick section revealed tracks in the depth region down to ~1.5 mm. The track densities in the solar flare region fell between 4×10^9 and $6 \times 10^6\ cm^{-2}$.

The remaining ~16 mm of the column were cut into 21 slices (perpendicular to the length of the column). Thin sections were prepared of those slices which contained substantial amounts of feldspar. Track densities were measured by transmission optical microscopy, but some difficulties were encountered due to the brecciated nature of the sample. Most polished surfaces became highly irregular after etching, presenting numerous dislocations and inclusions. Some thin sections proved unsuitable for track work, but four sections in the depth range between 5 mm and 20 mm had sufficiently clear crystals to give good counts. A 2μ counting criterion was used (i.e., any defect $<2\mu$ in length was not counted as a track).

RESULTS

A log–log plot of track density versus depth in 68815 is presented in Fig. 1. Note that SEM and optical counts are presented on the same graph. Usually, SEM counts are divided by two before comparison with optical measurements, in order to account for differences in track counts obtained by the two viewing techniques (Crozaz *et al.*, 1970). In this case, the effect was offset by the use of two different viewing geometries. The plane of observation for counts in the upper 5 mm was perpendicular to the nearest rock surface. Such counts are lower by a factor of two than counts made in the viewing plane used for the optical measurements below the 5 mm depth.

DISCUSSION

The three near-surface track density profiles shown in Fig. 1 have the general steeply falling character typified by solar flare tracks found in relatively fresh lunar rock surfaces (see for example Morrison and Zinner, 1975). We therefore have no reason to believe that radionuclide data would be disturbed by recent large scale chipping of the sample, either in handling and preparation or on the lunar surface. However, small scale chipping cannot be ruled out, and neither can the presence of a thin dust coating (a few tens of microns) distributed unevenly in pockets and depressions. It is possible that these factors may account for the variation of approximately a factor of two in absolute track densities between the three profiles.

The role of erosion in the history of this particular sample is somewhat difficult to define from the track standpoint. The slope of the track density profile in the solar flare region is 1.2 ± 0.1. This is somewhat steeper than the slope of ~ 1 found for many eroded lunar rocks (Crozaz *et al.*, 1971), yet less than the value of 1.7 to 1.8 found for relatively uneroded surfaces (Morrison and Zinner, 1975). The track profile is probably in a state of transition from an uneroded profile to one characteristic of erosion equilibrium.

Walker and Yuhas (1973) derived their empirical track production spectrum based on the measured galactic cosmic ray track densities in 68815 and the given 2.0 ± 0.2 m.y. exposure age. In the near surface region, the best fit to their track data was obtained by assuming an overall erosion rate of 1 mm/m.y. If this erosion rate is correct, we should expect erosional effects to be important for our sample at depths $\leqslant 2$ mm. For depths $\geqslant 2$ mm, erosional effects should become negligible as we approach the region in which galactic cosmic ray tracks predominate. For intermediate depths on the order of 2 mm we expect a transition region. Although there is some evidence for such a transition in the 1.5–3 mm range (see Fig. 1) as the slope of the profile flattens approaching the galactic region, the data in this depth range are very poor. The transition cannot be established with certainty. Track counts were extremely difficult in the 150μ–1.5 mm range, and impossible between 1.5 mm and about 5 mm, due to the shocked nature of this breccia and the scarcity of good feldspar crystals. Further observations of tracks in the 150μ–5 mm region would depend on analyzing samples with fewer shocked crystals in this depth range.

The previous track profile (Walker and Yuhas, 1973) had a data gap in the depth range between 5 mm and 2 cm. This range is covered in this work. Using the track production spectrum derived by Walker and Yuhas (1973), the appropriate rock geometry and viewing plane orientation, and the given exposure age of 2 m.y., we calculate track densities in complete agreement with the densities in the 5 mm to 2 cm range shown in Fig. 1. The empirical track production spectrum is thus confirmed. The agreement of this spectrum with that measured for contemporary cosmic rays (Yuhas and Walker, 1973) demonstrates again the long-term constancy of the galactic cosmic ray flux of very heavy ions.

Relationship of Track Studies to ^{53}Mn Measurements

Because of its brief and simple exposure history and the fact demonstrated above that the solar flare exposure for the surface is quite normal, measure-

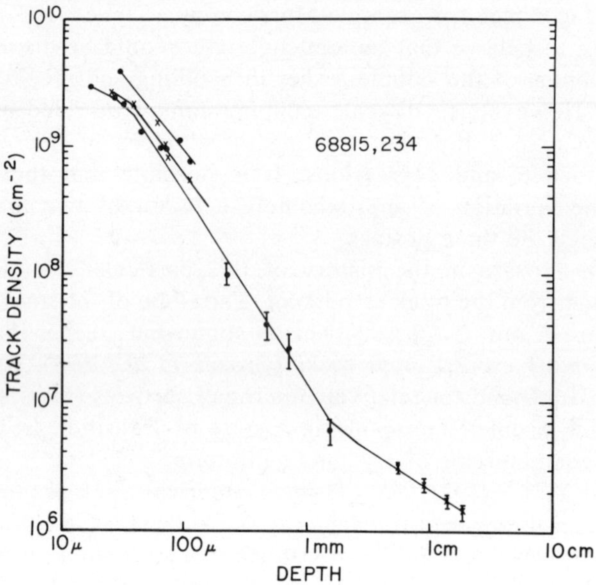

Fig. 1. Track densities as a function of depth in lunar rock 68815. The three near-surface profiles (depths <100μ) were obtained starting from different locations on the sample surface.

ments of the ^{53}Mn activity profile in 68815 provide a crucial test of the Reedy-Arnold model and of the constancy of solar flare protons and galactic cosmic rays.

In the near-surface regions, ^{53}Mn ($T_{1/2} = 3.7 \times 10^6$ yr) is produced primarily from iron by solar flare proton bombardment (see for example Imamura *et al.*, 1973). Production of ^{53}Mn from secondary neutrons is important at greater depths. As shown by Fruchter *et al.* (1977), the total galactic ^{53}Mn activity of 68815 is considerably lower than that for other lunar rocks, as would be expected from the brief exposure history. In recent work by Kohl *et al.* (1977), the Reedy-Arnold theoretical galactic cosmic ray (GCR) production rate of ^{53}Mn has been renormalized, raising the GCR production by 10% over that used by Nishiizumi *et al.* (1976). Kohl *et al.* (1977) obtain ^{53}Mn results at a single depth (>7 g/cm^2) in excellent agreement with the predictions of the Reedy-Arnold model, using the renormalized GCR component, the SCR flux calculated by Nishiizumi *et al.* (1976), and the 2 m.y. exposure age. This constitutes evidence both for the validity of the model and the constancy of the galactic cosmic ray flux over the time scale of 2 m.y. Fruchter *et al.* (1977) present results in good agreement with this, quoting a 1.9 m.y. ^{53}Mn exposure age for 68815. Detailed radionuclide measurements of the shallow regions of 68815 to establish the ^{53}Mn production rate by solar flare particles are still in progress (Kohl *et al.*, 1977).

Acknowledgments—We wish to thank Robert M. Walker and Ernst Zinner for their comments and suggestions and Diane Shuey for her expert help in preparation of the manuscript. This work was supported by NASA grant No. NGL-26-008-065.

REFERENCES

Behrmann C., Crozaz G., Drozd R., Hohenberg C., Ralston C., Walker R., and Yuhas D. (1973) Cosmic-ray exposure history of North Ray and South Ray material. *Proc. Lunar Sci. Conf. 4th*, p. 1957–1974.

Crozaz G., Haack U., Hair M., Maurette M., Walker R., and Woolum D. (1970) Nuclear track studies of ancient solar radiations and dynamic lunar surface processes. *Proc. Apollo 11 Lunar Sci. Conf.*, p. 2051–2080.

Crozaz G., Walker R., and Woolum D. (1971) Nuclear track studies of dynamic surface processes on the moon and the constancy of solar activity. *Proc. Lunar Sci. Conf. 2nd*, p. 2543–2558.

Crozaz G., Drozd R., Hohenberg C. M., Hoyt H. P., Jr., Ragan D., Walker R. M., and Yuhas D. (1972) Solar flare and galactic cosmic ray studies of Apollo 14 and 15 samples. *Proc. Lunar Sci. Conf. 3rd*, p. 2917–2931.

Fruchter J. S., Laul J. C., Rancitelli L. A., and Perkins R. W. (1977) Exposure age and other information on lunar surface material and meteorites from ^{53}Mn measurements (abstract). In *Lunar Science VIII*, p. 331–333. The Lunar Science Institute, Houston.

Imamura M., Finkel R. C., and Wahlen M. (1973) Depth profile of ^{53}Mn in the lunar surface. *Earth Planet. Sci. Lett.* **20**, 107–112.

Imamura M., Nishiizumi K., Honda M., Finkel R. C., Arnold J. R., and Kohl C. P. (1974) Depth profiles of ^{53}Mn in lunar rocks and soils. *Proc. Lunar Sci. Conf. 5th*, p. 2093–2103.

Kohl C. P., Russ G. P. III, Arnold J. R., Nishiizumi K., Imamura M., and Honda M. (1977) ^{53}Mn in lunar cores: Evidence for the time scale of surface gardening (abstract). In *Lunar Science VIII*, p. 552–554. The Lunar Science Institute, Houston.

Morrison D. A. and Zinner E. (1975) Studies of solar flares and impact craters in partially protected crystals. *Proc. Lunar Sci. Conf. 6th*, p. 3373–3390.

Nishiizumi K., Imamura M., Honda M., Russ G. P. III, Kohl C. P., and Arnold J. R. (1976) ^{53}Mn in the Apollo 15 and 16 drill stems: Evidence for surface mixing. *Proc. Lunar Sci. Conf. 7th*, p. 41–54.

Reedy R. C. and Arnold J. R. (1972) Interaction of solar and galactic cosmic-ray particles with the moon. *J. Geophys. Res.* **77**, 537–555.

Walker R. and Yuhas D. (1973) Cosmic ray track production rates in lunar materials. *Proc. Lunar Sci. Conf. 4th*, p. 2379–2389.

Yuhas D. and Walker R. (1973) Long term behavior of VH cosmic rays as observed in lunar rocks. Cosmic Ray Conference, 13th International, Denver, Conference Papers **2**, 1116–1121.

Proc. Lunar Sci. Conf. 8th (1977), p. 2321–2338.
Printed in the United States of America

In search of LKFM

ARCH M. REID, ANDREW R. DUNCAN, and STEPHEN H. RICHARDSON

School of Earth Sciences, University of Cape Town, Rondebosch, South Africa

Abstract—Low K Fra Mauro basalt (LKFM) was the name given to the composition of a group of glasses from Apollo 15 soils that are similar in composition to KREEP basalt but with a lower K content. This preferred glass composition occurs in several lunar soils and orbital X-ray data are consistent with the LKFM composition for many areas of the lunar surface. In addition rocks with LKFM bulk composition have been returned from the Apollo 15, 16, and 17 sites. The larger rocks with LKFM composition are impact produced and no primary igneous LKFM has been described. The origin of the LKFM glasses is by impact melting of pre-existing rocks and the immediate precursors are most probably the breccias, impact melts, and soils with LKFM composition. If the widespread occurrence of LKFM impact rocks can be related to a single major impact then the LKFM composition may have originated by mixing in the Imbrium event. Alternately, the abundance of "secondary" LKFM rocks over large portions of the lunar surface associated with several major impacts argues for a significant portion of the lower crust having LKFM composition early in lunar history.

PREAMBLE

THE TERM LOW K FRA MAURO BASALT (LKFM) was first used in a paper by Reid *et al.* (1972), dealing with glass compositions in the <1 mm fraction of three Apollo 15 soils. Over 500 glass particles in these soils were analysed for major elements by microprobe techniques. Eliminating glass compositions that indicate a substantial mare basalt component in the melt leaves a large number of glasses derived from non-mare* rocks with little admixture of mare material. Many factors contribute to producing a diversity of glass compositions but amid this spectrum certain preferred compositions can be discerned (the grouping of analytical data around certain preferred values was confirmed using cluster analysis techniques). Three major non-mare glass groups were recognised and named Highland basalt, Fra Mauro basalt (medium and high K), and low K Fra Mauro basalt. The LKFM glass component constitutes 9, 17, and 20% of the glasses in the three soils studied (15601, 15021, and 15101).

The glasses were interpreted by Reid *et al.* (1972), as melts of pre-existing rocks and the preferred compositions from the glass clusters were taken as indicative of the composition of locally or regionally abundant lunar rocks. The glasses retain only compositional evidence of the nature of the precursors which could thus be soil, breccia, or igneous rock. The LKFM composition was interpreted as being derived from an igneous parent because soils and breccias

*Non-mare is used as a compositional term to indicate the apparent absence of any component derived from high Fe (and high Ti) mare basalts. It is used here for convenience and we do not intend to imply that non-mare compositions in this sense could not have occupied mare basins.

seemed unlikely sources for preferred compositions and because the LKFM composition could be related to igneous KREEP basalt compositions by simple igneous processes.

Similar studies of soil from other lunar sampling sites have shown that LKFM glasses are common in several lunar soils. Table 1 shows the average composition for glasses that fall in this chemical group and the abundance of glasses of this type (relative to other glasses) for each of the sites studied. The data are all from the laboratory at the Johnson Space Center but other investigators have obtained essentially similar results.

If the glass data have been interpreted correctly then rocks of LKFM composition are a major component of the lunar crust. This interpretation is supported by the orbital X-ray and γ-ray data which are consistent with the LKFM composition for several areas of the lunar surface (Charette *et al.*, 1977). LKFM thus figures prominently in several lunar models and it has been suggested that it may make up almost a third of the lunar crust (Taylor *et al.*, 1973). Yet no primary igneous rock with LKFM composition has been described. Even in light of the limited extent of lunar sampling igneous LKFM does appear to be somewhat elusive for a rock type that is believed to be abundant. We report here on a search of Apollo 15 breccia samples for clasts that have igneous textures and are not obviously mare basalts. The objectives of this search were to assess the nature of non-mare igneous rocks in the Apollo 15 breccias and, in particular, to search for evidence of the elusive igneous LKFM. The data are presented and discussed briefly below with emphasis on the absence of manifestations of the aseity of igneous LKFM. The remainder of the paper is a discussion of the evidence of LKFM as a major lunar rock type and of the nature and origin of lunar rocks that have LKFM composition.

Table 1. Average composition in weight percent of LKFM glasses in soils from five lunar sites.

	Apollo 15	Apollo 16	Apollo 17	Luna 16	Luna 20
SiO_2	46.56 (1.78)	46.69 (2.07)	47.08 (2.25)	46.0	46.09 (2.10)
TiO_2	1.25 (0.45)	0.75 (0.45)	1.08 (0.92)	1.0	0.79 (0.56)
Al_2O_3	18.83 (1.85)	21.53 (2.47)	19.29 (2.65)	20.7	20.11 (3.14)
Cr_2O_3	0.20 (0.05)	0.11 (0.06)	0.23 (0.07)		0.19 (0.06)
FeO	9.67 (1.88)	6.89 (1.85)	8.76 (2.25)	9.8	9.36 (3.01)
MgO	11.04 (1.61)	10.36 (1.73)	11.50 (1.91)	8.8	10.27 (2.00)
CaO	11.60 (0.93)	12.86 (0.98)	11.76 (1.14)	13.0	12.60 (1.13)
Na_2O	0.37 (0.14)	0.31 (0.23)	0.27 (0.16)	0.4	0.37 (0.21)
K_2O	0.12 (0.07)	0.08 (0.09)	0.07 (0.04)	0.12	0.08 (0.07)
Percent abundance	15.0	9.2	5.5		16

Numbers in brackets are one standard deviation. Sources of data are Reid *et al.* (1972) and Ridley *et al.* (1973).

Methods and Results

The original description of the LKFM composition was based on the relative abundance of glass types in Apollo 15 soils (Reid *et al.*, 1972). From these data it was suggested that LKFM was a major contributor to the soil at the Apollo 15 site and that LKFM was abundant in the Apennine Front. For these reasons we chose to search the Apollo 15 breccias for clasts of igneous rocks of appropriate composition. The thin section collection in the Curator's Laboratory was surveyed and from the sections available at that particular time 36 sections from 10 rocks were selected (on the basis of their clast population) for more detailed study. Petrographic examination was then made of approximately 50 clasts that had igneous textures and were not mare basalts. The clasts can readily be classified into four major groups. The groups are (1) KREEP basalt clasts, (2) poikilitic clasts, (3) anorthositic clasts, and (4) troctolitic clasts. Clasts with poikilitic texture are only common in two of the rocks studied (15285, 15459) and troctolites are only common in 15445. The other two groups are much more common and are present in many of the sections studied.

Petrography

The KREEP basalt clasts are readily distinguished by their melt textures and by the common presence of zoned pyroxenes (orthopyroxene cores mantled by pigeonite) in subophitic intergrowth with plagioclase (Fig. 1a). The fragments have intergranular textures and there is a range in grain size from fine to medium grained. Evidence of deformation is absent or minor. The KREEP basalt clasts could have crystallised from impact-produced melts but there is no direct evidence of this. The absence of inclusions and of any textural indications that these are impact melts (e.g., Simonds *et al.*, 1973; Grieve *et al.*, 1974) argues for a primary igneous origin for the KREEP basalt clasts. The low trace siderophile element abundances in melt rocks that are petrographically similar also argues strongly for an igneous origin (e.g., Morgan *et al.*, 1973; Dowty *et al.*, 1976).

Feldspar and pyroxene in subequal abundance are the major minerals. Orthopyroxene is common in these clasts and the relative abundance of ortho- and clinopyroxene is different for different clasts. Clinopyroxene is predominantly pigeonite and occurs as rims to orthopyroxene or as separate grains or both. Cristobalite is prominent in about one third of the clasts studied. Ilmenite is the dominant opaque and the complex mesostasis, characteristic of KREEP basalts, is variable in abundance. These clasts may not be large enough to provide representative samples but most are fine grained and are obviously basaltic in composition. The apparent range in mineral abundances coupled to the measured range in Mg/Mg + Fe ratios in pyroxene cores indicates that there is a spectrum of KREEP basalts rather than a single KREEP basalt composition (see also Irving, 1977).

The various anorthositic, noritic, and troctolitic clasts all have textures that are dominated by secondary deformation. Many are cataclastic rocks in which

(a)

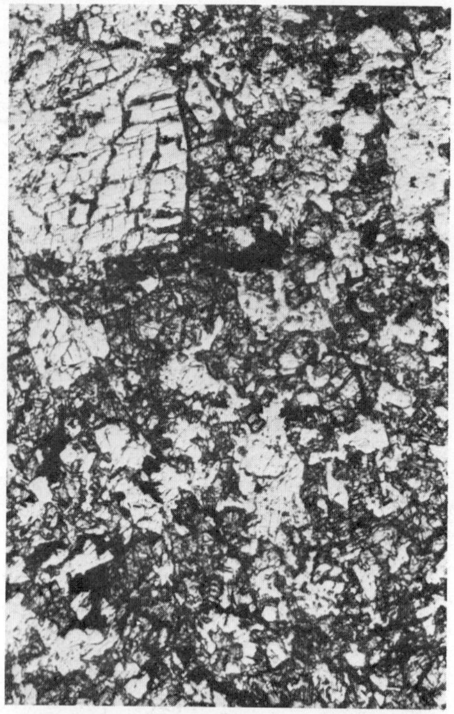

(b)

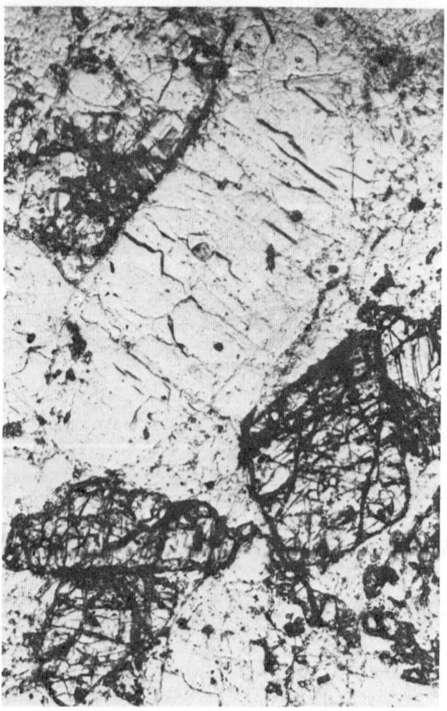

(c)

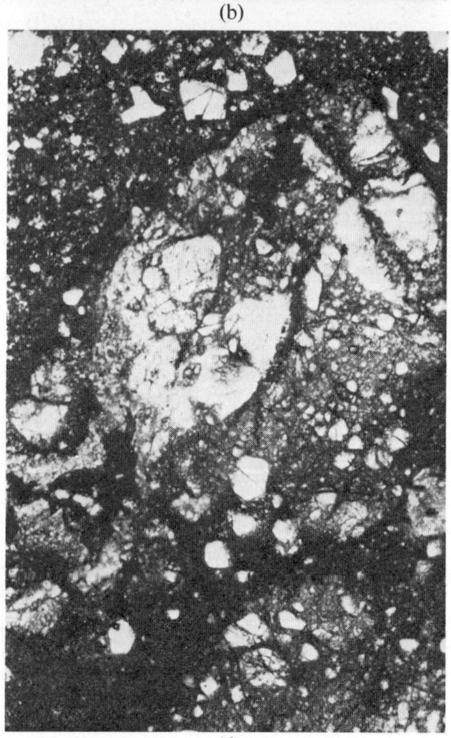

(d)

the original texture has been largely obliterated by secondary impact-induced deformation. Portions of several clasts retain some of the original coarse-grained texture (e.g., Fig. 1c) which is consistent with their origin as plutonic igneous rocks. With only partial preservation of these coarser-grained textures it is not possible to ascertain whether cumulate textures are present. The more feldspathic clasts range in composition, based on mineral abundances, from anorthosite to norite. Centimeter-sized clasts are not representative samples of rocks with grain sizes over 1 mm and the apparent mineral abundances, based on appearance in thin section, may be misleading. There is however a real range in mineral abundances with the more noritic types predominant. Feldspar is the major mineral with low-calcium pyroxene and olivine as the common mafic phases. The feldspar-olivine (troctolite) clasts commonly also carry a pink spinel.

The poikilitic clasts show a range of textures with low-calcium pyroxene as the dominant oikocryst. Feldspar is abundant as small subhedral to euhedral grains and in many clasts as large chadacrysts. Olivine is also present in most clasts as is ilmenite. Several authors (e.g., Bence *et al.*, 1973; Simonds *et al.*, 1973) have discussed the origin of poikilitic lunar rocks. The analogy to terrestrial impact-produced rocks (e.g., Grieve *et al.*, 1974; Grieve, 1975) leads to the conclusion that these rocks have originated as clast-laden impact melts. The similarities in texture among the various clasts with poikilitic texture indicate that they have all probably formed through impact mixing, heating, and melting. The range in mineral abundance and to some extent in texture implies that they are not all derived from the same precursor.

Mineral chemistry

Each petrographically distinguishable clast type has a characteristic and readily distinguishable set of major minerals. Feldspar is the single most abundant phase and the feldspar in the KREEP basalt clasts is more Na- and K-rich than in the other groups ($Or_1Ab_{16}An_{83}$–$Or_{0.5}Ab_{11}An_{88}$, see Fig. 2). The feldspar in these fine grained basalts also has higher contents of minor elements such as Mg and Fe (Fig. 3). The range in anorthite content at constant Fe/Fe + Mg is comparable to those for mare basalts and lower than the range in 14310 feldspars reported by Longhi *et al.* (1976) as evidence of mineralogical heterogeneity in an inhomogeneous impact melt. Feldspars in the KREEP basalts show normal

Fig. 1. Photomicrographs of clasts in Apollo 15 breccias. Plane polarised light. Scale bar is 0.3 mm (1b, 1c, 1d same scale). (1a) KREEP basalt clast in 15285,57. Major phases are plagioclase (An_{86}) and orthopyroxene ($Wo_{3.5}En_{79}Fs_{17.5}$) zoned to pigeonite rims. (1b) Portion of a poikilitic clast in 15459,123. Largest grain and light areas are feldspar. Plagioclase (An_{92}) and olivine (Fo_{70}) are poikilitically enclosed by low Ca pyroxene ($Wo_{5.5}En_{71}Fs_{23.5}$). (1c) Portion of an anorthositic norite clast in 15455,31. Plagioclase (An_{94}) and orthopyroxene ($Wo_{1.5}En_{81}Fs_{18}$) are relic grains in a clast with cataclastic texture. (1d) Spinel troctolite clast in 15445,148. Cataclastic texture with plagioclase (An_{96}), olivine (Fo_{89}), and pink spinel ($Mg_{.79}Fe_{.19}Cr_{.25}Al_{1.73}O_4$).

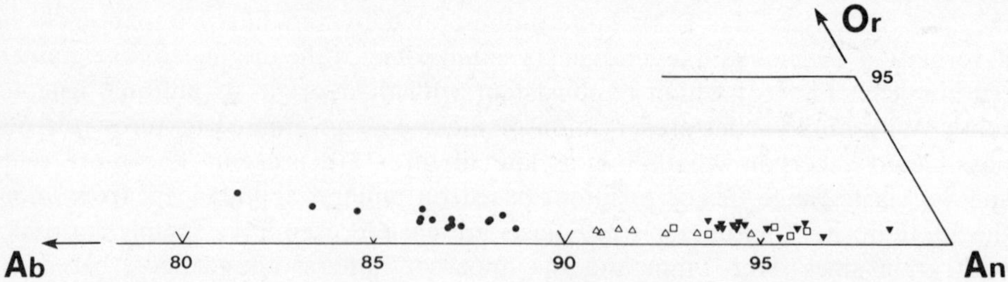

Fig. 2. Feldspar compositions in non-mare clasts from Apollo 15 breccias. ● KREEP basalt clasts, ▼ anorthositic clasts, △ poikilitic clasts, □ troctolite clasts.

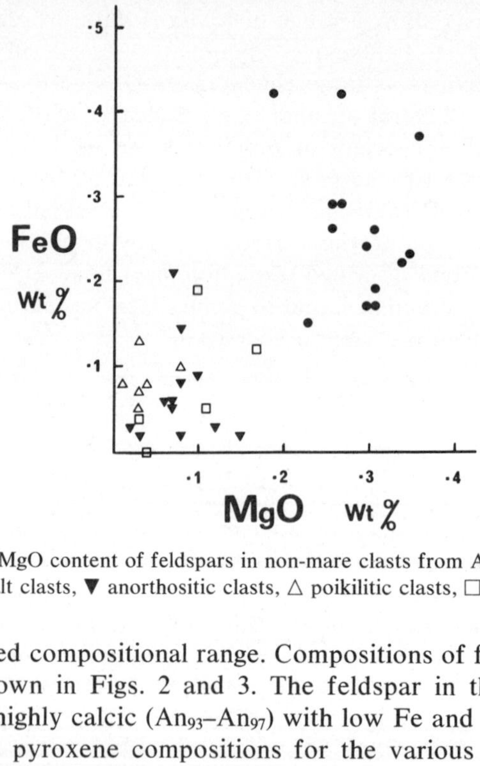

Fig. 3. FeO and MgO content of feldspars in non-mare clasts from Apollo 15 breccias. ● KREEP basalt clasts, ▼ anorthositic clasts, △ poikilitic clasts, □ troctolite clasts.

zoning with a limited compositional range. Compositions of feldspar in the other clast types are shown in Figs. 2 and 3. The feldspar in the anorthositic and troctolite clasts is highly calcic (An_{93}–An_{97}) with low Fe and Mg contents.

Figure 4 shows pyroxene compositions for the various clast types. Orthopyroxenes in KREEP basalts are as magnesian as En_{82} but even cores of larger grains show a range of Mg/Mg + Fe values ($Wo_3En_{82}Fs_{15}$–$Wo_8En_{59}Fs_{33}$). Extreme zoning is common with a normal zoned orthopyroxene core overlain by a more Fe- and Ca-rich pigeonite rim. The pyroxene trends are similar to those reported by several authors (e.g., Cameron *et al.*, 1973; Powell *et al.*, 1973; Dymek *et al.*, 1974; Irving, 1975; Basu and Bower, 1976; Meyer, 1977). The pyroxenes in the more anorthositic clasts are dominantly low Ca pyroxenes ($Wo_2En_{81}Fs_{17}$–$Wo_3En_{72}Fs_{25}$) and some clasts have minor amounts of diopsidic clinopyroxene.

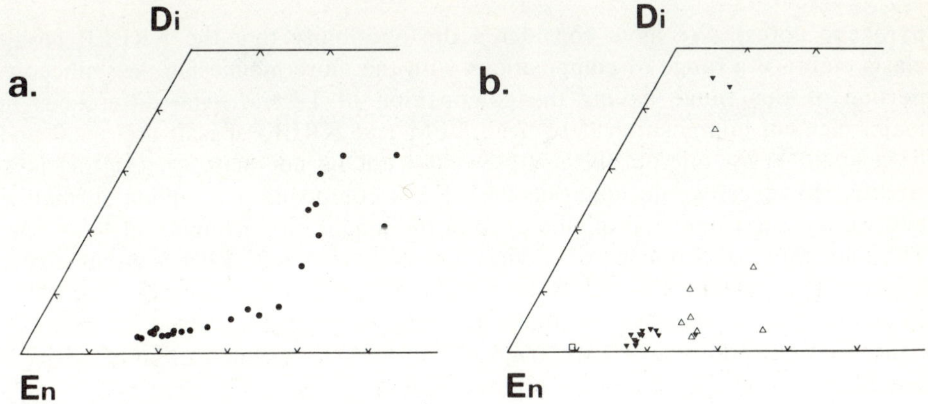

Fig. 4. Pyroxene compositions in non-mare clasts from Apollo 15 breccias. ● KREEP basalt clasts, ▼ anorthositic clasts, △ poikilitic clasts, □ troctolite clasts.

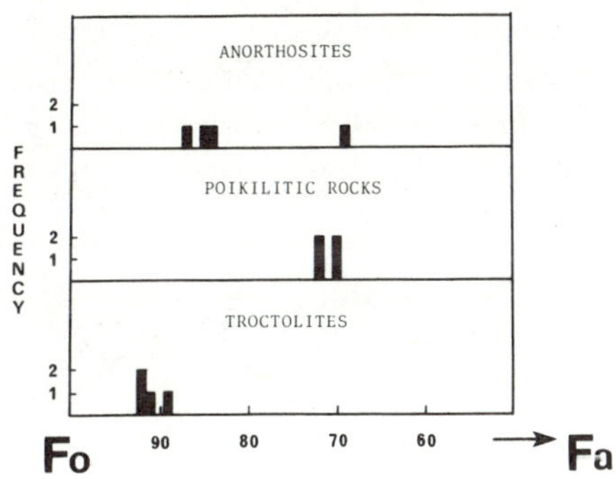

Fig. 5. Olivine compositions in non-mare clasts from Apollo 15 breccias.

Pyroxenes in the poikilitic clasts are predominantly low Ca types, both orthopyroxene and pigeonite ($Wo_2En_{73}Fs_{25}$–$Wo_{14}En_{58}Fs_{28}$).

Olivine is absent from the KREEP basalts, common in the more anorthositic clasts (mostly Fo_{84}–Fo_{87}) and most abundant and most magnesian (Fo_{89}–Fo_{92}) in the troctolites (Fig. 5). The less abundant phases have not yet been extensively analysed. Tabulated mineral compositional data are available from the authors.

Conclusions from these data are:

(1) The major non-mare igneous rocks contributing to the Apollo 15 breccias are KREEP basalts and plutonic anorthosites, norites, and troctolites.

(2) The KREEP basalt clasts are fine-grained melt rocks that show no evidence of being other than primary igneous basalts. They show little evidence of impact effects, in contrast to the clasts of the plutonic rocks. The mineral data suggest that the KREEP basalts do not all have the same composition but that a range of compositions exist (as reflected in the range of Mg/Mg + Fe values for

pyroxene cores). We have considered the possibility that the KREEP basalt clasts embrace a range of compositions with the more magnesian, less siliceous portion of this range having the composition of LKFM. The differences in major-element composition between LKFM and KREEP basalt are small and defocussed beam microprobe analyses may not be adequate to test this possibility. However, we do note that the LKFM composition is olivine normative and we have not detected olivine even in the magnesian members of this suite. The only three occurrences of olivine reported from KREEP basalts are from clasts in 15205 (Dymek *et al.*, 1974) and in 15404 (Ryder and Bower, 1976, 1977). In addition we note that other studies have demonstrated a comparable range in mineral compositions for rock fragments that have bulk composition of KREEP basalt (e.g., Irving, 1975, 1977; Basu and Bower, 1976).

(3) The highly feldspathic rocks are extensively deformed and obviously have a different source and history from the KREEP basalts. In many cases relict coarse-grained plutonic igneous textures can be discerned. The troctolite clasts have the most calcic feldspar and the most magnesian olivine. The other clasts range from anorthosite to anorthositic norite with the noritic end of the sequence being more common.

(4) The poikilitic clasts show a range in mineral assemblage and abundances. They probably derive in impact events from a variety of precursors.

(5) No igneous-textured clasts unequivocally equivalent to LKFM were found, even in this subjective search.

Occurrence of LKFM

We are thus faced with the fact that neither this study nor others on a variety of soils or breccias have discovered fragments of igneous rocks that match the LKFM composition. There are rocks with the composition of LKFM, but they are not igneous rocks. The best known example is probably the dark portion of 15455, the black and white breccia. Taylor *et al.* (1973) noted that the dark portion of the "black and white" breccia 15455 had essentially the major-element composition of LKFM. They analysed this portion and used these analyses to establish the trace-element characteristics of LKFM, since these data could not be obtained directly from the glass particles. The validity of using "15455 dark" to establish the trace-element characteristics of LKFM is critical to some models for the composition and petrochemical evolution of the lunar crust. Recent work by Meyer and Schonfeld (1977) on LKFM glasses shows that, for the few elements that have been determined in both LKFM glass and "15455 dark", the match for trace elements is reasonable. Figure 6 shows the dark portion of 15455, which is a microbreccia with few lithic clasts but a variety of mineral clasts with olivine and feldspar clasts predominating. Figure 7 is a histogram of olivine compositions from this material. Most fall in the range Fo_{79}–Fo_{88} but with a number of other grains extending the total spread from Fo_{64} to Fo_{92}. These data suggest that 15455 dark is indeed polymict but that a large number of the mineral clasts could be derived from a source with a limited compositional range.

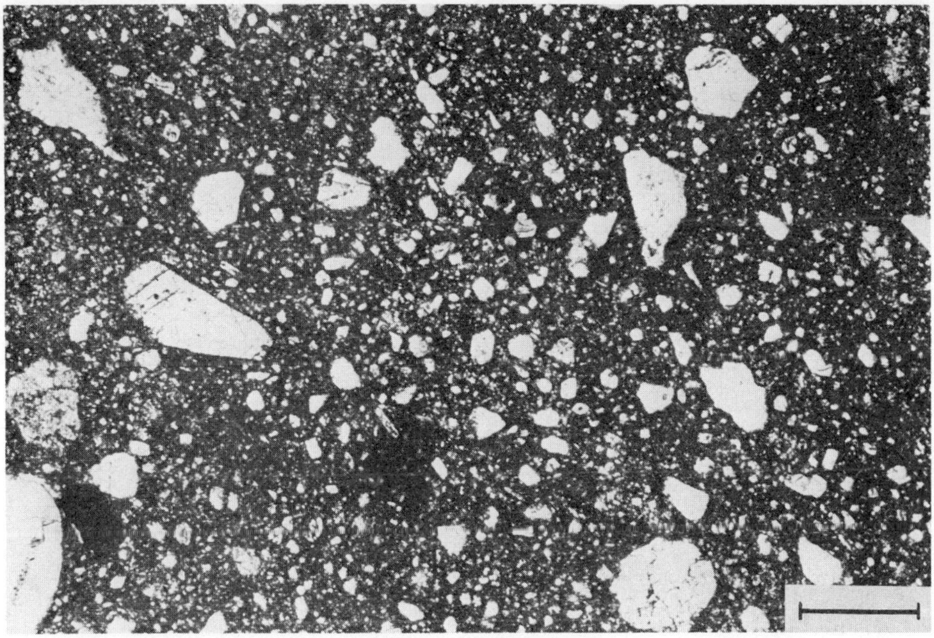

Fig. 6. Photomicrograph of a portion of "15455 dark" (in 15455,35) a microbreccia in which larger grains are single crystal fragments of plagioclase and olivine. Plane polarised light. Scale bar is 0.3 mm.

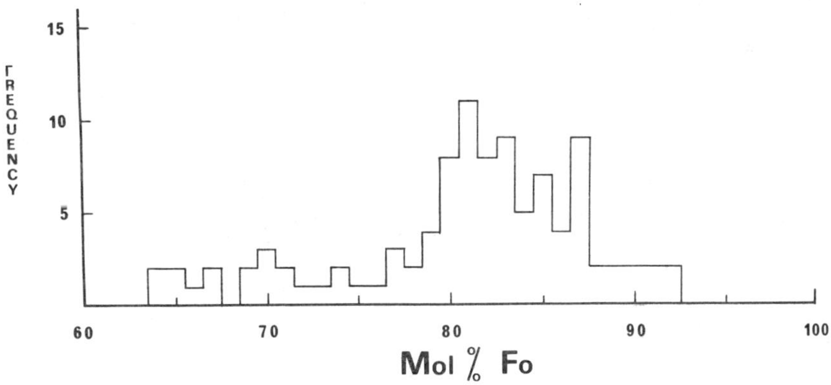

Fig. 7. Histogram of olivine compositions in a portion of "15455 dark".

Some of the brown matrix breccias at the Apollo 15 site also have bulk compositions that approach the LKFM composition as do some soils such as 15101.

Larger rocks with LKFM compositions also occur at the Apollo 16 site. Simonds *et al.* (1973) pointed out that several Apollo 16 poikilitic rocks approximate LKFM. These are impact-produced melts with a complex origin that probably includes a polymict predecessor. LKFM composition rocks also occur at the Apollo 17 site where several of the noritic breccias (LSPET, 1973b) have this approximate composition. James (1976) discusses the matrix and the aphanitic clasts in breccia 73215 which have compositions like LKFM. James also points out that a series of "fragment-laden" melt rocks, including Apollo 17 blue-gray breccias, the dark portions of Apollo 15 and 16 "black and white" rocks, dark clasts in Apollo 14 breccias, the dark portion of 12013, Apollo 17 green-gray breccias, and Apollo 16 poikilitic rocks all have a well defined and limited range in composition, falling along the olivine-plagioclase cotectic in the silica-olivine-anorthite pseudoternary. McKay and Weill (1976) also note the occurrence of lunar rocks with approximately this composition. From the entire range of highland rock types they find 33 samples that have compositions with the characteristics of potential parent materials for KREEP basalts: the average major-element composition of these samples is very close to LKFM. It is also true that these are not primary igneous rocks but impact-produced breccias. In fact the closest match to LKFM is in many cases with the breccia matrices, excluding clasts. These various impact breccias derive from a variety of sources and show evidence of being polymict. Why then, in impact-produced polymict rocks, should the LKFM composition be a common one? Three possible explanations are offered below.

ORIGIN OF LKFM

Hypothesis 1. LKFM is a myth.

In terms of this hypothesis the abundance of material of this specific composition on the lunar surface is more apparent than real. The original composition, derived from the glass data, is an average of many analyses that show a wide range of compositions (see Table 1 for standard deviations). In like manner the impact breccias and melts that have LKFM composition are not identical in composition. Ryder and Wood (1977) argue that the LKFM rocks associated with the Imbrium and Serenitatis events have compositions that differ significantly and may represent slightly different sampling levels in the crust. We are unfortunately not in a position to define a specific composition for LKFM or even the range of compositions that justify using the term LKFM. The concept is based originally on glass analyses that show considerable scatter. We are now using the term for compositions of impact rocks that by virtue of their origin and polymict nature also show a range of compositions. The validity of the LKFM composition rests on the common occurrence in glasses, lithic fragments and larger rocks of compositions that are broadly similar. It is also noted that all these samples plot at or close to the olivine-plagioclase cotectic in the system silica-olivine-anorthite as described by Walker *et al.* (1972).

Taylor *et al.* (1973) argue that the "average highlands" can be modelled by a mixture of 69% anorthositic gabbro and 31% LKFM and Charette *et al.* (1977) further show that the LKFM composition fits the rather limited compositional data available from orbiter experiments for large areas of the lunar surface. These various sources of data all suggest that the LKFM composition is common in the lunar crust, whatever its ultimate origin. This does not rule out the possibility that several chemically distinct rock types exist under the broad LKFM umbrella. Compositional differences do exist among the loosely-defined range of materials that broadly have the LKFM composition. The significance of such distinctions is however somewhat doubtful in light of the heterogeneous nature of many of these samples. We are led to the conclusion that LKFM is real: its nature and origin are further discussed below.

Hypothesis 2. LKFM is a mixture.

This hypothesis is undoubtedly correct for some if not all of the glasses and rocks of LKFM composition. Some of these are demonstrably polymict and are thus mixtures. We have repeated the work of Schonfeld (1974) using mixing model programs in order to get the best fit for LKFM using compositions of known lunar rocks (see Table 3). Modelling using only major elements does not adequately constrain the possible solutions and we have used the trace-element data of Taylor *et al.* (1973) on "15455 dark" for the LKFM composition. The data require that the LKFM mixture contains about 30% KREEP basalt. The source region must thus contain a large proportion of KREEP basalt but this must not be the major component. A large feldspathic component is required and reasonable fits are obtained with 45% anorthositic gabbro, using 15418 as the anorthositic gabbro component. The LKFM composition can thus be matched by a mixture of anorthositic gabbro and KREEP basalt. A close match however requires additional components. As Schonfeld (1974) noted an ultrabasic component must be added to boost the Mg content of the mixture. Using 72415 as an ultrabasic component, about 13% of this dunite must be added. This still leaves a deficiency of other elements, especially Ti, and the addition of 11% mare basalt aids the goodness of fit. The results of mixing model calculations are shown in Table 3.

We have also calculated mixing models for LKFM using a variety of components other than those given in Table 3. In particular we have utilised 15386 as a KREEP composition (but necessarily with estimates of certain trace-element concentrations) and a spinel troctolite as an alternative to the dunite. These alternative models show somewhat different mixing proportions, as would be expected, but do not provide a better fit to the LKFM composition.

There is no question that the components given in Table 3, or the alternative components discussed above, exist on the moon and could be mixed to produce LKFM. What is a little harder to accept is that this mixture was produced several times on the moon at several different impact sites. To explain this phenomenon two sub-hypotheses can be erected.

Table 2. Lunar samples with compositions approximating LKFM.

	15455[1]	60315	61156	62235	65467	64815	65015	72435	77135	73215[2]	73215[3]	76315
SiO_2	47.3	45.61	44.65	47.04	45.69	45.86	45.6	45.76	46.13	46.4	45.6	45.82
TiO_2	1.35	1.27	0.64	1.21	0.72	1.7	1.4	1.54	1.54	0.7	1.1	1.47
Al_2O_3	17.1	17.18	22.94	18.69	21.62	17.33	20.2	19.23	18.01	20.6	19.8	18.01
Cr_2O_3	0.26	0.21	0.14	0.20			0.2	0.20	0.20	0.25	0.25	0.19
FeO	8.79	10.53	7.75	9.45	7.08	9.5	9.5	8.7	9.11	7.30	8.13	8.94
MnO		0.12	0.12	0.11	0.07	0.12	0.1	0.11	0.13	0.104	0.106	0.11
MgO	13.3	13.15	9.60	10.14	11.50	11.83	9.9	11.63	12.63	11.6	12.0	12.41
CaO	10.6	10.41	13.34	11.52	12.52	12.05	11.9	11.72	11.03	11.9	11.5	11.06
Na_2O	0.58	0.56	0.39	0.48	0.42	0.47	0.3	0.52	0.53	0.50	0.52	0.57
K_2O	0.17	0.35	0.11	0.34	0.17	0.26	0.3	0.23	0.30	0.20	0.22	0.27
P_2O_5		0.45	0.22	0.41	0.19	0.29	0.3	0.27	0.28			0.29

[1]Black portion.
[2]Matrix.
[3]Aphanite clast.
Sources of data: Albee *et al.* (1973), Hubbard *et al.* (1973), LSPET (1973a,b), Taylor *et al.* (1973), James (1977).

Table 3. Mixing model calculation of LKFM composition.

	Mixing components					Mixing model				
	KREEP[1]	Anortho-sitic gabbro[2]	Dunite[3]	Mare basalt[4]	Mare basalt[5]	LKFM	LKFM (calculated)	Component	Percentage	S.D.
Weight percent										
SiO_2	48.13	44.97	39.93	37.84	47.69	46.73	44.34	KREEP[1]	29.1	1.6
TiO_2	1.90	0.27	0.03	13.03	1.92	1.26	1.29	Anorthositic		
Al_2O_3	17.50	26.73	1.53	3.70	9.60	18.90	18.37	Gabbro[2]	45.0	6.1
FeO	10.72	5.37	11.34	13.87	20.16	9.71	9.21	Dunite[3]	12.6	3.3
MgO	7.99	5.38	43.61	9.12	8.10	11.08	11.20	Mare basalt[4]	3.5	1.5
CaO	11.01	16.10	1.14	10.41	10.75	11.65	11.80	Mare basalt[5]	7.8	4.4
Na_2O	0.88	0.31	0.01	0.38	0.28	0.38	0.43	Total	98.0	8.5
K_2O	0.63	0.03	0.01	0.05	0.05	0.23	0.20			
P_2O_5	0.90	0.03	0.04	0.07	0.08	0.23	0.29			
Cr_2O_3	0.19	0.11	0.34	0.50	0.53	0.26	0.21			
ppm										
Nb	85	0.4	0.3	21	6	33	28			
Zr	1360	67	2.6	195	93	480	441			
Y	300	5.4	1.1	75	29	93	95			
Sr	200	152	11	147	120	141	143			
Rb	16.3	0.2	0.1	0.6	1	5	5			
Co	27	10	55	22	45	22	24			
V	50	18	50	90	130	39	42			
Ba	1030	20	4	95	63	370	318			

[1]KREEP—Schonfeld and Meyer (1972).
[2]Anorthositic gabbro (15418)—LSPET (1972), Taylor *et al.* (1973).
[3]Dunite (72415, 72417)—LSPET (1973b), Laul and Schmitt (1975).
[4]Average A-17 high-Ti basalt—LSPET (1973b), Rhodes *et al.* (1976), Duncan *et al.* (1974, 1976).
[5]Average A-15 quartz-normative basalt—LSPET (1972), Rhodes and Hubbard (1973), Willis *et al.* (1972), Duncan *et al.* (1976).

Hypothesis 2A. LKFM is a common mixture.

LKFM is a mixture and the components of the mixture exist in the proper amounts in large volumes of the crust. By this hypothesis any large impact can produce LKFM by wholesale mixing. If the average composition of the crust approximated LKFM then very large impacts would produce impact melts and breccias approximating LKFM. Dence *et al.* (1976) proposed this method to explain the Apollo 17 LKFM-like breccias. If these are produced by the Serenitatis event that homogenized the outer portion of the moon, down to 60 km according to Dence *et al.*, then the impact melts produced should yield an average composition for the upper 60 km. The ultramafic component in the mix is provided by incorporation of part of the upper mantle into the mixture. A shallower excavation that involves only crustal materials requires that dunite or some Mg-rich component be more abundant than most current lunar crustal models allow.

Ryder and Wood (1977) argue for a stratified crust in which about 18 km of approximately anorthositic gabbro average composition, overlies about 18 km approximating LKFM. Larger impacts such as Imbrium and Serenitatis excavate into the lower crust and deposit the deeper material (LKFM) selenographically separate from the shallower crustal material. They point out that a total crust of LKFM composition would be richer in heat-producing elements than the heat flow data allow.

If several large impacts have produced LKFM composition melt rocks and ejecta, then large volumes of the crust must be of LKFM composition. The LKFM mixture outlined above contains several rather disparate components. It is difficult to construct a model where these components are mixed in just these portions by several large impacts. The idea that LKFM is average crust conflicts with the heat flow data, with present crustal models, and with the idea suggested by several authors that anorthositic gabbro (or Highland basalt) is close to an average crust mixture. If several large impacts have produced LKFM composition melt rocks and ejecta then it is easier to model this process through a crust that contains large volumes of rock of LKFM composition. This does not solve the problem of the ultimate origin of LKFM but merely moves it one step back.

Hypothesis 2B. LKFM is a unique mixture.

The creation of the LKFM composition by mixing involves only known lunar components and runs into difficulty only because the nature of the mix (if we have modelled it correctly) is such that it is difficult to imagine several different events producing the same mixture. One can hypothesize that LKFM is the mixture produced in large volumes by the Imbrium event alone and that these Imbrian ejecta are abundant over much of the face of the moon. The Apollo 17 samples, including the very large boulders, would then represent, not Serenitatis ejecta, but Imbrian ejecta. The preponderance of 3.95 ages in highland samples may well support this idea (Schaeffer *et al.*, 1976). If the Apollo 14 samples are Imbrium ejecta then this model requires the generation of well-mixed ejecta at the Apollo 15, 16, and 17 sites, and KREEP-dominated poorly mixed ejecta covering the Apollo 14 site.

Hypothesis 3. LKFM is a primary igneous rock.

Reid *et al.* (1972) promoted LKFM as an igneous rock partly on a basis of an assumed relationship to KREEP but without any direct evidence. Walker *et al.* (1972) did provide indirect evidence for this idea by showing that LKFM (and medium and high KFM) fell along a cotectic in the system silica-olivine-anorthite. If these average compositions are meaningful then their position within this phase diagram is a strong argument for an origin controlled by crystal-liquid igneous processes. If LKFM is a mixture then either it is a mixture of components with compositions controlled by crystal-liquid processes or the fact that it plots close to the cotectic is purely fortuitous.

Schonfeld (1975) proposed a model for the generation of anorthositic gabbro as a major constituent of the lunar crust. In this model anorthositic gabbro is a mix of plagioclase and trapped liquid with the composition of LKFM. LKFM is by inference an igneous composition and is cited as a parent material for the formation of KREEP, by analogy with Gast's (1972) calculated precursor of KREEP.

McKay and Weill (1976, 1977) have argued for a parent material for KREEP that has many or all of the characteristics of LKFM. KREEP is then produced from primitive lunar material in at least a two-stage model with LKFM as the immediate precursor of KREEP. In the McKay and Weill model LKFM is a residual product from early lunar fractionation at around 4.4 b.y. Formation of igneous KREEP basalts by later low pressure fractionation of LKFM material is consistent with the experimental evidence and the major- and trace-element partitioning constraints. Hess *et al.* (1977) report experiments showing that LKFM fractionates to compositions more Fe-rich and Al-poor than KREEP basalt and also argue that the range in Ti/K values for LKFM, MKFM, and HKFM is inconsistent with low pressure fractionation of a common parent magma. The McKay and Weill calculations for the KREEP precursor require an LKFM-like rock early in lunar evolution. Their model requires LKFM as a forerunner to KREEP whereas the mixing model requires KREEP to be present as a component in order to generate LKFM.

However, a major difficulty with this hypothesis is the lack of any large rock, lithic fragment, or breccia clast that unambiguously combines the bulk composition of LKFM with an igneous texture. The limited number of returned lunar rocks contain representatives of all proposed crustal rocks except LKFM. In many samples the primary textures are largely obscured by later deformation but even deformed LKFM samples appear to be absent. KREEP basalt is not a dominant component of the lunar crust and in many models is much less abundant than LKFM. Yet KREEP basalt fragments with melt textures are not uncommon in soils or breccias. It is not too difficult to argue away the absence of large LKFM rocks with igneous textures because of the limited sample return. However, it is very difficult to explain the absence in soils and breccias of a rock type that may constitute 30% of the crust. Two explanations can be hazarded. (1) LKFM igneous fragments are not uncommon but we have not recognised them. The discussion earlier in this paper of LKFM as a possible Mg-rich member of the suite of KREEP basalt clasts does indicate that there is a need for further work on the range of compositions present. However, in light of the thorough studies that have been completed by several workers, this explanation seems highly unlikely. (2) LKFM is present but unrecognisable. This hypothesis requires special pleading that might be correct but cannot be substantiated. For example one can argue that LKFM is present as igneous rock at depth in the crust and is only excavated by large impacts. The nature of breccia samples preclude the preservation of very coarse-grained plutonic rocks except in very unusual circumstances. The process of bringing LKFM to the surface may destroy the original texture and only separate mineral grains may be preserved from the precursor material. Such arguments are plausible but not susceptible to testing. In contrast to the preservation of a 4.6 b.y. old dunite (72415) in the returned samples the absence of any igneous or pseudo-igneous fragments of LKFM is a strong argument against a primary igneous origin.

CONCLUSIONS

The results of sample studies coupled with orbiter data provide reasonable evidence (though not compelling evidence in view of the level of sampling and geologic control) that rocks with the composition of LKFM are a major constituent of the lunar crust. The LKFM composition glasses in lunar soils are derived by impact melting of precursors with LKFM composition. These precursors are mostly impact-produced breccias, impact melts and soils. The origin of these LKFM impact rocks is not clear. They may represent a unique composition generated by impact mixing and melting in the Imbrium event and present at several sampling sites as a partial overlay to earlier ejecta. Alternately they may represent ejecta from several major impacts in which case LKFM composition material must constitute a major portion of the lunar crust. In light of the apparent dominance of anorthositic gabbro in the upper crust (an interpretation reinforced by the ubiquitous presence in lunar soils of anorthositic gabbro or "Highland basalt" glasses) the most likely model appears to be one in which LKFM is a major constituent of the lower crust and is excavated by larger impact events. The nature of LKFM in the crust is not immediately apparent. It may be present as a series of discrete units or layers that are mixed by impact. In this case mixing model calculations can perhaps provide an insight into the nature of the lower crust. If LKFM is present in the lower crust as primary igneous material, then the composition is indeed controlled by crystal-liquid processes. LKFM may have formed very early in the evolution of the moon as part of an early large-scale differentiation of the surface regions. This igneous LKFM, formed around 4.4 b.y. or earlier, would then be a source material for the generation of KREEP basalt, as modelled by McKay and Weill (1976, 1977). The major fault in this hypothesis is the absence of samples of igneous LKFM. With the present data we are unable to discriminate among these alternatives other than by personal prejudice, and these vary among the authorship.

Acknowledgments—We thank K. Newton, J. McPherson, and C. Basson for their help in the preparation of this manuscript.

References

Albee A. L., Gancarz A. J., and Chodos A. A. (1973) Sanidinite metamorphism of Apollo 16 sample 65015 (abstract). In *Lunar Science IV*, p. 24–26. The Lunar Science Institute, Houston.

Basu A. and Bower J. F. (1976) Petrography of KREEP basalt fragments from Apollo 15 soils. *Proc. Lunar Sci. Conf. 7th*, p. 659–678.

Bence A. E., Papike J. J., Sueno S., and Delano J. W. (1973) Pyroxene poikiloblastic rocks from the lunar highlands. *Proc. Lunar Sci. Conf. 4th*, p. 597–611.

Cameron K. L., Delano J. W., Bence A. E., and Papike J. J. (1973) Petrology of the 2–4 mm soil fraction from the Hadley-Apennine region of the moon. *Earth Planet. Sci. Lett.* **19**, 9–21.

Charette M. P., Taylor S. R., Adams J. B., and McCord T. B. (1977) The detection of Fra Mauro basalts in the highlands by remote spectral reflectance techniques and implications for crustal stratigraphy (abstract). In *Lunar Science VIII*, p. 175–177. The Lunar Science Institute, Houston.

Dence M. R., Grieve R. A. F., and Plant A. G. (1976) Apollo 17 gray breccias and crustal composition in the Serenitatis Basin region. *Proc. Lunar Sci. Conf. 7th*, p. 1821–1832.

Dowty E., Keil K., Prinz M., Gros J., and Takahashi H. (1976) Meteorite-free Apollo 15 crystalline KREEP. *Proc. Lunar Sci. Conf. 7th*, p. 1833–1844.

Duncan A. R., Erlank A. J., Sher M. K., Abraham Y. C., Willis J. P., and Ahrens L. H. (1976) Some trace element constraints on lunar basalt genesis. *Proc. Lunar Sci. Conf. 7th*, p. 1659–1671.

Duncan A. R., Erlank A. J., Willis J. P., Sher M. K., and Ahrens L. H. (1974) Trace element evidence for a two-stage origin of some titaniferous mare basalts. *Proc. Lunar Sci. Conf. 5th*, p. 1147–1157.

Dymek R. F., Albee A. L., and Chodos A. A. (1974) Glass-coated soil breccia 15205: Selenologic history and petrologic constraints on the nature of its source region. *Proc. Lunar Sci. Conf. 5th*, p. 235–260.

Gast P. W. (1972) The chemical composition and structure of the moon. *The Moon* **5**, 121–148.

Grieve R. A. F. (1975) Petrology of the impact melt at Mistastin Lake Crater, Labrador. *Bull. Geol. Soc. Am.* **86**, 1617–1629.

Grieve R. A. F., Plant A. G., and Dence M. R. (1974) Lunar impact melts and terrestrial analogs: Their characteristics, formation and implications for lunar crustal evolution. *Proc. Lunar Sci. Conf. 5th*, p. 261–273.

Hess P. C., Rutherford M. J., Campbell H. W., and Dick P. A. (1977) Fractionation of lunar feldspathic magmas (abstract). In *Lunar Science VIII*, p. 433–435. The Lunar Science Institute, Houston.

Hubbard, N. J., Rhodes J. M., Gast P. W., Bansal B. M., Shih C.-Y., Weisman H., and Nyquist L. E. (1973) Lunar rock types: The role of plagioclase in non-mare and highland rock types. *Proc. Lunar Sci. Conf. 4th*, p. 1297–1312.

Irving A. J. (1975) Chemical, mineralogical and textural systematics of non-mare melt rocks: Implications for lunar impact and volcanic processes. *Proc. Lunar Sci. Conf. 6th*, p. 363–394.

Irving A. J. (1977) Chemical variation and fractionation of KREEP basalt magma. *Proc. Lunar Sci. Conf. 8th*. This volume.

James O. B. (1976) Petrology of aphanitic lithologies in consortium breccia 73215. *Proc. Lunar Sci. Conf. 7th*, p. 2145–2178.

Laul J. C. and Schmitt R. A. (1975) Dunite 72417: A chemical study and interpretation. *Proc. Lunar Sci. Conf. 6th*, p. 1231–1254.

Longhi J., Walker W., and Hays J. F. (1976) Fe and Mg in plagioclase. *Proc. Lunar Sci. Conf. 7th*, p. 1281–1300.

LPSET (1972) The Apollo 15 lunar samples: A preliminary description. *Science* **175**, 363–375.

LSPET (1973a) The Apollo 16 lunar samples: Petrographic and chemical description. *Science* **179**, 23–41.

LSPET (1973b) The Apollo 17 lunar samples: Chemical and petrographic description. *Science* **182**, 659–672.

McKay G. A. and Weill D. F. (1976) Petrogenesis of KREEP. *Proc. Lunar Sci. Conf. 7th*, p. 2427–2447.

McKay G. A. and Weill D. F. (1977) KREEP (abstract). In *Lunar Science VIII*, p. 649–651. The Lunar Science Institute, Houston.

Meyer C., Jr. (1977) Petrology, mineralogy and chemistry of KREEP basalt. *Phys. Chem. Earth.* **10**, 239–260.

Meyer C., Jr. and Schonfeld E. (1977) Ion microprobe study of glass particles from lunar sample 15101 (abstract). In *Lunar Science VIII*, p. 661–663. The Lunar Science Institute, Houston.

Morgan J. W., Krähenbühl U., Ganapathy R., Anders E., and Marvin U. B. (1973) Trace element abundances and petrology of separates from Apollo 15 soils. *Proc. Lunar Sci. Conf. 4th*, p. 1379–1398.

Powell B. N., Aitken F. K., and Weiblen P. W. (1973) Classification, distribution and origin of lithic fragments from the Hadley-Apennine region. *Proc. Lunar Sci. Conf. 4th*, p. 445–460.

Reid A. M., Warner J., Ridley W. I., and Brown R. W. (1972) Major element composition of glasses in three Apollo 15 soils. *Meteoritics* **7**, 395–415.

Rhodes J. M. and Hubbard N. J. (1973) Chemistry, classification and petrogenesis of Apollo 15 mare basalts. *Proc. Lunar Sci. Conf. 4th*, p. 1127–1148.

Ridley W. I., Reid A. M., Warner J. L., Brown R. W., Gooley R., and Donaldson C. (1973) Glass compositions in Apollo 16 soils 60501 and 61221. *Proc. Lunar Sci. Conf. 4th*, p. 309–321.

Ryder G. and Bower J. F. (1976) Sample 15405: Petrology. In *Interdisciplinary Studies by the Imbrium Consortium*, vol. 1 (J. A. Wood, ed.), p. 77–94.

Ryder G. and Bower J. F. (1977) Sample 15405: Petrology. In *Interdisciplinary Studies by the Imbrium Consortium*, vol. 2 (J. A. Wood, ed.), p. 19–26.

Ryder G. and Wood J. A. (1977) Serenitatis and Imbrium impact melts: Implications for lunar crustal composition and stratigraphy (abstract). In *Lunar Science VIII*, p. 826–828. The Lunar Science Institute, Houston.

Schaeffer O. A., Husain L., and Schaeffer G. A. (1976) Ages of highland rocks: The chronology of lunar basin formation revisited. *Proc. Lunar Sci. Conf. 7th*, p. 2067–2092.

Schonfeld E. (1974) The contamination of lunar highland rocks by KREEP: Interpretation by mixing models. *Proc. Lunar Sci. Conf. 5th*, p. 1269–1286.

Schonfeld E. (1975) A model for the lunar anorthositic gabbro. *Proc. Lunar Sci. Conf. 6th*, p. 1375–1384.

Schonfeld E. and Meyer C., Jr. (1972) The abundances of components of the lunar soils by a least-squares mixing model and the formation age of KREEP. *Proc. Lunar Sci. Conf. 3rd*, p. 1397–1420.

Simonds C. H., Warner J. L., and Phinney W. C. (1973) Petrology of Apollo 16 poikilitic rocks. *Proc. Lunar Sci. Conf. 4th*, p. 613–632.

Taylor S. R., Gorton M. P., Muir P., Nance W., Rudowski R., and Ware N. (1973) Lunar highlands composition: Apennine Front. *Proc. Lunar Sci. Conf. 4th*, p. 1445–1459.

Walker D., Longhi J., and Hays J. F. (1972) Experimental petrology and origin of Fra Mauro rocks and soil. *Proc. Lunar Sci. Conf. 3rd*, p. 797–817.

Willis J. P., Erlank A. J., Gurney J. J., and Ahrens L. H. (1972) Geochemical features of Apollo 15 materials. In *The Apollo 15 Lunar Samples* (J. W. Chamberlain and C. Watkins, eds.), p. 268–271. The Lunar Science Institute, Houston.

Proc. Lunar Sci. Conf. 8th (1977), p. 2339–2355.
Printed in the United States of America

KREEP petrogenesis revisited

G. A. McKay* and D. F. Weill

Department of Geology, University of Oregon, Eugene, Oregon 97403

Abstract—Solid/liquid distribution coefficients have been experimentally determined for Sr, Ba, Ce, Sm, Eu, and Yb in olivine and plagioclase at 1240°C, for Cr and Sc in olivine at 1240°C, and for Sc in low-Ca pyroxene at 1195°C. The equilibrium crystallization sequence for a synthetic Low-K Fra Mauro (LKFM) analog has been studied. Phase proportions from this crystallization sequence and crystal/liquid distribution coefficients are used to calculate the evolution of elemental abundances in residual liquids resulting from the differentiation of a LKFM parent composition. When the degree of differentiation is such that the calculated liquids have Sm contents similar to those of Apollo 15 igneous KREEP samples, the abundances of Mg, Sr, Eu, Ce, Yb, Ba, Rb, and K in these model liquids are also similar to the observed abundances of these elements in the Apollo 15 KREEP samples. However, the abundances of Ti and Fe in the model liquids are significantly greater than those of Apollo 15 KREEP. A revised model for the petrogenesis of the KREEP-rich samples is proposed in which material similar in composition to the LKFM breccias, but poorer in Ti and Fe, is the "proto-KREEP" from which more evolved KREEP-rich samples are derived.

The relative importance of physical mixing vs. igneous processes in determining the compositions of highland breccias is considered. It is concluded that the compositions of most samples richer in Sm than the LKFM breccias (~15 ppm) are dominated by igneous processes, while the compositions of most samples with lower Sm contents may be dominated by mixing. The KREEP-rich component in these possible mixtures is similar in composition to LKFM, rather than more highly evolved KREEP.

Introduction

Among the samples returned on the Apollo 12 mission is a group of soil fragments whose most distinguishing chemical feature is an unusual enrichment in the "excluded" minor and trace elements. These samples were given the acronym KREEP, after several of the elements enriched. Samples of similar composition have subsequently been found in the collections from all Apollo missions. The degree of enrichment in the excluded elements varies considerably from one sample to another, so that there is no single KREEP, but rather a continuum of compositions. In recognition of this fact, use of the noun KREEP has largely been replaced by use of the adjective KREEP-rich.

Walker *et al.* (1972, 1973) pointed out that the bulk compositions of the KREEP-rich samples, when plotted in the olivine-anorthite-silica (OL–AN–SI) pseudoternary low-pressure liquidus diagram, fall near the OL–PL cotectic curve, the OL–OPX–PL peritectic point, and the OPX–PL cotectic curve, suggesting that the samples may represent liquids which have equilibrated with plagioclase, olivine and/or pyroxene at relatively low pressure.

Most of the KREEP-rich samples are breccias or impact glasses, and we

*Present address: NASA Johnson Space Center, Houston, Texas 77058.

cannot be sure that their compositions truly represent igneous liquids, uncontaminated with other crustal materials. However, a number of KREEP-rich Apollo 15 samples have igneous textures and low abundances of elements associated with meteoritic contamination and with impact-generated melts (Irving, 1975; Dowty *et al.*, 1976). Most workers believe that these samples represent crystallized melts which were produced by endogenous igneous processes within the moon.

Trace-element modeling techniques have been used to constrain the nature of the source region for KREEP-rich partial melts (Hubbard and Gast, 1971; Helmke *et al.*, 1972). Weill and McKay (1975) demonstrated that KREEP-rich samples could not be generated by partial melting of material of proposed average lunar highlands composition (Taylor and Jakeš, 1974), but that they might be generated by low-pressure partial melting of bulk compositions roughly similar to that proposed for the whole moon (Taylor and Jakeš, 1974).

McKay and Weill (1976) proposed a model which relates the continuum of KREEP-rich samples in a scheme of igneous differentiation. Pseudoternary representations of several compositions which are important in this model are shown in Fig. 1. According to the model, liquid along the OL–PL cotectic (C in Fig. 1) was produced by fractionation of large amounts of olivine and lesser amounts of plagioclase (and possibly other minor phases) from undifferentiated

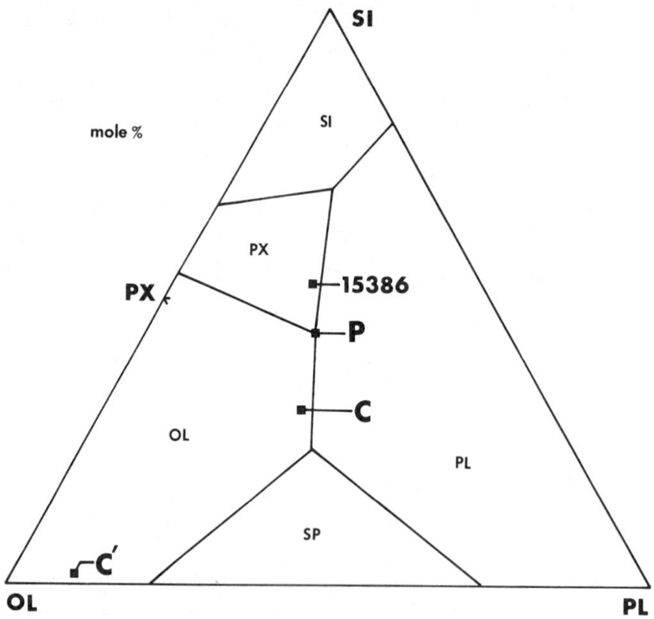

Fig. 1. Pseudoternary representations of important compositions in our model for the evolution of KREEP (McKay and Weill, 1976). Sample 15386 is a representative of Apollo 15 igneous KREEP. P is the peritectic composition which we proposed as the parent to 15386. C is the average composition of highlands breccias which we proposed as representatives of proto-KREEP, the parent to more evolved KREEP. C′ is the composition of a possible primitive parent to C.

lunar material (C′) at 4.4 G.y. This fractionation might have occurred during crystallization of the lunar magma ocean, or it might have been the result of partial melting of a relatively undifferentiated chilled magma ocean crust. The "proto-KREEP" produced in this event was similar in composition to the low-K Fra Mauro (LKFM) glasses of Reid *et al.* (1972) and is now represented by a number of highlands breccias (the LKFM breccias). After separation from its refractory solid assemblage, the proto-KREEP cotectic liquid differentiated, yielding a series of more evolved KREEP-rich residual liquids, including liquids of peritectic bulk composition (P in Fig. 1). Some of these liquids may have been extruded, but a large proportion of them crystallized within the lunar crust. In a later igneous event at 3.9 G.y., some of the more evolved KREEP-rich material was partially melted, producing a series of very highly evolved KREEP-rich liquids, including the Apollo 15 igneous KREEP (e.g., 15386, Fig. 1).

In developing the model which we proposed last year, we used the pseudoternary liquidus diagram, together with distribution coefficients for Fe and Mg, to estimate the compositions and relative proportions of residual liquid and each crystallizing phase involved in each fractionation step. We then used these estimated phase proportions, together with experimentally determined crystal/liquid distribution coefficients, to calculate the evolution of elemental abundances during the fractionation steps. Because compositions of natural samples are not completely represented in the pseudoternary system, the relative phase proportions calculated using the pseudoternary diagram, and in turn the elemental abundances calculated using these phase proportions, are subject to considerable uncertainty (cf. Hess *et al.*, 1977).

To minimize this uncertainty, we have now experimentally determined the crystallization sequence, including the variation of relative phase proportions during crystallization, for a composition resembling LKFM. In this paper we report the results of this study, apply them to the petrogenesis of Apollo 15 igneous KREEP, and present a somewhat revised model for the petrogenesis of these samples. We also consider the role which physical mixing plays in determining the compositions of highland samples, and the nature of the KREEP-rich end-member component in those samples which might be mixtures. Finally, we present recent determinations of solid/liquid distribution coefficients for Sr, Ba, and several of the REE in olivine and plagioclase, for Cr in olivine, and for Sc in olivine and low-Ca pyroxene.

Experimental Procedures

The crystallization sequence of a synthetic LKFM analog was determined by holding the bulk composition at temperature and oxygen fugacity (H_2 and CO_2 mixtures) for periods of 24–48 hr. Charges were held in Pt wire loops formed from 1.2×10^{-4} m diameter wire which had been previously held in contact with the same bulk composition for at least 48 hr at the temperature and oxygen fugacity of the run. Compositions of the individual phases are based on a minimum of 10 point analyses per phase. The modes were determined from the bulk composition and the individual phase compositions by multilinear regression analysis. Because of possible Fe exchange between charges and Pt loops, Fe was not included in the regression analysis. To monitor this exchange, the bulk Fe content of each charge, calculated from the mode and the phase compositions, was compared with the bulk Fe content of the starting material. Any charge indicating an apparent gain or loss of more than 7% Fe, relative to the amount in the starting material, was discarded. Bulk compositions to which 0.5–3 wt.% of a "trace" element had been added were held at T and f_{O_2} for 50–100 hr. Quenched crystals and glass were analyzed to determine crystal/liquid distribution coefficients. Phase compositions in these trace-element doped charges were very similar to the phase compositions in undoped charges run at the same temperature.

Experimental Results

The results of the experimental crystallization runs are given in Table 1. The variation of relative phase proportions with F, the weight fraction of residual liquid, is shown in Fig. 2. These results are in general agreement with results

Table 1. Experimental crystallization sequence of a synthetic LKFM analog. Mode and oxide concentrations are in wt.%.

T°C	$-\log f_{O_2}$	Phase	Mode	SiO_2	TiO_2	Al_2O_3	FeO	MgO	CaO	Na_2O
1300	11.9	GL	100	45.8	1.30	19.2	9.3	11.0	11.5	0.54
1250	12.5	GL	87	46.7	1.52	17.4	9.5	10.8	10.9	0.51
		OL	2	39.9	0.01	0.00	12.0	46.8	0.42	0.01
		PL	11	44.1	0.03	35.3	0.55	0.33	19.3	0.58
1245	12.5	GL	85	47.0	1.52	18.1	9.5	11.3	11.1	n.a.
		OL	3	40.5	0.06	0.20	12.4	45.9	0.33	n.a.
		PL	12	44.4	0.07	35.3	0.56	0.35	19.4	n.a.
1220	12.8	GL	82	47.6	1.56	17.6	10.3	10.0	11.3	0.52
		OL	6	39.8	0.01	0.04	13.6	45.0	0.42	0.00
		PL	12	46.2	0.02	34.1	0.53	0.43	18.4	0.81
1220	12.8	GL	80	47.5	1.64	17.4	9.8	10.2	11.4	0.49
		OL	6	40.0	0.02	0.02	13.3	45.3	0.42	0.00
		PL	14	45.8	0.04	34.2	0.48	0.43	18.4	0.89
1200	13.1	GL	52	49.6	2.41	15.3	11.6	8.8	10.4	n.a.
		OL	15	39.4	0.05	0.09	17.1	42.8	0.28	n.a.
		PL	33	44.7	0.11	34.9	0.86	0.56	19.0	n.a.
1180	13.3	GL	34	47.3	3.57	13.7	14.0	7.9	10.9	0.40
		OL	13	38.2	0.06	0.00	22.0	39.4	0.35	0.00
		PL	42	45.2	0.06	33.7	0.60	0.35	18.3	0.70
		PX	11	53.0	0.65	2.02	12.8	27.5	2.13	0.00
1180	13.3	GL	37	48.2	3.33	13.7	14.4	7.8	10.7	0.42
		OL	15	38.5	0.07	0.00	22.7	38.7	0.35	0.00
		PL	40	44.8	0.06	34.7	0.59	0.27	18.5	0.65
		PX	8	53.2	0.56	1.76	13.4	27.3	2.05	0.01
Av. LKFM breccia*				46.2	1.5	17.9	9.2	12.3	11.2	0.7

*Average of 13 breccias whose compositions resemble LKFM glasses of Reid *et al.* (1972). Data sources are given in Fig. 6.

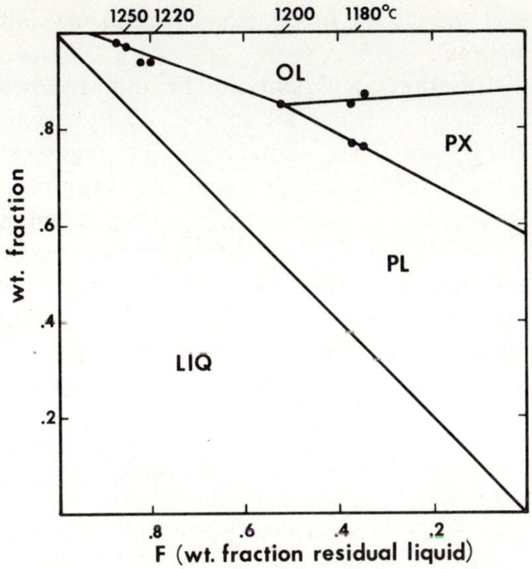

Fig. 2. Phase assemblages (weight fractions) during the equilibrium crystallization of a synthetic composition similar to the LKFM glasses of Reid *et al.* (1972).

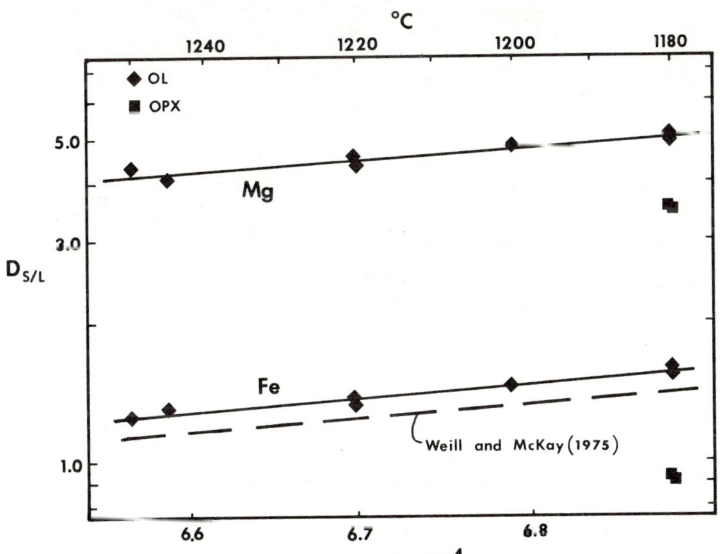

Fig. 3. Fe and Mg distribution coefficients (weight basis). Coefficients for the regression lines (solid lines) are given in Table 2. The regression line for $D^{Fe}_{OL/L}$ of Weill and McKay (1975) is shown with dashes. Their regression line for $D^{Mg}_{OL/L}$ nearly coincides with the one shown here.

obtained by Hess *et al.* (1977) in an experimental crystallization study of a similar bulk composition.

Olivine/liquid distribution coefficients for Fe and Mg for the charges listed in Table 1 are plotted against $1/T$ (°K) in Fig. 3. Pyroxene/liquid distribution coefficients at 1180°C are also shown. Linear regression coefficients for $\log_{10} D_{OL/L}$ vs. $1/T$ are given in Table 2, and the resulting regression lines are shown in Fig. 3. The regression line for $D^{Mg}_{OL/L}$ is in excellent agreement with that reported by Weill and McKay (1975). However, the regression line for $D^{Fe}_{OL/L}$ differs significantly from that which we previously reported (dashed line, Fig. 3). This difference is probably due to uncertainties in the mathematical correction for iron loss which we applied in calculating the previous distribution coefficients. By minimizing iron loss in the present series of experiments we

Table 2. Olivine/liquid distribution coefficients (weight basis) for Fe and Mg along the crystallization path for the synthetic LKFM composition. Coefficients in the equation $\log_{10} D = A + B/T$, where T is in °K. The Fe–Mg exchange coefficient, $D^{Fe-Mg}_{OL/L} = (X^{Fe}_{OL})(X_L{}^{Mg})/(X^{Fe}_L)(X^{Mg}_{OL})$, is equivalent to D^{Fe}/D^{Mg}. S is the standard error of estimate of $\log_{10} D$.

Element	*A*	*B*	*S*
Fe	−1.871	2999	.010
Mg	−1.004	2474	.012

Table 3. Crystal/liquid distribution coefficients. Uncertainties (1 σ) represent the standard deviation of the mean of 5–10 determinations. Phase compositions are similar to those given in Table 1 at similar temperatures.

	PL (1240°C)	OL (1240°C)	PX (1195°C)
Sr	1.87 ± .03	.003 ± .002	
Ce	.047 ± .003	.012 ± .003	
Sm	.033 ± .006	.010 ± .003	
Eu	1.20 ± .06	.010 ± .003	
Yb	.035 ± .006	.042 ± .005	
K	.20 ± .02		
Ba	.16 ± .01	.005 ± .003	
Rb	.08 ± .04		
Cr		1.08 ± .05	
Sc		.265 ± .003	1.4 ± .1

have eliminated this correction, and we believe that the current values are more reliable than our previous ones.

Crystal/liquid distribution coefficients for several trace elements are given in Table 3. These coefficients are consistent with those reported by McKay and Weill (1976). Distribution coefficients for both Eu and Cr are dependent on oxygen fugacity (Weill and McKay, 1975; McKay and Weill, 1976), and the values reported here were measured at oxygen fugacities appropriate for lunar magmatic conditions (Table 1). Our values for $D_{OL/L}^{Cr}$ are significantly higher than those reported by Akella *et al.* (1976) and by Huebner *et al.* (1976) for systems of different bulk composition. The difference is probably due to the strong compositional dependence of this coefficient (Schreiber and Haskin, 1976), and our values are appropriate only for the systems we consider in this paper.

Discussion

Suitability of LKFM as parent to highly evolved KREEP

We will now re-examine our model for the petrogenesis of the KREEP-rich samples using data from the experimentally determined LKFM crystallization sequence. The approach we will take is to use the phase proportions from the crystallization sequence (Table 1 and Fig. 2) and the crystal/liquid distribution coefficients reported here and in our previous work (McKay and Weill, 1976) to calculate the evolution of abundances of a number of key elements during the differentiation of the LKFM breccias (the proto-KREEP in our model), and then to compare the calculated abundances with those observed in a number of highly evolved KREEP-rich samples.

The abundances may be calculated using the mass balance equation for equilibrium crystal/liquid partitioning:

$$\bar{D} = \sum_{i=1}^{n} D_i W_i = (C_0 - C_L F)/(1 - F)C_L, \tag{1}$$

where $\bar{D}$ is the composite solid/liquid distribution coefficient and W_i is the weight fraction of mineral i ($\Sigma_{i=1}^{n} W_i \equiv 1$), while F is the weight fraction of the equilibrated liquid relative to the total weight of the system. C_L is the concentration of an element in the liquid and C_0 is the concentration in the total system. Values for C_0 are obtained by averaging abundances from a number of LKFM breccias (data sources are given in Fig. 6). Values for F and for the W's are taken from Fig. 2.

The elements we will consider are Ti and Sm, which are strongly excluded from all the major crystallizing phases, Fe and Mg, which are strongly excluded from plagioclase but not from the ferromagnesian minerals, and Sr, which is strongly excluded from the ferromagnesian minerals but not from plagioclase. We arbitrarily select the Sm abundance as an index of the degree of fractionation, against which we will plot the abundances of the other elements.

The results are shown in Fig. 4. Abundances for the LKFM breccias are

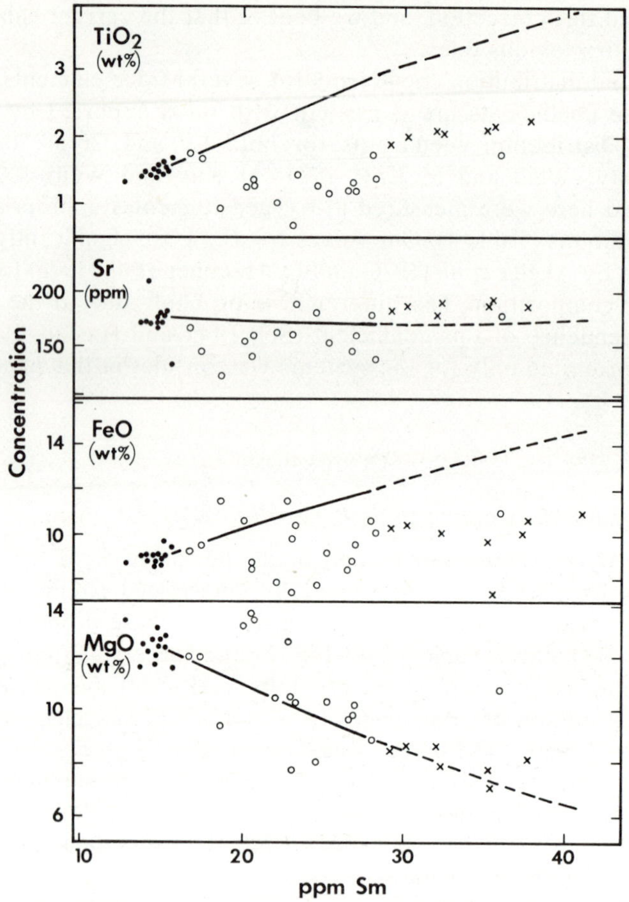

Fig. 4. Variation of the concentrations of TiO_2, Sr, FeO, and MgO with Sm content for a number of LKFM breccias (●), more evolved KREEP-rich breccias (○), and Apollo 15 igneous KREEP (×). Evolution trends, calculated as explained in the text, are shown for the differentiation according to the model of McKay and Weill (1976) of material having the average composition of the LKFM breccias (composition C in Fig. 1). The solid curves depict evolution trends during their proposed differentiation step C→P (cf. Fig. 1), and the dashed curves depict trends during the step P→Apollo 15 igneous KREEP (e.g., 15386). Data sources are given in Fig. 6.

shown as closed circles, abundances for other KREEP-rich breccias are shown as open circles, and abundances for Apollo 15 igneous KREEP are shown as *X*'s.

The solid curves show the calculated evolution of elemental abundances in residual liquids from the beginning of crystallization until the residual liquid reaches the OL–PL–PX peritectic. These are the evolution trends which our model predicts for residual liquids produced during the igneous event at 4.4 G.y.

Large fraction partial melting of these evolved near-peritectic compositions at 3.9 G.y., the origin we propose for Apollo 15 igneous KREEP, will produce liquids whose compositions lie along the PL–PX cotectic in the pseudoternary diagram. We again use Eq. (1) to calculate the evolution of elemental adundances during this partial melting event. In the absence of directly determined phase proportions, we use the PL–PX cotectic curve of Walker *et al.* (1973) to obtain the relative molar proportions of coexisting plagioclase, pyroxene, and liquid for a system of peritectic bulk composition, and convert these to weight fractions using $D_{PX/L}^{Fe-Mg}$ from the data in Table 1. Because the PL–PX cotectic curve of Walker *et al.* (1973) may not be strictly applicable to the bulk compositions which we consider (Hess *et al.*, 1977), the phase proportions obtained in this manner are subject to considerably more uncertainty than those determined directly. To emphasize this uncertainty, we plot the resulting evolution trends as dashed curves in Fig. 4.

If our model is correct, the calculated compositions of derivative liquids should match the observed compositions of the igneous KREEP samples. The curves for Sr and Mg fall very close to the *X*'s representing igneous KREEP. Calculated abundances for Eu, Ce, Yb, Ba, Rb, and K, not shown in Fig. 4, are also very similar to the abundances of these elements in igneous KREEP at equivalent Sm contents. In contrast, the evolution curves for Fe and Ti pass well above the igneous KREEP points in Fig. 4. Hess *et al.* (1977) also report large degrees of Fe enrichment in residual liquids produced during the experimental crystallization of LKFM.

One possible interpretation of these results is that material similar to LKFM in composition is not proto-KREEP. In this case, the similarity of the calculated abundances for Mg, Sr, Eu, Ce, Yb, Ba, Rb, and K to the abundances of these elements in Apollo 15 igneous KREEP at similar Sm contents would be fortuitous. We do not consider this very likely.

Another possible interpretation is that the LKFM breccias are samples of proto-KREEP, but that one or more Fe- and Ti-rich phases which do not appear in the experimental crystallization sequence were involved in the differentiation to more evolved KREEP-rich compositions on the moon. This could be the case only if our experiments and those of Hess *et al.* (1977) were conducted at oxygen fugacities or pressures inappropriate for lunar differentiation.

It is possible that lunar oxygen fugacities were somewhat lower than those in our experiments, resulting in crystallization of Fe metal or troilite. Fractionation of these phases would yield an evolution curve for Fe which is lower than the one shown in Fig. 4. However, Fe metal and troilite occur almost exclusively in mesostasis regions of Apollo 15 igneous KREEP samples (Basu and Bower, 1976), and it is therefore unlikely that these samples were saturated at their liquidi with these phases.

We consider crystallization of a Ti-rich phase to be unlikely. The KREEP-rich samples are very undersaturated with respect to any Ti-rich phase at low pressure and at the oxygen fugacities of our experiments. (We have observed spinel crystallization from experimental liquids similar in composition to KREEP-rich samples, but these spinels were poor in Ti.) We know of no evidence for a decrease in the solubility of any Ti-rich phase with increasing pressure or with changes in oxygen fugacity.

A third possible interpretation of the results in Fig. 4 is that the proto-KREEP which was ancestral to Apollo 15 igneous KREEP was similar in composition to the LKFM breccias, but had lower contents of TiO_2 (~1.0%) and FeO (~8%). This is the interpretation which we currently favor.

Rb–Sr isotopes

The arguments which we proposed last year concerning the times of the differentiation events leading to Apollo 15 igneous KREEP were based on the Rb–Sr isotopic compositions of the LKFM breccias (McKay and Weill, 1976). If it is true that these rocks do not accurately represent the precursors to Apollo 15 igneous KREEP, the arguments based on their isotopic compositions are no longer strictly valid. Nevertheless, our general conclusions still hold. The old Rb–Sr model ages of all KREEP-rich samples so far investigated indicate that the major fractionation of Rb/Sr from estimated whole-moon values to the much higher values typical of KREEP occurred prior to 4.3 G.y. The fact that the Apollo 15 igneous KREEP samples have crystallization ages (3.9 G.y.) which are much younger than their model ages (4.3 G.y., Nyquist *et al.*, 1973, 1974, 1975), implies that at least two episodes of igneous differentiation were involved in the formation of these samples. Furthermore, the old model ages and high Rb/Sr of five of the seven Apollo 15 igneous KREEP samples for which whole rock Rb–Sr isotopic data are available indicate that the precursors from which these five samples were derived at 3.9 G.y. had Rb/Sr values of at least .046 and were thus more evolved than the LKFM breccias (Rb/Sr = .038). Hence, although we can no longer precisely estimate the times of the differentiation events leading to Apollo 15 igneous KREEP, these times are still subject to important constraints.

Ti/Sm

The ratio of Ti to strongly excluded elements provides another important constraint on the origin of KREEP-rich materials (Duncan *et al.*, 1976). For example, Ti/Sm in Apollo 15 igneous KREEP is 4–6 times smaller than this ratio in chondritic or estimated whole-moon compositions. This means that either the estimated whole-moon Ti/Sm values are seriously in error, or else some phase which strongly incorporates Ti relative to Sm has crystallized during the formation of KREEP-rich samples from relatively undifferentiated material.

The major silicate phases which crystallize during the differentiation of estimated whole-moon compositions are olivine, pyroxene, plagioclase, and, at high pressures, garnet (Hodges and Kushiro, 1974), all of which strongly exclude Ti as well as Sm. Ilmenite crystallizes late in the sequence, and could be responsible for the reduction in Ti/Sm. However, ilmenite is far from the liquidus of KREEP-rich samples (e.g., Table 1; Hess *et al.*, 1977; Walker *et al.*, 1973).

We know of no simple mechanism by which an ilmenite-saturated liquid can evolve into a strongly ilmenite-undersaturated liquid. We are reluctantly forced to appeal to "unorthodox" mechanisms for the generation of proto-KREEP, such as magma hybridization (Hubbard and Minear, 1976) or assimilation of crustal materials by a superheated ilmenite-saturated magma. Such mechanisms are rather unconstrained, and we have not attempted to model them quantitatively.

A model for KREEP petrogenesis

Our revised model for the petrogenesis of the KREEP-rich samples is summarized in Fig. 5. It differs in several respects from the model which we proposed last year. We believe that the proto-KREEP which was ancestral to Apollo 15 igneous KREEP is similar in composition to the proto-KREEP of last year's model (the LKFM breccias), but is somewhat poorer in Ti and Fe. We no longer use the Rb–Sr isotopic compositions of the LKFM breccias to fix the time of differentiation of proto-KREEP from material with primitive Rb/Sr.

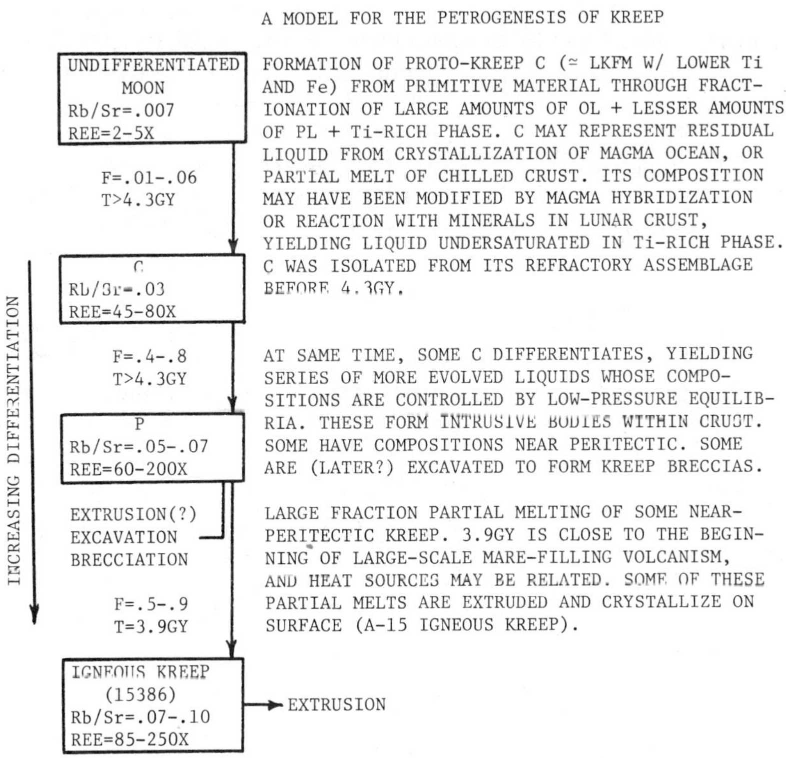

Fig. 5. Schematic representation of our revised model for the petrogenesis of the KREEP-rich samples. *F* is the fraction of liquid resulting from any differentiation step relative to the immediate parent material for that step.

However, the time of this differentiation is still constrained by the model ages of the KREEP-rich samples to be prior to 4.3 G.y. Finally, unless proposed whole-moon values of Ti/Sm are in error by more than a factor of 4, we must appeal to "unorthodox" processes such as magma hybridization or reaction of magma with crustal minerals to account for the lower values of Ti/Sm in KREEP-rich samples. Aside from these differences, our current model is similar to that which we proposed last year.

Relationship of KREEP to other highland rocks through mixing

A fundamental question in the petrogenesis of the highland breccias is whether chemical and petrologic trends established by primary igneous processes are still preserved in these samples, or whether such trends have been completely obscured by the processes of brecciation and physical mixing. Some workers have argued that the effects of igneous processes are still detectable in these samples (e.g., Wood, 1975), while other workers have argued that the compositions of most of these samples are totally dominated by physical mixing of a small number of end-member components (e.g., Wänke *et al.*, 1976).

Advocates of the mixing origin for highland breccias have proposed a variety of materials as the end-member components in the mixtures, but all proposed mixtures contain a KREEP-rich component. Occasionally, this component is comprised of less highly evolved KREEP. For example, Schonfeld (1974) used Apollo 17 LKFM breccias (Sm $\simeq$ 15 ppm) as the KREEP-rich end member for a mixing model analysis of the Apollo 17 soils. More frequently, very highly evolved KREEP is selected as the KREEP-rich component. In modeling a number of highland rocks (including Apollo 17 LKFM breccias) as mixtures, Schonfeld (1974) selected KREEP-rich glasses having Sm contents of 42 ppm (Schonfeld and Meyer, 1972) as the KREEP-rich end member. Wänke *et al.* (1976) suggest that the KREEP component in highland mixtures contains at least 32 ppm Sm, and possibly much more.

We now consider the questions of (1) the relative importance of igneous vs. mixing processes in determining the compositions of highland samples and (2) the nature of the KREEP-rich end member in any possible mixtures. Figure 6 shows the bulk compositions of over 100 highland rocks plotted in the OL–AN–SI pseudoternary diagram. These rocks include all the samples for which we could find published analyses of both major and trace elements for the same subsample. The rocks have been divided into groups based on their Sm contents, and each group is plotted separately. Samples with low Sm contents are dispersed throughout a wide region of the diagram. In contrast, samples with Sm contents of greater than 10 ppm lie close to the multiple saturation curves, suggesting igneous control of their compositions.

Figure 7 shows the variation of MgO content with Sm content for the same samples plotted in Fig. 6. The LKFM breccias are plotted as open circles, Apollo 15 igneous KREEP samples are plotted as *X*'s, and all other samples are plotted as filled circles. The curve is the calculated differentiation path for the LKFM

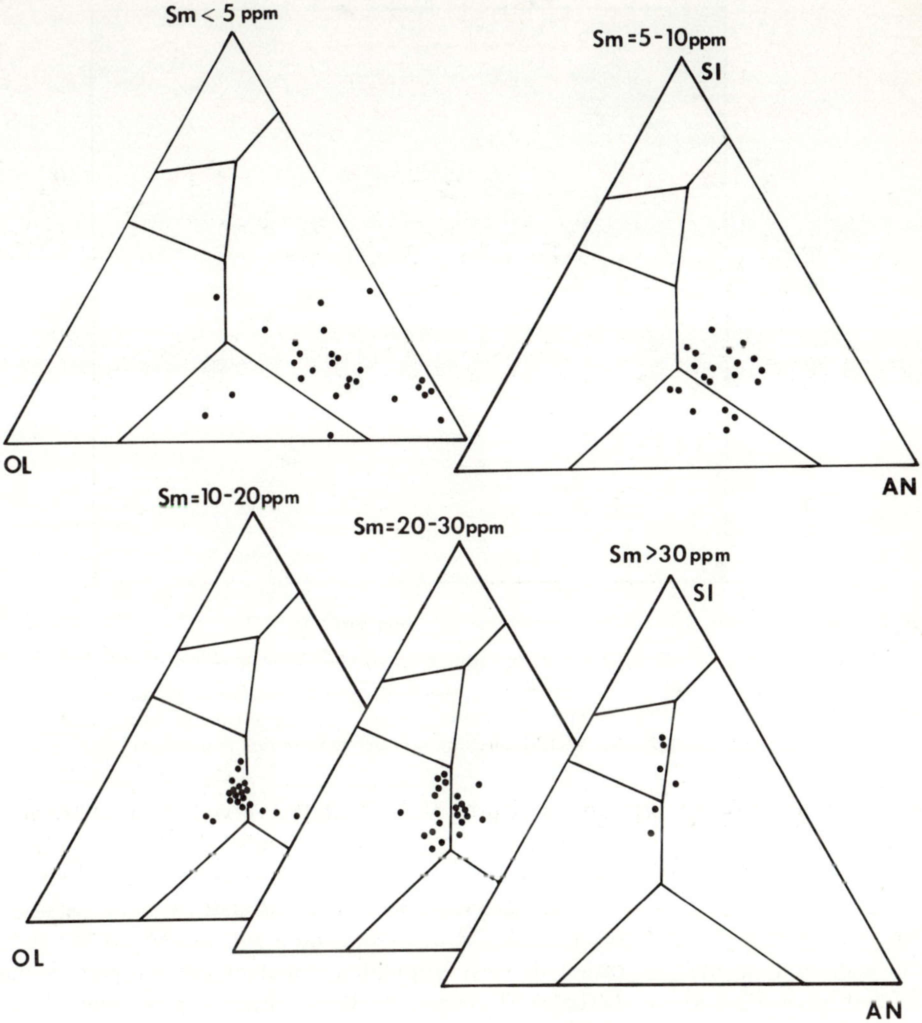

Fig. 6. Pseudoternary representations of the compositions of highlands samples, grouped according to Sm content. Data sources: Drake *et al.* (1973), Haskin *et al.* (1973), Hubbard *et al.* (1972, 1973), Laul *et al.* (1972), LSPET (1972, 1973), Nava (1975), Nakamura *et al.* (1973), Philpotts *et al.* (1972, 1974), Rhodes and Hubbard (1973), Rhodes *et al.* (1974), Rose *et al.* (1972), Taylor *et al.* (1972, 1973), Wänke *et al.* (1972, 1973, 1974, 1975, 1976), and Wiesmann and Hubbard (1975).

breccias, from Fig. 4. Most samples having Sm contents of greater than 15 ppm lie close to the calculated evolution curve. We infer from Figs. 6 and 7 that the compositions of most of these Sm-rich samples are dominated by igneous processes, although some mixing effects are probably superimposed.

There are few samples in the portion of Fig. 7 between the Apollo 15 igneous KREEP and the samples which are low in both MgO and Sm. If the KREEP-

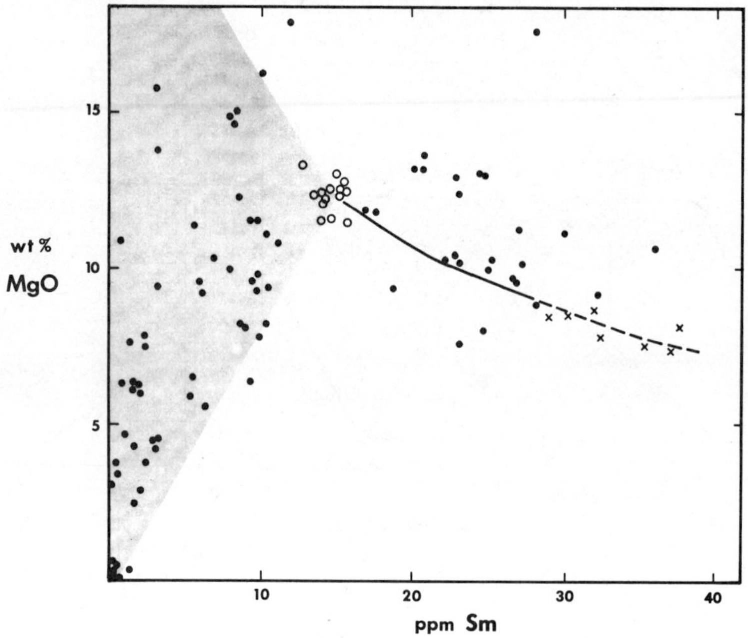

Fig. 7. MgO vs. Sm for LKFM breccias (○), Apollo 15 igneous KREEP (×), and other highland samples (●). The curve for the evolution of MgO and Sm is from Fig. 4. Samples in the shaded region have compositions which might be dominated by physical mixing of anorthosite, LKFM, and a poorly defined Mg-rich component.

rich end member of highland mixtures were highly evolved KREEP (Sm > 30 ppm), as suggested by Wänke *et al.* (1976) and Schonfeld (1974), we would expect to find this portion of the diagram as densely populated as any other. In contrast, the shaded region of the diagram, which is bounded by lines between the anorthosites (MgO < 1%), the LKFM breccias, and a poorly defined Mg-rich, Sm-poor end member, is rather densely populated. Most of the samples in the shaded region could be mixtures primarily of these three end-member components. Hence the data in Fig. 7 suggest that if the rocks having Sm contents of less than 15 ppm are indeed physical mixtures, then material similar in composition to the LKFM breccias, rather than more highly evolved KREEP, is the KREEP-rich end-member component of these mixtures.

SUMMARY

We have used an experimentally determined crystallization sequence for a composition similar to the LKFM breccias and experimentally determined crystal/liquid distribution coefficients to calculate the evolution of the abundances of several key elements during the crystallization of LKFM liquids. Results indicate that the LKFM breccias are not themselves precursors to Apollo 15 igneous KREEP, but that compositions slightly poorer in Ti and Fe might be. Consideration of the evolution of Ti and Sm abundances during the differentiation of proposed whole-moon compositions suggests that it may be necessary to appeal to "unorthodox" mechanisms such as magma hybridization or interaction of magma with genetically unrelated minerals within the lunar crust to produce KREEP-rich liquids having the observed Ti/Sm. We have presented a revised model for the petrogenesis of the KREEP-rich samples which is similar to the model we proposed last year, but which takes into account the new experimental data and the arguments involving Ti/Sm. Finally, we have considered the question of whether the present compositions of highland samples reflect mainly igneous processes, or whether they reflect primarily physical mixing. We conclude that the compositions of samples with Sm contents greater than those in LKFM (~15 ppm) are dominated by igneous processes, while samples with lower Sm contents may be mixtures of an anorthositic component, a Mg-rich Sm-poor component, and a component similar in composition to LKFM.

Acknowledgments—Our research is supported by NASA grants NGL-38-003-020 and NGL-38-003-022.

REFERENCES

Akella J., Williams R., and Mullins O. (1976) Solubility of Cr, Ti, and Al in coexisting olivine, spinel, and liquid at 1 atm. *Proc. Lunar Sci. Conf. 7th*, p. 1179–1194.

Basu A. and Bower J. (1976) Petrography of KREEP basalt fragments from Apollo 15 soils. *Proc. Lunar Sci. Conf. 7th*, p. 659–678.

Dowty E., Keil K., Prinz M., Gros J., and Takahashi H. (1976) Meteorite-free Apollo 15 crystalline KREEP. *Proc. Lunar Sci. Conf. 7th*, p. 1833–1844.

Drake M., Stoeser J., and Goles G. (1973) A unified approach to a fragmental problem: Petrological and geochemical studies of lithic fragments from Apollo 15 soils. *Earth Planet. Sci. Lett.* **20**, 425–439.

Duncan A., Erlank A., Scher M., Abraham Y., Willis J., and Ahrens L. (1976) Some trace element constraints on lunar basalt genesis. *Proc. Lunar Sci. Conf. 7th*, p. 1659–1671.

Haskin L., Helmke P., Blanchard D., Jacobs J., and Telander K. (1973) Major and trace element abundances in samples from the lunar highlands. *Proc. Lunar Sci. Conf. 4th*, p. 1275–1296.

Helmke P., Haskin L., Korotev R., and Ziege K. (1972) Rare earths and other trace elements in Apollo 14 samples. *Proc. Lunar Sci. Conf. 3rd*, p. 1275–1292.

Hess P., Rutherford M., Campbell H., and Dick P. (1977) Fractionation of lunar feldspathic magmas (abstract). In *Lunar Science VIII*, p. 433–435. The Lunar Science Institute, Houston.

Hodges F. and Kushiro I. (1974) Apollo 17 petrology and experimental determination of differentiation sequences in model moon compositions. *Proc. Lunar Sci. Conf. 5th*, p. 505–520.

Hubbard N. and Gast P. (1971) Chemical composition and origin of nonmare lunar basalts. *Proc. Lunar Sci. Conf. 2nd*, p. 999–1020.

Hubbard N., Gast P., Rhodes J., Bansal B., and Wiesmann H. (1972) Nonmare basalts: Part II. *Proc. Lunar Sci. Conf. 3rd*, p. 1161–1179.

Hubbard N. and Minear J. (1976) Petrogenesis in a modestly endowed moon. *Proc. Lunar Sci. Conf. 7th*, p. 3421–3435.

Hubbard N., Rhodes J., Gast P., Bansal B., Shih C., Wiesmann H., and Nyquist L. (1973) Lunar rock types. The role of plagioclase in nonmare highland rock types. *Proc. Lunar Sci. Conf. 4th*, p. 1297–1312.

Huebner J., Lipin B., and Wiggins L. (1976) Partitioning of chromium between silicate crystals and melts. *Proc. Lunar Sci. Conf. 7th*, p. 1195–1220.

Irving A. (1975) Chemical, mineralogical and textural systematics of non-mare melt rocks: Implications for lunar impact and volcanic processes. *Proc. Lunar Sci. Conf. 6th*, p. 363–394.

Laul J., Wakita H., Showalter D., Boynton W., and Schmitt R. (1972) Bulk, rare earth, and other trace elements in Apollo 14 and 15 and Luna 16 samples. *Proc. Lunar Sci. Conf. 3rd*, p. 1181–1200.

LSPET (1972) The Apollo 15 lunar samples: A preliminary description. *Science* **175**, 363–375.

LSPET (1973) The Apollo 16 lunar samples: Petrographic and chemical description. *Science* **179**, 23–34.

LSPET (1974) Apollo 17 lunar samples: Chemical and petrographic description. *Science* **182**, 659–690.

McKay G. and Weill D. (1976) The petrogenesis of KREEP. *Proc. Lunar Sci. Conf. 7th*, p. 2427–2447.

Nakamura N., Masuda A., Tanaka T., and Kurasawa H. (1973) Chemical compositions and rare-earth features of four Apollo 16 samples. *Proc. Lunar Sci. Conf. 4th*, p. 1407–1414.

Nava D. (1974) Chemical compositions of some soils and rock types from the Apollo 15, 16, and 17 lunar sites. *Proc. Lunar Sci. Conf. 5th*, p. 1087–1096.

Nyquist L., Bansal B. M., and Wiesmann H. (1975) Rb–Sr ages and initial $^{87}Sr/^{86}Sr$ for Apollo 17 basalts and KREEP basalt 15386. *Proc. Lunar Sci. Conf. 6th*, p. 1445–1465.

Nyquist L., Bansal B. M., Wiesmann H., and Jahn B.-M. (1974) Taurus-Littrow chronology: Some constraints on early lunar crustal development. *Proc. Lunar Sci. Conf. 5th*, p. 1515–1539.

Nyquist L., Hubbard N., and Gast P. (1973) Rb–Sr systematics for chemically defined Apollo 15 and 16 materials. *Proc. Lunar Sci. Conf. 4th*, p. 1823–1846.

Philpotts J., Schnetzler C., Nava D., Bottino M., Fullager P., Thomas H., Schuhmann S., and Kouns C. (1972) Apollo 14: Some geochemical aspects. *Proc. Lunar Sci. Conf. 3rd*, p. 1293–1305.

Philpotts J., Schuhmann S., Kouns C., Lum R., and Winzer S. (1974) Origin of Apollo 17 rocks and soils. *Proc. Lunar Sci. Conf. 5th*, p. 1255–1267.

Reid A., Warner J., Ridley W., and Brown R. (1972) Major element compositions of glasses in the three Apollo 15 soils. *Meteoritics* **7**, 395–415.

Rhodes J. and Hubbard N. (1973) Chemistry, classification, and petrogenesis of Apollo 15 mare basalts. *Proc. Lunar Sci. Conf. 4th*, p. 1127–1148.

Rhodes J., Rogers K., Shih C., Bansal B., Nyquist L., Wiesmann H., and Hubbard N. (1974) The relationships between geology and soil chemistry at the Apollo 17 landing site. *Proc. Lunar Sci. Conf. 5th*, p. 1097–1117.

Rose H., Cuttita F., Annell C., Curron M., Christian R., Dwornik E., Greenland L., and Ligon D. (1972) Compositional data for twenty-one Fra Mauro lunar materials. *Proc. Lunar Sci. Conf. 3rd*, p. 1215–1279.

Schonfeld E. (1974) The contamination of the lunar highlands by KREEP: Interpretation by mixing models. *Proc. Lunar Sci. Conf. 5th*, p. 1269–1286.

Schonfeld E. and Meyer C. (1972) The abundances of components of the lunar soils by a least-squares mixing model and the formation age of KREEP. *Proc. Lunar Sci. Conf. 3rd*, p. 1397–1420.

Schreiber H. and Haskin L. (1976) Chromium in basalts: Experimental determination of redox states and partitioning among synthetic silicate phases. *Proc. Lunar Sci. Conf. 7th*, p. 1221–1259.

Taylor S., Gorton M., Muir P., Nance W., Rudowski R., and Ware N. (1973) Lunar highlands composition: Apennine Front. *Proc. Lunar Sci. Conf. 4th*, p. 1445–1459.

Taylor S. and Jakeš P. (1974) The geochemical evolution of the moon. *Proc. Lunar Sci. Conf. 5th*, p. 1287-1305.

Taylor S., Kaye M., Muir P., Nance W., Rudowski R., and Ware N. (1972) Composition of the lunar uplands: Chemistry of Apollo 14 samples from Fra Mauro. *Proc. Lunar Sci. Conf. 3rd*, p. 1231–1249.

Walker D., Grove T., Longhi J., Stolper E., and Hays J. (1973) Origin of lunar feldspathic rocks. *Earth Planet. Sci. Lett.* **20**, 325–336.

Walker D., Longhi J., and Hays J. (1972) Experimental petrology and the origin of Fra Mauro rocks and soil. *Proc. Lunar Sci. Conf. 3rd*, p. 797–817.

Wänke H., Baddenhausen H., Balacescu A., Teschke F., Spettel B., Dreibus G., Palme H., Quijano-Rico M., Kruse H., Wlotzka F., and Begemann F. (1972) Multielement analysis of lunar samples and some implications of the results. *Proc. Lunar Sci. Conf. 3rd*, p. 1251–1268.

Wänke H., Baddenhausen H., Dreibus G., Jagoutz E., Kruse H., Palme H., Spettel B., and Teschke F. (1973) Multielement analyses of Apollo 15, 16, and 17 samples and the bulk composition of the moon. *Proc. Lunar Sci. Conf. 4th*, p. 1461–1481.

Wänke H., Palme H., Baddenhausen H., Dreibus G., Jagoutz E., Kruse H., Palme C., Spettel B., Teschke F., and Thacker R. (1975) New data on the chemistry of lunar samples: Primary matter in the lunar highlands and the bulk composition of the moon. *Proc. Lunar Sci. Conf. 6th*, p. 1313–1340.

Wänke H., Palme H., Baddenhausen H., Dreibus G., Jagoutz E., Kruse H., Spettel B., Teschke F., and Thacker R. (1974) Chemistry of Apollo 16 and 17 samples: Bulk composition, late stage accumulation and early differentiation of the moon. *Proc. Lunar Sci. Conf. 5th*, p. 1307–1335.

Wänke H., Palme H., Kruse H., Baddenhausen H., Cendales M., Dreibus G., Hofmeister H., Jagoutz E., Palme C., Spettel B., and Thacker R. (1976) Chemistry of lunar highland rocks: A refined evaluation of the composition of the primary matter. *Proc. Lunar Sci. Conf. 7h*, p. 3479–3499.

Weill D. and McKay G. (1975) The partitioning of Mg, Fe, Sr, Ce, Sm, Eu, and Yb in lunar igneous systems and a possible origin of KREEP by equilibrium partial melting. *Proc. Lunar Sci. Conf. 6th*, p. 1143–1158.

Wiesmann H. and Hubbard N. (1975) A compilation of the lunar sample data generated by the Gast, Nyquist and Hubbard lunar sample PI-ships. NASA Johnson Space Center, Houston, Texas. Unpublished manuscript.

Wood J. (1975) Lunar petrogenesis in a well-stirred magma ocean. *Proc. Lunar Sci. Conf. 6th*, p. 1087–1102.

Proc. Lunar Sci. Conf. 8th (1977), p. 2357–2373.
Printed in the United States of America

Origin and evolution of LKFM basalt

P. C. HESS, M. J. RUTHERFORD, and H. W. CAMPBELL

Geology Department, Brown University, Providence, Rhode Island 02912

Abstract—Although LKFM basalt is a major component in lunar soils and breccias, no igneous-textured clasts are found that are clearly identifiable with the LKFM compositions. Petrochemical studies of polymict breccias support the hypothesis that the LKFM composition is a mixture of various lithic components. However, crystallization experiments on a LKFM basalt composition show that both olivine and plagioclase are liquidus phases. The LKFM basalt is indeed a cotectic liquid and it strains credibility to explain these phase relations as the result of fortuitous and random mixing processes. We offer the hypothesis that LKFM basalt breccias are mixtures of cumulate rocks (anorthosite, troctolite, dunite) and residual liquid produced by the fractionation of a LKFM basalt.

An equilibrium liquid line of descent for a LKFM basalt produced FeO-enriched residual liquids which are compositionally equivalent to some M and HKFM basalt glasses found in the lunar soils. Residual liquids produced under more reducing conditions approximated the compositions of some Apollo 14 and 15 KREEP rocks. These experimental data support the model that LKFM basalt is parental to MKFM basalts.

LKFM basalt is believed to be an intercumulus component in the plagioclase-rich lunar crust, and therefore a relatively primitive liquid formed during the early crystallization of the moon. In order to explain the high trace-element abundances in LKFM basalt, two models are proposed. In a model of rapid accretion of the moon, LKFM basalt is a hybrid magma produced by the partial melting of the lunar crust by a KREEP-rich residual liquid. In a model of slow accretion, the outermost portions of the moon are only partially molten. Incompatible elements in the partially molten zone are concentrated into a smaller magma ocean which forms by gravity separation of solids and liquids. The magma ocean would then contain several times the accepted abundances of incompatibles and could account for the pattern of trace-element enrichments observed in lunar rocks.

INTRODUCTION

IT IS NOW WIDELY ACCEPTED that the lunar crust is a feldspathic cumulate layer formed by flotation of plagioclase within a fractionating magma of global dimensions. A corollary to this model is that the lunar crust must also include a significant fraction of intercumulus melt. The intercumulus melt is a direct sample of a magma on the liquid line of descent of the primitive moon, and would provide an important constraint to the modelling of lunar history. Unfortunately, it is not a straightforward exercise to identify this component. Most plutonic and volcanic non-mare rocks have been shocked, melted, and brecciated by meteorite impacts. Consequently, many non-mare rocks and glasses are mixtures of different rock types.

Averages of preferred glass compositions in the lunar soil provide information on the composition of lunar rock types (Ridley *et al.*, 1973; Reid *et al.*, 1972a,b). For example, the preferred glass compositions in the Apollo 11, 12, and 14 soils closely match rock compositions of high- and low-Ti mare basalts (Reid *et al.*, 1972a). One preferred glass composition, the Highland basalt, bears a strong resemblance to the average major-element composition of the upper 10–12 km of the lunar highlands (Taylor and Jakeš, 1974; Taylor and Bence, 1975). The average Highland crust composition is based on orbital XRF and gamma-ray data yielding Al/Si, Mg/Si, and Th values. These data are then reduced by assuming an average SiO_2 concentration (45%) and utilizing established inter-element relationships. The technique is empirical and is not dependent on mixing calculations using postulated Highland crustal rock types. Thus, the close correspondence to the Highland basalt composition is a powerful argument that the preferred glass compositions represent melted, homogenized upper crustal material. The other preferred glass composition, that of low-K Fra Mauro basalt (LKFM), is not found as clasts of pristine igneous rock in the returned lunar samples. One of the problems in lunar petrogenesis which has yet to be solved is that of the origin of the LKFM basalts.

OCCURRENCE OF LKFM BASALT

The average LKFM preferred glass compositions (Table 1) plot close to the low pressure olivine-plagioclase cotectic in the pseudoternary olivine-anorthite-silica phase diagram (Walker *et al.*, 1973a). The proximity of LKFM compositions to multisaturated liquids supports the view that these basalts are produced either by large degrees of partial melting of crustal rocks (Walker *et al.*, 1973a) or are the residual liquids to the fractionation of the outer moon (McKay and Weill, 1976, 1977). However, fragments or clasts of igneous-textured LKFM basalt have not been identified (Reid *et al.*, 1977; Meyer, in press) even though the LKFM composition does occur in many forms. Apollo 17 soil samples free of lithic mare contaminants are of LKFM composition (Table 1) and also plot close to the multisaturated liquidus curves (Bence *et al.*, 1974). Apollo 15 soils contain less Al_2O_3 and CaO (Table 1) but also resemble LKFM "basalts". Low KREEP poikilitic rocks, particularly those at the Apollo 16 site, have the LKFM chemistry (Table 1) (Simonds *et al.*, 1973). These poikilitic rocks contain clasts of anorthosite, troctolite, and non-mare basalt. The protolith most likely was a polymict breccia and/or lunar soil which was fused and metamorphosed by meteorite impacts (Albee *et al.*, 1973; Bence *et al.*, 1973; Simonds *et al.*, 1973). Clearly, these soils and breccias are not monolithic aggregates and do not simply represent degraded LKFM volcanics. The LKFM composition does appear to be a common one among the highland samples, however, and thus its origin is an important problem.

Dence *et al.* (1976) suggest that samples of LKFM represent homogenized samples of a shock-melted mixture of almost the entire 60 km crust. The LKFM composition is approximated by a mixture of 65% anorthosite and anorthosite gabbro, 17% gabbro, 17% troctolite, and 1% granite. The implication of this model is that the lunar crust is zoned from Highland basalt in the upper crust to less aluminous, more KREEP-enriched rocks at depth (see also Ryder and Wood, 1977). According to this model, the average crust contains about 2000–

Table 1.

	(1)	(2)	(3)	(4)	(5)	(6)	(7)	(8)
SiO_2	44.8	45.0	46.5	45.8	45.0	46.2	46.2	47.3
TiO_2	0.4	0.6	1.0	1.3	1.5	1.5	0.8	0.5
Al_2O_3	25.8	24.6	20.1	19.4	20.9	17.9	18.1	17.2
FeO	5.6	6.6	9.1	9.0	8.5	9.1	10.7	10.0
MgO	8.0	8.6	10.4	11.3	10.0	12.3	12.1	10.7
CaO	14.8	14.2	12.4	11.8	12.9	11.2	11.9	13.1
Na_2O	0.2	0.5	0.4	0.5	0.4	0.7	0.5	0.5
K_2O	0.0	0.1	0.1	0.3	0.2	0.3	0.2	—

(1) Average Highland basalt, soil glass (Ridley *et al.*, 1973). (2) Average Highland crust (Taylor and Bence (1975)). (3) Average LKFM basalt, soil glass (from Ridley *et al.*, 1973). (4) Average of light poikilitic rocks (Simonds *et al.*, 1973). (5) Apollo 17 soil, no mare lithic contaminants (Bence *et al.*, 1974). (6) Average of five matrix compositions to some clast-laden melt breccias (Simonds, 1975). (7) Average of 23 compositions of "cotectic liquids" obtained by subtracting plagioclase from lunar "cumulate" rocks. Range of compositions SiO_2 45–50, TiO_2 0.6–1.6, Al_2O_3 17–19, FeO 5–15, MgO 9–15, CaO 10–13, Na_2O 0.2–0.9, K_2O 0.1–0.5. (8) Residual liquid to Taylor–Jakeš (1974) moon at 10 kbar 1300°C (Ringwood, 1976).

3000 ppm K, and with the application of inter-element correlations, about 100× chondrite concentrations of incompatible elements. Recent thermal models of the moon indicate that the crust formed from a magma ocean about 200–300 km thick. Assuming this thickness and the Dence model, a mass balance calculation requires that the magma ocean would have contained incompatible elements at 25× chondritic abundances. These concentrations are not consistent with the recent low estimates (Taylor, 1976). In any case, it is clear that most LKFM compositions are obtained from mixtures of different rock types. One interpretation of this data is that the proximity of LKFM basalt to the olivine-plagioclase cotectic is fortuitous (Meyer, 1977).

Equilibrium Crystallization of LKFM Basalt

The apparent cotectic control of LKFM compositions in the olivine-anorthite-silica pseudoternary system does not require that these melts are indeed multisaturated. Components such as TiO_2, Na_2O, K_2O, P_2O_5, Cr_2O_3, and $CaSiO_3$ are excluded in this representation and could alter the phase boundaries. To examine this question, we have experimentally determined the low pressure liquidus for a LKFM composition (Table 3, column 1) very similar to the preferred LKFM glass composition in the Apollo 15 soils (Reid *et al.*, 1972). McKay and Weill (1976, 1977) proposed a model in which olivine-plagioclase cotectic liquids similar to LKFM basalt were parental to more evolved KREEP-rich compositions. Major- and trace-element partitioning data were used to calculate the composition of highland rocks that could be possible progenitors to KREEP-rich rocks. We have extended these studies by determining a low pressure equilibrium liquid line of descent for the LKFM composition.

Experiments were conducted in Mo foil containers sealed in evacuated glass tubes (Rutherford *et al.*, 1974). Starting materials were LKFM gels most of which were reduced to f_{O_2} values of the IW buffer. Some experiments were performed on LKFM gels reduced to f_{O_2} values 1 and 2 orders of magnitude below those determined by the IW buffer. The reduced gels were fused at 1325°C and quenched to a glass. These glasses were held at the desired run temperatures for 24–144 hr. FeO loss to the Mo container was minimal in runs where f_{O_2} was approximated by the IW buffer. Melts held at 1250°C for 48 hr contained virtually the same amount of FeO (9.6%) as the initial composition (FeO = 9.7%).

Results of these experiments are given in Tables 2 and 3. Runs obtained under f_{O_2} values of IW are most detailed and are considered first. Trace amounts of Cr-bearing aluminous spinel appear at 1265°C but are found only next to the Mo container walls. Microprobe analysis reveals that the spinels are zoned and contain significant concentrations of Mo (2–5%) in their cores. Therefore, 1265°C is not the true liquidus temperature because the Mo has probably stabilized the spinel at this and at lower temperatures. Rims of spinel are $(Mg_{.77}Fe_{.23})\times(Al_{1.6}Cr_{0.4})O_4$ and contain only about 1–2% Mo. These spinels (on Mo-free basis) are comparable to the more Cr-rich aluminous spinels found in Luna 20 spinel troctolites (Prinz *et al.*, 1973). Consequently, the LKFM basalt has a composition which is close to but not necessarily on the spinel liquidus. The true liquidus is between 1250°C and 1245°C where olivine (Fo_{86}) and plagioclase (An_{95}) crystallize together. All phase boundaries are based on reversals. Thus, the LKFM liquid is indeed a cotectic liquid saturated with both olivine and plagioclase. It strains credibility to explain these phase relations as the result of fortuitous and random mixing processes! A low calcium pyroxene $Wo_4En_{74}Fs_{32}$ joins olivine and plagioclase between 1190°C and 1175°C and these three phases coexist with liquid to at least 1150°C.

Compositions of representative residual liquids are given in Table 3. The equilibrium liquid line of descent is characterized by (1) a steady increase in FeO, TiO_2, K_2O, and to a modest extent, SiO_2; (2) a decrease in MgO and Al_2O_3; and (3) only small changes in Na_2O, CaO, and Cr_2O_3. Ti and K behave as incompatible elements and are quantitatively enriched in the residual liquids. The TiO_2/K_2O ratio must then remain nearly constant in the residual liquids. The FeO/TiO_2 ratio does not stay constant but decreases from a value of 7.9 to 3.5 at 1150°C.

The compositions of the residual liquids produced by more than 25% crystallization contain more than 11.8% FeO. The iron content is much higher than that observed in most KREEP rocks (Meyer, 1977). Compare the composition (Table 3, column 5) of the peritectic liquid residual to KREEP basalt 14310 (Walker *et al.*, 1973a; peritectic B) with that obtained from the LKFM basalt at nearly the same temperature (Table 3, column 3). The major-element compositions are similar excepting the FeO and K_2O contents. The discrepancy between K_2O contents is easily remedied if the starting LKFM contained $K_2O = 0.3$ rather than 0.1. Potassium contents as high as 0.3 are typical of many

Table 2.

Run	Starting material	T°C	Duration (hr)	Products
2	B	1265	36	G,S
3	A	1250	48	G,S
4	B	1245	84	G,O,P,S
5	A	1245	84	G,S
7	B	1235	84	G,O,P,S
8	A	1235	84	G,O,P,S
9	B	1225	24	G,O,P,S
10	A	1225	24	G,O,P,S
13	A	1200	61	G,O,P,S
14	A	1190	144	G,O,P,S
15	A	1175	68	G,O,P,Px,S
16	A	1150	144	G,O,P,Px,S
17	A	1200	84	G,O,P,S
18	A	1200	114	G,O,P
19	A	1200	114	G,O,P
20	C	1250	48	G,O,P
21	C	1180	48	G,O,P,Px

A—glass fused at 1320°C for 10–15 min.
B—chip from run 13.
C—peritectic composition "A" of Walker *et al.* (1973).
G = glass, O = olivine, P = plagioclase, Px = pyroxene, S = spinel.
Runs 18 and 19 were equilibrated at f_{O_2} 1 and 2 orders of magnitude respectively below IW.

Table 3.

	1	2	3	4	5	6	7	8	9	10
SiO_2	47.0	47.6	48.0	48.8	49.1	48.0	48.0	48.5	51.2	50.8
TiO_2	1.2	1.6	2.3	4.3	2.4	2.0	2.1	2.2	2.3	2.2
Al_2O_3	19.0	17.3	15.8	12.2	15.9	12.6	17.0	15.2	16.0	14.8
Cr_2O_3	0.2	0.2	0.2	0.2	0.3	0.3	—	—	0.3	—
FeO	9.7	11.8	12.5	15.3	9.3	11.9	10.9	11.7	10.8	10.6
MgO	11.3	11.0	8.5	6.7	9.2	8.5	8.7	8.4	10.0	8.2
CaO	11.8	11.0	10.8	11.3	12.1	15.2	10.7	10.7	11.1	9.7
Na_2O	0.4	0.4	0.3	0.4	0.7	0.6	0.7	0.7	0.4	0.7
K_2O	0.1	0.1	0.2	0.4	0.7	0.3	0.5	0.8	0.2	0.7

(1) Starting composition. (2) Residual liquid (RL) 1235°C. (3) RL 1190°C. (4) RL 1150°C. (5) RL to 14310 1190°C (Walker *et al.*, 1973a). (6) RL to composition A of Walker *et al.* (1973a), 1180°C. (7) MKFM glass in Apollo 14 soil (Ridley *et al.*, 1973). (8) M(H)KFM glass in Apollo 12 soil (Ridley *et al.*, 1973). (9) RL to starting composition (1) 1200°C for f_{O_2} 2 orders of magnitude below IW. (10) 15386,1 Rhodes and Hubbard (1973).

LKFM compositions (Table 1). The large difference of 3.2% FeO is not so easily dismissed. It is possible that cotectic melts of more primitive composition (i.e., lower FeO, higher MgO) are more suitable parents for KREEP rocks (see also McKay and Weill, 1977).

A possible parental liquid is composition A, the peritectic liquid coexisting with olivine, plagioclase, and spinel in experiments on 62295 (Walker *et al.*, 1973a). The liquid contains about 13% MgO and 7% FeO (compared to the LKFM basalt with 11% MgO and 10% FeO). An equilibrium liquid line of descent was determined for composition A using the same experimental procedures (including an f_{O_2} equivalent to the IW buffer). The residual liquid at 1180°C (Table 3, column 6) has FeO and MgO concentrations which are very similar to those obtained for the LKFM composition. Thus, iron-enriched KREEP liquids were produced with even a relatively primitive starting composition. Unfortunately, these conclusions are tempered by the observation that the 1180°C liquid contains 12.6% Al_2O_3 and 15.2% CaO. Lunar rocks of such low Al_2O_3 and high CaO contents are not known. It is concluded that the peritectic liquid used as a starting composition is not parental to any LKFM basalt and in fact, probably is not a valid primitive magma.

Cotectic crystallization of olivine and plagioclase will invariably lead to the production of iron-enriched liquids according to mass balance calculations which show that plagioclase and olivine crystallize in a weight ratio of about 2:1. This is verified by the results of the experiments. However, the pattern of iron enrichment is depressed under more strongly reducing conditions where Fe metal is a crystallizing phase. Equilibrium crystallization experiments were performed on the same LKFM gel reduced to oxygen fugacities 1 and 2 orders of magnitude below those of the IW buffer. Fe metal crystallized in both experiments. The residual liquid obtained at 1200°C and at two orders of magnitude below IW contains about 3.0% FeO less than that obtained at oxygen fugacities equivalent to the IW buffer. The liquid is a relatively good match to KREEP basalt 15386 (Table 3, cf., columns 9, 10). Again, the K_2O discrepancy is due to the very low K_2O starting composition. The liquid obtained at 1200°C and f_{O_2} conditions only 1 order of magnitude below IW contains about 1.0% less FeO than the liquid obtained under more oxidizing conditions.

The experiments described allow for the possibility that magmas similar to some Apollo 14 and 15 compositions could be obtained as residual liquids to LKFM basalt (containing about 0.2–0.3% K_2O) provided that the oxygen fugacities are about 2 orders of magnitude below the IW buffer. More FeO-enriched magmas similar to MKFM glasses in the lunar soils are obtained under more oxidizing conditions (Table 3, cf., column 2 with 7 and column 3 with 8), although it should be remembered that an exact match should not be expected, since the initial LKFM basalt composition and the glass soil compositions are each averages. Walker *et al.* (1973b), on the basis of low pressure experiments, conclude that some feldspathic magmas crystallized at oxygen fugacities about 1.5 orders of magnitude below the IW buffer. This conclusion was guided by the observation that some highland rocks do show evidence of Fe metal near their liquidi. However, KREEP basalts such as 14310 (Ridley *et al.*, 1972) and 15382 (Dowty *et al.*, 1976) do not contain Fe inclusions in the early crystallizing phases and apparently were not saturated with Fe metal near the liquidus. These observations appear to contradict the model proposed above for the origin of Apollo 14 and 15 KREEP rocks (Grove, pers. comm.).

We consider this question still unresolved. Fe-metal inclusions are not abundant in early crystallizing phases so that their absence in a particular rock cannot be taken as proof that these magmas did not have Fe metal on the liquidus. Moreover, Sato (1976, 1977) has shown that graphite-bearing melts will probably be reduced if brought to the surface of the moon. Thus, the occurrence of Fe-metal inclusions in early crystallizing phases does not require that these melts were saturated with Fe metal at depth. If Sato's ideas are correct, the same melt could crystallize under radically different oxidizing conditions. For example, a LKFM basalt might produce iron-enriched residual liquids at depth where the environment might be more oxidizing ($\sim\frac{1}{2}$ order below IW). Perhaps the preferred glass compositions of MKFM basalt in the lunar soils are representative of such magmas. The same magma when brought to the surface would crystallize Fe metal because of graphite oxidation and loss of CO [the f_{O_2} of the graphite-CO equilibrium decreases markedly with decreasing pressure according to French (1966, see Fig. 14)], and the result would be magmas more akin to the KREEP rock samples. Although there are many unresolved questions, the experiments described do indicate the possibility that the Fra Mauro basalts are genetically related. Since the experiments were in an equilibrium mode, the results apply equally well to the conditions of partial melting or crystallization of a LKFM composition.

Discussion

LKFM basalt—an igneous rock type

The conclusion that the soil, glass and poikilitic rock of LKFM composition are mixtures of different rock types is well documented; however, we do not agree with the argument that the apparent cotectic control of these compositions is fortuitous. Experiments on a LKFM composition, described in the previous section, do indeed demonstrate that olivine and plagioclase occur as liquidus phases! The two conclusions can be reconciled in several ways. Warner *et al.* (1974) propose that impact initiated partial melting of surficial material results in the production of multisaturated liquid. Segregation of this melt with some included mineral and lithic clasts would crystallize as a LKFM basalt whereas the more refractory residue would form some kind of breccia. This process is a viable one but, we believe, not the dominant one. Temperatures of the olivine-plagioclase cotectic are between 1190°C and 1280°C (Walker *et al.*, 1973). LKFM basalts have a liquidus of around 1250°C while the liquidus temperature of Highland basalt (=average upper highland crust) is somewhat higher than 1320°C (McKay and Weill, 1976). Because the liquidus temperatures for these compositions are so close together, it seems statistically unlikely that large volumes of LKFM basalt could be produced by impact-related partial melts. Melts produced by bulk melting of impacted material should be most abundant (Grieve, 1975). The fact that LKFM basalt compositions are abundant suggests that the basalt composition was already in existence as a target rock.

The target rocks could be flows of LKFM composition or shallow level plutonic rocks. Consider the evolution of a pluton containing a magma of LKFM composition. This magma crystallizes olivine and plagioclase in approximately a 1:2 weight ratio. This ratio is calculated from our experiments or can be estimated by extrapolating the metastable part of the cotectic to the olivine-plagioclase join. However, cumulates formed from these cotectic liquids wouldn't necessarily lie on this extension. The cumulate layers formed in the lower part of the chamber are enriched in settled olivine and will approach dunite in composition. Plagioclase will be floated toward the roof of the pluton producing anorthosites. Cumulate layers in the central region should be more troctolitic in character. Residual liquids will trend to the peritectic and beyond to the pyroxene-plagioclase cotectic, and are concentrated in the central to upper part of the pluton. A zoned pluton results that contains distinct lithic types such as dunites, troctolites, anorthosites, and KREEP basalt yet as a whole is of LKFM basalt composition.

If this pluton is impacted by a large meteorite, the impact melt generated by near total melting of a large fraction of the pluton must lie close to the cotectic. Clasts of troctolite, dunite, and anorthosite may be mixed with the magma. Thus, even a clast-laden melt breccia containing significant non-LKFM clasts will have a LKFM bulk composition because the included clast components are complementary products to the residual liquid. Smaller impacts that homogenize only the upper levels of the pluton would produce KREEP-rich norite breccias associated or mixed with anorthositic clasts. It is noteworthy that Schonfeld (1974) has shown that LKFM compositions can be approximated by a mixture of dunite, anorthosite, and KREEP.

LKFM basalt—an intercumulus magma

Our hypothesis is that LKFM basalt is a residual liquid to the primitive fractionation of the moon and occurs as an intercumulus component in plagioclase cumulate rocks. Plagioclase cumulate rocks of the Upper Border Group, UBG, in the Skaergaard (Wager and Brown, 1967) are assumed to be terrestrial analogues of the supposed lunar cumulate rocks. The UBG rocks differ from the contemporaneous layered series near the bottom of the Skaergaard in being free of phenocrysts of olivine and pyroxene and in having a greater proportion of intercumulus material. Equivalent lunar plagioclase-rich rocks of undoubtedly cumulate origin are rare because most samples are brecciated or melted by impact processes. We shall assume that anorthosites, gabbroic anorthosites, and anorthositic gabbros that are not polymict breccias are samples of plagioclase cumulates. The intercumulus component of these rocks is determined by sub-

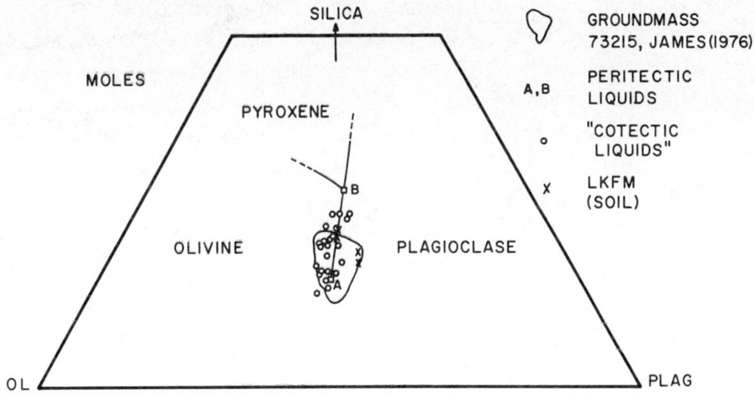

Fig. 1. Plagioclase-olivine-silica pseudoternary section. A composition is plotted by calculating a CIPW molecular norm and then using normative olivine, feldspar, and silica as the compositional coordinates. This diagram has the advantage that fractionation of plagioclase moves all compositions radial to the plagioclase apex and that all olivine-normative and all silica-normative compositions plot on the correct side of the pyroxene-plagioclase join. It has the deficiency in common with other diagrams, that all data are projected from the points $FeTiO_3$ and $CaSiO_3$ when, in fact, ilmenite and wollastonite are not liquidus phases. The peritectic liquids, A and B, are from Walker *et al.* (1973a). The LKFM soil compositions are from Reid *et al.* (1972a,b). The "cotectic liquids" are the calculated residual liquids to 23 anorthosites, gabbroic anorthosites, and anorthositic gabbros taken from the literature and chosen because of their apparent nonpolymict character.

tracting plagioclase, An_{95-98}, until the residual composition falls close to the olivine-plagioclase cotectic (Fig. 1). A similar scheme adopted by Schonfeld (1975) assumed that the intercumulus component was 50% of the rock. In contrast, the amount of residual liquid obtained from our technique varied from a low of 25% in the case of the more anorthositic samples (60017, 67075, 69935) to 56% in less plagioclase-rich rocks (15418, 76315, 78155). Anorthosites containing more than 32% Al_2O_3 were not used because of the large uncertainties involved in subtracting out large amounts of plagioclase.

The average composition of our calculated residual liquid is equivalent to a LKFM basalt (Table 1, column 7). In our opinion, the LKFM composition approximates an average intercumulus liquid that was the parent magma into which plagioclase was concentrated to form the lunar crust. Clearly, these are only rough numbers since not all rocks are necessarily cumulates to the same magma and moreover may contain some Fe–Mg cumulates as well. Nevertheless, we believe that the calculated LKFM composition places important constraints on the origin and evolution of the upper moon. Abundances of Sm and Eu were calculated for the average residual liquid using the distribution coefficients of Weill and McKay (1976), and REE data for the whole rock. Average chondrite normalized abundances are Sm = 20, Eu = 14, and Sm/Eu = 1.4. These data are only approximate and should simply be used in a qualitative way. However, if the initial magma ocean contained about 4–5× chondrite abundances of Sm and Eu, then plagioclase and the lunar crust began to accumulate after approximately 75% crystallization of the magma ocean. This is a reasonable figure. However, the LKFM composition (Table 1, column 7) also implies that the moon initially contained about 400–500 ppm K which is much higher than current estimates of 100 ppm (Taylor, 1976). A model for the origin of LKFM basalts must explain this discrepancy.

Origin of LKFM magma

The hypothesis that LKFM basalt originated as an intercumulus liquid within the lunar crust is not without problems. It is difficult to reconcile the high concentrations of incompatible elements in LKFM basalts with the conclusion that the basalts are relatively primitive magmas. Revised heat flow measurements indicate that the U abundances in the moon are 3–5 times chondritic (Langseth *et al.*, 1976). Similar enrichments are likely for other refractory incompatible elements. Volatile-involatile element ratios indicate K abundances of about 100 ppm (Taylor, 1976). Solomon and Chaiken (1976) show that the crystallization of an initially molten zone of 200 ± 100 km effectively balances the expansion of the moon due to internal warming. This satisfies the constraint that large-scale features indicative of compressive or tensional stresses are not observed in post-mare lunar terrains. Therefore, the crust represents from $\frac{1}{3}$ to $\frac{1}{4}$ of an initial magma ocean either 200 or 300 km thick. The intercumulus liquid in the upper crust should then contain about 300–400 ppm K and refractory incompatibles at concentrations of 15–20× chondrite.

In comparison, LKFM basalt contains 1000–3000 ppm K (Table 1) and REE concentrations enriched to about 75–100× chondrite (Blanchard *et al.*, 1976; Taylor *et al.*, 1973). It is possible that these trace-element enrichments are due to contamination with a KREEP-rich component. However, the high KREEP abundances are required in the initial LKFM composition if it is indeed a parental magma to KREEP basalts as is implied by our experimental data.

Accepting the hypothesis that LKFM basalt is indeed an intercumulus liquid, the enrichments of incompatible elements must be explained. Two models are proposed. Model one assumes a rapid accretion of the moon and the formation of an early magma ocean of global dimensions. Model two allows for a slower rate of accretion and the formation of a partially molten outer layer.

1. Rapid accretion model. In order to focus the discussion, a Taylor–Jakeš (T–J) lunar composition is assumed (Taylor and Jakeš, 1974).* The evolution of a T–J magma 300 km is obtained using the extensive crystallization experiments (Fig. 2) of Ringwood (1976). The temperature profile immediately following accretion is based roughly on the work of Walker *et al.* (1975). Olivine is the liquidus phase at all depths. Crystallization of olivine (Fo_{94}) is primarily from the

*More recent models tend to compositions lower in Al_2O_3 and CaO (Taylor and Jakeš, 1977).

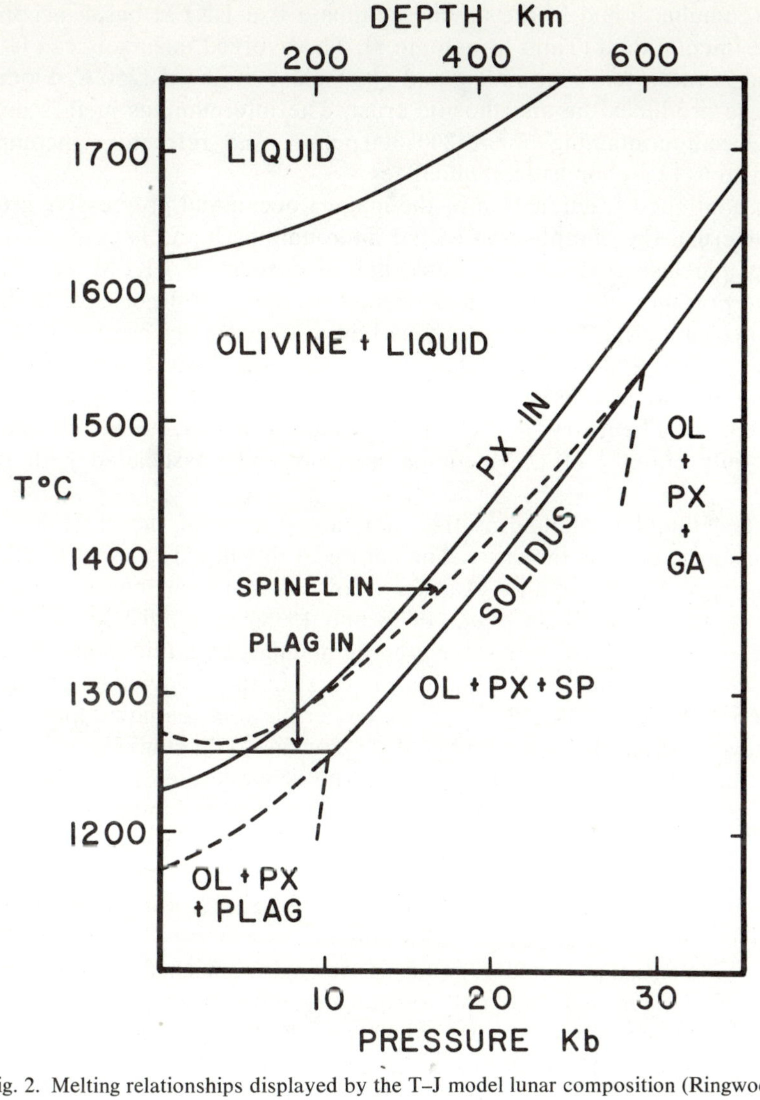

Fig. 2. Melting relationships displayed by the T–J model lunar composition (Ringwood, 1976).

top of the magma ocean (Solomon and Longhi, 1977). However, crystallization of olivine also occurs in the 200–300 km depth range where the melting point gradient exceeds the adiabatic gradient (Ringwood, 1976). Plagioclase or clinopyroxene + spinel join olivine after about 50% of the magma ocean is crystallized. The 300 km system now contains a 150 km base of cumulate olivine plus intercumulus liquid, a thin and probably ephemeral crust of quenched magma (Herbert *et al.*, 1977; Walker *et al.*, 1975) and an upper zone of convecting magma. The major-element composition of the remaining magma and the intercumulus liquid in the olivine cumulate is a LKFM basalt according to the experimental data (Table 1, column 8). The evolved magma ocean is 150 km thick and is saturated with olivine and plagioclase at about 1250°C. Flotation of plagioclase produces the anorthositic crust. The intercumulus melt is initially a LKFM basalt containing about 200–300 ppm K and refractory incompatibles enriched to 6–12× chondritic abundances.

With continued fractionation of the magma ocean and progressive growth of the lunar crust, the composition of the intercumulus liquid is controlled by the olivine-plagioclase cotectic. The liquid line of descent of LKFM magma in this moderate pressure range of 3–5 kbar should be similar but not exactly the same as obtained at 1 atm. The intercumulus liquid in the lower crust would contain more FeO, TiO_2, and incompatibles but less Al_2O_3 and MgO than LKFM basalt according to the low pressure experiments (see Table 3). It approaches the *major*-element chemistry of KREEP basalt. However, these liquids would contain only about $\frac{1}{5}$ of the incompatibles normally associated with KREEP rocks.

The last liquid remaining as the cumulate growth of the crust nears completion is at the base of the crust. The nature of this liquid cannot be established with certainty. The liquid lines of descent for non-mare magmas at 1 atm such as LKFM basalt, 14310 (Hess *et al.*, 1975) and 15386 (unpublished data) produce Fe-enriched magmas which exsolve granitic magmas by liquid immiscibility. It is not known whether immiscible liquids can exist in the 3–5 kbar range that would characterize the lower crust, nor can we be certain that the liquid line of descent would have reached the two liquid field. In any case, late-stage liquids residual to the fractionation of the outer moon should be of moderate SiO_2 (~50%), high in FeO (15–20%) and in TiO_2 (5–10%), and strongly enriched in incompatibles. This magma has the characteristics of the hypothetical KREEP component which Taylor and Bence (1976) suggest provides the KREEP but not the major components to KREEP rocks. The KREEP-rich liquid is not represented by any known igneous rock in the lunar collection.

This KREEP-rich residual liquid is strongly enriched in the radiogenic elements K, U, Th, and does not crystallize (Solomon and Longhi, 1977). One gram of liquid containing 10,000–20,000 ppm K and the appropriate concentrations of Th and U is capable of melting 8–16 g of rock in 100 m.y. The amount of rock melted depends on the temperature of the rock, its solidus and amount of minimum melting components in the rock, and the rate of heat loss by conduction at a given level within the lunar crust.

In thermal models (Hubbard and Minear, 1976; Solomon and Longhi, 1977), the KREEP residual liquid melts the underlying rocks and progresses down through the mafic cumulates toward the unfractionated moon. The efficiency of this process depends on how much intercumulus material resides in these cumulate rocks. Refractory cumulates have very high solidus temperatures and would prevent the downward advance of the melting zone (Solomon, pers. comm.). However, it is difficult to envision a totally refractory cumulate rock without some low melting component (see Ringwood, 1976, for contrary view).

The upward progression of the melt zone into the lunar crust is inhibited by the rapid loss of heat through the surface of the moon. However, the existence of a low melting, intercumulus component and high temperatures at the base of the crust provides an environment in which the residual liquid could eat its way into the lunar crust. This process must happen early in the lunar evolutionary history so that high temperatures are maintained in the lower crust and also to satisfy isotopic constraints that require a closed-system behavior after 4.4 AE (Hubbard and Minear, 1976).

It is suggested that LKFM basalt is a hybrid of two liquids which are products of the fractionation of the outer moon (see also Hubbard and Minear, 1976; Taylor and Jakeš, 1977). The major-element chemistry is dominated by the more primitive, intercumulus component and the incompatible elements by the KREEP magma. Since both components lie on a common liquid line of descent, the major-element composition of the hybrid liquids will not depart drastically from the compositions of true intermediate magmas.

2. Slow accretion model. Most treatments of the thermal and chemical evolution of the moon assume that gravitational energy of accretion was released at the surface and only by rapid accretion on a time scale of 10^3–10^4 yr are temperatures sufficiently high to produce an early magma ocean. Temperatures in excess of 1600°C are required to totally melt the outermost regions of a T–J moon (Fig. 2). However, thermal history calculations indicate that the source regions for mare basalts would be too hot to allow for closure of isotopic systems at 4.4 AE (Brett, 1977). The early moon must have been considerably cooler. Models of slow accretion would result in a cooler moon. In the Wetherill (1975, 1976) model, the moon is accreted on a time scale of 10^8 yr by impacting and burying large planetesimals of $r \geq 200$ km. Signficant fractions of the impact energy are trapped within the moon resulting in localized and heterogeneously distributed melted regions. At the end of accretion, the moon would be composed of large, overlapping, differentiated plutonic bodies (Wetherill, 1975). An enveloping magma ocean of lunar dimensions is not formed.

We propose a model developed by Alfvén and Arrhenius (1976) which is intermediate to those described above. Impacts by large planetesimals produce localized, near-surface differentiated plutons. These plutons may be impacted and partially or totally melted several times as the radius of the planet grows. This produces an "accretional front of hot spots", discontinuous in space and time which moves outward with the surface of the growing moon. The amount of melt occurring near the moon's surface at any given time depends on the rate of accretion. The "accretional front" produces magmas strongly enriched in incompatible elements because locally differentiated plutons will have formed with the more fractionated melts closer to the surface of the moon. This residual material is preferentially incorporated in the newly formed impact melt. The moon is thus zone-refined during accretion. As the moon approached its present size, accretion was more rapid and larger fractions of the outer moon were melted. It is possible that the thermal profile of the outer moon never exceeded the liquidus but was contained between the liquidus and the solidus. The outermost moon was then only partially molten and might have attained maximum initial temperatures some 200–300°C lower than usually assumed (e.g., see Walker *et al.*, 1975).

Most of the magma from the partially molten shell of the outer moon would segregate toward the surface possibly forming a relatively shallow magma ocean. The major-element composition of this magma would not be equal to the bulk composition of the moon. For a lunar composition similar to a T–J moon, it is possible, based on the available experimental data (Ringwood, 1975), that the melt would approximate a LKFM magma. However, the contents of incompatible elements might be enriched by several factors over that of the whole moon. The degree of enrichment depends on the efficiency of the zone refining mechanism.

The initial magma ocean produced in the zone refinement process is assumed to be enriched in trace elements (10–20× chondrite), Al_2O_3, and CaO relative to the whole moon. The lunar plagioclase-rich crust can therefore be formed from a small magma ocean not much larger than 100 km in depth. The intercumulus melt in this crust is even more enriched in trace elements. Total melting of a large fraction of the moon is not required to provide for the trace-element enrichments. The model provides for a cooler moon satisfying the constraints that isotopic systems were closed by 4.4 AE (Brett, 1977) and a shallow ocean as required by thermal calculations (Solomon and Chaiken, 1966). The partially molten moon also favors the formation and survival of an early, more mafic crust. At deeper levels within the moon, the mantle is much more heterogeneous and is dominated by locally differentiated igneous complexes (see Wetherill, 1975). Each of these plutons are complex miniatures of the upper moon. Cumulates in these plutons might then be the source rocks from which mare basalts are derived. Thus, non-mare and mare basalts need not be derived from products evolved from the same magma system.

CONCLUSIONS

(1) LKFM basalts are cotectic liquids containing both olivine and plagioclase as liquidus phases. This, plus the fact that they are major soil and rock components in the highlands, suggests that LKFM basalts are samples of a primitive magma which occurs as an intercumulus component in the anorthositic lunar crust. Polymict breccias of LKFM composition are thus explained as mixtures of cumulate rock(s) and residual liquids (rocks) produced by the fractionation of a LKFM magma.

(2) Residual liquids produced by either equilibrium crystallization or equilibrium partial melting of LKFM basalt are FeO enriched and are equivalent to some MKFM basaltic glasses which occur as preferred compositions in the lunar soils. Residual liquids produced under more reducing conditions; i.e. f_{O_2} 2 orders of magnitude below IW, precipitate

metallic Fe and are similar to the less FeO-enriched samples of KREEP rocks. LKFM basalt is then a possible parent to MKFM basalts.

(3) LKFM basalt could have originated (a) as a hybrid magma produced by partially melting of the lunar crust by a KREEP-rich magma or (b) as a magma residual to the crystallization of a primitive magma which was strongly enriched in incompatible elements. The enrichment in trace elements could have been accomplished by a mechanism involving zone refining of the outer moon.

Acknowledgments—The manuscript was greatly improved by the reviews of T. L. Grove and A. J. Irving.

References

Albee A. L., Gancarz A. J., and Chodos A. A. (1973) Metamorphism of Apollo 16 and 17 and Luna 20 metaclastic rocks at about 3.95 A.E.: Samples 61156, 64423, 14-2, 65015, 67483, 15-2, 76055, 22006, and 22007. *Proc. Lunar Sci. Conf. 4th*, p. 569–595.

Alfvén H. and Arrhenius G. (1976) Evolution of the solar system. NASA SP-345. 599 pp.

Bence A. E., Delano J. W., Papike J. J., and Cameron K. L. (1974) Petrology of the highland massifs at Taurus-Littrow: An analysis of the 2–4 mm soil fraction. *Proc. Lunar Sci. Conf. 5th*, p. 785–827.

Bence A. E., Papike J. J., Sueno S., and Delano J. W. (1973) Pyroxene poikiloblastic rocks from the lunar highlands. *Proc. Lunar Sci. Conf. 4th*, p. 597–611.

Blanchard D. P., Jacobs J. W., Brannon J. C., and Haskin L. A. (1976) Major and trace element compositions of matrix and aphanitic clasts from consortium breccia 73215. *Proc. Lunar Sci. Conf. 7th*, p. 2179–2187.

Brett R. (1977) The case against early melting of the bulk of the moon. *Geochim. Cosmochim. Acta* **41**, 443–445.

Dence M. R., Grieve R. A. F., and Plant A. G. (1976) Apollo 17 grey breccias and crustal composition in the Serenitatis Basin region. *Proc. Lunar Sci. Conf. 7th*, p. 1821–1832.

Dowty E., Keil K., Prinz M., Gros J., and Takahashi H. (1976) Meteorite-free Apollo 15 crystalline KREEP. *Proc. Lunar Sci. Conf. 7th*, p. 1833–1844.

French B. M. (1966) Some geological implications of equilibrium between graphite and a C–H–O gas at high temperatures and pressures. *Rev. Geophys.* **4**, 223–253.

Grieve R. A. F. (1975) Petrology and chemistry of the impact melt at Mistastin Lake Crater, Labrador. *Geol. Soc. Amer.* **86**, 1617–1629.

Herbert F., Drake M. J., Sonett C. P., Wiskerchen M. J. (1977) The thermal history of lunar magma ocean (abstract). In *Lunar Science VIII*, p. 424–426. The Lunar Science Institute, Houston.

Hess P. C., Rutherford M. J., Guillemette R. N., Ryerson F. J., and Tuchfeld H. A. (1975) Residual products of fractional crystallization of lunar magmas: An experimental study. *Proc. Lunar Sci. Conf. 6th*, p. 895–909.

Hubbard N. J. and Minear J. W. (1976) Petrogenesis in a modestly endowed moon. *Proc. Lunar Sci. Conf. 7th*, p. 3421–3435.

James O. B. (1976) Petrology of aphanitic lithologies in consortium breccia 73215. *Proc. Lunar Sci. Conf. 7th*, p. 2145–2178.

Langseth M. G., Keihm S. J., and Peters K. (1976) Revised heat-flow values. *Proc. Lunar Sci. Conf. 7th*, p. 3143–3171.

McKay G. A. and Weill D. F. (1976) Petrogenesis of KREEP. *Proc. Lunar Sci. Conf. 7th*, p. 2427–2447.

McKay G. A. and Weill D. F. (1977) KREEP (abstract). In *Lunar Science VIII*, p. 649–651. The Lunar Science Institute, Houston.

McKay G. A. and Weill D. F. (1977) KREEP petrogenesis revisited. *Proc. Lunar Sci. Conf. 8th*. This volume.

Meyer C., Jr. (1977) Petrology, mineralogy and chemistry of KREEP basalt (L. H. Ahrens, ed.) *Phys. Chem. Earth* **10**. In press.

Prinz M., Dowty E., Keil K., and Bunch T. E. (1973) Mineralogy, petrology and chemistry of lithic fragments from Luna 20 fines: Origin of cumulate ANT suite and its relationship to high alumina and mare basalts. *Geochim. Cosmochim. Acta* **37**, 979–1006.

Reid A. M., Warner J., Ridley W. I., Johnston D. A., Harmon R. S., Jakeš P., and Brown R. W. (1972a) The major element composition of lunar rocks as inferred from glass compositions in the lunar soils *Proc. Lunar Sci. Conf. 3rd*, p. 363–378.

Reid A. M., Warner J., Ridley W. I., and Brown R. W. (1972b) Major element composition of glasses in three Apollo 15 soils. *Meteoritics* **7**, 395–415.

Reid A. M., Richardson S. H., and Duncan A. R. (1977) Non-mare "igneous" clasts in Apollo 15 Breccias (abstract). In *Lunar Science VIII*, p. 801–803. The Lunar Science Institute, Houston.

Rhodes J. M. and Hubbard N. J. (1973) Chemistry, classification and petrogenesis of Apollo 15 mare basalts. *Proc. Lunar Sci. Conf. 4th*, p. 1127–1148.

Ridley W. I., Brett R., Williams R. J., Takeda H., and Brown R. W. (1972) Petrology of the Fra Mauro basalt 14310. *Proc. Lunar Sci. Conf. 3rd*, p. 159–170.

Ridley W. I., Reid A. M., Warner J. L., Brown R. W., Gooley R., and Donaldson C. (1973) Glass compositions in Apollo 16 soils 60501 and 61221. *Proc. Lunar Sci. Conf. 4th*, p. 309–321.

Ringwood A. E. (1976) Limits on the bulk composition of the moon. *Icarus* **28**, 325–349.

Rutherford M. J., Hess P. C., and Daniel G. H. (1974) Experimental liquid line of descent and liquid immiscibility for basalt 70017. *Proc. Lunar Sci. Conf. 5th*, p. 569–583.

Ryder G. and Wood J. A. (1977) Serenitatis and Imbrium impact melts: Implications for lunar crustal composition and stratigraphy (abstract). In *Lunar Science VIII*, p. 826–828. The Lunar Science Institute, Houston.

Sato M. (1976) Oxygen fugacity and other thermochemical parameters of Apollo 17 high-Ti basalts and their implications on the reduction mechanism. *Proc. Lunar Sci. Conf. 7th*, p. 1323–1344.

Sato M. (1977) The driving mechanism of lunar pyroclastic eruptions inferred from the oxygen fugacity behavior of Apollo 17 orange glass (abstract). *EOS* (*Trans. Amer. Geophys. Union*) **58**, 425.

Schonfeld E. (1974) The contamination of the lunar highlands rocks by KREEP: Interpretation of mixing models. *Proc. Lunar Sci. Conf. 5th*, p. 1269–1286.

Schonfeld E. (1975) A model for the lunar anorthositic gabbro. *Proc. Lunar Sci. Conf. 6th*, p. 1375–1384.

Simonds C. H., Warner J. L., and Phinney W. C. (1973) Petrology of Apollo 16 poikilitic rocks. *Proc. Lunar Sci. Conf. 4th*, p. 613–632.

Solomon S. C. and Chaiken J. (1976) Thermal expansion and thermal stress in the moon and terrestrial planets: Clues to early thermal history. *Proc. Lunar Sci. Conf. 7th*, p. 3229–3243.

Solomon S. C. and Longhi J. (1977) Magma oceanography: Thermal evolution (abstract). In *Lunar Science VIII*, p. 884–886. The Lunar Science Institute, Houston.

Taylor S. R. (1976) Geochemical constraints on the composition of the moon. *Proc. Lunar Sci. Conf. 7th*, p. 3461–3477.

Taylor S. R. and Bence A. E. (1975) Evolution of the lunar highland crust. *Proc. Lunar Sci. Conf. 6th*, p. 1121–1141.

Taylor S. R., Gorton M. P., Muir P., Nance W. B., Rudowski R., and Ware N. (1973) Composition of the Descartes region, lunar highlands. *Geochim. Cosmochim. Acta* **37**, 2665–2683.

Taylor S. R. and Jakeš P. (1974) The geochemical evolution of the moon. *Proc. Lunar Sci. Conf. 5th*, p. 1287–1305.

Taylor S. R. and Jakeš P. (1977) Geochemical evolution of the moon: Taylor–Jakeš–Bence models revisited (abstract). In *Lunar Science VIII*, p. 934–936. The Lunar Science Institute, Houston.

Wager L. R. and Brown G. M. (1967) Layered igneous rocks. Freeman, San Francisco. 588 pp.

Walker D., Grove T. L., Longhi J. Stolper E. M., and Hays J. F. (1973a) Origin of lunar feldspathic rocks. *Earth Planet. Sci. Lett.* **20**, 325–336.

Walker D., Longhi J., Grove T. L., Stolper E., and Hays J. F. (1973b) Experimental petrology and origin of rocks from the Descartes Highlands. *Proc. Lunar Sci. Conf. 4th*, p. 1013–1032.

Walker D., Longhi J., and Hays J. (1972) Experimental petrology and the origin of Frau Mauro rocks and soil. *Proc. Lunar Sci. Conf. 3rd*, p. 797–817.

Walker D., Longhi J., and Hays J. F. (1975) Differentiation of a very thick magma body and implications for the source regions of mare basalts. *Proc. Lunar Sci. Conf. 6th*, p. 1103–1120.

Warner J. L., Simonds C. H., and Phinney W. C. (1974) Impact-induced fractionation in the lunar highlands. *Proc. Lunar Sci. Conf. 5th*, p. 379–397.

Wetherill G. W. (1975) Possible slow accretion of the moon and its thermal and petrological consequences (abstract). In *Papers Presented to the Conference on the Origins of Mare Basalts and Their Implications for Lunar Evolution*, p. 184–188. The Lunar Science Institute, Houston.

Wetherill G. W. (1976) The role of large bodies in the formation of the earth and moon. *Proc. Lunar Sci. Conf. 7th*, p. 3245–3257.

Proc. Lunar Sci. Conf. 8th (1977), p. 2375–2401.
Printed in the United States of America

Origins of KREEP basalts

CHI-YU SHIH

Lockheed Electronics Company, Inc., Houston, Texas 77058

Abstract—A partial melting and a fractional crystallization model are proposed for the genesis of KREEP (or Fra Mauro) basalts. Both are consistent with the data of trace and major element chemistry, experimental phase relationships, and Rb–Sr isotopic geochemistry.

The partial melting model emphasizes that the anorthositic-noritic-troctolitic (ANT) rocks, especially those with high trace element abundances and high "mg" (atomic: MgO/(MgO + FeO)) values (~0.82), can be suitable parent sources for the KREEP basalts upon small degrees of internal melting at shallow depth at ~4.4 AE. However, the sources themselves were established at 4.6 AE.

The crystallization model requires that the outer portion of the moon be molten to a depth of at least 500 km at 4.6 AE. The crystallization of the magma body ended at 4.4 AE. KREEP basalts were the late-stage residual liquids highly enriched in trace elements. The negative Eu anomaly, low Ca/Al ratio, high "mg"-value and low Ti/Zr ratio, and slight enrichment of light REE over heavy REE in KREEP basalts indicate that they were formed from the magma by the separation of plagioclase, clinopyroxene, ilmenite, and garnet. These late-stage plagioclase-clinopyroxene-ilmenite separates constitute the sources for the high-Ti mare basalts.

In both models, the crystallization age of KREEP basalt at 3.9 AE could simply reflect the age of the impact-induced basin formation or of the terminal lunar cataclysm. The age for KREEP basalt is interpreted to be 4.4 AE. The chemical compositions of the basalts could be subject to slight modification at 4.1 AE but probably not at 3.9 AE.

The hypothetical "LKFM basalts" are probably mixtures of KREEP basalts, ANT series, and mare basalts and have no significance in the derivation of KREEP basalts.

1. INTRODUCTION

ONE OF THE most important highland samples returned from the Apollo missions is a basaltic rock of plagioclase-orthopyroxene-whitlockite assemblages with extremely high LIL element concentrations called KREEP basalts (Meyer and Hubbard 1970; Meyer, 1976), K, REE, P-rich norites (Marvin *et al.*, 1971), or Fra Mauro basalts (Reid *et al.*, 1972a; Taylor, 1973). It is an essential component of breccias and soils over the lunar highlands. No terrestrial analogue with this mineralogical and chemical (both major and LIL element) composition has ever been found. The uniqueness of the chemistry of KREEP basalts requires an unusual geological environment for their petrogenesis.

Partial melting models have been advanced for the origin of KREEP basalts (Hubbard and Gast, 1971; Gast, 1972) based on the high LIL element concentrations; however, the chemical and mineral compositions of the source material have not been specified in detail. An undifferentiated source was implied. Based on experimental petrology, MgO/(MgO + FeO) ratios and their geological abundance over the highlands, Walker *et al.* (1972, 1973) have proposed that the anorthositic norites could be suitable source materials for KREEP basalts, by a small degree of partial melting. LIL element chemistry has not been used or tested. However, Weill and McKay (1975) reached a very different result through experimental study of the behavior of the LIL and major elements in different source materials. They concluded that Apollo 16 KREEPy breccias could be generated by equilibrium partial melting of a ferromagnesian composition similar to the bulk moon composition of Ganapathy and Anders (1974) or Taylor and Jakeš (1974) but not to the highland composition. Taylor *et al.* (1973) have suggested a model for generating KREEP basalts by the mechanical mixing of anorthositic rocks with some undefined LIL element-rich residual liquid after fractional crystallization of the surface magma. It is difficult to justify this model because of the hypothetical nature of the end members.

In this paper I intend to investigate the possibility of producing KREEP basalts by two well known igneous processes, i.e., partial melting and fractional crystallization. In the former case, emphasis is placed on a combination of Rb–Sr systematics of the igneous KREEP rocks, the LIL element distribution law along with phase relationships developed by high pressure experimental petrology, and estimates of the LIL element composition as well as the major element composition of the source material for KREEP basalts. In the latter case, KREEP basalts are treated as late-stage residual liquids, a natural consequence of a molten, primitive upper portion of the moon undergoing slow cooling and magmatic differentiation under extremely reducing conditions. A similar two-stage model has been outlined earlier (Shih, 1976; Shih and Schonfeld, 1976). Recently, McKay and Weill (1976, 1977) have emphasized a multi-stage model for the generation of KREEP basalts through "LKFM" basalts which is worth exploring further here.

LIL element compositions of the non-mare igneous rocks used in the following discussion are mostly taken from the data derived by isotope dilution mass spectrometry at the Johnson Space Center (JSC) under the PI-ships of Gast, Hubbard, and Nyquist (Wiesmann and Hubbard, 1975). The high precision Sr isotopic data are taken from both the JSC (under Nyquist) and Caltech (under Wasserburg and Papanastassiou) laboratories for which the interlaboratory bias is known with confidence. The data cover a large variety of highland rock types and are fairly representative.

2. Chemical Composition of KREEP Basalts

Chemistry and its problem

KREEP basalts were first found as ropy and sculptured glasses or glassy matrix breccias in soil and breccia samples at the Apollo 12 site (Meyer *et al.*, 1970). Since then, KREEP basalts have been widely recognized, in varying abundances, in glass fragments in soil samples at every landing site (e.g., Apollo Soil Survey, 1971). However, they are probably highly concentrated at the Apollo 12, 14 and 15 sites. Sample 14310 is the only example of a hand specimen-sized KREEP basalt recovered, but its extremely high concentration of siderophile elements of meteoritic origin and other LIL element characteristics prompted some authors (Morgan *et al.*, 1972; Philpotts *et al.*, 1972) to conclude that it is a recrystallized surface soil or breccia. In addition, only small fragments of KREEP basalt showing igneous texture were reported at the Apollo 15 sites by Meyer (1972), Phinney *et al.* (1972), and Simonds *et al.* (1975). The limited number and amount of representative KREEP samples plus the extreme complexity of the impact history superimposed on the non-mare rock samples makes the study of the genesis of KREEP basalts difficult. Furthermore, the major element chemistry is based mainly on data from glasses of unknown origin or on less accurate, defocused electron beam analysis (DBA) of lithic fragments. The ranges of the compositions reported are shown in Fig. 1. The chemical differences between igneous KREEP basalt 15386 and KREEPy glasses are noted. Nevertheless, the general characteristics of KREEP basalts are: (a) they are saturated with SiO_2, generally have excess quartz in the CIPW normative mineralogy, and they are tholeiitic according to the classification of terrestrial volcanic rocks, and (b) "mg"-values (0.5–0.6) are the lowest of non-mare rock samples, with the exception of very rare granitic rock, and are slightly higher than those of the mare basalts.

Actually, KREEP basalts are named after their LIL element characteristics. They generally have light REE abundances around 200–300× chondrites, decreasing gradually to 100–150× chondrites at the heavy REE, and large negative Eu anomalies of only 40× chondrites. K, Rb, Ba, U, Th, and Zr are also high relative to other lunar rocks (Hubbard *et al.*, 1971). KREEP basalts and KREEPy breccias from different landing sites show a remarkably similar LIL element distribution pattern, although the absolute abundance varies by a factor of 4. There is an anti-correlation of the "mg"-value and LIL element abundance for the KREEP rocks. The variations are probably not primary features of the basalts at extrusion, but modifications after mixing and reworking with "mg"-rich, LIL element-poor ANT rocks by subsequent meteoritic bombardment. Apollo 15 samples with extremely low values of the meteoritic component are generally believed to be most representative (Dowty *et al.*, 1976). An igneous-textured, pigeonite basalt clast (Stoeser *et al.*, 1975; Blanchard *et al.*, 1975) is another rare example. Apollo 16 and 17 samples, showing the lowest LIL element contents are mixtures rather than primary KREEP basalts (Schonfeld, 1974).

Samples of igneous KREEP basalt are usually not large enough for both major and LIL element analysis. Correlation of major element chemistry from a suite of samples with LIL element chemistry from another suite of samples definitely introduces sampling errors. The igneous-textured sample 15386 is one of the rare exceptions, having reliable major and trace element data (Hubbard *et al.*, 1974; Rhodes and Hubbard, 1973). Its low siderophile element content indicates a lack of contamination (Irving, 1975). This sample will be emphasized in later discussions.

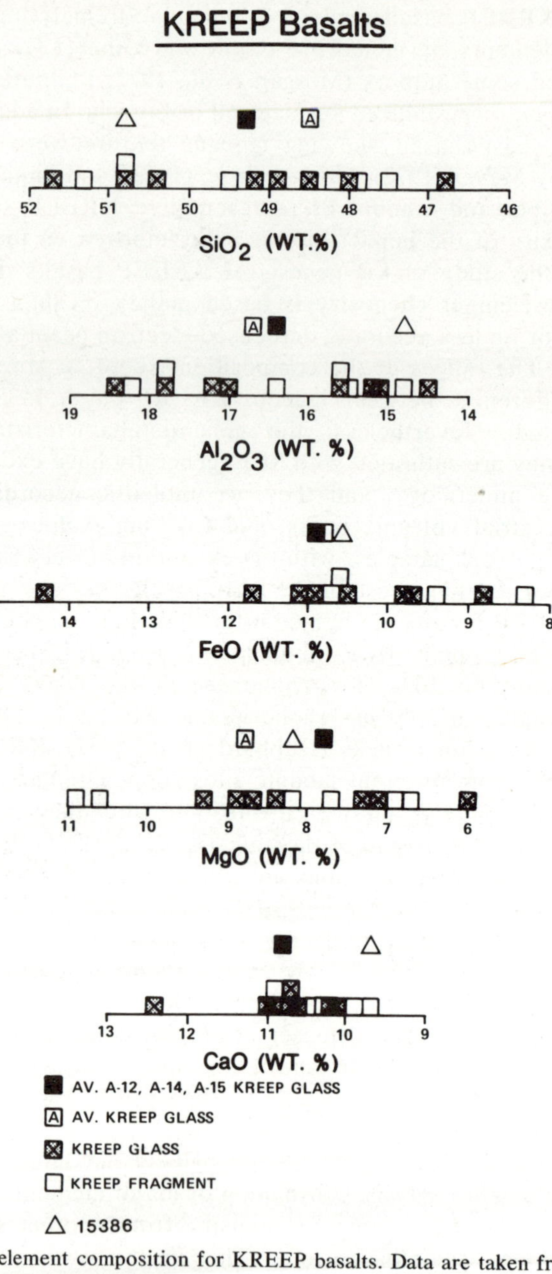

Fig. 1. Major element composition for KREEP basalts. Data are taken from: average Apollo 11, 12, 14, 15, 16, 17, Luna 16 and 20 KREEP glass (Reid *et al.*, 1972a, b; 1973; Ridley *et al.*, 1973a; Warner *et al.*, 1972; Jakeš *et al.*, 1972); Apollo 12 KREEP glass (Meyer *et al.*, 1971); Apollo 14 common KREEP basalts (Hubbard and Gast, 1971); high-KREEP norite 14303 (Wood, 1974); KREEP basalt 15386, 1 (Rhodes and Hubbard, 1973); average Apollo 15 KREEP fragments and pigeonite basaltic clast 72275 (Stoeser *et al.*, 1974).

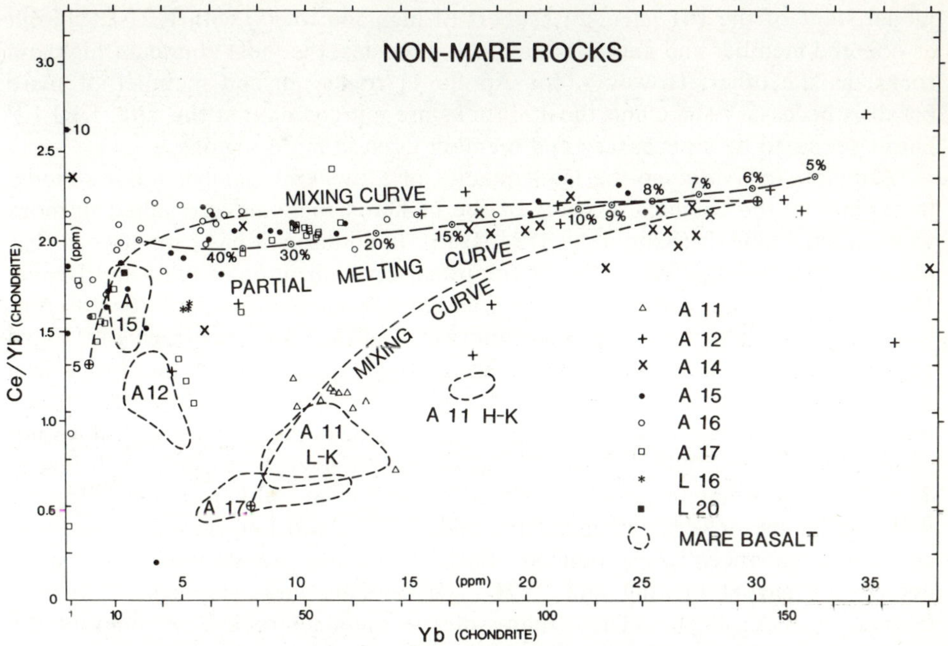

Fig. 2. Ce/Yb vs. Yb correlation for lunar highland rocks. Data are from Wiesmann and Hubbard (1975). Partial melting curve is calculated based on parameters listed in Fig. 6. Binary mixing curves are calculated based on two hypothetical end members shown in circled plus.

Parallelism of LIL element distribution pattern in highland rocks

One of the most distinct LIL element characteristics of the highland rocks is the parallel distribution of their chondrite-normalized distribution patterns (Fig. 2). The only discrepancies are the Eu and Sr concentrations which are strongly dependent on the plagioclase content of the sample. This parallelism can best be illustrated by a plot of the ratio of light REE (Ce) and heavy REE (Yb) vs. Yb (Fig. 2). For the vast majority of highland samples from anorthositic rocks to KREEP basalts, absolute values of LIL element concentrations vary by a factor of more than two orders of magnitude, but their relative LIL element distribution patterns, especially for the REE, are almost identical. The Ce/Yb ratio for most of the samples is relatively constant at around twice that of average chondrite, although it may be slightly lower for anorthositic rocks. This has been suggested as strong evidence of a mixing process for producing KREEPy rocks, with KREEP basalts as an essential end member. Because KREEP basalt has such a high concentration of LIL elements, it has a dominant role in controlling the LIL element chemistry (the Ce/Yb ratio) of the mixture. The identity of the LIL element poor end members is somewhat less certain, with anorthosite, anorthositic norite, dunite, mare basalts, and meteorites being probable candidates. For a simple mixing model with two end members, we can satisfactorily model some of the LIL element aspects of highland rocks, with KREEP basalt as one end member and anorthositic norite breccias, the most abundant highland rocks, as the other. However, for Apollo 11 rocks, an end member of mare basalt is necessary since anorthositic rocks are not common at this site. KREEP basalt seems to be a necessary end member even in mare regions.

Attempts to overcome the inadequacies of a two-end-member mixing model to explain all the chemical aspects of the highland rocks have resulted in more sophisticated models (Schonfeld, 1974, 1975). In addition, it has been argued that partial melting (Schonfeld, 1974) or fractional crystallization (Taylor and Bence, 1975) will produce liquids having fractionated LIL element patterns (or different Ce/Yb ratios). The observed constancy of Ce/Yb ratios for highland rocks suggests that they were not produced by such igneous differentiation processes. A partial melting model calculation of a most probable plagioclase-rich source material (see discussion later) for highland rocks (also plotted in Fig. 2) does not show a significant fractionation of Ce/Yb ratios over a reasonable range of degrees of partial melting, and as a result of the partial melting, a suite of liquids with similar major element chemistry and Ce/Yb ratio but very different LIL element abundances are produced. KREEP basalts are derived from small degrees of partial melting and "VHA" basalts by large degrees of melting (Hubbard *et al.*, 1973b). Thus, many of the highland rocks are undoubtedly produced through mixing of preexisting rocks (Dowty *et al.*, 1973) but partial melting may very likely produce not only KREEP basalt, an end member of the mixing model, but also a variety of liquids with chemical compositions similar to those observed for many highland rocks.

3. Geochronological Evolution of KREEP Basalts

KREEP basalts have probably undergone three thermal events. The earliest event at ~4.4 AE was proposed as the formation age of the KREEP basalts by Schonfeld and Meyer (1972) from a study of KREEP components in Apollo 12 soils and breccias. This age is supported by the following: (1) Experimental petrology indicates that all highland rocks are genetically related (Walker *et al.*, 1973). KREEP basalts could be the products of a major differentiation event occurring in the highlands as revealed by the upper intersect at ~4.4 AE on the concordia for all the highland rocks (Tera *et al.*, 1974); (2) Shih and Schonfeld (1976) and Wakita *et al.* (1975) provided chemical evidence that genetic relationships exist between KREEP basalts and the source of high-Ti mare basalts. The age for the source of mare basalts is also ~4.4 AE from the Rb–Sr and Sm–Nd whole rock isochrons of mare basalts (Shih and Schonfeld, 1976; Lugmair, 1975), and from the first event on the U, Th–Pb concordia of internal isochrons of mare basalts (Tera and Wasserburg, 1975, 1976; Papanastassiou *et al.*, 1977). The KREEP basalts and sources of mare basalts could be the end products of the global differentiation of the outer moon which ended at ~4.4 AE. The second episode is revealed by the whole rock isochron for Apollo 14 impact melt rocks and Apollo 15 igneous-texture KREEP basalts, defining a possible magmatic event at ~4.1 AE (Nyquist *et al.*, 1973; Papanastassiou and Wasserburg, 1976).

The latest event is indicated by the ~3.9 AE crystallization age of the igneous-textured KREEP basalts (Papanastassiou and Wasserburg, 1971, 1976; Wasserburg and Papanastassiou, 1971; Nyquist *et al.*, 1974, 1975). This age is consistent with the predominant age of highland rocks sampled from the ejecta blankets of the basins by the $^{40}Ar/^{39}Ar$ method (Turner *et al.*, 1973; Kirsten and Horn, 1974; Husain and Schaeffer, 1975; Schaeffer *et al.*, 1976). The 3.9 AE age has been suggested as the recrystallization or metamorphic age reset by the Imbrium basin-forming event (Tera *et al.*, 1974). It is very important to note that although the crystallization ages for these seven crystalline basalts are clustered at 3.9 AE, their initial $^{87}Sr/^{86}Sr$ ratios are distinctly different. At least three isotopically distinct KREEP basalts were established at 3.9 AE. In other words, the KREEP basalts could not be derived from a common source at 3.9 AE by any magmatic processes. The common source—probably there was one—could be established at 4.4 AE or as late as 4.1 AE.

A multi-stage evolution model for KREEP basalts is demonstrated by the age vs. $^{87}Sr/^{86}Sr$ diagram in Fig. 3. Lines correspond to the Sr isotopic evolution of the basalts. Slopes of lines are proportional to the Rb/Sr ratio of the rocks; a change in slope requires the fractionation of Rb/Sr ratio by igneous processes at that particular time. The most obvious time for change of slope is at 4.4 AE. Breaks at 4.1 and 3.9 AE do not change the Rb/Sr ratio significantly, if errors are considered. For simplicity, a simple two-stage model (one fractionation of Rb/Sr) is used in this discussion. It is assumed that at 4.4 AE KREEP basalts were formed from a low Rb/Sr source through magmatic processes. Subsequent events at 4.1 and 3.9 AE did not change the chemistry or Rb/Sr ratio appreciably, although the clock was reset. This explains the variable $^{87}Sr/^{86}Sr$ ratios for KREEP basalts at 3.9 AE. If the moon was isolated from a solar nebula at 4.6 AE with an initial $^{87}Sr/^{86}Sr$ ratio of LUNI (= 0.69903), then the Rb/Sr ratio variation F during such an isotopic equilibrium igneous process of generating KREEP basalts at 4.4 AE can be calculated according to the following equation (e.g., Schonfeld and Meyer, 1972; Shih *et al.*, 1975):

$$\begin{aligned} F &= \frac{(\mathrm{Rb/Sr})_{\mathrm{basalt},\,t}}{(\mathrm{Rb/Sr})_{\mathrm{source},\,t}} \\ &= \frac{[(^{87}\mathrm{Sr}/^{86}\mathrm{Sr})_{\mathrm{basalt,\,present}} - (^{87}\mathrm{Sr}/^{86}\mathrm{Sr})_{\mathrm{basalt},\,t}] \cdot e^{\lambda t}/(e^{\lambda t}-1)}{[(^{87}\mathrm{Sr}/^{86}\mathrm{Sr})_{\mathrm{source},\,t} - (^{87}\mathrm{Sr}/^{86}\mathrm{Sr})_{\mathrm{source},\,T}]/(e^{\lambda(T-t)}-1)} \\ &\sim \frac{I_{\mathrm{basalt,\,present}} - I_{\mathrm{basalt},\,t}}{I_{\mathrm{source},\,t} - I_{\mathrm{source},\,T}} \times \frac{T-t}{t}, \end{aligned} \tag{1}$$

where t is the presumed formation age of the KREEP basalts (or 4.4 AE) and T is the assumed age of the moon (or 4.6 AE) and λ is the decay constant of ^{87}Rb (or $1.39 \times 10^{-11}\ yr^{-1}$). The calculated F values for typical Apollo 15 igneous-

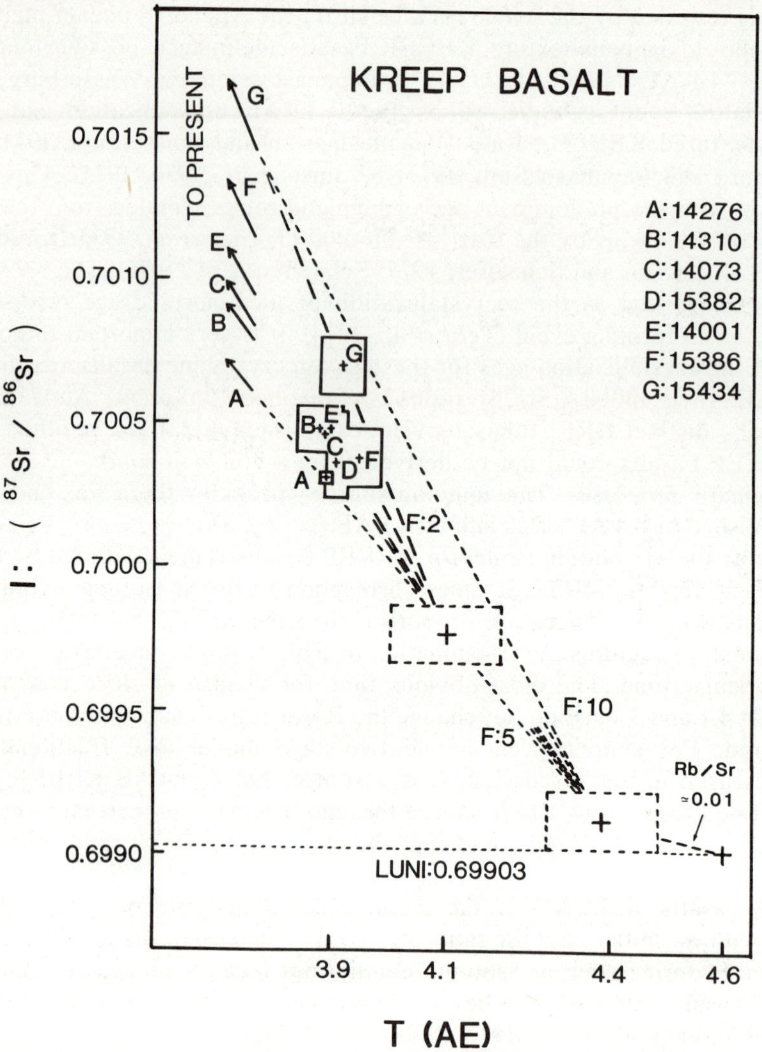

Fig. 3. *T–I* diagram for Apollo 14 and 15 crystalline KREEP basalts. Data are compiled from Papanastassiou and Wasserburg (1971; 1976), Wasserburg and Papanastassiou (1971), and Nyquist *et al.* (1974; 1975). 4.1 AE event is from Papanastassiou and Wasserburg (1976); 4.4 AE event is from Shih and Schonfeld (1976) and Shih (1976). All data are adjusted to JSC scale. Dotted rectangles are estimated uncertainties. Straight lines represent possible evolutions of $^{87}Sr/^{86}Sr$ in KREEP basalts. Slope of the growth line is proportional to the Rb/Sr ratio in the rock. Change of the slope demonstrates fractionation of Rb/Sr ratio at that particular time, e.g., considerable change occurred at 4.4 AE but insignificant change happened at 4.1 AE and 3.9 AE. Note that at least three isotopically distinct KREEP basalts were formed at 3.9 AE. Variations of Rb/Sr in the basalts must have been established at 4.4 or 4.1 AE but probably not at 3.9 AE.

textured KREEP basalts and "VHA basalts" of Hubbard *et al.* (1973) are ~10 and ~5, respectively. F values for the basalts cannot be determined better than within a factor of two because of the uncertainties of the differentiation time and initial $^{87}Sr/^{86}Sr$ ratios of the event. However, the T–I diagram places an important geochronological constraint on the LIL element fractionation (e.g., the Rb/Sr ratio) during the formation of these basalts. It should be noted that the T–I diagram provides evidence for changes in the Rb/Sr *ratio.* Two major igneous processes, partial melting and fractional crystallization, that will change Rb/Sr ratios are discussed in detail here.

4. Partial Melting Model for Generating KREEP Basalts

The equilibrium batch melting equation of Shaw (1971) has been especially useful for petrogenesis, namely

$$C_1/C_0 = 1/[D + X(1-P)], \tag{2}$$

where D and P are the bulk distribution coefficients for the source and melts, respectively, and X is the degree of partial melting.

The abundances and ratios of LIL elements in partial melts will generally change during melting. The magnitude of the variations is strongly dependent on their distribution coefficients as well as the degree of partial melting.

Minerals which are relevant to the genesis of KREEP basalts are plagioclase, low-Ca pyroxene (orthopyroxene) and olivine (Walker *et al.*, 1973). LIL element distribution coefficients for these minerals are very low; that is, LIL elements strongly favor the liquid upon melting, except Sr and Eu which are selectively retained in the residual plagioclase. The absolute distribution coefficients are difficult to estimate since they are highly dependent on physical and chemical parameters, not to mention the extremely complex nature of geological environments. However, some progress has been made recently, especially on the effect of temperature, chemical composition, and oxygen fugacity on LIL element distribution coefficients (Weill and McKay, 1975; Mysen, 1976). Unfortunately, Mysen (1976) found that the distribution coefficients determined by microprobe in an experimental system doped with high concentrations of the elements are probably not applicable to the natural system. Therefore, LIL element distribution coefficients used in the present calculations are taken from those measured in mineral-glass pairs of basalts (Philpotts and Schnetzler, 1970; Table 4 in Shih *et al.*, 1975). They are slightly different from the experimental measurements of Weill and McKay (1975).

(1) *Degree of partial melting*

The degree of partial melting, X, is a very important parameter for estimating the LIL element abundance of the source material and is also extremely difficult to calculate with accuracy. However, the two-stage Rb–Sr evolution model presented above provides a quantitative estimate of the Rb/Sr ratio variation in the partial melting event of the KREEP basalt genesis. The degree of partial melting X can also be related to the Rb/Sr fractionation during melting by the batch melting equation, namely

$$F = \frac{\left(\frac{\mathrm{Rb}}{\mathrm{Sr}}\right)_{\mathrm{KREEP}}}{\left(\frac{\mathrm{Rb}}{\mathrm{Sr}}\right)_{\mathrm{source}}} = \frac{D^{\mathrm{Sr}}_{\mathrm{source}} + X(1 - P^{\mathrm{Sr}}_{\mathrm{KREEP}})}{D^{\mathrm{Rb}}_{\mathrm{source}} + X(1 - P^{\mathrm{Rb}}_{\mathrm{KREEP}})} \tag{3}$$

as a function of source mineralogy (Fig. 4) assuming source is composed of three major minerals, plagioclase, orthopyroxene, and olivine. For the range of F calculated by the two-stage model earlier (Eq. 1), the degree of partial melting is less than 10% for an extremely plagioclase-rich source and becomes much less ($<1\%$) for a more mafic source. For a particular source mineralogy, the lower the degree of melting, the greater is the Rb/Sr ratio enrichment. It should be

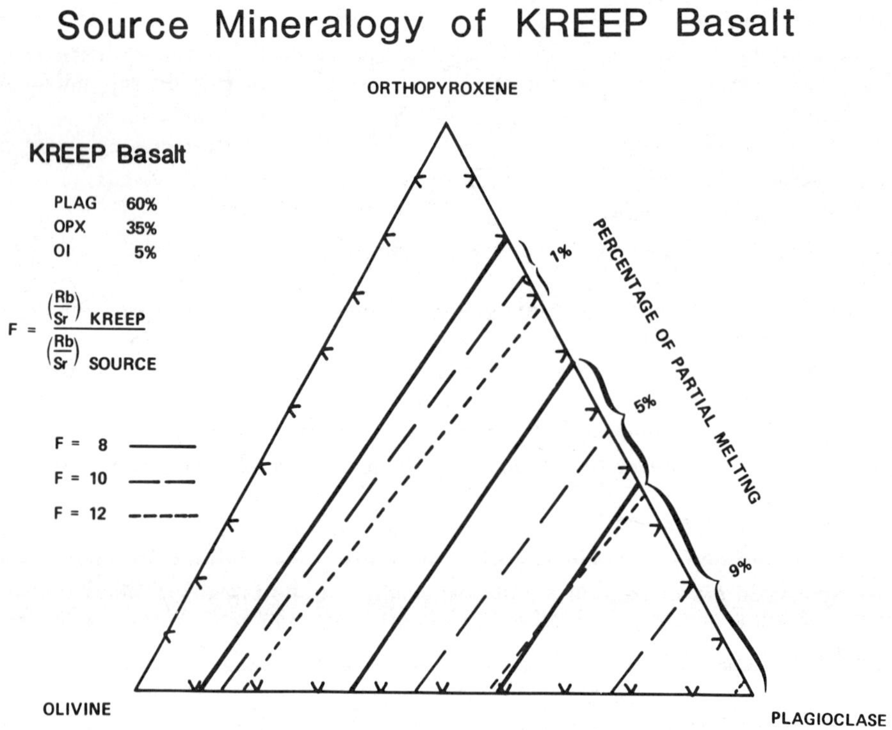

Fig. 4. Relationship between source mineral composition for KREEP basalts and degree of partial melting calculated from Eq. (3).

noted that the estimation is based strictly on Rb/Sr ratio constraints set by the two-stage evolution model and the assumed source mineralogy. Nevertheless, degree of melting cannot be calculated unless the source mineralogy can be quantitatively specified by other independent methods.

(2) *Source material for KREEP basalts*

Experimental petrology has revealed qualitatively that plagioclase, orthopyroxene, and olivine are the three essential phases saturated on the liquidus of KREEP basalt at shallow depth (Walker *et al.*, 1972, 1973). In addition, these authors found that KREEP basalts, including lithic fragments and glasses, are located around the peritectic of the SiO_2-olivine-anorthite pseudoternary diagram. Thus the liquids generated from any assemblages consisting of plagioclase-orthopyroxe-olivine will always be that of the peritectic composition until one phase is completely consumed. However, it is impossible to determine the exact proportion of each mineral phase in the source. To quantitatively estimate the mode of the source, a LIL element approach is attempted here. Sm, an arbitrarily chosen LIL element, is considered in the calculation. It is assumed that Sm in the primary KREEP basalts is ~175× chondritic abundance, the same as in the Apollo 15 igneous-textured KREEP basalts. The combination of the LIL element (Sm) and the Rb/Sr ratio constraints gives a quantitative relationship between source mode, source LIL element abundance and degree of partial melting which is shown in the triangular plots of plagioclase-olivine-orthopyroxene abundance (Figs. 5a, b, c). The LIL element chemistry of the source material and degree of partial melting can be simultaneously determined, if the mineral composition of the source material is known with accuracy. Figure 5 demonstrates that, for a plagioclase-rich source, around 15–20× chondrites of Sm (or LIL elements) is necessary, whereas for a mafic-rich source (ol–opx), the LIL elements abundance is much less.

(a) *Mineralogy.* If KREEP basalts were produced by partial melting of surface or near-surface highland materials which have been sampled, then it is possible to identify the sources from the returned highland rocks. CIPW norms of such samples including anorthosite, troctolite, norite, and anorthositic norite breccias are plotted in Fig. 5. For $F = 10$ the calculation requires >20× chondrites of Sm for anorthosite 15415 and >10× chondrites for both norite 78235 and troctolite 76535, if these are to be suitable sources of KREEP basalts. The published LIL element data are <1× chondrites for anorthosite 15415 (Hubbard *et al.*, 1971b), ~8× chondrites for norite 78235 (Winzer *et al.*, 1975), and ~5× chondrites for troctolite 76535 (Haskin *et al.*, 1974). All are at least a factor of two too low for the model; therefore, these rock types are probably not favorable sources for KREEP basalts. The only rock type which meets the requirement seems to be anorthositic norite breccias with ~18× chondrites of Sm, which have not only the proper mineralogy but also the correct LIL element

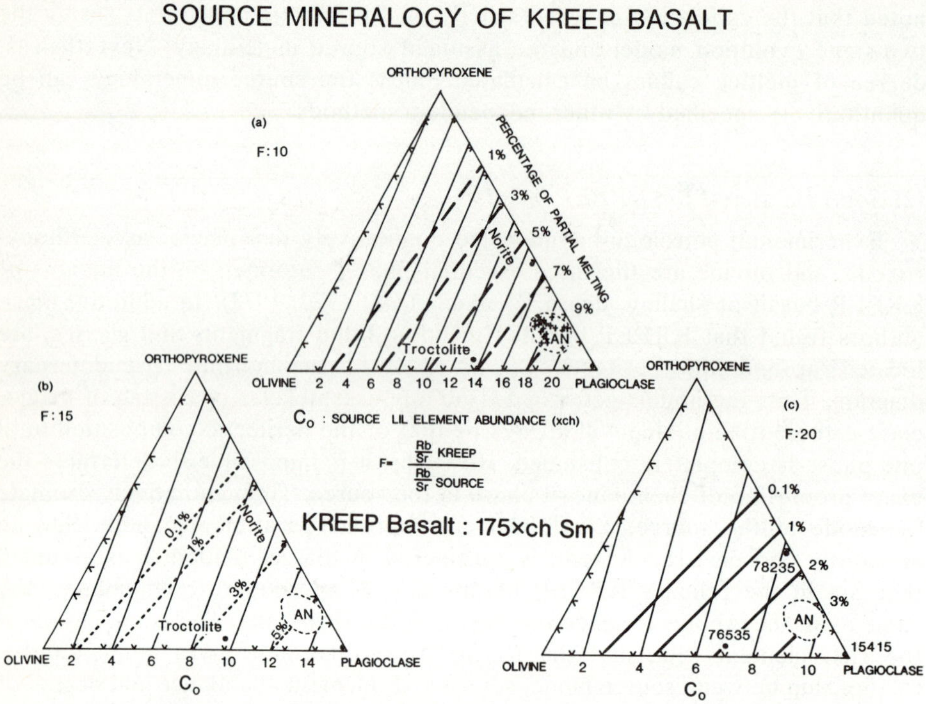

Fig. 5. Relationship among source mineral composition, source LIL element abundances and degree of partial melting calculated by simultaneously solving Eq. (2) and Eq. (3). Rock data are from Haskin *et al.* (1974) for troctolite 76535; Winzer *et al.* (1975) for norite 78235 and Hubbard *et al.* (1974) for anorthositic norite breccias (shown in +).

abundances. Anorthositic norite breccias are the most abundant rock type reported on the highland crust and are the possible source material for KREEP basalts suggested by Walker *et al.* (1973), based on the evidence of phase equilibrium studies. However, if $F = 15$–20, which is also within the uncertainty of the model, then norite 78235 and troctolite 76535 become eligible as a KREEP basalt source.

(b) *LIL element chemistry.* A suite of LIL elements including alkali, alkaline earths, and rare-earth elements with known distribution coefficients are used to test if their abundances in the source materials are consistent with those measured in anorthositic norite breccias. The source material composition is taken as 75% plagioclase, 15% orthopyroxene, and 10% olivine. The Eu distribution coefficient is highly dependent on oxygen fugacity. Based on direct oxygen fugacity measurement of a single highland KREEP basalt 14310 (Sato *et al.*, 1973), this rock seems to have been formed under slightly more oxidizing conditions, at 1200°C, than mare basalts. Furthermore, the measurement is probably an upper estimate because the basalt was contaminated with meteoritic Fe–Ni metals

(Axon and Goldstein, 1972) and has undergone oxygen loss by volatilization (Ford *et al.*, 1972). Thus, for plagioclase, the Eu distribution coefficient is taken as 0.8 at 1200°C in order to match the measures $\sim 10^{-11}$ atm oxygen fugacity for the calculation of KREEP basalt genesis. This value is slightly lower than that (1.0) used by Weill and McKay (1975). Comparable Eu distribution coefficients for olivine and orthopyroxene are also used. Assuming the Rb/Sr fractionation (F) is 10, the degree of melting is then set at 8% (Fig. 5). The calculated LIL element content of this hypothetical source material—both the absolute and relative abundances—falls within the ranges reported for anorthositic norite breccias (Fig. 6). The distribution pattern is similar to norite 78235 and troctolite 76535 but higher in abundances. Thus, a KREEP liquid could be derived by even a smaller degree ($\sim 1\%$ and $\sim 0.5\%$ respectively) of melting of these materials provided that $F = 15$–20. If the model is reasonably correct, it requires a source not only with rather high LIL element abundances but a very fractionated source; that is, one slightly enriched in light REE relative to heavy REE and a pronounced positive Eu anomaly. K and Rb would be highly depleted relative to the refractory LIL elements (Ba, Sr, and REE). This source is definitely not primitive material like chondritic meteorites, but rather a differentiated product of a major crystal-liquid fractionation process. It is very different from sources proposed earlier by Gast (1972) and Weill and McKay (1975) which are primitive undifferentiated materials with 5–10× chondritic abundances of the refractory LIL elements.

It is also interesting to note that the ratios of light REE to heavy REE do not change very much in the partial melting event according to the model (see Fig. 2). The chemistry of a liquid derived by 30% melting is in good agreement with that reported for "VHA basalts" (Hubbard *et al.*, 1973b). Hence, the model indicates the possibility of more extensive melting to produce "VHA basalts" from the common anorthositic norite source. A unique thermal regime must be invoked if the partial melting is to be limited only to a degree necessary for generating KREEP basalt without proceeding further to the production of other basalts of similar chemistry and mineralogy such as "VHA basalt".

(c) *Major element chemistry.* The batch melting equation is essentially a mass balance equation. It is used here to calculate the major element abundance for a KREEP basalt source assuming that the basalts are generated by 8% of partial melting at 1200°C. The major element distribution coefficients, strongly temperature dependent, are calculated from the compositions reported by Weill and McKay (1975) for liquid and coexisting minerals at the peritectic of the SiO_2-olivine-anorthite pseudoternary diagram. The concentrations of five major elements (SiO_2, Al_2O_3, CaO, MgO, FeO) in the source for KREEP basalts have been calculated; they are not remarkably different from those reported for the anorthositic noritic rocks, even considering the compositional range of KREEP basalts (Fig. 7). The serious discrepancy is the low FeO resulting in a high "mg"-value (0.85). With this exception, the major element as well as the LIL element chemistry seems to be consistent with partial melting of anorthositic

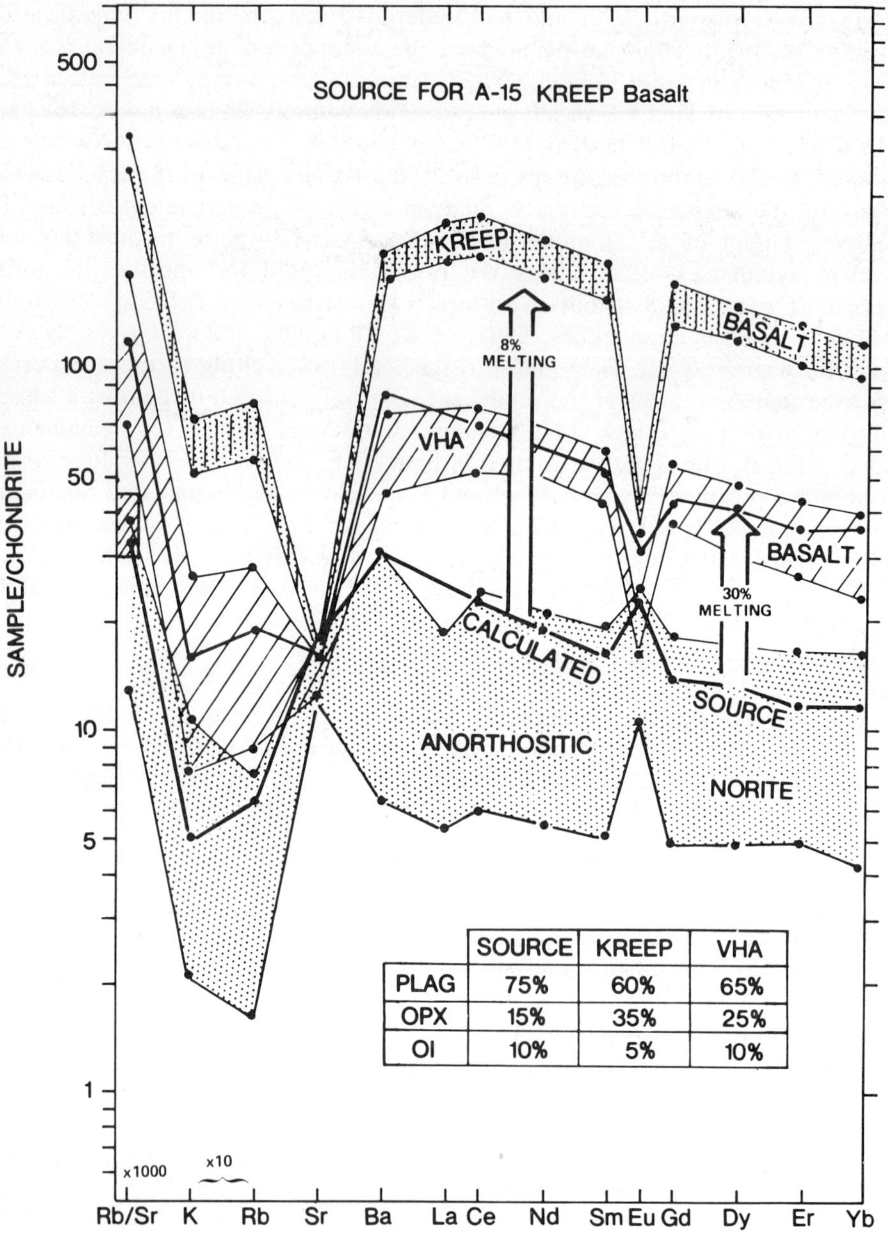

Fig. 6. Calculated LIL element distribution in the source material for Apollo 15 KREEP basalt.

norite to generate KREEP basalts. Weill and McKay (1975) also noted that the anorthositic norite rocks ("mg" = ~0.75) are probably too low in "mg" to be source materials, even though their conclusions are based heavily on the chemistry of Apollo 16 KREEPy breccias (impact melts) 64815, 62235, and 65015 which are probably not good representatives of KREEP basalts. These rocks are extremely high in meteorite-derived siderophile elements, low in overall LIL elements and high in "mg"-values, all suggesting an origin by mixing.

The constant relationship of Fe/Mg between olivine and liquid ($K_D = 0.3$) in most equilibrium experimental systems (Roeder and Emslie, 1970) is very useful to constrain the source mineralogy. A similar relationship exists for orthopyroxene ($K_D = 0.3$) and plagioclase ($K_D \cong 0.5$) (Weill and McKay, 1975; Longhi *et al.*, 1976). Since plagioclase does not contain significant amounts of MgO and FeO, a simplified equation can readily be used for estimating the "mg" of an olivine-orthopyroxene-plagioclase source providing the liquid in equilibrium with such a source is derived from a small degree of partial melting.

$$\text{“mg”}_{\text{source}} = \frac{1}{K_D\left(\dfrac{1}{\text{“mg”}_{\text{basalt}}} - 1\right) + 1} \qquad (4)$$

where $K_D = 0.3$ for an olivine-orthopyroxene-plagioclase source. The calculation gives only the upper limit to the source. For the Apollo 15 KREEP basalt ("mg" = 0.58), a maximum "mg" = 0.82 is required for the source. This is slightly higher than that of the anorthositic norite breccias ("mg" = 0.75), but is similar to that of norite 78235 ("mg" = 0.81), and troctolite 76535 ("mg" = 0.87). For Apollo 15 KREEP basalt, the "mg" of the source, with a small degree of partial melting (~5%), can be determined by a mass-balance equation similar to Eq. (3), or Eq. (3) of Shih and Schonfeld (1976) (shown in Fig. 8). At constant source mineralogy, the more "mg" the source, the lower the degree of melting required to produce KREEP basalt; at constant "mg" the more plagioclase in the source, the higher the degree of melting that can be accommodated. For sources similar to norite and troctolite, ~1% melting is needed to generate KREEP basalt, which is consistent with LIL element evidence derived earlier. These rocks are probably better candidates than anorthositic norite breccias. Additionally, the anorthositic norite breccias are too young (~3.9 AE) to be the source for KREEP basalts (> 4.4 AE).

Most of the samples returned from the lunar highlands are polymict breccias. Norite 78235 and troctolite 76535 are the few igneous rocks representing rocks at depth. These ANT rocks could be established at 4.6 AE as the result of a moon-wide chemical differentiation (Papanastassiou and Wasserburg, 1976). The source of thermal energy for melting is not clear, but it could be decay of radioactive elements since these rocks contained $\gtrsim 5\times$ chondritic abundances of U and Th. Internal partial melting of materials similar to these rocks at 0.2 AE later could give rise to KREEP basalts (Walker *et al.*, 1973; Delano, 1974; McCallum and Mathez, 1975).

The Apollo 15 KREEP basalt is located near the orthopyroxene and plagio-

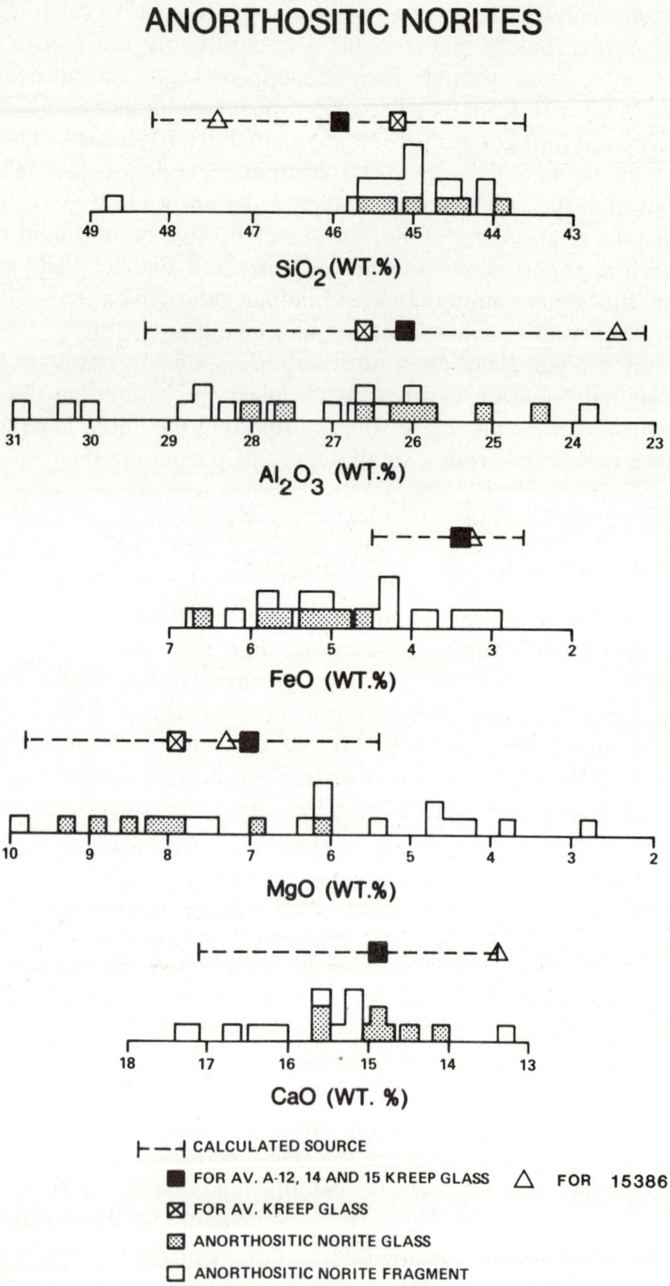

Fig. 7. Calculated major element composition for the source of Apollo 15 KREEP basalt by 8% melting. Anorthositic norite breccias are from data of Hubbard *et al.* (1974) and the Apollo Soil Survey (see references of Fig. 1). Dotted lines are the ranges calculated based on the ranges of KREEP basalts in Fig. 1. Solid squares are based on the average KREEP glass at all landing sites.

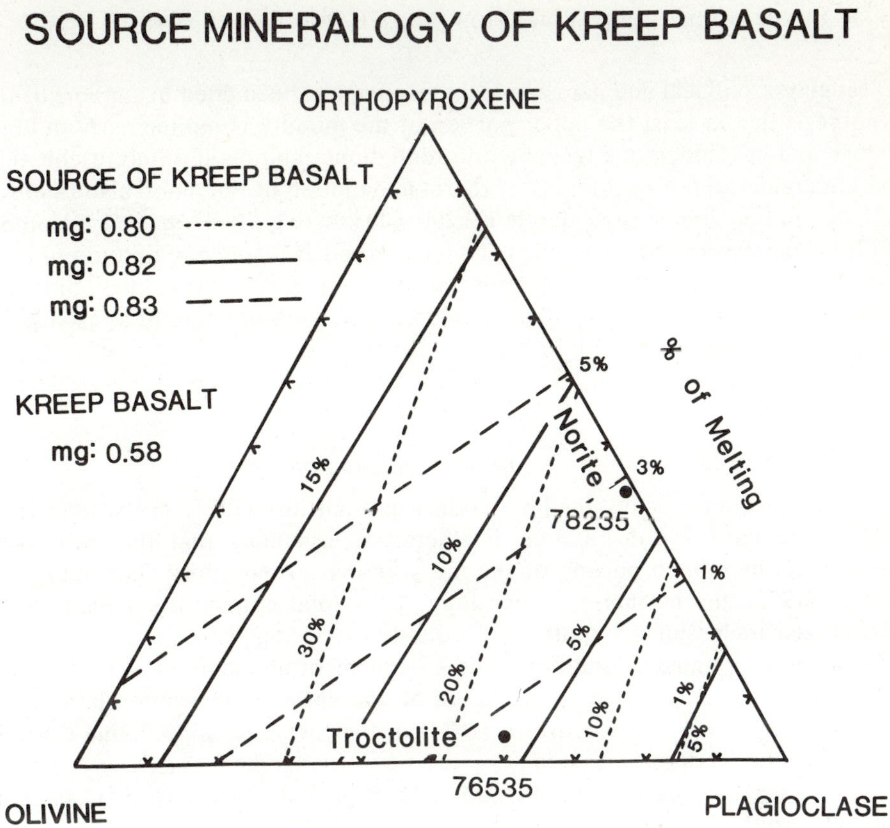

Fig. 8. Relationship between "mg" value of the source and degree of partial melting according to Eq. (3) of Shih and Schonfeld (1976) or similar to Eq. (3) by substituting Rb/Sr for Mg/Fe.

clase cotectic between point B and point C on the diagram (Walker *et al.*, 1973). It was probably evolved from point B by orthopyroxene and plagioclase fractionation (Walker *et al.*, 1973; McKay and Weill, 1976). If so, the event should have predated 3.9 AE, probably at 4.4 AE (partial melting and subsequent near-surface fractionation) and possibly at 4.1 AE (near-surface fractionation only) because KREEP basalts were isotopically distinct at 3.9 AE. The chemistry of the KREEP basalts remained unchanged but the Rb–Sr system was reequilibrated during the basin-forming event at 3.9 AE. Apollo 16, 17 KREEPy breccias and "LKFM basalts" could be formed by mixing pre-existing KREEP basalts and other non-mare rock types. Only small amounts of the uncontaminated KREEP basalts which have escaped the intensive meteoritic bombardment were sampled.

5. Fractional Crystallization Model for Generating KREEP Basalts

Much petrological and geochemical evidence has been cited in support of the hypothesis that at least the upper portion of the moon was molten early in lunar history and has undergone igneous differentiation resulting in different chemical and mineralogical layers. Models of thermal evolution of the moon also favor an initially molten moon to a depth of 200–600 km (e.g., Toksöz and Solomon, 1973; Strangway and Sharpe, 1975). To model the LIL element evolution of such a thick magma body through the entire process is possible. Previously, Shih and Schonfeld (1976) have satisfactorily demonstrated a similar kind of approach for the genesis of high-Ti mare basalts.

LIL element evolution in a differentiating magma body

Petrological models outlined by Hodges and Kushiro (1974) and Walker *et al.* (1975b) are used for discussion. Furthermore, assuming that the bulk composition of the molten portion of the moon is that proposed by Ganapathy and Anders (1974), the normative mineralogy is the total mineral assemblage to be crystallized under ideal hypothetical conditions. Using these petrological and mineralogical boundary conditions, the LIL element abundances can be readily calculated for each cotectic liquid, based on the phase equilibrium relationships developed by Hodges and Kushiro (1974) in the SiO_2-anorthite-olivine-Ca-rich-clinopyroxene quaternary system.

It is assumed that the moon had unfractionated refractory LIL element contents around 7× chondrites, Rb/Sr ratio of ~ 0.01, and chondritic K/Rb ratio. As the magma ocean cools slowly and crystallization proceeds, the LIL elements which are generally rejected by most of the common minerals are gradually concentrated in liquids (Fig. 9). At first, the liquids will start to show slight positive Eu anomalies until they reach saturation of plagioclase. Then the Eu anomalies become negative and with a pronounced increase because of the separation of plagioclase. Surprisingly, the light REE do not fractionate much over heavy REE. Walker *et al.* (1975b) have pointed out the dynamic importance of the sinking, early crystallized olivine and its reaction with the liquid to form orthopyroxene under a pressure gradient of ~ 20 kbar in a thick magma body. For an extreme case, if all the olivine reacted with the liquid to form orthopyroxene, the residual liquids would still not have REE fractionation patterns characteristic of most highland rocks (Fig. 9).

(a) *Garnet fractionation.* The distinctive light over heavy REE ratio for KREEP basalts of twice chondrite can only be accommodated if (1) the hypothetical starting materials were slightly enriched in light REE over heavy REE relative to chondritic meteorites (which is probably unlikely), or (2) minerals favoring heavy REE were crystallized somewhere in the differentiation

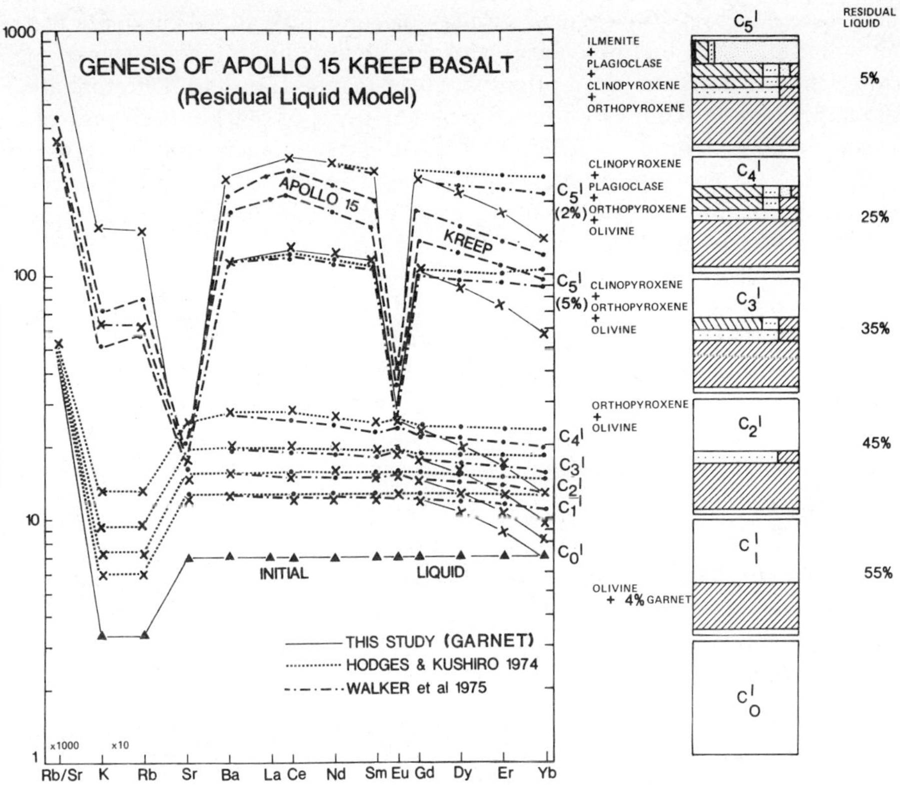

Fig. 9. Simplified model for LIL element fractionation after the initial accretional melting of total upper mantle (500 km thick) of the moon. In the molten layer, the refractory LIL element distribution is assumed to be chondritic, but ~7 times enriched, while the Rb/Sr ratio is ~0.01. The differentiation scheme is outlined by Hodges and Kushiro (1974) and Walker *et al.* (1975a,b). A small amount of garnet is expected to be present because of the 25 kbar pressure gradient. The last residual liquid subject to the separation of garnet, plagioclase, clinopyroxene, ilmenite bears the LIL element characteristics of KREEP basalt.

process. In the latter case, garnet is one probable candidate. Garnet is stable only at pressures greater than 25 kbar or at the bottom of a magma ocean of at least 500 km depth. The thickness of the initially molten layer is unknown. However, proposed thermal models speculate a layer ranging in thickness from 100–200 km (Strangway and Sharpe, 1975), 400–800 km (Toksöz and Solomon, 1973) to even whole moon (Smith, 1977). Recent seismic data from both surface and deep moonquake events (Dainty *et al.*, 1976; Goins *et al.*, 1977) characterized an upper mantle 500 km thick. They interpreted the sudden decrease in seismic velocities in the lower mantle as being due to substitution of Fe for Mg in silicate minerals. The seismic velocity discontinuity at 500 km depth can also be interpreted as the boundary between highly differentiated upper mantle (Mg-rich) and unfractionated primitive lower mantle (Fe-rich). Differentiation of the upper mantle requires an initially molten layer of 500 km thickness early in lunar evolution. Garnet is definitely a stable phase in such a thick magma body with a 25 kbar pressure gradient. The amount of garnet required is small (not more than a few percent), since it has very large heavy REE distribution coefficients resulting in the light REE being preferentially concentrated in the residual liquids (Fig. 10).

At present, seismic studies are not capable of detecting garnet (a layer of <35 km thickness for 5% garnet cumulated) at 500 km depth (Nakamura, 1975, pers. comm.). If garnets were disseminated with other minerals it is even more difficult to detect. Laul and Schmitt (1975) also noted the probable involvement of garnet in the magma from which dunite 72417 crystallized. Some indirect evidence for the existence of garnet in the lunar interior was reported by Ridley *et al.* (1973b).

(b) *Plagioclase fractionation.* As far as the LIL element concentrations are concerned, KREEP basalts can be achieved by late-stage residual liquid (less than 5% liquid or greater than 95% solidification). The size of negative Eu anomalies and also the high Rb/Sr ratios can also be satisfied by the separation of plagioclase.

(c) *Clinopyroxene fractionation.* Ca/Al ratio for KREEP basalt (0.82) is significantly lower than that for the meteorites (1.1). Separation of clinopyroxene prior to the formation of KREEP liquid in the differentiation process can account for the low Ca/Al ratio.

(d) *Ilmenite fractionation.* KREEP basalts have the lowest Ti/Zr ratio (~ 10) found in lunar rocks. Ilmenite separation has been suggested for considerably decreasing Ti/Zr ratios in a liquid from which to derive KREEP basalt by fractional crystallization (Duncan *et al.*, 1976; McCallum and Charette, 1977). It is interesting to note that TiO_2 concentration for ilmenite saturation in Fe-rich residual liquid was reported as low as 2.4% (Rutherford *et al.*, 1974; Hess *et al.*, 1975) which is similar to that for the KREEP basalts.

High "mg"-value in KREEP basalts

Gast (1972) has argued that KREEP basalts cannot be generated by extensive fractional crystallization processes because of their high "mg"-values ($0.50 \sim 0.60$). The magma becomes saturated with ilmenite as its TiO_2 concentration reaches $\sim 10\%$. Precipitation of ilmenite could deplete Fe from the liquid, thus increase "mg"-value of the liquid. In addition, it should be emphasized that for a highly reducing lunar magma with an oxygen fugacity several orders of magnitude lower than the terrestrial environment (Sato *et al.*, 1973), the late-stage liquids could become saturated with metallic Fe (Williams, 1975, pers. comm.). It is common to find metallic Fe in lunar basalts. Only a few percent of Fe metal need be removed to counterbalance the depletion of "mg"-values in the residual liquid.

Additionally, the "mg"-value in a late-stage liquid is not expected to be low if the process is not a *perfect* fractional crystallization, but involves crystallization within a *partially closed-system*. Walker *et al.* (1975a) have found that the last 10% of the residual liquid in equilibrium with plagioclase, clinopyroxene, ilmenite, and olivine in an experimental crystallization system would have an "mg"-value of about 0.50–0.55. It is interesting to note that KREEP basalts have similar "mg"-values. Heavy stirring of the magma ocean would also tend to reduce the Fe/Mg in late-stage liquids (Wood, 1975).

High Cr concentration in KREEP basalts

High Cr concentration (~ 1500 ppm) in KREEP basalts has also been used as an argument against the differentiation model (Gast, 1972). Using Cr distribution coefficients for different minerals measured under controlled reducing conditions (Ringwood, 1970; Akella *et al.*, 1976; Schreiber and Haskin, 1976; Huebner *et al.*, 1976), at least the chondritic amount of Cr (3000–4000 ppm) must be present in the molten portion of the moon—a factor of two higher than that proposed by Ganapathy and Anders (1974)—to generate a residual liquid having the observed Cr abundances of the KREEP basalts. Hess *et al.* (1975) suggested that the Cr would not be depleted in the residual liquid because of the resorption of early-crystallized spinel before it could be removed from the magma. It is also interesting to note that the white portion of the granitic rock 12013 (Ma *et al.*, 1977) has just about the same amount of Cr as the KREEP basalts. If the granitic rocks were formed from the late stage of fractional crystallization, then the process probably would not deplete Cr in the residual liquid as drastically as assumed from the terrestrial experience. Lunar rocks, especially mare basalts, seem to be highly enriched in Cr by terrestrial standards. Whether Cr is enriched intrinsically in the lunar material is open to further study. The calculated Cr abundance in a liquid coexisting with ilmenite-plagioclase clinopyroxene cumulates (high-Ti mare basalt) agrees very well with values measured for the KREEP basalts.

In conclusion, most of the chemical characteristics of the KREEP basalts can be satisfactorily explained as late-stage residual liquids which have undergone prior separation of garnet, plagioclase, clinopyroxene, and ilmenite. The late-stage cumulates, ilmenite-plagioclase-clinopyroxene, are the proper source of high-Ti mare basalt (Shih and Schonfeld, 1976). Theoretically, a bulk moon composition of Ganapathy and Anders (1974) which is olivine-normative should finally produce liquid of point B composition of the Walker diagram by *perfect* equilibrium crystallization. In order to derive the liquid composition of Apollo 15 KREEP basalts (quartz-normative) which are located between points B and C, *fractional* crystallization, which is more probable in nature, is required. Isotopic data are consistent with differentiation of a magma ocean starting at 4.6 AE and ending at 4.4 AE. KREEP basalts formed were subject to possible chemical modification at 4.1 AE and recrystallization at 3.9 AE.

Distribution of KREEP basalts

Taylor *et al.* (1972) and Gast (1972) argued that the uneven distribution of highly radioactive materials (KREEP rocks) over the lunar surface, as indicated from the gamma-ray experiments of Metzger *et al.* (1973), is evidence against the KREEP basalts being originated from whole moon differentiation. However, it is equally well-established that the KREEP basalts are a widely spread and important component of soils and breccias not only in the lunar highlands but also in mare regions. For example, (1) the KREEP basalts are found to be a necessary end-member for mixing models of soil and breccia formation in either highland or mare regions (Meyer *et al.*, 1971; Schonfeld and Meyer, 1972; Schonfeld, 1973, 1975). (2) The KREEP glasses were found at every landing site by the Apollo Soil Survey project (Reid, 1974), and a large KREEP fragment, Luny Rock 1, was reported at the Apollo 11 site (Albee, 1970). Wide distribution of KREEP basalts is certainly consistent with the model that they were produced by a moon-wide differentiation process. The localization of KREEP basalts at Apollo 12, 14, and 15 sites is probably related to variations of the thickness of the anorthositic crust.

"LKFM basalt"

Recently, much attention has been drawn to the "LKFM basalt" for its role in KREEP basalt genesis (McKay and Weill, 1976, 1977; Hess *et al.*, 1977). The name "LKFM basalt" was originally given to one of the most dominant glass compositions in highland soil by the Apollo Soil Survey (1971). It probably represents the composition of a major rock type. Hand specimens of Apollo 16 and 17 breccias are close to this in composition. However, it should be reemphasized that evidence for *an igneous rock type* is totally lacking. Chemical composition of the rock is not well known. Major element chemistry is based on glasses of unknown origin; LIL element chemistry is based solely on a black portion of 15455 (Taylor, 1973) which is a breccia with an unusually high amount of meteoritic component (Ganapathy *et al.*, 1973). The mixing model of Schonfeld (1974) has already demonstrated that "LKFM basalt" could be a mixture of predominant KREEP basalt, ANT rocks, and mare basalt. To treat "LKFM basalt" as an igneous rock type is probably unnecessary.

6. CONCLUSIONS

Constraints of major and LIL element chemistry, phase equilibria experiments and Rb–Sr isotopic data have been used to construct models of KREEP basalt genesis from which the following conclusions can be made:

(1) The partial melting model requires a high "mg"-value (~ 0.82) and high

LIL element abundances (5–15× chondrites) in the source for KREEP basalts. Norites and troctolites of the ANT series are probably better candidates than anorthositic norite breccias. The sources were established at ~4.6 AE. Generation of KREEP basalt by a small degree of internal melting occurred at ~4.4 AE.

(2) The crystallization model starts with melting of the total upper mantle (500 km thick) at ~4.6 AE. Differentiation of the thick magma ocean began immediately and ended at ~4.4 AE. KREEP basalts are the late-stage residual liquids highly enriched in LIL elements. Negative Eu anomalies, low Ca/Al ratios, high "mg"-values, low Ti/Zr ratios, and enrichment of light REE over heavy REE require the residual liquids to have undergone fractionation of plagioclase, clinopyroxene, ilmenite, and garnet in an extremely reducing environment. The late-stage plagioclase-clinopyroxene-ilmenite cumulates are suitable sources for high-Ti mare basalts. This model is favored because it can account for mare basalt genesis.

In both models, the formation age of KREEP basalts is interpreted to be ~4.4 AE. Their chemical composition had either been established at that time or had undergone minor modification at ~4.1 AE, but probably not at ~3.9 AE. The mineral isochron age of KREEP basalts of ~3.9 AE is probably a metamorphic age reset by the Imbrium event.

Acknowledgments—I am indebted to J. E. Wainwright, E. Schonfeld, L. A. Haskin, and L. E. Nyquist for their comments and assistance. Reviews by G. A. McKay, S. R. Winzer, C. H. Simonds, D. Walker, and R. A. Schmitt are gratefully acknowledged. I thank Ms. Lina Romero for typing the many drafts.

References

Akella J., Williams R. J., and Mullins O. (1976) Solubility of Cr, Ti, and Al in co-existing olivine, spinel, and liquid at 1 atm. *Proc. Lunar Sci. Conf. 7th*, p. 1179–1194.

Albee A. L. and Chodos A. A. (1970) Microprobe investigations on Apollo 11 samples. *Proc. Apollo 11 Lunar Sci. Conf.*, p. 135–157.

Apollo Soil Survey (1971) Apollo 14: Nature and origin or rock types in soil from the Fra Mauro Formation. *Earth Planet. Sci. Lett.* **12**, 49–54.

Axon H. J. and Goldstein J. I. (1972) Temperature-time relationships from lunar two phase metallic particles (14310, 14163, 14003). *Earth Planet. Sci. Lett.* **16**, 439–447.

Blanchard D. P., Haskin L. A., Jacobs J. W., Brannon J. C., and Korotev R. L. (1975) Major and trace element chemistry of Boulder 1 at Station 2, Apollo 17. *The Moon* **14**, 359–371.

Dainty A. M., Toksöz M. N., and Stein S. (1976) Seismic investigation of the lunar interior. *Proc. Lunar Sci. Conf. 7th*, p. 3057–3075.

Delano J. W. (1974) 76535: A possible source rock for lunar KREEP and VHA magmas (abstract). *EOS* (*Trans. Amer. Geophys. Union*) **55**, 324

Dowty E., Prinz M., and Keil K. (1973) "Very high alumina basalt": A mixture and not a magma type. *Science* **181**, 1214–1215.

Dowty E., Keil K., Prinz M., Gros J., and Takahashi H. (1976) Meteorite-free Apollo 15 crystalline KREEP. *Proc. Lunar Sci. Conf. 7th*, p. 1833–1844.

Duncan A. R., Erlank A. J., Sher M. K., Abraham Y. C., Willis J. P., and Ahrens L. H. (1976) Some trace element constraints on lunar basalt genesis. *Proc. Lunar Sci. Conf. 7th*, p. 1659–1671.

Ford C. E., Biggar G. M., Humphries D. J., Wilson G., Dixon D., and O'Hara M. J. (1972) Role of water in the evolution of the lunar crust; an experimental study of sample 14310; an indication of lunar calc-alkaline volcanism. *Proc. Lunar Sci. Conf. 3rd*, p. 207–229.

Ganapathy R. and Anders E. (1974) Bulk compositions of the moon and earth, estimated from meteorites. *Proc. Lunar Sci. Conf. 5th*, p. 1181–1206.

Ganapathy R., Morgan J. W., Krähenbühl U., and Anders E. (1973) Ancient meteoritic components in lunar highland rocks: Clues from trace elements in Apollo 15 and 16 samples. *Proc. Lunar Sci. Conf. 4th*, p. 1239–1261.

Gast P. W. (1972) The chemical composition and structure of the moon. *The Moon* **5**, 121–148.

Goins N. R., Dainty A. M., and Toksöz M. N. (1977) The structure of the lunar interior as determined from seismic data (abstract). In *Lunar Science VIII*, p. 354–356. The Lunar Science Institute, Houston.

Haskin L. A., Shih C. Y., Bansal B. M., Rhodes J. M., Wiesmann H., and Nyquist L. E. (1974) Chemical evidence for the origin of 76535 as a cumulate. *Proc. Lunar Sci. Conf. 5th*, 1213–1225.

Hess P. C., Rutherford M. J., Guillemette R. N., Ryerson F. J., and Tuchfeld H. A. (1975) Residual products of fractional crystallization of lunar magmas: An experimental study. *Proc. Lunar Sci. Conf. 6th*, p. 895–909.

Hess P. C., Rutherford M. J., Campbell H. W., and Dick P. A. (1977) Fractionation of lunar feldspathic magmas (abstract). In *Lunar Science VIII*, p. 433–435. The Lunar Science Institute, Houston.

Hodges F. N. and Kushiro I. (1974) Apollo 17 petrology and experimental determination of differentiation sequences in model moon compositions. *Proc. Lunar Sci. Conf. 5th*, p. 505–520.

Hubbard N. J. and Gast P. W. (1971) Chemical composition and origin of non-mare lunar basalts. *Proc. Lunar Sci. Conf. 2nd*, p. 999–1020.

Hubbard N. J., Meyer C., Jr., Gast P. W., and Wiesmann H. (1971a) The composition and derivation of Apollo 12 soils. *Earth Planet. Sci. Lett.* **10**, 341–350.

Hubbard N. J., Gast P. W., Meyer C., Nyquist L. E., Shih C., and Wiesmann H. (1971b) Chemical composition of lunar anorthosites and their parent liquids. *Earth Planet. Sci. Lett.* **10**, 341–350.

Hubbard N. J., Gast P. W., Rhodes J. M., Bansal B. M., Wiesmann H., and Church S. E. (1972) Non-mare basalts: Part II. *Proc. Lunar Sci. Conf. 2nd*, p. 1161–1179.

Hubbard N. J., Rhodes J. M., Gast P. W., Bansal B. M., Shih C. Y., Wiesmann H., and Nyquist L. E. (1973a) Lunar rock types: The role of plagioclase in non-mare and highland rock types. *Proc. Lunar Sci. Conf. 4th*, p. 1297–1312.

Hubbard N. J., Rhodes J. M., and Gast P. W. (1973b) Chemistry of lunar basalts with very high alumina contents. *Science* **181**, 339–342.

Hubbard N. J., Rhodes J. M., Wiesmann H., Shih C. Y., and Bansal B. M. (1974) The chemical definition and interpretation of rock types returned from the non-mare regions of the moon. *Proc. Lunar Sci. Conf. 5th*, p. 1227–1246.

Huebner J. S., Lipin B. R., and Wiggins L. B. (1976) Partitioning of chromium between silicate crystals and melts. *Proc. Lunar Sci. Conf. 7th*, p. 1195–1220.

Husain L. and Schaeffer O. A. (1975) Lunar evolution: The first 600 million years. *Geophys. Res. Lett.* **2**, 29–32.

Irving A. J. (1975) Chemical, mineralogical and textural systematics of non-mare melt rocks: Implications for lunar impact and volcanic processes. *Proc. Lunar Sci. Conf. 6th*, p. 363–394.

Jakeš P., Warner J., Ridley W. I., Reid A. M., Harmon R. S., Brett R., and Brown R. W. (1972) Petrology of a portion of the Mare Fecunditatis regolith. *Earth Planet. Sci. Lett.* **13**, 257–276.

Kirsten T. and Horn P. (1974) Chronology of the Taurus-Littrow region III: Ages of mare basalts and highland breccias and some remarks about the interpretation of lunar highland rock ages. *Proc. Lunar Sci. Conf. 5th*, p. 1451–1475.

Laul J. C. and Schmitt R. A. (1975) Dunite 72417: A chemical study and interpretation. *Proc. Lunar Sci. Conf. 6th*, p. 1231–1254.

Longhi J., Walker D., and Hays J. F. (1976) Fe and Mg in plagioclase. *Proc. Lunar Sci. Conf. 7th*, p. 1281–1300.

Lugmair G. W. (1975) Sm–Nd systematics of some Apollo 17 basalts. In *Papers Presented to the*

Conference on Origins of Mare Basalts and Their Implications for Lunar Evolution, p. 107–110. The Lunar Science Institute, Houston.

Ma M. S., Murali A. V., Schmitt R. A., Quick J. E., and Albee A. L. (1977) Major, minor and trace element compositions of two petrologically significant lithologics in 12013 (abstract). In *Lunar Science VIII*, p. 602–604. The Lunar Science Institute, Houston.

Marvin U. B., Wood J. A., Taylor G. J., Reid J. B., Jr., Powell B. N., Dickey J. S., Jr., and Bower J. F. (1971) Relative proportions and probable sources of rock fragments in Apollo 12 soil samples. *Proc. Lunar Sci. Conf. 2nd*, p. 679–700.

McCallum I. S. and Charette M. P. (1977) Partitioning of Zr between crystals and coexisting high-Ti mare basalt melt (abstract). In *Lunar Science VIII*, p. 637–639. The Lunar Science Institute, Houston.

McCallum I. S. and Mathez E. A. (1975) Petrology of noritic cumulates and a partial melting model for the genesis of Fra Mauro basalts. *Proc. Lunar Sci. Conf. 6th*, p. 395–414.

McKay G. A. and Weill D. F. (1976) Petrogenesis of KREEP. *Proc. Lunar Sci. Conf. 7th*, p. 2427–2447.

McKay G. A. and Weill D. F. (1977) KREEP (abstract). In *Lunar Science VIII*, p. 649–651. The Lunar Science Institute, Houston.

Metzger A. E., Trombka J. I., Peterson L. E., Reedy R. C., and Arnold J. R. (1973) Lunar surface radioactivity: Preliminary results of the Apollo 15 and Apollo 16 gamma-ray spectrometer experiments. *Science* **179**, 800–803.

Meyer C., Jr. (1972) Mineral assemblages and the origin of non-mare lunar rock type (abstract). In *Lunar Science III*, p. 542–544. The Lunar Science Institute, Houston.

Meyer C., Jr. (1976) Petrology, mineralogy and chemistry of KREEP basalt. *Phys. Chem. Earth* **10**. in press.

Meyer C., Jr. and Hubbard N. J. (1970) High potassium, high phosphorous glass as an important rock type in the Apollo 12 soil samples (abstract). *Meteoritics* **5**, 210–211.

Meyer C., Jr., Brett R., Hubbard N. J., Morrison D. A., McKay D. S., Aitken F. K., Takeda H., and Schonfeld E. (1971) Mineralogy, chemistry and origin of the KREEP component in soil samples from the Ocean of Storms. *Proc. Lunar Sci. Conf. 2nd*, p. 393–411.

Morgan J. W., Laul J. C., Krähenbühl U., Ganapathy R., and Anders E. (1972) Major impact on the Moon: Characterization from trace elements in Apollo 12 and 14 samples. *Proc. Lunar Sci. Conf. 3rd*, p. 1377–1395.

Mysen B. (1976) Partitioning of samarium and nickel between olivine, orthopyroxene, and liquid; preliminary data at 20 Kbar and 1025°C. *Earth Planet Sci. Lett.* **12**, 36–48.

Nyquist L. E., Bansal B. M., Wiesmann H., and Jahn B. M. (1974) Taurus-Littrow chronology: Some constraints on early lunar crustal development. *Proc. Lunar Sci. Conf. 5th*, p. 1515–1539.

Nyquist L. E., Bansal B. M., and Wiesmann H. (1975) Rb–Sr ages and initial $^{87}Sr/^{86}Sr$ for Apollo 17 basalts and KREEP basalt 15386. *Proc. Lunar Sci. Conf. 6th*, p. 1445–1465.

Papanastassiou D. A. and Wasserburg G. J. (1971) Rb–Sr ages of igneous rocks from the Apollo 14 mission and the age of the Fra Mauro Formation. *Earth Planet. Sci. Lett.* **12**, 36–48.

Papanastassiou D. A. and Wasserburg G. J. (1976) Rb–Sr age of troctolite 76535. *Proc. Lunar Sci. Conf. 7th*, p. 2035–2054.

Papanastassiou D. A., DePaolo D. J., Tera F., and Wasserburg G. J. (1977) An isotopic triptych on mare basalts: Rb–Sr, Sm–Nd, U–Pb (abstract). In *Lunar Science VIII*, p. 750–752. The Lunar Science Institute, Houston.

Philpotts J. A. and Schnetzler C. C. (1970) Phenocryst-matrix partition coefficients for K, Rb, Sr, and Ba, with application to anorthosite and basalt genesis. *Geochim. Cosmochim. Acta* **34**, 307–322.

Philpotts J. A., Schnetzler C. C., Nava D. F., Bottino M. L., Fullagar P. D., Thomas H. H., Schuhmann S., and Kouns C. W. (1972) Apollo 14: Some geochemical aspects. *Proc. Lunar Sci. Conf. 3rd*, p. 1293–1305.

Phinney W. C., Warner J. L., Simonds C. H., and Lofgren G. E. (1972) Classification and distribution of rock types at Spur Crater. In *The Apollo 15 Lunar Samples* (J. W. Chamberlain and C. Watkins, eds.), p. 149–152. The Lunar Science Institute, Houston.

Reid A. M. (1974) Rock types present in lunar highland soils. *The Moon* **9**, 141–146.

Reid A. M., Warner J., Ridley W. I., Johnston D. A., Harmon R. S., Jakeš P., and Brown R. W. (1972a) The major element compositions of lunar rocks as inferred from glass compositions in the lunar soils. *Proc. Lunar Sci. Conf. 3rd*, p. 363–378.

Reid A. M., Warner J., Ridley W. I., and Brown R. W. (1972b) Major element composition of glasses in three Apollo 15 soils. *Meteoritics* **7**, 395–415.

Reid A. M., Ridley W. I., Donaldson C., and Brown R. W. (1973) Glass compositions in the orange and gray soils from Shorty Crater, Apollo 17 (abstract). *EOS (Trans. Amer. Geophys. Union)* **54**, 607–609.

Rhodes J. M. and Hubbard N. J. (1973) Chemistry, classification, and petrogenesis of Apollo 15 mare basalts. *Proc. Lunar Sci. Conf. 4th*, p. 1127–1148.

Ridley W. I., Reid A. M., Warner J. L., Brown R. W., Gooley R., and Donaldson C. (1973a) Glass compositions in Apollo 16 soils 60501 and 61221. *Proc. Lunar Sci. Conf. 4th*, p. 309–321.

Ridley W. I., Hubbard N. J., Rhodes J. M., Wiesmann H., and Bansal B. M. (1973b) The petrology of lunar breccia 15445 and petrogenetic implications. *J. Geol.* **81**, 621–631.

Ringwood A. E. (1970) Petrogenesis of Apollo 11 basalts and implications for lunar origin. *J. Geophys. Res.* **75**, 6453–6479.

Roeder P. L. and Emslie R. F. (1970) Olivine-liquid equilibrium. *Contrib. Mineral. Petrol.* **29**, 275–289.

Rutherford M. J., Hess P. C., and Daniel G. H. (1974) Experimental liquid line of descent and liquid immiscibility for basalt 70017. *Proc. Lunar Sci. Conf. 5th*, p. 569–583.

Sato M., Hickling N. L., and McLane J. E. (1973) Oxygen fugacity values of Apollo 12, 14 and 15 lunar samples and reduced state of lunar magmas. *Proc. Lunar Sci. Conf. 4th*, p. 1061–1079.

Schonfeld E. (1974) The contamination of lunar highland rocks by KREEP: Interpretation by mixing models. *Proc. Lunar Sci. Conf. 5th*, p. 1269–1286.

Schonfeld E. (1975) Component abundances in Apollo 15 soils and breccias by the mixing model technique (abstract). In *Lunar Science VI*, p. 712–714. The Lunar Science Institute, Houston.

Schonfeld E. and Meyer C., Jr. (1972) The abundances of components of the lunar soils by a least-squares mixing model and the formation age of KREEP. *Proc. Lunar Sci. Conf. 3rd*, 1397–1420.

Schaeffer O. A., Husain L., and Schaeffer G. A. (1976) Ages of highland rocks: The chronology of lunar basin formation revisited. *Proc. Lunar Sci. Conf. 7th*, p. 2067–2092.

Schreiber H. D. and Haskin L. A. (1976) Chromium in basalts: Experimental determination of redox states and partitioning among synthetic silicate phases. *Proc. Lunar Sci. Conf. 7th*, p. 1221–1259.

Shaw D. M. (1970) Trace element fractionation during anatexis. *Geochim. Cosmochim. Acta* **34**, 237–243.

Shih C. (1976) On the Origin of KREEP basalts (abstract). In *Lunar Science VII*, p. 800–802. The Lunar Science Institute, Houston.

Shih C. and Schonfeld E. (1976) Mare basalt genesis: A cumulate-remelting model. *Proc. Lunar Sci. Conf. 7th*, p. 1757–1792.

Shih C., Haskin L. A., Wiesmann H., Bansal B. M., and Brannon J. C. (1975) On the origin of high-Ti mare basalts. *Proc. Lunar Sci. Conf. 6th*, p. 1255–2185.

Simonds C. H., Warner J. L., and Phinney W. C. (1975) The petrology of the Apennine Front revisited (abstract). In *Lunar Science VI*, p. 744–746. The Lunar Science Institute, Houston.

Smith J. V. (1977) Chemical evidence on origin of earth and moon (abstract). In *Lunar Science VIII*, p. 881–883. The Lunar Science Institute, Houston.

Stoeser D. B., Wolfe R. W., Wood J. A., and Bower J. F. (1974) Petrology. In *Interdisciplinary Studies of Samples from Boulder 1, Station 2, Apollo 17*, Vol. 1 (J. A. Wood, ed.). LSI Contr. No. 210D, p. 35–109. Smithsonian Astrophysical Observatory, Cambridge.

Strangway D. W. and Sharpe H. N. (1975) A model of lunar evolution. *The Moon* **12**, 369–397.

Taylor G. J., Marvin U. B., Reid J. B., Jr., and Wood J. A. (1972) Noritic fragments in the Apollo 14 and 12 soils and origin of Oceanus Procellarum. *Proc. Lunar Sci. Conf. 3rd*, p. 995–1014.

Taylor G. J., Drake M. J., Wood J. A., and Marvin U. B. (1973) The Luna 20 lithic fragments, and the composition and origin of the lunar highlands. *Geochim. Cosmochim. Acta* **37**, 1087–1106.

Taylor S. R. (1973) Geochemistry of the Lunar highlands. *The Moon* **7**, 181–195.

Taylor S. R. and Bence A. E. (1975) Evolution of the lunar highland crust. *Proc. Lunar Sci. Conf. 6th*, p. 1121–1141.

Taylor S. R. and Jakeš P. (1974) The geochemical evolution of the moon. *Proc. Lunar Sci. Conf. 5th*, p. 1287–1305.

Tera F. and Wasserburg G. J. (1975) The evolution and history of mare basalts as inferred from U–Th–Pb systematics (abstract). In *Lunar Science VI*, p. 807–809. The Lunar Science Institute, Houston.

Tera F. and Wasserburg G. J. (1976) Lunar ball games and other sports (abstract). In *Lunar Science VII*, p. 858–860. The Lunar Science Institute, Houston.

Tera F., Papanastassiou D. A., and Wasserburg G. J. (1974) Isotopic evidence for a terminal lunar cataclysm. *Earth Planet. Sci. Lett.* **22**, 1–21.

Toksöz M. N. and Solomon S. C. (1973) Thermal history and evolution of the Moon. *The Moon* **7**, 251–278.

Turner G., Cadogan P. H., and Yonge C. J. (1973) Argon selenochronology. *Proc. Lunar Sci. Conf. 4th*, p. 1889–1914.

Wakita H., Laul J. C., and Schmitt R. A. (1975) Some thoughts on the origin of lunar ANT-KREEP and mare basalts. *Geochem. J.* **9**, 25–41.

Walker D., Longhi J., and Hays J. F. (1972) Experimental petrology and origin of Fra Mauro rocks and soil. *Proc. Lunar Sci. Conf. 3rd*, p. 797–817.

Walker D., Grove T. L., Longhi J., Stolper E. M., and Hays J. F. (1973) Origin of lunar feldspathic rocks. *Earth Planet Sci. Lett.* **20**, 325–336.

Walker D., Longhi J., Grove T. L., Stolper E. M., and Hays J. F. (1975a) Origin of titaniferous lunar basalts. *Geochim. Cosmochim. Acta* **39**, 1219–1235.

Walker D., Longhi J., and Hays J. F. (1975b) Differentiation of a very thick magma body and implications for the source regions of mare basalts. *Proc. Lunar Sci. Conf. 6th*, p. 1103–1120.

Warner J., Reid A. M., Ridley W. I., and Brown R. W. (1972) Major element composition and Luna 20 glasses. *Earth Planet. Sci. Lett.* **17**, 7–12.

Wasserburg G. J. and Papanastassiou D. A. (1971) Age of an Apollo 15 mare basalt; Lunar crust and mantle evolution. *Earth Planet. Sci. Lett.* **13**, 97–104.

Weill D. F. and McKay G. A. (1972) The partitioning of Mg, Fe, Sr, Ce, Sm, Eu, and Yb in lunar igneous systems and a possible origin of KREEP by equilibrium partial melting. *Proc. Lunar Sci. Conf. 6th*, p. 1143–1158.

Wiesmann H. and Hubbard N. J. (1975) A compilation of the lunar sample data generated by the Bast, Nyquist and Hubbard Lunar Sample PI-Ships. NASA-Johnson Space Center, Houston. Unpublished manuscript.

Winzer S. R., Nava D. F., Lum R. K. L., Schuhmann S., Schuhmann P., and Philpotts J. A. (1975) Origin of 78235, a lunar norite cumulate. *Proc. Lunar Sci. Conf. 6th*, p. 1219–1229.

Wood J. A. (1974) A Survey of lunar rock types and comparison of the crusts of Earth and Moon. *Proc. Sov.–Amer. Conf. on Cosmochemistry of the Moon and Planets, Moscow*. NASA SP 370. In Press.

Wood J. A. (1975) Lunar petrogenesis in a well-stirred magma ocean. *Proc. Lunar Sci. Conf. 6th*, p. 1087–1102.

Proc. Lunar Sci. Conf. 8th (1977), p. 2403–2417.
Printed in the United States of America

Evolution of KREEP: Further petrologic evidence

MARIA LUISA CRAWFORD

Department of Geology, Bryn Mawr College, Bryn Mawr, Pennsylvania 19010

LINCOLN S. HOLLISTER

Department of Geological and Geophysical Sciences, Princeton University,
Princeton, New Jersey 08540

Abstract—The textures and the details of the chemical zoning of minerals in Apollo 15 KREEP samples 15382 and 15434,66, among others, support the contention that KREEP from the Apollo 15 site appears to be igneous. A comparison of the data from Apollo 15 KREEP to those of KREEP from other locations, particularly to 14310, shows that the similarities support an igneous origin for other samples of crystalline KREEP. The low siderophile element content and the lack of high pressure phenocrysts in Apollo 15 KREEP, in contrast to 14310, is explained by a model of slower rise of Apollo 15 KREEP melt to the surface. During ascent the melt may have crystallized, differentiated, and equilibrated to near-surface conditions prior to eruption. According to the model, early gravitational separation of Fe–Ni metal would deplete the final erupted melt in siderophile elements relative to the possibly more rapidly erupted 14310 liquid. Liquid immiscibility occurred during crystallization of these melts; the chemical effects of the separation into two melts can be identified in plagioclase and possibly also in pyroxene. The compositions of the two immiscible melts suggest KREEP may be the parent of Apollo 12 and 15 basalts on one hand, and of granitic rocks on the other, if KREEP melt crystallized to the point of becoming immiscible at depth within the moon, followed by segregation and subsequent eruption of the evolved magmas.

INTRODUCTION

THE ORIGIN AND EVOLUTION of feldspathic basalts enriched in LIL elements (KREEP basalts) continues to be controversial. A recent summary by Meyer (1977) presents the data and interpretations advanced by the many investigators interested in this problem. Recently, there seems to be general agreement among a number of authors that the KREEP fragments from the Apollo 15 site are igneous rocks originating in the lunar interior. The main reason for identifying these fragments as "primary" igneous KREEP is the low content of Au, Ir, Re, and Os in the basalts ($<.01$ ppm). 10^2–10^3 times higher concentrations of these siderophile elements have been interpreted as indicators of meteoritic contamination of lunar samples (Gros *et al.*, 1976, and earlier papers quoted therein). Dowty *et al.* (1976) and Meyer (1977) both point out, however, that the Rb/Sr isotopic and age data for Apollo 15 KREEP resembles that for KREEP from other sites.

If Apollo 15 KREEP melts were derived from the lunar interior and other crystalline KREEP samples were not, then we would expect to find some evidence for differences in the crystallization history in samples which have different origins. For example, a melt originating by partial melting in the lunar interior should have a bulk composition lying along a cotectic line or on a peritectic point of the appropriate phase diagram, and phenocrysts in such a rock should have chemical evidence of once being in equilibrium with the melt. On the other hand, a melt originating by impact would have the average composition of the impacted volume (Dence, 1971) and would therefore not be constrained to fall on a cotectic line or peritectic point in the appropriate phase diagram. Unmelted fragments in a rock originating from crystallization of an impact melt would not be constrained to be in chemical equilibrium with the impact melt (Simonds, 1975).

We have suggested that other KREEP samples, in particular 14310, are igneous rather than crystallized impact melts and that these melts started to crystallize prior to eruption onto the lunar surface (Crawford and Hollister, 1974). In the present paper we show that the mineral chemistry and crystallization sequence of samples of Apollo 15 KREEP resemble the patterns outlined for 14310 closely enough to suggest that radically different origins (i.e., melting due to impact versus eruption from within the moon) and crystallization histories for the two are unlikely. However, differences in the bulk chemistry and in the details of the crystallization histories require differences in the pre-eruption histories of the Apollo 15 KREEP and 14310.

PETROGRAPHY

Table 1 lists the Apollo 15 crystalline KREEP fragments which were included in this study. The main features of the mineralogy and phase chemistry of these samples have been described by a number of persons (for a complete list see Meyer, 1977). The grain size of individual samples varies from relatively coarse-grained (1–3 mm wide, 10 mm long) in sample 15434,47,128 to the more fine-grained vitrophyre 15434,66 with small (1 mm) skeletal crystals of pyroxene and thin plagioclase laths set in a predominantly glassy groundmass (Fig. 1). Many fragments contain long hollow crystals of pyroxene similar to those seen in mare basalts such as 12021. In some, e.g., 15434,47,148, plagioclase zoning patterns suggest the plagioclase laths were also initially hollow; others, among them 15382, have curved plagioclase laths (Fig. 1). These curved plagioclase grains appear to be unique to Apollo 15 samples. Aside from the fairly high percentage of glass or very fine-grained mesostasis the textures of these rocks

Table 1. Apollo 15 KREEP samples included in study.

15382,6	15434,37,109	15999,95
15382,7	15434,47,128	15999,96*
15382,17	15434,49,132	15999,97
15386,3		15999,98

*Sample 15434,66 is a small fragment in slide 15999,96.

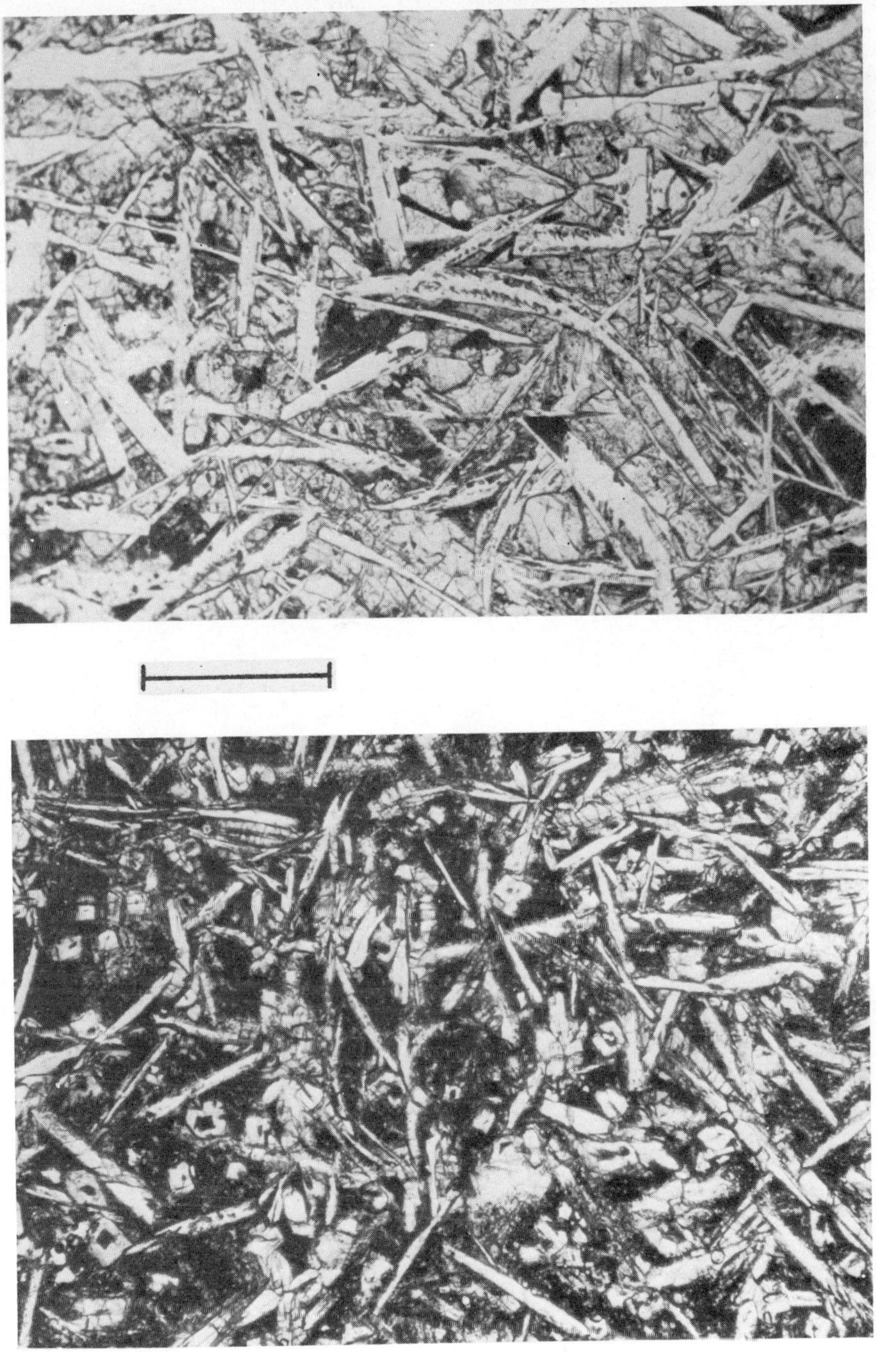

Fig. 1. General view of samples 15382,7 (1a) and 15434,66 (1b); plane light. Plagioclase laths (light) contain glass inclusions in 15382. 15382 also shows light gray pyroxene and minor amounts of dark glass matrix. 15434 has a higher proportion of matrix glass, pyroxene occurs as small hollow crystals, some oriented to show a square cross section with glass in the core. Scale bar = 500 μ.

bear a close resemblance to those of the mare basalts. If we can interpret lunar basaltic textures in terms of cooling histories, the resemblance of these textures to those of mare basalts and to some of the textures produced experimentally by Lofgren (1974) suggests rapid cooling.

Samples 15382 and 15434,66 have textures (Fig. 1) which suggest they represent two extremes of crystallization history for the Apollo 15 KREEP samples; therefore, they were selected for detailed study. 15382, described by Dowty *et al.* (1976), is largely crystalline and shows complex zoning in both pyroxene and plagioclase grains. Tabular plagioclase grains have a relatively homogeneous core which zones gradually and symmetrically outward from more calcic (An_{86}) to more sodic (An_{79}) composition. The core is surrounded by a zone of glass inclusions, a thin layer of higher An content (An_{84}), and finally a more sodic zoned outer rim (Fig. 3). Blocky grains, some of which appear to have

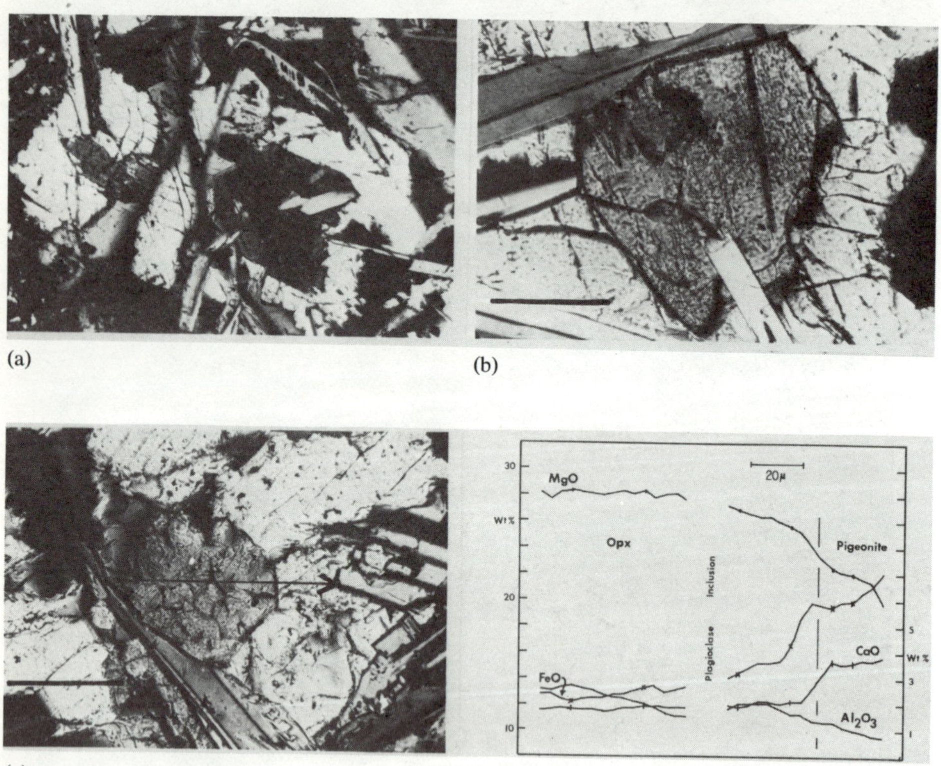

(a) (b) (c) (d)

Fig. 2a,b,c. Pyroxene grains in sample 15382, showing twinned plagioclase laths included in the core orthopyroxene (gray or black), and the sharp but sometimes ragged margin between orthopyroxene and pigeonite (white). Scale bar = 100 μ. (2d) Compositions measured along the traverse across pyroxene (2c) (marked on the photograph). The points indicated along the traverse correspond to points where complete analyses were calculated.

(a)

(b)

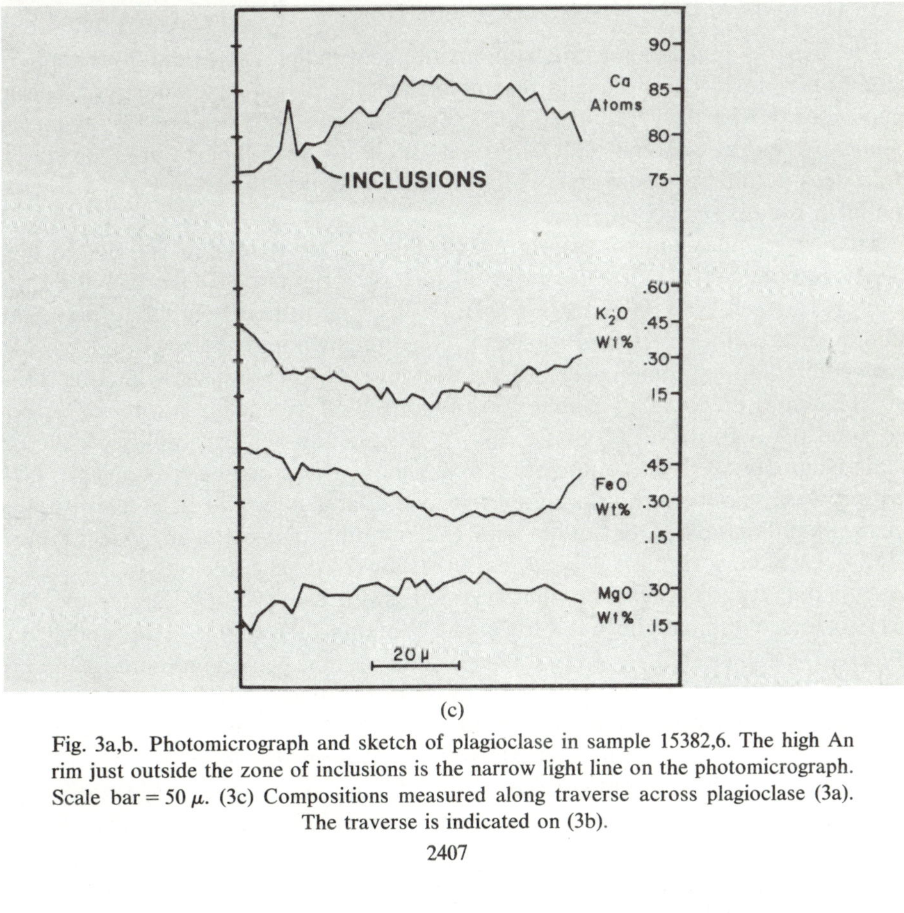

(c)

Fig. 3a,b. Photomicrograph and sketch of plagioclase in sample 15382,6. The high An rim just outside the zone of inclusions is the narrow light line on the photomicrograph. Scale bar = 50 μ. (3c) Compositions measured along traverse across plagioclase (3a). The traverse is indicated on (3b).

been broken, consist of a large volume of the more calcic core material whereas numerous long, often curved, laths containing glass inclusions are compositionally identical to the sodic rims. Many of the pyroxene grains consist of euhedral cores of orthopyroxene rimmed by pigeonite (Fig. 2). Plagioclase laths occur included within the pyroxene cores and are common in the rims. The mesostasis of this sample consists of glass and crystals of ilmenite, cristobalite, and minor amounts of accessory minerals.

15434,66 is the least crystallized of the samples studied. Figure 1 shows the hollow pyroxene grains and partly skeletal plagioclase laths set in a glassy matrix, a texture which suggests very rapid crystallization (i.e., quenching). The pyroxenes are considerably smaller than those in 15382 but the plagioclase laths are comparable in size. The matrix contains crystallites of plagioclase, ilmenite, crystalline SiO_2, FeS, zircon, and whitlockite plus glasses. The glass in the matrix of both 15434,66 and 15382 has separated into two immiscible fractions. The details of the occurrence and chemistry of these glasses are described by Hollister and Crawford (1977a,b).

COMPARISON OF MINERAL COMPOSITIONS IN KREEP BASALTS

The general features of the compositions of plagioclase and pyroxene in these Apollo 15 KREEP basalts resemble those of plagioclase and pyroxene in samples of KREEP from other sites, even though the Apollo 15 rocks are chemically more evolved (McKay and Weill, 1976). There are, however, differences in mineral chemistry which suggest a slightly different history for the Apollo 15 basalts.

First, in comparison to sample 14310, there is no evidence for the earliest crystals which, for 14310, were inferred to have originated at depth within the moon (Crawford and Hollister, 1974). 15382 and other coarsely crystalline samples of Apollo 15 KREEP do have euhedral orthopyroxene grains rimmed by pigeonite and, in some cases, augite. However, in 14310 it was demonstrated that the cores of some pyroxene grains consist of two generations of orthopyroxene (Crawford and Hollister, 1974). No unequivocal examples of such a relationship can be documented for 15382. The earliest orthopyroxene in 14310 was a low-Al variety in marked contrast to the later one, and it frequently was not in crystallographic continuity with the rimming high-Al type. One grain in 15382 does have a similar, low-Al, early orthopyroxene core, but it is possible that this one case represents a sector zoned grain. Dowty (1976) states that the {111} sectors of titanaugite have lower Al contents than {100}, {110}, and {010}. Our analyses show that the bulk of the orthopyroxene, pigeonite, and augite are compositionally indistinguishable from those in 14310 and other KREEP samples for both major and minor elements (see also Dowty *et al.*, 1976; Powell *et al.*, 1973).

For plagioclase, the earliest crystals in 14310 form the cores of large blocky grains and have a distinctly higher An content and lower Fe/(Fe + Mg) ratio than later laths (Fig. 4). No such grains were observed in the Apollo 15 KREEP

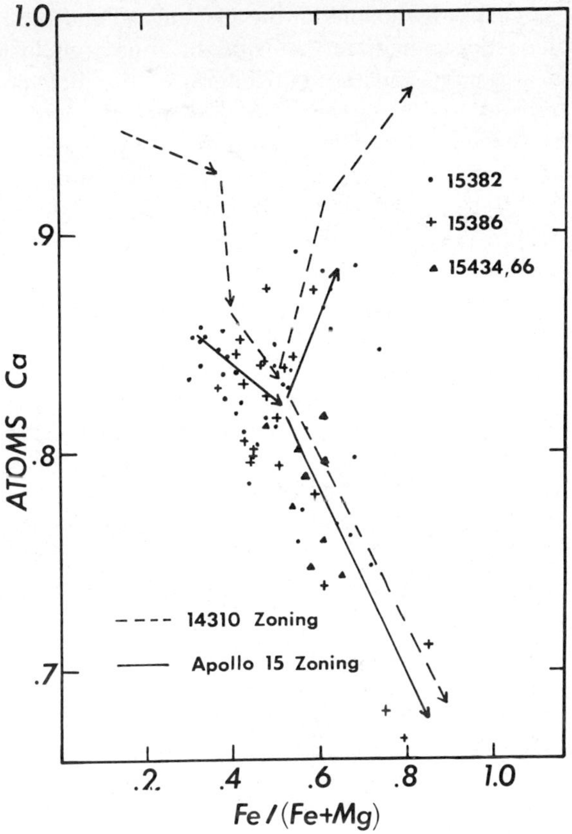

Fig. 4. Compositions of plagioclase grains in Apollo 15 KREEP and sample 14310. Fe/(Fe + Mg) is low in early formed plagioclase and increases as the melt becomes enriched in Fe relative to Mg during crystallization. 14310 plagioclases contain early calcic cores not seen in Apollo 15 samples. Other zoning trends in 14310 are closely paralleled by those of Apollo 15 plagioclases, including the division into sodic and calcic late laths and rims.

basalts. Figure 4 shows the striking similarity in composition of the lath-shaped grains of plagioclase in 14310 and those in 15382, 15386, and, with a more limited range of composition due to crystallization during quenching, of 15434,66. With the exception of the earliest feldspar, it would appear that the process of plagioclase crystallization has been very similar for all the KREEP basalts studied by us.

These observations on plagioclase and pyroxene would suggest that, in general, both the Apollo 15 KREEP and 14310 have broadly similar crystallization histories but that the more chemically evolved Apollo 15 KREEP lacks the evidence interpreted by us to suggest an origin deep within the moon.

A second difference between the samples described here and 14310 lies in the details of the plagioclase zoning in sample 15382. This rock is unique both for the large number of glass inclusions in the plagioclase and the rim of higher An plagioclase which occurs immediately outside the zone of inclusions (Fig. 3). The pattern of chemical zoning and the distribution of the inclusions suggests that the melt was trapped on plagioclase growth surfaces and in re-entrants in plagioclase grains possibly caused by resorption of the margins of crystals. The melt inclusions were trapped by crystallization of the more An-rich plagioclase zone. The zoning differs from 14310 in which glass inclusions do not occur in plagioclase and where plagioclase has a zone of lower, rather than higher, An content rimming the early crystallized cores. Glass inclusions do occur in the plagioclase of KREEP sample 62235, but they lie within a zone of lower An content and they are small, oval shapes rather than the large, often irregular, and interconnecting inclusions seen in 15382. Resorption of the plagioclase, followed by crystallization of more An-rich feldspar suggests a shift of crystal-melt equilibrium due either to a sudden drop in pressure in a volatile-free magma, an increase in temperature due to adiabatic crystallization, or a combination of both. In contrast to 14310 and 62235, the high An content plagioclase zone, preceded by resorption of the plagioclase, suggests a volatile-free crystallization history.

The third major distinction between the Apollo 14 and 16 KREEP studied by us and these Apollo 15 KREEP samples, is the evidence for reduction which accompanies the crystallization of 14310 and 62235, and the lack of that evidence in the Apollo 15 samples. We have previously commented that the separation of Fe–Ni globules from the melt of samples 14310 and 62235 suggests that reduction accompanied crystallization of that melt (Crawford and Hollister, 1974). Reverse zoning of pyroxene in 14310 (Hollister *et al*, 1972; James, 1973) and in 73217 (Crawford, 1975) may also be explained by removal of Fe metal from the melt during crystallization. We found no reverse zoned pyroxenes in the Apollo 15 samples; the amount of native iron crystallized from the melt is miniscule and is limited to the mesostasis. Presumably, therefore, the Apollo 15 KREEP melts were already adjusted to the f_{O_2} of the lunar surface prior to crystallization, whereas the evidence suggests that the Apollo 14 and 16 melts, and the minerals which melted during the formation of 73217, were not.

Discussion of Crystallization Histories: KREEP Basalts

(a) *Crystallization sequence*

The earliest crystals formed in sample 15382 are plagioclase and orthopyroxene. Some of the orthopyroxene includes plagioclase laths (Fig. 2) showing that plagioclase probably crystallized first, although both phases were present in the early melt. The orthopyroxene is separated from the pigeonite which encloses it by a sharp optical and chemical discontinuity (Fig. 2). Plagioclase laths nucleated abundantly at this boundary and are subsequently enveloped by pigeonite which has also nucleated on the orthopyroxene. The plagioclase followed by pigeonite sequence produces characteristic wedge-shaped ter-

Fig. 5. Sample 15434,66 showing the hollow pyroxene grains and the wedge-shaped plagioclase laths. Scale bar = 100 μ.

minations on the plagioclase laths, as is very clearly shown in sample 15434,66 (Fig. 5). Sample 15434,66 lacks the early phase of plagioclase and orthopyroxene crystallization; its crystallization begins with thin plagioclase laths, followed by co-precipitation of plagioclase and pigeonite. Crawford (1973) showed in mare basalt 12021 that rapid plagioclase growth forms very long, thin crystals which grow to more equant shapes by subsequent crystallization on the early cores. Inspection of Fig. 5 shows how this process will produce wedge-shaped ends on the plagioclase laths as plagioclase and pigeonite grow simultaneously. The nucleation of plagioclase on the orthopyroxene margin, combined with the abrupt change in pyroxene Ca content at the orthopyroxene-pigeonite boundary argues for a hiatus in pyroxene crystallization at this stage. Some pyroxenes (e.g., 62235, Crawford and Hollister, 1974) do not show a change in major-element chemistry at the orthopyroxene-pigeonite transition. These may have crystallized continuously without a break at the orthopyroxene-pigeonite transition. Some of the orthopyroxene cores appear to be corroded and irregular (Fig. 2), lending support to the hypothesis that pyroxene was not in equilibrium with the melt during the stage in the crystallization history represented by the growth of these plagioclase laths.

The plagioclase laths which nucleated on the orthopyroxene boundary grew and trapped melt inclusions before the laths were included in pigeonite. This leads to the further speculation that the hiatus in pyroxene crystallization

Table 2. Crystallization sequence of plagioclase and pyroxene in 15382 and 15434,66.

Core plagioclase—plagioclase laths ⁄ calcic late laths and rims
\+ ﹨ sodic late laths and rims
inclusions

Orthopyroxene pigeonite + augite

├──── 15382 ────┤

├── 15434,66 ──┤

accompanied an episode during which the original melt was subject to a change in physical conditions (emplacement at the surface?). Pyroxene reappeared as a crystallizing phase as a response by the liquid to the new surroundings following a crystallization stage involving only plagioclase. Table 2 summarizes the proposed crystallization history.

(b) *Effects of liquid immiscibility*

These Apollo 15 samples show evidence of silicate liquid immiscibility (Hollister and Crawford, 1977a). Immiscibility in KREEP-rich samples has also been documented for sample 14310 (Roedder and Weiblen, 1972), for 15272, (Switzer, 1972), and for lithic fragments in Apollo 16 soils (Powell *et al.*, 1975). In 15434,66 the basalt was quenched apparently at the point of separation of the two immiscible phases. However, the more slowly cooled basalts, such as 15382, continued to crystallize plagioclase and pyroxene, as well as late mesostasis minerals, following the separation of the two melt phases. Crawford and Hollister (1974) and Ridley *et al.* (1972) noted the connection between An content of the late crystallizing plagioclase in sample 14310 and the composition of the adjoining mosostasis. Silicate liquid immiscibility is probably the cause of the two distinct zoning trends in the later plagioclase (Fig. 4). The onset of silicate liquid immiscibility combined with rapid, probably disequilibrium crystallization may permit local control of the mineral zoning trends. The final cooling of these samples was evidently a rapid quench, as glass is preserved in the groundmass. Therefore, late crystallization was not necessarily an equilibrium process and the immiscible melts, once formed, need not have remained in equilibrium with each other or with crystals formed from those melts. Thus plagioclase which crystallized from high Fe melt became enriched in An relative to plagioclase crystallizing in contact with high silica liquids, as a consequence of the efficient partition of alkalies into the high silica melt phase. The split of late plagioclase into two different compositions (Table 2, Fig. 4) occurred for those portions of the grains outside the zone of glass inclusions in 15382. This suggests the melts were not immiscible until at least the stage in which melt was entrapped in the plagioclase grains.

We have not been able to determine whether there are measurable chemical effects of liquid immiscibility in pyroxene. The crystallization stage at which these effects might be found may nearly coincide with the change from orthopyroxene to pigeonite. Whether augite lamellae which form within the pigeonite in some grains close to the pigeonite-orthopyroxene boundary are related to crystallization from high iron rather than high silica melt is an intriguing speculation, but it cannot be proven. In any event, as outlined above, the boundary between the orthopyroxene and the pigeonite appears to represent an actual discontinuity in pyroxene crystallization, rather than simply the point during the crystallization history at which the composition of the melt crossed from the field of orthopyroxene to the field of pigeonite stability. Arguments proposed by Hollister and Crawford (1977b) suggest that immiscibility was immediately preceded by a period of crystallization of plagioclase only from the melt. The mineral intergrowth relationships would place this pre-immiscibility crystallization of plagioclase prior to the formation of pigeonite.

Modal analyses of samples 15382 and 15434,66 are presented in Table 3. If it is assumed that immiscibility in 15382 followed the formation of the core plagioclase (inside the zone of glass inclusions) and of the orthopyroxene cores, then immiscibility must have taken place in this rock when it was only 20% crystalline. Plagioclase and pyroxene in sample 15434,66 formed prior to immiscibility. This sample, therefore, was approximately 58% crystalline prior to immiscibility. Both of these figures document our contention that immiscibility occurred much earlier in these samples than has previously been suggested for lunar rocks.

In these samples the high iron immiscible glasses generally contain a significantly greater proportion of quench crystals than the high silica glasses. Thus crystallization of the high iron glass will produce areas enriched in plagioclase and pyroxene relative to the rest of the mesostasis and will also tend to deplete the overall amount of high iron melt remaining within the sample. The high silica melt is considerably more viscous and will probably not crystallize as readily. Modal estimates of the relative amounts of the two types of melt in samples which have experienced post-immiscibility disequilibrium crystallization should,

Table 3. Modes of samples 15382,6 and 15434,66.

15382,6	%	15434,66	%
Core plagioclase	8.7		
Rim plagioclase and laths	24.2	Plagioclase	20.7
Orthopyroxene	9.2		
Pigeonite	33.3	Pigeonite	35.4
Matrix minerals	10.5	Matrix minerals	3.5
Opaque	6.4	Opaque	7.2
Glass	7.7	Glass	31.8

therefore, result in higher proportions of high silica to high iron glass than would be the case at the onset of immiscibility. In sample 15434,66, which has not crystallized significantly following immiscibility, the high iron glass is the most abundant phase (Hollister and Crawford, 1977b).

(c) *Siderophile elements*

The miniscule amount of Fe–Ni metal and sulphide minerals and their very late crystallization, in contrast to 14310, suggest an alternative explanation to the one offered by Gros *et al.* (1976) for the low content of siderophile elements in Apollo 15 basalts. If these KREEP samples are derived from other more primitive melts, as suggested by McKay and Weill (1976), it is certainly plausible that separation of immiscible Fe metal and sulphide globules could happen early during that fractionation. In all but the Apollo 15 samples, early crystallization of Fe metal accompanies the crystallization of the earliest silicates.

It has also been demonstrated that green glass, believed by many to represent a true lunar magma (Heiken *et al.*, 1974; Meyer *et al.*, 1975) has fairly high contents of siderophile as well as volatile elements. Butler and Meyer (1976) note that Ni correlates with the high volatile element content of the surfaces of green glass spherules, and Wasson *et al.* (1976) document relatively high Au and Ir (both siderophile elements) concentrations on the surfaces of both green and orange glass spheres. Butler and Meyer (1976) and Wasson *et al.* (1976) ascribe the enrichment of these elements on the surfaces of the glass spheres to condensation of a vapor phase which accompanied the erupting magma. For this case, at least, there seems to be no argument that volatile and siderophile elements may be native to the lunar interior. Ashwal (1975) cites evidence that siderophile elements in plutonic rocks 67955 and 77017 also do not appear to be reasonably explained by a model of meteoritic contamination. Thus, we argue that 14310 is close to a KREEP primitive magma, which retained its siderophile and volatile elements until it erupted, whereas the Apollo 15 basalts have undergone fractional crystallization. During that crystallization, precipitation of an early Fe metal phase depleted the melt of its siderophile elements.

Origin and Evolution of KREEP

The near coincidence of the crystallization age of most KREEP samples (3.9 AE) to the similar dates from other samples which are interpreted to represent intense bombardment of the lunar surface suggests a cause and effect relationship. Whereas many models (e.g., Gast, 1972; Shih, 1976; McKay and Weill, 1976 and earlier papers) suggest that KREEP can be derived by partial melting of the lunar interior and provide trace-element partition data to support this model, no reason has been given for the melting episode at 3.9 AE. In addition, the chemistry of many KREEP samples shows that they cannot be directly related to low pressure multiply-saturated liquids on the pseudo-ternary liquidus diagrams of Walker *et al.* (1973). A model to explain the production of KREEP melts of various compositions which have all formed at approximately the same time, and consistent with the other data for these rocks, involves rapid crustal unloading by impact resulting in a decrease of the solidus temperature of material below the large impact sites (e.g., Imbrium) to below the ambient temperatures. This would result in rapid disequilibrium melting and extrusion of magma generated from the lowest-melting-point fraction of the underlying material. This model is consistent with the known occurrence of high KREEP rocks around the Imbrium and Procellarum basins. A similar model was proposed by Crawford (1975) in her interpretation of the partial remelting and subsequent rapid crystallization of sample 73217.

Chemical differences in KREEP samples may represent different degrees of partial melting and they may also be related to different extrusion histories. We interpret more "primitive" KREEP melts, like 14310, to have been generated at depth and to have risen rapidly to the surface with little interaction with surrounding rocks during their ascent. The more "chemically evolved" Apollo 15 KREEP has been interpreted by McKay and Weill (1976) to have originated by fractional crystallization from more primitive KREEP. We present evidence above to suggest that the f_{O_2} of these rocks equilibrated with lunar surface f_{O_2} values prior to crystallization. Fractional crystallization and equilibration to near-surface conditions would be expected in a magma which rose more slowly through the crust. A slower ascent would permit early crystallizing phases to separate from the rising magma; the evidence for a possible lower volatile content of the Apollo 15 KREEP could also result from gradual rather than rapid rise to the surface. Once the melt is actually emplaced at the surface, the outer parts of the flows will quench rapidly giving rise to vitrophyres like 15434,66. The interiors of the flows, preserving some of the crystals generated during the ascent, may heat up due to the heat of crystallization generated within the flow until the heat generated is balanced by heat loss to the surroundings. Both the ascent and the possible temperature increase could cause a hiatus in pyroxene crystallization, a resorption of plagioclase and precipitation of a more calcic plagioclase, before pyroxene crystallization resumes.

For all KREEP samples now observed at the surface, immiscibility occurred during crystallization following eruption. Because silicate liquid immiscibility results in the separation of K and P from each other, K going into Si-rich melt and P into the Fe-rich melt, creation and separation of immiscible silicate liquids cannot have happened prior to eruption. Creation of immiscible metal and/or sulphide liquids happens earlier in the crystallization histories of these rocks, however, and the Apollo 15 KREEP compositions (depleted in siderophile elements) suggest the immiscible metal liquids have separated from the silicate magma for some KREEP basalts. A melt with KREEP affinities which does not erupt to the surface, but crystallizes to the point of immiscibility within the lunar crust can result in separation of the Fe-rich melt from the Si-rich melt by gravitational settling of the dense (2.99 g/cc) Fe-rich melt and fractionation of the Si-rich component into the higher zones of the magma chamber. Hollister and Crawford (1977) have suggested that this process could result in the formation of the source region of the Apollo 12 and 15 basalts. This generation of secondary magmas, through silicate liquid immiscibility, from primary KREEP may have happened at any time during lunar crystal evolution, possibly as early as 4.4 AE, the earliest isotopic record of KREEP as a distinct magma type.

Acknowledgments—This study was supported by NASA grant NSF-9005. Comments and suggestions on the manuscript by B. N. Powell and W. I. Ridley are greatly appreciated.

REFERENCES

Ashwal L. D. (1975) Petrologic evidence for a plutonic igneous origin of anorthositic norite clasts in 67955 and 77017. *Proc. Lunar Sci. Conf. 6th*, p. 221–230.

Butler P., Jr. and Meyer C., Jr. (1976) Sulfur prevails in coatings on glass droplets: Apollo 15 green and brown glasses and Apollo 17 orange and black (devitrified) glasses. *Proc. Lunar Sci. Conf. 7th*, p. 1561–1581.

Crawford M. L. (1973) Crystallization of plagioclase in mare basalts. *Proc. Lunar Sci. Conf. 4th*, p. 705–717.

Crawford M. L. (1975) Magma genesis by *in situ* melting within the lunar crust. *Proc. Lunar Sci. Conf. 6th*, p. 249–261.

Crawford M. L. and Hollister L. L. (1974) KREEP basalt: A possible partial melt from the lunar interior. *Proc. Lunar Sci. Conf. 5th*, p. 399–419.

Dence M. R. (1971) Impact melts. *J. Geophys. Res.* **76**, 5552–5565.

Dowty E. (1976) Crystal structure and crystal growth: II. Sector zoning in minerals. *Amer. Mineral.* **61**, 460–469.

Dowty E., Keil K., Prinz M., Gros J., and Takahashi H. (1976) Meteorite-free Apollo 15 crystalline KREEP. *Proc. Lunar Sci. Conf. 7th*, p. 1833–1844.

Gast P. (1972) The chemical composition and structure of the moon. *The Moon* **5**, 121–148.

Gros J., Takahashi H., Hertogen J., Morgan J. W., and Anders E. (1976) Compositions of the projectiles that bombarded the lunar highlands. *Proc. Lunar Sci. Conf. 7th*, p. 2403–2425.

Heiken G. H., McKay D. S., and Brown R. W. (1974) Lunar deposits of possible pyroclastic origin. *Geochem. Cosmochim. Acta* **38**, 1703–1718.

Hollister L. S. and Crawford M. L. (1977a) Evidence for a direct genetic connection between Apollo 15 KREEP and Apollo 12 and 15 basalts (abstract). In *Lunar Science VIII*, p. 452–454. The Lunar Science Institute, Houston.

Hollister L. S. and Crawford M. L. (1977b) On the origin of high Fe mare basalts and KREEP basalts. *Proc. Lunar Sci. Conf. 8th*. This volume.

Hollister L. S., Trzcienski W. E., Jr., Dymek R., Kulick C. G., Weigand P. W., and Hargraves R. B. (1972) Igneous fragment 14310,21 and the origin of the mare basalts (abstract). In *Lunar Science VI*, p. 386–388. The Lunar Science Institute, Houston.

James O. B. (1973) Crystallization history of lunar feldspathic basalt 14310. *U.S. Geol. Survey Prof. Paper 84*, 29 pp.

Lofgren G. (1974) An experimental study of plagioclase crystal morphology: Isothermal crystallization. *Amer. J. Sci.* **274**, 243–273.

McKay G. A. and Weill D. F. (1976) Petrogenesis of KREEP. *Proc. Lunar Sci. Conf. 7th*, p. 2427–2447.

Meyer C., Jr. (1977) Petrology, mineralogy, and chemistry of KREEP basalt. *Phys. Chem. Earth* **10**. In press.

Meyer C., Jr., McKay D. S., Anderson D. H., and Butler P., Jr., (1975) The source of sublimates on the Apollo 15 green and Apollo 17 orange glass samples. *Proc. Lunar Sci. Conf. 6th*, p. 1673–1699.

Powell B. N., Aitken F. K., and Weiblen P. W. (1973) Classification, distribution, and origin of lithic fragments from the Hadley-Apennine region. *Proc. Lunar Sci. Conf. 4th*, p. 445–460.

Powell B. N., Dungan M. A., and Weiblen P. W. (1975) Apollo 16 feldspathic melt rocks: Clues to the magmatic history of the lunar crust. *Proc. Lunar Sci. Conf. 6th*, p. 415–433.

Ridley W. I., Brett R., Williams R. J., and Takeda H. (1972) Petrology of Fra Mauro basalt 14310. *Proc. Lunar Sci. Conf. 3rd*, p. 159–170.

Roedder E. and Weiblen P. W. (1972) Petrographic features and petrologic significance of melt inclusions in Apollo 14 and 15 rocks. *Proc. Lunar Sci. Conf. 3rd*, p. 251–279.

Rutherford M. J., Hess P. C., Ryerson F. J., Campbell H. W., and Dick P. A. (1976) The chemistry, origin and petrogenetic implications of lunar granite and monzonite. *Proc. Lunar Sci. Conf. 7th*, p. 1723–1740.

Shih C. (1976) On the origin of KREEP basalt (abstract). In *Lunar Science VII*, p. 800–802. The Lunar Science Institute, Houston.

Simonds C. (1975) Thermal regimes in impact melts and the petrology of the Apollo 17 Station 6 boulder. *Proc. Lunar Sci. Conf. 6th*, p. 641–672.

Switzer G. S. (1972) Composition of three glass phases in an Apollo 15 basalt fragment. *Smithsonian Contrib. Earth Sci.*, no. 14, p. 25–30.

Walker D., Grove T. L., Longhi J., Stolper E. M., and Hays J. F. (1973). Origin of lunar feldspathic rocks. *Earth Planet. Sci. Lett.* **20**, 325–336.

Wasson J. T., Boynton W. V., Kallemeyer G. W., Sandberg L. L., and Wai C. M. (1976) Volatile compounds released during lunar lava fountaining. *Proc. Lunar Sci. Conf. 7th*, p. 1583–1595.

Proc. Lunar Sci. Conf. 8th (1977), p. 2419–2432.
Printed in the United States of America

Melt immiscibility in Apollo 15 KREEP: Origin of Fe-rich mare basalts

LINCOLN S. HOLLISTER

Department of Geological and Geophysical Sciences, Princeton University, Princeton, New Jersey 08540

MARIA LUISA CRAWFORD

Department of Geology, Bryn Mawr College, Bryn Mawr, Pennsylvania 19010

Abstract—We describe the chemistry and occurrence of immiscible silicate liquids in two KREEP-rich Apollo 15 basalts. The two samples have different cooling histories but the same composition. In sample 15434,66, silicate liquid immiscibility (SLI) occurred prior to crystallization of 58% of the rock. In sample 15382, SLI occurred after 20% had crystallized. In order to enter the field of SLI, essentially plagioclase only had to crystallize during part of each rock's crystallization history, probably under disequilibrium conditions. Shortly after SLI had occurred in 15434, zircon, plagioclase, FeS, a high SiO_2 phase, merrillite (whitlockite), and ilmenite, but not pyroxene, were liquidus phases for each melt.

With fractionation, TiO_2, P_2O_5, and FeO/MgO increase in the Fe-rich immiscible melts. The earliest Fe-rich immiscible melts are chemically similar to the average compositions of several groups of Fe-rich Apollo 12 and 15 mare basalts. It is suggested that SLI may play an important role in the formation of the source regions of Fe-rich mare basalts.

INTRODUCTION

SILICATE LIQUID IMMISCIBILITY occurs in the late stages of crystallization of lunar basalts (Roedder and Weiblen, 1970, 1971, 1972; Powell *et al.*, 1975; Switzer, 1972; Weiblen and Roedder, 1973). Hess *et al.* (1975) and Rutherford *et al.* (1976) have attempted to evaluate the possible importance of this phenomenon through experimental investigation of the fractionation trends of late-stage immiscible melts in lunar basalts. They confirm that most, if not all, "granite" on the moon, including portions of rock 12013, originates as an immiscible melt phase. Quick and Albee (1976) and Dymek *et al.* (1977) have applied these results toward understanding the compositions of the light and dark regions of sample 12013. Hess *et al.* (1975) also noted the similarities of some Fe-rich immiscible liquids in feldspathic basalts to Ti-rich mare basalts and suggested a possible genetic connection.

The purpose of the present study was to determine, in KREEP-rich crystalline rocks, the earliest stage that silicate liquid immiscibility could be documented and to describe from direct observation the chemical trends of each melt with crystallization. We selected KREEP-rich rocks for this study because it was in such a rock (14310) that the largest amount of residual liquid showing silicate liquid immiscibility was documented previously (at least 12%; Roedder and Weiblen, 1972). We also wished to compare and contrast the crystallization history, including silicate liquid immiscibility, of Apollo 15 KREEP-rich rocks with our studies on 14310 and 62235 (Crawford and Hollister, 1974).

The crystallization history of the two Apollo 15 KREEP-rich rocks we have studied, 15382 and 15434,66, is given in Crawford and Hollister (1977). In the present paper, we describe the immiscible melt compositions, now glasses, with particular reference to the earliest Fe-rich immiscible melts, whose compositions are similar to some Fe-rich mare basalt compositions. We suggest that silicate liquid immiscibility may play an important role in the generation of Fe-rich Apollo 12 and 15 mare basalts, as well as "granitic" lunar rocks.

ANALYTICAL TECHNIQUE

All data on the glass inclusions, and the bulk rock data, were obtained with an automated energy-dispersive analysis (EDA) system with a 167 eV resolution, utilizing the electron gun and optics of an electron microprobe normally used for X-ray dispersive analysis (XDA). The EDA system is advantageous for broad beam analyses because the counting rate is not sensitive to the effects inherent in XDA analysis by departure of analysis area from the circumference of the Rowland circle. The constraints on operating conditions of EDA, which are necessary in order to minimize pulse pile-up effects and deadtime, are those which are optimal for obtaining a minimum beam diameter while maintaining stable electron gun operating conditions. A minimum beam diameter under stable conditions is required to obtain representative analyses of the inhomogeneous glass inclusions, many of which are less than 3 μm across on the exposed surface. Obtaining representative analyses using this technique involves some sacrifice in accuracy, however. No result on an individual element is accurate to better than 3% of the amount reported; uncertainties of some results, e.g., on phosphorous, potassium, and for the lower values of magnesium, exceed 5% but are probably better than 10%. Results are more inaccurate if they are less than 0.3 wt.%; Na_2O is also not reported because the accuracy of the detector at this atomic number is poor if Na_2O is less than 1 wt.%.

The analyses are corrected for absorption, atomic number, and fluorescence. Standards used are kaersutite for Fe, Ti, K, and Mg; bytownite for Ca, Al, and Si; and schreibersite for P. In general, the standards and unknowns had approximately similar total counting rates, making deadtime corrections unnecessary; the count rate was sufficiently low (1200 per sec) that pulse pile-up was not a significant factor even for Si in the high Si glasses. Counting times were approximately 100 sec in order to reduce errors from statistical counting fluctuations without significant contamination-spot buildup.

Representative analyses are given in Tables 1, 2, and 3A, B, and C. Typical compositions for coexisting glasses found as inclusions in feldspar are given in Table 3A, and for glasses in the interstices in Table 3B; the analyses of the earliest immiscible Fe-rich glasses are given in Table 3C. In total, we made 52 Fe-rich glass analyses, 25 Si-rich glass analyses, 5 broad beam glass matrix analyses, and 6 total rock analyses.

OCCURRENCE OF IMMISCIBLE MELTS IN 15382 AND 15434

The two samples, 15382 and 15434, have nearly identical bulk compositions (Table 1). Therefore, the obvious textural differences shown in Fig. 1 result from different cooling histories. As discussed in Crawford and Hollister (1977), 15434,66 probably cooled more rapidly and from a wholly liquid melt, whereas

Table 1. Major-element chemistry of KREEP basalts bulk analyses (wt.%).

	15382,6,7,17[1]	15434,63,66[1]	15382,7[2]		Mode (%) 15382,6	15434,66
MgO	7.01	6.75	7.1	Quench plag.	24.2	22.0
Al_2O_3	18.44	17.68	17.8	Core plag.	8.7	0
SiO_2	52.53	51.89	52.4	Quench px.	33.3	35.4
K_2O	.46	0.50	0.57	Core px.	9.2	0
CaO	10.30	10.29	9.9	Opaque	6.4	7.2
TiO_2	1.29	1.77	1.78	Matrix min.	10.5	3.5
FeO	8.27	9.42	8.6	Glass	7.7	31.8

[1] Average of three analyses each; normalized to 98.3% for purposes of comparison, to take account of approximately 1.7% Cr_2O_3, P_2O_5, and Na_2O (Dowty *et al.*, 1976). Analyses by broad beam EDA.
[2] Analysis by Dowty *et al.* (1976), by broad beam XDA.

Table 2. Analyses of matrix glasses, sample 15434,66.

	1	2	3	4	5	6
MgO	8.49	8.01	4.97	3.85	2.99	.71
Al_2O_3	4.92	4.70	7.52	10.87	10.40	11.98
SiO_2	48.78	45.50	45.93	47.80	48.97	76.07
P_2O_5	1.00	.08	2.30	2.68	4.92	—
K_2O	.30	.13	.95	1.08	1.54	4.81
CaO	10.26	11.41	11.11	10.47	11.89	1.75
TiO_2	1.77	3.13	4.84	4.58	3.21	.72
FeO	26.03	25.07	21.90	17.58	13.08	2.35
Total	101.55	101.03	99.52	98.91	97.00	98.39

Analyses 1–5, Fe-rich glass.
Analysis 6, Si-rich glass.

Table 3A. Analyses of coexisting glasses in inclusions in plagioclase, sample 15382,7.

	B8	B10	A5	A4	8-2	8-1	8-3	8-4
MgO	5.34	.76	4.02	—	2.86	.28	.38	.21
Al_2O_3	2.14	6.39	5.25	10.38	4.91	12.20	6.65	11.99
SiO_2	36.56	71.89	36.76	78.17	41.01	81.72	44.76	79.09
P_2O_5	2.03	—	7.15	—	9.60	.21	2.62	.07
K_2O	—	3.09	.66	6.26	2.53	3.91	1.43	5.95
CaO	10.16	3.38	15.93	.58	9.42	.85	8.47	.72
TiO_2	8.88	2.75	2.01	.71	7.04	.67	5.37	.82
FeO	34.15	9.28	27.95	.66	18.49	1.84	30.60	1.73
Total	99.26	97.54	99.73	96.76	95.86	101.47	100.28	100.51

15382 contained early crystals (core phases, Table 1) which acted as nucleation centers for later crystal growth.

In sample 15434,66, the glass matrix (Fig. 2) contains microcrystals of ilmenite, tiny rounded globs of FeS, phosphate (both rare-earth element bearing whitlockite and apatite are present), zircon (confirmed by analysis), a silica mineral, and plagioclase (An_{84}). There are also two glass phases, one high in Fe,

Table 3B. Analyses of glasses in interstices, sample 15382,7.

	A2	A1	A3	D1	D2	D3	B4	B5	A10
MgO	2.57	—	—	.50	—	.24	.39	—	—
Al_2O_3	1.11	13.36	11.46	3.77	8.41	10.53	3.65	14.50	11.16
SiO_2	33.62	72.88	67.50	23.55	80.44	67.96	31.57	70.99	77.00
P_2O_5	11.07	—	.77	9.03	—	.69	6.82	—	—
K_2O	—	6.87	5.96	.99	2.24	4.02	.73	5.43	5.88
CaO	13.69	1.75	2.94	10.29	1.96	3.48	4.48	2.60	.64
TiO_2	.72	.63	1.23	15.96	.87	1.65	10.16	1.49	.47
FeO	34.78	.80	7.28	36.92	.83	5.49	39.06	0.30	.96
Total	97.56	96.29	97.14	101.01	94.75	94.06	96.86	95.31	96.11

A2, D1, B4: Fe-rich blebs in interstial high Si glass.
A1, D2, B5, A10: Si-rich interstial glass.
A3, D3: broad beam glass analyses which include A2 + A1 and D1 + D2, respectively; also included are indeterminate amounts of microcrystalline phases.

Table 3C. Early Fe-rich immiscible melts.

Sample	15382,7				67942,1[1]	14310,5[2]	15272,11[3]
Analysis #	D9	C1	D5	A8	1619-11		3
MgO	9.04	8.23	5.05	7.05	9.4	6.3	4.7
Al_2O_3	4.06	2.30	9.99	2.18	5.6	3.9	8.0
SiO_2	44.82	37.60	40.05	41.19	41.1	37.6	43.4
P_2O_5	.74	< .05	1.86	1.48	3.5	3.1	0.8
K_2O	0	< .05	.08	.09	.04	.32	0.8
CaO	8.26	13.17	13.62	4.43	10.4	11.1	10.3
TiO_2	3.28	8.74	7.60	4.85	6.4	6.0	6.3
FeO	29.74	28.48	22.63	37.50	21.7	30.6	24.6
Total	99.94	98.52	100.88	98.77	98.14	98.92	98.9

[1] Powell *et al.* (1975).
[2] Roedder and Weiblen (1972); average of 6.
[3] Switzer (1972).

the other high in Si and K. The Fe-rich glass phase contains a lower fraction of the crystalline phases (Fig. 2) and is more abundant than the high Si and K glass. Pyroxene is not present as a crystalline phase in either glass. Analyses of both glass phases are given in Table 2, and illustrated in Fig. 3.

The more slowly cooled rock 15382 contains glassy interstitial matrix areas (interstices) and glass inclusions within the feldspar crystals. Both of these occurrences are illustrated in Fig. 4. The interstices contain mainly a Si-rich glass with small, rounded blebs of Fe-rich glass. This feature has been described in residual glasses of many lunar samples.

If a feldspar entraps melt inclusions, as it did in sample 15382, the inclusions need not represent the average melt composition in the matrix at the time of entrapment, especially if immiscibility of the melt has already occurred. The

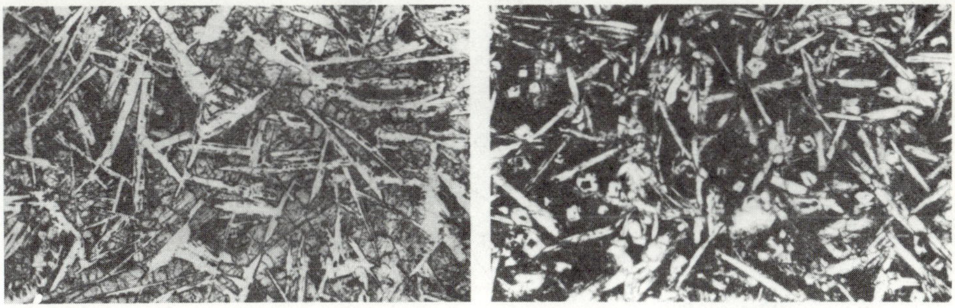

Fig. 1. Photomicrographs (plane light) of 15382,7 (left) and 15434,66 (right). Width of photos 1.5 mm. Region D, Fig. 4, is located in upper left center of 15382,7.

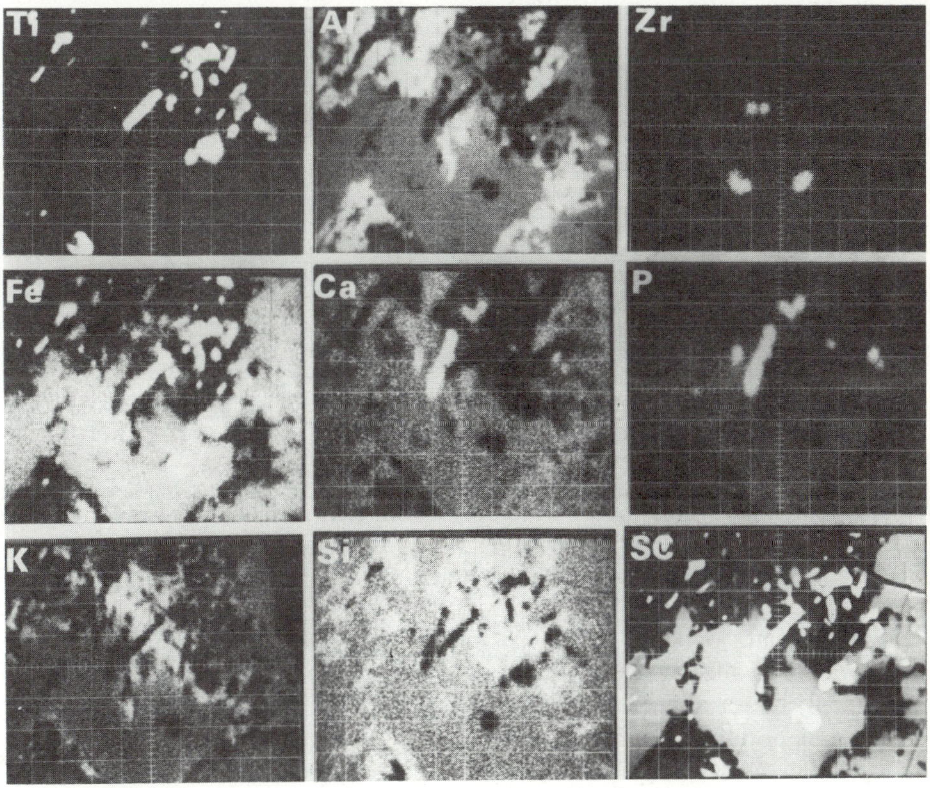

Fig. 2. X-ray area scans for Ti, Al, Zr, Fe, Ca, P, K, Si, and sample current display, of immiscible glasses and crystalline phases of matrix in sample 15434,66. Scale is 10 μm between grid lines. Al-rich phase, plagioclase; K (and Si)-rich phase, Si-rich glass; Si (only)-rich phase, probably cristobalite; Ca + P-rich phase, merrillite and apatite; Zr-rich phase, zircon; very Fe-rich phase, FeS; Fe-rich phase (and Si, K, P, Ca, Al, and Ti-bearing), Fe-rich glass; Ti-rich phase, ilmenite.

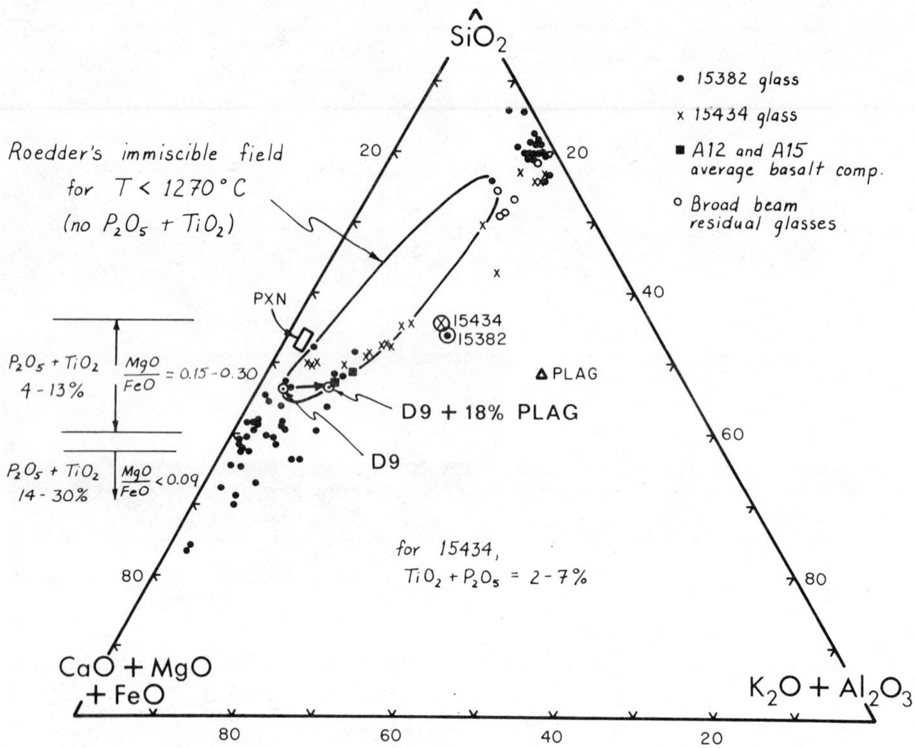

Fig. 3. Immiscible glass compositions in sample 15382 and 15434,66 plotted on diagram showing region of silicate liquid immiscibility (Weiblen and Roedder, 1973). (×), sample 15434; (●), sample 15382; (○), broad beam analyses of interstices, sample 15382. Bulk compositions of samples, plagioclase (Plag, An_{85}), and pyroxene (PXN) are shown. See text.

feldspar can conceivably incorporate only the Si-rich melt, only Fe-rich melt, or both melts simultaneously. Each liquid which is incorporated into the feldspar can, after incorporation, continue to separate into immiscible melts. Inclusions in the feldspar in sample 15382 tend to be either Fe-rich melt alone or Fe-rich melt plus subordinate amounts of Si-rich melt. This might be due to preferential incorporation of the Fe-rich melt, or, as is true for 15434, that the Fe-rich melt was initially more abundant. In some crystals (Fig. 5) sub-equal amounts of Fe-rich and Si-rich melts occur in the inclusions in the feldspar. The greater abundance of Si-rich glass in the matrix, in contrast to the relative proportions observed as inclusions, may reflect a greater tendency of the Fe-rich melt in the matrix to crystallize to pyroxene and ilmenite. It would clearly be difficult to estimate from the mode the relative amount of two immiscible melts if one of the melts had a greater tendency to crystallize for a given cooling rate.

In 15382, ilmenite, cristobalite, and pyroxene have crystallized in the matrix, reducing the amount of remaining residual glasses. Local crystallization has a

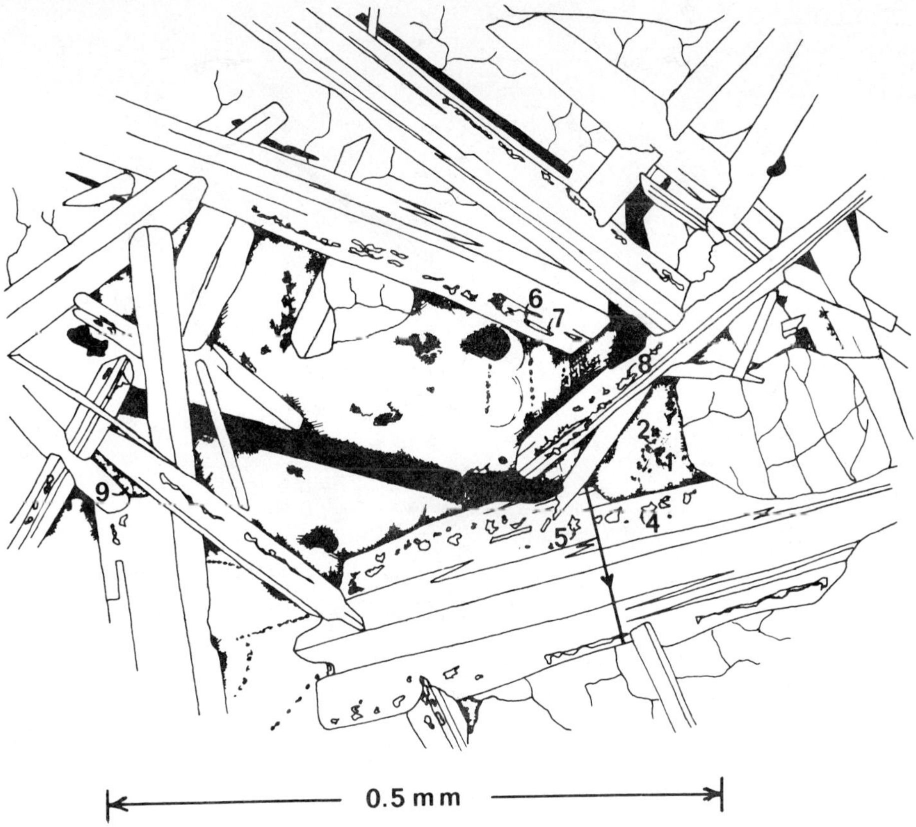

Fig. 4. Sketch of region D, sample 15382,7. Numbers refer to analysis points, some of which are given in Tables. Shown are plagioclase, pyroxene (cross-fractured), cristobalite (dendritic fracture pattern), ilmenite (large, tabular, black), and glass (dark = Fe-rich glass; light = Si-rich glass).

marked effect on the composition of a nearby glass; e.g., if ilmenite is present in glass, the glass is much lower in Ti than glass which does not contain ilmenite (compare analyses A2 and D1, Table 3B). Phases do appear to crystallize easily in both the immiscible melts even when very rapidly cooled. The X-ray area scans (Fig. 2) of a portion of 15434 clearly show the two immiscible melts plus several crystalline phases. The coexistence of each of these minerals with both glasses is apparent, as would be expected under equilibrium conditions at the point of silicate liquid immiscibility. The distribution of minerals in the X-ray area scans suggests nucleation of phosphate and ilmenite occurred near where the Si-rich melt formed; but most growth must be from the Fe-rich melt, which contains the major constituents of these phases (e.g., P_2O_5 and TiO_2).

As discussed in detail in Crawford and Hollister (1977), a minimum estimate of the amount of crystallization which took place prior to entrapment of the glass inclusions in 15382 is about 20%. This estimate is based on the modal

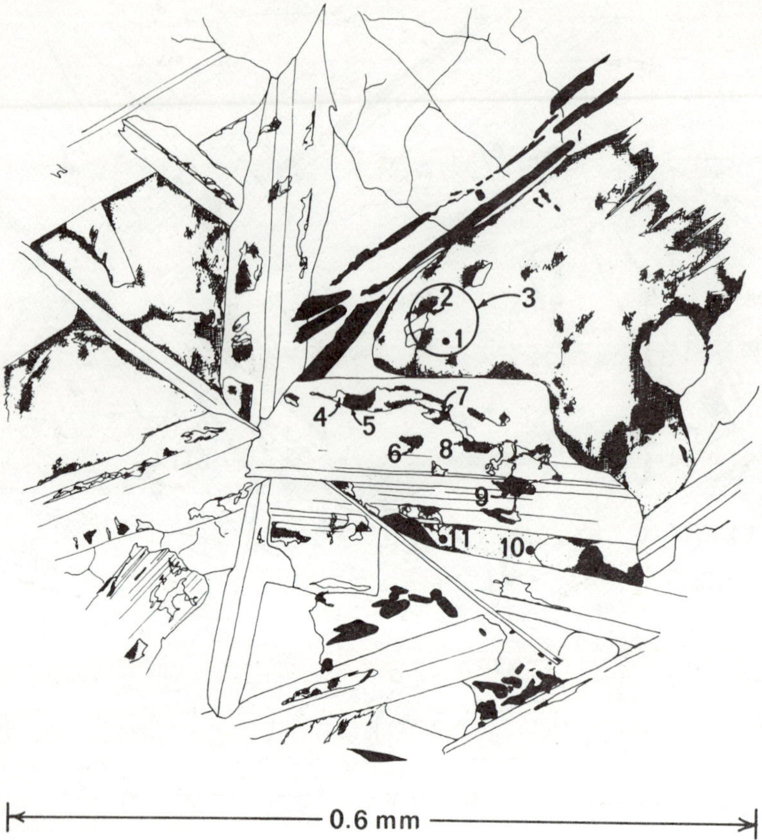

Fig. 5. Sketch of region A, sample 15382,7. Numbers refer to analysis points, some of which are given in Tables. Shown are plagioclase, pyroxene (cross-fractured), cristobalite (dentritic fracture pattern), ilmenite (large, tabular, black), and glass (dark = Fe-rich glass; light = Si-rich glass).

amount of plagioclase inside the zone of plagioclase containing melt inclusions, and the modal amount of orthopyroxene (Crawford and Hollister, 1977). An estimate for the maximum amount of crystallization which had occurred prior to the separation of the two melt phases, assuming plagioclase and pyroxene were the only crystalline phases on the liquidus, may be taken from the mode of sample 15434 (Table 1): about 58%. Based on their nearly identical compositions, it is concluded that immiscibility occurred in these samples after 20–58% of crystallization.

Results of Analyses

The data are plotted on a compositional triangle with SiO_2, $CaO + FeO + MgO$, and $K_2O + Al_2O_3$ at the corners (Fig. 3). This diagram, except for the absence of Na_2O from the $K_2O + Al_2O_3$ corner, is the same used by Weiblen and

Roedder (1973) to illustrate late-stage immiscible melts of the mare basalts. The diagram is adapted from Grieg (1927).

The ranges of $P_2O_5+TiO_2$ and of MgO/FeO for the Fe-rich immiscible glasses in 15382 are indicated on the left side of Fig. 3. Analyses of Fe-rich glass which plot within the triangle and below the lower horizontal line contain 14–30% combined $P_2O_5+TiO_2$, and have MgO/FeO less than 0.09. For analyses plotting between the other two horizontal lines, $P_2O_5+TiO_2$ ranges between 4 and 13%, and MgO/FeO is between 0.15 and 0.30. $TiO_2+P_2O_5$ for all glasses in 15434 is between 2 and 7%. Broad beam analysis of residual interstitial matrix areas are illustrated in Fig. 3 for sample 15382, and several are given in Table 3B. These analyses include both Si-rich glass, tiny globules of Fe-rich glass, and microcrystalline phases. Clearly, the interstitial matrix areas are predominately composed of Si-rich glass.

Although there is not a strictly monotonic increase of $TiO_2+P_2O_5$ as the relative content of CaO + MgO + FeO increases (Fig. 3) there is definitely an indication that the immiscibility gap is enlarged by addition of $P_2O_5+TiO_2$. The Fe-rich glasses show an increase in Fe/Mg with increase of $P_2O_5+TiO_2$, suggesting a fractionation trend toward increase of P_2O_5, TiO_2, and FeO/MgO with crystallization. This fractionation trend shows up in a general way on a plot of P_2O_5 vs. MgO/FeO (Fig. 6). The high P_2O_5 and low MgO/FeO values are mainly Fe-rich blebs in the interstitial matrix glass areas, or are connected to the matrix from within a feldspar, whereas the high MgO/FeO and low P_2/O_5 glasses occur mainly as inclusions in feldspar in sample 15382 or are the Fe-rich glasses in 15434,66. Except for 15434, the individual symbols refer to different regions in sample 15382; two such regions are illustrated in Fig. 4 and 5.

The lines visually fitted (Fig. 6) to the points for area D, 15382, and most of the points for 15434,66, do suggest fractionation trends. Differences from this trend for some individual regions may be accounted for by local crystallization of a phosphate phase (observed in one glass inclusion in 15382, and present in 15434). Crystallization of a phosphate phase would result in a drop of P_2O_5 in the residual melt without change of MgO/FeO. Ilmenite commonly occurs with the interstitial glasses (e.g., Figs. 4 and 5), resulting in greater scatter in plots of TiO_2 vs. MgO/FeO than for P_2O_5 vs. MgO/FeO.

Based on the apparent fractionation trend of P_2O_5 vs. MgO/FeO, it is reasonable to assume that the Fe-rich glasses with the highest MgO/FeO and lowest P_2O_5 are samples of the earliest immiscible Fe-rich glasses. The best analyses of these are labelled on Fig. 6 and tabulated in Table 3C. For comparison, analyses of Fe-rich immiscible glasses of other samples are also given in Table 3C. Glass D9 (analysis 9, region D), with the highest MgO/FeO ratio (0.30) and P_2O_5 content of only 0.4–0.74% (repeat analyses) is the most primitive of the analyzed Fe-rich glasses, including those of sample 15434,66. It also has no detectable K_2O, suggesting the maximum amount of extraction of Si-rich melt had already occurred when D9 became isolated from the remaining residual melt. The occurrence of this glass is illustrated in Fig. 4. Note that glass D5 (Fig. 4; Table 3C) is nearly as primitive.

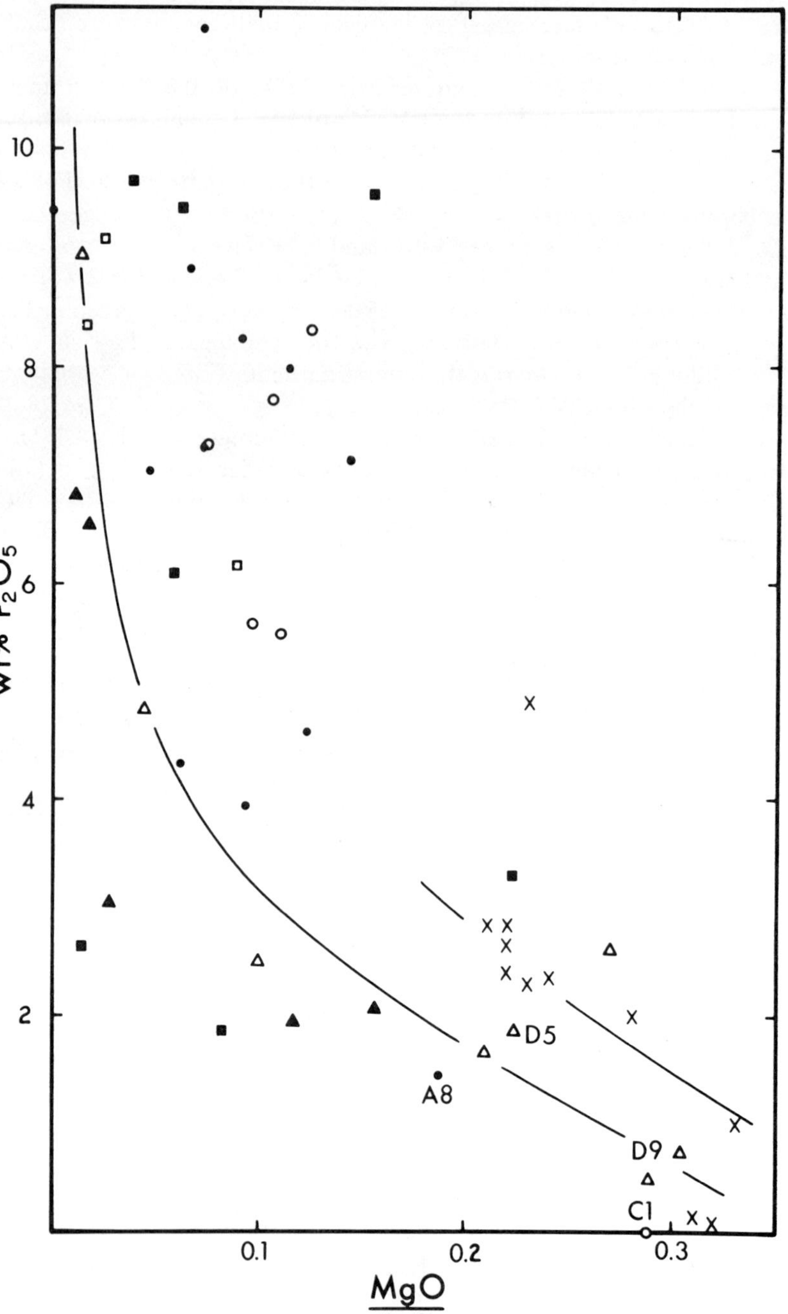

Fig. 6. P_2O_5 vs. MgO/FeO of Fe-rich glasses. (×), sample 15435; (△), area D, sample 15382,7; other symbols distinguish different regions (e.g., Fig. 4 and 5) in sample 15382. Lines visually fitted to some data from 15434 (×) and from region D, 15382 (△).

Plagioclase probably crystallized from the original entrapped melt onto the walls of the inclusion, which would act as a nucleus for further plagioclase growth. The fact that most of the Fe-rich glasses in 15382 plot further from the average plagioclase composition of this sample (An_{85}) on Fig. 3 than the glasses of sample 15434, all of which are part of the 42% glassy matrix, supports this view. It is difficult to specify precisely how much plagioclase component must be added to the Fe-rich glasses to obtain the composition of the original Fe-rich melt. However, a reasonable estimate can be based on the amount of Al_2O_3 in the Fe-rich matrix glasses of 15434,66.

Mixing 18% An_{85} with the composition of sample D9 (analyzed twice) results in Al_2O_3 values within the range of the Fe-rich matrix glasses of 15434,66, and results in a calculated composition remarkably similar to one group each of Apollo 12 and 15 basalts. The comparative analyses are given in Table 4. There are two points in this comparison that should be noted here. The first is that only the mare basalt compositions listed match closely the calculated Fe-rich immiscible melt composition (Table 4); the others which do not match the calculated analyses closely have higher silica. The second and most important point is that the calculated melt still has significant P_2O_5 (about 0.4%) whereas the mare basalts have negligible P_2O_5. Outside of these differences, the similarities of the analyses are striking, and this similarity is the subject of the discussion in the next section.

In order for the melt composition of 15382 or 15434 to enter the field of silicate liquid immiscibility (Fig. 3), plagioclase with the approximate composition (An_{85}) as marked on Fig. 3 must crystallize essentially alone for the early part of the crystallization history of the samples. This is difficult at equilibrium at low pressure, because the bulk compositions of both samples lie near the plagioclase-pyroxene cotectic (Meyer, 1977) of the silica-olivine-plagioclase pseudoternary (Walker *et al.*, 1973). The crystallization path necessary to reach the field of silicate liquid immiscibility (Fig. 3) therefore requires either early physical conditions where plagioclase only was a liquidus phase, or initial disequilibrium crystallization of only plagioclase during rapid cooling,

Table 4. Comparison of calculated immiscible glass and Fe-rich mare basalts.

	D9 + 18 % plag[1]	Group 2, Apollo 15[2]	Pigeonite, Apollo 12[3]
MgO	7.5	9.6	7.9
Al_2O_3	9.2	8.9	10.4
SiO_2	45.4	45.4	46.5
CaO	10.0	10.2	11.0
TiO_2	2.7	2.5	3.4
FeO	25.4	22.5	19.7

All analyses normalized to 100% for elements given.

[1] Average of 2.

[2] Butler (1976).

[3] Rhodes and Hubbard (1973).

followed by pyroxene crystallization from the high Fe residual melt. The latter process is supported by the textural relationships (Crawford and Hollister, 1977).

DISCUSSION

The striking similarity of the calculated early Fe-rich immiscible melts in KREEP basalts to Fe-rich mare basalts raises the question of whether there might be a direct genetic connection between the process of silicate liquid immiscibility and the generation of the Fe-rich mare basalts. If there is a connection, a mechanism must be found to remove the phosphorous from the Fe-rich immiscible melts, for example the 0.4% P_2O_5 in the D9 plus 18% plagioclase composition.

The presence of a phosphate phase, probably merrillite (REE-bearing, Dowty, 1977) which coexists with the Fe-rich immiscible melt of sample 15434 (Fig. 2) suggests the possibility that further crystallization of merrillite from the melt could deplete the melt of its phosphorous. Merrillite, with a density of 3.37 g/cc (Dowty, pers. comm.), would sink from the calculated Fe-rich melt which has a calculated density of 2.99 g/cc (Ashwal, pers. comm.). The Fe-rich melt itself would sink relative to feldspar and the high Si melt. Thus, in a favorable situation, the Fe-rich melt could be physically separated from the other phases of KREEP-rich rocks; and, as a cumulus phase, it could possibly be mixed with other cumulous phases such as olivine.

Extraction of a phosphate mineral is also required to remove trivalent trace elements from the Fe-rich immiscible melt to bring these element concentrations into the ranges measured for the Fe-rich mare basalts. It is well known that the KREEP-rich rocks are extremely enriched in these elements and that they would be even more enriched in any uncrystallized fraction of the KREEP-rich rocks. Clearly, K is enriched in the Si-rich immiscible melt and P in the Fe-rich melt (Fig. 3). Based on REE data on the dark and light portions of 12013 (Ma *et al.*, 1977), and on the experimental data of Watson (1976), the trivalent trace elements are probably enriched in the Fe-rich immiscible melt. The low SiO_2, Fe-rich Apollo 12 and 15 mare basalts have minimal Eu anomaly, but about the same Eu content as KREEP basalts (Schmitt and Laul, 1973). Removal of trivalent REE from the Fe-rich immiscible melt would result in an approach of the relative abundance of rare-earth elements in the Fe-rich immiscible melt to that in the mare basalts. Thus, if this immiscible melt is genetically related to the mare basalts, the trivalent rare-earth elements must be removed. This removal can be accomplished by removal of merrillite (whitlockite), because it is a major carrier of these elements (Griffin *et al.*, 1972). The merrillite in 15434,66 does have over 1.0% REE detected by the EDA technique.

Obviously, this study does not prove that Fe-rich mare basalts, at 3.3 AE, evolved from 3.9 AE KREEP. Only a process is demonstrated by the results of the present study. A model must be constructed to lead from melt which has undergone silicate liquid immiscibility to the actual generation of the mare basalts. A possible, but not necessarily unique, set of circumstances follows.

Assume, as many have, an early globe-encircling magma ocean. At some stage in its crystallization history, the stage depending on the initial bulk composition of the magma ocean, it enters a field of silicate liquid immiscibility. Earlier, dense phases such as olivine would certainly have settled from this melt. Plagioclase is on the liquidus; it may have been from the beginning, but this is not necessary to the model. The Fe-rich immiscible melt will sink from the plagioclase and the Si-rich melt. It would be a cumulate phase just as olivine would be, and probably would be associated with other cumulus phases. However, after sinking, crystallization of merrillite would result in a cumulus merrillite phase separating from the cumulus Fe-rich melt. The Fe-rich melt would, of course, subsequently crystallize; further cumulus phases are not required in our model.

The model now evolves, as do other models, to a remelting at 3.3 AE of the region of Fe-rich cumulate immiscible melt possibly mixed with other cumulus phases. However, in the remelting, we must presume the phosphate phase was sufficiently dense and refractory to not be incorporated into the melt at extrusion. A feature of this model is that the heat source for the remelting may be the phosphate itself, which also contains uranium and thorium.

Acknowledgments—C. Kulick assisted in the data collection, and L. Ashwal helped reduce the data. We thank E. Dowty for critical discussions. B. Murck drew the sketches of Fig. 4 and 5. E. Roedder and M. Dougan are thanked for their constructive reviews. Financial support provided by NASA grant NGL-31-001-283.

References

Butler J. C. (1976) Variations in chemical composition of Apollo 15 mare basalts. *Proc. Lunar Sci. Conf. 7th,* p. 1429–1447.

Crawford M. L. and Hollister L. S. (1974) Kreep basalt: A possible partial melt from the lunar interior. *Proc. Lunar Sci. Conf. 5th,* p. 399–419.

Crawford M. L. and Hollister L. S. (1977) Evolution of KREEP: Further petrologic evidences. *Proc. Lunar Sci. Conf. 8th,* Vol. 1.

Dowty E. (1977) Structure and composition of the $Ca_3(PO_4)_2$ minerals. *Earth Planet Sci. Lett.* In press.

Dowty E., Keil K., Prinz M., Gros J., and Takahashi H. (1976) Meteorite-free Apollo 15 crystalline KREEP. *Proc. Lunar Sci. Conf. 7th,* p. 1833–1844.

Dymek R. F., Quick J. E., and Albee A. L. (1977) The role of silicate liquid immiscibility: Constraints from a terrestrial example (abstract). In *Lunar Science VIII,* p. 272–274. The Lunar Science Institute, Houston.

Greig J. W. (1927) Immiscibility in silicate melts. *Amer. J. Sci.*, Series 5, **13**, 1–44.

Griffin W. L., Amli R., and Heier K. S. (1972) Whitlockite and apatite from lunar rock 14310 and from Odegarden, Norway. *Earth Planet. Sci. Lett.* **15**, 53–58.

Hess P. L., Rutherford M. H., Guillemette R. N., Ryerson F. J., and Tuchfeld H. A. (1975) Residual products of fractional crystallization of lunar magmas: An experimental study. *Proc. Lunar Sci. Conf. 6th,* p. 895–909.

Ma M.-S., Murali A. V., and Schmitt R. A. (1977) Major, minor and trace element compositions of two petrologically significant lithologies in 12013 (abstract). In *Lunar Science VIII*, p. 602–604. The Lunar Science Institute, Houston.

Meyer C., Jr. (1977) Petrology, mineralogy, and chemistry of KREEP basalt. *Phys. Chem. Earth* **10**. In press.

Powell B. N., Dungan M. A., and Weiblen P. W. (1975) Apollo 16 feldspathic melt rocks: Clues to the magmatic history of the lunar crust. *Proc. Lunar Sci. Conf. 6th,* p. 415–433.

Quick J. E. and Albee A. L. (1976) 12013 revisited—Two clast laden melts (abstract). In *Lunar Science VII,* p. 712–714. The Lunar Science Institute, Houston.

Rhodes J. M. and Hubbard N. J. (1973) Chemistry, classification, and petrogenesis of Apollo 12 mare basalts. *Proc. Lunar Sci. Conf. 4th.* p. 1127–1148.

Roedder E. and Weiblen P. W. (1970) Lunar petrology of silicate melt inclusions, Apollo 11 rocks. *Proc. Apollo 11 Lunar Sci. Conf.,* p. 801–837.

Roedder E. and Weiblen P. W. (1971) Petrology of silicate melt inclusions, Apollo 11 and Apollo 12, and terrestrial equivalents. *Proc. Lunar Sci. Conf. 2nd,* p. 507–528.

Roedder E. and Weiblen P. W. (1972) Petrographic features and petrologic significance of melt inclusions in Apollo 14 and 15 rocks. *Proc. Lunar Sci. Conf. 3rd,* p. 251–279.

Rutherford M. J., Hess P. C., Ryerson F. J., Campbell H. W., and Dick P. A. (1976) The chemistry, origin and petrogenetic implications of lunar granite and monzonite. *Proc. Lunar Sci. Conf. 7th,* p. 1723–1740.

Schmitt R. A. and Laul J. C. (1973) A survey of the selurochemistry of major, minor and trace elements. *The Moon* **8**, 182–209.

Switzer G. S. (1972) Composition of three glass phases in an Apollo 15 basalt fragment. *Smithson. Contrib. Earth Sci.* No. 14, p. 25–30.

Walker E., Grove T. L., Longhi J., Stolper E. M., and Hays J. F. (1973) Origin of lunar feldspathic rocks. *Earth Planet. Sci. Lett.* **20**, 325–336.

Watson E. B. (1976) Two-liquid partition coefficients: Experimental data and geochemical implications. *Contrib. Mineral. Petrol.* **56**, 119–134.

Weiblen P. W. and Roedder E. (1973) Petrology of melt inclusions in Apollo samples 15598 and 62295. *Proc. Lunar Sci. Conf. 4th,* p. 681–703.

Proc. Lunar Sci. Conf. 8th (1977), p. 2433–2448.
Printed in the United States of America

Chemical variation and fractionation of KREEP basalt magmas

ANTHONY J. IRVING

The Lunar Science Institute, 3303 NASA Road 1, Houston, Texas 77058

Abstract—A synthesis of available chemical data (including defocused beam microprobe analyses) for 53 Apollo 15 igneous KREEP basalts shows that their bulk compositional variation is much greater than previously recognized, especially in terms of 100 Mg/Mg + Fe which ranges from 73 to 35. Furthermore, there are systematic variations in K_2O and trace element abundances with Mg/Mg + Fe which are consistent with interpretation of the Apollo 15 KREEP basalts as a magma series generated by fractional crystallization processes. Experimental and chemical evidence indicate that this magma series results from low-pressure (subvolcanic?) fractional crystallization of a magnesian parental liquid (100 Mg/Mg + Fe ~ 72) by removal of low-Ca pyroxene and plagioclase, with eventual production of liquids similar in composition to 15405 quartz-monzodiorites. The Apollo 17 KREEPy basalts are probably also an igneous magma series which evolved by plagioclase-pyroxene fractionation, but from a different parental liquid. The postulated parental liquid for the Apollo 15 KREEP magma series is represented by at least one sample from the soils (SAO 465-11) and its composition is very similar to that of the low-pressure peritectic liquid in the olivine-anorthite-silica pseudoternary system. There is some evidence that this primitive liquid may be a direct partial melt of troctolitic materials in the deep lunar crust, but further experimental studies are needed to clarify this.

INTRODUCTION

THE APOLLO 15 AND 17 KREEP BASALTS are distinct in many respects from other lunar highland rocks with melt textures. As discussed by Irving (1975) these basalts have uniform melt textures, systematically lack the xenolithic debris characteristic of other melt rocks, contain metal with indigenous lunar rather than meteoritic contents of Ni and Co, and are uncontaminated by meteoritic siderophile trace elements. More recently, low siderophile element concentrations have been confirmed in KREEP basalt 15382 (Dowty *et al.*, 1976). These features overwhelmingly favor the interpretation that the Apollo 15 and 17 KREEP basalts are igneous lavas and not impact melts. It is probable that volcanic KREEP basalts are present in minor amounts at other sites, especially Apollos 14 and 12; however, until further studies are made, samples like 14310 must be excluded on the basis of the above criteria and are best interpreted as impact-melted mixtures of Apollo 15 volcanic KREEP basalt and minor ANT-suite rocks (see Irving, 1975).

One of the problems in treating the chemistry of Apollo 15 and 17 KREEP basalts is that all but two of the returned samples weigh less than 1 g. Rake samples 15386 (7.5 g) and 15382 (3.2 g) are the largest and best studied samples, but the remainder are soil particles and breccia clasts which are nearly all less than 100 mg in weight. These size limitations have not hindered precise trace element analysis, but they have severely limited major element analysis by more precise techniques (only *one* complete XRF analysis has been made). There exists, however, a large body of less precise and less accurate major element data obtained by defocused beam electron microprobe analysis (DBA) of small particles and clasts. Despite the limitations of these data, they are very valuable in defining the patterns of chemical variation of KREEP basalt magmas.

This paper examines the chemical variation of Apollo 15 and 17 KREEP basalts and shows that the compositional range is much more extensive than generally supposed. The two KREEP magma series are then interpreted as resulting from near-surface fractionation of volcanic KREEP basalt parent magmas derived by partial melting in the shallow lunar interior.

Data Sources and Uncertainties

The data set used in this synthesis includes 52 published major element analyses, 9 unpublished major element analyses and 19 trace element analyses of Apollo 15 and 17 KREEP basalts. These data include the analyses used by Irving (1975), with additional DBAs from Sewell *et al.* (1974), Ryder and Bower (1976), Basu and Bower (1976 and unpublished), and Ryder *et al.* (1977). Also included are high quality neutron activation analyses for trace elements and major elements (except Si and P) in seven KREEP basalt fragments from the Apollo 15 deep drill core (Lindstrom *et al.*, 1977). Sample numbers and references for all samples are given in the Appendix.

Of the 61 major element analyses used, 44 are by DBA, 1 by X-ray fluorescence spectrometry (XRF) and 3 by microprobe analysis of fused beads. A further 13 samples have been partially analyzed for major elements by neutron activation, atomic absorption, or colorimetric techniques. All but three of the DBAs were carried out at the Smithsonian Astrophysical Observatory or the University of Melbourne. Albee (1977) has discussed the systematic analytical errors in DBAs and suggests that uncertainties are typically about 5% relative if heterogeneous matrix corrections are applied. The Smithsonian data have been corrected in this way, but the University of Melbourne non-dispersive Si-detector data have not been corrected at all for matrix effects and may be of slightly lower accuracy. However, both of these laboratories have calibrated their techniques using well-analyzed fine-grained terrestrial basalts (Bower *et al.*, 1977; Gleadow *et al.*, 1974; D. A. Wark, pers. comm., 1977), and found overall uncertainties of around 5% or better for all major elements except FeO (which Bower *et al.* show can be up to 10% low).

Although the analytical errors associated with these DBAs appear to be sufficiently low to draw useful conclusions from large arrays of data, the sampling errors involved with small (1–10 mm) lithic fragments are potentially much greater. It is impossible to rigorously evaluate how representative a lithic fragment is of its parent rock, but there are several textural criteria which can be subjectively used to reject non-representative fragments. Photomicrographs of most of the fragments analyzed by DBA were used for this purpose, and data accepted only for fine-grained, even-textured samples lacking large phenocrysts (see Fig. 1). Over half of the University of Melbourne Apollo 15 KREEP basalt

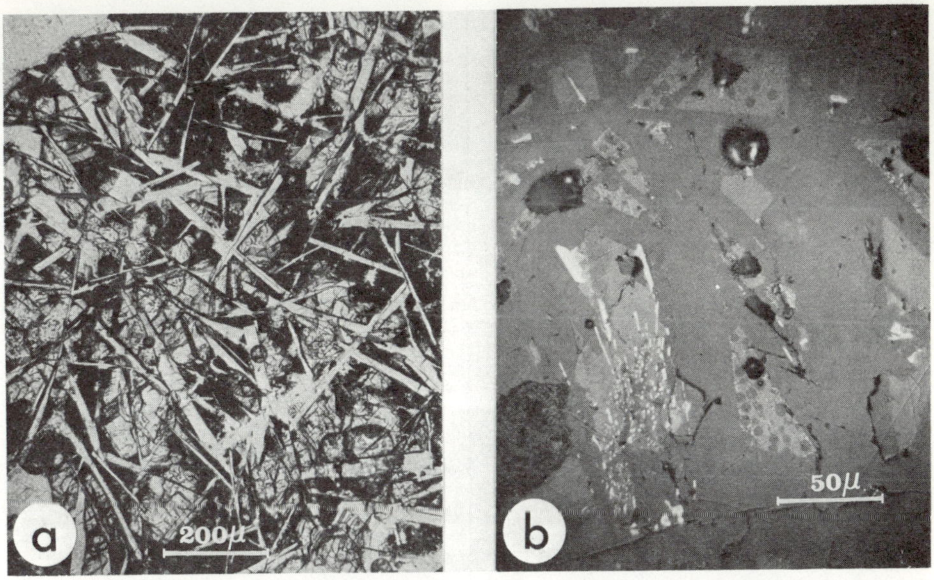

Fig. 1. Textural features of primitive Apollo 15 KREEP basalt sample SAO 465-11. (a) General view illustrating uniform melt (igneous) texture. Laths of plagioclase are enclosed subophitically by pyroxene with interstitial dark mesostasis. Transmitted light. (b) Detail of silicate immiscibility in mesostasis areas between plagioclase laths (gray). Dark gray globules of Fe-rich glass occur within light gray Si-rich glass. White bladed crystals at left are ilmenite. Reflected light.

clast analyses reported in Sewell *et al.* (1974) and discussed by Gleadow *et al.* (1974) were rejected on this basis, but 11 were judged acceptable. Five unpublished analyses previously excluded by Basu and Bower (1976) were also judged acceptable.

Chemical Variation of KREEP Basalts

The most significant major element variation in KREEP basalts is in MgO, FeO, and K_2O. Major element analyses illustrating the observed compositional range are given in Table 1. Figure 2 shows the variation in K_2O as a function of 100 Mg/Mg + Fe (henceforth Mg-value) for Apollo 15 and 17 KREEP basalts. Although there is some scatter in the data, the Apollo 15 KREEP basalts show a definite trend of increasing K_2O with decreasing Mg-value. Much of the scatter is probably analytical, but part may reflect real chemical differences among separate KREEP basalt flows sampled by cratering processes. The break in slope of the data at a Mg-value of 50 probably has no significance, given the analytical uncertainties.

The composition modally estimated by Taylor (1976) for a 15405 quartz-monzodiorite (QMD) clast also plots on this trend. These QMD clasts are coarser grained than the KREEP basalts with which they are spatially asso-

Table 1. Selected major and trace element data for Apollo 15 and 17 KREEP basalts and 15405 quartz-monzodiorite illustrating the ranges of compositional variation. All trace element abundances in ppm. Data sources are given in the Appendix. Experimentally inferred peritectic liquid compositions P (McKay and Weill, 1976) and B (Walker *et al.*, 1973b) are listed for comparison (see text).

	Apollo 15 KREEP basalts						Apollo 15 quartz-monzodiorite	Peritectic liquids		Apollo 17 KREEP basalts	
	SAO 465-11	15003,453	15386	15434,63(4)	SAO 490-65	15417B		P	B	72275,147	72275,135(C4)
SiO_2	49.2	(49.2)	50.83	52.1	51.6	53.4	57.4	49.1	49.1	50.1	49.4
TiO_2	1.2	2.6	2.23	2.3	2.6	3.5	1.1	2.3	2.4	0.83	0.92
Al_2O_3	15.5	15.7	14.77	17.0	12.9	11.2	13.2	15.9	15.9	14.1	18.1
Cr_2O_3	0.30	0.29	0.31	0.30	0.31	—	0.18	—	0.3	0.36	0.36
FeO	7.9	10.2	10.55	11.1	11.9	12.8	11.2	8.2	9.3	11.5	8.5
MnO	0.15	0.14	0.16	0.23	—	—	—	—	—	0.21	0.17
MgO	12.5	9.7	8.17	6.3	5.6	3.8	3.4	11.8	9.2	8.5	4.2
CaO	9.5	10.2	9.71	10.3	8.5	10.0	9.0	12.1	12.1	7.8	12.6
Na_2O	0.63	0.87	0.73	0.58	0.98	1.2	1.0	—	0.69	0.38	0.56
K_2O	0.48	0.55	0.67	0.73	1.7	1.7	1.9	—	0.67	0.37	0.32
P_2O_5	0.28	(0.6)	0.70	0.60	1.9	2.3	0.4	—	—	0.29	0.53
Sum	97.6	100.0	98.8	101.5	98.0	99.9	98.8	99.4	99.7	94.4	95.7
100 Mg/Mg + Fe	73.8	62.9	58.0	50.3	45.4	34.6	35.1	71.9	63.9	56.8	47.1

Method of analysis[1]	DBA	INAA	XRF	MFB	DBA	DBA	MODE	DBA	DBA
Type of sample	1–2 mm fines fragment	drill core fragment	rake sample	4–10 mm fines fragment	2–4 mm fines fragment	breccia clast	breccia clast	breccia clast	breccia clast
Sm (ppm)			29→42				92.5		23
Hf			19→40				51		18
Zr			630→1020				—		—
Ba			670→950				1490		—
Rb			13→18				39.8		8
U			3.1→4.0				11.5		1.5
Th			10→18				43		—
Eu			2.5–3.1				2.61		1.58
Sr			177–195				154		87
Sc			21→26				30.7		61
Cr			2870→1800				1220		3140
Co			18–23				7.8		37
Ni			18				≤2		43

[1] DBA = defocused beam microprobe analysis.
INAA = instrumental neutron activation analysis (P_2O_5 estimated, SiO_2 by difference).
MFB = fused bead microprobe analysis.
MODE = calculated from mode and mineral analyses.

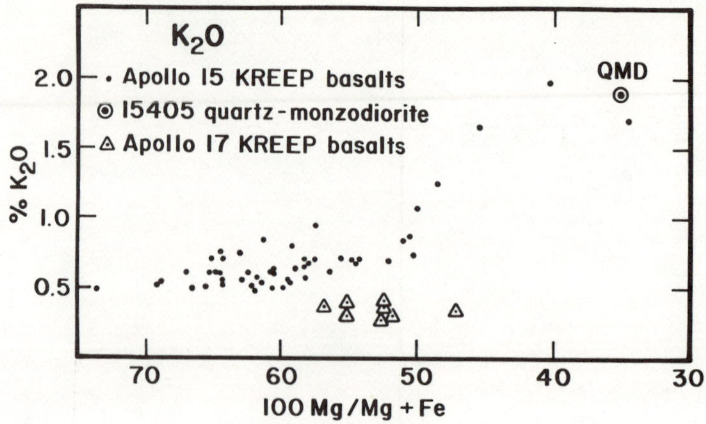

Fig. 2. Variation of K_2O with Mg-value for Apollo 15 and 17 KREEP basalts and 15405 quartz-monzodiorite.

ciated, but they share an igneous texture, low abundances of siderophile trace elements and other chemical features suggesting an origin by fractional crystallization of KREEP basalt magma (Ryder, 1976; Gros *et al.*, 1976; Nyquist *et al.*, 1977). The Apollo 17 data are too few to show a clear trend, but, as Ryder *et al.* (1977) have noted, these KREEPy basalts appear to be chemically quite distinct from the Apollo 15 samples.

Ranges of trace element abundances for KREEP basalts are given in Table 1. Figures 3 and 4 illustrate the variation in abundances of several key trace elements as a function of Mg-value (over the range 63–45 where data are available). Again the data show some scatter, but it is reasonably clear that for the Apollo 15 samples large ion lithophile (LIL) elements such as Ba, Th, Hf and the trivalent rare earths (exemplified by Sm) increase with decreasing Mg-value confirming the trend shown by K_2O. Trivalent rare-earth element patterns for Apollo 15 KREEP basalts are essentially parallel, with La ranging from 200 to 300× chondrites for Mg-values from 63 to 45 (see Irving, 1975, Fig. 4). In each case these trends in LIL elements are reinforced by the QMD data.

Other LIL elements such as Rb, U, and Zr show the same behavior (see Table 1), however Eu and Sr have a very different pattern of variation and are essentially constant with Mg-value, even in the more evolved QMD. Cr data show considerable scatter but a trend of decreasing Cr with Mg-value is consistent with the low Cr content of the QMD. Sc shows a small increase with decreasing Mg-value. Co is essentially constant in the KREEP basalts, but is lower in the QMD by a factor of 3. Trace element abundances for the single analyzed Apollo 17 KREEP basalt are very different from those for Apollo 15 KREEP basalts of corresponding Mg-value (e.g., lower LIL elements, higher Cr and Sc), thus confirming the existence of at least two separate igneous KREEP magma series (Blanchard *et al.*, 1975; Ryder *et al.*, 1977).

Variations in Al_2O_3, TiO_2, and P_2O_5 are difficult to assess properly with the existing data set since these elements are particularly subject to analytical error by DBA techniques (e.g., Albee, 1977). Al_2O_3, being the second most abundant major element, has the largest absolute error and is erroneously high in many DBAs (even if heterogeneous corrections have been applied). TiO_2 and P_2O_5 suffer from the added problem of being located mainly in accessory phases (ilmenite, apatite, whitlockite) which are subject to non-representative sampling during thin-section preparation and electron beam traverses. The limited amount of data by more precise methods indicates relatively small ranges in Al_2O_3 (15–17%) and TiO_2 (1.8–2.3%), but there is some suggestion of a decrease in Al_2O_3 and an increase in TiO_2 with Mg-value, a trend continued in the quartz-monzodiorite. A decrease in Al_2O_3 with increasing LIL element abundances (and hence decreasing Mg-value) for KREEP basalts was illustrated by Irving (1975, Figs. 1–3). There are few reliable P_2O_5 analyses but this element would be expected to follow K_2O. There is little systematic variation in CaO (7–10%), Na_2O (0.6–1.0%), and SiO_2 (49–52%), although there is some tendency for SiO_2 to increase with decreasing Mg-value.

Because of sample size limitations there are no data for Y and Nb in KREEP basalts, and only a small amount of Zr data of varying precision. Thus it is difficult to critically examine variations in all the diagnostic ratios of "incompatible" elements employed by Duncan *et al.* (1976). Nevertheless, K/Ba appears to be well determined at 6–7 for 12 Apollo 15 KREEP basalt samples, but is slightly higher (9.5) in 15405 quartz-monzodiorite (Nyquist *et al.*, 1977). The limited Zr data indicate Ba/Zr ~ 0.9, K/Zr ~ 5, and Ti/Zr ~ 15 for Apollo 15 KREEP basalts. All these ratios lie in the ranges found by Duncan *et al.* (1976) for Apollo 14 KREEP-rich breccias, and thus justify their assumption that the "incompatible" element abundances in the breccias are determined by a KREEP component.

Experimental Constraints on the Genesis of KREEP Basalts

The olivine-anorthite-silica pseudoternary liquidus diagram of Walker *et al.* (1973a,b) conveniently approximates the low-pressure ($\lesssim$2 kbar) phase relations for KREEP basalt magmas. Projected bulk compositions for 51 Apollo 15 KREEP basalts in part of the pseudoternary diagram are shown in Fig. 5. For nine of the plotted samples only partial major element analyses are available, so SiO_2 contents were calculated by difference after estimating P_2O_5 contents by analogy with other analyzed samples; two other samples could not be plotted because of insufficient major element data. Data for Apollo 17 KREEP basalts are not shown in Fig. 5 because their compositions are too different from those on which the projection is based. The plotted points for Apollo 15 KREEP basalts are concentrated along the pyroxene-plagioclase cotectic line from very close to peritectic point B toward cotectic point C, where Taylor's (1976) modally estimated QMD composition plots. The scatter in the data to the right of the cotectic line is almost entirely due to errors in Al_2O_3 determined by DBA (which is expressed as excess anorthite component). It is notable that all the

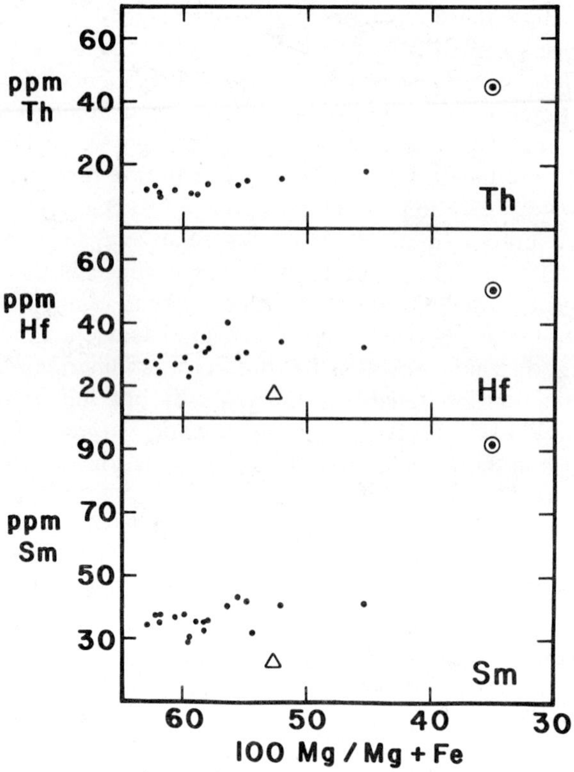

Fig. 3. Variation of Th, Hf, and Sm with Mg-value for Apollo 15 and 17 KREEP basalts and 15405 quartz-monzodiorite. Symbols as in Fig. 2.

compositions determined by more precise methods (including those for which SiO_2 was calculated by difference) plot very close to the pyroxene-plagioclase cotectic line.

The analysis which plots closest to point B (slightly within the olivine field) is for lithic fragment SAO 465-11 studied by Basu and Bower (1976), which is the most primitive (highest Mg-value) Apollo 15 KREEP basalt discovered to date. The textural features of this interesting sample are shown in Fig. 1. The textural uniformity and relatively fine grain size indicate that the published bulk composition should be among the more reliable determined by DBA.

Also plotted in Fig. 5 are DBAs of homogeneous mesostases from 11 Apollo 15 KREEP basalt samples. Their plotted positions within the silica field have no real significance in this projection (K_2O and Fe/Mg are too high), but it is perhaps surprising that mesostases from different samples have such similar bulk compositions. None of these mesostases show obvious silicate immiscibility. However, immiscibility is evident in SAO 465-11 (see Fig. 1b) and in a number of other Apollo 15 KREEP basalts (e.g., Meyer, 1972; Cameron *et al.*, 1973; Powell *et al.*, 1973; Switzer, 1975; Hollister and Crawford, 1977; Basu and Bower, 1977).

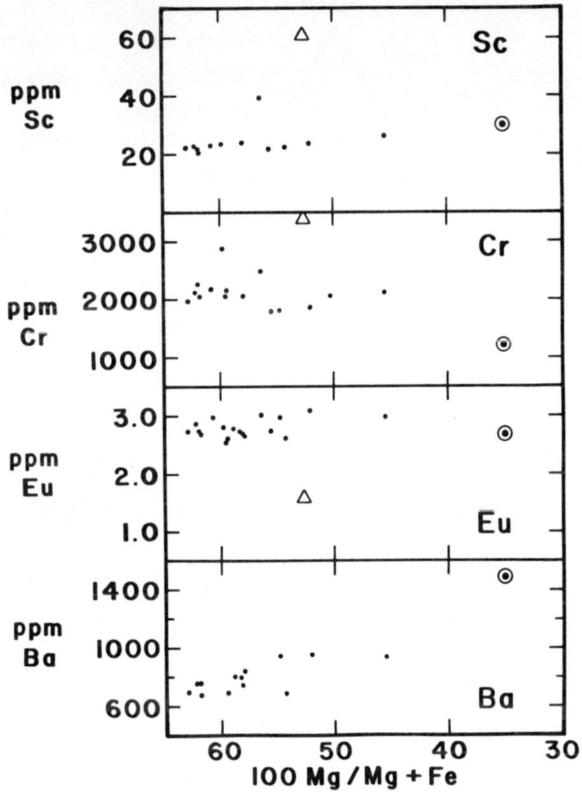

Fig. 4. Variation of Sc, Cr, Eu, and Ba with Mg-value for Apollo 15 and 17 KREEP basalts and 15405 quartz monzodiorite. Symbols as in Fig. 2.

An experimental study of phase relations of complex 12-component Apollo 15 and 17 KREEP basalt compositions (including 15386 and SAO 465-11) is in progress. Preliminary results (Irving, 1977 and unpublished data) indicate that plagioclase and low-Ca, low-Al pyroxene are the near-liquidus phases of both 15386 and an Apollo 17 KREEP basalt composition at pressures from 1 atm to 5 kbar.

Discussion

The observed chemical variation of KREEP basalt liquids is most readily interpreted in terms of a fractional crystallization model. The possibility of these liquids being instead a series of melts produced by fractional fusion of a suitable source is unlikely because of their relatively continuous compositional variation, which is expected in fractional crystallization but not in fractional fusion processes (Presnall, 1969). If alternatively these KREEP basalts were a series of equilibrium partial melts derived by different degrees of melting of one or more sources their observed relationship to the plagioclase-pyroxene cotectic in Fig. 5

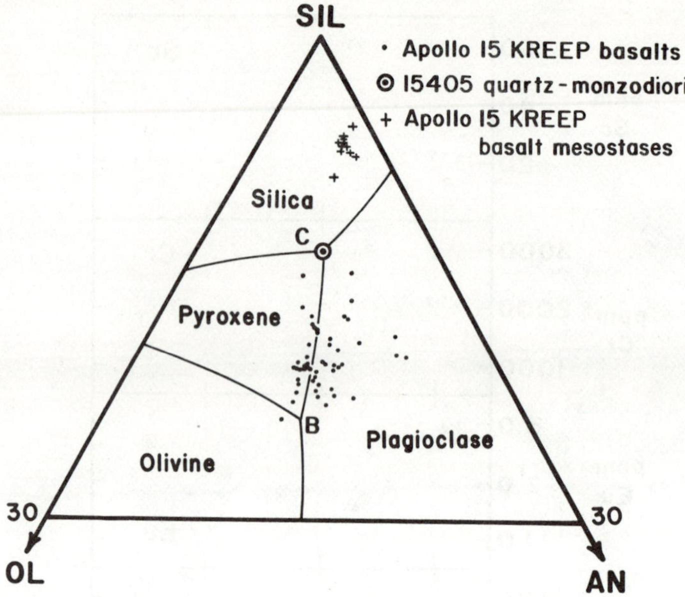

Fig. 5. Part of the olivine-anorthite-silica pseudoternary liquidus diagram (Walker *et al.*, 1973a,b) with data plotted for Apollo 15 KREEP basalts, 15405 quartz-monzodiorite, and Apollo 15 KREEP basalt mesostases (Phinney *et al.*, 1972; Drake *et al.*, 1973; Sewell *et al.*, 1974; Ryder, 1976).

would be extremely fortuitous. The experimental evidence both from Fig. 5 (based on Apollo 14 KREEP-rich compositions) and from studies of specific Apollo 15 and 17 KREEP basalt compositions strongly suggests low-pressure separation of plagioclase and low-Ca pyroxene as the controlling process in generating the observed spectrum of liquid compositions for both the Apollo 15 and 17 magma series.

Furthermore, separation of these phases (which are of course also the major phases of crystalline KREEP basalts) is entirely consistent with the observed trace element variation. The near constancy of Eu and Sr abundances with Mg-value and the concomitant increase in abundances of trivalent rare earths suggest removal of plagioclase, since it is the only common phase which discriminates so strongly among these elements (e.g., Hubbard *et al.*, 1973; McKay and Weill, 1976). Nyquist *et al.* (1977) have argued that the 15405 quartz-monzodiorites are an extreme product of this fractional crystallization process, involving also late removal of minor K–Ba feldspar.

The decrease in Cr and slight increase in Sc with decreasing Mg-value are also consistent with removal of pyroxene + plagioclase because of the difference in pyroxene/liquid partition coefficients (D) for these two elements. McKay and Weill (1976) determined a value of 5.2 for D_{Cr} between low-Ca pyroxene and model KREEP basaltic liquid at 1200°C and lunar f_{O_2}. McKay and Weill (1977) report a D_{Sc} value of 1.4 at 1195°C for low-Ca pyroxene crystallized from a low-K Fra

Mauro basalt bulk composition. Thus, with increasing fractional crystallization of pyroxene + plagioclase (i.e., decreasing Mg-value) Sc could be slightly enriched in the melt since it is strongly excluded from plagioclase. Although Cr is also excluded from plagioclase, Cr could be progressively depleted in the melt because of the dominant effect of pyroxene.

McKay and Weill (1976) have presented a similar model for KREEP basalt 15386. Using experimentally determined partition coefficients for selected major and trace elements they calculated the composition of the liquid at peritectic point B (their point P) which could yield 15386 by fractional crystallization of plagioclase and low-Ca pyroxene. Calculated liquid P differs in Mg/Fe from analyzed experimental liquid B of Walker *et al.* (1973b), but interestingly is more similar in major element composition to primitive KREEP basalt SAO 465-11 than is liquid B (see Table 1).

McKay and Weill also sought to relate peritectic liquid P by partial melting or fractional crystallization to a parental composition representing the average of those analyzed lunar highlands rocks which could yield liquid P without violating major and trace element partitioning constraints. The assumed parent composition is similar to the low-K Fra Mauro composition of Reid *et al.* (1972), however, Hess *et al.* (1977) have shown experimentally that at 1 atm the low-K Fra Mauro composition fractionates to a peritectic liquid significantly different from P (and B) in Al_2O_3 content and Mg/Fe ratio. Nevertheless, it seems clear that the source from which peritectic KREEP basalt liquids are derived must contain olivine and plagioclase (Weill and McKay, 1975), and it is intriguing that the Sm–Nd isotopic data of Lugmair *et al.* (1976) suggest a close genetic relationship between KREEP basalt 15382 and troctolite 76535.

The peritectic liquids too should be saturated with olivine as well as with pyroxene and plagioclase. Sample SAO 465-11 might be expected to contain olivine, but a careful thin-section search by G. Ryder revealed none. Olivine (Fo_{77-80}) has been observed in three KREEP basalt clasts in 15205 and 15405 (Dymek *et al.*, 1974; Ryder and Bower, 1976, 1977). None of these fragments has been analyzed, but the bulk Mg-values can be calculated (assuming olivine to be a liquidus phase) from the relation $X_{OL} = 100X_{LIQ} \div [X_{LIQ} + K_D(100 - X_{LIQ})]$, where X_{OL} is the Fo content of olivine, X_{LIQ} is the Mg-value of coexisting liquid, and K_D is the Fe/Mg distribution coefficient between olivine and liquid (cf., Irving and Green, 1976). Taking K_D to be 0.33 (Longhi *et al.*, 1977), these olivine-bearing KREEP basalts have Mg-values no greater than 56.9, significantly lower than the Mg-value of peritectic liquid (>70). Unless the olivines are crustal xenocrysts, their presence in these KREEP basalts must be a consequence of other unusual chemical features. Green *et al.* (1972) found olivine (Fo_{80}) as a liquidus phase of a KREEP-like composition, however, this composition is well outside the natural range, particularly because of low SiO_2 and high K_2O and Al_2O_3 relative to its Mg-value. Rutherford *et al.* (1976) also mention olivine in a disequilibrium cooling rate experiment on a 15405 KREEP basalt composition, but details are not yet available.

By analogy with the Apollo 15 rocks, the chemical variation observed for the

Apollo 17 KREEP basalts can also be interpreted as a fractional crystallization sequence. Because of the paucity of data it is difficult to speculate about a parent liquid composition, yet it seems clear that these Apollo 17 basalts are not related to the Apollo 15 KREEP basalts, but have evolved from a completely different parent liquid (cf., Ryder *et al.*, 1977). Experimental and petrographic evidence suggest that the parent liquid might have pyroxene, plagioclase, and chromite as near-liquidus phases. Olivine (Fo_{69}) occurs in two Apollo 17 KREEP basalt clasts (Ryder *et al.*, 1977), and again represents an enigma since it implies bulk Mg-values of 42.3, much lower than the range for analyzed samples.

The time at which fractionation of the two KREEP magma series took place is uncertain. The isotopic evolution of Apollo 15 KREEP basalts has been discussed by a number of workers (e.g., Nyquist *et al.*, 1975; Papanastassiou and Wasserburg, 1976; McKay and Weill, 1976; Shih, 1977). Rb–Sr internal isochron ages for three samples (15382, 15386, and 15434,73) are all ~3.93 AE, in agreement with two independent ^{40}Ar–^{39}Ar ages on 15382 (Stettler *et al.*, 1973; Turner *et al.*, 1973), but the $^{87}Sr/^{86}Sr$ initial ratios (adjusted for interlaboratory bias) have a wide range from 0.7004 to 0.7007. An internal Rb–Sr age of 4.01 AE was determined by Compston *et al.* (1975) for an Apollo 17 KREEP basalt. Evidence for older events in KREEP basalt genesis at ~4.1 and ~4.4 AE comes from Rb–Sr model ages, U–Pb systematics (Papanastassiou *et al.*, 1977) and fission track data (Haines *et al.*, 1975). These KREEP basalts were definitely liquids at 3.9–4.0 AE, and this seems like the most logical time for fractional crystallization, perhaps in subvolcanic magma chambers. Impact remelting at this time (e.g., Irving, 1975) may not be necessary if the source regions for the parental liquids were already differentiated (LIL element enriched) with high and locally variable $^{87}Sr/^{86}Sr$ ratios (cf., Papanastassiou and Wasserburg, 1976). At least for the Apollo 15 series, the extremely high Mg-value of the suggested parental (peritectic) liquid probably precludes its being a residual liquid from early lunar differentiation (e.g., Lipin, 1977). An origin involving partial melting in the deep lunar crust or uppermost mantle seems more likely. Perhaps, as suggested by Papanastassiou and Wasserburg (1976), this internal melting was triggered by large impacts such as the Imbrium event, which would explain the apparent preferential occurrence of KREEP-rich materials around the Imbrium Basin.

Acknowledgments—I am grateful to G. Ryder, D. A. Wark, M. J. Drake, A. Basu, and D. P. Blanchard for supplying unpublished chemical data or photomicrographs of analyzed samples. Reviews by J. L. Warner and G. A. McKay were very helpful. This research was supported by the Lunar Science Institute which is operated by the Universities Space Research Association under Contract No. NSR-09-051-001 with the National Aeronautics and Space Administration. This paper constitutes the Lunar Science Institute Contribution No. 284.

REFERENCES

Albee A. L., Quick J. E., and Chodos A. A. (1977) Source and magnitude of errors in "broad-beam analysis" (DBA) with the electron probe (abstract). In *Lunar Science VIII*, p. 7–9. The Lunar Science Institute, Houston.

Basu A. and Bower J. F. (1976) Petrography of KREEP basalt fragments from Apollo 15 soils. *Proc. Lunar Sci. Conf. 7th*, p. 659–678.

Basu A. and Bower J. F. (1977) Provenance of Apollo 15 deep drill core sediments. *Proc. Lunar Sci. Conf. 8th.* Vol. 1.

Blanchard D. P., Haskin L. A., Jacobs J. W., Brannon J. C., and Korotev R. L. (1975) Major and trace element chemistry of Boulder 1 at Station 2, Apollo 17. *The Moon* **14**, 359–371.

Bower J. F., Wood J. A., Richardson S. M., McSween H. Y., Jr., and Ryder G. (1977) Rock compositions by defocussed beam analysis (abstract). *8th Int. Conf. on X-ray Optics and Microanalysis and 12th Ann. Conf. of the Microbeam Analysis Soc.*, Boston, Mass., p. 182.

Cameron K. L., Delano J. W., Bence A. E., and Papike J. J. (1973) Petrology of the 2–4 mm soil fraction from the Hadley–Apennine region of the moon. *Earth Planet. Sci. Lett.* **19**, 9–21.

Church S. E., Bansal B. M., and Wiesmann H. (1972) The distribution of K, Ti, Zr, U, and Hf in Apollo 14 and 15 materials. In *The Apollo 15 Lunar Samples* (J. Chamberlain and C. Watkins, eds.), p. 210–213. The Lunar Science Institute, Houston.

Compston W., Foster J. J., and Gray C. M. (1975) Rb–Sr ages of clasts from within Boulder 1, Station 2, Apollo 17. *The Moon* **14**, 445–462.

Dowty E., Keil K., Prinz M., Gros J., and Takahashi H. (1976) Meteorite-free Apollo 15 crystalline KREEP. *Proc. Lunar Sci. Conf. 7th*, p. 1833–1844.

Drake M. J., Stoeser J. W., and Goles G. G. (1973) A unified approach to a fragmental problem: Petrological and geochemical studies of lithic fragments from Apollo 15 soils. *Earth Planet. Sci. Lett.* **20**, 425–439.

Duncan A. R., Erlank A. J., Sher M. K., Abraham Y. C., Willis J. P., and Ahrens L. H. (1976) Some trace element constraints on lunar basalt genesis. *Proc. Lunar Sci. Conf. 7th*, p. 1659–1671.

Dymek R. F., Albee A. L., and Chodos A. A. (1974) Glass-coated soil breccia 15205: Selenologic history and petrologic constraints on the nature of its source region. *Proc. Lunar Sci. Conf. 5th*, p. 235–260.

Gleadow A. J. W., Le Maitre R. W., Sewell D. K. B., and Lovering J. F. (1974) Chemical discrimination of petrographically defined clast groups in Apollo 14 and 15 breccias. *Chem. Geol.* **14**, 39–61.

Green D. H., Ringwood A. E., Ware N. G., and Hibberson W. O. (1972) Experimental petrology and petrogenesis of Apollo 14 basalts. *Proc. Lunar Sci. Conf. 3rd*, p. 197–206.

Gros J., Takahashi H., Hertogen J., Morgan J. W., and Anders E. (1976) Composition of the projectiles that bombarded the lunar highlands. *Proc. Lunar Sci. Conf. 7th*, p. 2403–2425.

Haines E. L., Hutcheon I. D., and Weiss J. R. (1975) The fission track record of Apennine Front KREEP basalts. *Proc. Lunar Sci. Conf. 6th*, p. 3527–3540.

Helmke P. A., Blanchard D. P., Haskin L. A., Telander K., Weiss C., and Jacobs J. W. (1973) Major and trace elements in igneous rocks from Apollo 15. *The Moon* **8**, 129–148.

Hess P. C., Rutherford M. J., and Campbell H. W. (1977) Origin and evolution of LKFM basalt. *Proc. Lunar Sci. Conf. 8th.* This volume.

Hlava P. F., Green J. A., Prinz M., Nehru C. E., Keil K., Dowty E., and Bunch T. E. (1973) Apollo 15 rake sample microbreccias and non-mare rocks: Bulk rock, mineral and glass electron microprobe analyses. Univ. New Mexico Inst. Meteoritics Spec. Publ. No. 11. 78 pp.

Hollister L. S. and Crawford M. L. (1977) Melt immiscibility in Apollo 15 KREEP: Origin of Fe mare basalts. *Proc. Lunar Sci. Conf. 8th.* This volume.

Hubbard N. J., Gast P. W., Rhodes J. M., Bansal B. M., and Wiesmann H. (1972) Nonmare basalts: Part II. *Proc. Lunar Sci. Conf. 3rd*, p. 1161–1179.

Hubbard N. J., Rhodes J. M., Gast P. W., Bansal B. M., Shih C.-Y., Wiesmann H., and Nyquist L. E. (1973) Lunar rock types: The role of plagioclase in non-mare and highland rock types. *Proc. Lunar Sci. Conf. 4th*, p. 1297–1312.

Hubbard N. J., Rhodes J. M., Wiesmann H., Shih C.-Y., and Bansal B. M. (1974) The chemical definition and interpretation of rock types returned from the non-mare regions of the moon. *Proc. Lunar Sci. Conf. 5th*, p. 1227–1246.

Irving A. J. (1975) Chemical, mineralogical and textural systematics of non-mare melt rocks: Implications for lunar impact and volcanic processes. *Proc. Lunar Sci. Conf. 6th*, p. 363–394.

Irving A. J. (1977) Chemical and experimental constraints on the genesis of Apollo 15 and Apollo 17 KREEP basalts (abstract). In *Lunar Science VIII*, p. 493–495. The Lunar Science Institute, Houston.

Irving A. J. and Green D. H. (1976) Geochemistry and petrogenesis of the Newer Basalts of Victoria and South Australia. *J. Geol. Soc. Australia* **23**, 45–66.

Lindstrom M. M., Nielsen R. L., and Drake M. J. (1977) Petrology and geochemistry of lithic fragments separated from the Apollo 15 deep drill core. *Proc. Lunar Sci. Conf. 8th*. Vol. 1.

Lipin B. R. (1977) The system Mg_2SiO_4–Fe_2SiO_4–$CaAl_2Si_2O_8$–SiO_2 and the origin of Fra Mauro basalts. *Amer. Mineral.* In press.

Longhi J., Walker D., and Hays J. F. (1977) The distribution of Fe and Mg between olivine and lunar basaltic liquids. *Geochim. Cosmochim. Acta.* In press.

Lugmair G. W., Marti K., Kurtz J. P., and Scheinin N. B. (1976) History and genesis of lunar troctolite 76535 or: How old is old? *Proc. Lunar Sci. Conf. 7th*, p. 2009–2033.

McKay G. A. and Weill D. F. (1976) Petrogenesis of KREEP. *Proc. Lunar Sci. Conf. 7th*, p. 2427–2447.

McKay G. A. and Weill D. F. (1977) KREEP petrogenesis revisited. *Proc. Lunar Sci. Conf. 8th*. This volume.

Meyer C., Jr. (1972) Mineral assemblages and the origin of non-mare lunar rock types (abstract). In *Lunar Science III*, p. 542–544. The Lunar Science Institute, Houston.

Morgan J. W., Higuchi H., and Anders E. (1975) Meteoritic material in a boulder from the Apollo 17 site: Implications for its origin. *The Moon* **14**, 373–383.

Nyquist L. E., Bansal B. M., and Wiesmann H. (1975) Rb–Sr ages and initial $^{87}Sr/^{86}Sr$ for Apollo 17 basalts and KREEP basalt 15386. *Proc. Lunar Sci. Conf. 6th*, p. 1445–1465.

Nyquist L. E., Wiesmann H., Shih C.-Y., and Bansal B. M. (1977) 15405 quartz-monzodiorite: Super KREEP. In *Interdisciplinary Studies by the Imbrium Consortium*, Vol. 2 (J. A. Wood, ed.). LSI Contr. No. 268D, p. 30–41. Center for Astrophysics, Cambridge.

Papanastassiou D. A., DePaolo D. J., Tera F., and Wasserburg, G. J. (1977) An isotopic tryptych on mare basalts (abstract). In *Lunar Science VIII*, p. 750–752. The Lunar Science Institute, Houston.

Papanastassiou D. A. and Wasserburg G. J. (1976) Rb–Sr age of troctolite 76535. *Proc. Lunar Sci. Conf. 7th*, p. 2035–2054.

Phinney W. C., Warner J. L., Simonds C. H., and Lofgren G. E. (1972) Classification and distribution of rock types at Spur Crater. In *The Apollo 15 Lunar Samples* (J. Chamberlain and C. Watkins, eds.), p. 149–152. The Lunar Science Institute, Houston.

Powell B. N., Aitken F. K., and Weiblen P. W. (1973) Classification, distribution and origin of lithic fragments from the Hadley–Apennine region. *Proc. Lunar Sci. Conf. 4th*, p. 445–460.

Presnall D. C. (1969) The geometrical analysis of partial fusion. *Amer. J. Sci.* **267**, 1178–1194.

Reid A. M., Warner J., Ridley W. I., and Brown R. W. (1972) Major element composition of glasses in three Apollo 15 soils. *Meteoritics* **7**, 395–415.

Rhodes J. M. and Hubbard N. J. (1973) Chemistry, classification and petrogenesis of Apollo 15 mare basalts. *Proc. Lunar Sci. Conf. 4th*, p. 1127–1148.

Rutherford M. J., Hess P. C., Ryerson F. J., Campbell H. W., and Dick P. A. (1976) The chemistry, origin and petrogenetic implications of lunar granite and monzonite. *Proc. Lunar Sci. Conf. 7th*, p. 1723–1740.

Ryder G. (1976) Lunar sample 15405: Remnant of a KREEP basalt-granite differentiated pluton. *Earth Planet. Sci. Lett.* **29**, 255–268.

Ryder G. and Bower J. F. (1976) Sample 15405: Petrology. In *Interdisciplinary Studies by the Imbrium Consortium*, Vol. 1 (J. A. Wood, ed.). LSI Contr. No. 267D, p. 77–94. Center for Astrophysics, Cambridge.

Ryder G. and Bower J. F. (1977) Sample 15405: Petrology. In *Interdisciplinary Studies by the*

Imbrium Consortium, Vol. 2 (J. A. Wood, ed.). LSI Contr. No. 268D, p. 19–26. Center for Astrophysics, Cambridge.

Ryder G., Stoeser D. B., and Wood J. A. (1977) Apollo 17 KREEPy basalt: A rock type intermediate between mare and KREEP basalts. *Earth Planet. Sci. Lett.* **35**, 1–13.

Schonfeld E., O'Kelley G. D. Eldridge J. S., and Northcutt K. J. (1972) K, U, and Th concentrations in rake sample 15382 by non-destructive gamma-ray spectroscopy. In *The Apollo 15 Lunar Samples* (J. Chamberlain and C. Watkins, eds.), p. 253–254. The Lunar Science Institute, Houston.

Sewell D. K. B., Gleadow A. J. W., Britten R., Cundari A., and Lovering J. F. (1974) Composition of rock clasts and their constituent minerals from Apollo 14 and 15 lunar breccias. Univ. Melbourne Dept. Geol., Publ. No. 2. 23 pp.

Shih C.-Y. (1977) Origins of KREEP basalts. *Proc. Lunar Sci. Conf. 8th.* This volume.

Stettler A., Eberhardt P., Geiss J., Grögler N., and Maurer P. (1973) Ar^{39}–Ar^{40} ages and Ar^{37}–Ar^{38} exposure ages of lunar rocks. *Proc. Lunar Sci. Conf. 4th*, p. 1865–1888.

Stoeser D. B., Wolfe R. W., Wood J. A., and Bower J. F. (1974) Petrology. In *Interdisciplinary Studies of Samples from Boulder 1, Station 2, Apollo 17*, Vol. 1 (J. A. Wood, ed.). LSI Contr. No. 210D, p. 35–109. Smithsonian Astrophysical Observatory, Cambridge.

Switzer G. S. (1975) Composition of three glass phases in an Apollo 15 basalt fragment. *Smithson. Contr. Earth Sci.* **15**, 25–30.

Taylor G. J. (1976) Sample 15405: Further petrological studies of the KREEP-rich quartz monzodiorite. In *Interdisciplinary Studies by the Imbrium Consortium*. Vol. 1 (J. A. Wood, ed.). LSI Contr. No. 267D, p. 94–95. Center for Astrophysics, Cambridge.

Turner G., Cadogan P. H., and Yonge C. J. (1973) Argon selenochronology. *Proc. Lunar Sci. Conf. 4th*, p. 1889–1914.

Walker D., Longhi J., Grove T. L., Stolper E. M., and Hays J. F. (1973a) Experimental petrology and origin of rocks from the Descartes Highlands. *Proc. Lunar Sci. Conf. 4th*, p. 1013–1032.

Walker D., Grove T. L., Longhi J., Stolper E. M., and Hays J. F. (1973b) Origin of lunar feldspathic rocks. *Earth Planet. Sci. Lett.* **20**, 325–336.

Weill D. F. and McKay G. A. (1975) The partitioning of Mg, Fe, Sr, Ce, Sm, Eu, and Yb in lunar ignous systems and a possible origin of KREEP by equilibrium partial melting. *Proc. Lunar Sci. Conf. 6th*, p. 1143–1158.

Wiesmann H. and Hubbard N. J. (1975) A compilation of lunar sample data generated by the Gast, Nyquist and Hubbard lunar sample PI-ships. Unpublished manuscript. NASA Johnson Space Center, Houston, Texas.

APPENDIX

Sample numbers and data sources used

Apollo 15 KREEP basalts. 15304,25; 15314,81 (Hubbard *et al.*, 1973; Wiesmann and Hubbard, 1975); 15023,2,5 (Hubbard *et al.*, 1972; Meyer, 1972); 15273,4,9; 15273,4,10 (Hubbard *et al.*, 1973; Church *et al.*, 1972); 15382 (Hubbard *et al.*, 1973; Schonfeld *et al.*, 1972; Church *et al.*, 1972; Dowty *et al.*, 1976); 15386 (Rhodes and Hubbard, 1973; Hubbard *et al.*, 1974; Wiesmann and Hubbard, 1975; D. P. Blanchard, unpublished data); 15564,17(5); 15434,67(7); 15434,63(4) (Helmke *et al.*, 1973; B. N. Powell, unpublished data); SAO 490-53; SAO 490-65 (Drake *et al.*, 1973); 15003,432; 15003,455; 15003,453; 15003,462; 15003,454; 15003,444; 15003,434 (Lindstrom *et al.*, 1977); SAO 411-3 (J. A. Wood, unpublished data); SAO 465-11; SAO 947-1; SAO 947-2; SAO 465-22; SAO 966-5; SAO 465-19; SAO 966-9; SAO 465-10; SAO 969-33 (Basu and Bower, 1976); SAO 929-1; SAO 465-5; SAO 465-12; SAO 965-12; SAO 953-5 (A. Basu and J. F. Bower, unpublished data); 15405,12(C-1); 15405,69(C-11); 15405,69(C-12); 15405,69(C-7); 15405,68(C-2); 15405,68(C-5) (Ryder and Bower, 1976); 15103,2,20 (Cameron *et al.*, 1973); clast in 15358 (Hlava *et al.*, 1973); 15565,87C; 15565,87D; 15565,87H; 15565,87M; 15565,87O; 15565,87P; 15565,87Q; 15015,19,62; 15015,19,502; 15285,18,29; 15417,4B (Sewell *et al.*, 1974).

Apollo 15 quartz-monzodiorites. 15405,51; 15405,85; 15405,86 (Taylor, 1976; Gros *et al.*, 1976; Nyquist *et al.*, 1977; D. P. Blanchard, unpublished data).

Apollo 17 KREEP basalts. 72275,128(C2); 72275,128(C19); 72275,135(C4); 72275,147 (Ryder *al.*, 1977); 72275,135(C1); 72275,135(C2); 72275,135(C3) (Stoeser *et al.*, 1974); 72275,91 (Blanchard *et al.*, 1975; Morgan *et al.*, 1975).

Proc. Lunar Sci. Conf. 8th (1977), p. 2449–2458.
Printed in the United States of America

Nearly pure Apollo 12 KREEP: Soil sample 12023

RICHARD V. MORRIS, JEFFREY L. WARNER, DAVID S. MCKAY
Code SN7, NASA Johnson Space Center, Houston, Texas 77058

ROY W. BROWN
Lockheed Electronics Company, Houston, Texas 77058

Abstract—Soil sample 12023, which was collected from a 20 cm trench on the east rim of Sharp Crater, is the first known example of a KREEP Apollo 12 soil. Its major element chemistry and FMR linewidth are indistinguishable from those of the Fra Mauro (Apollo 14) soils. The latter indicates that the soil matured on a KREEP regolith surface.

INTRODUCTION

PREVIOUS STUDIES (e.g., McKay *et al.*, 1971; Marvin *et al.*, 1971; Hubbard and Gast, 1971; Quaide *et al.*, 1971; Meyer *et al.*, 1971; Wasson and Baedecker, 1972) have shown that KREEP is an important chemical component of Apollo 12 soils. Mixing models (e.g., Schonfeld and Meyer, 1972; Wänke *et al.*, 1972) place the percentage of KREEP from ~20–30% for 12070 and 12037 to ~50–70% for 12032 and 12033. Meyer *et al.* (1971) point out that there are two morphological populations of Apollo 12 KREEP. One population, which dominates the KREEP component of 12032 and 12033, consists of KREEP glass particles (which we call ropy glass particles). The other population, which dominates the KREEP component of the other soils, consists of crystalline KREEP. The origin of KREEP at Apollo 12 is not agreed upon, but the alternatives suggested include ejecta of Copernicus or another major impact, transportation of Fra Mauro material from the Fra Mauro Formation or from local outcrops, and excavation of Fra Mauro material underlying the mare basalts (e.g., Meyer *et al.*, 1971; Marvin *et al.*, 1971; Wasson and Baedecker, 1972).

A number of soils from the Apollo 12 site have not been analyzed in any significant detail. One of them is 12023, which was collected from a 20 cm trench on the east rim of Sharp Crater (Shoemaker *et al.*, 1970). No major element or petrographic data are published for this soil. We shall show below that 12023 is a virtually pure KREEP soil, ~10% of which is KREEP ropy glass, and discuss the implications of this important soil for the evolution of the Apollo 12 site.

CHARACTERIZATION OF 12023

Major element chemistry

The <1 mm sieve fraction of 12023 was analyzed for major elements using a fused bead sample preparation technique (Brown, 1977) and an ARL microprobe. The average of two analyses is given in Table 1. For comparison, the average major element composition of KREEP ropy glass, gray-mottled KREEP, Apollo 14 soil, and mixing-model KREEP are also given. The KREEP ropy glass and gray-mottled KREEP compositions were determined from analyses of individual soil particles from Apollo 12 soils other than 12023. Mixing-model KREEP is the composition of Apollo 12 KREEP derived from mixing-model calculations. It is readily apparent from Table 1 that as far as major element data is concerned sample 12023 is pure KREEP. In fact, the major element compositions of 12023 and the average of Apollo 14 soils are indistinguishable within experimental error.

Ferromagnetic resonance (FMR) and magnetic studies

The value of the FMR linewidth (ΔH) for 12023 is 590 G. This value is in marked contrast to the 725–775 G range observed previously for all other Apollo 12 soils, and it falls in the 550–600 G range observed for Apollo 14 soils (Morris, 1976). Since ΔH co-varies with the concentration of FeO in a soil (Morris, 1976),

Table 1. Major element data for 12023 and representative KREEP compositions.

	12023,25	Average KREEP glass*	Average gray-mottled KREEP*	Average Apollo 14 soils	Mixing-model KREEP
SiO_2	47.6	48.2	49.5	47.9	46.2
TiO_2	1.7	2.0	1.4	1.7	—
Al_2O_3	18.3	15.5	18.2	17.6	17.0
Cr_2O_3	0.2	—	—	0.3	—
FeO	9.9	10.6	9.4	10.4	10.5
MgO	9.5	7.7	7.2	9.2	8.3
CaO	10.6	10.9	10.9	11.2	11.1
Na_2O	0.8	—	1.2	0.7	0.9
K_2O	0.6	1.1	0.9	0.6	0.6
P_2O_5	—	0.8	1.1	0.5	—
Total	99.2	96.8	99.8	100.1	—
Fe/(Fe + Mg)†	0.37	0.44	0.42	0.39	0.41
Preference	(a)	(b)	(b)	(c)	(d)

References: (a) This work; (b) Meyer (1977); (c) Rose *et al.* (1972); (d) Wänke *et al.* (1972).

*Measurements performed on KREEP particles in Apollo 12 soils other than 12023.

†Atomic ratio.

the lower linewidth of 12023 reflects its lower FeO concentration as compared to the rest of the Apollo 12 soils. It was this "anomalous" value of ΔH that first alerted us to the unique nature of sample 12023.

The value of the FMR maturity index, I_s/FeO (<250 μm), for 12023 is 55.0 units; soil 12023 is thus submature (Morris, 1976).

As measured magnetically, the iron metal concentration of 12023 is 0.51 wt.%. This value is typical for lunar soils (e.g., Pearce *et al.*, 1974; Cisowski *et al.*, 1974).

Grain-size analysis

The <1 mm bulk sample was sieved using standard procedures (McKay *et al.*, 1974) down to 20 μm. The sieve data are shown in Table 2. The 1 mm to 1 cm size distributions were obtained from the Curator. A histogram of the grain-size distribution is shown in Fig. 1 for all sieve fractions less than 1 cm. The graphic mean grain size (M_z) and inclusive graphic standard deviation (σ_I) for the <1 cm data are 4.01 ϕ (62 μm) and 2.17 ϕ (222 μm). For the <1 mm data, those parameters are 4.21 ϕ (54 μm) and 2.04 ϕ (243 μm). These grain-size parameters and the unimodal histogram shown in Fig. 1 are typical of submature to mature soils (McKay *et al.*, 1974).

Table 2. Sieve fraction data. The weights for sieve fractions greater than 1 mm were obtained from the Curator. Column 3 is the percent of soil in each sieve fraction for the entire soil (<1 cm).

Sieve fraction	Weight (g)	Percent < 1 cm soils
mm		
4–10	0.36	0.18
2–4	2.27	1.11
1–2	4.78	2.33
<1	197.88	96.39
Total	205.29	100.00
μm		
500–1000	0.02018	5.68
250–500	0.03436	9.66
150–250	0.03251	9.14
90–150	0.03938	11.08
75–90	0.01492	4.19
45–75	0.04806	13.51
20–45	0.06868	19.32
<20	0.08468	23.80
Total	0.34277	96.39

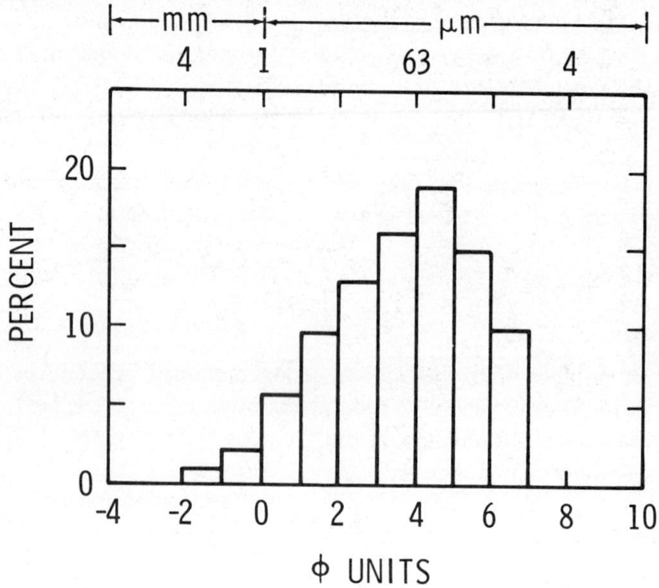

Fig. 1. Histogram of the grain-size distribution of 12023 for <1 cm sieve fraction data.

Petrography

Polished thin sections of grain mounts of the 90–150 and 250–500 μm sieve fractions were analyzed by petrographic techniques. The classification scheme adopted is given below, and the results are presented in Table 3.

Agglutinates. Irregular fragments that consist of a mixture of angular to subangular mineral and lithic debris with a light brown to very dark brown vesicular glass. The glass characteristically contains micrometer and smaller blebs of metal. The fragments commonly display banding due either to streamers of different colored glass or different amounts of metal blebs in the glass. Total abundance of debris is commonly over 50%, but in some fragments or portions of fragments the abundance of debris is as low as a few percent.

Vitric-matrix breccia. These fragments have been referred to by other authors as soil breccias and as glassy breccias. They are similar to the agglutinates in that they consist of a mixture of angular to subangular mineral and lithic debris that is held together by a matrix that contains glass fragments and has a brown to dark brown color. Whereas the agglutinate matrix consists of a solid mass of glass, the matrix of the vitric-matrix breccias consist of fragments of brown glass and mineral material.

Crystalline-matrix breccias. Fragments with a clast-matrix structure. The matrix is holocrystalline; it consists of intergrown plagioclase and pyroxene crystals 5–10 μm in size. The plagioclase-pyroxene network is speckled with micrometer-sized ilmenite crystals giving a salt-and-pepper appearance. The clasts consist of rounded to subrounded mineral and lithic grains.

Table 3. Modal analysis of the 90–150 and 250–500 μm sieve fractions of 12023,25. Numbers are percentages.

Component	90–150 μm	250–500 μm
Agglutinates	33.3	30.6
Breccias:		
Vitric-matrix br.	11.7	17.7
Crystalline-matrix br.	1.9	5.3
Gray-mottled frag.	18.1	23.0
Granulites	0.5	—
Basalt:		
Highland	4.0	4.5
KREEP	0.5	
Mare	2.4	
Undiff. + vitroph.	1.2	
Anorthosite	0.7	2.8
Feldspar	5.5	3.4
Pyroxene	2.6	0.8
Olivine	0.5	0.8
Metal	0.2	—
Glass		
Ropy	10.0	5.1
Sphere	0.2	1.1
Devitrified	1.7	0.8
Reddish brown	0.5	1.1
Orange + colorless	2.9	2.8
Heterogeneous	1.7	—
Total number of grains	420	356

Gray-mottled fragments. These are the fragments that were first described as annealed brecciated KREEP by Meyer *et al.* (1971) and gray-mottled fragments by Anderson and Smith (1971). They are holocrystalline and consist of plagioclase and low-Ca pyroxene with minor Cr-spinel, ilmenite, and metal arranged in an interlocking, but irregular texture. Set in this network are megacrysts (probably clasts) of plagioclase, low-Ca pyroxene, and rarely Cr-spinel. In some fragments the megacrysts are tabular. The texture, although irregular, resembles a mosaic or granular texture in some fragments and a subophitic texture in other fragments. These fragments probably result from the metamorphism of crushed or slightly crushed basaltic material. A few percent of these fragments contain an interstitial glass between the mineral crystals; this glass is similar to the reddish-brown glass described below. The association of holocrystalline fragments, fragments with some interstitial glass, and fragments that are mostly glass suggests that the glass is a partial melt of the holocrystalline fragments. A similar phenomenon has been described by Phinney *et al.* (1972) for material of similar texture and composition found on the Front at Apollo 15. Gray-mottled fragments are also equivalent to some of the noritic breccias (Taylor *et al.*, 1972) and metabreccias (Carr and Meyer, 1972) at the Apollo 14 site.

Granulite. Polymineralic fragments that consist of plagioclase and pyroxene and/or olivine with a granulitic texture. These fragments contain about 75% plagioclase. The texture is suggestive of solid-state annealing (metamorphism). These fragments have highland affinities and similar material has been described as breccias (Warner *et al.*, 1977).

Basalt. This category contains several types of holocrystalline materials with a "basaltic texture". That is, clast-free fragments that contain tabular-shaped plagioclase in an interlocking texture with mafic minerals. Highland basalts contain pale green to yellow pyroxenes that are not commonly zoned. KREEP basalts display plagioclase with small aspect ratios, bent plagioclase, abundant mesostasis commonly approaching 25%, and lack olivine. The KREEP basalts are similar to the feldspathic basalts described from Apollo 15 by Dymek *et al.* (1974) and the igneous KREEP 15382 and 15386. Mare basalts contain pinkish pyroxenes that commonly show zoning. The last category of undifferentiated basalts consists of those basaltic fragments that we could not assign to the above categories and several vitrophyres.

Anorthosite. Cataclastic anorthosites such as described in many studies (e.g., Warner *et al.*, 1973; Dymek *et al.*, 1975).

Mineral fragments—(feldspar, pyroxene, olivine, and metal). Simple fragments that each consists of one crystal. Some of the olivine crystals are coated with clear glass. Essentially all of these crystals have some agglutinitic glass adhering to their surfaces.

Ropy glass. Pale yellow-orange glass that contains micrometer-sized euhedral crystals and abundant mineral and lithic debris. The debris is confined to layers or bands within the fragments and to the surfaces of the fragments that are characteristically coated with debris.

Glass sphere. Complete to partial (broken) glass spheres consisting of optically homogeneous glass with much less than a percent of inclusions. Colors are pale yellow, pale orange, and cinnamon brown.

Devitrified glass. Irregular shapes or fragments of spheres that consist of radiating spherulites of plagioclase and a mafic mineral. The spherulites nucleate on the surfaces of the grains and each spherulite is about 100 μm across. Micrometer-sized blebs of spinel and/or metal are included in the spherulites.

Reddish-brown glass. This glass is reddish brown to orange, contains irregular fragments of plagioclase and pyroxene, but contains no lithic fragments. These fragments are transitional into gray-mottled fragments with interstitial glass of a similar color.

Orange and colorless glass. Irregular fragments of colorless to pale orange glass that contain less than 2% mineral and lithic debris.

Heterogeneous glass. This refers to glass fragments that are debris laden.

They form a transition group between orange and colorless glass, agglutinates, and ropy glass. If the fragments had vesicles and metal blebs they would be agglutinates, if they had crystallites and the debris was confined to bands they would be ropy, and if they did not have all the debris they would be orange and colorless.

Discussion

Since it is virtually pure KREEP, sample 12023 is obviously important for interpreting the geology of the Apollo 12 landing site. Figure 2 is a histogram of the total iron concentration for Apollo 12 soils, breccias, and mare basalts. It is readily apparent that, with respect to FeO, the soils and breccias can be considered to be appropriate combinations of 12023 and mare-basalt composition material. The close correspondence between the major element chemistry of 12023 and the major element chemistry of both KREEP soil particles in other Apollo 12 soils and the KREEP composition derived from mixing calculations support this interpretation (cf., Table 1).

As previously noted, the major element chemistry of 12023 is essentially identical to that of the average of Apollo 14 soils. The petrography is also very similar except that 12023 contains more ropy glass (~10%) in the 90–150 μm sieve fraction than do the Apollo 14 soils, which contain a maximum of 3% (McKay *et al.*, 1972).

The agglutinate particles contain all the other soil components, including ropy

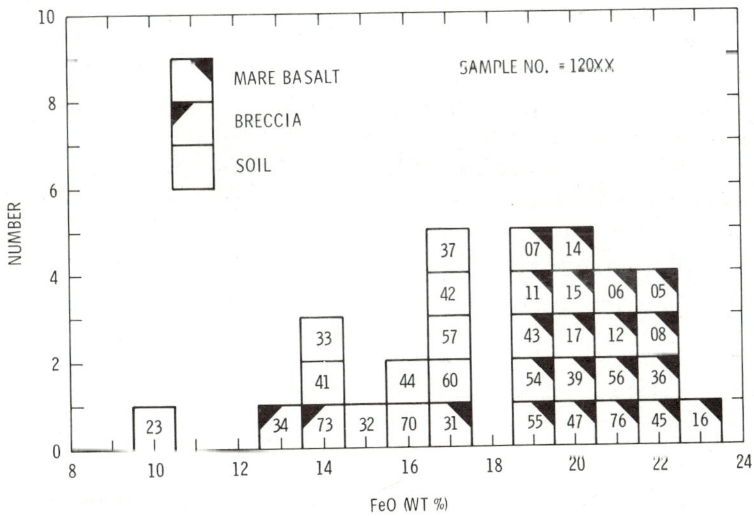

Fig. 2. Histogram of the concentration of FeO for Apollo 12 soils, breccias, and mare basalts. The mare basalt data are from Rhodes *et al.* (1977) and the other data from Warner (Curatorial data base).

glass, as fragments. This observation shows that sample 12023 underwent at least some maturation after all the components had been added.

Sample 12023 has survived significant mare contamination even though it has undergone maturation to a submature state, as shown by the major element chemistry and FMR linewidth (ΔH). If this maturation took place at the Apollo 12 site, there must be a significant amount of 12023-like material present to prevent admixture of a significant mare component. Evidence for this is the bright ray pattern around Sharp Crater (Shoemaker *et al.*, 1970), where 12023 was collected. This bright pattern may not be due as much to the freshness of the crater as to the ejection of high albedo KREEP onto the surface. In addition the Apollo 12 astronauts reported a change in ground firmness between Sharp and Bench craters (Bean *et al.*, 1970) which may be the edge of a KREEP deposit.

Whether such a large deposit of KREEP material can be part of a ray from Copernicus remains an open question. However, Oberbeck *et al.* (1974) have argued that rays are mostly excavated local material rather than primary crater ejecta. The source of 12023 may be more local, possibly ejecta from local outcrops of Fra Mauro material or Fra Mauro material excavated from beneath the mare basalts. These origins have been proposed before (e.g., Quaide *et al.*, 1971; Wasson and Baedecker, 1972; Marvin *et al.*, 1971; Meyer *et al.*, 1971). We do feel, as do others (e.g., Meyer *et al.*, 1971), that the ropy glasses, because they are impact melts, were added to the Apollo 12 site by a major impact, which may have been Copernicus, ~800 m.y. ago. The 800 m.y. formation age was derived most recently by Alexander *et al.* (1976) from sample 12033, whose KREEP component is mostly ropy glass (Meyer *et al.*, 1971; Marvin *et al.*, 1971).

SUMMARY

(1) Sample 12023 is virtually pure KREEP and is the first sample of its type known for the Apollo 12 landing site. The major element chemistry of 12023 is indistinguishable from that of Fra Mauro (Apollo 14) soils.

(2) The correspondence of the FMR linewidth (ΔH) of 12023 to those for the Apollo 14 soils shows that the reworking and maturation of 12023 occurred on a KREEP regolith surface; addition of mare material was negligible.

(3) The origin of 12023 may be ejecta from a local outcrop of Fra Mauro material or Fra Mauro material buried beneath the Apollo 12 landing site. Origin as ray material from Copernicus must also be considered.

(4) Undescribed samples may still be present in the lunar collection which have surprising and unexpected characteristics. An effort should be made to describe these remaining samples.

Acknowledgments—This paper has benefitted from discussions with Dr. C. Meyer, Jr. and the careful reviews of Drs. I. M. Steele and N. Z. Boctor. Carol Hardy is thanked for typing the manuscript.

References

Alexander E. C., Jr., Bates A., Coscio M. R., Jr., Dragon J. C., Murthy V. R., Pepin R. O., and Venkatesan T. R. (1976) K/Ar dating of lunar soils II. *Proc. Lunar Sci. Conf. 7th*, p. 625–648.

Anderson A. T., Jr. and Smith J. V. (1971) Nature, occurrence, and exotic origin of "gray-mottled" (Luny Rock) basalts in Apollo 12 soils and breccias. *Proc. Lunar Sci. Conf. 2nd*, p. 431–438.

Bean A. L., Conrad C., Jr., and Gordon R. F. (1970) Crew observations. In *Apollo 12 Prelim. Sci. Rep.*, NASA SP-235, p. 29–38.

Brown R. W. (1977) A sample fusion technique for whole rock analysis with the electron microprobe. *Geochim. Cosmochim. Acta* **41**, 435–438.

Carr M. H. and Meyer C. E. (1972) Chemical and petrographic characterization of Fra Mauro soils. *Proc. Lunar Sci. Conf. 3rd*, p. 1015–1027.

Cisowski C. S., Dunn J. R., Fuller M., Rose M. F., and Wasilewski P. J. (1974) Impact processes and lunar magnetism. *Proc. Lunar Sci. Conf. 5th*, p. 2841–2858.

Dymek R. F., Albee A. L., and Chodos A. A. (1974) Glass-coated soil breccia 15205: Selenologic history and petrologic constraints on the nature of its source region. *Proc. Lunar Sci. Conf. 5th*, p. 235–260.

Dymek R. F., Albee A. L., and Chodos A. A. (1975) Comparative petrology of lunar cumulate rocks of possible primary origin: Dunite 72415, troctolite 76535, norite 78235, and anorthosite 62237. *Proc. Lunar Sci. Conf. 6th*, p. 301–341.

Hubbard N. J. and Gast P. W. (1971) Chemical composition and origin of nonmare lunar basalts. *Proc. Lunar Sci. Conf. 2nd*, p. 999–1020.

Marvin U. B., Wood J. A., Taylor G. J., Reid J. B., Jr., Powell B. N., Dickey J. S., Jr., and Bower J. F. (1971) Relative proportions and probable sources of rock fragments in the Apollo 12 soil samples. *Proc. Lunar Sci. Conf. 2nd*, p. 679–699.

McKay D. S., Fruland R. M., and Heiken G. H. (1974) Grain size and the evolution of lunar soils. *Proc. Lunar Sci. Conf. 5th*, p. 887–906.

McKay D. S., Heiken G. H., Taylor R. M., Clanton U. S., and Morrison D. A. (1972) Apollo 14 soils: Size distribution and particle types. *Proc. Lunar Sci. Conf. 3rd*, p. 983–994.

McKay D. S., Morrison D. A., Clanton U. S., Ladle G. H., and Lindsay J. F. (1971) Apollo 12 soil and breccia. *Proc. Lunar Sci. Conf. 2nd*, p. 755–773.

Meyer C., Jr. (1977) Petrology, mineralogy and chemistry of KREEP basalt. *Phys. Chem. Earth* **10**. In press.

Meyer C., Jr., Brett R., Hubbard N. J., Morrison D. A., McKay D. S., Aitken F. K., Takeda H., Schonfeld E. (1971) Mineralogy, chemistry, and origin of the KREEP component in soil samples from the Ocean of Storms. *Proc. Lunar Sci. Conf. 2nd*, p. 393–411.

Morris R. V. (1976) Surface exposure indices of lunar soils: A comparative FMR study. *Proc. Lunar Sci. Conf. 7th*, p. 315–335.

Oberbeck V. R., Morrison R. H., Hörz F., Quaide W. L., and Gault D. E. (1974) Smooth plains and continuous deposits of craters and basins. *Proc. Lunar Sci. Conf. 5th*, p. 111–136.

Pearce G. W., Strangway D. W., and Gose W. A. (1974) Magnetic properties of Apollo samples and implications for regolith formation. *Proc. Lunar Sci. Conf. 5th*, p. 2815–2826.

Phinney W. C., Warner J. L., Simonds C. H., and Lofgren G. E. (1972) Classification and distribution of rock types at Spur Crater. In *The Apollo 15 Lunar Samples* (J. W. Chamberlain and C. Watkins, eds.), p. 149–153. The Lunar Science Institute, Houston.

Quaide W., Oberbeck V., Bunch T., and Polkowski G. (1971) Investigations of the natural history of the regolith at the Apollo 12 site. *Proc. Lunar Sci. Conf. 2nd*, p. 701–718.

Rhodes J. M., Blanchard D. P., Brannon J. C., Rodgers K. V., and Dungan M. (1977) Chemistry of Apollo 12 mare basalts: Magma types and fractionation processes. *Proc. Lunar Sci. Conf. 8th*. This volume.

Rose H. J., Jr., Cuttitta F., Annell C. S., Carron M. K., Christian R. P., Dwornik E. J., Greenland L. P., and Ligon D. T., Jr. (1972) Compositional data for twenty-one Fra Mauro lunar materials. *Proc. Lunar Sci. Conf. 3rd*, p. 1215–1229.

Schonfeld E. and Meyer C., Jr. (1972) The abundances of components of the lunar soils by a least-squares mixing model and the formation age of KREEP. *Proc. Lunar Sci. Conf. 3rd*, p. 1397–1420.

Shoemaker E. M., Batson R. M., Bean A. L., Conrad C., Jr., Dehlem D. H., Goddard E. N., Hait M. H., Larson K. B., Schaber G. G., Schleicher D. L., Sutton R. L., Swann G. A., and Waters A. C. (1970) Geology of the Apollo 12 landing site. In *Apollo 12 Prelim. Sci. Rep.*, NASA SP-235, p. 113–156.

Taylor G. J., Marvin U. B., Reid J. B., Jr., and Wood J. A. (1972) Noritic fragments in the Apollo 14 and 12 soils and the origin of Oceanus Procellarum. *Proc. Lunar Sci. Conf. 3rd*, p. 995–1014.

Wänke H., Baddenhausen H., Balacescu A., Teschke F., Spettel B., Dreibus G., Palme H., Quijano-Rico M., Kruse H., Wlotzka F., and Begemann F. (1972) Multielement analyses of lunar samples and some implications of the results. *Proc. Lunar Sci. Conf. 3rd*, p. 1251–1268.

Warner J. L., Simonds C. H., and Phinney W. C. (1973) Apollo 16 rocks: Classification and petrogenetic model. *Proc. Lunar Sci. Conf. 4th*, p. 481–504.

Warner J. L., Phinney W. C., Bickel C. E., and Simonds C. H. (1977) Feldspathic, granulitic impactites and pre-final bombardment lunar evolution. *Proc. Lunar Sci. Conf. 8th.* This volume.

Wasson J. T. and Baedecker P. A. (1972) Provenance of Apollo 12 KREEP. *Proc. Lunar Sci. Conf. 3rd*, p. 1315–1326.

Proc. Lunar Sci. Conf. 8th (1977), p. 2459–2494.
Printed in the United States of America

Petrology of four clasts from consortium breccia 73215

ODETTE B. JAMES and JANE G. HAMMARSTROM

U.S. Geological Survey, Reston, Virginia 22092

Abstract—Four clasts extracted from breccia 73215 for consortium study are: one of felsite ("granite"); and three of ANT-suite anorthositic gabbros. The felsite clast has two components: fragments of crystalline felsite; and veins and patches of felsic glass. The crystalline felsite consists largely of a vermicular intergrowth of quartz and Ba–K-feldspar; it also contains minor amounts of Ca–K-rich ternary plagioclase, Ca–Fe-rich pyroxene, Fe-rich olivine and ilmenite. Textures and mineralogy indicate it crystallized from a highly differentiated melt. The trace-element data indicate that immiscibility was probably involved in formation of the melt (Blanchard *et al.*, 1977); the Rb–Sr data suggest that igneous crystallization probably occurred between 3.90 and 4.05 b.y. (Compston *et al.*, 1977). The felsic glass component consists of crystallized brown and colorless glasses and uncrystallized colorless glass; all are K and Si rich. They contain relict fragments of quartz, olivine, plagioclase, pyroxene and other phases, and second-generation crystals of pyroxene and K-feldspar. Most of the glass was emplaced as dikelets of melt into fractures in the crystalline felsite. The clast was at a temperature ≥990°C when incorporated in the breccia, and its felsic glass component was mobile. During cooling olivine relics in the glass were enriched in Mg by reduction and by Fe–Mg exchange with breccia matrix. Textures of the brown glass and exsolution structures in the second-generation pyroxenes (Nord and James, 1977) suggest there was probably no appreciable reheating after the breccia-forming event. These data indicate that the 3.90–3.92 b.y. age of the clast (Compston *et al.*, 1977; Jessberger *et al.*, 1977) probably dates the breccia-forming event (possibly the South Serenitatis impact).

The three anorthositic gabbros have similar mineralogies and gradational textures. Plagioclase (An_{91-98}) is dominant; olivine (Fo_{71-80}) and orthopyroxene ($Wo_{3-4}En_{72-78}Fs_{19-24}$) are the major mafic minerals. One of the rocks has poikilitic texture, and the other two have both poikilitic and mosaic textures; in all three, the cores of large plagioclase grains represent preserved fragments of minerals from preexisting rocks. The petrologic data suggest that these anorthositic gabbros formed as heated, partly melted and/or recrystallized polymict breccias. The ~4.25 b.y. age obtained for all three rocks (Jessberger *et al.*, 1976; Müller *et al.*, 1977; Compston *et al.*, 1977) may be the date of the melting/recrystallization event.

INTRODUCTION

BRECCIA 73215 CONSISTS DOMINANTLY of a gray-to-black aphanitic rock. This aphanite has been termed matrix because it encloses physically separable clast material, but it itself consists of abundant small clasts set in a very fine-grained groundmass. Consortium studies (James *et al.*, 1975; James, 1976; James and Blanchard, 1976) have yielded the following interpretation of breccia origin. The aphanitic matrix formed as a mechanical mixture of hot silicate melt and cold crystalline clasts; the melt subsequently crystallized to form the groundmass. The breccia-forming event was probably a large impact, possibly the South Serenitatis basin-forming event (James *et al.*, 1975; James and Blanchard, 1976; Morgan *et al.*, 1976; Jessberger *et al.*, 1976). The clasts represent fragments of rocks granulated during the impact, and the melt likely represents impact melt; the mixing occurred as these materials were transported within, and ejected from the crater. The mixing, thermal equilibration between clasts and melt, melt crystallization, and subsequent cooling were all so rapid that the constituents of the breccia were not completely outgassed of radiogenic Ar or isotopically equilibrated (Jessberger *et al.*, 1976, 1977; Müller *et al.*, 1977; Compston *et al.*, 1977).

Four clasts have been extracted from the breccia and studied intensively by the consortium. Three of these are of anorthositic-noritic-troctolitic (ANT)-suite anorthositic gabbros that are typical of the dominant type of lithic clast in the breccia. The fourth clast is of felsite (fine-grained "granite"). Rocks of this type are present in only trace amounts in the breccia, but they are extremely important because their characteristics yield information on the genesis of lunar "granites". Locations of these four clasts relative to other samples studied by the consortium are given by James and Blanchard (1976).

FELSITE

This clast, taken from matrix subsample 43, had a total weight of ~20 mg. Slivers from its edges were made into thin sections for the petrologic and transmission electron microscopy (TEM) studies (Nord and James, 1977). The largest chip from the clast (13.6 mg) was irradiated and analyzed by instrumental neutron activation (INAA), and was then split for Rb–Sr and ^{40}Ar–^{39}Ar studies. The INAA, Rb–Sr, and ^{40}Ar–^{39}Ar data are reported by Blanchard *et al.* (1977), Compston *et al.* (1977), and Jessberger *et al.* (1977), respectively.

The clast consists of two felsic components: (1) fragments of crystalline felsite, and (2) veins and irregular patches of felsic glass. The crystalline felsite fragments are rounded to subangular, and the glass veins and patches cut and enclose them (Fig. 1a). (Felsic glass also is a minor constituent within the crystalline felsite.) Textures within the crystalline fragments do not match across the glass veins, but would match if the fragments had been dilated and laterally displaced about 0.2–0.3 mm. Locally veinlets of breccia groundmass cut the clast; these veinlets have intrusive contacts with crystalline felsite and gradational contacts with felsic glass.

Crystalline felsite

Texture. The bulk of this rock is a vermicular intergrowth of quartz and K-feldspar: worm-like blebs of quartz 10–20 μm across are poikilitically enclosed by ovoid grains of K-feldspar 0.1–0.8 mm across (Fig. 1b). Quartz forms ~40% and K-feldspar ~60% of this intergrowth. Minor plagioclase forms blebs associated with the quartz (Fig. 1c). The quartz and plagioclase blebs tend to have subparallel elongations (Fig. 1c), and to be optically continuous, within a given K-feldspar grain. Trace amounts of ilmenite, zircon, and olivine form 5–50 μm equant anhedral grains. Trace amounts of apatite and whitlockite form 5–50 μm equant grains and needles; these commonly are complex intergrowths of the two phases, with whitlockite being dominant.

The crystalline felsite contains sparse large grains of several minerals, and patches where intermediate-sized grains of two or three minerals are intergrown to form mosaics. The isolated large grains are of: irregular anhedral plagioclase—0.2–0.5 mm grains bordering K-feldspar; blocky plagioclase—a 0.5 mm grain with a 0.4 mm homogeneous core and a zoned overgrowth rim (Fig. 1d); blocky ilmenite—0.1 mm grains; and ovoid olivine—0.1 mm grains. In the mosaics grain size is about 50–75 μm, and the mineral assemblages are: quartz + plagioclase ± K-feldspar; quartz + olivine (Fig. 1b); quartz + pyroxene; and quartz + ilmenite. The quartz grains are equant and tend to be polygonal, the plagioclase grains are equant to elongate and irregular, the K-feldspar grains are generally quartz-free areas of grains that are part of the vermicular intergrowth, and the olivine and ilmenite grains are typically ovoid; where several olivine grains are present in a single mosaic, they commonly are optically continuous.

Mineralogy. The K-feldspar is a nearly pure Ba–K-feldspar (Figs. 2, 3). The plagioclase shows a considerably greater compositional range than the K-feldspar, with the dominant trend of variation being one of K enrichment and accompanying slight decrease in Na/Na + Ca (Figs. 2, 3). The bulk of the plagioclase is a unique ternary feldspar rich in both K and Ca; "normal" low-K plagioclase is present but only as the core of the 0.5 mm zoned grain mentioned above. Plagioclase composition is correlated with grain size and occurrence, as follows. The large zoned grain has a core of homogeneous $Or_2Ab_{23}An_{75}$; the rim of this grain first shows a trend of increasing Na/Na + Ca, but in a few microns this trend changes abruptly to one of rapidly increasing K and slightly decreasing Na/Na + Ca. The finest-grained plagioclase, that in the vermicular intergrowth, is the most K rich and averages $Or_{21}Ab_{21}An_{58}$; intermediate-sized plagioclase has intermediate compositions (Fig. 2). BaO does not increase with K_2O in the plagioclase, but instead remains fairly constant (Fig. 3). The olivine is very Fe rich, averaging $Fo_{19.4}$; CaO is ~0.25%, MnO ~0.70%, and Cr_2O_3 ~0.04%. The pyroxene is Ca–Fe rich, averaging $Wo_{36.9}En_{23.4}Fs_{39.7}$ (Fig. 4); TiO_2 is ~0.35%, Al_2O_3 is ~0.60%, Cr_2O_3 is ~0.25%, and MnO is ~0.30%. Compositions of the other phases are listed in Table 1.

Felsic glass

Texture. The glass is highly variable in vesicularity, color, and content of relict mineral fragments, second-generation crystals that precipitated from the melt as it was quenched, and minute crystallites (Fig. 1e). The vesicles are irregular, as much as 0.3 mm across, and occur in the larger patches and veins of glass (Fig. 1e). The bulk of the glass is brown and contains abundant minute needle-like crystallites with radiating or subparallel orientations; TEM studies have shown that these crystallites form only a small volume of the glass and that the bulk of this material is uncrystallized (Nord and James, 1977). The crystallites are too small to be identified optically, but their reflectivities suggest they are probably silica minerals and feldspars. Crystallites of mafic minerals are concentrated locally, and minute particles of Fe metal and/or troilite are ubiquitously abundant.

The remainder of the glass consists of two colorless varieties, one crystallized and the other uncrystallized. In the crystallized variety, the intergrowth is a mat of tiny (~1 μm) subparallel blades and locally has micrographic texture. The dominant phases appear to be quartz and K-feldspar; tiny blebs and prisms of pyroxene are locally concentrated in the intergrowth. The two varieties of colorless glass have different distributions. The crystallized variety occurs at the margins of and within the crystalline felsite fragments; it forms small irregular patches, bands ≤10 μm thick along grain boundaries, and irregular haloes around mafic mineral grains. The bulk of the uncrystallized colorless glass occurs within the veins and patches of felsic glass. It forms large irregular patches concentrated near the edge of the clast (Fig. 1e), and thin haloes surrounding mafic mineral grains and Fe-metal globules within brown glass. Uncrystallized colorless glass also is present within the crystalline felsite, as thin selvages marking grain boundaries (Fig. 1c; also see Nord and James, 1977, for TEM photographs). These selvages are thickest (up to 10 μm) surrounding ilmenite or mafic-mineral grains, and thinner at quartz-feldspar contacts (~1 μm). In crystalline felsite near the edge of the clast the selvages widen as the clast margin is approached; the clast-matrix contact is marked by a glass band 10–15 μm thick.

Fragments of relict mineral grains are scattered throughout the glass veins and patches (Fig. 1e). These are, in approximate order of abundance, of olivine, quartz, plagioclase, ilmenite, calcic pyroxene, phosphate minerals, zircon, pigeonite, and K-feldspar. Olivine relics are ovoid (Fig. 1e): some are small, have irregular shapes suggesting extensive resorption, and are encased in thick rims of pyroxene; others are larger, have more rounded outlines, and have thin pyroxene rims or are unrimmed. Quartz relics are ovoid to angular; they are cut by fractures filled by uncrystallized

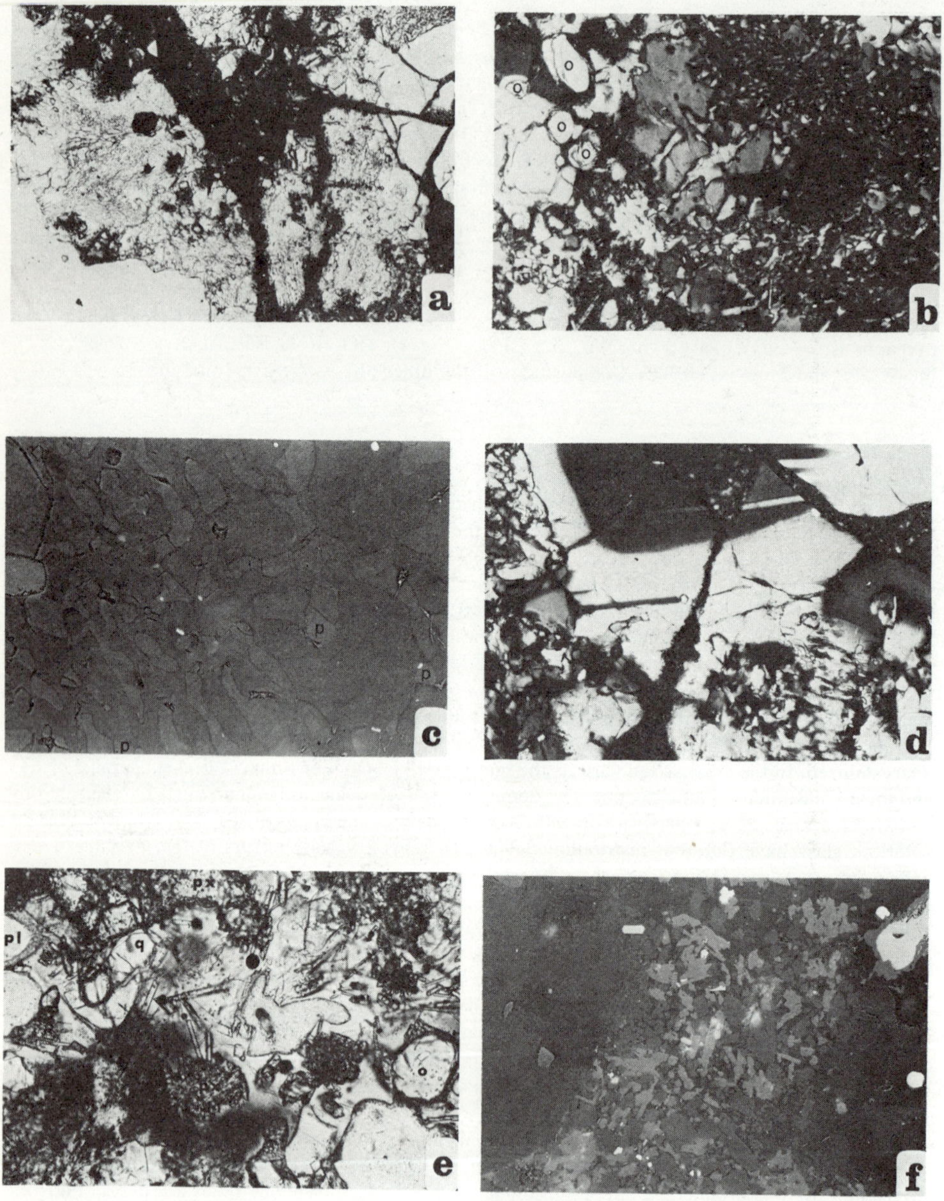
a
b
c
d
px
pl
q
e
f

colorless glass and are rimmed by similar glass. Plagioclase relics are ovoid to highly irregular (see Fig. 2 in Nord and James, 1977). They are cut and rimmed by 5–10 μm bands of reacted plagioclase that has slightly higher reflectivity than the unreacted mineral and is riddled with micron-sized K-feldspar inclusions (Nord and James, 1977). Sparse ovoid relics consist entirely of such reacted plagioclase. Ilmenite relics are ovoid to blocky, and some have rims of second-generation mafic minerals (Fig. 1f). Pyroxene relics have highly irregular shapes modified by overgrowth of skeletal second-generation pyroxene (Fig. 1e). Phosphate-mineral relics are ovoid to blocky, zircon relics are angular and internally fractured or equant and blocky, and K-feldspar relics are angular.

Second-generation mineral grains that precipitated from the felsic melt are abundant in the glass veins and patches. Most of these grains are of pyroxene and K-feldspar; a small number are of plagioclase, phosphate minerals, and olivine. The second-generation pyroxene rims pyroxene, olivine, and ilmenite relics as described above; it also forms prisms as much as 150 μm long (Fig. 1e) and 75 μm blocky skeletal grains. The K-feldspar forms ~15 μm euhedral laths, and skeletal rims on plagioclase relics. The phosphate minerals form needle-like grains, and the olivine rims ilmenite relics. The second-generation plagioclase, which is inclusion rich and very similar to the reacted plagioclase described above, forms rims on plagioclase relics (Fig. 1f); it also rims plagioclase and K-feldspar grains that occur within crystalline felsite but border felsic glass. The second-generation plagioclase is distinguished from reacted relict plagioclase as follows: (1) the inclusions it contains are typically of brown glass; (2) the inner parts of the rims it forms are much richer in inclusions than the outer parts, and the outer edges of the rims are marked by thin inclusion-free bands having euhedral terminations; and (3) it is abundant in parts of the clast near breccia matrix.

Fig. 1. Photomicrographs of felsite clast. (a) Low power view of entire clast. Plane polarized light; width of field of view is 2.2 mm. The clast consists of crystalline felsite (light areas) and felsic glass (dark areas). The black material at the lower right edge is breccia matrix. The bulk of the crystalline felsite is a vermicular intergrowth of quartz and K-feldspar; this intergrowth contains a large grain of zoned plagioclase (right edge). The felsic glass forms veinlets cutting the crystalline felsite (center of field) and irregular patches (bottom right). The large plagioclase grain is fractured; the fractures are filled by matrix-derived groundmass near the edge of the clast and by clast-derived felsic glass in the interior of the clast. (b) Vermicular intergrowth and mosaic patches in crystalline felsite. Crossed polarizers; width of field of view is 0.88 mm. The right half of the field shows the vermicular intergrowth; the small white blebs are of quartz and the light-gray, optically continuous grain enclosing them is of K-feldspar. The upper left corner of the field shows a mosaic patch that consists of olivine (o) and quartz. (c) Vermicular intergrowth in crystalline felsite. Reflected light; width of field of view is 0.22 mm. Quartz (light gray) is enclosed by K-feldspar (intermediate gray). Sparse plagioclase (p) is associated with the quartz. Quartz-feldspar boundaries are marked by thin septa of glass (dark gray). (d) Zoned plagioclase grain in crystalline felsite. Crossed polarizers; width of field of view is 0.88 mm. (e) Felsic glass patch. Plane polarized light; width of field of view is 0.55 mm. The glass ranges from clear and uncrystallized (center), to brown and clouded with minute crystallites (lower center). Irregular and ovoid vesicles are present (lower right, center). The glass contains relict fragments of olivine (o), pyroxene (px), quartz (q), and plagioclase (pl). Second-generation pyroxene forms rims on olivine relics, euhedral overgrowths on pyroxene relics, and prismatic crystals. (f) Zone of mixing of felsic glass and matrix mineral grains at edge of clast. Reflected light; width of field of view is 0.22 mm. The right, left, and top of the field are in the felsite clast, the bottom is in matrix, and the center is in the mixed material. The mixed material consists of matrix-derived subhedral grains of pyroxene (light gray) and plagioclase (intermediate gray), and clast-derived interstitial felsic glass (dark gray). At the left of the field plagioclase relics are overgrown by second-generation inclusion-rich plagioclase. At the top right of the field an ilmenite relic is rimmed by second-generation pyroxene.

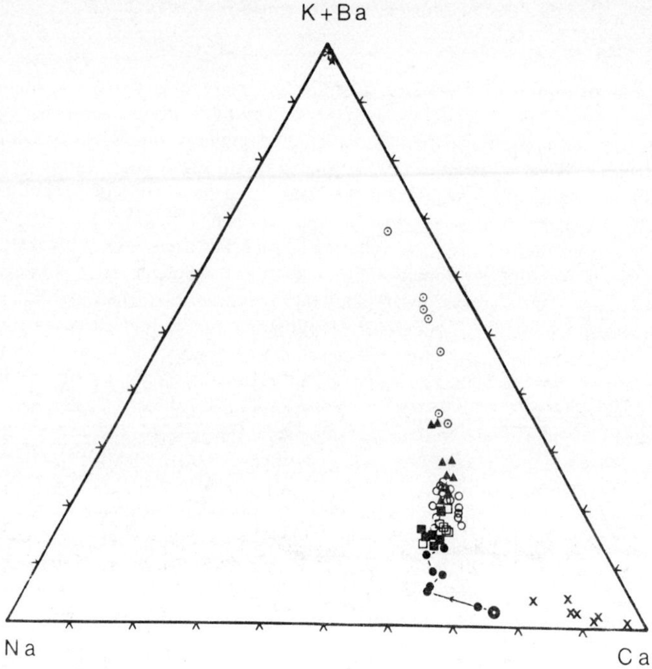

Fig. 2. K + Ba : Na : Ca contents of feldspars in felsite clast. Large circle, average of 14 analyses in core of zoned plagioclase grain (Fig. 1d); small filled circles, traverse from inner to outer boundary of rim of zoned plagioclase grain (3 μm steps, arrow indicates direction of traverse). Filled squares, points in equant 0.2 mm plagioclase grain. Open squares, plagioclase grains in mosaics. Small open circles, plagioclase grains in vermicular intergrowth. Filled triangles, centers of plagioclase relics in felsic glass. Small dotted circles, reacted and second-generation inclusion-rich plagioclase in felsic glass. Asterisk, average of 26 analyses of K-feldspar in vermicular intergrowth. Filled star, average of 11 analyses of second-generation K-feldspar in felsic glass. ×, small plagioclase grains in mixed material at edge of clast.

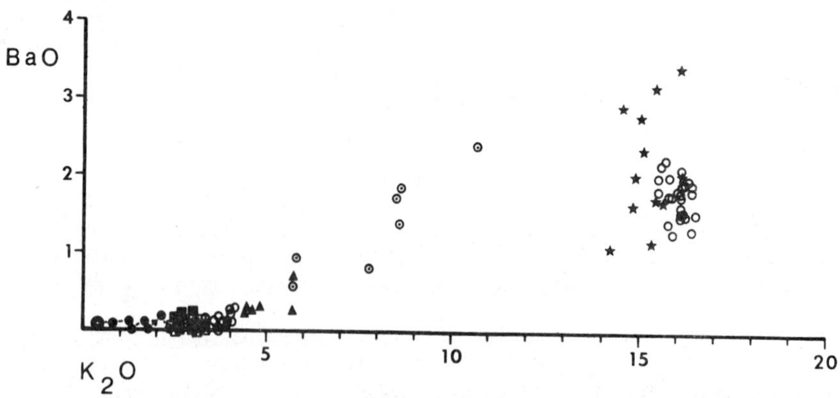

Fig. 3. K_2O vs. BaO contents in feldspars in felsite clast. Symbols are as in Fig. 2, except: open circles indicate all analyses of plagioclase and K-feldspar in vermicular intergrowth; filled stars indicate all analyses of second-generation K-feldspar in felsic glass; plagioclase in mixed material is not shown.

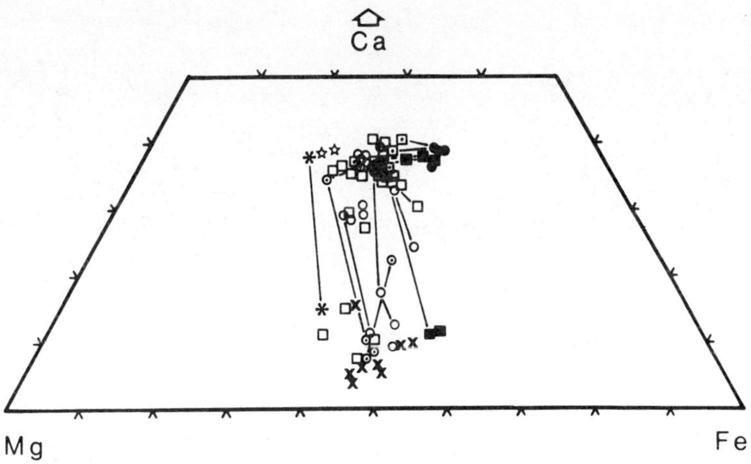

Fig. 4. Ca:Mg:Fe contents of pyroxenes in felsite clast. Filled circles, centers of pyroxene grains in crystalline felsite. Filled squares, centers of pyroxene relics in felsic glass. Open symbols are second-generation pyroxenes, as follows: dotted squares, edges of grains in crystalline felsite or edges of relics in felsic glass; circles, rims on olivine relics; squares, prismatic crystals; dotted circles, large blocky grains with associated Fe metal and troilite; stars, rims on ilmenite relics; asterisks, rim on olivine relic at edge of clast. ×, grains in mixed material at edge of clast. Tie lines connect compositions of points in the same grains.

Globules and irregular blebs of Fe metal are abundant in the glass veins and patches; sparse blebs of troilite are associated with these particles. These phases are concentrated in uncrystallized colorless glass and commonly are associated with second-generation pyroxene.

Locally material consisting of a mixture of clast-derived felsic glass and matrix-derived mineral grains forms: ovoid protrusions into the clast at its margin (Fig. 1f); irregular bands along the clast-matrix contact; or veinlets that cut crystalline felsite. The mixed material comprises 5–10 μm grains of matrix-derived plagioclase, pyroxene and olivine, and interstitial clast-derived felsic glass. The mixed material in contact with felsic glass contains ~25% interstitial glass; at increasing distances from this contact its content of felsic glass decreases, and the contact with matrix is gradational.

Locally felsic glass derived from the clast intrudes surrounding matrix, forming veinlets about 10 μm thick. These veinlets are mostly of uncrystallized colorless glass and contain sparse second-generation crystals of pyroxene and K-feldspar. Felsic glass also occurs filling minute interstices in the matrix as much as 100 μm from the edge of the clast.

Compositional data. Very finely crystallized brown glasses and uncrystallized colorless glasses have been analyzed (Table 1). All are rich in K_2O and SiO_2. The brown glasses average higher in K_2O and FeO, and lower in SiO_2 and Na_2O, than the colorless glasses.

The relict mineral fragments have compositions very similar to the compositions of the minerals in the crystalline felsite. (Pigeonite is an exception because this mineral was not found in the crystalline felsite examined; the relict pigeonite is very Fe rich, $Wo_{11.6}En_{35.9}Fs_{52.5}$, and contains 0.80% Al_2O_3, 0.52% TiO_2, 0.23% Cr_2O_3, and 0.62% MnO.) The compositions of some plagioclase, olivine, and pyroxene relics have been somewhat modified by reaction and diffusive exchange with surrounding felsic melt and with nearby matrix. Some plagioclase relics with rims of inclusion-rich plagioclase are significantly richer in K, slightly richer in Ba, and have lower Na/Na + Ca than plagioclase in the crystalline felsite (Figs. 2, 3). Olivine relics show increases in Mg/Mg + Fe correlated with extent of rimming by pyroxene, resorption, association with Fe-metal globules, and

Table 1. Compositions of minor phases and glasses in felsite (wt.%, electron microprobe analyses).

	(1)	(2)	(3)	(4)	(5)	(6)	(7)	(8)	(9)
SiO_2	0.22	0.28	—	—	74.2	78.2	76.7	76.8	78.1
TiO_2	50.7	1.79	—	—	0.73	0.36	0.65	0.28	0.49
Al_2O_3	0.13	22.2	—	—	13.4	12.9	13.1	13.4	12.6
FeO	45.4	31.8	1.81	0.34	1.24	0.61	0.81	0.38	0.40
MgO	1.15	1.89	2.29	0.07	0.04	0.01	0.04	0.03	0.03
CaO	0.09	0.09	38.7	53.4	0.62	0.64	0.79	0.75	0.77
Na_2O	—	—	—	—	0.10	0.14	0.21	0.17	0.30
K_2O	—	—	—	—	9.58	6.90	7.20	7.88	7.21
BaO	—	—	—	—	0.26	0.06	0.04	0.11	0.10
MnO	0.47	0.51	—	—	0.02	0.02	0.02	0.01	<0.01
Cr_2O_3	0.38	38.1	—	—	0.03	0.01	0.02	0.01	0.02
Y_2O_3	—	—	4.48	0.37	—	—	—	—	—
La_2O_3	—	—	1.07	0.13	—	—	—	—	—
Ce_2O_3	—	—	3.01	0.23	—	—	—	—	—
Nd_2O_3	—	—	1.63	0.19	—	—	—	—	—
Sm_2O_3	—	—	0.59	0.06	—	—	—	—	—
Gd_2O_3	—	—	1.37	0.20	—	—	—	—	—
P_2O_5	—	—	40.5	40.4	0.21	0.04	0.26	0.09	0.20
ZrO_2	—	—	—	—	0.11	0.13	0.22	0.04	0.07
F	—	—	0.43	2.88	—	—	—	—	—
Cl	—	—	0.29	0.89	—	—	—	—	—
Total	98.54	96.66	96.17	99.16	100.54	100.02	100.06	99.95	100.29

(1) Ilmenite: average of 12 analyses of grains in felsic glass and crystalline felsite.

(2) Aluminous chromite: inclusion in ilmenite grain in crystalline felsite (low total due to small size of grain).

(3) Whitlockite: average of three analyses of grains in crystalline felsite (low total due to destruction of grains during analysis).

(4) Apatite: average of eight analyses of grains in felsic glass and crystalline felsite.

(5) Brown glass: average of 20 analyses.

(6) Uncrystallized colorless glass in felsic glass veins and patches: average of ten analyses.

(7) Uncrystallized colorless glass haloes around second-generation and relict mafic mineral grains: average of 12 analyses.

(8) Uncrystallized colorless glass selvages in vermicular intergrowth in crystalline felsite: average of five analyses.

(9) Uncrystallized colorless glass bands at contact of crystalline felsite and matrix: average of five analyses.

proximity to matrix: relics with no pyroxene rims average $Fo_{22.8}$; relics with narrow rims average $Fo_{28.4}$; relics with wide rims and resorbed-appearing shapes average $Fo_{32.1}$; relics with wide rims and associated Fe-metal globules are $Fo_{39.3-50.1}$; and relics that have no pyroxene rims but are near matrix are $Fo_{30.4-35.8}$. Centers of Ca-rich pyroxene relics are slightly more magnesian than Ca-rich pyroxene in the crystalline felsite.

Second-generation pyroxenes have highly variable Mg/Mg + Fe and Ca/Ca + Mg + Fe, and they average more magnesian and less calcic than the pyroxene in the crystalline felsite (Fig. 4). The most magnesian second-generation pyroxenes occur in felsic glass near the edge of the clast, and in some cases have abundant associated Fe-metal globules. The grains appear optically homogeneous, but points analyzed within single grains commonly vary greatly in composition (Fig. 4), and the TEM studies have established that the grains consist of very finely exsolved mixtures of low-Ca and high-Ca pyroxene (Nord and James, 1977). TiO_2, Al_2O_3, Cr_2O_3, and MnO contents in the second-generation high-Ca pyroxenes average higher than in the high-Ca pyroxene of the crystalline felsite; contents of these oxides average 0.99, 1.49, 0.28, and 0.33%, respectively. Contents of these oxides in the second-generation low-Ca pyroxenes average 0.52, 0.81, 0.22, and 0.43%, respectively.

Second-generation K-feldspars have highly variable Ba contents, and Ba, Na, and Ca average higher than in the K-feldspar in the crystalline felsite (Figs. 2, 3). Analyses attempted of the reacted and second-generation plagioclase have yielded compositions that are mixtures of plagioclase and K-feldspar (Figs. 2, 3). The analyses establish that the plagioclase has lower Na/Na + Ca than any other plagioclase in the clast, and Na/Na + Ca appears negatively correlated with proximity to breccia matrix. Reflectivity of the plagioclase indicates it has lower K than unreacted relict plagioclase, but how much lower cannot be determined from the analyses.

The Fe-metal particles have very low Ni and Co contents. Ni ranges from 0.27 to 0.70% and averages 0.47%; Co ranges from 0.15 to 0.35% and averages 0.25%.

Compositions of matrix-derived grains in veinlets and in the mixed material at the clast margin reflect their matrix source. The plagioclase grains range from $Or_1Ab_3An_{96}$ to $Or_4Ab_{16}An_{80}$ (Fig. 2), the same range as found for small grains in breccia matrix (James, 1976). The pyroxene grains average about $Wo_8En_{46}Fs_{46}$ (Fig. 4), whereas small pyroxene grains in matrix average about $Wo_3En_{73.5}Fs_{23.5}$ (James, 1976); it appears that these grains were enriched in Fe due to exchange with the felsic material but retained the low Ca contents typical of matrix pyroxene. The olivine grains are $\sim Fo_{43}$, whereas small olivine grains in matrix are $\sim Fo_{70}$; thus olivine also apparently exchanged Fe–Mg with the felsic material.

Genesis and History of Felsite

Crystalline felsite

Grain shapes, spatial relations, and grain-size variations of its minerals indicate that the crystalline felsite crystallized from a melt; the compositional variations of the minerals are in accord with the paragenesis that is indicated by the textures. Crystallization took place as follows. The earliest phase to precipitate was "normal" plagioclase, $Or_2Ab_{23}An_{75}$, forming the largest grain in the clast. Plagioclase continued to precipitate but its composition changed rapidly, first following a trend of Na enrichment, and subsequently following a trend of K enrichment accompanied by decreasing Na/Na + Ca. Then plagioclase still richer in K co-precipitated with quartz, forming mosaic intergrowths of intermediate-sized grains. At this stage Ca–Fe-rich pyroxene, Fe-rich olivine, and ilmenite (and probably also Fe-rich pigeonite) were also crystallizing. The bulk of the mafic minerals appears to have crystallized at this time, but it is not clear from the textures whether or not these minerals were also crystallizing prior to the appearance of quartz. At the end of crystallization, fine-grained quartz and K-feldspar precipitated, together with minor K-rich plagioclase, and formed the vermicular intergrowth; minor ilmenite, zircon, phosphate minerals, olivine, and pyroxene were also precipitating. The order of appearance of the two feldspars and quartz, and their relative abundances in the various types of intergrowth, are as would be predicted for melt crystallization from the phase relations in the anhydrous system leucite-anorthite-silica (Schairer and Bowen, 1947).

The compositional variation trend in the plagioclases formed ternary feldspars with Ca–K-rich compositions unknown from natural terrestrial occurrences and laboratory crystallization experiments, but previously reported for feldspars from felsite clasts in breccias from Boulder 1, Station 2, Apollo 17 (Ryder *et al.*, 1975). The correlation of more K-enriched compositions with later crystallization time, and the fact that K enrichment was not accompanied by Ba enrichment (Fig. 3), suggest that these plagioclases crystallized from the melt as homogeneous feldspars with ternary compositions. The TEM studies have confirmed this supposition (Nord and James, 1977).

The high-Si, high-K melt that crystallized to form the felsite could have formed either as a partial melt or as a residual liquid in a differentiated igneous complex. The rock textures are similar to those of terrestrial granophyres, and favor the latter alternative. In particular, the presence of a large homogeneous plagioclase grain suggests that there was an earlier, slower stage of crystallization preceding crystallization of the bulk of the felsite; such a grain is an unlikely relic from partial melting.

Composition of the parent magma that differentiated to form the felsite melt, and whether or not separation of immiscible liquids occurred during the differentiation, are important points relevant to the broader question of the origin of lunar "granites". In considering these points, the major- and trace-element data provide the most critical information. The question of immiscibility is the easier to evaluate. During fractional crystallization, K, Rb, Ba, P, Zr, and REE are all strongly partitioned into the residual melt; if this melt splits into immiscible K–Si-rich and Fe-rich liquids, K, Rb, and Ba concentrate in the K–Si-rich liquid and the REE, P, and Zr concentrate in the Fe-rich liquid (Hess and Rutherford, 1974; Ryerson and Hess, 1975; Watson, 1976). Thus consideration of elemental ratios permits late-stage melts formed by immiscibility to be distinguished from those formed by fractional crystallization alone (Hess *et al.*, 1975). Using such criteria, Blanchard *et al.* (1977) conclude that separation of immiscible liquids was indeed involved in origin of the felsite melt. The question of composition of the parent magma that differentiated to form the felsite melt is considerably more difficult to evaluate. Blanchard *et al.* (1977) conclude that, as there is now very little data available on elemental partitioning during crystallization of late-stage immiscible melts, the differentiation process cannot yet be adequately modeled; thus the source of the felsite melt cannot yet be determined.

At present it is also impossible to make any definitive statement on the origin of lunar "granites" in general. Only a few other "granites" have been studied in detail, because such rocks are found in the returned samples primarily as very rare, <1 mm fragments in breccias and soils; thus isolation of material for analysis is very difficult. Various investigators have described the petrology of "granitic" fragments in soil and breccia samples (Wood, 1970; Marvin *et al.*, 1971; Anderson *et al.*, 1972; Brown *et al.*, 1972; Wilshire and Jackson, 1972; Roedder and Weiblen, 1972; Sclar *et al.*, 1974). Detailed petrologic studies were made of felsite clasts in Apollo 17 Boulder 1 at Station 2 (Ryder *et al.*, 1975), and matrix samples rich in such clasts were analyzed for Rb–Sr (Compston *et al.*, 1975) and for other trace elements (Jovanovic and Reed, 1975); no pure felsite was analyzed, however. Detailed petrologic studies were made of granite and related quartz monzodiorite clasts from breccia 15405 (Ryder, 1976); the quartz monzodiorite was analyzed and dated (Nyquist *et al.*, 1977a,b; Tatsumoto and Unruh, 1976; Bernatowicz *et al.*, 1977), but no similar studies were made of the granite. Felsite in breccia 12013 has also been intensively studied petrologically (Albee *et al.*, 1970; Drake *et al.*, 1970; James, 1970; Quick and Albee, 1976; Quick *et al.*, 1977). Subsamples from this breccia were analyzed and dated (see Earth and Planetary Science Letters, Vol. 9, No. 2, 1970, for results), but again none were of pure felsite; Quick *et al.* (1977), however, have used a combined petrologic-chemical study to determine major- and rare-earth element chemistry of the pure felsite. This lack of data makes it difficult to judge whether or not the 73215 felsite data are of general application.

The petrologic data indicate that most of the lunar "granites" that have been studied are igneous rocks. Although many of them show shock-induced deformation and vitrification produced by the events that incorporated them in their host breccias, they do not show evidence of the events of intense brecciation, high temperature metamorphism and remelting that most ANT-suite rocks appear to have experienced. Origin of the granitic melts, whether as residual liquids or partial melts, is in most cases impossible to determine because of the small sizes of the samples; where there is evidence on melt genesis, however, it usually suggests origin as a residual melt.

Whether or not separation of immiscible liquids was involved in the formation of lunar "granites" is a question that has been considered by many investigators. Most have favored origin of the "granites" by immiscibility (Roedder and Weiblen, 1970, 1972; Hess *et al.*, 1975; Rutherford and Hess, 1975; Rutherford *et al.*, 1976; Quick *et al.*, 1977; Blanchard *et al.*, 1977), and the arguments are well summarized by Hess *et al.* (1975), and Rutherford *et al.* (1976). In opposition to this view, Ryder (1976) and Nyquist *et al.* (1977a,b) have proposed that the granite and quartz monzodiorite they studied in 15405 formed by fractionation of a KREEP basalt magma without separation of immiscible liquids. The trace-element data do support such a relationship between KREEP basalt and the quartz monzodiorite, but the lack of trace-element data for the granite makes it difficult to evaluate the relationship of this rock to the quartz monzodiorite and to judge how it might be related to other lunar "granites". The major-element data suggest that the 15405 granite may indeed not be "typical"; its contents of FeO, MgO, and P_2O_5, and its Mg/Mg + Fe, are much higher than in the 73215 and 12013 felsites, and higher than in most small fragments of felsite and felsic glass that have been found in other breccias and soils.

The question of the composition of the parent magma that differentiated to form the granitic melts has also been considered by many investigators. Most have favored a KREEP basalt parent because there is a fairly consistent spatial association between KREEP basalts and "granites" at the collecting sites (Brown *et al.*, 1972; Meyer, 1972; Ryder *et al.*, 1975; Hess *et al.*, 1975; Nyquist *et al.*, 1977a,b; Ryder, 1976). The REE distribution patterns in the felsites, however, are very difficult to explain by such a relationship. Fractional crystallization of KREEP basalt magma followed by separation of late-stage immiscible liquids could not produce a granitic melt with the REE pattern observed in the felsites (Blanchard *et al.*, 1977; Quick *et al.*, 1977). It is further unlikely that the felsite REE pattern could have been produced by crystal fractionation within the granitic melt (i.e., by either addition or subtraction of crystals of a REE-bearing phase), because texture and grain-size relations in the 73215 felsite suggest that most of the REE-bearing phases precipitated relatively late, when physical movement of the crystals would have been impossible. It may be possible, however, to explain the felsite REE pattern by continued equilibration between the Fe-rich and K–Si-rich immiscible liquids during precipitation of their REE-bearing mineral phases. At present the feasibility of such a process cannot be evaluated because the necessary experimental data are not available, and because the REE patterns in the relevant minerals are poorly known in general. Thus it is not possible either to confirm a relationship between KREEP basalt and the lunar "granites", or to rule out such a relationship.

Felsic glass

The bulk of the felsic glass that forms veins and patches was emplaced as dikelets of melt into fractures in the crystalline felsite, and was not formed by *in situ* melting. The crystalline felsite was fractured and the melt dikelets were emplaced into it prior to incorporation of the clast in the 73215 breccia. Evidence leading to these conclusions is as follows. The walls of the glass veins are roughly parallel, indicating that they mark fractures. Crystalline felsite bordering the veins was apparently dilated and laterally displaced when the veins were emplaced, indicating intrusion. Thicknesses and distribution of the veins show no relationship to the clast margins, indicating that their formation predated incorporation of the clast in the breccia.

Bulk compositions of the glasses and crystalline felsite are similar; moreover, the glasses contain relict fragments of minerals from the crystalline felsite and no demonstrably exotic materials. Thus the parent rock for the felsic melt was almost certainly the same parent rock as for the crystalline felsite fragments.

There is no definitive evidence on the nature of the event that formed the felsic melt, but preliminary studies of the entire suite of felsite clasts in 73215 and in the related breccia 73255 (James and Marti, 1977) suggest that this event may have been an impact. Although the crystalline felsite of the clast described herein is unshocked (Nord and James, 1977), most similar clasts in 73255 and many in 73215 do show shock effects (their minerals show undulatory extinction, conversion to thetomorphic glass, and melting and vesiculation along grain boundaries). The fact that the felsic melt was emplaced into fractures suggests that the event was physically disruptive, additional evidence for an impact. Thus a possible interpretation of the clast history is that the felsic melt formed by shock melting of one part of a felsite parent body, and this melt was immediately injected into a fractured but unshocked part of the same body.

Although the felsic melt was injected into the crystalline felsite before the clast was incorporated in the breccia, the melt was mobile during breccia aggregation. The evidence is as follows: (1) clast-derived felsic glass and matrix-derived mineral grains are locally mixed at the edge of the clast, (2) veinlets of matrix-derived groundmass that cut crystalline felsite grade into veinlets of clast-derived felsic glass, (3) olivine relics in glass at the edge of the clast are enriched in Mg due to Fe–Mg exchange with adjacent matrix, and (4) clast-derived felsic glass intrudes and locally permeates matrix near the clast.

The bulk clast was quite hot when it was incorporated in the breccia, for some of the glass it contains formed at this time by *in situ* melting. The crystallized colorless glass within the crystalline felsite formed by *in situ* melting, but it is not clear if this melting occurred before or after the breccia-forming event. It is clear, however, that the uncrystallized colorless glass in the crystalline felsite (the selvages that mark grain boundaries and the bands at the clast-matrix contacts) formed after the breccia-forming event, for distribution of such glass is related to the clast margins. (TEM observations of the glass selvages by Nord and James (1977) also suggest origin by *in situ* melting.) This glass has the approximate composition of a K-feldspar-quartz eutectic melt. As the temperature of the anhydrous eutectic is 990°C (Schairer and Bowen, 1947), the bulk clast must have been at or above this temperature when incorporated.

After the clast was incorporated in the breccia, the felsic melt cooled. Olivine and plagioclase relics reacted with the melt and their compositions were modified by several processes. Olivine relics were enriched in Mg by Fe–Mg exchange with matrix, which has a much higher Mg/Mg + Fe than the clast (James, 1976), and by reduction of fayalite component to Fe metal plus silica (which combined with unreduced olivine to form second-generation pyroxene). Evidence for reduction is: Fe metal is very rare in the crystalline felsite but is abundant in the glass, and it has very low, nonmeteoritic, contents of Ni and Co; Fe metal is associated with partly resorbed olivine relics and with second-generation pyroxene; and Mg/Mg + Fe of reacted olivine is positively correlated with extent of reaction. Plagioclase relics were partly replaced by K-feldspar and were enriched in Ca relative to Ca + Na; it is not clear whether the Ca enrichment was produced by exchange with matrix, which has a higher Ca/Ca + Na than the clast (James, 1976), or whether it was produced by volatile loss of Na from the clast.

During the early stages of cooling of the felsic melt, second-generation crystals of pyroxene and K-feldspar precipitated. These grains are compositionally much more heterogeneous than those in the crystalline felsite, reflecting their rapid growth under nonequilibrium conditions. During later stages of cooling, crystallites of a silica mineral, feldspars, and mafic minerals formed. Most of these crystallites are very small, and the bulk of the glass did not crystallize (Nord and James, 1977). The second generation pyroxenes show exsolution textures like those observed in some mare basalts and in feldspathic basalt 14310, suggesting comparable subsolidus cooling histories (Nord and James, 1977). In sum, these features suggest that crystallization and exsolution occurred only during cooling following the breccia-forming event, and there was no later reheating of the clast (or the bulk breccia). (On the scale of observations reported herein and by Nord and James (1977), however, reheating at temperatures <500–600°C would probably produce no detectable effects.)

Age dating studies

Rb–Sr and ^{40}Ar–^{39}Ar studies of the felsite clast have yielded ages of 3.90 ± .05 and 3.92 ± .04 b.y., respectively (Compston *et al.*, 1977; Jessberger *et al.*, 1977). The Rb–Sr age is an internal isochron age based on analyses of separates of felsic glass and crystalline felsite; the ^{40}Ar–^{39}Ar age is a plateau age obtained by stepwise heating of a bulk chip. Laser-probe ^{40}Ar–^{39}Ar studies of a different felsic glass clast confirm a ~3.90 b.y. age for such material (Müller *et al.*, 1977). This ~3.90 b.y. date is at the least a lower limit to the age of the breccia-forming event, and it probably is the actual age of the event, as indicated by the following reasoning. The date can be no older than breccia aggregation because the clast was very hot when it was incorporated in the breccia; thus its Ar was probably outgassed and its Sr isotopes were probably reequilibrated at that time. The date is definitively only a lower limit to the time of breccia formation because reheating of the bulk rock cannot be precluded. Such reheating, however, would have had to reset both Rb–Sr and K–Ar systems in the clast to the same date, but not affect texture of the clast or the Rb–Sr and K–Ar systems in surrounding matrix; as such a cryptic and selective thermal event seems improbable, the ~3.90 b.y. age probably is the actual age of the breccia-forming event.

Age of the igneous crystallization of the granitic parent rock of the clast cannot so clearly be established. The Rb–Sr model-age data set a probable upper limit on crystallization age at 4.05 b.y. (Compston *et al.*, 1977). It thus appears that the felsite is a fairly "young" highlands rock, not a relic of the earliest lunar crustal differentiation.

Whether or not all lunar "granite" is as "young" as the 73215 felsite cannot be determined because so little data are currently available on this type of material. The other dated granitic samples have also been taken from breccias (12013, 15405), and in all cases the ages have been strongly affected by the breccia-forming events. In all the samples Ar seems to have been almost completely outgassed during breccia formation (Turner, 1971; Bernatowicz *et al.*, 1977; Jessberger *et al.*, 1977); the Pb–U–Th system seems to show the effects of multiple disturbances, both during and after the breccia-forming events (Tatsumoto, 1970; Tatsumoto and Unruh, 1976). The Rb–Sr system shows the effects of Sr isotopic reequilibration or Rb loss; this system, however, also seems to retain some record of events prior to breccia formation (Albee *et al.*, 1970; Nyquist *et al.*, 1977a,b; Compston *et al.*, 1977). Because the isotopic disturbances have in general been complex, none of the interpretations of original crystallization age for any of the granitic materials are unique. The 12013 granitic material was incorporated in the breccia as blobs of melt or fragments of glass and crystallized during breccia consolidation (James, 1970; Quick and Albee, 1976; Quick *et al.*, 1977), and it appears that there was extensive isotopic exchange between the various components of the breccia at that time (Albee *et al.*, 1970). On the basis of the Rb–Sr data, Albee *et al.* favor 4.5 b.y. as the date of crystallization of the parent body of the 12013 felsite. The 15405 quartz monzodiorite retained its primary igneous texture when incorporated in its host breccia, but it shows evidence of complex isotopic disturbances and Rb loss, possibly caused by more than one event (Nyquist *et al.*, 1977a,b; Tatsumoto and Unruh, 1976). The different investigators who have worked on 15405 have different interpretations of the crystallization age of the felsic clasts: 4.2–4.4 b.y. (Nyquist *et al.*, 1977a,b, from Rb–Sr data); 4.0 ± 0.1 b.y. (Tatsumoto and Unruh, 1976, from Pb–U–Th data); and 3.90–3.95 b.y. (Ryder, 1976, from Rb–Sr and geologic data). Of all the lunar granitic materials, the 73215 felsite described herein appears to have been least disturbed isotopically by the event that incorporated it in its host breccia. It apparently reequilibrated internally at the time of breccia aggregation, but it shows no evidence of Rb loss, complex or multiple isotopic disturbances, or large amounts of isotopic exchange with surrounding matrix. Thus the ~4.05 b.y. upper limit to igneous crystallization of the parent granitic body is a more firm estimate of crystallization age than are the age estimates for the other granitic materials. In fact, the data on the other materials will permit similar "young" crystallization ages for all, but to apply the 73215 age results to all lunar "granites" would be premature at this time.

ANORTHOSITIC GABBROS

The rock name "anorthositic gabbro" is used in this paper in a general sense and includes both "anorthositic gabbros" and "anorthositic norites" as defined by Prinz *et al.* (1973); by a more strict mineralogic definition all three clasts are actually anorthositic norites. Subsample numbers and total weights of the three clasts are as follows: 29,9—0.75 g; 46,25—2.3 g; and 46,33—0.53 g. Chips of all three clasts were made into thin sections, analyzed by INAA and atomic absorption spectrophotometry (AAS) (Blanchard *et al.*, 1977), and dated by $^{40}Ar–^{39}Ar$ stepwise heating (Jessberger *et al.*, 1976). In addition, chips of 29,9 were analyzed for Rb–Sr (Compston *et al.*, 1977) and dated by the laser-probe $^{40}Ar–^{39}Ar$ method (Müller *et al.*, 1977), and chips of 46,25 were analyzed for Rb–Sr (Compston *et al.*, 1977) and for siderophile and volatile trace elements (Morgan *et al.*, 1976). The three clasts appear to be genetically related, for they all have similar mineralogies and their textures form a gradational series. Texture and mineralogy of each clast are described in detail below.

Anorthositic gabbro 29,9

Texture. The rock is poikilitic, with blocky subhedral plagioclase grains and rounded irregular olivine grains enclosed by orthopyroxene oikocrysts (Fig. 5a). The plagioclase is inequigranular (0.03–1.0 mm), the olivine is comparatively equigranular (~0.2 mm), and the pyroxene oikocrysts are optically continuous over ~1 mm. Augite is a minor constituent, forming small globules and lamellae within orthopyroxene, thin bands at orthopyroxene-plagioclase contacts, and sparse small oikocrysts.

Many plagioclase grains contain "necklaces" of globular olivine inclusions (Fig. 5a). Such "necklaces" are present in all the largest grains, and in sparse grains as small as 0.15 mm across. The plagioclase cores within the "necklaces" are 0.1–0.5 mm across; many are inclusion free, many contain minute K-feldspar inclusions (Fig. 5a), and some contain streaks of tiny bubbles. In grains with inclusion-bearing cores, it can be seen that the edges of the cores are several tens of microns inside the olivine "necklaces". Many grains in the 0.1–0.2 mm size range also have cores containing abundant K-feldspar inclusions, but have no olivine "necklaces".

a
b
c
d
e
f

Many of the olivine grains in pyroxene oikocrysts contain globular inclusions of plagioclase. In most grains these inclusions are scattered unevenly throughout, but in sparse grains they tend to form "necklaces" like the olivine "necklaces" in the large plagioclase grains.

K-feldspar, K–Si-rich glass, apatite, whitlockite, ilmenite, chromite, Ni–Fe, troilite, and baddeleyite are trace constituents. The glass, K-feldspar and phosphate minerals fill interstices. Ilmenite and chromite form small grains at silicate-mineral contacts, and chromite also occurs as inclusions within olivine. Ni–Fe, along with minor troilite, forms irregular particles at grain boundaries and fills interstices. Baddeleyite forms tiny grains within troilite.

The rock is cut by a network of healed fractures that cross all mineral grains. Minute granules of troilite are concentrated along these fractures and locally also along grain boundaries.

Mineralogy. The compositions of all minerals in the clast have fairly restricted ranges. Plagioclase and olivine show compositional variations that are correlated with texture and occurrence, whereas augite and orthopyroxene do not. Plagioclase ranges from $An_{93.1}$ to $An_{96.5}$ (Fig. 6). The centers of subhedral grains in pyroxene oikocrysts, the centers of globules in olivine, and rims of all grains have the same compositions; overall average is $An_{95.3}$ (K_2O, BaO, and FeO average 0.08, 0.04, and 0.20%, respectively). Inclusion-free cores in grains with olivine "necklaces" average $An_{95.5}$; K-feldspar-bearing cores are slightly more sodic, averaging $An_{94.4}$. Electron microprobe traverses across grains that have K-feldspar-bearing cores and olivine "necklaces" show that the cores are uniform and relatively sodic, the inner rims are reversely zoned, the outer rims are uniform and relatively calcic, and the grain margins have normal progressive zoning. The extreme edges of grains are of the most sodic plagioclase in the rock, averaging $An_{94.0}$.

Olivine ranges from $Fo_{70.8}$ to $Fo_{75.3}$ (Fig. 7); CaO, MnO, and Cr_2O_3 average 0.37, 0.26, and 0.06%, respectively. Globules that form "necklaces" in plagioclase grains have centers with compositions covering nearly this entire range, but the globules in any given plagioclase are extremely uniform (e.g., four globules in one grain are $Fo_{74.2\pm0.2}$ and three in another are $Fo_{72.4\pm0.3}$). Grains in pyroxene oikocrysts

Fig. 5. Photomicrographs of anorthositic gabbros. (a) Sample 29,9. Crossed polarizers; width of field of view is 2.2 mm. Subhedral plagioclase and ovoid olivine (o) are enclosed by poikilitic orthopyroxene (dark gray). Locally olivine (om) is molded on subhedral plagioclase. The large plagioclase grains contain "necklaces" of olivine globules; the cores of some of these grains contain abundant tiny inclusions of K-feldspar (top left). (b) Sample 46,25. Crossed polarizers; width of field of view is 2.2 mm. Plagioclase and olivine form a mosaic of polygonal grains, and pyroxenes form oikocrysts scattered throughout this mosaic. The large plagioclase grains contain "necklaces" of olivine globules. (c) Mosaic texture in 46,25. Crossed polarizers; width of field of view is 0.88 mm. The grains are equant polygons with smoothly curving or planar boundaries, and three grains commonly meet at 120° junctions. Orthopyroxene oikocrysts (upper left, upper right, center left) include small globular plagioclase grains. (d) Large plagioclase grain in 46,25. Crossed polarizers; width of field of view is 0.70 mm. The grain has an olivine "necklace" and a core containing tiny bubbles. The bubbles form streaks that terminate at the core-rim boundary; this boundary is 10–20 μm inside the olivine "necklace" and is marked by an optical discontinuity. (e) Sample 46,33. Crossed polarizers; width of field of view is 2.2 mm. Mosaic intergrowths of plagioclase and olivine (lower center, top left edge, center right edge) and clusters of pyroxene oikocrysts (top right corner, center left edge) form distinct mosaic and poikilitic domains. Large grains of plagioclase and olivine are scattered throughout the rock, and many of those of plagioclase contain "necklaces" of olivine globules. (f) Contact between mosaic and poikilitic domains in 46,33. Plane polarized light; width of field of view is 0.88 mm. The right half of the field shows a mosaic domain, consisting of anhedral polygons of plagioclase and olivine. The left half of the field shows a poikilitic domain; many of the larger plagioclase grains in the oikocrysts are subhedral, whereas the smaller ones are virtually all anhedral and globular.

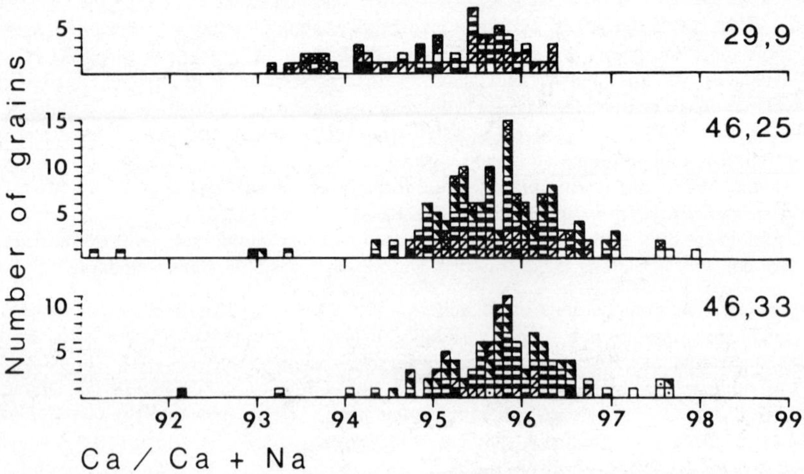

Fig. 6. Histograms of Ca/Ca + Na contents in plagioclase in anorthositic gabbros. Textures and occurrences of the individual grains analyzed are indicated by the following symbols in the boxes on the diagrams: filled, cores containing K-feldspar inclusions; empty, inclusion-free cores and cores containing bubbles; small dot, cores containing glass inclusions; diagonal lines, rims on large grains; filled lower half, grains in oikocrysts; filled upper right corner, grains in mosaics; filled upper half, small euhedral grains in oikocrysts; filled upper left corner, globules included in olivine (29,9) or deformed, recovered plagioclase in mosaics (46,33); crosses, grains in lathy aggregate (46,25) or grain margins (29,9).

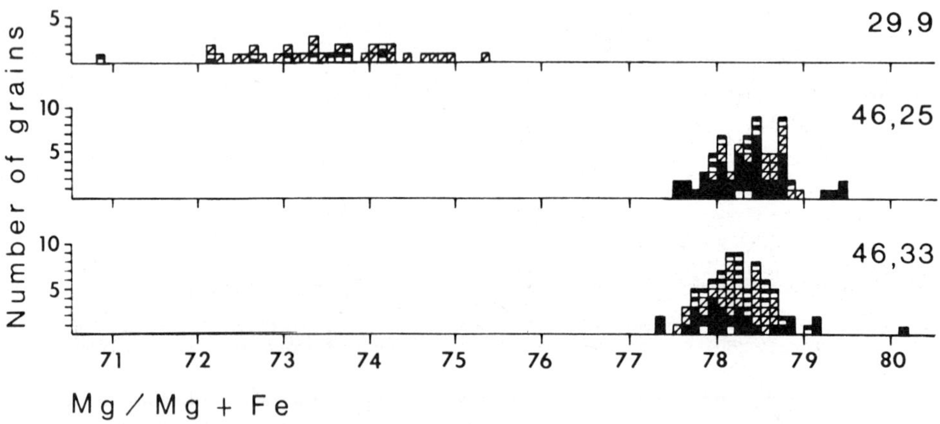

Fig. 7. Histograms of Mg/Mg + Fe contents in olivines in anorthositic gabbros. Textures and occurrences of the individual grains analyzed are indicated by the following symbols in the boxes on the diagrams: filled, grains in mosaics; empty, centers of large grains; diagonal lines, globules forming "necklaces" in plagioclase grains; filled upper half, grains in oikocrysts; filled lower half, grain margins.

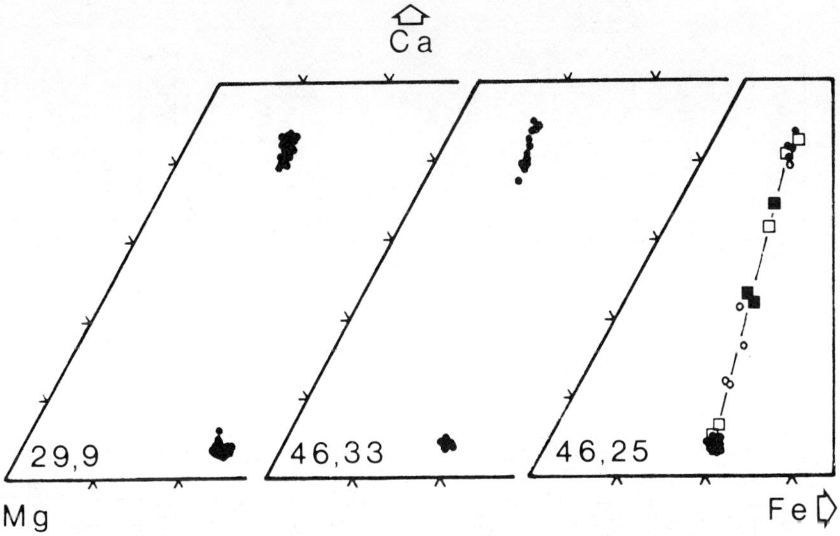

Fig. 8. Ca:Mg:Fe of pyroxenes in anorthositic gabbros. Filled circles, grains forming oikocrysts. Open circles, analyses attempted of exsolution lamellae. Squares indicate analyses of pyroxene globules included in plagioclase grains in 46,25: open squares are individual phases in these globules; filled squares are expanded-beam analyses of the bulk globules.

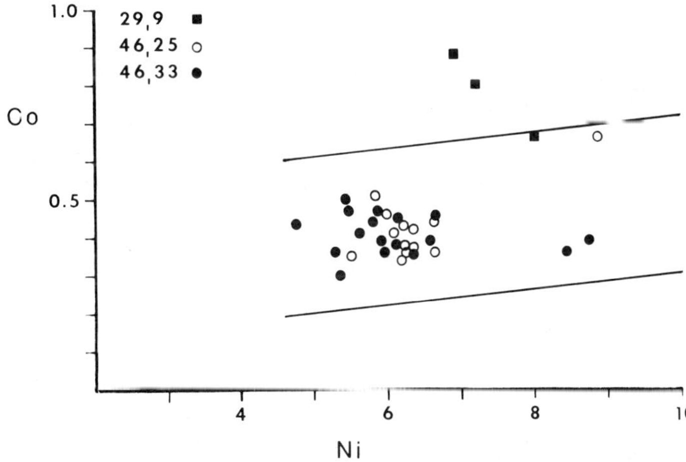

Fig. 9. Ni–Co contents (wt.%) in metal grains in anorthositic gabbros. The area within the lines is the range of compositions of "meteoritic" metal (Goldstein and Yakowitz, 1971).

Table 2. Compositions of pyroxene, trace minerals, and glasses in anorthositic gabbros (wt.%, electron microprobe analyses).

	29,9					46,25					46,33			
	(1)	(2)	(3)	(4)	(5)	(6)	(7)	(8)	(9)	(10)	(11)	(12)	(13)	(14)
SiO_2	0.20	0.19	0.39	0.34	79.4	52.4	0.19	<0.01	0.32	80.0	0.25	2.01	0.29	73.9
TiO_2	54.4	7.83	—	—	—	0.75	55.0	90.2	8.65	0.91	6.49	—	—	1.55
Al_2O_3	0.22	11.7	—	—	12.3	1.40	0.33	0.18	12.0	9.18	12.6	—	—	11.5
FeO	38.3	30.5	0.48	0.53	0.63	9.40	35.6	0.44	30.2	0.15	28.2	0.65	8.47	0.37
MgO	6.01	6.55	0.42	3.49	—	23.1	8.73	<0.01	5.46	0.02	6.58	6.04	18.9	0.12
CaO	0.46	0.26	53.2	40.4	0.84	11.5	0.33	<0.01	<0.01	0.57	0.48	40.9	24.1	0.79
Na_2O	—	—	—	—	0.25	—	—	—	—	0.46	—	—	—	0.34
K_2O	—	—	—	—	4.18	—	—	—	—	5.79	—	—	—	7.14
BaO	—	—	—	—	0.04	—	—	—	—	0.10	—	—	—	0.10
MnO	0.42	0.73	—	—	—	0.09	0.32	<0.01	—	<0.01	0.84	—	—	0.01
Cr_2O_3	0.38	41.6	—	—	—	0.42	0.76	1.81	42.5	<0.01	43.6	—	—	0.07
Y_2O_3	—	—	0.21	1.68	—	—	—	—	—	—	—	1.08	0.14	—
La_2O_3	—	—	0.12	0.95	—	—	—	—	—	—	—	0.25	0.05	—
Ce_2O_3	—	—	0.31	3.66	—	—	—	—	—	—	—	1.46	0.08	—
Nd_2O_3	—	—	0.25	1.92	—	—	—	—	—	—	—	1.16	0.13	—
Sm_2O_3	—	—	0.14	0.59	—	—	—	—	—	—	—	0.17	<0.01	—
Gd_2O_3	—	—	0.17	1.44	—	—	—	—	—	—	—	0.55	0.12	—
P_2O_5	—	—	40.3	42.6	—	—	—	—	—	0.02	—	42.4	46.7	0.18
ZrO_2	—	—	—	—	—	—	—	2.46	—	<0.01	—	—	—	0.09
Nb_2O_5	—	—	—	—	—	—	<0.01	4.30	<0.01	—	<0.01	—	—	—
F	—	—	2.50	0.21	—	—	—	—	—	—	—	0.24	0.39	—
Cl	—	—	0.49	0.05	—	—	—	—	—	—	—	0.03	0.07	—
Total	100.39	99.36	98.98	97.86	97.64	99.06	101.26	99.39	99.13	96.80	99.04	96.72	99.48	96.16

(1) Ilmenite: average of two analyses.
(2) Chromite.
(3) Apatite: average of five analyses.
(4) Whitlockite: average of three analyses.
(5) Large glassy inclusion in plagioclase.
(6) Pyroxene globule included in plagioclase: expanded beam bulk analysis.
(7) Ilmenite: average of two analyses.
(8) Rutile intergrown with ilmenite.
(9) Chromite.
(10) Large glassy inclusion in plagioclase.
(11) Chromite: average of three analyses.
(12) Whitlockite.
(13) Ca–Mg–Fe phosphate mineral.
(14) Small glassy inclusion in plagioclase core.

have centers with compositions in the middle of the observed range, and edges with compositions that tend to be at the Fe-rich end of the range.

Both orthopyroxene and augite have compositions that cluster tightly in Ca:Mg:Fe (Fig. 8); the slight variation in Ca content shown by the augite is probably due to fine-scale exsolution. Minor-element contents are also fairly uniform. In orthopyroxene, TiO_2, Al_2O_3, MnO, and Cr_2O_3 average 0.71, 1.16, 0.27, and 0.39%, respectively; in augite these oxides average 1.57, 2.24, 0.16, and 0.62%, respectively.

The K-feldspars fall into two compositional groups: (1) interstitial grains are fairly pure K-feldspar, with only 0.63% BaO, 0.86% CaO, and 0.03% Na_2O; (2) inclusions in plagioclase are poorer in BaO (0.13%), much richer in Na_2O (0.68%), and they may also contain significant CaO, but interference from surrounding plagioclase precludes accurate analyses for this oxide. Metal particles have Ni–Co contents that fall just outside the "meteoritic" range (Fig. 9). Compositions of other trace phases are listed in Table 2.

Anorthositic gabbro 46,25

Texture. Most of the plagioclase and olivine in the rock form a mosaic intergrowth (Fig. 5b,c). The grains in this mosaic are equant polygons, and grain junctions commonly form 120° angles. The plagioclase is inequigranular (5 μm–0.4 mm), and the olivine is more equigranular and averages much smaller (5 μm–0.1 mm). Pyroxene forms oikocrysts scattered throughout this mosaic. The oikocrysts contain small knots of mosaic plagioclase, sparse ovoid 10–25 μm olivine grains, and abundant ovoid 2–35 μm plagioclase grains. One orthopyroxene oikocryst has a 70 μm inclusion-free core and a zone surrounding this core that contains numerous small euhedral plagioclase grains (the only euhedral plagioclase observed in the rock). Most oikocrysts are of orthopyroxene, but a few are of augite; oikocrysts of both pyroxenes are scattered throughout the mosaic, and distribution of the two minerals appears unrelated. Oikocrysts of orthopyroxene and augite are optically continuous (though not physically continuous) for as much as 2 and 0.5 mm, respectively. Minor augite also forms tiny exsolved blebs and lamellae within orthopyroxene. In addition, sparse globules consisting of lamellar intergrowths of augite and orthopyroxene occur within plagioclase.

Sparse grains of olivine as much as 0.1 mm across occur in oikocrysts and in the mosaic; these grains include small globules of plagioclase. An intergrowth consisting entirely of lath-shaped plagioclase grains forms a patch ~0.3 mm across in the mosaic.

Many plagioclase grains >75 μm across contain "necklaces" of small olivine globules (Fig. 5d). The cores within the "necklaces" are as much as 75 μm across, and have rounded or angular outlines. Many cores are inclusion free, many include clouds and streaks of tiny bubbles (Fig. 5d), and a few include abundant minute particles of glass or K-feldspar. Sparse cores contain patches of discontinuous twin lamellae and recrystallized subgrains. In many grains, edges of the cores are marked by slight, but very sharp, discontinuities in extinction position (Fig. 5d). As in 29,9, the edges of these cores generally occur 10–20 μm inside the olivine "necklaces".

Trace constituents are chromite, rutile, zircon, Ni–Fe, and troilite. Ilmenite forms thin plates; it also occurs intergrown with rutile and chromite in anhedral interstitial grains. Zircon forms anhedral interstitial grains. Ni–Fe forms globules and irregular particles at grain boundaries. Troilite forms abundant micron-sized particles concentrated along grain boundaries and along sparse healed fractures that cut across silicate-mineral grains.

Mineralogy. Plagioclase is slightly more calcic, and the mafic minerals are appreciably more magnesian, than in 29,9. As in 29,9, however, all minerals have restricted ranges of composition. Plagioclase shows only minor variations in composition, but these variations are correlated with texture and appear to be significant. Grains in mosaics, grains within pyroxene oikocrysts, the rims of grains containing olivine "necklaces", and grains in the lathy intergrowth all have virtually identical compositions (Fig. 6); they average $An_{95.7}$, $An_{95.6}$, $An_{95.7}$, and $An_{95.8}$, respectively. K_2O, BaO, and FeO average 0.09, 0.03, and 0.15%, respectively, overall. The euhedral plagioclase grains are significantly more sodic, averaging $An_{92.7}$. The cores of grains containing olivine "necklaces" are also systematically different in composition from the bulk of the plagioclase: cores with glass or

K-feldspar inclusions are relatively sodic, averaging $An_{94.5}$; and cores that are inclusion free or contain tiny bubbles are relatively calcic, averaging $An_{97.1}$. Electron microprobe traverses across grains having cores that contain K-feldspar or glass inclusions show that: the cores are either uniform and relatively K and Na rich throughout, or they show abrupt reverse zoning at their edges; the rims are either uniform and relatively Ca rich throughout, or they are reversely zoned; and the core-rim transition is always abrupt. Electron microprobe traverses across grains having cores that are inclusion free or include bubbles show that: the cores are uniform and relatively Ca rich ($An_{96.4-97.9}$) at their centers, and generally show normal progressive zoning within 20–40 μm of their edges; and the rims are relatively sodic (about $An_{94.5}$) at the core-rim contacts and are reversely zoned.

Both olivine and pyroxene are extremely uniform in composition (Figs. 7, 8). In olivine CaO, MnO, and Cr_2O_3 average 0.30, 0.22, and 0.06%, respectively. In orthopyroxene TiO_2, Al_2O_3, MnO, and Cr_2O_3 average 0.80, 1.37, 0.24, and 0.49%, respectively; in augite these oxides average 1.74, 2.67, 0.15, and 0.83%, respectively. Analyses attempted of exsolved augite lamellae indicate that they are of the same composition as the augite that forms oikocrysts. Pyroxene globules included in plagioclase consist of intergrowths of orthopyroxene and augite that have compositions like the poikilitic pyroxenes, but in bulk composition these globules are subcalcic augites (Table 2, Fig. 8). Metal grains have Ni–Co contents that fall within the "meteoritic" range (Fig. 9). Compositions of other trace phases are listed in Table 2.

Anorthositic gabbro 46,33

Texture. This rock is similar to 46,25 in that most of the olivine and plagioclase grains form a mosaic intergrowth, and the pyroxene grains are poikilitic. In differs from 46,25 in that grain size is finer and the pyroxene oikocrysts are not scattered throughout the mosaic but are concentrated in clusters; thus the mosaic and poikilitic intergrowths form physically distinct domains (Fig. 5e,f). These domains are 0.2–1.0 mm across.

The mosaic intergrowth is like that in 46,25 but finer grained; plagioclase and olivine size ranges are 5 μm–0.3 mm and 2–70 μm, respectively. Unlike in 46,25, the plagioclase grains locally show textures indicating they have been deformed and subsequently recovered. These deformed grains consist of mosaics of subgrains, or they contain bands or wedges that differ slightly in extinction position from the rest of the grain. The deformation features extend to grain margins, and a few of the deformed grains include olivine globules; thus the deformation occurred *in situ.*

The poikilitic domains (Fig. 5f) consist of clusters of as many as six oikocrysts. These oikocrysts do not envelop knots of mosaic plagioclase as in 46,25, but they do include sparse ovoid olivine grains and abundant plagioclase grains; the larger plagioclase inclusions (25 μm–0.2 mm) are subhedral laths, and the smaller ones (5–25 μm) are anhedral globules (Fig. 5f). Most oikocrysts are of orthopyroxene; they average ~0.5 mm across and are physically as well as optically continuous over their entire extents. Oikocrysts of augite are ~0.3 mm across, and are concentrated at the edges of the poikilitic domains. Augite also forms sparse exsolution lamellae in orthopyroxene. Except for the finer grain size and the anhedral shapes of the smaller plagioclase inclusions, the texture is like the texture of 29,9.

The rock contains sparse large angular and rounded grains of plagioclase and olivine like those in 46,25 (Fig. 5e). The ones of plagioclase generally have "necklaces" of olivine globules, and their cores have variable contents of inclusions: many are inclusion free, some contain abundant tiny K-feldspar particles, and some contain glass globules. A few plagioclase cores show patchy extinction variations that indicate they have been deformed and later recovered. As in 46,25, core-rim contacts in some grains are marked by slight, but sharp, discontinuities in extinction position; such discontinuities and the edges of inclusion-rich cores typically are ~10 μm inside the olivine "necklaces".

Trace constituents are ilmenite, chromite, Ni–Fe, schreibersite, troilite, whitlockite, and an unidentified Ca–Mg–Fe phosphate mineral. The ilmenite, chromite, and Ni–Fe form small irregular grains that tend to be concentrated in the mosaic intergrowths. The schreibersite and a trace amount of troilite are associated with the metal grains, but the bulk of the troilite forms tiny particles concentrated along grain boundaries and along healed fractures.

Mineralogy. The mineral compositions are quite homogeneous and are virtually identical to those in 46,25. The bulk of the plagioclase is uniform: grains in mosaics, grains within pyroxene oikocrysts, rims of grains containing olivine "necklaces", and recovered grains in mosaics average $An_{95.7}$, $An_{95.7}$, $An_{95.8}$, and $An_{95.4}$, respectively (Fig. 6). K_2O, BaO, and FeO average 0.10, 0.04, and 0.23%, respectively, overall. As in 46,25, the cores of grains having olivine "necklaces" are slightly but systematically different in composition from the bulk of the plagioclase: inclusion-free cores average $An_{97.0}$, cores with glass inclusions average $An_{96.9}$, and cores with K-feldspar inclusions average $An_{94.6}$. Electron microprobe traverses across such plagioclase grains show several different patterns of compositional variation. Grains with cores that are inclusion free or contain glass inclusions show two types of pattern: (1) the center of the core is uniform and Ca rich, the edge of the core shows slight normal progressive zoning, the rim is fairly uniform and relatively Na rich, and the edge of the rim shows either reverse or normal progressive zoning; (2) the core is uniform and Ca rich throughout, and the rim shows smooth normal progressive zoning throughout. Grains with cores containing K-feldspar inclusions show the following pattern: the core is K and Na rich and uniform; the core-rim boundary is an abrupt discontinuity; and the rim is uniform and relatively Ca rich. The olivine is extremely uniform (Fig. 7); CaO, MnO, and Cr_2O_3 average 0.39, 0.22, and 0.08%, respectively. The poikilitic pyroxenes are also uniform, except that the augites show Ca variations due to very fine exsolution (Fig. 8). In orthopyroxene TiO_2, Al_2O_3, MnO, and Cr_2O_3 average 1.00, 1.58, 0.23, and 0.54%, respectively; in augite these oxides average 1.89, 2.52, 0.16, and 0.83%, respectively. The metal grains have Ni–Co contents identical to those in 46,25 (Fig. 9). Compositions of other trace phases are listed in Table 2.

Discussion of Anorthositic Gabbro Data

The similar mineral assemblages, similar mineral compositions, and gradational textures of these rocks suggest that they all formed by the same process. Unfortunately, however, not all of them display textures and compositional variations that yield unambiguous interpretations of their origins. For this reason, data from all three rocks must be considered together to arrive at an interpretation of genesis. In the following paragraphs, genesis of each of the three rocks is discussed separately. The sample with the simplest texture and least definitive characteristics (29,9) is discussed first, and the sample with the most complex texture and characteristics that set the strongest constraints on genesis (46,33) is discussed last.

29,9

The bulk of this rock crystallized from melt. Features diagnostic of melt crystallization are as follows: plagioclase grains are subhedral; some olivine grains are molded on subhedral plagioclase grains; olivine grains in pyroxene oikocrysts have shapes suggesting that they were partly resorbed; distribution of the augite oikocrysts suggests augite crystallized after orthopyroxene; and edges of plagioclase grains show normal progressive zoning. The paragenesis indicated by the textures is the same as would be predicted for crystallization of melt on the basis of experimental phase relations (Weill and McKay, 1975; Walker *et al.*, 1973).

The melt contained fragments of minerals derived from preexisting rocks; these were preserved as the cores of the plagioclase grains that have olivine "necklaces", K-feldspar-bearing cores in plagioclase grains that do not have olivine "necklaces", and cores of large olivine grains. Evidence that these were xenocrysts and not phenocrysts is as follows: their outlines are rounded rather than subhedral; some contain streaks of tiny bubbles that mark healed fractures; and the ones of plagioclase are of two types that differ in composition and in content of inclusions.

The characteristics of this rock unfortunately yield no definitive information on genesis of the melt it crystallized from. It could have been a fragment-laden melt that incorporated clasts by a mechanical mixing process, or it could have formed by melting of a preexisting breccia.

46,25

Grain growth in this rock was dominantly by solid-state crystallization. Diagnostic features are: polygonal outlines and common 120° triple junctions of grains in the mosaics, ovoid outlines of the plagioclase grains within pyroxene oikocrysts, distribution of augite and orthopyroxene oikocrysts that suggests both pyroxenes crystallized simultaneously, olivine and pyroxene compositions that are homogeneous in Fe/Fe + Mg within analytical error, and marginal reverse zoning in plagioclase grains.

As in 29,9, the cores of some plagioclase and olivine grains (and the core of one orthopyroxene oikocryst) appear to represent fragments of minerals from preexisting rocks. The textural and compositional diversity of these cores, their angular and rounded outlines, the presence in many of healed fractures, and the presence in some of deformation and recrystallization textures are all characteristics that indicate derivation from preexisting rocks.

A very small fraction of the rock appears to have crystallized from melt. The pyroxene globules in plagioclase apparently formed from melt, because their bulk compositions are those of metastable pyroxenes that probably could not have crystallized below the solidus. The sparse tiny grains of relatively sodic plagioclase that are found locally in pyroxene oikocrysts have euhedral shapes suggesting crystallization from melt. The plagioclase rims containing olivine "necklaces" probably also crystallized from melt; if they had grown by solid-state enlargement of the plagioclase cores, the core-rim boundaries should have been at the "necklaces" rather than well inside them, there probably should not be sharp compositional discontinuities at these boundaries, and the plagioclase of the inner rim should not be as sodic as it is (it should probably not be any more sodic than the average plagioclase of the rock).

The general characteristics of the rock do not yield much information on its nature prior to the episode of solid-state grain growth, but the detailed characteristics favor origin as a heated, partly melted and recrystallized breccia. Specifically, the fraction of the rock that crystallized from melt is a very low melting-temperature fraction, consisting of pyroxene and relatively sodic plagioclase; the most probable origin for a melt of this kind is by a low degree of partial melting of the bulk rock. The alternative interpretation of rock history, origin as a fragment-laden melt, is unlikely. If the rock had formed as a fragment-laden melt, only a small fraction had crystallized above the solidus, and the remainder had been quenched and crystallized below the solidus, then the fraction that crystallized from melt should have been a high melting-temperature rather than a low melting-temperature fraction. If the rock had formed as a fragment-laden melt, had crystallized, had later been deformed, and subsequently had been recrystallized so that all traces of igneous texture were destroyed, then the cores of the large plagioclase grains should show more evidence of deformation and recrystallization than they do.

46,33

A large fraction of the rock precipitated from melt (the poikilitic domains and the rims of the large plagioclase grains); the remainder crystallized by solid-state grain growth (the mosaic domains). Evidence for solid-state growth of the mosaics is the same as listed above for the 46,25 mosaic. Evidence for crystallization of the plagioclase rims from melt is the normal progressive zoning that they commonly show, and evidence for crystallization of the poikilitic domains from melt is as follows: the larger plagioclase grains in the oikocrysts are subhedral; distribution of the augite and orthopyroxene oikocrysts indicates that the augite precipitated later than the orthopyroxene; and pyroxene occurs only as oikocrysts, never as mosaic or relict grains. The end stages of crystallization of the oikocrysts may have occurred below the solidus, however, because the smaller plagioclase grains they contain are anhedral rather than subhedral.

As in 29,9 and 46,25, the cores of some plagioclase and olivine grains represent fragments of minerals from preexisting rocks; evidence for this conclusion is the same as given above for 46,25. Unlike in the other two rocks, however, local deformation occurred during solidification. The deformation was restricted to parts of the mosaic intergrowth, and the cores of large plagioclase grains and oikocrysts were not deformed; thus the deformation was limited in extent and had terminated before the poikilitic domains crystallized.

The evidence cited above strongly favors the interpretation that 46,33 is a heated breccia that partly melted and partly recrystallized. The fraction of the rock that crystallized from melt was, as in 46,25, a pyroxene-plagioclase mixture having a relatively low melting temperature; the rock textures are most easily explained if this melt formed by partial melting of the bulk rock. If instead 46,33 had formed as a fragment-laden melt and crystallized above the solidus, all of its olivine and much of its plagioclase should have precipitated before its pyroxene began to form; thus the olivine-plagioclase mosaics should show igneous texture rather than the metamorphic texture that is observed. If the rock had formed as a fragment-laden melt, had crystallized, and had been deformed midway in the course of its crystallization so that the igneous texture of all early-crystallized minerals was destroyed, then the cores of the plagioclase grains should show more evidence of deformation than they do.

Genesis

The evidence summarized above indicates that the most likely origin for the parent rocks of all three of the anorthositic gabbro clasts was as heated, partly melted and/or recrystallized breccias. For 46,33 the evidence is most definitive: in this rock an appreciable fraction of the breccia melted and crystallized with igneous texture, while the unmelted fraction recrystallized. For 46,25 the evidence is also fairly strong: the bulk of the rock recrystallized in the solid state, but a very small fraction partly melted. For 29,9 the evidence is not definitive, but the textural and mineralogic similarities between this rock and the other two indicate that genesis was probably similar: in this rock the bulk of the breccia melted and the melt crystallized with igneous texture. Textures suggest that the K-feldspar and glass inclusions in the preserved fragments of preexisting plagioclases precipitated from the plagioclases during the melting/recrystallization event.

The parent breccias were probably polymict rather than monomict. Each of the anorthositic gabbros contains preserved fragments of at least two kinds of plagioclase that differ in bulk composition, or nature of included phases, or both; thus it appears that at least two source rocks contributed fragments to the parent breccias. Moreover, each rock contains Ni–Fe particles with Ni–Co contents suitable for derivation from a meteoritic source; analyses of siderophile trace elements also suggest that the metal is meteoritic (Morgan *et al.*, 1976). These data indicate that the bulk compositions of the rocks cannot be regarded as those of pristine lunar highlands magmas or cumulates. The number of source rocks that supplied material to the breccias was apparently small, however, and it is likely that these breccias had not been subjected to much mixing or reworking before the heating event that produced their present textures and mineral assemblages.

All three of the anorthositic gabbros show healed fractures; these are marked by minute granules of troilite, and similar particles are also concentrated along grain boundaries. The event that produced the fractures occurred after the melting/recrystallization event because the fractures cut across all grains; subsequently temperatures and/or pressures were high enough that the fractures healed. Unfortunately the textures yield no information on timing of this episode, or on the temperatures and pressures that were involved. It is also not clear whether the troilite was introduced into the rocks in the fracturing event, or whether it was remobilized material that was already present.

Anorthositic gabbros having bulk compositions and textures like the clasts described herein are widespread on the lunar surface, and they apparently represent an important rock type in the highlands crust. Most of the returned samples of such rocks are small fragments in breccias and soils, but a few hand-specimen-sized samples have also been returned. Rocks of this type were first found at the Apollo 11 site, where they make up a very small fraction of the soil (Wood *et al.*, 1970). At the Apollo 15 site they have been reported as sparse small clasts in glassy breccias (Cameron *et al.*, 1973). They appear to be more abundant at the Apollo 16 site, where reported examples are: clasts in the Outhouse Rock breccia boulder (Hollister, 1973; Roedder and Weiblen, 1974; Ashwal, 1975; Nord *et al.*, 1975); rake samples (Dowty *et al.*, 1974); clasts in breccia 67435 ((Warner *et al.*, 1976); and sparse fragments in the 2–4 mm size fraction of the soils (Delano *et al.*, 1973; Taylor *et al.*, 1973). They are apparently most abundant at the Apollo 17 site, where they form: large and numerous clasts in individual breccia samples and in the boulders at Stations 2, 6, and 7 (Stoeser *et al.*, 1974; Ryder *et al.*, 1975; Dymek *et al.*, 1976; Simonds, 1975; Chao *et al.*, 1974); large rock samples 77017, 78155, and 79215; and fragments in the 2–4 mm size fractions of the massif and light-mantle soils (Bence *et al.*, 1974). Rocks having bulk compositions nearly identical to these anorthositic gabbros, but slightly different textures, have also been found in the Luna 20 soil (Prinz *et al.*, 1973). It does appear, however, that such rocks are not present or are present only in trace amounts at the Apollo 12 and 14 sites; no material that is clearly of this type has been described from either of these sites.

There has been a fair diversity of opinion published concerning the origin of these rocks. In studies of soil and breccia samples, fragments with textures like 46,25 or 46,33 have generally been interpreted as metamorphic rocks of unspecified origin or as recrystallized breccias; fragments with textures like 29,9 have been interpreted as metamorphic rocks by some investigators, and as products of melt crystallization by others. In the studies of the larger samples, the same spectrum of interpretations has been offered. Rocks with textures like 46,33 and 46,25 (78155, 79215, the matrix of 67955) have been interpreted as hornfelses (Nord *et al.*, 1975), as recrystallized monomict breccias (Bickel *et al.*, 1976) and as recrystallized polymict breccias (Bickel and Warner, 1977). Rocks with textures like 29,9 (77017, clasts in 67955 and in the Apollo 17 Station 6 boulder) have been interpreted as: the product of plutonic crystallization of highlands magma (Ashwal, 1975; Hodges and Kushiro, 1974), as a cumulate that was brecciated and partly melted, with the melt fraction subsequently crystallizing above the solidus (Helz and Appleman, 1974); as the product of subsolidus crystallization of a glass-rich monomict breccia that was originally a cumulate (McCallum *et al.*, 1974); or, tentatively, as the product of crystallization of impact melt (Simonds, 1975). Warner *et al.* (1977) have drawn together the data on all these samples and interpret all as recrystallized impact breccias. In spite of this diversity of opinion there is consensus on one point: the observed textures formed at temperatures near the solidus. Temperature estimates range from >900°C to 1100–1200°C (Hollister, 1973; McCallum *et al.*, 1974; Bickel and Warner, 1977; Bickel *et al.*, 1976).

The studies of the 73215 anorthositic gabbros, in conjunction with the studies of the other rocks cited above, permit some tentative statements to be made concerning genesis of the entire suite.

(1) The 73215 data, and observations made by Hollister (1973), Ashwal (1975), and Helz and Appleman (1974), indicate that crystallization from melt was indeed involved in formation of the textures of some of these rocks.
(2) It appears unlikely, in any of the rocks that did crystallize from melt, that the melt was a pristine highlands magma. Sample 29,9 clearly contained mineral fragments from preexisting rocks when it crystallized, and sample 77017, which is texturally similar to 29,9, contained lithic clasts (Helz and Appleman, 1974; McCallum *et al.*, 1974).
(3) The 73215 studies confirm that temperatures of recrystallization were very near solidus temperatures (about 1100–1150°C, on the basis of data given by Weill and McKay, 1975).
(4) It is likely that all the rocks represent reheated polymict breccias. The 73215 samples all contain mineral fragments from at least two different preexisting rocks, and 78155 and 77017 contain lithic fragments from several sources. In addition, all the samples contain material that is probably from a meteoritic source (Ganapathy *et al.*, 1974; Morgan *et al.*, 1974; Higuchi and Morgan, 1975). In general the source rocks seem to have been few in number, however, and mixing was not extreme.
(5) The dominant source rocks for the parent breccias appear to have been relatively coarse-grained, possibly plutonic, rocks, because most relict fragments are of single mineral grains. Source rocks that have been found as lithic clasts are: fine grained igneous anorthosites and medium-grained igneous and metamorphic anorthositic norites and noritic anorthosites (in 78155; Bickel and Warner, 1977); and mosaic-textured anorthosites and troctolitic anorthosites (in 77017; McCallum *et al.*, 1974; Helz and Appleman, 1974). All the source rocks were fairly feldspathic ANT-suite rocks, and none appear to have contained any appreciable KREEP component (Warner *et al.*, 1977).

In brief, the anorthositic gabbros discussed herein ("feldspathic, granulitic impactites" or "Early Impactites" of Warner *et al.*, 1977) record an episode of intense brecciation of the ANT-suite rocks of the highlands, accompanied by mixing in of metal of probable meteoritic origin. This brecciation was followed by prolonged heating at temperatures near or above the solidus, as a result of which the breccias recrystallized and/or melted, and the melts that were produced subsequently crystallized with igneous textures. The heating was apparently a static thermal heating, and not directly induced by shock. Equilibration of mineral compositions occurred over distances on the order of millimeters; these equilibration distances are intermediate between those observed in the Apollo 14 thermally metamorphosed breccias and in coarser-grained rocks like 15415 and 76535, indicating that annealing conditions were also intermediate (Warner *et al.*, 1977). The most likely mechanism for the brecciation was impact bombardment. The depth of burial and the heat source during the melting/recrystallization are somewhat more problematical. The rocks must have been buried deeply enough so that they were not excavated by impact for appreciable time periods; Warner *et al.* (1977) estimate depths on the order of several kilometers. The heat source could have been a high impact flux and high internal heat flow early in the history of the highlands crust (Warner *et al.*, 1977), or it might possibly have been emplacement of internally generated magma into the crust.

Age dating results

The $^{40}Ar/^{39}Ar$ stepwise heating data show double plateaus for all three of the 73215 clasts; intermediate-temperature plateau ages are 4.06–4.07 ± .04 b.y. and high-temperature plateau ages are 4.22–4.26 ± 0.04 b.y. (Jessberger *et al.*, 1976). The laser-probe $^{40}Ar/^{39}Ar$ data for 29,9 show considerable variations in "ages" of the materials studied, but as the data are relatively imprecise it is difficult to assess their chronologic significance. The cores of plagioclase grains with olivine "necklaces" give ages ranging from 4.16 to 4.37 b.y., with errors about ±0.1 b.y.; a composite of plagioclase grains that crystallized from the melt gives an age of 4.07 ± 0.05 b.y. (Müller *et al.*, 1977). The Rb–Sr internal isochron age of 29,9 is 4.24 ± 0.31 b.y.; the analyzed materials from 46,25 did not show a large enough range in Rb/Sr to define an internal isochron, but the data could fit the 29,9 isochron. Rb–Sr model ages for 29,9 and 46,25 are 4.30–4.35 ± 0.15 b.y. (Compston *et al.*, 1977). Thus the results of all dating methods consistently yield ages of 4.22–4.35 b.y., and the $^{40}Ar/^{39}Ar$ data additionally yield an age of 4.06–4.07 b.y.

Ages determined for the other anorthositic gabbros that are related to the 73215 clasts also show a tendency to fall into two ranges, i.e., a low range, around 4.0 b.y., and a high range, greater than 4.2 b.y. Sample 78155 has a $^{40}Ar/^{39}Ar$ plateau age of 4.22 ± 0.04 b.y. (Turner and Cadogan, 1975). A clast that is texturally like 29,9 but finer grained from the Apollo 17 Station 6 boulder shows a two-step $^{40}Ar/^{39}Ar$ age spectrum quite like the spectrum typical of the 73215 clasts: a well-defined intermediate-temperature plateau yields an age of 3.98 ± 0.05 b.y., and at higher temperatures the ages rise to 4.10 ± 0.05 b.y. (Turner and Cadogan, 1975). Sample 77017 has had considerable loss of gas due to a relatively recent shock event and injection of glass, but it shows a $^{40}Ar/^{39}Ar$ high-temperature plateau age of 4.05 ± 0.05 b.y. (Kirsten and Horn, 1974). Sample 76235, a clast in the Station 6 boulder having a texture somewhat like 46,33, has an $^{40}Ar/^{39}Ar$ plateau age of 3.94 ± 0.06 b.y. (Cadogan and Turner, 1976).

The ages in the lower range are the most easily interpreted. Jessberger *et al.* (1976) attributed the 4.06–4.07 b.y. ages shown by the 73215 clasts to partial outgassing of some components of these rocks when they were incorporated in the breccia. Cadogan and Turner (1976) attributed the 3.94–3.98 b.y. ages shown by the Boulder 6 clasts to total outgassing of these rocks at the time of the event that formed the boulder. On the basis of all the data, interpretations such as these seem to be the most likely.

The ages in the higher of the two ranges, 4.22–4.35 b.y., are not so simply interpreted. Other ANT-suite rocks from the Apollo 17 site, in addition to the 73215 clasts and 78155, also show ages in this range. Troctolite 76535 has a Sm–Nd age of 4.26 ± 0.06 b.y. (Lugmair *et al.*, 1976), and a K–Ar age of 4.23–4.34 b.y. (Huneke and Wasserburg, 1975; Husain and Schaeffer, 1975; Bogard *et al.*, 1974, 1975). Three 2–4 mm chips of anorthositic rocks from soils have $^{40}Ar/^{39}Ar$ ages of 4.23–4.28 b.y. (Schaeffer *et al.*, 1976; Kirsten and Horn, 1974; Schaeffer and Schaeffer, 1977). Moreover, a number of samples from the Apollo 16 site also show a similar ~4.25 b.y. age (Schaeffer and Husain, 1973). These data indicate that the ~4.25 b.y. age is the approximate time of an event in the lunar crust, but nature of this event is not clearly defined. Several authors have suggested that this event was excavation of material by a very large impact, and that prior to excavation the rocks were at depths where temperatures were high enough to permit continuous isotopic reequilibration and outgassing (Turner *et al.*, 1973; Schaeffer and Husain, 1974; Papanastassiou and Wasserburg, 1976). Other possible hypotheses are that the event was an episode of highlands magmatism, or an episode of high-temperature metamorphism.

The data for the 73215 anorthositic gabbros do not permit a definitive interpretation of the ~4.25 b.y. age, but they can be used to set some limits on possible interpretations. During the melting/recrystallization episode that formed their textures, the rocks were for significant periods of time at or near their solidus temperatures. Such heating, melting, and recrystallization would most likely have outgassed all constituents of the rocks of their radiogenic Ar and reequilibrated their Sr isotopes; thus it is unlikely that the K–Ar and Rb–Sr internal isochron age data reflect events prior to this episode. It is also unlikely that the late-stage fracturing event that affected the rocks is reflected in the age data, because this event produced only minor and local effects on rock texture (in addition the fact that the ~4.25 b.y. age is found in a significant number of rocks suggests it refers to a major rather than a minor event). Thus it is most likely that the ~4.25 b.y. age represents: (1) the date of the high-temperature melting/recrystallization event that formed the rock textures, or (2) the date of excavation of the rocks from depth. At present no definitive choice can be made between these alternatives, but the former hypothesis may be preferable to the latter because it indicates a simpler history for the rocks: if they were excavated at 4.25 b.y., they must have been stored in a location where they were unaffected by meteoritic bombardment and metamorphism for ~350 m.y. until they were incorporated in the breccias in which they are now found; if instead they were melted and recrystallized at 4.25 b.y. and excavated at 3.90 b.y., no intermediate storage is required.

Conclusions

Felsite clast

The parent rock of the felsite clast crystallized from a melt. This melt probably originated as a residual liquid in a differentiated igneous complex. Composition of the parent magma of the complex cannot yet be determined, but it is likely that the differentiation involved late-stage separation of immiscible liquids (Blanchard *et al.*, 1977). The Rb–Sr model-age data indicate that the time of this differentiation was probably between 3.90 and 4.05 b.y. (Compston *et al.*, 1977). Thus it appears that the felsite is a relatively "young" highlands rock; whether or not all lunar "granite" is similarly "young", however, cannot yet be determined because so few samples of this type of rock have been studied in detail.

The felsic glass in the clast was, for the most part, emplaced as dikelets of melt into fractures in the crystalline felsite. The parent rock for the felsic melt was probably the same parent rock as for the fragments of crystalline felsite. Fracturing of the felsite and emplacement of the melt dikelets occurred prior to incorporation of the clast in the 73215 breccia, and it is likely that the event that produced the fractures and the melt was an impact. The bulk clast was hot (≥990°C), and the felsic melt was mobile when the clast was incorporated in the breccia.

After incorporation of the clast in the breccia, the felsic melt cooled. Olivine relics in the melt were enriched in Mg by reduction and by Fe–Mg exchange with surrounding matrix. Plagioclase relics were depleted in Na and partly replaced by K-feldspar. During the early stages of cooling, second-generation crystals of pyroxene and K-feldspar crystallized from the melt. During later stages of cooling, minute crystallites of a silica mineral, feldspars, and mafic minerals precipitated. Textures formed during crystallization of the felsic melt, and exsolution structures in the second-generation pyroxene crystals (Nord and James, 1977) suggest that there was probably no significant reheating of the clast after the breccia-forming event.

Rb–Sr and ^{40}Ar–^{39}Ar ages of the felsite clast are 3.90–3.92 b.y. (Compston *et al.*, 1977; Jessberger *et al.*, 1977). The petrologic data outlined above indicate that this age probably dates the breccia-forming event. If the event was indeed the South Serenitatis basin-forming impact (James *et al.*, 1975; Morgan *et al.*, 1976; Jessberger *et al.*, 1976; Wood, 1975), then the time interval during which a large number of major lunar basins formed was very short, on the order of 100 m.y. This interpretation of the age data supports the concept of a "cataclysmic" episode of formation of lunar basins (Tera *et al.*, 1974).

Anorthositic gabbro clasts

The similar mineral assemblages, similar mineral compositions and gradational textures of the three anorthositic gabbro clasts suggest that their parent rocks all formed by the same process. Consideration of the petrologic data for all the clasts indicates that their most likely genesis was as heated, partly melted and/or recrystallized breccias. In 29,9 the bulk of the breccia melted and the melt crystallized with igneous texture. In 46,25 only a very small fraction of the breccia melted, and the bulk of the rock recrystallized in the solid state. In 46,33 an appreciable fraction of the breccia melted and crystallized with igneous texture, while the unmelted fraction recrystallized. The parent breccias were probably polymict rather than monomict; they apparently contained fragments from only a small number of source rocks, however, and they were probably not much mixed or reworked before they were heated and partly melted.

The age dating results yield two "ages" for each clast, one of about 4.25–4.35 b.y. and the other of 4.06–4.07 b.y. (Jessberger *et al.*, 1976; Müller *et al.*, 1977; Compston *et al.*, 1977). The younger age probably has no precise time significance but is instead the result of partial resetting of some components of the clasts after they were incorporated in the breccia (Jessberger *et al.*, 1976). The older age may be the date of the melting/recrystallization event that produced the textures of the rocks; alternatively it may be the date of excavation of the rocks from depth, with there having been continuous outgassing and isotopic reequilibration during the time interval between the melting/recrystallization event and excavation. The former hypothesis may be preferable to the latter because it indicates a simpler history for the rocks.

Acknowledgments—We would like to thank other members of the 73215 consortium, in particular D. P. Blanchard, W. Compston, E. K. Jessberger, T. Kirsten, H. W. Müller, G. L. Nord, Jr., T. Plieninger, and O. A. Schaeffer, for stimulating discussions and for the opportunity to participate in truly interactive, interdisciplinary research. We greatly appreciate D. P. Blanchard's promptness and care in handling the chores of sample splitting and distribution. Patricia Eck typed innumerable versions of this and previous manuscripts with unfailing good cheer; We are grateful for her helpfulness and patience. This research was done on behalf of NASA under contract T-3040C.

References

Albee A. L., Burnett D. S., Chodos A. A., Pananastassiou D. A., Podosek F. A., Russ G. P., and Wasserburg G. J. (1970) Mineralogic and isotopic investigations of lunar rock 12013. *Earth Planet. Sci. Lett.* **9**, 137–163.

Anderson A. T., Braziunas T. F., Jacoby J., and Smith J. V. (1972) Thermal and mechanical history of breccias 14306, 14063, 14270 and 14321. *Proc. Lunar Sci. Conf. 3rd*, p. 819–835.

Ashwal L. D. (1975) Petrologic evidence for a plutonic igneous origin of anorthositic norite clasts in 67955 and 77017. *Proc. Lunar Sci. Conf. 6th*, p. 221–230.

Bence A. E., Delano J. W., Papike J. J., and Cameron K. L. (1974) Petrology of the highlands massifs at Taurus-Littrow: An analysis of the 2–4 mm soil fraction. *Proc. Lunar Sci. Conf. 5th*, p. 785–827.

Bernatowicz T., Hohenberg C. M., Hudson B., Kennedy B. M., Morgan C. J., and Podosek F. A. (1977) Argon ages for Imbrium consortium samples 14064 and 15405 (abstract). In *Lunar Science VIII*, p. 97–99. The Lunar Science Institute, Houston.

Bickel C. E. and Warner J. L. (1977) Petrology of 78155: An early, thermally metamorphosed polymict breccia (abstract). In *Lunar Science VIII*, p. 109–111. The Lunar Science Institute, Houston.

Bickel C. E., Warner J. L., and Phinney W. C. (1976) Petrology of 79215: Brecciation of a lunar cumulate. *Proc. Lunar Sci. Conf. 7th*, p. 1793–1819.

Blanchard D. P., Jacobs J. W., and Brannon J. C. (1977) Chemistry of ANT-suite and felsite clasts from consortium breccia 73215, and of gabbroic anorthosite 79215. *Proc. Lunar Sci. Conf. 8th.* This volume.

Bogard D. D., Nyquist L. E., Bansal B. M., and Wiesmann H. (1974) 76535: An old lunar rock? (abstract). In *Lunar Science V*, p. 70–72. The Lunar Science Institute, Houston.

Bogard D. D., Nyquist L. E., Bansal B. M., Wiesmann H., and Shih C.-Y. (1975) 76535: An old lunar rock. *Earth Planet. Sci. Lett.* **26**, 69–80.

Brown G. M., Emeleus C. H., Holland J. G., Peckett A., and Phillips R. (1972) Mineral-chemical variations in Apollo 14 and Apollo 15 basalts and granitic fractions. *Proc. Lunar Sci. Conf. 3rd*, p. 141–157.

Cadogan P. H. and Turner G. (1976) The chronology of the Apollo 17 Station 6 boulder. *Proc. Lunar Sci. Conf. 7th*, p. 2267–2285.

Cameron K. L., Delano J. W., Bence A. E., and Papike J. J. (1973) Petrology of the 2–4 mm soil fraction from the Hadley–Appenine region of the moon. *Earth Planet. Sci. Lett.* **19**, 9–21.

Chao E. C. T., Minkin J. A., and Thompson C. L. (1974) Preliminary petrographic description and geologic implications of the Apollo 17 Station 7 boulder consortium samples. *Earth Planet Sci. Lett.* **23**, 413–428.

Compston W., Foster J. J., and Gray C. M. (1975) Rb–Sr ages of clasts from within Boulder 1, Station 2, Apollo 17. *The Moon* **14**, 445–462.

Compston W., Foster J. J., and Gray C. M. (1977) Rb–Sr systematics in clasts and aphanites from consortium breccia 73215. *Proc. Lunar Sci. Conf. 8th.* This volume.

Delano J. W., Bence A. E., Papike J. J., and Cameron K. L. (1973) Petrology of the 2–4 mm soil fraction from the Descartes region of the moon and stratigraphic implications. *Proc. Lunar Sci. Conf. 4th*, p. 537–551.

Dowty E., Prinz M., and Keil K. (1974) Ferroan anorthosite: A widespread and distinctive lunar rock type. *Earth Planet. Sci. Lett.* **24**, 15–25.

Drake M. J., McCallum I. S., McKay G. A., and Weill D. F. (1970) Mineralogy and petrology of Apollo 12 sample no. 12013: A progress report. *Earth Planet. Sci. Lett.* **9**, 103–123.

Dymek R. F., Albee A. L., and Chodos A. A. (1976) Petrology and origin of Boulders #2 and #3, Apollo 17 Station 2. *Proc. Lunar Sci. Conf. 7th*, p. 2335–2378.

Ganapathy R., Morgan J. W., Higuchi H., Anders E., and Anderson A. T. (1974) Meteoritic and volatile elements in Apollo 16 rocks and in separated phases from 14306. *Proc. Lunar Sci. Conf. 5th*, p. 1659–1683.

Goldstein J. I. and Yakowitz H. (1971) Metallic inclusions and metal particles in the Apollo 12 lunar soil. *Proc. Lunar Sci. Conf. 2nd*, p. 177–191.

Helz R. T. and Appleman D. E. (1974) Poikilitic and cumulate textures in rock 77017 (abstract). In *Lunar Science V*, p. 322–324. The Lunar Science Institute, Houston.

Hess P. C. and Rutherford M. J. (1974) Elemental fractionation between immiscible melts (abstract). In *Lunar Science V*, p. 328–330. The Lunar Science Institute, Houston.

Hess P. C., Rutherford M. J., Guillemette R. N., Ryerson F. J., and Tuchfeld H. A. (1975) Residual products of fractional crystallization of lunar magmas: An experimental study. *Proc. Lunar Sci. Conf. 6th*, p. 895–909.

Higuchi H. and Morgan J. W. (1975) Ancient meteoritic component in Apollo 17 boulders. *Proc. Lunar Sci. Conf. 6th*, p. 1625–1651.

Hodges F. N. and Kushiro I. (1974) Apollo 17 petrology and experimental determinations of differentiation sequences in model moon compositions. *Proc. Lunar Sci. Conf. 5th*, p. 505–520.

Hollister L. S. (1973) Sample 67955: A description and a problem. *Proc. Lunar Sci. Conf. 4th*, p. 633–641.

Huneke J. C. and Wasserburg G. J. (1975) Trapped ^{40}Ar in troctolite 76535 and evidence for enhanced ^{40}Ar–^{39}Ar age plateaus (abstract). In *Lunar Science VI*, p. 417–419. The Lunar Science Institute, Houston.

Husain L. and Schaeffer O. A. (1975) Lunar evolution: The first 600 million years. *Geophys. Res. Lett.* **2**, 29–32.

James O. B. (1970) Petrology of lunar microbreccia 12013,6. *U.S. Geol. Survey Interagency Report: Astrogeology* **23**.

James O. B. (1976) Petrology of aphanitic lithologies in consortium breccia 73215. *Proc. Lunar Sci. Conf. 7th*, p. 2145–2178.

James O. B. and Blanchard D. P. (1976) Consortium studies of light-gray breccia 73215: Introduction, subsample distribution data, and summary of results. *Proc. Lunar Sci. Conf. 7th*, p. 2131–2143.

James O. B. and Marti K. (1977) Consortium breccia 73255: Matrix petrography and exposure history (abstract). In *Lunar Science VIII*, p. 505–507. The Lunar Science Institute, Houston.

James O. B., Brecher A., Blanchard D. P., Jacobs J. W., Brannon J. C., Korotev R. L., Haskin L. A., Higuchi H., Morgan J. W., Anders E., Silver L. T., Marti K., Braddy D., Hutcheon I. D., Kirsten T., Kerridge J. F., Kaplan I. R., Pillinger C. T., and Gardiner L. R. (1975) Consortium studies of matrix of light gray breccia 73215. *Proc. Lunar Sci. Conf. 6th*, p. 547–577.

Jessberger E. K., Kirsten T., and Staudacher Th. (1977) One rock and many ages—further K–Ar data 73215. *Proc. Lunar Sci. Conf. 7th*, p. 2201–2215.

Jessberger E. K., Kirsten T., and Staudacher T. (1977) One rock and many ages—further K–Ar data on consortium breccia 73215. *Proc. Lunar Sci. Conf. 8th*. This volume.

Jovanovic S. and Reed G. W., Jr. (1975) History of Boulder 1 at Station 2, Apollo 17 based on trace element interrelationships. *The Moon* **14**, 385–393.

Kirsten T. and Horn P. (1974) Chronology of the Taurus-Littrow region III: Ages of mare basalts and highlands breccias and some remarks about the interpretation of lunar highland rock ages. *Proc. Lunar Sci. Conf. 5th*, p. 1451–1475.

Lugmair G. W., Marti K., Kurtz J. P., and Scheinin N. B. (1976) History and genesis of lunar troctolite 76535 or: How old is old? *Proc. Lunar Sci. Conf. 7th*, p. 2009–2033.

Marvin U. B., Wood J. A., Taylor G. J., Reid J. B., Jr., Powell B. N., Dickey J. S., Jr., and Bower J. F. (1971) Relative proportions and probable sources of rock fragments in the Apollo 12 soil samples. *Proc. Lunar Sci. Conf. 2nd*, p. 679–699.

McCallum I. S., Mathez E. A., Okamura F. P., and Ghose S. (1974) Petrology and crystal chemistry of poikilitic anorthositic gabbro 77017. *Proc. Lunar Sci. Conf. 5th*, p. 287–302.

Meyer C., Jr. (1972) Mineral assemblages and the origin of non-mare lunar rock types (abstract). In *Lunar Science III*, p. 542–544. The Lunar Science Institute, Houston.

Morgan J. W., Ganapathy R., Higuchi H., Krähenbühl U., and Anders E. (1974) Lunar basins: Tentative characterization of projectiles from meteoritic elements in Apollo 17 boulders. *Proc. Lunar Sci. Conf. 5th*, p. 1703–1736.

Morgan J. W., Gros J., Takahashi H., and Hertogen J. (1976) Lunar breccia 73215: Siderophile and volatile trace elements. *Proc. Lunar Sci. Conf. 7th*, p. 2189–2199.

Müller H. W., Plieninger T., James O. B., and Schaeffer O. A. (1977) Laser probe ^{39}Ar–^{40}Ar dating of materials from consortium breccia 73215. *Proc. Lunar Sci. Conf. 8th*. This volume.

Nord G. L., Jr., Christie J. M., Heuer A. H., and Lally J. S. (1975) North Ray crater breccias: An electron petrographic study. *Proc. Lunar Sci. Conf. 6th*, p. 779–797.

Nord G. L., Jr. and James O. B. (1977) Aphanitic matrix, an ANT-suite clast and a felsite clast in consortium breccia 73215: An electron petrographic study. *Proc. Lunar Sci. Conf. 8th*. This volume.

Nyquist L. E., Wiesmann H., Shih C.-Y., and Bansal B. M. (1977a) 15405 quartz-monzodiorite: Super KREEP. In *Interdisciplinary Studies by the Imbrium Consortium*, Vol. 2, (J. A. Wood, ed.). LSI Contr. No. 268D, p. 30–41. Center for Astrophysics, Cambridge.

Nyquist L. E., Wiesmann H., Shih C.-Y., and Bansal B. M. (1977b) REE and Rb–Sr analysis of 15405 quartz-monzodiorite (super KREEP) (abstract). In *Lunar Science VIII*, p. 738–740. The Lunar Science Institute, Houston.

Papanastassiou D. A. and Wasserburg G. J. (1976) Rb–Sr age of troctolite 76535. *Proc. Lunar Sci. Conf. 7th*, p. 2035–2054.

Prinz M., Dowty E., Keil K., and Bunch T. E. (1973) Mineralogy, petrology and chemistry of lithic fragments from Luna 20 fines: Origin of the cumulate ANT suite and its relationship to high-alumina and mare basalts. *Geochim. Cosmochim. Acta* **37**, 979–1006.

Quick J. E. and Albee A. L. (1976) 12013 revisited—two clast-laden melts (abstract). In *Lunar Science VII*, p. 712–714. The Lunar Science Institute, Houston.

Quick J. E., Albee A. L., Ma M.-S., Murali A. V., and Schmitt R. A. (1977) Chemical compositions and possible immiscibility of two silicate melts in 12013. *Proc. Lunar Sci. Conf. 8th*. This volume.

Roedder E. and Weiblen P. W. (1970) Lunar petrology of silicate melt inclusions, Apollo 11 rocks. *Proc. Apollo 11 Lunar Sci. Conf.*, p. 801–837.

Roedder E. and Weiblen P. W. (1972) Petrographic features and petrologic significance of melt inclusions in Apollo 14 and 15 rocks. *Proc. Lunar Sci. Conf. 3rd*, p. 251–279.

Roedder E. and Weiblen P. W. (1974) Petrology of clasts in lunar breccia 67915. *Proc. Lunar Sci. Conf. 5th*, p. 303–318.

Rutherford M. J. and Hess P. C. (1975) Origin of lunar granites as immiscible liquids (abstract). In *Lunar Science VI*, p. 696–698. The Lunar Science Institute, Houston.

Rutherford M. J., Hess P. C., Ryerson F. J., Campbell H. W., and Dick P. A. (1976) The chemistry, origin and petrogenetic implications of lunar granite and monzonite. *Proc. Lunar Sci. Conf. 7th*, p. 1723–1740.

Ryder G. (1976) Lunar sample 15405: Remnant of a KREEP basalt-granite differentiated pluton. *Earth Planet. Sci. Lett.* **29**, 255–268.

Ryder G., Stoeser D. B., Marvin U. B., and Bower J. F. (1975) Lunar granites with unique ternary feldspars. *Proc. Lunar Sci. Conf. 6th*, p. 435–449.

Ryerson F. J. and Hess P. C. (1975) The partioning of trace elements between immiscible silicate melts (abstract). *EOS* (*Trans. Amer. Geophys. Union*) **56**, 470.

Schaeffer O. A. and Husain L. (1973) Early lunar history: Ages of 2 to 4 mm soil fragments from the lunar highlands. *Proc. Lunar Sci. Conf. 4th*, p. 1847–1863.

Schaeffer O. A., Husain L., and Schaeffer G. A. (1976) Ages of highland rocks: The chronology of lunar basin formation revisited. *Proc. Lunar Sci. Conf. 7th*, p. 2067–2092.

Schaeffer O. A. and Schaeffer G. A. (1977) ^{39}Ar–^{40}Ar ages of lunar rocks (abstract). In *Lunar Science VIII*, p. 840–842. The Lunar Science Institute, Houston.

Schairer J. F. and Bowen N. L. (1947) The system anorthite-leucite-silica. *Soc. Geol. Finlande Bull.* **20**, 72–75.

Sclar C. B., Prinz M., and Keil K. (1974) Petrology of xenoliths of granitic composition in alkalic high-alumina basalt (KREEP) from Oceanus Procellarum (12057,22). *Geol. Soc. Amer. Abstracts with Programs* **6**, 945.

Simonds C. H. (1975) Thermal regimes in impact melts and the petrology of the Apollo 17 Station 6 boulder. *Proc. Lunar Sci. Conf. 6th*, p. 641–672.

Stoeser D. B., Marvin U. B., Wood J. A., Wolfe R. W., and Bower J. F. (1974) Petrology of a stratified boulder from South Massif, Taurus-Littrow. *Proc. Lunar Sci. Conf. 5th*, p. 355–377.

Tatsumoto M. (1970) U–Th–Pb age of Apollo 12 rock 12013. *Earth Planet. Sci. Lett.* **9**, 193–200.

Tatsumoto M. and Unruh D. M. (1976) KREEP basalt age: Grain by grain U–Th–Pb systematics study of the quartz monzodiorite clast 15405,88. *Proc. Lunar Sci. Conf. 7th*, p. 2107–2129.

Taylor G. J., Drake M. J., Hallam M. E., Marvin U. B., and Wood J. A. (1973) Apollo 16 stratigraphy: The ANT hills, the Cayley plains, and a pre-Imbrium regolith. *Proc. Lunar Sci. Conf. 4th*, p. 553–568.

Tera F., Papanastassiou D. A., and Wasserburg G. J. (1974) Isotopic evidence for a terminal lunar cataclysm. *Earth Planet. Sci. Lett.* **22**, 1–21.

Turner G. (1971) ^{40}Ar–^{39}Ar ages from the lunar maria. *Earth Planet Sci. Lett.* **11**, 169–191.

Turner G. and Cadogan P. H. (1975) The history of lunar bombardment inferred from ^{40}Ar–^{39}Ar dating of highland rocks. *Proc. Lunar Sci. Conf. 6th*, p. 1509–1538.

Turner G., Cadogan P. H., and Yonge C. J. (1973) Argon selenochronology. *Proc. Lunar Sci. Conf. 4th*, p. 1889–1914.

Walker D., Grove T. L., Longhi J., Stolper E. M., and Hays J. F. (1973) Origin of lunar feldspathic rocks. *Earth Planet. Sci. Lett.* **20**, 325–336.

Warner R. D., Planner H. N., Keil K., Murali A. V., Ma M.-S., Schmitt R. A., Ehmann W. D., James W. D., Jr., Clayton R. N., and Mayeda T. K. (1976) Consortium investigation of breccia 67435. *Proc. Lunar Sci. Conf. 7th*, p. 2379–2402.

Warner J. L., Bickel C. E., Phinney W. C., and Simonds C. H. (1977) Feldspathic granulitic impactites that predate the final lunar bombardment (abstract). In *Lunar Science VIII*, p. 979–981. The Lunar Science Institute, Houston.

Watson E. B. (1976) Two-liquid partition coefficients: Experimental data and geochemical implication. *Contrib. Mineral. Petrol.* **56**, 119–134.

Weill D. F. and McKay G. A. (1975) Major and trace element trends during minor experiments on the crystallization of lunar melts (abstract). In *Lunar Science VI*, p. 863–865. The Lunar Science Institute, Houston.

Wilshire H. G. and Jackson E. D. (1972) Petrology and stratigraphy of the Fra Mauro formation at the Apollo 14 site. *U.S. Geol. Survey Prof. Paper 785.*

Wood J. A. (1970) Petrology of the lunar soil and geophysical implications. *J. Geophys. Res.* **75**, 6497–6513.

Wood J. A. (1975) The nature and origin of Boulder 1, Station 2, Apollo 17. *The Moon* **14**, 505–517.

Wood J. A., Dickey J. S., Jr., Marvin U. B., and Powell B. N. (1970) Lunar anorthosites and a geophysical model of the moon. *Proc. Apollo 11 Lunar Sci. Conf.*, p. 965–988.

Proc. Lunar Sci. Conf. 8th (1977), p. 2495–2506.
Printed in the United States of America

Aphanitic matrix, an ANT-suite clast and a felsite clast in consortium breccia 73215: An electron petrographic study

GORDON L. NORD, JR. AND ODETTE B. JAMES

U.S. Geological Survey, Reston, Virginia 22092

Abstract—Light-gray breccia 73215 consists of physically separable clasts in an aphanitic matrix; the matrix itself is composed of small crystalline clasts set in a very fine-grained, texturally distinct groundmass. In this study, matrix areas and clasts first characterized optically and by electron microprobe analysis have been examined by transmission electron microscopy at 200 kV.

Grain size of the groundmass in the matrix is extremely fine (<10 μm); previous optical and SEM observations suggested that its texture is microsubophitic and that it crystallized from a melt. The TEM study confirms the presence of igneous textures in the groundmass and its origin as a melt. A small clast of ANT-suite rock shows microstructures indicating extensive shock, annealing, recovery, and recrystallization prior to its incorporation in the breccia.

The felsite clast examined has been the subject of intensive consortium study. It consists of crystalline felsite and felsic glass components. The crystalline felsite is mostly a vermicular intergrowth of quartz and K-feldspar, with minor Ca-K-rich ternary plagioclase. The ternary plagioclase is homogeneous optically; electron diffraction shows it has a single lattice with weak elongated (b) spots and a modulated "tweed" texture with a wavelength of 100 Å, which probably represents initial attempts at phase separation. The quartz is free of deformation features, but the K-feldspar has a moderate density of dislocations, subboundaries, and a few deformation bands.

The felsic glass component of the clast was a melt when the clast was incorporated in the breccia. It is in some places clear and free of crystallites, and in others brown and clouded with abundant small crystallites. It also contains prisms of pyroxene that precipitated as the melt cooled; exsolution structures in these crystals are like those seen in lunar basaltic pyroxenes. The nature of the exsolution in pyroxene and the degree of crystallinity of the glass are consistent with a single-stage crystallization and cooling of the felsic melt after incorporation of the clast in the breccia; no evidence for later reheating is observed. Thus it is likely that the 3.90–3.92 b.y. age obtained for this clast by Rb-Sr and $^{40}Ar/^{39}Ar$ methods dates the breccia-forming event (possibly the South Serenitatis basin-forming event).

INTRODUCTION

LUNAR BRECCIAS are extremely complex rocks, difficult to characterize petrologically because they commonly have very fine-grained matrices and each clast they contain has had its own unique history. A new and useful technique, which augments light optical microscopy in working out mineralogical and petrographic details and therefore geologic history, is transmission electron microscopy (TEM). In using this technique to study lunar breccias, individual clasts and small areas of matrix are first studied using the optical microscope and the electron microprobe in order to define problems concerning the petrologic interpretation of the breccia history. These areas are then thinned to less than 2 μm by ion bombardment so that they become electron transparent. By placing the sample on suitably shaped copper grids and substrates during the ion thinning, the position of the thinned area can be very precisely controlled. Previous TEM studies have established the usefulness of this technique in describing breccia formation mechanisms (Christie *et al.*, 1973) and delineating breccia thermal and mechanical histories (Lally *et al.*, 1976).

73215 is a light-gray breccia which has been shown by previous studies to have formed during a major lunar impact event as a fragment-laden melt (a mechanically mixed aggregate of crystalline clasts and melt). The breccia has been the subject of an intensive study by a consortium of investigators. This paper presents optical and TEM observations on the aphanitic matrix of the breccia, on one small clast of ANT-suite rock, and on a clast of felsite (lunar "granite"). Petrology of the bulk breccia was described by James *et al.* (1975), and petrology of the matrix was discussed in detail by James (1976). Petrology, bulk chemistry, Rb–Sr systematics and $^{40}Ar/^{39}Ar$ age dating results for the felsite clast are presented by James and Hammarstrom (1977), Blanchard *et al.* (1977), Compston *et al.* (1977) and Jessberger *et al.* (1977), respectively. The results of all the 73215 consortium studies as of May, 1976, were summarized by James and Blanchard (1976).

MATRIX

The matrix of the breccia is a gray-to-black aphanitic rock that consists of small mineral and lithic clasts set in a groundmass of very fine grain size ($\leq 10\,\mu m$). The groundmass is texturally distinct from the clasts it encloses. When viewed with the petrographic microscope, the groundmass texture appears to range from microsubophitic to microintergranular to microgranular (Fig. 1a), but the grain size is so fine that the textures are not always clearly resolvable and very small clasts cannot be distinguished from groundmass grains. When viewed with the scanning electron microscope, however, the groundmass intergrowth clearly shows lath-shaped plagioclase and microsubophitic texture. On the basis of these observations, James (1976) proposed that the groundmass crystallized from a melt. Study of the groundmass by TEM confirms the presence of textures indicating crystallization from melt. Many perfectly formed lath-shaped plagioclase grains are seen to be enclosed in pigeonitic pyroxene in a microsubophitic intergrowth (Figs. 1b,c). The plagioclase laths are commonly strongly twinned and have curved terminations; the textures are similar to textures seen in fine-grained areas which crystallized from a melt in the North Ray Crater breccias (Fig. 10b of Nord *et al.*, 1975). Locally orthopyroxene also is present in the groundmass, as euhedral grains; Fig. 1b shows one such grain that exhibits several sets of (100) stacking faults.

ANT-SUITE CLAST

A $200\,\mu m$ clast of fine-grained ($5–50\,\mu m$) hornfelsic norite was also examined. It consisted of a large ($50\,\mu m$) irregular plagioclase grain surrounded by an equigranular mosaic of $5\,\mu m$ grains of plagioclase and minor pyroxene.

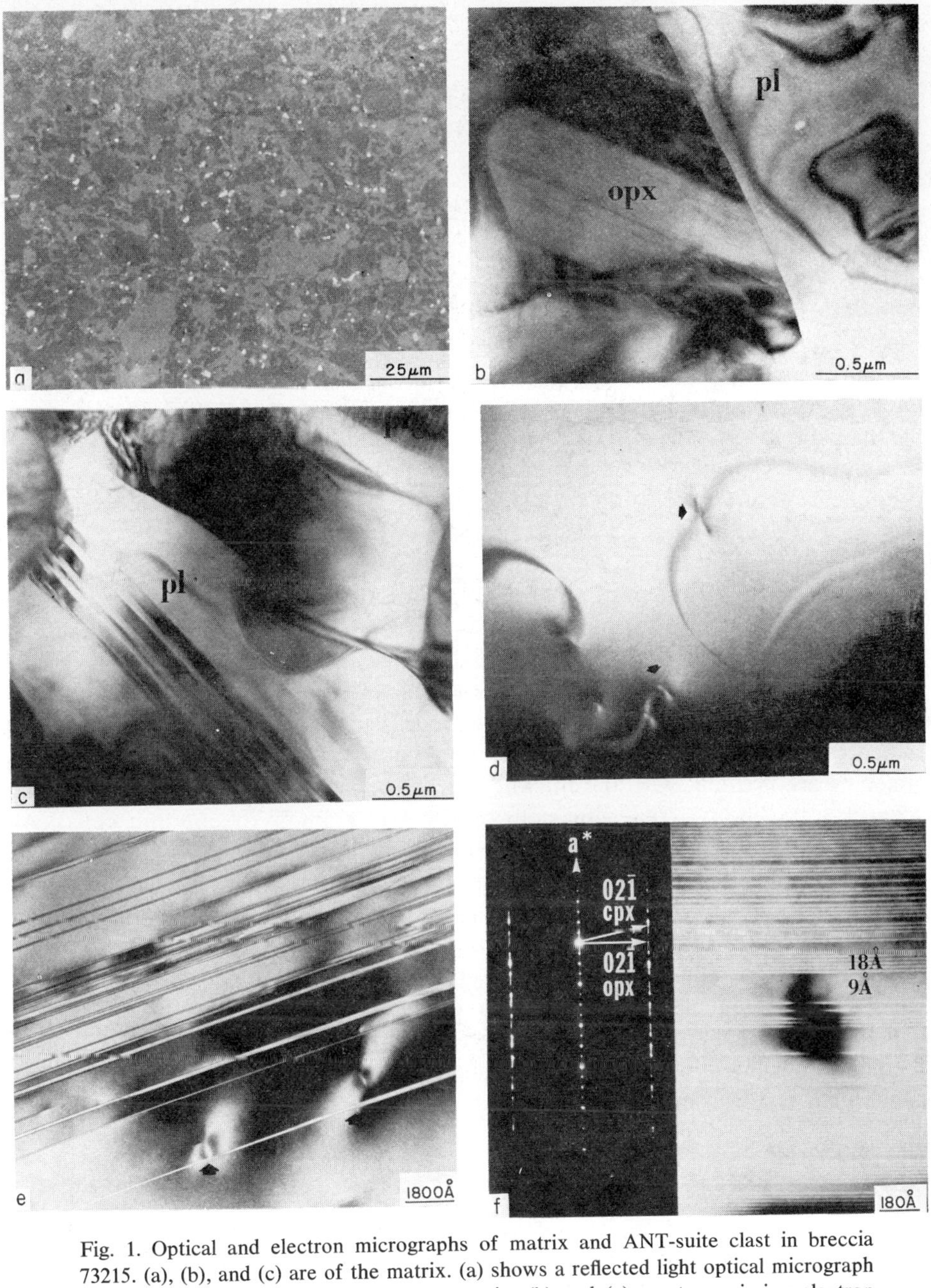

Fig. 1. Optical and electron micrographs of matrix and ANT-suite clast in breccia 73215. (a), (b), and (c) are of the matrix. (a) shows a reflected light optical micrograph of the fine-grained groundmass of the matrix. (b) and (c) are transmission electron micrographs of the groundmass, showing plagioclase (pl), orthopyroxene (opx), and pigeonite (pig). (d), (e), and (f) are of the ANT-suite clast. (d) is a dark field micrograph ($h + k$ odd, l odd = **g**) showing type (b) APB's in plagioclase terminated at dislocations (arrowed). (e) shows lamellae in orthopyroxene, with dislocations on the interfaces (arrowed). (f) shows the 18 Å lattice spacing of the host orthopyroxene and the 9 Å lattice spacing of the lamellae, imaged in dark field $\mathbf{g} = h2\bar{1}$; the indexed diffraction pattern shows reflections from both phases. All electron micrographs taken at 200 kV.

The texture suggests that the clast was strongly deformed and subsequently recrystallized on a fine scale prior to being incorporated in the breccia, but that some larger relict grains are preserved.

The TEM examination of the plagioclase confirms the presence of textures indicating extensive recrystallization. The small plagioclase grains commonly are bounded by straight contacts and triple junctions; such a texture is typical of metamorphic rocks. The plagioclase grains show type (b) antiphase domain boundaries (APB's), illustrated in Fig. 1d. This type of APB in plagioclase can be formed by two mechanisms. The first mechanism is domain growth during cooling through the $C\bar{1} \rightarrow I\bar{1}$ phase transition, and the APB's that are produced enclose domains and form a "foam" structure (Nord *et al.*, 1973). The second mechanism is growth mistakes that occur when crystallization or recrystallization takes place in the lower temperature $I\bar{1}$ field; the APB's that are produced commonly do not enclose domains. The APB's in the plagioclase in this clast are terminated by dislocations and do not enclose domains; thus they probably formed by the second mechanism, subsolidus recrystallization in the $I\bar{1}$ field, rather than by the first, a phase transition during cooling.

The TEM examination of the orthopyroxene in this clast shows that this mineral contains a moderate density of dislocations and a high density of linear defects (lamellae) that are parallel to (100). The lamellae are terminated by dislocations (Fig. 1e). A lattice image of the $h2\bar{1}$ row of diffraction spots (Fig. 1f) indicates that the lamellae are of clinopyroxene; areas of both 18 and 9 Å spacings are present, with the thicknesses of the areas with 9 Å spacing varying from 9 Å (1 clinopyroxene unit cell) to 126 Å (14 clinopyroxene unit cells). An electron diffraction pattern shows orthopyroxene lattice spots, and a set of clinopyroxene lattice spots in only one twin orientation. The $h00$ row spots coincide exactly, indicating that the clinopyroxene has a low calcium content; the presence of $h+k$ odd, diffraction spots in the clinopyroxene diffraction net prove that it has the primitive space group, $P2_1/c$. The data indicate that the clinopyroxene lamellae are clinohypersthene; they apparently formed by the ortho→clino transition, probably promoted by shock-induced shear. Similar formation of clinopyroxene by shock has been observed in meteoritic pyroxenes (Ashworth and Barber, 1975).

FELSITE CLAST

The felsite clast consists of fragments of crystalline felsite intruded and enclosed by felsic glass (James and Hammarstrom, 1977). The bulk of the crystalline felsite is a vermicular intergrowth of quartz and K-feldspar; this intergrowth contains minor amounts of Ca–K-rich ternary plagioclase, ilmenite, zircon, Fe-rich olivine, Ca–Fe-rich pyroxene, and phosphate minerals. The felsic glass ranges from clear and colorless, to clouded and brown. It contains second-generation crystals of pyroxene and K-feldspar, and relict fragments of olivine, pyroxene, quartz, ilmenite, phosphate minerals, zircon, and ternary plagioclase (Fig. 2a); the olivine, pyroxene, and ilmenite relics have rims of

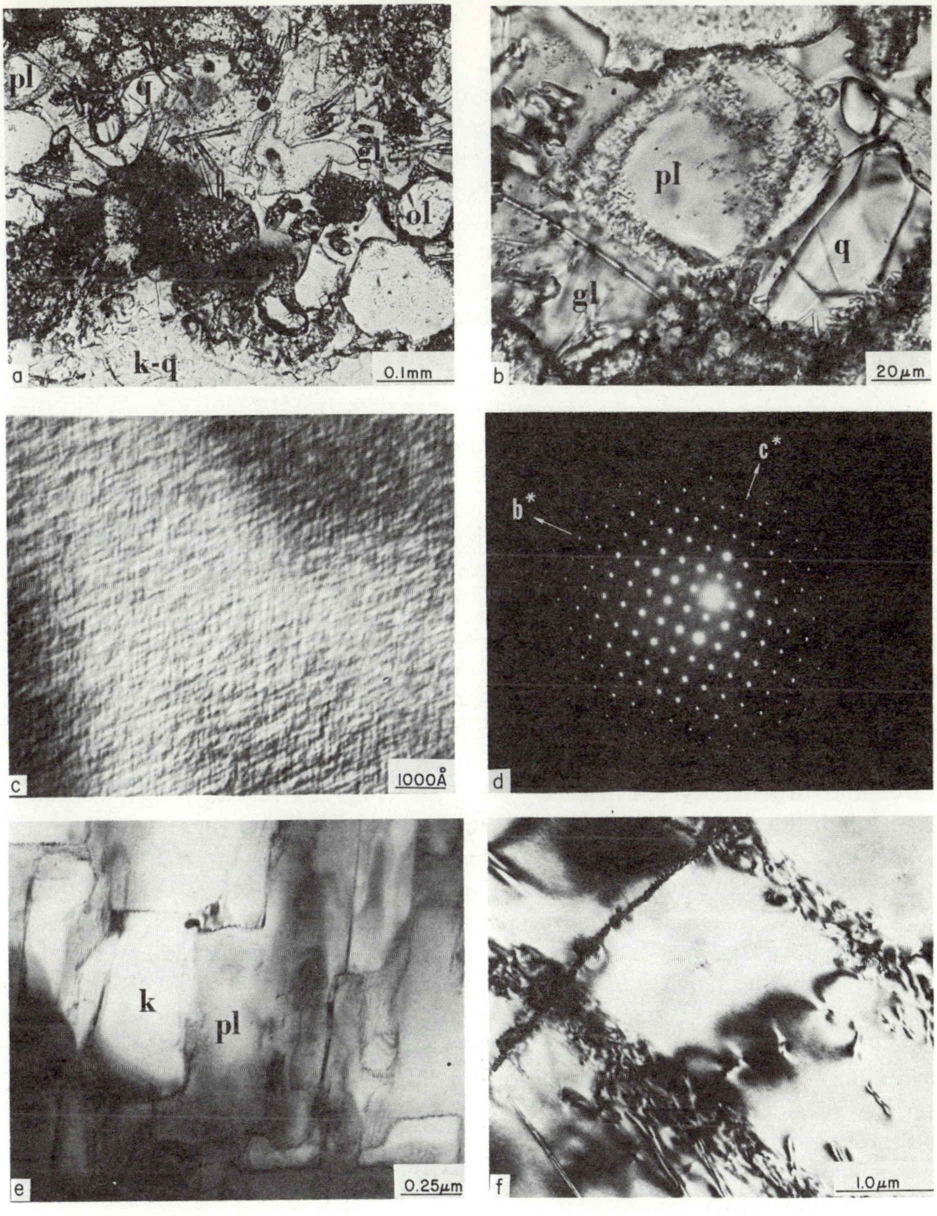

Fig. 2. Optical and electron micrographs of ternary plagioclase. (a) is a plane light optical micrograph of an area at the edge of the felsite clast. Ternary plagioclase (pl), olivine (ol), quartz (q), glass (gl), and the K-feldspar-quartz vermicular intergrowth (k-q) are shown. (b) shows relict fragments of quartz and ternary plagioclase with an inclusion-rich rim in felsic glass. (c) shows a dark field electron micrograph of the "tweed" texture of the ternary plagioclase in the core of the grain shown in (b). (d) is the electron diffraction pattern of the area shown in (c). (e) shows the blocky inclusions of K-feldspar (k) in the rim of the plagioclase grain. (f) is a micrograph of another ternary plagioclase grain showing deformation bands and a fine "tweed" texture. All electron micrographs taken at 200 kV.

second-generation pyroxene. A trace of felsic glass also occurs within the crystalline felsite, forming thin ribbons marking quartz-K-feldspar contacts in the vermicular intergrowth. The felsic glass is K- and Si-rich, containing 5–12% K_2O and 71–81% SiO_2. The glass ribbons in the crystalline felsite have compositions close to the composition of a K-feldspar-quartz eutectic melt.

The petrologic studies (James and Hammarstrom, 1977) indicate that the crystalline felsite crystallized from a melt that was probably a residual liquid in a differentiated igneous complex. The felsic glass was, for the most part, emplaced as dikelets of melt into fractures in the crystalline felsite. The felsic glass was still a melt when the clast was incorporated in the breccia.

Of particular interest in this clast is the Ca–K-rich ternary plagioclase. Plagioclase of this composition is unknown from natural terrestrial occurrences and laboratory crystallization experiments, but it has been found in felsite clasts in breccias from Boulder 1, Station 2, Apollo 17 (Ryder *et al.*, 1975). Several grains were examined in TEM, one of which was a relic within felsic glass (Fig. 2b). This grain is 80 μm across. It has an optically homogeneous core and a rim 10–30 μm thick that optically appears to consist of plagioclase riddled with minute inclusions of K-feldspar or glass. The core plagioclase has $Ab_{21}An_{57.5}Or_{21.5}$ composition. Analyses of the rim yield compositions with about 55 mol% Or; these clearly are mixtures of plagioclase and a K-rich phase that are intergrown on a scale too fine to be resolved with the electron microprobe. Both core and rim of this grain were examined in TEM. The core contains two directions of a modulated "tweed" structure (Fig. 2c), and this structure continues into the rim. The modulations are nearly parallel to (010) and (001) in the [100] zone orientation of the micrograph shown in Fig. 2c, and each has a wavelength of about 100 Å. This short wavelength produces two sets of streaks on every spot in the [100] electron diffraction pattern (Fig. 2d). In addition to type (a) reflections ($h+k$ even, l even), diffuse reflections elongated parallel to $01\bar{1}$ are present in the type (b) position ($h+k$ odd, l odd). The direction of elongation of these reflections is consistent with the direction of splitting of type (e) satellites, which are characteristic of the intermediate-plagioclase structure type (Smith, 1974). In sum, the electron diffraction pattern and image show that the ternary plagioclase is single phase, with a single lattice, modulated slightly in structure and probably also in composition. The structures observed in this grain are similar to those observed in intermediate plagioclases that have "normal" (low) contents of Or component.

The TEM studies of the rim of the plagioclase relic show that the K-rich inclusions are of K-feldspar. The inclusions become larger and more abundant as the edge of the grain is approached. Figure 2e shows an area of the rim in which fully half of the volume is K-feldspar. The plagioclase-K-feldspar interfaces are semi-coherent, as is demonstrated by the presence of a characteristic periodic set of interface dislocations. These dislocations relieve the interface strain induced by the large difference in lattice parameters between the phases ($\alpha_{plag} = 93° \ 10'$, $\alpha_{Kspar} = 90° \ 0'$). The 90° α angle observed for the K-feldspar suggests (within the $\pm 5'$ error of measurement) that it is monoclinic; there is no evidence of incipient

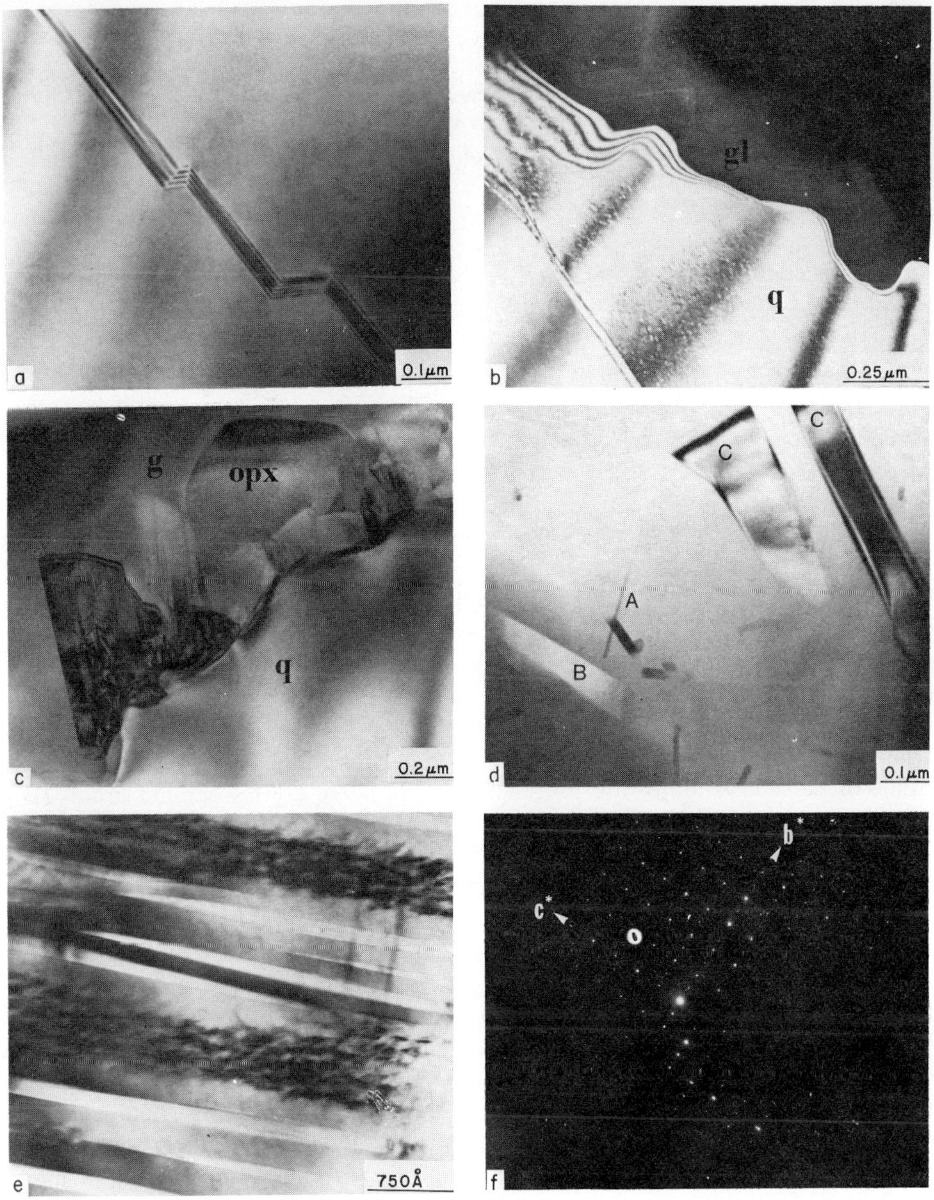

Fig. 3. Electron micrographs of quartz, clear and brown felsic glass, and prisms of second-generation pyroxene. (a) shows a Dauphiné twin boundary in quartz. (b) shows an irregular quartz (q)-glass (gl) interface in a dark field micrograph; the interface is marked by thickness fringes and changes its slope from steep to gentle across the micrograph. (c) shows the quartz-glass interface on which have grown abundant crystallites; some of these are orthopyroxene (opx). (d) shows the brown glass, which contains needle-like (A) and lath-shaped (B) crystallites; the two grains labeled (c) are in the same orientation but separated by glass, and they are part of a single skeletal crystal. (e) shows coarse and fine exsolution lamellae of augite in a prism of second-generation pigeonite. (f) shows the electron diffraction pattern for the two phases shown in (e). All micrographs taken at 200 kV.

inversion to the triclinic phase. The inclusions have parallel crystallographic axes and are probably interconnected in three dimensions; their crystallographic axes are within a few degrees of those of the host, indicating that the two phases either grew simultaneously or the inclusions formed by replacement. The ternary plagioclase relics are in general free of dislocations, but some areas do show significant numbers of dislocations (Fig. 2f).

Quartz relics in the felsic glass (Figs. 3a,b) were examined by TEM. The grains were identified as quartz by their optical reflectivity, their electron diffraction patterns and the presence of distinctive microstructures (such as Dauphiné twins). The quartz relics in glass have boundaries that are smoothly curved and highly irregular, and locally show deep embayments (Fig. 3b). Abundant small crystals of other minerals precipitated on the surfaces of many of these relics from the surrounding felsic glass (Fig. 3c). Some or all of these crystals are of orthopyroxene (identified by the 18 Å (100) spacing). The dislocation density in all the quartz relics examined is quite low. The most distinctive microstructures observed are Dauphiné twin boundaries (Fig. 3a); these twin boundaries arise from 180° parallel axis twinning, which usually occurs just prior to the $\beta \rightarrow \alpha$ transition (573°C). The quartz in the felsite clast may have cycled several times through this transition temperature, and the observed textures do not indicate during which cycle the twinning formed.

The clear colorless felsic glass is free of any crystallites or included particles, whereas the brown clouded felsic glass contains abundant crystallites (Fig. 3d). Some of these crystallites are long needles (A), some are flat tabular grains (B), and some are skeletal (C).

Microstructures in a $1 \times 10\ \mu m$ prism of second-generation pyroxene were examined. This crystal precipitated from the felsic melt after the clast was incorporated in the breccia (James and Hammarstorm, 1977). Although the early stages of cooling of the clast were rapid enough that the clear felsic glass was quenched, the later stages of cooling were slow enough that exsolution lamellae formed in the pyroxene (Fig. 3e,f). The bulk of the grain is pigeonite; it contains coarse (001) lamellae of augite, and between these lamellae finer (001) augite lamellae formed. The bimodal lamella size indicates that the overall cooling rate was more rapid than the growth rate for coarse lamellae; therefore small lamellae nucleated in the supersaturated regions which developed between the coarse lamellae. Identical microstructures are common in clinopyroxenes from mare basalts and feldspathic basalt 14310 (Lally *et al.*, 1975).

The K-feldspar-quartz vermicular intergrowth of the crystalline felsite (Fig. 4a) was examined for evidence of shock or deformation, and for evidence on the origin of the glass ribbons that mark grain boundaries. The TEM examination shows that the minerals in this intergrowth appear to be virtually unshocked and undeformed. The dislocation density in the quartz is very low, and that in the K-feldspar (Fig. 4c) is only slightly higher. No shock-produced thetomorphic glass was found. The glass ribbons were observed to occur at contacts between grains of K-feldspar and quartz (Fig. 4b); no glass was found within quartz or K-feldspar grains, or at boundaries between two grains of the same mineral. The

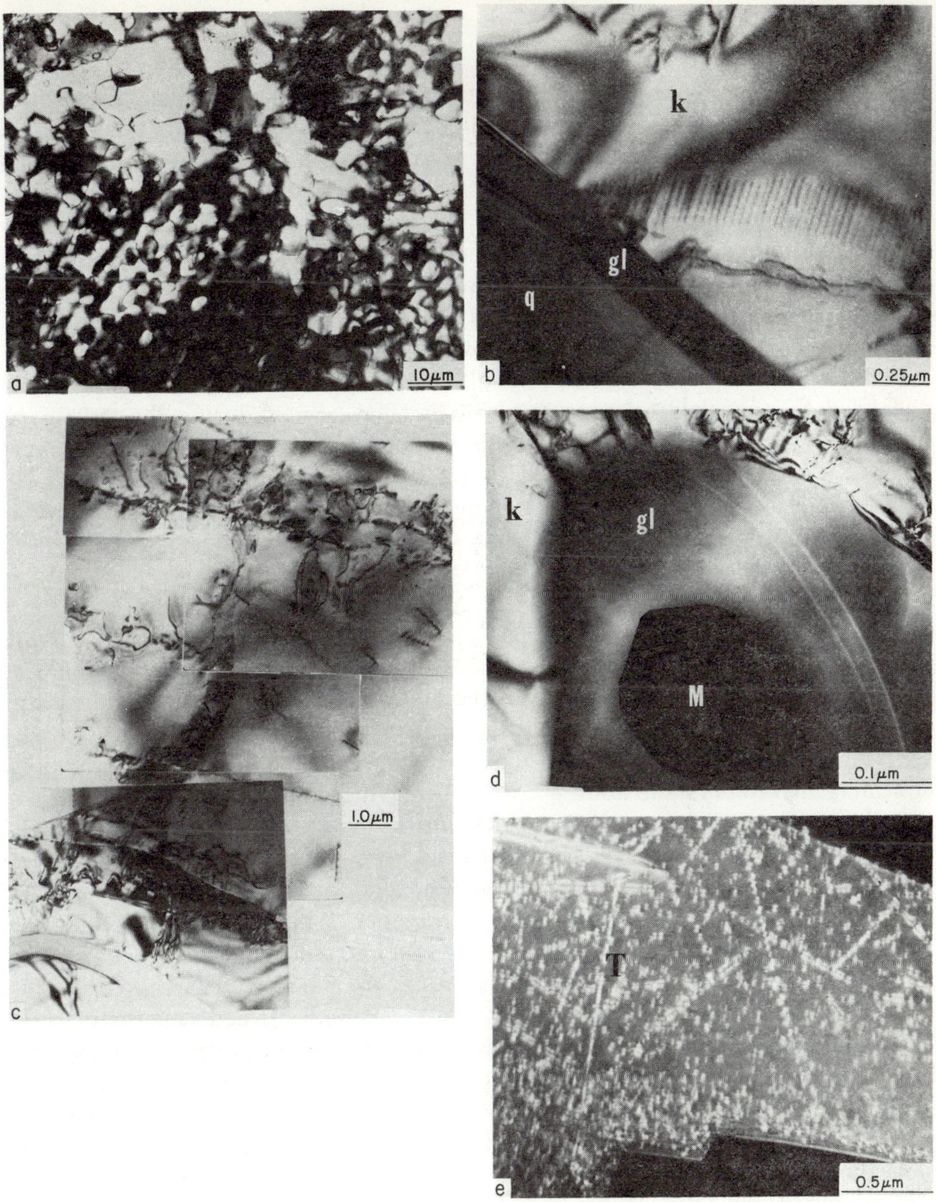

Fig. 4. Optical and electron micrographs of the K-feldspar-quartz vermicular intergrowth in the crystalline felsite. (a) is an optical micrograph that shows the intergrowth in cross-polarized light. (b) shows the intergrowth in a transmission electron micrograph; quartz (q)-K-feldspar (k) interfaces are marked by ribbons of glass (gl). (c) is a montage of transmission electron micrographs covering a large area of K-feldspar; a quartz-glass contact is visible in the lower left corner, and several bands of dislocations (healed fractures?) can be seen in the K-feldspar. (d) shows a slightly rounded metal particle (M) surrounded by glass (gl) in the K-feldspar (k)-quartz intergrowth. (e) shows a single Dauphiné twin of quartz in a dark field micrograph; cosmic ray tracks (T) are visible and have been enhanced by the spot radiation damage induced by the electron beam. All electron micrographs taken at 200 kV.

contacts between quartz and glass, and between K-feldspar and glass are smooth and relatively straight. Where the vermicular intergrowth contains other phases in addition to quartz and K-feldspar, larger amounts of glass are developed (Fig. 4d). In sum, these features indicate that the crystalline felsite experienced, at most, only slight deformation or shock. The glass ribbons along grain boundaries were not produced by shock-induced melting; they instead formed by *in situ* melting due to thermal heating. Where particles of other phases were present in the intergrowth, an increased amount of grain-boundary melt was produced.

One grain of quartz examined shows a charged particle track density of $> 10^8 T/cm^2$. The tracks, which are initially barely visible in dark-field images, are enhanced by the electron radiation damage that occurs during the course of observation. Close examination of other quartz grains in the clast, however, showed a lower track density consistent with that found by Hutcheon *et al.* (1974) of $5–3.7 \times 10^6 T/cm^2$. The local high density may be the result of irradiation by an adjacent U–Th-rich phase.

CONCLUSIONS

TEM studies demonstrate that microsubophitic texture is present in the groundmass of the aphanitic matrix of the breccia. This observation confirms the origin of the matrix as a fragment-laden melt.

A clast of ANT-suite rock studied had a history of shock-induced deformation, annealing and recrystallization prior to being incorporated in the breccia. It contains orthopyroxene that was partly transformed to clinopyroxene by shock.

Studies of the felsite clast clarify several aspects of the origin of the parent felsite, genesis of the felsic glass fraction, and thermal history of the clast during and after its incorporation in the breccia. First, the unique Ca–K-rich ternary plagioclase in the crystalline felsite precipitated as a homogeneous phase during the igneous crystallization of this rock; the TEM studies show that this plagioclase has a single lattice and a 100 Å wavelength "tweed" microstructure. Second, the event that produced the felsic melt component of the clast may have been a shock event, but if so, it did not significantly affect the crystalline felsite. Quartz in the crystalline felsite shows low dislocation densities, and K-feldspar shows low-to-moderate dislocation densities, suggesting that the maximum possible shock pressures experienced by this material were very low. Thus it is clear that the felsic melt cannot have been formed by *in situ* shock melting. Third, the observation of ribbons of eutectic glass along grain boundaries produced by *in situ* thermal melting of the crystalline felsite indicates that the temperature of the bulk clast when it was incorporated in the breccia was at or above the eutectic temperature for a quartz-K-feldspar mixture, 990°C (Schairer and Bowen, 1938). The petrologic studies (James and Hammarstrom, 1977) have shown that the injected felsic melt was still mobile when the clast was incorporated in the breccia. After the breccia-forming event, the felsic melts cooled rapidly enough to be quenched to glass; even the brown glass that contains abundant small crystallites is still largely glassy. A crystal of second-generation pigeonite that precipitated from the felsic melt as it cooled exsolved augite lamellae, however; width and spacing of the lamellae are comparable to those observed in pyroxenes in mare basalts and in feldspathic basalt 14310. These data are consistent with a single-stage crystallization and cooling of the felsic melt after the clast was incorporated in the breccia, and suggest that there was no appreciable later reheating. (On the scale of observations made in this study, however, a reheating to temperatures $\leqslant 500°C$ would produce no detectable effects, and it is possible that reheating to temperatures as high as 600°C would also not be detectable.) The thermal history outlined above indicates that the clast was probably outgassed of its radiogenic Ar at the time of the breccia-forming event, and its Sr isotopes probably reequilibrated. Thus the Rb–Sr internal isochron age for the clast of 3.90 ± 0.05 b.y. (Compston *et al.*, 1977) and the ^{40}Ar–^{39}Ar plateau age of 3.92 ± 0.04 b.y. (Jessberger *et al.*, 1977) can be no older than the breccia-forming event. As the results of the TEM studies and ^{40}Ar–^{39}Ar stepwise heating experiments suggest that the clast (and the bulk rock) was not reheated, this ~3.90 b.y. age is most likely the actual date of the breccia-forming event. Several authors (James *et al.*, 1975; Morgan *et al.*, 1976; Jessberger *et al.*, 1976; Wood, 1975) have proposed that this event was the South Serenitatis basin-forming impact. If the date of this event is indeed ~3.90 b.y., the age data on all the other Apollo 17 breccias will have to be reexamined in light of this fact. A ~3.90 b.y. age for the Serenitatis event would imply that a large number of the major lunar basins formed within a very short time span, on the order of 100 m.y.

Acknowledgments—We would like to thank Jane Hammarstrom for providing electron microprobe analyses of minerals studied in the felsite clast. This research was done on behalf of NASA, in part under contract T3040C.

References

Ashworth J. R. and Barber D. J. (1975) Shock effects in meteoritic pyroxene. *Proc. Electron Microscopy and Analysis Group*, p. 517–520. Inst. of Physics, Bristol.

Blanchard D. P., Jacobs J. W., and Brannon J. C. (1977) Chemistry of ANT-suite and felsite clasts from consortium breccia 73215, and of gabbroic anorthosite 79215. *Proc. Lunar Sci. Conf. 8th.* This volume.

Christie J. M., Griggs D. T., Heuer A. H., Nord G. L., Jr., Radcliffe S. V., Lally J. S., and Fisher R. M. (1973) Electron petrography of Apollo 14 and 15 breccias and shock-produced analogs. *Proc. Lunar Sci. Conf. 4th*, p. 365–382.

Compston W., Foster J. J., and Gray C. M. (1977) Rb–Sr systematics in clasts and aphanites from consortium breccia 73215. *Proc. Lunar Sci. Conf. 8th.* This volume.

Hutcheon I. D., MacDougall D., and Stevenson J. (1974) Apollo 17 particle track studies: Surface residence times and fission track ages for orange glass and large boulders. *Proc. Lunar Sci. Conf. 5th*, p. 2597–2608.

James O. B. (1976) Petrology of aphanitic lithologies in consortium breccia 73215. *Proc. Lunar Sci. Conf. 7th*, p. 2145–2178.

James O. B. and Blanchard D. P. (1976) Consortium studies of light-gray breccia 73215: Introduction, subsample distribution data, and summary of results. *Proc. Lunar Sci. Conf. 7th*, p. 2131–2143.

James O. B., Brecher A., Blanchard D. P., Jacobs J. W., Brannon J. C., Korotev R. L., Haskin L. A., Higuchi H., Morgan J. W., Anders E., Silver L. T., Marti K., Braddy D., Hutcheon I. D., Kirsten

T., Kerridge J. F., Kaplan I. R., Pillinger C. T., and Gardiner L. R. (1975) Consortium studies of matrix of light-gray breccia 73215. *Proc. Lunar Sci. Conf. 6th*, p. 547–577.

James O. B. and Hammarstrom J. G. (1977) Petrology of four clasts from consortium breccia 73215. *Proc. Lunar Sci. Conf. 8th.* This volume.

Jessberger E. K., Kirsten T. and Staudacher T. (1976) Argon-argon ages of consortium breccia 73215. *Proc. Lunar Sci. Conf. 7th*, p. 2201–2215.

Jessberger E. K., Kirsten T., and Staudacher T. (1977) One rock and many ages—further K–Ar data on consortium breccia 73215. *Proc. Lunar Sci. Conf. 8th.* This volume.

Lally J. S., Christie J. M., Nord G. L., Jr., and Heuer A. H. (1976) Deformation, recovery and recrystallization of lunar dunite 72417. *Proc. Lunar Sci. Conf. 7th*, p. 1845–1863.

Lally J. S., Heuer A. H., Nord G. L., Jr., and Christie J. M. (1975) Subsolidus reactions in lunar pyroxenes: An electron petrographic study. *Contrib. Mineral. Petrol.* **51**, 263–281.

MacDougall D., Rajan R. S., Hutcheon I. D., and Price P. B. (1973) Irradiation history and accretionary processes in lunar and meteoritic breccias. *Proc. Lunar Sci. Conf. 4th*, p. 2319–2336.

Morgan J. W., Gros J., Takahashi H., and Hertogen J. (1976) Lunar breccia 73215: Siderophile and volatile trace elements. *Proc. Lunar Sci. Conf. 7th*, p. 2189–2199.

Nord G. L., Jr., Christie J. M., Heuer A. H., and Lally J. S. (1975) North Ray Crater breccias: An electron petrographic study. *Proc. Lunar Sci. Conf. 6th*, p. 779–797.

Nord G. L., Jr., Lally J. S., Heuer A. H., Christie J. M., Radcliff S. V., Griggs D. T., and Fisher R. M. (1973) Petrologic study of igneous and metaigneous rocks from Apollo 15 and 16 using high voltage transmission electron microscopy. *Proc. Lunar Sci. Conf. 4th*, p. 953–970.

Ryder G., Stoeser D. B., Marvin U. B., and Bower J. F. (1975) Lunar granites with unique ternary feldspars. *Proc. Lunar Sci. Conf. 6th*, p. 435–449.

Schairer J. F. and Bowen N. L. (1938) The system leucite-diopside-silica. *Amer. J. Sci.* **35A**, 289–309.

Smith J. V. (1974) Feldspar Minerals, vol. 1, Crystal Structure and Physical Properties. Springer Verlag, New York. 627 pp.

Wood J. A. (1975) The nature and origin of Boulder 1, Station 2, Apollo 17. *The Moon* **14**, 505–517.

Proc. Lunar Sci. Conf. 8th (1977), p. 2507–2524.
Printed in the United States of America

Chemistry of ANT-suite and felsite clasts from consortium breccia 73215 and of gabbroic anorthosite 79215

DOUGLAS P. BLANCHARD

NASA/Johnson Space Center, Houston, Texas 77058

JEFFREY W. JACOBS* and JOYCE C. BRANNON*

Lockheed Electronics Corporation, 16811 El Camino Real, Houston, Texas 77058

Abstract—The bulk of 73215 consists of gray-to-black aphanite rock that formed as a mechanically mixed aggregate of crystalline clasts and silicate melt. In addition to this aphanitic matrix and cogenetic aphanitic clasts there are several distinct lithologies present as granulated material in bands and veins running through the breccia and as competent, separable clasts whose isotopic and compositional information has not been appreciably altered by the brecciation event that formed 73215.

We have analyzed these clasts and granulated materials from 73215 and samples from gabbroic anorthosite 79215 by instrumental neutron activation analysis for FeO, Na_2O, Cr_2O_3, Sc, Co, Ni, Hf, Ta, Th, La, Ce, Sm, Eu, Tb, Yb, and Lu and by atomic absorption spectrophotometry for SiO_2, TiO_2, Al_2O_3, MgO, CaO, and K_2O. Granulated feldspathic and troctolitic materials, occurring as veins and bands in the breccia, contain abundant aphanitic clasts, which appear to dominate REE abundances. Anorthositic gabbro clasts show no evidence of similar contamination by aphanitic materials. Petrologically, chronologically, and compositionally they resemble a group of rocks, which includes 79215, that predate the lunar cataclysm and thus preserve compositional and isotopic information about the early highland crust. Pink-spinel-bearing troctolitic basalt clasts bear no evidence of being disrupted by the consolidation event of 73215; however, we presently have too little supporting data to judge whether they too predate the lunar cataclysm.

The 13.6 mg clast of felsite from 73215 is one of very few separated clasts of lunar granite large enough to allow petrographic, compositional, and isotopic studies. It has a distinctive composition; it contains 7.0% K_2O, very low FeO, and has unusual REE contents with absolute abundances of about 130× chondrites and a bowl shaped relative pattern with HREE equally as enriched as LREE. In many respects it closely resembles felsite from 12013. The high K/La ratio (1690) for the 73215 felsite (versus 90 for normal lunar material) is compelling evidence that the felsite is the product of liquid immiscibility, but lack of quantitative data on elemental distributions between immiscible liquids and between each liquid and crystalline phases in equilibrium with it limit further characterization of possible parent compositions.

INTRODUCTION

BRECCIA 73215 IS a light-gray breccia collected from the light mantle regolith at Apollo 17, Station 3. This breccia was one of two chosen by consortium leader, O. B. James, for consortium research. Superficial examination suggested that the rock formed as an aggregate of melt and clasts produced during a major impact

*Present address: Department of Earth and Planetary Sciences, Washington University, St. Louis, Missouri 63130.

event and that it had experienced no significant post aggregation heating (James and Blanchard, 1976). The goals of the consortium were to date the impact event, to investigate the details of the consolidation of the breccia, and to look back through the breccia-forming event to infer characteristics of the preimpact lunar crust. Work on 73215 has continued for three years and various aspects of the work have already been reported. James and Blanchard (1976) have summarized the results of research as of March 1976.

Our previous work (James *et al.*, 1975; Blanchard *et al.*, 1976) centered on studies of the aphanitic matrix of the breccia and aphanitic clasts that are lithologically similar to the matrix. Our results indicated that all the 73215 aphanites are cogenetic: The clast-like bodies are not clasts of preexisting rock, but they instead formed at the same time as the matrix; the differences observed in occurrence, chemistry, and petrologic characteristics are probably the result of differences in conditions of aggregation.

Our recent work has centered on characteristics of the clast suite. In this paper we report data on the major types of lithic clasts in the breccia: Competent fine-grained anorthositic gabbros; granulated coarse-grained feldspathic and troctolitic rocks; competent fine-grained troctolitic basalts; and felsite (lunar "granite"). The results of other studies on some of the same clasts from 73215 that we have analyzed are reported by: James and Hammarstrom (1977)—petrology; Nord and James (1977)—transmission electron microscopy; Jessberger *et al.* (1976, 1977) and Müller *et al.* (1977)—^{39}Ar–^{40}Ar dating; Compston *et al.* (1977)—Rb–Sr isotopic systematics; and Morgan *et al.* (1976)—trace siderophile elements. Locations of the clasts within the 73215 breccia are given by James and Blanchard (1976).

At least some of the ANT-suite clasts from 73215, the anorthositic gabbros and the troctolitic basalts, appear to have survived breccia formation with only minor changes in chemical, isotopic, and petrologic characteristics. The granulated feldspathic and troctolitic materials, however, were apparently subjected to more severe reworking. If we can account for the effects of the breccia-forming event, these materials may offer us an excellent look at ancient precataclysm lunar crustal material. The felsite is also an extremely important sample, because it is one of the very few samples of lunar "granite" that have been isolated for detailed petrological, compositional, isotopic, and age studies.

We are also reporting results of analyses of six samples of breccia 79215. This rock is a holocrystalline, nearly monomict breccia which is thought to have been brecciated and annealed prior to the lunar cataclysm (Bickel *et al.*, 1976; Warner *et al.*, 1977). The rock has been well characterized petrographically (Bickel *et al.*, 1976). Petrologically and chemically it resembles the fine-grained anorthositic gabbro clasts from 73215. Discussion of this sample along with the 73215 samples allows us to put all the data into a somewhat larger perspective.

The rock name "anorthositic gabbro" for the fine-grained ANT-suite clasts is the designation for such rocks preferred by James and Hammarstrom (1977); similar samples from 79215 have been referred to as gabbroic anorthosites (Bickel *et al.*, 1976). The normative compositions of most of these samples lie near the dividing line between these groups as defined by Prinz *et al.* (1973). The distinctions are not critical to our present discussion, so we use both terms interchangeably without strictly adhering to definitions of Prinz *et al.* (1973).

METHODS

Sample aliquots of approximately 20 mg were analyzed for FeO, Na_2O, Cr_2O_3, Sc, Co, Ni, Hf, Ta, Th, and seven rare earth elements (La, Ce, Sm, Eu, Tb, Yb, Lu) by nondestructive instrumental neutron activation analysis (INAA) using our usual methods (Jacobs *et al.*, 1977). 10–20 mg aliquots (often the same used for INAA) were analyzed for major elements (SiO_2, TiO_2, Al_2O_3, FeO, MgO, CaO, K_2O) by atomic absorption spectrophotometry (AAS) (Blanchard, 1973).

Table 1 lists results of our analyses of the various samples from 73215. Averages are included for our previously published analyses of samples of breccia matrix, black aphanite clasts, and gray aphanite spheroids (James *et al.*, 1975; Blanchard *et al.*, 1976). Table 1 also includes results of our analyses by AAS and radiochemical neutron activation (against primary chemical standards) of USGS standard rocks BCR-1 and DTS-1, which we have used as irradiation monitors for INAA. The typical uncertainties listed in Table 1 are estimates of accuracy based on counting statistics (INAA), variance observed in series of replicate analyses (INAA and AAS), and the absolute uncertainties in our knowledge of the elemental concentrations in the irradiation monitors (INAA) (Jacobs *et al.*, 1977; Blanchard, 1973).

Table 2 lists results of our analyses of 79215. Typical uncertainties are the same as those in Table 1. Table 3 lists normative mineral assemblages calculated from major-element data in Tables 1 and 2.

RESULTS

Anorthositic gabbro clasts from 73215: 29,9; 46,25; and 46,33

Textures of these samples range from igneous poikilitic (29,9) to granulitic (46,25); textural and mineralogic evidence suggests that the samples represent reheated, partially melted, and/or recrystallized breccias (James and Hammarstrom, 1977). ^{40}Ar–^{39}Ar stepwise heating experiments on all three clasts showed double plateaus that yielded intermediate-temperature ages of 4.06–4.07 ± .05 b.y. and high-temperature ages of 4.24–4.26 ± .05 b.y. (Jessberger *et al.*, 1976). Jessberger *et al.* interpreted the older ages as dating the melt crystallization or metamorphism, and the intermediate ages as the products of partial outgassing of some components of the clasts when they were incorporated in the 73215 breccia. ^{40}Ar–^{39}Ar laser probe data on plagioclases in 29,9 yield ages similar to those obtained by stepwise heating, 4.07–4.37 b.y., but the chronologic significance of the results cannot be unambiguously determined (Müller *et al.*, 1977). The Rb–Sr internal isochron age for 29,9 is 4.24 ± .31 b.y. and data for 46,25 would fit the same line as 29,9; model ages are 4.30–4.35 ± .15 b.y. Compston *et al.* (1977) interpret the model ages as upper limits to the time of melt crystallization or metamorphism of these rocks. Siderophile-element ratios in 46,25 (the only sample analyzed for these elements) fall in group 3 of the groups defined by Higuchi and Morgan (1976), and are distinctly separated from group 2, the group in which most matrix material from Apollo 17 breccias (including 73215) plots (Morgan *et al.*, 1976). Morgan *et al.* interpret the group 2 siderophiles as the traces of the Serenitatis projectile and the group 3 sidero-

Table 1. Major- and trace-element compositions of samples of consortium breccia 73215.

	Average[(2)] matrix	Average[(3)] black aphanite clasts	Average[(4)] gray aphanite spheroid	Anorthositic gabbro clasts: Coarse grained 29,9	29,9	Fine grained 46,25	46,33	Granulated feldspathic materials 38,18	38,26	38,45	46,5	Granulated troctolite 38,5	Troctolitic basalts 38,23	38,33	170,5	Felsite 43,3	BCR-1
SiO_2	46.4	46.2	46.3	45.9	46.6	44.1		n.a.	45.6	46.8	44.9	42.2			42.8	n.a.	54.2
TiO_2	0.7	1.1	0.8	0.69	0.58	0.23		n.a.	0.48	0.37	1.13	0.29			0.47	n.a.	2.4
Al_2O_3	20.6	19.6	20.9	25.6	25.4	25.6		n.a.	25.4	30.6	26.3	6.69			12.6	n.a.	13.5
FeO[(1)]	7.3	8.26	6.98	4.44	5.14	5.82	5.1	5.44	5.14	1.79	5.23	10.1	8.36	7.22	8.6	2.98	12.2
MnO	0.104	0.113	0.097	0.062	0.077	0.067		n.a.	0.052	0.026	0.073	0.089			0.087	n.a.	0.87
MgO	11.6	12.1	11.6	8.36	8.42	9.41		n.a.	8.89	2.37	6.82	37.7			26.0	n.a.	3.47
CaO	11.9	11.4	11.8	13.9	14.1	13.8		n.a.	13.8	16.0	14.2	3.47			8.17	n.a.	7.03
Na_2O	0.50	0.52	0.52	0.430	0.403	0.356	0.375	0.495	0.483	0.951	0.494	0.127	0.320	0.47	0.311	0.194	3.29
K_2O	0.20	0.21	0.18	0.121	0.097	0.088		n.a.	0.109	0.197	0.126	0.038			0.060	7.0	1.71
Cr_2O_3	0.25	0.26	0.22	0.130	0.146	0.124	0.123	0.130	0.117	0.042	0.127	0.197	0.186	0.19	0.152	0.015	(0.669)[(5)]
Total				99.9	100.9	99.6			100.1	99.1	99.4	100.9			99.3		
Sc	14.7	16.1	12.4	7.16	9.04	7.12	8.2	7.83	7.55	2.66	9.71	3.64	4.09	10.1	3.7	4.8	31.6
Co	25.3	29.3	27.8	13.7	13.2	30.7	32.7	22.7	18.0	4.13	11.5	54.9	47.4	34.9	50.8	2.10	36.0
Ni	165	175	210	80	80	420	460	150	130		70	360	185	280	360	n.d.	(2390)[(5)]
Hf	9.5	10.0	8.4	3.1	2.2	1.5	1.4	4.7	4.0	0.94	3.8	1.67	0.82	6.4	0.9	25.6	5.2
Ta	1.4	1.6	1.4	0.3	0.2	0.3	0.16	0.6	0.44		0.70	0.17	—	0.8	—	5.4	0.9
Th	4.1	4.3	4.1	1.6	0.96	0.53	0.58	1.8	1.48	0.39	1.9	0.86	0.27	2.5	0.2	39.9	6.0
La	27.0	27.7	27.1	8.70	6.24	2.61	3.52	12.3	9.8	4.24	11.0	4.84	3.17	16.6	2.6	42.9	25.2
Ce	70	74	73	24.4	16.8	6.78	9.1	32.5	26.8	10.9	29.3	14.0	17.8	45.5	7.0	125	54.2
Sm	12.5	12.6	12.2	3.88	3.08	1.04	1.53	5.48	4.32	1.73	4.93	2.17	1.28	7.34	1.34	19.0	6.80
Eu	1.43	1.41	1.49	1.00	0.91	0.77	0.77	1.23	1.15	0.64	1.22	0.301	0.656	1.08	0.75	3.11	1.97
Tb	2.6	2.8	3.1	0.84	0.82	0.23	0.36	1.18	1.05	0.320	1.14	0.51	0.320	1.70	0.30	5.6	1.15
Yb	9.1	9.4	8.9	2.9	3.0	1.4	1.71	4.2	3.21	1.07	3.9	1.54	1.00	5.70	1.06	27.2	3.48
Lu	1.30	1.28	1.23	0.413	0.405	0.225	0.273	0.60	0.46	0.151	0.55	0.222	0.125	0.79	0.16	5.3	0.526

(1) Total Fe expressed as FeO.

(2) Average of 11 samples. For detailed discussion see James *et al.* (1975) and Blanchard *et al.* (1976).

(3) Average of 4 samples (Blanchard *et al.*, 1976).

(4) Average of 3 samples (Blanchard *et al.*, 1976).

(5) Values for USGS standard rock DTS-1.

Table 2. Major- and trace-element compositions of samples of granulitic gabbroic anorthosite 79215.

	21,1	22,1	24,1	27,1	38,1	39,1
SiO_2	39.4	43.2	44.8	43.5	44.4	44.8
TiO_2	0.3	0.3	0.3	0.4	0.3	0.5
Al_2O_3	10.4	25.8	27.6	26.7	27.9	27.4
$FeO^{(1)}$	18.2	4.96	4.15	4.91	3.40	4.86
MnO	0.170	0.058	0.054	0.063	0.045	0.064
MgO	24.2	8.51	5.84	7.33	6.18	7.40
CaO	6.17	15.6	16.3	15.6	16.0	14.4
Na_2O	0.222	0.616	0.549	0.557	0.616	0.580
K_2O	0.034	0.110	0.107	0.113	0.128	0.119
Cr_2O_3	0.238	0.069	0.098	0.115	0.541	0.108
Total	99.24	99.24	99.9	99.2	99.01	100.2
Sc	7.07	4.60	7.14	7.69	5.53	8.14
Co	71.2	17.6	18.8	16.6	7.3	18.9
Ni	680	152	215	110	—	126
Hf	0.6	0.56	1.34	1.1	0.45	1.2
Ta	—	—	—	0.14	—	—
Th	0.46	0.53	0.21	0.37	0.50	0.32
La	1.01	2.33	2.5	3.3	2.45	2.65
Ce	22.4	5.5	6.5	9.1	6.6	6.8
Sm	0.481	0.768	1.03	1.53	0.96	1.19
Eu	0.35	0.77	0.77	0.77	0.94	0.84
Tb	0.12	0.20	0.23	0.34	0.22	0.28
Yb	0.78	0.72	1.07	1.5	0.79	1.37
Lu	0.132	0.106	0.16	0.23	0.108	0.24

(1)Total Fe expressed as FeO.

philes as the traces of a projectile from a different, possibly earlier, large impact.

The two samples from clast 29,9 that we analyzed have somewhat different REE, FeO, and Sc abundances; the differences are qualitatively consistent with the small differences in diopside contents shown by the normative calculations (Table 3). Our samples from clasts 46,25 and 46,33 have nearly identical compositions and differ somewhat from those of the 29,9 samples (Tables 1,2, and 3). The REE patterns for 29,9 (Fig. 1) have negative slopes overall whereas the REE patterns for 46,25 and 46,33 have distinct breaks in slope, from a negative slope for the LREE (light rare earth elements) to a positive slope for the HREE (heavy rare earth elements). As we will discuss later, this break in slope distinguishes these clasts from the granulated feldspathic and troctolitic clast materials in 73215.

Gabbroic anorthosite 79215

This granulitic gabbroic anorthosite has been interpreted as a metamorphosed rock which originated as a plagioclase-olivine cumulate. Mineralogically it is

Table 3. Normative mineralogy of clasts from 73215 and of 79215.

	73215											79215					
				Anorthositic gabbros clasts													
	Average matrix	Average black aphanite clasts	Average gray aphanite spheroids	Coarse grained		Fine grained	Granulated feldspathic materials			Granulated troctolitic	Troctolitic basalt	Granulitic gabbroic anorthosite 79215					
				29,9	29,9	46,25	38,26	38,45	46,5	38,5	170,5	21,1	22,1	24,1	27,1	38,1	39,1
Orthoclase	1.2	1.2	1.1	0.7	0.6	0.5	0.6	1.2	0.7	0.2	0.4	0.2	0.7	0.6	0.7	0.8	0.7
Albite	4.2	4.4	4.4	3.6	3.4	3.0	4.1	8.1	4.2	1.1	2.6	0.9	2.0	4.7	4.3	5.2	4.9
Anorthite	53.4	50.5	54.2	67.6	67.2	68.0	66.8	78.7	69.2	17.2	32.8	27.3	67.3	72.5	70.0	73.0	71.4
Diopside	4.6	4.9	3.5	1.1	2.2	0.4	1.3	0.6	1.0		6.1	2.7	8.1	6.7	6.0	5.1	
En	(1.6)	(1.6)	(1.2)	(0.4)	(0.8)	(0.1)	(0.5)	(0.2)	(0.3)		(2.4)	(0.9)	(2.8)	(2.2)	(2.0)	(1.8)	
Fs	(0.7)	(0.7)	(0.5)	(0.1)	(0.3)	(0.1)	(0.2)	(0.1)	(0.2)		(0.6)	(0.5)	(1.1)	(1.1)	(0.9)	(0.6)	
Wo	(2.4)	(2.5)	(1.8)	(0.6)	(1.1)	(0.2)	(0.7)	(0.3)	(0.5)		(3.2)	(1.4)	(4.2)	(3.5)	(3.1)	(2.7)	
Hypersthene	23.1	23.6	23.4	19.5	20.7	10.1	15.7	8.3	15.9	5.8	4.5			4.3		3.2	9.9
En	(16.20)	(16.4)	(16.7)	(14.6)	(14.7)	(7.0)	(11.3)	(5.7)	(10.9)	(4.9)	(3.7)			(2.9)		(2.4)	(6.9)
Fs	(6.9)	(7.2)	(6.6)	(4.9)	(6.0)	(3.1)	(4.4)	(2.6)	(5.0)	(0.9)	(0.9)			(1.4)		(0.8)	(3.0)
Olivine	11.5	12.7	11.1	5.6	5.6	17.0	10.4		6.0	75.6	51.7	66.8	18.7	10.2	17.2	10.9	12.0
Fo	(7.8)	(8.5)	(7.7)	(4.1)	(3.9)	(11.4)	(7.3)		(4.0)	(62.4)	(41.2)	(41.6)	(12.9)	(6.6)	(11.4)	(7.8)	(8.1)
Fa	(3.6)	(4.2)	(3.4)	(1.5)	(1.7)	(5.5)	(3.1)		(2.0)	(13.2)	(10.5)	(25.2)	(5.8)	(3.6)	(5.8)	(3.0)	(3.9)
Chromite	0.4	0.4	0.3	0.2	0.2	0.2	0.2	0.1	0.2	0.3	0.2	0.4	0.1	0.1	0.2	0.8	0.2
Ilmenite	1.3	2.1	1.5	1.3	1.1	0.4	0.9	0.7	2.2	0.6	0.9	0.6	0.6	0.6	0.8	0.6	1.0
Quartz								1.63									
Corundum										0.1							0.1
Nepheline												0.5	0.2		0.2		

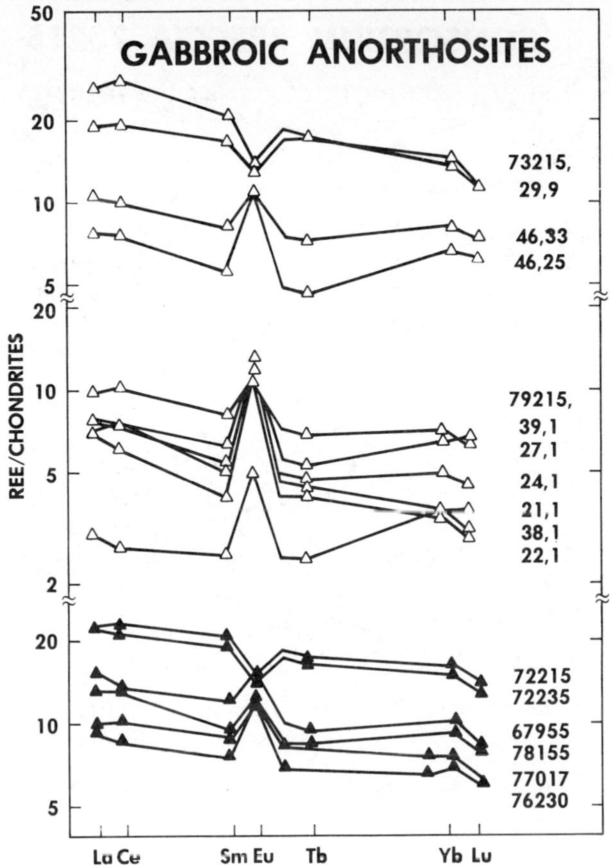

Fig. 1. Chondrite-normalized rare-earth-element abundances for anorthositic gabbroic clasts from 73215 (top); samples of 79215, a granulitic gabbroic anorthosite (middle); and other Apollo 16 and 17 gabbroic anorthosites (Blanchard *et al.*, 1975; Boynton *et al.*, 1976; Hubbard *et al.*, 1974; and Wänke *et al.*, 1976). Note in nearly all cases the break in slope between the light rare earths and the heavy rare earths.

distinguished from other ANT-suite rocks by the dominance of high-Ca pyroxene over low-Ca pyroxene and relatively high apatite content (Bickel *et al.*, 1976). Figure 1 shows the REE patterns for our samples of this rock. In all samples there is a break in the overall slope and in four of the six samples the slope of the HREE is positive. Sample 21,1 is clearly enriched in FeO, MgO, Co, and Ni, and depleted in SiO_2, Al_2O_3, CaO, alkalis and Eu relative to the other samples of 79215; these differences suggest a higher olivine to plagioclase ratio and the results of the normative calculations (Table 3) support this interpretation. The other five samples are nearly identical to each other and closely resemble the anorthositic gabbro from 73215, except that they have somewhat lower FeO, MgO, and higher Al_2O_3, CaO, and Na_2O; the differences are reflected in somewhat higher proportions of normative plagioclase in the 79215 samples.

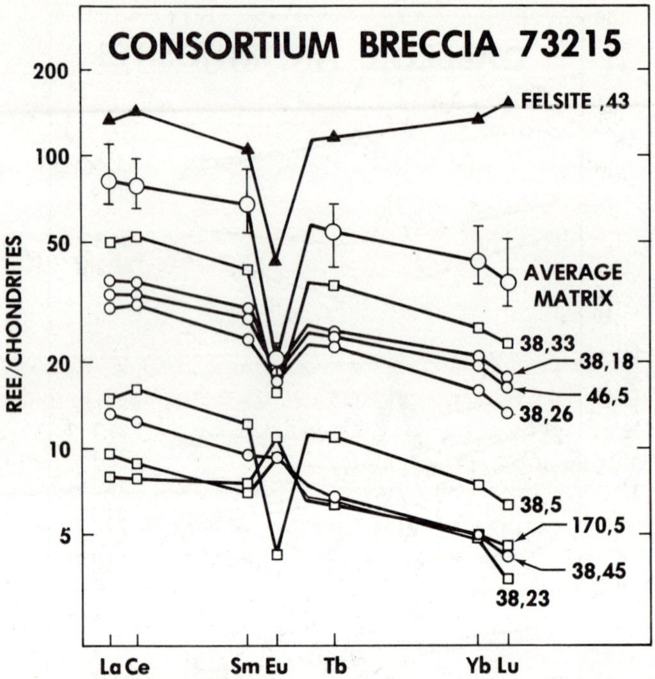

Fig. 2. Chondrite-normalized rare-earth-element abundances for feldspathic and troctolitic clast samples from 73215, and for felsite clast 73215,43. Also plotted are the average matrix composition and the range observed for separate samples of the matrix (James *et al.*, 1975). Aphanite clasts in this breccia all plot within the range of the matrix samples (Blanchard *et al.*, 1976). Note the unusual enrichment in heavy rare earths for 73215,43.

Granulated feldspathic materials from 73215: 38,18; 46,5; 38,26; and 38,45

These materials occur as incoherent bands and coatings surrounding the more competent clasts in 73215. We analyzed several samples attempting to characterize the composition of this rather abundant clastic constituent of the breccia. 73215,38,18 was a coating surrounding a gray aphanite spheriod. In thin section this material was found to be a brecciated mixture, about half of which is aphanite fragments and half is granulated clastic material; the granulated material consists mostly of fragments of single crystals of plagioclase with less abundant fragments of olivine and troctolitic basalt (James, pers. comm.). Sample 46,5 was carefully handpicked from a broad band of light gray to white friable material near the E end of the rock. In thin section this band is predominantly granulated clastic material with less than 25% small black aphanite clasts. The largest clasts are of coarse-grained anorthosite which bears evidence of deformation and recrystallization prior to incorporation into 73215. Mineral fragments are mostly plagioclase with minor amounts of olivine, pyroxene, and ilmenite (James, pers. comm.). Sample 38,26 was taken from stringers of "impure looking" feldspathic material. In thin section it resembles 38,18; 1/3 to 1/2 of the volume consists of aphanite fragments and the remainder is granulated clastic material consisting of mineral fragments and sparse clasts of troctolite basalt; there are also some fragments of large, undeformed plagioclase grains and grains consisting of blebby intergrowths of low-Ca and high-Ca pyroxene, suggesting an additional source rock of the 77215–76230 type (James, pers. comm.). Sample 38,45 was handpicked from stringers of "pure-looking" feldspathic material. In this section the granulated material consists almost entirely of fragments of feldspar and devitrified anorthositic glass; no lithic fragments are present in the small thin section of this material that we have (James, pers. comm.).

The samples of granulated feldspathic material that we have analyzed are clearly not monomict breccias nor are they crushed single lithologies; they are complex, polymict, brecciated materials. Their REE patterns (Fig. 2) are probably largely determined by the aphanitic material they contain; the "cleanest" of these samples, 38,45, shows the lowest REE abundances. We have clearly not yet analyzed 100% pure feldspathic component. While we are continuing our attempts to isolate this component, it is significant that all of the feldspathic materials we have sampled consist primarily of fragments of coarse-grained, highly feldspathic rocks, and also contain fragments of aphanite, troctolite basalt, and perhaps other source materials.

Rb–Sr data also suggest that the granulated feldspathic material is a mixture. Rb–Sr data (Compston *et al.*, 1977) show that separated feldspar grains from a sample corresponding to 46,5 have very low Rb/Sr and $^{87}Sr/^{86}Sr$ ratios; these grains lie on a 4.5 b.y. reference isochron. Rb–Sr data on the bulk sample of this lithology can be explained as a mixture of such feldspar with aphanite.

Granulated troctolite 73215,38,5

This material has a distribution and occurrence much like that of the granulated feldspathic materials; it forms thin veins and bands coating more competent clasts. Sample 38,5 was handpicked from material coating the surface of a gray aphanite spheroid, and was carefully cleaned of admixed dark material. The granulated troctolite consists of angular fragments of olivine and plagioclase, with minor fragments of spinel-pyroxene symplectite. The parent was relatively coarse-grained. Compositions of the feldspar and olivine, grain size, and the presence of symplectites suggest a possible relationship to rocks such as 76535 and dunite 72415–7 (James, pers. comm.). Granulated troctolite 73215,38,39, from the same material as 38,5 gave a single ^{39}Ar–^{40}Ar plateau age of $4.00 \pm .10$ b.y. which is interpreted either as representing substantial outgassing at that time or as an experimental artifact with no chronological significance (Jessberger *et al.*, 1976).

Sample 38,5 has low REE abundances (Fig. 2), Eu is severely depleted even relative to the low abundances of the other REE. The relative REE pattern is aphanite-like; perhaps a very small amount of REE-rich aphanite was incorporated into the granulated troctolite separated for analysis (35,5 is 92.8% normative olivine and plagioclase, Table 3).

Pink spinel-bearing troctolitic basalts from 73215: 170; 38,23; and 38;33

These are all fine-grained igneous textured rocks, and all show evidence of having been finely granulated and annealed after they crystallized (James, pers. comm.). The siderophilic elements in ,170 are indigenous with no detectable addition of meteoritic component (Morgan *et al.*, 1976). Sample 38,33 is a chip of a 1:1 mixture of troctolitic basalt and matrix aphanite; this explains the relatively high REE content of that sample.

Felsite 73215,43

This 13.6 mg clast consists of fragments of crystalline felsite intruded and enclosed by veinlets and patches of felsic glass. Most of the crystalline felsite is an intergrowth of quartz, K-feldspar, and minor plagioclase; the glass contains relict and second-generation grains of mafic minerals and feldspars and was clearly molten when the clast was incorporated in the breccia (James and Hammarstrom, 1977). Results of ^{39}Ar–^{40}Ar (Jessberger *et al.*, 1977) and Rb–Sr (Compston *et al.*, 1977) dating studies yield an age of about 3.90–3.92 b.y. for this material; this age is interpreted as the age of the breccia-forming event.

Clast 73215,43 is extremely rich in potassium (7.0%) and poor in iron (2.98%) and sodium (0.194). The REE abundances (Fig. 2) are high, but not extraordinarily so (La equals 130× chondrites). However, the relative abundances are unusual for lunar materials in that the HREE are not severely depleted but instead are enriched relative to Sm to the same extent that LREE are.

The lunar sample most nearly like the 13.6 mg chip of 73215 felsite that we have analyzed is the felsite component in breccia 12013. 12013 was the subject of a consortium study which resulted in an issue of *Earth and Planetary Science Letters* especially devoted to 12013 results (Sept., 1970). The reader is referred to that journal issue for the details of 12013. Recently an intensive effort (Quick and Albee, 1976; Ma *et al.*, 1977; Quick *et al.*, 1977) has been reinstituted to study the felsic portions of 12013. Figure 3 compares our analyses of felsite 73215,43 with pigeonite basalt, a candidate for a parental magma for KREEP basalts from boulder sample 72275 (Blanchard *et al.*, 1975) and with two analyses of gray breccia from 12013 (Ma *et al.*, 1977; Quick *et al.*, 1977).

Quite clearly the 73215 and 12013 felsites are very similar materials. Comparing the 73215 felsite clast with the felsite-rich gray breccia portions of 12013 (Quick *et al.*, 1977), we note that 73215 has less FeO (2.98% vs 9.6–14.0%), more K_2O (7.0% vs 2.6–3.74%), less Na_2O (0.194% vs 1.12–1.28%), less Sc (4.8 ppm vs 17–25 ppm), less Co (2.1 ppm vs 8.6–21 ppm), and more Hf (256 ppm vs 25–27 ppm). However, when the data for 73215 felsite are compared to the defocused beam microprobe analysis of only the felsite portions of the 12013 gray breccia, the materials appear to be more nearly identical; the felsite

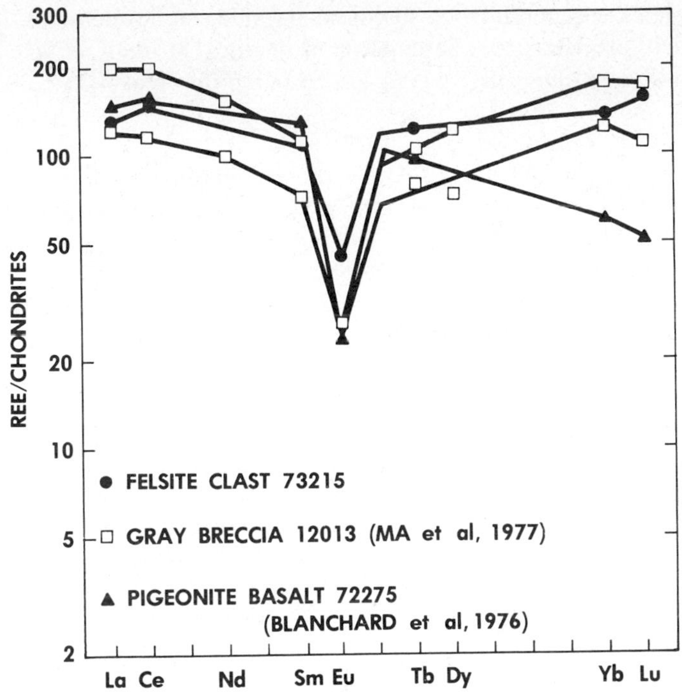

Fig. 3. Chondrite-normalized rare-earth-element abundances for felsite clast 73215,43 (filled circles), gray breccia samples from 12013 (open squares), and pigeonite basalt from 72275 (filled triangles). Note the similarity between 12013 and 73215 and the dissimilarity between them and the pigeonite basalt. 12013 data are from Ma *et al.* (1977); 72275 data are from Blanchard *et al.* (1975).

portions of 12013 have 1.4% FeO and 6.8% K_2O. Na_2O contents still differ by a factor of about seven, 1.4% for 12013 felsite vs 0.194% for 73215 felsite. We suspect the pure felsite in 12013 has a composition very similar to 73215,43.

Discussion

Warner *et al.* (1977) have outlined the characteristics of an especially significant group of lunar rocks, important in that they predate the lunar cataclysm and preserve information about the early history of the highland crust. The distinctive features of these rocks are their compositions which correspond to anorthositic norites and troctolites, granulated and annealed textures, homogeneous mineral compositions on the millimeter scale, unusual meteoritic component ratios, ages older than 4.2 b.y., and lithophile element abundances that are low and unfractionated relative to chondrites.

Data on anorthositic gabbro clasts from 73215 indicate that they also belong in this group of samples. They are annealed breccias with evidence of contributions from several source lithologies (James and Hammarstrom, 1977). The siderophile element ratios in the single anorthositic gabbro clast from 73215 so analyzed indicate that the rock acquired its siderophile elements from a projectile which predates the Serenitatis projectile (Morgan *et al.*, 1976). The isotopic data yield ages > 4.2 b.y.; a Rb–Sr isochron gives 4.24 ± .38 (Compston *et al.*, 1977) and the ^{39}Ar–^{40}Ar high temperature plateau gives a gas retention age of 4.23 ± .05 b.y. (Jessberger *et al.*, 1977).

REE abundances of the 73215 anorthositic gabbro clasts closely resemble those of the rocks in Warner's group. In Fig. 1, we have plotted REE data for 67955, 78155, 77017, 76230, and 79215 from Warner's group, 72215 and 72235, anorthositic gabbros from Boulder 1, Station 2, and compared them with REE data from anorthositic gabbros from 73215. All these samples have low, relatively unfractionated REE patterns without the KREEP-like depletion of the HREE.

The Co–Ni relationships (Fig. 4) among the rocks in Warner's group and clasts in 73215 are consistent with the proposed relationship among these samples. The Ni and Co in most of the clast and matrix of 73215 are dominated by their indigenous Co–Ni contents; the broken line in Fig. 4 joins the origin with 73215,170, a troctolite basalt which has no meteoritic component (Morgan *et al.*, 1976). The anorthositic gabbro clasts 46,25 and 46,33 are the only materials

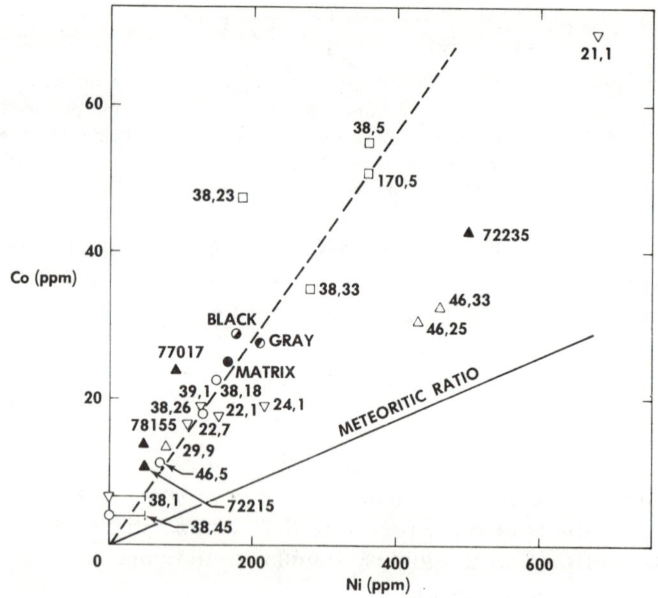

Fig. 4. Variation diagram of Co vs Ni. Open circles are granulated feldspathic samples; open squares are granulated troctolite samples; open triangles are gabbroic anorthosite clasts; filled circle is the average matrix samples; the half-filled circles are black and gray aphanite clasts, all from 73215. Inverted triangles correspond to 79215 samples. Filled triangles correspond to other Apollo 17 gabbroic anorthosites (Blanchard *et al.*, 1975; Boynton *et al.*, 1976; Wänke *et al.*, 1976). Meteoritic line corresponds to a Ni/Co ratio of 23.2. Dashed line passes through 170,5 and the origin; see text.

from 73215 that appear to have an observable component of meteoritic Ni and Co. We note that there is strong evidence which firmly establishes the relationship of clast 73215,29,9 with 73215,45,25, and 46,33; they are chemically and petrographically similar, yield nearly identical Rb–Sr data and give identical ^{39}Ar–^{40}Ar stepwise heating Ar release plateaus (James and Hammerstrom, 1977; Compston *et al.*, 1977; Jessberger *et al.*, 1977). A line joining these three clasts from 73215 would pass within error at most of the samples of anorthositic gabbros from 79215 and the other rocks for which we have data from the group defined by Warner *et al.* (1977); thus the Ni–Co data are consistent with the proposed association among these rocks.

We postulate that the gabbroic anorthosite materials represent lithologies from deep in the megaregolith which were excavated by the major impacts. The lack of evidence for melt contamination indicates that the REE abundances of the gabbroic anorthosite clasts from 73215 are representative of the REE abundances in the gabbroic anorthosite body from which they originated. They may have undergone metamorphic changes about 4.23 b.y. ago, but have since remained closed systems. 79215 may represent a somewhat more calcic sample of the sample crustal layer or similar body of gabbroic anorthosite and may have been sampled by a major impact event similar to that which produced 73215.

The granulated troctolitic and feldspathic materials in 73215 may represent either products of crushing previously uncrushed blocks of these materials in the megaregolith or the results of excavating fresh plutonic materials during the major impact that formed 73215. Unlike the anorthositic gabbros which show no evidence of contamination by the melt phase, these materials contain abundant aphanitic materials. Meager age data are consistent with a disruption of the granulated troctolite near the time of breccia formation; Jessberger *et al.* (1976, 1977) report a poorly defined plateau at $4.00 \pm .10$ b.y. for an olivine crystal separated from a troctolite vein. Our conclusions about the nature of the original troctolitic and feldspathic materials must be tentative as we have been unable to isolate samples of these materials free from admixed aphanite.

At present there is little data on the clasts of pink-spinel-bearing troctolitic basalt in 73215. On petrologic evidence, the clasts certainly predate the final consolidation of 73215 (James, pers. comm.), but there are no other age data available. This troctolite basalt is not the same troctolitic material that was granulated and mixed with aphanite to produce mixtures such as 38,5; they are petrographically distinct (James, pers. comm.) and a mix of troctolite basalts and aphanite would not result in the deep Eu anomaly observed in 38,5.

The question of the origin of felsite centers primarily on the question of the possible role of immiscibility. Hess *et al.* (1973) and Rutherford *et al.* (1976) argue that the felsites are derived from an immiscible liquid which fractionated, perhaps as a sizeable layer in the lunar crust. Ryder (1976) has taken the opposite viewpoint; he argues that lunar "granites" such as the 12013 felsite are related to the granitic mesostasis found in KREEP basalts, and that they are derived from KREEP basalt magmas by different degrees of extreme fractionation not involving immiscibility. Ryder cites the feldspathic nature of the basalts as well as the granites as evidence that they are not related by immiscibility.

The ratios of certain elements may be used to evaluate whether magmas have undergone separation of immiscible liquids. The K/La ratio in lunar rocks is not severely changed by processes of crystal fractionation. However, it has been shown experimentally (Hess *et al.*, 1974) that the K/La ratio can be changed by a factor of up to 20 by partitioning between immiscible liquids; the ratio is much larger in the Si-rich liquid than in the Fe-rich liquid. We note that the K/La ratio in the 73215 felsite is 1690, which is extremely high compared to a normal lunar ratio of 70 (Taylor and Jakeš, 1974). We feel that this extremely high K/La ratio is compelling evidence for an episode of immiscibility in the history of the 73215 felsite.

The REE patterns of both the 73215 and 12013 felsites show that these materials are unusually enriched in HREE and are somewhat similar to the relative patterns of some of the gabbroic anorthosites discussed earlier. Accounting for such unusual REE patterns at the high absolute REE abundances of the felsites is difficult. Experimental studies indicate the REE are not fractionated relative to each other by partitioning between immiscible liquids (Hess *et al.*, 1974; Hess, pers. comm.). They are partitioned as a group between the SiO_2-rich and FeO-rich melts in proportions of about 1:2. Consequently, both immiscible liquids, initially at least, should show REE patterns like those in the parent magma, but differing in absolute abundances. Because immiscibility itself does not change the relative REE patterns, explanation of the felsite REE pattern requires one of the following possibilities.

(1) The parent magma fractionated to an extreme degree prior to separation of the Fe-rich liquid from the Si-rich felsite liquid which then merely retained the extremely fractionated REE pattern of the parent magma.
(2) The immiscible liquids separated and initially had relative REE patterns which presumably were KREEP-like, similar to most of the other highly fractionated materials from the moon. Subsequent fractionation produced the bowl-shaped REE pattern in the felsite liquid. This fractionation could have been by one of the following mechanisms:
   (a) enrichments in the felsite liquid of a phase extremely rich in HREE, or
   (b) removal of a phase extremely rich in LREE and depleted in HREE, or
   (c) precipitation of a phase in the iron-rich liquid which would result in changes in the felsite melt as long as exchange was maintained between the liquids.

If the parent magma went to extreme degrees of fractionation prior to the onset of immiscibility (case 1), it would be reasonable to expect to find materials with REE patterns intermediate between those of the original parent magmas and those of the felsites. None have been found; the most fractionated materials instead are the most KREEP-like (e.g., the quartz monzodiorite of Nyquist *et al.*,

1977) with no indication of an evolution toward a felsite-like REE pattern. The alternative, fractionation of one or both of the separated liquids, should also result in compositions intermediate to the inferred starting compositions and the observed final product. In light of the few examples of felsite that have been analyzed for REE, however, this problem may simply be one of insufficient sampling.

If the felsite REE pattern results from inclusion of a HREE-rich phase, a candidate for that phase might be zircon. Nagasawa (1970) found that in acidic magmas zircon very effectively concentrates HREE ($Lu_{zircon}/Lu_{bulk\ rock} \cong 1000$) and shuns the LREE ($Ce_{zircon}/Ce_{bulk\ rock} \cong 0.2$). However, Zr should concentrate in the Fe-rich liquid, not the Si-rich liquid (Quick and Albee, 1976) and thus the bulk of the zircon should not be in the felsite liquid. Also, 12013 and 73215 felsites have very similar REE patterns but differ by an order of magnitude in Hf content and presumably also in Zr content. Accidental inclusion of zircon into the granites thus seems an unlikely means of explaining the REE patterns.

If the felsites have lost a phase rich in LREE and low in HREE, such a phase could be a phosphate mineral such as apatite or whitlockite. Apatite concentrates REE discriminating somewhat against the lightest and heaviest. The REE distribution coefficients for whitlockite have not yet been well documented except that this mineral seems to concentrate REE even more than apatite does. With the uncertainties in the distribution coefficients of apatite and whitlockite, quantitative calculations seem premature. Preliminary calculations indicate that if the parent magma had KREEP-like REE relative abundances, then removal of apatite alone would require that large amounts be removed and the result would still not be as relatively rich in HREE as the felsite is. If whitlockite is used as the removed phase we must infer that it discriminates more effectively against HREE than does apatite. With that assumption, smaller amounts of whitlockite would need to be removed for any given parent to produce the REE pattern of felsite. Precipitation of these phases in the Fe-rich liquid produces exactly the same effect as their removal from the Si-rich liquid as long as the exchange between the liquids is maintained. Unfortunately, until we have more information about *D* values for accessory phases and more experimental details about elemental fractionation during separation of immiscible liquids, discussion of parent magmas remains highly speculative and *ad hoc*.

Therefore while immiscibility seems to be required to explain the K/La ratios in the felsite, the nature of the parent liquid and the details of the fractionation processes remain problematical from the standpoint of accounting for the REE abundances.

Conclusions

Some of the clasts in consortium breccia 73215 contain information on the nature of the highlands crust prior to the lunar cataclysm. The anorthositic gabbro clasts clearly have preserved their compositional and isotopic identities since 4.23–4.24 b.y. ago. Pink-spinel-bearing troctolitic basalt clasts have not been disrupted by the consolidation of the 73215 breccia, but their history is not clear prior to that event. The granulated troctolitic and feldspathic materials were thoroughly mixed with aphanite material which was cogenetic with the melt phase in the breccia matrix. While analyses of pure troctolitic and feldspathic materials would be useful in characterizing ancient highlands materials, the presently available bulk samples of these materials are not.

The anorthositic gabbro clasts from 73215 should be included with the group of samples of ancient lunar highlands, which includes 79215, that have been identified by Warner *et al.* (1977). We postulate that these clasts represent material excavated from deep in the megaregolith by a major impact event. They have had a complex history showing evidence of brecciation and annealing prior to 4.23–4.24 b.y. ago, but have since remained closed systems with respect to compositional and isotopic alteration. Their low, relatively unfractionated REE abundances may result from the fact that they predate the appearance of KREEP basalt on the moon as suggested by Warner *et al.* (1977) or that they were merely spatially isolated from it; nevertheless, the REE patterns of these clasts are representative of an important rock type present in the early highlands crust.

The felsite clast was partially melted during consolidation of the breccia and thus, while it appears to have remained compositionally unaltered, it has lost isotopic information about its history prior to 3.9 b.y. ago. The history of the 73215 felsite clearly includes its separation as an immiscible liquid; this is the only viable explanation of its extremely fractionated K/La ratio. We suggest that the episode of immiscibility was not the last stage of the evolution of the felsite but that, following separation as an immiscible liquid, the felsite liquid further evolved by some manner of crystal fractionation which resulted in its unusual REE abundances.

Acknowledgments—We thank Tom Godsey and the Texas A&M reactor crew for irradiating the samples. We thank our consortium leader, O. B. James, for her exceptional efforts which have enabled this consortium to function successfully; this manuscript especially has benefitted from her comments and suggestions. We thank C. Hardy for preparing this manuscript.

REFERENCES

Bickel C. E., Warner J. L., and Phinney W. C. (1976) Petrology of 79215: Brecciation of a lunar cumulate. *Proc. Lunar Sci. Conf. 7th*, p. 1793–1819.

Blanchard D. P. (1973) The analysis of Apollo 15 rake sample walnuts and lunar breccias 15465 and 15565 for major and trace elements by atomic absorption spectrophotometry and neutron activation analysis. Ph.D. thesis. University of Wisconsin, Madison, Wisconsin.

Blanchard D. P., Haskin L. A., Jacobs J. W., Brannon J. C., and Korotev R. (1975) Major and trace element chemistry of Boulder 1 at Station 2, Apollo 17. *The Moon* **14**, 359–371.

Blanchard D. P., Jacobs J. W., Brannon J. C., and Haskin L. A. (1976) Major and trace element compositions of matrix and aphanitic clasts from consortium breccia 73215. *Proc. Lunar Sci. Conf. 7th*, p. 2179–2187.

Boynton W. V., Chou C-L., Robinson K. L., Warren P. H., and Wasson J. T. (1976) Lithophiles, siderophiles, and volatiles in Apollo 16 soils and rocks. *Proc. Lunar Sci. Conf. 7th*, p. 729–742.

Brecher A. (1976) The magnetic characteristics of consortium highland breccia 73215: Evidence for textural control of magnetism. *Proc. Lunar Sci. Conf. 7th*, p. 2217–2232.

Compston W., Foster J. J., and Gray C. M. (1977) Rb–Sr systematics in clasts and aphanites from consortium breccia 73215. *Proc. Lunar Sci. Conf. 8th.* This volume.

Floran R. J., Grieve R. A. F., Phinney W. C., Warner J. L., Simonds C. H., Blanchard D. P., and Dence M. R. (1977) Manicouagan impact melt, Quebec. Part I: Stratigraphy, petrology and chemistry. Submitted to *J. Geophys. Res.*

Hess P. C. and Rutherford M. J. (1974) Element fractionation between immiscible melts (abstract). In *Lunar Science V*, p. 328–330. The Lunar Science Institute, Houston.

Hess P. C., Rutherford M. J., Guillemette R. N., Ryerson F. J., and Tuchfeld H. A. (1975) Residual products of fractional crystallization of lunar magmas: An experimental study. *Proc. Lunar Sci. Conf. 6th*, p. 895–910.

Hubbard N. J., Rhodes J. M., Wiesmann H., Shih C-Y., and Bansal B. M. (1974) The chemical definition and interpretation of rock types returned from the non-mare regions of the moon. *Proc. Lunar Sci. Conf. 5th*, p. 1227–1246.

Jacobs J. W., Korotev R. L., Blanchard D. P., and Haskin L. A. (1977) A well tested procedure for instrumental neutron activation analysis of silicate rocks and minerals. *J. Radioanal. Chem.* In press.

James O. B. (1976) Petrology of aphanitic lithologies in consortium breccia 73215. *Proc. Lunar Sci. Conf. 7th*, p. 2145–2178.

James O. B. and Blanchard D. P. (1976) Consortium studies of light-gray breccia 73215: Introduction, subsample distribution data, and summary of results. *Proc. Lunar Sci. Conf. 7th*, p. 2131–2144.

James O. B., Brecher A., Marti K., Kirsten T., Braddy D., Hutcheon I. D., Blanchard D. P., Jacobs J. W., Brannon J. C., Korotev R. L., Haskin L. A., Silver L. T., Kerridge J. F., Kaplan I. R., Pillinger C. T., and Gardiner L. R. (1975) Consortium studies of matrix of light gray breccia 73215. *Proc. Lunar Sci. Conf. 6th*, p. 547–577.

James O. B. and Hammarstrom J. G. (1977) Petrology of four clasts from consortium breccia 73215. *Proc. Lunar Sci. Conf. 8th.* This volume.

Jessberger E., Kirsten T., and Staudacher Th. (1976) Argon-argon ages of consortium breccia 73215. *Proc. Lunar Sci. Conf. 7th*, p. 2201–2216.

Jessberger E. K., Kirsten T., and Staudacher Th. (1977) One rock and many ages—further K–Ar data on consortium breccia 73215: *Proc. Lunar Sci. Conf. 8th.* This volume.

Ma M-S., Murali A. V., Schmitt R. A., Quick J. E., and Albee A. L. (1977) Major, minor, and trace element compositions of two petrologically significant lithologies in 12013 (abstract). In *Lunar Science VIII*, p. 602–604. The Lunar Science Institute, Houston.

Morgan J. W., Gros J., Takahashi H., and Hertogen J. (1976) Lunar breccia 73215: Siderophile and volatile trace elements. *Proc. Lunar Sci. Conf. 7th*, p. 2189–2200.

Müller H. W., Plieninger T., James O. B., and Schaeffer O. A. (1977) Laser probe ^{39}Ar–^{40}Ar dating of materials from consortium breccia 73215. *Proc. Lunar Sci. Conf. 8th.* This volume.

Nagasawa H. (1970) Rare earth concentrations in zircons and apatites and their host dacites and granites. *Earth Planet. Sci. Lett.* **9**, 359–365.

Nord G. L., Jr. and James O. B. (1977) Aphanitic matrix, an ANT-suite clast and a felsite clast in consortium breccia 73215: An electron petrographic study. *Proc. Lunar Sci. Conf. 8th.* This volume.

Prinz M., Dowty E., and Keil K. (1973) Mineralogy, petrology and chemistry of lithic fragments from Luna 20 fines: Origin of the cumulate ANT suite and its relationship to high-alumina and mare basalts. *Geochim. Cosmochim. Acta* **37**, 979–1006.

Quick J. E. and Albee A. L. (1976) 12013 revisited—two clast-laden melts (abstract). In *Lunar Science VII*, p. 712–714. The Lunar Science Institute, Houston.

Quick J. E., Albee A. L., Ma M.-S., Murali A. V., and Schmitt R. A. (1977) Chemical compositions and possible immiscibility of two silicate melts in 12013. *Proc. Lunar Sci. Conf. 8th.* This volume.

Rutherford M. J., Hess P. C., Ryerson F. J., Campbell H. W., and Dick P. A. (1976) The chemistry, origin and petrogenetic implications of lunar granite and monzonite. *Proc. Lunar Sci. Conf. 7th*, p. 1723–1740.

Ryder G. (1976) Lunar sample 15405: Remanent of a KREEP basalt—granite differentiated pluton. *Earth Planet. Sci. Lett.* **9**, 359–365.

Ryder G., Stoeser D. B., Marvin U. B., Bower J. F., and Wood J. H. (1975) Boulder 1, Station 2, Apollo 17: Petrology and petrogenesis. *The Moon* **14**, 327–357.

Taylor S. R. and Jakeš P. (1974) The geochemical evolution of the Moon. *Proc. Lunar Sci. Conf. 5th*, p. 1287–1305.

Turner G. and Cadogan P. H. (1975) The history of lunar bombardment inferred from ^{40}Ar–^{39}Ar dating of highland rocks. *Proc. Lunar Sci. Conf. 6th*, p. 1509–1538.

Wänke H., Palme H., Kruse H., Baddenhaussen H., Cendales M., Dreibus G., Hofmeister H., Jagoutz E., Palme C., Spettel B., and Thacker R. (1976) Chemistry of highland rocks: A refined evaluation of the composition of the primary matter. *Proc. Lunar Sci. Conf. 7th*, p. 3479–3499.

Warner J. L., Bickel C. E., Phinney W. C., and Simonds C. H. (1977) Feldspathic, granulitic impactites that pre-date the final lunar bombardment (abstract). In *Lunar Science VIII*, p. 979–981. The Lunar Science Institute, Houston.

Proc. Lunar Sci. Conf. 8th (1977), p. 2525–2549.
Printed in the United States of America

Rb–Sr systematics in clasts and aphanites from consortium breccia 73215

W. COMPSTON and J. J. FOSTER

Research School of Earth Sciences, Australian National University, Canberra

C. M. GRAY

Department of Geology, LaTrobe University, Melbourne

Abstract—A small variable content of Rb-rich felsite clasts has generated Rb–Sr mixing lines in samples of aphanitic matrix and aphanite clasts from the light-gray breccias 73215 and Boulder 1 at Apollo 17, Station 2. These mixing lines can be mistaken for thermal disturbance of the Rb–Sr system during the breccia-forming event. Mineral and lithic fractions from a felsite clast give 3.90 ± 0.05 AE as the minimum age for primary crystallization, with high initial $^{87}Sr/^{86}Sr$ (0.7065); the Rb–Sr model age for a "total-rock" chip gives a maximum age estimate of 4.0 AE. Rb–Sr mixing is also evident with feldspathic clasts of the ANT-suite characterised by low initial $^{87}Sr/^{86}Sr$ ($\sim$0.6992), and with an unidentified component having a high initial $^{87}Sr/^{86}Sr$ ($>$0.701) that gives some of the aphanites model Rb–Sr ages that exceed 4.6 AE. The unidentified component could be a low abundance mineral or lithic clast, possibly plagioclase from the felsite, or it may be the groundmass melt itself. Volatile loss of Rb at $\sim$3.9 AE appears to be the best explanation of the old model ages. Clasts of anorthositic gabbro have a Rb–Sr age of $\sim$4.25 AE from imprecise internal isochrons, and model ages of 4.35 AE. If the clasts remained chemically closed during breccia formation, the model ages are older limits to their primary igneous or metamorphic ages. The felsite age of 3.90 ± 0.05 AE is equal to (or very slightly older) than the age of breccia aggregation. If breccia 73215 was formed by the Southern Serenitatis impact, as field and petrological evidence suggests, then several of the younger lunar basin-forming impacts occurred in a narrow time interval around 3.90 AE. The young age for the Serenitatis event also demonstrates that Ar was not totally degassed from most of the Apollo 17 breccias.

INTRODUCTION

WE DISCUSS HERE age determinations and Sr isotope geochemistry on the Apollo 17 consortium breccia 73215 (James *et al.*, 1975), a light-gray breccia collected from the light mantle avalanche deposit adjacent to South Massif. The breccia is very similar to samples from the nearby Boulder 1 at Apollo 17, Station 2, which we analyzed previously in a complementary study (Compston *et al.*, 1975). All these samples probably originated in the upper parts of the South Massif (Schmitt, 1973), which is thought to represent the ejecta blanket from the Southern Serenitatis impact (Wolfe and Reed, 1975).

The 73215 breccia consists of gray to black aphanitic rock and physically separable clast materials. The bulk of the aphanitic rock forms a "matrix" that encloses the clasts. This matrix is not clast free, however: it is composed of small lithic and mineral fragments set in a fine-grained intergrowth termed "groundmass" that crystallized from a melt. The breccia-forming process was one of mechanical mixing of melt and clasts during transport and ejection of debris from probably a single large impact event (James et al., 1975). The physically separable clast materials are fine-grained ANT-suite rocks, granulated feldspathic rocks and aphanites lithologically similar to the matrix aphanite.

In this study we have analyzed various bulk samples of matrix and clast aphanite, bulk samples and mineral separates from two ANT-suite clasts, and separates from a clast of felsite ("microgranite"). Our preliminary results were presented by James et al. (1976) and discussed by Morgan et al. (1976). Petrologic studies of these materials were presented by James et al. (1975), James (1976), and James and Hammarstrom (1977); major-, minor-, and trace-element chemistry was presented by James et al. (1975), Blanchard et al. (1976, 1977), and Morgan et al. (1976); results of ^{40}Ar–^{39}Ar age studies were given by Jessberger et al. (1976, 1977) and Müller et al. (1977). The detailed locations of our 73215 subsamples are given by James and Blanchard (1976). For brevity the subsample numbers will be used here without the rock number 73215 in most text discussion.

Our objectives in the present study were

(a) to test for strontium isotope equilibration and for redistribution of Rb and Sr in the matrix during breccia aggregation;
(b) to search for conclusive pre-4.0 AE lunar ages in separated clasts;
(c) to determine an older limit for the time of breccia formation from the age of its youngest clasts, and
(d) to search for mixing patterns in the collective Rb, Sr data from samples of aphanitic matrix. (Such patterns would be expected from the observed variations in clast contents of different types of matrix.)

In addition to the 73215 results, we report isotopic analyses of the meteorite Angra dos Reis and of NBS 987 standard strontium as reference points on our scale for normalised $^{87}Sr/^{86}Sr$. We also report an internal isochron for the feldspathic basalt 68416 to allow intercomparison of ages with other laboratories. The new data reported here and those from our earlier consortium study of Boulder 1 were produced by identical analytical methods and are strictly comparable. As the methods differ in some details from our earlier lunar work, they are described in Appendix 1. Ages are calculated using 1.39×10^{-11} yr^{-1} as the value of the ^{87}Rb decay constant.

AGES OF ANT-SUITE AND FELSITE CLASTS

ANT-suite anorthositic gabbro 29,9

This rock has a poikilitic texture (subhedral plagioclase and rounded olivine are enclosed by orthopyroxene) that indicates crystallization from a melt; minor numbers of xenocrysts of plagioclase and olivine were present in this melt (James and Hammarstrom, 1977). The melt probably represents a fused polymict breccia rather than a pristine magma from the lunar interior (James and Hammarstrom, 1977). The relatively coarse grain size and homogeneity of the mineral compositions indicate that the original body crystallized and cooled relatively slowly, so that Sr isotope homogenization between xenocrysts and melt probably occurred over the ~1 cm^3 volume of the actual clast.

The plagioclase and olivine I separates (Table 1) were handpicked and the total-rock samples were (non-representative) single chips from a 250 mg fragment of the clast. Olivine II was prepared at very fine grain size using redistilled heavy liquids, as there were insufficient single olivine grains for a duplicate handpicked sample. Unfortunately this concentrate contained enough composites with plagioclase to appreciably reduce its $^{87}Sr/^{86}Sr$ compared with olivine I, so the precision of the internal age, 4.24 ± 0.31 AE (2σ), is limited by the comparatively poor measurement for the latter. The age itself agrees with the more precise ^{40}Ar–^{39}Ar result of 4.24 ± 0.02 AE (2σ) for the high temperature plateau age of this clast (Jessberger *et al.*, 1976), but the large uncertainty in the Rb–Sr value precludes its use as independent corroboration of the Ar–Ar age.

Table 1. Rb–Sr data for clasts of anorthositic gabbro, granulated feldspathic material, and felsite from breccia 73215. A and B denote repeat analyses of similar but not necessarily identical samples, I and II denote essentially different mineral or rock fractions. The coefficient of variation for $^{87}Rb/^{86}Sr$ is 0.25% except for the felsite fractions of olivine I (0.5%). The uncertainties shown for $^{87}Sr/^{86}Sr$ refer to the last digit(s) and represent one standard error.

		Weight (mg)	Rb ppm	Sr ppm	$^{87}Rb/^{86}Sr$	$^{87}Sr/^{86}Sr$
A. 29,9	A	19.0	2.43	167.0	.0420	.70175 ± 5
1. total-rock	B	21.1	2.48	167.2	.0427	.70173 ± 5
2. plagioclase	A	5.0	1.45	208.6	.0201	.70038 ± 5
	B	6.2	1.63	185.9	.0253	.70078 ± 6
3. olivine	I	2.1	.40	15.45	.0755	.70406 ± 23
olivine	II	5.3	.70	39.2	.0510	.70220 ± 8
B. 46,25	A	16.8	2.21	143.4	.0444	.70185 ± 5
1. total-rock	B	21.1	2.31	143.5	.0465	.70195 ± 6
	A	5.6	3.55	193.1	.0531	.70238 ± 5
2. plagioclase	B	5.0	3.22	181.0	.0513	.70237 ± 8
C. 46,102 feldspathic	A	23.7	1.77	154.3	.03316	.70144 ± 5
material	B	25.9	1.76	154.8	.03278	.70139 ± 5
2. plagioclase	A	3.00	0.46	199.8	.00666	.69950 ± 5
	B	2.52	0.56	207.2	.00777	.69955 ± 5
D. 46,10 plagioclase fragment		4.1	1.55	202.4	.0221	.70043 ± 5
E. 43,IV felsite chip		1.62	255.5	158.0	4.666	.96616 ± 6
43,III glass concentrate		0.55	252.0	91.9	7.910	1.14662 ± 8
43,II grey fraction		1.19	290.0	156.4	5.350	1.00460 ± 8
43,I white fraction		0.84	342.8	213.4	4.634	.96423 ± 5

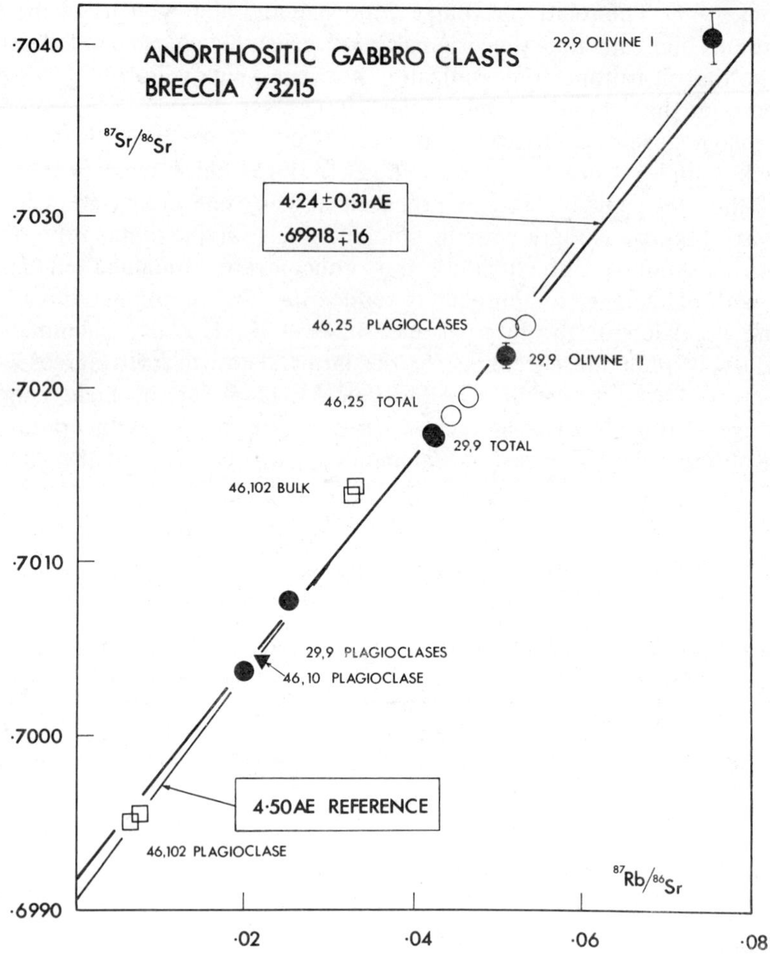

Fig. 1. Sr evolution diagram for gabbroic anorthosites 29,9 and 46,25. Jessberger *et al.* (1976) report high temperature plateau ages close to 4.25 AE for both clasts. The plagioclase and total-rock data for 46,25 will fit the internal isochron for 29,9 within experimental error to give a joint age of 4.31 ± .17 AE. However, other evidence (see text) argues against a strict cogenesis of the two rocks. Sample 46,102 is granulated feldspathic material of polymict origin, and the plagioclase fragment 46,10 was extracted from within a black aphanite clast.

The chronological significance of these data cannot be determined unambiguously, as it is possible that both the Rb–Sr and K–Ar systems might have been disturbed when the clast was incorporated in the breccia. However, the Rb–Sr data can be used to set a limit on the crystallization age of the clast. The total-rock model age is 4.35 AE, using 0.69912 as the initial $^{87}Sr/^{86}Sr$ (our estimate for BABI); error in the model age due to the measurements alone is about ±0.15 AE. This value represents a weighted mean age for the various components incorporated in the anorthositic gabbro melt, and it constitutes an older limit to the age of the melting event itself.

Despite the uncertainty in the Rb–Sr age, the initial $^{87}Sr/^{86}Sr$ for 29,9 is fairly well defined at 0.69918 ± 0.0016 (2σ). If the precise Ar–Ar age is assumed, the mean initial $^{87}Sr/^{86}Sr$ using all samples except olivine I becomes $0.69917 \pm .00006$, which we will employ later as characterizing this type of material within the aphanites.

Note that the above discussion assumes that the 29,9 clast was chemically closed during its emplacement in breccia 73215 and during older thermal events that the gabbroic anorthosite might have experienced. If the plagioclase exchanged with more-radiogenic Sr from the matrix, both the internal isochron and the total-rock model age would be reduced. From the results of Papanastassiou and Wasserburg (1975a, 1976a) for Rb,Sr-bearing inclusions within olivine, which are apparently protected from later Sr isotope exchange by the nature of the olivine crystal structure, it is possible that the best estimate for the ages of 29,9 is given by the olivine I analysis (Table 1). Olivine I must be employed rather than olivine II, which is known to contain plagioclase. The model age for olivine I is 4.56 ± 0.21 (σ) AE, so that on this hypothesis, we cannot exclude the possibility of a very early primary age for 29,9.

ANT-suite anorthositic gabbro 46,25

This rock, in contrast to 29,9, shows virtually no textural evidence for crystallization in contact with melt; instead, crystallization was dominantly subsolidus (James and Hammarstrom, 1977). Fragments of plagioclase and olivine derived from preexisting rocks are abundant. The high siderophile element content and siderophile element ratios suggest that the rock contains admixed meteoritic metal (Morgan *et al.*, 1976). It is most probable that it represents a metamorphosed polymict breccia (James and Hammarstrom, 1977).

Plagioclase was the only mineral that could be separated in sufficient quantity for analysis and this had a comparatively high Rb content (Table 1). As a result, the difference in Rb/Sr between the plagioclase and total rock is too small to define an internal isochron (Fig. 1) and we can report no independent age for this clast. However the model age is well defined at 4.30 ± 0.15 AE, and the initial $^{87}Sr/^{86}Sr$ is 0.69917 ∓ 0.00006 if the age is assumed to be 4.26 AE, as given by the high temperature Ar–Ar plateau age found by Jessberger *et al.* (1976). Thus both the model age and initial $^{87}Sr/^{86}Sr$ are indistinguishable from those of 29,9.

Granulated feldspathic material and plagioclase fragments

The granulated feldspathic material is a polymict lithology; it consists primarily of mineral fragments derived from relatively coarse-grained, highly feldspathic rocks and particles of black and gray aphanites. A bulk sample of this material, 46, 102, lies significantly above the anorthositic gabbro isochron, and also above the 4.50 AE reference isochron through BABI (Fig. 1). This may be related to its composite nature. The abundant small aphanitic clasts it contains could not be excluded from the "total-rock" sample, and the Rb–Sr parameters for a number of similar aphanites from 73215 make them suitable contaminants to displace the coordinates of the bulk 46,102 in Fig. 1 in the required direction. Such mixing will be discussed further below. In contrast to the bulk sample, plagioclase separated from 46,102 lies systematically below the anorthositic gabbro isochron (Fig. 1) and on the 4.5 AE reference line. This indicates that the parent rocks of the granulated feldspathic material were older than the anorthositic gabbro clasts, if the true ages of the latter are indeed ~4.25 AE.

The various aphanites contain sparse fragments of plagioclase crystals large enough to be physically separated, and the 46,10 plagioclase was such a fragment from black aphanite clast 46,10. This plagioclase has substantially higher Rb than the 46,102 plagioclase and in this respect strongly resembles the 29,9 plagioclases. It has slightly lower $^{87}Sr/^{86}Sr$ than the 29,9 plagioclase but the difference is not statistically significant.

Felsite, 43

This rock contains two components: A crystalline felsite and a felsite glass. The crystalline felsite consists mainly of a vermicular K-feldspar-quartz intergrowth with minor plagioclase, olivine, pyroxene, and accessory minerals. The felsic glass is similar to the crystalline felsite in bulk chemistry, and was injected into it as veins before incorporation of the clast in the 73215 breccia (James and Hammarstrom, 1977). Four fractions were handpicked from a coarsely crushed 9.1 mg fragment of the felsite clast (Table 1): I, white K-feldspar-quartz intergrowth as free as possible from coloured phases: II, gray material, representing composites of the white fraction with ilmenite and olivine; III, brown material, mainly composites of the brown glass with the other material; and IV, several comparatively large chips as a "total-rock" sample. These fractions define a perfectly fitted isochron (Fig. 2) of age 3.90 ± 0.05 AE (2σ) and initial $^{87}Sr/^{86}Sr$ 0.7065 ± 0.0040, but the slope of the isochron is controlled almost entirely by the brown glass point. Interpretation of the age is therefore critically dependent on the relationship of the brown glass to the other fractions, and we offer the following alternatives:

(1) If the brown glass formed by rapid melting within the same felsite body that was the source of the crystalline felsite, and if cooling within the clast after the melting episode was sufficiently fast, it is possible that no Sr isotope equilibration took place between the glass and the crystalline material. It is also possible that Rb/Sr of the glass is higher than that of the crystalline material because Rb/Sr in the particular part of the felsite body that was melted was higher than average. For this situation, the isochron would represent the original age of crystallization of the felsite, and the high initial $^{87}Sr/^{86}Sr$ would be a property of the original felsite body.

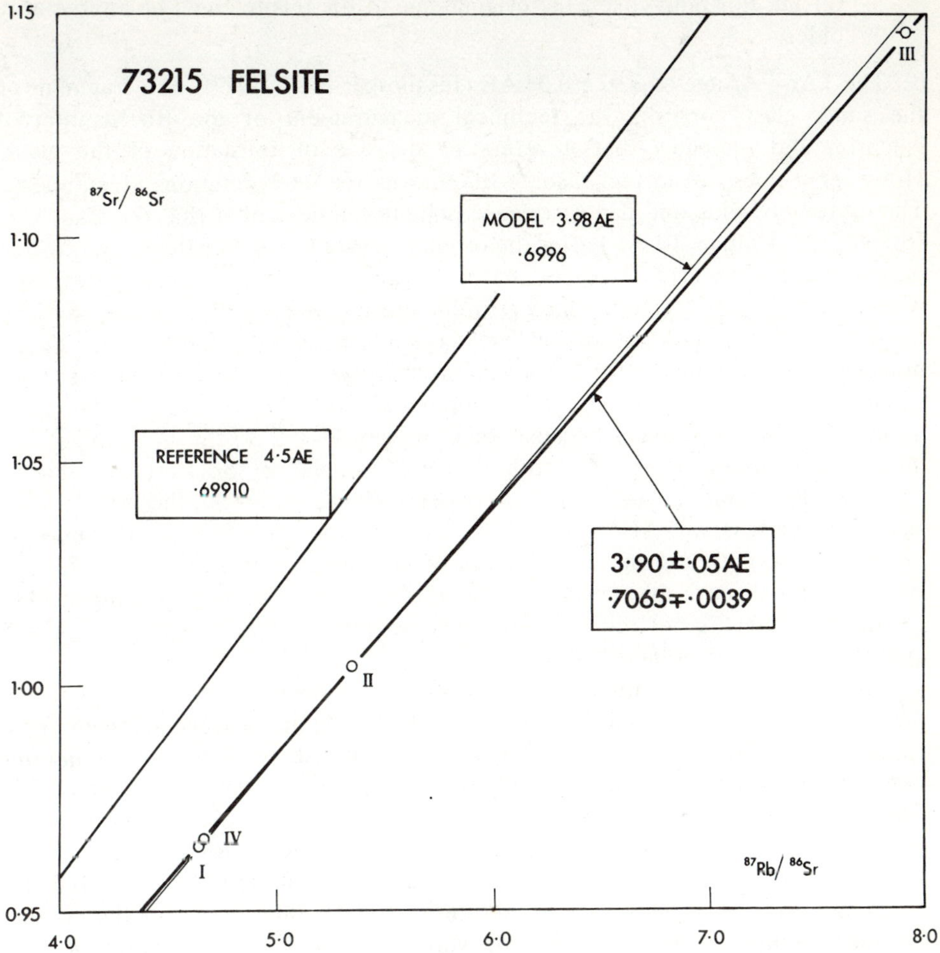

Fig. 2. Internal isochron for felsite clast 73215,43. The uncertainty in $^{87}Sr/^{86}Sr$ is not visible on the scale of the diagram but one standard error in $^{87}Rb/^{86}Sr$ is shown on each side of the points. The model age of 4.00 AE assumes the same initial $^{87}Sr/^{86}Sr$ as pigeonite basalt 72275,171 (Compston *et al.*, 1975) but its value is not sensitive to either of these parameters and is significantly greater than the internal age. The internal age is essentially controlled by the brown glass result.

(2) On the other hand, if (a) complete Sr isotope equilibration did occur between glass and the crystalline material in the clast, or if (b) the lower Sr in the glass reflects chemical fractionation during the melting event, then the isochron refers instead to the age of formation of the brown glass. For case (a) one or more minerals within the crystalline part of the clast would have a raised "initial" $^{87}Sr/^{86}Sr$ due to isotope exchange. This need not be so for case (b), in which plagioclase, for example, within the crystalline clast might preserve an original low initial $^{87}Sr/^{86}Sr$ for the

felsite. For both cases, the original age of the felsite could be appreciably older.

The ^{40}Ar–^{39}Ar age of 3.92 ± 0.04 AE (Jessberger *et al.*, 1977) for a fragment of the same clast confirms the technical measurement of the Rb–Sr internal isochron and indicates that it refers to the age of formation of the glass. However it does not in fact resolve the alternative interpretations given above. The extremely wide and flat Ar release plateau implies either that the glass was formed very soon ($\leq$.01 AE) after the original crystallization of the felsite, which permits alternative (1) (above), or that the felsite clast was completely degassed when the glass was formed, which permits alternative (2).

A limit may be set on the original crystallization age of the felsite by its model Rb–Sr age. Because the three points for crystalline fractions in Fig. 2 are so close together, their model ages are almost identical for a very wide choice of initial $^{87}Sr/^{86}Sr$. Moreover, because their present-day $^{87}Sr/^{86}Sr$ is so high, the model age changes only slowly with choice of initial $^{87}Sr/^{86}Sr$. For any value of "normal" lunar initial $^{87}Sr/^{86}Sr$, viz. between 0.6990 and 0.6996, the mean model age of the clast is 4.00 AE, and subject to one further qualification, this may be taken as a maximum estimate for the age of original crystallization of the felsite. We will later discuss the possibility that the very small total-rock sample, only 1.6 mg in weight, is not representative of the average felsite body, which will qualify the above argument.

The felsite is very similar in bulk chemistry, mineralogy, and occurrence to the microgranite described by Ryder *et al.* (1975) in breccia samples from Boulder 1, Apollo 17, Station 2. Its direct analysis here tests our predictions concerning the age and chemistry of the microgranite (Compston *et al.*, 1975). The measured model age of 4.00 AE is within 1% of the age calculation for the microgranite clasts in breccia 72215,104 but the predicted Rb content was too high by a factor of two. The latter was due to our assumption of K/Rb = 100, cf., ~225 as observed in the felsite. In addition, our estimate of 5% for the content of microgranite clasts in this sample was probably low by a factor of two, according to measurements of clast proportions by Stoeser *et al.* (1974), which would independently amend the content of Rb in the clasts to about 300 ppm.

Rb–Sr PATTERN IN 73215 APHANITES

Aphanites analyzed were six samples of matrix and five of clasts (Table 2). The matrix samples were from four units differentiated on the basis of variations in colour, friability, and clast assemblage. The clast samples were taken from the two most prominent types of aphanite clast: Gray spheroids that occur within matrix aphanites, and black clasts that occur within granulated feldspathic clast materials. The distinguishing characteristics of these various lithologies were detailed by James *et al.* (1975) and James (1976).

Although the noritic groundmass within all the above aphanites is fairly uniform chemically (James, 1976), chemical variations in the total sample can be expected when the proportion of groundmass to (micro)clasts varies and when the average composition of the clasts changes. Such changes will now be described in detail.

Felsite addition. The range in Rb/Sr for samples of matrix and aphanite clasts (Table 2, Fig. 3) is due principally to differences in their Rb contents. At least part of the reason for the latter is variation in the abundance of felsite clasts, which occur in all samples but are notably more abundant in the black matrix samples 38,49 and 36,3 which have the highest Rb contents. This control on Rb distribution is also pronounced in matrix samples such as 72215,104 from the Boulder 1 breccias (Compston *et al.*, 1975).

Table 2. Rb–Sr data for aphanites from breccia 73215, sampled as total-rock chips. 36,3 was divided into a black II, and a brown fraction, I, but the A,B fractions are splits of inhomogeneous coarse-grained chips. The coefficient of variation for $^{87}Rb/^{86}Sr$ is 0.25% except for 36,3 which is 0.5%. The uncertainties shown in $^{87}Sr/^{86}Sr$ refer to the last digit and represent one standard error.

			Rb	Sr	$^{87}Rb/^{86}Sr$	$^{87}Sr/^{86}Sr$
46,10.6 Black	A	22.2	4.83	137.1	.1016	.70551 ± 6
clast	B	10.1	5.02	143.7	.1007	.70551 ± 5
38,57.5	A	17.7	2.94	155.4	.0546	.70283 ± 8
Gray spheroid	B	20.0	3.02	150.4	.0579	.70308 ± 5
157 Heterogeneous	A	24.0	5.96	139.8	.1231	.70649 ± 5
black matrix	B	25.4	5.67	138.6	.1179	.70626 ± 5
258 Heterogeneous	A	43.1	3.01	145.8	.0595	.70315 ± 5
black matrix	B	28.8	2.88	141.4	.0588	.70300 ± 5
46,45 Light-	A	21.2	2.26	139.3	.0467	.70255 ± 6
gray matrix	B	16.9	2.33	140.4	.0478	.70243 ± 5
178 Schlieren-	A	25.3	3.46	138.9	.0719	.70390 ± 5
rich gray matrix	B	23.5	3.24	139.9	.0669	.70355 ± 5
38,49 Black	A	16.6	8.20	146.9	.1612	.70926 ± 7
matrix	B	15.4	8.32	148.5	.1618	.70919 ± 5
46,102 Gray aphanite clast	A	6.21	3.27	131.2	.0720	.70369 ± 5
46,102 Gray aphanite clast	B	6.77	3.45	133.3	.0747	.70392 ± 5
38,32 Gray	A	15.7	7.16	147.2	.1405	.70785 ± 5
spheroid	B	19.4	6.87	148.8	.1333	.70739 ± 5
46,19 Black aphanite clasts		25.6	3.65	136.1	.0773	.70427 ± 5
36,3 Black	I	22.2	58.2	165.4	1.0214	.75908 ± 5
matrix	II	21.4	10.53	173.5	0.1753	.70969 ± 5

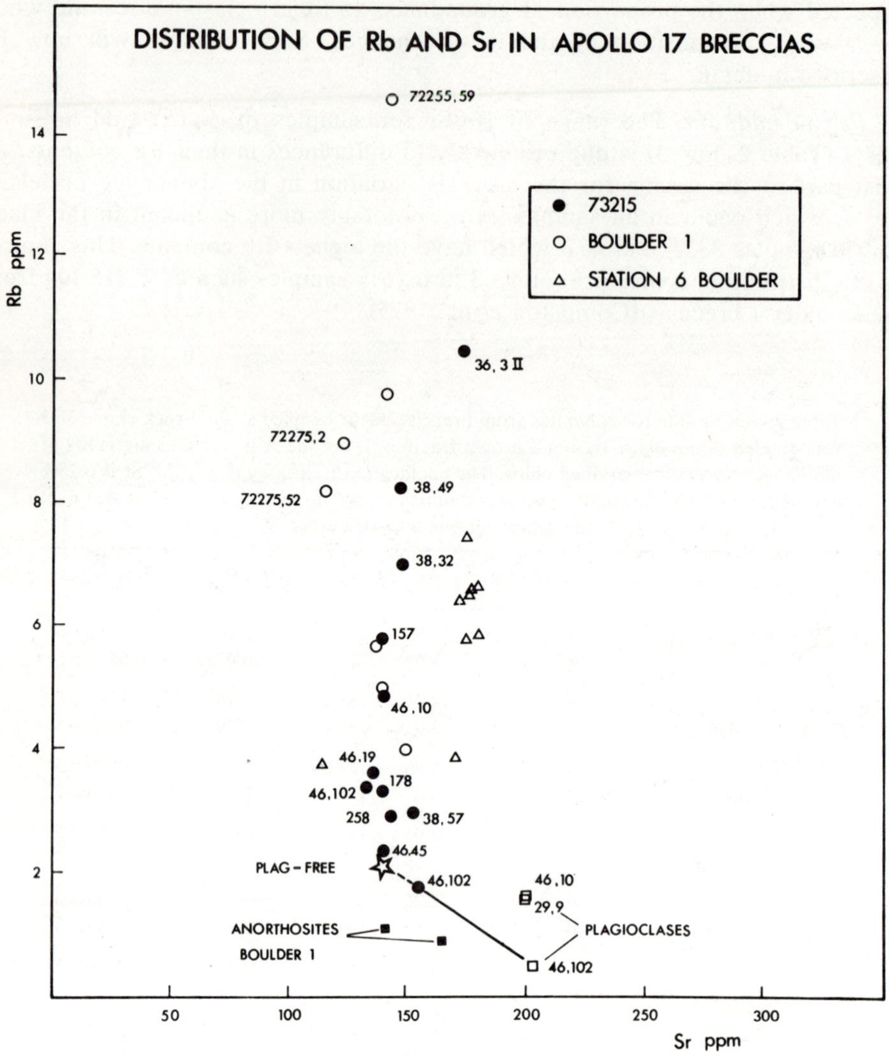

Fig. 3. The distribution pattern for Rb relative to Sr in aphanitic samples from Apollo 17 breccias. Sr is comparatively uniform in concentration whereas Rb ranges up to nearly 60 ppm (higher Rb samples not shown), which is due to the addition of high Rb felsite clasts. The data for the Station 6 Boulder are from Nyquist *et al.* (1974), and are included here as another coherent suite of breccia matrix samples.

The effect of the felsite clasts on the Rb–Sr pattern is shown by the modified Sr evolution diagram Fig. 4, in which radiogenic Sr grown in the samples during the past 3.90 AE has been subtracted from the observed $^{87}Sr/^{86}Sr$. The diagram gives the clearest visual resolution of small differences in $^{87}Sr/^{86}Sr$ between samples at 3.90 AE. The data for the felsite clast lie along a zero-slope isochron, corresponding to the 3.90 AE age of the glass. The data for the felsite-rich

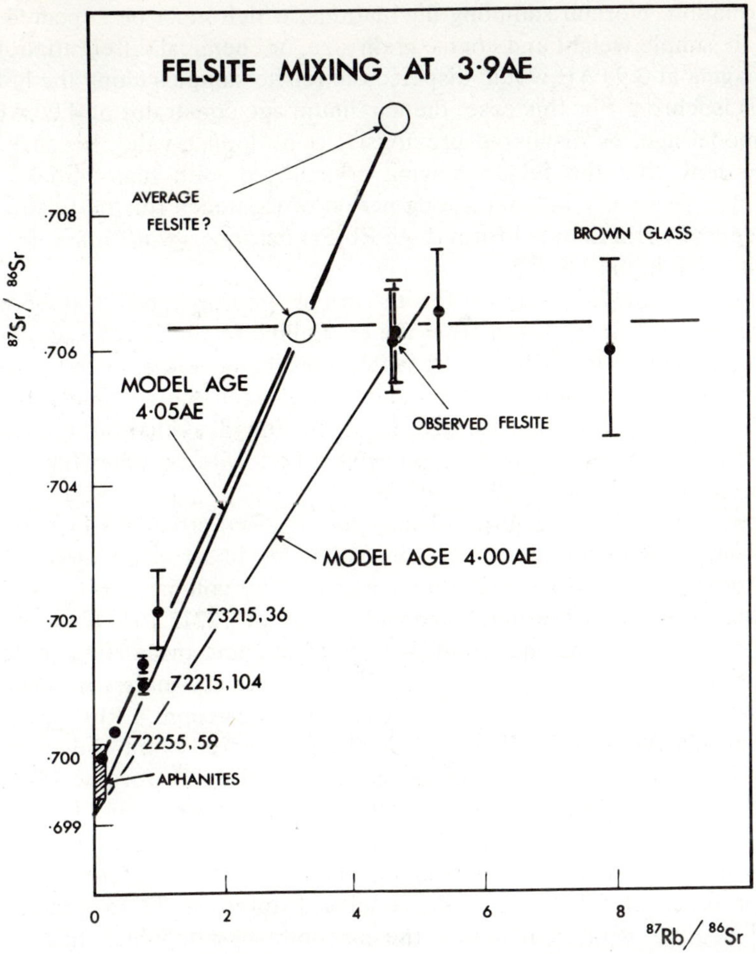

Fig. 4. Modified Sr evolution diagram for 73215 and Boulder 1 aphanites showing the felsite mixing line as it was 3.90 AE ago. Model ages are stated with respect to the present but slopes shown correspond to zero for 3.90 AE. The felsite mixing line as defined by the aphanite alignment does not pass exactly through the observed "bulk" felsite, which we suspect is biassed due to sampling fluctuations.

aphanites are well correlated along a mixing line that passes near, *but not exactly through*, the observed crystalline felsite. Because the fit of the measured felsite is not exact, we must modify our assessment of the Rb–Sr parameters of the original felsite body. We will now consider the possibility that the actual mean felsite that mixed into the aphanites was somewhere along the mixing line shown in Fig. 4, and probably between the two points indicated, rather than being represented by the measured felsite.

The lower of the two "average felsite" points in Fig. 4 corresponds to an original crystallization age of 3.90 AE for the felsite rather than only the age of glass formation. Normal sampling fluctuations, which must be expected for the very small sample weight and coarse grain size, or chemical differentiation of the felsite magma at 3.90 AE would disperse the felsite samples along the horizontal (3.90 AE) isochron. For this case, the maximum age constraint of 4.00 AE based on the model age, as discussed previously, is no longer valid, because we are assuming now that the felsite magma crystallized with high initial $^{87}Sr/^{86}Sr$. Instead, it is necessary to infer a long period of existence for the felsite magma in its chemically fractionated form (high Rb/Sr) *before crystallization*, in order to explain its high initial $^{87}Sr/^{86}Sr$.

The higher point for "average felsite" in Fig. 4 corresponds to the possibility that a net loss of ^{87}Sr occurred from the crystalline felsite in the clast without a significant change in its $^{87}Rb/^{86}Sr$, by isotopic exchange either with the intrusive glass or with formerly adjacent volumes of felsite in the original felsite body during the thermal event that formed the glass. For this situation, the maximum estimate for the original crystallization age of the felsite becomes the model age of the "average felsite", which is 4.05 AE.

Figure 5 illustrates the large change in $^{87}Rb/^{86}Sr$ produced in a particular matrix sample from Boulder 1 by subtracting its observed content of felsite clasts. Stoeser *et al.* (1974) counted felsite as 23% by volume of all clasts greater in size than 200 μm in the dark compact matrix of 72215,104. If we assume a total clast content in the matrix of 50% and the same proportion of felsite in clasts less than 200 μm, the total content of felsite in the matrix becomes about $11\frac{1}{2}$%. Taking the Rb content of "average felsite" as 280 ppm, $^{87}Rb/^{86}Sr$ as 3.2 and $^{87}Sr/^{86}Sr$ at 3.90 AE as 0.7065, we obtain the felsite-free point shown in Fig. 4. $^{87}Rb/^{86}Sr$ for this point is lower than the measured value for the matrix by a factor of more than 3.0. Such a large effect makes it reasonable to ascribe the variable location of other Boulder 1 samples, 72255,59 dark and light-gray matrix fractions and 72275,52, on the right-hand side of the 4.5 AE reference line to the known presence of felsite in these samples. [Moreover 72275 contains small clasts of another Rb-rich material, the pigeonite basalt, which has a crystallization age of 4.03 AE (Compston *et al.*, 1975)]. Similarly we suggest that other aphanites having Rb content greater than about 4 ppm (46,10; 157; 38,49; 38,32; 36,3 II in Fig. 3) owe their higher Rb to the presence of small variable amounts of felsite, and if this is so, the positions of the same samples in the isochron diagram Fig. 4 will be displaced along felsite mixing lines towards the felsite end member on the right-hand side. It is not possible to correct the data of these samples quantitatively to a felsite-free composition, as their felsite contents are too small for accurate visual estimation. The sensitivity of their Rb/Sr to felsite addition may be illustrated for sample 157: for a felsite content of 1%, the felsite-free Rb/Sr for this sample would just exceed half the observed Rb/Sr.

Anorthosite and ANT-suite plagioclase addition

Figure 3 shows that anorthosite clasts from within Boulder 1 aphanites (Compston *et al.*, 1975) and plagioclase from 73215 (Table 1) have similar Sr

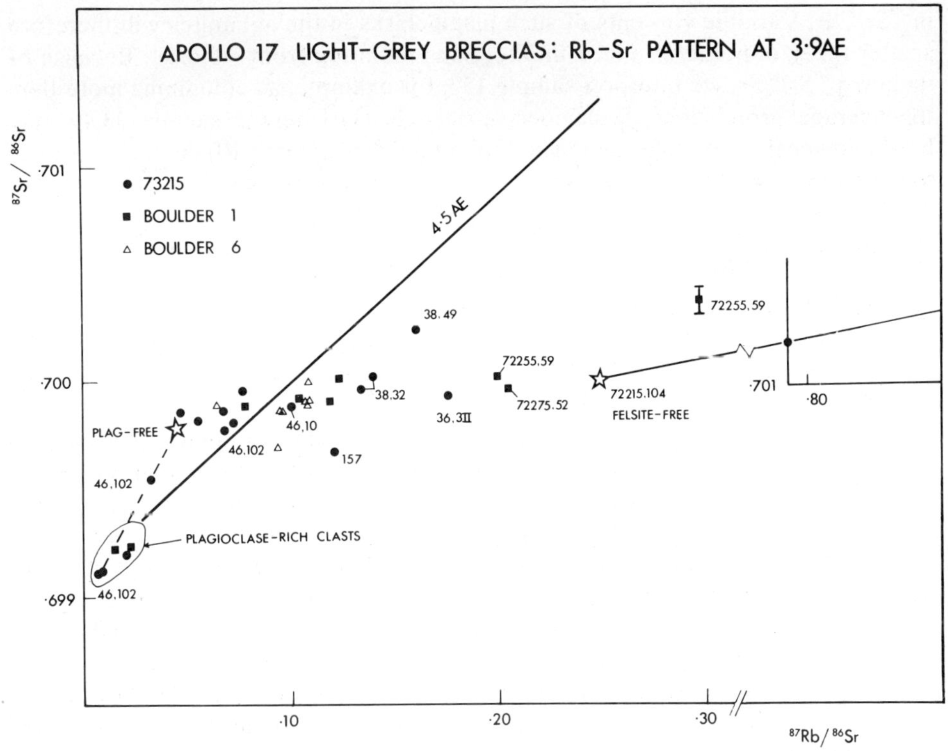

Fig. 5. Sr evolution diagram for 73215 aphanite clasts and aphanitic matrix samples after removal of radiogenic ^{87}Sr produced since 3.90 AE. Sample 47,102 is granulated feldspathic material containing small aphanite clasts. The diagram emphasizes differences in $^{87}Sr/^{86}Sr$ between samples immediately after aggregation of the breccia at 3.90 AE, demonstrating the lack of Sr isotope equilibration on a 1 cm to 1 mm scale. Aphanites with high Rb contents are not shown. The effect of felsite addition is illustrated by subtracting an estimated $11\frac{1}{2}\%$ of felsite from 73215,104 to give the felsite-free point. Other aphanites are affected by mixing with felsite to varying degrees.

contents to the 73215 aphanites but much lower Rb. Variable admixture of such clasts in the bulk aphanite samples therefore generates a range in their Rb/Sr, and mixing lines that have no direct time significance are produced also on the Sr evolution diagram, Fig. 5. The bulk sample of granulated feldspathic material 46,102 illustrates the process. In both Figs. 3 and 5, mixing lines are shown connecting the bulk sample with a separated plagioclase component, and notional locations of the mean plagioclase-free component(s) within the bulk sample are indicated along the extended mixing lines. The slope of the mixing line in Fig. 5, equivalent to 5.1 ± 0.3 AE ($2\,\sigma$), is impossibly old, and shows that 46,102 is an unequilibrated mixture of uncogenetic components. All other bulk aphanites will be located on other mixing lines which in general are different in slope but which have the same (or a similar) low Rb end member close to 0.6991 in $^{87}Sr/^{86}Sr$. Variable contents of such plagioclases in the aphanites will therefore scatter the data points in Fig. 5 in directions radiating from ~0.6991. Because of its lower $^{87}Sr/^{86}Sr$, we interpret sample 157, for example, as containing more than the average proportion of plagioclase-rich clasts, whereas sample 38,49 may have a lower than average proportion. (On the basis of their Rb contents, both of these samples also have detectably higher amounts of felsite.)

An "excess" model age component

The values of the Rb–Sr model age for the samples, calculated for an assumed initial $^{87}Sr/^{86}Sr$ of 0.6991, succinctly express their positions on the isochron diagram relative to the 4.50 AE reference line. The latter is a locus of particular single-stage Rb–Sr systems. Multi-stage systems may lie to the left, on, and to the right of the reference line, corresponding to model ages that are greater, equal, and less than 4.50 AE. Many lunar mineral phases that are low in Rb/Sr have model ages greater than 4.50 AE. Whole rock samples of low Rb/Sr sometimes have model ages greater than 4.50 AE, and those of high Rb/Sr almost never have model ages greater than 4.50 AE. For brevity, we will describe a model age that is greater than 4.50 AE as an "excess" model age, but it must be understood that in principle there is nothing strange about such values.

Figure 5 shows that many of the matrix aphanites and aphanite clasts lie on the left-hand side of the 4.5 AE reference isochron, in the field of "excess" single-stage Rb–Sr model ages. So too does the gray fraction of 72215,104 from Boulder 1, and if the felsite component could be accurately subtracted, it is likely that other Boulder 1 and 73215 aphanites would be found to lie in this field. The felsite-free composition of the dark gray fraction of 72215,104 does not do so (Fig. 5), as calculated on the basis previously discussed. If the actual felsite content of the sample amounted to 14½% rather than the 11½% estimated, the felsite-free point would lie on the 4.5 AE reference isochron. It is possible that the uncertainties in the calculation would allow the higher felsite content.

All of the aphanites shown in Fig. 5 contain a large proportion of ANT-suite clasts, some up to 50%, which must be subtracted from the composition of the bulk aphanites in order to specify the Rb–Sr coordinates of the excess-age component. This cannot be done quantitatively at the present time. However, it is plain that the excess-age component must lie *above* the measured point for each aphanite, along the extension of a mixing line drawn from the particular aphanite point and ~0.6991. The *minimum* value for $^{87}Sr/^{86}Sr$ in the excess-age component at 3.90 AE will be about 0.6998, which corresponds to the highest observed $^{87}Sr/^{86}Sr$ among the aphanites that have a definite excess model age (sample 46,45). If a 50% reduction in Rb/Sr is assumed for sample 38,49 to allow for its felsite content (based on reduction of its Rb content from 8.0 to 4.0 ppm), then this sample would assume an excess model age and the minimum $^{87}Sr/^{86}Sr$ for the excess-age component would become ~0.7001. A more realistic estimate comes from allowing for the known dilution by ANT-suite clasts. If the latter contribute 50% of the total Sr, then $^{87}Sr/^{86}Sr$ for the groundmass melt and excess-age component combined will be approximately .7008. This value remains a minimum estimate for the excess-age component because the groundmass melt itself probably contains a large component of (melted) ANT-suite material.

Discussion

Thermal redistribution of alkalis in ejecta sheets

A possible explanation of the "excess"-age component shown in Fig. 5 is the volatile loss of Rb at high temperature. Such loss might have occurred at 3.90 AE when the groundmass melt was formed. It might also have occurred earlier than 3.90 AE, if the source materials of the groundmass melt and/or the microclasts within it were subjected to high temperature events before they were aggregated into the particular breccia 73215. The effect of Rb loss at 3.90 AE is a horizontal displacement of the points in Fig. 5 to the left from original positions on or slightly to the right of the 4.5 AE reference line.

Gibson and Hubbard (1973) have demonstrated that appreciable amounts of both Rb and K can be rapidly removed from lunar rocks by vacuum evaporation at around 1000°C, whereas involatile elements such as Sr are not depleted. Nyquist *et al.* (1972, 1976) have suggested that lithic fragments from Apollo 14 soils and various fractions of Apollo 16 soils might owe their "excess" Rb–Sr model ages to Rb volatilization during impact events, but the low Rb/Sr of the particular samples makes the interpretation inconclusive because both igneous differentiation and Sr isotope exchange during metamorphism will also produce high model ages in such material. However Nyquist *et al.* (1976, 1977) have discovered model Rb–Sr ages of up to 5.4 AE in highly radiogenic clasts of quartz monzodiorite and adjacent matrix in breccia 15405, for which the volatile Rb loss explanation is much more attractive. The evidence is still not compelling, as in principle, the 15405 samples could have exchanged Sr isotopes with even more highly radiogenic Sr that might be nearby. For example, one mineral fraction from area #40, 12013,10 (Lunatic Asylum, 1970) is also highly radiogenic but its model age of 4.89 AE is best attributed to internal Sr isotope homogenization. Nevertheless, the 15405 data show that the possibility of volatile Rb loss as an important process in lunar impact events should be seriously considered.

Volatile loss of Rb from the 73215 samples would provide a third possible reason for the distribution pattern of Rb and Sr: samples with comparatively high Rb, such as gray spheroid 38,32 could be displaced downward parallel to the Rb axis in Fig. 3 to lower Rb, such as gray spheroid 38,57. This possibility cannot be distinguished in Fig. 3 from the effects of anorthosite addition which is in the same direction, and of felsite addition which is in the opposite. A problem is to identify an unaltered "starting" Rb content prior to the possible volatile loss.

The K contents of the above two gray spheroids cannot be used to support either alkali volatilization or felsite addition. The data of Gibson and Hubbard (1973) show that a decrease of ~900 ppm K would accompany a change from 7.0 ppm Rb (sample 38,32) to 2.9 ppm Rb (38,57) produced by alkali volatilization, but Blanchard *et al.* (1977) report a difference in K of only 250 ppm and Jessberger *et al.* (1977) find *no* significant difference in K between these two samples. Similarly, the observed K content of 38,32 is too low for felsite addition to be the sole explanation of its higher Rb content when compared with 38,57. Interelement comparisons such as the above are hampered by the strong heterogeneity within single clasts (e.g., 46,10 as discussed by Morgan *et al.*, 1976) and within single samples of matrix. The latter is particularly evident for 36,3 (Table 2) which was deliberately handpicked into two slightly different colour fractions, but in addition, the A and B portions of most samples in Table 2 differ by more than experimental precision.

Nyquist *et al.* (1974) attributed a 4.02 ± 0.10 AE Rb–Sr alignment observed in Apollo 17 noritic breccias to Rb volatilization during and immediately after the Serenitatis event. Their principal control for this alignment comes from sample 72275,2, a light-gray breccia from Boulder 1, for which Rb/Sr is about double the next highest value in the group. Figure 3 shows that the distribution of Rb in this sample and in other breccia-matrix samples analyzed by Nyquist *et al.* (1974) closely resembles the 73215 and Boulder 1 pattern. Rb is comparatively higher in 72275,2 and the sample is known to contain microclasts of felsite and of pigeonite basalt. We believe that the location of 72275,2 in particular on the 4.02 AE alignment should be assigned to felsite admixture rather than the addition of Rb vaporized from the low Rb breccias or to a variable Rb loss from all samples. We do not exclude the possibility that some thermal redistribution of Rb between the noritic breccias did occur, but we deny that the age of any such redistribution can be as well determined as 4.02 ± 0.1 AE using the present total-rock data.

In more general terms, it is puzzling in view of the apparent lunar evidence for Rb volatilization that Compston and Chapman (1969) could not detect this process in the Rb–Sr data for Australasian tektites.

The identity of the "excess"-age component

We do not yet know the identity of the excess age component in terms of a particular mineral or lithic phase. The fact that no unusual clasts have been observed in the samples concerned indicates that the component might be the groundmass melt itself. The simplest possibility would be the depletion of Rb relative to Sr in the groundmass melt by volatile loss at 3.90 AE, when the melt was formed. The question of volatile loss of alkalis has been discussed already and we cannot deny that this process may have operated for the low Rb samples.

A second possibility would be to view the groundmass melt as principally formed from a particular igneous suite in the target area that was enriched in a low Rb/Sr mineral such as plagioclase by normal crystal fractionation. The ANT suite and other clasts with the groundmass would be accidental xenoliths on this hypothesis, as their model ages are "normal". The fact that the chemistry of the matrix aphanites is so uniform (Blanchard *et al.*, 1976), independently of the excess model ages, shows that variation in the proportion of modal plagioclase is quite low and therefore insufficient to explain the variation in Rb/Sr.

A third candidate for the excess age component is plagioclase from the felsite, which might have an initial $^{87}Sr/^{86}Sr$ as high as 0.7065. Only very little of such a component would be needed to displace the aphanites into the excess model-age field, which would explain its apparent visual absence, and we would view the groundmass melt as almost entirely produced from ANT-suite materials that range in age from 4.2 AE to a maximum of 4.6 AE. Weaknesses of this proposal are (1) that the initial $^{87}Sr/^{86}Sr$ of plagioclase from the felsite has not yet been directly measured, and (2) that the mixing of felsitic material into the aphanites appears to be as lithic fragments rather than as separated mineral fragments.

Sr isotope equilibration during breccia aggregation

The internal scatter in $^{87}Sr/^{86}Sr$ at 3.90 AE between various 73215 aphanites (Fig. 5) shows that Sr isotope equilibration was not achieved over the ~1 cm scale of sampling during the impact event or during cooling in the ejecta sheet. Similarly, the differences in $^{87}Sr/^{86}Sr$ preserved between the 46,10 plagioclase fragment and 46,10 aphanite, and between the plagioclase and aphanite clasts and bulk aphanite for 46,102 (Fig. 5), show that isotope equilibration was not complete on a 1 mm scale. These observations support our previous conclusions for the light-gray breccias of Boulder 1, as well as the conclusions of Papanastassiou and Wasserburg (1975) for breccia 72435 and of Mark *et al.* (1975) for Apollo 14 breccias, that effective Sr isotope equilibration need not occur in ejecta sheets. On the other hand, the reaction rims and zoning of clasts described by James (1976) indicate that very local (~10 μm) Sr isotope exchange probably occurs in 73215.

The lack of Sr isotope equilibration within 73215 demonstrates that no single isochron or single mixing line can be defined for this rock with any precision. In this respect, the comparatively unannealed breccia 73215 differs from the more recrystallized 12013, in which a metamorphic event at 4.0 AE is clearly registered by mineral separates from within small volumes that have internally re-equilibrated their Sr isotopes.

The age of breccia aggregation

The intrusion of brown glass in the felsite preceded incorporation of the felsite as a clast within breccia 73215. However on the model that the glass was formed by the particular major impact that produced the light-gray and other Apollo 17 breccias (James and Hammarstrom, 1977), the age of the glass and that of the impact are effectively identical. Thus, the felsite isochron at 3.90 ±

0.05 AE measures the age of aggregation for 73215 and most probably for Boulder 1 at Station 2 also. If these breccias in turn are the single-stage products of the Southern Serenitatis impact, then the latter is dated at 3.90 ± 0.05 AE. This conclusion strongly reinforces the idea of a late heavy bombardment of the moon, as when combined with the mare basalt ages, it constrains the Serenitatis and Imbrium basin-forming impacts to a very short time period around 3.90 AE. It is possible that such a young age for Serenitatis is seriously in conflict with the ages of lunar surfaces given by the crater-count method (G. Neukum, pers. comm.). If so, the interpretation of breccia 73215 and Boulder 1 as undisturbed samples of the Serenitatis ejecta sheet should be critically reexamined. However it is quite conclusive that the age of aggregation of 73215 itself is 3.90 AE.

Breccia aggregation and Ar degassing

Independent ages for the felsite have been measured by Jessberger *et al.* (1977) by the ^{40}Ar–^{39}Ar plateau method and by Müller *et al.* (1977) by laser probe total ^{40}Ar–^{39}Ar dating. Their results confirm the Rb–Sr age, but substantially modify previous interpretations of ^{40}Ar–^{39}Ar results on lunar breccias. The essential conclusion is that total degassing of Ar in the light-gray breccias, even from the groundmass melt, did not occur during the impact event and the subsequent cooling in the ejecta blanket. For samples of aphanitic matrix and aphanite clasts, the apparent plateau age is affected by the amount and nature of microclasts, e.g., those in which most of the K is contributed by the 3.9 AE felsite show the youngest ages at ~4.0 AE whereas others show ages up to 4.2 AE. This is illustrated by Fig. 6.

Of the other dated Apollo 17 breccias, Müller *et al.* (1977) observe that the green-gray breccias which have the fewest clasts and were the most slowly cooled, also show the youngest ^{40}Ar–^{39}Ar ages (3.90–3.95 AE). This is consistent, as Müller *et al.* suggest, with the idea that the green-gray breccias are the most degassed of the Apollo 17 breccias, and that ~3.90 AE is the age of the degassing event. We suggested earlier (Compston *et al.*, 1975) that a group of isotopic ages around 3.90 AE from the Apollo 17, Station 7 Boulder might reflect the Serenitatis event, rather than the Ar age frequency peak at 3.98 AE emphasized by Turner and Cadogan (1975). It now appears that the 3.98 AE peak might be an artifact of sampling, and it certainly includes many breccias that we would not now regard as completely degassed.

If Rb volatilization is responsible for the excess Rb–Sr model ages, and if it occurred during breccia aggregation at 3.9 AE, it is likely that those samples which lost the most alkalis were also the more completely degassed. Such a process would generate an inverse relationship between Rb–Sr model ages and Ar–Ar ages. The light-gray matrix 46,45 has the oldest model age (5.1 AE) and assuming that it may be validly compared with the Jessberger *et al.* sample 46,44, its position below the mixing line in Fig. 6 would in fact correspond with significant degassing. By contrast, the gray spheroid 38,57, of model age 4.8 AE, does not fit this simple situation. The low Rb of 38,57 is consistent with

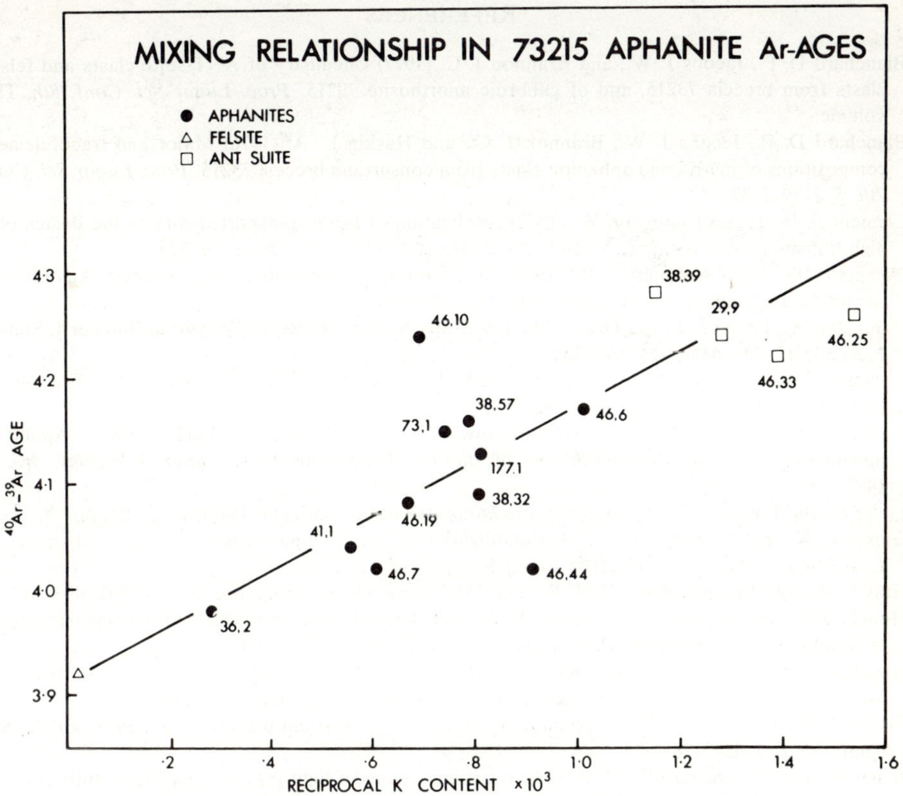

Fig. 6. The physical mixing of a high K, low age phase (felsite) with a low K, high age phase (ANT suite clasts) produces a linear correlation between age and the reciprocal of K content. Such a possible relationship is present in the Ar–Ar data of Jessberger *et al.* (1977). The effects of degassing of ~4.25 AE material at 3.90 AE would be superposed on this correlation.

substantial alkali loss compared with a second gray spheroid 38,32 but the Ar–Ar age of 38,57 is the greater of the two and somewhat above the proposed mixing line. The position of black clast 46,10 in Fig. 6 is also anomalous. If an appreciable amount of its K is contributed by 3.9 AE old felsite, it appears to have an older mixing component than the observed ANT-suite clasts. Overall we must view the Ar–Ar ages as complex products of partial degassing and the blending of multiple components of different ages.

Acknowledgments—We thank Dr. Odette B. James for invaluable aid throughout the course of the consortium project, R. Rudowski for some of the sample preparation and Odette B. James, D. A. Papanastassiou, and L. E. Nyquist for detailed reviews of the text.

REFERENCES

Blanchard D. P., Jacobs J. W., and Brannon J. C. (1977) Chemistry of ANT-suite clasts and felsite clasts from breccia 73215, and of gabbroic anorthosite 79215. *Proc. Lunar Sci. Conf. 8th.* This volume.

Blanchard D. P., Jacobs J. W., Brannon J. C., and Haskin L. A. (1976) Major and trace element compositions of matrix and aphanitic clasts from consortium breccia 73215. *Proc. Lunar Sci. Conf. 7th*, p. 2179–2187.

Clement S. W. J. and Compston W. (1972) Application of beam transport theory to the design of a high transmission mass spectrometer. *Int. J. Mass Spectrom. Ion Phys.* **10**, 323.

Compston W. and Chapman D. R. (1969) Sr isotope patterns within the Southeast Australasian strewn-field. *Geochim. Cosmochim. Acta* **33**, 1023–1036.

Compston W., Foster J. J., and Gray C. M. (1975) Rb–Sr ages of clasts from within Boulder I, Station 2, Apollo 17. *The Moon* **14**, 445–462.

Compston W., Vernon M. J., Berry H., and Rudowski R. (1971) The age of the Fra Mauro Formation. *Earth Planet. Sci. Lett.* **12**, 55–58.

Compston W., Vernon M. J., Berry H., Rudowski R., Gray C. M., and Ware N. (1972) Apollo 14 mineral ages and the thermal history of the Fra Formation. *Proc. Lunar Sci. Conf. 3rd*, p. 1487–1501.

Faure G. and Powell J. L. (1973) *Strontium Isotope Geology.* Springer, Heidelberg. 188 pp.

Gibson E. K. and Hubbard N. J. (1973) Volatilization studies on lunar soils and breccias (abstract). In *Lunar Science IV*, p. 286–288. The Lunar Science Institute, Houston.

Gray C. M. and Compston W. (1974) Excess ^{26}Mg in the Allende meteorite. *Nature* **251**, 495–497.

Huneke J. C., Jessberger E. K., Podosek F. A., and Wasserburg G. J. (1973) ^{40}Ar/^{39}Ar measurements in Apollo 16 and 17 samples. *Proc. Lunar Sci. Conf. 4th*, p. 1725–1756.

Husain L., Schaeffer O. A., Funkhouser J., and Sutter J. (1972) The ages of lunar material from Fra Mauro, Hadley Rille and Spur Crater. *Proc. Lunar Sci. Conf. 3rd*, p. 1557–1567.

James O. B. (1976) Petrology of aphanitic lithologies in consortium breccia 73215. *Proc. Lunar Sci. Conf. 7th*, p. 2145–2178.

James O. B. and Blanchard D. P. (1976) Consortium studies of light-gray breccia 73215: Introduction, subsample distribution data and summary of results. *Proc. Lunar Sci. Conf. 7th*, p. 2131–2143.

James O. B., Blanchard D. P., Jacobs J. W., Brannon J. C., Haskin L. A., Brecher A., Compston W., Marti K., Lugmair G. W., Gros J., Takahashi H., and Braddy D. (1976) Consortium studies of aphanitic lithologies and two anorthositic gabbro clasts in breccia 73215 (abstract). In *Lunar Science VII*, p. 423–425. The Lunar Science Institute, Houston.

James O. B., Brecher A., Blanchard D. P., Jacobs J. W., Brannon J. C., Korotev R. L., Haskin L. A., Higuchi H., Morgan J. W., Anders E., Silver L. T., Marti K., Braddy D., Hutcheon I. D., Kirsten T., Kerridge J. F., Kaplan I. R., Pillinger C. T., and Gardiner L. R. (1975) Consortium studies of matrix of light gray breccia 73215. *Proc. Lunar Sci. Conf. 6th*, p. 547–577.

James O. B. and Hammarstrom J. G. (1977) Petrology of four clasts from breccia 73215. *Proc. Lunar Sci. Conf. 8th.* This volume.

Jessberger E., Kirsten T., and Staudacher Th. (1976) Argon–argon ages of consortium breccia 73215. *Proc. Lunar Sci. Conf. 7th*, p. 2201–2215.

Jessberger E., Kirsten T., and Staudacher Th. (1977) One rock and many ages—further K–Ar data on Consortium breccia 73215. *Proc. Lunar Sci. Conf. 8th.* This volume.

Kirsten T., Horn, P., Heymann D., Hübner W., and Storzer D. (1973) Apollo 17 crystalline rocks and soils: Rare gases, ion tracks and ages (abstract). *EOS* (*Trans. Amer. Geophys. Union*) **54**, 595–596.

Lunatic Asylum (1970) Mineralogic and isotopic investigations on lunar rock 12013. *Earth Planet. Sci. Lett.* **9**, 137–163.

Mark R. K., Lee-Hu C., and Wetherill G. W. (1975) More on Rb–Sr in lunar breccia 14321. *Proc. Lunar Sci. Conf. 6th*, p. 1501–1507.

Morgan J. W., Gros J., Takahashi H., and Hertogen J. (1976) 73215: Siderophile and volatile trace elements. *Proc. Lunar Sci. Conf. 7th*, p. 2189–2199.

Müller H. W., Plieninger T., James O. B., and Schaeffer O. A. (1977) Laser probe ^{39}Ar–^{40}Ar dating of materials from consortium breccia 73215. *Proc. Lunar Sci. Conf. 8th.* This volume.

Murthy V. R., Evensen N. M., Jahn B. M., and Coscio M. R. (1972) Apollo 14 and 15 samples: Rb–Sr ages, trace elements and lunar evolution. *Proc. Lunar Sci. Conf. 3rd*, p. 1503–1514.

Nyquist L. E., Hubbard N. J., Gast P. W., Church S. E., Bansal B. M., and Wiesmann H. (1972) Rb–Sr systematics for chemically defined Apollo 14 breccias. *Proc. Lunar Sci. Conf. 3rd*, p. 1515–1530.

Nyquist L. E., Bansal B. M., Wiesmann H., and Jahn B. M. (1974) Taurus-Littrow chronology: Some constraints on early crustal development. *Proc. Lunar Sci. Conf. 5th*, p. 1515–1539.

Nyquist L. E., Bansal B. M., and Wiesmann H. (1976) Rb–Sr systematics of agglutinate fractions of Apollo 16 soils (abstract). In *Lunar Science VII*, p. 639–641. The Lunar Science Institute, Houston.

Nyquist L. E., Wiesmann H., Shih C.-Y., and Bansal B. M. (1977) REE and Rb–Sr analysis of 15405 quartz monzonite (abstract). In *Lunar Science VIII*, p. 738–740. The Lunar Science Institute, Houston.

Papanastassiou D. A. and Wasserburg G. J. (1975a) Rb–Sr study of a lunar dunite and evidence for early lunar differentiates. *Proc. Lunar Sci. Conf. 6th*, p. 1467–1489.

Papanastassiou D. A. and Wasserburg G. J. (1975b) A Rb–Sr study of Apollo 17 Boulder 3: Dunite clast, microclasts and matrix (abstract). In *Lunar Science VI*, p. 631–633. The Lunar Science Institute, Houston.

Papanastassiou D. A. and Wasserburg G. J. (1976a) Rb–Sr age of troctolite 76535. *Proc. Lunar Sci. Conf. 7th*, p. 2035–2054.

Papanastassiou D. A. and Wasserburg G. J. (1976b) Early lunar differentiates and lunar initial $^{87}Sr/^{86}Sr$ (abstract). In *Lunar Science VII*, p. 665–667. The Lunar Science Institute, Houston.

Ryder G., Stoeser D. B., Marvin U. B., and Bower J. F. (1975) Lunar granites with unique ternary feldspars. *Proc. Lunar Sci. Conf. 6th*, p. 435–449.

Schmitt H. H. (1973) Apollo 17 Report on the Valley of Taurus-Littrow. *Science* **182**, 681–690.

Stoeser D. B., Marvin U. B., and Bower J. F. (1974) Petrology and petrogenesis of Boulder 1. In *Interdisciplinary Studies of Samples from Boulder 1, Station 2, Apollo 17*, Vol. 2 (J. A. Wood, ed.). LSI Contr. No. 211D, p. III1–III51. Smithsonian Astrophysical Observatory, Cambridge, Mass.

Tatsumoto M., Hedge C. E., Doe B. R., and Unruh D. M. (1972) U–Th–Pb and Rb–Sr measurements on some Apollo 14 lunar samples. *Proc. Lunar Sci. Conf. 3rd*, p. 1531–1555.

Turner G. and Cadogan P. H. (1975) History of lunar bombardment inferred from ^{40}Ar–^{39}Ar dating of highland rocks. *Proc. Lunar Sci. Conf. 6th*, p. 1509–1538.

Wolfe E. W. and Reed V. S. (1974) Geological setting of the South Massif, Apollo 17 site. In *Interdiscriplinary Studies of Samples from Boulder 1, Station 2, Apollo 17*, Vol. 2 (J. A. Wood, ed.). LSI Contr. No. 211D, p. I1–I19. Smithsonian Astrophysical Observatory, Cambridge, Mass.

York D., Kenyon W. J., and Doyle R. J. (1972) ^{40}Ar–^{39}Ar ages of Apollo 14 and 15 samples. *Proc. Lunar Sci. Conf. 3rd*, p. 1613–1622.

Appendix 1: Experimental Methods

Samples are inspected, crushed, and mineral grains handpicked in a class 100 clean-air work station using stainless-steel tools. Aliquots from 0.5 to 30 mg in weight are dissolved chemically in 10 ml teflon beakers using minimal HF (up to 0.5 ml), $HClO_4$ (.05 ml) and 6 M HCl (.1 ml), dried down, and between 20 and 120 mg of mixed ^{85}Rb, ^{84}Sr spike as appropriate (1.9497 ppm ^{85}Rb, 0.4650 ppm ^{84}Sr) weighed in. Approximately 0.5 ml of 6 M HCl is then added, then 5 ml of water, and the solution again dried down to ensure equilibration. The residue is taken up in 1 ml of 0.25 M oxalic acid–0.1 M HCl solution, centrifuged and loaded onto a 2 g cation exchange column using Dowex 50 W × 8 resin. The oxalic acid complexes Fe, Al, and Ti, which are eluted by adding a further 5 ml oxalic HCl solution.

Rb is eluted using 0.5 M HCl and Sr using 2.5 M HCl. The Sr fraction is always processed twice as the production of thermal ions in the mass spectrometer is enhanced by sample purity. With Mg-rich samples, the Rb fraction is commonly diluted by Mg which inhibits Rb ionization and is physically bulky. Such fractions are transferred to silica centrifuge tubes, converted to nitrate, heated until the

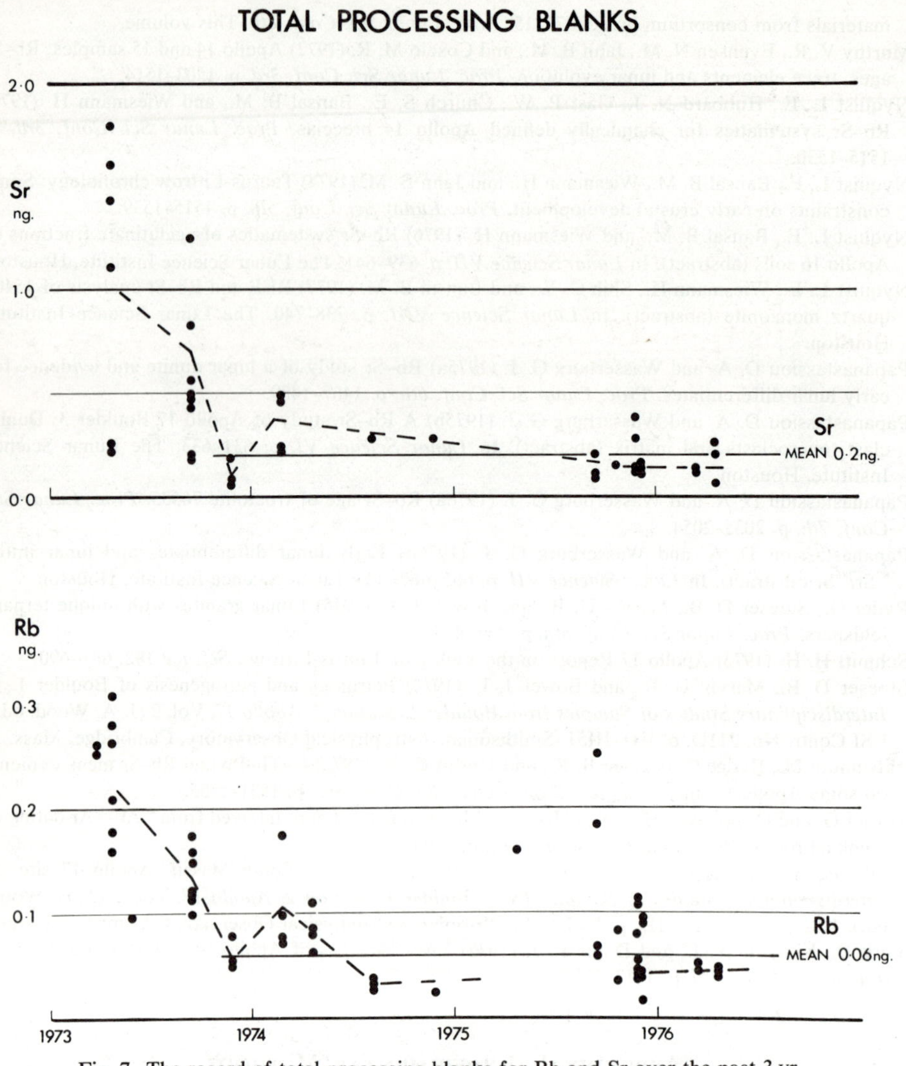

Fig. 7. The record of total processing blanks for Rb and Sr over the past 3 yr.

production of NO_2 at dull-red temperature signifies its decomposition to insoluble MgO, and leached with cold water to recover Rb. All evaporations are done in filtered N_2 in a clean-air hood. Both Rb and Sr are mounted as chlorides on Re side-filaments on a triple filament bead.

Figure 7 shows the total processing blanks observed since April 1973 using the above methods. The size and scatter of the blanks have been stable since late 1973, excepting four Rb values exceeding 0.15 ng that we choose to regard as unusual. The mean for 35 Rb blanks analyzed since late 1973 is 0.06 ± 0.03 ng (population standard error) and for 28 Sr blanks, 0.20 ± 0.11 ng. All lunar analyses are made in duplicate to avoid error in low Rb samples due to unusual blanks. For most samples, even the unusual blanks are negligible.

The mass spectrometer employed is a single-focusing 23 cm, 60° sector machine designed for high ion transmission (Clement and Compston, 1972), with magnetic field switching, Faraday cup collection, Cary model 36 electrometer, and Hewlett-Packard 5211B voltage-to-frequency converter, and modified counter for digital output. The digital system is on-line to a Hewlett-Packard 2116B computer (time-sharing operating system devised and written by Dr. P. A. Arriens) for real-time data reduction and analysis. Linearity of the ion-current measurement system, including the 5×10^{10} ohm input resistor has been checked using a ramp-generator and is limited by the v-f converter. It is better than 0.01% for signals greater than 10% of full-scale, and a fixed bias is applied to the input to the v-f converter to ensure this condition.

Each isotopic ratio is measured as the mean of a set of ten or more successive A-B-A type ion-current comparisons, e.g., ^{88}Sr–^{86}Sr–^{88}Sr etc., delaying one second for the response of the magnetic field and electrometer and integrating the ion-current for the following second. Sets of $^{84}Sr/^{86}Sr$ and $^{87}Sr/^{86}Sr$ ratios are alternated between $^{88}Sr/^{86}Sr$, the latter serving to monitor any variable mass discrimination. ^{85}Rb is measured before and after each set of $^{87}Sr/^{86}Sr$ and ^{87}Rb subtracted from the total mass 87 beam by time-interpolation using an assigned $^{87}Rb/^{85}Rb$. Observations of the zero point for the measurements are taken at mass 81.5 before each set, and allowance is made in data reduction for electrometer memory which adds 0.0042% of the preceding beam for the conditions described. Spectral cross-contamination is checked before and during the measurements and is usually $\leq$0.005% of mass 87. The within-set coefficient of variation for ratio measurements is 0.035% or less when the ^{87}Sr beam is 4.0×10^{-12} A or greater.

The measured isotopic ratios all change slowly between sets towards heavier values due to sample exhaustion. In addition they may be changed in either direction by adjustment of the electrostatic lenses in the ion-source. Comprehensive tests have shown recently that both the above changes in the various measured ratios for Sr are adequately corrected by *linear normalization* (Faure and Powell, 1973) to the standard value 8.3752 for $^{88}Sr/^{86}Sr$, rather than a more complex relationship such as observed for Mg (Gray and Compston, 1974). Most analyses show corrections of between 0.3 and 0.2% per mass unit.

Using the equipment and methods described above, the mean normalised $^{87}Sr/^{86}Sr$ for 29 different runs of unspiked NBS 987 reference strontium by different analysts between early 1973 and May 1976 is 0.71029, with the standard error for a single observation being 0.00006. For 11 runs since February 1975 of NBS 987 spiked with a mixed ^{85}Rb, ^{84}Sr tracer to simulate the lunar analyses, the mean normalised $^{87}Sr/^{86}Sr$ is 0.71031 and population standard error 0.00003. In addition, Table 3 shows analyses of a powdered fragment of the meteorite Angra dos Reis. The samples labelled ADR

Table 3. Isotopic analyses for the Angra dos Reis meteorite. Conditions for the ADR group were as described by Compston *et al.* (1971). Blanks for the ADRE group were lower at 0.06 ng Rb and 0.2 ng Sr, and the mass spectrometry was strictly as described in this paper.

	Rb ppm	Sr ppm	$^{87}Sr/^{86}Sr \pm$ s.e.	Initial $^{87}Sr/^{86}Sr$ at 4.60 AE
ADRE-1	.116	146.6	.69904 ± 5	
-3	.116	146.2	.69904 ± 5	
-4	.115	143.9	.69910 ± 5	
mean ADRE			.69906 ± 3	.69891
ADR-1	—	—	.69901 ± 10	
-2	—	—	.69896 ± 10	
-3	—	—	.69899 ± 7	
-4	—	—	.69898 ± 5	
-5	.106	144.4	.69904 ± 5	
-6	.175	142.2	.69908 ± 7	
-7	.909	144.7	.69908 ± 5	
-8	.118	144.9	—	
mean ADR			.69903 ± 2	.69888

were analyzed using a Cary model 31 electrometer for which the response-time corrections were greater, and an inferior v-f converter for which non-linearity corrections were applied following calibration using a ramp-generator. The results of the two different measurement systems are in adequate agreement, and we employ the mean initial ratio of 0.69890 to relate our $^{87}Sr/^{86}Sr$ scale to that of Papanastassiou and Wasserburg (1976) who report a mean value of 0.69881 for the same meteorite. The data for NBS 987 reported by Nyquist *et al.* (1974) show that there is no significant bias in $^{87}Sr/^{86}Sr$ between our measurements and those of Nyquist *et al.*

APPENDIX 2: ANALYSES OF FELDSPATHIC BASALT 68416

Technical biases in determining age are possible between different methods and different laboratories, and the question arises whether our 3.90 ± 0.05 AE result for the felsite is strictly comparable to the 3.92 ± 0.04 AE plateau age of Jessberger *et al.* (1977). Our internal isochron for the feldspathic basalt 68416 (Tables 4, 5, Fig. 8) at 3.79 ± 0.03 AE is in adequate agreement with

Table 4. Analytical data for feldspathic basalt 68416,36. Errors in $^{87}Sr/^{86}Sr$ and $^{87}Rb/^{86}Sr$ as for Tables 1 and 2, except that the plagioclase and mesostasis $^{87}Rb/^{86}Sr$ are subject to uncertainty in the blank correction (See Fig. 8).

	(mg) weight	ppm Rb	ppm total Sr	$^{87}Rb/^{86}Sr$	$^{87}Sr/^{86}Sr$
Plagioclase 1	3.8	.018	108.1	.00048	$.69947 \pm 5$
Plagioclase 2	3.6	.014	106.6	.00037	$.69943 \pm 5$
Plagioclase (fines)	3.1	.23	199.6	.00355	$.69957 \pm 5$
Total-rock 1	37.1	1.609	166.2	.02792	$.70082 \pm 5$
Total-rock 2	49.6	1.427	147.4	.02791	$.70086 \pm 5$
Mesostasis A1	0.9	8.03	74.2	.3214	$.71613 \pm 6$
Mesostasis A2	0.9	7.03	73.0	.2780	$.71444 \pm 8$
Mesostasis B1	1.2	9.96	148.4	.1938	$.70991 \pm 7$
Mesostasis B2	1.5	9.85	143.1	.1987	$.71021 \pm 5$

Table 5. Comparative age determinations by different laboratories on the feldspathic basalts 14310 and 68416.

	Ar–Ar age (AE)	Rb–Sr age (AE)
14310	$3.87 \pm .05$ (1)	$3.84 \pm .04$ (4)
	$3.79 \pm .03$ (2)	$3.93 \pm .06$ (5)
	$3.91 \pm .05$ (3)	$3.93 \pm .04$ (6)
		$3.87 \pm .04$ (7)
68416	$3.85 \pm .05$ (8)	$3.78 \pm .03$ (9)
		$3.84 \pm .02$ (10)

(1) Jessberger *et al.* (1973). (2) Husain *et al.* (1972). (3) York *et al.* (1972). (4) Tatsumoto *et al.* (1972). (5) Murthy *et al.* (1972). (6) Compston *et al.* (1972). (7) Papanastassiou and Wasserburg (1971). (8) Kirsten *et al.* (1973). (9) This paper. (10) Papanastassiou and Wasserburg (1975).

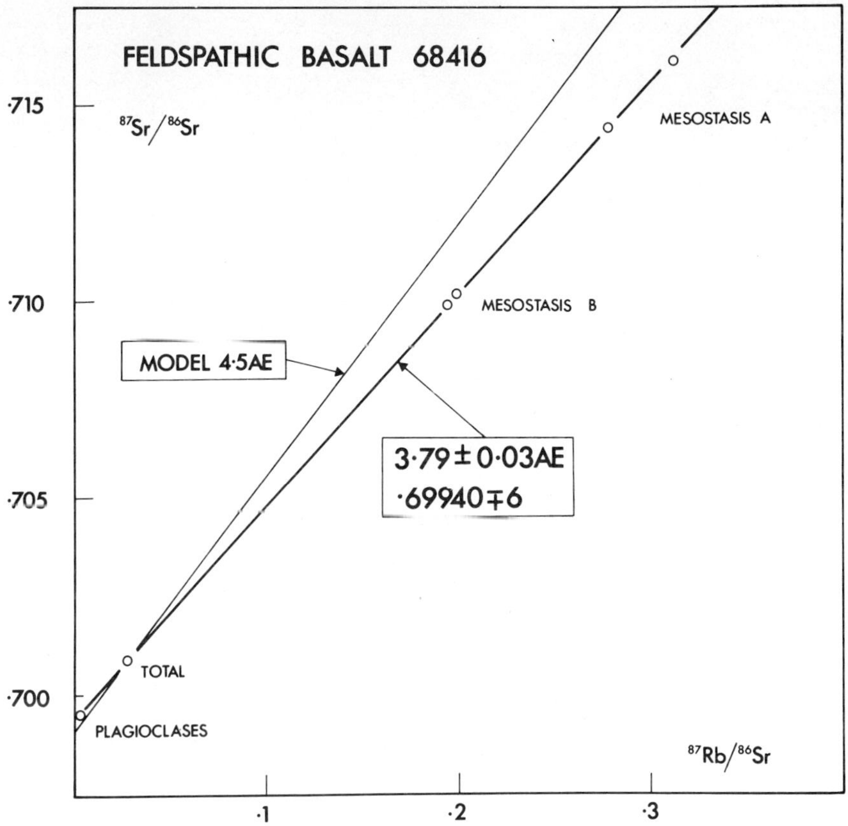

Fig. 8. Internal isochron for probable impact melt, 68416. The age is not sensitive to normal measurement errors in $^{87}Sr/^{86}Sr$ because of the comparatively wide range in the latter, but it is sensitive to the blank corrections. An additional ~0.1 ng Rb blank would increase the age by 1.8% to exact agreement with Kirsten *et al.* (1973) but for such an assumed increase, the amounts of Rb determined during the plagioclase measurements would be zero, which is unlikely. The best estimate for the age of 68416 (and 68415 from the same boulder) is probably the mean of Kirsten *et al.* (1973), Papanastassiou and Wasserburg (1976) and the above at 3.82 ± .02 AE.

3.85 ± 0.05 AE for the same rock by Kirsten *et al.* (1973) and with 3.84 ± 0.02 AE by Papanastassiou and Wasserburg (1975) for the combined data of samples 68415 and 68416 (fragments from the same rock). For the feldspathic basalt 14310, our result of 3.93 ± 0.04 AE compares with the mean value of 3.89 AE for three Ar–Ar and four Rb–Sr ages (Table 5), which show a population standard error of 0.06 AE (1.5%). These comparisons are based on the assumption that the lunar rocks concerned are ideal standard samples and it is acknowledged that some doubt exists for 68415, in this respect (Hunecke *et al.*, 1973). Unpublished internal isochrons for Apollo 17 mare basalts and new measurements on 15555 are in good agreement with various Ar–Ar and Rb–Sr ages from other laboratories, although such comparisons are less useful owing to the lower precision of the Rb–Sr results.

Our conclusion from the above is that biases between methods and between laboratories cannot be distinguished from random error and are probably less than 2%.

Proc. Lunar Sci. Conf. 8th (1977), p. 2551–2565.
Printed in the United States of America

Laser probe ^{39}Ar–^{40}Ar dating of materials from consortium breccia 73215

H. W. MÜLLER[1], T. PLIENINGER[2], O. B. JAMES[3], and O. A. SCHAEFFER[1]

[1]State University of New York at Stony Brook, Stony Brook, N.Y. 11794
[2]Max-Planck-Institut für Kernphysik, Heidelberg, Federal Republic of Germany
[3]U.S. Geological Survey, Reston, Virginia 22092

Abstract—The unique capacity of the laser probe ^{39}Ar–^{40}Ar method to determine ages of constituents in a rock sample which are only on the order of 100 μm apart has been applied to measure ages of the different constituents of consortium breccia 73215. Evidence is given that the age of the melt-derived groundmass, 4.01 ± 0.05 G.y., has no chronological significance. Instead, our best estimate for the age of the breccia-forming event, ~3.90 G.y., comes from the study of two clasts only a few hundred micrometers in size. The interpretation of this age as the age of the breccia-forming event is also supported by the studies of other consortium members and suggests a reinterpretation of the ages of most of the other Apollo 17 highland breccias.

INTRODUCTION

CONSORTIUM STUDIES of the light-gray breccia 73215 (James *et al.*, 1975; James and Blanchard, 1976; James, 1976) have demonstrated that this rock formed as a fragment-laden melt during a major lunar impact. The bulk of the breccia is a gray-to-black aphanitic rock termed matrix. This aphanite consists of small crystalline clasts set in a very fine-grained groundmass that crystallized from melt. The breccia-forming process was one of mechanical mixing of the clasts, which were relatively cold, with the melt, which may have been superheated. Mixing, thermal equilibration between clasts and melt, crystallization of the melt, and cooling of the bulk aggregate were probably all fairly rapid processes; there is no optically visible evidence of later reheating (James, 1976; James and Hammarstrom, 1977; Nord and James, 1977).

The breccia history outlined above suggested the possibility that some components of the rock might not have been completely outgassed of Ar in the breccia-forming and subsequent events. This supposition was confirmed by the results of ^{39}Ar–^{40}Ar stepwise heating experiments (Jessberger *et al.*, 1976, 1977). Clasts of anorthositic-noritic-troctolitic(ANT)-suite rocks extracted from the breccia showed double plateaus, with intermediate-temperature ages of 4.06–4.07 G.y. and high-temperature ages of 4.22–4.26 G.y., demonstrating that they were not completely outgassed during the breccia-forming event. Chips of matrix aphanite showed intermediate-temperature plateaus with ages ranging from 3.98 to 4.15 G.y. This spread in matrix ages was interpreted as the result of variations in degree of outgassing of the different components of the matrix, and variations in proportions of these components in the chips studied. Mixing of the matrix components is on a scale too fine to permit physical separation for stepwise heating experiments. For this reason the laser probe ^{39}Ar–^{40}Ar method (Megrue, 1973; Plieninger and Schaeffer, 1976), which has a spatial resolution as fine as a few tens of microns (Schaeffer *et al.*, 1977), has been used to attempt to determine the ages of the different components. In the laser probe study reported herein, it has been possible to date the groundmass and small clasts in the matrix separately, and to resolve age differences over distances on the order of 100 μm. A study has also been made of one of the ANT-suite clasts dated by stepwise heating in an attempt to elucidate its history.

Experimental Technique

In the laser probe method, the gas samples analyzed for Ar isotopes are obtained by melting of small volumes at rock surfaces with a laser beam. The beam is focussed through a petrographic microscope; thus location of the volume melted by each laser pulse can be very precisely controlled. Each gas sample analyzed consists of gas collected during about 50 laser pulses; the volumes melted by each pulse are very small, half spheres with diameters ranging from 10 to 100 μm. Thus it is possible to selectively melt individual phases in fine-grained rocks.

The 73215 chips studied range from 2 to 5 mm across. Each chip was potted in Lakeside and was ground flat and polished on two opposing surfaces; the Lakeside was then dissolved away with methanol. Plieninger and Schaeffer (1976) demonstrated that no detectable atmospheric Ar contamination is introduced by this polishing process. The chips were then irradiated, along with the NL-25 monitor (Schaeffer and Husain, 1973) in the core of the Brookhaven High Flux Beam Reactor. Total neutron dose was $\sim 1.6 \times 10^{19}$. A concise description of the neutron-irradiation procedure is given by Schaeffer and Schaeffer (1977).

For collection of the gas to be analyzed, the chips were mounted inside a small bakeable vacuum chamber with a silica glass port. The chamber was placed on the stage of a petrographic microscope equipped for vertical illumination. An additional light was placed adjacent to the microscope and was used to view the samples in oblique illumination; by this means it was possible to see a few tens of microns into the samples and avoid positioning the laser pulses in areas where the pits would penetrate into material different from that at the surface. The laser, a Biolaser (Control Data Corp.) with an output of $\sim 10^7$ W/cm^2, was mounted on the phototube of the microscope. The specimen chamber was connected to the gas purification section of a Nuclide (6-60SGA) static mass spectrometer. Most gas samples analyzed consisted of the gas released by 25–100 carefully positioned sequential laser pulses, each with a melted volume corresponding to a half sphere typically 30 μm in diameter. Hydrogen was gettered by a Ti–Zr getter at room temperature; other impurities were collected by charcoal maintained at the temperature of a dry ice-acetone mixture. After the gas was collected, it was admitted to the mass spectrometer and analyzed for Ar isotopes. Blank levels and other analytical details were given by Plieninger and Schaeffer (1976); procedures for calculation of the ages from the measured data are given by Schaeffer and Schaeffer (1977). (The J value for the irradiation of samples 177 and 29,9 was 0.09308 and for 131 it was 0.09422.)

The laser pulse produces a one-step, high-temperature heating that simultaneously outgasses sites varying in Ar retentivity. Thus if radiogenic Ar has been lost from less retentive sites by low-temperature diffusion, the ages determined from the laser-released gas samples may be appreciably lower than the actual ages. The stepwise heating data (Jessberger *et al.*, 1976, 1977) show that the 73215 matrix and clasts did not in general have much loss of Ar by low-temperature diffusion; the samples these authors studied had had in all but one case losses of radiogenic ^{40}Ar no greater than 5%, and in most cases Ar released at steps with temperatures above 550–650°C yielded plateau ages. Although these data suggest that prior gas loss should not greatly affect the 73215 ages measured by the laser method, we have employed a preheating technique in an attempt to further decrease the effects of gas loss of the measured ages. A few preliminary studies were made on two chips prior to preheating, but before most of the measurements reported in this study were made, all the chips were heated in vacuum at 550°C for $1\frac{1}{2}$–2 hr. The heating was done after irradiation, but prior to lasering; a temperature of 550°C was selected because the results of the stepwise heating studies indicated that at this temperature the sites that had had prior gas loss should be essentially outgassed. Two of the materials studied had characteristics suggesting that they might have been more susceptible to gas loss than the rest of the materials; therefore the chip containing these materials was repolished after the first analyses were made, repeat gas samples were collected, the chip was heated in vacuum at 750°C for 1 hr, and the two suspect materials were analyzed again.

In this work and in the work by Schaeffer *et al.* (1977) the technique of preheating the samples at different temperatures prior to lasering is applied for the first time. We feel it therefore appropriate to give some clarifying comments to make the underlying assumptions of this technique clear. In the normal ^{39}Ar–^{40}Ar stepwise heating technique, a differential gas release pattern is obtained; the basic premise of this technique is to deplete unretentive phases in the lower-temperature steps of the heating, and thus to single out the retentive phases which are thought to give the most chronologically meaningful ages. Exactly the same model assumption underlies our preheating technique. Whereas in the stepwise heating technique the gas released at a given temperature is analyzed, in our "preheating prior to lasering" technique the gas that is analyzed is instead the gas that remains in the sample after the heating. The preheating of the sample in different steps is done *after neutron irradiation* but prior to lasering; therefore, the preheating should not change the ^{39}Ar–^{40}Ar ratio in the more retentive sites, but should only outgas the less retentive sites of their remaining radiogenic ^{40}Ar and the ^{39}Ar produced by the irradiation. Thus the only difference between the stepwise heating technique and our "preheating prior to lasering" technique is that the stepwise heating technique gives a differential release pattern, and the "preheating prior to lasering" technique gives an integral pattern, in which each step integrates all the gas that would be released at temperatures higher than the preheating temperature. It has been convincingly shown by Schaeffer *et al.* (1977) that both the stepwise heating and our "preheating prior to lasering" techniques lead basically to the same results, and that the "preheating prior to lasering" technique is a valuable extension of the normal stepwise heating technique. Because only a tiny amount of material is needed for an age determination with the laser technique (Plieninger and Schaeffer, 1976), many repeat analyses can be made between sequential heating steps even on a single crystal. (For example, we may collect gas from a plagioclase crystal by lasering it, then repolish the crystal, then preheat it to a higher temperature, and then laser it again.)

Sample Description and Results

The materials studied were: Clasts and groundmass within two matrix chips (subsamples 177 and 131); and plagioclase grains within an ANT-suite clast (subsample 29,9). Subsample 177 is of schlieren-rich gray aphanitic matrix, and 131 is of gray aphanitic matrix (see James *et al.*, 1975; and James, 1976 for petrologic descriptions of these lithologies). Major-, minor-, and trace-element abundances and ^{39}Ar–^{40}Ar stepwise heating results are reported for materials of these lithologies by Blanchard *et al.* (1976) and Jessberger *et al.* (1976), respectively. These two papers include data on two other chips taken from the parent subsample 177. ANT-suite clast 29,9 is of anorthositic gabbro. Its petrology is described by James and Hammarstrom (1977), its composition is given by Blanchard *et al.* (1977), a Rb–Sr internal isochron is reported by Compston *et al.* (1977), and its ^{39}Ar–^{40}Ar age spectrum derived by stepwise heating was given by Jessberger *et al.* (1976). Locations of the 177 and 29,9 subsamples within the parent 73215 breccia were given by James and Blanchard (1976); subsample 131 was taken from the N_1 face of the slab adjacent to subsample 177. The specific

Table 1. Laser released Ar isotopes from various subsamples of breccia 73215 in 10^{-12} cm³ STP†.

Subsample	Lithology	^{40}Ar	$^{39}Ar^*$	$^{38}Ar^*$	^{37}Ar	$^{36}Ar^*$	100 × K/Ca	$^{40}Ar_K/^{39}Ar_K$	Age (G.y.)
177	Groundmass in[a] typical matrix	1541.7 ± 20.4	19.93 ± .29	10.55 ± .79	524.04 ± 14.78	19.23 ± 2.45	5.6 ± 2.4	76.99 ± 1.59	3.96 ± .04
177	Groundmass in[b] typical matrix	1645.1 ± 20.0	20.58 ± .99	8.05 ± 3.96	522.96 ± 45.32	13.61 ± 6.86	7.6 ± .8	79.71 ± 3.99	4.02 ± .07
131	Groundmass in[b] typical matrix	1271.4 ± 12.0	16.51 ± .19	9.02 ± 2.08	425.62 ± 4.93	16.67 ± 5.88	7.5 ± .2	76.62 ± 1.25	3.97 ± .03
131	Groundmass in[b] clast-poor matrix	3994.4 ± 50.0	50.09 ± .99	15.23 ± 3.17	894.48 ± 21.97	24.72 ± 6.86	10.9 ± .4	79.57 ± 1.88	4.04 ± .04
177	Large plagioclase[b] crystal fragment	475.49 ± 4.80	5.41 ± .32	5.65 ± 1.29	372.29 ± 46.11	6.63 ± 4.90	2.8 ± .3	87.54 ± 5.30	4.17 ± .09
177	Composite of[b] small plagioclase clasts	1190.68 ± 9.87	14.06 ± .34	12.62 ± 2.95	688.0 ± 112.04	24.54 ± 4.94	3.9 ± .6	83.99 ± 2.26	4.10 ± .05
177	Devitrified[b] maskelynite clast	1297.90 ± 21.36	15.08 ± .46	11.44 ± 2.68	698.38 ± 108.87	20.63 ± 6.59	4.2 ± .6	85.55 ± 3.05	4.13 ± .06
131	Granulated core of[b] anorthosite clast	2533.70 ± 53.69	35.44 ± .73	28.76 ± 3.17	2691.08 ± 8.41	37.29 ± 6.47	2.6 ± .07	71.18 ± 2.14	3.85 ± .05
131	Granulated core of[c] anorthosite clast	1381.50 ± 15.40	20.85 ± .58	18.91 ± 1.58	1406.6 ± 22.86	26.05 ± 2.75	2.9 ± .1	65.86 ± 2.02	3.72 ± .05
131	Granulated core of[d] anorthosite clast	1974.20 ± 8.9	26.54 ± .37	23.96 ± 1.44	2128.77 ± 32.12	26.96 ± 4.12	2.4 ± .07	74.14 ± 1.13	3.92 ± .03
131	Recrystallized rim[b] of anorthosite clast	1501.04 ± 12.90	19.07 ± .52	15.81 ± 2.61	1144.47 ± 8.54	26.29 ± 8.32	3.2 ± .1	78.21 ± 2.31	4.01 ± .05

131	Recrystallized(b) anorthosite clast	1347.87 ± 1[illegible].79	16.97 ± .63	13.22 ± 1.31	899.07 ± 8.37	21.09 ± 3.54	3.7 ± .1	78.99 ± 3.05	4.02 ± .06
131	Recrystallized(c) anorthosite clast	1033.40 ± 13.50	12.68 ± .31	9.64 ± 1.24	633.49 ± 31.82	17.55 ± 2.45	3.9 ± .2	81.01 ± 2.32	4.06 ± .05
131	Crystalline(b) felsite clast	785.46 ± 15.00	9.49 ± .59	3.07 ± 1.98	3.84 ± 3.84	9.02 ± 5.39	480. ± 480.	82.29 ± 5.49	4.09 ± .10
131	Felsic glass(b) clast	3118.22 ± 10.0	41.39 ± 1.00	5.03 ± 1.19	7.98 ± 2.76	14.73 ± 3.14	1008. ± 350.	75.16 ± 1.83	3.94 ± .04
131	Felsic glass(c) clast	22905.2 ± 178.8	327.59 ± 1.42	8.30 ± 0.79	98.52 ± 98.52	20.96 ± 2.16	647. ± 647.	69.89 ± .63	3.82 ± .03
131	Felsic glass(d) clast	13533.60 ± 83.30	185.04 ± .79	9.14 ± 1.49	59.11 ± 59.11	22.05 ± 1.75	609. ± 609.	73.09 ± .55	3.89 ± .03
29,9	Large plagioclase(a) xenocryst	3001.30 ± 100.0	32.86 ± 1.99	26.41 ± 4.95	1488.77 ± 14.78	51.42 ± 14.71	4.2 ± .2	90.73 ± 6.35	4.23 ± .10
29,9	Large plagioclase(b) xenocryst	941.97 ± 42.44	10.80 ± .42	8.91 ± 8.91	853.45 ± 124.69	3.82 ± 8.82	2.5 ± .4	87.02 ± 5.19	4.16 ± .09
29,9	Large plagioclase(b) xenocryst	1236.81 ± 65.50	12.49 ± 1.01	13.95 ± 5.06	974.47 ± 238.25	24.50 ± 15.29	2.5 ± .6	98.34 ± 9.61	4.37 ± .14
29,9	Large plagioclase(b) xenocryst	374.59 ± 4.50	3.81 ± .22	4.86 ± 1.49	285.42 ± 23.05	6.86 ± 2.94	2.6 ± .2	97.59 ± 5.81	4.36 ± .09
29,9	Composite of(b) intermediate-sized plagioclase xenocrysts	928.94 ± 3.20	9.65 ± .28	7.98 ± 3.97	503.48 ± 46.48	13.02 ± 9.80	3.7 ± .4	95.84 ± 3.01	4.33 ± .05
29,9	Composite of(b) small plagioclase grains	1387.52 ± 16.99	16.77 ± .36	15.47 ± 1.90	1127.96 ± 43.43	22.99 ± 4.64	2.9 ± .1	82.31 ± 2.10	4.07 ± .04

(a) = unheated. (b) = outgassed at 550°C. (c) = outgassed at 550°C and repolished. (d) = outgassed at 750°C.

*Corrected for n-induced contributions.

†The exact amount of material melted by the laser pulses cannot be determined, but the values in the table roughly correspond to units of 10^{-9} cm^3 STP/g, within a factor of about 5.

materials studied within each of the three chips are described below, and the ages derived from the Ar analyses are listed in Table 1.

Matrix subsamples 177 and 131

Groundmass in typical matrix. Typical matrix contains about 50% recognizable clasts (larger than 10 μm) and 50% melt-derived groundmass (James, 1976). The groundmass is a microsubophitic intergrowth of plagioclase and

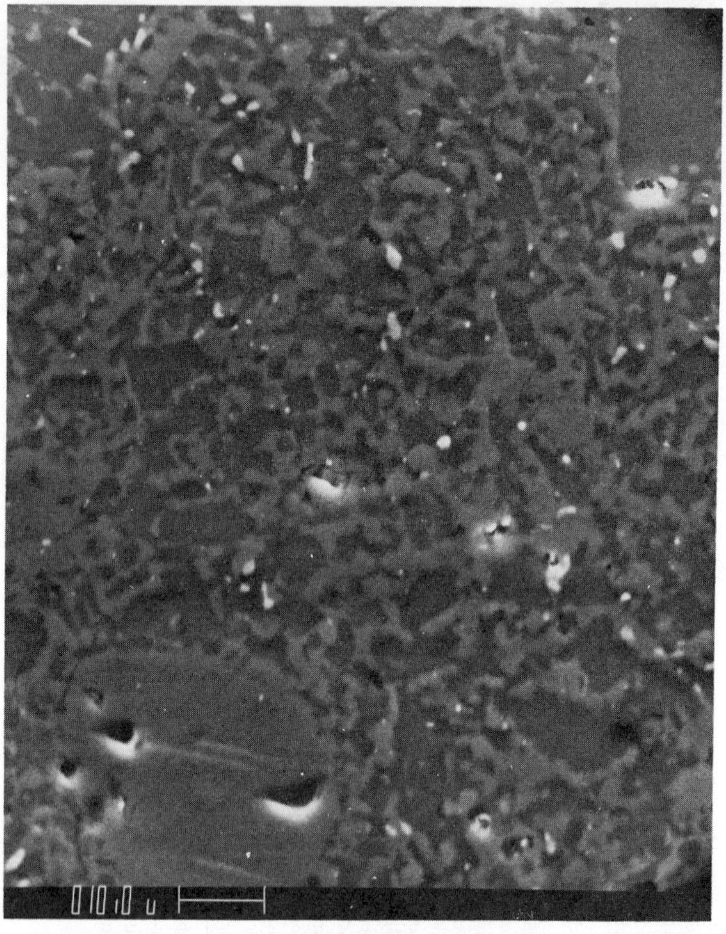

Fig. 1. Scanning electron microscope photograph of groundmass in typical matrix. The bar scale represents 10 μm. Plagioclase is dark gray, mafic minerals are light gray, and opaque minerals, metal and troilite grains, and dust specks are white. The large round pyroxene grain at the lower left and the angular plagioclase grain at the upper right are clasts. The remainder of the field is the groundmass intergrowth; 30 μm "spots" in areas identical to this one were laser melted for the analyses of groundmass reported in this study.

pyroxene; minor amounts of opaque oxides, Fe metal, and troilite are present (Fig. 1). Grain size of the intergrowth averages 5–10 μm, and some interstitial pore space is present. Bulk analyses indicate that phosphate minerals and K-rich phases (glass or feldspars) must also be present; as these are not visible optically, size of the grains must be <1 μm. Each gas sample analyzed was collected by laser melting of about 50 "spots" 25–30 μm across; these "spots" were free of recognizable clasts at their surfaces.

Groundmass in clast-poor matrix. This type of matrix forms irregular patches and schlieren within typical matrix; its clast content is <25%. Texture and grain size of the groundmass are the same as those in typical matrix, but porosity is essentially zero. The gas sample analyzed was collected in the same manner as were those from typical groundmass.

Large plagioclase crystal fragment. This plagioclase clast was 0.28 mm across; it was relatively transparent and showed traces of twin lamellae and cleavage, indicating it was a fragment of a single crystal. About 50 "spots" 30 μm across were melted by the laser; none of the "spots" were closer than 30 μm to the edge of the clast.

Composite of small plagioclase crystal fragments. The gas sample was collected from the centers of 29 fragments that ranged from 40 to 125 μm across. One to four 40 μm "spots" per fragment were melted, depending on clast size.

Clast of devitrified maskelynite. This clast was 0.24 mm across, and it showed the rounded outlines, cloudy appearance, and minute inclusions that typify devitrified maskelynite. The entire surface of the clast was covered with laser-melted 25–30 μm "spots".

Clasts of granulated, partly recrystallized anorthosite. Two clasts were studied. One was 0.9 mm across; it had a 0.4 mm porous core composed of angular plagioclase fragments, and 50–300 μm nonporous rim of recrystallized plagioclase (Fig. 2). A few percent minute grains of mafic minerals were present in both core and rim. The other clast was 0.4 mm across and consisted entirely of recrystallized plagioclase. Textures indicate recrystallization of these clasts occurred after they were incorporated in the breccia. From the core of the first clast three gas samples were collected and analyzed, in the following sequence: (1) after outgassing at 550°C, the entire polished surface of the core was covered with laser-melted "spots"; (2) after repolishing, half the exposed surface of the core was laser melted; and (3) after outgassing at 750°C, the remaining half of the core surface was laser melted. From the rim of the clast only one gas sample was collected; this was after the 550°C outgassing, and none of the laser-melted "spots" were any closer than 70 μm to the clast edge. From the second clast two gas samples were collected: The first was taken after the 550°C outgassing and the entire exposed surface of the clast was covered with laser-melted "spots"; the second was taken after repolishing and none of the "spots" were any closer than 30 μm to the surrounding groundmass.

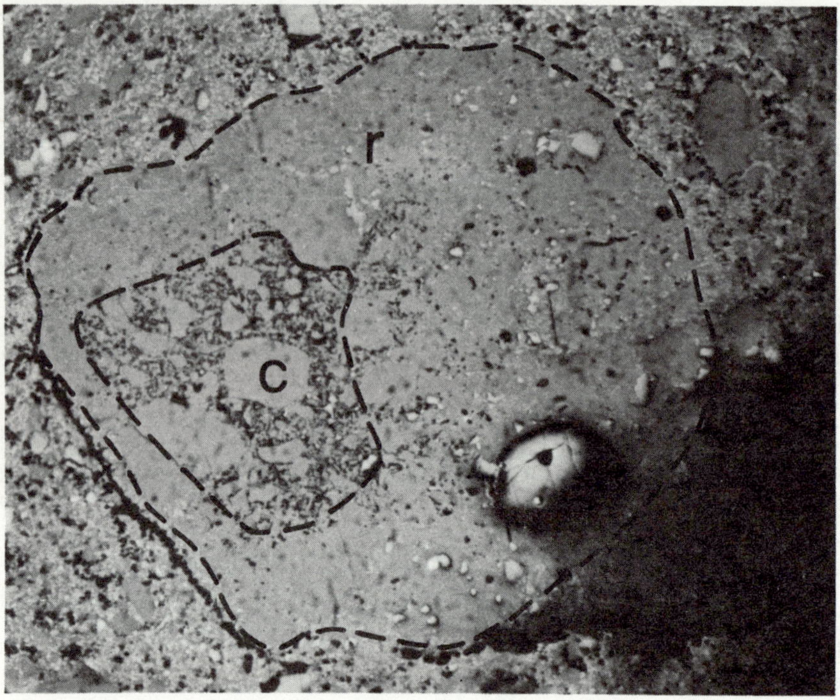

Fig. 2. Photomicrograph of clast of granulated, partly recrystallized anorthosite. Reflected light; width of the field of view is 1.65 mm. The clast has a recrystallized rim (r) and a porous unrecrystallized core (c).

Clast of crystalline felsite. This 0.15 mm clast consisted of a vermicular intergrowth of K-feldspar, quartz, and plagioclase; a single grain of a phosphate mineral was also present. Thin selvages of undevitrified glass marked grain boundaries and the clast-groundmass contact. The minerals were unusually transparent to the laser beam so that only a very small gas sample was obtained; about twenty five 10 μm spots, located within the plagioclase and K-feldspar grains and the glass selvages, were melted.

Clast of felsic glass. This 0.25 mm clast consisted almost entirely of barely devitrified felsic glass that contained sparse tiny euhedral pyroxene crystallites and globules of Fe metal and troilite. Minor plagioclase was also present, forming large angular grains concentrated along the clast margin. Textures of other such clasts in 73215 indicate that they were molten when incorporated in the breccia (James and Hammarstrom, 1977); thus they were probably totally outgassed of Ar at that time. Three gas samples were collected, in the following sequence: (1) after outgassing at 550°C, the entire exposed area of felsic glass was covered with laser-melted "spots"; (2) after repolishing, half the exposed area of glass was laser-melted; and (3) after outgassing at 750°C, the remaining exposed area of glass was melted.

Anorthositic gabbro clast 29,9

This clast consists of subhedral 0.1–1 mm plagioclase and ovoid 0.2 mm olivine grains enclosed by poikilitic pyroxene. Its texture indicates that the bulk of the rock crystallized from melt. Many large and intermediate-sized plagioclase grains contain cores outlined by "necklaces" of small globular inclusions of olivine; the petrologic data suggest that these cores represent xenocrysts incorporated in the melt (James and Hammarstrom, 1977).

Large plagioclase xenocrysts. Four large xenocrystic cores of plagioclase grains were studied. These were 0.3–0.5 mm across, and the rims that surrounded them averaged about 0.2 mm thick. Four gas samples were collected by laser melting the entire exposed surface of each core.

Composite of intermediate-sized plagioclase xenocrysts. Three smaller xenocrystic cores were also studied; these averaged about 0.2 mm across and had 0.3–0.1 mm rims. A composite gas sample was collected from all three cores; the entire exposed surface of each was laser melted.

Composite of small plagioclase grains. The centers of 48 subhedral plagioclase grains that crystallized from melt were laser melted; these grains ranged from 75 to 230 μm across.

DISCUSSION

For each radiometric dating method there are a number of physical and analytical factors that influence the accuracy and precision of the ages that may be determined. As the laser probe ^{39}Ar–^{40}Ar method is a fairly new technique, it has been necessary to investigate such factors to establish what limits they set on the use of the ages to derive precise chronological information. The factors that have been investigated are listed and discussed in the following paragraphs.

Knowledge of the spatial resolution of the technique is essential for evaluating the age data. The resolution is determined by the volume of rock that is outgassed during each laser pulse. Each pulse melts the rock within a half sphere and gas is freed from this melt. It is possible, however, that the material surrounding the half sphere might also be heated enough that gas could be freed from it. Schaeffer *et al.* (1977) investigated this question of the extent of gas release from material surrounding the laser pits in their study of Apollo 17 basalt samples. The pyroxene in these basalts contained inclusions of K-rich mesostasis several tens of microns across. Laser melting of pyroxene bordering the inclusions showed that no significant gas was released from them unless they were actually intersected by the laser pits. Thus the effective resolution of the technique is nearly identical to the dimensions of the laser pits that are produced, on the order of tens of microns. No systematic investigations were made, however, on the possible effects of mineralogy on resolution. Therefore, to insure that the gas samples collected from clasts were not affected by diffusion of gas from surrounding groundmass, these samples were taken wherever possible by laser melting parts of clasts more than one pit diameter from their edges.

An additional factor affecting resolution is the possibility that at depth in the sample a given laser pit might intersect material different from that at the surface. The use of oblique illumination in conjunction with optical illumination (as was done in the study reported herein) clearly shows the subsurface material, provided that the material at the surface has a grain size greater than about 20 μm, is not severely fractured, and is transparent to light. Thus the gas samples collected from plagioclase clasts can be confidently assumed to be entirely from the clasts, but there is more uncertainty in the source of the gas for the samples collected from groundmass.

An important limitation of the laser technique is the ^{39}Ar content of the gas sample collected, which is affected by both the transparency and the K content of the material being studied. If the material is transparent to the laser beam, it is not melted and no gas is released. The study of the felsite clast was severely hampered by this effect. If the material has a very low K content very little ^{39}Ar is freed. This effect was a problem in the study of the plagioclase clasts, most of which have K_2O contents less than 1000 ppm. Analysis of gas from clasts in which the ^{39}Ar levels were less than two times blank levels in general had unacceptably high errors, and the ages derived from them are not reported herein. The results of all other plagioclase analyses are reported.

The laser probe ^{39}Ar–^{40}Ar ages reported herein are total potassium argon ages, but the effects of low-temperature gas loss are assumed to have been essentially eliminated by outgassing of the samples at 550–750°C after the neutron irradiation and prior to lasering. Thus a direct comparison of the laser probe data with ^{39}Ar–^{40}Ar plateau ages from stepwise heating experiments can be made. Such a comparison is demonstrated to be basically valid by the more detailed outgassing prior to lasering experiments done on Apollo 17 basalt samples by Schaeffer *et al.* (1977).

The laser probe data for anorthositic gabbro 29,9 unfortunately do not yield a clear interpretation of the chronology of this clast. The ambiguities arise because of the relatively low precision of the ages, all determined on plagioclases, and because of the complex history of the rock. Textures of the rock (James and Hammarstrom, 1977) indicate four distinct events in its history that could be reflected in the age data: (1) formation of the parent rocks of the xenocrysts, (2) crystallization of the anorthositic gabbro melt, (3) late-stage fracturing and subsequent annealing, and (4) incorporation of the clast in the 73215 breccia. Stepwise heating experiments on a bulk chip yielded a double plateau: a high-temperature plateau at 4.24 G.y., and an intermediate-temperature plateau at 4.06 G.y. (Jessberger *et al.*, 1976). The high-temperature plateau was interpreted as crystallization age of the melt and the intermediate-temperature plateau was interpreted as the result of partial outgassing of some components of the clast when it was incorporated in the 73215 breccia. The laser probe data, however, do not confirm the presence of plagioclases with two distinct ages in this clast, but the errors in the measured ages are such that a bimodal distribution cannot be ruled out either; the laser probe ages on xenocrysts range from 4.16 to 4.37 G.y., with errors typically about ±0.1 G.y.; a composite of grains crystallized from the melt has an age 4.07 ± 0.04 G.y. The ~4.35 G.y. age clearly represents a lower limit to the age of the parent rock of the xenocrysts. Any of the ages between 4.35 and 4.07 G.y. could represent crystallization of the melt, or could reflect partial outgassing due to one of the later heating events.

The laser probe data for the matrix samples demonstrate clearly that during the breccia-forming event outgassing of Ar was incomplete on a scale of tens of microns. The three age determinations for groundmass yield the same result within experimental error; average is 4.01 G.y. Clasts of anorthosite that recrystallized after incorporation in the breccia yield essentially the same age as groundmass. Clasts of plagioclase crystals and devitrified maskelynite show consistently older ages, from 4.10 to 4.17 G.y. A clast of granulated anorthosite and a clast of felsic glass have younger ages, about 3.90 G.y. (It is not possible to assign precise ages to these materials with the present data. They are clearly younger than the other materials studied, but the measured ages show large and inexplicable variations. Studies are currently in progress of felsic clasts using a more detailed stepwise preheating procedure, and these studies are expected to clarify the chronologic interpretation of the data.)

The relatively old ages shown by the clasts of plagioclase and devitrified maskelynite represent lower limits on the times of the last significant heatings of these materials prior to their incorporation in the breccia. Two factors act to make these dates lower limits rather than actual ages. First, as discussed above, the gas released during lasering may have contained a small contribution from surrounding younger groundmass. Second, and more important, it is likely that these clasts were partly outgassed of Ar when they were incorporated in the breccia. Microprobe studies of fragments of olivine and shocked plagioclase in the matrix have demonstrated that heating of clasts on mixing with the groundmass melt was sufficient to permit appreciable diffusive exchange of cations between clasts and groundmass (James, 1976). Zones as wide as 50 μm at the edges of the olivine clasts show the effects of Fe–Mg exchange, and zones as wide as 20 μm at the edges of the shocked plagioclase clasts show the effects of Ca–Na–K exchange. Accompanying diffusive loss of Ar would probably be expected under such conditions. In fact, the analyses of plagioclase clasts reported herein appear to show a trend that may be the result of partial outgassing. The age data for the clasts in chip 73215,177 show a rough positive correlation of clast size with clast age (Fig. 3). In view of the small number of data points and the sizes of the errors, this trend cannot be considered as well established; if this trend is indeed real, however, it could have been produced by diffusive loss of Ar from the clasts after they were incorporated in the groundmass melt, assuming that they were all of about the same age prior to the breccia-forming event.

There are two possible interpretations for the relatively young ages, ~3.90 G.y., shown by two of the clasts studied. The first, and we feel most likely alternative, is that the ~3.90 G.y. age dates the breccia-forming event. The second alternative is that the date might reflect a post-consolidation heating

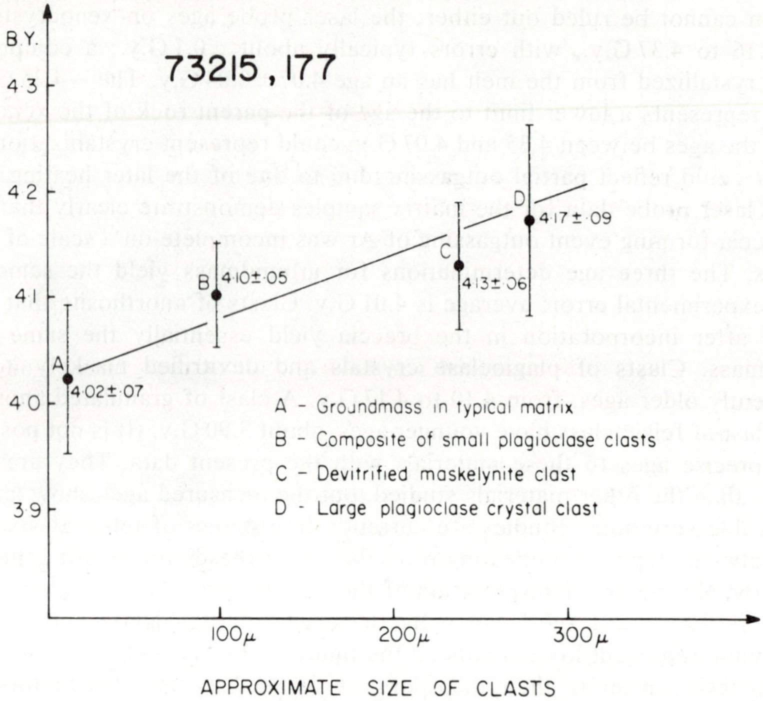

Fig. 3. Age to clast size correlation in 73215,177.

event. The first alternative is strongly supported by the results of studies done by other consortium members of a 20 mg clast of felsite extracted from the breccia. This clast contained both crystalline felsite and felsic glass components, having textures similar to the felsic clasts analyzed in this study. The clast shows a ^{39}Ar–^{40}Ar plateau age of 3.92 ± 0.04 G.y. (Jessberger *et al.*, 1977) and a Rb–Sr internal isochron age of 3.90 ± 0.05 G.y. (Compston *et al.*, 1977). The petrologic data (James and Hammarstrom, 1977) indicate that the clast was partly molten when incorporated in the breccia; thus its Sr isotopes probably reequilibrated, and it was probably outgassed of Ar at that time. The transmission electron microscopy observations indicate that there was probably no appreciable later reheating (Nord and James, 1977). Therefore, the ~3.90 G.y. age is best interpreted, in our view, as the date of the breccia-forming event. If the age were the date of a post-consolidation thermal event, it would be necessary that the rehating have reset both the Rb–Sr and K–Ar systems in the felsite equally, but left these systems virtually unaffected in the rest of the rock. Moreover, the rock textures would have had to be unaffected. A reheating with these characteristics seems improbable.

The significance of the ~4.01 G.y. age for groundmass is not entirely clear. The bulk of this material was a melt at the time of the breccia-forming event, but it may have contained appreciable numbers of plagioclase fragments $<10\ \mu m$ across. It is possible that these tiny clasts were not completely outgassed of their radiogenic Ar. It is also possible that the melt itself was not completely outgassed, or that it gained Ar during or after crystallization by incorporating gas that diffused out of the clasts. Some light is shed on this problem by the analyses of the clast of granulated anorthosite (Fig. 2). The granulated core of this clast gives the age we interpret as the age of the breccia-forming event, ~3.90 G.y., whereas the recrystallized rim shows essentially the same age as the surrounding groundmass, 4.01 ± 0.05 G.y. It would appear, therefore, that the recrystallized rim and the groundmass came to some equilibrium in their gas contents and did not outgas completely, whereas the granulated core was able to outgas completely. In any case, it does appear that the 4.01 G.y. age of the groundmass probably has no chronological significance.

The results presented herein are significant for the interpretation of stepwise heating data on lunar breccias. Stepwise heating experiments on samples of 73215 matrix show well-defined intermediate-temperature plateaus (Jessberger et al., 1976, 1977); such plateaus are generally interpreted as having age significance. The data reported herein, however, clearly indicate that matrix samples contain materials having a range of ages, and that the plateaus can only represent an average of the ages of the various components in any given chip. Thus the plateaus obtained with the stepwise heating technique do not necessarily date the formation of a breccia, but instead only set upper limits on the last significant outgassing event.

There is considerable evidence that the incomplete outgassing observed in 73215 may have been general in the Apollo 17 highland breccias. The laser probe results suggest a speculative reinterpretation of the breccia chronology at this site, as follows: The three major types of Apollo 17 highland breccias—light gray, blue gray, and green gray—all formed as fragment-laden melts (Simonds, 1975; Chao *et al.*, 1974; Dence *et al.*, 1976; Wood, 1975; G. Ryder, pers. comm.). They all originated as mechanically mixed aggregates of melt and clasts, and the mixing almost certainly took place in a major impact or several major impact events. Textures of the three types of breccia are gradational. The light-gray breccias (of which 73215 is a representative) incorporated the least melt and the most clasts, and were most quickly crystallized and cooled. The green-gray breccias incorporated the most melt and the fewest clasts, and were most slowly crystallized and cooled. The blue-gray breccias are intermediate in these characteristics. Bulk compositions of all these rocks are quite similar, and most have similar siderophile element ratios (Higuchi and Morgan, 1975). These relationships suggest a common origin for all the rocks. $^{39}Ar–^{40}Ar$ plateau ages range from 3.90 to 4.01 G.y. (Table 2), and there are rough correlations of age with breccia type: The green-gray breccias show the youngest ages, in the range 3.90–3.95 G.y.; the light-gray breccias show the oldest ages, >3.98 G.y.; and the blue-gray breccias for the most part have intermediate ages, 3.95–3.98 G.y. These apparent age relations are in conflict with those that would be predicted on the basis of the stratigraphy of the highland massifs from which the samples were derived: The light-gray breccias originated stop South Massif whereas the blue-gray and green-gray breccias probably came from considerably deeper

Table 2. ^{39}Ar–^{40}Ar plateau ages of Apollo 17 highland breccias.

Green-gray breccias		Blue-gray breccias		Light-gray breccias	
Age (G.y.)	Rock	Age (G.y.)	Rock	Age (G.y.)	Rock
3.90 ± 0.05	77135[1,5]	3.91 ± 0.04	77115[5]	4.01 ± 0.03	72255[2]
3.93 ± 0.04	76015[1]	3.95 ± 0.04	73235[3]	3.98–4.15 ± 0.04	73215[6]
3.94 ± 0.04	76215[1]	3.96 ± 0.04	76295[1]		
3.95 ± 0.04	73275[3]	3.98 ± 0.04	76315[1]		
		3.98 ± 0.03	77075[4]		

1. Cadogan and Turner (1976).
2. Leich *et al.* (1974).
3. Turner and Cadogan (1975).
4. Stettler *et al.* (1974).
5. Stettler *et al.* (1975).
6. Jessberger *et al.* (1976); Jessberger *et al.* (1977).

within the massifs, yet the light-gray breccias show consistently older ages. In sum, these data suggest the possibility that outgassing was incomplete in the majority of the Apollo 17 breccias, and that all formed in the same major impact event at ~3.90 G.y. The most slowly cooled (green-gray breccias) were most completely outgassed, and the least slowly cooled and stratigraphically higher (light-gray breccias) were least completely outgassed.

Acknowledgements—We want to express our gratitude to T. Ludkewycz for his skillful technical assistance. This work was greatly assisted by the glass blowing of R. Schlott. We would also like to thank J. C. Huneke and D. Phinney for constructive reviews of the manuscript. This research was supported by NASA grants T-3040C and NGL-33-015-174.

REFERENCES

Blanchard D. P., Jacobs J. W., Brannon J. C., and Haskin L. A. (1976) Major and trace element compositions of matrix and aphanitic clasts from consortium breccia 73215. *Proc. Lunar Sci. Conf. 7th*, p. 2179–2187.

Blanchard D. P., Jacobs J. W., and Brannon J. C. (1977) Chemistry of ANT-suite and felsite clasts from consortium breccia 73215 and gabbroic anorthosite 79215. *Proc. Lunar Sci. Conf. 8th.* This volume.

Cadogan P. H. and Turner G. (1976) The chronology of the Apollo 17 Station 6 boulder. *Proc. Lunar Sci. Conf. 7th*, p. 2267–2285.

Chao E. C. T., Minkin J. A., and Thompson C. L. (1974) Preliminary petrographic description and geologic implications of the Apollo 17 Station 7 boulder consortium samples. *Earth Planet. Sci. Lett.* **23**, 413–428.

Compston W., Foster J. J., and Gray C. M. (1977) Rb–Sr systematics in clasts and aphanites from consortium breccia 73215. *Proc. Lunar Sci. Conf. 8th.* This volume.

Dence M. R., Grieve R. A. F., and Plant A. G. (1976) Apollo 17 grey breccias and crustal composition in the Serenitatis Basin region. *Proc. Lunar Sci. Conf. 7th*, p. 1821–1832.

Higuchi H. and Morgan J. W. (1975) Ancient meteoritic component in Apollo 17 boulders. *Proc. Lunar Sci. Conf. 6th*, p. 1625–1651.

James O. B. (1976) Petrology of aphanitic lithologies in consortium breccia 73215. *Proc. Lunar Sci. Conf. 6th*, p. 2145–2178.

James O. B. and Blanchard D. P. (1976) Consortium studies of light-gray breccia 73215: Introduction, subsample distribution data, and summary of results. *Proc. Lunar Sci. Conf. 6th*, p. 2131–2143.

James O. B., Brecher A., Blanchard D. P., Jacobs J. W., Brannon J. C., Korotev R. L., Haskin L. A., Higuchi H., Morgan J. W., Anders E., Silver L. T., Marti K., Braddy D., Hutcheon I. D., Kirsten T., Kerridge J. F., Kaplan I. R., Pillinger C. T., and Gardiner L. R. (1975) Consortium studies of matrix of light gray breccia 73215. *Proc. Lunar Sci. Conf. 6th*, p. 547–577.

James O. B. and Hammarstrom J. G. (1977) Petrology of four clasts from consortium breccia 73215. *Proc. Lunar Sci. Conf. 8th*. This volume.

Jessberger E. K., Kirsten T., and Staudacher Th. (1976) Argon-argon ages of consortium breccia 73215. *Proc. Lunar Sci. Conf. 7th*, p. 2201–2215.

Jessberger E. K., Kirsten T., and Staudacher Th. (1977) One rock and many ages—Further K–Ar data on consortium breccia 73215. *Proc. Lunar Sci. Conf. 8th*. This volume.

Leich D. A., Kahl S. D., Kirschbaum A. R., Niemeyer S., and Phinney D. (1975) Rare gas constraints on the history of boulder 1, station 2, Apollo 17. *The Moon* **14**, 407–434.

Megrue G. (1973) Spatial distribution of $^{40}Ar/^{39}Ar$ ages in lunar breccia 14301. *J. Geophys. Res.* **78**, 3216–3221.

Nord G. L., Jr. and James O. B. (1977) Aphanitic matrix, an ANT-suite clast and a felsite clast in consortium breccia 73215: An electron petrographic study. *Proc. Lunar Sci. Conf. 8th*. This volume.

Plieninger T. and Schaeffer O. A. (1976) Laser probe ^{39}Ar–^{40}Ar ages of individual mineral grains in lunar basalt 15607 and lunar breccia 15465. *Proc. Lunar Sci. Conf. 7th*, p. 2055–2066.

Schaeffer G. A. and Schaeffer O. A. (1977) ^{39}Ar–^{40}Ar ages of lunar rocks. *Proc. Lunar Sci. Conf. 8th*. This volume.

Schaeffer O. A. and Husain L. (1973) Early lunar history: Ages of 2 to 4 mm soil fragments from the lunar highlands. *Proc. Lunar Sci. Conf. 4th*, p. 1847–1863.

Schaeffer O. A., Müller H. W., and Grove T. L. (1977) Laser ^{39}Ar–^{40}Ar study of Apollo 17 basalts. *Proc. Lunar Sci. Conf. 8th*. This volume.

Simonds C. H. (1976) Thermal regimes in impact melts and the petrology of the Apollo 17 Station 6 boulder. *Proc. Lunar Sci. Conf. 6th*, p. 641–672.

Stettler A., Eberhardt P., Geiss J., and Grögler N. (1974) ^{39}Ar–^{40}Ar ages of samples from the Apollo 17 Station 7 boulder and implications for its formation. *Earth Planet. Sci. Lett.* **23**, 453–461.

Stettler A., Eberhardt P., Geiss J., Grögler N., and Guggisberg S. (1975) Age sequence in the Apollo 17 Station 7 boulder (abstract). In *Lunar Science VI*, p. 771. The Lunar Science Institute, Houston.

Turner G. and Cadogan P. H. (1975) The history of lunar bombardment inferred from ^{40}Ar–^{39}Ar dating of highland rocks. *Proc. Lunar Sci. Conf. 6th*, p. 1509–1538.

Wood J. A. (1975) The nature and origin of boulder 1, Station 2, Apollo 17 period. *The Moon* **14**, 405–417.

[illegible] (1973) ^{40}Ar–^{39}Ar [illegible] samples from the Apollo [illegible] 22, [illegible]

[illegible]

[illegible]

[illegible]

Proc. Lunar Sci. Conf. 8th (1977), p. 2567–2580.
Printed in the United States of America

One rock and many ages—Further K–Ar data on consortium breccia 73215

E. K. Jessberger, T. Kirsten, and Th. Staudacher

Max-Planck-Institut für Kernphysik, P.O. Box 103980, 6900 Heidelberg/Germany

Abstract—^{40}Ar–^{39}Ar dating of six new subsamples of consortium breccia 73215 has confirmed and enlarged the age diversity found among the constituents of this breccia; this age diversity is especially striking because the breccia is the product of a one-stage event. An age range from 3.92 to 4.28 AE is now covered by 17 dated subsamples. This diversity clearly demonstrates the limited resetting power of basin-forming events. Only study of an exceptional rock type found as a clast within the breccia, the felsite, has made it possible to date the breccia-forming event; the date is 3.92 ± 0.04 AE. If 73215 is indeed consolidated Serenitatis ejecta, then all circular basins must have formed between 3.78 and 3.92 AE and the ages determined on most Apollo 17 highland rocks predate this period due to incomplete resetting.

Introduction

Promoted by multiple stimulations of the O. James Consortium, the particular features of light-gray breccia 73215 have motivated us to undertake what might well be the most extensive age-systematics investigation of any single lunar rock. We previously studied 11 subsamples of various clast and matrix lithologies; these samples showed a great range of ages, from 4.04 to 4.28 AE (Jessberger *et al.*, 1976a). We have argued that this age diversity is of exemplary significance for lunar highland rock chronology in general: It implies that even very small lithic clasts may have escaped total Ar loss during basin excavation and, as a consequence rocks of the pre-basin lunar crust are principally accessible for ^{40}Ar–^{39}Ar dating by analysis of clasts from basin ejecta (cf. Jessberger *et al.*, 1974). This statement is based on the fact that 73215 is not a regolith breccia composed of multiply reworked materials, but rather "formed as an aggregate of melt and clasts produced during a major (basin-forming) impact event, cooled rapidly after aggregation, and was never significantly reheated" (James and Blanchard, 1976). The implication is that it is extremely difficult to infer basin ages from the ages of breccias produced in basin-forming events; in spite of 11 age determinations, only an upper limit, 4.04 AE, could be established for the formation time of 73215. It is likely that the melt fraction of the breccia (the groundmass) was completely reset, but this melt was intimately mixed with abundant minute clasts which may have escaped complete outgassing. The breccia matrix is an aphanitic rock that consists of roughly 50% melt-derived groundmass and 50% small clasts (James, 1976), and a straightforward dating of the breccia-forming event by dating samples of such material is not possible if the clasts were not totally outgassed. In order to further investigate the effects of incomplete outgassing, we have dated five more aphanitic samples, two of matrix aphanites and three of aphanitic clasts lithologically similar to the matrix. In addition, we have dated a felsite clast which was partly melted (James and Hammarstrom, 1977) and presumably outgassed when the breccia formed: The age of this clast apparently yields the age of the breccia-forming event.

EXPERIMENTS

The two samples of aphanitic matrix studied are 46,44—light-gray matrix and 36,2—black matrix. The three samples of aphanitic clasts are: 38,32—gray spheroid from within matrix; 46,19 non-vesicular black clast from within granulated feldspathic clast material; and 46,7—vesicular black clast from within granulated feldspathic material. The petrologic nature of these rocks and their genetic context is described by James *et al.* (1975) and James (1976). The felsite clast is sample 43,3 and the chip we studied contained both crystalline and glassy fractions; the petrology of this clast is described in detail by James and Hammarstrom (1977). Data on aphanite samples comparable to the samples we analyzed have been reported by other members of the consortium: Blanchard *et al.* (1976), major, minor, and trace element analysis; Morgan *et al.* (1976), volatile and siderophile element analysis; Compston *et al.* (1977), Rb–Sr analysis. Data on the felsite clast are given in other papers in this volume: Blanchard *et al.* (1977), rare-earth element chemistry; Compston *et al.* (1977), Rb–Sr systematics; and Nord and James (1977), transmission electron microscopy. The preliminary report of the data given in this paper was presented by Staudacher *et al.* (1977).

The samples were irradiated in one irradiation (M1/76) in the BER-reactor at Mol/Belgium. Experimental procedures, corrections, monitors, and data reduction are the same as described in Jessberger *et al.* (1976a). The conversion factors for M1/76 are:

$$^{39}C_K = 4.48 \times 10^{-4}\ \text{cc STP}\ ^{39}\text{Ar/g K}$$

$$^{37}C_{Ca}/^{39}C_K = 0.51.$$

The flux gradient along the 3 cm capsule was monitored by three equal-spaced hornblende monitors NL-25. No gradient as measured by (^{40}Ar/^{39}Ar) is found outside the experimental error of 0.4%.

The data for the individual temperature fractions are listed in Table 1 at the same stage of data reduction as in Jessberger *et al.* (1976a). Subsequently, trapped and cosmogenic components are subtracted using $(^{36}\text{Ar}/^{38}\text{Ar})_C = 0.65 \pm 0.02$; $(^{40}\text{Ar}/^{38}\text{Ar})_C = 0.2 \pm 0.2$; $(^{36}\text{Ar}/^{38}\text{Ar})_T = 5.35$, and $(^{36}\text{Ar}/^{40}\text{Ar})_T = 1 \pm 1$. The resulting ^{40}Ar–^{39}Ar ages for each temperature fraction are also listed in Table 1, together with the total K–Ar age for each sample. Cosmic-ray exposure ages are calculated from the $^{38}\text{Ar}_C/^{37}\text{Ar}_{Ca}$ ratios using a production rate of $^{38}P_{Ca} = 1.4 \times 10^{-8}$ cc STP $^{38}\text{Ar}_C$/g Ca, m.y.

RESULTS

The spectra of apparent ^{40}Ar–^{39}Ar ages and the corresponding K/Ca ratios are plotted in Figs. 1 and 2 and all results are summarized in Fig. 3. All age spectra can be subdivided into three distinct regimes:

Regime 1. A low-temperature regime (typically 400–650°C) of monotonically increasing, reduced ages. These low ages are probably the product of classical volume diffusion losses. The $^{40}\text{Ar}_R$ losses calculated relative to the plateau ages given in Fig. 3 are generally less than 5% of the total $^{40}\text{Ar}_R$, except for the case of black matrix 36,2 where it is 12% of the total. Consequently, the total K–Ar ages (Table 1) are only ~2% lower than the plateau ages (4.5% for 36,2). Regime 1 is also characterized by a relatively sharp monotonic decrease of the K/Ca ratio, by factors between 2 and 4 (except for the felsite, 43,3).

Regime 2. An intermediate-temperature regime (typically 650–900°C) consisting of a plateau of at least five consecutive release fractions that have virtually identical ages; the Ar released in this range represents ~40% of the total ^{39}Ar. This regime is characterized by a rather constant K/Ca ratio with deviations from the mean of less than ±25%. As an exception, the felsite age plateau

Table 1. Argon isotope data of 73215 samples irradiated in M1/76.

Extraction temp [°C]	$^{40}Ar^{(a)}$ [10^{-8} cc STP/g]	$^{36}Ar/^{40}Ar^{(b)}$	$^{37}Ar/^{40}Ar^{(c)}$	$^{38}Ar/^{40}Ar^{(d)}$	$^{39}Ar/^{40}Ar^{(c,e)}$	Age[(f)] [AE]
38,32,5 (36.26 mg)		$\times 10^{-5}$	$\times 10^{-3}$	$\times 10^{-5}$	$\times 10^{-5}$	
410	47	162 ± 18	161 ± 6	176 ± 17	1450 ± 600	3.07 ± 0.06
480	113	117 ± 10	153 ± 3	172 ± 11	973 ± 19	3.70 ± 0.03
550	248	120 ± 4	157 ± 1	183 ± 4	820 ± 8	3.98 ± 0.02
600	349	124 ± 5	169 ± 1	213 ± 3	777 ± 7	4.07 ± 0.02
650	308	136 ± 5	188 ± 2	245 ± 4	750 ± 8	4.13 ± 0.02
700	468	171 ± 3	207 ± 1	277 ± 3	769 ± 5	4.09 ± 0.01
750	576	182 ± 3	214 ± 1	285 ± 3	769 ± 5	4.09 ± 0.01
800	644	183 ± 3	211 ± 1	281 ± 3	770 ± 4	4.08 ± 0.01
850	494	169 ± 5	204 ± 1	270 ± 2	765 ± 6	4.09 ± 0.01
900	422	183 ± 4	212 ± 2	292 ± 4	760 ± 6	4.11 ± 0.01
950	227	181 ± 7	241 ± 3	318 ± 7	757 ± 11	4.11 ± 0.02
1000	138	242 ± 11	290 ± 6	345 ± 10	810 ± 18	4.00 ± 0.04
1050	313	326 ± 5	350 ± 3	488 ± 6	863 ± 9	3.89 ± 0.02
1100	132	381 ± 12	416 ± 9	595 ± 14	863 ± 21	3.89 ± 0.04
1150	804	284 ± 3	285 ± 1	417 ± 3	826 ± 5	3.97 ± 0.01
1200	525	322 ± 3	319 ± 2	471 ± 4	805 ± 6	4.01 ± 0.01
1300	435	339 ± 5	342 ± 2	500 ± 4	774 ± 7	4.07 ± 0.01
1450	616	304 ± 4	327 ± 2	454 ± 4	782 ± 6	4.06 ± 0.01
1540	30	40 ± 70	20 ± 1	80 ± 30	720 ± 130	4.19 ± 0.31
Total	6889	226 ± 1	252 ± 1	346 ± 1	795 ± 2	4.03 ± 0.01
46,19,5 (57.19 mg)		$\times 10^{-5}$	$\times 10^{-3}$	$\times 10^{-5}$	$\times 10^{-5}$	
400	21	190 ± 29	143 ± 7	150 ± 50	1640 ± 90	2.89 ± 0.08
450	55	241 ± 10	104 ± 2	138 ± 11	971 ± 36	3.70 ± 0.06
500	125	169 ± 6	111 ± 1	139 ± 4	874 ± 13	3.87 ± 0.02
550	230	97 ± 4	114 ± 1	139 ± 5	790 ± 6	4.04 ± 0.01
600	273	98 ± 4	119 ± 1	153 ± 4	786 ± 5	4.05 ± 0.01
650	409	136 ± 2	140 ± 1	191 ± 2	771 ± 5	4.08 ± 0.01
700	424	158 ± 2	152 ± 1	208 ± 2	763 ± 4	4.10 ± 0.01
750	561	177 ± 3	159 ± 1	218 ± 3	768 ± 3	4.09 ± 0.01
800	552	183 ± 2	156 ± 1	225 ± 2	766 ± 3	4.09 ± 0.01
850	501	176 ± 2	153 ± 1	214 ± 2	766 ± 4	4.09 ± 0.01
900	472	236 ± 3	167 ± 1	245 ± 3	772 ± 4	4.08 ± 0.01
950	206	181 ± 5	196 ± 2	277 ± 5	785 ± 9	4.05 ± 0.02
1000	107	260 ± 9	242 ± 4	318 ± 9	787 ± 14	4.05 ± 0.03
1060	173	292 ± 7	279 ± 3	391 ± 5	808 ± 10	4.00 ± 0.02
1100	76	282 ± 13	324 ± 8	433 ± 14	784 ± 19	4.05 ± 0.04
1150	1476	222 ± 2	219 ± 1	317 ± 2	818 ± 3	3.98 ± 0.01
1200	1355	232 ± 2	229 ± 1	331 ± 2	822 ± 2	3.97 ± 0.01
1300	1170	281 ± 3	277 ± 1	409 ± 2	810 ± 3	4.00 ± 0.01
1450	161	393 ± 13	434 ± 10	581 ± 15	710 ± 17	4.22 ± 0.04
Total	8347	213 ± 1	204 ± 1	289 ± 1	798 ± 1	4.02 ± 0.01

Table 1. (*Continued*).

Extraction temp [°C]	$^{40}Ar^{(a)}$ [10^{-8} cc STP/g]	$^{36}Ar/^{40}Ar^{(b)}$	$^{37}Ar/^{40}Ar^{(c)}$	$^{38}Ar/^{40}Ar^{(d)}$	$^{39}Ar/^{40}Ar^{(c,e)}$	Age[(f)] [AE]
46,7,3 (12.73 mg)		$\times 10^{-5}$	$\times 10^{-4}$	$\times 10^{-5}$	$\times 10^{-5}$	
440	56	125 ± 40	847 ± 104	240 ± 65	1910 ± 290	2.66 ± 0.21
500	86	70 ± 26	781 ± 69	118 ± 43	1280 ± 160	3.26 ± 0.19
560	169	84 ± 13	845 ± 38	139 ± 20	957 ± 75	3.73 ± 0.13
630	309	83 ± 7	862 ± 21	134 ± 11	908 ± 38	3.81 ± 0.07
680	346	104 ± 7	1086 ± 23	164 ± 10	795 ± 40	4.03 ± 0.08
740	391	129 ± 7	1365 ± 27	202 ± 10	783 ± 32	4.05 ± 0.07
790	356	140 ± 8	1506 ± 34	220 ± 11	790 ± 36	4.04 ± 0.08
830	350	156 ± 9	1687 ± 40	241 ± 11	775 ± 34	4.07 ± 0.07
880	434	139 ± 7	1488 ± 29	217 ± 9	767 ± 28	4.09 ± 0.06
930	594	150 ± 17	1522 ± 21	226 ± 27	783 ± 18	4.06 ± 0.04
980	796	186 ± 17	1909 ± 21	284 ± 5	820 ± 13	3.98 ± 0.03
1000	411	204 ± 9	2128 ± 46	312 ± 12	809 ± 31	4.00 ± 0.06
1050	1399	213 ± 3	2166 ± 15	324 ± 3	822 ± 9	3.97 ± 0.02
1100	1191	239 ± 4	2395 ± 22	354 ± 5	790 ± 12	4.04 ± 0.02
1150	587	215 ± 7	2103 ± 36	323 ± 29	807 ± 17	4.00 ± 0.04
1200	671	189 ± 7	1801 ± 33	278 ± 7	802 ± 19	4.02 ± 0.04
1250	264	214 ± 30	2389 ± 179	372 ± 30	768 ± 66	4.09 ± 0.14
1310	184	237 ± 62	2855 ± 427	461 ± 71	943 ± 148	3.75 ± 0.25
1450	150	265 ± 112	3424 ± 895	521 ± 138	1070 ± 290	3.55 ± 0.42
Total	8745	183 ± 4	1887 ± 21	284 ± 5	825 ± 9	3.97 ± 0.02
46,44 (88.0 mg)		$\times 10^{-5}$	$\times 10^{-3}$	$\times 10^{-5}$	$\times 10^{-5}$	
390	17	110 ± 30	225 ± 9	203 ± 16	2290 ± 100	2.41 ± 0.06
460	58	101 ± 7	148 ± 2	154 ± 6	1330 ± 20	3.20 ± 0.02
510	165	118 ± 4	159 ± 1	202 ± 3	975 ± 6	3.70 ± 0.01
550	200	127 ± 3	161 ± 1	208 ± 3	875 ± 5	3.87 ± 0.01
600	336	146 ± 3	175 ± 1	229 ± 2	827 ± 4	3.97 ± 0.01
650	401	168 ± 2	196 ± 1	268 ± 2	820 ± 4	3.98 ± 0.01
700	466	191 ± 2	218 ± 1	296 ± 2	809 ± 3	4.00 ± 0.01
750	521	202 ± 2	231 ± 1	313 ± 2	805 ± 2	4.01 ± 0.01
800	748	204 ± 2	233 ± 1	314 ± 2	796 ± 2	4.03 ± 0.01
850	451	196 ± 2	223 ± 1	302 ± 2	793 ± 4	4.03 ± 0.01
910	358	203 ± 2	231 ± 1	311 ± 2	803 ± 3	4.01 ± 0.01
960	239	241 ± 4	274 ± 1	374 ± 4	833 ± 5	3.95 ± 0.01
1010	186	315 ± 4	342 ± 2	480 ± 4	876 ± 7	3.87 ± 0.01
1060	346	365 ± 3	386 ± 2	553 ± 3	873 ± 4	3.87 ± 0.01
1110	156	593 ± 7	587 ± 5	868 ± 8	910 ± 9	3.80 ± 0.02
1160	88	644 ± 12	609 ± 8	937 ± 14	872 ± 15	3.87 ± 0.03
1210	401	387 ± 3	380 ± 1	564 ± 3	850 ± 3	3.92 ± 0.01
1300	441	467 ± 4	466 ± 2	682 ± 4	814 ± 4	3.99 ± 0.01
1500	192	519 ± 8	572 ± 7	786 ± 10	735 ± 10	4.16 ± 0.02
Total	5770	265 ± 1	288 ± 1	402 ± 1	835 ± 2	3.95 ± 0.01

[(a)]Absolute amounts are uncertain to ±5%. Corrected for $^{40}Ar_K$ and system blank.

[(b)]Corrected for $^{36}Ar_{Ca}$ and system blank. Error figures are statistical variations in the measured isotope ratios and include also the uncertainties of the correction ratios and a 50% uncertainty in system blank.

[(c)]Decay-corrected.

[(d)]Corrected for $^{38}Ar_{Ca}$ and system blank.

[(e)]Corrected for $^{39}Ar_{Ca}$ and mass spectrometer background.

[(f)]The error in the apparent ages reflects the uncertainties as given in footnote (b). $\lambda_\beta + \lambda_\epsilon = 5.305 \times 10^{-10}\ yr^{-1}$, $^{40}K/K = 0.0119$ atom %.

Table 1. (*Continued*).

Extraction temp [°C]	$^{40}Ar^{(a)}$ [10^{-8} cc STP/g]	$^{36}Ar/^{40}Ar^{(b)}$	$^{37}Ar/^{40}Ar^{(c)}$	$^{38}Ar/^{40}Ar^{(d)}$	$^{39}Ar/^{40}Ar^{(c,e)}$	Age[(f)] [AE]
36,2 (73.74 mg)		$\times 10^{-5}$	$\times 10^{-3}$	$\times 10^{-5}$	$\times 10^{-5}$	
460	563	24 ± 1	20 ± 1	20 ± 2	2073 ± 4	2.55 ± 0.01
510	646	23 ± 1	21 ± 1	29 ± 1	1268 ± 4	3.28 ± 0.01
560	951	26 ± 3	27 ± 1	38 ± 1	1004 ± 2	3.65 ± 0.01
610	1288	30 ± 1	28 ± 1	42 ± 1	894 ± 2	3.84 ± 0.01
650	1678	35 ± 1	34 ± 1	50 ± 1	846 ± 2	3.93 ± 0.01
700	1524	45 ± 1	47 ± 1	67 ± 1	826 ± 2	3.97 ± 0.01
750	1182	63 ± 1	67 ± 1	93 ± 1	820 ± 2	3.98 ± 0.01
800	1154	76 ± 1	80 ± 1	109 ± 1	819 ± 2	3.98 ± 0.01
850	1329	59 ± 1	62 ± 1	89 ± 2	820 ± 3	3.98 ± 0.01
900	880	58 ± 1	58 ± 1	81 ± 1	836 ± 2	3.95 ± 0.01
960	740	78 ± 1	77 ± 1	107 ± 1	859 ± 2	3.90 ± 0.01
900	24	100 ± 20	100 ± 6	157 ± 10	883 ± 50	3.86 ± 0.09
950	68	139 ± 7	132 ± 3	196 ± 5	851 ± 18	3.92 ± 0.03
1000	106	144 ± 5	138 ± 2	205 ± 3	843 ± 12	3.93 ± 0.02
1050	201	191 ± 3	189 ± 1	278 ± 3	839 ± 7	3.94 ± 0.01
1100	442	157 ± 2	151 ± 1	223 ± 1	826 ± 3	3.97 ± 0.01
1150	949	90 ± 1	81 ± 1	125 ± 1	830 ± 2	3.96 ± 0.01
1200	1143	158 ± 1	154 ± 1	225 ± 1	836 ± 2	3.95 ± 0.01
1260	790	201 ± 2	202 ± 5	292 ± 2	830 ± 4	3.96 ± 0.01
1340	558	223 ± 4	227 ± 2	332 ± 3	867 ± 8	3.89 ± 0.01
1450	1049	223 ± 3	228 ± 2	325 ± 2	822 ± 6	3.97 ± 0.01
1500	109	207 ± 30	235 ± 18	330 ± 26	855 ± 67	3.91 ± 0.13
Total	17,380	87 ± 1	87 ± 1	125 ± 1	903 ± 1	3.82 ± 0.01
43,3 (4.42 mg)	[10^{-6} cc STP/g]	$\times 10^{-6}$	$\times 10^{-5}$	$\times 10^{-6}$	$\times 10^{-5}$	
350	2.0	—	340 ± 90	290 ± 260	4180 ± 280	1.67 ± 0.07
400	8.9	—	130 ± 30	—	2440 ± 40	2.33 ± 0.02
450	32.0	50 ± 20	50 ± 10	30 ± 30	1392 ± 7	3.13 ± 0.01
500	71.6	40 ± 10	50 ± 4	30 ± 10	1010 ± 4	3.64 ± 0.01
550	128.7	38 ± 6	40 ± 1	60 ± 10	885 ± 2	3.85 ± 0.01
600	191.8	63 ± 5	48 ± 1	60 ± 9	858 ± 2	3.90 ± 0.01
640	301.2	55 ± 3	47 ± 1	65 ± 9	849 ± 1	3.92 ± 0.01
680	351.8	56 ± 3	45 ± 1	77 ± 9	844 ± 1	3.93 ± 0.01
720	319.5	59 ± 3	48 ± 1	72 ± 9	847 ± 2	3.93 ± 0.01
760	168.3	70 ± 5	70 ± 2	60 ± 10	854 ± 2	3.91 ± 0.01
800	120.0	72 ± 7	130 ± 4	82 ± 9	853 ± 3	3.91 ± 0.01
840	119.6	87 ± 8	165 ± 3	77 ± 10	853 ± 3	3.91 ± 0.01
880	109.4	73 ± 8	143 ± 3	50 ± 10	852 ± 3	3.92 ± 0.01
920	103.9	55 ± 8	132 ± 4	50 ± 10	845 ± 3	3.93 ± 0.01
970	94.2	50 ± 10	110 ± 3	60 ± 10	849 ± 3	3.92 ± 0.01
1020	60.1	40 ± 10	53 ± 2	76 ± 9	836 ± 4	3.95 ± 0.01
1070	67.6	40 ± 20	42 ± 6	79 ± 10	833 ± 4	3.95 ± 0.01
1170	109.8	85 ± 9	60 ± 2	63 ± 10	841 ± 3	3.94 ± 0.01
1220	8.3	—	90 ± 20	—	754 ± 23	4.12 ± 0.05
1300	1.9	—	—	—	702 ± 86	4.24 ± 0.21
1450	4.2	—	60 ± 40	—	587 ± 38	4.54 ± 0.11
Total	2375	58 ± 2	70 ± 1	65 ± 3	870 ± 1	3.88 ± 0.01

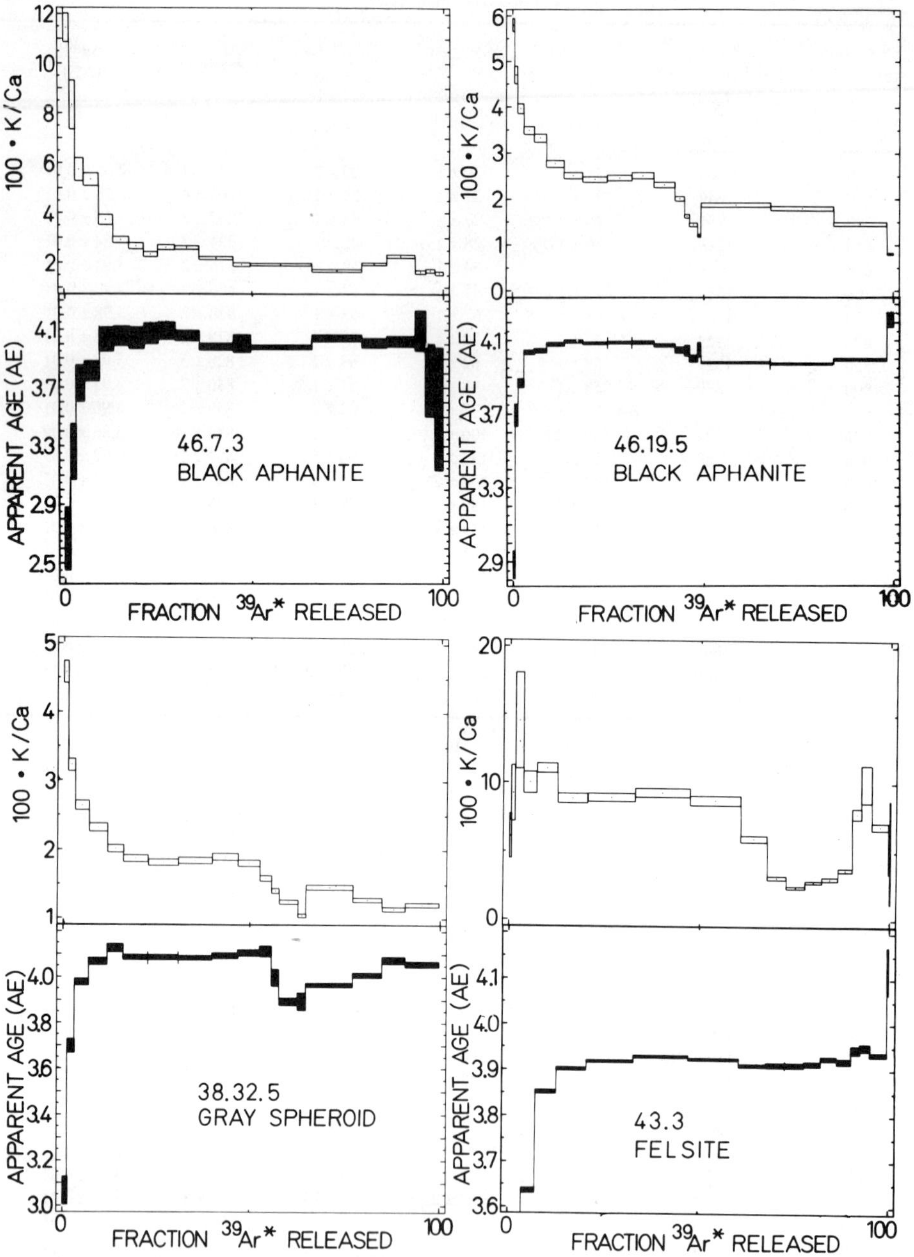

Fig. 1. ^{40}Ar–^{39}Ar age spectra and associated K/Ca ratios versus the fraction of total ^{39}Ar released for three samples of aphanitic clasts and the felsite clast from 73215. Note that the expanded age scales and linear K/Ca scales are different for the four samples.

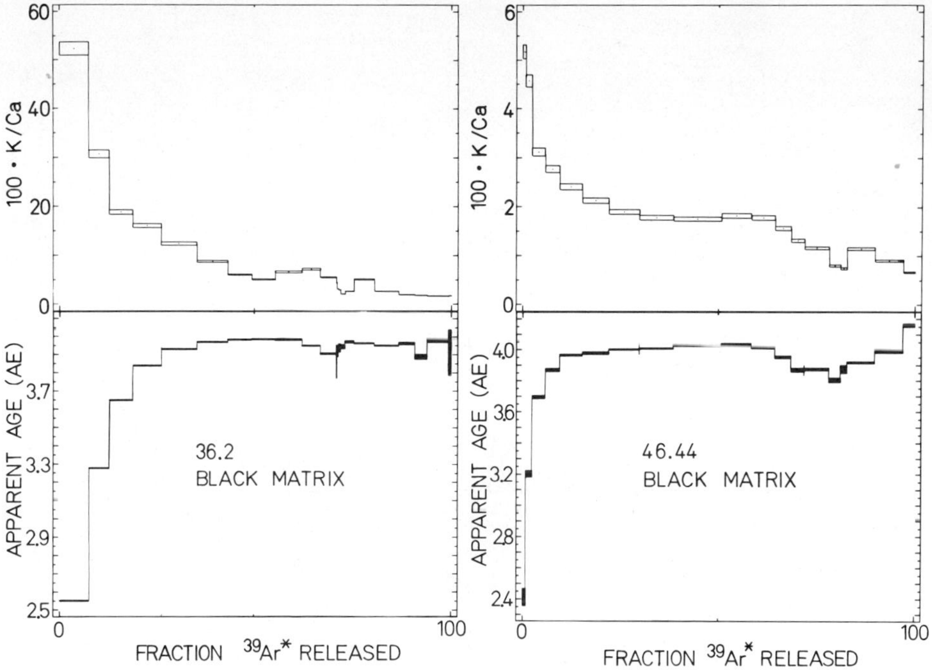

Fig. 2. Apparent ^{40}Ar–^{39}Ar ages and K/Ca ratios versus the fraction of ^{39}Ar released for two samples of aphanitic matrix from 73215. Note that the expanded age scales and linear K/Ca scales are different for the two samples.

extends up to the 1170°C fraction, and the K/Ca ratio is depressed as much as a factor of 3 between 760 and 970°C.

Regime 3. A high-temperature regime of reduced ages. The beginning of this regime is marked by a moderate age depression of 3–5%, and there is a tendency for partial recovery at higher temperatures; in general, there is more scatter of apparent ages than in Regime 2, even though the total scatter is still rather restricted. The onset of the age reduction is accompanied by a slight decrease of the K/Ca ratios. These ratios also display slightly more variation than in the plateau regime.

The question arises whether chronologically significant ages should be based on Regime 2 alone or whether Regimes 2 and 3 should be combined. If the deviations in Regime 3 are merely a result of an *internal* redistribution of the Ar isotopes within the various phases of the sample, the latter approach would be appropriate. On the contrary, if these deviations are due to recoil losses or other unspecified loss mechanisms (e.g., devitrification), Regime 2 alone should be used to infer the plateau age. A definite decision between these two alternatives cannot be made, but we prefer to base our plateau ages on Regime 2 alone; only in Regime 2 is the plateau truly "well defined" and paralleled by a constant K/Ca

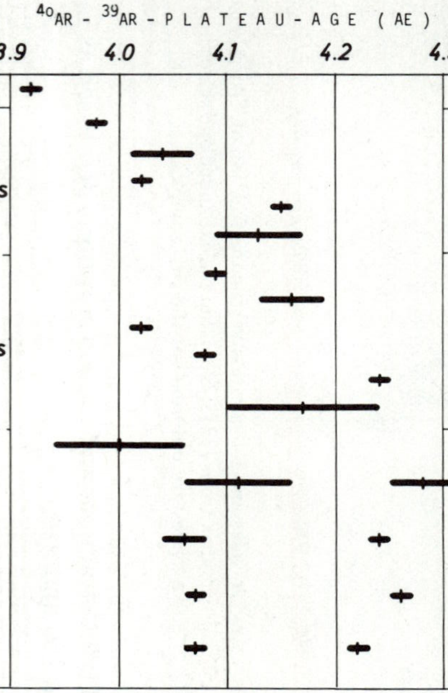

	SUBSAMPLE NUMBER	LITHOLOGY		^{40}AR - ^{39}AR - PLATEAU-AGE (AE)	K (ppm)	Ca (%)	EXPOSURE-AGE (m.y.)
X	43,3	felsite clast		3.92 ± .01	47100	1.8	-
X	36,2	black	matrix aphanites	3.98 ± .01	3550	6.7	232 ± 9
	41,1	black		4.04 ± .03	1800	7.1	254 ± 9
X	46,44	light gray		4.02 ± .01	1090	7.4	250 ± 10
	73,1	gray		4.15 ± .01	1350	7.5	238 ± 12
	177,1	schlieren rich gray		4.13 ± .04	1230	6.8	257 ± 11
X	38,32,5	gray spheroid	clast aphanites	4.09 ± .01	1240	7.7	230 ± 10
	38,57,4	gray spheroid		4.16 ± .03	1270	7.1	259 ± 12
X	46,7,3	vesicular black clast		4.02 ± .01	1630	7.4	235 ± 15
X	46,19,5	nonvesicular black clast		4.08 ± .01	1490	7.5	230 ± 15
	46,10,7	nonvesicular black clast		4.24 ± .01	1440	6.4	252 ± 22
	46,6,1	dark gray clast		4.17 ± .07	990	6.6	256 ± 24
	38,39,1,1	troctolite vein		4.00 ± .06	150	1.2	247 ± 62
	38,39,1,2	feldspathic clast		4.11 ± .05 4.28 ± .03	870	9.8	228 ± 40
	29,9,6	anorthositic gabbro clasts		4.06 ± .02 4.24 ± .01	780	10.3	268 ± 6
	46,25,5			4.07 ± .01 4.26 ± .01	650	9.2	250 ± 12
	46,33,4			4.07 ± .01 4.22 ± .01	720	8.1	244 ± 10

Fig. 3. Summary of ^{40}Ar–^{39}Ar dating results for 73215 matrix and clast samples. The table includes the results of the previous (Jessberger *et al.*, 1976a) and the present (marked by *X*) study. The quoted errors are the unweighted average quadratic deviations from the plateau values and do not include the absolute uncertainty (for the plateau age this is ±0.04 AE). The anorthositic gabbro clasts and the feldspathic clast 38,39,1,2 showed spectra with two-step plateaux, in which each plateau has chronological significance (Jessberger *et al.*, 1976a).

ratio, which indicates that a chemically homogeneous phase is responsible for the gas release. Fortunately, the choice between the two alternatives is not a very critical one. If we were to take the alternative approach, the resulting ages would be only 0.03 to 0.05 AE lower than the plateau ages based on Regime 2 alone. For the especially important felsite sample, there would be no difference at all. In the study of 73215 we have applied a very restrictive "plateau" definition for which the exceptional constancy of the K/Ca ratios in Regime 2 provides the basis. We note, however, that the majority of ^{40}Ar–^{39}Ar ages of highland and mare whole rock samples apparently are meaningful in spite of usually greater K/Ca variations than in 73215. The plateau ages compiled in Fig. 3 are the weighted averages of the plateau fractions in Regime 2. The stated error is the unweighted standard deviation of the plateau fractions and the absolute error of 0.04 AE is not included. For purposes of comparison, Fig. 3 also lists our previous results for other 73215 subsamples. The samples analyzed in this work are marked by an *X*.

Age spectra of individual samples

For the gray aphanite spheroid 38,32, seven plateau fractions define an age of 4.09 ± 0.01 AE. The form of the age spectrum is quite similar to that of gray spheroid 38,57 (Jessberger *et al.*, 1976a), which has a plateau age of 4.16 ± 0.03 AE. The two black aphanite clasts 46,19 and 46,7 give plateau ages of 4.08 ± 0.01 and 4.02 ± 0.01 AE, respectively, and the previously dated black clast 46,10 has an age of 4.24 ± 0.01 AE. In spite of the different absolute ages, the structure of the age spectra is similar for all three samples. Quite clearly, real diversity of plateau ages exists even among petrologically very similar materials. The same conclusion can be drawn from a comparison of light-gray matrix 46,44 (plateau age 4.02 ± 0.01 AE) with gray matrix 73,1 (4.15 ± 0.01 AE, Jessberger *et al.*, 1976a).

The black matrix 36,2 has a plateau age of 3.98 ± 0.01 AE. In this case the distinctions between Regimes 2 and 3, as well as the corresponding K/Ca variations, are not as apparent as in the other cases, so that one could equally well justify a plateau age of 3.95 ± 0.01 AE based on the combined data from Regimes 2 and 3. Another black matrix sample 41,1 gave a plateau age of 4.04 ± 0.03 AE (Jessberger *et al.*, 1976a), but it should be noted that judging from the K concentration these two samples are chemically different.

The lowest of the plateau ages is obtained for the extremely K-rich felsite clast 43,3. The sample exhibits an extraordinarily well-defined plateau that yields an age of 3.92 ± 0.01 AE (Fig. 1). This age is in excellent agreement with the 3.90 ± 0.05 AE Rb–Sr internal isochron age for felsite reported by Compston *et al.* (1977).

K and Ca concentrations; Ca–^{38}Ar exposure ages

Figure 3 lists K and Ca concentrations inferred from the total quantities of ^{39}Ar and ^{37}Ar released. The values are accurate to within ±5%. Black matrix 36,2 has an exceptionally high concentration of K, 0.355%, probably due to an unusually large proportion of small felsite clasts in this sample (James, pers. comm.). The Ca concentrations of all the aphanitic samples are uniform and range from 6.4 to 7.7%. Compared with the chemical analyses of Blanchard *et al.* (1976) our Ca data are systematically lower by 8–10%. For K, differences range up to 25%, but they are not systematic. As an example, for 46,7 our 1630 ppm K is to be compared with 1660 ppm from Blanchard *et al.* Our 4.4 mg felsite sample contains 4.71% K; Blanchard *et al.* (1977) report 5.8% K for the bulk 13.6 mg clast from which our sample was taken.

The Ca–^{38}Ar exposure ages of the samples analyzed in this work are, within the limits of error, consistent with our previously determined values. Taking into account the new data, the total mean value is reduced from 250 ± 11 m.y. (Jessberger *et al.*, 1976a) to 244 ± 9 m.y., which closely agrees with the Kr–Kr age of 243 ± 7 m.y. (James *et al.*, 1975). Because of its high K/Ca ratio no exposure age was determined for the felsite. Instead, we have used the data to estimate the production rate of $^{38}Ar_C$ from potassium by assuming an exposure age of 244 m.y. for the clast:

$$^{38}P_K = 0.70 \times 10^{-8} \text{ cc STP } ^{38}Ar_c/\text{g K, m.y.}$$

Reported literature values (0.54: Turner *et al.*, 1973; 1.4: Turner *et al.*, 1971) are less accurate since they are deduced from less K-rich samples.

DISCUSSION

The newly acquired data have confirmed the age diversity of the constituents of breccia 73215 found by Jessberger *et al.* (1976a) and extended the range of observed ages. The total range of matrix ages is now 3.98–4.15 AE, as compared to 4.04–4.15 AE from the previous data, and the total range of aphanitic clast ages is now 4.02–4.24 AE, compared to 4.16–4.24 AE found previously. In addition, an important minimum age of 3.92 AE has been obtained from the felsite clast. Obviously, the breccia components were in general not isotopically equilibrated during breccia formation, and even the lowest of the ages determined in the matrix samples must still be regarded as only an upper limit for the time of the event in which the breccia was assembled. This is also supported by the findings of Müller *et al.* (1977). The felsite clast, on the other hand, is a component which was completely reset when the breccia formed. This is inferred from the combination of three observations:

(1) The well-defined age plateau indicates identical ages for the crystalline and the glassy components in the clast.
(2) The felsic glass was a melt when the clast was incorporated in the breccia and the clast temperature was $\geqslant$990°C (James and Hammarstrom, 1977; Nord and James, 1977).
(3) The transmission electron microscopy studies indicate that probably no

significant (>500°C) reheating of the clast occurred after the breccia-forming event (Nord and James, 1977). Consequently, the date of the 73215 breccia-forming event is 3.92 ± 0.04 AE (absolute error included). This age for the breccia-forming event is in agreement, within error limits, of the date for this event given by Müller *et al.* (1977) and Compston *et al.* (1977).

A 3.92 AE age of breccia 73215 has severe implications for lunar basin chronology, since it is generally believed that this as well as other Apollo 17 breccias are consolidated ejecta of the southern Serenitatis basin-forming impact (e.g., Reed and Wolfe, 1975; Morgan *et al.*, 1976). This implies:

(1) Since the youngest circular basin, Orientale, is certainly older than 3.78 AE (Kirsten and Horn, 1974), and since Serenitatis is stratigraphically the oldest circular basin, it follows that at least 13 circular basins (Stuart-Alexander and Howard, 1970; Hartmann and Wood, 1971) were formed within at most 140 m.y. only, between 3.92 and 3.78 AE (rather than between 4.0 and 3.78 AE, as was argued previously by Kirsten and Horn, 1974; Horn and Kirsten, 1977; Jessberger *et al.*, 1974; Turner, 1977).

(2) Almost none of the Apollo 17 highland breccias were completely reset in the event that formed them, since their radiometric ages cluster around 4.0 AE rather than 3.92 AE (Horn and Kirsten, 1977; Kirsten, 1977). Some of the other Apollo 17 breccias do also clearly record the ~3.9 AE event, however (77135, Tatsumoto *et al.*, 1974; 77135, Stettler *et al.*, 1975). One would have to conclude that isotopic equilibration of basin ejecta is the exception rather than the rule. [^{40}Ar–^{39}Ar dating of hornblende separates from the Nördlinger Ries provided further evidence for the stability of the K–Ar system in an impact event (Jessberger *et al.*, 1976b).]

We concede that there is a general problem in determining which lunar event formed any given lunar breccia, and one cannot be absolutely certain that 73215 is a sample of consolidated Serenitatis ejecta. On the other hand, the interpretation that the typical Apollo 17 highland rock ages are all older than the date(s) of the event(s) that formed them is a logical consequence of the K–Ar data on 73215 [including the data on ANT-suite clasts presented by Jessberger *et al.* (1976a)]. All matrix samples contain clasts which were not completely outgassed when the breccia formed and the melt-derived groundmass crystallized. Incomplete outgassing is also demonstrated by the results of laser-probe ^{40}Ar–^{39}Ar dating of 73215 samples. The laser-probe data show that clasts as small as 100 μm across were not completely outgassed (Müller *et al.*, 1977). The laser data for the groundmass itself yield an age of 3.97 ± 0.03 AE (since all groundmass melt formed contemporaneously, we take the lowest groundmass age reported by Müller *et al.* (1977) rather than the average, 4.01 AE); this age is in marginal agreement with the felsite age of 3.92 ± 0.04 AE.

To refine our estimate of the age of the Serenitatis Basin, we shall consider the following relevant age figures:

73215 felsite Rb–Sr internal isochron	3.90 ± 0.05 AE (Compston *et al.*, 1977)
73215 felsite ^{39}Ar–^{40}Ar isochron age	3.92 ± 0.04 AE (this work)
77135 Rb–Sr and ^{39}Ar–^{40}Ar isochron ages	3.90 ± 0.03 AE (Tatsumoto *et al.*, 1974; Stettler *et al.*, 1975)
77115,75 ^{39}Ar–^{40}Ar matrix age	3.91 ± 0.04 AE (Stettler *et al.*, 1975)
73215,131 melt-derived groundmass ^{39}Ar–^{40}Ar laser age	3.97 ± 0.03 AE (Müller *et al.*, 1977)
73215,36,2 lowest ^{39}Ar–^{40}Ar age of matrix	3.98 ± 0.04 AE (this work)
Apollo 17 highland breccias mean age	3.98 AE (e.g., Kirsten, 1977; Schaeffer and Schaeffer, 1977)

Considering the error limits associated with the ages enumerated above, a safe conclusion is certainly that Serenitatis formed 3.95 ± 0.05 AE ago. A more precise estimate depends on evaluation of the errors. One could argue that the actual age is near 3.95 AE and there is virtual agreement of the felsite age with the majority of Apollo 17 rock ages. We feel, however, that the 73215 results of our laboratory promote the alternative interpretation which is that the felsite age is in truth lower than the ages of about 3.98 AE, and that the samples showing the older ages were not completely reset.

The foregoing discussion centered around the question whether there exist at least some matrix samples which were completely equilibrated when the breccia formed. However, this discussion should not obscure the fact that the general pattern is that most matrix samples were *definitely not* reset on this occasion but are distinctly older than the felsite. This demonstrates in any case the limited resetting power of basin-forming events. The implications of this observation have been discussed previously (Jessberger *et al.*, 1976a) and shall not be repeated here. Additional evidence has been obtained by dating the Apolllo 16 breccia 67435, in which 4.4 AE old millimeter-sized clasts survived the breccia-forming event (Jessberger *et al.*, 1977). We admit, however, that the quick cooling experienced by 73215 [subsolidus cooling ~60 days (James, 1976)] may not be generalized for other breccias and other conditions could apply in thick hot ejecta blankets.

The age variability of the aphanitic samples can be explained if they are considered as mixtures of low K inherited clasts which retain Pre-Serenitatis ages, a relatively low-K phase (groundmass?) that was largely or completely degassed in the breccia-forming event, and felsite that was completely degassed. This leads to a rough anticorrelation of K content and apparent age in the more felsite-rich samples. For example, if we compare the K and age "balance" for black matrix samples 41,1 and 36,2 and the felsic clast 43,3, both K contents and apparent ages would be consistent with 4% more felsite in 36,2 than in 41,1. In contrast, the age variations shown by the less felsite-rich aphanites and ANT-suite clasts cannot be explained in this way. Instead, their age diversity must date back to the pre-excavation stratification, that is, mixing in of clasts from preexisting rocks of different ages.

A final point of concern is the question of the age relationship between aphanitic clasts and aphanitic matrix. The aphanitic clasts are lithologically like the matrix aphanites (i.e., they consist of mixtures of small clasts and a groundmass that crystallized from melt), but they form distinct clast-like bodies within the breccia. It is possible that these bodies actually do represent clasts of preexisting rock; alternatively, they could be co-genetic with the matrix aphanites but formed under slightly different conditions. James (1976) compared the petrologic characteristics of these clasts with those of matrix aphanites, and James and Blanchard (1976) summarized the data obtained by various consortium members in an attempt to resolve the question of genesis. In brief, most of this evidence favors the interpretation that all types of aphanite are co-genetic, but strong evidence to the contrary came from the previous ^{40}Ar–^{39}Ar stepwise heating results (Jessberger *et al.*, 1976a): These results showed that the clast aphanites tended to have older plateau ages than the matrix aphanites. Even with the better statistics achieved in this work, there is still a tendency for aphanitic clasts to be older than matrix, in spite of similar "miniclasts"/groundmass proportions. However, the new age for 46,7 (4.02 AE old clast) has increased the age range overlap (matrix 3.98–4.15 AE; aphanitic clasts 4.02–4.24 AE). Possibly the difference is in part a statistical phenomenon. On the other hand, it is still conceivable that some aphanitic clasts represent preexisting impact breccias even though the petrological and chemical evidence does not favor such interpretation.

Acknowledgment—We wish to thank O. James for many valuable suggestions and discussions.

References

Blanchard D. P., Jacobs J. W., Brannon J. C., and Haskin L. A. (1976) Major and trace element compositions of matrix and aphanitic clasts from consortium breccia 73215. *Proc. Lunar Sci. Conf. 7th*, p. 2179–2187.

Blanchard D. P., Jacobs J. W., and Brannon J. C. (1977) Chemistry of ANT-suite and felsite clasts from consortium breccia 73215 and of gabbroic anorthosite 79215. *Proc. Lunar Sci. Conf. 8th*. This volume.

Compston W., Foster J. J., and Gray C. M. (1977) Rb–Sr systematics in clasts and aphanites from consortium breccia 73215. *Proc. Lunar Sci. Conf. 8th*. This volume.

Hartmann W. K. and Wood C. A. (1971) Moon: Origin and evolution of multi-ring basins. *The Moon* **3**, 3–78.

Horn P. and Kirsten T. (1977) Lunar highland stratigraphy and radiometric dating. In *The Moon, A New Appraisal from Space Missions and Laboratory Analyses*. In press. Also published in *Phil. Trans. Roy. Soc. London* **A285**, 145–150.

James O. B. (1976) Petrology of aphanitic lithologies in consortium breccia 73215. *Proc. Lunar Sci. Conf. 7th*, p. 2145–2178.

James O. B. and Blanchard D. P. (1976) Subsamples of consortium breccia 73215: Spatial relations and lithologic correlation. *Proc. Lunar Sci. Conf. 7th*, p. 2131–2143.

James O. B. and Hammarstrom J. G. (1977) Petrology of four clasts from consortium breccia 73215. *Proc. Lunar Sci. Conf. 8th.* This volume.

James O. B., Brecher A., Blanchard D. P., Jacobs J. W., Brannon J. C., Korotev R. L., Haskin L. A., Higuchi H., Morgan J. W., Anders E., Silver L. T., Marti K., Braddy D., Hutcheon I. D., Kirsten T., Kerridge J. F., Kaplan J. R., Pillinger C. T., and Gardiner L. R. (1975) Consortium studies of matrix of light-gray breccia 73215. *Proc. Lunar Sci. Conf. 6th*, p. 547–577.

Jessberger E. K., Dominik B., Kirsten T., and Staudacher Th. (1977) New ^{40}Ar–^{39}Ar ages of Apollo 16 breccias and 4.42 AE old anorthosites (abstract). In *Lunar Science VIII*, p. 511–513. The Lunar Science Institute, Houston.

Jessberger E. K., Huneke J. C., Podosek F. A., and Wasserburg G. J. (1974) High resolution argon analysis of neutron-irradiated Apollo 16 rocks and separated minerals. *Proc. Lunar Sci. Conf. 5th*, p. 1419–1449.

Jessberger E. K., Kirsten T., and Staudacher Th. (1976a) Argon–argon ages of consortium breccia 73215. *Proc. Lunar Sci. Conf. 7th*, p. 2201–2215.

Jessberger E. K., Dence M., Dominik B., Hartung J., Kirsten T., and Schaeffer O. A. (1976b) Crater chronology: Resetting of K–Ar ages by craters of 1–20 km size from studies on the Ries and Brent Craters (abstract). In *Papers Presented to the Conference on Comparisons of Mercury and the Moon*, p. 19. The Lunar Science Institute, Houston.

Kirsten T. (1977) Lunar highland chronology. Submitted to *Proceedings of the 2nd Symposium on the Origin and Distribution of the Elements.* Paris, UNESCO (IAGC).

Kirsten T. and Horn P. (1974) Chronology of the Taurus-Littrow Region III: Ages of mare basalts and highland breccias and some remarks about the interpretation of lunar highland rock ages. *Proc. Lunar Sci. Conf. 5th*, p. 1451–1475.

Morgan J. W., Gros J., Takahashi H., and Hertogen J. (1976) Lunar breccia 73215: Siderophile and volatile trace elements. *Proc. Lunar Sci. Conf. 7th*, p. 2189–2199.

Müller H. W., Plieninger T., James O. B., and Schaeffer O. A. (1977) Laser-probe ^{39}Ar–^{40}Ar dating of materials from consortium breccia 73215. *Proc. Lunar Sci. Conf. 8th.* This volume.

Nord G. L. and James O. B. (1977) Aphanitic matrix, an ANT-suite clast and a felsite clast in consortium breccia 73215: An electron petrographic study. *Proc. Lunar Sci. Conf. 8th.* This volume.

Reed V. and Wolfe E. (1975) Origin of the Taurus-Littrow massifs. *Proc. Lunar Sci. Conf. 6th*, p. 2443–2461.

Schaeffer O. A. and Schaeffer G. A. (1977) ^{39}Ar–^{40}Ar ages of lunar rocks (abstract). In *Lunar Science VIII*, p. 840–842. The Lunar Science Institute, Houston.

Stettler A., Eberhardt P., Geiss J., Grögler N., and Guggisberg S. (1975) Age sequence in the Apollo 17 Station 7 boulder (abstract). In *Lunar Science VI*, p. 771–773. The Lunar Science Institute, Houston.

Staudacher Th., Jessberger E. K., and Kirsten T. (1977) ^{40}Ar–^{39}Ar age systematics of consortium breccia 73215. II. (abstract). In *Lunar Science VIII*, p. 896–898. The Lunar Science Institute, Houston.

Stuart-Alexander D. E. and Howard K. A. (1970) Lunar maria and circular basins—A review. *Icarus* **12**, 440.

Tatsumoto M., Nunes P. D., Knight R. J., and Unruh D. M. (1974) Rb–Sr and U–Th systematics of boulders 1 and 7, Apollo 17 (abstract). In *Lunar Science V*, p. 774–776. The Lunar Science Institute, Houston.

Turner G. (1977) Potassium-argon chronology of the moon. *Phys. Chem. Earth* **10**, 145–195.

Turner G., Cadogan P., and Yonge C. (1973) Argon selenochronology. *Proc. Lunar Sci. Conf. 4th*, p. 1889–1914.

Turner G., Huneke J., Podosek F., and Wasserburg G. (1971) ^{40}Ar–^{39}Ar ages and cosmic ray exposure ages of Apollo 14 samples. *Earth Planet. Sci. Lett.* **12**, 19–35.

Proc. Lunar Sci. Conf. 8th (1977), p. 2581–2592.
Printed in the United States of America

Matrix glass vs. intruded glass in lunar breccia 15286

C.A. HANDWERKER, L. C. KLEIN, P. I. K. ONORATO, and D. R. UHLMANN

Department of Materials Science and Engineering, Massachusetts Institute of Technology, Cambridge, Massachusetts

Abstract—The viscous flow and crystallization behavior of the matrix composition of breccia 15286 have been determined over a wide range of temperature. The temperature ranges covered by the viscous flow data are 1159–1307°C and 644–770°C. The variation of viscosity with temperature is closely described by the semi-empirical model of Bottinga and Weill (1972), and even more closely by the model of Shaw (1972). The crystal growth rate has been determined over the temperature range 821–1185°C and has a maximum of about 10^{-2} cm min^{-1} at 1090°C.

Time-temperature-transformation (TTT), logarithmic cooling, and continuous cooling (CT) curves were constructed from the measured viscosities and crystal growth rates assuming a nucleation barrier of 60 kT at a relative undercooling, $\Delta T/T_E$, of 0.2. From the TTT and CT curves, it is estimated that the matrix of the 15286 breccia cooled at a rate greater than 80°K min^{-1} through the glass transition region. Annealing of the matrix glass in the breccia increases its density, which suggests that the breccia formed by cooling from the molten state rather than by direct shock-induced transition from crystal to glass.

An analysis of viscous sintering under continuous cooling and logarithmic cooling conditions indicates that significant sintering will be achieved for particles which come into contact at any temperature about or somewhat below the liquidus temperature.

I. INTRODUCTION

IN A PREVIOUS STUDY of the glassy intrusion of the 15286 breccia (Uhlmann and Klein, 1976) it was suggested that the glass was formed by the intrusion of molten material into cold rock. The cooling rate estimated from a heat flow analysis was consistent with (exceeded) the minimum for forming this composition as a glass estimated using a kinetic analysis developed earlier (Uhlmann *et al.*, 1975). These results suggested that a study be undertaken of the matrix glass of the breccia to determine the conditions of its formation for comparison with the glassy intrusion.

The kinetic analysis used a combined treatment of the crystallization and sintering processes. The analysis is based on the construction of continuous cooling (CT) curves which are obtained from time-temperature-transformation (TTT) curves for an indicated degree of crystallinity.

The TTT and CT curves corresponding to a just-detectable volume fraction crystallized ($V_c/V = 10^{-6}$) are shown in Fig. 1 for the glassy intrusion in the 15286 breccia. The continuous cooling curve indicates that it would be necessary to cool this composition at a rate of 120°K min^{-1} or higher to produce the observed glassy intrusion.

The present paper reports the application of the kinetic analysis to the matrix composition of the 15286 breccia. In applying this analysis, measurements have

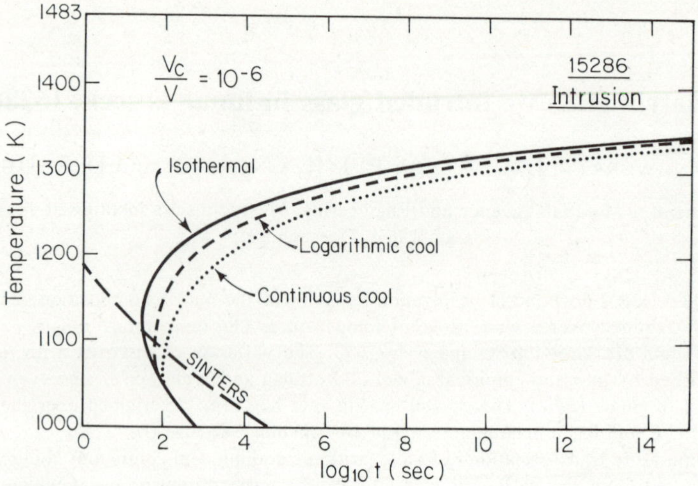

Fig. 1. Isothermal time-temperature-transformation, logarithmic cooling and constant-rate continuous cooling relations for the glassy intrusion in lunar breccia 15286. Volume fraction crystallized (V_c/V) = 10^{-6}. After Uhlmann and Klein (1976).

been made of the viscous flow and crystallization behavior of the matrix; and these results are compared with corresponding data on the glassy intrusion. The conclusions of the kinetic analysis are viewed in light of heat flow calculations for various geometries of interest.

The 15286 sample is a vesicular, vitric-matrix microbreccia weighing 34.6 g which was collected at Station 6 on the Apennine Front. Its dimensions are $5.5 \times 3.5 \times 3$ cm, and appears to have broken from a larger rock. The formation of the glassy intrusion and the formation of the matrix glass are treated as two separate events. This is in accord with the sharp breccia-intrusion interfaces reported by Wosinski *et al.* (1973). The grayish-black glassy intrusion, which makes up about 60% of sample 15286, is approximately 1 cm in thickness. The brown matrix glass, which makes up about 38% of the sample, exhibits clear evidence of devitrification, primarily around vesicles. Wosinski *et al.* (1973) also report the presence of 3–25 μm metallic inclusions in the areas of devitrified matrix glass, as compared to less than 1 μm inclusions in the intruded glass. The presence of the metallic particles which are largely Fe and Ni was associated with subsolidus reduction.

The compositions of the intrusion and the matrix are compared in Table 1. Both compositions were determined by electron microprobe analyses of representative areas of the lunar samples. As shown there, the compositions of the homogeneous glass intrusion and the matrix of the breccia are similar.

II. Experimental Procedure

The materials studied in this investigation were prepared from reagent grade chemicals. The melting procedure, sample preparation method and measuring techniques for the viscous flow study

Table 1. Compositions (wt.%) of intrusion and matrix of breccia 15286.

Intrusion		Matrix
46.1	SiO_2	47.6
1.6	TiO_2	1.2
14.3	Al_2O_3	13.3
14.2	FeO	13.6
12.5	MgO	13.6
10.4	CaO	9.8
0.8	Na_2O	0.7
0.1	K_2O	0.3

have been described previously (Cukierman *et al.*, 1972). The melting was carried out in multi-kilogram batches in a large-volume, Mo-wound furnace. This promotes homogeneity in composition and avoids many problems associated with the volatility of constituents in the batch. Viscosities in the molten range were determined using a rotating-cylinder viscosimeter, while viscosities greater than about 10^8 poise were determined using a bending-beam viscosimeter. Both viscosimeters were provided with an atmosphere at 5% H_2-forming gas; both were calibrated using NBS reference materials.

The measurements of crystal growth rates were carried out on small glass specimens cut from the material used in the viscous flow study. The oxygen fugacity was controlled to the conditions suggested by Sato *et al.* (1973). A solid ceramic eletrolyte cell was used to monitor the oxygen fugacity continuously in the furnace (Williams, 1972).

III. Viscous Flow and Crystallization Behavior

The variations of viscosity with temperature for the matrix and inclusion compositions are shown in Fig. 2. The dashed interpolations of each curve represent regions of temperature for which reliable flow data could not be obtained because of crystallization. Data in the high-temperature range below the liquids were obtained by superheating the specimens well above the respective liquidus temperatures prior to cooling.

The glass-transition temperature ($\eta = 10^{13}$ poise) for the intrusion is 650°C, while that for the matrix composition is about 644°C. The apparent activation energies for flow in the high-temperature region are about 60 kcal mole^{-1} for the intrusion and 135 kcal mole^{-1} for matrix composition; the corresponding values in the low-temperature region above the glass transition are about 180 kcal mole^{-1} for the glassy intrusion and 140 kcal mole^{-1} for the matrix composition.

The melting conditions used in preparing the present samples are somewhat more reducing than those generally estimated for the lunar environment (Sato *et al.*, 1973). It has been reported, however, that the 15286 sample contains metallic inclusions (Wosinski *et al.*, 1973), which are believed to be the products of subsolidus reduction. In any case, differences in the state of reduction over the range of concern for the moon have been shown to have an insignificant effect on liquid viscosity in the high-temperature region for lunar composition 15555,

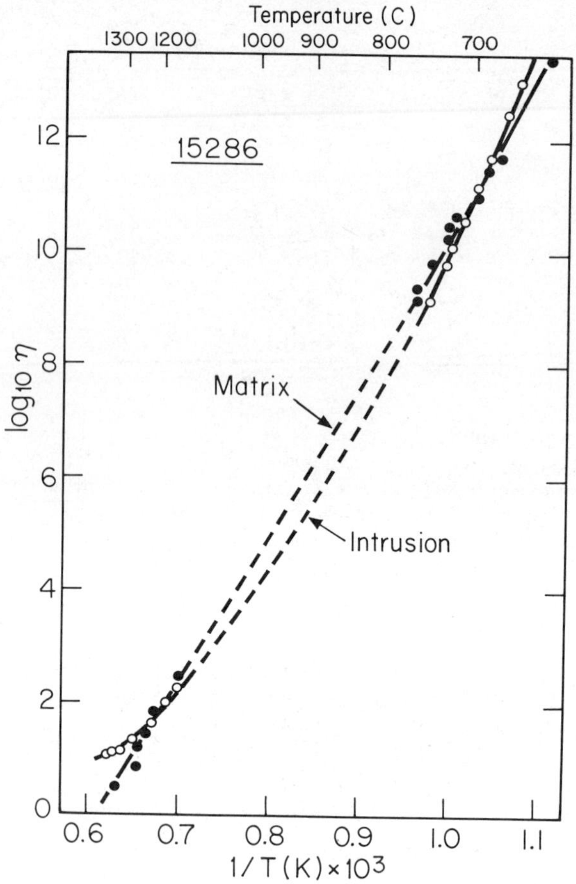

Fig. 2. Viscosity vs. temperature relations for the matrix and intrusion compositions in lunar breccia 15286.

which contains 22.5 wt.% FeO (Cukierman and Uhlmann, 1974). Since the present compositions contain only 14.2 wt. % FeO (intrusion) and 13.6 wt.% FeO (matrix), the effects of the state of reduction on melt viscosity are expected to be minimal.

For both the matrix and inclusion liquids, the high-temperature flow data are predicted within 0.2 in $\log_{10} \eta$ by the semi-empirical model of Bottinga and Weill; and the predictions of the semi-empirical model of Shaw (1972) provide even better agreement with the experimental data.

The overall variation of viscosity with temperature for the 15286 matrix composition falls within the range of viscosity-temperature relations found for other lunar materials. In detail, however, the apparent activation energy for flow on the high-temperature region is appreciably higher than that for other lunar compositions; and the change in apparent activation energy with temperature is correspondingly smaller. The origin of this difference in flow behavior is not clear.

The matrix and intrusion compositions both have moderate-to-high SiO_2 + Al_2O_3 contents—about 60.9 and 60.4 wt.%, respectively. Both show strong temperature dependences in their viscous flow behavior; they have fairly high viscosities at their respective liquidus temperatures and show rapid increases in viscosity with falling temperature below the liquidus. These characteristics make both compositions moderately good glass formers (Uhlmann *et al.*, 1974); but by no means are they comparable with commercial glasses or with lunar composition 67975 in glass-forming ability.

At all temperatures, the crystallization of both matrix and intrusion compositions proceeded uniformly from the external surfaces of the samples, and the thickness of the crystal layer increased linearly with time at all temperatures. Growth rates were determined from the slopes of least-squares lines drawn through the thickness vs. time-at-temperature plots. The growth rates of the two compositions vary with temperature as shown in Fig. 3. The maximum growth rate of both compositions is about 1.1×10^{-2} cm min^{-1}; it occurs at an undercooling of about 120°C for the intrusion ($T_{\mathrm{LIQUIDUS}} = 1210°C \pm 10°C$), and at an undercooling of about 180°C for the matrix ($T_{\mathrm{LIQUIDUS}} = 1270°C \pm 10°C$).

The forms of the growth rate vs. temperature relations shown in Fig. 3 are similar to those generally found for glass-forming materials. The maximum reflects a competition between the drive force for crystallization (which increases with increasing undercooling) and the molecular mobility (which decreases with increasing undercooling). The magnitudes of the maximum growth rates for the matrix and intrusion compositions are similar to those

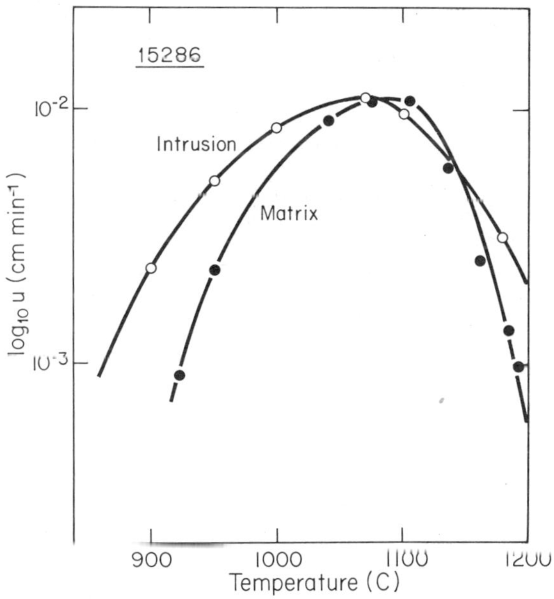

Fig. 3. Crystal growth rates vs. temperature for matrix and inclusion compositions in lunar breccia 15286.

measured for a variety of other lunar compositions (e.g., Uhlmann *et al.*, 1974).The similarity in the maximum growth rate for matrix and intrusion very likely reflects the higher liquidus temperature of the matrix being compensated (at least in part) by its higher viscosity at a given temperature in the range of rapid crystallization.

IV. Breccia Formation: Cooling Rate, Coalescence, and Heat Flow

The kinetic model for breccia formation (Uhlmann *et al.*, 1975) can be used to describe the conditions under which the 15286 breccia formed. First, a time-temperature-transformation (TTT) curve was constructed for the volume fraction of crystals (V_c/V) observed in the matrix of the breccia ($V_c/V = 10^{-2}$). The TTT curve was constructed from the data on growth rate and viscosity, together with an assumed barrier for homogeneous nucleation of 60 kT at a relative undercooling ($\Delta T/T_E$) of 0.2. Here ΔT is the undercooling and T_E is the liquidus temperature. This assumption is in accord with observations on the glass-forming tendencies of a number of other lunar compositions (Uhlmann *et al.*, 1974) as well as experimental results on the 70019 composition (Klein and Uhlmann, 1976).

Continuous cooling (CT) curves, which provide a description of crystallization under various conditions of continuous cooling, are constructed from the corresponding isothermal TTT curve following the approach of Grange and Kiefer (1941) as elaborated and applied to the problem of glass formation by Uhlmann and Onorato (1976). CT curves for logarithmic and constant-rate cooling for the matrix composition of the 15286 breccia, together with the TTT curve from which they were derived, are shown in Fig. 4. From the CT curve for constant-rate cooling, it is estimated that the matrix of the 15286 breccia cooled through the temperature range below the liquidus at a rate of about 80°K min^{-1}. This is typical of those cooling rates estimated for other Apollo 15 samples studied in our laboratory (15286 intrusion, 2.4×10^2 °K min^{-1}; 15498, 1.3×10^2 °K min^{-1}).

The cooling rate required to form a glass of the 15286 matrix composition has also been evaluated using the analysis of crystallization statistics (Hopper *et al.*, 1974). This analysis uses a crystal distribution function, $\psi(r, t, R)$, which contains complete statistical information about the state of crystallinity in the body. The crystal distribution function, ψ, has been related to the thermal history as a function of time and position, the nucleation frequency, and the growth rate. This function is defined such that the number of crystals at time t in a volume element dy at r having radii between R and $R + dR$ is

$$N = \int \psi(r, t, R)\, dv\, dR. \tag{1}$$

With this type of kinetic analysis, it is possible to describe crystallization which takes place at temperatures below the nose of the CT curve, an area which is neglected in estimating cooling rates from CT curves. Using the cooling rate

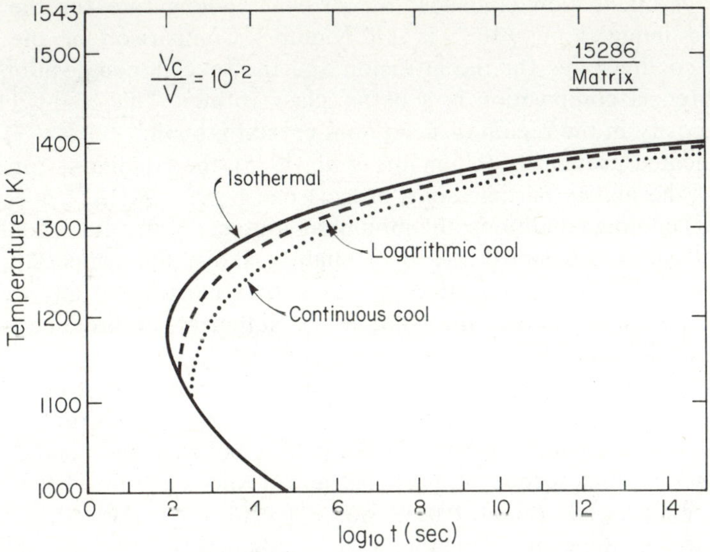

Fig. 4. Isothermal TTT, logarithmic cooling CT and constant-rate cooling CT curves for the matrix composition of lunar breccia 15286. Volume fraction crystallized = 10^{-2}.

estimated for the matrix composition from the CT analysis (80°K min^{-1}), the distribution function and the volume fraction crystallized were calculated for various temperatures during cooling. The relationship between the minimum cooling rates calculated from the CT analysis and from crystallization statistics are presented in a companion paper in this volume (Uhlmann *et al.*, 1977). For the matrix composition of 15286, the analysis of crystallization statistics indicates a fraction crystallized of about 8.1×10^{-4} at the temperature of the nose of the CT curve (1105°K), while the CT analysis indicates a fraction crystallized of 10^{-2}. This difference between the CT analysis and the analysis of crystallization statistics is larger than that found for the 67975 composition (Uhlmann *et al.* 1977), and suggests that the approximations used in the CT analysis are more appropriate for the 67975 material, and the variation of its crystallization kinetics with temperature.

Crystallization statistics also indicate that significant crystallization takes place at temperatures between the nose of the CT curve (T_N) and the glass transition (T_g). For example, for the 15286 composition cooled at 80° K min^{-1}, the volume fraction crystallized at T_g is about 9.6×10^{-4}. The increase by a factor of about 1.2 in volume fraction crystallized between T_N and T_g is found over a wide range of cooling rates for this composition. In general, the CT analysis overestimates the critical cooling rate for the 15286 composition by a factor of about 1.9. The cooling rate estimated by the analysis of crystallization statistics is about 42°K min^{-1}; and this value will be taken as the critical cooling rate of the 15286 breccia.

Using crystallization statistics, the critical cooling rate for the 15286 intrusion, assuming $V_c/V = 10^{-6}$, is 120°K min^{-1}. Comparison of the minimum estimated cooling rates for the intrusion and the breccia composition indicates that the breccia composition is a better glass former. This is reflected in the higher viscosity in the region of maximum crystal growth.

As indicated previously (Uhlmann *et al.*, 1975), the conditions for significant welding of the matrix particles in a breccia can be calculated. Under constant-rate (linear) cooling conditions, the minimum contact temperatures for significant welding of the 15286 matrix are only slightly above the temperatures of the curve. It is concluded that contact of the matrix particles at any temperature above or somewhat below the liquidus is sufficient to produce substantial sintering of the bodies for cooling rates of interest.

The present discussion of the 15286 breccia has been based on the assumption that this breccia formed by viscous sintering in a stress-free environment, rather than by shock lithification. To differentiate between glass bodies produced by the viscous coalescence of particles and a shock-produced glass, a small section of matrix glass (about 0.01 g) was extracted from the 15286 sample and subjected to an annealing schedule (1 hr at the glass transition temperature, followed by an overnight furnace cool to ambient). The index of refraction was determined by the Central Illumination Method (Wahlstrom, 1969) both prior to and after annealing. The index was observed to increase upon annealing, by about 0.004, indicating relaxation of the glass to a more dense state during the anneal.

An increase in the index of refraction upon annealing is characteristic of glasses initially cooled through the glass transition more rapidly than during the annealing treatment. Under these conditions, the glass will densify upon subsequent heat treatment in the vicinity of the glass transition. The increase in the index of refraction is contrary, however, to the annealing behavior of glasses produced directly by shock. Diaplectic glasses generally relax to a less dense state upon annealing and exhibit a corresponding decrease in index of refraction (e.g., Arndt, 1971). In contrast, investigations of shock-produced Apollo 11 soil and breccias, terrestrial maskelynite, as well as shocked quartz and plagioclase indicate an increase in index of refraction upon annealing (von Engelhardt, *et al.* 1970); Stöffler, 1974; Bunch *et al.*, 1967; Bunch *et al.*, 1968). In each case, the increase in refractive index was associated with the occurrence of crystallization upon annealing. As the annealing time increased, the index of refraction of the samples approached the index of the crystalline phase.

In the present study of the 15286 material, no detectable crystallization took place during annealing. The annealing treatment of the 15286 material was initiated at the glass transition temperature. For such an initial temperature of heat treatment, and for the annealing times used in the present work, the extent of crystallization is expected to be negligible—an expectation in accord with observation. The annealing results thus provide strong evidence that the matrix glass in the 15286 breccia formed by the cooling of molten material, rather than by a shock-induced crystal-glass transition.

V. Heat Flow Analyses

The thickness of the 15286 matrix material which can be obtained with the observed degree of crystallinity is estimated to order-of-magnitude accuracy as about 3 cm. This estimate is obtained from the relation:

$$y \approx (D_{TH}\,\tau_N)^{1/2}, \tag{2}$$

where the thermal diffusivity, D_{TH}, has been taken as 4×10^{-3} cm^2 sec^{-1}; y is the thickness; and τ_N is the time at the nose of the TTT curve. The lunar breccia 15286 has dimensions about $5.5 \times 3.5 \times 2$ cm; the glassy intrusion in the breccia is about $3.5 \times 2 \times 1$ cm. The thickness estimated from Eq. (2) is approximately equal to that of the lunar sample.

More precise estimates of the size of a body which can formed with the observed degree of crystallinity can be obtained from the two-stage cooling model suggested by Simonds (1975) and applied to the Apollo 17, Station 6 boulder by Onorato *et al.* (1976). This model treats the cooling behavior of a clast-laden melt in terms of two distinct stages, the first involving rapid heat flow from hot melt to cold clasts, and the second involving heat flow from the melt to its surroundings.

Calculations show that the temperature of thermal equilibration of the melt-clast system is above the nose of the CT curve for the 15286 matrix composition for any combination of initial clast content and initial temperature which gives the final volume fraction of clasts observed in the lunar sample.

This result indicates that for the partially crystalline 15286 matrix, heat flow to the surroundings is the primary mechanism of cooling through the region of the nose of the CT curve (cooling to sufficiently low temperatures that the rate of crystallization is insignificant). Three models have been employed in treating this cooling: (1) cooling of the melt sheet by conduction to a cold basement below and by radiation from the top surface; (2) cooling by conduction to a cold basement below and to a similar cold layer on top (corresponding to the melt sheet being covered by a cold blanket); and (3) cooling by radiation as a single body, modelled as a sphere. The last problem has been treated by Hopper *et al.* (1974a).

For a melt sheet cooled by conduction below and radiation above, it is found that the critical cooling rate (42°K min^{-1}) would be achieved at the slowest-cooling part of a body about 3.5 cm thick. For a melt sheet cooled by conduction at both surfaces, the critical cooling rate would be reached at the slowest-cooling part (the center) of a body about 3 cm thick; and for a spherical body cooling by radiation, the critical rate would be achieved at the center of a body between 5 and 10 cm in diameter.

It has been suggested that the 15286 breccia formed as part of a larger body. However, the heat flow analysis indicates that the center of any body of this composition which is much larger than the lunar sample could not have cooled from its liquidus temperature at the indicated rate (~42°K min^{-1}). Hence if the breccia did form as part of a larger body, it must have come from the peripheral regions of such a body.

The results reported previously for the glassy intrusion in the 15286 breccia suggested that molten material intruded into cold rock. This suggestion that the glassy intrusion was formed in a separate event from that which formed the breccia is in accord with the sharp interfaces between the breccia and the glass intrusion reported by Wosinski *et al.* (1973). These authors also reported the presence of sharp breccia-devitrified glass interfaces. The results presented here are in agreement with the suggestion of Wosinski *et al.* The liquidus temperature of the 15286 breccia composition (1270°C ± 10°C) is higher than the liquidus temperature of the glassy intrusion (1210°C ± 10°C). The difference in liquidus temperatures, along with the presence of sharp breccia-intrusion interfaces, suggests that the initial temperature of the molten material intruded into the cold breccia was greater than 1210°C but not much greater than 1270°C.

Crystallization statistics can also describe the state of crystallinity in a body which has been cooled down and subsequently reheated. The effect on the degree of crystallinity of intruding molten material into a previously cooled breccia can be estimated using the analysis, which has been applied to the 15286 matrix composition. If the breccia is reheated at a rate about 1–5°C sec^{-1}, which is appropriate for the heating of 1 cm body in contact with an intrusion at a temperature greater than 1210°C, the breccia would develop significant crystallinity at temperatures greater than about 825°C. The fact that most of the breccia did not devitrify indicates that the maximum average temperature to which the matrix was raised is less than 825°C. Evidence of devitrification near the matrix-intrusion interface indicates that locally the temperature was raised above 825°C.

VI. DISCUSSION AND CONCLUSIONS

The kinetic analysis of breccia formation by viscous sintering (Uhlmann *et al.*, 1975) can be used with the combined data on crystallization kinetics and viscosity to evaluate the cooling rates at which breccias formed. The increase in the index of refraction of the glassy matrix after annealing suggests that the 15286 breccia formed by viscous coalescence of clast-laden melt, rather than by shock lithification. The kinetic analysis can, therefore, be well applied to the 15286 breccia.

The minimum critical cooling rate has been estimated both from the construction of CT curves and from crystallization statistics. A comparison of the results of these two analyses indicates that the critical cooling rate calculated from the CT analysis overestimates the volume fraction crystallized. According to crystallization statistics, the minimum critical cooling rate for the 15286 breccia is about 42°K min^{-1}. This cooling rate is consistent with the breccia having formed in its present size. The heat flow analysis indicates that the breccia could not have cooled at a rate as high as 42°K min^{-1} if it came from the center of a body appreciably larger than 5 cm of a larger body.

The analysis of crystallization statistics applied to the matrix of the 15286 breccia indicates that the matrix would develop significant crystallinity if reheated to temperatures higher than 825°C. The small degree of crystallinity observed in the matrix indicates that the maximum average temperature to which the breccia was reheated on intrusion of the molten material is less than 825°C. Evidence of devitrification near the matrix-intrusion interface suggests that locally the temperature was raised to higher temperatures.

Acknowledgments—Financial support for the present work was provided by the National Aeronautics and Space Administration under grant NGR-22-009-646, and by Owens-Illinois Inc., who provided one of the authors (P.I.K.O.) with the Owens-Illinois Fellowship in Materials Science. This support is gratefully acknowledged.

REFERENCES

Arndt J., Hornemann U., and Müller W. F. (1971) Shock-wave densification of silica glass. *Phys. Chem. Glasses* **12**, 1–7.

Bottinga Y. and Weill D. F. (1972) The viscosity of magmatic silicate liquids: A model for calculation. *Amer. J. Sci.* **272**, 438–475.

Bunch T. E., Cohen A. J., and Dence M. R. (1967) Natural terrestrial maskelynite. *Amer. Mineral.* **52**, 244–253.

Bunch T. E., Cohen A. J., and Dence M. R. (1968) Shock-induced structural disorder in plagioclase and quartz. In *Shock Metamorphism of Natural Materials.* (B. M. French and N. M. Short eds.), p. 509–518. Mono Book Corp., Baltimore, Maryland.

Cukierman M., Tutts P. M., and Uhlmann D. R. (1972) Viscous flow behavior of lunar compositions 14259 and 14310. *Proc. Lunar Sci. Conf. 3rd*, p. 2619–2625.

Cukierman M. and Uhlmann D. R. (1974) Effects of iron oxidation state on viscosity, lunar composition 15555. *J. Geophys. Res.* **79**, 1594–1598.

Grange R. A. and Kiefer J. M. (1941) Transformation of austenite on continuous cooling and its relation to transformation at constant temperature. *Trans. ASM* **29**, 85–115.

Hopper R. W., Onorato P. I. K., and Uhlmann D. R. (1974a) Thermal histories and crystal distributions in partly devitrified lunar glasses cooled by radiation. *Proc. Lunar Sci. Conf. 5th*, p. 2257–2273.

Hopper R. W., Scherer G., and Uhlmann D. R. (1974b) Crystallization statistics, thermal history and glass formation. *J. Non-Cryst. Solids* **15**, 45–62.

Onorato P. I. K. and Uhlmann D. R. (1976) Nucleating heterogenities and glass formation. *J. Non-Cryst. Solids* **22**, 367–378.

Onorato P. I. K., Uhlmann D. R., and Simonds C. H. (1976) Heat flow in impact melts: Apollo 17 Station 6 Boulder and some applications to other breccias and xenolith laden melts. *Proc. Lunar Sci. Conf. 7th*, p. 2449–2467.

Sato M., Hickling N. L., and McLane J. E. (1973) Oxygen fugacity values of Apollo 12, 14, and 15 lunar samples and reduced state of lunar magmas. *Proc. Lunar Sci. Conf. 4th*, p. 1061–1079.

Shaw H. R. (1972) Viscosities of magmatic silicate liquids: An empirical method of prediction. *Amer. J. Sci.* **272**, 870–893.

Simonds C. H. (1975) Thermal regimes in impact melts and the petrology of the Apollo 17 Station 6 boulder. *Proc. Lunar Sci. Conf. 6th*, p. 641–672.

Stöffler D. (1974) Deformation and transformation of rock-forming minerals by natural and experimental shock processes. *Fortschr. Miner.* **51**, 256–289.

Uhlmann D. R. and Klein L. C. (1976) Crystallization kinetics, viscous flow and thermal histories of lunar breccias 15286 and 15298. *Proc. Lunar Sci. Conf. 7th*, p. 2529–2540.

Uhlmann D. R., Klein L. C., and Handwerker C. A. (1977) Crystallization kinetics, viscous flow and thermal history of lunar breccia 67975. *Proc. Lunar Sci. Conf. 8th*. This volume.

Uhlmann D. R., Klein L. C., Kritchevsky G., and Hopper R. W. (1974) The formation of lunar glasses. *Proc. Lunar Sci. Conf. 5th*, p. 2317–2331.

Uhlmann D. R., Klein L. C., Onorato P. I. K., and Hopper R. W. (1975) The formation of lunar breccias: Sintering and crystallization kinetics. *Proc. Lunar Sci. Conf. 6th*, p. 693–705.

von Engelhardt W., Arndt J., Müller W. F., and Stöffler D. (1970) Shock metamorphism of lunar rocks and origin of the regolith at the Apollo 11 landing site. *Proc. Apollo 11 Lunar Sci. Conf.*, p. 363–384.

Wahlstrom E. E. (1969) *Optical Crystallography*, p. 112–114. John Wiley and Sons, New York.

Williams R. J. (1972) A solid ceramic electrolyte system for measuring redox conditions in high temperature gas mixing studies. NASA TM X-58105. 16pp.

Wosinski J. F., Williams J. P., Korda E. J., and Geiger G. A. (1973) Glass and devitrified zones on lunar breccia 15286,3. *Proc. Lunar Sci. Conf. 4th*, p. 383–388.

Wosinski J. F., Williams J. P., Korda E. J., and Geiger G. A. (1973) Glass and devitrified zones in lunar breccia 15286,3 (abstract). In *Lunar Science IV*, p. 796–797. The Lunar Science Institute, Houston.

Proc. Lunar Sci. Conf. 8th (1977), p. 2593–2615.
Printed in the United States of America

Shock glass veins in some lunar and meteoritic samples—Their nature and possible origin

EDWIN ROEDDER

959 National Center, U.S. Geological Survey, Reston, Virginia 22092

PAUL WEIBLEN

Department of Geology and Geophysics, University of Minnesota, Minneapolis, Minnesota 55455

Abstract—Glassy veinlets, 15 μm to 1 mm thick and now in part devitrified, cut through breccias 67915, 67936, and 67955 from Outhouse Rock, and similar veins have been found in several other lunar samples and in meteorites. The veins have features suggesting injection of extremely hot material, possibly at several thousands degrees of superheat, presumably from an impact event. The vein compositions are very similar but not identical to the bulk compositions of the host rocks. Two mechanisms of injection are considered feasible—injection of hot liquid and injection of hot particulate matter (resembling a "fluidized bed"), but neither mechanism fully explains all details of the occurrences.

INTRODUCTION

THE PROCESSES THAT OCCUR at the moment of hypervelocity impact are complex and yield a wide spectrum of products. Melting of rock is common, as shown by the presence of glassy quenched material that forms droplets, coatings, and crosscutting glass veins in many lunar samples, as well as in some materials from terrestrial impact craters and certain meteorites. Several theories have been suggested for the mechanism of formation of these veins (e.g., Wilshire, 1971; Fredriksson *et al.*, 1963; Jackson *et al.*, 1975). We presented a brief description of some very small glass veins in Apollo 16 breccia sample 67915 from Outhouse Rock (Weiblen and Roedder, 1973) that have features that are difficult to explain by any of these theories. In this paper we give a fuller description of these veins, as well as possibly related veins in other samples, and explore the possible modes of origin.

PHYSICAL NATURE OF VEINS

Samples studied

The veins studied here occur in many parts of sample 67915 (subsamples 3, 12, 24, 45, 81, 83, and 84); 67936 (subsamples 3, 4, 5, and 13); and 67955,53. Samples 67915 and 67936 are dark-matrix breccias, and 67955 is a light-matrix breccia (Ulrich, 1973), all three samples are from Outhouse Rock on the edge of North Ray Crater in the Descartes region. Possibly similar veins, but of very different composition, were studied in 79155,66. For comparison, cursory optical examinations were made of somewhat similar veins in the meteorites Ibitira, Parandiba, Chantonnay, Cachari, Wethersfield, and Girgenti.

Size, shape, and orientation of the veins

Most of the veins in the lunar samples are <50 μm thick, and seldom are >100 μm, except in 67936, where anastomosing masses of glassy material several millimeters wide are common. Some veins are only ~15 μm thick. Individual veins change thickness as much as tenfold along their length, and such changes can take place very abruptly, within a few micrometers. Some branch and reunite around individual mineral clasts, and occasionally the whole vein changes direction as much as 90° to follow around an individual clast. When several veins occur in the same section, they are commonly but not always subparallel, and some veins obviously cut other, earlier veins (e.g., 67915,24-4, P2[1]).

Vein terminations are not commonly observed because most veins are truncated by the edges of the section. Some are observed to feather out, however, and several terminate rather abruptly, with blunt ends (67915,3-3, P1—see Fig. 1; 67915,3-3, P4; 67955,53, P10). Some short segments of veins that cut clasts in 67915 end abruptly at the clast boundary, and others are offset within single clasts (Fig. 2), indicating several stages of breccia formation. There is an obvious offset of the host rock of ~0.5 mm on opposite sides of one vein (Weiblen and Roedder, 1973, their Fig. 8), but such offsets do not appear to be general. A very few smooth-walled subspherical vesicles were found in the thickest parts of the largest veins in 67915 and 67936,13.

Petrography of the veins and host rocks

The veins in the lunar samples are various shades of yellow or brown, except those in 79155, which are red. They are generally zoned roughly parallel to the vein walls; the zones consist of trains of minute opaque spheres ≤~1 μm, or gross differences in opacity, from nearly opaque to colorless and nearly transparent (Fig. 3). The larger opaque spheres are globules of metal and/or sulfide. The index of refraction of the glass shows minor variations across a given vein, outlining striae or schlieren that sometimes also differ in color. These structures generally parallel the vein edge.

Some veins have a ≤5 μm thick marginal zone of pyroxene (see below), and veins both with and without such pyroxene zones are found in some samples (e.g., 67915,81). In one example (67915,3-3, P8a) the pyroxene appears euhedral against the plagioclase of the host rock.

The glass of the veins is isotopic or shows low birefringence and wavy extinction, suggesting devitrification. The devitrified glass is generally at the

[1]The numbers immediately following the dash in sample designations throughout this report refer to slides prepared in our laboratory; the numbers prefixed "P" are locating photograph numbers. Both slides and photographs will eventually be placed in NASA archives.

margins, and locally the extinction parallels that of plagioclase in the walls. Short segments of older veins within breccia clasts have higher birefringence (~0.01) and uniform, but not sharp, extinction over the entire vein, as though recrystallized to a single subparallel crystal group. Others have crystallized to a mass of random 1–2 μm crystals.

Angular to rounded 10–20 μm fragments of unshocked to weakly shocked pyroxene and olivine crystals (i.e., with reduced birefringence, undulatory extinction, and fine scale fracturing), generally <1%, are scattered through the glass of some of the veins (Fig. 5). The only plagioclase clasts within the veins are rare "islands" of polycrystalline plagioclase enclosed by the two arms of a branching vein. These various clasts have sharp contacts with the glass of the veins. No change in index, color, or composition of the host glass was detected adjacent to them, except for several small clasts of ilmenite that had short "tails" of reddish-brown glass and a higher index (67915,45-1c, P5 and Fig. 4). Schaal and Hörz (1977) report plagioclase clasts, maskelynite clasts, and colorless plagioclase schlieren in the glass veins in another sample of 79155 (slide 60).

The contacts of the veins with the walls are optically sharp to the resolution of the optical microscope. However, locally some of the wall materials adjacent to the veins have been partly melted. Olivine and pyroxene are unaffected, but plagioclase is commonly converted to glass. This glass is optically homogeneous and completely isotropic, and forms a zone as much as several times wider than the thickness of the vein. The contact of the outer edge of the isotropic zone with the adjacent plagioclase within a given crystal is sharp but irregular. All plagioclase in the zone has been isotropized; no differences were noted between the behavior of plagioclase in large single crystal clasts, in clasts of recrystallized maskelynite, or in the granular matrix of metamorphic rock clasts. Within the isotropic zone, whether it be in a single crystal or across a polycrystalline clast, all minute cracks and index variations in the plagioclase produced by previous shock deformation have disappeared, but small pyroxene or olivine crystals embedded in the plagioclase seem completely unchanged in position, shape, and birefringence (see Fig. 8) so we presume the plagioclase glass has not flowed. A similar band of clear plagioclase adjacent to a vein in 60015 is shown by Sclar and Bauer (1974; their Fig. 4), but it is wider and has apparently devitrified completely.

In a very few places there is some evidence of dragging of wall material along the length of the vein. Thus, in 67915,45-1d, P5 there are several iron sulfide globules in the vein within a few micrometers of a mass of sulfide plus ilmenite in the wall.

As glass veins have been reported in numerous meteorites (e.g., Fredriksson *et al.*, 1963; Fredriksson and Kraut, 1967), slides of a series of meteorites known to contain veins were also examined, including the following (USNM numbers): Ibitira (5280); Parandiba (2336); Chantonnay (176); Cachari (3209 and 3258); Wethersfield (5596); and Girgenti (2799). Although some of these are difficult to study because of weathering, etc., veins of isotropic glassy material visually similar to those in the lunar samples were found in all. These veins were in

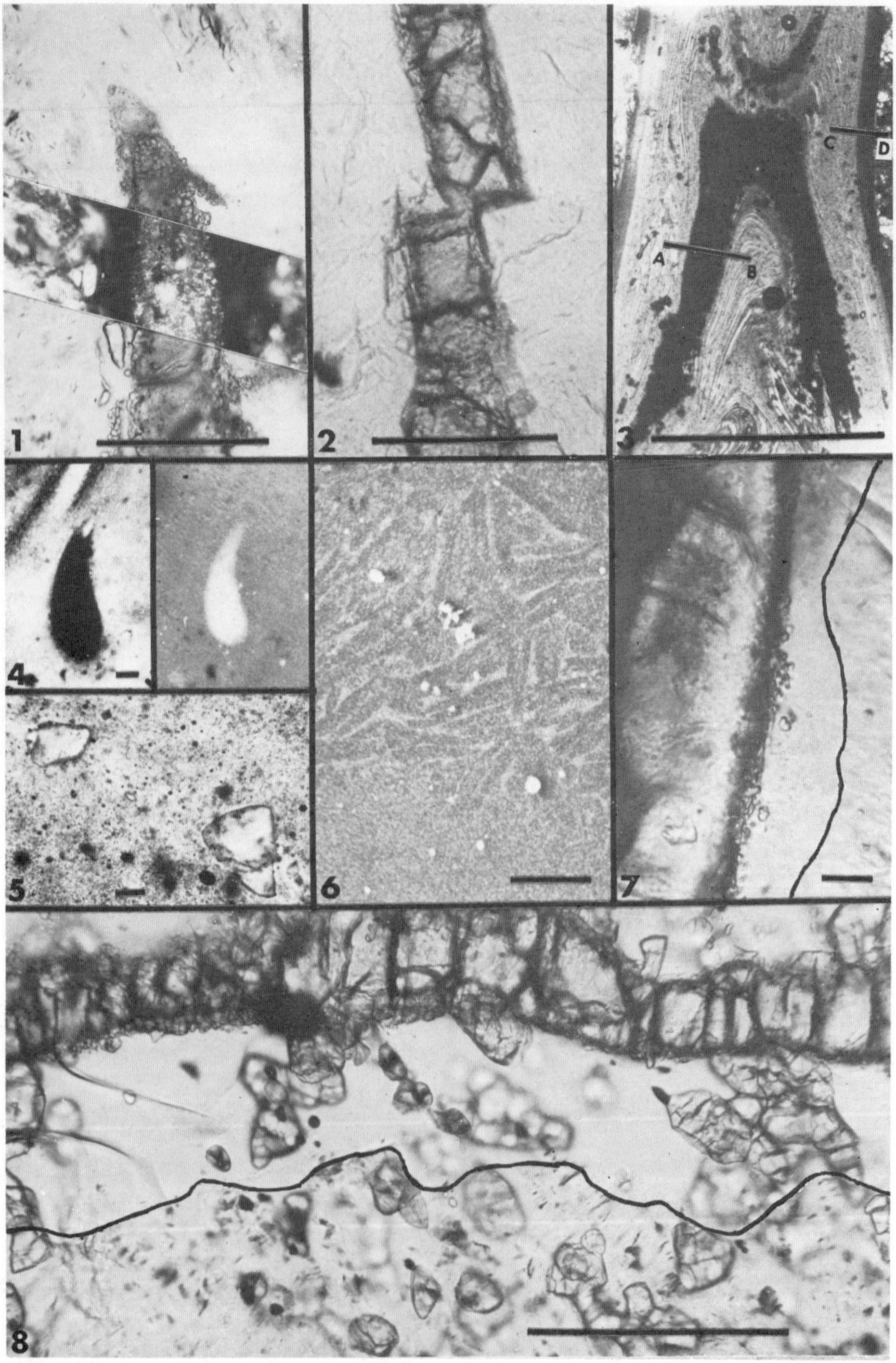
1
2
3
A
B
C
D
4
5
6
7
8

contact with plagioclase in three slides only, one from Ibitira and two from Cachari. Only one of these (Cachari 3209) showed any isotropic zone in the feldspar adjacent to the vein.

Chemical composition of the veins

Bulk composition of glassy part. Tables 1–3 present analyses of glass veins (and their devitrified equivalents and associated phases) from 67915, 67936, 67955, and 79155. These analyses were made with a 20 kV, 0.015 μA beam of ~20 μm diameter to minimize the effects of microheterogeneities. Since these are mostly ≤1 μm, they should introduce no serious errors (Albee *et al.*, 1977); mineral clasts and larger color inhomogeneities were avoided. The data were reduced using the method described in Roedder and Weiblen (1970, p. 821). All analyses of 67915 glass veins (Table 1, analyses 1–8) are very similar to those obtained earlier from

---

Fig. 1. Blunt ended vein in partially isotropized plagioclase; 67915,3-3, P1. Plane polarized transmitted light except for center band (crossed polars). Bar = 100 μm.

Fig. 2. Recrystallized and offset vein in plagioclase clast; in transmitted plane polarized light; 67915,81-3. Bar = 100μm.

Fig. 3. Zoned and striated vein in plane polarized transmitted light; 67936,5. Electron microprobe traverse A–B showed a somewhat higher and more variable concentration of Fe, Ti, and Mg in the dark zone (see Fig. 6), but traverse C–D showed an essentially uniform concentration in both the clear glass and the opaque zone. Bar = 1 mm. Clear zone in center—analysis 12.

Fig. 4. Partly dissolved ilmenite (?) clast in vein glass, in plane polarized transmitted (left) and reflected (right) light; 67915,45-1c, P5. The concentrations of Fe and Ti in the object are high and in about the correct ratio for ilmenite. Bar = 10 μm.

Fig. 5. Subangular crystal clasts in glass in plane polarized transmitted light; 67915,45-1d, P4. Bar = 10 μm.

Fig. 6. Scanning electron microscope image of polished surface at the edge of the dark zone near traverse A–B (Fig. 3), taken at 1200×. Bar = 10 μm. Dark zone (upper part) consists of prisms of lower average atomic weight (plagioclase?) with interstitial pyroxene (?). Note tiny spheres of metal.

Fig. 7. Vein-plagioclase contact at electron microprobe traverse Fig. 9, in plane polarized transmitted light, showing (from left to right): almost opaque isotropic vein, transparent devitrified vein border, granules of pyroxene at vein wall, isotropized feldspar zone, and unaffected birefringent, recrystallized maskelynite clast. The approximate boundary between the last two zones (at the slide surface) is indicated. 67915,84-17. Bar = 10 μm.

Fig. 8. Photomicrograph, in transmitted plane polarized light, showing zone of isotropized plagioclase at contact of vein (dark band at top) and anorthositic clast; 67915,76. The contact between isotropized and birefringent plagioclase has been drawn in. Note that although the pyroxene crystals and opaque grains in the isotropized zone appear unaffected, almost all minute cracks and lamellae (shock-induced?) that occur uniformly throughout the plagioclase of this ~1 mm clast are absent in the isotropic zone. Bar = 100 μm.

Table 1. Microprobe analyses and CIPW norms (wt.%) of shock glass veins (Paul W. Weiblen, analyst).

	67915								67936					67955				79155		
Anal. No.	(1)	(2)	(3)	(4)	(5)	(6)	(7)	(8)	(9)	(10)	(11)	(12)	(13)	(14)	(15)	(16)	(17)	(18)	(19)	(20)
Slide	3-1a-2	45-1	3-3	3-3	24-4	24-4	3-3	Avg.	3	3	4	5	Avg.	53	53	53	Avg.	66	66	Avg.
Lab. No.	12	34	L57	L58	L38	L37	L54	of	M177	M178	M187	M184	of	L67	L68	L69	of	M144	M146	of
Area	1-2	P7-4	P8	P8	P3	P3	P7	15	P0-1	P0-2	P1-2	P0-2	4	P3	P4	P5	3	P12-1	P13-2	2
SiO_2	43.6	48.1	44.3	43.5	43.6	45.6	48.6	45.71	46.6	46.2	46.0	45.5	46.08	46.0	46.5	46.5	46.33	44.4	42.7	43.6
Al_2O_3	31.5	27.8	32.5	31.2	32.0	33.0	27.7	30.51	27.4	23.8	24.3	25.0	25.13	29.0	30.0	30.5	29.83	12.0	12.4	12.2
FeO	2.80	3.48	1.08	6.56	2.48	1.22	2.84	2.69	4.30	6.32	5.92	6.44	5.74	3.24	2.80	2.82	2.95	16.5	19.5	18.0
MgO	2.17	2.13	3.89	3.15	3.86	2.30	2.62	2.86	5.04	9.26	6.97	7.07	7.08	5.51	4.69	4.37	4.86	7.81	8.04	7.92
CaO	17.5	16.8	18.5	15.9	17.3	18.3	17.1	17.47	16.4	14.3	15.0	15.4	15.28	15.7	14.9	15.9	15.51	10.6	10.6	10.6
Na_2O	0.27	0.56	<0.05	0.34	0.62	0.39	0.32	0.39	<0.05	<0.05	0.52	0.32	0.21	0.17	0.21	0.31	0.23	<0.05	0.24	0.12
K_2O	0.03	0.17	0.05	<0.05	0.05	0.05	0.10	0.06	0.03	0.05	0.02	0.02	0.03	0.06	0.07	0.06	0.06	<0.05	<0.05	<0.05
TiO_2	0.20	0.52	0.06	0.43	0.15	0.11	0.33	0.25	0.19	0.27	0.20	0.12	0.20	0.21	0.21	0.20	0.21	8.88	7.96	8.42
P_2O_5	—	—	—	<0.05	—	—	0.09	0.03	<0.05	<0.05	<0.05	<0.05	—	<0.05	<0.05	<0.05	<0.05	<0.05	<0.05	<0.05
MnO	0.03	0.03	<0.05	0.05	<0.05	<0.05	<0.05	0.06	<0.05	<0.05	0.06	<0.05	0.02	<0.05	<0.05	0.08	<0.05	0.23	0.20	0.22
Cr_2O_3	0.03	—	—	<0.05	<0.05	<0.05	<0.05	0.03	<0.05	<0.05	<0.05	<0.05	—	<0.05	—	—	<0.05	0.13	0.07	0.10
NiO	<0.05	—	—	<0.05	—	—	<0.05	—	—	—	—	—	—	—	—	—	—	—	—	—
Total	98.13	99.59	100.38	101.13	100.06	100.97	99.70	100.06	99.96	100.20	98.99	99.87	99.77	99.89	99.38	100.74	99.98	100.55	101.71	101.18
Norm																				
Q	—	4.8	—	—	—	0.1	6.2	0.6	1.8	—	—	—	—	0.4	3.9	1.7	2.0	7.2	0.6	3.9
C	—	—	—	1.7	1.4	—	—	—	—	—	—	—	—	0.1	2.5	1.0	1.2	—	—	—
or	0.2	1.0	0.3	—	0.3	0.3	0.6	0.4	0.2	0.3	0.1	0.1	0.2	0.4	0.4	0.4	0.4	—	—	—
ab	2.3	4.8	—	2.8	5.2	3.3	2.7	3.3	—	—	4.4	2.7	1.8	1.4	1.8	2.6	1.9	—	2.0	1.0
an	86.4	73.1	88.3	78.1	80.6	87.3	74.2	81.4	74.6	64.7	64.8	66.9	67.6	78.0	74.5	78.3	76.9	32.6	32.2	32.5
wo	0.9	4.4	1.3	—	2.2	1.1	4.3	2.1	2.8	2.6	4.4	4.0	3.5	—	—	—	—	8.3	8.2	8.2
en	4.3	5.3	5.3	4.0	—	5.7	6.5	7.1	12.6	17.7	12.5	10.5	14.4	13.7	11.7	10.8	12.1	19.4	19.7	19.6
fs	3.9	5.6	1.0	5.7	—	2.0	4.7	4.5	7.6	8.5	7.6	7.0	8.3	5.6	4.8	4.8	5.1	15.5	22.3	19.0
fo	0.8	—	3.0	2.7	6.7	—	—	—	—	3.7	3.5	5.0	2.3	—	—	—	—	—	—	—
fa	0.8	—	0.7	4.2	3.3	—	—	—	—	2.0	2.3	3.6	1.5	—	—	—	—	—	—	—
il	0.4	1.0	0.1	0.8	0.3	0.2	0.6	0.5	0.4	0.5	0.4	0.2	0.4	0.4	0.4	0.4	0.4	16.8	14.9	15.8
cm	—	—	—	—	—	—	—	—	—	—	—	—	—	—	—	—	—	0.2	0.1	—
ap	—	—	—	—	—	—	0.2	0.1	—	—	—	—	—	—	—	—	—	—	—	—
Total	100.0	100.0	100.0	100.0	100.0	100.0	100.0	100.0	100.0	100.0	100.0	100.0	100.0	100.0	100.0	100.0	100.0	100.0	100.0	100.0

(1) Mass of fine feathery crystals in breccia; probably a fragment of devitrified vein.
(2) Glassy vein near clasts of Fig. 4 and plagioclase of analysis 24. See Fig. 10.
(3) New, glassy vein, alongside old vein (analysis 4). See Fig. 10.
(4) Old, crystalline vein, alongside new vein (analysis 3).
(5) New, glassy vein. Feldspar is equiv. to Ab_6An_{94}.
(6) Old, crystalline vein, parallel and adjacent to No. 5. Feldspar is equiv. to Ab_4An_{96}.
(7) Center of vein—see traverses, Fig. 11.
(8) Avg. of 15 vein analyses (analyses 1–7 from this table and eight analyses from Weiblen and Roedder, 1973).
(9) Clear glass part of composite vein.
(10) Dark zone at edge of clear vein.
(11) Clear glass vein, 20 μm beam.
(12) Clear center of glass vein, 20 μm beam. See Fig. 3.
(13) Avg. of analyses 9–12.
(14) Clear glass center of vein.
(15) Clear glass center of vein, near analysis 16. Feldspar is equiv. to Ab_3An_{97}.
(16) Center of vein.
(17) Avg. of analyses 14–16.
(18) Vein at traverse—clear part.
(19) Vein near traverse—clear part.
(20) Avg. of analyses 18–19.

other sections (Weiblen and Roedder, 1973, analyses 107–112, 115, 116); they all have 43.5–48.6% SiO_2 and about 85% normative plagioclase ($\sim An_{96}$)[2]; pyroxene makes up most of the remainder of the norm. Some are quartz normative and others are olivine normative.

Four analyses of veins from 67936 (analyses 9–12) show little variation among themselves, but are significantly higher in Fe, Mg, and Mg/Fe, and lower in Al and Ca, than those in 67915. Three analyses of veins from 67955 (analyses 14–16) are intermediate in composition between those in 67915 and in 67936, but show very little overlap with them.

Two analyses of a vein in 79155 show much higher Fe and Ti, as might be expected from its red color; these have a normative composition of a quartz-bearing ilmenite basalt.

Composition of non-glassy parts. Table 2 presents analyses of mineral clasts from the veins. Most are olivine or pyroxene, but a few are clasts of chromite and spinel (analyses 27, 31, 32), silica minerals, and ilmenite (Fig. 4). One small spinel had apparent compositional gradients across its width, but this could be an artifact from the strong relief of the grain in polished section. Olivines in veins from 67915 range from Fo_{73} to Fo_{89} (Roedder and Weiblen, 1974, analyses 44–46 and other, partial analyses), and pyroxenes include both high- and low-calcium types (e.g., $En_{71}Fs_{25}Wo_4$, Roedder and Weiblen, 1974, analysis 47). A similar assemblage, but having a wider range of olivines (Fo_{64-80}), is present in veins in 67936. Both olivine and pyroxene are relatively low in NiO ($\leq 0.06\%$).

The tiny spherical globules of metal and/or iron sulfide that occur throughout much but not all the glass are mostly $<1\ \mu m$ in diameter and hence are too small to analyze by electron microprobe. The one metal globule that could be analyzed (Table 2, analysis 34) contains 6.14% Ni and 1.71% Co. Although these globules appear in the transmitted light photographs to be very abundant, their true abundance, as seen in reflected light, varies widely, rarely exceeds 1%, and is generally $<0.1\%$. Scanning electron microscope (SEM) studies of polished surfaces at magnifications up to 6000$\times$ show that the larger of these spheres (up to 15 μm; 67955,53,P5) are frequently a composite of iron metal and sulfide, with appreciable nickel. The smaller are mainly iron metal, the proportions of Ni and S generally decreasing with sphere size. Thus in 67936,5,2–7 μm spheres contained 7% Ni and 5% S, plus minor P; spheres 0.1–0.2 μm in diameter had similar Ni but almost no S; and spheres of $<0.1\ \mu m$ had only $\sim 3\%$ Ni.

Some dark to opaque zones in the veins were also examined by SEM. Those in 67936,5 (Fig. 6), consist of crudely lathlike crystals $\sim 1 \times 8\ \mu m$, presumably of plagioclase, in a matrix that was higher in Fe, Mg, and Ti, and hence richer in pyroxene component. These same zones generally had a higher total Fe, Mg, and Ti. Some of the lighter brown veins, as in 67955,53, had the reverse texture

[2]All mineral element ratio and formula compositions in this report are on an atomic or molecular basis; all CIPW norms are on a weight basis.

under the SEM—roughly rectangular crystals, 0.5–1 μm in diameter, of apparently higher average atomic weight (pyroxene?) in a matrix of lower average weight. In both samples the compositional contrast was small, so presumably these crystals are skeletal and include much glass.

Variation in composition of individual veins. The striae in the veins appear as minor compositional oscillations on microprobe traverses (Figs. 9–11). In spite of these local variations, however, the bulk analyses are remarkably uniform from one place to another (Table 1). No significant variation was recognized along the length of any given vein, and the optically visible variations across the width of the veins are reflected in surprisingly small compositional variations (e.g., compare analyses 14 and 15, and Figs. 9–11), even for the almost opaque zones in clear glass, as in 67936. Some of these very dark zones are somewhat higher in Fe and Mg than adjoining clear glass, and one thin dark band (67936,3, P0) is much higher (analyses 9 and 10), but in some (e.g., 67955,53, P3) the dark border is slightly lower in Fe. Those veins that appear completely crystallized have overall compositions generally very similar to the glassy ones (e.g., analyses 3–4; 5–6).

The most striking compositional discontinuity is the pyroxene zone at the edge of some veins (Figs. 1, 7, and 9–10; Roedder and Weiblen, 1974). Only two of these were thick enough to analyze (Table 2, analysis 26, and Weiblen and Roedder, 1973, analysis 117). Many of the veins show no pyroxene zone; in these, the composition of the vein is uniform right up to the contact, within the resolution of the optical microscope and electron microprobe. The composition of the glassy part of the veins is essentially independent of the nature of the immediately surrounding clast—except within a few micrometers, no change is detectable in the composition of a given vein where it cuts either plagioclase or pyroxene in the host rock.

Vein/host-rock interface relationships

The interface between the vein and the host rock was examined in some detail in several samples to see what constraints it might provide on the conditions of formation and subsequent cooling. In sample 67955,53, P4, several olivine crystals in direct contact with a thick (~270 μm) isotropic vein were studied. Analysis of one of these two crystals verified that they were the same composition as the other olivines in this rock (~Fo_{80}). Inspection with the microscope showed no optically visible contact effects in the adjacent glass, but electron microprobe step scanning (1 μm steps) across the boundary revealed a zone about 6 μm wide in the glass at the crystal interface that was depleted in iron relative to the glass toward the center of the vein. The lowest value was about half the normal amount, which we assume to be 2.80% FeO (analysis 15). The actual width of the depleted zone is probably <6 μm, because of edge effects. SEM examination of this boundary at high magnification showed no details indicative of either solution or crystallization of olivine.

Although fine feathery devitrification of the vein material is common,

Table 2. Microprobe analyses (wt.%) and atomic ratios of host rock minerals and of mineral clasts within shock glass veins (Paul W. Weiblen, analyst).

	67915						67955,33	67936							79155,66					
Anal. No.	(21)	(22)	(23)	(24)	(25)	(26)	(27)	(28)	(29)	(30)	(31)	(32)	(33)	(34)	(35)	(36)	(37)	(38)	(39)	(40)
Slide	45-1d	3-1a-1	45-1d	45-1d	3-3	3-3	53	3	3	3	3	4	5	5	66	66	66	66	66	66
Lab. No.	35	70	23	24	L55	L56	L66	M179	M180	M181	M182	M186	M183	M188b	M153	M154	M155	M145	M152	—
Area	P7-3	3	P3-1	P7-1	P7	P8a	P5	P0-3	P0-4	P0-5	P0-6	P1-1	P0-1	P0-4	P12-4	P12-5	P12-6	P13-1	P12-3	P12-2
SiO_2	44.2	45.9	44.3	53.2	43.9	52.7	1.05	39.6	48.4	38.8	0.72	0.51	36.8		50.7	49.9	45.6	51.3	48.0	—
Al_2O_3	36.0	33.0	35.7	29.7	34.8	1.12	17.0	0.24	32.3	0.31	15.4	61.0	0.46	Al 0.11	2.31	0.92	32.8	1.08	3.97	—
FeO	0.11	0.65	0.10	0.31	0.07	21.6	27.9	18.8	0.33	20.2	28.1	18.8	31.6	Fe 88.55	14.7	19.1	0.55	13.2	9.40	48.1
MgO	<0.05	1.30	0.10	<0.05	<0.05	22.2	3.48	41.0	0.07	37.8	3.84	14.0	31.2	Mg 1.49	15.7	20.0	0.33	16.8	14.8	1.2
CaO	19.4	18.0	19.2	12.2	19.5	1.57	0.21	0.10	17.3	0.39	0.32	0.10	0.12	Ca 0.41	14.5	7.07	17.5	15.4	20.0	—
Na_2O	0.65	0.34	0.26	3.15	0.28	—	—	—	0.61	—	—	—	—	—	<0.05	<0.05	1.86	<0.05	<0.05	—
K_2O	<0.05	0.02	<0.05	0.07	<0.05	—	—	—	0.03	—	—	—	—	—	—	—	<0.05	—	—	—
TiO_2	<0.05	<0.05	<0.05	0.02	—	0.21	4.41	<0.05	<0.05	<0.05	1.97	0.19	<0.05	Ti 0.06	0.87	0.58	<0.05	0.97	1.90	50.4
P_2O_5	—	<0.05	—	—	—	—	—	—	—	—	—	—	—	—	—	—	—	—	—	—
MnO	<0.05	<0.05	<0.05	<0.05	—	0.43	8.95	0.11	<0.05	0.20	0.59	0.07	0.11	—	0.39	0.46	<0.05	0.20	0.21	—
BaO	—	—	—	—	—	—	—	—	—	—	—	—	—	Co 1.71	—	—	—	—	—	—
Cr_2O_3	—	<0.05	—	—	<0.05	0.21	44.4	0.04	<0.05	<0.05	49.2	6.22	<0.05	—	0.10	0.08	<0.05	0.07	0.51	—
NiO	—	<0.05	—	—	—	<0.05	—	0.06	—	0.03	—	—	<0.05	Ni 6.14	—	—	—	—	—	—
S	—	<0.05	—	—	—	—	—	—	—	—	—	—	—	S 0.03	—	—	—	—	—	—
Total	100.36	99.21	99.66	98.65	98.55	100.04	99.40	99.95	99.04	97.73	100.14	100.89	100.29	98.50	99.27	98.11	98.64	99.02	98.79	99.7
Atoms																				
O	8.000	8.000	8.000	8.000	8.000	6.000	4.000	4.000	8.000	4.000	4.000	4.000	4.000		6.000	6.000	8.000	6.000	6.000	
Si	2.038	2.137	2.052	2.427	2.059	1.962	0.035	1.011	2.234	1.021	0.024	0.013	0.998		1.920	1.917	2.138	1.939	1.823	
Ti	0.00	—	0.00	0.001	—	0.006	0.110	—	—	—	0.050	0.004	—		0.025	0.017	—	0.028	0.054	
Al	1.958	1.812	1.950	1.598	1.926	0.049	0.663	0.007	1.757	0.010	0.605	1.878	0.015		0.103	0.041	1.813	0.048	0.177	
Cr	—	—	—	—	0.00	0.006	1.161	0.001	—	—	1.296	0.129	—		0.003	0.002	—	0.002	0.015	
Fe	0.004	0.026	0.004	0.012	0.003	0.672	0.772	0.402	0.013	0.445	0.784	0.411	0.716		0.465	0.614	0.022	0.417	0.299	
Mn	0.00	—	0.00	—	—	0.014	0.026	0.002	—	0.004	0.017	0.001	0.003		0.013	0.015	—	0.006	0.007	
Mg	0.00	0.073	0.006	—	0.00	1.232	0.172	1.560	0.005	1.483	0.190	0.545	1.261		0.886	1.145	0.023	0.946	0.837	
Ca	0.958	0.898	0.953	0.597	0.981	0.063	0.008	0.003	0.855	0.011	0.011	0.003	0.004		0.588	0.290	0.879	0.624	0.814	
Ba	—	—	—	—	—	—	—	—	—	—	—	—	—		—	—	—	—	—	
Na	0.058	0.031	0.023	0.279	0.026	—	—	—	0.054	—	—	—	—		—	—	0.170	—	—	
K	0.00	0.001	0.00	0.004	0.00	—	—	—	0.001	—	—	—	—		—	—	—	—	—	

(21) Melted plagioclase near vein. Equiv. to Ab_6An_{94}.
(22) Vein between two olivine clasts. Essentially melted plagioclase Ab_3An_{97}.
(23) Core of plagioclase clast with glassy but partly recrystallized edge. Equiv. to Ab_2An_{98}.
(24) Core of plagioclase clast with glassy edge and no recrystallization. Equiv. to $Ab_{32}An_{68}$.
(25) Plagioclase crystal (birefringent) adjacent to vein at analysis 7. Equiv. to Ab_3An_{97}.
(26) Pyroxene crystal at edge of vein. Equiv. to $En_{63}Fs_{34}Wo_3$. See Fig. 10.
(27) Chromite clast (~10 μm diam.) in vein.
(28) Olivine in breccia host—$Fo_{80}Fa_{20}$.
(29) Plagioclase in breccia host—Ab_6An_{94}.
(30) Olivine clast $Fo_{77}Fa_{23}$ in vein (analyses 9 and 10)
(31) Chromite clast in vein.
(32) Spinel clast in vein.
(33) Olivine clast in host rock—$Fo_{64}Fa_{36}$.
(34) Metal globule in vein after correction for 3.74% silicate.
(35) Host-rock pyroxene crystal cut by vein—$En_{46}Fs_{24}Wo_{30}$.
(36) Low-calcium pyroxene crystal cut by vein—$En_{56}Fs_{30}Wo_{14}$.
(37) Host-rock plagioclase near vein—$Ab_{16}An_{84}$.
(38) Host-rock pyroxene cut by vein—$En_{48}Fs_{21}Wo_{31}$.
(39) Low-birefringence host-rock pyroxene cut by vein—$En_{43}Fs_{15}Wo_{42}$.
(40) Ilmenite clast in vein.

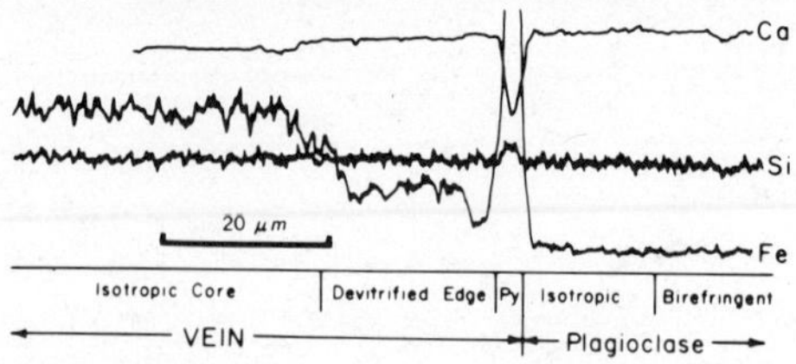

Fig. 9. Electron microprobe traces across vein in 67915,84-17, showing a devitrified border zone and thin layer of pyroxene crystals at the contact. The plagioclase at the right is a previously shocked and recrystallized clast that has an isotropic zone at the contact with the vein. The devitrified edge of the vein is more transparent than the isotropic core (which presumably is dark from incipient devitrification.) See Fig. 7 and Weiblen and Roedder (1973, analyses 115–119). These traverses were made with a focused 20 kV, 0.015 μm, ~0.5 μm beam, in a continuous mode, with advance at 20 μm/min. The counting rates were 1500, 750, and 1000 cps, and the line-to-background ratios were 1000, 500, and 750, for Ca, Si, and Fe, respectively, on the isotropic core.

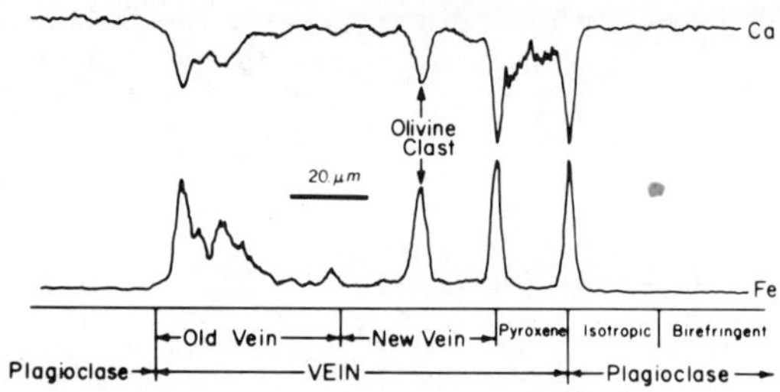

Fig. 10. Electron microprobe traces across composite vein in 67915,3-3, P8, between two plagioclase clasts, showing two relatively large pyroxene crystals ($En_{63}Fs_{34}Wo_3$; analysis 26) at contact of new vein (analysis 3) with old vein (analysis 4); the latter now consists of 1–2 μm crystals. Analytical conditions as in Fig. 9.

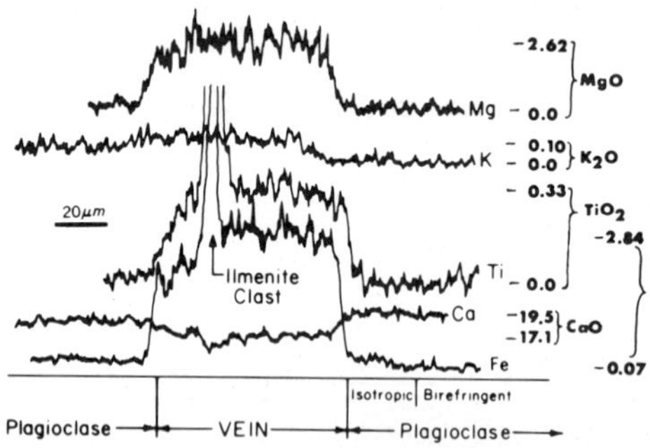

Fig. 11. Electron microprobe traces across vein in 67915,3-3, P7, between two plagioclase clasts of different composition. Plagioclase on left averages 17.1% CaO; that on right, with isotropized zone, averages 19.5% CaO. Central area of vein, analysis 7. Analytical conditions as in Fig. 9, with counting rates of 200, 100, and 150, and line-to-background ratios of 10, 20, and 15 for Mg, K, and Ti, respectively, on the central vein. The Mg, Ti, Ca, and Fe data were reduced with a pyroxene standard (Kakanui augite), with background corrections only, and K with an orthoclase standard.

apparently nucleated by crystals in the walls, only one example was found of crystallization occurring in the isotropized zone of the plagioclase, in a clast in 67915,45-1d, P3. This was a very calcic clast (An_{98}; analysis 23), and had a ~50 μm isotropic zone adjacent to a 250 μm wide vein. The contact between isotropic and birefringent plagioclase consists of a row of subparallel euhedral 5 μm prisms, up to 30 μm long, arranged perpendicular to the contact and in optical continuity with the unaffected plagioclase clast. The larger thickness of vein here no doubt aided this crystallization by slowing the cooling rate. However, a nearby clast that is much more sodic (An_{68}; analysis 24), shows no trace of crystallization into its isotropic zone, presumably because of the composition. The amount of crystal growth found is appropriate to cooling times in the range of one second (Uhlmann *et al.*, 1974).

Composition relative to host rock

The mineral clasts in the veins have compositions similar to those occurring in the host rock, but not necessarily like those in the immediately adjacent host rock or clast (for 67915, see particularly Table 2; Weiblen and Roedder, 1973, analyses 107–119 and 136–167; and Roedder and Weiblen, 1974, analyses 1–27 and 31–47; for 67936 see Table 2). Breccia 67936 contains 2.9% normative quartz, but the veins in it average 3.8% normative olivine. Twenty-five analyses of clasts in vein and host rock in 67955 and 79155 are available (Table 2; Roedder and Weiblen, 1974, 1975).

Although there is a gross similarity between the composition of the veins in the three breccia samples and the host breccias, and between the vein in basalt 79155 and its host, the compositions do not correlate in detail. The largest deviations for the three breccia samples are in Mg, Fe, and Mg/Fe (Table 3). Owing to the problems of sampling breccias, it is not possible to get truly valid averages, but it should be noted that the veins with the highest average FeO (5.74%; 67936) occur in the rock with the lowest FeO (2.77%). Unfortunately this latter number is based on only a single defocused beam analysis of the matrix of this breccia (Albee *et al.*, 1977). Also, the veins in both 67915 and 67955 have lower FeO than the minimum value given by Warner *et al.* (1974) for Apollo 16 breccia analyses. The only element showing a consistent pattern is sodium, which is lower in the four veins than in the corresponding host rocks. All other elements for which we have data show values both above and below those of the host rock.

As there are no recorded bulk analyses of 79155, we can only compare the vein in 79155,66 with another, probably very similar basalt, 70017 (Table 3); the differences in four major constituents, (SiO_2, Al_2O_3, MgO, and TiO_2; 2–5 wt.% each) are appreciable. Mao *et al.* (1974) reported an analysis of a thick red glass coating on 79155,25, which differs from both 70017 and our vein analyses (Table 3). Glass veins in another slide of 79155 (no. 60) were found to have a wide range of compositions (Schaal and Hörz, 1977). The pyroxenes adjacent to the vein in 79155, and throughout it, are all ≥15 mol.% Fs, and range from 14 to 42% Wo

Table 3. Average compositions of veins and host rocks.

Anal. No.	(I)	(II)	(III)	(IV)	(V)	(VI)	(VII)	(VIII)	(IX)
Sample	67915	67915	67936	67936,5	67955,53	67955,8	79155,66	70017,30	79155,25
Type	Veins	Rock	Veins	Breccia matrix	Veins	Rock	Veins	Rock	Glass coating
Avg. of	15	2	4	—	3	—	2	—	—
SiO_2	45.71	44.95	46.08	46.4	46.33	45.01	43.6	38.80	40.85
Al_2O_3	30.51	29.43	25.13	31.5	29.83	27.68	12.2	8.54	8.80
FeO	2.69	3.58	5.74	2.77	2.95	3.84	18.0	18.12	16.68
MgO	2.86	4.97	7.08	2.11	4.86	7.69	7.92	10.16	11.20
CaO	17.47	16.42	15.28	16.7	15.51	15.54	10.6	10.56	10.45
Na_2O	0.39	0.42	0.21	0.39	0.23	0.40	0.12	0.33	0.23
K_2O	0.06	0.063	0.03	<0.05	0.06	0.05	<0.05	0.07	0.
TiO_2	0.25	0.32	0.20	<0.05	0.21	0.27	8.42	12.84	7.57
P_2O_5	0.03	0.072	—	<0.05	<0.05	0.03	<0.05	0.04	—
MnO	0.06	0.047	0.02	<0.05	<0.05	0.05	0.22	0.24	0.35
Cr_2O_3	0.03	—	—	<0.05	<0.05	0.01	0.10	0.49	0.5
NiO	—	—	—	—	—	—	—	—	0.
S	—	0.008	—	—	—	0.01	—	—	—
Total	100.06	100.280	99.77	99.87	99.98	100.58	101.13	100.19	96.63
Norm									
Q	0.6	—	—	2.9	2.0	—	3.9	0.2	—
C	—	—	—	0.5	1.2	—	—	—	—
or	0.4	0.4	0.2	—	0.4	0.3	—	0.4	—
ab	3.3	3.5	1.8	3.3	1.9	3.4	1.0	2.8	2.9
an	81.4	78.1	67.6	82.9	76.9	73.1	32.5	21.6	23.9
wo	2.1	1.1	3.5	—	—	1.4	8.2	12.7	12.5
en	7.1	6.8	14.4	5.3	12.1	7.3	19.6	25.4	25.7
fs	4.5	3.3	8.3	5.1	5.1	2.5	19.0	11.6	16.3
fo	—	3.9	2.3	—	—	8.3	—	—	2.3
fa	—	2.1	1.5	—	—	3.1	—	—	1.6
il	0.5	0.6	0.4	—	0.4	0.5	15.8	24.5	14.9
ru	—	—	—	—	—	—	—	—	—
cm	—	—	—	—	—	—	—	0.7	0.8
ap	0.1	0.2	—	—	—	0.1	—	0.1	—
Total	100.0	100.0	100.0	100.0	100.0	100.0	100.0	100.0	100.0

(I) Average of 15 vein analyses (analysis 8, Table 1).
(II) Average of analyses of light and dark parts of bulk rock (Duncan *et al.*, 1973, p. 1098).
(III) Average of four vein analyses (analysis 13, Table 1).
(IV) Defocused beam analysis of breccia matrix (Weiblen and Roedder, 1973, analysis 185).
(V) Average of three vein analyses (analysis 17, Table 1).
(VI) LSPET (1973).
(VII) Average of two vein analyses (analysis 20, Table 1).
(VIII) Rose *et al.* (1974).
(IX) Red glass coating, away from contact with rock (Mao *et al.*, 1974).

(Table 2, analyses 35–36, 38–39; and Roedder and Weiblen, 1975, analyses 195–198). The one analyzed ilmenite clast within the vein in 79155 (analysis 40) was similar to that found in this rock by Roedder and Weiblen (1975, analyses 183–192).

Except for sodium, most of the differences in gross composition between veins and host breccias are sufficiently small that they may represent merely sampling and instrumental problems. Thus, in 67915, Weiblen and Roedder (1973) reported that the veins have about twice as much normative orthoclase (0.4%) as was found in the lithic clasts analyzed from this sample, but the difference is not present when the veins are compared with the whole rock (see Table 3, analyses I and II). Strontium isotope and rare-earth studies may eventually help to clarify the source of the material for these veins.

Origin of the Veins

Source of the vein material

As the veins in both the lunar breccias and the igneous rock clast 67955 have gross compositions similar to their host rocks, it is logical to assume that the veins were formed from local material by the cratering event that formed Outhouse Rock. The presence of several generations of adjacent veins with different degrees of crystallinity would seem to require several different events, rather than a sequence of stages in a single event, but whether these events were separated by seconds or eons is unknown.

If the vein material represents melted and injected regolith, the process might have been fast enough to preserve some solar wind component in the resulting quenched glass in spite of the high temperatures. Sample 67936 was the only one available in which the glass veins were thick enough to yield an adequate sample for analysis of solar wind gases. Two samples were handpicked, a 7 mg sample of glass vein, essentially free of host-rock contamination (except for fragments rafted within the glass), and 25 mg of adjacent matrix breccia, essentially free of vein material. These were submitted to E. Calvin Alexander, Jr., for rare gas analysis. His results are given in Table 4.

Table 4. Gas analyses of glass veins and host matrix, from sample 67936 (E. Calvin Alexander, Jr., analyst). Concentrations in cc/g.

	^{4}He	^{40}Ar	^{84}Kr	^{132}Xe	$^{40}Ar/^{36}Ar$
Vein[1]	4.809×10^{-5}	6.459×10^{-5}	6.448×10^{-10}	1.43×10^{-10}	220.2
Matrix[2]	10.99×10^{-5}	5.017×10^{-5}	24.2×10^{-10}	5.84×10^{-10}	379.2

[1]NASA number 9001, 7.302 mg, a composite of vein matter from subsamples 67936,3; 67936,4; and 67936,5.

[2]NASA number 9002, 25.235 mg, a composite of adjacent matrix from the same three subsamples.

As the vein sample has *less* He, Kr, and Xe than the matrix, and regolith samples generally have two or three orders of magnitude *more* He, Kr, and Xe than was found in either the veins or the matrix, there is no indication of the presence of solar wind gases in the vein. Furthermore, the ratio $^{40}Ar/^{36}Ar$ in soils is ~1.0, whereas the veins had a ratio of 220, i.e., essentially pure radiogenic argon. The lower concentration of gases in the veins than in the matrix is expectable if matrix-type material were heated to form the veins and hence partly outgassed, but the degree of outgassing that would occur in such a transient event is difficult to assess (E. Calvin Alexander, Jr., pers. comm.).

Possible mechanisms of formation

Four different mechanisms have been proposed for such glass veins:

1. *Intense mylonitization.* Glassy or devitrified glass veins, called pseudotachylite, occur in several places on earth, most notably in the Vredefort Dome in South Africa (e.g., Bisschoff, 1962; Wilshire, 1971). In most of these occurrences, the glassy material, plus large amounts of host-rock fragments, constitute anastomosing dikes or veins separating larger breccia fragments, as much as meters in size. Some are "barely perceptible to the naked eye" (Bisschoff, 1962, p. 208). The thickness varies widely and abruptly in any given vein, and some end abruptly. In this and several other respects, they resemble the lunar veins examined.

The origin of the Vredefort pseudotachylite veins has been discussed at length, and some authors have attributed them to the heat of friction in a mylonitization process, particularly of hot rocks (Philpotts, 1964; Higgins, 1971). Several features of the present veins are inexplicable by such a mylonitization mechanism: (1) in general, the composition of the veins is not exactly that of the host rocks, so at least some material had to be moved in or out. In detail, even the thinnest veins do not change composition on traversing clasts of grossly different composition (e.g., plagioclase and pyroxene) as they do in Vredefort (Wilshire, 1971, p. 199); (2) the zone of isotropized plagioclase requires unrealistically high local frictional heating on the shear plane, particularly in view of the lack of evidence of parallel shears in the immediately adjacent host rock; and (3) the abrupt pinch-offs and the sudden changes in vein direction around some clasts.

2. *Shock-induced local melting.* Fredriksson *et al.* (1963) and Fredriksson and Kraut (1967) proposed that some apparently similar glass veins in certain eucritic and chondritic meteorites represented the locus of intersecting shock waves within the meteorites, where the energy dissipation was adequate to cause comminution and melting. This mechanism should yield melts chemically similar to the immediately adjacent host rock, which our veins definitely are not, and, just as above, the overall composition of the veins should closely match that of the rock, which ours do not.

At first, the isotropized zone of plagioclase adjacent to the veins might seem appropriate for a vein origin by shock. However, the sharp interface between completely isotropic plagioclase glass and plagioclase with normal birefringence within a single crystal seems to preclude an origin for this isotropization by shock. There are no zones of plagioclase showing intermediate shock effects (fracturing, irregular extinction, deformation lamellae, decreased birefringence, etc.; Chao, 1968; Chao *et al.*, 1970). Also, some of the veins seem too thin to be from shock wave interaction. Most important, however, is the range of shock effects in the host rock. There are some severely shocked clasts in the host breccias, but most of the rock shows little or no shock effects. Although individual minerals differ widely, and plagioclase melts by shock more easily than pyroxene or olivine, pressures in the range of a megabar are needed for extensive melting (Kieffer *et al.*, 1976). Hence the presence of essentially unshocked plagioclase only a few micrometers from a glass vein would require rather extreme degrees of local amplification or attenuation of shock pressures if the mechanism proposed by Fredriksson and co-workers were correct. In some veins this attenuation would have to be ~100,000 bars/μm (and perhaps 200°C/μm?). This might occur adjacent to a microcrater (Schaal *et al.*, 1976), but is not feasible along the length of a vein. Schaal and Hörz (1977) suggest both *in situ* shock melting and injection of an allochthonous melt for the glasses they found in 79155,60.

3. *Injection of superheated liquid.* This is the most direct and obvious mechanism for the formation of such veins, corresponding to the vein networks of Wilshire and Moore (1974). This mechanism has also been suggested for somewhat different veins in sample 15205 (Dymek *et al.*, 1974). However, several features of the samples examined are difficult to explain by this mechanism. The liquid would have to be sufficiently low in viscosity to permit it to be forced into very tiny cracks in a very short time. The major parameters involved in injection are the temperature of the liquid, the pressure difference, the liquid viscosity and its change with temperature and pressure, the temperature of the host rock, the dimensions and configuration of the crack, and the thermal transmissivity of liquid and host rock by both conductive and radiative processes. Additional parameters include the specific heat and compressibility of the host rock and of the liquid, frictional heat generated during injection, etc. It may not be possible to place significant quantitative restraints on all these parameters, but it is obvious that during the injection the pressure on the fluid flowing into the crack must be high enough to overcome viscous drag and that this flow must be fast enough to permit injection into the entire length of the crack before the rise in viscosity of the cooling flow front stops the flow.[3] There must also be sufficient

[3]No data seem to be available on the viscosity of pertinent silicate liquids above about 1700°C, but the drop in viscosity with increase in temperature, although exceedingly rapid at low temperatures (Morey, 1954), becomes much less at high temperatures (i.e., ~1700°C, where viscosities are in the range 1–10 poise; Uhlmann *et al.*, 1974). The exact formulation of the rate of cooling of the veins would lead to a nonlinear integrodifferential equation of formidable difficulty (Hopper *et al.*, 1974).

heat flow from the vein filling into the walls during and after the injection to isotropize the plagioclase in an amount of host rock $\leqslant\sim 5$ times the volume of the final vein filling.

An additional problem with simple liquid injection, without intervention of shock effects, involves the nature of the mineral clasts present in the veins. Fudali (1974) pointed out that mafic and ultramafic rocks melt much more rapidly than granitic rocks held at the same temperature, even though the granitic rocks are ultimately less refractory in terms of liquidus temperatures (presumably this effect stems from the orders-of-magnitude difference in viscosity). Yet in our veins, we find mineral clasts of pyroxene and olivine, and none of feldspar.

The presence of both rounded and sharply angular mineral clasts in the veins could be a result of differing histories. Those that are rounded could have been partly dissolved in the embedding liquid, and the angular ones could be crystals picked up (or broken) near to where they are now found. If the rounded ones got that way by solution, there should be higher index tails around them, but such evidence is limited to a few ilmenite crystals that have tails only a few micrometers long (Fig. 4).

Veins of glass in some lunar samples have occasionally been found to be directly connected to thick glass coatings on the outside (e.g., Grieve and Plant, 1973; Jackson *et al.*, 1975, their Fig. 6). Such connections are expectable, to be sure, but these veins are generally one or more millimeters thick, and hence may have formed by a different process than the thin veins we describe. Some of these also have compositions differing greatly from that of the host rock, and represent merely liquid splashed onto and into the lunar samples.

We have no data on the type of heating cycle needed to cause isotropization of calcic plagioclase. If, however, we assume that the host-rock temperature before impact was 600°C[4], and that 1600°C is necessary to cause melting, then each volume of host rock adjacent to a vein that is brought up to that temperature requires ~1000°C of superheat (i.e., degrees above the liquidus) in the vein filling. Thus, if the isotropic zone on each side is twice the width of the vein, the vein material would have had to come in at $1600°C + (4 \times 1000°C) = 5600°C$. No allowance has been made for heats of melting or differences in specific heats of liquid and solid because these uncertainties are minor compared with those in some other parameters. Thus there is no assurance that even 1600°C will be adequate to cause melting of these An_{96} plagioclase crystals in the time frame involved (probably well under 1 sec for the small veins; Uhlmann *et al.*, 1974; Hopper *et al.*, 1974), in view of the classic studies of the amazingly slow isotropization (i.e., melting) of admittedly much more alkali-rich and hence viscous feldspars by Day *et al.* (1905). There is also a certainly positive and possibly large contribution of frictional heat to the walls, and there must be heat flow into the walls during the injection; both these effects would tend to

[4]This is an assumed metamorphic temperature, and even this may be too high, as many of the glass veins have remained isotropic after impact, and Outhouse Rock is large enough to have cooled very slowly. Thus, even more superheat may be required.

decrease the initial vein-matter temperature needed to achieve a given width of isotropic zone. Irregularities in the flow paths taken could possibly explain the great range of widths of the isotropic zone in any given vein.

4. *Injection of "superheated dust"*. In order to minimize the problem of injection of silicate liquids having relatively high viscosities, even at very high temperatures, Weiblen and Roedder (1973, p. 701) suggested the possibility that instead of a liquid silicate melt, the injected material might have been very fine-grained particulate "material", at well above its melting point. As such, this is merely a small scale, but very hot version of the injection of "unmelted crushed debris" described by Moore (1969), Wilshire *et al.* (1971, 1973), and Wilshire (1971). It is even possible that the difference between solid and liquid is inconsequential. Not only do some crystals take a finite time to melt, but highly viscous liquids (e.g., tar) will break or even powder rather than flow as a liquid if the stress rate is sufficiently high. Hence it seems at least possible that during impact, a strongly shocked, finely crushed, and very hot "dust" of plagioclase-rich composition could have formed in some areas of the impact area. The plagioclase in this dust is assumed to have lost all its crystalline structure during the process, and then to have been crushed into smaller grains than the pyroxene and olivine, some grains of which still persist. (An alternative hypothesis is that the mineral clasts found in the veins were picked up during transport, and all plagioclase clasts picked up this way have lost their crystallinity and mixed in.)

Once this speculative postulate is accepted, it is easy to visualize such a powder flowing into a crack under pressure. On earth, certain powders flow very easily under trivial pressure gradients, indicating very low apparent viscosities. Such "fluidized bed" flow has been proposed for various geological environments (Reynolds, 1954), including the pseudotachylite veins in the Vredefort Dome (Bisschoff, 1962). This behavior requires the presence of gas between the particles. Whether similar flow could occur in a vacuum is not known, but under the extreme temperature conditions of impact on the moon, it is likely that at least some gas phase (alkalies, sulfur, etc.) would be present. (This might explain the consistently lower sodium values in the veins.) The presence of sharply angular mineral clasts in the glass is concordant with this mode of origin. The rounded nature of some mineral clasts could be the result of mechanical abrasion during injection, and perhaps this is the stage in the process at which plagioclase clasts in the vein material were destroyed. Presumably, once the powder stopped moving, the superheat of the particles was sufficient to cause them to fuse together to a glass, and the heat flow into the host rock sufficient to cause the isotropization of the plagioclase near the vein. The total heating and cooling cycle was of too short duration to permit any significant diffusion of constituents, except perhaps to homogenize on a microscopic scale, but during cooling some very fine devitrification of the vein material occurred where the host rock provided nuclei. Shrinkage on cooling and crystallization resulted in formation of spherical gas bubbles, particularly in the larger bulges in the vein where the material stayed molten for the longest time. Minor amounts of flow in the melt in the vein just before solidification, perhaps in response to post-shock volume changes (Ahrens and Rosenberg, 1968; Ahrens *et al.*, 1973) yielded the striae and other evidence of fluid flow.

This mechanism also is not free of difficulties. Thus we have no proof that the hot dust grains, once formed, would not stick together on contact rather than flow while being pushed into a crack under high pressure. The absence of plagioclase clasts in the veins presents another problem. These must certainly have been picked up during injection, since the host rock is 70–80% plagioclase, yet they are not present in the veins either as birefringent crystals or as isotropic clear areas. This would seem to require that these plagioclase grains have been melted and mixed so thoroughly that they now exist only as the relatively minor striae visible under strongly collimated lighting (except in 79155; Schaal and Hörz, 1977). A third problem involves the lack of any evidence of vein fillings of unfused dust. The first material flowing into a crack will be cooled most rapidly and might be too cool to even weld when it finally comes to rest. We have not seen such dust-filled veins, but their preservation in the returned samples would be very unlikely.

Discussion

We favor the last two hypotheses, but we do not have adequate evidence to decide between injection of hot liquid and injection of hot "dust" as the probable mechanism of formation. In either case, there are some rather inexplicable features in these veins:

(1) The thin pyroxene surface film on some veins: Weiblen and Roedder (1973) proposed that these formed from the condensation of a vapor phase injected ahead of the main silicate, but this is strictly an *ad hoc* hypothesis.

(2) The band of devitrification: Devitrification along the margins of the veins appears to be a result of nucleation from the crystalline walls, but the pyroxene and olivine clasts scattered throughout the veins have no dendrites at their edges as might be expected, since they crystallize much more readily, particularly in the middle of the vein where cooling rates were lowest. It is possible that the devitrification is only from crystallization of plagioclase and since all plagioclase in the vein has been melted, there are no nuclei there. The black devitrified bands in sample 67936,3, P0 could represent an exception in which there was sufficient FeO and MgO present (perhaps also a slower cooling) to cause nucleation and growth of minerals other than plagioclase.

(3) Variations in the width of the isotropic zone: The zone width varies irregularly along the length of the vein, and does not seem to be correlative with the variations in vein thickness.

(4) The sharp edge of the outer margin of the isotropic zone: This margin, although irregular, seems to be very sharp, since no gradation is recognizable with the light microscope. Perhaps this margin indicates simply

the particular time-temperature combination at which melting occurs, but, if so, one would expect at least some gradation due to minor compositional variations in the feldspars.

(5) Assuming that the vein material came from an impact event, there may be some problems in the distance between the event and the emplacement of the veins: Stöffler (1971) showed that peak shock pressures of >800 kbar are needed to convert quartzofeldspathic rocks to inhomogeneous glasses with post shock temperatures of >3000°C. Yet, the relatively unshocked clasts that constitute the bulk of the host rock for the veins shows that there has been no severe shock for the whole assemblage in that area. The problem lies in the distance that the hot liquid (or "dust") must be transported beyond the zone of intense shock, and yet arrive at the site of the vein still under high enough pressure and hot enough to cause injection, and the isotropization of plagioclase. In this connection, Higuchi *et al.* (1976) have proposed that the injection of vein material in the ureilites coincided with the shock event that produced the diamonds in these veins.

Conclusions

Thin veins of glass or devitrified glass occur in several Apollo 16 lunar breccia samples from Outhouse Rock. The vein features suggest injection of extremely hot material, presumably from an impact event, into fractures in relatively cool host rock. In composition, the veins are very similar but not identical to the bulk composition of the host rocks, suggesting their derivation from a local source. Although most of the host rocks are very high in plagioclase, and the veins are similarly high in normative plagioclase, the only mineral clasts found in the veins are pyroxene and olivine, suggesting a selective destruction of plagioclase. Apparent thermal effects of the veins on adjoining host rock suggest a large amount of superheat. Four possible modes of origin are suggested. Physical and chemical features of the veins seem to preclude two origins: intense mylonitization, and shock-induced local melting. Available data are inadequate to choose between two other suggested origins: Injection of superheated liquid, and injection of superheated "dust". There are several features of the veins that are difficult to explain by either of these suggested origins.

Acknowledgments—We thank Robin Brett and Odette James for thoughtful reviews of the manuscript, H. E. Belkin for sample preparation, and E. C. Alexander, Jr., for the rare gas analysis data. Support for this study from National Aeronautics and Space Administration grants T-1170-A and NSG-9050 is gratefully acknowledged.

References

Ahrens T. J., O'Keefe J. D., and Gibbons R. V. (1973) Shock compression of a crystallized anorthositic rock from Apollo 15. *Proc. Lunar Sci. Conf. 4th*, p. 2575–2590.

Ahrens T. J. and Rosenberg J. T. (1968) Shock metamorphism: Experiments on quartz and plagioclase. In *Shock Metamorphism of Natural Materials* (B. M. French and N. M. Short, eds.), p. 59–81. Mono Book Corp., Baltimore, Maryland.

Albee A. L., Quick J. E., and Chodos A. A. (1977) Source and magnitude of errors in "broad-beam analysis" (DBA) with the electron probe (abstract). In *Lunar Science VIII*, p. 7–9. The Lunar Science Institute, Houston.

Bisschoff A. A. (1962) The pseudotachylite of the Vredefort Dome. *Geol. Soc. South Africa, Trans.* **65**, 207–230.

Chao E. C. T. (1968) Pressure and temperature histories of impact metamorphosed rocks—based on petrographic observations. In *Shock Metamorphism of Natural Materials* (B. M. French and N. M. Short, eds.), p. 135–158. Mono Book Corp., Baltimore, Maryland.

Chao E. C. T., James O. B., Minkin J. A., and Boreman J. A. (1970) Petrology of unshocked crystalline rocks and evidence of impact metamorphism in Apollo 11 returned lunar sample. *Proc. Apollo 11 Lunar Sci. Conf.*, p. 287–314.

Day A. L., Allen E. T., and Iddings J. P. (1905) The isomorphism and thermal properties of the feldspars. *Carnegie Inst. Wash., Pub. 31.* 121 pp.

Duncan A. R., Erlank A. J., Willis J. P., and Ahrens L. H. (1973) Composition and inter-relationships of some Apollo 16 samples. *Proc. Lunar Sci. Conf. 4th*, p. 1097–1113.

Dymek R. F., Albee A. L., and Chodos A. A. (1974) Glass-coated soil breccia 15205: Selenologic history and petrologic constraints on the nature of its source region. *Proc. Lunar Sci. Conf. 5th*, p. 235–260.

Fredriksson K., DeCarli P. S., and Aarmäe A. (1963) Shock-induced veins in chondrites. In *Space Research III, Proc. 3rd Internat. Space Sci. Symp.* (W. Priester, ed.), p. 974–983. North-Holland Pub. Co., Amsterdam.

Fredriksson K. and Kraut F. O. (1967) Impact glass in the Cachari eucrite. *Geochim. Cosmochim. Acta* **31**, 1701–1704.

Fudali R. F. (1974) Genesis of the melt rocks at Tenoumer Crater, Mauritania. *J. Geophys. Res.* **79**, 2115–2121.

Grieve R. A. F. and Plant A. G. (1973) Partial melting on the lunar surface, as observed in glass coated Apollo 16 samples. *Proc. Lunar Sci. Conf. 4th*, p. 667–679.

Higgins M. W. (1971) Cataclastic rocks. *U.S. Geol. Survey Prof. Paper, 687.* 97 pp.

Higuchi H., Morgan J. W., Ganapathy R., and Anders E. (1976) Chemical fractionations in meteorites—X. Ureilites. *Geochim. Cosmochim. Acta* **40**, 1563–1571.

Hopper R. W., Onorato P., and Uhlmann D. R. (1974) Thermal histories and crystal distributions in partly devitrified lunar glasses cooled by radiation. *Proc. Lunar Sci. Conf. 5th*, p. 2257–2273.

Jackson E. D., Sutton R. L., and Wilshire H. G. (1975) Structure and petrology of a cumulus norite boulder sampled by Apollo 17 in Taurus-Littrow Valley, the Moon. *Bull. Geol. Soc. Amer.* **86,** 433–442.

Kieffer S. W., Schaal R. B., Gibbons R., Hörz F., Milton D. J., and Dube A., (1976) Shocked basalt from Lonar Impact Crater, India, and experimental analogues. *Proc. Lunar Sci. Conf. 7th*, p. 1391–1412.

LSPET (1973) The Apollo 16 lunar samples: Petrographic and chemical description. *Science* **179**, 23–34.

Mao H. K., El Goresy A., and Bell P. M. (1974) Evidence of extensive chemical reduction in lunar regolith samples from the Apollo 17 site. *Proc. Lunar Sci. Conf. 5th*, p. 673–683.

Moore H. G. (1969) Subsurface deformation resulting from missile impact. In Geological Survey Research, 1969, p. B107–B112. *U.S. Geological Survey Prof. Paper 650B.*

Morey G. W. (1954) *The Properties of Glass.* Reinhold Pub. Corp., New York. 591 pp.

Philpotts A. R. (1964) Origin of pseudotachylites. *Amer. J. Sci.* **262**, 1008–1035.

Reynolds D. L. (1954) Fluidization as a geological process and its bearing on the problem of intrusive granites. *Amer. J. Sci.* **252**, 577–641.

Roedder E. and Weiblen P. W. (1970) Lunar petrology of silicate melt inclusions, Apollo 11 rocks. *Proc. Apollo 11 Lunar Sci. Conf.*, p. 801–837.

Roedder E. and Weiblen P. W. (1974) Petrology of clasts in lunar breccia 67915. *Proc. Lunar Sci. Conf. 5th*, p. 303–318.

Roedder E. and Weiblen P. W. (1975) Anomalous low-K silicate melt inclusions in ilmenite from Apollo 17 basalts. *Proc. Lunar Sci. Conf. 5th*, p. 147–164.

Rose H. J., Jr., Cuttitta F., Berman S., Brown F. W., Carron M. K., Christian R. P., Dwornik E. J., and Greenland L. P. (1974) Chemical composition of rocks and soils at Taurus-Littrow. *Proc. Lunar Sci. Conf. 5th*, p. 1119–1133.

Schaal R. B., and Hörz F. (1977) Shock effects in some lunar basalts (abstract). In *Lunar Science VIII*, p. 832–834. The Lunar Science Institute, Houston.

Schaal R. B., Hörz F., and Gibbons R. V. (1976) Shock metamorphic effects in lunar microcraters. *Proc. Lunar Sci. Conf. 7th*, p. 1039–1054.

Sclar C. B. and Bauer J. F. (1974) Shock-induced melting in anorthositic rock 60015 and a fragment of anorthositic breccia from the "picking pot" (70052). *Proc. Lunar Sci. Conf. 5th*, p. 319–336.

Stöffler D. (1971) Progressive metamorphism and classification of shocked and brecciated crystalline rocks at impact craters. *J. Geophys. Res.* **76**, 5541–5551.

Uhlmann D. R., Klein L., Kritchevsky G., and Hopper R. W. (1974) The formation of lunar glasses. *Proc. Lunar Sci. Conf. 5th*, p. 2317–2331.

Ulrich G. E. (1973) A geologic model for North Ray Crater and stratigraphic implications for the Descartes region. *Proc. Lunar Sci. Conf. 4th*, p. 27–39.

Warner J. L., Simonds C. H., and Phinney W. C. (1974) Impact-induced fractionation in the lunar highlands. *Proc. Lunar Sci. Conf. 5th*, p. 379–397.

Weiblen P. W. and Roedder E. (1973) Petrology of melt inclusions in Apollo samples 15598 and 62295, and of clasts in 67915 and several lunar soils. *Proc. Lunar Sci. Conf. 4th*, p. 681–703.

Wilshire H. G. (1971) Pseudotachylite from the Vredefort Ring, South Africa. *J. Geol.* **79**, 195–206.

Wilshire H. G., Howard K. A., and Offield T. W. (1971) Impact breccias in carbonate rocks, Sierra Madera, Texas. *Bull. Geol. Soc. Amer.* **82,** 1009–1018.

Wilshire H. G. and Moore H. J. (1974) Glass-coated lunar rock fragments. *J. Geol.* **82**, 403–417.

Wilshire H. G., Stuart-Alexander D. E., and Jackson E. D. (1973) Apollo 16 rocks: Petrology and classification. *J. Geophys. Res.* **78**, 2379–2392.

Proc. Lunar Sci. Conf. 8th (1977), p. 2617–2626.
Printed in the United States of America

Inverted pigeonites from lunar breccia 76255 and pyroxene-crystallization trends in lunar and achondritic crusts

HIROSHI TAKEDA

Mineralogical Institute, Faculty of Science, University of Tokyo, Hongo, Tokyo 113, Japan

MASAMICHI MIYAMOTO

Department of Earth Sciences, Kobe University, Nada-ku, Kobe 657, Japan

Abstract—Single crystal X-ray diffraction studies and electron microprobe analyses of pyroxenes in breccia 76255 have indicated that they are inverted pigeonites with exsolution lamellae of augite about 15 μm thick and developed in a primary pigeonite with (001) in common. The host orthopyroxene contains blebby augite inclusions similar to those found in the most magnesian eucrite, Binda. The orientation of orthopyroxenes with respect to the (001) augite differs from grain to grain in 76255 but retains some geometrical relationships. A simple computer simulation of the growth rates of lamellae in this pyroxene suggests that the observed texture was developed at a depth of a few kilometers in the lunar crust. Exsolution textures of some cumulate eucrites are estimated to be produced at greater depth. The Yamato 75032 achondrite contains the most iron-rich diogenitic orthopyroxene with Fe/(Mg + Fe) close to that of Binda. These pyroxenes delineate the crystallization trend in an achondrite crust, which is simpler than that of the lunar trends.

INTRODUCTION

SINGLE CRYSTAL X-RAY DIFFRACTION and electron microprobe studies of pyroxene sample 76255,61 have been carried out as a part of consortium studies of Apollo 17, Station 6 boulder headed by W. C. Phinney. The PET thin section of sample 76250 (chipped from the boulder at Station 6) contains pyroxenes which have the coarsest exsolution textures yet found on the moon, thereby suggesting that 76250 is a slowly-cooled, plutonic rock (LSPET, 1973). The detailed petrology of the breccia matrix and igneous clasts of 76255 has been reported by Warner *et al.* (1976). Since it is proposed that these plutonic clasts may be products of the primitive lunar differentiation, the crystallographic data of the pyroxenes which relate to the cooling history should be useful in characterizing such crusts. The results have been compared with those of the achondritic pyroxenes of similar composition to deduce the pyroxene crystallization trends of the lunar and achondritic crusts.

EXPERIMENTAL TECHNIQUES

Sample 76255,61 is a collection of five coarse pyroxene grains (No. 1 to 5) from unit 4 of rock 76255 (W. C. Phinney, pers. comm., 1975). The pyroxenes are pale yellowish-brown crystals 0.5–2 mm in diameter with exsolution lamellae visible under a binocular. Each crystal was broken into several fragments. One crystal was mounted on a glass fiber for the X-ray examination and the other was mounted in araldite for the microprobe analysis.

In order to obtain the orientational relationship between the exsolved augite and the host orthopyroxene inverted from pigeonite, the crystals were mounted nearly perpendicular to the augite plate and aligned with spindle axis parallel to the c^* direction of the augite. Then, precession photographs of $h0l$ and $0kl$ nets of augite and those containing each of a^*, b^*, and c^* of the orthopyroxene were taken by rotating the dial of the camera, using Zr-filtered Mo$K\alpha$ radiation. The $h0l$ and $0kl$ nets of the orthopyroxene were photographed after remounting and readjustments.

Polished grain mounts were prepared so that the coarse platy augite lamellae were nearly perpendicular to the thin section. The quantitative chemical analyses of the augite lamellae and the host orthopyroxenes were made with a JEOL JXA-5 electron probe X-ray microanalyzer with a 40° take off angle. The method was the same as that described by Nakamura and Kushiro (1970). The bulk composition of the entire crystal including augite lamellae was computed from the composition of the lamellae and the host phase and the width of the lamellae and their intervals obtained by microprobe traverses. Electron microprobe techniques have also been utilized to study pyroxene clasts in unfinished sections 76250,6 and 76255,83. Thin section 76250,5 has been studied for comparison.

Pyroxenes in the Yamato 75032 meteorite which was found by the Japanese Antarctic Research Expedition field party in 1975 (Matsumoto, 1977) were also studied by the above methods (Miyamoto *et al.*, 1977; Takeda *et al.*, 1977).

RESULTS

Pyroxenes in sample 76255,61 and pyroxene clasts in 76250,5 display two sets of exsolved augites. The coarser, presumably first, set of platy lamellae are assumed to be exsolved on (001) of the original pigeonite. For brevity this augite will be referred to as "remnant (001) augite" through the rest of this paper. The second, and presumably later, set of augite is blebby and occurs in the orthopyroxene host between the first set of lamellae. Many augite blebs are elongate in the c direction of the host orthopyroxene. This host orthopyroxene has been considered to be inverted from pigeonite.

The orientation of the host orthopyroxene with respect to the remnant (001) augite differs from grain to grain. The crystallographic orientations between orthopyroxene and augite are still maintained in some cases as is shown in the stereographic projection (Fig. 1); grain No. 2 shows characteristic diffraction pattern which can be explained by a mechanism proposed by Poldervaart and Hess (1951) for orthopyroxene inverted from pigeonite from the Stillwater complex. They maintain that during the inversion from pigeonite to orthopyroxene, the pigeonite b and c crystallographic axes (or a^*) are retained in the orthopyroxene. Therefore, the remnant (001) augite and the host orthopyroxene thus produced have common direction of neither a^* nor c^*. Only one inverted pigeonite found in Moama, among many pyroxenes thus far studied, shows such orientational relationships (Takeda *et al.*, 1976). Weak augite reflections with (100) in common with the orthopyroxene were observed in the host orthopyroxene of inverted pigeonite grain No. 2. The a^* direction of augite in grain No. 4 is nearly parallel to [031]* and is contained in the $h{\cdot}3l{\cdot}l$ net of orthopyroxene, and the diffraction spots of the augite lamellae are sharp, whereas those of the host orthopyroxene are more misoriented with respect to the augite spots than those of other grains. In grain No. 5, c^* of augite is contained in the $h0l$ net of the orthopyroxene. The results are consistent with those obtained previously for typical inverted pigeonites from the Stillwater complex (Takeda, 1973).

The host pyroxene (No. 4) between the thick (001) augite lamellae, is now composed of orthopyroxene with considerable amounts of thinner augite blebs with common (100). These blebby augites are similar to those found in the inverted pigeonite in the Binda eucrite (Takeda *et al.*, 1976). The intensity of the $\overline{10}{\cdot}0{\cdot}2$ reflection of the (100) augite, is intermediate between those of the $\overline{17}{\cdot}0{\cdot}2$ and $\overline{18}{\cdot}0{\cdot}2$ reflections of the orthopyroxene, and is much stronger than those of augite exsolved from the primary orthopyroxene. The intensity of the 802 augite reflection is weaker than that of $\overline{10}{\cdot}0{\cdot}2$. Because this intensity relationship is the same as that observed for an augite single crystal, the

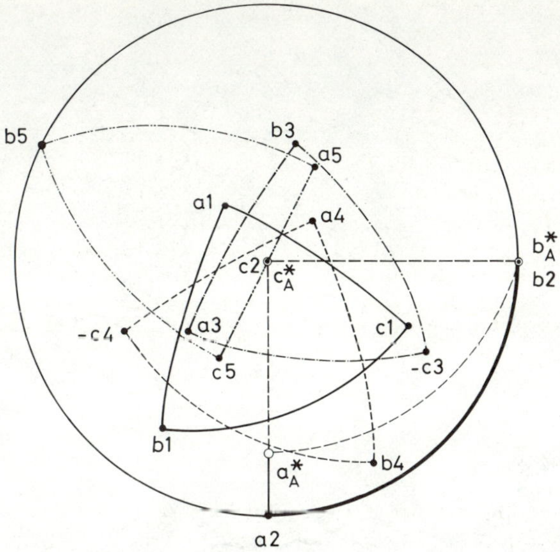

Fig. 1. Stereographic projection of the orientation of *a*, *b*, and *c* axes of orthopyroxenes (inverted pigeonites) with respect to the remnant (001) augite lamellae. The direction of a^*, b^*, and c^* axes of augites (A) is plotted common for all five grains (1 to 5).

comparison of their intensities indicates that the majority of the augite is not twinned on (100). These phenomena are different from the regular and platy exsolution lamellae on (100) in the primary orthopyroxene.

An inverted pigeonite grain contained in thin section 76250,5 has four thick (001) augite lamellae (Fig. 2) and is separated into two parts with different orientation by one of the lamellae. The orientation of orthopyroxene of the lower part with respect to the (001) augite is the same as what was found in grain No. 2 (76255,61), whereas the orthopyroxene of the upper part has an apparently random orientation. In the host orthopyroxene, there is a second set of augite blebs elongate in the *c* direction of orthopyroxene (Fig. 2).

Microprobe traverses perpendicular to the lamellae (Ca$K\alpha$ radiation) showed that coarse lamellae ranging from 5 to 15 μm in width (ave. 11 μm for No. 1, 8 μm for No. 3) are regularly separated by about 40–95 μm, and between each pair of the coarse augite lamellae there are one to ten thinner lamellae ranging from 1.5 to 4 μm. The thinner lamellae in some crystals (e.g. No. 4) tend to be rather blebby. The compositions of the thick augite lamellae, $Ca_{44}Mg_{42}Fe_{14}$, are uniform. The host orthopyroxenes range from $Ca_2Mg_{66}Fe_{32}$ (No. 1, 4) to $Ca_2Mg_{63}Fe_{35}$ (No. 5) (Fig. 3), and are also the same as those of inverted pigeonites in the norite clast from this specimen (Warner *et al.*, 1976). The composition of orthopyroxene No. 2 is $Ca_2Mg_{65}Fe_{33}$.

The bulk composition of the host between the thick (001) augite lamellae (Table 1, low bulk: $Ca_6Mg_{64}Fe_{30}$) is considerably more Ca-rich than that of common plutonic orthopyroxene. The bulk composition of the entire crystal computed from the host/lamellae proportion is that of pigeonite, $Ca_{12}Mg_{58}Fe_{30}$.

The composition of the Yamato 75032 pyroxene with Fe/(Mg + Fe) ratio, similar to that of 76255 and a little smaller than that of the Binda low-Ca pigeonite, is given in Table 1 for comparison (Takeda *et al.*, 1977), and is plotted in Fig. 4. This pyroxene-rich achondrite is a monomict breccia with orthopyroxene and minor augite. Small fragments of calcic plagioclase ($Ca_{91}Na_9$) and minute grains of troilite and chromite are present in the matrix (Takeda *et al.*, 1977). The precession photograph of an orthopyroxene with high bulk calcium content (9 in Table 1) and with blebby

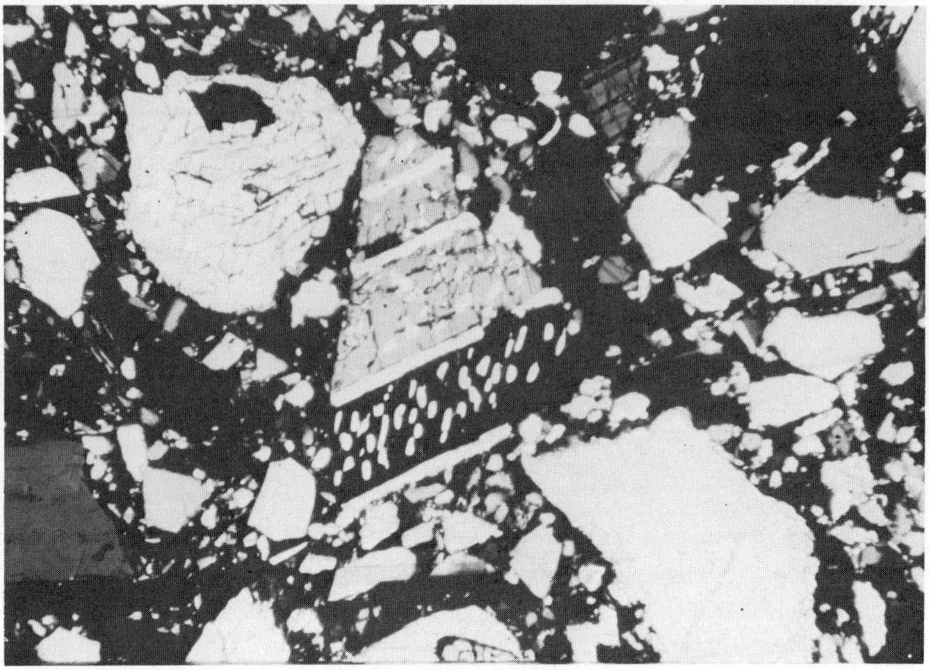

Fig. 2. Photomicrograph (cross nicols) of an inverted pigeonite in thin section 76250,5. Width is 1.2 mm. There are four thick (001) augite lamellae oriented 2–8 o'clock within a grain in the center of the photograph. The portion below the third lamella is in the extinction position. The second set of blebby augite is seen in the host orthopyroxene.

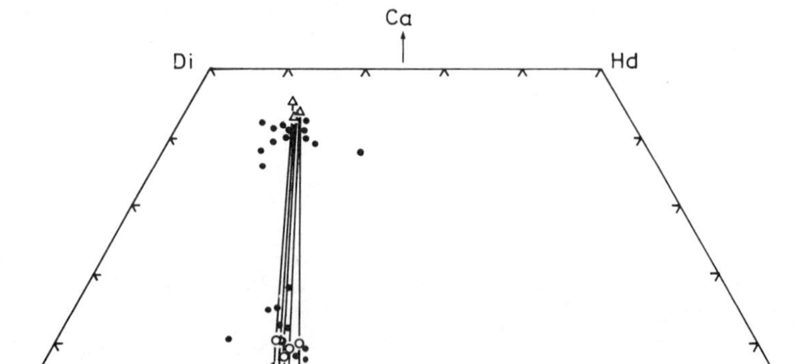

Fig. 3. Pyroxene compositions of five inverted pigeonites of sample 76255,61. Large open circle: bulk composition, large solid circle: host orthopyroxene, and triangle: (001) augite lamellae. Small dots are compositions of pyroxene clasts in section 76255,83 and 76250,6. Many dots overlap with large solid circles.

Table 1. Selected pyroxene compositions of lunar breccia (L) 76255,61 and Yamato achondrite (Y).

Sample	L76255,61							Y75032			
no.	1	1	3	3	3	4	4	8	8	9	9
phase	opx.	low	opx.	aug.	pig.	aug.	aug.	opx.	aug.	opx.	aug.
remark	host	bulk	host	lam.	bulk	blebs	lam.	host	bulk	bulk	blebs
SiO_2	52.3	52.1	52.0	51.5	51.6	51.5	51.7	52.2	51.9	52.3	52.8
Al_2O_3	0.80	0.87	0.71	1.61	0.90	1.45	1.39	0.65	1.15	0.61	0.83
TiO_2	0.51	0.60	0.56	1.19	0.72	1.14	1.20	0.26	0.61	0.36	0.58
Cr_2O_3	0.34	0.38	0.27	0.53	0.33	0.53	0.51	0.25	0.46	0.66	0.35
FeO	20.0	19.30	21.8	8.91	18.67	8.20	8.71	21.9	8.96	20.12	7.71
MnO	0.32	0.31	0.31	0.20	0.27	0.15	0.17	0.69	0.31	0.65	0.31
MgO	23.6	23.0	22.2	14.48	20.3	14.79	14.55	23.2	15.12	22.6	15.47
CaO	1.07	2.86	1.22	21.5	6.09	21.9	21.5	0.83	21.5	2.62	22.5
Na_2O	0.01	0.03	0.00	0.20	0.06	0.20	0.19	0.02	0.12	0.05	0.14
Total	98.95	99.45	99.07	100.12	98.94	99.86	99.92	100.00	100.13	99.97	100.69
Ca*	2.2	5.7	2.5	44.3	12.5	44.8	44.3	1.7	43.4	5.3	44.9
Mg	66.3	64.1	62.8	41.4	57.7	42.1	41.7	64.3	42.5	63.2	43.0
Fe	31.5	30.2	34.7	14.3	29.8	13.1	14.0	34.0	14.1	31.6	12.0

*Atomic %.
opx: orthopyroxene, pig: pigeonite, aug: augite, and lam: lamella.

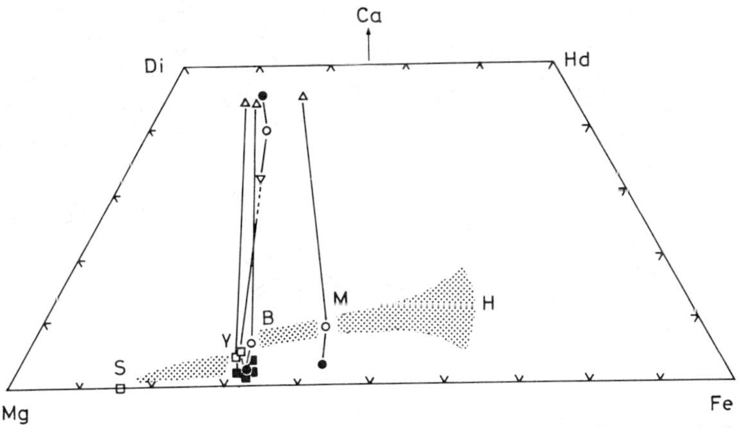

Fig. 4. Pyroxene crystallization trend in a model layered crust of a parent body of achondrites (from S to H) and pyroxene compositions of the Yamato 75032 meteorite (Y: squares). S: Steinbach, B: Binda, and M: Moama. Open symbol: bulk, filled: host, and triangle: lamellae.

inclusions of augite shows reflections of augite with (100) in common. This calcic orthopyroxene resembles the low-Ca pigeonite in the Binda eucrite, though the number of augite blebs per unit area is less than in Binda. The occurrence of such calcic orthopyroxene, and compositions of plagioclase and chromite indicate some resemblance of the Yamato 75032 meteorite to the Mg-rich eucrites. Although we cannot exclude the possibility that this calcic orthopyroxene was originally a low-calcium pigeonite, there is another orthopyroxene with very low bulk calcium content in contact with a primary augite (8 in Table 1). 75032 may be the most Fe-rich diogenite.

ESTIMATION OF COOLING RATES

An attempt has been made to calculate the cooling rate from the width of exsolution lamellae based on certain assumptions feasible for the present purpose of the computer simulation. By taking into account temperature dependence of the atomic diffusion coefficient of Ca in pigeonite, we integrated the equation derived from diffusion equation, with respect to temperature (1200–850°C) by the numerical method for the observed width of augite lamella and bulk Ca content in pigeonite.

A similar approach has been made by Huebner *et al.* (1975); however, we have used different boundary conditions. Atomic diffusion coefficient (D) was derived from McCallister and Yund's experiments (1974), and Fernández-Morán *et al.*'s observation (1971), by employing a solution of the diffusion equation. We adopted as D (at 1125°C) of 4.5×10^{-19} cm^2/sec with an activation energy of 28 kcal/mol (McCallister, 1974) for this simulation. The calcium contents of the augite and the host pigeonite at each temperature during cooling have been estimated from the shape of the pigeonite-augite solvus (Kushiro, 1972; Ross *et al.*, 1973). The temperature-time variation within a lunar crust 200 km thick and a parent body of 1000 km in diameter was estimated by the equation of Carslow and Jaeger (1959). Because the temperature-time variation curves at any depth beneath the surface for the above model are given, the time necessary to develop the lamellae can be converted into the depth. The detailed method of the computer simulation is given in a separate paper (Miyamoto and Takeda, 1977).

A computer simulation of lamellae growth rates of 76255 pyroxene suggests that the observed (001) augite exsolution lamellae in the pigeonite would develop if the cooling rate from 1200°C to 900°C were *ca.* 18°C/10^4 yr. Such conditions could occur about 2.5 km beneath the lunar surface if the original crust was 200 km thick. For a parent body of 1000 km in diameter, the depth estimated for the Moama eucrite is 12 km beneath the surface, Moore County 9 km, Juvinas 1.7 km, and Mt. Padbury 0.5 km. An absolute cooling rate for the first two meteorites is about 1°C/10^4 yr. The eucrite crust may be thinner than 15 km. The results were compared with those of the lunar crust.

DISCUSSION

All the crystallographic and chemical characteristics of the 76255,61 pyroxenes described in the previous sections indicate that they are inverted pigeonites first described in the Stillwater complex by Poldervaart and Hess (1951). Such inverted pigeonite with the remnant (001) augite lamellae has been called "orthopyroxene of the Stillwater type". The mechanism to develop this exsolution texture has been proposed by many investigators (e.g., Poldervaart and Hess, 1951; Bonnichsen, 1969; Ishii and Takeda, 1974). Discussion on Bonnichsen's findings and mechanism is given in our previous paper on a lunar inverted pigeonite (Takeda, 1973). Bonnichsen proposed a mechanism to produce random orientation between the pigeonite and orthopyroxene axes. We found more close orientational relationships in the 76255 pyroxenes by the X-ray method.

The orientational relationship between the thick augite lamellae and the host orthopyroxene of the 76255 inverted pigeonites can be explained by the nucleation and growth mechanism, a modification of Bonnichsen's proposal. The orientation of the orthopyroxene nucleus epitaxially grown at the augite-pigeonite interface may determine the final orientation when the nucleation rate is very slow. Two orthopyroxene regions in different orientation bound by the thick augite lamella within the originally homogeneous pigeonite crystal found in thin section 76250,5 suggest two nucleation sites in one crystal and support the above mechanism (Fig. 2).

A process that may account for the above exsolution pattern, is the following. The calcic pigeonite first crystallized with augite in cotectic relation at high temperature, about 1140°C (Ishii *et al.*, 1976) 10–50°C higher than the lowest stability limit of the pigeonite or pigeonite eutectoid reaction (PER) line, *ca.* 1110°C (Ishii, 1975) then upon passing into the pigeonite + augite immiscibility gap stabley exsolves augite oriented with (001) as the common plane. The augite lamellae might have continued to grow along the metastable extension of the pigeonite solvus slightly below the PER line. The degree of such growth may be influenced by inversion of the host pigeonite to orthopyroxene and amount of supercooling. The high calcium contents of the host phase between the thick lamellae indicate that the exsolution took place at a higher temperature than for common inverted pigeonites. The composition of the augite-orthopyroxene pair suggests that the last equilibrated temperature is about 970°C.

Common inverted pigeonites from other localities have fine regular platy augite lamellae in the host orthopyroxene with (100) in common. However, the second set of augite produced between the thick (001) augite lamellae of the 76255 pyroxenes has a blebby shape. The texture resembles augite blebs found in pyroxenes from the Binda meteorite (Takeda *et al.*, 1976). The bulk chemical composition of the matrix pyroxene between the (001) augite lamellae of 76255 is close to that of the Binda pyroxene. Their calcium contents are lower than that of pigeonite at the PER line. Within one crystal of these low-Ca pigeonites, the most augite blebs are oriented in the same direction sharing common (100) plane with the host orthopyroxene.

Comparisons of the texture and crystallographic characteristics of these pyroxenes with similar Fe/(Mg + Fe) ratios and different calcium contents in lunar sample 76255, Yamato 75032, and Binda will be useful in deducing the effect of Ca contents on the process of inversion and exsolution. It has been proposed that inversion or decomposition of the pigeonite producing the blebby augite in the host orthopyroxene with (100) in common took place on or not far below the PER line, where three phases, orthopyroxene, augite, and pigeonite, coexist (Ishii and Takeda, 1974). The above comparison suggests that the (100) blebby augites in the host orthopyroxenes are characteristic of inverted or decomposed pigeonites with calcium content lower than that of pigeonite at the PER line and at an unmixing or decomposition temperature on or lower than that of the PER line. Nakazawa and Hafner (1976) proposed that the (100) augite exsolution in pigeonite might have been produced at lower temperature than the (001) exsolution. We also proposed that the (100) exsolution is common for a low-Ca pigeonite (Takeda *et al.*, 1975).

If one accepts the above proposal that a pyroxene which crystallized on or a little below the PER line and cooled slowly will show this characteristic unmixing texture, such texture can be used as a signature of a pyroxene-crystallization trend, and the minimum temperature of crystallization of such pigeonite can be estimated by the temperature-composition diagram of Ross and Huebner (1975) and Ishii (1975).

To our knowledge, the 76255 pyroxene is one of the most Mg-rich Stillwater-type inverted pigeonites reported on the moon, suggesting the high temperature of crystallization, 10–50°C above the PER line. The inverted pigeonites in 15459,38 (Takeda, 1973) and 77215 (Huebner *et al.*, 1975) do not contain the thick remnant (001) augite. Other inverted pigeonites in 14082 (Papike and Bence, 1972), 67075 (McCallum *et al.*, 1975), and 60025 (Takeda *et al.*, 1976) are more Fe-rich. Many of these pyroxenes are not simply related to a differentiation sequence in a model moon. The temperature-composition variation line of these pyroxene crystallization trends crossed the PER line at different temperatures (Ishii *et al.*, 1976). For example above *ca.* 1150°C for 15459,38 and 1050°C for 60025.

In contrast with the lunar crustal pyroxene trends, the crystallization trend of the pyroxenes in the primitive crust of the achondrite parent body (Takeda *et al.*, 1976) shows simple sequential features which can be interpreted as a product of a single step differentiation of a thin crust formed early in the history of the solar system. The pyroxene compositions of the achondritic trends deduced from pyroxenes in polymict (genomict) surface breccia, howardites, and in single meteorites range approximately from that of Steinbach (S in Fig. 4) to that of Haraiya (H). The Binda pyroxene, which has the same composition as that of 76255 inverted pigeonite, is the most magnesian pigeonite to crystallize at the highest temperature in the trend (Fig. 4). The temperature of the point at which the trend reaches the PER line is estimated to be above 1100°C, which is lower than the crystallization temperature of the 76255 pyroxene. There has been a compositional gap between the Binda pigeonite and the most calcic and Fe-rich pyroxene found in diogenites. The composition of the Yamato 75032 pyroxene (Y in Fig. 4) is slightly more Mg-rich than Binda, and it appears to represent the last orthopyroxene crystallized in the trend (Fig. 4), supporting our differentiation model.

Some models proposed for achondrite parent body or bodies have dealt with only eucrites (Stolper, 1975; Drake and Consolmagno, 1977). However, close occurrence of diogenites and eucrites in many howardites and the oxygen isotope data, all indicate that they are genetically related. Our new finding of a diogenite-like meteorite with pyroxene composition close to that of Binda emphasizes the important genetic relationship between eucrites and diogenites. As far as this pyroxene crystallization trend is concerned, there is no apparent gap between them.

The result of a simplified computer simulation on lamellae growth of the 76255 pyroxene is consistent with their plutonic origin and with Huebner *et al.*'s view (1975, p. 529) that "the presence of coarse pyroxene exsolution lamellae does not necessarily indicate extensive periods of annealing which can only occur at depth within the lunar crust". The depth estimated for some cumulate eucrites is deeper than that of the lunar pyroxene. This is primarily because we assumed a smaller eucrite parent body, which cools more rapidly than in the lunar case. Although there are some uncertainties of the parameters used in the simulation, the relative depths may not change.

In conclusion, the 76255 pyroxene represents one of the most magnesian inverted pigeonites with the coarse remnant (001) augite lamellae, indicating a higher temperature of crystallization and slow cooling. The result is consistent with its plutonic origin, but pyroxenes in some cumulate eucrites show exsolution textures which were developed much deeper within the crust than the lunar ones. Pyroxene crystallization trends in the crust of a parent body of some achondrites is much simpler than that in the lunar crust. The finding of the most iron-rich diogenite with composition close to that of Binda suggests that diogenites and eucrites are closely related as far as their pyroxene crystallization trend is concerned.

Acknowledgments—We are indebted to Dr. W. C. Phinney of NASA-JSC for access to sample 76255, and to Prof. T. Nagata and Dr. K. Yanai of the National Institute for Polar Research and Prof. Y. Matsumoto for access to the Yamato meteorites. We thank Mrs. M. Hatano and Miss M. Sato for technical assistances. A part of this study has been supported by Itoh Science Foundation grant and a Fund for Scientific Research of the Ministry of Education. We thank Drs. W. C. Phinney, J. L. Warner, C. H. Simonds, and M. B. Duke for discussion and Drs. R. J. Williams and J. R. Smyth for critical reading of the manuscript.

References

Bonnichsen B. (1969) Metamorphic pyroxenes and amphiboles in the Biwabik iron formation, Dunka River area, Minnesota. *Mineral Soc. Amer. Spec. Pap.* **2**, 217–239.

Carslow H. S. and Jaeger J. C. (1959) *Conduction of Heat in Solids*, 2nd ed. Oxford Univ. Press, New York. 510 pp.

Drake M. J. and Consolmagno G. J. (1977) Asteroid 4 Vesta: Possible bulk composition deduced from geochemistry of eucrites (abstract). In *Lunar Science VIII*, p. 248–250. The Lunar Science Institute, Houston.

Fernández-Morán H., Ohtsuki M., and Hibino A. (1971) Correlated electron microscopy and diffraction of lunar clinopyroxenes from Apollo 12 samples. *Proc. Lunar Sci. Conf. 2nd*, p. 109–116.

Huebner J. S., Ross M., and Hickling N. (1975) Significance of exsolved pyroxenes from lunar breccia 77215. *Proc. Lunar Sci. Conf. 6th*, p. 529–546.

Ishii T. (1975) The relation between temperature and composition of pigeonite in some lavas and their application to geothermometry. *Min. Journ. (Japan)* **8**, 48–57.

Ishii T., Miyamoto M., and Takeda H. (1976) Pyroxene geothermometry and crystallization, subsolidus equilibration temperatures of lunar and achondritic pyroxenes (abstract). In *Lunar Science VII*, p. 408–410. The Lunar Science Institute, Houston.

Ishii T. and Takeda H. (1974) Inversion, decomposition and exsolution phenomena of terrestrial and extraterrestrial pigeonites. *Mem. Geol. Soc. Japan* **11**, 19–36.

Kushiro I. (1972) Determination of liquidus relation in synthetic silicate systems with electron probe analysis: The system forsterite-diopside-silica at 1 atomosphere. *Amer. Mineral.* **57**, 1260–1271.

LSPET (1973) Apollo 17 lunar samples: Chemical and petrographic description. *Science* **182**, 659–672.

McCallister R. H. (1974) Kinetics of enstatite exsolution from supersaturated diopsides. In *Geochemical Transport and Kinetics* (A. W. Hofmann, B. J. Giletti, H. S. Yoder, Jr., and R. A. Yund, eds.), p. 195–203. Carnegie Inst. Wash. Publication 634. Washington, D.C.

McCallister R. H. and Yund R. A. (1974) Kinetics and microstructure of pyroxene exsolution. *Carnegie Inst. Wash. Yearb.* **74**, 433–436.

McCallum I. S., Okamura F. P., and Ghose S. (1975) Mineralogy and petrology of sample 67075 and the origin of lunar anorthosites. *Earth Planet. Sci. Lett.* **26**, 36–53.

Matsumoto Y. (1977) On the collection of the Yamato-75 meteorites. *Abstr. 2nd Symp. on Yamato Meteorites*, Natnl. Inst. Polar Res., Tokyo.

Miyamoto M. and Takeda H. (1977) Evaluation of a crust model of eucrites from the width of exsolved pyroxenes. *Geochem. J.* **11**, 161–169.

Miyamoto M., Yanai K., and Takeda H. (1977) Movements of materials on the surface of an achondrite parent body as inferred from its crust model and the Yamato achondrites (abstract). In *Lunar Science VIII*, p. 670–672. The Lunar Science Institute, Houston.

Nakamura Y. and Kushiro I. (1970) Compositional relations of coexisting orthopyroxene, pigeonite and augite in a tholeiitic andesite from Hakone volcano. *Contrib. Mineral. Petrol.* **26**, 265–275.

Nakazawa H. and Hafner S. S. (1976) X-ray diffraction profiles and exsolution history of pigeonite. *Proc. Lunar Sci. Conf. 7th*, p. 1865–1873.

Papike J. J. and Bence A. E. (1972) Apollo 14 inverted pigeonites: Possible samples of lunar plutonic rocks. *Earth Planet. Sci. Lett.* **14**, 176–182.

Poldervaart A. and Hess H. H. (1951) Pyroxenes in the crystallization of basaltic magma. *J. Geol.* **59**, 472–489.

Ross M. and Huebner J. S. (1975) A pyroxene geothermometer based on composition-temperature relationships of naturally occurring orthopyroxene, pigeonite and augite (abstract). *Int. Conf. Geothermometry Geobarometry*, p. 173–176. Penn. State Univ.

Ross M., Huebner J. S., and Dowty E. (1973) Delineation of the one atomosphere augite-pigeonite miscibility gap for pyroxenes from lunar basalt 12021. *Amer. Mineral.* **58**, 619–635.

Stolper E. (1975) Experimental petrology of the Juvinas eucrite (abstract). *Meteoritics* **10**, 498–499.

Takeda H. (1973) Inverted pigeonites from a clast of rock 15459 and basaltic achondrites. *Proc. Lunar Sci. Conf. 4th*, p. 875–885.

Takeda H., Miyamoto M., Ishii T., and Lofgren G. E. (1975) Relative cooling rates of mare basalts at the Apollo 12 and 15 sites as estimated from pyroxene exsolution data. *Proc. Lunar Sci. Conf. 6th*, p. 987–996.

Takeda H., Miyamoto M., Ishii T., and Reid A. M. (1976) Characterization of crust formation on a parent body of achondrites and the moon by pyroxene crystallography and chemistry. *Proc. Lunar Sci. Conf. 7th*, p. 3535–3548.

Takeda H., Miyamoto M., and Yanai K. (1977) Achondrites in the Yamato meteorites. *Abstr. 2nd Symp. on Yamato Meteorites*. Natnl. Inst. Polar Res., Tokyo.

Warner J. L., Simonds C. H., and Phinney W. C. (1976) Apollo 17, Station 6 boulder sample 76255: Absolute petrology of breccia matrix and igneous clasts. *Proc. Lunar Sci. Conf. 7th*, p. 2233–2250.

Proc. Lunar Sci. Conf. 8th (1977), p. 2627–2640.
Printed in the United States of America

Effects of thermal metamorphism on compositions of lunar spinels

K. C. MISRA and L. A. TAYLOR

Department of Geological Sciences, The University of Tennessee, Knoxville, Tennessee 37916

Abstract – The use of spinel compositions and textures in interpreting the crystallization history of the mare basalts implies the assumption that the spinels have retained their signature of original crystallization despite subsequent subsolidus equilibration. We have performed a heating experiment on olivine basalt, 12018, 43, containing spinels of a wide range of compositions, to evaluate the changes that may be produced in lunar spinels by later thermal metamorphism. The experiment was conducted at 950°C at an estimated f_{O_2} of about $10^{-16.08}$ atm for a period of 10 days. With a special mounting technique (for EMP analysis) that did not require an epoxy base, it was possible to study identical polished surfaces of the rock and compare directly the textural and compositional features in corresponding grains before and after the heating experiment.

In the unheated sample, the spinels belong to the ulvöspinel–chromite (hercynite) solid solution series. Compositionally, they fall into three gradational subgroups: low-Ti spinels, intermediate-Ti spinels, and high-Ti spinels, but the compositional distribution is essentially biomodal. The low-Ti spinels show pronounced compositional zoning, from Cr-rich cores to Ti-enriched rims. The high-Ti spinels show unsymmetrical to oscillatory zoning.

Compositional changes in the high-Ti and intermediate -Ti spinels in the heated sample are fairly pronounced; these include an increase in TiO_2 and FeO, with concommitant decrease in Cr_2O_3, Al_2O_3, and MgO. Compositional changes in the heated low-Ti spinels are less pronounced and less systematic; as a group they seem to show an increase in Cr_2O_3 accompanied by a corresponding decrease in FeO and TiO_2. The ilmenites in the heated sample show an increase in TiO_2 (about 2%) and a decrease in FeO (about 1 to 2%). The pyroxene grains adjacent to high-Ti spinels show a decrease in FeO and an increase in MgO and Al_2O3 at the interface.

The nature of the compositional changes in the heated spinels cannot be explained by intragranular adjustments alone. Vapor transport of some of the oxide elements as their respective carbonyls may account for the non-isochemical changes in the heated spinels and ilmenites.

The heating experiment demonstrates that thermal metamorphism of lunar basalts is likely to modify the compositions of their pre-existing spinels (and ilmenites). These possible modifications must be taken into account if spinel compositions are used as indicators of lunar petrogenesis.

INTRODUCTION

SPINELS IN MARE BASALTS essentially belong to the ulvöspinel (Fe_2TiO_4)–chromite ($FeCr_2O_4$)–hercynite ($FeAl_2O_4$) solid solution series. Compositional variations in lunar spinels may be considered to be of two kinds: (1) an overall variation which is reflected in the markedly bimodal distribution of the spinel compositions (titanian chromite and chromian ulvöspinel) in lunar samples; and (2) a finer scale, gradational to stepped variation across individual spinel grains in a particular lunar sample. The compositional gap between Ti-chromite and Cr-ulvöspinel has been interpreted to indicate two stages of spinel crystallization (Cameron, 1971; Kushiro *et al.*, 1971; Bushe *et al.*, 1972). Recently, compositional zoning in individual spinel grains have been used to reconstruct the crystallization sequence of mare basalts (El Goresy *et al.*, 1976; El Goresy and

Ramdohr, 1977). The use of spinel compositions and textures as indicators of the crystallization history of the mare basalts is conditioned by the underlying assumption that these spinels have retained their signature of the original crystallization through later subsolidus equilibration. However, subsolidus reduction, especially of the Cr-ulvöspinel, is rather widespread in the mare basalts (e. g., Haggerty, 1971, 1972b; El Goresy *et al.*, 1972; El Goresy and Ramdohr, 1975) suggesting that solid-state diffusion and reequilibration have taken place in many of these assemblages. Of particular importance is the observation of Haggerty (1972b) that high intensity reduction of Cr-ulvöspinel may produce a spinel (Al-Ti chromite) which is compositionally similar to the intermediate primary precipitates. Moreover, Taylor *et al.* (1976) have demonstrated by heating experiments that subsolidus annealing, such as might occur in an ejecta blanket, can significantly modify the textures and compositions of native FeNi metal in lunar rocks. Therefore, it is reasonable to consider if the spinels in lunar rocks may respond to similar modifications during subsolidus annealing. The present study is an attempt to evaluate experimentally the compositional and textural changes that are likely to be observed in the Fe-Ti-Cr spinels of lunar igneous rocks by subsequent thermal metamorphism.

METHOD OF STUDY

Sample

The Apollo 12 igneous rock, 12018,43, was chosen for this investigation because an earlier study of another fraction of this rock by El Goresy *et al.* (1971) has reported an almost continuous trend of spinel compositions between Cr-ulvöspinel (Usp_{81}–Chr_{19}) and Ti-chromite (Usp_{10}–Chr_{90}), including several compositions within the so-called "Apollo 12 Compositional Gap" referred to by Haggerty and Meyer (1970).

The rock 12018,43 is an olivine basalt (Warner, 1971; James and Wright, 1972), and is characterized by phenocrysts of rounded olivine and zoned accicular clinopyroxene, laths of plagioclase, and subordinate amounts of opaque phases: ilmenite, Fe-Ti-Cr spinels, native FeNi, and troilite. Native copper, mackinawite, and chalcopyrrhotite (?) also have been reported from 12018,49 (El Goresy *et al.*, 1971), but were not observed in the present sample. According to El Goresy *et al.* (1971), the presence of mackinawite is an evidence of reequilibration of this rock during cooling down to temperatures well below 200°C.

Experimental

A fundamental limitation of our earlier heating experiment studies (Taylor *et al.*, 1976) has been the fact that the polished surfaces of a sample examined before and after the heating experiment are similar, but not identical. Although the textural and compositional changes inferred in such studies are statistically valid, they are not correlatable to specific grains in the sample. A highlight of the present study is the formulation of a special mounting technique (for EMP analysis) that allowed us to examine specific grains on identical surfaces of the sample before and after heating.

The parent chip of the rock 12018,43, as received from NASA, had a maximum dimension of the order of 8.0 mm and it weighed 0.875 g. An approximate cube (edge length about 4 mm; weight 0.212 g) was cut from this chip and three adjacent surfaces of the cube were polished for reflected-light examination and EMP analyses both before and after the heating experiment. For the purpose of EMP analysis of each surface, the cube was mounted in a cylindrical specimen holder with a crude assortment of aluminum foils, *without* the use of any epoxy base. Particular care was taken to wrap the edges and corners of the exposed surface with strips of aluminum foil to protect them from carbon coating. This procedure was repeated for each of the three polished surfaces. After completion of the EMP analyses, the aluminum foil wrappings were removed and the carbon coating on the sample was thoroughly cleaned before it was subjected to the heating experiment. After the heating experiment, the three polished surfaces were buffed with 0.05 micron alumina powder. For EMP analysis, each surface was mounted in the same orientation using the same mounting technique as for the unheated sample.

The heating experiment was conducted in an evacuated, sealed, silica-glass tube. A small amount (0.136 g) of pelletized, synthetic ulvöspinel, physically separated from the rock sample, was also placed in the tube. The synthetic ulvöspinel was used with two objectives in view: (a) reduction of the ulvöspinel would serve as an indicator of the oxygen fugacity conditions during the heating experiment, and (b) any carbon released from the unremoved carbon coating on the sample during heating would attack the pure ulvöspinel in preference to the Cr-ulvöspinel grains in the rock sample, so that Cr-ulvöspinel would be retained in the annealed assemblage. The silica tube was heated for 10 days at 950°C, and was quenched in cold water. Examination of the quenched tube with a binocular microscope showed no signs of breakage; the lack of any leakage was confirmed also by weighing the tube. However, the synthetic ulvöspinel in the quenched tube was found to have suffered a loss of about 10 wt.%. Subsequent examination of polished sections of the synthetic ulvöspinel showed that this weight loss could be tied to partial reduction of the ulvöspinel to ilmenite + Fe. The quenched silica tube contained some dust-like black particles of undetermined composition adhering to its inner walls.

The oxygen fugacity over the charge was not controlled. However, the partial reduction of the synthetic ulvöspinel to ilmenite but lack of ilmenite reduction to rutile suggested that the oxygen fugacity of the run was close to the ulvöspinel/ilmenite + Fe buffer. At 950°C, this corresponds to an oxygen fugacity of $10^{-16.08}$ atm, at the approximate position of the Usp/Il + Fe buffer curve determined by Taylor *et al.* (1972).

Although the carbon coated surfaces of the rock chip were thoroughly cleaned before it was heated, a small amount of carbon was probably retained in the pores of the chip. Rough calculations based on the area and porosity of the polished surfaces of the chip showed that the carbon that was likely to be retained in the pores of the chip would amount to 10–30 ppm considering the entire mass of the chip. Similar C contents have been reported for the lunar igneous rocks. This coupled with the partial reduction of the synthetic ulvöspinel (in the quenched charge) to ilmenite + Fe indicated that the reducing conditions under which the lunar igneous rocks may be metamorphosed were approximately simulated in our heating experiment.

Analytical

All chemical analyses were performed using an automated, three-channel, MAC-400S electron microprobe analyzer with the following operating conditions: 15 KV, beam current of about 0.05 μA on benitoite, and a spot diameter of 1–2 μm. The spinels and ilmenites in the three polished surfaces were analyzed for Ti, Cr, V, Al, Fe, Mg, and Ca using appropriate synthetic and natural standards. Many pyroxene and olivine grains in contact with the spinel grains were also analyzed for the above elements and Si, using suitable synthetic pyroxene and olivine standards for Fe, Mg, Al, and Si. Matrix corrections for both the oxides and silicates were applied according to Bence and Albee (1968) and Albee and Ray (1970).

For the unheated sample, the analyses on each ilmenite and spinel grain (and adjacent silicate grains in some cases) were done essentially along several N–S and E–W profiles, at spots generally 5–10 μm apart; the analyzed points were mapped on polaroid photographs. Since the orientation of each polished surface for EMP analysis was kept the same after the heating experiment, it was possible to analyze the same spots (or, at least points along the same profiles) in the heated sample as had been done in the unheated one. This enabled us to compare *directly* the compositional and textural features along selected profiles in corresponding grains before and after the heating experiment. Some representative profiles are presented in Figs. 2 and 3. A few typical analyses of ilmenite and the different compositional types of spinels are given in Table 2; compositional ranges of these phases are summarized in Table 1.

RESULTS OF THE HEATING EXPERIMENT

Unheated sample

The spinels in this sample belong to the ulvöspinel–chromite (hercynite) solid solution series (Fig. 1). Compositionally they fall into three gradational subgroups (Table 1): (1) Low-Ti spinels (Ti-Al chromites), (2) intermediate-Ti spinels (Al-Cr ulvöspinels), and (3) high-Ti spinels (Cr-ulvöspinels). The so-called "Apollo 12 Compositional Gap" is not present in this sample, but the compositional distribution is bimodal in that the great majority of the compositions are either high-Ti spinels (containing > 19 wt.% TiO_2) or low-Ti spinels (containing < 13 wt.% TiO_2), with only a few (less than 10% of the analyzed compositions) plotting in the intermediate range.

The low-Ti and high-Ti spinels in the present sample occur as discrete grains—the former typically as euhedral grains, and the latter invariably as anhedral grains. Almost all the spinels show discernible compositional zoning, but the zoning is gradational and optically indistinct. The most striking and systematic zoning is seen in the low-Ti spinels, from Cr-rich cores to relatively Ti- and Fe-enriched rims (Fig. 2). The maximum compositional variation observed across a single low-Ti spinel grain was as follows (wt.%): TiO_2, 4.2–11.6; Cr_2O_3, 46.9–35.5; Al_2O_3, 12.4–10.3; FeO, 30.6–36.7; and MgO, 4.8–5.0. Even the most Ti-rich compositions at the outer rims of these low-Ti spinel grains are Ti-chromites rather than Al-Cr- or Cr-ulvöspinels. Spinel grains with a low-Ti (chromite) core grading sharply to a high-Ti (Cr-ulvöspinel) mantle, as has been

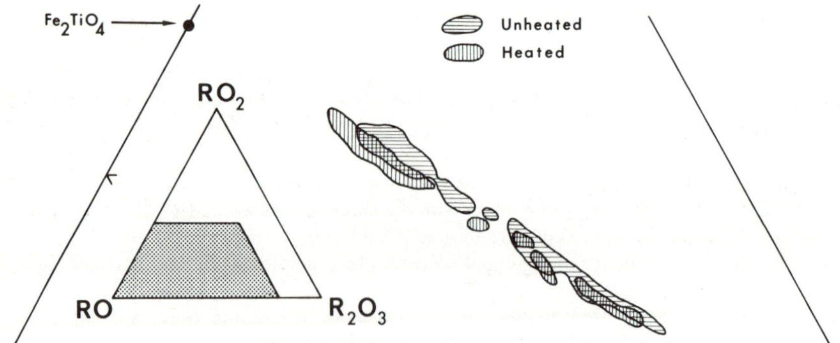

Fig. 1. RO_2–R_2O_3–RO (wt.%) diagram showing the compositional fields of the spinels analyzed in the unheated and heated rock sample 12018,43 (about 200 analyses for each).

Table 1. Compositional ranges of Spinels and Ilmenites in 12018,43.

	TiO_2	Cr_2O_3	V_2O_3	Al_2O_3	FeO	MgO
Ilmenite	52.0-53.5	0.2-0.6	0.0	0.1-0.3	44.0-47.0	0.4-1.5
High-Ti Spinel	19.0-25.0	19.0-26.0	0.5-0.7	3.5-7.5	43.0-51.0	2.0-5.0
Intermed.-Ti Spinel	16.0-20.0	25.0-32.0	0.5-0.7	6.0-7.5	40.0-44.0	3.5-5.0
Low-Ti Spinel - Core	4.0-6.5	43.0-47.0	0.6-1.0	9.0-12.5	31.0-32.0	4.5-5.5
Rim	7.0-12.0	35.0-42.0	0.6-1.0	9.0-12.5	33.0-36.0	4.5-5.5

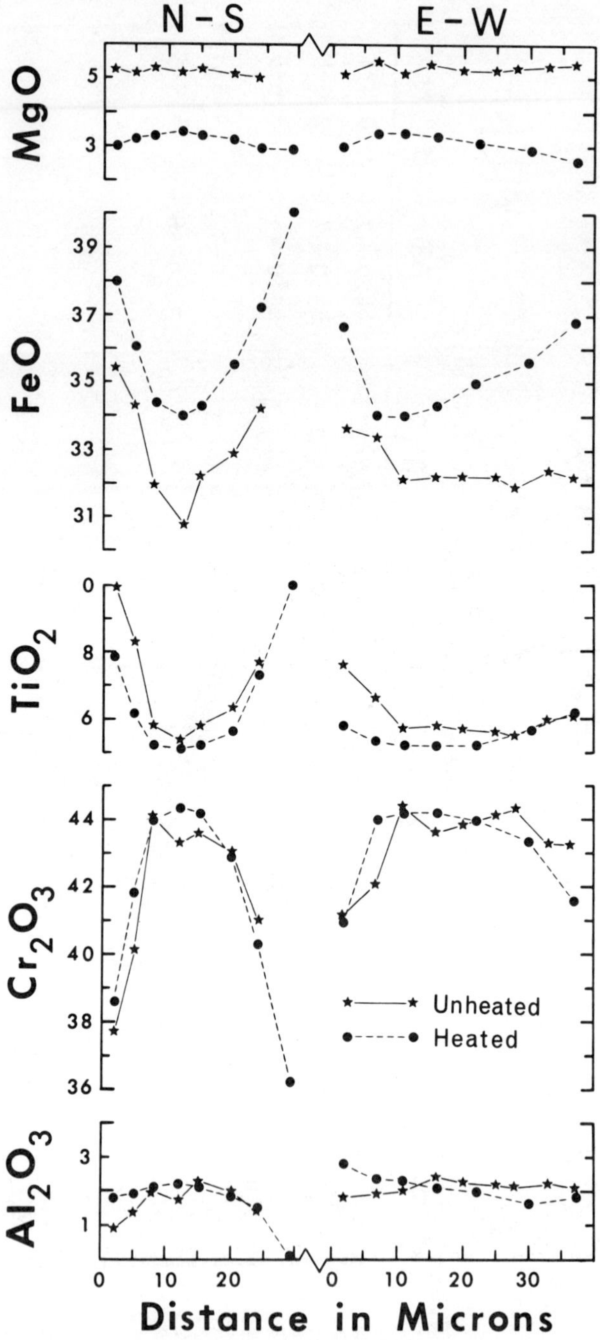

Fig. 2. Compositional (wt.%) profiles across a low-Ti spinel grain in 12018,43 before and after heating. The compositional zonings seen in this diagram are typical of the low-Ti spinels in the sample.

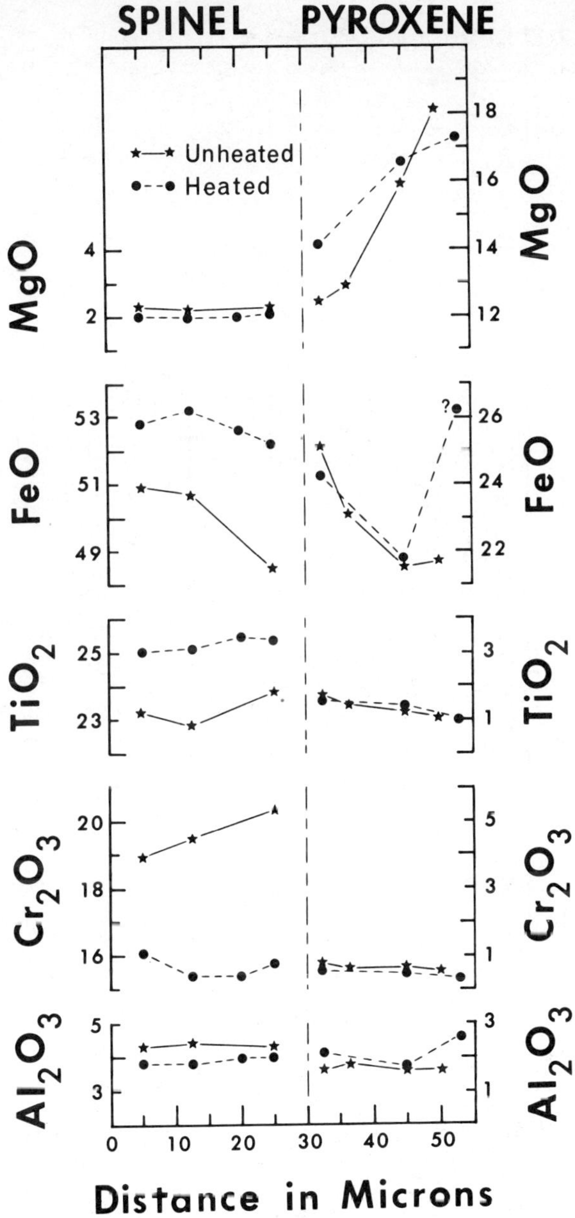

Fig. 3. Compositional (wt.%) profiles across a high-Ti spinel grain and an adjacent pyroxene grain in 12018,43 before and after heating. The trends of compositional changes due to heating shown in this diagram are typical of the high-Ti spinels in the sample, although magnitudes vary considerably between different high-Ti spinel grains.

Table 2. Selected analyses of corresponding High-Ti, Intermediate-Ti, and Low-Ti spinels in unheated (U) and heated (H) Apollo 12 sample 12018,43.

	1		2		3		4		5		6		7		8	
	U	H	U	H	U	H	U	H	U	H	U	H	U	H	U	H
TiO_2	22.75	23.89	20.42	23.17	20.05	21.28	19.53	20.86	17.16	19.66	11.38	14.58	10.66	7.78	7.55	5.93
Al_2O_3	5.18	4.25	6.52	5.01	5.52	5.29	6.07	5.18	7.24	6.20	10.11	8.27	11.06	11.04	9.30	11.12
Cr_2O_3	20.43	16.11	23.68	17.70	23.99	19.66	24.69	20.53	27.39	21.70	34.83	29.31	36.31	39.18	41.53	42.20
V_2O_3	0.48	0.41	0.45	0.24	0.57	0.58	0.49	0.46	0.62	0.49	0.61	0.53	0.73	0.66	0.71	0.80
FeO	46.02	51.04	43.93	48.99	46.09	49.17	44.45	48.39	42.05	46.22	35.94	42.62	35.42	36.51	35.81	36.26
MgO	4.18	2.51	4.14	3.11	2.32	2.12	3.81	2.33	4.02	2.88	4.90	2.83	4.94	2.92	2.82	2.17
CaO	0.18	0.07	0.32	0.11	0.18	0.11	0.20	0.10	n.d.	0.11	0.25	0.16	0.23	0.08	0.28	0.25
Total	99.22	98.28	99.46	98.33	98.72	98.20	99.35	97.85	98.48	97.26	98.02	98.30	99.35	98.17	98.00	98.73
	Atomic Proportions Based on 32 Oxygen Atoms															
Ti	4.92	5.25	4.31	5.04	4.34	4.67	4.15	4.58	3.65	4.30	2.39	3.13	2.20	1.65	1.62	1.26
Al	1.68	1.46	2.15	1.71	1.87	1.82	2.02	1.78	2.42	2.13	3.33	2.79	3.57	3.67	3.12	3.70
Cr	4.54	3.72	5.25	4.04	5.46	4.53	5.52	4.74	6.13	4.99	7.69	6.62	7.87	8.74	9.35	9.41
V	0.12	0.10	0.10	0.06	0.13	0.14	0.11	0.11	0.14	0.11	0.14	0.12	0.16	0.15	0.16	0.18
Fe	10.88	12.47	10.30	11.84	11.09	11.99	10.54	11.83	9.96	11.25	8.39	10.19	8.12	8.61	8.53	8.56
Mg	1.71	1.09	1.73	1.34	0.99	0.92	1.61	1.02	1.70	1.25	2.04	1.21	2.02	1.23	1.20	0.91
Ca	0.06	0.02	0.10	0.03	0.06	0.03	0.06	0.03	n.d.	0.03	0.07	0.05	0.07	0.02	0.09	0.08
Total	23.91	24.11	23.94	24.06	23.93	24.09	24.02	24.10	24.00	24.08	24.04	24.10	24.00	24.07	24.06	24.09

described from other Apollo 12 basalts, were not seen in the present sample. The high-Ti spinels, which have been interpreted in other similar samples to represent a later period of crystallization, show unsymmetrical (to occasionally oscillatory) compositional zoning (Fig. 3). The maximum compositional variation observed across a single high-Ti spinel grain was as follows (wt.%): TiO_2, 19.7–23.2; Cr_2O_3, 24.9–20.7; Al_2O_3, 6.1–5.1; FeO, 44.8–46.3; and MgO, 3.8–4.0.

Some typical analyses of the three groups of spinels are presented in Table 2. Computation of atomic proportions on the basis of 32 oxygen atoms show no charge imbalance. This suggests that Fe^{+3}, Ti^{+3}, and Cr^{+2}, if present, are quantitatively insignificant. Also, the data presented in Table 2 indicates that the compositional variations are systematic so that stoichiometry is maintained from the most Ti-poor (chromites) to the most Ti-rich (ulvöspinels) spinels. The systematic nature of the compositional variations is also discernible in Figs. 2 and 3.

The spinels show a strong Al^{+3}–Cr^{+3} positive correlation (Fig. 4); this is to be expected because both of these ions have high B^{VI} site preference energies so that R^{+3} substitution is restricted to B^{VI} sites in the spinel structure. The Ti^{+4} vs. $(Cr + Al + V)^{+3}$ plot for these spinels (Fig. 5) shows that almost all the analyzed points plot on or very close to the line representing the theoretical concentration

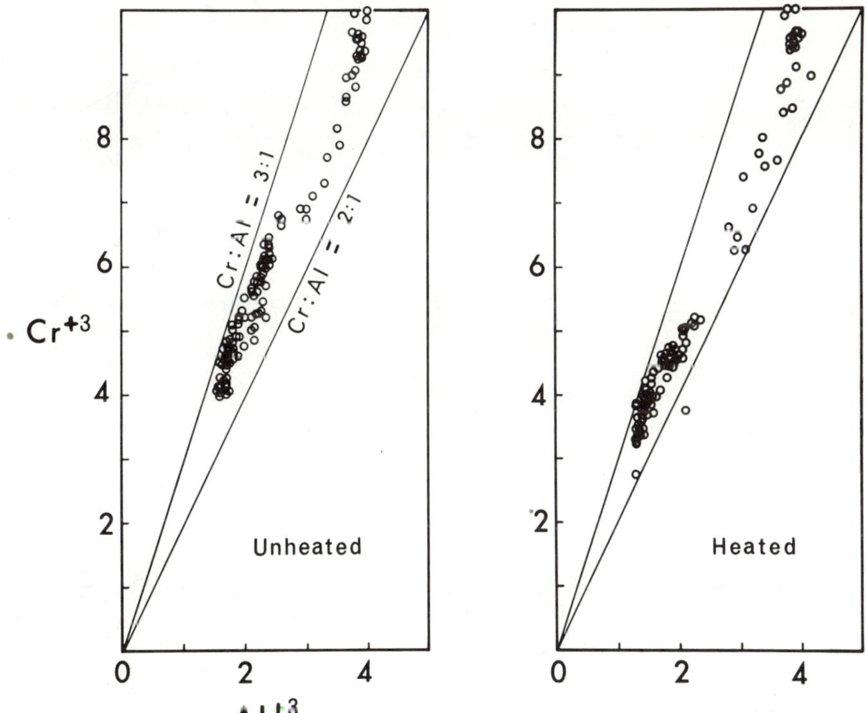

Fig. 4. Cr^{+3}–Al^{+3} substitution in spinels in unheated and heated 12018,43 (Ions based on 32 oxygen atoms).

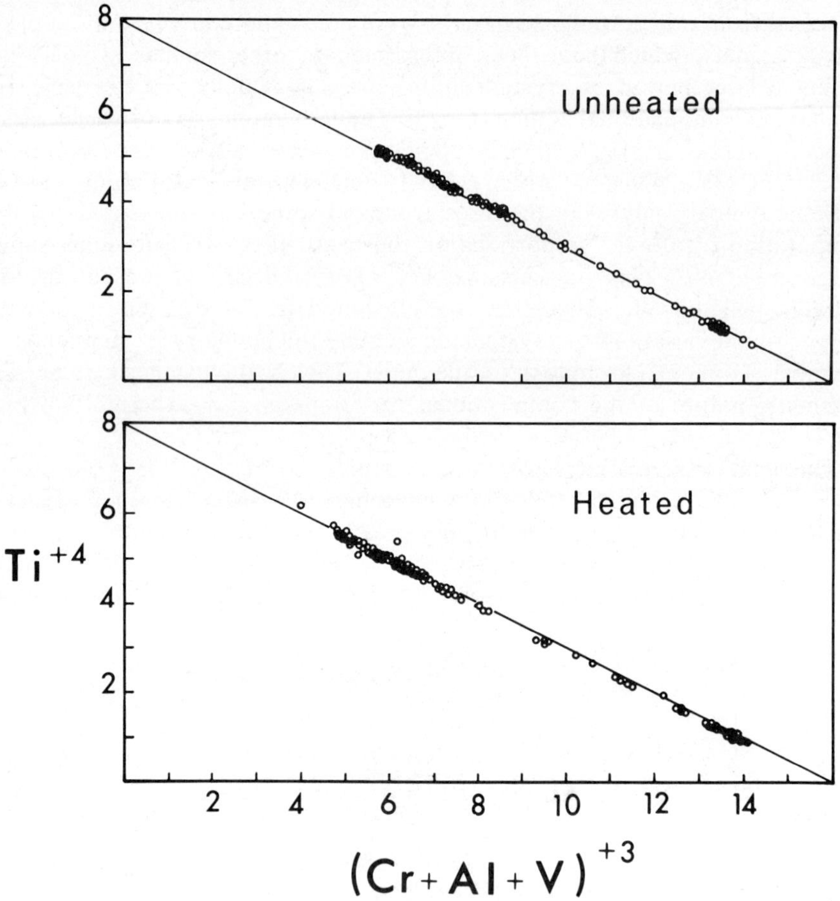

Fig. 5. Ti^{+4}–$(Cr + Al + V)^{+3}$ substitution in spinels in unheated and heated 12018,43 (Ions based on 32 oxygen atoms).

of these elements in the pure "normal" (chromite-hercynite) and "inverse" (ulvöspinel) end members of the spinel series (the "Octahedral line" of Haggerty, 1972b). This suggests that the Ti^{+4} in the spinel substitutes for $(Al, Cr)^{+3}$ in the B^{VI} sites. Thus, as suggested by Gibbs *et al.* (1970), $(Fe, Mg)^{+2} + Ti^{+4} \rightleftharpoons 2(Cr, Al)^{+3}$ appears to be the substitutional mechanism that enables the spinel stoichiometry to be maintained through the normal-inverse transition in the spinel structure. Evidently, the chemistry of the spinels in the present sample is very similar to that of the Apollo 15 spinels described in detail by Haggerty (1972b).

Heated sample

Heating of the sample did not produce much textural changes except for the appearance of $<5\,\mu m$ wide blebs of Cr-bearing high-Ti spinel (checked by semi-quantitative EMP analysis) along the cracks of an ilmenite grain, and homogenization of micron-size high-Ti blebs (ilmenite?) in some high-Ti spinel grains. However, the compositional changes observed in the corresponding spinel (and ilmenite) grains are significant and systematic relative to their initial compositions (Figs. 1, 2, 3, and 6).

The compositional changes observed in the high-Ti and intermediate-Ti spinels are fairly pronounced; these include an increase in TiO_2 and FeO contents in the heated sample, with a concommitant decrease in Cr_2O_3, Al_2O_3, and MgO (Figs. 3 and 6). This is typical of all high-Ti and intermediate-Ti spinels analyzed in the present study. The compositional changes observed in the low-Ti spinels are not as systematic or conspicuous. For the low-Ti spinel grain represented in Fig. 2, heating has produced a significant increase in FeO and a corresponding decrease in MgO and TiO_2, whereas no clear trend is discernible for Cr_2O_3 and Al_2O_3. However, a few low-Ti spinel grains show a reverse trend (increase in TiO_2 and decrease in FeO and Cr_2O_3—see Table 2, Analysis 6); in some others the changes produced by heating are not significant. As a group the low-Ti spinels are judged to show on heating an increase in Cr_2O_3 (accompanied by a corresponding decrease in FeO and TiO_2).

Although no compositional profiles for ilmenites are being included here, it may be pointed out that the ilmenites in the heated sample show an increase in TiO_2 (about 2 wt.%) and a decrease in FeO (about 1–2 wt.%), probably accompanied by a very small increase (0.1–0.2 wt.%) in Cr_2O_3, Al_2O_3, and MgO.

Some typical analyses of the three types of spinels in the heated sample (along with their unheated compositions) are given in Table 2. The Cr^{+3}–Al^{+3} plot (Fig. 4) and the Ti^{+4}–$(Cr + Al + V)^{+3}$ plot (Fig. 5) for the spinels in the heated sample are essentially the same as for the corresponding spinel grains in the unheated sample—except for the shifts in the absolute positions of the data points consequent upon the relative changes in elemental composition produced

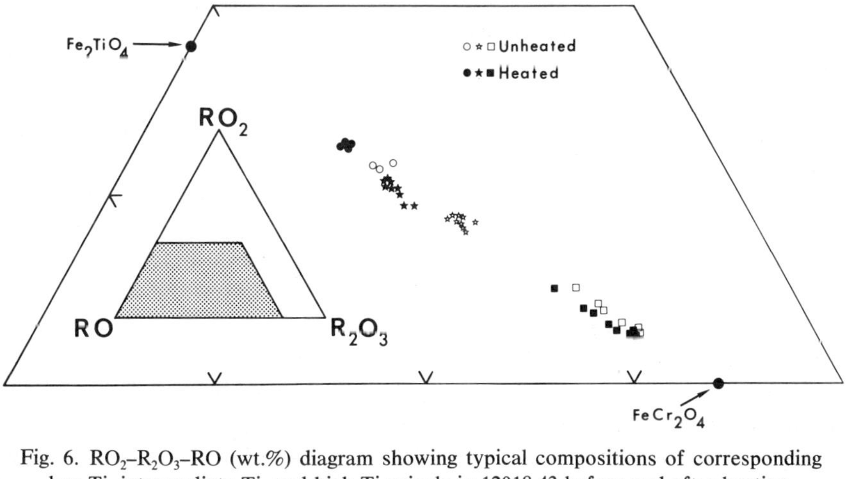

Fig. 6. RO_2–R_2O_3–RO (wt.%) diagram showing typical compositions of corresponding low-Ti, intermediate-Ti, and high-Ti spinels in 12018,43 before and after heating.

by heating. These data suggest that despite the highly variable and often large compositional changes in the spinels produced by heating, the stoichiometry is still maintained in the normal-inverse spinel series by the same substitutional mechanism discussed earlier.

DISCUSSION OF RESULTS

Our data from the present study do not allow us to evaluate all the different hypotheses proposed for the bimodal distribution of spinel compositions in lunar basalts. However, the marked tendency of the high-Ti and intermediate-Ti spinels to shift toward ulvöspinel composition is compatible with the suggestion of Haggerty (1972a) that the intermediate members of the normal-inverse spinel series are structurally metastable which may account for their less common occurrence in the lunar basalts. The lack of a clear trend for the low-Ti spinels to shift toward the chromite (-hercynite) composition is related possibly to slower kinetics due to the relatively smaller departure of these compositions from $Fe(Cr, Al)_2O_4$ end member composition.

The nature of the compositional changes in the heated spinels (and ilmenites) is such that they cannot be explained by intragranular adjustments alone. In the heated sample, pyroxene grains adjacent to the high-Ti spinels show a decrease in FeO, and an increase in MgO and Al_2O_3 at the interface (Fig. 3); this may account for the MgO depletion and a part of the FeO enrichment in the high-Ti spinels. A part of the FeO enrichment in the heated high-Ti spinels may be correlated also with the decrease in FeO in the ilmenites. The increase in TiO_2 and decrease Cr_2O_3 in high-Ti spinels may be tied to the reverse trends observed in some of the lower-Ti spinels. However, judging from the relative abundance of the relevant mineral phases and the magnitude of compositional changes in the heated sample, it is difficult to reconcile the bulk changes in spinel compositions entirely by isochemical subsolidus reequilibration.

Much of the native FeNi metal present on the polished surfaces of the original sample was absent in the heated sample. It would appear that the metal grains were lost by evaporation during the heating experiment. As mentioned earlier, although the rock chip was thoroughly cleaned before it was heated, it was anticipated that a small amount of carbon would be retained in the pores of the chip, simulating the conditions under which lunar basalts might be thermally metamorphosed. Therefore, it is possible that the native FeNi was vaporized and the metals transported as a carbonyl (e.g., $Fe(CO)_5$), similar to the mechanism proposed by El Goresy and Ramdohr (1975). The black coloring on the sides of the quenched silica tube may represent carbon residue from the carbonyl. Vapor transport of some of the oxide elements as their respective carbonyls could account for the non-isochemical changes in the heated spinels and ilmenites.

Admittedly, we have no definite explanation at present for the observed compositional changes in the spinels in the heated sample. Also, the persistence of compositional inhomogeneities and gradients in the heated spinels strongly suggest the possibility that the spinels (and ilmenites and the associated silicates) did not attain their equilibrium compositions during the heating experiment, presumably due to kinetic factors. However, we believe that our heating experiment does demonstrate that *thermal metamorphism of lunar basalts is likely to modify the compositions of their pre-existing spinels* (*and ilmenites*), as demonstrated earlier by Taylor *et al.* (1976) for the FeNi metals in Apollo 16 breccias and melt rocks. These compositional changes may not be drastic enough to invalidate paragenetic relationships deduced for the distinctly different compositional types of spinels (e.g., Ti-Chromite vs. Cr-ulvöspinel), but application of compositional zoning in individual spinel grains to trace the progressive changes in the composition of the crystallizing lunar melt may be misleading.

Acknowledgments—We would like to express our gratitude to Drs. A. El Goresy, J. I. Goldstein, and B. Walker for their helpful discussions and suggestions. Drs. J. Friel and T. M. Usselman are acknowledged for the review of the manuscript. We also thank the University of Tennessee for providing substantial support for the purchase of the electron microprobe. The research presented in this paper was supported by NASA grant NGL 43-001-140.

References

Albee A. L. and Ray L. (1970) Correction factors for electron probe microanalysis of silicates, oxides, carbonates, phosphates, and sulfates. *Anal. Chem.* **42**, 1408–1414.

Bence A. E. and Albee A. L. (1968) Empirical correction factors for the electron microanalysis of silicates and oxides. *J. Geol.* **76**, 382–403.

Bushe F. D., Prinz M., Keil K., and Bunch T. E. (1972) Spinels and the petrogenesis of some Apollo 12 igneous rocks. *Amer. Mineral.* **57**, 1729–1747.

Cameron E. (1971) Opaque minerals in certain lunar rocks from Apollo 12. *Proc. Lunar Sci. Conf. 2nd*, p. 193–206.

El Goresy A., Prinz M., and Ramdohr P. (1976) Zoning in spinels as an indicator of the crystallization histories of mare basalts. *Proc. Lunar Sci. Conf. 7th*, p. 1261–1279.

El Goresy A. and Ramdohr P. (1977) Apollo 17 TiO_2-rich basalts: Spinel chemical bimodality in the two major basalt types and genetic significance of inverted zoning of chromian ulvospinel (abstract). In *Lunar Science VIII*, p. 281–283. The Lunar Science Institute, Houston.

El Goresy A. and Ramdohr P. (1975) Subsolidus reduction of lunar opaque oxides: Textures, assemblages, geochemistry, and evidence for a late-stage endogenic gaseous mixture. *Proc. Lunar Sci. Conf. 6th*, p. 729–745.

El Goresy A., Ramdohr P., and Taylor L. A. (1971) The opaque minerals in the lunar rocks from Oceanus Procellarum. *Proc. Lunar Sci. Conf. 2nd.*, p. 219–235.

El Goresy A., Ramdohr P., and Taylor L. A. (1972) Frau Mauro crystalline rocks: Mineralogy, geochemistry and subsolidus reduction of the opaque minerals. *Proc. Lunar Sci. Conf. 3rd*, p. 333–349.

Gibb F. G. F., Stumpfl E. F., and Zussman J. (1970) Opaque minerals in an Apollo 12 rock. *Earth Planet. Sci. Lett.* **9**, 217–224.

Haggerty S. E. (1971) Subsolidus reduction of lunar spinels. *Nature Phys. Sci.* **234**, 113–117.

Haggerty S. E. (1972a) Solid solution, subsolidus reduction and compositional characteristics of spinels in some Apollo 15 basalts. *Meteoritics* **7**, 353–370.

Haggerty S. E. (1972b) Apollo 14: Subsolidus reduction and compositional variations of spinels. *Proc. Lunar Sci. Conf. 3rd*, p. 305–332.

Haggerty S. E. and Meyer H. O. A. (1970) Apollo 12: Opaque oxides. *Earth Planet. Sci. Lett.* **9**,

James O. B. and Wright T. L. (1972) Apollo 11 and 12 mare basalts and gabbros: Classification, compositional variations, and possible petrogenetic relations. *Bull. Geol. Soc. Am.* **83**, 2357–2382.

Kushiro I., Nakamura Y., Kitayama K., and Akimoto S. (1971) Petrology of some Apollo 12 crystalline rocks. *Proc. Lunar Sci. Conf. 2nd,* p. 481–495.

Muan A., Hauck J., and Lofall T. (1972) Equilibrium studies with a bearing on lunar rocks. *Proc. Lunar Sci. Conf. 3rd,* p. 185–196.

Taylor L. A., Williams R. J., and McCallister R. (1972) Stability relations of ilmenite and ulvospinel in the Fe-Ti-O system and application of these data to lunar assemblages. *Earth Planet. Sci. Lett.* **23**, 220–228.

Taylor L. A., Misra K. C., and Walker B. M. (1976) Subsolidus reequilibration, grain growth, and compositional changes of native FeNi metal in lunar rocks. *Proc. Lunar Sci. Conf. 7th,* p. 837–856.

Warner J. L. (1971) Lunar crystalline rocks: Petrology and geology. *Proc. Lunar Sci. Conf. 2nd,* p. 469–480.

Proc. Lunar Sci. Conf. 8th (1977), p. 2641–2654.
Printed in the United States of America

Barred olivine "chondrules" in lunar spinel troctolite 62295

EDWIN ROEDDER

U.S. Geological Survey, 959 National Center, Reston, Virginia 22092

PAUL W. WEIBLEN

Department of Geology and Geophysics, University of Minnesota, Minneapolis, Minnesota 55455

Abstract—Several objects have been found in sections of lunar igneous spinel troctolite 62295 that resemble certain meteoritic barred olivine chondrules. Each consists of an apparently spherical single crystal of Fo_{90} olivine, ~0.6–0.8 mm in diameter, containing a set of ~30–40 subparallel stringers of An_{95} plagioclase, whereas the stringers in ordinary meteoritic chondrules consist of glass. The olivine of the 62295 chondrules is also more magnesian, and is radially zoned, having a relatively iron-rich core and rim and an iron-poor intermediate zone. Several possible origins are proposed: impact-generated melt globules solidified in flight, spherical phenocrysts, and meteoritic chondrules, but none of these seems adequate to explain the detailed observations.

INTRODUCTION

CHONDRULES, AS DEFINED by Fredriksson *et al.* (1973b, p. 476), are "once wholly or partly liquid, individual bodies which cooled, solidified or crystallized rapidly". In spite of extensive studies, the enigma of chondrules in meteorites has been with us for a long time. A variety of theories has been proposed to explain the origin of these little (~1 mm) spheroids and the manner in which they have been incorporated in the chondritic meteorites, yet many questions remain unanswered. The presence of spherical objects that strongly resemble meteoritic chondrules in non-meteoritic samples provides a new approach and possibly new insight into the problems of their origins. Some globules of molten silicate have formed by meteorite impact on the earth (Fredriksson *et al.*, 1973a), and on the moon, where they are found in the soils and breccias. These have crystallized to yield textures similar to some meteoritic chondrules, as pointed out by several authors (Haggerty, 1974; Roedder and Weiblen, 1973a,b; Kurat *et al.*, 1972; Bunch *et al.*, 1972; Fredriksson *et al.*, 1970a, 1973b; Ivanov *et al.*, 1976). Except for one "chondrule" illustrated by Fredriksson *et al.* (1973b, their Fig. 1), these various "chondrules" are not "barred" (i.e., consisting of parallel bars of an optically continuous crystal separated by bars of another composition), and even that one differs in several respects from typical meteoritic chondrules. One important difference, for example, is the lack of the generally characteristic outer continuous single crystal rim (Van Schmus, 1969). Also, the compositions are not generally similar to those of chondrules from meteorites. In this paper we discuss an occurrence of apparent barred olivine "chondrules" in a lunar igneous rock, 62295, in which both the texture and composition are similar to those of meteoritic chondrules. These "chondrules", which might be xenoliths, differ from all other xenoliths we have seen and present several unsolved problems concerning their origin, as well as that of the host rock. Unfortunately the occurrence raises more questions than it answers on the problem of the origin of meteoritic chondrules.

Host Rock 62295

General petrography

Rock 62295 is a small sample of an apparently igneous rock collected during the Apollo 16 mission. Although it is unique in composition and texture among lunar samples of this size, many small clasts of grossly similar composition have been found in various lunar soils and breccias, so the rock type is not truly rare on the moon. Modally, the rock is a spinel troctolite with a few percent of lithic debris (Wilshire *et al.*, 1973) and presumed xenocrysts (Agrell *et al.*, 1972; Walker *et al.*, 1973). Agrell *et al.* (1972) report modes in the following ranges: plagioclase 48–57; mesostasis 15–20; olivine 1; spinel 3–6; ilmenite 1; metal and sulfide 1 (*sic.*; total of their reported mode is low), and call the rock spinel allivalite. Texturally, the host rock is porphyritic and has been described as a variolitic and intersertal spinel olivine basalt (Ganapathy *et al.*, 1974). The lithic fragments and xenocrysts are assumed by some to be either unmelted relicts from the fused source materials, or solids that were later incorporated into a lava produced by either normal igneous processes or impact.

The inclusions in the rock are possibly pertinent to the problem of the origin of the "chondrules". Although the rock weighed only 251 g, rather gross differences exist between the various sections available, particularly in the nature of the xenoliths or xenocrysts. One of the xenolith types consists of an oval mass of very fine-grained pure anorthite, $An_{99.6}$, approximately 200 μm long, that presumably was present as a solid and acted as a nucleus for the crystallization of the An_{91-95} laths that make up the bulk of the surrounding rock. This may be a devitrified fragment of anorthitic maskelynite, from an earlier impact event, incorporated into the 62295 melt. In addition to this type of plagioclase, single crystals of plagioclase, olivine (in part shocked?), spinel, and metal are among the possible xenocrysts. Sharply euhedral, in part skeletal, phenocrysts of olivine are also present (Weiblen and Roedder, 1973); these show *normal* zoning from Fo_{95} in the core to Fo_{80} in the rims and are unique among lunar olivines in their low content of Cr_2O_3 (Brown *et al.*, 1973). Some compositionally homogeneous plagioclase xenoliths are very well rounded ((Nord *et al.*, 1973, Fig. 5a).

Composition of 62295 and its minerals

Details on the bulk composition of 62295, plus 20 analyses of minerals in it, and nine analyses of silicate-melt inclusions trapped in these various minerals were given by Weiblen and Roedder (1973) and need not be repeated here. The five olivines that were analyzed were all hosts for silicate-melt inclusions and ranged from $Fo_{85.6}$ to $Fo_{93.4}$. CaO in these olivines averaged 0.13%, but this average includes one analysis yielding 0.33% CaO. NiO was $\leq 0.05\%$. Five plagioclase crystals analyzed by Weiblen and Roedder (1973) ranged from An_{91} to An_{97}, with the average $An_{93.6}Ab_{6.1}Or_{0.3}$, but Agrell *et al.* (1972) reported $An_{90}Ab_{9.2}Or_{0.8}$.

"Chondrules"

Physical features

The only complete "chondrule" found is a single crystal of olivine, almost perfectly circular in section ($623 \times 660\ \mu m$), but jagged in detail, that contains 30–40 bifurcating but subparallel stringers of plagioclase separating the olivine into a series of bars (Figs. 1 and 2). All the olivine goes to extinction essentially simultaneously ($\pm 0.5°$). In several areas within the "chondrule", groups of adjoining stringers of plagioclase extinguish simultaneously and hence, each group is presumed to comprise a single crystal that was connected in the third dimension before sectioning. As these extinction positions bear no systematic relation to those of the host olivine, presumably no regular crystallographic relationship exists between the two, even though the texture would suggest it. Near the edge of the "chondrule" about half of these stringers end abruptly, leaving a $20\ \mu m$ band of olivine relatively free of stringers. Some of the plagioclase stringers that do extend through this band to the outside are crystallographically parallel with plagioclase crystals in the host rock. In addition, there are many small, irregular areas of olivine in the surrounding rock that are physically and optically continuous with the "chondrule" olivine crystal, and others within $\sim 100\ \mu m$ that extinguish with it and hence are presumably connected (Fig. 2). These features suggest that the "chondrule" was present during crystallization of the host and acted as a nucleus, but do not prove that the band necessarily grew during the crystallization of the host rock. Other than the similar crystallographic orientations of plagioclase and olivine crystals across the "chondrule" edge, no effects on the host-rock texture or composition near the "chondrule" are noticeable.

Tiny opaque particles (~ 1–$2\ \mu m$) are present in the plagioclase stringers and make up $\sim 0.1\%$ of the "chondrule". Many silicate-melt inclusions, each now consisting apparently of glass with an approximately 10 vol.% vapor bubble, are present in the olivine. Most are only a few micrometers in length, much smaller than those in the euhedral phenocrysts of olivine in this same rock that were analyzed earlier (Weiblen and Roedder, 1973). A few of the tiny melt inclusions contain opaque grains, possibly daughter crystals. Some inclusions occur in crude rows approximately on the centerline of olivine bars. These could be simply glass of plagioclase composition that has failed to nucleate and crystallize; such metastability is common in terrestrial silicate-melt inclusions (Roedder, 1969, 1971).

Although the "chondrule" illustrated in Fig. 1 is the only complete section of

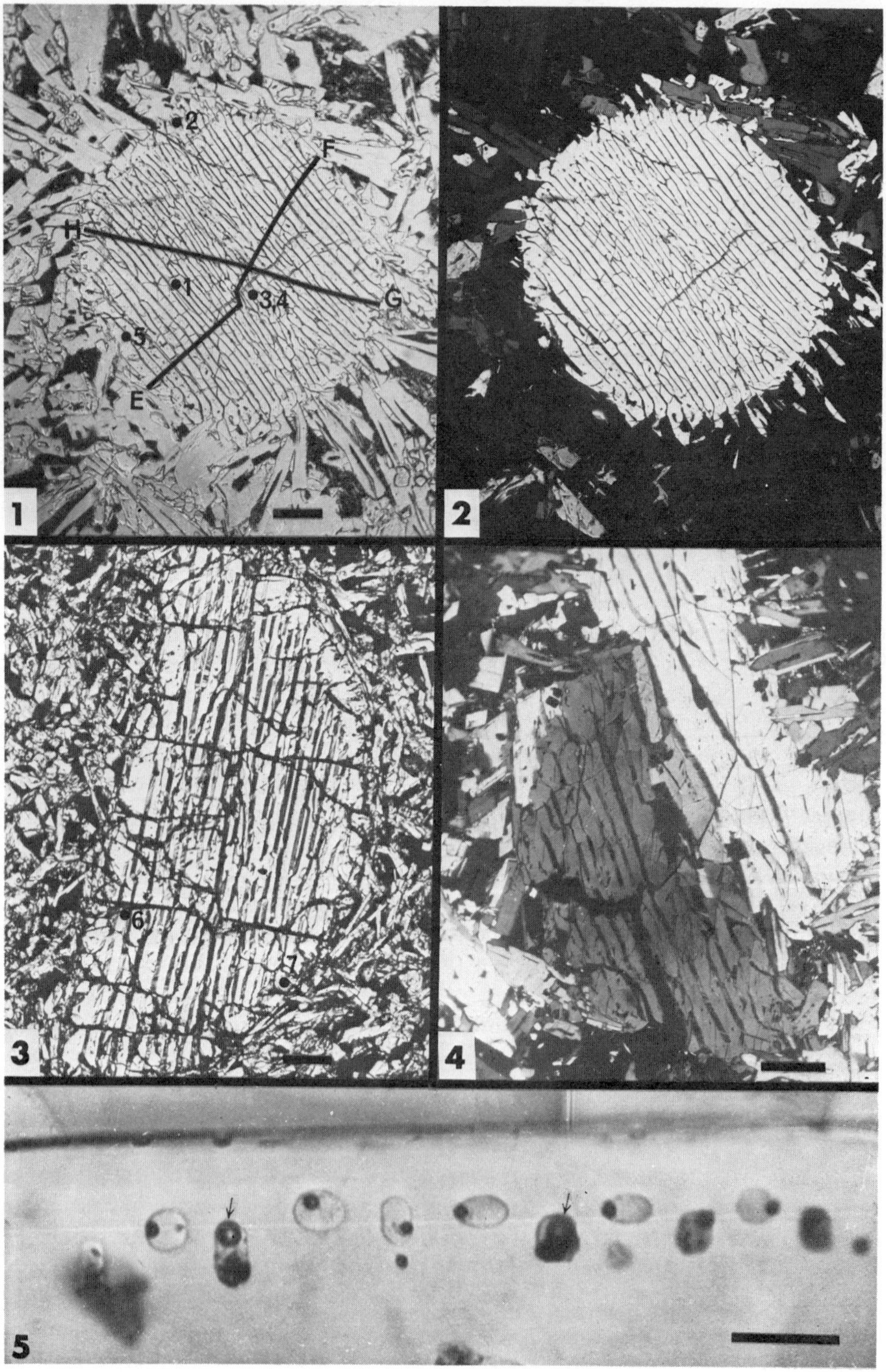
1
2
F
H
1
3,4
G
5
E
2
3
6
7
4
5

a "chondrule" we have found, examination of all other available slides revealed several chord-shaped fragments of similar apparent chondrules. One of these (Fig. 3) was found by Nord *et al.* (1973) in 62295 and was examined by them. They reported that the bars appeared to consist of silicate glass but we found them to be birefringent. As all these "chondrules" had essentially parallel bars, it seems unlikely that they are merely the result of a special angle of cut through a radial-type chondrule, as was suggested by Levi–Donati (1970) for other apparent lunar barred chondrules. We do not know whether any of these sections cut through the actual center of a "chondrule".

Composition

The mode of the "chondrule" illustrated in Fig. 1 is olivine—91.2, plagioclase—8.8 wt.%. It was determined from a series of microscope traverses across the surface at 1200×, in reflected light; linear intercept distances were measured by use of a filar micrometer eyepiece. The linear and hence volume ratio in this section was 89.6/10.4.

The olivine is not uniform in composition but is zoned from Fo_{87} to Fo_{94} (Table 1) in a peculiar manner, in that the core and rim of the "chondrule" are relatively iron-rich ($Fo_{88.5}$ and Fo_{87}, respectively) and the intermediate zone is iron-poor (Fo_{92}). This is illustrated in Fig. 6 on which is plotted the FeO content of the olivine in each bar along two traverses, which were inclined ~60° to each other (Fig. 1) and which were made with some slight offsets to avoid cracks and pits. Calcium, chromium, and aluminum vary sympathetically with iron (0.1–0.3, 0.06–0.12, and 0.1–0.2%, respectively). Although some of the smaller compositional excursions on the traces in Fig. 6 may be spurious, the gross zoning picture is certainly valid. Each analysis was made on the center of a bar in order to avoid contamination with stringer material, and represents essentially pure olivine. If any compositional zoning across individual bars exists, it must be

Figs. 1, 2. Barred olivine "chondrule" from 62295,2, in transmitted plain light (Fig. 1) and with crossed polarizers (Fig. 2). Bar = 100 μm. Electron microprobe analysis points 1–5 (see Table 1) and step scan traverses E–F and G–H (see Fig. 6) are shown on Fig. 1. (Offsets on E–F were made to avoid pits and cracks.)

Fig. 3. Fragment of barred olivine "chondrule" from 62295 32A-3, in plain light. Bar = 100 μm. Electron microprobe analyses made on points 6 and 7 (see Table 1). Photo courtesy of G. L. Nord, Jr.

Fig. 4. One of a series of phenocrysts having barred texture and tiny silicate-melt inclusions, from 62295,2 (see, e.g., Table 1, anal. 10). Partly crossed polarizers. Most such phenocrysts are single crystals but this one is a pair. Bar = 100 μm.

Fig. 5. Silicate-melt inclusions in rim of sharply euhedral olivine phenocryst containing glass stringers resembling bars, from 62295,83a (see Weiblen and Roedder, 1973, Fig. 4). Bar = 10 μm. The stringers (and probably these melt inclusions also) consist of basaltic glass (Weiblen and Roedder, 1973, Table 5, anal. 75–78). Small normal shrinkage bubbles occur in six inclusions, but several inclusions containing opaque daughter minerals (?) have anomalously large bubbles (arrows).

Table 1. Electron microprobe analyses of barred olivine chondrules and associated phases in 62295 (P. W. Weiblen analyst).

Anal. ≠	(1)	(2)	(3)	(4)	(5)	(6)	(7)	(8)	(9)	(10)	(11)	(12)	(13)	(14)	(15)
Slide	2	2	2	2	2	32 A3	32 A3	38-83B	83A	2	2	38-83C	2	32 A3	2
Lab no.	−2	−1	P-3	P-3	P-0	P2-13	P2-19	P1	P1	P4-6	P1-18	P-1	P0	P3-1	P1A
SiO_2	42.00	41.30	44.80	46.09	47.10	40.60	41.20	42.00	39.60	40.30	40.10	40.00	46.60	45.30	.74
Al_2O_3	.10	.21	30.40	36.30	31.30	.09	0	−.05	−.05	.05	.39	−.05	34.20	34.20	63.10
FeO	6.25	10.60	1.94	.36	.37	6.90	10.50	5.15	10.90	8.93	17.10	9.60	.25	.37	12.60
MgO	52.10	48.80	7.81	0	1.18	51.30	47.70	53.30	47.30	50.40	42.60	50.30	.54	.21	19.60
CaO	.13	.10	14.00	16.70	18.30	.17	.29	.07	.28	.09	.20	.13	18.50	19.60	−.05
Na_2O	0	0	.35	.42	.51	0	0	0	0	0	0	0	.36	−.10	0
K_2O	0	0	.06	.07	.10	0	0	0	0	0	0	0	.07	.06	0
TiO_2	0	0	.05	.06	.07	−.05	−.05	−.05	−.05	−.05	.07	−.05	.38	−.05	.10
MnO	.08	.15	−.05	0	−.05	.10	.15	.06	.17	.09	.24	.09	−.05	−.05	.20
Cr_2O_3	.07	.06	−.05	0	−.05	.10	.15	0	−.05	−.05	−.05	−.05	−.05	−.05	3.77
NiO	−.02	−.02	0	0	0	−.05	−.05	−.05	−.05	−.05	−.05	0	−0	0	0
Total	100.73	101.22	99.41	100.00	98.93	99.26	99.99	100.58	98.25	99.86	100.70	100.12	100.90	99.74	100.11
Cation proportions															
O	4.000	4.000	8.000	8.000	8.000	4.000	4.000	4.000	4.000	4.000	4.000	4.000	8.000	8.000	4.000
Si	1.004	1.002	2.083	2.106	2.192	.990	1.012	1.001	.995	.986	1.007	.980	2.124	2.098	.019
Ti	0	0	.002	.002	.002	0	0	0	0	0	.001	0	.013	0	.002
Al	.003	.006	1.666	1.955	1.717	.003	0	0	0	.001	.012	0	1.838	1.867	1.892
Cr	.001	.001	0	0	0	.002	.003	0	0	0	0	0	0	0	.069
Fe	.125	.215	.075	.014	.014	.141	.216	.103	.229	.183	.359	.197	.010	.014	.268
Mn	.002	.003	0	0	0	.002	.003	.001	.004	.002	.005	.002	0	0	.004
Mg	1.856	1.765	.541	0	.082	1.865	1.746	1.893	1.771	1.839	1.595	1.837	.037	.014	.743

Ca	.003	.003	.697	.818	.913	004	.008	.002	.008	.002	.005	.003	.904	.973	0
Na	0	0	.032	.037	.046	0	0	0	0	0	0	0	.032	0	0
K	0	0	.004	.004	.006	0	0	0	0	0	0	0	.004	.004	0
Total	2.994	2.995	5.100	4.935	4.973	3.007	2.987	2.999	3.005	3.013	2.985	3.020	4.961	4.970	2.998
Mineral compositions															
Fo	93.69	89.14	0	0	0	92.98	89.01	94.86	88.55	90.96	81.62	90.33	0	0	73.49
Fa	6.31	10.86	0	0	0	7.02	10.99	5.14	11.45	9.04	18.38	9.67	0	0	26.51
An	0	0	95.21	95.19	94.61	0	0	0	0	0	0	0	96.18	99.64	0
Ab	0	0	4.31	4.33	4.77	0	0	0	0	0	0	0	3.39	0	0
Or	0	0	.49	.48	.62	0	0	0	0	0	0	0	.43	.36	0

A negative sign before an analytical value signifies less than.

A zero in the right-hand column of any analysis signifies not analyzed.

(1) A point about two thirds of the way in from rim to core of olivine "chondrule". See Fig. 1.

(2) One of the "prominences" on the olivine "chondrule". The rim under these shows 7.77% FeO and 0.35% CaO. See Fig. 1.

(3) Wide part in plagioclase stringer in interior of "chondrule"; includes some olivine (see anal. 4 and Fig. 1).

(4) Analysis 3 minus 15.9% olivine of local composition.

(5) Thick plagioclase stringer near edge of "chondrule".

(6) A point about half way in from rim to core of broken olivine "chondrule" (see Fig. 3).

(7) A point on the rim of broken olivine "chondrule" (see Fig. 3).

(8) Core of olivine phenocryst similar to that shown in Fig. 4; three more analyses out to rim show FeO 5.23, 7.99, and 9.01%. Stringers are high in Ca and Al.

(9) Intermediate position in square olivine phenocryst containing glass stringers resembling bars (Weiblen and Roedder, 1973, Fig. 4). The FeO contents (in percent), from core to rim, were 6.2, 7.0, 7.5, 7.8, 9.0, 10.9, 11.0, 11.5, and 13.0.

(10) Barred olivine phenocryst similar to that shown in Fig. 4.

(11) Small olivine grain in matrix.

(12) Shocked olivine xenocryst. Three other points in same grain gave FeO 10.2, 10.2, and 10.5%.

(13) Plagioclase crystal just outside full "chondrule".

(14) Nearby plagioclase xenolith.

(15) Small clear spinel octahedron.

small. The small olivine crystal extensions or "prominences" on the "chondrule" out into the surrounding rock are essentially the same iron-rich olivine as the outer edge of the "chondrule" proper. The olivine of the "chondrule" is essentially similar to that present in the bulk of the rock in both major and minor elements (Table 1, anal. 10; see also Table 5 in Weiblen and Roedder, 1973), but more magnesian than that in most meteoritic chondrules.

The stringers are plagioclase An_{95}, on the basis of electron microprobe analyses, corrected for olivine contamination (Table 1, anals. 3–5). The opaque particles in the stringers consist of troilite plus nickel iron with some cobalt, but are too small for quantitative analysis by electron microprobe.

Several other olivine crystals in this rock, some sharply euhedral and some rounded to irregular, have roughly parallel lamellar inclusions (Fig. 4). These, however, are thicker than those in the "chondrules", are much more widely spaced, and consist of glass containing only 63–86% normative plagioclase; similar compositions are found for the glass inclusions in the phenocrysts of olivine from this rock (Fig. 5, and Weiblen and Roedder, 1973).

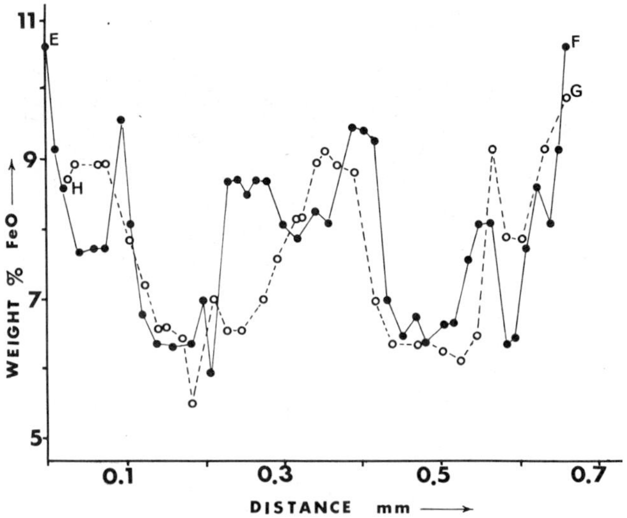

Fig. 6. Electron microprobe step scan analyses of olivine bars along two traverses shown in Fig. 1. The lines connecting points are shown for visual clarity only. These traverses were made by centering the beam with the viewing microscope on the electron microprobe on the centers of individual olivine bars. The olivine bars were oriented in the direction of the analyzing crystal to minimize edge effects. Analyses 1 and 2 (Table 1) were made by the same procedure and indicate that the edge effects were minimal and that the bars are olivine. The analyses were made at 20 kV, with a 1 μm beam diameter, at 0.015 μA and a 10 sec counting interval. The counting rate on the olivine was approximately 100 cps with a background of 10 cps. The total counting error and error due to edge effects is estimated to be within the range of 10% of the amount present. A similar traverse on the portion of a "chondrule" shown in Fig. 3 showed a similar variation in iron and a sympathetic variation in magnesium, but this section does not include a central core.

The melt inclusions in the "chondrule" represent a sample of the liquid at the growing crystal face, minus whatever olivine may have crystallized on the walls after trapping. This crystallization may have major effects on the composition of the remaining melt, as well as on the vapor/melt ratio (Roedder, 1969). Unfortunately, none of these inclusions at the surface of the "chondrule" is large enough for analysis and comparison with the much larger melt inclusions in the phenocrysts of olivine in this same rock (Fig. 5) that were analyzed by Weiblen and Roedder (1973).

Possible Origins of "Chondrules" in 62295

Three possible modes of origin can be postulated but none seem completely adequate to explain the details observed in the "chondrules" on 62295. These three modes are: (1) crystallization of globules of an impact-generated melt on the moon, (2) formation of roughly spherical phenocrysts, and (3) incorporation of preexisting meteoritic chondrules.

Impact-generated melt globules

In this mode of origin, the "chondrules" would result from crystallization of an impact-generated melt that was blasted out into space, crystallized, and fell back to the moon to be incorporated into a magma of 62295 composition. (Formation and incorporation may have been one continuous event or two completely independent events.) As a result of its composition, a "chondrule" of this type would be very resistant to melting and dissolution (G. Kurat, pers. comm., 1974). The distribution of silicate-melt inclusions along the centerlines and the barred structure suggest rapid feathery growth of olivine. Barred and feathery textures in synthetic chondrules as described by Nelson *et al.* (1972) and in lunar and meteoritic chondrules as described by Fredriksson *et al.* (1973b) are generally considered to be a result of extremely rapid crystallization from strongly supercooled melts, as indicated by the work of Kurat (1967). The circular cross section seen here suggests that the "chondrule" originated from crystallization of a spherical droplet in free fall or zero gravity, as meteoritic chondrules are generally assumed to originate (Wood, 1963). Somewhat similar barred "chondrules" of olivine containing stringers of *glass* have been described by Fredriksson *et al.* (1973b) and Kurat *et al.* (1972) from lunar samples, and we have reported them in orange soil 74220 (Roedder and Weiblen, 1973a). The stringers of plagioclase rather than of glass in the present sample may merely reflect the gross composition of the original melt. They may well have been trapped as stringers of glass and crystallized to plagioclase later. However, the gross composition of the "chondrule", which contains 91% olivine, contrasts strongly with the bulk composition of 62295, which contains only ~25% olivine, raising a problem of the origin of such a magnesian melt droplet. Fuchs *et al.* (1973) reported plagioclase-rich glass inclusions in the olivine of some meteoritic chondrites, and Blander *et al.* (1976) obtained highly siliceous residual glasses in their synthetic chondrules. If a troctolitic liquid were trapped as an inclusion in olivine, crystallization on the walls would yield a plagioclase-rich residue. Such a process does not explain the textural features of the "chondrule", however.

The rather unusual nature of some of the inclusions in rock 62295, such as the mass of almost pure anorthite, suggests that they may have come from an impact on a coarse-grained rock. If this rock also contained olivine, it would also permit formation of an olivine-plagioclase melt to yield the "chondrule". However, if this were the origin for these two, the rock would have to be rather unusual to provide both plagioclase of $An_{99.6}$ and olivine of Fo_{90}.

Spherical phenocrysts

A second possibility is that the "chondrule" grew in 62295 magma (whatever its origin) similar to the almost spherical plagioclase phenocryst noted by Nord *et al.* (1973). If so, it is difficult to explain the circular shape and the lack of intermediate types between this and the euhedral olivine phenocrysts in this rock (Weiblen and Roedder, 1973, their Fig. 4), but it is true that the circular outline and symmetry of the "chondrule" could be from a fortuitous section through a corner of a rounded and zoned phenocryst.

Many phenocrysts apparently start by rapid, feathery crystallization, presumably under conditions of strong supercooling prior to nucleation. Once the local supersaturation is eliminated, subsequent growth is slower and the feathery aggregate becomes covered with solid growth. In support of this possibility is the presence in sample 62295 of other olivine crystals containing roughly parallel lamellae. These lamellae, however, differ from those in the "chondrule" in many details, particularly in composition, since they consist of basaltic glass (Fig. 5; Weiblen and Roedder, 1973, Table 5, anals. 75–78). Furthermore, these olivine phenocrysts are either sharply euhedral or irregular (Fig. 4). Irregularly shaped phenocrysts would presumably become more spherical on partial dissolution, as suggested by E. C. T. Chao (pers. comm., 1973), but because the olivine phenocrysts in 62295 have only normal zoning (e.g., Table 1, footnote to anal. 9), this explanation seems invalid. The fact that sharply euhedral and apparently spherical crystals of olivine occur together in the same rock, however, suggests two different origins or at least different times of origin.

Meteoritic chondrules

The third possible mode of origin for the "chondrules" is that they are meteoritic xenoliths, i.e., that they are actually chondrules in the original sense. This would require that they have survived disaggregation on impact (except for those that are apparently fragments of chondrules), before incorporation in the 62295 magma. At least some meteoritic chondrites have been subjected to shock. Although most lunar meteorite impacts apparently result in complete disaggregation, fusion, or even partial vaporization of the projectile, it is possible that a special combination of low speed and low angle of the impact might result in incomplete destruction of a part of the projectile (e.g., Fredriksson *et al.*, 1970b). Considerable evidence exists for the presence of small grains of meteoritic material in the lunar soils (e.g., Brecher *et al.*, 1973; Ganapathy *et al.*, 1974), but the rarity of fragments of material on the moon larger than a few tens of micrometers that are believed to be meteoritic in origin is an expectable consequence of the rarity of such a special type of impact.

It is odd, however, that meteoritic chondrules have not been found in the lunar soils or breccias. Roedder and Weiblen (1974, Fig. 13) described one 250 μm, roughly circular object from breccia 67915 that looks like a barred chondrule but consists of an optically single crystal of pyroxene ($En_{39.0}Wo_{21.0}Fs_{40.0}$), low in Ti, Ni, and Cr, separated by ~20 bars of siliceous material comprising ~15 vol.%. These bars consist of 92.6% SiO_2 and 0.9% TiO_2, and CaO, MgO, and FeO make up most of the balance. To our knowledge, such compositions have not been reported for meteoritic chondrules. We are also not aware of any published report of meteoritic chondrules with the unusual zoning pattern found in those from 62295, but G. Kurat has found similar zoning in the Mezo–Madaras and Ticschitz meteorites (pers. comm., 1974). Fredriksson (pers. comm., 1977) reports that some polycrystalline chondrule-like objects from C3 or C4 chondrites have higher iron olivine grains in the cores, and Fredriksson *et al.* (1969) reported the presence of separate, new, low-iron olivine crystals in the outer parts of some chondrules from the Sharps chondrite, which they ascribed to a two-stage process with more reducing conditions in the latter stage. Others (e.g., Levi–Donati and Jarosevich, 1972) have described barred chondrules having metallic iron rims.

Keil and Fredriksson (1964) and Van Schmus (1969) have examined the olivine of many meteoritic chondrules. Van Schmus (1969) showed that the olivine of the chondrites is uniform in composition (i.e., unzoned) and that a trimodal distribution of these compositions exists, such that individual classes of chondrites can be recognized by their olivine compositions. The values for mole% fayalite are 16–20, 22–24, and 27–31; complete gaps exist between the groups. The olivine in the 62295 "chondrule" contains only 7–13 mole% fayalite. However, Olsen and Jarosevich (1971) reported olivine of 14 mole% fayalite among the rare first chondrules found in an iron meteorite. The compositional similarity in 62295 of the "chondrule" olivine to the olivine of the surrounding rock is unusually fortuitous, unless 62295 itself is a fused meteorite of unusual composition.

It is also important to note that the interstitial material in meteoritic barred olivine chondrules normally consists of a soda-alumina-silica glass (e.g., in Karkh, Dube *et al.*, 1977) or microcrystalline material (Van Schmus, 1969), whereas that in 62295 is essentially pure single crystal plagioclase, An_{95}. A clear glass of composition An_{85} was reported in chondrules from Boredy (Graham *et al.*, 1976) but this was in orthopyroxene-rich chondrules. Larimer (1966) suggested that partial melting of olivine with separation of the liquid phase along (010) cleavage planes can give the barred olivine texture. It seems unlikely, however, that an originally homogeneous crystal of olivine could have held as much as 8.8 wt.% plagioclase components in solution. Fredriksson (pers. comm., 1977) has suggested that some of the glass stringers in meteoritic chondrules might represent former crystalline plagioclase stringers that have been shocked.

Wood (1963, Appendix, p. 176) has made an important point of the fact that meteoritic chondrites do not have bubbles. Although we have no reason to doubt this statement as it applies to the major melt (glassy) parts of chondrules, once such melt becomes enclosed within a rigid single crystal as a melt inclusion, further cooling can and does form shrinkage bubbles in these inclusions, in both meteoritic chondrules and the 62295 "chondrules", wherever the mass of melt is large enough to result in "vapor" nucleation during quenching (Roedder, 1969; Roedder and Weiblen, 1973b, p. 1048).

A variant on the meteoritic mode would have the "chondrules" in 62295 as part of the original lunar material under the assumption that the moon has formed by the same processes as the chondrites (Ganapathy and Anders, 1974).

DISCUSSION

The unusual zoning poses problems for all three possible modes of origin. Metastable equilibria were probably involved in the crystallization of chondrules (Blander *et al.*, 1976), and this may hold the explanation for the zoning. Furthermore, the crystals in meteoritic barred chondrules are usually assumed to have grown from a nucleus on the outer surface of the droplet, which would not yield the observed concentric zoning. Fuchs *et al.* (1973), Dodd (1971), and Hutchison and Symes (1972) have reported iron-rich rims on some meteoritic olivine crystals and chondrules, possibly in part from equilibration with the matrix, and some barred chondrules even have rims of metallic iron, but we know of none that have iron-rich rims *and* cores.

It is possible to set up *ad hoc* explanations of the observed zoning, for example, by recourse to rapidly changing f_{O_2} and/or temperature of crystallization from the heat of crystallization released during rapid crystallization of strongly supercooled drops [as shown by Nelson *et al.* (1972) and Keil *et al.* (1973)], or by taking cognizance of the two distinctly different environments implicit in all three proposed modes of origin, i.e., the free fall in space followed by re-equilibration in 62295 magma for the meteoritic and droplet modes, or deep and shallow crystallization for a phenocryst mode of origin.

CONCLUSIONS

The barred olivine "chondrules" in 62295 are similar to meteoritic chondrules in several respects, but have a lower Fa content, have Fe-rich cores *and* rims, and have single crystal plagioclase as interbar stringers, rather than glass. We hoped that the discovery of these "chondrules" in a lunar igneous rock might shed some light on the whole enigma of the origin of meteoritic chondrules. Unfortunately, however, study of these lunar "chondrules" has not permitted us to select from three possible origins: impact-generated melt globules solidified in flight, spherical phenocrysts, and preexisting meteoritic chondrules, as each proposed origin requires unlikely *ad hoc* conditions. In any of these modes, a two-stage process seems to be needed to form the record of a reversal in the activity ratio between olivine components.

Acknowledgments—We thank S. V. Radcliffe for the transfer of his samples of 62295, and H. E. Belkin for sample preparation and the modal analysis of the "chondrule". The manuscript has profitted from discussions with many individuals and from the reviews of—Gordon L. Nord, Jr., J. Stephen Huebner, Kurt Fredrikkson, and Ursula Marvin. Support for this study from National Aeronautics and Space Administration grants T-1170-A and NSG-9050 is gratefully acknowledged.

References

Agrell S. O., Agrell J. E., Arnold A. R., and Long J. V. P. (1972) Some observations on rock 62295 (abstract). In *Lunar Science IV*, p. 15–17. The Lunar Science Institute, Houston.

Blander M., Planner H. N., Keil K., Nelson L. S., and Richardson N. L. (1976) The origin of chondrules: Experimental investigation of metastable liquids in the system Mg_2SiO_4–SiO_2. *Geochim. Cosmochim. Acta* **40**, 889–896.

Brecher A., Vaughan D. J., Burns R. G., and Morash K. R. (1973) Magnetic and Mössbauer studies of Apollo 16 rock chips 60315,51 and 62295,27. *Proc. Lunar Sci. Conf. 4th*, p. 2991–3002.

Brown G. M., Peckett A., Phillips R., and Emelius C. H. (1973) Mineral-chemical variations in the Apollo 16 magnesio-feldspathic highland rocks. *Proc. Lunar Sci. Conf. 4th*, p. 505–518.

Bunch T. E., Quade W., Prinz M., Keil K., and Dowty E. (1972) Lunar ultramafic glasses, chondrules and rocks. *Nature Phys. Sci.* **239**, 57–59.

Dodd R. T. (1971) The petrology of chondrules in the Sharps meteorite. *Contrib. Mineral Petrol* **31**, 201–227.

Dube A., Fredriksson B. J., Jarosewich E., Nelen J. A., Noonan A. F., O'Keefe J., and Fredrikkson K. (1977) Eight L-group chondrites: A comparative study. *Mineral Sciences Investigations 1974–1975. Smithson. Contrib. Earth Sci.* no. 19, p. 71–82.

Fredriksson K., Jarosevich E., and Nelen J. (1969) The Sharps chondrite—new evidence on the origin of chondrules and chondrites. In *Meteorite Research* (P. M. Millman, ed.), p. 155–165. Reidel Pub. Co., Dordrecht, Netherlands.

Fredriksson K., Nelen J., and Melson W. G. (1970a) Petrography and origin of lunar breccias and glasses. *Proc. Apollo 11 Lunar Sci. Conf.*, p. 419–432.

Fredriksson K., Nelen J., Melson W. G., Henderson E. P., and Andersen C. A. (1970b) Lunar glasses and micro-breccias: Properties and origin. *Science* **167**, 664–666.

Fredriksson K., Dube A., Milton D. J., and Balasundaram M. S. (1973a) Lonar Lake, India: An impact crater in basalt. *Science* **180**, 862–864.

Fredriksson K., Noonan A., and Nelen J. (1973b) Meteoritic, lunar and Lonar impact chondrules. *The Moon* **7**, 475–482.

Fuchs L. H., Olsen E., and Jensen K. J. (1973) Mineralogy, mineral-chemistry, and composition of the Murchison (C2) meteorite. *Smithson. Contrib. Earth Sci.* **10**, 33 pp.

Ganapathy R. and Anders E. (1974) Bulk compositions of the Moon and Earth, estimated from meteorites. *Proc. Lunar Sci. Conf. 5th*, p. 1181–1206.

Ganapathy R., Morgan J. W., Higuchi H., Anders E., and Anderson A. T. (1974) Meteoritic and volatile elements in Apollo 16 rocks and in separated phases from 14306. *Proc. Lunar Sci. Conf. 5th*, p. 1659–1683.

Graham A. L., Easton A. J., Hutchinson R., and Jérome D. Y. (1976) The Boredy meteorite; mineral chemistry and origin of its Ca-rich glass inclusions. *Geochim. Cosmochim. Acta* **40**, 529–535.

Haggerty S. E. (1974) Apollo 17 orange glass: Textural and morphological characteristics of devitrification. *Proc. Lunar Sci. Conf. 5th*, p. 193–205.

Hutchison R. and Symes R. F. (1972) Calcium variation in olivines of the Murchison and Vigarano meteorites. *Meteoritics* **7**, 23–29.

Ivanov A. V., Nazarov M. A., Rode O. D., and Shevaleevsky I. D. (1976) Chondrule-like particles from Luna 16 and Luna 20 regolith samples. *Proc. Lunar Sci. Conf. 7th*, p. 743–757.

Keil K. and Fredriksson K. (1964) The iron, magnesium, and calcium distribution in coexisting olivines and rhombic pyroxenes of chondrites. *J. Geophys. Res.* **69**(16), 3487–3515.

Keil K., Prinz M., Planner H. N., Skaggs S. R., Dowty E., Nelson L. S., Richardson N. L., and Blander M. (1973) A qualitative comparison of textures in lunar chondrules and CO_2 laser-formed synthetic chondrule-like spherules. University of New Mexico, Inst. Meteoritics Spec. Publ. 7. 16 pp.

Kurat G. (1967) Zur Entstehung der Chondren. *Geochim. Cosmochim. Acta* **31**, 491–502.

Kurat G., Keil K., Prinz M., and Nehru C. E. (1972) Chondrules of lunar origin. *Proc. Lunar Sci. Conf. 3rd*, p. 707–721.

Larimer J. W. (1966) The petrology of chondritic meteorites in the light of experimental studies. Ph.D. Dissertation, Lehigh University, Lehigh, Pennsylvania. 185 pp.

Levi-Donati G. R. (1970) Microchondrules. *Meteoritics* **5**, 33–42.

Levi-Donati G. R. and Jarosevich Eugene (1972) Mineralogical and chemical researches on L-chondrites: Girgenti. *Meteoritics* **7**, 109–125.

Nelson L. S., Blander M., Skaggs S. R., and Keil K. (1972) Use of a CO_2 laser to prepare chondrule-like spherules from supercooled molten oxide and silicate droplets. *Earth Planet. Sci. Lett.* **14**, 338–344.

Nord G. L., Jr., Lally J. S., Heuer A. H., Christie J. M., Radcliffe S. V., Griggs D. T., and Fisher R. M. (1973) Petrologic study of igneous and metaigneous rocks from Apollo 15 and 16 using high voltage transmission electron microscopy. *Proc. Lunar Sci. Conf. 4th*, p. 953–970.

Olsen E. and Jarosevich E. (1971) Chondrules: First occurrence in an iron meteorite. *Science* **174**, 583–585.

Roedder E. (1969) Interpretation of silicate melt inclusions in lunar, terrestrial, and meteoritic minerals (abstract). *Fluid Inclusion Research, Proceedings of COFFI* (Commission on Forming in Fluids and Inclusions) Vol. 2, p. 21–22. Published by author.

Roedder E. (1971) Metastability in fluid inclusions. In *IMA-IAGOD, IAGOD Volume.* (Society of Mining Geologists of Japan, Special issue No. 3). (T. Tatsumi, ed.), p. 327–334. Asashi Evening News, Tokyo.

Roedder E. and Weiblen P. W. (1973a) Apollo 17 "orange soil"—a result of meteorite impact on liquid lava? *Nature* **244**, 210–212.

Roedder E. and Weiblen P. W. (1973b) Petrology of some lithic fragments from Luna 20. *Geochim. Cosmochim. Acta* **37**, 1031–1052.

Roedder E. and Weiblen P. W. (1974) Petrology of clasts in lunar breccia 67915. *Proc. Lunar Sci. Conf. 5th*, p. 303–318.

Van Schmus W. R. (1969) The mineralogy and petrology of chondritic meteorites. *Earth Sci. Rev.* **5**, 145–184.

Walker D., Longhi J., Grove T. L., Stolper E., and Hays J. F. (1973) Experimental petrology and origin of rocks from the Descartes Highlands. *Proc. Lunar Sci. Conf. 4th*, p. 1013–1032.

Weiblen P. W. and Roedder E. (1973) Petrology of melt inclusions in Apollo samples 15598 and 62295, and of clasts in 67915 and several lunar soils. *Proc. Lunar Sci. Conf. 4th*, p. 681–703.

Whipple F. L. (1972) On certain aerodynamic processes for asteroids and comets. In *Nobel Symposium 21, from Plasma to Planet* (A. Elvius, ed.), p. 211–232. Almqvist and Wiksell, Stockholm, Sweden.

Wilshire H. G., Stuart-Alexander D. E., and Jackson E. D. (1973) Apollo 16 rocks: petrology and classification. *J. Geophys. Res.* **78**, 2379–2392.

Wood J. A. (1963) On the origin of chondrules and chondrites. *Icarus* **2**, 152–180.

Lunar Sample Index

Pages 1–1303: Volume 1, The Moon and the Inner Solar System

Pages 1305–2654: Volume 2, Petrogenetic Studies of Mare and Highland Rocks

Pages 2655–3965: Volume 3, Planetary and Lunar Surfaces

(Index entries were compiled from information supplied by the authors, and refer only to opening pages of articles.)

LUNA SAMPLES

Subject Index

Pages 1–1303: Volume 1, The Moon and the Inner Solar System
Pages 1305–2654: Volume 2, Petrogenetic Studies of Mare and Highland Rocks
Pages 2655–3965: Volume 3, Planetary and Lunar Surfaces

(Index entries were compiled from key words supplied by the authors, and refer only to opening pages of articles.)

Author Index

Pages 1–1303: Volume 1, The Moon and the Inner Solar System
Pages 1305–2654: Volume 2, Petrogenetic Studies of Mare and Highland Rocks
Pages 2655–3965: Volume 3, Planetary and Lunar Surfaces